American College of Veterinary Internal Medicine (ACVIM) Forum 2016

Denver, Colorado, USA
8 - 11 June 2016

Volume 1 of 2

ISBN: 978-1-5108-2540-6

Printed from e-media with permission by:
Curran Associates, Inc.
57 Morehouse Lane
Red Hook, NY 12571

Some format issues inherent in the e-media version may also appear in this print version.

Printed by Curran Associates, Inc. (2016)

Address correspondence to:
AMERICAN COLLEGE OF VETERINARY INTERNAL MEDICINE (ACVIM)
1997 WADSWORTH BLVD.
LAKEWOOD, CO 80214-5293
www.ACVIM.org

Additional copies of this publication are available from:

Curran Associates, Inc.
57 Morehouse Lane
Red Hook, NY 12571 USA
Phone: 845-758-0400
Fax: 845-758-2633
Email: curran@proceedings.com
Web: www.proceedings.com

TABLE OF CONTENTS

VOLUME 1

CARDIOLOGY

ONCOLOGY

NEUROLOGY

SMALL ANIMAL-ENDOCRINOLOGY

SMALL ANIMAL-GASTROENTEROLOGY

SMALL ANIMAL-HEMATOLOGY

SMALL ANIMAL-HEPATOLOGY

SMALL ANIMAL-INFECTIOUS DISEASE

SMALL ANIMAL-NUTRITION/METABOLISM

SMALL ANIMAL-NEPHROLOGY/UROLOGY

SMALL ANIMAL-OTHER

SMALL ANIMAL-PHARMACOLOGY

SMALL ANIMAL-RESPIRATORY

EQUINE

FOOD ANIMAL

AMERICAN ASSOCIATION OF VETERINARY NUTRITION/AMERICAN COLLEGE OF VETERINARY NUTRITION

AMERICAN ASSOCIATION OF VETERINARY NUTRITION/AMERICAN COLLEGE OF VETERINARY NUTRITION FOOD ANIMAL

AMERICAN COLLEGE OF VETERINARY CLINICAL PHARMACOLOGY/AMERICAN ACADEMY OF VETERINARY PHARMACOLOGY AND THERAPEUTICS

AMERICAN SOCIETY FOR VETERINARY CLINICAL PATHOLOGY

AMERICAN SOCIETY OF VETERINARY NEPHROLOGY & UROLOGY

ASSOCIATION OF VETERINARY HEMATOLOGY AND TRANSFUSION MEDICINE

BUSINESS MANAGEMENT

COMPANION ANIMAL PARASITE COUNCIL

COMPARATIVE GASTROENTEROLOGY SOCIETY

VETERINARY EMERGENCY AND CRITICAL CARE SOCIETY

CARDIOLOGY

NEUROLOGY

VOLUME 2

ONCOLOGY

SMALL ANIMAL INTERNAL MEDICINE

EQUINE

FOOD ANIMAL

LARGE ANIMAL INTERNAL MEDICINE

TECHNICIAN PROGRAM

ADDITIONAL PAPER

C01

SINUS RATE APPROXIMATION WITH VVI, VVIR, AND VDD IN DOGS WITH THIRD DEGREE ATRIOVENTRICULAR BLOCK

Bryan Eason[1], Dan Hogan[1], Ryan Baumwart[2]

[1]*Purdue University, West Lafayette, IN, USA,* [2]*Oklahoma State University, Stillwater, OK, USA*

The purpose of this study was to investigate which pacing modality (VVI, VVIR, or VDD) most closely approximates the intrinsic sinus rate in dogs with third degree AV block (3AVB) over a range of typical daily activities. A secondary aim of the study was to determine if echocardiographic parameters were different between pacing modalities. We hypothesized that VDD would most closely track the intrinsic sinus rate for all activities and that echocardiographic dimensions would be different between AV asynchronous (VVI, VVIR) and AV synchronous (VDD) pacing modalities.

Dogs were prospectively recruited following implantation of a St. Jude dual-chamber pacemaker system for 3AVB; dogs with evidence of sinus node dysfunction or clinically important arrhythmias other than 3AVB were excluded. All dogs were programmed to VDD at study entry. The average, lowest and highest heart rates during VDD pacing were determined; the average heart rate was programmed as the base rate for VVI and VVIR. The highest heart rate was programmed as the upper tracking rate for VVIR with a threshold of 2.5, slope of 16, and fast reaction and recovery times. The rest function was used for all pacing modalities and was programmed to the lowest heart rate during VDD pacing; if the lowest heart rate was at the lower rate limit, then the rest rate was programmed 20% slower. Dogs were randomly assigned to VVI or VVIR after initial VDD interrogation and echocardiography (2D, Doppler, TDI). Ambulatory ECGs were recorded for 36 hours and the owners were instructed to record the time and duration of 5 specific activities: 1) urination/defecation, 2) eating, 3) sleeping, 4) walking, and 5) play. After three months, echocardiography was repeated, the dog was crossed over to VVI or VVIR pacing, and ambulatory ECG monitoring was repeated. After three additional months, echocardiography was repeated, the pacing modality was programmed to VDD and ambulatory ECG monitoring was repeated. Atrial rate, ventricular rate, and atrial-to-ventricular ratio were determined for each of the 5 activities after an initial 12 hours acclimatization period.

This is an ongoing study and currently 6 dogs are enrolled where 5 dogs have completed the VVI phase, 4 dogs have completed the VVIR phase, 4 dogs have completed both the VVI and VVIR phases, and 1 dog has completed all VVI, VVIR, and VDD phases. Atrial rates and atrial-to-ventricular ratios for VVI, VVIR, and VDD at each activity are as follows: urination/defecation (156.0 ± 42.9, 2.01 ± 0.39),(134.4 ± 24.0, 0.95 ± 0.14), (105.7, 1.0); eating (116.4 ± 20.8, 1.51 ± 0.23),(126.2 ± 11.7, 0.97 ± 0.22), (95.7, 1.0); sleeping (99.1 ± 38.3, 1.32 ± 0.62),(79.8 ± 15.7, 0.88 ± 0.30), (51.5, 0.76); walking (183.2 ± 24.7, 2.4 ± 0.41),(157.4 ± 38.9, 1.09 ± 0.12), (91.4, 1.0); play (125.4 ± 38.3, 1.66 ± 0.61),(105.8 ± 17.0, 0.83 ± 0.07), (108.9, 1.0). Echocardiographic data is available for all 6 dogs in VDD, 4 dogs in VVIR, and 2 dogs in VVI. The left ventricle appears to be larger in diastole and systole for VVI (LVAd/Ao= 6.85 ± 0.76, EDV/Ao= 26.08 ± 2.80, LVAs/Ao= 2.74 ± 0.19, ESV/Ao= 9.23 ± 0.08) and VVIR (LVAd/Ao= 5.24 ± 0.57, EDV/Ao= 24.78 ± 3.04, LVAs/Ao= 2.38 ± 0.47, ESV/Ao= 10.65 ± 3.21) than VDD pacing (LVAd/Ao= 5.14 ± 1.38, EDV/Ao= 23.13 ± 7.46, LVAs/Ao= 2.30 ± 0.84, ESV/Ao= 7.43 ± 4.27).

As the study is ongoing no final conclusions can be made although synchronous VDD pacing appears to more closely approximate sinus rate and have slower atrial rates than asynchronous VVI or VVIR pacing. All dogs will have completed the study within the next two months at which time final conclusions can be made.

C02

VALIDATION OF A METHOD FOR QUANTITATION OF CLOPIDOGREL AND CLOPIDOGREL ACTIVE METABOLITE IN FELINE PLASMA

Janne G Lyngby, Michael H Court, Pamela M Lee

Program in Individualized Medicine (PrlMe), Department of Veterinary Clinical Sciences, College of Veterinary Medicine, Washington State University, Pullman, WA, USA

Clopidogrel is considered mainstay therapy for the prevention of cardiogenic thromboemboli in cats and has been shown to be superior to aspirin. Clopidogrel is a pro-drug and requires hepatic bioactivation through cytochrome p450 for formation of the clopidogrel active metabolite (CAM), which is unstable and challenging to quantitate. After bioactivation, CAM binds to the ADP-receptor on platelets and inhibits platelet aggregation. Recently, a method to stabilize CAM in human plasma has been developed by adding 2-bromo-3'methoxyacetophenone (BMAP) to blood tubes. Once stabilized, the derivatized CAM (CAM-D) can be quantitated using high-performance liquid chromatography and tandem mass spectrometry (HPLC-MS/MS).

Our aim is to validate HPLC-MS/MS for the detection of clopidogrel, CAM-D, and other metabolites in feline plasma. We hypothesized that HPLC-MS/MS is both accurate and precise. The quantitation of CAM is the first step in determining if cats can consistently metabolize clopidogrel or if significant interindividual variation exists as in humans with genetic polymorphisms of hepatic cytochrome P450. Furthermore, having an accurate and precise method for measuring the plasma concentration of CAM will allow safer use of clopidogrel by potentially establishing a therapeutic index and by evaluating whether coadministration of drugs like omeprazole inhibit CAM formation in cats.

FDA guidelines to assess accuracy and precision were followed. The *in vitro* stability of clopidogrel, CAM-D, and clopidogrel acid in both citrate and EDTA feline plasma was evaluated at room temperature, 4°C, -20°C and -80°C at 0, 6, 12, 24, 48, 72 hours, and 7 days. Linear regression was performed on an 8 point standard curve (clopidogrel and CAM-D at 1-100ng/mL, clopidogrel acid at 0.2-20ug/mL). To evaluate clinical application, two healthy cats were given clopidogrel (Plavix) at the clinically used dose of 18.75 mg orally once daily. Blood was taken prior to clopidogrel administration and at 20 min, 40 min, 60 min, 90 min, 4 hr, 6 hr, 8 hr, and 12 hr post clopidogrel administration. All blood samples were collected aseptically into 2 mL cryotubes containing 10 uL of 500 mM BMAP and 20 uL of 500 mM EDTA and then centrifuged. The supernatant was then transferred to new cryotubes and stored at -80°C. All samples were analyzed using HPLC-MS/MS in duplicate with a standard curve.

In vitro samples were stable at all temperatures for a minimum of 72 hours. No difference was appreciated in the stability of clopidogrel and its metabolites between EDTA and citrated plasma. Standard curves showed linearity for clopidogrel, CAM-D, and clopidogrel acid (r=0.99). The assay measured clopidogrel, CAM-D, and clopidogrel acid with an accuracy of <±18.7% and a precision within runs of <±7%. Plasma concentrations of clopidogrel, CAM-D, and clopidogrel acid were readily detected in plasma from healthy cats receiving clopidogrel.

This is the first simultaneous quantitation of clopidogrel, CAM-D, and clopidogrel acid in feline plasma. HPLC-MS/MS is an accurate and precise method for simultaneous quantitation of clopidogrel and metabolites in feline plasma both *in vitro* and from healthy cats receiving oral clopidogrel.

C03

EFFICACY OF BRONCHIAL STENTING IN DOGS WITH MYXOMATOUS MITRAL VALVE DISEASE AND BRONCHIAL COLLAPSE

Dar Ozer[1], Samantha Siess[2], Brienne Williams[1], Nikki Gaudette[1], George Kramer[1]

[1]*Atlantic Coast Veterinary Specialists, Bohemia, NY, USA,* [2]*Stony Brook University, Stony Brook, NY, USA*

Chronic airway disease and myxomatous mitral valve disease (MMVD) are frequent comorbidities in small breed dogs. Bronchial collapse can cause coughing, tachypnea and hypoxemia. Pulmonary hypertension can be seen with both MMVD and chronic airway disease. The purpose of this study was to review outcomes in dogs that had bronchial stents placed due to bronchial collapse. It was hypothesized that moderate to severe MMVD or the presence of pulmonary hypertension would not have a negative effect on lifespan after stent placement.

Medical records of 18 small breed dogs that had bronchial stent placement for chronic coughing secondary to bronchial collapse were reviewed. A hierarchical multiple linear regression analysis was conducted to predict lifespan after bronchial stent placement based on age at time of stent placement, severity of MMVD and the presence of pulmonary hypertension.

Eighteen dogs had bronchial stents placed over a period of 7 years. Age at the time of stent placement ranged from 6.5 years to 14 years of age (M = 10.47 ± 1.85). Breeds represented included Cavalier King Charles spaniel (2), beagle (2), Chihuahua (4), Pomeranian (3), toy poodle (2), Yorkshire terrier (2), Maltese (1), Coton de Tulear (1) and shih tzu (1). There were 11 males and 7 females. Twelve dogs (66.67%) had evidence of moderate to severe MMVD and 4 (22.22%) had evidence of pulmonary hypertension. Six dogs (33.33%) had CHF prior to stent placement and 6 dogs (33.33%) had CHF after stent placement. Syncope was reported in 6 dogs prior to stent placement and 5 dogs after stent placement. The average lifespan after stent placement was 203.56 ± 250.72 days. Three dogs are currently alive post-stent placement (1013, 559 and 411 days).

A hierarchical multiple linear regression was calculated to predict lifespan after placement of a bronchial stent based on age at the time of stent placement, the presence of pulmonary hypertension, and severity of MMVD. In stage one, age at the time of stent placement significantly predicted lifespan after placement of a bronchial stent ($\beta = -7.10$, $p < .02$); with lifespan decreasing by 7.10 days for each additional month of age at the time of stent placement. The presence of pulmonary hypertension did not significantly predict lifespan ($\beta = 235.32$, $p = .10$). Severity of MMVD did not contribute significantly to the model in stage two ($\beta = 101.38$, $p = .43$).

Results from this study indicate that the severity of MMVD or the presence of pulmonary hypertension did not negatively affect lifespan after bronchial stent placement. As such, moderate to severe MMVD and the presence of pulmonary hypertension should not be viewed as exclusion criteria when assessing candidates for possible bronchial stenting. Prospective studies should be conducted to further investigate the clinical benefit of bronchial stenting in dogs with severe bronchial collapse.

C04

STUDY OF ECHOCARDIOGRAPHIC VARIABILITY IN ESTIMATING PULMONARY ARTERY PRESSURE AND PULMONARY VASCULAR RESISTANCE IN DOGS.

JD Rhinehart, JD Bonagura, BA Scansen, KE Schober

College of Veterinary Medicine, The Ohio State University, Columbus, OH, USA

Pulmonary hypertension (PH) is important in clinical practice and is related to clinical symptoms and prognosis. We hypothesized that Doppler echocardiographic (DE) indices of PH and pulmonary vascular resistance (PVR) are influenced by a variety of independent factors leading to clinically important variability of DE-based pressure estimates in dogs.

Dogs with naturally acquired tricuspid regurgitation (TR) were studied prospectively. All dogs had degenerative valve disease. Target variables were acquired during 4 study periods (dogs in lateral recumbency or standing, after a 6-minute walk test [6-MWT] and after sedation [0.25mg/kg butorphanol, IM]), and by two different observers. Heart rate (HR), TR flow velocity (TRFV), PVR, estimates of right atrial pressure, stroke volume, cardiac output, and 23 other variables were quantified. Statistical methods included repeated-measures ANOVA and mixed model analysis. $P<0.05$ was considered significant.

Thirty-two dogs of 10 small breeds with varying TRFVs (1.9-5.28 m/s) and PVR (2.9-33.0 WU) were studied. There was a significant effect of observer ($P<0.05$), respiration ($P<0.01$), and image quality ($P<0.05$) on outcome variables whereas imaging position (dogs lying vs. standing) did not cause differences ($P>0.20$). Sedation significantly increased mean TRFV ($P<0.05$, from 3.15 ± 0.71 m/s to 3.51 ± 0.83 m/s; increase observed in 55% dogs) whereas 6-MWT had no effect (3.16 ± 0.75 m/s, $P>0.05$). HR declined after sedation (mean reduction 13 ± 19 bpm; $P<0.05$) and increased after 6MWT (mean increase 30 ± 28 bpm; $P<0.05$) but had no correlation to TVFV or PVR. 6-MWT did not influence TRFV and PVR.

These data document relevant variability of DE estimates of PH that deserve clinical consideration.

C05

CHANGES IN NT-PROBNP ASSOCIATED WITH TREATMENT AND SURVIVAL TIME IN CATS WITH CONGESTIVE HEART FAILURE

Kursten Roderick, John Rush, Lisa Freeman, Vicky Yang, Suzanne Cunningham

Cummings School of Veterinary Medicine at Tufts University, North Grafton, MA, USA

Survival time in cats with congestive heart failure (CHF) is variable, and predictors of outcome would be useful to help guide treatment and patient assessment. In people with CHF, the change in NT-proBNP concentrations after initiating treatment and/or targeting NT-proBNP concentrations below an absolute level provide valuable information that can be used to help guide treatment. The change in NT-proBNP concentrations in cats with CHF after treatment and their association with survival have not been reported. Therefore, the purpose of this study was to measure the change in NT-proBNP in cats with CHF after treatment and to determine whether serial NT-proBNP measurements provide prognostic information in this population.

Client-owned cats presenting with new onset CHF secondary to cardiomyopathy were eligible for enrollment. Blood was collected for NT-proBNP analysis within 4 hours of admission, on the day of discharge, and at the re-evaluation 7-10 days later.

Thirty-two cats were enrolled (25 male, 7 female; all neutered), with a median age of 9.6 years (range 1.1 to 18.0 years). Thirty-one (97%) cats survived to discharge and 28 cats (88%) survived to the 7-10 day re-evaluation. Median NT-proBNP concentrations at admission and hospital discharge were 1,713 pmol/L (range, 160 to 3,784 pmol/L) and 902 pmol/L (range, 147 to 3,223 pmol/L), respectively. Median NT-proBNP concentration at re-evaluation was 1,124 pmol/L (range, 111 to 2,727 pmol/L). Median NT-proBNP concentrations decreased significantly from admission to discharge ($p<0.001$) and from admission to re-evaluation ($p<0.001$), but not from discharge to re-evaluation ($p=0.28$). Median survival time was 109 days (range, 1 to 606 days), with 6 cats (19%) still alive at the time of analysis. Cats that received pimobendan had a significantly greater decrease in NT-proBNP concentrations [-821 pmol/L (range, -2,163 to -267 pmol/L) from admission to discharge compared to cats that did not receive pimobendan [-561 pmol/L (range, -2,137 to 889 pmol/L); $p=0.045$]. Cats whose owners had difficulty administering medications (n=11) had a shorter survival time [median 60 days (range, 2 to 234 days)] compared to owners who did not report difficulty [n=21, median 115 days (range, 1 to 606 days)]; however this was not statistically significant ($p=0.12$).

The association between pimobendan and a larger decrease in NT-proBNP concentration warrants additional research. Larger studies are needed to determine whether NT-proBNP can be used to help guide treatment in cats with CHF.

C06

RIGHT VENTRICULAR OUTFLOW TRACT OBSTRUCTION AND CORONARY ANATOMY IN FRENCH AND ENGLISH BULLDOGS WITH PULMONIC STENOSIS

Courtney Smith, Katherine Scollan, David Sisson, Nicole LeBlanc

Oregon State University, Corvallis, OR, USA

Pulmonic stenosis (PS) is one of the most common congenital cardiac defects in dogs. Stenotic lesions described range from a normal pulmonic annulus with domed leaflets to a hypoplastic pulmonary annulus with thickened leaflets. Additionally, anomalous coronary artery anatomy (commonly described as an R2A) has been reported impinging on the right ventricular outflow tract (RVOT) in English Bulldogs and a Boxer dog with PS. The aim of this study was to investigate the stenotic lesion morphology and coronary artery anatomy in French Bulldogs in comparison to English Bulldogs to assess similarities and/or differences between the breeds.

Medical records were reviewed for French and English Bulldogs presented to Oregon State University for heart murmur evaluation or imaging prior to balloon valvuloplasty (BVP). Dogs were included if they had severe PS (pressure gradient >80 mmHg estimated via echocardiography) and confirmation of their coronary artery anatomy by either angiography or computed tomography (CT). Echocardiographic, angiographic, and CT images were reviewed. The pulmonic annulus (PA), aortic annulus (Ao), PA:Ao ratio, valvular anatomy, and coronary anatomy were assessed on each imaging modality by three observers (CS, KS, DDS) and averaged. Tricuspid regurgitation (TR) was assessed on echocardiographic and angiographic studies. Breed medians or means were compared by Mann-Whitney U test or unpaired t-tests as indicated. In the dogs that had both CT and angiographic studies, values were compared by paired t-tests.

Twenty-six dogs were reviewed and 6 dogs (all English Bulldogs) were excluded due to lack of definitive coronary artery anatomy imaging. Of the 20 included dogs, 7 were English (7 male) and 13 were French Bulldogs (4 female and 9 male). Mean weight was 13.8 ±4.5 kg in the English and 8.9 ±3.9 kg in the French Bulldogs (p=0.02). CT and angiographic imaging was performed on 6/7 and 2/7 English, and 9/13 and 12/13 French Bulldogs, respectively. An R2A coronary artery anomaly was definitively identified in 6/7 English Bulldogs and 0/13 French Bulldogs. Median PA:Ao annulus ratio was not statistically different between the English (0.88, range 0.64-1.0) and French Bulldogs (0.89, range 0.85-1.1; p= 0.37). In addition, no difference was found between CT and angiographic measurements of the PA:Ao ratio (p= 0.38). The mean PA annulus indexed to body weight to the 1/3 power was not different between English (0.54 ±0.07) and French Bulldogs (0.53 ±0.04; p=0.73). TR was noted in 9/13 French Bulldogs (4 mild, 2 moderate, 3 severe) while 3/7 English Bulldogs had TR (1 mild, 1 moderate, 1 severe).

Anomalous development of the coronary arteries was not observed in any of the French Bulldogs evaluated in this study, whereas the R2A coronary anomaly was identified in 6/7 English Bulldogs with a RVOT obstruction. Marked thickening of the pulmonic valve leaflets with markedly reduced leaflet mobility was observed in all of the French Bulldogs and annular hypoplasia was not present or was very modest in both breeds. In the dogs with R2A anomalies, the curvature and degree of encroachment from the left coronary artery on the RVOT was markedly variable. A combination of valvular and supravalvular obstruction was identified in the only English Bulldog with normal coronary anatomy.

C07

CARDIAC FUNCTION AND METABOLIC PARAMETERS IN OBESE DOGS

Melissa Tropf, O. Lynne Nelson, Pamela Lee

Washington State University, Pullman, WA, USA

In people, obesity is an independent risk factor for the development of obesity related cardiac dysfunction (ORCD) characterized by systolic dysfunction, diastolic dysfunction, and/or vascular endothelial dysfunction. Components of metabolic syndrome including dyslipidemias, elevated inflammatory markers, and insulin resistance are believed to play an important role in the pathophysiology of ORCD. Obesity is reaching epidemic proportions in dogs and it is estimated that approximately 34% of dogs are overweight and 5% are obese in the United States. Recent studies performed in dogs suggest that metabolic syndrome also exists in obese dogs, warranting further investigation into the risk for ORCD.

We hypothesized that obese dogs have cardiac dysfunction and have elevations in inflammatory markers, altered insulin sensitivity, and dyslipidemia consistent with metabolic syndrome when compared with non-obese dogs. The aims of this study were first, to assess the presence of cardiac dysfunction in obese dogs compared to dogs with ideal body composition and second, to compare metabolic parameters including lipid analysis and inflammatory markers in obese versus normal dogs.

Healthy dogs weighing less than 25 pounds with ideal body condition (n = 17) and obese body condition (n = 29) were enrolled. All dogs had normal physical exam and thoracic auscultation with no evidence of systemic illness on routine laboratory diagnostics. A body composition using a validated 9 point scale, systolic arterial Doppler blood pressure, a 6-lead electrocardiogram, and an echocardiogram (M-mode, B-mode, Doppler and strain imaging) were obtained. Finally, metabolic parameters including a lipid profile (cholesterol, triglyceride, HDL, LDL), insulin, adiponectin, and a canine inflammatory marker ELISA panel (GM-CSF, IL2-18, TNF-alpha, INF-gamma, TNF-gamma CRP, MCP-1, IP-10, KC-like) were evaluated. Paired T tests were performed to compare the study groups and a p-value of less than 0.05 considered statistically significant.

Compared to dogs with ideal body condition scores, obese dogs had significantly increased septal wall thickness, increased systolic function measures, and significantly reduced diastolic function measures. Obese dogs had significantly increased insulin, insulin:glucose ratio, cholesterol, triglyceride, and HDL levels. The remaining metabolic parameters including adiponectin were not significantly different between the groups. IL-8 and KC-like inflammatory markers were significantly higher in the obese group. While the mean blood pressure was higher in obese patients, the difference was not statistically significant. A number of obese dogs (n = 6) were confirmed to have systemic hypertension on subsequent examinations.

In conclusion, obese dogs do have alterations in cardiac structure and function as well as dyslipidaemia, altered insulin sensitivity, and elevated inflammatory markers. These findings warrant additional studies to investigate inflammation, dyslipidaemia, or possibly systemic hypertension as potential etiologies of altered cardiac function.

C08

COMPARISON OF FUROSEMIDE INFUSION DILUTED WITH 2.4% HYPERTONIC SALINE VS. DEXTROSE 5% IN WATER (D5W)

Darcy Adin[1], Teresa DeFrancesco[1], Clarke Atkins[1], Kari Kurtz[1], Manoela Penteado[2]

[1]North Carolina State University, Raleigh, NC, USA, [2]2University of Sao Paulo, Sao Paulo, Brazil

Continuous rate infusion (CRI) of furosemide causes more diuresis than intermittent bolus administration, but can exacerbate dehydration and renal dysfunction. The co-administration of hypertonic saline (HS) may mitigate this, however, volume expansion could be detrimental in heart failure.

Furosemide CRI with D5W (FCRI-D5W) was compared to furosemide CRI with 2.4% hypertonic saline (FCRI-HS) in 6 normal dogs, in a randomized, blinded, crossover study. Furosemide was diluted to 2.2% with either 1.5 ml/kg D5W for the FCRI-D5W treatment or with 1.0 ml/kg D5W and 0.5 ml/kg of 7.2% hypertonic saline for the FCRI-HS treatment. The furosemide CRI rate was 0.66 mg/kg/hr for 5 hrs in equal volume for both treatments and was preceded by a 0.66 mg/kg furosemide IV bolus.

Mean PCV and %increase in BUN were less for FCRI-HS vs FCRI-D5W ($P=0.046$ both). Mean CVP decreased over time ($P<0.001$), with no treatment difference ($P=0.4$). Urine output, water intake, and creatinine increased significantly, while serum osmolality, serum/urine electrolytes, urine furosemide, and body weight decreased significantly, without differences between treatments. The %decrease in serum potassium ($P=0.09$) and %increase in urinary sodium ($P=0.08$) trended higher in FCRI-HS. Plasma furosemide was stable and not different between treatments. Urinary neutrophil gelatinase-associated lipocalin (NGAL) was unchanged, while both treatments doubled urinary aldosterone excretion suggesting similar activation of RAAS.

Diuresis was similar for both treatments however, FCRI-HS appeared to produce less dehydration (lower PCV, smaller %increase in BUN). Absence of intravascular volume expansion, based on CVP, suggests FCRI-HS may be safe in heart failure.

C09

MEDICAL MANAGEMENT AND SURVIVAL TIME ASSOCIATED WITH CONGESTIVE HEART FAILURE STAGE D: A RETROSPECTIVE STUDY

Amelie Beaumier, John E. Rush, Lisa M. Freeman

Tufts Cummings School of Veterinary Medicine, North Grafton, MA, USA

Degenerative mitral valve disease (DMVD) is the most common acquired heart disease in dogs. The ACVIM consensus statement for dogs with DMVD provided recommendations for diagnosis and management of Stages A through D. Recommendations for Stage D were limited by the paucity of published information for dogs with advanced congestive heart failure (CHF).

The objective of this retrospective study was to investigate the survival and treatment of dogs diagnosed with stage D CHF. For the purpose of this study, we defined stage D as the persistence of CHF despite receiving the label-approved dose of pimobendan, an angiotensin-converting-enzyme inhibitor (ACEI), and furosemide $\geq$4 mg/kg/day. Dogs receiving a dosage of furosemide <4 mg/kg/day were enrolled only if the dose was increased to $\geq$4 mg/kg/day and if a new medication to control CHF was introduced at the same visit.

Forty-three dogs (27 male and 16 female, all neutered) met eligibility criteria and were included in the study. Mean age at the time of diagnosis of Stage D CHF was 10.1 ± 1.8 years. The most common breeds represented were Cavalier King Charles spaniel (n=9), Chihuahua (n=4), Dachshund (n=4), and Shih Tzu (n=3). Twelve dogs were hospitalized on the day of diagnosis of stage D CHF [median duration of hospitalization=1 day (range, 1-3 days)]. The median duration between the first diagnosis of CHF and diagnosis of stage D was 157 days (range, 11-743 days). At the time of diagnosis of stage D, 29 dogs had medications added [1 medication added (n=26), >1 medication added (n=3)], the diuretic dose was escalated in 27 dogs, and the pimobendan dose was increased to >0.6 mg/kg/day in 20 dogs.

Thirty-two (74%) dogs had subsequent additional changes in medications during the course of stage D CHF. Most dogs (n=31; 72%) dogs were ultimately receiving at least 5 cardiac medications for the management of stage D CHF. Ultimate medication dosages during Stage D were: Furosemide (7.4 ± 2.7 mg/kg/day), pimobendan (1.0 ± 0.3 mg/kg/day), and ACEI (0.9 ± 0.2 mg/kg/day). Additional medications administered to these dogs included spironolactone (n=31), sildenafil (n=21), torsemide (n=16), digoxin (n=8), and hydrochlorothiazide with spironolactone (n=8). Median survival time after onset of Stage D CHF was 311 days (range, 9-658 days), with 14 dogs still alive at the time of analysis.

Multiple medication adjustments were necessary to manage dogs with stage D CHF. However, the survival time may be encouraging to the dedicated owner.

C10

MRI 7T RESONANCE SPECTROSCOPY (MRS) PREDICTS CARDIAC ENERGETIC RESERVES IN DOGS WITH PRECLINICAL MITRAL INSUFFICIENCY

A Ray Dillon[1], Tom Denney[2], Seungwoo Jung[1], Randolph Winter[1], Juming Zhong[1], Michael Tillson[1], Xiaoxia Zhang[2], Nouha Salibi[3], Sharron Barney[1]

[1]College of Veterinary Medicine, Auburn University, Auburn, AL, USA, [2]College of Engineering, MRI Center, Auburn University, Auburn, AL, USA, [3]Siemens Healthcare, MR Research and Development, MRI Center, Auburn University, Auburn, AL, USA

Early stages of heart disease are characterized by reduction in energy reserves with impairment of myocardial contractility and relaxation. However, these alterations are difficult to measure with existing clinical imaging modalities such as echocardiography. Recent advances in phosphorous (^{31}P) magnetic resonance spectroscopy (MRS) enable changes in energy reserves to be measured non-invasively.

The creatine kinase (CK) reaction is the primary energy reservoir of the heart. The CK reaction reversibly converts adenosine diphosphate (ADP) and phosphocreatine (PCr) to ATP and creatine. In heart failure, the flux through the CK reaction decreases as both creatine concentration (Cr) and CK activity decreases. Unlike ATP, which can be generated by de novo synthesis pathways in the myocyte, Cr is not produced in excitable tissues. Therefore, in heart failure the decrease in Cr occurs earlier and is faster than the decrease in ATP. In this study, the PCr/ATP ratio was used as a marker of cardiac energetics. The hypothesis was that changes in cardiac energetic reserves measured by MRS precede global MRI, echocardiographic, and biochemical markers in volume overload heart failure.

This hypothesis was evaluated in a canine model of preclinical isolated mitral regurgitation (MR) via chordal rupture. In each group, 2 dogs had minimal and 2 dogs had more severe MR at induction. One group of dogs (4) was evaluated at baseline and data collected at 2, 6, 12, 16 and 18 weeks after induction. A second group of dogs (4) was evaluated at baseline and 2, 4, and 6 weeks after induction. At these parallel time points, cardiac evaluation, thoracic radiographs, echocardiography, and MRS and standard 7T MRI were collected. In addition, peripheral blood samples were collected for microRNA and biomarker evaluation (NTproBNP). Baseline and endpoint cardiac catheterization were recorded for cardiac output, right and left ventricular function and vascular resistance. At the conclusion of the study, myocardial function was assessed by measuring isolated myocyte morphology, size, and myocyte contractility and Ca flux. Left ventricle and mitral valve tissues were collected for molecular studies of mRNA and miRNA.

Utilizing a 7 Tesla (T) MRI scanner, ^{31}P magnetic resonance spectra were collected with an ungated, free breathing three-dimensional chemical shift imaging sequence. In each dog, three 2.8cc voxels on the septal wall near midventricle were selected for analysis. The PCr/ATP ratio was computed without saturation or blood correction in each voxel and averaged.

Although there was a range in severity of MR in each group, all dogs were asymptomatic, untreated and stage B1 or B2 during the observation period. All data were collected at each time point. Using each dog as its own control for echocardiographic evaluation, the end-point LVDd increased in diameter (35.1% in 18wk dogs, 19.9% in 6 wk dogs) and left atria dimensions increased (27.3% in 18 wk dogs and 31.2% in 6 wk dogs) and the fractional shortening increased (22.6% in 18 wk dogs, 24.3% in 6 wk dogs). Compared to baseline, the LVDs dimension increased (18.7% at 18 wks in the 18 wk dogs, and 8.6% in the 6 wk dogs). At endpoint catheterization, the forward LV stoke volume decreased (35.2% in 18wk dogs, 22.2% in 6 wk dogs) in all dogs

The normal PCr/ATP ratio without saturation or blood correction was determined at baseline for normal dogs (1.455 mean, 0.097 SEM). In the unloaded LV at 2 weeks, the PCr/ATP ratio was increased in 6 of the 8 dogs. Over time with dysfunctional remodelling in MR dogs, the PCr/ATP ratio decreased as myocytes energy reserves were depleted especially in the dogs with systolic dysfunction.

PCr/ATP ratio would appear to be a measure of cardiac energetics early in the remodelling phase of volume overload MR and allows for the linear measure of the same dog over time to determine disease progression. MRS may provide a useful tool in the evaluation of drug therapy on the cardiac energetics of MR in dogs.

C11

IMMUNOHISTOCHEMICAL STUDY OF THE PRO-NATRIURETIC PEPTIDE CONVERTASE CORIN IN SEVERE CANINE MYXOMATOUS MITRAL VALVE DISEASE

Alex Sahagian[2], Chris Lam[1], Bob Bao[2], Sanaz Maleki[2], Brett Hambly[2], Niek Beijerink[1]

[1]*University Veterinary Teaching Hospital, Faculty of Veterinary Science, Sydney University, Sydney, NSW, Australia,* [2]*Discipline of Pathology, Bosch Institute and School of Medical Sciences, Sydney University, Sydney, NSW, Australia*

Corin is expressed by cardiomyocytes and converts pro-natriuretic peptides to their mature forms, which play a key role in fluid homeostasis and cardiac remodelling. In mice and human with failing hearts, myocardial corin protein levels have been reported to be increased, however without a concomitant increase in corin activity. These findings can partly explain the clear paradox in congestive heart failure (CHF) in that the natriuretic peptides are elevated, yet the body is fluid retaining (ultimately resulting in CHF). Interestingly in the Cavalier King Charles Spaniel, 2 loci associated with development of myxomatous mitral valve disease (MMVD) have been identified. Corin is on one of these loci on the latest dog assembly. This further raises the question whether deficiencies in myocardial corin expression or activity could be a mechanism explaining the progression of heart disease and CHF in dogs as well. The purpose of this study therefore was to characterize the immunohistochemical expression of corin in myocardial tissue in normal dogs and in dogs with severe MMVD.

Two groups of dogs of a variety of breeds were recruited including 5 control dogs (euthanized for severe non-cardiac disease; age range 4 months-12 years), and 6 dogs with severe MMVD (euthanized for refractory CHF; age range 9-14 years). Samples were collected from the right atrium (RA), right ventricle (RV), left atrium (LA), left ventricle (LV) and interventricular septum (IVS). Immunohistochemical staining of all tissues with a polyclonal rabbit anti-human antibody against corin was carried out, and computational quantification was used to compare the staining density between the two groups and between cardiac regions.

Positive transmembrane and intracellular immunostaining for corin was widely observed in all parts of the myocardium in both groups. Corin expression was not up- or downregulated in any of the cardiac chambers of the dogs with severe MMVD (figure 1), nor was any difference seen between the chambers within each group.

Our results indicate that corin expression might be a rate-limiting step in dogs with CHF due to MMVD. A lack of an increase in corin expression or activation can be postulated to be a possible important pathophysiological mechanism responsible for the natriuretic peptide paradox in dogs with CHF, and warrants further investigation.

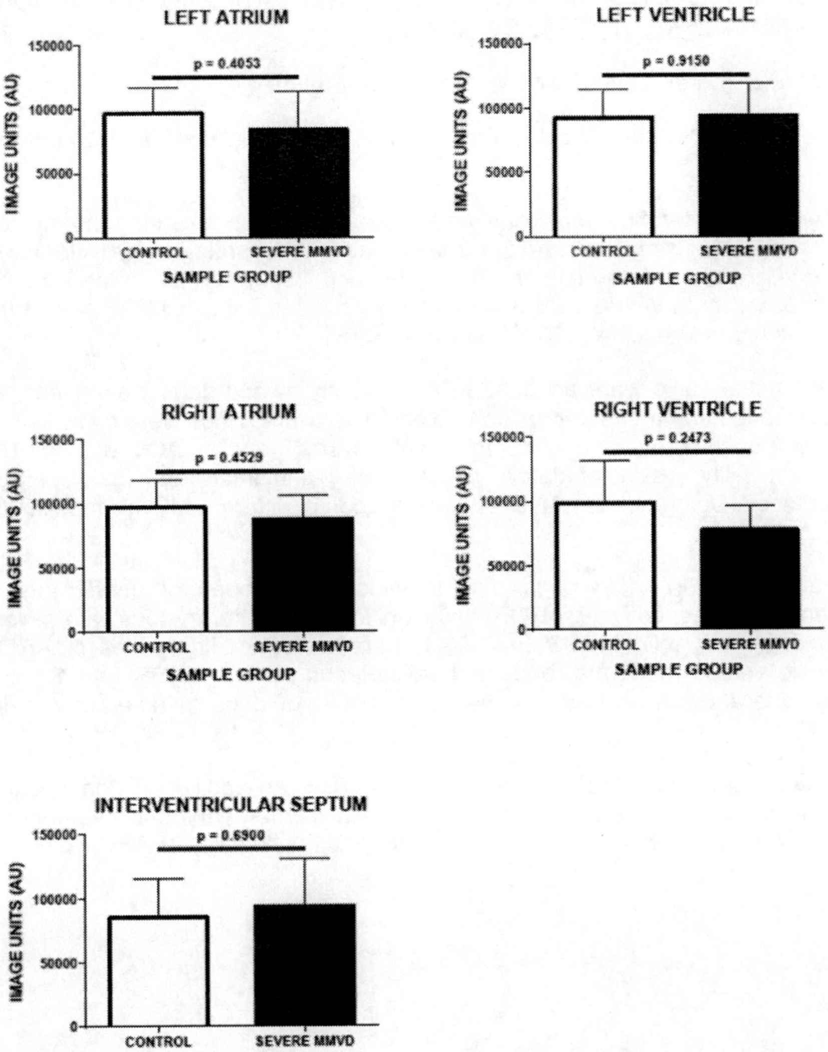

Figure 1: Quantitative corin expression in myocardial tissue in control dogs, and in dogs with congestive heart failure secondary to myxomatous mitral valve disease

C12

ANATOMIC REGURGITANT ORIFICE AREA USING 3D-ECHOCARDIOGRAPHY IN DOGS WITH MYXOMATOUS MITRAL VALVE DISEASE (MMVD)

Giulio Menciotti[1], Michele Borgarelli[1], Sandra Müller[2], Jonathan Abbott[1]

[1]*VA-MD College of Veterinary Medicine, Blacksburg, VA, USA,* [2]*Utrecht University, Utrecht, The Netherlands*

Aims of this study were to determine feasibility and repeatability of measuring anatomical regurgitant orifice area (AROA) using real-time 3D transthoracic echocardiography (RT3DE) in dogs affected by MMVD, and to determine if AROA obtained using RT3DE relates to the severity of MMVD assessed both with an echocardiographic scoring system (MRSS) and the American College of Veterinary Medicine (ACVIM) staging system.

RT3DE datasets of the mitral valve were acquired in 81 privately owned dogs diagnosed with MMVD and AROA was traced and measured using dedicated software. Dogs were classified as mild, moderate or severe according to an echocardiographic MRSS, and as ACVIM Stage B1, B2 or C. Technique feasibility (as % of dataset measurable) and inter- and intra-operator repeatability were assessed. Differences in AROA between dogs in different MRSS and ACVIM stages were investigated.

AROA was measurable in 60 (74.1%) dogs. The inter- and intra-operator coefficients of variation were 26% and 21%, respectively. AROA was significantly greater in dogs with severe MRSS compared to dogs with mild MRSS (p=0.04). There were no differences in AROA between dogs with mild versus moderate, and moderate versus severe MRSS (p=0.84 and p=0.28, respectively). There was no difference between AROA of dogs in different ACVIM clinical stages (p=0.17)

In conclusion obtaining AROA using RT3DE is feasible and offers an additional, non-invasive technique to stratify MR severity in dogs with MMVD. The relatively low repeatability suggests standardization of the technique is desirable. Diagnostic and prognostic utility of AROA deserve further investigation.

C13

ANALYSIS OF MITRAL VALVE MORPHOLOGY WITH REAL-TIME 3-DIMENSIONAL ECHOCARDIOGRAPHY IN DOGS UNDERGOING MITRAL VALVE REPAIR

Takeshi Mizuno, Masashi Mizuno, Kayoko Harada, Hiroshi Takano, Arane Takahashi, Kazuya Mamada, Junsoek Lee, Sayaka Takeuchi, Ayaka Chen, Tamotsu Sawada, Takahiro Mizukoshi, Asako Shinoda, Shuhei Uchida, Takurou Mori, Kazuki Takamura, Junichirou Takeuchi, Masami Uechi

JASMINE Veterinary Cardiovascular Medical Center, Yokohama, Kanagawa, Japan

Three-dimensional (3D) saddle shape of the mitral valve annulus was first reported in 1987 by Levine and colleagues. Previous studies showed annulus height to commissural width ratio (AHCWR), an index parameter for saddle-shape of mitral annulus, is preserved across mammalian species (15-20%). This suggests that nature conserves the saddle-shaped annulus for mechanical benefit. Menciotti et al described 3D transthoracic echocardiography (TTE) of normal dogs and AHCWR was 21±0.05%. There are no reports on 3D transesophageal echocardiography (TEE) studies on dogs with myxomatous mitral valve disease (MMVD). This study aims to evaluate mitral valve annulus morphology using 3D TEE in dogs with MMVD.

3D TEE was performed in dogs weighting more than 5kg during mitral valve repair surgery. The 3D geometry of mitral valve apparatus was measured with dedicated quantification software. The classification of MMVD dogs used in this study was the basis of ACVIM staging. The 3D TEE measurements were compared between each group of dogs.

Twenty-six dogs were included in the study, 2 dogs were excluded from the study due to the uninterruptable echocardiographic images. Thus, a total of 24 dogs were evaluated in this study. The study group included 7 dogs with ACVIM Stage B2, 10 dogs with Stage C, and 7 dogs with Stage D. Characteristics of each group were summarized below:

	B2 (n=7)	C (n=10)	D (n=7)
Age (month)	101 ± 16	131 ± 14*	123 ± 21
Body Weight (kg)	6.84 ± 2.64	7.61 ± 1.33	7.6 ± 2.43
NT-proBNP (pg/ml)	2448 ± 1016	4786 ± 2893	3650 ± 2277
VHS (v)	12.1 ± 0.85	12.9 ± 1.3	13.2 ± 2.43
LA/Ao	1.85 ± 0.24	2.45 ± 0.34*	2.81 ± 0.25*
LVEDDN ($cm/kg^{0.294}$)	2.29 ± 0.22	2.32 ± 0.27	2.53 ± 0.30

Significant difference compared with B2 group, P < 0.05 (Tukey's multiple comparison test)

AHCWR decreased as the disease progressed (Stage B2; 14.2 ± 4.2%, Stage C; 9.0 ± 4.2 5, Stage D; 9.5 ± 3.3%, Figure). Stage D dogs showed significantly lower values (P<0.05) compared to Stage B2 dogs. Non-planarity angle measurement tends to increase in value as the severity of MMVD worsened, however, there were no significant differences between the groups (B2; 157.7 ± 8.1°, C; 167.9 ± 9.2°, D; 164.4 ± 8.5°, Figure). There were no significant differences between the groups for annular circumference and annular area values.

Results

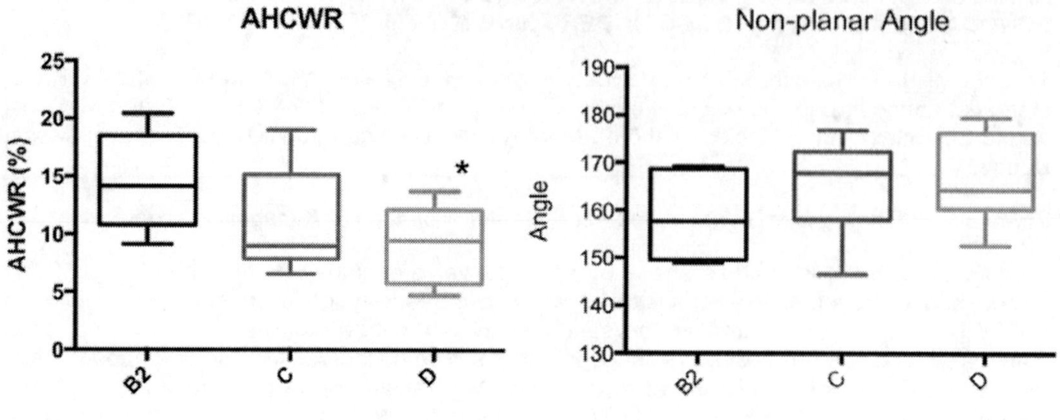

In conclusions, the mitral valve of dogs with MMVD loses the saddle shape as the mitral valve disease progresses. The results of this study suggested that as the severity of MMDV worsened, more mechanical stress would be placed on the mitral valve, which could increase the progression of mitral regurgitation. To the best of our knowledge, this is the first report about the morphological changes of mitral valve annulus according to disease severity in dogs with MMVD.

C14

LOW-DENSITY LIPOPROTEIN OXIDATION IS BREED AND GENDER DEPENDENT IN DOGS WITH MYXOMATOUS MITRAL VALVE DISEASE

Maria Josefine Reimann[1], Jens Häggström[2], Jacob Eifer Møller[3], Jens Lykkesfeldt[1], Torkel Falk[4], Lisbeth Høier Olsen[1]

[1]University of Copenhagen, Frederiksberg C, Denmark, [2]Swedish University of Agricultural Sciences, Uppsala, Sweden, [3]Odense University Hospital, Odense, Denmark, [4]Din Veterinär, Helsingborg, Sweden

Oxidative stress has been associated cardiovascular disease and suggested to contribute to cardiac remodeling in human patients with mitral regurgitation. Few previous studies have evaluated the relationship between oxidative stress and myxomatous mitral valve disease (MMVD) in dogs. The objective of this study was to determine plasma concentrations of certain markers of oxidative stress and investigate if they were associated with MMVD severity, selected clinical variables and disease progression in dogs with no or different severities of MMVD.

The study included 89 privately-owned dogs: 14 healthy control Beagles, 59 cavalier King Charles spaniels (CKCS) with different severities of MMVD, and 16 dogs of different breeds with clinical signs of congestive heart failure due to MMVD.

Markers of oxidative stress including malondialdehyde (MDA), oxidized low-density lipoprotein (oxLDL) and vitamin E (α-tocopherol and γ-tocopherol) were measured in plasma and their influence on 2-year progression of MMVD was determined.

Plasma oxLDL concentration was significantly lower in control Beagles compared to CKCS (P=0.0005) and females had a significantly lower oxLDL plasma concentration compared to males (P=0.007). Plasma vitamin E concentrations were associated with body condition score (BCS) (α-tocopherol, P=0.002; γ-tocopherol, P=0.01). All the markers of oxidative stress (MDA, oxLDL, vitamin E) were positively associated with serum cholesterol ($P \leq 0.02$) but none of the markers were associated with severity or progression of MMVD. The effect of BCS disappeared when adjusting for serum cholesterol concentration.

In conclusion, Beagles and female dogs appear to have lower plasma concentrations of oxLDL. The results cannot confirm a role of oxidative stress in dogs with MMVD.

C15

CARDIAC OUTPUT MEASURED BY ECHOCARDIOGRAPHY AND CARDIAC-GATED COMPUTED TOMOGRAPHY COMPARED TO THERMODILUTION

Nicole LeBlanc, Katherine Scollan

Oregon State University, College of Veterinary Medicine, Corvallis, OR, USA

The accurate measurement of stroke volume and cardiac output (CO) has important applications in both clinical and research arenas. Invasively measured CO by thermodilution (TD) remains the gold standard method. To date, there are very few comparative studies evaluating echocardiographic CO in dogs using noninvasive imaging modalities. There are no direct comparisons between noninvasive CO estimation using 64-slice multidetector computed tomography (MDCT) and TD in dogs. The aim of this study was to evaluate the accuracy of CO estimation by noninvasive methods (MDCT, 1-, 2-, and 3-dimensional echocardiography) compared to invasively measured CO obtained by TD.

Six intact hound cross dogs weighing between 19.5-23.8 kg were anesthetized using a standardized protocol and spontaneous ventilation. Each dog underwent MDCT, echocardiography, and TD. Dogs were randomized to the order of MDCT and echocardiography, and TD was performed last. CO using TD was measured in triplicate with <10% variation and minimal changes in heart rate. For imaging methods, CO was calculated by volumetric measurements of SV multiplied by the average heart rate. Echocardiographic left ventricular (LV) end-diastolic volumes (EDV) and end-systolic volumes (ESV) were measured on right long axis 1D images (Teichholz) and right long axis and left apical 2D images (single plane Simpson's method of disks, MOD) on 3 averaged consecutive beats. Real-time 3-dimensional echocardiographic (RT3DE) LV volumes were measured using 3D functional analysis software from both right long axis and left apical imaging planes. MDCT EDV and ESV were measured using specific LV functional analysis software on ECG-gated studies.

Excellent correlation was found between CO values measured by TD and 2D MOD (r=0.98) from the left apical window, as well as 2D MOD (r=0.90) and RT3DE (r=0.92) from the right long axis window. A less robust correlation was found between RT3DE values obtained from the left apical window (r=0.68) and TD. MDCT-derived CO were also strongly correlated (r=0.80) with TD. CO estimation by Teichholz formula (r=-0.05) had no significant correlation with TD. Bland-Altman analysis showed a systemic underestimation of CO values derived both from echocardiographic and MDCT measurements. The bias for RT3DE CO obtained from the right long axis window was -0.23 L/min (95% confidence interval (CI) -0.56 to 0.09 L/min), and for 2D from the left apical window was -0.66 L/min (95% CI -0.91 to -0.42 L/min). The bias for CO values obtained from MDCT was -0.57 L/min (95% CI -0.99 to -0.15 L/min).

These results suggest certain echocardiographic techniques available for CO measurement are comparable to invasively measured CO. Specifically, 2D MOD using the right long axis and left apical 4-chamber views, or RT3DE from the right long axis view, had excellent agreement with TD. The excellent correlation between methods indicates a close relationship between select techniques, although they are not interchangeable. CO is underestimated by the noninvasive techniques studied here compared to TD. The differences between TD and noninvasive measures may be due to errors of the method, patient stroke volume variability, heart rate fluctuation, or inaccuracies of TD. One of the major limitations of this study is the small number of patients, which allows preliminary conclusions only. A larger prospective study is needed to delineate the benefits and constraints of these methods.

C16

ANGIOTENSIN CONVERTING ENZYME ACTIVITY AND RESPONSE TO ENALAPRIL IN DOGS WITH AN ACE GENE POLYMORPHISM

Kathryn Meurs[1], Darcy Adin[1], Brent Aona[1], Teresa DeFrancesco[1], Bruce Keene[1], Yamir Reina[2], Josh Stern[1], Sandy Tou[3], Jess Ward[1], Kate Woodruff[1], Clarke Atkins[1]

[1]*North Carolina State University, Raleigh, NC, USA,* [2]*University of California, Davis, Davis, CA, USA,* [3]*Iowa State University, Ames, IA, USA*

Myxomatous mitral valve disease (MMVD) is the most common heart disease in the dog. Angiotensin converting enzyme (ACE) inhibitors are frequently recommended for management of dogs with MMVD and asymptomatic cardiac enlargement. However, the benefit of ACE inhibition in dogs before the onset of congestive heart failure (CHF) is controversial, with different studies showing conflicting results. A variable response to ACE inhibitor therapy has also been observed in human beings with heart disease and in some cases this has been attributed to a polymorphism in the ACE gene. We have previously demonstrated a polymorphism in the canine ACE gene, although the clinical significance of this finding is unknown.

We hypothesized that dogs with MMVD, cardiac enlargement and the ACE gene polymorphism would be less responsive to ACE inhibition with enalapril, as measured by plasma ACE activity, than dogs without the polymorphism.

We evaluated dogs with pre-CHF MMVD and a vertebral heart score of ≥ 11, presenting to the NCSU CVM in the last 18 months. Dogs were genotyped for the ACE polymorphism. Serum samples were collected and analyzed for ACE activity measurement by radioimmunoassay. Dogs were prescribed enalapril, 0.5 mg/kg orally twice a day, and revaluated in 14-21 days with a second measurement of ACE activity. A t test was used to compare ACE activity in the wildtype (normal canine sequence) and DNA variant groups at baseline and after therapy. A paired t test was used to compare each genotype's ACE activity before and after therapy.

Thirty-one dogs were evaluated. Genotypes included 12 homozygous for the DNA variant and 19 homozygous for the wild type. Median baseline ACE activity was significantly lower for dogs homozygous for the DNA variant (18.05 U/L) than for dogs with the wildtype sequence (27.4 U/L) ($p=0.01$). Twenty-seven dogs returned for re-evaluation in 14-21 days. Both genotypes showed a significant suppression of ACE activity on enalapril, ($p= 0.0065$ DNA variant dogs; $p < 0.0001$ wildtype dogs). Median post therapy ACE Activity was not different for the two genotypes (DNA variant, < 5.0 U/L; wildtype dogs <5.0 U/L) ($p=0.24$)).

We conclude that dogs that are homozygous for the ACE variant have lower baseline levels of ACE activity but still demonstrate ACE inhibition in response to enalapril. Further study is warranted to evaluate the clinical importance of these findings.

C17

ANGIOTENSIN CONVERTING ENZYME ACTIVITY IN CAVALIER KING CHARLES SPANIELS WITH AN ACE GENE POLYMORPHISM

Kathryn Meurs[1], Maria Josefine Reimann[2], Lisbeth Heier Olsen[2]

[1]North Carolina State University, Raleigh, NC, USA, [2]University of Copenhagen, Copenhagen, Denmark

The Cavalier King Charles Spaniel is one of the most commonly reported breeds of dogs affected with myxomatous mitral valve disease (MMVD). Angiotensin converting enzyme (ACE) inhibitors are frequently recommended for management of dogs with MMVD. However, the benefit of ACE inhibition in MMVD before congestive heart failure has developed is controversial with different studies showing conflicting results. A variable response to ACE inhibitor therapy has also been observed in human beings with heart disease and has been attributed to a polymorphism in the ACE gene. We have previously demonstrated a polymorphism in the canine ACE gene in small breed dogs.

We hypothesized that the polymorphism would be common in CKCS dog and that this would impact ACE activity.

We collected DNA and plasma samples from CKCS with MMVD. All dogs were genotyped for the known ACE polymorphism. Plasma samples were collected and analyzed for ACE activity measurement with a radioimmunoassay. A t test was used to compare ACE activity in the wildtype (control sequence) and DNA variant groups

Sixty-two dogs were evaluated. Genotypes of 38 (61%) dogs were homozygous for the DNA variant, 3 (5%) were heterozygous and 21 (34%) were homozygous for the wild type (normal canine sequence). Samples from 40 dogs were submitted for ACE analysis including 20 homozygous for the wild type and 20 homozygous for the variant. Median baseline ACE activity was significantly lower for dogs homozygous for the DNA variant (25.0 U/L) than for dogs with the wildtype sequence (31.0 U/L) (p=0.02).

We conclude that the ACE polymorphism appears to be common in the CKCS and dogs that are homozygous for the ACE variant have lower baseline levels of ACE activity. Further study is warranted to evaluate the clinical importance of these findings.

C18

PREVALENCE OF *DIROFILARIA IMMITIS* ANTIGEN IN CLIENT-OWNED PET DOGS BEFORE AND AFTER SERUM HEAT TREATMENT

Laura Nafe, Susan Little, Paul DeMars, Ryan Baumwart, Nalani Yamada, Eileen Johnson

Oklahoma State University, Stillwater, OK, USA

Pet dogs are routinely screened for heartworm disease as a component of annual wellness evaluation. Recent evidence suggests that antigen-antibody complexes in canine serum may preclude accurate diagnosis of heartworm disease with commercially available *Dirofilaria immitis* antigen assays. The objective of this study was to determine whether normal dogs presenting to the Oklahoma State Veterinary Teaching Hospital Community Practice Service for annual veterinary care warrant heat treatment of serum to confirm a negative heartworm antigen test by 4DX Snap (Idexx). Serum and EDTA samples were collected from 201 client-owned pet dogs between December 2014 and October 2015. Dogs ranged in age from 10 months to 15 years (median 4 yrs) and included 108 males and 93 females. Complete survey results (available for 194/201 dogs) revealed 139 dogs (72%) received heartworm preventative regularly over the past year. All EDTA samples were evaluated for microfilaria and serum samples for *D. immitis* antigen before and after heat treatment using a commericial assay (DiroCHEK, Zoetis). Four dogs (2%) tested positive for *D. immitis* antigen on 4DX Snap Test (Idexx) and 2 were microfilaremic. On DiroCHEK antigen testing 2 dogs were positive before and after heat treatment, 1 dog was positive after heat treatment only, and 1 dog was negative before and after heat treatment despite being microfilaremic. One dog received heartworm preventative (Heartgard; Merial) regularly (2/12 doses missed). Heat treatment does not appear to improve detection of *D. immitis* antigen in asymptomatic pet dogs undergoing routine heartworm disease testing with 4DX Snap Test (Idexx).

C19

CARDIAC BIOMARKERS TROPONIN I AND N-TERMINAL PRO-B-TYPE NATRIURETIC PEPTIDE IN CANINE CHRONIC KIDNEY DISEASE PATIENTS

Lena Pelander, Jens Häggström, Ingrid Ljungvall

Department of small animal clinical sciences, Swedish university of agrisultural sciences, Uppsala, Sweden

Increased concentrations of N-terminal pro brain natriuretic peptide (NT-proBNP) and cardiac Troponin I (cTnI) in azotemic dogs, cats and people have been repeatedly reported. Knowledge of the mechanisms behind increased concentrations of these cardiac biomarkers in azotemic dogs is warranted for correct interpretation of test results. The aim of this study was to investigate the association between measured glomerular filtration rate (GFR) and plasma concentrations of cTnI and NT-proBNP in stable canine CKD patients.

Fifty client-owned dogs were prospectively included into the study: Twenty-seven dogs with a diagnosis of CKD (structural and/or functional abnormalities of one or both kidneys that had persisted for at least 3 months) and without signs of other significant disease and twenty-three healthy dogs were included as controls. All dogs were investigated by a physical examination, repeated blood pressure measurements, a complete urinalysis, a full hematology and biochemistry panel, an echocardiographic examination, an ultrasound examination of the entire urinary tract, and a scintigraphic examination for estimation of kidney glomerular filtration rate (GFR). Plasma from each dog was stored in -70°C and assayed in batch.

Median age of included dogs was 6 years (range 0.5-14 years). Median bodyweight was 19.5 kg (range 2-42 kg). There was no difference in age or weight between healthy and CKD dogs. Among dogs with CKD, 16 had a decreased GFR and 11 had a GFR within the normal range. Dogs with CKD and a GFR in the normal range had other evidence of kidney disease such as persistent renal proteinuria or morphological abnormalities.

Univariate analyses were performed to evaluate associations between dog characteristics and cTnI and NT-proBNP, respectively. Age ($r= 0.59$; $p<0.0001$), body weight (BW; $r=0.29$; $p=0.04$), creatinine ($r=0.29$; $p=0.04$) and systolic blood pressure (SBP; $r=0.42$; $p=0.06$) were associated with cTnI concentration. Creatinine ($r=0.46$; $p=0.001$), GFR ($r= -0.39$; $p=0.0063$), albumin ($r= -0.35$; $p=0.0141$), and erythrocyte volume fraction (EVF; $r= -0.35$; $p=0.0072$) were associated with NT-proBNP concentration.

All variables that were associated with cTnI or NT-proBNP concentration, respectively, with a p-value of <0.2 were included in the multiple regression analysis (except creatinine because of covariance with GFR). Age, BW and SBP were the variables remaining in the final model for cTnI (adjusted R-square 0.50, $p<0.0001$). GFR, BW and serum albumin concentration were the variables remaining in the final model for NT-proBNP (adjusted R-square 0.29, $p<0.0001$).

In conclusion, plasma concentration of cTnI in dogs with stable CKD was not influenced by a decreased GFR. Plasma concentration of NT-proBNP in stable canine CKD patients on the other hand, appeared to be influenced by GFR.

C20

ESTABLISHING NORMAL 24 HOUR HOLTER MONITOR VALUES IN HEALTHY PUPPIES

Rebecca Tracey[1], Dongyun Wang[2], Pamela Lee[1]

[1]Washington State University, Pullman, WA, USA, [2]University of Idaho, Moscow, ID, USA

Currently, published literature focuses on the normal 24 hour Holter monitor values of adult dogs greater than 1 year of age. Studies have shown that normal healthy adult dogs may have occasional ventricular premature complexes (VPCs) and that more frequent or sequential VPCs on Holter monitoring may indicate cardiac or clinical disease. This poses a problem for cardiac evaluation of puppies for veterinarians. This lack of information inhibits the ability to screen puppies for indicators of cardiac disease. This is especially important in breeds that tend to have inherited cardiac arrhythmias such as German Shepherd puppies. The purpose of this study is to establish normal ambulatory ECG values using Holter monitors in puppies in an effort to enhance the ability to screen puppies for cardiac disease.

The primary objective was to determine the minimum, average, and maximum heart rate and incidence of arrhythmias for healthy puppies. The secondary objective was to evaluate the correlation between Holter heart rate and age, weight or breed. We hypothesized that healthy puppies will have less arrhythmias and higher Holter minimum, average, and maximum heart rates than healthy adult dogs.

Eligible puppies had to be healthy and between the ages of 12-51 weeks. History, physical exam, and thorough auscultation were performed. An echocardiogram was performed on any puppy with a heart murmur on auscultation. Puppies with physiologic heart murmurs were included in the study while puppies with structural heart disease were excluded. A 24 hour 3-channel Holter monitor was placed on all puppies included in the study. Holter placement followed established methods from the WSU Veterinary Holter Service.

A total of 44 puppies representing 20 breeds were included in this study. Minimum, average, and maximum Holter heart rates were higher in puppies compared to healthy adult dogs. The number of supraventricular complexes seen in puppies were less than in healthy adult dogs. The range of ventricular complexes in puppies were higher than in healthy adult dogs. Minimum and average Holter heart rate decreased with increasing age while maximum Holter heart rate was not correlated with age. No correlation between minimum, average, and maximum Holter heart rates and weight was appreciated. There was a statistically significant difference between maximum Holter heart rate and different breed classifications by size (small vs. medium vs. large vs. giant breed). No correlation between minimum and average Holter heart rate and breed classifications was identified in healthy puppies.

This study is the first study to evaluate normal ambulatory electrocardiographic values in healthy puppies. This information can be used to potentially screen puppies at an earlier age for cardiac disease. The number of ventricular arrhythmias considered normal for healthy adult dogs is thought to vary with dog breed. Whether or not this breed variation is true in puppies is unknown. Now that a normal baseline has been established for puppies, further studies can be conducted to evaluate for breed variation.

C21

DISTRIBUTION OF ALVEOLAR-INTERSTITIAL SYNDROME IN DYSPNEIC VETERINARY PATIENTS ASSESSED BY LUNG ULTRASOUND VERSUS THORACIC RADIOGRAPHS

Jessica Ward[1], Gregory Lisciandro[3], Teresa DeFrancesco[2]

[1]*Iowa State University, Ames, IA, USA,* [2]*North Carolina State University, Raleigh, NC, USA,* [3]*Hill Country Veterinary Specialists, San Antonio, TX, USA*

Lung ultrasound (LUS) is a point-of-care imaging technique that identifies alveolar-interstitial syndrome (AIS) through the observation of ultrasound artifacts called B-lines. This study assessed distribution of AIS detected by LUS compared to thoracic radiographs (TXR) in a population of 100 acutely dyspneic veterinary patients (76 dogs and 24 cats).

Patients underwent LUS and TXR within 6 hours. LUS images were scored for presence of B-lines at 4 sites on each hemithorax. An individual LUS site was scored positive if >3 B-lines were observed within a single intercostal space at that site. TXR were scored for presence of alveolar-interstitial infiltrates at 4 sites on each hemithorax, analogous to LUS. An individual TXR site was scored positive if infiltrate was present in least 25% of the site. Medical records were evaluated for final diagnosis. Agreement in distribution of positive sites between LUS and TXR was compared using a Cohen's Kappa coefficient. Patterns of distribution of AIS among different final diagnoses were compared using Fisher's exact tests.

When considered site-by-site, agreement in distribution of AIS between LUS and TXR was poor to fair (K = 0.25-0.5). However, when considering larger spatial quadrants, agreement between LUS and TXR was good (K = 0.45-0.65). Distribution of AIS differed significantly based on final diagnosis for both cardiogenic and noncardiogenic causes of dyspnea. Dogs with mitral valve disease were more likely to have a caudal distribution of AIS ($p = 0.0031$), while dogs with dilated cardiomyopathy were more likely to have diffuse AIS ($p = 0.0008$). Patients with airway disease were more likely to have absence of AIS in all sites ($p = 0.0006$), while patients with pneumonia were more likely to have unilateral AIS ($p = 0.0055$).

LUS was useful for detecting AIS apparent on TXR. Agreement between the two modalities improved when considering larger spatial quadrants rather than smaller focal sites. Distribution of AIS differed based on final diagnosis, suggesting that a pattern-based LUS approach may prove diagnostically useful.

C22

CIRCULATING PLASMA MIRNA AS NOVEL MOLECULAR BIOMERKERS IN CONGESTIVE HEART FAILURE

SeungWoo Jung, Amy Bohan

Auburn University, Auburn, AL, USA

Congestive heart failure (CHF) carries poor prognosis in dogs with myxomatous mitral valve disease (MMVD). Early identification of dogs predisposed to developing CHF is critical to improve long term survival outcome. The aim of this study is to characterize plasma molecular profiles in dogs with CHF secondary to MMVD and to develop novel molecular biomarkers via liquid biopsy. MicroRNAs (miRNAs) are small, non-coding RNAs and have been known to regulate expression levels of target genes involved in human CHF. Regulatory networks of miRNAs in dogs with MMVD need to be established to further enhance our understanding of CHF pathogenesis. The hypothesis tested in the study was that plasma miRNAs may be differentially expressed in dogs with CHF secondary to MMVD.

A minimum of seven dogs in each group (normal vs CHF) were needed to demonstrate the statistical power with a probability of 0.9 and type one error of $p<0.05$ based on previous publications. Blood was collected in EDTA tubes from 9 normal geriatric dogs (free of heart murmur and absent history of CHF) and 8 dogs with CHF secondary to MMVD (confirmed by clinical signs, chest x-ray and echocardiography). Plasma samples were subjected to isolation of miRNAs with a miRNeasy Plasma Kit (Qiagen) and subsequently reverse-transcription reaction with a miScript II RT Kit (Qiagen). Validated canine primers of 12 candidate miRNAs and miR-39 (spike-in control) were employed to quantify relative expression levels of each miRNA between normal and CHF dogs via real time PCR (miScript SYBR Green PCR Kit, Qiagen).

Canine specific miRNAs were successfully isolated from plasma and amplified. Plasma miR-21 ($p=0.015$), miR-133 ($p=0.01$), and miR-1 ($p=0.01$) were differentially expressed in CHF dogs with statistical significance when compared to normal dogs. These miRNAs are known to be involved in extracellular matrix remodeling and the regulation of fibrotic genes such as TGF-β and Smad. The results suggest that alterations in the interaction between miRNAs expression profiles and their downstream target genes may play a role in cardiac progression to CHF in dogs with MMVD. Further investigation remains warranted to determine their diagnostic potential as molecular biomarkers for early detection of CHF in dogs with MMVD.

C23

A MULTICENTER EVALUATION OF SIGNALMENT AND COMORBID CONDITIONS ASSOCIATED WITH AORTIC THROMBOSIS IN 291 DOGS

Randolph L. Winter, Christine M. Budke

[1]*Auburn University College of Veterinary Medicine, Auburn, AL, USA,* [2]*Texas A&M University Department of Veterinary Integrative Biosciences, College Station, TX, USA*

Aortic thrombotic disease is a primary disease or secondary complication of a systemic disease state with devastating effects to both dogs and cats. Cats with aortic thromboembolism often have significant concurrent cardiac disease. While there is a substantial amount of information available about cats with this disease process, aortic thrombosis (AT) is not well characterized in dogs.

In order to assess signalment and concurrent disease processes in dogs diagnosed with aortic thrombotic disease, a retrospective review was performed of records available through the Veterinary Medical Database of dogs examined between 1985 and 2011. Information on age, sex, breed, weight, and concurrent disease processes in dogs reported with AT were reviewed. Five control dogs without a diagnosis of AT were selected for each AT case.

AT was diagnosed in 291 of 984,973 dogs included in the VMDB from 1985 to 2011. The odds of a dog having AT did not differ significantly by sex (p=0.0593), age (p=0.9833), or weight (p=0.7766). Compared to mixed breed dogs, Shetland sheepdogs had a significantly higher odds (OR=2.44, p=0.0027) of being diagnosed with AT. Protein-losing nephropathy was the most commonly reported disease occurring concurrently in dogs with AT (64/291).

Signalment data including age, sex, and weight were not significantly different in dogs diagnosed with AT compared to dogs not diagnosed with AT. Dogs with AT were commonly diagnosed concurrently with PLN. Contrary to previous reports, cardiac disease was not a common concurrent diagnosis in the dogs with AT included in this study.

C24

RIGHT VENTRICULAR SYSTOLIC FUNCTION IN DOGS WITH POST-CAPILLARY PULMONARY HYPERTENSION

Amalia Turner Giannico[1], Gustavo Dittrich[1], Bruna Cristina Bruler[1], Tilde Rodrigues Froes[1], Aparecido Antonio Camacho[2], Marlos Goncalves Sousa[1]

[1]federal University of Parana, Curitiba, PR, Brazil, [2]Sao Paulo State University, Jaboticabal, SP, Brazil

The function of right ventricle (RV) has gained increased recognition over the past years. In pulmonary hypertension (PH) patients, this led to the reconceptualization of the RV as part of right ventricular pulmonic circulation unit and the acknowledgement that RV function is a major determinant of prognosis in pulmonary arterial hypertension. With this in mind, we sought to investigate the RV systolic function in dogs with post-capillary PH ascribed to mitral valve disease (MVD). Twenty nine dogs with PH (2.5-17.2 kg; 7-15y) and 51 dogs without PH (1.5-32.0 kg; 2-16y) were recruited into this prospective cross-sectional observational study. The PH group was subdivided into mild PH (n=19), moderate PA (n=6), and severe PH (n=4). Several breeds were represented. All animals underwent a complete echocardiogram, which included the measurement of RV fractional area change (RV FAC); tricuspid annular plane systolic excursion (TAPSE); RV free-wall longitudinal systolic velocity (tricuspid annular S′ wave), and RV fractional shortening (RV $FS_{\%}$). For improved statistical analysis and normalization of data according to animal size, a body weight-indexed TAPSE ($TAPSE_{BW\text{-}indexed}$) was created. Some standard echocardiographic indices of congestion and LV function, including early-to-late LV filling velocities ratio (E/A), isovolumic relaxation time (IVRT), E-to-IVRT ratio (E/IVRT), left atrium-to-aorta ratio (LA/Ao), and LV fractional shortening (LV $FS_{\%}$) were recorded as well. The data was not normally distributed. The Mann-Whitney test showed no differences ($P>0.05$) to exist for RV FAC, RV $FS_{\%}$, TAPSE, $TAPSE_{BW\text{-}index}$, and tricuspid annular S′ wave between dogs with PH and those without PH. Also, the analysis of the subdivided PH group showed no statistical difference between the three degrees of PH. Spearman correlation coefficients between RV systolic data and LV indices of congestion and function showed a significant positive correlation between TAPSE and E/IVRT (R=0.3917, P=0.0003), TAPSE and LA/Ao (R=0.3298, P=0.0029), and $TAPSE_{BW\text{-}indexed}$ and LV FS% (R=0.2757, P=0.0166). Finally, a significant positive correlation was documented between RV FAC and RV $FS_{\%}$ (R=0.3507, P=0.0100), as well as between TAPSE and tricuspid annular S′ wave (R=0.4575, $P<0.0001$). Because only a few animals with severe PH were recruited into this study, the absence of association between the severity of PH and the identification of RV systolic impairment needs further investigation. Nevertheless, we observed a downward trend for RV FS%, TAPSE, $TAPSE_{BW\text{-}indexed}$, and tricuspid annular S′ wave along with the aggravation of PH. Our study found correlations between a few congestion and LV function parameters, which may prove useful in monitoring RV function in dogs with severe MVD and overt signs of post-capillary PH. Since the majority of PH cases were mild, it is likely that only a few had reactive PH, which may explain why RV function of either group behaved very similarly. Also, it is not clear whether RV remodeling and impairment would be stronger if pre-capillary PH dogs had been included in this investigation.

C25

VASOVAGAL TONUS INDEX IN DOGS WITH MYXOMATOUS MITRAL VALVE DISEASE

Bruna Cristina Bruler[1], Gustavo Dittrich[1], Amalia Turner Giannico[1], Tilde Rodrigues Froes[1], Aparecido Antonio Camacho[2], Marlos Goncalves Sousa[1]

[1]*Federal University of Parana, Curitiba, PR, Brazil,* [2]*Sao Paulo State University, Jaboticabal, SP, Brazil*

Decreased heart rate variability (HRV) has been found in dogs in congestive heart failure (CHF) and previous studies have acknowledged the use of HRV indices as prognostic indicators in patients with mitral valve disease (MVD). The vasovagal tonus index (VVTI) is an unconventional time domain indicator of HRV, which is mainly influenced by cardiac parasympathetic tone. In this cross-sectional observational study, we sought to investigate the VVTI in dogs with MVD. Electrocardiographic recordings of 30 dogs (7-16y, 3.5-15.5kg) previously classified into ACVIM stages A (controls), B, or C (10 dogs each group) were used to calculate the VVTI. For this, 20 consecutive R-R intervals were measured from each ECG recording (R-R1 to R-R20), and the index was obtained from the formula VVTI = LN [VAR (R-R1 – R-R20)], where LN: natural logarithm and VAR: variance. Bad quality ECG tracings and recordings from dogs with non-sinus rhythms or animals undergoing anti-arrhythmic treatment were not included in this investigation. Also, we recorded the BW-indexed LV in diastole and systole, wall stress index in diastole and systole, fractional shortening, left atrium-to-aorta ratio, mitral E wave, mitral E/A, isovolumetric relaxation time, and the E-to-IVRT ratio. All data underwent the Shapiro-Wilk test to check for normal distribution, while ANOVA, followed by Tukey's test, were used to compare the VVTI between groups. Pearson's test was used to search for linear correlations between the VVTI and the echocardiographic data. The mean values (with lower and upper 95% CI of mean) of VVTI for dogs in stages A, B, and C were, respectively: 8.45 (7.36 – 9.54), 6.09 (4.47 – 7.71) and 6.34 (5.01 – 7.68). A significant difference was found between groups (P=0.0189), with the mean VVTI being significantly higher in control animals as compared to dogs with stage B MVD (P<0.05). When it comes to the relationship between VVTI and cardiac rhythms, a significant difference existed between animals in sinus rhythm (SR), sinus arrhythmia (SA), and sinus tachycardia (ST) (P=0.0083). The lowest VVTI was documented for dogs in ST (5.82; 95% CI 4.14 – 7.49), while the higher was found for animals presenting SR (8.27; 95% CI 7.40 – 9.15). Significant negative correlations were found between VVTI and LA/Ao (R=-0.3699; P=0.0443), as well as between VVTI and heart rate (R= -0.4864; P=0.0064). Although no correlation existed between body weight and the VVTI, age and VVTI attained a significant negative correlation (R=- 0.3827; P=0.0369). The negative correlation between VVTI and heart rate is likely ascribed to the role played by the parasympathetic tone in VVTI, therefore producing higher values when slower rates and irregular rhythms are present. Even though uncontrolled conditions during ECG recording, including stress and agitation, may increase HR, the lower VVTI found in animals exhibiting sinus tachycardia may suggest a sustained sympathetic activation. Although further investigation is warranted, the confidence intervals of this study point to the VVTI <5 being a potential prognostic indicator for CHF, whereas a VVTI >7.7 is likely suggestive of reduced risk for congestion. This is supported by the correlation between VVTI and LA/Ao, as well as the difference between the means of control dogs (stage A) and stage B MVD animals.

C26

SICK SINUS SYNDROME AND SINUS NODE DYSFUNCTION: CASE SERIES REPORTS OF FMVZ-USP CARDIOLOGY SERVICE (2007-2015)

Rebecca Bastos Pessoa[1], Guilherme Teixeira Goldfeder[1], Patricia Pereira Costa Chamas[2], Denise Saretta Schwartz[1], Paula Staudacher Leal de Carvalho[1], Maria Helena Matiko Akao Larsson[1]

[1]*School of Veterinary Medicine and Animal Science - University of Sao Paulo (FMVZ-USP), São Paulo, SP, Brazil,* [2]*Paulista University, São Paulo, SP, Brazil*

Sick Sinus Syndrome is a complex disease characterized by an abnormal cardiac impulse formation on the sinus node. The term Sinus Node Dysfunction refers to the variety of electrocardiographic features involved in this condition. When Sinus Node Dysfunction is associated with clinical manifestations it is referred to as the Sick Sinus Syndrome. A survey of affected dogs attended at the Cardiology Service in the Department of Internal Medicine of the Veterinary Teaching Hospital (HOVET) of the School of Veterinary Medicine and Animal Science - University of Sao Paulo (FMVZ-USP) from 2007 to 2015 was performed. Seventeen cases were reviewed. Breeds involved were Schnauzer (71%), Cocker Spaniel (6%), Poodle (6%) and Dachshund (6%), besides a mongrel dog (6%). There was a predominance of female dogs (71%). The most common clinical manifestations observed were syncope (71%), presyncope (24%), dyspnea (18%) and exercise intolerance (12%). Twelve percent of the dogs were asymptomatic. The most frequent electrocardiographic changes found were sinus arrest (75%), junctional escape rhythm (73%), supraventricular tachycardia (53%) and sinus arrhytmia (53%). Forty-seven percent of the dogs presented tachycardia-bradycardia syndrome. Only 24% of the animals received pacemaker implants, given the high percentage of animals unable to go through chirurgical/anesthetic procedures (29%), and the presence of asymptomatic animals. The dogs that received pacemaker implants lived for 20.4 months (mean value), whereas the ones that were treated conservatively lived for 10.8 months (mean value). These values showed benefit to those patients who received pacemaker implants, but was hindered by the high percentage of comorbidities in older animals, besides the small number of patients enrolled in this retrospective study.

C27

THE UTILITY OF SPECTRAL DOPPLER OF THE HEPATIC VEINS IN DOGS WITH TRICUSPID REGURGITATION

Kensuke Nakamura, Tatsuyuki Osuga, Tomoya Morita, Keitaro Morishita, Noboru Sasaki, Hiroshi Ohta, Mitsuyoshi Takiguchi

Hokkaido University, Sapporo, Japan

Echocardiographic assessment of the severity of tricuspid regurgitation (TR) is less reliable compared to mitral regurgitation due to the complicated geometry and function of right ventricle. Spectral Doppler of hepatic veins (SDHV) has emerged as a useful examination to evaluate the severity of TR in humans. The hepatic vein blood flow waveform is multiphasic and typically consisted with four peaks. A small retrograde a-wave occurring after the P-wave of the electrocardiographic is created by retrograde blood flow during atrial contraction. A large anterograde S-wave occurring immediately after the QRX complex is created by blood flow toward the heart during ventricular systole. A small retrograde v-wave occurs after ventricular systole. A large anterograde D-wave occurring after T-wave is created by ventricular diastole. The objective of the present study was to investigate the correlation between the presence of ascites and the findings of SDHV in dogs with TR. This is a clinical cohort study including twenty-seven client-owned dogs with TR. Dogs were divided into two groups based on the presence of ascites. Physical examination, SDHV and echocardiographic variables were compared between the groups. For the assessment of the comparative accuracy in identifying patients with ascites, receiver operating characteristic (ROC) curves was used. Peak velocity of v-wave and D-wave in dogs with ascites (n=11) was significantly higher than dogs without ascites (n=16). The highest accuracy was obtained for v-wave, with an area under the ROC curve (AUC) of 0.99, followed by D-wave, with an AUC of 0.95; late diastolic velocity of the septal mitral annulus (Am), with an AUC of 0.907. TR peak velocity had no significant difference between two groups. In conclusion, v-wave and D-wave obtained by SDHV has strong correlation with the presence of ascites in dogs with TR.

C28

CUTANEOUS USE OF AN IMPLANTABLE LOOP RECORDER (Reveal LINQ™) FOR CARDIAC EVENT MONITORING IN DOGS

Emily Menzen, Amara Estrada, Brandy Winter, Melanie Powell

University of Florida College of Veterinary Medicine, Gainesville, Florida, USA

Event monitors are a useful diagnostic tool for evaluating patients with episodic cardiac events. The ideal event monitor for use in veterinary practice would be non-invasive, affordable, accurate and reusable. Such devices are becoming increasing difficult to obtain. This study aimed to prospectively evaluate the feasibility of using the Medtronic Reveal LINQ™ human implantable cardiac monitor (ICM) for cutaneous event monitoring for up to 5 days in 14 clinically healthy canines (21.5 kg, range 13.4-31.2kg; 4.5 years range 1.5-13 years old) and two canines with reported collapse episodes (4.7 and 6.4kg, 2.5 and 11 years old). The study consisted of developing a method to secure the low-profile device without causing discomfort at the location generating the strongest signal, adjusting the auto activation parameters, activating the device manually, and assessing the recordings for accuracy.

Reveal LINQ™ wireless cardiac device were obtained via donation following explantation from human patients and the battery life assessed prior to use in each canine patient using a Medtronic Carelink® programmer with appropriate software loaded. An area on the left lateral thorax was clipped and alcohol was used to remove oils. Adhesive electro-conductive gel pads were applied underneath the device and it was placed within the 4th intercostal space parallel to the ribs and then slowly moved in a grid pattern to find the most consistent electrocardiogram (ECG) on the programmer screen. The device was then secured to the thorax using Tegaderm™ film, gauze, and Elastikon®. It was then taped around the thorax with 2 inch white tape and Vetrap™ and the patients were fitted with a Holter monitor vest to reduce likelihood of device tampering or consumption of bandage materials and device. The device was programmed with the patient's information and the pre-programmed manufacturer heart rate parameters Patients wore the ICM for 4 to 5 days and owners were asked to activate the monitor manually for 4 total recordings during different activities and to keep a diary. Accuracy was assessed based on quality of ECG tracings (consistent ability to measure heart rate, visualization of entire p-QRS-T waves) and presence of a recording during manual activation. Two dogs were fitted with 24 hour Holter monitors as a control

Of the sixteen patients, two were unable to be fitted with a device due to a weak signal, one client terminated the study after 1 day, and the device moved out of place and stopped recording on one dog. The remaining twelve dogs had at least 3 out of 4 interpretable tracings from manually activated recordings. The auto-activated recordings triggered by pre-set parameters were of good quality. After each set of auto-activated recordings were reviewed, the parameters were adjusted to define the outer range of normal and recorded rhythms were reclassified. Over the course of the study patient heart rates did not exceed 222 bpm for 32 beats or reach 30 bpm for 12 consecutive beats. Fewer auto-activations occurred per dog when the pause parameter was set at 4.5 seconds than 3 seconds. One symptomatic patient had three episodes while wearing the device however there was no indication of a rhythm abnormality in the manually activated recordings. The other symptomatic patient did not have an episode while wearing the monitor and the device incorrectly logged an atrial fibrillation episode. During this time period, the ECG from the ICM was not of sufficient quality to identify p waves but the control Holter recording clearly showed a sinus arrhythmia was present. The device was still recording at time of removal in all but one dog. Five dogs had self-limiting, mild skin irritation after removal of the device.

The Medtronic Reveal LINQ™ ICM is a viable option for cutaneous cardiac monitoring in dogs for up to 5 days. Veterinary parameters for auto-activation of the device should be programmed at the upper and lower limits of the auto-activated parameters to reduce the number of incorrectly labelled recordings and all recordings should be reviewed by a veterinarian familiar with normal cardiac rhythms.

C29

AORTOSEPTAL ANGLE AND RESPONSE TO BALLOON VALVULOPLASTY IN DOGS AFFECTED WITH SEVERE SUBAORTIC STENOSIS

Lilian Shen[1], Amara Estrada[1], Etienne Côté[2], Brandy Winter[1], Kenneth Lamb[3]

[1]*University of Florida, Gainesville, Florida, USA,* [2]*university of Prince Edward Island, Charlottetown, Prince Edward Island, Canada,* [3]*Lamb Consulting, West St. Paul, Minnesota, USA*

Subaortic stenosis (SAS) is the most common congenital cardiac defect identified in dogs. While it is not difficult to diagnose severe forms of this disease, treatment and decisions on when to consider interventional therapy remain a challenge. It has been suggested that SAS, in some breeds, may be related to an abnormally steep aortoseptal angle (AoSA) - a quality which may limit efficacy of ballooning methods. If an abnormally steep AoSA contributes to SAS and is not remedied by ballooning, it is possible that the magnitude of the AoSA may serve as a useful criterion for determining which dogs will respond better to balloon valvuloplasty and are therefore better candidates for the procedure.

This study investigated the relationship between the measured AoSA and response to combined cutting and high pressure balloon valvuloplasty (CB/HPBV), in dogs with severe SAS. Retrospective evaluation of angiographic and echocardiographic video loops of 22 client-owned dogs of various breeds affected with severe SAS (PG mean 143.36 mmHg; range 80.49-322.5 mmHg) who underwent CB/HPBV were evaluated at the baseline time point immediately prior to the procedure. Two board-certified veterinary cardiologists and 1 novice evaluated these video loops to identify still frame images to be used to measure the AoSA. Each evaluator was blinded to the other 2 evaluators' selected images and subsequent measurements of the AoSA. All evaluators were blinded to initial PG and patient outcomes following the procedure. The percentage decrease in PG at 24 hours, 6 months, and 12 months after ballooning was then calculated and correlated to the measured AoSA at each recheck and for each imaging plane.

The inter-observer variability analysis suggested significant mean differences for the novice observer when compared to the 2 cardiologists for all angle types. Significant differences were observed between the cardiologists for left apical echocardiographic (P = 0.0163) AoSA measurements, while no significant differences were observed for angiographic (P = 0.1042) or right sided long axis LVOT echocardiographic AoSA (P = 0.7058) measurements or for all measurements of AoSA combined combined (P = 0.1937).

Based on right sided long axis LVOT echocardiographic AoSA measurements made by the cardiologists, greater mean differences of PG reduction ratio were observed for angles >160° compared to angles <160°. At 24 hours, significantly greater (P = 0.003) mean differences of PG ratio were observed in the >160° (54.45 s.E. +/- 3.8) compared to the mean in the <160° (39.88 s.E. +/- 2.09). This difference persisted at 6 months (>160° mean 57.73 s.E. +/- 10.9; <160° mean 28.22 s.E. +/- 3.42) and at 12 months (>160° mean 76.11 s.E. +/- 17.5; <160° mean 27.61 s.E. +/- 6.44).

The findings of this study suggest that dogs with severe SAS with an AoSA >160° as measured on right sided long axis LVOT echocardiographic imaging respond to CB/HPBV better than dogs with an AoSA <160°. The difference in response to the same procedure by the two sets of dogs suggests that the magnitude of the AoSA does play a role in SAS and its treatment, and it is possible that measurement of the AoSA may be helpful in the screening process and determination of prognosis following CB/HPBV. However, further investigation of the effect of LVOT PG reduction on survival times and quality of life is warranted in order to determine if AoSA is truly a suitable criterion for selecting candidates for CB/HPBV.

C30

ECHOCARDIOGRAPHIC EVALUATION OF PULMONARY HYPERTENSION IN CATS WITH CARDIAC DISEASE AND PLEURAL EFFUSION VERSUS PULMONARY EDEMA

Melissa Wilson, Katherine Scollan, Nicole LeBlanc

Oregon State University, Corvallis, OR, USA

Cardiomyopathies, including hypertrophic and restrictive, are the most common acquired cardiac diseases in cats. Many affected cats develop pulmonary edema secondary to left-sided congestive heart failure (L-CHF). Alternatively, cats with primarily left-sided cardiomyopathy can develop pleural effusion without the presence of pulmonary edema. It is well known, but not well understood, why cats with left-sided CHF can develop both pulmonary edema and pleural effusion. The development of pulmonary hypertension could lead to right-sided heart failure and potentially cause pleural effusion in cats with predominately left-sided cardiac disease. The aim of this study was to evaluate for echocardiographic evidence of pulmonary hypertension in cats in L-CHF with pleural effusion in comparison to cats with pulmonary edema, other pulmonary disease, and healthy cats.

Medical records and echocardiographic studies from 2008-2015 were reviewed and measured. Cats were categorized into four groups: 1) L-CHF cats with pleural effusion, with or without pulmonary edema, 2) L-CHF cats with pulmonary edema only, 3) cats with pulmonary disease without evidence of cardiomyopathy, and 4) healthy cats. Pulsed wave Doppler recordings of pulmonic outflow were used to measure acceleration time (AT), ejection time (ET), AT:ET ratio, and maximum velocity through the pulmonic valve (PA Vmax). The main pulmonary artery diameter (MPA), aortic diameter (Ao), MPA:Ao ratio, left atrial diameter (LA), and LA:Ao ratio were measured from right parasternal short axis images, while right atrial (RA) maximum diameter, left atrial maximal diameter (LA_{Long}), RA:LA_{Long} ratio, and RA:Ao ratio were measured from right parasternal long axis images. Studies were reviewed for the presence of tricuspid regurgitation (TR) and the maximum TR velocity (TR Vmax) was measured when available. Groups were compared by one-way ANOVA or Kruskal Wallis tests where appropriate.

Ninety-three cats were included in the study, 31 cats with L-CHF and pleural effusion, 25 cats with L-CHF and pulmonary edema, 15 cats with pulmonary disease without evidence of cardiomyopathy, and 21 healthy cats. The median RA:Ao of the pleural effusion group (1.54, range 0.85-2.70) was significantly larger (p=0.0004) than all other groups. The mean RA:LA ratio of the pleural effusion (0.73 ±0.2) and pulmonary disease (0.75 ±0.1) groups were different from the pulmonary edema (0.56 ±0.2) and healthy (0.95 ±0.1) groups ($p < 0.0001$), though not different from each other. Comparison of pulsed wave Doppler measurements of the pulmonic outflow revealed no significant difference in AT, ET, AT:ET, or PA Vmax between any groups. TR Vmax was measured in 7 pleural effusion cats (median 2.2 m/s, range 1.2-3.2) and 1 healthy cat (1.7 m/s).

The results of our study indicate that cats that developed pleural effusion had significantly more right atrial enlargement in addition to left atrial enlargement than cats that developed only pulmonary edema. No echocardiographic evidence of pulmonary hypertension was found in any of the four groups of cats. The potential causes for right atrial enlargement in the cats with pleural effusion include concurrent right-sided cardiac disease or pulmonary hypertension that was not detected by these echocardiographic parameters. Additional studies determining the degree of RA enlargement associated with the development of pleural effusion could be useful in predicting CHF manifestation in cats with cardiomyopathy.

C31

POST-OPERATIVE DYSRHYTHMIAS AFTER CARDIAC SURGERY UNDER CARDIOPULMONARY BYPASS

Junseok Lee, Takeshi Mizuno, Masashi Mizuno, Kayoko Harada, Hiroshi Takano, Arane Takahashi, Kazushi Mamada, Sayaka Takeuchi, Ayaka Chen, Tamotsu Sawada, Takahiro Mizukoshi, Aasako Shinoda, Shuhei Uchida, Takurou Mori, Kazuki Takamura, Junichirou Takeuchi, Masami Uechi

Japan Animal Specialty Medical Institute. JASMINE Veterinary Cardiovascular Medical Center, Yokohama, Kanagawa, Japan

In human, post-operative dysrhythmia is an important complication after cardiac surgery, a major cause of increasing mortality and lengthening hospitalization. In veterinary field, however, there is little knowledge on rhythmic disturbances complicated by cardiac surgery. The aim of this study is to characterize post-operative arrhythmias and to determine peri-operative predisposing factors to arrhythmias in dogs with open-heart surgery under cardiopulmonary bypass (CPB).

This is a prospective and observational study of 70 consecutive dogs undergoing mitral valve repair surgery. The incidence and the types of post-operative arrhythmias were determined by continuously monitoring electrocardiogram for the first 7 days during hospitalization. Occasional premature atrial or ventricular beats were considered insignificant. Serologic and blood-gas parameters measured at serial time points during CPB were normalized and compared between two groups with non-arrhythmic and arrhythmic episodes. All patients showed normal sinus rhythm before surgery.

32 patients (45.7%) had a total of 55 episodes of arrhythmia after cardiac surgery, where ventricular arrhythmias were the most frequently identified (26 of 55; 47.3%). Supraventricular, nodal arrhythmias and conduction blocks accounted for 23.6%, 14.5% and 14.5%, respectively. 53.2% of patients with arrhythmic episodes were recovered to normal sinus rhythm within the first 2 days after surgery. Patients with persistent arrhythmia until discharge were 18.8% (6 of 32). There was no case that permanent pacemaker should be implanted. Although intravenous bolus injection of lidocaine (2mg/kg) was applied to 2 dogs with polymorphic ventricular tachycardia lasted until 7 days, no improvement was found. No hospital deaths were caused by post-operative arrhythmias (deaths in hospital; 1 respiratory failure, 2 embolic events). Furthermore, the arrhythmic group showed longer duration of CPB (107.2±33.6 min, $P<0.05$) and aortic cross-clamping time (59.9±13.3 min, $P<0.01$), and lower glucose level normalized by CPB duration (82.6±38.9 mg/dL/CPB time, $P<0.01$), compared to the non-arrhythmic group (CPB time, 90.3±25.7 min; cross-clamping time, 50.1±20.6 min; normalized glucose level, 126.8±66.7 mg/dL/CPB time). Also, logistic regression model for the probability of post-operative arrhythmias revealed significant effects of the aortic cross-clamping time and the normalized glucose level (adjusted R^2=0.30, $P<0.001$).

The incidence of cardiac arrhythmias is quite high after open-heart surgery under CPB in dogs. However, their clinical significance as a complication of cardiac surgery may be minimal. In addition, hypoglycemic or myocardial ischemic environment during CPB may predispose to post-operative cardiac dysrhythmias.

O01

THE ASSOCIATION OF ENDOTHELIN-1 SIGNALING WITH BONE ALKALINE PHOSPHATASE EXPRESSION AND PRO-TUMORIGENIC ACTIVITIES IN CANINE OSTEOSARCOMA
(*VCS Award Winner*)

Neumann ZL[1], Pondenis HC[1], Masyr A[1], Byrum ML[1], Wycislo KL[2], Fan TM[1]

[1]*Department of Veterinary Clinical Medicine, University of Illinois, Urbana, IL, USA,* [2]*Department of Pathobiology, University of Illinois, Urbana, IL, USA*

Background- Canine osteosarcoma (OS) is an aggressive sarcoma characterized by pathologic skeletal resorption and pulmonary metastases. A limited number of negative prognostic factors, including bone alkaline phosphatase, have been identified in dogs diagnosed with OS; however, the biologic underpinnings for such observations have not been thoroughly investigated. Endothelin-1 mediated signaling is active during bone repair, and is responsible for osteoblast migration, proliferation, and bone alkaline phosphatase expression.

Hypothesis- The endothelin-1 signaling axis is active in canine OS cells, and this pathway is utilized by malignant osteoblasts for promoting cellular migration, proliferation, and bone alkaline phosphatase activities.

Animals- 45 dogs with appendicular OS.

Methods- The expressions of endothelin-1 and endothelin A receptor were studied in OS cell lines and spontaneous tumor samples. The biologic activities mediated by endothelin-1 signaling were investigated by characterizing responses in 3 OS cell lines. In 45 dogs with OS, bone alkaline phosphatase concentrations were correlated with primary tumor osteoproductivity.

Results- Canine OS cells express endothelin-1 and endothelin A receptor, and this signaling axis is functional and mediates OS migration, proliferation, and bone alkaline phosphatase activities. In OS-bearing dogs, circulating bone alkaline phosphatase activities were positively correlated with primary tumor relative bone mineral densities.

Conclusions and clinical importance- Canine OS cells express endothelin-1 and functional endothelin A receptors; with the potential for an autocrine or paracrine pro-tumorigenic signaling loop. Elevations of bone alkaline phosphatase are associated with osteoblastic OS lesions, and might be an epiphenomenon of active endothelin-1 signaling or excessive osteoproduction within the localized bone microenvironment.

O02

SAFETY EVALUATION OF COMBINATION DOXORUBICIN AND TOCERANIB PHOSPHATE (PALLADIA$^{(R)}$) IN TUMOR BEARING DOGS: A PHASE I DOSE FINDING STUDY
(*VCS Award Winner*)

Pellin MA[1], Wouda RM[3], Robinson K[1], Tsimbas K[1], Kurzman ID[1], and Vail DM[1,2]

[1]Department of Medical Sciences, School of Veterinary Medicine, University of Wisconsin-Madison, Madison, WI, USA; [2]Carbone Cancer Center, University of Wisconsin-Madison, Madison, WI, USA, [3]Kansas State University, College of Veterinary Medicine, Manhattan, KS, USA

Introduction
Combination chemotherapy holds the promise of improved outcome in canine malignancy, but care must be taken to avoid overlapping toxicity.

Methods
Dogs with cytologically or histologically diagnosed tumors of any histology (excluding mast cell tumors) and any stage were allowed. A standard, open-label, phase I, 3+3 dose-cohort escalation design was employed. The toceranib dose remained constant at 2.75 mg/kg administered orally every other day. Doxorubicin was initiated at 20mg/m2 administered IV every 21 days for four doses. The doxorubicin dose was escalated by 5 mg/m2 each cohort until MTD or 30mg/m2 was reached. DLT was defined as any grade 3 adverse event with the exception of hematopoietic AEs for which a grade 4 AE was considered dose limiting. Any grade AE that was refractory to supportive care or persisted beyond seven days was also considered dose-limiting.

Results
Twenty-one patients were enrolled. Represented histologies included hemangiosarcoma, multicentric lymphoma, anal sac adenocarcinoma, epitheliotropic lymphoma, and prostatic carcinoma. The DLT was neutropenia occurring with doxorubicin 30 mg/m2. The MTD of the combination was determined to be doxorubicin 25 mg/m2 and toceranib 2.75 mg/kg. No grade 3 or 4 gastrointestinal or novel AEs were observed. Additionally, anti-tumor activity was observed in several cases.

Conclusion
The combination of doxorubicin 25 mg/m2 IV q21 days and toceranib 2.75 mg/kg PO EOD was well tolerated. Efficacy and longer term toxicity of this combination should be further investigated in phase II/III trials.

O03

COMPARISON OF SERUM CYTOKINE LEVELS BETWEEN DOGS WITH MULTICENTRIC LYMPHOMA AND HEALTHY DOGS

Jerome Calvalido, Paul Woods, Geoffrey Wood, Anthony Mutsaers, Darren Wood, William Sears

University of Guelph, Guelph, Ontario, Canada

In people, changes in the cytokine environment have been linked to the development of lymphoma, and multiple circulating cytokines have been shown to be relevant biomarkers for response to chemotherapy and prognosis. Therefore, the purpose of this study was to measure the serum levels of 13 cytokines in dogs with multicentric lymphoma and compare them to a population of healthy dogs. It was hypothesized there would be a significant difference in cytokine levels between dogs with lymphoma and healthy dogs, and between dogs with B cell and T cell lymphoma.

Thirty one dogs cytologically or histopathologically diagnosed with multicentric lymphoma were prospectively enrolled in the lymphoma group. CBC and biochemistry were performed in all dogs. Immunophenotype was determined by flow cytometry in all dogs, separating them into 2 groups: B cell ($n = 21$) and T cell ($n = 10$). Nineteen healthy dogs were enrolled in the control group. All dogs had similar exclusion criteria including previous administration of chemotherapy, systemic immunomodulatory medication (including corticosteroids and NSAID) within the last two weeks, and vaccination within the last four weeks.

T-test and Chi-square test were used to compare and match continuous variables (age, weight) and categorical variables (sex), respectively, between lymphoma and control groups. The cytokine concentrations were measured using a commercial canine multiplex magnetic bead-based (Luminex)® assay which measured IL-2, IL-6, IL-7, IL-8, IL-10, IL-15, IL-18, GM-CSF, TNF-α, IFN-γ, IP-10, KC-like, and MCP-1. The serum levels of each cytokine were compared amongst the three groups (B cell lymphoma, T cell lymphoma, healthy control) using ANOVA or Kruskal-Wallis tests, as appropriate. Statistical significance was set at $p < 0.05$.

There was no significant difference between the lymphoma and healthy control groups regarding sex, age and weight. IL-6, IL-10, KC-like, and MCP-1 were significantly higher in dogs with lymphoma compared to healthy dogs ($p = 0.01$, $p = 0.03$, $p = 0.01$, $p < 0.01$, respectively). IL-10, KC-like, and MCP-1 were significantly higher in the B cell lymphoma group than in the healthy group ($p = 0.01$, $p < 0.01$, $p = 0.01$, respectively). MCP-1, IL-6 and IL-15 levels were significantly higher in the T cell lymphoma group than in the healthy group ($p = 0.02$, $p < 0.01$, $p < 0.01$, respectively). Finally, IL-6 and IL-15 were significantly higher in the T cell lymphoma group than in the B cell lymphoma group ($p = 0.03$. $p = 0.02$, respectively). There were no significant differences in serum IL-2, IL-7, IL-8, IL-18, GM-CSF, TNF-α, IFN-γ, and IP-10 between the lymphoma and control groups.

In conclusion, this study found different serum cytokine profiles between dogs with lymphoma and healthy dogs, and also different serum cytokine profiles between dogs with B cell and T-cell lymphoma. Further study is necessary (and underway) to investigate the role of these cytokines in lymphoma pathogenesis, response to treatment, and prognosis

O04

POST-CHEMOTHERAPY PERFORATION IN CATS WITH A DIAGNOSIS OF INTERMEDIATE OR HIGH GRADE GASTROINTESTINAL LYMPHOMA

Zachary Crouse[1], Brenda Phillips[3], Andi Flory[3], Jennifer Mahoney[1], Keith Richter[3], Linda Kidd[2]

[1]Angell Animal Medical Center, Boston, Massachusetts, USA, [2]Western University of Health Sciences, Pomona, California, USA, [3]Veterinary Specialty Hospital of San Diego, San Diego, California, USA

Whether solitary gastrointestinal (GIT) mass lesions caused by intermediate or large cell lymphoma in cats should be treated initially with surgical excision or chemotherapy is debated. There is a theoretical risk of perforation due to rapid cell death associated with chemotherapy. The prevalence of, and clinical findings associated with perforation have not been described.

The objective of this study was to document the prevalence and timing of post-chemotherapy perforation in cats with GIT masses caused by intermediate or large cell lymphoma. We hypothesized that tumor size, presence of hypoproteinemia, or suppurative inflammation within the mass at the time of diagnosis would be associated with perforation.

Cats with a diagnosis of intermediate to large cell lymphoma based on cytologic examination of a mass lesion of the GIT, and treated with variable chemotherapy were identified by searching the patient record database of two referral hospitals. Gastrointestinal perforation was confirmed with histopathology or using ultrasound and analysis of abdominal fluid.

Twenty-three cats with intermediate (n=3) or large cell (n=20) lymphoma were identified. Gastrointestinal perforation was confirmed in 4 of 23 cats (17.4%). Perforation occurred at 23, 56, 59, and 87 days after induction. There was no association between tumor size, the presence of hypoproteinemia, or suppurative inflammation within the mass at the time of diagnosis, and subsequent perforation.

Perforation is relatively common in cats with GIT mass lesions caused by intermediate or large cell lymphoma. However, the findings of this exploratory study suggest that acute perforation after induction of chemotherapy is not common.

O05

A RETROSPECTIVE STUDY ON THE INCIDENCE OF PROTEINURIA ASSOCIATED WITH THE USE OF TOCERANIB PHOSPHATE

Sindy Piscoya, Cheryl Balkman, Kelly Hume

Cornell University, Ithaca, NY, USA

Tyrosine kinase inhibitors (TKIs) are a commonly used therapy in both human and veterinary medicine. The incidence of proteinuria in humans receiving VEGFR-TKIs is reported to be as high as 18.7%. In veterinary medicine, the TKI masitinib mesylate has been shown to induce a protein-losing nephropathy in a subset of patients. Documented reports of proteinuria with the use of toceranib phosphate are scarce. The purpose of this study was to describe the overall incidence of proteinuria associated with the use of toceranib phosphate. Medical records were retrospectively reviewed from September 2009 through October 2014 for canine patients with any type of malignancy receiving toceranib phosphate for a minimum of 3 weeks and a urinalysis and urine protein creatinine ratio (UPC) when indicated. Fifty-five cases met the inclusion criteria. The median dose of toceranib phosphate was 2.6 mg/kg (range, 2.1-2.8 mg/kg) given on Monday/Wednesday/Friday. Prednisone was used in conjunction with toceranib phosphate in 26 dogs with a median dose of 0.5 mg/kg (range 0.25-1.5 mg/kg) given on Tuesday/Thursday/Saturday/Sunday. In the toceranib phosphate plus prednisone group, proteinuria developed in 7 dogs at a median of 69 days (range, 15-122 days) with a median UPC of 0.8 (range, 0.6-1.1). In the toceranib phosphate alone group, 5 dogs developed proteinuria at a median of 47 days (range, 19-626 days) with a median UPC of 0.7 (range, 0.6-4.9). The use of toceranib phosphate was associated with development of proteinuria in a subset of patients and careful monitoring including frequent serial urine protein-creatinine ratios is recommended.

O06

GROWTH PATHWAYS IN FELINE ORAL SQUAMOUS CELL CARCINOMA

Krystal Harris, Howard Gelberg, Stuart Helfand

Oregon State University, Corvallis, OR, USA

The present study was designed to reach a better understanding of the biologic behavior of feline oral squamous cell carcinoma (FOSCC) by evaluating a variety of growth factor pathways known to contribute to aggressive behavior in many cancers. In addition, we evaluated a novel transcription factor, Bcl11b, a biomarker for head and neck squamous cell carcinoma (HNSCC).

FOSCC biopsies were graded and immunohistochemistry used to quantify relative expression of Bcl11b, EGFR, Ki67, and VEGF-D. Chi-square tests determined if Bcl11b expression correlated with expression of other receptors and growth factors. Western blots on FOSCC cell lines (SCCF1, SCCF2, SCCF3) were employed to verify expression of Bcl11b, EGFR, Src, ERK, and VEGF-D. Western blot on FOSCC cell lines determined phosphorylation of Bcl11b and the effect of a src inhibitor, dasatinib, on this activation. An MTT assay was applied on FOSCC cell lines following treatment with dasatinib. ELISA quantified VEGF production by FOSCC cells with and without dasatinib treatment.

Bcl11b staining intensity was moderate to strong in 79.4% (27/34) of FOSCC's. FOSCC biopsies with intense Bcl11b staining had the highest number of Bcl11b positive cells ($P<0.001$). All biopsies expressed VEGF-D and 85.7% (18/21) expressed EGFR at varying intensities. A correlation between increasing Bcl11b and EGFR expression approached significance ($P=0.05598$). All FOSCC cell lines expressed pBcl11b, pEGFR, pSrc, pERK, and VEGF-D. Dasatinib treatment lowered cell viability, inhibited activation of Bcl11b, and suppressed VEGF production. Ligand stimulation of EGFR resulted in increased activation of EGFR and ERK, increased VEGF/VEGF-D production, and partially rescued FOSCC cells from the effects of dasatinib.

This study represents the first description of Bcl11b expression in the cat and in FOSCC. We demonstrated activation of the Src pathway by FOSCC cells making Src a strategic therapeutic target that could be used in a clinical setting. Exposing FOSCC cells to dasatinib has inhibitory effects on cell proliferation, EGFR, and consistently decreased production of VEGF. Based on these results, dasatinib may have both anti-proliferative and anti-angiogenic effects in FOSCC. Since EGF is able to partially rescue cells from the effects of dasatinib, future studies should evaluate dual target inhibition of Src and EGFR, EGFR and VEGFR, as well as combining Src inhibition with standard chemotherapy agents.

O07

A RETROSPECTIVE ANALYSIS OF MULTIMODALITY TREATMENT FOR CANINE ORAL MELANOMA: 126 CASES

Michelle Turek[1], Tracy LaDue[2], Jayme Looper[3], Koichi Nagata[4], Keijiro Shiomitsu[5], Michele Keyerleber[6], Julia Buchholz[7], Tracy Gieger[5], David Vail[1]

[1]University of Wisconsin-Madison School of Veterinary Medicine, Madison, WI, USA, [2]Southeast Veterinary Oncology and Internal Medicine, Orange Park, FL, USA, [3]Chicago Veterinary Specialty Group, Chicago, IL, USA, [4]University of Georgia College of Veterinary Medicine, Athens, GA, USA, [5]Louisiana State University School of Veterinary Medicine, Baton Rouge, LA, USA, [6]Tufts University Cummings School of Veterinary Medicine, North Grafton, MA, USA, [7]Animal Oncology and Imaging Center, Hunenberg, Switzerland

Oral melanoma (OM) has a high propensity to metastasize to regional lymph nodes (LN) and lungs. Surgery and/or radiotherapy (RT) are effective local treatments, however most dogs succumb to distant metastasis. Immunotherapy represents an attractive strategy for this potentially immunogenic tumor. The objective of this multi-institutional retrospective study was to examine the clinical outcome of dogs with OM treated with ONCEPT™ melanoma vaccine, +/- surgery and/or RT.

Medical records of dogs with OM treated with ONCEPT™ were reviewed from seven veterinary radiotherapy facilities. Many dogs underwent concurrent surgery and/or RT (8 Gray X 4 weekly fractions). Dogs with distant metastasis and those receiving concurrent chemotherapy were excluded.

One hundred and twenty six dogs were included. All received ONCEPT™. Sixty had adequate local control (ALC; complete excision or irradiation of microscopic disease). Fifteen were treated in the microscopic setting. Fifty-one were treated in the gross disease setting, of which 39 underwent RT. Median time to progression, median progression-free survival and median overall survival were 304, 244 and 617 days, respectively. Dogs with ALC had improved clinical outcomes. The following also correlated with favorable clinical outcomes: rostral location, stage 1, absence of LN metastasis, low mitotic index, absence of bony lysis, absence of gross disease.

This is the largest reported series of dogs with OM treated with ONCEPT™. Clinical outcomes are similar to those reported recently for dogs treated with surgery and/or RT alone. Several prognostic indicators were confirmed. A prospective, randomized, controlled study is needed to determine the clinical benefit of ONCEPT™.

O08

ENHANCEMENT OF DOXORUBICIN EFFECTIVENESS WHEN COMBINED WITH SALINOMYCIN IN FISS CELL LINES

Lucia Borlle[1], Brittany Zumbo[2], Kelly R. Hume[1]

[1]*Cornell University, Ithaca, New York, USA,* [2]*Michigan State University, Michigan, USA*

Feline injection site sarcoma (FISS) is an aggressive neoplasia that remains a challenge for clinicians. Radical surgery is the first choice treatment but is not always feasible. Chemotherapeutic responses have been documented but are generally of short duration; tumors often develop resistance quickly. Salinomycin is an ionophore that has been shown to inhibit cancer stem cells and to increase the sensitivity of human soft tissue sarcoma cells to doxorubicin chemotherapy. The aim of our study was to determine if salinomycin could be used to increase the sensitivity of FISS cell lines to doxorubicin. Two primary cell lines cultured from feline patients with histologically confirmed FISS were used for this purpose. Since the threshold response to doxorubicin differs, we ranked the cells according to the sensitivity to doxorubicin and used one sensitive line (C10) and one resistant line (B4). Cell viability assays (MTT) were used to determine the response to single agent and combination therapy. Three independent assays were performed, with samples evaluated in triplicate in each assay. Student's t-tests were employed to compare the results of individual doxorubicin concentrations with and without salinomycin combination therapy. Addition of 5 uM salinomycin resulted in a statistically significant ($p < 0.05$) reduction in cell viability compared to single agent doxorubicin at multiple concentrations for both cell lines. Our results highlight a potential chemotherapy protocol for the treatment of FISS. Further investigations need to be pursued to define the effect of this novel therapeutic combination and to validate our results *in vivo*.

O09

CLINICAL ADVANCEMENT OF A RNA-TRANSFECTED CD40-B CELL VACCINE FOR THE TREATMENT OF CANINE NON-HODGKIN'S LYMPHOMA

Nicola Mason, Martha MaloneyHuss, Josephine Gnanandarajah

University of Pennsylvania, Philadelphia, PA, USA

Canine Non-Hodgkin's lymphoma (NHL) occurs in approximately 30-100/100,000 dogs per year and accounts for up to 83% of canine hematopoietic cancers. Standard–of-care treatment for canine NHL comprises multi-drug chemotherapeutic protocols. 60-85% of dogs enter a complete clinical remission following induction chemotherapy, and median first-remission durations range from 140 to 385 days. Resistance to chemotherapy increases with each subsequent relapse and less than 10% of patients survive longer than 2 years from diagnosis. To prolong overall survival, more effective therapies are required to prevent relapse following successful induction chemotherapy.

Directing the immune system to specifically target malignant cells is a powerful strategy in the treatment of cancer. We have previously developed a vaccine strategy that utilizes autologous CD40-activated B cells as antigen-presenting cells electroporated with autologous tumor RNA as the antigenic payload to stimulate anti-tumor immunity in dogs with NHL. In our first clinical trial, we showed that dogs vaccinated three times at three week intervals following successful induction chemotherapy had a tendency towards prolonged remission times and overall survival when compared to historical controls. Furthermore, vaccinated dogs that did relapse showed a statistically significant increase in overall survival following rescue chemotherapy. These results coupled with immune analysis from vaccinated patients suggested that while RNA loaded CD40-activated B cells primed an anti-tumor immune response, it was insufficient to maintain remission. Furthermore, we hypothesized that rescue chemotherapy boosted vaccine primed tumor-specific immunity and lead to prolonged second remission times. We therefore hypothesized that repeat administrations of RNA-loaded CD40-activated B cells given every 2 months following an initial priming course would maintain anti-tumor immunity and prevent relapse.

To investigate this hypothesis we are currently performing a second phase I/II clinical trial using RNA-loaded CD40-activated B cells. Fifteen dogs will be recruited to this study. Eligibility criteria include a recent diagnosis of NHL, no prior treatment with chemotherapy or steroids, and no evidence of circulating blasts. At the time of diagnosis, peripheral blood mononuclear cells and malignant lymph node tissue are harvested to generate CD40-activated B cells and whole tumor RNA respectively. Dogs then undergo a 19-week CHOP-based chemotherapy protocol. Only dogs in complete clinical remission following induction chemotherapy are eligible to receive the autologous vaccine. Dogs are vaccinated every three weeks for three doses, and then every two months until clinical relapse. Induction and maintenance of anti-tumor immunity is being determined using an IFN-γ ELISpot assay. The primary endpoint of this trial is time to progression, and the secondary endpoints include duration of anti-tumor immunity, overall survival and lymphoma-specific survival.

To date, ten dogs have been recruited and CD40-activated B cells have been successfully generated for nine dogs. Briefly, PBMCs are co-cultured with feeder cells consisting of irradiated K562 cells transfected with human CD40L (KtCD40L) in the presence of cyclosporine and IL-4, to generate CD40-activated B cells. CD40-B cells are then electroporated with RNA extracted from autologous malignant lymph node tissue or with canine distemper virus Hemagglutinin antigen (CDV-HA) mRNA as a control vaccine. At the time of writing, 2/10 dogs have completed their induction chemotherapy and have received their initial priming vaccinations. No adverse events have occurred following vaccine administration, and neither dog has yet relapsed. The trial is ongoing and the preliminary data confirm the feasibility of autologous CD40-B cell vaccine generation and safety of administration. ELISpot analysis at baseline and at each vaccine time point are being performed to elucidate whether repeat vaccinations effectively maintain anti-tumor immunity.

O10

VITAMIN D STATUS AND ACUTE PHASE PROTEIN CONCENTRATIONS IN CANINE CANCER PATIENTS

Nicole Weidner[1], J. Paul Woods[2], Anthony J. Mutsaers[2], Julie Bayle[3], Adronie Verbrugghe[2]

[1]Department of Biomedical Sciences, Ontario Veterinary College, University of Guelph, Guelph, ON, Canada, [2]Department of Clinical Studies, Ontario Veterinary College, University of Guelph, Guelph, ON, Canada, [3]Royal Canin Research Center, Aimargues, France

Circulating concentrations of acute phase proteins (APP) have been used as biomarkers in canine oncology. Increased concentrations have been observed in cancer patients and decreased concentrations are linked to patient remission. In humans, increased APP concentrations have been linked with low vitamin D status in several disease states. Low vitamin D status has also been observed in human and canine cancer patients. Therefore, there is the possibility that increased APP concentrations will be linked with decreased vitamin D status in canine cancer patients. The objective of this study was to examine the relationship between blood vitamin D and APP concentrations in healthy dogs and dogs with cancer.

Dogs with lymphoma (n=34), osteosarcoma (n=21) and mast cell tumors (n=26) diagnosed by cytology and/or histology presenting to one referral hospital for treatment, as well as client-owned healthy control dogs (n=25), were enrolled. Blood samples were collected from all dogs prior to any treatment and analyzed for plasma 25-hydroxyvitamin D (25(OH)D) using radioimmunoassays, and haptoglobin (Hp), alpha-1-acid glycoprotein (AAG) and serum amyloid A (SAA) using ELISA kits. Statistical analyses were completed with SAS 9.3. Vitamin D and APP concentrations were compared between groups using ANOVAs followed by Tukey-Kramer post-hoc analyses. Pearson correlation coefficients were performed to determine relationships between vitamin D and APP concentrations within each group.

Mean±SD plasma 25(OH)D concentrations were significantly lower in dogs with lymphoma (102±33 nmol/L) than in healthy dogs (129±41 nmol/L, $p=0.007$). Mean±SD plasma Hp and AAG concentrations were significantly higher in dogs with lymphoma (9.6±7.3 ng/ml and 92.4±71.0 ng/ml, respectively) than in healthy dogs (3.0±1.7 ng/ml, $p<0.001$ and 43.7 ± 30.8 ng/ml, $p<0.001$, respectively). There were no significant differences in blood SAA levels among groups. Pearson correlation coefficients revealed a moderate inverse relationship between plasma AAG and 25(OH)D concentrations ($R= -0.39$, $p = 0.024$) in lymphoma patients only.

The lower plasma 25(OH)D and higher Hp and AAG concentrations in canine lymphoma patients compared to healthy dogs is consistent with previous reports. However, this study is the first to report these observations within the same population. Further research is needed to investigate whether decreased plasma 25(OH)D concentration is a factor in cancer development, or a consequence of cancer, and to determine the clinical relevance of these findings.

O11

RETROSPECTIVE EVALUATION OF METRONOMIC CYCLOPHOSPHAMIDE IN EPITHELIAL AND MESENCHYMAL MALIGNANT TUMOURS

J Ellis, R D Foale

Dick White Referrals, Six Mile Bottom, Cambridgeshire, UK

Continuous low dose chemotherapy administration has been reported to have anti-angiogenic effects with efficacy against multiple tumour types. This study aimed to retrospectively assess the use of metronomic cyclophosphamide for histopathologically confirmed solid malignant canine tumours.

Patient records from 2009 to 2014 were reviewed at a multi-disciplinary referral hospital. Dogs without histological confirmation of diagnosis were excluded. Thirty-seven cases of epithelial or mesenchymal malignancy in dogs treated with 10mg cyclophosphamide daily/every other day were identified. This included 14 high grade soft tissue sarcomas, 5 soft tissue sarcomas of undefined grade, 4 intermediate grade soft tissue sarcomas, 3 splenic haemangiosarcomas, 3 incompletely excised low grade soft tissue sarcomas, 2 thyroid carcinomas, and 1 each of adenocarcinoma, bladder haemangiosarcoma, splenic fibrosarcoma, apocrine gland carcinoma, metastatic prostate carcinoma, and metastatic carcinoma of unknown origin. Two were lost to follow up. Twenty-two had at least one attempt at surgical excision and three underwent radiotherapy prior to initiation of chemotherapy.

Three of the dogs (8.6%) developed haematuria, but no other adverse effects were reported. The haematuria resolved in all three dogs after cessation of cyclophosphamide.

74% of dogs had stable disease or partial/complete remission >100days with cyclophosphamide, including two dogs that had failed previous chemotherapy protocols and one that had received radiotherapy. Four dogs did not have resection of the primary tumour, and three of these remained clinically well for >100 days. 20 of 35 dogs (57%) had DFI >6 months with 13 (37%) still alive over one year later.

Of the dogs with intermediate/high grade sarcomas (n=18) the median survival time (MST) was 362 days, with one dog lost to follow up and four still alive.

Of the dogs with carcinomas (n=6) one was lost to follow up and two are still alive >1 year later with no clinical signs. One dog had a reduced rate of progression and stable disease for >100 days with no clinical signs of disease. Two dogs with thyroid carcinoma were still alive three years later. This is in contrast to the reported MST for invasive thyroid carcinoma treated with cisplatin of 98 days.

This study adds to the evidence that metronomic cyclophosphamide may have some use in the treatment of malignant mesenchymal and epithelial tumours. However, it is important to note that even at low doses side effects may occur.

O12

EVALUATION OF TOCERANIB PHOSPHATE (PALLADIA®) IN THE TREATMENT OF FELINE MAST CELL NEOPLASIA: 53 CASES

Erika Berger[1], Chad Johannes[1], Gerald Post[2], Gillian Rothchild[2], Kai-Biu Shiu[3], Sarah Wetzel[3], Chelsea Tripp[4], Theresa Arteaga[5], Martin Crawford-Jukubiak[6], Anna Rita Serras[7], Sarah Gillings[8], Christine Swanson[9], Shawna Greene[10], Kelly O'Neill[11], Michelle Pressel[12]

[1]Iowa State University College of Veterinary Medicine, Ames, IA, USA, [2]The Veterinary Cancer Center, Norwalk, CT, USA, [3]Veterinary Specialty Center, Madison, WI, USA, [4]Veterinary Cancer Specialty Center, Seattle, WA, USA, [5]Animal Cancer Center, Monterey, CA, USA, [6]Sage Centers for Veterinary Specialty and Emergency Care, Concord, CA, USA, [7]Animal Health Trust, Newmarket, Suffolk, UK, [8]Summit Veterinary Referral Center, Tacoma, WA, USA, [9]Blue Pearl Specialty and Emergency Medicine for Pets, Grand Rapids, MI, USA, [10]Iowa Veterinary Specialties, Des Moines, IA, USA, [11]Saint Francis Veterinary Specialists, Decatur, GA, USA, [12]Pacific Veterinary Specialists, Capitola, CA, USA

The purpose of this study was to solicit and compile data from practicing veterinary specialists regarding their use of toceranib in cats with mast cell neoplasia and to provide initial assessment of possible clinical benefit and toxicity.

The American College of Veterinary Internal Medicine Oncology and Small Animal Internal Medicine listserves were used to solicit data pertaining to cases in which toceranib was used in the treatment of feline mast cell neoplasia. Cases were included if the following data were received: signalment (age, gender, breed), anatomic classification of disease (cutaneous or visceral), previous and concurrent treatment, toceranib dose (mg/kg) and schedule, duration of therapy, best response, and documentation of adverse events.

Case data from 53 cats with cutaneous (n=25) or visceral (n=28) mast cell neoplasia were received. Clinical benefit (CB) was seen in 81% (43/53), including 88% (22/25) with cutaneous (8CR, 12PR, 2SD) and 75% (21/28) with visceral (5CR, 13PR, 3SD) involvement. Most cats (n=49) did not receive concurrent radiation or chemotherapy, though a majority (n=38) received glucocorticoids during toceranib treatment. Mean duration of toceranib treatment in cats experiencing CB was 37 weeks (4-106) and 48 weeks (10-199) for cutaneous and visceral cases, respectively. Toceranib was administered at a median dose of 2.5 mg/kg in cats with CB; in 91% (39/43) the drug was given 3 times per week. Treatment was generally well tolerated with 57% (30/53) cats experiencing adverse effects at some point during toceranib treatment. The majority of these signs were low-grade (Grade 1 or 2) gastrointestinal or hematologic toxicities that resolved with treatment break and/or dose adjustment.

Toceranib appears to be well-tolerated in feline patients with mast cell neoplasia. Biologic activity of this drug is evident in the studied cats, however further prospective studies are needed to fully elucidate its role in treatment of this condition.

O13

EXPRESSION OF P GLYCOPROTEIN (ABCB1) IN CATS WITH T-CELL LYMPHOCYTIC GASTROINTESTINAL LYMPHOMA

Valter de Medeiros Winkel, Bruno Cogliati, Archivaldo Reche Jr, Ana Luiza Nairismagi Alves, Sílvia Regina Ricci Lucas

University of Sao Paulo, Sao Paulo, SP, Brazil

Gastrointestinal is the most common anatomical form of lymphoma in cats and systemic chemotherapy is indicated for the treatment. Although most of cats can achieve a complete remission with chemotherapy, relapse can occurs and may be associated to the multidrug resistance (MDR). Cell mechanisms of MDR include activation of transmembrane proteins, that reduce intracellular concentrations of different chemical compounds, and alterations in drug target. The main protein related to MDR phenotype is P-glycoprotein. Its overexpression results in reduced concentration of certain drugs within the cell, which is linked to the resistance. The aim of this study was to investigate P glycoprotein (P-gp) expression in cats with T-cell lymphocytic gastrointestinal lymphoma. Immunohistochemistry was performed in 40 samples using the monoclonal antibodies mouse anti-PAX-5 (PAX5-Invitrogen), polyclonal rabbit anti-CD3 (DAKO) and mouse anti-P-glycoprotein (clone C494-Enzo Life Sciences). P-gp immunoreactivity was analysed by degree of staining intensity, between 0 (no expression) to 3 (strong expression). Lymphoma was considered to be positive for P-gp when more than 10% of the neoplastic population expressed the protein. Three cats had a strong and 7 had a moderate expression, comprising 25% of the cases. The median of overall survival time was 24 months for these 10 cats; 60% had complete and 40% partial response to the treatment, and it was the same median observed for all cats studied. In conclusion, the expression of P-gp was observed in 25% of the cats with T-cell lymphocytic gastrointestinal lymphoma and this expression was not related to the treatment response or overall survival time.

O14

EARLY EXPERIENCES WITH STEREOTACTIC RADIATION THERAPY FOR THE TREATMENT OF CANINE NON-LYMPHOMATOUS NASAL TUMORS

Tracy Gieger, Michael Nolan

Department of Clinical Sciences (College of Veterinary Medicine), and Comparative Medicine Institute, Raleigh NC, USA

The purpose of this study was to describe initial experiences with a stereotactic radiation therapy (SRT) protocol (10 Gy x 3 daily fractions delivered with 6 MV photons via a linear accelerator) used as treatment for non-lymphomatous nasal tumors in dogs. A retrospective analysis of cases treated from August 2013 to October 2015 at NC State Veterinary Hospital was performed. Dogs were included if they had biopsy-confirmed non-lymphomatous nasal tumors. Dogs treated with chemotherapy were excluded. The gross tumor volume (GTV) was contoured using pre- and post-contrast 1-2 mm slice thickness CT image sets. A clinical target volume (CTV) expansion was used to include the ipsilateral nasal cavity and sinuses as well as 0.5-1 cm of the contralateral nasal cavity. A PTV expansion was not used (CTV = PTV). Thirty Gray (Gy) was prescribed to 99% of the GTV and > 95% of the PTV with a planned simultaneous boost to the center of the GTV. Cone beam CT, indexed bite-block system, and a treatment couch with 6 degrees of freedom were used for set up and delivery. Cases were reviewed and information related to patient demographics, tumor characteristics, and radiation plan dosimetry was collected.

Twenty-five cases were included. Stages included T1 (n=1), T2 (n=10), T3 (n=2), T4 (n=12). None had evidence of distant metastasis. Nineteen had regional lymph node aspirates and metastasis was present in one. Tumor types included adenocarcinomas (n=15), chondrosarcomas (n=5), transitional nasal carcinomas (n=3), intranasal squamous cell carcinoma (n=1), and neuroesthesioblastoma (n=1). Clinical signs improved in all cases. Twelve cases had recheck CT scans 3-4 months post-SRT and partial or complete tumor response was seen in all cases. Minimal acute toxicity was detected. The only late effect noted to date is oronasal fistula development in one dog 4 months post-SRT (this was anticipated due to tumor invasion into the palate at diagnosis). The median disease progression-free survival time has not been reached, and 52% of dogs are progression-free at 1 year. Nine dogs are deceased (median survival time 375 days) and 40% are alive 600 days post SRT.

Three fraction SRT has been used to treat non-lymphomatous nasal tumors in 25 dogs with limited adverse events. Continued accrual and follow-up will be necessary to confirm low toxicity, further elucidate effects of stage on outcome, and characterize clinical efficacy.

O15

TREATMENT OF CANINE APPENDICULAR OSTEOSARCOMA WITH AMPUTATION, CARBOPLATIN, AND TOCERANIB PHOSPHATE

Tracy Gieger[1], Dawn Clarke[2], Julie Nettifee-Osborne[1], Chad Johannes[4], Brianna Hallman[1], Michael Nolan[1], Laurel Williams[3]

[1]North Carolina State University, Raleigh NC, USA, [2]University of Georgia College of veterinary Medicine, Athens GA, USA, [3]Veterinary Specialty Hospital of the Carolinas, Cary NC, USA, [4]Iowa State University, Ames IA, USA

Osteosarcoma (OSA) is locally invasive and highly metastatic. We hypothesize that standard-of-care carboplatin chemotherapy followed by continued administration of toceranib phosphate (TP) will result in improved survival in dogs with OSA as compared to historical controls that did not receive TP. In addition, we hypothesize that TP will decrease circulating levels of pro-angiogenic MMP-9, a downstream factor and surrogate marker for VEGF receptor inhibition.

Ten dogs with histologically confirmed OSA with no evidence of thoracic metastasis at diagnosis were treated with amputation followed by 4 doses of carboplatin chemotherapy (300 mg/m2 IV q 3 weeks). Fourteen days after chemotherapy was completed, TP was initiated at a dose of 2.75 mg/kg PO q 48 hours. Serum samples for VEGF and MMP-9, bloodwork to evaluate toxicity, screening for metastasis with thoracic radiographs, and quality-of-life (QOL) scores monitored via owner surveys were performed monthly. The Kaplan-Meier method was used to calculate OSA-free survival and overall survival (OS). A Friedman's test was used to compare VEGF and MMP values at baseline, pre-TP, and end of study. A Kruskal-Wallis test of repeated measures was used to compare platelet-corrected VEGF values at baseline, pre-TP, and end of study. P values of <0.1 were considered significant.

Three dogs required dose reductions of carboplatin secondary to neutropenia, and two required drug holidays during TP treatment. No dogs were withdrawn from the study due to toxicity. Seven dogs developed local recurrence or metastasis as the first event. One dog was withdrawn from the study due to owner wishes, one was euthanized due to osteoarthritis, and one developed lymphoma with no evidence of OSA metastasis. There was no significant difference in MMP or VEGF (absolute or platelet-corrected) at any time point, and there was no correlation of these values with OSA-free survival or OS. Median OSA-free survival was 183 days (range, 55-576) and OS was 250 days (range, 80-882).

Although the combination of carboplatin followed by TP was well-tolerated in dogs with appendicular OSA, survival times did not exceed previously published data from dogs treated with amputation followed by chemotherapy. Serum VEGF and MMP-9 levels remained unchanged from baseline despite the continued administration of TP. The results of this study do not support ongoing TP administration post amputation and standard-of care chemotherapy in dogs with appendicular OSA.

O16

SAFETY ASSESSMENT OF A NOVEL ONCOLYTIC MARABA VIRUS IN CATS

J. Paul Woods[1], Byram Bridle[1], Dorothee Bienzle[1], Josepha Delay[2], Annette Morrison[3], Michelle Cieplak[3], Jeff Hummel[4], Brian Lichty[4]

[1]Ontario Veterinary College, University of Guelph, Guelph, ON, Canada, [2]Animal Health Laboratory, University of Guelph, Guelph, ON, Canada, [3]Campus Animal Facilities, University of Guelph, Guelph, ON, Canada, [4]McMaster Immunology Research Centre, McMaster University, Hamilton, ON, Canada

Oncolytic viruses are viruses whose replication is restricted to tumour cells leading to cytolysis. A growing number of viruses are being developed as oncolytic viruses and various strategies have been utilized to target tumour cells. Our group developed a biotherapeutic that combines two emerging cancer treatment modalities, tumour vaccination and oncolytic viruses. We recently determined that oncolytic viruses expressing tumour antigen transgenes are an excellent means to boost and enhance tumour vaccination. In this preliminary study, cats were first vaccinated with an adenovirus expressing a cancer gene, and then boosted with another virus (Maraba) also expressing this cancer gene. This vaccination protocol should elicit strong immunity to cancer cells, but should not cause any ill effects to the cats. The purpose of this study was to determine the safety (or toxicity) of a genetically altered oncolytic Maraba virus in cats.

Five young male (6-12 months old) purpose-bred cats were vaccinated intramuscularly with 9 x 10^9 PFU of a non-replicating human serotype 5 adenoviral vaccine vector (prime). Three weeks later the cats were administered 2.64 x 10^{11} PFU of MG1 Maraba virus intravenously (boost). Cats were monitored with physical exams observing for clinical signs, and by complete blood counts (CBC). Peripheral blood, saliva, urine, and feces were collected for virus isolation on days 1, 2, 3, 4 and 7 post- Maraba virus infusion.

There were no severe adverse events. Two cats had mild transient pyrexia (40 - 41.4 C) between 2 and 5 hours post Maraba virus infusion (typical of cytokine induction). The CBCs revealed a transient mild neutrophilia in 2 cats one day after Maraba virus infusion (14.92 and 8.73; reference interval 2.1-8.3 x 10^9 /L). Thrombocytopenia of 27 x 10^9/L (reference range 93-514 x 10^9/L) was noted in one cat 2 days post Maraba virus infusion, and in another cat thrombocytopenia of 9 x 10^9/L and 33 x 10^9/L on days 3 and 7, respectively, post Maraba virus infusion.

Viral genomes but not replication-competent viruses were detected in secondary lymphoid tissues (i.e. spleen) on day 39. Replication-competent virus was not detected in peripheral blood, saliva, urine, or feces.

Viral genomes in secondary lymphoid tissues without concurrent replication-competent virus is a finding similar to that in mice and primates, and reflects the propensity of Maraba virus to be taken up by antigen-presenting cells when delivered systemically, making it an ideal vector for boosting immune responses.

The lack of replication-competent virus in saliva, urine, feces or peripheral blood satisfied Canadian Food Inspection Agency (CFIA) concerns regarding potential shedding, and enabled initiation of a clinical trial utilizing Maraba virus to treat cats with mammary cancer. In this trial a heterologous prime:boost strategy targeting tumour antigens will be tested.

O17

ROLE OF MONOCYTE RECRUITMENT IN HEMANGIOSARCOMA METASTASIS IN DOGS (*VCS Award Winner*)

Regan, Daniel[1]; Escaffi, Andrea[1]; Coy, Jonathan[1]; Kurihara, Jade[1]; Dow, Steven[1]

[1] *Flint Animal Cancer Center, Dept. of Clinical Sciences, Colorado State University, Fort Collins, CO, USA*

INTRODUCTION

Canine hemangiosarcoma is a highly malignant tumor, which is associated with poor long-term survival due to the development of early and widespread metastatic disease. Currently, little is known regarding the biology of canine hemangiosarcoma, and the mechanisms accounting for the highly metastatic nature of the tumor are poorly understood. In humans and rodents, monocytes have been shown to play key roles in metastasis through promotion of tumor cell extravasation, seeding, growth, and angiogenesis, as well as suppression of anti-tumor immunity. However, there has been little investigation into the role of monocytes in canine tumor metastasis. Therefore, we investigated the potential role of monocyte infiltration in the regulation of tumor metastasis in dogs.

METHODS

To address this question, we initially performed immunohistochemistry for CD18 to determine the degree of monocyte infiltration in necropsy samples obtained from several common metastatic tumors of dogs, including hemangiosarcoma, osteosarcoma, and various carcinomas.

RESULTS

We found that compared to other tumor types, hemangiosarcoma metastases had significantly greater infiltration of CD18+ monocytes. Next, migration assays were used to compare the ability of tumor cell lines to stimulate monocyte migration in vitro. Hemangiosarcoma cell lines were among the strongest at stimulating monocyte migration, and were also found to be the highest producers of the monocyte chemoattractant CCL2. In addition, hemangiosarcoma metastases in vivo were found to produce large amounts of CCL2, compared to other tumor metastases.

CONCLUSION

These results are consistent therefore with the hypothesis that overexpression of CCL2 and recruitment of large numbers of monocytes may explain in part the aggressive metastatic nature of canine hemangiosarcoma. Moreover, these findings suggest that immunotherapeutic interventions designed to block monocyte recruitment or mobilization may be an effective adjuvant strategy for suppressing tumor metastasis in dogs with hemangiosarcoma.

O18

PHASE I CLINICAL TRIAL OF THE TARGETED CHEMOTHERAPEUTIC DRUG, FOLATE-TUBULYSIN, IN DOGS WITH URINARY TRACT TRANSITIONAL CELL CARCINOMA (*VCS Award Winner*)

Szigetvari, Nicholas[1]; Dhawan, Deepika[1]; Ramos-Vara, José A. [1]; Leamon, Christopher P.[2]; Klein, Patrick J. [2]; Heng, Hock Gan[1]; Vlahov, Iontcho R. [2]; Low, Philip S[2]; Fourez, Lindsey M. [1]; Knapp, Deborah W. [1]

[1] *Purdue University College of Veterinary Medicine, West Lafayette, IN, USA,* [2] *Purdue University College of Chemistry, West Lafayette, IN, USA*

Introduction: Targeted chemotherapy can include repurposing drugs previously considered clinically irrelevant due to marked toxicity and low therapeutic index. Tubulysin is one such drug with compelling antiproliferative activity *in vitro*, but poor tolerability *in vivo*. High affinity folate receptor (FR) expression has previously been found in canine transitional cell carcinoma (TCC) and serve as a druggable target. A phase I clinical trial of Folate-Tubulysin was conducted in dogs with urinary tract TCC.

Methods: With PACUC approval, eligible dogs were enrolled in a 3+3 dose escalating cohort study of Folate-Tubulysin (Endocyte, West Lafayette, IN). Inclusion criteria were histologic diagnosis of TCC, presence of FRs on the TCC cells detected by immunohistochemistry and/or folate uptake in the TCC detected by nuclear scintigraphy, and informed pet owner consent. The initial dose was 0.2 mg/kg given intravenously every two weeks, with intra- and inter-cohort escalation by 0.02 mg/kg after at least two consecutive treatments. Escalation continued until dose limiting toxicity (DLT) was observed and the maximum tolerated dose (MTD) determined.

Results: Eleven dogs were enrolled to obtain the maximum tolerated dose (MTD). A total of twenty seven dogs were enrolled to further investigate toxicity and tumor response. The Folate-Tubulysin MTD was 0.26 mg/kg administered every two weeks. The most common DLTs were neutropenia, anorexia, and/or lethargy. Resolution of side effects occurred with minimal supportive care and in some dogs treatment delay. Early categorical tumor responses have included 3 partial remissions, 16 stable disease, 8 progressive disease.

Conclusion: Folate-Tubulysin was generally well tolerated when given at or below the MTD. Conjugation of Tubulysin to folate allows use of this compound that otherwise would not be possible due to toxicity.

This work was supported by the Animal Cancer Foundation and Endocyte.

N01

KINEMATIC MAGNETIC RESONANCE IMAGING FOR EVALUATION OF DISC-ASSOCIATED CERVICAL SPONDYLOMYELOPATHY IN DOBERMAN PINSCHERS

Michele Provencher, Amy Habing, Laurie Cook, Sarah Moore, Gary Phillips, Ronaldo da Costa

The Ohio State University, Columbus, OH, USA

Disc-associated cervical spondylomyelopathy (DA-CSM) is a common condition in Doberman Pinschers (DPs) that is similar in pathogenesis to cervical spondylotic myelopathy in humans (hCSM). In patients with hCSM, kinematic MRI (kMRI) is used to evaluate the cervical vertebral column in flexion and extension. The purpose of this study was to evaluate kMRI in DPs with DA-CSM using a novel positioning device that allowed controlled flexion and extension of the cervical vertebral column. We hypothesized that kMRI would cause worsening of compressive lesions and the development of new compressions, especially in extension.

Nine client-owned DPs that had a neurologic examination consistent with a cervical myelopathy were prospectively evaluated. All dogs underwent standard MRI in a neutral position using a 3.0 T magnet. Following standard MRI, the patients were placed in right lateral recumbency on a positioning device and additional MRI sequences were obtained with the cervical vertebral column positioned in flexion and extension. Morphologic and morphometric assessments were performed using the neutral, traction, flexion and extension images. Morphologic analysis included a modified spinal cord compression score (previously published), direction of spinal cord compression, signal intensity changes within the spinal cord, and worst site of spinal cord compression. Morphometric assessment included spinal cord height, intervertebral disc width, spinal cord width, spinal cord area, and vertebral canal height. A Fischer's exact test was used to evaluate the morphologic data and mixed-effects linear regression was used to analyse the morphometric data. Interobserver and intraobserver analyses were performed.

Due to space constraints, only part of the results is presented. With neutral positioning, 14 total compressions were noted; 5/9 patients had 1 compression, 3 patients had 2 compressions, and 1 patient had 3 compressions. With flexion, 12 total compressions were noted; 1 patient had no compressions, 5 patients had 1 compression, 2 patients had 2 compressions, and 1 patient had 3 compressions. With extension, 26 total compressions were noted; 1 patient had 1 compression, 6 patients had 2 compressions, 3 patients had 3 compressions, and 1 patient had 4 compressions. In neutral, the majority of lesions were mild (11/14) and located ventrally (9/14). Extension was associated with worsening of compressions in 7 out of 14 sites and the presence of severe compressions (4/26). It was also associated with multiple directions of compression; there was ventral and dorsal compression at 12/26 sites with extension. Morphometric results revealed a decrease in spinal cord height at C6-C7 (midsagittal images) from neutral to extension (p=0.003) and from flexion to extension (p=0.001).

This study supports the use of kMRI to evaluate the dynamic component of DA-CSM. Results suggest that extension is most useful; not only did it demonstrate worsening of multiple compressions but it was also able to detect new compressions not visualized in a neutral position.

N02

3D SOLITARY SEQUENCE MAGNETIC RESONANCE IMAGING OF THE DOG SPINE IN HANSEN TYPE I INTERVERTEBRAL DISK DISEASE

Melissa Carpentier-Anderson[1], Fred Wininger[2], Shannon Holmes[3], Daniela Mauler[2]

[1]*Veterinary Specialty Servies, Manchester, Missouri, USA,* [2]*University of Missouri Veterinary Health Center, Columbia, Missouri, USA,* [3]*University of Georgia College of Veterinary Medicine, Athens, Georgia, USA*

SPACE (Sampling Perfection with Application optimized Contrasts using different flip angle Evolutions) is a T2-weighted three-dimensional magnetic resonance imaging sequence. Currently, 2 dimensional T2-weighted spin echo imaging is the gold standard diagnostic imaging for thoracolumbar intervertebral disk disease. SPACE is potentially advantageous to standard MRI in that it is a single acquisition, faster, technically easier modality with higher resolution and the potential to reconstruct the image set in any plane of section. The purpose of this study is to determine the agreement between SPACE and T2-weighted standard spin echo imaging sequences in the presence of Hansen Type I intervertebral disk disease in the thoracolumbar spine of dogs to determine if SPACE is a viable imaging sequence in the canine patient. Our hypothesis was that the MRI sequence SPACE is able to positively identify Hansen Type 1 intervertebral disk disease with equal accuracy to standard T2-weighted spin echo imaging.

This was a double blinded controlled study of dogs with an acute onset T3-L3 compressive myelopathy secondary to Hansen Type I intervertebral disk disease. All patients had standard pre-anesthetic screening and underwent magnetic resonance imaging of the T3-L3 spine using a Siemens 1.5T scanner. Images were obtained, in at least, the following sequences: standard T2-W sagittal and transverse, STIR dorsal, and SPACE. All dogs underwent a decompressive hemilaminectomy based on the imaging findings. The MRI sequences were randomized and three blinded individuals interpreted the MR images a total of three times. The evaluators were of different skill levels, including one ECVN diplomate, one ACVR diplomate, and a third year neurology resident. The standard T2-weighted and SPACE sequences were individually evaluated for site, side, and degree of compression as well as surgical recommendation and subjective confidence of imaging findings.

23 dogs met the inclusion criteria. There was statistically significant agreement for both site and side (P-Value <0.001) between the standard T2-weighted spin echo imaging and SPACE. There was complete agreement by all three researchers concerning both the site and side of disk herniation (comparing standard T2-weighted standard spin echo imaging and SPACE) for 20 of the 23 dogs, and partial agreement for the other 3 dogs. Compressive Length Ratio (CLR) estimates by the two methods were similar, with more variability due to evaluator variability than to differences between the two systems, and similar internal (intra-rater) variabilities between the two systems. The degree of compression, as measured by the standard mild/medium/severe rating protocol shows significantly positive association between the two methods, although not as strong as that between the two methods for the CLR measurements.

The MRI sequence SPACE was able to positively identify Hansen Type I Intervertebral disk disease in the canine thoracolumbar spine as confirmed via surgery with equal accuracy to standard T2-weighted imaging.

N03

CLINICAL CHARACTERISTICS OF DOGS WITH PROGRESSIVE MYELOMALACIA FOLLOWING ACUTE INTERVERTEBRAL DISC HERNIATION

Aude Castel, Natasha Olby

North Carolina State University, College of Veterinary School, Raleigh, NC, USA

Progressive myelomalacia (PMM) is a catastrophic disease associated with acute intervertebral disc herniation (IVDH). Published data on the clinical characteristics of this disease are limited. The aim of this retrospective study was to describe the onset and progression of clinical signs of PMM in a large case cohort.

Dogs with confirmed IVDH and either a histopathologic diagnosis of PMM or a high clinical suspicion (acute paraplegia, loss of nociception, loss of patellar reflex and cranial advancement of the cutaneous trunci muscle reflex inconsistent with location of disc herniation) were identified by medical record search.

Forty-eight cases were identified, 16 confirmed via histopathology, and 32 suspected based on clinical signs and imaging. The majority of dogs were chondrodystrophic, including 24 Dachshunds. Onset of PMM signs ranged from present at first evaluation (n=16) to 5 days after presentation, with 43/48 developing signs within 3 days. Progression of signs from onset to euthanasia, excluding 10 cases euthanized at presentation, ranged from 1 to 13 days with 32 being euthanized within 72 hours. Systemic signs (depression, anorexia, abnormal temperature and diffuse pain) were documented in 34 dogs. Focal disc herniation was present in 35 dogs, while 7 dogs had extensive herniations.

In summary, the majority of dogs that develop PMM do so within 3 days of presentation and progress to euthanasia within another 3 days. However, onset can be delayed up to 5 days after presentation with progression to death taking as long as 2 weeks. Non specific systemic signs are commonly reported.

N04

IN VITRO ANTI-TUBULIN EFFECTS OF BENZIMIDAZOLE ANTHELMINTICS MEBENDAZOLE AND FENBENDAZOLE ON CANINE GLIOBLASTOMA CELLS

Serene Lai[1], Jennifer Koehler[2]

[1]*Auburn University College of Veterinary Medicine, Dept of Clinical Sciences, Auburn, AL, USA,*
[2]*Auburn University College of Veterinary Medicine, Dept of Pathobiology, Auburn, AL, USA*

Benzimidazole anthelmintic drugs have been reported to have antiproliferative effects on several cancers both *in vitro* and *in vivo*, with reduced off-target toxicity as compared to other microtubule-disrupting drugs. In addition, these small, lipophilic drugs readily cross the blood-brain barrier. The purpose of this study was to evaluate the *in vitro* chemosensitivity of canine glioblastoma J3T cells to mebendazole (MBZ) and fenbendazole (FBZ).

Cells were exposed to drugs for 72 hours, and cell viability was evaluated using the MTT assay. The half-maximal inhibitory concentration (IC_{50}) of MBZ and FBZ was calculated using a four-parameter variable slope curve fit nonlinear regression analysis. The IC_{50} was compared between the two drugs. Western blot was used to compare the ratio of polymerized to depolymerized tubulin between drug-treated cells and untreated controls. Immunocytochemistry was performed on treated and untreated cells to visualize effects of treatment on tubulin.

The mean IC_{50} of MBZ and FBZ were 0.026 μM +/- 0.03 and 0.55 μM +/- 0.02 respectively. Treatment with MBZ and FBZ resulted in increased depolymerization of tubulin compared to the untreated control. Immunocytochemical studies showed disruption of tubulin and evidence of apoptotic bodies after only 2 hours of treatment.

Our *in vitro* data suggest that MBZ and FBZ may be good candidates for treatment of canine glioblastomas. Further *in vivo* studies are required.

N05

SUBARACHNOID-SUBARACHNOID SHUNTING FOR TREATMENT OF SUBARACHNOID CEREBROSPINAL FLUID FLOW OBSTRUCTION IN 9 DOGS

Ilyssa Meren, Jessica Chavera, Nick Jeffery

Iowa State University, Ames, IA, USA

Cerebrospinal fluid (CSF) flow in the subarachnoid space can be impeded by developmental anomalies of the subarachnoid space (i.e. diverticula) or acquired adhesions in older animals. Developmental lesions can occur in any type of dog and clinical signs typically present in animals less than one year old; acquired lesions appear to be most common in middle-aged and older small dogs, notably pugs. Clinical signs include progressive proprioceptive ataxia and, at later stages, fecal and urinary incontinence. Traditional surgical treatment of acquired subarachnoid adhesions (constrictive myelopathy) is associated with unreliable outcomes. Here we report on the surgical procedure and outcome using subarachnoid-subarachnoid shunting as a means of bridging the area of adhesion.

Nine dogs were treated by this technique at Iowa State University between November 2013 and August 2015. The spinal cord was exposed via routine dorsal laminectomy in 8 cases (thoracolumbar spinal cord) and via C2 laminotomy in one. A durotomy was made over the lesion site using a hypodermic needle bevel and expanded laterally to disrupt adhesions between the pia mater and arachnoid mater. A shunt tube consisting of the ventricular catheter from a ventriculoperitoneal shunt kit or a red rubber catheter of appropriate diameter was trimmed to span the lesion plus ~5mm on either side of the adhesion and secured within the subarachnoid space using 5/0 non-absorbable (polypropylene [Prolene]) sutures. The dura was then closed over the shunt.

One patient had a developmental subarachnoid diverticulum (the patient with a C2 laminotomy), and the remaining 8 had an acquired adhesion (constrictive myelopathy) documented at surgery. Four patients showed improvement of clinical signs, of which three showed complete recovery. The one patient that has not shown a complete recovery in this group was the young patient with a developmental subarachnoid diverticulum, but is continuing to improve at 3 months post-operatively. One patient had mixed success: improvement of fecal control, but deterioration in gait (although with subsequent gradual improvement). Three patients did not show major improvement after surgery. Of these, one showed mild improvement after surgery but acute progression of paraparesis at 10 months prompted euthanasia (cause of acute paraparesis was not ascertained because of lack of post mortem examination). The other two cases showed unchanged clinical signs at 7 and 24 months post operatively. The last remaining case improved immediately post-operatively, but was lost to long-term follow-up.

Obstruction of CSF flow in the subarachnoid space can be associated with severe and life-threatening (fecal incontinence is a common reason for euthanasia) neurologic deficits. Our results suggest that subarachnoid-subarachnoid shunting can be an effective surgical technique to ameliorate or prevent progression of associated clinical signs.

N06

11-DEHYDRO THROMBOXANE B2 AS A BIOMARKER FOR INTRACRANIAL NEOPLASIA IN DOGS.

Jessica A. Rivera[1], Elizabeth Hurst[2], James L. Mobley[2], Peter J. Dickinson[1]

[1]*University of California, Davis, Davis, California, USA,* [2]*Cayman Chemical Company, Ann Arbor, MI, USA*

Meningiomas and gliomas are the most common intracranial neoplasms in dogs, and definitive diagnosis is often problematic due to the cost and invasiveness of an intracranial biopsy. Biomarkers in serum, urine, and/or CSF can provide adjunctive diagnostic data in conjunction with standard techniques such as magnetic resonance imaging. The cycloogygenase pathway has been identified as a key player in cancer progression, and several pathway products including thromboxane A_2 (TxA_2) are over expressed in a variety of human cancers, including gliomas.

The purpose of this pilot study was to determine levels of 11-dehydro-thromboxane B_2 ($11TxB_2$) (a stable hydrolysis product of TxA_2) in dogs with a variety of neurological and non-neurological diseases, using an enzyme immunoassay. Urine, serum, and CSF samples were stored immediately following collection at -20°C (or at 4°C for up to 7 days prior to -20°C storage).

$11TxB_2$ was reliably detected in urine, but not in serum or CSF. Storage at 4°C up to 7 days did not significantly affect urine TxB_2 levels, and concurrent corticosteroid or non-steroidal anti-inflammatory medication did not apparently affect levels within the study group as a whole. Overlap of $11TxB_2$ urine levels was seen between clinically normal and neurological animal groups. Although some of the highest $11TxB_2$ values were seen in animals with intracranial neoplasia, overlap with other neurological disease groups was present. Assay of $11TxB_2$ is feasible; however, additional analysis of a large study population will be necessary to determine whether a threshold urine $11TxB_2$ level may be useful as a diagnostic biomarker.

N07

LUMBAR FRACTALKINE AND M2 MICROGLIA INCREASE THROUGHOUT DISEASE PROGRESSSION IN CANINE DEGENERATIVE MYELOPATHY

Christine Sibigtroth[1], Maria Jones[2], Virginia Garcia[2], Gayle Johnson[3], Joan Coates[1], Eric Villalón[2], Michael Garcia[2]

[1]*Department of Veterinary Medicine and Surgery, College of Veterinary Medicine, University of Missouri, Columbia, MO, USA,* [2]*Division of Biological Sciences, College of Arts and Sciences, University of Missouri, Columbia, MO, USA,* [3]*Department of Veterinary Pathobiology, College of Veterinary Medicine, Univeristy of Missouri, Columbia, MO, USA*

Canine degenerative myelopathy (DM) is an adult-onset neurodegenerative disorder. Clinical progression is homogeneous within and across breeds, thus four distinct disease stages have been identified (Kanazono et al., 2013; Coates and Wininger, 2010). DM shares similarities with some forms of amyotrophic lateral sclerosis (ALS), including underlying mutations in *superoxide dismutase-1* (*SOD1*). SOD1-mutant microglia play a central role in ALS progression in rodent models. Microglia behavior is complex, as expression patterns of classic M1 and M2 molecules are variable throughout disease and between species studied. The mechanism(s) underlying microglial phenotype determination within ALS is not known. Fractalkine, a neuronally produced chemokine, has been shown to suppress *SOD1* mutant microglial-mediated neurotoxicity in vitro. Thus, fractalkine is a possible contributor to microglial phenotype determination in ALS. Increased spinal cord microglia have been documented in end-stage DM-affected dogs, however, their response and phenotype throughout disease progression is unknown. **The goals of this study were to 1) quantify and phenotype microglial cells in close proximity to lumbar spinal cord motor neurons of DM-affected dogs at each disease stage and compare these findings to age-matched, neurologically normal dogs and 2) correlate microglial phenotype with fractalkine protein levels in lumbar spinal cord throughout disease progression.**

Spinal cord tissue was collected post-mortem from client-owned dogs donated to the University of Missouri for research purposes. Lumbar spinal cord microglia analyses were conducted with immunofluorescence and confocal microscopy. To identify microglia in close proximity to motor neurons (MNs), concentric circles with increasing radii of 6μM were applied to MNs of DM-affected dogs at each disease stage (stage 1 n=120; stage 2 n=118; late DM (3&4) n=149) and age-matched control dogs (n=145). Cells with positive immunoreactivity for Iba-1/iNOS (M1 microglia), or Iba-1/arginase-1 (M2 microglia), were quantified in a double-blind manner. Data were analyzed via Kruskal-Wallis ANOVA on ranks with *post-hoc* Dunn's method. Fractalkine total protein was quantified via western blot analysis of lumbar spinal cord from DM-affected dogs (stage 1 n=2; stage 2 n=4; late DM n=3) and age-matched control dogs (n=3). Relative optical densities were normalized to control dogs. Data were analyzed via ANOVA with *post-hoc* Holm-Sidak method.

A progressive increase in total microglia closely associated with lumbar motor neurons was observed in DM-affected dogs (stage1: 19±0.7; stage 2: 25±0.9; late DM: 28±0.9; $p<0.05$) compared to age-matched controls (13±0.4; $p<0.05$). Total M2 microglia were increased in stage 2 (7.5±0.5) and late DM-affected dogs (6.2±0.5) compared to age-matched controls (3.3±0.7; $p<0.05$). No differences were observed in total M1 microglia throughout disease progression. Lumbar spinal cord fractalkine protein was increased in stage 1 (2.2±0.2) and late DM-affected dogs (2.2±0.3) compared to age-matched controls (1±0.1; $p<0.05$).

These data suggest that M2, but not M1, microglia are progressively recruited to motor neurons during disease progression. Increased lumbar spinal cord fractalkine correlates with progressive M2 microglia in DM-affected dogs.

N08

INCREASED PHOSPHORYLATED NEUROFILAMENT HEAVY IN CEREBROSPINAL FLUID AS A DISEASE MARKER OF CANINE DEGENERATIVE MYELOPATHY

Christine Sibigtroth[1], Virginia Garcia[2], Gerry Shaw[3], Joan Coates[1], Patrick Pithua[1], Michael Garcia[2]

[1]Department of Veterinary Medicine and Surgery, College of Veterinary Medicine, University of Missouri, Columbia, MO, USA, [2]Division of Biological Sciences, College of Arts and Sciences, University of Missouri, Columbia, MO, USA, [3]EnCor Biotechnology, Gainesville, FL, USA

Canine degenerative myelopathy (DM) is an adult-onset neurodegenerative disorder with four defined disease stages. DM initially manifests as spastic upper motor neuron paraparesis and general proprioceptive ataxia (stage 1). Progressive neurodegeneration results in non-ambulatory paraparesis/paraplegia (stage 2) and thoracic limb paresis (stage 3). End-stage disease culminates in flaccid tetraplegia, widespread muscle atrophy and signs of brainstem dysfunction (stage 4). The clinical spectrum of DM is homogeneous within and across breeds. Phosphorylated neurofilament heavy (pNF-H), a major structural protein of myelinated motor axons, has shown promise as a prognostic biomarker in other diseases of the nervous system such as amyotrophic lateral sclerosis (ALS). **The goals of this study were to 1) quantify the concentration of pNF-H in the CSF of DM-affected dogs throughout disease progression and compare these finding to age-matched, neurologically normal dogs with and without *SOD1* mutations and 2) compare CSF concentrations of pNF-H between DM-affected and dogs with other chronic spinal cord disease that mimic DM.**

Using an enzyme-linked immunosorbent assay (ELISA), we measured pNF-H concentration in CSF from aged control dogs (n=10) and DM-affected dogs at all stages (stage 1 n=9; stage 2 n=6; stage 3 n=6; stage 4 n=3). Data were analyzed using Kruskal-Wallis ANOVA on ranks with *post-hoc* Dunn's method. Additionally, CSF pNF-H concentration was compared between DM-affected dogs (n=24) and dogs with other chronic spinal cord disease (n=7). Those data were analyzed by the student's t test. To evaluate diagnostic performance, a receiver operating characteristic curve was generated.

CSF concentrations of pNF-H were increased in DM-affected dogs at all disease stages compared to aged control dogs ($p = 0.002$); pNF-H concentrations did not differ between DM disease stages ($p = 0.704$). DM-affected dogs had increased CSF pNF-H compared to dogs with other chronic spinal disease ($p < 0.001$). Using a cut-off concentration of 1.5ng/ml, sensitivity and specificity for diagnosis of DM were 0.83 and 0.95, respectively. Area under the receiver operator characteristic curve was 0.911.

These preliminary findings suggest that pNF-H in CSF is a sensitive and specific disease marker for DM. These data warrant further study of pNF-H in CSF and serum for diagnosis of disease and longitudinal monitoring of therapeutic efficacy in DM.

N09

ETIOLOGY OF FELINE JUVENILE ONSET SEIZURES

Muna Qahwash, Heidi Barnes Heller

University Of Wisconsin, School of Veterinary Medicine, Madison, WI, USA

Seizure etiology are well documented in adult cats. Despite this awareness, little is published about the etiology and outcome of juvenile onset seizures in cats. Juvenile onset seizures are defined as seizures starting prior to 12 months of age for this study. The aims of this study were to 1) identify the common etiology and 2) describe the outcome (dead or alive) in cats with juvenile onset seizures. Inclusion criteria were 1) a confirmed or suspected seizure before 12 months of age and 2) complete medical records including a final diagnosis. The diagnosis was reclassified for each cat according to the 2010 ILAE guidelines. Fifteen cats met the inclusion criteria. Median age at onset was 24 weeks (range, 0.4 – 40 weeks). Six cats (40%) were diagnosed with structural epilepsy, 5 cats (33%) were diagnosed with unknown epilepsy, and 4 cats (26%) were diagnosed with reactive seizures. Generalized seizures were documented in 9 cats, 4 cats had focal seizures and 2 cats had both focal and generalized seizures. Six cats (40%) had cluster seizures, and status epilepticus was documented for 1 cat. Overall 8 cats (53%) were alive, 6 cats were dead and 1 cat was lost to follow up. (Follow up 16 – 173 months after diagnosis) Three cats died, or were euthanized, secondary to seizures or the underlying diagnosis and 3 cats died or were euthanized unrelated to their seizures or seizure diagnosis and 1 cat died due to unknown causes.

N10

COMPARISON OF SERUM TRACE NUTRIENT CONCENTRATION IN DOGS WITH PRIMARY GENETIC EPILEPSY VERSUS HEALTHY DOGS

Samantha Vitale, Devon Hague, Laura Selmic

University of Illinois Veterinary Clinical Medicine, Urbana, Illinois, USA

Primary genetic epilepsy is the most common cause of seizures in dogs, reportedly affecting 0.5-5.7% of the population. There have been several investigations regarding the serum concentrations of trace nutrients, including copper, selenium, zinc, manganese, and iron in human epileptics and animal models. However, no research of this nature is available in dogs with primary genetic epilepsy. This is a prospective pilot study, designed to compare the serum concentrations of several trace nutrients in non-epileptic dogs compared to dogs with epilepsy.

Blood samples were collected from 7 healthy control dogs and 8 dogs with primary genetic epilepsy. Serum was evaluated for concentrations of copper, selenium, zinc, cobalt, manganese, molybdenum, and iron using inductively coupled plasma mass spectroscopy. Mann U Whitney tests were used to compare normal and epileptic dog serum levels of the trace nutrient. A significantly higher level of copper ($p=0.006$), selenium ($p=0.018$), and manganese ($p=0.017$) were found in epileptic dogs compared to normal dogs. No significant differences were found in levels of zinc ($p=0.64$), cobalt ($p=0.82$), molybdenum ($p=0.60$), or iron ($p=0.95$). These results suggest that there are differences in the concentrations of some trace nutrients in epileptic dogs compared to normal dogs, which warrants further investigation.

N11

SERUM LACTATE AS A PROGNOSTIC FACTOR IN DOGS AND CATS WITH HEAD TRAUMA: 93 CASES (2003-2014)

Dana Sengewald, Devon Hague, Jason Tarricone, Maureen McMichael, Laura Selmic

University of Illinois Veterinary Clinical Medicine, Urbana, Illinois, USA

Previous studies identified associations of Modified Glasgow Coma Scale (MGCS) and hyperglycemia with survival in veterinary patients with head trauma. Lactate is the product of anaerobic glycolysis, and hyperlactatemia has classically been associated with decreased tissue perfusion. Some studies in dogs with various diseases, including trauma, have showed that higher initial lactate concentration in dogs is associated with increased risk of mortality. The main aim of this study is to determine if elevated serum lactate following head trauma in dogs and cats is associated with mortality and whether the degree of hyperlactatemia corresponds with clinical outcome.

Medical records for dogs and cats diagnosed with head trauma from the VetCOT Trauma Registry and searching the electronic medical record at the veterinary teaching hospital from 2003-2014 were examined. Cases were included if a diagnosis of head trauma was made and if blood lactate concentration was measured on the day of presentation. For each animal with head trauma included in the study, information about signalment, vital parameters, diagnostic testing, treatment and outcome was recorded. Based on information from the medical records, a MGCS was assigned to patients. Univariable logistic regression was used to evaluate association with survival between signalment, examination parameters, laboratory testing, and administration of treatment.

A total of 93 animals (78 dogs and 15 cats) met the inclusion criteria for this study. There was no evidence that hyperlactatemia (P=0.38) or hyperglycemia (P=0.69) were associated with mortality. The MGCS Score at initial evaluation was the only variable significantly associated with outcome (P=0.011). Dogs and cats with head trauma may have elevated serum lactate levels, but the degree of hyperlactatemia is not associated with severity of head trauma or with outcome.

N12

COMPARISON OF FECAL MICROBIOMES BETWEEN DOGS WITH MENINGOENCEPHALOMYELITIS OF UNKNOWN ORIGIN AND CONTROLS

Nick Jeffery[1], Jan Suchodolski[2], Al Jergens[1], Jon Levine[2]

[1]Iowa State University, Iowa, IA, USA, [2]Texas A&M University, College Station, TX, USA

Recent experiments in laboratory rodents have suggested that the constituents of the fecal microbiome influence the activity of the immune system. Specifically, it has been shown that gnotobiotic rats are resistant to development of experimental autoimmune encephalitis (EAE), an immune-mediated disease of the CNS that mimics aspects of multiple sclerosis in humans and meningoencephalomyelitis of unknown origin (MUO) in dogs. Further experiments have shown that restoration of a normal microbiome abrogates this resistance to EAE.

In this study we investigated whether a similar susceptibility to MUO might exist in dogs that have specific microbiome constituents. Previous published work in other species suggests that *Faecalibacterium* spp, specifically *F. prausnitzii,* may modulate the immune system to reduce risk of inflammatory and immune-mediated diseases. Therefore, in this study we tested the hypothesis that a relative deficiency of *Faecalibacterium* spp, especially *F. prausnitzii,* would be associated with development of MUO.

We obtained fecal samples from dogs diagnosed with MUO and from controls, many of which were matched for breed, age and gender. Dogs that had been treated with antibiotics within the previous 4 weeks were excluded. The fecal microbiome was characterized using 16S rRNA sequencing by Illumina and qPCR assays targeting *Faecalibacterium* spp, and abundances were compared between affected dogs and controls.

In analysis of the entire sampled population there was a significant reduction (of log DNA/gram feces) in *Faecalibacterium* spp in affected dogs compared to controls (median 6.5 versus 3.8; Mann-Whitney test p=0.012). However when affected dogs were compared only with matched controls, the median population of *Faecalibacterium* spp was smaller in the affected dogs (3.9 versus 6.4) but the Wilcoxon signed rank p = 0.074. Moreover, in 6/19 pairs the affected dogs had a smaller population of *Faecalibacterium* spp than the unaffected dogs. Similarly, although the population proportion of *F. prausnitzii* was smaller in affected dogs than controls (0.01% versus 0.1%), comparison using the Mann-Whitney test revealed p = 0.089. Examination of paired data again revealed lower median proportion in the affected dogs (0.03% versus 0.1%) but a Wilcoxon paired rank p = 0.15.

This study suggests that there is some, albeit limited, evidence to suggest that abundance of *Faecalibacterium* spp may influence the susceptibility of individual dogs to developing MUO. One possible explanation for our results is that the well-recognized breed susceptibility to MUO may be mediated, at least partially, through effects on gut microbiome. Here we examined just one highly specific hypothesis regarding gut microbiota in this disease; further analysis of the extensive fecal microbiome results available from this study may generate hypotheses that can be tested in future studies.

N13

CHARACTERIZATION OF IATROGENIC BLOOD CONTAMINATION ON LACTATE DEHYDROGENASE AND CREATINE KINASE IN CANINE CEREBROSPINAL FLUID

Jourdan Brune, Annie Chen-Allen

Washington State University, Pullman, WA, USA

Cerebrospinal fluid (CSF) analysis is regularly performed in patients with neurological disease but iatrogenic blood contamination can distort results. The effects of iatrogenic blood contamination on canine CSF lactate dehydrogenase (LD) and creatine kinase (CK) total and isoenzyme activity have not been previously evaluated. CK exists as 3 isoenzymes (CK-MM, CK-MB, CK-BB) and LD exists as 5 isoenzymes (LD1-5) in both serum and CSF. Investigating the effects of blood contamination on total LD and CK and their isoenzyme activity could allow clinicians to better interpret the diagnostic and prognostic value of LD and CK isoenzymes in canine neurologic patients.

CSF samples were contaminated with incremental amount of whole blood via serial dilution and LD and CK total and isoenzyme activity values were compared to an uncontaminated CSF sample and the established reference interval. Total LD and CK values were determined spectrophotometrically using a Sieman Dimension Xpand and isoenzyme activity was determined using gel electrophoresis and densitometric scanning. In contaminated samples, total LD and CK values were increased compared to the uncontaminated sample after contamination with > 2000 and > 4000 RBC/ul, respectively. Additionally, CSF isoenzyme activity began to emulate serum isoenzyme activity, where CK-BB and LD-5 banding became more prominent, when CSF was contaminated with > 4000 RBC/ul.

These findings suggest there is a threshold where blood contamination can artificially distort CSF LD and CK total and isoenzyme analysis. Therefore, these values will need to be interpreted with caution when there is iatrogenic blood contamination in CSF.

N14

OUTCOME COMPARISON IN DOGS WITH THORACOLUMBAR ACUTE NON-COMPRESSIVE NUCLEUS PULPOSUS EXTRUSION OR FIBROCARTILAGINOUS EMBOLIC MYELOPATHY

Lorenzo Mari[1], Anita Shea[1], Philippa Johnson[2], Luisa De Risio[1]

[1]*Neurology and neurosurgery service, Centre for Small Animal Studies, Animal Health Trust,, Newmarket, UK,* [2]*Section of diagnostic imaging, Cornell College of Veterinary Medicine, Ithaca, USA*

The objective of the study was to compare the clinical outcome between dogs with acute non-compressive nucleus pulposus extrusion (ANNPE) or presumptive fibrocartilaginous embolic myelopathy (FCEM) affecting the thoracolumbar spinal cord segments.

The medical records of dogs presented to the Animal Health Trust between May 2006 and April 2015 and diagnosed with ANNPE or presumptive FCEM were retrospectively reviewed. A randomized list was generated of the MRI studies of 87 cases matching the inclusion criteria and submitted to two board-certified neurologists (LDR and AS) and one board-certified radiologist (PJ), blinded to any clinical data. Each observer was requested to make a diagnosis of ANNPE or FCEM based on previously published diagnostic criteria. Inter-observer agreement was evaluated with Kappa statistic. The clinical outcome of the consensus diagnoses was compared. Recovery of motor and autonomic function was evaluated via medical records and a telephone questionnaire with the owners. Clinical re-examinations were performed when possible.

Based on the review of the MRI studies, 72 cases were diagnosed with ANNPE and 15 with presumptive FCEM. The level of agreement between LDR-AS (95.3%; $\kappa = 0.853$) or AS-PJ (95.3%; $\kappa = 0.819$) was almost perfect while that between LDR-PJ (90.6%; $\kappa = 0.68$) was substantial.

Impaired urinary and/or fecal continence was reported only in the ANNPE group in 20/71 dogs (28%, P=0.033); all these dogs had concurrent persistent motor deficits. Four dogs with impaired urinary and/or fecal continence were ambulatory at presentation. Inability to control the urge to defecate was the most frequently reported deficit in faecal continence.

All but one dog became or remained ambulatory without assistance. One dog diagnosed with presumptive FCEM presented paraplegic without nociception, did not recover an ambulatory status after 310 days and was euthanized. Only 20% dogs with ANNPE and 14% dogs with presumptive FCEM recovered a completely normal gait according to their owners, including 29% of the grade 2/5 dogs, 12.5% of the grade 3/5 dogs and none of the dogs that were plegic at presentation. The percentage of dogs with persistent motor deficits was not statistically different between the two groups. The time to recovery of some motor function in the most affected limb when presented with mono- or paraplegia was significantly longer in dogs with presumptive FCEM compared to dogs with ANNPE (P=0.025); however only two cases with this variable were present in the FCEM group. Dogs diagnosed with presumptive FCEM were significantly younger (P=0.025) and lighter (P=0.037) than dogs with ANNPE and had a longer intramedullary hyperintense lesion in sagittal T2-weighted MRI images (P=0.001). Dogs with ANNPE were more frequently reported to be performing physical activity (P=0.001) and to vocalise (P<0.0001) at the time of onset of the clinical signs and more frequently received anti-inflammatory drugs after diagnosis (P=0.012) compared to dogs with presumptive FCEM.

In conclusion, dogs diagnosed with ANNPE affecting the thoracolumbar spinal cord segment developed impaired urinary/fecal continence significantly more frequently than dogs with presumptive FCEM. Impaired urinary/fecal continence can occur also in dogs that are ambulatory at presentation. The presence of persistent motor deficits is extremely common, particularly in dogs that presented non-ambulatory; however, no statistically significant difference was detected in the degree of recovery of motor function between the two groups.

N15

MORPHOLOGY OF THE CAUDAL FOSSA IN MESATICEPHALIC AND BRACHYCEPHALIC CATS AND ASSOCIATED CLINICAL SIGNS

Georgios (Yorgo) Varotsis[1], Tobias Schwarz[1], Rodolfo Cappello[2], Danielle Gunn-Moore[1], Katia Marioni-Henry[1]

[1]*Hospital for Small Animals, R(D)SVS and the Roslin Institute, University of Edinburgh, Edinburgh, Midlothian, UK,* [2]*North Downs Specialist Referrals, Bletchingley, Surrey, UK*

Morphological abnormalities of the caudal fossa are increasingly recognized as a cause of morbidity in many brachycephalic dogs but there is little information about these conditions in cats. The objective of this study was to investigate presence of similar morphological abnormalities of the caudal fossa of mesaticephalic and brachycephalic cats and possible association with clinical signs and final diagnosis.

Feline brain MRI studies at two referral centres were reviewed for cases without visible disease that could alter anatomic landmarks or raise intracranial pressure. T2 weighted median plane images of the brain and if available spinal cord were reviewed for the presence of caudal cerebellar indentation, coning, herniation, syringohydromyelia, and fluid accumulation in the middle ear. Area of the cerebral hemispheres and cerebellum was also measured Blind measurements were taken by two observers and reviewed.

There were 53 feline MRI studies meeting the inclusion criteria. Cerebellar herniation was found in 48% of mesaticephalic and 78% of brachycephalic cats. Syringohydromyelia was not identified in any of the MR studies including only the first two cervical spinal cord segments (36 cats) or more regions of the spinal cord (17 cats). High signal on T2-weighted transverse images of the middle ear was detected in 8% of mesaticephalic and 17% of brachycephalic cat. Information on presenting complaint, results of neurological examination and final diagnosis in cats with and without cerebellar herniation was reviewed.

Mesaticephalic breed of cats show indentation of the caudal aspect of the cerebellum and caudal cerebellar coning similar to brachycephalic breeds of cats. The dimensions of the foramen magnum planum and the area of the cerebral hemispheres and cerebellum measured on a T2W sagittal MR image of the brain were also similar between the two groups. However, brachycephalic cats had a higher percentage of herniation of the cerebellar vermis through the foramen magnum. Herniation of the cerebellar vermis does not appear to be associated with syringomyelia in cats.

N16

DIAGNOSTIC INVESTIGATION IN 13 CATS WITH SUSPECTED FELINE HYPERESTHESIA SYNDROME

Pablo Amengual[1], Clare Rusbridge[2], Tim Nuttall[1], Sarah Heath[3], Katia Marioni-Henry[1]

[1]Hospital for Small Animals, Royal (Dick) School of Veterinary Studies, University of Edinburgh, Roslin, Scotland, UK, [2]School of Veterinary Medicine, Faculty of Health & Medical Sciences, University of Surrey, Guildford, England, UK, [3]Behavioural Referrals Veterinary Practice, Chester, England, UK

Feline hyperesthesia syndrome (FHS) was first reported in 1979 and described as an accentuation of an otherwise normal behavior in cats consisting in episodes of tail chasing, biting or licking the lumbar area, flank, anal area or tail; skin rippling and muscle spasms; excessive and unusual vocalizations, wild and uncontrolled jumping and running and presumed hallucinations. FHS has been anecdotally reported in association with dermatological, behavioural, orthopaedic and neurological conditions. Treatment with phenobarbital, gabapentin, prednisolone, amitriptyline, fluoxetine, clomipramine and multivitamin supplements have been used with inconsistent responses. However, we are not aware of any clinical scientific investigation on Feline Hyperesthesia Syndrome.

We retrospectively collected data on the history, signalment, diagnostic workup and treatment from three referral veterinary centers. The inclusion criteria were a history of attacking or over-grooming the tail, flank or perineal area associated with either vocalization or lumbar hyperesthesia, manifested as rippling of the thoracolumbar skin occurring spontaneously or induced by gentle touch. Information on physical and neurological examination, and diagnostic work up were also necessary for inclusion in the study.

Thirteen cats were included in the study. The median age of presentation was 1 year (1-7 years). Eleven cats were attacking or over-grooming their tails and two cats their flank or perineal area. Tail mutilation was reported in 9/13 cats, rippling of the thoracolumbar skin in 10/13 cats and 6 cats presented with unusual vocalization during the episodes. The episodes occurred multiple times per day in 12/13 of the cases, and multiple times per week in 1/13. No consistent triggers were reported. Results of CBC and biochemistry were available for 12/13 cats, spinal radiographs for 7/13, magnetic resonance imaging for 6/13, cerebrospinal fluid analysis for 3/13, electromyography for 3/13, dermatological assessment for 7/13, and joint fluid analysis for one cat. The diagnostic work up led to the diagnosis of allergic dermatitis in 2 cases and immune-mediated polyarthritis in one case, but a definitive diagnosis was not reached in the remaining 10 cases. Treatment included corticosteroids in 7/13 of the cats, gabapentin in 8/13, clomipramine in 4/13, topiramate in 2/13, phenobarbital in 2/13 and cyclosporine in 6/13. The majority of the cases (10/13) received a combination of 2 or more drugs.

Our study confirmed previous anecdotal reports of FHS affecting young cats and the possibility of that FHS signs are secondary to inflammatory skin conditions. A definitive diagnosis was not reached in 10 cases. However, not all the cases had the same diagnostic work up and not all had a full behavioral and dermatological assessment. Based on the results of this retrospective study, specialists in neurology, behavior and dermatology have formulated a standardized questionnaire and proposed a rational diagnostic approach for investigation and treatment of this constellation of clinical signs in cats and as a platform for future prospective studies.

N17

CYSTOMETRIC CHARACTERIZATION OF URINARY BLADDER DYSFUNCTION IN CHRONICALLY PARALYZED DOGS

Hilary Hu, Nick Jeffery, Chelsea Marko, Victoria Kichler

Iowa State University, Ames, USA

Increasing numbers of pet owners elect to keep dogs that have suffered permanent thoracolumbar spinal cord injury as pets. Although pelvic limb paralysis can be partially circumvented by using carts to allow greater mobility, care of the urinary bladder is often more problematic. In addition to urinary incontinence and recurrent urinary tract infections, it is possible that many such affected animals can suffer from detrusor hyper-reflexia. This is a risk factor for chronic elevation of bladder pressure and secondary renal damage in both human and canine spinal cord injury patients.

Here we describe cytometric quantification of bladder dysfunction in chronic paraplegic pet dogs. These dogs were <20kg and had sustained severe T3-L3 spinal cord injury resulting in inability to walk or control urination persisting for at least three months post-injury. A dual-lumen sterile urinary catheter was used for cystometry: one lumen was used to infuse 0.9% sterile saline into the bladder while the other measured intravesicular pressure. Infusion was terminated when: (1) leaking was apparent at the external urethral orifice; or, (2) when the physiologic maximum holding capacity of 20ml/kg was reached; or, (3) when the 50mmHg intra-vesicular pressure threshold was reached.

Several cystometric metrics were measured in this study: (1) intravesicular pressure at leak; (2) bladder holding capacity expressed as a percentage of the physiologic maximum holding capacity of 20ml/kg; and (3) bladder compliance, which was determined as change in bladder volume divided by change in intra-vesicular pressure. Two standard points were used for the calculation of bladder compliance: (1) the intra-vesicular pressure and bladder volume at the start of bladder filling (both were zero); and (2) the intravesicular pressure and bladder volume at cystometric capacity or immediately before the start of any detrusor contraction that resulted in urine leakage.

Cystometric data was recorded in 42 dogs with chronic spinal cord injury. Intravesicular pressure at leak ranged from 1mmHg to 50mmHg with a mean of 18.6mmHg and median of 17mmHg. In 2 (5%) patients, the intravesicular pressure reached the 50mmHg threshold. The two additional parameters used to quantify detrusor hyper-reflexia were both highly variable within the population as a whole: (1) bladder compliance (mean = 27.3ml/mmHg, median = 5.1ml/mmHg, range = 0.4-180ml/mmHg); and, (2) bladder holding capacity (mean = 68%, median = 68%, range = 9-100%).

We conclude that there is evidence that detrusor hyper-reflexia occurs in a small proportion of pet dogs paralyzed because of severe T3-L3 spinal cord injury. It is characterized by elevation in intravesicular pressure and reduction in bladder compliance and bladder holding capacity. This identification of a specific type of neurogenic bladder dysfunction permits timely interventions to mitigate renal damage. Potential therapies for treating detrusor hyper-reflexia can now be evaluated using these quantifiable bladder functional parameters derived from cystometry.

N18

EVALUATION OF KINEMATIC MAGNETIC RESONANCE IMAGING IN DOGS WITH OSSEOUS-ASSOCIATED CERVICAL SPONDYLOMYELOPATHY

Michele Provencher, Amy Habing, Sarah Moore, Laurie Cook, Gary Phillips, Ronaldo da Costa

The Ohio State University, Columbus, OH, USA

Osseous-associated cervical spondylomyelopathy (OA-CSM) is a condition characterized by static and dynamic spinal cord compressions. The dynamic component of cervical spondylotic myelopathy in humans (hCSM) is evaluated with kinematic MRI (kMRI). The purpose of this study was to evaluate kMRI in dogs with OA-CSM using a positioning device that allowed controlled flexion and extension of the cervical vertebral column. Our hypothesis was that kMRI would reveal new compressive lesions not identified with standard positioning.

Ten Great Danes and 2 Doberman Pinschers with neurologic examination findings consistent with a cervical myelopathy were prospectively evaluated. All dogs underwent standard MRI in a neutral position (sagittal and transverse T1 and T2 weighted images) using a 3.0 T magnet. The patients were then placed in right lateral recumbency on a positioning device and the cervical vertebral column was first flexed and then extended. Sagittal and transverse T2 weighted images were acquired in both flexion and extension. Morphologic and morphometric assessments of the cervical vertebral column in neutral, flexion and extension were performed. Morphologic analysis included a modified spinal cord compression score (previously published), direction of spinal cord compression, signal intensity changes within the spinal cord, and worst site of spinal cord compression. Morphometric assessment included spinal cord height, intervertebral disc width, spinal cord width, spinal cord area, and vertebral canal height. A Fischer's exact test was used to evaluate the morphologic data and mixed-effects linear regression was used to analyze the morphometric data. Interobserver and intraobserver analyses were performed.

Given space constraints, only a portion of the results is reported. In neutral 4 patients had 1 compression, 4 patients had 2 compressions, 3 patients had 3 compressions, and 1 patient had 4 compressions (25 total compressions). In flexion 1 patient had 0 compressions, 4 patients had 1 compression, 4 patients had 2 compressions, 2 patients had 3 compressions, and 1 patient had 4 compressions (22 total compressions). In extension 2 patients had 1 compression, 1 patient had 2 compressions, 8 patients had 3 compressions, and 1 patient had 4 compressions (32 total compressions). Extension was associated with mild compression at C4-C5 (p=0.02) that was not noted in neutral or flexion, and extension caused worsening of 28% (7/25 compressions) of the compressions that were present in neutral. There were 11/32 compressions (34%) in extension that were not present in neutral and 10/11 (91%) of these compressions had a dorsal component; the presence of dorsal compression with extension was significant at C4-C5 (p=0.01). Flexion was associated with an increase in intervertebral disc width at C3-C4 (p=0.02) and extension caused widening of the intervertebral disc at C6-C7; no other significant morphometric findings were noted. In 1 patient, a synovial cyst that was not identified in neutral was noted to cause mild spinal cord compression in extension.

Our results support the use of kMRI in patients with OA-CSM to reveal new compressive sites, dorsal compressions and to enhance visualization of extradural compressive lesions, such as synovial cysts.

N19

BARTONELLA SPP. PCR ASSAY RESULTS USING CEREBROSPINAL FLUID OF DOGS WITH CENTRAL NERVOUS SYSTEM DISEASE

Lisa Bartner[1], Stephanie McGrath[1], Adam Drury[1], Annie Chen[2], Arianne Morris[1], Melissa Brewer[1], Meri Hall[2], Michael Lappin[1]

[1]*Colorado State University, Fort Collins, CO, USA,* [2]*Washington State University, Pullman WA, USA*

A multitude of infectious agents are on the differential list for dogs with focal or multifocal neurological dysfunction. A number of *Bartonella* spp., including *B. henselae, B. vinsonii* subsp. *berkhoffii*, and *B. clarridgeaie,* are known to infect dogs. While multiple human cases of neurobartonellosis have recently been described, the role these organisms play in clinical diseases of the central nervous system of dogs has not been widely explored. The purpose of this study was to use polymerase chain reaction (PCR) to amplify *Bartonella* spp. DNA from cerebrospinal fluid (CSF) of naturally exposed dogs in endemic areas meeting criteria for inflammatory central nervous system (CNS) disease.

CSF samples from 175 pure or mixed breed dogs were submitted to Colorado State University (www.dlab.colostate.edu) from either the Washington State University Neurology Department between January 2012 and September 2014 or the Colorado State University Veterinary Teaching Hospital between January 2012 and September 2015. The CSF samples were stored at -80C until evaluated in this study. Dogs with neurologic examinations consistent with focal and multifocal neurologic dysfunction and CSF pleocytosis (total nucleated cell count >5 nucleated cells/µl and red blood cell <4,000 cells/µl) were included. Animals with normal neurologic examinations were also included if their CSF met our criteria. The CSF was thawed and centrifuged at 10,000 X g for 15 minutes. The supernatant was removed and the pellet assayed in a previously published PCR assay that targets the 16S-23S rRNA intergenic region. All positive amplicons were sequenced to determine the infective *Bartonella* spp.

A total of 67 dogs met the inclusion criteria, none of which were positive for *Bartonella* spp. DNA in CSF. Of the other 108 CSF samples, one was positive for *B. henselae* DNA. The CSF from this dog contained red blood cells (94 RBC/µl).

As *Bartonella* spp. have an intra-erythrocytic phase, we speculate that minimal peripheral RBC contamination in the CSF of dogs with systemic *Bartonella* spp. infection may lead to positive *Bartonella* PCR assay results in the absence of a CNS disease association. Thus, positive PCR for *Bartonella* spp. DNA in the CSF of dogs must be interpreted in light of number of RBCs within the sample as well as the presence of inflammation or systemic infection. Failure to amplify *Bartonella* spp. DNA from the CSF of the 67 dogs with inflammatory disease studied suggests the organism was not involved, the organism was in CNS tissues but not in the CSF, or the organism was present but in quantities undetectable by this PCR assay. The combination of PCR and culture has been found to be the most sensitive way to detect *Bartonella* spp. in samples from dogs and humans and the use of that technique should be considered in future studies.

N20

PREICTAL, POSTICTAL AND INTERICTAL BEHAVIORAL CHANGES IN DOGS WITH GENETIC EPILEPSY COMPARED TO CONTROL DOGS

Hilary Levitin[1], Devon Hague[1], Kelly Ballantyne[2], Laura Selmic[1]

[1]*University of Illinois Veterinary Clinical Medicine, Urbana, Illinois, USA,* [2]*Medical District Veterinary Clinic at Illinois, Chicago, Illinois, USA*

While anxiety related behaviors have been reported in humans diagnosed with epilepsy, only one veterinary study has shown an increased frequency of anxiety behaviors in epileptic dogs compared to other medical populations. The purpose of this study was to determine the presence and severity of anxiety related behaviors in dogs with genetic epilepsy when compared to control patients. The study's other aims were to investigate if an association exists between increased anxiety related behaviors in the preictal and postictal phases of a seizure episode compared to the interictal period and whether dogs with refractory epilepsy have more severe seizures, which could account for the increased anxiety related behavioral side effects as reported in humans and dogs.

This multi-institutional study used a questionnaire based on the Canine Behavioral Assessment and Research Questionnaire (CBARQ), that has been previously validated for its ability to analyze canine behavior. Owners of dogs that had been diagnosed with genetic epilepsy were asked to complete the questionnaire and a 10-point seizure severity scale. Sections of the CBARQ relating to separation, attachment and attention seeking, and obedience and training behaviors were modified to include additional parameters for detecting anxiety based behavioral changes associated with different phases of the seizure episode. The questionnaire was also distributed to primary care clients presenting dogs for wellness examinations and owners of dogs with intervertebral disc disease (IVDD). Veterinarians of the dogs in the study also completed a brief questionnaire as well to identify and verify the diagnoses of each dog that participated in this study. Fischer's exact tests were utilized, to assess for differences in behavioral traits exhibited between the epileptic, IVDD and primary care groups. If a significant difference between the groups was detected, then post hoc testing with Fischer's exact tests for pairwise comparisons was performed and a Bonferroni correction was applied to account for multiple comparisons.

Dogs in the epilepsy and IVDD groups were more likely to be fearful/anxious when approached by an unfamiliar dog than those dogs in the wellness group (P= 0.0011). Epileptic dogs with increasing seizure severity were significantly more likely to show fear and anxiety when approached by an unfamiliar dog (p= 0.0129) and showed significantly less excitement just before just before being taken on a car trip (P= 0.0126). Epileptic dogs receiving polytherapy had a significant decrease in excitement just before taking a walk (P= 0.0007) and just before being taken on a car trip when compared to dogs on monotherapy (P= 0.027). Dogs receiving polytherapy had significantly increased fear and anxiety when groomed or bathed by a household member (P= 0.0197), and had an increase in shaking, shivering, or trembling when left or are about to be left on their own (P= 0.0004).

Dogs receiving polytherapy had a significant increase in becoming agitated when their owner or others show affection for another person during preictal (P= 0.005), postictal (P= 0.001) and interictal (P_r= 0.0083) period and when their owner or others show affection for another dog or animal during preictal (P= 0.0068), postictal (P= 0.0068) and interictal (P= 0.02) period when compared to dogs on monotherapy. The finding of behavioral changes present throughout different seizure phases in this study raises further questions about how epilepsy affects canine patient's quality of life.

EN01

GLUCOSE HOMEOSTASIS DETERIORATES MORE RAPIDLY WITH AGE IN BURMESE CATS COMPARED TO NON-BURMESE

Margareth Rose Lederer[1], Jacquie Rand[2], Nubia Lopes[2], John Morton[3], Nick Jonsson[1]

[1]University of Glasgow, Glasgow, Scotland, UK, [2]University of Queensland, Gatton, Queensland, Australia, [3]Jemora Pty Ltd., Geelong, Victoria, Australia

Burmese cats are predisposed to diabetes mellitus in Australia, New Zealand and United Kingdom, and the disease is 3 to 4 times more prevalent in this breed compared to domestic cats. Mechanisms predisposing Burmese cats to diabetes remain unclear. This study compared hormonal and biochemical variables associated with glucose and lipid metabolism between mixed-age, lean to overweight, clinically healthy Burmese and non-Burmese cats. Burmese cats older than 3.5 years had higher fasting glucose compared to non-Burmese ($P = 0.02$), with 1.3 mmol/L higher at 3.5 years and 3.5 mmol/L higher at 10 years of age ($P<0.001$), but not at 2 years of age. Mean and 2-hr glucose concentrations during a glucose tolerance test were higher in Burmese than non-Burmese cats ($P <0.006$ to 0.031), indicating relative glucose intolerance. Lean Burmese cats had 4.8 μU/mL higher fasting insulin concentrations than non-Burmese, suggesting they are insulin resistant ($P= 0.031$). At time 10 min during the GTT, Burmese cats had lower insulin concentration compared to non-Burmese, but were higher at 120 minutes ($P=0.066$ and $p=0.046$). This is consistent with diminished first phase, and increased and prolonged second phase insulin release, a pattern of insulin secretion described in prediabetic, diabetic, obese humans and obese cats. Fasting triglyceride concentration was 0.2 mmol/L higher ($P=0.015$) and fasting NEFA concentration was 0.18 mmol/L higher ($P=0.016$) in Burmese compared to non-Burmese. These results demonstrate that Burmese cats have more rapid deterioration of β-cell function with age, and impaired glucose homeostasis and lipid metabolism occur concurrently in these cats.

EN02

PERTURBATIONS IN SERUM FRUCTOSAMINE LEVEL IN DIABETIC HYPERTHYROID CATS – A RETROSPECTIVE STUDY

Arnon Gal[1], Brie Trusiano[3], Adrienne French[2], Nicolas Lopez-Villalobos[1], Amy MacNeill[3]

[1]Institute of Veterinary, Animal and Biomedical Sciences, Palmerston North, New Zealand, [2]New Zealand Veterinary Pathology, Palmerston North, New Zealand, [3]Colorado State University, Ft. Collins, USA

Previous studies indicated that serum fructosamine level is decreased in hyperthyroid cats; however, its clinical utility in diabetic hyperthyroid cats was not reported. We hypothesized that hyperthyroidism in diabetic cats will result in a clinically significant decrease in serum fructosamine level compared to euthyroid diabetic cats. Data of serum total thyroxine (TT4) and fructosamine of hyperthyroid/euthyroid diabetic/nondiabetic cats from the New Zealand Veterinary Pathology and Colorado State University were retrieved and statistically analyzed. Serum fructosamine was significantly lower in hyperthyroid diabetic than euthyroid diabetic cats (mean 332μmol/L 95% CI 291-379, n=18 versus mean 527μmol/L 95% CI 515-553, n=186) while not different between hyperthyroid diabetic and euthyroid nondiabetic cats (mean 332μmol/L 95% CI 291-379, n=18 versus mean 321μmol/L 95% CI 296-345, n=128). There was a significant negative correlation between serum TT4 concentration and serum fructosamine and glucose concentrations (n=659, $p<0.01$, and n=297, $p<0.01$, respectively). Hyperthyroid cats (diabetic or not) had significantly ($p<0.05$) lower serum glucose than euthyroid cats (diabetic or not). Diabetes and random variability accounted for 49.7% and 45.4% of the change in serum fructosamine; whereas, age (3.1%) and population (1.8%) had minimal impact on serum fructosamine. In conclusion, in the current study, serum fructosamine concentration was significantly decreased in diabetic hyperthyroid cats and was similar to the fructosamine concentration in euthyroid nondiabetic cats. In addition, approximately 45% of the change in serum fructosamine was independent of diabetes. Future studies should be undertaken to explore specific factors that have an impact on fructosamine level.

EN03

PHARMACODYNAMIC AND PHARMACOKINETIC PROPERTIES OF INSULIN ASPART FOLLOWING SUBCUTANEOUS AND INTRAMUSCULAR INJECTION IN CATS

Hannah Pipe-Martin[1], Jon Fletcher[1], Chen Gilor[2], Michael Kearney[3]

[1]Department of Veterinary Clinical Sciences, School of Veterinary Medicine, Louisiana State University, Baton Rouge, LA, USA, [2]Department of Veterinary Clinical Sciences, College of Veterinary Medicine, The Ohio State University, Columbus, OH, USA, [3]Department of Pathobiological Sciences, School of Veterinary Medicine, Louisiana State University, Baton Rouge, LA, USA

Insulin aspart (Aspart) is a rapid-acting insulin analog that has been shown to be an effective treatment for diabetic ketoacidosis (DKA) when administered by intermittent subcutaneous (SC) injection in humans. In cats, rapid-acting insulin analogs have not yet been evaluated, and the most commonly used protocols for DKA involve delivery of regular insulin via constant intravenous infusion, which can be technically challenging. The aim of this study was to determine the pharmacodynamics (PD) and pharmacokinetics (PK) of Aspart following SC and intramuscular (IM) injections in healthy cats.

Using the isoglycemic clamp method, time-action profiles were generated in eight healthy cats to determine pharmacodynamics of Aspart following SC and IM injection. Pharmacokinetics were determined by measuring the plasma Aspart concentration every 15 minutes during the clamp using the Mercodia® Iso-insulin ELISA kit.

Mean ± SEM time to onset of action and duration of action did not differ significantly between SC (11.12 ± 0.97 min and 154 ± 14.8 min) and IM injection (13.75 ± 3.1 min and 176 ± 19.8 min), and both methods of administration resulted in acceptable plasma concentrations and blood glucose lowering effect. Subcutaneous injection resulted in greater plasma Aspart concentration at 15 and 30 min and greater peak concentrations when compared to IM injection ($p < 0.0001$ and $p = 0.0005$ respectively).

Aspart is rapidly absorbed following SC and IM injection in healthy cats. The effective and predictable glucose lowering effect of insulin aspart may be advantageous in the treatment of DKA in cats.

EN04

THE EFFECT OF TETRA-HYDROXYLATED BILE ACID ON ADIPOCYTE SIZE AND INSULIN SENSITIVITY IN HEALTHY CATS

Isabelle Rast[1], Elena Salesov[1], Thomas Lutz[2], Eric Zini[1], Christian Wolfrum[3], Claudia Reusch[1]

[1]Clinic for Small Animal Internal Medicine, Vetsuisse Faculty, University of Zurich, Zurich, Switzerland, [2]Institute of Veterinary Physiology, Vetsuisse Faculty, University of Zurich, Zurich, Switzerland, [3]Laboratory of Translational Nutrition Biology, Swiss Federal Institute of Technology Zurich, Schwerzenbach, Switzerland

Obesity is a major risk factor for the development of diabetes in cats. Recently, retinoid-related orphan receptor gamma (RORγ) has been identified as an important transcription factor in the development of large insulin-resistant adipocytes. RORγ can be inhibited by its ligand tetra-hydroxylated bile acid (THBA). Oral supplementation of THBA reduces adipocyte size, improves insulin sensitivity and prevents hyperglycemia in obese mice. The purposes of this study were to investigate palatability and possible side effects of THBA, and the effects of THBA supplementation on adipocyte size and insulin sensitivity in healthy cats.

Six healthy purpose-bred cats were fed 5 mg/kg/day of THBA for 8 weeks. THBA was given in a small amount of food to each cat individually before the daily ration was offered. Complete blood count, biochemical profile and insulin tolerance test (ITT; 0.1 IU/kg insulin aspart IV) were performed at weeks 0 and 8. After 12-18 hours of fasting blood glucose concentration was recorded and the rate constant for the disappearance of glucose (K_{ITT}) was calculated. Subcutaneous fat tissue samples were taken in the umbilical area by a 4 mm biopsy punch, and adipocyte size was determined by histological and image based analysis of at least 10 sections per animal. Data were analyzed by parametric tests ($p < 0.05$).

Oral supplementation of THBA was well accepted in all cats, and no side effects were noted. None of the cats displayed abnormalities on complete blood count and serum biochemical profile. Fasting blood glucose concentration and average size of white subcutaneous adipocytes were significantly lower after 8 weeks of THBA administration, whereas K_{ITT} did not differ.

The reduction of adipocyte size in response to THBA treatment is in accordance with data from mice and underscores the possible role of RORγ signaling in the control of adipose tissue metabolism. While we observed a significant reduction in fasting glucose, overall insulin sensitivity did not improve. These findings might be explained by the fact that insulin sensitivity in normal weight subjects is mainly determined by liver and muscle cell function, while altered adipocyte size would not contribute significantly to the ITT. Future studies in obese cats and altered dosing regimes are required to assess whether THBA improves insulin resistance in cats, in addition to direct effects on adipocyte size.

EN05

THE EFFECT OF ADIPOSITY AND DIET ON SECRETION OF INCRETIN HORMONES IN CATS

Chen Gilor, Katie McCool, Adam Rudinsky, Valerie Parker, Prosper Boyaka

The Ohio State University, Columbus, OH, USA

Degree of adiposity and dietary macronutrient composition affect the secretion of incretin hormones but little is known about their effect in cats.

In this study, 8 overweight cats were fed a maintenance diet (MD) for 3 weeks followed by a weight control diet (WCD, lower fat, higher in carbohydrates and fiber). Cats were fed ad libitum initially and then food was restricted to achieve 1-2% loss of body weight weekly (11 weeks). When lean, cats were fed MD for 2 additional weeks. A standardized meal test (SMT) using a third diet was performed after at least 7 days on each diet, before and after weight loss (4 SMT's total). Glucose, insulin, glucagon-like peptide-1 (GLP-1) and glucose-dependent insulinotropic peptide (GIP) concentrations were measured immediately before and over 6 hours after feeding the SMT. Area under the concentrations curve (AUC) was compared using non-parametric tests.

Glucose, insulin and GLP-1 concentrations did not differ between SMT's at the 4 time points. Fasting GIP concentrations also did not differ but post-prandial GIP secretion was affected by degree of adiposity (P=0.025) with a trend towards an effect of diet (P=0.085). GIP secretion in lean cats fed WCD was higher than in obese fed MD (P=0.0062).

In conclusion, compared to the overweight state, GIP secretion was increased in cats in response to SMT in the lean state. On-going exposure to WCD increased GIP responses to SMT, suggesting that dietary macronutrient content is important in determining GIP responses not only acutely but also on a long-term basis.

EN06

BREED, COAT COLOUR AND HAIR LENGTH AS RISK FACTORS FOR FELINE HYPERTHYROIDISM

Victoria J Crossley, Yu-Mei Chang, Robert C Fowkes, Jonathan Elliott, Harriet M Syme

Royal Veterinary College, London, UK

Hyperthyroidism commonly affects geriatric cats, but its aetiology is poorly understood. Previous studies have shown Siamese, Himalayan and Burmese cats to have reduced risk of developing hyperthyroidism. These breeds have colourpoint coats as a result of temperature-sensitive mutations in the tyrosinase gene, which limit conversion of tyrosine to melanin pigment. Tyrosine is essential for synthesis of both melanin and thyroid hormones and, as such, coat pigmentation may impact relative tyrosine availability for thyroid hormone production. This study aimed to identify potential associations between coat phenotype and development of hyperthyroidism, by investigation of breed, coat colour and hair length as risk factors for feline hyperthyroidism.

This retrospective study of cats aged ≥10 years, referred to a single UK veterinary teaching hospital (2006-2014), used electronic patient records to obtain owner-reported data regarding breed, coat colour, age and sex, and clinical data to classify cats as hyperthyroid/euthyroid. Bayesian multivariable logistic regression was used to evaluate breed, age and sex as risk factors for hyperthyroid status in all the cats (purebred and non-purebred). To avoid interactions with coat length or colour in purebreds, the effect of coat colour/pattern, colour dilution, base pigment, white markings and hair length were assessed only in non-purebred cats in a separate analysis. Variables with P<0.2 in univariate analyses were evaluated in multivariable models, with final variables significant at P≤0.05. Risk factors are quantified as odds ratio ((OR) P-value).

Of 3934 cats included in the final analysis, 885 were hyperthyroid and 3049 euthyroid. Multivariable results showed Burmese (OR 0.01, P=0.005), Persians (OR 0.20, P<0.001), Siamese (OR 0.32, P=0.006), Abyssinians (OR 0.04, P=0.035) and British shorthairs (OR 0.52, P=0.014) had reduced risk of hyperthyroidism compared to domestic shorthairs. Domestic longhairs (OR 1.34, P=0.015) showed increased risk of hyperthyroidism compared to domestic shorthairs, but coat colour, colour dilution, base pigment and white markings did not have a significant effect. Overall, females (OR 1.39, P<0.001) had increased risk of hyperthyroidism compared to males, and, compared to 10 year olds, increased risk was found in cats aged 11-12 years (OR 1.45, P=0.007), 13-14 years (1.83, P<0.001) and 15-17 years (OR 1.50, P=0.004).

To authors' knowledge this is the first study to report increased risk of hyperthyroidism in long haired, non-purebred, domestic cats. Consistent with a number of previous studies, increased risk of hyperthyroidism was found with increased age and female sex, and reduced risk in Burmese, Siamese and Persian breeds. However, this study also newly identified two further purebreds, Abyssinian and British shorthair, at reduced risk of hyperthyroidism, neither of which were exclusively colourpoint. Coat colouration was not found to be associated with risk of hyperthyroidism in the analysis of domestic cats, however, reliance on secondary data may have resulted in misclassification errors and coat phenotype may be a poor surrogate marker for melanin concentrations. Further studies are required to evaluate tyrosine availability as a potential factor in the aetiology of feline hyperthyroidism, and to investigate the apparent protective effect in certain purebreds and associations between hyperthyroidism and hair length.

EN07

DOES A LIMITED IODINE DIET AFFECT THE RESPONSE TO RADIOACTIVE IODINE THERAPY IN HYPERTHYROID CATS?

Allison Rowland, Karyn Harrell, Katharine Lunn

North Carolina State University, Raleigh, NC, USA

A retrospective study was performed to determine the outcome of radioactive iodine (131-I) therapy in cats previously treated with a limited iodine diet (Hill's Prescription Diet y/d®).

Medical records were searched for cats receiving 131-I therapy after documented feeding of the prescription diet for at least 14 days. Cats were excluded if the prescription diet had been supplemented with other foods, or if methimazole had been administered concurrently with, or subsequent to, the prescription diet. Owners were contacted directly to provide missing dietary information.

Nine cats fulfilled the study criteria. The prescription diet was fed for a median of 91 days (range: 19-371). The interval between discontinuing the diet and administering 131-I was 0 days in 3 cats, 4-5 days in 3 cats and 16-24 days in 3 cats. Highest serum total thyroxine (T4) without therapy was used in calculating 131-I dose; median highest T4 was 10.3 mcg/dl (range: 4.3-23.8 mcg/dl; reference range: 0.78-4.27 mcg/dl). The median dose of 131-I administered was 3.64 mCi (range: 3.06-5.26 mCi). The cats were initially evaluated 22-57 days after 131-I therapy and hyperthyroidism had resolved in all 9 cats, with a median post-therapy total T4 of 1.3 mcg/dl (range: 0.5-1.8 mcg/dl). Total T4 was below the reference range in 2 cats (0.5 and 0.55 mcg/dl) at the initial post-therapy evaluation, but was normal 22 and 34 days later (0.9 and 1.6 mcg/dl respectively).

In this small group of cats, pre-treatment with a limited iodine diet did not adversely affect the response to 131-I therapy.

EN08

EFFECTS OF LEVOTHYROXINE ADMINSTRATION AND WITHDRAWAL ON THE HYPOTHALAMIC-PITUITARY-THYROID AXIS IN EUTHYROID DOGS

Vincent Ziglioli, David Panciera, Edward Monroe, Gregory Troy, Katie Boes

Virginia Maryland College of Veterinary Medicine, Blacksburg, VA, USA

Many dogs are misdiagnosed with hypothyroidism because of the vague presenting signs and limitations of thyroid function tests, leading to inappropriate treatment with levothyroxine. During chronic supplementation, the hypothalamic-pituitary-thyroid axis (HPTA) is suppressed and may make it difficult to accurately determine thyroid function following withdrawal of levothyroxine. We sought to determine if the HPTA is suppressed following levothyroxine administration in euthyroid dogs and the time required for resolution of any suppression. We hypothesized that levothyroxine administration would suppress the HPTA in euthyroid dogs and that the HPTA would recover within 8 weeks in all dogs, regardless of the duration of treatment.

Twenty-eight healthy euthyroid dogs were administered levothyroxine (mean dose 0.024mg/kg every 24 hours) for either 8 weeks (Group 1) or 16 weeks (Group 2). The dose of levothyroxine was adjusted weekly until a target total thyroxine (T_4) serum concentration of 40-70nmol/L was obtained 4-6 hours post pill. Serum concentrations of T_4, free thyroxine (fT_4) by equilibrium dialysis, thyrotropin (TSH), and 3,5,3'-triiodothyronine (T_3) were measured every 4 weeks during supplementation and for 4 months after levothyroxine was discontinued. A mixed model ANOVA followed by Tukey's procedure for multiple comparisons was used to compare thyroid function tests both within and between groups. The level of significance was set at <0.05.

Mean serum concentrations of T_4 and fT_4 were significantly higher and TSH was significantly lower during levothyroxine administration compared to baseline in both groups. Mean serum concentrations of T_4, fT_4 and TSH beginning 1 week after levothyroxine was discontinued were significantly different compared to values during levothyroxine administration but not compared to baseline values. When groups 1 and 2 were compared, there was no difference in mean T_4, fT_4, and T_3 concentrations between groups during or after levothyroxine supplementation. Suppression of the HPTA occurred during levothyroxine supplementation, with mean serum T_4, fT_4 and TSH concentrations returning to the reference interval by 1 week after discontinuation in both groups. It appears that assessing thyroid function tests 1 week after long-term levothyroxine supplementation may reliably establish dogs as being euthyroid.

EN09

VARIABILITY OF P450SCC AUTOANTIBODY PERSISTENCE IN DOGS AFFECTED WITH HYPOADRENOCORTICISM

Alisdair Boag[1], Michael Christie[3], Kerry McLaughlin[5], Peter Graham[4], Harriet Syme[2], Brian Catchpole[2]

[1]Royal (Dick) School of Veterinary Studies, Edinburgh, UK, [2]Royal Veterinary College, London, UK, [3]University of Lincoln, Lincoln, UK, [4]University of Nottingham, Nottingham, UK, [5]King's College, London, UK

Autoantibodies directed against the P450 side chain cleavage enzyme (P450scc) have been recently described in a proportion of dogs affected with hypoadrenocorticism, consistent with an immune-mediated disease process. In humans affected with Addison's disease, autoantibodies can have a predictive value, being detected prior to clinical signs developing, and have been shown to persist post-diagnosis. Furthermore, an autoantibody positive status post-diagnosis has been associated with successful remission of Addison's disease following B-cell depletion with rituximab, suggesting active pathology in these cases.

The current study explored changes in serum P450scc autoantibody status over time in dogs diagnosed with spontaneous hypoadrenocorticism. P450scc autoantibodies were measured by radioimmunoassay in an initial cohort of 213 dogs, indicating a prevalence of 24%. Thirty two of these dogs had repeat samples (n = 80 in total) available for analysis. Between two and six samples were available for each dog, obtained between two and 787 days apart.

Five dogs were consistently autoantibody positive in all samples, up to 425 days between sampling times, these comprised one male and four female dogs, with three crossbreeds, a Dogo Argentino and an English springer spaniel. Three dogs were initially autoantibody positive, then became negative at later time points, a 3 y 11 month old male entire bull terrier, a 4 y 5 month old female neutered beagle and a 2 y female entire crossbreed dog. One dog, a 1 y 8 month old female entire standard poodle, initially negative for P450scc autoantibodies, seroconverted 18 months after diagnosis. The remaining 23 dogs with multiple samples available were consistently autoantibody negative. Persistence was not associated with sex ($p = 0.673$).

This study demonstrates persistence of P450scc autoantibodies in a subset of dogs affected with hypoadrenocorticism and seroconversion over one year post-diagnosis. Whilst P450scc autoantibody positivity has been associated with sex, with females having a higher prevalence, there was no sex difference in persistence demonstrated. P450scc expression in the ovary may act as an additional source of antigenic stimulation in female dogs, explaining the later seroconversion observed in one individual. However, paralleling humans with Addison's disease, antibody persistence in dogs with hypoadrenocorticism might represent persistent pathology, due to residual antigenic stimulation and autoimmune inflammation in the adrenal gland.

EN10

JUVENILE HYPOADRENOCORTICISM IN THE NOVA SCOTIA DUCK TOLLING RETRIEVER: A RECESSIVE MONOGENIC AUTOIMMUNE DISEASE

Emily Brown[1], Amy Young[2], Zena Wolf[3], Claire Wade[4], Angela Hughes[5], Oded Foreman[6], Kartika Jayashankar[1], Anita Oberbauer[2], Noa Safra[1], Kirstin Linblad-Toh[7], Shelley Burton[8], Danika Bannasch[1]

[1]*University of California, Davis School of Veterinary Medicine, Davis, California, USA,* [2]*University of California, Davis, Davis, California, USA,* [3]*Brigham and Women's Hospital and Harvard Medical School, Boston, Massachusetts, USA,* [4]*University of Sydney, Sydney, New South Wales, Australia,* [5]*Mars Veterinary, Germantown, Maryland, USA,* [6]*Genentech, Inc., South San Francisco, California, USA,* [7]*Broad Institute of Harvard and Massachusetts Institute of Technology, Cambridge, Massachusetts, USA,* [8]*Atlantic Veterinary College, University of Prince Edward Island, Charlottetown, Prince Edward Island, Canada*

While hypoadrenocorticism, or Addison's disease, can occur in a dog of any breed at any age, certain breeds, including Nova Scotia Duck Tolling Retrievers (NSDTRs), have an increased incidence of the disease. Specifically in the NSDTR, there appears to be at least two forms of Addison's disease in the breed: a juvenile onset form (JADD) occurring in dogs under 1 year of age, and an adult onset form occurring in dogs at 4.5 years of age on average. JADD differs from adult onset Addison's disease, as it can be a multisystemic illness and appears to involve an autoimmune component. This is supported by the identification of a CD3+ T cell infiltrate in histologic sections of a JADD affected adrenal gland sample and that many cases are often affected with other autoimmune diseases, including immune-mediated hemolytic anemia and thrombocytopenia, immune-mediated polyarthritis, and hypothyroidism. To identify a genetic basis of JADD, a genome-wide association study was performed using the Illumina Canine HD 173,000 SNP array with 14 NSDTRs diagnosed with Addison's disease less than 1 year of age and 33 healthy control NSDTRs over 6 years of age. All cases were definitively diagnosed through an adrenocorticotrophic hormone stimulation test. Genome-wide association analysis identified a 1.7Mb associated haplotype on chromosome 27. Whole genome sequencing was performed on 2 NSDTRs with the associated haplotype, as well as 6 unaffected NSDTRs and 11 other dogs from 5 different breeds. Analysis of variants yielded 5 segregating SNPs in the associated region: 1 intergenic, 2 intronic, 1 coding, and 1 in the 3' untranslated region of a gene. The only coding variant, a missense mutation causing an amino acid change from a proline to a leucine, occurs in a currently uncharacterized gene, but is predicted to be damaging and deleterious based on sequence conservation. In summary, genome-wide association and whole genome sequencing analysis has identified a novel gene and mutation implicated in multi-organ autoimmunity and juvenile onset Addison's disease in the NSDTR.

EN11

VARIABILITY IN POST ACTH STIMULATION SERUM CORTISOL FOLLOWING ADMINISTRATION OF CORTISONE ACETATE IN HEALTHY DOGS

Arnon Gal, Karin Weidgraaf, James Bowden, Nicolas Lopez-Villalobos, Nick Cave

Institute of Veterinary, Animal and Biomedical Sciences, Massey University, Palmerston North, New Zealand

The adrenocorticotropic hormone stimulation test (AST), often used for the diagnosis of canine hyperadrenocorticism, has a wide range of sensitivity and specificity. Thus, we hypothesized that standardized negative feedback on the hypothalamic-pituitary-adrenal-axis will result in large variability in the post AST serum cortisol concentrations. Therefore, this study aimed to evaluate the variability in the post AST serum cortisol in healthy dogs and to statistically examine parameters that may influence cortisol responses.

Fourteen healthy Harrier Hound dogs (mean age 7.1y±3.2; mean body weight 26.5kg±2.7) were randomized by a Latin square design to receive 5 treatments [placebo, 50mg, 37.5mg, 25mg, and 12.5mg of cortisone acetate (CA)] in each of the five 7-day treatment periods, which were separated by 14-day washout periods. An AST was performed 24h after the last dose of CA for each of the 5 treatments and serum cortisol was analyzed by an electrochemiluminescence immunoassay.

A significant dose dependent decrease in the mean (±SEM) post AST serum cortisol was observed (placebo 365±13nmol/L, CA 12.5mg 347±13nmol/L, CA 25mg 335±13nmol/L, CA 37.5mg 326±13nmol/L, CA 50mg 308±13nmol/L). There was a significant ($p<0.05$) between-week variability in post AST serum cortisol in the placebo and 25mg and 37.5mg CA groups. The overall variability in the post AST serum cortisol concentrations was attributed to the individual dog (31%), random unexplained variability (31%), dose of CA (16%), age (15%), body weight (4%), and timing of treatment (3%).

The wide range in the published specificity of the AST may be related to large variations in cortisol responses.

EN12

IONIZED HYPERCALCEMIA IN CATS: ETIOLOGIES AND ASSOCIATED CLINICAL SIGNS

Muzzammil Sayyid, Chen Gilor, Valerie Parker, Adam Rudinsky, Dennis Chew

The Ohio State University, Columbus, OH, USA

The etiologies, concurrent diseases, and clinical signs associated with hypercalcemia in cats have been reported in the past based only on measurement of total calcium and excluding cases of idiopathic hypercalcemia (IHC). Here we report on the relative frequency of various diagnoses and concurrent clinical signs associated with ionized hypercalcemia in cats. The medical records of an academic institution that treats primary and secondary cases was searched for cases treated between the years 2003-2013. Cases were included if serum or plasma ionized calcium concentrations were above the upper limit of the reference interval at least twice. Cats were excluded if ionized hypercalcemia was not present on follow-up examination or if a final diagnosis was not made.

Sixty-nine cases were identified. The most common diagnoses were IHC (48%) and chronic kidney disease (35%). Urolithiasis was diagnosed in 14% (43% calcium oxalate), neoplasia in 13% (46% lymphoma) and primary hyperparathyroidism in 3%. Other diagnoses included hyperthyroidism (8%), hypertrophic cardiomyopathy (4%), feline idiopathic cystitis (2%), and diabetes mellitus (2%). The average age at diagnosis of hypercalcemia was 10.0±4.61 years. Most common clinical signs reported in the IHC group vs. other diagnoses (in which clinical signs could be related to the primary disease, and not necessarily to hypercalcemia), were anorexia (21.7% vs. 31.9%), vomiting (14.5% vs. 26.1%), lethargy (10.1% vs. 7.2%), and constipation (10.1% vs. 1.4%).

In conclusion, ionized hypercalcemia in cats is frequently idiopathic. The association between feline ionized hypercalcemia and its etiologies and clinical signs should be further explored in case-controlled studies.

EN13

EVALUATION OF BASAL CORTISOL CONCENTRATIONS FOR THE DIAGNOSIS OF HYPOADRENOCORTICISM IN DOGS

Ashley Gold, Joe Hauptman, Daniel Langlois

College of Veterinary Medicine, Michigan State University, East Lansing, MI, USA

Canine hypoadrenocorticism is a potentially life-threatening endocrine disorder characterized by glucocorticoid deficiency, with or without concurrent mineralocorticoid deficiency. The current standard for diagnosing hypoadrenocorticism is the ACTH stimulation test, but a simpler and less expensive test with similar accuracy would be desirable. Previous studies regarding the diagnostic utility of basal plasma or serum cortisol concentrations are limited by low numbers of affected dogs, and receiver operating characteristics (ROC) methods were not utilized to identify optimal cut-off points. The purpose of our study was to evaluate the diagnostic utility of basal cortisol concentrations for the diagnosis of hypoadrenocorticism in a large number of dogs by constructing an ROC curve to identify optimal cut-offs for sensitivity and specificity. Furthermore, the diagnostic utility was evaluated in two sub-populations of dogs: those with electrolyte abnormalities (Na:K ratio ≤28) and those without electrolyte abnormalities (Na:K ratio >28).

Medical records for dogs in which an ACTH stimulation test had been performed for a clinical suspicion of hypoadrenocorticism between 2005 and 2015 were retrospectively reviewed. Dogs that had received treatment for hyperadrenocorticism or recent glucocorticoid therapy were excluded. Five hundred and fourteen dogs met inclusion criteria, including 163 dogs with hypoadrenocorticism (post-ACTH stimulated cortisol concentration ≤55 nmol/L). An ROC curve was constructed for basal cortisol concentrations using commercially available software. The area under the ROC curve was 0.991. The ROC curves were nearly identical when evaluating dogs with and without electrolyte abnormalities. Accuracy was optimal at 22 nmol/L, which provided a sensitivity and specificity of 96.9% and 95.7%, respectively. At cut-offs of 55 nmol/L and 4 nmol/L, sensitivity and specificity were 99.4% and 99.7%, respectively. Using a previously estimated prevalence of 15%, cut points of 55 nmol/L and 4 nmol/L resulted in negative and positive predictive values of 99.8% and 97.9%, respectively. Even at lower prevalence rates, positive predictive values remained high at low cut points. Results of this study reaffirm the high sensitivity and negative predictive value of basal cortisol concentrations for the diagnosis of hypoadrenocorticism. However, the high specificity and positive predictive values at low cut points could be of major importance for veterinary practitioners. Prospective studies are needed to further substantiate these findings.

EN14

THYROID PROFILES IN HEALTHY KITTENS AGED TWO TO SIXTEEN WEEKS OF AGE

Christina Marino, Margret Casal

University of Pennsylvania, Matthew J. Ryan Veterinary Hospital, Philadelphia, PA, USA

Congenital hypothyroidism (CH) is associated with significant morbidity and mortality in pediatric felines. Thyroid hormones (TH) are essential for normal development of the nervous and skeletal systems. The paucity of information on TH levels and the inability to identify early histopathologic changes has led to CH underdiagnosis. The literature on TH levels in kittens is sparse and incongruent, comprised of mostly case reports. References state kitten total thyroxine (TT4) levels are 2 to 3 times higher than the adult cat TT4 level. They also make the assumption free thyroxine (fT4) levels are higher than the adult cat. An abstract was previously presented showing kitten TH levels did not exceed the adult cat normals. These values were measured with assays that are no longer used. The normal pediatric TH levels must be evaluated with the newer and currently available assays. The purpose of this study was to determine the TT4, fT4, total tri-idiothyronine (T3), free tri-idiothyronine (fT3) and thyroid stimulating hormone (TSH) levels in healthy kittens aged 2 through 16 weeks measuring levels at weekly intervals using the current available diagnostic methods.

Normal, healthy kittens from a research colony in an approved facility at the University of Pennsylvania, School of Veterinary Medicine were used. One half to one milliliter of whole blood based on body weight was drawn by jugular venipuncture each week. Serum was immediately obtained and frozen until analysis. The samples were submitted to the Michigan State University Diagnostic Laboratory for Animal and Population Health. All TH measurements were performed on serum. The TSH was measured using a canine chemiluminescent assay. The TT4, TT3 and fT3 were measured with radioimmunoassays and the fT4 was measured using equilibrium dialysis. All assays have been previously validated in cats.

Samples were collected at the same time in the morning at weekly intervals. All values are reported as the mean. The TT4 was initially above the adult cat reference range (ACRR; 10-47 nmol/L) at 2 weeks and peaked at 6 weeks (50.9 nmol/L and 64.6 nmol/L, respectively). The TT4 returned to within the ACRR at 8 weeks, was elevated again at 10 weeks (50.0 nmol/L), returned within the ACRR at 13 weeks and remained stable through 16 weeks. The fT4 remained within the ACRR (10-53 pmol/L) through all 16 weeks, but showed a similar peak at 6 and 10 weeks (50 pmol/L and 39.8 pmol/L, respectively). The TSH was within the ACRR (0-0.38 ng/ml) throughout the 16 weeks. The TSH remained stable at 0.16 ng/ml from 2 through 4 weeks, decreased within the ACRR at 8 weeks, peaked at 10 weeks (0.18 ng/ml) and then continuously decreased to 0.06 ng/ml at 16 weeks. The TT3 and fT3 remained within the ACRR through all 16 weeks. Both hormones peaked at 6 weeks and remained at that level through 16 weeks.

The TH levels in pediatric kittens were measured with the newer assays and with the exception of TT4, all TH levels were within the ACRR. The TT4 did not elevate more than twice the upper limit of the ACRR at any time point and was consistently within the ACRR after 12 weeks of age. A peak in TT4, fT4, TT3 and fT3 was seen at 6 weeks and a peak in TT4 and fT4 was also seen at 10 weeks. TSH levels using the canine TSH assay have been validated for use in adult cats, but these values have not yet been reported in kittens. The TSH levels remained within the ACRR and it showed a similar peak at 10 weeks. These patterns likely occur due to an increased growth and development at 6 and 10 weeks.

EN15

SPECTROPHOTOMETRY AND ULTRACENTRIFUGATION FOR MEASUREMENT OF PLASMA LIPIDS IN DOGS WITH DIABETES MELLITUS

Eileen Seage, Kenneth Drobatz, Rebecka Hess

Matthew J. Ryan Veterinary Hospital of the University of Pennsylvania, Philadelphia, PA, USA

Conflicting results have been reported regarding lipid concentrations in dogs with diabetes mellitus (DM). Some studies have reported that low-density lipoprotein-cholesterol (LDL-C) is not significantly different in dogs with and without DM. The goal of this study was to report total cholesterol (TC), total triglycerides (TG), high-density lipoprotein-cholesterol (HDL-C), very low-density lipoprotein-cholesterol (VLDL-C), and LDL-C in fasted dogs with and without DM.

Twenty-two dogs with DM and nine healthy control dogs with normocholesterolemia were prospectively enrolled following a minimum eight-hour fast. Plasma lipid fractions were measured by ultracentrifugation (TC, HDL-C, VLDL-C, and LDL-C) or direct sample analysis (TG) coupled with enzymatic reactions and spectrophotometric end points (Roche COBAS c311). Median lipid concentrations in dogs with and without DM were compared using a two-sample Wilcoxon rank-sum test. A p-value <0.05 was considered significant.

In dogs with DM, median TC (343mg/dl), TG (98 mg/dl), HDL-C (196mg/dl), VLDL-C (39.5mg/dl), and LDL-C (67.5mg/dl) concentrations were significantly higher than concentrations in healthy dogs (197mg/dl, 57mg/dl, 168mg/dl, 12mg/dl, 16mg/dl, respectively, $p<.05$ for all comparisons). The greatest difference was measured in LDL-C concentration, which was 4.2 times higher in dogs with DM compared to controls. VLDL-C, TC, and TG concentrations were 3.3, 1.7, and 1.7 times higher, respectively, in dogs with DM compared to healthy dogs. The smallest difference was noted in HDL-C concentrations.

It is concluded that the increase in LDL-C is an important component of dyslipidemia in dogs with DM. Future studies of therapeutics focused at decreasing LDL-C in dogs with DM may be warranted.

EN16

EFFECT OF HYDROCORTISONE ADMINISTRATION ON LEPTIN AND ADIPONECTIN SYNTHESIS IN HEALTHY DOGS

Hye-Ryung Choo, Changhwan Ahn, Woon-Bum Baek, Bohye Shin, Hakhyun Kim, Ji-Houn Kang, Eui-Bae Jeung, Mhan-Pyo Yang

Chungbuk National University College of Veterinary Medicine, Cheongju, Chungbuk, Republic of Korea

The objective of the present study was to examine whether circulating concentrations of leptin and adiponectin, distribution of abdominal fat, and mRNA expressions of leptin and adiponectin in abdominal adipose tissue were affected by hydrocortisone administration. Six laboratory dogs each received hydrocortisone (8.5 mg/kg) orally every 12 hours for 90 days. We measured the serum concentrations of leptin and adiponectin using canine-specific ELISA kits. We quantified visceral fat using CT scanning, and we analyzed the mRNA expressions of leptin and adiponectin in abdominal fat using real-time PCR.

Hydrocortisone administration resulted in increased abdominal visceral fat mass on days 30, 60, and 90 compared with fat mass before administration (day 0). Additionally, visceral fat area at the L3 level also was increased during hydrocortisone administration. Serum leptin concentrations began to increase on day 1 of hydrocortisone administration and were significantly increased on days 1, 3, 7, 30, 60, and 90 compared with concentrations on day 0. Serum adiponectin concentrations on days 1, 3, 7, 30, 60, and 90 were decreased compared with concentrations on day 0. The mRNA expressions of leptin and adiponectin in the abdominal fat were increased on day 30 compared with expressions on day 0, but the expressions were decreased on days 60 and 90 compared with those on day 30. We identified a correlation between serum leptin and adiponectin concentrations and visceral fat distribution. This study showed that hydrocortisone administration affected visceral fat distribution and serum concentrations of leptin and adiponectin by dysregulating leptin and adiponectin expression.

EN17

USING PURIFIED FELINE INSULIN TO EVALUATE CROSS-REACTIVITY WITH A HUMAN INSULIN ANALOG ELISA

Jon Fletcher, Hannah Pipe-Martin, Chin-chi Liu

Department of Veterinary Clinical Sciences, School of Veterinary Medicine, Louisiana State University, Baton Rouge, LA, USA

Recombinant DNA technology has been used to modify the structure of human insulin to yield analogs that are more rapidly-acting than regular insulin or have a longer duration of action than NPH insulin. The long-acting insulin analogs have been studied in cats and are routinely used to manage feline diabetics. More recently, the isoglycemic clamp method has been used to study the metabolic effect of a rapid-acting insulin analog in healthy cats. Generating time-action profiles to study the pharmacodynamic properties of insulin formulations does not require measurement of plasma insulin concentrations, but if measured, they provide useful pharmacokinetic data that can be used to validate the time-action profiles. The Iso-Insulin ELISA (Mercodia AB, Uppsala, Sweden) is used to measure the concentration of insulin analogs in plasma. A previous study using actual feline serum samples with a wide range of insulin concentrations measured by a commercial ELISA failed to detect cross-reactivity of feline insulin with the Iso-Insulin ELISA.

The purpose of this study was to evaluate the cross-reactivity of feline insulin with the Iso-Insulin ELISA using purified feline insulin standards. Five purified feline insulin standards with known concentrations ranging from 9.13 ng/L to 702 ng/L that are used as calibrators for the commercial feline insulin ELISA (Mercodia AB, Uppsala, Sweden) were used. The ELISAs were performed in duplicate according to the manufacturer's recommendations. The mean absorbance values of the feline insulin ELISA calibrators were below the mean absorbance value of the lowest Iso-Insulin calibrator (0.087) except for the first calibrator, which had a slightly higher mean absorbance of 0.099. The slightly higher absorbance for the first calibrator was likely the result of a minor technical issue with this very sensitive ELISA since all calibrators with higher insulin concentrations consistently had absorbances at or below the lowest Iso-insulin calibrator.

These results demonstrate that purified feline insulin concentrations up to 702 ng/L are not detected by the Iso-Insulin ELISA, and therefore will not interfere with insulin analog measurements in feline plasma.

EN18

STEROID HORMONE PANEL TEST AND TRILOSTANE AND MELATONIN THERAPY IN POMERANIAN DOGS WITH ALOPECIA X

Melissa Sanches[1], Marcia Jericó[2], Eric Januário[1]

[1]Pet Care Centro Veterinário, Sao Paulo, SP, Brazil, [2]Universidade Anhembi Morumbi and Clinica Veterinária Alto da Lapa, Sao Paulo, SP, Brazi

Alopecia X is a skin disease of unknown etiology characterized by non-inflammatory, bilateral and symmetric alopecia, with hyperpigmentation and absence of pruritus. This condition is most frequently observed in young dogs of Nordic breeds, and is thought to be related to abnormal adrenal gland function, with trilostane and melatonin having been proposed as possible treatment alternatives. This study aims to evaluate the comprehensive steroid panel and therapeutic response of Pomeranian dogs diagnosed with Alopecia X. Eighteen Pomeranian dogs admitted at veterinary clinics (São Paulo, Brazil) with a chief complaint of symmetric and bilateral alopecia were studied. These included 12 males and 6 females, 16 of which were neutered, aging from 2 to 5 years. All animals were subjected to thorough laboratory screening: complete blood count with differential leukocyte count, liver function enzymes, urea, creatinine, triglycerides, cholesterol, blood glucose, and thyroid panel, as well as urinalysis and abdominal ultrasonography; all exam results were within the reference values. Subsequently, blood samples were collected to quantify by radioimmunoassay at Provet Laboratories (Brazil) and Tennessee University the following hormones, both before and after synthetic ACTH stimulation (single dose 0.05 mg/kg/EV): 17-hydroxiprogesterone, aldosterone, androstenedione, cortisol, estradiol, progesterone and testosterone. Eight dogs were then subjected treatment A (trilostane 1mg/kg PO BID), seven dogs were subjected to treatment B (trilostane 1mg/kg PO BID and melatonin 3 mg/dog PO BID) and two dogs received treatment C (melatonin 3 mg/dog PO BID); three animals were lost to follow-up. In comparison to reference values established by the laboratories for both neutered and non-neutered dogs, the animals with Alopecia X presented higher levels in three of the seven hormones both before and after the ACTH stimulation test. Abnormally elevated hormone levels were observed prior to and following ACTH stimulation in respectively: 3 (17%) and 16 dogs (89%) for 17-hydroxiprogesterone, 9 (50%) and 10 dogs (55%) for androstenedione, and 2 (11%) and 16 dogs (89%) for progesterone. Also, the abnormally elevated testosterone levels were observed in 6 dogs (33%) prior to ACTH stimulation and 8 dogs (44%) following ACTH stimulation. Two dogs presented abnormally decreased estradiol levels, both prior to and following ACTH stimulation. On the other hand, abnormally elevated cortisol levels were observed in 3 dogs (17%) prior to ACTH stimulation, and these elevated levels were maintained by only one dog (6%) following ACTH administration. In contrast, all dogs had aldosterone levels within reference ranges, with the exception of one dog with slightly elevated aldosterone levels prior to ACTH stimulation. Of the dogs subjected to treatment A, four (50%) had complete hair regrowth and three (37.5%) had partial regrowth after three months of treatment. Of the dogs subjected to treatment B, one (14%) had complete hair regrowth, four (57%) had partial regrowth, and one (14%) did not show signs of regrowth. Treatment for one of the two dogs subjected to treatment C had to be withdrawn due to emesis. Pomeranian dogs affected by Alopecia X presented increased levels of 17-hydroxiprogesterone, androstenedione, testosterone and progesterone, both prior to and following stimulation with synthetic ACTH, suggesting an adrenal gland increased function in the sex hormones production. Due this, trilostane, as a steroidogenesis blocker, was found to be a positive therapeutic option for Alopecia X, with or without the association of melatonin.

EN19

METABOLITE DIFFERENCES BETWEEN HEALTHY SENIOR BURMESE AND NON-BURMESE CATS, AND ASSOCIATIONS BETWEEN METABOLITES AND MEASURES OF GLUCOSE METABOLISM

Mia Reeve-Johnson[1], Jacquie Rand[1], Stephen Anderson[2], Diane Vankan[1], Daniel Dias[3], Berin Boughton[3], Alysha De Livera[3], Katsumi Ishioka[4], Ute Roessner[4]

[1]School of Veterinary Science, University of Queensland, Brisbane, Australia, [2]School of Biomedical Sciences, University of Queensland, Brisbane, Australia, [3]Centre for Molecular, Environmental, Genetic and Analytic (MEGA) Epidemiology, Melbourne School of Population and Global Health, The University of Melbourne, Melbourne, Australia, [4]School of Veterinary Nursing & Technology, Faculty of Veterinary Science, Nippon Veterinary and Life Science University, Tokyo, Japan

Phenotypic identification of cats at risk of developing diabetes has yet to be defined. Burmese cats are reported to have a dyslipidemia and are 3 to 4 times more likely to develop diabetes than other breeds.

To identify plasma metabolite differences between healthy senior Burmese and matched non-Burmese cats ranging from lean to obese, and to determine if significant correlations exist between these metabolites and measures of glucose metabolism, hormonal and biochemical related to diabetes.

Of the 69 cats tested, 49 non-Burmese; 20 lean (BCS 4-5/9), 12 overweight (BCS 6-7/9) and 17 obese (BCS 8-9/9) and 20 were Burmese (6 lean, 12 overweight and 2 obese).

Food was withheld 18 - 24 hr, and a cephalic vein catheter implanted. A venous blood sample was collected and the serum and plasma separated and frozen at -80 °C. A glucose tolerance test (glucose 0.5 g/kg) was performed. Biochemical (screening, fasting and 2- hr blood glucose in the glucose tolerance test, triglyceride), hormonal (leptin, adiponectin, leptin:adiponectin ratio,insulin, glucose:insulin ratio) and enzymatic (fPLI, MCP-1) analyses were performed on plasma and serum samples. After extraction, analyses using GC-MS was performed. Linear modeling was used to identify statistically significant (adjusted p-value<0.05) metabolites associated with obesity and whether these responses were affected by sex and age. Pearson correlations were used to identify relationships between metabolites and other parameters.

GC-MS analyses identified 25 metabolites significantly (p<0.05) different between Burmese and non-Burmese cats. In Burmese cats, metabolites that are involved in amino acid metabolism were either increased (phenylalanine, tyrosine, valine, serine, threonine, cysteine, proline and arginine) or decreased (cysteine, alanine and b-alanine) compared to non-Burmese cats. Some metabolites in carbohydrate metabolism were decreased (glucose, fructose, lactic acid), as were some involved in lipid metabolism (glucose 3 phosphate, phosphoric acid, stearic acid, cholesterol). When these metabolites were correlated with known measures of glucose metabolism, numerous amino acids were positively correlated with insulin and triglyceride (tyrosine, phenylalanine cysteine, proline, glutamine and arginine were all positively correlated with insulin, and tyrosine and phenylalanine were also correlated with triglyceride). These results agreed with human studies where future risk of developing diabetes was found to be associated with these changes in metabolites. The amino acid metabolites alanine, threonine and tyrosine were negatively correlated with adiponectin and glucose and fructose were positively associated with 2-hr blood glucose following a glucose tolerance test. However, some metabolites considered discriminators of impaired fasting glucose and increased in humans at risk of diabetes (phosphoric acid, alanine and stearic acid) were decreased in Burmese cats, as were metabolites of glycerolipid metabolism (glycerate, glycerol 3 phosphate, glucose and fructose).

This study is the first to report metabolite differences between healthy non-Burmese and Burmese cats and identifies some biomarkers involved in amino acid, carbohydrate and lipid metabolism that are altered in Burmese cats in a similar pattern to those identified in humans at risk of diabetes. Further research is required to determine if these are useful as markers of metabolic dysfunction in cats at risk of developing diabetes.

EN20

A GENOME-WIDE ASSOCIATION STUDY IDENTIFIES NOVEL CANDIDATE GENES FOR SUSCEPTIBILITY TO DIABETES MELLITUS IN DSH CATS
(*ESVE Award Winner*)

Y. Forcada[1],M. Boursnell[2], B. Catchpole[1], D.B. Church[1]

[1]*The Royal Veterinary College, North Mymms, UK,* [2]*Animal Health Trust, Newmarket, UK*

Diabetes mellitus (DM) is a common feline endocrinopathy and pathophysiologically similar to human type 2 diabetes (T2DM). T2DM occurs due to a combination of insulin resistance and β-cell dysfunction. Several studies have identified environmental and genetic susceptibility factors for T2DM. In cats, environmental factors such as obesity and physical inactivity have been linked with DM; however, identification of genetic factors has been challenging. To date, MC4R is the only gene shown to be associated with increased susceptibility to DM in overweight domestic short hair (DSH) cats. The aim of the present study was to perform a genome-wide association study (GWAS) to identify loci associated with DM in lean DSH cats.

Illumina Infinium 63k iSelect DNA arrays were used to interrogate genomic DNA samples from 200 lean diabetic DSH cats from the Royal Veterinary College Feline DM Archive and 400 control DSH cats. The data was analysed using PLINK whole genome data analysis toolset. Significance was established at p < 1 x 10- 5. SNPs with a minor allele frequency below 0.05 and a call rate below 95% and individuals with a genotyping rate < 90% were excluded from analysis.

A total of 49,930 SNPs were available for analysis. After excluding cats with low genotypic rate, 389 control DSH and 192 lean diabetic DSH cats were evaluated. Diabetic cats had a mean (SD) age of 11.62 (3.44) years; 123 (63%) were male, 71 (37%) female. Nondiabetic cats had a mean (SD) age of 14.83 (2.06) years; 216 (54%) were female, 183 (46%) male. Control cats were significantly older than diabetic cats (p < 0.0001; t-test). Five significant SNPs were identified: chrA2.4150731 (p = 1.4 x 10- 7); chrUn17.115508 (p = 7 x 10- 7); chrUn17.394136 (p = 3 x 10- 7); chrUn17.314128 (p = 3 x 10- 7) and chrUn17.7283 (p = 9 x 10- 6). The first SNP is located within chromosome A2; the others are located within a 0.8 Mb region towards the end of chromosome A3. The SNP in chromosome A2 is located 3kb upstream of dipeptidyl-peptidase-9 (DPP9), a peptidase similar to DPP-4, involved in incretin inactivation. Within the identified region of chromosome A3, genes of interest include TMEM18 and ACP1; both have been associated with T2DM in humans, most likely causing insulin resistance. This is the first GWAS of DM in cats. A number of significant SNPs have been identified, some of which are located in proximity to genes that have been associated with T2DM in humans; others could be involved in pathophysiology related to DM. Further investigation of these candidate genes is warranted.

DISCLOSURES
Disclosures to report.
SNP chips for the GWAS were provided by the Morris Animal Foundation.

GI01

SERUM PANCREATIC LIPASE IMMUNOREACTIVITY CONCENTRATIONS IN DOGS WITH GASTROINTESTINAL FOREIGN BODIES

LAUREN COCHRAN[1], STEVE HILL[1], JAN SUCHODOLSKI[2], JONATHAN LIDBURY[2], JOERG STEINER[2]

[1]VETERINARY SPECIALTY HOSPITAL OF SAN DIEGO, SAN DIEGO, CALIFORNIA, USA, [2]COLLEGE OF VETERINARY MEDICINE, TEXAS A&M UNIVERSITY, COLLEGE STATION, TEXAS, USA

Elevated serum canine pancreatic lipase immunoreactivity (cPLI) concentrations are highly specific for pancreatitis. Additionally, some studies in dogs with primary gastrointestinal (GI) disease have suggested a negative outcome in dogs with increased serum cPLI concentrations. GI disease, including foreign bodies (FBs) can mimic clinical signs of pancreatitis. GI FBs have also been proposed as a risk factor for the development of pancreatitis in dogs. To date the prevalence of pancreatitis in dogs with GI FBs is unknown. The primary aim of this study was to determine the prevalence of elevated cPLI concentrations (as measured by Spec cPL®) in dogs with GI FBs, and to correlate Spec cPL concentrations with patient survival. A secondary aim was to compare Spec cPL concentrations among dogs based on age, FB type, FB location, and type of removal procedure.

Serum Spec cPL was measured in 49 dogs prospectively enrolled with endoscopically or surgically confirmed GI FBs from May 2014 to May 2015. Samples were obtained the day of the procedure prior to anesthetic induction. Dogs were retrospectively grouped based on age (≤4 years, n=22; 5-8 years, n=11; ≥9 years, n=16), FB type (linear, n=14; discrete, n=35), FB location (stomach, n=21; duodenum, n=3; jejunum, n=12; involving multiple segments, n=13) and type of removal procedure (endoscopy, n=19; surgery, n=30). The prevalence of a serum Spec cPL above the upper limit of the reference interval (≥ 200 µg/L) or the suggested cut-off value for pancreatitis (≥ 400 µg/L) was determined in all dogs. Serum Spec cPL concentrations were compared amongst groups using Kruskal-Wallis or Mann-Whitney U tests, as appropriate. Statistical significance was set at $P < 0.05$.

A Spec cPL≥ 200 µg/L was identified in 12/49 dogs (24.5%), 9 of which (18.4%) had a Spec cPL≥ 400 µg/L. Spec cPL (median, range) was higher in dogs 5-8 years (125 ug/L, <30-2885 µg/L) and ≥9 years (233 ug/L, 43-2816 µg/L) of age than in dogs ≤4 years (59 ug/L, <30-246 µg/L; P <0.0001) of age. Spec cPL was higher in dogs with linear FBs (376 ug/L, <30-2885 µg/L) than in dogs with discrete FBs (68 ug/L, <30-837 µg/L; P = 0.0038). Spec cPL was higher in dogs with FBs affecting multiple GI segments (569 ug/L, <30-2885 µg/L) than in dogs with gastric FBs (60 ug/L, <30-185 µg/L; P = 0.0159). Spec cPL was higher in dogs undergoing surgery (108.5 ug/L, <30-2885 µg/L) than in dogs undergoing endoscopy (63 ug/L, <30-185 µg/L; P = 0.0322). Two of 49 dogs (4%) were euthanized during the course of the study. Both dogs had evidence of septic peritonitis at the time of euthanasia. In both dogs, Spec cPL concentrations were elevated (1292 ug/L and 2816 ug/L, respectively).

The overall prevalence of Spec cPL ≥ 200 ug/L and ≥ 400 ug/L in dogs with GI FBs was relatively low (24.5% and 18.4%, respectively). Spec cPL was higher in middle aged and older dogs, dogs with linear FBs, and dogs that needed to undergo surgery for FB removal. A Spec cPL≥ 400 ug/L was noted in both dogs that were euthanized. Due to this small number, additional studies would be needed to confirm if an elevated Spec cPL is associated with a negative outcome in dogs with GI FBs.

GI02

SERUM PANCREATIC LIPASE IMMUNOREACTIVITY CONCENTRATIONS AFTER CHRONIC ADMINISTRATION OF SUPRAPHYSIOLOGIC DOSES OF GLUCOCORTICOIDS TO DOGS

Sarah Cocker[1], Keith Richter[1], Joerg Steiner[2], Jonathan Lidbury[2], Jan Suchodolski[2]

[1]*Veterinary Specialty Hospital, San Diego, CA, USA,* [2]*Gastrointestinal Laboratory, Department of Small Animal Medicine and Surgery, College of Veterinary Medicine, Texas A&M University, College Station, TX, USA*

Prednisone and other glucocorticoids (GCs) are commonly used as long-term medications to treat a multitude of immune mediated, inflammatory, and neoplastic diseases in dogs. Patients with many of these diseases may have clinical signs that overlap with those of patients with pancreatitis. The effect of long-term GC administration on canine serum Spec cPL concentrations has not been adequately assessed. Therefore, the aim of this study was to determine if long-term administration of GCs at supraphysiologic doses affects serum concentrations of Spec cPL in dogs with a variety of different diseases.

Dogs were prospectively enrolled at the Veterinary Specialty Hospital of San Diego between February 2015 and October 2015. Inclusion criteria for enrollment included prescribed long-term (≥ 3 weeks) administration of at least 0.5 mg/kg/day prednisone-equivalent of GCs. Serum Spec cPL concentration was measured prior to initiation of GC therapy so each dog served as its own control. A second measurement of serum Spec cPL concentration was obtained in all dogs after at least 3 weeks of treatment with GCs. Only dogs that had no dosage change within one week of sampling were evaluated. The dose of GCs recorded was the dose that was being administered at the time that the second serum sample was collected. Cases were excluded if there was poor owner compliance with GC administration (defined as missing more than two consecutive doses of GCs) or if there was administration of other medications known or suspected to increase serum Spec cPL concentrations (e.g. potassium bromide or phenobarbital) or GC administration within the 3 months prior to enrollment. Comparison of serum Spec cPL between pre GC administration and ≥ 3 weeks post GC administration was performed using Wilcoxon matched-pairs signed rank test. Statistical significance was set as $P < 0.05$.

A total of 23 dogs were enrolled. Eight cases were excluded due to euthanasia (n=3), lack of follow up for second sample (n=3), GC dose <0.5 mg/kg/day at time of second sample collection (n=1), and change in GC dose < 1week before second sample collection (n=1). Fifteen cases met the inclusion criteria. Of the 15 cases enrolled reasons for chronic GC administration included: immune mediated thrombocytopenia (n=4), immune mediated hemolytic anemia (n=4), protein losing enteropathy (n=3), histiocytic sarcoma (n=1), suspected immune mediated pancytopenia (n=1), myelofibrosis (n=1), and mast cell tumor (n=1). The median Spec cPL concentrations for pre and post GC administration were 68 µg/L (30 – 1,997 µg/L) and 165 µg/L (30 – 1,628 µg/L), respectively. There was no statistically significant difference of serum Spec cPL concentrations between baseline and post GC therapy of ≥ 3 weeks in this group of patients ($P = 0.19$; Figure 1).

The results of this study did not reveal a statistically significant difference in SPEC cPL concentrations in dogs prior to administration of GC and ≥ 3 weeks after administration of supraphysiologic doses of GCs. Thus, the administration of GCs should not compromise the clinician's ability to interpret results of SPEC cPL concentrations.

GI03

SERUM IL-2, IL-6, IL-8, AND TNF-α CONCENTRATIONS IN DOGS WITH INCREASED SERUM SPEC cPL® CONCENTRATIONS

Agostino Buono[1], Andrea Petrelli[2], Jonathan Lidbury[1], Jan Suchodolski[1], Joerg Steiner[1]

[1]*Gastrointestinal Laboratory, Department of Small Animal Clinical Sciences, Texas A&M University, College Station, Texas, USA,* [2]*Department of Veterinary Medical Sciences, University of Bologna, Ozzano dell'Emilia, Italy*

The role of cytokines in pancreatitis is well characterized in human medicine. However, limited information is available about their role in dogs with pancreatitis. The aim of this study was to describe serum concentrations of interleukin (IL) -2, IL-6, IL-8, and tumor necrosis factor-alpha (TNF-α) in dogs with increased serum concentrations of pancreatic lipase immunoreactivity (cPLI; as measured as Spec cPL®).

Forty-nine surplus serum samples from dogs with a serum cPLI concentration >1000 µg/L were used for this study. Ten healthy dogs with serum cPLI concentration <200 µg/L were used as controls. Serum cytokine concentrations were measured using previously analytically validated electrochemiluminescence immunoassays. Data were analyzed using nonparametric statistics. Significance was set at $p<0.05$.

The cytokine concentrations were significantly increased in dogs with serum cPLI concentrations >1000 µg/L when compared to healthy controls, (increased cPLI group median [minimum-maximum] vs. healthy group median [minimum-maximum]): IL-2 (18.9 pg/ml [0-559.2] vs. 7.4 [1.7-56.8], $p=0.0167$); IL-6 (17.0 pg/ml [0-1085] vs. 4.5 [0.6-88] $p=0.0003$); IL-8 (6054.4 pg/ml [667.9-19664.9] vs. 2849.8 [1538.5-4247.5], $p=0.0029$); TNF-α (1.2 pg/ml [0–115.1 pg/ml] vs. 0.3 [0.1-0.6], $p=0.0014$). No correlation was found between age, sex, or serum cPLI concentration and any of the serum cytokine concentrations.

In conclusion, serum IL-2, IL-6, IL-8, and TNF-α concentrations were significantly increased in dogs with serum cPLI concentrations >1,000 µg/L. Prospective studies are needed to assess the utility of these cytokines as early diagnostic markers or prognostic markers in dogs with pancreatitis.

GI04

FISHHOOK FOREIGN BODIES IN DOGS AND CATS: 107 CASES (2004-2015)

Austin Hardegree[1], James Barr[1], Micah Bishop[2], Medora Pashmakova[1]

[1]*Texas A&M University, College Station, TX, USA,* [2]*Animal Specialty Hospital, Naples, FL, USA*

Fishhook foreign bodies are a common presentation in certain geographic areas and often require emergent removal from the gastrointestinal tract or skin. In the authors' institution, endoscopic retrieval is considered highly successful in contrast to previous literature. The objective of this study was to describe location, retrieval, complications, and outcomes associated with fishhooks.

Electronic medical records of cats and dogs admitted between 2004 and 2015 for fishhook foreign bodies were searched for radiographic location of fishhook, retrieval method, and documentation of complications and outcome. Data was tested for normality. A Mann-Whitney test or odds ratios were constructed to compare association between different variables.

One hundred and seven cases (5 cats and 102 dogs) with a total of 111 fishhooks were included. Single shanked hooks comprised 52/111 (46.85%) and multiple shanked hooks comprised 59/111 (53.15%) of cases. Most common locations were the pharyngeal/oral region 38/111 (34.26%)), cervical esophagus (25/111 (22.52%)), and stomach (27/111 (24.32 %)). Three hooks were cutaneously embedded. Hooks with multiple shanks were 2.314 times more likely to be associated with esophageal damage than single shanked hooks (95% CI: 1.038-5.159; P=0.0478). Endoscopic retrieval was successful in 60/69 (87.0%) cases in which it was attempted. Esophageal perforation occurred in 2 endoscopic cases and a total of 9 cases necessitated surgery. Survival was 100%.

Our study supports that endoscopic retrieval is highly successful and survival is 100%, even in cases requiring surgery. While multiple shanked fishhooks are more likely to cause esophageal damage, neither type of hook nor location affected outcome.

GI05

PERTURBATIONS OF THE INTESTINAL MICROBIOTA AND BILE ACID METABOLISM IN DOGS WITH DIABETES MELLITUS

Alana Redfern-Allen[1], Blake Guard[2], Jan Suchodolski[2], Joerg Steiner[2], Jonathan Lidbury[2], Albert Jergens[1]

[1]*Iowa State University, Ames, IA, USA,* [2]*Texas A&M University, College Station, TX, USA*

The intestinal microbiota has been recently linked to the development of human metabolic diseases including diabetes mellitus (DM), due to its role in energy balance, gut permeability, and host inflammatory state. Microbial imbalances may also affect bile acid (BA) metabolism through altered profiles of primary and secondary BA, which modulate signaling pathways involved in insulin resistance via GLP-1 induction. Characterization of the intestinal microbiota and BA in dogs with DM has not been systematically evaluated. Therefore, the aim of this study was to investigate the fecal microbiota and BA composition of healthy dogs and dogs with DM.

Fecal samples were collected from 27 heathy control dogs and 23 diabetic dogs. The fecal microbiota was evaluated prior to intervention using quantitative polymerase chain reaction (qPCR) for *E. coli*, *Streptococcus*, *Turicibacter*, *Faecalibacterium*, *Fusobacterium*, and *Blautia*. The fecal concentrations of primary (i.e., cholic and chenodeoxycholic acid) and secondary (i.e., litho-, deoxy-, and ursodeoxycholic acid) BA were measured using gas chromatography with mass spectrometry. Non-parametric Mann-Whitney *U* tests were used to compare these parameters between healthy dogs and dogs with DM. Statistical significance was set at $P<0.05$.

Results indicate that DM dogs exhibit imbalances in their fecal microbiota characterized by decreased numbers of *Turicibacter* ($P=0.026$) and *Faecalibacterium* (P=0.002) as compared to healthy dogs. Perturbations in BA metabolism were observed in the DM cohort and involved an increase in primary BAs and a decrease in secondary BAs ($P<0.001$ for both).

We conclude that dogs with DM have significant imbalances in intestinal microbiota and also altered bile acid metabolism, which may contribute to disease pathogenesis and morbidity. Further studies are warranted to elucidate mechanisms behind these alterations, and whether they are primarily due to disease or secondary to environmental factors including diet and antibiotics.

GI06

VARIATION OF THE MICROBIOME AND METABOLOME ALONG THE CANINE GASTROINTESTINAL TRACT

JB Honneffer, JM Steiner, JA Lidbury, JS Suchodolski

Gastrointestinal Laboratory, Texas A&M University, College Station, TX, USA

The gastrointestinal tract (GIT) of the dog is a complex ecosystem and its biological and biochemical compositions vary along the length of the GIT. The analysis of fecal samples is frequently used as a surrogate indicator for gastrointestinal disease occurring in more proximal intestinal sections. While the canine gastrointestinal biogeography of the microbiome has previously been described, corresponding metabolite profiles at various points along the GIT of healthy dogs have not been characterized to our knowledge. Therefore, the aim of this study was to perform 16s rRNA gene sequencing and untargeted metabolomics on samples from multiple sites to characterize the metabolome and microbiome along the GIT of healthy dogs.

Samples of the duodenal, ileal, colonic, and rectal contents were collected from six adult dogs immediately after humane euthanasia for another study. The microbiota was characterized using Illumina sequencing of the 16S rRNA gene. The metabolome was characterized by untargeted mass spectrometry-based methods. Inter-animal variation was statistically addressed by blocking in the software JMP, with subsequent Wilcoxon tests and Benjamini-Hochberg correction for multiple comparisons. Statistical significance was set at $p<0.05$.

Predominant phyla throughout the samples were Proteobacteria, Firmicutes, Fusobacteria, and Bacteroidetes. Percentages of reads identified as Proteobacteria decreased aborally ($p=0.017$; duodenum, 62%; ileum, 48%; colon, 9%; rectum, 6%), while those of Firmicutes increased ($p=0.032$; 20%, 43%, 68%, 69%). On the gas chromatography platform, 530 unique compounds were detected, and 199 of these were assigned to named metabolites. 134 named metabolites had at least one site-pair comparison reaching statistical significance. Key fermentation compounds such as sugars and amino acids had varied site distributions: glucose increased progressively along the GIT along with 6-deoxyglucose ($p=0.047$ and 0.034, respectively). Nearly all amino acids exhibited a decrease between the ileum and colon (*e.g.*, alanine, citrulline, glutamine, glycine, lysine, methionine, tryptophan, and valine, all $p<0.03$).

In conclusion, the metabolome and microbiome vary along the canine GIT and further studies are needed to explore the effect of local disease at specific sites, to determine the host versus microbial origin of these metabolites, and possibly to enhance extrapolation from fecal assay results to improve our understanding of GI diseases.

GI07

EFFECTS OF A HYDROLYZED PROTEIN DIET AND METRONIDAZOLE ON THE FECAL MICROBIOME AND METABOLOME IN HEALTHY DOGS

Jan Suchodolski[1], Erin Olson[2], Julia Honneffer[1], Blake Guard[1], Amanda Blake[1], AlShawaqfeh Mustafa[1], Jörg Steiner[1], James Barr[1], Frederic Gaschen[2]

[1]Gastrointestinal Laboratory, Texas A&M University, College Station, TX, USA, [2]Louisiana State University, Baton Rouge, LA, USA

The abnormal interaction between the GI microbiome and the immune system is a major contributing factor to IBD. Treatment often includes the use of antimicrobials like metronidazole or the use of an elimination diet (i.e., hydrolyzed protein diet), but their effects on the intestinal microbiota are not well studied. Recent epidemiological studies in humans have linked gut dysbiosis due to antibiotics as risk factor for various chronic diseases, such as obesity, diabetes, and asthma. A better understanding of the functional effects of antibiotics and diets on the microbiome is needed. The aim of this study was to prospectively evaluate the impact of metronidazole administration vs. hydrolyzed protein diet on the metabolome of healthy dogs.

Twenty-four healthy pet dogs were assigned to one of 3 groups (8 dogs each): control dogs with no intervention (group 1, G1), dogs treated with a soy-based hydrolyzed diet (Purina HA) for 12 weeks with administration of metronidazole 15 mg/kg PO q12h during weeks 6 to 8 (G2), and dogs maintained on their usual diet and administered the same dose of metronidazole for 2 weeks (G3). Feces and blood were collected at day (D) 0, 21 and 42 in G1; and D0, 14, 28 and 42 days in G2 and G3. Microbial communities were analyzed by Illumina sequencing of 16S rRNA genes, and the software PICRUSt was used to predict functional gene families. The serum and fecal metabolome was assessed by an untargeted approach combining various mass spectrometry platforms. All assessed parameters were adjusted for multiple comparisons using the Benjamin Hochberg adjustment, and an adjusted $p<0.05$ was considered significant.

No significant differences in any of the evaluated parameters were observed in G1 (control group). Also, in G2, the dietary switch to the hydrolyzed diet (between baseline and week 6) did not lead to any significant changes in any of the evaluated parameters. However, when dogs in G2 and the dogs in G3 received metronidazole, significant changes were observed. Because the responses in these 2 groups were similar, they were combined for statistical analysis. Microbiome structure and diversity were significantly altered between baseline and time of completion of metronidazole treatment, and also 4 weeks after end of metronidazole treatment (ANOSIM: $p=0.002$). The most significant changes were increases in *E. coli* and decreases in Firmicutes. Functional gene families found to be significantly more abundant during antibiotic administration included: lipopolysaccharide biosynthesis, glutathione metabolism, recombination and repair proteins, tryptophan metabolism, and fatty acid metabolism ($p<0.0001$ for all). At cessation of antibiotic administration (D14), functional gene families found to be significantly more abundant included: bile secretion, xylene degradation, dioxin degradation, and signal transduction mechanisms ($p<0.0001$ for all). Metronidazole lead to alteration in 98 out of 469 measured fecal metabolites. Major changes observed were reduction in secondary bile acids ($P<0.001$), increases in oxidative stress pathways ($P<0.0001$), and changes in tryptophan-indole pathways ($P<0.001$). While most of the changes were reversed 14 days after the end of antibiotic administration, some of the evaluated bacterial taxa and metabolites remained significantly altered up to 4 weeks after end of administration (end of study).

In conclusion, metronidazole has a profound effect on the fecal microbiome and metabolome, which only partially resolves after 4 weeks. Further studies are warranted to assess whether the dysbiosis and these disrupted metabolic pathways due to metronidazole may pose a risk factor for development of some chronic diseases. In contrast, the hydrolyzed diet had no noticeable effect on the microbiome and metabolome of healthy dogs.

GI08

ALTERED FECAL BILE ACID METABOLISM IN DOGS WITH CHRONIC ENTEROPATHY

Blake Guard[1], Linda Toresson[2], Julia Honneffer[1], Amanda Blake[1], Yuri Lawrence[1], Jonathan Lidbury[1], Joerg Steiner[1], Jan Suchodolski[1]

[1]*Texas A&M University, College Station, TX, USA,* [2]*Evidensia Specialist Animal Hospital/Helsinki University, Helsingborg,Sweden/Helsinki, Finland*

Chronic enteropathy in dogs has been characterized as a multi-factorial disease that is thought to involve inappropriate and ongoing activation of the mucosal immune system in addition to a microbial dysbiosis. Recently, studies in humans have suggested that in some patients with inflammatory bowel disease, microbial dysbiosis drives bile acid (BA) dysmetabolism. Primary BAs are synthesized by the liver and secreted into the intestinal lumen, where enzymatic reactions catalyzed by bacteria lead to BA biotransformation. Additionally, BAs participate in gastrointestinal mucosal defense. For example, secondary BAs have been shown to be anti-inflammatory as they have the potential to decrease the synthesis of proinflammatory cytokines such as TNF-α. Therefore, the aim of this study was to investigate the fecal BA profiles of dogs with chronic enteropathy.

Fecal samples were collected from healthy dogs (n=13) and dogs with chronic enteropathy with histologically confirmed inflammation (n=13). Fecal bile acids were analyzed by gas chromatography coupled with mass spectrometry. A Mann-Whitney *U* test was used to compare concentrations of BAs between groups. Cholic acid and chenodeoxycholic acid were among the primary bile acids measured, while deoxycholic acid, lithocholic acid, and ursodeoxycholic acid were among the secondary bile acids measured. Statistical significance was set at $p<0.05$.

No significant differences in fecal concentrations of primary bile acids were found between healthy dogs and dogs with chronic enteropathy. Deoxycholic acid was significantly decreased in dogs with chronic enteropathy (median [min-max]: 0.26 µg/mg [0.11-13.18 µg/mg]) compared to healthy dogs (median [min-max]: 4.64 µg/mg [0.11-18.94 µg/mg]; $p=0.0240$). Lithocholic acid was also significantly decreased in the dogs with chronic enteropathy (median [min-max]: 0.07 µg/mg [0.01-2.94 µg/mg]) compared to healthy dogs (median [min-max]: 1.05 ug/mg [0.03-3.21 µg/mg]; $p=0.0159$).

In conclusion, this study revealed a decrease in fecal secondary bile acids in dogs with chronic enteropathy. Further studies are needed to investigate the interplay between specific bacterial groups and altered bile acid metabolism as well as their relationship with intestinal inflammation in dogs with chronic enteropathy.

GI09

ALTERED FECAL STEROL PROFILES IN DOGS WITH CHRONIC INFLAMMATORY ENTEROPATHY

JB Honneffer[1], L Toresson[2], BC Guard[1], R Lopes[1], AB Blake[1], JA Lidbury[1], JM Steiner[1], JS Suchodolski[1]

[1]*Gastrointestinal Laboratory, Texas A&M University, College Station, TX, USA,* [2]*Evidensia Specialist Animal Hospital, Helsingborg, Sweden and University of Helsinki, Finland*

Diseases of chronic inflammation of the gastrointestinal tract, including idiopathic inflammatory bowel disease (IBD), are thought to be perpetuated by bacterial dysbiosis and dysregulation of the mucosal immune system in both humans and animals. Gastrointestinal (GI) absorption of metabolites is altered by inflammation, with concurrent changes in microbial metabolism within the GI tract. Cholesterol is the primary sterol in mammals; it is incorporated into cell membranes throughout the body and used as a building block for endogenous hormones, steroids, and bile acids. Some phytosterols have shown anti-inflammatory properties and have been shown to decrease disease severity in a murine colitis model. Therefore, the aim of this study was to compare fecal sterol profiles of dogs with chronic enteropathy to those of healthy dogs.

Fecal samples were collected from healthy dogs (n=13) and dogs with chronic enteropathy (n=13) with biopsy-confirmed inflammation. Sterols in lyophilized feces were subjected to trimethylsilyl ether derivatization and analyzed by gas chromatography-mass spectrometry (GC/MS) operating in Single Ion Monitoring (SIM) mode. Target analytes included cholesterol, cholestanol, β-sitosterol, sitostanol, fucosterol, stigmaterol, and campesterol. Fecal concentrations of sterols were expressed as µg/mg of lyophilized feces. A Mann-Whitney *U* test was used for comparison between animal groups with statistical significance set at $p<0.05$.

Fecal cholesterol and cholestanol were not significantly altered between healthy dogs and dogs with chronic enteropathy. However, phytosterols were significantly decreased in dogs with chronic enteropathy (healthy group median [min-max] vs. chronic enteropathy group median [min-max]): β-sitosterol (p=0.0066, 2.53 µg/mg [1.41-4.62] vs. 1.34 [0.084-3.70]); sitostanol (p=0.0001, 0.54 µg/mg [0.17-1.14] vs. 0.08 [0.01-0.45]); fucosterol (p=0.0138, 0.23 µg/mg [0.10-0.33] vs. 0.09 [0.01-0.80]); stigmasterol (p=0.0138, 0.46 µg/mg [0.23-0.66] vs. 0.30 [0.01-0.52]); and campesterol (p=0.0313, 0.84 µg/mg [0.45-1.76] vs. 0.53 [0.14-1.79]).

In conclusion, this study suggests that the fecal sterol profile in dogs with chronic enteropathy is characterized by decreased phytosterols. Additional studies are required to explore the effect of dietary supplementation of phytosterols and to investigate how these sterols are related to specific bacterial groups that digest dietary plant material in these patients.

GI10

THE FECAL MICROBIOME OF DOGS WITH EXOCRINE PANCREATIC INSUFFICIENCY.

Anitha Isaiah, Joseph Cyrus Parambeth, Jonathan A Lidbury, Joerg M Steiner, Jan S Suchodolski

Gastrointestinal Laboratory, Texas A&M University, College Station, Texas, USA

Exocrine pancreatic insufficiency (EPI) in dogs is a syndrome of inadequate synthesis and secretion of pancreatic enzymes, which leads to clinical signs of maldigestion. Previous reports show that small intestinal bacterial dysbiosis occurs in dogs with EPI and is reversed with pancreatic enzyme therapy. However, there are no studies evaluating the fecal microbiome of dogs with EPI. The objective of this study was to compare the fecal microbiome of healthy dogs (n=18), untreated (n=7) dogs with EPI, and dogs with EPI treated with enzyme replacement therapy (n=19).

To be included into the study, the dogs had to be at least 1 year of age, have clinical signs of EPI, a serum cTLI concentration ≤2.5 µg/L, and be free from any other concurrent disease. Three naturally voided fecal samples collected over three consecutive days were frozen immediately after collection and pooled. Fecal samples were collected in a similar manner from healthy dogs without any clinical signs of gastrointestinal disease. Extracted DNA from fecal samples was used for Illumina sequencing of the bacterial 16S rRNA gene and analyzed using Quantitative Insights Into Microbial Ecology (QIIME). The analysis of similarities (ANOSIM) function in the statistical software package PRIMER 6 (PRIMER-E Ltd., Luton, UK) was used on the unweighted UniFrac distance matrix to determine if any groups of samples contained significantly different bacterial communities. There was a significant difference in fecal microbial communities when healthy dogs were compared to treated (p=0.001) and untreated (p=0.001) dogs with EPI. Quantitative Insights Into Microbial Ecology (QIIME). The analysis of similarities (ANOSIM) function in the statistical software package PRIMER 6 (PRIMER-E Ltd., Luton, UK) was used on the unweighted UniFrac distance matrix to determine if any groups of samples contained significantly different bacterial communities. There was a significant difference in fecal microbial communities when healthy dogs were compared to treated (p=0.001) and untreated (p=0.001) dogs with EPI. Alpha diversity was significantly decreased in untreated and treated EPI dogs when compared to the healthy dogs (p<0.01). The families Bifidobacteriaceae (p=0.006), Enterococcaceae (p=0.035), and Lactobacillaceae (p=0.001) were significantly increased in the untreated and treated dogs with EPI when compared to healthy dogs. In contrast, Lachnospiraceae (p<0.001), and *Blautia* (p<0.001) were significantly decreased in dogs with EPI.

In conclusion, this study suggests that the fecal microbiome of dogs with EPI (both treated and untreated) is different from that of healthy dogs.

GI11

EFFECT OF WEIGHT LOSS AND DIET ON FECAL MICROBIOTA AND FECAL METABOLOMICS IN CATS

Adam Rudinsky[1], Katie McCool[1], Valerie Parker[1], Prosper Boyaka[1], Josh Daniels[1], Rosana Lopes[2], Joerg Steiner[2], Jan Suchodolski[2], Chen Gilor[1]

[1]The Ohio State University College of Veterinary Medicine, Columbus, Ohio, USA, [2]Texas A&M University College of Veterinary Medicine, College Station, Texas, USA

Intestinal microbiota and short chain fatty acids (SCFA) are important in a variety of metabolic disorders. We hypothesized that obesity and diet affect intestinal microbiota and SCFA in cats.

Eight cats were used in a repeated-measures study. Cats were fed a maintenance diet (M-diet) from study days -21-0 then switched to a weight-loss diet (WL-diet) on day 0 without calorie restriction until day 7. On days 8-84, daily intake was restricted to achieve 1-2% weight-loss per week. On days 84-98, feeding the M-diet resumed while maintaining stable body condition score (BCS). Fecal samples were collected and stored at -80°C until analysis. Samples were collected on day -7 and 0 (Obese/M-diet), 5 and 7 (Obese/WL-diet), 77 and 84 (Lean/WL-diet) and 86, 89, 91 and 98 (Lean/M-diet) for microbiota and SCFA profiles. Body weight, BCS, muscle condition score, and body fat percentage were assessed throughout the study.

Microbiota and SCFA profiles revealed changes associated with diet and BCS. Microbiota diversity and species richness alterations were noted. Acetic and butyric acid production was significantly different between obese and lean cats. Diet and BCS had significant interaction in the statistical model for remaining SCFA. On post-hoc analysis, isobutyric, valeric and isovaleric acids were significantly different between obese and lean cats on the maintenance diet. Valeric and isovaleric were different between the M-diet and WL-diet in obese cats.

Microbiota and SCFA profile alterations were noted based on body condition. Further study is warranted to determine the impact of these changes on inflammation and metabolic disorders.

GI12

PROBIOTIC MIXTURE VSL#3 INCREASES BENEFICIAL FECAL AND MUCOSAL MICROBIOTA IN CANINE INFLAMMATORY BOWEL DISEASE

Jan Suchodolski[2], Robin White[1], Blake Guard[2], Todd Atherly[3], Peter Sciabarra[1], Steve Hill[5], Craig Webb[4], Curtis Mosher[1], Giacomo Rossi[6], Albert Jergens[1]

[1]*Iowa State University, Ames, IA, USA,* [2]*Texas A&M University, College Station, TX, USA,* [3]*USDA-ARS, Ames, IA, USA,* [4]*Colorado State University, Fort Collins, CO, USA,* [5]*Veterinary Specialty Hospital, San Diego, CA, USA,* [6]*University of Camerino, Camerino, Italy*

Canine inflammatory bowel disease (IBD) is an immune-mediated enteropathy likely triggered by environmental and immunoregulatory factors in genetically susceptible dogs. Previous studies suggest a pivotal role for intestinal bacteria in disease pathogenesis since luminal microbial composition is markedly altered at diagnosis. It has been suggested that probiotic bacteria are effective and promising agents for the treatment of IBD in humans and animals. However, the effects of probiotic bacteria that directly interact with the host's resident intestinal microbiota are poorly understood. The aim of the present study was to investigate the impact of multi-strain probiotic VSL#3 supplementation (Visbiome; Exegi Pharma) at a dose of 50 billion bacteria/kg/day on the fecal and intestinal mucosal microbiota in dogs with IBD. Nineteen dogs diagnosed with moderate-to-severe IBD (CIBDAI $\geq$ 6) were randomized to receive standard IBD therapy (ie, elimination diet and prednisone) with or without probiotic VSL#3 for 8 weeks. Dogs were evaluated clinically at 0, 3, and 8 weeks and endoscopically and histopathologically at 0 and 8 weeks. The fecal and mucosal microbiota were investigated before and during treatment using 16S rRNA based quantitative polymerase chain reaction (qPCR) and fluorescence in situ hybridization (FISH), respectively. Both cohorts of IBD dogs showed a reduction in GI signs after 8 weeks of therapy compared to baseline CIBDAI scores (P<0.05). Using qPCR, the placebo group failed to show significant changes in any of the bacterial groups analyzed. In contrast, VSL#3 treated dogs showed a significant (P=0.001) increase in total lactic acid bacteria (*Bifidobacteria, Streptococci, Lactobacilli*) at treatment weeks 3 and 8 versus baseline values. *Bifidobacteria* were also significantly (P=0.018) increased at 3 and 8 weeks post-VSL#3 treatment. FISH analysis showed that VSL#3 altered the number and spatial distribution of most colonic mucosal bacterial groups including total bacteria and the number of *Bifidobacteria, Fecalibacteria, Lactobacilli, Streptococci,* and *Enterobacteriaceae* (P<0.05 for each). The number of mucus-laden and attaching *Bifidobacteria* and *Streptococci* were also significantly (P<0.05) increased in dogs receiving VSL#3. In conclusion, probiotic therapy of IBD dogs with VSL#3 increases the number of fecal and mucosal probiotic genera, especially the lactic acid bacteria. IBD dogs receiving standard therapy (placebo) showed little change in the composition of microbial communities despite exhibiting clinical improvement.

GI13

CHANGES IN INTESTINAL MACROPHAGE POPULATIONS FOLLOWING CLINICAL RESOLUTION IN DOGS WITH CHRONIC ENTEROPATHY

JULIEN R. S. DANDRIEUX[1], WAYNE KIMPTON[1], BARBARA BACCI[2], KARIN ALLENSPACH[3], ALBERT E. JERGENS[4], CAROLINE MANSFIELD[1]

[1]UNIVERSITY OF MELBOURNE, WERRIBEE, VICTORIA, AUSTRALIA, [2]UNIVERSITY OF SURREY, GUILDFORD, SURREY, UK, [3]ROYAL VETERINARY COLLEGE, LONDON, UK, [4]IOWA STATE UNIVERSITY, AMES, IA, USA

Chronic enteropathy (CE) is a multi-factorial disease, which involves aberrant immune responses to commensal bacteria or dietary antigens. Although macrophages have been shown to play an important role in human disease very little information is available in dogs. CD163 is a scavenger receptor specific for macrophages, and has been detected in canine duodenum. The aim of this study was to determine if intestinal CD163 positive cell ($CD163^+$) populations alter after successful treatment of CE. Calprotectin was used as an additional marker for macrophages.

Dogs were prospectively enrolled and underwent standard diagnosis and treatment trial for CE. Duodenal biopsies were taken before and after long-term resolution of clinical signs ($\geq$2 months). Serial histologic sections were stained for CD163 (using AM-3K) and calprotectin. Positively stained cells in five villous and crypt regions were manually counted. The percentages of positively stained cells from the total nucleated cells per 10,000 [illegible]^2 before and after treatment were compared with repeated-measure ANOVA.

Five dogs had dietary-responsive and three steroid-responsive CE. $CD163^+$ significantly increased following treatment in the villi (from 4.8% [0-12.6] to 6.2% [0.0-12.4], P=0.043), whereas $calprotectin^+$ significantly decreased (from 3.7 [0.0-11.7] to 2.0 [0.0-9.7]). Increased $CD163^+$ (5.3 [1.0-9.7] to 6.2 [0.8-12.5]) and decreased $calprotectin^+$ (1.3 [0.0-4.2] to 0.9 [0.0-4.5]) was also observed following treatment in the crypts, although this was not significant. Similar changes were seen in food and steroid responsive dogs.

These results suggest that macrophages may play a role in canine CE. Further characterization of macrophages is warranted to understand their role in CE.

GI14

HISTOPATHOLOGIC CHARACTERISTICS OF INTESTINAL BIOPSIES FROM DOGS WITH CHRONIC ENTEROPATHY WITH AND WITHOUT HYPOALBUMINEMIA

Sara Wennogle[1], Craig Webb[1], Simon Priestnall[2]

[1]*Colorado State University, Fort Collins CO, USA,* [2]*Royal Veterinary College, Hertfordshire, UK*

The term protein-losing enteropathy (PLE) is often applied to cases of chronic enteropathy (CE) when the disease process results in hypoalbuminemia. Multiple previous studies have identified hypoalbuminemia as a risk factor for a negative outcome in dogs with chronic enteropathy, yet it has not been determined whether the histopathology differs between CE dogs with and without hypoalbuminemia. The goal of this retrospective study was to compare histopathologic findings (as determined by the WSAVA scoring system) in dogs with CE with and without hypoalbuminemia.

Electronic medical records at Colorado State University were reviewed for dogs having intestinal biopsy performed between January 2010-July 2015. Included dogs had clinical signs compatible with CE of at least 3 weeks duration. Dogs with significant concurrent disease, potential non-GI causes of hypoalbuminemia, intestinal lymphoma, or for which a complete medical record could not be obtained were excluded. 83 cases were suitable for inclusion. Dogs were placed in either the CE (albumin ≥ 3.0 gm/dL), or CE w/ PLE (albumin <3.0 gm/dL) group. Recorded data included signalment, presenting complaint, clinical signs, additional clinicopathologic abnormalities, tissue types available, and method of biopsy procurement. A blinded pathologist (SLP) was asked to review all biopsies and apply the WSAVA scoring system to each intestinal biopsy. In cases where more than one tissue type was available (13/83), the tissue type with the highest total WSAVA score was used in the analysis.

Results are summarized in the table below. The most notable discrepancies between groups occurred in the categories of crypt distension (CD), lacteal dilation (LD), villous stunting (VS), and lamina propria neutrophils (LP N), with a greater proportion of dogs in the CE w/ PLE exhibiting those changes on histopathology. Villous epithelial injury (VEI) and intraepithelial lymphocytes (IEL) were also appreciably more common in dogs in the CE w/ PLE group. Lamina propria lymphocytes and plasma cells (LP L/PC) and lamina propria eosinophils (LP E) were observed in both groups. Mucosal fibrosis (MF) was uncommon in both groups.

Dogs with PLE secondary to CE had a greater percentage of lacteal dilation, injury to the villi, crypt distension, and lamina propria neutrophils than dogs with CE that had normal albumin.

Furthermore, although lacteal dilation was noted in 76% of PLE cases, it was listed as a primary diagnosis in only 5 cases. Further investigation is required to determine if these differences represent a difference in underlying etiology, or should alter treatment strategy in specific cases.

	LP L/PC	LP E	LP N	IEL	LD	VS	VEI	CD	MF
CE	22/46 (48%)	12/46 (26%)	7/46 (15%)	1/46 (2%)	16/46 (35%)	23/46 (50%)	4/46 (9%)	5/46 (11%)	0/37 (0%)
CE w/ PLE	23/37 (62%)	11/37 (30%)	17/37 (46%)	11/37 (30%)	28/37 (76%)	35/37 (95%)	14/37 (38%)	24/37 (65%)	2/37 (5%)
Total	45/83 (54%)	23/83 (28%)	24/83 (29%)	12/83 (14%)	44/83 (53%)	58/83 (70%)	18/83 (22%)	29/83 (35%)	2/83 (2%)

GI15

HUMAN GRANULOCYTE IMMUNOFLUORESCENCE ASSAY FOR ANTI-NEUTROPHIL ANTIBODIES SHOWS STRONG ASSOCIATION WITH CANINE FOOD RESPONSIVE DIARRHEA

Karin Allenspach[1], Shayna Streu[2], Valerie DiMuro[3], Anna Riddle[1], Joseph Kirk[1], Lucrezia Perazzotti[4], Regina Wagner[5], Elisbabeth Mueller[5], Jessica Florey[1]

[1]*Royal Veterinary College, London, UK,* [2]*Michigan State University, East Lansing, MI, USA,* [3]*Arcadia University, Philadelphia, PA, USA,* [4]*Euroimmun UK Ltd, Wimbledon, UK,* [5]*Laboklin AG, Bad Kissingen, Germany*

Perinuclear anti-neutrophilic antibodies (pANCA) have previously been shown to be serum markers in canine IBD, and more likely to be associated with food responsive diarrhea (FRD) cases than steroid responsive disease (SRD). The indirect immunofluorescence assay used in previous publications is based on isolation of canine granulocytes from donor dogs, which are subsequently fixed in ethanol on microscope slides, and then incubated with the patient serum and secondary antibody. This assay is time consuming and has been shown to have low inter-observer agreement. In this pilot study we therefore sought to investigate whether a commercially available human granulocyte indirect immunofluorescence assay could be used to detect pANCA in dogs and to evaluate preliminary sensitivity and specificity of the assay for groups of dogs with FRD *versus* SRD.

Forty-four dogs with FRD and 20 dogs with SRD were included in the study. Granulocyte Mosaic Biochip slides (EUROIMMUN AG) containing human granulocytes fixed in ethanol, human granulocytes fixed in formalin, purified myeloperoxidase protein (MPO), purified proteinase-3 protein (PR-3), and HEp-2 cells were used for the assay. These biochips allow detection of pANCA, cytoplasmic antineutrophil antibodies (cANCA), as well as anti-nuclear antibodies (ANA) on the same slide. The assay was performed according to manufacturer's instructions. Briefly, biochips were incubated with 30ul of 1:10 and 1:100 diluted serum samples for 30 Min, then washed and incubated with secondary anti-dog IgG FITC-labeled antibody for 30 Min. After washing, the slides were embedded in medium, covered and evaluated under fluorescence light at 20-40x magnification by two independent observers.

The number of pANCA or cANCA positive results in the group of FRD dogs was 27/44 (61%). To the contrary, none of the dogs in the SRD group had evidence of pANCA or cANCA seroreactivity (0/20). Consequently, sensitivity of the new assay to predict FRD amongst a group of dogs with chronic enteropathies was 0.61 (95% CI 0.45-0.75); and specificity was 1.0 (95% CI 0.89-1.0). This is comparable to previous studies using the canine granulocyte method (sensitivity 0.51; specificity 0.95). Fishers exact test revealed a statistically significant association of a positive pANCA or cANCA test with FRD ($p<0.0001$). Inter-observer agreement for the assay was good (kappa 0.75). Furthermore, in 8 of the ANCA positive cases, it was possible to identify either MPO or PR-3 as antigens on the biochips.

In conclusion, the human granulocyte immunofluorescence assay used in this study was able to identify sero-positivity for ANCA in a high percentage of dogs with FRD. The data suggests that this assay could be used as an alternative to the canine assay for detection of ANCA in dogs.

GI16

SERUM CITRULLINE CONCENTRATIONS IN DOGS WITH CHRONIC ENTEROPATHY

Rosana Lopes, Yuri Lawrence, Robert K. Phillips, Jonathan A. Lidbury, Jan S. Suchodolski, Jörg M. Steiner

Gastrointestinal Laboratory, Texas A&M University, College Station, TX, USA

Citrulline, a non-proteinogenic amino acid, has recently been described as a biomarker of enterocyte function in humans. Decreased concentrations of plasma citrulline have been reported in patients with short bowel syndrome, Crohn's disease, viral enteritis, or in critically ill patients. In dogs, the relationship between serum citrulline concentration and small intestinal dysfunction has not been well characterized. Citrulline is mainly synthesized from glutamine in enterocytes and is metabolized in the duodenum and jejunum. Therefore, intestinal injury can lead to decreased serum citrulline concentrations. The aims of this study were to analytically validate a GC/MS method for the measurement of citrulline in canine serum and to assess the concentration of this amino acid in dogs with chronic enteropathy.

Surplus serum samples from 40 dogs were used for analytical validation of the citrulline assay. Thirty-one surplus serum samples from healthy dogs and dogs with chronic enteropathy were used to evaluate possible alterations in serum citrulline concentrations in dogs with chronic enteropathy. Validation variables included the lower limit of detection, linearity, spiking recovery, and precision. Serum citrulline concentrations from 19 dogs with chronic enteropathy were compared to those from 12 healthy dogs. A Mann-Whitney *U* test was performed to compare serum citrulline concentrations between healthy and diseased dogs. Statistical significance was set as $p < 0.05$.

The lower limit of detection for citrulline in serum was 4.9 µmol/L. Observed-to-expected ratios for citrulline dilutional parallelism (n = 6) were 83.2% to 116.7%. Recoveries for spiking concentrations of citrulline in serum samples (n = 5) were between 84.9% to 119.4%. Mean intra- and inter-assay coefficients of variation (%CVs) for the citrulline assay were 5.4% and 10.9%, respectively. Dogs with chronic enteropathy had significantly lower serum citrulline concentrations (median [minimum - maximum]: 10.5 µmol/L [2.8 - 46.0]) than healthy dogs (median [minimum - maximum] 32.5 µmol/L [20.0 - 44.0]; $p < 0.0001$).

The assay evaluated here was shown to be linear, precise, accurate, and reproducible. Dogs with chronic enteropathy had significantly lower serum citrulline concentrations than healthy dogs. Further studies are warranted to assess the clinical utility of the serum citrulline as a biomarker for enterocyte function in dogs.

GI17

ORAL VERSUS PARENTERAL COBALAMIN SUPPLEMENTATION IN DOGS WITH CHRONIC ENTEROPATHIES AND HYPOCOBALAMINEMIA

Linda Toresson[3], Joerg Steiner[2], Jan Suchodolski[2], Thomas Spillmann[1]

[1]Helsinki University, Helsinki, Finland, [2]Gastrointestinal Laboratory, Texas A&M University, College Station, TX, USA, [3]Evidensia Specialist Animal Hospital/Helsinki University, Helsingborg, Sweden/Helsinki, Finland, Sweden

Hypocobalaminemia is a common sequel to chronic enteropathies (CE) in dogs and has been reported to be associated with a negative outcome. Current supplementation protocols for cobalamin (cbl) in dogs call for parenteral (PE) supplementation, although no protocol has been validated. In humans, several studies have reported equal efficacy of per oral (PO) and PE administration of cbl. Further, this group has previously reported about successful PO cbl supplementation in dogs with CE and hypocobalaminemia in a retrospective study. Thus, the purpose of this prospective randomized study was to validate the current PE and a new PO cbl supplementation protocol and compare the efficacy of PO versus PE cbl supplementation in dogs with hypocobalaminemia.

Dogs with symptoms of CE and a serum cbl below 285 ng/L (reference interval: 244-959 ng/L) were included from Evidensia Specialist Animal Hospital, Helsingborg, Sweden, Helsinki University Hospital, Finland, and from two other Swedish Veterinary Clinics. Dogs were randomized to treatment with either daily PO cbl tablets during the whole study period (cyanocobalamin (Behepan) 1 mg/tablet; dogs up to 10 kg received ¼ tablet, dogs 10 to 20 kg ½ tablet and dogs ≥ 20 kg 1 tablet/day) or PE cbl according to a protocol currently suggested by a diagnostic laboratory (http://vetmed.tamu.edu/gilab/research/cobalamin-information#dosing). A block-randomized schedule was performed by an external statistician prior to the start of the study in March 2014. The study was approved by the Swedish Board of Agriculture and the Viikki Campus Research Ethics Committee, Finland. Serum cbl for follow-up was analyzed 28 ± 5 days and 90 ± 15 days after initiation of supplementation using an automated chemiluminescence immunoassay (Immulite 2000, Siemens Healthcare Diagnostics).

41 dogs of 22 breeds were included with a median age of 6.2 (range 1.7-13.1) All of the dogs had symptoms of CE. Concurrent medical treatment and diet was given based on clinical judgement. Intestinal biopsies from the small and large intestine confirming chronic inflammation were available from 25/41 dogs. There was no statistically significant difference in serum cbl concentrations at baseline between the two groups ($p=0.25$, Mann Whitney test). Serum cbl increased in all dogs after supplementation. In the group receiving PO supplementation, median serum cbl concentrations was 245 ng/l (150-285) at inclusion ($n=22$), 970 ng/L (564-2026) after 28 days ($n=22$), and 1334 ng/L (768-3921) after 3 months ($n=20$). The difference between baseline and 28 days was statistically significant ($p<0.0001$; Wilcoxon matched-pairs signed rank test) as well as the difference between 28 days and 3 months (<0.006; Wilcoxon matched-pairs signed rank test). In the group receiving PE cbl, mean cbl concentration was 221 ± 43 ng/L at inclusion ($n=19$), 1686 ± 757 ng/L after 28 days ($n=19$), and 803 ± 275 ng/L ($n=13$) after 3 months. The difference between baseline and 28 days was statistically significant ($p<0.0001$; paired t-test) as well as the difference between 28 days and 3 months ($p=0.0003$, paired t-test). Comparing the increase in serum cbl concentrations at 28 days to baseline, the increase was significantly higher in the PE group than the PO group (mean increase 1465 ± 745 ng/L and 888 ± 449 ng/L, respectively, $p=0.004$, unpaired t-test). However, after 3 months, the median (range) increase in serum cbl concentrations compared to baseline group were significantly higher in the PO than the PE group (970 ng/L (538-3663) and 608 ng/L (38-997), respectively, $p=0.0001$, Mann Whitney test).

We conclude that both PO and PE cbl supplementation appears to effectively increase serum cbl concentrations in dogs with CE and hypocobalaminemia. Further studies comparing cellular cobalamin status after PO or PE supplementation are warranted.

GI18

PLASMA ESSENTIAL TRACE ELEMENT CONCENTRATIONS IN DOGS WITH CHRONIC ENTEROPATHY

Nozomu Yokoyama, Hiroshi Ohta, Hazuki Mizukawa, Yumiko Kagawa, Shouta Nakayama, Noboru Sasaki, Keitaro Morishita, Kensuke Nakamura, Yoshinori Ikenaka, Mayumi Ishizuka, Mistuyoshi Takiguchi

Department of Veterinary Clinical Sciences, Graduate School of Veterinary Medicine, Hokkaido University, Sapporo, Hokkaido, Japan

Chronic enteropathy (CE) is a common cause of malnutrition in dogs. Essential trace elements (ETE) are defined as micronutrients that are required for proper biochemical processes in the body. To date, little is known about plasma ETE status in dogs with CE. The aim of this study was to determine whether ETE deficiency occurred in dogs with CE.

Dogs with CE that were presented to Hokkaido University Veterinary Teaching Hospital between July 2014 and June 2015 were recruited. Fourteen laboratory beagles were used as controls. Plasma ETE concentrations (Chromium, manganese, iron, cobalt, copper, zinc, selenium, molybdenum) were measured by inductively coupled plasma-mass spectrometry. Plasma ETE concentrations in CE dogs were compared with those of controls. In CE dogs, correlation between plasma ETE concentrations and canine chronic enteropathy activity index (CCECAI), plasma albumin concentration and histopathologic score calculated according to the World Small Animal Veterinary Association (WSAVA) criteria.

Twenty-two dogs with CE were enrolled in this study. There were no significant differences between plasma ETE concentrations of CE dogs and those of controls, except for molybdenum. Plasma molybdenum concentration in CE dogs (median 6.9 ng/ml, range 0.38-60 ng/ml) was higher than that of controls (median 3.2 ng/ml, range 0.72-4.7 ng/ml) (P = .015). However, plasma molybdenum concentration didn't correlate with CCECAI, albumin concentration and histopathologic score.

These results indicate that there might be no obvious plasma ETE deficiency in dogs with CE. Further study is necessary to clarify the reason for increased plasma molybdenum concentration in dogs with CE.

GI19

VALIDATION OF ULTRASONOGRAPHIC MEASUREMENT OF GASTRIC EMPTYING TIME IN HEALTHY CATS USING RADIONUCLIDE SCINTIGRAPHY

Roman Husnik, Jon Fletcher, Lorrie Gaschen, Frederic Gaschen

Department of Veterinary Clinical Sciences, School of Veterinary Medicine, Louisiana State University, Baton Rouge, LA, USA

The prevalence of gastric emptying (GE) disorders in cats is unknown due to lack of clinically applicable diagnostic test. This study assesses agreement for GE times between ultrasonography and the gold standard in 7 healthy cats.

For scintigraphy, 60-second static acquisitions were performed in right and left lateral recumbency at 0, 5, 15, 30, 45, and 60 minutes following a 4 mCi Tc99M-mebrofenin solid test meal (20% of daily energy requirements), and at 30-minute intervals thereafter. Mean total counts from a gastric region of interest were measured from both images. Geometric means of the decay-corrected counts were calculated, and the amount of gastric radioactivity was expressed as a percentage of the initial count. Using nonlinear regression analysis, 25%, 50%, 75% and 90% GE times were determined. Sonographic assessment of GE was repeated three times after the same meal (non-labeled). Using the same time points, mean cross-sectional area of transverse images of the relaxed antrum was obtained and expressed as a percentage of the maximal area. The area under the curve was calculated and 25%, 50%, 75% and 90% GE times determined.

Good correlation was found between scintigraphy and sonography for the 75% (287 ± 67 [mean ± SD] and 306 ± 28 minutes respectively, $r = 0.85$, $p < 0.02$) and 90% GE times (340 ± 77 and 381 ± 29 minutes respectively, $r = 0.85$; $p < 0.02$). Ultrasound evaluation of antral cross-sectional area is a valid alternative to scintigraphy in the late phase of GE in healthy cats.

GI20

EFFECT OF MOSAPRIDE ON POSTPRANDIAL GALLBLADDER MOTILITY AND PLASMA LEVELS OF MOTILIN IN DOGS

Toshiaki Kakimoto, Hideyuki Kanemoto, Kenjiro Fukushima, Koichi Ohno, Hajime Tsujimoto

The University of Tokyo, Tokyo, Japan

Impairment of gallbladder motility might be responsible for the pathogenesis of gallbladder mucocele and biliary sludge in dogs and the improvement of the gallbladder motility has a possibility to exert the preventive effects on the progression of the gallbladder diseases. However, little has been studied on the treatments to enhance the gallbladder motility in dogs. It was reported that 5-HT_4 receptor agonist improves gallbladder emptying in human gallstone patients, and that mosapride, which is a selective 5-HT_4 receptor agonist practically used for prokinetics of canine gastrointestinal disease, induces the increase of motilin, which is a physiological inducer of gallbladder motor activity. In this study, we investigated the effects of mosapride on the postprandial gallbladder motility and the plasma levels of motilin in dogs with or without biliary sludge. Furthermore, we examined the expression of 5-HT_4 receptor using gallbladder tissue of a normal dog.

Nine apparently healthy beagles, owned by the University of Tokyo, were enrolled in this study. Of these, 5 did not have any gallbladder abnormality (group N), and 4 had biliary sludge (group S) based on the assessment by ultrasonography. Gallbladder ejection fractions (EF) were compared between dogs that were given diet with and without simultaneous single oral administration of mosapride 1 mg/kg. EF was estimated by ultrasonography, as described in the previous study. Briefly, dogs were fasted at least 12 hours before examination. Once the fasting gallbladder volume was ultrasonographically measured, they were given a test diet with and without mosapride. Postprandial gallbladder volume was recorded at an hour after the feeding. Before each ultrasonography, blood was collected for the measurement of plasma motilin levels. Experiments and animal care complied with the policies outlined in the Guide to Animal Use and Care of the University of Tokyo. The expression of 5-HT_4 receptor was evaluated by Western blotting using an archived canine gallbladder.

In the group N, median EF of the dogs treated with mosapride was significantly higher than those without the treatment [46.1% (range: 44.1 to 60.6) vs. 38.1% (24.2 to 43.5%), respectively, $P < 0.05$]. Such tendency was also observed in the group S, although the difference was not significant [treatment: 43.3% (33.9 to 63.2%) vs. without treatment: 34.8% (23.6 to 37.7%), $P = 0.06$]. However, the median degree of the increase of the plasma motilin concentrations after feeding was not significantly different between the dogs with and without mosapride in either group [Group N: mosaporide: -1.6 pg / ml (range: - 123.3 to 168.3 pg / ml) vs. feeding only: 65.0 pg /ml (- 25.0 to 415.0 pg / ml), Group S: mosaporide: 37.0 pg / ml (range: - 273.0 to 95.0 pg / ml) vs. feeding only: 173.3 pg /ml (5.0 to 363.3 pg / ml)]. In Western blotting, an immunoreactive band was observed at the position of 45 kDa using the protein purified from the archived gallbladder, suggesting that 5-HT_4 is expressed in the canine gallbladder.

These results suggested that mosapride enhances postprandial gallbladder motility. Although the mechanisms of this effect are still to be clarified, our results suggested that mosapride enhances gallbladder motility possibly via gallbladder 5-HT_4 receptor and motilin does not play a role in this effect. Further studies are needed to support our results.

GI21

EVALUATION OF THE PRESENCE AND ROLE OF CYSTEINE PROTEASE 30 IN FELINE *T. FOETUS*

M. Katherine Tolbert, Emily Gould, Mabre Brand, Stephen Kania

University of Tennessee, Knoxville, TN, USA

Tritrichomonas foetus (Tf) is a flagellated protozoan parasite that is recognized as a significant cause of diarrhea in domestic cats with a prevalence rate as high as 30%. No drugs have been shown to consistently eliminate Tf infection in all cats. Cysteine proteases (CPs) have been identified as mediators of Tf-induced adhesion-dependent cytotoxicity to the intestinal epithelium. Thus, CPs represent novel targets for the treatment of feline trichomonosis. However, cats also produce CPs that are part of life-critical systems. Thus, parasitic CPs need to be selectively targeted to reduce the potential for host toxicity. Previous studies have demonstrated the importance of a specific CP, CP30, in mediating bovine and human trichomonad cytopathogenicity. Thus, the aims of this study were to characterize the presence of CP30 in feline Tf isolates and to evaluate the effect of targeted inhibition of CP30 on feline Tf-induced adhesion dependent cytotoxicity.

The presence of CP30 in 4 different feline Tf isolates was evaluated by indirect immunofluorescence and flow cytometry using a rabbit polyclonal antibody that targets bovine Tf CP30 (α-CP30). Feline Tf treated with rabbit IgG and/or secondary Ab in the absence of α-CP30 Ab and bovine Tf were used as negative and positive controls, respectively. The effect of inhibition of CP30 activity on Tf adhesion-dependent cytopathogenicity was determined using CFSE-labeled feline Tf that were allowed to adhere to monolayers of porcine intestinal epithelial cells (IPEC) in co-culture. The effect of CP30 inhibition on feline Tf adhesion and cytotoxicity was evaluated by immunofluorescence quantification assays and crystal violet spectrophotometric analysis. A minimum of 6 replicates were performed for each co-culture experiment. Data were analyzed using Systat software ($p<0.05$).

CP30 expression was identified in all feline Tf isolates tested by indirect immunofluorescence and flow cytometry. Treatment of feline isolates with α-CP30 Ab resulted in significantly decreased Tf adhesion to and cytotoxicity towards IPEC-J2 monolayers compared to rabbit IgG-treated Tf isolates.

These studies establish that CP30 is expressed by feline Tf isolates and may be an important virulence factor in the cytopathogenicity of feline Tf. The results of these studies provide strong evidence-based justification for investigation of CP30 as a novel target for the prevention and/or treatment of feline trichomonosis.

GI22

EVALUATION OF THE EFFECT OF OMEPRAZOLE ON SERUM CALCIUM, MAGNESIUM, GASTRIN AND BONE IN CATS

Emily Gould[1], Casey Clements[1], Ann Reed[1], Luca Giori[1], Joerg M. Steiner[2], Jonathan A. Lidbury[2], Jan S. Suchodolski[2], Mabre Brand[1], Tamberlyn Moyers[1], Lee Emery[1], M. Katherine Tolbert[1]

[1]University of Tennessee College of Veterinary Medicine, Knoxville, TN, USA, [2]Texas A&M University College of Veterinary Medicine & Diagnostic Sciences, College Station, TX, USA

Chronic proton pump inhibitor administration has been associated with electrolyte imbalances and cobalamin deficiency, disrupted bone homeostasis, serum hypergastrinemia and rebound acid hypersecretion in humans. These side effects could have important clinical implications in cats. However, no studies have explored if cats receiving prolonged oral omeprazole (e.g., > 4 weeks in duration) experience similar side effects. We hypothesized that prolonged oral omeprazole administration would result in altered bone homeostasis and/or serum calcium, magnesium, cobalamin and/or gastrin concentrations in healthy cats.

In a within subjects, pre to post design, 6 adult healthy colony cats received either omeprazole (0.83-1.6 mg/kg PO q12 hrs) or placebo (250 mg lactulose PO q12 hrs) for 60 days. Analyses of serum calcium, magnesium, cobalamin and gastrin concentrations were performed on days 0, 30, and 60, with dual energy x-ray absorptiometry performed at days 0 and 60 for each treatment. On day 60 of omeprazole treatment, intragastric pH capsules were placed in two of the cats to measure the effects of rapid omeprazole cessation. Continuous data were analysed using a two-way ANOVA and tested to account for the effects of phase of treatment, time of measurement and the interaction of phase and time. Log transformation of serum gastrin, bone mineral density (BMD) and bone mineral content (BMC) were required for the data to meet ANOVA normality assumptions. To reduce the possibility of a Type I error due to analysis of multiple study variables, the Bonferroni correction was applied ($\alpha = 0.006$). SAS 9.4 was used to perform all data analyses and generate all descriptive statistics.

No significant changes were observed between treatment and placebo for any parameters, except serum gastrin, which was significantly higher during omeprazole treatment in comparison to placebo ($p < 0.002$). Although the change was not significant, all cats experienced a decrease in BMC by the end of the omeprazole treatment arm. Evidence of gastric hyperacidity was seen in both cats following cessation of omeprazole.

In conclusion, the results of this study provide evidence that omeprazole may be relatively safe when administered to cats for 60 days or less, but slow withdrawal of the medication is recommended based on the hypergastrinemia and rebound hyperacidity demonstrated in this pilot trial. However, further studies involving more cats are needed to confirm these results.

GI23

GRAVITY-ASSISTED ESOPHAGEAL TRANSIT CHARACTERISTICS IN DOGS WITH MEGAESOPHAGUS

Jillian Haines[1], Alison Khoo[3], John Thomason[2], Erin Brinkman[2], Andrew Mackin[2]

[1]Washington State University, Pullman, WA, USA, [2]Mississippi State University, Mississippi State, MS, USA, [3]Western College of Veterinary Medicine, Saskatoon, Saskatchewan, Canada

Treatment of megaesophagus (ME) often relies on feedings in a vertical position. Little is known about the behavior of food material in the esophagus during transit to the stomach in dogs with ME. The aims of our project were to standardize a contrast video fluoroscopy technique to evaluate esophageal transit time in dogs with ME while sitting in a vertical position, and to use this information to determine an appropriate duration an individual patient with ME should remain vertical after drinking and feeding.

Twelve dogs with ME were included in the study. Six dogs had congenital ME, 4 dogs had idiopathic acquired ME, and 2 dogs had ME secondary to myasthenia gravis. All dogs were placed in a vertical position using adjustable size-appropriate Bailey chairs. Bailey chairs were constructed from thin wood which allowed them to be radiolucent. In three different feeding sessions, dogs were given standardized amounts of liquid barium, canned food meatballs mixed with barium, or their normal diet also mixed with barium (if the diet was a different consistency than meatballs, generally a slurry). Fluoroscopy was performed in 5 minute increments to visualize transit of food or barium from time to ingestion until full clearance of material from the esophagus, or for a maximum of 20 minutes for liquid barium and 30 minutes for food material.

In dogs that were accustomed to the Bailey chair, performance of contrast video fluoroscopy was technically straightforward. Significant individual variation was seen between dogs for transit time of all three ingesta types. Ingesta consistency affected transit time, but the consistency that allowed the most rapid transit time varied by individual. Mean transit time for liquid was 15.4 mins with 58% (7/12) of dogs having no clearance of liquid by 20 mins. Mean transit time for meatballs was 24.6 mins with 67% (8/12) having incomplete clearance of meatballs at 30 minutes. Mean transit time for the slurry consistency was 26.3 mins with 75% (6/8) of tested dogs having incomplete clearance of food at 30 mins. Three dogs were identified as having gastroesophageal reflux and one dog was found to have a concurrent swallowing disorder. Based on individual imaging results, recommendations were made to alter management strategies for 8/12 dogs; change food consistency in 50%, change time kept upright in 25% of dogs, and alter pharmacologic therapy in 17%.

Fluoroscopy performed utilizing vertical positioning in modified Bailey chairs in dogs with ME provided clear information that allowed for individual evaluation of the most ideal food consistency, anticipated behavior of ingested water, and overall esophageal transit time for ingested food material. This information permitted informed decision-making with regards to what food type appeared to be best for individual dogs, whether providing drinking water at times separate to feeding was likely to be well tolerated, and how long the dog would need to remain in a vertical position after eating or drinking. The imaging also allowed for identification of concurrent disorders affecting food tolerance and allowed for more specific management strategies in the majority of dogs.

GI24

FEASIBILITY OF MEASURING GASTROINTESTINAL TRANSIT TIME IN HEALTHY DOGS USING ALICAM

Jill S. Pomrantz[1], Jonathan A. Lidbury[2], Brian T. Hardy[3], Jörg M. Steiner[2], Jan S. Suchodolski[2], Brigitte B. Mcatee[2], Jeffrey A. Solomon[1]

[1]Infiniti Medical, LLC, Menlo Park, CA, USA, [2]Texas A&M University, College Station, TX, USA, [3]university of California, Davis, Davis, CA, USA

The lack of a simple and widely available test for assessing gastrointestinal motility in dogs can make diagnosing disorders difficult. Ambulatory light-based imaging is an easy to use, novel gastrointestinal imaging technique performed by oral administration of a fully automated capsule-sized camera device (ALICAM) that is propelled by natural peristalsis. The time required for passage of the capsule through the stomach and small intestine can be determined based on correlation of images with an internal clock. The aim of this study was to evaluate the feasibility of using ALICAM to measure gastric transit time (GTT) and small intestinal transit time (STT) in healthy dogs.

The study subjects were 10 clinically healthy dogs between 3 and 8 years of age that weighed 17-32 kg. Dogs were fasted for 16-24 hours before and 8 hours after capsule administration. Images were downloaded and independently reviewed by 3 board-certified internists. Each reviewer determined GTT and STT. The reproducibility of the results was compared between reviewers.

The median (min-max) GTT was 86 (13-218) minutes. The median (min-max) STT was 124 (34-168) minutes. There was agreement among the reviewers with an intra-class correlation coefficient of >0.999 and 0.969 for GTT and STT respectively. The upper limits of the 95% confidence intervals for GTT and STT in healthy dogs were 243 and 179 minutes, respectively.

Ambulatory light-based imaging can be used to assess gastrointestinal transit time in healthy dogs. Further studies are needed to determine its role in dogs with gastrointestinal disease.

GI25

COMPARISON OF GASTRIC TRANSIT TIME IN HEALTHY DOGS AND DOGS WITH SIGNS OF GASTRIC HYPOMOTILITY

Jill S. Pomrantz[1], Jonathan A. Lidbury[2], Brian T. Hardy[3], Jörg M. Steiner[2], Jan S. Suchodolski[2], Brigitte B. Mcatee[2], Jeffrey A. Solomon[1]

[1]*Infiniti Medical, LLC, Menlo Park, CA, USA,* [2]*Texas A&M University, College Station, TX, USA,* [3]*University of California, Davis, Davis, CA, USA*

Ambulatory light-based imaging is a gastrointestinal imaging technique performed by oral administration of a fully automated camera device contained in a capsule (ALICAM) that is propelled by natural peristalsis. Gastric transit time (GTT) can be determined by visualization of the capsule's passage from the stomach into the duodenum and correlating images with the clock function of the device. The aim of this study was to use the ALICAM system to compare the GTT in healthy dogs with the GTT in sick dogs with clinical signs that could be consistent with gastric hypomotility.

The study population consisted of a group of 10 clinically healthy dogs and a group of 26 dogs with retching, regurgitation, and/or vomiting. All dogs were fasted for 16-24 hours before and 8 hours after capsule administration. Studies were reviewed by a board-certified internist. For each case, GTT was determined. A Wilcoxin rank sum test was used to compare the two groups.

The median GTT (min-max) in healthy dogs was 86 (13-218) minutes. The median (min-max) GTT in sick dogs was 187 (2-1405) minutes. In 7 sick dogs, the capsule remained in the stomach for the duration of the study (435-1405 minutes) and the recorded time represented the minimum GTT. The GTT for sick dogs was significantly longer than for healthy dogs ($p=0.046$).

The results of this study suggest that gastric hypomotility occurs in a subset of dogs presenting with retching, regurgitation, and/or vomiting. Further studies are needed to determine if these dogs would benefit from prokinetic therapy.

GI26

FEASIBILITY OF A NOVEL GASTROINTESTINAL IMAGING DEVICE FOR USE IN DOGS

Jill S. Pomrantz[1], Brian T. Hardy[2], Ajay Sharma[3], Jeffrey A. Solomon[1]

[1]infiniti Medical, Llc, Menlo Park, Ca, Usa, [2]university Of California, Davis, Davis, Ca, Usa, [3]university Of Georgia, Athens, GA, USA

The purpose of this study was to test the safety and feasibility of a disposable and fully automated ingestible camera system that images the gastrointestinal tract in ambulatory dogs.

Five ambulatory light-based imaging (ALI) devices were constructed and contained within a translucent 11x31mm capsule. The key components were a battery, a light source, 4 auto-focusing cameras, an internal memory system, and a microprocessor with an accelerometer that synchronized camera activity with device motion.

Five client-owned dogs were food restricted for 24 hours before and 8 hours after capsule administration. Capsules were administered using a direct pilling technique. Normal activity and access to water were permitted throughout the study. After recovery, capsules were inspected for damage. Images were then downloaded and independently reviewed by three board-certified internists.

Capsules were successfully administered to 5/5 dogs and were recovered within 24 to 36 hours. There were no adverse events. Median (min-max) study duration was 16 (8-18) hours. Capsules recorded a median (min-max) of 19713 (8572-51683) images. Median (min-max) gastric and small intestinal transit times were 89 (10-110) minutes and 134 (68-177) minutes respectively. Visualization of the mucosa and image quality were described as excellent for the stomach and small bowel. The colon was poorly imaged due to retained feces.

ALI is safe and feasible in dogs. It produced high quality images throughout the majority of the gastrointestinal tract. Visualization of the colon might be feasible with additional preparation. This non-invasive technique may expand the role of imaging in dogs with gastrointestinal signs.

GI27

CORRELATION BETWEEN NOVEL IMMUNOHISTOCHEMICAL MARKERS, HISTOPATHOLOGY AND CLINICAL SCORES IN CATS WITH CHRONIC ENTEROPATHY

Giacomo Rossi[3], Alison Manchester[1], Jan Suchodolski[2], Craig Webb[1]

[1]*Colorado State University, Fort Collins, CO, USA,* [2]*Texas A&M University, College Station, TX, USA,* [3]*University of Camerino, Matelica, Italy*

A number of studies have demonstrated a poor correlation between histopathology and the clinical severity of chronic enteropathy in dogs, as well as the potential contribution of various immunohistochemical (IHC) markers to enhance interpretation of intestinal biopsies. Although a standardized scoring system has been developed for feline enteropathy using clinical parameters and histopathologic severity, a similar investigation, including the utility of IHC markers, has not yet been performed in cats.The purpose of this study was to identify correlations between histopathologic severity, immuno-histochemical markers, and clinical scoring of cats with chronic enteropathy.

A modified feline chronic enteropathy activity index score (FCEAI), a standard panel of gastrointestinal serum parameters, a standardized quantification of histopathology, a quantitative fecal score, and a group of immunohistochemical (IHC) markers were determined in 29 cats presented for signs consistent with chronic enteropathy. Study participants were from a client-owned population of cats experiencing diarrhea and/or vomiting of at least 3-weeks duration; had no evidence of a non-gastrointestinal disease based on history, physical examination, biochemical profile, complete blood count, urinalysis, total T4, and fPLI and fTLI; and had endoscopy with biopsies and histopathology of the small intestine as part of their diagnostic work-up. Serum cobalamin and folate concentrations were also determined. Histopathologic grading and IHC quantification was performed by a single, board certified pathologist blinded to any additional information regarding the case. Immunohistochemistry was performed to identify the prevalence of cellular staining for TGF-β, CD3, Foxp3, and MAC387. The degree of cellular staining was measured at 3 random and separate areas and those 3 values were averaged to give the final immunohistochemical score for each particular marker. Biochemical parameters, FCEAI scores, and histopathologic grading scores were tested for normality using the Kolmogorov-Smirnov test, and since the vast majority of data was not normally distributed, non-parametric statistics were used for analysis.

Nonlinear regression analysis demonstrated a significant positive correlation between the FCEAI score and all 4 IHC markers: TGF-beta (r^2 0.24, p=0.008), CD3 (r^2 0.24, p=0.011), Foxp3 (r^2 0.17, p=0.026), and MAC 387 (r^2 0.25, p=.006). The histopathologic severity score was found to have a significant negative correlation with the IHC for TGF-beta (r^2 0.29, p=0.004) and Fox p3 (r^2 0.25, p=0.007) and a significant positive correlation with CD3 (r^2 0.40, p=0.0007), but not MAC 387. This study failed to demonstrate a significant correlation between the histopathologic severity score and either the fecal score or the modified FCEAI scoring index.

As with dogs, there appears to be an absence of a significant correlation between histologic severity and clinical scorings such as a clinical severity index or fecal score in cats with chronic enteropathy. A significant correlation was, however, identified between a novel set of IHC markers and both the histopathology and clinical severity scores. The pathophysiologic relevance and clinical utility of these markers remains to be determined.

GI28

FECAL D-/L-LACTATE CONCENTRATIONS AND ABUNDANCE OF LACTIC ACID BACTERIA IN DOGS WITH EXOCRINE PANCREATIC INSUFFICIENCY

Amanda Blake, Joseph Cyrus Parambeth, Anitha Isaiah, Blake Guard, Jonathan Lidbury, Jorg Steiner, Jan Suchodolski

Gastrointestinal Laboratory, Texas A&M University, College Station, TX, USA

Exocrine pancreatic insufficiency (EPI) in dogs is characterized by maldigestion as a result of inadequate synthesis and secretion of pancreatic enzymes. Previous studies have shown that dogs with EPI have small intestinal dysbiosis (SID) and may require concurrent antibiotic therapy in addition to pancreatic enzyme replacement therapy. In humans, patients with short bowel syndrome, which is also characterized by maldigestion and SID, have been reported to have increased fecal lactate concentrations. Therefore, the aim of this study was to compare the abundance of lactic acid-producing bacteria and D- and L-lactate concentrations in feces from healthy dogs (n=18), dogs undergoing enzyme replacement therapy for EPI (n=17), and untreated dogs with EPI (n=5).

Inclusion criteria for dogs with EPI were: a serum cTLI concentration ≤ 2.5 µg/L, age ≥ 1 year, clinical signs of EPI, and no other concurrent disease. Healthy dogs had no clinical signs of gastrointestinal disease. Fecal samples were obtained for three consecutive days, pooled, and DNA was extracted using the PowerSoil®DNA Isolation Kit (MoBio Laboratories) for analysis of lactic acid producing bacteria by quantitative PCR. Fecal samples were deproteinized and fecal D- and L-lactate were measured by enzymatic methods (R-Biopharm, D-/L-lactic acid kit). A Kruskal-Wallis test followed by Dunn's post-test was used to compare lactate concentrations and bacterial abundances between groups. Statistical significance was set at $p < 0.05$.

D-lactate was significantly increased in both the treated (median [min-max]: 4.34 mM [0.16-20.77 mM]; p=0.0081) and untreated (11.45 mM [5.65-26.49 mM]; p=0.0013) dogs with EPI when compared to the healthy controls (0.35 mM [0.13-0.74 mM]). L-lactate was significantly increased in both the treated (median [min-max]: 18.25 mM [0.29-32.61 mM]; p=0.0023) and untreated (23.31 mM [13.28-24.76 mM]; p=0.0019) dogs with EPI when compared to the healthy controls (0.47 mM [0.18-2.79 mM]). *Lactobacillus* and *Bifidobacterium* were significantly increased in both treated (p=0.0007and p=0.0019, respectively) and untreated dogs (p=0.0120 and p=0.0076, respectively) with EPI when compared to healthy controls.

In conclusion, this study suggests that fecal D- and L-lactate are increased in dogs with exocrine pancreatic insufficiency, regardless of enzyme replacement treatment status. Also, lactate-producing bacterial groups were altered in dogs with EPI when compared to healthy control dogs.

GI29

EVALUATION OF SERUM BETA-HYDROXYBUTYRATE CONCENTRATIONS IN DOGS WITH CHRONIC ENTEROPATHIES

G Esposito[1], K Allenspach[2], JM Steiner[1], JA Lidbury[1], JS Suchodolski[1]

[1]*Gastrointestinal Laboratory, College of Veterinary Medicine Texas A&M University, College Station, TX, USA,* [2]*Royal Veterinary College, London, UK*

Beta-hydroxybutyrate (BHB), a ketone body, is synthesized in the liver from acetyl CoA and is utilized as an energy source when glucose supplies are limited. BHB has been demonstrated to play a role in the oxidative stress response by the binding of histone deacetylase and in the suppression of inflammatory disease by antagonizing the cryopyrin inflammasome (NALP3). Previous untargeted metabolomics studies of both human and canine patients with idiopathic inflammatory bowel disease have shown significantly higher serum concentrations of BHB in diseased patients than in healthy controls, suggesting that BHB may be useful as a biomarker for inflammatory bowel disease. In this study, we investigated serum BHB concentrations in healthy dogs and dogs with two types of chronic enteropathy: food responsive diarrhea and steroid-responsive diarrhea

Serum samples from dogs with food responsive diarrhea (FRD, n=30), steroid responsive diarrhea (SRD, n=14), and a group of healthy control dogs (n=30) were used. All samples were surplus material from a previous study. Dogs were diagnosed as having food responsive or steroid responsive diarrhea based on response to treatment. Serum concentrations of BHB were measured in serum samples using a colorimetric assay kit (beta-hydroxybutyrate colorimetric assay kit, Cayman Chemicals). A Kruskal-Wallis was performed to compare serum concentrations between groups.

Median serum BHB concentrations were (minimum – maximum): 0.45 mmol/L (0.02 – 0.11) for healthy dogs, 0.45 mmol/L (0.00 – 0.27) for dogs with SRD, and 0.03 mmol/L (0.01 – 0.13) for dogs with FRD. There was no significant different between the three groups ($p = 0.328$).

These results suggest that serum concentrations of BHB are not significantly increased in dogs with chronic enteropathies compared to healthy dogs.

GI30

EFFECT OF SERUM CREATININE ON FELINE SERUM DGGR-LIPASE AND SERUM PANCREATIC LIPASE IMMUNOREACTIVITY

Ellen Everson, Anthony Abrams-Ogg, Helen Kocmarek, Kristiina Ruotsalo

University of Guelph, Guelph, Ontario, Canada

The objective of this study was to examine the influence of kidney function (indicated by serum creatinine level) on serum 1,2-o-dilauryl-rac-glycero-3-glutaric acid-(6'methylresorufin) ester (DGGR) lipase and pancreatic lipase immunoreactivity (PLI) in cats.

Using electronic record searches, all cats were identified that had had a serum biochemistry profile ordered at the Ontario Veterinary College between January 1, 2010 and April 14, 2015. Serum creatinine had been measured at the Animal Health Laboratory (AHL), University of Guelph, by enzymatic UV method (Randox) on a Cobas 6000 c501 analyzer (Roche Diagnostics). Serum lipase had been measured at AHL by the DGGR colorimetric assay (Cobas) on the same analyzer. In some cases, an aliquot of the same serum sample had been submitted for Spec fPL® at IDEXX laboratories, Markham, ON. Laboratory results were retrieved and corresponding medical records reviewed. Data obtained from the records included, when available, hydration status and body condition score as assessed by the attending clinician.

Data were first assessed as independent events by analyzing only the results of the first serum biochemistry profile obtained for each cat (n=1314). The median lipase value in the group of cats with creatinine values above reference interval (>190 µmol/L, n=162) was 18.5 U/L (range 7-594 U/L), which was higher (p<0.001) than the median lipase value of 14 U/L (range 5-747 U/L) in the group of cats with creatinine values ≤190 (n=1152), although ranges overlapped substantially. The same trend (p=0.06) was observed for PLI in these respective groups: median was 16.6 µg/L (range 1.5-50 µg/L) for cats with creatinine >190 µmol/L (n=9); median was 3.85 µg/L (range 0.5-50 µg/L) for cats with creatinine ≤190 µm/L (n=55). There was a significant, but weak, correlation between creatinine and lipase values across all cats (rho=0.2, p<0.001, n=1314), but no correlation was present for the group of cats with creatinine values >190 µmol/L (n=162). There was no correlation between creatinine and PLI. Multivariate linear regression was used to assess the influence of creatinine, hydration, and body condition on lipase over all cats for which this information was available (n=313); none of these variables demonstrated a significant effect (p=0.09, 0.22, and 0.97, for creatinine, hydration, and body condition, respectively). The data were next assessed by MANOVA for variation over time within individual cats for which repeat serum biochemistries and medical record data were available (n=279). Hydration demonstrated a significant effect on creatinine (p<0.001), and a trend towards effect on lipase (p=0.07). Creatinine did not demonstrate any further relationship with lipase (p=0.39). PLI was repeated in 15 cats; values moved in the same direction as creatinine in only 4 cases.

The data indicate that kidney function has a mild effect on DGGR-lipase level in cats, which may in part be due to hydration status. The study may be underpowered to show this effect in all analyses and on PLI. Creatinine may underestimate decreased kidney function in underweight cats, affecting results, but the overall weak associations between creatinine and lipase, and creatinine and PLI, suggest that elevated lipase or PLI should not be attributed solely to decreased kidney function in azotemic cats. This may be because the magnitude of increases in lipase and PLI in pancreatic disorders overwhelms the magnitude of effect of kidney function.

GI31

PREVALENCE OF *HELICOBACTER* SPECIES AND THEIR ASSOCIATION WITH GASTRIC PATHOLOGY IN DOGS WITH GASTROINTESTINAL DISEASE

Roman Husnik[1], Jiri Klimes[2], Simona Kralova-Kovarikova[3], Petr Fictum[4], Michal Kolorz[5], Frederic Gaschen[1]

[1]Department of Veterinary Clinical Sciences, School of Veterinary Medicine, Louisiana State University, Baton Rouge, LA, USA, [2]Department of Biology and Wildlife Diseases, Faculty of Veterinary Hygiene and Ecology, University of Veterinary and Pharmaceutical Sciences Brno, Brno, Czech Republic, [3]Department of Small Animal Internal Medicine, Faculty of Veterinary Medicine, University of Veterinary and Pharmaceutical Sciences Brno, Brno, Czech Republic, [4]department of Pathological Morphology And Parasitology, Faculty of Veterinary Medicine, University of Veterinary and Pharmaceutical Sciences Brno, Brno, Czech Republic, [5]Department of Human Pharmacology and Toxicology, Faculty of Pharmacy, University of Veterinary and Pharmaceutical Sciences Brno, Brno, Czech Republic

The prevalence and pathogenic role of different species of *Helicobacter* in the dog have not been definitely documented to date. This prospective study included 84 privately owned dogs for which the prevalence of gastric *Helicobacter* species and associated gastric lesions was determined, and the accuracy of different diagnostic techniques was compared.

Inclusion criteria consisted of chronic GI clinical signs (vomiting, diarrhea, anorexia, weight loss), exclusion of non-GI diseases and discontinuation of all medications for at least 2 weeks. Biopsy specimens were obtained by gastroduodenoscopy. The presence of *Helicobacter* spp. was detected using genus- and species-specific PCR (investigated species were *H. heilmannii* sensu stricto [HH s.s.], *H. bizzozeronii* [HB], *H. felis* [HF], *H. salomonis* [HS] and *H. pylori* [HP]), cytology, rapid urease test, and histologic examination with hematoxylin-eosin and silver staining of gastric biopsy samples.

PCR detected *Helicobacter* DNA in 60 dogs (71.4%). Comparing the diagnostic value of cytology, rapid urease test and histology to PCR, cytology yielded a sensitivity of 88.3% and a specificity of 91.7%. The rapid urease test had a sensitivity of 85% and a specificity of 91.7% and histology had a sensitivity of 81.7% and a specificity of 83.3%. Using a combination of diagnostic tests for *Helicobacter* spp. including rapid urease test, cytology, histology and PCR, 65 dogs (77.4%) had one or more positive result. For 61 dogs (72.6%) diagnostic tests were concordant. Single infections with HH s.s. was detected by PCR in 23.3%, HB in 15%, HS in 6.7% and HF in 1.7% of 60 PCR positive dogs. Mixed infection of HH s.s. and HB was found in 38.3% and combination of HH s.s. and HF was detected in 1.7% of the dogs. *Helicobacter* species remained undetermined in 13.3% of the patients. No HP was found in animals examined. In most dogs, the endoscopic appearance of the gastric mucosa was normal. Overall, the most common endoscopic lesions were erosions or ulcers and increased granularity of gastric mucosa. Gastritis was diagnosed histologically in 36.7% of dogs positive for *Helicobacter* spp. and in 41.7% of dogs without evidence of *Helicobacter* colonization. Severe lymphocytic-plasmacytic gastritis (11.1%) and acute purulent-necrotic gastritis (1.5%) were found in single HB infections and moderate lymphocytic-plasmacytic gastritis (14.3%) in single HH s.s. infections. An association between the presence of mixed infection (including HH s.s. and HB) and the lower severity of gastric inflammation was detected when compared to single *Helicobacter* species infection.

This is a large study evaluating association of the individual *Helicobacter* species with histologically confirmed gastritis in dogs. Dogs are frequently colonized by *Helicobacter* spp. other than HP, and may represent a reservoir for human infection with non-HP *Helicobacter* spp. The combination of cytology and rapid urease test performed on gastric biopsy samples represents a reliable diagnostic method suitable for clinical practice. Competitive inhibition may occur when 2 *Helicobacter* species are simultaneously detected in gastric biopsies, and the presence of one *Helicobacter* species may interfere with the virulence of other *Helicobacter* species.

GI32

NORMAL AND ABNORMAL FINDINGS IN THE CANINE GASTROINTESTINAL TRACT USING AMBULATORY LIGHT-BASED IMAGING

Jill S. Pomrantz[1], Brian T. Hardy[2], Jeffrey A. Solomon[1]

[1]*Infiniti Medical, LLC, Menlo Park, CA, USA,* [2]*University of California, Davis, Davis, CA, USA*

Ambulatory light-based imaging (ALI) is a new imaging modality that allows for non-invasive endoluminal visualization of the gastrointestinal mucosa. ALI is performed by oral administration of a capsule containing a fully automated camera (ALICAM) that is propelled by natural peristalsis. The capsule is retrieved from the dogs' feces and images are downloaded. Familiarity with normal and abnormal findings is essential to accurate interpretation of these studies, but these have yet to be documented in veterinary patients. The aim of this study was to present a gallery of normal and abnormal images of the canine gastrointestinal tract using ALICAM.

ALICAM studies from 65 dogs were evaluated retrospectively by a board-certified internist. The average study consisted of 21003 images obtained over 16 hours. Patients evaluated included 55 dogs with clinical signs of gastrointestinal disease and 10 asymptomatic dogs. The patients ranged in age from 4.5 months to 13.7 years old and in weight from 7.8 to 72kg. Images representative of normal mucosa and common lesions were saved and logged.

The most common reasons for ALICAM administration were vomiting, diarrhea, signs of gastrointestinal bleeding, or a combination of several gastrointestinal signs. Gastric lesions identified in dogs with gastrointestinal signs included irregular mucosa, polyps, ulcers, erosions, masses and foreign material. Small intestinal lesions identified included irregular mucosa, dilated lacteals, ulcers, erosions, masses, and parasites (tapeworms and hookworms). Colonic lesions included masses and erosions.

Establishing characteristics of normal and abnormal findings is an important step in the evolution of ALI as a diagnostic test.

GI33

FECAL α_1-PROTEINASE INHIBITOR CONCENTRATIONS IN DOGS WITH CARDIAC DISEASE

Joseph Cyrus Parambeth, Jordon P Vitt, Jan S Suchodolski, Jonathan A Lidbury, Ashley B Saunders, Joerg M Steiner

Department of Small Animal Clinical Sciences, College Station, TX, USA

Protein-losing enteropathy (PLE) has been documented in various cardiac diseases (i.e., congestive heart failure, constrictive pericarditis, and congenital heart diseases) typically related to increased interstitial pressure and intestinal lymphangiectasia in humans. Fecal α_1-proteinase inhibitor (fα_1-PI) is a stable protein lost into the gastrointestinal tract at a rate comparable to that of albumin and therefore serves as an endogenous marker of gastrointestinal protein loss. While PLE has been suspected in dogs with cardiac disease, no published studies are available. The aim of this study was to evaluate fα_1-PI concentrations in dogs with variety of cardiac diseases.

Single, naturally voided fecal samples from 27 adult dogs (≥1 year of age) and 15 puppies (<1 year of age) were collected from dogs with congenital and acquired cardiac diseases. Two dogs, one puppy and one adult, had serial fecal evaluations (i.e., before and after occlusion of their patent ductus arteriosus (PDA) with an Amplatz® canine ductal occluder (ACDO) device). Each dog in the study had an echocardiogram performed by a board certified cardiologist or a resident under the supervision of a board certified cardiologist. Fecal samples were frozen upon collection and fα_1-PI concentration was measured using an in-house radioimmunoassay and was compared with an established reference interval for healthy adult dogs. A fα_1-PI concentration of ≥21.0 µg/g feces was considered abnormal.

Three of 27 adult dogs (11%) all with congenital cardiac disease had an increased fα_1-PI concentration (median [minimum - maximum]: 10.1 µg/g feces [1.8 - 45.6]). Also, nine of 15 (60%) of the puppies, all with congenital heart diease, had an increased fα_1-PI concentration (median [minimum - maximum]: 24.8 µg/g feces [6.6 - 79.1]). Some of the dogs had multiple congenital heart defects on echocardiographic evaluation. Also, the severity of cardiac disease varied in each patient. For the two PDA dogs for which samples from two time points were available, there was a reduction in fα_1-PI concentration post coil embolization; from 79.1 to 21.5 µg/g feces in the puppy, and from 36.5 to 14.1 µg/g feces in the adult. The increased fα_1-PI concentrations in puppies have to be interpreted with caution as higher fα_1-PI concentrations have been reported previously in puppies and a specific reference interval for puppies has not been established.

Limitations of this study include the small number of dogs, the lack of intestinal biopsies to rule out any other causes of increased gastrointestinal protein loss, the lack of multiple fecal samples to account for day to day variability in fα_1-PI concentrations reported earlier, the variety of cardiac diseases, and lack of complete serum chemistry profiles for all dogs. However, this study shows that fα_1-PI concentrations may be increased in a subset of dogs with cardiac disease. Prospective studies are needed to establish if increased fα_1-PI concentrations in dogs occur consistently with congenital heart diseases, are increased in dogs with acquired heart diseases that increase interstitial pressure, and if elevations have any influence on treatment or prognosis.

GI34

EVALUATION OF THE EFFECTS OF PRE-CONDITIONING ON FEMALE CANINE ADIPOSE-DERIVED MESENCHYMAL STEM CELL CYTOKINE PRODUCTION

Rebecca Timmons, Stephanie Smith, Craig Webb, Tracy Webb

Colorado State University, Fort Collins, CO, USA

Clinical use of mesenchymal stem cell (MSC) therapy may be optimized through pre-conditioning methods that would enhance their immunomodulatory functions. We investigated the effects of pre-conditioning strategies on cytokine production of female canine adipose-derived MSC that may optimize clinical effects of MSC application in this species. Adipose-derived MSC were generated from healthy adult female dogs and subjected to 6 conditions. Supernatants were harvested after 1, 6, 12, 18, and 24 hours, and levels of cytokines present in the supernatants were determined by ELISA. IL-10 and TGFβ were not released in measurable levels within a 24 hour period under any of the tested culture conditions. Large amounts of MCP-1 (mean=10.2ng/ml) and VEGF (mean=2.9ng/ml) and very small amounts of IL-8 (mean=40.9pg/ml) were constitutively produced by the cells at 24 hours. All of the pre-conditioning strategies except TGFβ altered measured cytokine production in the MSC from control levels at 24 hours: poly I:C increased IL-8 and MCP-1 secretion; IFNγ increased MCP-1 and decreased VEGF and IL-8 secretion; serum-free decreased IL-8, MCP-1, and VEGF secretion; and the mixed hypoxic environment inhibited MCP-1, VEGF, and IL-8 secretion. All changes in cytokine secretion were evident at the 6 hour time point with most at maximal levels by 12 hours suggesting that 12 hour culture periods are sufficient to induce relevant changes in MSC cytokine secretion. This initial study shows that specific pre-conditioning methods hold promise for tailoring and augmenting canine MSC immunomodulatory properties to treat specific diseases, although further study is necessary and currently on going.

HM01

REPRODUCIBILITY, STABILITY AND BIOLOGICAL VARIABILITY OF THROMBIN GENERATION USING CALIBRATED AUTOMATED THROMBOGRAPHY IN HEALTHY DOGS

Benoît Cuq[1], Shauna Blois[1], Darren Wood[1], Anthony Abrams-Ogg[1], Christian Bédard[2]

[1]*Ontatio Veterinary College, University of Guelph, Guelph, Ontario, Canada,* [2]*Faculté de Médecine Vétérinaire, Université de Montréal, St Hyacinthe, Quebec, Canada*

Thrombin plays a central role in hemostasis and thrombosis, and is involved in all *in vivo* hemostatic pathways. Calibrated automated thrombography (CAT), a thrombin generation assay, may therefore be a valid physiological test for hemostatic disorders in dogs. Our objectives were to establish reference intervals for CAT thrombin generation, and assess the effects of pre-analytical and biological variability on this assay in healthy dogs.

Lag time (lag), time to peak (ttpeak), peak thrombin generation (peak), and endogenous thrombin potential (ETP) were measured in 41 clinically healthy dogs. Both direct jugular venipuncture and winged-needle catheter-assisted saphenous venipuncture were used to collect samples from each dog. Storage stability at -80°C was assessed over 2 months in a subset of samples. Biological variability of CAT was assessed via nested ANOVA using samples obtained weekly from 9 dogs for 4 consecutive weeks.

Samples for CAT analysis were stable at -80°C for up to 2 months. Samples collected via winged-needle catheter venipuncture showed poor repeatability compared to direct venipuncture samples; there was also poor agreement between the two sampling methods. Intra-individual variability of CAT variables was low (≤7.8%); inter-individual variability ranged from 4.8-76.8%. Analytic precision goals were met for ttpeak, peak and ETP. Indices of individuality ranged from 0 to 0.12.

In conclusions, thrombin generation seems to be stable and repeatable in healthy dogs. Direct venipuncture sampling is recommended for CAT. Indices of individuality were low for all CAT variables, suggesting limited utility of population-based reference intervals for this assay.

HM02

DEVELOPMENT AND VALIDATION OF A NOVEL CANINE IMMUNE THROMBOCYTOPENIA BLEEDING SCORE

KELLY MAKIELSKI[1], MARJORY BROOKS[2], CHONG WANG[1], JONAH CULLEN[1], ANNETTE O'CONNOR[1], DANA LEVINE[1]

[1]*IOWA STATE UNIVERSITY, COLLEGE OF VETERINARY MEDICINE, AMES, IOWA, USA,* [2]*CORNELL UNIVERSITY, COLLEGE OF VETERINARY MEDICINE, ITHACA, NEW YORK, USA*

A method of objective quantification of clinical bleeding in canine immune-mediated thrombocytopenia (ITP) is needed because ITP patients have variable bleeding tendencies that do not directly relate to severity of thrombocytopenia. A consistent description of their bleeding would allow more direct comparisons of this bleeding heterogeneity in multi-institution studies. Application of a standardized bleeding assessment tool would help stratify ITP patients in clinical trials based on their bleeding severity to provide an objective comparison of patient outcomes among treatment protocols. In this study we aimed to develop and validate a daily canine ITP bleeding assessment tool (DOGiBAT) and to evaluate the utility of a training course for improving accurate implementation of the DOGiBAT.

A novel bleeding assessment tool, DOGiBAT, was developed for canine ITP comprising nine different anatomic site-specific bleeding grades ranging from 0 (none) to 2 (severe). An online training course was developed for application of the DOGiBAT to score clinical cases. A case-based quiz set of still images was used to assess the trainees' ability to apply the DOGiBAT tool. Additionally, to assess the efficacy of the training course, 70 veterinary student volunteers were randomized to take the quiz with (n=35) or without (n=35) the training. All students scored all sites from all cases and site scores were considered correct if they agreed with the investigators' pre-specified classification. The frequency of correct scores was compared between trained and untrained students. A logistic regression model was used to assess the association between training and score, while adjusting for correlated responses from sites within cases.

Clinicians (n=13) and technicians (n=3) taking the quiz following training were able to correctly apply the DOGiBAT, scoring 100% of responses correctly. 86.1% of responses from students that received training were correctly scored, compared to only 78.2% of responses from students that did not receive training. The odds of trained students giving correct answers were higher than untrained students ($p < 0.0001$). The anatomic site for which training was most important for correct scoring was oral bleeding (82.9% with training, 47.6% without; $p < 0.0001$). Similarly, scoring of ocular (75.2% with training, 58.1% without; $p=0.0097$) and cutaneous bleeding (84.8% with training, 66.7% without; $p=0.003$) were improved by the training course. Students with more clinical experience were better able to apply the DOGiBAT correctly (fourth years, 84.2%; third years, 80.1%; $p < 0.0149$).

Our study suggests that the DOGiBAT is simple scoring tool for classification of bleeding severity in dogs with ITP. A short online training course was shown to improve the accuracy and consistency of DOGiBAT scoring. Adoption of the DOGiBAT scoring system for future clinical trials of canine ITP patients will facilitate more rigorous assessment of treatment-effect based on standardized bleeding outcomes.

HM03

EFFECTS OF ASPIRIN DOSE ESCALATION ON CANINE PLATELET FUNCTION AND URINARY THROMBOXANE AND PROSTACYCLIN LEVELS

Natalie McLewee, Todd Archer, Robert Wills, Andrew Mackin, John Thomason

Mississippi State University, Mississippi State, MS, USA

Aspirin is commonly used in an effort to prevent thrombus formation in dogs. High-dose aspirin (10 mg/kg, PO, q 12h) reliably inhibits platelet function and thromboxane A_2 (TXA_2) synthesis, but can be associated with unacceptable side effects, including inhibition of prostacyclin synthesis, which counteracts the effects of inhibition of TXA_2. Lower doses of aspirin, in contrast, inhibit platelet function and TXA_2 synthesis, have few side effects, and allow ongoing prostacyclin synthesis. However, low-dose aspirin therapy at standard dosage (0.5-1 mg/kg, PO, q 24h), does not reliably inhibit platelet function, a phenomenon known as "aspirin resistance". Aspirin-associated platelet dysfunction has been shown to be highly dose-dependent, suggesting that traditional "low-doses" of aspirin may be under-dosing canine patients. Our study used incremental increases in dosages to determine the dose of aspirin that consistently inhibited platelet function and TXA_2 synthesis without inhibiting prostacyclin synthesis.

Eight healthy adult research dogs were separated into one of 5 groups, with each group given different oral doses of aspirin: Dose A, 0.5 mg/kg, q24h; Dose B, 1 mg/kg, q24h; Dose C, 2 mg/kg, q24h; Dose D, 4 mg/kg, q24h or Dose E, 10 mg/kg, q12h. Prior to aspirin therapy (Day 0) and again after one week (Day 7) of drug administration, blood and urine were collected to assess platelet function (turbidimetric aggregometry and PFA-100® (closure time)) and urine 11-dehydro-thromboxane B_2 (11-dTXB_2, a stable metabolite of TXA_2) and 6-keto-prostaglandin $F_{1\alpha}$ (6-keto-PGF1_α, a stable metabolite of prostacyclin) synthesis. A dog was considered to be an aspirin responder if there was >25% reduction in maximal amplitude on aggregometry, or the Day 7 PFA-100® closure time was >300 seconds.

On aggregometry, for Doses A-E, there was a mean decrease in maximum amplitude of 3.8%, 41.3%, 81.3%, 75.9%, and 87.8%, respectively. Compared to Day 0, there was a significant decrease on Day 7 for Doses B-E. The maximum amplitude was decreased >25% (indicating aspirin responsiveness) in all dogs receiving Dose C or greater. Similar to aggregometry, when the PFA-100® closure times on Day 0 were compared to Day 7, there was a significant increase for Doses B-E. On Day 7, for Doses A-E, mean PFA-100® closure time was >300 seconds (indicating aspirin responsiveness) in 0%, 25%, 62.5%, 87.5%, and 87.5%, respectively. Compared to Day 0, there was a significant decrease on Day 7 for the concentrations of 11-dTXB_2 ($p=0.0003$) and 6-keto-PGF1_α ($p=0.0075$). However, the changes in 11-dTXB_2 and 6-keto-PGF1_α were not dependent on aspirin dose.

The results of our study suggest that an aspirin dose of 2 mg/kg, PO, q24h reliably inhibits platelet function in normal dogs based on the established "gold standard" assay, platelet aggregometry. Additionally, even though all aspirin doses inhibited prostacyclin synthesis to some degree, there was no difference in prostacyclin concentration between a dose of 2 mg/kg, PO, q24h and the traditionally used lower doses. Based on our results, an increased standard aspirin "low-dose" of 2mg/kg, PO, q24h may minimize the risk of aspirin resistance in dogs.

HM04

B CELL ACTIVATING FACTOR AS A BIOMARKER IN DOGS WITH PRIMARY IMMUNE MEDIATED THROMBOCYTOPENIA

Jessica Pritchard[1], Michael Wood[2], Adam Birkenheuer[1], Henry Marr[1]

[1]*Department of Clinical Sciences, College of Veterinary Medicine, North Carolina State University, Raleigh, NC, USA,* [2]*Department of Medical Sciences, School of Veterinary Medicine, University of Wisconsin - Madison, Madison, WI, USA*

There is a critical need for a biomarker that differentiates active primary immune mediated thrombocytopenia (pITP) from secondary causes of thrombocytopenia, and the concentration of which correlates with disease status in dogs. Such a biomarker would provide a valuable diagnostic tool to modulate therapy, evaluate risk of relapse, and reduce patient exposure to potentially harmful immunosuppressive medications. B cell activating factor (BAFF) is a cytokine within the tumor necrosis factor family that plays a crucial role in B cell maturation, survival, and class switching. Increased concentrations of BAFF in mice and people have been linked to immune-mediated diseases such as systemic lupus erythematosus, rheumatoid arthritis, and ITP. Our central hypothesis is that BAFF production will be significantly and specifically increased in dogs with untreated pITP.

The purpose of this pilot study was to quantify and compare the transcription of BAFF in the blood of dogs with untreated pITP, dogs with secondary thrombocytopenia, and healthy control dogs.

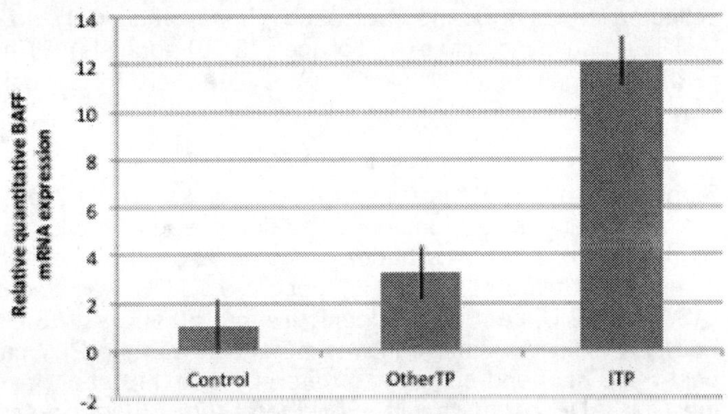

Figure 1. Levels of plasma BAFF mRNA in healthy dogs (control), dogs with secondary thrombocytopenia (otherTP), and dogs with ITP. The ratios of BAFF mRNA in dogs with ITP compared to those with secondary thrombocytopenia and healthy dogs are 3.25 and 12.13, respectively.

Quantitative real-time PCR (qRT-PCR) was used to assess BAFF mRNA expression. Total RNA was isolated from 6 dogs with untreated pITP, 5 healthy dogs, and 15 dogs with secondary thrombocytopenia. pITP was defined as dogs that had platelet counts <20,000 cells/ml, were PCR and serology negative for *Babesia*, *Ehrlichia*, and *Anaplasma*, had no evidence of neoplasia on abdominal and thoracic imaging, and no history of drug administration (antibiotics, corticosteroids, vaccines, or chemotherapeutics). Secondary thrombocytopenic control dogs had platelet counts below the reference interval due to diagnoses other than pITP, and healthy dogs presented for wellness exams and had normal platelet counts. qRT-PCR was performed using intron-spanning primers to amplify a 129 base pair region of the mRNA. β-actin was used as a reference gene.

The relative mRNA expression of BAFF in pITP dogs was increased 12.13-fold and 3.25-fold compared to healthy control dogs and dogs with secondary thrombocytopenia, respectively ($p<0.01$) (Figure 1).

These results indicate that relative expression of BAFF mRNA is increased significantly and specifically in dogs with pITP compared to both healthy dogs and dogs with other causes of thrombocytopenia, similar to previous findings in people. Thus, BAFF may be a useful biomarker to distinguish dogs with pITP from those with other causes of thrombocytopenia. Using these primers our group has also identified a canine macrophage cell line (030-D) that expresses BAFF and is utilizing them for protein identification and quantitation as well as further characterization of BAFF mRNA splice variants.

HM05

COMPARISON OF MULTIPLATE, PLATELET FUNCTION ANALYZER-200, AND PLATELETWORKS IN DOGS TREATED WITH ASPIRIN AND CLOPIDOGREL

Sophie Saati, Anthony Abrams-Ogg, Shauna Blois, Darren Wood

University of Guelph, Guelph, Ontario, Canada

In dogs, the antiplatelet drugs aspirin (ASA) and clopidogrel may be used in an effort to prevent thromboembolism. Response to these drugs might be variable as in humans, therefore platelet function testing may be warranted to monitor treatment effect. This study aimed to evaluate the effects of ASA, clopidogrel, or combination therapy, in healthy laboratory Beagles using three platelet function tests: Multiplate Analyzer (MP), Platelet Function Analyzer-200 (PFA), and Plateletworks (PW). Secondary goals were comparison of PW using two hematology analyzers: Vetscan HM5 (using impedance) and ADVIA 2120 (using flow cytometry); and evaluation of the novel INNOVANCE P2Y cartridge for PFA to evaluate clopidogrel-induced platelet inhibition.

Six dogs were given ASA 1 mg/kg/day and placebo, clopidogrel 2 mg/kg/day and placebo, or a combination of both drugs, for 1 week each. Drugs were given in a randomized crossover design (Latin square). Each dog was randomized to one of the three treatments in three different phases such that each dog received each treatment. Investigators were blinded to the treatments. A washout period of 2 weeks occurred between phases. Blood samples were collected on days 0 and 7 of each phase, and analyzed using MP (adenosine diphosphate [ADP], arachidonic acid [AA], collagen [COL] agonists), PFA (P2Y, COL-ADP, COL-Epinephrine [EPI] cartridges), and PW (ADP, AA, COL agonist tubes) using the two hematology analyzers.

Significant ($p<0.05$) changes in mean values pre and post treatments were: For MP - Area under the curve (AUC) units (U) was decreased with COL only with combination therapy (pre 19 U, post 6 U). AUC was decreased with AA with either clopidogrel (pre 41 U, post 19 U) or combination therapy (pre 44 U, post 9 U). AUC was decreased with ADP with all treatments (ASA pre 62 U, post 52 U; clopidogrel pre 59 U, post 13 U; combination pre 58 U, post 12 U). For PFA – closure time (CT) was increased with the P2Y cartridge with clopidogrel (pre 95 sec, post >300 sec) and combination therapy (pre 119 sec, post >300 sec). CT was increased with the COL-ADP cartridge with clopidogrel (pre 65 sec, post 185 sec) and combination therapy (pre 68 sec, post 215 sec). CT was increased with the COL-EPI cartridge only with ASA (pre 167 sec, post 235 sec). For PW using the HM5 analyzer - % aggregation (% Agg) with AA was decreased for all treatments (ASA pre 80%, post 7%; clopidogrel pre 90%, post 4%; combination pre 78%, post 12%). With ADP, % Agg was decreased for clopidogrel (pre 45%, post 9%) and combination therapy (pre 47%, post 9%). PW results for HM5 and ADVIA 2120 showed almost perfect agreement (kappa 0.83-0.98). Although the results above revealed changes in mean values with treatments, results were variable for individual dogs. The only tests for which every dog's individual values followed the changes in mean values were: MP with COL with combination therapy; PFA with the P2Y cartridge with clopidogrel or combination therapy; PW with ADP for clopidogrel or combination therapy, and PW with AA for all treatments.

All three platelet function tests detected ASA and clopidogrel effects in some dogs and may have utility for monitoring antiplatelet therapy. The P2Y cartridge was superior to the COL-ADP cartridge for detection of clopidogrel effect with PFA. PW can be performed with either Vetscan HM5 or ADVIA 2120, and PW using AA was the only test that detected all drug effects.

HM06

EVALUATION OF THE RISK OF RELAPSE OF CANINE IMMUNE-MEDIATED THROMBOCYTOPENIA AFTER ROUTINE VACCINATION

Jenny Ellis, Patricia M Ward, Robert D Foale

Dick White Referrals, Six Mile Bottom, Cambridgeshire, UK

Vaccination has been suggested as a cause of initial development and relapse of immune-mediated thrombocytopenia (ITP), although this relationship has not been definitively established in dogs. This study aimed to identify cases of canine primary ITP that had been vaccinated after cessation of treatment and to assess if any of these vaccinated cases showed signs of relapse.

Records of animals diagnosed with ITP between 2004 and 2014 were retrospectively reviewed at a multi-disciplinary hospital. Cases were included if a diagnosis of presumptive primary ITP had been made with a platelet count of below $50x10^9/l$ (ref 200 – 500). Dogs diagnosed with secondary ITP were excluded, as were cases that did not have assessment for underlying disease, with minimum investigation required comprising full haematological and biochemical assessment including blood film evaluation, abdominal ultrasonography and thoracic radiography. Infectious disease PCR tests were also performed in some cases at the discretion of the clinician involved. Referring veterinary practices were contacted for follow-up information regarding length of treatment, ongoing vaccination protocols including dates and vaccination administered, and survival data as required.

One hundred and forty-six thrombocytopenic dogs were identified, with 32 dogs not surviving to discharge. Sixty-eight dogs were excluded due to detection of an underlying infectious or neoplastic disease. Of the dogs with presumptive ITP, 11 were lost to follow-up and 14 dogs were still receiving prednisolone or other immunosuppressive agents. Four of these dogs remained on long-term prednisolone due to relapse when the dose was tapered.

Of the 21 dogs that were in remission with all immunosuppressive treatment withdrawn, 9 did not receive any vaccinations after diagnosis. None of these dogs showed signs of relapse during the follow-up period. The remaining 12 dogs received at least one vaccination a minimum of one month after discontinuing all immunosuppressive treatments. One of these dogs had initially relapsed early in the treatment course, prompting a longer treatment protocol before withdrawal of treatment. None of the dogs receiving vaccinations once immunosuppressive treatment was withdrawn showed signs of relapse, including those receiving multiple vaccinations over several years. Vaccinations administered included a combination of core boosters, intranasal vaccination for *Bordetella bronchiseptica*, and one dog received a Rabies booster for travel.

These results failed to show an association between relapse of presumptive primary ITP in dogs and vaccination, although the possibility cannot be excluded. Future studies assessing higher numbers of dogs would be required to further assess a potential association.

HM07

PREDICTING IN VIVO RESPONSE TO LOW-DOSE ASPIRIN IN HEALTHY DOGS USING IN VITRO PLATELET AGGREGOMETRY

Rachel Hegedus, Pamela Lee, Michael Court, Jillian Haines

Washington State University, Pullman, WA, USA

Aspirin, or acetylsalicylic acid (ASA), is a cyclooxygenase inhibitor that prevents the synthesis of hemostatically active prostaglandins. At low doses aspirin is used to inhibit platelet function and ultimately thrombus formation. Some human patients appear non-responsive to these antiplatelet effects, termed "aspirin resistance." This same phenomenon has been identified in dogs and could be a contributing factor to the formation of thromboemboli in some at risk patients. Accurately predicting aspirin response prior to treatment could prevent inappropriate drug therapy and devastating thrombus formation in otherwise manageable diseases.

The objective of our study was to determine whether the ASA concentration needed to cause 50% inhibition of platelet aggregation (EC50) measured using dog blood collected before ASA treatment (i.e. the *in vitro* EC50) could be used to predict the EC50 value measured in the same dogs after administration of low-dose oral aspirin (i.e. the *in vivo* EC50).

Twenty client-owned healthy dogs were utilized for this study. *In vitro* testing was performed on two separate occasions at least 1 week apart to determine individual repeatability. Blood was incubated with six concentrations of aspirin and aggregation was measured using a multiple electrode impedance aggregometer (Multiplate instrument). *In vivo* testing was performed at least 1 week after *in* vitro testing. The same dogs received a single dose of 1 mg/kg aspirin orally and aggregation was assessed at baseline prior to administration, then 20 minutes, 40 minutes, and 180 minutes after administration. Additionally, plasma concentrations of ASA and salicylic acid (SA) were measured at each time point using liquid chromatography–mass spectrometry (LC-MS). SA is the major ASA metabolite and may be a better indicator of ASA exposure (bioavailability) than ASA.

Calculation of an *in vitro* EC50 allowed dogs to be separated into four different response groups. Sixteen of the 20 dogs (80%) had *in vitro* EC50 values below 250 umol/L consistent with an appropriate aspirin response with mean +/- SD (range) values of 84.5 +/-36.0 (34.8-151.4) umol/L. When these values were evaluated for repeatability a mean difference of 44.9% was found with the highest difference between independent measurements for each dog of 95.1%. One dog showed a consistently poor response with *in vitro* EC50 values between 250-500 umol/L at both times measured. Two dogs had substantial variation in their *in vitro* EC50 values between the two time points and were classified as variable responders. One additional dog showed no significant inhibition of platelet aggregation at any time up to an ASA concentration of 3300 umol/L and was classified as a non-responder. *In vivo* aspirin dosing at 1 mg/kg did not cause measureable inhibition of platelet aggregation in any of the dogs after a single dose at any time point. An *in vivo* EC50 could not be established based on these findings as 50% inhibition was never achieved at this dose. Plasma ASA and SA concentrations were available for 18 of 20 dogs. The mean +/- SD (range) maximum plasma concentrations (Cmax) for ASA were 0.23 +/- 0.21 (0 - 0.77) umol/L and for SA were 21 +/- 5 (14 - 34) umol/L. Two dogs did not have ASA concentrations above the detection limit (0.03 umol/L).

This study suggests individual variation in aspirin response does exist in dogs as assessed by multiple electrode aggregometry. A single low dose (1 mg/kg) of aspirin was not sufficient to cause measurable inhibition of platelet aggregation in healthy dogs and resulted in ASA and SA concentrations that were substantially less than the *in vitro* EC50. Low-dose (1 mg/kg) aspirin therapy may not be a reliable antithrombotic therapy after only a single dose thus necessitating the potential need for additional therapies during the first 24 hours. Further research looking at higher oral aspirin doses or multiple day dosing is warranted to better establish an *in vivo* EC50 and determine the predictive value of *in vitro* EC50 determination prior to commencement of therapy.

HM08

RETROSPECTIVE EVALUATION OF ANEMIA AND ERYTHROCYTE MORPHOLOGY IN DOGS WITH LYMPHOMA AND INFLAMMATORY BOWEL DISEASE

Cyril Parachini-Winter, Lisa Carioto, Carolyn Gara-Boivin

School of Veterinary Medicine of the University of Montreal, Saint-Hyacinthe, Canada

The differentiation between intestinal lymphoma and inflammatory bowel disease (IBD) can be diagnostically challenging. Anemia is a frequent paraneoplastic syndrome in dogs with lymphoma, and eccentrocytes have also been reported in canine lymphoma. The purpose of this study was to assess the prevalence of anemia and various RBC morphologic anomalies in the peripheral blood of dogs with a diagnosis of lymphoma or IBD, to determine if there is an association with the specific disease and with the severity of lymphoma.

Medical records of 32 dogs diagnosed with all forms of lymphoma, 23 dogs diagnosed with IBD and 28 control dogs, presented to the Centre Hospitalier Universitaire Vétérinaire between 2006 and 2014 were retrospectively reviewed. Each blood smear was blindly re-assessed by a board-certified pathologist. A comparison between the following 3 groups: healthy dogs, dogs with IBD and dogs with lymphoma, as well as within the lymphoma group (stages 1,2,3 vs. 4/5, sub-stage a vs. b, and small vs. intermediate/large cells) was performed for the following hematologic parameters: hematocrit, hemoglobin, erythrocyte count, mean corpuscular volume, and mean corpuscular hemoglobin concentration. The total numbers and each specific RBC anomaly were compared between these 3 groups. RBC morphologic anomalies were also categorized by the underlying pathologic process (regenerative anemia, oxidative stress, or other morphologic changes) and compared between the 3 groups.

Prevalence of anemia was significantly higher in dogs with lymphoma than dogs with IBD ($p=0.02$) and healthy dogs ($p=0.0001$), and significantly higher in stages 4 and 5 lymphoma as compared to stage 1, 2 or 3 ($p=0.045$). There was a significantly higher total number of RBC morphological anomalies in the peripheral blood of dogs with lymphoma as compared to dogs with IBD ($p=0.02$) and healthy dogs ($p<0.0001$). Dogs with sub-stage b lymphoma also had a higher prevalence of total RBC morphologic anomalies as compared to sub-stage a ($p=0.04$). When categorized by pathologic etiology, dogs with lymphoma had a significantly higher number of RBC anomalies associated with regenerative anemia ($p<0.0001$), oxidative stress ($p=0.002$), and other RBC anomalies ($p<0.0001$) when compared to healthy dogs, but no significant difference when compared to dogs with IBD. The prevalence of eccentrocytes was significantly higher in dogs with lymphoma compared to IBD ($p=0.03$), as well as healthy dogs ($p=0.002$). No differences were noted between small, intermediate and large cell lymphoma for all parameters studied.

In conclusion, observation of anemia, of an increased total number of RBC morphological anomalies on blood smear, and specifically the presence of eccentrocytes could increase the clinical suspicion of lymphoma when compared to IBD in dogs with a suggestive clinical presentation.

HM09

AN IN-VITRO ASSESSMENT OF CANINE TO FELINE RED BLOOD CELL XENOTRANSFUSION

Michele Wilkinson, Ian McClure, Patricia Kaufman

Animal Blood Resources International, Stockbridge, MI, USA

Canine to feline xenotransfusions are increasingly reported in the veterinary literature as a temporary but questionably effective means of addressing life threatening anemia in recipient cats. This study was conducted to determine the in-vitro compatibility of feline red blood cells (RBCs) with canine sera and conversely canine RBCs with feline sera. Each cat and dog had a saline assessment initially conducted to determine if pre-existing auto-agglutination existed. Both major and minor cross matches were performed, with major defined as feline sera against canine RBCs and minor defined as feline RBCs against canine sera. All cross matches were performed via manual technique at 4, 25 and 37 degrees Centigrade. The end assessment of incompatibility was based on scored assessments of macroscopic or microscopic agglutination and/or hemolysis of RBCs. The first group of cross matches consisted of 19 blood Type A cats against five dogs. The dogs were selected to comprise a broad spectrum of the Dog Erythrocyte Antigen (DEA) possibilities. Each of the 19 cats had a major and minor cross match with each of the five dogs at all three temperatures for a total of 570 data points. None of the cross matches were compatible (100% incompatibility). To expand the database of canine bloodtypes, a second novel group of seven blood Type A cats were major and minor cross matched against ten dogs, five dogs of which were cohorts from the first set of cross match data. This yielded 420 data points for assessment of compatibility. Similar to the first data set, none of the cross matches were compatible (100% incompatibility). This data suggests there is an inherent incompatibility between dog and cat blood regardless of components tested. Clinicians contemplating canine to feline xenotransfusions should be cautious of utilizing this technique even in the presence of life threatening anemia.

HM10

CHARACTERIZATION OF THE IMMUNOPHENOTYPE OF DOGS WITH IMMUNE-MEDIATED HEMOLYTIC ANEMIA

James Swann, Julia Wu, Barbara Glanemann, Oliver Garden

Royal Veterinary College, London, UK

Whereas the clinical features of immune-mediated hemolytic anemia (IMHA) have been described extensively in dogs, the immunologic phenotype of this disease has yet to be fully characterized. We hypothesized that this phenotype would differ between dogs with primary IMHA and dogs with inflammatory diseases that did not have an immune-mediated etiology.

Serum and EDTA-anticoagulated blood samples were collected from healthy control dogs (HC), dogs with primary IMHA (IMHA), and dogs with inflammatory diseases (INF). Dogs were considered to have primary IMHA if they were anemic (with hematocrit less than 35%) and fulfilled one of the following criteria: significant spherocytosis on fresh blood smear, persistent agglutination of erythrocytes after dilution in saline, or a titer of greater than 1:16 in a direct antiglobulin test. Dogs were excluded if any underlying cause of disease was detected by serum biochemical testing, imaging of the thorax and abdomen and testing for infectious agents. Dogs in the INF group had diseases that were not considered to have an immune-mediated etiology and had no previous history of autoimmune disease. Dogs in the HC group had no previous history of autoimmune disease and had unremarkable physical examinations. Owners of dogs consented for collection and use of samples, and the study was approved by an institutional ethical review committee.

After lysis of red blood cells, peripheral blood mononuclear cells (PBMCs) were stained with fluorophore-conjugated antibodies specific for CD4, CD5, CD8, CD79b and FoxP3 and analyzed by flow cytometry to identify major lymphocyte subsets. Total RNA was extracted from PBMCs for quantitative reverse transcriptase polymerase chain reactions using primers for three reference genes and for interferon gamma (IFNγ) and interleukin(IL)-10. The serum concentrations of five cytokines (IL-2, IL-6, IL-8, IL-10 and TNFα) were measured using enhanced chemiluminescent assays. Results of complete blood cell counts and serum biochemical profiles were recorded for all INF and IMHA cases.

Samples were obtained from 25 IMHA dogs, in addition to 20 HC and 11 INF dogs; these groups had similar age and sex distributions. There was no difference in the proportion of regulatory T cells, helper T cells or cytotoxic T cells between groups ($p=0.226$), but the serum concentrations of IL-2 ($p<0.001$) and TNFα ($p<0.001$) were significantly greater in IMHA dogs compared to either of the other groups. The serum concentrations of IL-6 ($p=0.001$) and IL-8 ($p=0.001$) were significantly greater in IMHA dogs compared to HC, but there was no difference compared to the INF group. There was no difference between groups in the serum concentration of IL-10 ($p=0.731$), nor in the relative expression of IL-10 ($p=0.805$) or IFNγ ($p=0.670$). Principal component analysis revealed three major factors underlying the variability in the dataset, of which one, positively correlated with IL-2, IL-8, TNFα, blood monocyte count, serum total bilirubin concentration and negatively correlated with hematocrit, differed between IMHA and INF dogs ($p=0.001$).

Immunophenotyping of dogs with IMHA revealed increased concentrations of some pro-inflammatory cytokines compared to HC and INF dogs. Characterization of factors explaining variability in all parameters suggested development of different underlying inflammatory processes in IMHA and INF groups. Further work is required to determine whether these changes are associated with survival and whether interventions that ameliorate the effects of IL-2 and TNFα might improve outcome in dogs with IMHA.

HM11

A RETROSPECTIVE STUDY ON USE OF LEFLUNOMIDE IN DOGS WITH IMMUNE MEDIATED DISEASES

Masahiko Sato, Julia Veir, Marie Legare, Michael Lappin

Colorado State University, Fort Collins, CO, USA

Leflunomide is an immunomodulating agent labeled for the treatment of rheumatoid arthritis and psoriatic arthritis in humans. The drug inhibits the synthesis of ribonucleotide uridine monophosphate pyrimidine which ultimately lessens lymphocyte replication. This effect could be of benefit in the treatment of a variety of immune mediated diseases in dogs. The pharmacokinetics of leflunomide are known in dogs and the drug can suppress lymphocyte proliferation in this species. However, little clinical information concerning use of the drug in dogs is available. The purpose of this retrospective study was to report safety and efficacy of leflunomide for the treatment of naturally occurring immune mediated diseases in dogs.

Medical records from 1995-2014 at a USA Veterinary Teaching Hospital were retrospectively searched for dogs with immune mediated diseases administered leflunomide. Data that were extracted from the medical records included signalment, bodyweight, underlying indication for leflunomide, dose of leflunomide, treatment duration, concurrent medications, treatment response, and adverse events. Dogs with immune mediated polyarthritis (IMPA) were assessed for response to treatment on the basis of clinical signs of IMPA and physical examination findings. Dogs with immune mediated thrombocytopenia (IMTP) were assessed for response to treatment based on platelet count. A dog with cutaneous histiocytosis (CH) was assessed for treatment response on the basis of gross appearance of the skin lesion.

A total of 92 cases were included. The mean dose of leflunomide was 1.79 ± 0.8 mg/kg/day. The median duration of the use of leflunomide was 23.5 weeks. Adverse events which could be related to leflunomide administration included diarrhea (3 of 92, 3.3%), lethargy (2 of 92, 2.2%), suspect blood dyscrasia (3 of 92, 3.3%), thrombocytopenia (2 of 31, 6.5%) and increased liver enzyme activities (1 of 16, 6.3%). Leflunomide was discontinued due to the possible adverse events in 6 dogs (6.5%). Significant dose differences between dogs with adverse events (n=11, 2.6 ± 0.8 mg/kg/day) and dogs without adverse events (n=81, 1.7 ± 0.8 mg/kg/day) were found ($P<0.001$). The treatment response could be evaluated in 9 dogs with suspected IMPA, 7 dogs with IMTP, and 1 dog with CH. All of the 9 dogs with IMPA, 1 dog with IMTP and 1 dog with CH received leflunomide without administration of concurrent immunosuppressant drugs. Six of the dogs with IMTP received leflunomide with concurrent immunosuppressant drugs, the doses of which were unchanged when leflunomide was added to the treatment. Of these 17 dogs, 12 dogs (70.5%; 7 dogs with IMPA, 4 dogs with IMTP, 1 dog with CH) had an apparent positive response to the use of leflunomide. There was no significant difference ($P=0.18$) in doses between dogs that responded to leflunomide (2.0 ± 0.8 mg/kg/day) and those that did not respond (1.5 ± 0.4 mg/kg/day).

A lower starting dose of leflunomide of 2 mg/kg/day than the current suggested dose of leflunomide of 3-4 mg/kg/day is recommended based on the findings in this study.

HM12

EFFECTS OF LEUKOREDUCTION AND STORAGE ON EICOSANOID CONCENTRATIONS IN UNITS OF CANINE PACKED RED CELLS

Samantha Muro, Jung Hwa Lee, Matthew Ross, Robert Wills, Todd Archer, Andrew Mackin, John Thomason

Mississippi State University College of Veterinary Medicine, Starkville, Mississippi, USA

Storage of blood products creates an unnatural environment that can accelerate product degradation and pose risk to the recipient. Eicosanoids are arachidonic acid-derived bioactive molecules that are involved in inflammation and can accumulate inside units of packed red blood cells (pRBCs). Since leukocytes and platelets are the primary sources of eicosanoid synthesis within units of pRBCs, removal of these cells via leukoreduction may decrease eicosanoid synthesis and potentially reduce the risk of transfusion-related complications.

In this crossover study, units of blood were collected from 8 healthy dogs. In half of the units, leukocytes and platelets were removed via leukoreduction. Units of pRBCs were then created via centrifugation and stored for 10 or 21 days. Initial baseline plasma samples were collected after processing and before storage (Day 0). Additional samples were collected following storage for 10 or 21 days, both after removal from refrigeration and again after 5 hours at room temperature (to simulate transfusion conditions). Concentrations (ng/mL) of arachidonic acid (AA), prostaglandin E_2 (PGE_2), prostaglandin D_2 (PGD_2), prostaglandin $F_{2\alpha}$ ($PGF_{2\alpha}$), thromboxane B_2 (TXB_2, stable metabolite of TXA_2), 6-keto-prostaglandin $F_{1\alpha}$ (6-keto-$PGF_{1\alpha}$, stable metabolite of prostacyclin) and leukotriene B_4 (LTB_4) were quantified by liquid chromatography-mass spectrometry.

There was no difference in AA, PGE_2, PGD_2, and LTB_4 concentrations at any time point, and there was no difference in the concentrations between leukoreduced and non-leukoreduced units.

Prior to storage (Day 0), there were significantly greater concentrations of TXB_2 in units of leukoreduced pRBCs compared to non-leukoreduced units. In all leukoreduced samples, except for Day 10 post-transfusion, the TXB_2 concentrations decreased compared to Day 0. Except for the pre-storage samples, there was no difference in TXB_2 concentrations between leukoreduced and non-leukoreduced units. On both Days 10 and 21, there was a significant increase in 6-keto-$PGF_{1\alpha}$ concentrations for both leukoreduced and non-leukoreduced units. There was no difference in the 6-keto-$PGF_{1\alpha}$ concentrations between leukoreduced and non-leukoreduced units, except for the Day 10 post transfusion samples, in which the leukoreduced units were increased compared to the non-leukoreduced units.

Prior to storage, the $PGF_{2\alpha}$ concentrations were significantly greater in units of leukoreduced pRBCs compared to non-leukoreduced units. However, for the remaining time points, the $PGF_{2\alpha}$ concentrations decreased compared to Day 0, and there was no difference between leukoreduced and non-leukoreduced units.

In units of pRBCs, leukoreduction markedly influenced some eicosanoid concentrations. In general, removal of platelets and white cells tended to decrease production of most eicosanoids during storage. For some eicosanoids, however, this effect was counteracted by the passage of blood through the leukoreduction filter, which lead to marked increases in eicosanoid levels in units of pRBCs prior to storage. Further studies will be needed to determine what impact the complex effects of leukoreduction on eicosanoids in stored pRBCs will have on the incidence of transfusion reactions.

HM13

INCIDENCE OF DEA 5 IN CANINE POPULATION USING NOVEL CANINE ANTISERA

Ian McClure, Michele Wilkinson, Patricia Kaufman

Animal Blood Resources International, Dixon, California, USA

Dog Erythrocyte Antigen (DEA) 5 is a red cell membrane receptor, and is one of the six blood groups that have been described having clinical implications[1, 2]. Up until recently, blood typing was unavailable for this antigen. Previous supplies of the DEA 5 antisera had been exhausted, and the ability to create more was seriously hindered. A novel antisera was recently discovered, bringing back the ability to accurately blood type for the DEA 5 antigen.

A proprietary testing methodology was developed for this antisera using known facts about the DEA 5 antigen. The antisera was tested against canines previously known to be positive for the DEA 5 antigen. The antisera proved to identify the DEA 5 antigen with 100 percent positivity. Blinded in-house trials were performed as well, and technicians could correctly identify canines with the DEA 5 antigen.

Once the testing methodology was codified, and a Standard Operating Procedure (SOP) written, testing for the DEA 5 antigen moved into phase two, blind clinical testing. Two closed canine colonies were tested for the DEA 5 antigen, one comprised entirely of unrelated greyhounds, the other of mixed canines. Previously, a portion of these mixed canines had been typed for the DEA 5 antigen, so the sample set had known positives mixed with unknowns. Samples were blinded via identification number. All canines were infectious disease screened, up to date on vaccinations and physical examinations. Testing on the two canine colonies revealed an interesting find. Previously, it was thought that greyhounds had up to a 30% incidence rate of the DEA 5 antigen[3]. However, in testing of the closed greyhound colony, the incidence rate proved to be zero. In a mixed breed closed canine colony, the incidence of the DEA 5 antigen proved to be 21.6%. All previously known DEA 5 antigen positive mixed canines were correctly identified by laboratory technicians.

In phase three of the testing methodology, clinical samples sent to the laboratory were typed for the DEA 5 antigen to establish an incidence rate. Over a period of two months, the laboratory used convenience sampling to test for the DEA 5 antigen. At the closing of phase three, a total of 777 random canine samples were tested, with a total of 63 canines testing positive for the DEA 5 antigen, resulting in an incidence rate of 8.1 %, consistent with previously known incidence rates. Therefore, at the conclusion of phase three, it was determined the novel antisera was valid for the detection of the DEA 5 antigen.

With the re-introduction of a novel anti-DEA 5 reagent, a more complete profile of canine blood types can be achieved. This can result in better clinical decision making regarding transfusion in the canine patient, and therefore better outcomes, including the avoidance of delayed transfusion reactions when a DEA 5 incompatibility exists.

1. Colling, DT, Saison R: Canine Blood Groups: II. Description of a new allele in the Tr Blood Group System. Animal Genetics 11:13-20, 1980.
2. Symons M, Bell K: Canine blood groups: Descriptions of 20 specificities. Animal Genetics 23:509-515, 1992.
3. Hale, Anne: Canine Blood Groups and Their Importance. Transfusion Medicine25:6 1995.

HM14

PROSPECTIVE STUDY IN THE TREATMENT OF NONREGENERATIVE IMMUNE-MEDIATED ANEMIA IN 8 DOGS

Keitaro Morishita, Kyosuke Takeuchi, Shunsuke Mori, Sayuri Hayakawa, Kensuke Nakamura, Noboru Sasaki, Hiroshi Ohta, Mitsuyoshi Takiguchi

Hokkaido university, Sapporo, Japan

Nonregenerative immune-mediated anemia (NRIMA) is a common hematologic disorder in dogs. Because the bone marrow of the affected dogs shows erythroid hyperplasia and/or erythroid maturation arrest, it has been recognized as immune-mediated destruction of erythroid precursor cells. A variety of immunosuppressive agents have been used in the treatment of this disease, but information regarding the therapeutic efficacy is limited due to the lack of prospective research. The aim of this pilot study was to assess the treatment outcome to the pre designed immunosuppressive regimen including corticosteroids, cyclosporine and mycophenolate mofetil.

8 dogs with NRIMA referred to Hokkaido university teaching hospital between 2012 and 2015 were included. Inclusion criteria in this research were a minimum 5-day history of anemia, a severe nonregenerative anemia (Hct < 20%) with absolute reticulocyte count < 60×10^3 /μl, and absence of any identifiable underlying disease which could induce severe anemia. The following data were collected for each dog: age, breed, sex, results of hematological analyses performed on admission, results of bone marrow examination. First treatment regimen included prednisolone (2 mg/kg/day) and cyclosporine (10 mg/kg/day) for 8 weeks. Good response were defined as an absolute reticulocyte count > 60×10^3 /μl or increasing Hct. Dogs that responded within 8 weeks continued to be treated with gradual tapering of dose, and re-evaluated after 6 month. Dogs that did not respond to the first regimen proceeded to second regimen including prednisolone and mycophenolate mofetil (15 mg/kg, twice a day). Similarly, good response for second regimen was determined as explained above and dogs that did not respond to both regimens were defined as poor response.

Dogs ranged 5 to 9 years old (median age, 9.5 years), and included 5 castrated males, 2 intact females and 1 spayed female. There were 6 miniature dachshunds, 1 Shih-tzu and 1 Papillon. Median Hct and absolute reticulocyte count were 16.3% and 59,640 /μl, respectively. Bone marrow erythroid series were evaluated as hyperplasia in 5 dogs, euplasia in 2 dogs and hypoplasia in 1 dog. Erythroid maturation arrest was observed in all dogs. 6 of 8 dogs underwent core biopsy and 2 dogs showed myelofibrosis. 5 of 8 dogs responded to the first regimen and time to response was between 28 and 52 days. 3 dogs stopped receiving the first regimen due to anorexia that might be induced by Cyclosporine. 6 month prognoses of the other 2 dogs were complete remission (CR: Hct increased to reference range) and partial remission (PR: Hct did not reach the reference range), respectively. 3 dogs that did not respond to first regimen were proceeded to the second regimen. 1 of 3 dogs showed response after 29 days and 6 month prognosis was PR. The other 3 dogs that discontinued receiving Cyclosporine were also proceeded to the second regimen, and 6 month prognoses were CR in 2 dogs and PR in 1 dog. 2 dogs that did not show any response to both treatments had myelofibrosis.

In this research, response to both regimens could be evaluated in 6 dogs and median response time was 35 days (range 28-52). These results may be valuable for practitioners in consideration of changing immunosuppressive agent.

HM15

VALIDATION OF RAPID THROMBOELASTOGRAPHIC ANALYSIS ON CITRATED AND NATIVE WHOLE BLOOD FROM HEALTHY DOGS

Hye-Bin Wang[1], Hyeri Shin[1], Jin-Young Chung[2], Kun-Ho Song[1], Kyoung-Won Seo[1]

[1]*Chungnam National University, Daejeon, Republic of Korea,* [2]*Kangwon National University, Chuncheon, Republic of Korea*

Thromboelastography (TEG) is rapid and non-invasive diagnostic instrument that has been used in veterinary medicine for evaluation of haemostatic function in whole blood (WB). Recently, rapid thromboelastography (r-TEG) has been recognized as an overall evaluation of coagulation abnormalities in human medicine and become available faster than conventional coagulation tests (CCTs). Although r-TEG is becoming more generally used in human medicine, there are no studies describing the use of r-TEG in veterinary medicine. The objectives of this study were to 1) validation of r-TEG values with native and citrated WB samples from healthy dogs, and to compare stability and repeatability between two types of samples, and 2) examine whether the use of r-TEG could accelerate the coagulation pathway and decrease in taken time in comparison to kaolin activated TEG. Blood was collected from 16 clinically healthy Beagle dogs. TEG analysis was performed on kaolin-activated, citrated r-TEG, and native r-TEG blood samples. TEG activated clotting time (ACT), kappa value (k), alpha angle (α), and maximum amplitude (MA) were recorded. Coefficients of variation (CV) of native r-TEG and citrate r-TEG for TEG ACT, K, α angle, and MA were 13.4/18.8%, 11.1/16.6%, 4.2/5.1% and 10.0/10/0%, respectively. The mean time for the citrated and native r-TEG were 1313.9 ± 250.9/1351.3 ± 264.6 seconds (mean ± SD) respectively, in comparison with 1779.9 ± 197.0 seconds for kaolin activated TEG.

The native and citrated r-TEG tests were taken significantly less time than the kaolin activated TEG test ($p < 0.05$). Based on the results of this study, the r-TEG can evaluate haemostatic status rapidly than other coagulation tests and it is expected to be usefully used in critical setting in veterinary medicine.

HM16

SERIAL EVALUATION OF THROMBOELASTOGRAPHY AND PLATELET AGGREGOMETRY IN HEALTHY DOGS

Sarah Shropshire, Christine Olver, Michael Lappin

Colorado State University, Fort Collins, CO, USA

The coagulation system is comprised of many complex interactions where a balance between the pro-thrombotic and fibrinolytic arms is necessary to maintain normal hemostatic function. Thromboelastography (TEG) and platelet aggregometry are specialized assays that help to evaluate the viscoelastic properties of blood clotting and platelet function, respectively. If interindividual variability is high, serial evaluation within an individual rather than single measurements may provide greater sensitivity to detect clinically significant changes. If the intraindividual variability is greater than the interindividual variability however, then serial evaluation would not offer increased sensitivity. The purpose of this study was to compare interindividual and intraindividual variability over time for TEG and platelet aggregometry variables in healthy dogs.

Tissue factor activated TEG was performed in six specific pathogen free (SPF) sex and age-matched healthy Beagles at three different time points where the variables reaction time (R), clotting time (K), rate of clot formation (α), and maximum amplitude (MA) were recorded. Adenosine diphosphate (ADP)-induced and arachidonic acid (AA)-induced platelet aggregometry in addition to platelet count, hematocrit, and fibrinogen were also performed concurrently. APTT, OSPT, antithrombin activity, and D-dimer concentrations were measured on the first day of the study. A one-way random effects model was used to analyze the following variables; R, K, α, MA, area under the curve for ADP (AUC_{ADP}), and area under the curve for AA (AUC_{AA}). Variance components were created from the random effects model and used to calculate coefficients of variation for interindividual variability (CV_G) and intraindividual variability (CV_I). A repeated measures ANOVA was performed on the platelet count, hematocrit, and fibrinogen over the three time points with $p < 0.05$ considered significant.

The intraindividual variability was lower than the interindividual variability over time for MA, AUC_{ADP}, and AUC_{AA} however the intraindividual variability was higher than the interindividual variability for the TEG variables R, K, and α. There were no statistical differences in the platelet count, hematocrit, and fibrinogen measurements over time.

Based on the results of this study, serial measurements for ADP and AA-induced platelet aggregometry and the TEG variable MA may provide a more sensitive method to detect relevant changes when monitoring patients. However, due to the high intraindividual variability for the TEG variables R, K, and α, serial measurements may not be more beneficial than a single measurement and comparing it to population based reference intervals.

HP01

FLUORESCENCE IN SITU HYBRIDIZATION IDENTIFIES OCCULT BACTERIAL INFECTION IN GALLBLADDER MUCOCELES

Sara Wennogle[1], David Twedt[1], Kenneth Simpson[2]

[1]*Colorado State University, Fort Collins CO, USA,* [2]*Cornell University, Ithaca NY, USA*

Gallbladder mucocele is a commonly recognized form of gallbladder disease in the dog that is characterized by cystic mucinous hyperplasia and insidious accumulation of viscous bile and mucus within the gallbladder. Although several predisposing factors have been proposed, the underlying etiology of gallbladder mucocele formation is unknown. Previous studies have found limited associations with cholecystitis or bacterial infection. The aim of this study was to utilize culture-independent, fluorescence in situ hybridization (FISH) to determine the presence of bacteria in gallbladder mucoceles with and without concurrent cholecystitis.

Electronic medical records at Colorado State University were reviewed for cases of histopathologically confirmed gallbladder mucocele between December 2010 and January 2015. 29 cases were available and suitable for evaluation. Case data included patient signalment, clinical signs, clinicopathological abnormalities, ultrasonographic findings, histopathology, and bacterial culture of bile and liver. Formalin-fixed, paraffin-embedded gallbladder sections were mounted on charged glass slides and evaluated by FISH using a eubacterial probe (5Cy3-EUB-338) concurrently with a control probe (56 FAM-Non-EUB-338). Sections were examined on a BX51 epifluorescence microscope, and images were captured with an Olympus DP-7 camera.

Dogs undergoing cholecystectomy for treatment of gallbladder mucocele ranged in age from 6-14 years old. The most common breeds were mixed (n=10), Shetland Sheepdog (n=3), Miniature Schnauzer (n=2), Pomeranian (n=2), and Miniature Dachshund (n=2). Ultrasound was performed in all 29 cases, with 22/29 (76%) exams concluding the appearance of the gallbladder was consistent with a mucocele. Histopathology revealed cystic mucinous hyperplasia in all cases, with associated cholecystitis noted in 13/29(45%) cases. Bacteria were detected by eubacterial FISH in 8/29 (28%) cases. Bacteria were visualized as multifocal clusters of short rods within the mucus (n=1), adherent to the wall (n=1), or within the parenchyma (n=4) in 6/29 (21%) gallbladders. In 2 cases, a single bacterium was observed within the gallbladder parenchyma. Of the 6 cases with multifocal bacteria, 4 had concurrent cholecystitis. Bacterial culture of mucocele contents was positive in only 1/24 cases, yielding moderate growth of *Escherichia coli*. FISH analysis of this case showed diffuse colonization of mucocele contents and mucosa by E.coli.

FISH detected the presence of bacteria in 28% of mucoceles evaluated. FISH was more sensitive than bacterial culture, which was positive in only 1/8 cases with visible bacteria. Our results reveal that bacterial infection is more common than previously thought, and support the need for further studies to examine the relationship between gallbladder mucoceles, cholecystitis, and bacterial infection.

HP02

ASSOCIATION OF GALLBLADDER MUCOCELE FORMATION WITH OCCULT HYPOTHYROIDISM IN DOGS: A MATCHED CASE-CONTROLLED STUDY

Kathleen Aicher, John Cullen, Gabriela Seiler, Katharine Lunn, Kyle Mathews, Maria Correa, Jody Gookin

North Carolina State University, Raleigh, NC, USA

Gallbladder mucocele (GBM) formation results from accumulation of viscous bile that obstructs or ruptures the gallbladder. After cholecystectomy a median of 27% of dogs die before hospital discharge. The cause of GBM formation is unknown. Retrospective studies identify 14-17% of dogs with GBM have a diagnosis of hypothyroidism compared to 5-9% of normal dogs. Whether hypothyroidism is an independent risk factor for GBM formation or the consequence of an underlying metabolic syndrome is unknown. This distinction has important implications regarding the cause of GBM formation and may establish a rationale for diagnosis or treatment of hypothyroidism in these dogs. We undertook a prospective case controlled study of dogs newly diagnosed with GBM formation and lacking clinical signs of hypothyroidism to test a hypothesis that occult hypothyroidism exists at a high prevalence in these dogs and that the cause may differ from classic lymphocytic thyroiditis. Dogs diagnosed with GBM by ultrasound and healthy, age, breed and sex-matched dogs with normal gallbladder ultrasound were enrolled. Exclusion criteria included clinical evidence, prior diagnosis or treatment for hypothyroidism; drugs interfering with thyroid function, intact reproductive status, or emergent disease. Testing included a CBC, chem panel, UA, thyroid function testing including serum total T4, total T3, free thyroxine by equilibrium dialysis (FT_4), free triiodothyronine (FT_3), thyrotropin (TSH) and autoantibodies to T4AA, T3AA and TGAA. Dogs were stratified by disease severity into 1) no clinical signs, 2) mild clinical disease, 3) moderate to require hospitalization, and 4) severe requiring intensive care (incl. emergent surgery). Thyroid histology of GBM dogs and controls was examined by a pathologist blinded to the diagnosis. Parametric data were analyzed by t-test and non-parametric data by Mann-Whitney U test. $P < 0.05$ was considered significant. Of 106 screened dogs, 38 with GBM met study inclusion criteria and ranged in age from 6-16 yrs (median 10.5 yrs). Breeds included 11 Shetland sheepdogs, 4 Cocker Spaniels, 3 Chihuahuas, 3 Beagles, and 16 other pure breeds. Illness severity scores were: 37% absent; 21% mild; 21% moderate; and 21% severe. Of 18 healthy dogs screened, 12 were enrolled. Dogs with GBM had significantly lower serum T_4 (median 11, range 0-44 nM), T_3 (median 0.7, range 0-1.7 nM), FT_4 (median 15, range 1-27 pM), and FT_3 (median 3, range 0-10.5 pM) compared to control dogs. Relative to control dogs, dogs with GBM frequently had severely low individual thyroid hormone concentrations. Serum T4 was below reference range (RR) in 50% of GBM dogs compared to 8% controls. Four (10.5%) GBM dogs had a T4=0. Serum T3 was below RR in 47% of GBM dogs compared to 25% controls. Three (8%) GBM dogs had a T3=0. Serum FT4 was below RR in 12% of GBM dogs compared to 0% controls. Serum FT3 was below RR in 17% of GBM dogs compared to 0% controls. Significant differences in T4AA, T3AA or mean TSH were not identified. However 23.5% of GBM dogs and only 8% control dogs had an elevated TSH. Serum TGAA was elevated in 17% of control dogs and 0% of GBM dogs. No inflammation was detected in the thyroids of any of 12 dogs with GBM undergoing autopsy. These results identify a high prevalence of occult, often severe, decreases in individual serum thyroid hormone concentrations in GBM dogs. Influence of non-thyroidal illness must be considered for some of these abnormalities. Lack of serum TGAA and absent histologic inflammation in thyroids of GBM dogs is noteworthy. These findings suggest that an alternative mechanism for abnormal thyroid function in dogs with GBM should be considered.

HP03

INTEROBSERVER AGREEMENT FOR HISTOLOGICAL SCORING OF CANINE HEPATIC FIBROSIS

Jonathan Lidbury[1], Aline Rodrigues Hoffmann[1], Fabiano Olivera[2], Brian Porter[1], Thomas Van Winkle[3], Jan Suchodolski[1], Guy Grinwis[4], Renata Ivanek[1], John Cullen[5], Joerg Steiner[1]

[1]*Texas A&M University, College Station, Texas, USA,* [2]*Antech Diagnostics, College Station, Texas, USA,* [3]*University of Pennsylvania, Philadelphia, Pennsylvania, USA,* [4]*Utrecht University, Utrecht, The Netherlands,* [5]*North Carolina State University, Raleigh, North Carolina, USA*

Hepatic fibrosis is an important complication of canine chronic hepatitis. To the authors' knowledge, the interobserver agreement associated with the histological scoring of fibrosis from canine hepatic biopsy specimens has not previously been reported. Therefore, the aim of this study was to evaluate the interobserver agreement for the histological scoring of canine hepatic fibrosis.

Paraffin-embedded liver specimens from 50 dogs were selected from the tissue archives at Texas A&M University and North Carolina State University. Two contiguous sections were cut from each specimen, one was stained with hematoxylin and eosin (H&E) and one with picrosirius red (PR). Six board-certified veterinary pathologists assigned a fibrosis score (0: absent, 1: mild, 2: moderate, 3: marked, or 4: very marked) to each section using a scoring system adapted from the Ishak scoring system that is used in humans. Interobserver agreement for the H&E and PR stained sections was evaluated using the kappa and weighted-kappa statistics. The median fibrosis score assigned to H&E and PR stained sections from each dog were compared using the Wilcoxon signed-rank test.

Multiuser kappa (95% CI) for H&E stained sections was 0.35 (0.26 − 0.44) and multiuser weighted-kappa was 0.59 (0.50 − 0.70). Pathologists were in complete agreement 49% of the time, differed by one score level 41% of the time, and differed by more than 1 score level 11% of the time. Multiuser kappa (95% CI) for PR stained sections was 0.39 (0.30 − 0.49) and multiuser weighted-kappa was 0.64 (0.55 − 0.73). Pathologists were in complete agreement 53% of the time, differed by one score level 42% of the time, and differed by more than one score level 5% of the time. There was no significant difference in the median fibrosis scores assigned by the 6 pathologists between H&E and PR stained sections (both medians were 2; $P = 0.248$).

There was only fair interobserver agreement when veterinary pathologists assessed canine hepatic fibrosis from H&E or PR stained sections. However, because there was frequent partial agreement between pathologists, this scoring system may be acceptable for clinical use. A clear benefit of the routine staining with PR for the assessment of hepatic fibrosis was not demonstrated in this study.

HP04

EMBOLIZATION OF INTRATRAHEPATIC PORTOSYSTEMIC SHUNTS IN DOGS WITH A PROTOTYPE COIL

Matthias Schneider, Andreas Stosic, Stephan Bayer

Small Animal Clinic, Giessen, Germany

Surgical treatment of intrahepatic portosystemic shunts (iPSS) is challenging. Frequently, multiple interventions are needed to avoid portal hypertension and to achieve a complete closure. Complications seen in catheter-assisted embolization with commercial coils include coil migration, portal hypertension (rapid thrombotic occlusion), and incomplete occlusion. Attenuation of hepatic vein by means of stent-supported coil embolization may lead to the development of intrahepatic venous collaterals. In the present study we postulated that implantation of a single prototype coil combined with immediate unfractionated heparin therapy (UHT) is an effective treatment for iPSS in dogs and has fewer complications than reported with open surgery.

Dogs with single iPSS with or without intrahepatic venous collaterals were included. A percutaneous access to the right jugular vein was used for retrograde portal vein catheterization. Portal vein pressure and iPSS diameter were measured unblocked and blocked (balloon catheter: 8F wedge catheter or 20/27mm occlusion catheter). The prototype coil with polyester fibers was selected to be either at least 1 mm larger in diameter than the blocked iPSS or twice the diameter of the unblocked iPSS, and was implanted using an 8F guiding catheter. Coagulation times were assessed by activated coagulation time (ACT) at baseline and during UHT. Heparin was started before coil implantation with a bolus of 100 IU/kg (injected into the portal vein), followed by continuous rate infusion (25 IU/kg/h). Postoperatively, UHT was administered 150 IU/kg q6h subcutaneously and intermittent 50 IU/kg intravenously to achieve an ACT of 150 to 200 seconds, and then adjusted based on sonographic proven development/resolution of ascites. Functional closure was judged according to the result of the oral ammonia tolerance test. Descriptive statistics and Kaplan-Meier analysis was used.

The prototype coil was implanted in a total of 25 dogs. Most common breeds were Hovawart (n=8) and Golden Retriever (n=5). Median age was 10 months (range 6 – 35) and median body weight 26 kg (range 12 – 48). iPSS originated from the right lateral (n=1), right medial (n=7) or left portal vein branch (n=17). 19/25 dogs showed portal hypertension during shunt blocking. Unblocked and blocked shunt diameter were 5.2 - 13.0 mm (median 8.3; n=25) and 9.0 - 21.0 (median 13.3; n=24), respectively. Intervention was successful at the first (n=24) or second (n=1) intervention. Major complications developed in 5 dogs leading to death in 2 dogs. Minor complications were common and were mostly associated with bleeding problems. One dog was lost to follow-up after three months. Of the remaining 22 dogs, 5 experienced complications associated with change to a non-protein restricted diet: severe neurological abnormalities leading to euthanasia (n=1), moderate symptoms leading to re-intervention (n=2), mild symptoms which could be treated successfully by changing to a moderately protein restricted diet (n=2). All dogs without portal hypertension showed functional closure at 3 months follow-up. In dogs with portal hypertension the cumulative proportion of functional closure reached 81% at 24 months follow-up.

In conclusion, stiffness and size of the prototype coil were responsible for the high procedural success and functional closure rates. Unfractionated heparin treatment needs optimization to reduce complications, but the general treatment regimen seems promising.

HP05

INVESTIGATION OF HEPATIC COPPER ACCUMULATION IN DOGS FROM TWO TIME PERIODS (1982-1988 AND 2009-2015)

Ryan Schultz, Rebecca Smedley, John Buchweitz, Katie Barnes, Sarah Abood, Daniel Langlois

Michigan State University College of Veterinary Medicine, East Lansing, MI, USA

Canine copper-associated hepatitis (CCAH) has become an increasingly recognized cause of progressive and potentially fatal liver disease in dogs; however, the exact etiology remains unclear. To date, no broad epidemiologic investigations quantifying hepatic copper concentrations over time in both predisposed (to CCAH) and non-predisposed breeds have been conducted. The purpose of our study was to determine and compare quantitative hepatic copper concentrations in breeds that are, or are not, predisposed to CCAH from two study periods. We hypothesized that hepatic copper concentrations have increased for all dogs over time, but the relative increase will be greater in predisposed breeds. A retrospective search of the histopathology database at the Michigan State University Diagnostic Center for Population and Animal Health was conducted for two study periods: 1982 through 1988, and 2009 through 2015. Doberman Pinschers, Labrador Retrievers, and West Highland White Terriers were considered predisposed to CCAH. Dalmatians, Skye Terriers, and Bedlington Terriers were excluded from analysis due to a rarity of cases, limited information on predisposition, or a known genetic etiology, respectively. Cases with archived liver tissue available were randomly selected, reviewed, and then classified by breed and presence or absence of inflammatory liver disease. Hepatic copper concentrations (parts per million, dry weight basis) were measured in all specimens using inductively coupled plasma mass spectrometry. Comparisons were made within and across groupings using commercially available statistical software.

A total of 373 cases were evaluated, including 207 dogs predisposed to CCAH and 166 non-predisposed dogs. Independent of breeds groupings and inflammatory activity, mean hepatic copper concentrations in the recent time period were 225 ppm greater than those in the historical time period ($P < 0.05$). This effect was predominantly due an increase of 386 ppm over time in predisposed breeds ($P < 0.001$). The mean hepatic copper concentration of 338 ppm in non-predisposed breeds in recent years was greater than the mean hepatic copper concentration of 271 ppm in non-predisposed breeds in the historical period, but this trend did not reach significance ($P = 0.08$). In the recent period, the mean hepatic copper concentration of 519 ppm in predisposed breeds with non-inflammatory liver histology was greater than the mean hepatic copper concentration of 297 ppm in non-predisposed breeds with non-inflammatory liver histology ($P < 0.001$). This difference was not observed in the historical time period. In both periods, dogs with hepatitis had higher hepatic copper concentrations than those without hepatitis, but this difference was magnified in recent years. In the aggregate, these findings confirm that hepatic copper concentrations have changed over time. The etiology of these changes is unknown, but suggestive of changes in environmental copper exposure. Detailed investigations of potential sources of copper exposure, namely drinking water and diet, are needed to further elucidate a potential environmental etiology. Additional investigations also are needed to determine if these changes are important for all breeds of dog.

ID01

EVIDENCE FOR GENETIC PREDISPOSITON TO *BORRELIA BURGDORFERI* INFECTION IN PURPOSE BRED BEAGLES

Elena Contreras[1], Scott Moroff[2], Michael Lappin[1]

[1]Colorado State University Veterinary Teaching Hospital, Dept of Clinical Sciences, Fort Collins, Colorado, USA, [2]Antech Diagnostics, Lake Success, New York, USA

Borrelia burgdorferi (BB) and *Anaplasma phagocytophilum* (AP) are transmitted by *Ixodes* spp., and seroprevalence rates in endemic areas can be greater than 15% in dogs. A genetic predisposition to infection (Bernese mountain dogs) or clinical illness associated with BB has been suggested, particularly in people and dogs with Lyme nephritis (Retriever breeds). In vaccine studies, wild caught *I. scapularis* adults placed in feeding chambers and allowed to feed for seven days usually results in at least a 75% BB infection rate. Recently, an *I. scapularis* infestation study using a previously reported model and ticks with a similar BB carriage rate from the same collection area previously used, resulted in only a 29.2% infection rate. The purpose of this report is to describe the genetic relatedness of the dogs used in that study.

A total of 10 female and 14 male purpose-bred beagles, 9-12 months of age, and negative for antibodies against *A. phagocytophilum*, *B. burgdorferi*, and *E. canis* (Accuplex 4 BioCD system; Antech Diagnostics), were infested for 7 days with 13 female and 12 male wild-caught *I. scapularis* with an estimated 48% BB infection rate and a 15% AP infection rate. Blood was collected from all dogs prior to tick attachment and then on Weeks 1-12 and Week 18 after tick attachment and screened for BB and AP antibodies as described. Dogs that seroconverted were assumed to have been infected. Statistical analyses were performed with commercial software; $p \leq 0.05$ was considered statistically significant.

Antibodies against BB were detected in 7 of 24 dogs (29.2%) between 4-8 weeks post-infestation; all positive dogs were females ($p=0.0003$). When positive and negative dogs were compared, there were no significant differences in the ages, number of ticks recovered, or number of fed ticks. Antibodies against AP were detected in the serum of 10 of 24 dogs (41.7%). Results using the same source of dogs, ticks, and model in previous publications were 77.8% and 55.5%% for BB and AP, respectively. The differences in infection rates between the current and historical studies were similar for AP but significantly lower in the current study for BB ($p = 0.004$). The most recent common ancestor (MRCA) for 6 of the 7 (86%) BB-positive dogs was either a sibling or parent. Dogs A and B were siblings and had the same father as Dog C. The mother of Dog D was the maternal grandmother of Dogs A and B. The father of Dog E was the paternal great grandfather of Dogs A, B, C. The mother of Dog F was the maternal grandmother of Dog E. The MRCA for 8 of the 17 (47.1%) BB-negative dogs was either a sibling or parent. None of the BB positive dogs had a MRCA sibling or parent in common with any of the BB negative dogs.

These findings provide further evidence that a genetic predisposition to *B. burgdorferi* infection exists. As studies have investigated common genotypes and alleles among humans with Lyme borreliosis and arthritis, this should also be considered in future studies of canine Lyme borreliosis.

ID02

RAPID DIAGNOSIS OF *BABESIA GIBSONI* USING POINT-OF-CARE INSULATED ISOTHERMIC POLYMERASE CHAIN REACTION ASSAY

Kirsten Cooke[1], Patrick Frenzer[1], Julie K. Levy[1], Sylvia Tucker[1], P. Cynda Crawford[1], Sarah Kirk[2]

[1]University of Florida, Gainesville, FL, USA, [2]The American Society for the Prevention of Cruelty to Animals, New York, NY, USA

Babesia gibsoni is commonly identified in pit bull-type dogs, particularly those involved in dogfights. Dogs seized from fighting operations undergo health and behavior evaluations to assess suitability for adoption. Adoption carries the risk for sending dogs harboring infectious diseases to unsuspecting owners or to regions where the infections are not endemic. Screening large numbers of dogs for infectious diseases is expensive and it may be days before results are available. A more rapid and cost-effective diagnostic test would make identification and treatment of *B. gibsoni* infected dogs more feasible for rescue organizations. Recently, the development of insulated isothermic PCR (iiPCR) has made mobile PCR a reality. The purpose of this study was to determine performance of the POCKIT® iiPCR system for diagnosis of *B. gibsoni* in dogs rescued from fighting organizations.

Whole blood was collected from 260 pit bull-type dogs rescued during a federal dogfighting investigation. Real-time PCR was performed for *Babesia* species, and positive samples underwent a second round of species-specific PCR in a commercial diagnostic laboratory (IDEXX Laboratories, Inc.). Residual whole blood samples were frozen and stored at -80 C pending subsequent analysis. POCKIT® iiPCR was performed twice on each sample, once with a kit specific for *B. gibsoni* and once with a kit that detects both *B. gibsoni* and *Babesia canis* (Babesiosis). The sensitivity and specificity of each test were calculated using the commercial test as the gold standard. Positive (PPV) and negative (NPV) predictive values were calculated for theoretical 5%, 15%, and 30% prevalences. The study was approved by the UF IACUC.

A total of 80 dogs (31%) were positive for *B. gibsoni* by the gold standard commercial PCR test. Of these, 72 were identified as positive by the POCKIT® iiPCR *B. gibsoni* kit, and 70 were identified as positive by the POCKIT® iiPCR *Babesiosis* kit. Sensitivity and specificity of the *B. gibsoni* kit were 90% and 99%, respectively. Sensitivity and specificity of the *Babesiosis* kit were 88% and 98%, respectively. The PPVs of the B. gibsoni kit at 5%, 15%, and 30% prevalence were 82%, 94%, and 97% respectively. The NPVs of the *B. gibsoni* kit were 99%, 98%, and 96% respectively. The PPVs of the *Babesiosis* kit were 70%, 89%, and 96% respectively. The NPVs of the *Babesiosis* kit were 99%, 98%, and 95% respectively.

The POCKIT® iiPCR system was quick to learn, portable, and had a short run time of approximately 1 hour. There were few false-positive results, indicating that positive results are likely to represent true infections when tested in high-risk animals. Approximately 10-15% of infected dogs would be missed by the POCKIT® iiPCR system. However, the portability and speed with which results can be obtained may result in more infected dogs being diagnosed than in the current situation in which testing is rarely performed in dog-fighting cases due to cost and logistics of sending samples to outside laboratories. This system may also be used for *B. gibsoni* treatment monitoring using a hybrid approach. For example, following initial diagnosis with either the POCKIT® system or a commercial laboratory, the POCKIT® system could be used for testing during treatment. Once a negative result was obtained, a sample could be submitted to a commercial laboratory to confirm treatment efficacy. Although this study focused on mass screening for *B. gibsoni*, the portable iiPCR platform has potential to aid in rapid detection of a variety of infections under field conditions.

ID03

PREVENTION OF *BORRELIA BURGDORFERI* AND *ANAPLASMA PHAGOCYTOPHILUM* TRANSMISSION FROM *IXODES SCAPULARIS* TO DOGS BY SAROLANER

Nicole Honsberger, Robert Six, Thomas Heinz, Angela Weber, Sean Mahabir, Thomas Berg

Zoetis, Kalamazoo, MI, USA

The efficacy of sarolaner (Simparica™, Zoetis) to prevent transmission of *Borrelia burgdorferi* and *Anaplasma phagocytophilum* from infected wild-caught *Ixodes scapularis* to dogs was evaluated in a well-controlled laboratory study. Twenty-four purpose-bred laboratory Beagle dogs seronegative for *B. burgdorferi* and *A. phagocytophilum* antibody were allocated randomly to one of three oral treatment groups: placebo administered on Days 0 and 7, sarolaner administered on Day 0 (28 days prior to tick infestation), or sarolaner administered on Day 7 (21 days prior to tick infestation). Sarolaner tablets were shaved and/or sanded based on each dog's individual bodyweight to provide a dosage of 2 mg/kg. On Day 28 each dog was infested with approximately 25 female and 25 male wild caught adult *I. scapularis* that were determined by random sampling to have infection rates of 57% for *B. burgdorferi* and 6.7% for *A. phagocytophilum* by PCR. *In situ* tick counts were conducted on Days 29 and 30. On Day 33, all ticks were counted and removed. Blood samples collected from each dog on Days 27, 49, 63, 77, 91 and 104 were tested for the presence of *B. burgdorferi* and *A. phagocytophilum* antibodies using the SNAP® 4Dx® Plus Test, and quantitatively assayed for *B. burgdorferi* antibodies using an ELISA test. Skin biopsies collected on Day 104 were tested for the presence of *B. burgdorferi* by bacterial culture and PCR. Acaricidal efficacy was calculated based on the reduction of geometric mean live tick counts in the sarolaner-treated groups compared to the placebo-treated group for each tick count.

Geometric mean live tick counts for placebo-treated dogs were 14.8, 12.8 and 19.1 on Days 29, 30 and 33, respectively. For the group treated with sarolaner 21 days prior to infestation, the percent reductions in mean live tick counts on Days 29, 30, and 33 were 86.3%, 100%, and 100%, and for the group treated with sarolaner 28 days prior to infestation were 90.9%, 97.1%, and 100%. Geometric mean live tick counts for both sarolaner-treated groups were significantly lower than those for the placebo group on all count days (P<0.0001).

Successful transmission of *B. burgdorferi* to all eight placebo-treated dogs was confirmed by positive antibody (6 of 8 dogs), PCR (7 of 8 dogs), and/or culture result (7 of 8 dogs). Similarly, transmission of *A. phagocytophilum* was confirmed by the presence of antibodies in four placebo-treated dogs. In contrast, a single oral dose of sarolaner prevented transmission of *B. burgdorferi* from infected ticks to dogs infested 21 or 28 days after treatment as demonstrated by negative antibody, PCR, and culture results. Prevention of *A. phagocytophilum* transmission was also demonstrated by negative antibody in all sarolaner-treated dogs. There were no adverse reactions to treatment with sarolaner.

ID04

MOLECULAR ANALYSIS ON CANINE BABESIA PREVALENCES AND EXPOSURE TO ANAPLASMA, EHRLICHIA, BORRELIA AND DIROFILARIA IN MOST POPULATED SERBIAN REGION – BELGRADE AREA

Milica Kovacevic Filipovic[1], Anja Ilic[1], Zorana Milanovic[1], Dragan Bogdanovic[4], Vladimir Radonjic[2], Jelena Francuski Andric[1], Nenad Andric[1], Brendon Thatcher[3], Christian Leutenegger[3], Phyllis Tyrrell[3], Antony Mestek[3], Chandrashekar Ramaswamy[3]

[1]Faculty of Veterinary Medicine, University of Belgrade, Belgrade, Serbia, [2]Private Veterinary Practice Petrovac", Belgrade, Serbia, [3]IDEXX Laboratories, Westbrook, Maine, USA, [4]State University of Novi Pazar, Novi Pazar, Serbia

Babesiosis is endemic in many parts of Serbia. Other tick borne diseases are less commonly diagnosed because their clinical signs are nonspecific and there is no published data indicating their prevalence. The aim of this study was to determine molecular prevalence of *Babesia canis canis* and *B. gibsoni*, molecular and seroprevalence of *Anaplasma phagocytophilum*, *A. platys*, *Ehrlichia canis*, *E. ewingii*, *E. chaffeensis* and *Borrelia burgdorferi* and *Dirofilaria sp.* microfilariae in different outdoor canine populations.

Blood samples from apparently healthy stray (n=38), shelter (n=39) and hunting dogs (n=35) and outdoor pet dogs suspected to have tick borne diseases (n=50) were tested using real time PCR and serological assays. Real-Time PCR assays were used to amplify *Babesia*, *Anaplasma* and *Ehrlichia* DNA from blood samples. An experimental SNAP® Multi-Analyte Assay (SNAP® M-A) (IDEXX Laboratories, Inc. Westbrook, Maine, USA) was used to screen all sera for antibodies to *Anaplasma* and *Ehrlichia* genus peptides and *A. phagocytophilum*, *A. platys*, *B. burgdorferi*, *E. canis*, *E. ewingii*, and *E. chaffeensis* species-specific peptides. Modified Knott test was used to identify dogs with *Dirofilaria sp.* microfilaria presence.

In total, prevalence of *Babesia sp.* infections among apparently healthy dogs was 16% and among sick pet dogs, was 65%. Prevalence of *B. canis* was the highest in shelter (25%) and the lowest in stray dogs (8%). *B. gibsoni* was rare among shelter, stray and pet dogs, while it was not detected among hunting dogs. Molecular analysis did not reveal the presence of dogs with active anaplasmosis, ehrlichiosis or borreliosis. However, 32% of stray and shelter dogs, 12% of hunting dogs and 4% of sick pet dogs had antibodies to *A. phagocytophilum*. Only one dog (hunting group) was seropositive to *A. platys*, one dog to *E. ewingii* (stray group) and two dogs to *B. burgdorferi* (stray group). None of the dogs was seropositive to *E. canis* and *E. chaffeensis*. Also, 33% of pet, 18% of hunting, 11% of stray and 3% of shelter dogs had microfilariae in the blood. Number of dogs exposed to two and three pathogens was the highest among shelter dogs (31% and 10%, respectively), and number of dogs exposed to four pathogens was the highest among stray dogs (5%).

In Belgrade area, *B. canis* and *A. phagocytophilum* are the most prevalent vector-borne pathogens in apparently healthy outdoor dog population, while *B. canis* and *Dirofilaria sp.* are the most important pathogens among sick pet dogs.

ID05

Enhanced serologic surveillance to detect prevalence of canine vector-borne infections on the island of St. Kitts, West Indies

Diana Scorpio[1], Jennifer Ketzis[1], Linda Shell[1], Brendon Thatcher[2], Ramaswamy Chandrashekar[2]

[1]*Ross University School of Veterinary Medicine, Basseterre, Saint Kitts and Nevis,* [2]*Idexx Laboratories, Westbrook, Maine, USA*

The island of St. Kitts (West Indies) has a large feral and semi-feral dog population. Semi-feral dogs present a unique opportunity for study since they are mostly outdoors but are comparatively well taken care of compared to feral dogs. Care tends to range from minimal to adequate, though many owners use the Ministry of Agriculture Veterinary Services for preventative medicine rather than a veterinary hospital setting. Historically, dogs on the island are endemically infected and infested with internal and ectoparasites, which are also associated with a number of vector-borne diseases. Due to the high prevalence of the vector Rhipicephalus sanguineus, it is not unusual to see high prevalence of Ehrlichia canis and *Anaplasma platys*. Samples collected from an island-wide field study from last year revealed an abundance of serologic data. These samples have been further tested utilizing a research-based serology platform (IDEXX Laboratories) and additional canine vector-borne infections data has been collected from these analyses.

Over a period of 3 months, veterinarians and students from RUSVM visited all 9 parishes of St. Kitts, covering a diverse geographic area with a wide diversity in ecosystem landscape. One hundred and eleven dogs were identified and visited, and were given a complete physical exam, then blood and feces were collected. Demographic data was collected, including breed, age, housing, other animals in household, evidence of ticks or fleas, deworming and vaccine administration. Blood was analyzed using a research-based broader canine vector-borne infection serologic platform (SNAP® M-A, IDEXX Laboratories), which tests for specific antibodies against *E. ewingii, E. canis, E. chaffeensis, Anaplasma platys, A. phagocytophilum,* and *Borrelia burgdorferi.* This testing platform is also based on a rapid ELISA SNAP® technology.

Analyses using the new SNAP® M-A serologic platform revealed no evidence of exposure to *E. ewingii, E. chaffeensis, or Borrelia burgdorferi.* However, as expected due to the endemic nature of the tick vector, the platform identified *A. platys* exposures of approximately 20.7%. *E. canis* exposures were detected at a prevalence of approximately 32%. Of interest is that the SNAP®

M-A test also identified 19 dogs (17%) with antibodies to *A. phagocytophilum*, not previously described on St. Kitts due to the absence of *Ixodes* tick vectors.

Using the SNAP® M-A serologic platform, there was a robust capacity to detect up to 2 *Anaplasma* species, in addition to having a sensitivity which corroborates previously published prevalence rates of *E. canis.* Of note is the presence of *A. phagocytophilum* exposures, and additional studies are warranted to understand and explain these findings. The data collected in this study supports the importance of including an exhaustive list of differentials when suspecting canine vector-borne infections, in addition to aggressive and consistent use of environmental and topical parasiticides for tick control.

ID06

PERFORMANCE OF POINT-OF-CARE ASSAYS FOR FELV AND FIV

Julie Levy

Maddie's Shelter Medicine Program, University of Florida, Gainesville, Florida, USA

At least 5% of cats in the US are infected with FeLV or FIV, contagious retroviruses that predispose to bone marrow suppression, cancer, chronic inflammatory conditions, immune dysfunction, and wasting syndromes. This equates to more than 5 million cats affected by these preventable viruses and at risk for disease and premature death. Although vaccination plays a role in prevention, the cornerstone of control is identification and segregation of infected cats. Recently, new point-of-care screening tests have become available, but independent comparison of test performance is needed. The purpose of this study was to determine the performance of currently available FeLV antigen/FIV antibody combination test kits.

For assessment of FeLV tests, serum or EDTA whole blood samples were collected from animal shelters or reference laboratory and tested with two microtiter plate ELISAs for the detection of FeLV p27 antigen (ViraCHEK®/FeLV and PetChek® FeLV Antigen Test) to establish a gold standard. Only samples with concordant results were included in the study. For assessment of FIV tests, plasma samples previously collected from naturally or experimentally infected cats and verified by virus culture and from uninfected SPF cats were included. None of the cats had been vaccinated against FIV. The final study set included 146 FeLV+, 154 FeLV-, 94 FIV+, and 97 FIV- samples. Point-of-care tests evaluated in this study included SNAP® Combo FeLV Ag/FIV Ab Test, WITNESS® FeLV-FIV Test Kit, Anigen® Rapid FIV Ab/FeLV Ag Test Kit, VetScan® Feline FeLV/FIV Rapid Test. All test results were visually read by two blinded observers.

	SNAP®	**WITNESS®**	**ANIGEN®**	**VETSCAN®**
FeLV Sensitivity (95% CI)	**100%** (96.9-100)	**89.0%** (82.8-93.2)	**91.8%** (86.1-95.4)	**85.6%** (78.9-90.5)
FeLV Specificity (95% CI)	**100%** (97.1-100)	**95.5%** (90.8-98.0)	**95.5%** (90.8-98.0)	**85.7%** (79.3-90.48)
FIV Sensitivity (95% CI)	**97.9%** (92.1-99.9)	**94.7%** (87.9-98.0)	**96.8%** (90.6-99.3)	**91.5%** (83.9-95.8)
FIV Specificity (95% CI)	**99%** (93.8-100)	**100%** (95.4-100)	**99%** (93.8-100)	**99%** (93.8-100)

The SNAP® test had the overall best performance. In high-stakes testing, test kits should be selected for both high sensitivity and specificity. When tests lack sensitivity, infected cats may escape detection and remain at risk for infecting other cats. When specificity falls, uninfected cats may be unnecessarily segregated or even euthanized. In low infection prevalence, such as observed in FeLV and FIV, decreased specificity has the largest impact on erroneous test results, leading to decreased positive predictive value (increased false-positives).

ID07

DOES A DIVA TEST EXIST FOR DIFFERENTIATING FIV INFECTION FROM FIV VACCINATION?

Cynda Crawford

Maddie's Shelter Medicine Program, University of Florida, Gainesville, Florida, USA

Diagnosis of FIV infection is based on detection of circulating FIV antibodies. Introduction of FIV vaccines created a diagnostic dilemma because vaccine antibodies were indistinguishable from those induced by infection using existing serological assays. New point-of-care tests for FIV antibody are now available to practitioners. The purpose of this study was to determine the accuracy of FIV diagnostic tests in FIV-vaccinated cats and whether any of the tests could serve as a DIVA test to differentiate infected from vaccinated animals.

Plasma samples were collected from 104 uninfected SPF cats vaccinated 3 times with a killed dual-subtype FIV vaccine (Fel-O-Vax, Boehringer Ingelheim) according to manufacturer instructions. Age at vaccination ranged from 2.5 to 13 months. The interval between vaccination and sample collection ranged from 1.75 to 14 months. The FIV-free infection status of all cats was confirmed by virus culture. Three FIV-infected cats were similarly vaccinated and tested. The plasma samples were tested in 4 commercial point-of-care assays: SNAP® Feline Combo FeLV Ag/FIV Ab Test, WITNESS® FeLV-FIV Test, Anigen® Rapid FIV Ab/FeLV Ag Test Kit, and VetScan® Feline FeLV/FIV Rapid Test. All testing was performed by personnel blinded to sample status. Each test result was independently interpreted by 2 personnel.

The SNAP® and VetScan® tests detected FIV antibodies in 102/104 and 88/104 uninfected vaccinated cats, respectively. This would lead to misclassification of most of the vaccinated cats as infected. The WITNESS® and Anigen® tests had very high specificity, indicating that nearly all vaccinated cats were correctly identified as uninfected. All 4 tests correctly identified the 3 FIV-infected cats that were vaccinated against FIV.

Specificity and 95% Confidence Intervals for 4 Point-of-Care Tests for FIV Infection Performed in 104 Uninfected Cats that were Vaccinated against FIV

WITNESS®	ANIGEN®	VETSCAN®	SNAP®
98.1% (92.8-99.9)	**98.1%** (92.8-99.9)	**21.2%** (14.3-30.0)	**1.9%** (0.1-7.2)

Accurate detection of FIV-infected cats is essential to providing appropriate care and segregation from uninfected cats. Since euthanasia is a tool employed in the control of FIV, especially in animal shelters where the previous vaccination history of cats is usually unknown, misidentification of vaccinated cats as being infected may have unwarranted fatal consequences. Based on this study, the WITNESS® FeLV-FIV Test and the Anigen® FIV Ab/FeLV Ag Test are the most accurate for differentiating FIV-infected from FIV-vaccinated cats and meet the criteria for DIVA tests.

ID08

AN EPIDEMIOLOGICAL STUDY OF GAMMAHERPESVIRUSES IN DOMESTIC CATS IN JAPAN

Morihiro TATENO, Yasuyuki ENDO, Masashi TAKAHASHI

Kagoshima University, Kagoshima, Japan

Gammaherpesviruses (GHVs) are members of an emerging subfamily of the Herpesvividae. The two major human GHVs, Epstein-Barr virus (EBV) and Kaposi's sarcoma-associated herpesvirus (KSHV) are clinically important pathogens because those viruses possess tumorigenesis potential especially in immune-compromised patients. In feline practice, we often experience tumor cases, especially lymphomas. Feline leukemia virus (FeLV) was a major cause for the development of lymphoma in the past days, however, the majority of lymphoma cases are shifting to FeLV-free cats. But the mechanism of tumorigenesis in these "spontaneously" developed lymphomas has not been well understood. Under this situation, GHV became a candidate which explains the lymphomagenesis in domestic cats. Recent study identified a novel GHV in domestic cats (Felis catus GHV, FcaGHV), and epidemiological surveys found that FcaGHVs are distributing in several countries including the United States, Australia, Singapore, Germany and Austria. In the present study, a molecular epidemiological study was conducted to investigate the prevalence of GHVs in Japan using large numbers of samples which were previously used to a nation-wide epidemiological survey for two feline retroviruses in Japan (Nakamura Y et al., J Vet Med Sci, 72:1051-1056, 2010).

GHV-derived DNA in blood samples was detected by degenerate nested pan-GHV PCR targeting gB gene as previously reported. Nucleotide sequences of obtained DNA fragments were determined and BLAST analysis was carried out. Phylogenic analysis was also conducted. Risk factors relating to GHV infection on clinical parameters, including age, sex, status of retrovirus infections, bite wound history and clinical signs, were statistically analyzed using univariable (Fisher's exact test) and multivariable (Logistic regression model) analyses.

GHV-derived DNA was detected in 2.6% of Japanese domestic cats. BLAST analysis revealed that all of them showed high similarity with FcaGHVs which were isolated from domestic cats in the United States. Phylogenic analysis also revealed that all Japanese isolates formed one cluster with FcaGHV. Age (> 2 years old) (Odds ratio [OR], 6.02; 95% confidential interval [CI], 0.79-46.0; P=0.05) and feline immunodeficiency virus (FIV) infection (OR, 5.41; 95% CI, 1.93-15.1; P<0.01) were found to be a risk factor in univariable analysis. In multivariable analysis, only FIV infection (OR, 3.74; 95% CI, 1.28-10.94; P=0.016) was a significant risk factor.

In this study, the prevalence of GHVs in Japanese domestic cats was surveyed and risk factor for GHV infection was determined. FcaGHV prevalence in Japan (2.6%) was lower than previously reported prevalence in other countries and areas (US, 16%; Europe, 16.2%). It was unable to determine the cause of this difference. However, Japanese isolates were identified to have close genetic relationship to domestic cat derived FcaGHV. This finding suggested that the virus is highly conserved in each animal species and GHV is harboring in host specific manner as well as other herpesviruses. The risk factor for GHV infection in Japanese cats was FIV infection and it was same as previously reported. But we have to clarify how FIV enhances the GHV infection and/or viremia. The clinical importance of GHVs infection in feline practice is still unknown. Further investigation will be necessary to identify the pathogenicity of GHVs in domestic cats.

ID09

CANINE INFLUENZA H3N2 INFECTION IN FOUR DOGS

Anne Cohen[1], Jill Richardson[2], Amy Glaser[3], Edward Dubovi[3], Nyssa Reine-Salz[2]

[1]Chicago Veterinary Emergency and Specialty, Chicago, USA, [2]Merck Animal Health, Madison, USA, [3]Cornell University, Ithaca, USA

The H3N2 influenza virus is of avian origin and was first isolated from clinically ill dogs in China in 2006 and South Korea in 2007. [1] Canine H3N2 influenza virus has been associated with severe respiratory signs and other clinical signs such as fever, reduced body weight, and interstitial pneumonia. [1] This case report presents four confirmed, naturally infected clinical cases of Canine Influenza H3N2 in the United States. The cases presented were all pet dogs that had histories of recent exposure to other dogs in a social setting including dog parks and doggie day cares. None of the dogs originated from, or had traveled to, typical regions endemic for this viral disease. All dogs were presented at different stages of illness attributed to infectious respiratory disease.

One dog , a 1.5 year old Corgi, was evaluated for lethargy, anorexia, and loose stool for 2 days. He had also developed a cough on the day of presentation and had mild increases in respiratory rate. On physical examination, he had a fever with mildly increased respiratory sounds. The patient was re-presented the next day for worsened signs of coughing, lethargy and hyporexia. Thoracic radiographs were performed which showed severe bilateral bronchopneumonia. Based on his poor response to supportive care, the owners elected humane euthanasia. Necropsy was performed at the New York State Veterinary Diagnostic Laboratory at Cornell University. Necropsy revealed Severe, acute, locally extensive necrohemorrhagic interstitial pneumonia with epithelial necrosis.

Another dog, a 8 year 10 month old Miniature Pincher, was evaluated after a 9 day history of coughing, wheezing, weakness, ataxia, hyporexia and one day of clear nasal discharge. On physical examination, the patient was weak, ~7% dehydrated, had serous nasal discharge, muffled heart sounds, poor peripheral femoral pulses, and harsh bilateral thoracic sounds with crackles and wheezes. Thoracic radiographs were performed which showed right caudodorsal and left cranial bronchoalveolar bronchopneumonia. The patient experienced cardiopulmonary arrest and upon return of spontaneous circulation the owners elected humane euthanasia.. Necropsy results were severe, acute, multifocal to coalescing necrohemorrhagic pneumonia with hyaline membrane formation and epithelial necrosis. Enteric lesions were also noted with moderate, diffuse, chronic lymphoplasmacytic enteritis with multifocal crypt necrosis.

Another dog, a 4 year old Greater Swiss Mountain dog, was referred for hospitalization due to right cranial and middle lung lobe consolidation, fever, dyspnea, gagging with phlegm production, exercise intolerance and hyporexia. On physical examination, he was quiet, febrile at 105.1 degrees Fahrenheit, had an increased respiratory rate of 60 breaths per minute with harsh bronchovesicular lung sounds. He became oxygen dependent with a pulse oximetry of 89-91%. Recheck thoracic radiographs were performed on the fourth day of hospitalization which showed a worsened alveolar pattern in the right and left cranial thorax. He was hospitalized for 7 days, making gradual improvements until discharge. H3N2 was confirmed via PCR at Cornell University.

The fourth case, a 10 year old terrier mix breed, was presented for evaluation of a productive cough for two days and one day of lethargy, anorexia and weakness. On physical examination, harsh lung sounds with mild increased respiratory rate was appreciated. The patient vomited a large volume of bile material after examination. A severe fever of 107.1 degrees Fahrenheit was appreciated with a respiratory rate of 48 breaths per minute and ~5% dehydration. Thoracic radiographs were performed which showed a mild to moderate bronchointerstitial pattern most concentrated in the caudodorsal lung fields. The patient was hospitalized for 4 days and was eventually tapered off of oxygen and sent home on oral antibiotic therapy. A respiratory panel was sent to IDEXX Reference Laboratory. The patient was positive for an Influenza A Assay, negative for an H3N8 PCR and positive for H3N2 PCR.

All four of the dogs were diagnosed with Canine Influenza H3N2 through PCR, three through the New York State Veterinary Diagnostic Laboratory and one through Idexx Diagnostic Laboratory. In patients with characteristic clinical features, Canine Influenza H3N2 virus infection should still be considered as differential diagnosis.

ID10

DEMONSTRATION OF PROTECTION AGAINST CANINE INFLUENZA VIRUS H3N2 INFECTION FOLLOWING VACCINATION WITH INACTIVATED CIV H3N2

Rhonda LaFleur[1], Tamara Davis[1], Patrick Tuma[1], Huchappa Jayappa[1], Mike Francis[2], Ian Tarpey[2]

[1]Merck Animal Health, Elkhorn, NE, USA, [2]MSD Animal Health, Milton Keynes, UK

A study was conducted in dogs to evaluate the efficacy of an inactivated Canine Influenza Virus (CIV) H3N2 vaccine following experimental challenge with a virulent heterologous strain of CIV H3N2, isolated from the recent CIV H3N2 outbreak in the United States. Eleven dogs, 7-8 weeks of age, were vaccinated with 2 doses of an inactivated CIV H3N2 vaccine, 3 weeks apart, and 19 dogs were vaccinated with a placebo. Two weeks after the second vaccination, all dogs were challenged intranasally with virulent CIV H3N2 and then monitored daily for 10 days for clinical signs including fever, nasal discharge, sneezing, coughing, depression, and dyspnea. Nasal swabs were collected to evaluate viral shedding, and serum samples were collected at various time points to determine antibody titers. At necropsy, lungs were scored for consolidation. Following the booster vaccination, the placebo-vaccinated control dogs remained seronegative (<10) to CIV H3N2, while 10 of the 11 vaccinated dogs developed an antibody titer to CIV H3N2 (GMT = <80; Range = <10 - 320). Antibody titers in dogs from both treatment groups increased following challenge, but the increase was greater in the vaccinated dogs (GMT = >1452; Range = 40 - >10240). Following challenge, 8 (42%) of the 19 placebo-vaccinated control dogs were euthanized prior to the 10-day post-challenge observation period due to severe clinical signs, including difficulty breathing, depression, fever, and severe coughing with retching; whereas, none (0%) of the vaccinates had to be euthanized ($p = 0.014$). Clinical signs were evaluated based on a weighted scoring system. The mean clinical score for the placebo-vaccinated control group was 24.9, compared to only 8.7 for the vaccinated group ($p = 0.036$). The placebo-vaccinated control group shed CIV H3N2 virus for a mean of 1.9 days, compared to 1.4 days for the vaccinated group ($p = 0.507$). The median lung consolidation score for the placebo-vaccinated control group was 7.4, compared to 0.0 for the vaccinated group ($p = 0.026$). Results of this study demonstrate that this inactivated CIV H3N2 vaccine significantly protects dogs against severe clinical disease and lung consolidation associated with a virulent CIV H3N2 infection.

ID11

PREVALENCE OF CANINE INFECTIOUS RESPIRATORY DISEASE COMPLEX PATHOGENS IN DOGS IN GEORGIA AND NORTH CAROLINA

Jill Richardson[1], Amy Glaser[2], Edward Dubovi[2], Nyssa Reine-Salz[1]

[1]*Merck Animal Health, Madison, NJ, USA,* [2]*Cornell University, Ithaca, NY, USA*

Canine infectious respiratory disease complex (CIRDC) is caused by many different viruses and bacteria. In June 2015, veterinarians around Atlanta, Georgia and also western areas of North Carolina began to notice an unusual increase in dogs presenting to their clinics with signs of infectious respiratory disease. Nasal and pharyngeal swabs from dogs exhibiting clinical signs were submitted to the Animal Health Diagnostic Center (AHDC) at Cornell University. A canine respiratory polymerase chain reaction (PCR) screening panel was utilized which allows identification of the following CIRDC pathogens: *Bordetella bronchiseptica*, *Mycoplasma cynos,* canine adenovirus 1 and 2, canine distemper, canine influenza A, H3N8 and H3N2, parainfluenza virus 5, pneumovirus and respiratory coronavirus.

Between June 1 and August 7, 2015, Merck Animal Health's Diagnostic Support Program tested over 175 samples from dogs with clinical signs of respiratory disease from Georgia and North Carolina. Of those dogs, 51 tested positive for Canine Influenza virus H3N2 - 29 of these cases were identified from Georgia and 22 cases were from North Carolina. None of the tested dogs were confirmed to have the H3N8 canine influenza strain.

Results of the testing in these regions also identified 38 cases of parainfluenza virus, 10 cases of *B. bronchiseptica*, four cases of adenovirus type 2, 29 cases of pneumovirus, 33 cases of respiratory coronavirus, and one case of canine distemper. Further evaluation of the positive cases of parainfluenza showed that 34 of the 38 had received parainfluenza vaccination with the injectable distemper, adenovirus type 2, parvovirus and parainfluenza virus combination vaccine. Also, of the 38 parainfluenza positive dogs, 21 had been vaccinated with a monovalent *B. bronchiseptica* vaccine (oral, injectable or intranasal). Seven of these dogs had been vaccinated with a *B. bronchiseptica* and parainfluenza combination intranasal product, of which two had been recently vaccinated and the parainfluenza sampled may have been of vaccine origin.

The information gathered from this testing program documents the spread of Canine Influenza virus H3N2 to two new locations in the US. In addition, data collected supports the role of parainfluenza virus as a major preventable pathogen in CIRDC and route of vaccination should be considered in vaccination protocols.

ID12

FREQUENCY, BENEFITS AND HEALTH RISKS OF ANIMALS IN NURSING HOMES: CROSS-SECTIONAL STUDY OF OHIO FACILITIES

Jason Stull, Timothy Landers

The Ohio State University, Columbus, OH, USA

Animals are argued to play an important role in nursing homes, providing health benefits, but may also be a source of infections and injuries for animals and patients. To-date little is known of the involvement of animals in nursing homes, practices in place to protect pet and resident health, or specific health benefits and risks. The objective of this study was to determine the frequency and avenues in which animals are utilized in the nursing home environment, practices to safeguard health, and benefits and risks of resident-animal interactions. An anonymous 15-minute online questionnaire was developed and delivered to a convenience sample of nursing home administrators in Ohio, USA from November 2014 through March 2015.

Responses from 95 facilities (approximately 5% of all registered facilities in OH) of varying sizes and geographic locations were received and adequately completed to allow for analysis. Nearly all respondents (94; 99%) allowed animals in their facilities (live-in, visiting, or both), with dogs being the most frequent animal involved with facility activities (99%), followed by cats (89%), birds (74%), fish (57%), and rodents (28%). High-risk species such as reptiles/amphibians and "petting zoo farm animals (e.g., pigs, goats, chickens)" were often permitted in facilities (27% and 12%, respectively). Few facilities had restrictions targeting particular species (n=9) or age (n=1) of animals. Animals were reported to visit most (71%) facilities weekly or more often. One respondent (1%) reported an animal (cat) was mistreated by a resident or staff in the preceding 12 months. No known or suspected animal-associated infections were reported for the preceding 12 months. Perceived benefits of animals in facilities were high (e.g., all participants agreed/strongly agreed that "residents appeared to be happier after interacting with an animal" and "animals were useful for calming agitated residents").

These results highlight the frequent use and perceived benefits of involving animals in nursing homes, but also common gaps in established policies and best practices important for reducing resident and animal health risks. Findings can assist in the development of directed guidance documents for nursing homes to better facilitate safe interactions between residents and animals.

ID13

EFFECT OF QUORUM QUENCHING WITH AZITHROMYCIN ON *PSEUDOMONAS AERUGINOSA* ASSOCIATED OTITIS EXTERNA/MEDIA IN DOGS

Modest Vengust[1], Javid Hosseini[1], Darja Kusar[1], Gabrijela Tavcar Kalcher[1], Irena Zdovc[1], Rok Blagus[2], Tina Kotnik[1]

[1]*University of Ljubljana, Veterinary Faculty, Ljubljana, Slovenia,* [2]*University of Ljubljana, Medical Faculty, Ljubljana, Slovenia*

Many bacteria can express their virulence via quorum sensing (QS); in *Pseudomonas aeruginosa* (PA) the QS signal molecules are N-acyl-homoserine lactones (PA-AHLs). The use of QS blockers or quorum quenching (QQ) to attenuate bacterial growth and virulence is a new strategy for treatment of resistant bacterial pathogens. Azithromycin (AZ) expresses in-vitro and in-vivo QQ effect, and is used clinically to reduce PA virulence. Hypotheses were tested that topical treatment with AZ will: 1) Attenuate the formation of most common PA-AHLs: N-butyryl-L-homoserine lactone (BHL), N-decanoyl-L- homoserine lactone (HHL) and N-(3-oxo-dodecanoyl)-L-homoserine-lactone (OdDHL); 2) Enhance PA clearance from the ear canal; 3) Have a positive effect on ear canal sample cytology; and 4) Reduce the severity of clinical signs related to PA associated otitis externa/media (PAOEM) in dogs.

This was a randomized double blind placebo controlled study. Samples were collected from 21 dogs with PAOEM, which were treated with topical AZ (1ml AZ solution at the concentration of 50µg/ml, SID) or placebo (1ml sterile 0.9% saline) for 56 days. All dogs were prescribed standard PAOEM treatment and regularly evaluated by their veterinarians. For the purpose of this study dogs were clinically evaluated and ear canal samples taken on days 0, 7, 14, 21, 28, 42 and 56. Samples were obtained by rinsing ear canals with 30ml of sterile saline. The concentration of BHL, HHL and OdDHL in ng/ml was determined by liquid chromatography coupled with tandem mass spectrometry. Bacteriological and cytological analyses were performed. Clinical signs (ear canal edema, pain, stenosis and erythema) were graded from 0-4; grades were then summarized to form a single clinical sign grade. Data were analyzed using linear mixed effects model for continuous outcomes and general linear mixed effects model for the binary outcomes; a P-value of <0.05 was considered significant.

The concentration of BHL, HHL and OdDHL were not affected by time (P≥0.1) or AZ (P≥0.2). The presence of PA in ear canal samples was affected by time (P=0.002) but not AZ (P=0.06). The presence of rod shaped bacteria, cocci and neutrophils in the sample were affected by time (P≤0.02) but not AZ (P≥0.08). Macrophages were not affected by time or AZ. Clinical signs and, therefore, the duration of the disease was influenced by time (P<0.0001). Azithromycin had no effect on clinical signs and the duration of the disease (P=0.5).

Detection of PA-AHLs in this study showed that QS acts as a promoter of PA virulence in otitis externa/media. PAOEM in dogs is, therefore, an appropriate disease model to study the effects of QQ on PA. Neither of our hypotheses was verified. Azithromycin had no effect on PA-AHLs and the elimination of PA from the ear canal, nor had any effect on cytology of the sample or the outcome of the disease. Only a weak tendency towards quicker elimination of PA from the ear canal was noted in this study.

ID14

COMPARATIVE EVALUATION OF FIVE IN-CLINIC RAPID TESTS FOR FELINE LEUKEMIA VIRUS INFECTION

Jiayou Liu[1], Melissa Beall[1], Tom O'Connor[1], Ramaswamy Chandrashekar[1], Mike Lappin[2]

[1]IDEXX Laboratories, Westbrook, ME, USA, [2]Colorado State University, Fort Collins, CO, USA

The feline leukemia virus is a potentially life-threatening oncogenic retrovirus that is transmitted horizontally by close contact or vertically by in-utero transmission. The p27 viral core protein is produced by the virus in infected cells and is found in the cytoplasm in several blood cells as well as free in the serum and plasma of infected cats. Reference laboratories and veterinary clinics commonly use some form of enzyme-linked immunoassay (ELISA) or particle-based immunoassay to detect the presence of the p27 protein antigen in samples obtained from blood. The purpose of this study was to compare the performance of several in-clinic tests: the SNAP® Feline Triple® Test (IDEXX), the VetScan® FeLV/FIV Rapid Test (Abaxis), the Witness® FeLV-FIV Test (Zoetis), Anigen® Rapid FIV/FeLV Test (BioNote), Speed Duo® FeLV/FIV Test (Virbac).

The sample population (84 positive and 101 negative) consisted of serum and plasma samples submitted to IDEXX's worldwide Reference Laboratory for feline retrovirus testing. Virus isolation and reverse transcriptase polymerase chain reaction results were not available and so samples were judged to be positive or negative based on the results of the ViraCHEK® FeLV (Zoetis) microtiter plate assay. ViraCHEK® was used as the reference test because it was reported in a published study to have high sensitivity and specificity (Hartmann et al. 2007) and it detects the same FeLV antigen as does each of the in-clinic tests evaluated in this study.

When compared to the ViraCHEK® microtiter plate assay results, the SNAP® Feline Triple® Test had the highest sensitivity among all in-clinic tests evaluated (Table below).

In-clinic Test	Comparative Sensitivity (95% CI)	Comparative Specificity (95% CI)
SNAP® Feline Triple FeLV Test	97.6% (91.1-99.8)	100% (95.5-100)
WITNESS® FeLV Test	76.2% (66.0-84.0)	97.0% (91.2-99.3)
VetScan® FeLV Test	69.0% (58.5-77.9)	97.0% (91.2-99.3)
Anigen® FeLV Test	66.7% (56.0-75.8)	97.0% (91.2-99.3)
Speed Duo® FeLV Test	51.2% (40.7-61.6)	99.0% (94.0-100)

These findings are important because in a clinical setting positive test results may be the only mechanism to identify infected cats – false negative results could delay supportive care and result in transmission to causal-contact naïve cats. Additional studies are needed to determine the clinical sensitivity and specificity of these tests using samples from cats characterized by PCR and/or viral culture.

ID15

DETECTION OF GIARDIASIS IN DOGS:COMPARISON OF THREE RAPID DIAGNOSTIC TESTS

Hannah Bewsey, Jan Drexel, Jiayou Liu, Brendon Thatcher, Tom O'Connor, Ramaswamy Chandrashekar

IDEXX Laboratories, Westbrook, ME, USA

Giardiasis is one of the most common intestinal parasitic infections affecting both dogs and cats. Infections may be subclinical or cause severe diarrhea. Dogs become infected by ingestion of *Giardia* cysts in contaminated water or feces. A single fecal flotation has a diagnostic sensitivity of approximately 70% because cysts can be few in number, are shed intermittently, and can be difficult to distinguish from other fecal matter by inexperienced observers. Fecal flotation combined with fecal antigen tests that detect soluble cyst wall antigen results in a diagnostic sensitivity of approximately 97%. While new *Giardia* antigen assays are now available for veterinary use, comparison studies are for the most part lacking.

The present study compared the performance of three in-clinic tests, the SNAP® *Giardia* Test, the VetScan® Canine *Giardia* Rapid Test (Abaxis), and the Anigen® Rapid CPV/CCV/*Giardia* Antigen Test (BioNote). Canine fecal samples used were submitted to IDEXX Reference Laboratories for ova and parasite testing by centrifugal flotation. After testing, the samples were stored at -20°C. The samples were then evaluated for soluble cyst wall antigen using the Thermo Scientific™ ProSpecT™ *Giardia* Microplate Assay. Positive samples were defined as those testing positive by both reference methods while negative samples were defined as those testing negative on both reference methods. A total of 95 positive samples and 81 negative samples were identified and tested on all three in-clinic tests. Six samples were excluded from sensitivity/specificity analysis because they failed to produce positive control lines on the VetScan® Test, despite producing valid results on the other two in-clinic tests. Results were compared to the consensus reference method results for calculation of sensitivity and specificity.

In-clinic Test	Comparative Sensitivity (95% CI)	Comparative Specificity (95% CI)
SNAP® Giardia Test	89.2% (81.1-94.2)	100.0% (94.2-100.0)
VetScan® Giardia Test	71.0% (61.0-79.2)	83.1% (73.0-89.9)
Anigen® Giardia Test	78.5% (69.0-85.7)	70.1% (59.1-79.2)

Using these criteria, the SNAP *Giardia* Test had numerically greater sensitivity and specificity than the other 2 tests. This study used another antigen test in the definition of true positive and true negative results since the samples were not fresh fecals; and not tested using a different test method like immunofluorescent antibody assay.

ID16

SEROPREVALENCE OF ANTIBODIES TO ANAPLASMA PHAGOCYTOPHILUM AND BORRELIA BURGDORFERI IN DOMESTIC CATS

Jill Saucier[1], Bryan Leonard[1], Kristine Hoyt[2], Michael R Lappin[3], Ramaswamy Chandrashekar[1], Jesse S Buch[1]

[1]*IDEXX Laboratories, Inc., Westbrook, Maine, USA,* [2]*Cats on Call Clinic, Scarborough, Maine, USA,* [3]*Colorado State University, Fort Collins, Colorado, USA*

Anaplasma phagocytophilum (*Aph*) and *Borrelia burgdorferi* (*Bb*) are pathogenic bacteria vectored by *Ixodes* spp. and are known to infect cats. While associations between *Bb* infections and feline clinical syndromes are unclear, studies have demonstrated *Aph* infection to be associated with fever, severe lethargy, and thrombocytopenia in cats. The present study aimed to evaluate the seroprevalence of antibodies to *Aph* and *Bb* in a large population of domestic cats from across the United States to better understand their risk of exposure to these agents.

A total of 5,416 samples submitted between September 2014 and February 2015 for Complete Blood Count (CBC)-inclusive test profiles were obtained from IDEXX Reference Laboratories located in *Ixodes* spp.-endemic areas. Samples originated from 26 states with California (24%), Illinois (19%), and Massachusetts (13%) contributing most to the study population. The population had a median age of 12.0 years (range: 1 month – 25 years) with similar representation between males (51%) and females (49%). All samples were screened by *Aph*-specific and *Bb*-specific peptide immunoassays and the results were evaluated based on various test and demographic parameters.

Overall, 9.7% of the feline sera tested in this study population contained antibodies to *Aph* while 2.8% had antibodies to *Bb*. Additionally, 1.2% of cats tested had antibodies to both *Aph* and *Bb*. As observed in dogs and humans, antibody prevalence rates among domestic cats varied by geography. The highest positive rates for *Aph* and *Bb* were observed in Connecticut (20.2%, n=337) and Maine (14.0%, n=50), respectively, while the lowest seroprevalance for both *Aph* and *Bb* was observed in Nevada (*Aph* - 2.1%, *Bb* = 0.0%, n=47).

These results corroborate findings in previous studies which indicate that cats can be infected with *Aph* and *Bb* and suggest that tick prevention is an important consideration in cats as well as dogs.

ID17

VALIDATION OF A HIGH-THROUGHPUT SEROLOGICAL ELISA METHOD FOR FELV P27 ANTIGEN DETECTION

Genevieve Clark, Roberta Cahill, Brendon Thatcher, Peter F Smith, Thomas P O'Connor, Ramaswamy Chandrashekar, Melissa J Beall, Jesse S Buch

IDEXX Laboratories, Inc., Westbrook, Maine, USA

Feline leukemia virus (FeLV) is a globally-distributed oncogenic retrovirus that can cause fatal disease in cats. Immunoassays that detect FeLV p27 antigen and are readily used to aid in the diagnosis of FeLV have been commercially-available for more than two decades. While single-sample rapid test formats are preferred for in-clinic use, the microtiter plate format enzyme-linked immunosorbent assay (ELISA) is the methodology most commonly employed in reference laboratory settings to facilitate batch testing. However, current commercially-available microtiter plate format ELISAs for FeLV are not well suited for this purpose because they rely on manual wash steps, drop-wise addition of reagents from reagent bottles, and visual interpretation of results. The purpose of this study was to validate a microtiter plate format ELISA designed specifically to maintain performance while enhancing throughput in a reference laboratory setting.

The FeLV microtiter plate format ELISA method validated in this study facilitates high throughput operation by combining parallel fluid transfer processes with automated data acquisition and results interpretation. The assay method is based on sequential orthogonal screening and confirmatory protocols. The screening protocol utilizes two distinct anti-FeLV p27 mouse monoclonal antibodies that produce a colorimetric response in samples containing FeLV p27 antigen. The confirmatory protocol requires neutralization of positive samples in a separate set of controlled assays using an anti-FeLV polyclonal antibody that blocks binding of the mouse monoclonal antibodies to FeLV p27 antigen and thereby inhibits color generation. The confirmatory step offers increased assurance of specificity by enabling discrimination between infected true positives and false positives associated with several factors including anti-mouse heterophilic antibodies in patient samples.

Precision of the screening assay was determined using 3 samples (negative, low positive, and high positive). The intra-assay coefficient of variation (CV) ranged from 3.9% to 7.9% while the inter-assay CV ranged from 6.0% to 8.6%. For the confirmatory assay, the same low positive and high positives samples were used demonstrating an intra-assay CV which ranged from 3.0% to 4.7% and an inter-assay CV between 7.4% and 9.7%. No interference was observed for either protocol at bilirubin, hemoglobin, and lipid concentrations up to 14.4 mg/dl, 550 mg/dl, and 6.4 OD_{660nm}, respectively. The analytical sensitivity for FeLV p27 antigen was established at 1.7 ng/mL for inactivated whole FeLV and at 1.0 ng/mL for purified recombinant FeLV p27. Discriminating analytical specificity was demonstrated based on the absence of cross-reactivity to related retrovirus antigen from murine leukemia virus and Feline RD-114. Overall, the assay exhibited 100% diagnostic accuracy in 83 patient samples (n=63 negative; n=20 positive) with FeLV status defined on the combined basis of PCR (FeLV RealPCR; IDEXX Laboratories, Inc.) and independent ELISA (ViraCHEK FeLV; Zoetis, Inc.).

Results of this study validate that the new microtiter plate format ELISA is a highly specific, precise, and accurate method for detection of FeLV p27 antigen in feline patient samples and is appropriate for high-throughput applications in a reference laboratory setting.

ID18

INVESTIGATION OF WHETHER LEPTOSPIRA VACCINAL ANTIBODIES REACT WITH BORRELIA PEPTIDES USED IN A COMMERCIAL ASSAY

Amber Caress[1], Scott Moroff[2], Michael Lappin[1]

[1]*Colorado State University, Fort Collins, CO, USA,* [2]*Antech Diagnostics, Lake Success, NY, USA*

In small animal veterinary medicine, two of the most common pathogenic spirochete genera are *Borrelia* spp. and *Leptospira* spp.. While antibodies that develop against these organisms over the course of infection in dogs are thought to be genus specific, studies evaluating for cross-reactive antibodies against individual *Borrelia* spp. or *Leptospira* spp. peptides are for the most part lacking. The purpose of this study was to determine if *Leptospira* spp. antibodies induced by administration of one of four commercial vaccines would cross-react with the *B. burgdorferi* antigens used in a commercially available assay.

Staff and student owned dogs were recruited at a Veterinary Teaching Hospital in a *B. burgdorferi* non-endemic area. Inclusion criteria stipulated that dogs were apparently healthy, weighed at least 10 kg, and had not been administered a *Leptospira* spp. vaccine within the past year. The dogs were randomized to be administered one of four commercially available *Leptospira* spp. vaccines that all contained serovars *Canicola, Gryppotyphosa, Icterohemorrhagiae,* and *Pomona.* Blood was collected prior to vaccine administration on Week 0 and Week 3 and then again on Weeks 4, 8, and 12. *Leptospira* spp. microagglutination titers (MAT) were determined using a commercially available service. After confirming the maximal *Leptospira* spp. titers occurred on Week 4, an aliquot of this sera was shipped to Antech Diagnostics for analysis of *B. burgdorferi* antibodies against OspA, OspC, and OspF with the Accuplex 4 BioCD system.

The Week 4 sera from all 31 dogs had an MAT titer of 1:100 for at least 1 *Leptospira* spp. serovar. MAT titers of 1:800 or greater were detected against serovar *Gryppotyphosa* and serovar *Pomona* in 27 dogs. None of these 31 samples contained antibodies against the *B. burgdorferi* OspA, OspC, and OspF peptides used in the commercially available assay.

In conclusion, the *B. burgdorferi* peptides used in the Accuplex 4 BioCD system do not recognize antibodies induced by the commercially available *Leptospira* spp. vaccines administered in this study. However, lateral flow device manufacturers and laboratories providing *B. burgdorferi* serological assays have specific sources of peptides that may vary antigenically from those used here and so the results of this study may not be the same for all laboratories.

ID19

CLINICAL AND LABORATORY FINDINGS IN DOGS WITH *IXODES*-INDUCED CHRONIC *ANAPLASMA PHAGOCYTOPHILUM* INFECTION AFTER PREDNISOLONE ADMINISTRATION

Elena Contreras[1], Kristy Dowers[1], Scott Moroff[2], Michael Lappin[1]

[1]*Colorado State University Veterinary Teaching Hospital, Dept of Clinical Sciences, Ft Collins, Colorado, USA,* [2]*Antech Diagnostics, Lake Success, New York, USA*

Anaplasma phagocytophilum (AP) is transmitted by *Ixodes* spp. and is the cause of granulocytic anaplasmosis in many species including dogs, cats, horses, and people. In endemic areas, antibodies against AP are detected in the serum of up to 40% of dogs and some healthy dogs are AP PCR positive. The effect of prednisolone administration on dogs infected with AP after *I. scapularis* infestation has not been determined.

A total of six young laboratory-reared beagles that were infected by AP, but no other vector borne agents, as evidenced by detection of AP DNA in blood by PCR assay (FastPanel, Antech Diagnostics) and AP antibodies in serum (Accuplex 4 BioCD system; Antech Diagnostics) after infestation with wild-caught *I. scapularis* were selected for study. On Week 20 after *I. scapularis* infestation, the dogs were placed into two Groups, each containing one AP PCR positive, AP antibody positive dog and two AP PCR negative, AP antibody positive dogs. Group 1 was administered doxycycline at 5 mg/kg, PO, twice daily for 4 weeks. After sample collection the morning of Week 24, all dogs were administered prednisolone at 2 mg/kg, PO, daily for 2 weeks. The dogs were evaluated daily for clinical abnormalities and by CBC, AP PCR and AP antibody assay Weeks 24-28. Sera from 5 dogs was assayed for AP antibodies on Week 48.

Clinical abnormalities were not detected over the course of the study. The only CBC abnormalities noted were anemia, thrombocytosis and neutropenia in one Group 1 dog and one Group 2 dog on Week 24, prior to prednisolone administration. After Week 20, only one dog was positive for AP DNA in blood (Group 2; Week 24 and Week 25). One Group 1 dog was negative for AP antibodies in all samples collected after Week 20, and the other two Group 1 dogs were negative for AP antibodies Week 48. In contrast, all dogs in Group 2 were AP antibody positive on Week 48.

The 2 dogs with anemia, thrombocytosis and neutropenia on the day prednisolone was initiated were clinically normal, AP PCR negative, and were normal in all other CBC evaluations, suggesting these results were spurious. The one dog in Group 2 that was positive for AP DNA on Week 25, was also positive on Week 24 prior to prednisolone administration. None of the other samples assessed by CBC and AP PCR assay after initiation of prednisolone administration showed evidence for activation of AP infection, suggesting that this protocol is unlikely to adversely affect dogs with chronic AP infection in the field. Whether other immune suppressive protocols would activate chronic AP infection remains to be proven. The only dogs that became AP seronegative by Week 48 were administered doxycycline, suggesting treatment may aid in the elimination of chronic AP infection, even after prednisolone administration.

ID20

SERUM NEUTRALIZATION OF US FELINE CALICIVIRUS ISOLATES FOLLOWING VACCINATION WITH A NON-ADJUVANTED, KILLED, FCV-BIVALENT VACCINE

LeMac Morris[2], Jennifer Hess[2], Michael Lappin[1]

[1]*Colorado State University, Fort Collins, Colorado, USA,* [2]*Boehringer Ingelheim, St. Joseph, Kansas, USA*

The objective of this study was to determine the proportion of feline calicivirus (FCV) field isolates that would be neutralized in vitro by sera from cats that were routinely vaccinated with a product containing two caliciviruses. Serum from cats hyperinoculated against 2 feline caliciviruses (FCV) has been shown to cross neutralize a greater proportion (90.2%) of 61 FCV field isolates from the United States when compared to serum from cats hyperinoculated with 1 FCV isolate (23.0%)

At 8 and 11 weeks of age, each of the 44 kittens was administered an experimental vaccine containing modified-live panleukopenia, modified-live feline herpesvirus 1, and non-adjuvanted inactivated calicivirus strains FCV-255 and FCV-DD1. Serum was collected from each kitten 7 and 14 days after the second dose of vaccine and pooled by day.

The ability of the pooled sera to neutralize a total of 33 FCV strains collected from cats in the USA was determined. The specific FCV serum neutralization titer for each FCV was defined as the serum dilution causing a 50% inhibition of virus replication. Using the Day 7 post-vaccine pooled sera, 24 of 33 (72.7%) FCV isolates were neutralized. Using the Day 14 post-vaccine pooled sera, 20 of 33 (60.6%) FCV isolates were neutralized. The placebo serum pool had undetectable antibody titers against all 33 FCV field strains that were screened.

Cross-neutralization of USA field strains of FCV was > 60% using pooled sera from kittens that were only administered 2 doses of an inactivated non-adjuvant vaccine containing FCV-DD1 and FCV-255 strains. These results suggest that cats vaccinated with products containing more than one FCV isolate will likely have broad heterologous protection if exposed to field strains of FCV.

ID21

RISK FACTORS ASSOCIATED WITH *GIARDIA* AND *CRYPTOSPORIDIUM* IN PET DOGS AND CATS IN THE USA

Hanaa Thigeel[1], Francisco Olea-Popelka[1], Valeria Scorza[1], David Aucoin[2], Michael Lappin[1]

[1]*Colorado State University, Fort Collins, CO, USA,* [2]*Antech Diagnostics, Irvine, CA, USA*

Giardia spp. and *Cryptosporidium* spp. are protozoans that colonize and reproduce in the intestines of many domesticated animals, including dogs and cats. Outcomes range from subclinical infection to severe diarrhea. *Cryptosporidium* spp. infection rates in dogs and cats are largely unknown as sensitive diagnostic procedures were not previously available. Polymerase chain reaction (PCR) assays are now available to amplify *Giardia* spp. and *Cryptosporidium* spp. DNA from feces. The purpose of this study was to report fecal PCR assay results for *Giardia* or *Cryptosporidium* in a sample set of dogs and cats in the USA.

All fecal samples that were submitted to a commercial laboratory (ANTECH® Diagnostics) between 2010 and 2015 for evaluation with a PCR panel of assays that amply DNA of a select group of infectious agents including *Giardia* spp. (dogs and cats), *Cryptosporidium* spp. (dogs and cats), and *Cryptosporidium felis* (cats only) were analyzed. Descriptive, univariable, and multivariable logistic regression analyses were conducted to assess associations amongst age, sex, region, and season and the probability of testing positive to either *Giardia* spp. or *Cryptosporidium* spp. in pet dogs and cats.

Giardia spp. and *Cryptosporidium* spp. fecal PCR assay results were available for 23,042 dogs and 16,326 cats.

Species	Pathogen	Percent	95% CI
Dog	*Giardia* spp.	7.7	7.3-8.0
	Cryptosporidium spp.	5.4	5.1-5.7
Cat	*Giardia* spp.	5.2	4.9-5.5
	Cryptosporidium spp.	7.4	7.0-7.9
	Cryptosporidium felis	5.1	4.7-5.4

Cryptosporidium spp. DNA was amplified from feces of 336 of 1,762 *Giardia* spp. DNA positive dogs (19.1%) and 132 of 843 *Giardia* spp. DNA positive cats (15.7%). Of the 843 *Giardia* spp. DNA positive cats, 97 (11.8%) were positive for *Cryptosporidium felis* DNA. Logistic regression models results showed that age (young) and region were associated with presence of *Giardia* spp. DNA and *Cryptosporidium* spp. DNA in feces of dogs and cats. In addition, presence of *Giardia* spp. DNA in feces of dogs varied by season.

The results indicate that *Giardia* spp. DNA, *Cryptosporidium* spp. DNA, and *Cryptosporidium felis* DNA are commonly amplified from feces of dogs and cats, coinfections are common, and positive rates can vary by age, season, and region. Either of the agents should be suspected more highly in young animals. Additional studies will be required to evaluate for associations of positive test results with clinical findings and to determine the likelihood dogs or cats are carrying zoonotic *Giardia* spp. or *Cryptosporidium* spp.

ID22

CLINICALLY SIGNIFICANT DIFFERENCES IN SENSITIVITY LEVELS OF IN-CLINIC RAPID TESTS FOR ANTIBODIES TO *ANAPLASMA* SPP. IN DOGS

Jiayou Liu[1], Ramaswamy Chandrashekar[1], Mathew Eberts[2], Mary Menard[3], Katie Erswell[4]

[1]IDEXX Laboratories, Westbrook, ME, USA, [2]Lakeland Veterinary Clinic, Baxter, MN, USA, [3]Borador Animal Hospital, Salem, NY, USA, [4]Pine Point Animal Hospital, Scarborough, ME, USA

Tick-borne diseases, including Lyme disease, ehrlichiosis, and anaplasmosis, are becoming increasingly prevalent as tick distributions expand through climate change, wildlife migration, and increased relocation of companion animals. Granulocytic anaplasmosis is caused by the obligate intracellular bacterium *Anaplasma phagocytophilum*, which is transmitted by *Ixodes* spp. of ticks. Both dogs and humans are susceptible to this infection. In dogs common signs of the infection, such as lethargy, anorexia, lameness and fever, are non-specific. Differential diagnosis is supported by evidence of morulae or DNA, especially in acute cases, and by detection of specific antibodies using IFA or ELISA-based tests.[1] A related but distinct bacterium, *A. platys*, also infects dogs and causes infectious cyclic thrombocytopenia.[2]

The present study compared the performance of two in-clinic rapid tests, the SNAP® 4Dx Plus® Test (IDEXX), and the VetScan® Canine *Anaplasma* Test (Abaxis). Random canine samples sourced from endemic regions for *A. phagocytophilum* were characterized by *A. phagocytophilum* Immunoflourescence assay (IFA) which was performed by a commercial reference laboratory. *A. platys*-specific samples were sourced from dogs living in the southwest, an *A. platys*-endemic region. These samples were characterized using A. *platys*-specific peptide ELISA.[3] The characterized samples were then blinded and randomized before testing with the two in-clinic rapid tests. Results are shown in the table below.

		Relative Sensitivity		Relative Specificity	
Organism	**Reference Test (sample size - pos/neg)**	**SNAP 4Dx Plus**	**Abaxis VetScan Test**	**SNAP 4Dx Plus**	**Abaxis VetScan Test**
A. phagocytophilum	IFA (143/90)	86.0%	32.2%	96.7%	88.9%
A. platys	ELISA (68/147)	85.3%	63.2%	98.0%	89.1%

Notably, the VetScan® Test failed to detect more than 60% of the medium to high titer (1:400 and above) *A. phagocytophilum* samples.

Independent evaluation of SNAP 4Dx Plus and VetScan *Anaplasma* Tests was performed in three different veterinary practices (Maine, New York and Minnesota) using samples obtained from pet dogs. Similar results were obtained - the VetScan Test failed to detect over half of the IFA positive dogs detected by the SNAP 4Dx Plus Test in each of the three clinics.

These results demonstrate the SNAP 4Dx Plus Test has better sensitivity for the detection of antibodies to both *A. phagocytophilum* and *A. platys* when compared to the Abaxis VetScan *Anaplasma* Test.

1. Carrade et al. J Vet Intern Med. 2009, 23:1129
2. Bowman DD, Topics in Companion Anim Med. 2011, 26:173
3. Qurollo BA et al. Infect Ecol Epidemiol. 2014, 4: 24699

ID23

PREVALENCE OF SELECT INFECTIOUS DISEASE AGENTS IN CLIENT OWNED CATS IN MOSCOW, RUSSIA

Katie MacMillan[1], Natalia Volgina[2], Michael Lappin[1]

[1]*Colorado State University, Fort Collins Colorado, USA,* [2]*Private Practice, Moscow, Russia*

Client-owned cats in Moscow, Russia can be exposed to fleas and ticks and some develop fever, hemolytic anemia, thrombocytopenia, gingivitis, and other clinical problems noted in cats of other countries. However, limited data are available concerning prevalence rates for infectious agents common to eastern and western Europe. The overall objective of this study was to determine the estimated prevalence of select feline infectious disease agents in a convenience sample population of cats from Moscow, Russia.

Remnant blood in EDTA and serum from client owned cats in Moscow that were evaluated for reasons independent to the study were collected with permission and stored a 4C until evaluated. Using previously validated assays, sera were assayed for *Toxoplasma gondii* IgG (Toxo), *Bartonella* spp. IgG (Bart), FeLV antigen (Ag), and FIV antibody (Ab). Total DNA was extracted from blood samples and DNA of *Anaplasma* spp., *Bartonella* spp., *Ehrlichia* spp., and haemoplasmas were amplified by previously validated PCR assays with *Mycoplasma hemofelis* (Mhf) amplicons confirmed by sequencing.

None of the cats were positive for DNA of *Anaplasma* spp. or *Ehrlichia* spp. but many of the cats showed evidence of exposure or current infection with one of more of the other agents. *Bartonella* spp. antibodies were most common in serum and '*Candidatus* M. hemominutum (Mhm) DNA was most common in blood.

Serology (positive/tested [%])				PCR (positive/tested [%])			
Toxo IgG	Bart igG	FeLV Ag	FIV Ab	Mhm alone	Mhf alone	Mhm/Mhf	Bart PCR
24/96 (25%)	30/96 (30%)	4/93 (4.3%)	1/93 (1.1%)	14/100 (14%)	2/100 (2%)	1/100 (1%)	2/100 (2%)

Toxoplasma gondii IgG and *Bartonella* spp. IgG titers ranged from 1:64 to 1:2048. Of the 81 cat samples evaluated in all assays, single, dual, triple, and quadruplicate infections were documented in 31, 11, 2, and 2 samples, respectively. Using this subset of samples, 57.8% of the cats had evidence of infection or exposure to one or more of the select agents.

The results from this sample population suggest cats in Moscow are exposed to intermediate hosts or undercooked meat (*T. gondii*), fleas (*Bartonella* spp. and hemoplasmas), and come in direct contact with infected cats (hemoplasmas, FeLV and FIV). These infectious agents should be on appropriate clinical differential lists, processed foods should be fed, ectoparasite control should be maintained, and feline-feline contact should be avoided when possible.

ID24

COXIELLA BURNETII DNA NOT IDENTIFIED IN FLEAS FROM DOMESTIC CATS IN AUSTRALIA AND THE USA

Alison Manchester[1], Jennifer Hawley[1], Julia Beatty[2], Vanessa Barrs[2], Michael Lappin[1]

[1]*Colorado State University, Fort Collins, CO, USA,* [2]*Valentine Charlton Cat Centre, University of Sydney, Sydney, Australia*

Coxiella burnetii is a rickettsial pathogen with serious zoonotic implications (Q fever) transmitted by direct contact and vectored by numerous tick species. Cats infected *C. burnetii* have been implicated in human infections in Australia and the United States but minimal information exists concerning routes of feline infection. Recently, *C. burnetii* DNA was isolated from *Ctenocephalides felis* collected from wildlife in Cyprus; this flea commonly infests cats. The purpose of this study was to evaluate select groups of fleas from cats in the Australia and United States for the presence of *C. burnetii* DNA using a previously optimized PCR assay.

The DNA samples utilized in the study had been extracted from fleas infesting 96 cats in previously published *Bartonella* spp., hemoplasmas, and *Rickettsia felis* studies. They were maintained at -80°C until the present study. The fleas were collected from cats in eastern Australia (86) and the southern United States (4 Alabama, 6 Florida) and pooled in groups comprised of a maximum of 5 fleas before DNA extraction. A previously reported conventional PCR assay utilizing primers pairs targeting the IS-1111 (IS-5, IS-9, IS-14 and IS-20) insertion sequences transposase elements in the *C. burnetii* genome was used. If positive amplicon was detected, genetic sequencing would be performed to confirm the presence of the pathogen.

For this study, fleas were not speciated before DNA extraction. However, previous studies in both countries suggest *C. felis* was most likely. The detection sensitivity of the assay was shown to be 3.14 ng total *C. burnetii* genomic DNA per PCR reaction utilized in all 4 primer sets. While all positive and negative controls performed as expected, none of the 96 pooled flea extracts were positive for *C. burnetii* DNA.

Coxiella burnetii DNA was not recovered from fleas collected from cats in eastern Australia or the southern United States. The results suggest that fleas infesting domestic cats in these regions are not important vectors for *C. burnetii* and are unlikely to play a role in the transmission of the organism to humans with Q fever.

ID25

RISK OF HEARTWORM INFECTION IN DOMESTIC CANINES OF NORTHWESTERN OREGON

Kirk Miller[1], William Rausch[2]

[1]*Oregon State University College of Veterinary Medicine, Corvallis, OR, USA,* [2]*Portland Veterinary Cardiology, Portland, OR, USA*

Canine infection with *Dirofilaria immitis*, the parasitic agent responsible for causing heartworm disease, is easily and safely prevented via the periodic administration of one of a number of preventative agents. Many veterinarians in Oregon do not routinely recommend the use of heartworm prophylactics; therefore, dogs are commonly not on a preventative. Confusion regarding the use of heartworm preventatives in Oregon is understandable, given the paucity of incidence data in our region. The decision between a veterinarian and a client on whether or not to start a preventative is often made based on a cost-benefit analysis, where only the cost is known. The purpose of this study was to evaluate the risk of developing heartworm infection amongst unprotected dogs in Northwestern Oregon. This information will enable veterinarians and their clients to make more informed decisions regarding heartworm prevention.

Between March 2010 and August 2015 owners of all dogs surrendered to the Oregon Humane Society filled out a short questionnaire relating to their dogs' histories. This questionnaire was used to determine the eligibility of each dog for this study. Criteria for inclusion in the study included that the dogs: 1) had been in the possession of the surrendering party since they were 6 months of age or younger; 2) were at least 12 months of age; 3) had no history of travel outside of the state of Oregon; and 4) were never administered heartworm preventative. Dogs that did not meet all four of these criteria were excluded from the study. Additional information gathered from the records included breed, age, sex and spay/neuter status, location (by zip code), and the average number of hours the dog spent outdoors. If a dog did meet the study criteria then blood was drawn and tested for the presence of *D. immitis* using a commercially available ELISA test (IDEXX Snap 4Dx Plus).

A total of 5,103 dogs were surrendered to the Oregon Humane Society during this period. Of these, 298 met the inclusion criteria and were included in the study. None of these 298 dogs tested positive for *Dirofilaria immitis*. The tested dogs had spent a median of 2 years with their previous owners (range 6 months to 16 years) for a combined total of 1300.45 years of exposure. Using a likelihood-based method, the maximum likelihood estimate (MLE) for the risk is 0 (95% CI: [0, 0.0023]) per year.

The findings of this study suggest that the likelihood of unprotected Northwestern Oregon dogs developing heartworm infection is currently very low. Heartworm transmission prediction models based on climate data (Knight/Lok) have shown that the climate conditions in the area can support heartworm transmission. Data compiled by the Companion Animal Parasite Council (Capcvet.org) confirm that *Dirofilaria immitis* has been diagnosed within the state of Oregon, although only 0.18% of all positive cases of heartworm in the U.S. are in the state of Oregon. It is important to consider that these data reflect a point in time. As the population of Oregon continues to grow, more heartworm infected dogs will invariably arrive. Changes in climate patterns will also influence the transmission of this disease. Continued monitoring and vigilance are recommended.

ID26

ECTOPARASITES AND VECTOR-BORNE PATHOGENS OF DOGS IN BAJA CALIFORNIA SUR

Cody Minor, Dana Hill, Danielle Straatmann, Michael Lappin

Colorado State University, Fort Collins, Colorado, USA

In the last several years, clinical illness in a number of people in Baja California Sur, México, has been suspected to be associated with rickettsial disease agents such as *Rickettsia rickettsii*, *Rickettsia felis*, and/or *Ehrlichia* species. The objective of this study is to determine the prevalence of select vectors and vector-borne disease agents carried by companion dogs of the region.

During planned community outreach sterilization and vaccination clinics for local dogs with consenting owners, samples of ectoparasites were collected from infested dogs and preserved in alcohol for identification, and blood samples were collected in EDTA and non-additive tubes retained for testing. Sera were assayed for antigen of *Dirofilaria immitis* and antibodies against *Ehrlichia canis/E. ewingii*, *Anaplasma phagocytophilum/A. platys*, and *Borrelia burgdorferi* using a commercial kit (SNAP® 4DXPlus®, IDEXX Laboratories). Sera were also assayed for antibodies against *Rickettsia* species by indirect fluorescent antibody (IFA) testing (*R. rickettsia* antigen source). Total DNA was extracted from blood and evaluated with conventional polymerase chain reaction (PCR) assays to amplify DNA of *Anaplasma* species, *Babesia* species, *Bartonella* species, *Ehrlichia* species, *Rickettsia* species, and hemotropic *Mycoplasma* species (Hemoplasmas). Positive amplicons were sequenced to ascertain the species.

Samples from 67 dogs were available. From these, ectoparasites *Rhipicephalus sanguineus* (33 dogs; 49.3%), *Ctenocephalides felis* (11 dogs; 16.4%), and *Pulex irritans/simulans* (four dogs; 6.0%) were identified. *Dirofilaria immitis* antigen (one dog; 1.5%) and antibodies against *A. phagocytophilum/A. platys* (10 dogs; 14.9%) and *E. canis/E. ewingii* (31 dogs; 46.3%) were detected in the serum of some dogs. Two samples had probable antibodies against *R. rickettsii* through IFA testing and are being confirmed. Evaluation of DNA from 41 dogs has been completed to date; *E. canis* DNA was amplified from eight dogs (19.5%), *A. platys* DNA from six dogs (14.6%), *Babesia canis vogeli* DNA from three dogs (7.3%), and *Mycoplasma haemocanis* DNA from three dogs (7.3%). Coinfections were confirmed in four dogs (9.8%) and consisted of *D. immitis/E. canis* (one dog), *E. canis/A. platys* (one dog), *E.* canis/M. *haemocanis* (one dog), and *E. canis/A. platys/M. haemocanis* (one dog).

Vector-borne agents detected to date likely reflect common exposure to *R. sanguineus*, as this tick vectors each of the PCR-confirmed agents. Further information will be gained by completion of the PCR assay analysis of the blood, fleas, and ticks.

ID27

IDENTIFYING AGREEMENT AND BARRIERS TO PROPOSED CANINE INFECTIOUS DISEASE GUIDELINES FOR DOG GROUP SETTINGS

Jason Stull[1], Michelle Evason[2], Jennifer Kasten[1]

[1]*Ohio State University, Columbus, Ohio, USA,* [2]*Rayne Clinical Nutrition Canada, Burnaby, BC, Canada*

Canine group settings, locations or events where dogs temporarily come together in a shared environment (e.g., shows, sporting events, dog parks) pose an increased risk for infectious disease transmission. Despite this increased risk, few guidelines exist to provide recommendations for reducing disease risk in these settings. During 2014-2015 a panel of canine infectious disease experts reviewed the current literature and drafted a set of 44 evidence-based recommendations for prevention of infectious diseases for dogs in group settings. In August 2015 a survey of attendees at the AKC Canine Health Foundation Parent Organization conference was completed to determine agreement with and perceived barriers to these recommendations. The 15-minute self-administered survey was provided to 238 Conference attendees and consisted of a series of Likert-type and open-ended questions (online and paper format) seeking feedback on 22 of the recommendations. The survey was completed by 185 individuals (78%), and all responses were reviewed, summarized, and open-ended comments categorized by theme. Respondents self-identified as: participants, judges, and breeders in a variety of local, national, and international canine group events. Most respondents (> 40%) agreed with all but three of the panel's recommendations, yet a majority of respondents stated the recommendations would be difficult or very difficult to implement in their setting (primarily dog shows). Common survey result themes related to difficulty of implementation included: administrative concerns (cost, human resources/manpower), enforcement issues, ethical concerns, privacy concerns, and strong need for official outreach to promote awareness and education related to canine infectious diseases. Survey responses identified needs for: further refinement of recommendations to aid comprehension and clarity (especially around ecto- and endoparasite control), and education to promote culture changes related to disease risk prevention. In order to raise awareness of canine infectious disease in group settings amoungst event participants, attendees, and organizers; an online freely available canine infectious disease risk calculator tool is being developed.

ID28

CORRELATION OF *MYCOPLASMA* QUANTITATIVE PCR TO SEVERITY OF CONJUNCTIVITIS IN CATS

Alexis Dubin, Jennifer Hawley, Cynthia Powell, Michael Lappin, Julia Veir

Colorado State University, Center for Companion Animal Studies, Fort Collins, CO, USA

Mycoplasma species are one of the most common infectious causes of conjunctivitis in cats. *Mycoplasma felis* is commonly implicated as a primary pathogen, but other *Mycoplasma* species have also been detected in clinically ill cats. Findings from previous studies using conventional PCR (cPCR) to investigate the role of *Mycoplasma* species in causation of feline conjunctivitis have been mixed as *Mycoplasma* can be carried by apparently normal cats. Therefore, the purpose of this study was to determine if increasing severity of conjunctivitis in cats correlates with higher *Mycoplasma* species copy numbers using qPCR.

A total of 77 conjunctival swabs collected from 29 shelter cats with conjunctivitis and confirmed to contain *Mycoplasma* species DNA using cPCR were selected for study. The severity of conjunctivitis at the time the samples were acquired was determined using a grading scheme from 0 - 9. The samples were evaluated using a previously validated qPCR to determine the *Mycoplasma* copy number. Statistical methods consisted of using the Spearman's rho test to determine if severity of conjunctivitis was correlated to qPCR *Mycoplasma* species copy number.

The results revealed the severity of conjunctivitis significantly correlated to qPCR *Mycoplasma* copy number (Spearman's correlation coefficient -0.32, P=0.0042), however, the strength of this correlation was only mild to moderate.

Based on the results of this study, future investigation of the impact of *Mycoplasma* species other than *M. felis* on the correlation of qPCR and severity of conjunctivitis in cats should be performed.

NM01

EFFECT OF THE HYPER-IMMUNE EGG YOLK SUPPLEMENTATION ON WEIGHT GAIN IN NEONATE PUPPIES

Hanna Mila[1], Alexandre Feugier[2], Claire Mariani[2], Aurelien Grellet[2], Sylvie Chastant-Maillard[1]

[1]*École nationale vétérinaire de Toulouse, UMR 1225 Interactions Hôte-Pathogènes, INP, Toulouse, France,* [2]*Royal Canin, Aimargues, France*

Colostrum provides puppies with most of their passive immune transfer, as in dogs only 5% of immunoglobulin G (IgG) is acquired via transplacental transfer. Inadequate colostrum intake during the first day of life will deprive puppies not only of immunoglobulins, but also from many hormones, growth factors and nutrients. Hence, it increases the risk for neonatal morbidity and mortality. Supplementation during the first hours of life with canine serum or plasma increased blood IgG concentration[1] in colostrum deprived puppies as well as improved their growth during the entire neonatal period (0-3 weeks)[2]. This study aimed to evaluate the effect of exogenous specific antibodies administrated via egg yolk before the intestinal barrier closure on growth in pre-weaning puppies.

Specific antibodies against canine parvovirus type 2 (CPV2) and *E.coli* were obtained in eggs from hens vaccinated separately against one of the mentioned agents[3]. Egg yolk was flash-dried and tested for the presence of CPV2 and *E.coli*-specific antibodies. Hyper-immune solution was then prepared by mixing egg powder with a commercial milk replacer (Babydog Milk, Royal Canin, Aimargues, France; 1g of egg powder with CPV2 antibodies and 1g of egg powder with *E.coli* antibodies with 12ml of reconstituted milk). A total of 334 puppies from 16 different breeds, enrolled in one breeding kennel, were included in the study. Depending on the expected adult body weight, puppies were classified into small breed dogs (S; adult weight < 25 kg), and large breed dogs (L; > 25 kg). Within each litter and taking into account the birth weight, puppies were assigned into supplemented or control group. Each puppy from the supplemented group received orally 1.5 ml/100g bw of hyper-immune solution at once within the first 8 hours after birth. Puppies from the control group received at the same dose (1.5 ml / 100g) and time after birth (< 8h) the milk replacer only. All puppies were weighed at birth and at 7, 14, and 21 days of life. Linear mixed models (MIXED procedure; SAS Institute Inc., Cary, NC, USA) with litter modeled as a random effect were performed to determine the variables affecting birth weight and weight gain during the neonatal period: breed size (small; large), age (0-7; 7-14; 14-21 days), supplementation (supplemented or control group). All the interactions between mentioned fixed effects were also tested.

L represent 38.3% (128/334) of the included puppies. Among L and S, 65 (50.8%) and 104 (50.5%) were supplemented, respectively. The weight at birth was significantly higher in L compared with S (median: 370g [interquartile range: 325; 408g] vs. 200g [156; 248g]; $p<0.001$). Birth weight was found not different between supplemented and control puppies whatever the breed size ($p=0.14$): supplemented L 366g [330; 412g] vs. 375g [322; 405g] in controls; supplemented S 199g [155; 247g] vs. 201g [159; 248g] in controls. Weight gain during the neonatal period was influenced by time ($p<0.001$), supplementation ($p=0.031$) and the interactions between the breed size and supplementation ($p=0.027$) and time and breed size ($p=0.001$). L gained 176g [67; 294g] during the 1st week, 223g [158; 324g] during the 2nd week and 260g [160; 382g] during the 3rd week of life, with significantly greater weight gain compared with S only during the 1st week (S=116g [66; 172g]; $p=0.001$). Whatever the period concerned, supplemented L gained more weight during the entire neonatal period than the controls (841g [485; 1087g] vs. 623g [436; 858g]; $p=0.048$). No difference was evidenced between supplemented and control S ($p=1$).

In our study large breed puppies supplemented at birth with the hyper-immune egg yolk had greater weight gain during the entire neonatal period. Retarded growth at the early stage of life increases the risk of morbidity and mortality in puppies[4]. Thus it could be hypothesized that better growth in supplemented puppies reflects a better health. Nevertheless, further studies are needed in order to confirm our findings in other breeding kennels and on large number of individuals.

[1]Poffenbarger et al., 1991; [2]"Canine health product containing antibodies against canine parvovirus type 2" WO2015004181 A1.; [3]Van Nguyen et al., 2006; [4]Mila et al., 2012.

NM02

EFFECTS OF DIETARY MEDIUM CHAIN TRIGLYCERIDES ON VOLUNTARY ACTIVITY IN Dogs And Cats

Yuanlong Pan, Hui Xu, Sandeep Bhatnagar, Janet Jackson

Nestle Purina Research, St Louis, MO, USA

Decline in cerebral glucose metabolism is a common feature of aging in people and animals including rats, dogs, and monkeys, and it is closely associated with age-dependent cognitive decline. We had confirmed in a previous study that dietary medium chain triglycerides (MCTs) can enhance cognitive functions in dogs by providing the brain with an alternative energy source. In this study, we investigated the effects of dietary MCTs on voluntary activity in dogs and cats.

In the cat study, sixteen middle-aged and senior cats were fed 100% of their maintenance energy requirements (MERs) with a control diet for one week, and then were switched to the MCT-containing diet for 28 days. Their daily activity was monitored during the feeding trial with activity monitors. In the dog study, twenty senior dogs were fed 100% of their MERs with a control diet for 6 days, and then were assigned to one of 2 MCT-containing diets with 10 dogs per diet for 27 days. Then the dogs were switched to the alternative MCT diet for 32 days. Daily activity was recorded with activity monitors during the feeding trial. In cats, the MCT-containing diet significantly increased both daytime (17381.23±2551.18 vs 22628.55±2790.61, $p<0.0001$) and night time activity (4415.56±654.71 vs 5926.66±786.88, $p=0.002$). Interestingly, in dogs, both MCT diets significantly increased only daytime activity (80349.46±9703.34 vs 99254.33±12163.83; 80349.46±9703.34 vs 105120.38±11550.91, $p<0.05$). The results show that both dogs and cats became more active when fed a diet containing MCTs

NM03

EFFECTS OF DIETARY MACRONUTRIENT CONTENT AND FEEDING PATTERN ON LEPTIN CONCENTRATIONS IN LEAN HEALTHY CATS

DAGMAR TARKOSOVA[1], JACQUIE RAND[2], HEIDI FARROW[2], MARCIA CORADINI[2], JOHN MORTON[3]

[1]*UNIVERSITY OF VETERINARY AND PHARMACEUTICAL SCIENCES, BRNO, CZECH REPUBLIC,* [2]*THE UNIVERSITY OF QUEENSLAND, BRISBANE, AUSTRALIA,* [3]*JEMORA PTY LTD, GEELONG, AUSTRALIA*

Excessive weight gain is associated with metabolic and hormonal changes that predispose cats to insulin resistance and other disorders. The adipokine leptin is an important regulator of energy metabolism, and circulating leptin concentrations correlate with fat mass in cats and other species. In cats, leptin is associated with insulin resistance, independent of bodyweight. Although dietary factors such as energy density, fat content and feeding pattern increase the risk of obesity, there is limited information on how dietary factors such as feeding, fasting, and macronutrient content influence leptin concentrations in cats. The aims of this study were to compare leptin concentrations in lean, healthy cats fed diets high in protein, fat and carbohydrate, to assess associations between leptin concentrations and insulin, glucose, and NEFA concentrations, and to determine the effects of feeding pattern on baseline and postprandial leptin concentrations.

A controlled trial was conducted with clinically healthy cats (n=24). Mean body weight was 4.9 kg and all cats had ideal body condition score of 3 on a 5-point scale. All cats were fed a "washout" diet (commercial feline diet) for five weeks, and were then fed one of three test diets high in one of protein, fat or carbohydrate for five weeks. Diets were dry extruded formulations, and each test diet provided approximately 50% of energy from the test macronutrient, and 25% of energy from each of the other two test macronutrients. Leptin concentrations were measured during two feeding patterns: a meal-feeding test (once daily feeding) and an *ad libitum*-feeding test, conducted 4 weeks after commencement of the test diets. During the meal-feeding test, cats ate 90 - 100% of the 12 hour *ad libitum* intake as a single meal in 0.5 hour. Blood samples (4 mL), were collected over 24 hours in the meal feeding test and 12 hours in the *ad libitum* test.

Distributions of leptin variables in the *ad libitum* and meal-feeding tests were similar across dietary groups after consumption of the washout diet for 5 weeks. Mean baseline concentrations (average of -30 and -5 min values), mean concentrations over 24 h (mean 24-h), and peak leptin concentrations in the meal feeding test varied significantly by diet (overall $P<0.001$ to 0.027). Baseline, mean 24-h and peak leptin concentrations for the high fat diet were significantly higher than for the high protein diet, and for baseline and mean 24-h, for the high carbohydrate diet. A similar pattern was observed in the *ad libitum* feeding test with highest leptin concentrations in cats consuming the high fat diet. In general, leptin concentrations after consuming the high carbohydrate diet were not significantly different from the high protein diet, but if they differed, leptin concentration for the high carbohydrate diet was higher compared to the high protein. There was no significant effect of diet on the time to peak leptin concentration in the meal-feeding test (overall $P=0.855$), and median times to peak leptin concentration were 15 hours for all diets. In the meal-feeding test, only 3/8 cats significantly exceeded their baseline leptin concentration for each of the high protein and high fat diets, but 7/8 did so for the high carbohydrate diet. In general, leptin concentrations decreased significantly approximately 2 hr after eating in the meal feeding test, but remained relatively constant during *ad libitum* feeding. When data for all diets were combined, leptin was significantly positively correlated with insulin concentrations during *ad libitum* feeding ($P=0.008$ to 0.036), but only at baseline and at 24hrs in the meal feeding test ($P=0.025$ and <0.001, respectively). There were no correlations between leptin and either glucose or NEFA across all diets. Glucose, insulin and NEFA did not account for the decline in leptin soon after feeding in the meal feeding test.

In conclusion, independent of feeding pattern, leptin concentrations tend to be highest in lean cats consuming a diet with 50% of energy from fat, compared to diets high in protein or carbohydrate. Leptin concentrations vary minimally over 12 hr of *ad libitum* feeding, whereas leptin decreases approximately 2 hrs after feeding following a fast, and this decrease is not accounted for by changes in insulin, glucose or NEFA concentrations. Further investigation is needed to understand the interactions between hormones associated with satiety, dietary factors and weight gain in cats.

NM04

EFFECT OF HIGH SODIUM DIET ON BLOOD PRESSURE AND CARDIAC FUNCTION IN HEALTHY ADULT DOGS

Hui Xu, Dottie Laflamme, Sandeep Bhatnagar, Xuemei Si, Grace Long

Nestle Purina PetCare Research, St. Louis, Missouri, USA

Promoting water intake is recommended for managing dogs with lower urinary tract disease. Increased dietary salt has been used for this purpose in veterinary therapeutic diets. The objective of this study was to evaluate potential adverse effects of dietary salt (sodium chloride) on blood pressure and cardiac function in dogs.

Following a two week baseline period where dogs were fed a Control Diet containing 0.13 g sodium/100 kcal ME, twenty healthy dogs (6-10 years old) were allocated to two groups and fed diets differing only in total sodium chloride content: CONTROL (CON) = 0.13 g sodium /100 kcal ME, or High Sodium Diet (HNA) = 0.41 g sodium/100 kcal ME. Dogs were fed their respective diet for 6 months. During this period, indirect systolic blood pressure was recorded monthly. Overall health, cardiac function evaluated by echocardiography (performed by boarded veterinary cardiologist), plasma B-type natriuretic peptide (BNP), serum biochemistry and urinalyses were assessed at 0 and 6 months on test. Body composition was analyzed by dual energy X-ray absorptiometry (DEXA) at 0, 3 and 6 months. Treatment effects were analyzed using t-test and Generalized Linear Mixed Models.

Overall, dietary sodium level had no significant effect on blood pressure, cardiac function, plasma BNP, serum biochemical profiles or urinalyses. There were no dietary effects or diet by time interactions on bone mineral content or body composition. In conclusion, consumption of a diet containing 0.41 g sodium/100 kcal ME had no adverse effects on blood pressure or cardiac function in healthy adult dogs over a 6-month period.

NM05

DETERMINING THE LACTATE AND GLUCOSE THRESHOLDS AND THE ACID-BASE IMBALANCES IN BEAGLES DOGS

Alejandro Zamora Restan, Evandro Zacché, Jaislaine Braz, Raphaela Marques Canola, Samara Beretta, Antonio Queiroz-Neto, Guilherme Camargo Ferraz, Aparecido Antonio Camacho

Universidade Estadual Paulista – UNESP, Jaboticabal, Brazil

The determination of lactate and glucose threshold in dogs submitted to incremental exercise test (IET) is scarce. This study aimed (1) to determine the lactate and glucose thresholds (2) to investigate the acid-base responses of untrained dogs submitted to incremental exercise test. Fourteen Beagle dogs, 1-2 years old, healthy and untrained were used for determination of lactate and glucose thresholds and submitted to IET using treadmill, with slop from 5 to 7.5% and increasing speeds of 0.5 m/s every 5 minutes. The test was concluded when dogs presented fatigue signals. Blood samples were collected after 90 seconds of each effort step, though venous catheter placed in the jugular vein. The samples used for determination of acid-base alterations were collected in the moments before of IET, immediately and five minutes after the fatigue. Data were submitted to Shapiro-Wilk test, and the Pearson was used to correlate the running velocities corresponding to the lactate and glucose threshold and analysis between moments of assessment was performed by using repeated measures one-way ANOVA ($p < 0.05$). It was found high correlation between lactate and glucose threshold velocities, $r = 0.84$ ($p < 0.01$), and for pH, PCO_2, HCO_3^- strong ions difference (SID) were found moment effect ($p < 0.001$). The basal values of pH increased ($p < 0.001$) in fatigue (7.341 ± 0.02 to 7.390 ± 0.04, respectively), and the PCO_2 presented reduction ($p < 0.001$) in the fatigue and 5 minutes after fatigue (31.74 ± 5.9, and 37.01 ± 5.8), in comparison with basal value (41.4 ± 2.2). The HCO_3^- decreased in fatigue and 5 minutes after IET (18.2 ± 2.3 and 19.1 ± 2.5) when compared with basal (21.9 ± 1.1). The values of SID showed decreased in fatigue compared to values before exercise (36,2 ± 1,7 e 36,6±2.1) respectively. Running velocities corresponding to the lactate and glucose threshold showed accordance between them, what has been already shown in horses and humans, besides results showed respiratory alkalosis and metabolic acidosis, which is considered a change due to hyperventilation and strenuous anaerobic activity, decreasing PCO_2 concentrations, which leads to increase pH through the increase of the relation of HCO_3^- and PCO_2. SID showed a metabolic acidosis possibly from lactate values increase. Results of this study suggest that glucose threshold can be used as a marker of aerobics capacity in dogs as it is used in humans and horses, besides, maximum exercises in dogs cause acid basic alterations.

NM06

WHEN FED FOODS WITH SIMILAR PALATABILITY, CATS CHOOSE 30%, DOGS 23% OF CALORIES AS PROTEIN

Dennis Jewell[1], Jodi Vondran[1], Melissa Vanchina[1], Jean Hall[2]

[1]*Hill's Pet Nutrition, Inc, Topeka, KS, USA,* [2]*Oregon State University, Corvallis OR, USA*

Dogs and cats have specific taste preferences that are influenced by macronutrient composition and physical characteristics of food, as well as presence or absence of specific taste imparting compounds (palatability enhancers). This study investigated the relationship of food choice on macronutrient composition and that choice on subsequent plasma amino acid concentrations. Experimental foods for dogs and cats were individually prepared to have similar palatability through manipulations of factors independent of macronutrients. Foods with similar palatability preference for their respective populations as a whole were then investigated for intake choice using 20 dogs with varying body fat composition (mean, 3.9 kg; SD, 0.9 kg; range, 2.4-6.2 kg;) and 27 cats (mean body fat composition, 1.4 kg; SD, 0.9 kg; range 0.3-2.2 kg). Subsequent effects of food choice on serum amino acids were also determined. Four completely balanced foods were available to dogs and cats at all times. The four varied in macronutrient content: Food 1: high protein (33% of calories from protein for dogs; 42% for cats); Food 2: high fat (54% of calories from fat for dogs; 44% for cats); Food 3: high carbohydrates (56% of calories from carbohydrate for dogs; 52% for cats); and Food 4: blended macronutrients. Dogs on average chose to consume 23.0% calories from protein (SD, 1.5%; range 20-26%) whereas cats chose 30.3% of calories from protein (SD, 4%; range 24-38%). Body composition influenced their choice of calories from protein. In dogs, there was a positive relationship between body fat and percent calories consumed as protein (r=0.56; P=0.02). There was no relationship between lean body mass and percent calories consumed as protein (r=0.06; P=0.82). Overall, dogs with high body fat showed the highest preference for dietary calories from protein. In cats, both body fat and lean body mass were negatively associated with calories consumed as protein (P=0.04; P=0.02, respectively), with a positive association for the interaction (P=0.02). Overall, cats with high body fat and high lean body mass showed the highest preference for dietary calories from protein. Dogs had higher circulating amino acid concentrations of threonine, methionine, isoleucine, and valine (P<0.05) compared with cats, the later which had higher circulating amino acid concentrations of isoleucine, valine, tyrosine, phenylalanine, and histidine (P<0.05). In both dogs and cats there was no effect of protein intake on circulating essential amino acid concentrations. Increased intake of calories as protein was associated with increasing plasma ornithine (dogs only) and citrulline (dogs and cats) concentrations (P<0.05). In summary, given the opportunity to choose between foods with similar palatability, cats chose to consume 30.3% and dogs 23.0% of their calories as protein. Although there were species differences between circulating amino acid concentrations indicating metabolic shifts associated with protein intake, there were no changes in circulating essential amino acid concentrations.

NM07

A DOUBLE MASKED CLINICAL TRIAL OF A THERAPEUTIC FOOD IN THE MANAGEMENT OF CANINE ATOPY

Jennifer Macleay, Heidi Schiefelbein, Kathy Gross

Hill's Pet Nutrition, INC, Topeka, KS, USA

The purpose of this double masked, controlled multicenter clinical trial was to determine the impact of a food with ingredients designed to have skin and coat health benefits and immune modulating effects on clinical signs of seasonal atopic dermatitis versus a control food. Dogs with a history of seasonal dermatitis, but without current clinical signs consistent with atopy (CADESI and pruritis scores = 0) were recruited in the spring from 11 general practices in the United States. Dogs currently on foods designed for adverse food reactions were excluded. Eligible dogs were randomly assigned to Test or Control groups and evaluated by their veterinarian at 0, 4, 8, 12 and 16 weeks. Consistent with accepted standards of care, treatment was not withheld, therefore, prescribed medications and dosing information was collected; only oclacitinib which was not uniformly available to all practices was disallowed. Veterinarians graded skin lesions using a modified CADESI (0:absent - 4:severe for erythema, alopecia, excoriations and lichenification for 27 body sites, max score of 432) and a pruritis score (0:absent - 4:severe).

Forty-four adult dogs (22 Test, 22 Control) were enrolled and completed the study. During the course of the study 18% of all dogs had no reported dermatological clinical signs whereas 82% developed a positive CADESI and/or pruritis score. CADESI and pruritis scores rose significantly from 0 at all subsequent time points but were generally low. Mean ± SEM CADESI scores at weeks 4, 8, 12 and 16 were; 2.6 ± 0.98, 5.8 ± 2.47, 8.6 ± 3.31 and 6.6 ± 1.91 respectively. Mean ± SEM pruritis scores were 0.5 ± 0.16 at week 4 and were highest at week 16; 1.2 ± 0.16. Incidence, location and types of clinical signs manifested were not different between groups. Medications were only dispensed to dogs with a positive CADESI and/or pruritis score, which was 66% of all dogs (73% of Control vs. 59% of Test). Medications dispensed were categorized as a topical containing a glucocorticoid, a systemic antihistamine and/or a systemically administered glucocorticoid. An event history analysis revealed that time until prescribing topical medications was highly similar between groups (Test 98 ± 7.0 vs. Control 100 ± 5.2 days) but was less so for antihistamines (Test 62 ± 4.1 vs. Control 55 ± 3.3 days) and systemic glucocorticoids (Test 105 ± 4.4 vs. Control 90 ± 6.4 days).

This study is the first to describe the recurrence rate of seasonal atopy in dogs from general practices as part of a grade 1, multicenter clinical trial. Onset of clinical signs resulted in medical intervention in the majority of dogs. Statistical power in this study was likely influenced by the generally low CADESI and pruritis scores combined with medical intervention. In addition, the interval to medical intervention was likely influenced by the study schedule which promoted frequent evaluations and case management. This data suggests that dogs with seasonal atopy my benefit from frequent evaluations. This data also suggests altered management of dogs fed the Test food as medication use was lower and the interval prior to the onset of systemic antihistamines and/or glucocorticoids was longer. A food designed for skin and coat health benefits and immune modulating effects can be part of a proactive atopy management program and may result in less medication needed.

NU01

LONGITUDINAL EVALUATION OF SERUM SYMMETRIC DIMETHYLARGININE (SDMA) AND CREATININE (sCr) IN DOGS WITH EARLY CKD

Sarah Guess[1], Maha Yerramilli[2], Edward Obare[2], Greg Grauer[0]

[1]*Kansas State University, Manhattan, KS, USA,* [2]*IDEXX Laboratories, Westbrook, ME, USA*

Retrospective feline studies have shown SDMA is more sensitive than sCr for early CKD detection. Our objective was to compare SDMA and sCr for detection of early CKD in a prospective, longitudinal study of older dogs.

SDMA and traditional serum and urine clinicopathologic tests were measured biannually for four years in 43 dogs with an initial median age of 8.9 years. Persistent increases in UPC, sCr, or SDMA or decreases in USG triggered renal ultrasound (US) and GFR (plasma iohexol clearance) evaluations. Necropsies were performed whenever possible when dogs died or were euthanized.

CKD was documented in 23 dogs (53%) by US abnormalities (n=13), decreased GFR (> 40% reduction) (n=13), persistent renal proteinuria (UPC ≥ 0.5) (n=6), or renal histology (n=6) (12 dogs had multiple abnormalities).

9 of 23 dogs had increased SDMA (≥ 14μg/dl) without hypersthenuria concurrent or subsequent to CKD diagnosis. One dog had a single (endpoint) increased SDMA and 8 had persistent/multiple SDMA increases. Conversely, only 2 of the 23 dogs had increased sCr (> 1.8mg/dl) at any point and both of these dogs had concurrent/prior SDMA increases. In the 13 dogs with decreased GFR, sCr and SDMA were increased in 1 and 7 dogs, respectively. There were no persistent increases in SDMA without hypersthenuria without CKD. Increased SDMA without hypersthenuria had 39% sensitivity and 100% specificity for CKD whereas increased sCr without hypersthenuria had 9% sensitivity but 100% specificity (p=0.024).

SDMA is a sensitive biomarker for early canine CKD.

NU02

SDMA CORRELATES BETTER WITH CREATININE THAN HIGH THROUGHPUT IMMUNOTURBIDOMETRIC CYSTATIN C ASSAY IN FELINE SERUM.

Polina Prusevich, Julie Cross

IDEXX Laboratories, Westbrook, ME, USA

The purpose of this study was to determine the correlation of SDMA and Cystatin C with creatinine in felines, to assess the utility of these markers together in diagnosing chronic kidney disease (CKD).

Fifty-six feline serum samples were sourced from samples submitted to IDEXX Reference Laboratories for general chemistry testing. Cystatin C, SDMA and creatinine were assayed using a Beckman AU5800 clinical chemistry analyzer. All three methods were calibrated appropriately on the day of assay, and QC was performed prior to analysis of the feline serum samples. Fifty-six feline serum samples were assayed in duplicate for all three analytes on the same day.

SDMA is validated in cats, with values above 14 μg/dL indicating that CKD is likely present. In this study, SDMA correlated well with serum creatinine. All but one of the samples above the feline creatinine cut-off of 2.5 mg/dL was also above the SDMA cut-off (19 of the 20 elevated creatinine samples). This one sample was very near the cut-offs for both SDMA and creatinine. Samples with normal creatinine values were distributed between normal SDMA values (22 of 36 samples with normal creatinine values) and elevated SDMA values (14 of 36 samples with normal creatinine values) consistent with previous data showing that SDMA can be elevated earlier than creatinine in some cases of CKD.

However, Cystatin C did not effectively differentiate between felines with normal or abnormal creatinine levels, with 25% (5 of 20 samples) of the feline samples with abnormal creatinine levels showing normal Cystatin C, using the previously published Cystatin C cut-off of 1.95 mg/dL, which was determined in cats. For cats with normal creatinine levels, half of the Cystatin C results were discrepant, with 50% (18 of 36 samples) having abnormal Cystatin C levels.

In this study, there was no alternative cut-off value that could be determined for this Cystatin C assay, which would significantly increase its diagnostic value.

SDMA has previously been shown to correlate well to GFR, and can be elevated prior to creatinine in some animals, making it an excellent marker for CKD in felines. However, this high-throughput immunoturbidometric Cystatin C assay did not differentiate between animals with elevated creatinine, and cannot be recommended for aid in diagnosing CKD in felines.

NU03

FIBROBLAST GROWTH FACTOR-23 IN CANINE CHRONIC KIDNEY DISEASE

Laura Harjes[1], Valerie Parker[1], Katarzyna Dembek[1], Luciano Henrique Giovaninni[2], Márcia Mery Kogika[2], Dennis Chew[1], Ramiro Toribio[1]

[1]The Ohio State University, Columbus, OH, USA, [2]University of São Paulo, São Paulo, Brazil

In humans and cats with chronic kidney disease (CKD), fibroblast growth factor-23 (FGF-23) has been associated with disease progression and development of renal secondary hyperparathyroidism (RHPT). In humans, elevations in FGF-23 often precede hyperphosphatemia and RHPT. The objectives of this study were to measure plasma FGF-23 concentrations in healthy dogs and dogs with CKD and to determine its association with serum creatinine, phosphorus, and parathyroid hormone (PTH) concentrations.

Thirty-four dogs with CKD and 10 healthy dogs were included. Dogs with CKD were staged according to International Renal Interest Society (IRIS) guidelines. A human FGF-23 ELISA was validated for canine samples, showing linearity and intra- and interassay coefficients of variation <7%. Values are presented as median (range). Plasma FGF-23 concentrations in healthy dogs and dogs with IRIS stages 3 and 4 CKD were 315 pg/mL (211-449), 2,302 (455-24,409) and 7,733 (2,520-41,265), respectively ($P < 0.01$).

Plasma FGF-23 concentrations were positively correlated with creatinine ($r = 0.86$, $P < 0.01$), phosphorus ($r = 0.69$, $P < 0.01$), and PTH ($r = 0.72$, $P < 0.01$) concentrations in all study dogs. Nineteen (56%) CKD dogs had FGF-23 concentrations above the upper range of normal dogs. Based on healthy dog phosphorus concentrations, 11 (32%) CKD dogs had elevated phosphorus concentrations. Only eight (24%) CKD dogs had hyperparathyroidism.

These findings suggest that FGF-23 concentrations rise in canine CKD before other traditional markers of RHPT do. Consequently, FGF-23 concentrations should be considered as an early indicator of disease progression, RHPT, and potential target for therapeutic intervention.

NU04

THE USE OF DARBEPOETIN ALFA TO STIMULATE ERYTHROPOIESIS IN DOGS WITH CHRONIC KIDNEY DISEASE

E. Hathaway Fiocchi[1], Dorothy Brown[1], Larry Cowgill[2], Samuel Tucker[3], Jessica Markovich[3], Mary Ann Labato[3], Mary Beth Callan[1]

[1]University of Pennsylvania School of Veterinary Medicine, Philadelphia, PA, USA, [2]University of California Davis School of Veterinary Medicine, Davis, CA, USA, [3]Tufts University School of Veterinary Medicine, North Grafton, MA, USA

Erythropoiesis-stimulating agents are currently recommended for veterinary patients with anemia secondary to chronic kidney disease (CKD). Darbepoetin alfa (darbepoetin) has replaced epoetin due to a three-fold longer half-life allowing for less frequent dosing. Also, darbepoetin is perceived to be less likely to elicit formation of anti-erythropoietin antibodies resulting in pure red cell aplasia (PRCA). A previous study examined the efficacy of darbepoetin in cats with CKD; 14 of 25 cats responded, with adverse events observed in 12 cats, including two with possible PRCA. This retrospective study evaluated response to darbepoetin therapy, protocols for administration, and potential adverse events in dogs with CKD.

Thirty-four dogs with IRIS stage 3 or 4 CKD met the inclusion criteria and their medical records were reviewed. Data recorded included CBC, serum biochemistry panel, reticulocyte count, serum iron parameters, systolic blood pressure, packed red blood cell (pRBC) transfusion, iron supplementation, darbepoetin dose and frequency of administration, comorbidities, adverse events, and survival.

Starting HCT ranged from 9.6-28.9% (mean of 19.7%). The starting dose of darbepoetin ranged from 0.4 - 2.08 mcg/kg (median 0.5 mcg/kg). The initial dosing frequency was every 7 days in 31/34 dogs. Nine of 34 dogs received a pRBC transfusion and 27/34 dogs received iron supplementation in the form of a pRBC transfusion, intramuscular iron dextran, oral supplementation, or a combination thereof at the start of darbepoetin therapy. Sixteen of 34 dogs received on-going iron supplementation throughout treatment.

Survival following initiation of darbepoetin therapy ranged from 30 – 546 days (median 118 days). An estimated 442 doses of darbepoetin were administered over that course of treatment; the range of doses administered per dog was 2 - 61 (median 11.5 doses per dog). Dogs were considered responders if they achieved a HCT of 30% over the course of treatment. Twenty-nine of 34 dogs were responders. Time to response ranged from 6 - 106 days (median 28 days). There was no association between starting HCT, dose or frequency of administration of darbepoetin, or initial iron supplementation between responders and non-responders. Dogs that received a pRBC transfusion at the onset of treatment were more likely to be non-responders ($P = 0.013$). Twenty-two dogs were transitioned to an extended dosing interval following response, and 15 of those dogs subsequently required the dosing interval to be shortened to maintain target HCT. No dog sustained a response at a dosing interval greater than every 21 days. Potential adverse events were documented in 19/34 dogs and included increased systolic blood pressure requiring treatment (n=12), seizures (n=5), vomiting (n=3), diarrhea (n=3), hypersensitivity reaction to iron injection (n=1), and possible PRCA (1). Development of potential adverse events resulted in discontinuation of darbepoetin only with respect to possible PRCA; in all other cases, adverse events were more likely attributed to underlying disease.

Darbepoetin is an effective treatment for anemia secondary to CKD. A dosing interval of >21 days is not effective at maintaining a response to therapy. Further studies are needed to define the most effective dose. PRCA was a possible adverse event in 1/34 dogs (2.9%).

NU05

VITAMIN D-BINDING PROTEIN – EARLY MARKER OF TUBULAR INJURY IN DOGS WITH CHRONIC KIDNEY DISEASE

Fernanda Chacar[1], Márcia Kogika[1], Talita Sanches[2], Douglas Caragelasco[1], Cinthia Martorelli[1], Camila Rodrigues[2], José Capcha[2], Lúcia Andrade[2]

[1]*School of Veterinary Medicine and Animal Science, University of Sao Paulo, Sao Paulo, Sao Paulo, Brazil,* [2]*School of Medicine, University of Sao Paulo, Sao Paulo, Sao Paulo, Brazil*

Vitamin D-binding protein (VDBP) is the main carrier of vitamin D metabolites into circulation, and it is freely filtrated through glomerulus, and the complex VDBP-25-OH-vitamin D is uptaken by megalin-cubilin receptors in proximal tubules. Once the presence of VDBP in urine, it may indicate proximal tubular dysfunction/injury as well as interstitial tubular fibrosis. Also it still remains to be proven whether loss of VDBP in urine may lead to vitamin D deficiency and its contribution to the progression of chronic kidney disease (CKD) in dogs. The hypothesis and the aim of this study were to investigate whether VDBP could be previously detected in urine of dogs with CKD regardless of urinary-to-protein ratio (UPC) values, and it will be an early and specific marker of proximal tubular injury. Forty dogs according to IRIS staging of CKD (stage 1 n= 10; stage 2 n= 10; stage 3 n= 10; stage 4 n= 10) were enrolled, various breeds and age, and with no proteinuric-associated disease. Nine clinically healthy dogs, different breeds and age, composed the control group. Dogs with CKD were classified according to IRIS guidelines based on UPC as non-proteinuric (UPC<0.2), borderline proteinuric (UPC= 0.2 to 0.5) and proteinuric (UPC>0.5). Western blotting was performed to investigate VDBP in canine urine (Anti-Vitamin D Binding Protein antibody, ab95469, Abcam© 1:500). No urinary VDBP was detected in control dogs and UPC was 0.14±0.04 (mean±SEM) and min= 0.028 and max= 0.41. In CKD dogs, urinary VDBP was observed since the early stages of CKD, stages 1 and 2, and the UPC was 0.39±0.29; 0.02-3.01 and 0.33±0.12; 0.14-1.14 (mean±SEM; min - max), respectively. In stage 1 CKD dogs, urinary VDBP was detected in 7 out of 10 dogs and in 4 of those 7 dogs, UPC was into the non-proteinuric range, 2 dogs in borderline range and only one was proteinuric. In stage 2 CKD dogs, VDBP in urine was noticed in 9 out of 10 CKD dogs and 5 of them had UPC < 0.2, one dog was in borderline (UPC= 0.42) and 3 dogs were slightly proteinuric (UPC of 0.67, 0.72 and 1.14). In CKD dogs in stage 3, UPC was 1.51±0.54 (0.07–4.57) and VDBP in urine was immunodetected in 8 out of 10 dogs and in 3 of those 8 dogs, UPC was ≤ 0.28 and the left 5 dogs were proteinuric. All CKD dogs in stage 4 were proteinuric (UPC = 4.37±0.47; 1.4–6.94) and showed VDBP in urine. In conclusion, the immunodetection of VDBP in urine noticed in the early stages of CKD in dogs, mainly in stages 1 or 2, associated with UPC into non or borderline proteinuric range, it reinforces that urinary VDBP could be a potential early marker of kidney injury as well as to detect the specific segment (proximal) of the nephron affected.

NU06

ASSESSMENT OF REPEATED ADMINISTRATION OF A FELINE FVRCP VACCINE AS A MODEL FOR INTERSTITIAL NEPHRITIS

Stacie Summers, Shannon McLeland, Jennifer Hawley, Jessica Quimby, Randall Basaraba, Catriona MacPhail, Michael Lappin

Colorado State University, Fort Collins, CO, USA

Progressive interstitial nephritis (IN) is the primary cause of feline chronic kidney disease which is considered to be the leading cause of death in adult cats. The Crandell Rees feline kidney (CRFK) cell line is commonly used to grow feline herpesvirus 1 (FHV-1), calicivirus, and panleukopenia virus used in (FVRCP) vaccine production. Previous studies have shown that cats administered FVRCP vaccines parenterally develop antibodies against CRFK lysates and alpha enolase (glycolytic pathway enzyme in all mammalian cells) and these antibodies can bind to feline renal cell lysates. In addition, three of six cats that were hyperinoculated (11 injections) with CRFK lysates over two years had IN on renal biopsy collected two weeks after the last booster. The primary objective of this study was to determine whether IN could be induced over 16 weeks by repeatedly administering a market leading parenteral FVRCP vaccine to potentially use as a short term model to study biomarkers associated with IN in cats.

A total of six (three male; three female) one-year-old purpose bred cats were included in the IACUC approved study. The cats were previously maintained as unvaccinated controls in a FHV-1 study at five months of age and none of the cats had clinical signs of FHV-1 when enrolled in the current study. On Week 0, blood, serum, urine were collected for biomarker assays and a wedge kidney biopsy for histopathological evaluation and alpha-enolase immunohistochemistry was obtained. After sample collection on Week 0 and again on Weeks 2, 4, 6, 8, 10, 12, and 14, all cats were administered a commercially available FVRCP vaccine shown previously to induce anti-CRFK and anti-enolase antibodies and samples were collected for biomarker assays. After sample collection on Week 16, similarly sized renal biopsies were made, avoiding the previous biopsy sites. Haematoxylin and eosin stained sections were provided to two board-certified pathologists that were masked to the timing of the biopsies. Anti-CRFK and anti-enolase antibodies levels in serum were determined by ELISAs.

All 6 cats developed progressively increasing anti-enolase and anti-CRFK serum antibodies.

Mean and SD of ELISA absorbance values of 6 cats administered an FVRCP vaccine					
Antibody	Week 0	Week 4	Week 8	Week 12	Week 16
Enolase	0.57 ± 0.16	2.00 ± 0.74	3.48 ± 0.72	3.82 ± 0.47	3.88 ± 0.43
CRFK	0.28 ± 0.03	0.66 ± 0.23	1.38 ± 0.39	1.63 ± 0.37	1.81 ± 0.4

Histological evidence of interstitial nephritis was not detected by light microscopy in any of the tissue biopsies. Significant biochemical or urinalysis changes during the study were not detected and the cats were adopted to private homes for long term monitoring.

While anti-CRFK and anti-enolase antibodies were induced in this model, biochemical or histopathology abnormalities were not detected. Results of selected biomarkers and enolase immunohistochemical staining will be used to further evaluate this potential IN model.

NU07

COMPARISON OF VISUAL AND AUTOMATED INTERPRETATION OF URINARY DIPSTICKS WITH GLUCOSE:CREATININE RATIO AND GLUCOSE CONCENTRATION

Caroline Aldridge[1], Ellen N. Behrend[1], Jo Smith[2], Elizabeth G. Welles[1], Hollie P. Lee[1]

[1]Auburn University, Auburn, AL, USA, [2]University of Georgia, Athens, GA, USA

The accuracy of Bayer Multistix with visual reading for determination of the glucose concentration in canine urine is only 59% (Behrend et al, J Vet Int Med, 2009). The correlation between visual and automated assessment has not been investigated. Furthermore, urine glucose:creatinine ratio (UGCR) may be a useful method in quantifying glucosuria similar to urine protein:creatinine ratio but has not been assessed. The purpose of this study was to compare automated readings, visual readings and UGCR for quantification of glucosuria in canine urine.

Urine samples submitted to the Auburn University Clinical Pathology Laboratory for urinalysis and that were naturally glucosuric (n=39) were included. Multistix (Bayer) were used according to manufacturer instructions and were read visually by a single trained investigator ("visual reading") and by the CLINITEK 50 analyzer (Bayer; "automated reading"). Urine glucose and urine creatinine concentrations were measured using a Hitachi 911 Chemistry Analyzer (Boehringer Mannheim Corp.); the glucose measurements were considered to be the true concentration (gold standard). Spearman's rank order correlation was used to evaluate the relationship between: 1. glucose concentration and automated readings; 2. glucose concentration and visual reading; 3. automated and visual readings; 4. automated readings and UGCR; and 5. visual readings and UGCR. Significance was set at $p<0.05$. Correlation was classified as excellent for r=0.93-0.99, good for r = 0.8-0.92, fair for r = 0.59-0.79 and poor for r < 0.59. Reference intervals that correlated with the colors on the test strip were devised based on the package insert (trace=76-175 mg/dL; 1+ = 176-375 mg/dL; 2+ = 376-750 mg/dL; 3+ = 751-1500 mg/dL; 4+ = >1500 mg/dL).

Significant correlation was detected between both automated and visual readings and glucose concentration ($p<0.0001$). The correlation for automated readings was good (r=0.834) but fair for visual readings (r=0.761). Overall, the underestimations and overestimations of glucose concentration for the automated readings was 56.4% and 15.4%, respectively. For the visual readings, they were 13.2% and 36.8%, respectively. Significant correlation was detected between both automated and visual readings and UGCR ($p<0.0001$). However, correlation for automated readings was classified as good (r=0.801) but for visual readings was fair (r=0.719). Furthermore, automated and visual readings had a significant ($p<0.0001$) and good correlation (r=0.817).

Our results indicate that despite a good correlation between automated and visual reading of Bayer Multistix, automated reading may be better. Interestingly the automated readings tend to underestimate glucose concentration while with visual reading, overestimations are more common. The UGCR is a valid way to assess urinary glucose excretion; however, it offers no advantage over measurement of glucose concentration in urine.

NU08

IMPACT OF CANINE PANCREAS-SPECIFIC LIPASE ON THE OUTCOME OF DOGS WITH HEMODIALYSIS-DEPENDENT ACUTE KIDNEY INJURY

Kanae Takada, Carrie Palm, Steven Epstein, Larry Cowgill

University of California-Davis Veterinary Medical Teaching Hospital, Davis, CA, USA

Acute pancreatitis (AP) and acute kidney injury (AKI) are common co-morbidities in both dogs and humans. In human patients the presence of AP with AKI is a negative prognostic factor; however, a similar relationship has not been evaluated in dogs. When considering hemodialysis for management of AKI in dogs, this prognostic information could be important for therapeutic decisions. Canine pancreas-specific lipase (SPEC) is used commonly to support the diagnosis of AP in dogs. The purpose of this study was to investigate elevated SPEC on the outcome of dogs with AKI treated with hemodialysis.

Medical records were evaluated retrospectively from August 2011 to June 2015 to identify dogs presented to the UC Davis VMTH that received intermittent hemodialysis for management of AKI and also had a SPEC (Spec cPL™, IDEXX Laboratories) measured during the course of therapy. Outcome was assessed at 30 days from admission, and dogs were classified as died/euthanized or alive. Surviving dogs were stratified further into non dialysis-dependent or dialysis-dependent. An elevated SPEC was defined as ≥ 400 mcg/L. Median values of SPEC were compared with a Mann-Whitney U test for each outcome category, above. Categorical data were compared with a Fisher's exact test. A *P*-value of <0.05 was considered significant. Data are presented as median [range].

Forty-three client-owned dogs were included. Median initial serum creatinine (iSCr) was 8.5 [3.6 - 16.6] mg/dl and median age at presentation was 5.6 [0.70 - 13.5] years. Eighteen were male neutered, 14 were female spayed, 4 were intact male, and 7 were intact female dogs. At 30 days, 30/43 dogs (70%) were alive (iSCr, 8.6 [4.7 - 16.6] mg/dl) and 13/43 (30%) had died or been euthanized (iSCr, 8.0 [3.6 - 15.3] mg/dl). Nine of 30 dogs (30%) were still dialysis dependent at 30 days (iSCr, 10.9 [6.1 - 15.4] mg/dl), and 21/30 (70%) were non dialysis-dependent (iSCr, 8.3 [4.7 - 16.6] mg/dl). SPEC was not significantly different between dogs alive at 30 days and dogs that had died or were euthanized (SPEC 627.5 [29 - 2001] mcg/L vs 455.0 [30 - 2001] mcg/L, $P = 0.50$). SPEC in dialysis-dependent dogs was significantly higher than non dialysis-dependent dogs (SPEC 1001.0 [177 - 2001] mcg/L vs 379.0 [29 - 1001] mcg/L, $P = 0.008$). The proportion of dogs alive at 30 days with SPEC ≥ 400 mcg/L during the treatment course was not different from alive dogs with SPEC < 400 mcg/L (71% vs 68%, $P = 1.0$). In addition, the proportion of the non dialysis-dependent dogs with SPEC ≥ 400 mcg/L during the treatment course was not different from non dialysis-dependent dogs with SPEC < 400 mcg/L (42% vs 58%, $P = 0.36$).

These results suggest the presence of pancreatitis assessed as SPEC ≥ 400 mcg/L did not affect survival outcome at 30 days. However, pancreatitis may be a more common co-morbidity in dogs with more severe (dialysis-dependent) AKI and it may influence renal recovery.

NU09

CHARACTERIZATION OF SUBCLINICAL BACTERIURIA, URINARY TRACT INFECTION, AND PYELONEPHRITIS IN DOGS WITH CHRONIC KIDNEY DISEASE

Jonathan Foster, Harathi Krishnan, Stephen Cole

University of Pennsylvania School of Veterinary Medicine, Philadelphia, PA, USA

Bacterial urinary tract infection (UTI) is common in dogs, however the prevalence of subclinical bacteriuria (SBU) in healthy dogs has recently been demonstrated to be substantially lower. Following standards created by human infectious disease specialists, there has been more initiative put towards identifying and characterizing SBU in dogs. Although it has been considered a risk factor for UTI, the prevalence of UTI and SBU in dogs with chronic kidney disease (CKD) is not known. The objectives of this study were to retrospectively evaluate the prevalence of positive urine cultures in dogs with CKD, and to further categorize these patients as having SBU, clinical UTI, or pyelonephritis.

Medical records were reviewed retrospectively from 1/2010-7/2015 for dogs with a diagnosis of CKD who had a urine culture submitted. The diagnosis of CKD was based on International Renal Interest Society (IRIS) guidelines. Patients were excluded if they had another reason to develop UTI, namely endocrine disease (hyperadrenocorticism, diabetes mellitus), urolithiasis, urinary incontinence, urinary bladder neoplasia, had a nidus for infection such as ureteral stent or subcutaneous ureteral bypass, or were receiving systemic immunosuppression. Based on previously reported definitions, patients were placed into one of the following three categories based on the review of their medial record: SBU, pyelonephritis, or UTI. The frequency of negative culture, SBU, pyelonephritis, and UTI were compared across IRIS stages using Fisher's exact analysis. $P < 0.05$ was considered significant for all comparisons.

Two hundred and eighty-two cultures were submitted on 195 patients during the study time period. Forty-seven cultures performed on 13 patients were excluded due to the presence of ureteral stents (5 patients), chemotherapy (2 patients), diabetes mellitus (2 patients), transitional cell carcinoma (2 patients), urinary incontinence and cystic calculi (1 patient each). A total of 235 cultures submitted on 182 patients were included in the final analysis. There were 8 intact males, 52 castrated males, 5 intact females, and 117 spayed females. There were 40 positive urine cultures (17.0% of all cultures) obtained on 33 patients (18.1% of all dogs). Of all positive cultures, the most frequently determined diagnosis was SBU (18 cultures, 45%), followed by pyelonephritis (16 cultures, 40%), then UTI (6 cultures, 15%). There were no statistically significant observed relationships between any IRIS stage and diagnosis ($p=0.635$). The number of positive cultures, regardless of patient diagnosis, was not statistically significant between IRIS stages ($p=0.432$). *Escherichia coli* was the most frequently observed isolate (29/40 cultures, 72.5%).

In this population of dogs with CKD the prevalence of dogs with positive urine culture was 18.1%. The most frequently observed diagnosis associated with a positive culture was SBU. Increasing IRIS stage of CKD was not found to be associated with a higher frequency of any particular diagnosis. *E. coli* was the most commonly observed isolate. Despite the rigorous criteria for defining SBU, pyelonephritis, and UTI in the current study, it is possible that some patients were incorrectly categorized. Prospective research is needed to determine if antibiotic therapy is indicated in dogs with CKD and SBU. Additionally, renal biomarkers may allow for better characterization of patients with SBU and occult pyelonephritis and improve antibiotic stewardship.

NU10

INITIAL OUTCOMES AND COMPLICATIONS OF THE SUBCUTANEOUS URETERAL BYPASS PROCEDURE AT TWO UNIVERSITY HOSPITALS (2012-2015)

Ewan D. S. Wolff[1], Roswitha Dorsch[2], Julia Knebel[2], Daniel J. Duffy[1], Lynetta J. Freeman[1], Lynn Guptill[1], Larry G. Adams[1]

[1]*Purdue University, West Lafayette, IN, USA,* [2]*Ludwig Maximilians University Munchen, Munich, Bavaria, Germany*

The aim of this study was to report combined experience of two institutions placing subcutaneous ureteral bypass systems (SUBs). We compared our data to historical data from cats with ureteral stents for ureteral obstruction. Hypotheses included 1) Patients with SUBs would have the same or better 7-14 day and 3-4 month post-operative creatinine as historic ureteral stent patients, 2) SUB patients would experience the same or improved peri-operative, post-operative and recurrent urinary tract infection rates as ureteral stent cats, 3) clinical signs related to implants would be similar or better with the SUB system vs. ureteral stenting.

Adult cats were enrolled retrospectively, subsequent to the 2012 alteration to SUB placement procedure. All cats were managed per clinician preference pre-operatively. Flushes of SUB systems were performed every 3-6 months for the majority of patients.

Records from 19 cats from Purdue and Munich Universities were reviewed. Median age was 9 years at Purdue and 7.7 years at Munich. There were 12 spayed female and 7 castrated male cats; 13 domestic short-haired cats, 2 domestic longhaired cats, and three other breeds. 17% of cats had preexisting IRIS Stage II CKD. Lethargy, vomiting and inappetence were the most common clinical signs (>25% of patients). Median renal pelvic diameter in affected kidneys was >0.5cm at both institutions. Median creatinine immediately post-operatively was 2.6 mg/dL (0.9-2.3 mg/dL) at Purdue and 1.43 mg/dL at Munich. Major peri-operative complications were uncommon; one cat had cardiac arrest and another was transfusion dependent. Post-operative complications included 2 cats with fluid overload and pleural effusion. 8/19 cats required a blood transfusion in the peri- or post-operative period. The median creatinine was 3.6 mg/dL at 3 months and 1.8 mg/dL at 6 months. 3 patients had repeat surgeries; median time to repeat surgery was 225 days. Causes of repeated surgery were SUB port obstruction or encrustation of SUB tubing. Total mortality was 21% with a median disease free interval of 180 days and a 1 year survival of 83%.

No cats had active urine sediment or culture prior to SUB placement, but post-operative and follow up urinary tract infections occurred. 21% of patients had urinary tract infection within 10 days of surgery. Two patients had recurrent multi-drug resistant *E. coli* infection, one was cultured as a biofilm. Another patient developed encrustation 10 months post-procedure. Mandelic acid, EDTA, antibiotic infusions into SUBs failed to resolve UTI in 2 cats.

Comparing SUB outcomes to ureteral stenting, 6 month post-procedural creatinine was better in the SUB group (1.8 mg/dL) than that reported for stents (2.6 mg/dL). One year survival was similar for stent cats (86%) and SUB cats (83%). Long-term infection rates were similar (around 31%). Only one SUB patient experienced dysuria beyond the post-operative period whereas this is more common in cats with stents.

This study suggests that the SUB system carries similar potential complications of re-obstruction, encrustation and infection to ureteral stenting. Recurrent and resistant infections and the potential for biofilms is a significant concern in SUB patients. Protocols for treatment of resistant UTIs utilizing local instillation and indwelling bladder treatment should be developed in a prospective fashion. Further work with more cases could explore risk factors for UTI and obstruction.

NU11

OCCULT URINARY TRACT INFECTION IN CATS: PREVALENCE AND FINDINGS ON CONTEMPORANEOUS URINALYSIS

Melanie Puchot[1], Audrey Cook[1], Caitlin Pohlit[2]

[1]*Texas A&M University, College Station, TX, USA,* [2]*Michigan State University, East Lansing, MI, USA*

Cats are frequently screened for bacterial urinary tract infection (UTI) in the absence of clinical signs of lower urinary tract disease (LUTD) when managing conditions such as chronic kidney disease and diabetes mellitus. The purposes of this study were to determine the prevalence of occult UTI in cats and to describe results of the contemporaneous urinalysis (UA). We hypothesized that a positive quantified urine culture (QUC) could be predicted by findings on microscopic sediment examination (MSE) such as bacturia, pyuria (≥3 white blood cells/hpf) and hematuria (≥10 red blood cells/hpf).

An electronic search identified all QUCs performed on cats at a university teaching hospital between 2009 and 2015. Results were included in the study if there were no signs of LUTD and the QUC was not performed as follow up to a previous UTI. A QUC was classified as positive if growth >1,000 cfu/ml was reported. For each positive, five controls were randomly selected from samples with a negative QUC. Medical records for both groups were reviewed.

In all, 31/500 (6.2%) samples were positive. Most infections were a single organism (n=27); four contained multiple organisms. *E. coli* was the most common species (58% of isolates) followed by *Enterococcus* species (25% of isolates). Positive specimens were more likely to be from female cats (n=24) versus male (n=7) ($p = 0.0054$). Patient age and body weight were not associated with increased risk of occult UTI. Median urine specific gravity was similar for the positive group (1.022) and controls (1.024). The prevalence of proteinuria on dipstick testing was also similar (84% v 80%). Occult UTIs were strongly associated with bacteriuria (60% v 6%; odds ratio 24.3 [CI: 9.0-65.7]; $p < 0.0001$) or pyuria (67 v 19%; odds ratio 8.7 [CI: 3.7-20.5]; $p < 0.0001$) but not microscopic hematuria. Positive specimens were significantly more likely to have an abnormal sediment exam (odds ratio 13.5 [CI 3.9-45.5]; $p < 0.0001$).

Occult UTI appears to be uncommon in cats and can usually be predicted on the basis on an abnormal MSE (bacturia, pyuria). Based on this study, there is little indication to perform a QUC in a cat with no clinical signs of LUTD and an unremarkable MSE.

NU12

DIAGNOSTIC PERFORMANCE OF URINARY CANINE CALGRANULINS IN DOGS WITH LOWER URINARY TRACT CARCINOMA

Romy Heilmann[1], Elizabeth McNeil[2], David Lanerie[1], Niels Grützner[3], Jan Suchodolski[1], Joerg Steiner[1]

[1]*Gastrointestinal Laboratory, Texas A&M University, College Station, TX, USA,* [2]*Cummings School of Veterinary Medicine, Tufts University, Boston, MA, USA,* [3]*Farm Animal Clinic, Vetsuisse University Bern, Bern, BE, Switzerland*

Aside from its association with several inflammatory diseases, the S100A8/A9 calgranulin complex was shown to be expressed by epithelial cells after malignant transformation. In humans, S100A9 or S100A8 have been reported to be up-regulated in patients with prostate carcinoma (PCA) or transitional cell carcinoma (TCC), respectively. Results of the expression of the S100A12 calgranulin in TCC/PCA in humans are ambiguous. Pilot data by our group suggest that the urine canine S100A8/A9-to-S100A12 ratio (uCalR) may have potential as a marker for TCC/PCA in dogs. Further evaluation of urinary S100A8/A9, S100A12, and uCalR in dogs is needed to determine its specificity for a diagnosis of TCC/PCA. Thus, this study investigated urinary S100A8/A9, S100A12, and uCalR in dogs with TCC/PCA, other urinary tract diseases, or other malignancies.

Urine samples from dogs with TCC/PCA that were treatment-naïve (n=22) or under treatment (n=40), dogs with non-neoplastic diseases of the urinary tract (n=22) or other neoplasms (n=35), dogs with a urinary tract infection (UTI, n=45), and a group of healthy control dogs (n=75) were used. Urinary canine S100A8/A9 and S100A12 were measured in all samples using in-house radioimmunoassays and the uCalR was calculated. All parameters were tested among the different groups of dogs using non-parametric multiple-/two-group comparisons. Sensitivities and specificities were determined for the optimal cut-off values, and positive (PPV) and negative predictive values (NPV) were calculated. Significance was set at $p<0.05$.

Dogs with TCC/PCA had significantly higher urinary S100A8/A9 and S100A12 concentrations than all other groups of dogs (all $p<0.016$) except for dogs with UTI, whereas uCalR was significantly higher in TCC/PCA dogs compared to only those with a UTI ($p<0.001$). Treatment-naïve TCC/PCA patients and UTI dogs were best distinguished by a uCalR of ≥9.1 (sensitivity: 91%, specificity: 60%). A urine canine S100A8/A9 concentration (normalized against the urine specific gravity [USG]) of ≥109.9 had the highest sensitivity (96%) and specificity (66%) to diagnose treatment-naïve TCC/PCA dogs from all other groups of dogs, and the specificity increased to 75% if a UTI had been excluded. Using a urine S100A8/A9 concentration (normalized against USG) of ≥109.9 to screen dogs ≥6 years of age for TCC/PCA (estimated prevalence: <1%) yielded a PPV of 4% and a NPV of 100%.

The results of this study show that urine S100A8/A9 and the uCalR have utility for the diagnosis of TCC/PCA in dogs. The high sensitivity and NPV of urine S100A8/A9 also suggest that S100A8/A9 may be a good screening test for TCC/PCA in dogs. Further studies are warranted to validate our findings in a larger cohort of dogs and to evaluate the source of S100A8/A9 and S100A12 expression in canine TCC/PCA.

NU13

SURVEY OF SUBCUTANEOUS FLUID PRACTICES IN CATS WITH CHRONIC KIDNEY DISEASE

Crystal Cooley[1], Sarah Caney[2], Liberty Sieberg[1], Jessica Quimby[1]

[1]*Colorado State University, Fort Collins, CO, USA,* [2]*Vet Professionals, Roslin, Midlothian, UK*

Chronic kidney disease (CKD) is common in elderly cats. The purpose of this study was to describe subcutaneous fluid (SQ) administration practices of owners of CKD cats to help more owners successfully give SQ fluids to their cat.

An anonymous web-based survey was advertised via list serves, websites and social media. Owners of 468 cats with CKD participated. 87% of cats were 10 years or older. Cats were IRIS stage I (1%), II (20%), III (37%), IV (17%), and unknown (25%).

95% of owners said they discussed giving fluids with their veterinarian. 399 respondents stated they gave SQ fluids, 57 did not, and 12 tried but could not. Only 42% of owners were given additional educational resources. 79% said the process was ok/easy to learn. Once experienced, 15% said it was still somewhat/highly stressful on them, and 11% said it was somewhat/highly stressful for the cat. To improve tolerance 57% used food for positive reinforcement with 59% stating this improved tolerance, 60% warmed the fluids and 83% felt warming fluids increased tolerance. 74% felt that length of time it took to administer fluids affected tolerance. 82% said needle size affected tolerance. 40% of owners checked hydration status daily or twice daily and 18% of owners did not know how. 43% said they skipped/added fluids based on hydration assessment.

The majority of owners were successful in administering fluids but additional education materials could be provided. Variables such as needle size, warming fluids, and length of time of administration may improve tolerance.

NU14

QUANTIFYING URINE ELIMINATION BEHAVIORS IN CATS USING A VIDEO RECORDING SYSTEM

Ryan Dulaney, John Kruger, Marie Hopfensperger, Rob Malinowski, Joe Hauptman, Brenna Lashbrook

Michigan State University, East Lansing, MI, USA

Feline lower urinary tract disorders often rely on subjective caregiver quantification of clinical signs to establish a diagnosis and monitor therapeutic outcomes. The objective of this study was to investigate use of a video recording system (VRS) to more reliably assess and quantify urination behaviors in cats. Litter box urination behaviors were quantified in 11 healthy cats and 8 cats with abnormal urination patterns using a VRS for 14 days and compared to daily caregiver observations. Video recordings were analyzed using a behavior analysis software program.

The mean number of urinations per day detected by VRS (2.5±0.2) was significantly greater compared to caregiver observations (0.6±0.2; P<0.0001). Five of 19 cats were never observed by their caregivers in the litter box. The mean number of urinations per day detected by VRS was significantly higher for abnormal cats (2.9±0.3) compared to healthy cats (2.1±0.2; P=0.02); there were no apparent differences in frequency between healthy and abnormal cats reported by caregivers (0.7±0.3 and 0.5±0.3 respectively). There were no differences in mean urination time between normal and abnormal cats determined by VRS or caregivers. Mean cover-up time determined by VRS was significantly longer in normal cats (23±4 seconds/urination compared to abnormal cats (9±5 seconds/urination; P=0.03); differences in cover-up time were not detected by caregiver observations.

In conclusion, there was considerable disparity between caregiver and video observations with caregivers commonly underestimating urination frequency. VRS appears to facilitate objective assessment of urination behavior and could be of value in future clinical studies of feline urinary disorders.

NU15

EVALUATION OF THE EFFECT OF URINE DIP VERSUS URINE DRIP ON MULTITEST STRIP RESULTS

Alisdair Boag, Craig Breheny, Adam Gow

Royal (Dick) School of Veterinary Studies, Edinburgh, UK

Standard operating procedures (SOPs) have been established for in-clinic laboratory tests, such as the World Health Organisation guidelines for packed cell volume. No independent evidence based guidelines exist for dipstick urinalysis, however manufacturer's instructions state to dip the stick into urine. In veterinary medicine, occasionally urine samples of such small volume are obtained that this is impractical and dripping urine on the stick from a pipette or syringe is performed. This also has the advantage of conserving the sample for additional tests.

The hypothesis of this study was that the method of drip or drip would have no effect on the test results. Siemens Multistix 10SG strips were used. To standardise the method of strip analysis, a Siemens Clinitek Status+ analyser was used to read the strips; controlling time of reading the results and user variability in interpretation of colourimetric changes. Three individuals tested 25 aliquots of urine from dogs with a range of disease processes by both the dip and drip method. Results were compared for each variable between drip and dip using Wilcoxon-signed rank test.

Across the nine variables assessed (SG was not), a significant difference between methods was found for one variable, bilirubin ($p = 0.046$). This difference was accounted for by one dog who tested negative for bilirubin for all three users using a dip method and had a trace value for all three users using a drip methodology. No other significant differences were found between variables.

This study demonstrated a significant difference between dipping and dripping for bilirubin, caused by one dog with the same level of variability for all three users. This difference is, in the authors' opinion, clinically insignificant. This study therefore supports the use of either a dipping or dripping approach to urinalysis. Further studies are required to more fully assess the role of individual variation when performing the test including level of experience.

NU16

POSITIVE IMPACT OF NUTRITIONAL INTERVENTIONS IN CLIENT-OWNED DOGS WITH IRIS STAGE-1 CHRONIC KIDNEY DISEASE

Jean Hall[1], Dale Fritsch[2], Maha Yerramilli[3], Edward Obare[3], Murthy Yerramilli[3], Dennis Jewell[2]

[1]Oregon State University, Corvallis, OR, USA, [2]Hill's Pet Nutrition, Inc., Topeka, KS, USA, [3]IDEXX Laboratories, Inc, Westbrook, ME, USA

A prospective 12-month clinical trial was performed in client-owned dogs with IRIS stage 1 chronic kidney disease (CKD) to measure their ease of transition to a commercial renal protective food and to assess the effects of food on renal biomarkers and quality of life attributes. Dogs with IRIS stage 1 CKD (n=36) were transitioned over 1 week from various grocery brand foods to a modified protein, low phosphorus, antioxidant enriched test food (Prescription Diet® k/d®, Hill's Pet Nutrition, Inc.). At months 0, 3, 6, 9, and 12 owners completed a questionnaire to assess their pet's acceptance of the test food and their dog's quality of life. Renal biomarkers, including serum creatinine (Cr), blood urea nitrogen (BUN), and symmetric dimethylarginine (SDMA), and urinalysis parameters, including urine specific gravity (USG) and urine protein:creatinine ratio (UPC), were measured. Of the 36 dogs initially enrolled, 35 (97%) transitioned to the test food. Pets moderately or extremely liked the test food 88% of the time, ate most or all of the food 84% of the time, and were moderately or extremely enthusiastic while eating 76% of the time. Dogs consuming test food showed a decrease in serum Cr and BUN concentrations across time (both P=0.01) and a decrease in serum SDMA concentration and USG across time (both P=0.09). All serum renal biomarkers (Cr, BUN, SDMA) were decreased (P≤0.05) from baseline at 3 months, and remained decreased from baseline at 12 months in dogs completing the study (n=20). This study shows that dogs with IRIS stage 1 CKD readily transition to a commercial renal food, and decreasing serum biomarker concentrations suggest improvement in kidney function. In addition, owners reported improvement in a consortium of overall health and quality of life attributes (P<0.01) and hair and coat quality attributes (P<0.01).

NU17

FIBROBLAST GROWTH FACTOR 23 (FGF-23) IN DOGS WITH NATURALLY OCCURRING CHRONIC KIDNEY DISEASE

Marcia Kogika[1], Cinthia Martorelli[1], Douglas Caragelasco[1], Fernanda Chacar[1], Luciano Giovaninni[2], Dennis Chew[3], Ramiro Toribio[3], Katarzyna Dembek[3], Camila Rodrigues[4], Talita Sanches[4], Lucia Andrade[4]

[1]School of Veterinary Medicine, University of São Paulo, Sao Paulo, Sao Paulo, Brazil, [2]Sao Paulo, Sao Paulo, Sao Paulo, Brazil, [3]College of Veterinary Medicine, The Ohio State University, Columbus, Ohio, USA, [4]School of Medicine, University of São Paulo, Sao Paulo, Sao Paulo, Brazil

Decreased glomerular filtration rate in chronic kidney disease (CKD) reduces renal phosphorus excretion, leading to increased total body phosphorus retention and eventually hyperphosphatemia. In addition of promoting parathyroid hormone synthesis, hyperphosphatemia stimulates the synthesis of fibroblast growth factor 23 (FGF-23), a phosphaturic/hypophosphatemic hormone. There are no reports on serum FGF-23 in healthy dogs or in dogs with CKD. The goal of this study was to measure serum FGF-23 concentrations in healthy dogs and in dogs with CKD, and to determine its association with serum phosphorus concentrations and severity of renal disease based on IRIS staging of CKD. We hypothesized that serum FGF-23 will be increased in dogs with CKD, that its concentrations will be proportional to CKD severity, and that it will be an early marker of altered phosphorus metabolism.

A total of 17 dogs of various breeds and age with CKD (n=6 in IRIS stage 2; n=11 in IRIS stage 3) that were being fed a renal diet, and 15 healthy dogs of different breeds and age, being fed a maintenance diet, were included. Dogs with CKD were followed for up to 12 months or until death. A human-specific FGF-23 ELISA was validated for this study, showing linearity and 2.8% and 8.8% intra and inter-assay coefficients of variation, respectively. Serum FGF-23 concentrations were higher in CKD dogs in stages 2 and 3 on presentation (2,605.4±872.0 pg/mL; mean±SEM) compared to healthy dogs (258.7±21.2 pg/mL); significant difference was detected between healthy vs CKD dogs in stage 3 ($P < 0.01$;one-way ANOVA), and stage 2 vs stage 3 ($P < 0.05$). On presentation, all dogs in stage 2 CKD were normophosphatemic (3.8±0.2 mg/dL), serum FGF-23 was 499.1±135.8 and 3/6 dogs showed increased FGF-23 concentrations (505.5 – 1,065.5 pg/mL), and there was no correlation between FGF-23 and phosphorus concentrations on admission and follow up ($r=0.286$; $P= 0.1185$).

In stage 3 CKD dogs, on presentation, 7 were normophosphatemic (2.5 – 5.0 mg/dL) and 4 were hyperphosphatemic (5.3 – 6.6 mg/dL), but all (n=11) had increased serum FGF-23 concentrations (3,754.3±1,225.2 pg/mL; 957.6-14,193.0 pg/mL) that remained elevated or increased during follow up (843.9-184,581.0 pg/mL). Of the normophosphatemic dogs in stage 3 CKD (n=7), 5 developed hyperphosphatemia (5.6 – 16.0 mg/dL) after 6 months of follow up. Serum phosphorus was positively correlated with FGF-23 in these dogs ($r= 0.71$; $P <0.0001$). In the study period, 7/11 dogs in stage 3 died.

In conclusion, serum FGF-23 could be used as an early marker of phosphorus dysregulation in dogs with CKD as increased FGF-23 concentrations were found in normophosphatemic CKD dogs in stages 2 and 3. More marked elevated serum levels of FGF-23 were found in advanced renal disease – IRIS stage 3. It is likely that increased concentrations of FGF-23 contribute to disturbed calcium and phosphorus homeostasis and altered vitamin D metabolism and parathyroid gland function. Further studies at targeted control of serum phosphorous and FGF23 concentrations in dogs with CKD are warranted.

NU18

SERUM CYTOKINE PROFILES OF CATS WITH IDIOPATHIC CYSTITIS

Maciej Parys, Vilma Yuzbasiyan-Gurkan, John Kruger

Department of Small Animal Clinical Sciences, Michigan State University, East Lansing, USA

Idiopathic cystitis (IC) is a common lower urinary tract disorder in cats that resembles painful bladder syndrome/interstitial cystitis in humans. In both disorders, the etiology is unknown and specific therapy is not available. Diagnosis is based on clinical signs and exclusion of other disorders due to lack of characteristic pathologic findings, well-defined phenotypes, or objective biomarkers. Cytokines can serve as noninvasive biomarkers to define the presence, severity, and progression of disease, and response to treatment. The objective of this pilot study was to determine concentrations of selected cytokines in serum from healthy cats and cats with IC. Thirteen serum samples from healthy cats and 13 from cats with IC were evaluated. Multiplex analysis of 19 cytokines (sFas, TNF-α, SDF-1, SCF, RANTES, PDGF-BB, MCP-1, KC, IL-18, IL-13, IL-12 (p40), IL-8, IL-6, IL-4, IL-2, IL-1β, IFN-γ, GM-CSF, Flt-3L) was performed with a multiple antigen Luminex bead-based assay according to manufacturer recommendations. Results were then analyzed for normality of data with D'Agostino-Pearson test and groups were compared using Mann-Whitney test.

Significantly increased mean concentrations of IL12(p40) (1191±484.9 vs 219.6±393.3 pg/mL; $p<0.0001$), SDF1 (2015±1131 vs 528.7±1002 pg/mL; $p=0.002$), and IL18 (483.8±468.8 vs 92.03±272.6 pg/mL; $p=0.02$) were detected in the serum of cats with IC compared to unaffected cats. Further studies are necessary to evaluate the diagnostic and prognostic utility of serum cytokine biomarkers in cats with IC.

NU19

CYSTOLITH DISSOLUTION IN CATS USING A COMMERCIALLY AVAILABLE DIET

Camille Torres-Henderson[2], Jamie Bunkers[2], Elena Contreras[2], Deborah Greco[1], Emily Cross[1], Michael Lappin[2]

[1]Nestle Purina PetCare, St. Louis, MO, USA, [2]Colorado State University, Fort Collins, CO, USA

The most common radiopaque cystoliths in cats are composed of struvite or calcium oxalate minerals. Struvite cystoliths have been shown to dissolve with dietary management whereas calcium oxalate cystoliths must be removed surgically. The commercially available diet, Purina® Pro Plan® Veterinary Diets UR Urinary® St/Ox®, has been formulated for the dissolution of struvite cystoliths and to lessen the recurrence of both struvite and oxalate cystoliths. The purpose of this study was to describe the clinical and laboratory findings in cats with radiopaque cystoliths fed this commercially available diet.

The study was announced to northern Colorado animals shelters by list serve. Cats with clinical signs of lower urinary tract disease and cystoliths confirmed via radiographs were enrolled in this IACUC approved study. Qualifying cats were transferred and maintained at the research study site to assure that only the test diet was fed. Complete blood cell count, serum biochemistry profile, abdominal ultrasound, and urinalysis with aerobic bacterial culture and antimicrobial sensitivity were performed on entry to and exit from the study. The cats were housed in a gang room, fed the study diet ad libitum, and assessed by abdominal radiographs weekly. Cats with cystoliths that resolved based on two sequential weekly radiographs and confirmatory ultrasound examination were considered diet successes. Cats with no change in cystolith size after four weeks underwent cystotomy for stone removal, aerobic culture and antimicrobial susceptibility testing, and analysis at the University of Minnesota Urolith Center. The cats were then adopted to private homes and when possible, the study diet was continued and abdominal radiographs were obtained to evaluate for recurrence.

To date, five cats between the ages of four and eight years of age have completed the study. All cats accepted the study diet, weight loss was not noted over the course of the study in any cat, serum biochemical data were normal for all cats, and urine culture was negative for all cats when exiting the study. Total cystolith dissolution was achieved by Week 2 for three cats and two cats still had radiographic evidence of cystoliths by Week 4. One cat with persistent cystoliths had a single ammonium urate cystolith with a struvite nidus (normal pre and post-bile acids) and the other cat had multiple calcium oxalate cystoliths; cultures were negative for both cats. Recheck radiographs at two months for two cats with stone resolution on diet and the cat with calcium oxalate cystoliths (three month post-cystotomy) showed no evidence of cystolith recurrence.

While larger case numbers are needed, the preliminary results suggest that feeding Purina Veterinary Diet UR Urinary® St/Ox® can successfully dissolve cystoliths that are likely struvite and may lessen risk of recurrence of struvite and calcium oxalate cystoliths.

OT01

EFFECT OF THREE RESUSCITATIVE FLUID PROTOCOLS ON N-TERMINAL PROHORMONE BRAIN NATRIURETIC PEPTIDE IN HEALTHY DOGS

Alison Khoo, Matthew Gaunt, Anthony Carr, Cheryl Waldner

University of Saskatchewan, Saskatoon, Saskatchewan, Canada

The purpose of this crossover design study was to determine if 3 different resuscitative fluid therapy protocols would increase N-terminal proBNP (NT proBNP) levels in healthy dogs. The protocols included a balanced isotonic replacement crystalloid solution (Normosol R) administered at 90ml/kg/hr over one hour, a 7.2% hypertonic saline bolus administered at 5ml/kg over 15 minutes, and a 6% hydroxyethyl starch fluid bolus (Voluven ®) administered at 20ml/kg over 30 minutes. Serum NT-proBNP concentrations were measured in 6 healthy purpose bred dogs prior to fluid administration and then 0.5, 1, 1.5, 2, 4, 8, 12, 18, 24, 36 and 48 hours after receiving one of the three fluid therapy protocols. Each dog was randomly assigned to receive a different fluid protocol each week for 3 weeks, allowing a 5-day washout period between treatment protocols.

Serum NT-proBNP concentration was significantly more elevated from baseline in dogs receiving 90ml/kg/hr of crystalloid solution compared to dogs receiving hypertonic saline or hydroxyethyl starch boluses. None of the fluid therapy protocols caused elevations of serum NT-proBNP above the previously reported cut-off concentration used to discriminate between dogs with congestive heart failure and dogs with primary respiratory disease.

Resuscitative fluid therapy does increase serum NT-proBNP concentrations but does not impact the ability to use serum NT-proBNP concentration to discriminate between cardiac failure and respiratory disease in healthy dogs.

OT02

DEXMEDETOMIDINE OROMUCOSAL GEL FOR ALLEVIATION OF ACUTE ANXIETY AND FEAR ASSOCIATED WITH NOISE IN DOGS

Mira Korpivaara[1], Kaisa Laapas[1], Mirja Huhtinen[1], Barbara Schöning[3], Karen Overall[2]

[1]Orion Corporation, Espoo, Finland, [2]Biology Department, University of Pennsylvania, Philadelphia, PA, USA, [3]Veterinary Specialty Practice for Behavior, Hamburg, Germany

Distress, fear, and anxiety during exposure to loud noises (e.g., storms, fireworks) are behavioral and welfare concerns for dogs, and affect up to 50% of dogs over their lifetime. Our objective was to confirm that a sub-sedative dose of dexmedetomidine, an alpha-2 adrenoceptor agonist, alleviates acute anxiety and fear associated with noise in dogs.

This randomized, double-blind, placebo-controlled, parallel-group study was conducted at 17 sites in 2 European countries on New Year's Eve. 182 dogs with a history of anxiety or fear due to fireworks received 125 $\mu g/m^2$ dexmedetomidine as an oromucosal gel (89 dogs) or placebo (93 dogs). Owners applied the first dose before or at the onset of distress associated with fireworks. Dosing could be repeated up to 4 times, every 2 hours. Owners assessed treatment effect and dog's behavior. They also reported oral mucosal color, observational and functional alertness, adverse events, and ease of use.

An excellent or good treatment effect was seen in 64 (72%) dexmedetomidine-treated dogs and in 34 (37%) placebo-treated dogs. The overall treatment effect was superior with dexmedetomidine ($P<0.0001$). Dogs on dexmedetomidine displayed less panting, trembling, and hiding. Transient pale mucous membranes were observed more frequently with dexmedetomidine, but no mucosal irritation was reported. Most dogs (82/89, 92%) remained functional and no adverse events of concern were observed. Most owners (154/182, 85%) considered the product very easy or quite easy to use.

This study confirms that oromucosal dexmedetomidine at a sub-sedative dose alleviates acute anxiety and fear associated with noise in dogs.

OT03

PROGNOSTIC INDICATORS IN CATS WITH SEPTIC PERITONITIS: 44 CASES (2002-2015)

Katherine Scotti[1], James Barr[1], Micah Bishop[3], Michael Kato[2], Matthew Mellema[2], Medora Pashmakova[1]

[1]*Texas A&M, College Station, TX, USA,* [2]*UC Davis, Davis, CA, USA,* [3]*Animal Specialty Hospital of Florida, Naples, FL, USA*

Septic peritonitis (SP) in cats carries significant mortality (12-70%) and etiologies may be indeterminate. The aim of this study was to identify etiologies and prognostic factors in cats with SP.

Medical records of cats hospitalized for SP between 2002 and 2015 at Texas A&M University and University of California (Davis) were reviewed. Septic peritonitis was diagnosed by intracellular bacteria on cytology or positive bacterial culture. Data was checked for normality and compared using an unpaired t-test and a Mann-Whitney test. Fisher's exact test and Odds ratios were used to compare variables. Alpha was set at 0.05.

Seventy-nine cats were identified and 55 met inclusion criteria. The source of SP was undetermined in 28/55 cats. Gastrointestinal perforation was the cause of SP in 16/55 cats. Thirty of 44 (68.18%) cats that were surgically managed and 5/11 (45.45%) that were medically managed survived to discharge. In this population, there was no association between leukocytosis, leukopenia, anemia, hypoalbuminemia, hyperlactatemia, ionized hypocalcemia, or band neutrophilia and survival. Cats with increased creatinine were 4.68 times more likely to die (95%CI:1.40-15.69; P=0.012). Hypothermic cats were 4.67 more likely to die (95% CI: 1.36-15.98; P=0.014), while febrile cats were 5.409 times more likely to live (95% CI: 1.58-18.58; P= 0.0063).

The origin of SP was undetermined in half of the cats in this study. Survival was in agreement with previous reports. Azotemia and hypothermia, which may reflect persistence of shock in this species, should be closely monitored in septic cats.

OT04

RELIABILITY OF THERMOMETER PROTECTIVE SHEATHS FOR MEASUREMENT OF RECTAL TEMPERATURE IN DOGS

Franck Jolivet[1], Maud Pic[1], Didier Concordet[2], Olivier Dossin[3]

[1]Department of Small Animal Clinical Sciences, INP - National Veterinary School and Federal University of Toulouse Midi Pyrénées, Toulouse, France, [2]UMR1331 Toxalim, INP - National Veterinary School and Federal University of Toulouse Midi Pyrénées, Toulouse, France, [3]Department of Small Animal Clinical Sciences and IRSD INSERM 1220, INP - National Veterinary School and Federal University of Toulouse Midi Pyrénées, Toulouse, France

Rectal temperature is commonly used in clinical evaluation of canine patients. Single-use lubricated protective thermometer sheaths are largely used in human medicine but this practice is not routinely reported in veterinary medicine. Therefore the aim of our study was to compare rectal temperature measured with the same thermometer with and without protective sheaths in a large canine population.

Three hundred and eighty two dogs were included in the study. Rectal temperature was assessed in a standardized manner by two persons (FJ and MP) with two identical digital rectal thermometers. The thermometers were validated for accuracy and repeatability (within-day variability) with a high precision thermometer. The thermometers validation was performed with and without protective sheaths. The rectal temperature was measured consecutively in each dog with and without protective sheaths in a randomized order. The difference of temperature measured with the two methods was analyzed with a generalized linear model with order of temperature measurement, age, body condition score (underweight, normal and overweight), format (stratified as mini, medium and large) and conscience level (conscious, sedated and anesthetized) as explanatory variables. Results are presented as mean ± SD.

The accuracy of the study thermometers was excellent with average temperature differences of 0.067 ± 0.064 and 0.066 ± 0.054 and ranging from -0.28 to +0.15 and from -0.22 to +0.1 with and without sheath respectively. In the field study, the difference of the temperatures measured with and without sheaths were within the -0.1 and +0.1 °C range for 80% of the measurements. Moreover, these differences were ≤ -0.5°C and ≥ +0.5°C for 3 and 5 dogs respectively. Except for the body condition score none of the explanatory variables studied were significantly associated with the differences measured suggesting that the differences were more important in overweight dogs that in normal dogs with average differences of -0.087 and -0.009°C respectively ($p=0.013$).

This study shows that thermometer protective sheaths can be used reliably to measure rectal temperature in dogs in a clinical setting. Therefore, it is strongly advised to use this hygienic protective measure in daily clinical practice because even in overweight dogs the differences between the two methods are clinically non significant.

OT05

ISOLATION AND IDENTIFICATION OF MICRORNA FROM THE MATURE FELINE ERYTHROCYTE: A PILOT STUDY

Katrina Stewart, Pierre Deshuillers, Joanne Messick

Purdue University, West Lafayette, IN, USA

MicroRNAs have been identified as sensitive, non-invasive diagnostic markers for early detection of human malignancies. To date there have been very few studies of microRNA expression in the serum of domestic animals with no known analysis of the microRNA present in mature erythrocytes. Feline erythrocytes are known to be sensitive to oxidative damage and we are interested in their potential use as a model for the study of systemic oxidative stress. Identification of specific microRNA which are dysregulated under these conditions may be sensitive biomarkers of oxidative stress.

This pilot study was aimed at identifying the microRNAs expressed in mature feline erythrocytes. This specifically involved isolation of a purified population of mature erythrocytes from cat blood; extraction of high quality microRNA from the purified population of mature erythrocytes and sequencing of the microRNA population to form a library from which to facilitate further study.

We successfully purified mature feline erythrocytes from whole blood with a combination of centrifugation and leukoreduction with a specialized filter. This removed a majority of the reticulocytes, white blood cells and platelets, however a final purification step utilizing magnetic beads coated with antibodies was subsequently used to ensure a pure mature population of erythrocytes. For this immune-depletion step, antibodies were initially tested for reactivity in the feline, then biotinylated for use within the columns. The antibodies used in this study were: anti-CD71 (anti-transferrin receptor to remove reticulocytes); anti-CD18 (anti-beta2 integrin to remove leukocytes) and anti-CD61 (anti-beta 3 integrin to remove platelets). Assessment of the mature red blood cell preparation purity was performed with an automated flow cytometry-based hematology analyzer to identify the presence of leukocytes, platelets and reticulocytes.

Isolation of total RNA was performed with a modified TRIzol phase separation method with concentration and quality of the RNA assessed prior to library construction. A small RNA library was constructed with cluster generation and sequencing followed by quality trimming and mapping of reads to the *Felis catus* genome.

With this study we have successfully isolated mature feline erythrocytes from whole blood samples, extracted quality microRNA from the isolated erythrocytes and developed a library of microRNA present in these cells in a normal population of cats. Further work will establish comparison of the microRNA profile between this normal population of cats and those at risk of or with evidence of oxidative damage. This will help to identify new and potentially informative biomarkers for systemic redox status and complications associated with oxidative stress in cats.

OT06

EFFECTS OF SEDATION AND ANESTHESIA ON CANINE HEMATOLOGIC AND SERUM BIOCHEMICAL ANALYSES DURING PREVENTIVE HEALTHCARE

Laura McPhee, Xuemei Si, Carolyn Cupp

Nestle Purina Research, St. Louis, MO, USA

This study aims to determine the effects of sedation and anesthesia on complete blood count (CBC) and serum chemistry results in dogs undergoing routine health and wellness care. Ten healthy adult dogs were selected based on physical exams and the need to undergo anesthesia for routine dental prophylaxis. Blood was collected by venipuncture at a series of timepoints: a) before receiving a premedication, b) prior to induction of anesthesia, c) 5 minutes after induction of anesthesia, d) 15 minutes after induction of anesthesia, and e) 45 minutes after induction of anesthesia. Premedication was a combination of acepromazine (0.025 mg/kg) and butorphanol (0.2 mg/kg) given as a single subcutaneous injection 20 minutes prior to induction with propofol. Propofol was calculated at 4 mg/kg and administered intravenously to effect. Anesthesia was maintained with inhalation of isoflurane gas via endotracheal intubation. Generalized linear mixed models were conducted to compare parameters between various time points to account for the correlated data (the same dog participated in multiple time points). P-values less than 0.05 are considered as statistically significant. When compared to blood results prior to premedication, statistically significant changes were observed in average hematocrit, hemoglobin, mean cell volume, red blood cell, albumin, cholesterol, and protein values after premedication alone. After induction with propofol, variation was additionally observed with white blood cell count, alkaline phosphatase, calcium, globulin, potassium, triglycerides, and whole blood DNA concentration. After 15 minutes of maintenance on isoflurane gas, mean cell hemoglobin concentration, monocytes, phosphorus, and bilirubin values were additionally affected. This study highlights the significant effect that sedation and anesthesia have on blood analyses and subsequent interpretation and clinical relevance.

OT07

DOES FINE NEEDLE ASPIRATION AFFECT MANAGEMENT OF DOGS WITH INCIDENTAL SPLENIC NODULES OR HETEROGENEOUS PARENCHYMA?

Igor Yankin, Sarah Nemanic, Silvia Funes, Helio de Morais

Oregon State University, Corvallis, Oregon, USA

Splenic nodules and heterogeneous splenic parenchyma are relatively common findings in older dogs undergoing abdominal ultrasound. The impact of fine needle aspiration and cytology in the management of those patients is not clear.

The aim of this retrospective study was to determine how often results of ultrasound-guided splenic fine-needle aspirates impact case management in adult dogs with splenic nodules or heterogeneous splenic parenchyma.

Medical records from adult dogs in which ultrasound-guided splenic fine-needle aspirates were performed at the Oregon State University Veterinary Teaching Hospital from June 2011 through July 2015 were retrospectively reviewed. Dogs were included if splenic nodules or heterogeneous splenic parenchyma was incidentally found under ultrasonographic examination. Only patients with splenic fine-needle aspirates performed on the same day of the ultrasonographic examination were included. Dogs were not included if a disease associated with splenic abnormalities was known to be present before the ultrasound.

One hundred and twenty six adult client-owned dogs met the inclusion criteria. Fine-needle aspiration changed the clinical management of 25 of 126 dogs (20%). Those included patients diagnosed with malignant neoplasia or suppurative inflammation: lymphoma (n=10), sarcoma (n=3), malignant histiocytosis (n=3), suppurative inflammation (n=3), epithelial neoplasm (n=2), melanoma (n=1), hemangiosarcoma (n=1), carcinoma (n=1), and spindle cell tumor (n=1). In the remaining 101 dogs (80%), fine-needle aspiration of the spleen did not change the management of the case. Those include patients with extramedullary hematopoiesis (n=79), lymphoid reactivity/hyperplasia (n=11), normal spleen (n=9), hyperplastic nodule (n=1), and non-diagnostic sample (n=1).

Splenic fine-needle aspirate is a minimally invasive diagnostic modality providing information that changed case management in 1/5 adult dogs with splenic nodules or heterogeneous splenic parenchyma, and should be considered in this population of patients.

P01

ITRACONAZOLE ABSORPTION FROM PROPRIETARY AND COMPOUNDED FORMULATIONS IN HEALTHY CATS

Dianne Mawby[1], Leanne Fowler[1], Mark Papich[2], Jacqueline Whittemore[1]

[1]University of Tennessee, Knoxville, TN, USA, [2]North Carolina State University, Raleigh, NC, USA

Itraconazole is available in FDA-approved oral capsules and solution (Sporanox®), as well as compounded formulations. The purpose of this study was to determine absorption of proprietary and compounded itraconazole in cats.

After a 12 hour fast, 8 cats received 50mg (~12.7mg/kg) of itraconazole (Sporanox® capsule, Sporanox® solution, compounded capsule, and compounded suspension) in a randomized crossover with a 21-day washout. Capsules were administered with a small meal whereas solution was not. Plasma was collected at pre-determined intervals for analysis using high pressure liquid chromatography. Compartmental and non-compartmental pharmacokinetic analyses were performed using Phoenix® software (Certara) on the proprietary Sporanox® and compounded formulation data, respectively.

For the Sporanox® capsule, the peak concentration (C_{MAX}) and half-life (t½) were 0.54µg/mL (37% CV) and 18 hours (21%CV), respectively. These values for Sporanox® solution were 1.82µg/mL (50% CV) and 23.7 hours (42%CV), respectively. Absorption of the Sporanox® solution was 4.2 times greater than for the Sporanox® capsule based on AUC ratios, and C_{MAX} was 3.4 times higher for the solution. Compounded formulations were absorbed inconsistently, commonly resulting in undetectable concentrations. Complete pharmacokinetic results were obtained from only 3 and 1 cats in the compounded capsule and suspension groups, respectively. Relative absorption of compounded itraconazole was 8% from capsule and 2% from suspension.

The long half-life of itraconazole in cats allows once daily oral dosing. Due to much higher absorption of Sporanox® solution, versus capsules, the oral solution dose can be substantially reduced. Compounded itraconazole should not be used in cats because of poor oral absorption.

P02

BIOAVAILABILITY OF A NOVEL FORMULATION OF S-ADENOSYLMETHIOINE GIVEN WITH FOOD IN BEAGLE DOGS

David Griffin, Carolyn Warner

Nutramax Laboratories Veterinary Sciences, Inc., Lancaster, SC, USA

S-adenosylmethionine (SAMe) is used to support liver function both in companion animals and humans, and is also used in humans for mental and joint health. Previous studies have shown that when SAMe is administered with food there is a substantial decrease in SAMe bioavability (Table 1). In a blinded crossover study a new Denamarin® (Nutramax Laboratories Veterinary Sciences, Inc.) chewable tablet containing a novel formulation of SAMe salt (NMXSS75A™) was administered to beagle dogs to assess its bioavailability relative to administration with food. Six beagles were dosed after an overnight fast with 225 mg SAMe ion (mean 21.7 mg/kg) in one formulation of the supplement, followed by the other formulation 7 days later administered with food. Plasma values for SAMe before dosing and at 0.5, 1, 2, 4, 6, 8, and 24 hrs were determined using an HPLC-ESI-MS/MS assay. With both supplements SAMe plasma levels increased substantially above baseline values (21-38 ng/ml) and returned to baseline after 24 hrs. Mean values for C_{max}, T_{max}, AUC_{0-24} and $t_{1/2}$ were comparable for both formulations with no statistically significant differences (Table 2). The data indicate that this novel SAMe salt may be used in dogs with equivalent bioavailability when administered with food which may increase owner compliance.

Table 1: Effects of Food Consumption on SAMe Tosylate Absorption.

Grams of Food	Cmax	± Std
0	1,736	907
100	713	350
225	562	512

Table 2: Effects of Novel SAMe Salt with Food Consumption Compared to SAMe Tosylate fasted.

	SAMe Tosylate Chewable Tablet (Mean ± Std)	Novel Salt NMXSS75A™ Chewable Tablet (Mean ± Std)
	Fasted (n=6)	**Fed** (n=6)
C_{max} (ng/ml)	1,202 ± 365	1,157 ± 683
T_{max} (hr)	1.5 ± 0.5	1.9 ± 2.0
AUC_{0-24} (ng•hr/ml)	7,060 ± 1,431	4,579 ± 1,454
$t_{1/2}$ (hr)	9.03 ± 3.39	10.78 ± 4.57

P03

POSACONAZOLE PHARMACOKINETICS IN CATS AFTER ORAL AND IV ADMINISTRATION

Dianne Mawby[1], Leanne Fowler[1], Mark Papich[2], Jacqueline Whittemore[1]

[1]University of Tennessee, Knoxville, TN, USA, [2]North Carolina State University, Raleigh, NC, USA

Posaconazole is the most active available azole antifungal drug, but absorption and pharmacokinetics are not available to guide dosing regimens in cats. The purpose of this study was to determine pharmacokinetics of oral suspension and IV formulations of posaconazole in cats to determine the feasibility of oral administration for treatment of fungal infections.

After a 12 hour fast, each of 6 healthy cats received 15 mg/kg of posaconasole (Noxafil®) oral suspension with food. Four cats also received 3 mg/kg IV posaconazole after a 7-day washout period. Plasma was collected at predetermined intervals for analysis using high pressure liquid chromatography (HPLC). Concentration data were analyzed using a two-compartment pharmacokinetic analysis for IV data and a one-compartment analysis with first-order input for oral data using Phoenix® software (Certara).

After IV dosing, volume of distribution (V_{SS}) was 1.9 (0.3) L/kg (mean, std. deviation), terminal half-life (t½) was 39.4 (15.9) hours, and clearance was 28.1 (17.3) mL/kg/hr. After oral dosing, peak concentration (C_{MAX}) was 1.2 (0.5) µg/mL and terminal t½ was 38.1 (15.0) hours. Bioavailability from oral administration was 15.9% (8.6). No adverse effects were observed from either route of administration.

Despite low oral absorption, these data allow for simulation of oral dosage regimens that could be explored in clinical studies. Two regimens can be considered to maintain targeted trough concentrations of 0.5-0.7 µg/mL: 1) 30 mg/kg PO loading dose followed by 15 mg/kg q 48 hours, or 2) 15 mg/kg PO loading dose followed by 7.5 mg/kg q 24 hours.

P04

EVALUATION OF INTRAVENOUS MYCOPHENOLATE MOFETIL USE IN HEALTHY CATS

J.E. Slovak, S.M. Rivera, J.K. Hwang, M.H. Court, N. Villarino

Washington State University, Pullman, WA, USA

Mycophenolic acid (MPA) is the active moiety of the prodrug mycophenolate mofetil (MMF). It is a selective non-competitive inhibitor of inosine-5'-monophosphate dehydrogenase. MPA is an attractive immunosuppressant option in veterinary medicine due to its rapid onset and commercial availability in multiple formulations. Although MPA is used routinely in humans and canines, there is a paucity of literature supporting MPA's use in feline patients, likely from cats' limited ability to glucuronidate certain medications. Our goal for this study was to evaluate the pharmacokinetics, pharmacodynamics, and safety of MPA in 9 healthy cats.

The pharmacokinetics of MPA was evaluated following an intravenous infusion of 5 mg/kg (n=2) and 10 mg/kg (n=1) of MMF (CellCept® intravenous, Roche Lab Inc.) Surprisingly, the plasma concentration of MPA in cats at these doses were remarkably low compared with the plasma levels consistent with immunosuppressive levels in humans. Therefore, we evaluated the PK of MPA using 20 mg/kg of MMF administered twice 12 hs apart (n=6). Following the infusion, all cats metabolized MMF into MPA. The maximum observed MPA plasma concentration ranged between 3.4 and 22 µg/mL after the first dose and between 6.7 and 20 µg/mL, following the second dose. The evaluation of the concentration time profile of MPA in plasma revealed additional peaks of MPA which could correspond to enterohepatic recycling of MPA.

Mycophenolic acid is eliminated by liver biotransformation. Unexpectedly, the concentration of MPA in plasma decayed quickly (terminal half-life ~1.5 hs) in all the cats, suggesting a fast rate of biotransformation, explaining the higher dose of MMF. The metabolite MPA phenol glucuronide was quantifiable in 2 out of 6 cats while acyl glucuronide was not detected in any of the cats. In contrast, the metabolite MPA phenol glucoside was quantifiable in plasma from all the cats, suggesting that glucosidation of MPA is a key route for MPA elimination in cats.

The administration of MMF showed a rapid effect on immune cells. After the second dose of MMF, there was a reduction of the peripheral blood mononuclear cells (PBMC) by ~24%. The PBMCs recovered to baseline levels ~36 hs after stopping the infusion of MMF. This indicates that repeated administrations of MMF are required to maintain a constant suppression of immune cells in cats. Importantly, there were neither serum biochemistry changes nor significant adverse events associated with the administration of MMF in any of the cats.

This study is the first step towards improving treatment of various autoimmune diseases in cats. The results of this study suggest that the infusion of MMF once at 20 mg/kg 12 hs apart, provides a safe and rapid immunosuppressive effect in cats. Future studies will evaluate the pharmacokinetics and the effect of MPA on target lymphocyte subpopulations ($CD25^{+}CD8^{+}$ and $CD25^{+}CD4^{+}$ T cells) following multi-day intravenous and oral dosing of MMF.

P05

THE ACCURACY, PRECISION AND STABILITY OF COMPOUNDED MILBEMYCIN OXIME IN AQUEOUS SUSPENSION

Zachary Cochrane, Darren Berger, Dwayne Schrunk, Johann (Hans) Coetzee

Iowa State University College of Veterinary Medicine, Ames, IA, USA

Production of a veterinary proprietary product solely containing milbemycin oxime (MO) for use in the United States was temporally ceased in 2011, resulting in use of compounded MO formulations for the treatment of canine generalized demodicosis. Despite the return of a labeled single agent product to the veterinary marketplace, compounded formulations are still available through national compounding pharmacies. This study describes the accuracy, precision and stability of MO strength when compounded as an aqueous suspension (20 mg/mL).

Preparation choice reflected current prescribing practices and routinely utilized concentration. Samples were acquired by prescription from two national veterinary compounding pharmacies at three time points spanning a 30-day period. Two different storage conditions were evaluated and sampled at four time points from the order date (Day 7, 14, 21, and 28). MO recovery was performed by solid-phase extraction and concentration strength measured via high-performance liquid chromatography.

Overall, accuracy ranged from 67.5 to 135.5% with pharmacy A having a range of 67.5 to 91.5% and pharmacy B 78.5 to 135.5%. Of the initial Day 7 samples, only 3 of 12 were within the United States Pharmacopeia's guideline defining acceptable range of 90 to 110% compared with labeled strength. The precision for pharmacy A was 1.0 to 16.9% while for pharmacy B it was 3.9 to 32.8%. Storage condition did not affect stability in the tested aqueous suspensions. The average decrease in MO concentration in pharmacy A's samples was 21.8% (13.3 to 26.1%) while from pharmacy B it was 18.6% (9.6 to 27.7%).

Compounded MO suspension deviated by more than 10% from their labeled strength, which could lead to potential over-dosage and severe neurologic toxicity in predisposed breeds. Bioavailability and clinical efficacy of compounded MO suspensions still remain unknown, and the use of these products should be discouraged at this time.

P06

PHARMACOKINETICS OF INTRAVENOUS AND SUBCUTANEOUS DOLASETRON AND PHARMACODYNAMICS OF SUBCUTANEOUS DOLASETRON IN PURPOSE-BRED CATS

Andrea Herndon, Liberty Sieberg, Leigh Davis, Amber Caress, Ryan Hansen, Luke Wittenburg, Daniel Gustafson, Jessica Quimby

Colorado State University, Fort Collins, CO, USA

Dolasetron is a 5-HT_3 receptor antagonist anti-emetic and is typically dosed at 0.5-1 mg/kg IV or SQ once daily. Pharmacokinetic (PK) and pharmacodynamic (PD) studies in cats have not been previously performed. The purpose of this study was to evaluate these parameters in purpose-bred cats.

Five purpose-bred cats with unremarkable complete blood count, serum biochemistry and urinalyses were utilized. PK study: Each cat received 0.8mg/kg subcutaneous (SQ) and intravenous (IV) dolasetron in a cross-over manner. Serum samples were obtained via jugular catheter at 0, 0.25, 0.5, 1, 2, 4, 8, 12, 24, 36 and 48 hours after administration of dolasetron. Dolasetron and the active metabolite hydrodolasetron were measured using liquid chromatography/tandem mass spectrometry. Non-compartmental pharmacokinetic analysis was performed. PD study: Subcutaneous dolasetron (0.8mg/kg or 1.0mg/kg) or placebo was administered 30 minutes prior to intramuscular administration of 0.44 mg/kg xylazine in a randomized crossover manner. Number of emetic events, lip licks, time to onset of vomiting or visual nausea score were scored by a blinded observer.

PK: Dolasetron was quickly metabolized to hydrodolasetron with only two cats in each administration route having measurable concentrations beyond 2 hours, limiting the assessment of dolasetron PK parameters. There was no significant difference in exposure to the active metabolite between the two routes of administration. Pharmacokinetic parameters are summarized below.

	IV administration		SQ administration	
Parameter	Dolasetron	Hydrodolasetron	Dolasetron	Hydrodolasetron
C_{max} (ng/ml)	761.4 ± 518.9	140.6 ± 106.0	196.6 ± 253.1	78.3 ± 26.1
T_{max} (hr)	0.3 ± 0.1	0.4 ± 0.1	0.3 ± 0.1	0.63 ± 0.3
T_{last} (hr)	5 ± 4.8	13.6 ± 6.1	3.8 ± 3.1	18 ± 6.9
$T_{1/2}$ λ (hr)	NP	4.2 ± 1.8	NP	4.0 ± 1.0
AUC $_{0-T}$ (hr*ng/mL)	325.0 ± 210.0	314.0 ± 130.9	92.0 ± 86.7	429.4 ± 165.4

PD: When dolasetron or placebo was administered prior to xylazine, there was no significant difference in the mean number of emetic events, lip licks, time to onset of vomiting or visual nausea score when compared to placebo.

At 0.8 mg/kg dolasetron does not maintain serum concentrations for 24 hours and does not adequately control xylazine-induced vomiting when given SQ even at 1 mg/kg. Additional dose studies are needed to determine if a higher dose is more effective.

P07

INVESTIGATION OF THE PHARMACOKINETICS OF TRANSDERMAL ONDANSETRON IN NORMAL PURPOSE-BRED CATS

Lara Zajic, Andrea Herndon, Liberty Sieberg, Amber Caress, Ryan Hansen, Daniel Gustafson, Jessica Quimby

Colorado State University, Fort Collins, CO, USA

Ondansetron is a 5-HT3 receptor antagonist used as an anti-emetic in ill cats. Ondansetron can be dosed orally, IV, or subcutaneously but has previously been demonstrated to have poor oral bioavailability and a short elimination half-life requiring frequent dosing (every 6-8 hours). Because of the impracticality of dosing (>3 doses/day) at home, it is often resigned to in-hospital administration. Ondansetron is a candidate for a transdermal medication because it is small in size (294 Daltons) and is moderately lipophilic (log p ~ 2.1). The purpose of this study was to assess the pharmacokinetics of transdermal ondansetron administration in healthy, purpose-bred cats.

Five purpose-bred cats with unremarkable CBC, biochemistry and urinalysis were utilized. 2 mg transdermal ondansetron Lipoderm gel was applied once to the internal ear pinna (total volume of 0.1ml). Blood samples were collected via jugular catheter over a 48 hour period following administration (0, 15min, 30min, 1hr, 2hr, 4hr, 8hr, 12hr, 24hr and 48hr). Serum was separated and frozen prior to analysis. Ondansetron was measured via liquid chromatography coupled to tandem mass spectrometry.

Analysis revealed no appreciable drug was present in serum after transdermal administration of 2 mg ondansetron, indicating that this is not an acceptable method of drug delivery despite the characteristics of the drug that imply that it would adequately pass the skin barrier.

This study highlights the importance of assessing the potential of each medication for transdermal administration.

R01

COMPARISON OF BRONCHOALVEOLAR LAVAGE TECHNIQUES FOR SAMPLING LOWER AIRWAYS IN CATS

Kimberly Hooi, Alice Defarges, Andrea Sanchez, Stephanie Nykamp, Anthony Abrams-Ogg, Scott Weese, Dorothee Bienzle

Ontario Veterinary College, University of Guelph, Guelph, ON, Canada

The purpose of this study was to compare bronchoscopic bronchoalveolar lavage (B-BAL) and non-bronchoscopic bronchoalveolar lavage (NB-BAL) in clinically healthy cats.

Two bronchoalveolar lavages (BAL) were performed 2 weeks apart in twelve cats free of respiratory disease. BALs were performed in a randomized order under general anesthesia by wedging a 2.9mm fiber-optic bronchoscope (B-BAL) or an 8Fr red rubber catheter (NB-BAL). Ten mL aliquots of saline were infused into the left and right caudal lung fields and aspirated with a 20mL syringe. Proportion of fluid retrieved, depth of wedging and anesthetic complications were recorded. Total nucleated cell count, differential cell count and semi-quantitative evaluation of cytologic quality were performed on all BALF samples. Results were compared using ANOVA and Wilcoxon signed-rank tests.

Proportion of BALF retrieved and depth of wedging was significantly greater for B-BAL than NB-BAL. Differential cell counts and cytologic quality did not differ significantly between techniques. Results were [median (range)]:

		NB-BAL	B-BAL	P-value
Proportion of BALF retrieved (%)		55.0 (20.0-80.0)	70.0 (10.0-100.0)	0.008*
Depth of wedging (mm)		81.6 (67.0-93.0)	90.3 (72.8-122.0)	<0.001*
Differential cell count (%)	Macrophages	68.5 (24.0-90.0)	70.0 (41.0-91.0)	0.851
	Eosinophils	13.0 (0.0-73.0)	10.0 (0.0-28.0)	0.451
	Lymphocytes	5.0 (1.0-17.0)	5.5 (1.0-23.0)	0.853
	Neutrophils	8.5 (0.0-36.0)	9.5 (2.0-36.0)	0.270

* p<0.05

Complications included transient hemoglobin desaturation (24/24 BALs) and prolonged anesthetic recovery time (4/24 BALs). However, anesthetic recovery scores did not differ significantly between techniques.

The results suggest that NB-BAL procedure provided BALF samples of equivalent quality to B-BAL procedure in cats clinically free of respiratory disease.

R02

CHARACTERIZATION OF THE FELINE RESPIRATORY MICROBIOME

Aida I. Vientos-Plotts[1], Aaron Ericsson[2], Hans Rindt[1], Megan Grobman[1], Amber Graham[1], Kaitlin Bishop[1], Leah Cohn[1], Carol Reinero[1]

[1]*University of Missouri College of Veterinary Medicine Comparative Internal Medicine Laboratory, Columbia, MO, USA,* [2]*University of Missouri Metagenomics Center, Columbia, MO, USA*

Although lower airways have been considered sterile by culture, sequencing 16S rRNA amplicons has revealed diverse bacterial species in human lungs. Using healthy research cats in a stable environment, we hypothesized that 16S rRNA amplicon sequencing of DNA from bronchoalveolar lavage (BAL), oropharyngeal and fecal swabs, and blood, would reveal distinct microbiomes from each site.

Six healthy research cats <1 year old had BAL, oropharyngeal and fecal samples collected at day 0, week 3 and 8; and blood week 8. Extracted DNA underwent PCR of the 16S rRNA gene. Once sequenced, number of reads/sample, richness and relative abundance of representative operational taxonomic units (OTUs) were determined. Principle component analysis (PCA) described relatedness of samples.

Fecal and oropharyngeal swabs yielded a mean±SD of 65,652±26,069 and 20,632±18,499 reads/sample, respectively. BAL and blood yielded lower coverage at 1,489±1,822 and 269±43 reads/sample, respectively. Oropharyngeal and fecal swabs were significantly richer than BAL (mean number OTUs 93, 88 and 36, respectively; $p<0.001$) with no significant difference ($p=0.180$) in richness between time points. *Fusobacterium sp.*, *Campylobacter sp.*, and *Bacteroides sp.* were predominant fecal taxa. *Pseudomonas sp.* was most abundant in the oropharynx (day 0) and *Pasteurellaceae* and *Porphyromonas sp.* were most abundant weeks 3&8. In BAL, *Pseudomonas sp.*, *Bradyrhizobiaceae*, and *Sphingobacteriaceae* predominated day 0 and week 3 with abundance of *Pseudomonas sp.* declining week 8. *Bradyrhizobiaceae* and *Sphingobacteriaceae* were predominant taxa in blood. PCA revealed distinct microbiomes in each site; however, blood had marked compositional similarity with BAL. Samples clustered more by time than by individual, with oropharyngeal swabs having subjectively greater variation than other samples.

Healthy cats have a rich and distinct airway microbiome with dynamic bacterial populations likely depending on age and environment. Further data is necessary to determine how the distinct feline microbiomes from the upper and lower airways, feces and blood are established and evolve. These data are relevant for comparisons between healthy cats and cats with respiratory disease.

E01

PULMONARY DISPOSITION AND PHARMACOKINETICS OF ORAL MINOCYCLINE IN THE ADULT HORSE

Kate Echeverria[1], Kara Lascola[1], Steve Giguere[2], Jon Foreman[1], Scott Austin[1], Pamela Wilkins[1]

[1]University of Illinois College of Veterinary Medicine, Urbana, IL, USA, [2]University of Georgia College of Veterinary Medicine, Athens, GA, USA

The objective of this study was to determine the pharmacokinetics and pulmonary disposition of minocycline in adult horses after multiple intragastric doses.

Minocycline (4 mg/kg) was administered via the intragastric route to 6 healthy adult horses every 12 hours for a total of 5 doses. Minocycline concentrations were measured in plasma, pulmonary epithelial lining fluid (PELF), and bronchoalveolar lavage (BAL) cells via liquid chromatography mass spectrometry at multiple time points over a 96h time period. Pharmacokinetic variables were determined after the last dose using a noncompartmental approach. Comparison of minocycline concentrations between sampling sites was done using a Friedman repeated measures ANOVA and significance was set at $P < 0.05$.

After the last dose, the mean + s.d. maximum plasma concentration (Cmax) of minocycline was 2.3 + 1.3 µg/mL and the terminal half-life (T1/2) was 11.8 + 0.5 hrs. The median (25th and 75th percentiles) time to peak concentration (Tmax) was 1.3 (1.0 – 1.5) hrs. The Cmax and Tmax of minocycline in the PELF were 10.5 +12.8 µg/mL and 9.0 (5.5 – 12.0) hrs, respectively. The Cmax and Tmax for BAL cells were 0.24 + 0.1 µg/mL and 6.0 (0.0 – 6.0) hrs, respectively. Peak and trough minocycline concentrations in PELF were significantly higher than in plasma whereas concentrations in BAL cells were significantly lower.

Orally administered minocycline distributes into the PELF and BAL cells of the horse. Knowledge of specific minimum inhibitory concentrations is required to predict efficacy of minocycline for the treatment of bacterial pneumonia in adult horses.

E02

INTRAVENOUS ADMINISTRATION OF COBALT CHLORIDE IS ASSOCIATED WITH HEMODYNAMIC ALTERATIONS IN HORSES

Teresa Burns, Kasia Dembek, Laura Dunbar, Sara Brewington, Linda Bednarski, Carl O'Brien, Turi Aarnes, Brice Dooley, Jeffrey Lakritz, Ramiro Toribio

The Ohio State University, Columbus, OH, USA

Cobalt is a substance of abuse in humans and animals performing in strenuous athletic competitions. When administered at pharmacologic doses, it has been associated with induction of erythropoietin release, increased hematopoiesis, which is thought to confer competitive advantage. Cobalt chloride ($CoCl_2$) is reportedly given to racehorses to enhance performance, and recently, allowable limits for cobalt have been set by several racing jurisdictions for post-race illicit substance testing of blood and urine. While preliminary single-dose pharmacokinetic data have been published for $CoCl_2$ in horses, information regarding the effects of repeated dosing (which is how the substance is reportedly used illicitly in performance horses) is unavailable. Even fewer data have been published describing the pharmacodynamic effects of $CoCl_2$ administration in horses, particularly at high doses. The purpose of this pilot study was to describe the physiologic and biochemical effects of weekly intravenous doses of $CoCl_2$ to Standardbred horses. This report describes the hemodynamic effects of $CoCl_2$ in a dose escalation study.

Five Standardbred mares (12-13 years-old; 460-530 kg) were randomly assigned to receive one of 5 doses of $CoCl_2$ (4, 2, 1, 0.5, or 0.25 mg/kg) as an intravenous bolus (infused over 1 minute) once weekly for 5 weeks. Prior to each dose, animals were instrumented with pulmonary artery and right atrial catheters, a transverse facial artery catheter, two external jugular venous catheters, an indwelling urinary catheter, and electrocardiography leads. Physical examination parameters, blood pressure (systolic arterial pressure [SAP], diastolic arterial pressure [DAP], and mean arterial pressure [MAP]), cardiac output, and qualitative ECG assessment were evaluated every 5-10 minutes for 4 hours immediately after administration of the first and fifth weekly doses of $CoCl_2$.

All mares were subjectively anxious (nostril flaring, muscular tremor/fasciculation, pawing, straining) by 5 minutes following the $CoCl_2$ infusion; this persisted for ~60 minutes in mares receiving higher doses (4, 2, and 1 mg/kg). Mares receiving 4, 2, or 1 mg/kg doses developed tachycardia immediately after dosing (HR 60-126 bpm), but this was not observed in mares receiving 0.5 or 0.25 mg/kg (HR 36-52). Paroxysmal ventricular tachycardia was noted in the first 10 minutes post-administration in the mare receiving the 4 mg/kg dose. Elevations in SAP, DAP, and MAP were noted following drug administration at most doses; while profound hypertension was observed at 4 mg/kg (SAP/DAP, MAP [mmHg] = 291-300/163-213, 218-279), all mares became hypertensive in the 30-45 minutes following $CoCl_2$ administration. Mares receiving 4 and 2 mg/kg developed conspicuous oral mucous membrane congestion that persisted for 20 minutes post-dosing and subsequently resolved. At all doses, cardiovascular parameters returned to baseline by 1-2 hours post-administration.

Results of this preliminary study document significant, repeatable hemodynamic alterations associated with intravenous $CoCl_2$ administration to horses. Further, the degree of hypertension observed following infusion raises humane and human safety concerns if doses of >1 mg/kg are used.

E03

THE PHARMACOKINETICS OF INTRAVENOUS GENTAMICIN IN HEALTHY YOUNG-ADULT VERSUS GERIATRIC HORSES

Andrew Gestrich[1], Daniela Bedenice[1], Michelle Ceresia[2], Iman Zaghloul[2], Nisanne Ghonem[2], Mary Rose Paradis[1]

[1]Cummings School of Veterinary Medicine at Tufts University, North Grafton, MA, USA, [2]School of Pharmacy, MCPHS University, Boston, MA, USA

The study objective was to evaluate the effects of patient age on aminoglycoside pharmacokinetics in young-adult (5-10 years) versus geriatric (≥25 years) healthy horses, receiving a single 6.6mg/kg intravenous injection of gentamicin (Henry Schein®). Heparinized plasma samples were obtained at designated time-points following drug administration and frozen at -80˚C until assayed by a validated immunoassay (QMS®). Gentamicin plasma concentrations versus time plots were analyzed using a two-compartment model and commercial software (WinNonlin-v6.4). Data were compared between age-groups using independent samples t-test; $P<0.05$ considered significant.

Baseline physical examination and hematological parameters were not significantly different between young (median age [range]: 7 [6-8] years) and aged (26 [25-41] years) horses, aside from a lower mean bodyweight in the geriatric group (402±56 vs. 477±38 kg). Similarly, pharmacokinetic parameters were not statistically different between age groups (mean ±SD; $P>0.05$):

	C_0 (μg/mL)	$AUC_{0\text{-}\infty}$ (μg*h/mL)	$T_{1/2\alpha}$ (h)	$T_{1/2\beta}$ (h)	V_{ss} (L)	Cl (mL/h/kg)
Young-adult (n=8)	86.2±25.9	75.6±12.9	0.1±0.03	1.22±0.2	69.2±12.9	89.3 ±13.8
Geriatric (n=8)	86.6±21.8	82.7±10.3	0.11±0.04	1.42±0.33	60.2±12.2	80.8 ±9.4

C_0–Initial plasma concentration; AUC–Area under the curve; $T_{1/2}$ α and β–Distribution and elimination half-life; V_{ss}–Volume of distribution at steady state; Cl–Clearance

Peak plasma concentrations (C_{pk}) at the end of the distribution phase were 31.7±4.4μg/ml in young and 32±2.2μg/ml in aged horses (*P=0.901*), while gentamicin trough levels reached 2μg/ml (common MIC for equine pathogens) at 6.1±1.2 and 7.3±1.5 hours, respectively (*P=0.09*). Considering an expected post-antibiotic effect of ≤8hrs, dosing adjustments are likely required to meet therapeutic needs in both study groups.

E04

ENDOTOXIN-INDUCED MICRORNA EXPRESSION IN EQUINE PERIPHERAL BLOOD MONONUCLEAR CELLS

Nicholas Parkinson, Virginia Buechner-Maxwell, Sharon Witonsky, R. Scott Pleasant, Ansar Ahmed

Virginia-Maryland College of Veterinary Medicine, Blacksburg, Virginia, USA

The innate immune response to bacterial lipopolysaccharide (LPS) mediated by toll-like receptor 4 (TLR4) contributes substantially to the morbidity of a number of important equine disease syndromes including gastrointestinal disease and neonatal sepsis, and contributes to other chronic pro-inflammatory states. MicroRNAs (miRNAs) are small non-coding RNA molecules 17-25 nucleotides in length that function as post-transcriptional regulators of gene expression. miRNAs have recently been found to have a central role in the modulation of the TLR4 signaling cascade in other species, and altered expression has been implicated in the pathogenesis of inflammatory diseases. Limited information is available on miRNA expression in horses, and there is no published data on expression in equine inflammatory cells. This objective of this study was to use next generation sequencing to characterize the basal miRNA transcriptome in isolated equine peripheral blood mononuclear cells, and to test the hypothesis that LPS would induce differential expression of miRNAs.

Peripheral blood mononuclear cells were isolated from four healthy adult horses. Cells were cultured with LPS (10 ng/mL) or control for 0, 2 and 4 hours. Tumor necrosis factor alpha (TNFα) was measured in culture supernatants by ELISA. RNA was extracted from the cells and the miRNA transcriptome was sequenced using an Illumina HiSeq 2500 analyzer after size selection. High quality sequence data were mapped to the equine genome and annotated using the miRNA database miRBase 21. Differential expression was analyzed with generalized linear models, using the Benjamini-Hochberg procedure to control the false discovery rate at 0.1.

327 mature miRNAs were detected, with 191 present in all samples. Basal expression was highly consistent between horses. The most abundant miRNAs in baseline samples were miR-21, let-7g and miR-150, miRNAs associated with cell cycle regulation and both innate and adaptive immunity in other species. TNFα expression was significantly higher in the supernatants from LPS-treated cells than controls at both 2 and 4 hours ($p=0.016$ and 0.0003 respectively). After correction for multiple comparisons, only miR-155 was significantly upregulated by LPS ($p=0.00018$, 1.5 to 1.6 fold change versus controls). 9 miRNAs showed statistically significant expression changes with time. These included miR-146a and miR-146b, which are induced by LPS in other species but had non-significant upregulation by LPS here. MiR-155 expression was significantly correlated to supernatant TNFα ($R^2=0.78$).

The basal expression characterized by these data provide a foundation for future research into the roles of specific miRNAs in equine inflammatory responses. miR-155 is the principal LPS-induced miRNA in horses, as in humans and mice. This miRNA has documented roles in TLR4 signaling regulation, including enhancement of TNFα translation and suppression of intermediate signal transduction proteins. It is thus likely to influence the magnitude and nature of the acute inflammatory response to LPS in the horse, and could be a target for immune modulatory interventions. It may also prove a useful marker of TLR activation. Further research will be necessary to validate expression changes in a wider sample set, investigate targets and functions of these miRNAs in horses, and to establish roles in naturally occurring disease.

E05

SEVERE EQUINE ASTHMA (HEAVES) IS ASSOCIATED WITH AN INCREASED NUMBER OF CIRCULATING LOW-DENSITY GRANULOCYTES

Nicolas Herteman, Amandine Vargas, Jean-Pierre Lavoie

Department of Clinical Sciences, Faculty of Veterinary Medicine, University of Montreal, Saint-Hyacinthe, Québec, Canada

Airway neutrophilia is a characteristic finding of severe equine asthma (heaves), a common and incurable condition of horses. However, how polymorphonuclear neutrophils (PMNs) contribute to the disease remains poorly understood. “Low-Density Granulocytes” (LDGs), a subpopulation of PMNs, have been suggested to have strong pro-inflammatory properties, which can promote the recruitment of PMNs in an autocrine manner and thus amplifying the inflammatory response. We therefore hypothesized that LDGs are increased in the blood of heaves-affected horses compared to healthy subjects.

We studied 8 horses with heaves during disease exacerbation and remission, and 6 healthy animals. LDGs and normal-density neutrophils were isolated from peripheral blood using Ficoll. Quantification and morphological characterization of LDGs were performed blindly using an optical microscope. LDGs were further characterized by flow cytometry using anti-CD90 and anti-myeloperoxidase (MPO) antibodies.

The number of LDGs was significantly elevated in horses with heaves during both exacerbation and remission compared to healthy horses ($p=0.005$ and $p=0.029$, respectively). No significant difference was found between disease exacerbation and remission ($p>0.1$). LDGs displayed a significantly greater percentage of normally segmented than hyposegmented nuclei in heaves-affected and control horses ($p=0.0002$ and $p=0.0022$, respectively). MPO's mean fluorescence intensity of LDGs was significantly lower during heaves exacerbation when compared to control horses ($p=0.0442$). PMNs counts in BAL and LDG levels in blood of horses with heaves were not correlated ($p>0.1$ for both disease exacerbation and remission).

This study describes for the first time an association between an increased subpopulation of PMNs, namely the LDGs, and an inflammatory disease in horses. LDGs appear to be mostly constituted of mature cells in both healthy and heaves-affected horses. The significance of these findings remains to be ascertained.

E06

CAN LEVAMISOLE UPREGULATE THE EQUINE CELL-MEDIATED IMMUNE RESPONSE IN VITRO?

Amy Santonastaso[1], Bettina Wagner[2], Siobhan Ellison[3], Virginia Maxwell[1], Scott Pleasant[1], David Lindsay[4], Sharon Witonsky[1]

[1]Virginia-Maryland Regional College of Veterinary Medicine, Blacksburg, VA, USA, [2]College of Veterinary Medicine, Cornell University, Ithaca, NY, USA, [3]Pathogenes, Inc, Reddick, FL, USA, [4]Department of Biomedical Sciences and Pathobiology, Virginia Tech, Blacksburg, VA, USA

Equine Protozoal Myeloencephalitis (EPM) is arguably the most common and costly equine neurologic diseases in the United States. The national seroprevalence is >50%, but only 0.5-1% of all horses develops disease during their lifetimes. Some EPM affected horses have decreased immune responses. A cell-mediated immune response appears to be protective for development of EPM after infection with *S. neurona* in mouse models. Therefore, in addition to specific anti-protozoal medications, levamisole has been proposed as an adjunctive therapy to upregulate the immune response in EPM affected horses. Studies have been performed in other species to address whether levamisole alters immune function in vitro or in vivo, but there are very limited studies in equids. We hypothesized that We hypothesized that levamisole can upregulate a cell mediated macrophage (M1) dendritic cell (DC1) CD4 T-helper 1 (Th1) CD8 Tc1 immune response in vitro.

The first aim was to determine both the optimal conditions and effects of levamisole on cellular proliferation. Equine PBMCs were harvested from ten horses seronegative for *S. neurona*. Cells were cultured alone, or with each of the following mitogens: concanavalin A (conA), phorbol 12-myristate 13-acetate and ionomycin (PMA/I), or with a combination of the above mitogens and levamisole at several conditions. Cellular proliferation was assessed using a colorimetric bromodeoxyuridine ELISA assay (Roche Life Sciences, 11647229001).

The second aim was to determine the ability of levamisole, under optimized conditions, to upregulate the M1 DC1 CD4Th1 CD8 Tc1 response in vitro, based on activation and function.

PBMCs from the same 10 horses were cultured with each of the following, no stimulation, conA, and levamisole with and without conA. To determine proliferation of each specific subset, cells were labelled with a fluorescent dye (CellTrace Violet, ThermoFisher Scientific, C34557). Proliferation of each subset was determined based on dye dilution using flow cytometry (FacsAria Flow Cytometer). To determine the ability of levamisole to upregulate or alter the immune response, immune subsets were identified (CD4, CD8, CD21, CD172a, CD14) using fluorescent labelled antibodies. Activation was assessed for macrophages and DCs using MHC class II and CD86 expression. Induction of T-regs was based on CD4, foxp3 expression. Specific immune phenotypes were determined based on intracellular cytokine expression of specific subsets (M1 DC1 CD4 Th1 CD8: IFN-gamma) vs. (M2 DC2 CD4 Th2: IL4) vs (CD4 T-reg: IL-10). Significant differences in response were determined using a mixed model ANOVA with significance set at $p<0.05$.

Study results indicated that cells cultured with levamisole alone did not alter PBMC proliferation compared to the response of unstimulated cells. Cells cultured with either conA or PMA/I resulted in a statistically significant ($p<.05$) increase in proliferation compared to unstimulated cells. Cells cultured with conA and levamisole at 1µg/mL resulted in a significant ($p<.05$) decrease in proliferation compared with cells cultured with conA alone. Flow cytometry data analysis results to assess how levamisole directs the immune phenotype are currently pending.

These data demonstrate that, under these conditions, levamisole downregulates conA stimulated PBMC proliferation. Based on these in vitro results, further studies to determine the effectiveness of levamisole on modulating the equine immune system in vivo are warranted.

E07

NOVEL PHARMACOLOGICAL TREATMENT REGIMES IN EQUINE ATRIAL FIBRILLATION

Helena Carstensen[1], Eva Z. Hesselkilde[1], Jørgen K. Kanters[2], Steen Pehrson[3], Maria Mathilde Haugaard[1], Thomas Jespersen[2], Rikke Buhl[1]

[1]Department of Large Animal Sciences, Faculty of Health and Medical Sciences, University of Copenhagen, Taastrup, Denmark, [2]Department of Biomedical Sciences, Faculty of Health and Medical Sciences, University of Copenhagen, Copenhagen, Denmark, [3]Department of Cardiology, The Heart Centre, Copenhagen University Hospital, Copenhagen, Denmark

Atrial fibrillation (AF) is the most common pathological arrhythmia diagnosed in horses. Due to the adverse effects encountered when treating with quinidine sulphate we wanted to improve the medical treatment of AF. In this study we aimed to evaluate the effects of the antiarrhythmic compounds dofetilide and ranolazine in the equine heart. Furthermore, we hypothesized that the use of a combination of two antiarrhythmic compounds with distinct mechanisms would allow us to use lower doses of each compound and thereby minimize the risk of adverse effects.

Three horses were included in a safety study testing administration of intravenous dofetilide (8.0µg/kg BWT), ranolazine (2.4mg/kg BWT) and a combination of the two using similar doses. The horses were monitored clinically for adverse reactions and Holter monitor ECG recordings, as well as blood pressure measurements, were obtained throughout the study. Following the safety study, nine horses were included in an electrophysiological study testing the effects of dofetilide and ranolazine on electrical induced AF. By placing two intraatrial electrodes we were able to record atrial electrograms and perform atrial pacing for measurements of atrial effective refractory periods (aERPs) and induction of AF. All horses went through four procedures (dofetilide and ranolazine respectively, combination of the two drugs using lower dosages and a control procedure where only saline was administered) with 7 days wash out periods between treatments.

All horses tolerated both drugs and the combination well without development of any clinical signs of discomfort or pain and neither of the drugs resulted in alterations in blood pressures. Plasma levels of cardiac Troponin I and creatinine kinase fraction MB did not increase due to intraatrial pacing or any of the treatments given. Ranolazine alone and the combination therapy terminated AF faster than in control procedures ($p=0.0178$). The combination therapy resulted in decreased AF duration ($p=0.0469$) and AF vulnerability ($p=0.0213$) but neither of the treatments altered the aERP significantly. Dofetilide significantly increased the QT_c-interval at multiple time points. This was, however, not the case when lower doses of dofetilide were administered in the combination procedures.

In this study dofetilide and ranolazine were safe to administer in healthy horses and none of the compounds or the electrophysiological procedure caused myocardial damage in the horses. When dofetilide was given at a high dose the QT- interval prolonged indicating altered ventricular repolarization. Ranolazine showed an antiarrhythmic effect against acutely induced AF in horses. By combining dofetilide and ranolazine antiarrhythmic effects were obtained at low doses that would have been ineffective when administered separately. These findings lead us to believe that combination therapy may be a future treatment regime in equine AF.

E08

CAN EXERCISING ELECTROCARDIOGRAPHY PREDICT PERFORMANCE IN YOUNG STANDARDBRED HORSES AT THE START OF TRAINING?

Linda Frellstedt, Gareth Fitch, Chris Rogers

Massey University, Palmerston North, New Zealand

The aim of this study was to describe the occurrence of cardiac arrhythmias in young Standardbred horses that have recently entered training, to determine whether cardiac arrhythmias are consistently present during several training sessions and to evaluate if performance can be predicted based on the type and frequency of observed arrhythmias.

Fifty-one Standardbred horses (aged 2-4 years, 16 F, 32 G, 2 M) were examined during multiple training sessions over a three month period. Resting and exercising electrocardiograms (Holter monitor, Televet®) as well as exercising speeds (GPS, Polar®) were recorded. Descriptive statistics were used to analyze the data.

In total 120 training sessions were recorded. Thirty horses completed 3 training sessions, 8 horses completed 2 training sessions and 13 horses completed one training session. Single or multiple arrhythmias were observed during 42/120 (35%) sessions in a total of 30/51 horses. Secondary AV-block at rest was observed during 15/120 sessions. Ventricular premature contractions (VPCs) were recorded during 20/120, supraventricular premature contractions (SVPCs) during 4/120, and sinus arrhythmias (SAs) during 9/120 sessions. Thirteen horses dropped out of training after their first training session, 7/13 (54%) of these had arrhythmias, 3/7 had VPCs, 2/7 has SVPCs and 2/7 has SAs. Three horses dropped out of training after two sessions, all of them had arrhythmias (100%), 2/3 had VPCs and 1/3 had SA. In 5/51 (9.8%) horses, arrhythmias were consistently recorded during several training sessions. Horses that dropped out of training had >5 VPCs per session (mean of 11.14±7.66).

Cardiac arrhythmias are commonly observed in young Standardbred horses that have recently entered training. Horses with cardiac arrhythmias observed during multiple training sessions and with more significant abnormalities may be less likely to continue training and to reach their first trials.

E09

IS ECG IN HORSES ONLY FOR DYSRHYTHMIA DIAGNOSIS? INTRODUCING A NEW METHOD FOR 12-LEAD ECG

Eva Z. Hesselkilde[1], Helena Carstensen[1], Bettina Vandsø[1], Thomas Jespersen[2], Jørgen Kanters[2], Rikke Buhl[1]

[1]Department of Large Animal Sciences, Faculty of Health and Medical Sciences, University of Copenhagen, Copenhagen, Denmark, [2]Department of Biomedical Sciences, Faculty of Health and Medical Sciences, University of Copenhagen, Copenhagen, Denmark

ECG is an invaluable diagnostic tool in human cardiology where conditions such as left ventricular hypertrophy (LVH) and bundle branch block can be diagnosed based on the 12-lead ECG. Guidelines for 12-lead ECG in horses exist. These guidelines have imitated the human electrode placement with the limb leads recorded from the extremities and the precordial leads on the thorax. However, the different orientation of the equine heart alters Einthoven's triangle and Wilson's Central Terminal (WCT) and thereby changing the basis of the recordings.

The aim of this study was to evaluate if a different electrode placement based on the physiological ideas of the human Einthoven's triangle and WCT may provide better information about the equine atria and ventricles compared to the standard extremity leads.

Development of the new method for 12-lead ECG recordings was performed by placing the limb electrodes at right and left scapula and in the midline caudal to the xiphoid process. This will place WCT in the middle of the heart and Einthoven's triangle will be nearly perpendicular to the main cardiac depolarization vector. By placing the precordial leads between the superficial pectoral muscles (V1), on the ventral part of both the left and right triceps muscle (V2 and V6), in the 6st intercostal space, in level of the shoulder joint of the left side (V3) and in level of the elbow joint on both sides (V4 and V5), information from all heart chambers would be recorded. 35 Standardbred horses (14 mares, 15 geldings and 6 stallions. Mean age: 5.8 years range 2-14) had ECGs recorded using both the existing method and the new modified method. All horses had an echocardiographic examination performed to measure atrial (left atrium area (LAA) and left atrium diameter (LAD)) and ventricular size (including calculation of LV mass). All waves from each lead were measured and all data were blinded before analyses.

Comparing the various leads of the ECGs from the two methods showed a significant increase in both P wave amplitude and duration in the new modified method ($P<0.0005$). Significant correlation between P wave duration and LAA and LAD was found for the new modified method ($P<0.0005$, $R^2=0.32$ and $P<0.005$ $R^2=0.27$, respectively). When amplitudes of R and S wave was added, significant correlation to LV mass was found for the new modified method ($P=0.0005$, $R^2=0.31$) while the existing method did not ($P=0.08$, $R^2=0.01$). Also, significant correlation between mean QRS duration and LV mass were found for the new modified method ($P<0.005$, $R^2=0.22$) but not for the existing method ($P=0.54$, $R^2=0.01$).

This study shows that the new method were better correlated with left atrial size and LV mass suggesting that the correction of Einthoven's triangle WCT gives better results when obtaining 12-lead ECGs in horses.

E10

HUMORAL HYPERCALCEMIA OF MALIGNANCY IN HORSES: A RETROSPECTIVE STUDY (2010-2015)

Kate Hepworth-Warren[2], Stephanie Caston[1]

[1]Iowa State University, Ames, Iowa, USA, [2]Rancocas Veterinary Associates, Mount Holly, NJ, USA

The reported frequency of neoplasia in the horse is low but is likely under-reported due to the difficulty of establishing a definitive diagnosis. Clinical signs may be vague, and neoplasia is often identified only at post-mortem examination. Delays in diagnosis may lead to increased morbidity whereas definitive diagnoses in terminal cases provide objective evidence in electing for humane euthanasia. Humoral hypercalcemia of malignancy is a common paraneoplastic syndrome in other species, and could be utilized to aid in the identification of equine neoplasia. The objectives of this study were to identify the prevalence of hypercalcemia among horses with neoplasia, and to compare calcium, albumin, and the relationship between these values in horses with neoplasia, systemically ill horses without neoplasia, and healthy animals.

An electronic medical record search was performed to identify all horses over 6 months of age on which serum total calcium and albumin were measured. Eligible cases grouped by diagnosis: neoplasia (n=43), sick non-neoplasia (n=466), and healthy/orthopedic (n=55). The sick non-neoplasia group was further sub-divided based on the body system affected by disease for additional comparisons. Exclusion criteria included lack of definitive diagnosis or a diagnosis of renal disease or rhabdomyolysis.

Preliminary data identified higher calcium and lower albumin in the neoplasia group when compared to the sick non-neoplasia group. The calcium:albumin ratio was higher in the neoplasia group than in the healthy/orthopedic group, and lower than the sick non-neoplasia group. This data suggests that hypercalcemia may support a diagnosis of neoplasia in equine patients.

E11

DEVELOPMENT OF A TECHNIQUE FOR DETERMINATION OF PULMONARY ARTERY PULSE WAVE VELOCITY IN HORSES

Gonçalo Silva[1], Bruce Guest[2], Diego Gomez[1], Martine McGregor[2], John Runciman[2], Laurent Viel[1], Luis Arroyo[1]

[1]Departement of Clinical Studies, University of Guelph, Guelph, Ontario, Canada, [2]School of Engineering, University of Guelph, Guelph, Ontario, Canada

Calcification of the tunica media of the main pulmonary arteries has been observed in a large proportion of young racehorses. In humans, medial calcification is the most important cause of increased arterial stiffness, and has been implicated in the pathogenesis of cerebral and renal microvascular diseases. Pulse wave velocity (PWV) is a marker of arterial stiffness. This study aimed to develop a technique for determination of pulse wave velocity of the main pulmonary arteries of horses.

A convenience sample of 9 adult horses with no history of respiratory or cardiovascular disease were used. The horses were sedated for catheter placement, and continuously monitored with electrocardiography during the procedure. The pulmonary artery (PA) trunk was cannulated via right heart catheterization, with a catheter introducer sheath (9Fr x 100cm). Introducer placement was guided with echocardiography. A custom-made dual pressure sensor catheter (PSC) (7Fr x 170cm) was inserted through the introducer sheath, and into one of the main branches of the PA. The position of the PSC in one of the main branches of the PA was confirmed with thoracic radiography. The catheter was withdrawn in 5cm steps and pressure measurements were recorded at each location. The time delay of the pulse waves between the two sensors was used to calculate PWV. Histology of the PA trunk and main branches was performed to investigate the presence of medial lesions and calcification.

The PSC placement was successfully achieved in all horses (9/9), without significant complications, aside from transient arrhythmias. The catheter was more commonly located on the left PA (8/9). At the time of pressure measurements, the level of sedation was variable between horses. The mean (±SD) PWV, was 2.3±0.7m/s in the proximal PA trunk and 1.1±0.1m/s further distal (20cm), in a main PA branch. The mean (±SD) of mean arterial pressures in the proximal PA trunk was 30.1±5.2mmHg, and 22.0±6.0mmHg further distal (20cm), in a main PA branch. The mean (±SD) pulse pressure in the proximal PA trunk was 15.0±4.7mmHg, and 13.5±3.3mmHg further distal (20cm), in a main PA branch. Five out of nine horses presented moderate to severe lesions and calcification of the tunica media of the PA trunk and/or main branches.

This study demonstrated the feasibility of a technique to determine PA-PWV in standing horses. The technique developed may allow further investigation of the effect of calcification of large pulmonary arteries in the development of pulmonary microvascular disorders in horses.

E12

REPEATED MEASUREMENTS OF AUTONOMIC TONE MARKERS OVER A TRAINING SEASON IN EVENTING AND ENDURANCE HORSES

Olivia Lorello[1], Alessandra Ramseyer[1], Dominik Burger[1], Vinzenz Gerber[1], Rupert Bruckmaier[1], Han Van der Kolk[1], Cristobal Navas de Solis[2]

[1]University of Bern, Bern, Switzerland, [2]Texas A&M University, College Station, TX, USA

The purpose of this study was to describe normal values and changes over a competition season of markers of autonomic tone in competing eventing (EV) and endurance (EN) horses, in conjunction with previously reported variables to monitor training and compare these to non-competitive breed-matched controls (EVc and ENc).

26 EV, 13 EVc, 11 EN and 7ENc started the project. Heart rate variability (HRV), non-invasive blood pressure (NIBP), splenic volume, pre- and post-exercise hematocrit and cortisol, standardized exercise tests (SETs) and muscle enzyme activities, were measured pre-season (T1), mid-season (T2) and at the peak/end of the competition season (T3).

HRV was lower ($p<0.05$) in EV than in EVc at all times and post-exercise cortisol lower at T3. There were no significant differences in EN versus ENc. HRV and post-exercise cortisol did not change over the season in any group. EV had higher fitness markers during SETs compared to EVc, but there was no difference between EN and ENc. NIBP, splenic volume, hematocrit, pre-exercise cortisol, muscle enzyme activities, and weight were not significantly different between groups and did not change over the season.

In conclusion, competing eventing horses showed higher sympathetic tone and lower post-exercise cortisol than controls. NIBP, splenic volume and resting or post-exercise hematocrit did not detect changes in autonomic tone in this population. The studied markers of autonomic tone did not change significantly throughout the season in any group and did not detect changes in competing endurance horses compared to controls.

E13

EPIDEMIOLOGICAL CHARACTERISTICS OF HORSES WITH HYPERINSULINEMIA IN A LARGE POPULATION OF HORSES

Steven Grubbs[1], Dwana Neal[1], Thomas Keefe[2]

[1]Boehringer Ingelheim Vetmedica, Inc., St. Joseph, MO, USA, [2]Colorado State University, Fort Collins, CO, USA

Pituitary pars intermedia dysfunction (PPID) and Equine Metabolic Syndrome (EMS) have been described as the most common endocrinologic disorders in horses. Few large studies exist that describe the epidemiological characteristics of horses with these endocrine diseases. The purpose of this study was to obtain information that included age, breed, sex, clinical signs, and ACTH/insulin/glucose status at initial diagnosis of potential endocrine cases (new cases) to determine the epidemiological characteristics of horses with hyperinsulinemia.

Horses of any age, breed, and sex from the continental US were eligible for study enrollment as long as they were documented to be exhibiting one or more of the following clinical signs: generalized or regional hypertrichosis, muscle wasting, abnormal fat distribution, lethargy, laminitis (unknown etiology), polyuria, polydipsia, susceptibility to infections, abnormal sweating, and/or inappropriate lactation. Normal horses were excluded from the study. At initial visit, demographic data, signalment, a physical examination was conducted. Clinical signs were documented and blood was drawn for basal ACTH, fasting insulin, and glucose. Blood samples were processed and shipped overnight to the Animal Health Diagnostic Center, Cornell University, Ithaca, NY for analysis. The association between hyperinsulinemia (based on fasting insulin results), clinical signs and glucose were statistically evaluated individually using the Pearson chi-square test. Odds ratios for significant predictors of hyperinsulinemia were computed using corresponding 95% confidence intervals when applying multiple logistic regression analysis.

Four-hundred eighteen of nine-hundred eighty two horses with complete epidemiological information were included in the final data analysis. Of the 418 horses, 221 (52.8%) were HI^+. Further, horses were stratified into 1 of 3 groups based on ACTH and insulin laboratory results ($PPID^+/HI^+$), ($PPID^-/HI^+$), and ($PPID^-/HI^-$). Of the 418 horses, 115 (27.5%) were $PPID^+/HI^+$, 106 (25.3%) were $PPID^-/HI^+$, and 197 (47.1%) were ($PPID^-/HI^-$). Of the 418 horses evaluated, 9.6% of horses <15 years of age were $PPID^+HI^+$, 23.5% of horses 15 to 19.9 years of age were $PPID^+HI^+$, 27.8% of horses 20 to 24.9 years of age were $PPID^+HI^+$, and 39.1% of horses >25 years of age were $PPID^+HI^+$. Further, 47.3% of horses <15 years of age were HI^+, 61.2% of horses 15 to 19.9 years of age were HI^+, 46.7% of horses 20 to 24.9 years of age were HI^+, and 44.8% of horses >25 years of age were HI^+. One-hundred ninety seven of 418 horses were $PPID^-HI^-$.

Analysis for PPID+/HI+ and PPID-/HI+ horses was evaluated. Based on the data for clinical signs only in the PPID-/HI+ horses, HI was significantly ($p<0.000$) greater among horses with cresty neck and laminitis present. Although not statistically significant ($P>0.025$), HI was greater in horses with 2 other clinical signs; abnormal sweating and pot belly/weight gain. Based on clinical signs data, the odds of HI in horses with cresty neck was more than twice (2.3X) that for horses without this clinical sign. Similarly for horses with laminitis, the odds ratio of HI was more than twice (2.2X) that for horses without this clinical sign.

Of the 418 enrolled horses, 52.8% were diagnosed with hyperinsulinemia. Based on the combined data of clinical signs (11 interpretable), glucose levels and fasting status, cresty neck, laminitis and high glucose were significant predictors of HI. Specifically, horses with high glucose levels were seen to have almost four times the odds (3.5X) of HI compared to those with normal glucose levels.Therefore, when evaluating horses with suspected endocrine disease, at a minimum, ACTH, insulin and glucose should be evaluated. Long term studies need to be conducted in large populations of horses to further evaluate endocrinopathies in horses.

E14

MANAGEMENT OF EARLY PPID IN HORSES

John Haffner[1], Christine Cocquyt[2], Dwana Neal[3], Steven Grubbs[3], Thomas Keefe[4]

[1]Middle Tennessee State University, Murfreesboro, TN, USA, [2]Tennessee Equine Hospital, Thompson Station, TN, USA, [3]Boehringer Ingelheim Vetmedica, Inc, St. Joseph, MO, USA, [4]Colorado State University, Fort Collins, CO, USA

Pituitary Pars Intermedia Dysfunction (PPID) has previously been diagnosed and managed as an older horse disease with advanced clinical signs. However, in early PPID, the clinical signs are typically more subtle. Clinical signs of early PPID include: decreased athletic performance, change in attitude/lethargy, delayed hair coat shedding (subtle), regional hypertrichosis, loss of epaxial muscle mass (topline), regional adiposity, and laminitis. One of the major limitations of diagnostic testing for PPID is the sensitivity of available diagnostic assays. Thyrotropin-releasing hormone (TRH) stimulation test has been shown to have an increased sensitivity of detecting horses with early PPID compared to resting ACTH. Few studies exist concerning endocrinologic testing and follow-up monitoring the clinical progression and diagnostic assay results following treatment in horses with early PPID. The purpose of this study was to identify cases of early PPID based on clinical signs and laboratory testing then determine improvements in ACTH levels using resting ACTH, TRH stimulation test, and clinical signs over time.

Sixteen horses were enrolled (January) with at least one clinical sign of early PPID listed above and diagnostically confirmed by the TRH stimulation test measuring ACTH at 0 (T0ACTH; pre-TRH) and 10 (T10ACTH) min following 1mg i.v. TRH administration. Fasting insulin and glucose was also determined pre-treatment and at each follow-up visit. Each horse was administered Prascend® (pergolide mesylate) orally at a starting dosage of 2 mcg/kg once daily, dosage adjusted to effect if required, but not to exceed 4 mcg/kg daily, per manufacturers label recommendations. A history, physical examination, and TRH stimulation test evaluating T0ACTH and T10ACTH was conducted for 5 months (February through June) following initial treatment administration. The comparative changes to the baseline ACTH levels over time were evaluated using descriptive statistics. The statistical significance of the change in ACTH levels to be evaluated using a confidence interval of 95%.

At initial examination, the most common clinical signs at initial diagnosis were regional hypertrichosis (9/16), delayed shedding (5/16) and muscle wasting (4/16). The baseline (arithmetic mean) for T0ACTH and T10ACTH for all horses was 44.6pg/mL and 360.5pg/mL, respectively. Following six-months of treatment, the resting T0ACTH in 12 of 16 (75%) horses were less than normal reference range (35pg/mL), whereas 7 of 16 (44%) of horses had T10ACTH less than the recommended T10ACTH reference range (110 pg/mL). In June, the decrease from baseline for T0ACTH was 7.1%; whereas the decrease from baseline for T10ACTH was 49.1%. All horses with regional hypertrichosis and delayed shedding were considered improved or resolved, whereas body condition and muscle wasting was slightly improved. Four adverse events not related to treatment were reported and considered resolved within 24 hours.

All horses had improvements in clinical signs following treatment. No significant changes from baseline were observed in insulin or glucose parameters. Following 5 months of treatment, 12 of 16 (75%) horses T0ACTH decreased below normal range compared to 7/16 (44%) T10ACTH that decreased below the recommended reference range. Although all horses responded clinically, the T10ACTH in 9/16 (54%) horses remained positive. Even though T10ACTH did not decrease below reference range, the option to hold the treatment dosage at the same level is justifiable based on clinical sign improvement. Further long term studies in large numbers of horses should be conducted in horses with early PPID.

E15

DEVELOPMENT OF AN OCTREOTIDE RESPONSE TEST FOR DETECTION OF INSULIN DYSREGULATION IN HORSES

Nicholas Frank[1], Pilar Hermida[1], Alfredo Sanchez-Londono[1], Cassandra Uricchio[2], Ranee Singh[3]

[1]*Cummings School of Veterinary Medicine at Tufts University, North Grafton, MA, USA,* [2]*Stockbridge School of Agriculture, University of Massachusetts, Amherst, MA, USA,* [3]*Veterinary Medicine Faculty, Khon Kaen University, Khon Kaen, Thailand*

Octreotide is a somatostatin analog that suppresses insulin secretion and we hypothesized that an octreotide response test (ORT) could be developed to detect insulin dysregulation (ID).

Twelve Morgan horses were included and insulin status was assessed with the oral sugar test (OST). An ORT was then performed by fasting horses overnight and administering octreotide IV at a dosage of 1 µg/kg. Blood was collected at 0, 5, 10, 15, 20, 25, 30, 45, 60, 75, and 90 min, and 2h, 3h, 4h, 6h, 8h, 12h, and 24h.

Mean AUCg did not differ significantly between normal (n=5) and ID (n=7) groups, but mean glucose concentrations at 75, 180, 240, and 360 min were significantly ($P < 0.05$) higher in the ID group, when compared with t-tests. Mean AUCi was significantly higher ($P = 0.003$) in the ID group when compared with the normal group. Mean plasma insulin concentrations differed significantly between groups at 180, 240, 360, and 480 min. A significant time effect ($P < 0.001$) was detected and there was a trend towards higher insulin concentrations over time (group × time; $P = 0.091$).

Results show that insulin concentrations decrease within 60 min of octreotide administration and hyperglycemia develops as a result. As the effects of octreotide wane, insulin concentrations increase above the pre-injection baseline as more insulin is secreted to regain glycemic control. The magnitude of the increase in insulin concentrations reflects insulin sensitivity, as evidenced by the higher AUCi values detected in horses from the ID group.

E16

INSULIN AND INCRETIN HORMONE CONCENTRATIONS IN HORSES DURING AN ORAL SUGAR TEST AND PASTURE CHALLENGE

Nicholas Frank[1], Pilar Hermida[1], Alfredo Sanchez-Londono[1], Cassandra Uricchio[2]

[1]*Cummings School of Veterinary Medicine at Tufts University, North Grafton, MA, USA,* [2]*Stockbridge School of Agriculture, University of Massachusetts, Amherst, MA, USA*

An oral sugar test (OST) is used to assess insulin status in horses and this study was undertaken to test the hypothesis that test results are closely correlated with blood insulin responses to grazing on pasture grass. Correlations among insulin and incretin hormone concentrations were also assessed to better understand mechanisms of hyperinsulinemia in horses.

Eight Morgan horses from the University of Massachusetts herd were fasted overnight and then subjected to an OST over 180 min, immediately followed by turnout on pasture grass. Blood samples were collected at 0, 30, 60, 75, 90, 120, 150, 180, 240, 300, 360, 420, and 540 min. Plasma active and total glucagon-like peptide-1 (aGLP-1 and tGLP-1, respectively) and glucose-dependent insulinotropic polypeptide (GIP) concentrations were measured, along with glucose and insulin concentrations. Peak concentrations and area under the curve values were compared between the OST (0 to 180 min) and pasture challenge (180 to 540 min).

Peak insulin concentration (r_s = 0.83; P = 0.015) and area under the insulin curve values (r_s = 0.95; P = 0.001) were positively correlated between the OST and pasture challenge. Area under the tGLP-1 curve (0-540 min) and area under the GIP curve were correlated with one another (r_s = 0.76; P = 0.037), but did not correlate with area under the insulin curve. Active GLP-1 concentrations are pending.

Our hypothesis was supported and we conclude that the OST can be used to predict insulin responses to grazing on pasture grass.

E17

ASSOCIATION OF ANDROGENS AND PREGNANES RESPONSE TO ACTH STIMULATION WITH ADRENAL DYSFUNCTION IN HOSPITALIZED FOALS

Katarzyna Dembek[1], Jillian Minuto[1], Teresa Burns[1], Bonnie Barr[2], Nathan Slovis[3], Ramiro Toribio[1]

[1]The Ohio State University, College of Veterinary Medicine, Columbus, USA, [2]Rood and Riddle Equine Hospital, Lexington, USA, [3]Hagyard Equine Medical Institute, Lexington, USA

Sepsis continues to cause high mortality in human and equine neonates. The hypothalamic-pituitary-adrenal axis is activated during sepsis to increase cortisol concentrations; however, the secretion of cortisol can be insufficient for the severity of disease in a number of sick foals. Our group and others have shown that relative adrenal insufficiency (RAI), characterized by a poor cortisol response to adrenocorticotropin (ACTH), is associated with mortality and severity of disease in foals. Most studies investigating RAI in equine patients have been focused on cortisol, while other adrenocortical steroids have been overlooked.

We hypothesized that a high dose ACTH stimulation test will stimulate the release of multiple adrenocortical steroids; however, RAI in septic foals will be characterized by a weak glucocorticoid and mineralocorticoid response, but normal androgen and pregnane response when compared to healthy foals and sick foals with normal adrenocortical function. We also proposed that the steroid response to ACTH will be associated with severity of disease.

Foals <1 day of age (n=25) were categorized into 3 groups: septic (n=5), sick non-septic (SNS; n=10) and healthy (n=10). After baseline blood sample collection on admission (time 0), foals received ACTH (100 µg, IV). Additional samples were collected at 30 and 90 min post-ACTH. Hormone concentrations were determined by immunoassays. The delta hormone was defined as the percent concentration change between time 0 and 30 min ($Delta_{0-30}$).

Septic and SNS foals had higher pregnenolone, progesterone, 17α-OH-progesterone and estradiol compared to healthy foals at all 3 time points ($P<0.05$). Cortisol, aldosterone and androstenedione concentrations were higher in septic and SNS than healthy foals at time 0 and 90 min ($P<0.05$). The $Delta_{0-30}$ was lower in septic compared to healthy foals for cortisol and aldosterone ($P<0.05$). Septic foals had higher dehydroepiandrosterone (DHEA) $Delta_{0-30}$ than healthy foals ($P<0.05$).

This study shows that an impaired response to ACTH in septic foals involves multiple adrenocortical steroids. The increased DHEA response to ACTH stimulation might be a good indicator of RAI in septic foals.

E18

ASSOCIATION OF OXYTOCIN AND NEUROSTEROIDS WITH NEONATAL MALADJUSTMENT SYNDROME (NMS) IN HOSPITALIZED FOALS

KATARZYNA DEMBEK[1], CAROLINE BROWN[1], MARGARET MUDGE[1], STEVEN REED[2], BARRY DAVID[3], RAMIRO TORIBIO[1]

[1]THE OHIO STATE UNIVERSITY, COLLEGE OF VETERINARY MEDICINE, COLUMBUS, USA, [2]ROOD AND RIDDLE EQUINE HOSPITAL, LEXINGTON, USA, [3]HAGYARD EQUINE MEDICAL INSTITUTE, LEXINGTON, USA

Neonatal maladjustment syndrome (NMS) is a common illness of newborn foals. Neuroactive steroids (pregnanes, androgens) participate in neurological development; imbalances have been implicated in human disorders, and more recently in the pathogenesis of NMS. Oxytocin regulates uterine contraction, lactation, and social functioning. Oxytocin modulates neurosteroid actions and neurosteroids (pregnanes) can stimulate oxytocin release in growing animals. Oxytocin release may be altered in human patients with a variety of neurological disorders; however, the role of oxytocin in pathologies of newborn foals, including sepsis and NMS, has not been investigated.

We hypothesized that foals with NMS and sepsis will have higher oxytocin and neurosteroids concentrations compared to healthy foals. We also proposed that concentrations of neuroactive steroids and oxytocin will be associated with outcome.

Blood samples were collected on admission from 23 foals with NMS, 37 foals with other neonatal diseases and 15 healthy foals of <3 days of age. Blood concentrations of steroids and oxytocin were determined by immunoassays.

Oxytocin concentrations were higher in foals with NMS compared to healthy foals but lower in foals with other disorders including sepsis compared to healthy and NMS foals ($P<0.05$). Sick foals (NMS and other diseases) had increased cortisol, aldosterone, 17α-OH-progesterone, progesterone and pregnenolone concentrations compared to healthy foals ($P<0.05$). There were no differences in steroid concentrations between foals with NMS and other diseases. Progesterone concentration was higher in sick non-surviving foals compared to survivors ($P<0.05$). Androstenedione and dehydepiandrosterone concentrations were not different between groups of foals.

Due to its neuronal modulatory actions, our study suggests that oxytocin concentration may be involved in the pathogenesis of NMS and other disorders of the equine neonate. Hyperprogesteronemia was associated with NMS, sepsis, and poor prognosis for survival in hospitalized foals.

E19

IDENTIFICATION OF GENETIC LOCI UNDERLYING EQUINE METABOLIC SYNDROME AND LAMINITIS RISK IN WELSH PONIES

Elaine Norton[1], Nichol Schultz[1], James Mickelson[1], Ray Geor[2], Molly McCue[1]
[1]University of Minnesota, Saint Paul, MN, USA, [2]Massey University Colllege of Science, Palmerston North, New Zealand

Equine metabolic syndrome (EMS), a clustering of clinical signs including insulin resistance and dyslipidemia, is the most common cause of laminitis, a painful and life-threatening disease of the horse's hoof. In a large across-breeds study of metabolic variation in horses, our lab demonstrated that EMS phenotypic variability in metabolic traits is influenced by genetic and environmental factors. Further, we demonstrated that certain features of the EMS phenotype are different between breeds. The objective of this project was to identify genomic regions contributing to EMS by performing a genome-wide association study (GWAS) in a cohort of 232 Welsh ponies phenotyped for 11 metabolic traits. Individuals were genotyped on one of two SNP arrays (670,000 to 1,800,000 SNPs); the software program Beagle was used to generate a uniform set of makers across both platforms (~1.8 million SNPs). GWAS was performed using a mixed linear regression model that included a random polygenic term determined from a genomic relationship matrix calculated from select trait associated SNPs, random herd effect, and fixed covariates sex and age. Significant loci were identified for several EMS traits. Specific examples include loci on ECA1 (p=5.04e-08) and ECA15 (p=5.43e-12) for fasting adiponectin and triglyceride levels, respectively. For fasting insulin levels, significant loci were identified on ECA18 (p=4.77e-10) and ECA6 (p=4.33e-09). Candidate genes in these regions include a gene associated with obesity in humans on ECA18, and a gene associated with height and insulin resistance on ECA6. Future directions include haplotype analysis and interrogation of these regions through whole genome sequencing.

E20

THE FIBROBLAST GROWTH FACTOR-23/KLOTHO AXIS IN HEALTHY AND HOSPITALIZED FOALS

Ahmed Kamr[1], Katarzyna Dembek[1], Blake Hildreth III[1], Stephen Reed[2], Nathan Slovis[3], Bonnie Barr[2], Ahmed Zaghawa[4], Ramiro Toribio[1]

[1]*College of Veterinary Medicine, The Ohio State University, Columbus, Ohio, USA,* [2]*Rood and Riddle Equine Hospital, Lexington, Kentucky, USA,* [3]*Hagyard Equine Medical Institute, Lexington, Kentucky, USA,* [4]*Faculty of Veterinary Medicine, University of Sadat City, Sadat City, Egypt*

Hyperphosphatemia, hypocalcemia, increased parathyroid hormone (PTH), and reduced vitamin D metabolite concentrations were recently documented in critically ill foals by our group. However, information on fibroblast growth factor-23 (FGF-23) and klotho, two factors that regulate vitamin D activation and PTH secretion, and their association with calcium, phosphorus, PTH, and vitamin D concentrations in healthy and hospitalized foals is lacking. FGF-23 is released by osteocytes in response to increased $1,25(OH)_2D_3$, PTH, and phosphorus concentrations. FGF-23 suppresses renal 1α-hydroxylase activity and PTH synthesis. Klotho is secreted by the kidneys and acts as a co-receptor for FGF-23. The goal of this study was to determine the relevance of FGF-23 and klotho in calcium, phosphorus, PTH, and vitamin D regulation, including their association with clinical and laboratorial findings, disease severity and outcome in hospitalized foals. We hypothesized that elevated FGF-23 and reduced klotho concentrations will be frequent, associated with calcium and phosphorus dysregulation, PTH and vitamin D metabolite ($25(OH)D_3$, $1,25(OH)_2D_3$) concentrations, severity of illness, and mortality in critically ill foals.

One hundred newborn foals ≤ 3 days old divided into hospitalized (n=83; 59 septic, 24 sick non-septic [SNS]) and healthy (n=17) groups were included. Blood samples were collected on admission. Serum FGF-23, klotho, PTH, and vitamin D metabolites were measured by immunoassays. Data were analyzed by non-parametric methods and logistic regression.

Serum FGF-23 concentrations were significantly higher while klotho, $25(OH)D_3$, and $1,25(OH)_2D_3$ concentrations were significantly lower in septic and SNS compared to healthy foals ($P < 0.01$). Septic foals had higher phosphorus and PTH, and low calcium concentrations than SNS and healthy foals ($P < 0.05$). In hospitalized and septic foals, serum FGF-23 concentrations were associated with phosphorus and PTH ($P < 0.05$), but not with calcium and vitamin D metabolite concentrations ($P > 0.05$). In septic foals, serum klotho concentrations were positively associated with $1,25(OH)_2D_3$ concentrations (rs =0.42; P = 0.01). Hospitalized foals with the highest FGF-23 and lowest klotho concentrations were more likely to die (OR= 3.1; 95% CI = 1.2-9.1; OR=3.6; 95% CI=1.4-10.0, respectively).

High FGF-23 and low klotho concentrations in the presence of elevated phosphorus and PTH concentrations suggests that FGF-23 resistance may be involved in the pathogenesis of mineral and endocrine disorders of critically ill foals.

E21

EFFECT OF RRR-ALPHA-TOCOPHEROL FORMULATION ON SERUM AND CSF ALPHA-TOCOPHEROL CONCENTRATIONS IN VITAMIN E DEFICIENT HORSES

Jennifer C. Brown[1], Stephanie J. Valberg[1], Mattie Hogg[1], Carrie J. Finno[2]

[1]*University of Minnesota, College of Veterinary Medicine, St Paul, Minnesota, USA,* [2]*University of California, School of Veterinary Medicine, Davis, California, USA*

Alpha-tocopherol (α-TP) deficiency in horses can lead to nutritional myodegeneration, neuraxonal dystrophy, vitamin E deficient myopathy or motor neuron disease depending on its temporal occurrence and individual susceptibility. In the absence of adequate grass pasture, α-TP supplementation is becoming increasingly important and, due to better bioavailability, the 'natural' isoform of α-TP, RRR-α-TP, is increasingly being used. RRR-α-TP is available in two formulations, a water-dispersible liquid and an acetate powder, with a more rapid increase in serum α-TP occurring with the more expensive liquid RRR-α-TP. RRR-α-TP acetate is often used for long-term supplementation however, due to the higher cost of liquid RRR-α-T.

The objective of this eight-week study was to compare serum and cerebrospinal fluid (CSF) α-TP concentrations in three groups of horses that had serum α-TP concentrations of <2 μg/mL. Groups were 1) no α-TP (n=5), 2) 5000IU daily of RRR-α-TP acetate (n= 7) and 3) 5000 IU daily of RRR-α-TP liquid, transitioning to 5000 IU daily of acetate (n=7). The transition was; liquid RRR- α-TP (2 weeks), transition (4 weeks) and RRR-α-TP acetate (2 weeks). Horses appeared healthy and were fed the same diet of hay cubes in a dry lot. Serum samples were obtained weekly and atlanto-occipital CSF fluid was obtained before and after the eight-week supplementation period. A GLM ANOVA with post hoc Bonferroni pair wise comparisons (P<0.05) was performed.

Beginning in week 1, serum α-TP was significantly higher in the liquid to acetate RRR-α-TP transition group than in the acetate and the no supplement groups. Serum α-TP increased more gradually in the RRR-α-TP acetate group and was significantly higher than in the no supplement group. CSF α-TP increased significantly between pre and post supplementation in the liquid to acetate RRR-α-TP group only.

In conclusion, for horses deficient in α-TP, the fastest and highest increase in serum α-TP was obtained using water-dispersible RRR-α-TP. A regimen of RRR-α-TP water-dispersible product, tapered to RRR-α-TP acetate, provided an effective means to rapidly and sustainably increase α-TP in serum and CSF over an eight-week period.

E22

DIFFERENTIAL GENE EXPRESSION IN EQUINE SUBCUTANEOUS AND INTERNAL ADIPOSE DEPOTS

Ann Kemper, James Mickelson, Molly McCue

University of Minnesota, St Paul, MN, USA

Equine metabolic syndrome is characterized by regional adiposity, insulin resistance, predisposition to laminitis, and activated inflammation. Regional adiposity in horses is characterized by subcutaneous fat deposits at the nuchal ligament, behind the shoulder, and at the tail head, and increased neck circumference is associated with insulin resistance. In humans and mice, visceral fat is associated with insulin resistance and development of diabetes, and is more metabolically active than other fat depots. In contrast, candidate gene expression studies in horses have suggested that nuchal fat may be more metabolically active than visceral fat. We have collected samples from six adipose tissue depots within the horse; nuchal, shoulder, tail head, (subcutaneous, SQ) and visceral, omental, and retroperitoneal (internal) in six horses of varying age, breed, and sex. To quantify the differences in gene expression in the various depots, an average of 19,085,946 paired-end RNA sequencing reads per sample were generated on the Illumina HiSeq platform. Limma-voom and linear modeling identified 4,252 genes that were differentially expressed between SQ and internal adipose tissue depots. Interestingly, several pathways, including PPAR and PTEN, associated with metabolically active beige/brown fat were highly expressed in SQ compared to internal fat depots. Further, SHOX2 (p=1.54E-10), a marker for beige fat whose subcutaneous expression levels are also positively correlated with central obesity in humans, was overexpressed in SQ. TBX15 (p=4.72E-8) and ZIC1 (p=5.03E-5), genes expressed in brown adipose tissue, were also overexpressed in subcutaneous tissue. Several differences in SQ gene expression were driven by differences in the nuchal fat.

E23

EFFECTS OF A COMMERCIAL ANIONIC SUPPLEMENT ON URINARY ACIDIFICATION IN HORSES

Elizabeth Nelson, Martha Mallicote, Lori Warren, Ariel Robelen, Sarah Reuss

University of Florida, Gainesville, Florida, USA

There are multiple clinical applications for the acidification of equine urine, including prevention of reoccurrence of urolithiasis. This study evaluated the efficacy of a commercial anionic supplement (SoyChlor®) for urinary acidification without altering systemic pH. Eight horses were administered SoyChlor® to achieve a DCAD of 0mEq/kg or -40mEq/kg or fed a control diet for 14 days in a randomized crossover trial. Urine and plasma pH, plasma strong ion difference, plasma anion gap and fractional excretion of calcium were measured and compared between treatment groups. Mean urine pH across each collection period was significantly greater in control horses compared to either treatment group but there was no difference between treatment groups. Blood pH and anion gap were not affected by treatment, but blood pH decreased and anion gap increased over time. Mean bicarbonate concentration was lower in the 0 mEq/kg group than the control group, but did not differ between the 0 mEq and -40 mEq/kg treatments. Strong ion difference of the 0 mEq treatment was lower than the control group. The mean fractional excretion of calcium over all time points did not differ between treatment groups but there was an effect of time and treatment. Both treatment groups had increased fractional excretion of calcium relative to the control group at trial day 7 only. This study shows SoyChlor supplementation can decrease equine urinary pH. Systemic pH was not affected but an increase in fractional excretion of calcium did occur.

E24

EFFECTS OF ABRUPT CONCENTRATE INCREASE AND PREBIOTIC SUPPLEMENTATION ON EQUINE CECAL PH AND LACTATE

Katherine Williamson[1], Amanda Reeg[2], Teresa Douthit[2], Murali Raghavendra Rao[1], Mary Beth Gordon[1]

[1]*Purina Animal Nutrition, LLC, Gray Summit, Missouri, USA,* [2]*Kansas State University, Manhattan, Kansas, USA*

Abrupt changes in diet have been shown to cause gastrointestinal upset in horses. In particular, sudden large increases in concentrate intake can lead to hindgut disruption due to acidosis. Various dietary additives have been studied to determine if this response may be ameliorated or blunted by exerting a prebiotic effect. The objective of this study was to test the hypothesis that the supplementation of a proprietary yeast fermentation product would ameliorate the disruptive effects of a sudden increase in concentrate intake on the cecum.

Nine cecally-cannulated Quarter horses (6-7 y, 4 mares, 5 geldings, approximately 500 kg) were divided into control (n = 4) and treatment (n = 5) groups based on gender and body weight and acclimated for 19 d to the diets and housing. The control diet consisted of 1.5% BW/d prairie grass hay (divided into 2 feedings at 0700 & 1900) and 0.5% BW/d of a 12% protein commercially-available concentrate (Omolene 200®, Purina Animal Nutrition) top-dressed with 10g/45.5 kg BW alfalfa pellet placebo (1 feeding at 0700). The treatment diet was identical but concentrate was top-dressed with 10 g/45.5 kg BW of a pelleted proprietary yeast fermentation product. On d 19, cecal fluid samples were collected 1 h before and after feeding (-1 h, +1 h) and at + 4 h, +8 h, +12 h, +16 h, +20 h and +24 h for 2 additional days to establish baseline levels. On d 22, the morning meal consisted of 1.25% BW concentrate (challenge) followed by 0.75% BW hay. Cecal samples were collected as on previous days. Cecal fluid pH was measured in samples at the time of collection. Lactate was later assessed using colorimetric analysis. A completely randomized split-split-plot design was used to test the effects of the yeast fermentation product. Analysis of variance was done with mixed models (SAS 9.3, SAS Institute Inc., Cary, NC, USA) and least squares means were compared using Fisher's least significant difference ($P < 0.05$).

No adverse clinical effects were noted following dietary challenge. Mean cecal pH was decreased at +8 h and +16 h ($P = 0.017$ & 0.001 respectively) and mean cecal lactate was increased at +1 h, +8 h and +24 h ($P = 0.046$, <0.0001 & 0.031 respectively) after challenge as compared to baseline regardless of treatment. Treatment group horses had increased mean cecal pH at +4 h ($P = 0.042$) and decreased mean cecal lactate at +16 h ($P = 0.026$) compared to control group. There were no significant differences between control and treatment horses' mean cecal pH and lactate at other sampling times.

The sudden increase in concentrate intake, from 0.5% BW to 1.25% BW, induced a significant, transient decrease in cecal pH and increase in cecal lactate which was inconsistently blunted by consumption of a proprietary yeast fermentation product.

E25

EFFECTS OF A SUPPLEMENT (ALFA-LOX FORAGE®) ON EQUINE GASTRIC ULCER SCORES AND GASTRIC JUICE PH

Frank Andrews[1], Pilar Camacho-Luna[1], Kelsey Bailey[1], Isabelle Nesen[2], Michael Keowen[1], Frank Garza, Jr.[1], Chin-Chi Liu[1]

[1]*Equine Health Studies Program, Louisiana State University School of Veterinary Medicine, Baton Rouge, LA, USA,* [2]*École Nationale Vétérinaire, Toulouse, France*

Gastric ulcers are common in stall-confined horses under intermittent feeding conditions. Alfalfa hay feeding has been shown to buffer stomach contents and decrease gastric ulcer scores. The purpose of this study was to determine the effects of a supplement (ALF; Alfa-Lox Forage, Triple Crown Nutrition, Inc., Wayzata, MN), containing chopped alfalfa hay, mannanoligosaccharides, Omega 3 Fatty Acids, and L-Carnitine, top-dressed on grain on nonglandular gastric ulcer scores and gastric juice pH in horses after omeprazole treatment and after intermittent feeding. Eight healthy Thoroughbreds (7 geldings and 1 mare; ages 2-12 yrs.) were used in a 2-period crossover study, where horses were fed ALF (2.0 lbs. [0.91 Kg], twice daily, for 56 days) top dressed on grain (2.5 Kg; Omelene® 100, Purina Animal Nutrition, LLC., Grey Summit, MO) or control (grain alone). While being treated the horses had their stomach scoped on day 0 (before treatment with ALF or grain only) and the on days 14, 28, 42, 49 and 56. In addition horses were weighed, body condition score (BCS) recorded and gastric juice pH measured. During the 56 day periods, from days 14 to 28 horses were treated with omeprazole paste (OME; 4.0 mg/kg, P.O., Q.D.; GastroGard®, Merial Ltd., Duluth, GA) and from days 42 to 49 horses underwent an intermittent feeding model.[1] Stomach nonglandular gastric ulcer number (NGN) and severity (NGS) scores were assigned at each scoping by the PI (FMA), who was masked to treatment. Previous forage analysis data was used to compare nutritional quality between the two periods. Data were analyzed statistically using an ANOVA for repeated measures and when significant differences were found in the main model, a post-hoc Least Squares Means test was performed to determine significant differences at $P<0.05$. Alfa-Lox Forage® top-dressed on grain was readily eaten by all horses, and did not result in any adverse effects. When period data were pooled, Alfa-Lox Forage® did not have a treatment by day effect on gastric ulcer scores or gastric juice pH during the study period. However, there was an overall period effect on treatment, in that BCS was higher and gastric ulcer scores were lower in the AFL-treated horses during the first period of the study, but no treatment by day effect was seen. Forage analysis showed lower digestible energy in the first period of the study. Other significant effects seen during the study included OME-treatment significantly decreased gastric ulcer scores and increased gastric juice pH in both ALF-treated and control horses and gastric ulcers healed in 6/8 (75%) horses after 14 days of treatment. In addition, gastric ulcer scores significantly increased 2 weeks after OME treatment and after feed-deprivation regardless of treatment. Alfa-Lox Forage® had no effect on gastric ulcers scores after OME treatment or during feed-deprivation when compared to untreated controls. However, when forage had lower digestible energy and higher fiber content, Alfa-Lox® Forage-fed horses had better BCS and fewer ulcers. Thus, Alfa-Lox Forage® might provide gastric support when forage has marginal nutritional value. It should be pointed out that the quantity of Alfa-Lox Forage® fed may not have been adequate to affect gastric ulcer scores or gastric juice pH in horses in this study, but other nutritional benefits might result.

[1]Murray MJ and Eichorn ES. (1996) *Am J Vet Res.* 57(11):1599-1603.

E26

CHANGES OF THE EQUINE NEONATAL INTESTINAL MICROBIOTA ASSOCIATED WITH AGE AND DIARRHEA

Angelika Schoster[1], Henry Staempfli[2], Mohammad Jalali[3], Luca Guardabassi[4], Scott Weese[3]

[1]University of Zurich, Vetsuisse faculty, Clinic of Equine Internal Medicine, Zurich, Switzerland, [2]University of Guelph, Ontario Veterinary College, Departemnt of Clinical Science, Guelph, ON, Canada, [3]University of Guelph, Ontario Veterinary College, Departmnet of Pathobiology, Guelph, ON, Canada, [4]University of Copenhagen, Faculty of Health and Medical Sciences, Departement of Veterinary Disease Biology, Frederiksberg, Copenhagen, Denmark

Neonatal diarrhea is an important disease complex in foals. Diarrhea is associated with significant changes in composition and structure of the microbiota in adult horses. The changes associated with diarrhea in neonatal foals are unknown.

The objective was to investigate changes of the equine neonatal microbiota associated with age and occurrence of diarrhea in a field trial.

Fecal samples were collected from 20 foals at 2, 4 and 6 weeks of age and assessed via metagenomic sequencing. Foals were monitored and occurrence of diarrhea was recorded. The Wilcoxon and Steelman-Dwass tests were used to assess differences between age groups and between foals with diarrhea and healthy foals within an age group.

Age had a significant effect on relative abundance of bacterial taxa on the phylum, order and class level. There was a significant increase in richness over time (p=0.02). 9/20 (45%) foals developed diarrhea. The structure of the microbiota was significantly different at Week 2 compared to Week 4 (p=0.09) and Week 6 (p=0.02). Diarrhea had a significant effect on relative abundance of orders and classes. Foals with diarrhea had reduced bacterial richness at four weeks of age (p=0.04). Diarrhea did not have a significant effect on the beta diversity in any age group (all p>0.36).

The composition and richness of the equine neonatal microbiota is changing rapidly after birth. Similar to adults, foals with diarrhea have changes in composition and richness of the microbiota. Further studies are needed to assess whether these changes are the cause or effect of diarrhea.

E27

COMPARISON OF TUBE, GEL, AND IMMUNOCHROMATOGRAPHIC STRIP METHODS FOR EVALUATION OF EQUINE BLOOD TRANSFUSION COMPATIBILITY

Daniela Luethy[1], Sean Owens[2], Urs Giger[3], Darko Stefanovski[1], Rose Nolen-Walston[1]

[1]School of Veterinary Medicine, University of Pennsylvania, Clinical Studies-New Bolton Center, Kennett Square, PA, USA, [2]University of California, Davis, School of Veterinary Medicine, Department of Pathology, Microbiology and Immunology, Davis, CA, USA, [3]School of Veterinary Medicine, University of Pennsylvania, Ryan Small Animal Hospital, Philadelphia, PA, USA

Assessment of compatibility is recommended prior to blood transfusion in horses, typically using gross or microscopic tube agglutination and hemolysin crossmatch, or less commonly, as predicted by blood type and alloantibody assay. In contrast, gel column and immunochromatographic strip methods are the preferred techniques for compatibility testing in small animals. The purpose of this study was to determine agreement between novel and standard crossmatch and typing methods. 33 blood typed and alloantibody screened horses were evaluated. TUBE and GEL crossmatches were performed on 144 recipient-donor pairs and compared using non-parametric area under the curve (AUC) receiver operating characteristic (ROC) analysis. With TUBE assigned as the reference variable, TUBE and GEL had excellent agreement for agglutination (AUC ROC = 0.87), but borderline for hemolysis (AUC ROC = 0.64). There was outstanding agreement between TUBE (gross) and TUBE (microscopic) scores for agglutination (AUC ROC = 0.91). The predicted crossmatch compatibility based on blood type and alloantibody assay showed excellent agreement with TUBE and GEL (AUC ROC = 0.84 and 0.89, respectively). However, there were more recipient-donor pairs identified as incompatible by both TUBE and GEL than predicted by blood type and antibody screen, suggesting unidentified alloantibodies. A blood typing STRIP assay was performed on each aliquot and exhibited 100% sensitivity and specificity for 30 Ca+ and 3 Ca- horses. These findings show that gel column crossmatch techniques appear to be a practical and accurate equine blood compatibility test method, and that immunochromatographic strips are accurate at identifying the Ca blood type.

E28

INFLUENZA-SPECIFIC IMMUNE RESPONSES TO A COMBINATION VACCINE IN NAÏVE PONIES

Amanda Adams[1], Stephanie Reedy[1], Melissa Siard[1], Sarah Elzinga[1], Thomas Chambers[1], Steven Grubbs[2]

[1]*Gluck Equine Research Center, Department of Veterinary Science, University of Kentucky, Lexington, Kentucky, USA,* [2]*Boehringer Ingelheim Vetmedica, Inc, St. Joseph, MO, USA*

Equids, like humans and swine, are recognized as natural hosts for a variety of common viral pathogens, in particular equine influenza virus (EIV). EIV has resulted in worldwide disease outbreaks. Vaccination is essential to prevent or limit disease spread. Many commercially available vaccines are sold as single pathogen vaccines, however nowadays several are available as multi-pathogen or combination vaccines. A significant amount of equine research has been conducted to measure immune responses to single pathogen vaccines however data is lacking to determine if combination vaccines stimulate the immune system in particular cell-mediated immunity. Thus, the objective of this study was to characterize the cell-mediated immunity (CMI) and humoral immune responses to a combination vaccine in influenza naïve animals. A total of 16 influenza naïve yearling ponies were randomly assigned to receive one of two treatment groups: (Group 1) (n=8) Vetera® EWT + EIV/EHV (BIVI) vaccination and (Group 2) (n=8) controls to receive a saline vaccination. All vaccinate ponies received a primary vaccination on day 0, followed by a booster vaccination on day 28. Peripheral blood was collected from all ponies prior to vaccination on day 0 and every 2 weeks following for an 8 week period. Serum and peripheral blood mononuclear cells were isolated to measure influenza specific antibody and cell-mediated immune responses by the following assays: HI, ELISA, interferon-gamma intracellular staining by flow cytometry, and Real-Time PCR. Statistical analysis was performed using SIGMA Plotv12.0™ (Systat Inc., Richmond, CA) with significance determined at the level of ($P \leq 0.05$). Non-normally distributed data was log transformed prior to statistical analysis. For each variable measured (cell-mediated immunity, etc.), a mixed model two way ANOVA with repeated measures was used. In measuring the humoral immune response to the EIV component of the vaccine, results showed an overall significant ($P<0.05$) increase in the change of HI antibody titers in vaccinated ponies compared to non-vaccinated saline controls. Further, there was an increase in EIV-specific IgGa and IgGb antibody titers in the vaccinated group of ponies compared to the saline controls, there was no change in IgGT responses in the vaccinated ponies. Measurement of CMI responses by flow cytometry showed a significant increase in EIV-specific interferon-gamma being produced per cell from the vaccinated ponies compared to the saline controls. Measurement of EIV specific CMI responses by Real-Time PCR showed a significant increase in EIV-specific interferon gamma and granzyme B gene expression in the vaccinated ponies compared to controls. There was no significant difference in the gene expression of perforin, IL-2 or IL-18 when comparing the vaccinated ponies to the controls. Overall, the combination killed vaccine induced significant humoral and cell-mediated responses in naïve animals.

E29

WNV-SPECIFIC IMMUNE RESPONSES TO A COMBINATION VACCINE IN NAÏVE PONIES

Amanda Adams[1], Alex Betancourt[1], Day Barker[1], Melissa Siard[1], Sarah Elzinga[1], Stephanie Reedy[1], Steven Grubbs[2]

[1]*M.H. Gluck Equine Research Center, Department of Veterinary Science, University of Kentucky, Lexington, Kentucky, USA,* [2]*Boehringer Ingelheim Vetmedica, Inc., St. Joseph, MO, USA*

West Nile virus (WNV) mosquito-borne flavivirus and is the leading cause of arbovirus encephalitis in horses in the United States. WNV infection can cause severe acute illness, with devastating clinical signs of illness affecting gait and behavioural abnormalities often times resulting in a high case fatality rate. Thus, WNV vaccination is a recommended core vaccine and is an essential standard of care for all horses in North America as recommended by the AAEP. Many commercially available vaccines are sold as single pathogen vaccines, however nowadays several are available as multi-pathogen or combination vaccines. Little data exists to characterize the immune response to the WNV antigen following a combination vaccination. We hypothesize that a multi-pathogen "combination" vaccine will elicit significant cell-mediated and humoral immune responses in naïve ponies. A total of 16 influenza naïve yearling ponies were randomly assigned to receive one of two treatment groups: (Group 1) (n=8) Vetera® EWT + EIV + WNV (BIVI) vaccination and (Group 2) (n=8) controls to receive a saline vaccination. All vaccinate ponies received a primary vaccination on day 0, followed by a booster vaccination on day 28. Peripheral blood was collected from all ponies prior to vaccination on day 0 and every 2 weeks following for an 8 week period. Serum and peripheral blood mononuclear cells were isolated to measure WNV-specific antibody and cell-mediated immune responses by the following assays: ELISA, interferon-gamma intracellular staining by flow cytometry, and Real-Time PCR. Statistical analysis was performed using SIGMA Plotv12.0™ (Systat Inc., Richmond, CA) with significance determined at the level of ($P \leq 0.05$). Non-normally distributed data were log transformed prior to statistical analysis. For each variable measured (cell-mediated immunity, etc.), a mixed model two way ANOVA with repeated measures was used. In measuring the humoral immune response to the WNV component of the vaccine, results showed a significant ($P<0.05$) increase in WNV IgG titers in vaccinated ponies compared to non-vaccinated saline controls. Measurement of CMI responses by flow cytometry showed a significant ($P<0.05$) increase in the percentage of WNV-specific interferon-gamma producing cells from the vaccinated ponies compared to the saline controls. Measurement of WNV-specific CMI responses by RT-PCR showed a significant ($P<0.05$) increase in WNV-specific interferon gamma and IL-2 gene expression in the vaccinated ponies compared to controls. There was no significant ($P>0.05$) difference in the gene expression of GrzB, IL-6 or IL-18 when comparing the vaccinated ponies to the controls. Overall, the combination killed vaccine induced significant WNV-specific humoral and cell-mediated immune responses in naïve animals.

E30

CHARACTERISTICS OF INFECTION CONTROL PRACTICES AT NORTH AMERICAN VETERINARY TEACHING HOSPITALS

Cristine S. De La Hoz Ulloa[1], Katharine M. Benedict[2], Paul S. Morley[2], Brandy A. Burgess[1]

[1]*Virginia Tech, Blacksburg, VA, USA,* [2]*Colorado State University, Fort Collins, CO, USA*

Infection control is crucial in the operation of all veterinary hospitals to not only protect the patients and hospital, but to protect personnel as those in the veterinary field have an increased risk of occupational exposure to zoonotic diseases. In a 2008 study, 50% of Veterinary Teaching Hospitals (VTHs) surveyed reported significant health problems due to zoonotic infections among hospital personnel. *Cryptosporidium parvum* infections accounted for 68% of these infections. The objective of this study was to characterize current infection control practices in place for the prevention of zoonotic disease infection, specifically infection of *C. parvum,* in VTHs.

All VTHs located in North America (n=35) that had been operational for at least one year were eligible to participate in this study. Phone surveys of biosecurity experts were conducted from July - October 2015 which addressed policies for hygiene, surveillance, patient contact, education, awareness, and enteric infectious disease control.

Among participating VTHs (n=20/35; 57.1%), greater than half reported significant outbreaks of disease among hospitalized patients in the previous 5 years; most commonly due to *Salmonella enterica* or equine herpesvirus-1. In addition, 50% reported significant health problems in personnel, in the previous 2 years, that most likely resulted from zoonotic infection. Of these, *Cryptosporidium* was identified as the most common agent. The majority of VTHs surveyed had a committee that oversaw biosecurity program activities as well as written biosecurity policy documents. However, only half conducted mandatory training on the biosecurity and infectious disease control program.

The results of this study will help to improve strategies for preventing healthcare-associated infection, including zoonotic diseases, among patients and veterinary personnel; and allow for targeted educational tools to promote a safety culture in veterinary medicine.

E31

ANTI-ENDOTOXIC PROPERTIES OF KETOROLAC TROMETHAMINE AND FLUNIXIN MEGLUMINE IN HORSES

Alexandra Bianco, George Moore, Sandra Taylor

Purdue University, West Lafayette, IN, USA

Non-steroidal anti-inflammatory drugs (NSAIDs) play an integral role in equine medicine due to their combined analgesic and anti-inflammatory properties. Despite their widespread use, there are limited NSAIDs available that demonstrate variable adverse effects and safety. While several NSAIDs have been proven to be adequate analgesics, few have undergone rigorous evaluation for anti-inflammatory efficacy. Ketorolac tromethamine (KT) is a non-selective cyclooxygenase inhibitor that has been used in human patients since 1989. The pharmacokinetic properties of KT have recently been determined in the horse, and have been previously determined in several other species. However, there have been only two previous studies examining KT's anti-inflammatory properties in animals, both of which were *in vivo.* There have been no published studies evaluating the *in vitro* anti-inflammatory effects of KT in any veterinary species. The purpose of this study was to evaluate the anti-inflammatory effects of KT compared to flunixin meglumine using an *in vitro* model of LPS-stimulated equine monocytes.

Equine monocytes were isolated from whole blood from a single horse and incubated with either KT or flunixin meglumine at six concentrations ranging from 2.5 µg/mL to 80 µg/mL for 1 hour. After the initial incubation, *E. coli* 055:B5 LPS was added to each well at a concentration of 1 µg/mL, which has been shown to consistently activate monocytes. The wells were incubated at 37°C and 5% CO_2 for 4, 8, 12, and 24 hours. Samples were collected at each time point and stored at -80°C until analysis. Equine-specific ELISAs were used to measure the eicosanoids PGE_2 and TXB_2 as well as the cytokines TNF_{α}, IL-6, and IL-8.

Preliminary results demonstrated that flunixin meglumine suppressed PGE_2 production up to 12 hours at concentrations ≥5 µg/mL and up to 24 hours at concentrations ≥40 µg/mL. Flunixin meglumine suppressed TXB_2 production up to 12 hours at all concentrations and up to 24 hours at concentrations ≥20 µg/mL. Ketorolac tromethamine suppressed PGE_2 production up to 12 hours at all concentrations and up to 24 hours at concentrations ≥20 µg/mL. Ketorolac tromethamine also suppressed TXB_2 production up to 12 hours at all concentrations but did not suppress production up to 24 hours at any concentration. Peak eicosanoid concentration in the non-treated samples occurred at 4 hours for PGE_2 and 12 hours for TXB_2.

While *in vitro* results cannot be directly correlated to *in vivo* efficacy, the results thus far indicate both drugs effectively suppress eicosanoid production after LPS stimulation, with an effect of both time and drug concentration. Based on the results of this study, a therapeutic dose of 2.5 µg/mL KT would effectively suppress eicosanoid production in cases of endotoxemia. Further research is needed to correlate *in vitro* results with *in vivo* efficacy.

E32

EFFECTS ON SWEATING OF CHLORAMPHENICOL AND THE MACROLIDE GAMITHROMYCIN: COMPARISON WITH ERYTHROMYCIN

Amy Stieler[2], Chris Sanchez[1], Martha Mallicote[1], Amy Smith[1], Jim Burrow[1], Rob MacKay[1]

[1]*Department of Large Animal Clinical Sciences, University of Florida, Gainesville, FL, USA,* [2]*University of Georgia, Athens, GA, USA*

Hyperthermia in foals treated with erythromycin (ERY), clarithromycin, or azithromycin is associated with reduction of sweating by an unknown mechanism. The aims of this study were to evaluate effects on sweating of (1) chloramphenicol (CHL) which, like macrolides, binds to the 50S subunit of the bacterial ribosome and (2) gamithromycin (GAM), a macrolide sub-class 7a azalide which is not yet licensed for treatment of foals. Over 3 experimental periods, 12 foals (1 to 3 months old) were treated with ERY (25 mg/kg PO, 3 times daily for 5 days), GAM (8 mg/kg IM once) and CHL (50 mg/kg PO, 3 times daily for 5 days) according to a masked, duplicated, fully counterbalanced design. Quantitative intradermal terbutaline sweat tests were performed on 3 successive days before treatment (baseline) and on days 1, 2, 5, 9, 24, and 39 after treatment began. Data were analyzed by 1-, 2-, and 3-factor repeated measures ANOVA with *post hoc* pairwise analyses of significant tests. There was significant (P <0. 05) suppression of sweating by ERY but not CHL or GAM. Compared with sweating at baseline, values when foals were given ERY were significantly lower on all test days except day 39. Sweating responses in foals given ERY were significantly lower than when given GAM or CHL on days 2, 5 and 9 at every terbutaline concentration except 100 μg/mL. In contrast, there was no difference between CHL and GAM. Results show that (1) ability to bind the bacterial 50S ribosome does not by itself confer potential to induce anhidrosis and (2) unlike the macrolides ERY and clarithromycin and the 9a-azalide azithromycin, the 7a-azalide GAM did not suppress sweating responses.

E33

COMPARISON OF THE CLINICOPATHOLOGIC SIGNATURES OF EQUINE CORONAVIRUS AND *SALMONELLA* ENTEROCOLITIS

Arlie Manship, Johanna Elfenbein

North Carolina State University College of Veterinary Medicine, Raleigh, NC, USA

Recent worldwide outbreaks of Equine Corona Virus (ECoV) have led to an increased awareness of this pathogen as a cause of colitis in adult horses. The objective of this study was to determine whether ECoV enterocolitis has a distinct clinical and/or clinicopathologic signature that can distinguish it from enteric salmonellosis or unknown causes of fever and neutropenia.

Data were collected from medical records from North Carolina State University (2003-2015). Horses >1 year old were divided into 3 groups based on diagnosis: ECoV (fecal PCR), enteric salmonellosis (fecal culture or PCR), or unknown diagnosis by ruling out ECoV and *Salmonella* (3 negative *Salmonella* cultures, negative ECoV and *Salmonella* PCR). Clinical features, diagnostic test results, and outcome were compared between groups. Statistical significance was determined using a Mann-Whitney U test with $p<0.05$.

Data were obtained from 7 horses with ECoV, 11 horses with *Salmonella*, and 11 horses without a diagnosis. ECoV cases had significantly fewer neutrophils (median 569/uL; range 105-1,029/uL) compared with all other diagnoses (1,310/uL; 337-9,720/uL) but there was no difference in neutrophil count when compared with *Salmonella* or unknown diagnosis. There were no differences in temperature, lymphocyte or platelet counts, fibrinogen, or any biochemical parameter between groups.

The results suggest that disease with ECoV and Salmonella share clinical features and should be included together on a differential list for fever and neutropenia. Further investigation is warranted to determine whether there is an unknown pathogen as the cause of the fever and leukopenia in the group without a diagnosis.

E34

Phenotypic Characterization of *Sarcocystis neurona* Lesions in Gravely Affected Horses

JunJie Liu[1], Micheal Dark[1], Lisa Farina[1], Robert MacKay[2], Nancy Denslow[3], Lei Zhou[4], Maureen Long[1]

[1]*University of Florida Department of Infectious Diseases and Pathology, Gainesville, FL, USA,* [2]*University of Florida Department of Large Animal Clinical Sciences, Gainesville, FL, USA,* [3]*University of Florida Department of Physiology, Gainesville, FL, USA,* [4]*University of Florida Department of, Gainesville, FL, USA*

The primary goal of this research was to contribute new knowledge to the understanding of *Sarcocystis neurona* (SN) infection in horses. Our goal was to identify SN infected horses and to use different markers to investigate and characterize the infiltration of inflammatory cells and resident cellular changes in EPM-diseased horses by genetic (PCR) and immunohistochemical (IHC) techniques. The hypothesis investigated was that phenotypic characterization the CNS of gravely SN affected horses could define disease pathogenesis. Tissues from 2 groups of humanely euthanized horses with either grave or no neurological symptoms were collected under University of Florida IACUC protocols 3984, 4109, and 4559. Sets of CNS tissues from all levels of brain and spinal cord were fixed for pathology and immunohistochemistry and frozen for nucleic acid extraction. Histopathologically, encephalomyelitis and normal CNS was confirmed in affected and unaffected horses, respectively. SN, *Neospora hughesi*, EHV-1, WNV, and EEEV virus PCR was performed to confirm neuropathogen status in all horses. For RNA expression analysis, the tissues in which SN was detected, the corresponding section of frozen tissue was cut into 30-50 30 μg sized cubes. Relative parasite burden was calculated based on the standard curve generated by the use of cultured SN merozoites. 160 samples were tested and the correlation coefficient (R, Psych package) between $CD3^+$ and TNF-α expression was 0.874. The correlation coefficient between $CD3^+$ expression and parasite burden was 0.8304 (186 samples) and the correlation coefficient between TNF-α and parasite burden was 0.7757 (101 samples). Immunohistochemical phenotypic analysis was conducted for T cells, macrophages, B cells, microglia and astrocytes for C3+ T cells, B cells, macrophage cells, glial cells, and astrocytes and this was performed on 12/14 EPM cases that had SN+ and adjacent SN- tissue formalin fixed tissue (FFPE) blocks. Eleven normal control horses were used and sections were location matched to lesion sections. All sections were compared for significant differences by ANOVA ($P<0.05$). The number of $CD3^+$ T cells and microglia/macrophage cells of SN+ tissues were significantly increased compared with its adjacent section and the sections from the normal horses. Although present in extremely limited numbers (< 10 cells per site), macrophage only staining cells had statistically higher cell counts in affected horses compared to normal horses. Because CD3+ cells were the most numerous, differentiation of CD4+ and CD8+ cells was attempted using IHC by several markers, none of which worked on FFPE sections. CD4+ and CD8+ relative expression was investigated by quantitative PCR and the most significant change was that of CD4+ cell marker expression at 144.6 fold (log_2) over controls. The fold change of CD8+ between the lesion group and the normal group was 10.8. In conclusion, the presence of SN was only present within inflammatory nodules and the presence of parasite burden was associated with the presence of CD3+CD4+T cell markers compared to that of CD8+ T cells. Based on the presence of inflammatory markers, the presence of SN was not detected outside of aggregates of either resident or infiltrating inflammatory cells.

E35

Normal Ultrasonographic Pleural Thickness in Clinically Healthy Adult Horses

Breanna Sheahan, Kara Lascola, Scott Austin

University of Illinois Veterinary Teaching Hospital, Urbana, IL, USA

The objectives of this study were to establish reference ranges for normal pleural thickness measurements in healthy adult horses and to compare measurements between horses < 15 years of age (YOUNG) and ≥ 15 years of age (OLD). Pleural thickness is assessed as part of a complete thoracic ultrasonographic examination but no normal measurements have previously been reported. We hypothesized that measurements of pleural thickness would be < 2 mm.

18 clinically healthy adult horses (11 YOUNG; 7 OLD) were examined with a curvilinear 3.5-8 mHz probe in eight thoracic locations (four per side): cranioventral (intercostal space (ICS) 6), mid-thoracic (15 cm dorsal to point of shoulder (POS) in ICS 10), caudoventral (ICS 10 at POS), and caudodorsal (ICS 14). All measurements were performed post hoc by two observers on three recorded images from each site. Differences among locations were evaluated with Kruskal-Wallis. Difference between age groups was evaluated with Mann Whitney U. Inter-observer reliability was evaluated using intraclass correlation (ICC). Significance was set at $P < 0.05$.

Pleural thickness median (range) was 1.5 (1.11-2.33) mm in all horses, 1.47 (1.11-2.09) mm in YOUNG horses and 1.55 (1.18-2.33) mm in OLD horses. Significant differences among locations and between age groups were not noted. Mild pleural irregularities were subjectively more common in older horses. ICC was significant for measurements at each location.

Older horses did not have significantly greater pleural thickness measurements. Measurements were reproducible. Measurements above the normal variation should be cause for further investigation of respiratory disease.

E36

SYRINGE VERSUS MECHANICAL SUCTION WITH N-BUTYLSCOPOLAMMONIUM EFFECTS ON BAL PARAMETERS IN HORSES WITH PASTURE RAO

Jacquelyn Bowser, Lais Costa, Christine Lopp, Rachel Wilson, Rebecca Byrne, Melissa Steichen, Santosh Kumar TK, Alba Rodil, Michael Brashier, Cyprianna Swiderski

Mississippi State University College of Veterinary Medicine, Mississippi State, MS, USA

Bronchoalveolar lavage (BAL) is frequently employed to evaluate lower airway inflammation in horses with RAO. In these horses, bronchial collapse often limits retrieval of BALF. N-butylscopolammonium bromide (Buscopan®), a parasympatholygic agent, safely produces short term bronchodilation, reversing airway obstruction RAO-affected horses.

Based upon the hypothesis that syringe aspiration would improve BALF retrieval, we performed BAL in the left and right lungs of 8 sedated summer pasture associated RAO-affected horses using a flexible endoscope. Saline (300cc) was instilled via a polyurethane tube inserted through the biopsy port using a 60 cc syringe and aspirated using gentle manual suction (SS) with a 60cc syringe or with mechanical suction (MS) using a pressure regulated suction pump. Pressures of 15-75 mm H_2O retrieved BALF without airway collapse. Bilateral BAL was performed twice in each horse with or without Buscopan® (0.3mg/kg IV) in a crossover fashion with a 21 day washout period. The effects of suction type and Buscopan® administration on volume yield, cellularlity and cell viability were determined.

BAL performed using SS demonstrated higher volume yield, with lower RBC and lower nucleated cell counts. There was a tendency for improved viability with MS. Differences in yield, nucleated cell count and red blood cell count were not identified with Buscopan®, with the exception of horses demonstrating profound collapse.

Use of a modified BAL technique that includes manual syringe suction with Buscopan® administration may provide an advantage in horses with significant airway hyper-reactivity by limiting airway collapse and enhancing the diagnostic utility of BALF obtained.

E37

PREVALENCE OF FUNGI IN RESPIRATORY SAMPLES OF HORSES WITH INFLAMMATORY AIRWAY DISEASE

Julie Dauvillier, Emmanuelle van Erck

Equine Sports Medicine Practice, Waterloo, Belgium

Inflammatory Airway Disease (IAD) is a common cause of respiratory problems and poor performance in horses. It has recently been suggested that IAD and Recurrent Airway Obstruction (RAO) could be two forms of a same inflammatory process and recent evidence tends to establish a close parallel between these equine diseases and asthma in humans. Fungi have been shown to contribute to the inflammatory response of lungs in RAO horses and in some forms of asthma in humans. The role of fungi in IAD has not yet been assessed to the knowledge of the authors.

The purpose of this prospective observational clinical study was (1) to evaluate the prevalence of fungal isolates in the respiratory samples of horses diagnosed with IAD, (2) to describe clinical signs associated with the presence of fungi in respiratory samples of horses diagnosed with IAD and (3) to assess the risk factors associated with IAD and positive fungal culture and proliferation. The study was performed in a population of working sport, race and leisure horses based in France, Belgium and the Netherlands.

A total of 482 cases horses, referred to a specialized ambulatory internal medicine practice in Europe, were assessed between June 2013 and December 2014. The horses were referred either for respiratory problems or loss of performance. For each case, an environmental evaluation, a clinical examination, an airway endoscopy, a tracheal wash (TW) and a bronchoalveolar lavage (BAL) were performed. The TW and the BAL underwent cytologic evaluation and the TW was submitted for bacteriology and mycology. Diagnosis of IAD was established based on BAL fluid cytology, according to the ACVIM consensus statement definition. On both cytologic examinations, the presence of fungal elements (spores, conidiophores, hyphae) was recorded, as well as signs of active proliferation.

In the population studied, a diagnosis of IAD was established in 84% of cases. A positive fungal culture was obtained in 49% of IAD+ cases and in 44% of IAD- cases. The most commonly isolated fungus were *Aspergillus sp, Penicillium* and *Rhizomucor*. No relationship could be established between positive mycology and cytologic evidence of fungal elements or active proliferation of fungi in the airways. However horses with fungal elements on the TW cytology had 3.8 more chances of having IAD than horses with no fungi (OR=3.8; 95% CI 1.8-7.8; p=0.0003). Twenty IAD horses had signs of fungal proliferation in their TW and/or BAL, but all mycologies remained negative in these cases. Distinctive respiratory clinical signs such as cough, breathlessness or nasal discharge showed low sensitivity and high specificity for IAD diagnosis. Concomitant bacterial infection did not increase the risk of fungal isolation. The degree of inflammation (% inflammatory cells in BAL) was higher when horses were housed indoors, bedded on straw vs. shavings (Mann-Whitney test, 25.53 +/-17.14 vs. 19.9+/-17.5; p<0.001) or fed dry hay vs. steamed hay (23.2 +/- 17.5 vs. 18.8 +/-18.9; p<0.001).

The prevalence of fungal contamination in IAD cases is high in Europe. Horses inhaling aerosolized fungal particles are at a significantly higher risk of developing IAD. The type of bedding and forage as well as other environmental conditions represent significant risk factors for IAD and fungal contamination of equine airways. The identification of proliferating fungi in the airways is a sign of fungal infection and warrants further investigation. In the current study, mycology culture seems to lack sensitivity in the detection of fungi in respiratory samples.

E38

INTRAVENOUS MAGNESIUM SULFATE AS A RESCUE THERAPEUTIC FOR BRONCHOCONSTRICTION IN HORSES

Jacquelyn Bowser, Caitlin Wenzel, Robert Wills, Noel Bondi, Rachel Wilson, Rebecca Byrne, Cyprianna Swiderski

Mississippi State University College of Veterinary Medicine, Mississippi State, MS, USA

Horses with RAO exhibit exacerbations of acute airway obstruction that cause respiratory distress resulting from bronchospasm, mucus accumulation, and airway inflammation. Emergency therapeutic options for control and reversal of airway bronchoconstriction in equine airways include parasympatholytic agents which inhibit gastrointestinal motility, β2-adrenergic agonists, as well as systemic corticosteroids, which require 4-6 hours for full effect. Meta-analysis of emergency intervention for asthmatic adults found that adjunctive infusion of magnesium sulfate ($MgSO_4$) improves lung function and reduces hospital admissions. This study examined utility and effectiveness of intravenous $MgSO_4$ as a novel, inexpensive, easily administered bronchodilatory agent for horses with a naturally occurring pasture-associated form of RAO (SPARAO).

Six SPARAO-affected horses exhibiting naturally occurring disease exacerbation (CSRE $\geq$5, ΔPpl_{max} > 15 cm H_2O) were treated with three doubling dosages of $MgSO_4$ by rapid IV infusion using three way cross-over design. Horses were instrumented for continuous ECG and blood gas monitoring. Tidal volume (TV), pulmonary resistance (R_L), dynamic compliance (C_{dyn}), and maximum change in pleural pressure (ΔPpl_{max}) were monitored at 5 minute intervals using conventional pulmonary mechanics.

Mixed models were fit for percent change from baseline values for each outcome using PROC MIXED (SAS, SAS Institute. Cary, NC). Significant decreases in ΔPpl_{max}, R_L and significant increases in C_{dyn} were identified at all dosages and were accompanied by increased P_aO_2 and decreased P_aCO_2, substantiating improved ventilation. We conclude that intravenous infusion of $MgSO_4$ has immediate bronchodilatory effects in horses making this therapy a valuable adjunct in horses with signs of respiratory distress referable to bronchoconstriction.

E39

INVESTIGATION OF MISOPROSTOL AS A NOVEL ANTI-INFLAMMATORY IN EQUINE LEUKOCYTES

Emily Medlin Martin, Samuel Jones

North Carolina State University College of Veterinary Medicine, Raleigh, NC, USA

E-type prostaglandins (PGEs) play diverse roles throughout the body, including regulation of gastrointestinal (GI) homeostasis and modulation of inflammation. To treat inflammation, anti-inflammatory medications including nonsteroidal anti-inflammatory drugs (NSAIDs) are designed to inhibit prostaglandin production. Unfortunately, potential adverse effects of NSAID use in horses include gastric ulceration and right dorsal colitis. To prevent and treat NSAID-associated GI injury, equine practitioners commonly administer the PGE_1 analog, misoprostol. In addition to serving as a gastroprotectant, misoprostol has been shown to have anti-inflammatory effects in cell models, potentially by increasing intracellular cyclic AMP (cAMP). Elevated cAMP is known to dampen leukocyte effector functions that can damage host tissues, including cellular adhesion, tissue infiltration, and cytokine and superoxide production. The effects of misoprostol on equine leukocyte effector functions are undetermined. Thus, the purpose of this study was to determine the effects of misoprostol on equine leukocyte pro-inflammatory cytokine and superoxide production *in vitro*. Equine leukocytes were isolated from whole blood and stimulated with lipopolysaccharide (LPS) in the presence or absence of misoprostol. Misoprostol at concentrations of 1-10uM significantly inhibited tumor necrosis factor-α, interleukin-1β, and interleukin-6 mRNA production, as measured by qPCR. Misoprostol also elicited concentration-dependent decreases in neutrophil superoxide production in response to granulocyte-monocyte colony-stimulating factor priming and LPS stimulation, as measured by luminol-enhanced chemiluminescence. From this data we conclude that misoprostol produces anti-inflammatory effects in equine leukocytes *in vitro*, paving the way for *in vivo* studies investigating misoprostol as an anti-inflammatory treatment in horses.

E40

DO ENDOCRINE DISRUPTING CHEMICALS PLAY A ROLE IN HORSES WITH EQUINE METABOLIC SYNDROME?

Sian Durward-Akhurst[1], Nichol Schulz[1], Elaine Norton[1], Raymond Geor[2], James Mickelson[1], Molly Mccue[1]

[1]University of Minnesota, Minnesota, USA, Mmassey University, Palmerston North, New Zealand

Equine Metabolic Syndrome (EMS) is characterized by abnormalities in insulin regulation, increased adiposity and laminitis. In a large across-breed study (610 individuals) of 11 morphometric and biochemical metabolic traits, our lab has demonstrated that 51-77% of the phenotypic variability is due to individual factors including age, breed, sex and genetics and 23-49% of the phenotypic variability is the result of shared environment. Despite up to 49% of the variability in EMS phenotype being related to shared environment; only 4-18% of this variability is explained by diet, amount and intensity of exercise and season, suggesting that other environmental factors play a role in EMS development. Recent work has identified associations between Endocrine Disrupting Chemicals (EDCs) and human metabolic syndrome and other endocrine abnormalities. Our preliminary data demonstrated that horses from farms within 30 miles of EDC disposal sites (EPA 'Superfund sites') were more likely to have had laminitis ($p=0.002$) and had higher post oral sugar challenge insulin concentrations (OST INS) ($p=0.00005$), suggesting EDC exposure is an EMS risk factor. The objective of this study was to determine if plasma EDC concentration is correlated to metabolic measurements. Plasma EDC concentrations were measured using the CALUX-DR bioassay in 158 Morgans and 137 Welsh Ponies from 32 farms. Grams fat extracted from plasma and sex (female) were negatively correlated with EDC concentration. EDC concentration was positively associated with OST INS ($p=0.03$). The results suggest that some of the unexplained environmental variance identified in individuals with EMS is due to exposure to EDCs.

E41

EVALUATION OF AN ORAL SUGAR TEST FOR DYNAMIC ASSESSMENT OF FIVE EQUINE BREEDS' INSULIN RESPONSE/SENSITIVITY

Jane Manfredi[1], Raymond Geor[2], Patty Sue Weber[1], Bo Norby[1], L Jill McCutcheon[2]

[1]Michigan State University, East Lansing, MI, USA, [2]Massey University, Palmerston North, New Zealand

Veterinarians have identified equine metabolic syndrome (EMS) as the most common cause of laminitis within equine practice. An oral sugar test (OST) has been used clinically to identify "at risk" horses although it has not been rigorously compared to a "gold standard" frequently sampled intravenous glucose tolerance test (FSIGTT) in a large number of horses or across different breeds.

Eighty-nine horses/ponies of five different breeds (Quarter Horses [QH], Arabians, Morgans, Welsh Ponies, and Thoroughbreds) had both an OST (0.25 ml/kg Karo syrup) and an FSIGTT performed at least 24 hours apart. Optimal time for sampling was determined with a repeated measures ANOVA. A ROC curve used to identify an optimal threshold insulin value for insulin resistant (IR) equids (insulin sensitivity [SI] < 1). Spearman correlations were utilized to examine possible correlations between OST outcome measures (area under the curve for insulin [AUCi] and peak insulin concentration) and SI and the acute insulin response to glucose (AIRg) (the latter two determined by Minimal Model analysis of the FSIGTT).

IR horses had OST insulin but not glucose concentrations from 60-120 minutes that were significantly higher than insulin sensitive horses. In this cohort, a single time point insulin concentration of ≥30.2 IU/mL was indicative of insulin resistance. Moderate correlations (*rho*=-0.6) to SI and strong correlations to AIRg (*rho*≥0.74) were evident for AUCi and peak insulin concentration. Weak correlations existed between glucose concentrations and SI and AIRg. QHs have significantly lower OST AUCi than all breeds except Morgans.

Different from what is currently recommended in the literature, we advise measuring the insulin concentration of a single blood sample at 60, 75, 90, or 120 minutes post–Karo syrup administration and using a ≥30.2 IU/mL threshold. Practitioners should be aware that breed may play a role in insulin concentrations that are achieved during the OST.

E42

BASAL INSULIN CONCENTRATION IN COMPETITION DRAFT HORSES

Harold Schott[1], Lisanne Gallant[1], Elizabeth Tadros[1], Sarah Jacob[1], Jane Woodrow[2], Melissa Hines[2], Susan Ewart[1]

[1]*Michigan State University, East Lansing, MI, USA,* [2]*University of Tennessee, Knoxville, TN, USA*

Measurement of basal insulin (INS) concentration is currently advocated as the endocrine test of choice for screening horses for insulin dysregulation. Values >20 µU/mL after overnight fasting and >50 µU/mL in hay fed horses are considered supportive of insulin dysregulation. Although it is well recognized that INS increases with body condition (fat mass), recent evidence suggests that there may also be age and breed differences with regard to basal insulin concentrations. However, little data exists for draft breeds. Consequently, we tested the hypothesis that INS differs in various breeds of draft horses, independent of age and body condition. Age was obtained from owners/trainers and body condition score (BCS) was assessed and basal (resting but not fasting) INS was measured in 164 draft horses participating in an international competition, including 48 Belgians, 52 Clydesdales, and 64 Percherons. INS was measured using a radioimmunoassay validated for equine serum (with average intra- and inter-assays CVs of 3.4% & 6.9% and 4.7% & 7.2% for low and high equine control sera, respectively). None of the horses examined had a history or current evidence of laminitis. Age, BCS, and INS were not normally distributed; thus, median values and 25% and 75% confidence intervals were determined (**Table**). Data were analyzed using ANOVA on ranks, Chi square analysis, and Spearman rank order correlation.

	Belgian	**Clydesdale**	**Percheron**	**P value**
Age	5 (3-7.5)	6 (3-8.5)	6 (4-9.5)	0.33
BCS	6 (6-7)[a]	6 (5-6)[b]	6 (6-7)[a,b]	**0.007**
Insulin	37.1 (29.5-72.0)	36.7 (23.4-53.5)	38.0 (29.2-62.5)	0.30
# (%) >50 µU/mL	17 (35%)	15 (29%)	39 (47%)	0.07

Median age was not different between horses sampled for each breed. Although the Clydesdale breed had a lower BCS as compared to the Belgian breed, there was no difference in median INS between breeds or the proportion of horses with an elevated INS (>50 µU/mL). Further, there were significant, albeit weak, positive correlations between age and INS ($R=0.32$, $p<0.01$), BCS and INS ($R=0.19$, $p<0.02$), and age and BCS ($R=0.36$, $p<0.01$). However, of greatest interest, one-third to nearly half (varying with breed) of all horses tested had an elevated INS, supportive of insulin dysregulation despite horses being trained for competition and without evidence of laminitis. In conclusion, this convenience sampling study of draft horses provides support that INS may vary between light breeds and draft breeds and that breed needs to be considered when interpreting INS results.

E43

MATRIX METALLOPROTEINASE-2 AND -9 LEVELS IN HORSES WITH EXPERIMENTAL SMALL COLON INTRALUMINAL OBSTRUCTION

Kamila Gravena[1], Beatriz de Assis Pimenta[1], Lina Maria Wehrle Gomide[2], Vinicius Athaydes Canello[1], Nara Saraiva Bernardi[1], Amanda Festa Sabes[1], Guilherme Dias Melo[2], Paulo Ricardo Dell'Armelina Rocha[2], Gisele Fabrino Machado[2], José Corrêa Lacerda-Neto[1]

[1]FCAV - Univ. Estadual Paulista - Unesp, Jaboticabal, São Paulo, Brazil, [2]FMVA - Univ. Estadual Paulista, Araçatuba, São Paulo, Brazil

Among different causes of colic, intestinal obstruction has been identified as the major cause of hospitalization and death in horses around the world. In small colon occurs mainly due to fecalomas or enteroliths. During ischemic process, decreases in oxygen tension inside the cell occur, reducing the generation of adenosine triphosphate (ATP) and resulting in cellular homeostasis disorders. Calcium accumulation inside the cell produces significant increase in the phospholipase A_2 activity, which directly damage cell membrane through increased production of arachidonic acid, lysophosphatidylcholine and platelet aggregating factor. The platelet aggregation factor is a potent inflammatory mediator, which has an important effect on vascular permeability and on neutrophil activity. Neutrophils adhesion to endothelium may also cause damage to endothelial cells, thus amplifying the reaction. This occurs due to release of reactive oxygen species (ROS), proteoglycans enzymes such as matrix metalloproteinases (MMPs) and elastase from lysosomes during cell degeneration. Therefore, the aim of this study was to evaluate neutrophil migration and matrix metalloproteinase-2 and -9 activity in horses with experimental intraluminal obstruction.

For this purpose, eight healthy adult mixed-breed horses were subjected to small colon distension using a surgically implanted latex ball in the lumen (Ethical approval: CEBEA-Protocol #007568-09). Blood and peritoneal fluid samples were obtained before intestinal distension (M0), after 4 hours of distension (M4 - when the ball was deflated and removed) and 72 hours after decompression (M72). Twelve hours after the ball removal was also collected peritoneal fluid sample for neutrophils count. Data were submitted to Friedman's test followed by post-hoc Dunn's Multiple Comparison Test.

This model induced alterations in neutrophils migration into peritoneal fluid, the concentration ($x10^3/\mu L$) at M12 and M72 was significantly higher when compared to M0 (**M0**=0.55±0.14; **M4**=3.21±1.24; **M12**=133.18±35.80; **M72**=35.07±6.08). There was also alterations in plasma pro-enzyme form of MMP-2 (pro-MMP-2) at M4 and in peritoneal fluid pro-MMP-2 at M72 and pro-MMP-9 was significantly increase in M4 when compared to M0. Mean (± standard error) in arbitrary units of plasma pro-MMP-2 was **M0**=94.25±6.14; **M4**=63.72±4.57 and **M72**=89.52±6.70. Peritoneal fluid Pro-MMP-2 was **M0**=53.07±10.90; **M4**=89.42±6.23 and **M72**=128.13±10.86. Plasma Pro-MMP-9 was **M0**=46.75±6.77; **M4**=50.65±6.45 and **M72**=44.92±4.74. Peritoneal fluid Pro-MMP-9 was **M0**=16.70±5.10; **M4**=155.77±5.77 and **M72**=90.21±13.14.

Knowing that multiple proteinases are present in the granules of neutrophils, it was conclude that increase in pro-enzyme form after small colon decompression may be associated with neutrophil migration.

E44

EFFECTS OF COLLAGEN HYDROLYSATES ON EQUINE GASTRIC ULCER SCORES AND GASTRIC JUICE PH IN HORSES

Michael Keowen[1], Pilar Camacho-Luna[1], Lisa Micheau[2], Frank Garza, Jr.[1], Jos Olijve[3], Brian Lamp[3], Chin-Chi Liu[1], Frank Andrews[1]

[1]Equine Health Studies Program, Louisiana State University School of Veterinary Medicine, Baton Rouge, LA, USA, [2]École Nationale Vétérinaire, Toulouse, France, [3]Darling Ingredients International Holding B.V., Kanaaldij Noord 20-21, 5691 NM Son, The Netherlands

Nonglandular (NG) gastric ulcers, as part of the equine gastric ulcer syndrome (EGUS), are common in performance horses and the current FDA-approved pharmaceutical agent, omeprazole is effective in treatment, but expensive and alters gastric juice pH. Recently, there has been interest in natural feed supplements that improve stomach health and can be used in horses competing in drug-free competitions. The purpose of this study was to evaluate the effect of a feed supplement containing porcine collagen hydrolysates on gastric ulcer scores and gastric juice pH in stall-confined horses, treated with omeprazole and undergoing intermittent feeding. Ten Thoroughbreds were used in a two-period crossover study. The supplement (CHL; collagen hydrolysates, 45 grams) was mixed with sweet feed (Omelene 100, Purina Animal Nutrition, LLC, Gray Summit, MO) or sweet feed only (Control group) twice daily for two 56-day periods. From days 14 to 28, both treatment groups were administered omeprazole (OME; 4.0 mg/kg, P.O., Q.D.; Gastrogard®, Merial Limited, Duluth, GA) and from days 42 to 49, horses underwent an alternating feed-deprivation period, to induce or worsen ulcers.[1] Gastroscopies were performed on all horses on days -1 (before treatment) and days14, 28, 42, 49 and 56. Gastric juice was aspirated pH measured at each scoping. Nonglandular gastric ulcer number (NGN) and severity (NGS) scores were assigned at each examination by the PI (FMA), who was masked to treatment. Data were analyzed using an ANOVA for repeated measures and when significant differences were found in the main model, a post-hoc Tukey's test was used to determine significant differences of $P < 0.05$. The CHL-supplement in powder form, mixed with the grain, was readily consumed by all horses and no adverse effects were seen. Mean NGN and NGS gastric ulcer scores were lower at each gastroscopy examination and a significant treatment effect was seen on day 56 of the study. In addition, gastric ulcer scores were significantly lower in both groups on day 28, compared to the other days, due to omeprazole treatment. By day 42, two weeks after omeprazole was discontinued, ulcers reoccurred in both groups of horses. On day 49, gastric ulcer scores were significantly increased due to intermittent feeding, but no treatment effect was observed. Gastric juice pH values were low and variable throughout the trial period, except on day 28 when pH was significantly higher in CHL-supplement treated group compared to control group. Omeprazole treatment was effective in increasing gastric juice pH and significantly decreasing gastric ulcers scores. In addition, the collagen hydrolysate supplement significantly increased gastric juice pH while horses were on omeprazole treatment.

Conclusion: Collagen hydrolysates ameliorated the severity of gastric ulcers in stall-confined horses undergoing feed stress after 56 days of feeding.

[1]Murray MJ and Eichorn ES. (1996) Am J Vet Res. 57(11):1599-1603.

E45

SAFETY, HUMORAL IMMUNE RESPONSE AND FECAL SHEDDING OF MODIFIED-LIVE BOVINE CORONAVIRUS VACCINEs GIVEN TO HORSES

James Prutton[2], Samantha Mapes[1], Nicola Pusterla[1]

[1]*University of California, Davis, USA,* [2]*Liphook Equine Hospital, Liphook, UK*

Equine coronavirus (ECoV) is considered an emerging enteric virus with reported morbidity rates ranging from 10-83% and a fatality rates ranging from 7-27% in adult horses.

The objectives of the study were to assess the safety, humoral response and viral shedding of a commercially available modified-live bovine coronavirus (BCoV) vaccine given to healthy adult horses. Twelve confirmed healthy adult horses were vaccinated twice, 3 weeks apart, either orally, intra-nasally or intra-rectally. Two healthy unvaccinated horses served as sentinel controls. Following each vaccine administration, horses were monitored daily for physical abnormalities and onset and duration of BCoV shedding determined by quantitative PCR (qPCR) in nasal secretions and feces. Whole blood was collected every 3 weeks to determine BCoV-specific antibodies.

With the exception of transient and self-limiting changes in fecal character observed in 7 vaccinated and 1 control horse, no additional abnormal clinical findings were found in the study horses. Following the first and second vaccine administration, two and one horse, respectively, tested qPCR-positive for BCoV in nasal secretions 1 d post intra-nasal vaccination. No vaccinated horses tested qPCR positive for BCoV in feces following each vaccine administration. One of the two horses that shed BCoV seroconverted to BCoV after the first vaccine administration and an additional 2 vaccinated horses (oral and intra-rectal) seroconverted to BCoV after the second vaccine administration.

In conclusion, the results show that the modified-live BCoV is safe to administer to horses via various routes, causes minimal shedding and detectable antibodies to BCoV in 25% of the vaccinates.

E46

DO HORSES WITH EQUINE METABOLIC SYNDROME HAVE REDUCED IMMUNE RESPONSES TO VACCINATION?

Sarah Elzinga, Stephanie Reedy, Day Barker, Tom Chambers, Amanda Adams

University of Kentucky, Lexington, KY, USA

Recent reports indicate the percentage of overweight horses may range anywhere from 20.6-41%. Obesity is a key component of Equine Metabolic Syndrome (EMS), which is typically defined by criteria set in a 2010 ACVIM consensus statement as; regional (neck crest, rump, etc.) or general obesity, hyperinsulinemia or insulin resistance, and a history of or predisposition towards laminitis. Obese humans and diet-induced obese mice have been shown to have reduced immune responses to vaccination. However, the immune response of EMS horses to vaccination is unknown. Therefore, the objective of this study was to determine the effect of EMS on immune responses to routine vaccination. We hypothesized that, similar to other species, EMS horses would have a reduced response to a routine influenza vaccine compared to non-EMS age matched controls.

To establish this, 25 horses from a pre-existing herd were used for this study. Of these, 13 were classified as EMS by the 2010 AVCIM consensus statement criteria, and 12 were non-EMS age matched controls. Within each group, 4 served as saline controls and the remaining horses served as vaccinates. A low-dose dexamethasone suppression test and TRH stimulation test were performed on all horses prior to the start of the study to insure no horses tested positive for pituitary pars intermedia dysfunction (PPID). Venous blood was drawn weekly for 6 consecutive weeks, with vaccination or saline administered intramuscularly on weeks 0 and 3. Horses were vaccinated with an inactivated influenza vaccine (Fluvac Innovator®, Zoetis). Blood samples were used to determine humoral and cell mediated immune responses. Humoral immune responses against KY/97 were measured by hemagglutination assay for total Ig antibody responses and ELISA for IgGa, IgGb, and IgGt antibody isotype responses. Cell mediated immune measures were determined via PBMC isolation, EIV-stimulation, and subsequent flow cytometry and RT-PCR analysis for cytokine protein and gene expression, respectively. Humoral immune data were log transformed and analyzed using a repeated measures design with protected LSD, adjusting for age. All analyses were completed using PROC MIXED, SAS 9.4 (SAS Institute Inc., Cary, NC, USA). Results are expressed as least square means $\pm$ standard error of the mean. Data were considered statistically significant when $P \leq 0.05$.

Overall, all horses responded to vaccination with a significant increase ($p < 0.05$) in HI titers following initial vaccination compared to saline controls. Significant increases ($p < 0.001$) in IgGa and IgGb, but not IgGt EIV specific antibody responses to vaccination were also observed. EMS did not significantly affect ($p > 0.05$) humoral immune responses as measured by HI titers or IgG isotypes to influenza vaccination, however more work is needed to explore possible differences in CMI responses. It is possible that obesity as compared to insulin resistance plays a larger role in modulating the immune process considering that while EMS horses were insulin resistant and had significantly greater regional obesity (cresty neck score) compared to non-EMS controls, they had moderate general obesity (body condition score), which may have not been sufficient to impact immune response to vaccination.

E47

IMMUNOLOGICAL COMPARISONS OF AGED HORSES WITH VS. WITHOUT PITUITARY PARS INTERMEDIA DYSFUNCTION.

Melissa H. Siard, Virgina D. Barker, Stephanie E. Reedy, Amanda A. Adams

Gluck Equine Research Center, Lexington, KY, USA

Pituitary pars intermedia dysfunction (PPID) is an endocrinopathy affecting 15-30% of the aged equid population. PPID is characterized by various clinical manifestations, the hallmark of which is hypertrichosis. Additionally, PPID horses exhibit immunosuppression, making them more susceptible to opportunistic infections. Our lab has established that aged horses exhibit both immunosenescence and inflamm-aging, the low-grade chronic inflammation occurring systemically with aging. To determine whether these age-related changes in cell-mediated immune responses may be affected by the presence of PPID, peripheral blood mononuclear cell (PBMC) proliferation and cytokine production were compared in age-matched PPID and non-PPID horses. PPID status was determined using thyrotropin releasing hormone (TRH) stimulation testing.

Sixteen aged horses (18-27 yrs), including n=8 PPID horses (mean = 24.5±2.2 yrs) and n=8 non-PPID horses (mean = 23.1±3.1 yrs) of mixed breeds and sex, were sampled to determine immune function. Heparinized blood was collected aseptically, and PBMCs were isolated, stimulated with Phorbol 12-myristate 13-acetate (PMA), and stained intracellularly for pro-inflammatory cytokines interferon-γ (IFNγ) and tumor necrosis factor-α (TNFα). Flow cytometry was performed to determine the percent of lymphocytes producing IFNγ and TNFα. Real-time polymerase chain reaction (RT-PCR) was used to determine PBMC gene expression of cytokines interleukin (IL)-2, IL-4, IL-6, IFNγ, and TNFα. PBMC proliferation was also determined using carboxyfluorescein succinimidyl ester (CFSE) staining and stimulation with various amount of concanavalin A (Con A; 2.5, 5, and 10 L/mL). TRH stimulation testing was performed, in which adrenocorticotropin hormone (ACTH) levels were measured in plasma 10 minutes post (T-10) intravenous administration of TRH (1mg/mL saline/horse).

Data was analyzed using the statistical software program SigmaPlot version 12.3. To analyze flow cytometry and RT-PCR data for differences between PPID and non-PPID horses, t-tests were performed. To analyze lymphocyte proliferation data at various concentrations of Con A stimulation, two-way analysis of variance with repeated measures was performed. No significant differences between PPID and non-PPID horses were found for any of the immune measures examined including flow cytometry, RT-PCR, and proliferation data ($p>0.05$). These results indicate that the immunosuppression associated with PPID does not appear to be due to differences in PBMC proliferation or cytokine production when comparing aged PPID horses with aged non-PPID horses. Thus, more research is warranted to further understand the immunological mechanisms responsible for immunosuppression and susceptibility to opportunistic infections in the PPID horse.

E48

CORYNEBACTERIUM PSEUDOTUBERCULOSIS ANTIBODY DETECTION IN HORSES: SYNERGISTIC HEMOLYSIS INHIBITION TEST AND SMALL RUMINANT ELISA COMPARISON

Marta Barba[1], Anne A. Wooldridge[1], Robert Glass[2], Thomas Passler[1], Allison J. Stewart[1]

[1]*Auburn University, Auburn, AL, USA,* [2]*Pan American Veterinary Laboratories, Hutto, TX, USA*

False positive results may occur when testing horses for antibodies against *Corynebacterium pseudotuberculosis* by synergistic hemolysis inhibition (SHI). In small ruminants, ELISA tests based on exotoxin and cell wall antigens have greater accuracy than SHI in diagnosis of caseous lymphadenitis. The purpose of this study was to compare the detection of *C. pseudotuberculosis* antibodies in equine serum by SHI and a small ruminant ELISA test that uses exotoxin and cell wall antigens. Sera from 7 ponies experimentally infected with *C. pseudotuberculosis* were analyzed by both tests. Correlation and agreement were calculated by Pearson and Kappa coefficients, respectively. Receiver operating characteristic analysis was used to obtain the optimal cut-off value for the calculated ELISA score (Patient optical density / Negative control optical density x 100). Antigen reactivity to the sera was evaluated by immunoblotting. When SHI titers ≥1:128 were considered a positive result, the optimal ELISA score cut-off to determine positive status was 106%, with 73% sensitivity and 72% specificity with respect to SHI. Correlation between both tests was strong [r=0.904; (95%CI 0.472-0.986, P=0.005)]. Agreement in determining positive status was poor (Kappa=0.439; (95%CI 0.226-0.652). Correlation and agreement were strong in 3/7 ponies, but weak in 4/7. Immunoblot showed a band 13.4 times more intense in infected compared to pre-infected sera corresponding with the exotoxin antigen, but only non-specific reactivity with the cell wall antigen. The use of this ELISA test in horses is not recommended. Development of an ELISA test with specific antigens from *C. pseudotuberculosis* biovar *equi* is needed.

E49

EFFECTS OF ERYTHROMYCIN ON RESPONSES INDUCED IN FOALS BY INTRAVENOUS EPINEPHRINE.

Amy Stieler[2], Chris Sanchez[1], Martha Mallicote[1], Ashley Coxen[1], Jim Burrow[1], Rob MacKay[1]

[1]*Department of Large Animal Clinical Sciences, University of Florida, Gainesville, FL, USA,* [2]*University of Georgia, Athens, GA, USA*

Macrolide-induced anhidrosis is a concern in foals treated for *Rhodococcus equi* pneumonia. Because sweating in horses is caused by activation of β2-adrenergic receptors on sweat glands, it was of interest to evaluate responses to IV epinephrine in foals given erythromycin. Four pony-cross foals were treated for 3 days prior to epinephrine infusion with either erythromycin (25 mg/kg orally, 3 times daily) or a control preparation according to a 2-period randomized masked design. Epinephrine was given by CRI over 10 minutes at 0.5 μg/kg/minute and the following data were collected before and at intervals for 180 minutes after onset of infusion: sweat (absorbed into absorbent pads), heart and respiratory rates, systemic arterial BP, pupil size, PCV, plasma protein, and blood glucose concentration. Cardiac rhythm was monitored continuously by ECG. Univariate and 2-factor repeated measures ANOVA were used to analyze data. Friedman tests were used for data sets that failed assumptions for ANOVA. *Post hoc* pairwise analyses were performed after ANOVA indicated significant ($P<0.05$) effect. In foals given either treatment, there were significant ($P<0.05$) effects of epinephrine on heart rate, pupil size, PCV and blood glucose. Sweating and arterial BP also increased significantly in control- but not erythromycin-treated foals. Significant treatment x time interaction was found only for sweating; pairwise comparisons were significant at every point after time 0. These results suggest that erythromycin does not cause generalized suppression of α- and β-adrenergic responses. Anhidrosis may be a tissue-specific receptor or post-receptor effect of erythromycin and other macrolides.

E50

EQUINE VECTOR-BORNE DISEASES DETERMINED BY SEROLOGICAL AND MOLECULAR DIAGNSOTICS

Barbara Qurollo[1], Barbara Hegarty[1], Jeffrey Tyrrell[1], Edward Breitschwerdt[1], Susan Tornquist[2], Kathryn Schlaich[2], Jennifer Kelsey[2], Melissa Beall[3], Ramaswamy Chandrashekar[3]

[1]Intracelluar Pathogens Research Laboratory, College of Veterinary Medicine, North Carolina State University, Raleigh, North Carolina, USA, [2]Oregon State University, College of Veterinary Medicine, Corvallis, OR, USA, [3]IDEXX Laboratories, Inc, Westbrook, Maine, USA

Vector-borne disease impacts the health and welfare of horses, worldwide. Recognized vector-borne diseases among horses include Anaplasmosis (*Anaplasma phagocytophilum*), Lyme Disease (*Borrelia burgdorferi*), Potomac Horse Fever (*Neorickettsia risticii*) and Piroplasmosis (*Babesia caballi /Theleria equi*). Additionally, *Bartonella*, *Ehrlichia*, *Rickettsia*, and hemotropic *Mycoplasma* species have been detected in horses with clinical disease spectrums ranging from nonclinical to fatal. The goal of this study is to investigate serological and molecular (PCR) tests that will detect vector-borne disease (VBD) pathogens in horses and apply a broad diagnostic panel to gain a more definitive picture of vector-borne equine pathogens, including novel species and co-infections. Three populations of horses were used to test serological and molecular diagnostic tests. Group I (n=16): EDTA-whole blood (WB) and serum (S) sets from clinically healthy horses pastured in North Carolina with no clinical suspicion of VBD; Group II (n=20): WB and S sets from horses submitted for diagnostic testing to the VBDDL, where presumably the clinician suspected vector-borne disease; Group III (n=15): WB from horses in Mérida, Nicaragua with a high degree of tick exposure. Serologic and PCR testing were performed retrospectively. DNA was extracted from WB, and qPCR assays were performed to test samples for infection by bacterial species within the following genera: *Anaplasma, Babesia, Bartonella, Ehrlichia, Mycoplasma, Neorickettsia, Rickettsia* and *Theilera*. For samples with positive results, the infectious agent was speciated by species-specific qPCR assay of the sample and/or DNA sequencing of the PCR amplicon. Seroreactivity to *Babesia*, *Bartonella*, *Neorickettsia* and *Rickettsia* antigens was determined using IFA, and seroreactivity to 3 *Ehrlichia species*, 2 *Anaplasma species* and *B. burgdorferi* antigens was determined using both a commercially available SNAP®4DX®Plus ELISA and a research-based SNAP® MA ELISA. The combined use of serological and molecular assays in this study gave evidence of infection (by PCR) or exposure (by seroreactivity) to *Anaplasma*, *Babesia*, *Bartonella*, *Borrelia*, *Ehrlichia spp.*, *Theileria,* and *Rickettsia* spp. None of the samples contained DNA from Mycoplasma spp. and two samples were positive for infection by *Rickettsia felis*, constituting the first reported instances of equine infection with *R. felis* detected by PCR. A novel Equine *Ehrlichia* species, reported in horses from Nicaragua, was detected in additional Nicaraguan horse samples from this study by both PCR, IFA and ELISA. This study highlights the wide-range of exposure to and infection with vector-borne pathogens in horses.

E51

SELENIUM DEFICIENCY ASSOCIATED WITH THE DEATHS OF FIFTEEN ADULT HORSES

Andrew Allen, Debra Sellon, Fairfield Bain, Danielle Nelson

Washington State University, Pullman, WA, USA

Nutritional myopathy due to selenium (Se) deficiency typically presents itself in young rapidly growing calves, lambs, kids, and foals. Reports of myodegeneration secondary to selenium deficiency in adult horses are less frequent. We investigated 2 herds with nutritional myopathy with generalized skeletal muscle degeneration and significant death loss in adult horses. In 2014, two American Paint horses from the same farm presented to our diagnostic laboratory for post mortem evaluation. Significant necropsy and histopathology findings included myonecrosis of the cardiac and skeletal muscles, widespread subcutaneous edema, pleural effusion and ulcerative glossitis. An on farm investigation was performed where it was discovered that 13 horses, including the two presented horses, had died on this farm with similar clinical signs over the last year. Complete blood counts and serum chemistry panels were performed on 4 of the 12 remaining horses and physical examinations were performed on the entire herd. All four horses had elevated creatinine kinase activity (CK) and aspartate aminotransferase (AST) levels, the herd had an overall low body condition score, 3 horses had elevated heart rates, and one horse had pitting ventral edema. All of the horses had extremely low blood selenium levels, no box elder trees were found on the farm and gastric contents of the dead horses and feed supplements were found to be free of ionophores. An on farm echocardiographic study was performed 6 months post initial investigation on three of the most severely affected horses. All findings appeared within normal limits. In 2015 a second investigation of two dead horses revealed a very similar scenario and herd wide myodegeneration secondary to selenium deficiency. Although unusual, selenium deficiency should be considered a rule out with adult horses showing signs of ventral edema, weakness, and abrupt death.

E52

EFFECTS OF ALFA-LOX FORAGE® ON BLOOD GLUCOSE AND INSULIN ACTIVITY AFTER GRAIN FEEDING IN HORSES

Frank Garza, Jr.[1], Pilar Camacho-Luna[1], Kelsey Bailey[1], Isabelle Nesen[2], Michael Keowen[1], Liu Chin-Chi[1], Frank Andrews[1]

[1]Equine Health Studies Program, Louisiana State University School of Veterinary Medicine, Baton Rouge, LA, USA, [2]Ecole Nationale Veterinaire, Toulouse, France

Processed concentrate feeds contain a high concentration of sugars and in some cases non-structural carbohydrates (NSC) exceed 35%. These feeds are commonly fed to stall-confined horses, especially those in competition. Grains, high in NSC, can overwhelm glucose receptors in the proximal small intestine and increase delivery of sugars to the hindgut leading to fermentation and colonic acidosis. Feed supplements, that might enhance glucose absorption in the small intestine, would provide more energy substrate for exercise and decrease delivery of sugars to the hindgut. The purpose of this study was to determine effects of a feed supplement (ALF; Alfa-Lox Forage, Triple Crown Nutrition, Inc., Wayzata, MN), containing chopped alfalfa hay, mannanoligosaccharides, Omega 3 Fatty Acids, and L-Carnitine, top-dressed on grain, on plasma glucose concentration and serum insulin activity in horses before and after grain feeding. Eight healthy Thoroughbreds (7 geldings and 1 mare; ages 2-12 yrs.) were used in a 2-period cross over study, where horses were fed ALF (2.0 lbs. [0.91 Kg], twice daily) top dressed on a sweet feed grain (2.5 Kg; Omelene® 100, Purina Animal Nutrition, LLC., Grey Summit, MO) or control (grain alone) for 56 days, as part of another study. On Days 41 and 48, after withdrawing feed overnight, blood samples were drawn at 8 am, before the morning grain feeding, with or without ALF, and then again 2 hours after ALF was top-dressed on grain (treated) or grain alone (control) was fed. Plasma and serum were separated within 30 minutes of collection and samples frozen at -70 C for shipment to the Diagnostic Center for Population Animal Health (DCPAH; Michigan State University, East Lansing, MI). Plasma glucose and serum insulin were analyzed using an automated chemistry analyzer. Data were analyzed statistically using an ANOVA (PROC Mixed; SAS, Cary, NC) for repeated measures and when significant differences ($P < 0.05$) were found in the main model, post-hoc t-tests were performed to determine significant differences. Alfa-Lox Forage® top-dressed on grain was readily eaten by all horses, and did not result in any adverse effects. When period data were pooled, on day 42, fasting mean (SEM) serum insulin activity significantly increased from 65.1 (3.5) pmol/L and 61.9 (5.1) pmol/L in control and ALF-treated horses, respectively, to 126.3 (24.2) pmol/L and 203.1 (30.1) pmol/L, respectively, 2 hours after feeding sweet feed. In addition, on day 48, mean serum insulin activity significantly increased from mean (SEM) serum insulin concentrations of 76.3 (4.3) pmol/L and 69.9 (3.8) pmol/L in control and ALF-treated horses, respectively to 194.6 pmol/L and 290.0 pmol/L, respectively, two hours after feeding sweet feed. Mean serum insulin activity was significantly higher in ALF-treated horses compared to control horses 2 hours after feeding sweet feed. Mean plasma glucose concentrations did not significantly increase 2 hours after feeding. Alfa-Lox® Forage top-dressed on grain resulted in higher blood insulin values, but not significantly increased blood glucose concentrations 2 hours after feeding sweet feed. Alfa-Lox Forage® resulted in increased absorption of sugars from the small intestine and more efficient utilization for glucose for energy, which likely decreased the amount of sugars delivered to the large colon.

E53

VENOUS BLOOD GAS, ELECTROLYTE AND METABOLITE FINDINGS IN HEALTHY NEONATAL FOALS RECEIVING SODIUM LACTATE INFUSIONS

Breanna Sheahan[1], Pamela Wilkins[1], Ray Boston[2], Kara Lascola[1], Igor Canisso[1], Kevin Kline[3], Scott Austin[1], Levent Dirikolu[1], Brian Aldridge[1]

[1]*University of Illinois College of Veterinary Medicine, Urbana, IL, USA,* [2]*University of Pennsylvania School of Veterinary Medicine, New Bolton Center, Kennett Square, USA,* [3]*University of Illinois Department of Animal Science, Urbana, IL, USA*

Exogenous sodium L-lactate (LAC) (1 mmol/kg BWT over 15 min) was administered to 6 healthy neonatal foals to determine LAC clearance variables at 12 hours (12Hr) and again at 60 hours (60Hr) of age. We hypothesized there would be differences in some venous blood gas (VBG) variables associated with both age (12Hr vs. 60Hr) and sample time.

Venous blood was collected into heparinized syringes via a catheter in the right jugular vein prior to lactate infusion for blood gas (BG), electrolyte and metabolite analyses and repeated at 0, 15, and 60 minutes after infusion completion using the NOVA Critical Care Express. Statistical comparisons were by Friedman analysis between time points of the same infusion, and Wilcoxon Rank Sum between age groups, $p < 0.05$. Two-way ANOVA by general linear model was also used to assess differences and interactions associated with age at testing and sample time.

There were no appreciable deleterious clinical effects of infusion. Sodium LAC infusion in normal neonatal foals resulted in clinically unimportant changes in VBG measures. HCT, O2CT, Cl-, LAC, creatinine and BUN were significantly greater in 12Hr foals. pH, O2SAT, HCO3-, BEecf, K+ and Ca++ were significantly greater in 60Hr foals. Only LAC was different related to sample time, being largest at end infusion. No differences were found for pCO2, pO2, Mg++, glucose, TCO2, Gap, or osmolarity.

Infusion of sodium LAC does not importantly impact VBG, electrolyte or metabolite measures in healthy 12Hr and 60Hr foals, with the identified differences primarily due to age.

E54

GROWTH AND FUNCTION OF EQUINE ENDOTHELIAL PROGENITOR CELLS LABELED WITH SEMICONDUCTOR QUANTUM DOTS

Randolph L. Winter[1], Wen Seeto[2], Yuan Tian[2], Fred J. Caldwell[1], Elizabeth A. Lipke[2], Anne. A Wooldridge[1]

[1]Auburn University College of Veterinary Medicine, Department of Clinical Sciences, Auburn, AL, USA, [2]Auburn University, Department of Chemical Engineering, Auburn, AL, USA

Endothelial progenitor cells (EPCs) contribute to neovascularization and vascular repair in vivo and are attractive for clinical use in ischemic disease. EPC tracking is essential to determine engraftment after administration. Semiconductor quantum dots (QD) are promising for cell labeling due to ease of uptake by many cell lines and their continued presence after many cell generations. The purpose of this study was to evaluate function and growth of equine EPCs after QD label.

Equine EPCs were incubated with QD (2 – 20nM) for 12-hr or 24-hr, and intensity of label was assessed with fluorescent microscopy. Cell proliferation of EPCs labeled with QD for optimum time and concentration was then assessed by comparing the number of cell doublings per day (NCD) and the population doubling time (PDT) in labeled and unlabeled cells. EPC function was assessed by comparing uptake of acetylated low-density lipoprotein (DiO-Ac-LDL) and in vitro tubule formation in labeled and unlabeled cells.

Equine EPCs readily labeled with QD, showing maximum fluorescence using 20nM QD, with a 24-hr incubation time. NCD and PDT were unchanged with QD label ($p=0.95$, $p=0.99$). Uptake of DiO-Ac-LDL by EPCs was not affected by the presence of QD label (97.9% labeled cells; 97.0% unlabeled cells). Tubule formation on Matrigel was not affected by the presence of QD label.

Equine EPCs are effectively labeled with QD, and QD concentrations up to 20nM do not affect cell growth or function. The use of QD labeling with equine EPCs may help track engraftment of EPCs in clinical applications.

F01

PREVALENCE OF COAGULASE-NEGATIVE STAPHYLOCOCCI SPECIES IN INTRAMAMMARY INFECTION IN DAIRY GOATS

Véronique Bernier Gosselin, Pamela R. F. Adkins, John R. Middleton

University of Missouri, Columbia, Missouri, USA

Coagulase negative staphylococci (CNS) are the most prevalent pathogens isolated from intramammary infection in goats, with more than 15 different species being identified in goat milk. Unlike cows, the relationship between CNS infection, somatic cell count and milk loss in goats is less clear. The conflicting results between studies on CNS mastitis in goats may be explained by the variation between studies in regard to distribution of CNS species, varying pathogenicity between species, the presence of confounding factors affecting outcome measures, and variation between studies in regard to speciation methods. Phenotypic speciation was used in most studies; however genotyping is considered more reliable than phenotyping for identification of CNS species in goats. The objective of this study was to estimate the prevalence of species-specific CNS intramammary infections in a large dairy goat herd in Missouri.

Composite milk samples were aseptically collected from all 940 lactating goats and aerobic bacteriological culture was performed. A culture yielding 100 CFU/ml CNS was considered positive. Isolates from milk samples yielding a single colony morphology identified as CNS were used to prepare lysates for PCR amplification. Species identification was based on PCR amplification and sequencing of either the *rpoB*, *tuf* gene, or 16S rRNA. Sequences were compared with the GenBank database using the NCBI Nucleotide-BLAST algorithm. For *rpoB*, species identification was assigned with ≥ 97% identity and ≥ 5% separation between different species. For *tuf*, species identification was assigned with ≥ 98% identity and > 0.8% separation between different species. For 16S, species identification was assigned with ≥ 99% identity and > 0.8% separation between different species. PCR amplification and sequencing of *rpoB*, *tuf* and 16S were performed sequentially until identification was assigned. For instance, *tuf* was performed on isolates for which *rpoB* either yielded unsuccessful amplification, or criterion for identification with sequence analysis were not met, followed by 16S if *tuf* amplification or sequence analysis failed.

From 940 milk samples cultured, 214 yielded growth of CNS of a single colony morphology. Of these 214 isolates, 184 were successfully speciated, of which 157 were identified with *rpoB*, 26 with *tuf*, and 1 with 16S. The remaining 30 isolates were unable to be speciated based on the criteria listed above. Overall, 13 different species were identified (Table 1). Among isolates that could not be assigned an identification, 3 met the criterion for identity but insufficient separation between the species (*S. warneri* and *S. pasteuri*) with both *rpoB* and *tuf* sequencing.

Table 1. Species-specific prevalence among 184 CNS isolates and number of isolates successfully speciated by each gene, per species.

Species	Number of isolates	%	*rpoB*	*tuf*	16S
S. simulans	69	37.5	65	4	0
S. caprae	31	16.8	30	1	0
S. chromogenes	26	14.1	25	0	1

S. epidermidis	19	10.3	17	2	0
S. xylosus	16	8.7	12	4	0
S. equorum	7	3.8	0	7	0
S. cohnii	6	3.3	3	3	0
S. lentus	5	2.7	1	4	0
S. arlettae	1	0.5	1	0	0
S. aureus	1	0.5	1	0	0
S. auricularis	1	0.5	1	0	0
S. gallinarum	1	0.5	1	0	0
S. sciuri	1	0.5	0	1	0
Total	184	100	157	26	0

Typeability of CNS isolates from goat milk in this study using gene sequencing was 86%. Typeability with *rpoB* gene sequencing alone was 73%, which was lower than expected. This may result from underrepresentation of isolates of caprine origin in the reference database. Unsuccessful PCR amplification in a proportion of isolates could reflect imperfect match between the primer and the target sequence. Relative prevalence of different species was similar to that of previous studies using genotyping or phenotyping.

F02

ASSESSMENT OF AN ANTIMICROBIAL-USE ALGORITHM FOR TREATMENT OF DIARRHEA IN DAIRY CALVES

Diego E. Gomez[1], Luis G. Arroyo[2], Laurent Viel[2], J. Scott Weese[1]

[1]Department of Pathobiology, Ontario Veterinary College, University of Guelph., Guelph, Ontario., Canada, [2]Department of Clinical Studies, Ontario Veterinary College, University of Guelph., Guelph, Ontario., Canada

The objective of this study was to evaluate the impact of implementing an antimicrobial-use algorithm on antimicrobial (AB) treatment and mortality rates of diarrheic calves.

Three commercial dairy farms (F1, F2, F3) were enrolled. Farm records were obtained to collect data about the incidence of diarrhea, AB treatment and mortality rates for the preceding 18 months. Then, an antimicrobial-use algorithm for farm personnel based on the presence of three clinical signs (diarrhea, fever and decreased milk intake) was developed to direct AB therapy in diarrheic calves (<30 day-old) and the same farm data were collected from the same records systems for 5, 2 and 2 months for F1, F2 and F3, respectively.

Treatment records of 1208 (F1), 1051 (F2) and 296 (F3) calves were available for the pre-intervention period. The incidence of diarrhea was 86, 90 and 57%, AB were administered to 99, 95 and 33% of diarrheic calves and mortality rates attributed to diarrhea were 3, 3 and 2% in F1, F2 and F3, respectively. After the algorithm implementation, data were available for 105 (F1), 282 (F2) and 31(F3) calves. Diarrhea developed in 74, 89 and 67% calves on the three farms, but treatment rates were significantly lower; 39, 14 and 0% ($P<0.0001$), with no impact on mortality (1.3, 0.8 and 0%, respectively, $P>0.05$).

The use of a simple and practical antimicrobial-use algorithm can provided a significant reduction in the use of antimicrobials, decreasing treatment costs and improving antimicrobial stewardship with no negative impacts on calf health.

F03

PREVALENCE OF BOVINE CORONAVIRUS IN FECES OF HEALTHY AND DIARRHEIC CALVES

Diego E. Gomez[1], Luis G. Arroyo[2], Laurent Viel[2], J. Scott Weese[1]

[1]Department of Pathobiology, Ontario Veterinary College, University of Guelph., Guelph, Ontario, Canada, [2]department of Clinical Studies, Ontario Veterinary College, University of Guelph., Guelph, Ontario, Canada

Bovine coronavirus (BoCV) is an endemic pathogen that can be associated with enteric disease, especially in calves. This study was conducted to assess the prevalence of BCoV in healthy and diarrheic calves and to determine whether the presence of BCoV in feces is associated with the age and health status of the calf.

This was a prospective case-control study. Fecal samples from 89 diarrheic calves from three different commercial dairy farms were tested for BCoV using a PCR assay. For each study case one age-matched control calf from the same farm was enrolled. Categorical variables were analyzed using X^2 and univariable logistic regression analyses. Multivariable analysis using a mixed effect logistic regression model (with farm as a random effect) was performed to determine significant association between testing positive for BCoV in feces with the age and the presence of diarrhea. Age of the calf was categorized using a cut-off of 10 days.

The age of healthy and diarrheic calves was 8.5±3 and 9.4±3.9 days, respectively (P=0.06). There was a significant difference in the prevalence of BCoV between healthy (54%) and diarrheic (69%) calves (P=0.039). Multivariable logistic regression revealed that calves positive for BCoV were more likely to be older than 10 days of age (OR: 2.2, CI, 1.21 to 3.95; P=0.009) and have diarrhea (OR: 1.73, CI, 1.07 to 2.81; P=0.025).

This study determined that BCoV can be detected in both healthy and diarrheic calves; however, the likelihood of BCoV is higher in diarrheic calves. Future studies investigating whether BCoV strains detected in healthy and diarrheic calves are genetically different are warranted.

F04

USE OF AN ALIVECOR HEART MONITOR FOR HEART RATE AND RHYTHM EVALUATION IN DOMESTIC GOATS

Joe Smith[1], *Meera Heller*[2], *Fauna Smith*[2], *Mikaela Mueller*[3], *Jessica Ward*[1]

[1]Iowa State University, Ames, Iowa, USA, [2]University of California Davis, Davis, California, USA, [3]Blue Pearl Veterinary Partners, Seattle, Washinton, USA

The AliveCor Heart Monitor, a smartphone-based electrocardiogram (ECG) recording device, is a promising new diagnostic tool in veterinary medicine. While this device has been validated for canine, feline, and equine species, its clinical utility in domestic goats has not yet been evaluated.

Eight healthy adult dairy breed and eight healthy adult meat breed goats had simultaneous AliveCor and standard base apex ECGs performed in a standing position. The AliveCor ECG was recorded using two different device positions (over the left cardiac apex and over the brisket region). Paper speed was 25 mm/sec and amplitude was 10 mm/mV. All ECGs were evaluated unpaired and independently by a data-blinded veterinary cardiologist. The average heart rate was determined using 6 seconds of ECG recording, rhythm diagnosis was recorded for each tracing, and each ECG was given a quality score (0-3). The findings were then reviewed to determine if the AliveCor device was comparable to the base apex ECG for rate and rhythm diagnosis via a paired t-test. Differences in quality scores between the readings for each goat were assessed via a Mann-Whitney U-test. Quality score differences were assessed with a Mann-Whitney U-test between breeds and location of AliveCor ECG. Rhythm agreement was assessed by comparing the blinded rhythm diagnoses made by the cardiologist.

No significant differences were found between heart rates collected via AliveCor or standard base apex ECGs ($P = 0.8151$). Quality scores were significantly higher for AliveCor tracings (1.85 +/- 0.66) compared to standard base apex tracings (0.94 +/- 0.78) ($P < 0.0001$). No significant difference in AliveCor quality score was noted between meat and dairy breeds ($P = 0.094$). No significant differences were noted between quality scores of AliveCor readings taken from the left cardiac apex or sternal region ($P = 1.000$). Normal sinus rhythm was present in all goats, and could be diagnosed via the AliveCor tracings in 93.75% of cases. Based on this study, the AliveCor ECG was clinically useful for heart rate determination and diagnosis of normal sinus rhythm in goats. Interestingly, AliveCor ECGs had higher quality scores compared to standard base apex ECGs.

F05

LEFT DISPLACEMENT OF THE ABOMASUM IN FOUR BEEF CALVES

Rachel Oman[1], Robert Streeter[1], Emily Reppert[2], Clemence Chako[3]

[1]Oklahoma State University, Stillwater, OK, USA, [2]Kansas State University, Manhattan, KS, USA, [3]Midwestern University, Glendale, AZ, USA

Left displaced abomasum (LDA) is an uncommon condition in beef cattle compared to dairy cattle and is rarely described in sucking beef calves. The true prevalence of LDA in beef calves is unknown and difficult to gauge based on the few reported cases. There remains a paucity of information regarding the diagnosis and treatment of LDA in beef calves.

Four beef calves (two mixed breed beef, one Beefmaster, and one Angus) between two weeks and three months of age were treated for LDA between November 2013 and March 2015. All four calves had a history of decreased appetite and left-sided abdominal distention; two had recently been treated for necrotic laryngitis and one calf was being treated for suspected clostridial abomasitis. A high-pitched ping and splash were ausculted with percussion and succussion of the left abdomen. Ultrasonography confirmed the abomasum to be displaced between the rumen and the left body wall in all four calves. Blood gas analysis revealed hypokalemic, hypochloremic metabolic alkalosis in three of the calves.

All four calves were initially treated by rolling to correct the left abomasal displacement, and facilitate correction of metabolic abnormalities. Following rolling, the abomasum re-displaced in three of the calves within one hour to 11 days; one calf became acutely colicky as a result of a mesenteric volvulus. A right paramedian abomasopexy was preformed in all cases. Three of four calves grew well and remained in the herd six to 18 months later; one calf was euthanized 10 days after surgery due to complications associated with necrotic laryngitis.

Based on these findings, LDA is likely more common in beef calves than many veterinarians recognize and it should be considered as a differential for calves that present with abdominal distention. Concurrent laryngeal dysfunction has not been previously described in beef calves with LDA and may be a contributing factor in the pathogenesis of this disorder. Ultrasonography is a widely available chute-side diagnostic technique that can rapidly confirm the diagnosis in beef calves. Rolling is a useful adjunctive measure for allowing temporary improvement in cases with severe metabolic alterations, but may be associated with catastrophic intestinal complications in calves as has been reported in adult cows. Therefore, surgical correction is recommended, and right paramedian abomasopexy under sedation and local block is a feasible method of correction of LDA in beef calves.

F06

APPARENT EFFICIENCY OF COLOSTRAL IMMUNOGLOBULIN ABSORPTION IN HOLSTEIN HEIFERS

Jennifer Halleran, Derek Foster

[1]*Oklahoma State University, Stillwater, OK, USA,* [2]*North Carolina State University, Raleigh, NC, USA*

Adequate absorption of bovine colostrum correlates with improved neonatal health. The apparent efficiency of absorption (AEA) of immunoglobulins can be measured using a mathematical equation based on serum and colostral IgG concentration levels, as well as the volume of colostrum being fed and the body weight of the calf. While commonly measured in research projects, little information is available on the normal AEA across a large group of healthy calves on multiple farms. The purpose of this study was to observe how contributing factors (volume of feeding, birthweight and time of feeding) can alter AEA, and to establish a normal range of AEA in healthy calves.

One hundred and two Holstein heifer calves from 5 different dairies in North Carolina and Colorado were sampled. After a normal calving, an aliquot of colostrum being fed to the heifer was saved and the heifer received either 4 liters or 5.6 liters of colostrum within 6 hours of birth and a blood sample was collected between 24 and 36 hours after birth. Birthweights were measured using the same weight tape on each farm. Radial immunodiffusion assay was performed to obtain IgG concentrations in the colostrum and serum samples. From this data, the AEA was calculated.

AEA ranged from 7.7% to 59.9% with mean of 28.1% +/- 9.5 and median of 27.5%. AEA of 69% (70/102) of the calves fell between 21-40%. There was a significant correlation ($p=0.033$, $r=0.21$) between birthweight and AEA and volume fed and AEA ($p=0.030$, $r=0.22$). There was no correlation between time of feeding and AEA or total IgG mass fed and AEA.

AEA varies widely between calves even when feeding is standardized. There appears to be opportunities to improve the serum IgG concentration in calves by feeding increased volumes of colostrum, as in this study we were unable to saturate the absorption process. With the wide range of AEA values obtained, there may be opportunities to improve AEA through genetic selection.

F07

CONCENTRATIONS OF CHLORTETRACYCLINE IN FETAL TISSUES FOLLOWING ORAL ADMINISTRATION IN THE EWE

Kevin Washburn[1], Virginia Fajt[1], Hans Coetzee[2], Shannon Washburn[1]

[1]Texas A&M University, College Station, USA, [2]Iowa State University, Ames, Iova, USA

The objective of this study was to determine the disposition of chlortetracycline (CTC) in pregnant ewes and their unborn fetuses following oral administration of 500 mg/head/day during the last trimester of gestation.

Five pregnant sheep were administered 500 mg of CTC in divided doses for 8 days. On day 7, the fetus was surgically implanted with venous lines for sampling, and placenta and amniotic fluid were harvested. Fetal plasma was sampled for 36 hours after the last dose of CTC. Subsequently, ewes were sacrificed and samples of amniotic fluid, placenta, fetal kidneys, fetal liver, and fetal stomach contents were collected.

Observed CTC concentrations were below the limit of quantification for all fetal tissues with the exception of liver and kidney in all but two fetuses. Fetal liver concentrations ranged from 4.6 to125.9 ng/ml, while fetal kidney concentrations ranged from 8.1 to 17.4 ng/ml 36 hr after the last dose of CTC.

Although our study indicated that CTC crosses the placental barrier based on accumulation in the fetal kidney and liver, concentrations achieved were far below the minimum inhibitory concentrations (MIC) reported in the literature for isolates from abortion outbreaks and below our limit of quantification in amniotic fluid, fetal stomach contents and placenta. No studies demonstrate the concentrations of CTC that are preventative for abortion from *Campylobacter* spp.; our findings suggest that either the dosage used in this study is not high enough or that the pharmacodynamic parameter relating preventive dose to pathogen MICs is yet to be determined.

F08

HYPERKETONEMIA AS A TOOL TO PREDICT MORTALITY IN DAIRY GOAT DURING LAST MONTH OF PREGNANCY

VINCENT DORE[1], JOCELYN DUBUC[2], ANNE-MARIE BELANGER[2], SEBASTIEN BUCZINSKI[2]

[1]NORTH CAROLINA STATE UNIVERSITY, RALEIGH, NORTH CAROLINA, USA, [2]UNIVERSITE DE MONTREAL, SAINT-HYACINTHE, QUEBEC, CANADA

Increased ketone bodies concentration in blood has been used to monitor goat during last month of pregnancy to diagnose animal with subclinical and/or clinical cases pregnancy toxemia (PT). Since clinical cases of PT are highly fatal, early diagnostic tool to predict mortality on farm would be useful to producer to save money on animal with lower chance of survival or to initiate more aggressive therapy depending on the value of the animal and its fetuses.

The objective of this study was to determine if increased ketone bodies concentration were associated with mortality in goats from herds with a history of pregnancy toxemia.

A prospective cohort study was performed on 1242 lactating goats from 10 commercial herds in Québec, Canada. Each goat was followed weekly during the 5 last weeks of pregnancy or until parturition. During each farm visit all pregnant goats were sampled until more than 95% of the group had kidded. Blood samples from jugular vessel were collected and were analyzed directly on farm using an electronic on-farm test for the quantification of blood β-hydroxybutyrate acid (BHBA) concentration (Precision Xtra, Abbott Diabetes Care, Saint-Laurent, Canada). Producers evaluated each goat for presence of PT based on a 4 degrees scale (absence, low, moderate, or strong suspicion of PT) using a standardized definition. Number of fetuses, mortality, difficulty of kidding, and presence of treatment, anorexia, lethargy and isolation from the rest of the herd were evaluated during the first week following parturition.

Pregnancy toxemia was diagnosis in 108 goats. Prevalence of mortality was 5.0% during prepartum and first week of lactation period. There were 26 cases of mortality during prepartum period (96.2% associated with PT) and 28 cases of mortality during the first week of lactation (71.4% associated with PT). Critical cutpoints for predicting mortality during last month of pregnancy and first week of lactation were determined based on the highest sum of sensitivity and specificity every week. Blood concentration above 0.6mmol/L during week 4 before kidding (sensitivity (Se): 44.7%; specificity (Sp): 86.7%), 3 (Se: 65.4% ; Sp: 82.7%), and 2 (Se: 70.8%; Sp: 76.1%) before kidding, above 1.4mmol/L during week 1 before kidding (Se: 51.3% ; Sp: 93.5%) and above 1.7 for the first week of lactation (Se: 28.6% ; Sp: 95.0%) were selected as the optimal cutpoints in this study. Presence of treatment during the last month of pregnancy was associated with increased mortality in dairy goat (OR: 11.79, $p<0.01$)

In conclusion, even if hyperketonemia is associated with low to moderate sensitivity from week to week, the test shows good specificity with relatively low prevalence of mortality. For this reason hyperketonemia appeared as a cheap test to assess the risk of mortality during the last month of pregnancy. However, others parameters should be evaluated to access prognosis in goat with PT.

F09

EFFICACY AND PHARMACOKINETICS OF INTRAVENOUS FAMOTIDINE IN ADULT STEERS

Christie Balcomb[1], Meera Heller[1], Heather Knych[1], Munashe Chigerwe[1], Allison Meyer[2]

[1]University of California, Davis, Davis, CA, USA, [2]University of Missouri, Columbia, Missouri, USA

The objectives were to measure the pH of abomasal outflow fluid as an indicator of abomasal pH in adult cattle after administration of the H_2-receptor agonist famotidine as a single dose of 0.4 mg/kg IV or as three doses of 0.4 mg/kg IV every eight hours to mimic clinical application and secondly to describe the pharmacokinetic behavior of intravenous famotidine in adult cattle.

A prospective random two-way crossover study design was implemented using 4 healthy adult Angus steers previously fitted with duodenal cannulae located orad to the bile and pancreatic ducts. The steers were fed alfalfa hay *ad-lib* and whole corn and soybean hull meal pellets at 4 lbs/head 12 hours apart. On Day 1, steers were administered either 0.4 mg/kg famotidine or an equal volume of saline via jugular catheters. Whole blood and abomasal outflow samples were obtained at intervals for a 12-hour period. On Day 2, steers were administered either 0.4 mg/kg famotidine or saline IV at time 0, 8, and 16 hours. After a 24-hour washout period, the opposite treatments were administered and the study repeated. Whole blood and abomasal outflow samples were obtained at intervals for a 24-hour period. Abomasal outflow fluid was tested for pH using a commercial bench top pH meter within 5 minutes of sampling. Fluid pH was analysed using a mixed model in SAS 9.4 with treatment, sampling hour, and treatment*hour as fixed effects and steer and treatment period as random effects. Least square means were separated using least significant difference. Serum was separated and stored at -20° C until pharmacokinetic analysis. Serum famotidine analysis was performed using tandem liquid chromatography mass spectrometry and pharmacokinetic analysis was performed using commercial software.

The treatment*sampling hour interaction affected abomasal outflow pH for both a single dose and multiple doses of famotidine ($P < 0.001$). A single dose of intravenous famotidine resulted in a significant increase of abomasal outflow pH as compared to the control group for 4 hours after administration ($P < 0.05$). When administered every 8 hours, the abomasal outflow pH of the famotidine treated group was greater than the control group for 4 hours after the first dose ($P < 0.05$), 2 hours after the second dose, and 1 hour after the third dose ($P < 0.05$). Pharmacokinetic analysis demonstrated that famotidine had a terminal half-life of 3.33 (3.21 – 3.54; median and range) hours, a volume of distribution of 0.042 (0.014 – 1.89) L/kg and a clearance of 1.26 (0.625 – 11.5) mL/min/kg.

In conclusion, an H_2-antagonist, famotidine increases gastric pH by reducing acid production by parietal cells. This study showed that a single intravenous dose of famotidine administered at 0.4 mg/kg is effective at increasing the abomasal outflow pH of adult cattle for up to 4 hours as compared to saline control. When administered every 8 hours, to mimic clinical therapeutic recommendation, the abomasal outflow pH was significantly higher for 4 hours after administration of the first dose; however, the effect seemed to decrease after each subsequent dose. Results of this study indicate that famotidine administered at 0.4 mg/kg IV is effective at raising the abomasal pH and may be an effective treatment for abomasal ulceration in adult cattle. However, the efficacy of treatment may decrease over time or may require more frequent dosing to sustain elevated abomasal pH.

F10

THE PHARMACOKINETICS OF INTRAVENOUS GENTAMICIN IN HEALTHY YOUNG-ADULT VERSUS AGED ALPACAS

Andrew Gestrich[1], Daniela Bedenice[1], Michelle Ceresia[2], Cheryl Stockman[1], Iman Zaghloul[2]

[1]*Cummings School of Veterinary Medicine at Tufts University, North Grafton, MA, USA,* [2]*School of Pharmacy, MCPHS University, Boston, MA, USA*

The study objective was to evaluate the effects of patient age on aminoglycoside pharmacokinetics in young-adult (≤5 years) and aged (≥14 years) healthy alpacas, receiving a single 6.6mg/kg intravenous injection of gentamicin (Henry Schein®). Heparinized plasma samples were obtained at designated time-points following drug administration and frozen at -80˚C until assayed by a validated immunoassay (QMS®). Gentamicin plasma concentrations versus time plots were analyzed using a two-compartment model and commercial software (WinNonlin-v6.4). Data were compared between age-groups using independent samples t-test; *P<0.05* considered significant.

Baseline physical and hematological parameters were not significantly different between young (median age [range]: 2.6 [1.7-3.7] years) and aged (15.5 [14.6-18] years) alpacas, aside from a lower mean heart rate in older animals (50±13 vs. 79±15 beats/minute). Laboratory results did not differ pre and post-treatment.

Comparative pharmacokinetic parameters are displayed below (mean ±SD; $^{\dagger}P<0.05$):

	C_0 (µg/mL)	$AUC_{0-\infty}$ † (µg*h/mL)	$T_{1/2\alpha}$ (h)	$T_{1/2\beta}$ (h)	V_{ss} (L/kg)	Cl † (mL/h/kg)
Young-adult (n=8)	54.6±6.5	70.4±10.5	0.13±0.02	1.7±0.02	0.21±0.02	95.5±13.4
Aged (n=8)	65.2±13.2	90.4±17.6	0.12±0.04	2.1±0.5	0.21±0.02	75.6±16.1
C_0–Initial plasma concentration; AUC–Area under the curve; $T_{1/2}$ α and β–Distribution and elimination half-life; V_{ss}–Volume of distribution at steady state; Cl–Clearance						

Both the peak gentamicin concentration at the end of the distribution phase (23.8±2.1 vs. 26.1±2 µg/ml, *P=0.04*) and the mean time-point at which gentamicin trough levels reached 2µg/ml (common MIC for large-animal pathogens) were significantly lower in young-adult versus older animals (6.7±1 versus 8.7±1.9 hours, *P=0.03*). These data indicate reduced renal drug clearance in aged alpacas.

F11

PROGNOSIS ASSOCIATED WITH CEREBROSPINAL FLUID ANALYSIS RESULTS IN RECUMBENT DAIRY CATTLE : RETROSPECTIVE STUDY (2006-2014)

Marie-Ève Bilodeau, David Francoz, Damien Achard, André Desrochers, Sylvain Nichols, Marie Babkine, Hélène Lardé, Carolyn Grimes, Gilles Fecteau

Université de Montréal, Saint-Hyacinthe, Québec, Canada

In recumbent dairy cows, the analysis of the cerebrospinal fluid (CSF) is an ancillary test frequently used at our referral center. Interpretation of the CSF analysis results in cattle is challenging, in part, because reported intervals for this species differ. In a previous retrospective study, we demonstrated that recumbent cows with a CSF total nucleated cells count (TNCC) ≥ 4.5 cells/μL and/or a protein concentration ≥ 0.40 g/L had a spinal cord lesion on post mortem examination. However, CSF results below those threshold values did not exclude the presence of such a lesion.

The objective of this study was to determine the short-term and long-term prognosis in recumbent adult dairy cows who underwent CSF analysis. The previously mentioned threshold values for TNCC and protein concentration were used. We hypothesized that recumbent cows with CSF values higher than the pre-determined threshold values would have a less favorable short-term and long-term prognosis. All dairy cattle, of 2 years or more of age, that were recumbent and who have had a CSF analysis were included in the study (n=214 cases). CSF analysis was performed by board certified pathologists, and only CSF analysis which contained both the TNCC and protein concentration were included in the study. A successful short-term prognosis was based on the return to the farm. A successful long-term prognosis was based on the completion of at least one lactation using the Canadian Dairy Network database. Statistical analysis was performed using the Chi-squared test to compare numbers of cows in each group.

Age of patients ranged from 2.0 to 14.5 years (median 5.6 years). Two hundred and eight cows were Holstein, 4 were Ayrshire and 2 were Jersey. The duration of hospitalization ranged from 1 to 48 days (median 7 days). In cows for which the information was available, days in milk ranged from 1 to 516 days (median 25 days) and 18% of the cows (23/130) were dried. Of the 214 cases studied, 167 cows had CSF values for TNCC and protein concentration below the threshold values, and 47 cows had CSF values above the pre-determined threshold values.

The results showed that TNCC ≥ 4.5 cells/μL and/or a protein concentration ≥ 0.40 g/L was associated with a worse short-term prognosis in recumbent dairy cattle. Indeed, 53% of cows with CSF values below the pre-determined threshold values were discharged from the hospital, which is comparable to the survival rate of all recumbent adult dairy cows that present to our referral center (survival rate of 48% between 1994 and 2014, n=1271). The short-term survival rate was 30% in cows where only the TNCC value was above the threshold ($p = 0.035$), 28% if only the protein concentration was above the threshold ($p=0.012$), and 34% if both CSF parameters were above the thresholds ($p=0.024$). However, in cows discharged from the hospital (n=104), the long-term prognosis was comparable amongst groups; 60% (53/88) of cows with CSF values below the thresholds, and 75% (12/16) of cows with one or more CSF value above the thresholds completed at least one lactation after their return to the farm ($p = 0.37$).

Based on this study, we conclude that the short-term prognosis of recumbent adult dairy cattle with CSF TNCC ≥ 4.5 cells/μL and/or a protein concentration ≥0.40 g/L is lower than that of cows with CSF values below those thresholds. However, some cattle with CSF values above the thresholds survive and their long-term prognosis seems to be acceptable suggesting that these cows may have a spinal cord lesion that is reversible.

The Premiumization of Dog Foods –What's Real and What's Not?

Lisa P. Weeth, DVM, DACVN
Edinburgh, UK

INTRODUCTION

The field of veterinary nutrition is often being pulled between the science of nutrition and the psychology of marketing and nowhere is that more evident than in the rise of premiumization in dog foods. In the early post-World War II years of the pet food industry, convenience and ease of preparation were the driving forces behind the pet food market. People's attitudes towards a particular dog food were dependent on whether the food was nutritionally balanced and how rapidly the dog finished its meal.[1] Today, meeting nutritional requirements and being palatable are minimal expectations. With advances in manufacturing technology, growth in the field of nutrigenomics, and identifications of breed specific nutritional requirements, pet owners are seeking diets that optimize health and wellness in companion dogs and are willing to pay more for ones they feel provide these benefits.

Many dog owners purchase "premium" brands as an expression of their level of commitment to their dog, and these dog food choices help shape their personal, emotional and social identities.[1] Pet food companies on the other hand have realized that positioning a dog food as a "premium" product is good for business. Pet food sales in the United States rose from $12.2 billion in 2000 to $21.4 billion in 2014,[2] with the "premium" market segment accounting for 42% of all pet food sales.[3] Current human food trends, marketing efforts by manufacturers, and recommendations promoted by well-intended but often uninformed pet food advocacy groups have left pet owners confused and concerned about what to feed their companion animals. Owners then turn to veterinarians for guidance when making these decisions and it is important for veterinary professionals (both doctors and support staff) to understand not only basic nutrition concepts, but the pet food marketplace as well.

PERCEPTION VS. REALITY OF "PREMIUM" DOG FOOD

The word "premiumization" was first introduced into the marketing lexicon through the liquor industry and referred to the concept that consumers were willing to spend more money for an item that was perceived to be of a higher quality, whether it was simply a rebranded mass market product or one that was truly unique and different.[4] Premiumization of dog foods follows a similar pricing structure with "premium" pet foods being defined as those that are priced 10-20% over the category average while "superpremium" are those priced >20% over the category average.[5] Aside from price positioning though there is little consistency and no regulation that defines what "premium" means for the product itself.

While convenience and cost are still considerations for pet owners, the marketing of premium foods has been very effective. Many pet owners now believe that by spending more money on a particular food they are getting a higher-quality, healthier, more nutritious diet for their dogs, and that they are better "dog parents" because of it.[1] Marketing experts have recognized that the emotional connections a person has with their dog will often dictate how they shop for dog food[6] and that many owners will prioritize their dog's food over their own.[7] Dog food marketing studies[1,6,7] have also found that pet owners who treat their dogs as human equivalents or view their dog as an extension of their self-identity can help drive financial growth in the premium pet food market.

Pet food companies have been quick to capitalize on this anthropomorphism of dogs. The marketing of "premium" foods and treats appeals to the ethical eating standards of pet owners by using terms like "human-grade", "natural", and "organic", and the advertising targets their emotions by refer to pet owners as "pet parents" and dogs as their "fur babies". The problem is that the terms "premium" and "superpremium" are based on price points, not ingredients or manufacturer quality; "human-grade" has no regulatory definition; "natural" refers to how ingredients are derived, and "organic" refers to a husbandry/agricultural practices, not the nutritional adequacy of the raw materials or finished product.[8]

UNDERSTAND THE DIFFERENCE, OR LACK THEREOF

Edible products for dogs and cat (foods, treats, supplements, and chews) are classified as animal feed by the US government and are therefore under the authority of the Food and Drug Administration (FDA) Center for Veterinary Medicine (CVM). The FDA-CVM is responsible for overseeing ingredients, manufacturing, and distribution of products for interstate sales, as well as advising state regulators on issues of adulterated or misbranded foods sold within a given state.[9] This is in contrast to the role of the Association of American Feed Control Officials (AAFCO), which is comprised of state, federal, and foreign officials, that serves a more advisory function. AAFCO provides a Model Bill and Regulation that can be adopted by individual states and territories and also publishes an annual Official Publication on topics such as ingredient definitions and adequacy standards.[8] AAFCO has no regulator or judicial oversight and cannot enforce laws; enforcement of pet food regulations is instead made by the individual state governments. Neither of these entities, FDA-CVM or AAFCO, have defined what constitutes a "premium" pet food or has address whether they are any better (or worse) than non-premium or mass marketed pet foods.

Commercial dog foods that are marketed as complete and balanced will provide all of the essential nutrients that are required by dogs, irrespective of price point. More economically priced dog food will often contain larger proportions of plant-based proteins, such as soy and wheat, which are less expensive than animal proteins; will use more rendered or dehydrated animals protein (e.g., "chicken meals") rather than fresh animal-proteins, which again are more expensive to purchase, ship, and store; and may include more generic categories of ingredients, such as "poultry by-product", which is a combination of chicken, duck and/or turkey internal organs, head and/or feet. Dog foods that use these types of least cost formulation will still meet the essential nutritional requirements of dogs in general, though may not be optimally meet the needs of a given individual. For example, plant-based proteins can be less digestible than animal-based,[10] which would be contraindicated for a dog requiring a highly digestible diet. Additionally, the use of generic categories ingredients (e.g., "poultry by-product") can lead to ingredient variability in each production batch, which could be problematic for animals with certain medical condition.

Diets targeting "premium" pet food shoppers tend to use more fixed diet formulations and higher proportions of animal-proteins, which will increase the cost of manufacturing, though whether the raw materials are truly a higher quality and in the proportions listed on the labels are dependent on the manufacturer integrity. Recent publications have highlighted the potential for undeclared ingredients in over-the-counter diets[11-13] and a major marketer of "premium" pet foods settled a class action lawsuit in December 2015 for false advertising and inclusion of lower cost ingredients not reported on the ingredient list.[14]

THE PET FOOD LABEL

While the majority of a pet food label is taken up with color graphics and marketing terms, there are a few sources of nutritional information that will help consumers evaluate pet food across market segments: Nutritional Adequacy Statement, Ingredient List, and Guaranteed Analysis.

Nutritional Adequacy Statement

All pet foods and treats sold in the United States must carry a Nutritional Adequacy Statement. These statements can be written either as "intended for intermittent or supplemental feeding" or "complete and balanced" for a particular life-stage based on AAFCO nutrient guidelines. Dog foods labeled as "complete and balanced" are verified for completeness either by computer formulation or animal feeding trials. Those that are intended for intermittent or supplemental feeding do not provide all essential nutrients for dog and should not be fed as a sole source of nutrition. Products carrying adequacy statements that reference an entity other than AAFCO may be in violation of state pet food label regulations. While feeding trials only include a small number of healthy individuals (8 adult dogs in an adult maintenance trial) and run for a limited amount of time (6 months for adult maintenance) these trials will give basic information on the longer-term acceptance of a diet (i.e., more than a week or two), effects on stool quality and flatulence production, quality of skin and hair coat, and reveal any glaring adverse ingredient or nutrient interactions.

Ingredients

The AAFCO Model Bill and Regulations outline ingredients that can and cannot be included in commercial pet foods and treats. This list of ingredients is required on all pet food labels and must be ordered by weight of raw material. The challenge for veterinarians and caregivers is that the AAFCO definitions do not address differences in manufacturing processes or variation in the quality of raw materials available for use by pet food manufacturers. As such, ingredient comparisons between foods or brands may not be an accurate metric to evaluate diet quality. Variations in manufacturing processes and standards, as well as storage and product handling, will impact diet digestibility and can diminish or enhance the bioavailability of nutrients in a particular ingredient or ingredient combination. Additionally, ingredients not found on a pet food label may be included in trace amounts through the use of shared manufacturing equipment. Cross-contamination of ingredients can occur during normal manufacturing processes, but unlike human foods that must comply with the Food Allergen Labeling and Consumer Protection Act of 2004 and disclose any potential food allergens in a finished product, there are no allergen label requirements for pet foods. Surveys of over-the-counter diets[11-13] have found trace amounts of ingredients not reported on the labels of a number of dry and canned diets analyzed; it is not clear if these finding represent unintentional or intentional adulteration of pet foods, but they highlight that fact that over-the-counter diets must be used with caution in dogs with known or suspected food hypersensitivities.

Guaranteed Analysis

A Guaranteed Analysis including percentages of Crude Protein, Crude Fat, Moisture, and Crude Fiber is also required on pet food and treat labels. Anything else that is reported has been added for marketing affect and to draw the pet owner's attention to key nutrient(s) the manufacturer would like to highlight in the product, such as glucosamine or taurine. Guaranteed Analysis values are derived from a chemical or thermal analysis of the food and do not indicate quality or type of protein, fat or fiber included. Crude Protein, for example, measures total nitrogen content of the finished product and uses this result as a surrogate for the level of protein in the diet. It does not distinguish plant- from animal-based protein, nor does it indicate whether the protein is from a highly digestible or a poorly digestible source. Guaranteed Analysis values are reported as either minimums (for protein and fat) or maximums (for moisture and fiber) and may or may not represent the average or typical analysis of the food. The largest deviation between what is reported on a pet food label and what is actually provided by the diet may come from Crude Fiber. Crude Fiber primarily measures the insoluble fiber content of a food and significantly underestimates the total dietary fiber levels in diets containing high levels of soluble, fermentable fibers such as fructo-oligosaccharides (FOS) and inulin.[15]

SUMMARY

There is no one commercial diet that can be used to provide optimal nutrient intake for all dogs and while some higher priced foods may use higher quality ingredients and incorporate advanced in pet food science and technology, others may simply be taking advantage of the emotional bond between caregiver and companion dog to claim a piece of the multibillion dollar pet food market. Individual animal differences as well as variability in digestibility and bioavailability of nutrients in commercial pet foods will influence "performance" of a diet and there are good quality diets found in every market segment. Individual dog response to a particular diet, such as changes in stool quality or skin and coat health, should influence diet selection more than current food fads and marketing strategies.

REFERENCES

1. Boya UO, et al **Int J Cons Stud** 2015;39(1):74.
2. **Pet Food Institute** Accessed January 2016 http://www.petfoodinstitute.org/?page=PetfoodSales
3. Sprinkle D. **Petfood Industry** Jan 2015; http://www.petfoodindustry.com/articles/4849-pet-humanization-and-product-premiumization
4. Ritson M. **Marketing Magazine** Apr 2008 http://www.marketingmagazine.co.uk/article/805488/mark-ritson-branding-why-brands-premium
5. Philips-Donaldson D. **Petfood Industry** Jan 2016; http://www.petfoodindustry.com/blogs/7-adventures-in-pet-food/post/5599-premium-pet-food-is-industrys-lifeblood-a-dying-breed
6. Boya UO, et al **J Targ Meas Anal Mark** 2012;20:133.
7. Dotson MJ and Hyatt EM. **J Busin Res** 2008;61:457.
8. Official Publication of the Association of American Feed Control Officials 2016. Oxford, IN: Association of American Feed Control Officials 2016.
9. FDA's regulation of pet food. Accessed January 2016 at fda.gov/animalveterinary/products/animalfoodfeeds/petfood/default.htm
10. Bednar GE, et al **Arch Tierernahr** 2000;53(3)127.
11. Okuma TA and Hellberg RS. **Food Control** 2015;50:9.
12. Maine IR, et al **Acta Vet Scand** 2015;10;57:7.
13. Raditic DM, et al **JAPAN** 2010; 95: 90.
14. **Petfood Industry** Dec 2015; http://www.petfoodindustry.com/articles/5549-blue-buffalo-settles-class-action-lawsuit-over-ingredients
15. Farcas AK, et al **JAVMA** 2013;242(7):936.

Handling Nontraditional Dietary Requests

Lisa P. Weeth, DVM, DACVN
Edinburgh, UK

INTRODUCTION

Feeding a complete and balanced diet is known to promote wellness in dogs and cats, but a number of highly publicized pet food recalls, as well as a growing appreciation for the role of diet in health and disease, has raised questions about the use of commercial pet foods to meet this end. Additionally, concerns over specific ingredients, such as corn or wheat, or the use of nebulous ingredient names, such as "by-product", have been raised by veterinary advice columns and pet food marketing companies and have prompted owners to become leery of manufacturers that incorporate these ingredients into their foods. Many owners, and some veterinarians, also advocate feeding dogs and cats home-prepared foods exclusively (raw or cooked, or both) and either cite perceived health benefits or a general mistrust of the pet food industry. At its most basic concept, food is a means of getting essential nutrients into the body. How these nutrients are delivered or which specific ingredients are being used (barring ingredient sensitivity) is much less important than ensuring that the optimal amounts are provided on a regular basis. Whatever the reason for a particular diet strategy, it is important for veterinary professionals (both doctors and support staff) to understand the motivations, risks and benefits of that diet type to ensure that the nutritional needs of the individual animal are being met.

WHAT IS A "TRADITIONAL" DIET?

The commercial pet food industry as we know it is a somewhat young field relative to the duration of time humans have been living with companion animals. Prior to the early 1900s, dogs and cats were fed household leftovers and human food production offal (i.e., "by-product"). The first pet foods to be marketed to owners were dog biscuits at the turn of the twentieth century, followed by canned meat dog foods and dry dog meals in the 1920s, and dry expanded kibble in the 1950s.[1] The convenience and ease of feeding dry extruded diets were attractive features for pet owners especially those in busy households in the booming post-World War II economy. The rise in popularity (and market share) of dry extruded kibble over the last 60 years saw pet food sales generated a total of $21.4 billion in the United States in 2014 with 60% ($13 billion) in dry pet food sales alone.[2] Nutritional and nutritional marketing trends in pet food have mirrored those in the human food market with pet owners now demanding the same dietary standards for everyone in the household, human and humanized pet alike. Many of today's pet owners are not just looking for convenience and ease of feeding, they want a diet that mimics the food choices they make for themselves; their choice in feeding strategy reinforces the emotional bond they have with their dog or cat.[3]

TYPES OF NONTRADITIONAL DIETS

Nontraditional diets are those that vary from the established norm created over the last 60 years of pet ownership and pet food production. This can be as simple as changing from a mass market pet food to a boutique brand of food, or a complicated as a complete eschewing of the commercial industry and preparing all food and treats at home using ingredients intended for human consumption.

Boutique or "Premium" Pet Foods

Boutique or "premium" wet and dry pet foods do not differ significantly from traditional diets and are instead identified by inclusion or exclusion on their ingredient lists. Boutique and "premium" pet food may use terms like "organic", "natural", or "human-grade" to promote their products and may avoid ingredients found in lower cost and mass market pet foods, such as corn or animal by-products. While the terms "organic" and "natural" have specific regulatory definitions,[4] "human-grade" does not, and none of these terms denotes the quality of a particular ingredients or product. There are no established health benefits to avoiding common or lower cost pet food ingredients unless the dog or cat has a specific sensitivity to a particular ingredient.

Alternative Commercial Diets –Raw Mat-Based Diets

The Association of American Feed Control Officials (AAFCO) has established model guidelines for the states regarding pet food labels, ingredient definitions, what can and cannot go into pet foods, and levels of specific essential nutrients required for a given life-stage.[4] Any pet food with an AAFCO label of adequacy must have met these guidelines. Commercially-prepared pet foods, whether cooked or raw meat-based, sold in the United States must carry a Nutritional Adequacy Statement that the give diet is either "complete and balanced" for long-term feeding or intended as "supplemental feeding only".

Raw meat-based commercial diets can be frozen, dehydrated, or fermented. They are similar to boutique and "premium" pet foods in that they are distinguished by what they include or exclude from the ingredient list. Specifically raw meat-based commercial diets include uncooked animal proteins that have not undergone thermal changes to alter the protein structure (or kill potentially pathogenic bacteria). Many commercial raw meat-based diets are labeled for long-term feeding and have been formulated to AAFCO nutritional profiles, with a few even undergone AAFCO feeding trials.

Nutritional adequacy is typically not the concern with complete and balanced commercial raw meat-based diet, food safety for pets and people is. Any raw meat ingredients can be a potential source of foodborne pathogens, including *Salmonella*, *E. coli*, *Campylobacter*, and *Listeria.* While healthy adult dogs and cats may resist disease induced by these pathogens, young or immune-compromised animals and people in the household can be at risk of illness and death with exposure. Otherwise healthy adult animals fed raw meat diets can serve as sources of contamination to people and any other pets in the household,[5,6] though admittedly contaminated dry commercial foods are also not without risk.[7] One survey of commercial raw meat-based diets found that 64% of the diets sampled tested positive for *E. coli*, 20% had positive *Salmonella* cultures, and 20% were positive for *Clostridium perfringens*.[8] Another looking at both commercial raw meat-based diets and more traditional wet and dry pet foods found that 5.9% of the raw meat diets tested positive for *Salmonella*, while none of the traditional pet food did.[9] A more recent bacterial analysis of dry, semimoist, and raw meat-based diets and treats found that 2 out of the 480 traditional diets tested positive for potentially pathogenic bacteria (1 for *Salmonella* and 1 for *Listeria*), while 81 out of the 576 raw meat-based diets and treats had positive bacterial assay results (15 for *Salmonella* and 66 for *Listeria*).[10]

Alternative Commercial Diets –Premixes

Commercial premixes are combinations of carbohydrates, vegetables, and vitamin and mineral supplements that are marketed for feeding to dogs. They are not complete and balanced when fed alone, but are intended to be mixed with fresh cooked or raw animal proteins prior to feeding. There are no legal requirements for a commercial diet to be complete and balanced unless it makes a label claim as such; commercial premixes are labeled for "supplemental feeding only" and are unlikely to meet AAFCO nutritional profile recommendations. Commercial premixes are similar to home-prepared diets in that it allows owners flexibility in animal protein ingredients but these types of diets are not suitable to feeding to growing animals[11] and may not be suitable for long-term feeding to adult dogs as the sole source of nutrition.

Home-Prepared Diets – Cooked & Raw Meat

Home-prepared diets have grown in popularity over the last few years. For some pet owners it is in response to concerns about the production of commercial diets, for others feeding home-prepared foods reinforces the human-animal bond, and for still others a home-prepared diet is required to manage a medical condition.[12-14] Proponents of home-prepared diets claim that these foods are a safe and natural way to feed animals.[15,16] It is true that fresh meat, whether it is fed raw or cooked, is palatable to most dogs and cats; can be highly digestible;[17] and depending on the cut of meat selected is higher in fat than most dry kibbles. This leads to the general perception that animals eating home-prepared diets will more readily eat their food, have lower stool volumes than when fed traditional commercial dry pet foods, and can have shinier coats. These positives attributes are held up as proof of nutritional superiority to commercial foods, while ignoring the root cause of these beneficial changes (higher meat and fat content) and potential negative consequences of this feeding strategy (bacterial and parasitic contamination, dietary imbalance).

Home-prepared raw meat diets carry the same risk of bacterial contamination as commercial raw meat-based diet. Raw and cooked bones specifically carry a risk of gastrointestinal obstruction/perforation and oral trauma[18,19] and are a poor source of essential minerals (such as calcium, phosphorus and magnesium) due to the poor digestibility of larger bones within the canine and feline digestive tract.[20] Ultimately, the animal's acceptance of a home-prepared diet does not change significantly when the meat is cooked or when more bioavailable sources of nutrients are used.

Published reviews of the nutritional adequacy of home-prepared (cooked and raw) diet recipes in recent years found that the majority of recipes used by the pet owners do not provide a complete and balanced source of essential nutrients.[21-25] Unless someone with veterinary nutrition training developed the diet, there are a few consistent deficiencies found in most home-prepared recipes irrespective of the ingredients used: they lack a sufficient source of calcium; trace minerals (such as iodine, selenium, copper, and zinc); linoleic acid; and essential fat-soluble and water-soluble vitamins. While the perceived benefits of home-prepared diets are reinforced daily to the owner, nutrient deficiencies or toxicities in adult dogs and cats are insidious and can lead to long-term complications, which can vary from poor skin and coat health to chronic diarrhea, poor reproductive performance, osteopenia, anemia, altered drug metabolism, and hepatic lipidosis depending on the specific nutrients imbalance in the diet.[26-29]

DIETARY WANTS VS. NEEDS

Companion animals have become members of the family with many owners prioritizing their pet's food over their own.[30] Owners that select nontraditional diets are doing so because they believe that these types of foods are higher-quality, healthier, more nutritious for their dog or cat, and that they are better "pet parents" because of it.[4] The best way to handle nontraditional diet requests is by acknowledging the emotional bond between caregiver and pet and focusing the diet discussion on the nutritional needs of the individual dog or cat. Is the diet providing adequate calories to maintain an optimal body weight and condition? Is the diet complete and balanced either by formulation/feeding trials to meet the AAFCO nutrition requirements, or formulation by a trained Veterinary Nutrition Specialist? Does any aspect of the diet or individuals ingredient pose a health risk to animals or people in the household? Are there any concurrent conditions that may affect nutritional requirements? The role of the veterinarians it to ensure that the diet the owners elect to feed is healthy for the dog or cat and safe for all members of the family.

MONITORING FOR EFFECT

Individual animals may vary in their response to any given diet (commercially- or home-prepared) and there is no one diet-that-fits-all. If an owner elects to feed a raw meat-based diet or a home-prepared diet (raw or cooked) they should be counseled on the risks of this feeding strategy and cautioned that nutritionally-related disease can mimic other forms of chronic illness. Any animal eating a home-prepared diet should have at least an annual physical exam and health screening, including serum biochemistry (with T4), hematology, and urinalysis profiles. While blood work and urinalysis results will give you a general overview of the animal's health status, it will not pick-out specific deficiencies or excesses. A complete diet history (all foods and supplements) should be collected from the owner at each visit. Any home-prepared diet recipes should be obtained from a reputable, trained source. Additional resources, including a list of Board-certified Veterinary Nutritionists, can be found through the American College of Veterinary Nutrition (www.acvn.org).

REFERENCES

1. **Pet Food Institute** Accessed January 2016 http://www.petfoodinstitute.org/?page=HistoryofPetFood
2. **Pet Food Institute** Accessed January 2016 http://www.petfoodinstitute.org/?page=PetfoodSales
3. Boya UO, et al **Int J Cons Stud** 2015;39(1):74.
4. Official Publication of the Association of American Feed Control Officials 2016. Oxford, IN: Association of American Feed Control Officials 2016.
5. Joffe DJ and Schlesinger DP. **Can Vet J** 2002; 43:441.
6. Lejeune JT and Hancock DD. **JAVMA** 2001; 219(9):1222.
7. Imanishi M, et al **JAVMA** 2014;244(5):545.
8. Weese JS, et al **Can Vet J** 2005; 46:513-516.
9. Strohmeyer RA, et al **JAVMA** 2006; 228(4):537.
10. Nemser SM, et al **Foodbourne Pathol Dis** 2014;11(9):706.
11. Taylor MB, et al **JAVMA** 2009;234(8):1041.
12. Willoughby KN, et al **JAPAN** 2005; 89:428.
13. Roudebush P and Cowell CS. **Vet Dermatol** 1992; 3:23.
14. Wakefield LA, et al **JAVMA** 2006; 229:70.
15. Billinghurst I. Give your dog a bone. The practical and common sense way to feed your dog. Alexandria (Australia): Bridge Printery; 1993.
16. Schultze KR. The ultimate diet. Descanso (CA): Affenbar Ink; 1998.
17. Vester BM, et al **Zoo Biology** 2010;29:676.
18. Thompson HC, et al **JVECCS** 2012;22:253.
19. Frowde PE, et al **J Fel Med Surg** 2011;13:564.
20. Delay J and Laing J. **AHL Newsletter.** 2002;6(2):23.
21. Stockman J, et al **JAVMA** 2013;242(11):1500.
22. Freeman LM and Michel KE. **JAVMA** 2001;218(5):705.
23. Gray CM, et al **JAVMA** 2004;225(11):1670.
24. Larsen JA, et al **JAVMA** 2012;240:532.
25. Heinze CR, et al **JAVMA** 2012;241:1453.
26. Becker N, et al **Tier Prax Ausg K Klein** 2012;40(2):52.
27. De Fornel-Thibaud P, et al **JAAHA** 2007;43(1):52.
28. Sontas BH, et al **Can Vet J** 2014;55(8):78.
29. Kohler B, et al **JSAP** 2012;53(3):182.
30. Dotson MJ and Hyatt EM. **J Busin Res** 2008;61:457.

Micronutrient Nutrition in Fetal Health and Development

Robert J. Van Saun, DVM, MS, PhD, DACT, DACVN
University Park, PA

INTRODUCTION

Significant economic losses occur in agricultural enterprises as a result of prenatal and natal deaths and neonatal disease processes. Diagnosis of abortion losses still remains below 45% and a significant portion of these "idiopathic" abortions have been hypothesized to be related to nutritional causes. In a path model of risk factors for natal and prenatal deaths in beef herds, nutritional deficiencies and toxins were linked with every primary factor identified in causing these deaths as well as factors leading to inadequate passive transfer and neonatal disease.[1] Hepatic trace mineral concentrations were significantly lower in aborted fetuses compared to "control" slaughterhouse fetal specimens.[2] However, it has not been determined as to whether this finding suggests a direct role of trace mineral deficiency in abortion or whether it is a consequence of the abortion. Many specific trace mineral deficiencies can result in abortion, stillbirths or weak neonates.[3,4]

Assessing trace mineral and vitamin status of the fetus or neonate may provide critical information as to the underlying cause of a given diagnostic dilemma. Nutrient concentration within fetal or neonatal liver specimens or blood can be just as easily determined as is routinely completed for adults. Of concern however, is how one can interpret these concentrations relative to mineral or vitamin status. Can we assume metabolism is the same between fetal, neonatal and adult animals? Is it appropriate to use current criteria for adult animals on fetal and neonatal samples? If the answer to these questions are "no", then we may be mis-diagnosing many perinatal nutritional problems. The objective of this presentation is to discuss the dynamics of vitamins and trace mineral metabolism in fetal development and subsequent neonatal health. Findings from a recent survey of stillborn calves relative to potential nutritional factors will be highlighted in addressing recommendations for assessment and supplementation.

VITAMIN and MINERAL METABOLIC DYNAMICS

Trace minerals are indirectly or directly associated with a tremendous variety of metabolic processes in animals of all ages. Deficiency diseases affect almost every physiologic and metabolic function and include immune dysfunction (Cu, Zn, Se); developmental abnormalities (Cu, Mn, I); abortion (Cu, I, Se); retained placenta (Cu, Se, I); metabolic disturbances (Co, Fe, Zn, I); and poor growth (Co, Cu, Fe, I, Se, Zn).[3,4] In addition and more importantly, subclinical disease resulting in reduced productive efficiency (reproduction, growth, lactation) and increased disease susceptibility is a more economically important problem associated with marginal mineral deficiencies. Subclinical disease is often difficult for the producer to identify within the herd without appropriate records evaluation and production bench marking. Presence of trace mineral deficiencies, clinical or subclinical, seem prevalent within the beef and dairy industries. Fat-soluble vitamins (ie, A, D and E) are associated with specific clinical disease manifestations, but much less is known relative to subclinical disease concerns.

Trace minerals and fat-soluble vitamins are not homeostatically regulated, but more controlled through movement between pools. Microminerals can be found in the body as a component to one or more metalloenzymes (biochemical function pool), transported on carrier proteins (transport mineral pool) or stored as a metal complex (storage mineral pool). Fat soluble vitamins are managed in a similar way, though they exert their biological effects as hormones affecting gene expression (A, D) or direct action (A, E). The body makes every effort to maintain a necessary level of activity in the biochemical pool to ensure normal function. The storage pool holds a reserve and is sensitive to nutritional status. If nutrient intake is in excess of requirements, excess intake will be stored until other regulatory processes, reduced absorptive efficiency or increased renal excretion, modify net mineral retention back into balance. Transport pool is dynamic in reflecting changes relative to either deficient or sufficient nutrient state. In a situation of nutritional inadequacy, hepatic storage will be mobilized and used to maintain biochemical pool activity until absorptive efficiency, reduced excretion or both can be enacted to raise net retention.

Using these concepts of mineral pools within the body, Suttle has described four progressive phases of mineral deficiency disease.[4,5] Phases of mineral status move from depletion, loss of mineral in storage; deficiency, loss of mineral in transport pool; dysfunction, compromise of the function pool and finally disease (clinical signs associated with critically reduced function of a specific metalloenzyme). For example, copper deficiency reduces tyrosinase activity, which then decreases production of melanin pigment resulting in clinical signs of achromotrichia. Subclinical disease occurs during deficiency and dysfunction phases, often defined as impaired immune function, reduced growth rate or reproductive efficiency or other non-specific declines in productive efficiency. Disease due to deficiency of fat soluble vitamins would follow this same pattern of change in respective biologic pools.

MATERNAL-FETAL-NEONATAL INTERRELATIONSHIPS

The developing fetus is totally dependent upon availability of essential nutrients from placental transfer from maternal blood. As a result, fetal nutrient status is reflective of maternal nutrient status. Maternal nutrients available to the fetus would include those from the consumed diet as well as mobilized reserves, if needed. Swenson showed a decline in maternal liver copper concentration during late pregnancy, which would be consistent with maternal transfer of mineral to fetus.[6] Numerous studies have observed a fetal liver concentrating ability for minerals in finding fetal hepatic mineral concentrations to exceed maternal values.[2,7-14] In many situations, fetal mineral concentrations are twice that of their dams.

In contrast to hepatic mineral concentrations, mean maternal serum selenium concentration was twice that of her fetus.[12] However, whole blood and erythrocyte selenium concentrations were not different between fetus and dam. Mean whole blood glutathione peroxidase activity was only slightly greater in the fetus compared to the dam. Whole blood and erythrocyte Se concentrations as well as whole blood glutathione peroxidase activity represent the functional pool of selenium as an antioxidant and these data suggest an approximately equal requirement for both fetus and dam. Other studies have also shown minimal differences between non-hepatic fetal tissue mineral concentrations and maternal tissue mineral concentrations.[9,13]

Observed differences between serum and liver selenium concentrations suggest altered transport and storage pools between dam and fetus. Higher maternal serum selenium concentration provides a substantial concentration gradient necessary for efficient placental selenium transport. Higher fetal liver Se concentrations infers a preferential storage of excess Se by the fetal liver over and above tissue requirements. Although similar relationships have not been completely identified for the other trace elements, similar mechanisms are thought to be in place. Given these relationships, fetal liver and serum mineral concentrations must be interpreted differently from adult values. Nutrient-specific enzyme activity most likely is similar between fetus and adult animals.

Fat-soluble vitamins A, D and E do not appreciably cross the placenta as evidenced by much lower serum and liver fat-soluble vitamin concentrations in fetal samples compared to adult cattle.[15,16] Mechanisms for placental transport of fat-soluble vitamins exist, most likely to ensure sufficient amounts to meet fetal metabolic needs. Amounts necessary to support fetal needs are not defined and why concentrations decline with age is undetermined. The neonate's primary source of fat-soluble vitamins comes via colostrum ingestion supplied from an adequately supplemented dam. Maintenance of neonatal vitamin status will come from milk consumption.

During the early postnatal period, almost all essential nutrients are adequately provided for by milk consumption. However, a number of critical micronutrients, namely Cu, Fe, Zn and Se, are insufficiently to marginally provided by milk consumption alone, thus requiring additional sources to meet daily needs. Milk will contain some fat-soluble vitamins, but this will depend upon maternal supplementation. Fetal hepatic nutrient reserves play a critical role in maintaining adequate micromineral concentrations to support daily nutrient requirements in the milk-fed postnatal animal. Hepatic mineral reserves are augmented by consumption of colostrum, a highly concentrated source of most essential minerals and fat-soluble vitamins, which is dependent upon maternal nutrient status.

ROLE IN PERINATAL DISEASE

Clinical disease associated with specific trace minerals or fat-soluble vitamins has been described,[4] but these situations are not generally prevalent unless serious dietary issues are present. Marginal deficiencies of trace minerals have been clinically implicated in prenatal and postnatal disease issues leading to abortion, stillbirth, weak neonates and impaired immune response, but a definitive cause and effect has not been established through controlled studies.[1,3,4,17] The role of fat-soluble vitamins is less well defined. Survey studies have associated low hepatic copper, selenium, manganese and zinc concentrations with abortion with and without an identified infectious agent.[7,18] In a recent survey of stillborn beef and dairy calves, 78% of cases had low vitamin A and one or more trace mineral deficiencies (Van Saun, unpublished data, 2015). This observed role of vitamin A is consistent with other observations of low vitamin A status being associated with weak beef calves.[19] Trace minerals and vitamins may play an important role in maintaining antioxidant status of the fetus during development as well as during the process of parturition.[20,21] Inadequate antioxidant capacity could result in demise of the early embryo, fetus and placenta, thus accounting for pregnancy wastage issues. The adaptation of the fetus from a low oxygen environment to a high oxygen environment increases reactive oxygen species and need for antioxidants and may account for disease issues occurring at or soon after birth.

Adequacy of neonatal nutrient reserves might explain differences in time frame and severity of specific nutrient deficiency disease occurrence. If a pregnant dam is severely deficient, mineral transfer to the fetus may be so limiting as to compromise normal functions, resulting in fetal death and abortion. If the deficiency is lessened but still serious, the fetus may die during parturition or soon thereafter. If mineral status is sufficient to maintain fetal development, hepatic reserves may be limited to various degrees. This then may result in clinical deficiency signs in the neonate within a week or two of birth. In other neonates where hepatic mineral reserves were slightly better, one might see clinical signs at one month or later or may not see clinical signs at all, but rather poor growth and performance. At this time, we do not know what mineral storage amount is necessary in the neonatal liver to minimize clinical and subclinical problems. Much more research in this area is needed.

DIAGNOSTIC EVALUATION

A better database of adequate fetal and neonatal trace mineral and vitamin concentrations is required for proper diagnosis and monitoring. In attempting to determine mineral status, one needs to consider what question is being asked. First if one is interested in determining cause-effect relationship between a mineral deficiency and specific pathologic lesion, then one needs to look at the physiologic or biochemical role of mineral relative to biochemical function. On the other hand, what is most often asked is: What is the nutritional status of the animal? This is entirely different question and reflects the status of a different nutrient pool, the storage pool. Unfortunately collection of serum or whole blood is not the preferred specimen for determining nutritional status. Interpretation of this pool is difficult in many circumstances due to dynamic circumstances of nutrient flux through this pool. As a result of these relationships, collection of a liver biopsy specimen is the preferred sample to determine nutritional status of the animal. Liver samples can be obtained from any portion of the liver. No data has shown any evidence of mineral concentration variation within the liver.[22]

Given the described differences between fetal and maternal mineral metabolism, and understanding that neonatal mineral metabolism is a gradual progression from fetal to adult metabolic patterns, it seems obvious that adult-based diagnostic criteria cannot be used for either fetal or neonatal evaluations. Some diagnostic laboratories have recognized these differences and have established age-based criteria. At Michigan State University's Clinical Nutrition laboratory diagnostic criteria have been estimated for fetal, newborn (1-9 days), infant (10-29 days), juvenile (30-300 days), yearling (301-700 days) and adult (>700 days) age categories. This laboratory has a tremendously large database by which these criteria were empirically derived for vitamins A and E and selenium in serum and liver samples. Without age-based criteria, all younger animals and fetuses would be considered deficient in most trace minerals and vitamins. While this approach is a tremendous move forward in diagnostic capabilities, more data is needed to refine these diagnostic criteria. Our ability to make diagnostic interpretations from fetal and neonatal liver mineral concentrations may be improved if evaluations are based on age and hepatic dry matter content. More controlled research is needed to specifically determine adequate hepatic mineral concentrations in bovine fetus and neonate.

SUPPLEMENTATION APPROACHES

It is absolutely essential that the pregnant animal receive an adequate amount of all minerals and fat-soluble vitamins to support both maternal maintenance and conceptus development throughout the duration of gestation to minimize deficiency disease problems of either the dam or neonate. House and Bell have suggested NRC mineral requirements were sufficient to support pregnancy,[14] but vitamins A and E were increased with the later NRC publication.[23] In beef cattle it is suggested to increase NRC mineral requirements during pregnancy by 125% to ensure sufficient fetal transfer and maintain maternal status. Dietary supplementation is more physiologic and should be maintained throughout the gestation feeding period. Challenges occur with the rumen in microbial degradation of fat-soluble vitamins (A predominately) and production of interfering agents. Often a recommendation to include between 25 and 30% mineral supplement from chelated or organic forms in made.

Injectable minerals and vitamins have been used to correct diagnosed deficiencies or in the place of dietary supplementation programs. Injectable minerals and vitamins are more biologically available, but increasing serum concentrations may result in a greater percent of the injected dose being excreted. Selenium deficient beef heifers administered a label dose of injectable sodium selenite excreted nearly 25% of the injected dose within 24 hours and selenium status adequacy was not achieved.[24] More recent work has shown some beneficial effects on cow performance with injectable minerals even when the diet fed was within recommendations.[25,26] A better understanding of how supplemental nutrition can influence immune response is necessary to better define mode and rate of trace mineral and vitamin supplementation to enhance cow and calf health and performance.

REFERENCES

[1]Toombs RE, et al. Vet Clinics NA: Food Animal Pract 1994;10(1):137. [2]Graham TW, et al. J. Vet Diagn Invest 1994;6:77. [3]Hostetler CE, et al. Vet J 2003;166:125. [4]Suttle NF. The Mineral Nutrition of Livestock, 4th ed., New York, NY: CABI Publishing, 2010. [5]Suttle NF. In Practice 2004; 26(10):553. [6]Patterson J, et al. Proc Mid-South Ruminant Nutrition Conference, Dallas-Ft. Worth, Tx, 1999, pp. 1-12. [7]Abdelrahman MM, Kincaid RL. J. Dairy Sci. 1993;76:3588. [8]House WA, Bell AW. J. Dairy Sci. 1994;77:1860. [9]Hidiroglou M, Knipfel JE. J. Dairy Sci. 1981;64:1637. [10]Gooneratne SR, Christensen DA. Can. J. Anim. Sci. 1989a;69:141. [11]Gooneratne SR, Christensen DA. Can. J. Anim. Sci. 1989b;69:151. [12]Van Saun RJ, et al. J. Nutr. 1989a; 119:1128. [13]Hidiroglou M. Can Vet J 1980;21:328. [14]House WA, Bell AW. J. Dairy Sci. 1993;76:2999. [15]Njeru CA, et al. J Anim Sci 1994; 72:1636-1640. [16]Van Saun RJ, et al. J Nutr 1989b; 119:1156. [17]Enjalbert F, et al. JAPAN 2006; 90:459-466. [18]Van Saun RJ. *Proc 14th International Congress of Production Diseases in Farm Animals*, Ghent, Belgium, June 20-24, pg. 167-168, 2010. [19]Waldner CL, Blakley B. J Vet Diag Invest 2014; 26:376. [20]Mutinati M, et al. Reprod Dom Anim 2013;48:353. 21Mutinati M, et al. Reprod Dom Anim 2014;49:7. [22]Abdelrahim AI, et al. Res. Vet. Sci. 1986;40:209. [23]National Research Council. Nutrient Requirements of Dairy Cattle, Washington, DC, National Academy Press, 2001. [24]Maas J, et al. J Vet Intern Med 1993; 7:342-348. [25]Machado VS, et al. Vet J 2014; 200:299. [26]Machado VS, et al. Vet J 2013; 197:451.

Manganese Deficiency in Cattle

Thomas H. Herdt, DVM, MS, DACVIM (LAIM), DACVN
East Lansing, MI

INTRODUCTION

Manganese is a metal that exists in relative abundance in the earth's surface. It is, however, one of the least abundant metals in the animal body. It exists in a variety of oxidation states, which probably explains its usefulness in many biological systems. It has been recognized as an essential nutrient in animals since the 1930s. Recently there has been renewed interest in manganese nutrition in cattle. This interest appears to have been stimulated by a number of factors including controversy over the recommended dietary Mn concentration for dairy cattle [1], concern over the potential role glyphosate (Roundup) may play in the Mn concentration in feedstuffs, and a perceived increase in the incidence of congenital malformations of calves that may be related to prenatal Mn supply.

PHYSIOLOGICAL ROLE

Manganese plays a role in the function of numerous body enzymes, including pyruvate carboxylase, superoxide dismutase (MnSOD), and glycosyltransferase. Pyruvate carboxylase is a critical enzyme in carbohydrate oxidation and physiological energy generation. Clinical manifestations of manganese deficiency due to reduced activity of this enzyme are not described. Manganese-dependent superoxide dismutase is an intramitochondrial enzyme critical in protecting the animal against the effects of free radicals, particularly superoxide. In clinical situations or on diets prone to the induction of oxidative stress, manganese deficiency might result in adverse clinical effects via increased susceptibility to free radical damage. However, defined clinical syndromes associated with this mechanism are not described. The use of tissue MnSOD activity, or the abundance of its RNA message hold promise as diagnostic methods in the diagnosis of Mn deficiency, but such tests are not currently available for practical application in veterinary medicine.

Glycosyltransferase is essential for the normal formation of bone and other connective tissue. Impaired glycosyltransferase activity reduces the synthesis of glycosaminoglycan and oligosaccharide side chains in animals deprived of manganese, resulting in altered bone formation. It is this reaction that is responsible for the most recognized and well-described manifestations of Mn deficiency in animals.

Based on variation in tissue Mn concentrations among cycling and anestrous ewes, there has been speculation that Mn is important for corpora luteal function and sex steroid production. This observation has not been supported by more recent work showing normal cycling and conception rates in heifers deprived of Mn to the extent that they delivered malformed calves[2].

DIETARY MANGANESE SUPPLY AND AVAILABILITY - CAUSES FOR CURRENT CONCERN

The manganese supply from natural feed ingredients varies widely. Concentrates, especially those containing the bran portion of grains, generally are richer in Mn than forages, although this is not always the case. Increases in soil pH markedly decrease the uptake of Mn by plants. The general comment can be made that corn, either as silage or corn grain, is a relatively poor source of Mn, compared to other common feed sources.

There is some indication that glyphosate herbicide (Roundup) reduces the uptake of Mn by plants[3], although this is not the mechanism of herbicidal action of this compound. This has caused concern that residual glyphosate in the soil (it has a relatively long half-life in soil) may be reducing the concentration of Mn in forage plants. However, there is little indication that the Mn concentration of plants has diminished since the introduction of glyphosate and this potential does not appear to be of practical concern [4]. The Mn concentration of natural feedstuffs should not be relied upon for animal Mn supply without chemical analysis.

The efficiency of absorption of dietary Mn, like many other minerals, is inversely proportional to its concentration in the diet and the Mn nutritional status of the animal [5]. This suggests a physiological mechanism of Mn homeostasis, which probably involves expression of the divalent metal cation transporter in the intestinal epithelium. The NRC (2001) estimated the efficiency of Mn absorption on typical cattle diets at 6 to 7%, but this probably varies widely based on dietary Mn sources and the Mn status of the animals. Phytate, fiber, and calcium have been suggested to alter the availability of dietary Mn, but the role of these factors in ruminant diets is unknown and probably not of practical significance on well balanced diets. Inorganic sources of Mn including manganese sulfate and manganese chloride are more readily available than manganous oxide and the organic forms of Mn present in natural feedstuffs.

Dietary iron and the iron status of the animal appear to be among the most influential of extraneous nutritional factors affecting the availability of dietary Mn. As dietary iron concentration increases, the proportion of dietary Mn absorbed diminishes[6] and this is probably true in cattle as it is in other species [7]. The iron concentrations of forages in the US are typically high, and high hepatic iron stores are a frequent finding in US cattle (Herdt, MSU DCPAH data). Thus the influence of high iron intake should be considered, in addition to other factors, when Mn deficiency is suspected.

CLINICAL MANEFESTATIONS OF MANGANESE DEFICIENCY

Manganese has multiple functions in the body, but the function most associated with clinical manifestations of deficiency is its role in glycosyltransferase activity. Impaired glycosyltransferase activity reduces the synthesis of glycosaminoglycan and oligosaccharide side chains in animals deprived of manganese resulting in aberrations of skeletal and connective tissue development. The most well recognized Mn deficiency syndromes in animals are perosis and nutritional chondrodystrophy of poultry. Similar skeletal malformations occur in manganese-deficient bovine fetuses [8-10]. The fetus is completely dependent on the dam for all nutrients including trace elements, thus Mn deficiency manifestations in the fetus reflect the Mn status of the dam. Dr. Jeremy Schefers of the Minnesota Veterinary Diagnostic Laboratory reviewed a series of ten cases of bovine fetal or congenital abnormalities that were apparently associated with Mn deficiency. These cases occurred in 2010. All cases had one or more of the following abnormalities: twisted or short limbs with swollen joints, dwarfism, and/or shortened nasomaxillary bones. Live calves were reported to be weak and unsteady. Mean hepatic manganese concentration of these fetuses on a wet tissue basis was 0.88 ug/g, which compared to a 1.2 ug/g mean hepatic Mn concentration of bovine fetuses seen in Dr. Schefers' laboratory that were not affected with skeletal abnormalities. No infectious or hereditary causes of these abnormalities could be identified. All ten herds from which the affected fetuses originated fed diets with less than 50 mg Mn/kg dry matter. Six herds fed manganese oxide as a supplement; manganese from this source is poorly available. One herd fed a diet entirely of corn forage and byproducts, which have inherently low Mn concentrations. Increasing the Mn intake in these herds did not prevent further problems during the current gestation, but after supplementation no similar problems were seen in the subsequent breeding year in these herds.

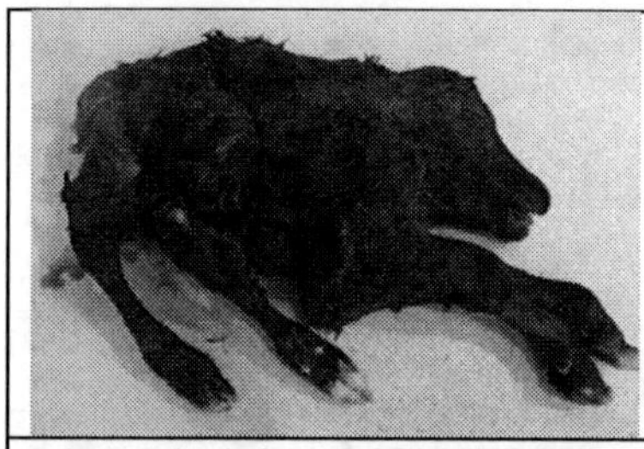

Extremely deformed fetus apparently affected by Mn deficiency. Photos courtesy of Dr. Jeremy Schefers, University of Minnesota

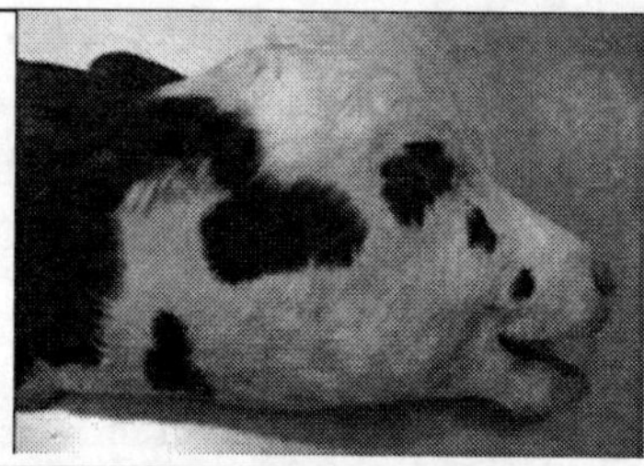

Very typical head and facial configuration of Mn deficient fetus. Note shortened nasomaxillary bones and domed cranium.

DIAGNOSIS OF MANGANESE DEFICIENCY

Manganese deficiency is probably most definitively diagnosed by demonstrating a reduction in the activity of MnSOD. When fresh necropsy tissue is available, heart MsSOD activity appears to be a sensitive indicator of Mn intake, at least in lambs [11] Analysis of the activity of this enzyme in leukocytes may provide the most practical approach to this assay in clinical application. Such "functional" measures of trace element nutriture are generally recognized as the "gold standard" for identifying deficiency. However, such tests are typically impractical in veterinary medicine because they have limited availability or are prohibitively expensive, are dependent on rigorous sample handling protocols, and because reference ranges are not available. Furthermore, while they may clearly identify cases of deficiency, they may be insensitive to more marginal depletion of body trace element stores, which would occur prior to clinical deficiency. Therefore, analysis of Mn concentrations in body tissues and fluids holds the most promise as a practical means of evaluating Mn nutriture in veterinary patients.

Blood is the most readily available sample for Mn evaluation. The Mn concentration in erythrocytes is ten to twenty times higher than in plasma or serum, making whole blood Mn concentrations substantially higher than serum or plasma concentrations, thus care must be taken in determining that the reference range employed is for the appropriate sample. Limited information suggests that bovine whole-blood Mn concentrations greater than 20 ng/mL indicate adequate Mn nutriture [12]. Furthermore, it appears, based on data from human subjects, that whole-blood Mn concentrations are more stable within day and more reflective of Mn exposure than are serum Mn concentrations [13].

Blood serum and plasma are unreliable samples for evaluation of Mn status. Typical Mn concentrations in serum or plasma are very low, requiring very sensitive assays for determination. Such assays may not always be available. A major drawback related to serum or plasma Mn determination is that Mn from the erythrocytes appears to leach from the cells if the serum remains in contact with the clot, or plasma with the red cells. This occurs within a matter of a few hours and is present even in the absence of visible hemolysis. This can increase the serum Mn concentration several fold and make interpretation impossible. As mentioned above [13], serum Mn concentrations tend to be variable within day and not good indicators of Mn exposure.

Attempts to use hair as an index of Mn status generally yield highly variable results within animal and do not appear reliable.

Liver is the tissue with highest Mn concentration and liver tissue from biopsy samples or necropsy specimens is frequently used for Mn evaluation. Relative to necropsy specimens, however, heart Mn concentration may be a more sensitive reflection of Mn status than is liver. Antemortem, liver biopsy specimens appear to be the most practical means of assessing Mn status, although they appear to be insensitive indicators with relatively severe Mn deficiency necessary to reduce hepatic Mn concentrations [14].

Current interpretation of bovine hepatic manganese concentrations, Diagnostic Center for Population and Animal Health, Michigan State University – Values are in ug/g of dry tissue.

	Critical low	Low reference range	High reference range
Fetal and neonatal samples	1.7	2.7	14
Older calves and adults	3.1	5.5	15

REFERENCES

1. Weiss, W.P. and M.T. Socha, J Dairy Sci, 2005. **88**(7): p. 2517-23.
2. Hansen, S.L., et al., J Dairy Sci, 2006. **89**(11): p. 4305-11.
3. Eker, S., et al., J Agric Food Chem, 2006. **54**(26): p. 10019-25.
4. Camberato, J., K. Wise, and B. Johnson. 2010; Available from: https://www.btny.purdue.edu/WeedScience/2010/GlyphosateMn.pdf.
5. Jenkins, K.J. and M. Hidiroglou, J Dairy Sci, 1991. **74**(3): p. 1047-53.
6. Davis, C.D., T.L. Wolf, and J.L. Greger, J Nutr, 1992. **122**(6): p. 1300-8.
7. Ho, S.Y., et al., J Dairy Sci, 1984. **67**(7): p. 1489-95.
8. Rojas, M.A., I.A. Dyer, and W.A. Cassatt, J Anim Sci, 1965. **24**: p. 664-7.
9. Hansen, S.L., et al., J Dairy Science, 2006. **89**(11): p. 4305-4311.
10. Anon, Vet Rec, 2013. **172**(15): p. 389-92.
11. Masters, D., et al., Aust J Agr Res. 1988. **39**: p. 517-524.
12. Hidiroglou, M., et al., Can J Comp Med, 1978. **42**(1): p. 100-7.
13. Baker, M.G., et al., J Trace Elem Med Biol, 2015. **29**: p. 123-9.
14. Ahola, J.K., et al., J Anim Sci, 2004. **82**(8): p. 2375-83.

Adipose Tissue Metabolism in Relation to Transition Cow Disease – A New Look at Fat Cows

G. Andres Contreras, DVM MS PhD
East Lansing, MI

INTRODUCTION

During the transition period, dairy cows mobilize triglyceride reserves stored in adipose depots to offset the negative energy balance (NEB) induced by fetal requirements, the onset of lactation, and limited dry matter intake. Through lipolysis, nonesterified fatty acids (NEFA) are released from the glycerol molecule to be used as energy substrate in different organs. In modern dairy cows, the lipolytic process is exacerbated because of their homeorhetic drive to meet genetic potential for high lactation efficiency. Excessive rates of lipolysis, usually measured as plasma NEFA, are associated with increased morbidity and mortality rates and are a major risk factor for the development of displaced abomasum (DA), ketosis, and metritis. These transition cow disorders have high economic impact on dairy producers, for example in the United States, costs associated with treating clinical cases of metritis, displaced abomasum (DA) and ketosis range between $145 and $340. Actual losses are higher as these values do not include revenue reductions related to poor lactation performance and increased cull rates. Remarkably, transition cattle with high body condition scores are more susceptible to periparturient disorders. Higher risk for clinical disease in these cows is linked with enhanced lipolysis rates, pronounced NEB after calving, and the development of a low grade inflammatory state that is accompanied by oxidative stress [1,2]. Although the mechanisms leading to this state are not well characterized, lipolysis induced changes in adipose tissue function and structure may play an important role in altering systemic immune function, especially in fat cows. This manuscript will summarize recent studies on adipose tissue metabolic function during NEB in monogastric and ruminant animals focusing on the effects on immune cell populations.

LIPOLYSIS AND ADIPOSE TISSUE REMODELING

In non-ruminants, lipolysis induces a remodeling process within the adipose tissue that activates an inflammatory response with enhanced macrophage infiltration [3,4]. Macrophages are part of the stromal vascular cellular fraction (SVF) of adipose tissues, which also contains fibroblasts, endothelial cells, blood cells, pericytes, preadipocytes, and mesenchymal stem cells, but not adipocytes. Macrophages are the most abundant immune cell in fat comprising ~5% of SVF and more than half of the leukocytes [4]. Upon migration to the adipose, macrophages acquire functional phenotypes in a process known as polarization. Classically activated adipose tissue macrophages (ATM), also termed as M1, are proflammatory. In contrast, M2 are considered to be alternatively activated and anti-inflammatory. During short term periods of NEB, lipolysis induces ATM infiltration with a predominant M2 phenotype [5]. In contrast, extended periods of lipolysis such as those observed in human obesity and metabolic syndrome enhance ATM infiltration and polarization to the M1 phenotype. Pro-inflammatory macrophages tend to aggregate into multicellular structures, known as crown like structures, that surround necrotic adipocytes or lipid droplets remnants. The presence of M1 aggregations is strongly associated with the development of local adipose inflammation and systemic insulin resistance (IR) in rodents and humans [6].

In transition dairy cattle, there is strong evidence indicating a similar lipolysis induced remodeling process that is directly associated with NEB. First, intense lipolytic rates during the first weeks of lactation enhance adipose gene transcription of pro- inflammatory cytokines including TNFα, Interleukin 1β, and Interleukin 6 as well as important components of the pro-inflammatory pathway NFκB [7,8]. Second, around parturition dairy cows exhibit adipose tissue macrophage (ATM) infiltration in omental and subcutaneous adipose tissues. This increased flux of immune cells was described in a study that compared ATM populations in transition cows with high lipolytic rates that developed DA with a group on non-lactating cows in positive energy balance [9]. ATM comprised ~9% of the cells in SVF of omental and subcutaneous fat depots of non-lactating cows, compared to 20% in transition cows with DA. Analysis of the inflammatory phenotype of infiltrating ATM in DA cows indicated that the majority were of M1 phenotype and formed multi-cellular aggregates. Although some of the DA cows in this study were over-conditioned prior to calving, it is currently unknown if ATM infiltration is exacerbated in obese dairy cows during intense lipolytic stimuli such as occurs at the onset of lactation. In summary, findings from studies in monogastric species and dairy cattle indicate that ATM infiltration exacerbates adipose inflammatory responses possibly extending and enhancing lipolysis and increasing disease risk in cows during early lactation [9].

ADIPOSE TISSUE INSULIN RESISTANCE

Modern dairy cows ensure sufficient glucose flow to the insulin-independent mammary gland by developing a transient IR state that limits glucose uptake in adipose tissue and skeletal muscle and stimulates adipose lipolysis [10]. IR refers to a reduced biological response to normal insulin concentrations that in transition dairy cows is further aggravated by hypoinsulinemia during the first week after parturition. Adipocytes respond to insulin stimulation by up-regulating glucose transport and lipogenesis, enhancing the uptake of NEFA from systemic circulation, and inhibiting lipolysis [11]. Extended periods of IR perpetuate the lipolytic stimuli releasing more NEFA into circulation leading to dyslipidemias such as high plasma NEFA [12]. Adipose tissue IR, specifically, is observed in cows that exhibit high plasma NEFA and extreme weight loss during the first weeks of lactation, such as obese cows [13]. The effects of periparturient-associated IR on immune function include enhanced synthesis of pro-inflammatory cytokines TNFα, interleukin 1β, and interleukin 6 by adipocytes and immune cells and restricted energy availability to immune cells [11]. The autocrine effect of the aforementioned cytokines further promotes lipolysis, thus increasing the concentration of NEFA in the adipose milieu and systemically[14]. Adipose tissue remodeling together with the development of IR enhance the low grade inflammation state that promotes excessive inflammatory responses in non-adipose organs and predisposes to clinical diseases [15].

ADIPOKINES AND IMMUNE FUNCTION

Besides providing NEFA as an alternate energy substrate during the transition period, adipose tissue exerts an autocrine and endocrine control of transition cattle metabolic function through the expression of adipocyte derived peptides (i.e. adipokines) such as adiponectin, leptin, and resistin. Adiponectin enhances insulin sensitivity in adipose, liver, and muscle. At the same time, this adipokine promotes fatty acid β-oxidation in hepatocytes and myocytes and increases dry matter intake. In dairy cows, plasma adiponectin concentrations are reduced during the first week after parturition compared to levels observed during the dry period and peak lactation [16]. In addition to metabolic effects, adiponectin modulates the inflammatory responses of human and bovine macrophages by reducing their expression and secretion of TNF alpha and other pro-inflammatory cytokines [16,17]. Adiponectin is also an important modulator of adaptive immunity as it is required for dendritic cell activation and T-cell polarization [18]. The role of adiponectin in modulating the inflammatory phenotype of ATM and other immune cells during periparturient inflammatory responses remains unclear.

Leptin is another important adipokine because it reduces dry matter intake, enhances adipose tissue lipolysis, and inhibits adipocyte and hepatocyte lipogenesis. In liver and skeletal muscle, leptin promotes fatty acid β-oxidation. Leptin also modifies metabolic function indirectly by reducing insulin secretion and stimulating growth hormone expression [19]. Importantly, leptin is a regulator of immune function. Hypoleptinemia impairs the efficacy of T cell immune responses by promoting a shift in the phenotype of these cells from type 1 (pro-inflammatory) to a T2 helper. Leptin is also necessary for adequate maturation and inflammatory responses in dendritic cells. In macrophages and polymorphonuclear cells, leptin signaling is required for phagocytosis in response to toll-like-receptor activation [20]. Similar to adiponectin, leptin reaches its nadir during the first week after calving, while the highest plasma concentrations are observed early during the dry period [19].

A third adipokine with important effects on immune function is resistin. The dynamics of expression and secretion of resistin in dairy cows were described recently. In contrast to leptin and adiponectin, resistin plasma concentrations spike during the first week after calving and return to pre-partum levels by 5 weeks into lactation [21]. In humans and rodents, resistin expression in adipocytes is stimulated by Il-6, hyperglycemia and growth hormone. Resistin impairs insulin signaling in adipocytes and is characterized as a pro- inflammatory adipokine [22]. In bovines, resistin promotes lipolysis in adipocytes but its effects on immune cells are currently unknown.

ASSESSING ADIPOSE TISSUE FUNCTION DURING THE TRANSITION PERIOD

Transition cow management programs often include routine measures of clinical and production parameters that can directly or indirectly evaluate adipose tissue function. Body condition score (BCS) is a good measure of subcutaneous adiposity and the dynamics of BCS changes around parturition subjectively describe lipolysis rates. Alternatively, the use of image biomarkers obtained during ultrasound examination of adipose tissues provides an objective evaluation of BCS avoiding the variability associated with subjective visual measurements. Subcutaneous adipose tissue depth is strongly correlated with BCS evaluation when measured by trained personnel and is highly sensitive and specific in predicting plasma NEFA values at close-up and calving in dairy cattle [23]. If using subjective BCS assessment, mature cows should approach calving with a BCS of 3.0 and heifers with 3.25 as excessively thin or over- conditioned cows are more susceptible to disease.

Currently the most common direct measure of lipolysis is plasma NEFA. In preventive herd medicine, pre and post calving plasma NEFA values are used as early lactation disease predictors (Table 1). Similar to plasma NEFA, post-partum plasma BHB indicate NEB and predict disease risk in early lactation [24]. Lipolysis can also be evaluated at the group or individual animal level using the milk fat to milk protein percentage ratio (Table 1). Milk fat increases as plasma NEFA rise. Cows with milk fat to milk protein ratio values higher than 2 during the first week after calving are at a higher risk for developing retained fetal membranes, DA, clinical endometritis, and being culled before the end of lactation [25].

Single biomarkers do not provide enough information to support management decisions during the transition period. However, when multiple biomarkers are analyzed together and combined with health, production, nutritional, and environmental data, biomarkers become essential for identifying metabolic problems related to extended periods of intense lipolysis.

CONCLUSIONS AND FUTURE DIRECTIONS

The impact of intense and extended adipose lipolysis rates on periparturient disease susceptibility and poor lactation performance may be directly related to the development of a low grade systemic pro-inflammatory state that alters immune function[2]. The mechanisms for this cause effect relationship may include lipolysis induced adipose remodeling processes, alterations in the expression of adipokines, and the development of insulin resistance. Certainly the focus on adipose tissue biology research given by human obesity and diabetes epidemics in western countries has expanded our understanding of the role of lipolysis in metabolic and immune function. However, recent studies highlight marked differences between ruminant and monogastric adipose organ physiology, demonstrating that focused research is required on specific inflammatory and metabolic pathways linking adipocyte and immune cells in dairy cattle [26]. Given their intense lipolytic responses, over-conditioned transition cows may serve as a model in which to study the impact of lipolysis on inflammatory responses of circulating and tissue resident immune cells.

To improve transition cow performance and minimize periparturient disease in commercial dairy operations, it is necessary to expand the discovery and application of lipolysis and functional adipose tissue biomarkers. Although current biomarkers such as BCS and plasma NEFA and BHB adequately describe lipolysis rates, they lack specificity with regards to disease diagnosis, are retrospective in nature, and are poor predictors of disease during the early dry period a time when the success of nutritional or pharmaceutical interventions is often greatest. Understanding the inherent characteristics of adipose tissue function in transition dairy cattle will reveal more sensitive and specific biomarkers of disease and productivity, allowing for improved herd health and profitability.

Table 1. Selected biomarkers of transition cow health and adipose tissue function.

BIOMARKER	Reference Values
NEFA[27]	<0.50 mEq/L
BHB [27]	12.5 mg/dL Subclinical Ketosis; 14.5 mg/dL Clinical Ketosis
Cholesterol[28]	1.7 - 4.3 mmol prepartum; 2.7 - 5.3 mmol pospartum
Triglycerides [29]	0.12-0.65 mmol
GOT/AST [29]	46.5–103 IU/L
GGT [29]	21-37 IU/L
Acetoacetate/acetone [30]	> 10 mg/dL of acetoacetate and acetone.
Albumin	3.2 – 3.7 g/dL
Milk fat: protein ratio [25]	1.00 – 1.25
NEFA:cholesterol [28]	<0.2

NEFA=nonesterified fatty acids; BHB=beta-hydroxybutirate; GOT/AST= glutamic-oxaloacetic transaminase/aspartate aminotransferase; GGT=gamma-glutamyl transferase

REFERENCES

[1] O'Boyle N, et al. Vet Immunol Immunopathol. 2006;113:297-304. [2] Contreras GA, et al. Comp Immunol Microbiol Infect Dis 2011;34:281-289. [3] Granneman JG, et al. Am J Physiol Endocrinol Metab 2005;289:E608-E616. [4] Kosteli A, et al. J Clin Invest 2010;120:3466-3479. [5] Lee Y-H, et al. Biochim Biophys Acta 2014;1842:358-369. [6] Cinti S, et al. J Lipid Res 2005;46:2347-2355.[7] Schmitt E, et al. J Dairy Sci 2011;94:5913-5925.[8] Saremi B, et al. PLoS ONE 2014;9:e86211.[9] Contreras GA, et al. J Dairy Sci 2015;98:6176-6187. [10] Bell AW, et al. Adaptations of glucose metabolism during pregnancy and lactation. J Mammary Gland Biol Neoplasia 1997;2:265-278.[11] De Koster JD, et al. Vet Clin North Am Food Anim Pract 2013;29:299-322. [12] Pires JAA, et al. J Dairy Sci 2007;90:2735-2744. [13] Zachut M, et al. J Dairy Sci 2013;96:5656–5669. [14] Chen X, et al. Cell Biochem Funct 2009;27:407-416. [15] Sordillo LM, et al. Vet Clin North Am Food Anim Pract 2013;29:267-278. [16] Kabara E, et al. Comp Immunol Microbiol Infect Dis 2014;37:49-58. [17] Folco EJ, et al. J Biol Chem 2009;284:25569-25575. [18] Jung MY, et al. J Immunol 2012;188:2592-2601. [19] Chilliard Y, et al. Domest Anim Endocrinol 2005;29:3-22. [20] Naylor C, et al. Trends Mol Med 2016;22:88-98. [21] Reverchon M, et al. PLoS ONE 2014;9:e93198. [22] AL-Suhaimi E, et al. Eur J Med Res 2013;18:12. [23] Strieder-Barboza C, et al. Livestock Sci 2015;177:159-164. [24] Ospina PA, et al. Vet Clin North Am Food Anim Pract 2013;29:387-412. [25] Toni F, et al. J Dairy Sci 2011;94:1772-1783. [26] Laliotis GP, et al. Curr Genom 2010;11:168-183. [27] Ospina PA, et al. J Dairy Sci 2010;93:546-554. [28] Kaneene JB, et al. Prev Vet Med 1997;31:59-72. [29] Bertoni G,et al. Vet Clin North Am Food Anim Pract 2013;29:413-431. [30] Krogh MA, et al. J Dairy Sci 2011;94:2360-2367.

Clinical Effects of Subclinical Hypocalcaemia in Peripartum Dairy Cows

Vengai Mavangira, BVSc, DACVIM (LAIM)
East Lansing, MI

INTRODUCTION

The importance of calcium metabolism in lactating dairy cattle has been appreciated for several decades. Efficient homeostatic mechanisms that maintain normal blood calcium concentrations are critical to meet calcium demands in the periparturient dairy cow. Calcium demands are greatest in the late transition phase when high amounts are required for colostrum and milk synthesis. Blood calcium represents the immediate calcium pool available for use in meeting the lactation demands, in addition to maintaining normal physiological homeostasis. Failure to replenish the blood calcium pool will result in a state of negative calcium balance. The decrease in blood calcium concentrations immediately postpartum associated with clinical signs, such as recumbency, is known as milk fever. In contrast, subclinical hypocalcemia (SCH) develops when blood calcium concentrations decrease with no observable clinical signs. The SCH is more costly than milk fever because it contributes to negative health and performance outcomes in a greater proportion of cows.[1]

The reported prevalence of milk fever in US dairy herds is around 5%[2] but can reach up to 10%.[3] Milk fever is generally controllable through dietary modifications including manipulation of the dietary cationic-anionic difference (DCAD) or managing dietary potassium (K) content. Conversely, SCH has a greater prevalence and affects cows of all parities with incidences ranging from 25 - 54%.[4] Use of DCAD diets decreases rates of SCH to between 15% - 25 %.[5] Managing dietary K content decreases the incidence of SCH, however, a 60% incidence was reported in cows ≥ 3rd lactation in herds that did not use DCAD diets.[6] These observations suggest that management of SCH to optimize calcium homeostasis needs to be expanded beyond dietary manipulation. This is important considering that dairy cattle experience negative calcium balance for periods of 1 to several weeks post-partum.[7,8] The period of negative calcium balance is coincident with the post-partum period with highest rates of common diseases of dairy cows.

Overview of pathophysiology of subclinical hypocalcemia (SCH)

The pathophysiology of hypocalcemia and strategies for limiting the occurrence of milk fever and SCH in dairy cattle were reviewed extensively in literature.[2,9-11] Colostrum and milk synthesis in the periparturient dairy cow results in losses of much as 20-30 g/day[9] which can reach up to 50 g/day according to some estimates.[10] Early postpartum calcium loss cannot be met by dietary sources alone, rather, endogenous mechanisms must be activated to mobilize calcium from bone to replace the serum concentrations depleted by milk production.[9] These mechanisms involve parathyroid hormone (PTH), whose release is stimulated by the decrease in blood calcium. Subsequently, PTH acts on the kidneys to enhance the production of 1,25-dihydroxyvitamin D (the active form of vitamin D) and to increase renal reabsorption of calcium. In turn, the formation of 1,25-dihydroxyvitamin D enhances the efficient absorption of calcium from the gut. Cows are at increased risk of negative health outcomes when the mechanisms for enhancing calcium mobilization from bone, absorption from the gut, and renal excretion are inadequate to replace the deficit placed by hypocalcemia.

How calcium is linked to disease pathophysiology

The link between development of negative health outcomes and post-partum hypocalcemia was shown in early studies. For example, the occurrence of diseases including dystocia, retained fetal membranes, ketosis, and mastitis was significantly associated with parturient hypocalcemia.[12,13] The association of SCH with health disorders results from perturbed cellular functions that require calcium. Calcium mediates intracellular signaling for many cellular pathways and is also essential for normal functioning of nervous and muscle tissue. Optimal functioning of leukocytes including neutrophils and monocytes/macrophages also requires calcium. These leukocytes are crucial first lines of defense during infections such as mastitis, metritis and endometritis.[14,15] In peripheral blood mononuclear cells (PBMCs) from cows with SCH, intracellular calcium was decreased resulting in poor responses to stimulation with CD3+ antibody.[16] Similarly, neutrophils from cows with SCH had reduced intracellular calcium and diminished respiratory burst capacity.[17] Thus, SCH contributes to the immunosuppression that occurs in parturient dairy cattle.
Concentrations of total blood calcium for defining SCH and its association with postpartum health problems are considered to be between 5.5 – 8.0 mg/dL [1.38 – 2.0 mmol/L].[18] Specific cutoff values vary among studies; however, a common finding in recent studies is that blood calcium concentrations significantly associated with negative health outcomes are higher than previously reported.[17] The observation that higher blood calcium concentrations for defining SCH fall in the traditionally normocalcemic range (serum calcium concentrations of 8.02 – 11.22 mg/dL [2.0 – 2.8 mmol/L]), raises the question of whether it is physiologically possible to raise blood calcium concentrations in post-partum dairy cows beyond that achieved by proven dietary interventions, such as use of DCAD diets.

SCH on infectious conditions

Occurrence of infectious diseases such as mastitis and metritis is highest during the periparturient period. Epidemiologic studies showed that neutrophils from dairy cattle with metritis had diminished respiratory burst capacity when stimulated with E. coli.[17] Neutrophils in mid-lactation cows with induced SCH showed decreased cell counts, decreased intracellular calcium, and reduced respiratory burst activity.[19] In addition, the decreased intracellular calcium concentrations was also shown in PBMCs which exhibited poor responses to immune stimulation.[16] Taken together, the decrease in intracellular calcium could explain, in part, the immune suppression that occurs around parturition, thus predisposing the cow to post-partum diseases.

Mastitis

The occurrence of mastitis in cows with milk fever was shown in previous reports.[12] Explanations for this relationship include reduction in teat sphincter smooth muscle function and higher levels of cortisol in SCH cows.[3,20] Additionally, using the cut-off point of serum calcium ≤ 1.8 mmol/L [7.21 mg/dL], SCH-affected cows spent 2.7 hours less time standing on first day after parturition,[21] suggesting that increased contact with the environmental sources of bacteria causing mastitis could put cows at increased risk for mastitis. Specific cut-off values have not been defined in literature for SCH associated with increased incidence of mastitis; however, the increased severity of mastitis in cows with decreased cell mediated and humoral immune responses[22,23] suggests a potential role for SCH.

Retained placenta and uterine infection

Serum calcium concentrations < 8.0 mg/dL [2.0 mmol/L] were suggested to increase the risk for retained fetal membranes (RFM) in some studies.[20] In one study, marginal calcium concentrations of < 7.5 mg/dL [1.87 mmol/L] were associated with RFM in cows in their second lactation or greater.[24] Another study evaluated the occurrence of RFM, metritis, endometritis, and puerperal metritis and showed that the severity of uterine infection was related to the degree of SCH. In that study, the SCH (Ca < 8.0 mg/dL) affected a greater proportion of cows (93 %) for an extended period of 6 weeks that spanned the sampling period.[7] Martinez et al also showed that cows with calving disorders (dystocia, twins, stillbirths, and retained placenta) had significantly higher rates of metritis associated with higher cut-off values for SCH (serum calcium ≤ 8.59 mg/dL [2.14 mmol/L] than previously reported.[17] In addition, SCH in the study by Martinez et al was associated with higher NEFA and BHBA suggesting that calcium homeostasis and energy balance could both influence the occurrence of metritis. Using the same cut-off point, SCH in grazing dairy cows had higher rates of uterine infection, increased NEFA, and higher rates of subclinical ketosis.[25] Significant associations were shown between decreased calcium of 8.28 mg/dL [2.07 mmol/L] and 7.99 mg/dL [1.99 mmol/L] with RFM and metritis, repectively.[26]

SCH on Metabolic Diseases

Displaced abomasum

The occurrence of displaced abomasum (DA) in postpartum dairy cows with SCH can be partly explained by decreased abomasal motility. Several studies showed increased risk for DA using varying cut-offs for defining SCH ranging from ≤ 2.1 mmol/L (8.42 mg/dL) up to ≤ 2.3 mmol/L.[27-29] Higher odds of developing DA were observed for much lower calcium concentrations, however, the lower cutoffs had poor specificity and sensitivity.[29] It is noteworthy that other studies did not find an association between calcium and DA even when SCH was defined at lower concentrations of < 1.8, < 1.9 and < 2.0 mmol/L at days -1, 0 and 1 relative to calving day, respectively.[30] The authors suggested that SCH was not the cause of DA occurrence.

Hepatic lipidosis and ketosis

The degree of fatty infiltration of the liver and the development of hepatic lipidosis in dairy cows is known to be a risk factor for post-partum health problems, most of which are also associated with SCH such as metritis, DA and ketosis. Using a cutoff of < 1.0 mmol/L for ionized serum calcium to define SCH, affected cows demonstrated greater plasma NEFA and hepatic lipid content, but no significant differences in disease occurrence were detected in comparison to normocalcemic cows.[31] Cows with SCH (< 2.14 mmol/L) had greater rates of elevated NEFA and subclinical ketosis.[25] These data suggest a relationship between energy balance and calcium metabolism. In grazing dairy cows, restricted feed intake decreased NEFA concentrations while increasing postpartum calcium concentrations. In that study, using a cutoff of < 2mmol/L, the proportion of cows with SCH were lowest, intermediate, and highest for cows fed at 75%, 100% and 125% of their close-up energy requirements, respectively.[32]

SCH on culling rates and other diseases

Culling rates for dairy cattle are highest in the first 2 months post-partum and the majority are due to negative heath events.[33] The specific association of calcium is noteworthy because cutoffs of ≤ 2.2 mmol/L [8.82 mg/dL] and ≤ 2.3 [9.22 mg/dL] were associated with increased culling odds of 2.4 and 5.3, respectively.[29] The associations suggest a contributory role of SCH on several of the negative health events. Indeed, in grazing dairy cattle, a mean calcium of 2.08 ± 0.06 mmol/L [8.34 mg/dL] was associated with a significant risk for developing multiple clinical events including RFM, clinical hypocalcemia, clinical mastitis, and mild or severe metrtitis.[34]

Variability in blood calcium concentrations

Thresholds for SCH and contrasting findings among research studies relating SCH to various postpartum disorders likely reflect herd to herd differences. The range of calcium concentrations in apparently healthy lactating dairy cattle is 6.57 – 10.46 mg/dL [1.64 – 2.61 mmol/L][35] for total calcium and 3.79 – 5.25 mg/dL [0.95 – 1.31 mmol/L][36] for ionized calcium. Ionized calcium concentrations in dairy cattle are approximately half of those of total calcium concentrations. The wide calcium concentrations range may explain, in part, the lack of significant relationships reported in some studies between SCH and some postpartum diseases.[30,31] Variations in calcium concentrations also depend on farm-specific nutritional strategies for controlling hypocalcemia such as inclusion of DCAD diets or by limiting dietary cations (mostly K). Feed restriction strategies in the transition period to minimize the extent and impact of NEB may affect the incidence of hypocalcemia. Restricting energy intake in grazing dairy cattle was associated with better postpartum energy balance and lower proportion of cows with SCH.[32] Another source of variation may relate to differences in sampling times among studies. Typically calcium concentrations fall to their nadir within 12 hours of calving and return to pre-calving concentrations by 24 hours in healthy cows.[16] Cows developing health problems in which SCH plays a significant role experience a lower calcium nadir which they recover from at a slower pace than their healthy counter parts.[17]

Conclusions and Future directions

The SCH thresholds in recent studies (calcium ≤ 8.59 mg/dL[17] and ≤ 8.59 mg/dL[29]) fall within the historically normocalcemic range.[17] Whether calcium concentrations can be improved beyond the higher SCH thresholds remains to be determined. Possible options for raising calcium concentrations could be extended DCAD supplementation strategies pre-partum compared to the traditional 21 day supplementation. For example, a 42-day DCAD supplementation strategy tended to have greater total calcium concentrations (7.5 mg/dL) than the 21-day strategy (6.8 mg/dL), $P = 0.06$.[37] However, this strategy does not appear to adequately raise calcium concentrations above recent high SCH thresholds. Another approach for augmenting calcium blood concentrations in the postpartum period is oral administration of commercial calcium products to individual cows. This approach was reviewed recently as beneficial and cost-effective when used in a targeted approach.[1] Further studies are required to explore the practical utility of this approach and its influence on SCH and post-partum disorders.

The variability in calcium concentrations from herd-to-herd can be explained by farm-specific practices that influence the occurrence of SCH. Other than practices (including controlled DMI, dietary K and DCAD manipulations, and macro-mineral supplementation) for enhancing calcium homeostatic responses around parturition, a recently reported potential link between oxidant status and calcium metabolism should be evaluated. Recently, the development of milk fever in dairy cattle was associated with decreased concentrations of selenium and decreased antioxidant capacity.[38] Low blood selenium and low serum calcium were previously associated with retained placenta and metritis. These observations would suggest that oxidant status plays a role in calcium metabolism and warrants further investigation.

In conclusion, the effect of SCH in predisposing dairy cattle to peripartum disorders is probably greater than previously reported as recent studies suggest that cutoff values for SCH are much higher. In addition to the proven methods for reducing the incidence and effects of hypocalcemia in dairy cattle (DCAD diets, dietary K, dietary magnesium, oral calcium supplementation) further research is required to elucidate the reasons why SCH continues to occur at high levels. These areas include energy metabolism and the changes in oxidant status of dairy cattle in the periparturient period.

REFERENCES

1. Oetzel GR. Vet Clin North Am Food Anim Pract 2013;29:447.
2. Goff JP. Vet J 2008;176:50.
3. Mulligan F, et al. Ir Vet J 2006;59:697.
4. Reinhardt TA, et al. Vet J 2011;188:122.
5. Oetzel GR. Vet Clin North Am Food Anim Pract 2004;20:651.
6. Caixeta LS, et al. Vet J 2015;204:150.
7. Mateus L, et al. Revista Portuguesa de Ciencias Veterinarias 2002;97:35.
8. Goff JP, et al. J Dairy Sci 1996;79:378.
9. Goff JP. Vet Clin North Am Food Anim Pract 2014;30:359.
10. DeGaris PJ et al. Vet J 2008;176:58.
11. Martin-Tereso J, et al. Vet Clin North Am Food Anim Pract 2014;30:643.
12. Curtis CR, et al. J Am Vet Med Assoc 1983;183:559.
13. Curtis CR, et al. J Dairy Sci 1985;68:2347.
14. Galvao KN, et al. Theriogenology 2011;76:290.
15. Galvao KN, et al. Theriogenology 2012;77:356.
16. Kimura K, et al. J Dairy Sci 2006;89:2588.
17. Martinez N, et al. J Dairy Sci 2012;95:7158.
18. Horst R, et al. In: Production Diseases in Farm Animals: 12th International Conference 2006;215.
19. Martinez N, et al. J Dairy Sci 2014;97:874.
20. Goff J. Recent Research on Hypocalcemia and Immunity. 2013.
21. Benzaquen M, et al. Vet J 2015;204:186.
22. Thompson-Crispi KA, et al. J Dairy Sci 2012;95:3888.
23. Thompson-Crispi KA, et al. Clin Vaccine Immunol 2013;20:106.
24. Melendez P, et al. Am J Vet Res 2004;65:1071.
25. Ribeiro ES, et al. J Dairy Sci 2013;96:5682.
26. Bicalho MLS, et al. J Dairy Sci 2014;97:4281.
27. Chapinal N, et al. J Dairy Sci 2012;95:5676.
28. Chapinal N, et al. J Dairy Sci 2011;94:4897.

29. Seifi HA, et al. Vet J 2011;188:216.
30. LeBlanc SJ, et al. J Dairy Sci 2005;88:159.
31. Chamberlin WG, et al. J Dairy Sci 2013;96:7001.
32. Roche JR, et al. J Dairy Sci 2015;98:7164.
33. Roberts T, et al. J Dairy Sci 2012;95:3057.
34. Sepulveda-Varas P, et al. PLoS One 2015;10.
35. Quiroz-Rocha GF, et al. Can J Vet Res 2009;50:383.
36. Lincoln S, et al. J Am Vet Med Assoc 1990;197:1471.
37. Weich W, et al. J Dairy Sci 2013;96:5780.
38. Heilig M, et al. Tierarztl Prax Ausg G Grosstiere Nutztiere 2014;42:261.

Choosing The Best Anti-Epileptic Drug For Your Patient

Heidi Barnes Heller, DVM, DACVIM (Neurology)
Madison, WI

INTRODUCTION

Numerous anti-epileptic drugs (AED) are available for use therefore it is important to choose the appropriate one carefully. Factors including seizure etiology, species specific metabolism, administration route and ease of administration should be considered.

SEIZURE GOALS

Seizures are among the most difficult neurologic conditions for owners cope with. Therefore setting up appropriate goals and expectations up front is paramount to successful seizure management. Often the stated **goals of the owner** for their pet are 1) no seizures, 2) no side effects on medications and 3) few, or infrequent drug administrations. These goals are in contrast to **our goals** which are 1) 1 seizure or less every 4-6 weeks, each seizure less than 5 minutes in duration, minimal side effects and regular administration of medication(s). During the initial consultation, we discuss these goals with the owner therefore attempting to limit unfair therapeutic expectations by the client.

INITIATION OF MEDICATION

Anti-epileptic drugs (AED) are recommended for dogs and cats if they have greater than 1 seizure every 4-6 weeks, have any 1 seizure lasting longer than 5 minutes, or if there is aggression during the post ictal phase. Aggression may not be managed with the AED but a reduction in seizure frequency limits exposure to other people and animals to the aggressive behavior.

SEIZURE ETIOLOGY

Identification of the underlying cause of seizures can be an important factor when deciding to start anti epileptic drugs (AED). Classification of etiology based on the International League Against Epilepsy (ILAE) classification scheme from 2010 has been applied to dogs and cats. [1,2] (See Table 1). Knowledge of the seizure etiology may result in avoidance of an AED that requires hepatic metabolism or renal excretion in animals with hepatic or renal disease, for example. Additionally, animals with structural epilepsy may be more likely to have AED drug resistance, as defined by the ILAE. Drug resistant epilepsy is defined as "failure of adequate trials of two tolerated and appropriately chosen and used AED schedules (whether as monotherapies or in combination) to achieve sustained seizure freedom."[3] Therefore, initiation of treatment earlier or with more than 1 AED may be considered following a diagnosis of structural epilepsy in humans[4] and possibly veterinary species as well.

Table 1: Classification, inclusion criteria as applied to veterinary patients and an example of appropriate patient inclusion modified from the ILAE 2010 consensus report on classification and terminology. [5]

Current classification	Definition	Case example for inclusion	Previous classification
Genetic epilepsy	An intracranial disorder that has a confirmed genetic predisposition in the breed or animal	A Labrador retriever dog with seizure onset between 1-5 years old, with a familial history of seizures	Idiopathic epilepsy, true epilepsy, primary epilepsy
Unknown epilepsy	No known genetic predisposition but no identifiable cause on diagnostic testing (MRI, CSF and laboratory testing)	A cat with a seizure disorder with no evidence of disease on MRI, CSF or laboratory testing and no familial history of seizures	Idiopathic epilepsy, probably symptomatic epilepsy, cryptogenic epilepsy
Structural epilepsy	An identifiable cause other than metabolic disease, on diagnostic testing (MRI, CSF)	A Boxer dog with evidence of an intra-axial mass lesion.	Symptomatic epilepsy, secondary epilepsy
Metabolic seizures	NOT epilepsy, but a seizure disorder caused by a metabolic or biochemical abnormality	A young Maltese dog with evidence of hepatic dysfunction on biochemical analysis	Extracranial disorder, reactive epilepsy.

Table 2: Mechanism of action (MOA), metabolism and recommended use for the 4 most commonly prescribed AED in small animals.

AED	MOA	Metabolism	Recommended species use
Phenobarbital	Enhances inhibitory mechanism in GABA-A receptors: Increases Cl- permeability. Hyperpolarizes post-synaptic neuronal membrane, raising the seizure threshold.	Hepatic with auto induction of hepatic metabolism	Dog, cat
Bromide	Negatively charged ion, which readily diffuses across membranes to result in hyperpolarization.	Renal excretion with glomerular resorption	Dog
Levetiracetam	Reversibly bind to the synaptic vesicle 2A (SV2A) on the presynaptic membrane, resulting in reduced or absent release of acetylcholine into the neuromuscular junction.	Renal excretion of unchanged drug molecule	Dog, cat
Zonisamide	Blocks sodium channels (?)	Mostly renal excretion with some hepatic metabolism occurs	Dog, cat

Table 3: Recommended doses, routes of administration, monitoring (if available) and reported side effects for the 4 most commonly prescribed AED in small animals. D = dog; C = cat; α= not routinely evaluated in clinical patients; β = author starts with 22 mg/kg q12h, however most published doses are q24h; PU/PD/PP = polyuria, polydipsia, polyphagia.

AED	Recommended dosage	Route of administration	Therapeutic monitoring range	Reported side effects
Phenobarbital	2-5 mg/kg q12h (D,C)[6] 3-5 mg/kg (D, C)[7] 9 mg/kg q12h (C)[8]	PO IV Transdermal (C) in Lipoderm ®	15-45 ug/ml (typical goal 20-25 ug/ml)	Sedation, paresis, PU/PD/PP.
Bromide	20-40 mg/kg q12-24h[6,9], [β]	PO	1-3 mg/ ml	Sedation, ataxia, paresis, PU/PD/PP
Levetiracetam - **Standard release**	20 mg/kg q8h (D, C)[10,11] 30 – 60 mg/kg once (D)[12] 40 mg/kg once (D)[13]	PO IV Rectal	5-45 ug/mL[α]	Sedation, ataxia, GI
- **Extended release**	30 m/kg q12h (D)[14] ? (C)	PO PO		Possible GI or mild sedation
Zonisamide	5-10 mg/kg q12 (D)[15,16]	PO	10-40 ug/mL	Sulfa allergic dogs should not receive this drug. Sedation, ataxia, anorexia

REFERENCES

1. Finnerty KE, Barnes Heller HL, Mercier MN, Giovanella CJ, Lau VW, Rylander H. System for Seizures in Cats : 30 Cases (2004 – 2013). *J Am Vet Med Assoc* 2013: 195-199.
2. Mariani CL. Terminology and classification of seizures and epilepsy in veterinary patients. *Top Companion Anim Med.* 2013;28(2):34-41.
3. Kwan P, Arzimanoglou A, Berg AT, et al. Definition of drug resistant epilepsy: Consensus proposal by the ad hoc Task Force of the ILAE Commission on Therapeutic Strategies. *Epilepsia.* 2010;51(6):1069-1077.
4. Téllez-Zenteno JF, Hernández-Ronquillo L, Buckley S, Zahagun R, Rizvi S. A validation of the new definition of drug-resistant epilepsy by the International League Against Epilepsy. *Epilepsia.* 2014;55(6):829-834.
5. Berg AT, Berkovic SF, Brodie MJ, et al. Revised terminology and concepts for organization of seizures and epilepsies: Report of the ILAE Commission on Classification and Terminology, 2005-2009. *Epilepsia.* 2010;51(4):676-685.
6. Thomas WB. Idiopathic Epilepsy in Dogs and Cats. *Vet Clin North Am Small Anim Pract.* 2010;40(1):161-179.
7. Patterson ENE. Status Epilepticus and Cluster Seizures. *Vet Clin North Am Small Anim Pract.* 2014;44(6):1103-1112.
8. Delamaide Gasper J a, Barnes Heller HL, Robertson M, Trepanier L a. Therapeutic serum phenobarbital concentrations obtained using chronic transdermal administration of phenobarbital in healthy cats. *J Feline Med Surg.* 2014:1-5.
9. Gindiciosi B, Palus V, Eminaga S, Villiers E, Bruto Cherubini G. Serum bromide concentrations following loading dose in epileptic dogs. *J Small Anim Pract.* 2014;55(February):108-111.
10. Carnes MB, Axlund TW, Boothe DM. Pharmacokinetics of levetiracetam after oral and intravenous administration of a single dose to clinically normal cats. *Am J Vet Res.* 2011;72(9):7-12.
11. Volk H a., Matiasek L a., Feliu-Pascual AL, Platt SR, Chandler KE. The efficacy and tolerability of levetiracetam in pharmacoresistant epileptic dogs. *Vet J.* 2008;176(3):310-319.
12. Hardy BT, Patterson EE, Cloyd JM, Hardy RM, Leppik IE. Double-Masked, Placebo-Controlled Study of Intravenous Levetiracetam for the Treatment of Status Epilepticus and Acute Repetitive Seizures in Dogs. *J Vet Intern Med.* 2012;26(2):334-340..
13. Peters RK, Schubert T, Clemmons R, Vickroy T. Levetiracetam rectal administration in healthy dogs. *J Vet Intern Med.* 2014;28(2):504-509.
14. Beasley MJ, Boothe DM. Disposition of Extended Release Levetiracetam in Normal Healthy Dogs After Single Oral Dosing. *J Vet Intern Med.* 2015: 29: 1348-1353.
15. Von Klopmann T, Rambeck B, Tipold a. Prospective study of zonisamide therapy for refractory idiopathic epilepsy in dogs: Paper. *J Small Anim Pract.* 2007;48(3):134-138..
16. Chung J, Hwang C, Chae J, et al. Zonisamide monotherapy for idiopathic epilepsy in dogs. *N Z Vet J.* 2012;60(March 2015):357-359.

Choosing Appropriate Acid Suppressant Therapy

Stanley L. Marks, BVSc, PhD, DACVIM (Small Animal Internal Medicine, Oncology), DACVN
Davis, CA

Gastric acid secretion is under neural and hormonal control of parietal cells located in the gastric glands lining the fundus and the body of the stomach. This physiologic process is controlled by a number of redundant second messenger pathways activated as a result of the binding of gastrin, acetylcholine, histamine, and prostaglandins to specific receptors on the basolateral surface of parietal cells. An increase in cytosolic calcium mediates the stimulatory effect of acetylcholine and gastrin, whereas the activation of adenylate cyclase with generation of cyclic AMP (cAMP) mediates the stimulatory effect of histamine. Strong potentiation between histamine and either gastrin or acetylcholine reflects post-receptor interaction between the pathways as well as the ability of acetylcholine and gastrin to release histamine from mucosal enterochromaffin-like (ECL) cells.[1] Although gastrin is the major circulating stimulus of acid secretion, it is likely that gastrin does not stimulate the parietal cells directly, but mobilizes histamine from the ECL cells in the oxyntic mucosa. Gastrin acts on CCK2 receptors to control the synthesis of ECL cell histamine, accelerating the formation of the histamine-forming enzyme, histidine decarboxylase (HDC). Stimulation of the proton pump (H^+, K^+-ATPase) to secrete hydrogen ions into the gastric lumen in exchange for potassium ions is the ultimate factor in acid secretion. A feedback system terminates gastric acid secretion after sufficient acid secretion has occurred. A decrease of intragastric pH stimulates somatostatin (SST) release from antral D cells. This peptide inhibits acid secretion and also blunts gastrin release and its stimulatory effect on CCK2 receptors located on parietal and ECL cells. In addition, acidification of the duodenum triggers secretin release, which also inhibits gastric acid secretion.[2] The inhibitory effects of SST and endogenous prostaglandins (EPs) on acid secretion are mediated by receptors (SST2 and EP3, respectively) coupled by guanine nucleotide-binding proteins to inhibition of adenylate cyclase activity. All the pathways (both stimulatory and inhibitory) converge on and modulate the activity of the luminal enzyme (i.e. H^+, K^+ -ATPase).[3]

PPIs are the most potent inhibitors of gastric acid secretion available and are most effective when the parietal cell is stimulated to secrete acid postprandially. Chemically, all the available proton pump inhibitors (PPIs; e.g., omeprazole, pantoprazole, esomeprazole, rabeprazole, landoprazole) consist of a benzimidazole ring and a pyridine ring, but vary in the specific side ring substitution.[4] The PPIs are weak bases that are unprotonated and act as pro-drugs in the physiologic pH of the blood. Once the PPI diffuses and accumulates in the acidic environment of the secretory canaliculi of the active parietal cell, the drug becomes protonated and trapped where it forms disulfide bonds with cysteine residues on the alpha-subunit of the H^+-K^+-ATPase enzymes resulting in prolonged to irreversible enzyme inactivation.[5] Acid secretion resumes only after new proton pumps are synthesized in the parietal cells resulting in a prolonged effect after administration. Previously dormant parietal cells are activated following initial PPI administration. Thus, inhibition of acid secretion is approximately 30% of maximal on day 1 of PPI administration because not all H^+-K^+-ATPase enzymes are bound by drug. Maximal inhibitory effect is achieved within approximately 2 to 4 days of PPI administration.[6] Esomeprazole administered intravenously to 41 dogs on 2 separate occasions 18-24 hours apart was associated with a mean gastric pH of 7.7 ± 1.7 compared to a placebo group (pH 3.9 ± 1.9) underscoring the relatively rapid onset of action of the PPIs in dogs.[7] In addition, omeprazole administered orally to cats also has a relatively rapid onset of action based upon ongoing studies in the author's laboratory. Omeprazole administered twice PO to 10 cats (18-24 hours apart) was associated with a mean gastric pH of 7.2 ± 0.4 compared to a placebo group in which the gastric pH was 2.8 ± 1.0. Because the amount of H^+, K^+ -ATPase present in the parietal cell is greatest after a prolonged fast, PPIs should be administered 30-60 minutes before the first meal of the day.

PPIs are effective for treatment of all acid-related disorders, i.e. peptic ulcer (PU) and eradication of *Helicobacter pylori*, Zollinger-Ellison syndrome (ZES), gastroesophageal reflux disease (GERD) and its complications, nonsteroidal anti-inflammatory drugs (NSAID)-associated gastroduodenal ulcers, as well as PU bleeding. In a landmark study, Burget et al. found a highly significant correlation between ulcer healing and the degree of acid suppression, the duration of acid suppression, and the length of therapy.[8] The shape of the contour expression of these relationships showed that healing rate increased as the duration of suppression increased and as gastric pH increased. Being more potent and longer lasting than H_2RAs, PPIs are therefore more effective in healing both gastric and duodenal ulcers and this observation has been confirmed in several meta-analyses.[9, 10] Once-daily PPI dosing inhibits maximal acid output by about 66% after 5 days in people. Thus, the occasional use of a PPI taken on an 'as needed' basis would not be expected to reliably provide adequate acid inhibition and would be unlikely to produce a consistent or satisfactory clinical response Clear instructions in human practice concerning empirical therapy of PPIs are lacking. That PPIs are often taken inappropriately is revealed by a US survey that showed that only 27% of GERD patients dosed their PPI correctly (i.e. up to 60 min before any meal of the day).[11]

H_2RAs inhibit acid secretion by competitively blocking histamine type-2 receptors on the parietal cell, thus reducing basal and stimulated gastric acid secretion. Pepsin secretion is also reduced, which results in decreased peptic activity. The use of H_2RAs can suppress gastric acid production by 37 to 68% over 24 hours in people and there does appear to be a dose-response relationship between H_2RAs and gastric ulcer prevention and healing. Both a meta-analysis and a Cochrane review found that standard doses of H_2RAs were effective at reducing the risk of duodenal but not gastric NSAID-associated ulcers.[12, 13] The same Cochrane review concluded that high dose (double-dose) H_2RAs were effective at preventing chronic NSAID-related duodenal and gastric ulcers and also significantly reduced symptoms of abdominal pain compared with placebo.[13] There are currently three combination products aimed at decreasing the risk of NSAID-associated GI complications that are approved for human use in the US. Each of these combinations – diclofenac and misoprostol, naproxen and esomeprazole, or ibuprofen and famotidine have been shown to lower the incidence of GI toxicity compared to NSAID use alone.[14]

CLINICAL APPLICATIONS OF ACID SUPPRESSANTS IN DOGS

Unfortunately, critical assessment of gastric acid suppressants in experimental animal models or dogs with spontaneous disease is sparse, and most of the studies evaluating the efficacy of acid suppressants have been completed in healthy animals. In people, optimal gastroduodenal ulcer healing and treatment of gastroesophageal reflux disease occur at an intragastric pH of ≥ 3 for approximately 75% of the day, and a pH of ≥ 4 for approximately 67% of the day, respectively.[15] In dogs, effective clinical doses of anti-secretory drugs have not been established, but clinical experience suggests that the recommended oral omeprazole dose (0.7-1.0 mg/kg q 24h) may not suppress acid secretion adequately. This impression is underscored by the findings of Bersenas et al. who determined the degree of gastric acid suppression associated with the administration of 4 gastric acid suppressants (ranitidine 2 mg/kg IV q 12h; famotidine 0.5 mg/kg IV q 12h; pantoprazole 1 mg/kg IV q 12h; and omeprazole 1 mg/kg PO q 24h or 1 mg/kg PO q 12h) compared to saline solution in 12 healthy Beagles.[16] Famotidine, pantoprazole, and omeprazole significantly suppressed gastric acid secretion compared to saline (placebo); however, ranitidine did not. In addition, administration of famotidine q 8h instead of q 12h did not affect gastric pH significantly, suggesting that increased frequency of administration of this H_2RA to three times daily is of no additional benefit. Omeprazole administered q 12h instead of q 24h had a significant effect on gastric acid secretion and was the only regimen tested that approached the potential therapeutic efficacy for acid-related disease when assessed by criteria used for human patients.

The ability to accurately and noninvasively measure intragastric pH using catheterless radiotelemetric pH monitoring devices in dogs and cats since 2002 has resulted in renewed understanding of the effects of various antacids on intragastric pH in animals. Several important studies have been published utilizing the Bravo system for assessment of intragastric pH in healthy dogs and cats following administration of PPIs and H_2RAs. Comparison of the gastric acid suppressing effects of an equine omeprazole paste (Gastrogard®) with omeprazole tablets and famotidine tablets (1.0 – 1.3 mg/kg q 12h) was performed in 6 healthy adult dogs.[17] The dose of omeprazole administered (1.5-2.6 mg/kg q 24h) was slightly higher than the recommended dose, and both omeprazole formulations provided superior acid suppression in dogs compared to famotidine or placebo. Mean ± SD % of time the intragastric pH was ≥ 3 was 62.5 ± 6.1% for omeprazole tablet, 54.1 ± 8.3% for omeprazole paste, 21.5 ± 13.6% for famotidine, and 6.1 ± 16.8% for placebo. Famotidine and placebo were not significantly different from each other in this study.

The potential benefits of co-administering a more rapidly acting H_2RA (famotidine) with a PPI (pantoprazole) intravenously versus pantoprazole alone in 12 healthy dogs has also been investigated.[18] Many veterinarians administer the rapidly acting famotidine in the first few days of treatment in combination with the slower acting, but more potent PPIs to critically ill dogs or animals at high risk for GI bleeding. The potential concern with this practice however is that the simultaneous treatment with an H_2RA might decrease the efficacy of PPIs because of the requirement for PPI activation and trapping in the acidic canalicular environment of the parietal cell. There were no significant differences in the mean % time intragastric pH was ≥ 3 and 4 between the two groups, and pantoprazole administered alone achieved pH goals established for humans with acid-related disorders. These results should be interpreted cautiously because there is evidence for the benefits of combining an H_2RA at night with PPI therapy for people with acid reflux experiencing nocturnal acid breakthrough.[19]

The efficacy of omeprazole (0.85 mg/kg q 24h) and standard dose famotidine (1mg/kg q 24h) and high-dose famotidine (2 mg/kg q 12h) administered orally was evaluated in racing Alaskan sled dogs that have been shown to be at high risk for gastric erosions and ulcerations.[20,21] The three most important findings in the studies were that a) administration of famotidine at the standard dose of 1mg/kg q 24h was not as effective in lessening the severity of gastric lesions in dogs running 330 miles as it was in dogs running 100 miles; b) administration of high-dose famotidine (2 mg/kg q 12h) was roughly similar in its effects in lessening the severity of gastric lesions as was the standard dose of famotidine; and c) administration of omeprazole at the recommended dose was superior to both doses of famotidine in lessening the severity of gastric lesions.

The benefits of PPIs in people with moderate to severe gastroesophageal reflux disease (GERD) have been well documented and several studies have investigated the effects of PPIs and H_2RAs on esophageal pH in dogs in an effort to manage or prevent reflux esophagitis and esophageal stricture formation associated with general anesthesia.[7] Administration of esomeprazole to 22 dogs undergoing elective orthopedic procedures was associated with a significant increase in gastric and esophageal pH throughout the surgery procedure but did not have an impact on the number of reflux events. Nevertheless, the maintenance of an esophageal pH > 4 in all of the dogs treated probably minimized the likelihood of reflux esophagitis and stricture formation in the 36% of dogs that had ≥ 1 reflux event because the proteolytic pH range for the conversion of pepsinogen to pepsin is between 1.5 and 3.5,[22] and esophageal acid clearance does not occur during sleep or anesthesia. There is compelling in vivo evidence in experimental animal models of esophagitis induced in rabbits, dogs, and cats that the dripping or infusion of an acidic solution of hydrochloric acid alone (pH 1.6–2.0) on the esophagus over a period of 30–60 minutes is associated with minimal to no damage to the esophagus, but the addition of pepsin to the hydrochloric acid is associated with both macroscopic (eg, edema, hyperemia, hemorrhage) and microscopic (eg, epithelial loss, erosions, inflammation, mucosal, and submucosal hemorrhage) evidence of acute esophagitis.[23] Further studies are warranted to determine the benefits of pre-anesthetic administration of PPIs to prevent esophagitis and stricture formation in dogs.

Misoprostol, a synthetic prostaglandin E1 analog, has been shown to significantly reduce endoscopically detectable aspirin-induced gastroduodenal mucosal lesions in healthy dogs compared to a placebo group.[24] Misoprostol was administered orally at 3 µg/kg q 8h and aspirin was administered at 35 mg/kg q 8h and lesions were scored on days 5, 14, and 30 of the study. In contrast to the previous study, misoprostol administered at 2 µg/kg q 8h had no effect in preventing gastric lesions in dogs with acute intervertebral disc disease following administration of a single dose of dexamethasone at 2 mg/kg followed by a tapering regime of prednisolone (2 mg/kg to 0.5 mg/kg over the course of 5-6 days).[25]

CLINICAL APPLICATIONS OF ACID SUPPRESSANTS IN CATS

There are fewer studies critically evaluating the efficacy of PPIs and H_2RAs in cats compared to dogs. Dosages for acid suppressants in cats have been extrapolated from results of studies in dogs. The Bravo system was used by Parkinson et al. for the wireless measurement of intragrastric pH following twice daily administration of omeprazole (fractionated tablet or reformulated paste) versus standard doses of famotidine administered q 12h.[26] The mean percentage time ± SD that intragastric pH was ≥ 3 was 68.4 ± 35.0% for the fractionated omeprazole tablet, 73.9 ± 23.2% for the reformulated omeprazole paste, 42.8 ± 18.6% for famotidine, and 16.0 ± 14.2% for placebo. Importantly, there were no significant differences in the plasma concentrations between the 2 omeprazole formulations and both were in a range associated with acid suppression. This is the first study in small animals to show that fractionated enteric-coated omeprazole is an effective acid suppressant despite disruption of the enteric coating. In addition, the acid suppressive effects of famotidine were significantly weaker than that of both omeprazole formulations.

ACID SUPPRESSION FOR ACID RELATED AND NON-ACID RELATED DISORDERS

Gastritis

Despite the frequent administration of gastric acid suppressants for the management of suspected or confirmed gastritis in dogs and cats, there is no evidence to date documenting the benefit of this application. It is likely that this approach has been extrapolated from the human literature in which gastric acid suppression with H_2RAs or PPIs represents a well-established practice in addition to antimicrobial therapy for the management of *Helicobacter pylori* gastritis. Acid suppression with famotidine (0.5 mg/kg SID) did not affect treatment efficacy or frequency of clinical signs in 23 dogs with histologic evidence of gastritis and spiral bacteria in gastric mucosal biopsy samples.[27]

Hepatic Disease

Despite the apparent association between hepatic dysfunction and gastroduodenal ulceration in dogs, there is a dearth of literature proving this association. The pathogenesis of GI ulcer formation secondary to hepatic dysfunction include portal hypertension, hypergastrinemia, and hyperhistaminemia; however, serum gastrin concentrations did not correlate with hepatic disease in dogs in 2 studies.[28, 29] In humans, gastric acid suppression is generally not considered effective at reducing bleeding from portal hypertensive gastropathy because these patients already have hypochlorhydria.[30] However, proton pump inhibitors may indirectly stop bleeding from the stomach by raising intraluminal gastric pH and thereby stabilizing blood clots.[31] The efficacy of acid suppressants for ulcerative gastrointestinal disease in cats and dogs with liver disease requires further study.

Pancreatitis

Gastric acid suppressants are commonly administered to dogs and cats with acute pancreatitis in an effort to prevent or minimize gastric ulceration associated with hypovolemia and ischemia and focal peritonitis. Despite this practice, there is no evidence documenting any benefit for the administration of PPIs (pantoprazole) in a placebo-controlled study in humans with acute pancreatitis.[32] In addition, there have been no studies in veterinary medicine critically assessing the potential association between pancreatitis and upper GI bleeding in dogs or cats.

Renal Disease

Hypergastrinemia, gastritis, GI bleeding, and peptic ulceration are common complications of end-stage renal disease in human patients,[33,34] although gastric acid secretion in humans may be influenced by *H. pylori* infection status.[35] There are no recommendations for the use of prophylactic acid suppressant therapy in human patients with renal disease who do not have ulcer bleeding; however, acid suppressants are generally recommended if other risk factors (e.g., NSAID and/or corticosteroid therapy) for ulcer development are present. There are no published studies evaluating the gastric acid secretion profile in dogs or cats with renal disease, and the contribution of hypergastrinemia and gastric hyperacidity (if present) to the development of gastric mineralization has not been determined. The deleterious effects of chronic acid suppressant administration on calcium metabolism and bone remodeling in dogs and cats with CKD could lead to alterations in calcium and PTH metabolism and increased risk of osteoporosis, and potentially pathologic bone fractures in at-risk populations. Gastric acid suppressants should be used judiciously in dogs and cats with renal disease, and should be restricted to animals with additional risk factors for ulceration or when concern for GI bleeding (e.g. melena, severe iron deficiency anemia) or vomiting-induced esophagitis exists.

ADVERSE EFFECTS OF CHRONIC PPI AND H_2RA ADMINISTRATION

The long-term effects of PPI and H2RA administration have not been well studied in dogs and cats to date and effects have been extrapolated from human studies. The most commonly reported adverse effects associated with chronic PPI therapy include vitamin B12 deficiency, diarrhea, enteric infections, particularly *Clostridium difficile* infection, increased risk of hospital acquired pneumonia, and the potential increased risk of fractures in geriatric people.[36] Diarrhea is the most common adverse effect reported in association with PPI administration in dogs.[21] This side effect has not been frequently reported in cats. A recent study evaluated the adverse effects of a longer course (60 days) of omeprazole in a reformulated capsule at 1mg/kg q 12h in 6 healthy cats (Gould E, personal communication). There were no significant alterations in hematological or biochemical parameters, including cobalamin and methylmalonic acid; however, bone mineral density (measured via DEXA) was decreased by > 8% in 4/6 cats. Measurement of gastric pH via Bravo pH capsule technology in 2 cats following abrupt cessation of PPI therapy after 60 days showed significant rebound acid hypersecretion with both cats having gastric pH < 1.5 for several days. Proton pump inhibitors should be gradually tapered after their prolonged use to minimize rebound acid hypersecretion.

The effect of omeprazole administration on the gastric and duodenal microbiota was evaluated in 8 healthy dogs administered omeprazole at 1.1 mg/kg q 12h for 15 days.[37] Gastric fluid pH was significantly increased at the end of the treatment period (median: 7.1; range: 4.2-8.1); however, omeprazole did not induce major changes in the microbiome in either the stomach or the duodenum. It should be emphasized that the approach used allowed only the assessment of relative changes in the proportions of bacterial groups within the total population and significant changes in the total number of bacteria in response to omeprazole treatment may not have been discovered.

In conclusion, there is a strong need for veterinarians to better define the clinical applications of PPIs and H_2RAs in small animals in an effort to minimize the injudicious administration of these drugs. The evidence is fairly compelling that PPIs administered twice daily provide the most significant acid 24-hour suppression, and PPIs administered once daily or H2RAs administered twice daily are suboptimal. Many animals are administered omeprazole, famotidine and sucralfate concurrently when there is little to no benefit documenting the additive effects of co-administering these drugs. The long-term effects of these drugs must be better investigated for animals requiring chronic acid suppressant therapy, and the potential concerns of H_2RA tolerance must be studied. Finally, the effect of canine breed on the variation in PPI efficacy warrants further investigation given the genetic polymorphism in cytochrome P450 genotypes that is a well-established cause for the large inter-individual and inter-ethnic variation in the pharmacokinetics and pharmacodynamics of PPIs in humans.[38]

REFERENCES

[1]Hersey SJ, et al. Physiol Rev 1995;75:155-189; [2]Chey WY, et al. J Physiol Pharmacol 2003,54(suppl 4).1005-112; [3]Aihara T, et al. Pharmacol Ther 2003; 98:109–127; [4]Stedman CAM, et al. Aliment Pharmacol Ther 2000;14:963-978; [5]Fellenius E, et al. Nature 1981;290:159-161; [6]Kromer W, et al. Pharmacol 2000;60:179-187; [7]Zacuto A, et al. J Vet Int Med 2012;26:518-525; [8]Burget DW, et al. Gastroenterol 1990;99:345–351; [9]Poynard T, et al. Eur J Gastroenterol Hepatol 1995;7:661-665; [10]Salas M, et al. BMC Gastroenterol 2002;2:17-23; [11]Pezanoski J, et al. Gastroenterol 2003;124(suppl):A228; [12]Koch M, et al. Arch Intern Med 1996;156:2321-2332; [13]Rostom A, et al. Cochrane Database Syst Rev 2002;4:CD002296; [14]Schiff M, et al. Expert Rev Gastroenterol Hepatol 2012;6:25-35; [15]Bell NJ, et al. Digestion 1992;51(suppl 1):59-67; [16]Bersenas AM, et al. Am J Vet Res 2005;66:425-431; [17]Tolbert K, et al. J Vet Int Med 2011;25:47-54; [18]Tolbert K, et al. J Vet Int Med 2015;29:556-560; [19]Mainie I, et al. J Clin Gastroenterol 2008;42:676-9; [20]Williamson KK, et al. J Vet Int Med 2007;21:924-927; [21]Williamson KK, et al. J Vet Int Med 2010;24:285-288; [22]Hirschowitz BI. Yale J Biol Med 1999;72:133–143; [23]Lanas A, et al. Gastroenterol 1999;116:97–107; [24]Johnston SA, et al. J Vet Int Med 1995;9:32-38; [25]Neiger R, et al. J Vet Int Med 2000;14:33-36; [26]Parkinson S, et al. J Vet Int Med 2015;29:104-12; [27]Leib MS, et al. J Vet Intern Med 2007;21:1185-1192; [28]Boothe, DM. J Vet Intern Med 1990;4,122 (abstract); [29]Mazaki-Tovi M, et al. Vet Rec 2012;171:19; [30]Gjeorgjievski M, et al. World J Hepatol 2016;8:231-262; [31]Zhou Y, et al. J Gastroenterol Hepatol 2002;17:973-979; [32]Yoo JH, et al. Korean J Gastroenterol 2012;60:362-367; [33]Falcao HA, et al. J Surg Res 1975;18:107-111; [34]Chalasani N, et al. Am J Gastroenterol 1996;91:2329-2332; [35]Chang SS, et al. World J Gastrointestinal Pharmacol Therap 2015;6:127-136; [36]Sheen E, et al. Dig Dis Sci 2011;56:931-950; [37]Garcia-Mazcorro JF, et al. FEMS Microbiol Ecol 2012;80:624-636; [38]Egan L et al. Digest Dis 2000;18:58-63.

Evidence-based Insulin Therapy in Dogs and Cats

Melissa Clark, DVM, DACVCP
New York, NY

INTRODUCTION

Evidence-based medicine has been defined as "the conscientious, explicit and judicious use of current best evidence in making decisions about the care of the individual patient".[1] With specific reference to diabetes in dogs and cats, this involves evaluating available literature describing a growing variety of insulin preparations and protocols, and integrating this information with clinical expertise and client/patient factors to determine which regimen might be best suited to accomplish treatment goals. In general, the goals of treatment in diabetic dogs are to control clinical signs of diabetes (which is facilitated by maintenance of a blood glucose concentration below the renal threshold for the majority of the dosing period) without inducing hypoglycemia. This requires an insulin preparation with an appropriate duration of action to provide optimal control but avoid overlap of insulin action between one dosing period and the next. Some degree of consistency in absorption of an insulin preparation is also desirable, to allow predictable effect among or within individuals. In cats, an additional major goal of treatment is induction of diabetic remission, which appears more likely to be achieved when blood glucose concentrations are well-controlled throughout the dosing period.

WHAT TYPES OF EVIDENCE ARE AVAILABLE TO GUIDE INSULIN TREATMENT IN DOGS AND CATS?

Maintenance insulin preparations currently in routine use in veterinary medicine include neutral protamine Hagedorn (NPH), lente, protamine zinc (PZI), glargine, and detemir insulin. For particular insulin products used in dogs and cats, the types of published evidence available for assessment of suitability for achieving clinical goals include 1) pharmacokinetic/pharmacodynamic (PK/PD) studies in healthy or diabetic animals, 2) retrospective or observational studies describing the use of the insulin in a clinical situation, 3) small clinical trials involving a single treatment regimen, 4) clinical trials in larger numbers of animals (including pivotal trials conducted under the auspices of the drug sponsor) and 5) comparative studies of different insulin treatment regimens. Randomized controlled trials, although they form the foundation for evidence-based insulin therapy in human medicine, are rare in veterinary medical literature describing treatment of diabetes, and no large-scale prospective comparisons of different insulin products and treatment regimens in dogs and cats have been reported. Nonetheless, some useful information regarding insulin therapy in dogs and cats can be gained by reviewing available descriptive studies, clinical trials, and small comparative studies.

INSULIN THERAPY IN DOGS

NPH

Efficacy of NPH in dogs is supported largely by expert opinion and clinical experience, and by a prospective study in 54 dogs.[2] That study found resolution of clinical signs by 2 months in approximately 95% of dogs receiving recombinant human NPH twice a day, with 91-94% achieving good glycemic control. However, assessment of glycemic control in that study was based on morning blood and urine glucose only, and the proportion of dogs achieving good glycemic control with NPH does not appear to have been reported in other investigations. In a recent PK/PD study, the duration of action of NPH was 5.5 h (range 4-8) in four dogs, but >10h in another four, suggesting that the duration of action may be too short for optimal control in some dogs.[3] The authors of a retrospective study of involving treatment with several insulin types, but predominantly NPH, found a duration of optimal action (hours BG <200) of >12 h in only 4% of 74 dogs treated with NPH at the time of clinical control, supporting twice a day dosing with this formulation.[4]

Lente

Lente insulin appears to have a slightly more consistent time to nadir and longer duration of action (median 14 h, range 10 - >24h) than NPH in dogs.[5,6] Despite the longer duration of action, reduction of blood glucose (BG) was noted for ≤12 h in 5/8 dogs in one study, and the duration of optimal action (maintenance of BG 90-234 mg/dL) was <13 h in 5/8 dogs in another, leading the authors of both studies to conclude that twice a day administration would be appropriate in most dogs.[7,8] In a prospective investigation of efficacy of lente insulin in 53 diabetic dogs (submitted as part of the US approval package for Vetsulin), good glycemic control was achieved in approximately 75% of diabetic dogs by 2 months of treatment, with twice a day dosing ultimately being necessary in 66%. Owner-reported improvement in clinical signs ranged from 80-96%, and most episodes of suspected clinical hypoglycemia (which occurred in 38% of dogs) developed during the dose adjustment period. Biochemical hypoglycemia (BG <60 mg/dL) was noted in 36% of dogs and 15% of BG curves.[9]

PZI

PZI has a substantially longer duration with more variable absorption than intermediate-acting insulins in dogs. A study of the bovine product reported a duration of action of >24 hours in 8 diabetic dogs[5]; for the recently available recombinant human PZI (rhPZI), a PK/PD study in healthy dogs also found a duration of action of >24 hours (range 16->24h) [10], and in both studies, there was substantial between-dog variability in the time course of insulin action. A prospective clinical trial in 17 diabetic dogs treated with rhPZI twice daily found that the time to glucose nadir was late (median 8-10h) and variable, and occurred at the beginning or end of a 10-hour sampling period in 94% of the dogs (54% of curves). The magnitude of the nadir was also variable and unpredictable.

Nonetheless, by 60 days, 82% of dogs had improved clinical signs, 76% of dogs had a mean BG of <250 mg/dL, and 50% had fructosamine <450 umol/L (the study criteria for good glycemic control). Biochemical hypoglycemia (defined in this study as BG < 80 mg/dL) was reported in 12% of dogs and 6% of curves, with clinical signs of hypoglycemia in 18% of dogs.[11] The authors concluded that overlap of insulin action between one dosing period and the next (with concern about a Somogyi response) is likely when PZI is dosed twice a day in dogs, but that it may be appropriate in dogs with too short a duration of action of NPH or lente.

Glargine and detemir

Glargine and detemir are both human insulin analogues and the PK/PD have been minimally studied in dogs. One abstract described a duration of action of 13.2 ± 1.9 h for glargine in 9 healthy dogs, with minimal glucose-lowering effect in 7 of the dogs.[a] Two other studies found a longer duration of action (≥24 h) for both glargine and detemir in groups of 3 healthy dogs based on a glucose clamp; the same group administered the insulins twice a day to diabetic dogs, and after 7 days of treatment, several dogs in each group had preinsulin BG ≤80 mg/dL.[12-13]

Thus far, only very small studies (10-12 subjects) are available for evaluation of the clinical efficacy of glargine or detemir in dogs, but it appears possible to achieve good glycemic control with either in 50-60% of dogs, and owners reported improved clinical signs in 90-100% of cases.[14-16] As is the case for PZI, variability in time to nadir and possible insulin overlap have been reported with these formulations, with time to nadir ranging from the first to last measured time points of serial glucose curves. Biochemical hypoglycemia (BG <60 mg/dL) was noted in 50% of dogs and 5% of glucose curves for glargine in one study[14], and in 30% of dogs and 9% of curves for detemir.[16] Although clinical hypoglycemia appears relatively infrequent with glargine (0-20% of dogs), it may be problematic with detemir treatment in small dogs. The likely reason for this is that the myristic acid moiety conjugated to B29 in detemir does not cause steric hindrance and reduced receptor binding in dogs as it does in humans, and the more concentrated product marketed for humans may be more difficult to dose precisely in smaller dogs. The median dose of detemir at the conclusion of the single available clinical trial was 0.12 U/kg every 12 hours.[16]

In summary, available data demonstrate that clinical control can be achieved in a proportion of diabetic dogs with any of the insulin formulations mentioned above. Overall, direct comparisons between insulin protocols are hampered by variability between studies in study duration, inclusion criteria, definitions of diabetic control, and outcome measures reported, and no prospective comparative studies of different maintenance insulins in dogs are available.

INSULIN THERAPY IN CATS

NPH

Published data regarding treatment of cats with NPH insulin are very limited and are confined to descriptions involving the historically available porcine product. In healthy cats, insulin concentrations were reported to peak early (1.5 h; range 1-2h) and return to basal relatively quickly (mean of 7.7 ± 0.8 h; range 6-10h).[17] In 6 diabetic cats, glucose nadir occurred at a median of 6 hours post-injection (range, 3-8h), but a duration of action was not calculated as sampling continued only for 8 h in some cats. In the two cats for which glucose curves were displayed, glucose concentration was precipitously increasing after the 6 hour nadir (although one of these cats may have been experiencing a Somogyi response). Treatment with NPH once or twice daily in seven cats reportedly resulted in two diabetic remissions (one in a cat previously treated with megestrol), three cases of "satisfactory" control, and two cases of poor control after development of concurrent hyperthyroidism or renal disease.[18] Since these reports were published, they have been cited numerous times as support for statements that the duration of action of NPH is too short in most cats. As duration of action was not actually reported, such statements may also reflect subsequent clinical experience not subjected to formal analysis.

Lente

The same porcine lente product approved for dogs is licensed for use in cats in the USA, Europe, and Australia. In 25 diabetic cats, lente was reported to have a mean duration of action of 9.3 ± 0.5 h, and all cats in this study eventually needed twice-a-day administration for control.[19] Efficacy of porcine lente insulin has been studied in 46 cats, and in a different population of 77 cats, with naturally occurring DM. In the first study, 15% of cats went into diabetic remission and 60% were considered "clinically stable" (interactive at home with subjectively normal appetite and water consumption, and stable body weight) and had improved glycemic control within 4 months of starting twice daily insulin treatment. In this study, diabetic remission was defined as persistent normoglycemia 1 week after insulin discontinuation, and though inclusion criteria relating to diagnosis of DM were defined, some cats included in the study did not meet all criteria.[20] In the second study (part of the US approval package for Vetsulin), all cats except one were treated with lente insulin twice a day. By day 60 of treatment, 75% of cats had a mean blood glucose <300 mg/dL, and 72.4% had a glucose nadir <200; percentages were similar at day 180. Mean scores on a visual analogue scale of diabetic control (incorporating clinical signs, body weight, and blood glucose results, with a score of 100 representing no control) improved from 94 at study initiation to 31 at 2 months and 25 at 6 months of follow-up. Diabetic remission was achieved in 5% of cats at 4 months, with 7% of the remaining cats achieving remission during an extended use period of 1-83 weeks (overall remission rate of 10% by 72 weeks). Blood glucose <50 mg/dL was documented in 79% of cats, and confirmed clinical hypoglycemia occurred in 17%.[21]

PZI

As is the case in dogs, PZI has a longer duration of action and more variable PK/PD than less complex insulins in cats. In healthy cats administered beef-pork PZI, peak insulin concentration occurred at approximately 4 h (range 1-8)[17], time to glucose nadir was 3-4 h[22,23] and duration of action was 20.9 h (range 9->24).[23] A study of 67 naïve and previously treated diabetic cats given twice-daily beef-pork PZI reported a slightly longer time to nadir (5-7 h); in that study, by 45 days of treatment, 85% of the cats had moderate to good glycemic control (fructosamine <500 umol/L, mean 9-h BG <300 mg/dL, and a nadir BG of <200 mg/dL), with slightly lower percentages in the subgroup of previously treated cats, and 90% of owners reported improvement in clinical signs.[24] The largest efficacy study of a particular insulin type in dogs or cats to date is a clinical trial of PZIR in 133 diabetic cats. Based on owner-assessed improvement in PU/PD, presence of appropriate body condition, and improvement in one category of glycemic control (with good, moderate, and poor control defined as fructosamine <450, 450-500, and >500 umol/L, and 9-h mean BG <200, 200-300, and >300, respectively), 85% of cats were categorized as treatment successes by the study endpoint of 45 days. Blood glucose <80 mg/dL was documented in 22% of glucose curves and 64% of cats, and clinical signs of hypoglycemia were reported in 21% of cats (2 confirmed cases). Diabetic remission was not reported in any of the cats in this study, but the authors speculated that some of the cats with low 9-hour mean BG and fructosamine concentrations were in the process of remission.[25]

Glargine

Following the introduction of glargine onto the human market as a basal insulin, it became the subject of numerous investigations in diabetic cats, with a focus on its potential to induce diabetic remission. Pharmacokinetic studies in 9 healthy cats suggested a long duration of action (mean 22.2h, range of 12->24 h).[26] In 6 healthy cats, glucose concentration was still suppressed below baseline at 24 h in 50% of cats after once or twice a day administration, and the second nadir with twice a day administration was lower than the first (suggesting some carryover of the previous dose).[27] Most studies evaluating the efficacy of glargine in diabetic cats have involved dietary modification and/or specific intensive glucose monitoring protocols. Remission rates with twice-daily glargine and a low-carbohydrate, high-protein (LCHP) diet have been reported as 17, 40, 50, or 100%.[28-31] The study with 100% remission, which included newly diagnosed cats only, found remission rates of 38% and 25%, respectively, for PZI and lente by the study endpoint of 4 months, but cats in this study were not randomized to treatment groups, and sample size was small (n=8 per group).[31] The only other prospective study (a randomized clinical trial) to compare insulin types between groups found remission rates of 17% for once-a-day glargine and 43% for twice-a-day lente (not significantly different) in a similarly small number of cats, all transitioned to high-protein diets at study enrollment.[32] A retrospective study of factors affecting diabetic remission in cats, in which diet was not standardized, found a remission rate of 72% in cats treated with glargine (n=29) and 56% in those treated with lente insulin (n=41). This difference was significant in univariate analysis, but a multivariate analysis incorporating insulin type was not possible because of the number of cats that did not survive to discharge and therefore did not receive either insulin type.[33]

Three studies involving glargine and specific protocols targeting euglycemia are available. In a prospective, uncontrolled study of 55 cats managed with twice-daily glargine, LCHP diet, and intensive home blood glucose monitoring (minimum of three daily glucose measurements, target daily glucose concentrations between 50 and 200 mg/dL) remission rate was 64% within 10 months. However, 47% of the cats achieving remission had been receiving corticosteroid treatment in the preceding 6 months, and 26% later relapsed. Biochemical hypoglycemia (BG <50 mg/dL) occurred frequently (93% of cats) but clinical hypoglycemia was reportedly rare (1.8%). Of cats not achieving remission, 70% were considered well-regulated (mean daily blood glucose ≤ 150 mg/dL).[34] A retrospective study compared near-euglycemic and traditional management paradigms in newly diagnosed diabetic cats treated with twice-daily glargine and LCHP and found that in the near-euglycemic group (NE; blood glucose evaluated at least every 2 weeks, with a target range of 60-160 mg/dL; n=18), remission rate was 78%, vs. 14% in the traditionally managed group (TM; insulin dose adjusted based primarily on owner-reported clinical signs; n=36). Biochemical and clinical hypoglycemia were noted in 44% and 11% of cats managed with the NE paradigm and 33% and 13% of those managed traditionally.[35] Lastly, a prospective, randomized, controlled trial assessed the effect of early intensive insulin therapy in cats with newly diagnosed diabetes. In 15 cats receiving an infusion of insulin aspart to maintain BG between 90 and 180 mg/dL for 6 days, followed by treatment with twice-daily subcutaneous glargine and a LCHP diet, remission rate by 4 months was 67%, with a 10% relapse rate. Of cats that continued to require insulin, 60% had good metabolic control (absence of clinical signs, blood glucose values between 90 and 270 mg/dL during curves, fructosamine <400 umol/L) by the end of the 6-month study period. For 15 cats treated only with twice-daily subcutaneous glargine and the diet, remission rate was 50% with a 14% relapse rate, and only 14% of cats still requiring insulin had good metabolic control at the end of the study. These differences between groups were not significant (possibly as a consequence of sample size), but cats that received initial intensive insulin treatment required less insulin during the subsequent 6 months.[30]

A systematic review has analyzed publications describing treatment or prognostic factors associated with diabetic remission in cats. Most studies relevant to the effects of different insulin protocols on diabetic remission were nonrandomized trials or follow-up studies (prospective single-group cohort studies or trials) considered to have moderate or large risk of bias. Concerns about study design included lack of blinding, potentially inadequate follow-up time to allow remission, risk of including cats with acromegaly, small sample size with lack of power calculation beforehand, and low external validity. Populations were diverse and definitions of remission were variable among studies. A mean remission rate of 47.6% for glargine was calculated from all included papers, although 95% confidence intervals for remission rate in each paper were wide. The authors concluded that remission could occur with a variety of protocols, that study design and reporting could be improved, and that blinded, randomized trials were needed for comparison among protocols.[36]

Detemir

Detemir had a duration of action of 13.5 ± 3.5 h in one study involving an isoglycemic clamp in healthy cats.[37] It has been evaluated in conjunction with a high-protein diet and intensive blood glucose monitoring in a small number of diabetic cats (n=18), as described previously for glargine. In this setting, remission rate was 67% (with 25% relapse) within 5.3 months, and good (83%; mean BG ≤150) or moderate (17%; mean BG ≤200) glycemic control was achieved in the remaining cats. Also similar to glargine, detemir was associated with BG<50 mg/dL in 10% of BG curves but only 1 mild episode of clinical hypoglycemia.[38] The efficacy of detemir has also been reported in a retrospective study of 14 cats on twice daily insulin therapy. Clinical signs were resolved in 71% of cats, and 21% of cats achieved remission, within the first 3 months. By this time, 54% of cats had a mean BG ≤180 mg/dL, and 23% had mean BG of 180-270 mg/dL (classified as good or moderate glycemic control, respectively). When glycemic control was based on fructosamine, the corresponding percentages were lower: 30% had good control (fructosamine <450 umol/L), 0% moderate control, and 70% poor control (>550 umol/L). Subclinical biochemical hypoglycemia (BG <45) and clinical signs of hypoglycemia were each reported in 14% of cats.[39]

In general, as in dogs, direct comparisons between insulin protocols have rarely been made in cats, although there is evidence that most maintenance insulins discussed (with the exception of NPH) can provide good clinical and glycemic control. Some evidence suggests glargine is more likely to induce remission than PZI or lente; however, this conclusion is based primarily on nonrandomized studies with small sample size or retrospective evaluations. Clinical hypoglycemia may be less common with glargine or detemir, although subtle clinical signs of hypoglycemia may be difficult to recognize in cats.

CONCLUSION

Evidence to support different insulin regimens in dogs and cats consists primarily of descriptive studies (PK/PD or observational), and small nonrandomized, noncontrolled clinical trials, with larger noncontrolled clinical trials available for those preparations with a veterinary sponsor. Data from these sources indicates that good diabetic control can be achieved with a variety of protocols. Based on PK/PD, intermediate-acting insulins may be more suitable as a first choice for dogs (with probable need for twice-a-day administration in most dogs), and there is minimal evidence to support efficacy of NPH in cats. However, because of nonuniform study conditions and the lack of large randomized trials, further direct comparison among different regimens is difficult. Some evidence suggests that glargine insulin may be associated with a higher rate of remission in diabetic cats, but studies supporting this conclusion have been prone to bias for various reasons (retrospective nature, small sample size, lack of randomization) and superiority of glargine over other insulins for this purpose has not yet been conclusively demonstrated. As mentioned by the authors of a systematic review of factors affecting diabetic remission in cats, there is a need for blinded, randomized, controlled clinical trials comparing different insulin treatment regimens in veterinary medicine, although appropriate sample size may be an obstacle.

REFERENCES

a. Stenner VJ *et al. J Vet Intern Med* 2004; 18:444 (abstract).

1. Sackett D, *et al. BMJ* 1996; 312.
2. Lorenzen F. *Acta Vet Scand* 1992; 33(3): 219.
3. Palm C, *et al. J Vet Intern Med* 2009; 23:50.
4. Hess RS, *et al. J Am Vet Med Assoc* 2000; 216:217.
5. Church DB. *J Small Animal Pract* 1981; 22:301-310.
6. Graham PA, *et al. J Small Anim Pract* 1997; 38(10):434.
7. Horn B, *et al. Aust Vet J* 2000; 78:831.
8. Fleeman L, *et al. Vet Rec* 2009; 164:232.
9. Monroe W, *et al., J Vet Intern Med* 2005; 19(5):675.
10. Clark M, *et al. J Vet Pharm Ther* 2011; 35:342.
11. Della Maggiore A, *et al. J Vet Intern Med* 2012; 26:109.
12. Mori A, *et al., Vet Res Commun* 2008; 32:563.
13. Sako T, *et al. Res Vet Sci* 2011; 90:396.
14. Hess RS, *et al. J Am Vet Med Assoc* 2013; 243:1154.
15. Fracassi F., *et al. Vet Rec* 2012; 170:52.
16. Fracassi F., *et al. J Am Vet Med Assoc* 2015; 246:74.
17. Wallace MS, *et al. Domest Anim Endocrinol* 1990; 7:509.
18. Moise NS, *et al. J Am Vet Med Assoc* 1983; 182:158.
19. Martin GJ *et al. J Feline Med Surg* 2001; 3:23.
20. Michiels L, *et al. J Feline Med Surg* 2008; 10:439.

21. Intervet, Vetsulin Freedom of Information Summary, 2008.
22. Broussard JD, *et al. Am J Vet Res* 1994; 55:127.
23. Marshall RD, *et al. J Vet Pharmacol Ther* 2008; 31:205-212.
24. Nelson RW, *et al. J Am Vet Med Assoc* 2001; 218:38-42.
25. Nelson RW, *et al. J Vet Intern Med* 2009; 23:787-793.
26. Marshall RD, *et al. J Vet Pharmacol Therap* 2008; 31:205.
27. Marshall RD, *et al. J Feline Med Surg* 2008; 10:488.
28. Hall TD, *et al. J Feline Med Surg* 2009; 11:125.
29. Boari A, *et al. Vet Res Commun* 2008; 32:S243.
30. Hafner M, *et al. J Vet Intern Med* 2014; 28:1753.
31. Marshall RD, *et al. J Feline Med Surg* 2009; 11:683.
32. Weaver KE, *et al., J Vet Intern Med* 2006; 20:234.
33. Zini E, *et al. J Vet Intern Med* 2010; 24:1314.
34. Roomp K, *et al. J Feline Med Surg* 2009; 11:668.
35. Nack R, *et al. Vet Quart* 2014; 34:132.
36. Gostelow R, *et al. Vet J* 2014; 202:208.
37. Gilor C, *et al. J Vet Intern Med* 2010; 24:870.
38. Roomp K, *et al. J Feline Med Surg* 2012; 14:566.
39. Hoelmkjaar KM, *et al. J Feline Med Surg* 2015; 17:144.

Equine Drug Testing in the 21st Century

Mary A. Robinson, VMD, PhD
Kennett Square, PA

INTRODUCTION

The Penn Vet Equine Pharmacology Laboratory's mission is *To Promote the Welfare of the Equine Athlete while maintaining the Integrity of Equine Sport through Pharmacological and Forensic Research* (1). The Pennsylvania Equine Toxicology and Research Laboratory's mission is *To Perform Equine Drug Testing for the Commonwealth of Pennsylvania's Six Racetracks.* I consider it a privilege to be the Director and Acting Director, respectively, of these laboratories, and I'd like to start with acknowledging all of the wonderful people who work in the laboratories, my current and past mentors, the members of the Pennsylvania Racing Commissions, and many of the racing industry stakeholders who share my passion for ensuring the welfare of the horse while maintaining the integrity of equine sport.

Funding for these laboratories is provided primarily by the Pennsylvania Department of Agriculture, Horse and Harness Racing Commissions (2). Additional monies and/or equipment have been donated to the Penn Vet Equine Pharmacology Laboratory by the Pennsylvania Harness Horsemen Association at Pocono Downs and Chester Downs, the Meadows Standardbred Owners Association, and the Horsemen Benevolent and Protective Association at Penn National and Presque Isles Downs. I have no commercial financial disclosures to declare.

The practice of racing horses has existed for centuries as the "sport of kings". Race training requires an intense exercise program and an exceptional nutritional program to optimize the development and performance of the horse's biological machinery. Especially in the last 100 years, the serendipitous discovery and the targeted development of pharmaceutical agents that can clearly enhance and/or modify the development and performance of the body has created an ethical dilemma for competitors, regulatory bodies, and fans of all sports. The intended purpose of such agents is to relieve suffering, to aid the healing process, and/or to maintain the welfare of the patient. However, the use of these agents in equine athletes training for competition, and/or during competition – termed "doping" – can give the user an unfair advantage over other competitors who have trained without the use of performance altering or enhancing drugs, and some drugs can compromise the safety and welfare of the competitors. In an effort to maintain the integrity of the sport and to ensure the welfare of the horse, the use of all drugs was banned by regulatory agencies, and legislatively by many states within the USA, and these states and some countries around the world, still operate legally under a "zero tolerance" policy when it comes to drug use in equine athletes.

DRUG TESTING PROCEDURES

To enforce the ban on pharmaceutical assistance during competition, drug testing procedures and laboratories were established. In PA, testing began at each individual race track in 1964, and was centralized at the Pennsylvania Equine Toxicology and Research Laboratory (PETRL) in 1986 in West Chester PA. Modern day procedures require blood and urine samples to be collected from the winner and one or two other competitors (i.e. "specials"). Horses must report directly to the test barn from the race track at the conclusion of the race for sample collection, and only authorized personnel are allowed to enter this area. In addition, horses are selected randomly for pre-race testing to prevent the use of alkalizing agents (i.e. "milkshaking" with sodium bicarbonate).

Samples are collected and packaged by the Commission's veterinarian and/or their designee, and all samples are assigned a barcode and entered into a Laboratory Information Management System (LIMS). The samples are barcoded so that the laboratory personnel cannot identify the horse from which the sample came or the owner and trainer of the horse, and this sample ID number is also used to track the tests performed on each sample at the laboratory. The appropriate packaging of the samples with the corresponding paperwork is very important because a continuous "chain of custody" must be documented for each sample from the time the sample is collected to the time the sample is delivered to the testing laboratory and throughout the testing process at the laboratory.

Following sample collection at the racetrack, part of the sample is sent to PETRL for testing and the other part of the sample is kept in a double-locked freezer at the race track (i.e. two locks). One lock can only be opened by the Commission staff and the other can only be opened by a representative from the horseman's association, ensuring both representatives are present for sample addition/removal. If a drug is reported by PETRL, the trainer can request the "split sample" (i.e. the sample maintained at the race track) to be sent to another comparable drug testing laboratory for testing; a list of acceptable laboratories is provided by the Commissions. The second laboratory must confirm PETRL's reported finding in order for penalties to be assigned for the finding.

The testing procedures are similar for the Fédération Equestre Internationale (FEI), which is the regulatory body that oversees Olympic Equestrian Sports (3). The USA National Federation is the United States Equestrian Federation (www.usef.org)(4). Samples are collected by FEI Official Veterinarians; veterinarians interested in helping to collect samples for testing can apply to become first a Permitted Treating Veterinarian and then an Official Veterinarian at FEI level events. See the FEI website's Anti-Doping and Veterinary Matters Tab for more information on the FEI Anti-Doping effort and on becoming an FEI-approved veterinarian (5).

ANALYTICAL METHODOLOGIES

PETRL receives approximately 40,000 samples per year for drug testing from the six racetracks in PA. During the summer racing season, PETRL receives as many as 1000 samples per week. Blood and urine samples are routinely used to test for approximately 600 drugs using a multitude of technologies. The samples are first analyzed with several screening methods (i.e. methods which look for a large number of drugs but do not provide quantitative information). If a drug is suspected in a sample, the sample is then subjected to a specific confirmatory test. The confirmatory test confirms the presence of the suspected drug by comparison of the sample's results to the result obtained for a certified reference standard (positive control); this test also provides quantitative information. The tests performed for each sample are recorded on paper and tracked electronically using LIMS.

Drug testing was originally performed using thin layer chromatography, however this technology is now obsolete in most drug testing laboratories due to the development of faster, more specific, and more sensitive methods: enzyme linked immuno-sorbent assay (ELISA) and mass spectrometry (MS). Mass spectrometry is coupled with a variety of separation and/or ionization techniques including liquid chromatography (LC), gas chromatography (GC), and inductively coupled plasma (ICP). In addition, DNA testing can now be performed to confirm the identity of the horse from which the sample was collected if the chain of custody is questioned. Hair testing is also a new method being implemented by some laboratories.

ELISA is a relatively cheap and sensitive method that uses antibodies to detect the presence of drugs. A good explanation of how the ELISA works can be found on Neogen's website, a manufacturer of many drug testing ELISA kits for horses (6). Because antibodies often will bind to multiple members of the same drug family, ELISA can only be used as a screening method to suggest the presence of a prohibited drug. A confirmatory method must then be performed to identify which drug is present and to accurately quantify the amount present.

Mass spectrometry is the preferred method for confirming the presence of a drug, and this technology can also be used for screening samples for multiple drugs (7). It is more specific than ELISA because the drug is processed by several steps that provide identifying information. First, an extraction step uses the chemical nature of the drug to group drugs with similar properties together (e.g. basic drugs and acidic drugs are grouped together). Second, chromatography physically separates the drugs from each other by passing them through a solid column filled with a material that will bind to the drugs. The length of time a drug remains inside the column (i.e. retention time) will depend on the size of the drug, how well it binds to the substance in the column, and on the solvents flowing through the column. By using different types of columns and solvents, a particular drug can be isolated from the other components of the sample. The eluent from the column flows directly into the mass spectrometer. There are different types of mass spectrometers, but all of them measure the mass and charge of the drug which is the m/z ratio. Sometimes a drug is also subjected to a process inside the mass spectrometer to create fragments of the drug. The fragmentation pattern (i.e. m/z ratio of the fragments produced) and the ion ratio (i.e. the proportion of those fragments to each other) provide additional identifying information. If the retention time, m/z ratio, fragmentation pattern, and ion ratio match between the sample and the reference standard, according to the criteria set forth by the Association of Official Racing Chemists (9), then the drug has been confirmed in the sample.

Ongoing technological developments in mass spectrometry and data acquisition software have significantly increased the ability of drug testing laboratories to identify and quantify drugs. Within the last 5 years, the sensitivity of our analytical methods have improved in some cases by 1000 fold e.g. drugs that were accurately quantified at plasma concentrations ranging between 1 µg/mL (0.001 mg/mL) and 1 ng/mL (0.000001 mg/mL) can now be accurately quantified at 1 pg/mL (0.000000001 mg/mL). This is the equivalent of being able to accurately measure the concentration of dissolving 4 grains of salt (assuming 0.7 mg/grain) in an Olympic size swimming pool (2,500,000 L)! This dramatic increase in sensitivity is an advantage for detecting substances of abuse that should never be found in the horse e.g. dermorphin – a synthetic opioid agonist (8), but it has created a nightmare with regard to therapeutic drugs used by practicing veterinarians, which in some cases (e.g. sotolol) can now be detected days to weeks after there is no longer a measureable pharmacological effect.

MEDICATION RULES

Who makes the rules for Olympic equestrian sports?

Organized anti-doping efforts for human and equine sport have been in existence since the beginning of the 20th century. However, human deaths and scandals due to doping prompted the formation of the World Anti-Doping Agency in 1999, which now creates the rules and regulations for human Olympic sports (10). Dr. Arne Ljungqvist, who was vice-president of WADA from 2008 to 2013, served as a consultant to the FEI from 2008 to 2010 in order to develop a similar approach for Olympic equestrian sports (11). These rules include a list of prohibited drugs (i.e. The Prohibited List) and this list is updated annually by the FEI Prohibited Substances List Expert Group. This group is composed of sport veterinarians, pharmacology and toxicology specialists, and researchers (3). Anyone can submit suggestions for consideration by using a form found on the FEI website under the Anti-Doping and Controlled Medication tab.

Who makes the rules for racing?

In the United States, each state is responsible for regulating racing. The members of the Racing Commissions, which are appointed by the governor, are responsible for developing and enforcing the rules and regulations in their state, including those pertaining to drug testing. They receive guidance from a number of sources. The Association of Racing Commissioners International

(ARCI) was established in 1934 to try to unify rules and regulations across state lines, and many Commissioners are members of this organization (12). In 2003, racing industry representatives supported the formation of the Racing Medication and Testing Consortium (RMTC) so that it could provide guidance to ARCI and others specifically regarding drug testing (13). The RMTC Scientific Advisory Committee, of which I am a member, meets approximately every 6 months to discuss and formulate recommendations pertaining to medications and best laboratory practices. The horsemen and women associations (e.g. The Horsemen's Benevolent and Protective Association) also provide input to the Racing Commissions regarding their interests with respect to drug policies and testing. Some states have an equine medical director, typically a veterinarian, employed by the Racing Commission to provide them guidance. And some states (e.g. PA) support a research laboratory so that drug questions specific to their jurisdiction can be immediately and confidentially addressed with targeted research in pharmacology and analytical chemistry.

Which drugs are prohibited for racing?

The first list of prohibited drugs for racing was generated by ARCI in 1991, and includes a classification scheme with 5 categories (12). Class 1 drugs are those with the highest potential to compromise the integrity of the sport or the welfare of the horse and include DEA scheduled drugs (e.g. morphine). Class 2 includes drugs that could be used therapeutically, but also have a high potential to compromise the integrity of the sport and/or welfare of the horse (e.g. anesthetics including local blocking agents). Class 3 drugs are those with moderate potential (e.g. bronchodilators and anabolic steroids). Class 4 drugs are those with some potential but less potential than Class 3 drugs (e.g. nonsteroidal anti-inflammatories and corticosteroids), and Class 5 are those with essentially no potential (e.g. anti-ulcer medications) but are prohibited to be given in many jurisdictions within 24 hours (e.g. PA) or 48 hours (e.g. CA) of the race. ARCI also recommends penalties appropriate for the classified drugs. With regard to "zero tolerance" ARCI states: "Found substances or drugs not included in these guidelines should be treated as Class 1 violations warranting a Class A penalty unless otherwise advised by consultation with Racing Commissioners International (RCI) and/or the Racing Medication and Testing Consortium (RMTC)." (12) In addition, as of 2012, the Thoroughbred Owners and Breeders Association annually issues a list of drugs that must be included in the drug testing program for Graded and Listed Stakes races.

Every year, more drugs are added to the lists of prohibited drugs. However, the speed at which approved medications and illegal substances have been discovered, developed, and made available for purchase has consistently surpassed the ability of forensic chemists to develop analytical methods for identifying individuals who have administered these substances to their horse. It was estimated in 2011 that 8,969 unique compounds are approved for therapeutic or experimental use in humans and animals, and synthetic chemists have created libraries of over 100,000 compounds (14). Many of these substances do have the potential to compromise the integrity of the sport and/or the welfare of the animal, however, many do not. It is up to the researchers and research sponsors, in collaboration with private investigators, to prioritize which drugs are studied first. Historically, much of the focus of this work has been only on the identification of these drugs with little regard to the concentration measured – zero tolerance means that legally any measureable amount is a violation. The dramatic increase in sensitivity that has been achieved over the last 5 to 10 years, however, has forced a re-evaluation of this philosophy for drugs routinely used by practicing veterinarians to treat disease.

THRESHOLDS AND WITHDRAWAL TIMES

A threshold is the concentration below which a drug will be ignored in a sample (i.e. thresholds allow exceptions to the zero tolerance rule). The first drug to be assigned a threshold for racing horses was phenylbutazone (threshold = 2 μg/mL), the most commonly used non-steroidal anti-inflammatory. A threshold can be set in plasma or urine. A threshold in plasma is recommended if the drug can be confirmed and quantified in plasma after it is no longer having a measureable pharmacological effect (e.g. flunixin meglumine). A threshold in urine is recommended if the pharmacological effect is measureable after the drug can no longer be confirmed in plasma (e.g. clenbuterol). For short acting drugs, the threshold is set to correlate with 24 or 48 hours to ensure no drugs have been administered during this time frame – this assures compliance with the "no drug administration" rule stated earlier.

A withdrawal time is the recommended length of time required for the concentration to fall below the threshold (or limit of confirmation if there is no threshold). A withdrawal time can be estimated for a particular drug only if the threshold/limit of confirmation is known and pharmacokinetic data are available, however it can never be guaranteed. These studies are typically performed in a relatively small number of healthy animals (6 to 20) and the data are extrapolated to the entire racing population. Ideally population pharmacokinetic studies would be performed to ensure the pharmacokinetic model generated with the small number of horses applies to the population, however the resources to do these intensive studies are limited. Thus, a safety factor is applied to try to account for the limitations of the data at hand. The RMTC calculates the upper 95/95 tolerance limit to provide withdrawal guidance, and more information describing this calculation can be found on their website (13).

Thresholds and associated withdrawal times have now been proposed for 26 medications by the RMTC, and the RMTC scientific advisory committee is working with the American Association of Equine Practitioners (AAEP) Racing Committee (15) to identify which drugs should be prioritized for research efforts going forward. Because the Center for Veterinary Medicine (16, 17), the branch of the Food and Drug Administration responsible for the approval of veterinary drugs, does not require pharmacokinetic data for the approval of equine drugs, and because of extra-label drug use, these data have primarily been generated in academic research laboratories.

WHAT IS APPROPRIATE DRUG USE?

This question has yet to be answered and is the subject of much debate in the United States and globally. Many feel that no drugs should be used. Many others consider it cruel to prohibit the use of drugs that have a medical benefit, even if they compromise integrity. Currently, it is up to racing commissioners to decide the answer to this question for race horses.

RESOURCES AND REFERENCES

All website links were accessed on February 2nd, 2016 to confirm they were intact. (1) Penn Vet Equine Pharmacology Laboratory www.vet.upenn.edu (2) PA Department of Agriculture Horse and Harness Racing Commissions www.agriculture.pa.gov (3) FEI Anti-Doping Rules http://www.fei.org/content/anti-doping-rules (4) United States Equestrian Federation Rules and Governance: Drugs and Medications http://www.usef.org/_IFrames/Drugs/Default.aspx (5)FEI Official Veterinarian Program http://www.fei.org/fei/horse-health-and-welfare/vet-matters (6) Neogen http://neogen.com/Toxicology/Assay_Principle.html (7) Liu Y, et. al Anal Chem 2011;83(17):6834. (8) Robinson MA, et. al J Vet Pharmacol Therap 2014;38:321. (9) Association of Official Racing Chemists http://www.aorc-online.org/home (10) World Anti-Doping Agency https://www.wada-ama.org/en (11) Professor Arne Ljungqvist Anti-Doping Foundation http://www.arneljungqvist.com (12) Association for Racing Commissioners International http://www.arci.com (13) Racing Medication and Testing Consortium http://www.rmtcnet.com (14) Huang R, et. al Sci Tran Med 2011;3(80):80ps16. (15) American Association of Equine Practitioners Whitepapers http://www.aaep.org/info/white-papers (16) FDA Center for Veterinary Medicine http://www.fda.gov/AnimalVeterinary/default.htm/ (17) Animal Drugs@FDA http://www.accessdata.fda.gov/scripts/animaldrugsatfda/

Impact of Disease on Drug Pharmacokinetics in Food Animals

Johann F. Coetzee, BVSc, Cert CHP, PhD, DACVCP, DACAW
Locke Karriker, DVM, MS, DACVPM
Ames, IA

INTRODUCTION

Antimicrobials approved for use in food animals are studied in healthy animals to determine their pharmacokinetic (PK) profile. There are few studies evaluating the impact of disease on pharmacokinetics and the implications on modifying treatment regimens. Managing the effective use of antimicrobials for treatment of disease is an important responsibility of veterinarians, which involves detailed knowledge of these agents. Pharmacokinetic studies in diseased animals better represent the circumstances where most drugs are applied in practice and will allow for the modification of treatment regimens in consideration of the altered physiologic status of diseased animals. In this session we will examine the results of recent studies that have investigated the impact of disease on the pharmacokinetics of drugs in pigs and cattle.

IMPACT OF DISEASE ON PHARMACOKINETICS IN SWINE

There are relatively few studies evaluating the impact of disease on pharmacokinetics of antimicrobials for use in swine. A study by Tantituvanont and colleagues evaluated the pharmacokinetics of ceftiofur hydrochloride in 4 week old pigs infected with Porcine Reproductive Respiratory Virus (PRRSV).[1] Pigs were administered a 3 mg/kg IM dose of ceftiofur hydrochloride at 7 days post infection with a Thai PRRSV isolate. Plasma ceftiofur concentrations were measured using HPLC and pharmacokinetic parameters were analyzed using a non-compartmental model. Pharmacokinetic analysis revealed significant differences ($P < 0.01$) in PK parameters between control and PRRSV infected pigs. PRRSV infected pigs had a 116% increase in Vz/F (volume of distribution per fraction of dose absorbed) and a 234% increase in Cl/F (plasma clearance per fraction of dose absorbed) compared to non-infected pigs. Additionally, the C_{MAX} and AUC of infected pigs decreased by 54% and 70% when compared to non-infected pigs, and the half-life was reduced from 21.0 hours to 16.0 hours on average in infected pigs. The conclusions of this study suggested pharmacokinetic parameters of ceftiofur hydrochloride were altered as a consequence of PRRSV infection. The authors proposed decreased plasma protein binding resulted in increased plasma clearance, increased volume of distribution, and shortened half-life, although changes in total protein or protein binding were not measured. Reduced protein binding would lead to increased fractions of free desfuroylceftiofur (DFC). Free DFC is more readily exposed to renal elimination resulting in increased ceftiofur clearance and decreased half-life. A greater portion of free ceftiofur is also available for tissue deposition leading to an increased volume of distribution. The authors further suggested alterations in dosage regimens may be warranted to achieve maximum efficacy of ceftiofur hydrochloride treatment against secondary bacterial pathogens in PRRSV infected pigs.

The impact of an experimental Porcine Reproductive and Respiratory Syndrome Virus (PRRSV) and *Streptococcus suis* coinfection on the pharmacokinetic (PK) profile of ceftiofur hydrochloride in pigs after intramuscular (IM) injection was recently investigated.[2] Eighteen clinically normal crossbred gilts, 7 wk of age, were sourced from a PRRSV negative herd with no clinical signs of *Streptococcus suis* infection. Animals were ranked and randomly assigned by weight into a challenge group (ten pigs) and control group (eight pigs) which were housed separately. The first PK assessment occurred in both groups before disease challenge. Pigs received a single IM injection of ceftiofur hydrochloride (Excenel RTU Sterile Suspension, Zoetis, Kalamazoo, MI) at a 5 mg/kg BW dose. Blood samples were collected from both groups at 0, 10, 20, 40, 60 min and at 2, 4, 8, 16, 24, 36, and 48 h post injection to characterize the plasma concentration curve. After a 10 day drug washout period, the challenge group was inoculated with 2 mL of PRRSV isolate VR-2385 ($10^{5.75}$ 50% tissue culture infective doses per mL) intranasally (IN). Eight days after PRRSV inoculation, the same pigs were inoculated with 2 mL of *Streptococcus suis* serotype 2 (1.23×10^9 cfu/mL) IN. When clinical disease was evident in greater than 50% of the pigs in the challenge group, the second PK assessment began in both challenge and control groups using the same collection schedules. Ceftiofur and related metabolites were measured using a liquid chromatography-mass spectrometry technique. Plasma concentration-time curves were analyzed using a non-compartmental model. A mixed model ANOVA was used to compare PK parameters. Coinfected pigs demonstrated lower values of area under the curve (AUC) (P=0.0050) and maximum serum concentration (C_{MAX}) (P=0.0069), but higher values of plasma clearance per fraction of dose absorbed (Cl/F) (P=0.0019) and volume of distribution per fraction of dose absorbed (Vz/F) (P=0.0054) indicating drug kinetics were altered by infection. Comparisons completed on the same animals reduced biological variation and increased statistical power as indicated by C_{MAX} which was not statistically significant when comparing across control and challenge groups at the second PK (P=0.4682) assessment, yet was significant when comparing the same pigs before and after challenge (P=0.0069). The data from this study has implications on ceftiofur treatment regimens in diseased pigs.

Given that infection with wildtype PRRS virus has been shown to alter the pharmacokinetics of ceftiofur in ways that predict a negative impact on efficacy and extra-label modification of dose to compensate for these impacts is not permitted for ceftiofur, the potential for viral vaccination to preserve antimicrobial drug efficacy was investigated.[3] Thirty-eight high-health, PRRSv naïve barrows were allotted to one of four groups A) Control, n=8; B) Vaccinated Only, n=10; C) Challenged Only, n=10; D) Vaccinated followed by challenged, n=10. Groups B and D were administered a 2mL intramuscular injection of Ingelvac® PRRS MLV vaccine (Boehringer Ingelheim Vetmedica). Subsequently, groups C and D were intranasally challenged with PRRSv VR-2385. After challenge and vaccination events, all groups were intramuscularly administered ceftiofur crystalline free acid (CCFA, Excede for Swine®, Zoetis) at a 5 mg/kg dosage. Blood was sampled 0 (pre-treatment), 0.25, 0.5, 1, 6, 12, 24, 48, 96,144,192, and 240 hours post-injection. Plasma was analyzed using liquid chromatography coupled with mass spectrometry to determine concentrations of ceftiofur and related metabolites. For each pig, non-compartmental PK modeling (WinNonlin 5.2) of plasma drug concentrations was used to determine the following parameters: area under curve time 0 to last time point (AUC_{last}), maximum concentration (Cmax), time to maximum concentration (Tmax), terminal half-life ($T_{½}\lambda_z$), clearance per fraction of dose absorbed (Cl/F), volume of distribution per fraction of dose absorbed (Vz/F). Mixed model ANOVA was used to compare PK parameters between groups using Tukey's test to determine significant mean differences, $P \leq 0.05$ (SAS 9.3). Group A and Group B were not significantly different on any evaluated PK parameter. When compared to Group C, Group D had significantly higher AUC_{last}, lower Cl/F, lower $T_{½}\lambda_z$, and lower Vz/F. When considered alone, vaccination with Ingelvac® PRRS MLV does not influence PK parameters of CCFA. When faced with PRRSv challenge, previous vaccination with Ingelvac® PRRS MLV, appeared to increase the absorption of CCFA, indicated by a larger AUC_{last}. $T_{½}\lambda_z$, Vz/F, and Cl/F were lower in pigs that were vaccinated before challenge compared to those only challenged; this could reflect changes in intramuscular bioavailability (F). The results of this study suggest that vaccination preserved the pharmacokinetic parameters that are used to predict the efficacy of ceftiofur when vaccinated pigs are subsequently exposed to PRRS virus.

The pharmacokinetics of amoxicillin in healthy and respiratory diseased pigs was examined after ad libitum administration in medicated feed.[4] Diseased pigs had a naturally occurring respiratory infection with PRRSV in addition to secondary bacterial agents isolated from lung tissue at the conclusion of the study. Isolated bacterial agents included *Pasturella multocida, Bordetella bronchiseptica*, and *Streptococcus suis.* Diseased pigs received amoxicillin in medicated feed at a mean dose of 9 mg/kg/day for five consecutive days. Plasma concentrations of amoxicillin were determined using HPLC and pharmacokinetic analysis utilized a non-compartmental approach. Diseased animals compared to healthy animals had larger values at steady state of AUC, C_{MAX}, and C_{MIN}, as well as longer absorptive periods (4-6 hours versus 2 hours). The authors attributed these differences to the significantly higher bioavailability (14.1% to 44.7%) in diseased pigs which may occur from reduced intestinal transit rate increasing the extent of drug absorption. Intestinal transit rate was not directly measured in this study. The results of this research suggest the pharmacokinetic behavior of amoxicillin after oral administration is modified in the disease state and influenced by animal feeding behavior. Dosage regimens, when using medicated feed, should consider these alterations in PK parameters and the potential effects on gastrointestinal transit time to optimize treatment efficacy.

The effect of salmonellosis in swine on the pharmacokinetics of antimicrobials has been evaluated in two different studies. Salmonella infections can be asymptomatic or cause disease manifested as septicemia or enterocolitis. The impact of *Salmonella typhimurium* infection on the pharmacokinetics of danofloxacin, a fluoroquinolone, administered IV at a dose of 2.4 mg/kg BW was evaluated in a study by Lindecrona and colleagues.[5] A total of 24 male pigs were utilized for the study which were evenly divided into a control and disease group. Inoculation of the disease group was performed intragastrically via a stomach tube with 5.5×10^{10}CFU of *Salmonella typhimurium*. Twenty-four hours post inoculation the PK assessment occurred in both inoculated and control animals. Danofloxacin concentrations in plasma were determined using HPLC and a non-compartmental model was utilized for pharmacokinetic analysis. In Salmonella infected pigs, the mean elimination half-life was increased from 6.7 to 9.4 hours resulting from decreased body clearance in infected pigs. Diseased animals also had a significant reduction in the volume of the peripheral compartment. These changes in PK parameters lead to an increase in the area under the curve. The authors concluded prolonged half-life and reduced body clearance may be attributed to a reduced metabolic capacity occurring during infection.

The pharmacokinetic impact of *Salmonella typhimurium* was again investigated, but in pigs administered amoxicillin IM at 15 mg/kg BW dose.[6] A total of 24 healthy gilts were utilized and divided evenly into a control and challenge group. Salmonella inoculation was performed intragastrically by use of a stomach tube with 6×10^{10}CFU of Salmonella strain NVL810. Amoxicillin was administered 24 hours after inoculation and blood samples collected in both diseased and healthy pigs. Plasma amoxicillin concentrations were determined by HPLC and pharmacokinetic analysis conducted using a one-compartment model with first order absorption. Salmonella-inoculated pigs experienced changes in the plasma concentration-time profile. Diseased animals had an increase in the absorption rate constant, a prolonged elimination half-life, and an increased volume of distribution. The authors attributed the increased absorption rate constant in infected pigs to the increased body temperature and accompanying muscle shivering which would lead to increased blood flow at the site of drug injection. The prolonged half-life in inoculated animals was explained by the increased volume of distribution which may result from alteration in blood flow as a consequence of infection. This study showed *Salmonella typhimurium* inoculation altered the pharmacokinetics of amoxicillin. The authors concluded inflammation alone may not be responsible for changes in the kinetics of a drug, but alterations in blood flow may play an important role.

Impact of Disease on Pharmacokinetics in cattle

Recently a study was conducted at North Carolina State University to compare the pharmacokinetics and milk elimination of flunixin and 5-hydroxy flunixin between healthy and mastitic cows.[7] Twenty lactating Holstein cows with mastitis and matched control cows received flunixin IV, ceftiofur IM, and cephapirin or ceftiofur, intramammary. Blood samples were collected before (time 0) and 0.25, 0.5, 1, 2, 4, 8, 12, 24, and 36 hours after flunixin administration. Composite milk samples were collected at 0, 2, 12, 24, 36, 48, 60, 72, 84, and 96 hours. Plasma and milk samples were analyzed by use of ultra-high-performance liquid chromatography with mass spectrometric detection. For flunixin in plasma samples, differences in area under the concentration-time curve and clearance were detected between groups. Differences in flunixin and 5-hydroxy flunixin concentrations in milk were also detected at various time points. At 36 hours after flunixin administration (milk withdrawal time), 8 cows with mastitis had 5-

hydroxy flunixin concentrations higher than the tolerance limit (ie, residues). Flunixin residues persisted in milk up to 60 hours after administration in 3 of 10 mastitic cows. It was concluded that the pharmacokinetics and elimination of flunixin and 5-hydroxy flunixin in milk differed between mastitic and healthy cows, potentially resulting in violative residues. The authors hypothesized that this may partially explain the high number of flunixin residues reported in beef and dairy cattle.

In a recent study published by our group at Iowa State University, we compared the plasma PK of CEF in healthy cows with those with severe endotoxic mastitis.[8] Eight cows with naturally occurring mastitis and 8 clinically healthy cows were treated with 2.2 mg of CEF per kilogram of body weight once daily for 5 d via the intramuscular route. Blood was collected at 0, 0.33, 0.67, 1, 1.5, 2, 3, 4, 8, 16, and 24 h after the first CEF administration and every 8 h thereafter until 120 h after the final dose. Plasma samples were analyzed for CEF concentrations using liquid chromatography coupled with mass spectrometry. With the exception of time 0, CEF was detected at all time points. The disease group had a significantly higher plasma CEF concentration at $t = 3$ h after the first injection and a significantly lower plasma concentration from 40 to 152 h following the first injection, with the exception of the $t = 64$ h time point. Data following the first injection (time 0–24 h) were fit to a single-dose, noncompartmental PK model. This model indicated that the disease group had a shorter plasma half-life. A multidose, noncompartmental model was used to determine steady-state PK (**Figure 1**). Compared with control cows, the disease group had an initially higher peak concentration and a higher volume of distribution and drug clearance rates. The disease group also had a lower area under the curve per dosing interval, steady-state concentration maximum, and dose-adjusted peak steady-state concentration. All other PK parameters were not different between the 2 groups. Altered PK, as suggested by this trial, may contribute to an increased risk for the development of a violative residue in meat. Further research is needed to more completely characterize drug distribution in diseased cattle and to study the effect of coadministration of other drugs on drug distribution.

Figure 1.

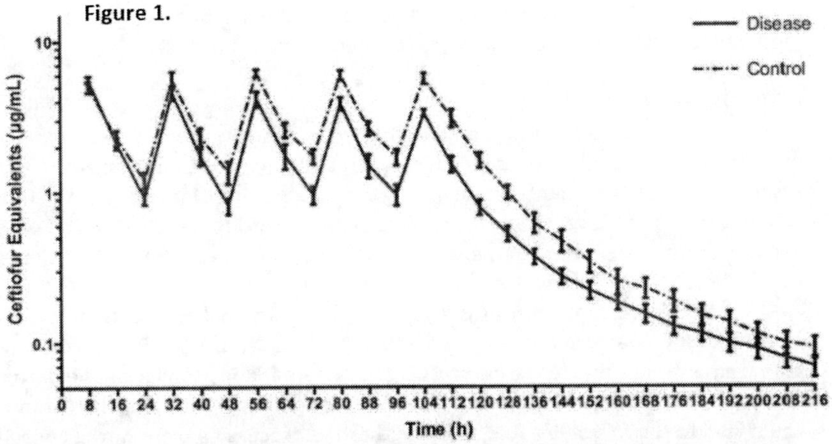

CONCLUSION

The few studies examining the impact of disease on pharmacokinetics evaluated a limited number of drugs in few species. The same pharmacokinetic parameters were not uniformly affected between studies, however, a commonality exists that disease impacts pharmacokinetics. It is challenging to draw conclusions across different studies about specific changes in PK parameters resulting from experimental infection, as there are many factors involved including: the disease model, inoculum concentration, antimicrobial class evaluated, species, and the animal's response to infection. Studies utilized different routes of infection and incubation periods, which likely contributed to the variation in study results. Additionally, the different PK parameters affected between studies and ranges in magnitude of effect indicate this relationship is likely disease and antimicrobial dependent. This supports the need for additional pharmacokinetic studies examining multiple antimicrobial classes and disease challenge models. The effects on pharmacokinetics as a consequence of disease also illustrate the need for PK studies completed in both diseased and healthy animals to be included in the drug approval process.

REFERENCES

1. Tantituvanont, A., W. et al. 2009. J. Antimicrob. Chemother. 63(2):369-373.
2. Day, D.N.et al. 2015. *Journal of Veterinary Pharmacology and Therapeutics*. 38(5): 475–481.
3. Sparks J, et al. 2014. Proceeding of the Conference of Research Workers in Animal Diseases. Chicago, Illinois. December 2014. Abstract number 138.
4. Godoy, C., G. et al. 2010. J. Vet. Pharmacol. Therap. 34:265-276.
5. Lindecrona, R.H.et al. 2000. Res. Vet. Sci. 68(3):211-216.
6. Agerso, H., C. et al. 2000. Am. J. Vet. Res. 61(8):992-996.
7. Kissell LW, et al. 2015. J Am Vet Med Assoc;246 (1):118-25
8. Gorden, P.J. et al. 2016. *Journal of Dairy Science*. 99 (1): 505–514

RESIDUES FOLLOWING ORGANIC THERAPIES IN LIVESTOCK.

Ronald E. Baynes, DVM, PhD; Keena A. E. Mullen, PhD; Sharon E. Mason, DVM, PhD;
Kevin L. Anderson, DVM, PhD; DABVP, Steven P. Washburn, PhD
Raleigh, NC

ABSTRACT

There is increasing demand in the U.S. that livestock be raised without antibiotics to be certified as organic. This presentation will focus on our recent studies that demonstrated meat and milk residues in dairy cattle following exposure to two organic products used to **treat** mastitis on organic dairy farms. We hope to use the data generated and novel pharmacometric methods to estimate withdrawal times for these organic products.

INTRODUCTION

In spite of the public demand for organic foods, there is limited availability of scientifically tested **organic therapeutics**, often in the form of plant essential oils, for treating mastitis in dairy cows. There is also limited anecdotal evidence that these plant essential oils for treatment of mastitis in dairy cattle are efficacious and there is no information or data describing the chemical residues from these organic products once administered to dairy cows. The potential mechanisms of action of these essential oils in mastitis therapy have not been evaluated. It should be stressed that for organic dairy cattle to maintain their organic status in the U.S., antibiotic use is prohibited.

There are some concerns about organic dairy milk production as we know little about milk quality. As a prelude to the current study, our co-authors (Anderson, Mullen, and Washburn) recently demonstrated that despite differences in herd management, milk quality based on SCC can be remarkably similar between the organic and conventional dairies compared for this study (Mullen et al., 2013, 2014a). However, residues of plant-based treatments might cause off-flavors or other concerns which could lead to the milk being rejected by either the processing plant or the consumers.

The veterinary community has been slow to adopt organic mastitis products because of the potential risk of residues entering the milk supply and the fact that these products are not FDA-approved. The ingredients in these organic products are generally recognized as safe (GRAS). U.S. FDA regulations state that GRAS substances are safe to be used as human food additives (U.S. FDA, 2016). The reader should be aware that GRAS substances in some products are not U.S. FDA approved products and therefore the detection of any chemical residues from these products in meat or dairy products is a residue violation. The Animal Medicinal Drug Use and Clarification Act (AMDUCA) allows for the extralabel use of drugs with very strict allowances and one of them being that the product needs to be an approved drug. Therefore veterinarians should be aware that use of organic therapies in food animals are not covered by AMDUCA (U.S. FDA, 2012). For this reason, we developed a very sensitive assay which is briefly described below that is capable of detecting the major ingredient chemicals in these organic therapies. As these products are not approved, there are no established U.S. safe levels or tolerances, EU maximum residue levels (MRLs), or Codex MRLs. Therefore, future computation of a safe withholding time for these products will require the use of pharmacokinetic data and the most sensitive limits of detect (LOD) and limits of quantification (LOQ) from validated analytical chemistry methods briefly described below. Several authors (Anderson, Baynes, Mason) of this paper have recently published in Baynes and Riviere (2014) strategies for minimizing drug and chemical residues in food animals, which could be applied in this situation.

An essential oil-based botanical intramammary infusion, Phyto-Mast® (CowMaster LLC, Narvon, PA), contains thymol (5,500 ppm) and carvacrol (10,200 ppm) among other ingredients and has been used with some effectiveness as a **dry treatment** in mature cows (Mullen et al., 2014a). One component of this botanical infusion, thyme essential oil, has documented evidence of antibacterial activity (Mullen et al., 2014b). In fact, the efficacy from the organic therapy was comparable to conventional therapies and there were no adverse effects from the organic therapies. Another product, Uddersol™ (Ralco Animal Health, Marshall, MN), is applied topically to the affected udder and contains only carvacrol (40,000 ppm) which is believed to be absorbed across the udder of the skin and has antibacterial activity. Thymol and carvacrol are two monoterpene phenols (Cowan, 1999) Thymol can be found in essential oils of thyme, *Thymus vulgaris* or *Thymus zygis*. Carvacrol occurs in many essential oils of the family Labiatae, including *Origanum, Satureja, Thymbra, Thymus,* and *Corodothymus* species. There is some evidence that thymol and carvacrol may have some additional useful biological activities, as antioxidant, antifungal property and effective against mites [Burt et al., 2004; Cowan, 1999; Tabari et al., 2015; Vale-Silva et al., 2010). One recent study in our laboratory described the milk and plasma pharmacokinetics of Phyto-Mast administered to lactating dairy goats (McPhee et al., 2011). That study demonstrated that thymol can persist in milk for a few milkings and therefore a zero milk withholding time will not be appropriate for lactating dairy goats and our hypothesis was that this would be the case for lactating dairy cows similarly treated. This was part of the stimulus for the work described below.

METHODS

Experimental Design

Organic dairy cows were recruited from several certified organic dairy farms in North Carolina and West Virginia. These were healthy mid-lactation Jersey, Holstein, or crossbred predominantly Holstein or Jersey dairy cows. A randomized study design was developed such that there were four treatment groups consisting of 10 dairy cows per treatment group. Two groups were treated with either 1X or 2X of the recommended dose of Phyto-Mast and two groups were treated with 1X or 2X of the recommended dose of Uddersol. This presentation will only describe the 2X recommended dose. For Phyto-Mast treated cows, each cow received two 12 ml tubes intramammary per udder quarter applied to two contralateral quarters every 12 hours for 4 doses and for Uddersol treated cows, each cow received 25 ml topically per quarter applied to two contralateral quarters every 12 hours for 4 doses. For both treatments, the udder was evenly massaged by the same person following product administration. Plasma and milk samples were collected periodically for up to 5 days after the last dose. Tissue samples were also collected to assess residue depletion in liver, kidney, muscle and fat.

Analytical method

Thymol and carvacrol are both semi-volatile compounds, which facilitated analysis by headspace solid-phase microextraction coupled with gas chromatography-mass spectrometry (SPME-GC/MS/MS), as described in our one of recent publications which describes the method for these two chemicals in more detail (Armorini et al., 2016).

RESULTS

Analytical Method

Thymol and carvacrol showed an excellent linearity over the concentration range 0.0005-1.0 ppm for all the matrices tested, except for fat tissue, with a limit of determination (LOD) equal to 0.0005 ppm and a limit of quantification (LOQ) of 0.001 ppm. For fat tissue, the LOQ was 0.010 ppm and LOD was 0.005 ppm. Inter- and intra-day precision and accuracy values were satisfactory in all matrices (Armorini et al., 2016). This finding is important as are the LOD and LOQ values that will be used to establish safe withdrawal periods for these products based on the presence of thymol and carvacrol alone. It should be noted that we have previously detected other chemicals from these products in the milk and plasma of dairy goats (McPhee et al., 2011). Current work in our laboratory is focussed on probing for these other chemical components in Phyto-Mast and Uddersol. Although we suspect that they may contribute only a very small percentage of the chemical residues present in meat and milk products, they may very well be below LOD and LOQs in the open literature.

Chemical Depletion in Plasma

Thymol concentrations in plasma ranged from 0.005 to 0.05 ppm during the first hour after the last Phyto-Mast IMM infusion and then exponentially declined to near the LOD (0.0005 ppm) at 8 hours. A similar trend was observed for carvacrol depletion following the Phtyo-Mast IMM infusion except there was more inter-animal variability for carvacrol plasma concentrations than with thymol plasma concentrations. Topical application of Uddersol resulted in a very different plasma concentration – time profile for carvacrol. There was an obvious absorption phase with maximum plasma concentrations ranging from 0.0007 – 0.0042 ppm within the first 2.0 hours post administration followed by a steady linear decline to just above 0.0005 ppm at 12 hours. There was considerable variability in carvacrol concentrations during the early time points and for the maximum plasma concentrations. In summary, plasma concentration profiles suggest that Phyto-Mast IMM infusion resulted in comparable plasma concentrations – time profiles of thymol and carvacrol during the first 8 hours after the last IMM dose. However, Uddersol topical application resulted in higher carvacrol concentrations and area under the curve (AUC) in plasma compared to carvacrol plasma concentrations following IMM Phyto-Mast.

Chemical Depletion in Milk

Thymol concentrations in milk ranged from 2.0 to 9.0 ppm during the four hour sample after Phyto-Mast IMM infusion and then declined exponentially to just above LOD levels at 12 hours. A similar trend was observed for carvacrol depletion following the Phtyo-Mast IMM infusion although there was greater variability in carvacrol milk concentrations across all time points, but carvacrol was detected in milk at 24 hours in several cows. As with the plasma concentration – time profiles, topical application of Uddersol resulted in a very different milk concentration – time profile. There was an obvious absorption phase with maximum milk concentrations ranging from 0.05 – 0.35 ppm within 4- 8 hours post topical application followed by a steady linear decline to levels just above the LOD of 0.0005 ppm at 12 hours for most cows but also out to 36 hours for several cows. In summary, milk concentration profiles suggests that Phyto-Mast IMM infusion results in comparable concentrations profiles of thymol and carvacrol during the first 24 hours after the last IMM dose. However, Uddersol application resulted in lower carvacrol concentrations and area under the curve (AUC) in milk compared to carvacrol concentrations following IMM Phyto-Mast. It should be noted that carvacrol concentrations in milk were lower following Uddersol topical application albeit carvacrol concentrations in the Uddersol dose were four times greater than for the IMM Phyto-Mast dose.

Tissue Residues

The table below depicts the presence of thymol and carvacrol in two of four tissues at different slaughter times following IMM infusion of Phyto-Mast at 2X the recommended dose. Neither of the two chemicals were detected in any of the four tissues at 72 hours after the last dose, with the slowest depleting tissues appearing to be liver. When Uddersol was administered topically, a similar pattern of tissue residues depletion was observed (data not shown), although there was a tendency for greater carvacrol residues with Uddersol application.

	Average Thymol Concentrations (ppm)				Average Carvacrol Concentrations (ppm)			
Time	Muscle	Kidney	Liver	Fat	Muscle	Kidney	Liver	Fat
12 Hours	<LOD	<LOD	0.003	<LOD	<LOD	<LOD	0.0033	<LOD
24 Hours	<LOD	0.0006	0.0019	<LOD	<LOD	0.0008	0.0027	<LOD
48 Hours	<LOD	0.0013	0.002	<LOD	<LOD	0.0008	0.002	<LOD
72 Hours	<LOD	<LOD	<LOD	<LOD	<LOD	<LOD	<LOD	<LOD

CONCLUSIONS

Plasma concentrations of target molecules were below the limit of detection by 12 hours (Uddersol topical at 2x dose) and 8 hours (Phyto-Mast IMM at 2x dose) after the last dose. This was not the case for chemical residues in various tissues where chemical residues persisted for as long as 48 hours after the last dose of Phyto-Mast and Uddersol. Based on these preliminary findings, meat withholding times greater than 48 hours would be appropriate. The reader should be aware that we have not utilized the tolerance limit method which accounts for 99th percentile of the population and the 95th confidence interval (U.S. FDA, 2006). Therefore a meat withholding time could very likely extend to as much as much as 3 to 4 days at 2X recommended dose for these products.

Milk concentrations of target molecules were below the limit of detection by 24 to 36 hours after the last dose was administered for both products. This does not mean that the milk withholding time would be 36 hours but could be as long as 48 or 96 hours pending further analysis and using the U.S. FDA tolerance limit method. Current research is focused on determining the influence of mastitis on plasma and milk concentration profiles. Previous research in our laboratory (Kissell et al., 2015) demonstrated that mastitis can influence the clearance of veterinary approved drugs and therefore the withholding time for milk may need to be increased for diseased animals.

ACKNOWLEDGMENTS

This material is based upon work that is supported by the National Institute of Food and Agriculture, U. S. Department of Agriculture, under award number 2014-51300-22234. Any opinions, findings, conclusions, or recommendations expressed in this publication are those of the author(s) and do not necessarily reflect the view of the U. S. Department of Agriculture.

REFERENCES

Armorini, S., Yeatts, J., Mullen, K., Mehmeti, E., Mason, S., Anderson, K., Washburn, S., and Baynes, RE. (2016). Development and validation of HS-SPME-GC/MS/MS method for the quantification of thymol and carvacrol in bovine matrices. (to be submitted to *J. Chromatog. B*).

Baynes RE, Riviere JE. (2014). Strategies for Reducing Drug and Chemical Residues in Food Animals: International Approaches to Residue Avoidance, Management and Testing. John Wiley & Sons, Inc, Hoboken, NJ.

Burt, S. (2004). Essential oils: their antibacterial properties and potential applications in foods-a review*, Int. J. Food Microbial.* 94: 223.

Kissell, L., Leavens, T., Baynes, RE., Riviere, JE., Smith, G (2015). Comparison of flunixin pharmacokinetics and milk elimination in healthy cows and cows with mastitis. *J. Amer. Vet. Med Assoc.* 246(1):118-125.

Cowan, M. M. 1999. Plant products as antimicrobial agents. Clin. Microbiol. Rev. 12:564- 582

McPhee, C., Anderson, KL., Barlow, B., Yeatts., J., Mason, SE., and Baynes, RE. (2011). Milk and plasma disposition following intramammary administration of a phytoceutical mastitis treatment. *J. Dairy. Sci*. 94(4): 1738-1743.

Mullen, KAE, LG Sparks, RL Lyman, SP Washburn, and KL Anderson. (2013). Comparisons of milk quality on North Carolina organic and conventional dairies. *J Dairy Sci* 96: 6753-6762.

Mullen KAE, KL Anderson, SP Washburn. (2014a). Effect of 2 herbal intramammary products on milk quantity and quality compared with conventional and no dry cow therapy. *J Dairy Sci* 97(6): 3509-3522.

Mullen, KAE, AR Lee, RL Lyman, SE Mason, SP Washburn, KL Anderson (2014b). Short Communication: An in vitro assessment of the antibacterial activity of plant-derived oils. *J Dairy Sci*, 97: 5587-5591.

Tabari, MA., Youssefi, M.R., Barimani, A., and Araghi, A. (2015). Carvacrol as a potent natural acaricide against *Dermanyssusgallinae, Parasitol. Res* 114: 3801.

U.S FDA (2006). Guidance for Industry. General Principles for Evaluating the Safety of Compounds Used in Food-Producing Animals. U.S. Department of Health and Human Services Center for Veterinary Medicine July 25, 2006 Food and Drug Administration. Rockville, MD. USA.

U.S. FDA (2016). Title 21: Food and Drugs Subpart A – General Provisions. Accessed Feb. 1, 2016. http://www.ecfr.gov/cgi-bin/text-idx?SID=239c3d298515e58327158692df09eea1&mc=true&node=sp21.6.582.a&rgn=div6

Vale-Silva, L.A., Gonçalves, M.J., Cavaleiro, C., Salgueiro, L., and Pinto, E. (2010). Antifungal Activity of the Essential Oil of Thymus x viciosoi against Candida, Cryptococcus, Aspergillus, and dermatophyte Species, *Planta. Med*. 76: 882.

Treatment of Lameness in Cattle: Evidence and Residue Considerations

Sarah A. Wagner, DVM, PhD, DACVCP
Fargo, North Dakota

Introduction

Lameness occurs commonly in dairy cows. Lameness decreases cow productivity, reproductive performance, and well-being, and increases the risk of culling. Consequently, there is motivation to find effective prevention and treatment strategies for lameness in cattle. Despite the widespread and serious negative impact of lameness on dairy cows and in dairy herds, the number of published clinical trials for treatments of lameness is low. The objective of this manuscript is to provide a brief overview of research that has been published regarding lameness treatment, including the drug residue considerations that go along with lameness therapies. Some commentary is also included regarding preventive measures and potential avenues for future research.

Digital Dermatitis

Digital dermatitis (DD), which is also known by other names including hairy heel wart, is the most common infectious cause of lameness in dairy herds in the United States and Canada. Digital dermatitis causes painful ulcerative lesions, most commonly on the heels of the rear feet. DD can also occur in beef herds, but it is more common on dairies (Brown et al, 2000).

The etiology of DD is multifactorial and the pathology is complex. The occurrence and resolution of the disease are associated with farm factors such as management and hygiene, and animal factors such as parity and stage of lactation. It is widely believed that the disease is caused by infection with a spirochete organism, specifically a treponeme, but the relationship between the organism and disease occurrence is not straightforward. The disease does have an infectious component, as evidenced by its rapid spread through herds after an affected animal is introduced and by the improvement of clinical lesions following treatment with antimicrobial drugs. Refaai et al published an excellent review of the pathogenesis, risk factors, and principles of treatment of infectious causes of lameness in 2013 (Refaai et al, 2013).

Properly designed and implemented footbaths, used frequently, can help to prevent and control DD. Formalin and copper sulfate have demonstrated evidence of efficacy as footbath ingredients. (Hoffman, 2012) Unfortunately, some contamination of the environment is inevitable with footbath use; consequently, these compounds cannot be used without concern, and at some point regulatory agencies may curtail or prohibit their use. Footbaths are not effective for producing whole-herd cures of DD.

In a review of all aspects of DD, Evans et al (2015) describe the treponemes associated with DD as most susceptible to penicillins and macrolides, yet tetracyclines are most commonly used for topical treatment, even though the susceptibility of the organisms to tetracyclines is described as "intermediate". The cure rate for DD treated with a topical tetracycline is typically 60-70% (Holzhauer 2001, Berry 2010), but recurrence has been observed to be about 50% within 12 months (Evans, 2015). In a recent review of treatments for infectious causes of lameness in cattle, Apley (2015), reported a lack of well-executed published clinical trials of systemic antimicrobial drug therapy for the treatment of DD. Evans et al suggest that efficacy could be improved, and recurrence rates reduced, by extending the duration of treatment with penicillins to 7-10 days. Whether farms would choose to extend therapy, which would require also extending associated withholding times, in return for increased cures and fewer recurrences, has not been examined.

Therapeutic hoof trimming also appears to be beneficial in the treatment of cows with DD; Ando et al (2009) found that augmenting antimicrobial therapy with hoof trimming lowered serum cortisol, considered a marker of stress, in animals treated with drugs and hoof trimming, when compared to cows treated with drugs only.

As no drug is approved for the topical treatment of DD, drug residues following such treatment must be avoided. Regarding residues, cows that had topical oxytetracycline applied as a spray or on a bandage did not achieve violative residues of the drug in milk (Britt, 1999).

In summary, published clinical trial results for therapies of this painful condition are limited. Teteracyclines appear to be somewhat effective, with minimal residue risks, but the rate of recurrence after treatment is high. Longer duration, systemic therapy with a macrolide or penicillin drug, used in conjunction with therapeutic hoof trimming, might be more effective, but published information about the actual clinical efficacy of this approach is lacking.

Claw horn lesions

The other common causes of lameness in dairy cows are claw horn lesions such as sole ulcers and white line disease. As with DD, the etiology of claw horn lesions is multifactorial; it involves the interplay of nutrition, body condition, endocrine and inflammatory factors, and facilities and management. These factors and their effects on the development of claw horn lesions have been well described by Shearer, Plummer, and Schleining (2015).

The most common approach to the treatment of claw horn lesions in dairy cows is corrective hoof trimming directed at restoring normal claw shape, removing abnormal and damaged hoof tissue, facilitating the drainage of abscessed areas, and redistributing weight-bearing. Unfortunately, the results of clinical trials evaluating the effects of corrective hoof trimming on the healing of claw lesions could not be found. The conduct of such trials would be complicated by the ethical issues involved with enrolling an untreated control group for these painful hoof conditions.

In addition to corrective trimming, a block is commonly applied to the weight-bearing surface of the healthier claw on the same limb, in order to lift the more affected claw off the ground and reduce its weight-bearing load. As reported by Shearer et al (2015), these blocks have been found to be effective in accelerating the healing of solar lesions.

Treatment of claw horn lesions with the systemic administration of antimicrobial drugs does not appear to be common or indicated. In the highly contaminated environment of a dairy farm, such compounds are not necessary if proper drainage of abscessed areas is established by corrective hoof trimming, and such drugs are unlikely to be beneficial if adequate drainage is not established.

Topical drug therapy of claw horn lesions is sometimes undertaken. Topical application of caustic or irritating compounds such as copper sulfate or tetracyclines is likely to delay healing, as these irritating compounds promote the production of excessive granulation tissue. Furthermore, the topical application of such compounds causes cows to display increased pain behaviors after treatment. (Shearer, 2015).

Similarly to studies investigating the presence of residues in milk after topical application of oxytetracycline to treat DD, plasma levels of tetracyclines after topical administration to solar lesions were detectable but not high enough to trigger regulatory action (Shearer, 2015)

Although antimicrobial drugs are not known to benefit healing of claw horn lesions, anti-inflammatory/analgesic drugs have demonstrated evidence of improving healing. Thomas et al (2014) reported that cows treated with corrective trimming, a block on the less affected claw, and the NSAID ketoprofen, had better locomotion scores 35 days after treatment when compared to cows treated with trimming only. Neither ketoprofen nor any other NSAID drug is approved for the treatment of pain or lameness in cattle in the United States; consequently such use is extra-label and therefore must be prescribed or administered by a veterinarian who assigns meat and milk withholding times. Fortunately, data about the pharmacokinetics of NSAIDs in cattle is available and can be used as a basis for assigning such withholding times. Veterinarians in the US are encouraged to contact the Food Animal Residue Avoidance Databank at 1-800-USFARAD or www.farad.org for guidance.

REFERENCES

Ando, T., Fujiwara, H., Kohiruimaki, M., Hayashi, T., Ohtsuka, H., Watanabe, D., ... & Koiwa, M. (2009). Peripheral blood leukocyte subpopulation of dairy cows with digital dermatitis and effect of hoof trimming with antibiotic treatment. *Journal of Veterinary Medical Science, 71*(4), 391-395.

Apley, M. D. (2015). Clinical evidence for individual animal therapy for papillomatous digital dermatitis (hairy heel wart) and infectious bovine pododermatitis (foot rot). *Veterinary Clinics of North America: Food Animal Practice, 31*(1), 81-95.

Britt, J. S., Carson, M. C., Von Bredow, J. D., & Condon, R. J. (1999). Antibiotic residues in milk samples obtained from cows after treatment for papillomatous digital dermatitis. *Journal of the American Veterinary Medical Association, 215*(6), 833-836.

Brown, C. C., Kilgo, P. D., & Jacobsen, K. L. (2000). Prevalence of papillomatous digital dermatitis among culled adult cattle in the southeastern United States. *American journal of veterinary research, 61*(8), 928-930.

Hoffman, A. Footbaths for the Treatment or Control of Hairy Heel Warts (Digital Dermatitis) in Dairy Herds: Summary of Seven Studies (2012) Washington State University Veterinary Medicine Extension, http://vetextension.wsu.edu/wpcontent/uploads/sites/8/2015/03/Footbaths20122.pdf, Accessed March 4, 2016

Refaai, W., Van Aert, M., El-Aal, A. A., Behery, A. E., & Opsomer, G. (2013). Infectious diseases causing lameness in cattle with a main emphasis on digital dermatitis (Mortellaro disease). *Livestock science, 156*(1), 53-63.

Shearer, J.K., Plummer, P. J., and Schleining, J.A., (2015). Perspectives on the treatment of claw lesions in cattle. *Veterinary Medicine: Research and Reports, 6*, 273-292

Thomas, H. J., Miguel-Pacheco, G. G., Bollard, N. J., Archer, S. C., Bell, N. J., Mason, C., ... & Huxley, J. N. (2015). Evaluation of treatments for claw horn lesions in dairy cows in a randomized controlled trial. *Journal of dairy science, 98*(7), 4477-4486.

How Flow Cytometry is Expanding Our Understanding of Leukemia

Paul R. Avery, VMD, PhD, DACVP
Fort Collins, CO

INTRODUCTION

Flow cytometry has long been an integral tool in the early assessment of human leukemias and is now commonly used in veterinary medicine as well. While human leukemias have been very specifically sub-classified using flow cytometry, we are still expanding our understanding of canine and feline leukemias as reagents become available. Identifying phenotypic features of canine and feline leukemia that help predict disease course and response to therapy would greatly aid in making rationale treatment decisions. A major focus of our laboratory is to extend the current knowledge of the range of phenotypes seen in canine and feline leukemia and to determine if certain phenotypic markers predict treatment response rates and/or survival times.

ACUTE LEUKEMIAS

Acute leukemias tend to be highly aggressive neoplasms of the hematopoietic system in both dogs and cats (Williams 2008, Novaco 2015, Mylonakis 2008). Acute leukemias can arise from both myeloid and lymphoid lineages and, while cellular morphology can be helpful in some cases, lineage assignment can be difficult to determine in those case with minimal differentiation (Novaco 2015, Graves 2007). This distinction is generally accomplished through the application of detailed flow cytometric phenotype analysis in people. Additional difficulties arise in distinguishing acute lymphoblastic leukemia from advanced stage V lymphoma in veterinary patients. This is often done based on somewhat subjective assessment of the magnitude of peripheral lymph node involvement. This classification issue in people has largely been addressed through the discovery of common phenotypic expression patterns and nomenclature that classifies both forms as a spectrum of one disease; ie. precursor T acute lymphoblastic leukemia (ALL)/ lymphoblastic lymphoma (Han 2001). The expression of the stem cell marker CD34 is an objective marker of an acute leukemia but there is evidence in people and animals that not all acute leukemias express this antigen (Villiers 2006, Creutig 1995). Because targeted treatment modalities are utilized in people based on the specific type of acute leukemia, it is imperative that we begin to dissect acute leukemias in more detail in our veterinary patients.

ACUTE LEUKEMIA DATA TO BE DISCUSSED

We have recently retrospectively examined the case records from 50 dogs seen at CSU with either 1. Greater than 1,000/uL circulating CD34+/MHCII- cells and/or 2. Greater than 20% circulating blast cells in the blood or bone marrow with concurrent peripheral bi or pancytopenias. This analysis allows the comparison of clinical presentation and treatment outcome between molecularly and clinically diagnosed acute leukemias. Additionally, in order to better define the patient characteristics, potential lineage assignment and outcome of CD34+ acute leukemias we have collected data on a total of 192 flow-defined cases of CD34+/MHCII- canine leukemia with clinical follow-up on 59 of these dogs. Three major phenotypes including myeloid, T lymphoid and undetermined emerged which differed in presenting cell counts, breed distribution and age range. Importantly, there were no B cell cases in this series of CD34+/MHCII- leukemias based on flow and PCR for antigen receptor rearrangements (PARR). Median overall survival was short for all subtypes consistent with the widely held clinical impression of this disease. Interestingly, the only dogs surviving beyond 60 days (n=5) were in the myeloid category.

CHRONIC LYMPHOCYTIC LEUKEMIA

Chronic leukemias are typically far easier to classify based on morphology as they demonstrate recognizable differentiation. While both myeloid and lymphoid forms are seen in dogs and cats, chronic lymphocytic leukemia (CLL) is far more common than chronic myeloid leukemias (CML) (Adam 2009, Workman 2003). Published data suggests that T cell CLL is more common than B-CLL in dogs and prognostic features have been described between and within phenotypes (Tasca 2009, Williams 2008, Comazzi 2011). Relatively little has been published describing feline CLL but it has been suggested that T cell CLL predominates, it occurs more commonly in older cats and may be amenable to treatment with oral chlorambucil and prednisolone (Campbell 2013, Workman 2003). CLL in dogs can be associated with peripheral and visceral lymphadenopathy and splenomegaly similar to the spectrum of disease reported in human B cell chronic lymphocytic lymphoma/small lymphocytic lymphoma (SLL) (Williams 2008, Batata 1992). It is unclear how, if at all, CLL in cats is related to small cell lymphoma of the feline GI tract.

CLL DATA TO BE DISCUSSED

Recently, 491 canine cases that met the inclusion criteria for B-CLL of having >5,000 lymphocytes/uL in the peripheral blood with >60% of the lymphocytes being small CD21+ lymphocytes were examined. As expected, peripheral cytopenias were quite rare yet significant proportions of the dogs had peripheral lymphadenopathy, splenomegaly and visceral lymphadenopathy supporting the use of the terminology B cell chronic lymphocytic leukemia/small cell lymphoma. Certain breeds, particularly small breed dogs, had increased odds of B-CLL whereas certain breeds such as the Golden retriever were very rarely represented. Phenotypic characteristics of the B cells such as CD25 and MHCII expression appeared distinctive in English Bulldogs suggesting that there may be more than one variant of canine B-CLL.

Cell phenotype, clinical follow-up and outcome data has been obtained from 98 cats with a persistent (present in 2 or more CBCs) lymphocytosis of >6,000 cells/uL. A CD4+ T cell phenotype was the most frequent phenotype seen with far fewer CD8+ cases and CD5+CD4/8- cases reported. An isolated increase in B cells was uncommon and none of the samples tested by PARR were B cell positive. Unlike dogs, cats were more likely to have elevations in more than one lymphocyte subset with PARR data suggesting that in some cases this was a reactive process. The prolonged survival times of the CD4+ and CD8+ T cell cases (median=752 days and 1217 days) is comparable to that reported in a small case series of feline CLL (Campbell 2013) and also with similarly treated feline small cell GI lymphoma cases (Stein 2010, Kiselow 2008). This, coupled with the large percentage of cats in our study that had associated gastrointestinal signs, opens the possibility that we are describing a subset of GI lymphoma cases with significant peripheral lymphocytosis. If corroborated in future studies, this would lend further credence to using phenotype to capture feline CLL/SLL as spectrums of the same disease.

REFERENCES

Adam, et al. Vet Comp Oncol. 2009;7(3):181
Batata, et al. Cancer 1992, 70(3):625
Campbell, et al. Vet Comp Oncol 2013 11(4):256
Comazzi, et al. J Vet Intern Med 2011 25(1):100
Creutig, et al. Blood. 1995;86(8):3097
Graves, et. al. JAAHA 1997;33(1)
Han, et al. Am J Clin Pathol 2001 127(4):528
Kiselow, et al. J Am Vet Med Assoc 2008;232(3):405
Mylonakis, et al. Aust Vet J. 2008;86(6):22
Novaco, et al. Vet Comp Oncol online 2015
Stein, et al. JAAHA 2010 46(6):413
Tasca, et al. Vet Clin Path 2009 38(1)2
Villiers, et al. Vet Clin Path 2006;35:55
Williams, et al. J Vet Intern Med 2008;22(3):596
Workman, et.al. Vet Clin Small Anim 2003;33:1379

Laboratory Diagnosis and Classification Of Anemia

Carolyn Grimes, DVM, DACVP (Clinical Pathology)
Montréal, QC, Canada

INTRODUCTION

For most species, laboratory hematologic evaluation (i.e., the hemogram) consists of an automated complete blood count (CBC), microhematocrit tube evaluation, refractometric evaluation of plasma protein, and microscopic examination of a blood smear. Additional tests are used to supplement this information in some species (e.g., fibrinogen precipitation test in large animals) or when specific pathologies are suspected (e.g., methemoglobinemia "spot test", saline test, new methylene blue staining). For complete evaluation of the hematopoietic system, clinical findings as well as cytopathologic and histopathologic evaluation of bone marrow and other ancillary tests (e.g., iron panel) should be considered. The goal of this session is to review traditional classification schemes of anemia and to discuss the value of blood smear exam findings and hematology analyzer output.

LABORATORY CLASSIFICATION OF ANEMIA

Two laboratory classification schemes are used to describe anemia in veterinary patients: classification by red blood cell (RBC) indices and classification by bone marrow (BM) responsiveness. Classification by RBC indices uses specific RBC parameters (the Wintrobe indices: mean corpuscular volume (MCV), mean corpuscular hemoglobin concentration (MCHC), and mean cell hemoglobin (MCH)) to describe trends in red cell size and hemoglobin concentration. According to this scheme, an anemia may be described as macrocytic and hypochromic, normocytic and normochromic, etc. Classification by BM responsiveness is based on evidence of BM compensation for anemia, either in the form of reticulocytosis (or aggregate reticulocytosis in the cat), polychromasia, or erythroid hyperplasia in the BM. In most species, a reticulocyte count is considered the easiest, most reliable measure of marrow responsiveness (in the absence of a reticulocyte count, the number of polychromatophils serves as a crude estimate of the BM response). A notable exception is the horse, which releases few to no immature RBCs into circulation; a BM sample must be examined to determine the erythroid response in this species. These two laboratory classification schemes help clinicians refine the list of differential diagnoses in order to narrow the cause of anemia. The two schemes are often used in tandem (e.g., a microcytic, hypochromic, nonregenerative anemia).

BLOOD SMEAR EXAMINATION

Microscopic examination of a blood smear is an essential part of the hemogram as it can provide vital diagnostic information that is not present in the CBC. The morphology, distribution, and quantity of all three cell lines (RBCs, white blood cells (WBCs), and platelets) should be evaluated. However, RBC morphology (not including erythroparasites) is the focus of this section.

A survey done by Newman et al.[1] examined agreement between clinical pathologists and clinicians regarding RBC morphology terms and their clinical value. The results pointed to important shortcomings in how these two groups communicate caused by different understanding of certain RBC morphology terms and, possibly, suboptimal reporting practices. Most of the RBC morphologic terms used by clinical pathologists are presented in **Table 1**. Images of the changes and comprehensive explanations of their clinical significance are available in various hematology texts[2]. As a note, the term **poikilocyte** refers to any abnormally shaped RBC and includes those with specific morphologies (e.g., acanthocytes, spherocytes) and abnormally shaped RBCs that do not fit into a specific category.

Although some morphologic changes are of limited clinical utility, some provide very important diagnostic information. Selected profiles of morphologic changes and their clinical significance are presented here:

Table 1. RBC morphology terms

Acanthocyte	Leptocyte*
Basophilic stippling	Macrocyte/microcyte*
Cabot rings*	Nucleated RBC (metarubricyte, rubricyte)
Codocyte (target cell)	Ovalocyte/Elliptocyte*
Crystallized hemoglobin*	Polychromatophil
Dacryocyte*	Poikilocyte
Drepanocyte*	Reticulocyte (NMB stain only)
Eccentrocyte	Schizocyte (schistocyte)
Echinocyte	Siderotic inclusions*
Erythroid loop*	Spherocyte
Ghost cell	Stomatocyte
Heinz body	
Howell-Jolly body	**Descriptors for RBC populations**
Hypochromic cell	Anisocytosis
Keratocyte	Agglutination
Knizocyte*	Rouleaux

*indicates infrequently observed cell or infrequently used term

- **Polychromasia, anisocytosis, rubricytosis, Howell-Jolly bodies, stomatocytes, basophilic stippling:** Often seen in animals with regenerative anemia. However, species differences exist (e.g., basophilic stippling is frequently seen in ruminants, but not small animals; horses typically do not have circulating evidence of regeneration; stomatocytes are most frequently seen in dogs).
- **Schizocytes, acanthocytes**: Associated with erythrocyte fragmentation (mechanical damage) secondary to microangiopathy, as may be seen secondary to neoplasia (in particular, hemangiosarcoma), disseminated intravascular coagulopathy, glomerulonephritis, and vasculitis. Acanthocytes may also be seen in dogs with liver disease or lipid metabolopathy.
- **Spherocytes, agglutination**: May be seen in animals with immune-mediated hemolytic anemia (IMHA).

- **Heinz bodies, eccentrocytes (and, rarely, pykocytes):** Evidence of oxidative damage to RBCs.
- **Hypochromasia with associated poikilocytosis** (including microcytes, schizocytes, keratocytes, acanthocytes, codocytes): Often seen in animals with iron deficiency.
- **Ghost cells**: If not lipemic "smudge" cells (seen in lipemic samples, resembling and sometimes described as ghost cells), an indicator of intravascular hemolysis (sometimes seen in IMHA or Heinz body hemolytic anemia).
- **Ovalocytosis with macrocytosis:** In dogs with persistent, nonregenerative anemia, a strong indicator of myelofibrosis[3].
- **Reticulocytosis in the absence of anemia**: Although traditionally thought to indicate recovery from anemia or non-anemic chronic hypoxia (e.g., cardiovascular disease, pulmonary disease), the prevalence of reticulocytosis in the absence of anemia appears to be increasing in dogs[4]. Several risk factors (or their cumulative effects) may be involved, including osteoarthritis and their various associated treatments, e.g., anti-inflammatories, nutraceuticals.
- **Inappropriate rubricytosis** (increased nucleated RBCs in the absence of a regenerative anemia): Seen in heat stroke, septicemia, and lead poisoning. May also be an indicator of dyserythropoiesis.
- **Basophilic stippling and rubricytosis in the absence of regenerative anemia**: Seen in lead poisoning in some species.

HEMATOLOGY ANALYZERS

Introduction: Impedance versus flow cytometry analyzers

Most of the automated hematology analyzers commonly used in veterinary medicine are either impedance counters, flow cytometers, or a combination of the two. In impedance counters, particles (cells) in a blood sample are suspended in an electrolyte solution and passed through a microchannel that connects two chambers, one containing a positive electrode, and one containing a negative electrode. Passage of a particle through the channel invokes a brief change in the electrical impedance between the electrodes, which is proportional to the size of the particle. Particles are then classified as cells according to their apparent size. When measuring RBCs, all cells pass through the channel and are classified as either a RBC or a platelet, based on its size (WBCs are counted, but the number is usually too small to significantly affect the RBC count). When measuring WBCs, the cells are lysed and the nuclei are measured. Free hemoglobin (from the lysed RBCs and the small amount that was already present in the plasma) is measured by spectrophotometry. Measured values from impedance counters include RBC count ([RBC]), MCV, platelet count ([plt]), hemoglobin concentration ([Hgb]), WBC count ([WBC]), and a 3- to 5- part WBC differential. From the measured values, additional values, notably, MCHC and hematocrit (HCT), are calculated.

Flow cytometers depend on light scatter principles to evaluate cells. Cells are differentiated based on their size, internal structure, and granularity, as these properties determine the different kinds of scatter that will be produced when laser light hits a cell. In addition to the parameters measured by impedance counters, flow cytometers also measure cellular hemoglobin concentration mean (CHCM). As in impedance counters, MCHC and HCT are calculated. Although MCHC (provided by both impedance and flow methods) and CHCM (flow only) are similar, CHCM is a more reliable indicator of a RBC's hemoglobin concentration. Since MCHC is calculated using [Hgb], [RBC], and MCV values, erroneous measurement of any of these values will alter MCHC. Discrepancies between MCHC and CHCM can thus potentially alert the clinician to an error in [Hgb], [RBC], or MCV. HCT on automated analyzers is calculated from MCV and [RBC]. A discrepancy between the calculated HCT and the hematocrit measured directly using a capillary tube (i.e., microhematocrit) often indicates an erroneous MCV or [RBC].

Quality assurance

The importance of a comprehensive quality assurance (QA) program for all laboratory equipment and procedures cannot be understated. Unreliable results or a lack of awareness of the limitations of the methods being used can lead to errors in interpretation and compromised medical decision making. Individual veterinary labs vary in terms of their investment in and commitment to QA. Resources are available (e.g., the American Association of Veterinary Laboratory Diagnosticians (AAVLD)[5], the American Society for Veterinary Clinical Pathology (ASVCP)[6]) to help individual laboratories launch and maintain a QA program. That said, even when machines are functioning optimally, the clinician interpreting the data must be aware of machine limitations and potential artifacts.

Manufacturer-provided manuals should be consulted for particularities of individual machines. However, the technology common to impedance and flow cytometry analyzers can cause errors in these machines. Limitations of selected RBC-related parameters are presented here:

- **Prolonged storage** prior to analysis (usually when samples are transported to external labs) may cause *in vitro* RBC swelling, resulting in spuriously increased MCV and spuriously decreased MCHC and CHCM. The effect is more likely when samples are stored at room temperature.
- As [Hgb] is measured spectrophotometrically, any substance that alters the turbidity of the solution may falsely increase [Hgb]. **Lipemia, Heinz bodies, icterus, excess numbers of nuclei (e.g., severe leukocytosis), cold agglutinins, and other particles** may cause this. If [Hgb] is increased, MCHC will also be increased, as [Hgb] is used to calculate this value (CHCM would not be affected).

- **Hemolysis** (either *in vitro* or intravascular *in vivo*) causes an increase in the amount of free hemoglobin present in blood. While [Hgb] is tehcnically accurate in such cases, MCHC will be spuriously increased, as MCHC is calculated from [Hgb], [RBC], and MCV (basically, the machine "assumes" that all of the hemoglobin present was liberated from intact cells that were lysed for analysis). Again, CHCM is not affected. The discrepancy between the two values may help the clinician identify hemolysis (if other causes of spuriously increased [Hgb], e.g., lipemia, are excluded).
- **Blood clots** remove RBCs (and other cells and protein) from suspension, decreasing their counts. Clots may also interfere with optimal functioning of some analyzers.
- **Very high [WBC]** may spuriously increase [RBC]. RBCs and WBCs are usually counted together during enumeration of RBCs. In most circumstances, relative WBC numbers are too low to cause a significant change in [RBC].
- **Large platelets**, in particular in patients with relatively small RBCs (cats, goats, sheep, animals with iron deficiency), may be counted as RBCs and vice versa. This may alter both [RBC] and [plt]. Usually, the change to [RBC] is insignificant, but may be noticeable when high numbers of large platelets are present.
- **Microcytic RBCs** (seen in some species, certain dog breeds, animals with iron deficiency and dogs with portovascular anomalies) may be too small to be recognized as RBCs and thus erroneously counted as platelets, affecting both [RBC] and [plt].
- **Agglutinated RBCs** can spuriously increase MCV and simultaneously decrease [RBC] when an aggregate of agglutinated RBCs is counted as a single, large, RBC. Very large aggregates are usually not recognized (and thus not counted) as RBCs.
- **Severe, chronic hypoosmolality or hyperosmolality** (e.g., secondary to severe, persistent electrolyte abnormalities) alters the osmolality of RBCs which, when subject to analyzer diluents, may swell or shrink, thereby increasing or decreasing their MCV.
- **Anything that spuriously affects [RBC] or MCV** will in turn spuriously affect MCHC and HCT. **Anything that spuriously affects [Hgb]** will also affect MCHC.

Other information provided by analyzers

RBC histograms and cytograms provide a visual representation of RBC data generated by the analyzer. These can be useful in characterizing RBC populations and subpopulations, particularly in cases where the numeric data does not "tell the whole story" or is potentially erroneous. RBC morphology flags and reticulocyte indices (generated by some flow cytometers) also provide additional diagnostic information. Most reference laboratories do not routinely include histograms, cytograms, morphology flagging results, and reticulocyte indices in the hematology report. However, they should be readily accessible to clinicians using in-house analyzers and can be requested from reference laboratories. RBC data is the focus of this session. However, resources are available[7,8,9] regarding histograms, cytograms, and additional indices that characterize the WBC and platelet portions of the hemogram.

The RBC volume histogram shows the distribution of RBCs according to their size. Evaluation of the histogram is helpful in several ways: (1) The visual appearance of the histogram demonstrates the extent of variation in size amongst the RBC population (the red cell distribution width (RDW) provides corollary information, as it is the coefficient of variation of RBC volume). (2) When large platelets and/or small RBCs are misclassified, overlap of the volumes of the RBC and platelet populations can often be identified. (3) Two distinct RBC populations that have different volumes (e.g., a normocytic population and a macrocytic population) can be visualized in some cases when one or more pathologic subpopulations are not evident in the numerical data (i.e., MCV).

The RBC scatter cytogram (produced by flow cytometers) is a scatter plot of RBCs (or particles assumed to be RBCs) based on each RBC's volume (y-axis, low-angle scatter) and hemoglobin concentration (x-axis, high angle scatter). Similar to the histogram, the scatter cytogram allows for identification of RBC subpopulations that are not reflected in the numerical data that is routinely reported. RBCs that have subnormal or supranormal volume, hemoglobin concentration, or both, can be readily identified. In fact, the percentage of cells that are abnormal (e.g., percentage of RBCs with subnormal volume, percentage of RBCs with supranormal hemoglobin concentration) is also reported. In some cases, percentages of abnormal RBCs may be significant enough to generate a morphology flag (e.g., microcytosis ++, hyperchromasia +++). The default thresholds used to define normal values are likely not optimal for all species[10]; however, they may provide a more sensitive means of detecting RBC subpopulations (e.g., microcytic, hypochromic cells in dogs with iron deficiency anemia (IDA)) than routine hematologic parameters[10,11].

Some flow cytometers are capable of measuring reticulocytes and various reticulocyte indices (e.g., reticulocyte MCV (MCVr), reticulocyte hemoglobin concentration (CHCMr), reticulocyte hemoglobin content (CHr)). In human medicine, reticulocyte indices are used as markers of iron availability and engraftment and for monitoring the therapeutic response of anemia[12]. In veterinary medicine, a few studies have shown that evaluation of reticulocyte indices may be useful in dogs with iron deficiency anemia (IDA), anemia of inflammation, and breed-associated microcytosis[11,13,14] and that these parameters may be more sensitive than traditional RBC parameters, in particular for diagnosing IDA[11,13]. Finally, Foy et al.[15] described reticulocyte parameters in canine serial blood donors. In this study, dogs which donate blood >5 times annually showed reticulocyte parameters compatible with iron-limited erythropoiesis, putting them at risk for IDA.

The diagnostic potential of histograms, scattergrams, morphology flags, and reticulocyte indices for characterizing RBCs in veterinary patients is not yet known. Clinicians and clinical pathologists alike are encouraged to evaluate this data in both clinical and research contexts, so that normal and abnormal patterns for different species and different pathologies may be better characterized.

REFERENCES

1. Newman AW, et al. Vet Clin Pathol 2014;43(4):487.; 2. Harvey JW. Veterinary Hematology: A Diagnostic Guide and Color Atlas. 2012; Saunders.; 3. Hollinger CH, et al. Vet Clin Pathol 2011;40(4);595.; 4. Pattullo KM, et al. Vet Clin Pathol 2015; 44(1):26.; 5. The American Association of Veterinary Laboratory Diagnosticians (AAVLD), www.aavld.org; 6. The American Society of Veterinary Clinical Pathology (ASVCP), http://www.asvcp.org/pubs/qas/index.cfm; 7. Gibbs G. Advia Hematology Systems, 2nd ed. 2014; Lulu Press, Inc.; 8. Weiss DJ, et al. Schalm's Veterinary Hematology, 6th ed. 2010.; 9. Wilkerson MJ. Vet Clin North Am Small Anim Pract 2012;42(1):53.; 10. Grimes CN, et al. Can Vet J 2014;55(12):1173.; 11. Schaefer DM, et al. Vet Clin Pathol 2015;44(1):109; 12. Piva E, et al. Clin Lab Med 2015;35(1):133.; 13. Fry MM, et al. Vet Clin Pathol 2006;35(2):172.; 14. Meléndez-Lazo A, et al. J Comp Pathol 2015;152(4):304.; 15. Foy DS, et al. J Vet Int Med 2015;29(5):1376.

Nonregenerative Anemia: Mechanisms of Decreased or Ineffective Erythropoiesis

Carolyn Grimes, DVM, DACVP (Clinical Pathology)
Montréal, QC, Canada

INTRODUCTION

Anemia without an appropriate compensatory hematopoietic response is termed nonregenerative. Nonregenerative (NR) anemia is common in veterinary species and occurs in association with many different disease processes and organ system pathologies. The goal of this session is to review the different factors that contribute to the development of NR anemia in patients with various pathologies, as familiarity with these processes is important when developing diagnostic and treatment plans and may encourage further research investigations in this field. Known mechanisms of NR anemia in veterinary patients, as well as selected mechanisms of NR anemia in human patients that may have veterinary applications, are covered.

NORMAL ERYTHROPOEISIS

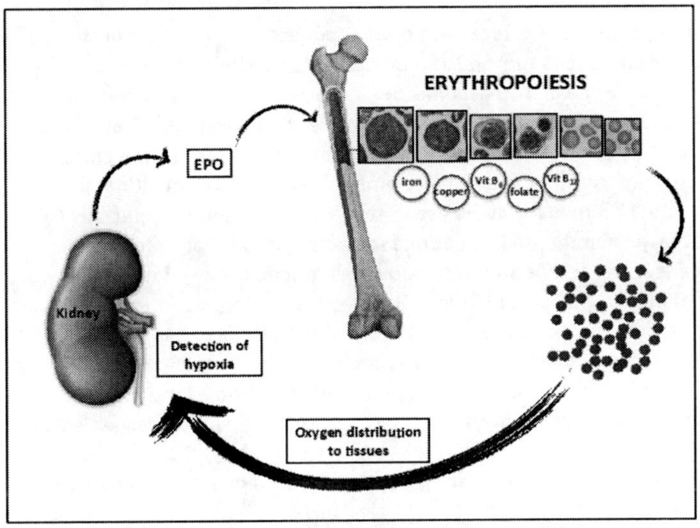

Figure 1. Basic processes of erythropoiesis. EPO: erythropoietin.
Note: Photomicrographs courtesy of Dr. Mike Fry, University of Tennessee

The term erythropoiesis refers to the formation and development of red blood cells (RBCs). It is a complex, highly regulated process that, in healthy adult animals, takes place primarily in the bone marrow. The basic process is illustrated in Figure 1. The most important physiologic trigger for erythropoiesis is hypoxia, which is detected by the kidneys. When hypoxic conditions are present, renal production and secretion of erythropoietin (EPO) is up-regulated, which stimulates the proliferation and differentiation of erythroid precursors. Healthy maturation of RBCs depends, in part, on the availability copper, iron, folate, cobalt, vitamin B_6, and vitamin B_{12}. Also, like other blood cells, the health of the hematopoietic microenvironment and stem cells, as well as the interplay of various growth factors and cytokines plays an important role in RBC development. Pathology that affects any of these elements may compromise erythropoiesis and lead to anemia.

NONREGENERATIVE ANEMIA ASSOCIATED WITH VARIOUS DISEASE PROCESSES

Inflammation

Anemia of inflammation (AI) a common cause of NR anemia in humans and animals. Although many factors contribute to AI, increased hepatic expression of hepcidin triggered by inflammatory cytokines is considered the most important. Circulating hepcidin binds to ferroportin, a membrane protein that normally enables efflux of iron stored from macrophages, enterocytes, and other cells, into the blood. When hepcidin binds ferroportin, ferroportin is internalized and degraded in the cell, leaving intracellular (stored) iron without an exit port. This causes a "functional" iron deficiency; although total body iron content is unchanged, plasma iron concentrations decrease and are unavailable to developing erythroid cells. Inflammatory cytokines may also play a role in dysmetabolism of iron through increased ferritin synthesis, decreased transferrin receptor (TfR) expression, and disruption of transferrin binding to TfR.

Although interfering with iron availability is the principal mechanism in AI, inflammatory cytokines may contribute to anemia in various other ways, such as irreversible inhibition of erythroid progenitors, decreased hemoglobin synthesis, inhibition of EPO gene expression and production, inhibition of EPO-mediated erythroid differentiation, EPO resistance, and decreased RBC survival time.

Immune-mediated disease

Although immune-mediated hemolytic anemia (IMHA) typically causes a regenerative anemia, NR IMHA has been described in dogs, cats, and cattle. An inadequate regenerative response in patients with IMHA may occur because of underlying disease that inhibits erythropoiesis (e.g., concurrent renal failure) or because of targeted destruction of developing erythroid cells.

Hemophagocytic syndrome is a condition in which excessive stimulation and proliferation of non-neoplastic macrophages results in unrestrained phagocytosis of RBCs and other blood cells. It has been reported in several veterinary species, including dogs, cats, cattle, chickens, and a macaque[1].

Bovine neonatal pancytopenia (BNP), also called "bleeding calf syndrome", causes, among other problems, a severe hemorrhagic anemia in calves that have ingested colostrum containing Bovine Viral Diarrhea vaccine-induced maternal alloantibodies. Although principally recognized in Europe, similar cases have been reported in Canada[2] and Japan[3]. BNP is characterized by pancytopenia and bone marrow aplasia in neonatal calves that often present with signs related to hemorrhagic diathesis. A heritable trait of the dam that favors development of the alloantibodies may be implicated[4]. A BNP-like syndrome has also been described in dromedary camels[5].

Neoplasia

Anemia is common in human and veterinary oncology patients, although the prevalence and severity of anemia depends on the type and extent of cancer. Cancer-related anemia may be regenerative or nonregenerative, is often multifactorial, and may include both direct and indirect effects of tumors and treatments.

Inflammation is generally considered the most common cause of anemia in cancer patients. Inflammation may play a role in the development of some cancers and cancer frequently causes secondary inflammation.

Myelophthisis (replacement of hematopoietic tissue), myelonecrosis, and myelofibrosis can occur secondary to neoplastic diseases. Both primary and secondary tumors may proliferate in the marrow, causing myelophthisis, destruction of hematopoietic cells, and disruption of the marrow microenvironment through dysregulation of growth factors and cytokines. Myelonecrosis may be caused by direct toxic injury and/or local ischemia. Myelofibrosis may occur secondary to necrosis caused by tumors or due to induction of cytokines that induce fibroblast proliferation in the marrow.

Neoplastic cells may also produce other hormones and factors that interfere with erythropoiesis. Production of estrogen or estrogen-related compounds, as reported in Sertoli cell tumors, granulosa cell tumors, and interstitial cell tumors, may cause suppression of hematopoietic tissue in humans and animals. Also, pure red cell aplasia has been reported in a variety of cancer types in humans, notably thymoma, and excess production of hepcidin by hepatic adenomas in humans has also been described[6].

Nutritional deficiencies in cancer patients may also inhibit erythropoiesis. Potential contributing causes include chronic tumor-associated hemorrhage (which may cause iron deficiency), inadequate dietary intake, and impaired gastrointestinal (GI) absorption of erythropoietic nutrients. Decreased serum cobalamin and folate may be associated with lymphoma (especially GI lymphoma) in cats[7] and hypocobalaminemia in dogs with multicentric lymphoma has been associated with a poor outcome[8].

The kidneys may be damaged as a result of cancer or cancer treatment and impaired renal function may contribute to anemia. Primary and metastatic renal tumors can significantly damage renal parenchyma. Paraneoplastic production of hormones or proteins may lead to biochemical (e.g., hypercalcemia) or hemodynamic (e.g., hyperproteinemia) disturbances that may also contribute to renal damage. Various malignancies in humans have been associated with lesions such as membranous nephropathy, membranoproliferative glomerulonephritis, glomerulosclerosis, renal vasculitis, and hemolytic-uremic syndrome, and have been rarely reported in veterinary patients[9]. Some cancer therapies may also have renal-damaging side effects. Specific mechanisms of anemia associated with renal disease are discussed later.

Chemotherapeutic agents, radiation therapy, and even erythropoiesis-stimulating agents may also cause or exacerbate anemia in some cases.

Infectious disease

Anemia secondary to infection may be regenerative or nonregenerative and often occurs because of associated inflammation, organ damage, blood loss, or a misdirected immune response. However, some infectious agents cause anemia more directly, for example, by infecting erythroid cells or producing erythropoiesis-inhibiting substances. Selected infectious agents that demonstrate particular anemogenic mechanisms are presented here.

Viral infections can have various hematologic effects, including nonselective infection of hematopoietic cells (feline leukemia virus (FeLV), various parvoviruses), damage to the hematopoietic microenvironment (e.g., feline immunodeficiency virus), selective erythrocytopathic effects (e.g., FeLV-subgroup C, chicken anemia virus, and simian, human, and possibly canine parvoviruses), GI hemorrhage and inflammation (e.g., canine parvovirus), dysregulation of inflammatory cytokines (equine infectious anemia, various human viruses), and induction of hemophagocytic syndrome (several human viruses and possibly a simian retrovirus[10]).

Hematoprotozoal infections often cause hemolytic, regenerative anemia. However, a few exceptions exist. *Theileira parva* (the causative agent of East Coast Fever in cattle in Africa) can cause destruction of hematopoietic tissue, resulting in severe pancytopenia, including NR anemia. *Cytauxzoon felis* infection in cats is also often associated with NR anemia. The lack of regeneration may be due to rapid progression of disease (preregenerative anemia) and/or cytokine-induced marrow suppression. *Plasmodium* spp. also demonstrate unique (and interesting!) anemogenic mechanisms related to inhibition of erythropoiesis and promotion of dyserythropoiesis and erythrophagocytosis. One study suggested that *Plasmodium falciparum* synthesizes a parasite-specific transferrin receptor which, when localized to the erythrocyte surface, enables extraction of iron from the host[11], aggravating the already iron-limited state of erythropoiesis.

Certain fungal organisms and mycotoxins may cause or contribute to anemia in various species. *Stachybotrys* spp. infection in sheep, tricothecenes (a mycotoxin produced by *Fusarium* spp. and other fungi) in cattle and cats, and diatcetoxyscirpenol (another *Fusarium*-associated toxin) in pigs, calves, and dogs, are among those reported. Bone marrow damage secondary to disseminated fungal disease and granulomatous inflammation is also possible.

Toxicity

Some therapeutic or environmental agents may damage erythroid precursor cells or the hematopoietic environment, causing a NR anemia. The presence and severity of anemia may be idiosyncratic or dose-dependent and patient factors, such as species or breed characteristics, genetic susceptibility, and pre-existing conditions, may influence the development of anemia. A list of therapeutic and environmental agents that have been associated with the development of NR anemia in veterinary species is provided elsewhere[12].

Congenital disorders

Congenital dyserythropoiesis has been described in English springer spaniels and was associated with polymyopathy, megaesophagus, and cardiac disease[13]. Congenital dyserythropoiesis associated with alopecia, dyskeratosis, and, often, death occurs in some polled Hereford cattle[14]. Inherited disorders causing defective cobalamin absorption, hypocobalaminemia, and dyserythropoiesis have been reported in dogs[15] and a cat[16].

Renal disease

Anemia is very common in veterinary patients with chronic renal failure (CRF). Although both regenerative and nonregenerative mechanisms may contribute, the anemia of CRF is most commonly associated with decreased erythropoiesis. Cats and dogs with anemia and CRF typically have insufficient EPO production. This is generally thought to be a result of damage to EPO-producing renal peritubular cells, however some authors challenge this notion, suggesting that decreased EPO production may be part of a physiologic response to other processes involved in CRF (e.g., dysregulation of the renin-angiotensin-aldosterone system (RAAS))[17]. Other mechanisms that may contribute to NR anemia in CRF patients include neocytolysis (selective lysis of young RBCs) secondary to EPO suppression; an inflammatory state associated with CRF or uremia; inhibition of EPO, erythroid development, and heme synthesis by uremic toxins; alterations in calcium phosphorous homeostasis (e.g., increased parathyroid hormone (PTH)); nutritional deficiencies; chronic hemorrhage; and certain treatments (e.g., the use of recombinant EPO, complications of dialysis).

Heart failure and cardio-renal syndrome

In humans, anemia is a common complication of heart failure (HF). The pathophysiology of anemia in these patients is not entirely clear, but proposed mechanisms are not unique to human heart disease, and include inflammation, GI dysfunction, RAAS activation, as well as certain RAAS-targeted therapies. Concurrent renal disease (often referred to as cardio-renal syndrome in the human medical literature) is common in human and veterinary HF patients and may also contribute to anemia. Chronic renal hypoxia likely plays a central role in progression of renal disease in HF patients.

Endocrine disease

NR anemia is a common finding in many patients with various endocrine disorders. As thyroid hormones stimulate erythropoiesis both directly and indirectly, hypothyroid dogs often develop NR anemia. Decreased stimulation of erythropoiesis similarly occurs in dogs with hypoadrenocorticism, due to deficiencies in cortisol and androgens, although anemia may be masked in these patients because of dehydration. Anemia is a relatively common finding in diabetic dogs and cats. Although not extensively studied in animals, proposed mechanisms for anemia in diabetic humans include include inflammation (associated with obesity, microangiopathy, and other co-morbidities), GI disease, renal injury, autonomic neuropathy (dysfunction of the anemia-sensing mechanism), decreased stabilization of hypoxia-inducible factor-1a, and decreased iron-binding capacity.

Hepatic disease

As with many types of non-hepatic disease, NR anemia may develop secondary to inflammatory, malnutritive, and/or immune-mediated processes in veterinary patients with liver disease. Portosystemic vascular anomalies (PSVA) in dogs and cats are frequently associated with microcytic, normochromic, NR anemia. Studies looking at iron parameters in dogs with PSVA suggest that the pathophysiology may be related to iron sequestration (a functional iron deficiency, similar to that which occurs in inflammation). However, it is not clear how portal blood flow contributes to this process. Chronic GI hemorrhage secondary to portal hypertension can also cause or contribute to anemia via blood loss and iron deficiency.

Nutritional deficiency and gastrointestinal disease

Iron deficiency in veterinary species is most often caused by chronic blood loss, with GI blood loss being the most heavily implicated. Neoplastic and non-neoplastic GI ulceration and intestinal parasites (e.g., trichuriasis) are common culprits. Copper deficiency can lead to secondary iron deficiency (as copper is required for normal iron transport) and copper deficiency-associated anemia has been reported in dogs, pigs, ruminants, and humans. A relatively high prevalence of cobalamin deficiency in dogs[18] and cats[19] with GI disease has also been reported. Although hypocobalaminemia is associated with anemia in human patients, the relationship between cobalamin deficiency secondary to GI disease and anemia in veterinary patients is not clear. Finally, a strong link between *Helicobacter pylori* infection and iron deficiency anemia has been established in humans and may have implications for veterinary species. However, the pathogenic role of *Helicobacter* spp. infection in gastric disease in veterinary patients is not entirely clear.

PHYSIOLOGIC AND ENVIRONMENTAL FACTORS

Healthy humans develop increased EPO concentrations as they age. Absence of this rise may play a role in "unexplained anemia" in elderly patients. The physiologic erythrokinetic changes of geriatric animals are not well characterized.

Anemia associated with pregnancy in humans is highly prevalent and may be caused by nutritional deficiencies, infectious agents, hormonal imbalances, and other problems. Pregnancy-associated anemia has not been as extensively studied in veterinary species. However, a physiologic decrease in hematocrit during late pregnancy and early lactation is seen in some species (e.g., dogs, cats, pigs) and may be significant enough to cause a mild NR anemia.

Intense exercise may also be anemogenic. Although some mechanisms of anemia associated with strenuous exercise can cause regenerative anemia (e.g., hematuria, GI hemorrhage, hemolysis), iron loss and inflammation can result in NR anemia. Strenuous exercise in rats induced inflammation that was associated with increased expression of inflammatory cytokines, increased hepcidin expression, and anemia[20].

Neocytolysis may occur when EPO secretion is inhibited, as occurs in individuals descending from high altitude, newborns emerging from the hypoxic intrauterine environment, and astronauts during space travel (central pooling of blood in microgravity is thought to increase the body's "central" hematocrit, causing suppression of EPO).

REFERENCES

1. Controneo TM, et al. Vet Pathol 2011;48(6):1138; 2. Gosselin VB, et al. Can Vet J 2011;52(7):788.; 3. Fukunaka M, et al. J Vet Med Sci 2010;72(12):1655.; 4. Benedictus L, et al. Vet Res 2014;45(6):129.; 5. Wernery U. Vet Rec 2011;169(8):214.; 6. Yu A, et al. Faseb Journal 2008;22(1191):12.; 7. Simpson KW, et al. J Vet Intern Med 2001;15(1):26.; 8. Cook AK, et al. J Am Vet Med Assoc 2009;235(12):1437.; 9. Page RL, et al. J Am Vet Med Assoc 1990;196(4):620.; 10. O'Sullivan MG, et al. J Clin Invest 1994;93(4):1571.; 11. Rodriguez MH, et al. Nature 1986;325(6102):374.; 12. Grimes CN, et al. Vet Pathol 2015;52(2):298.; 13. Holland CT, et al. J Vet Intern Med 1991;5(3):151.; 14. Steffen DJ, et al. J Hered 1993;84(4):263.; 15. Fyfe JC, et al. Pediatr Res 1991;29(1):24.; 16. Vaden SL, et al. J Am Vet Med Assoc 1992;200(8):1101.; 17. Dunn A, et al. Am J Med Sci 2007;334(1):65.; 18. Berghoff N, et al. Am J Vet Res 2013;74():84.; 19. Simpson KW, et al. J Vet Intern Med 2001;15():26.; 20. Liu YQ, et al. Int J Sport Nutr Exerc Metab 2011;21(1):19.

Hereditary Erythrocyte Defects Causing Anemia in Dogs and Cats

Urs Giger, DACVIM-SA, DECVCP, Osheiza Abdulmalik, Polly Fourman, Guldal Inal Gultekin,
Sarah Lehman, Keijiro Mizukami, Karthik Raj, Claudia Tritschler
Philadelphia, PA

Anemia is one of the most common clinical signs and abnormal laboratory test results in companion animals. Although acquired conditions such as infections, immune disorders, intoxication, blood loss, and chronic organ failure represent the main causes of anemia, hereditary erythrocyte defects leading to anemia are becoming increasingly important in clinical practice. Several hereditary erythrocyte defects have been clinically reported in companion animals, and much new information, including the molecular basis of some disorders, has emerged. Frequently they are only considered after treatments for immune and infectious diseases have failed or the anemia is recurring, despite the fact that in some breeds these conditions are common and well recognized.

Canine and Feline Erythrocytes

Although the major structural features and functions of erythrocytes are similar among mammals, some noticeable differences exist between canine and feline erythrocytes. Feline erythrocytes are smaller than canine erythrocytes. Normal variations observed in many Akitas and Miniature Poodles are erythrocyte microcytosis and macrocytosis, respectively. The normal life spans of erythrocytes in dogs and humans are similar (100-120 days), but erythrocyte life span is only 70-75 days in cats.

Devoid of a nucleus and mitochondria, erythrocytes have a limited and specialized metabolism that enables them to survive in circulation and to adequately transport oxygen. Energy is generated almost exclusively through anaerobic glycolysis, also known as the Embden-Meyerhof pathway. The hexose monophosphate shunt reduces pyridine nucleotides and glutathione, which are necessary for degradation of oxidants, thereby preventing membrane damage and hemoglobin denaturation. In addition, the methemoglobin or cytochrome-b5 reductase system reduces heme iron from the ferric (Fe^{3+}) to the ferrous (Fe^{2+}) form. The Rapoport-Luebering pathway is responsible for the synthesis of 2,3-diphosphoglycerate (DPG), which influences the oxygen affinity of canine - but not feline - hemoglobin. The erythrocytic DPG concentration is similar in dogs and humans but very low in cats.

Dogs and cats apparently have embryonic, but no fetal hemoglobin. With the exception of some Japanese breeds, which have two hemoglobins (HbA and HbB), only one adult hemoglobin has been found in dogs. However, further studies are needed to characterize canine hemoglobin. In cats, one α-globin and at least six different β-globins have been found. With each cat having one to four different β-globins, many hemoglobin patterns are recognized.

The erythrocyte membrane consists of a lipid bilayer, affixed to a membrane skeleton, which determines cell shape and deformability. Various transmembrane glycoproteins function as receptors or transporters. Canine and feline erythrocytes lose their Na^+,K^+-ATPase during late maturation in the bone marrow, owing to proteolysis (with the exception of erythrocytes from some Japanese dogs). Therefore, erythrocytes' high sodium and low potassium concentrations are similar to those of serum electrolytes. Consequently, hyperkalemia generally does not occur after intravascular hemolysis or storage of unseparated clotted blood unless stress reticulocytes or high-potassium erythrocytes from Japanese dogs are lysed. In fact, Akitas' erythrocytes are "leaky" in vitro, and pseudohyperkalemia has been observed in serum that has not been separated immediately from its clot. Finally, canine erythrocytes have been noted to be uniquely fragile under alkaline conditions, compared with erythrocytes from cats and other species, presumably because of a facilitated calcium entry under these conditions. This pH sensitivity may explain the tendency of canine erythrocytes to lyse in uncapped tubes in the laboratory.

Classification of Erythrocyte Defects

Hereditary erythrocyte defects form a large and clinically heterogeneous group of diseases. Each erythrocyte disorder is observed only rarely, although some enzyme deficiencies occur frequently within certain breeds. The mode of inheritance is autosomal recessive, with the exception of feline porphyria, which can be inherited as a dominant or recessive trait. While the degree of characterization is varied, many seem to be homologous to hereditary diseases in humans. These disorders have been classified into four groups: (1) heme defects and hemoglobinopathies, (2) cytosolic enzyme deficiencies, (3) membrane abnormalities, and (4) production and maturation defects; a few specific disorders are discussed in more detail.

Although the signalment, type, and severity of the anemia and the pleiotropic effects observed may provide clues to an inherited erythrocyte defect, a full laboratory evaluation is essential in validating the diagnosis or discovering new inherited disorders. In a few breeds, more than one erythrocyte disorder has been recognized. Routine hematological laboratory tests are used to detect hematological abnormalities and to rule out acquired anemia. An inherited erythrocyte defect should be considered in animals with Coombs-negative hemolytic anemia, a lack of toxin exposure or infection, and adequate kidney as well as liver function. A careful examination of a peripheral blood smear is pivotal in recognizing any poikilocytosis, such as elliptocytosis, spherocytosis and stomatocytosis, although most erythrocyte defects cause no change in cell shape and are historically known as non-spherocytic hemolytic anemias. The degree of reticulocytosis is often marked, even with mild anemia and in the case of canine PK deficiency with osteosclerosis, and is generally proportional to the shortened survival of defective erythrocytes. Thus, a bone marrow examination rarely provides new information. The signs of hemolysis may be mild owing to the chronicity, strong compensation, and low-grade hemolysis. Hyperbilirubinuria and -emia are generally noted, but may be mild due to adaptation. Low serum haptoglobin concentrations, hemoglobinemia, or hemoglobinuria, which all indicate intravascular hemolysis, have been reported in PFK deficiency but are unlikely to be associated with other defects. Some defective erythrocytes appear extremely fragile in vitro, resulting in artificial in vitro lysis in blood tubes. Animals with methemoglobinemia show cyanosis (even when oxygen). Furthermore, cats with porphyria exhibit erythrodontia and pigmenturia.

Special laboratory tests for defining the nature of an intrinsic erythrocyte defect can be divided into general screening tests, used to characterize unknown erythrocyte disorders, and specific screening tests for known defects in certain breeds. Both are performed only in specialized laboratories such as the PennGen laboratory at the School of Veterinary Medicine, University of Pennsylvania, Philadelphia (http://www.vet.upenn.edu/penngen which also offers a listing of all available DNA tests worldwide) and several veterinary DNA testing laboratories.

Hemolysis resulting from erythrocyte defects may be well compensated for by marked erythropoiesis, thereby causing no or only minimal clinical signs except during crises and allowing the animal to have a normal life expectancy. Furthermore, affected animals may have adapted well to the chronic anemia. In contrast, some disorders are associated with severe hemolytic crises for which animals may need to receive supportive therapy, including blood-typed and, if previously transfused, crossmatch-compatible transfusions. Cats with increased erythrocytic osmotic fragility or PK deficiency develop marked splenomegaly and may be helped with splenectomy by removing a major site of erythrocyte destruction; however, it appears that PK- and PFK-deficient dogs have not benefited from this procedure. Finally, affected and carrier animals should not be used for breeding in order to prevent the further spread of these disorders. However, in case of a narrow gene pool or due to many desirable traits, an asymptomatic carrier may be bred to a clear animal, as long as all offspring considered to be used for breeding are also screened.

Hemoglobin Defects including Methemoglobinemia, Porphyrias and Iron-Resistant Iron Deficiency Anemia

In contrast to the common occurrence of thalassemia and sickle cell anemia in people, no hemoglobinopathies have been documented in dogs and cats. Isolated cases of **methemoglobinemia** associated with cytochrome-b5 reductase deficiency were found among dogs of various breeds and domestic shorthair cats. Note, hereditary methemoglobinopathies are associated with cyanosis and frequently erythrocytosis rather than anemia. The molecular basis has been characterized in one dog.

Porphyrias are a group of inborn errors resulting from accumulation of porphyrins due to deficient activities of specific enzymes in heme biosynthesis seen thus far only in cats but not dogs. In humans, they are clinically classified as either erythroid with cutaneous involvement or hepatic with acute neurovisceral attacks. All reported porphyric cats had erythrodontia (brown discolored teeth which fluoresced pink), porphyrinuria, and a mild compensated hemolytic disorder, but no evidence of acute life-threatening neurovisceral attacks or cutaneous lesions (with near normal life-expectancies). Urinary uroporphyrin I and coproporphyrin I concentrations were increased in the porphyric cats. However, cats with **acute intermittent porphyria (AIP)** had high urinary 5-aminolevulinic acid and porphobilinogen concentrations, while the cat with **congenital erythroid porphyria (CEP)** had normal or low levels of these porphyrin precursors. Cats with AIP had <50% of normal hydroxymethylbilane synthase (HMBS) activity in erythrocytes and tissues. In contrast, the erythrocyte and hepatic uroporphyrinogen III-synthase (UROS) activity was markedly decreased in the cat with CEP. Sequencing the feline *HMBS* gene revealed different mutations in each of the 8 lines of cats with AIP including duplications, deletions, and missense mutations, making it thus far the most mutated gene in companion animals. Interestingly, 7 cats were heterozygous for the *HMBS* mutations, while one was homozygous for a mild missense mutation. The cat with CEP was homozygous for a missense mutation in *UROS* resulting in drastically reduced enzyme activity and stability. Thus, cats with discolored teeth and normal or mild hemolysis may have either deficient HMBS or UROS activity and represent homologues of human AIP or CEP, respectively. The biochemical and molecular characterization facilitates screening of affected cats which is offered through the PennGen laboratory.

A severely microcytic-hypochromic anemia with low serum iron and unresponsive to oral iron supplementation has been observed in a Cocker spaniel and a few other dogs. This **Iron-Resistant Iron Deficiency Anemia (IRIDA)** was found due to a defect in *TMPRSS6* (matriptase-2) gene which regulates with hepcidin iron absorption and bioavailability.

Membrane Defects including Stomatocytosis and Increased Osmotic Fragility

Elliptocytosis and **spherocytosis** resulting from a deficiency of the protein band 4.1 and spectrin of the cytoskeleton have been characterized at the molecular level in mongrel dogs. Other presumed membrane abnormalities include stomatocytosis in Alaskan malamutes and in Miniature and Middle Schnauzers, spherocytosis in Golden Retrievers with gastritis, nonspherocytic anemia in Beagles, and erythrocytes with increased osmotic fragility in Abyssinian, Somali and non purpose-bred cats.

Increased osmotic fragility of erythrocytes suggests a membrane or ion transport defect. Chondrodysplastic Alaskan malamute dwarfs with **stomatocytosis** were the first dogs described with fragile erythrocytes, but the exact mechanism is still unknown. Miniature and Middle Schnauzers with stomatocytosis had no skeletal abnormalities, and they had only mild anemia based on hemoglobin measures. Interestingly while the macrocytosis is severe, the anemia is only mild.

A marked **osmotic fragility** of erythrocytes associated with intermittent anemia, severe splenomegaly, and hyperglobulinemia has been observed in Abyssinian, Somali and non pupose-bred cats. These erythrocytes are macrocytic, extremely fragile in vitro, and can lyse when stored overnight in the refrigerator. Although the cause has not been identified, a membrane defect is suspected. Affected cats with marked splenomegaly may benefit from anti-inflammatory doses of prednisolone. If anemia is severe and frequently recurring, and if there is massive splenomagly, splenectomy may help;

splenectomized animals are particularly prone to develop sepsis for the first month post-operatively. Osmotic fragility testing is offered through the PennGen laboratory.

Erythroenzymopathies including PK and PFK Deficiency

Deficiencies of the two key regulatory glycolytic enzymes result in distinctly different forms of hemolytic anemia in several breeds of dogs, while only pyruvate kinase (PK) deficiency has been seen in many breeds of cats. Although PK deficiency was first, and best, characterized in the Basenji breed more than 4 decades ago, the typical clinical features and biochemical abnormalities appear identical in other canine breeds. In contrast PK-deficient cats seem to have intermittent anemia which is more similar to phosphofructokinase (PFK) deficiency in dogs. Many animals are presumptively treated for immune-mediated hemolytic anemia for months to years prior to diagnosis and thereby are subjected to unnecessary diagnostics and potentially harmful treatments.

Phosphofructokinase (PFK) Deficiency - Despite its discovery 3 decades ago and the availability of enzymatic and DNA screening tests, this glycolytic enzyme deficiency is still seen in field-trial English Springer spaniels in the United States and Europe, but has also been reported in some bench/show dogs as well as Cocker spaniels, Whippets, Wachtelhunds, and mixed-breed dogs. PFK deficiency is caused by a single missense mutation of the muscle-type PFK that results in a truncation and instability of the PFK enzyme protein in all breeds reported, except for the Wachtelhund which has another PFK mutation.

The disorder is characterized by hemolytic crises and exertional myopathy. S poradic dark pigmenturia resulting from severe hemoglobinuria and hyperbilirubinuria is a key feature and commonly develops after episodes of excessive panting and barking, intensive exercise, and fever or high environmental temperature. Consequently, hyperventilation-induced alkalemia results in intravascular lysis of PFK-deficient erythrocytes. During these crises, affected dogs may become severely anemic and icteric, and show fever, lethargy, and anorexia, which usually resolve within days. Furthermore, affected dogs totally lack PFK activity in muscle; therefore, they have a metabolic myopathy characterized by exercise intolerance, occasional muscle cramps, and mildly increased serum creatine kinase activity. As they cannot run hard and fast for any length, they will perform poorly as field-trial dogs.

Simple mutation-specific tests are offered to accurately diagnose PFK-deficient and carrier dogs. For those breeds without a known PFK mutation an enzyme activity (low) or hemoglobin-oxygen dissociation curve (high affinity) test can be helpful. Situations triggering hemolytic crises should be avoided such as excessive panting, barking, exercise, and heat. When experiencing a crisis, resting will help them, but they may require supportive and transfusion therapy. PFK-deficient dogs may have a normal life span, but have persistent hyperbilirubinuria and reticulocytosis, despite a near normal hematocrit because of the high hemoglobin-oxygen affinity of PFK-deficient erythrocytes.

Pyruvate Kinase (PK) Deficiency - Despite the severity and persistency of the anemia in PK-deficient dogs, the clinical signs, except for pallor, are mild but triggered hemolytic crisis can occur at any age, which may be fatal. The anemia is highly regenerative, with numerous circulating metarubricytes and reticulocyte counts as high as 90%. An unexplained progressive myelofibrosis and osteosclerosis of the bone marrow and generalized hemosiderosis/-chromatosis with associated hepatic failure develop, causing death usually before 6 years of age, though some PK-deficient West Highland White and Cairn Terriers and Beagles have reached 10 years of age. The molecular genetic basis of PK deficiency has been identified in Basenjis, Beagles, Labrador Retrievers, Pugs, and Cairn and West Highland White Terriers, and mutation-specific DNA tests are available for these breeds, but not for others. PK deficiency has also been reported in Miniature Poodles, Eskimo Toys, Dachshunds, and Chihuahuas. It appears probable that the previously described non-spherocytic hemolytic anemia and osteosclerosis in Poodles was caused by a PK deficiency. In these breeds, a cumbersome PK-enzyme test with isozyme characterization is required to define R-PK deficiency due to anomalous M-PK activity in vitro. Carriers do not express the M-PK form and have half the normal PK activity; however, differentiation between carriers and homozygous normal dogs based on enzyme activity can be difficult. Clinical signs are mild, likely due to chronic adaptation to severe anemia and high erythrocytic DPG concentrations facilitating oxygen delivery (low hemoglobin-oxygen affinity). Hepatosplenomegaly may result from severe extravascular hemolysis, extramedullary hematopoiesis and hemosiderosis/-chromatosis. Iron chelation has been proposed but not yet shown to be effective and safe. Splenectomy has not been shown to be effective in dogs. Bone marrow transplantation has been experimentally proven effective for both enzymopathies in dogs, but because of the likely lack of a major histocompatibility complex compatible donor and the need for severe bone marrow suppression, this treatment is not clinically offered.

In cats, PK deficiency causes intermittent anemia with a mild to moderate regenerative response, but cats do not develop osteosclerosis. Affected cats may develop bilirubin calculi in gall bladder, hepatic failure, and mild splenomegaly. Anti-inflammatory doses of prednisolone and in severe cases splenectomy appear to ameliorate the clinical signs of intermittent anemia, with the oldest such cat reaching 11 years of age. Erythrocyte PK activity is severely reduced, and there is no M-type PK expression, thereby simplifying the biochemical diagnosis. In the late 1990s, we documented that PK deficiency is caused by a single splicing defect resulting in a 13 base deletion and a DNA screening test for PK deficiency in Abyssinian, Somali, and several other breeds of cats as well as domestic shorthair cats and is seen on different continents. Testing has been available since then from PennGen and more recently many other laboratories. Any cat with unexplained persistent or recurring anemia following screening for toxic, infectious and immune-mediated causess should be screened for PK deficiency, as this is a much more likely cause of anemia than.

Reduced Erythropoiesis including Hereditary Cobalamin Malabsorption

Whereas the previous defects resulted in shortened erythrocyte survival and regenerative anemias, the production and maturation disorders are reflected not only in a non-regenerative anemia but also in changes in other bone marrow-derived cells. Selective cobalamin (vitamin B_{12}) malabsorption, also known as Imerslund-Gräsbeck syndrome, resulting from the ileal intrinsic cobalamin receptor defect has been reported in several breeds of dogs. Giant Schnauzers and Australian shepherds have mutations in the *AMN* gene and Beagles and Border collies have mutations in *CUBN*. Affected animals fail to thrive, and have varied degrees of cachexia, neurologic signs, leukopenia, thrombocytopenia, anemia, low serum cobalamin levels and a methylmalonic aciduria. They respond completely to parenteral cobalamin administration every 2-4 weeks. A detailed description is given in another presentation.

REFERENCES

Available from authors.

SUPPORT

Supported in part by a grant from the NIH OD 010939.

Blood Typing and Blood Compatibility Testing in Dogs

Urs Giger, DACVIM-SA, DECVIM-CA, DECVCP, Michelle M. Acierno, Catharina C. Euler, Hee Y. Kim, JaeHo Lee, Keijiro Mizukami, Klaudia Polak, Karthik Raj
Philadelphia, PA

Introduction

Veterinary clinicians and clinical pathologists play a key role in providing safe and effective transfusion therapy. Blood typing is clinically important to ensure blood compatibility and therefore is recommended for any dog in need of a transfusion or considered to become a blood donor. Moreover, previously transfused dogs also should be crossmatched. Unless blood typing is performed regularly in practice, blood should be sent to a clinical pathology laboratory for typing. Different viewpoints exist regarding the extent and methods used for compatibility testing, and various techniques for laboratory and point-of-care use have been applied or are being developed. This presentation will present the current knowledge of the canine blood types and their clinical importance, review the different typing and crossmatching techniques for the in-practice and clinical pathology laboratory, and illustrate this with case examples.

Canine Blood Types

Blood types are genetic markers on erythrocyte surfaces that are antigenic and species specific. A set of blood types of two or more alleles makes up a blood group system. Dogs have likely more than a dozen blood group systems mostly known as dog erythrocyte antigens (DEA). However, there is no DEA 2 blood group and some may be rather labeled high frequency or common red blood cell (RBC) antigens (e.g. DEA 4) and some have not yet received a DEA designation (e.g. Dal). Canine erythrocytes are either positive or negative for a blood type (e.g., DEA 4+ or DEA 4-), and these blood types are likely codominantly inherited. The DEA 1 system was thought to be an exception with DEA 1.1 (A1), DEA 1.2 (A2) and potentially DEA 1.3 (A3) being allelic. Thus, a dog could apparently be DEA 1.1+ or DEA 1.1- and DEA 1.1- dogs can be DEA 1.2+ or DEA 1.2-. However, these studies were based upon weak polyclonal antibodies (DEA 1.1 and 1.X) requiring Coombs' reagents. Recent studies with a monoclonal antibody showed that the DEA 1 blood group is a continuum from DEA 1- to weakly to strongly DEA 1+; hence DEA 1.2 typing is no longer offered. The degree of DEA 1 expression is constant and DEA 1+ appears to be dominantly inherited. A recent survey in North America indicates that most dogs are either DEA 1- or strongly DEA 1+ with fewer dogs being weakly to moderately DEA 1+. The biochemical structure of the DEA 1 remains still unknown, but a genome wide association study has identified a single locus.

Recent surveys revealed that the Dal- type is not restricted to Dalmatians but is also seen in Beagles, Doberman Pinschers, and Lhasa Apsos and thus typing for this blood type is becoming more important particularly for those requiring multiple transfusions. In a related study dogs from North America were screened for two new blood types, preliminarily called Kai 1 and Kai 2. Most dogs were Kai 1+ and only few dogs were Kai 2+ or Kai 1-/Kai2-. The clinical importance is yet to be determined albeit anecdotally dogs can develop anti-Kai 1 alloantibodies. The PennGen Laboratory currently offers Dal and Kai 1/2 typing.

The clinically most important canine blood type is DEA 1, which elicits a strong alloantibody response after sensitization of a DEA 1- dog by a transfusion and thus can be responsible for a transfusion reaction in a DEA 1- dog previously transfused with DEA 1+ blood. It is currently unknown if DEA 1- dogs are equally sensitized by weakly to strongly DEA 1+ blood or if weakly DEA 1+ dogs are sensitized by strongly DEA 1+ blood. Furthermore, transfusion reactions against other blood types or common antigens have rarely been observed and reported. They include reactions against the DEA 4, Dal, Kai 1 and another common RBC antigens, and other clinically important blood types may be found in the future. No reagents currently are available against several antigens or are only available on a limited basis, and additional blood types continue to be recognized. Only limited surveys on the frequency of these blood types have been reported, which suggest possible geographic and breed-associated differences.

Strongly antigenic blood types are of great clinical importance because they can elicit a potent alloantibody response. These alloantibodies may be of the immunoglobulin G (IgG) or IgM class and may be hemagglutinins or hemolysins. Based upon experimental and clinical data, dogs can become sensitized after receiving a mismatched transfusion (i.e., a blood unit positive for one or more blood types not found on the recipient's RBCs). There are no clinically important, naturally occurring alloantibodies (also known as isoantibodies) present before sensitization of a dog with a transfusion. Sensitizing dogs in experimental studies in the 1950s led to the documentation of some transfusion reactions caused by blood group incompatibilities and to the characterization of new blood types.

Clinically the most antigenic blood type in dogs is the DEA 1. Transfusion of DEA 1+ RBCs to a DEA 1- dog invariably elicits a strong alloantibody response. Following a first transfusion, anti-DEA 1 antibodies develop after more than 4 days and may cause a delayed transfusion reaction (rarely clinically documented). However, a previously sensitized DEA 1- dog develops an acute hemolytic reaction after transfusion of DEA 1+ blood. Transfusion reactions also may occur after a sensitized dog receives blood that is mismatched for a RBC antigen other than DEA 1 (e.g. DEA 4 and Dal). However, in most cases the incompatible blood type has not been determined. Because administration of a small (<1 ml) amount of incompatible blood can result in life-threatening reactions, the practice of giving small "test volumes" of donor blood to assess blood-type compatibilities is unacceptable. In contrast, pregnancy does not cause sensitization in dogs because of a complete placenta and does not induce alloantibody production; thus dogs with prior pregnancies can be used safely as blood donors.

Canine Blood-Typing Procedures

Because of the strong antigenicity of DEA 1, typing of donors for DEA 1 is recommended. Whenever possible, the recipient also should be typed to allow the use of DEA 1+ blood for DEA 1+ recipients. Canine blood typing are generally based on serologic identification by agglutination reactions but chromatographic strip methods are also offered. Originally serum from sensitized dogs has been used for typing, but such polyvalent alloantibodies vary from batch to batch, may require Coombs' reagent to enhance agglutination, and may not be always available and are therefore not optimal. Monoclonal antibodies against DEA 1 have been developed at Kansas State University and at the University of Lyon. The gel column technology, widely used in human blood banking, was found to be an excellent standardized laboratory method (DiaMed), but is unfortunately no longer commercially available. A blood typing card has been available with modifications since the mid-1990s as a simple in-practice kit to classify dogs as DEA 1- or DEA 1+ (degree of reaction can vary). a standardized simple immunochromatographic technique became available in the mid-2000s from Alvedia. Another cartridge with a similar strip technique was introduced by DMS/AgroLabo, but has not been evaluated. Moreover, a third cartridge method in which blood flows through the cartridge is also available (DMS/Abaxis) but was producing inconsistent results.

Polyclonal reagents against other DEA types are currently only available on a limited bases for DEA 3, 4 and 7 from Animal Blood Resource International (prior Michigan state University and Midwest Blood Services). And only limited anti-Dal reagents from sensitized dogs are currently available in a couple of laboratories like Montreal University and PennGen, monoclonal anti-Kai 1 and anti-Kai 2 alloantibodies have been developed in South Korea. DEA 1 typed and matched patients in need of a transfusion may be typed for DEA 4, Dal and Kai 1/2, which may then permit the localization of a type matched donor dog.

Caution should be exercised whenever the patient's blood is autoagglutinating or has a low hematocrit (<10%). If autoagglutination is not too severe, it does not appear to affect the Alvedia strip technique because only free RBCs are moving up the strip. Clinicians and technicians should check for autoagglutination of blood with buffer/saline on a slide or the card. Autoagglutinating blood may be first washed three times with ample physiological saline to overcome the apparent autoagglutination similar to what is done for the Coombs' and crossmatch testing. However, if autoagglutination after three washes persists at more than 1+, it is considered to reflect true autoagglutination, which may preclude typing (as well as Coombs' testing and crossmatching, because it always looks like DEA 1+ blood). In such circumstances, DEA 1- blood should be used, until the patient does not agglutinate anymore and can be retyped. DEA 1+ blood from severely anemic animals may not agglutinate when exposed to the anti-DEA 1 or other reagents because of a prozone effect. In these cases some of the patient's plasma may be discarded before applying a drop of blood onto the card. Finally, recently transfused dogs may display a mixed field reaction, with only the transfused or recipient cells agglutinating if they were DEA 1 mismatched.

Blood Crossmatching Test

Whereas blood typing tests reveal the blood group antigens on the red blood cell surface, blood crossmatching tests assess the serologic compatibility or incompatibility between donor and recipient. Thus the crossmatch test checks for the presence or absence of naturally occurring and induced alloantibodies in serum (or plasma) without determining the blood type and thus does not replace blood typing. These antibodies may be hemagglutinins and/or hemolysins and can be directed against known blood groups or other RBC surface antigens. Many laboratories commonly use a standardized tube crossmatching procedure, but the interpretation of the agglutination reaction is highly variable. The crossmatching test requires some technical expertise, may be accomplished through a veterinary laboratory along with blood typing, and is done with washed EDTA-anticoagulated blood from recipient and potential donor(s). The DiaMed gel column technique and more recently the in-clinic DMS gel tube assay have been evaluated and were found to be simple, sensitive, and standardized methods to crossmatch dogs and cats. In addition, Alvedia introduced a simple strip crossmatch test with a Coombs' phase.

The major crossmatch tests search for alloantibodies in the recipient's plasma against donor cells, whereas the minor crossmatch test looks for alloantibodies in the donor's plasma against the recipient's RBCs. Generally tube segments from collection bags are used for this purpose in dogs. The presence of autoagglutination or severe hemolysis may preclude the crossmatch testing. A major crossmatch incompatibility is of greatest importance, because it predicts that the transfused donor cells will be attacked by the patient's plasma, thereby causing a potentially life-threatening acute hemolytic transfusion reaction. Because fatal reactions may occur with less than 1 ml of incompatible blood, compatibility testing by administering a small amount of blood is not appropriate; this has been shown in experimental studies to potentially result in fatal reactions. A minor crossmatch incompatibility should not occur in dogs if canine donors have not been transfused previously and is of lesser concern because donor's plasma volume is small, particularly with packed red cell products, and is diluted markedly in the patient. Do not use previously used dogs as donors.

The initial blood crossmatch between two dogs that have never before received a transfusion should be compatible, because dogs do not have naturally occurring alloantibodies. Therefore, a crossmatch may be omitted before the first transfusion in clinical practice for dogs. Because the crossmatch does not determine the blood type of the patient and donor, a compatible crossmatch does not prevent sensitization of the patient against donor cells within 1 to 2 weeks. Thus, previously transfused dogs should always be crossmatched, even when receiving again blood from the same donor. The time span between the initial transfusion and incompatibility reactions may be as short as 4 days and the induced alloantibody can last for many months to years (i.e., years after the last transfusion alloantibodies may be present). Again, a blood donor never should have received a blood transfusion to avoid sensitization. The practice of transfusing patients with the least compatible unit does not have any scientific basis. Nevertheless, some minor agglutination results in crossmatching a patient may be unrelated to alloantibodies and unspecific (e.g., patient's RBC damage by uremia and other illnesses, donor cells after extended storage of unit in the refrigerator). Of course, any patient with true/persistent autoagglutination may not be matched to any donor.

Although transfusion of blood and its components is usually a safe and temporarily effective form of therapy, there is always a risk for potential hazards. Adverse reactions usually occur during or shortly after the transfusion and can be due to any component of whole blood. Most transfusion reactions can be avoided by carefully selecting only healthy donors; using appropriate collection, storage, and administration techniques; performing blood typing and crossmatching; and administering only the needed blood components.

SELECTED RECENT REFERENCES

Acierno MA, Raj K, Giger U (2014). Varied DEA 1 expression on canine erythrocytes analyzed by immunochromatographic and flow cytometric techniques. J Vet Intern Med, in 28: 592–598.

Blais MC, Berman L, Oakley DA, Giger U (2007). Canine Dal blood type: a red cell antigen lacking in some Dalmatians, J Vet Intern Med 21: 281-286.

Blais MC, Rozanski EA, Hale AS, Shaw SP, Cotter SM (2009). Lack of evidence of pregnancy-induced alloantibodies in dogs. J Vet Intern Med. 23:462-465.

Blois SL, Richardson DM, Abrams-Ogg AC (2013). Comparison of a gel column blood typing method and a point-of-care cartridge for dog erythrocyte antigen 1.1.J Vet Emerg Crit Care. 23:340-343

Ergul Ekiz E, Arslan M, Ozcan M, Inal Gultekin G, T Kirmizibayrak, Giger U (2011): Prevalence of dog erythrocyte antigen DEA 1.1 in 4 breeds native to different areas in Turkey. Vet Clin Pathol, 40:518–523.

Euler CC, Mizukami K, Raj K, Murray L, Chen CY, Mackin A, Giger U (2016): Xenotransfusion of anemic cats with blood compatibility issues: Pre- and post-transfusion laboratory diagnostic and crossmatching studies. Submitted.

Euler CC, Lee JH, Kim HY, Raj K, Mizukami K, Urs Giger U (2016): Survey of two new (Kai 1 and Kai 2) and other blood groups in dogs of North America. Sbmitted.

Ferreira RR, Gopegui RR, Matos AJ (2011). Frequency of dog erythrocyte antigen 1.1 expression in dogs from Portugal. Vet Clin Pathol. 40:198-201.

Giger U, Stieger K, Palos H (2005): Comparison of various canine bloodtyping methods, Am J Vet Res 66:1386-90.

Giger U (2014). Blood-typing and crossmatching to ensure blood compatibility. Chapter 26 in Kirk's Current Veterianry Therapy. Online publication.

Hohenhaus AE (2004): Importance of blood groups and blood group antibodies in companion animals , Transfus Med Rev 18 : 117-22.

Iazbik MC, O'Donnell M, Marin L, Zaldivar S, Hudson D, Couto CG (2010). Prevalence of dog erythrocyte antigens in retired racing Greyhounds. Vet Clin Pathol.39:433-435.

Kessler RJ, Reese J, Chang D, Seth M, Hale AS, Giger U (2010): Dog erythrocyte antigens 1.1, 1.2, 3, 4, 7, and Dal blood typing and cross-matching by gel column technique. Vet Clin Pathol, 39:306-16.

Kohn B, Classe G, Weingart C (2012). Clinical evaluation of the QuickVet/ RapidVet canine dog erythrocyte antigen 1.1 blood-typing test. J Vet Diagn Invest. 24:539-45.

Lucidi C, Takahira RK, Gerlach JA, Davis JM, Schwartz KA, Scott MA (2011). Flow cytometric assessment of canine erythrocytes and platelets for dog erythrocyte antigen 1.1. Vet Clin Pathol. 40:435-43.

Patterson J, Rousseau A, Kessler RJ, Giger U (2011): In vitro lysis and acute transfusion reactions with hemolysis caused by inappropriate storage of canine red blood cell products. J Vet Intern Med. 25:927-933.

Polak K, Acierno MM, Raj K, Mizukami K, Siegel DL, Giger U (2015): Dog erythrocyte antigen 1: Mode of inheritance and initial characterization. Vet Clin Pathol. 44: 369-379.

Riond B, Schuler E, Rogg E, Hofmann-Lehmann R, Lutz H (2011). Prevalence of dog erythrocyte antigen 1.1 in dogs in Switzerland evaluated with the gel column technique. Schweiz Arch Tierheilkd. 153:369-74.

Seth M, Jackson K,V, Winzelberg S, Giger U (2012): Comparison of gel column, card and cartridge techniques for Dog Erythrocyte Antigen. Am J Vet Res, 73:213–219.

SUPPORT

Supported in part by a grant from the NIH (OD 010939).

CONFLICT OF INTEREST

The authors' laboratory is offering quantitative DEA 1, Dal and Kai typing. Alvedia and DMS Laboratories kindly provided reagents and kits for the authors' studies.

What is the Role of Protein Restriction in Feline CKD? Perspective: Internist

David J. Polzin, DVM, PhD, DACVIM
St. Paul, MN

INTRODUCTION

"Renal diets" have been considered to be "standard therapy" for managing cats with chronic kidney disease (CKD) for many years.[1] These diets continue to be recommended by the IRIS Board for therapy of cats with IRIS CKD Stages 2-4.[2] However, recently some feline practitioners have begun recommending feeding diets containing higher levels of dietary protein; in some instances clearly excessive levels of dietary protein. While the focus of recommendations proposed by feline practitioners are intended to specifically increase dietary protein intake, in reality, most if not all aspects of the "renal diet" have been altered because specific dietary modifications other than protein are not defined. As a consequence, a recommendation to feed "high(er?) protein diets" to cats with IRIS CKD Stages 2-4 fails to achieve many of the basic metabolic intentions on which the medical considerations of feline renal diets are formulated and recommended.[3]

There is evidence that poor nutrition may be associated with poorer outcomes in CKD.[4] While some cats with CKD fed renal diets may become nutritionally depleted, it has not been confirmed that the protein content of these diets is consistently the mechanism underlying the nutritional depletion. Further, it has been shown that adequate nutrition can be sustained in cats fed renal diets.[5] However, the ideal dietary protein content to provide to cats with CKD is unknown. Studies on the medical efficacy of renal diets in cats were performed using the protein content provided by the manufacturers, and the stated goals of these studies did not seek to determine the optimum protein content for these products. While small differences in the protein content exist among commercial products, studies directly comparing the efficacy of different commercial products have not been published.

While a consensus on whether renal diets contain the optimum protein content for cats with CKD has not been achieved, there is ample scientific evidence that excessive dietary protein content is potentially harmful in cats with CKD.[6] Simply choosing to feed a cat with CKD a higher protein diet is not the same as choosing a higher protein "renal diet." Although commonly called "low-protein" diets, feline renal diets are not actually low in protein and they do contain many other important modifications including reduced phosphorus and sodium content, increased B-vitamins, potassium and caloric density, added soluble fiber, a neutral effect on acid-base balance, and supplementation of omega-3-polyunsaturated fatty acids and antioxidants. As a consequence, feeding more protein potentially involves multiple errors in feeding cats with respect to phosphorus, potassium, sodium, fatty acids, antioxidants, fiber, and alkalinizing effect. While the evidence supporting these dietary components varies, it is generally accepted at least that phosphorus, fatty acids, antioxidants and alkalinizing impact may influence progressive loss of kidney function.[1,3]

Evidence Supporting Efficacy of Feline Renal Diets

There are three published studies that address the efficacy of feline renal diets.[5,7,8] The first study compared a manufactured renal diet (RD) to no diet change.[7] This study was neither randomized nor blinded. Cats fed RD (mean survival time = 633 days) survived substantially longer than cats consuming their regular diet (mean survival time = 264 days). In addition, BUN, serum phosphorus, and parathyroid hormone concentrations were reduced in cats consuming RD. The second study was a randomized, masked clinical trial with 22 cats randomized to a manufactured renal diet (RD) and 23 cats were randomized to a prototypical maintenance diet (MD).[5] Cats in this study had serum creatinine values ranging from 2.0 mg/dl to 4.5 mg/dl (IRIS CKD stages 2 & 3). Risks of uremic crises and renal deaths were significantly reduced when RD was fed. Among the 22 cats fed RD, there were no uremic crises or renal deaths over two years of study. In contrast, among the 23 cats consuming MD, 6 developed uremic crises, 5 died of renal causes. The third study was a retrospective study performed in 31 first-opinion veterinary practices in The Netherlands and compared survival times for cats fed one or more of 7 commercial feline renal diets to cats not fed a renal diet.[8] Median survival time for cats fed a renal diet was 16 months compared to 7 months for cats fed their usual diet. The consistent findings in these three studies using different diets and methodologies and performed in 3 different countries and by 3 different groups of researchers appear to strongly support that renal diets favor better clinical outcomes (survival and reduced uremic crises).

Evidence Supporting Adverse Effects of Feeding High Protein Diets in Cats

Studies in cats with induced CKD indicate that feeding higher protein diets to cats with CKD promote glomerular hyperfiltration, proteinuria and glomerular sclerosis.[6] Proteinuria has been associated with increased mortality in cats with spontaneous CKD, even at borderline proteinuria.[6] The suggestion that the decrease in serum creatinine concentration that follows feeding high protein diets is a favorable response is unsubstantiated. In fact, the reduction in creatinine values is likely to be accompanied by increased proteinuria and glomerular lesions. While a limited increase in protein intake might have a favorable nutritional benefit, the impact on proteinuria and progression of CKD should be balanced against the impact on nutrition.

Is the Purported Nutritional Impairment Due to Protein Content of Renal Diets or Failure to Assure Adequate Food Intake?

The 2-year randomized study of renal diets in feline CKD described above reported the nutritional characteristics of cats on the renal diet and maintenance diet.[6] The nutritional status of the two diets were not statistically significantly different. The nutritional status of cats fed the renal diet did not develop significant deterioration of nutritional measures over two years of study. This finding suggests that the purported malnutrition associated with feeding renal diets to cats may in fact be consequent to failure to assure that cats are consuming adequate quantities of the renal diet. Feeding adequate calories is essential when feeding renal diets. When necessary, efforts to increase food intake may be a more prudent approach to promoting adequate nutrition, particularly given the potential adverse effects of high protein intake on the diseased kidneys.

Recommendations

Feeding cats with CKD renal diets are recommended for cats with IRIS CKD Stages 2-4. Nutrition status should be closely monitored as should food intake be monitored. Optimum nutritional intake may be assured by: 1) assuring adequate calories are consumed (by assuring food presentation is optimized and hydration is maintained, pharmaceutical appetite stimulation and/or tube feeding) 2) appropriate titrated supplementation of protein or amino acids, and 3) monitoring for evidence of potentially adverse effects of excessive protein supplementation (proteinuria, marked increase in BUN, signs of uremia).

REFERENCES

1. Polzin DJ. A Step-Wise Approach to Managing Chronic Kidney Disease: Application of Evidence Based Medicine. J Vet Emerg Crit Care 2012; 205-15, 2013.
2. www.iris-kidney.com
3. Roudebush P, Polzin DJ, Towell TL, Adams LG and Forrester SD: An evidence-based review of therapies for canine chronic kidney disease. J Small Anim Pract 51(5):244-252, 2010.
4. Parker VJ and Freeman LM: Association between body condition and survival in dogs with acquired chronic kidney disease. J Vet Intern Med 25:1306-1311, 2011.
5. Ross SJ, Osborne CA, Kirk CA, et al. Clinical evaluation of dietary modification for treatment of spontaneous chronic kidney disease in cats. J Am Vet Med Assoc 2006; 229:949-957.
6. Adams L, Polzin D, Osborne C, et al: Influence of dietary protein/calorie intake on renal morphology and function in cats with 5/6 nephrectomy. Laboratory Investigation 1994; 70:347-357.
7. Elliott J, Rawlings JM, Markwell PJ, et al. Survival of cats with naturally occurring chronic renal failure: effect of dietary management. *J Small Anim Pract* 2000;41:235–242.
8. Plantinga EA, Everts H, Kastelein AMC, Beynen AC: Retrospective study of the survival of cats with acquired chronic renal insufficiency offered different commercial diets. Vet Rec 2005. 157:185-7.

Should Protein Always Be Restricted in Feline Chronic Kidney Diseases?

Margie Scherk, DVM, DABVP (Feline Practice)
Vancouver, Canada

INTRODUCTION

Since the 1950's, dietary therapy has been the cornerstone for the treatment of chronic kidney diseases (CKD) in dogs and cats. It continues to be the most common recommendation made for affected feline patients and is credited with the highest strength and quality of evidence.[1] In addition to protein restriction, dietary modifications include reduced phosphorus and sodium, increased omega-3 polyunsaturated fatty acids, adaptations to promote neutral acid-base balance and possibly increased caloric density, B-vitamins, and fiber. Combined, they provide a beneficial "renal effect", however evaluation of the effect of individual components remains untested. Assessing these components individually remains an important goal in order to optimize outcome and reduce detrimental effects.

EVALUATING THE EVIDENCE FOR PROTEIN RESTRICTION

Conclusions regarding dietary therapy for feline CKD come from: 1) pathophysiologic reasoning (e.g., inflammatory attributes of protein in renal tubules[2]); 2) extrapolation from other species (rodent or canine models and human clinical trials, e.g., phosphate restriction prevented renal mineralization in nephrectomized rats[3]); 3) observations from renal ablation studies in cats (e.g., caloric rather than protein restriction resulted in fewer nonglomerular lesions[4]); and 4) trials in naturally occurring CKD in cats (e.g., cats fed renal diet have fewer uremic episodes and renal related deaths compared to cats fed maintenance diet[5]). Results from these different studies have not however been consistent or have been limited by study design or size.

Critical to this presentation is that there are no clinical studies in cats with naturally occurring disease selectively comparing protein levels. A retrospective study (321 cats) comparing cats fed a variety of renal diets to those fed uncontrolled and varied maintenance diets found longer survival in the group fed any renal diet.[6] A non-randomized prospective study (50 cats) compared 29 that willingly ate a dry or wet renal diet to those that wouldn't and were fed an uncontrolled variety of diets. This study found a survival benefit in the renal diet group. It must also be noted that the renal group received phosphorus binders to control hyperphosphatemia as needed; this group had lower serum phosphorus and parathyroid hormone levels.[7]

Two randomized, controlled prospective trials have been conducted. In one, 74% of 35 cats were fed a renal diet with the balance being fed a control maintenance diet for 24 weeks.[8] Those fed the renal diet had less clinical deterioration and less increase in serum creatinine and urea nitrogen (SUN). In the other study, 45 cats (IRIS Stage 2 or 3) were randomly assigned to eat a renal or control maintenance diet. Results revealed fewer uremic episodes and renal related deaths in those eating the renal diet.[5] Despite these studies, because protein has not been parsed out specifically and the diets fed are too dissimilar, it remains unclear what if any role protein restriction plays in the composite effect of dietary therapy.

Similarly, optimal levels of dietary protein in treatment of proteinuria in cats is unknown. While proteinuria is recognized as a negative prognostic factor, dietary restriction of this nutrient has not been shown to be beneficial either in angiotensin converting enzyme inhibitor (ACE-I) or angiotensin receptor blocker trials in naturally occurring disease.[9-11] In the nephrectomy model, proteinuria appeared to be associated with hypokalemia[12] and did not develop in cats fed a high protein, potassium replete diet.[4] In the randomized, controlled prospective study of Stage 2 and 3 cats, dietary protein level did not result in a difference in proteinuria.[5]

As obligate carnivores, cats have evolved to manage a high dietary protein load[13-15] but adapt to various intake as long as their minimum needs are met.[16] Dietary protein is not toxic to kidneys. Creatinine and urea result from metabolizing ingested protein as well as from turnover of endogenous stores in health or in protein deficiency. In excess of 60 purported uremic toxins are reported.[17] Interactions between many endogenous metabolites[18] resulting from inflammation, malnutrition, increased concentrations of protein-bound solutes and hypoalbuminemia as well as non-nutritional toxins contribute to the clinical spectrum of uremia.[17,19,20] Thus, while uremic toxins can result in malnutrition, malnutrition itself results in inflammation, morbidity and mortality in human patients with CKD.[20]

Despite having greater resources and the ability to perform significant meta-analysis, the question of whether or not protein restriction is beneficial in human CKD remains controversial.[21-23] Similar to cats, protein malnutrition, sarcopenia, and iron deficiency are clinical problems in human CKD patients following low protein regimes. A prospective study in human CKD patients contrasting protein and phosphorus restriction concluded that the risks associated with protein restriction may outweigh the benefits from phosphorus restriction, and suggested that protein restriction as a means to control phosphorus intake is not justified.[24] A meta-analysis regarding the effects of dietary protein restriction in human CKD concludes that the relatively weak effect on progression of disease suggests that better therapies are needed.[25]

EVALUATING THE EVIDENCE FOR OTHER ASPECTS OF RENAL DIETS

Phosphorus

Numerous studies using the remnant kidney model have verified that controlling serum phosphorus levels with resultant modulation of parathyroid hormone (PTH) and fibroblast growth factor-23 (FGF-23), results in reduced renal pathology, morbidity and mortality in rats as well as in cats. Lumlertgul and Koizumi both showed that restriction of phosphorus arrested progression of CKD in nephrectomized rats independent of dietary protein intake.[26,27] Several studies in naturally occurring feline CKD have similarly shown that when phosphorus was controlled in addition to feeding a protein and phosphorus restricted diet, reductions in serum phosphorus,[5] PTH,[6,7,28] and FGF-23[29] are seen along with increased survival.[30-32] Lacking are studies that isolate phosphorus restriction from protein content in naturally occurring feline disease.

Acidosis

Acidosis stimulates numerous endocrine responses[33] contributing to protein catabolism, decreased protein synthesis, loss of lean body mass (LBM),[34] increased calcium and potassium excretion. Correction of acid-base imbalance results in improvement in clinical signs that include nausea, anorexia, vomiting, weight loss, weakness, and lethargy. Additionally, it reduces the loss of LBM and may slow the progression of CKD. In retrospective studies of humans with CKD on dialysis, as well as a prospective controlled clinical trial of non-dialysis CKD patients, correction of acidosis not only normalized protein catabolism, but also improved markers of nutritional status, and resulted in less deterioration of creatinine clearance compared to the control group.[35] Net nitrogen balance, SUN and total body protein synthesis are positively affected by ensuring a neutral serum pH.[36]

Potassium

Hypokalemia is both a cause, as well as a result, of CKD. Metabolic acidosis results in an extracellular shift of potassium that masks the severity of the intracellular and total body potassium level. Independent of acid-base, potassium supplementation has been shown to reduce serum creatinine.[37,38] In another small study, potassium supplementation in normokalemic cats with CKD, resulted in improvement in creatinine, SUN and increased, or no decline, in GFR, when compared to cats supplemented with sodium gluconate.[39] These findings lack power to recommend potassium supplementation, per se, however provision of adequate potassium in a non-acidifying diet does appear to be relevant.

EVALUATING THE EVIDENCE FOR MAINTAINING BODY CONDITION

As cats age, they tend to lose weight and LBM. In part this is associated with decreased digestion of fat and protein,[40] in part due to increased requirements for protein and calories that may not be met.[40-44] Maintenance of weight and condition prolongs life.[45] Given that cats with CKD often live for a long time,[10,31] the nutritional needs of older cats should be taken into consideration in treatment of those with CKD.

Weight loss with poor muscle condition is common among cats with CKD. Weight loss begins before 1 to 3 years before a diagnosis of CKD has been made[46] and is associated with decreased survival.[31,46] A decrease in muscle mass is associated with increased morbidity and mortality in human CKD patients.[47] Protein synthesis is reduced in these individuals compared to healthy human patients. While inadequate caloric intake contributes to weight, fat and LBM loss, reduction in dietary protein exacerbates the CKD-related muscle loss[47] in attempts to meet the needs for protein turnover and ongoing metabolic needs. Increasing dietary protein helps to regulate acid-base balance through excretion of hydrogen ions associated with ammonium.[48]

SUMMARY

While feeding a renal diet is accepted as a core part of therapeutic management, each patient's response to the disease and to nutritional intervention varies dramatically, and individualizing therapy is required, especially as different pathologic processes may underlie the CKD. At the present time, there are no studies that evaluate the independent effects of restricting protein, controlling acidosis, potassium, or reducing phosphorus. There is, however, sufficient evidence showing that low weight and reduction in LBM is detrimental to quality of life, progression and survival of cats with CKD.

In order to ascertain the role (benefit or not) of protein restriction, randomized, controlled prospective studies need to be performed in cats with naturally occurring renal disease comparing diets that are as identical as possible with one having a higher proportion of calories from protein than the other. Until such time, we need to consider that not all cats from IRIS Stage 2 onward be offered a renal diet and focus on the risks of inadequate dietary protein. It is critical to monitor every patient closely for the subtle changes associated with protein deficiency that can shorten life.

(Adapted from chapter in press by Scherk MA, Laflamme DP: Renal Diets are Indicated for Cats with IRIS CKD Stages 2-4: The Con View. Vet Clin N Am Small)

REFERENCES

[1]Roudebush P, et al. J Feline Med Surg 2009; 11: 195.[2]Lawson J, et al. Vet J 2015; 203: 18.[3]Ibels LS, et al. N Engl J Med 1976; 298: 122.[4]Finco DR, et al. Am J Vet Res 1998; 59: 575.[5]Ross SJ, et al. J Am Vet Med Assoc 2006; 229: 949.[6]Plantinga EA, et al. Vet Rec 2005; 157: 185.[7]Elliott J, et al. J Small Anim Pract 2000; 41: 235.[8]Harte JG, et al. J Nutr 1994; 124: 26660S.[9]King JN, et al. J Vet Intern Med 2006; 20: 1054.[10]King JN, et al. J Vet Intern Med 2007; 21: 906.[11]Sent U, et al. J Vet Intern Med. 2015; 29: 1479.[12]Adams LG, et al. Lab Invest 1994; 70: 347.[13]Hewson-Hughes AK, et al. J Expl Bio. 2011; 214: 1039.[14]Hewson-Hughes AK, et al. J Comp Physiol B. 2013; 183: 525.[15]Plantinga EA, et al. Br J Nutr. 2011; 106: S35.[16]Green AS, et al. J. Nutr. 2008; 138: 1053.[17]Vanholder R, et al. Kidney Int 2003; 63: S6.[18]Lisowaska-Myjak B. Nephron Clin Pract 2014; 128: 303.[19]Stenvinkel P, et al. Nephrol Dial Transplant 2000; 15: 953.[20]Vanholder R, et al. Semin Dial 2002; 15: 311.[21]Levey AS, et al. J Am Soc Nephrol 1999; 10: 2426. [22]Levey A, et al. Am J Kidney Dis 2006; 48: 879. [23]Kopple JD, et al. Kidney Int Suppl 2005; 95: S21.[24]Shinaberger CS, et al. Am J Clin Nutr 2008; 88: 1511.[25]Kasiske

BL, et al. Am J Kidney Dis 1998; 31: 954. [26]Lumlertgul D, et al. Kidney Int 1986; 29: 658. [27]Koizumi T, et al. Biochem Bioph Res Co 2002; 295: 917. [28]Barber PJ, et al. J Small Anim Pract 1999; 40: 62.[29]Geddes RF, et al. J Vet Intern Med 2013; 27: 1354. [30]Kuwahara Y, et al. J Small Anim Pract 2006; 47: 446.[31]Boyd LM, et al. J Vet Intern Med 2008; 22: 1111. [32]Chakrabarti S, et al. J Vet Intern Med 2012; 26: 275.[33]Franch HA, et al. J Am Soc Nephrol 1998; 9: S78. [34]Franch HA, et al. Am J Physiol Renal Physiol 2004; 287: F700.[35]De Brito-Ashurst I, et al. J Am Soc Nephrol 2009; 20: 2075.[36]Mehrotra R, et al. Am J Clin Nutr 2009; 90: 1532.[37]Dow SW, et al. J Am Vet Med Assoc 1987; 191: 1569.[38]Dow SW, et al. J Nutr 1990; 120: 569.[39]Theisen SK, et al. J Vet Intern Med 1997; 11: 212.[40]Perez-Camargo G. Comp Cont Edu Pract. 2004; 26 (suppl 2A): 5.[41]Harper EJ. J Nutr 1998; 128: 2627S. [42]Laflamme DP, et al. Vet Clin N Am Small 2014; 44: 761.[43]Laflamme DP, et al. J Fel Med Surg: In press.[44]Villaverde C, et al. Vet Clin N Am Small 2014; 44: 699.[45]Cupp CJ, et al. Proc Nestlé Purina Comp Anim Nutr Summit 2010: 36. [46]Freeman LM, et al. J Vet Intern Med 2015; 29: 935. [47]Wang XH, et al. Nat Rev Nephrol 2014; 10: 504.[48]Remer T. Eur J Nutr 2001; 40: 214.

What is the Role of Protein Restriction in Feline CKD? Perspective: Nutritionist

Valerie J. Parker, DVM, DACVIM (SAIM), DACVN
Columbus, OH

How much dietary protein is recommended for a healthy cat?

Minimal and (some) maximal nutrient recommendations for cat food are provided by the Association of American Feed Control Officials (AAFCO) and the National Research Council (NRC). On a caloric basis, the AAFCO minimum protein recommendation for feline adult maintenance is 65 grams per 1000 kcal (Mcal) metabolizable energy (ME) or 6.5 g per 100 kcal. The minimum protein recommendation for feline growth and reproduction is 75 g per Mcal (7.5 g per 100 kcal).

Most diets sold over-the-counter should meet AAFCO requirements; however, veterinary therapeutic diets may be specifically modified in order to provide certain nutrients at concentrations less than AAFCO minimums. The average protein content of over-the-counter commercial cat foods in the USA is approximately 8.5-9.0 grams per 100 kcal, though some contain up to 18.0 g per 100 kcal. Veterinary therapeutic diets specifically marketed for chronic kidney disease (CKD) provide an average of 6.9 g protein per 100 kcal. In contrast to veterinary therapeutic renal diets designed for dogs that typically provide less protein than AAFCO minimum, most of these diets exceed AAFCO minimum recommendation for protein.

Is there anything unique about cats and their protein requirement?

Cats have several unique nutritional requirements, including a high protein requirement. While growing kittens need about 1.5 times a much protein as other species (e.g., chicks, pigs), this protein requirement does not seem to be due to higher amino acid requirements; it may be due to a higher nitrogen requirement. One of the reasons that cats (and other carnivores) have this higher nitrogen requirement is that they have little ability to adapt aminotransferases involved in amino acid catabolism. Even when fed low-protein diets, cats do not effectively reduce urea cycle enzyme activity; thus, ammonia is continually converted to urea rather than providing nitrogen for synthesis of amino acids. One benefit to cats regarding this lack of enzyme regulation is the ability to use amino acids as a source of energy and for gluconeogenesis.[1]

What is the evidence for reducing dietary protein?

Feeding therapeutic renal diets has been shown to reduce the risk of uremic episodes and improve survival in cats with naturally-occurring CKD.[2,3] However, it is difficult to determine which specific nutritional modification(s) (e.g., reduced phosphorus, reduced protein, etc.) affected these findings. In dogs with experimentally-induced CKD, dietary phosphorus intake seemed to impact survival to a greater extent than dietary protein intake.[4] Additionally, protein intake was not associated with progression of kidney disease in cats with experimentally-induced CKD.[5]

Dietary protein intake should be reduced if a cat has been diagnosed with significant proteinuria as reducing dietary protein is the nutritional mainstay of treatment of protein-losing nephropathy.[6,7] If dietary therapy is considered as part of proteinuria management, a complete diet history, including all protein sources, should be obtained to ascertain the cat's starting dietary protein intake.[8] Medical management with ACE inhibitors and angiotensin receptor blockers should be considered.[9]

Why might dietary protein restriction be contraindicated?

One of the most compelling reasons to avoid protein restriction in cats with CKD is to reduce the risk of protein-energy wasting (PEW). Protein-energy wasting is defined as "a state of nutritional and metabolic derangements... characterized by simultaneous loss of systematic body protein and energy stores, leading ultimately to loss of muscle and fat mass and cachexia."[10] The criteria considered when considering a diagnosis of PEW include: 1) biochemical criteria (e.g., serum albumin); 2) body mass (e.g., unintentional weight loss); 3) decreased muscle mass; 4) dietary intake (e.g., low protein or energy intake).[11] Several factors influence the development of muscle loss including increased inflammatory cytokines and metabolic acidosis; thus, increasing dietary protein alone will not necessarily improve muscle condition. Figure 1 (used with permission) illustrates the concept of PEW.

In a recent study evaluating the effects of feeding diets with variable dietary protein concentrations to healthy adult cats, those fed low and moderate protein diets (5.6 grams and 7.3 grams protein per 100 kcal, respectively) lost lean body mass. The authors suggested that adult cats require 5.2 grams protein per kg body weight (7.8 g/kg$^{0.75}$), which is higher than current AAFCO recommendations of approximately 3.3-3.9 grams protein per kg body weight.[12]

Does the source of protein matter?

In people, there is increasing evidence that the source of dietary protein may play a significant role in the progression of CKD.[13] Metabolic acidosis is a major consequence of CKD that contributes to protein catabolism via the ATP-dependent ubiquitin-proteasome system. Meat-based diets are typically acid-producing and are more likely to induce metabolic acidosis. A recent meta-analysis in people with CKD investigated the effects of soy protein containing isoflavones. The soy protein significantly reduced serum creatinine, phosphorus, C-reactive protein and proteinuria.[14]

It can be challenging to formulate a vegetarian diet for cats since vegetarian protein sources do not typically provide appropriate amino acid profiles and are often poor sources of certain essential nutrients (e.g., vitamin D, vitamin A, cobalamin, arachidonic acid, etc.) A recent study that assessed protein and amino acid concentrations of 24 commercial vegetarian diets for dogs and cats found several diets were deficient in essential amino acids, including taurine and its precursors methionine and cystine.[15] It is possible to formulate predominately vegetarian cat foods with the use of proper vitamin/mineral supplementation and purified essential amino acids; however, this highlights the importance of recommending specific reputable diets carefully. Since renal diets are formulated to promote a neutral urine pH, it may not be as important to consider a vegetarian diet for CKD cats.

What about the relationship between protein and phosphorus?

There is a strong correlation between dietary protein and phosphorus; dietary phosphorus intake can be predicted based on dietary protein intake in people with CKD.[16] It can be difficult, although not impossible, to reduce dietary phosphorus intake without concurrently reducing dietary protein. Knowing how to interpret nutrient profiles can help veterinarians more critically evaluate various diet options when making recommendations for their patients. For example, Hill's Science Diet Mature Adult Tender Tuna Dinner (canned) provides 10.5 grams protein and 151 mg phosphorus per 100 kcal while Royal Canin Veterinary Diet Senior Consult (dry) provides 7.4 grams protein and 150 mg phosphorus per 100 kcal. It is important to realize that dry and canned varieties of the same diet, as well as various flavors, may have significantly different nutrient profiles. For example, Purina Fancy Feast Gourmet Cat Food with Savory Chicken & Turkey provides 271 mg phosphorus per 100 kcal while Purina Fancy Feast Chunky Turkey Feast provides 474 mg phosphorus per 100 kcal.

Many pets with CKD receive people food, with some eating home-cooked diets exclusively. Different methods of handing meat can drastically alter nutrient profiles. The effect of various conditions (e.g., boiling, direction of cut) on protein and phosphorus concentrations of beef was assessed. Slicing (i.e., cutting the muscle fiber vertically) and boiling the meat for 10 minutes reduced phosphorus concentration by 50% while protein retention remained > 90% (compared to raw).[17] More information about nutrient profiles for specific people food can be accessed on the USDA National Nutrient Database website: http://ndb.nal.usda.gov/.

So what should I feed my feline CKD patients?!

It depends! Each patient should be considered as an individual. A nutritional assessment should be performed, including a complete diet history and body composition assessment. The animal's comorbidities and potential other nutrients of concern should be considered. The animal's CKD should be staged and substaged according to the International Renal Interest Society (IRIS) guidelines. Based on that information, taking pet and owner preferences into account, a nutritional recommendation should be made, realizing that one or more modifications may be necessary based on the cat's response and course of disease progression.

D Fouque et al.: **Protein-energy wasting in kidney disease** meeting report

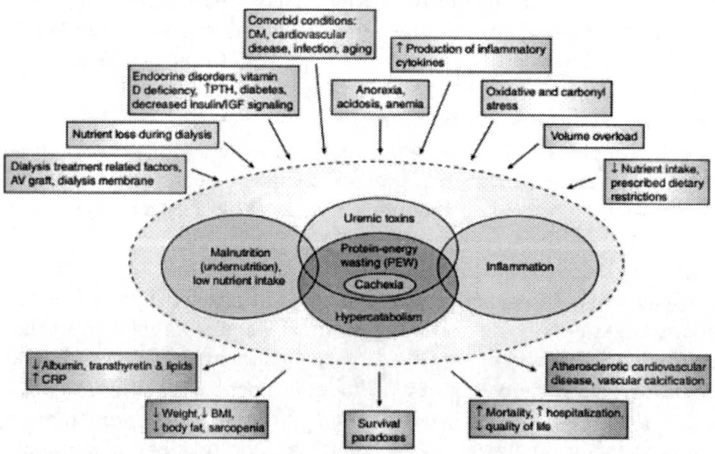

Figure 1 | Schematic representation of the causes and manifestations of the protein-energy wasting syndrome in kidney disease.

REFERENCES

1. Morris JG. Nutr Res Rev 2002;15:153-168.
2. Elliott J, et al. J Small Anim Pract 2000;41:235-242.
3. Ross SJ, et al. J Am Vet Med Assoc 2006;229:949-957.
4. Finco DR, et al. Am J Vet Res 1992;53:2264-2271.
5. Adams LG, et al. Am J Vet Res 1993;54:1653-1662.
6. Adams LG, et al.Lab Invest 1994;70:347-357.
7. Littman MP. Vet Clin N Am - Small 2011;41:31-62.
8. Parker VJ, et al. Compendium 2012;July:E1-E5.
9. Sent U, et al. J Vet Intern Med 2015.
10. Obi Y, et al. Curr Opin Clin Nutr 2015;18:254-262.
11. Fouque D, et al. Kidney Int 2008;73:391-398.
12. Laflamme DP, et al. J Fel Med Surg 2013;15:691-697.

13. Goraya N, et al. Curr Opin Nephrol Hy 2015;24:505-510.
14. Jing Z, et al. Clin Nutr 2015.
15. Kanakubo et al. J Am Vet Med Assoc 2015;247:385-392.
16. Boaz M, et al. J Am Diet Assoc 1996;96:1268-1270.
17. Ando S, et al. J Renal Nutr 2015;25:504-509.

Utilization of Feeding Tubes in the Management of CKD
Sheri Ross, DVM, PhD, DACVIM (Small Animal Internal Medicine)
San Diego, CA

The long-term use of esophagostomy feeding tubes provides the means to maintain hydration, maintain adequate nutrition, and facilitate medication administration in patients with chronic kidney disease. This talk will present the associated logistics, clinical outcomes and complications of patients with IRIS stages 2-4 CKD that have had esophagostomy tubes placed as part of their overall management plan.

Appropriate nutrition is critical in the management of CKD

Dietary therapy is an extremely important component in the management of patients with chronic kidney disease. Proper nutrition may forestall many of the complications associated with advancing renal disease and has been shown to improve the patient's quality and quantity of life. [1,2] Compared to adult maintenance diets, diets formulated specifically for patients with chronic kidney disease typically have reduced protein, phosphorus, and sodium content; increased potassium, B-vitamin content and caloric density; a neutral effect on acid-base balance; and an increased omega-3/omega-6 polyunsaturated fatty acid (PUFA) ratio. Currently there is grade I evidence supporting the recommendation that cats with serum creatinine concentrations in excess of 2.0 mg/dl (IRIS CKD Stages mid-2 through 4) be fed a diet formulated specifically for kidney disease.[1] Similar results were observed in an earlier non-randomized clinical trial in cats with naturally occurring CKD, where the median survival time for cats consuming the renal diet was 633 days compared to 264 days for cats that did not switch to a renal diet. [3]

Anorexia in CKD is multifactorial

Anorexia secondary to CKD is usually encountered in the advanced stages of the disease and is multifactorial in origin. Contributing factors are believed to include: hypergastrinemia, stomatitis, gastritis, altered taste, and stimulation of the chemoreceptor trigger zone by circulating uremic toxins. Uremic anorexia may also be aggravated by management problems, including forced feeding, sudden changes in diets, intolerance to medications, and other factors associated with food aversion. A variety of metabolic causes may be associated with poor appetite in patients with renal insufficiency including: 1) anemia, 2) uremic gastritis, 3) dehydration, 4) metabolic acidosis, 5) hypokalemia, and 6) renal secondary hyperparathyroidism. Most of these conditions can be managed with appropriate therapy. Providing frequent small meals may be helpful in increasing calorie intake in patients that are partially anorexic. In addition, anorexia may be ameliorated by modifying feeding schedules, the use of pharmacologic appetite stimulants, and/or use of assisted enteral nutrition. The use of maropitant in patients with CKD has been very promising. A clinical trial demonstrated that use of maropitant resulted in a clinically significant reduction in vomiting when used in cats with CKD. [4] Anecdotally, the chronic use of maropitant may result in increased food intake in patients that are not obviously nauseated. Mirtazapine is a commonly used appetite stimulant in patients with partial anorexia. The appetite stimulating properties and pharmacokinetics of mirtazapine have been established for cats with CKD. [5,6] Mirtazapine appears to be a promising adjunctive therapy for the nutritional management of cats with CKD, especially when nutritional support is only needed short-term and/or in cases where feeding tube placement is not an option. Despite these and other medical interventions, many patients show a progressive decline in body condition.

Enteral feeding tubes are an option for many patients

In desperation to see their pet eat, many owners will offer anything to their pets in an effort to increase their food intake. Although the commonly touted phrase "they have to eat something" is certainly true, the quality and quantity of a patient's life will be prolonged if they consume an adequate amount of a diet designed for patients with kidney disease. (Ross 2006, Jacob 2003) At the first signs of weight loss, owners should be encourage to allow placement of an esophagostomy tube for long term nutritional support. Proactive intervention with enteral feeding tubes combats catabolism and helps prevent loss of lean body mass. Esophagostomy tubes will also allow for long-term provision of nutrition, hydration, and/or medications.

Assisted enteral feeding, using either a gastrostomy or esophagostomy tubes, has been shown to decrease morbidity and mortality in both dogs and cats. Placement techniques along with associated complications and limitations have also been described for both tube types. [7,8,9] Traditionally, gastrostomy tubes have been preferred for chronic nutritional management, while esophagostomy tubes have been used for shorter periods (weeks), and usually only in the management of hepatic or gastrointestinal disorders. The chronic use of assisted feeding in the management of chronic kidney disease has been described in dogs.[7] In this group of dogs, nutritional support via gastrostomy tube was successful in maintaining adequate nutritional plane. There is a paucity of information in the literature evaluating the long-term management of CKD using esophagostomy tubes.

We have been aggressive with the use of E-tubes in the management of CKD in both cats and dogs over the past 10 years. The majority of E-tubes are placed in patients with IRIS stage 3 CKD or above due to the associated metabolic complications seen at this level of disease. Patients that received an E-tube either maintained or improved their body condition score and body weight. In the subset of cats with CKD, the average body weight at the time of death was greater than the average body weight at the time of tube placement. Patients with an E-tube rarely required hospitalization for IV fluid diuresis, and supplemental hydration via the use of subcutaneous fluids at home was rarely required. Dietary and medication compliance for patients with E-tubes were also improved. In the review of our cases, cats had the E-tubes in place longer than the dogs. This is likely due to the usual prolonged course of feline CKD compared to canine CKD and the resulting longer survival times. Feeding tubes were removed in less than 20% of patients, and only after the patient demonstrated that they were able to maintain an acceptable plane of nutrition. Within our patient group, the longest recorded continuous use of an esophagostomy tube in the management of CKD was in a cat, whose tube was in place for 1882 days. This cat had regular (q 3 month) exchanges of a 14 French red rubber tube, using the original stoma for the entire period.

Complications associated with feeding tubes

The most commonly encountered complication in our patient group was cellulitis or abscess formation at the insertion site. This was seen in less than 30% of patients (in some sub-groups, less than 5%), and in almost all cases responded readily to 1-2 weeks of oral antibiotic therapy and hot-packing the site 2-3 times daily. Although placement of an E-tube is not a sterile procedure, every effort should be made to keep the procedure as clean as possible. Problems during the initial placement of the tube were rare in our patient population, but have been reported previously.[8-11] Bleeding from the insertion site was seen in a few cases, but this quickly resolved without significant intervention. The second most commonly encountered complication within our group of patients was inadvertent tube removal at home. In one subset of 35 CKD cats, 8 cats presented for tube removal at home, with 2 of these being repeat offenders. In all but one case, the tube was replaced in the existing stoma without the need for anesthesia. Clogging of the tube was also encountered, and seemed to be a repeatable problem for certain patients. This was almost always corrected with husbandry changes such as more care in preparing medications for tube administration, different approach to blending the food, or in 2 cases, straining the food prior to feeding.

Summary

Esophagostomy feeding tubes are useful in the management of patients with advanced chronic kidney disease. Although they are often not considered as a long-term management option, we have had many patients with advanced CKD maintain a good quality of life through the use of feeding tubes. Esophagostomy tubes allow patients to stay hydrated, avoiding episodes of hospitalization for diuresis. In addition tubes facilitate the administration of medications, which helps to mitigate the metabolic problems seen with advanced chronic kidney disease. And finally, feeding tubes provide a means to maintain an adequate nutritional plane in a patient that otherwise would be subject to chronic wasting.

REFERENCES

1. Ross SJ, et al. JAVMA : 2006 :229, 949-957
2. Jacob F, et al JAVMA 2003, 220:1163-70
3. Elliott J, et al. JSAP 2000 41:235-242
4. Quimby JM, et al. J Feline Med Surg. 2015;17(8):692-7
5. Quimby JM, et al. JVIM. 2011 Sep-Oct; 25(5):985-9
6. Quimby JM, et al. Vet J. 2013 Sep; 197(3):651-5
7. Elliott DA, et al. JAVMA 2000 Nov 1; 217(9):1337-42
8. Ireland LM. Et al. JAHA 2003: 39:3, 241-246.
9. Levine PB, et al. JAHA 1997, 33:5, 405-410.
10. Crowe DT, et al. JAHA 1997, 33: 393-403
11. Devitt CM et al. JAHA 1997, 33: 55-60

Understanding the Limitations of Phosphate Binders

Michael W. Wood DVM, PhD, DACVIM (SAIM)
Madison, WI

INTRODUCTION

Increased serum phosphate concentrations have been linked to higher mortality in dogs, cats, and people with chronic kidney disease (CKD). The observed hyperphosphatemia occurs secondary to an imbalance in gastrointestinal phosphate absorption, bone formation/resorption, and renal phosphate clearance. In veterinary patients, limiting serum phosphorus concentrations is primarily achieved by restricting the dietary consumption of phosphorus and the administration of oral phosphate binders. In the last 15 years a number of new phosphate binders have been approved for use in human medicine. Each of these binders has variable potency, side effects, and cost allowing for greater flexibility and individualization of phosphate binding therapy. The purpose of this lecture is to provide an overview of available phosphate binders highlighting their mechanism of action, efficacy, and secondary effects that should be considered when creating a phosphate binder treatment plan. The lecture will also review ongoing work developing alternative therapeutics to manage persistent hyperphosphatemia.

PHOSPHATE BINDER THEORY

Phosphate binders primarily exert their therapeutic effect by ionic binding or ion exchange. The former class of phosphate binders is primarily metal-based and contains two elements that disassociate within the gastrointestinal tract. One of the elements binds phosphate with high affinity to create an insoluble complex that is not absorbed by the body. The freed second element is then available to bind cations within the intestine such as hydrogen and frequently acts as an acid buffer thereby exerting a secondary effect. For most ionic binders a portion of disassociated elements fails to bind phosphorus and is systemically absorbed potentially eliciting a tertiary effect. Examples of ionic binding phosphate binders include aluminum hydroxide, calcium carbonate, calcium acetate, magnesium carbonate, lanthanum carbonate, ferric citrate, and sucroferric oxyhydroxide. Alternatively, ion exchange phosphate binders bind phosphate to a large cationic polymer by exchanging an ion such as chloride or bicarbonate for phosphorus. The insoluble polymer thus makes any bound phosphorus unavailable for absorption into circulation. Examples of ion exchange phosphate binders include sevelamer hydrochloride, sevelamer carbonate, and chitosan.

PHOSPHATE BINDER OVERVIEW

IONIC BINDING PHOSPHATE BINDERS

Aluminum salts

In human medicine the use of aluminum salts for the treatment of hyperphosphatemia has been largely abandoned except for short-term salvage therapy to rapidly reduce serum phosphorus concentrations. The side effects commonly observed in people with long-term aluminum consumption including cognitive dysfunction, osteomalacia, and bone marrow suppression, are deemed unacceptable given the alternative treatments available. In veterinary medicine these side effects are observed at higher dosages, however, at the current time the potential benefits of this drug outweigh the side effects. The most important benefit may be the high potency of aluminum salts. Aluminum hydroxide binds phosphorus with a higher affinity than the calcium salts and in general is considered one of the most potent phosphate binders available. It is soluble at a wide range of pH allowing disassociation throughout greater lengths of the gastrointestinal tract compared with other drugs. The disassociation creates an aluminum ion capable of binding phosphorus as well as a hydroxide ion capable of reducing the body's acid load thereby allowing aluminum hydroxide to have a dual benefit in patients with CKD. The medication is also inexpensive when compared to the newer phosphate binders available on the market.

Calcium salts

Calcium carbonate was the second phosphate binder widely used in clinical human medicine. Like the hydroxide in aluminum hydroxide, the carbonate in calcium carbonate provides some buffering capacity however calcium carbonate's phosphorus binding capacity is less than that of aluminum. Potential explanations for this reduction are that calcium carbonate has reduced efficacy at a low pH because hydrogen ions effectively compete with calcium for phosphate binding. Unfortunately, the calcium carbonate tablets dissolve best in an acidic pH resulting in greatest drug availability in an environment suboptimal to maximize phosphate binding. This also means that availability of calcium carbonate may be reduced in patients that receive proton pump inhibitors or H2 blockers. An alternative treatment is calcium acetate (Phoslo). This binder is soluble at a wider range of pH and hence may be a better choice particularly in patients receiving acid suppressors. It is slightly more expensive than calcium carbonate but very affordable when compared to newer binders.

While calcium salts are an affordable, effective treatment for hyperphosphatemia, the greatest obstacle to their use is the potential induction of hypercalcemia. Up to 50% of people taking calcium-based binders will develop hypercalcemia. This side effect is exacerbated by the co-administration of vitamin D analogues. In a recent meta-analysis, the use of calcium-based phosphate binders increased mortality by 22% at 24 months as compared to people taking sevelamer or lanthanum carbonate. Similar studies in veterinary medicine do not exist although it is accepted that calcium salts should not be administered to hypercalcemic patients. The use of calcium salts in combination with aluminum hydroxide is an alternative that may reduce the dosage of both drugs thereby limiting the potential side effects.

Lanthanum carbonate

Lanthanum carbonate (Fosrenol) is a relatively new calcium-free phosphate binder on the U.S. market. It has the advantage of binding phosphorus within a wide pH range and is effective throughout the gastrointestinal tract. In animal studies lanthanum was comparable to aluminum salts in phosphate binding capacity. Largely insoluble, 93.4% of the drug is excreted in the feces of dogs after oral administration. Despite this, accumulation of the metal in the bones, brain, and liver has been detected with unknown consequences. The greatest argument against the use of lanthanum carbonate is its high cost with limited proven survival benefit over less expense alternatives. Evidence does exist demonstrating that lanthanum reduces aortocoronary calcification when compared to calcium-based phosphorus binders and its use should be considered in patients with current or historical hypercalcemia particularly when a second agent is necessary to control hyperphosphatemia.

Iron salts

The FDA approved the use of ferric citrate (Auryxia) as a phosphate binder in human dialysis patients in 2014. Given its composition of ferric ions and citrate this drug may potentially act as a phosphate binder, supplier of elemental iron, and as an acid buffer. Of these, only the two former attributes have been demonstrated in clinical studies. Ferric citrate can be titrated to achieve phosphorus control similar to other phosphate binders, however at a similar dose calcium carbonate was shown to be more effective at binding phosphorus. Ferric citrate has the added benefit of increasing plasma iron. After 1 year of administration human patients benefited from increased plasma iron, transferrin saturation, and serum ferritin such that erythropoietin usage was reduced by 25% and intravenous iron administration was reduced by 50%. Iron overload is the greatest risk in administering ferric citrate and caution should be used when administering ferric citrate in patients that are not iron deficient. In veterinary patients where the use of aluminum hydroxide is common, caution must be used when considering adding ferric citrate as a secondary agent. The release of citrate may enhance gastrointestinal absorption of aluminum increasing the risk of aluminum toxicity. No clinical studies exist examining the use of ferric citrate as a phosphate binder in veterinary medicine, and its very high cost may limit its use in most instances.

Sucroferric oxyhydroxide (Velphoro) is a second iron-based phosphate binder however the iron in this compound is not systemically absorbed. The drug is a mixture of sucrose, starch, water, and iron(III)-oxyhydroxide. The hydroxide groups are exchanged with phosphate throughout the gastrointestinal tract to elicit a similar serum phosphorus lowering effect when compared to sevelamer carbonate albeit at a much lower daily dose. Given the insolubility of the iron molecules this binder carries a low risk of iron overload with no evidence of iron accumulation after 1 year of use. Again in veterinary medicine there are not any clinical studies examining sucroferric oxyhydroxide's use as a phosphate binder, and its high cost may limit its utility.

Magnesium salts

Traditionally magnesium salts have not been used as phosphate binders because the development of hypermagnesemia. In limited clinical studies there is evidence that a formulation of magnesium carbonate and calcium carbonate may be a more effective phosphate binder than calcium carbonate alone with the added benefits of a decreased incidence of hypercalcemia and gastrointestinal side effects. Despite the lack of concrete evidence supporting its use, the low cost and limited gastrointestinal side effects observed may result in more widespread availability and use in the coming years.

ION EXCHANGE PHOSPHATE BINDERS

Sevelamer

Sevelamer is a nonabsorbable cationic polymer that exists in two forms as sevelamer hydrochloride (Renagel) and sevelamer carbonate (Renvela). The hydrochloride formulation has been removed from the U.S. market because the ion exchange of phosphorus for chloride created hydrochloric acid in the gastrointestinal tract and potentiated metabolic acidosis. To overcome this problem sevelamer carbonate was created where the chloride ions are replaced with bicarbonate thereby creating a dual effect of phosphate binding and bicarbonate production. Sevelamer's peak binding pH is 7 and it has less phosphate binding capacity per gram than most other phosphate binders. This reduced potency is related to a number of factors. First, sevelamer is a non-specific ion exchanger hence as it moves through the GI tract phosphate can be displaced by other anions. Second, within the duodenum sevelamer couples with bile salts limiting its ability to bind phosphorus. An additional side effect of the bile salt binding is that sevelamer can limit the absorption of lipid soluble vitamins and drugs such as vitamin D, ciprofloxacin, and mycophenolate. Overall this requires a higher dose of sevelamer to achieve similar serum phosphorus reductions in CKD patients and also higher costs.

Chitosan

Chitosan is a glucosamine polymer similar to sevelamer in that the compound contains positively charged amine moieties capable of binding phosphorus. Given that chitosan has limited amine groups (estimated to be less than ½) when compared to sevelamer, chitosan is an inferior phosphate binder at the same dosage. In addition, since chitosan is a non-specific ion exchanger, it will bind other ions throughout the gastrointestinal tract thereby further limiting its efficacy.

PHOSPHATE BINDER FAILURE AND FUTURE DIRECTIONS

Despite the use of phosphate binders, hyperphophatemia persists in many of our patients. There are many factors that may perpetuate this hyperphosphatemia including inadequate dietary phosphorus restriction and sub-optimal administration of phosphate binders. An additional factor may include the up regulation of NaPi-2b intestinal co-transporters in response to low phosphorus concentrations that occur secondary to reduced dietary phosphorus and phosphate binder use. Increased expression of sodium phosphate intestinal co-transporters may also help explain why long-term phosphate binder use fails to consistently reduce PTH and FGF-23 serum concentrations in clinical trials. In research studies NaPi-2b blockade in combination with phosphate binders improved phosphate control in uremic animals. Currently, the NaPi-2b inhibitor nicotinamide has been used in a number of small human clinical studies with some success in reducing plasma phosphate concentrations. Side effects reported included vomiting, diarrhea, significant thrombocytopenia, and one case of serve hepatotoxicity. Additional studies assessing the benefits/risks (or development of better tolerated alternatives) are necessary before nicotinamide is used more regularly in human medicine. In veterinary medicine nicotinamide has been used in dogs to treat immune mediated skin disease, however its value as an adjunctive treatment to treat hyperphosphatemia has not been reported.

REFERENCES

1. Damment S, et.al Clin Pharmacokinet 2008;47:553.
2. Froment D, et.al Kidney Int 1989;36:978.
3. Greig S, et.al Drugs 2015;75:533-542.
4. Jamal S, et.al Lancet 2013;382:1268.
5. Kays M, et.al Am J Kidney Dis 2003;42:1253.
6. Lenglet A, et.al Drugs 2013;13:165.
7. Lewis J, et.al J Am Soc Nephrol 2014;26:493.
8. Oh M, et.al Clin J Am Soc Nephrol 2014;9(5):967.
9. Pieper A, et.al Nephrol Dial Tranplant 2004;19:2630.
10. Schaefer K, et.al Am J Kidney Dis 1992;19:460.

Microanatomy of Feline Nephrolithiasis

Jody Lulich, DVM, PhD, DACVIM (Small Animal Internal Medicine)
St. Paul, MN

The occurrence of upper tract uroliths composed of CaOx pose unique management problems in part because a method to promote medical dissolution of CaOx uroliths has not yet been developed. The problem is magnified in cats with chronic renal failure since partial or total obstruction of just one ureter with a urolith may precipitate an acute onset of a uremic crisis (so-called "acute-on-chronic" renal failure). Results of a recent retrospective study concluded that medical and surgical management of ureteroliths in cats are associated with high morbidity and mortality. Because of the complex nature of this clinical syndrome, no routine approach is appropriate for every patient. Development of effective prevention and treatment strategies require an understating of how nephroureteroliths form. This article will present new information on the microanatomy of feline kidneys in cats with nephroureteroliths.

WHAT WE KNOW

1. The incidence of nephroureterolithiasis is increasing. Between 1980 and 1999, a 10 fold increase in the frequency of upper tract uroliths was extracted from the Veterinary Medical Data Base (Lekaroensuk 2005). In 1980, the hospital proportional morbidity rate of cats with upper tract uroliths was three cases per 10,000 cats. By 1999, the morbidity rate of cats with upper tract uroliths was 35 cases per 10,000 cats. Incidence rates during the past 15 years have not been reported.

2. Nephroureterolithiais is common in cats with chronic kidney disease (CKD). Of 88 cats with CKD entering a clinical trial, 47% (n=41 cats) had nephroureteroliths (Ross 2007).

3. Calcium oxalate is the most common upper urinary tract stone. In 2014, the Minnesota Urolith Center analyzed uroliths 16,056 cats. Four hundred and fourteen submissions were uroliths removed from the kidneys and/or ureters. Eight-three percent were composed primarily of calcium oxalate (figure 1).

Figure 1. Mineral composition of Feline Nephroureteroliths

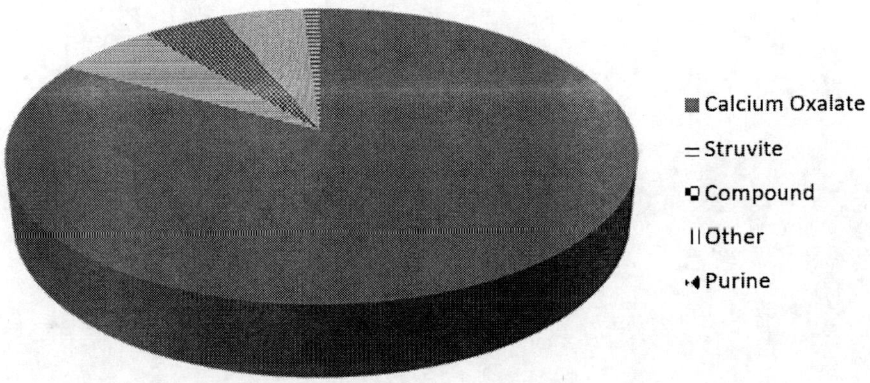

4. Obstructive ureterolithiasis is a cause of acute kidney injury.

WHAT WE DO NOT KNOW

1. The sequence of events, etiologies and risk factors leading up to calcium oxalate stone formation in the upper or lower tract of cats is poorly understood.

2. If nephrolithiasis is a consequence of renal disease or if renal disease is a consequence of nephrolithiasis.

3. Intermittent or persistent idiopathic hypercalcemia occurs in up to one third of cats with calcium oxalate nephroliths (Aronson 2006). The role of hypercalcemia in the formation of calcium oxalate uroliths seems obvious; however, it is not a requirement for most stone formers. Absence of evidence is not evidence of absence. In other words, is hypercalcemia not commonly associated with calcium oxalate urolithiasis because of its intermittent nature and therefore not commonly recognized?

WHAT WE ASSUME

Nephroliths form in the urine as a consequence of increasing supersaturation of urine for calcium oxalate. This assumption makes scientific sense. Increasing saturation promotes precipitation of minerals from their dissolved state favoring crystal formation and their growth into macroscopic uroliths. Most stones appear to reside in the urinary space also supports the idea that stone formation is primarily governed by the concentrations and functions of the constituents in urine. Because most stones are recognized free in the urinary bladder, this phenomenon is also likely occurring in the kidney. Therefore it is logical that therapy designed to prevent urolith recurrence lower saturation of mineral species in the urine.

Several questions are fundamental to insure that our preventative strategies are effectively managing the disease processes that we believe are underlying this disease. 1. Do nephroliths from in the urinary space or renal parenchyma? 2. Is the composition of the mineral deposits in the kidney identical to the composition of uroliths in the urinary space? To begin to answer these questions and others we looked at the microanatomy of the kidney.

Figure 2. Images of the right kidney from a 12-yr-old male-neutered cat with calcium oxalate nephroliths. A) Bisected right kidney with two stones in the renal pelvis of a (arrows). B) Micro CT demonstrating the two stones and mineralization in the renal parenchyma. C) Micro CT after subtraction of tissue revealing the two pelvic stones and multiple smaller stones in the renal parenchyma (arrow). D) Light-microscopic image of renal tissue near the renal papilla; the sites of mineral deposits (arrows) were stained brown by von Kassa methodology.

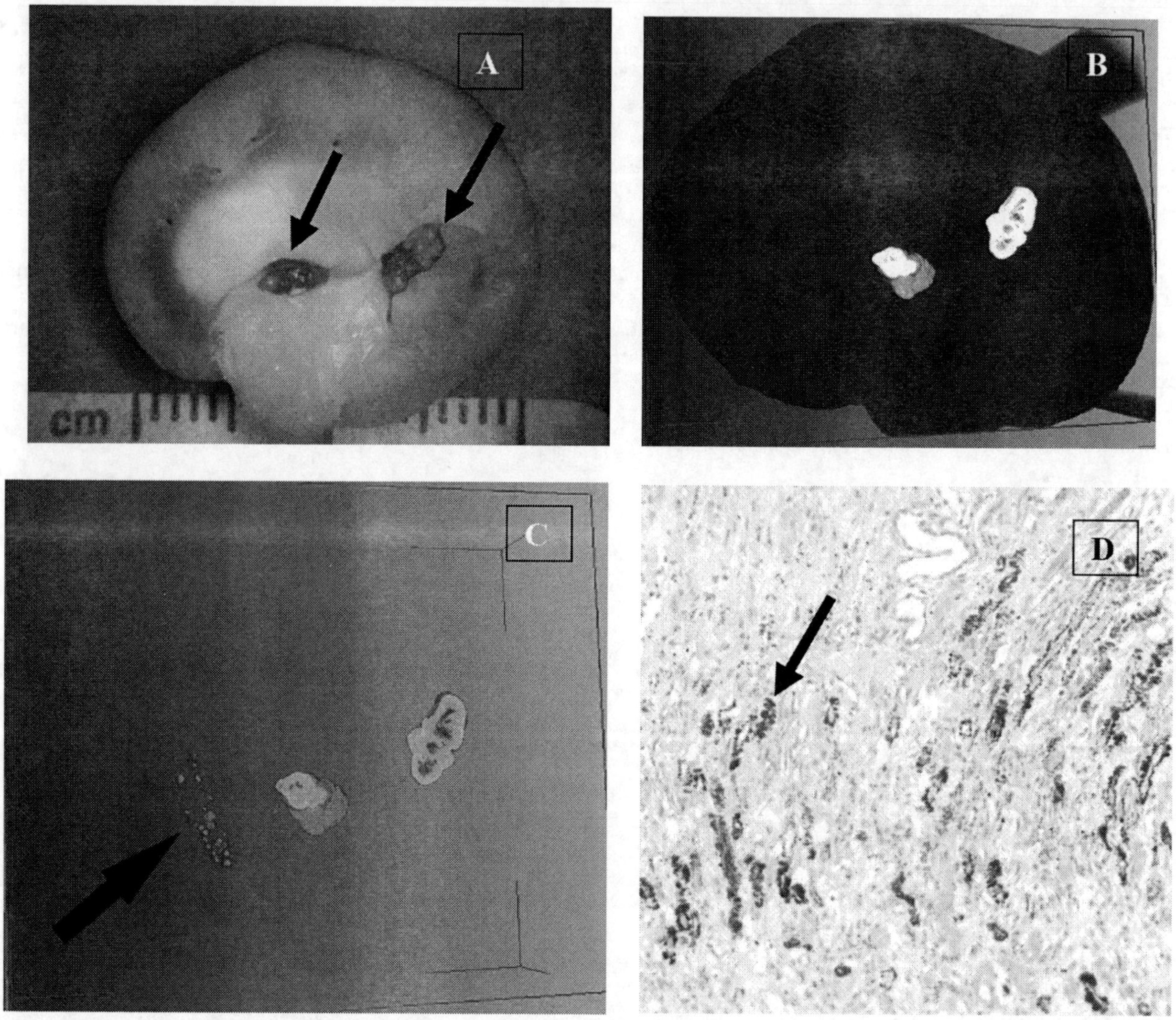

These findings indicate that calcium oxalate urolithiasis is a disease that involves the renal parenchyma and urinary space. Therapy only directed at urine saturation for calcium oxalate may not effectively prevent stones forming in the renal parenchyma.

REFERENCES

Lekcharoensuk C, et al. J Amer Anim Hosp Association, 2005;41:39

Ross SJ, et al. J Am Vet Med Assoc. 2007;230:1854

Aronson LR. et. al. J Am Vet Med Assoc 2006;228:743

Ureteral Obstructions: Stents Versus Subs-The Last 10 Years of Experience

Dr. Allyson Berent, DVM, DACVIM (SAIM)
New York, NY

Introduction

An increasing incidence of problematic ureteroliths, ureteral strictures, obstructive pyonephrosis, and obstructive neoplasia has been seen in veterinary practice over the past decade. The invasiveness and morbidity associated with traditional surgical techniques make these conditions a management dilemma for most veterinarians. Interventional radiologic (IR) and interventional endoscopic (IE) techniques have enabled clinicians to simultaneously diagnose and treat upper urinary tract obstructions more effective, more successful, and significantly less invasive than traditional surgical techniques. The technology requires the use of fluoroscopy, often with flexible and/or rigid endoscopy, and various wires, catheters, and/or stents. The development and improvements in endourology in human medicine (like ureteroscopy, ureteral stenting, ESWL, laser lithotripsy, laparoscopy, and percutaneous nephroureterolithotomy [PCNUL]) have almost eradicated the need for open surgery for upper urinary obstructive disease. The same trend has occurred in veterinary medicine. This talk will focus on the experience over the past 10 years we have encountered using internvetional techniquies for the management of ureteral obstructions in dogs and cats, with a focus on the over 400 treated feline cases and 150 canine cases to date, in the speakers practice. Please note that this lecture will expand upon the speaker's personal experiences, and will also address many of the controversies that exist in veterinary medicine in making medical, interventional and surgical decisions, both diagnostically and therapeutically. Other lectures in this conference will focus strictly on complications associated with some of these interventional techniques.

Traditional therapeutic options will only be briefly reviewed, and the speaker refers the audience to the published literature for more details.

ETIOLOGY

Greater than 90% of feline, and 50% of canine, upper tract stones are reported to have a composition of calcium oxalate, meaning that they will not dissolve medically. They either need to pass spontaneously, be removed, or the urine needs to be diverted when an obstruction is present. Additionally, obstructions can occur for other reasons beyond stone disease, like dried solidified blood stones (cats), strictures (25% or more of cats; <10% in dogs), and tumors (<5% of dogs or cats with ureteral obstructions). For each of these conditions different options exist and should be considered.

Regardless of the cause of the obstruction treatment needs to be considered in an expedited manner and should not be delayed. The physiologic response to a ureteral obstruction is very complex. In dog models it is shown that after a complete ureteral obstruction an immediate increase in renal pelvic pressure occurs and the renal blood flow diminishes by 60% over the first 24 hours, and 80% within 2 weeks. This excessive pressure decreases the glomerular filtration rate (GFR). The contralateral kidney will increase its GFR in response, as long as that kidney is normal and has the potential for hypertrophic compensation. In has been shown that the longer the ureter remains obstructed, the more progressive the damage (after 7 days the GFR permanently diminished by 35%, and after 2 weeks by 54%). Since this research model was performed in normal dogs with a complete ureteral obstruction, without pre-existing azotemia, chronic interstitial nephritis, fibrosis, or chronic obstruction, an extrapolation of a worse outcome might be expected in a clinical patient once the hypertrophy mechanisms are exhausted. Additionally, since over 30% of cats will develop renal azotemia as an adult, leaving less than 25% of renal function, any preventable loss in GFR should be avoided. This encourages aggressive and timely intervention whenever possible. Interestingly, in contrast to the irreversibility of a complete obstruction, partial obstructions have been shown to result in less severe and slower nephron destruction, giving more time for intervention when needed. It is estimated in the speaker's practice that over 75% of feline ureteral obstruction and 40-50% of canine ureteral obstructions are considered partial, based on ureteropyelography. In a dog model it was found that after decompression the GFR returned to normal after 4 weeks of a partial obstruction. Knowing that many patients are partially obstructed, with concurrent renal compromise, aggressive diagnosis, management, and obstruction relief is recommended whenever possible.

It is now clear that partial obstructions result in permanent renal damage and treatment should always be considered, regardless of the degree of pelvic dilation noted on ultrasound, or the suspected chronicity of the condition. Knowing the cause of the obstruction is very important when deciding on traditional surgical fixation methods as different approaches will be taken for a tumor, stricture, or stone induced obstruction. For the newer interventional treatment options (double pigtail ureteral stents or a subcutaneous ureteral bypass device), these details are slightly less important, as all can safely be used to treat most causes of a ureteral obstruction (stone, stricture, or tumor), regardless of obstruction cause, location, stone number, or the presence of concurrent nephrolithiasis.

TREATMENT OF URETERAL OBSTRUCTIONS

Once medical management fails to encourage stone passage, which is the case is approximately 90% of veterinary patients, (traditionally: intravenous fluid therapy, mannitol CRI and alpha-adrenergic blockade 24-48 hours) the immediate resolution of the obstruction should be considered. It is important to realize that over 25% of feline ureteral obstructions are reported to be associated with strictures (+/-stones), where medical management and traditional surgery are not expected to be highly effective.

Once medical management has been exhausted than emergency decompression should be considered to avoid permanent damage. For cases that are oliguric, anuric, hyperkalemic, or overhydrated, than immediate intervention or intermittent hemodialysis therapy should be considered. The main interventional options considered in the author's practice include: ureteral stenting (canine), and the placement of a subcutaneous ureteral bypass (SUB) device (feline). Decompression of the renal pelvis will decrease the hydrostatic pressure, and stop the permanent damage that is occurring to the renal parenchyma, allowing for the possibility of the most immediate renal recovery.

Traditional Surgery

In the past, traditional intervention for ureteral obstructions has been accomplished surgically via ureterotomy, neoureterocystostomy, ureteronephrectomy, and renal transplantation. Kyles et al. in 2005 reported two retrospective studies in a large number of cats. There were high procedure associated complications (over 30%) and perioperative mortality rates (21%) documented. This study was from two universities that have extensive expertise with ureteral surgery. The morbidity and mortality may be higher in environments where operating microscopes and microsurgical experience is not as readily available as these two institutions. Many of the associated complications with surgery are due to site edema, recurrence of obstruction from stones that pass from the renal pelvis to the surgery site, stricture formation, and ureterotomy-associated or nephrostomy tube-associated urine leakage.. Of the cats that had long term imaging follow-up, 40% had evidence of ureteral obstruction recurrence, 85% percent of which had evidence of ipsilateral nephrolithiasis at the time of the first ureteral surgery. The number of animals that did not have stone recurrence with prior nephrolithiasis was not evident in that study. Another study, by Roberts et al, documented the same perioperative mortality rate after ureterotomy of 21%. Chronic kidney disease was found to be common at the time of diagnosis (>75% were azotemic with a unilateral obstruction), and persistent azotemia was commonly seen after a successful surgery (over 50-80% of cats). Though, with all of the surgical concerns, the survival rates were dramatically higher for cats that had surgical intervention performed when compared to those treated with medical management alone (33% mortality prior to discharge; 87% failure to see renal function improvement). This was a large series was in cats in which surgery was performed or attempted, hence excluding patients that were not considered good surgical candidates, those with strictures, too many stones to attempt multiple ureterotomies, etc. Over the past 10 years the number of stones found in the ureter and kidney have been much greater (median of 4 stones per ureter in a recent study of 79 ureteral obstructions in cats, and 57 ureteral obstructions in dogs).

Ureteral strictures were reported in 10 cats in 2011 by Zaid et al, and have since been reported in a larger series of 79 feline ureteral obstruction (Berent, JAVMA 2014), in a series of 22 cats with circumcaval ureters (Steinhaus, abstract, JVIM 2013), and in 5 of 57 dog ureters (Pavia, 2015). Ureteral strictures can occur for various reasons, and are most commonly seen secondary to a previous surgery of the ureter for stone disease, from a stone getting embedded in the ureteral mucosa, as a congenital abnormality, associated with idiopathic renal hematuria, as an idiopathic process, and possibly associated with a circumcaval (retrocaval) ureter. Most ureteral strictures were found to occur in the very proximal ureter (<2.5 cm from the renal pelvis), in which traditional surgical approaches were difficult requiring re-implantation with renal desensus and psoas cystopexy or a ureteral resection and anastomosis. With a cat ureter being 0.3-0.4 mm in luminal diameter, this can be a very challenging situation, particularly in a potentially unstable patient.

Finding an alternative that results in immediate decompression and stabilization of the kidney, while concurrently allowing patency to be established quickly and effectively, for any cause of a ureteral obstruction (stone, numerous stones, stricture, debris, blood clots, purulent debris, blood stones, etc), which can decrease post-operative complications, mortality and long-term re-obstruction is ideal. The use of interventional radiologic techniques, either minimally invasively, or with surgical assistance, has aided in circumventing the complications of ureteral surgery (peri-operative leakage, stricture, re-obstruction, long anesthesia times, worsening renal function, etc), particularly in cases that are not considered good surgical candidates. IR techniques have allowed successful and efficient stabilization of the patient regardless of the cause of ureteral obstruction, while decreasing renal pelvic pressure and stopping the cycle of pressure induced nephron damage and loss.

Interventional Options

The Subcutaneous Ureteral Bypass Device (SUB)

The SUB device (Norfolk Vet Products) was initially created for feline patients as an alternative to ureteral stents when either a stricture was present (higher rate of stent occlusion [>50%]), or a stent was unable to be successfully placed due to excessive stones, very narrow lumen, or patient stability. Over the course of the following 7 years, the long-term outcomes documented with the SUB device, as far as long-term complications and patient comfort (primarily stent occlusion and recurrent dysuria), were found to be superior to that of stents in cats. Due to these findings, the trend in the author's practice has moved toward using the SUB device as a primary treatment option of feline ureteral obstructions and ureteral stents for the treatment in dogs. It is very important to understand the risks and benefits of traditional surgery, ureteral stenting and SUB device placement device prior to deciding the best treatment option for your patient, all of which should be carefully discussed with the pet owner. The SUB device has been used in a small subset of dogs, but stents have been so successful, placed endoscopically, on an outpatient basis, and with excellent long-term result, that the author does not prefer the SUB in dogs, unless a stent fails or is contraindicated.

The placement of nephrostomy catheters in veterinary medicine has been met with excellent success when the appropriate device is used, but the limitation is the externalized drainage, which is not a long-term fixation, requiring an additional surgery to something more definitive. The development of an indwelling SUB device using a combination locking-loop nephrostomy catheter and cystostomy catheter connected to a subcutaneous shunting port has been highly successful for the treatment of all causes of ureteral obstructions (stones, stricture, tumors and obstructive pyelonephritis). This device reduces the complications associated with externalized nephrostomy tubes and ureteral stents, and have been shown to be associated with lower mortality and long-term complications than both traditional surgery and ureteral stents in cats. This presentation will expand upon the author's outcome of using a SUB device in 175 patients with both short and long-term outcomes.

The use of the feline SUB device has recently been described in various abstract and peer-reviewed formats. To date, in the authors practice, over 225 of these devices have been placed over a 7 year period with good short and long-term success. Updated statistics will be presented in this larger group of cats during this lecture. In a previous abstract on 61 SUBs the median pre-operative creatinine was 6.8 mg/dL and 2.4 mg/dL post-SUB placement. The peri-operative mortality after surgery was 5.6%, with 1 death associated with a procedure related complication early in the course of the study. This compares well to traditional surgery (21% peri-operative mortality) and ureteral stenting (7.5%). The median followup time is 16 months (range 6-95 months), with a majority of patients still alive. Peri-operative complications are typically rare and include leakage (at the port or nephrostomy tube insertion site) (<5%), kinking of the catheters (or failure of the renal function to recover (<3%), and blockage of the system with either blood clots, debris, purulent material, or future stones (~15-20%). Dysuria is rarely seen in cats with a SUB device (<5%) compared to the ureteral stents (38%). Chronic urinary tract infections is also uncommonly seen in SUB cats, and are more typically associated with cat that had infections prior to SUB placement or those that had externalized urinary catheters during or after SUB placement. The SUB device is considered a functional option and is the author's treatment of choice in cats with ureteral obstructions.

Ureteral Stenting

Ureteral stenting has been performed for a variety of disorders in both dogs and cats. The goal of using a feline double pigtail ureteral stent (Infiniti Medical, LLC) is to divert urine from the renal pelvis into the urinary bladder during a ureteral obstruction (ureterolithiasis, obstructive neoplasia, ureteral stricture/stenosis, dried solidified blood clots or severe obstructive pyelonephritis or pyonephrosis). Stents also encourage passive ureteral dilation (for ureteral stenosis/strictures or future ureteroscopy, extracorporeal shockwave lithotripsy [ESWL], etc) in the ureter which was recently proven in dogs after endoscopic stent placement over a 10 week period. The double pigtail stent, which is the method of choice for decompression in dogs, is completely intracorporeal, and can remain in place long-term maintaining ureteral patency and passive ureteral dilation (a 0.4 mm ureter was documented to dilate to 1.5-2.0 mm within 3-7 days in over 90% of cats, and a1-1.5 mm ureter in dogs dilated to 1.5-2.5 mm). Each loop of the pigtail is curled (one in the bladder and one in the renal pelvis) with the shaft of the stent travelling within the ureteral lumen, allowing for direct urinary diversion from the kidney to the urinary bladder, around the stones, or through the stricture/tumor.

Double pigtail ureteral stents have many benefits: (1) immediate decompression of the renal pelvis, (2) passive ureteral dilation for urine and stone passage, (3) prevention of strictures and leakage associated with surgical or non-surgical procedures, and (4) prevention of a ureteral obstruction after extracorporeal shock-wave lithotripsy (ESWL). All of these benefits have been clearly documented in human medicine and, more recently, in veterinary medicine, with the placement of ureteral stents in more than 125 feline and 150 canine ureters for various indications in the author's practice. The use of double pigtail ureteral stents in dogs and cats has been investigated as a potentially less invasive alternative to traditional surgery, resulting in immediate ureteral decompression, stabilization of the associated azotemia, a decreased risk of ureteral stricture or ureteral leakage/uroabdomen, and a decreased rate of ureteral obstruction recurrence. Placement of a double-pigtail ureteral stent, via either minimally invasive techniques (endoscopy and interventional fluoroscopy in most dogs) or with surgical assistance using fluoroscopic guidance (nephrostomy access in most cats), seem to have minimized the complications of surgery alone (e.g., leakage, stricture, re-obstruction), and has resulted in expedited stabilization. There are currently veterinary ureteral stents available, which were designed for use in dogs and cats. Prior to considering this option, the operator must be appropriately trained in stent placement, the risks of stent placement needs to be understood, and owners must be educated on these risks. The operator must understanding the nuances of stent placement, which can be very difficult, and these outcomes reported are all from the same operators, where over 250 stents have been placed to date. Once experience is obtained, than stent placement is highly successful. The main concerns with stents are typically seen months to years after placement. They are not usually life threatening, and are relatively easy to address on an outpatient basis. These include: dysuria, stent migration, stent occlusion (seen most commonly with ureteral strictures), and stent encrustation. These problems are all more common in cats than in dogs, which has pushed the move toward using the subcutaneous ureteral bypass (SUB) device preferentially in cats and ureteral stents in dogs.

In dogs a recent report by Pavia et al in 44 dogs with 57 ureteral obstrluctions caused by ureterolithiasis (84%) or strictures (17%), ureteral stenting was highly successful. It is most commmonly performed with endoscopic and fluoroscopic guidance in a retrograde manner, and is considered an outpatient procedure. This has been successful over 90% of dogs to date in the author's practice for all causes (stones, stricture and tumors). The complications observed in dogs are far fewer than that seen in cats in all time periods (peri-operative, short and long-term). The mortality rate was 2%. Comlications included recurrent urinary tract infections (~25%, with approximately 60% having infections pre-stent), proliferative tissue growth around the distal loop of the stent at the ureterovesicular junction (~15%), stent migration (<5%), stent occlusion (<10%), and encrustation (<10%). Dysuria is much less common in dogs than in cats after stent placement (<2%), and both species are typically glucocorticoid responsive if it does not resolve spontaneously. The median followup time was 1158 days.

A recent study evaluating 69 feline patients (79 ureters) with ureteral obstructions treated with a ureteral stent had a 96% placement success, 14% were bilaterally obstructed, 28% were associated with strictures (+/- concurrent ureteroliths), and 85% of cases had concurrent nephroliths. The median number of stones in the obstructed ureter was 4. Approximately 1/3 of cats had documentation of a urinary tract infection prior to stent placement, and 30% of cats had at least 1 positive culture in their lifetime post-stent placement. The median creatinine was 5.3 mg/dL pre- and 2.1 mg/dL post-stent placement. The peri-operative mortality after surgery was 7.5%, with no death associated with a ureteral obstruction or the procedure. This compares to 21% mortality rate with traditional surgery. The most common long-term (>1 month) complications after ureteral stent placement was dysuria (38%, but only persistent in 1.7%), chronic hematuria (18%), stent occlusion (19%), stent migration (6%), and ureteritis/mucosal proliferation (4%). The dysuria typically resolved with short-term steroid therapy. A ureteral stent exchange with either a bigger stent or a SUB device was needed in 27% of cats for either occlusion or migration. This complication was most commonly associated with ureteral strictures (over 50% of strictured ureters experienced reocclusion), or an occlusion at a previous ureterotomy site. Ureters remained patent long-term in most cats with the longest stent in place and functional for over 5 years. The median survival time was 498 days. The median survival time when renal cause of death was over 1250 days, with only 21% of cats ultimately dying of suspected CKD over the course of the entire followup period. Another study evaluating pre-operative prognostic factors for short- and long-term survival and renal recovery did not identify any clinical, biochemical or imaging findings to suggest or predict overall survival times in feline patients This preliminary data suggests that ureteral stenting, in both dogs and cats, is safe and effective, resulting in immediate decompression of the renal collection system. Few major procedural or peri-operative complications occurred, particularly in dogs, but the learning curve was steep. In cats long-term stent exchange (~20%) or manipulation may be necessary if dysuria, a ureteral reaction, stricture-induced stent occlusion, (18.5%) or stent migration (6%) occurs. Because of the management issues associated with feline ureteral stents, which are tolerable but can be expensive and frustrating, the author currently recommends the subcutaneous ureteral bypass device (SUB) for cats with ureteral obstruction and ureteral stents for dogs.

Conclusion

Ureteral obstructions are frustrating cases on both dogs and cats, but with the new technology and intense training available, successful treatments are becoming more available to specialists. With the recent advances in veterinary interventional endourologic techniques, diagnosis and treatment has become less invasive, more effective, and safer. Proper training and the availability of specialized equipment are needed to help these procedures become more available in the future. With these new devices in veterinary medicine, we are hoping to find better alternatives for these problematic conditions, as we have seen in our human counterparts. These techniques offer an alternative option to traditional surgery, and as the data is published the veterinary community can decide if, and when, using interventional techniques will replaced traditional surgery, as is the case in human medicine.

REFERENCES

Available upon request.

Conflict of Interest statement

The author is a consultant for Infiniti Medical, LLC and Norfolk Vet Products.

Leukoreduction: Disadvantages from the Perspective of the Blood Bank

Anne S. Hale, DVM
Albuquerque, NM

Universal leukoreduction of blood products is performed in most industrialized countries except the United States. Historical debate regarding the advantages and disadvantages of leukoreduction for blood products has been well documented in the human transfusion literature. With the progression of veterinary transfusion medicine and hematology, the question of leukoreduction has been raised. While the advantage of leukoreduction is well documented, there are significant reasons why this practice is not recommended in the veterinary blood banking community. Simply stated, leukoreduction requires additional cost and the logistics of processing blood products with existing closed system devices are difficult.

Veterinary blood banking involves three tiers of production: nationwide, regional and local. The scope of these production programs vary in product produced, species provided, annual production and donor source. Successful conversion to universal leukoreduction requires standardized production protocols using existing closed system devices. Closed systems do not exist for most domestic species including the cat, horse, cow and llama. Use of sterile docking systems necessary for modifying collection systems for these species adds $10,000.00-$20,000.00 to the annual cost of good. It is estimated that 40-50% of veterinary blood products used in transfusion practice is provided by regional or local blood banks. Feasibility for the regional or local program is impractical due to their small annual production volumes.

Nationwide programs would experience an estimated increase of $40.00/unit in order to provide leukoreduced products. All blood bank personnel would have to be trained in the appropriate use of leukoreduction filter systems incurring additional labor cost for each program. Device costs for changes in processing as well as individual filter costs are also part of this cost increase. In veterinary medicine, the client still pays directly for veterinary services with few third party pay systems in place. Increasing the cost of a blood product by $40.00 in this system would limit blood product use due to limited financial resource in many cases.

Leukoreduction costs remain high in the United States by comparison to other countries. Universal leukoreduction in human blood banking is the driving factor in cost savings for this technique. Leukoreduction is unlikely to become universal in the United States in the near future. Therefore, cost of leukoreduction will remain high until this policy is changed.

Logistically, universal leukoreduction of veterinary blood products will not occur.
Currently, the industry lacks standards and protocols vary widely from program to program. Advanced leukoreduction filters use both physical and chemical features of blood to successfully remove the white blood cells. Variance in blood cells, proteins and chemical processing between species would require extensive research and validation studies to document leukoreduction. As previously mentioned, few veterinary species can be drawn economically and safely using closed system human devices. Further variance in end product would be introduced with modification of the systems to accommodate individual species requirements.

Leukoreduction filters also cause the loss of 20-30% of the end product due to processing consumption. These filters are designed around high volume processing for multiple units of 400-500ml. This process can be used for dogs, sheep and goats using the current systems; however, due to volume variance in processing, cats, horses, cattle and llamas would require significant modifications of the available closed systems. Furthermore, due their small volume, feline units could not be processed as single donor units with available commercial filters.

"Fourth generation" leukoreduction filters that are commercially available have been optimized to filter human leukocytes from whole blood. Materials have been selected not only for their filtration ability but also because they select by "charge" the leukocytes as they pass through the filter. Species variance in veterinary medicine is significant enough to lose the optimization of the filter mechanism. This means that filters would have to be evaluated for each species and validated before use. Primary grants and research dollars for this type of validation are hard to achieve as they do not provide a model for human disease.

Cost makes universal leukoreduction of veterinary blood products unlikely. The logistics of implementing leukoreduction of veterinary blood products in any species make the practice difficult at best. Although the advantages of leukoreduction provide a convincing argument for its use in veterinary medicine, the likelihood of implementation using currently available commercial systems is low. Further development of closed systems that can easily incorporate species specific leukoreduction filters would be needed for universal use.

REFERENCES

1. Brownlee L, et al JVIM2000 Jul-Aug;14(4):412-7.
2. Slichter S, et al BLOOD, 15Jan2005:;vol 105(2): 847-854.
3. Hod EA, et al Transfus Clin Biol 2012 June;19(3):84-9.
4. McMicheal MA, et al JVIM2010 Sep-Oct;24(5): 1131-7.
5. Lacerda LA, et al Vet Clin Pathol 2014 Sep;43(3): 362-70.
6. Obrador R, et al JVECC(San Antonio) 2015 Mar-Apr;25(2):187-99.
7. Smith SA, et al AJVR 2015 Jan;76(1):35-41.

The Case for Anticoagulant Therapy in Canine IMHA

Rita M. Hanel, DVM, DACVIM (SAIM), DACVECC
Raleigh, NC

INTRODUCTION

In order to prevent thrombosis, one has to know or predict when or if it will occur. How to do this remains an enigma in both human and veterinary medicine. With the advent of the cell-based model of coagulation, it has become increasingly apparent that the hemostatic system is reliant on both plasmatic and cell based factors, as well as the endothelium and the microenvironment in which it lives. Any test interrogating just one portion of this system at a single point in time is likely to fail in its ability to predict how the system will behave as a whole. The recent exploitation of viscoelastic monitors as a means of global hemostatic evaluation has enabled veterinarians to document hypercoagulable states in patients with variable diseases, including: protein losing nephropathy, protein losing enteropathy, neoplasia, disseminated intravascular coagulation, parvovirus, hepatic disease, immune mediated hemolytic anemia or thrombocytopenia. Although a hypercoagulable tracing may signify a risk for thrombosis, it still represents only one arm of the so-called Virchow's triad; which includes stasis, vascular endothelial injury, and a hypercoagulable state. Thus, the optimal strategy to prevent thromboembolic disease remains elusive. This session will focus on the evidence supporting the use of anticoagulants to prevent thrombosis, with a focus on canine patients with immune mediated hemolytic anemia (IMHA). This will be broken into three sections: 1) the distribution and type of thrombus formation, 2) documented success and failure of anticoagulants in the veterinary literature, 3) and possible rationale for these outcomes.

Distribution and Nature of Thrombi

Thromboemboli are classified as arterial or venous in origin, which are often referred to as "white" or "red" clots, respectively. Arterial thrombi generally occur in high flow environments, rely on platelet activation and are rich in platelets and fibrin. Venous thrombi generally form in low flow environments, whereby they accumulate large amounts of fibrin and trap red blood cells. IMHA, the most common immune-mediated disease in the dog, has reported mortality rates of 30 to 70%, and thromboembolism is identified on post-mortem in up to 100% of these patients.[1-6] Pulmonary thromboemboli are commonly reported, but embolism been has been documented to occur in other various locations, including the heart,[7] within the arterial (i.e. splenic, renal, iliac, and mesenteric) and venous systems (i.e. portal, splenic, and vena cava).[3,4,8-10] According to American College of Chest Physicians (ACCP) guidelines, immediate therapeutic anticoagulation is recommended in people in whom deep venous thrombosis or pulmonary embolism (grade 1B) is suspected.[11] Low molecular weight heparin or unfractionated heparin is used and either is typically combined with an oral anticoagulant. Efficacy depends on achieving rapid therapeutic levels of the chosen strategy within the first 24 hours of treatment. Based on the aforementioned, anticoagulation appears to be a rational approach for canine patients with IMHA. However, given the lack of standardization concerning strategies and what represents "therapeutic" in veterinary medicine, comparisons between published outcomes and approaches present a challenge.

Anticoagulants in the Veterinary Literature

Studies evaluating the efficacy of anticoagulants are lacking, and several involve the use of fixed dosages of unfractionated heparin (UFH). One prospective study of the use of UFH in 18 dogs with IMHA evaluated the effect of fixed dosages of UFH (300 IU/kg subcutaneously every 6 hours) on activated partial thromboplastin time (aPTT) as well as factor Xa inhibition (e.g. anti-Xa or AXa activity).[12] During the initial 40 hours, only eight of 18 dogs attained therapeutic AXa activities, and 15 out of 18 dogs survived to discharge. In a second prospective report of hemostatic abnormalities in 20 dogs with IMHA, 17 of 20 dogs were treated with heparin at a starting dose of 100 IU/kg body weight subcutaneously every 6 hours, for a range of 3 to 28 days.[8] The UFH dose was adjusted in order to prolong aPTT by 1.5 to 2.0 times above reference range. Four dogs died during hospitalization and two additional died during the first month after discharge. Thromboembolism was the most common finding on necropsy. Further information regarding UFH therapy, such as time to therapeutic concentration, was not mentioned. A recent randomized, prospective, controlled clinical trial in dogs with IMHA showed an 85% reduction in 180 day mortality when UFH anticoagulant therapy was adjusted based on AXa activity versus standardized or fixed dose administration.[13] Seven out of 8 dogs with individually adjusted heparin survived, whereas only one out of seven survived in the fixed dosage cohort.

Low molecular weight heparins (LMWH) are a preferred parenteral anticoagulant in human medicine, owing to their predictable pharmacokinetics and pharmacodynamics. Based on their predictability, LMWH is not typically titrated and monitored but administered as a fixed dosage. One retrospective study examined a LMWH, dalteparin, in 38 critically ill dogs, and found that fixed dosages were often not therapeutic and dose adjustment in response to AXa activity interpretation inconsistently resulted in subsequent attainment of the target AXa range.[14] Also, dogs that appeared more hypercoagulable on pre-treatment thromboelastography were significantly less likely to achieve the target AXa activity. Thus, once cost and the aforementioned results are taken into consideration, the benefits of LMWH in veterinary medicine compared to UFH are not as clear.

Rationale

The high mortality and shortcomings of the aforementioned anticoagulation strategies in dogs with IMHA may be related to our gaps in knowledge surrounding the pathogenesis of thromboembolism or our inability to appropriately titrate therapy. Several studies have documented alterations in platelet activation, likely to be discussed in the other half of this session. Factors contributing to the initiation of coagulation include, but are not limited to: elevated levels of fibrinogen, exposure to red cell membrane components, circulating microparticles, endothelial activation, the presence of neutrophil extracellular traps, and intravascular expression of tissue factor. In the author's opinion, the likelihood that multiple of the aforementioned elements contribute to thrombus formation is high. As such, the blockage of any one mechanism or receptor is unlikely to effect resolution. Furthermore, platelets are consumed in this thrombotic, inflammatory condition, and in the face of thrombocytopenia, mortality remains high.[3,8,12] Yet, even if the use of anticoagulants appears rational, veterinary medicine is still lacking a consensus regarding how to target therapy.

Intravenous UFH use has declined as newer drugs have been introduced. However, in the setting of the critically ill or inpatient, it will likely continue to be used due to its short half-life, ability to be reversed, and non-renal route of elimination. As such, it is prudent to be familiar with the currently available methods of monitoring. There are three main categories of available monitoring: global clotting assays measuring primarily the intrinsic and common pathways of coagulation, global assays utilizing viscoelastic technology, and methods of measuring heparin concentration. These three approaches will be briefly discussed during the session and preliminary results from an ongoing study at the author's institution will be presented. Factors affecting the usage of aPTT and AXa activity have been summarized in Table 1.

Summary

In summary, there is strong rationale for the use of anticoagulants in canine patients with IMHA. Heparins are the most commonly used anticoagulant and measurement of the functional activity of heparin, via AXa activity, is considered an optimal method for monitoring and is recommended by the American College of Chest Physicians (ACCP) and the College of American Pathologists (CAP). If this is not readily available, it is recommended to establish institution specific references and nomograms for therapeutic aPTT ranges corresponding to heparin concentrations, based on AXa activity or protamine titration. This is due to the marked variability in aPTT results by different reagent-instrument combinations, which has been demonstrated by multiple studies in humans as well as in dogs.[15] Although a canine specific target therapeutic concentration has not been established, it appears reasonable to initially adopt human standards based on the results from Helmond et al.[13] Studies are needed to evaluate the utility of novel oral anticoagulants in this setting.

Table 1: Factors known to influence activated partial thromboplastin time (APTT) and anti-Factor Xa (AXa) activity in humans when used in the monitoring of unfractionated heparin (UFH). I=increase in laboratory result; D=decrease in laboratory result; N=minimal effect on laboratory result. * Indicates that dose adjustment based on result may be incorrect.

PREANALYTICAL AND ANALYTICAL FACTORS	APTT	AXA ACTIVITY
Blood sampling in the morning/evening (diurnal variations)	D/I*	N
Underfilled tubes / too much citrate	I*	N
Sampling too close to site/improper flushing; heparin contamination	I*	I*
Delay in analysis / processing sample (> 1 hour)	D*	D*
Inadequate centrifugation (platelet contamination; heparin neutralization by platelet factor 4)	D*	D*
Hemolysis	N	D*
Change in reagents; necessitates re-calibration	I or D	I or D
Change in machine/coagulometer; necessitates re-calibration	I or D	I or D
BIOLOGICAL FACTORS	**APTT**	**AXA ACTIVITY**
Hyperbilirubinemia (> 6.6 mg/dl in humans)	N	D*
Hyperlipidemia	N	I*
Antithrombin deficiency; congenital or acquired form of "heparin resistance"	D	D or N (1 or 2 stage)
Increased acute phase reactants (fVIII, fibrinogen, etc...); mimics "heparin resistance" if using aPTT	D*	N
Increased heparin binding proteins (infection, inflammation, malignancy); common cause of "heparin resistance" requiring higher heparin dosages	D	D
Obesity (increased volume of distribution)	D	D
Impaired renal function (decreased UFH elimination)	I	I
Liver dysfunction/disease or Vitamin K antagonism (decreased factor production)	I*	N
Consumptive coagulopathy	I*	N
Clotting factor deficiencies (fXII, VIII, IX, etc...)	I*	N

REFERENCES

1. Reimer ME, et.al J Am Anim Hosp Assoc 1999; 35: 384.
2. Weinkle TK, et.al J Am Vet Med Assoc 2005; 226: 1869.
3. Carr A, et.al J Vet Intern Med 2002; 16:504.
4. McManus PM, et.al J Am Vet Med Assoc 2001; 218: 1308.
5. Mason N, et.al J Vet Intern Med 2003;17(2):206.
6. Lunsford KV, et.al Vet Clin North Am Small Anim Pract 2007;37(3):579.
7. Orcutt ES, et.al J Vet Emerg Crit Care 2010;20(3):338.
8. Scott-Moncreiff JC, et.al J Am Anim Hosp Assoc 2001;37(3):220.
9. Klein MK, et.al J Am Vet Med Assoc 1989;195(2):246.
10. Johnson LR, et.al J Vet Intern Med. 1999 Jul-Aug;13(4):338.
11. Guyatt GH, et.al Chest 2012 Feb. 141(2 Suppl):7S.
12. Bruehl EL, et.al J Am Anim Hosp Assoc 2009;45:125.
13. Helmond SE, et.al J Vet Intern Med 2010; 24(3): 597.
14. Lynch A, et.al J Vet Emerg Crit Care 2014; 24(4): 421.
15. Mischke RH, et.al Am J Vet Res 2001; 62(12): 1887.

The Case for Anti-Platelet Therapy in Canine IMHA

Kari V. Lunsford, DVM, DACVIM (Small Animal Internal Medicine)
Starkville, MS

INTRODUCTION

Targeted thromboprophylactic therapy for dogs with IMHA remains a challenge. In order to provide targeted therapy we must predict the location or microenvironment in which a thrombus may occur. With currently available measures of global hemostatic evaluation, we are able to detect and predict which disease states may impart a greater risk for thrombus formation and many, if not all dogs with IMHA are hypercoagulable on presentation. According to Virchow's triad, the hypercoagulable state is one of the three possible criteria for thrombus formation. It is created by both plasmic factors and cellular factors and is best described by the cell based model of coagulation which highlights the role cellular factors, including platelet factors, in all phases of hemostasis. This is however only one arm of the traditional Virchow's triad. The second and third arms are blood stasis or alterations in blood flow and vascular damage respectively. The location of thrombus formation largely determines the blood flow conditions which may range from blood stasis in the venous circulation which may activate plasmic coagulation factors to high shear conditions in the arterial circulation which may activate platelets. Finally, the presence of vascular damage reflects a state in which disruption of vascular integrity alters the microenvironment in which the thrombus may form. Within this microenvironment we find the vascular endothelial cells which interact with platelets and are important players in the hemostatic system. Therefore both plasmic factors and cellular factors play a role in thrombus formation to different degrees depending on location and microenvironment. The role each of these factors play in initiating and propagating thrombus formation may also vary over time. In dogs with IMHA, all of these factors seem to vary from patient to patient.

PLATELET INVOLVEMENT IN THROMBUS FORMATION

The current classification of thromboemboli is dependent on location. Thrombi that form in the arterial circulation result in "white clots" which are composed primarily of platelets and fibrin while thrombi that form in the venous circulation result in "red clots" composed primarily of fibrin and entrapped erythrocytes. It is therefore reasoned that the platelet rich thrombi in the arterial circulation are more likely to be initiated by platelet activation and thus, prevention of these thrombi is best provided by anti-platelet therapy such as aspirin. By the same reasoning, fibrin rich thrombi in the venous circulation, it seems, would be best prevented by anti-coagulant drugs such as the heparins. For canine IMHA, most available evidence pertaining to thromboembolic complications, is focused on pulmonary thromboembolism (PTE). The typical human model of PTE consists of pulmonary emboli broken from thrombi that have formed in the deep leg veins, or deep vein thrombosis (DVT). The initiating factors for these thrombi have long been thought to be blood stasis in the deep vein, usually secondary to reduced mobility, and a hypercoagulable state brought about by some other illness, inherited susceptibility, or surgery. In this scenario, the most appropriate treatment would appear to be anti-coagulants such as the heparins. Recently however, the platelet has also been implicated in DVT formation in people. The thrombi tend to form behind the cusps of poorly functional valves in the venous vasculature. Not only is there blood stasis but the endothelium in these areas become hypoxic. This in turn activates endothelial cells which in turn activate platelets. In fact, some physicians are now recommending anti-platelet therapy in patients at risk for DVT. Patients undergoing some routine orthopedic procedures are often now placed on anti-platelet therapy prior to the procedure. The argument is being made that the assumption that thrombi are initiated by/or the result of primarily plasmic factors is erroneous and needs revision based on this new information.[1] Dogs, of course, do not typically form thrombi in leg veins and the origin of the pulmonary embolus is often unclear and likely varies from patient to patient. The origin may in fact be the pulmonary circulation itself. As the thrombus initiating factors are difficult to predict in the dog, and canine IMHA patients have been found to have circulating activated platelets, prophylactic therapy that includes anti-platelet drugs such as aspirin is likely beneficial.

ASPIRIN IN THE VETERINARY LITERATURE

There are few studies in the veterinary literature evaluating the efficacy of thromboprophylactic aspirin therapy in dogs with IMHA. A retrospective study of prognostic factors, survival rates and treatment protocols in 151 dogs with IMHA found that treatment including ultra low dose aspirin (ULDA - 0.5 mg/kg) produced survival rates at discharge, one month and one year of 88%, 82% and 69% respectively.[2] This compares to dogs treated without ULDA with survival rates of 74%, 57% and 45% at discharge, one month and one year respectively.[2] Dogs whose treatment included ULDA and heparin had discharge, one month and one year survival rates of 70% 67% and 64%.[2] This same study showed that protocols that included heparin as the sole thromboprophylactic resulted in discharge, one month and one year survival rates of 23%, 17% and 17% respectively although a conservative fixed dose of heparin was used (75 – 125 U/kg q6hr).[2] In the context of this study there was a clear survival advantage for dogs with treatment protocols that included aspirin over those without aspirin. Even in more severely affected patients, when aspirin was included, survival times were significantly longer. In another study, the anti-platelet drug Clopidogrel was evaluated in comparison with ULDA and ULDA in combination with Clopidogrel.[3] No significant differences were found between groups. 7/8 dogs in the Clopidogrel group and in the ULDA group were alive at discharge and 8/8 in the group given both drugs. At 90 days 6/8, 6/8 and 7/8 were alive in the Clopidogrel alone, the ULDA alone and the two drugs combined groups respectively. There was no evidence of bleeding in any group. Thrombosis occurred in one dog in each of the single drug groups but not in the group given both drugs. A third study evaluated mycophenolate mofetil in five dogs with IMHA.[4] All of the dogs received low dose aspirin (1 mg/kg) and although only three of the five survived for more than a year, none of the dogs developed thrombosis.

RATIONALE

In spite of therapy with aspirin or an anti-coagulant drugs, dogs with IMHA continue to have thromboembolic complications. Treatment failure in dogs treated with aspirin may be due to an insufficient dose. Recent studies in normal dogs have shown that higher doses of aspirin may be needed to impair platelet function, at least in some dogs, with dose recommendations as as high as 2 mg/kg (unpublished data). In normal dogs there is considerable variability in response to aspirin, at least when platelet function is measured with available assays. There is also concern that these measures of platelet function do not necessarily represent platelet function *in-vivo* or in the clinical patient. Aspirin therapy may fail when thrombi are initiated primarily by plasmic factors. These patients may have been better protected by an anti-coagulant drugs such as heparin. Clinical experience, the few studies in the veterinary literature, and extrapolation from human medicine show that aspirin may be viable choice for thromboprophylaxis in dogs with IMHA. Ongoing work also demonstrates that further studies more clearly defining the optimum dose, or establishing protocols for identifying the optimum dose in individual dogs are certainly warranted. There are very likely a subset of dogs with IMHA that would respond better to anti-coagulant therapy or some that require combination therapy. Unfortunately we currently have no way to predict which dogs need aspirin, anti-coagulants, or both. One approach may be to routinely use both aspirin and an anti-coagulant but the optimum doses for this combination remains elusive and the risk for bleeding complications in clinical patients remains undefined.

SUMMARY

The rationale for the use of anti-platelet drugs for thromboprophylactic therapy in dogs with IMHA is relatively strong. Aspirin therapy makes physiologic sense, in terms of both the traditional Virchow's triad and the cell based model of coagulation. Aspirin therapy is supported by the few veterinary studies available. And finally, aspirin therapy is supported by extrapolation from human medicine. Aspirin is an attractive choice because of its low cost, ease of use and relative safety in dogs at recommended doses. Unfortunately, some dogs develop thrombi in spite of aspirin therapy. This may be the result of an insufficient aspirin dose, or may be due to plasmic factors that remained unchecked in the absence of anti-coagulants. The optimum dose of aspirin in clinical IMHA patients is not defined and the dose requirements may in fact vary from patient to patient. It is also impossible at this point to predict which patients may need anti-coagulant therapy instead of, or in addition to, aspirin therapy. The best approach therefore may be to combine anti-platelet and anti-coagulant therapies although this may increase the risk for bleeding and these patients need to be monitored closely.

REFERENCES

1. Jameson SS, et.al. Bone Joint Res 2014; 3:146
2. Weinkle TK, et.al. JAVMA 2005; 226(11):1869
3. Mellett AM, et.al. J Vet Intern Med 2011; 25:71
4. West LD, et.al. JVECC 2014; 24(2):226

Antifibrinolytics as a Treatment for Bleeding: Benefits and Risks

Benjamin Brainard, VMD, DACVAA, DACVECC and Dana N. LeVine, DVM, PhD, DACVIM (Small Animal Internal Medicine)
Athens, GA and Ames, IA

INTRODUCTION

The use and availability of antifibrinolytic agents can save lives and decrease morbidity and mortality in a wide variety of bleeding patients. Fibrinolysis occurs as a natural progression of clot formation and breakdown in the body, and allows restoration of blood flow to areas, following repair of vascular endothelial injury. Components of fibrinolysis are integrated into the fibrin matrix, and allow precise orchestration of the dissolution of the clot. Excessive or inappropriate fibrinolysis can lead to hemorrhage, and inhibition of fibrinolysis can be used strategically to stabilize clots.

Fibrinolysis occurs when circulating plasminogen binds to a clot, where it is activated to plasmin by clot-bound tissue-type plasminogen activator (tPA). Urokinase-type plasminogen activator (uPA) is another serine protease that can activate plasminogen. tPA has a high affinity for fibrin, and low activity when not bound to fibrin.[1] tPA is secreted from endothelial cells (EC) in response to fibrin, thrombin, or by the effects of vessel occlusion from the clot.[1] The formation of fibrin strands exposes lysine-rich binding sites for plasminogen which will integrate it into the forming clot, where it will also be somewhat shielded from inhibitor enzymes.[2] tPA also binds to fibrin via a lysine-binding site mechanism, which brings the two molecules in close apposition.[3] Plasmin, once activated from plasminogen, is a protease that will cause proteolysis of the clot.[2]

Thrombin-activatable fibrinolysis inhibitor (TAFI) is activated by the thrombin-thrombomodulin complex and slows fibrinolysis by cleaving exposed lysine residues, decreasing the positive feedback mechanism that converts additional plasminogen to plasmin via tPA.[1] Because of the requirement for thrombin to support this modality, TAFI sits at the intersection between coagulation and fibrinolysis. Patients with pathologic conditions associated with decreased thrombin formation (e.g. hemophilia) may also generate weaker clots when they make clots, as they produce less thrombin and thus have less TAFI to stabilize the clot. α2-antiplasmin (also called α2-plasmin inhibitor) is present in circulation and impedes the association of plasminogen with fibrin (by binding to the lysine residues of plasminogen) and decreases fibrinolysis. Plasminogen activator inhibitor (PAI-1) is released by platelets and endothelial cells, and also acts to decrease the initiation of fibrinolysis. Excessive release of plasminogen activator from EC leading to excessive fibrinolysis may occur with pathologic conditions including trauma, neoplasia, heatstroke, and other inflammatory diseases.

Currently available antifibrinolytic drugs are lysine analogs such as ε-aminocaproic acid (EACA) and tranexamic acid (TXA). These compounds are widely used in human and veterinary medicine to decrease fibrinolysis in diverse applications. In general, they work by reversibly binding to the lysine binding sites of plasminogen, preventing its association with the lysine residues on fibrin, and preventing eventual activation of plasminogen by tPA. Both drugs are FDA approved for use in humans.

INDICTIONS FOR INHIBITION OF FIBRINOLYSIS IN HUMAN MEDICINE

TXA has been used since the 1970s for prevention of cyclic heavy menstrual bleeding and for short term use to prevent bleeding in people with hemophilia during dental procedures. Both TXA and EACA have been used to decrease hemorrhage during and after surgery. In orthopedic surgeries, the drugs have been used both topically as a lavage solution,[4] as well as intravenously.[5,6] Use has been reported in cardiovascular procedures,[7,8] including pediatric cardiothoracic procedures.[9,10] The use of antifibrinolytic agents has resulted in a decrease in transfusion requirements, an increased postoperative hematocrit, and decreased blood product usage. There is some evidence that the use of antifibrinolytic agents may also attenuate inflammatory cytokine production in the context of cardiopulmonary bypass.[11]

The CRASH-2 (Clinical Randomisation of an Antifibrinolytic in Significant Haemorrhage 2) trial provided significant evidence to support the use of TXA in human victims of trauma.[12] Recruiting over 20,000 patients, this study randomized subjects to within 8 hours of injury to receive either TXA or placebo therapy. TXA administration resulted in a decrease in all-cause mortality (14.5% vs. 16%) as well as the risk of death due to bleeding (4.9% vs. 5.7%). Further analysis of the data supported the use of TXA as early as possible in the course of injury (*i.e.* less than 3 hours after injury).[13] Those patients receiving TXA less than 1 hour after injury had a relative risk of mortality of 0.68 (95% CI 0.57-0.82) compared to those in the placebo group. A Cochrane database review also supported the use of antifibrinolytic agents following trauma, suggesting a 10% decrease in all-cause mortality in patients receiving TXA (although the majority of data for this analysis came from the CRASH-2 data).[14] No adverse effects, including thrombotic events, were increased in the patients receiving antifibrinolytic agents.

Antifibrinolytics have been used to limit hemorrhage in patients with hereditary bleeding disorders including vonWillebrand disease (VWD) and hemophilia. This ranges from oral mucosal and nasal bleeding[15] to supplementing the use of coagulation factor concentrates for more invasive surgical procedures.[16] TXA has been used as a mouthwash following oral surgery in susceptible individuals with good success.[17] There is also evidence that EACA may reduce thrombocytopenic bleeding and transfusion dependency.[18] Patients that have been treated with various oral anticoagulant regimens can also benefit from the inclusion of antifibrinolytics in the perioperative period.

RISKS OF ANTIFIBRINOLYTIC THERAPY

Although antifibrinolytics are clearly effective in limiting hemorrhage, their use is not entirely without risk. In some patients, especially those with underlying procoagulant disease, antifibrinolytic therapy can increase the risk for thrombosis.[19] Several trauma studies since the CRASH-2 trial have found a significant increase in thromboembolic events in patients receiving TXA, and the CRASH-2 study has been criticized for not actively searching for thrombotic complications and for having a study population at relatively low risk for thromboembolic complications.[20-23] A retrospective observational study, Military Application of Tranexamic Acid in Trauma Emergency Resuscitation (MATTERs) comparing TXA administration in combat injury patients demonstrated a 9 fold increase in the rate of pulmonary embolism and a 12 fold increase in deep vein thrombosis rate in patients who received TXA compared to those who did not.[24] Indeed, a canine model of chronic pulmonary thromboembolism employed TXA to prevent thrombolysis and maintain emboli following their induction; TXA alone did not induce thrombi.[25] Despite these, most evaluations of TXA in a variety of disease states find no increased risk of thrombosis with its use.[12,26,27]

TXA can also cause seizures by antagonizing inhibitory central nervous system gamma-aminobutyric acid (GABA) and glycine receptors.[28-30] In one retrospective study of over 11,000 cardiac surgery patients TXA was a strong independent predictor of seizures (OR 14.3, 95% CI 5.5-36.7) and those patients with convulsive seizures had 2.5 times higher in-hospital mortality than those without seizures.[31] EACA also inhibits glycine receptors, though for unknown reasons, seizures have not been reported with this drug.[29]

Another side effect of TXA, hypotension with rapid infusion, may have contributed to significantly increased mortality with TXA observed in one retrospective study of critically injured patients with severe traumatic shock.[32] Hypotension resulted in higher crystalloid requirements in TXA treated trauma patients, leading to worse outcomes.[32]

Given they are not without risk, one of the biggest questions relating to use of antifibrinolytics is when are they truly indicated? There has been significant criticism of the CRASH-2 trial, which would imply that every trauma patient should receive TXA, since it was performed predominantly in low to moderate income countries without advanced prehospital care or critical care. The reported beneficial effects might not translate to more mature trauma systems.[20] Further studies are needed to determine which patients will clearly benefit from antifibrinolytic therapy, and which are at highest risk of adverse events. Documenting hyperfibrinolysis prior to TXA therapy rather than administering TXA to all trauma patients may be ideal. Currently a large prospective, randomized, placebo-controlled multicenter trial in Australia and New Zealand, the Pre-hospital Anti-fibrinolytics for Trauma Coagulopathy and Haemorrhage (PATCH) study, aims to examine the impact of prehospital administration of TXA on mortality in a controlled environment with standardized transfusion protocols and assessment of TXA on coagulation, fibrinolysis, and vascular occlusive complications.[23]

ANTIFIBRINOLYTIC USE IN VETERINARY MEDICINE

Antifibrinolytics have seen perioperative use in horses undergoing sinusotomy or other surgeries associated with significant blood loss,[33,34] but until recently have not been reported for use in small animal patients. Investigating delayed postoperative bleeding in greyhounds, a group at Ohio State University hypothesized that these dogs had a hyperfibrinolytic condition[35] and successfully minimized delayed postoperative bleeding in a population of greyhounds following limb amputation for osteosarcoma using aminocaproic acid.[36] Other hyperfibrinolytic states have been identified in dogs, including hemoperitoneum caused by a ruptured abdominal tumor,[37] and in dogs following trauma.[38] Anti-fibrinolytic agents may be indicated for use in both of these populations. In addition, canine patients at risk for hemorrhage following routine or emergent surgery may benefit from the use of antifibrinolytic agents. TXA was administered to dogs with Scott syndrome after ovariohysterectomy and castration and EACA was given to a Scott syndrome dog with a pelvic limb hematoma.[39] Treatment trials will be need to determine the efficacy of antifibrinolytics to decrease transfusion requirements and prevent rebleeding in such dogs with platelet defects and perhaps thrombocytopenic patients.

Recent studies have attempted to verify the therapeutic levels for TXA and EACA in dogs[40] and horses,[41] and have supported the concept that TXA is approximately 10 times more potent than EACA. Using a TEG assay modified to simulate hyperfibrinolysis, these studies determined the plasma drug concentration that was associated with complete inhibition of fibrinolysis. A subsequent *ex vivo* study of EACA in dogs was unable to achieve the recommended plasma levels in the *in vitro* study but did show a significant decrease in fibrinolysis using the hyperfibrinolytic TEG model when EACA was administered orally at 100 mg/kg, using either the injectable formulation or the pill formulation.[42] TXA has been studied as an emetic in dogs, perhaps working through neurokinin receptors.[43] Healthy adult dogs vomited following high IV doses of TXA (≥15 mg/kg) administered rapidly, but did not demonstrate other serious adverse effects.[44] Further studies in clinical patients with hyperfibrinolytic conditions are necessary to assess both the efficacy and the safety profile of the antifibrinolytic agents in small animals and to determine which of our patients would benefit most from these therapies.

REFERENCES

1. Marder In: Hemostasis and thrombosis, 5th ed. Philadelphia PA: Lippincott Williams and Wilkins. 2001. Pp. 350
2. Hantgan RR, Lord ST. In: Hemostasis and thrombosis, 5th ed. Philadelphia PA: Lippincott Williams and Wilkins. 2001. Pp. 285
3. Tengborn L, et al. Thromb Res. 2015; 135 (2): 231.
4. Panteli M, et al. Knee. 2013;20(5):300.
5. Wang H, et al. Knee. 2014;21(6):987

6. Boyle JA, et al. J Pharm Pract. 2015 pii: 0897190015621813.
7. Sigaut S, et al. Anesthesiology. 2014;120(3):590.
8. Du Y, et al. J Cardiothorac Vasc Anesth. 2014;28(5):1233
9. Eaton MP, et al. Anesthesiology. 2015;122(5):1002.
10. Wesley MC, et al. Anesthesiology. 2015;122(4):746
11. Greilich PE, et al. J Thorac Cardiovasc Surg. 2003; 126(5):1498.
12. CRASH-2 trial collaborators. Lancet. 2010; 376(9734):23.
13. CRASH-2 trial collaborators. Lancet. 2011; 377(9771):1096
14. Ker K, et al. Cochrane Database Syst Rev. 2015; 5:CD004896. doi: 10.1002/14651858.CD004896.pub4.
15. Pasi KJ, et al. Haemophilia. 2004; 10(3):218
16. Löfqvist T, et al. Acta Orthop Scand. 1996; 67(4):321
17. Sindet-Pedersen S. J Clin Pharm, 1987; 27: 1005
18. Kalmadi S, et al. Cancer. 2006; 107(1): 136.
19. Mutter WP, et al. Am J Kidney Dis. 2009; 53(2): 346.
20. Binz S, et al. J Blood Transfus. 2015; 2015: 874920.
21. Swendsen H, et al. J Trauma Treat. 2013; 2 (4): 1.
22. Cole E, et al. Ann Surg. 2015; 261(2): 390.
23. Gruen RL, et al. Med J Aust. 2013; 199(5): 310.
24. Morrison JJ, et al. Ann Surg. 2012; 147(2): 113.
25. Moser KM, et al. Circulation. 1991; 83(4): 1371.
26. Lukes AS, et al. Am J Obstet Gynecol. 2010; 116(4): 865.
27. Huang F, et al. Hip Int. 2015; 25(6): 502-9.
28. Furtmüller R, et al. J Pharmacol Exp Ther. 2002; 301 (1): 168.
29. Lecker I, et al. J Clin Invest. 2012; 122(12): 4654.
30. Martin K, et al. J Cardiothorac Vasc Anesth. 2011; 25(1): 20.
31. Sharma V, et al. Anaesthesia. 2013; 69(2): 124.
32. Valle EJ, et al. J Trauma Acute Care Surg. 2014; 76(6): 1373.
33. Ross J, et al. Am J Vet Res. 2007; 68(9):1016
34. Heidmann P, et al. Am J Vet Res. 2005; 66(2):313-8.
35. Lara-García A, et al J Vet Intern Med. 2008; 22(3):525.
36. Marin LM, et al. J Vet Emerg Crit Care. 2012; 22(3):332
37. Fletcher DJ, et al. J Vet Emerg Crit Care. 2016; 26(1):41.
38. Holowaychuk MK, et al. J Vet Emerg Crit Care. 2014; 24(1):93
39. Jandry KE, et al. J Vet Intern Med. 2012; 26(6): 1402.
40. Fletcher DJ, et al. Am J Vet Res. 2014; 75(8):731.
41. Fletcher DJ, et al. J Vet Intern Med. 2013; 27(6):1589
42. Brown JA, et al. A J Vet Res. 2016; in press.
43. Kakiuchi H, et al. Am J Vet Res. 2014;75(12):1099
44. Kelmer E, et al. J Vet Emerg Crit Care. 2015;25(4):495

Leading Through Adversity: Connecting with Your Team

Bonnie K. Lefbom, DVM, DACVIM (Cardiology)
Vienna, VA

INTRODUCTION

As veterinarians, we are tasked with being the leader in our office whether we want to or not -even when we experience tragedy in our personal or professional lives. How can we lead and support each other when we lose someone we love or a team member has a life threatening injury or illness? If we allow ourselves to demonstrate empathy and compassion, we become authentic leaders who not only serve the team but lead in the face of whatever comes our way.
Sometimes, it is OK to cry in front of everyone.

CONNECTION, RECOVERY AND LEADERSHIP

Allow life's sorrows to break our hearts and its joys to make us whole again.

It turns out the loss of loved ones, medical disasters, car crashes and cancer are all a part of the human condition. The best way to get through any of these disasters is to allow ourselves to talk about it. Many veterinarians are perfectionists and as such feel a need to appear stoic or unaffected by tragedy. Others hide their feelings behind a wall of disconnection under the guise of "getting through the day". Letting that wall down is difficult but we grow exponentially when we accept that we are vulnerable to all life throws our way and become open enough to share our feelings with our team. As we allow ourselves to not only feel the emotions of the surrounding tragedy, articulating those emotions evolves into deeper connection with our team. In these conversations, we learn about the tragedies others have endured and this humanizes not only the doctors in the room but develops deeper mutual respect for everyone. "Me, too" moments are powerful and bring us closer to whole-hearted living. In a practice setting, allowing team members to feel safe sharing their emotions leads to improved engagement for the tasks of the day. Team members are then freed from trying to hide their upset and can focus their energy and compassion on client service and the tasks at hand.

CHOOSE OUR RESPONSE

Accepting that we can not control a lot of what happens around us can be difficult for everyone, particularly veterinarians. We prefer to be in charge, in control and want to feel like what we do influences the outcome of events. It doesn't always work out that way. Being able to choose our response to tragedy and demonstrating emotional intelligence surrounding these events (identifying our feelings, letting them all the way in, then letting the feelings go) is crucial to authentic leadership. Our job is to convey empathy not only to our clients, but also to ourselves and our team.

POW-WOW IN THE AM

At the start of every work day, whether surrounding a tragedy or not, a great habit to develop is to make time to meet and review personal as well as business concerns. This simple act humanizes the team and frees everyone's mind to be their most productive self. It is ideal to have a joke of the day or at least to inject some humor into the conversation. Joyfulness is the key – happiness is conditional, joyfulness is unconditional.

PHILANTHROPY AND HOSPITAL WIDE OUTREACH

In times of tragedy, the act of giving or performing community service can have a tremendous healing effect. Consider donating to a favored charity and even more powerful, setting up a day of volunteering for the whole office to participate in the giving.

PRACTICAL PERSONAL TIPS TO PREPARE FOR LEADING IN TIMES OF ADVERSITY

Develop your emotional intelligence – invest the time and energy
Empathize with the upset - yours, the team members, clients
Be your whole self - discuss your feelings, sadness, concerns
Begin each day with positive energy and manage your emotional "wake"
Connect with your team – be a good listener, encourage others to talk about their feelings
Listen to comedy on the way to the office

PRACTICAL TIPS FOR THE PRACTICE TO WORK THROUGH TIMES OF ADVERSITY

Hospital wide meeting as soon as possible - empathize with each other
Share facts, not rumors
Have a consistent media message
Bring in professional counselors – human behavior is a specialty
Allow for/expect variation in response

RECOMMENDED READING/REFERENCES

1. Arbinger Institute. *Leadership and Self-Deception: Getting Out of the Box.* 7th edition. San Francisco, CA: Berrett-Koehler, 2000.
2. Bradberry, T, etal. *Emotional Intelligence 2.0.* np TalentSmart, 2009.
3. Brown, B. *Daring Greatly: How the Courage to Be Vulnerable Transforms the Way We Live, Love, Parent, and Lead.* New York, NY: Gotham, 2012.
4. Carnegie D. *How to Win friends and influence people.* NY, NY: Gallery Books; 1981
5. Collins, James C. *Good to Great: Why Some Companies Make the Leap--and Others Don't.* New York, NY: HarperBusiness, 2001
6. Frankl, Viktor E. *Man's Search for Meaning.* Boston: Beacon, 2006.
7. George, Bill. *Authentic Leadership: Rediscovering the Secrets to Creating Lasting Value.* San Francisco: Jossey-Bass, 2003
8. Olson, Jeff, and John David. Mann. *The Slight Edge: Turning Simple Disciplines into Massive Success and Happiness.* 8th edition, Austin, TX: Greenleaf Book Group, 2013.
9. Steil, L K, Bommelje RK. *Listening Leaders: The Ten Golden Rules to Listen, Lead & Succeed.* Edina, MN.: Beaver's Pond, 2004.

Battling Burnout: Keeping the Flame Alive

Bonnie Lefbom, DVM, DACVIM (Cardiology)
Vienna, VA

INTRODUCTION

Is it harder and harder to get out of bed and go to work? Know anyone who is grumpy in the morning until they finish a large cup of stimulant? Have you overheard team members talk down to each other or make snide comments about other veterinarians? Are you exhausted at the end of every day? You may be struggling with compassion fatigue or burnout ... we are all at risk.

BURNOUT & COMPASSION FATIGUE

By now, most of us are aware of the concept of compassion fatigue. It's a specific type of burnout experienced by medical professionals of all varieties, that is physically, emotionally and spiritually exhausting. Veterinarians are particularly susceptible to this because of the depth of our compassion and the sincere connections we make with our clients. We become preoccupied with the suffering of both our clients and patients which leads to a form of secondary traumatic stress. We continue to give all that we have and end up unable to balance empathy and objectivity, generally losing the ability to convey the compassion that we feel so deeply. At that point, we are burned out.

RECOGNIZING THE SIGNS OF COMPASSION FATIGUE

When we start to feel overwhelmed by the tasks of the day or unable to accomplish what we think we "should" be able to do, we start to either shut down emotionally or lash out at those closest to us. Listening to patterns of speech for negative generalizations such as: Why won't they...?; How could they...?; They are so... are all signs of compassion wearing down. Other signs include blaming others for our discontent, not caring about personal appearance or clothes, inability to articulate and/or feel emotions, and loss of joy or pleasure in life. Denial is common and one of the most detrimental symptoms because it prevents people from seeking help. Becoming self aware is a journey worth taking and allows us to notice when our energy is fading, ideally far ahead of a burnout.

RECOVERY IS ACHIEVABLE

Every one of us is unique so techniques that resonates with one may be a total disconnect for the next. The place to start for most of us is with self-reflection - identify what's truly important and live in a way that reflects it. Develop principles for life — guidelines of personal integrity that articulate personal values. Commit to live and work within these principles. Recall the passion that drove us to be veterinarians in the first place and hold on to a little bit of that each day. Choose a path for recovery that includes gratitude and joy. For some, recovering the joy of living includes self affirmation in the morning mirror, for others prayers of thanksgiving at night, others develop a regular exercise program or enroll in group support. For everyone, changing our habits is difficult but can absolutely be done. Neuroscience has proven we can re-train our brains and we are not hard wired to maintain old habits and negative self talk.

PRACTICAL TIPS FOR THOSE ON THE BRINK:

Take action to change – you are the only one that can make it better
Be kind to yourself – shut down the mean, judgmental voice in your head
Develop gratitude for all that is good – in your life and the world
Exercise and get enough sleep – really
Find a way to laugh, every single day – smiles are contagious
Choose positive friends, dump the draining ones – consider/develop a support group
Find your voice - articulate what you need, be specific
Develop outside interests – your passion doesn't have to be your job
Delegate as many things as possible – allow for less than perfect
Find meaning outside of work – family, religion, volunteerism

ASSUMPTIONS TO LIVE BY: —STEPHEN COVEY

Body – assume you've had a heart attack; live accordingly
Mind – assume the half-life of your profession is two years; prepare accordingly
Heart—assume everything you say about another, they can overhear; speak accordingly
Spirit—assume you have a one-on-one visit with your Creator every quarter; live accordingly

RECOMMENDED READING/REFERENCES

1. Arbinger Institute. *Leadership and Self-Deception: Getting Out of the Box.* 7th edition. San Francisco, CA: Berrett-Koehler, 2000.
2. Bradberry, T, et al. *Emotional Intelligence 2.0.* np TalentSmart, 2009.
3. Brown, B. *Daring Greatly: How the Courage to Be Vulnerable Transforms the Way We Live, Love, Parent, and Lead.* New York, NY: Gotham, 2012.
4. Brown, B. *The Gifts of Imperfection: Let Go of Who You Think You're Supposed to Be and Embrace Who You Are.* Center City, MN: Hazelden Publishing; 2010
5. Covey, S. The 8th Habit: From Effectiveness to Greatness. New York:Free, 2005.
6. Frankl, Viktor E. *Man's Search for Meaning.* Boston: Beacon, 2006.
7. Marsh, Nigel. *Overworked and Underlaid: A Seriously Funny Guide to Life.* Crows Nest, N.S.W.: Allen & Unwin, 2009.
8. Olson, Jeff, et al *The Slight Edge: Turning Simple Disciplines into Massive Success and Happiness.* 8th edition, Austin, TX: Greenleaf Book Group, 2013.
9. Pink, Daniel H. *A Whole New Mind: Moving from the Information Age to the Conceptual Age.* New York: Riverhead, 2005.
10. Steil, L K, et al. *Listening Leaders: The Ten Golden Rules to Listen, Lead & Succeed.* Edina, MN.: Beaver's Pond, 2004.
11. Ziegler Z. A view from the top. NY NY: Simon & Schuster Audio/Nightingale-Conant; 2002

The Profitability of Co-Managed Care

Bonnie Lefbom, DVM, DACVIM (Cardiology)
Vienna, VA

INTRODUCTION

We continue to hear rumblings about competition for cases between primary care veterinarians(pcDVM) and specialists. This is a tragedy for the profession. As specialists, it is crucial that we develop and nurture respectful relationships with primary care veterinarians and we must be able to demonstrate the value of our service in providing optimal care for companion animals.

GOOD MEDICINE IS GOOD BUSINESS

Co-managed care provides optimal care for pets. As veterinarians, we know this statement is true yet we struggle in our role as personal counselors for our clients – sometimes offering less than ideal options in the spirit of saving the clients money. The reality is, we have no business presuming another person's financial ability and we have no idea how much or how little each client values a particular animal. US pet owners spent over $60 Billion on pets in 2015, with only 27% of that on veterinary care. It is crucial that we improve our ability to demonstrate the value of our service to both the pet owning public and primary care veterinarians. In a recent study, our practice demonstrated that small breed dogs with congestive heart failure live an average of 75% longer when co-managed with both a pcDVM and in-person board certified veterinary cardiologist when compared to dogs managed with the pcDVM alone. We also demonstrated that the pcDVM generates 22% more income when cases were co-managed compared to those managed alone. We need more studies like this that demonstrate both the outcome and income benefit of co-managed care.

SUPPORTING THE pcDVM WITH CLIENTS

One of the most impactful practices we can develop as specialists is to support the pcDVM at every possible moment in the client experience - from the time the client walks in the door of the specialist's office to the time they leave. We can praise the pcDVM to the owner for many reasons – initially for simply sending the client to us, then for providing records with past history and labwork, for providing temporary and sometimes life saving treatment, and so on. Another crucial component to creating trust and support for co-managed care is to send clients back to the pcDVM for follow up at very specific times. It is ideal to set up the medical records system to send clients reminders to see their pcDVM for follow up care. These reminders help bond both the client and the pcDVM to our office. We can also be clear with our verbal instructions to include the pcDVM in the long term care plan. Recommending seeing the pcDVM to discuss nutrition and weight loss plans, explaining that the three of us (client, pcDVM and specialist) will be managing the case together, outlining wellness care plans with specific times(3 months, 6 months, 12 months) are all ways to support and include the pcDVM in the overall case management.

RELATIONSHIP BUILDING WITH pcDVMs

Getting our reports out as soon as possible, ideally the same day, to all interested parties is crucial. We need to be sure to question clients about what other specialists are involved, including other departments in house that may not share the medical record system, as well as ancillary services such as holistic or re-habilitation. In our practice, almost 12% of cases are referred by other specialists. Getting out and visiting with or offering small group CE can be huge for relationship building with pcDVMs. When we compare referrals from the six months before to the 6 months after, we documented an 18% increase in referrals from practices that we provided lunch and learn CE and a 14% increase in referral for practices that received practice visits. It is worth your time! A crucial and often overlooked component of practice visitation however is to select carefully who will be the contact – sometimes the specialist is not the right person for an event. Another component of small group or practice visits is to include information on how the pcDVMs can build their own practice - examples for cardiology include using labwork to check for cardiac disease with NT proBNP testing, monitoring blood pressure as a routine component of wellness visits, discussing why early intervention is the key to better outcomes along with the economics behind the lifetime value of each pet.

SUPPORT THE pcDVM WITH OUR PRACTICE CULTURE

It's up to us to provide optimal care - to clients, patients and pcDVMs. The culture of the practice has to reflect that commitment with zero tolerance for negativity when it comes to complaining about anything related to the referral of a case. They got in the door and that by itself is great! Providing clear values that the whole team can embrace as well as having specialist veterinarians demonstrating those values is a huge "lead by example" opportunity. We constantly bring the conversation back to the concept that we are all veterinarians in it together – working toward optimal care.

REFERENCES

American Pet Products Association – *US Pet Industry Spending Figures and Future Outlook 2015.* www.americanpetproducts.org/press_industrytrends.asp. Visited Mar 12, 2016

Lefbom B, Peckens N. *Impact of collaborative care on survival time for dogs with CHF and revenue for pcDVMs.* in press JAVMA. 2016.

Veterinary Referral Dynamics and Behaviors – Findings of the VetSOAP White Paper

Saundra E. Willis, DVM, DACVIM (SAIM), Bonnie K. Lefbom, DVM, DACVIM (Cardiology)
Tacoma, WA

INTRODUCTION

VetSOAP (Veterinary Specialists Outreach and Awareness Project) is a nonprofit organization focused on creating a culture of collaboration between primary care veterinarians and veterinary specialists. Building a better understanding of the referral process, from what's working to what needs work, is the foundation that will drive the development of VetSOAP's education and outreach plan.

In 2014, VetSOAP embarked on a journey to learn everything possible about referral dynamics in veterinary medicine as a means to identify gaps hindering optimal collaboration. VetSOAP views this research process as one which is no different than how one would proceed when presented with a diseased patient. Therefore, Phase I is the history and physical examination (i.e. the White Paper), Phase II is the diagnostic testing to fill in the gaps (i.e. Phase IIa/IIb will be the qualitative and quantitative research, respectively), and Phase III is the diagnosis and treatment.

In this presentation the results of Phase I, the White Paper, will be presented and discussed. The entire White Paper, including a list of references sourced for the paper, is available at www.vetsoap.org.

SUMMARY OF KEY FINDINGS OF THE REPORT

Collaborative care between the primary care DVM (pcDVM) and the specialist, including early referral, can benefit all.
pcDVM' s base their referral to a specialist on trust, communication, and strong relationships.
Clients and pcDVM's both want clear recommendations and discussion of all options, including costs, by specialists.
The clients' reality and the pcDVM's perception on client willingness to spend are often incongruous and this creates a barrier to quality of care and referral.
Increased insurance penetration benefits all and would drive client spending for both the pcDVM and specialists.
Independent public awareness of referral options by clients remains low unless discussed by the pcDVM.
Income pressures related to decreased client visits and increased competition limits the referral tendencies of pcDVM's.[1]

EXECUTIVE SUMMARY OF THE WHITE PAPER

Veterinary Specialists Outreach and Awareness Project Literature Review: Referral Relationships in Companion Animal Veterinary Medicine.

Executive Summary

The landscape of veterinary medicine is every changing, but slow to adopt a clinically-based referral system. In an effort to support evolution of care for companion animals and identify needs gaps to ease referral barriers. VetSOAP completed the first step in analyzing referral dynamics through an intensive literature review. This literature review summarizes existing information regarding collaborative referral dynamics between the Primary Care Veterinarian (pcDVM) and veterinary specialists. The study also compares this information to available information for the human health and human dentistry arenas. The review uncovered knowledge and process gaps that exist and supports continued areas of study. Over one hundred published sources involving referral data across multiple disciplines of veterinary and human medicine as well as dentistry were analyzed and reviewed for this study.

Summary of Information Gathered

Market Dynamics

Factors impacting the veterinary profession overall can influence the increase and/or decrease of referral frequency. This literature review surfaced marked dynamics and possible trends to be considered for tracking and evaluation for impact on referral dynamics. One example is market dynamics that could impact DVM competition. Visits to the pcDVM have been in decline to slight growth. There is a reported increased number of specialists as well as an increased number of new graduates which are reported less likely to refer.

pcDVM Referral Dynamics

Cost, relationship and communication appear as frequent barriers to referral.

Current indications point to a difference between the pcDVM's perception of the client's willingness to pay and the client's actual willingness to pay. These pcDVM perceptions affect their willingness to refer.

A common driver for a client's selection of a specialist is recommendation by pcDVM, a recommendation which relies on a trusted relationship built on good communication between professionals.

Prompt communication methodologies between pcDVM and specialists are needed and not routinely available.

Clients and pcDVM's want clear recommendations along with all of the options and full explanations of costs.

Pet Owner Dynamics

pcDVM referral remains the primary source of pet owner awareness with the pcDVM indicating that pet owner request is not common. Pet owner awareness of specialists remains low. Referred clients place high value and regard toward the pcDVM who makes the referral.

Parallels between Veterinary and Human Disciplines

There are significant and noteworthy similarities between veterinary medicine and the human dental profession in regards to referral relationships. However, these professions currently have divergent referral methodologies involving clinical referral, standard business practices and client expectations.

Dentists see referral as part of their practice rather than a potential loss of revenue.

Few clients assume that their general dentist has all of the expertise to serve their dental needs.

The relationship between primary care and specialty human dentistry is comfortable and symbiotic.

Continued Areas of Study

While the in-depth literature review provides perspective, additional research and data is required to truly understand gaps and identify potential solutions to improve the referral dynamic and collaborative patient care in veterinary medicine. As a result, VetSOAP is continuing research efforts into understanding specific gaps in the pcDVM/specialist referral dynamic, client satisfaction and patient outcomes. VetSOAP" has prioritized these areas/topics.

Financial Impact of the pcDVM referral

Does a robust referral dynamic result in a healthier/more successful pcDVM practice and by what measure?

Define the metrics to employ for defining a healthier/more successful pcDVM practice (include growth, revenue, client loyalty, community reputation, etc.)

Define indications for referral by representative clinical case types. Measure frequency of referral occurrence by representative clinical case types and stage of disease presentation.

What percentage of patients are referred from healthy, successful practices compared to clinics not typically referring? Is the difference minimal or remarkable?

pcDVM Referral Dynamics

Benchmark successful pcDVM/specialist relationships.

What are the characteristics of a successful pcDVM and specialist relationship?

What do pcDVM's and specialists expect from each other and what factors remain to be optimized in the relationship? What are the obstacles to prompt communication methodologies?

Research is required to determine the current magnitude of the differential (and presence of the differential) between pcDVM's perception of the willingness to pay and the client's willingness to pay. In addition, research is required to determine the impact of these perceptions on the pcDVM's willingness to refer.

Client Satisfaction and Patient Outcomes

What is the pet owner's awareness of, understanding of, and expectations related to veterinary specialists? What is their perception of the referral process and care outcome?

How does the referral experience impact the client's perception of the pcDVM and the veterinary practice?

Do companion animals with a variety of conditions lead healthier and longer lives with early referral and extended collaborative care?

Summary

Many of us in the veterinary community share these goals: defining and creating a path forward to advance veterinary patient care, providing longer, healthier patient lives and creating overall economic progress for our profession. It is evident from this extensive review of the present state of referral dynamics that shifting toward adoption of a true symbiotic and collaborative healthcare approach for the patient is part of this critical path forward. VetSOAP is committed to continuing qualitative and quantitative studies toward understanding and achieving these common goals in support of the veterinary profession.

REFERENCES

1. Murtaugh R, et.al Clinician's Brief, in press.

Systems that Ensure Consistency of Care
Peter A. Weinstein, DVM, MBA
Irvine, CA

INTRODUCTION

Consistency of care means providing a predictable and repetitive approach to clinical cases in an effort to ensure best outcomes for patients and clients and referring veterinarians. The provision of exceptional care starts with client service and doesn't end even after the client and patient have returned home. To be able to deliver at this high level, a practice must develop a system that includes the individual processes that are needed to have the entire team working together and delivering as a unit. If you can do this, you will have a business that exceeds those in your community and acts as a standard for others to emulate.

WHAT ARE SYSTEMS

People and animals are perfect examples of integrated systems working together to survive and thrive. Whether it is the respiratory or cardiovascular system, the renal or urogenital system, the endocrine or neurological system, we are all alive today because of our functioning systems. Within each of these systems are organs-heart, lungs, kidneys, pancreas, liver, etc. And each of these organs has a function or functions that work either dependently or independently of other organs or other systems. Within each organ, there are very specific cells that have functions. And the system, organ, cell relationship works with checks and balances and feedback to keep the animal alive.

Small businesses are no different.

And the provision of healthcare is no different.

The simplest way to think of systems in a business or clinical setting is a set of things, actions, ideas and information that interact with each other, and in doing so alter other systems. You will have a client service system, an exam room technician system, an animal care system, a bookkeeping system, etc. These systems may have subsystems and they are composed of processes.

Processes are a list of steps taken in order to achieve a specific goal that supports the system. There may be the new client telephone answering process that is a part of the telephone subsystem of the client service system.

And to make sure the right work is done in the right order each time and every time, you have these processes and they may be integrated into a checklist that acts as a memory jog to ensure the perfect outcome.

Simple stuff, right?

WHAT IS CONSISTENT CARE?

80% of what we see in practice, at any level, presents itself with a predictable set of signs and symptoms that IF you identify them direct you to a series of diagnostic tools. These tools give you information that helps you come closer to a diagnosis OR direct you to run additional tests. The combination of physical examination, diagnostic testing, knowledge and experience allow you set up an appropriate treatment plan.

What happens if you skip a step? The physical exam, for example. You go right to testing without making some preliminary assessments. Could you choose the wrong tests? Could you misinterpret results that lead you down the wrong path and possibly the wrong treatment? And what happens when you skip the testing and go right to treatment based solely upon your examination. And while doing your testing or treatment, without a specific process in place, could you forget to dilute a product appropriately; or forget to do a surgical prep before you do a procedure.

Consistent care is based upon knowing what to do; in what order; and doing that time after time. From taking a history to releasing a pet to go home, consistent care can be systematized to ensure you don't miss a beat. And thus help to create more predictable outcomes.

I'M STILL CONFUSED...

Let's look at a case from start to finish.

A dog with polyuria, polydipsia, polyphagia.

The case will most likely start as a referral from another veterinarian. Without a system, processes and checklists in place, you are depending upon your client service team to: get a completed referral form; get copies of medical records; get copies of related laboratory testing; get copies of radiographs; schedule an appointment for an appropriately allotted time; and ensure that the client knows where to go; where to park; where to check in; etc. And they have to do all of this without forgetting any one step.

Upon arrival, what does your front office team do? You don't know? Well neither will they!! How about filling out paperwork, collecting medical records and radiographs (if they hadn't been sent already), offering coffee, making the client feel more comfortable and safe, make the pet feel more comfortable and safe. Are you consistent in the manner by which this is handled? Consistency of care is NOT just the medical components, it is the entire experience.

The patient/client experience continues in the examination/consultation room. This is one of the areas where having a good system and processes are imperative. You need to conduct a thorough history of each pet. To do this having a Client/Patient History process that takes you step by step through the information you want to gather will insure you never miss the important questions—how long has this been going on? What have you tried at home to treat this? Besides dog food, what are feeding? What did Dr. Google suggest might be going on? There may even be separate history processes for the exam room nurse and the veterinarian AND even a separate history for referred cases including complicated diabetic, uncomplicated diabetic, chronic diarrhea dog, chronic diarrhea cat, etc.

Having completed the history, the physical examination is itself a process. Have you ever forgotten to check an entire organ system because you were distracted by the 5 year old drawing on the exam room walls? The physical exam process is frequently monitored using your practice management software and PE templates. If you are using an exam room nurse, they can help ensure that you follow the process by asking about specific areas of the physical exam you haven't noted or documented. Could there be disease specific physical examinations? Absolutely, especially for cardiovascular diseases or cancers, among others.

Usually after the examination is completed, there is a discussion of what testing needs to be done to confirm a diagnosis or to get additional information. For almost every presenting set of signs or symptoms, there is a traditional work up that is done. From CBC, Chemistry and Urinanalysis to radiographs, EKG, ultrasound, blood pressure, there is a minimum database that you need. The Work Up System would include your PU/PD/PP process for working up the case. And this process will also help populate the treatment plan and the healthcare plan (estimate). The importance of the history and physical examination processes and systems is second only to the work up systems and processes. It is hard to explain or reconcile a failure to collect a component of the minimum database of a work up.

Besides the Basic Work Up System, it would be beneficial to have an Advanced Work Up System. This may be the ultrasound and radiographs or additional testing needed after the Basic Work Up. Your PU/PD/PP Work Up would have a Basic and Advanced and maybe even another level for a process to further confirm diagnosis. In each process there may be sub-processes as well. If you are doing any stimulation testing, what is the protocol? This is a process. This process can be referred to during another process to avoid mistakes in dilution, time between before and after, sample handling, etc.

Other areas for diagnostic systems and processes in the diagnostic area include the cardiac ultrasound, the colonoscopy, the rhinoscopy, the abdominal ultrasound, and the list goes on. The goal of the processes here are to ensue proper patient preparation, proper equipment preparation, proper procedure from start to finish, and proper equipment handling after the procedure, just to name a few.

Finally, with the patient admitted, examined, tested, poked, prodded, etc., there is the Patient AFTER CARE SYSTEM and associated processes. This will include post visit aftercare; ensuring proper medication dispensing; follow up phone calls; follow up evaluating and testing systems; and communication with the referring veterinarian to ensure needed follow up. Have you ever done a great job with the patient and the client but completely forgot to send home medications or written the referring veterinarian about the case?

Does this all make sense? Do you see how having systems and processes in place will help you deliver a consistent level of service to your clients and to your patients? Do you see how having systems and processes in place will help your staff succeed by setting them up with a step by step manner to avoid making mistakes? Do you see how consistent delivery of patient care will help create better patient outcomes and more profit (you won't forget to do things or charge for them)?

BUT HOW?

The 'How' is actually the easy part...that is except for the time that is needed to do the 'How'. As noted earlier, almost 80% of what you do each day is repetitive. From phone to clean up and everything in between is pretty much the same from day to day. Start with the simple systems and processes. Identify the general areas of the hospital, let's call them zones. Client Service Zone. Exam Room Zone. Treatment Area Zone. Animal Care Zone. Management Zone. Are there others?

Each zone will have systems within it. For example:
Client Service Zone has:
The Telephone Answering System. The Reception Room Comfort System. The Reception Area Maintenance System. The Computer System. The Money System. Etc.

Exam Room Zone:
The Exam Room Client System. The Exam Patient System. The Exam Room Comfort System. The Exam Room Maintenance System. The Exam Room Computer System. Etc.

Treatment Room Zone:
The Treatment Room System. The Treatment Room Comfort System. The Treatment Maintenance System. The Ultrasound System. The Endoscopy System.

And that is just scratching the service.

To identify all of the actions/systems that happen during the day, week, month, give everybody a small notebook and have them Identify everything they do during the day. I mean everything! Answer the phone. Walk the dogs. Pay bills. Prep ultrasound unit. This is the general overview notebook. By the time the week is over you should have about 80 % of what you do daily noted and identified. You now have a 35,000 foot overview of your practice operating system.

The 5 foot view

From 35000 feet we need to get down to 5 feet. For each general system there are subsystems.

The Telephone Answering System will include subsystems for: New client/New patient; Old client/New patient; Recheck; Emergency Referral; Rx Refill, etc.

The Ultrasound System will include subsystems for: cardiac ultrasound; abdominal ultrasound; recheck ultrasound; kidney biopsy ultrasound; liver biopsy ultrasound.

The Bathroom Maintenance System will includes subsystems for: Toilet cleaning; floor cleaning; sink cleaning etc

To get from 35000 feet to 5 feet, you give each employee a new note book and ask questions like: how is it that you answer the phone; how do you use the ultrasound; how do you clean the bathroom.

You will have an even more in depth view of how your practice operates by doing this. But that isn't fine-tuned enough. We need to get to the molecular level

The Molecular Level

Just when you thought it couldn't get any more granular, it can!!! You know what you do. It is time to learn how. Step by step in the most detail possible. What do you do when the phone rings? How do you answer the phone? What do you say? How do you ask? What are the settings on the ultrasound set to? How do you prep the patient for a chest tap? What is the preparation for an endoscopy?

The Third Notebook gets you molecular. For every subsystem, there are some very specific procedures that need to be documented. Step by step. In as great as detail as possible. In fact in so much detail that ANYBODY with or without experience could pick up the description and perform the process!!! From start to finish. Without question. Without fail.

FINISHING UP

Lots of work, right? Yes, but once you do it right the first time, you have a training manual, an accountability program, a system of checks and balance to ensure the delivery of a consistent patient, client and staff experience. And it is in the "HOW WE DO IT HERE MANUAL".

Securing consistent patient care with systems will give your practice peace of mind. It will help to minimize errors of omission and commission. And with better patient and client care supported with systems, the profitability will also improve.

Creating Checklists to Ensure Consistent Patient Care

Peter A. Weinstein, DVM, MBA
Irvine, CA

INTRODUCTION

With systems created and processes identified, you need a method to reduce failure by compensating for potential limits of human memory and attention. Checklists help to ensure consistency and completeness in carrying out a task. This could be as simple as a bathroom maintenance checklist or as complex as an upper gastrointestinal endoscopy checklist or specific chemotherapeutic agent checklist that ensure safety and efficacy of the staff and for the patient.

ISN'T A CHECKLIST BUSY WORK?

In many practices, I have seen checklists used as punishing documents. They are not memory jogs as much as termination jogs. That is not what checklists are for.

If you have ever been on an airplane you wouldn't want to be a passenger if the pilot didn't go through their routine checklist to ensure you got their safely.

If you have ever built a house or hospital, you wouldn't want a contractor who didn't have a series of checklists that they went through to build the house and ensure you had electricity and water.

If you went to the hospital, you wouldn't want surgery done on the wrong body part because the nurse didn't follow the checklist that has you initial the correct knee.

Checklists should not be busy work. They should be quality control components of your practice to ensure the consistent client, patient, staff and ownership system that makes a business fun to work at and profitable.

HOW DO CHECKLISTS ENSURE CONSISTENT PATIENT CARE?

Veterinary medicine has so many moving parts. From the client service interactions through technician involvement to the doctor's diagnosis, treatment plan, and subsequent care, there are SO many opportunities for something to slip through the cracks. Some simple examples that could have been avoided with a checklist:

Forgetting to tell a client to fast a patient
Forgetting to get an estimate signed and deposit collected
Not appropriately diluting an injectable product
Not changing the Baralyme on the anesthetic machine
Failing to run quality control checks on in house lab equipment
Neglecting to get a clotting time before doing liver biopsies during an ultrasound
Not sending home a prescription with a patient upon release
Charging for one unit of an item when 60 units were used
And the list goes on...

By utilizing a checklist as a memory jog those things that have become mundane are monitored while those things that might be life threatening if ignored are insured the attention they require. Consistent patient care comes by doing procedures the same correct way, each time, every time, in the same order, without fail UNTIL a better way of doing things is identified. The vast majority of the procedures we do in practice are done multiple times per month or per week or even per day. These MUST be done the same way each time to ensure accuracy, dependability and predictability. Checklists are a system for ensuring this desired outcome.

WHAT ARE SOME EXAMPLES OF CHECKLISTS WE SHOULD HAVE?

As noted, veterinary medicine is pretty repetitive on daily basis. This repetition lends itself to complacency and with complacency comes loss of attention to detail. Some of the daily routines that SHOULD have a checklist include:

Patient Drop Off Checklist
Outpatient Appointment Checklist
Patient Release Checklist
Abdominal Ultrasound Checklist
Colonoscopy Checklist
Feline Chemotherapy Checklist
MRI Checklist
End of Day Close Out Checklist

In looking at the list above, which is by no means complete, can you think of ONE thing that if you forgot could be damaging to the patient or even fatal? That is what a checklist is for.

WHAT GOES INTO A CHECKLIST? HOW DO I DECIDE WHAT TO INCLUDE?

I don't see checklists as a step-by-step guideline as much as general outlines of processes that need to be completed. In fact the processes themselves (essentially the content of your Operations Manual) might be the items of a checklist.

To review, the veterinary business is composed of systems and subsystems that allow them to operate effectively and efficiently. Within each of these are a number of processes and the associated steps needed to accomplish the requisite tasks. The New Client Appointment Process and the steps to do so. The Chest Tube Insertion Process and its associated steps. The Post Chemotherapy Release Process and its associated steps. The Referring Veterinarian Communication Process and its related steps.

Checklists are composed of the necessary processes to complete the task. The processes have been trained, approved and documented in the Operations Manuals. Besides the list of processes, I would include ANY and ALL steps that if missed could have serious repercussions to the patient and the relationship with the client.

If the checklists you create is pages long, it is too long. Most checklists will be 7-12 items in length max. If they are longer than that the detail may be too great and will make the checklist too daunting to deal with. To create a checklist, start by including everything that you think is important. Everything!! With this initial list in hand, see which items are actually included in a more global process that already exists. Also, checklists must be in chronological order that must be adhered to. Baking the cake before you mix the ingredients will not get you the desired dessert.

PUTTING IT ALL TOGETHER AS A CHECKLIST

With the list of items that you think you should have In the checklist collected, make sure they are in the correct order. How many items do you have in the list? Are some of the items actually steps in a more global process? Are the critical checklist items identified?

What are critical items? They fit into multiple categories but they all include patient and staff safety. Fluid additives at the correct dilution. Anesthetic agents to the correct dilution. Safe handling of chemotherapeutic agents. Accurate placement of an IV catheter for administration of caustic agents. Having a contact number for the owner in case of emergency or to reach with results.

The steps/process for a cardiac ultrasound might include:

1. Cardiac Ultrasound appointment process
2. Cardiac Ultrasound admission process
    a. Was the patient fasted?
3. Cardiac Ultrasound calibration process
4. Cardiac Ultrasound prep process
5. Cardiac Ultrasound process
6. Cardiac Ultrasound Data Collection process
    a. Is there sufficient memory to capture and store images
7. Cardiac Ultrasound Results Reporting process
    a. Copy of report to referring vet
8. Cardiac Ultrasound Release Process
    a. Medications to go home process
    b. Follow up appointment process

WHO DOES WHAT?

In the sample of a Cardiac Ultrasound noted above, there are various team members involved including client service, imaging technicians, veterinarians, and possibly animal caretakers, and other technicians. Each of the 8 line items noted could have a different person involved. It doesn't matter how many individuals are involved only that the outcome is perfect and that there is someone who has ultimate oversight of the entire checklist. And, by the way, in very well organized practice, it is NOT a person's name that is associated with a line item, it is a job description/title. For example:

1. Cardiac Ultrasound appointment process (CSR)
2. Cardiac Ultrasound admission process (Outpatient Nurse)
    a. Was the patient fasted? (OPN)
3. Cardiac Ultrasound calibration process (Imaging Tech)
4. Cardiac Ultrasound prep process (Technician Assistant)
5. Cardiac Ultrasound process (DVM and Technician)
6. Cardiac Ultrasound Data Collection process (DVM)
    a. Is there sufficient memory to capture and store images (Imaging Tech)
7. Cardiac Ultrasound Results Reporting process (Technician or DVM)
    a. Copy of report to referring vet (CSR)
8. Cardiac Ultrasound Release Process (Outpatient Nurse)
    a. Medications to go home process (Outpatient Nurse)
    b. Follow up appointment process (CSR)

DO YOU EXPECT SOMEONE TO INITIAL AND COMPLETE A CHECKLIST FOR EVERYTING WE DO?

Yes

And No.

Initially, while the checklists are becoming part of the day-to-day operations, it would be beneficial to manually check off the line items as they are completed. For any new hires going through training, it would be beneficial to manually check off the line items as they are completed. Anytime that there is a major change to a process or to a checklist, it would be beneficial to manually check off the line items as they are completed.

Once the checklists are well engrained into your practice's structure, checklist oversight should be assigned to an individual in that zone, for example a receptionist will handle the End of Day Checklist; an imaging technician will handle the Abdominal Ultrasound Checklist; a lab technician may handle the Cushing's Disease Work Up Checklist.

The person in charge of the checklist is ultimately accountable that all of the line items are completed whether they themselves actually perform the action or not. They have oversight of the outcome by making sure that others take care of their respective roles. They can do this simply by asking, "Kathy, did you calibrate the ultrasound this morning?" "Becky, are the endoscopes clean?" "Jack, is the chemistry analyzer warmed up?"

Over time, the checklists will become second nature but SOMEBODY must go through each step to make sure nothing was missed and especially critical items.

SUMMARY

Unlike human healthcare where MDs are so specialized that they focus so much on one part of one body that they don't have to even think of the patient as a whole, veterinary practices have way too many moving parts and our patients can't tell us what's going on. We have to depend upon the pet owner/client during veterinary interactions, which adds another variable to everything we need to do. The goal of systems, processes, and checklists is to control the variables as much as possible to minimize mistakes, ensure consistent client and patient care, and optimize outcomes.

In "The Checklist Manifesto", Dr. Atul Gawande describes how one small change and the integration of checklists decreased morbidity and mortality for something so seemingly simple as a central IV line. In extrapolating this to the veterinary field, we have so many different opportunities to improve our outcomes by simply introducing these systems of checks and balances and minimizing human error with the application of checklists to the multitude of tasks that we do on a daily basis.

There is no question that this will take time that is a rare commodity in veterinary practices. However, the time invested NOW will return to you in multiples of minutes and dollars down the road.

Teamwork It Makes the Dreamwork

Peter A. Weinstein, DVM, MBA
Irvine, CA

INTRODUCTION

Veterinary medicine is a team sport. The people on your team will ultimately determine the success of your practice. Putting this team together and getting them to perform at the highest level is one of the great challenges of any small business and definitely in a veterinary hospital where human resource management is still an evolving practice. Your dream business requires hiring correctly; training correctly; motivating correctly; and, making sure you have a truly engaged group. Teamwork thus makes your dream work.

WHY IS TEAMWORK SO IMPORTANT

Have you ever watched a well-developed sports team perform? It is truly magical to watch a perfectly trained offense or defense and their game plan come together because the people are teammates sacrificing for the greatness of the group. On the other hand, have you ever seen a group of players lead by one superstar whose ego and self-image are more important to them than the ultimate outcome of the team. The statistics are more important to that person than winning a championship. In a veterinary practice, the same issues arise. Whether it is the doctor (or doctor-owner), a technician, receptionist, etc., there is no room for ego in veterinary practice. With a practice focused on patient advocacy and client advocacy, everybody must be committed to the greater good. The client service team must be sensitive to the needs of the technician who are sensitive to the needs of the doctor team who are sensitive to the needs of the animal care team. The well-managed veterinary practice is magic to watch as they orchestrate the client and patient experience. On the other hand, when teamwork is not evident clients wait extended periods of time; mistakes are made in patient care; and the practice is dirty because there is no pride of ownership. Teamwork in veterinary medicine is not different than teamwork on football, basketball, baseball or hockey; it needs role players and leaders all pulling for each other and for the practice.

GETTING ORGANIZED

Jim Collins in his obligatory read, Good to Great, writes about the most successful businesses and how they got that way. One of the principles that he noted is that the great organizations 'start by getting the right people on the bus, the wrong people off the bus, and the right people in the right seats'. What are the right seats? Who are the wrong people? Who are the right people? Answering those questions will get your practice closer to the well-oiled machine it needs to be deliver your promises to the pets and clients.

The Right Seats

You can call them job descriptions or position statements. What are the roles and responsibilities that your practice and business need to be successful? Start by defining what you need first. You need to create the position statements or job descriptions for what you need and want, NOT for the people you already have. By clearly defining expectations, duties, tasks, objectives, and accountabilities, you will define what the 'seats' look like and then, and only then, can you find the right person for that seat.

The Right Person

The right people for your practice aren't necessarily the ones with the most skills or knowledge; they are the ones who share your organization's core values and mission. Only after you assess cultural fit and a team fit, do you look to see if they have the abilities needed to fill one of the seats. Hiring people for skills and without a team fit will ultimately disrupt the teamwork. Hiring people for cultural fit and then providing the training and coaching necessary to be successful in one of the seats will always create a team victory. Finally, the right people 'get it'. They don't need to be micro-managed. They are fully engaged. They accept responsibility and accountability and deliver what is asked of them. The right people get it, want it and have the capacity to deliver it. They are self-motivated and eager to be a part of a team that is expected to deliver great results.

The Wrong People

There have been numerous examples in the world of athletics where a superb player is added to an already successful team with great expectations for performance only to see the team implode and neither the superb player nor the supporting cast is able to deliver. The people you put on the team will determine how smoothly the bus ride will be. The wrong person will steer the team in the wrong direction because they focus specifically on their needs NOT on the greater needs of the organization. With the wrong people on the bus no matter what you do, the ride will be rocky and greatness will be hard to achieve.

You can look at this step as getting your Organization Chart in order. Have you created an organization chart representing whom reports to whom and who is accountable to whom? Within the org chart are specific roles or jobs. There are client service representatives, licensed technicians, exam room technicians, imaging technicians, animal caretakers, etc. For each of these jobs, you need to clearly define the responsibilities, expectations, and accountabilities.

With a clearly defined organization chart, you can place the people in their seats. Do you have the right people already? Great! Are some of your great people in the wrong seats based upon their skills? Maybe you need to move some people around the bus. Once you have the seats filled, do you have some outliers that have metaphorically 'missed the bus'? Driving around with people who shouldn't be on the bus will not drive you to success.

STEPS FOR SUCCESSFUL TEAM WORK

With members of the team in place and in the right seats on the bus, and the bus moving forward with great speed and smoothness how do you keep things working. Here are some thoughts:

1. Fire fast, Hire Slow

To keep the right people-right seats working, you need to take your time to identify people with a cultural fit. Using a Hiring Team composed of different staff within the practice-technician, reception, animal care- to evaluate the people skills of an applicant will be more effective than any questionnaire or even a one-on-one interview. The Hiring Team can write the job listing, screen the resumes, and even do some initial telephone screening (with an appropriate script). After the first cuts, the Hiring Team can conduct an initial group face to face interview and monitor during a working interview. After the Hiring Team has made its list of the Top Three candidates, the management team can get involved to help the Hiring Team choose amongst the Top Three.

The Hiring Team will be looking for those people, like them, who are committed to the team and the work ethic. And after the top candidate is hired, the Hiring Team will work to ensure their success because they don't want the 'wrong person on the bus'. This process may take longer than you are used to but taking your time hiring and also hiring in anticipation of needs rather than in response to needs will create a much better team fit.

After a candidate is hired, it shouldn't take very long to determine that somebody isn't going to work. How long are you going to give somebody the benefit of the doubt? Two weeks? Two months? Two Years? Within 60 to 90 days, you should be able to know if a new hire is going to be a team player or disruptive factor. Don't wait until good people get off the bus to get rid of a bad hire.

2. Set People up for success, not failure

With the optimal candidate chosen, a formal process for ONBOARDING AND TRAINING should be in place. You have the best person for the team; NOW, make sure you give them the tools for success. After a formal onboarding schedule to take care of legal requirements, a formal training and integration schedule should exist. Checklists, oversight, and monitoring are all a part of this process. Daily, weekly, and monthly benchmarks of success need to be reviewed. A good cultural fit without a formal training program may get along with the team but not be a productive member of the practice. Even someone who comes in with years of experience in the field should be put through a formal training program for your practice. Remember, there is no ego in a team based veterinary practice. Skills are only as good as the ability to work with others and to do things in the manner that your practice finds successful. Create an Onboarding and Training Program that clearly lines out daily, weekly, and monthly steps so that by 90 days you can determine: Do they understand the responsibilities of the job? Can they do the job? Do they want the job? Do they fit the team? If not...FIRE FAST.

3. Communication, Communication

When you watch sports on television, you will hear the announcers talk about the chemistry of the team and how well they communicate with one another. Whether it is the pitcher and catcher in baseball, the quarterback and receivers in football, or the point guard and forwards in basketball, there are both spoken and unspoken communication that leads to victory. The same should hold true in a practice situation. Doctor-technician, doctor-reception, Technician-reception, animal care-technician/reception, management to everybody, it doesn't matter what the dynamic is, there must be communication. This communication must be both verbal and, because of the healthcare nature of veterinary medicine, it must be written in the medical record. As much as we would love to have a team full of mind readers, that is not the way things work. There may come a time when your team can predict the needs of their co-workers and especially of the doctors, but until that time comes, the true test of a successful team is how well they and how clearly they communicate with one another.

There is very clear de-militarized zone that exists between the treatment room and the client service area where neither group wants to cross into the other's zone. If this zone exists, it must be removed. The client service team needs to know what is going on in the treatment area, the treatment area needs to know how smoothly things are going in the exam rooms, the doctors need to know what is happening in the cases that are active in the hospital and those that have gone home. Whether it is Rounds, Huddles, or Time Outs, consider creating a formal communication program during the day to ensure everybody is on the program. Taking the time to communicate with one another will communicate to the clients that you are all part of the same team and the patient care will also be at a higher level.

Besides daily communication, schedule formal department and staff meetings to work on education and communication on the global needs of the practice. Using staff meetings to teach and role-play and problem solve between and among zones will again add to the team feel.

The Hand-Off is one of the more common sources for fumbles in football. The same exists in a veterinary practice. When does a hand-off occur in veterinary practice? Every client or patient is transferred from the care of one person to another!! A phone call being handed from a receptionist to a technician. A patient's care being transferred from one doctor to another. A client being transferred from release instructions to collecting money. A patient's care being transferred from a doctor to a technician. And in specialty setting, a patient going from a referring DVM to a boarded specialist and back. This area of communication can very easily lead to malpractice or negligence issues. Work on this area of communication to minimize fumbles. A fumble in a veterinary practice can lead to the loss of a client or more significantly the loss of a patient.

4. It's All About the Trophy

In team sports, the salaries can be great. The year-end awards can be rewarding. The accolades for individual performance are headlines. However, it is winning the Trophy at the end of the year that results from the team working together, training, and performing day in and day out. The pain, the endless hours, and the sacrifice are all worth it for the Trophy represents the success of all the members of the team working together. What is your trophy? What is your measure of success? And how do you recognize the team effort that goes into 'winning the trophy'.

In a practice situation, the owner (as in a sports team) is the primary recipient of the trophy if money is the measuring stick. The owner gets the profits. But it is the team that helped to create the profits. The team needs to be acknowledged as well. Team based rewards recognize this fact. Everybody from the youngest animal caretaker to the most senior doctor helps in every case. There is no ONE person. Somebody walked the dog or cleaned the litter box. Somebody answered the phone call and made the appointment. Somebody monitored anesthesia or took radiographs. The outcome comes from everybody pulling for each other. The rewards should recognize the fact that the outcome is what you strive for.

Identify your trophy. Identify the rewards. Recognize the team. Teamwork will create more victories when the TEAM is the ultimate victor.

HOW DO YOU KNOW IT'S WORKING?

There a few metrics you can look like that indicate that the team is working. They include:

Client satisfaction surveys
Staff longevity
Profitability
Staff Engagement levels
Patient referrals

If you have a revolving door of staff and clients-it's not working
If you have low profitability—it's not working
If you have poor client satisfaction survey results—it's not working
If you are getting fewer and fewer referrals—it's not working

But the greatest way that you know it's working is how you feel at the end of the day; the end of the week; the end of the month; when the feeling in the pit of your stomach is gone and your ability to smile has returned.

Retention Marketing for Specialty Practices

Peter A. Weinstein, DVM, MBA
Irvine, CA

INTRODUCTION

Specialty practices have two primary populations that they must grab hold of and hang onto: pet owners and referring veterinarians. You need to focus on both. Throughout the discussion I will be addressing both of your target audiences. Which is more important to target? They are both equally important. However, there is a much higher return when you have advocacy from your local veterinary population as they have contact with many, many pet owners during the course of the year. Advocacy from pet owners that you have made happy works in multiple directions. They will seek your services when their pets have needs you can help them with; they will advise their friends of your services; and they will let their referring veterinarian know of their satisfaction with your care. This triad of the referring veterinarian-pet /owner – and specialist is a triad that provides optimal patient care, client satisfaction and best outcomes for the clinicians. Retention marketing is everything that you do maintain that triad intact.

WHAT IS RETENTION MARKETING?

Retention Marketing is directed at strengthening the already existing relationship that you have with your pet owning clients and rDVM (referring DVM clients) so that they have absolutely no desire to leave the warm, cozy comfort provided by you, your practice team, and your practice. It is based upon ongoing communication and education with a WE CARE attitude.

Through the use of various communication tools and education, you strengthen your existing client bonds as well as the bond with your rDVMs. Retention marketing places emphasis on personal media such as letters, e-mails, telephone, and postcards. It may also include newsletters and health alerts. In addition, with rDVMs it also includes personal visits to their practices as well as inviting them to your practice for visits, tours, and case discussions. The financial return on these education and communication outreach efforts is significantly higher than in acquisition marketing, sometimes reaching a $10 return for every dollar invested. And this return solidifies the relationship that you have especially with the rDVMs.

Reports indicate that it is about 10 times more expensive to acquire a client than to retain those that you already have; this is a general business benchmark. I would suggest it is even higher with getting referrals from local DVMs. Thus, acquisition marketing has a much higher price tag than just communicating, educating, and relationship building with the pet parents and rDVMs that you already have. Doesn't it make sense to do everything you possibly can to keep, preserve, and retain, that which is rightfully, dutifully, and already YOURS? Shouldn't the bulk of your advertising or marketing budget be directed toward retaining your already satisfied rDVMs and their (and your) clients? Thus, the term RETENTION MARKETING is used to describe any and all communications, promotions, marketing developed to maintain, keep or retain clients that are already part of your hospital family.

WHAT IS INCLUDED IN RETENTION MARKETING?

Retention marketing includes controlling all of the contact points with a client. These may include the times when the client contacts your practice or when you communicate with a client. (Remember client in this case is both the pet owner and the rDVM.)

INREACH: the client contacts you. Anytime a client calls or visits your practice, they develop an image. If the image is cold, careless, and isolated, the relationship that forms with your practice is weakened. If the image is warm, caring, comforting, and concerned, the relationship is strengthened.

IMAGE: It is imperative to work on your image every day. You should have a vision of how you want to relate to your clients. How do you want to answer the phones? How do you want to welcome clients when they come into the practice? What are you doing to make your existing clients feel welcome and part of the family?

APPEARANCE: Even the exterior and interior appearances of your practice are part of retention marketing. How warm is the atmosphere? What colors did you use to create a welcoming feel? Does your staff have a dress code or appearance? What about the nature of the seats in the reception area—are they comfortable or stiff? Pictures on the walls also emit a message. You have to think about all the details as your clients would. What a healthcare professional perceives as warm and comforting may not be what makes a consumer warm and comfortable. Seek professional assistance in developing your image and appearance. If your clients are comfortable, you will know because they are in no rush to leave and will hang around chit-chatting. That is a good sign.

OUTREACH: this is YOU communicating with the client. It is the mainstay of RETENTION Marketing. General practice communication, in the "good old" days, was written communication primarily in the form of reminder cards. These were sterile notes designed just to get a client to take a specific action. There was very little in the way of relationship built with reminder cards. And virtually NO education was included on the cards. Subsequently, newsletters became a popular form of communication. They allowed for more personalized approach to inform clients about the new products or services available in your veterinary practices and to provide information about medical conditions that pets might develop. Newsletters definitely allowed for more relationship building.

As the cost of care escalates, so has the need for more intensive marketing programs that focus on communication, education, and calls to action. The goal of 21st century Retention Marketing is to educate, communicate caring and needs, and encourage action.

As a specialty practice, you need to tap into everything that the rDVMs do PLUS you have to develop programs that are specifically designed to retain your rDVMs. Your educational pieces for the pet owner and the educational pieces for the rDVMs are going to be different. You have two populations to outreach to and target.

RETENTION MARKETING TECHNIQUES FOR PET OWNERS

As a specialty practice, you are frequently asked to perform procedures or techniques that your rDVM doesn't have the skills or technology to perform. You may be asked to take over a case that the rDVM hasn't been able to handle. Or you may be referred directly because the rDVM realized that, in the pet's best interest and the client's best interest, a referral was best.

After you have taken over the case and started to engage with the client, you move to client retention mode. However, in contrast to a rDVM that is working towards keeping a client from going to a competitive rDVM, your goal is multi-factorial. You want the pet owner to be satisfied enough to come back for ongoing care of this case (if needed); you want the pet owner satisfied enough to go back to their rDVM for ongoing care (if desired); you want the pet owner to think of you for this pet or their other pets if a challenging situation develops in the future; AND you want to maintain the referral relationship with the rDVM and make sure that they are comfortable.

Pet owners can never get enough education about responsible pet care. The best place to get his information is NOT Dr. Google but from veterinarians. Retention marketing means taking the time to fully educate your pet owners about the condition that is under treatment. Whether it is handouts or referring them to your website for additional information, the more information you can provide, the more educated your client is about the condition, the more compliant a client will be when it comes to treatment at home, follow up testing, and the longer the pet will live. Professionally created handouts on the major conditions that you see have a great value and will help communicate more effectively to the visual learners that we have. Creating simple and short videos on these same conditions can help communicate with audio or visual or kinesthetic learners.

Think of it this way...Teach to Treat. If you can teach the pet owners about their pet's condition the greater the likelihood you can treat the condition.

By the way, these same resources should be shared with your rDVM population. I know that I handed out hundreds of client education pieces from specialists while sending the pet owner out the door to see that very same specialist.

After the care has ended and the pet has been referred back to the rDVM should you do anything to communicate with the pet owner? YES. However, your goal is simply to remain top of mind to the pet owner. Following up by email or mail, days, weeks or even months down the road offering updates in the research on a condition is an inexpensive but very valuable way of staying top of mind. Remember, you don't want to be their primary care veterinarian you want to be their choice when the care is beyond that of the scope of the primary care veterinarian. The best way to do that is stay in touch and show you care about them and their pet.

Consider creating a quarterly electronic newsletter for pet owners that discusses new technologies, new specialists, or new thoughts on cases that you focus on. This is an eighth grade level, graphic heavy, pet picture heavy educational piece that can easily be forwarded to friends or family members that might have a need. Again, show you care; stay top of mind; be a resource.

Encourage pet owners to use Facebook as a way to let other pet owners know of your skills and expertise. And concurrently post success stories of pets on your Facebook page (get a release before doing so).

Retention of pet owners is important and doing a great job with them is integral for retention marketing of rDVMS. The manner in which you handle the pet owner is very important but even more important on case by case basis is how you communicate and handle the rDVM.

RETENTION MARKETING TO rDVMS

Having initiated care with a pet owner that has a primary care veterinarian, you enter into Retention Marketing with the rDVM. This is even while the case is still active. More important than the ultimate outcome to the rDVM is how you handle his client and even more important than that is how you communicate with the rDVM about what you are doing; why are you doing what you are doing; and what will they (the rDVM) be doing to help manage the case after you refer it back to them.

And by the way the greatest form of retention marketing for rDVMs is to make sure you refer a case back to them and work with them to manage the case. One of the greatest issues that rDVMs have is to never see a case again after it is referred. So, if you really want to impress your rDVM database, work with them!!!

Having done a great job on the case in question for a rDVM and also referring it back for after care, what more can you do to Retain that Referral Relationship?

As a former rDVM, I appreciated every effort by a specialist to help me become a better clinician and to be able to take better care of my clients and their pets. As noted above, the client education handouts that the specialists were using for their clients were a great resource for my practice and me and we handed out significant numbers of those to client as we were referring them for specialist care.

Many of the specialty practices now have case based e-newsletters that are sent to their rDVMs. These are great as they are adding to the educational level. Most of the articles are based upon cases from colleagues in the community and help the rDVM see what cases are being referred and how the referral relationship works.

An integral part of the retention marketing is something that is probably considered trite to the specialist but huge to the rDVM-the Referral Letter. A thorough explanation of how the case was managed by the specialist, the results of the tests run, the prognosis given to the client, the medications sent home, the expectations for needed after care AND what role the rDVM will have in that after care. This letter is HUGE (capitals intentional).

Communication between a specialist and rDVM can be challenging because everybody is busy. However, the ability to get on the phone with the specialist was a deal maker for me and a deal breaker at times. If the rDVM respects your time and your expertise as a specialist the willingness to get on the phone and talk about an active case, a potential case or just to go over some test results is Retention Marketing at its finest

Tell me you love me!! I loved it when specialists or their representatives would stop by and introduce themselves to the staff. This face-to-face meeting was a factor in whether our client service or technician team would advocate for referral to a particular specialty practice. You don't need to bring anything more than a smile (although donuts, bagels or cookies are great for staff's memory). Once or twice a year road trips show me that the relationship is more than just one way.

Finally, invite me to your house. Metaphorically. Invite your local referral base to the practice for CE programs. Encourage them to come by and talk to you about the case they referred. Encourage them to come by with radiographs or lab results. Invite your local referral base's technicians to your practice for CE by your technicians. Have an open house. The more comfortable we are with each other the more likely that there will be referrals.

READ THIS—for the last few years, a local veterinary specialty practice has been running hands on wet labs for their local practitioners. This is retention marketing at its finest. 6 hours of continuing education taught by specialist with a focus on practical and useful techniques that can be brought back to the practice an implemented. 10 stations with no more than 10 people per station with the ability to visit 10 different stations. It is more complex than space allows but consider opening your practice, sharing you knowledge and shaking the hands of your rDVMs.

SUMMARY

The specialty practice world is getting more and more competitive. Whether it is more boarded people or more generalists doing more and referring less, the need to develop and hold onto relationships is more important than ever. It is imperative to take the time to keep in touch with the pet owners after you have finished caring for their pets and just let them know you care.

More importantly, it is integral to strengthen the relationship with your rDVM population using impersonal communications such as e-newsletters and educational handouts. From a personal standpoint, go above and beyond with your referral letters, your availability on the telephone, your willingness to visit your referral base, and to invite your referral base to visit you.

Veterinary medicine is all about building relationships and keeping them. It is how well that you do this that will ultimately define your success.

The Genus *Rickettsia*: Review and Update

Dwight D. Bowman, MS, PhD, DACVM (Parasitology, Honorary)
Ithaca, NY
Edward B. Breitschwerdt, DVM, DACVIM (Small Animal Internal Medicine)
Raleigh, NC

INTRODUCTION

The genus *Rickettsia*, best known in North America for containing the organism causing Rocky Mountain spotted fever, *Rickettsia rickettsii*, which is a member of the larger group known as the Rickettsiales that contains two families important to veterinary medicine, i.e., the Rickettsiaceae and the Anaplasmataceae.[1] The Anaplasmataceae contains three genera that are disease agents in domestic animals, e.g., *Ehrlichia*, Anaplas*ma*, *Neorickettsia*, and the genus *Wolbachia* that has been recognized as a symbiont in arthropods (e.g., mosquitoes) and some of the nematodes (e.g., *Dirofilaria immitis*) vectored by these arthropods.[1] *Rickettsia* and *Orientia* are the two genera in the family Rickettsiaceae.[2] This well-known family is closely related to another lesser-known small Rickettsiales family, the Midichloriaceae with the single genus, *Midichloria*, that is symbiotic in *Ixodes ricinus* ticks.[3] Within the Rickettsiaceae, *Orientia* has only a single species, *Orientia tsutsugamushi*. *O. tsutsugamushi* is a zoonotic disease agent of humans causing scrub typhus or tsutsugamushi fever throughout southeastern Asia and the western Pacific region.[4] This disease is transmitted by chigger mites of the genus *Leptotrombidium* with the mites being the reservoir and vector with infections in the mites being maintained in the environment by transovarial transmission of the agent within the mites. The genus *Rickettsia* contains some 25 species which are divided into two major groups, the Typhus Group and the Spotted Fever Group along with a third group the "ancestral" group. The Ancestral Group consists of two species that are endosymbionts of ticks.[1] The Typhus Group contains two species: *Rickettsia prowazekii* transmitted by lice (epidemic typhus) and *Rickettsia typhi* (endemic typhus) transmitted by fleas.[1] The Spotted Fever Group contains species transmitted by ticks throughout the world. The two species that seem to stand apart a bit from other groups are *R. felis* that is seemingly transmitted perhaps by fleas, ticks, and mites and *Rickettsia akari*, the causative agent of mite-borne rickettsialpox.[1]

THE TYPHUS GROUP

The typhus group rickettsia, which includes *Rickettsia typhi* and *Rickettsia prowazekii*, have not been implicated as a cause of illness in domestic or wild animals. Experimental infection of dogs with typhus group rickettsiae in the laboratory at North Carolina State University did not result in illness.[5] Both agents do however, sporadically infect people in the United States.[6-8]

Rickettsia prowazekii has caused disease at levels of world changing proportions from the times of ancient Greece through World War II.[9,10] The control of human body lice and the introduction of antibiotics have in recent years pretty much eradicated disease outbreaks or controlled any flare ups that have occurred during war or famine. In the United States, there are rare human cases of sylvatic epidemic typhus (sporadic epidemic typhus) that are associated with the reservoir flying squirrels (*Glaucomys volans*). Transmission among squirrels is associated with the flea *Orchopeas howardi* and squirrel lice with transmission to humans by inhalation of aerosolized flea feces from nests in attics or eaves.[6]

Rickettsia typhi is an agent transmitted to people by rodent flea feces that are inoculated into the bite site by scratching of the wound. Cases still occur on occasion in the United States in people living around domestic rodents. Cases have occurred amongst the homeless in Houston, Texas,[11] with occasional cases being reported annually in humans in California and Hawaii.[7,8]

RICKETTSIA AKARI

The agent of Rickettsialpox was first recognized in 1946 in an outbreak in New York City.[12] The house mouse, *Mus musculus*, is the major reservoir, and it is transmitted to people by the house mouse mite *Liponyssoides sanguineus*. Long considered a member of the Spotted Fever Group, recent molecular characterization has aligned this agent with the *Orientia* Group;[2] another agent that has a mite-like vector. Rickettsialpox is usually a nonfatal illness that begins with a papule at each infectious bite site. Initial lesions can progress to an eschar. There are systemic symptoms that appear at the time of eschar formation, about 2 weeks after the initial bite that include headache and muscle aches. Most cases resolve without treatment, but antibiotic treatment is curative.

RICKETTSIA FELIS

Rickettsia felis has been found in domestic cats, dogs and opossums (*Didelphis virginiana*) in the United States and other countries and to be transmitted through the feces of the cat flea, *Ctenocephalides felis*. *Rickettsia felis* has been passed for 12 generations in cat fleas via transovarial transmission.[13] There are also reports of the finding of this agent in ticks and mites from around the world, and there are even suggestions that it might be vectored by mosquitoes.[2] Recent molecular work has suggested that *R. felis* is associated with the Ancestral group of non-pathogenic vector symbionts – one with a wide host range.[2] Despite the seemingly symbiotic relationship with multiple vectors, *R. felis* causes acute-onset, febrile illness in people around the world. Concern has been expressed about a number of cases in people from whom DNA has been recovered but who have never seroconverted to *R. felis*.[14] In Australia, *R. felis* seroprevalence was 50.7%, and rickettsemia was documented in 2.3% of dogs from indigenous communities, 9% of pound dogs and 19.8% of fleas collected from cats.[15] Although *Rickettsia felis* rickettsemia has been documented in healthy and sick dogs, the potential clinical importance of this species is yet to be determined. If dogs are a competent reservoir host for *R. felis*, as occurs with dogs, *Rickettsia conorii* and *Rhipicephalus sanguineus* in Europe, Asia and Africa,[16] peridomestic flea infestations on dogs could result on zoonotic human infection. *Rickettsia asembonensis*, initially identified molecularly in fleas collected in 2009 from Asembo, Kenya and subsequently in *Ctenocephalides felis* fleas removed from dogs in Ecuador.[17-19] The pathogenic relevance for this novel flea-associated *Rickettsia* sp. is unknown.

SPOTTED FEVER GROUP

Spotted fever group (SFG) rickettsiae have been described from all continents. In North America, *Rickettsia rickettsii* is the most important SFG rickettsiae because this tick-transmitted organism can cause serious or fatal illness in dogs and people. The SFG group includes numerous closely related species including *R. rickettsii* (the type species), *R. africae*, *R. australis*, *R. conorii*, *R. felis*, *R. montana*, *R. parkeri*, *R. rhipicephali* and *R. sibirica*, although many other SFG rickettsia have been described. Throughout various regions of the world, spotted fever group rickettsiae are transmitted by *Dermacentor*, *Rhipicephalus*, *Haemaphysalis* and *Amblyomma* tick species. Regardless of the strain or species of SFG rickettsiae, these organisms generally induce an acute febrile illness secondary to endothelial cell damage, which results in vasculitis, altered vascular permeability, edema and necrosis.[20] To date, chronic infection with a SFG rickettsiae has not been confirmed in dogs or human beings in North America. Although it seems likely that other SFG rickettsiae could induce disease in dogs, to date only *R. rickettsii* in North America and *R. conorii* in southern Europe have been documented as canine pathogens.[21] *Rickettsia parkeri* is a recently documented pathogenic species in humans in the eastern US and is found in 29% of *Amblyomma maculatum* in North Carolina, the state with the highest annual incidence of RMSF.[22,23] *Candidatus Rickettsia amblyommii* was pathogenic in a Guinea pig model and prior infection with this species was protective when Guinea pigs were challenged with *R. rickettsii*.[24] In Oklahoma, *R. montanensis* and *R. amblyommii* DNA were PCR amplified from the blood of naïve dogs naturally-exposed to ticks (90% *Ambylomma americanum*, and 10% *Dermacentor variabilis* and *A. maculatum*).[25] No dog developed signs of illness and no hematological abnormalities were found when complete blood counts were performed every two weeks during the 8 week tick infestation period.

ROCKY MOUNTAIN SPOTTED FEVER

Rocky Mountain spotted fever (RMSF) is a tick-transmitted rickettsial disease of dogs, people, and potentially other vertebrate species. The causative agent is *Rickettsia rickettsii*. In the context of morbidity, mortality, and severity of disease, RMSF is the most important tick-borne infection of dogs in the United States.[26-28] Due to variation in the severity and location of vascular lesions among different patients, veterinarians should anticipate a spectrum of disease manifestations following naturally-occurring infection with *R. rickettsii*.[27-34] Much of the United States is considered endemic for the ticks (*Dermacentor variabilis*, *Dermacentor andersoni*, *Rhipicephalus sanguineus*, *Amblyomma americanum*) that transmit *R. rickettsii*. *Rickettsia rickettsii* infection ("Rocky Mountain spotted fever") also occurs in areas of Central and South America, where peri-urban disease out breaks have been associated with brown dog tick (*R. sanguineus*) infestations. Clinical abnormalities associated with RMSF include fever, anorexia, depression, mucopurulent ocular discharge, scleral injection, tachypnea, coughing, vomiting, diarrhea, muscle pain, neutrophilic polyarthritis, and a diverse group of neurologic signs including hyperesthesia, ataxia, vestibular signs, stupor, seizures, and coma. In some dogs weight loss is very severe, considering the short duration of illness. Poorly localizing joint, muscle and/or neurologic pain suggestive of polyarthritis, polymyositis, or meningitis may represent the only or most prominent clinical finding. Retinal hemorrhages are a consistent finding, but may be absent early in the course of the disease.[30] Epistaxis, melena, hematuria, and petechial to ecchymotic hemorrhages occur in some dogs, but may not develop unless diagnosis and treatment are delayed for five or more days after the onset of clinical signs. Scrotal edema, hyperemia, hemorrhage, and epididymal pain are frequently observed in male dogs. Signs associated with cardiovascular collapse, oliguric renal failure or brain death can develop in the terminal stages of the disease. Gangrene affecting the distal extremities, scrotum, mammary glands, nose or lips is associated with severe vascular obstruction and can induce substantial tissue loss, necessitating reconstructive surgery.[31] Clinical manifestations in dogs are identical in most instances to disease manifestations reported in human patients.[26-28] From a public health perspective, it is important that veterinarians recognize and accurately diagnose RMSF, so as to educate the client as to the potential risk to the family.

Seasonal occurrence, history of tick infestation, fever, and combinations of the previously described clinical findings should suggest the possibility of RMSF.[26] Due to marked variation in clinical presentation, confirmation of the diagnosis of RMSF is also important for the clinician to gain familiarity with the disease. In addition, confirmation of RMSF in household pets, warrants owner education as to the enhanced risk associated with tick exposure in the local environment, as transovarial transmission of R. *rickettsii* from female ticks to progeny results in focal geographic locations in which a high percentage of ticks are capable of *R. rickettsii* transmission. Thrombocytopenia, generally mild in degree, is the most consistent hematologic finding associated with RMSF in dogs and people. Leukopenia generally occurs during the early stages of infection (first 24 to 48 hours), followed by progressive leukocytosis, which increases in proportion to the severity of vascular injury.[26-27] Toxic granulation of neutrophils, metamyelocytes, eosinopenia, lymphopenia, and monocytosis may accompany the more typical changes in platelet and neutrophil number. A mild anemia may occur. Severe anemia, leukopenia, and thrombocytopenia can accompany severe or fatal cases of canine Rocky Mountain spotted fever. In severe cases, co-infection with other tick-borne organisms should be pursued diagnostically. Biochemical abnormalities reflect the effects of generalized vascular damage and vary with the severity and duration of infection. Hypoproteinemia, hypoalbuminemia, azotemia, hyponatremia, hypocalcemia, bilirubinuria and bilirubinemia and increased liver enzymes (serum alkaline phosphatase, alanine aminotransferase) may occur in dogs with RMSF.[26,27] Synovial or cerebrospinal fluid analysis will generally reflect a mild increase in protein and cells, which, in the early disease process, are composed of a high percentage of neutrophils and later, mononuclear cells.

Diagnostic conformation of RMSF requires either direct immunofluorescent testing for *R. rickettsii* antigen in tissue biopsy or necropsy specimens, documentation of seroconversion, or PCR amplification of rickettsial DNA from blood, CSF, or tissue samples.[33] Utilizing the indirect fluorescent antibody test, documentation of a fourfold or greater increase in antibody titer between acute and convalescent sera confirms a diagnosis of RMSF.[29] To facilitate the accurate interpretation of serologic results, it is important to obtain the acute phase sample as early in the disease course as possible and the convalescent sample two to three weeks thereafter. If acute phase sera is obtained several (i.e. 5-7 days or longer) days after the onset of clinical signs, a high *R. rickettsii* antibody titer may be found. Although early initiation of anti-rickettsial antibiotic therapy will slightly decrease the intensity of the humoral immune response to *R. rickettsii* antigen, experimental studies indicate that collection of appropriately timed serum samples should reflect seroconversion and thereby facilitate an accurate diagnosis of RMSF in dogs.[29,32] Direct immunofluorescent testing or PCR amplification of *Rickettsia* DNA provides the opportunity for rapid diagnosis by demonstrating *R. rickettsii* in skin biopsies or by detecting DNA in blood or other diagnostic specimens. Unfortunately, direct FA testing is not readily available, however, PCR testing is now widely available for use by practicing veterinarians. Importantly, different commercial diagnostic laboratories use different PCR gene targets that vary in sensitivity. Despite the severity of RMSF, the low number of *Rickettsia* in systemic circulation (organism causes endothelial infection and damage), limits the diagnostic sensitivity of PCR. A positive PCR result confirms the diagnosis, a negative PCR result does not rule out RMSF. Prior initiation of antibiotic therapy can cause a false-negative test results when using direct immunofluorescent testing or PCR amplification of *R. rickettsii* DNA.

Tetracycline (22 mg/kg TID for 14 days) or doxycycline (5 mg/kg BID for 14 days) is effective for treatment of SFG rickettsiae infection.[26-28,32] Fluoroquinolones, such as enrofloxacin, appear to be equally effective for treating RMSF, but are not effective for treatment of ehrlichiosis. Following initiation of treatment, a rapid clinical response occurs in dogs without severe vascular damage or neurologic sequela. Defervescence should be anticipated within 24 hours after the initiation of antibiotics.[34] Delay in diagnosis, or the use of antibiotics that lack anti-rickettsial spectrums, such as penicillin, cephalosporin derivatives, or aminoglycosides, may result in increased morbidity or mortality. Due to severe vascular damage, fluid therapy should be utilized with extreme caution to avoid iatrogenic induction of cerebral edema.[34] Prednisolone administered at anti-inflammatory or immunosuppressive dosages in conjunction with doxycycline does not potentiate the severity of R. rickettsii infection in experimentally-infected dogs.[35]

Immunity following natural infection is probably life-long. It is possible that undiagnosed mild or asymptomatic *R. rickettsii* infections, or repeated exposure to nonpathogenic spotted fever group rickettsiae (such as *Rickettsia montanensis* or *Rickettsia rhipicephali*) can also contribute to naturally-induced immunity and therefore the prevention of severe RMSF in dogs with prior tick exposure in endemic regions. Following infection with *R. rickettsii*, the duration of rickettsemia in dogs is brief, approximately 5-14 days.[32,35] Therefore, infected dogs do not play an important reservoir role, and pose a minimal zoonotic threat to humans. In contrast, if RMSF is diagnosed in a household pet, the dog serves as a valuable environmental sentinel for potential human exposure to infected ticks. Care should be exercised in removing ticks, so as not to contaminate one's hands with infective hemolymph from ticks. Although the risk of inadvertent transmission is small, contact with rickettsemic blood during intravenous catheter placement, blood collection procedures, or laboratory sample analysis should be avoided. Zoonotic warning labels are recommended for blood samples obtained from dogs in which a differential diagnosis of RMSF is being considered.

FLEA AND TICK PREVENTION

Minimizing tick exposure, the use of acaricidal preparations, and routine, rapid removal of ticks from dogs represent the most effective means of prevention for RMSF. There is a broad spectrum of acaricides currently available in the markets of the United States. There is a large number topically applied products in many formulations and newly developed long-lasting tick collars that prevent tick attachment and feeding. Also, now becoming available are orally administered acaricides of the isoxazoline class that kill ticks soon after they begin to feed on a dog.[36] Restating the text above: "In the context of morbidity, mortality, and severity of disease, RMSF is the most important tick-borne infection of dogs in the United States." Dogs that go outside, especially in the eastern United States where RMSF is most common, need to be protected against tick attack. They should be protected against ticks year round.

REFERENCES

[1]Little SE ***Parasites for Veterinarians***, 10th ed, Elsevier, 2014; 246. [2]Shpynov S et al **Microb Infect** 2015; 17:839. [3]Sassera D et al **Int J Syst Evol Microbiol** 2006; 56: 2535. [4]Traub R, Wisseman CL **J Med Entomol** 1974; 11:237. [5]Breitschwedt EB et al **J Am Vet Med Assoc** 1995; 207:58. [6]Prusinki MA et al. **Vector-Borne Zoon Dis** 2014; 14:240. [7]Eremeeva ME et al **J Med Entomol** 2012; 1485. [8]Hoskinson S et al **MMWR** 2002; 52:1224. [9]Zinsser H. ***Rats, Lice, and History*** NY, 1935; Bantom. [10]Stapleton DH. **Technol Culture** 2008 46:513. [11]Reeves WK et al. **J Vector Ecol** 2008; 33:205. [12]Huebner et al. **Pub Hlth Rep** 1946; 61:1677. [13]Wedincamp J, Foil LD. **J Vector Ecol** 2002; 27:96; [14]Labruna MG, Walker DH **Emerg Inf Dis** 2014; 1768. [15]Hii SF et al **Parasit Vectors** 2013; 6:159. [16]Ionita M et al. **Ticks Tick-Borne Dis** 2015; 7:150. [17]Jiang J et al **Vector Borne Zoon Dis** 2013; 13:550. [18]Oteo JA et al. **Parasit Vectors** 2014 7:455. [19]Luce-Fedrow A et al **Vector Borne Zoon Dis** 2015 15:268. [20]Davidson MG et al **Am J Vet Res** 1990; 51:165. [21]Gasser AM et al **J Am An Hosp Assoc** 2001; 37 41. [22]Kidd L et al **Vet Microbiol** 2008; 129:294. [23]Varela-Stokes AS et al **Emerg Inf Dis** 2011; 17:2350. [24]Rivas JJ et al **Ticks Tick-Borne Dis** 2015; 6:805. [25]Barrett A et al **Vector-Borne Zoon Dis** 2014; 14:20. [26]Kidd L, Breitschwerdt EB **NAVC Clin Brief** 2009, April 9. [27]Greene CE et al., **J Am Vet Med Assoc** 1985; 186:465. [28]Breitschwerdt EB et al **Am J. Vet Res** 1985; 46:2124. [29]Breitschwerdt EB et al **Am J Vet Res** 1990; 51:1312. [30]Davidson M et al **J Am Vet Med Assoc** 1989; 194:777. [31]Weiser I, Greene CE. **J Am Vet Med Assoc** 1989; 195:1756. [32]Breitschwerdt EB et al **Animicrob Agents Chemother** 1991; 35:2375. [33]Davidson MG et al **J Vet Inter Med** 1989; 3:8. [34]Gasser AM, et al **J Am An Hosp Assoc** 2001; 37:41. [35]Breitschwerdt EB et al **Antimicrob Agents Chemother** 1997; 48:141. [36]Ozoe Y et al **Biochem Biophys Res Commun** 2010 391:744.

The Genus *Bartonella*: Review and Update

Dwight D. Bowman, MS, PhD, DACVM (Parasitology, Honorary)
Ithaca, NY
Edward B. Breitschwerdt, DVM, DACVIM (Small Animal Internal Medicine)
Raleigh, NC

INTRODUCTION

Bartonella species are fastidious Gram-negative bacteria that are highly adapted to a mammalian reservoir host and within which the bacteria usually cause a long-lasting intraerythrocytic bacteremia.[1-3] These facts are of particular importance to veterinarians and physicians, as an increasing number of animal reservoir hosts have been identified for various *Bartonella* species. Among numerous other examples, *Bartonella henselae* has co-evolved with cats, *Bartonella vinsonii* subsp. *berkhoffii* has co-evolved with dogs and wild canines, and *Bartonella bovis* has co-evolved with cattle. Importantly, the list of reservoir-adapted *Bartonella* species, including a large number of rodent and bat species continues to grow exponentially, as new *Bartonella* spp. are discovered.[2-3] Prior to 1990, there were only two named *Bartonella* species, whereas there are now at least 36 named and 16 unnamed or *candidatus* species, based upon deposited GenBank sequences or preliminary reports, respectively.

HISTORY

The agent of Carrión's Disease (Oroya fever or verruga peruana) was the first member of this genus described.[4] It was well known from work on the disease by a medical student (Daniel Carrión) in Peru for whom the disease is named. He died of febrile anemia after purposely infecting himself with the agent inoculated from a patient's skin lesion. The agent was described as *Bartonella bacilliformis* in 1913. This is the only species of *Bartonella* known to be transmitted by a sand fly, *Lutzomyia verrucarum*, and until recently it was thought this was the only species to vector the infection. However recent cases of the disease have been identified in patients outside the "verruga zone., where the sandfly is not found."

Trench Fever (5-day fever), caused by *Bartonella quintana*, was a major disease amongst the armies of both sides in World War I.[4] It was found in the blood and described as *Rickettsia quintana* by Schminke in 1917.[5] The disease is spread through the posterior station method, i.e., with the organisms being passed in the feces of the infected louse, and in the dried feces the agent can remain infective for several months. People are infected when the louse feces are rubbed into the bite wound. Infections have occurred in the homeless in the USA, and in immunocompromised individuals *B. quintana* has been found in lesions of bacillary angiomatosis.[4]

Bartonella henselae the causative agent of cat-scratch disease had been known as a disease for nearly 100 years before the bacterium was initially isolated in 1992, not from a case of cat-scratch disease but form an immune-impaired person with recurrent fever.[4] After being recovered from this HIV infected human, the agent has subsequently been found in cats, dogs, horses, marine mammals and other small terrestrial mammals. *Bartonella clarridgeiae* was recovered from a cat owned by a person septic with B. *henselae.*[6] *B. clarridgeiae* has since been shown to be present in cats around the world,[7-9] and has been identified in veterinarian who was bitten on the finger by an adopted 6-week-old cat with a flea infestation.[10] *Bartonella vinsonii* subsp. *berkhoffii* was initially isolated from a dog with endocarditis in the Rickettsiology Laboratory at the College of Veterinary Medicine at North Carolina University in 1993,[11] and subsequently, from cats, coyotes and human patients. Retrospectively, long-term administration of immunosuppressive doses of corticosteroids for a presumptive diagnosis of systemic lupus erythematosus with cutaneous vasculitis may have facilitated the isolation of the original type strain of *B. vinsonii (berkhoffii)* from this dog that subsequently developed endocarditis.

EPIDEMIOLOGY

In cats, *B. henselae* and *B. clarridgeiae* have been amplified or grown most frequently from cats and the fleas collected from cats. Flea-associated transmission has been well documented amongst cats. In flea endemic areas, *Bartonella* spp. seroprevalence rates in cats can be greater than 90% and bacteremia rates can be greater than 50%. Granulomatous myocarditis has been reported in cats, naturally or experimentally-infected with *B. henselae. B. vinsonii* (*berkhoffii)* caused recurrent osteomyelitis in a cat.[3,11]

Due to the relatively recent recognition that dogs can be infected with *B. vinsonii* (*berkhoffii*), *B. henselae* and potentially other *Bartonella* spp., seroprevalence data is somewhat limited.[3] Seroprevalence was determined in 1,920 sick dogs from North Carolina or surrounding states that were evaluated at a veterinary teaching hospital. Using a reciprocal titer of >32, only 3.6% of sick dogs had antibodies to *B. vinsonii (berkhoffii).* Risk factors that could be associated with seroreactivity included: heavy tick exposure (Odds ratio 14.2), cattle exposure (OR 9.3), rural vs. urban environment (OR 7.1) and heavy flea exposure (OR 5.6). Experimental flea transmission of *B. henselae* to dogs has now been confirmed in the laboratory (Lappin and Breitschwerdt, unpublished data). Also, using sera from dogs experimentally infected with *R. rickettsii* or *Ehrlichia canis*, cross reactivity to *Bartonella* antigens was not detected. However, 36% of serum samples derived from dogs naturally infected with *E. canis* were reactive to *B. vinsonii* antigens. As *E. canis* is transmitted by *Rhipicephalus sanguineous*, this tick may be involved in the transmission of *B. vinsonii*. The possibility of tick transmission was further supported by two additional studies involving dogs infected with one or more *Ehrlichia* spp. from the same geographic region, in which seroreactivity to *B. vinsonii (berkhoffii)* antigens was 30% and 89%, respectively. Seroprevalence, using *B. vinsonii (berkhoffii)* antigens, was 10% (4/40 dogs) in dogs with suspected tick-borne illness from Israel and 36% in dogs with fever and thrombocytopenia from Thailand. Using an ELISA assay, 35% of 869 samples, derived from coyotes in California, contained antibodies to *B. vinsonii (berkhoffii)* antigens. Current data indicates that canine and human exposure to *B. henselae* and *B. vinsonii (berkhoffii)* can be found throughout much of the United States and most tropical and subtropical regions of the world.

Studies from Hawaii, the United Kingdom and Japan identified *B. henselae* seroprevalences of 6.5% (2/31dogs), 3.0% (3/100 dogs) and 7.7% (4/52). In the Rickettsiology Lab at North Carolina, *B. henselae* is the most common *Bartonella* species found in sick dogs using the BAPGM enrichment blood culture platform. The pathogenicity of all *Bartonella* spp. in dogs is poorly understood; however, like humans, dogs can be infected with numerous *Bartonella* spp. *B. henselae* was amplified and sequenced from the liver of a dog with peliosis hepatis, a unique pathological lesion also reported in *B. henselae* infected people. *B. henselae* DNA was also amplified from a dog with granulomatous hepatitis, a histopathological lesion reported with some frequency in *B. henselae*-infected children and adults. Similarly, *B. clarridgeae* DNA has been amplified and sequenced from the liver of a Doberman pincher with copper storage disease and from the aortic valve of a dog with vegetative valvular endocarditis. *Bartonella elizabethae*, a species that infects rodents, was found in a dog that had experienced chronic weight loss culminating in sudden unexplained death. Based upon a large seroepidemiological, controlled study from the University of California (Davis), dogs that were seroreactive to either *B. henselae, Bartonella clarridgeae* or *B. vinsonii (berkhoffii)* were referred for evaluation of lameness, neutrophilic polyarthritis, nasal discharge, epistaxis and splenomegaly.[12]

PATHOGENESIS

Although as yet unproven, *B. vinsonii (berkhoffii)* is presumably transmitted to dogs by the bite of an infected tick. Based upon anecdotal evidence, dogs may become infected with *B. henselae* by a cat bite or scratch, analogous to cat scratch disease in people. *B. vinsonii* appears to cause chronic intraerythrocytic and endothelial cell infections in dogs for extended periods of time, potentially resulting in vasoproliferative pathologies. Similar to other highly adapted intracellular vector-transmitted pathogens, the factors that ultimately result in these bacteria causing disease manifestations are yet to be determined. If similar to babesiosis, another intraerythrocytic pathogen, stress, hard work, parturition, concurrent infection with other organisms or therapeutic immunosuppression may contribute to the development of pathology. Following experimental inoculation of SPF dogs with culture grown *B. vinsonii (berkhoffii)*, there was sustained suppression of peripheral blood CD8+ lymphocytes, accompanied by an altered cell surface phenotype and an increase in CD4+ lymphocytes in the peripheral lymph nodes.[12] Therefore, infection with *B. vinsonii (berkhoffii)* appears to induce a degree of chronic immunosuppression that might predispose dogs to other infectious agents, resulting in a wide array of clinical manifestations in naturally-infected dogs. Less is known about pathogenesis of disease in cats but some mechanisms are likely similar. Cats infected with *Bartonella* spp. are commonly co-infected with hemoplasmas and at times, more than one *Bartonella* sp. However, whether co-infections magnify disease manifestations of either genera is unclear and in most studies co-infections did not appear to potentiate illness. From an evolutionary perspective, it is obvious that vectors, vector-borne organisms, and animal and human hosts have developed a highly adapted form of interaction. These factors serve to illustrate the potential difficulty in establishing causation in cats, dogs or people co-infected with multiple tick-transmitted pathogens. Recently, we proposed an addition to Koch's postulates entitled the Postulate of Comparative Infectious Disease Causation.[12] By satisfying this postulate, *Bartonella* species appear to be able to cause, endocarditis, granulomatous inflammatory diseases, particularly involving heart, liver, lymph nodes, and spleen, persistent intravascular infections and the induction of vasoproliferative tumors.[3,13-16]

CLINICAL FINDINGS

The spectrum of disease associated with *Bartonella* infection in dogs and most other animal species is currently unknown. Endocarditis, has been reported in cats, dogs and humans, infected with a spectrum of *Bartonella* spp. In some dogs, intermittent lameness, bone pain, epistasis or fever of unknown origin can precede the diagnosis of endocarditis for several months, whereas other dogs will present with an acute history of cardiopulmonary decompensation.[3,13] Cardiac arrhythmias secondary to myocarditis can be detected in dogs without echocardiographic evidence of endocarditis. Granulomatous lymphadenitis has been associated with *B. vinsonii (berkhoffii)* and *B. henselae* in dogs. *B. vinsonii (berkhoffii)* and other *Bartonella* species appear to contribute to the development of dermatologic lesions indicative of a cutaneous vasculitis, panniculitis, as well as anterior uveitis, polyarthritis, meningoencephalitis and immune-mediated hemolytic anemia.[3,14-18] Additional research efforts, using carefully designed case controlled studies are necessary to establish the frequency and extent to which *Bartonella* spp. contribute to dermatological, ocular, orthopedic, neurological or hematological abnormalities in dogs (and humans).

Clinically, many disease manifestations have also been attributed to *Bartonella* spp. infections in cats.[3] However, it is very difficult to prove disease associations in cats in the field because of the high prevalence rates in non-clinical carriers. In research cats that are infected by exposure to *C. felis*, fever, endocarditis and myocarditis are the most common disease manifestations. As discussed for dogs, additional case controlled studies are needed in cats.

DIAGNOSIS

In the natural reservoir host, chronic bacteremia with a *Bartonella* species can frequently be detected by blood culture or PCR in outwardly healthy individuals. In contrast, the diagnostic detection of a *Bartonella* spp. in a non-reservoir adapted host can be extremely difficult. Most, although not all diseases caused by *Bartonella* spp., occur in accidental hosts and these organisms are being increasingly implicated as a cause of zoonotic infections.[19-23] Until recently, mechanisms that facilitate persistent *Bartonella* bacteremia in mammals were not well understood. Recent reports have identified an intra-endothelial and intra-erythrocytic localization for these bacteria, which represents a unique strategy for bacterial persistence.[2,3] Non-hemolytic intracellular colonization of erythrocytes and endothelial cells would preserve the organisms for efficient vector transmission, protect *Bartonella* from the host immune response, and potentially contribute to decreased antimicrobial efficacy. Other *in vitro* studies indicate that *Bartonella* spp. can infect dendritic cells, microglial cells, pericytes, monocytes and CD34+ bone marrow progenitor cells.

Thrombocytopenia, anemia, which frequently can be immune-mediated, and neutropenia or neutrophilic leukocytosis are the hematological abnormalities in dogs that are seroreactive or BAPGM enrichment blood culture/PCR positive.[15,18] Thrombocytopenia is found in approximately half, eosinophilia approximately one third of infected dogs and monocytosis frequently occurs in Bartonella endocarditis. Hematological abnormalities have been rarely reported in cats, but similar to dogs, a subset of Bartonella-infected cats are neutropenic. Serum biochemical abnormalities are usually very mild or nonexistent in both cats and dogs. In cats, *Bartonella* spp. antibodies have correlated with polyclonal hyperglobulinemia and hypoglycemia.[24]

As *B. vinsonii (berkhoffii)* antibodies are infrequently detected (<4%) in sick or healthy (<1%) dog populations in endemic regions, detection of *B. vinsonii (berkhoffii)* antibodies in a sick dog provides strong clinical evidence for prior exposure and potentially active infection. For this reason, treatment of seroreactive dogs or dogs from which any *Bartonella* spp. DNA is detected in blood or tissue samples would be recommended.

ISOLATION AND MOLECULAR DETECTION OF *BARTONELLA* SPECIES

Because conventional microbiological isolation techniques lack sensitivity, bartonellosis is usually diagnosed by PCR amplification of organism specific DNA sequences and/or through serological testing. Recently, the development of a more sensitive isolation approach, using BAPGM (*Bartonella* alpha *Proteobacteria* growth medium) followed by real time PCR has greatly facilitated the molecular detection or isolation of *Bartonella* species from the blood of sick or healthy animals, including dogs and human beings.[11,20,21] Obviously, the relative sensitivity of the diagnostic methods used to detect *Bartonella* species infection greatly influences an investigator's ability to establish disease causation or a clinicians ability to initiate appropriate treatment. The use of this optimized microbiological approach has facilitated the recognition of blood-borne *Bartonella* spp. infections in dogs, horses, human beings and porpoises.[11] Diagnostic testing (animals and humans) for *Bartonella* species (serology, PCR and BAPGM Enrichment Blood Culture/PCR) is available through Galaxy Diagnostics, Inc. (contact@galaxydx.com). In cats, serology, PCR or culture combined with serology is recommended and can be procured at Galaxy Diagnostics Inc. and Colorado State University (www.dlab.colostate.edu)

PATHOLOGIC FINDINGS

In dogs, pathologic findings associated with *Bartonella* spp. infection include endocarditis, myocarditis, granulomatous lymphadenitis, granulomatous hepatitis, osteomyelitis, bacillary angiomatosis and peliosis hepatitis.[3,13,15,16] Multifocal areas of severe myocardial inflammation can be found in dogs with *B. vinsonii (berkhoffii)* endocarditis. Although not specific for *Bartonella* infections, organisms can be detected in diseased tissues using silver stains, particularly in acute *Bartonella* infections, such as acute regional lymphadenitis (cat scratch disease). During chronic infections, organisms are often too few in number to be detected in tissues by silver staining, unless a fulminate infection is localized to heart valves. The cardiac abnormalities noted in cats to date are similar to those described for dogs.[13] It seems likely that the spleen plays an important immunomodulatory role in controlling persistent *Bartonella* spp. bacteremia in animals and people.[16] The extent to which *Bartonella* spp. induce splenic pathology deserves additional research consideration.

THERAPY

To date, an optimal protocol has not been established for the treatment of *Bartonella* infections in cats, dogs, or people.[3,11] Regardless of the antibiotic(s) that is used for treatment, a long duration of antibiotic administration (4-6 weeks) may be necessary to eliminate the infection. Due to the rapid development of resistance, macrolides (azithromycin) I no longer recommend these antibiotics for treating *Bartonella* infections. Fluoroquinolones in combination with doxycycline are currently being used by the author to treat clinical cases of bartonellosis, while exploring efficacy through experimental infection studies. Doxycycline alone does not appear to eliminate *B. vinsonii (berkhoffii), B. henselae* or *B. clarridgeae* in cats, dogs or other animal species. Serum antibody titers decrease rapidly (3-6 months) and are generally no longer detectable in dogs that recover following antimicrobial therapy. Therefore, post-treatment serology may be a useful adjunct to BAPGM/PCR to determine if therapeutic elimination of Bartonella infections has been achieved. Whether there is clinical benefit to follow serologic or molecular assay results in cats has not been widely studied, but most treated cats do not become seronegative in the short term. However, bacteremia can resolve after treatment or resolve spontaneously in some cats, whereas other cats remain bacteremic despite four to six weeks of antibiotic (documented for several antibiotic regimens) administration, and despite resolution of clinical abnormalities (such as lethargy, inappetence and fever).

PREVENTION

Minimizing or eliminating flea and tick exposure is perhaps of greater veterinary and public health importance today, than during any previous time in history. Currently the role of the common "cat flea" that can also infest dogs, coyotes, foxes, raccoons and opossums represents an underappreciated vector for disease transmission to animals and their owners. When rigorous flea and tick control measures are instituted, it is highly probable that transmission of *Bartonella* species to pets and their owners will be greatly reduced or eliminated.[25]

PUBLIC AND OCCUPATIONAL HEALTH CONSIDERATIONS

There is increasing evidence to support an important role for *Bartonella* species as a cause of a spectrum of disease manifestations in human patients.[11,19,26-32] CDC has added a page on the Bartonella website with information for veterinarians: http://www.cdc.gov/bartonella/veterinarians/ Due to extensive contact with a variety of animal species, veterinary professionals appear to be at occupational risk for infection because of frequent exposure to *Bartonella* spp.; therefore, these individuals should exercise increased precautions to avoid arthropod bites, arthropod feces (i.e. fleas and lice), animal bites or scratches and direct contact with bodily fluids from sick animals.[33] As *Bartonella* spp. have been isolated from cat, dog or human blood, cerebrospinal fluid, joint fluid, aqueous fluid, seroma fluid and from pleural, pericardial and abdominal effusions, a substantial number of diagnostic biological samples collected on a daily basis in veterinary practices could contain viable bacteria. The increasing number of defined *Bartonella* spp., in conjunction with the high level of bacteremia found in reservoir-adapted hosts, which represent the veterinary patient population, ensures that all veterinary professionals will experience frequent and repeated exposure to animals harboring these bacteria. Therefore, personal protective equipment, frequent hand washing and avoiding cuts and needle sticks have become more important as our knowledge of this genus has improved and various modes of transmission have been defined. Physicians should be educated as to the large number of *Bartonella* spp. in nature, the extensive spectrum of animal reservoir hosts, and the diversity of confirmed and potential arthropod vectors, current limitations associated with diagnosis and treatment efficacy, and the ecological and the medical complexities induced by these highly evolved intravascular, endotheliotropic bacteria.

REFERENCES

[1]Breitschwerdt EB, et al **Clin Microbiol Rev** 2000;13:428. [2]Chomel BB, et al **Ann N Y Acad Sci** 2003;990:1. [3]Breitschwerdt EB, et al **J Vet Emerg Crit Care** 2010;20:8. [4]Karem KL et al **Microbes Inf** 2000;2:1193. [5]Schmincke A **Munsch Med Wochschr** 1917;64:91. [6]Lawson PA, Collins MD. **Med Microbiol Let** 1996;5:64. [7]Braga IA et al **Braz J Inf Dis** 2015;19:451. [8]Fleischman DA et al **Appl Environ Microbiol** 2015;81:6446. [9]Mokhtar AS, Tay ST. **Am J Trop Med Hyg** 2011;85:931. [10]Kordick DL et al **J Clin Microbiol** 1997;35:1813. [11]Breitschwerdt EB **ILAR J** 2014;55:46. [12]Breitschwerdt EB, et al **J Comp Pathol** 2013;1:1. [13]Chomel BB, et al **Ann N Y Acad Sci** 2009;1166:120. [14]Henn JB, et al **Am J Vet Res** 2005;66:688. [15]Perez C, et al **J Vet Intern Med** 2011;25:805. [16]Varanat M, et al **J Vet Intern Med** 2011;25:1284. [17]Beerlage C, et al **Med Microbiol Immunol** 2012: 319. [18]Pérez Veraa C, et al **Comp Immunol Microbiol Infect Dis** 2013;36:481. [19]Breitschwerdt EB, et al **J Neuroparasitology** 2012;3:1. [20]Mascarelli PE, et al **Parasit Vectors** 2013;6:98. [21]Maggi RG, et al **Parasit Vectors** 2013;6:101. [22]Maggi RG, et al **Parsit Vectors** 2013;6:106. [23]Balakrishnan N, et al **J Med Microbiol** 2013;62:338. [16]Whittemore et al **J Vet Intern Med** 2012;26:639. [18]Lappin MR, et al **Parasit Vectors** 2013;25:26. [24]Breitschwerdt EB, et al **J Clin Microbiol** 2011;49:3415. [25]Cherry NA, et al **Vector Borne Zoonotic Dis** 2011;11:1425. [26]Maggi RG, et al **Diagn Microbiol Infect Dis** 2011;71:43. [27]Breitschwerdt EB, et al **J Clin Microbiol** 2007;45:270. [28]Breitschwerdt EB, et al **J Clin Microbiol** 2010;48:2289. [29]Oliveira AM, et al **J Vet Intern Med** 2010;24:1229. [30]Rossi MA, et al **Vet Dermatol** 2015;26(1):60-e22. [31]Balakrishnan N, et al **J Clin Microbiol** 2014;52:3447.

Fecal Microbiota Transplantation in Veterinary Medicine: From Bench to Cage Side

Stanley L. Marks, BVSc, PhD, DACVIM (Small Animal Internal Medicine, Oncology), DACVN and
J. Scott Weese, DVM, DVSc, DACVIM (Small Animal Internal Medicine)
Davis, CA and Guelph, ON

The mammalian intestinal tract contains a complex, dynamic, and diverse population of predominantly non-pathogenic bacteria that play equally complex and poorly understood roles in the maintenance of health. Researchers have estimated that the average human contains 100 trillion microbes in the gut, which is 10 times more than the cells of the human body.[1] The intestinal microbiota influences the health of the host by providing nutritional substrates, modulating the immune system, and providing a support in the defence against intestinal pathogens.[2] The term *microbiota* refers to the microbial population present in different ecosystems in the body while *microbiome* refers to the total number of microorganisms and their genetic material.

Previous approaches to gastrointestinal disease tended to focus on the actions of single pathogens on the gastrointestinal tract or body as a whole. While this provides some understanding of pathophysiology of disease, it overlooks the response of the microbiota to pathogens and other influences, and the potential role of an altered microbiota (often referred to as *dysbiosis*) in disease. With recent technological advances, particularly next generation sequencing, in-depth investigation of the complex gastrointestinal microbiota has become relatively easy and economical, so recent attention has focused on the indigenous non-pathogenic microbiota and the ways in which it may benefit the host.

Initial colonization of the sterile newborn intestine occurs with maternal vaginal and fecal bacterial microbiota. The first colonizers have a high reduction potential and include species such as *Enterobacter*, *Streptococcus*, and *Staphylococcus*.[3] These bacteria metabolize oxygen, favouring the growth of anaerobic bacteria, including lactobacilli and bifidobacteria. Colonization with these bacteria is significantly delayed in caesarean deliveries,[4,5] leading to delayed activation of the efferent limb of the mucosal immune response. Additional beneficial effects of developing a normal bacterial flora are seen in germ free mice that have small intestines that weigh less than their healthy counterparts. This effect occurs partly due to underdevelopment of lymphoid constituents, with a lack of plasma cells in the lamina propria and Peyer's patches, and subsequent reduction in IgA production. Exposure to bacteria results in a reversal of this phenomenon within 28 days of exposure.[6]

Exposure to the intestinal microbiota can both produce and prevent disease, depending on the situation. Alterations in the intestinal microbiota have been associated with Crohn's disease and ulcerative colitis, as well as irritable bowel syndrome in humans.[7-9] In addition, the intestinal microbiota has also been implicated in the pathogenesis of various canine GI disorders, either associated with the presence of specific enteropathogens such as *Salmonella*, *Clostridium perfringens*, and viruses in acute episodes of diarrhea or a non-specific dysbiosis precipitating inflammatory bowel disease (IBD).[10,11] Molecular –phylogenetic studies have revealed a bacterial and/or fungal dysbiosis in the duodenum of dogs with IBD. A common trend observed in dogs and other species with GI disturbances has been profound shifts in some major members of the microbiota. Often, this involves a decrease in the proportion of Clostridiales and an increase in Proteobacteria,[11] and Proteobacteria over-abundance may be a general indicator of intestinal microbiota dysbiosis.

Only a few studies have described the fecal microbiota of dogs with acute and chronic GI disorders. Dogs with acute diarrhea, particularly those with acute hemorrhagic diarrhea have the most profound alterations in their microbiota characterized by decreases in Blautia, Ruminococcaceae including *Faecalibacterium* and *Turicibacter* spp, and significant increases in genus *Sutterella* and *Clostridium perfringens* compared to healthy dogs. Dogs with clinically active IBD had decreased *Faecalibacterium* spp. and Fusobacteria that increased during resolution of the IBD.[11] The bacterial species that are commonly decreased during diarrhea are thought to be important short-chain fatty acid producers and could promote intestinal health. A deeper understanding of the gut microbiome will provide reference values for healthy populations and assist in diagnosing and treating diseased animals. In addition, manipulation of the intestinal microbiome via dietary intervention, administration of probiotics, prebiotics, or synbiotics, and fecal transplantation is currently being performed to maintain gut health in people and companion animals.

Thus, while understanding of the canine and feline gut microbiota is still very much in its infancy, data from dogs, cats and other species clearly supports the notion that the gut microbiota is a critical factor in gastrointestinal and systemic health. Accordingly, this suggests that manipulation of the intestinal microbiota might be an important therapeutic tool. This should come as no surprise since various methods of microbiota manipulation have been widely employed for the treatment of GI disease in dogs and cast, such as the use of antimicrobials (e.g. tylosin, metronidazole). While often effective, this is a rather crude and untargeted approach that carries potential risks. Other approaches have been employed with varying success. Manipulation of canine GI microbiota to improve health via dietary intervention did not begin until the early 1990's. Dietary fiber, prebiotics and probiotics have been the major nutritional strategies studied to modulate the canine and feline GI microbiota. Unfortunately, most of the research studies published to date have evaluated the effects of dietary manipulation of the GI microbiota in clinically healthy dogs and cats, and many of these studies have used traditional plating techniques or qPCR to quantify a limited number of bacteria (e.g., *Lactobacillus*, *Bifidobacteria*, *C. perfringens*, and *E. coli*) to assess efficacy. While subtle shifts in certain microbial components can sometimes be found, clinical relevance of these has been unclear and clinical studies of probiotics and prebiotics have largely been disappointing. One likely problem is the composition of the probiotic, with commercial probiotics using relatively small numbers (in proportion to the existing gut population) or a small number of bacteria that comprise a distinct minority of the microbiota of healthy animals. Measures to provide larger numbers of organisms that specifically target certain health or disease characteristics would be ideal, and a 'synthetic stool' approach is being investigated in humans.[12] However, this is not currently an option for animals.

Fecal Microbiotia Transplantation (FMT)

Fecal microbiotal transplantation or infusion of a fecal suspension from a healthy individual into the gastrointestinal tract of another person to cure a specific disease was first performed in the 4th Century when people with severe diarrhea ingested a fecal suspension (yellow soup). Fecal transplantation was performed in horses with diarrhea in the 17th century, and rumen transfaunation has represented a similar approach in cattle. While specific reasons were not known, these approaches were based on the assumption that 'good' bacteria were present in the donor sample and outcompeted or replaced the abnormal components of the microbiota in an individual with disease. Use of FMT in humans has been more limited, probably more because of aversion to the procedure rather than any inherent difference between disease or potential response to treatment between species. The procedure gained increasing interest in 1958 when a fecal enema was administered to 4 human patients with severe pseudomembranous colitis secondary to *Clostridium difficile*. The patients had a dramatic response to the procedure within 48 hours. However, its use was, at best, rare and the medical community largely saw no reason to embrace such a treatment in an era where 'modern' medicine was the focus. However, profound increases or recurrent and debilitating *C. difficile* infection (CDI) in humans has lead to renewed interest in FMT. Data from various human studies have reported astounding cure rates for recurrent CDI, which tends to carry a poor prognosis. This has led to expansion of study into other conditions such as ulcerative colitis (UC) and Crohn's disease (CD).[13-15] In a recently published randomized controlled trial in patients with UC, FMT (one enema once weekly for 6 weeks) was statistically superior to placebo to induce remission at week 7; however, one of the donors was associated with better clinical outcome, suggesting that donor characteristics are the driver of the potential beneficial effect of FMT in IBD.[14] In an effort to facilitate use of this treatment, frozen donor stool has been studied, with no difference in outcome compared to fresh stool.[16] While issues remain with regulatory approval for the process, standardization of donor testing and a few other procedure or regulatory aspects, there is much interest in this minimally invasive approach to some frustrating GI diseases.

FMT has also been used in animals, with abundant anecdotal information but limited objective data. FMT has been performed in dogs and cats with a variety of chronic enteropathies,[17] and one of the authors (SLM) is currently completing a clinical trial evaluating the efficacy of FMT in macaques with chronic colitis. There are various difficulties in performing and interpreting clinical trials for chronic diarrhea in dogs and cats, particularly the multifactorial nature of disease. Regardless, FMT is of increasing interest for treatment of chronic enteropathies as an apparently safe, effective, drug-free and relatively low cost approach. However, much needs to be defined and refined for successful and safe broader application of this approach. Objective information for most aspects of FMT in dogs and cats is lacking but extrapolation from humans and knowledge of infectious disease can lead to reasonable preliminary recommendations.

Selection of the donor

In humans, donors are extensively screened for enteropathogens and bloodborne pathogens.[18] Additional restrictions typically involve recent antimicrobial usage, immunosuppressive administration, GI disease and other comorbidities. Guidelines for dogs and cats are lacking and probably vary depending on the species, pathogen prevalence in the region, risk aversion and cost. Ideally, donor fecal samples are screened for relevant bacterial, viral, and parasitic enteropathogens. Questions such as what to screen for, how frequently to screen and how to test can be controversial or at least lacking in consensus. Preferably, all donor dogs undergo a comprehensive evaluation including physical examination, CBC, chemistry panel, urinalysis, serum B12/folate, serum TLI, Spec cPL (dogs), fecal centrifugation flotation for ova, cysts, and oocysts, fecal PCR testing for the following enteropathogens (Parvovirus, Circovirus, *Campylobacter jejuni, C. coli; Salmonella* spp., enterotoxigenic *Clostridium perfringens, Giardia* spp., and *Cryptosporidium* spp). This can be cost-prohibitive and use of a narrower selection of tests based on a risk assessment can be justified. Limited screening can perhaps be justified if donors are from the same household, based on the notion that there is probably already ample cross-exposure to potential pathogens. Regardless of testing, exclusion of animals with a recent history (e.g. past 3 months) of antimicrobial therapy, immunosuppressive therapy or enteric disease is reasonable. Patients with IBD, atopy, GI malignancy, obesity, and chronic diarrhea should also be excluded from being donors.

Sample collection

Time delays presumably result in progressive loss of some components of the microbiota. Whether short-term delays are of clinical relevance is unknown, in large part because the specific components of stool that are responsible for clinical cure are not well defined. While strict guidelines are not possible at this time, it is logical to recommend treatment occur as soon as reasonably possible after stool collection, ideally within 4 hours.

Donor stool storage: fresh vs frozen vs freeze dried feces?

Frozen-and-thawed feces has been successfully used in humans for the management of recurrent CDI, and a recent study reported lack of inferiority of frozen stool compared to fresh stool.[16] The ability to use frozen stool eliminates many of the logistical burdens inherent to FMT, because stool collection and processing need not be tied to the procedure date and time. The use of centralized stool banks is rapidly gaining traction and has removed barriers for FMT by making available to clinicians safe, screened stool that can be shipped and stored frozen and thawed for use as needed.

Other storage approaches might also be feasible, such as freeze-drying. Freeze dried canine feces has been used by one of the authors (JSW) with clinical success, but objective data are lacking.

Administration protocols

Despite the positive clinical results obtained through formal studies and anecdotes, standard treatment regimens are lacking in human and veterinary medicine. Questions pertaining to important subjects such as route of administration (upper vs. lower gastrointestinal tract vs. both), method of administration (e.g. colonoscope vs. enema), peri-procedure administration or withdrawal of antimicrobials, colon cleansing, amount of fecal material infused, volume of dilution, and nature of the diluent (e.g. saline vs. water) remain unanswered. The amount of donor stool used has varied; however, in a recent review, relapse was four-fold greater when < 50 g of stool was used in people with RCDI.[19] Stool is most commonly suspended in nonbacteriostatic saline; however, other diluents (e.g., yoghurt, milk, bottled water) have been successfully used. The donor stool is mixed with diluent to a consistency that can be injected via the biopsy channel of a colonoscope or administered via red rubber tube or Foley catheter. The suspension should be filtered through gauze pads or strainer to remove large particulate matter before aspiration into the syringe. The volume of stool suspension that is deposited in the colon varies tremendously. Dosing regimens such as 10 ml/kg or less specific volumes such as 300-500 mL/patient have been used. A larger volume may be beneficial by allowing the clinician to deposit aliquots of 90-100 mL into multiple locations within the intestinal tract, including the jejunum, ileum, ascending colon, transverse colon, and upper descending colon via endoscope. Large volumes could correspondingly increase the likelihood of straining and rapid defecation post-treatment.

The timing of FMT is another pivotal factor that has been poorly studied to date. It is plausible that the effect of FMT in a recipient that achieved remission after conventional treatment might be associated with fewer side effects and different clinical outcome than in a patient with active inflammation. Another vital question pertains to the characteristics of the donor that are "predictors" for a favourable outcome. A recent study by Vermeire and colleagues observed that donors from whom the stools resulted in a successful FMT were characterized by a significantly higher bacterial richness compared to donors from whom stools did not result in a successful FMT. In addition, the diversity of the post-FMT recipients microbiotia were associated with improved outcome suggesting that this parameter could be utilized to better predict the clinical effect of FMT.[13] Other features of the donor's microbiota that should be studied include abundance in anti-inflammatory taxa like *Faecalibacterium prausnitzii* or in butyrate producers.

Conclusions and future directions

The current collective understanding of the microbiome remains incomplete, but given demonstrated associations between the gut microbiota and conditions such as obesity, inflammatory bowel disease, diabetes, and colon cancer, there is increased interest in expanding the therapeutic scope of FMT. Defining a role for FMT in dogs and cats will require completion of prospective, randomized, placebo-controlled studies that rely on clinically relevant end points related to a particular physiological or pathological condition, and should involve assessment of "predictors" in the donor that are associated with improved outcome.

REFERENCES

[1] Varga C, et al. ***BMC public health*** 2012;12:866.
[2] Hooper LV, et al. ***Science (New York, NY)*** 2001;292:1115-1118.
[3] Insoft RM, et al. ***Pediatric clinics of North America*** 1996;43:551-571.
[4] Murgas Torrazza R, et al. ***J Perinatol*** 2011;31 Suppl 1:S29-34.
[5] Dominguez-Bello MG, et al. ***Proc Natl Acad Sci USA*** 2010;107:11971-11975.
[6] McCracken VJ, et al. ***Cellular microbiology*** 2001;3:1-11.
[7] Seksik P, et al. ***Gut*** 2003;52:237-242.
[8] Sokol H, et al. ***Inflammatory bowel diseases*** 2006;12:106-111.
[9] Nobaek S, et al. ***The American journal of gastroenterology*** 2000;95:1231-1238.
[10] Xenoulis PG, et al. ***FEMS Microbiology Ecology*** 2008;66:579-589.
[11] Suchodolski JS, et al. ***PLoS ONE*** 2012;7:e51907.
[12] Petrof EO, et al. ***Microbiome*** 2013;27:457-462.
[13] Vermeire S, et al. ***Journal of Crohn's colitis*** 2015.
[14] Moayyedi P, et al. ***Gastroenterology*** 2015;149:102-109.e106.
[15] Kunde S, et al. ***Journal of pediatric gastroenterology and nutrition*** 2013;56:597-601.
[16] Lee CH, et al. ***JAMA*** 2016;315:142-149.
[17] Weese JS, et al. ***American College of Veterinary Internal Medicine Forum***. Seattle, Wa, 2013.
[18] Hamilton MJ, et al. ***The American journal of gastroenterology*** 2012;107:761-767.
[19] Gough E, et al. ***Clin Infect Dis*** 2010;53:994-1002.

Antimicrobial Use Guidelines for Treatment of Respiratory Tract Disease in Dogs and Cats: Antimicrobial Guidelines Working Group of the International Society for Companion Animal Infectious Diseases

Michael R. Lappin, DVM, PhD, DACVIM (Small Animal Internal Medicine)
Fort Collins, CO

The Antimicrobial Guidelines Working Group of the International Society for Companion Animal Infectious Diseases (www.iscaid.org) developed guidelines for treatment of bacterial causes of feline upper respiratory tract disease, canine infectious respiratory disease complex (CIRDC; previously known as canine infectious tracheobronchitis or kennel cough complex), bronchitis, pneumonia, and pyothorax. The document was based on review of the current literature including other veterinary recommendations on antimicrobial therapy (1-4) and corresponding guidelines for human medicine (5,6).

After the review of the literature, the first meeting of the Working Group were in person and led to the initial draft of the guidelines. A multitude of revisions were then made electronically in an attempt to develop wording of the recommendations that would lead to consensus within the Working Group in regards to each recommendation. During the development of the guidelines, The Working Group unanimously agreed that there were many significant limitations in objective, published information relevant to the treatment of bacterial canine and feline respiratory diseases. Accordingly, a modification of the Delhi method for consensus building was used in the development of the guidelines (7). While the preliminary final draft of the document was complete, each recommendation was provided to all Working Group members who were asked to individually list whether they agreed, were neutral, or disagreed with a recommendation. The document was then updated with the Working Group responses and the document provided to 6 expert volunteers in the field who were asked to rate each recommendation using the same system. For those recommendations that received any "disagree" votes from the 17 total reviewers (Working Group and outside reviewers), the percentage distribution of all reviewers and appropriate comments are presented throughout the document. After completion of the final draft, the manuscript was submitted for peer review as a primary publication.

Each section of the document was subdivided into 1. Definitions and causes, 2. Diagnosis, 3. Treatment, and 4. Monitoring. The document contains two Tables that are similar to those used in the Antimicrobial Use Guidelines for Treatment of Urinary Tract Disease in Dogs and Cats (8). Table 1, adapted for this proceedings, is a listing of the first-line antimicrobial recommendations for each bacterial respiratory syndrome in the dog and cat. For each recommendation, extensive justifications with references are provided in the text. In Table 2 (not shown), the primary drugs used in veterinary medicine for the treatment of bacterial respiratory infections in dogs and cats are listed, showing the doses recommended by the Working Group.

In the ACVIM presentation corresponding to this proceedings, the recommendations from the Working Group will be presented for each section with a listing of the final voting results and associated comments. In the published version of the document, over 100 references were utilized.

The Working Group consisted of Joseph Blondeau, Dawn Booth, Edward Breitschwerdt, Luca Guardabassi, Michael Lappin, David Lloyd, Mark G. Papich, Shelley Rankin, Jane Sykes, John Turnidge, and Scott Weese. The Working Group give special thanks to Leah Cohn, Joshua Daniels, Eleanor Hawkins, Steve Holloway, Lynelle Johnson, and Carol Reneiro for provided external votes prior to submission of the manuscript.

REFERENCES

1. Morley, P.S., et al, 2005. Antimicrobial Drug Use in Veterinary Medicine. ACVIM Consensus Statement. Journal of Veterinary Internal Medicine 19, 617–629.
2. Guardabassi, L., et al, 2008. Guidelines to antimicrobial use in dogs and cats. In Antimicrobial use in animals. Blackwell Publishing 183-206.
3. Swedish Veterinary Association, 2009. Guidelines for the clinical use of antimicrobials in the treatment of dogs and cats.
4. Danish Small Animal Veterinary Association (SvHKS), 2013. Antimicrobial Use Guidelines for Companion Animal Practice. Danish Small Animal Veterinary Association.
5. Bradley, J.S., et al., 2011. The management of community-acquired pneumonia in infants and children older than 3 months of age: Clinical Practice Guidelines by the Pediatric Infectious Diseases Society and the Infectious Diseases Society of America. Clinical Infectious Diseases 53, 325-76
6. Lionel, A., et al., 2007. Infectious Diseases Society of America/American Thoracic Society Consensus Guidelines on the Management of Community-Acquired Pneumonia in Adults. Clinical Infectious Diseases 44, 27–72.
7. Murphy, M.K., et al., 1998. Consensus development methods, and their use in clinical guideline development. Health Technology Asses 2,1-88.
8. Weese, J.S, et al., 2011. Antimicrobial Use Guidelines for Treatment of Urinary Tract Disease in Dogs and Cats: Antimicrobial Guidelines: Working Group of the International Society for Companion Animal Infectious Diseases. Vet Med International Volume 2011, Article ID 263768, 9 pages doi:10.4061/2011/263768

Table 1. First-line antimicrobial options for bacterial respiratory infections in the dog and cat

Infection Type	**First-line drug options**
Feline acute bacterial URI	Doxycycline or amoxicillin per os (PO)
Feline chronic bacterial URI	Doxycycline or amoxicillin PO Base the choice on culture and antimicrobial susceptibility (C&S) testing if available
Canine infectious respiratory disease complex (bacterial component)	Doxycycline or amoxicillin-clavulanate PO
Bacterial bronchitis (dogs or cats)	Doxycycline PO Base changes if needed on clinical responses and C&S if available
Pneumonia in animals with extensive contact with other animals that have no systemic manifestations of disease (i.e. fever, lethargy, dehydration)	Doxycycline PO Base changes if needed on clinical responses and C&S if available
Pneumonia with or without clinical evidence of sepsis [c]	Parenteral administration of a fluoroquinolone and a penicillin or clindamycin[e] initially. Base oral drug choices to follow on clinical responses and C&S results if available.
Pyothorax (dogs or cats)	Parenteral administration of a fluoroquinolone and a penicillin or clindamycin initially combined with therapeutic lavage initially. Base oral drug choices to follow on clinical responses and C&S results if available.

ISCAID Consensus Statement: Antimicrobial Guidelines for the Treatment of Urinary Tract Infections in Dogs and Cats

J. Scott Weese[1], DVM, DVSc, DACVIM; Joseph M. Blondeau[2]; Dawn Boothe[3], DVM, MS, PhD; Luca Guardabassi[4], DVM, PhD; Nigel Gumley[5], DVM, MS, DABVP (C/F); Michael Lappin[6], DVM, PhD, DACVIM; Mark G. Papich[7], DVM, MS; Shelley C. Rankin[8], PhD, DACVCP; Jane E. Sykes[9], BVSc, PhD, DACVIM

[1]Guelph, ON, Canada; [2]Saskatoon, SK, Canada; [3]Auburn, AL; [4]Basseterra, St. Kitts and Nevis, West Indies; [5]Ottawa, ON, Canada; [6]Fort Collins, CO; [7]Raleigh, NC; [8]Philadelphia, PA; [9]Davis, CA

INTRODUCTION

Urinary tract disease is frequently encountered in dogs and cats, and is a common reason for antimicrobial therapy. In humans, concerted efforts are underway to improve and limit antimicrobial use because of concerns regarding overuse, misuse and ineffective use.[1-3] Some of these same issues have been identified in veterinary medicine, leading to development of clinical guidelines for diagnosis and management of urinary tract infections in dogs and cats.[4] As the field has advanced, more companion animal-specific research has been published.[5-7] Feedback was also obtained from users. It became clear that these guidelines needed to evolve; therefore, the process of revising of the 2011 ISCAID guidelines was initiated in 2015. As with any clinical guideline, these guidelines are designed to provide 'guidance' not dictate a standard of care that must apply to all situations. They are based on evidence, whenever applicable, but the paucity of high level evidence in veterinary medicine results in the need for extrapolation from human medicine and the use of personal opinions from an expert panel.

SCOPE

The revised guidelines expand on the 2011 version both in terms of specific recommendations, as well as in scope and depth. It was realized that defining of some patient/disease categories was difficult (or difficult to justify) and that extrapolation of some definitions from human medicine was problematic. For example, 'uncomplicated' or 'simple uncomplicated UTI' are common categorizations in humans.[8] While it is based on the presence of UTI in an individual with a normal urogenital tract, specific testing is rarely performed and this population consists predominantly of young, otherwise healthy sexually active women. Therefore, this is not analogous to most UTIs in female dogs. Infections in men are typically considered complicated, something that would reasonably apply to intact male dogs; however, intact males constitute a great minority of dogs in many regions. Recurrent disease is a common problem in companion animals, and the reasons can be diverse. Differentiating a new infection caused by an unaddressed predisposing factor (e.g. anatomical abnormality, immune suppression, neurogenic bladder) from an infection that was not completely eliminated (e.g. urolith, bladder wall mass, poor owner compliance with antimicrobials) is important, and detailed consideration of what 'recurrent' means is required to properly diagnose and manage these complicated cases. There are also important subpopulations that have specific facets that necessitate additional or different approaches to diagnosis and treatment, including pregnant bitches, intact male dogs, and individuals with urologic implants, urinary catheter or uroliths. Further, advances in interest and study of a range of alternative or complementary approaches to the management of UTIs, particularly recurrent UTIs, necessitate specific consideration of these non-antimicrobial therapies

As such, the following conditions or topics have been addressed, attempting to separate topics with important clinical differences and improve on existing (and often unclear) definitions:

- First episode urinary tract infection in the female and neutered male dogs
- First episode urinary tract infection in the cat
- First episode urinary tract infection in the intact male dog
- Recurrent bacterial urinary tract infection
- Pyelonephritis
- Bacterial prostatitis
- Bacterial UTI in the pregnant bitch
- Subclinical bacteriuria
- Bacteriuria, infection and management of urinary catheters
- Peri-operative prophylaxis for urological surgery and minimally invasive urological procedures
- Prophylaxis and treatment of UTI/bacteriuria in animals with urologic implants
- Medical management of infection-induced uroliths
- Non-antimicrobial therapies
- Interpretation of culture and susceptibility tests for treating urinary tract infections

Discussion of these topics is well beyond what can be presented in these proceedings; however, general drug dosing regimens are presented in Table 1.

DRUG TREATMENT TABLE

Table 1: Antimicrobial recommendations for treatment of urinary tract infections in dogs and cats.

Drug	Dose	Comments
Amoxicillin	11-15 mg/kg PO q12h	Good first-line option for bacterial cystitis in the absence significant tissue involvement (e.g. prostatitis, pyelonephritis). Excreted in urine predominantly in active form if normal kidney function is present.
Amikacin	Dogs: 15-30 mg/kg IV/IM/SC q24h Cats: 10-14 mg/kg IV/IM/SC q24h	Not recommended for routine use but may be useful for treatment of multidrug resistant organisms. Potentially nephrotoxic. Avoid in animals with chronic kidney disease. Other factors (eg, low pH) can affect aminoglycoside activity, which should be considered.
Amoxicillin / clavulanate	12.5-25 mg/kg POq12h (dose based on combination of amoxicillin + clavulanate	First-line option for bacterial cystitis in the absence significant tissue involvement (e.g. prostatitis, pyelonephritis).
Ampicillin		Not recommended because of poor oral bioavailability. Amoxicillin is preferred.
Cephalexin, Cefadroxil	12-25 mg/kg PO q12h	Enterococci are resistant. Limited activity against Enterobacteriaceae.
Cefovecin	8 mg/kg single SC injection. Can be repeated once after 7-14 days.	Duration is longer than is typically needed. Should only be used in situations where oral treatment is problematic. Enterococci are resistant. Approved for treatment of cystitis in some countries, but limited efficacy against Enterobacteriaceae apart from cystitis. Lack of clinical breakpoints hampers assessment of susceptibility, especially in tissue-associated infections.
Cefpodoxime proxetil	Dogs: 5 to 10 mg/kg q24h PO Cats: no dose established.	More active than cephalexin or cefadroxil against Enterobacteriaceae. Enterococci are resistant.
Ceftiofur	Dogs: 2 mg/kg q12-24h SC Cats: no dose established.	Approved for treatment of UTIs in dogs in some regions. Enterococci are resistant
Chloramphenicol	Dogs: 40-50 mg/kg PO q8h Cats: 12.5-20 mg/kg PO q12h	Reserved for rare situations with multidrug resistant infections having few other treatment options. Myelosuppression can occur, particularly with long-term (e.g. >28d) therapy. Inhibits metabolism of some drugs. Avoid contact by humans because of rare idiosyncratic aplastic anemia.
Ciprofloxacin	30 mg/kg PO q24h	Difficult to justify over approved fluoroquinolones. Lower and more variable oral bioavailability than approved veterinary fluoroquinolones. Sometimes used because of lower cost than enrofloxacin. Dosing recommendations are empirical and based on limited pharmacokinetic studies. Human breakpoints do not apply: no Interpretive Criteria are available for testing isolates from animals.
Doxycycline	5 mg/kg PO q12h	Reserved for infections caused by pathogens that are resistant to drugs that are actively excreted in urine in active form.
Enrofloxacin	5-20 mg/kg q24h (dogs)	Excreted in urine predominantly in active form. Reserve for documented resistant cystitis but good first line choice for pyelonephritis in dogs at the higher end of the dosing range. Considered first-line choice for infections that involve the

		prostate. Dosing should consider the bacterium and degree of tissue involvement. Lower doses should be avoided in complicated or tissue-associated infections and with certain pathogens (e.g. *Pseudomonas*). Not recommended for enterococci. Not recommended for cats because of the risk of retinopathy at high doses and resistance emergence at lower doses.
Imipenem-cilastatin	5 mg/kg IV/IM q6-8h	Reserve for treatment of multidrug resistant infections, particularly those caused by ESBL-producing Enterobacteriaceae or *Pseudomonas aeruginosa*. Recommend consultation with a urinary or infectious disease veterinary specialist or veterinary clinical pharmacologist prior to use.
Marbofloxacin	2.7-5.5 mg/kg PO q24h	Excreted in urine predominantly in active form. Reserve for documented resistant cystitis but good first line choice for pyelonephritis at the higher end of the dosing range. Considered first-line choice for infections that involve the prostate. Dosing should consider the bacterium and degree of tissue involvement. Lower doses should be avoided in complicated or tissue-associated infections and with certain pathogens (e.g. *Pseudomonas*). Not recommended for enterococci.
Meropenem	Dogs: 8.5 mg/kg SC/IV q 12 (SC) or 8 (IV) h Cats: 10 mg/kg q12h, IV, SC, IM.	Reserve for treatment of multidrug resistant infections, particularly those caused by ESBL-producing Enterobacteriaceae or *Pseudomonas aeruginosa*. Recommend consultation with a urinary or infectious disease veterinary specialist or veterinary clinical pharmacologist prior to use.
Nitrofurantoin	4.4-5 mg/kg PO q8	An option for some cases of cystitis, particularly when multidrug resistant pathogens are involved. Not appropriate for tissue-associated infections.
Orbifloxacin	Tablets: 2.5-7.5 mg/kg PO once daily Suspension (cats): 7.5 mg/kg once daily.	Excreted in urine predominantly in active form. Reserve for documented resistant cystitis but good first line choice for pyelonephritis at the higher end of the dosing range. Considered first-line choice for infections that involve the prostate. Dosing should consider the bacterium and degree of tissue involvement. Lower doses should be avoided in complicated or tissue-associated infections and with certain pathogens (e.g. *Pseudomonas*). Not recommended for enterococci.
Trimethoprim-sulfadiazine/ Trimethoprim-sulfamethoxazole/Ormetoprim-sulfadimethoxine	15-30 mg/kg PO q12h Note: dosing is based on total trimethoprim + sulfadiazine concentration	Concerns regarding idiosyncratic and immune-mediated adverse effects in some patients, especially dogs and with prolonged therapy. If prolonged (>7d) therapy is anticipated, baseline Schirmer's tear testing is recommended, with periodic re-evaluation and owner monitoring for ocular discharge. Avoid in dogs that may be sensitive to potential adverse effects such as KCS, hepatopathy, hypersensitivity and skin eruptions. Activity against *Enterococcus* in urine is controversial and should be avoided. Trimethoprim can be considered a treatment choice for prostate infections in male dogs. Susceptibility testing is based on trimethoprim-sulfamethoxazole and breakpoints have not been established for dogs or cats.
Fosfomycin	40 mg/kg PO (with food) q12h	Should be reserved for multidrug resistant infections. Not appropriate for tissue-associated infections. Efficacy in dogs or cats is undetermined.

Pradofloxacin	Dogs: 3-5 mg/kg PO once daily. Cats: 3-5 mg/kg once daily (tablets) or 5-7.5 mg/kg once daily (suspension)	Excreted in urine predominantly in active form. Reserve for documented resistant cystitis but good first line choice for pyelonephritis at the higher end of the dosing range. Considered first-line choice for infections that involve the prostate. Dosing should consider the bacterium and degree of tissue involvement. Lower doses should be avoided in complicated or tissue-associated infections and with certain pathogens (e.g. *Pseudomonas*). Not recommended for enterococci. Greater activity against some bacteria than older fluoroquinolones (enrofloxacin, marbofloxacin, orbifloxacin). Approved for dogs and cats in Europe and Canada. Approved only for cats in the U.S. (liquid suspension). Published evidence for efficacy treating UTI in dogs and cats.
Minocycline	Cats: 8.8 mg/kg PO q24, (or 50 mg per cat). Dogs: 5 mg/kg PO q12h.	Reserved for infections caused by pathogens that are resistant to drugs that are actively excreted in urine in active form. Can be considered an alternative to doxycycline when doxycycline is not available. Some doxycycline-resistant staphylococci are susceptible. Higher doses (10 mg/kg) can be used in dogs but is more likely to cause vomiting.

SUBCLINICAL BACTERIURIA

Another important topic is subclinical bacteriuria, one that is receiving increasing attention because of overtreatment in humans and animals. Subclinical bacteriuria is defined as the presence of bacteria in urine as determined by positive bacterial culture from a properly collected urine sample, in the absence of clinical evidence of signs of lower urinary tract disease. Some studies in the veterinary literature use the term 'urinary tract infection' or 'occult infections' in reference to animals with positive bacterial cultures but no clinical signs, this terminology should be avoided. The term bacteriuria has been used to describe cases where bacteria are visible cytologically, irrespective of culture results; however, diagnosis should be based on culture.[4] Cytological evaluation is an important part of urinalysis in animals with suspected urinary tract disease and results must be considered in the context of other data. An increased urine sediment white blood cell count has been associated with increased odds of a positive culture,[9,10] but this has not been a consistent finding.[11] Poor agreement between cytological detection of bacteria and positive urine culture has been reported in dogs.[12,13] Increased urine sediment red blood cell count is also not predictive of positive cultures.[9,10] Thus, cytological data are useful adjunctive data to assess animals with potential urinary tract disease but may not be highly predictive of culture results or infectious disease.

Subclinical bacteriuria is not uncommon, even in individuals with no known predisposing factors. Rates of 2.1-8.9% have been reported in healthy dogs,[10,12-15] with higher rates (15-31%) in groups such as dogs with diabetes, morbidly obese dogs, puppies with parvoviral enteritis and dogs treated with cyclosporine or glucocorticoids.[11,12,16-18] Study of subclinical bacteriuria has been limited in cats and the prevalence may be lower than reported in dogs, as one study identified bacteriuria in only 0.9% of healthy cats.[19] No evidence of an association between subclinical bacteriuria and risk of development of UTI or other infectious complications has been reported in dogs or cats. A study of healthy female dogs identified bacteriuria in 8.9% and found no association with subsequent UTI development.[15]

In humans, there is abundant support for not treating asymptomatic bacteria (the human analogue of subclinical bacteriuria), even in most compromised patients. While bacteriuria rates are high in various populations (e.g. diabetics, the elderly, patients with paralysis), treatment guidelines such as Infectious Diseases Society of America Guidelines for the Diagnosis and Treatment of Asymptomatic Bacteriuria in Adults do not recommend treating asymptomatic bacteriuria in almost all patient groups.[20] Exceptions are patients undergoing transurethral resection of the prostate and patients that will be undergoing urologic procedures that result in mucosal bleeding. Screening and treatment of pregnant women is recommended; however, this has recently been questioned because while an association between untreated bacteriuria and pyelonephritis was identified, the low burden of pyelonephritis and potential adverse effects of antimicrobials may not justify universal treatment.[21] Treatment is specifically not recommended for pre-menopausal, non-pregnant women, those with diabetes, older individuals in the community, elderly institutionalized individuals or individuals with spinal cord injuries.[4] Thus, even in what would be considered high-risk populations, treatment of asymptomatic bacteriuria is discouraged and intensive measures are used to reduce the treatment of asymptomatic bacteriuria. These efforts are typically focused around antimicrobial stewardship from an antimicrobial resistance standpoint, but reduction in unnecessary treatment is also desirable because of cost, adverse effects of antimicrobials, and lack of evidence that treatment improves outcome in almost all patient groups. While treatment might eliminate the current bacteriuria event, recolonization often follows.[22] A systematic review concluded that while bacteriuria may be eliminated in the short-term, the effect is not sustained and re-colonization is common, leading to no impact on overall morbidity or mortality.[23] Further, two studies have reported significantly higher bacteriuria recurrence rates in women treated for asymptomatic bacteriuria compared to untreated controls.[24,25] Treated women also had higher rates of antimicrobial resistance in *E. coli* isolated from subsequent UTIs.[25]

Summary of Preliminary Recommendations

Diagnosis

1) **Subclinical bacteriuria should be based on identification of bacteria by culture of urine collected via cystocentesis in an animal without clinical signs attributable to UTI. Confirmation of bacteriuria through isolation of the same bacterium in two subsequent samples is a requirement for diagnosis in humans[20] and would be ideal in animals.**
2) **Urine collected via catheter should not be used for diagnosis of subclinical bacteriuria unless there are contraindications to cystocentesis. If urine collected via catheter is used, isolation of the same bacterial species from two sequential samples collected at least one week apart is required.**
3) **Subclinical bacteriuria should never be diagnosed from free-catch urine samples.**
4) **Quantitative culture results cannot differentiate subclinical bacteriuria from UTI.** Heavy growth on quantitative culture data (e.g. >100 000 CFU/ml) can be present in animals with subclinical bacteriuria[9] and there is no evidence that high CFU counts indicate a greater risk of disease development.

Treatment

1) **Treatment of bacteriuria is rarely indicated in animals that have no clinical signs of UTI (e.g. stranguria, hematuria, pollakiuria, dysuria).**
2) **Treatment of animals with pyuria or other cytological abnormalities but no clinical abnormalities is not recommended.** Previous guidelines supported treatment of animals with no clinical signs but cytological evidence of inflammation (pyuria). However, there is currently no evidence in veterinary medicine that treatment of pyuria is beneficial or that pyuria in the absence of clinical signs represents a relevant disease state, and treatment of pyuria in humans in the absence of clinical evidence of UTI is not recommended.[20]
3) **The presence of a multidrug bacterium should not affect decisions about whether to treat bacteriuria.** Antimicrobial resistance genes are not virulence factors and resistant organisms are not more likely to cause disease than their susceptible counterparts. Anecdotal information suggests that multidrug resistant organisms will sometimes be replaced with susceptible organisms if treatment is withheld, and then treatment with routine antimicrobials may be more practical if decolonization is desired or if clinical disease develops.
4) **Treatment of bacteriuria caused by multidrug resistant pathogens for infection control purposes (e.g. to eliminate urine shedding of a possible pathogen) is not recommended.** It is reasonable to assume that the bacterial strain in the bladder is also present in the gastrointestinal tract; therefore, even if bacteria are eliminated from the bladder with antimicrobials, it would likely have limited impact on the overall risk posed by the patient.
5) **Culture of urine from clinically normal animals should not be performed when there would be no indication to treat based on a positive culture result.**
6) **In some circumstances, treatment may be considered if there is concern that there is a particularly high risk of ascending or systemic infection (e.g. immunocompromised patients, patients with underlying renal disease) or that the bladder may be a focus of extra-urinary infection.** Good data are lacking to define high-risk populations and it is likely that there are few animals that fall into this category. Therefore, treatment on the basis of suspected high risk of complications should be used sparingly.
7) **In patients that are unable to display clinical signs of UTI (e.g. spinal injury), a clinical judgement must be made, ensuring that consideration of the need and potential adverse impacts (e.g. adverse effects, antimicrobial resistance) are balanced.** Patients that are unable to display clinical signs of UTI complicate decision-making as determining what represents 'subclinical' is difficult. In those patients, the presence of systemic signs (e.g. fever) and changes in urine appearance and odour (e.g. gross discoloration, malodor) can likely help differentiate infection from subclinical bacteriuria, but clear guidance is lacking.
8) **Treatment of subclinical bacteriuria caused by *Corynebacterium urealyticum* should be considered because of its association with encrusting cystitis**, a condition that is difficult to treat and evidence from humans indicating an increased risk of UTI and obstructive uropathy in transplant patients with *C. urealyticum* bacteriuria.[26]
9) **There is currently no evidence that screening bacterial isolates for urovirulence factors should impact decision-making for subclinical bacteria.**

REFERENCES

[1] Kim M, et al. *Infection* 2015;43:89-94.
[2] Lee MJ, et al. *BMC infectious diseases* 2015;15:289.
[3] Trautner BW, et al. *JAMA internal medicine* 2015;175:1120-1127.
[4] Weese JS, et al. *J Vet Int Med* 2015;29:487-498.
[5] Jessen LR, et al. *Vet J* 2015;203:270-277.
[6] Westropp J, et al. *J Vet Int Med* 2012;26:506-512.

[7] Clare S, et al. *J Vet Int Med* 2014;28:818-826.
[8] Warren JW, et al. *Clin Infect Dis* 1999;29:745-758.
[9] Forrester SD, et al. *J Vet Intern Med* 1999;13:557-560.
[10] O'Neil E, et al. *Can Vet J* 2013;54:1061-1066.
[11] McGuire NC, et al. *J Am Anim Hosp Assoc* 2002;38:541-544.
[12] Peterson AL, et al. *Vet Dermatol* 2012;23:201-e243.
[13] McGhie JA, et al. *Aust Vet J* 2014;92:33-37.
[14] Way LI, et al. *J Vet Emerg Crit Care* 2013;23:23-28.
[15] Wan SY, et al. *J Am Vet Med Assoc* 2014;245:106-112.
[16] Torres SMF, et al. *J Am Vet Med Assoc* 2005;227:239-243.
[17] Lusby AL, et al. Prevalence of asymptomatic bacterial urinary tract infections in morbidly obese dogs. *ACVIM Forum*. Denver, CO, 2011.
[18] Koutinas AF, et al. *Vet Microbiol* 1998;63:109-116.
[19] Eggertsdóttir AV, et al. *J Feline Med Surg* 2011;13:800-803.
[20] Nicolle LE, et al. *Clin Infect Dis* 2005;40:643-654.
[21] Nicolle LE. *The Lancet infectious diseases* 2015.
[22] Dalal S, et al. *Clin Infect Dis* 2009;49:491-497.
[23] Dull RB, et al. *Pharmacotherapy* 2014;34:941-960.
[24] Cai T, et al. *Clin Infect Dis* 2012;55:771-777.
[25] Cai T, et al. *Clin Infect Dis* 2015;61:1655-1661.
[26] López-Medrano F, et al. *Clin Infect Dis* 2008;46:825-830.

Echinococcus multilocularis: An Emerging Zoonotic Disease

Andrew S. Peregrine, BVMS, PhD, DVM, DEVPC, DACVM (Parasitology); Benoît Cuq, DMV
Guelph, ON, Canada

INTRODUCTION

Echinococcus multilocularis is a zoonotic tapeworm that occurs in central Europe, much of northern, central and eastern Eurasia, and parts of North America[1]. Adult parasites reside within the small intestine of definitive hosts, which are primarily wild canids (e.g. foxes, coyotes, wolves). However, in some areas, domestic dogs, and to a lesser extent cats, may act as definitive hosts. Eggs shed in the feces of these species are morphologically indistinguishable from *Taenia*-type eggs and are immediately infective for intermediate hosts which include various species of wild rodents[2]. Subsequent to ingestion, a hexacanth embryo is released from the egg, travels to the liver via the hepatic portal circulation and develops into alveolar hydatid cysts[1]. This larval stage of the parasite is comprised of numerous small vesicles lined with a germinal epithelium from which multiple protoscolices develop, undergoes exogenous budding, and behaves like an invasive tumour[1]. The resultant disease, alveolar echinococcosis (AE), is associated with extensive damage to the liver and occasionally spread to other locations within the intermediate host. When an infected intermediate host is ingested by a definitive host the life cycle is completed; development to the mature tapeworm takes approximately 4-5 weeks[3]. Lastly, ingestion of eggs by people may also result in AE, a potentially severe, fatal disease that is currently an emerging issue in parts of central Europe[4].

In North America, *E. multilocularis* has historically been considered to occur in two distinct geographic regions: the Northern Tundra Zone that extends from the west coast of Alaska to parts of the Canadian arctic[5], and the North Central Region that includes the southern parts of 3 Canadian provinces (Alberta, Saskatchewan, Manitoba) and 13 neighbouring USA states (North/South Dakota, Iowa, Minnesota, Montana, Wyoming, Nebraska, Illinois, Wisconsin, Indiana, Ohio, Missouri, Michigan)[5,6]. Within this latter region, red foxes and coyotes are considered the primary definitive hosts; the most significant intermediate hosts include voles and mice[1,7]. Strangely, despite this large geographic distribution, there are very few reports of *E. multilocularis* intestinal infections in dogs or cats in the USA and Canada. Outside Alaska, patent intestinal infections in dogs appear to have been described only once; using molecular methods, *E. multilocularis* was identified in the feces of 1/218 dogs in 2014 in Alberta[8]. Similarly, only 5 cats have been reported with intestinal infections; 3 in 1971 in Saskatchewan[9], and 2 in 1972 in North Dakota[10]. Likewise, only 3 cases of alveolar echinococcosis in people are described that appear to have been acquired in the USA or Canada; one in 1937 in Manitoba[11], one in 1979 in Minnesota[12], and one in 2014 in Alberta[8].

Information concerning dogs as definitive hosts for *E. multilocularis* is contained in veterinary parasitology textbooks. However, what is missing in almost every book is the fact that since the late 1980s, cases of AE, primarily involving the liver, have been described in dogs in Switzerland, Germany, France and Belgium[13]. These cases are thought to occur as a result of either ingestion of large numbers of eggs or by autoinfection in association with the presence of adult tapeworms in the small intestine[14].

CASES OF CANINE ALVEOLAR ECHINOCOCCOSIS IN NORTH AMERICA

Prior to 2009, *E. multilocularis* had never been diagnosed in a dog in Canada. However, in that year hepatic AE was diagnosed in a 3-year old dog that had lived permanently in British Columbia; *E. multilocularis* was identified on histology and direct immunofluorescence of hepatic tissue; sequence data for the mitochondrial 12S rRNA gene, and RFLP analysis of the mitochondrial NADH dehydrogenase 1 and 12S rRNA genes, confirmed the diagnosis[15]. In 2012, a second case was similarly diagnosed in a 2-year old dog that resided in southern Ontario[16]; a third case was diagnosed in a 4-year old dog that lived in Alberta and Manitoba[17]. Between 2013 and 2015, three additional cases were diagnosed in southern Ontario[18]. None of the 6 dogs were related; all 6 dogs had never travelled outside Canada; 4 of the dogs had lived their entire lives in provinces where *E. multilocularis* had never been diagnosed prior to the occurrence of these cases.

CLINICAL PRESENTATION AND DIAGNOSIS

Corsini and others recently published a retrospective analysis of twenty confirmed (n = 18) or probable (n = 2) cases of canine AE in Europe[14]. The median age at diagnosis was 3.1 years (range = 1.1-10.7 years) and the most common clinical signs included abdominal distension, lethargy, anorexia and vomiting; less common clinical signs included diarrhea, weight loss, polypnea, fever and polyphagia. Most dogs had lived all their adult lives in areas where foxes occurred and were exercised outdoors.

Abdominal ultrasound and clinical pathology

In dogs with clinical AE, abdominal ultrasound typically reveals the presence of multiple, large, heterogeneous, poorly limited, cavitated hepatic masses. Usually, there is a hyperechoic thick periphery, an irregular inner surface, and a centrally located cavity filled with corpuscular fluid[14]. Extension to neighbouring organs is observed in approximately one third of cases; free abdominal fluid is present in a similar proportion[14]. In many dogs, ultrasound-guided fine-needle aspirates of intralesional or peritoneal fluid generates samples with cytological morphology consistent with AE; calcareous corpuscles and folded membranous structures[18]. While the former indicates the presence of any cestode, the latter has only been described with *E. multilocularis*.

Histopathology

In dogs with hepatic AE, there is typically extensive involvement of the liver with a multinodular firm pale-tan mass that grossly has the appearance of a neoplasm. Histological examination of hepatic biopsies is extremely helpful as the morphology is characteristic for *E. multilocularis*[13,15]; hepatic architecture is replaced by multi-loculated coalescing cystic structures surrounded by fibrosis. Lining individual cysts is a hyaline membrane ("laminated layer") that stains with Periodic acid-Schiff (PAS) stain. The inner lining of the hyaline membrane typically comprises a basophilic matrix that contains occasional calcareous corpuscles. Intraluminal protoscolices are sometimes observed. A prominent chronic eosinophilic and granulomatous inflammation is also typically present.

Serology and PCR

Positive serology to the *E. multilocularis* Em2-antigen is observed in most dogs with AE[14,19]. However, at the present time this diagnostic method is only available at the University of Bern, Switzerland. In addition, some dogs will seroconvert, but not develop AE due to abortive infections[20,21]. As a result, examination of intralesional or abdominal fluid, or hepatic tissue, using an *E. multilocularis*-specific PCR provides confirmatory diagnostic information[22-24].

MANAGEMENT

Historically, dogs with AE have been managed with surgery and/or treatment with albendazole at 10 mg/kg bodyweight, daily, for life. In the aforementioned case series from Europe, dogs that received any intervention (i.e. only albendazole, or surgery and albendazole combined) were shown to survive significantly longer than dogs with no intervention. However, no significant difference in survival time was observed between these two intervention groups[14]. While low numbers of dogs limited the statistical power of the study, the authors concluded that there was no evidence that debulking surgery (cytoreduction) followed by treatment with albendazole resulted in a superior outcome compared to treatment with albendazole alone[14]. As a result, debulking surgery could not be recommended at the present time; as in people, surgery should only be attempted if complete resection is possible[14]. Thus, ideally, investigation of potential cases requires advanced imaging (CT-Scan) to assess whether surgical excision is possible. In non-resectable cases with a centrally located fluid cavity, a pigtail retention catheter can be placed attached to a sealed collection system to provide temporary relief. Lastly, it has historically been assumed that medical management of dogs with AE is associated with short-term survival. However, 4 of 6 dogs managed with only daily albendazole were alive 0.5, 0.5, 2.6 and 9.5 years after the initial diagnosis and were considered to be in remission[14]. Medical management of dogs with AE can therefore be associated with long-term survival; however, daily treatment with albendazole is required.

PUBLIC HEALTH CONCERNS

The intermediate (larval) stage of *E. multilocularis* in the abdomen of dogs constitutes no risk to public health. However, some dogs with hepatic AE may also have patent intestinal infections and therefore may constitute a zoonotic threat. Thus, as soon as a presumptive diagnosis of hepatic AE is made, dogs should be treated with praziquantel at 5 mg/kg bodyweight to eliminate intestinal *E. multilocularis* infection[25]; treatment should be repeated 24 hours later to ensure the correct dosage is administered. In areas endemic for *E. multilocularis*, monthly administration of praziquantel at 5 mg/kg bodyweight is required to prevent patent intestinal infections, and may be effective at preventing canine AE[19,26]; a risk assessment should be carried out to determine if such preventive treatment is necessary.

If exposure of people to *E. multilocularis* eggs is a concern, serological testing at 3, 6 and 12 months following exposure using three different ELISAs (EgHF-ELISA, Em2-ELISA and Em18-ELISA) is recommended[27,28].

CONCLUSION

In dogs with hepatic masses, travel to areas endemic for *E. multilocularis* should be evaluated. If hepatic lesions are first detected during an exploratory laparotomy, it should be recognised that the gross appearance of hepatic AE in dogs is very similar to that of hepatic neoplasia. Thus, histological examination of hepatic biopsies should be carried out to confirm tentative gross diagnoses, particularly in young dogs. Management of AE involves daily treatment with albendazole for life, with or without surgery; due to the late stage at which infections are diagnosed, the prognosis for long-term survival is typically poor.

References

[1]Eckert J & Deplazes P Clin Microbiol Rev 2004;17:107. [2]Taylor M, et.al Veterinary Parasitology. 3rd ed. Oxford, UK: Blackwell Publishing 2007;377. [3]Deplazes P, et.al Vet Parasitol 2011;182:41. [4]Gottstein B, et.al Trends Parasitol 2015;31:407. [5]Eckert J, et.al Int J Parasitol 2000;30:1283. [6]Kazacos K Suppl Compend Contin Educ Pract Vet 2003;25:16. [7]Kapel C, et.al Int J Parasitol 2006;36:79. [8]Massolo A, et.al Parasite 2014;21;73. [9]Wobeser G Can Vet J 1971;12:65. [10]Leiby P & Kritsky D J Parasitol 1972;58:1213. [11]James E & Boyd W Can Med Assoc J 1937;36:354. [12]Gamble W, et.al J Am Med Assoc 1979;241:904. [13]Deplazes P & Eckert J Vet Parasitol 2001;98:65. [14]Corsini M, et.al Vet Rec 2015;177:569. [15]Peregrine A, et.al Can Vet J 2012;53:870. [16]Skelding A, et.al Can Vet J 2014;55:551. [17]Peregrine A, et.al WAAVP Proceedings, 2013:345. [18]Oscos-Snowball A, et.al Vet Clin Path 2015;44:167. [19]Staebler S, et.al Vet Parasitol 2006;141:243. [20]Rausch R, et.al Am J Trop Med Hyg 1987;36:576. [21]Gottstein B, et.al Parasite 2014;21:63. [22]Diebold-Berger S, et.al Acta Cytol 1997;41:1381. [23]Dinkel A, et.al J Clin Microbiol 1998;36:1871. [24]Trachsel D, et.al Parasitology 2007;134:911. [25]Panel on Animal Health and Welfare EFSA Journal 2015;13:4373. [26]Behrens S, et.al Kleintiermedizin 2011;6:300.

[27]Gottstein B, et.al J Clin Microbiol 1993;31:373. [28]Gottstein B, personal communication.

Resistant Gram-Negative Infections

Mary F. Thompson, BVSc(Hons), DACVIM (SAIM), MANZCVS (Assoc)
Murdoch, Western Australia

INTRODUCTION

Resistant Gram-negative bacteria, specifically multidrug-resistant (MDR) bacteria, represent a considerable challenge in both human and small animal medicine. Although resistance in bacteria in human medicine has increased dramatically in the last two decades, in particular in Gram-negative bacteria, development of new antimicrobials has effectively stalled.[1] In the past twenty-five years, only two new classes of antimicrobials have been developed and both target Gram-positive bacteria. As such, no new antimicrobials have been recently developed for treatment of resistant Gram-negative infections and there is little financial incentive for drug companies to contribute in this arena. As veterinarians, we are not only confronted with the individual animal and its difficult-to-treat infection, we must also bear some responsibility for the global picture of human and animal health, recognizing that our antimicrobial choices could well have implications beyond the individual, and even beyond our clinics and countries.

MECHANISMS OF RESISTANCE

There are several key mechanisms of resistance in Gram-negative bacteria:

- β–lactamases e.g. CTX-M-15
- AmpC enzymes e.g. CMY-2
- Metallo-β-lactamases e.g. NDM-1
- Molecular class A carbapenem-hydrolysing enzymes e.g. KPC-type β-lactamase
- Fluoroquinolone resistance acquisition through mutation or plasmid mediated transfer

INCREASING RESISTANCE IN BACTERIAL ISOLATES

Increasing resistance on a large scale can emerge as a result of transfer of resistance genes (largely carried on plasmids) from one bacterial species to another or may reflect expansion of successful MDR clonal groups.[2] A prominent example of large-scale acquisition of resistance via plasmid transfer is the New Delhi metallo-β-lactamase (NDM-1) carbapenemase, which was initially linked to human patients who had travelled to India and Pakistan but was subsequently found in individuals devoid of a travel history.[3] Some isolates were clonally related but many were not. The plasmid is most often found in *E. coli* and *Klebsiella pneumoniae* isolates and confers resistance to all antimicrobials apart from tigecycline (member of new tetracycline class) and colistin. The concern in human medicine is that ongoing spread of the NDM-1 plasmid, which occurs by stealth because bacteria containing the plasmid are carried in the gut in the normal flora and are only detected when extraintestinal infection occurs, could result in large scale occurrence of carbapenem-resistant isolates in clinical infections. Companion animals have not been implicated as a major source of NDM-1 plasmids but their role as a potential reservoir is largely unknown. NDM-1 plasmids have been reported in six non-clonal canine *E. coli* isolates in the USA,[4] indicating that they may play some epidemiological part.

The best example of worldwide expansion of a successful clonal bacterial strain is arguably that of *E. coli* sequence type 131 (ST131), which was unrecognized prior to 2008 and is now the dominant extraintestinal pathogenic (ExPEC) *E. coli* strain in humans globally, frequently bearing the CTX-M-15 β-lactamase, and in doing so increasing the selection pressure for carbapenem resistance, for example via NDM-1 plasmid-bearing bacteria as discussed above.[5] ST131 strains are most frequently associated with urinary tract infection (UTI) but also with bacteriaemia and other infections.[6] The role of companion animals as reservoirs of *E. coli* ST131 has undergone some examination and it seems unlikely that they are a major source of the organism for human infections. Fluoroquinolone-resistant *E. coli* isolates in humans in Australia had a greater proportion of ST131 isolates (202/585, 35%) in comparison to companion animal isolates (9/125, 7.2%) indicating that both humans and companion animals are likely involved in the dissemination of the clone but that companion animals are probably not a major source for human infections.[7]

Another important clonal group in human medicine is *Klebsiella pneumoniae* ST258, which carry the Klebsiella Pneumoniae Carbapenemase (KPC), facilitating an increase in carbapenem resistance in many countries.[2]

Most studies regarding resistant Gram-negative organisms in canine and feline medicine concern collections of isolates, typically *E. coli*, with certain resistance characteristics, thus information on prevalence of MDR bacteria is scarce. Use of molecular characterization tools to examine these isolate collections, however, has allowed some insight into human-associated clonal groups and plasmid-associated resistance genes that can be found in companion animals. Whilst there are some reports of clonal spread of resistant isolates in companion animal medicine,[8,9], most reports describe collections of more distantly related isolates.

Recent characterization of drug resistance in Enterobacteriaceae in diseased dogs and cats (no antimicrobial exposure in month prior) in Europe found that 22/608 (3.6%) were resistant to cefotaxime (all 22 were sensitive to carbapenems).[10] Ten isolates were extended spectrum β-lactamase (ESBL) producers and 12 were AmpC-producing. Two CTX-M-15-producing isolates belonged to *E. coli* clonal group ST131 and similarities to isolates of human origin were suggestive of the possibility of exchange of isolates between humans and animals, as has been previously suggested.[11] Fluoroquinolone resistance was characterised in a similar isolate group; 69/604 (11%) were FQ-resistant, 17% with at least one plasmid-mediated quinolone resistance (PMQR) gene and the majority of the remainder with mutations in the Quinolone Resistance Determining Region (QRDR), indicating that this may be a more important contributor to fluoroquinolone resistance in Enterobacteriaceae from dogs and cats.[12]

RISK FACTORS FOR MDR INFECTIONS IN DOGS AND CATS

There are few studies that examine the risk of MDR infections in dogs and cats, again because many focus on isolate collections rather than case descriptions. In humans, risk factors for infection with ESBL-producing Enterobacteriaceae in the community include recent antimicrobial use, recent hospitalization, advanced age, and male gender.[13] Hypotheses for risk factors were generated for 37 dogs acquiring extraintestinal infection by MDR *E. coli* and *Enterobacter* spp. (62% of which were UTI).[14] Most (97%) of dogs had an underlying disease, received prior antimicrobial treatment (87%), were hospitalized ≥ 3 days (82%) and had a surgical intervention (57%). Risk factors for MDR *E. coli* UTI in cats have also been examined; exposure to a greater number of antibiotic groups in the three months prior increased the risk of MDR UTI but association between number of days of hospitalization in the prior three months and increased risk of MDR *E. coli* UTI did not reach significance.[15] A significant association between number of days of hospitalization and rectal colonisation of dogs with antimicrobial resistant *E. coli* has previously been reported.[16] In particular, dogs that had been treated with fluoroquinolones had a greatly increased risk of rectal colonisation with FQ-resistant *E. coli* whereas prior treatment with β-lactam antimicrobials did not predict development of antimicrobial resistant fecal strains. In another study, fluoroquinolone resistant isolates were found in the feces of 23/123 (18.7%) hospitalized dogs, with profiling suggesting a potential common source of acquisition. Most belonged to clonally related lineages of high virulence potential, with five isolates belonging to the human-associated *E. coli* clonal group ST131.[17]

VIRULENCE OF MDR ISOLATES

Several reports suggest decreased virulence in drug resistant Enterobacteriaceae e.g. FQ-resistant *E. coli* isolates,[18-20] but with the expansion of human-associated successful clonal strains, the relationship is less predictable. For example, ST131 isolates belong to the more pathogenic B2 phylogenetic group of *E. coli* and are thus typically both virulent and MDR. *E. coli* isolates with broad-spectrum-cephalosporin resistance recovered from companion animals in the USA mostly belonged to the more virulent phylogroups D (39%) and B2 (30%), as did fluoroquinolone-resistant *E. coli* isolates from companion animals in Australia (D 59%, B2 13%).[21,4]

There is scant information in the veterinary literature with regard to outcomes associated with MDR Gram-negative infections in dogs and cats. In an Australian study concerning MDR *E. coli* and *Enterobacter* spp. in 37 dogs, only one dog was euthanized with the remaining dogs treated with a variety of antimicrobials. The most successful treatment regimen, used for 14 UTI cases, was arguably high dose co-amoxyclavulanate treatment, even though the isolates showed *in vitro* resistance. With regard to highly successful clonal strains such as *E. coli* ST131 in humans, mortality is not necessarily higher than that associated with non-ST131 MDR strains.[22]

APPROACH TO TREATMENT

Treatment of dogs and cats with MDR infections requires both a global and individual approach. As with every case from which bacteria are isolated, the first question is whether treatment with antimicrobials is actually required. MDR bacteria are isolated from the urine of dogs and cats with some frequency but do not always cause clinical signs i.e. subclinical bacteriuria, and thus likely do not always require treatment.[23] Additionally, examination and elimination of underlying causes for infection is key for a successful outcome. In one human study to evaluating a variety of antimicrobials to treat ESBL-producing organisms, the value of ancillary treatments such as urinary drainage, abscess drainage and mucolytic therapy was recognised in expediting successful treatment. [24] With regard to empirical treatment for serious infections, local knowledge of resistant pathogens is paramount.

Key features of appropriate treatment of MDR include (from Abraham et al, 2014) :

- short high-dose courses of appropriate antimicrobials
- use of carbapenems only in cases that lack other alternatives based on culture and sensitivity
- MIC-based susceptibility testing for recurrent or persistent UTI

Finally, the transfer of resistant pathogens or resistance elements between infected animals and humans (and vice versa) should be considered a possibility, particularly for those who are very young or immunosuppressed, and nursing practices and advice to clients should reflect this.

REFERENCES

[1]Sabtu N, et.al Br Med Bull 2015;116:105. [2]Canton R, et al Clin Microbiol Infect 2012;18:413. [3]Kumarasamy K, et al Lancet Infect Dis 2010;10. [4]Shaheen B, et al Antimicrob Agents Chemother 2013;57:2902. [5]Price L, et al. mBio 2013;4:e00377. [6]Rogers B, et al J Antimicrob Chemother 2011;66:1. [7]Platell J, et al J Antimicrob Chemother 2010;65:1936. [8]Haenni M, et al J Antimicrob Chemother 2012;67:770. [9]Stolle I, et al. J Antimicrob Chemother 2013;68:2802. [10]Bogaerts P, et al Microb Drug Resist 2015;21:643. [11]Platell J, et al Antimicrob Agents and Chemother 2011;55:3782. [12]Guillard T, et al. Vet Microbiol 2015 (in press). [13]Ben-Ami R, et al. Clin Infect Dis 2009;49:682. [14]Gibson J, et al J Vet Intern Med 2008;22:844. [15]Hernandez J, et al. J Feline Med Surg 2014;16:75. [16]Ogeer-Gyles J, et al J Am Vet Med Assoc 2006;229:694. [17]Guo S, et al. J Antimicrob Chemother 2013;68:1025. [18]Horcajada J, et al J. Clin. Microbiol 2005;43:2962. [19]Moreno E, et al J Antimicrob Chemother 2006;57:204. [20]Johnson J, et al Vet Microbiol 2009;136:108. [21]Platell J, et al J Antimicrob Chemother 2010;65:1936. [22]Cho S, et al Microb Drug Resist

2015;21:463. [23]Weese J, et al. Vet Med Int. 2011:263768. [24]Bin C, et al Diagn Microbiol Infect Dis 2006;56:351. [25]Abraham S, et al J Antimicrob Chemother 2014;69:1155.

New Methods of Testing for Infectious Agents

Jane E. Sykes, BVSc(Hons), PhD, DACVIM (SAIM)
Davis, CA

INTRODUCTION

Over the past 15 years, there has been increasing clinical application of nucleic acid-detection assays to rapidly detect pathogens, such as polymerase chain reaction (PCR)-based assays and fluorescent in situ hybridization (FISH). Refinement of molecular diagnostic assays has continued to occur in veterinary diagnostic laboratories, with improved standardization and quality assurance. There have also been advances in creation of nucleic acid-based assays that can be performed in the clinic. However, as with other assays, it remains important that practitioners' understand the limitations of these assays and how to properly interpret test results. In addition to molecular based assays has come the introduction of flow cytometry and mass spectrometry instruments into diagnostic laboratories. Mass spectrometry instruments can rapidly and accurately identify pathogens that are grown in culture, bypassing the need for time-consuming biochemical testing or sequence analysis. Some assays have somewhat lost favor due to recognition of poor specificity, e.g. use of direct immunofluorescent antibody on smears or cytospin preparations for diagnosis of canine distemper. The purpose of this talk is to provide an update on novel techniques that are now being used in medical and veterinary medical laboratories and clinics to detect pathogens, and to discuss their clinical relevance and their limitations. The focus will be on nucleic acid detection techniques, flow cytometry, and mass spectrometry.

LABORATORY-BASED NUCLEIC ACID-BASED PATHOGEN DETECTION AND IDENTIFICATION TECHNIQUES

Real-time Polymerase Chain Reaction

The use of nucleic acid-based assays such as PCR for diagnosis has increased dramatically in veterinary medicine over the past 10 years. Nucleic acid detection methods are ideally suited to detection of organisms that are not easily found using cytology or histopathology, are slow-growing, or are difficult to culture, as well as when a rapid (<12 hours) diagnosis is required.

Because of the high sensitivity of some nucleic-acid detection assays, especially those that involve DNA amplification, positive results may reflect contamination that occurs in the laboratory. Contamination problems have decreased with increased automation and the use of fluorescence-based real-time PCR assays (instead of conventional PCR and nested PCR, which require agarose gel electrophoresis for detection of the PCR product). Fluorescence-based PCR assays used in veterinary medicine employ either 1) a fluorescent DNA probe in addition to primers that bind to a specific pathogen DNA sequence (TaqMan PCR), or 2) a fluorescent dye that binds only to double stranded DNA (SYBR green assays). Fluorescence is detected by a spectrophotometer as amplification occurs (i.e., in real-time), without any need to open a tube and contaminate the laboratory environment with PCR amplicons. Nevertheless, false positive results can still occur, and the detection of a pathogen from a clinical specimen does not always imply its involvement in disease, and so test results always need to be interpreted in light of the overall clinical picture.

It is important to recognize that not all PCR assays for a specific pathogen are equal, because the nucleic acid primers and probes and the equipment used to run each assay may be different. So a PCR assay for *Leptospira* performed in one laboratory may perform differently than an assay used in another laboratory.

The amount of fluorescence emitted during real-time PCR is proportional to the number of copies of template DNA in the specimen, so real-time PCR assays can be quantitative. A plot of PCR cycle number against fluorescence is generated by an attached computer. The number of cycles that elapse before fluorescence is detectable above a certain threshold value is the CT (cycle threshold) value. Most laboratories run 40 cycles, so specimens that generate CT values less than 40 are positive. The lower the CT value, the more target DNA is present in the sample, because fewer cycles elapse before fluorescence is detected. Because specimen size can vary, the CT value may be normalized to the number of host cells present in the sample. Knowledge of organism load may have prognostic value for some infections, can allow infection to be monitored over time with treatment, and may be useful to differentiate low-level colonization from active infection causing clinical signs. Organism load is being increasingly reported by veterinary diagnostic laboratories performing real-time PCR, requiring interpretation by the clinic.

The presence of substances that inhibit a PCR assay can be identified in the laboratory using a *housekeeping gene* control PCR assay. This is performed on a separate aliquot of the specimen at the same time as the assay for the target pathogen. Housekeeping genes are host (dog or cat) genes that are expected to be present in the specimen. Examples of commonly used housekeeping gene PCR assays are those for canine or feline glyceraldehyde-3-phosphate dehydrogenase (GAPDH), histone, or 18S DNA. A negative housekeeping gene PCR assay result (or a high CT value for the assay, which is equivalent to a weak positive result) suggests that inhibitors may be present in the specimen, or the DNA in the submitted specimen may have degraded during storage or transport. Some PCR assays use an *internal housekeeping gene control*, in which a housekeeping gene PCR is multiplexed with the PCR assay for the target microorganism. Knowledge of the CT value for the housekeeping gene control impacts interpretation of the PCR test result for the pathogen of interest. For example, if the PCR result for the pathogen of interest is negative, but the CT value for the housekeeping gene control is high (e.g., 38), a false negative result may be possible due to inadequate amounts of DNA in the extracted sample.

DNA Microarrays

One of the downsides of culture is its inability to detect a target pathogen when it is combined with a large number of other pathogens or commensal organisms. In DNA microarrays, the products of PCR amplification of a universal bacterial gene (such as bacterial 16S rRNA) are hybridized onto a solid platform that contains thousands to more than a million oligonucleotide probes, which either hybridize to bacterial 16S rRNA genes; genus, species, or strain-specific genes; genes that specify resistance to antimicrobials; or toxin genes. The products of amplification are labeled with fluorescent dyes and the amplicon-probe interactions are measured using a laser scanner and analyzed by a computer. Examples are PhyloChip (Second Generation Solutions), which has been widely used to analyze microbial communities in gastrointestinal flora using 16S bacterial rRNA probes[1]; and Prove-It Sepsis (Mobidiag, Finland), which detects 60 species of bacteria and 13 fungi, as well as the *mecA* resistance gene, in blood specimens within a 3.5–hour time frame.[2,3] Although not always as sensitive as culture, when combined with culture, overall sensitivity is improved because of detection of pathogens that may be unculturable.

Isothermal PCR and In-Clinic PCR

Isothermal PCR is a process that amplifies nucleic acids at a constant temperature, and has been the basis for in-clinic PCR assay development, because an expensive thermocycler is not required. Dozens of complex different isothermal amplification methods have been published, and some have reached the practice environment in human medicine (e.g., Loopamp MALARIA)[4] and most recently in veterinary medicine (BioGal PCRun). Examples include rolling-circle amplification and loop-mediated isothermal amplification, which can reach an amplification efficiency comparable to that of PCR. In-clinic assays involve a DNA extraction step, followed by addition of reagents required for amplification; incubation in a heat block; then detection of the PCR product. The PCRun assay involves incubation at 60°C for one hour. Detection of the PCR product either involves the use of a lateral-flow DNA detection device (into which the tube containing the amplified DNA is inserted), or a small benchtop bioilluminator device that can semi-quantify the amount of DNA present in the reaction. BioGal has developed assays for *Leptospira, Ehrlichia canis, Anaplasma platys, Leishmania, Babesia canis*, canine parvovirus, feline panleukopenia virus, and *Mycoplasma haemofelis*, which allow detection of a PCR product within 75 minutes. The clinical sensitivity and specificity of these assays have not been published, and the assays are currently only available for research purposes in the United States. In addition, these assays currently do not incorporate a negative extraction control.

URINE FLOW CYTOMETRY

In human clinical microbiology laboratories, flow cytometry has been increasingly used to detect bacteria in urine specimens, without the need for prior culture. Urine flow cytometers (e.g., UF1000i, Sysmex) are able to detect particles in urine, and are useful for rapidly ruling out the presence of infection. Although the manufacturer has incorporated software to distinguish between cocci and rods, recent studies have shown that while 90% of the time the cytometer accurately identified rods, only 29% of the time did it accurately identify spherical-shaped bacteria.[5]

MATRIX ABSORPTION LASER DEIONIZATION TIME OF FLIGHT MASS SPECTROMETRY (MALDI TOF-MS)

Matrix absorption laser deionization time of flight mass spectrometry (MALDI TOF-MS) is a technique used to rapidly identify bacterial and fungal pathogens based on their protein composition. The availability of instrumentation such as the Bruker Biotyper and VITEK MS systems has revolutionized bacterial identification in the clinical microbiology laboratory. These systems use mass spectrometry to rapidly (on average in 5 to 8 minutes) and accurately (often >95%) identify organisms to the species level based on their chemical composition, provided a reference spectrum for the organism is available in the database used for comparison. This compares with days to weeks for conventional bacterial and fungal identification using phenotypic methods. Although the instrumentation is expensive, the ultimate cost is lower than with conventional biochemical methods, because of reduction in labor and reagent costs. Once the organism has been cultured, a small amount of a colony is spotted onto a MALDI target plate, which is then inserted into the spectrometer. A matrix that consists of α-cyano-4-hydroxycinnamic acid together with a solvent (acetonitrile and trifluoroacetic acid) is added. This lyses bacterial cell walls, and extracts and separates proteins. Subsequently, the mixture is subjected to laser light pulses, which are absorbed by the matrix molecules and convert them to an excited state. Because the matrix is acidic, it donates positive charge to the analytes (organism proteins). Localized heating of the material then causes micro-explosions and the ions desorb from the target surface, drifting through a vertical free field towards a detector plate. The speed that the ions travel (time-of-flight) is proportional to the ion's mass, the smaller ions reaching the detector first. This leads to generation of a molecular fingerprint that is unique for a pathogen species (and sometimes even molecular type). If the fingerprint is available in a library of fingerprints for comparison by the computer software, the organism can be accurately identified. Different extraction methods have been used to generate fingerprints for specific types of pathogens. For example, extraction of mycolic acids has been used as a means to type mycobacteria, with high sensitivity and specificity for tuberculosis diagnosis in humans.[6]

Advantages of rapid bacterial and fungal identification using MALDI-TOF include improved antimicrobial selection, early recognition of zoonotic pathogens, improved hospital biosecurity operations, improved patient care and shorter hospital stays, higher throughput due to improved automation, decreased medical waste and decreased send-out sequencing costs. Over several years, costs of identification are reduced (and therefore costs of culture are reduced) because of the cost of biochemical identification reagents and labor. Currently, susceptibility testing must still be performed using routine broth microdilution or diffusion methods. Methods have been developed to identify resistance to antibiotics by hydrolysis (e.g. beta-lactamase production) by incubating bacteria with an antibiotic. If the antibiotic is hydrolyzed, there is a shift in the mass spectrometry profile.[7]

REFERENCES

[1]Liu J, et al. BMC Complement Altern Med 2015;15:279. [2]Hinic V, et al. J Microbiol Methods 2014;107:38. [3]Ljungstrom L, et al. BMC Infect Dis 2015;15:199. [4]Oriero EC, et al. Malar J 2015;14:396. [5]Geerts N, et al. Clin Chim Acta 2015;448:86. [6]Cheng K, et al. Proteomics Clin Appl 2016. [7]Oviano M, et al. J Clin Microbiol 2015.

Emerging Resistance to Antifungal Drugs

Vanessa R. Barrs, BVSc(hons) PhD, MVetClinStud, FANZCVS (Feline Medicine)
Sydney, NSW, Australia

INTRODUCTION

Antifungal drugs can be grouped into four major classes based on their mechanism of action: (1) alteration of membrane function (polyenes - amphotericin B), (2) inhibition of nucleic acid synthesis (flucytosine), (3) inhibition of ergosterol synthesis (azoles and allylamines), and (4) inhibition of glucan synthesis (echinocandins – e.g. caspofungin). Mechanisms and prevalence of antifungal resistance have been studied extensively in clinical isolates of *Candida* from human patients with invasive infections, and, to a lesser extent, in human and veterinary isolates of *Aspergillus* and *Cryptococcus* spp. By contrast, little is known about resistance to antifungal drugs among other important fungal pathogens including *Coccidioides* spp. and *Histoplasma capsulatum*.

Of greatest concern currently is the increasing detection of *Aspergillus fumigatus* isolates with acquired resistance to azole drugs. This is a serious and rapidly emerging global problem. Mortality rates in humans with invasive aspergillosis involving azole-resistant strains approach 90%[1]. The major selection pressures driving azole resistance are the widespread use of azole fungicides in agriculture and previous azole therapy [1,2].

Further, some *Aspergillus* species are intrinsically less susceptible (i.e. have primary resistance) than *A. fumigatus* to some classes of antifungal drugs, especially the polyenes and azoles. For fungal pathogens of cats and dogs including *A. felis, A. udagawae, A. lentulus* and *A. fischeri*, the minimum inhibitory concentrations (MICs) of one or more azoles, amphotericin B and/or echinocandins are frequently elevated [3-6]. The different antifungal susceptibility profiles of these species compared to *A. fumigatus* highlights the need for accurate identification of *Aspergillus* spp from clinical samples. Most of these species cannot be reliably distinguished from *A. fumigatus* using fungal culture micro- and macro-morphology and hence are called "cryptic' species. There are more than 50 different cryptic species in the *A. fumigatus* complex, and most have pathogenic potential. Their definitive identification requires comparative sequence analysis of partial gene regions such as *ß*-tubulin or calmodulin [7]. Where this is not readily available, demonstration of thermotolerance, i.e. ability to grow at 50°C is highly supportive of *A. fumigatus,* since cryptic species do not usually grow at this temperature[3,8]

WHEN IS AN ISOLATE RESISTANT?

Two reference methods are available to determine antifungal susceptibility *in vitro*; the Clinical and Laboratory Standards Institute (CLSI) and European Committee on Antimicrobial Susceptiblity Testing (ECAST) methods. Both utilize the same 96-well broth microdilution format, broth media (RPMI 1640) and minimum inhibitory concentration (MIC) endpoint (complete inhibition of growth determined by visual inspection). The Sensititre YeastOne (SYO) system (TREK Diagnostic Systems, Cleveland, OH), which has wide application for *Candida, Cryptococcus* and *Aspergillus* spp is a non-reference method that is commercially available. This system, which utilizes azole and echinocandin antifungals in serial 2-fold dilutions on a 96-well dried, colorimetric broth microdilution panel has high interlaboratory reproducibility and correlates well with the CLSI reference method for susceptiblity testing of *Candida, Aspergillus fumigatus* and *Cryptococcus* spp. against azole antifungal drugs [9-11].

Importantly, clinical MIC breakpoints, which correlate *in vitro* data including susceptibility testing results and drug pharmacokinetics/pharmacodynamics with clinical trial outcomes in animals and humans, have not yet been defined for *Aspergillus* spp, although MIC breakpoints (EUCAST method) for itraconazole (ITC) and voriconazole (VRC) (> 2 µg/mL), and posaconazole (POS) (> 0.5 µg/mL) to define resistance in *A. fumigatus* have been proposed [12]. Epidemiological cutoff values (ECVs) of MICs, using CLSI (M38-A2 protocol) or EUCAST methods, are being increasingly used to detect decreased susceptibility or resistance *in vitro*. ECVs are upper limit values set by determining the distribution of MICs for wild-type (WT) isolates. The WT population is most commonly defined as those isolates around the modal MIC ± one 2-fold dilution, which encompasses >95% of strains. Although the clinical relevance of ECVs is not yet fully determined, they are used as a practical and standardized tool for the detection and monitoring of antifungal resistance [13]. More recently, *Cyp51A* gene sequencing data for *A. fumigatus* isolates was included with susceptibility data to determine ECVs of triazoles for *A. fumigatus,* which for ITC and VRC were 1 µg/mL, and for POS were 0.125 µg/mL[14].

MOLECULAR MECHANISMS OF AZOLE RESISTANCE

Molecular mechanisms of azole resistance in pathogenic fungi including *Aspergillus fumigatus*, *Cryptococcus* spp and *Candida* spp are below. For *Candida* spp. azole resistance results from a combination of these mechanisms, whereas single resistance mechanisms are more common among *A. fumigatus* and *Cryptococcus* spp isolates:

Mutations or overexpression of the azole target (*CYP51A/ERG11*)

Azole drugs target ergosterol biosynthesis by binding with the cytochrome P450-dependent enzyme14α-lanosterol demethylase, resulting in defective fungal cell wall formation. In *A. fumigatus* this enzyme is encoded by two gene homologues, *CYP51A* and *CYP51B*, while for yeasts, including *Candida* and *Cryptococcus,* it is encoded by *ERG11.* More than 140 *ERG11* amino acid substitutions have been reported among azole-resistant strains of *Candida* and multiple substitutions can have an additive effect[15]. By contrast, *ERG11* mutations are less common among resistant strains of *C. neoformans*. Overexpression of *ERG11,* another common mechanism of azole resistance in *Candida* spp involves multiple pathways including partial or complete duplication of the chromosome in which *ERG11* resides, as well as mutations in the gene encoding the transcription factor Upc2 (*UPC2)*[16].

Cyp51A and Cyp51B are protein paralogues. In experimental "knock out" models deletion of both *CYP51A* and *CYP51B* is lethal, while deletion of either gene is not. Since *CYP51* mutations occur preferentially in the *CYP51A* gene, it has been postulated that this gene encodes the major 14α-lanosterol demethylase required for growth [17,18].

Over 30 different non-synonymous single nucleotide polymorphisms (SNPs) in the *CYP51A* gene of *A. fumigatus* confer azole resistance. The most common "hot-spots" are G54, G138, M220 and G448 [17]. In *A. fumigatus,* SNPs accompanied by tandem repeats (TR) in the *CYP51A* promoter region are currently recognised as the most common molecular mechanism causing acquired azole resistance. Resistant isolates usually harbor one of two genetic signatures, TR_{34}/L98H or TR_{46}/Y121F/T289. The former is associated with high level ITC resistance or intermediate-level VRC and/or POS resistance, and the latter with high-level VRC resistance[19]. In isolates with these combined mutations, the mechanism of azole resistance is upregulation and overexpression of *CYP51A*, by up to 8-times compared to wild-type strains of *A. fumigatus* [20]. The combination of both modifications (TR and L98H or Y121F/T289) is necessary to confer azole resistance in these strains [21]. Over-expression of *CYP51B* has been documented in some azole resistant isolates of *A. fumigatus,* in the absence of *CYP51A* modifications, but is a less well-understood mechanism of azole resistance in *A. fumigatus.*

Activation of drug efflux pumps

Over-expression of plasma membrane proteins that reduce the accumulation of drugs intracellularly by pumping them out of the cell is a major mechanism of azole resistance for both *Candida* and *Cryptococcus*. The two major types of efflux pump are ATP-binding cassette (ABC) transporters and the major facilitator superfamily (MFS). *Candida* Drug Resistance Proteins (Cdr1p and Cdr2p) are major ABC multidrug transporters in *C. albicans* encoded by *CDR1* and *CDR2*. Overexpression of *CDR1* or *CDR2* is associated with azole resistance due to reduced drug accumulation as a result of rapid efflux [22]. Induction of *CDR* gene-encoded efflux pumps causes multiazole drug resistance. Similarly, overexpression of *CaMDR1,* a major MFS multidrug transporter of *C. albicans*, is linked to azole resistance, but is selective for fluconazole. Mutations in the transcription factors *TAC1* and *MRR1* (multi-drug resistance regulator 1) leading to overexpression of efflux pumps are also described in azole resistant *Candia* spp. In *C. neoformans* and *C. gattii*, overexpression of genes encoding ABC transporters, including *AFR1, MDR1* and *AFR2* is associated with azole resistance [23].

The significance of efflux pump overexpression as a mechanism of azole resistance in *A. fumigatus* is less well understood than for yeasts. Although overexpression of a number of efflux pumps has been described *in vitro* among azole resistant strains of *A. fumigatus* their clinical significance for the most-part remains unclear[17]. Of several ABC transporter genes in *A. fumigatus* shown to be up-regulated in response to azole exposure, to date, only one of these (*AfuMDR1*, which has been renamed *CDR1B*), is directly implicated in azole resistance [24].

Modifications of the cellular stress response pathway – stress adaptation mechanisms

Another mechanism by which azole resistance can develop is by modification of intracellular signaling pathways involved in stress responses including [10,17]:

(i) Unfolded protein response (UPR).
Aberrantly folded or unfolded proteins can accumulate in the endoplasmic reticulum (ER) of *Aspergillus, Cryptococcus* and *Candida* spp. The UPR is a protective cellular mechanism by which protein folding homeostasis is regained by increasing the folding capacity of the ER and regulating disposal of damaged proteins by ER-associated degradation.

(ii) Oxidative stress response.

(iii) Membrane and cell stress responses

Chromosomal abnormalities

Genomic alterations associated with azole resistance in *Candida* and *Cryptococcus* spp include loss of heterozygosity of specific genomic regions, increase of chromosome copy number and segmental or chromosomal aneuploides[16]

EMERGENCE OF AZOLE RESISTANCE IN *ASPERGILLUS FUMIGATUS*

Lanosterol 14α-demethylase is the rate-limiting enzyme in the ergosterol biosynthesis pathway. Azoles that inhibit C-14 demethylation of precursor sterols include imidazoles and triazoles, collectively referred to as sterol demethylation inhibitors (DMIs). For over 30 years DMIs have been effectively used in agricultural crop protection worldwide. Triazole antifungal drugs (including ITC, VRC, POS and most recently, isavuconazole) are also used as first line drugs for treatment of all forms of aspergillosis in humans, for invasive forms of aspergillosis in cats and dogs, and for aspergillosis prophylaxis in at-risk human patients.

Most fungicide DMIs share a core structure with the medical triazole antifungals; propriconazole and bromuconazole have a similar core structure to ITC and POS, while tebuconazole and epoxiconazole are similar to voriconazole. All four fungicide DMIs have no activity *in vitro* against isolates of *A. fumigatus* with the TR_{34}/L98H mutation. All four were first used in the Netherlands between 1990 and 1996, preceding identification of the first clinical azole-resistant *A. fumigatus* strains with this mutation in patients from the year 2000 onwards [2,21]. The prevalence of azole resistant strains of *A. fumigatus* isolated from human patients varies with geographic region. In a UK study in 2009 azole resistant strains were detected in 20% of patients with *A. fumigatus* infections[25] Resistance rates in Europe are the highest in the world with rates of 24 % detected in a multicentre study in the Netherlands in 2015[26]. In the US 10% of *A. fumigatus* isolates were azole resistant and had itraconazole MICs of ≥ 16 µg/mL and most of these had the TR_{34}/L98H or TR_{46}/Y121F/T289 mutations that have been linked with fungicide usage[27]. The overall prevalence of azole resistance in a prospective multicentre international study (SCARE study) involving almost 4000 *Aspergillus* isolates from 19 countries was 3.2%[28].

Surveillance for azole resistance among veterinary clinical isolates of *A. fumigatus* is limited to one study from Australia[8], where the prevalence of azole resistance among human isolates is reportedly < 1%[28,29]. Fifty isolates of *A. fumigatus* from dogs and cats with sino-nasal aspergillosis collected between 1988 and 2014 were screened for susceptibility to ITC, POS and VRC using the Sensititre YeastOne broth microdilution method, and tested for susceptibility to enilconazole and clotrimazole using the CLSI reference method (M38-A2). Only one canine isolate from 1992 was azole-resistant[8]. Ongoing international surveillance is warranted.

AZOLE RESISTANCE IN *CRYPTOCOCCUS*

The major mechanisms of azole resistance identified in *C. neoformans* and *C. gattii* are mutations or overexpression of *ERG11*, overexpression of ABC transporters and chromosomal abnormalities[23]. Azole resistance in *C. neoformans* is associated with disomies of chromosomes 1 and 4. Chromosome 1 contains *ERG11* and *AFR1*[16]. Changes in copy number of chromosomes 1 and 4 is associated with heteroresistance, where azole resistance in subpopulations within a clone varies based on frequent loss and gain of chromosomes in response to selection[16].

Similar to *Aspergillus* spp, clinical MIC breakpoints have not been established for *Cryptococcus* spp. Antifungal susceptibility of 42 *C. neoformans* and *C. gattii* isolates from cats and dogs from California was determined using the Sensititre YeastOne system[30]. A target fluconazole area-under-the-curve (AUC)/MIC ratio (CLSI method) has been previously established in a mouse meningoencephalitis model to prevent progressive fungal growth in the CNS. FLC AUC/MIC ratios did not exceed that target in 41 of the 42 isolates. It was concluded that similar to human patients, FLC may be inappropriate for treatment of canine and feline cryptococcal meningoencephalitis as single agent therapy, and results may explain poor treatment responses encountered when FLC MICs are relatively low. Although not directly comparable due to the methodology used and the small sample size, ECVs of azoles were 1- to 2-fold higher than ECVs established for wild type isolates. Similar to other studies, susceptibility profiles differed with molecular type of isolates. A *C. neoformans* isolate from a cat in Japan was found to have high MICs of fluconazole (128 µg/mL) using another commercially available non-reference method (E-test) for determining antifungal susceptibility. No mutations were identified in *ERG11* on sequencing; however, there was overexpression of both *ERG11* and *AFR1*[31].

RESISTANCE TO OTHER ANTIFUNGAL DRUGS

Acquired resistance can be rapid and is frequent, when flucytosine is used as a single agent to treat cryptococcosis precluding its use in monotherapy. Acquired resistance to echinocandins, the newest class of antifungal drugs (US FDA approved since 2001) is being increasingly reported in human patients with invasive candidiasis, both in patients previously exposed and unexposed to this drug class. Of concern is that many *Candida* isolates show cross-resistance to azole antifungal drugs. Resistance mechanisms identified include mutations in the *FKS1* gene, which encode for ß-1,3-glucan synthase (Fks1) and increased chitin synthesis[32]. Acquired resistance of *A. fumigatus* to echinocandins is currently uncommon, probably reflecting more restricted usage of this drug in patients with aspergillosis, compared to those with invasive *Candida* infections. Acquired resistance to amphotericin B is rare among clinical isolates of *Aspergillus, Candida* and *Cryptococcus* spp[33]

REFERENCES

1. van der Linden JWM, et al EID 2011;17:1846. 2. Snelders E, et al PLOS Med 2008;5:1629. 3. Barrs VR, et al PLOS One 2013;8:e64871. 4. Barrs VR & Talbot JJ Vet Clin North Am 2014;44:51. 5. Kano R, et al. J Vet Med Sci 2015;77:1183. 6. Krockenberger MB, et al Aust Vet J 2011;89:9. 7. Samson RA, et al. Stud Mycol 2007:147. 8. Talbot JJ, et al CIMID 2015;42:37. 9. Pfaller MA, et al Diag Micro Inf Dis 2012;73:365. 10. Guinea J, et al Diag Micro Inf Dis 2006;56:53-55. 11. Torres-Rodriguez JM & Alvarado-Ramirez E.J Antimicrob Chemo 2007;60:658. 12. Verweij PE, et al Drug Resist Updates 2009;12:141. 13. Pfaller MA, et al. J Clin Microbiol 2009;47:3142. 14. Meletiadis J, et al. Antimicrob Ag Chemo 2012;56:2524. 15. Morio F, et al Diag Micro Inf Dis 2010;66:373. 16. Cowen LE, et al Cold Spring Harbor Perspectives in Medicine 2015;5. 17. Wei X & Zhang Y, Lu L. J Microbiol 2015;53:91. 18. Chowdhary A, et al. Fut Microbiol 2014;9:697. 19. Verweij PE, et al. Clin Inf Dis 2016;62:362. 20. Cuenca-Estrella M, Clin Microbiol Inf 2014;20:54. 21. Mellado E, et al Antimicrob Ag Chemo 2007;51:1897. 22. Prasad R, et al . Adv Exptl Med Biol 2016;892:351. 23. Basso LR, et al J Antimicrob Chem 2015;70:1396. 24. Fraczek MG, et al J Antimicrob Chem 2013;68:1486. 25. Bueid A et al J Antimicrob Chem 2010;65:2116. 26. Ingen Jv, et al J Antimicrob Chem 2015;70:178. 27. Wiederhold NP, et al J Clin Microbiol 2016;54:168. 28. van der Linden JWM, et al. EID 2015;21:104. 29. Kidd SE, et al. Mycoses 2015;58:350. 30. Singer LM, et al J Clin Microbiol 2014;52:2061. 31. Kano R, et al Mycopathologia 2015;180:427. 32. Perlin DS. Clin Infect Dis 2015;61:S612. 33. Smith KD, et al Antimicrob Ag Chemo 2015;59:7197.

The Role of Inflammation in Hepatic Encephalopathy

Adam G. Gow, PhD, BVM&S, DSAM, DECVIM-CA, MRCVS
Edinburgh, UK

INTRODUCTION

Hepatic encephalopathy (HE) is defined as a neuropsychiatric disorder which develops as a result of hepatic dysfunction. HE may be seen with acute liver failure and chronic liver failure (with acquired portosystemic shunting) but is most commonly recognised in association with congenital portosystemic shunting. The spectrum of clinical signs are extremely wide, ranging from just perceptible alterations in behaviour through to coma. The severity of the signs can be graded on a scale modified from human medicine (1-4). In addition to overt HE, in human medicine, covert HE (cHE, also known as minimal HE) is recognised which can only be diagnosed with psychometric testing. By definition, this is not recognised in veterinary medicine.

HISTORY

It is irrefutable that ammonia plays a crucial role in the syndrome. Seminal work by Pavlov and Nencki in the 1890's elegantly demonstrated this fact. Surgically creating a portosystemic shunt caused clinical signs of HE when meat was fed, this was described as "meat intoxication syndrome." Further studies demonstrated that nitrogen-rich meals or ammonia salts produced the same effect, and that urine ammonia concentrations increased, and urea decreased compared to control animals. Furthermore, in control animals, portal vein ammonia concentrations were 3-4 times higher than in the systemic vasculature but in portocaval shunted animals systemic concentrations approached that of the portal vein and crucially, brain ammonia was four times that of the control animals, leading to the hypothesis that ammonia was causing the neurological signs. Since this time, much further work has been performed to define the mechanism of ammonia on the CNS.

PATHOPHYSIOLOGY -AMMONIA

In post-mortem examinations from cases of HE, a consistent finding is swelling of astrocytes and the pathophysiology of HE centres on the astrocyte, a subpopulation of glial cells. These were initially thought to be "gap-filler" cells in the CNS but are now recognised to play a multitude of crucial functions including: metabolic support of neurones (providing glucose from glycogen reserves and lactate), transmitter uptake and release, maintenance of the blood-brain barrier and regulation of extracellular ions. As a result, astrocyte dysfunction has serious consequences for neurological function. Of key relevance is the astrocytic neuroprotective function of buffering high ammonia concentrations in the CNS. Astrocytes contain glutamine synthetase an enzyme which converts glutamate to glutamine thereby consuming one ammonia molecule. One mechanism of dysfunction is that glutamine is osmotically active and, as it accumulates, cell swelling occurs due to osmotic stress- the "osmotic gliopathy" theory.[1] In acute liver failure and HE (thankfully a less common presentation in veterinary medicine), life-threatening cerebral oedema may result. In chronic HE cases, astrocytes compensate for increased concentrations of intra-cytoplasmic glutamine by decreasing concentrations of other osmoles, specifically myo-inositol. Histologically, astrocyte swelling is recognised by the presence of Alzheimer type II astrocytes which is documented in post-mortem examination of chronic HE cases. Decreased concentrations of glutamate inside the astrocyte stimulates production by amination of α-ketoglutarate, removed from the TCA cycle. Ammonia in itself inhibits TCS cycle enzymes, both factors favouring lactate production.

Glutamate/glutamine conversion is not a stand-alone process but part of neurotransmitter shuttling between astrocytes and neurones (Figure 1). Astrocytic glutamine is exported extracellularly via specific glutamine transporter protein (SNAT3) where it is then taken up by neurones and converted to glutamate for neurotransmission. Conversely, once glutamate has been released by the neurone into the synaptic cleft, although it may be scavenged by presynaptic and postsynaptic neurones, the majority of glutamate is removed by astrocytes. Increasing ammonia concentrations has been shown to decrease SNAT3 activity, compounding the increasing concentrations of glutamine within the astrocyte and also disrupting glutamate/glutamine shuttling and therefore neurotransmission.

A further theory is that glutamine acts as a "Trojan horse", being transported into astrocyte mitochondria whereby it is converted back to glutamate by phosphate-activated glutaminase thereby releasing ammonia again.[2] Ammonia within the mitochondria causes oxidative stress and membrane potential collapse within the mitochondria resulting in dysfunction, energy production failure and further production of free radicals. It is hypothesised that this energy failure and oxidative damage then results in the astrocytic swelling seen. Indeed, experimentally, anti-oxidants have been shown to reduce swelling. To tie these two theories together, swelling of astrocytes in itself produced free radicals, making it possible that both mechanisms are playing a role. Although it has been over 100 years since ammonia was implicated in HE pathogenesis, the exact details are still under much debate and subject to active research.

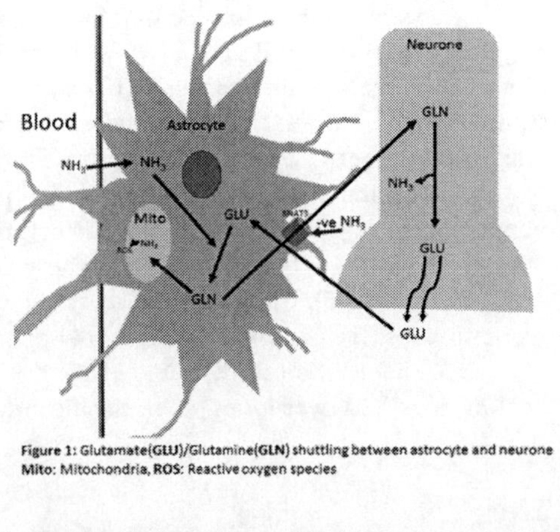

Figure 1: Glutamate(GLU)/Glutamine(GLN) shuttling between astrocyte and neurone
Mito: Mitochondria, ROS: Reactive oxygen species

Despite this consensus on astrocytic dysfunction as a result of hyperammonemia, it is also clear from clinical studies that although in groups of patients/animals there is correlation between the degree of HE and ammonia concentrations, in the individual, this is not the case. It has been demonstrated that overt HE may coincide with blood ammonia concentrations within the reference range and conversely, extremely high blood ammonia concentrations can be found in asymptomatic individuals. Therefore this has led to the concept of synergism. In human medicine, factors such as manganese, hyponatremia, hypokalemia, alkalosis, and inflammation have been implicated. There is now a large body of evidence that inflammation plays a major role in HE pathogenesis.

PATHOPHYSIOLOGY -INFECTION

Infection and neurological dysfunction has been described since the time of Hippocrates, and it is now recognised as "septic encephalopathy." In human medicine and also experimentally, the hypothesis of synergy between ammonia and inflammation resulting in encephalopathy has been explored. Clinically in human medicine, sepsis has been demonstrated to produce a rapid deterioration in HE clinical score in both acute liver failure and also cirrhotic patients.[3] In one study of human cirrhotic patients an association was demonstrated between infection/SIRS and HE score whereas, no association was demonstrated between arterial ammonia and HE score.[4] Laboratory studies with hyperammonemic mice have demonstrated that injection with endotoxin to produce a pro-inflammatory cytokine response produced greater encephalopathy than hyperammonemia in isolation.[5] A further study demonstrated that induction of hyperammonemia produced a more marked deterioration of cHE in cirrhotic patients which demonstrated signs of inflammation (these were stable patients, none were overtly septic) compared to those who did not, despite both groups developing similar ammonia concentrations.[6] Interestingly, in a further laboratory animal study, rats with HE due to a surgically created porto-systemic shunt demonstrated restored learning ability with the administration of non-steroidal anti-inflammatory medication.[7]

Pro-inflammatory cytokines have been implicated in driving this synergism, specifically IL6, TNFα. One immediate question is how systemic inflammation produces an effect in the CNS as the blood brain barrier prevents direct diffusion or transport. Various mechanisms to produce a CNS effect have been demonstrated. Peripheral inflammatory mediators can act via vagal afferents to induce cytokine production in the CNS, CNS endothelial cells have cytokine receptors which will increase cerebral prostanoids and nitric oxide.[2] Finally, the circumventricular organs lack a blood brain barrier and cytokines may have direct access via this route. Microglial activation is another feature that has been demonstrated in laboratory models of HE and also human patients with HE.

As a further mechanism of action, pro-inflammatory cytokines increase CNS endothelial cell permeability to ammonia, which may potentiate any given concentration of ammonia.

As an interesting twist, neutrophil dysfunction has been documented in human patients with cirrhosis causing reduced phagocytic ability therefore preventing effective bacterial and cell debris clearance as well as increased oxidative burst, which may cause collateral damage to surrounding tissue, resulting in more inflammation. This dysfunction has been replicated in vitro where clinically relevant ammonia concentrations have been shown to cause neutrophil dysfunction.[8] Therefore it appears that hyperammonemia will compromise the innate immune system, making infection more likely in these case.

Although the initiating source of the resultant inflammatory response is irrelevant for the consequences, in human medicine, a common source would be translocation of bacteria or endotoxins through the gastrointestinal wall. Cirrhosis causes portal and mesenteric hypertension which in turn increases gut wall permeability seeding through the portal vein in to the systemic circulation or direct translocation causing bacterial peritonitis.

EVIDENCE IN VETERINARY MEDICINE

This is all very well but there are profound differences between human and veterinary medical populations which suffer from HE. By far the most common presentation in veterinary medicine is as a result of congenital portosystemic shunting (cPSS), a condition which is exceptionally rare in human medicine. With this condition, these is no portal/ mesenteric vein hypertension and therefore no reason to assume an effect of gastrointestinal permeability. Conversely, HE in the human population is most commonly seen in end stage chronic liver failure i.e. cirrhosis, often as a result of viral infection. The analogous disease in veterinary medicine is canine chronic hepatitis which, although by definition inflammation is present, no infectious cause has yet been documented which may further drive the inflammatory process. Another obvious difference between the two populations is the availability of euthanasia in veterinary medicine compared to palliative care in human medicine and as a consequence of this, our population may not reach the same stage of terminal disease and be at lower risk of SIRS/sepsis as well as less profound portal hypertension and risk of bacterial translocation.

So what is the evidence that inflammation can have a role to play in HE seen in veterinary medicine? Certainly in regard to correlation of the clinical signs of HE with ammonia concentrations it is recognised that although broadly there is correlation in groups of animals, in the individual, the ammonia concentration has been stated to be an unreliable indicator of HE. Again in large study of dogs with cPSS, ammonia did not correlate with HE on an individual basis.[12] Therefore in this regard, it would appear that it is likely that other factors are playing a role in pathogenesis.

An initial study examined the concentration of C-reactive protein, a sensitive but non-specific marker of inflammation in a population of dogs with a congenital portosystemic shunt.[9] A total of 46 dogs with cPSS were included as well as 30 clinically healthy dogs. A significant difference was found between healthy dogs and dogs with HE ($p<0.001$) and also between cPSS dogs with and without HE ($p<0.05$). This was the first study to find an association between canine HE and inflammation and importantly, with this group the possible confounding factor of hepatitis is not present making a clean model of HE.

As a follow on study, a panel of cytokines; IL-2, IL-6, IL-8 and TNF-α were measured in 36 dogs with a cPSS compared to healthy dogs.[10] IL6 was found to be significantly increased compared to healthy dogs. The majority of dogs were displaying overt HE at the time of sampling (29/36), thus comparing sub-divisions was underpowered. IL-6 has been shown to be increased in human patients with HE and correlate with disease severity. The majority of these dogs were grade I/II again making sub-group analysis impossible.

A large study involving 120 dogs with a cPSS was undertaken and a large number of clinical, haematological and biochemical variables were used to analyse their ability to predict HE both in isolation, and also as part of a multivariable model.[8] This study demonstrated that both ammonia and a high SIRS score (3/4) predicted the presence of HE independently. Interestingly, and additive or synergistic effect between these two parameters could not be demonstrated. In a sub-population of 30 of these dogs where follow-up was available post-surgical attenuation, 25 were asymptomatic. Of the asymptomatic dogs, where measured (n=18), plasma ammonia had reduced significantly but SIRS scoring was not significantly different post-attenuation (n=14). One noted limitation of the study was that by using SIRS scoring, all the individual parameters making up the score may have reduced post-attenuation but not enough to reduce the SIRS category and also the low number of dogs with post-attenuation SIRS score. A further retrospective study including all dogs with HE presenting to a referral hospital did not demonstrate an association with HE grade and SIRS, although SIRS definition differed between studies and the latter's temporal association between blood sampling and documenting the clinical signs was not as tight.[11]

A further study of 140 dogs investigated markers of inflammation before and after surgical correction of the cPSS.[12] A greater proportion of dogs exhibiting HE at the time of sampling had a monocyte count above the reference range compared to asymptomatic dogs. Interestingly, at follow-up after surgical correction, the dogs which were asymptomatic prior and also after surgery demonstrated a significant decrease in neutrophil and monocyte counts, as did the dogs which had HE at initial evaluation and were subsequently asymptomatic after surgery. C-reactive protein before and after was available in a sub-set of dogs, again, CRP significantly decreased. This suggests that systemic delivery of portal blood potentiates a pro-inflammatory state and is consistent with a study demonstrating increased systemic lipopolysaccharide in dogs with PSS.[13]

CONCLUSION

From these veterinary studies it appears that dogs with cPSS have increased systemic LPS which pre-disposes them to a pro-inflammatory state and that markers of inflammation are associated with the presence of HE. Although at present we have an index of suspicion, association is not causation. For that leap we need to rely on experimental animal model studies and the provocative testing that has been done in human clinical patients. In light of data implicating ammonia in producing innate immune dysfunction, this intimate association will also have to be unpicked. Despite the obvious between veterinary and human medical populations with regard to the most common cause of HE (i.e. congenital versus acquired hepatic disease), it is looking likely that inflammation is a key component of canine HE. If this is confirmed then, as with the situation in human medicine, this opens up the opportunity to investigate modulating inflammation as a method of controlling HE in our patients. Methods of HE management in our patients are all directed at lowering ammonia, sometimes without sufficient improvement in quality of life, therefore an additional therapeutic angle is to be welcomed.

REFERENCES

[1] Brusilow S, Neurotherapeutics 2010; 7(4) 452. [2] Albrecht J, et. al Hepatology 2006; 44(4):788. [3] Aldridge et. al J Clin Exp Hep 2015; 5(S1) S7. [4] Shawcross et. al J Hepatol 2011; 54(4) 640. [5] Marini J et. al Mol Genet Metabol 2006 88(2) 131. [6] Shawcross et. Al Metab Brain Dis 2007; 22(1):125. [7] Cauli et. al 2007; Hepatology 46(2):514. [8] Rajkovic et. Al Clin Sci Lon 1985; 68(3) 247. [9] Gow A, et. al Metab Brain Dis 2012 27(2): 227. [10] Kilpatrick S et. al Vet J 2014; 200(1); 197. [11] Lidbury J et. al J Am Vet Med Assoc 2015; 247(2); 176. [12] Tivers M et. al PLoS One 2015; 10(2). [13] Tivers M et. al Vet J 2015; 206(3) 404.

Treatment of Acute and Chronic Hepatic Encephalopathy in Dogs and Cats

Adam G. Gow, PhD, BVM&S, DSAM, DECVIM-CA, MRCVS
Edinburgh, UK

DEFINING HEPATIC ENCEPHALOPATHY

Hepatic encephalopathy (HE) is defined as a neuropsychiatric disorder which results from liver disease. Immediately this definition causes some difficulty in veterinary medicine. The West Haven criteria developed for human medicine has been modified for use in veterinary medicine (Grades 1-4).[1] These classifications are for overt HE. In human medicine an additional classification is described; minimal HE. Individuals with minimal HE are by definition clinically asymptomatic but have subtle cognitive difficulties such as decision making or reduced psychomotor speed. Minimal HE is uncovered by psychometric testing in human medicine, using a combination of reaction time tests and paper and pencil tests. It has been stated that a diagnosis of Grade 1 HE using the West Haven criteria is challenging and that minimal HE and Grade 1 be combined under the term "covert HE". In the veterinary medicine grading scheme, grade 2 and above are sufficiently coarse that correct classification should be straightforward. In veterinary medicine we have the same difficulty as in human medicine; in accurately assigning grade 1 to our patients. In unfamiliar surroundings and the restricted time of an examination room, it is sometimes difficult to identify if behaviour is appropriate. This is also a small snap-shot in time and the animal may exhibit encephalopathy at other times. This would then reply on the client identifying the syndrome. We have all seen owners with obviously encephalopathic puppies that are blissfully unaware that they are not normal, therefore client history may be unrewarding. In dogs with chronic liver disease, the clinical signs of HE may be ascribed to "old age" and missed. These reasons may wrongly categorise the animal as asymptomatic. Equally, another pathology such as canine cognitive dysfunction or CNS neoplasia in animals with acquired liver disease, may cause an animal to be incorrectly categorised as having grade 1 HE.

Obviously, there is a balance to be struck; our companion animals may function well with subtle cognitive defects, therefore our cut-off for diagnosing HE may not have to be as stringent as in human medicine. However as there is no easy sensitive and specific test for HE, it is my impression that many cases, especially with chronic liver disease may have HE with a significant impact on the animal's quality of life that may go undetected.

PATHOPHYSIOLOGY OF HEPATIC ENCEPHALOPATHY

The role of ammonia and inflammation have been discussed in the earlier talk. It is clear that the syndrome of HE is a consequence of potentially multiple compounds and predisposing factors summating or acting synergistically to cause neurological dysfunction. Many putative factors have been appropriated from human medicine and it is not until recently we have begun to build-up any evidence of their prevalence in, and association with HE in our veterinary patients. Figure 1 depicts putative factors pre-disposing to HE in veterinary patients with an indication of evidence base for this.

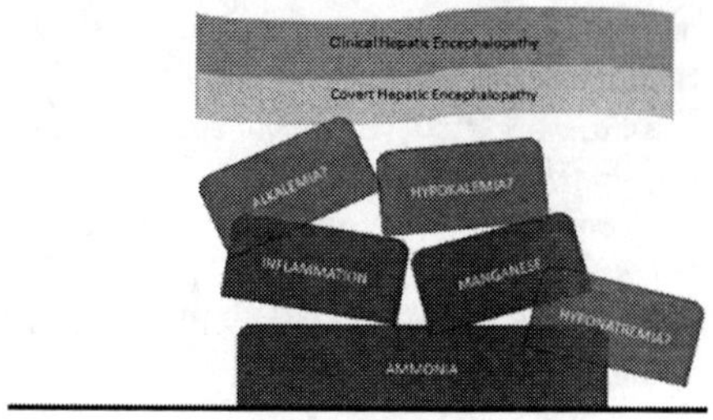

Figure 1: Building blocks of HE. Darker blocks have the most evidence in veterinary medicine

CAUSES OF HEPATIC ENCEPHALOPATHY

Again, broad categories have been taken from the human literature for use in veterinary medicine. The main causes of HE have been categorised as Type A: Acute liver failure without pre-existing disease, Type B: portosystemic shunting without intrinsic hepatocellular disease and Type C: Cirrhosis, portal hypertension and/or acquired shunt. In veterinary medicine, Type C has been expanded to include all acquired intrinsic hepatocellular disease.[2] HE as a consequence of portosystemic shunting is most commonly seen in veterinary practice; in acquired disease and especially the congenital form. The main difference between these three categories is that with Type A (Acute Liver Failure), acute onset HE is frequently seen whereas chronic HE is seen with cPSS and acquired shunting. This has pathophysiological and also treatment implications. Acute-on-chronic HE may be seen in type B & C if there is a sudden increase in precipitating factors. Thankfully, as the liver has a large reserve capacity, acute onset HE due to acute liver failure is encountered rarely in canine veterinary medicine, but most commonly due to infection or toxin ingestion.

ACUTE HEPATIC ENCEPHALOPATHY (Type A)

The rapidity of onset means that there is insufficient time for equilibration of osmolality, glutamine accumulates inside the astrocytes without commensurate reduction in myo-inositol. As a result, cellular swelling is more pronounced and as astrocytes account for 25-50% of brain volume this results in cerebral edema, increased intracranial pressure (ICP), potentially progressing to cerebral herniation and death if left untreated. Additionally there is loss of cerebral blood flow regulation resulting in increased flow, increasing hydrostatic pressure and potentiating cerebral edema and also episodes of hypotension and hypoxia. Evidence base for treating veterinary patients with acute HE is lacking and treatment recommendations are based on human medical data and standard of care in treating cerebral edema in veterinary medicine. In the human literature it is stated that traditional signs of ICP are unreliable in acute onset HE.[3] Avoiding procedures which increase ICP such as jugular sampling as well as keeping the head raised at a 30° angle to improve venous flow. More specific details are out-with the scope of this talk and reference should be made to critical care literature for further details of managing increased ICP and avoiding brain herniation as well as managing the other complications of acute liver failure. Of note is that hyperventilation to produce a respiratory alkalosis, which is recommended as a strategy to reduce ICP will favor formation of ammonia rather than the ammonium ion, which has been shown to be the state most favourable to cross the blood-brain barrier thus potentiating neurotoxicity. Other interventions are directed at dealing with pre-disposing factors in a similar manner to the more chronic or acute-on-chronic forms.

CHRONIC HEPATIC ENCEPHALOPATHY (Type B & C)

At present, the standard of care is directed at lowering ammonia concentrations however, as was discussed previously not only is hyperammonemia not the sole cause of HE, but a significant proportion of cases will have plasma ammonia concentrations within the reference range. Most cases present with an exacerbation of HE and consideration should be given to other potential causes of the neurological signs, specifically in cPSS and chronic liver disease, hypoglycaemia is a significant possibility. This should be part of the minimum database at presentation. Hyponatremia, hypokalemia and alkalosis are known precipitating factors for HE in humans.[4] The significance of these as regards HE in our veterinary patients is as yet unknown. One study examining multiple variables in HE pathogenesis in canine cPSS found no association with sodium or potassium, and a further study examining these variables, and alkalosis in canine HE from multiple causes also found no association with the clinical signs.[2,5] This difference may in part be due to the difference in main cause of HE between human and veterinary cases, although a recent study of canine chronic hepatitis cases found hyponatremia in only 1/55 cases, hypokalemia in 3/55 cases and these were all extremely mild.[6] This may reflect the difference in severity of the populations. Nonetheless, as these are known precipitators in human medicine, these avenues should be investigated and any derangements corrected.

For acute exacerbations, management will also involve attending to the recumbent/stuporous/comatose patient, again, out-with the scope of this talk. At present in veterinary medicine, methods to lower ammonia concentrations involve reducing production and increasing excretion.

DIETARY MANAGEMENT

As a large proportion of ammonia production occurs due to protein catabolism, historic recommendations were to restrict dietary protein. There are significant problems with overzealous dietary protein restriction. This may exacerbate hypoalbuminemia; in chronic hepatitis cases this reduction in oncotic pressure will further drive ascites formation. Body protein catabolism will release ammonia and finally, muscle is a significant reservoir for ammonia detoxification, converting glutamate to glutamine via glutamine synthetase. Any reduction in muscle mass reduces this alternative detoxification process, potentiating hyperammonemia. Protein should be highly digestible and of high biological value thus reducing the potential for catabolism, nitrogen and ammonia production. As regards quantity of protein, there is a paucity of evidence in veterinary medicine at present, however feeding an age-appropriate normal protein amount should be the aim (There is some evidence that chronic hepatitis and portosystemic shunt cases may have increased protein requirement). It must be remembered that young animals with a congenital shunt have high protein requirements due to their rapid growth, and proprietary "hepatic diets" will be excessively protein restricted, requiring the addition of extra protein; commonly cottage cheese. One study has demonstrated that dogs with cPSS on a soy-based diet had lower ammonia concentrations compared to a chicken-based diet.[7]

LACTULOSE

This non-absorbable disaccharide which passes through the small intestine intact. In the colon it is degraded by bacteria to short-chain fatty acids including acetic and lactic acid. This process decreases the luminal pH tending to trap nitrogen at the less absorbable NH_4^+. It may also provide an alternative energy source for bacteria. A laxative effect is also seen, causing increased frequency of defecation therefore increasing nitrogen excretion. Lastly, lactulose is a prebiotic which has been shown to modulate gut flora in humans, and experimentally, increasing *Lactobacilli* and this may reduce intestinal ammonia production. Although lactulose remains a standard of care for dogs with HE, and anecdotally appears effective, no studies have explicitly examined it's efficacy in controlling HE signs. Lactitol in powder form, where available, may be better tolerated especially by cats. Dosing is to effect, aiming for two to three soft but formed stools per day. Side-effects of both lactulose and lactitol include diarrhoea, abdominal discomfort and nausea.

ANTIBIOTICS

The main mechanism of action is to reduce intestinal bacterial load. This may reduce ammoniogenic bacteria as well as reduce a source of inflammation. Amoxycillin (10mg/kg three times daily), metronidazole (7.5mg/kg, twice daily – reduced dose due to hepatic excretion) and neomycin are commonly used. Neomycin (20mg/kg three times daily) is not abdorbed systemically, therefore if there is any suspicion of systemic infection then an alternative should be used. If there is any systemic absorbtion of neomycin then ototoxicity and renal toxicity may occur.

MANAGEMENT OF SEVERE HE

As above, appropriate consideration of managing the recumbent/comatose patient and also correcting any other metabolic perturbations should be given. Intravenous fluids will stimulate diuresis and renal ammonia excretion. Solutions containing lactate should be avoided as this requires hepatic metabolism. Acid/base status and electrolytes should be monitored regularly. Warm water enemata will evacuate the colon then a retention enema using a Foley catheter administered (3 parts lactulose to 7 parts water for 20 minutes, 20ml/kg, every 8 hours as required). Other additives such as povidine-iodine, neomycin and metronidazole have been described. A thorough evaluation for precipitating factors should be undertaken and these addressed. As inflammation is potentially a major factor, this in particular should be investigated and managed. Gastrointestinal haemorrhage has the potential to precipitate severe HE with the components of high protein content of blood, increased intestinal permeability for bacteria and endotoxins as well as a focus of inflammation.

FUTURE HE THERAPIES

Alternative ammonia lowering strategies: These include *L-Ornithine L-Aspartate* (LOLA) which are substrates for glutamate production in skeletal muscle allowing production of glutamine. Administration is intravenous, limiting use and there are concerns that a rebound hyperammonemia may occur on stopping treatment and some human trials have not demonstrated a beneficial effect.[8] There is one case series of LOLA administration in dogs with HE.[9] *L-Ornithine phenylacetate* (OP) is another compound used in human HE that appears a promising candidate. *L-ornithine* is used as before, *phenylacetate* then traps glutamine forming phenylacetylglutamine which is then excreted by the kidney.[8]

Chelation therapy to reduce Manganese (Mn): Hypermangaesemia and effect on astrocytes and microglial cells has been demonstrated experimentally and in humans to be implicated in HE pathogenesis. Hypermangasemia has been demonstrated in both dogs with cPSS and acquired liver disease.[10,11] In addition, manganese deposition in the CNS has been demonstrated in dogs with cPSS. Chelation therapy has been used in humans to produce a clinical effect.[10]

Modulating inflammation: Experimentally, NSAID's have been shown to improve HE, although not without significant risk of other adverse effects. *N-Acetylcystine* has also been used experimentally as an anti-oxidant to beneficial effect. Rifaximin-α is a synthetic non-absorbable antibiotic which has been shown to reduce the risk of HE recurrence in humans with cirrhosis. The mechanism(s) of action are still being elucidated; it may modulate intestinal flora composition and/or functionality as well as increasing systemic anti-inflammatory cytokines.[12]

Probiotics: An attractive target is modulating gastrointestinal flora to a less pro-inflammatory pattern and potentially less ammoniogenic species. In cirrhotic patients, probiotics have lowered IL-10 and improved neutrophilic phagocytic function.[12]

CONCLUSIONS

There is a difficulty in accurately defining and diagnosing low grade HE in veterinary medicine. In addition, as clinical signs will wax and wane, symptomatic animals may be missed at the point of consultation. A high index of suspicion should be maintained in animals with diseases which may precipitate HE. At present, our therapeutic strategies are based on mechanisms to lower systemic ammonia concentrations yet our evidence base even for these interventions is weak. There is increasing evidence that known precipitating factors in human medicine also play a role in our patients. There should be caution in assuming that HE pathogenesis in veterinary medicine is directly comparable to that in human medicine, although from the initial studies there appears to be remarkable similarities. If this is the case, then it is hoped that many novel therapeutic options will be on the horizon.

REFERENCES

[1] Rothuizen J, Vet Clin N Amer. 2009; 39(3): 419. [2]Lidbury J et. al J Am Vet Med Assoc 2015; 247(2); 176. [4] Sheasgreen, C. Inflammopharm. 2014; 22(6); 319. [5]Tivers M et. al PLoS One 2015; 10(2). [6] Kilpatrick, S. PLoS One 2016 11(1) [7]Proot s, J Vet Int Med. 2009 23(4); 794. Jalan, R. Med Hypoth. 2007 69(5); 1064. [9] Ahn, J. J. Vet Sci. 2015 Dec. [10] Gow A, J Vet Int Med 2010 24(1); 90. [11]Kilpatrick S, J Small Anim Prac. 2014 55(5); 241. [12]Aldridge D, J Clin Exp Hep. 2015 5(S1) S7

Climate Change: A Veterinary Problem?

Angela E Frimberger, VMD, DACVIM (Oncology), MANZCVS
Wauchope, NSW, Australia

Climate change (also known as global warming) raises significant veterinary issues, including domestic and wild animal welfare, infectious disease, food security, agricultural practice, and more. Veterinarians not only need to be aware of these, but should have an understanding of the science around climate change and potential solutions. This session will discuss these concepts.

Talking about climate change is not unlike talking with a client about their pet's cancer... "Good to meet you. Come in, please have a seat. I'm afraid I have some bad news. I know this is upsetting, but we do have some options to help control the condition and they're not as bad as you may think. So, first I'll tell you more about the condition and what can be expected if we do nothing; and then we'll talk about what we can do to help..."

THE PROBLEM

Two years ago, Working Group II of the Intergovernmental Panel on Climate Change (IPCC) issued a report titled Climate Change 2014: Impacts, Adaptation, and Vulnerability, stating that the effects of climate change are already occurring around the world. These changes are resulting in a variety of risks which will ultimately be difficult to manage if the levels of warming become high. The report was compiled by over 700 authors and editors from 70 countries and over 1700 reviewers. The press release for the report states that "The nature of the risks of climate change is increasingly clear, though climate change will also continue to produce surprises." "We live in an era of man-made climate change," said Vicente Barros, Co-Chair of Working Group II. "In many cases, we are not prepared for the ... risks that we already face."

What is meant by "if the levels of warming become high"? A general consensus of mainstream climate science suggests that many of the impacts of climate change might be manageable if long term warming were capped at 2.0°C compared with pre-industrial levels. More recently, at the Paris Climate Summit in December 2015, agreement was reached to strive to cap warming at 1.5°C compared with pre-industrial levels in order to improve the likelihood that most long-term impacts will be manageable and to reduce the risk of catastrophic impacts on the most vulnerable human communities, such as low-lying island nations. Presently the average global temperature is approximately 1.0°C higher than in 1880, and since 1975 has been increasing at approximately 0.15-0.20°C per decade. 2015 was the hottest year on record globally, by the highest margin yet.

It is well understood that climate change is caused by the accumulation "greenhouse gases" in our atmosphere, and that CO_2 is chief among these, although by far not the only contributor. The effect is not perfectly linear because of interacting factors and the potential for "tipping point" events; however it gives rise to the concept of a "carbon budget" based on how much CO_2 can probably be safely emitted in total before warming becomes unmanageable, how much has already been emitted and how much can still be released. By far the major source of CO_2 emissions is fossil fuel production and combustion, mostly for energy in the form of electricity and transportation. Under a "business as usual" scenario, CO_2 emissions would continue to rise at an accelerating rate because of population growth, increasing standard of living in developing countries and increasing consumption in developed countries. Thus international climate negotiations have focused on reducing fossil fuel use and the need to develop alternative / renewable energy sources. These negotiations are not ignorant to other contributing factors in global warming but focus on CO_2 because, as noted above, it is a major culprit; and also because, as difficult as it may seem to control CO_2 emissions and reduce fossil fuel use for electricity production, it is actually one of the more easily addressed causes due to the ability to control emissions with reasonable policy measures and the availability of technically viable alternative forms of electricity production.

Veterinary Impacts of Climate Change

This is certainly compelling - but is it a veterinary issue? Veterinarians are charged with the protection of animal health and welfare, the prevention and relief of animal suffering, and the promotion of public health. Climate change threatens animal health, welfare and public health, and risks domestic and wild animal suffering.

There is an obvious risk of suffering to wild animal species whose habitat is changing faster or farther than the species can adjust, or that is exposed to floods or wildfires (climate change is not a direct cause of either floods or wildfires; but the frequency and severity of both types of events is increased by climate change). As examples from Australia, both koalas and platypus are vulnerable to heat stress, and neither is able to extend its range substantially on an individual level to adjust. Koalas also suffer because hot growing conditions reduces the nutritional content of their sole food source of eucalyptus leaves, and when they travel in search of better quality food or relief from hot weather they become vulnerable to predators and car trauma.

There is also a risk of overt suffering to both pet domestic animals and production animals in heatwaves, floods and wildfires. It is estimated that over 1,000,000 animals lost their lives in Australia's Black Saturday bushfires. Moreover, a common reason that people put themselves in dangerous situations in wildfires is hesitation at leaving their animals.

Less dramatic and immediate threats to animal health and welfare also exist, and these overlap with threats to public health. For the most part these centre on infectious disease and food production. The World Health Organization predicts changes in infectious disease transmission patterns as a likely major consequence of climate change, partly as a direct result of changing temperature and water distribution, and partly as a result of increased range of arthropod vectors. These trends have been extensively studied and many have already been documented. Diseases expected to become more widespread include Lyme

Borreliosis, Dirofilariasis, Leptospirosis, bluetongue, blowfly strike, Rift Valley fever, West Nile virus, Dengue fever, malaria, Schistosomiasis, Leishmaniasis and cholera.

Food pressure is another climate change-related stress on domestic animals, especially production animals. It has already been shown that the productivity and nutritional quality of key food crops including corn and wheat is reduced by heat stress. Food plants are also vulnerable to an increase in pests and diseases with warmer temperatures. Likewise, water pressure is another major stress on both domestic and wild animals in drought conditions, which are clearly exacerbated by climate change.

Finally, climate change-related severe weather events such as droughts, extreme storms, floods and bushfires not only cause immediate suffering and loss of life but also result in substantial economic losses to individuals and communities. A healthy and well maintained pet population depends on a reasonably prosperous society, so in the face of economic stress the pet population is at risk for reduced quality of husbandry and health care. This is turn also means that the economic viability of veterinary medicine for pet animals as a profession is also at risk.

SOLUTIONS

Clearly, the need for action to control global warming is both important and urgent.

"Solutions" to climate change fall into two categories. Adaptation means accepting some degree of change and devising strategies for managing the impacts (so in fact it is more of a response than a solution). Mitigation means continued effort at reducing the degree that will occur. Because some degree of change is inevitable, adaptation is important; but from a veterinary perspective, adaptation is very limited as a solution because it is by definition anthropocentric and does not at all address the many impacts on wildlife. An excessive focus on adaptation can also receive the perceived importance of mitigation.

Individuals are well aware of the "right things to do" at home and in the office, that each help a bit in reducing our "carbon footprint". These include building energy efficient dwellings and practices, choosing the most fuel efficient vehicle for the task, powering appliances down completely when not in use, and recycling and using recycled where possible. But still the climate keeps changing, agencies announce that these steps are not enough, and these hard-working individuals may become frustrated.

In reality the concern and efforts made on an individual and a grassroots level are indeed important and necessary, and should be continued; but they are not sufficient to contain the problem because to date they have not been matched by concern and efforts on the part of industry and government. This is beginning to change strongly in both industry and government, with advancements in renewable energy technology and uptake, and the recent Paris climate summit outcome being a prime example of nascent ambition among governments. However in order to bring these agreements to fruition governments will need continued support or pressure from their electorate to take the steps necessary to fulfil their commitments.

The Potential Role of Veterinarians

The British Veterinary Association states "The effects (of climate change) on the weather patterns of our planet impact on agriculture, animal and human health and the economy. Veterinary surgeons will play an important role in responding to and mitigating the effects of climate change."

What, then, can veterinarians do to reduce the risks of climate change to our patients, their human families, and our profession? Minimizing our personal "carbon footprint" is important and should continue. One Health initiatives like this one are important in raising awareness, and some veterinary school have begun integrating climate change in various course content or even offering training specifically in climate change. Some veterinary organizations have included action on climate change as part of their mission. There may be a role for an organization that can offer a unified veterinary voice on climate change.

As individuals in our communities, veterinarians are respected opinion leaders and we should not flinch from utilizing this to support and encourage action to mitigate climate change as it is a major problem for our patients, our clients and our profession.

REFERENCES

Smith J B and Tirpak D A, US EPA report for Congress 1989 Appendix G. Klinedinst P L et al, Climat Change 1993;23(1):21. Wittman E J and Baylis M, Vet J 2000;160(2):107. Marcogliese D J, Can J Zool 2001;79(8):1331. Erasmus B N F et al, Glob Change Biol 2002;8(7):679. White N et al, Climat Change 2003;61:157. Brownstein J S et al, EcoHealth 2005;2(1):38. Learmonth J A et al, Oceanog Marine Biol Ann Rev 2006;44:431. Black PF et al, Rev Sci Tech 2008;27(2):563. de La Rocque S et al, Rev Sci Tech. 2008;27(2):339. Forman S et al, Rev Sci Tech 2008;27(2):1. Mas-Coma S et al, Rev Sci Tech 2008;27(2):443. Zhou X-N et al, Am J Trop Med Hyg 2008;78(2)188. Gale P et al, J Appl Microbiol 2009;106(5):1409. Schlenker, M J et al, Proc. Natl. Acad. Sci. USA 2009;10:1073. Wall R and Morgan E (Eds), Vet Parasitol Special Section 2009;163(4):263. McIntyre K M et al, Vet Rec 2010;167(15):586. Sachan and Singh, Vet World, 2010;3(11):519. van Dijk J et al, Animal 2010;4(3):377. Adams-Hosking C et al, Wildl Res 2011;38(2):122. Rose H and Wall R, Int J Parasitol 2011;41(7):739. Cox R et al, PLoS One 2012;7(7):e41590. Esbaugh A J et al, J Comp Physiol B 2012;182(7):921. Guis H et al, J R Soc Interface 2012;9(67):339. Yatoo M I et al, Int J Livest Res 2012;2(3):15. Altize S et al, Science 2013;341:514. Myers S S et al, Nature 2014;510:139. Nidumolu U et al, Int J Biometeorol 2014;58(6):1095. Rose H et al, Int J Parasitol Parasites Wildl.2014;3(2):209. Cooper K M et al, Trends Food Sci Tech 2015;44(1):21. Multiple authors, Lancet 2015;June special section. http://earthobservatory.nasa.gov/Features/WorldOfChange/decadaltemp.php. http://www.theage.com.au/victoria/black-saturday-cost-44-billion-20100801-11116.html. http://www.who.int/bulletin/volumes/92/8/14-143891/en/. http://www.bva.co.uk/atoz/Climatechangetheroleofvets.aspx.

ONE HEALTH VIEW OF VETERINARY DISASTER MANAGEMENT

Sebastian E. Heath, Vet MB, PhD, DACVIM, DACVPM
Washington, DC

Throughout the centuries veterinarians have played critical interdisciplinary roles in disasters. After several decades of large scale disasters that had animals as a focus, such as the recent events in Japan, Haiti, the global Foot and Mouth Disease pandemic, hurricanes and oil spills in the U.S., and many other less publicized events it is timely to review critical roles veterinarians play in disasters.

In a generic sense disasters are shocks to social and economic systems, where, temporarily, the need for resources outstrips their supply; if people or animals are not affected we usually do not think of the event as disastrous. In most natural and technological disasters the shock is typically the result of a sudden destructive event, such as earthquake, inclement weather, or exposure to an infectious disease agent, hazardous material or radiation. For these types of disasters the scope of impact can usually be defined quickly and the response is rapid supporting the desire to return to status quo as quickly as possible. In biological disasters, such as an outbreak of contagious disease, the scope of the event is often not known at the outset and the immediate response is often clouded by debates over how best to respond which can be perceived as inaction.

Through a One Health lens veterinarians have and will continue to play a critical interdisciplinary role in support of animal wellbeing, public health, the environment and the economy. Veterinarians have supported the response to natural and technological disasters as an extension of their practice and have effectively supported the response to disease outbreak by setting policy and enforcing regulations. Future opportunities for veterinarians in disasters lie in using scientific principles of One Health to shape how we, as a society, want to live with animals.

The Environment as a Source of Infectious Diseases in Humans and Animals

Lucas Pantaleon, DVM, MS, DACVIM, (LAIM), MBA
Versailles, KY

INTRODUCTION

One health is a concept that relates to the intricate relationship between animal, human and environment with regards to health and disease. Furthermore this concept highlights the collaboration of multiple disciplines at different levels in order to attain optimal health. [1] Biosecurity has been defined as work of strategy, efforts and planning to protect human, animal and environmental health against biological threats. [2] The FAO in 2003 adopted biosecurity as an all-inclusive term that includes policy and regulation to protect agriculture, food and the environment from biological risk. [3] The three pillars of biosecurity (human, animal and environment) are shared by the One Health concept and biosecurity should be implemented with One Health in mind, especially when it comes to collaboration and cross-learning from different disciplines in order to control the spread of pathogenic microorganisms. The importance of the interrelation between animals (colonized or infected), people and environment and their intricate interaction with microbes is an important aspect of infection control.

INFECTIOUS DISEASES AND HEALTH

Biosecurity must be an essential part of patient care and animal welfare. Excellent medical care is not possible without taking measures aimed at preventing and or minimizing the spread of contagious infectious diseases amongst patients. Similarly implementing measures aiming at limiting the introduction and spread of pathogens is paramount in livestock animal production systems.

Infectious pathogens are a serious threat to human and animal health. In human medicine the improvements in medicine, public health and social standards have led to a paradoxical increase in exposure and susceptibility to pathogens. [4] It can be argued that the same trend is occurring in veterinary medicine with more intensive production systems, increase in the number of specialized veterinary hospitals that treat sicker patients, increase in the case load, etc. There is consensus that owing to globalization and the 'four Ts' (trade, travel, transportation and tourism) biosecurity problems are worsening. [5] Moreover antibiotics used in human and in veterinary medicine has added to the selective pressures for development of microbial resistance and emergence of multi-resistant microorganisms. [4,6] According to a recent report two main trends are contributing to the increase of antibiotic use that is, raising income leading to increased access to antibiotics and a raise in demand for animal protein leading to increased use of antibiotics in agriculture. [7]

Biosecurity practices, antibiotic stewardship and innovation of drugs and therapies are important to combat and or control the emergence of pathogens resistance and contagious infectious diseases. [6] However the limited prospect for newer, safer and affordable antimicrobials and vaccines to cure and prevent disease, along with the emergence of multi-resistant microbes; makes biosecurity strategies even more vital.[4,6] Therefore implementing biosecurity and antibiotic stewardship with a One Health concept in mind would lead to better understanding and collaboration leading to increased success, thus reducing the need of antibiotics use, decreasing costs and saving lives.

BIOSECURITY IMPLEMENTATION

Properly implemented, biosecurity, along with antibiotic stewardship, can decrease or eliminate the use of antibiotics and prevent or limit the spread of disease. [6] Biosecurity should be proactive rather than reactive and involve all levels in the organization so everyone has a sense of ownership and everyone is accountable for implementing such measures. This is true at different organizational levels (institution, state, national and international) where a more risk-based approach focused on prevention and anticipation of disease emergence has been adopted. [5] For example at the international level the OIE has adopted a single list for animal diseases and includes notification requirements as well as patterns of emerging disease with significant morbidity/mortality or zoonotic potential. [5] "Buy in" from leadership is key to assure that the implementation of a successful organization cultural change, such as institutions of infectious disease control protocols, succeeds. Furthermore personnel and leadership education is paramount for biosecurity compliance to be achieved.

RECENT BIOSECURITY OUTBREAKS

Pandemics caused by severe acute respiratory syndrome coronavirus (SARS-CoV), Middle East respiratory syndrome coronavirus (MERS-CoV), avian influenza and swine influenza, although contained, showed how easily viruses can mutate and jump from one specie to another. [8,9]

The later outbreak of porcine epidemic diarrhea (PED), exotic in the US until May 2013, spread rapidly through swine farms causing high mortality in young pigs. [9] In 2015 the poultry industry experienced the devastating impact of a highly pathogenic avian influenza virus (HPAI).

PED was introduced in the US from China, likely via the use of Flexible Intermediate Bulk Containers (FIBC). [9,10] As with the apparent PED source in the US, accidental human assisted movement of pathogens across borders appears to be the major cause for animal biosecurity problems. [5] In the case of HPAI, migratory wild birds were responsible for the introduction of the virus in the commercial flocks, however transmission from farm to farm likely also occurred. [11] Biosecurity failures appeared to have contributed to the spread of HPAI between farms, in a report from USDA-APHIS it was found that some farms shared equipment, personnel movement between affected and non-affected premises occurred and there also was an improper cleaning and disinfection of transport vehicles. [11] As demonstrated by the HPAI outbreak, the movement of pathogens between wildlife and domestic animals has been a major driver for new disease emergence. [5] Also it is important to highlight the human factor that contributed to biosecurity failures as stated in the HPAI outbreak.

ECONOMICS OF BIOSECURITY OUTBREAKS

The costs of an infectious outbreak can be divided into direct fixed costs (remodeling of the facility, labor, equipment, utilities), direct variable costs (medications, treatments, procedures, supplies, diagnostic testing), indirect costs (morbidity, mortality, affected athletic performance potential, decreased worker productivity, liability) and intangible (bad publicity, loss of clients, loss of teaching opportunities, decreased morale, etc.). [12,13]

Health-care associated infections (HAIs) in hospitals impose significant economic consequences to the health care system. [13] In the US over 1.7 million patients per year suffer HAIs and one-third or more are believed to be preventable. [14] The annual direct hospital costs of treating HAIs in the US ranges from $28.4 billion to $45 billion. [13] Prevention of HAI is an example of value based medicine, where the goals of improving quality and decreasing cost are synergistic. [14,15] In the US HAIs are subject to mandatory reporting and are linked to reimbursement, thus reducing payment for HAIs which in turns incentivizes providers to improve infection prevention measures. [14] Furthermore HAI prevention may also contribute to the reduction of antimicrobial resistant infections. [14]

Outbreaks in veterinary teaching hospitals reported loses from thousands of dollars to $ 4.12 million per outbreak, these costs were likely conservative and much greater in the long term. [12,16] A recent survey of veterinary teaching hospitals in North America revealed that 82 % of the institutions had reported an outbreak of an infectious disease. *Salmonella enterica* was the leading cause of outbreaks (65 %) followed by methicillin resistant *Staphylococcus aureus* (MRSA) (42 %). Fifty-eight percent of these institutions had to restrict patient admissions while 32 % had to close in order to control the outbreak. *Salmonella enterica* has been the agent most often (77 %) involved in restricting admissions, followed by *Equine Herpes Virus – type 1* (EHV-1) (14 %). [17] The importance of nosocomial Salmonellosis lays on the fact of its high mortality rate of 30 to 60 %. [18]

In recent years livestock biosecurity problems have emerged in different parts of the world, characterized for their high costs and public profile. [5] The classic swine fever outbreak in 1997 cost The Netherlands approximately £2.4bn. The 2001 foot and mouth disease (FMD) outbreak cost the UK £7bn. The 2003 appearance of bovine spongiform encephalopathy (BSE) cost US and Canada $3 to 4bn in lost trade. [5] The PED outbreak in the US cost the swine industry $900 Million to $1.8bn. [19] The HPAI had a significant economic impact to the poultry, turkey and egg industry due to the loss of animals and negative trade impact. Just in the State of Iowa the outbreak was estimated to cost $ 1.2bn due to the death of 30 million hens and 1.5 million turkeys. [20]
Contagious infectious diseases in a hospital and livestock environment not only are a risk for the animal resident population but some are zoonotic diseases posing a risk for the human caregivers. In a study 50% of the surveyed institutions reported a zoonotic infection and attack rates during outbreak can be as high as 20 to 50 %. [17,18] MRSA is an important emerging zoonotic infectious disease with prevalence in the horse population in one teaching hospital of 1.7%. [21] A recent report highlights the fact that the human and animal relationships will continue to intensify due to animal husbandry practices, the growth of the companion animal market, climate change and ecosystem disruption, anthropogenic development of habitats and global travel and commerce. [22] Thus as the animal – human interactions rise, so does the potential for zoonosis and reverse zoonosis to occur.

Ambulatory practices for the most part deal with healthy populations, in which vaccination is considered to be the main tool for preventing the spread of infectious diseases. A good vaccination plan should be a key element in prevention of infectious diseases, but other forgotten practices such as proper environmental disinfection, isolation, hand hygiene, equipment disinfection and education, must also be part of a comprehensive protocol for prevention and control of pathogens. As it is the author's experience, a late trend in the equine industry has been to towards treating sick animals at the farm, in these cases it is paramount to educate farm personnel and implement strong biosecurity measures in order to prevent the spread of potential contagious diseases to the rest of the population. Ambulatory practitioners could also carry contagious infectious diseases to a large number of farms and for long distances, if biosecurity is not practiced. Therefore proper hygiene, protective clothing and disinfection should be part of the tools that ambulatory practitioners use to prevent infectious diseases spreading. It is important for ambulatory veterinarians to exercise good biosecurity in order to deliver high standards of care to their patients and clients. Biosecurity does not stop with us veterinarians: we must also educate clients and producers about the importance of infection prevention practices and the costs of doing nothing. Having in place a proper biosecurity plan and personnel education would help to prevent or minimize the consequences of disease outbreaks, thus saving money and lives.

KEY POINTS

In a recent consensus statement it was concluded that veterinarians and managers need to be aware that there is a recognizable standard of care with regards to infection control. Meaning that measures geared towards minimizing the spread of contagious infectious diseases and education must be part of the care provided by veterinarians to their patients and clients. Not meeting these standards constitutes malpractice and represents a failure to meet the ethical responsibilities to patients and clients. [18,23]

Biosecurity protocols need to be tailored to each individual situation. However some common facts for such a program are: surveillance, personnel training and education, leadership involvement, control of direct and indirect contact and optimizing personal and environmental hygiene. [23] These points can also be adjusted and applied to the hospital and field situations.

Infectious control needs to be a proactive rather than a reactive practice. Neglecting infectious disease control is unethical, constitutes malpractice and can be very costly in the short and long term of the organization.

REFERENCES

1. King L. One Health: A New Professional Imperative In. American Veterinary Medical Association American Veterinary Medical Association 2008:1-76.
2. Dunowska M. Biosecurity In: Sellon DC, Long MT, eds. Equine Infectious Diseases. St Louis, MO: Saunders; 2007:528-539.
3. FAO. Biosecurity in Food and Agriculture In: Committee in Agriculture 17th ed. Rome: FAO; 2003.
4. Sattar SA. Allen Denver Russell Memorial Lecture, 2006. The use of microbicides in infection control: a critical look at safety, testing and applications. Journal of applied microbiology 2006;101:743-753.
5. Waage JK, Mumford JD. Agricultural biosecurity. Philosophical transactions of the Royal Society of London Series B, Biological sciences 2008;363:863-876.
6. Spellberg B, Bartlett JG, Gilbert DN. The future of antibiotics and resistance. The New England journal of medicine 2013;368:299-302.
7. Gelband H. The State of the World's Antibiotics 2015 In: Center for Disease Dynamics, Economics & Policy 2015:1-84.
8. L. H. The Dangers of Hubris on Human Health In: Global risks 2013, 8th edWorld Economic Forum 2013:28-33.
9. Huang YW. Origin, evolution, and genotyping of emergent porcine epidemic diarrhea virus strains in the United States. mBio 2013;4:737-713.
10. USDA-APHIS. Swine Enteric Coronavirus Introduction to the United States: Root Cause In: USDA- APHIS 2015:1-53.
11. USDA-APHIS. Epidemiologic and Other Analyses of HPAI-Affected Poultry Flocks: June 15, 2015 Report In: USDA-APHIS; 2015:1-38.
12. Morley PS. Biosecurity of veterinary practices. The Veterinary clinics of North America Food animal practice 2002;18:133-155, vii.
13. Scott D. The Direct Medical Costs of Healthcare-Associated Infections in the US Hospitals and the Benefits of Prevention In. Division of Healthcare Quality Promotion National Center for Preparedness, Detection, and Control of Infectious Diseases Coordinating Center for Infectious Diseases Centers for Disease Control and Prevention: 2009:1-16.
14. Roberts RR. Costs attributable to healthcare-acquired infection in hospitalized adults and a comparison of economic methods. Medical care 2010;48:1026-1035.
15. Porter M, Lee, TH. The Strategy That Will Fix Health Care. Harvard business review 2013:1-19.
16. Dallap Schaer BL. Outbreak of salmonellosis caused by Salmonella enterica serovar Newport MDR-AmpC in a large animal veterinary teaching hospital. Journal of veterinary internal medicine / American College of Veterinary Internal Medicine 2010;24:1138-1146.
17. Benedict KM. Characteristics of biosecurity and infection control programs at veterinary teaching hospitals. Journal of the American Veterinary Medical Association 2008;233:767-773.
18. Morley PS. Evidence-based infection control in clinical practice: if you buy clothes for the emperor, will he wear them? Journal of veterinary internal medicine / American College of Veterinary Internal Medicine 2013;27:430-438.
19. Paarlberg P. Updated Estimated Economic Welfare Impacts of Porcine Epidemic Diarrhea Virus (PEDV). In. Dept. Agricultural Economics Purdue University 2014:1-38.
20. Steimel D. Study: Avian flu cost Iowa dearly in jobs, economic output. In: Iowa Farm Bureau, 8/17/2015 ed2015.
21. Stull JW, et.al. Methicillin-Resistant Staphylococcus aureus (MRSA) Surveillance in Horses at a Veterinary Teaching Hospital. Journal of Equine Veterinary Science 2012;32:S3 - S4.
22. Messenger AM, Barnes AN, Gray GC. Reverse zoonotic disease transmission (zooanthroponosis): a systematic review of seldom-documented human biological threats to animals. PloS one 2014;9:e89055.
23. Morley PS, Anderson ME, Burgess BA, et al. Report of the third Havemeyer workshop on infection control in equine populations. Equine veterinary journal 2013;45:131-136.

Interprofessional Education and One Health Initiatives – What is the Difference and Why Should We Recognize This?

Amara H. Estrada, DVM, DACVIM (Cardiology)
Gainesville, FL

INTRODUCTION

Proposed over a decade ago as a core competency for all health professions education by the Institute of Medicine[1], interprofessional education (IPE)—commonly defined as "learning about, from and with members of other health professions to improve collaboration and the quality of care"[2] —has become an increasing imperative as more health professions educational programs[3-5] including veterinary medicine[6], incorporate IPE requirements within accreditation standards. While IPE has had a long history, characterized by periods of waxing and waning[3], IPE's presence within health professions education programs is more pronounced than ever before. The goals of IPE are to develop learners' interprofessional skills as well as improve patient oriented care and community health outcomes through enhanced teamwork. Such teamwork is accomplished through interactive effort, a high level of communication, mutual planning, collective decision making, and shared responsibilities.[7] One Health initiatives have also become integral parts of core competencies for veterinary curricular development. One Health has been defined as "the collaborative effort of multiple disciplines working locally, nationally, and globally to attain optimal health for people, animals and our environment".[8] While the overall objectives of an IPE program are similar to those of a One Health initiative, they are not identical. There are unique differences in expectations and outcomes for an IPE program. Powerful experiential learning can occur for both the veterinary students and other health profession students involved. A recent scoping review article[9] suggested that there is some type of IPE program in place at many veterinary colleges but that these programs are diverse in their structure, delivery, objectives and outcomes across institutions. Putting veterinary and other health related professional students together at the beginning of their curriculum to work together as an interprofessional team at the human, animal, and environmental interface is a critical first step to successfully achieving the full potential benefits of IPE and One Health curriculum.

IMPORTANCE

One Health initiatives aim to address issues such as the impact of zoonotic diseases, water and food security, the relationship of plant and animal agriculture on human nutrition, and the impact of climate and land-use changes on human and animal health. Despite the recognized importance of the involvement of veterinarians, professional students may have very different attitudes and impressions regarding veterinary roles in these initiatives. Human health profession students may interpret One Health initiatives to be lessons learned from animal experimentation for human benefit. Similarly, veterinary students may misunderstand their role as care givers for animals who are involved in the research. Early in their careers, neither group seems to fully grasp the breadth of skills, perspective, training or possibilities that veterinarians could bring to help solve complex community and health issues. As specialists in veterinary medicine, all of us have inevitably had interactions with our human counterparts where we hear the phrase "you can do that in veterinary medicine?" or "I didn't know veterinarians could specialize". These types of conversations illustrate the misunderstanding of our profession but should also serve as recognition of an extremely important opportunity for us to help educate our colleagues while serving our mission and dedication to our goals of One Health. In order for our profession to be able to recognize the full potential of a One Health model, we must start with our veterinary and health profession students as well as the faculty who help to train and mentor them. Creating a program designed to bring veterinary and human health students and faculty to work together provides a platform for fostering this understanding and collaboration between professions. Because of the team work and interdisciplinary approach used in an IPE program, other critical competencies that are addressed include communication skills, collaboration, leadership development, awareness and application of an understanding of diversity and necessity to adapt to changing environments.

ACCOMPLISHING AND INITIATING INTERPROFESSIONAL EDUCATION

Interprofessional educational programs in the health science field vary widely, mostly dependent on the make-up of the institution health science center campus and ability for students to interact. The extent and nature of these programs range from single to multiple sessions, online offerings, or entire MPH and One Health programs. The introduction of the Canadian Interprofessional Health Collaborative (CIHC) and Interprofessional Education Collaborative (IPEC) competences in 2010 and 2011, respectively, provide educators a common language and framework to guide implementation of IPE learning experiences.[10,11] Given the multiple forces currently accelerating the need for IPE within health professions education programs, commentary and perspective on successful practices for program design and implementation is valuable to institutions initiating this work. Several reports in the literature discuss challenges and associated recommendations with IPE program design. These are based upon institutional experiences[12-17] or are framed around existing literature.[18-21] Uniformly reported is the logistical challenges of bringing together learners and faculty from multiple, separate professional programs and departments, each with their own calendar, curriculum and assessment goals. At some institutions, these logistical challenges are compounded by physical barriers of professions with separate office space, libraries and classrooms. Logistics are especially hard to work out when trying to create a common framework for interprofessional learning across clinical practice sites both in the institution and in outside placements across the community. Another key logistical challenge is identifying and integrating the content relevant to IPE and practice that already occurs in separate professional courses. That is, many learners are exposed to IPE content and practical experiences that are not marked as such. Identifying the many different places where IPE-relevant learning is already occurring throughout a large, complex educational institution (such as a large teaching hospital), with its many different curricula and training programs, is a complex task in itself. Many of these challenges can be met through working "person to person" across the institution to build a shared conceptual framework and sense of mission and commitment for IPE. Identifying dedicated local champions within each program involved who work closely together to establish and maintain an IPE that is relevant and successful for each program is key to establishing and accomplishing the learning activities. Lastly, as with any curriculum, when constructing an IPE program, a framework and method of evaluating student learning and whether the IPE program is meeting the intended learning outcomes is important. Assessment of individual learners and teams of learners in IPE is an area where concepts, approaches and standards are all experimental and under development.[19,22-26] As IPE programs within our veterinary curricula continue to develop and expand in scope, attention to assessment/evaluation approaches will become increasingly important as we seek evidence that graduates are able to demonstrate the interprofessional collaborative competencies that we have set as goals.

REFERENCES

[1] IOM. H Prof Ed 2003. [2] H Prof Network, W. H. O. http://www.who.int/hrh/resources/framework_action/en/ 2010.[3] Cerra F et al. Health Affairs 31(11), 2485-2492, 2011; [4] Foundation J.M.J http://macyfoundation.org/publications, 2013; [5]Zorek J et al, J Interprof Care 27(2), 123-130, 2013. [6]NAVMEC Consortium, http://aavmc.org/data/files/navmec/navmec_roadmapreport. [7]Nancarrow SA et al. H Res Health 11(19), 2013. [8]American Veterinary Medical Association, 2008.[9]Courtney M et al. J Interprof Care 28(4), 323-330, 2014. [10]Collaborative, C.C.I.H., Can Interprof Health Collab, 2010. [11]IOM, H Prof Ed proceedings Wash DC, Nat Acad Press 2003. [12]Mitchell et al., JAMA 81(10), 891-896, 2006. [13]Ho et al. Academic Medicine 83, 934-940, 2008. [14]Blue et al., JAAMC 85(8), 1290-1295, 2010. [15]Blue A et al, J Interprof Care 29(3):253-5, 2015. [16]Shrader S et al. J Interprof Care 29(2):165-7, 2015. [17]Buff SM et al, J Interprof Care 29(2):159-61, 2015. [18]Fook et al., J Interprof Care 27(4):286-91, 2013 [19]Abu-Rish et al. J Interprof Care, 26(6), 444-451, 2012;.[20]Clark PG et al, J Interprof Care 25(5), 321-327, 2011. [21]Reeves S et al. J Allied Health 36,231-235, 2007. [22]Parsell G et al. Med Educ 33(2); 95-100, 1999. [23]Luecht RM et al. J Allied Health 19(2), 181-191, 1990. [24]Sargeant J et al. J Contin Educ Health Prof 30 (2), 126-131, 2010. [25]PIPEs http://umanitoba.ca/programs/interprofessional/media/PIPEs, 2009. 25APTR, http://www.aptrweb.org, 2013. [26]Oandasan I et al. IP-COMPASS, https://www.mededportal.org/publication/9257, 2012.

Genetics of Endocrine Disease in Dogs and Cats

Lucy J. Davison, MA, VetMB, PhD, DSAM, DECVIM-CA, MRCVS
Cambridge, UK

INTRODUCTION

Observations that certain endocrine diseases occur more commonly in certain breeds of dog and cat have long suggested that genetic pre-dispositions to disease exist. Examples of this include diabetes mellitus in the Burmese cat, hyperadrenocorticism in the miniature poodle or hyperparathyroidism in the Keeshond. Understanding the genetic basis of endocrine disease is an exciting challenge and promises to deliver novel insights into pathophysiological mechanisms. Work is already underway to try to dissect the genetic basis of many human and veterinary endocrine diseases with the hope that such work may reveal new pathways and therapeutic or preventative opportunities. This presentation will be divided into several sections, beginning with an exploration of the way in which genes can affect disease risk, moving on to the techniques available for studying genotype-phenotype relationships and concluding with some examples of genetic studies in canine and feline endocrine disease.

HOW DO GENES IMPACT ON RISK OF ENDOCRINE DISEASE?

There are many ways in which disease risk can be influenced by genetics. The genome of an individual is unique (other than in identical twins) and contains millions of small variations, some of which may have a functional impact on the protein they encode. The variations may be inherited (somatic) and present in every cell in the body or, in the case of functional endocrine tumours for example, acquired mutations may alter the behaviour of a subset of cells in an endocrine gland, leading to failure of feedback mechanisms and / or hyper functionality. When a collection of variants (alleles) is inherited together, this is known as a haplotype.

Broadly, the most common types of genetic variations are:

1. **Single nucleotide polymorphisms (SNPs)** – these are changes in the A,T,G,C nucleotide code at a single site and may be found in the coding sequence (exons) of a gene, regions between coding sequences (introns) or inter-genic regions

If the SNP falls in a coding sequence changes the amino acid sequence of a protein it is known as a non-synonymous SNP (nsSNP), but if it does not change the amino acid sequence it is a synonymous SNP (sSNP). It is straightforward to see how nsSNPs may impact on disease risk because the amino acid change can alter the function of a protein. However, sSNPs and SNPs in non-coding regions may also have an effect on disease risk by changing the function of a regulatory region and affecting factors such as gene transcription, gene translation or transcript stability. This effect may only be apparent in certain tissues or at certain times in development so may be very challenging to identify. Many SNPs in the dog and cat genome have now been described and knowledge of these SNP maps has been crucial in the science of Genome-wide Association Studies, discussed later.

2. **Indels (insertions and / or deletions)** – these are insertions or deletions in the genome and can be more difficult to discover by some techniques. The consequence of indels depends on their size and location, but generally insertions and deletions can impact on gene function in a similar way to SNPs.

3. **Copy number variants (CNVs)** – sometimes there is more than one copy of a gene or part of a gene on a chromosome, (something which can be challenging to detect without sequencing long regions of the genome). Gene duplication can impact on the amount of the transcript and protein that is made from that gene.

4. **Inversions and chromosomal rearrangements** – larger rearrangements of whole sections of chromosome are also possible, although these are not usually recognised as being responsible for endocrine diseases.

CHALLENGES IN INVESTIGATING THE GENETIC RISK OF ENDOCRINE DISEASE IN DOGS AND CATS

Most genetic studies rely on examining the differences between a case population, suffering from the disease in question, and a control population, in which the disease is absent. Before discussing the techniques used in genetic studies, it is important to review briefly some challenges in designing studies to examine the genetics of endocrine disease in dogs and cats.

1. **Complexity of genetic pre-disposition** - Unlike monogenic diseases, caused by a mutation in a single gene, most endocrine conditions are considered to be complex diseases. Exceptions to this are endocrine diseases which occur very early in life (e.g. congenital hypothyroidism) and some other breed-associated conditions such as primary hyperparathyroidism of the Keeshond (see later). The inheritance of disease risk in common complex endocrine diseases is usually the result of small variations in a large number of genes. The fact that multiple genes play a role in disease susceptibility can make it more challenging to find all the genes involved in a particular disease.

2. **The role of the environment** (e.g. dietary factors, infectious diseases). Environmental triggers of certain diseases are highlighted by the fact that genetically identical human beings are not always concordant for endocrine diseases. For

example if one twin suffers from type 1 diabetes, the other twin, whilst having an increased risk of the disease, is not guaranteed to develop it. The power of genetic studies is reduced by the influences of the environment, since cases and controls may be genetically very similar but have simply differed in the environmental factors they have been exposed to. A disease where this might be particularly important is diabetes mellitus in cats, where overfeeding and lack of activity lease to obesity, which increases disease risk.

3. **Phenocopies of disease** – It sounds obvious, but it is crucially important to ensure that all 'case' animals within a particular study actually have the same disease. Phenocopies are disease cases which converge on a similar set of clinical signs but actually have different diseases, each with a different underlying pathogenesis and genetic basis. This is particularly true of some diseases where multiple mechanisms may result in the clinical diagnosis e.g. diabetes mellitus in dogs caused by several mechanisms including pancreatitis, insulin resistance due to hyperadrenocorticism or potentially related to pancreatic abiotrophy in juvenile dogs. For genetic study purposes, these dogs should be treated as separate endocrine disease populations. Similarly, for the purposes of genetic work, pituitary dependent and adrenal dependent hyperadrenocorticism should be treated as (at least) two separate entities.

4. **Age-matched healthy control samples for case-control studies** - As well as avoiding phenocopies, it is imperative to ensure that the control population is not contaminated with unrecognised / undiagnosed cases, or animals which have not yet reached an age at which clinical signs of the disease in question might appear. For example in feline hyperthyroidism, it is important that young cats are not used as a control population since most cats do not develop hyperthyroidism until they are over the age of 10.

5. **Population structure** – Population structure describes the overall genetic architecture of the population being studied (1). In dogs, the limited variation within breeds contrasts with the wide variation between breeds, and this can be advantageous when designing genetic studies. When comparing cases and controls, population structure should also be considered carefully and ideally cases and controls should come from the same geographical area / population, otherwise any genetic differences found might be, for example, geographical in origin and simply relate to the population as a whole rather than the disease.

6. **The concept of 'fixed disease risk' within certain breeds or populations** Recognised dog and cat breeds have already undergone a high degree of genetic selection, effectively being pushed through a "bottleneck" and limiting diversity within the breed to achieve certain characteristics. In cases of endocrine disease where a whole breed is known to be at increased risk, for example hypoadrenocorticism in the Bearded collie, it is important to remember that this risk may affect the entire breed equally, with mainly environmental factors actually being responsible for whether the animal gets a disease or not. If this is the case, then no genetic differences between cases and controls would be seen if they are compared, despite there being a strong genetic risk factor within the breed as a whole.

7. **Knowledge of normal variation in the canine and feline genome** - Any genetic variations discovered in an individual with a disease must be placed into the context of known variations in the rest of the population. Much progress has been made in recent years in identifying and recording SNPs in the canine (1) and feline genome (2) but these maps are not yet complete (particularly in the cat), so it is not always immediately possible to determine which variations in a particular patient are most significant. Finding or mapping a region associated with disease is not the same as finding the associated gene and its mechanism for increasing disease risk.

8. **Epigenetic influences** - It is becoming clear that it is not just the genome sequence that is important in conferring risk but that there are also heritable epigenetic "marks" on the genome such as gene methylation (3). These add an extra layer of complexity to the analysis because epigenetic features can also influence gene activity and disease risk, for example by switching genes on and off. These marks are not detected by standard sequencing techniques.

TECHNIQUES FOR IDENTIFICATION OF GENES INVOLVED IN ENDOCRINE DISEASE

In recent years, techniques for studying the canine and feline genome have evolved to become far more efficient, not only with respect to cost, but also in the time taken to generate sequence data. There are several possible types of study that can be performed to analyse the genetics of endocrine diseases:

- **Candidate gene study** – This involves sequencing genes known to be involved in the biology of the diseased organ to look for mutations, although this technique is inherently biased as it necessarily relies on previous knowledge of the disease.
- **Linkage analysis** – This technique is most useful for simple monogenic diseases where large family pedigrees are known, so is not a particularly effective approach to complex diseases in dogs and cats.

- **Genome-wide Association Study (GWAS)** – Our improved knowledge of the range of variation in the canine and feline genome has allowed canine and feline SNP-chips to be developed, allowing an animal to be genotyped for thousands of SNPs at a time, using just a small DNA sample. GWAS involves analysing these SNPs across the whole genome to determine whether a particular allele of any SNP is present at a higher frequency in the diseased or the control population. This technique identifies a region of the genome as being associated with a disease, but a lot more work is then needed to "fine-map" the signal and to determine which gene in the region is responsible for conferring disease risk and how it does so.
- **Next generation sequencing technology** – The first human genome more than $3 billion and took many years to sequence. Now it is possible for a scientist to sequence a whole genome in a few weeks for a few thousand dollars and to sequence an exome (the coding sequence of the genome) for around a thousand dollars, using next generating sequencing (NGS). As well as comparing the sequence of cases to controls, it is also possible to use NGS to look at gene expression for example to compare normal and hyperplastic thyroid tissue or to look for genomic DNA sequence differences between normal adrenal and adrenal tumour tissue. However, the amount of data generated from such studies requires specialist bioinformatic and statistical expertise full analysis as well as further follow up studies to prove disease association and determine the functional impact of any identified mutations.

EXAMPLES OF GENETIC DISCOVERIES IN ENDOCRINE DISEASE

Known variants in human endocrine deficiency diseases

In humans, many of the autoimmune endocrine (hormone deficiency) diseases share a number of common variants which all impact on disease risk, reflecting an underlying defect in immune tolerance (4). The most striking example of this is in the locus encoding the highly polymorphic major histocompatibility complex (MHC), involved in antigen presentation. Certain MHC alleles are associated with endocrine diseases including for example type 1 diabetes and hypoadrenocorticism, making this a popular region for canine and feline candidate gene studies.

Known variants in human endocrine diseases of hormone excess

In contrast, in endocrine conditions where there is a hormone excess, such as hyperthyroidism or hyperadrenocorticism, the underlying genetic predisposition to the disease is often less clear. Multifactorial analysis (including exome sequencing, gene transcription analysis, methylated gene analysis) of human adrenal tumours has led to the discovery of variations in many different genes "driving" excess hormone production by, for example, alteration of G-protein signalling or the cyclic-AMP / Protein Kinase A pathway. There are also Multiple Endocrine Neoplasia syndromes in humans, some with a very well defined genetic basis (5).

Canine primary hyperparathyroidism

One of the earliest genetic endocrine diseases in a single breed to receive specific attention was primary hyperparathyroidism in the Keeshond (6). Here, a candidate gene approach was employed, as well as studies of segregation of the disease in certain families. Initial work ruled out certain mutations commonly associated with the disease in humans and also documented autosomal dominant inheritance with age-related expression of the disease (penetrance). This work, followed by further mapping studies, eventually led to discovery of a positional candidate for the disease, from which a commercially available genetic test was developed.

Canine hypothyroidism

Breed-association studies have long suggested a role for genetics in susceptibility to canine hypothyroidism, with the Boxer, Doberman Pinscher, Gordon Setter and Beagle amongst those breeds commonly affected. Unlike Keeshond hyperparathyroidism however, this disease is complex and heterogeneous, with some dogs demonstrating lymphocytic thyroid infiltrates and others thyroid atrophy, making it challenging to study. Nonetheless, the first genetic region associated with disease risk was that containing the MHC genes (known as Dog Leukocyte Antigen *DLA* in dogs) (7), although this association was only evident in certain breeds. Another candidate region was the region encoding thyroid peroxidase or *TPO*, in which disease-associated mutations were found in congenital hypothyroidism of Tenterfield and Toy Fox terriers (8). A GWAS and meta-analysis in the Gordon setter, Hovawart and Rhodesian ridgeback also demonstrated a novel region on canine chromosome 12 associated with canine hypothyroidism. The exact causal variant has not been identified but there are three genes in the region, *LHFPL5, SRPK1* and *SLC26A8* requiring further study (9).

Feline hyperthyroidism

The availability of the feline genome sequence has enabled study of feline genetics to make tremendous advances in recent years, although the availability of data regarding normal feline variation is still more limited than such information in canines. Although feline hyperthyroidism is a very common disease, any underlying genetic cause remains elusive. Several candidate gene studies have attempted to examine mutations thyroid-related genes in feline thyroid adenomas, with limited success. In some thyroid nodules, mutations with mis-sense mutations in the TSH receptor gene and / or the Gs alpha gene have been reported but none can explain the disease completely (10).

Canine diabetes mellitus

At present, all published canine genetic findings in diabetes mellitus are based on candidate gene studies but some GWAS work, yet to be published, has also been undertaken. Canine DM is particularly challenging to study because of the number of disease phenocopies which exist (11). Within some breeds, certain *DLA* haplotypes are associated with risk of canine DM, implicating antigen presentation and the adaptive immune response in the pathogenesis of canine DM in these breeds (12). More recently, again using SNP discovery and a candidate gene approach, polymorphisms in several cytokine and immune response genes have been implicated in DM risk in certain breeds, include those encoding IFN-gamma, IL-10, IL-12 beta, IL-6, IL-4, PTPN22, TNF alpha (13). Here, associations are most common in terriers and spaniels, but there are clear differences across breeds. Current work is also using next generation sequencing to investigate some potential monogenic causes of diabetes in younger dogs.

Feline diabetes mellitus and feline acromegaly

Although the majority of feline acromegalic patients are diabetic, acromegalic cats are generally considered separately (for genetic purposes) from non-acromegalic cats with diabetes. A candidate gene study in feline acromegaly, based on the fact that up to 20% of human familial isolated pituitary adenomas are caused by germline mutations of the aryl-hydrocarbon-receptor interacting protein encoding gene (AIP) did find some nsSNPs in the gene, but no SNP segregated entirely with the disease (14). A further challenge in non-acromegalic diabetic cats is to account for the potential environmental influence of obesity on the development of the diabetic phenotype. An early gene associated with diabetes in overweight cats, by candidate gene study, was the melanocortin 4 receptor gene, also associated with obesity in humans. In addition, GWAS technology has been applied to the comparison of lean cats with obese diabetic cats, with several interesting candidate regions identified for further study (15). A recent GWAS in diabetic and non-diabetic Burmese cats has also reported two regions of potential interest for follow-up (16).

Canine Hypoadrenocorticism

One study in canine hypoadrenocorticism demonstrated a *DLA*-association in Nova Scotia Duck-tolling retrievers and a more recent analysis revealed further disease-associated *DLA* haplotypes in Cocker spaniels, Springer spaniels and Standard poodles. Further candidate gene studies also point to the role of the immune system in pathogenesis of disease, with variations in the T cell signalling pathway genes *CTLA4* and *PTPN22* also associated with risk of Hypoadrenocorticism in Springer and Cocker Spaniels respectively (17).

Canine Hyperadrenocorticism

Genotype-specific steroid profiles have been associated with adrenal adenomas in humans. Candidate gene studies of canine cortisol-secreting adrenal tumours have identified that approximately one third of tumours have an activating mutation of the *GNAS* gene, associated with constitutive activation of cAMP signalling (18). However, molecular screening for somatic mutations in corticotroph adenomas of dogs with pituitary dependent hyperadrenocorticism has been less rewarding (19), although similar studies using next generation sequencing technology have provided insight into disease pathogenesis in humans.

FUTURE PERSPECTIVES

Although the canine and feline genome sequences are now available and advances in technology have made large scale genetic studies more affordable, there is still much work to be done in understanding the role of genes and environment in canine and feline endocrine diseases. Such exciting insights may also offer a valuable perspective on the genetics of human endocrine disease.

REFERENCES AND RECOMMENDED FURTHER READING

1. Lindblad-Toh, K et al. (2005) Nature 438(7069):803-19; 2. Faillot S, Assie G. (2015) Journal of Feline Medicine and Surgery 17 : 405-15; 3. Yan H et al (2016) Am J Epidemiol. Jan 15;183(2):96-109; 4. Tait KF, Gough, SCL (2003) Clin. Endo. 59:1-22; 5. Norton JA (2015) Surg Oncol Clin N Am. Oct;24(4):795-832; 6. Goldstein RE, et al. (2007) J Vet Intern Med;21:199-203. 7. Kennedy, L J et al. (2006) Tissue Antigens 67 (1) : 53-6; 8.Dodgson S.E. (2012) J Vet Intern Med. 2012 Nov-Dec;26(6):1350-7; 9. Bianchi, M et al. (2015) PLoS ONE DOI : 10.1371/journal.pone.0134720; 10. Watson SG et al. (2005) Journal of Endocrinology 186: 523037; 11. Catchpole, B., et al.,. J Small Anim Pract, 2008. 49(1): p. 4-10; 12. Kennedy, LJ, et al., (2006). Tissue Antigens, 68(6): p. 467-76.; 13. Short, AD, et al., (2007) J Hered,. 98(5): p. 518-25.; 14. Scudder et al ACVIM Proceedings 2015; 15. Forcada, Y et al. (2014) J Vet Intern Med. 2014 Mar-Apr;28(2):458-64; 16. Samaha,G et al. Proceedings of Tufts Canine and Feline Breeding and Genetics Conference 2015; 17. Boag AM et al. (2014) Topics in Compan An Med 29 : 96- 101; 18. Kool, MMJ et al. (2013) J Vet Intern Med 27:1486-92; 19. Van Wijk P A (1997) J. Endocrinol Invest 20:1-7.

Etiopathogenesis of Feline Hyperthyroidism

Harriet M. Syme, BSc, BVetMed, PhD, FHEA, MRCVS, DACVIM, DECVIM-CA
London, England

Hyperthyroidism is the most common endocrinopathy of middle-aged and older cats. Although it was not recognised as a clinical entity until the late 1970s it is now very common world-wide (or at least in the industrialised nations where it has been studied), although in some countries it is likely that it remains under-diagnosed.[1,2] It is unclear whether feline hyperthyroidism existed as a clinical entity prior to this; histopathological studies had sometimes recognised thyroid nodules but these were apparently not associated with signs of thyrotoxicosis. A number of factors could have contributed to the emergence of this clinical condition including aging of the feline population, and an increasing willingness of owners to pay for veterinary care, widespread availability of diagnostic tests, and increased veterinary recognition, but it is generally believed that this is unlikely to solely account for the rapid and dramatic increase in disease prevalence.

Thyroid Pathology

Feline hyperthyroidism most often results from benign adenomatous nodules or hyperplasia of the thyroid gland and is clinically and pathologically similar to toxic multinodular goitre (MNG) in humans. The thyroid glands of hyperthyroid cats contain single, or more commonly multiple, adenomatous nodules with heterogeneous appearance. The nodules vary in size and in follicular appearance. The intensity of PAS-staining of thyroglobulin-containing colloid within the follicles is generally less than that of the surrounding, non-adenomatous tissue. The epithelial cells lining the hyperplastic areas and adenomatous nodules are cuboidal to columnar, while those in the surrounding normal tissue are quite flat, consistent with endocrine-suppression due to the absence of TSH.

Thyroid carcinoma does occur in the cat but is probably only responsible for about 2% of cases of hyperthyroidism; benign disease is much more common. However, given that thyroid carcinomas in cats are relatively well differentiated, with low metastatic potential, and often respond well to conventional treatments for hyperthyroidism, it is possible that this malignancy is not always recognised since tissue from hyperthyroid cats is rarely submitted for histopathology.

Grave's Disease is not the Underlying Mechanism for Hyperthyroidism in Cats

The most common cause of hyperthyroidism in humans, at least in iodine replete areas, is Graves' disease. Graves' disease is an autoimmune disorder where autoantibodies bind to the TSH receptor within thyroid tissue resulting in thyroid hormone production and secretion. The thyroid stimulating immunoglobulins (TSIs) that are produced in Graves' disease would be expected to stimulate all thyrocytes within the tissue resulting in diffuse hyperplasia, not the multi-nodular appearance found in the goitres of hyperthyroid cats. The serum from cats with hyperthyroidism has also not been consistently shown to contain TSIs (although there were some initial reports that suggested that TSIs were present, this was not replicated in later studies); serum from hyperthyroid cats did not stimulate cAMP production by cultured thyroid cells.[3,4] However, since the cultured cells were from rats it was suggested that this could be due to species-differences in receptor structure. Subsequent experiments, however, cloned and expressed the feline TSH receptor and still failed to demonstrate the presence of any TSIs within feline serum, while serum samples from humans with Graves' disease did demonstrate agonistic activity.[5] Taken together these studies seem to provide conclusive evidence that Graves' disease is not the underlying cause of hyperthyroidism in cats.

Isolated studies in both cats and humans have suggested that other autoantibodies could be present in patients with toxic multinodular goitre with the antibodies acting to stimulate thyroid growth but not hormone production.[6] It is postulated that this could result in the transformation of normal thyroid tissues into multinodular goitre. Although never entirely disproven this hypothesis has not found widespread support.

Thyroid autonomy

Thyroid tissue from cats with hyperthyroidism continues to grow in a nodular histological pattern and to produce hormone following transplantation into nude mice. This process occurs regardless of whether mice are supplemented with thyroxine (to suppress TSH) or given methimazole (to increase TSH).[7] Consistent with these xenotransplantation studies adenomatous cells from hyperthyroid cats continue to grow and function when cultured *ex vivo* in TSH-free medium. This contrasts with normal thyroid tissue for which significant incorporation of iodine only occurs after stimulation of the cells with TSH.[8] These results are similar to those reported from experiments with adenomatous thyroid tissue obtained from humans.

Sub-Clinical Hyperthyroidism

Not all patients with MNG are clinically or biochemically hyperthyroid ('toxic'). Both euthyroid and toxic MNG are characterised by a mixture of hypo-functioning, normal-functioning and hyper-functioning nodules within the thyroid; in humans this classification has been based on the pattern of iodine uptake on scintigraphy. In general terms about 85% of nodules are 'cold', 10% are 'normal' and 5% are 'hot', although of course this balance is shifted towards 'hot' nodules in patients that are hyperthyroid. Autoradiographs of thyroid tissue obtained from hyperthyroid cats treated by thyroidectomy shortly following injection of ^{125}I show that the radioiodine incorporation into nodules varies considerably suggesting that their activity is similarly heterogeneous.[7] There are also several clinical reports of cats with palpable goiter,[9,10] or histological changes in their thyroid glands,[9,11] that are not (or not yet) hyperthyroid.

The occurrence of a clinical syndrome of sub-clinical hyperthyroidism is a well-accepted in human endocrinology, although whether or not these patients should be widely screened for, and/or treated, remains controversial.[12] Sub-clinical hyperthyroidism is a biochemical diagnosis applied to patients with normal (free) T4 concentration but sub-normal TSH. In spite of the name these patients are not always symptomless; sub-clinical hyperthyroidism has been found to increase the risk of cardiac arrhythmias and osteoporosis, as well as development of overt hyperthyroidism over time.[13] Sub-clinical hyperthyroidism due to toxic MNG is particularly prevalent in elderly people living in iodine deficient regions.

There is some evidence for a state of sub-clinical hyperthyroidism in cats. In one study histological features of MNG were scored in thyroid glands collected from 4 groups of cats at post-mortem; young cats and hyperthyroid cats provided negative and positive control groups respectively, and euthyroid senior (>7years) cats were divided into 2 groups, those with undetectable TSH (akin to a state of sub-clinical hyperthyroidism) and those with detectable TSH (euthyroid group). Cats in the group with undetectable TSH had a higher frequency of nodules, a greater percentage of abnormal thyroid tissue, and a higher overall histopathological grade than cats with detectable TSH.[11] It has also been shown that cats with undetectable TSH are more likely to develop overt hyperthyroidism during follow-up than cats with detectable TSH. When a group of 104 senior (≥ 9 years), euthyroid cats were followed for a year hyperthyroidism was diagnosed in 40% of those with undetectable TSH, compared with only 1.4% of those with detectable TSH, at the study outset.[14] There was no difference in the total T4 measurements in the two groups at baseline.

Abnormalities in Signalling Pathways

Constitutive activation of the cAMP signalling pathway is widely accepted as the biochemical driving force in the development of most cases of thyroid autonomy in humans. This occurs most frequently due to somatic mutations within either the TSH receptor itself, or less frequently within the G-protein alpha (Gαs) subunit that stimulates adenylate cyclase. These mutations are found in the hyper-functional nodules (either adenomas or within MNG) and are associated with 'toxicity' rather than the hyperplastic changes; mutations are not usually identified in non-functional nodules.[15] Mutations have also been identified within the thyroids of cats with hyperthyroidism. In one study where DNA was extracted from 134 nodules in 50 cats, 56% of the cats had at least one mutation in the TSH receptor and 34% of cats had 2 or more mutations identified in different nodules.[16] The functionality of the individual nodules was not tested. In addition, a separate study identified mutations in Gαs subunits in 4/10 hyperthyroid cats.[17]

Expression of G-proteins in tissues from euthyroid and hyperthyroid cats has also been studied. In hyperthyroid tissue it was found that while expression of G_s proteins was not altered, there was decreased expression of G_i protein,[18] specifically of the $G_{i\alpha2}$ isoform.[19] This may result in decreased inhibition of adenylate cyclase activity with a consequent increase in hormone production and/or stimulation of nodule growth. Other work by the same group has shown that TSH binding stimulates GTP binding in a similar concentration-dependent manner in euthyroid and hyperthyroid cats although the basal activity of the hyperthyroid cell membranes was greater.[20] It was not determined whether any of the components of the signalling pathway were mutated in these expression studies.

The frequent occurrence of somatic mutations in thyroid tissue, particularly hyperplastic thyroid tissue, is thought to relate to the vulnerability of these cells to mutagenesis due to the highly oxidising environment created by the generation of hydrogen peroxide and reactive oxygen species in the process of iodinating thyroglobulin. This oxidative stress, together with the generally long life-span of thyroid cells, results in a high rate of acquired mutations, not all of which are of clinical consequence. Overall the mutation rate of thyroid tissue is thought to be 8-10 times greater than that of other tissues.[21]

Thyroid hyperplasia

It is important to appreciate that thyroid autonomy and thyroid hyperplasia are inter-related, in that they commonly occur in the same individual, but they are not the same thing. In humans it is generally believed that somatic mutations conferring thyroid autonomy occur in tissue that is already hyperplastic. The best characterised mechanism for development of thyroid hyperplasia in humans is iodine deficiency. The potential role of iodine deficiency in the etiopathogenesis of feline hyperthyroidism will be considered in a companion lecture (Ingrid van Hoek: 'Dietary links to feline hyperthyroidism') so will not be discussed in detail here. Regardless, it seems likely that some factor(s) results in widespread thyroid hyperplasia in cats and iodine deficiency is only one potential mechanism; other putative goitrogens are endocrine disruptors such as polybrominated diphenyl ethers (PBDEs),[22-25] dietary flavonoids,[20] and bisphenol A.[26] Cats have significant differences in their xenobiotic detoxification mechanisms compared to other mammals which may make them particularly vulnerable when exposed to some of these products.

PBDEs are used in the manufacture of numerous products but in particular they are added to fabrics intended for domestic use because of their flame retardant properties. With wear PBDEs may be leached from the surface of these materials into house dust. Cats have been demonstrated to have 10-100-fold higher concentrations of PBDEs in their serum than humans.[23,24] It is thought this is because cats ingest these chemicals from house dust while grooming their fur. In support of this idea the concentrations of PBDEs have been shown to be higher in pet cats than feral cats.[25] Additionally, the concentration of PBDEs in cat food has been shown to be relatively low.[25] PBDEs have been implicated in the pathogenesis of other hyperplastic feline endocrinopathies, most notably acromegaly.[27]

An alternative mechanism for thyroid hyperplasia that does not appear to have been given great consideration is the production of a TSH-mimetic. This could not only result in thyroid hyperplasia but would also explain the apparently low prevailing concentration of endogenous TSH in most cats which results in its limited clinical utility as a diagnostic test, at least when used in isolation. Of course, an alternative explanation for the low measured concentrations of TSH in cats is that the (canine) assay performance is poor when used with feline samples.[28]

Development of MNG in humans does not appear to be entirely due to environmental factors although these are undoubtedly important; genetics also plays a role. There is familial clustering of goitres in both iodine replete and deficient regions and monozygotic twins have a higher concordance rate than dizygotic twins.[29] Studies of the genetic basis for hyperthyroidism in cats have not been published but several epidemiological studies have found that certain breeds (Siamese and Burmese for example) are relatively protected from the development of hyperthyroidism.[30-32]

Epidemiological Studies of Feline Hyperthyroidism

There have been several epidemiological studies of feline hyperthyroidism from many different geographic regions. These studies could give insights into disease mechanisms. In all these studies hyperthyroidism has been found more commonly in old cats (typically over 10 years).[30,33] This is in keeping with what has been found with the development of MNG in humans. In some studies a slight female predisposition has been noted,[31,34] but this has not been a consistent finding.[30,31] Dietary risk factors for development of hyperthyroidism have also been identified but these will be discussed in the companion lecture. Other possible risk factors are related to the cats' lifestyle and include factors such as litter box use, living indoors and exposure to insecticides.[30,32,35] Associations with these factors has been inconsistent, perhaps due to differing relevance of the risk-factors in different geographical locations.

One factor that has been shown to be relatively protective against the development of hyperthyroidism in several studies is being pure-bred,[30,34] or more specifically being of the Siamese breed.[31,32,34] A recent, large, study that was predominantly based on cats diagnosed in first opinion practice was able to also show protection in the Burmese and Persian breeds.[31] A common feature of the breeds that have been found to be at reduced risk of hyperthyroidism is that they are color-pointed. It has been proposed that a potential mechanism for protection could be that cats of these breeds utilise less tyrosine (or its precursor phenylalanine) in making melanin for their coats, so more is available for the manufacture of thyroid hormones.[35] If this were found to be true it would then point to a relative deficiency in these amino-acids as a mechanism for development of thyroid hyperplasia. There is evidence that these amino-acids are lacking in some cat foods, resulting in lightening of cats' coats.[36] A preliminary study comparing concentrations of these amino-acids in hyperthyroid and euthyroid cats, and also in color-point and non-color-point animals did not find a difference, however, although this could have been because the study was relatively under-powered.[35]

REFERENCES

1. Wakeling J, Congress Proceedings: 15th ECVIM-CA Congress; 2005:220. 2. Dias Neves R, Horspool LJI. Congress Proceedings. 24th ECVIM-CA Congress; 2014. 3. Peterson ME, *et al.* Vet Immunol Immunopathol 1987;16:277-282. 4. Kennedy RL, Thoday KL. Autoimmunity 1989;3:317-318. 5. Nguyen LQ, *et al.* Endocrinology 2002;143:395-402. 6. Brown RS, *et al.* Thyroid 1992;2:125-130. 7 Peter HJ, *et al.* J Clin Invest 1987;80:491-498. 8. Peter HJ, *et al.* Thyroid 1991;1:331-338. 9 Norsworthy GD, *et al.* J Feline Med Surg 2002;4:139-143. 10. Paepe D, *et al.* J Feline Med Surg 2008;10:558-565. 11. Wakeling J, *et al.* Thyroid 2007;17:1201-1209. 12. Biondi B, *et al.* European *thyroid* journal 2015;4:149-163. 13. Das G, *et al.* Clinical Endocrinology 2012;77:146-151. 14. Wakeling J, *et al.* J Vet Intern Med 2011;25:1057-1065. 15. Derwahl M, Studer H. Trends in Endocrinology & Metabolism 2002;13:23-28. 16. Watson SG, *et al.* J Endocrinol 2005;186:523-537. 17. Peeters ME, *et al.* Thyroid 2002;12:571-575. 18. Hammer KB, *et al.* Am J Vet Res 2000;61:874-879. 19. Ward CR, *et al.* Am J Vet Res 2005;66:1478-1482. 20. Ward CR, *et al.* Am J Vet Res 2010;71:643-648. 21. Maier J, *et al.* Endocrinology 2006;147:3391-3397. 22. Chow K, *et al.* Vet Rec 2014;175:433-434. 23. Chow K, *et al.* Environmental research 2015;136:173-179. 24. Dye JA, *et al.* Environ Sci Technol 2007;41:6350-6356. 25. Mensching DA, *et al.* Journal of toxicology and environmental health 2012;75:201-212. 26. Kang JH, Kondo F. Res Vet Sci 2002;73:177-182. 27. Dirtu AC, *et al.* Environment international 2013;57-58:60-67. 28. Rayalam S, *et al.* Dom Ani Endocrinol 2006;30:185-202. 29. Hansen PS, *et al.* J Clin Endocrinol Metab 2004;89:2071-2077. 30. Wakeling J, *et al.* JSAP 2009;50:406-414. 31. Stephens MJ, *et al.* Vet Rec 2014;175:458. 32. Kass PH, *et al.* J Vet Intern Med 1999;13:323-329. 33. Edinboro CH, *et al.* J Am Vet Med Assoc 2004;224:879-886. 34. Olczak J, *et al.* Multivariate analysis of risk factors for feline hyperthyroidism in New Zealand. N Z Vet J 2005;53:53-58. 36. Sabatino BR, *et al.* J Vet Intern Med 2013;27:1049-1055. 37. Yu S, *et al.* JSAP 2001;42:176-180.

Dietary Links to Feline Hyperthyroidism

Ingrid van Hoek, DVM, PhD
Aimargues, France

INTRODUCTION

Excessive production and secretion of thyroid hormones in feline hyperthyroidism (HT) is caused in 98% of cases by benign adenomatous hyperplasia of the thyroid gland.[1] The role of diet in etiology and treatment of feline hyperthyroidism should be regarded differently.

DIET IN ETIOLOGY OF FELINE HYPERTHYROIDISM

The etiology of HT is likely to be multifactorial and since the first description of HT in 1979,[2] several studies were published reporting food-associated risk factors for HT[3-11]. Prevalence of HT differs among different regions[4,12] and the clinical presentation of HT has shifted over the years. This might suggest different etiology between regions and over time.

Available Scientific Evidence for Food-Associated Factors

Scientific literature comprises 9 publications[4-11,13] investigating food-associated factors in HT. To date, no prospective study on HT has been performed in cats, and therefore no true causative factor has been identified. All studies investigating risk factors for HT are retrospective case-control studies with data collection mostly performed using a questionnaire. Case-control studies are prone to bias, and the exposure in the control group should be representative of the population. Retrospective studies are less likely to identify etiological factors in diseases with a low (< 10%) prevalence like is the case for HT.

Three publications only qualitatively investigated preferred food flavors instead of quantifying specific food-associated factors.[8,9,13] One publication found the risk for feeding canned food to be non-significant.[5] Five publications associated the quantitative aspect of feeding canned food with increased risk for HT,[4,6,7,10,11] but those studies showed important limitations in their design: controls included sick cats,[4,6,11] cases outnumbered controls,[7,10] cases and controls differed in age,[6,10,11] or diet information did not reflect diet fed at time HT developed[7,11].

The causal component of an association is enforced by meeting more criteria established by Bradford-Hill: strength, temporality, consistency, specificity, dose-response, plausibility, coherence, experiment, and analogy.[14] An increased risk from feeding canned diet on HT meets only 3 of those criteria (strength, temporality, consistency).

Hypothesized Dietary Factors

From the available publications there is insufficient evidence to conclude canned diet is a food-associated factor in etiology of HT. Nonetheless, this has led to hypotheses to explain diet in etiology of HT. For instance bisphenol A (BPA), polybrominated diphenyl ethers (PBDEs), isoflavones, selenium, and iodine contents are suggested to cause a decreased serum thyroid hormone concentration and stimulate secretion of thyroid-stimulating hormone (TSH), leading to thyroid overstimulation and hypertrophy of thyroid follicular cells.[15] Neither this mechanism nor parts of it have been investigated or proven to be present in cats developing HT.

Bisphenol A (BPA)

Bisphenol A (BPA) is a monomer that is used to make polycarbonates in plastics. BADGE is the diglycidyl ether of BPA. BADGE epoxy resins are used in epoxy lacquer-coated food cans. BPA has structural similarity to thyroid hormones and might act as a thyroid hormone receptor antagonist.[16] To date, no study has been carried out to show BPA in blood or tissue of HT cats. BPA has been detected in canned cat food in Japan[17] and in the USA[18] at a concentration ranging from below the detection limit to well below the current EU maximum.

Polybrominated Diphenyl Ethers (PBDE)

Polybrominated diphenyl ethers (PBDEs) are flame retardants and known endocrine disruptors due to the structural similarities of various PBDE isomers with thyroid hormone. No case-control studies have been carried out to show an association between PBDE exposure from diet and HT. It is suggested that ingestion of household dust is the primary source for cats. Dust PBDE, and not serum PBDE, was significantly correlated to serum T4.[19] A recent study showed that PBDE is higher in serum of HT cats than healthy cats, but the study had important limitations in the control group because of an age difference and unclear health status.[20]

Isoflavones

In vitro studies demonstrated that isoflavones from soy can act as alternative substrate for iodination.[21-23] To date, there is only one short-term study in healthy cats evaluating the effect of soy in diet on thyroid function.[24] A soybean meal and soy hydrolysates-containing diet caused increased total T4 and free T4 concentrations compared to a control diet. However, the study had several limitations like a difference in content of iodine between diets, the isoflavone concentration in the soy-containing diet was much higher than levels reported in commercial feline diets,[25,26] and the iodine content of the soy-containing diet was higher than the legal maximum concentration. When purified sources of soy, like soy isolates and soy hydrolysates, are used, they contain limited levels of natural isoflavones. Data suggest that a goitrogenic effect of soy is possible when the diet is deficient in dietary iodine.

Selenium

Selenium is known to be implicated in thyroid hormone metabolism, and selenium deficiency is hypothesized to cause increased synthesis of T3 and T4 by the thyroid gland and to decrease conversion of T4 to T3 in peripheral tissues and increased TSH stimulation. Only 2 studies have investigated the effects of selenium in the diet on thyroid function in cats. A significant increase in serum total T4 and decrease in serum total T3 in kittens fed a low-selenium diet for 8 weeks was found,[27] but another study found no significant difference between cats from different regions, or between euthyroid and hyperthyroid cats from regions with high incidence of HT. As for flavonoids, it remains unclear what the role of selenium is in the multifactorial etiopathology of HT.[12]

Dietary Iodine

Dietary iodine has been proposed as a causal factor of HT in a variety of ways: fluctuant iodine intake as well as insufficient or excessive iodine intake would contribute to the development of thyroid disorders.

Iodine recommended intake has varied over the years[28] and wide ranges in iodine content have been found in commercial cat foods[3,29-31]. The recommendations from FEDIAF were below NRC until 2012, when FEDIAF increased the minimum recommended iodine concentration to levels equal to the NRC recommendations.[32,33] No safe upper limit has been defined by the NRC however. European authorities established a legal maximum for iodine when supplemented in pet foods.

A recent retrospective case-control study showed that cats fed commercial diets, without iodine supplementation according to listed ingredients, were 4 times more likely to develop HT compared to cats eating iodine-supplemented cat foods.[34] Humans with endemic goiter due to iodine deficiency, will develop increased thyroid hormone synthesis in the autonomous nodules (iodine-induced hyperthyroidism) when small increases in iodine intake occur (Jod-Basedow effect).[35] Further research is needed to investigate whether long-term inadequate iodine intake can cause pathological changes in the thyroid gland due to chronic stimulation by TSH, with the development of HT after switching to iodine-rich diets, as is the case in humans.

High iodine ingestion has also been suggested as a cause of HT, from extrapolation in humans. When diets with different levels of iodine were fed to healthy cats for 2 weeks, serum thyroid hormone concentrations responded inversely to dietary iodine;[30,36] however, when diets with low and high iodine were fed for 5 months no significant difference in serum free T4 was noted[37]. An increased risk for fish-flavored food has been suggested in epidemiological studies.[4,8] A recent study showed that median iodine intake on a dry matter basis and estimated daily iodine intake were not significantly different for foods with fish ingredients compared to foods without them.[31]

DIET FOR CATS WITH FELINE HYPERTHYROIDISM

The role of diet during treatment of HT can be on one hand directed at treating the disease (Table 1), and on the other hand having a supportive role. Once a cat is treated reversibly by anti-thyroid medication, or irreversibly by thyroidectomy or the gold standard treatment of radioactive iodine, the effects of hyperthyroidism on the metabolic state of the cat can be of less importance and a diet formulated for senior cats or diets addressing specific dietary needs like in the case of chronic renal failure are suited.

Table 1. Overview of clinical studies testing efficacy of a diet with restricted iodine content

References	[38]	[39]	[40]	[41]
Study design	Open label, no control	Prospective, blinded, controlled	Retrospective, no control, based on medical records	Prospective, no control
I level in diet	0.19 ppm DM dry diet 0.14 ppm DM wet diet	Mean 0.36 ppm DM (0.27–0.6) dry diet Mean 0.21 ppm DM (0.14–0.27) wet diet	0.2 ppm dry and wet diet	0.2 ppm
Number of cats included (n cats finalising)	225 (68)	18 (12) test diet, 15 (10) control diet	49 (47)	8 (8)
Weeks of follow-up	0-4-8	0-3-6-12	9 (60 days)–26 (180 days)	4–8
% of cats in reference range of TT4 at follow-up	8 weeks: 75%	12 weeks: 50%	60 days: 42% 180 days: 83%	8 weeks: 87.5%
Bodyweight	NA	NA	No change	No change
Body condition score	No change	no change	NA	NA
Heart rate	No change	NA	No change	NA
Clinical symptoms	Resolved	NA	NA	Resolved
Non-thyroidal illness (NTI)		11% (2 of 18)	25% (12 of 47)	
Palatability	Score 3.7/5 (4w), 3.79/5 (8w)	NA	NA	NA
Compliance	Score 2.75/4 (4w), 2.58/4 (8w)	Poor for 2 of 18 cats	Poor for 1 of 49 cats	Poor for 1 of 8 cats
Study limitations	- Different analytical methods for T4	- Prototype diet with differing levels of iodine	- Cats with NTI - Different analytical methods	

	- Open label - Data collected by owner and veterinarian		for T4 with single reference range - Different number of measures for cases	

NA: Not assessed.

CONCLUSION

Epidemiological studies investigating environmental and nutritional exposures suggest HT to be a complex multifactorial disease. Diet and feeding canned food have been associated with HT in retrospective studies which have their limitations in design. No hypothesized risk factors to explain etiology of diet in HT have been confirmed to induce HT yet. Lifelong prospective longitudinal studies based on the strongest epidemiological evidence to date are needed to investigate the role of food-associated risk factors in the development of HT.

REFERENCES

1. Meeking SA. *Vet Clin North Am Small Anim Pract*. 2005;35(3):635–653.
2. Peterson ME, Johnson JG, Andrews LK. In: Proceedings of the American College of Veterinary Internal Medicine Forum;1979:108.
3. Mumma RO, *et al. Am J Vet Res*. 1986;47(7):1633–1637.
4. Scarlett JM, Moise NS, Rayl J. *Prev Vet Med*. 1988;6:295–309.
5. De Wet CS, *et al. J Feline Med Surg*. 2009;11(4):315–321.
6. Edinboro CH, *et al. J Am Vet Med Assoc*. 2004;224(6):879-886.
7. Kass PH, *et al. J Vet Intern Med*. 1999;13(4):323–329.
8. Martin KM, *et al. J Am Vet Med Assoc*. 2000;217(6):853–856.
9. Olczak J, *et al. N Z Vet J*. 2005;53(1):53–58.
10. Wakeling J *et al. J Small Anim Pract*. 2009;50(8):406–414.
11. Gojska-Zygner O, Lechowski R, Zygner W. *Bull Vet Inst Pulawy*. 2014;58:267–271.
12. Foster DJ, *et al. Am J Vet Res*. 2001;62(6):934–937.
13. Gallagher B, Mooney CT. In: Proceedings of the American College of Veterinary Internal Medicine forum;2013. Seattle, WA.
14. Thygesen LC, Andersen GS, Andersen H. *J Epidemiol Community Health*. 2005;59(6):512–516.
15. Peterson M. *J Feline Med Surg*. 2012;14(11):804–818.
16. Moriyama K, *et al. J Clin Endocrinol Metab*. 2002;87(11):5185–5190.
17. Kang JH, Kondo F. *Res Vet Sci*. 2002;73(2):177–182.
18. Schecter A, *et al. Environ Sci Technol*. 2010;44(24):9425–9430.
19. Mensching DA, *et al. J Toxicol Environ Health A*. 2012;75(4):201–212.
20. Norrgran J, *et al. Environ Sci Technol*. 2015;49(8):5107–5114.
21. Divi RL, *et al. Biochem Pharmacol*. 1997;54(10):1087–1096.
22. Divi RL, Doerge DR. *Chem Res Toxicol*. 1996;9(1):16–23.
23. Doerge DR, Sheehan DM. *Environ Health Perspect*. 2002;110(Suppl 3):349–353.
24. White HL, *et al. Am J Vet Res*. 2004;65(5):586–591.
25. Bell KM, *et al. N Z Vet J*. 2006;54(3):103–108.
26. Court MH, *et al. Am J Vet Res*. 2002;63(2):181–185.
27. Yu S, *et al. J Anim Physiol Anim Nutr (Berl)*. 2002;86(1–2):36–41.
28. Edinboro CH, *et al. J Feline Med Surg*. 2010;12(9):672–679.
29. Johnson LA, *et al. N Z Vet J*. 1992;40(1):18–20.
30. Ranz D, *et al. J Nutr*. 2002;132(6 Suppl 2):1751S–1753S.
31. Edinboro CH, *et al. J Feline Med Surg*. 2013;15(8):717–724.
32. *Nutrient Requirements of Dogs and Cats*. 2006, Washington: National Research Council.
33. *The Reference Document on Cat and Dog Nutrition*. 2012, Bruxelles: FEDIAF.
34. Edinboro CH, *et al. Thyroid*. 2004;14:722.
35. Woeber KA. *Medical Clinics of North America*. 1991;75(1):169–178.
36. Tarttelin MF, *et al. N Z Vet J*. 1992;40(2):66–68.
37. Tarttelin MF, *et al. J Nutr*. 1994;124(12 Suppl):2577S–2578S.
38. van der Kooij M, *et al. J Feline Med Surg*. 2013;16(6):491–498.
39. Fritsch DA, *et al. Int J Appl Res Vet Med*. 2014;12(1):24–32.
40. Hui TY, *et al. J Vet Intern Med*. 2015;29(4):1063–1068.
41. Scott-Moncrieff JC, *et al. J Vet Intern Med*. 2015;29(5):1322–1326.

Gene Therapy for Lung Disease

Alisha M. Gruntman, DVM, ACVIM (LAIM)
Worcester MA

INTRODUCTION

After many decades of slogging along with minimal successes and many set-backs, gene therapy has finally reached a tipping point in which a successful therapy was just licensed in Europe and several are on the brink of clinical use within the US. While gene therapy approaches are rapidly progressing for many diseases affecting multiple body systems, gene therapy for lung diseases still remain a significant challenge. In this proceedings I will outline the basics of gene therapy approaches and elucidate what makes the lung such a difficult organ to target.

The overall premise behind gene therapy is relatively simple, although the development of a clinical gene therapy product can be incredibly complex. Gene therapy often employs viral vectors to deliver a therapeutic gene to a cell-type of interest. The vector is able to use the cellular machinery within the target cell to induce transcription of that gene product and, where applicable, translation of a therapeutic protein. In order to make a virus all or nearly all encoding viral genes are removed so that infected cells do not produce viral proteins that could elicit an immune response or cause disease (**Figure 1**). This approach for therapeutic gene delivery can be applied both when a therapeutic protein is needed (a protein coding gene is delivered) as well as when a therapeutic nucleotide is needed (such as a micro-RNA).

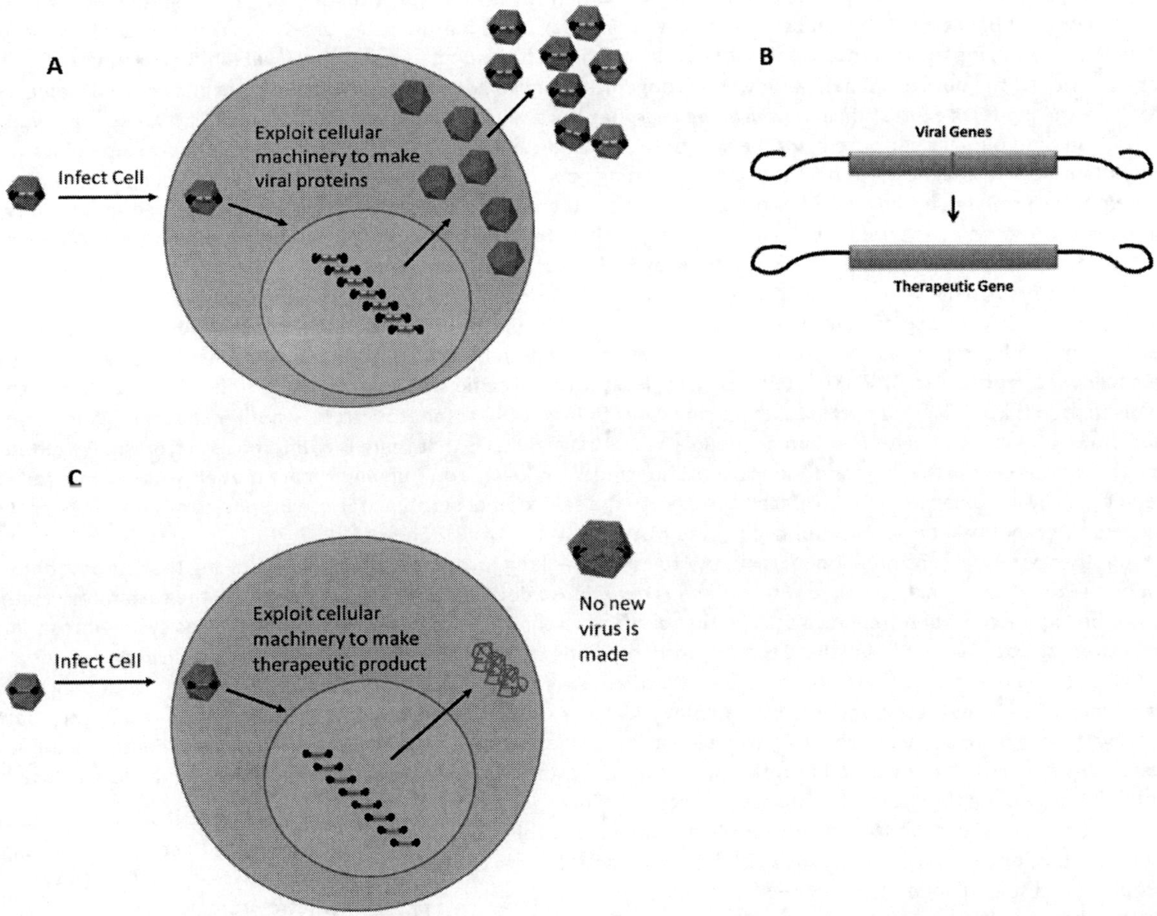

FIGURE 1. A) Wild type viruses exploit the cellular machinery to copy their genomes, make viral proteins, package the genomes in capsid then exit the cell. **B)** In order to make a viral vector the viral genes are removed from the viral genome and replaced by a therapeutic gene of interest. **C)** Vector viruses use the cellular machinery to express the therapeutic gene without making any viral proteins

Several factors are critical to a successful gene therapy. The disease or condition must have a genetic target (preferably a single target) that can either be augmented or knocked down/silenced. A good example of this is cystic fibrosis, in which the disease is caused by mutations in the CFTR gene. This is a disease in which augmentation via production of a normal CFTR protein would be therapeutic. Diseases with multiple potential gene targets, such as inflammatory conditions like sepsis or asthma, would be significantly more challenging to treat via gene therapy. The second factor necessary for successful viral mediated gene therapy is the availability of a viral vector and a route by which that vector can deliver the therapeutic gene to the cell-type or organ of interest. There a several viruses that have been employed for use in gene therapy. The most common are adeno-associated viruses (AAV), retroviruses, and adenoviruses. In order to be successful in treating a specific cell type or organ, the virus has to not only transduce (infect) but also persist for the desired length of time. This can be particularly challenging in dividing cell types such as lung epithelium where long term expression is often needed but not obtained if the therapeutic gene of interest is not appropriately integrated into the cellular genome following vector delivery. The final main factor in developing a gene therapy is an *in-vitro*, or preferably, and *in-vivo* disease model in order to provide proof of concept of both efficacy and safety prior to proceeding to patients.

Factors Necessary for a Successful Viral Gene Therapy

- Single gene disorder
- Available viral vector and delivery route to target the target cell or tissue
- An in-vitro or in-vivo model of disease for therapeutic testing

Canine and feline naturally occurring animal models of human diseases have often been employed in final stage (after cell and mouse work) gene therapy testing prior to proceeding to human patients. These studies have provided strong proof-of-concept evidence that gene therapy strategies in veterinary species are not only feasible but an effective and long-term therapy for previously fatal genetic conditions.

LUNG GENE THERAPY

ADENO-ASSOCIATED VIRUS (AAV)

The most commonly used viral vector for clinical gene therapy is AAV. This virus was originally discovered as a contaminant of adenoviral preps, but is actually a member of the *Parvovirinae* family. It has many factors that make it an ideal vector for gene therapy delivery, including the fact that it only integrates into the host genome at a very low level (limiting chances of turning on an oncogene or altering the host germline), has low immunogenicity (unlike adenovirus), is able to transduce both dividing and non-diving cells, with long-term expression in non-dividing cells, has the ability to target a wide range of tissues, and has a very strong safety profile in both humans and veterinary species (both as a wild-type virus and a vector). Unfortunately some of its strengths are also some of its biggest weakness, for one, the non-integrating nature makes long-term expression in dividing cells challenging and while the immune profile is low compared to other viral vectors there is still exist pre-existing immunity to some capsid types within both humans and veterinary species that can make dosing challenging. The other challenging factor when using AAV for gene therapy is its small size, which limits the gene size that can be packaged and delivered.

AAV was first used in a human clinical trial to treat cystic fibrosis in 1996[1]. This trial demonstrated safety but the problem of sustained expression in the face of a dividing cell population was encountered. Again, this is because AAVs don't integrate into the host genome, but rather persist as an extrachromosomal element within the cell[2]. Because a robust neutralizing antibody response develops following exposure to AAV, repeated dosing (at least with the same serotype) is not feasible. The inability to express long-term in dividing cells is the biggest hurdle to employing AAVs to target the respiratory track. Another challenge, not unique to the lung, but showing that the value of non-human models has its limitations, is that there is a difference in tropism for different AAV serotypes to various cells in the respiratory track in murine and even lower non-human primate models when compared to what has been seen clinically in human patients[3]. This can complicate the selection of serotype for use in species other than mice. Currently the AAV serotypes that best target the murine lung are rAAV5, rAAV6, rAAV6.2 and rAAV9[4].

Another potential consideration in using AAV to target the lung is recent work in mice showing that current or recent infection with *Bordetella bronchiseptica* or *Pseudomonas aeruginosa* decreased AAV gene transfer to the respiratory epithelium[5]. This should be considered when treating a disease that can lead to chronic respiratory infections, such as cystic fibrosis. In order to overcome these shortcomings of AAV, there is much work on-going to improve long-term expression in dividing cells or allow for repeat dosing. In the mean-time there are still situations where AAV targeting of the respiratory epithelium with only short-term expression might be an advantage. The first is to employ AAVs as a means of delivering an antibody to the respiratory tract that could protect from a respiratory disease. This method was used to express an antibody known to broadly neutralize influenza via intranasal delivery of rAAV9 to mice and ferrets[6]. The results indicated complete protection against a median lethal dose of five different influenza isolates in mice and complete protection following a lethal dose of two strains of influenza in ferrets. The authors propose that this may be a means of developing a vaccination quickly in the face of an influenza or other respiratory disease pandemic.

Another potential use of short-term expressing AAV gene therapy is in the face of an acute respiratory insult, such as that seen in acute lung injury (ALI). It has recently been shown that it is possible to attenuate the ALI induced by LPS via inhibition of nuclear factor kappa betta transcription factor within the lung[7]. They surmise that this approach might also be applicable to other disorders such as COPD, asthma, or ARDS.

Viral Gene Therapy Vectors for the Lung

- Adeno-Associate Virus (AAV)
- Retroviruses
  - Lentivirus
    - HIV
    - FIV
- Adenovirus

RETROVIRUSES

Retroviruses are RNA viruses that use cellular reverse transcriptase to produce a DNA genome that is then integrated into the host genome. This integration allows for long-term gene expression in dividing cell populations such as hematopoietic and epithelial cells. One of the first applications of a retrovirus was in the use of a gamma-retrovirus to treat children with X-linked severe combined immunodeficiency (SCID). While the gene therapy was successful in curing the T-cell deficiency and improving the NK and B-cell deficiency, 5/20 patients developed leukemia as a results of the retrovirus integrating in an oncogene. The immune correction however was sustained in 17 of the patients through the 13 year follow-up[8]. A subsequent trial modified the vector to decrease the dangerous integration patterns seen in the first trial and no cases of leukemia were seen. One of the downfalls of retroviruses in the lung is their poor ability to transduce terminally differentiated non-dividing cells, such a lung epithelium. The use of lentiviruses has improved transduction of terminally differentiated cells over previously used retroviruses. Both modified forms of human immunodeficiency virus (HIV) and feline immunodeficiency virus (FIV) have been used to target the lung.

ADENOVIRUS

Adenoviruses have been used extensively in the lung. Adenovirus is also non-integrating and will transduce both dividing and non-dividing cells, similar to AAV. An advantage of adenovirus over AAV is that is has a much larger gene packaging capacity, allowing packaging and delivery of much large genes. However, adenovirus has historically had limited use as a gene therapy vector because of the strong immune response to the vector that is elicited in the host. This can cause a severe immune reaction that clears the virally infected cells and terminates expression of the therapeutic gene. Adenovirus is also less efficient at targeting the respiratory track than other viral vectors.

SUMMARY

While many obstacles to successful pulmonary gene therapy have been overcome, there is still a long road ahead before we will be able to treat lung diseases successfully in both veterinary and human patients. Some of the short-acting AAV gene therapy applications, such as expressing antibodies in the airway as a fast acting vaccination may be applicable in the near future for veterinary use, particularly in the face of an outbreak situation. The use of a temporary expression of an inflammatory repressor locally in the lung may also have veterinary applications in the future.

REFERENCES

[1]Flotte TR, et.al Hum Gene Ther 1996;7:1145-1159. [2]Afione SA, et.al J Virol 1996;70:3235-3241. [3]Flotte TR, et.al Mol Ther 2010;18:594-600. [4]Gruntman AM et.al Current Protocols in Micro 2012;14D.2.1. [5]Myint M. et.al Hum Gene Ther 2014;25:966-976. [6]Limberis MP, et.al Sci Trans Med 2013;5(187). [7]MacLoughlin RJ, Hum Gene Ther 2015;26:36-46. [8]Cavazzana M, et.al Hum Gen Ther 2016.

Identification of Aeroallergens from Hay Associated with Recurrent Airway Obstruction Using Immunoblotting

Laurent L. Couetil, DVM, PhD, DACVIM (LAIM)
West Lafayette, IN

Recurrent airway obstruction or severe equine asthma is believed to result from an allergic reaction to inhaled molds present in moldy hay. Various species of fungi and fungal-like organisms have been implicated but most of the research has been focused on a handful of fungi abundant in moldy hay such as *Aspergillus fumigatus* and *Faenia rectivirgula*. However, whether these fungi or other species are responsible for disease exacerbation in naturally occurring disease is currently unknown. Furthermore, the difference in antigenic exposure between horses fed moldy versus good quality hay has not been reported. Exposure to potential respiratory irritants is most accurately estimated using personal sampling strategies in the breathing zone to avoid underestimation routinely associated with area sampling. We developed an immunoblotting technique to identify aeroallergens in hay dust collected in the breathing zone of horses in order to compare immunogenicity of individual fungi isolated from inhaled dust. We identified 31 different mold species in hay dust and most were found both in good quality hay as well as moldy hay. Only 4 fungal species resulted in unique immunogenic reactions in asthmatic horses.

We concluded that horses fed good quality or moldy hay are exposed to similar fungal species in their breathing zone but only a few fungal species result in distinctive immunogenic reactions in susceptible asthmatic horses.

Presence of Streptococcus in the Lower Airways

Sanni Hansen, DVM, PhD;
Mette Christoffersen, DVM, PhD; Julie Fjeldborg, DVM, PhD; Keith E. Baptiste, BVMS, PhD, DACVIM, DECEIM; Morten R. Petersen, DVM, PhD, DACT and Anders M. Bojesen DVM, PhD, ECPVS
Copenhagen, Denmark

INTRODUCTION

Equine respiratory diseases represent the second most common reason for lost training days in athletic horses, and consequently the second most common reason for usage of antimicrobials in horses (Traub-Dargatz and Dargatz, 2009). Typically, these lower airway infections occur within the first two years of the horse's life.

Many different bacterial types are responsible for the high incidence of equine lower airway diseases. However, Streptococcus equi subsp. zooepidemicus (S. zooepidemicus) a mucosal commensal organism, typically residing in lingual and palatine tonsular-like tissue, is the bacterium most frequently isolated from cases of equine lower airway infections and pleuropneumonia (Raphel and Beech, 1982; Sweeney et al., 1985, 1991; Yoshikawa et al., 2003). Generally S. zooepidemicus has been regarded as the most important secondary pathogen of the respiratory tract of horses following primary viral infections (Wood and Chanter, 1994; Allpress, 1997). Epidemiological studies have demonstrated significant associations between lower airway infections with S. zooepidemicus in racehorses and both inflammatory airway disease (Wood et al., 1993; Chapman et al., 2000; Wood et al., 2005) and coughing (Christley et al., 2001). Furthermore, S. zooepidemicus is also responsible for a variety of infections in several different species including humans, pigs and dogs (Chalker et al., 2003; Salasia et al., 2004; Eyre et al., 2010).

PERSISTENT INFECTION

Clinical lower airway disease is often sporadic and of short duration, but S. zooepidemicus has been associated with persistent infections (Salasia et al., 2004). However, it is not clear how the prolonged and repeated nature of respiratory disease in young horses could relate to S. zooepidemicus lower airway infections. It is a generel assumption is that young horses recover completely from S. zooepidemicus lower airway infections through immunological clearance in combination with veterinary treatments. However, could this assumption be challenged? Could S. zooepidermicus persist after lower airway infections in a ´dormant´state? Such assumptions are being proposed in human medicine. For example, persistence of group A streptococci have been found in cultured human respiratory cells (Spinaci et al., 2006). This could contribute to clinical problems (e.g. micro-pustule formation) such as relapsing infections and persistent carriage despite prolonged antimicrobial treatment.

Often several serovars of S. zooepidemicus can be isolated from the same horse (e.g. in tonsular-like tissue), but often only one clone of S. zooepidemicus are isolated from lower airway infections (Anzai et al., 2000). The mechanism by which S. zooepidemicus form and persist as invasive clones are yet to be found, but possible mechanisms such as mutation or regulation in virulence genes, gene deletion or acquisition of virulence factors have been suggested. Examples include surface anchored proteins such as SzP, a surface M-like protein with anti-phagocytic effect and the fibrinogen and plasminogen binding SzM (Velineni and Timoney, 2013).

In the endometrium of mares S. zooepidemicus can establish as a chronic subclinical latent infection by residing within the endometrial tissue, leading to infertility issues in mares as well as resistance to standard antimicrobial treatment (Petersen et al., 2015). Furthermore, strains of S. zooepidemicus isolated from the endometrium of mares with endometritis belonged to the same distinct subpopulation of S. zooepidemicus.

PERSISTER CELLS

Many bacteria are capable of forming persister cells. Persister cells are characterized as slow-growing or dormant sub-population of cells that are able to survive antimicrobial treatment, immune responses and other environmental stress factors (Maisonneuve and Gerdes, 2014). Persister cells resume growth and produce progeny cells genetically identical to the original cell. Persister cells together with planktonic cells can form biofilm; biofilm are defined by Burmøller et al. (2010) as "A coherent cluster of bacterial cells imbedded in a biopolymer matrix, which, compared with planktonic cells, shows increased tolerance to antimicrobials and resists the antimicrobial properties of the host defense". A recent study has shown that S. zooepidemicus are able to form biofilm in vitro and further up- and down-regulate certain protein encoding glycolysis and fermentation (Li et al., 2014). Persistence of S. zooepidemicus infection in the lower airway of the horse as a chronic infection, as persister cells and eventually as biofilm could go on to cause further problems later in the horse life, including poor performance and inflammatory airway disease. Therefore, the purpose of this study was to investigate if sub-epithelial S. zooepidemicus are present in lung biopsies from horses.

STUDY SUMMARY

Bronchiolar biopsies were obtained, under sterile conditions, from newly euthanized horses (n=4). Two horses were prior to euthanasia diagnosed with recurrent airway obstruction (RAO) and two horses were euthanized for unknown reasons. Samples were obtained from four different parts of the lungs, including the left and right cranio-ventral lung lobes and the left and right caudo-dorsal lung lobe.

The presence of group C streptococci was determined based on two methods: 1) Culture of biopsies on blood agar: B Biopsy material was streaked on blood agar plates and incubated at 37°C for 48-72 h. Colonies of S. zooepidemicus were identified by β-haemolysis, colony morphology (mucoid and non-pigmented) and Gram stain. 2) Immunohistochemistry (IHC): Formalin fixated lung biopsies (4%) were embedded in paraffin and cut into 5-μm-thick sections. Slides were permeabilized with tris buffered saline (TBS) and blocked with hydrogenperoxide (3%) and goat serum (5%) and washed with TBS and subsequently stained with primary antibody (Lancefield Group C Streptococci antiserum, Statens serum Institute, SSI Diagnostica, Copenhagen). The Dako EnVision®+ system-HRP (AEC+) kit (Dako) was used for detection and visualisation of primary antibody. Counter-staining was done with hematoxylin.

Using IHC, S. zooepidemicus could be visualized in the lung from the biopsies collected from the cranio-ventral parts of the lungs as small colonies of the bacteria, without changes in tissue morphology. In these likely chronic cases, S. zooepidemicus were localized primarily in distinct foci just below the luminal epithelium. Examination of lung tissue inoculated with S. zooepidemicus, strain A31S1 (positive control) revealed a similar bacteria appearance. Negative control, without strep-ab did not result in any signal.

Neutrophilic asthma in humans that responds poorly to treatment with glucocorticosteroids has been linked to chronic colonization by bacteria although the mechanism behind is lacking (Essilfie et al., 2012; Beigelman et al., 2014). Asthma in horses is a common disease among adult and old horses and is characterized by a neutrophilic inflammation in airway sputum. Further studies are needed in order to address if S. zooepidermicus can play a role in equine asthma.

REFERENCES

[1]Traub-Dargatz, J.L, Dargatz, D.A, Eq Vet Edu 2009;21,49.[2]Raphel C.F, Beech J, J Am Vet Med Ass 1982;181, 808. [3]Sweeney C.R, et al J Am Vet Med Ass 1985;187, 721. [4]Sweeney C.R, et al J Am Vet Med Ass 1991:198, 39. [5]Yoshikawa H, et al J Vet Med Sci 2003;65, 787. [6]Wood J.L.N, Chanter N, Eq Vet Edu 1994;28,177. [7]Allpress RG, Eq Vet Edu 1997; 9,68. [8]Wood J.L.N, et al Eq Vet J 1993;25,314. [9]Wood J.L.N, et al J Clin Microbiology 2005; 43,120. [10]Christley R.M, et al Vet Record 2001;148 99. [11]Chalker et al Vet Microbiol 2003;95(1-2),149. [12]Salasia et al J Vet Sci 2004;5(3),263. [13]Eyre et al Eur J Clin Microbiol Infect Dis 2010;29(12),1459. [14]Spinaci C, et al Pediatr Infect Dis J. 2006;25(10):880.[15]Anzai T et al Am J Vet Res 2000;61(2),162.[16]Velineni S, Timoney J.F, Clin Vaccine Immunol 2013;20(8),1181.[17]Petersen M.R, et al Vet Microbiol 2015;179,119.[18]Maisonneuve E, Gerdes K, Cell 2014;157(3),539.[19]Burmølle M et al FEMS Immunol Med Microbiol 2010 59(3), 324.[20]Li Yi et al Curr Microbiol 2014;69,227. [21]Beigelman A, et al Curr Opin Allergy Clin Immunol 2014;14.137. [22]Essilfie P.G, et al Thorax 2012;67,588.

How to Get the Most from Your Equine Bronchoalveolar Lavage (BAL) Sample

Julie Fjeldborg, DVM, PhD; Christina Birkholm Madsen, DVM; Charlotta Elinor Nevander, DVM; Keith E. Baptiste, BVMS, PhD, DACVIM, DECEIM and Sanni Hansen, DVM, PhD
Copenhagen, Denmark

INTRODUCTION

Bronchoalveolar lavage (BAL) has been for used for basic diagnostics in horses for many years but no international standardized operating procedure has been established in relation to performing the BAL, staining, cytospin contra pellet-smear, skilled versus non-skilled laboratory evaluators.

By comparison to human medicine, recommendations are established for standardization of all steps of the BAL procedure, which have been evaluated and published by the European Respiratory Society in 1999 (Haslam and Baughman, 1999) and the British Thoracic Society in 2001(British Thoracic Society Bronchoscopy Guidelines Committee, 2001).

This presentation aims to outline considerations for a standardized BAL procedure in horses for cytology in relation to: Inter – and intra-observer agreement, mast cell coloring, cytospin versus pellet-smear, temperature and storage time.

INTER- AND INTRA-OBSERVER AGREEMENT

Bronchoalveolar lavage cytology has always been thought of as an easy task, but several studies including inexperienced and experienced observer of BAL differential cell counts found variation in the inter-and intra-observer agreement. High agreement is important when results from the BAL cytology are interpreted as the level of laboratory experience can affect the final results of the differential cell count.

A study performed in our group including two experienced and three inexperienced observers evaluating BAL cytology slides (n=50) counting 200 cells showed high intra-observer agreement for the two experienced observers and adequate to poor intra-observer agreement for the three inexperienced observers. The observer agreement showed high inter-observer agreement between experienced observers, except for mast cells, and adequate inter-observer agreement between inexperienced observers.

An equine study found a high agreement among neutrophils, lymphocytes and macrophages and adequate for mast cells and eosinophils (Fernandez et al., 2013) and a human study (De Brauwer et al., 2002) also found a poor agreement for mast cells. These study shows that laboratory experience is an important factor when the counting is performed.

MAST CELL IDENTIFICATION

Toludine blue is the superior cell stain for accurate mast cell identification compared to other staining methods (Xaubet et al., 1991, Leclere et al., 2006). However, Toluidine blue only stains the metachromatic granules and does not stain other nucleated cells, indicating that it cannot stand alone as a cell stain for a complete differential cell count. The May-Grünwald Giemsa staining can be used for a full differential cell count as it stains both the metachromatic granules and the other nucleated cells.

A study from our group with the aim to evaluate the efficiency of two staining methods, May-Grünwald Giemsa and Toluidine Blue, on the percentage of mast cells was conducted on sixty-eight BAL samples and the results showed a significantly higher mean percentage of mast cells in the May-Grünwald Giemsa compared to the Toluidine Blue stain.

CYTOSPIN VERSUS PELLET-SMEAR

The cytospin technique has been the most common method used when BAL fluid is processed (De Klech and Pohl, 1989, Brauwer et al., 2000, Pickles et al., 2002b). The pellet-smear technique is a cheap and an easy alternative to the cytospin technique, which represents an alternative laboratory technique to diagnosing lower airway diseases in equine practice. Previous research on a smaller population of horses has shown that there is a difference in the differential cell count between cytospin preparations and pellet-smears, and there is no complete agreement among the studies (Pickles et al. 2002b, Dehard et al., 2008, Viscardi et al., 2013). Some of the important significant differences are a low lymphocyte cell count from cytospin preparations compared to smears (Pickles et al. 2002b) and a significantly higher neutrophil, mast cell and eosinophil counts from cytospin preparations compared to smears (Pickles et al., 2002b, Wenisch et al., 2001). One study found no significant difference between the two methods (Viscardi et al. 2013). The cell morphology has also been shown to be different where cells in smear preparations are more difficult to identify compared to cytospin (Dehard et al., 2008, Pickles et al., 2002b).

A new study from our group found a significant difference between cytospin preparations and pellet-smears. A significantly lower number of lymphocytes and a significantly higher number of macrophages, neutrophils, mast cells and eosinophils were found in cytospin preparations compared to pellet-smears. The cytospin technique is recommended over the pellet-smear technique due to a higher detection of neutrophils and mast cells and furthermore the cell morphology in cytospin preparations is better. The cytospin technique also has its biases considering the lower lymphocyte cell count that is important to remember when the results are interpreted.

EFFECT OF TEMPERATURE AND STORAGE TIME

The storage temperature and time of the BAL sample before analysis is important to take into considerations when differential cell counts are performed on BAL samples. This is due to possible cell morphological deterioration of the cells which might affect the final results of the differential cell count.

A study performed in our group investigated the effect of storage and time before analysis of the BAL sample. The study examined the effect at time (hours) zero, four, twelve and twenty-four, fourty-eight, and seventy-two hours after collection and the samples were stored by 6°C and 21°C. The differential cell count was stable within twenty-four hours when the sample was cooled. The most significant changes were found in mast cells where a decrease was detected when the sample was stored at room temperature for twenty-four hours. A significant cell morphological deterioration of the cells was found when BAL samples were stored at room temperature for twenty-four hours. BAL samples stored at 6°C showed significant changes in neutrophils and mast cells after twenty-four hours.

In conclusion, the study and other studies clarify that BAL samples should be cooled immediately after collection and analyzed within twenty-four hours (McGorum and Dixon 1994, Pickles et al., 2002a).

REFERENCES

[1]Haslam P.L, Baughman R.P Eur Respir J 1999;14, 245.[2]British Thoracic Society Bronchoscopy Guidelines Committee, Thorax 2001;56,Suppl 1, i1-21.[3]Fernandez N.J, et al Vet Clin Pat 2013;42(1), 92.[4]De Brauwer E, et al Anal Quant Cytol Histol 2002;24(6),337.

[5]Xaubet A, et al Allergy 1991;46(3), 222.[6]Leclere M, et al J Vet Int Med 2006, 20(2),377. [7]Klech H, Pohl W, Eur Respir J 1989;2(6) 561. [8]De Brauwer E, et al. AQCH 2000;22(5),416.[9]Pickles K, et al 2002b. Eq Vet J 2002b,34(3),292.[10]Dehard S, et al. J Am Anim Hosp Assoc 2008;44(6)285.[11]Viscardi V, et al. Braz. J. Vet. Res. Anim. Sci 2013;50(1) 68.[12]Wenisch T, et al Tierarztl Prax Ausg G 2001;29(4), 249. [13]Dehard S, et al J Am Anim Hosp Assoc 2008;44(6), 285.[14]McGorum B.C,, Dixon P.M, Eq Vet Edu 1994;6(4), 203.[15]Pickles K, et al Eq Vet J 2002a;34(3), 297.

Air Quality and Indoor Environmental Exposures: Clinical Impacts
Janice A. Dye, DVM, MS, PhD, DACVIM (SAIM)
Research Triangle Park, NC

INTRODUCTION

Indoor air quality (IAQ) is a term which refers to the air quality within and around buildings and homes as it relates to the health and comfort of the occupants. Many ambient (outdoor) air pollutants readily permeate indoor spaces. Because indoor air can be considerably more polluted than ambient air, the USEPA lists poor IAQ as a major environmental concern (www.epa.gov/iaq/pubs/). In the sections that follow, health effects associated with commonly encountered ambient air pollutants and indoor contaminants will be broken down by agent class. In some cases, exposure may be acute, with one or more pets (and owners) experiencing signs within a relatively short period. However, most exposures are episodic or chronic, making it difficult to definitively link poor IAQ to respiratory or other adverse health outcomes. Age or underlying immunologic, cardiac, or respiratory disease may further complicate the clinical picture, as those patients may be more sensitive to (and affected by) lower concentrations than prove problematic for healthy housemates. Because pets, like their owners, spend most of their lives indoors, we will discuss how certain home conditions can worsen indoor air quality and will briefly discuss measures to improve IAQ for owners and their pets (1).

BIOLOGIC CONTAMINANTS

Dampness and Mold

Over the last 40 years, buildings have become increasingly airtight as a response to stricter energy efficiency requirements. At the same time, changes are occurring in the ways in which household energy is supplied, distributed and used. These changes are having important impacts on exposure to indoor air pollutants in residential buildings. For example, as modern homes recirculate more of the indoor air to increase efficiency of heating and cooling, the volume of outdoor air entering the home is significantly reduced. This can lead to stagnant air and increased humidity. So called **tight building syndrome** is especially problematic if the home's heating ventilation and air conditioning (HVAC) system is not functioning optimally. Such effects are further compounded by water intrusion (e.g., flooding, appliance malfunction, or roof leaks). Excess moisture promotes growth of mold, fungi, and bacteria - which emit spores, hyphae fragments, microbial volatile organic compounds (VOCs), and/or endotoxin into the indoor air.

Definitive linkage of biocontaminant exposure to specific respiratory or neurologic syndromes remains elusive. Microbes can release VOCs, which are known for their ocular and upper respiratory tract irritant properties. Select biocontaminants are known to cause respiratory or systemic infection (i.e., "Legionnaire's Disease"). Increasingly, biocontaminants are implicated in hypersensitivity responses (e.g., pneumonitis, chronic rhinosinusitis), as well as exacerbation of allergies / asthma, and poorly defined but serious chronic neurologic conditions. Cats, not unlike young children, may be at greater risk for exposure to such agents because they typically inhale plus ingest greater quantities of microbe-laden house dust (cats due to extensive grooming and kids due to increased floor contact time and "mouthing" behaviors). In a case report describing two cats that developed acute pulmonary hemorrhage and died, signs were consistent with exposure to the so-called toxic black mold, *Stachybotrys chartarum* (2). Satrotoxin G adducts were detected in archived serum samples. Findings highlight the potential risk of acute pulmonary hemorrhage in animals living in an environment contaminated with mold following flood damage. In the cats (and similarly in human infant case reports), concurrent exposure to other respiratory irritants (e.g., cigarette smoke and inhalant anesthesia) may have contributed to the overall lung insult. Efforts to better define and quantify indoor biocontaminants are ongoing and include exposure assessments such as detection of mycotoxins in urine. Much remains to be established as to how damp indoor environments are associated with poor health.

Allergens

Data linking respiratory disease in companion animals to allergen exposure is quite limited. Supportive evidence comes from clinical improvement after allergen avoidance trials. In addition, cats with bronchial disease appear to have significantly greater positive intradermal skin test (IDST) reactions to the common indoor aeroallergen, house dust mite, than do healthy cats. Serum allergen-specific IgE in cats with bronchial disease were strongly reactive to dust mite allergen and select regional outdoor aeroallergens.

CHEMICAL TOXICANTS

Combustion-Derived Products

Second-hand smoke (SHS) contains gases and respirable particles. In poorly ventilated air spaces, it can cause ocular and respiratory tract irritation, especially in children. The Surgeon General concluded that there is no risk-free level of exposure to SHS. A smoke-free environment is considered the only way to fully protect nonsmokers from its hazardous effects. Pets may be at greater risk from SHS than humans because they spend more time in the home and closer to carpets where carcinogenic particles can linger. By extension, pets with pre-existing upper or lower respiratory disease may be especially sensitive to the irritating effects of SHS. Based on limited studies, SHS exposure in dogs is associated with decreased lung functional reserve, increased risk of lung cancer (in brachy- and mesocephalic dogs) and nasal cancer (in dolichocephalic breeds). In cats, SHS exposure is associated with increased risk of malignant lymphoma and oral squamous cell carcinoma, likely owing to inhalation plus extensive oral (via grooming) exposure routes. Nicotine toxicity can occur in pets ingesting cigarette butts, and may prove fatal in birds and small dogs. Recent upticks in calls concerning cases of nicotine poisoning in pets relate to ingestion of e-cigarettes or liquid nicotine refill solution. While dogs account for the majority of cases, nicotine in e-cigarettes and liquid refill solution is toxic to cats as well.

Ambient Air Pollutants

Historically, catastrophic air pollution episodes have occurred and were infamous for their toll on human life (e.g., the London Fog of 1952). They also caused considerable morbidity and mortality in animals. In North American, industrial accidents in combination with thermal weather inversions include Donora, PA (1948), killing 20 people, 10 dogs, 3 cats, and 40% of poultry in four nearby flocks; and in Poza Rica, Mexico (1950) where hydrogen sulfide release during a thermal inversion led to the death of 22 people and vast numbers of dogs, swine, cattle, and nearly 100% of the canaries in the area. Modern day pollution regulations and control technologies have greatly reduced contemporary air pollution levels. Still, air pollution remains associated with significant economic toll owing to health care costs, and lost work and school productivity. Individual air pollutants and their associated health effects will be discussed in brief below. It should be noted that health associations are inferred from epidemiologic associations in humans (3) in combination with data derived from relatively short-term, high-dose experimental exposures in laboratory animals (4). Episodic accidental leaks or chemical releases (e.g., methane) and increasingly common wildfires impact humans and pets alike. Thus, any evacuation efforts should allow for removal of pets.

Ambient particulate matter (PM) is a major outdoor environmental health concern in people, especially near large roadways. Fine ($PM_{2.5}$) and coarse (PM_{10}) particles are complex mixtures of microscopic carbon-based particles, with adsorbed materials that can be inhaled deep into the lungs. Health associations include: increased respiratory signs, decreased lung function, asthma and COPD exacerbation, arrhythmias, cardiac ischemic events, and premature death, especially in people with pre-existing heart or lung disease (3). Ambient gaseous pollutants include **nitrogen dioxide,** a poisonous, corrosive, brown gas, which can be both a major indoor and outdoor air pollutant. Its primary indoor sources relate to poorly ventilated biomass-burning. In pets, nitrogen dioxide exposure may result in increased frequency and severity of respiratory infection. It may also contribute to chronic bronchial and lung pathologic changes. **Sulfur dioxide** is another ambient air pollutant, and is the major precursor to so-called acid rain. Veterinary concerns regarding this pollution primarily relate to aquatic and wildlife health. One important exception is that of unvented kerosene space heaters, which when used in closed-space situations, can lead to exorbitant total airborne particle concentrations and associated acid aerosols. **Ozone,** is a potent oxidant gas and is one of the most directly injurious of the outdoor air pollutants. Exposure can induce respiratory tract irritation, injury, and inflammation (4). In some instances, ozone is emitted directly indoors by (e.g., electric generators, photocopiers and laser printers). Ironically, *ionizing* **air-purifiers** intentionally generate ozone to react with and remove odorants from the indoor air. Their use, however, can result in indoor ozone concentrations that exceed levels found in smog. **Carbon monoxide** (CO) is a colorless, odorless gas. Unlike most air pollutants, a single (likely to be encountered) exposure to CO may prove lethal. Significant exposure can occur during use of charcoal grills or unvented kerosene heaters in relatively closed air spaces (garages), or when chimneys or venting ducts of central heating systems are obstructed. Birds are especially sensitive to CO, as evidenced by their use as sentinels in coal mine shafts. CO poisoning is also an important cause of the morbidity and mortality in pets rescued from burning buildings. Having preset clinic protocols in place for managing animal victims of smoke inhalation can help to ensure a successful outcome.

OFF-GASSING EMISSIONS

Formaldehyde is an overpowering, highly water soluble, reactive gas. It can cause ocular and upper respiratory tract irritation. Due to off-gassing from resins in particleboard, wood paneling, foam insulation, carpeting, upholstery, and draperies, indoor levels can be quite high, especially in new homes and mobile homes (which are made of many of these materials). Excessive acute exposure can be associated with headache, nausea, dizziness, and disorientation. **Volatile organic compounds (VOCs)** can easily become vapors or gases and are present in a wide variety of household products and aerosol sprays numbering in the thousands. Concentrations of many VOCs are consistently higher indoors (up to ten times higher than outdoor levels).

Importantly, VOCs can act as sensory irritants and may have short- and long-term adverse health effects. Greater, prolonged, or combined exposures may lead to dizziness, headache, nausea, and even allergic-like reactions. Not uncommonly, owners fail to use these products as per warning label instructions, making VOC exposures in pets problematic. Thus, for pets presenting with acute onset of eye, skin, and/or respiratory signs, one should consider recent VOC exposure. Avian veterinarians should be alert that, like CO, pet birds are very sensitive to aerosolized VOCs. Fumes from cooking pans, self-cleaning ovens, or heat lamp bulbs lined with polytetrafluoroethylene (PTFE; Teflon® or Silverstone®) have also been associated with acute and fatal toxicity in birds.

FIBERS, PARTICLES, AND HOUSE DUST

Indoor particles represent both intruded outdoor PM, plus particles from indoor SHS, cooking, heating (fireplaces) and so forth. Indoor cats are further exposed to particles in litter (e.g., silicates or dusty clay), which anecdotally may be associated with respiratory distress in cats with feline bronchial disease. **Asbestos fibers** are naturally-occurring mineral silicates that were used extensively in friction products and as a noncombustible insulation. Asbestos fibers are small, readily inhaled, and difficult to clear from the lungs. Asbestos-associated disease (i.e., mesothelioma) has been reported in dogs. Very long latency periods (up to 30 years) are known to occur in humans, and are associated with development of interstitial lung fibrosis, carcinoma, and mesothelioma.

House dust is a mixture of lint, particles, and adsorbed substances from household products. Dust may be laden with metals, pesticides, semivolatile persistent organic pollutants (POPs) such as polybrominated diphenyl ethers (PBDEs, a type of flame retardant), or perfluoroalkyl substances (PFAS, used in anti-stain applications and nonstick coatings). Pets are continually exposed to these contaminants through dust exposure. We were one of the first to demonstrate that house dust represents a major source of PBDE exposure for indoor cats, likely due to the time they spend on/near treated carpeting, furniture, and electronics as well as to grooming activities (5). We have also shown that indoor cats have higher serum levels of PFAS (6). Recent efforts in California to eliminate PBDE flame retardant usage in household materials has led to decreased serum levels in Californian cats (7). Importantly, certain POPs purportedly act as endocrine disruptors (i.e., thyroid effects) and as obesogens (i.e., metabolic effects) (8). In the U.S., obesity and type II diabetes are a growing problems in both people and pets.

IMPROVING INDOOR AIR QUALITY

Key elements of improved air quality are source control, adequate ventilation (especially in shower, laundry and cooking areas), maintaining relative humidity between 30–50%, changing air filters on a regular basis, and air duct cleaning as needed. To limit dust accumulation, use a quality vacuum with a high efficiency filter weekly. Water leaks should be promptly repaired, and after any flooding, areas should be thoroughly cleaned and dried. If materials cannot be dried promptly, they should be replaced. CO monitors should be installed in the home. All gas appliances should be properly functioning, or inspected and adequately cleaned and repaired. Gas stoves and heaters should be vented to the outside of the home. Owners should be reminded that gas ranges should *never* be used as a heat source. Smoke detectors, along with the enforcement of improved building codes are effective in reducing fire-related deaths. Owners should be cautioned about so-called air purifiers that use ionization as a means of "clearing the air", but actually increase indoor ozone concentrations, at times to levels well in excess of that considered safe. Kerosene heaters should be used only as indicated by the manufacturer's instructions, and should be refueled outdoors using specially manufactured low-sulfur fuel. Cars, lawn mowers, etc., should never be left running inside a garage or shed, especially if the space is ever used to house a pet.

Pet owners should similarly be cautioned to follow the manufacturer's directions during use of common household cleaners, especially the aerosol-type products. Similar to recommendations for using pest control products, the owner should ensure that there is adequate ventilation. One common mistake is for an owner to mix bleach (sodium hypochlorite) containing products with ammonia-based cleaning products. This is a recipe for serious problems because together, these compounds can generate toxic (e.g., chloramine) fumes. Pet owners, especially of birds, should minimize use of all VOC-containing products. Encourage use of unscented and non-aerosol alternatives. Many newer and so-called "green" substitutes are designed to have much lower content of known irritants (e.g., low-VOC paints). Despite these precautions, if the owner perceives that use of any cleaning product, paint, new carpeting, and like agent is associated with ocular irritation, sore throat, headache or an offensive odor, the owner should ensure that neither they nor their pet maintain prolonged exposure in the area until odors dissipate. If this is not possible, efforts to improve ventilation are vital. Again, the Surgeon General has concluded that only by eliminating cigarette smoking in indoor spaces, can one fully protect nonsmokers from SHS exposure. Relatedly, a recent pilot study on owner attitudes toward smoking showed that by informing pet owners of the risks of SHS for their pets, the owner may be more motivated to quit smoking altogether. For young puppies and kittens, geriatric pets, and especially pets with known respiratory or heart disease, the above precautions are especially important due to increased risk of exposure exacerbating any underlying conditions.

REFERENCES AND HEALTHY HOME WEBSITES

1. Dye JA. Respiratory toxicants of interest to pet owners. In: Bonagura JD, Twedt D, eds. Kirk's Current Veterinary Therapy XV. Elsevier Science (2015).
2. Mader DR, et al. Acute pulmonary hemorrhage during isoflurane anesthesia in two cats exposed to toxic black mold. *J Am Vet Med Assoc.* 2007;231:731–5.
3. Brook RD, et al. Particulate matter air pollution and cardiovascular disease: An update to the scientific statement from the American Heart Association. Circulation. 2010;121(21):2331-78.
4. Dye JA, et al. Executive Summary: variation in susceptibility to ozone-induced health effects in rodent models of cardiometabolic disease. Inhal Toxicol. 2015 Suppl 1:105-15
5. Dye JA, et al. Elevated PBDE levels in pet cats: sentinels for humans? Environ Sci Technol. 2007; 41:6350–6.
6. Dye JA, et al. Perfluorinated compounds (PFCs) in serum of cats – linkage to indoor exposures. 31th ACVIM Forum Proceedings, Seattle, WA, 2013.
7. Guo W, et al. Temporal changes in PBDE levels in California house cats and a link to cat hyperthyroidism. Environ Sci Technol. 2016; 50(3):1510-8.
8. Janesick AS, Blumberg B. Obesogens: an emerging threat to public health. Am J Obstet Gynecol. 2016. S0002-9378(16)00232-5.

EPA's publication A Brief Guide to Mold, Moisture, and Your Home at http://www.epa.gov/mold/moldguide.html
CDC at http://www.cdc.gov/healthyhomes/bytopic/airquality.html

NIEHS at https://www.niehs.nih.gov/health/materials/index.cfm

DISCLAIMER

Proceedings do not reflect USEPA policy.

TRACHEAL AND BRONCHOMALACIA: WHERE WE ARE IN 2016 – Diagnostic Advances

Lynelle R. Johnson DVM, MS, PhD, DACVIM (Small Animal Internal Medicine)
Davis, CA

INTRODUCTION

Tracheal and bronchial collapse are common findings in dogs with cough. Tracheal collapse can affect the cervical tracheal, the intrathoracic region, or both segments and is a clinical entity in small breed dogs. In some studies, bronchomalacia commonly complicates tracheal collapse although in certain dogs, only tracheal collapse is present. The severity of tracheal collapse has traditionally been based on the percentage reduction in luminal diameter, with 25% collapse identified as Grade 1, 50% as Grade 2, 75% as grade 3, and 90% collapse or a double lumen "W" shaped trachea classified as grade 4. Grades 1 and 2 tracheal collapse are likely of little clinical consequence, and normal dogs can have variation in luminal dimensions up to 25% between inspiration and expiration. It has been suggested that tracheal and bronchial collapse are progressive in nature, although very young animals can present with the most severe form of disease (grade 4/4 tracheal collapse).

Bronchial collapse occurs in both large and small breed dogs. It has been suggested that bronchial collapse should be classified by severity and type to allow better characterization of disease. Malacic airway disease in people has been described by bronchoscopic visualization of >50% narrowing of the bronchial lumen, and this designation could be used to reflect grade 1 collapse. Grade 2 would indicate increased severity with grade 3 as >75% collapse. It has been proposed that this should be further characterized as static or dynamic to allow improved documentation of disease and comparison of conditions among dogs.

Bronchomalacia can be found alone or in association with infectious or inflammatory airway disease (chronic bronchitis or eosinophilic lung disease), although the relationship between airway collapse and inflammatory airway disease is somewhat unclear. In dogs examined bronchoscopically, inflammatory airway disease was found in equal numbers of dogs with and without airway collapse. Interestingly, normal bronchoalveolar lavage cytology was more common in the group of dogs with airway collapse than in the group without collapse, perhaps because lavage fluid was collected distal to the site of collapse. Infection is rarely a substantial problem in dogs with airway collapse, although laryngeal paresis or paralysis resulting in low-grade aspiration injury might contribute to the waxing and waning of clinical signs. Concurrent bronchiectasis can worsen cough in some cases, although there is no direct link between the presence airway collapse and development of bronchiectasis.

Although the clinical diagnosis of tracheal or airway collapse can be relatively easy based on history and physical examination, confirmation of the extent of disease can require multiple complementary tests. Identification of various complications or trigger events is important in individualizing therapy.

CLINICAL FINDINGS

The most common clinical complaint in dogs with airway collapse is a paroxysmal cough that waxes and wanes in intensity and duration. Dogs that have severe tracheal collapse can also present with respiratory distress, with or without cough. Syncope might be reported in dogs that cough severely and trigger a vagal reflex leading to collapse ("cough drop" or perhaps reflex or vasovagal syncope). Syncope could also be an indicator that pulmonary hypertension has developed.

Physical examination generally reveals moderate to several tracheal sensitivity and palpation can induce a constant honking cough on inspiration and expiration. Stridor can be auscultated over the upper airway in some dogs due to severe narrowing of the tracheal but could also represent concurrent laryngeal paralysis. An end expiratory snap is suggestive of bronchial collapse while an expiratory push might be indicative of generalized smaller airway collapse or concurrent bronchitis. Ballooning of the thoracic inlet during expiration or coughing suggests cranial lung herniation, a common finding in dogs with tracheobronchomalacia. The presence of crackles over the thorax could reflect airway opening and closure due to airway collapse or could indicate concurrent bronchitis, although lung sounds are often difficult to assess due to tachypnea, obesity, or referred upper airway sounds. Many dogs with chronic lower airway disease have concurrent mitral insufficiency and careful cardiac auscultation is advised. A right-sided murmur should raise suspicion for pulmonary hypertension developing as a consequence of chronic tracheobronchial disease.

DIAGNOSTIC TESTING

Cervical and thoracic radiographs are indicated to initiate the diagnosis of airway collapse and to identify concurrent conditions contributing to cough. Cervical radiographs can highlight caudal retraction of the larynx associated with inspiratory effort or might aid in the diagnosis of upper airway obstructive disease due to elongation of the soft palate, a hyperplastic epiglottis, or laryngeal edema. In addition, obtaining a cervical radiograph along with lateral views increases the chance of obtaining images during various phases of respiration, and therefore potentially increases the likelihood that cervical, intrathoracic, or bronchial collapse will be identified. Collapse of the cervical trachea down to the level of the thoracic inlet should be visualized during inspiration, while intrathoracic and potentially bronchial collapse would be observed on expiration. The right middle and left cranial lobar bronchi are most commonly affected by collapse, highlighting the necessity to obtain both lateral views.

Given the immediate availability of digital radiographic images in comparison to standard film radiography, it is easier to obtain images during various phases of respiration with little extra effort or cough involved. However, static radiographs likely over-estimate the presence of tracheal collapse in the cervical region. Firstly, the dorsal opacity in the tracheal lumen does not necessarily represent an elongated dorsal tracheal membrane from tracheal collapse and can lead to a false diagnosis. Either left or right lateral radiographs can show increased opacity along the dorsal margin of the trachea in dogs with and without tracheal collapse. This could be related to obliquity in the radiograph or super-imposition of structures in the region. Secondly, in dogs with fluoroscopically confirmed tracheal collapse, lateral radiographs were positive for collapse in over 90% of cases, however the location of collapse identified using the 2 imaging modalities agreed only 50% of the time. This is because collapse in the cervical trachea is easier to identify due to the increased size of the air column in that region. Bronchial collapse is more difficult to see because of the smaller luminal diameter as well as because of overlapping structures. Finally, the dynamic nature of respiration and coughing as well as positioning for imaging procedures can influence the detection of airway collapse, making advanced imaging important in the completing the diagnosis.

Fluoroscopy has poorer image resolution than radiography and also potentially results in more radiation exposure to both technicians and animals but has the advantage of providing dynamic evaluation of the respiratory tract during tidal respiration and during cough. Given the multiple locations of bronchial collapse that have been reported, fluoroscopic assessment in left and right lateral positioning as well as in sternal recumbency might be required to obtain a complete evaluation of the airways. As with videofluoroscopic swallowing studies, it is likely that a better study could be obtained with the animal in a standing position, however this would not obviate the need for multiple views.

Bronchoscopy can document tracheal collapse when radiographs or fluoroscopy are inconclusive but it does require general anesthesia. Bronchoscopy also provides the following benefits: laryngeal structure (edema, saccular eversion) and intrinsic motion or function can be evaluated; the location and degree of tracheal ring flattening can be graded; protrusion of the dorsal tracheal membrane into the tracheal lumen can be accurately, though subjectively, assessed; the extent of tracheal collapse can be defined for surgical intervention; tracheal inflammation or irritation can be visualized; intrathoracic collapse can be confirmed or denied; both dynamic and structural changes in the lower airways can be visualized; and the contribution of small airway disease to cough can be defined. Obtaining bronchoalveolar lavage fluid for cytologic and microbiologic analysis can be worthwhile for guiding therapy. Infection can be documented by finding septic suppurative inflammation on cytology, although up to 25% of dogs with lower respiratory tract infection lack such a finding. It is unclear whether extracellular bacteria or growth of small numbers of oropharyngeal flora are of any consequence in dogs with airway collapse because bacteria are found at the carina of healthy dogs. Similarly, identification of neutrophilic inflammation can be difficult to interpret, because it is possible that healthy, non-coughing dogs might have minor amounts of inflammation present. Chronic cough likely leads to hyperemia and mucus due to mechanical trauma alone. If this is isolated to the large airway, it may or may not result in abnormal findings in bronchoalveolar lavage fluid. The role of endobronchial brush cytology in documenting clinically significant inflammation requires further study. Taken together, clinical judgment must be used to determine when the value of bronchoscopy outweighs the risks of anesthesia.

RECENT ADVANCES

Computed tomography is commonly used to identify changes in airway dimensions in human patients, and dynamic expiratory CT can show dramatic airway narrowing in chronic inflammatory conditions as well as in primary tracheobronchomalacia. CT in non-anesthetized dogs has proven useful for documentation of obstructive lesions, however accurate measurements of tracheal dimensions typically require positive pressure ventilation under anesthesia. In a study of normal dogs of unknown breed, radiographs underestimated tracheal height by 1mm in comparison to CT, which could impact stent lengthening post-deployment. A preliminary study in dogs with tracheal collapse that compared CT to fluoroscopic measurements demonstrated that the lateral diameter of the trachea is substantially greater than the dorsoventral diameter, which could lead to a significant error in stent sizing. It is clear that CT provides three-dimensional imaging that can assist with diagnosis of the grade of tracheal collapse and likely would provide valuable information on tracheal distensibility. However, further evaluation of the use of CT in the diagnosis and utilization of this information for improved treatment strategies is required.

Tidal breathing flow-volume loops represent a non-invasive assessment of pulmonary function that can be used for identification of airway obstruction because flow limitation through the cervical trachea results in prolongation of inspiratory time and reduction in inspiratory flow. In one study, calculation of multiple flow-volume loop parameters allowed differentiation of normal dogs from dogs with grade 3 tracheal collapse. However, the limited availability of the equipment needed to perform such analysis hampers widespread utilization.

REFERENCES

Bottero E et al, J Vet Int Med 2013, 27: 840, Bylicki BJ et al, Vet Rad US 2015; 56: 602, Johnson LR & Pollard RE, J Vet Int Med 2010; 24(2): 298, Johnson LR et al, J Vet Int Med 2015; 29(6): 1619, Hawkins EC et al, Am J Vet Res 2007; 68: 435. Leonard CD et al, Am J Vet Res 2009; 70(8): 986, Macready DM et al, J Am Vet Med Assoc 2007; 230(12): 1870, Montgomery JE et al, Vet Surg 2015; 44: 115, Nafe LA et al, Can Vet J 2013; 54(10): 955, Pardali D et al, J Vet Int Med 2010; 24: 832, Singh MK et al, J Vet Int Med 2012; 26(2): 312.

Tracheal and Bronchomalacia: Where We Are in 2016 – Management Strategies

Eleanor C. Hawkins, DVM, DACVIM (SAIM)
Raleigh, NC

INTRODUCTION

Tracheal collapse is the historically, and commonly, used term for dogs with tracheomalacia. So, why the terms tracheomalacia and bronchomalacia (or tracheobronchomalacia)? First, many dogs with tracheal collapse have bronchial collapse as well. Second, including the term "malacia," as is used in people, is a reminder that tracheobronchomalacia is a description of pathologically soft large airways with many clinical manifestations rather than one stereotypical disease. Dogs with tracheobronchomalacia can have: various clinical signs; severity of signs that range from none (incidental finding) to life-threatening; and, one or more predisposing, concurrent, or complicating problems. As such, there is no simple treatment plan that can be recommended for the management of tracheobronchomalacia in dogs. In adult people, tracheobronchomalacia may be present concurrently with COPD and management is initially aimed at controlling this (or any other concurrent or underlying) disease.[1] I believe this to be similarly occurring in many adult dogs, especially those presenting with cough as the primary complaint. In young dogs, signs may be more often related to airway obstruction. Further, each dog will likely require re-assessment and adjustments in management throughout the remainder of their life because of the progressive nature of the process and the occurrence of exacerbations. Given these many factors to consider in developing a treatment plan, specific treatment options will be reviewed individually (the tools in our toolbox for managing these dogs), including their rationale and indications for use. As there is little objective data on which to base decisions, these recommendations are based largely on pathophysiology, medical practice in people, and personal experience.

TOOLS IN THE TOOLBOX

There are no tools in the toolbox that are known to strengthen malacic airway cartilage *in vivo*. If malacia is sufficiently severe to prevent effective ventilation because of tracheal obstruction, we rely on tracheal stenting. Fortunately, cough, rather than overt airway obstruction, is the primary complaint in the majority of cases and can most often be managed effectively by medical means. Dogs can also be presented with early signs of obstruction, such as stertorous breathing with worsening during excitement, which may also be amenable to medical management for some period of time.

Ideally, treatments are selected after a thorough diagnostic evaluation has been performed so that the most directly applicable treatments are possible. Unfortunately, even in cases with a thorough evaluation, no specific etiology may be identified for direct treatment. Further, infections with organisms such as *Mycoplasma* cannot always be effectively ruled out regardless of efforts. In the absence of a clear primary process (such as bacterial or eosinophilic bronchitis) in patients that are clinically stable, my general preference is to add one treatment at a time to assess its effect. Using this strategy, cough suppressants are withheld initially, as they are generally effective in decreasing cough even when there are underlying infections, inflammation, or other problems. While multiple treatments may be needed for full control, assessing response to one change in treatment at a time may avoid long-term continuation of ineffective treatments, and their accompanying adverse effects. In patients with severe signs, it may be necessary to make these determinations in reverse order; that is, treating broadly, then eliminating treatments to see if control is maintained. Universally beneficial, long-term, management strategies such as improving air quality and weight loss, can be initiated at any time. Regardless, effective management requires close communication with clients. Clients should be encouraged to keep a diary of treatments given (including frequency of administration) and a daily subjective assessment of frequency, duration, and severity of cough, stertor, or breathing efforts, as appropriate to the individual case.

Corticosteroids

In theory, tracheobronchomalcia does not have to be inflammatory. In practice, dogs that are symptomatic for tracheal collapse usually have some degree of airway inflammation. In the early histopathology study identifying cartilage abnormalities in a series of 10 dogs with tracheal collapse, intraluminal exudate was seen in the region of collapse, there was considerable neutrophilic infiltration of the submucosa, and sporadic lymphoplasmacytic infiltrates were seen.[2] The sequence of malacia and inflammation may be different from one dog to the next. Small breed dogs may have congenitally abnormal cartilage leading to chronic cough and airway collapse, which in turns leads to inflammation of the airways. Alternatively (or possibly concurrently), primary inflammatory airway disease such as with chronic or eosinophilic bronchitis, may lead to airway malacia due to collagenases, proteases and other damaging products along with relatively wide intra-thoracic pressure gradients that are generated during cough. Corticosteroids may also provide benefit to dogs with extra-thoracic airway obstruction and concurrent laryngeal inflammation or collapse by decreasing inflammation and edema.

When using steroids to treat chronic inflammatory disease I use an approach similar to those used in immune-mediated diseases, although beginning at lower dosages. Oral prednisone or prednisolone is used initially to gauge response to treatment, eliminating the added variables associated with use of an inhaler. Oral steroids are also a less expensive treatment trial than inhaled steroids. In patients with specific concerns for systemic steroid use, inhalers can certainly be used instead. Beginning with an initial dose of 0.5 – 1 mg/kg q12h, a beneficial response is expected to be noted within a week. If a beneficial response is seen, steroids are continued at that initial dosage for a week or two beyond the maximum effect that is achieved, and then the dosage slowly decreased to the least effective alternate day dose that maintains that effect. A typical tapering regimen is a 20-30% decrease every 2 weeks. If no beneficial response is seen, steroids are discontinued. A lack of any positive response should prompt reconsideration of the prior diagnostic evaluation for possible missed diagnoses, as it is unusual (in my experience) for steroids to be completely ineffective. In dogs that are responsive, a transition to inhaled steroids is discussed with the clients. Delivering steroids by inhalation rather than orally should minimize, if not eliminate, systemic signs. Some systemic side-effects of oral steroids, such as polyphagia and weight gain, work against the overall management plan.

Cough suppressants

The most effective cough suppressants seem to be narcotics. They act directly on the cough center in the medulla, possibly through either mu or kappa opiate receptors.[3] My preference is hydrocodone (0.25 mg/kg q6-8h, initially) for cough that is difficult to control, although butorphanol has the advantage of being licensed for use in dogs and may be as or more effective. If this dosage fails to achieve a sufficient response, and other potential contributing factors have been addressed fully, the clients are carefully questioned about any transient benefit. Based on this information, adjustments in frequency or dose are made to achieve a better effect. The primary limiting factor of these drugs is excess sedation. For dogs that seem intolerant of narcotics, maropitant can be tried. Maropitant (2 mg/kg q48h for 14 days) was reported by abstract to reduce client perception of cough, but not airway inflammation, in 7 dogs with chronic bronchitis.[4]

With increased scrutiny of narcotics, physicians are increasingly prescribing benzonatate (Tessalon Perles®) for cough control, however its use carries significant risk and I do not recommend it. Benzonatate is related to tetracaine and is thought to act on stretch receptors in the lung to dampen the stimulation of cough. The danger with this product is that it is formulated in small gel caps that if dissolved or punctured in the oral cavity can cause laryngospasm, bronchospasm, and aspiration. Adverse effects of proper dosing include a feeling of numbness, mental confusion, a burning sensation of the eyes, and visual hallucinations. Overdosage can cause seizures, cardiac arrhythmias and death.[5] The FDA has warned against its use in children under 10 years of age, and dogs seem quite at risk of having drug released directly into their mouth by clients not getting the capsule far enough down the throat or inadvertent biting of the capsule.

The goal of cough suppressants varies with the individual patient. If the dog is coughing in severe paroxysms throughout the day, hydrocodone may be prescribed for consistent administration for a week or more (along with other measures) to attempt to help break the "cough-cough" cycle. In mild cases, the emphasis is on effecting a response as needed to allow sleep or improve quality of life.

Bronchodilators

In theory, bronchodilators may not be indicated for the treatment of malacia. However, treatment with bronchodilators alone can result in marked decrease in coughing in some dogs. Presumably these dogs have a component of bronchitis, but even so, bronchospasm is rare in dogs. Theophyllines, in particular, may work through mechanisms such as dampening of inflammation, enhancing the effects of corticosteroids, and/or improved mucociliary clearance. Lower concentrations (5-10 ug/ml) of drug may be needed for anti-inflammatory effects, if they indeed occur in dogs.[6] The initial dosage for oral extended-release theophylline is 10 mg/kg q12h.[7] If treating concurrently with a fluoroquinolone, increase the dosage interval to once daily to avoid possible side-effects from delayed clearance of theophylline.

In people theophyllines are not prescribed as first line therapy for bronchitis because of side-effects, particularly nausea or jitteriness, and the availability of alternatives. However, the benefits of these drugs were sufficient that pharmaceutical companies have been attempting to develop phosphodiesterase-4 inhibitors with fewer side effects. In dogs, side-effects are infrequently reported by clients. Any adverse signs generally respond rapidly to discontinuation of drug, with reinstitution at a lower dose after signs have completely resolved. Assays are readily available to measure serum concentrations of drug should such information be useful in confirming cause of adverse effects or adequacy of dosing.

A concern with bronchodilators as a sole treatment for dogs with tracheobronchomalacia is that any anti-inflammatory effect may be insufficient to minimize ongoing damage to the airways. No studies are available to provide guidance in this regard, and there are associated risks with the use of corticosteroids. Regardless, it is unlikely that bronchodilators alone would be effective in controlling cough except in patients with mild clinical signs.

Antibiotics

Antibiotics are best selected based on the findings of airway washing cytology and culture obtained prior to initiating antibiotic therapy. Although infection is not present in the majority of cases, it is present in some. Studies published from referral institutions have the challenging limitation that their patients have often been receiving antibiotics in the recent past, which means reports in the literature may be under-estimating the incidence of infection. Infection may represent a primary infection, such as with *Bordetella bronchiseptica* or *Mycoplasma cynos*, or opportunistic infections from oral flora. The combination of poor oral health and reduced airway clearance would seem to favor the latter. Trial therapy with doxycycline may be useful because methods for identifying *Mycoplasma* (culture and PCR) can miss infections, and organisms are not visible in cytologic specimens. If aiming for bacteria originating from the oral cavity, broader spectrum drugs should be considered. Because of the chronic nature of these cases, and impaired airway defenses, I generally continue administration of antibiotics for a total of 4-6 weeks IF they have resulted in a reduction in signs within 1-2 weeks of administration. If signs worsen dramatically within a 1-2 weeks of discontinuing the drug, reinstitute treatment with the same antibiotic.

Anti-anxiety drugs

Some dogs with cough or signs of airway obstruction and tracheal collapse are minimally symptomatic (with or without other treatments) except during periods of great excitement or stress. These dogs may benefit from anti-anxiety drugs, which many minimize periods of excess excitement with less sedative effects than would be seen by increasing the dose or frequency of a narcotic drug. Trazadone can be effective in this setting, and is administered at a dosage of 2-5 mg/kg orally, divided every 8-24 hours.[8] Clinical experience suggests that this drug is most effective if given at least daily, with an increase in dose prior to a known stressful event. That is, administering the drug only intermittently, in anticipation of a stressful event, is less effective.

Surgical management of concurrent airway disease

Occasionally, dogs with tracheal collapse have primary or concurrent airway disease rostral to the trachea. Addressing abnormalities such as pharyngeal polyps, elongated soft palate, and laryngeal collapse can result in dramatic improvement in clinical signs in these patients by decreasing airway pressure gradients.

Weight loss counseling

Obesity can contribute to signs of tracheal collapse through numerous potential mechanisms. Most obvious would be the overall increased work of breathing and narrowing of the airway, particularly in the pharynx, associated with increased fat deposition. Specific guidance regarding diet and recommendations for controlled exercise are indicated for overweight dogs.

Dental prophylaxis

Oral flora gains access to large airways in normal people through ongoing aspiration of small quantities of saliva, particularly during sleep.[9] In people, there is an association between gum disease and COPD, speculated to be related to oral flora.[10] As the majority of dogs with tracheal collapse are of the breeds commonly plagued by chronic dental disease, careful attention to oral health is warranted. Because of the progressive nature of the airway disease, it is ideal to perform a thorough, careful dental prophylaxis early in the course of disease and to be particularly engaged in helping the client incorporate daily dental care into the home routine.

Cardiac drugs

Moderate to severely enlarged left atria resulting from myxomatous mitral valve degeneration can be seen to impinge on the left mainstem bronchi. Additional insult to the large airways could potentially occur as a result of vibrations from the regurgitant jet of blood impacting the atrial wall opposite the mitral valve. A study of 16 dogs with chronic cough, 10 of which had moderate to severely enlarged left atria, found that all dogs had collapse of the left cranial lobar bronchus, most dogs had collapse of other bronchi, and all dogs had bronchoscopic and cytologic evidence of inflammation.[11] Although this study failed to find an association in any of the examined signs and left atrial enlargement, more investigation is needed.

Improving indoor air quality

Triggers for cough should be avoided, and inflamed airways are particularly sensitive to inhaled irritants or allergens. The American Lung Association (www.lung.org) has a Healthy Air initiative with easily accessible information on indoor air quality at their website. Non-proprietary sources, such as these, are preferred to commercial sources as they are more likely to provide unbiased information. Certain air cleaning devices can potentially increase ozone concentrations in the home, which can contribute to or exacerbate respiratory signs.

Reduction of external pressure: finding the right harness

Using a harness for restraint, rather than a collar, is a long-time standard to counter neck pressure as an exacerbating factor for cough or obstruction. Design of harness and conformation of dog influence how well pressure on the neck is minimized. It is helpful for the client to be educated about the cervical anatomy to assist them in making a wise selection.

Drugs to improve cartilage strength

Only one study has been published showing improvement in grade of tracheal collapse with medical treatment.[12] Stanzolol, an anabolic steroid, was administered in a prospective placebo controlled clinical trial.[12] Improvement in clinical signs and tracheoscopy findings was noted following 75 days of treatment. Eighteen of the twenty-two dogs studied had minimal collapse (Grade 1 or Grade 2). All 4 dogs with grade 3 tracheal collapse had improvement in grade of collapse following treatment, but none of these dogs were given placebo to insure blindedness. Further reports or anecdotal experience are lacking, and drug availability is a limiting factor. There are no studies supporting the use of Adequan® (Novartis) or oral glycosaminoglycan to strengthen tracheal cartilage. Clearly more research is needed to understand the pathophysiology of this disease and in what circumstances these drugs, or novel treatments, would be of clinical benefit.

Tracheal stents

Tracheal stenting is recommended for patients who have failed aggressive medical management and have signs of significant large airway obstruction, such as labored breathing, cyanosis, or collapse associated with inspiratory stertor or stridor and/or expiratory wheeze. Differential diagnoses other than tracheal obstruction should be considered carefully to avoid placing a stent unnecessarily. Stents are sometimes placed in dogs with incessant coughing that is unresponsive to intense medical management, but in my experience with a limited number of cases, stents are of limited value in such patients. In fact, even for dogs with significant large airway obstruction, when cough is present concurrently, clients must be warned that ongoing management of cough will need to be continued. Cough also likely contributes to fatigue of the stent and may contribute to excessive granulation tissue formation.

The most commonly used stents are self-expanding, intra-luminal stents made of nickel-titanium alloy.[13,14] In experienced hands, they are simple to place during a short period of anesthesia using fluoroscopic or bronchoscopic guidance. There is minimal morbidity associated with stent placement, and for dogs whose clinical signs are predominately those of tracheal obstruction, response can be immediate. Complications of intra-luminal stents include bacterial infections, obstruction due to excess granulation tissue, stent fracture, and stent migration. Tapered stents are now available that will better match the anatomy of the trachea in many dogs, where the extra-thoracic trachea is of greater circumference than the intra-thoracic trachea. In theory, a better fit should decrease complications and improve the life-span of the stent. In the 2008 report by Sura, et al., survival times ranged from 1-48 months, with 7 of 12 dogs surviving greater than 2 years.[13]

External stenting remains favored by some surgeons, using prosthetic rings in the cervical region. Median survival was not influenced by the presence of intra-thoracic airway collapse in dogs treated with extra-luminal cervical rings based on statistical comparison of data in a study by Becker, et al.[15] The median lifespan of dogs without intra-thoracic airway collapse was >2500 days (median not reached) compared with 1500 days for dogs with concurrent intra-thoracic collapse. The major limitation to external stenting remains the high rate of significant complications. In the Becker study, laryngeal paralysis requiring tie-back immediately postoperatively occurred in 7 of 33 dogs (21%), laryngeal collapse occurred in 8 dogs (including 5 of those requiring the tie-back), and tracheostomy was needed in 1 dog. Three dogs (9%) died within 1 week of surgery. Chisnell, et al. reported good success with extra-luminal rings in 23 dogs.[16] For the 22 dogs surviving to discharge, the time from surgery to last evaluation or owner communication was a median of 50.5 months (approximately 1500 days) with a range of 10-103 months. The complication rate was 35%. Interestingly, only one dog had mainstem bronchial collapse and it was assessed as mild.

A comparison between outcomes of dogs treated with extra-luminal rings and intra-luminal stents for tracheal collapse was recently reported by Tinga, et al.[17] It is a retrospective study, as were all others, and numerous limitations are acknowledged by the investigators. Dogs with extra-luminal rings placed were younger and less likely to have main-stem bronchial collapse compared with dogs with intra-luminal stents. However, regardless of stent type, younger dogs and dogs without main-stem bronchial collapse had longer survival times. Remembering that difference in patient population, survival time was longer for dogs with extra-luminal rings (median 1,460 days; 95% CI: 840-1,866 days) than dogs with intra-luminal stents (median 365 days; 95% CI: 183-1,134 days). The median survival of all dogs without mainstem bronchial collapse was 1,600 days, compared with 325 days for dogs with mainstem bronchial collapse. Both types of stents were associated with major complications in nearly half of cases (42 and 43%), such as death, laryngeal paralysis, and the need for additional surgery for continued collapse.

In summary, tracheal stents can be life-saving for dogs with significant limitation of airflow through the trachea and may prolong life for years. Potential complications for both intra- and extra- luminal stents are not trivial. Dogs with cough as a predominant clinical sign (along with airway obstruction) and dogs with concurrent mainstem bronchial collapse may have relatively poorer outcomes. Differences in patient characteristics, bias in case selection, and clinician experience inherently influence the outcome of clinical studies.

REFERENCES

[1]Kandaswamy C, et al. Curr Opin Pulm Med 2009;15:113. [2]Done SH, et al. J Small Anim Pract 1976;17:783. [3]Takahama, et al. Cough 2007;3:1. [4]Grobman M, et al. (abstract) J Vet Intern Med 2015; 29:1227. [5]Abramowicz, M (ed). The Medical Letter 2011;53:9. [6]Barnes PJ. Am J Resp Crit Care Med 2003;167:813. [7]Bach JE, et al. J Am Vet Med Assoc 2004;224:1113. [8]Gruen, ME, et al. J Am Vet Med Assoc 2008;233:1902. [9]Gleeson K, et al. Chest 1997;111:1266. [10]Real FG, et al. PLoS ONE 2016;11:e0147518. [11]Singh MK, J Vet Intern Med 2012;26:312. [12]Adamama-Moraitou KK, et al. Intl J Immunopathol Pharmacol 2011;24:111. [13]Sura PA, et al. J Am Vet Med Assoc 2008;232:228. [14]Durant AM, et al. Vet Surg 2012;41:807. [15]Becker WM, et al. Vet Surg 2012;41:501. [16]Chisnell HK, et al. Vet Surg 2015;44:103. [17]Tinga S, et al. Vet Surg 2015;44:858.

The Respiratory Microbiome in Dogs and Cats

Carol Reinero, DVM, DACVIM, PhD
Columbia, MO

REVISITING THE DOGMA OF STERILITY OF THE LUNG

What have we been taught? The lung has long been described as a sterile organ, in large part based on the inability to routinely culture bacteria from healthy lung specimens (e.g., bronchial brushings, bronchoalveolar lavage, and/or lung biopsy). Defense mechanisms to prevent microbial colonization in the respiratory tract include, but are not limited to, reflexes (cough and sneeze), the mucociliary apparatus, a host of anti-microbial chemical compounds, and innate and adaptive immune responses. When bacteria have been cultured from the lung, this is perceived as a pathogenic state resulting from loss of or ineffective normal immune defense mechanisms, or when there is introduction of foreign material into the deep lung. Under these conditions, bacteria and other microorganisms like fungi and viruses are allowed to proliferate, creating inflammation and other pathogenic sequelae. Thus the role of microbes in pulmonary disorders has focused on infection by pathogenic bacteria leading to development, persistence, exacerbation, and/or progression of respiratory diseases.[1]

What is the current line of thinking? The reason the lung has been thought to be sterile is because standard laboratory-based culture techniques have failed to grow bacteria. However, fewer than 1% of all bacteria can be cultured in the laboratory[2], making the "gold standard" a very insensitive diagnostic. Many species while well adapted to grow in selective body sites *in vivo* will not grow *in vitro* laboratory conditions. As bacteria have been identified in all extremes of environmental temperature, pH, salinity, humidity, etc., it is somewhat absurd to think lung tissue would somehow be protected against bacterial colonization. There is direct contact with the environment when air, which is neither sterile nor particularly pure, is inhaled, and unlike the gastrointestinal tract where there are sphincters separating the different gut compartments, no physical barrier exists between the larynx and alveoli.

How can microbes be identified in the lung? As stated above, laboratory based culture techniques are notoriously insensitive, allowing only a small percentage of all bacteria to be grown in the laboratory. A switch in technique from culture-based to sequencing-based molecular tools to identify microbes has revealed the presence of bacteria where they were previously thought to be devoid of microbial life.[3] Sequencing involves identification of an appropriate marker gene of microbes such as the 16S ribosomal RNA (rRNA) gene and using it to identify and discriminate different microbes in a sample.

The endogenous microbiota is best studied at another mucosal site, namely the gastrointestinal tract, and is known to play a fundamental role in establishing health versus diseased states. The microbiota has a symbiotic relationship with the host providing a barrier function against pathogenic species, regulating cell growth, promoting healing, and modulating immune responses.[4] Exposure of the immune system to a diverse bacterial environment is critical for immunologic development.[5] A similar role for microbial populations in the lung is postulated, although this is a newer field of study.

LET'S TALK TERMINOLOGY

As the microbial communities inhabiting any site are complex, and as the terms used to describe these populations can be confusing, an overview is warranted. All living organisms are hierarchically classified into eight levels starting with the broadest: domain, kingdom, phylum, class, order, family, genus, and species. This classification scheme in part helps to describe the genetic and evolutionary relationships between organisms. While the most precise information may be derived from studies at the species level, it is not always possible to identify the species. That is, new species are continually being discovered with powerful sequencing analysis. Bacterial DNA sequences are thus used as a proxy to identify organisms present in a sample. The most commonly used marker gene is the 16S rRNA gene which is present in all Bacteria and Archeae; it has both highly conserved and hypervariable regions, with the latter allowing taxonomic identification of bacteria. Very briefly, after sample collection bacterial DNA is extracted, the 16S rRNA gene is sequenced (at a depth to accurately reflect the members of the community), the sequences are identified and statistical analysis is performed to describe the microbiome. Sequences obtained from a run can be identified as a specific type of organism or clustered into operational taxonomic units (OTUs) which have a pre-specified degree of similarity (e.g. 97% similarity is commonly used). These OTUs can then be assigned a taxonomy based on comparisons of known sequences present in a reference database. The reference databases have large collections of 16S rRNA sequences from cultured and uncultured microbes, with the latter population having inferred taxonomy based on similarity to known organisms. Populations can be described according to their richness (number of unique organisms or OTUs in a sample) and relative abundance (proportion of the unique organisms or OTUs in relation to all other the others in a sample). Sample diversity takes into consideration species richness and abundance to measure the microbial variability within a sample ("alpha diversity") or between samples ("beta diversity").

DO HEALTHY DOG AND CAT AIRWAYS HARBOR RESIDENT BACTERIAL POPULATIONS?

In this lecture, results of a study of 16 healthy dogs sampled from the upper airways (nasal swabs and oropharyngeal swabs) and lower airways (bronchoalveolar lavage fluid) will be described. Specifically, a summary of the sequence coverage, microbial richness between sites and the complex microbiota specific to each sample site will be presented. Collectively the results document that healthy dogs have identifiable microbial communities throughout the respiratory tract with a degree of regional continuity with deeper descent into the airways. Additionally, using the Phylogenetic Investigations into Communities by

Reconstruction of Unobserved States (PICRUSt) software, gene content of the respiratory microbiota was analyzed providing insight into metabolic pathways used by bacteria colonizing nutrient-poor, low-throughput environments.

Another study serially investigated oropharyngeal and bronchoalveolar lavage fluid samples at three time points from 6 healthy cats. Results of the coverage, richness, abundance and diversity will be discussed. This study illustrated that in healthy cats as has been demonstrated in dogs, there exists a respiratory microbiome.

DOES THE CANINE AND FELINE RESPIRATORY MICROBIOME CHANGE IN THE PRESENCE OF INFLAMMATORY AIRWAY DISEASE?

Chronic bronchitis (CB) is a common and irreversible inflammatory airway disorder of dogs. There are many possible causes of CB including infection and damaging irritants, toxicants, and gases. Infection may also episodically exacerbate CB. In humans, the composition of the airway microbiota impacts health versus disease, with deviations from a core airway microbiota predisposing to development, persistence, exacerbation, and/or progression of respiratory diseases similar to CB such as asthma and chronic obstructive pulmonary disease (COPD). Thus it is plausible that dysbiosis in the canine airway similarly contributes to chronic bronchitis. We will present the results of a study characterizing the respiratory microbiome in pet dogs with chronic bronchitis.

Feline asthma has been estimated to affect between 1-5% of the pet cat population and leads to substantial morbidity and occasional mortality. Allergic asthma is orchestrated by allergen-specific T helper 2 cells that, in genetically susceptible individuals with certain environmental influences, trigger and amplify a local inflammatory response. There are many environmental influences which could tip the balance from a tolerogenic to inflammatory environment and/or sustain airway inflammation. The presence of and role of the respiratory microbiome has not yet been studied to date in asthmatic cats. Our laboratory has developed and characterized an experimental model of feline allergic asthma which closely mimics the spontaneous pet cat and the human disease. Our laboratory has previously used this model to understand more about the pathogenesis of allergic asthma as well as to develop novel diagnostics and therapeutics and have undertaken a study to characterize the endogenous bacterial populations in these experimentally asthmatic cats.

CONCLUSIONS

It is clear from the pilot studies in our laboratory that both upper and lower airways of healthy dogs and cats as well as dogs with chronic bronchitis and cats with asthma are comprised of rich microbial populations that can be identified using sequencing of bacterial 16S rRNA genes. Future studies will be needed to determine if modulation of the microbial populations in inflammatory airway conditions will lead to attenuation of disease.

REFERENCES

1. Beigelman A, et al. *Curr Opin Allergy Clin Immunol* 2014;14:137-142.
2. Hilty M, et al. *PLoS ONE* 2010;5:0008578.
3. Salipante SJ, et al. *PLoS ONE* 2013;8
4. Neish AS. *Ann Am Thorac Soc* 2014;11:201306-201161MG.
5. Segal LN, et al. *Ann Am Thorac Soc* 2014;11:108-116.

Venous Function in Critical Illness

Bernie Hansen, DVM, MS, DACVECC, DACVIM (Small Animal Internal Medicine)
Raleigh, North Carolina

INTRODUCTION

Clinicians treating animals with critical illness recognize the need to monitor and address left heart function, arterial blood pressure, and arterial blood flow. Critical care specialists often bring to bear the technology of arterial hemodynamic monitoring such as echocardiography, direct blood pressure measurement, and measurement of arterial blood flow (cardiac output). In contrast to the perceived primacy of these "glam" monitored variables, veins are considered to be a means to an end: catheters are inserted into them for administration of fluid and drugs, but the conceptual end-point of therapy is support of arterial pressure and blood flow, not optimization of venous function or control of edema accumulation. So why can't the venous system get any respect?

VENOUS RETURN = CARDIAC OUTPUT

In humans about 64% of the estimated total blood volume (V_t) is contained within systemic veins; the remainder is in systemic arteries (13%), the lungs (9%), heart (7%), and capillaries (7%).[1] Veins comprise a low-pressure system; the intraluminal pressure (measured via an indwelling catheter) is not much higher than the pressure of the tissue surrounding the vein, and the pressure gradient across the vessel wall (*transmural* pressure) is very low. In fact, veins are the most compliant structures in the vascular system, and can accommodate large changes in blood volume with little-to-no change in transmural pressure. This property makes the venous system an ideal reservoir for blood, allowing it to provide a relatively constant supply of blood to the heart despite large changes in blood volume.

Venous return

Venous return (VR) must equal cardiac output; there are only slight differences between the two at any moment in time. Under most conditions cardiac output is primarily *determined* by VR to the heart. The driving force that returns blood to the heart is the difference between right atrial intraluminal pressure and the pressure upstream. The relevant upstream pressure is the intraluminal pressure in the venules immediately after the capillary beds, and the VR is proportional to that pressure difference divided by the resistance to flow in the venous system as described by the biological application of Hagen-Pouseulle's law:

$$\mathbf{VR = (P_{ms} - P_{ra})/R_v}$$

where R_v is venous resistance, P_{ra} = right atrial pressure and P_{ms} is the *mean systemic filling pressure*. The mean systemic filling pressure, which can't be measured clinically in the live patient, is the average pressure present throughout the circulation when flow = 0, that is, under conditions of cardiac arrest. In the moments after the heart is stopped arterial pressure falls dramatically and venous pressures rise slightly as the pressure equilibrates across capillary beds. During this process the postcapillary venule pressure does not change; the pressure there has been called the circulatory *pivot pressure* and it is nearly identical with the mean systemic filling pressure. Note that the only involvement of the heart in the VR equation is indirect, as reflected by right atrial pressure. The operating pressure range for the venous circulation is low; 8-12 mm Hg in the venules and 1-2 mm Hg in the vena cava and right atrium. Thus, small changes in P_{ra} can produce dramatic change in the pressure gradient and VR.

Venous response to blood loss

Venous blood can be conceptualized as existing in two compartments, *stressed* and *unstressed* volume (Figure 1). Unstressed blood volume is the amount of blood needed to fill the vascular system to capacity with no increase in vein transmural pressure, and stressed volume is the additional amount required to reach a normal total blood volume and generate normal transmural venous pressure. Consider the blood loss that would occur during passive exsanguination of an animal in lateral recumbency via an intravenous catheter open to the atmosphere at the level of the uppermost surface of the body. Blood would continue to flow from the catheter until the intravascular pressure equilibrated with atmospheric pressure and the transmural pressure reached 0. In the absence of any autonomic response, the blood remaining in the animal is the unstressed volume, and the amount shed through the catheter is the stressed volume.

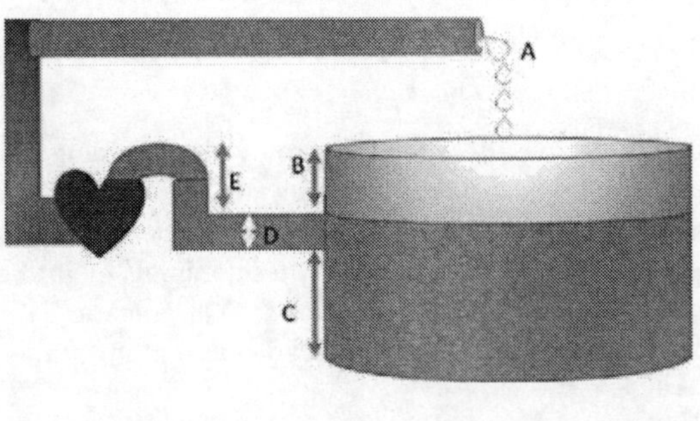

Figure 1: stressed and unstressed volumes. Lowering the tank's drain pipe is the schematic equivalent of shifting blood from V_o to V_s.

A. Arterial blood flow
B. Stressed volume, V_s
C. Unstressed volume, V_o
D. Venous resistance, R_v
E. Central venous pressure, CVP

In addition to the functional division of venous blood into stressed and unstressed compartments, an anatomical division exists between the splanchnic circulation and systemic circulation. The splanchnic circulation includes a very compliant venous system and an arterial supply that is under regional autonomic control that is somewhat isolated from other organ systems. When cardiac output falls, sympathetic activation to the mesenteric arterial supply reduces blood flow and the comparatively very compliant venous system empties blood into the systemic circulation. This shift of blood is rapid; when 90-130 ml of blood was collected over 2 minutes from experimental dogs averaging 14.6 kg in weight, an average of 56 ml of blood shifted from the splanchnic circulation to systemic by the end of the collection.[2]

Other mechanisms serve to maintain VR during hemorrhage. The autonomic response includes rapid constriction of systemic arterioles to maintain BP and venules/veins to shift blood from V_o to V_s; this latter effect serves to maintain the driving pressure for VR. A more gradual response is a shift of fluid from the interstitial compartment to the vasculature as capillary pressure falls. This response is slow (requiring 2-3 days) but is another mechanism to increase V_t and improve VR.

Venous function in distributive shock

Distributive shock is characterized by maldistribution of blood flow due to excessive dilation of regional arterioles and venules. The most common cause in companion animals is severe sepsis and septic shock. Early in the course of severe sepsis, P_{ms} falls secondary to a fall in V_o as plasma fluid shifts to the interstitium secondary to increased capillary permeability and uncoiling of interstitial glycoproteins[3], and blood shifts from V_s to V_o secondary to venule dilation.[4] Although hemoconcentration increases blood viscosity and tends to increase R_v, this is offset by a reduction in R_v by dilation of large veins. The combination of lower R_v and shunting of arterial blood through low-resistance arteriole tend to increase VR/CO, an effect that is markedly enhanced by increases in V_o from fluid administration as described below.

WAYS TO INCREASE VENOUS RETURN

Increase driving pressure

VR may be increased by increasing the driving pressure ($P_{ms} - P_{ra}$), reducing blood viscosity, or increasing the diameter of large veins (these last two both reduce R_v). Clinically, the only situations where driving pressure may be increased via reduction of P_{ra} is when P_{ra} is pathologically high secondary to pericardial tamponade, pulmonary embolism/hypertension, or right heart failure, and interventions such as pericardiocentesis, anticoagulation, or administration of pulmonary vasodilators are applied to reduce it. In most situations driving pressure is increased therapeutically by increasing P_{ms} relative to P_{ra} (and CVP). This is routinely accomplished by administration of enough intravenous fluids to effect an increase in V_s. Because P_{ra} and CVP are *barriers* to VR, increasing the CVP with fluid therapy will be associated with increased VR and CO *only* if the associated increase in P_{ms} is greater than the increase in CVP. Furthermore, the heart must be operating at a relatively steep portion of the Frank-Starling curve to respond to an increase in preload with a clinically significant increase in CO. Significant right and/or left ventricular dysfunction will blunt the response to increases in driving pressure. Cardiac reflexes are designed to hold CVP (and right heart filling pressure) constant within a broad physiological range – that's why the driving pressure and VR are increased in fluid-responsive patients. Administering fluids until the CVP rises substantially almost guarantees that P_{ms} will be high enough to favor generalized edema.

Lower viscosity and/or R_v

Transfusion of whole blood with viscosity identical to the recipient blood will increase V_t, V_s, and P_{ms} with no change in R_v; VR is increased solely by the increase in driving pressure. In addition to increasing the driving pressure for VR, administration of crystalloid fluids also reduces Rv by reducing blood viscosity and has greater impact on VR per ml of increase in V_t. Vasoactive drugs have varied effects on VR. Pure vasoconstrictors increase systemic vascular resistance and constrict large veins, both of which tend to reduce VR/CO by increasing afterload and R_v. However, these drugs (for example, phenylephrine) also constrict postcapillary venules, shifting more blood to V_s and increasing P_{ms}, an effect that counters some of the reduction in VR/CO. The net result is an increase in CVP/P_{ra} with a reduction in VR/CO. Vasopressors that also stimulate the heart (e.g. dopamine, norepinephrine) combine the venous effects of pure vasopressors with inotropic properties that on balance usually increase VR/CO. Inodilators (e.g., milrinone, pimobendan, dobutamine) reduce afterload and increase myocardial contractility at the same time the drugs reduce P_{ms} and R_v; the net effect is a larger increase in VR/CO than seen with dopamine. For any of these 3 classes of drugs the effect in sick animals will depend on the underlying disease and on the baseline conditions of V_s, V_o, R_v, and cardiac function.

PREDICTING FLUID RESPONSIVENESS

Once the underlying cause of hypovolemic shock from hemorrhage or dehydration is identified and any active bleeding controlled, resuscitative fluid therapy is relatively straightforward and patient response is usually unambiguous and quantifiable based on physical signs alone. In contrast, the response to fluids in animals with distributive or cardiogenic shock may be ambiguous and the demarcation between benefit and harm from fluid resuscitation is blurred. Supportive treatment of these animals might be substantially improved if clinicians could accurately predict whether administration of fluids might achieve clinically important improvement in VR/CO with minimal accumulation of edema. Central venous pressure has been used for decades as a monitored variable and continues to be widely used despite accumulating evidence that as a single measurement CVP has essentially no relationship with circulating blood volume or fluid responsiveness when the baseline value is within a wide physiological range.[5] That CVP has no bearing on fluid responsiveness should come as no surprise when one realizes that the cardiovascular system is *designed* to prevent fluctuations in CVP across a large range of blood volume and heart function; it does not change unless alterations in either of those variables is extreme. Whereas animals with distributive shock routinely partially respond to initial fluid therapy, most have "normal" CVP at presentation and immediately after initial fluid administration, and persistently normal CVP despite continued signs of shock. Similarly, in one structured review baseline pulmonary artery occlusion pressure did not predict the response to fluids and an increase in PAOP following fluid administration had no relationship with change in cardiac output in human ICU patients (who had presumably already been fluid resuscitated prior to ICU admission).[6] In that same review, other static measurements that are surrogates for pressure (right ventricular end-diastolic volume, left ventricular end-diastolic area) fared no better.

Because of the potential for harm from excessive fluid administration – especially in immobile, inflamed ICU patients - many investigators have studied alternative methods of predicting fluid responsiveness. Most studies in humans define fluid responsiveness as an increase in cardiac index > 15% or improvement in some other surrogate measure following rapid administration of 500-1000 ml of crystalloid fluid. Although no single measurement is useful in every patient, some dynamic measurements do predict fluid responsiveness in many. Examples of these include provocative tests of venous return (passive leg raises), transesophageal echocardiographic (TEE) measurement of changes in the cyclical diameter of the superior (cranial) or inferior (caudal) vena cavae induced by positive pressure ventilation, and TEE measurement of peak aortic flow velocity in response to the same maneuvers. Passive leg raises (lifting the legs to a 45° angle relative to the horizontal head and torso) abruptly increase VR without the administration of fluids, and response is measured by echocardiography, changes in pulse pressure, or other surrogates for CO. Because no fluids are administered, failure to respond is not complicated by unproductive aggravation of edema. If CVP is used to guide fluid therapy, a few suggestions are offered:

- You're wasting your time if you use a water column manometer; electronic monitoring is essential for accurate interpretation.
- Baseline values don't mean much unless they are at either extreme: subatmospheric or > 10-12 mm Hg
    - Also: If you needed CVP to tell you that your patient was hypovolemic, someone else should be doing your physical exams for you!
- Using CVP to guide "chronic" (more than an hour) fluid therapy is worthless. Worse, targeting a mild elevation of CVP to guide fluid therapy for AKI support or sepsis or other conditions guarantees edema (and potentially worsening of the AKI)
- CVP should be used as a guide for when to *stop* fluid resuscitation based on change following rapid (1-5 minutes) administration of a considered volume of fluid.
- Don't feel inferior for using CVP instead of a surrogate static measure like echocardiographic estimate of EDD: they have identical limitations can be equally bad predictors of fluid responsiveness.

REFERENCES

1 Milnor WR. The circulatory system. In: Milnor WR, editor. Cardiovascular physiology.New York: Oxford University Press; 1990. p. 3-28.
2 Brooksby GA, Donald DE. Dynamic changes in splanchnic blood flow and blood volume in dogs during activation of sympathetic nerves. Circ Res 1971 Sep;29(3):227-38.
3 Lucas CE. The water of life: a century of confusion. J Am Coll Surg 2001 Jan;192(1):86-93.
4 Magder S, Vanelli G. Circuit factors in the high cardiac output of sepsis. J Crit Care 1996 Dec;11(4):155-66.
5 Marik PE, Cavallazzi R. Does the central venous pressure predict fluid responsiveness? An updated meta-analysis and a plea for some common sense. Crit Care Med 2013 Jul;41(7):1774-81.
6 Michard F, Teboul JL. Predicting fluid responsiveness in ICU patients: a critical analysis of the evidence. Chest 2002 Jun;121(6):2000-8.

Arterial Function in Critical Illness
Stephanie Istvan, VMD, DACVECC
San Diego, CA, USA

INTRODUCTION

The critically ill patient suffers from a diverse range of cardiovascular compromise including hypovolemic, cardiogenic and distributive shock states. Clinically, an understanding of the effects of critical illness on both the cardiac and vascular elements together is useful in identification and resuscitation of cardiovascular dysfunction to reduce end-organ damage and mortality. Although improvements in resuscitation and post-resuscitation care have been made, organ dysfunction, increased length of hospitalization and mortality rates are associated with both lack of aggressive resuscitation and over resuscitation.[1,2]

ARTERIAL FUNCTION

The arterial system has two closely related hemodynamic functions. First, the arterial system delivers and distributes blood from the left ventricle to peripheral tissues to maintain the optimal relationship between oxygen supply and demand. Tissue demands for oxygen and carbon dioxide exchange contribute to autoregulation. Autoregulation is the ability of organs to maintain constant blood flow despite changes in blood pressure and direct blood flow distribution from large arteries to arterioles. Precapillary sphincters located at the root of each arteriole control blood flow into the tissue capillary bed. These sphincters are affected by local mediators (hydrogen ions, oxygen, and carbon dioxide) and the sympathetic nervous system.[3] The important network of capillaries between arterioles and venules is known as the microcirculation. The microcirculation regulates hemodynamics and supplies oxygen and nutrients to tissues.

The second major function of the arterial system, the "dampening function", diminishes blood flow and pressure oscillations resulting from variations in the pulsatile ejection of blood from the left ventricle. This function allows peripheral organ perfusion through a steady flow and pressure.[4]

Collectively, the arterial circulation is the effective circulating volume. Effective circulating volume depends upon an efficient cardiac pump, vascular integrity and adequate blood pressure (which is determined by cardiac output, vasomotor tone and perfusion). If any of these components are inadequate, a neurohumoral response augments these components to restore effective circulation in the arterial system.

RESPONSE OF THE CARDIOVASCULAR SYSTEM TO SHOCK

The cardiac and vascular elements can respond appropriately or inappropriately to neurohumoral activity. Response is related to the underlying cause of the illness, and when there is an inappropriate response, advanced pharmacologic intervention is required to restore effective circulation.

Neurohurmoral response is mediated by receptors in the carotid sinus and aortic arch. These receptors sense a decrease in effective circulating volume via a decrease in vessel wall stretch, thus activating the sympathetic nervous system and adrenal cortex. Receptors in the kidney, including the granular cells in the juxtaglomerular apparatus in the afferent arterial wall, also sense changes in effective circulating volume. An appropriate response ultimately leads to increased renin release, which results in increased angiotensin II. Increased angiotensin II leads to vasoconstriction, enhanced tubuloglomerular feedback and decreased GFR, all of which have a direct effect to increase sodium and water reabsorption and stimulation of ADH release. Angiotensin II stimulates release of aldosterone from glomerulosa cells in the adrenal cortex promoting sodium and water reabsorption in the principal cells of the distal tubule.[5] Common shock states with appropriate compensation include hemorrhagic shock, heart failure and hypovolemia.

Conversely, anaphylaxis, systemic inflammatory response syndrome (SIRS), sepsis, and severe or end-stage shock of any category make up a group of shock states with inappropriate response to decreased effective circulating volume. In these states, vascular hyporeactivity and hypotension persist after correction of the hypovolemic component due to vasodilation. Hypotension associated with vascular hyporeactivity is correlated with a high mortality rate.[6]

Specifically, the arterial effects of septic shock are complex and incompletely understood. Multiple proinflammatory mediators (interleukin-1β, interleukin-6, tumor necrosis factor alpha) result in inappropriate peripheral vasodilation, altered distribution of blood flow, and myocardial dysfunction. The following are purported mechanisms of vasodilation in these pathologic states:

- Potassium channels (K_{ATP}) in the arterial system are activated with septic shock (and with lactic acidosis) allowing for potassium to exit the cell, further hyperpolarizing the cell membrane. Hyperpolarized cell membranes lead to inhibition of cytoplasmic calcium influx by inactivation of the voltage-gated calcium channel, which is required for vasoconstriction.[7]
- High levels of nitric oxide result from increased inducible nitric oxide synthase in the vascular smooth muscle and endothelial cells. Nitric oxide activates the myosin light-chain phosphatase and plays a role in activation of the potassium channel that is sensitive to cytosolic calcium (K_{Ca}).
- α-adrenergic receptor internalization, uncoupling of receptors from secondary messengers and reduced catecholamine availability
- Depletion and decreased action of vasopressin[8]

- Myocardial depression (myocardial depressant factor) associated with systemic inflammation or sepsis results in decreased preload and cardiac output, and peripheral vasoconstriction. Dysfunction is complicated and involves down-regulation of β-1 adrenergic receptors and decreased sensitivity to calcium.
- Critical illness-related corticosteroid insufficiency results from inadequate corticosteroid release, peripheral resistance to corticosteroids, and alteration in glucocorticoid receptors.

EVALUATION OF ARTERIAL FUNCTION

Evaluation of arterial function includes physical examination, assessing urine output, arterial blood pressure measurement, blood lactate concentration, cardiac output measurement, ultrasound imaging, and assessment of central venous oxygen saturation. Additionally, determining cardiac filling pressures via cardiac catheterization, and microcirculatory imaging (near-infrared spectroscopy, sidestream dark field microscopy) can be utilized. Many of these techniques offer advanced evaluation of tissue perfusion and blood flow. Given that the latter are less routinely available in practice, physical examination, measurement of blood lactate concentrations, and arterial blood pressure evaluation are more routinely used to evaluate the cardiovascular system.

Arterial blood pressure monitoring provides evaluation of cardiac output (amount of blood in the system) and systemic vascular resistance (precapillary vascular tone). Non-invasive monitoring with oscillometric or Doppler methodology provides intermittent readings, while invasive blood pressure monitoring allows for continuous evaluation of arterial pressure waveforms and blood pressure. Systolic pressure is primary determined by stroke volume and arterial tone, while diastolic pressure is determined by vascular tone and heart rate.

Of note, pulse pressure variation can be utilized to assess hypovolemia in an intubated, manually ventilated patient. During a positive pressure breath, the positive intrathoracic pressure reduces left ventricular afterload and left ventricular preload is augmented. The maximum pulse pressure will occur during positive pressure inspiration (PP_{max}) and will be at a minimum value (PP_{min}) during expiration. Over a single breath, pulse pressure variation can be calculated. A value of > 13% correlates with volume responsiveness.

TREATMENT

Treatment should be directed towards the underlying disease process. Intravenous fluid administration is one of the first interventions in resuscitating the hemodynamically unstable patient. Once intravascular volume is deemed appropriate and cardiovascular dysfunction persists, vasopressor therapy should be initiated.

Norepinephrine is a potent α-1 agonist-mediated vasoconstrictor, and is a commonly used first-line agent for refractory hypotension in sepsis.[9] Infusions are titrated to achieve desired blood pressure with a dose range of 0.1mcg/kg/min – 1mcg/kg/min.

Epinephrine has α-1 and β-1 agonist properties that are most commonly used for anaphylaxis or cardiopulmonary resuscitation. Epinephrine increases inotropy and chronotropy as well as acts as a potent vasoconstrictor. Epinephrine can be administered intermittently for anaphylaxis but is most often administered as a continuous rate infusion for refractory hypotension with dose ranges of 0.005mcg/kg/min – 1mcg/kg/min. Epinephrine may have deleterious effects in dogs with septic shock.

Dobutamine, a β-1 agonist, is a positive inotropic agent used in shock states to increase cardiac output in patients with persistently low cardiac output. Dobutamine can be administered initially at an infusion rate of 5mcg/kg/min, and titrated up in increments of 2.5-5mcg/kg/min to 15mcg/kg/min in dogs. Dobutamine should be use with caution in cats due to potential central nervous system complication (seizures).

Dopamine, an epinephrine and norepinephrine precursor, generally has variable effects depending on the dose. At low doses (2.5-5mcg/kg/min), dopamine stimulates DA_1 and DA_2 receptors and has been advocated for renal protection. However, a recent meta-analysis no longer recommends its use for renal protection.[10] At doses between 5 and 10 mcg/kg/min, dopamine has β-1 adrenergic effects, resulting in positive inotropy, thus increasing cardiac output. At doses greater than 10mcg/kg/min, there are α-adrenergic effects, which result in vasoconstriction. Dopamine is no longer recommended as a first line vasopressor in severe sepsis.[9]

Vasopressin (antidiuretic hormone, ADH) exerts effects predominantly through the V_1 receptor in vascular smooth muscle (skin, skeletal muscle, adipose, bladder, myometrium, liver, spleen, pancreas, and thyroid gland) leading to vasoconstriction via the G_q protein cell-signaling cascade. Vasopressin concentrations initially rise but eventually become depleted in progressive shock states.[8] A benefit has been shown for vasopressin administration in these cases. In a study with five dogs, there was an elevation in mean arterial pressures after administration of vasopressin to treat dopamine-resistant hypotension.[11] Vasopressin can also be utilized to take advantage of its permissive effects on vascular response to catecholamines, such that low dose vasopressin infusion may restore vascular responsiveness and improve the hemodynamic response to therapy with catecholamines. The dose in dogs was extrapolated from humans and ranges from 0.05-4 milliunits/kg/min.

CONCLUSION

Critically ill patients with evidence of arterial dysfunction demand aggressive treatment to prevent multiple organ damage. Vasopressor agents are often required in severe shock states. An understanding of cardiac and vascular dysfunction in the critically ill will aid in personalizing management of shock and appropriate treatment selection.

REFERENCES

1. Rivers E, et. al New Eng J Med. 2001;354:1368.
2. Silverstein DC, et. al J Vet Emerg Crit Care. 2012;22(6):666.
3. Cunningham JG, et. al Textbook of Veterinary Physiology, 4th Ed. Saunders, Stl Louis, 2007
4. Nichols WW, et. al McDonald's Blood Flow in Arteries: theoretical, experimental and clinical principles. 6th Ed. Boca Raton, FL, 2011:77
5. Rose B, Clinical Physiology of Acid-Base and Electrolyte Disorders. 5 ed. New York: McGraw-Hill, 2001;258.
6. Bernardin, et. al Intensive Care Med. 1996; 22:17.
7. Landry DW, et. al New Eng J Med. 2001;345(8):588.
8. Blanchard T, et. al Crit Care Med. 2003;31:1752.
9. Dellinger RP, et. al Crit Care Med. 2013;41(2):580.
10. Kellum, et. al Crit Care Med. 2001;29:1526.
11. Silverstein DC, et. al J Vet Emerg Crit Care. 2007;17(4):399.

Vascular Access

Elizabeth Rozanski DVM, DACVIM(SAIM), DACVECC
North Grafton, MS

Vascular access is essential for care of ill animals; dogs and cats with critical illness are often hospitalized for an extended period of time while animals with life-threatening emergencies will usually present to the Emergency service without the presence of a pre-existing IV catheter. In both scenarios, it will be essential to gain vascular access. The goal of this session is to discuss the pros and cons for vascular access in small animals in a variety of settings.

TYPES OF VASCULAR ACCESS

1) Peripheral catheter – This is the most commonly placed type of catheter. These catheters are available in a number of sizes (gauge) with common sizes of 16, 18, 20, 22 and 24 ga, with higher numbers representing smaller sizes.
2) Central venous catheter- Catheters may be preferentially inserted into a larger vein, ideally the vena cava. This permits the delivery of hyperosmolar fluids (eg. High concentrations of dextrose, KCl or parenteral nutrition) as well as aspiration of blood. These are particularly useful in dogs and catheter with an anticipated long course in the hospital. Central venous catheters can have a single lumen or multiple lumens.
3) Intraosseous catheter- IO catheters or needles a placed into the marrow cavity of a long bone (eg. Femur or humerus).

They are almost always used during resuscitative efforts, or in very small patients. May be considered a central line. Easy to place with automated devices (EZ-IO catheter). They may be be hard to keep in place, and painful in conscious patients. However, they are often very useful in rapid access. A recent study documented that EZ-IO catheters are faster to place for almost all individuals that IV catheters.

Vein choices

Conventional vein choices include the following: In animals that have had limited catheters placed in the past, any of the following veins are usually successful

1) Cephalic
2) Saphenous
3) Jugular
4) Medial femoral (cats)
5) Umbilical veins (neonates)

Placement technique

1) Percutaneous-
2) Cut-down/Mini-cut down
3) Ultrasound guided

THE EMERGENCY PET WITH COLLAPSED(HYPOVOLEMIC) VEINS.

For collapsed dogs in lateral recumbency, the DOWN cephalic vein is much more easily catheterized, due to an easier angle, and also some internal occlusion by the body weight which helps to distend the vein. The UP saphenous vein is also easier to hit. In collapsed cats, the veins of choice are the medial saphenous vein and the jugular vein. In some primary care practices, jugular venipuncture is less commonly pursued; if it is possible to use the jugular vein more routinely, it is often easier to successfully catheterize this vein the pet with "collapsed" veins. Place a towel under the neck to better raise the vein. Catheter brands may have different characteristics in tissue handling and ability to show a "flash". Recall that in pets with lower venous pressure (eg. Hypovolemia), it may be hard to get a flash, or may be relatively "slow" compared to normovolemic pets. The size of the catheter placed should be a large as you can comfortably place, but it may be required to go a size or two smaller. In most cats, a 20 or 22 gauge catheter can be placed, and in most dogs, an 18 or 20 ga catheter. In collapsed neonates or pediatric patients, a jugular catheter can often be relatively easily placed, although it should be recalled that the vein will be quite superficial. A 20 ga needle may also be placed as an IO catheter, most easily down the shaft of the femur. Intraosseous catheters may also be placed quickly in larger or adult animals, using either a bone marrow needle, or a commercially available device called an EZ-IO gun. The EZ-IO gun is quite simple to use, even with minimal training. Individual needles are expensive, but potentially may be autoclaved and re-used. The final option for vascular access is a cut down, which may be performed either as a mini-cut down, or as a full cut down. A mini-cut down is performed by nicking the skin with a scalpel blade or needle, to simply try to reduce the likelihood that the catheter will burr while a full cut down is performed by making a incision parallel to the vein, and then carefully isolating the vein and introducing a catheter. This technique can be very effective although it should be noted it is easy to go through the back of the vein.

THE ICU PET WITH "NO VEINS LEFT"

ICU pets often have significant challenges with vascular access. Useful tips for placement of catheters in this population include the following.

1) Use of "secret" veins, these include metatarsal and metacarpal vessels, and "new" branches of the cephalic vein. The ear veins may be used as well.
2) Ultrasound assisted access is very common in people, and may be useful in animals as well
3) A smaller catheter can be traded out for a larger catheter using a guide-wire exchange; this is very useful in situations were it is not clear there are other good options.
4) In dogs with severe edema, wrapping the leg tightly for a few minutes and milking the edema away from the site commonly reveal a useful vein in the anatomically correct position.
5) Clotted catheters may be saved via infusion of TPA (1 mg total)
6) In cases where palpation and venous pressure prevents ease of "bleed back" of the catheter, ultrasound evaluation of the right heart can support or refute if the catheter is in the vein. Similarly, contrast material may be used if fluoroscopy if available. (Figure 1)

Brachycephalics/Chondrodystrophic dogs may be particularly challenging to place catheters in. In the front legs of chondrodystrophc dogs, it is often most rewarding to place the catheter almost perpendicular to the long access of the limb, where the vein crosses the carpus, which results in the catheter running at an "odd" angle.

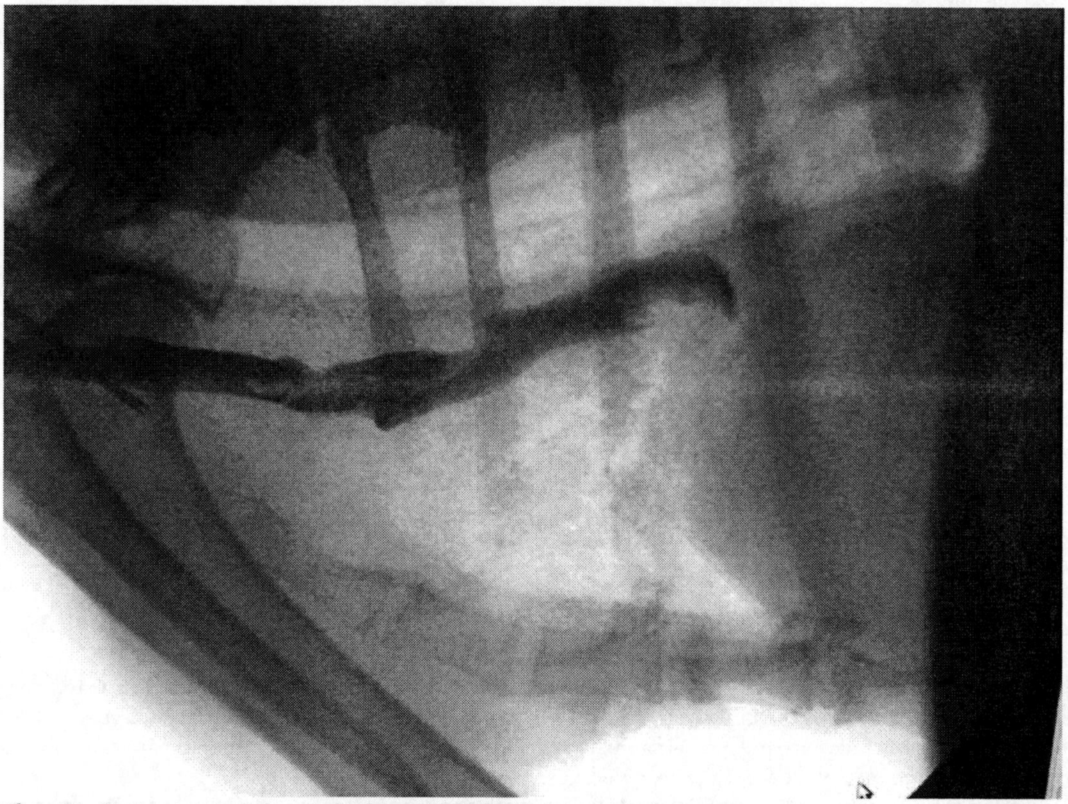

Figure 1. Contrast material may also be used to confirm correct placement.

Myths of catheter placements

1) *Automatic exchanges*: In some practices, it is an automatic nursing order to change a catheter every 3-5 days. There is no evidence in people or in animals that this is required and in fact in people there is evidence that it is harmful. Catheter wraps should be inspected daily, and broken down and re-wrapped if there is evidence of a problem or the bandage becomes soiled.
2) *Vein valves*: While veins have valve within them, they are not the reason a catheter won't feed; the catheter that won't feed well after apparently successful venipuncture, is almost invariable "burred" so the stylet is within the vein, but the catheter part is not.
3) *Bigger is better:* The size (ga) of the catheter being placed should be a combination of the flow rate required, and the size of the vein of the patient. The small collapsed cat with congestive heart failure, does not require a 20 ga catheter to give furosemide, and beagle getting ivf at 40 ml/hr does not benefit from an 18 ga catheter. While large bore catheters are very useful for bolusing large volumes of fluid, if they are not required, a larger catheter is more likely to result in thrombophlebitis and vasculitis, as well as discomfort to the patient during placement.

CUT DOWN CATHETERS

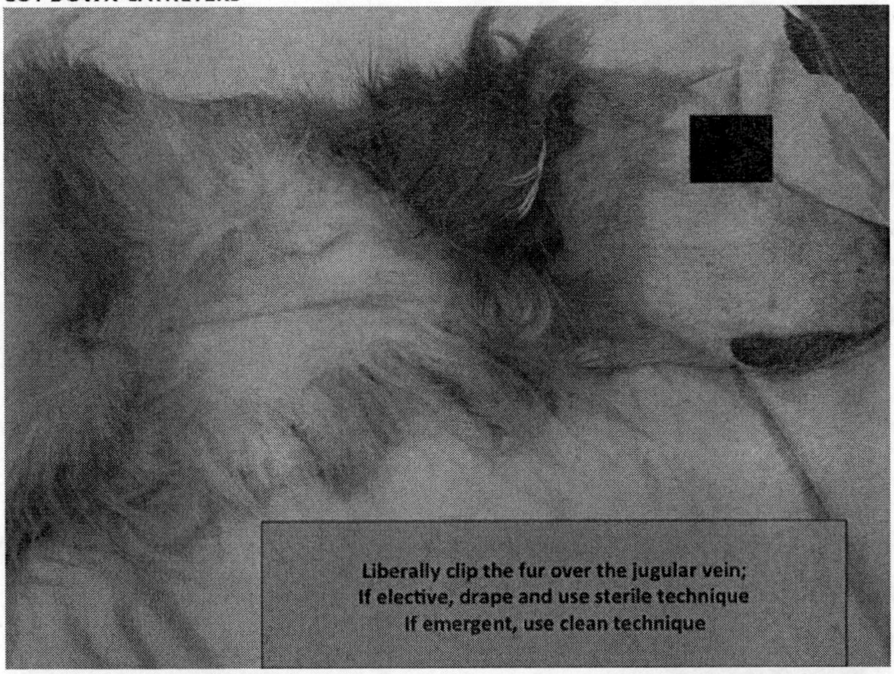

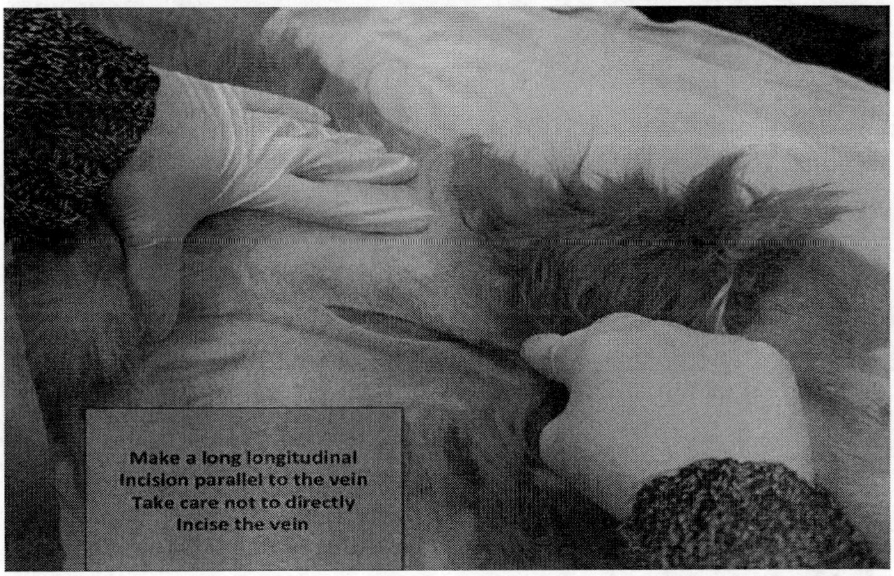

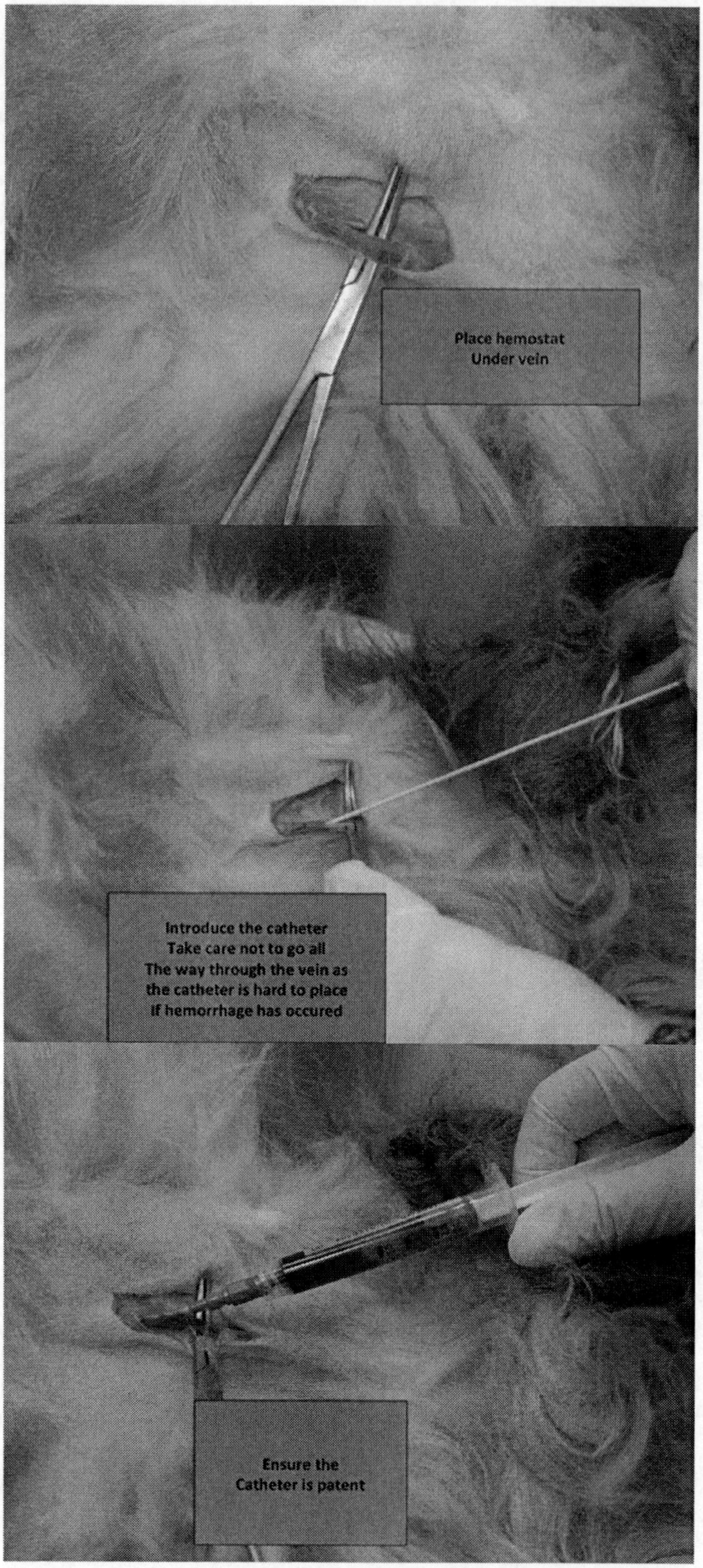

REFERENCES

Allukian et al. JVECC, in press, Chamberlain et al, JVECC 2013m 23(5) 498-503 Cahalane AK, J Am Vet Med Assoc. 2007 Nov 1;231(9):1354-60.Beal MW, Hughes D.Clin Tech Small Anim Pract. 2000 May;15(2):101-9.

Early Goal Directed Therapy: Updates and Practices

Lauren A. Sullivan, DVM, MS, DACVECC
Fort Collins, Colorado

Optimizing outcomes during the resuscitation of critically ill patients requires prompt recognition of the underlying condition, administering appropriate interventions, and monitoring the individual's therapeutic response using both quantitative and qualitative markers. The concept of "early goal directed therapy" (EGDT) specifically focuses on those patients at an increased risk of morbidity and mortality, who are likely to require aggressive management of their hemodynamic status due to a high degree of systemic illness. Those patients experiencing severe sepsis or septic shock are the representative population who receives EGDT as soon as feasibly possible. The impact of EGDT on outcome in people with severe sepsis or septic shock was first reported in a landmark study published by Rivers et al in 2001. The authors found that targeting specific hemodynamic goals during the first 6 hours of resuscitation lead to improved hemodynamic parameters, lower clinical severity scores, and less severe organ dysfunction. Their algorithm for resuscitation of people in severe sepsis or septic shock involved targeted central venous pressure (CVP, 8-12 mmHg), mean arterial pressure (MAP, >65 mmHg), central venous oxygen saturation (ScvO2, $\geq$ 70%), and hematocrit (Hct, $\geq$ 30%).

In addition to the previously mentioned benefits, the Rivers study also identified a significant difference in mortality when comparing the EGDT to a standard group. Those received EGDT had lower mortality rates in-hospital, as well as at the 30-day and 60-day time points. The overall conclusion was that EGDT therapy, provided at the earliest stages of severe sepsis and septic shock, has significant short-term and long-term benefits. EGDT was subsequently incorporated into the 6-hour resuscitation bundle of the Surviving Sepsis Campaign (2008, 2008 and 2012), placing a focus on aggressive fluid resuscitation to obtain a CVP > 8mmHg, use of vasoactive agents to achieve a MAP $\geq$ 65mmHg, transfusion of red blood cells to maintain Hct $\geq$ 30%, and inotropic agents to achieve a ScvO2 $\geq$ 70%.

The benefits of early goal-directed therapy in terms of outcome are thought to be multifactorial. In the Rivers study, the standard therapy group was twice as likely to die due to sudden cardiovascular collapse, suggesting an abrupt transition to severe disease as an important cause of death. Early identification of patients at high risk for cardiovascular collapse, as well as early therapeutic intervention to restore a balance between oxygen delivery and oxygen demand, may help decrease mortality in this patient population. Close monitoring of hemodynamic status (CVP, BP, ScvO2) and adjustment of interventions to ensure adequate oxygen delivery is the crux of EGDT. Prevention of hypoxemia through the use of EGDT may also mitigate other downstream effects of severe sepsis. Global tissue hypoxia is a stimulus of the systemic inflammatory response syndrome, contributing to endothelial activation and disruption of the homeostatic balance among coagulation, vascular permeability, and vascular tone. Prolonged or severe imbalances in oxygen delivery and utilization lead to microcirculatory failure, intracellular hypoxia and organ dysfunction. For these reasons, rapid disease identification and intervention in those with sepsis could prevent downstream sequelae that often contribute to mortality.

Following the Rivers study, EGDT became the standard of care when treating those with severe sepsis or septic shock. Further studies evaluating EGDT were largely observational or "before/after" studies. Several meta-analyses were published utilizing these studies to further demonstrate the benefit of EGDT. Barochia et al (*Crit Care Med* 2010) evaluated eight unblinded trials: one randomized clinical trial and seven observational studies with historical controls. This meta-analysis found that use of the sepsis bundles was associated with a consistent and significant increase in survival (odds ratio 1.91). A second meta-analysis (Chamberlain et al, 2011) evaluated both the 6-hour resuscitation bundle as well as 24-hour sepsis management bundle. Details of those bundles were as follows:

Sepsis Resuscitation 6-hour bundle	**Sepsis Management 24-hour bundle**
• Obtain microbiology samples and lactate measurement	• Administer low-dose corticosteroids, based on hospital policy
• Administer appropriate antibiotics	• Administer recombinant human activated protein C, based on hospital policy
• Administer early resuscitation therapy, including: **1.** Fluids to achieve a CVP of 8-12 mmHg **2.** Vasopressors to achieve a MAP > 65 mmHg **3.** Maintain ScvO2 $\geq$ 70% with packed red blood cells or inotrope therapy	• Maintain glycemic control (120-150 mg/dL) or (6.8-8.4 mmol/L)
	• Maintain plateau airway pressures $\leq$ 30 cmH20 in mechanically ventilated patients

Pooled results from this meta-analysis showed that the 6-hour resuscitation bundle achieved the greatest survival benefit (odds ratio 2.12), but survival benefits were also observed with the 24-hour management bundle (odds ratio 1.65). Both bundles together achieved a combined survival benefit (odds ratio 1.74). Additionally, this study reported a positive association in individual bundle components (targets of ScvO2 and blood glucose management). Through 2011, individual studies as well as meta-analyses overwhelmingly supported EGDT as a means to improve outcome in severe sepsis and septic shock. Limitations of these studies, though, were noted by several authors. Most studies were not methodological rigorous in design (e.g., lack of blinding, before/after study designs, retrospectively identified historical controls, potential selection bias, and completeness of data collection). There was also variation in participating healthcare workers and patients studied. It was believed this most recent meta-analysis brought together a large enough sample (23,438 patients) to outweigh these issues.

Conclusions reached by Rivers and other investigators went largely unchallenged until the publication of three additional studies. These large, randomized controlled trials extended across three continents and were identified as follows: (1) the ARISE study (Australasian Resuscitation in Sepsis Evaluation; Australia; *N Engl J Med* 2014), (2) the ProCESS study (Protocolized Care for Early Septic Shock; USA, *N Engl J Med* 2014), and (3) the ProMISe study (Protocolised Management In Sepsis; United Kingdom; *Health Technol Assess* 2015). All three studies failed to identify an outcome benefit when EGDT was performed, when compared to a control group receiving standard resuscitation. The ARISE study (Australia) enrolled 1600 patients, assigning 7696 to the EGDT group and 804 to the usual-care group. Patients in the EGDT group received a larger volume of fluids and they were more likely to receive vasopressors, red-cell transfusions, and dobutamine. Mortaility was 18.6% and 18.8% for the EGDT and usual-care groups, respectively. There was no significant difference in survival time, in-hospital mortality, duration of organ support, or length of hospital stay. The ProCESS study (USA) enrolled 1341 patients, of whom 439 were randomly assigned to protocol-based EGDT, 446 to protocol-based standard therapy, and 456 to usual care. Resuscitation strategies differed significantly with respect to monitoring of ScvO2 and the use of intravenous fluids, vasopressors, inotropes, and blood transfusions. By 60 days, mortality was 21%, 18.6% and 18.8% for each respective group. The ProMISE study (UK) tested both the clinical effectiveness and cost-effectiveness of EGDT. This study enrolled 623 patients in the EGDT group and 620 in the usual-care group. Similar to the other two studies, there was no difference in mortality (29.5% versus 29.2%) at 90 days. The ProMISe study also identified an increased cost associated with EGDT.

The most recent meta-analysis evaluating EGDT, (Zhang et al, *BMC Medicine* 2015), which included the ARISE and ProCESS trials, also failed to identify a survival benefit with EGDT. This meta-analysis found that those within the EGDT group had a similar length in ICU stay, in-hospital stay, ventilation rate and vasopressor support when compared to the control group. Results from these latter studies call the benefits of EGDT into question, indicating that patient outcome may not be improved using current resuscitation endpoints. Possible explanations for why existing EGDT recommendations may not be improving outcome are as follows:

- Central venous pressure (CVP): This is the first endpoint of the 6-hour resuscitation bundle. However, CVP does not reliably predict volume responsiveness and its utility during fluid resuscitation has come under heavy criticism. Elevated CVP (> 8-12 mmHg) has been associated with fluid overload, impaired microcirculatory perfusion and increased organ dysfunction. Kelm et al (*Shock* 2015) found that, in septic patients undergoing EGDT therapy, fluid overload was present in 27% of patients at 24 hours, and 48% of patients had fluid overload at 72 hours. In several subsets of critically ill people, fluid overload has been identified as an independent predictor of mortality. Therefore, using CVP to determine adequacy of volume resuscitation, or need for additional intravenous fluids, likely leads to fluid overload and its associated consequences.
- Blood transfusions and ScvO2: Blood transfusions are not without their inherent risks or side effects, and may further contribute to morbidity in critically ill patients. Providing additional red blood cells to a patient with a Hgb > 7 g/dl will not increase ScvO2 or improve blood lactate concentration, and blood transfusions have not been shown to positively influence outcome in septic shock. The Transfusion Requirements in Septic Shock (TRISS trial, *N Engl J Med* 2014) did not identify an advantage to targeting a Hgb concentration of 9 g/dl, when compared to a more conservative Hgb concentration (7 g/dl). The ProCESS study did not require placement of a central venous catheter and monitoring of ScvO2 in the control group; instead this group was automatically transfused once Hgb < 7.5 g/dl. Given a similar outcome in both groups, placing a central line and monitoring ScvO2 may not be warranted for all patients. Additionally, targeting higher Hgb concentrations has not demonstrated a survival advantage in septic people.
- Positive inotropes: Use of agents like dobutamine assist with increasing cardiac output, but positive intotropes may not improve perfusion parameters. A 2013 study in septic shock patients found that dobutamine failed to improve sublingual microcirculatory, metabolic hepatosplanchnic, or peripheral perfusion parameters despite inducing a significant increase in systemic hemodynamic variables.

Another criticism of the Rivers study was an inability to determine which EGDT intervention was providing an outcome advantage. It is difficult to discern elements that could be helping versus harming due to the study design of most EGDT trials. Some EGDT elements, as outlined above, may present deleterious effects to the patient. Further research is needed to determine which EGDT are the most helpful when treating sepsis, or if alternative definitions of EGDT are warranted. These recent studies suggest that aggressive and invasive monitoring (continuous ScvO2) may not be necessary for all individuals with severe sepsis. A critical understanding from these trial results is that bundled sepsis care should not be abandoned completely. All three author groups strongly endorse the use of sepsis algorithms and care bundles. How these recommendations will change remains to be seen; a meta-analysis of all 3 trials has yet to be published but would likely yield meaningful information regarding individual EGDT elements.

It is also important to note a difference in mortality between the Rivers study and the ARISE/ProCESS/ProMISE studies. Improved outcome in the latter studies is likely due to the development of other processes outside of EGDT. Examples of how standard-of-care for the treatment of sepsis has advanced over the past 15 years include the following:

- Consistent use of earlier and appropriate antibiotics. A landmark article published in 2006 found that administration of an effective antimicrobial within the first hour of documented hypotension was associated with a survival rate of 79.9% in adult patients with septic shock. Furthermore, this study found that each hour of delay in antimicrobial administration over the ensuing 6 hours was associated with an average decrease in survival of 7.6%. At that time, only 50% of septic shock patients were receiving effective antimicrobial therapy within 6 hours of documented hypotension. Increased awareness regarding the importance of timely antimicrobial administration has made a positive impact on overall outcome in recent years.
- Earlier recognition of patients requiring surgical intervention. The Surviving Sepsis Campaign (SSC) Guidelines urge clinicians to identify and treat the source of sepsis as soon as possible after initial resuscitation (except in cases of peripancreatic necrosis, to allow for adequate demarcation of viable and nonviable tissue). Expedited surgical care leads to earlier source control, which decreases morbidity and improves overall outcome.
- More readily available non-bundled therapies. Respiratory support has advanced and recommendations now exist for mechanically ventilating septic people with acute respiratory distress syndrome (ARDS). Similarly, the use of extracorporeal therapies to aid in the treatment of acute kidney injury and fluid overload may also provide an outcome advantage in select cases of severe sepsis and septic shock.
- Additional education or treatment aids. The SSC Guidelines have organized the care and management of a complex disease into instructions the healthcare team can effectively use. These guidelines include evidence-based recommendations for many components of sepsis care, including but not exclusive to EGDT.

The Surviving Sepsis Campaign (www.survivingsepsis.org) published a six-hour revised bundle in 2015, based upon results of the ARISE/ProCESS/ProMISE studies. The SSC states that, with publication of 3 trials that do not demonstrate superiority of required use of a central venous catheter to monitor CVP and ScvO2 in all patients with septic shock who have received timely antibiotics and fluid resuscitation, SSC Executive Committee has revised the improvement bundles as follows:

To be completed within **3 hours** of time of presentation:	To be completed within **6 hours** of time of presentation:
• Measure lactate level	• Apply vasopressors (for hypotension that does not respond to initial fluid resuscitation) to maintain a mean arterial pressure (MAP) ≥65mmHg
• Obtain blood cultures prior to administration of antibiotics	• In the event of persistent hypotension after initial fluid administration (MAP < 65 mm Hg) or if initial lactate was ≥4 mmol/L, re-assess volume status and tissue perfusion and document findings
• Administer broad spectrum antibiotics	• Re-measure lactate if initial lactate elevated
• Administer 30 ml/kg crystalloid for hypotension or lactate ≥ 4 mmol/L	

The SCC now recommends documenting the reassessment of volume status and tissue perfusion with either: (1) Repeat focused exam (after initial fluid resuscitation) by an independent practitioner including vital signs, cardiopulmonary, capillary refill, pulse, and skin findings, OR (2) Two of the following: measure CVP, measure ScvO2, bedside cardiovascular ultrasound, dynamic assessment of fluid responsiveness with passive leg raise or fluid challenge. Of note, only the 6-hour bundle has been updated (the 3-hour SSC bundle is unchanged).

It is likely that other markers of volume status will be integrated into sepsis care and EGDT. Variations in stroke volume and pulse pressure can be measured in mechanically ventilated patients. The inferior vena cava diameter can be measured in spontaneously breathing patients by ultrasound, although this only provides an indirect assessment of CVP. Real-time monitoring of cardiac output and bioimpedance analysis hold promise for noninvasively measuring volume status in critically ill patients. Markers of tissue perfusion and oxygenation may add information on oxygen delivery/utilization that blood lactate measurement does not provide. Additional recommendations might include earlier use of vasopressors, to prevent the development of fluid overload and its associated complications. The future of EGDT remains somewhat unknown at this point in time; as scientific progress continues and more information is provided, it is likely to change sepsis resuscitation and EGDT as many have come to understand it.

REFERENCES

ARISE investigators. Goal-directed resuscitation for patients with early septic shock. N Engl J Med 2014; 371:1496-506.

Barochia AV, Cui X, Vitberg D, et al. Bundled care for septic shock: an analysis of clinical trials. Crit Care Med 2010; 38:668-78.
Boyd JH, Forbes J, Nakada T, et al. Fluid resuscitation in septic shock: a positive fluid balance and elevated central venous pressure increase mortality. Crit Care Med 2011; 39:259-65.
Chamberlain DJ, Willis EM, Bersten AB. The severe sepsis bundles as processes of care: a meta-analysis. Aust Crit Care 2011; 24(4):229-43.
Dellinger RP, Levy MM, Rhodes A, et al. Surviving Sepsis Campaign: International guidelines for management of severe sepsis and septic shock: 2012. Crit Care Med 2013; 41:580–637.
Drumheller BC, Agarwal A, Mikkelsen ME, et al. Risk factors for mortality despite early protocolized resuscitation for severe sepsis and septic shock in the emergency department. J Crit Care 2016;31(1):13-20.
Fuller BM, Gajera M, Schorr C. The impact of packed red blood cell transfusion on clinical outcomes in patients with septic shock treated with early goal directed therapy. Indian J Crit Care Med 2010; 14:165-9.
Gupta RG, Hartigan SM, Kashiouris MG, et al. Early goal-directed resuscitation of patients with septic shock: current evidence and future directions. Critical Care 2015; 19:286-296.
Hernandez G, Bruhn A, Luengo C. Effects of dobutamine on systemic, regional and microcirculatory perfusion parameters in septic shock: a randomized, placebo-controlled double-blind crossover study. Intensive Care Med 2013; 39:1435-1443.
Holst LB, Haase N, Wetterslev J, et al. Lower versus higher hemoglobin threshold for transfusion in septic shock. N Engl J med 2014; 371:1381-91.
Joosten A, Alexander B, Cannesson M. Defining goals of resuscitation in the critically ill patient. Crit Care Clin 2015; 31(1):113-132.
Jones AE, Brown MD, Trzeciak S. The effects of a quantitative resuscitation strategy on mortality in patients with sepsis: a meta-analysis. Crit Care Med 2008; 36(10): 2734-9.
Kelm DJ, Perrin JT, Cartin-Ceba R, et al. Fluid overload in patients with severe sepsis and septic shock treated with early-goal directed therapy is associated with increased acute need for fluid-related medical interventions and hospital death. Shock 2015; 43(1):68-73.
Kumar A, Roberts D, Wood Ke, et al. Duration of hypotension before initiation of effective antimicrobial therapy is a critical determinant of survival in human septic shock. Crit Care Med 2006; 34:1589-96.
Marik PE. The demise of early goal-directed therapy for severe sepsis and septic shock. Acta Anaesth Scand 2015; 59:561-567.
Marik PE, Cavallazi R. Does the central venous pressure (CVP) predict fluid responsiveness: an update meta-analysis and plea for some common sense. Crit Care med 2013; 41:1774-81.
Marik PDE. Early management of severe sepsis: current concepts and controversies. Chest 2014; 145:1407-18.
Mouncey PR, Osborn TM, Power GS. Protocolised management of sepsis (ProMISe): a multicentre randomized controlled trial of the clinical effectiveness of early, goal-directed, protocolised resuscitation for emerging septic shock. Health Technol Assess 2015; 19(97):1-150.
Ripolles-Melchor J, Espinosa A, Martinez-Hurtado E, et al. Perioperative goal-directed hemodynamic therapy in noncardiac surgery: a systematic review and meta-analysis. J Clin Anesth 2015; Epub doi: 10.1016/j.jclinane.2015.08.004
Rivers E, Nguyen B, Havstad S, et al. Early goal-directed therapy in the treatment of severe sepsis and septic shock. N Engl J Med 2001;345(19)1368-77.
Sharif S, Owen JJ, Upadhye S. The end of early-goal directed therapy? Am J Emerg Med 2015; Epub doi: 10.1016/j.ajem.2015.10.039
The ProCESS Investigators. A randomized trial of protocol-based care for early septic shock. N Engl J Med 2014; 370(18):1683-93.
Vellinga NA, Ince C, Boerma EC. Elevated central venous pressure is associated with impairment of microcirculatory blood flow in sepsis. BMC Anesthesiol 2013;13:17.
Waldron NH, Miller TE, Gan TJ. Perioperative goal-directed therapy. J Cardiothorac Vasc Anesth 2014; 28(6):1635-41.

Disorders of Osmolality: Hypo/Hypernatremia

Stephanie Istvan, VMD, DACVECC
San Diego, CA

INTRODUCTION

The challenge of recognizing and providing treatment to patients with sodium disorders is driven by the complex relationships between extracellular fluid volume, osmolality, sodium and water. A recent study reported the incidence of hypernatremia of dogs and cats[1] (5.7%, 8% respectively) and hyponatremia (25.5% and 49.5%)[2]. Mortality rates for patients with both sodium disorders, especially hypernatremia (20.6% and 28.1%) were higher than patients with normal values. The increase in fatalities with disorders of sodium mirrors reports from human medicine and confirms the importance of management.

Sodium is the major extracellular osmole and is a primary determinant of serum osmolality and extracellular fluid volume. Total body sodium content (absolute amount of sodium and water) influences extracellular fluid volume, while osmolality and serum sodium concentration (ratio of sodium to water) are determined by water balance. Serum sodium concentration values do not provide direct information about total body sodium content. The kidneys play a key role in maintaining sodium and water homeostasis.

Two primary mechanisms for osmolality control are antidiuretic hormone (ADH) and thirst. Osmoreceptors located in the anterior wall of the third ventricle of the hypothalamus rapidly detect alterations in osmolality. A 1-2% plasma osmolality increase results in osmoreceptor shrinkage and increased ADH release. ADH is synthesized by neurons with cell bodies in the supraoptic and paraventricular nuclei of the hypothalamus, then transported to the posterior lobe of the pituitary by the supraopticohypophyseal tract for release into circulation.[3] ADH binds to V_2 receptors on the basolateral membrane of the renal collecting duct principal cells. Activation of the V_2 receptor increases intracellular cyclic AMP resulting in phosphorylation of serine residues to allow for movement of aquaporin-2 channels to the luminal membrane for passive water reabsorption. In addition, ADH increases sodium reabsorption in the collecting duct and independent urea reabsorption for additional water reabsorption.

Effective circulating volume (ECV) is the intravascular volume in the arterial system perfusing tissues. Under normal conditions, the renin-angiotensin-aldosterone system regulates ECV. Because sodium is the major determinant of extracellular volume, sodium balance is directly tied to ECV, however ECV is prioritized over plasma osmolality in states of decreased circulating volume. Ultimately, baroreceptors in the carotid bodies/aortic arch and mechanoreceptors in the cardiac atria/pulmonary vessels sense ECV and influence ADH release.[4] Additional stretch receptors located in the renal macula densa and juxtaglomerular apparatus of the afferent arteriole detect decreased renal perfusion. Sodium handling and reabsorption in the kidney depend on intrarenal feedback mechanisms (glomerulotubular balance), aldosterone, angiotensin, ADH, sympathetic nervous system, natriuretic peptides and serum sodium concentration.

HYPERNATREMIA

Hypernatremia is a rise in sodium concentration > 155 mEq/L in dogs and > 162 mEq/L in cats resulting in increased extracellular fluid tonicity.[5] Hypernatremia induces water movement across cell membranes from low to high osmolality. Rapidly increasing serum sodium concentrations can result in muscle tremors, rigidity, altered mentation and seizures. Clinical signs related to hypernatremia result from intracellular dehydration and decreased cell volume. Cell shrinkage may result in vascular stretching and meningeal vessel rupture. Chronic or slowly developing hypernatremia allows for electrolyte shifts and adaptive intracellular solutes to normalize brain cell volume. Idiogenic osmoles, e.g. sorbitol, inositol and the amino acids glutamate, glutamine, taurine, are major components of the adaptation.[6]

Causes of hypernatremia can be classified as below:

- <u>Hypervolemic hypernatremia:</u> Sodium gain result in an expanded extracellular fluid volume.
    - Salt ingestion (seawater, salt-flour dough mixtures)
    - Hypertonic saline or sodium bicarbonate infusions
    - Hyperaldosteronism
- <u>Euvolemic hypernatremia:</u> Pure water loss can result in euvolemic hypernatremia.
    - Inadequate intake (inadequate water access, hypodipsia, adipsia)
    - Insensible losses (fever, respiratory tract, mechanical ventilation)
    - Renal loss (central diabetes insipidus, nephrogenic diabetes insipidus e.g. hypercalcemia, neoplasia, renal injury, head trauma, drug-induced including aminoglycosides, amphotericin B)
- <u>Hypovolemic hypernatremia</u>: Water deficit with hypotonic fluid losses that are associated with hypovolemia. Hypovolemic hypernatremia is likely the most commonly encountered cause of hypernatremia in veterinary medicine.
    - Extrarenal: GI (vomiting, diarrhea, lactulose), third spacing, burns
    - Renal
        - Loop diuretics (furosemide)
        - Osmotic diuresis (mannitol, diabetes mellitus). Mannitol and glucose initially cause hyponatremia due to increased extracellular fluid compartment water. The osmotic diuresis that ultimately develops

results in free water loss, in addition to the solute replacing sodium in urine contributing to hypernatremia.

- Induced diuresis (post-obstructive diuresis, chronic or acute polyuric kidney injury)

Treatment of Hypernatremia

The primary goals of treatment are to correct the underlying disease process and replace the free water deficit without exacerbating clinical signs. To minimize complications during correction, sodium concentration should be lowered at a maximum rate of 0.5 - 1 mEq/L/hr, unless neurologic signs consistent with rapid increase are present. Hypernatremia in acute scenarios may require rapid sodium correction of 1 mEq/L/hr initially.[7]

Fluid therapy should consist of initial restoration of intravascular fluid volume in a patient with hypovolemic hypernatremia. Resuscitation should be performed with a similar sodium concentration to the patient's serum sodium concentration to minimize a rapid decrease in sodium. This can be achieved with 0.9% saline, but ideally a calculated number of mEq (e.g., 7.2% hypertonic saline contains 1.232 mEq/mL sodium) can be added to 1L of 0.9% NaCl during resuscitation to match the patient's serum sodium concentration. Following correction of hypovolemia, the serum sodium concentration should be rechecked prior to further calculations. Ongoing fluid therapy will consist of correction of the free water deficit and replacement of ongoing losses.

$$\text{Free water deficit (L)} = 0.6 \times \text{body weight (Kg)} \times \left(\frac{\text{sodium concentration of patient}}{\text{desired sodium concentration}} - 1\right)$$

$$\text{Change in serum sodium (per L)}^{8,9} = \frac{\text{fluid sodium} - \text{serum sodium}}{0.6 \times \text{body weight (Kg)} + 1}$$

Free water deficit replacement through the oral route is preferred. Patients with contraindications to enteral water intake, require parenteral replacement. Any type of hypotonic fluid may be used including 0.45% NaCl, Ringer's lactate, 5% dextrose in water (D5W) for correction. When D5W is used, the free water deficit calculation can be used exclusively. If another hypotonic fluid (e.g. 0.45% NaCl) is used, the expected serum sodium concentration change per liter administered of any fluid can be calculated.[8,9]

Regardless of calculation, each calculation assumes that hypernatremia is related to pure water loss without other solute loss. Thus the calculation may underestimate or overestimate. These equations should be used as a guide. Sodium concentrations require evaluation every 2-4 hours to adjust therapy. Development of cerebral edema due to overzealous correction is treated with intravenous mannitol (0.5-1 g/kg IV over 20-30 minutes) or hypertonic saline (7.2% sodium chloride 2-4 ml/kg IV over 20 minutes). Alternative therapy options include synthetic analogs of desmopressin in severe cases of central diabetes insipidus.

HYPONATREMIA

Hyponatremia is defined as serum sodium concentration < 140mEq/L in dogs and < 149 mEq/L in cats.[5] Hyponatremia induces water shifts from the extracellular space to the intracellular space, resulting in cellular edema. Clinical signs of brain swelling include dull mentation, seizures and head pressing. Unlike hypernatremia, the presence of hyponatremia does not always indicate hypoosmolality. Approaches to patients with hyponatremia are dependent upon the plasma osmolality and volume status.

Normal Plasma Osmolality

Hyponatremia with a normal plasma osmolality is due to hyperlipidemia or hyperproteinemia. These large molecules contribute little to plasma osmolality. Serum sodium is underestimated when flame photometry methodology is used, but can be avoided with ion specific electrodes.

Increased Plasma Osmolality

The addition of impermeable solutes with osmotic activity, such as glucose increases water movement into extracellular fluid. This movement results in decreased serum sodium concentration with hyperosmolar plasma. Each 100 mg/dL increase in glucose results in a 2.4 mEq/L serum sodium decrease.

Decreased Plasma Osmolality

Determination of fluid volume status is important to evaluate the causes of hypoosmolar hyponatremia.

Hypovolemia: In hypovolemic, hypoosmolar hyponatremia, decreased ECV and total body sodium deficit result in ADH release, increased sodium and water reabsorption in the proximal tubule and decreased water excretion. Renin and angiotensin release inhibit water excretion by encouraging reabsorption at the collecting tubules. Thirst stimulates patients to continue drinking water. The combined mechanisms result in increased water retention and a dilution effect on the remaining bodily fluids. The most common causes are gastrointestinal fluid loss, third spacing, cutaneous loss and hypoadrenocorticism.

Hypervolemia: In hypervolemic hyponatremia, the extracellular fluid volume and total body sodium are increased, yet hyponatremia can occur. Hepatic failure, nephrotic syndrome, congestive heart failure and pregnancy have variable mechanisms of decreased ECV. Each results in impaired water excretion and sodium retention from activation of the renin-angiotensin-aldosterone system and inappropriate ADH release.

Euvolemia: Euvolemic hyponatremia is characterized by mild increase in extracellular fluid volume. Psychogenic polydipsia, syndrome of inappropriate secretion of ADH (SIADH), drug administration (nitrous oxide, vincristine, barbiturates), hypotonic fluid infusion and myxedema coma are known causes of euvolemic hyponatremia. SIADH can be caused by general anesthesia, pain, nausea, hemorrhage or thrombosis of the central nervous system, and opiate derivatives in the critically ill.

Treatment of Hyponatremia

As with hypernatremia, treatment is directed at the underlying disease. Sodium concentration should rise at a maximum rate of 0.5 – 1 mEq/L/hr, unless neurologic signs consistent with acute hyponatremia are present. Conservative therapeutic goals include raising sodium concentration by 8mEq/L in 24 hours, 14mEq/L in 48hrs and 16 mEq/L in 72 hours.[10] Rarely, acute neurologic signs from chronic severe hyponatremia require treatment with an infusion of 3% NaCl to raise the sodium by 4-6 mEq/L in the first 2-4 hours or until abnormal neurologic activity resolves.

Fluid therapy in hypovolemic hyponatremia should consist of initial restoration of intravascular fluid volume. Resuscitation should be performed with a similar sodium concentration to the patient to minimize complications. A crystalloid with a lower sodium concentration (e.g. Ringer's lactate) may better choice than 0.9% NaCl and a designer fluid with a sodium concentration matching the patient in severe cases should be considered. Replacing an amount of the isotonic crystalloid fluid with sterile water USP creates a patient-specific low sodium fluid (i.e. replacing 150 mL of 1L LRS with sterile water yields a sodium content of approximately 110mEq/L). Once the hypovolemia-induced stimulus for ADH release is corrected, free water will be excreted more effectively leading to increasing sodium values. Similarly, correction of hypokalemia can contribute to sodium correction.[11] The remainder of the sodium deficit and the expected change in serum sodium equations can be used to select a fluid and rate, adjusting for ongoing losses to safely correct hyponatremia.

Sodium deficit (mEq) = 0.6 x body weight (Kg) x (desired sodium – patient sodium)

Sodium administration should be avoided in hypervolemic hyponatremia due to congestive heart failure. Improving circulation is the key to treatment. In other hypervolemic and euvolemic patients, treatments aim to remove water through restriction, loop diuretics or ADH receptor antagonists.

Overcorrection can lead to osmotic demyelination syndrome of the central thalamus, subcortical white matter and the pons (ODS, central pontine myelinolysis).[12] ODS is a syndrome of delayed neurologic deterioration 1-4 days after correction of hyponatremia. Clinical signs include depression, encephalopathy, dysphagia, tremors, weakness, ataxia, myoclonus or coma. A proposed mechanism is that overcorrection leads to axon shrinkage and disruption of tight cell junctions. Desmopressin analogues are being investigated for use with concurrent sodium infusions to prevent ODS formation.

CONCLUSION

Patients with sodium disturbances present a unique challenge. The formulas presented can be used for calculating free water or sodium deficits and are simply guides. Other than prevention, serial monitoring is essential to safely correct hyponatremia or hypernatremia without causing harm to the patient.

REFERENCES

1. Ueda Y, et. al J Vet Intern Med. 2015;29:794.
2. Ueda Y, et. al J Vet Intern Med. 2015;29:801.
3. Rose BD, Clinical Physiology of Acid-Base and Electrolyte Disorders. 5 ed. New York: McGraw-Hill, 2001;168.
4. Rose B ,Clinical Physiology of Acid-Base and Electrolyte Disorders. 5 ed. New York: McGraw-Hill, 2001;259.
5. Dibartola SP, Fluid, Electrolyte and Acid-base Disorders in Small Animal Practice. 4 ed. Saunders/Elsevier, 2012;45.
6. Lien YH, et. al J Clin Invest. 1990;85:1427.
7. Rose BD, Clinical Physiology of Acid-Base and Electrolyte Disorders. 5 ed. New York: McGraw-Hill, 2001;746.
8. Adrogué HJ, et. al New Engl J Med. 2000;342(20):1493.
9. Adrogué HJ, et. al New Engl J Med. 2000;342(21):1581.
10. Sterns RH, et. al Semin Nephrol. 2009;29:282.
11. Heng AE, et. al Clin Nephrol. 2007;67:345.
12. Murase T, et. al Am J Med. 2006;119:S69.

Fluid Overload in the Critically Ill: Consequences and Strategies

Lauren A. Sullivan, DVM, MS, DACVECC
Fort Collins, Colorado

Fluid administration is cornerstone of therapy in emergency and critical care medicine. Common indications for fluids include provision of water and electrolytes as "maintenance fluids," parenteral nutrition, for dilution of medications, or to protect the kidneys from endogenous and exogenous substances (e.g., iodinated contrast, hemoglobin or myoglobin). Patients presenting in shock states may require large fluid volumes to quickly optimize hemodynamics. Sepsis, burns, pancreatitis, and trauma constitute common diseases that may require intravascular volume resuscitation due to hypovolemia. The body's physiological response to these inflammatory diseases is dualistic, first termed in the 1940's as the "ebb and flow" phases of shock. The "ebb phase" represents distributive shock and is characterized by arterial vasodilation, transcapillary leakage of albumin and fluid, and decreased plasma oncotic pressure. Hypovolemia decreases blood volume, subsequently decreasing the gradient between mean systemic filling pressure and central venous pressure (CVP). Correction of an intravascular volume deficit is necessary to normalize venous return, cardiac output and oxygen delivery. Fluid therapy during this stage allows for restoration of intravascular volume and optimization of oxygen delivery. Following hemodynamic stabilization, the "flow phase" of shock ensues. During this phase, mobilization of extravascular fluid and volume diuresis through the kidneys returns fluid balance to a neutral state.

Previous recommendations for management of systemic inflammatory disease include aggressive fluid resuscitation during the first 24 hours of hospital admission. Justifications for liberal fluid administration include facilitation of hemodynamic status into the "flow phase" of shock, improve preload and cardiac output, provide oxygen and nutrients to peripheral tissues and reduce the incidence of organ dysfunction. Despite reasonable justifications for aggressive fluid therapy, less than 50% of hemodynamically unstable patients clinically respond to fluids, defined as an increase in stroke volume of 10-15% following a fluid challenge. A 2012 study in dogs (Silverstein et al, *J Vet Emerg Crit Care*) found that 65% of dogs showed normalization of blood pressure (defined as a Doppler blood pressure >90 mmHg) following bolus IV fluid therapy within 1 hour of presentation. Explanations for a lack of response are several-fold. Routine hemodynamic targets of resuscitation (blood pressure, heart rate, CVP, blood lactate concentration) can be poorly predictive of cardiac output and adequate tissue perfusion. Additionally, the cardiovascular response to fluids is modified in acute illness due to changes in myocardial function, arterial resistance, and venous capacitance. This response is further complicated by chronic disease and/or concurrent drug therapy. For these reasons, the very patients that may initially benefit from fluid therapy are also at risk for developing complications associated with fluid overload.

Fluid overload is defined as pathologic accumulation of water and dissolved electrolytes in the body beyond that observed in healthy individuals. More completely referred to as "hydroelectrolytic overload," this condition leads to physiological derangements and poor outcome in critically ill people. It is primarily characterized by a combination of increased fluid intake and reduced urine output can result in microscopic fluid accumulation within the interstitial compartment. Fluid also accumulates macroscopically within body cavities as pleural effusion or ascites. A positive fluid balance in these patients during acute illness may be secondary to various factors, depending upon the period of hospitalization (O'Connor et al, *Crit Care Clinics* 2015):

	Early in Admission to ICU	**Later in Admission to ICU**
Excessive intake of fluid	• Need for blood products • Intravenous fluid resuscitation	• Obligate daily fluid needs in terms of drug therapies and nutrition • Clinical events requiring fluid resuscitation
Inadequate fluid elimination	• Acute or chronic kidney disease • Acute hemodynamic instability • Fluid losses from circulation to interstitial space	• Ongoing renal impairment • Cardiac and liver dysfunction • Sequestration of fluid in the interstitium and body cavities.

Fluid resuscitation is commonly performed using isotonic crystalloids. The tendency for these fluids to move from the intravascular to the interstitial space has been well documented. In states of health, 85% of an infused bolus moves to the interstitium within 4 hours of administration. Transcapillary fluid dynamics change dramatically in critically ill patients, where less than 5% of a fluid bolus remains within the intravascular space after 90 minutes. Fluid dynamics across the capillary wall has been traditionally described using Starling's hypothesis. This hypothesis states that fluid movement across a capillary wall is dependent on the balance between the hydrostatic pressure gradient and the oncotic pressure gradient across the capillary. Starling's equation is as follows:

$$J_v = K_f([P_c - P_i] - \sigma[\pi_c - \pi_i])$$

In the above equation, J_v is defined as the net fluid movement between compartments, and $K_f([P_c-P_i] - \sigma[\pi_c-\pi_i])$ is the net driving force between the intravascular and interstitial space. Components of the net driving force are further described as follows: filtration coefficient (K_f), capillary hydrostatic pressure (P_c), interstitial hydrostatic pressure (P_i), reflection coefficient (σ), capillary oncotic pressure (π_c) and interstitial oncotic pressure (π_i).

Recent insights regarding vascular biology have added another crucial participant to transcapillary fluid dynamics: the glycocalyx. Located on the apical surface of vascular endothelial cells, the glycocalyx lines the capillary lumen and is a key structure in maintaining vascular integrity and intravascular volume. It includes of a complex network of cell-bound proteoglycans, glycosaminoglycan side chains, and sialo-proteins. The glycocalyx participates in numerous physiologic functions including regulation of vascular permeability, storage of non-circulating plasma volume, and modulation of inflammation and hemostasis. Regarding fluid balance, the glycocalyx binds plasma proteins and excludes them from the subendothelial cleft, creating a local oncotic gradient within the capillary endothelial lining. The difference in osmotic pressure between the plasma and the subendothelial cleft ultimately opposes plasma water efflux from the vascular space. At the same time, the subendothelial cleft also limits resorption of fluid from the interstitium. Pathophysiological implications of the glycocalyx relative to fluid overload include the following:

- During acute inflammatory states (e.g., sepsis, major surgery, or trauma), increased capillary permeability is largely due to disruption of the glycocalyx. The disruption is a physical process that is not easily reversible, subsequently placing critically ill patients at ongoing risk for interstitial fluid accumulation.
- The most important determinant of capillary fluid loss is not plasma oncotic pressure, as previously thought. Instead, capillary hydrostatic pressure is primarily responsible for the development of interstitial edema. Increased hydrostatic pressure may be secondary to (1) increased venous pressure or (2) precapillary arteriolar vasodilation. Within the available literature, provision of colloidal support does not significantly resolve interstitial edema or its associated clinical signs. The modified Starling hypothesis provides a physiologic explanation behind the equivocal use of colloids during resuscitation.
- Fluid flux across the capillary wall is generally small and unidirectional (from the capillary towards the interstitum); consequently, almost all fluid removal from the interstitium occurs via lymphatic drainage. This translates to a slow removal of fluid during interstitial edema, as well as slow vascular refilling. In critically ill patients, rapid rates of fluid removal are poorly tolerated for these reasons. Additionally, increasing colloid-osmotic pressure is unlikely to significantly improve transcapillary hemodynamics during rapid fluid removal.

In states of systemic inflammation, the integrity of the glycocalyx is altered due to the effects of ischemia, oxidative stress, circulating toxins (LPS, TNF-α), acute hyperglycemia, or the release of atrial natriuretic peptide [(ANP); secondary to fluid resuscitation and maximization of cardiac output]. As a result, impaired glycocalyx function can occur during varying states of critical illness. Loss of the glycocalyx layer ultimately leads to plasma water and protein extravasation, further potentiating interstitial edema. Damage to the glycocalyx also exposes endothelial cell adhesion molecules, augmenting the systemic inflammatory response. The ongoing cycle between glycocalyx damage and progressive inflammation lead to further loss of intravascular volume, clinical signs of hypovolemia, and worsening interstitial edema. In states of health, fluid loss from capillaries increases interstitial pressure and opposes trans-capillary ultrafiltration. This encourages lymphatic drainage in an effort to re-establish a neutral fluid balance. In states of disease, inflammation within the interstitial space results in *decreased* pressure within the extracellular matrix. Local tissue injury or exposure to inflammatory mediators further modify this response, with decreases in interstitial pressure causing retention of fluid and impairment of lymphatic drainage from the interstitial space.

The major consequence of fluid overload during critical illness is described as a syndrome of interstitial edema and total of body sodium excess (hydroelectrolytic overload). Reduced urine output and impaired renal electrolyte excretion contribute to fluid overload during critically illness. Oliguria is often interpreted as a marker of poor cardiac output during initial stabilization of hemodynamics. This typically results in aggressive intravenous fluid administration to increase preload and enhance kidney perfusion. Overzealous fluid administration coupled with impaired renal excretion may result in a "kidney paradox," where acute kidney injury (AKI) is further potentiated by a positive fluid balance. Aggressive intravenous fluids following adequate volume resuscitation may lead to increased venous pressure, extrarenal compression, and increased intra-renal interstitial pressure. These forces act to oppose renal blood flow and reduce glomerular filtration. Additional volume is not well tolerated in encapsulated organs (e.g., kidney or liver) without an increase in insterstitial pressure, causing further impairment of organ perfusion and function. Edema within the renal interstitium leads to lower perfusion pressure, oliguria and AKI. Furthermore, increased interstitial pressures may also lead to impaired sodium excretion due to decreased renal blood flow and glomerular filtration rate.

Hypernatremia, hyperchloremic acidosis, and progressive fluid accumulation individually contribute to mortality in the ICU setting. Therefore, electrolyte composition of the fluids being administered should be considered when prescribing fluid therapy. Sodium is the main cation provided within resuscitation fluids. Hypernatremia is present at hospital admission and at ICU admission in 2% and 7% of patients, respectively. Up to 27% of human ICU patients develop hypernatremia during their hospital stay. Sodium overload is associated with disease severity, kidney injury and dysfunction, duration of mechanical ventilation and length of ICU stay. Hypernatremia is also associated with higher in-hospital mortality in people. A retrospective study in small animals (Ueda et al, *J Vet Intern Med* 2015) identified hypernatremia in 5.7% of dogs and 8.0% of cats within a university teaching hospital. The magnitude of hypernatremia was linearly associated with higher mortality rates. This study suggests that hypernatremia may be significantly linked to prognosis in small animals. Correcting hypernatremia may ultimately result in better survival, which has been documented in hospitalized people.

Chloride is the main anion provided within resuscitation fluids. Additional chloride can amplify systemic inflammation and further contribute to metabolic acidosis due to an elevation of the strong ion difference (SID). An SID acidosis has been associated with higher mortality in septic patients. In a canine model of endotoxemia, only 1/3 of the post-resuscitation SID acidosis could be explained by exogenous chloride administration. Beyond fluid therapy, chloride elevation in patients with systemic inflammation may occur secondary to transvascular shifts. Chloride overload can influence renal blood flow through its effects on afferent and intra-renal arterial vessels. A biphasic response following administration of 0.9% NaCl has been observed in denervated canine kidneys. Initially, hyperosmolality leads to renal vasodilation and increased renal blood flow. After several minutes, elevated tubular chloride concentrations stimulate ATP release from the macula densa and initiate the tubuloglomerular feedback response. Afferent arteriolar resistance increases and reduces glomerular filtration, negatively affecting kidney function.

Progressive fluid accumulation leads to worsening organ function within multiple body systems. Interstitial edema causes impaired oxygen and metabolite diffusion, distorted tissue architecture, obstruction of organ perfusion, venous outflow and lymphatic drainage, and disturbed cell-cell interactions. Fluid overload also increases hydrostatic pressure, generating mechanical stress against capillary walls. This stress causes ultrastructural capillary damage, altering permeability to proteins and activating the inflammatory response. Increases in systemic venous pressure impair lymphatic drainage and further contribute to interstitial edema. The lungs are most affected by these changes, as interstitial edema leaks into the bronchioles and then accumulates within alveolar spaces. Pulmonary edema compromises gas exchange, reduces pulmonary compliance and increases work of breathing. Hypoxemia from impaired gas exchange results in pulmonary hypoxic vasoconstriction and shunting of regional blood flow to non-dependent regions. This inverted pulmonary perfusion pattern further contributes to impaired gas exchange and tissue hypoxia.

Fluid overload adversely impacts almost every body system during critical illness. Interstitial edema delays healing of skin and soft tissue wounds, increases surgical complications, and contributes to pressure ulceration during the perioperative period. Manifestations of fluid overload within the gastrointestinal system include malabsorption, ileus (with an associated delay in providing enteral nutrition) and anastomosis failure. Ischemic hepatitis, cholestasis, impaired synthetic function may all occur secondary to fluid overload within the hepatic system. Cardiac consequences of fluid overload include myocardial edema, poor ventricular function, decreased cardiac conduction and hypertension. Visceral edema contributes to abdominal compartment syndrome, which in turn worsens organ dysfunction. Increased incidence of delirium and impaired cognition may occur within the central nervous system.

Many conditions in which fluid overload has been documented share a common feature: the presence of systemic inflammation (septic or non-septic in origin) and a high risk of organ dysfunction. Those that are severely affected may be more likely to develop fluid overload as a consequence of their cardiovascular instability and need for additional fluids. It is possible that positive fluid balance is a marker for illness severity, rather than an independent predictor of outcome. Proving that fluid overload is causation for organ dysfunction or death has not been feasible using observational studies alone. The landmark Fluid and Catheter Treatment Trial (FACTT, *N Engl J Med* 2006) was the first multicenter, prospective randomized controlled study examining fluid management strategies in the ICU. The FACTT trial focused on those with severe lung injury and it included two different protocols for the fluid management of patients with ARDS. Fluid restriction and diuretics were used to maintain lower CVPs and pulmonary artery occlusion pressures in the conservative arm of the study. Those in the conservative fluid group achieved a negative fluid balance throughout hospitalization, whereas patients in the liberal fluid group progressively accumulated more fluid during their ICU stay. Those in the conservative group achieved a higher number of ventilator-free days (as well as ICU-free days), although mortality was not different between groups. A follow-up retrospective study using the FACTT trial data (Semler et al, *Crit Care Med* 2016) examined the relationship between initial central venous pressure, fluid strategy, and 60-day mortality in those presenting without evidence of shock. In 609 patients, those with initial central venous pressure >8 mmHg experienced similar mortality with conservative and liberal fluid management (18% vs 18%; $p = 0.928$), whereas those with central venous pressure of 8 mmHg or less experienced lower mortality with a conservative strategy (17% vs 36%; $p = 0.005$).

These observational and randomized clinical trials support the concern that fluid overload is deleterious, particularly later in the course of disease. Achieving a net neutral or negative fluid balance during hospitalization may lead to better outcomes. The first step in managing fluid balance is determination of percent fluid overload, which has traditionally been calculated using the following equation:

$$\%\ \text{Fluid Overload} = \frac{\text{Fluid intake (L)} - \text{Fluid output (L)}}{\text{ICU admission weight (kg)}} \times 100$$

The above formula provides a feasible assessment of relative fluid overload and has been performed in a small animal ICU setting (Mobley et al, *J Vet Emerg Crit Care*). This formula does require urinary catheterization to fully account for sensible losses (e.g., urine output). It does not account for insensible losses or muscle mass lost during extended hospitalization. Studies in pediatric critical care reveal that >10-20% fluid overload prior to initiation of renal replacement therapy confers a three- to eight-fold increased odds for mortality, after adjustment for illness severity and multi-organ failure. Data from adult studies have shown similar results. Recommendations for management of fluid therapy in human medicine include the following:

- Following initial fluid resuscitation, ongoing assessment of daily fluid balance and presence of fluid accumulation should occur. Fluid balance should be viewed as a correlative "vital sign" during critical illness, requiring serial evaluation throughout hospitalization.
- Revisit the "ebb and flow" model of fluid management to understand when a patient may require fluid loading, versus when they may be at risk of fluid overload or require fluid removal. Fluid removal is recommended when fluid accumulation contributes (or is likely to contribute) to patient morbidity. Examples include delayed weaning from mechanical ventilation, impaired organ recovery, the development of AKI, or disrupted wound healing.
- Fluid removal occurs in the "de-escalation" phase once the patient is deemed hemodynamically stable and individual resuscitation endpoints have been reached.
    - Begin judicious fluid management by minimizing all non-essential fluid therapy, while considering active pharmacologic fluid removal. Similar to fluid resuscitation, goals for fluid removal should be tailored to the individual patient (e.g. improve oxygenation and facilitate ventilator weaning).
    - Consider loop diuretics to achieve a negative fluid balance, either through continuous or intermittent bolus infusion. A recent study (Yeh et al, *J Emerg Trauma Shock* 2015) suggests that goal-directed diuresis by continuous furosemide infusion is more successful in achieving net negative fluid balance in patients with fluid overload when compared to intermittent bolus injection.
    - Newer drugs, such as natriuretics or aquaretics, may be added in select circumstances. A recent study evaluating initial experiences using tolvaptan in critically ill patients with fluid overload (Ruiz-Ramos et al, *J Pharm Clin Therap* 2015) showed a good response-to-treatment in patients that did not show a clinical response to furosemide.
    - Early renal replacement therapy may be indicated in cases of life-threatening fluid overload, refractory oliguria, the need for solute removal, or in cases that are refractory to pharmacological management of fluid overload. Prescription of mechanical fluid removal should include consideration of expected fluid inputs and losses, the expected speed to vascular refilling, and the patient's physiological tolerance to a transient reduction in intravascular volume. Ultrafiltration is the primary modality for fluid removal during extracorporeal therapy.

Mechanisms to assess volume status are critical to determining the presence of volume overload, interstitial edema and need for therapeutic intervention. Methods to help assess fluid balance in critically ill patients include the following (Rosner et al, *Brit J Anaesth 2014*):

Clinical variables	Laboratory variables	Static hemodynamic variables	Dynamic hemodynamic variables
• Serial body weight	• Blood lactate and lactate clearance	• Central venous and pulmonary artery pressure measurements	• Stroke volume or pulse pressure variation
• "Ins and outs"	• Central or mixed venous oxygen saturation	• Echocardiographic variables (ventricular sizes and diastolic volume)	• Dynamic changes in central venous or pulmonary artery occlusion pressure
• Vital signs	• Urinary biochemistry		• Esophageal Doppler-corrected aortic flow time
• Urine output	• Bioelectrical impedance and vector analysis		• Changes in vena cava diameter in response to PPV
• Physical examination and historical findings			• End-expiratory occlusion induced changes in hemodynamic variables
• Chest radiograph			• Echocardiographic variables (ejection fraction, fractional shortening)

Bedside measurement of extravascular lung water (EVLW) and bio-electrical impedance analysis (BIA) have recently gained attention for the identification and management of fluid overload. EVLW increases either because of increased lung permeability or because of increased hydrostatic pressure in the pulmonary capillaries. Transpulmonary thermodilution provides a relatively easy bedside measurement of EVLW in critically ill people. When EVLW is indexed to predicted body weight, it allows for improved correlation with lung injury scores, oxygenation and mortality. Transpulmonary thermodilution also provides the pulmonary vascular permeability index (PVPI), which is an indirect reflection of the alveolocapillary barrier integrity. A recent review (Jozwiak et al, *Ann Intensive Care* 2015) recommends measurement of EVLW in the critical care setting to assess pulmonary edema due to increased lung permeability and/or hydrostatic pressure. BIA emits an electrical current that travels through the body, and that current is impeded based upon the length and specific resistivity of the tissue. New BIA techniques allow for measurement of total body water with separation into intracellular and extracellular water. More research is needed in critically ill patients before widespread use of BIA can be suggested, due to measurement variability in the clinical setting.

In summary, fluid overload is a serious complication of critical illness and its presence independently increases the risk of morbidity and mortality in the ICU setting. When performing fluid resuscitation and subsequent maintenance fluid therapy, one suggested model identifies five distinct stages of fluid administration: Resuscitation, Optimization, Stabilization, and Evacuation (ROSE), followed by a potential risk for Hypoperfusion (Malbrain et al, *Anaesth Inten Ther* 2014). These describe the five different clinical phases of fluid therapy, occurring over the time course when patients experience a different impact on end-organ function. Modifying the fluid plan to meet specific goals during the course of therapy (**R**esuscitation-correcting shock, **O**ptimization-maintain tissue perfusion, **S**tabilization-aim for zero or negative fluid balance, **E**vacuation-mobilize fluid accumulation) will help provide the necessary balance between optimization of preload and oxygen delivery, while avoiding the deleterious effects of interstitial edema and organ dysfunction.

REFERENCES

Basso F, Berdin G, Virzi GM, et al. Fluid management in the intensive care unit: bioelectrical impedance vector analysis as a tool to assess hydration status and optimal fluid balance in critically ill patients Blood Purif 2013; 36:192-199.

Bensen BA, Gobatto AL, Melro LM, et al. Fluid and electrolyte overload in critically ill patients: An overview. World J Crit Care Med 2015; 4(2): 116-129.

Bouchard J, Mehta RL. Fluid balance issues in the critically ill patient. Contrib Nephrol 2010; 164:69-78.

Child DL, Cao Z, Seiberlich LE, et al. The costs of fluid overload in the adult intensive care unit: is a small-volume infusion model a proactive solution? Clinicoeconom Outcomes Res 2014; 7:1-8.

Cordemans C, De laet I, Van Regenmortel N, et al. Fluid management in critically ill patients: the role of extravascular lung water, abdominal hypertension, capillary leak, and fluid balance. Ann Intensive Care 2012; 2(suppl 1):S1-12.

Godin M, Bouchard J, Mehta R. Fluid balance in patients with acute kidney injury: emerging concepts. Nephron Clin Pract 2013;123:238-245.

Goldstein S, Bagshaw S, Cecconi M, et al. Pharmacological management of fluid overload. Br J Anaesth 2014; 113(5):756-63.

Goldstein SL. Fluid management in acute kidney injury. J Intensive Care Med 2014; 29(4):183-189.

Jozwiak M, Teboul J, Monnet X. Extravascular lung water in critical care: recent advances and clinical applications. Ann Intensive Care 2015;5:38.

Kellum JA, Bellomo R, Kramer DJ, et al. Etiology of metabolic acidosis during saline resuscitation in endotoxemia. Shock 1998; 9:364-368.

Malbrain M, Marik PE, Witters I, et al. Fluid overload, de-resuscitation, and outcomes in critically ill or injured patients: a systematic review with suggestions for clinical practice. Anaesth Inten Ther 2014; 46(5):361-380.

Malbrain M, Huygh J, Dabrowski W, et al. The use of bio-electrical impedance analysis (BIA) to guide fluid management, resuscitation and deresuscitation in critically ill patients: a bench-to-bedside review. Anaesth Inten Ther 2014; 46(5):381-391.

McDermid RC, Raghunathan K, Romanovsky A, et al. Controversies in fluid therapy: Type, dose and toxicity. World J Crit Care Med 2014; 3(1):24-33.

Mobley A, Sullivan LA. Retrospective determination of fluid overload in critically ill dogs. J Vet Emerg Crit Care 2012; S(8).

National Heart, Lung and Blood Institute Acute Respiratory Distress Syndrome (ARDS) Clinical Trials Network, Wiedemann HP, Wheeler HL, et al. Comparison of two fluid-management strategies in acute lung injury. N Engl J Med 2006; 354(24):2564-75.

O'Connor ME, Prowle JR. Fluid overload. Crit Care Clin 2015; 31:803-821.

Prowle JR, Echeverri JE, Ligabo EV. Fluid balance and acute kidney injury. Nat Rev Nephrol 2010; 6:107-115.

Rosner MH, Ostermann M, Murugan R, et al. Indications and management of mechanical fluid removal in critical illness. Br J Anaesth 2014; 113(5):746-71.

Ruiz-Ramos J, Gordon M, Cortes A, et al. Initial experience of use of tolvaptan in critically ill patients with fluid overload. J Clin Pharm Therap 2015; 40:339-341.

Semler MW, Wheeler AP, Thompson BT, et al. Impact of initial central venous pressure on outcomes of conservative versus liberal fluid management in acute respiratory distress syndrome. Crit Care Med 2016; DOI: 10.1097/CCM.0000000000001555.

Silverstein DC, Kleiner J, Drobatz KJ. Effectiveness of intravenous fluid resuscitation in the emergency room for treatment of hypotension in dogs: 35 cases (2000-2010) J Vet Emerg Crit Care 2012; 22(6):666-673.

Ueda Y, Hopper K, Epstein SE. Incidence, severity and prognosis associated with hypernatremia in dogs and cats. J Vet Intern Med 2015; 29(3):794-800.

Wang N, Jiang L, Wen Y, et al. Fluid balance and mortality in critically ill patients with acute kindey injury: a multicenter prospective epidemiological study. Critical Care 2015;19:371.

Wilcox CS. Regulation of renal blood flow by plasma chloride. J Clin Invest 1983; 71:726-735.
Woodcock TE, Woodcock TM. Revised Starling equation and the glycocalyx model of transvascular fluid exchange: an improved paradigm for prescribing intravenous fluid therapy. Br J Anaesth 2012; 108(3):384-94.
Yeh DD, Van der Wilden GM, Cropano C, et al. Goal-directed diuresis: A case-control study of continuous furosemide infusion in critically ill trauma patients. J Emerg Trauma Shock 2015; 8(1):34-38.

Quantitative Acid-Base Assessment

Bernie Hansen, DVM, MS, DACVECC, DACVIM (Small Animal Internal Medicine)
Raleigh, North Carolina

Blood gas analysis and assessment of electrolytes and acid-base balance provide a broad view of your patient's metabolic 'landscape', giving you insight into lung function and primary and secondary disturbances of regulation of many substances in the body. The gas pressure of oxygen (pO_2) in the arterial blood tells you about lung function and the pO_2 in venous blood tells you something about oxygen delivery to the tissues served by that vein. The gas pressure of carbon dioxide (pCO_2) defines ventilation. Analysis of electrolyte concentrations and hydrogen activity (aH^+) provide insight into many different primary disorders, and their concentrations have direct effects on metabolic function.

WHAT DETERMINES ACID-BASE BALANCE?

Three things:

1) Ventilation via the respiratory system
2) The concentration of "strong" ions (Na, Cl, K, Ca, lactate, and others)
3) The concentration of "weak" ions, which in biology are weak acids, and tend not to dissociate as strongly as the "strong" guys at physiologically relevant aH^+. In healthy animals this is mostly protein (70% of the activity is from albumin) and phosphate).

Some details about these 3 actors: Although it was widely known that ions in water affected acid-base chemistry for decades, it was Peter Stewart who in the 1970's separated these three independent factors that determine hydrogen activity from dependent variables and modelled them mathematically. Respiratory system function - ventilation - is *defined* by pCO_2. The effect of carbon dioxide gas pressure on acid-base balance is direct, consistent, linear, and straightforward. When dogs have a primary metabolic (nonrespiratory) disturbance in hydrogen ion, they should react by increasing or decreasing their ventilation to adjust hydrogen ion concentration back towards normal, and they do this by a reasonably predictable amount (cats are less predictable). "Strong "ions are the charged atoms or molecules that result from dissolving salts in water. These particles are charged AND they aren't reactive – that is, they do not bind with oppositely charged things and exist all alone because at physiological aH^+ they are highly dissociated and they carry about their business hydrated by surrounding water molecules. In biology most of them are also electrolytes, salts that conduct electricity in solution. "Weak ions" are mostly weak acids whose pK (the pH at which the compound is 50% dissociated) is within a single pH unit of 7.4. So, at clinically relevant pH (the inverse log of aH^+) the ability of these compounds to carry or release a proton isn't constant and varies with the aH^+ of the solution (body fluid). Hydrogen activity is provided by the dissociation of water into H+ and OH-, and the concentration of these two is a function of the relative concentration of all the other charged ions in the solution. Stated another way, the dissociation of water into H+ and OH- is *determined* by the relative concentrations of cations and anions in that solution.

Evaluating ion activity: the anion gap

By the law of electroneutrality, the concentration of positively charged ions in solution in water has to equal the concentration of negatively charged ions. Because it is the relative concentration of these that determines hydrogen activity, we can use the readings from chemistry panel to get some insights into the cause of acid-base disturbances! So, in extracellular fluid (sampled as plasma):

[all the cations] = [all the anions]

We commonly measure some of these and the formula can be expanded to:

[Na] + [K] + [all the *other* cations] = [Cl] + [HCO3] + [all the *other* anions]

If you switch things around a bit we get:

[Na] + [K] - [Cl] - [HCO3] = [all the *other* anions] - [all the *other* cations]

...and this is the classic formula for the anion gap. So the anion gap is the equivalent of subtracting the "unmeasured" cation activity from "unmeasured" anion activity, arrived at by subtracting the concentration of chloride and bicarbonate ions from the cations sodium and potassium. The word "unmeasured" is in quotes because some of them are in fact measured on the chemistry profile (calcium, phosphate, magnesium, and lactate) or could be measured if specially requested (sulphate, urate, and beta-hydroxybutyrate). Some – albumin and total protein- are always measured but we didn't know what to do with them because their ionic activity is complex and was comparatively unknown until recently.

Laboratories establish a large range of normal values for the anion gap. The reason for this is that the concentration of chloride relative to sodium is affected by water balance, and also because there is a wide range of normal concentration for protein and phosphorus, two anions with more complex charge behavior that depends on the current pH. As a consequence of this large normal range, major changes in the system like the addition of excess organic anions can be easily hidden by low or lowish albumin and/or phosphorus, and major deviations in chloride relative to sodium happen secondary to changes in water balance. In the last

20 years several methods have been introduced in an attempt to "tighten up" the anion gap concept to make it a more sensitive instrument for detecting the presence of ions that shouldn't be there.

Quantitative Chemistry – modifications of the Anion Gap

We now have some means to factor in these "unmeasured" ions into the anion gap to look for evidence that they are in fact players for individual patients with metabolic acid-base disorders. The anion gap formula can be adjusted for albumin or total protein, for example in dogs{Constable, 2005 12691 /id}:

Albumin-adjusted AG: AG + 0.42 X (37.7 - [albumin]) *Total protein–adjusted AG:* AG + 0.25 X (63.7 - [total protein]),

where albumin or total protein are expressed as grams/liter (multiply the reported gm/dl values by 10). The estimate can be further refined by adding in phosphate:

Albumin-adjusted AG: AG + 0.42 X (37.7 - [albumin]) + 1.4 X (1.8 – [phosphate])

TP–adjusted AG: AG + .25 X (63.7 - [total protein]) + 1.4 X (1.8 – [phosphate]);

where phosphate is expressed as mmols/liter (to get mmols/liter, multiply the mg/dl value by .3229).

An adjusted anion gap that lies outside of the laboratory reference range for normal dogs provides more specific evidence that unmeasured anions are present when albumin or phosphate is abnormal. However, this approach does not take into account the effects of water balance or the effect of pH on the ionic activities of protein and phosphorus. Those factors are at lease partially addressed by the...

Base Excess Gap

Another approach has been to combine the approaches of Siggaard-Andersen (base excess) with Stewart (quantification of strong ions) by partitioning the base excess measurement into four 'physical chemical' components. These are known as the Fencl corrections to the base excess:

1) Free water effect:
   $(Na_{patient} - Na_{normal})$ x .24 (dog) or .22 (cat)
2) Plasma [Cl] effect:
   $Cl_{normal} - Cl_{corr}$, where $Cl_{corr} = Cl_{patient} \times Na_{normal}/Na_{patient}$
3) Plasma albumin effect:
   =0.37 x $(Albumin_{normal} - Albumin_{patient})$
4) Serum phosphate effect:
   0.58 $(PO4_{normal} - PO4_{measured})$, in mg/dl
5) Unmeasured anions:
   Standard BE – #1 - #2 - #3 - #4

In each of these corrections, "normal" may be the midpoint normal for the laboratory used or may be the mean normal value for that laboratory. Although this approach uses a fixed, linear approximation of the activity of albumin it does allow for more sensitive identification of contributors to a base excess or deficit.

The Strong Ion Gap (SIG)

McCullough[1] and Constable[2] went a step further than this by using a quantitative chemistry approach to determine the presence of unmeasured anions in cats and dogs, respectively. From their work, simplified formulas to identify the presence of unmeasured ions are based on the concept that the net positive charge derived from strong ions (Na + K + 2Ca + 2Mg – Cl – Lactate) is balanced electrically by the anion activity of albumin, total protein, and phosphorus. If phosphorus concentration is not markedly abnormal, convenient (well, convenient if you have a scientific calculator!) formulas to estimate SIG based only on the pH-dependent anion activity of albumin and the results of a standard chemistry profile are, for the dog:

SIG = [albumin] X (0.348 + 0.469/{1 + 10^$^{7.77\text{-pH}}$}) – AG (suggested normal range = 0 +/- 5) or

SIG = [TP] x (0.206 + 0.273/1 + 10^$^{7.77\text{-pH}}$}) – AG (suggested normal = 0 +/- 6)

and for cats:

SIG = [albumin] X (0.76/{1 + 10$^{\wedge 7.17\text{-pH}}$}) – AG + 9 (suggested normal = 0 +/- 5).

For all of these, the SIG is expressed in mEq/L, the albumin or TP concentration is in grams/L, and the AG is in mEq/L. Values that lie below the lower limit of the suggested normal range (- 5 or -6 depending on whether using albumin or TP) suggests the presence of excess unmeasured anions, and values > than the upper limits (+5 or +6) suggest the presence of unmeasured cations, a relatively rare event. The presence of unmeasured anion in the absence of obvious cause (shock, ketoacidosis, uremia) should prompt further investigation into the cause.

QUANTITATIVE ACID-BASE ANALYSIS AND THE ACID BASE CALCULATOR

A major barrier to clinical use of Stewart's approach to acid-base analysis has been that the relationships between the independent determinants was described by a 4th-order polynomial equation ($ax^4 + bx^3 + cx^2 + dx + e = 0$) that can only be solved with an iterative approach, not possible with calculators or (until recently) spreadsheets. The other obstacle to its use was that Stewart did not have measures of physiological constants. Peter Constable developed two approaches to Stewart's work that brought it into the realm of clinical use. One was simplification of the polynomial model to a quadratic equation by both ignoring aH+ behavior outside of the narrow range of plasma and ignoring micromolar and nanomolar concentrations of relevant species when following the law of electrical neutrality. The other was experimental determination of the equilibrium constants for total weak anion activity (A_{tot}) and net protein charge by CO2 titration studies of plasma from horses[1], humans[2], cats[3], and dogs[4]. These advances are codified in the simplified strong ion equation (SSIE):

$$aH = \frac{K_1'S_{CO2}P_{CO2} + K_a[A_{tot}] - K_a[SID] + \sqrt{(K_1'S_{CO2}P_{CO2} + K_a[SID] + K_a[A_{tot}])^2 - 4K_a^2[SID][A_{tot}]}}{2[SID]}$$

Despite its apparent complexity, the formula is fundamentally similar to the Henderson equation for acid-base. Once the net protein charge and equilibrium constants were developed, creation of spreadsheet-based, species-specific calculators for quantitative acid-base analysis became possible. This was first performed by the late Peter Lloyd for humans[5], and in this presentation I will demonstrate a calculator for dogs and cats that is patterned on his work. Using species-specific constants, customized midpoint normal values, and solving the SSIE for each independent variable a comprehensive assessment of acid-base balance is presented. A sample of the analysis output is illustrated in Figure 1 below.

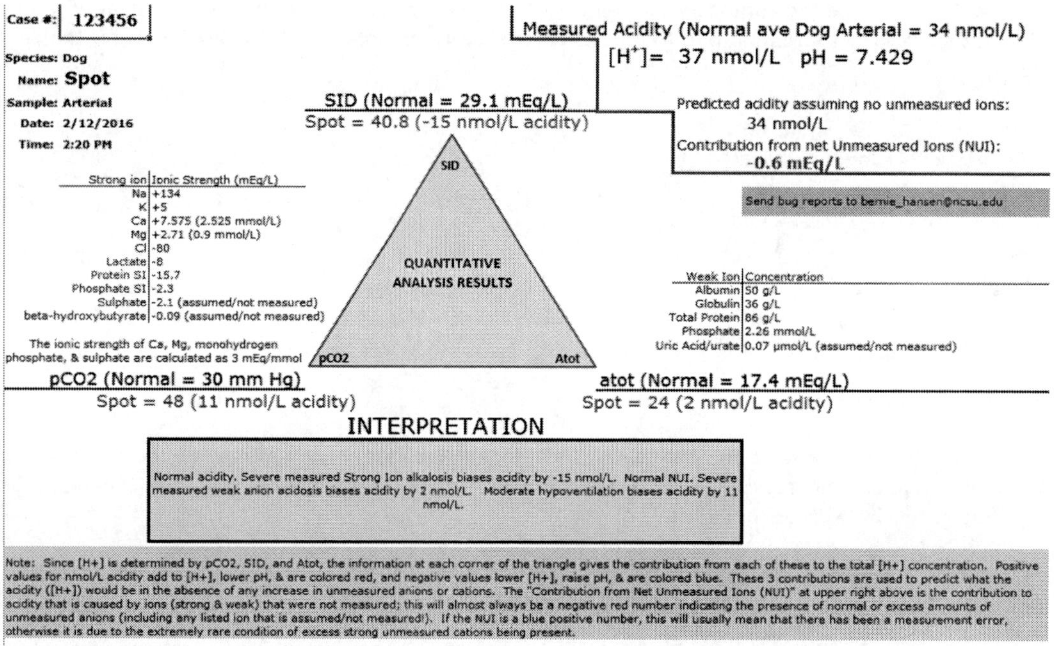

Figure 1: Sample output from the Acid-Base Calculator using data obtained from a dog.

REFERENCES

1 Constable PD. A simplified strong ion model for acid-base equilibria: application to horse plasma. J Appl Physiol (1985) 1997 Jul;83(1):297-311.

2 Constable PD. Total weak acid concentration and effective dissociation constant of nonvolatile buffers in human plasma. J Appl Physiol (1985) 2001 Sep;91(3):1364-71.

3 McCullough SM, Constable PD. Calculation of the total plasma concentration of nonvolatile weak acids and the effective dissociation constant of nonvolatile buffers in plasma for use in the strong ion approach to acid-base balance in cats. Am J Vet Res 2003 Aug;64(8):1047-51.

4 Constable PD, Stampfli HR. Experimental determination of net protein charge and A(tot) and K(a) of nonvolatile buffers in canine plasma. J Vet Intern Med 2005 Jul;19(4):507-14.

5 Lloyd P. Strong ion calculator--a practical bedside application of modern quantitative acid-base physiology. Crit Care Resusc 2004 Dec;6(4):285-94.

Chronic Medicine Cases in My ER

Elizabeth Rozanski, DACVIM (SAIM), DACVECC
North Grafton, MA

The appearance of a "chronic" medicine or surgery cases in the Emergency Service may have the perception of irritation to Emergency Clinicians. It can be hard to balance out the needs of the patient/client, internal medicine clinician, and ER clinician. The following concepts may be helpful to consider when considering these patients.

1) Relationship of the Emergency Service to the Internal Medicine services- 24 hour specialty hospitals that house both an Emergency Service and an Internal Medicine service often share cases; in this situation it is easier to get records and it may even be there is some familiarity with the patient before re-presentation to the ER. In night/weekend hospitals, or where the internist does not practice at the same site it can be much more challenging to determine what is the intended course of action might be. For ERs that commonly receive patients from other (non-internal DACVIMs), there should be regular discussion and bidirectional feedback on case approaches.
2) Personality of the ER and IM doctors-Some internists are very keen to be actively involved in all aspects of patient care, while others are more comfortable with sharing case management For internally shared patients, it should be clear when "Dr X" should be called and Dr. X should be happy to be called if she says to call her. Similarly, some ER clinicians enjoy medicine cases, while others are not a fan of long-term conditions. Importantly, animals that are being treated for long-standing medicine conditions often have remarkably committed and well-educated owners and may be clear plans as to the perceived next action step and this should be followed as warranted by the case.
3) Last medicine visit- A patient is not a "medicine" case, if it was seen once 5 years ago, for a problem that has long since resolved. Similarly, it is wise to recognize, particularly in smaller facilities, that it may be frustrating for clients to be told over the phone that their internist who discharged the dog yesterday is not available, only to run into him in the lobby.
4) Current presenting complaint/ Relationship of chronic disease to presenting complaint -If the case has a progressive decline of the chronic condition, then further investigation of that specific disease is warranted. A new or unrelated disease or trauma may also occur, and should be considered. While it is often more likely that progression of a single disease is occurring, it is not impossible for a new disease to develop or for a peripherally related complication to arise (eg. Pancreatitis in a dog with epilepsy being treated with anti-convulsants)
5) Maximizing communication; Recall that true medicine cases have owners that are not "rookies"; they are often knowledgeable and committed as well as having the resources to support patient care. Adequate resources should not be taken as a green-light to run every test known to man, especially if the patient is returning to the care of the internist the following am. Additionally, understanding and having a mutually agreeable clinical course (eg. Start amikacin if hospital – acquired infection) is ideal.

SPECIFIC DISEASES

The other important major issues are disease-specific; while any disease may be chronic, the specific list of concerning diseases includes the following:

1)*Addisons*

Dogs with Addison's disease are rarely sick if ever from their Addison's disease unless their supplementation has been inadvertently stopped. However, Addison's that are sick with something else will almost always need some extra glucocorticoids, depending on what disease or injury they have. Anesthesia is also a stressor. There is never a need to repeat an ACTH stimulation test on an emergency basis in a dog that has been previously identified as Addisonian. Dogs that are receiving daily mineralocorticoid supplementation should their electrolytes closely followed, and oral medication administered as soon as possible. Dogs on monthly DOCP are not likely to experience electrolyte disturbances.

2) *Cushings* disease

Cushingoid dogs are most likely to become sick during induction with Lysodren; this is far less frequent now that trilostane is more commonly used. Sick Cushing's patients should be evaluated for hypocortisolism; this is an important disease to detect. If possible, the ER clinician should obtain careful history as to the dose of medication, any medication changes, and the date of the last ACTH stimulation test. As an internist, patients with Cushing diseases should be advise to keep a log book of significant dates in the management of their dog. Dogs with Cushings may also develop other complications, particularly if untreated, including diabetes or diabetic ketoacidosis, pancreatitis, or urinary tract infections

3) *Diabetes mellitus*

Diabetes is a common endocrinopathy; dogs and cats are frequently presented out of hours with diabetic complications, including hypoglycemia, and hyperglycemia/ketosis. Hypoglycemia should be promptly treated, and the cause investigated; don't forget to offer a patient food as hypoglycemia is a potent appetite stimulant. Causes may be simple (double dosing) or more complex. A pet with a hypoglycemic crisis should be discharged on no or a significantly decreased dose of insulin. Hyperglycemia should be treated, and the cause investigated, insulin type is depends on clinician preference.

4) Diabetics sick with other diseases

Diabetic animals sick with something else (eg. cataracts, trauma, seizures etc) are also commonly treated in the ICU, and depending on the hospital dynamics, may be treated by critical care clinicians. As most diabetic patients are only moderately regulated in comparison to human standards, the typical approach to hospitalization of the diabetic pet includes administering half the normal daily dose until a patient is discharged or otherwise returns to normal appetite and activity. This approach may be adequate for day cases, or animals with good underlying appetite and health; however, in dogs and cats prone to complicated diabetes (eg. Cushings, Brittle diabetes) inadequate daily insulin may be associated with perpetuating anorexia and hyperglycemia. These patients in particularly may benefit from continuing a near normal dose (eg. 80%) of long-acting insulin, and monitoring for hyper or hypoglycemia. All hospitalized diabetic patients should have a dedicated clinician or nurse with special interest/training in diabetes overseeing their care to prevent inadvertent patient care misadventures.

5) *Critically ill animals with hyperglycemia*

Hyperglycemia in people and dogs has been associated with a worse outcome in critically ill people. One study in people supported a better outcome with tight glycemic control, while later studies have not supported that finding. Unexplained hyperglycemia should be carefully monitored and consideration for treating animals with marked elevations (>300 mg/dl).However, transient hyperglycemia should NOT be treated. Hyperglycemia will commonly accompany near arrest (due to epinephrine release), and as hyperglycemia of several hours duration is not harmful, watchful waiting is indicated in this population.

6) *Chronic renal failure*

CRF patients are commonly represented to the Emergency service with progressive failure to do well, vomiting or dehydration. Significant useful facts to recall include that CRF patients will have tremendous fluid requirements (eg. 120 ml/kg/day will not work!). Additionally, secondary urinary tract infection is common, as are symptomatic anemia. Normal SQ fluid dosing should be recorded, and even in light of apparently inactive urine sediment, a urine culture should be considered. Antiemetics are beneficial. Some internists (and owners) are very vested in the creatinine, and it important to recognize that different machines may have different numbers.

7)*Oncology*

Patients undergoing chemotherapy often present to the ER; it is wise to sort out if they have a chemotherapy complications (eg. Sepsis/gi upset) or progression of disease (eg. Recurrent hemoabdomen) or unrelated problem. Evaluation for fever, and neutropenia is warranted. Owners again tend to be quite dedicated, and a strong plan of supportive care is advised.

8) *Congestive heart failure*

Patients will present to the ER either in heart failure (needing more medications) or from excessive dehydration (eg.too much Lasix). In general, dogs with MVD will tolerate low dose to moderate IV therapy for rehydration, while cats and DCM dogs are perhaps slightly less tolerant. However, stopping the Lasix and encouraging rehydration is often useful. The pet should be discharged on a lower dose of diuretic after hydration is restored. Second line diuretics (eg. Chlorothiazide and spironolactone) may be associated with marked electrolyte disturbances. In animals with active heart failure, further diuretics, oxygen and vasodilators are indicted. Cavity fluid accumulation should be drained. Recheck echocardiograms within a few weeks or months of the last echo rarely provide significant new information.

PU/PD diseases (eg renal failure, diabetics, Cushing, and diuretic receiving patients) will have more challenging fluid requirements. Specifically, in cases of PU/PD, animal may be drinking large amounts (> 50 ml/kg/day) of fluids prior to illness, and after dehydration develops, they will need much higher than anticipated fluid replacement in order to be rehydrated.

It may be helpful to recall the common diseases that may result in abnormal urine concentrating ability. While intrinsic renal disease results in dilute urine, other common conditions that result in dilute urine include unregulated or poorly regulated diabetes mellitus, Cushing's syndrome or exogenous glucocorticoid therapy, phenobarbital and diuretic therapy. Of specific note, it is difficult to know the pre-renal component of azotemia in a clinically dehydrated patient who has been receiving furosemide therapy. Additionally, post–obstructive diuresis, such as that which may accompany the urethral obstruction, may also result in large volumes of dilute urine.

Chronic medicine patients with a history of PU/PD should be carefully monitored. The majority of the time, a higher than "maintenance or 2x maintenance" fluid rate is required to rehydrate the patient and some patients may require a very large volume of fluids to restore hydration.

Dogs that pant or vocalize excessively may also have increased fluid needs; this may be under-appreciated at time. Of interest, these patients will lose more free water with panting and vocalizing and may risk the development of hypernatremia.

REFERENCES

There are no specific references regarding this topic, we had planned this to try to bring together divergent opinions, and provide better care for our cases after hours.

EXPRESSION OF MYOCARDIAL REMODELING MARKERS: GENDER, AGE AND CARDIAC DISEASE ASSOCIATED VARIATIONS IN CATS

Sonja Fonfara[1], Sarah Kitz[2], Udo Hetzel[2], Anja Kipar[2]

[1]*University of Guelph, Guelph, Ontario, Canada,* [2]*University of Zurich, Zurich, Switzerland*

Age and gender associated myocardial remodeling has not been studied in cats. The aim of the present study was to investigate the influence of age, gender and cardiac disease on the transcription of relevant cytokines and extracellular matrix (ECM) remodeling enzymes.

The study was performed on 48 cats (age: 1.5 – 19 y) that had been euthanized for non-cardiac disease or old age (n=31; 20 female, 1 entire; 11 male, 2 entire) or hypertrophic cardiomyopathy (n=17; 14 male neutered; 3 female, 1 entire). Institutional ethical approval was obtained. All hearts were assessed for gross and histopathological changes, and the myocardium was analyzed for interleukin (IL)-1, -2, -4, -6, -18, tumor necrosis factor (TNF)-α, interferon (IFN)-γ, transforming growth factor (TGF)-β, matrix metalloproteinase (MMP)-2, -3, -13, tissue inhibitor of MMP (TIMP)-1, -2 and -3 transcription using cat specific quantitative RT-PCR assays. Data were normalized and cardiac regions, gender and groups compared using 1-way ANOVA and unpaired t-tests as appropriate. Linear regression analysis served to explore the relationship between marker expression levels and age.

Constitutive transcription of cytokines and ECM remodeling enzymes was detected, with generally higher levels in atria than in ventricles. Female cats showed overall lower transcription levels than male cats. However, with age, myocardial TGF-β transcription increased in both genders. Female cats exhibited a progressive decrease in the transcription of almost all markers, whereas male cats showed an increase in Th1 cytokines IL-2 and IFN-γ and TIMP-3 mRNA levels. Interestingly, the male cats with cardiac disease showed lower myocardial inflammatory cytokine, Th2 cytokine IL-4, TIMP-1 and -2 and increased MMP-2 expression than their counterpart with non-cardiac diseases.

The constitutive transcription of ECM remodeling enzymes suggests continuous myocardial remodeling throughout the entire life of cats. The observed changes during ageing indicate a modified reactivity of the myocardium with age, which differs between male and female cats. Gender differences in myocardial inflammation and ECM remodeling might influence the ability for repair, cardiac function, and development of disease. Interestingly, cardiac disease was associated with a reduction of inflammatory cytokine and pro-fibrotic TIMP-1 and -2 transcription, which might be consistent with impaired myocardial repair and progression of disease.

Despite the majority of cats being neutered, gender associated differences in myocardial reactivity were observed suggesting that non-hormonal factors are involved in myocardial remodeling and cardiac function. In future, more detailed investigations into the influence of gender into the ageing process of the feline myocardium, and on the influence of cardiac diseases on the myocardium of female cats would be of interest.

ECHOCARDIOGRAPHIC ASSESSMENT OF THE RIGHT VENTRICLE IN CATS WITH HYPERTROPHIC CARDIOMYOPATHY

Lance Visser, Joshua Stern

University of California, Davis, Davis, CA, USA

The purpose of this study was to compare right ventricular (RV) size, wall thickness, and function between cats with hypertrophic cardiomyopathy (HCM) ± heart failure (HF) and healthy cats (controls).

Clinical and echocardiographic data from control (n=20), preclinical HCM (n=24), and HCM + HF (n=21) cats were obtained. In addition to standard left-sided echocardiographic measurements, right atrial (RA) size and RV internal dimensions (RVID), freewall thickness (RVFW), and function, as determined by fractional area change (FAC) and tricuspid annular plane systolic excursion (TAPSE), were measured by 2D echocardiography in long axis. Differences among the 3 groups were determined via one-way ANOVA and Tukey's test.

Body weight, heart rate, age, and gender were not significantly different among the groups. Mean (± SD) maximum RA diameter in HF cats (13.8 ± 2.7 mm) was significantly larger than control (11.3 ± 1.4 mm) and preclinical HCM (11.1 ± 1.2 mm) cats. Diastolic RVFW thickness was significantly increased in preclinical HCM (3.1 ± 0.7 mm) and HF (3.5 ± 0.9 mm) cats compared to controls (2.3 ± 0.4 mm). Diastolic RVID was significantly greater in HF (8.1 ± 1.6 mm) compared to preclinical HCM (6.5 ± 1.3 mm) and control (6.7 ± 1.3 mm) cats. TAPSE and FAC were significantly decreased in HF (6.7 ± 1.8 mm; 51.8 ± 14.6%) compared to preclinical HCM (8.4 ± 1.0 mm; 63.3 ± 10.9%) and control (9.1 ± 1.5 mm; 62.9 ± 6.2%) cats.

These results suggest RV size and function are altered in cats with HCM.

PREDICTION OF LEFT VENTRICULAR VOLUMES FROM M-MODE DIMENSIONS IN THE DOG

Eric de Madron, Jérôme del Castillo

[1]*Centre Vétérinaire DMV, Montreal, QC, Canada,* [2]*University of Montreal, St-Hyacinthe, QC, Canada*

In human medicine, the Teichholz formula V = D^3 * 7/(2.4+D) has been widely used to estimate left ventricular volume from its M-mode short axis diameter. However, this formula has been shown to overestimate volumes in the dog, due to differences in the left ventricular sphericity index between the human and the dog. Echocardiograms from 28 dogs with or without heart disease were examined. M-mode measurements included short axis diameters (D) of the left ventricle and aorta (Ao). Bidimensional measurements included systolic and diastolic left ventricular lengths and volumes, obtained by apical 4 chambers views planimetry (Simpson's monoplane method). The relationships between the diastolic and systolic internal short axis diameters and lengths were analyzed, a geometrical model of the canine left ventricle chosen, and various modifications of the human Teichholz formula using aortic ratios (D/Ao): (V = D^3 * a/(b +c(D/Ao)) analyzed in order to determine which formula would best predict left ventricular volumes measured by planimetry. We also assessed with nonlinear regression, residual analysis, and the Akaike goodness-of-fit information criterion the type of relationship between measured ventricular volume and short axis diameter. We found that a power model of D best fitted our data. We also examined whether these new models would represent an improvement over the human formula. We conclude that modifications of the classical Teichholz formula using aortic ratios can mitigate somewhat the overestimation of measured left ventricular volumes in the dog. However, a model based on V = aD^b predicts these volumes more accurately and precisely.

AN INVESTIGATION OF ULTRA-WIDEBAND RADAR TECHNOLOGY TO EVALUATE CANINE HEART RATE AND RESPIRATORY RATE.

Amanda Landis-Hanna[1], Joe Wakshlag[2], Marc Krauss[2], Paul Tupin[1], Albert Goldfain[3]

[1]i4C Innovations, Chantilly, VA, USA, [2]University of PennsylvaniaSmall Animal Teaching Hosptial, Philadelphia, PA, USA, [3]Blue Highway, Syracuse, NY, USA

Ultra-wideband (UWB) technology has been used to characterize complex microwave systems since 1969. In the last ten years, UWB has been applied to communications and imaging, such as human fetal monitoring and detecting stroke volume. UWB technology is now leveraged in a proprietary health monitor known as "Voyce" which is worn around the canine neck. The Voyce Health Monitor(TM) is designed to collect physiological information by utilizing variations in the dielectric properties of tissues. The monitor has no wires, probes, or chest attachments, making it a non-invasive device. Voyce collects data and remotely transmits it to a computer or mobile device so the information can be reviewed anytime. This investigation evaluated the accuracy of the monitor as compared to gold standard technology (Televet 100 EKG) and manual readings (auscultation and palpation).

The study was designed in two stages. Stage 1 consisted of investigatory testing of the device using 30 canine subjects at the Cornell University Small Animal Teaching Hospital. Stage 2 consisted of testing of 70 canine subjects, in over 1500 test cases, in Chantilly, VA. In both stages, a variety of ages, sizes, and breeds were tested, to provide a representative population. Subjects wore a 4-lead Televet system, monitoring R wave to R wave (R-R) intervals. The subjects also wore the Voyce monitor, which recorded raw UWB data. That data was then converted to heart rate (HR) and respiratory rate (RR) data using proprietary data-processing algorithms.

In both study locations, significant accuracy was found for both the HR and RR. At Cornell and Chantilly, accuracy of HR and RR was found to be greater than 86%. Algorithm refinement and additional data collection is expected to allow improved accuracy throughout 2016. These tests suggest that Voyce can be utilized to obtain objective HR and RR data in a non-clinical setting.

HIGH SENSIVITY CARDIAC TROPONIN IN 60 BULLDOGS WITH OR WITHOUT PS AND ABERRANT CORONARY ANATOMY

Brian Scansen

Colorado State University, Fort Collins, Colorado, USA

High sensitivity cardiac troponin I (HScTnI) is a marker of myocardial damage. Bulldogs have high incidence of pulmonary valve stenosis (PS) and coronary artery (CA) anomalies that limit intervention. Diagnosis of CA anomalies requires cardiac catheterization or advanced imaging. This study compared HScTnI in bulldogs with PS to those without and evaluated the effect of CA anomalies on HScTnI.

Between July 2011 and September 2015, 52 English and 8 French bulldogs with (N = 33) and without (N = 27) PS had samples prospectively acquired. Frozen samples (58 serum, 2 plasma) were batch analyzed simultaneously through the Gastrointestinal Laboratory at Texas A&M University. One dog in congestive heart failure with a value of 1.331 ng/mL was excluded from analysis as an outlier; no other dogs were in heart failure.

HScTnI was significantly different (P = 0.0014) between dogs with PS (median = 0.155 ng/mL; range = 0.055 – 0.658 ng/mL) and those without PS (0.076; 0.024 – 0.363 ng/mL). Among dogs with PS and confirmed CA anatomy (N = 22), HScTnI was significantly (P = 0.009) higher in dogs with PS and CA anomaly (0.300; 0.094 – 0.658 ng/mL) compared to dogs with PS and normal CA anatomy (0.119; 0.055 – 0.247 ng/mL). Among dogs with PS (N = 32), there was no correlation between HScTnI and maximal transpulmonary pressure gradient (P = 0.45).

These results suggest myocardial damage in bulldogs with PS. HScTnI may help to distinguish dogs with PS and CA anomalies from normal CA anatomy, though further study is required.

EFFECTS OF A SMALL MOLECULE MODULATOR OF SARCOMERE CONTRACTILITY IN CATS WITH HYPERTROPHIC CARDIOMYOPATHY

Joshua Stern[1], Svetlana Markova[2], Yu Ueda[1], Jae Kim[2], Peter Pascoe[3], Marc Evanchik[2], Eric Green[2], Samantha Harris[4]

[1]Department of Medicine & Epidemiology; School of Veterinary Medicine; University of California Davis, Davis, CA, USA, [2]Myokardia, Inc., South San Francisco, CA, USA, [3]Department of Surgical and Radiological Sciences; School of Veterinary Medicine; University of California Davis, Davis, CA, USA, [4]Department of Cellular and Molecular Medicine; College of Medicine; Sarver Heart Center; University of Arizona, Tucson, AZ, USA

Feline hypertrophic cardiomyopathy (HCM) has an estimated prevalence of 14.5% with few established treatment options. A recently described small molecule modulator of sarcomere contractility (MYK-461) shows promise for prevention of hypertrophy in a mouse model of HCM and is now in clinical trials for human HCM. The purpose of this study was to identify the effects of MYK-461 in cats with subclinical HCM and left ventricular outflow tract obstruction (LVOTO). We hypothesized that MYK-461 would relieve LVOTO and reduce hyperdynamic function associated with HCM.

Five cats with HCM and dynamic LVOTO were identified from a research colony. An IV drug infusion and echocardiography protocol was developed that provided serial assessments of LV function and outflow tract gradient at multiple MYK-461 concentrations. Isoproterenol was used to increase heart rate and produce obstruction under anesthesia. Seven echocardiographic assessments were obtained: baseline, isoproterenol only, isoproterenol with increasing concentrations of MYK-461, and finally MYK-461 alone. Three cats served as controls after a washout period and followed the same protocol receiving only vehicle without MYK-461.

MYK-461 successfully reduced LVOTO gradients and eliminated systolic anterior motion of the mitral valve in all cats. A linear correlation was observed between reduction of fractional shortening (FS) and MYK-461 plasma concentration ($r=0.85$). All cats receiving MYK-461 had significant reduction in FS ($p=0.01$), which was not observed in control cats.

MYK-461 can relieve LVOTO and generate an exposure-dependent reduction in LV contractility in cats. Clinical studies of inhibitors of sarcomere contractility in feline HCM warrant consideration.

Problems and Pitfalls of Recording Exercising ECGs in Horses

Gayle D. Hallowell, MA, VetMB, PhD, CertVA, DACVIM, DACVECC, PFHEA, MRCVS
Nottingham, UK

INTRODUCTION

Electrocardiograms (ECGs) performed at exercise are used to evaluate cases that present with poor performance as well as to evaluate safety for ridden work that have cardiac murmurs and secondary chamber enlargement or that have rhythm disturbances present at rest such as atrial fibrillation. There are however several challenges to obtaining appropriate and diagnostic ECGs in these cases for a number of reasons that will be discussed to include patient and equipment factors.

Clinical factors

In order to increase the chances of obtaining a diagnostic ECG, a detailed history is required. It is well established that in horses that present with poor performance[1], cardiac dysrhythmias are often found concurrently with other more common underlying causes including lameness and airway disease, particularly of the upper airway. The link between airway disease and cardiac dysrhythmias is attributed to myocardial hypoxia. As such, if the horse has a history of a cough or noise at exercise or an increased resting respiratory rate, it would be prudent to perform tracheal wash cytology and/or dynamic airway endoscopy – the exercising ECG can thus be performed concurrently with the latter test. This will prevent over-interpretation of rhythm disturbances identified at exercise, that otherwise may be career ending, but completely reversible if the exercise-induced hypoxia is reversed with airway surgery or improved airway management.

The second factor that may influence results from an exercising ECG is the horse being concurrently lame. How lame is too lame? It is felt that increased sympathetic activity[2] could result in an increased number of aberrant beats during an exercise test. As such, significant lameness should be ruled out before an exercise test is undertaken to include examination in a straight line and on circles in order to avoid missing bilateral disease.

On a similar theme to the lameness issue, if the horse seems unduly stressed in the environment it is being examined, it may be prudent to perform the exercising ECG in a more familiar or less stressful environment. This again should minimize the risk of obtaining 'false positive' diagnoses. This is highlighted in a recent publication, where a very large proportion of horses with atrial fibrillation were found to have concurrent ventricular ectopics or aberrant idioventricular conduction both at rest and during exercise when examined in a busy hospital and exposed to a very stressful stimulus[3]. The accompanying editorial questions the validity and clinical relevance of these findings[2]. However, some clinicians feel that exercise testing should include some artificial stimulus, as this may mimic similar stressful situations that horses may encounter during regular use.

Another challenge we are often faced with is the request to examine horses that have been partially or totally de-trained. This again produces potential to obtain both false positive and negative results. It is accepted that the exercise test that horses undertake should be of a similar or greater intensity to what they do on a day-to-day basis[4]. If this test is performed at the exercise intensity that they will do when trained, there is the risk of musculoskeletal injury and again an activated sympathetic response. If it is performed at the intensity that the horse is currently fit to perform at, it may not reveal abnormalities that would be present at faster gaits and higher heart rates. There may be liability issues with injuring horses by asking them to perform above their normal exercise intensity or if they are unfit.

Exercise test suitability

For exercising ECGs to be diagnostic, they need to represent the type of the exercise that the horse normally undertakes and be performed in a horse that is fit enough to undertake the exercise that is required of it, as discussed above. Another decision is whether it is appropriate to perform the exercise test on a lunge, on a treadmill or whether the horse should be ridden. As a general rule, horses do not work as hard on a lunge as they do when they are ridden. But this has to be balanced with the liability factor, considering the risk that the horse may collapse. Treadmill exercise tests have the advantage that they are very controlled. However as discussed above, horses may find this type of exercise unusual and quite stressful and many horses after two or three training sessions have sore feet and thus may result in false positive results. It is inappropriate to perform exercising ECGs on a lunge in an event horse or racehorse, although we are sometimes pressured to do so. In addition, the ECG needs to be recorded for a long enough period to obtain heart rates representative of what the horse does and ideally should mimic normal exercise conditions. As such, it may be more appropriate to allow an exercising ECG in an endurance horse to be recorded over a long training session over a relevant distance. School work however is generally appropriate for dressage and showjumping horses, whereas event horses and racehorses should be exercised in a larger space at gallop.

Another relevant issue relating to exercising ECGs is repeatability. Does an isolated exercising ECG, even if performed in 'ideal' conditions represent what happens every day the horse is ridden? Unpublished work suggests that horses that have aberrant beats usually will have them on multiple recorded ECGs. It should be noted that normal horses commonly have aberrant beats at exercise[1,5], thus making it challenging sometimes to discern clinically relevant rhythm abnormalities from irrelevant ones (see Jo Ann Slack's talk).

Patient preparation and equipment factors

In order to obtain the best quality ECG recordings, good quality, equine specific (which contain more gel and are more adhesive) pads should be used. Ideally patients should not be clipped prior to placement of the self-adhesive pads, although very thick coats may need trimming to allow the center of the pad making contact with the skin. Alcohol-based hand gels or ultrasound coupling gel may be applied in small quantities in the center of the pad to improve electrical contact. If too much is applied, the pad will not adequately stick. Clipping also results in poor adhesion, particularly if the horse sweats. Some people use superglue or tissue glue to improve adhesiveness of the pads to the skin. However, in a small proportion of horses, growth of circular white hairs in these regions has been reported and thus should be used with caution as in some horse types could affect re-sale value and lead to litigation issues. The author has also had issues with surgical spirit resulting in severe areas of dermatitis. One group has also successfully used a girth with implantable rubber electrodes to record diagnostic exercising ECGs[6] and these may be the future.

Optimal electrode placement is another source of debate. The aim of the exercising ECG is to be able to make a definitive rhythm diagnosis at peak exercise. At these speeds is when we are most likely to encounter artifacts and excessive signal-to-noise ratio. Trying to optimize ECG quality whilst minimizing artifacts is an art. One of the main reasons for a poor quality ECG is the rider and the movement caused by their seat and legs. This 'artifact' is often less with professional riders and when jockeys are largely off the horses' back. Some believe that electrodes should be placed beneath tack and others as far away from tack as possible. Other causes of artifact include muscle activity, fat movement or movement of any part of the recording device. Data by the author has shown that most artifact regarding recording devices comes from movement of the electrodes and then movement at the electrode-wire connection. Movement of the ECG machine and wires should also be minimized and attached such that it is still and away from the rider. These artifacts are the main limitation of this technique and makes recording consistently diagnostic ECGs in all horse types significantly challenging.

Electrode position needs to be such that all sections of the ECG are present and diagnostic and ideally different leads can be assessed. In an ideal world this would result in electrodes being placed in a base-apex or modified base-apex position with electrodes being on both the left and right sides of the horse allowing more than one lead to be recorded. However these positions can be problematic regarding muscle, limb and rider movement. Six or 12 lead ECGs result in a poor quality trace due to frequent shifts in limb position and muscle tremors[7]. It is documented that the base-apex lead is sufficient to detect rhythm disturbances in the horse[4,7-9]. The base-apex lead configuration consists of the positive electrode located over the left apex of the heart and the negative electrode located in the right jugular groove[9]. This results in a positive P wave, large, negative QRS (although the Q wave is often not visible) and a positive or negative T when examined in lead I. Multiple leads may allow more accurate differentiation between normal and abnormal complexes[10], which can be very valuable in some cases. Alternative methods may include positioning electrodes under the saddle, although different algorithms may be required to optimize the trace obtained.

Another challenge faced in exercising ECG assessment is lack of automated software to read exercising ECGs although one system can fairly accurately detect R-R intervals, even at exercise. However, this frequently means that these ECGs need detailed, manual examination, bringing in the human error factor to diagnosis. The better the quality of the ECG, the less likely mistakes will happen.

The future

Significant improvements are required if we are to record consistent, diagnostic ECGs. In order to do this, we need to overcome movement artifact when recording ECGs, both from equipment, rider and patient factors. Perhaps implantable ECGs will be the future if cost and altered algorithms allow them to be implanted distant to the heart (and tack position). Adapted algorithms and superior sampling frequency of ECG machines may allow electrodes to be placed in positions distant to the heart thus reducing artifacts and improving the signal-to-noise ratio and improving what is actually recorded.

SUMMARY

In summary, to ensure the diagnostic recording of ECGs at rest and exercise, a detailed and targeted history for a particular horse needs to be obtained in order that the most appropriate ECG is recorded and combined with any other appropriate diagnostic tests. Placement of electrodes should be based upon the shape and body condition score of the horse, type of rider and saddle and position in the saddle. Being flexible with electrode placement and adapting to different situations, may improve the exercising ECG recordings you obtain.

REFERENCES

1.Martin BB, Jr., et al. *J Am Vet Med Assoc* 2000;216:554-558; 2.Physick-Sheard PW. *Equine Vet J* 2013;45:270-272; 3.Verheyen T, et al. *Equine Vet J* 2013;45:309-314; 4.Reef VB, et al. *J Vet Intern Med* 2014;28:749-761; 5.Ryan N et al *Equine Vet J* 2005;37:265-268; 6. Physick-Sheard PW, McGurrin MK. *J Vet Intern Med* 2010;24:1158-1166; 7.Robertson SA. *In Practice* 1990;12:59-67; 8. Marr CM, Bowen M. *Cardiology of the Horse*: Saunders Elsevier, 1999; 9. Reed SM, Bayly WM, Sellon DC. *Equine Internal Medicine*. 2nd ed: Saunders, 2004; 10. Reimer JM et al. *J Am Vet Med Assoc* 1992;201:1237-1243.

Exercising Arrhythmias and Sudden Death During Sports in Horses

Cris Navas, LV, PhD, DACVIM
College Station, TX

INTRODUCTION

Arrhythmias during or immediately after exercise are common in equine athletes. Many of these arrhythmias are clinically irrelevant while others can cause poor performance or rarely sudden cardiac death (SCD). This situation can have catastrophic consequences for the health of the horse, the safety of the human partner and the public perception of welfare during equestrian sports. A review of literature about SCD and exercising arrhythmias in human being provides an interesting background. Direct extrapolation can, however, not be made due to the many differences in causes and consequences of the arrhythmias and the different diagnostic tests available for human and equine athletes. In this session a short a review the current knowledge in equine exercise arrhythmias and its implication in SCD from a comparative perspective is presented. The two talks about exercising arrhythmias presented during the same session will give more insight into the importance and problems of acquiring and interpreting exercising electrocardiograms in horses.

EPIDEMIOLOGY

The risk of SCD in human athletes is higher than in the general population and the incidence approximately 0.5-4/100,000 athletes/year.[1-4] The incidence of sudden death (SD) in horses is not well established but literature and lay sources can be used to obtain rough estimations.[5] Lyle reported 28.7 sudden deaths/100,000 starts in Thoroughbred horses in the UK. If we estimate that a racehorse starts an average of 5-6 starts/year, we could propose an approximate number of 150 SD/100,000 Thoroughbred racehorses/year. This includes deaths related to exercise excluding the ones related to musculoskeletal injuries. It is not possible to make direct comparisons between human SCD and equine SD as a large percentage of cases the precise cause of death remains unknown. The numbers stated above for horses are for deaths associated with exercise excluding musculoskeletal injuries and therefore not specific for SCD.

In the last 6 years (2009-2014) the FEI reports 98,690 three-day eventing starts. In the same period, at least 20 press reports mentioned a horse collapsed and died during or shortly after competition. If we estimate that each three-day-event horse competes 6-10 times/year the approximate number would also be 150 sudden deaths/100,000 eventing horses/year.

Sudden death compared to total number of deaths is approximately 19% in flat races, 3.5% in jump races and probably around 40% in three day eventers.[6] The fact that the actual reason for sudden death is rarely known complicates the estimation of the risk of SCD. These numbers only account for deaths during competition and the total number is likely higher if deaths during training could be included. It appears that horses have several-fold higher risk, perhaps 25-100 fold, of SD during sports when compared to humans. Please consider these calculations rough estimates. The numbers in eventers are likely less accurate than the presented for racehorses.

PATHOPHYSIOLOGY

Exercise has effects on cardiac electrophysiology that may result in a tendency for arrhythmia development.[7] Acute changes associated with exercise include hyperkalemia, acidosis, raise in catecholamine plasma concentration and increase in the activity of the sympathetic nervous system.[6,8] Chronic changes derived from intense physical activity can mimic the changes in diseased hearts. Atrial or ventricular hypertrophy,[9] increased vagal tone, increased sympathetic activation/stimulation,[10,11] or fibrosis formation are associated with arrhythmia development in human athletes. The influence of the autonomic nervous system in arrhythmogenesis is several-fold and deserves special attention. Interestingly, vagal rebound is considered the most important arrhythmogenic factor in post-exercise arrhythmias in horses[12] while in humans has been suggested in humans to minimize the potentially pro-arrhythmic.[13,14]

ARRHYTHMIAS IN EQUINE VERSUS HUMAN ATHLETES

Both bradyarrhythmias and tachyarrhythmias can lead to inadequate cerebral perfusion, causing syncope.[15] Bradyarrhythmias that can induce syncope include sick sinus syndrome, high grade second degree AV block and third degree AV block. These are rare and more easily detectable than intermittent tachyarrhythmias.[17] The difference between pathologic and physiologic second degree AV block can be sometimes difficult to define.

Arrhythmias in athletic and sedentary humans are common if continuous ECGs are obtained. The following numbers may give an idea of their prevalence. Up to 40-80% of highly trained athletes have abnormal baseline ECGs. Vagally mediated bradyarrhythmias or isolated premature depolarizations (PDs) are the most common arrhythmias.[18,19] Ventricular premature depolarizations (VPDs) are also common in trained athletes (6-75%)[7] and complex (repetitive, polymorphic, R-on-T) VPDs occur in 1% to 6% of healthy persons.[20,21] Supraventricular (SV) arrhythmias are relatively rare during exercise when compared to ventricular arrhythmias and appear to be benign except for a possible relationship to atrial fibrillation.[7] The quantitative importance of this risk is unknown. Arrhythmias in the absence of structural heart disease or clinical signs (syncope, dizziness, palpitations, chest pain etc.) seem to have little clinical relevance,[13,18,19,21] but it is not always possible to differentiate physiological from malignant arrhythmias.

Several studies have reported the presence of arrhythmias during exercise and immediately afterwards in healthy and poorly performing horses.[22-28] Supraventricular arrhythmias occur more frequently than ventricular arrhythmias during exercise. Up to 50% of horses can have some form of supraventricular complexes during exercise and up to 5% premature ventricular beats. Ventricular ectopy is more common after than during exercise, with the more severe arrhythmias being reported during fast heart rate deceleration in racehorses. It is important to note that the study groups, methods and definitions of certain arrhythmias vary between studies and that the occurrence of arrhythmias varies depending on the type of exercise or discipline, age, intensity of exercise etc.

Occasional single VPDs and SVPDs and accelerated idioventricular rhythms in the absence of structural heart disease are currently accepted as clinically irrelevant in horses without underlying heart disease by most authors. A specific cut-off for how many PDs are acceptable or concerning is not available.[28]

Atrial fibrillation is the most common clinically relevant arrhythmia in humans and horses and deserves special consideration. There is debate about the safety/risk for collapse in horses with AF[17,28,29] with some authors suggesting that disproportionate tachycardia, QRS broadening and R-on-T morphology occur frequently during exercise in these horses potentially predisposing them to collapse.[30] Other authors have suggested that 'there appears to be no immediate danger in lone equine AF unless it is associated with underlying disease'.[29] Current practice in horses with AF is to convert the horse to sinus rhythm before exercising it. If this is not possible or other circumstances prevent it, an echocardiogram and an exercising electrocardiogram are recommended to help determine if the horse is allowed to exercise.[28]

DIAGNOSTIC TESTS

History, physical examination, echocardiograms, resting ECGs, exercising ECGs, 24-hour continuous Holter ECG recordings, and measurements of cardiac biomarkers (cTnI/cTnT) are the available tools for the equine veterinarian evaluating a horse for exercising arrhythmias and underlying heart disease. These were discussed in a consensus statement published jointly by the ACVIM and ECEIM about cardiovascular disease in equine athletes.[28] In humans 12-lead ECGs, electrophysiology testing, gadolinium enhanced cardiac MRI, coronary angiography, genetic testing or myocardial biopsy are also considered,[19,31] but these techniques are not possible or frequently performed in horses.

The above mentioned consensus statement has recommended echocardiograms for horses with murmurs suggesting significant valvular or congenital disease[28]. However the sensitivity of an echocardiogram to detect at risk horses is, to the best of our knowledge, unknown. Plasma cardiac troponin concentrations can increase moderately in normal horses during prolonged or strenuous exercise.[32,33] A significant increase in troponins in the face of exercising arrhythmias would appear intuitively a negative sign and this consideration has been recommended in cases of ventricular arrhythmias in sports horses by a panel of experts. However, the predictive value of cardiac biomarkers for detecting relevant cardiac disease in horses with exercising arrhythmias is unknown.

PREPARTICIPATION SCREENING

Which athletes should be banned from competitive sports, whether or not preparticipation screening is best practice for human competitive athletes and which tests should be included in this screening is a constant matter of debate. Different sets of guidelines exist in Europe (European Society of Cardiology) and North America (Bethesda Conference) and these guidelines are based mainly on expert opinion. Physical examination, history and in some instance a 12-lead ECG are the most commonly recommended tests in humans. All other tests are considered if relevant abnormalities are detected in these. Some consider that there is evidence to support that preparticipation screenings save lives. The potential for a high number of false positives in a problem with a low prevalence is the argument against preparticipation screening.

To the best of my knowledge no official rules exist for preparticipation screening in any of the equine sports disciplines besides the cardiac auscultation performed as the part of the general physical examination in horse shows or in 'vet checks' of endurance and eventing horses. The ACVIM/ECEIM consensus statement about cardiovascular disease in sports horses gives recommendations for which horses should be considered less safe to ride than their peers.[28] In some instances, and particularly in the case of exercising arrhythmias, these recommendations are non-specific due to the gaps in the knowledge about the significance of some rhythm disturbances and other cardiovascular problems.

CONCLUSION

Equestrian sports have inherent risks and no horse is completely safe to ride.[29] However, the goal of the clinician evaluating an equine athlete with heart disease should be to determine if the risk is increased in a particular horse when compared to its peers. The veterinarian is often asked to advice riders or appropriate authorities so that the best decision can be made. This is a difficult decision that needs to balance the risk of under-diagnosing a problem that could trigger dangerous events versus the risk of over-diagnosis that may result in unjustified alarm or disqualification from sports.[21] In the future, the veterinary community (riders, trainers, regulatory organs, clinicians and researchers) will need to work collaboratively to gather enough information to solve all the questions and uncertainties related to equine exercising arrhythmias.

REFERENCES

1. Mont L. Arrhythmias and sport practice. Heart. 2010;96(5):398-405. 2. Kane SF, Oriscello RG, Wenzel RB. Electrocardiographic findings in sports medicine: normal variants and the ones that should not be missed. Curr Sports Med Rep. 2005;4(2):68-75. 3. Corrado D, Basso C, Schiavon M. et al. Pre-participation screening of young competitive athletes for prevention of sudden cardiac death. J Am Coll Cardiol. 2008;52(24):1981-9. 4. Alattar A, Maffulli N. The Validity of Adding ECG to the Preparticipation Screening of Athletes An Evidence Based Literature Review. Transl Med UniSa 2014;19(11):2-13. 5. Lyle CH,

Blissitt KJ, Kennedy RN et al. Risk factors for race-associated sudden death in Thoroughbred racehorses in the UK (2000-2007).Equine Vet J. 2012;44(4):459-65. 6. Lyle CH, Uzal FA, McGorum BC et al. Sudden death in racing Thoroughbred horses: an international multicentre study of post mortem findings, Equine Vet J. 2011;43(3):324-31. 7. Beckerman J, Wu T, Jones S et al. Exercise test-induced arrhythmias. Cardiovasc Dis. 2005;47(4):285-305. 8. Paterson DJ. Antiarrhythmic mechanisms during exercise. J. Appl Physiol 1996;80(6):1853-62. 9. Ormaetxe JM, Martínez Alday JD, Almendral J et al. Prognostic significance of ventricular arrhythmias in the presence of pathological left ventricular hypertrophy. Eur Heart J. 1993;14 Suppl J:73-5. 10. La Gerche A. Exercisisng arrhythmias in endurance athletes Curr Sports Med Rep. 2013;12(2):63-9. 11. Benito B, Gay-Jordi G, Serrano-Mollar A et al. Cardiac arrhythmogenic remodeling in a rat model of long-term intensive exercise training. Circulation. 2011;123(1):13-22. 12. Physick-Sheard PW, McGurrin MK. Ventricular arrhythmias during race recovery in Standardbred Racehorses and associations with autonomic activity. J Vet Intern Med. 2010;24(5):1158-66. 13. Stoebner R, Bellin DA, Haigney MC. Cardiac electrophysiology and the athlete: a primer for the sports clinician. Curr Sports Med Rep. 2012;11(2):70-7. 14. Scheinman MM, Lam J. Exercise-induced ventricular arrhythmias in patients with no structural cardiac disease. Annu Rev Med. 2006;57:473-84. 15. Walker J, Calkins H, Nazarian S. Evaluation of cardiac arrhythmia among athletes. Am J Med. 2010;123(12):1075-81. 16. Lyle CH, Keen JA. Episodic collapse in the horse Equine vet. Educ. 2010;22(11):576-586. 17. Sanchez LD, Pereira J, Berkoff DJ. The evaluation of cardiac complaints in marathon runners. J Emerg Med. 36(4):369-76. 18. Giada F, Conte R, Pescatore V et al. Sports and arrhythmias. Minerva Med. 2011;102(3):239-47. 19. Lampert R. Evaluation and management of arrhythmia in the athletic patient. Prog Cardiovasc Dis. 2012;54(5):423-31. 20. Messineo FC. Ventricular ectopic activity: prevalence and risk. Am J Cardiol. 1989;64(20):53J-56J. 21. Fragakis N, Pagourelias ED, Koskinas KC et al. Arrhythmias in athletes: evidence Based strategies and challenges for diagnosis, management, and sports eligibility. Cardiol Rev. 2013;21(5):229-38. 22. Jose-Cunilleras E, Young LE, Newton JR et al. Cardiac arrhythmias during and after treadmill exercise in poorly performing thoroughbred racehorses. Equine Vet J Suppl. 2006;(36):163-70. 23. Ryan N, Marr CM, McGladdery AJ. Survey of cardiac arrhythmias during submaximal and maximal exercise in Thoroughbred racehorses. Equine Vet J. 2005;37(3):265-8. 24. Barbesgaard L, Buhl R, Meldgaard C. Prevalence of exercise-associated arrhythmias in normal performing dressage horses. Equine Vet J Suppl. 2010;(38):202-7. 25. Buhl R, Meldgaard C, Barbesgaard L. Cardiac arrhythmias in clinically healthy showjumping horses. Equine Vet J Suppl. 2010;(38):196-201. 26. Martin BB, Reef VB, Parente et al. Causes of poor performance of horses during training, racing, or showing: 348 cases (1992-1996). J Am Vet Med Assoc. 2000;216(4):554-8. 27. Slack J, Boston RC, Soma LR, Reef VB. Occurrence of cardiac arrhythmias in Standardbred racehorses. Equine Vet J. 2015:47(4):398-404. Reef VB, Bonagura J, Buhl R et al. Recommendations for management of equine athletes with cardiovascular abnormalities. J Vet Intern Med. 2014;28(3):749-61. 29. Physick-Sheard PW. Seek and ye shall find: Cardiac arrhythmias in the horse. Equine Vet. J. 2013;45:270–272. 30. Verheyen T, Decloedt A, van der Vekens N et al. Ventricular response during lungeing exercise in horses with lone atrial fibrillation. Equine Vet J. 2013;45(3):309-14. 31. Sen-Chowdhry S, McKenna WJ. Sudden cardiac death in the young: a strategy for prevention by targeted evaluation. Cardiology. 2006;105(4):196-206. 32. Holbrook TC, Birks EK, Sleeper MM et al. Endurance exercise is associated with increased plasma cardiac troponin I in horses. Equine Vet J Suppl. 2006;(36):27-31. 33. Durando MM, Reef VB, Kline K et al. Acute effects of short duration, maximal exercise on cardiac troponin I in healthy horses. Equine and Comparative Exercise Physiology 2006;3(4):217–223.

EPIC Trial Results

Adrian Boswood[1], DVC, MA, MRCVS, VetMB; Jens Häggström[2], DVM, DECVIM (Cardiology); Sonya Gordon[3], DVM, DVSc, DACVIM (Cardiology)
[1] Hertfordshire, UK, [2]Uppsala, Sweden, [3]College Station, TX

A potential benefit of pimobendan therapy in delaying or preventing the onset of clinical signs in dogs with preclinical MMVD has not previously been evaluated. The combination of preload and afterload reduction in combination with inotropic support may result in a reduction in cardiac size and filling pressures in dogs with remodelling secondary to MMVD. These effects could delay the onset of CHF.

The objective of the EPIC study (Evaluation of PImobendan in dogs with Cardiomegaly caused by preclinical mitral valve disease) was, to determine whether chronic oral administration of pimobendan (Vetmedin® chewable tablets) in dogs with evidence of increased heart size secondary to preclinical MMVD, could delay the onset of CHF. Inclusion criteria: Dogs of at least 6 years of age, with a body weight ≥ 4.1 and ≤ 15 kg and ACVIM Stage B2 MMVD. Exclusion criteria: severe systemic disease, clinically significant supraventricular and/or ventricular tachyarrhythmias, clinically relevant pulmonary hypertension (RA:RV gradient > 65 mmHg), pre-treatment with cardioactive medication (e.g. ACEi, furosemide etc.)

The trial was designed as a prospective double-blind, randomized, placebo-controlled, international, multi-center clinical study. Dogs were enrolled by specialized veterinary cardiologists at 36 trial centres in Australia, Canada, France, Germany, Italy, Japan, The Netherlands, Spain, Sweden, United Kingdom, and the United States of America. The 360 canine patients enrolled in the trial were randomly allocated to either a pimobendan or a placebo treatment group, with 180 dogs in each group.

The study was started in June 2010. A planned interim analysis was conducted with a number of pre-specified stopping criteria, agreed upon in consultation with an independent statistical consultant, and was concluded on the 16th of February 2015 by the independent interim analysis committee. The interim analysis indicated that there was clear evidence of benefit of the administration of pimobendan in prolonging the time to the primary endpoint of the study, which was a composite of the development of left-sided congestive heart failure, or death presumed to be cardiac in origin. The interim analysis did not raise any concern over the safety of pimobendan administration. Based on these results the interim analysis committee recommended that the study be stopped and the lead investigators ended the study as of March 1, 2015.

The full and final results of the study will be presented for the first time during this session.

Evaluation of Pimobendan in Dogs with Cardiomegaly Caused by Preclinical Mitral Valve Disease: EPIC Trial Discussion

Adrian Boswood[1], DVC, MA, MRCVS, VetMB; Jens Häggström[2], DVM, DECVIM (Cardiology); Sonya Gordon[3], DVM, DVSc, DACVIM (Cardiology)
[1] Hertfordshire, UK, [2]Uppsala, Sweden, [3]College Station, TX

A potential benefit of pimobendan therapy in delaying or preventing the onset of clinical signs in dogs with preclinical MMVD has not previously been evaluated. The combination of preload and afterload reduction in combination with inotropic support may result in a reduction in cardiac size and filling pressures in dogs with remodelling secondary to MMVD. These effects could delay the onset of CHF.

The objective of the EPIC study (Evaluation of Pimobendan In dogs with Cardiomegaly caused by preclinical mitral valve disease) was, to determine whether chronic oral administration of pimobendan (Vetmedin® chewable tablets) in dogs with evidence of increased heart size secondary to preclinical MMVD, would delay the onset of CHF.

The full and final results of the study will be presented for the first time during the preceding session. This session will be an open question forum in which the lead investigators will answer questions that arise relating to the study design, results and conclusions of the study.

Computed Tomography to Assess the Bracycephalic Airway
Brian A. Scansen, DVM, MS, DACVIM (Cardiology)
Fort Collins, CO

INTRODUCTION

The brachycephalic airway syndrome (BAS) is a disease of the nasal cavity, nasopharynx, and larynx related to early ankylosis of the basicranial epiphyseal cartilage. As a result of this early fusion, the longitudinal axis of the skull is shortened, particularly the upper jaw and upper respiratory tract. With a shortened nasal cavity, the nares and nasal cavity become occluded by growth of tissue and nasal conchae that have nowhere else to develop resulting in intranasal airflow obstruction. The soft palate extends caudal to the epiglottis due to the shortened upper jaw and results in stertorous breathing. Chronic increases in negative pressure within airway results in pharyngeal edema and laryngeal inversion. The disease occurs in the popular brachycephalic breeds including Pugs, English & French Bulldogs, Boston Terriers, Pekingese, Shih Tzu, etc. It is important to remember that BAS also occurs in feline breeds – in particular the Persian. In cats, the BAS has also been shown to impair nasolacrimal drainage resulting in ocular disease.[1]

Traditional evaluation of dogs with BAS involved direct visualization of the upper airway – nares, soft palate, and larynx. However, evaluation of the canine brachycephalic airway syndrome (BAS) by direct visualization is subjective and operator dependent. Not surprisingly, current surgical therapies show variable success. Recently, computed tomographic techniques to evaluate BAS in dogs have been published.[2-5] These reports have involved dogs under general anesthesia and either intubated or with the mouth fixed open. This lecture will review CT findings in dogs with BAS and present an approach for CT evaluation of dogs with BAS under conscious sedation, which avoids distortion of the complex 3D morphology of BAS. Furthermore, advanced techniques to quantitate airflow resistance will be introduced, which allows for objective assessment of medical therapy or surgery on BAS severity.

COMPONENTS AND CLASSIFICATION

Historical classification systems grouped BAS lesions into primary and secondary components. The primary components being stenotic nares, elongated soft palate, deviated nasal septum, nasopharyngeal turbinates, macroglossia, and redundant pharyngeal folds; the secondary components being everted laryngeal saccules, laryngeal collapse, tonsilar eversion, and mucosal edema. In a review of 90 dogs with BAS, 94% had an elongated soft palate, 77% had stenotic nares, 66% had eversion of the laryngeal saccules, and 56% had tonsilar eversion.[6] In another study, stenotic nares were present in 43% of dogs, elongated soft palate in 86%, and laryngeal collapse in 53% of dogs with BAS.[7]

While the above abnormalities are typically considered the primary components of BAS, work by Gerhard Oechtering suggests that the traditional scheme is superficial as the areas addressed by conventional surgery are before and after the real problem – too small a nasal cavity resulting in intranasal obstruction.[5] Similarly, mucosal contact points within the nasal cavity reduce the lumen for air passage and result in further intranasal obstruction.[8] Complete evaluation of the degree of intranasal obstruction requires computed tomography or rhinoscopy. A report evaluated functional assessment of the degree of intranasal airway resistance with impulse oscillometry,[9] though this is likely only applicable in the research setting and has not been adopted in clinical cases. Treating the nares rostral to the nasal cavity by alarplasty and the palate caudal to the nasal cavity by staphylectomy may not be addressing the primary site of the problem that is obstruction within the nasal cavity. Laser-assisted turbinectomy has been described to address intranasal obstruction with six-month nasal patency reported at 84%.[10]

Tracheal hypoplasia is commonly considered in the context of BAS, though is primarily a condition of the English Bulldog. More recent work suggests that the tracheal diameter of Bulldog puppies may grow/improve with age, while this increase was not seen in puppies of other breeds.[11] In addition to tracheal abnormalities in BAS, bronchial collapse also appears to be a co-finding in dogs with BAS.[12] In a study of 40 dogs with BAS, fixed bronchial collapse was noted in 35 of the dogs and the degree of laryngeal collapse was positively correlated with severity of the bronchial collapse.[12] While elongation of the soft palate is commonly considered in the evaluation of BAS, the presence of palate thickening has also been shown to occur in these dogs.[2] Whether the thickening is primary or secondary to mucosal edema and swelling from the high intranasal pressures is unknown.

The presence of nasopharyngeal turbinates was originally reported in ~20% of dogs and cats with BAS as observed by retroflex endoscopy,[13] though recent reports suggest two-thirds of dogs with BAS have caudal aberrant turbinates.[5] As noted above, this relates to expansion of normal structures into atypical locations due to the smaller nasal cavity. The association of nasopharyngeal turbinates and clinical disease is not proven, but they likely play a role in airway obstruction and the clinical syndrome. Intranasal epidermoid cysts have also been reported in dogs with BAS, exacerbating upper airway obstruction.[14] Last, gastrointestinal abnormalities appear commonly in dogs with BAS. A study of 73 dogs with BAS confirmed the presence of significant GI pathology in 74% of animals.[15] Findings included esophageal and gastric inflammation, gastro-esophageal reflux, gastric stasis, and pyloric stenosis.[15] Another study evaluating 50 dogs with BAS reported gastrointestinal signs in 88%.[16] Therapy for GI disorders, together with surgical amelioration of BAS, has been shown to improve both respiratory and GI symptoms in dogs with BAS.[17]

COMPUTED TOMOGRAPHY FOR BAS

Several reports have described the use of CT for evaluation of BAS in dogs. A report in 2007 of 23 dogs with BAS described two forms of aberrant conchae – rostral conchae obstructing the nasal passage and caudal conchae obstructing the choanae.[18] In a subsequent series,[2] 26 dogs were evaluated with differing severity of respiratory symptoms compared to 5 control dogs. Dogs with severe BAS had thicker soft palates as compared to mildly affected dogs and controls; however, no differences were found between groups for soft palate length.[2] Minimal cross-sectional area of the nasopharynx was slightly larger for less affected dogs, though significant differences between groups for normalized cross-sectional area were not found.[2] A more recent study used CT to compare nasopharyngeal dimensions between Pugs and French Bulldogs.[4] In this comparison, Pugs had a smaller cross-sectional area within the nasopharynx compared to the French Bulldogs; additionally, the soft palates of the Pugs were shorter and thinner than the French Bulldogs.[4]

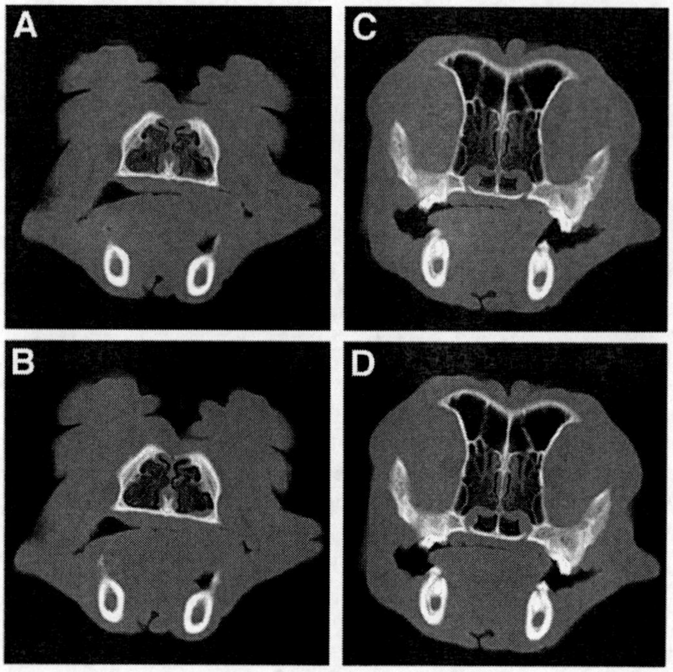

Figure 1. Axial images of the rostral nasal cavity (A, B) and at the level of the choanae (C, D) of an English bulldog with BAS before (A, C) and after (B, D) administration of oxymetazoline (Afrin). Images were acquired at the identical site before and 15min after the decongestant; note the reduction in soft tissue opacity lining the nasal turbinates and choanae, reflecting reduced mucosal swelling and improved cross-sectional dimension of the air passages.

The largest series of CT findings in dogs with BAS evaluated 132 dogs comprised of 66 Pugs, 55 French Bulldogs, and 11 English Bulldogs.[5] Notable findings included abnormal nasal conchae in all dogs, resulting in variable degrees of intranasal obstruction; interconchal and intraconchal mucosal contact points were apparent in 92% of dogs.[5] These same authors reported clinical improvement in these dogs when laser-assisted turbinectomy was performed to remove the aberrant conchae and open the nasal cavity.[10]

Data will be presented of CT findings in 23 dogs with BAS performed under conscious sedation without endotracheal intubation and with the mouth closed, representing non-distorted airway passages. Additionally, data will be presented from CT findings before and after conventional surgery for BAS in six dogs. Last, the effect of 0.05% oxymetazoline (Afrin) on CT findings will be shown, indicating that atomization of this decongestant into the nasal cavity of BAS is both tolerated and reduces mucosal swelling and contact points (**Figure 1**).

QUANTITATIVE EVALUATION OF BAS SEVERITY

Current evaluation of dogs with BAS relies on history (presence of clinical signs) and physical examination (visual inspection of stenotic nares, soft palate, and larynx). In light of the spectrum of findings noted above and differing surgical approaches to this disease, it seems plausible to pursue a more quantitative assessment of the disease to determine both the need for intervention as well as to measure the success of a given intervention. A few approaches have been reported to accomplish this quantitative assessment – impulse oscillometry[9] and whole body barometric plethysmography.[19] Both techniques require advanced equipment and impulse oscillometry requires the patient to be anesthetized. Whole body plethysmography does not yield a direct or linear measure of airway resistance; it has been reported to derive a predictive index of BAS severity based on several respiratory parameters derived from breathing curves. However, whole body plethysmography has the advantage of being measurable on the awake, non-sedated dog.

The author will present data on a novel approach to quantitation of airway resistance in dogs with BAS: computational fluid dynamic modeling of the nasal passage. Computational fluid dynamics (CFD) is a computer-based methodology that uses an imaging dataset to construct an anatomically accurate geometry of the upper respiratory tract, which can be extrapolated to a three-dimensional meshwork and allow mathematical simulation of airflow within this geometry (**Figure 2**). We are then able to use these CFD models to estimate airflow resistance within the airway under differing conditions.

Computed tomography datasets from brachycephalic dogs were used to investigate the feasibility of three-dimensional reconstruction and CFD modeling of airflow resistance. The transverse, bone algorithm datasets were imported into the imaging software for evaluation and generation of a volume meshwork. Semi-automated segmentation of the nasal passageway from the nares to the nasopharynx at the level of the caudal tip of the soft palate was achieved. Volumes of the airway were constructed into a three-dimensional meshwork (**Figure 2**). The volumetric dataset was then saved as a stereolithography Computer Aided Design file and imported into the modeling software. The models were evaluated in a CFD software package and airflow simulated by applying a pressure gradient from the nasal passage to the pharynx. The CFD simulations demonstrated a significant increase in airflow and a significant decrease in overall airflow resistance after administration of the decongestant oxymetazoline (Afrin); analysis of the before and after surgery datasets is ongoing.

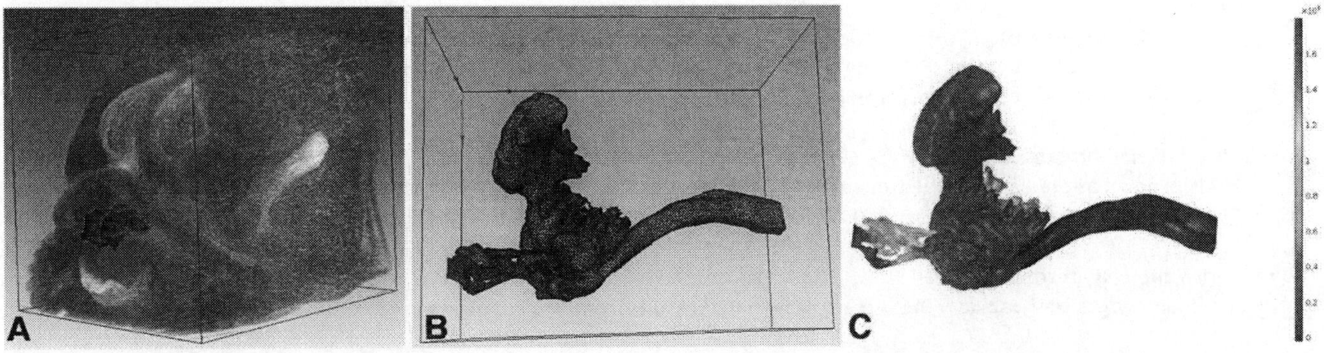

Figure 2. Airflow modeling from an English Bulldog. The CT data is manipulated to define the air-filled spaces of the nasal passage (A). The air passages are then extracted and modeled as a 3-dimensional surface geometry (B). Computer modeling of the air passage geometry is then performed and resistance calculated along the nasal passage with gradations in color signifying a change in resistance (C). In this dog, the greatest step-up in resistance occurs in the nasal turbinates (to the left of the image).

SUMMARY

Generations of selective breeding have led to brachycephalic dog breeds characterized by a shortened craniofacial conformation and the clinical syndrome of BAS. Cross-sectional imaging such as computed tomography has enabled better depiction of complex anatomy that is grossly inaccessible due to physical barriers. Computed tomography and magnetic resonance imaging are the standard of care for evaluation of intranasal obstruction in humans. Similarly, CT evaluation of BAS can provide additional insight into the morphology and severity of the obstructive process in dogs. We have utilized CFD modeling derived from CT scans of dogs with BAS to construct an anatomically accurate geometry of the upper respiratory tract via generation of a three-dimensional meshwork. Further work will assess the utility of this method to quantify airflow resistance in affected dogs and determine optimal therapeutic strategies.

REFERENCES

[1]Schlueter C, et al. J Feline Med Surg 2009;11:891. [2]Grand JG, et al. J Small Anim Pract 2011;52:232. [3]Kaye BM, et al. Vet Radiol Ultrasound 2015;56:609. [4]Heidenreich D, et al. Vet Surg 2016;45:83. [5]Oechtering GU, et al. Vet Surg 2016;45:165. [6]Fasanella FJ, et al. J Am Vet Med Assoc 2010;237:1048. [7]Torrez CV, et al. J Small Anim Pract 2006;47:150. [8]Schuenemann R, et al. J Am Anim Hosp Assoc 2014;50:237. [9]Lippert JP, et al. Pneumologie 2010;64:452. [10]Oechtering GU, et al. Vet Surg 2016;45:173. [11]Clarke DL, et al. J Am Anim Hosp Assoc 2011;47:329. [12]De Lorenzi D, et al. J Am Vet Med Assoc 2009;235:835. [13]Ginn JA, et al. J Am Anim Hosp Assoc 2008;44:243. [14]Murgia D, et al. J Small Anim Pract 2014;55:431. [15]Poncet CM, et al. J Small Anim Pract 2005;46:273-9. [16]Planellas M, et al. BMC Vet Res 2012;8:152. [17]Poncet CM, et al. J Small Anim Pract 2006;47:137. [18]Oechtering TH, et al. Tierarztl Prax 2007;35:177. [19]Liu NC, et al. PloS One 2015;10:e0130741.

Pulmonary Hypertension: Review and Update

Heidi B. Kellihan, DVM, DACVIM (Cardiology)
Madison, WI

DEFINITIONS

Pulmonary hypertension (PH) is defined as a mean pulmonary artery pressure (PAP) >25mmHg. In dogs, PH is defined as an estimated systolic PAP of >30mmHg. Pulmonary artery pressure is influenced by: 1. pulmonary blood flow, 2. pulmonary vascular resistance and 3. pulmonary venous pressure.

Pulmonary Arterial Hypertension (PAH)

Also known as pre-capillary PH. An increase in PAP results from an increase in pulmonary vascular resistance (PVR) and a normal pulmonary capillary wedge pressure (PCWP).

Pulmonary Venous Hypertension (PVH)

Also known as post-capillary PH. An increase in PAP results from and increase in the PCWP and a normal PVR.

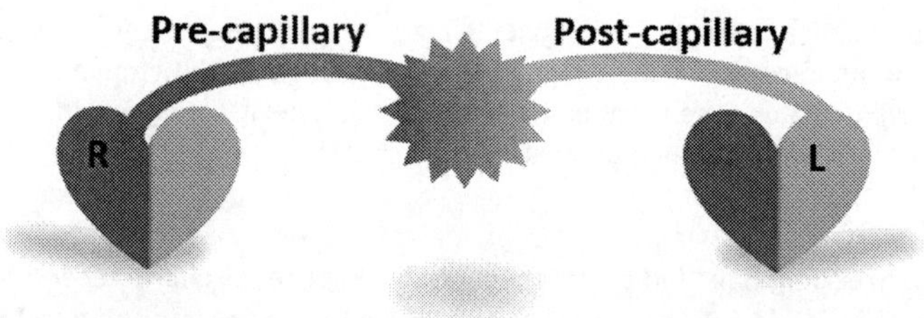

Reactive Pulmonary Hypertension

In this situation, the increase in PAP is a result of increased PCWP and increased PVR. The increased PVR can be transient and reversible from PA vasoconstriction secondary to hypoxia, as with pulmonary edema, or permanent and irreversible secondary to chronic vascular remodeling secondary to hypoxia. Disproportionate reactive PH is seen when the mPAP is disproportionately increased above what the PCWP would be with left-sided heart disease alone. This implies that there is intrinsic disease of the pulmonary arterial system from chronic remodeling changes secondary to chronic left-heart disease or from others diseases resulting in pulmonary arterial hypertension.

ETIOLOGIES OF PULMONARY HYPERTENSION

Pulmonary arterial hypertension	Congenital left-to-right shunt	Atrial septal defect Ventricular septal defect Patent ductus arteriosus Arteriovenous malformation
	Pulmonary disease	Chronic obstructive pulmonary disease Interstitial pulmonary fibrosis Neoplasia Reactive pulmonary artery vasoconstriction
	Thrombolic or embolic disease	Neoplasia Immune-mediated hemolytic anemia Protein-losing disease Hyperadrenocoticism Disseminated intravascular coagulation Sepsis Trauma Cardiac disease Recent surgery Heartworm disease
	High altitude disease	
	Parasitic disease	*Dirofilaria immitis* *Angiostrongylus vasorum*
	Necrotizing vasculitis/arteritis	
	Idiopathic	

Pulmonary venous hypertension	Left heart disease	Mitral valve disease Left ventricular failure Pulmonary venous stenosis Myocardial disease

PATHOLOGY AND PATHOPHYSIOLOGY

General Pulmonary Hypertension Pathology

The normal pulmonary circulation is comprised of a vast network of thin-walled arteries, veins and capillaries characterized by a very low vascular resistance, low pressure and high capacitance. Pulmonary hypertension occurs secondary to an imbalance of endogenous and exogenous pulmonary arterial vasoconstrictors and vasodilators resulting in pulmonary arterial intimal proliferation, medial hypertrophy, vasoconstriction and thrombosis.

In response to hypoxia (global or local), pulmonary arteries vasoconstrict. This is very different from the systemic vasculature, which vasodilates in response to hypoxia. In the face of hypoxia, PH can worsen.

Endothelin Pathway

One of the major mediators of PH is endothelin-1 (ET-1) which is released by pulmonary vascular endothelium resulting in potent pulmonary artery (PA) and vein vasoconstriction, smooth muscle cell proliferation and increased collagen synthesis. A primary stimulus for increased ET-1 concentrations is heart failure due to increased local production, decreased pulmonary clearance and upregulation of endothelin-A receptors (ETA). In addition to powerful affects influencing PH, ET-1 also results in sodium and water retention, potentiation of other neurohormonal systems and promotion of myocardial hypertrophy, fibrosis and remodeling.

Prostanoid Pathway

Prostacyclin and thromboxane A2 are endogenous arachidonic acid metabolites produces from PA endothelial cells and platelets involved in maintenance of vascular tone in normal physiologic and pathyphysiologic states. Prostacyclin is a vasodilator, inhibitor of platelet activation and has antiproliferative effects on the PA. In contrast, thromboxane A2 is a potent vasoconstrictor and promotor of platelet activation. In cases of pulmonary hypertension, the balance of protacyclin and thromboxane A2 formation favors thromboxane A2 leading to vasoconstriction, cellular proliferation and thrombosis.

Nitric Oxide Pathway

Nitric oxide (NO) is synthesized endogenously from L-arginine and oxygen by nitric oxide synthase isoenzymes in the PA vascular endothelium. Nitric oxide activates guanylate cyclase and subsequently cyclic guanosine monophosphate (cGMP) which results in a reduction of intracellular calcium concentration. The NO pathway enhances smooth muscle cell relaxation, inhibits smooth muscle cell proliferation and hypertrophy, and inhibits platelet aggregation and adhesion. cGMP is rapidly inactivated by phosphodiesterase (PDE), particularly PDE-5 isoenzymes. In experimental models and clinical studies of heart failure in people, a deficiency in basal NO synthesis as well as a decrease in endothelial release and response to NO has been found, suggesting that the loss of NO-dependent vasodilation and anti-cellular proliferation may contribute to the production of PH through vascular remodeling.

Natriuretic Pathway

Natriuretic peptides increase cGMP, independent of NO-cGMP. Natriuretic peptides result in PA vasodilation and they inhibit mitogenesis. In heart failure, there is down-regulation of natriuretic peptides and Upregulation of PDE-5. NP-cGMP mediated vasodilation is attenuated.

Miscellaneous Mediators

Miscellaneous mediators involved with PH include, but are not limited to: serotonin, vascular endothelial growth factor (VEGF) and bone morphogenic protein receptor (BMPR). These mediators result in cellular proliferation and vascular remodeling.

In PH, the endothelin pathway is upregulated, the NO pathway is downregulated, the prostanoid pathway is downregulated.

Intimal Proliferation

Intimal lesions associated with PH consist of eccentric intimal thickening and concentric fibrotic plexiform lesions comprised of myofibroblasts or endothelial cell in an "onion-skin" morphology that account for most of the reduction in luminal area of the small pulmonary arteries. Deposition of an acellular extracellular matrix in the lumen of pulmonary arteries also contributes to luminal obstruction and an increase in PVR. The pathogenesis of these intimal lesions is largely undetermined, although it is thought that these lesions may result from endothelial injury, smooth muscle cell migration, extracellular matrix deposition and endothelial proliferation. Microthrombi may also serve as the scaffolding for theses lesions.

Medial Hypertrophy

The pathologic feature of medial smooth muscle hypertrophy is characteristic of PH and involves muscularization of pulmonary arteries. Normally, the medial smooth muscle layer of pulmonary arteries represents 10-15% of the total outside vessel diameter. In PH, medial hypertrophy can result in an increase in pulmonary artery outside vessel diameter of up to 30-60%. The increase in the tunica media layer occurs as a result of both smooth muscle cell hypertrophy and hyperplasia. The mechanism by which this hypertrophy and hyperplasia occurs is thought to be primarily a result of cell proliferation and also inhibition of cell apoptosis.

Adventitial Remodeling

More recently, the fibroblast-rich adventitia has been postulated to play a role in PH by regulating vascular function from the "outside-in". In idiopathic PH in people, the adventitial layer, which is normally 15% of the external diameter of pulmonary arteries larger than 50 microns in diameter has been shown to be increased to 28% of the artery diameter because of collagen deposition. Inflammatory cells are present in the adventitial layer and may affect fibroblasts contributing to PH when vascular stress, hypoxia, high flow states and toxins (i.e. monocrotaline) are present.

Right Ventricular Pathology

The right ventricle's (RV) response to PH is RV concentric hypertrophy, when compensated, and RV eccentric hypertrophy, when decompensated. The morbidity and mortality associated with PH results from RV failure, in the form of ascites, low output signs and death.

CLINICAL SIGNS AND PHYSICAL EXAMINATION

Patients with PH can present with a variety of clinical signs, some resulting from the primary etiology of PH or directly from the PH. These signs include: cough, respiratory distress, lethargy, syncope and exercise intolerance.[1]

On physical examination, PH patients may have mitral murmurs, tricuspid murmurs, mitral and tricuspid murmurs, a split second heart sounds, pulmonary crackles, cyanosis and ascites. Again, these physical examination findings may be a result of the primary underlying disease or directly a result of the PH.

DIAGNOSTICS

Invasive Hemodynamic Assessment

Right heart catheterization can provide multiple hemodynamic parameters that aide in the diagnosis and etiologic classification of PH. Right heart catheterization provides hemodynamic information regarding the right atrial, right ventricular, pulmonary artery and pulmonary capillary wedge pressures. Cardiac output measurements can also be calculated, which can be used to calculate the PVR.

Thoracic Radiographs

Thoracic radiographs are not diagnostic of PH but they can, firstly, help determine the etiology of PH (i.e. left-heart disease, interstitial pulmonary fibrosis, neoplasia, PDA) and, secondly, help support a diagnosis of PH (i.e. right heart enlargement, main pulmonary artery enlargement, branch PA enlargement, tortuous and blunted PAs). Non-cardiogenic pulmonary edema can also form as a result of PH. The distribution of edema, in these cases, is typically patchy and often associated with dyspnea and syncope.

Echocardiography

Echocardiography can estimate the systolic, diastolic and mean PAP. The peak systolic TR velocity and gradient reflects the estimated peak systolic right ventricular pressure and is used to classify PH as mild (≥ 2.8 to <3.5 m/s, ≥ 31.4 to <50 mmHg), moderate (3.5-4.3 m/s, 50-75 mmHg) or severe (>4.3 m/s, >75 mmHg). There can be problems with the estimation of the PAP based on TR gradients and this measurement should be used in conjunction this other echo parameters of PH. Other parameters that echo can evaluate include: RV morphology, septal flattening, pulmonary artery systolic flow profiles, PA systolic time intervals, Tei index of myocardial performance, systolic time intervals, RV tissue Doppler, tricuspid annular plane systolic excursion and the right PA relative area change/distensibility index.

Cardiac biomarkers

NT-proBNP has been shown to be elevated in dogs in the presence of PAH and PVH. NT-proBNP can be helpful in supporting a diagnosis of PH in a dog with pre-capillary PH, especially when an estimated PAP is difficult to interpret.

TREATMENT

Pulmonary hypertension treatment focuses on targeting the derangements that result in medial hypertrophy, intimal proliferation and decreased vascular compliance. Pulmonary vasodilating drugs currently in use target the pathophysiologic abnormalities associated with the pulmonary arterial endothelin pathway (endothelin receptor antagonists), prostanoid pathway (prostacyclin analogues), and nitric oxide pathway (specific or non-specific phosphodiesterase inhibitors). The goals of treatment are to: 1. Treat the underlying disease 2. Promote PA vasodilation 3. Suppress cellular proliferation 4. Improve cardiac output by decreasing PVR and 5. Enhance RV inotropy. The ideal PH therapy would decrease PVR, spare the systemic circulation and increase RV inotropy.

PDE-5 Inhibitors

Sildenafil (Viagra®, Revatio®), tadalafil, vardenafil, udenafil, avanafil are orally active, highly selective PDE-5 inhibitors. Multiple studies have demonstrated the benefits of sildenafil in people and dogs with PH. Sildenafil appears to produce beneficial effects in PH by multiple mechanisms (i.e. improvement of: RV and LV function, ventricular interdependence, pulmonary lymphatic drainage) in pulmonary, but the primary mechanism operative in PH patients appears to be direct pulmonary artery vasodilation. Sildenafil should be used cautiously in cases of left-heart disease. If the PA pressure falls below the left atrial pressure, with sildenafil, forward blood flow from the right heart to the left heart can become severely impaired or the pulmonary capillaries can flood because of higher pulmonary venous pressures, resulting in pulmonary edema. Tadalafil has been described in the clinical and experimental settings.

Endothelin Antagonists

Bosentan, sitaxsentan, ambrisentan are endothelin antagonists that are given orally, yet have historically been cost prohibitive (i.e. > $40,000/year).

Prostacyclin Analogs

Epoprostenol, treprostinil, iloprost are prostacyclin analogs that are administered as continuous rate infusions, frequent SQ injections or frequent inhalations (i.e. 6-12 x/day).

Pimobendan

Pimobendan and levosimendan are calcium sensitizing agents and PDE-3 inhibitors. PDE-3 has activity at the level of large and small (resistance)PAs, whereas PDE-5 exerts its activity in primarily large pulmonary arteries. PDE-3 inhibitors promote PA vasodilation via enhancement of cAMP-dependent/adrenergic relaxation. The dual action of PDE-3 inhibition and the positive inotropic effects of calcium sensitization may provide some attenuation of PH, especially in PH secondary to left-sided heart disease. Ultimately, the RV needs to be supported if it fails and pimobendan may be beneficial as a RV positive inotrope.

L-Arginine

L-arginine is an amino acid that results in NO and L-ciltrulline production. There is the potential for L-arginine supplementation to improve PH by increasing NO bioavailability and potentiating the effects of PDE-5 inhibition.

REFERENCES

1. Kellum HB, et.al. J Vet Intern Med. 2007;21(6):1258.
2. Kellihan HB, et.al. Veterinary Clinics of North America. 2010;40(4):623.
3. Kellihan HB, et. al. J Vet Cardiol. 2012;14(1):149.
4. Kellihan HB, et. al. J Vet Cardiol. 2011;13(3):171.
5. Kellihan, et. al. J Vet Cardiol. 2015;17(3):182.
6. Soydan LC, et. al. J Vet Cardiol. 2015;17(1):13-24.
7. Tian L, et. al. J of Biomechanics. 2014;47(12):2904.
8. Johnson L, et. al. J Vet Intern Med. 1999;13(5):440.
9. Schober KE, et. al. J Vet Intern Med 2006;20(4):912.
10. Simonneau G, et. al. J Am Coll Cardiol 2004;43(12 Suppl S):5S.
11. Serres F, et. al. J Vet Intern Med. 2007;21(6):1280.
12. Serres FJ, et. al. J Am Vet Med Assoc 2006;229(11): 1772.
13. Atkinson KJ. et. al. J Vet Intern Med 2009;23(6):1190.
14. McLaughlin VV, et. al. Circulation 2009;119(16):2250.
15. Visser LC, et. al. ACVIM Forum Proceedings Abstract. 2015.
16. Visser LC, et. al. J Vet Cardiol. 2015;17(3):161.
17.Visser LC. Et. al. J Vet Cardiol. 2015;17(2):83.

Pulmonary Thromboembolism

Etienne Côté, DVM, DACVIM (Cardiology, Small Animal Internal Medicine)
Charlottetown, PE, Canada

INTRODUCTION

Pulmonary thromboembolism (PTE), which is the embedding and/or local formation of blood clots in the pulmonary arterial circulation, is a potentially devastating disorder in dogs, cats, and humans.[1-7] Its management is complicated by imprecision of current diagnostic tests and limited efficacy of existing treatments. Nevertheless, the potentially fatal outcome of PTE, and the recognized association between it and certain diseases, mean that recognition and treatment are important in veterinary medicine.

ANATOMY AND PATHOPHYSIOLOGY

In many animal species, including dogs, cats, horses, cattle, and humans, the lungs have a dual blood supply.[8,9] Systemic venous blood that returns to the right side of the heart is pumped by the right ventricle to the pulmonary arteries and through the pulmonary capillaries, where gas exchange occurs, then to the left side of the heart for systemic distribution. Thus, blood in the pulmonary arteries travels through, rather than to, the lungs. Very small branches of the descending aorta, the bronchial arteries, perfuse the bronchial walls and, in some species, the lungs.[8] The relevance to the current topic is that PTE can be expected to increase resistance to outflow of the right ventricle, and with massive PTE, severe cor pulmonale and oxygen desaturation are possible, but infarction is less common because the airways are perfused separately.[8,9]

The classic mechanism of PTE involves formation of a blood clot in the venous system, movement of this clot with systemic venous return through the right heart, and lodging of the clot in the pulmonary arteries or arterioles, when the clot reaches a blood vessel that is smaller in diameter than the clot. At such a point, while endogenous fibrinolytic mechanisms can be activated to begin to lyse the clot, the pathologic hypercoagulable mechanisms that led to the clot's initial formation can be expected to remain active, leading to clot expansion. This antagonistic dynamic between clot formation and clot lysis is pivotal in determining whether the patient recovers from PTE or is harmed by it. Logically, the clinician's goal is to promote clot lysis, reduce clot formation, and reduce the formation and propagation of new clots.

The image of a single clot in a pulmonary artery acting as one cork in a bottle likely is incomplete, inasmuch as histologic evidence often reveals a multifocal process consistent with recurrent embolic showering. This is an important nuance because physical removal of clots (thrombectomy) can address single large clots ("central," or Vanderbilt type I in humans) but not numerous (e.g., potentially hundreds of) small ones.

Virchow's triad of collagen/subendothelial exposure, a hypercoagulable state, and blood stasis explains the factors that can predispose to PTE. However, these neatly described categories often can overlap in disease processes, and thrombus formation can be favored through various mechanisms, including collagen exposure within a neoplasm, platelet hyperaggregability,[10] disruption of blood flow, and elaboration of paraneoplastic substances (neoplasia); endothelial splitting and sluggish blood flow (heart disease); autoagglutination (immune-mediated hemolytic anemia); and increased concentrations of coagulation factors and thrombin-antithrombin complexes[11] and nonspecific hemostatic alterations[12,13] (hyperadrenocorticism), among others. Therefore, the clinician's goals should include an understanding of the ways in which a particular disease can induce one or more of the components of Virchow's triad, and then to pursue means by which these changes can be suppressed and the normal state restored.

DIAGNOSIS

"The greatest problem diagnosing acute pulmonary embolism is omitting it from consideration as a possible explanation for a patient's presentation."[14]

Clinical suspicion

PTE is found consistently in association with other disorders that can predispose a patient to thromboembolic disease. Meta-analysis of eight retrospective studies totaling 120 dogs and 46 cats with PTE reveals that PTE most often coexists with neoplasia (27% of dogs with PTE, 35% of cats with PTE), heart disease (16% [dogs], 28% [cats]), sepsis (19% [dogs], 4% [cats]), hyperadrenocorticism (14% [dogs]), disseminated intravascular coagulation (9% [dogs], 11% [cats]), immune-mediated hemolytic anemia (IHA; 14% [dogs], 4% [cats]), protein-losing nephropathy (12% [dogs], 11% [cats]), and protein-losing enteropathy (9% [cats]).[1-5,15-17] Presence of these diseases in any patient warrants the consideration of PTE as a complicating factor, i.e., "guilt by association." The likelihood of PTE is increased if clinical signs are observed that reflect circulatory embarrassment and/or respiratory system dysfunction in patients with these disorders. Physical abnormalities in PTE patients include tachypnea/dyspnea (65% of cases) and tachycardia (42%) in dogs[1,3,15] and tachypnea/dyspnea (57%), lethargy (46%), dehydration (28%), hypothermia (17%), and icterus (11%) in cats.[2,4]

A "snowball effect" of rapidly expanding thrombus and recurrent emboli can be expected in the more severe cases of PTE, which lends validity to early suspicion and management as a way of mitigating the consequences of PTE.

Practical diagnostic approach

Algorithmic analysis allows the clinician to combine test modalities in a decision-tree format. Such algorithms have not been validated or assessed formally in veterinary medicine. The author's clinical approach, based on available evidence and practical limitations, is shown in the figure, below.

Bloodborne markers of hemostasis

Measuring an analyte on a blood sample carries the enormous advantages of being negligibly invasive and involving minimal patient restraint. The main drawbacks are limited sensitivity and specificity, and the need for an on-site diagnostic analyzer. D-dimers, which are a specific subgroup of fibrin degradation products, are released during clot lysis. High concentrations of D-dimers are noted in humans with PTE[7,14] and some dogs with PTE.[5,18] In two studies of dogs with PTE that was identified via computed tomographic (CT) angiography, D-dimers were not diagnostically helpful in one case series,[18] whereas in another, a high value was nonspecific but a value <103 ng/mL was 100% specific for ruling out PTE.[5]

A circumstantial association exists between PTE and general markers of hypercoagulability, as would be expected. For example, thromboemboli have been found in 50-80% of cases of IHA,[19-21] and dogs with IHA repeatably have shown thromboelastographic (TEG) profiles consistent with hypercoagulability.[22,23] However, the prospective value of such testing has been called into question because TEG results consistent with hypercoagulability have not been found consistently in dogs with thromboembolic disease. For example, a recent study of 39 critically ill dogs with necropsy-confirmed thrombosis found no correlation between TEG results and the presence or absence of thrombi.[24]

Short (rapid) prothrombin times (PT) and activated partial thromboplastin times (aPTT) could be thought to indicate hypercoagulability, since these changes represent the opposite of hypocoagulability, which these tests are designed to identify. However, a short PT and/or aPTT result has no association with hypercoagulability and is not diagnostically useful.

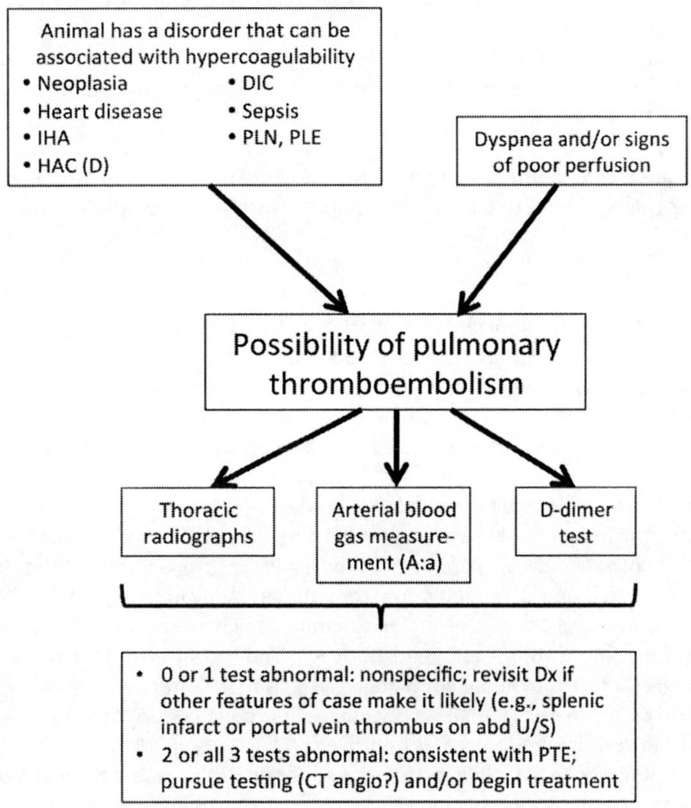

Abd U/S, abdominal ultrasound; CT angio, computed tomographic angiogram; DIC, disseminated intravascular coagulation; D, dog; Dx, diagnosis; HAC, hyperadrenocorticism; IHA, immune-mediated hemolytic anemia; PLE, protein-losing enteropathy; PLN, protein-losing nephropathy

Bloodborne assessment of gas exchange

Arterial blood gas results in dogs with PTE typically reveal mild hypoxemia (e.g., 58.8 +/- 10.0 mm Hg, normal range 67–92 mm Hg; room air at 1500 m elevation) and normo- or hypocarbia.[3] The alveolar to arterial ratio (A:a) represents the difference between oxygen content of systemic arterial blood and of gas in the alveoli, and normally should be < 15 mm Hg. Clinically, A:a is calculated using the formula A:a = $Fi_{O2} \times (P_b - P_{H2O}) - (Pa_{CO2}/R) - Pa_{O2}$, where F_{iO2} is the fraction of inspired oxygen, expressed as a decimal (e.g., 20% in room air, i.e., 0.2), P_b is the barometric pressure (627-643 mm Hg; exact value for day and geographic location available at weather.com), P_{H2O} = is the partial pressure of water (47 mm Hg), R is a constant (0.8), and P_{aO2} is the arterial oxygen content, which is displayed on the arterial blood gas result. A:a values ≥ 15 mm Hg suggest incomplete transfer of oxygen from alveoli to blood, as occurs with PTE and many other pulmonary disorders. In two case series totaling 20 dogs with PTE, the mean A:a was 35.9 +/- 12.2 (range, 18.0–44.3) mm Hg,[3] and 42.2 +/- 10.95 mm Hg,[15] and suggesting that this is a sensitive, if not specific, test for PTE in dogs.

Diagnostic imaging

Plain thoracic radiographs reveal variable, often overlapping, findings with PTE. In dogs, a pulmonary interstitial pattern (35% of cases), a mixed interstitial and alveolar pattern (15%), an alveolar pattern (17%), vascular changes (engorgement; pruning; loss of lobar artery, vein, or both; 28%), pleural effusion (34%), and normal radiographs (13%) have been reported in retrospective studies.[1,3,15,16] Pleural effusion is associated with pulmonary infarction at necropsy: in one detailed radiographic case series, half of dogs with pleural effusion had histologic evidence of lung infarction, compared to 0% in dogs without pleural effusion.[16] In cats, radiographic findings of PTE include vascular changes (asymmetric vessel size or distribution among lung lobes; 63% of cases), pleural effusion (58%), peripheral lung lobe consolidation (32%), hypolucent pulmonary regions (21%), atelectasis (16%), thoracic mass(es) (16%), and an alveolar pattern (11%).[2,4]

Computed tomographic angiography can be performed with conscious sedation in small animals suspected of having PTE.[5,18,25] In a series of 12 dogs suspected of being hypercoagulable due to naturally-occurring IHA, four (33%) were classified as having PTE, 3 (25%) were suspected of having PTE, and 5 (42%) did not have CT evidence of PTE. A gold standard was not available in this study but findings were superior to echocardiographic and hemostatic test results.[18] In an experimental study of induced PTE in dogs, the sensitivity for detecting pulmonary emboli via CT angiography was 64-76%, and specificity was >99%, with necropsy confirmation of the emboli detected on CT.[25] This sensitivity was greater than that of magnetic resonance imaging angiography performed on the same dogs.

Ventilation-perfusion (V-Q) scanning, long considered the most definitive test for a clinical diagnosis of PTE, is hampered by limited sensitivity and the need for specialized equipment. It is used rarely, if at all, for the diagnosis of PTE in veterinary medicine.

TREATMENT

Pulmonary thromboembolism carries the unfortunate combination of being confirmed uncommonly enough that clinical trials of treatments are difficult to conduct, yet being devastating enough that the information such trials would provide would be extremely valuable.

Inciting cause

Intuitively, the most important intervention to help the patient is to remove inciting or exacerbating causes. Intravenous catheters almost always are present in hospitalized patients with severe illness, and they should be removed or changed in PTE patients, especially jugular catheters, since these offer a surface for platelet adhesion and clot formation. Reduction of tumor burden similarly can be expected to reduce the propensity for progression of PTE when hypercoagulability was a paraneoplastic phenomenon. Elimination of adult heartworms from the pulmonary arteries is a third example.

Prevention of clot worsening and thromboprophylaxis

Medications that prevent the growth of existing blood clots and reduce the likelihood of formation of new ones are antiplatelet or anticoagulant drugs. Aspirin (5-10 mg/kg PO q 12h), an antiplatelet drug, has enjoyed a long history of use in small animal thrombus prevention via extrapolation from human medicine. Regrettably, the most recent work shows that aspirin's effects on platelet function are negligible.[26] It is still used empirically by some clinicians when heparins would be felt to be ineffective (e.g., protein-losing states, where antithrombin might be lost and heparin would therefore be ineffective).

Clopidogrel, a thienopyridine antiplatelet drug, is more effective than aspirin for prevention of recurrence of aortic thromboemboli in cats.[27] In dogs with IHA, clopidogrel has been administered at 2-3 mg/kg PO q 24h, with no convincing survival benefit.[28] Efficacy in veterinary PTE cases is unproven.

Warfarin (0.2 mg/kg PO q 24h, titrated to maintain prothrombin time at 1.5-2.5 times the upper limit)[29] subjectively is a more effective antithrombotic drug. Its requirement for close supervision for precise dosage titration, and risk of excessive anticoagulation, are limiting factors.

Low molecular weight heparin (e.g., dalteparin 20-40 IU/kg IV once, then 16.7-33 IU/kh/h IV CRI), initially thought to be suitable for long interdose intervals and a more predictable effect than unfractionated heparin, remains of limited efficacy and predictability. This dosage range produced anti-factor-Xa activity levels that were subtherapeutic (low end of the dosage range) or that were in the range but with unproven safety (higher end).[30]

Rivaroxaban, an orally-administered factor Xa inhibitor, has shown efficacy in the management of PTE in humans.[31] In dogs with PTE, clinical trials have not yet been conducted but an in vitro study using canine plasma has demonstrated that the drug has a dose-dependent anticoagulant effect.[32]

Thrombolysis

Streptokinase,[33,34] urokinase, and tissue plasminogen activator have been given clinically to dogs and cats to lyse arterial thrombi. Adverse effects have been frequent, and appear to offset benefits of such treatment compared to the more conservative approach of preventing clot worsening and allowing endogenous fibrinolysis to reduce clot volume. Currently, only tissue plasminogen activator is available for (human) medical use in the United States.

Clot removal

Physical removal of clots via catheter-based techniques or open surgical removal is effective in humans with PTE.[7,35] While such attempts have not been reported to be successful in dogs or cats, a similar technique was used in 4 cats with distal aortic thromboembolism.[36] Trials comparing this form of treatment to non-interventional means, be they thrombolytic or simply preventing clot expansion, have not been undertaken in small animal medicine.

PREVENTION

Thromboprophylaxis is provided anecdotally in patients at risk for PTE, and in patients who have experienced PTE and recovered from it if the underlying cause cannot be eliminated. In both groups, single-agent therapy with antiplatelet drugs (e.g., clopidogrel 2-3 mg/kg PO q 24h) is used most often. Other medications listed above (under Prevention of clot worsening) can be considered.

FOLLOW-UP AND OUTCOME

A successful response to treatment for PTE involves a patient whose clinical signs plateau and gradually improve. A dramatic change for the better is less likely than a slow improvement over many days, and prolonged, or indefinite, treatment is expected, particularly if the disease that is responsible for hypercoagulability cannot be eliminated completely.

SUMMARY

The presence of neoplasia, heart disease, IHA, or other diseases that can be associated with PTE should bring PTE to the forefront of the clinician's mind. The suspicion of PTE is reinforced if the patient shows lower respiratory signs, especially if these are new in onset. Confirmation is challenging in the clinical setting, and D-dimer concentrations, an abnormal alveolar to arterial ratio, or both, make PTE more likely but are not conclusive; conclusive diagnostic tests (angiography, ventilation-perfusion scanning) are complex, of limited precision, and often not justifiable. Treatment of PTE is warranted based on clinical suspicion and corroborative evidence as just described, and involves anticoagulation, antiplatelet drugs, and/or fibrinolysis; physical extraction of the clot is rarely performed. The patient's outcome is determined by the clinician's ability to control both the underlying disease and PTE.

REFERENCES

[1]LaRue MJ, et al. J Am Vet Med Assoc 1990;197:1368. [2]Norris CR, et al. J Am Vet Med Assoc 1999;215:1650. [3]Johnson LR, et al. J Vet Internal Med 1999;13:338. [4]Schermerhorn T, et al. J Vet Intern Med 2004;18:533. [5]Epstein SE, et al. J Vet Intern Med 2013;27:1646. [6]Williams JG. In Ettinger SJ, et al., eds. Textbook of Veterinary Internal Medicine, 8th ed (St. Louis: Elsevier Saunders, 2017): in press. [7]Konstantinides SV, et al. Eur Heart J 2014;35:3033. [8]McLaughlin RF, et al. J Am Med Assoc 1961;175:694. [9]Dyce KM, et al. Textbook of Veterinary Anatomy, 4th ed (St. Louis: Elsevier Saunders, 2010), 248. [10]Thomas JS, et al. J Vet Internal Med 1999;13:319. [11]Jacoby RC, et al. Arch Surg 2001;136:1003. [12]Pace SL. J Vet Intern Med 2013;27:1113. [13]Park FM. J Vet Intern Med. 2013;27:1136. [14]Goldhaber SZ. In Mann DL, et al., eds. Braunwald's Heart Disease, 10th ed (Philadelphia: Elsevier Saunders, 2015), 1664. [15]Klein MK, et al. J Am Vet Med Assoc 1989; 195, 246. [16]Fluckinger MA, et al. Vet Radiol 1984; 25:124. [17]Burns MG, et al. J Am Vet Med Assoc 1981;178:388. [18]Goggs R, et al. J Small Anim Pract 2014;55:190. [19]McManus PM, et al. J Am Vet Med Assoc 2001;218:1308. [20]Scott-Moncrieff C, et al. J Am Anim Hosp Assoc 2001; 37:11. [21]Carr AP, et al. J Vet Intern Med 2002;16:504. [22]Sinnott VB, et al. J Vet Emerg Crit Care 2009; 19:484. [23]Fenty RK, et al. J Am Vet Med Assoc 2011; 238:463. [24]Thawley VJ, et al. J Vet Emerg Crit Care 2016 Jan 8 (e-pub ahead of print). doi: 10.1111/vec.12441. [25]Hurst DR, et al. J Vasc Investig Radiol 1999;10:309. [26]Blois SL, et al. Am J Vet Res 2010; 71:349. [27]Hogan DF, et al. J Vet Cardiol 2015;17:S306. [28]Mellett AM, et al. J Vet Intern Med 2011;25:71. [29]Monnet E, et al. Am J Vet Res 2000;61:48. [30]Scott KC, et al. J Vet Emerg Crit Care 2009;19:74. [31]EINSTEIN-PE Investigators. N Engl J Med 2012;366:1287. [32]Conversy B, et al. Vet J 2013;198;437. [33]Moore KE, et al. J Vet Emerg Crit Care 2000; 10: 245. [34]Ramsey CC, et al. J Am Vet Med Assoc 1996;209:780. [35]Neely RC, et al. Ann Thor Surg 2015;100:1245. [36]Reimer SB, et al. J Vet Intern Med 2006;20:290.

The Importance of Body Size and Composition in Small Animal Heart Disease

Lisa M. Freeman, DVM, PhD, DACVN
North Grafton, MA

INTRODUCTION

Traditionally, nutritional modifications have been considered only in dogs and cats with congestive heart failure (CHF) and few, if any, nutritional recommendations have been made for animals in earlier stages of heart disease. However, if one waits to address nutrition until this later stage of disease, the benefits of optimizing nutrition during the longer, and more common, earlier stages are lost. In addition, there is some evidence for a role of diet in the development of heart disease. Therefore, attention to body composition appears to be increasingly important in both the development and progression of heart disease in dogs and cats.

CACHEXIA

Cachexia, a loss of muscle, occurs in a variety of acute and chronic diseases in people and companion animals, including cancer (cancer cachexia), heart failure (cardiac cachexia), chronic kidney disease (renal cachexia), chronic obstructive pulmonary disease, and a variety of acute illnesses and injuries seen in the intensive care unit (Freeman, 2012). The weight loss that occurs in cachexia is unlike that seen in a healthy dog, cat, or person that loses weight. In a healthy animal that is receiving insufficient calories to meet requirements, metabolic adaptations allow fat to be used as the primary fuel source, thus preserving lean body mass. Conversely, the primary fuel source in animal with chronic disease is amino acids from muscle so that these animals quickly catabolize muscle and lean body mass. Therefore, loss of lean body mass is a hallmark of cachexia. Fat and bone also are lost to lesser degrees although in advanced cachexia, depletion of all compartments is apparent. In addition to quantitative changes in muscle, there are also qualitative changes in muscle function. Loss of lean body mass and functional muscle changes have direct and deleterious effects on strength, immune function, and survival. In people, cachexia is an independent predictor of survival.

Severe weight loss in a patient with advanced CHF or cancer represents a classical picture of cachexia. However, cachexia actually begins gradually and is initially more subtle. It is at these early stages, when muscle loss is only mild, when identification and treatment is most effective. Therefore, developing clinically applicable techniques to identify lean body mass loss at an early stage is important because it is at this stage that treatments are much more likely to be successful. Although numerous research techniques are available for assessing lean body mass (e.g., dual-energy x-ray absorptiometry, MRI), they have inherent assumptions that affect their interpretation and are not readily available for clinical patients. The author is currently validating an ultrasound-based technique for quantifying muscle loss. In the meantime, an easy and clinically-applicable method of assessing muscle is the muscle condition score (MCS) which should be evaluated in every patient at every visit, in addition to body weight, body condition score (BCS), and diet history (WSAVA Nutritional Assessment Guidelines).

One of the forms of cachexia that has been studied in dogs and cats is cardiac cachexia (Freeman et al, 2012). In one study of dogs with CHF, over 50% of patients had cachexia (Freeman et al, 1998). Cardiac cachexia is a multifactorial process caused by the reduced calorie intake, increased energy requirements, and metabolic alterations. Appetite is commonly affected in animals with CHF and can present as anorexia (complete loss of appetite), hyporexia (reduced appetite), or, most commonly, dysrexia (changes in appetite and food preferences). However, it is inflammatory cytokines, such as tumor necrosis factor (TNF) and interleukin-1 (IL-1), that are primary mediators of cachexia (Freeman et al, 1998; Freeman et al, 2012). These inflammatory cytokines are known to directly decrease appetite, to increase energy requirements, and to increase the catabolism of lean body mass. Of particular pertinence to heart disease, TNF and IL-1 also cause cardiac myocyte hypertrophy and fibrosis and have negative inotropic effects.

SARCOPENIA

Sarcopenia is similar to cachexia in that it is characterized by muscle loss, but sarcopenia is a syndrome seen during aging in the absence of disease (although cachexia and sarcopenia can occur concurrently). In sarcopenia, the loss of lean body mass often is accompanied by an increase in fat mass so the total weight may not change (or may even increase), thus masking the sarcopenia. Like cachexia, sarcopenia has important clinical implications because it is associated with increased mortality. The mechanisms of sarcopenia involve physical inactivity, elevated cytokine production, decreased concentrations of growth hormone and testosterone, changes in type II muscle fibers, insulin resistance, and decreased protein synthesis. There is more limited research on sarcopenia in dogs and cats compared to humans but data suggest that dogs and cats also lose lean body mass during aging. Because heart disease typically affects older dogs and cats, sarcopenia (and cachexia) can complicate the treatment of patients with heart disease.

OBESITY

Approximately half of dogs and cats in the US are overweight or obese. Although pathophysiologic changes associated with obesity (e.g., increased C-reactive protein, hormonal changes, blood pressure) could alter cardiovascular structure and function, the role of obesity in the development of heart disease in dogs is unknown. It is often reported that one epidemiologic study from 1986 showed an association between obesity and cardiovascular disease (Edney and Smith, 1986), but it is difficult to determine from that paper whether this is a true association. More recent studies have identified echocardiographic changes in obese dogs and cats so obesity could play a role in the development and progression of cardiac disease. In one case-control study of Maine Coon cats, obesity was significantly associated with hypertrophic cardiomyopathy (HCM; Freeman et al, 2013), but in a case-control study of cats of other breeds, cats with HCM were not more likely to be overweight or obese compared to cats without the disease (Yang et al, 2008). Further research is needed to better understand the potential role of obesity in the development and progression of heart disease in dogs and cats.

OBESITY PARADOX

While additional research is needed to determine whether obesity is a risk factor for heart disease in dogs and cats, obesity is a well-known risk factor for the development of heart disease in people. However, once CHF is present, obesity may actually be associated with a protective effect - this is known as the obesity paradox and has been documented in numerous human studies. Given the adverse effects of cachexia, the association between obesity and improved survival in heart failure appears to be due to a *lack* of cachexia, rather than to the obesity *per se*. This is due to the increased reserve of lean body mass in overweight and obese people (when one gains weight, the weight gained is typically composed of 75% fat and 25% lean). The obesity paradox also has been demonstrated in dogs and cats with CHF (Slupe et al, 2008; Finn et al, 2010). These data emphasize the importance of avoiding weight (and muscle) loss in the patient with CHF.

***IN UTERO* AND EARLY GROWTH IN CATS**

Because heart disease is so common in cats, it is important to find ways to reduce the incidence or to slow progression. While it is not possible to identify cats at risk (other than predisposed breeds, such as the Maine Coon cat or Ragdoll, in which genetic testing can be done), the changes that *may* help reduce the phenotype of cardiac disease are easily achieved and are good general recommendations for feeding cats in general.

One of the challenges in heart disease is the wide phenotypic variation among cats with cardiomyopathies – some cats have asymptomatic HCM which never progresses, while others have rapid progression of disease with early death. This phenotypic variation also occurs in people where genetic mutations have been identified in half of people with HCM, but the phenotypic expression of the disease still is highly variable, even in people within the same family with the same genetic mutation. The cause for this phenotypic disparity is unknown but environmental modifiers, especially gene-nutrient interactions, may play a major role. Nutrient deficiencies and excesses during the *in utero* and early postnatal period influence the cardiovascular system through alterations in the growth hormone/insulin-like growth factor-1 (IGF-1) system and glucose metabolism. These gene-nutrient interactions suggest that the phenotype of genetically-based heart diseases, such as HCM, may be modifiable depending upon early nutrition or other environmental factors.

Our group has shown that cats with HCM are significantly larger (i.e., body weight, head length and width, and humeral and vertebral length) than healthy controls (Yang et al, 2008; Freeman et al, 2013). In addition, cats with HCM had higher IGF-1 concentrations compared to healthy controls (Freeman et al, 2013). Obesity also was significantly associated with HCM in a case-control study of Maine Coon cats (Freeman et al, 2013), but not in a study of cats of other breeds (Yang et al, 2008). Also, in the study of Maine Coon cats, body weight at six and 12 months was significantly higher in cats with HCM compared to controls, suggesting that the larger size was already developing at an early age. Finally, in a recent study, head width, body weight, NT-proBNP concentrations, and IGF-1 concentrations were significantly associated with left ventricular hypertrophy in a colony of cats. These data support the hypothesis of a relationship between early nutrition, body size, and HCM.

Much more information is needed in this area, however, until then, general recommendations can be made for young cats and dogs (see below). While these recommendations may not prevent heart disease, they are good general recommendations for young animals, and may also help to reduce the risk for other medical conditions (e.g., obesity, diabetes mellitus, osteoarthritis)

PRACTICAL IMPLICATIONS FOR DOGS AND CATS

Key nutritional issues that are important for optimal care of dogs and cats with heart disease include:

Assess diet history

A complete diet history includes the pet food (commercial or home-prepared, including the exact brand, product, and flavor for commercial foods and exact recipe for home-prepared foods), treats, table food, dietary supplements, and foods used for medication administration. This can help to determine if the diet is contributing to suboptimal body weight or body composition:

- Overall nutritional balance of the diet: Unless formulated by a veterinary nutritionist, nearly all home-prepared diets are nutritionally unbalanced. Even commercial diets, if made by a small company with inadequate nutritional knowledge and/or quality control, can have nutritional deficiencies or excesses. Over-the-counter diets labeled as "for intermittent or supplemental use" will be deficient in one or more nutrients and should not be fed as a significant portion of the diet.
- Calories: The diet(s), treats, and other foods, may provide too many or not enough calories to maintain ideal body weight and BCS in an individual animal.
- Protein: Diets vary tremendously in their protein content so the diet history can identify whether protein intake is appropriate.
- Sodium: The sodium content of the diet, treats, table food, and foods used for medication administration should be addressed for the individual animal.
- Other nutrients: Deficiencies or excesses of other nutrients also can occur, particularly if the owner is feeding an unbalanced diet, giving large amounts of table food or treats (or foods used to administer medications), or dietary supplements.

Physical examination

Alterations in body composition are common and important issues in dogs and cats with heart disease or CHF and can negatively affect the outcome and quality of life.

- Weight loss: Use a consistent scale to assess weight at each visit to identify subtle changes or trends.
- BCS: Use a consistent method to determine BCS to assess current status and changes over time (either a 9-point scale or a 5-point scale). The BCS evaluates body fat, with a goal for healthy dogs and cats of 4-5 on a 9-point scale. However, in dogs and cats with CHF, a BCS of 5-7 may be more desirable.
- Muscle loss: The MCS differs from the BCS in that the MCS evaluates muscle mass. Dogs and cats can be very obese and yet have severe muscle loss and conversely, cats can be thin but have normal muscle mass. Evaluation of muscle mass includes visual examination and palpation over the temporal bones, scapulae, lumbar vertebrae, and pelvic bones. Assessing MCS is important as dogs and cats with CHF lose primarily muscle, compared to a healthy animal that would lose fat. Early identification of muscle loss, at the "mild muscle loss" stage is valuable for successful intervention. Muscle condition scoring systems have been developed for both dogs and cats (see the WSAVA Nutrition Toolkit).

Make specific and individualized nutritional recommendations

Always make a specific recommendation for diet – this may be only to say that the diet the animal is currently eating is a good choice and that the owner is feeding an appropriate amount to keep the dog or cat in ideal body condition. It is important to reinforce good nutritional choices that owners are making, since there are so many myths about pet nutrition. It also is important to make recommendations specific for an individual animal based not only on the stage of disease, but also on the animal's nutritional assessment, physical examination, laboratory results, individual food preferences, and concurrent diseases. Ensuring adequate intake of protein and other nutrients is critical to maintain lean body mass. In complicated cases, consider referring the patient to a board-certified veterinary nutritionist (www.acvn.org or http://www.esvcn.eu/college).

Omega-3 fatty acid supplementation may provide benefits in dogs and cats with heart disease. The omega-3 fatty acids, eicosapentaenoic acid (EPA) and docosahexaenoic acid (DHA), are normally are found in very low concentrations in cell membranes compared to the omega-6 fatty acids, but they can be increased by a food or supplement enriched in omega-3 fatty acids. Omega-3 fatty acids reduce the production of the more inflammatory eicosanoids and of the inflammatory cytokines, TNF and IL-1. The anti-inflammatory, antiarrhythmic, and antithrombotic effects of omega-3 fatty acids have many potential benefits in dogs and cats with heart disease (e.g., reduced muscle loss, improved appetite, association with longer survival; Slupe et al, 2008; Freeman et al, 2010).

Finally, during growth (particularly for cats, but also for dogs), ensure that the owner is feeding a food made by a well-known, reputable manufacturer throughout the growth period (until one year of age). Food during this period should be nutritionally complete and balanced for growth or all life stages (if formulated to meet the Association of American Feed Control Officials profiles) or for growth (if tested by feeding trials, which is ideal). Tips during growth:

- Keep kittens and puppies trim from weaning through adulthood (and beyond; 4 to 5 on a 9-point BCS scale).
- Encourage clients to meal feed (feeding a measured amount of food two to three times a day), rather than feeding *ad libitum*. Few puppies and kittens can eat *ad libitum* and remain trim.
- Reduce the growing animal's calorie intake at time of neutering (energy requirements will decrease 15-20% at this time).
- Perform a nutritional assessment at each visit (i.e., complete diet history, BCS, MCS, body weight) and make specific nutritional recommendations.
- If the kitten or puppy is becoming overweight during the growth phase, do not change to an "adult" cat or dog food before the animal is one year of age. Instead, switch to a lower calorie density food (kitten/puppy foods range from <300 kcals/cup to >600 kcals/cup!).

REFERENCES

1. Edney ATB, Smith PM. Study of obesity in dogs visiting veterinary practices in the United Kingdom. Vet Rec 1986;118:391-396.
2. Finn E, Freeman LM, Rush JE, et al. The relationship between body weight, body condition, and survival in cats with heart failure. J Vet Intern Med 2010; 24: 1369-1374.
3. Freeman LM, Rush JE, Kehayias JJ, et al. Nutritional alterations and the effect of fish oil supplementation in dogs with heart failure. J Vet Intern Med 1998; 12: 440-448.
4. Freeman LM. Beneficial effects of omega-3 fatty acids in cardiovascular disease. J Small Anim Pract 2010; 51: 462-470.
5. Freeman LM. Cachexia and sarcopenia: Emerging syndromes of importance in dogs and cats. J Vet Intern Med 2012; 26: 3-17.
6. Freeman LM, Rush JE, Meurs KM, et al. Body size and metabolic differences in Maine Coon cats with and without hypertrophic cardiomyopathy. J Feline Med Surg 2013;15:74-80.
7. Slupe JL, Freeman LM, Rush JE. The relationship between body weight, body condition, and survival in dogs with heart failure. J Vet Intern Med 2008; 22: 561-565.
8. Tufts HeartSmart website [information on heart disease for pet owners, including nutrition (eg, low sodium diets for dogs and cats, treats, omega-3 fatty acid supplements)]. Available at: www.tufts.edu/vet/heartsmart January 31, 2016.
9. World Small Animal Veterinary Association Global Nutrition Committee. Muscle condition score chart, body condition score chart, diet history form, Nutrition on the Internet, Recommendations for Selecting a Pet Food, and other nutritional tools. http://www.wsava.org/nutrition-toolkit. Accessed January 31, 2016.
10. WSAVA Nutritional Assessment Guidelines Taskforce: Freeman L, Becvarova I, Cave N, et al. 2011 Nutritional Assessment Guidelines. J Small Anim Pract 2011; 52: 385-396.
11. Yang VK, Freeman LM, Rush JE. Morphometric measurements and insulin-like growth factor in normal cats and cats with hypertrophic cardiomyopathy. Am J Vet Res 2008; 69: 1061-1066.

Top 10 Tips for Cats with Heart Disease

Lisa M. Freeman, DVM, PhD, DACVN
North Grafton, MA

Despite the increasingly common diagnosis of cardiac disease in cats, managing cardiac disease in this species can be particularly challenging. Successful management requires careful selection of treatment strategies, regular monitoring, clear client communication, and careful attention to nutrition. The author has identified a number of nutritional tips that can be helpful in the care of feline cardiac patients.

TOP TEN TIPS

1. **Perform a complete nutritional assessment at every visit.** This includes body weight, body condition score, muscle condition score, and diet history. A complete diet history includes the main cat food being provided, but also treats, table food, dietary supplements, and foods used for medication administration. This can provide important information on whether the diet is optimized, whether it may be contributing to the underlying disease or clinical signs, and whether changes in diet would be beneficial. Tools are available to make this assessment quick and easy (WSAVA Nutrition Toolkit).
2. **Both body condition AND muscle condition scores should be determined for every patient with heart disease at every visit.** Body weight and body condition score, which primarily assesses body fat, are important because they have been associated with overall survival. Cats that are underweight and cats that are significantly overweight have the shortest survival while cats in moderate body condition (4-6/9) had the longest survival (Finn et al, 2010). However, it also is important to assess muscle condition at every visit, particularly in cats with cardiac disease. The muscle condition score is a subjective score which assesses muscle condition in the areas of the epaxial, gluteal, scapula, and temporal areas and graded as normal muscle condition or mild, moderate, or severe muscle loss. Because of increased production of inflammatory cytokines in congestive heart failure (CHF), muscle is lost preferentially to fat. Therefore, even in overweight cats, significant muscle wasting can be present. This muscle loss has important clinical implications as it negatively impacts strength, immune function, wound healing, and survival. By assessing muscle condition at every visit, muscle loss can be identified at its early stages when intervention is more likely to be successful. Optimizing diet (e.g., ensuring a nutritionally balanced diet with adequate calorie and protein intake) and omega-3 fatty acid supplementation can help to reduce muscle loss.
3. **Make individualized nutritional recommendations.** Always make a specific recommendation for diet which can be as simple as telling the owner that the diet the cat is currently eating is a good choice and that the owner is feeding an appropriate amount to keep the cat in ideal body condition. It is important to reinforce good nutritional choices that owners are making, since there are so many myths about cat nutrition. It also is important to develop nutritional recommendations specifically for an individual cat. Just as one wouldn't recommend an angiotensin converting enzyme inhibitor or furosemide for every cat with heart disease, the same nutritional recommendations are not ideal for all cats. There is usually not a single "best" diet for any cat – it depends on factors such as the severity of disease, clinical signs, physical examination findings, laboratory values, individual preferences of the cat and the owner, and concurrent diseases [which affected 56% of cats with heart disease in one study (Torin et al, 2007]. The author typically determines several diets that would be appropriate for an individual cat based on the cat's individual clinical situation. These diets are offered as choices for the owner and for the cat. Once they determine which diet the cat prefers, this diet can be used. Be sure to make specific diet recommendations, not just "a low sodium diet" or "a senior diet." It is especially important with respect to senior diets as the sodium content can vary dramatically (Freeman, unpublished data). Also, even if the cat is discharged with an optimal diet plan, don't assume that the owner is following these recommendations at subsequent visits; ask exactly what the owner is feeding at each visit. In complicated cases, consider referring the cat to a board-certified veterinary nutritionist (www.acvn.org or http://www.esvcn.eu/college).
4. **Address all components of the diet.** Be sure to address all components of the diet – not only the cat food, but also the treats, table food, and foods used for medication administration. In some cases, cats may be eating an ideal cat food but are getting large amounts of sodium from treats or table food [>30% of cats with heart disease received treats (Torin et al, 2007)]. Therefore, a diet must be selected that has the desired nutritional properties and palatability, but it also is important to devise an overall dietary plan that includes appropriate treats, if desired by the owner (and treats that should not be fed). Foods to be avoided include most commercial cat treats (unless specifically determined to be low in sodium), baby food, lunch meats and cold cuts, canned fish, and most cheeses. Acceptable treats (in limited quantities) include cat treats that are determined to be low in sodium (less than 5 mg sodium/treat). Note that even low-sodium treats and foods can provide large doses of sodium if fed in large quantities.
5. **Be sure to specifically discuss supplements.** Ask at each visit if the owner is administering dietary supplements. If so, ensure that the supplements are safe, are not interacting with the diet or medications, and are being administered at an appropriate dose. Fewer cats than dogs with cardiac disease receive dietary supplements (13% of cats vs. 31% of dogs; Torin et al, 2007; Freeman et al, 2003) but addressing this issue with the owner is important as "Internet surfing" for alternative treatments of cardiac disease is common. In addition to safety and efficacy issues, there are significant concerns about the quality control of dietary supplements since they do not require proof of safety, efficacy, or quality control to be marketed.

Therefore careful selection of type, dose, and brand is important to avoid toxicities or complete lack of efficacy. Therefore, veterinarians should consider recommending specific brands of dietary supplements that bear the logo of the United States Pharmacopeia Dietary Supplement Verification Program (http://www.usp.org/usp-verification-services), which tests human dietary supplements for ingredients, concentrations, dissolvability, and contaminants. Another good resource is www.Consumerlab.com, which performs independent testing of dietary supplements (primarily human supplements but also some pet products).

6. **Don't make major dietary changes while the cat is hospitalized.** Most cats with acute CHF will not eat but should begin eating when the CHF is better controlled. While hospitalized, avoid diets that are high in sodium but don't make major dietary changes during hospitalization. Once the cat is home, feeling better, and stabilized on medications, a gradual change to a new and more optimal diet can be made. Forced dietary changes when the cat is sick can induce food aversions. Knowing what the cat eats at home can help to know what food preferences will be (e.g., canned vs dry, flavor, kibble shape) since some cats have very distinct preferences. Make all dietary changes gradually over a period of three to seven days. It also is important to instruct the owner to notify the veterinarian if the cat does not eat adequate amounts of the new food so that other options can be devised. Some cats transition to new foods best by mixing the old food and the new food in varying proportions over the course of three to seven days (gradually reducing the amount of the old food and gradually increasing the amount of the new food, until the cat has completely transitioned to the new food). However, for other cats, a preferable way to transition is to offer the two foods (i.e., the old food and the new food) side-by-side for several days to ensure the cat is eating the new food well before removing the old food.

7. **Ensure clear client communication**. Good client communication is important for achieving desired outcomes, particularly in cats with cardiac disease. Discussion of diet, treats, table food, dietary supplements, and effective medication administration is beneficial for both the owner's and the cat's quality of life. Try to engage the client in decision making and defining expectations, so that recommendations can address the cat's preferences, as well as the client's time, lifestyle, and financial limitations. Owners appear much more likely to change dietary factors (e.g., diet, treats, supplements) than they would a medication. Therefore, it is relatively common to have various aspects of the diet change from one visit to the next and important to assess diet and to make specific recommendations at every visit (even if just to say that what the owner doing is ideal). Demonstrating and teaching the client to effectively administer medications and to evaluate the body weight, body condition score, and muscle condition score is beneficial for engaging the client in their cat's care.

8. **Provide appropriate and effective methods for administering medications.** Specifically discuss (and demonstrate, if necessary) appropriate methods for medication administration as cats are difficult to pill and many common foods used to administer medications are high in sodium. Administering medications was very difficult for many owners in one study (median=3, range 1-10) and had negative effects on the cats' quality of life [median=2, range 1-10 (Reynolds et al, 2010)]. The authors also found that the owners' concern increased with increasing number and frequency of medications. Cats with heart disease can have many factors that reduce their quality of life, including respiratory distress, weakness after arterial embolism, drug administration issues and drug side effects, trips to the veterinarian, and altered interactions with owners or other animals in the household. Identification of these factors, and determining which are most important for an individual cat and cat owner, can help limit the impact of heart disease on the quality of life. Reducing frequency of medication administration, and considering other alterations in drug administration method can maintain the bond between cat and owner. Owners can be taught to administer the pill without using foods (either by hand or using a device designed for this purpose, such as a pet piller or pet pill gun). Alternatively, foods such as low-sodium canned cat food or home-cooked meat (cooked without salt, not lunch meats) can be used. Many owners find that reserving a tasty, low-sodium canned food only for administering medications can make life easier. Pill administration treats also can be used but be sure to check the individual brand for content of sodium and other nutrients of concern for an individual patient. This becomes particularly important if many of these treats per day are used for medication administration. Finally, a compounded liquid medication can be considered although the pharmacokinetics of medications may be significantly altered by compounding.

9. **Help owners to use the Internet more effectively (and objectively).** There is a great deal of myth and misinformation about cat nutrition and cat foods. Provide owners with skills for searching the Internet about nutrition topics and useful websites to help them avoid the many common myths about cat nutrition (see reference section below for examples).

10. **Warn owners of cyclical/variable appetite in cats with CHF.** Cats with CHF can have anorexia (complete loss of appetite), hyporexia (reduced appetite), or dysrexia (changes in appetite and food preferences). Dysrexia is very common (i.e., the cat will eat one diet well for a few days but then stop eating it although it will eat a different food. However, they'll often eat a different food for a few days or more). It is helpful to warn the owner about these common appetite changes and provide multiple appropriate diet options for the owner to feed (i.e., they can feed one diet for several days until the cat no longer has interest for that food and then switch to a different food. They'll often be willing to eat the original food again so owners can use a "rotation" of foods to keep the cat eating). In addition to having different food options, other tips to increase appetite include:
    - Warm the food to body temperature.
    - Feed several small frequent meals during the day, rather than just one or two meals.
    - Feed the cat from a different type of dish (e.g., a new cat food dish or a human dinner plate).

- Try feeding in a different location in the house.
- Add homemade chicken, beef, or fish broth to the food (even low sodium store-bought broths are too high in sodium).
- Add a small amount (1-2 teaspoons) of cooked meat (hamburger, chicken, or fish) to the food. Be sure to instruct the owner not to use any prepared foods, such as rotisserie chicken, deli meats, or canned meats or fish due to their high sodium content.
- Fish oil, which is high in omega-3 fatty acids, reduces inflammatory cytokines and may have modest benefits for appetite. Fish oil (rather than flax or cod liver oil) should be used and a good quality brand should be used since there is little regulation of dietary supplements. The dose used by the authors is enough fish oil to provide 40 mg/kg eicosapentaenoic acid (EPA) and 25 mg/kg docosahexaenoic acid (DHA). For cats, a one gram capsule containing 180 mg EPA and 120 mg DHA can be easily used to provide this dose when given as one capsule per cat per day. Specific brands can be found on the Tufts HeartSmart website (see below)

One note of caution: If the cat also is acting lethargic, is tachypneic, won't eat *any* foods or if the cat has been eating very well and then suddenly stops eating, this usually indicates the need for reassessment and adjustment of medications.

REFERENCES

1. Finn E, Freeman LM, Rush JE, Lee Y. The relationship between body weight, body condition, and survival in cats with heart failure. J Vet Intern Med 2010; 24: 1369-1374
2. Freeman LM. Cachexia and sarcopenia: Emerging syndromes of importance in dogs and cats. J Vet Intern Med 2012;26:3-17.
3. Freeman LM. Beneficial effects of omega-3 fatty acids in cardiovascular disease. J Small Anim Pract 2010; 51: 462-470.
4. Freeman LM. Nutritional management of heart disease and Implementing a nutritional program for your cardiac patient. Cardiac Education Group website. http://www.cardiaceducationgroup.org/learning-center/circulations-articles. Accessed January 31, 2016.
5. Freeman LM. Nutritional management of heart disease. In: Little SP. August's consultations in feline internal medicine, 7th ed, 2015. St Louis: Elsevier: 403-411.
6. Freeman LM, Rush JE. Nutrition and cardiovascular disease. In: Tilley L, Smith FWK, Oyama M, Sleeper M. Canine and feline cardiology, 5th ed. St Louis: Elsevier: 394-404.
7. Pet Nutrition Alliance. www.petnutritionalliance.org. Accessed January 31, 2016.

1. Reynolds CA, Oyama MA, Rush JE, et al. Perceptions of quality of life and priorities of owners of cats with heart disease. J Vet Intern Med 2010;24:1421-1426.

8. Torin DS, Freeman LM, Rush JE. Dietary patterns of cats with cardiac disease. J Am Vet Med Assoc 2007; 230: 862-867.
9. Tufts HeartSmart website [information on heart disease for pet owners, including nutrition (eg, low sodium diets for dogs and cats, treats, omega-3 fatty acid supplements)]. Available at: www.tufts.edu/vet/heartsmart January 31, 2016.
10. World Small Animal Veterinary Association Global Nutrition Committee. Muscle condition score chart, body condition score chart, diet history form, Nutrition on the Internet, Recommendations for Selecting a Pet Food, and other nutritional tools. http://www.wsava.org/nutrition-toolkit. Accessed January 31, 2016.
11. WSAVA Nutritional Assessment Guidelines Taskforce: Freeman L, Becvarova I, Cave N, MacKay C, Nguyen P, Rama B, Takashima G, Tiffin R, Van Beukelen P, Yathiraj S. 2011 Nutritional Assessment Guidelines. J Small Anim Pract 2011; 52: 385-396.

Review of Cardiac Surgery in Dogs

Augusta Pelosi, DVM, DACVS, DACVIM (Cardiology)
East Lansing, MI

The history of cardiac surgery is one of the most fascinating and intriguing pieces of medical history. Around ---1950, brave pioneers initiated the era of human cardiac surgery at a time when cardiac surgery was considered off-limits. Despite the strong general opposition and the unlikely chances of success with cardiac surgery, cardiac surgery started to be explored in different venues and modalities. John Lewis used hypothermia to do cardiac procedures at the University of Minnesota and Drs. Blalock and Taussig created extra cardiac shunts to address "blue-baby" syndrome.

The second-world war was a determining factor in the development of this specialty. It opened the door to new opportunities and a new approach to surgery. Despite the devastation and death, it represented one of the most powerful boosts for cardiac surgery. The large number of casualties from fire arms and the devastating damage of thoracic wounds pushed the field surgeons to become braver in addressing conditions that once were considered untreatable.

The first bypass machine was built in 1948 by Dr. John H. Gibbon, of the Jefferson Medical Center, as a response to observing one of his patients dying because of pulmonary hypertension. In the spring of 1953, he performed a successful closure of a septum secundum atrial septal defect (ASD) in an18-year-old woman, using cardiopulmonary bypass. The new idea of using the cardiopulmonary bypass machine in order to operate inside the heart while having a dry field and archiving full control of the hemodynamics of the patients represented a major achievement. Following a series of surgical deaths under bypass, the concept of cross circulation by Dr. Lillehei was used for cardiac surgery instead. However, the return of the bypass machine was inevitable, considered the overall high mortality rate and the poor control of the patient's circulation while performing cross-circulation techniques. The era of intracardiac surgery had begun.

The specialty of human cardiac surgery has just celebrated its 60th birthday, while a full appreciation of this era in veterinary medicine has still to be explored. Veterinary pioneers of cardiac surgery had introduced this specialty to our surgical world, but only to be sporadically followed over the years. The action and courage of these men lead to some of the first veterinary interventions: first pacemaker (Dr. Jim Buchanan), VSD closure (Dr. Gene Breznock), ASD closure (Dr. George Eyster). Dr. Christopher Orton has significantly contributed to the development and expansion of cardiac surgery with several publications and study series, and in the more recent years several Japanese groups (Drs. Kanemoto and Uechi) have focused their attention on mitral valve surgical techniques, particularly in small dogs. However, several other manuscripts and small studies have been published looking at the many cardiac diseases occurring in dogs and have significantly contributed in the past years to the development of cardiac surgery. The contribution of their effort and passion to the cause of developing cardiac surgery was the pushing motivation to initiate this investigation. The primary goal was to review the published data and evaluate survival: operative, postoperative, and after 12 months. We also wanted to investigate factors affecting survival. Lastly, within the group of dogs receiving mitral valve surgery, we compared survival of mechanical valves to bioprostetic valves, and mitral valve repair (MVRep) techniques to mitral valve replacement (MVR), in the canine population.

Emerging Interventions for Mitral Regurgitation
Christopher Orton, DVM, PhD, DACVS
Fort Collins, CO

INTRODUCTION

Mitral regurgitation (MR) is classified etiologically as Primary (Organic) or Functional (Secondary). Primary MR is defined as MR caused by a primary leaflet abnormality. Causes of primary MR include degenerative (myxomatous) disease, endocarditis, congenital defects, rheumatic heart disease, and certain serotoninergic drugs. Functional MR (FMR) results from left ventricular remodeling and/or dysfunction associated with primary cardiomyopathies. In dilated forms of primary cardiomyopathy (idiopathic, ischemic), FMR results from the combined effects of displacement of the papillary apparatus leading to restrictive leaflet motion and dilation of the mitral annulus. For purposes of surgical mitral valve repair, MR has been classified functionally by Carpentier as Type I (annular dilation), Type II (leaflet prolapse), and Type III (restrictive leaflet motion) .[1] For chronic severe degenerative MR, the functional classification would include both a Type I and Type II defect. Interventional therapies for correction of MR take into consideration both its etiologic and functional classification.

OPEN MITRAL VALVE REPAIR

Open mitral valve repair is the standard of care for degenerative MR in humans.[2, 3] Open mitral repair is preferred over valve replacement for human patients and carries a Class I treatment effect (Benefit>>>Risk) based on Level B evidence. Open mitral repair in dogs was first reported in a case series of 18 dogs.[4] Repair techniques included ring annuloplasty, artificial chordae, and edge-to-edge repair. In this series, operative survival was 66% with long-term palliation in most dogs surviving surgery. More recently a Japanese group reported improved results (93% survival) for small dogs (2 – 5 kg) using artificial chordae and annuloplasty for open mitral valve repair .[5] These results demonstrate the feasibility of open mitral valve repair in small dogs, however the challenges of widespread application of open heart surgery in dogs remain significant, especially small breed dogs.

EXPTERNAL ANNULOPLASTY

Dilation of the mitral valve annulus (Type I defect) contributes to the severity and progression of all types of MR, including degenerative MR. External annuloplasty involves the surgical placement of an external plicating suture on beating heart in an effort to decrease the effective office area of the mitral annulus to decrease MR. Buchanan et al. reported results of 15 dogs undergoing a complete circumferential external annuloplasty.[6] They demonstrated long-term palliation of MR in 3 dogs, but the complication rate due primarily to hemorrhage was unacceptability high. A modified partial external mitral annuloplasty techniques has been reported in 9 dogs with degenerative MR.[7] Palliation of MR was demonstrated in 2 of 5 dogs surviving surgery. Based on high perioperative mortality and lack of palliation in most dogs, external surgical annuloplasty by suture is not viable option for dogs with degenerative MR. An external annuloplasty device for use in humans (Mitral-Touch®) or dogs (Mitrex®) has been described. The device was shown to be effective for reducing FMR in dogs with pacing-induced cardiomyopathy.[8] Clinical results in humans or dogs have not been reported.

TRANSCATHETER MITRAL ANNULOPLASTY

Several transcatheter devices for mitral annuloplasty in various stages of development have been described.[9] These devices are classified as indirect and direct. Indirect devices exploit the proximity of the coronary sinus to mitral annulus and include the Monarc® and CARILLON® devices. The Monarc® device has undergone a Phase I study in 72 humans with FMR.[10] The device was successfully implanted in 59 patients (82%) and decreased MR ≥ 1 grade in 50% of patients. Compression of coronary arteries was a significant complication. The CARILLON device has received CE Mark approval and is currently undergoing a Phase II clinical trial with comparison to optimized medical therapy for patients with FMR.[11] Mitralign® and Annucinch® are direct transcatheter annuloplasty devices that are applied to the ventricular side of the mitral annulus. Cardioband® is a direct annuloplasty system applied to the atrial side of the annulus using a transseptal approach. First-in-human results have been reported for the Mitralign® and Cardioband® devices.[12, 13]

TRANSCATHETER MITRAL VALVE REPAIR: EDGE-TO-EDGE

MitraClip® (Abbot Vascular) is a transcatheter-delivered clip that attaches the anterior and posterior mitral leaflets at the point of leaflet prolapse. It is based on surgical edge-to-edge valve repair technique first described by Alfieri.[14] Deployment of MitraClip® is by percutaneous transvenous access and transseptal puncture of the atrial septum and is performed under transesophageal echocardiography (TEE) guidance. The MitraClip® system uses a 22Fr steerable guide catheter for transseptal puncture, a steerable delivery catheter, and an implantable clip. The procedure has specific anatomic requirements for leaflet coaptation length, coaptation depth, flail gap, and flail width; and thus must be deployed before significant annular dilation (Type I defect) occurs. The EVEREST II clinical trial demonstrated MitraClip® repair to be safer but less effective than surgical repair in human patients selected using standard AHA/ACC recommended guidelines for surgery.[15] Currently MitraClip® is approved for patients with degenerative MR who are poor surgical risks. FMR is an evolving indication for MitraClip® device in humans. Application of transcatheter edge-to-edge mitral repair in dogs would require considerable down scaling of the current human device and delivery system. A hybrid transapical approach might increase feasibility of a canine device Early intervention would likely be necessary to meet the anatomic requirements for leaflet coaptation and flail size.

TRANSCATHETER MITRAL VALVE REPAIR: ARTIFICIAL CHORDAE

At least two systems for transcatheter implantation of artificial chordae for treatment of MR secondary to leaflet prolapse (Type II defect) are in development. Both utilize a hybrid transapical approach for implantation. Artificial chordae are externalized through the LV apex and can be adjusted in the beating heart under TEE guidance to optimize length. First-in-human results have been reported for the NeoChord® system.[16] Acute reduction of residual MR to < 2+ was achieved in all 49 patients reported. Three-month survival was 98% and freedom from reoperation was 92%. An early feasibility first-in-human study for the Haroon Medical device is currently underway.

TRANSCATHETER MITRAL VALVE IMPLANTATION

Several devices and delivery systems for transcatheter mitral valve implantation (TMVI) are currently in development (**Table 1**). First-in-human results have been reported for 4 of these devices[17-21]. The CardiAQ device has been deployed using a completely percutaneous approach.[17] All devices utilize a self-expanding Nitinol frame and bioprosthetic leaflets (bovine or porcine pericardium).

Table 1: Transcatheter Mitral Valve Implantation (TMVI) Devices

HUMAN	Design-Fixation	Approach	Delivery	First-in-Human
CardiAQ® (CardiAQ)	Oval, Trileaflet, Atrial & Ventricular leaflet anchors	Transapical (hybrid) Transfemoral-Transseptal	32 F	Yes [17, 18]
Tiara® (Neovasc)	D-shaped, Trileaflet, Atrial Cuff, 3 Ventricular leaflet anchors	Transapical (hybrid)	32 F	Yes [19]
Tendyne® (Abbot)	D-shaped, Trileaflet, Atrial cuff, Apical tether with button	Transapical (hybrid)	34 F	Yes [20]
FORTIS® (Edwards)	Oval, Trileaflet, Atrial Cuff, 2 Ventricular leaflet paddles	Transapical (hybrid)	42 F	Yes [21]
Medtronic-TMV® (Medtronic)	Oval, Trileaflet, Atrial cuff, Ventricular leaflet anchors	Transatrial (hybrid) Transseptal		
Cardiovalve® (Valtech)	Two-piece device, Oval, Trileaflet, Atrial Cuff, Ventricular leaflet anchors	Transfemoral-Transseptal	26 F	
HighLife-TMV® (Highlife)	Oval, Trileaflet, Atrial Cuff, Ventricular subannular band	Transfemoral-Transaortic & Transatrial (hybrid)		
Endovalve® (MID)	Oval, Trileaflet, Ventricular leaflet grippers	Transapical (hybrid)		
Gorman-TMV® (Gorman)	Oval, Trileaflet, Annular and ventricular leaflet anchors	Transatrial (hybrid)	30 F	
MitrAssist® (Misgav)	Asymmetric oval, Bileaflet	Transfemoral-Transaortic	18 F	
CANINE				
MitralSeal® (Avalon)	Oval, Trileaflet, Atrial cuff 3 Apical tethers	Transapical (hybrid)	19 F/22 F	

MitralSeal (Avalon Medical) is transcatheter mitral valve prosthesis being developed for dogs. The prosthesis is deployed into a beating heart through the left ventricular apex under fluoroscopic and TEE guidance. The prosthesis is a self-expanding bioprosthesis with an atrial cuff and 3 tethers that secure the prosthesis to the ventricular apex. Access to the left ventricular apex is gained through a small (minimally-invasive) left thoracotomy. The anchoring tethers are externalized through the LV apex and fixed to the outside of the apex. The first-generation prosthesis consisted of a woven Nitinol wire stent and porcine pericardium bioprosthetic leaflets. The leaflets are suspended above the atrial cuff ("atrialized" design) to shorten the stent body and prevent interference with the LVOT. The first-generation delivery system was a two-step system consisting of a delivery catheter and dilating obturator for entry through the LV apex without an introducer; and a loading/transfer catheter containing the prosthesis. A second-generation prosthesis consisting of a laser-cut electropolished stent has been developed to address issues with thrombogenicity. A second-generation delivery system that "preloads" the prosthesis in the end delivery catheter and allows delivery using a single-step has also been developed. Additional modifications to the delivery protocol have been made to assure accurate delivery and alignment of the prosthesis.

The first-generation MitralSeal device has been implanted in 5 dogs with Stage D MR under an early feasibility clinical trial protocol. The early feasibility trial follows FDA guidelines for early feasibility/first-in-human clinical testing which allows modifications to the device based on clinical results. The prosthesis was successfully delivered in 4 of 5 dogs. One dog did not survive surgery because of damage to the prosthesis during prosthesis transfer and delivery. Short-term hemodynamic response to

TMVI was very favorable in the remaining dogs despite the advanced disease stage in these dogs. Thrombogenicity of the first-generation prosthesis was the major issue limiting long-term survival. Implantations of the second-generation prosthesis and using the new delivery system are scheduled for March 2016.

REFERENCES

1. Carpentier A. Cardiac valve surgery--the "French correction". *J Thorac Cardiovasc Surg*. 1983;86:323-37.

2. Bortolotti U, Milano AD and Frater RW. Mitral valve repair with artificial chordae: a review of its history, technical details, long-term results, and pathology. *The Annals of thoracic surgery*. 2012;93:684-91.

3. Bonow RO, Carabello BA, Chatterjee K, de Leon AC, Jr., Faxon DP, Freed MD, Gaasch WH, Lytle BW, Nishimura RA, O'Gara PT, O'Rourke RA, Otto CM, Shah PM and Shanewise JS. 2008 Focused update incorporated into the ACC/AHA 2006 guidelines for the management of patients with valvular heart disease. *Circulation*. 2008;118:e523-661.

4. Griffiths LG, Orton EC and Boon JA. Evaluation of techniques and outcomes of mitral valve repair in dogs. *Journal of the American Veterinary Medical Association*. 2004;224:1941-5.

5. Uechi M, Mizukoshi T, Mizuno T, Mizuno M, Harada K, Ebisawa T, Takeuchi J, Sawada T, Uchida S, Shinoda A, Kasuya A, Endo M, Nishida M, Kono S, Fujiwara M and Nakamura T. Mitral valve repair under cardiopulmonary bypass in small-breed dogs: 48 cases (2006-2009). *Journal of the American Veterinary Medical Association*. 2012;240:1194-201.

6. Buchanan JW and Sammarco CD. Circumferential suture of the mitral annulus for correction of mitral regurgitation in dogs. *Veterinary surgery : VS*. 1998;27:182-93.

7. de Andrade JN, Orton EC, Boon J, Nishimori CT, Olivaes C and Camacho AA. Partial external mitral annuloplasty in dogs with myxomatous mitral valve degeneration and congestive heart failure: outcome in 9 cases. *Journal of veterinary cardiology : the official journal of the European Society of Veterinary Cardiology*. 2011;13:197-201.

8. Takaseya T, Shiose A, Saraiva RM, Fumoto H, Arakawa Y, Juravic M, Lombardi P and Fukamachi K. Novel epicardial off-pump device for mitral regurgitation: acute evaluation. *European journal of cardio-thoracic surgery : official journal of the European Association for Cardio-thoracic Surgery*. 2010;37:1291-6.

9. Chiam PT and Ruiz CE. Percutaneous transcatheter mitral valve repair: a classification of the technology. *JACC Cardiovascular interventions*. 2011;4:1-13.

10. Harnek J, Webb JG, Kuck KH, Tschope C, Vahanian A, Buller CE, James SK, Tiefenbacher CP and Stone GW. Transcatheter implantation of the MONARC coronary sinus device for mitral regurgitation: 1-year results from the EVOLUTION phase I study (Clinical Evaluation of the Edwards Lifesciences Percutaneous Mitral Annuloplasty System for the Treatment of Mitral Regurgitation). *JACC Cardiovascular interventions*. 2011;4:115-22.

11. Goldberg SL, Lipiecki J and Sievert H. The CARILLON Mitral Contour transcatheter indirect mitral valve annuloplasty system. *EuroIntervention : journal of EuroPCR in collaboration with the Working Group on Interventional Cardiology of the European Society of Cardiology*. 2015;11 Suppl W:W64-6.

12. Siminiak T, Dankowski R, Baszko A, Lee C, Firek L, Kalmucki P, Szyszka A and Groothuis A. Percutaneous direct mitral annuloplasty using the Mitralign Bident system: description of the method and a case report. *Kardiologia polska*. 2013;71:1287-92.

13. Maisano F, Taramasso M, Nickenig G, Hammerstingl C, Vahanian A, Messika-Zeitoun D, Baldus S, Huntgeburth M, Alfieri O, Colombo A, La Canna G, Agricola E, Zuber M, Tanner FC, Topilsky Y, Kreidel F and Kuck KH. Cardioband, a transcatheter surgical-like direct mitral valve annuloplasty system: early results of the feasibility trial. *European heart journal*. 2015.

14. Alegria-Barrero E, Chan PH, Paulo M, Duncan A, Price S, Moat N, Di Mario C, Serra A, Garcia E and Franzen O. Edge-to-edge percutaneous repair of severe mitral regurgitation--state-of-the-art for Mitraclip(R) implantation. *Circ J*. 2012;76:801-8.

15. Feldman T, Foster E, Glower DD, Kar S, Rinaldi MJ, Fail PS, Smalling RW, Siegel R, Rose GA, Engeron E, Loghin C, Trento A, Skipper ER, Fudge T, Letsou GV, Massaro JM and Mauri L. Percutaneous repair or surgery for mitral regurgitation. *N Engl J Med*. 2011;364:1395-406.

16. Colli A, Manzan E, Zucchetta F, Bizzotto E, Besola L, Bagozzi L, Bellu R, Sarais C, Pittarello D and Gerosa G. Transapical off-pump mitral valve repair with Neochord implantation: Early clinical results. *International journal of cardiology*. 2016;204:23-8.

17. Ussia GP, Quadri A, Cammalleri V, De Vico P, Muscoli S, Marchei M, Ruvolo G, Sondergaard L and Romeo F. Percutaneous transfemoral-transseptal implantation of a second-generation CardiAQ mitral valve bioprosthesis: first procedure description and 30-day follow-up. *EuroIntervention : journal of EuroPCR in collaboration with the Working Group on Interventional Cardiology of the European Society of Cardiology*. 2015;11.

18. Sondergaard L, De Backer O, Franzen OW, Holme SJ, Ihlemann N, Vejlstrup NG, Hansen PB and Quadri A. First-in-Human Case of Transfemoral CardiAQ Mitral Valve Implantation. *Circulation Cardiovascular interventions*. 2015;8:e002135.

19. Cheung A, Webb J, Verheye S, Moss R, Boone R, Leipsic J, Ree R and Banai S. Short-term results of transapical transcatheter mitral valve implantation for mitral regurgitation. *Journal of the American College of Cardiology*. 2014;64:1814-9.

20. Perpetua EM and Reisman M. The Tendyne transcatheter mitral valve implantation system. *EuroIntervention : journal of EuroPCR in collaboration with the Working Group on Interventional Cardiology of the European Society of Cardiology*. 2015; Suppl W:W78-9.

21. Abdul-Jawad Altisent O, Dumont E, Dagenais F, Senechal M, Bernier M, O'Connor K, Paradis JM, Bilodeau S, Pasian S and Rodes-Cabau J. Transcatheter Mitral Valve Implantation With the FORTIS Device: Insights Into the Evaluation of Device Success. *JACC Cardiovascular interventions*. 2015;8:994-5.

Research Update on MVD in Dogs

Lisbeth H. Olsen, DVM, DVSc
Frederiksberg, Denmark

INTRODUCTION

Myxomatous mitral valve disease (MVD) is the most common heart disease in dogs. The mitral valve becomes thickened and/or enlarged leading to mitral regurgitation. The prevalence of the disease increases with advancing age and the disease is most common seen in small to medium-size dog breeds, especially Cavalier King Charles Spaniels (CKCS). The progression of the disease can be divided into 3 stages: 1. Early signs of mitral valve prolapse (MVP) and/or thickening of the mitral valve, combined with no or minimal mitral regurgitation, 2. Presence of mild to severe mitral regurgitation, but without clinical signs of congestive heart failure (CHF) and 3. Presence of severe mitral regurgitation and clinical signs of CHF. Many dogs with MVD do not develop clinical signs of CHF; therefore MVD is seen as relative benign condition.[1] However, it is challenging to identify MVD dogs at high risk of developing CHF. New research aimed reducing the prevalence of MVD in dogs, improving diagnosis and increasing the understanding of the disease mechanisms has been published.

GENETICS AND BREEDING RESTRICTIONS

A genetic background of MVD in dogs has been suggested for long time since a higher prevalence of the MVD appears in certain dog breeds. Studies reported strong genetic background with polygenic inheritance. In CKCS, a positive association between parental heart status and mitral regurgitation murmur intensity in 5 year-old offspring has been reported.[2] Moreover, a positive association between the degree of MVP in parents and offspring has been reported in Dachshunds.[3] A high heritability for MVD in CKCS has been estimated from data obtained in a United Kingdom (UK) breeding scheme.[4] The heritability of having a mitral regurgitation murmur and of the grade of murmur intensity was estimated to 0.33 and 0.67, respectively. Genome-wide association studies have been performed. Interestingly, two gene loci on chromosomes 13 and 14 were found to differ between CKCS with early and late onset of MMVD.[5] However, another similar genome-wide association study, including a lower number of CKCS, did not find any association between gene loci and MMVD disease severity.[6] In Whippet dogs, another dog breed predisposed to MVD, chromosome 15 was reported to be associated with MVD.[7] Inconsistency between genome wide association studies in dogs with MVD may be due to differences between dog breeds and/or dog populations.

Breeding restrictions or recommendations aimed at decreasing the prevalence of MVD in CKCS have been developed in breeding clubs in different countries. The high heritability of the disease indicates that selection against the disease could be successful.[4] Some ongoing breeding schemes are mandatory, where others are voluntary, and the schemes include cardiac auscultation or both cardiac auscultation and echocardiography. Parent cardiac information is included in some breeding schemes. Moreover, examination intervals differ between breeding schemes. Some schemes include annual cardiac examinations where as others are based on a more limited number of cardiac examinations during the dog's lifetime. In some breeding clubs, the cardiac examination results are open to the public, in others the results are only known by the dog owner. It is still unknown how much these above mentioned factors influence the efficacy of a breeding scheme aimed at reducing the prevalence of MVD.

Preliminary findings from a British voluntary breeding scheme based on auscultation reported an effect in a subgroup of dogs auscultated by general practitioners but only a trend was reported when the dogs were auscultated by a veterinary cardiologist.[8] A mandatory breeding scheme in Sweden including annual cardiac auscultation and parent information did not result in significant effect of the scheme, however the follow-up period was relative short (from 2007 to 2009).[9] A mandatory Danish breeding scheme based on auscultation and echocardiography recently reported a decrease the prevalence of MVD in CKCS over an 8- to 10-year time period.[10] The breeding scheme estimated the degree of mitral regurgitation using auscultation (grade 1-6) and degree of mitral valve prolapse (MVP) by echocardiography (grade 0-3). Risk of having a mitral regurgitation murmur or MVP>1 in 2010-2011 compared with 2002-2003 was evaluated for dogs that were products of the breeding scheme (defined as dogs with both parents approved by the breeding scheme before breeding) and non-products of the breeding scheme (defined as dogs with at least 1 parent with unknown cardiac status before breeding). In 2010–2011, the odds of having mitral regurgitation murmur were 0.27 if dogs were a product of the breeding scheme, reflecting a 73% decreased risk over an 8- to 10-year time period. No significant risk reduction was found for the dogs that were non-products of the breeding scheme. These findings indicate that it is possible to reduce the prevalence of MVD in CKCS over an 8- to 10-year period by a mandatory breeding scheme based on auscultation and echocardiography.

LOCAL VALVULAR DISEASE MECHANISMS

Alterations in valvular extracellular matrix including deposition of mucopolyschaccarides and changes in collagen structure are central in the valvular pathogenesis of MVD. The changes seem to be mediated by activation of normally quiescent valvular interstitial cells (VIC).[11,12] The exact molecular mechanisms are not understood and several local mediators are involved. Recent studies suggest that changes in serotonin (5-hydroxytryptamine, 5-HT) signaling pathways play a major role in the pathogenesis of MVD.[13-15] Studies evaluating gene expression reported increased transcription of components of serotonin signaling pathways in canine mitral valves affected by MVD.[14,16] The findings have been supported by a number of studies based on immunohistochemistry and immunoblotting of serotonin signaling pathways. These studies reports increased expression of the serotonin receptor 2B in myxomatous mitral valves from dogs,[14,16,17] and tryptophan hydroxylase-1 (TPH1) (it is the rate limiting enzyme in serotonin synthesis).[18] Moreover a recent study analyzed tissue serotonin content in mitral valve tissue high-performance liquid chromatography (HPLC).[19] The study revealed higher levels of serotonin in mitral valves from dogs with MVD compared to control dogs euthanized due to non-cardiac reasons. Interestingly, it has be hypothesized that serotonin is involved in the activation of VIC in MVD. It has been suggested that serotonin induces transformation of valvular interstitial cells (VIC) into active myofibroblasts via transforming growth factor-beta (TGF-β).[17,20,21] Moreover, serotonin 2A and 2B receptors have been related to VIC activation.[18,20]

CIRCULATING BIOMARKERS

Higher levels of serum serotonin are reported CKCS with MVD compared with healthy dogs of other breeds.[22,23] A study including CKCS with different severities of MVD reported that the serum serotonin concentration decreased with increasing MMVD severity.[23] This study suggested that if serotonin plays a role in the pathophysiology of MVD it may mainly be in the early stages of the disease. Most of the circulating serotonin is contained within blood platelets, and it is released into serum during platelet aggregation and activation. Levels of free circulating serotonin estimated by measuring serotonin concentration in platelet poor plasma were not found be higher in dogs with MVD, however in agreement with serum serotonin findings, dogs with MVD had a higher content of serotonin in blood platelets compared to healthy non-CKCS dogs.[19] However, another platelet study could not confirm that the amount of serotonin stored in platelets of dogs with MMVD was different from healthy dogs.[24]

Circulating biomarkers of inflammation and oxidative stress have recently been evaluated in dogs with MVD. Circulating C-reactive protein (CRP), an acute-phase protein that increases in response to inflammation, is found mildly increased in dogs with CHF due to MVD. [25,26] The disease MVD is generally considered as a non-inflammatory condition since inflammatory cells in the mitral valves or myocardium are uncommon findings.[27,28] However, studies report increased myocardial cytokine expression in the atria and ventricles in dogs with CHF due cardiac diseases including MVD.[14,29] Furthermore, increased circulating concentrations of different inflammatory cytokine have also been reported in dogs with CHF due to MVD.[30] Few studies have suggested altered oxidative status in dogs with MVD.[31,32] Higher plasma tetrahydrobiopterin (BH4) and dihydrobiopterin (BH2) concentrations are reported dogs in CHF due to MVD compared to dogs without CHF.[32] Plasma concentrations of BH2 and BH4 may reflect levels of oxidative stress and endothelial dysfunction. Additional studies are needed to clarify the clinical implications of the findings.

MYOCARDIAL CHANGES

Myocardial fibrosis and intramural arterial changes in the left ventricle have been reported in dogs with CHF due to MVD compared to dogs euthanized due to non-cardiac reasons.[33,34] Increased myocardial fibrosis and narrowing of intra myocardial vessels in the left ventricle were associated with echocardiographic measurements of deceased systolic myocardial function (decreased fraction shortening) and decreased survival time.[34] New studies found that circulating levels of myocardial markers are associated with myocardial fibrosis in dogs with MVD. Increased myocardial fibrosis are reported to be associated with higher levels of circulating troponin I, a biomarker of myocardial damage.[35] Furthermore, a study evaluated galectin-3, a cardiac fibrosis marker, in dogs with CHF due to MVD.[36] Associations between myocardial fibrosis, myocardial expression of galectin and circulating plasma galectin were reported. Another recent study studied myocardial changes in dogs with different degrees of MVD.[37] Myocardial biopsies were collected during mitral valve repair surgery. Severe myocardial changes were reported in all included disease stages of MVD (International Small Animal Cardiac Health Council (ISACHC) classes I-III).

ECHOCARDIOGRAPHY

Global pump function of the myocardium is apparently maintained for a long time in most dogs with CHF due to MVD, despite of myocardial fibrotic and vascular changes that have been reported.[38-40] Conventional echocardiographic measurements of LV function such as fractional shortening are highly influenced by change in hemodynamic load. Therefore myocardial dysfunction may be masked by changed hemodynamic loading conditions associated with the presence of mitral regurgitation. Advanced echocardiographic methods, including speckle-tracking echocardiography, have been developed for evaluating left ventricular function and some of these methods are reported to be lesser load dependent. Some speckle-derived variables were found to be decreased in dogs with severe MVD. These studies suggested left ventricular dysfunction in dogs with MVD.[40]

Other new studies regarding echocardiography and MVD in dogs have been published. Studies reported that pulmonary hypertension estimated using echocardiography is common in dogs with severe MVD.[40,41] Moreover pulmonary hypertension is associated with a poor prognosis in dogs with severe MVD.[41] An echocardiographic study reported that R-R interval variations may affect the estimation of mitral regurgitation severity using colour Doppler echocardiography.[43] Mitral regurgitation severity was found to increase in long R-R intervals when these follow a short R-R interval. This phenomenon is only described in CKCS. The pathophysiology behind the finding and the prognostic significance remain to be elucidated. Finally, a recent interesting study compared mitral regurgitation severity measured using cardiac magnetic resonance imaging and echocardiography.[44] This study reported that echocardiographic vena contracta/aortic diameter-ratio and E-wave velocity correlated most significant with the cardiac magnetic resonance imaging estimated mitral regurgitation severity. These results are promising; however the study was limited by a relative low sample size. Further studies are warranted to validate echocardiographic estimation of mitral regurgitation in dogs with MVD.

CONCLUSIONS AND PERSPECTIVES

New knowledge regarding MVD in dogs have been published, however there are still a need for studies to further clarify the complex pathophysiology of MVD. Hopefully research with in MVD leads to improved prognostic, diagnostic and therapeutic possibilities for dogs with MVD in the future.

REFERENCES

[1]Borgarelli M, et al. J Vet Intern Med 2012;26(1):69. [2]Swenson L, et al. J Am Vet Med Assoc 1996;208(12):2009. [3]Olsen LH, et al. J Vet Intern Med 1999;13(5):448. [4]Lewis T, et al. Vet J 2011;188(1):73. [5]Madsen MB, et al. J Hered 2011;102(suppl 1):S62. [6]French et al. Vet J 2012;193(1):283. [7]Stern et al. PLoS One 2015;10(10):e0141234.[8]Swift S, Cripps P. Proceeding ACVIM meeting 2013. [9]Lundin et al. Acta Vet Scand 2010;52:54. [10]Birkegård et al. J Vet Intern Med 2016;30(1):63. [11]Liu et al Am J Path 2007;171,1407. [12] Rabkin et al. Circulation 2001;104:2525. [13]Orton et al. J Vet Cardiol 2012;14:7. [14]Oyama and Chittur. Am J Vet Res 2006;67:1307. [15]Oyama et al. J Vet Intern Med 2010;24:27. [16]Cremer et al. Res Vet Sci 2015;00:197. [17]Disatian and Orton 2009. J Heart Valve Disease 2009;18,44. [18]Disatian et al. J Heart Valve Disease 2010;19,71. [19]Cremer et al. J Vet Intern Med 2014;28(5):1534.[20]Connolly et al. Am J Pathol 2009;175,988. [21]Hutcheson et al. J Mol Cell Cardiol 2012;53,707. [22]Arndt et al. J Vet Intern Med 2009;23:1208. [23]Ljungvall et al. J Vet Intern Med 2013;27:1105. [24]Mangklabruks and Surachetpong. J Vet Cardiol 2014;16(3):155. [25]Rush et al. J Vet Intern Med 2006;20: 635. [26]Reimann et al. Vet J 2016 (in press). [27]Buchanan. Adv Vet Sci Comp Med 1977;21:75. [28]Aupperle and Disatian. J Vet Cardiol 2012;14,59. [29]Fonfara et al. Am J Vet Res 2013;74,408. [30]Zois et al. Vet J 2012;192(1):106. [31]Freeman et al, J Vet Intern Med 2005;19(4):537. [32]Reimann et al. J Vet Intern Med 2014;28(5):1520.[33]Jonsson. Acta Vet Scan 1972;Supp 38, 1.[34]Falk et al. Vet J 2010;185:68. [35]Falk et al. J Vet Intern Med 2013;27:500. [36]Sakarin et al. J Vet Cardiol 2016 (in press). [37]Lee et al. J Vet Intern Med 2015;29:1313. [38]Smith et al. J Vet Cardiol 2012;14:231. [39]Zois el al. J Vet Intern Med 2012;26:1309. [40]Zois et al. J Vet Intern Med 2013;27:875. [41]Tidholm et al. J Vet Intern Med 2015;29:1510. [42]Borgarelli et al. J Vet Intern Med 2015;29:569. [43]Reimann et al. Vet J 2014;199: 348. [44]Sargent et al. J Small Anim Pract 2015;56:641.

Micrornas: Novel Biomarkers and Therapeutic Targets in MMVD

Brendan M. Corcoran, MVB, DipPharm, PhD, MRCVS
Edinburgh, Scotland, UK

INTRODUCTION

Myxomatous mitral valve disease (MMVD) is the most important cardiac disease of the dog. Much is known about the disease, but there are no treatments to slow, arrest or reverse valve changes. While the end-stage pathology of myxomatous mitral valve disease (MMVD) has been described in detail, the molecular processes initiating and perpetuating disease pathogenesis are only partially understood. There is detachment and loss of endothelial cells, followed by a loss of collagen and proliferation of α-smooth muscle actin (α-SMA) and embryonic smooth muscle myosin (Smemb) positive cells adjacent to the valve endothelium. These α-SMA+ve and/or Smemb+ve cells represent a differentiated population of valvular interstitial cells (VIC) that become activated and destroy the valve stroma through a process of aberrant remodelling. While this process is likely to involve the TGF-β superfamily and 5-HT, a much wider range of less well understood mechanisms and pathways are also involved. Whatever signalling pathways are important they are likely to be driven in part by endothelial changes coupled with re-activation of endothelial-to-mesenchymal transition (EndoMT). There is an increasing body of evidence that endothelial dysfunction is likely the major driver of MMVD pathogenesis, and that examining microRNA (miR) expression will identify novel biomarkers and drug discovery targets.

MICRORNA EXPRESSION IN NORMAL AND DISEASED MITRAL VALVES

The activity and effects of microRNAs (miRs) are of particular interest, not least because miRs have potential as biomarkers, prognostic indicators and novel therapeutic targets[1,2,3,4]. MiRs are short sections of DNA that control gene function. While the functions of canine miRs remain largely unproven, target prediction software and likely conservation across species give clues to their probable role in controlling gene expression[5] (Figure 1). A variety of interactions between miRs and endothelin function have been identified, and we have identified changes in miR expression in canine valve tissue. MiR-21 is involved in TGF-β–induced EndoMT via a PTEN/Akt-dependent pathway and exposure of ECs to TGF-β directly up-regulates miR-21[3]. ET-1 up-regulates miR-23 which protects against oxygen radical formation and TGF-β induced EndoMT, while MiR-125 prevents endothelin production, and both are down regulated in MMVD[6]. ET-1 prevents the expression of miR-15, which acts to suppress cell proliferation and enhances apoptosis, and is also down-regulated in MMVD. MiR-218 has the highest signal intensity in canine mitral valve tissue and is significantly down-regulated, with matched changes in SLIT2 expression (miR-218 is encoded in an intron of SLIT2), which is important in EC function and migration. We have identified changes in several other canine mitral valve miRs that would suggest dysregulation of endothelial function and involving effects as varied as redox reactions, EndoMT, inflammation and fibrosis.

MiRs are also important in EndoMT and miR-29 and miR-30, both significantly down-regulated in MMVD, are associated with SNAI1 (up-regulated in MMVD), a key inhibitor of E-cadherin expression, which is fundamental to EndoMT. The changes in expression of miR-29 family members are particularly interesting, with at least 16 confirmed ECM genes regulated by miR-29, including several that are necessary for EndoMT, including cathepsins, hyaluronan synthases (Has2) and laminin, all of which are significantly altered in the MMVD transcriptome. Other targets include VEGFA (vascular endothelial cell growth factor A) and roundabout receptors (SLIT-ROBO signalling) which are key regulators of endothelial cell function and migration. Dysregulation of miR-218 in MMVD would suggest there is aberrant SLIT-ROBO signalling leading to endothelial dysfunction and migration, excessive and disorganized extracellular matrix production and enhanced myofibroblastic differentiation.

From studies of other chronic degenerative diseases, there is sufficient evidence to suggest a role for miRs, endothelial dysfunction and EndoMT in disease pathogenesis, and emerging evidence of the therapeutic potential of miR-mimcs and antagonists in disease management and as disease biomarkers[1,2,3]. For example; injection of stable, non-toxic anti-miR-21 has anti-fibrotic effects, Miraversin (miR-122 antagonist) reduces rates of hepatitis C infection and miR-29a acts as a biomarker for human hypertrophic cardiomyopathy[3,7,8].

MICRORNAS AS DISEASE BIOMARKERS

There is a particular interest across many diseases in human and veterinary medicine in the use of circulating miRs as biomarkers of disease and disease severity, and by extension of disease progression, prognosis and effectiveness of treatment. Changes in circulating miRs have been identified in MMVD using RT-PCR to look at 277 miRs in one study, with changes in 11 key miRs noted between mild/moderate and severe cases compared to normal, and in another study of 9 specific miRs in Dachshunds[9,10]. While there are thousands of miRs so far identified there is only a likely important role for a limited number (hundreds) and many are expressed at such low levels as to be of little importance. In the Dachshunds there was significant down-regulation of miR-30b (ACVIM Stage B) and 133b (ACVIM Stage C), and a trend to downregulation for 5 others (125, 126, 21, 29b and 30b) in ACVIM Stage C[10]. In the study using various breeds 7 miRs (302d, 380, 874, 582, 490, 329b and 487b) were down-regulated and four (103, 98, let-7b and let-7c) up-regulated[9]. Several of these miRs are known to affect TGF-β activity and collagen synthesis, but also can be changed where there is myocardial stress and myocardial infarction. More importantly is the change in expression of -29, -30 and let- miR family members, as changes are also found with these classes in diseased canine valve tissue[11]. This would suggest that at least some circulating miRs can be indicative of severity of valve pathology and not just general markers of disease. A complimentary study of circulating miRs in Doberman Pinschers with DCM failed to identify any significant changes[12]. A possible speculative interpretation at this stage would be that congestive heart failure alone does not appreciably change miR expression and that changes in MMVD indeed do reflect disease pathology.

EXOSOMES, MIRS AND BIOMARKERS

Exosomes are lipid-bilayer enclosed extracellular vesicles secreted by cells into the circulation, with some degree of disease specificity, that carry small numbers of miRs and allow cell-to-cell signalling locally and at a distance. It is likely their miR content can affect gene expression in other cells. There is increasing evidence that manipulating miRs and exosomes can have therapeutic benefit in a range of diseases, and exosomes themselves have potential application as biomarkers of disease[14]. There are now commercially available exosome kits that simplify procedures for obtaining purified exosome preparations and their identity can be determined by detecting the small numbers of miRs they contain and members of the tetraspanin and heat shock (Hsp60, 70 & 90) protein families. Considering the possible direct link between circulating miRs and pathology it is exosomal miR expression that might be more sensitive as a general disease biomarker and marker of pathology stage.

MICRORNAS AND THERAPY

The potential to pharmacologically manipulate global endothelial dysfunction and valve endothelial and interstitial cell function in MMVD by targeting endothelium-derived microRNAs (miRs) and exosomes could provide an avenue for manipulation of disease pathogenesis and end-stage pathology. While much of this is hypothetical there is evidence from other diseases that this might be a successful strategy. However it will be some time before we can successfully modify miR function in the clinical setting. Studies are needed to examine further the expression of canine miRs associated with damage to valves to allow identification of relevant predicted targets for pharmacological or genomic manipulation. For the former, this could involve screening a range of readily available drugs (particularly those affecting endothelial function). For the latter, extensive laboratory based work is needed. This will likely involve introduction of RNA oligonucleotides of interest into VEC cells by transfection and miR activity blocking using locked nucleic acid (LAN) technology to see if there is a clear biological effect that can be translated to clinical practice. Lastly, studies will be needed to examine the effects of selected drugs on exosomes and their miR expression. This strategy will allow us to clarify the role of miRs and exosomes in MMVD and how they can be manipulated using readily available drugs or newly designed miR-mics and mir-antagonists.

REFERENCES

[1]Condorelli G, et al JACC 2014; 63: 2177. [2]Fiedler J, et al Herz 2014; 39: 154. [3]Roncarati R, et al JACC 2014; 63: 920. [4]Van Rooij E, et al NatRevDrugDisc 2012; 11: 860. [5]Kozomara A, et al NucAcRes 2014; 42: D68. [6]Lagendijk AK, et al CircRes 2011; 109: 649. [7]Chau BN, et al SciTransMed 2012; 4: 121. [8]Jansen HLA, et al NewEngJM 2013; 358: 1685. [9]Li J, et al Neph 2015; 15: 507. [10]Hulanicka M, et al BMC VetRes 2014; 10: 205. [11]Lu C-C, et al TVJ 2015; 204: 23. [12]Steudemann C, et al BMC Vet Res 2013; 9: 12. [14]Melo SA, et al Nature 2015; 523: 177.

Figure 1. Clustering of miRNAs based on their expression profiles comparing CKCSs and normal dogs using *Biolayout Express 3D*. Each cluster shows the predicted biological effect or context of each miR. The closer the miR nodes are to each other the closer the biological links and interplay in gene function control. This way the complex manner in which multiple miRs might control single biological functions can be understood.

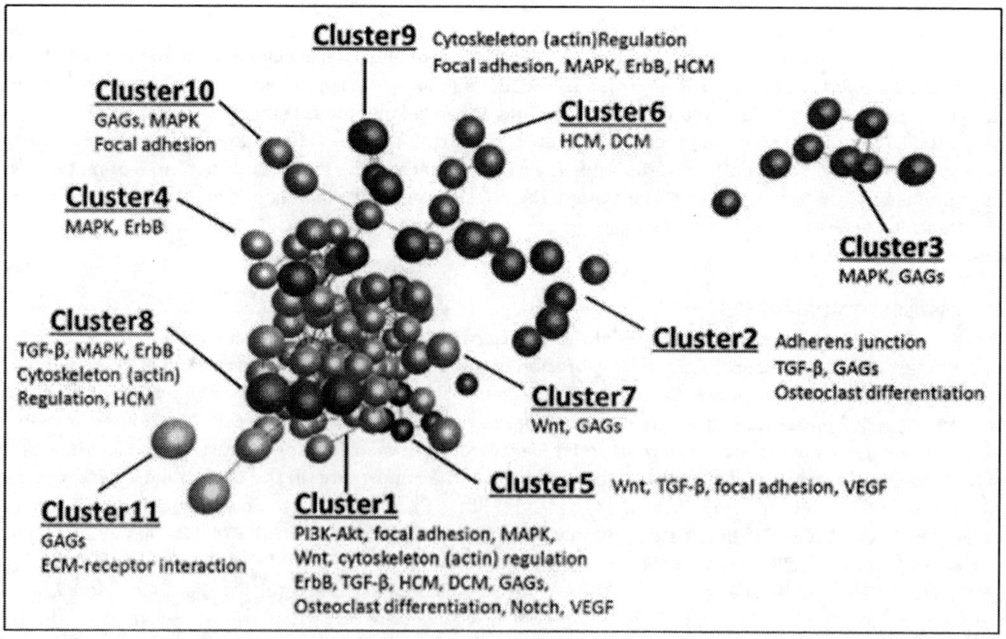

Calcium Mismanagement in Arrhythmogenesis and Heart Failure

Elizabeth Bode, BVSc, PhD, MRCVS
Easter Bush, Midlothian, UK

Introduction

Excitation-contraction (EC) coupling is the mechanism by which the heart converts electrical activity into mechanical force. This is a highly coordinated and regulated physiological process and results in the heart contracting as a syncytium, pumping blood around the systemic and pulmonary circulations. Alterations to this pathway can have serious consequences in terms of cardiac function and sudden death. Derangements in EC coupling have been well documented in heart failure [1] and abnormalities of both systolic and diastolic function result in worsening cardiac performance. Furthermore, many human heart failure patients die from sudden onset cardiac arrhythmias. In the majority of patients, these arrhythmias are due to abnormal automaticity or triggered activity, i.e. delayed after depolarisations (DADs) or early after depolarisations (EADs). Triggered arrhythmias, particularly DADs and under certain conditions EADs, are the result of abnormal Ca^{2+} handling [1].

Excitation-contraction Coupling

In order for cardiac contraction and relaxation to occur, intracellular calcium levels must rise and fall. Changes in intracellular calcium are brought about by the physiological process of EC coupling (figure 1). Initially, an action potential depolarises the cell membrane, penetrating deep inside the cell via the membranous invagination known as the transverse tubule. This depolarisation opens the L-type calcium channels causing a small amount of calcium to enter the cytosol from the extracellular milieu. This calcium then results in the release of a larger amount of calcium from the intracellular calcium store, the sarcoplasmic reticulum, in a process known as calcium-induced calcium-release. Movement of calcium from the sarcoplasmic reticulum is a passive process that involves the calcium release channel or ryanodine receptor. Calcium-induced calcium-release gives rise to the systolic calcium transient. This elevation in intracellular calcium results in calcium binding to troponin C. Binding of calcium to troponin C strengthens the interaction of troponin C with troponin I and destabilises the relationship between troponin I and actin. This allows the troponin T–tropomyosin complex to roll deeper into the actin groove. This movement of the troponin T–tropomyosin complex exposes the sites on the actin filaments where myosin interacts, enabling cross-bridge formation and sarcomeric shortening.

The heart then needs to relax and refill before it can contract again. Free intracellular calcium has to be removed from the cytosol after it dissociates from the myofilaments. There are several pathways by which this occurs. The dominant mechanism is the sarco(endo)plasmic reticulum calcium-ATPase (SERCA2a), which takes calcium back up into the sarcoplasmic reticulum. SERCA removes 70-90% of calcium from the cytosol [2,3], replenishing the intracellular calcium store ready for the next contraction. The sarcolemmal sodium/ calcium exchanger accounts for between 10-30% of calcium removal from the cytosol [2,3], transporting calcium back in to the extracellular space. Two minor mechanisms known to remove calcium from the cytosol include the plasma membrane calcium-ATPase transporter and the mitochondrial calcium uniport. These transport a very small amount of calcium (1-3%) either out of the cell or into the mitochondria respectively [2,3].

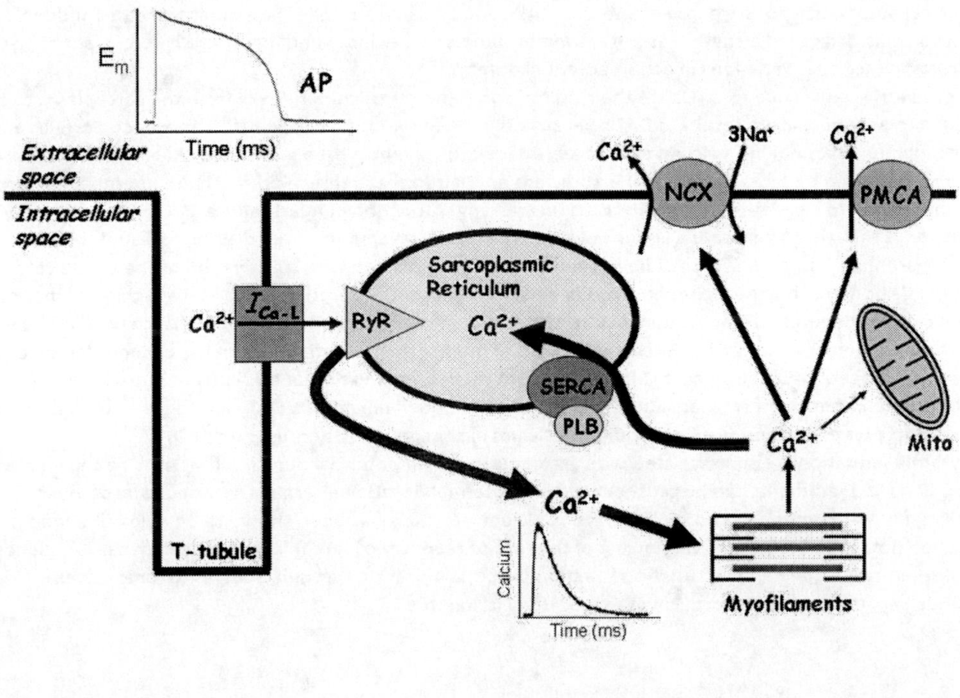

Figure 1. The excitation-contraction Coupling Pathway
Action potential (AP); I_{Ca-L} (L-type calcium current); RyR (Ryanodine receptor); T-tubule (Transverse tubule); SERCA (Sarco(endo)plasmic reticulum calcium ATP-ase pump); PLB (Phospholamban); NCX (Sodium/calcium exchanger); PMCA (Plasma membrane calcium ATP-ase pump); Mito (Mitochondria); Ca^{2+} (Calcium). Arrows represent the path travelled by calcium, with arrow thickness representing the proportion of calcium taken up by each removal pathway.

β-adrenergic Stimulation

Phosphorylation in response to β-adrenergic receptor (β-AR) stimulation plays a key role in normal EC coupling within the heart. In response to sympathetic stimulation, protein kinase A and calcium/calmodulin dependent protein kinase (CAMKII) both phosphorylate multiple targets within the cardiac myocyte, including the L-type calcium channel, phospholamban, ryanodine receptor, troponin-I and possibly the sodium/ calcium exchanger [4]. In most situations β-AR stimulation augments the actions of these cellular targets, resulting in positive inotropy, chronotropy and lusitropy. Perturbations in the β-adrenergic cascade, or in the responsiveness of its receptors, results in a heart that cannot adapt appropriately to sympathetic stimulation.

Excitation-contraction Coupling in Heart Failure

Heart failure is the end result of many disease processes such as hypertension, myocardial ischaemia and cardiac myopathies. Common to these aetiologies are alterations to calcium handling and ionic currents within cardiac myocytes (for general review see [1]). Shifts in the mechanical properties of the myocardium are also observed, largely due to geometric and structural disturbances [5]. These alterations result in systolic and/ or diastolic dysfunction and arrhythmogenesis or a combination of these.

Contractile force and, therefore, systolic function is decreased in heart failure. It is the systolic calcium transient that is crucial for initiating contraction and as such a ubiquitous finding amongst various models of heart failure is one of diminished calcium transient amplitude [6-8]. A reduction in the systolic calcium transient amplitude may be the result of either decreased intracellular calcium stores and/ or altered L-type calcium channel properties. Most studies have found that there is little change to the peak L-type calcium current density in heart failure [7,9]. Therefore, other factors must cause the diminished calcium transient amplitude and decreased levels of calcium stored within the sarcoplasmic reticulum are the main culprit [7,9]. Decreased sarcoplasmic reticulum content may be multifactorial in origin. In heart failure, a general observation is down-regulation of SERCA activity and up-regulation of sodium/ calcium exchanger function [6,7,9]. These alterations work synergistically to lower sarcoplasmic reticulum calcium levels by removing more calcium from the cell than is taken back up into the sarcoplasmic reticulum via SERCA. A lower SERCA activity can also cause diastolic dysfunction due to a slower rate of decline of intracellular calcium. However, up-regulation of the sodium/ calcium exchanger function may help to normalise this by removing a greater quantity of calcium at each beat [10].

Enhanced calcium leak from the sarcoplasmic reticulum in heart failure has also been observed [11]. This leak results in calcium leaving the sarcoplasmic reticulum during diastole, lowering its calcium content.

Arrhythmogenic Mechanisms in Heart Failure

Disturbances to calcium handling will directly affect contraction and relaxation, but may also lead to arrhythmias [12]. Many heart failure patients die from sudden onset cardiac arrhythmias. In the majority of patients these arrhythmias are caused by either abnormal automaticity or triggered events, i.e. DADs and EADs. In a smaller percentage of cases sudden death is due to re-entrant mechanisms. Triggered arrhythmias are caused by abnormal calcium handling whereas re-entrant arrhythmias are generally the result of regional variations in action potential duration.

Triggered arrhythmias due to DADs are the result of spontaneous calcium release from the sarcoplasmic reticulum [1]. This spontaneous release of calcium results in DADs because the calcium that is released from the sarcoplasmic reticulum is removed by the electrogenic sodium/ calcium exchanger. An inward current is then generated, which can depolarise the cell membrane. If this change in membrane potential is sufficient enough to reach threshold an action potential is triggered. Furthermore, there is an up-regulation of sodium/ calcium exchanger function in heart failure [7,9,10]. Enhanced sodium/ calcium exchanger function is thought to generate a larger inward current for a given amount of diastolic calcium release, facilitating a triggered action potential [10]. In contrast, EADs have been classically associated with reactivation of the L-type calcium current [13]. The generation of EADs in this manner relies on a prolonged action potential duration and an L-type calcium channel that has recovered from calcium-dependent inactivation. During the lengthened plateau phase the membrane potential is within the range of the L-type calcium channel's window current (range of membrane potentials where the channel can be both activated and inactivated). This together with a smaller calcium transient makes reactivation of this current more likely. EADs can also be observed during the later phases of repolarisation. These EADs have been linked to a DAD-like origin [14], where release of calcium from the sarcoplasmic reticulum and resultant sodium/ calcium exchanger activity causes an EAD.

Arrhythmia formation is also associated with gradients in action potential duration that provide the ideal substrate for re-entry [15]. Repolarisation gradients can be caused by action potential duration alternans, which occurs at much slower heart rates in the failing myocardium than in normal hearts [16]. Calcium alternans has been shown to be a possible mechanism for the induction of action potential alternans [16] via slowing of the rate of recovery of sarcoplasmic reticulum calcium release [17]. Other mechanisms of re-entry include regional variation in action potential duration through changes to ionic currents [15] and also structural remodelling, e.g. that which occurs with myocardial ischaemia.

Novel Therapeutic Strategies

Many drugs have been developed that target the excitation-contraction coupling pathway. Istaroxime increases SERCA activity and given intravenously may be of benefit in acute heart failure, reducing pulmonary capillary wedge pressures [18]. It appears to work via uncoupling phospholamban from SERCA resulting in both positive lusitropy and inotropy [19]. It also appears to affect the Na/K ATPase, similar to digoxin, but does not appear to have the same safety issues. Omecamtiv mecarbil, a direct activator of cardiac myosin, has also shown promise in terms of treatment of heart failure with reduced ejection fraction [20]. Recently, interest in gene therapy for heart failure has resulted in clinical trials involving the transfer of SERCA2a via adeno-associated virus type 1 (AAV1) to diseased myocardium. Clinical trials have been started both in USA and Europe.

The calcium upregulation by percutaneous administration of gene therapy in cardiac disease (CUPID) trial, performed in the USA, has completed both phase 1 and phase 2 clinical trials[21,22]. AAV1/SERCA2a were delivered to the myocardium via intracoronary perfusion over a 10 minute period to patients with New York Heart Association Class III/IV heart disease. Patients that received a high dose of AAVI/SERCA2a met all primary endpoints with reductions in both left ventricular end systolic and diastolic volume [22]. Benefits of high dose transfer persisted over a 3 year period [23]. In the UK, recruitment has now stopped for evaluation of AAVI/SERCA2a transfer to patients with left ventricular assist devices and analysis is underway.

REFERENCES

1. Bers DM. Altered cardiac myocyte Ca regulation in heart failure. Physiology 2006;21:380-387.
2. Bassani JW, Bassani RA, Bers DM. Relaxation in rabbit and rat cardiac cells: species-dependent differences in cellular mechanisms. The Journal of Physiology 1994;476:279-293.
3. Puglisi JL, Bassani RA, Bassani JW, et al. Temperature and relative contributions of Ca transport systems in cardiac myocyte relaxation. The American Journal of Physiology 1996;270:H1772-1778.
4. Bers DM. Excitation-contraction coupling and cardiac contractile force, 2nd ed. Dordrecht ; Boston: Kluwer Academic Publishers; 2001
5. Kass DA, Bronzwaer JG, Paulus WJ. What mechanisms underlie diastolic dysfunction in heart failure? Circulation Research 2004;94:1533-1542.
6. Briston SJ, Caldwell JL, Horn MA, et al. Impaired beta-adrenergic responsiveness accentuates dysfunctional excitation-contraction coupling in an ovine model of tachypacing-induced heart failure. The Journal of Physiology 2011;589:1367-1382.
7. Piacentino V, 3rd, Weber CR, Chen X, et al. Cellular basis of abnormal calcium transients of failing human ventricular myocytes. Circulation Research 2003;92:651-658.
8. Pogwizd SM, Schlotthauer K, Li L, et al. Arrhythmogenesis and contractile dysfunction in heart failure: Roles of sodium-calcium exchange, inward rectifier potassium current, and residual beta-adrenergic responsiveness. Circulation Research 2001;88:1159-1167.
9. Diaz ME, Graham HK, Trafford AW. Enhanced sarcolemmal Ca2+ efflux reduces sarcoplasmic reticulum calcium content and systolic calcium in cardiac hypertrophy. Cardiovascular Research 2004;62:538-547.
10. Pogwizd SM, Qi M, Yuan W, et al. Upregulation of Na(+)/Ca(2+) exchanger expression and function in an arrhythmogenic rabbit model of heart failure. Circulation Research 1999;85:1009-1019.
11. Shannon TR, Pogwizd SM, Bers DM. Elevated sarcoplasmic reticulum calcium leak in intact ventricular myocytes from rabbits in heart failure. Circulation Research 2003;93:592-594.
12. Aistrup GL, Balke CW, Wasserstrom JA. Arrhythmia triggers in heart failure: the smoking gun of [Ca2+]i dysregulation. Heart rhythm : the official journal of the Heart Rhythm Society 2011;8:1804-1808.
13. January CT, Riddle JM. Early afterdepolarizations: mechanism of induction and block. A role for L-type calcium current. Circulation Research 1989;64:977-990.
14. Volders PG, Kulcsar A, Vos MA, et al. Similarities between early and delayed afterdepolarizations induced by isoproterenol in canine ventricular myocytes. Cardiovascular Research 1997;34:348-359.
15. Antzelevitch C. Cardiac repolarization. The long and short of it. Europace : European pacing, arrhythmias, and cardiac electrophysiology 2005;7 Suppl 2:3-9.
16. Wilson LD, Jeyaraj D, Wan X, et al. Heart failure enhances susceptibility to arrhythmogenic cardiac alternans. Heart rhythm 2009;6:251-259.
17. Wasserstrom JA, Sharma R, Kapur S, et al. Multiple defects in intracellular calcium cycling in whole failing rat heart. Circulation Heart Failure 2009;2:223-232.
18. Shah SJ, Blair JE, Filippatos GS, et al. Effects of istaroxime on diastolic stiffness in acute heart failure syndromes: results from the Hemodynamic, Echocardiographic, and Neurohormonal Effects of Istaroxime, a Novel Intravenous Inotropic and Lusitropic Agent: a Randomized Controlled Trial in Patients Hospitalized with Heart Failure (HORIZON-HF) trial. American Heart Journal 2009;157:1035-1041.
19. Ferrandi M, Barassi P, Tadini-Buoninsegni F, et al. Istaroxime stimulates SERCA2a and accelerates calcium cycling in heart failure by relieving phospholamban inhibition. British Journal of Pharmacology 2013;169:1849-1861.
20. Cleland JG, Teerlink JR, Senior R, et al. The effects of the cardiac myosin activator, omecamtiv mecarbil, on cardiac function in systolic heart failure: a double-blind, placebo-controlled, crossover, dose-ranging phase 2 trial. Lancet 2011;378:676-683.
21. Jaski BE, Jessup ML, Mancini DM, et al. Calcium upregulation by percutaneous administration of gene therapy in cardiac disease (CUPID Trial), a first-in-human phase 1/2 clinical trial. Journal of Cardiac Failure 2009;15:171-181.
22. Jessup M, Greenberg B, Mancini D, et al. Calcium Upregulation by Percutaneous Administration of Gene Therapy in Cardiac Disease (CUPID): a phase 2 trial of intracoronary gene therapy of sarcoplasmic reticulum Ca2+-ATPase in patients with advanced heart failure. Circulation 2011;124:304-313.

23. Zsebo K, Yaroshinsky A, Rudy JJ, et al. Long-term effects of AAV1/SERCA2a gene transfer in patients with severe heart failure: analysis of recurrent cardiovascular events and mortality. Circulation Research 2014;114:101-108.

Cardiac CT and MRI: Visualizing the Heart in 3 Dimensions

Brian A. Scansen, DVM, MS, DACVIM (Cardiology)
Fort Collins, CO

INTRODUCTION

The heart is a 3-dimensional structure, yet nearly all cardiac imaging performed in animals relies on 2-dimensional imaging techniques such as radiography, fluoroscopy, and ultrasonography. The advancement of 3-dimensional imaging modalities over the past decade has been dramatic and techniques such as 3-dimensional echocardiography, cardiac computed tomography (cCT), and cardiac magnetic resonance imaging (CMR) allow an understanding of cardiac anatomy and an evaluation of cardiac function previously impossible to achieve. Within the vasculature, cross-sectional imaging is similarly transformative with computed tomography angiography (CTA) and magnetic resonance angiography (MRA) allowing detailed characterization of vascular malformations such as portocaval communications,[1] vascular ring anomalies,[2] and arterial-venous fistulae.[3] The focus of this lecture is to discuss the technical aspects of cCT, CTA, CMR, and MRA with an intent to highlight current and future capabilities of these modalities and to inform the audience on how to achieve optimal cross-sectional imaging of the heart and vasculature. Clinical cases will be used as examples.

CARDIAC CT

Cardiac CT involves the timed injection of iodinated contrast into a peripheral vein, followed by image acquisition when the contrast reaches the cardiac structures of interest. In a high motion organ, such as the heart, many technical factors play an important role in optimizing image quality.[4] These factors include the type of CT scanner, spatial and temporal resolution, gating and acquisition mode, and contrast administration.

CT Detectors

Early generation CT scanners lacked sufficient spatial and temporal resolution to characterize intracardiac anatomy. The development of multi-detector CT (MDCT), also known as multi-slice CT, and cardiac gating improved both spatial and temporal resolution such that cCT is now clinically feasible. A detector is the portion of the scanner that receives the photons generated by the X-rays passing through the patient. A single detector is comprised of many elements along its circumference, but is limited to a narrow width in order to resolve structures of interest. Multiple detectors allow the radiation effect to be collected along a broader slab width, known as the z-axis, which increases the slab thickness that can be scanned at one time. The first generation MDCT scanners used 4 detectors to improve z-axis thickness, with subsequent generations increasing to 8, 16, 64, 128, 320, and even 600 detectors. A MDCT with 320 detectors can typically cover the z-axis of a human heart allowing the entire heart to be evaluated in one rotation of the CT gantry. It should be noted that the length of the z-axis that can be imaged by a given scanner is dependent on both the number of detectors as well as the width of each detector such that two different 64-slice scanners may not have the same z-axis coverage. As an example, a 64-slice scanner with detectors of 1.25mm width could cover a z-axis thickness of 80mm while a 64-slice scanner from another manufacturer with detectors of 1.0mm width would yield a z-axis thickness of 60mm with each gantry rotation. Both the width in the z-axis that can be covered by each rotation as well as the speed of gantry rotation impact the final image quality of the scanner, as will be discussed below.

CT Resolution

Spatial resolution is defined as the ability to resolve structures that lie next to one another on an image. Spatial resolution is predominately affected by the size of the detector and the spacing between detectors; the detector must be smaller than an object to resolve it and detectors must be sufficiently close together to resolve two objects that lie close to one another in the patient. In cCT, the z-axis spatial resolution is equally important to allow reconstruction of the image in all planes without distortion, known as isotropic resolution, and for this reason MDCT scanners with a high number of detectors are desired to optimize cCT imaging. Spatial resolution is also improved by advancements in image detector technology and reconstruction software. New high-definition scanners are available that can achieve spatial resolution in the x- and y-axes to 0.23mm using novel detector material and advanced reconstruction algorithms. As dogs and cats have smaller hearts than humans, spatial resolution of small structures (e.g., coronary vessels) remains a challenge in these species with all but the newest generation of scanners.

Temporal resolution describes the ability of the scanner to resolve a structure moving in time and is best measured as the time required to acquire successive images. Temporal resolution is dependent on gantry rotation speed, mode of image acquisition, gating, and pitch. The fastest scanners currently on the market can perform a full gantry rotation in ~250msec, which means that they can resolve an image in roughly 125ms since a CT image is created/extrapolated from ½ of a gantry rotation, as the x-ray source moves from 0 degrees to 180 degrees. Adding an additional X-ray source placed 90 degrees to the other is one strategy to minimize the time required to circumferentially rotate around the patient and improve temporal resolution. These dual-source CT scanners have two unique x-ray sources (often of different energy spectra) as well as 2 sets of detectors and need to only rotate 90 degrees in order to acquire a half gantry rotation, reducing temporal resolution down to 66msec. The mode of acquisition refers to prospective or retrospective image acquisition/reconstruction and will be discussed in the section on gating. Pitch is defined as the distance the table advances per gantry rotation divided by the total thickness of the acquired slab (in the z-axis). With MDCT, the thickness of the

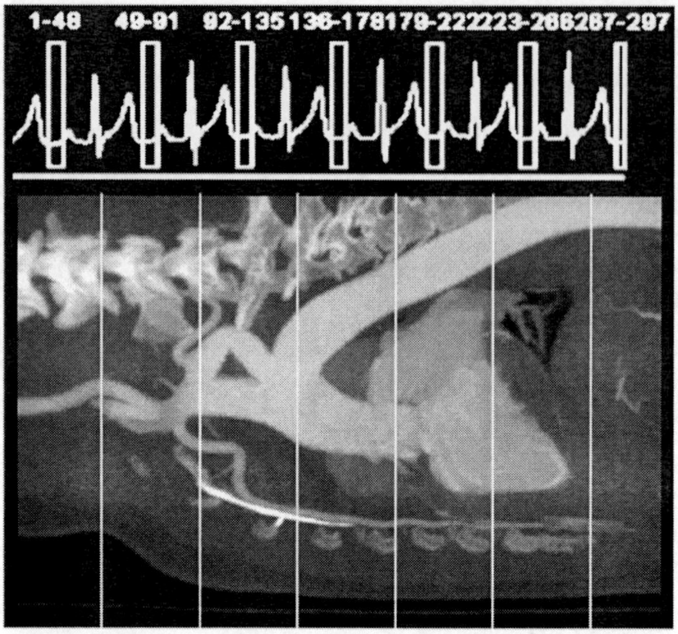

Figure 1. ECG-gated cardiac CT. The dog's ECG is seen along the top of the image with the numbers and boxes representing the slices acquired during that cardiac cycle. Retrospective gating allows the images to be acquired at the same point in each cardiac cycle (here in early diastole). The image is constructed in the z-axis by stitching together each slab of data, ~45 slices per slab with the 7 slabs together creating an image of the entire heart.

slab is the sum of the width of all detectors activated during acquisition. A pitch greater than 1 implies gaps in the z-axis coverage as the table moved too rapidly to allow the scanner to image every slab in the field of view. Some degree of overlap in the z-axis is preferred to adequately reconstruct images of the heart. The rapid heart rates of dogs and cats require excellent temporal resolution to achieve high-quality cCT imaging.

Gating

Gating is used to time the acquisition sequences to the cardiac cycle, thereby freezing cardiac motion in the acquired image series (**Figure 1**). The animal's ECG is used to reconstruct (retrospective) or trigger (prospective) the scan. With retrospective gating, the entire cardiac cycle is scanned (usually several cardiac cycles) and the desired image reconstructed from the raw dataset based on the simultaneous ECG. This allows for reconstruction of different phases of the cardiac cycle and for dynamic analysis (**Figure 2**). However, retrospective gating increases radiation exposure to the patient. Prospective gating triggers the scanner to acquire images only during a portion of each cardiac cycle, typically when cardiac motion is least apparent such as mid-diastole or end-systole (for higher heart rates). This is known as a move-and-shoot technique; the scanner is only activated for small windows of time and a single portion of the cardiac cycle is recorded, limiting radiation exposure but preventing dynamic analysis. Respiratory gating, or subjecting the animal to a breath-hold or apneic period during the scan, is also frequently employed to avoid movement during scan acquisition. The ability of the scanner to appropriately gate the study relates to the temporal resolution of the scanner – at fast heart rates the scanner may not be able to acquire the images rapidly enough to adequately resolve the desired image.

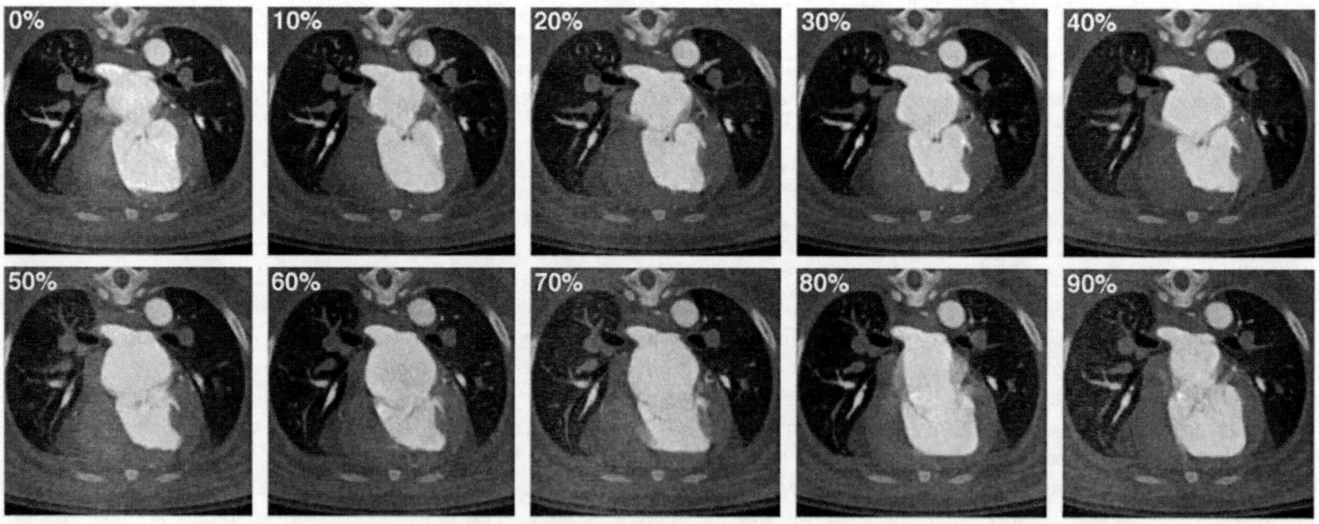

Figure 2. Dynamic cardiac CT with retrospective gating of the left atrium and left ventricle from a dog with mitral valve dysplasia. Note the change in chamber dimensions throughout the cardiac cycle, divided into 10% increments from the start of one R wave to the next.

CARDIAC MRI

Cardiac MRI shares many similarities with cCT in that both allow cross-sectional imaging and reconstruction of the heart in multiple planes. The advantages of CMR over cCT relate to the lack of ionizing radiation and the ability to view the blood pool without intravenous contrast administration. Additionally, tissue characteristics can be obtained by CMR (e.g., fibrosis, fat infiltration) that cannot be detected by cCT. The downside to CMR relates to the lower spatial and temporal resolution as compared to new generation cCT scanners, as well as the requirement for general anesthesia and longer scanning times for the patient. The cost and technical expertise required to have a proficient CMR program also exceeds that for cCT.

Technical Aspects of CMR in Animals

The physics of CMR are different than that for cCT. In brief, radiofrequency (RF) pulses are transmitted into the body to alter the state of protons that are aligned to a strong magnetic field. The RF pulses change the alignment and phase of the spinning protons within the body; as they return to their base state after the RF pulse passes, they emit small bursts of RF energy that are received by coils placed around the body. These signals are received by the coils and used to generate an image; because the rate of relaxation and phase realignment differs depending on the tissue type, images of the patient's tissue composition can be generated.

Cardiac MRI is most often performed with magnets of 1.5 Tesla and 3 Tesla field strengths with phased array coils placed around the thorax to both transmit and receive the RF pulses that will generate the image. Most CMR has been performed on 1.5 T magnets, though higher field strength magnets are becoming more popular for neurologic and musculoskeletal imaging. While 3 T magnets are the "new generation", the transition to 3 T imaging is not comparable to the advancements in cCT – newer is not always better. The higher field strength of a 3 T magnet provides a better signal-to-noise ratio and quicker scan times, but increased artifacts and more challenging sequence optimization are difficulties for 3 T CMR imaging.[5]

The standard CMR image sequence for determining cardiac anatomy is 3-dimensional steady-state free precession (SSFP) imaging.[6] This sequence provides a bright blood pool with good endocardial border detection and can be acquired throughout the cardiac cycle to allow cine imaging and full chamber volume characterization. This can either be performed during induced apnea since the animal is under general anesthesia, or there are respiratory gating mechanisms that can monitor the diaphragm and control for respiratory motion. Similar to cCT, ECG-gating can be performed to acquire slices of the heart throughout the cardiac cycle. Cardiac MRI also brings novel ability to characterize tissues, with sequences for determination of tissue fibrosis (late gadolinium enhancement or T1 mapping) as well as tissue edema (T2 mapping).[6] Phase-contrast sequences can be used to evaluated flow and pressure gradients[6] – in the author's experience, these are challenging to perform accurately in animals. Last, the stiffness of the myocardium can be evaluated by magnetic resonance elastography – where sound waves are introduced into the patient and their propagation monitored and analyzed by the MR system.[7] Examples will be provided from clinical canine cases.

CT AND MR ANGIOGRAPHY

Both CTA and MRA have been employed to evaluate vascular structures in animals. The benefit of cross-sectional imaging such as CTA or MRA is the avoidance of selective central catheterization as only peripheral venous access is required for these techniques. Most of the published literature on these modalities in animals has focused on anatomic delineation of portosystemic shunts, though they can be used to investigate any vascular structure. Timing of the contrast administration relative to scan acquisition of the structure of interest is both challenging and paramount to obtaining a useful imaging study. Various protocols for contrast timing are available, utilizing either a test bolus scan to precisely measure the time that contrast arrives at the structure of interest and then adjusting scan time to this interval, or using a machine-specific monitoring scan that initiates the full scan once an increase in radiodensity is noted in the structure being monitored (e.g., bolus-tracking). Unique to MRA, imaging techniques are available to visualize blood clearly on the image without the administration of contrast. Contrast agents may still be used and are preferred for some studies (such as perfusion scans), but may not be necessary for anatomic delineation.

ROTATIONAL ANGIOGRAPHY

Unrelated to cCT or CMR, a new cross-sectional imaging modality is rotational angiography. Rotational angiography is performed in the catheterization laboratory with new generation fluoroscopy systems that can rapidly rotate around the patient during contrast administration and create a 3-dimensional (CT-like) image of the patient's anatomy.[8] The benefits of this technique are that injections from multiple image planes need not be recorded, thereby reducing contrast load, and improved anatomic delineation of the structure of interest can be viewed during the interventional procedure – optimizing image guidance.

3-DIMENSIONAL ASSESSMENT

Once cross-sectional imaging of the heart is obtained, either by cCT or CMR, numerous software packages are available to reconstruct the anatomy in nearly any conceivable plane or as a 3-dimensional projection. This dramatically improves understanding of spatial relationships or intracardiac defects and can be useful for planning surgical or transcatheter interventions. Additionally, these datasets can be transferred to a 3-D printer to provide hands-on objects for student teaching, resident training, or research endeavors. While echocardiography will never be replaced by these new 3D technologies, they provide the veterinary cardiologist with novel ways to assess, evaluate, and understand the hearts of animals.

REFERENCES

[1]Zwingenberger A. Vet Clin North Am 2009;39:783. [2]Henjes CR, et al. BMC Vet Res 2011;7:57. [3]Zwingenberger AL, et al. Vet Radiol Ultrasound 2005;46:472. [4]Drees R, et al. Vet Radiol Ultrasound 2014;55:229. [5]Rajiah P, et al. Radiographics 2014;34:1612. [6]Bonnemains L, et al. Arch Cardiovasc Dis 2016; E-pub. [7]da Silveira JS, et al. Magn Reson Imaging 2016;34:26. [8]Pedra CA, et al. Curr Opin Cardiol 2011;26:86.

3D Modeling from CT Angiography in Congenital Heart Disease

Ashley B. Saunders, DVM, DACVIM (Cardiology); Sonya G. Gordon, DVM, DVSc, DACVIM (Cardiology)
College Station, TX

Procedural planning has classically relied on echocardiography and angiography to evaluate cardiac anatomy.[1] A detailed knowledge of the anatomy is critical to procedural success. Computed tomography (CT) angiography provides high-resolution images of the cardiac structures, and reconstructed models can be rendered using a variety of software applications to produce three-dimensional (3D) images. From image processing applications, the images are flattened into a 2D image that represents a 3D structure. Applying color, for example, to a tissue density or Hounsfield unit creates a medical illustration of 3D. From CT studies, the images can be manipulated to position the structures in a specific orientation that highlights the area of interest (i.e. oblique view) and the model rendered and viewed in a controlled manner. The authors will present a virtual reality system where a CT study can be imported minutes after it is acquired to view an interactive True 3D representation. A clipping plane is applied to view and measure internal structures such as patent ductus arteriosus pulmonary ostia or septal defects.

Advanced medical imaging software (Mimics®, Materialise, Belgium) allows users to develop wire mesh and 3D anatomic models from CT data sets.[2] Polygonal modeling of structures creates a 3D model from specific points called vertices that define a shape. These wireframe models approximate the surface of a structure and allow inner structures to be viewed and measured in cross section or volume estimated. The data sets are loaded and density thresholds applied to the tissues of interest to create and export a 3D model as an STL file that can be 3D printed. Standard threshold definition values are applied then customized depending on the scan.

These models have applications for surgical procedural planning,[3,4] for the engineering of new devices, to enhance visuospatial understanding of anatomy at all levels of training, and can be utilized to educate owners. This session will emphasize practical application of this technology in a variety of clinical cases.

REFERENCES

[1] Saunders AB, Miller MW, Gordon SG, Bahr A. Echocardiographic and angiographic comparison of minimal ductal diameter in dogs with patent ductus arteriosus. J Vet Intern Med 2007;21:68-75.

[2] Jacobs S, Grunert R, Mohr FW, et al. 3D-Imaging of cardiac structures using 3D heart models for planning in heart surgery: a preliminary study. Interact Cardiovasc Thorac Surg 2008;7:6-9.

[3] Ryan JR, Moe TG, Richardson R, et al. A novel approach to neonatal management of tetralogy of Fallot, with pulmonary atresia, and multiple aortopulmonary collaterals. J Am Coll Cardiol: Cardiovascular Imaging 2015;8:103-110.

[4] Saunders AB, Birch SA. Three-dimensional modeling of a patent ductus arteriosus in a cat. J Vet Cardiol 2015;17:S349-S353.

SURGERY AND VACCINE-BASED IMMUNOTHERAPY FOR CANINE GLIOMA

G. Elizabeth Pluhar, Matthew Hunt, John Ohlfest, Michael Olin

Department of Veterinary Clinical Sciences, College of Veterinary Medicine, University of Minnesota, Saint Paul, MN, USA

Glioma is the most common primary intra-axial tumor in dogs and people carrying a poor prognosis. Our goal is to develop effective immunotherapies with minimal adverse effects. We are conducting clinical trials on dogs with glioma using tumor lysate-based vaccines after surgery to stimulate a tumor specific immune response to kill residual tumor.

Dogs with MRI-confirmed solitary intra-axial masses underwent surgical debulking and randomization into 5 treatment groups: [1] temozolomide (TMZ) (n=4), [2] OX40L+TMZ (n=4), [3] vaccines (vax)+TMZ (n=8), [4] vax+OX40L (n-8), [5] vax+OX40L+TMZ (n=8). Follow-up was serial neurological examinations and MRIs. Outcome measures were adverse events, and progression-free (PFS) and overall survival (OS) times. Survival curves were compared using log-rank testing; significance level $P<0.05$.

Vaccine reactions were mild inflammation at injection sites and malaise. Two dogs weighing < 5 kg developed anemia and thrombocytopenia from TMZ. Median PFS and OS were: [1] 270 and 301 days, [2] 276 and 276 days, [3] 240 and 331 days, [4] 239 and 246 days, and [5] 239 and 246 days. There were no significant differences in PFS or OS among the treatment groups ($P>0.5$). Recurrent tumor was found in 26 of 28 brains with postmortem examination. We treated dogs with glioma using surgery and immunotherapy, extending survival over palliative care with few adverse effects and good quality of life. The lack of significant differences in survival among groups was likely due to type II error from small sample sizes. We are currently working on new immunotherapy combinations to improve our clinical response.

DOES A TH17 SKEWED IMMUNE RESPONSE IN STEROID-RESPONSIVE MENINGITIS-ARTERITIS EXIST?

Andrea Tipold, Jessica Freundt-Revilla

Klinik für Kleintiere, Stiftung Tierärztliche Hochschule Hannover, Hannover, Germany

Previous studies evaluating the pathogenesis of Steroid-Responsive Meningitis-Arteritis (SRMA) have shown increased levels of IL-6 and TGF-$ß_1$ in cerebrospinal fluid (CSF) samples. In the presence of these cytokines, naïve $CD4^+$ cells may differentiate into Th17 lymphocytes synthetizing IL-17. IL-17 induces and mediates inflammatory responses and plays an important role in autoimmune diseases and recruitment of neutrophils.

The purpose of this study was to confirm the hypothesis of a Th17 skewed immune response in dogs with SRMA by evaluating IL-17 using enzyme-linked immunosorbent assays (ELISA) and enzyme-linked immunoSpot assays (ELISpot). The cytokine was determined in eighty-four CSF and eighty-eight serum samples and in peripheral blood mononuclear cells of seventy one selected patients to identify and enumerate IL-17 producing cells at the single cell level. Different groups were compared: dogs suffering from SRMA in the acute stage (SRMA A), under treatment with glucocorticosteroids (SRMA T), during recurrence of the disease (SRMA R) and with other disorders and in healthy dogs (animal experiment 05-14A453).

The highest levels of IL-17 were found in CSF samples of dogs with SRMA A (median 901 pg/mL) and SRMA R (median 533 pg/mL). Both groups showed significant differences compared with all other groups - SRMA T (30 pg/mL), other disorders and healthy dogs ($p<0.0001$ with SRMA A; $p<0.005$ with SRMA R). Serum IL-17 levels were lower than CSF values, the highest levels were again measured in SRMA A and SRMA R. ELISpot assays confirmed IL-17 production at the single cell level. Furthermore, CSF concentrations of IL-17 showed a strong positive correlation with the degree of pleocytosis (rSpear= 0.8924; $p<0.0001$).

These results imply that Th17 cells are involved in the pathogenesis of SRMA and highly likely induce an autoimmune response. Furthermore, the development of severe neutrophilic pleocytosis can be explained considering a Th17 skewed immune response as well as the reduced neutrophil numbers and IL-17 levels in CSF under treatment with glucocorticosteroids. The investigation of the role of IL-17 in SRMA adds to the knowledge of pathophysiological mechanisms in SRMA and opens the discussion about new therapeutic strategies such as the application of agents influencing IL-17 or the respective receptor.

IMEPITOIN: FIELD OBSERVATIONS

Andrea Tipold, Jasmin Nessler

Klinik für Kleintiere, Stiftung Tierärztliche Hochschule Hannover, Hannover, Germany

To treat idiopathic epilepsy, the most frequently occurring chronic intracranial disease in dogs, different anti-epileptic drugs are used. Imepitoin was detected, when evaluating different substances for their anti-epileptic properties and was found to have useful and predictable pharmacokinetics in dogs. A rapid onset of effect from immediate release followed by a prolonged absorption was described.

Before imepitoin was available on the market in Europe several clinical studies were performed. In a multicentre field efficacy study efficacy of imepitoin was compared with the efficacy of phenobarbital in a blinded parallel group design. The oral administration of imepitoin demonstrated comparable efficacy to phenobarbital in controlling generalised seizures in dogs. The frequency of adverse events including somnolence/sedation, polydipsia and increased appetite was significantly higher in the phenobarbital group, in the group receiving imepitoin more hyperactive dogs were observed.

Furthermore, a randomized double blinded study was performed to demonstrate superior anticonvulsant activity of a high dose over a low dose of imepitoin in dogs with newly diagnosed idiopathic epilepsy. Indeed, treatment with the high dosage resulted in a significantly higher reduction in monthly frequency of generalised seizures compared to the low dosage. An effect on cluster seizures could not be demonstrated in this study.

Since imepitoin was licensed in Germany cases were anecdotally reported with the finding that imepitoin resulted in cluster seizures after add-on application to phenobarbital. Therefore, in an open label trial three groups of dogs not responding to the primary antiepileptic drug were compared: imepitoin was given add-on to phenobarbital in a dosage of either 10 mg/kg or 5 mg/kg twice daily or phenobarbital was given add-on to imepitoin. Add-on treatment of imepitoin in a dosage of 5 mg/kg twice daily and phenobarbital in a dosage of 0.5 mg/kg twice daily did not result in adverse effects such as the occurrence of cluster seizures. A reduction of seizure frequency was observed in part of the dogs as expected.

DOES A PANEL OF CSF BIOMARKERS ENHANCE THE PROGNOSTIC VALUE IN CANINE SPINAL CORD INJURY?

Stefanie Wicha, Regina Carlson, Andrea Tipold, Veronika Stein

University of Veterinary Medicine Hannover, Hannover, Lower Saxony, Germany

The prediction of a reliable prognosis can be challenging in paraplegic dogs with spinal cord injury (SCI), particularly when nociception is lost. Cerebrospinal fluid (CSF) analysis has been proposed as a minimally invasive method to assess inflammatory responses following SCI that could provide support to estimate the prognosis. The value of various substances as CSF biomarkers should be assessed for establishing a prognosis in paraplegic dogs with SCI.

Tau protein, macrophage inflammatory protein 3 beta (MIP-3ß) and glial fibrillary acidic protein (GFAP) are potential markers of spinal cord tissue destruction. Concentrations of tau protein, MIP-3ß and GFAP were measured in cisternal and lumbar CSF samples using enzyme-linked immunosorbent assay (ELISA) in 36 dogs with acute/subacute (≤28 days) and 13 dogs with chronic (>28 days) SCI. The dogs were classified as grade 4 or grade 5 according to the presence (n = 21) or absence (n = 28) of deep pain perception (Sharp and Wheeler, 2005). Seven healthy dogs served as controls. Outcome of dogs with acute/subacute SCI was monitored by neurological control exams and was defined to be successful when an improvement of at least one grade according to Sharp and Wheeler (2005) was noted within four weeks after surgery.

Paraplegic dogs with acute/subacute SCI had significantly ($p<0.05$) higher tau protein and MIP-3ß concentrations compared to dogs with chronic paraplegia and to the control dogs. GFAP values in cisternal CSF of dogs with acute/subacute SCI, grade 5, were significantly increased compared to the control dogs. In chronic SCI significantly elevated tau protein and MIP-3ß concentrations were detected in lumbar CSF compared to control dogs. Dogs with neurological improvement had significantly lower cisternal tau protein values.

In conclusion, SCI leads to a release of tau protein, MIP-3ß, and GFAP in CSF with lower values in dogs with chronic paraplegia. Cisternal tau protein concentrations may serve as a prognostic indicator for neurological improvement of paraplegic dogs. The combination of the concentrations of these three biomarkers together could not increase the predictive value for an improvement of neurological signs.

HOPE FOR TREATING KRABBE DISEASE

Charles H. Vite

Department of Clinical Studies, School of Veterinary Medicine, University of Pennsylvania, Philadelphia, PA, USA

Globoid cell leukodystrophy (GLD), or Krabbe disease, results from a deficiency in the hydrolytic enzyme galactosylceramidase (GALC), which is responsible for degrading the central and peripheral nervous system myelin lipids galactosylceramide and galactosylsphingosine (psychosine).

Combination gene therapy has been evaluated in GLD affected dogs using AAVrh10 encoding canine *GALC* (AAVrh10-cGALC). Two GLD dogs received a low dose of AAVrh10-cGALC, 1.2E12, by combination intravenous (IV) and intracerebroventricular (ICV) injection routes and displayed a modest increase in survival to 17.9 and 22.1 weeks of age. Two additional GLD dogs were treated with an increased dose of AAVrh10-cGALC, 1.9E13, and survival was further increased to 30.3 and 43.1 weeks of age. Both low and high dose combination IV and ICV therapy delayed the onset of neurological signs and prevented the onset of tremors. High dose AAVrh10-cGALC, but not low dose, resulted in normalization of pelvic and thoracic limb nerve conduction velocity and near normal sensory nerve conduction velocity. Combination therapy of either dose had negligible effect on the auditory system. After high dose treatment with AAVrh10-cGALC, GALC activity reached near normal levels in the cerebellum and sciatic nerve. GLD dogs treated with high dose AAV had CSF psychosine concentrations lower than untreated and low dose AAV-treated GLD dogs, despite being substantially older.

Efficacy of hematopoietic stem cell transplant alone and in combination with IV AAVrh10-cGALC is being evaluated in GLD dogs. Ongoing studies suggest that addition of IV infusion substantially increases survival as compared to HSCT alone.

MICRORNAS MIR-21 AND MIR-181C IN CEREBROSPINAL FLUID AND SERUM IN CANINE MENINGOENCEPHALOMYELITIS OF UNKNOWN ORIGIN

Luis Gaitero, Stewart Russell, Gabrielle Monteith, Jonathan Lamarre

Ontario Veterinary College, University of Guelph, Guelph, Ontario, Canada

The purpose of this study was to determine and compare the expression of microRNAs miR-21 and miR-181c in cerebrospinal fluid (CSF) and serum of dogs with and without meningoencephalomyelitis of unknown origin (MUO).

CSF and serum samples were obtained from ten dogs with MUO (MUO group) and from eight dogs with non-inflammatory neurological diseases with a normal CSF nucleated cell count (NCC) that were selected as control group (NINDC). Serum from four healthy dogs was used as second control group (HC). Relative expression of miR-21 and miR-181c was determined by qRT-PCR using the ΔΔCt method after adding cel-miR-39 as a spike-in control. Results were compared between groups using a general linear model. Pearson correlation analysis assessed association between parameters.

MiR-21 and miR-181c were significantly increased in CSF from the MUO group compared to controls ($P<0.002$ and $P<0.007$ respectively) but not in serum ($P=0.149$ and $P=0.172$ respectively). A strong correlation was present between NCC and miR-21 in CSF ($r = 0.891$; $P<0.001$). No correlation was observed between microRNA levels in CSF and serum in the MUO group, suggesting intra-thecal origin.

In conclusion, this is the first study to evaluate the expression of inflammation-associated miRNAs in CSF and serum in dogs with MUO in which an increase in these biomarkers in CSF was identified. The data support additional studies on the potential value of microRNAs as biomarkers for MUO, particularly miR-181c, since the absence of a correlation with NCC suggests that it may be an independent diagnostic and prognostic marker.

C-REACTIVE PROTEIN IN DIAGNOSIS OF DISCOSPONDYLITIS

William Bush, Sarah Trub, Dan Cuff, David Brewer, Joli Jarboe, Martin Young, Michael Higginbotham, Jessica Barker, Casey Neary, Lisa Lipitz

Bush Veterinary Neurology Service, Leesburg, Virginia, USA

C-reactive protein (CRP) is an acute-phase protein often elevated in inflammatory disease conditions. CRP elevations have been studied in veterinary neurology with significant elevations noted with steroid responsive meningitis arteritis (SRMA) but not meningoencephalitis, intervertebral disk disease, CNS tumors or idiopathic epilepsy. In human medicine verterbral osteomyelitis is the equivalent to discospondylitis and CRP elevations aid in the diagnosis, guide treatment efficacy, and predict outcome. To the author's knowledge there are no veterinary publications investigating CRP elevations and discospondylitis. The purpose of this study was to define the incidence of CRP elevations compared to historically used clinicopathologic and radiographic findings in a population of dogs diagnosed with discospondylitis using MRI.

Dogs were included if they had CBC, Chemistry, CRP, neurological examination, and MRI consistent with discospondylitis. In this study discospondylitis often manifested on MRI as T1-hypointensity, STIR hyperintensity, and contrast enhancement of the intervertebral disk, vertebral bodies, paravertebral soft tissues and epidural space. Radiographs of the MRI identified lesion were also reviewed when available.

Nineteen dogs were included in analysis. Eighteen dogs were large breed and 1 dogs was a beagle. Mean weight of these patients was 35.6 kg. Mean age was 5.4 years and 50% were male. All patients had spinal pain on examination, 10 had paresis and 2 were non-ambulatory. Six patients presented with fever. Two patients had leukocytosis, 7 neutrophilia, 6 hyperglobulinemia, 12 had elevated CRP. Twelve patients has spinal radiographs with 3 having abnormal radiographs consistent with discospondylitis. Among the 19 dogs, 41 MRI lesions consistent with discospondylitis were identified with L7-S1 being the most commonly affected disk space. A causative organism identified in 7/19 patients.

In the commonly utilized tests in patients with suspected discospondylitis, fever, CBC abnormalities hyperglobulinemia and radiographic lesions were only noted in about 25% of patients. CRP was elevated in 63% of cases and should be considered a useful adjunctive test in suspected cases of discospondylitis.

DO DOGS SPINAL WALK? ELECTROPHYSIOLOGIC CHARACTERIZATION OF LONG TRACT INTEGRITY IN CANINE SPINAL CORD INJURY

Melissa Lewis, Natasha Olby

North Carolina State University, Raleigh, NC, USA

Severe, acute thoracolumbar spinal cord injury (TLSCI) may leave dogs with permanent loss of pain perception. Despite this, 20-40% recover the ability to walk in the year following injury. It is unclear whether this recovery requires descending control or occurs through local circuitry. The aim of this study was to characterize the electrophysiologic status of motor and sensory tracts in dogs with chronic motor and sensory deficits due to acute TLSCI and to correlate findings to gait.

Nineteen dogs that failed to recover pelvic limb pain perception following acute TLSCI suffered at least 3 months previously were enrolled. Electrophysiologic testing included transcranial magnetic motor evoked potentials (TMMEPs) and somatosensory spinal evoked potentials (SSEPs). Gait was categorized as ambulatory (yes or no) and quantified using a published 0-12 open field gait scale (OFS). Association between walking and presence of TMMEP or SSEP was evaluated using Chi square analysis.

TMMEPs were present in the pelvic limbs of 4 dogs and cortical SSEPs were not detected following tibial stimulation in any dog. Four dogs were ambulatory with a median OFS of 6.5 (6-9). The remaining non-ambulatory dogs had a median OFS of 1 (0-4). Of the 4 dogs with pelvic limb MEPs, 3 were ambulatory with a median OFS of 6.5 (4-9). Ambulatory dogs were significantly more likely to have detectable trans-lesional spinal cord conduction ($p=0.0157$).

In summary, these findings suggest that some dogs with persistent deficits in pelvic limb pain perception have incomplete injuries and volitional motor control.

FRAMELESS STEREOTACTIC RADIOTHERAPY ALONE AND COMBINED WITH TEMOZOLOMIDE IN CANINE GLIOMAS

Mario Dolera, Luca Malfassi

La Cittadina Fondazione Studi e Ricerche Veterinarie, Romanengo, CR, Italy

Brain gliomas are diagnosed in an increased number of dogs. Few data regarding the best treatment regimen are available. The primary aim of this work was to evaluate the feasibility and efficacy of high dose hypo fractionated volume modulated arc radiotherapy (VMAT RT) in canine gliomas. The secondary aim was to assess the efficacy and toxicity of combination of radiotherapy with temozolomide.

72 client owned dogs with imaging based or histologically confirmed diagnosis of brain gliomas were enrolled. LINAC based frameless stereotactic radiotherapy was offered to all dogs. Dogs whose owners refused RT were palliatively treated (palliation arm). The schedule of treatment has ranged between 33 Gy/5 fx and 42 Gy/10 fx according the volumetric dose constraints of the brain (RT arm). During the second phase of the study, the same RT treatment was offered combined with temozolomide (TMZ) 65 mg/m^2 orally administered 6 hours prior each fraction and then for 5 days monthly for 6 cycles (RT+TMZ arm). Regular clinical examinations were performed during and after irradiation time, with regard to mentation, deambulation, cranial nerve dysfunction and seizures. Serial magnetic resonance imaging exams were done 2, 4, 6, 12, 18, 24 months after irradiation. Volumetric disease response and imaging findings were implemented with clinical neurological systematic evaluation to assess the course of the disease. Multivariate analysis to assess any prognostic significance was performed. Overall and disease specific survival were estimated using the Kaplan Meier curves.

30 dogs were palliated, 22 dogs were treated with RT, 20 were treated with RT+TMZ. Among RT arm, 7 dogs were treated with 42 Gy/10 fx, 15 dogs with 35 Gy/5 fx. Among the RT+TMZ arm, 10 dogs were treated with 42 Gy/10 fx, 2 dogs with 38 Gy/5 fx, 4 dogs with 35 Gy/5 fx, 4 dogs with 33 Gy/5 fx. 1 grade II radiotoxicity was observed. Median overall survival in palliated, RT and RT+TMZ arm was 94 days, 383 days, 420 days. 11/42 dogs have died from the brain tumor, 8/11 due to recurrent disease within the irradiation field, 3 due to spinal seeding. No statistically significant difference in survival was evident for the different grade of tumors (GII-GIII p=0.10; GIII-GIV p=0.68), presence of oedema, mass effect and between RT and RT+TMZ (p=0.61), whereas RT and RT+TMX were different from palliated (p=0.00089; p=0.000061). The statistically significant prognostic factors were the relative volume of the tumor <5% (p=0.013) and the clinical presentation with no alteration mental status (p=0.032).

VMAT RT is feasible and effective for canine brain gliomas. Small tumors and normal mental status are positive prognostic factors. The combination with TMZ at the used dose don't elicited any additional improvement in survival rate. Dose escalation in TMZ combination could be evaluated in further research.

OLIGODENDROCYTE PROGENITOR CELLS: TARGETS TO IMPROVE CNS FUNCTION AND REPAIR IN AGEING AND DISEASE

Abbe Crawford[1], Richa Tripathi[2], William Richardson[2], Robin Franklin[1]

[1]University of Cambridge, Cambridge, UK, [2]University College London, London, UK, [3]Royal Veterinary College, London, UK

Pathological destruction of the oligodendrocyte and its protective myelin sheath leaves neuronal axons exposed and susceptible to degeneration, ultimately resulting in progressive neurological decline. A population of oligodendrocyte progenitor cells (OPCs) exists within the central nervous system. In response to demyelination OPCs become activated and migrate into the injured area where they are capable of generating new myelin sheaths. Thus OPCs can restore the cytoarchitecture of the damaged tissue and so orchestrate the highly efficient regenerative process of remyelination. OPCs are additionally able to differentiate into Schwann cells, astrocytes and potentially neurons, and can generate action potentials; a property previously thought to be the sole preserve of neurons. However, they are a highly heterogeneous population of cells, which show failure in ageing and disease. Finding means to maintain OPC proliferation and new myelin production in ageing and disease could promote the long-term maintenance of white matter integrity, improve motor skills learning, provide treatment tools for neurodegenerative diseases and slow down age-related cognitive decline.

In this project, we are working to understand the diversity of the OPC, particularly focusing on the relevance of the different OPC populations that emerge in the developing nervous system. Use of a transgenic mouse line allows the different developmental populations to be traced due to their differential expression of fluorescent proteins, and their contribution to repair has been studied.

We have identified one population of OPCs that play a major role in repair. These cells shown enhanced migration abilities, as well as increased efficiency of differentiation into mature, myelin-forming oligodendrocytes. By studying their detailed biology we hope to identify new targets for their therapeutic manipulation to ultimately overcome the age- and disease-associated failure and so enhance regeneration and remyelination within the central nervous system.

FELINE ANTI-NERVE GROWTH FACTOR ANTIBODY IMPROVES MOBILITY IN CATS WITH DEGENERATIVE JOINT DISEASE-ASSOCIATED PAIN

Margaret Gruen[1], Andrea Thomson[1], Emily Griffith[1], Hayley Paradise[1], David Gearing[2], Duncan Lascelles[1]

[1]*North Carolina State University, Raleigh, NC, USA,* [2]*Nexvet Biopharma, Melbourne, Australia*

There is a critical need for proven drugs for treatment of degenerative joint disease (DJD) pain in cats. In the United States, there are no approved drugs for the treatment of musculoskeletal pain, and only a single NSAID is approved in other parts of the world. Additionally, there are concerns with the use of NSAIDs in cats because chronic kidney disease is so common. Neutralizing antibodies against Nerve Growth Factor (NGF) are analgesic in rodent models and in humans with chronic pain, although none are currently approved for use in humans. Using a proprietary technique for inter-species conversion of antibodies based on expressed cDNA sequence analysis (PETization™) Nexvet Biopharma developed a novel therapeutic fully-felinized anti-NGF mAb (named NV-02) for the alleviation of pain in cats. This pilot study evaluated the efficacy of a fully felinized anti-NGF antibody (NV-02) for the treatment of DJD pain and mobility impairment in cats.

In a placebo-controlled, pilot, masked clinical study, 34 cats with DJD-associated pain and mobility impairment were randomized to a single treatment with NV-02 (400 mcg/kg SC [n=11] or 800mcg/kg SC [n=12]) or placebo (saline, SC [n=11]). Cats wore an accelerometer for the duration of the study from screening onwards. In addition to objective accelerometry, owners completed clinical metrology instruments (CMIs) (Client-Specific Outcome Measures [CSOM] and Feline Musculoskeletal Pain Index [FMPI]) on days 0 (screening), 14 (baseline/treatment), 35, 56 and 77. Data were evaluated within and between groups using appropriate non-parametric methods. Activity data were summarized as mean counts per minute over each week of the study, and the percentage change from baseline (first 2 weeks) calculated for each cat and used in the analysis. A repeated measures least squares mean model was used to evaluate the objective activity data, with contrasts used to determine effects at each week. At the end of the study owners were asked if they thought their cat had received activity treatment or placebo and the data were analyzed using Chi square analysis.

All groups showed significantly improved owner assessments at all time points across both the clinical metrology instruments ($P<0.01$). CSOM scores ($P=0.035$) and pain over the last 3 weeks ($P=0.024$) were significantly improved in the combined treatment group compared to the placebo group 3 weeks following treatment; similar for the FMPI data ($P=0.061$). Objectively measured activity significantly increased in the NV-02 group compared to the placebo group between 2 and 6 weeks following the single treatment with NV-02 ($P=0.017$ overall), with approximately a 10% increase in measured activity over baseline. Activity increased following treatment and was significantly greater than in the placebo group at 2 ($P=0.035$), 3 ($P=0.007$), 4 ($P=0.006$), 5 ($P=0.007$) and 6 ($P=0.017$) weeks following treatment. In the placebo group, 45% of owners correctly identified the treatment administered during study compared to 83% of the owners of cats in the treatment group ($P=0.013$).

These pilot data demonstrate a positive analgesic effect of significant duration of this fully felinized anti-NGF antibody in cats suffering from chronic pain. The magnitude of the effect was greater than that seen with an NSAID.

THE PHARMACOKINETICS AND SAFETY OF GLIAL GROWTH FACTOR 2 IN DOGS

Natasha Olby, Ji-Hey Lim, Kimberly Williams

[1]*North Carolina State University College of Veterinary Medicine, Raleigh, NC, USA,* [2]*Comparative Medicine Institute, Raleigh, NC, USA*

Glial growth factor 2 (GGF2) is a neuregulin vital to nervous system development. It improves outcome in rodent spinal cord injury (SCI) with a long therapeutic window. This study aimed to examine the pharmacokinetics (PK) and safety of GGF2 administered to dogs using doses extrapolated from rodent studies.

Human recombinant GGF2 was administered to 6 dogs subcutaneously (SC) once daily for 7 days at a dose of 0.24mg/kg. Serial blood samples were taken for PK studies on days 1 and 7 (following intravenous administration on these 2 days). Serum chemistry and blood cell counts were performed at baseline and 3, 7 and 21 days post dosing. After a 4-week wash out, the protocol was repeated at a reduced dose of 0.08mg/kg (due to injection site pain). The 0.08mg/kg dose was administered to 10 additional normal dogs and 5 dogs with clinically complete SCI due to intervertebral disc herniation; 5 SCI dogs received vehicle only.

Dogs developed injection site pain at 0.24mg/kg but not 0.08mg/kg but signs of hypersensitivity occurred in the second week of dosing. Half-life was 13.8 +/- 4.2 minutes. Pain and hypersensitivity were not seen when the additional 10 normal and 5 SCI dogs were dosed at 0.08mg/kg SC once daily for 7 days.

We conclude that repeated rounds of dosing with recombinant GGF2 causes sensitization in dogs, but a single course of 7 days daily dosing at 0.08mg/kg SC is well tolerated in normal and SCI dogs. Efficacy using this protocol can be evaluated in further trials.

PHARMACOKINETICS AND PHARMACODYNAMICS OF IV DIAZEPAM AND IV TOPIRAMATE IN DOGS WITH AND WITHOUT EPILEPSY

Edward Patterson[1], Lisa Coles[2], Ilo Leppik[3], Gregory Worrell[4], Daniel Crepeau[4], Irene Vuu[2], Usha Mishra[2], Andrea Eckert[1], Patricia Maglalang[2], James Cloyd[2]

[1]University of Minnesota College of Veterinar Medicine, Saint Paul, MN, USA, [2]University of Minnesota College of Pharmacy, Minneapolis, MN, USA, [3]UMP MINCEP Epilepsy Care, Minneapolis, MN, USA, [4]mayo Clinic, Rochester, MN, USA

The development of new therapies for status epilepticus is hindered by the limitations of efficiently and successfully translating results from rodents. Our research group is studying dogs with epilepsy to evaluate the safety and efficacy of novel compounds and also to inform the design of canine and human trials. The objective of this study was to characterize the pharmacokinetics (PK) and pharmacodynamics (PD) of diazepam (DZP) and topiramate (TPM) in dogs with and without epilepsy.

Two with epilepsy and two dogs control dogs each received an investigational formulation of TPM (10 mg/mL in Captisol®) as a single 10 mg/kg and 20 mg/kg intravenous dose infused over 5 minutes with a minimum washout period of 14 days. Blood samples were collected at scheduled times and plasma TPM concentrations were measured using a validated HPLC-MS method.) from EEGs with 16 channels which were continuously recorded from which alpha and beta frequency amplitudes were determined and used as PD markers. As a positive control, intravenous DZP (5 mg/kg) was also studied. Compartmental PK and PD Emax models were evaluated to relate the drug concentration with changes in iEEG.

Both the TPM and DZP plasma concentrations were best fit by a two-compartment PK model. TPM clearance was ~4-fold greater and elimination half-life ~4-fold shorter in the dogs with epilepsy that are on maintenance phenobarbital therapy compared with dogs not on co-medications. Both DZP and TPM dosing resulted in significantly greater beta phase frequency amplitude compared with the hour prior to dosing while the alpha phase frequency was reduced. For DZP, the PK/PD data were best fit by PK-linked, sigmoidal Emax PD model with rapid (within one minute) onset of action.

From this work, we have shown that both BZD and TPM produce significant changes in the EEG and oscillatory activity in the brain. This effect is similar to what has been shown with benzodiazepines in humans with the largest effect at high frequency beta (18 - 30 Hz) range. This suggests that TPM may play a role in acute management of seizures and characterizing the PK and PD in dogs with epilepsy can be used to further guide AED development.

CORRELATING HEAD AND NECK PAIN WITH INTRACRANIAL DISEASE

Nicholas De Pompa, Jill Narak

Auburn University, Auburn, AL, USA

The purpose of this prospective clinical study was to determine the association of head and/or neck pain with evidence of intracranial pathology identified via magnetic resonance imaging (MRI) and/or cerebrospinal fluid (CSF) analysis.

85 dogs and 2 cats received brain MRI and/or CSF analysis as part of a routine neurologic work-up. MRI sequences included, at minimum, T2-weighted sagittal, T2-weighted transverse, and T1-weighted pre- and post-intravenous contrast with gadolinium. CSF collected from the cisterna magna or lumbar cistern was analyzed for color, opacity, total nucleated cell count, red blood cell count, protein, and cytologic evaluation. Each patient was also assigned an intracranial pain score at the time of admission. The Intracranial Pain Score (IPS) was based on history, posture, and observation of behavioral signs of pain on cervical range of motion and direct palpation of the head and neck.

70 dogs and 2 cats were included in the final analysis. The IPS had a sensitivity of 75.47% (95% CI = 61.7-86.23%) and specificity of 89.47% (95% CI = 66.82-98.39%) for detecting the presence of intracranial disease, with a positive predictive value of 95.24% and a negative predictive value of 56.67%. The mean IPS for structural brain disease, including vascular disease (2.5, $p < 0.001$), neoplasia (2.24, $p < 0.001$), and encephalitis (2.6, $p < 0.001$) was significantly increased compared to the mean IPS (0.15) of patients with normal brain MRI and/or CSF.

Clinical evidence of head and/or neck pain, quantified by the IPS, is correlated with the presence of intracranial pathology.

3D PRINTING APPLICATIONS FOR THE SPINE

Fred Wininger

University of Missouri-College of Veterinary Medicine, Saint Louis, MO, MO USA

3D printing or additive manufacturing (AM) is gaining popularity for the synthesis of anatomic modeling and custom prosthetics. AM may be of particular value in veterinary medicine because of the anatomic variability in size and morphometry of domestic animals. In the spine addition applications include implant guidance and stabilizing/fusion apparatuses.

Spinal structures are converted into digital surface 3D models from high-resolution tomographic data sets that acquire small isotropic voxels. CT is often preferable to MR imaging because of the superior bone contrast resolution. Stereolithography (STL) files can be generated from surface models available in common imaging platforms. The files are then corrected for errors in surface mesh and imported into slicer programs for the AM process. Once the file is generated, it can be digitally sent and printed virtually anywhere.

Because of their availability and cost, extrusion techniques such as fused deposition modeling (FDM) are the most common encountered in non-commercial use. The technique is valuable for creating anatomic models. FDM's lower resolution and thermoplastics materials may limit its utility in implantable prosthetics. FDM anatomic models have been reported to aid in surgical planning and pre-operative molding of metallic implants.

Light polymerized techniques such as stereolithography solidify liquid resin into complex polymers with light of specific wavelengths. These techniques have higher resolution, are recently cost effective and have the potential for biocompatibility, as the polymers are methacrylate based. However, their biomechanical properties are unclear, they have more complex post-processing and are prone to printing error. In addition to anatomic modeling they are useful in custom jig manufacturing for implant guidance as well as casting molds for metallic fabrication.

Binding techniques such as selective laser sintering are attractive because of their high resolution and biocompatible/biomechanical materials such as polyether ether ketone (PEEK) and titanium. Though too expensive for on-site printing, private companies will custom print pieces from validated STL files. These techniques may have the greatest potential for custom implantable prosthetics capable of vertebral fusion

The scope of AM is only limited to the modeling and materials available. The customizable nature makes the applications numerous but the biomechanical validation difficult. As the field grows, the safety and efficacy of AM warrants further study.

CEREBRAL MICROBLEEDS IN DOGS: A RETROSPECTIVE STUDY OF DEMOGRAPHICS, CLINICAL ASSOCIATIONS AND PATIENT OUTCOME

Sharon Kerwin, C. Elizabeth Boudreau, Christine Budke, John Griffin 4th, Jonathan Levine

Texas A&M College of Medicine and Biomedical Sciences, College Station, TX, USA

Cerebral microbleeds (CMBs) are small, well-defined, intraparenchymal susceptibility artifacts seen via magnetic resonance imaging (MRI). They are thought to represent previous hemorrhage. CMBs are recognized in canine brain MRIs, but their clinical importance is unknown.

Our study reports the prevalence of CMBs in dogs, and examines associations between CMBs and demographic and clinical variables, including survival time.

We performed a case-control retrospective study of dogs having a brain MRI at our institution between 2011 and 2016. The occurrence of CMBs as a function of age, weight, breed and gender was assessed within this sample population. An age- and breed-matched control subset was then generated to examine the relationship between CMBs and clinical variables, including survival time.

In our initial sample population of 579 dogs, 53 (9.2%) had CMBs. There was a strong relationship between age and CMBs, with an ~66% increase in the risk of CMBs for every year of life. Smaller dogs were more likely to have CMBs, with an ~3% decrease in risk with every kilogram of body weight. There was no effect of gender.

Within our age- and breed-matched control group, we used univariate analysis to identify candidate variables for a multivariate regression model. In our final model, CMBs were significantly associated with signs of vestibular dysfunction and the presence of brain atrophy on MRI. There was no significant effect of concurrent metabolic disease, though renal disease approached significance (p=0.08). Despite this, survival time was shorter for dogs with CMBs than for their matched counterparts.

Fundamentals of Neurologic Examination Neurologic Examination Made Easy!

Ronaldo C. da Costa, DMV, MSc, PhD, DACVIM (Neurology)
Columbus, OH

The neurologic examination is the most important tool in the diagnostic evaluation of patients with neurologic disease. A good neurologic examination allows lesion localization, which is key for a successful neurologic work up. The main components of the neurologic exam are evaluation of mental status and behavior, gait and posture, postural reactions, cranial nerves, spinal reflexes, palpation, and pain perception.

Before handling the patient, let the patient explore the examination room, if ambulatory, and observe the patient's reaction to the surroundings. Mental status should be evaluated both in terms of level and content of consciousness. Abnormal behavior is identified by comparing the patient's behavior to expected behavior for animals of a similar breed and age.

Gait and posture evaluation

When evaluating the gait, we need to ask 3 questions:

- Is the gait normal or abnormal?
- If it is abnormal, which limbs are affected? All 4 limbs, pelvic limbs only, or just one limb?
- What is wrong with the gait?

There are 3 main types of gait abnormalities, ataxia, paresis or paralysis and lameness. Lameness is typically caused by musculoskeletal diseases, though it can be seen with neuropathic (nerves and nerve roots) diseases. Ataxia, paresis, and paralysis are clinical signs commonly seen in various diseases of the brain, spinal cord, and peripheral nerves. The main difference between ataxia and paresis is that ataxia affects coordination without affecting strength, while paresis affects only strength. These clinical signs provide fundamental information for lesion localization. There are three types of ataxia, namely *proprioceptive, cerebellar,* and *vestibular.*

Vestibular ataxia is the easiest to recognize. Vestibular ataxia is characterized predominantly by a head tilt. Usually the side of the head tilt indicates the side of the lesion.

Cerebellar ataxia is characterized by dysmetria (inability to control the rate and range of stepping movements), which is usually manifested by hypermetria (exaggerated step). It is normally easier to recognize a hypermetric gait in the thoracic limbs.

Proprioceptive ataxia is the type primarily related to spinal cord diseases. This ataxia can be differentiated from vestibular and cerebellar ataxias by the absence of head involvement. Proprioceptive ataxia is a phenomenon of the spinal cord's white matter, reflecting a dysfunction of the sensory tracts carrying unconscious proprioception (dorsal, ventral, and cranial spinocerebellar tracts, as well as the cuneocerebellar tract). Clinical signs seen with proprioceptive ataxia are truncal sway (wobbliness) and abnormal limb stance and gait such as circumduction, abduction, or adduction with the limbs crossing with each other as the animal walks. Some animals also display a delay in initiating the protraction phase of the gait with a slight hyperflexion of the limb and a longer stride than normal.

Paresis means partial loss of motor function, which is usually manifested as weakness. Paralysis (plegia) refers to the complete loss of motor function. The terms *plegia* or *paresis* can be used in association with a prefix to specify which limb(s) are involved. Tetra-, para-, hemi-, or mono-paresis/-plegia refers to involvement of all four limbs, pelvic limbs, ipsilateral limbs, or a single limb, respectively. It is important to make the distinction between ambulatory and non-ambulatory paresis and plegia. A "down" dog can have non-ambulatory paraparesis and if appropriate treatment is established, dogs with paresis will typically recover faster than those with paralysis (plegia). Paresis and proprioceptive ataxia are common signs in patients with spinal cord diseases. The more severe the spinal cord involvement, the weaker the patient becomes, until the point of paralysis.

Cranial nerves

There are 12 pairs of cranial nerves, but not all of them are tested. We use reflexes and reactions to assess the most important functions of cranial nerves. Only the tests routinely performed are listed.

The first test is the menace response. It evaluates the ipsilateral optic and facial cranial nerves, as well as the ipsilateral cerebellum and the contralateral forebrain (thalamocortex or cerebrum). The pupillary light reflex tests the sensory function of the optic nerve and the parasympathetic function of the oculomotor nerve. Both eyes need to be tested and the direct and consensual reflexes are evaluated. The vestibulo-ocular or oculocephalic reflex evaluates the sensory function of the vestibulocochlear nerve, and the motor function of cranial nerves III, IV, and VI (oculomotor, trochlear and abducent, respectively). The eyes are also evaluated for nystagmus or strabismus. The trigeminal nerve is tested with the palpebral reflex (sensory function of the trigeminal nerve and motor function of facial nerve), palpation of the masticatory muscles (motor function of the trigeminal nerve), and nasal sensation, that tests the sensory function of the trigeminal nerve, as well as the thalamocortical response to the noxious stimulus. Cranial nerve VII (facial nerve) is evaluated by assessing the muscles of facial expression, looking for evidence of asymmetry ("droopy" facial expression). The facial nerve is also tested in the menace response and palpebral reflex. The vestibulocochlear nerve is assessed by looking at the posture of the head. A head tilt is a consistent sign of vestibular disease. Other signs of vestibular disease are nystagmus and positional strabismus. Cranial nerves glossopharyngeal and vagus are evaluated indirectly, investigating if the patient has signs of dysphonia, dysphagia or stertorous breathing. Unilateral atrophy of the tongue is a sign of hypoglossal nerve dysfunction.

Postural reactions

Postural reactions test the same neurologic pathways involved in gait, namely the proprioceptive and motor systems. Their main value is to detect subtle deficits or inconspicuous asymmetry that may not be obvious during the observation of gait. They are very valuable to confirm neurologic disease but by themselves cannot localize a lesion since patients can have postural reaction deficits with lesions affecting the forebrain, brainstem, any region of the spinal cord region, as well as the peripheral nerves. Thus, the one must use the other components of the neurologic exam to "solve the puzzle" and localize the lesion. Postural reactions are also useful in discriminating between orthopedic and neurologic disorders, though cannot replace the gait exam, which is more reliable for this distinction because some cases with chronic proprioceptive ataxia may have normal postural reactions. Frequently it is only necessary to perform two postural reaction tests, proprioceptive positioning and hopping. It is very important to properly support the patient's weight when performing it to detect proprioceptive deficits.

Spinal reflexes

Spinal reflexes assess the integrity of the sensory and motor components of the reflex arc and the influence of descending upper motor neuron pathways. Several reflexes have been reported but many are unreliable. The following spinal reflexes are consistently observed in most dogs and cats and should always be tested: muscle tone, patellar, flexor, perineal and cutaneous trunci reflexes. Lesion localization can be effectively achieved performing only these reflexes.

It is important to only perform tests that can perceived as uncomfortable to the patient last, as it is important to have their cooperation to properly perform testing of spinal reflexes. It is recommended to start with assessment of muscle tone, proceed with myotatic reflexes next (patellar reflex), and leave the flexor (withdrawal) reflexes last. It is also recommended to begin testing the pelvic limbs before testing the thoracic limbs. The perineal reflex and cutaneous trunci reflex are the last reflexes tested.

Palpation and sensory evaluation

Light palpation helps detect swelling or atrophy. Light palpation is also useful to evaluate the vertebral column for areas of luxation or crepitus. Deep palpation and manipulation detect painful regions. If crying, whimpering, or muscle tensing occur on palpation, more vigorous maneuvers, such as manipulation, are unnecessary and may be dangerous in patients with unstable fractures or luxations.

It is important, especially in nonambulatory patients, to determine whether or not noxious stimuli applied to the limbs are traversing the damaged segment of spinal cord to reach the brain for conscious perception. The presence or absence of deep pain perception (nociception) is important in assessing prognosis for recovery. It is critical not to confuse reflex withdrawal with conscious perception. As a rule of thumb, perform this test watching the head of the patient for reaction. As long as the patient does not have a LMN injury, a flexor reflex will always be present. The key point is to see a conscious behavioral response involving head movement or vocalization.

The Key to a Successful Neurologic Work – Lesion Localization!

Ronaldo C. da Costa, DMV, MSc, PhD, DACVIM (Neurology)
Columbus, OH

The main reason why neurology is considered intimidating is because it requires precise lesion localization for a successful work-up. A "down" or paralyzed patient can have a lesion in the spinal cord or somewhere along the peripheral nervous system. An MRI examination of the spine would be useless if the lesion is in the peripheral nervous system instead of the spinal cord!

When localizing lesions it is helpful to start with "big regions" first and then progressively narrow it down. This process can start with defining whether the lesion is in the central or peripheral nervous system (CNS or PNS). Then if it is within the CNS, localize it to the brain or spinal cord. Considering the high prevalence of spinal cord disorders, the text will discuss primarily lesion localization within the spinal cord.

It is important to localize the spinal lesion as closely as possible to concentrate diagnostic efforts on the affected region. The principles of upper motor neuron (UMN) and lower motor neuron (LMN) are used to localize lesions within the spinal cord. *Upper motor neuron* is a term used to designate a group of motor tracts originating from the brain and terminating within the spinal cord. The most important tracts forming the UMN are the corticospinal, rubrospinal, reticulospinal, and vestibulospinal tracts. Basically, the combined function of the UMN is to facilitate gait in animals, inhibiting extensor muscles of the limbs while facilitating flexor muscles. As such, when the animal has an UMN spinal cord lesion, there is paresis or paralysis with increased extensor tone of the limbs (due to the lack of UMN inhibition). The hallmark of UMN signs are then paralysis or paresis with increased extensor tone (spasticity or hypertonus), normal to increased spinal reflexes (hyperreflexia), and slowly progressive muscle atrophy from disuse. On the other hand, LMN is formed by a group of neurons that originate in the ventral grey horn of the spinal cord or in a nuclei of the brainstem, to give origin to peripheral or cranial nerves that then innervate the target muscle(s).

Lower motor neuron is also known as the final common pathway because any motor activity to be displayed has to go through the LMN. When the patient has a lesion somewhere along the LMN pathway, such as in the spinal cord enlargement (ventral grey horn), nerve roots, spinal nerve, peripheral nerve, or neuromuscular junction, LMN dysfunction occurs. Clinical signs of LMN dysfunction are basically the opposite as those seen with UMN lesions: paresis or paralysis with absent to decreased extensor tone (flaccidity or hypotonus), decreased or absent spinal reflexes (hyporeflexia or areflexia, respectively), and rapid and severe (neurogenic) muscle atrophy. The presence of UMN or LMN signs dictates the location of the lesion within the spinal cord.

Clinically, the spinal cord can be divided into four regions, cervical (C1-5), cervicothoracic (C6-T2), thoracolumbar (T3-L3), and lumbosacral (L4-S3). Involvement of all four limbs suggests a lesion at either C1-C5 or C6-T2. Assuming increased tone and reflexes in the pelvic limbs, normal to increased tone and reflexes (UMN signs) in the thoracic limbs indicate a lesion at C1-5, while decreased to absent tone and reflexes (LMN signs) suggest a lesion at C6-T2.

Paraplegia or paraparesis indicates a lesion caudal to T2. Normal to increased reflexes and muscle tone points to a T3-L3 lesion, the most common site for spinal lesions. More specific lesion localization within the segment T3-L3 can be achieved using the cutaneous trunci reflex and spinal palpation. Decreased tone and reflexes suggests a L4-S3 lesion. It is important to note that the lumbosacral spinal cord segments do not match with their corresponding vertebrae. The entire lumbar enlargement (L4-S3) is located within the vertebrae L4-L5 in most dogs, and L5-L6 in cats. Lesions involving the vertebrae L6, L7, and S1 in dogs affect the nerve roots for the pelvic limbs, perineal region, sphincters and tail (reflecting involvement of sciatic, pudendal, pelvic and caudal nerves). The femoral nerve (which innervates the quadriceps muscle, the most important extensor muscle of the pelvic limb) is spared because it emerges before the lumbosacral region, therefore dogs with lumbosacral disease are not paralyzed. Also, as there is no spinal cord in this region, no proprioceptive ataxia is observed (although proprioceptive positioning deficits may be seen). Common gait signs are lameness and mild weakness (paraparesis) associated with spinal pain. The flexor and perineal reflexes may display signs of LMN (depending on the involvement of these nerves).

Differential Diagnosis of Spinal Diseases

Ronaldo C. da Costa, DMV, MSc, PhD, DACVIM (Neurology)
Columbus, OH

When approaching a case with spinal problems, the simplest approach, once the lesion is localized, is to consider differential diagnoses using the acronym lists based on pathophysiologic mechanisms. These acronyms are called either VITAMIN-D or DAMNIT-V and are very useful and practical ways to approach neurological diseases.

When using this acronym, it is useful to consider the signalment and history to develop appropriate differential diagnoses for the patient. For example, even though intervertebral disc disease is the most common spinal disease of dogs, it is not a reasonable differential diagnosis for a 6-month old dog with chronic paraparesis. Some generalities should be considered when using the VITAMIN-D acronym. Young dogs are more likely to have congenital or inflammatory conditions. Acute presentations are usually caused by vascular or traumatic conditions. Chronic presentations are usually seen with degenerative or neoplastic processes. Another way to approach patients with spinal diseases is to develop a list of diseases that are known to affect specific spinal regions. This is useful because although many diseases affect several spinal regions (e.g. intervertebral disc disease, discospondylitis, fibrocartilaginous embolic myelopathy), many are region specific (e.g. atlantoaxial instability, cervical spondylomyelopathy, degenerative myelopathy in the earlier phases).

Once the list of differential diagnoses is prepared for the patient, the most probably causes should be ruled-in or ruled-out based on appropriate diagnostic tests. It is recommended to always perform survey radiographs (always obtaining orthogonal views). If nothing is found on radiographs, then magnetic resonance imaging, computed tomography, myelography or cerebrospinal fluid should be used to rule-in or out the differential diagnoses.

Differentials according to specific regions

C1 - C5 spinal cord segments / C1 – mid C5 vertebrae

Common diseases affecting the C1-5 spinal cord segments in small breeds are atlantoaxial subluxation and cervical intervertebral disc disease. Spinal pain is often present with these diseases. Primary differentials for large breed dogs with C1 to C5 lesions are intervertebral disc disease (IVDD), cervical spondylomyelopathy, and spinal neoplasia, mainly meningiomas. Trauma is also common and affects both small and large breeds. Cervical pain without neurological deficits affecting the C1 to C5 regions is usually caused by steroid responsive meningitis-arteritis, cervical intervertebral disc disease, or discospondylitis.

C6-T2 spinal cord segments (cervical enlargement) / vertebrae C5 to T1.

Frequent conditions seen at this region are cervical intervertebral disc disease in large and small breed dogs, and cervical spondylomyelopathy in large and giant breed dogs. Neoplasia, discospondylitis, osteomyelitis, trauma, and fibrocartilaginous embolic myelopathy can also occur in this region.

T3-L3 spinal cord segments / T2 – L3 vertebrae

Most spinal diseases in dogs and cats affect the T3-L3 spinal segments. Intervertebral disc disease (either extrusion or protrusion) is very common in this location. Other common diseases are degenerative myelopathy, spinal trauma, neoplasia, and fibrocartilaginous embolic myelopathy. If the lesion is localized to the mid to cranial thoracic region, between T2 to T10 vertebrae (based on the cut-off of the cutaneous trunci reflex and/or spinal pain), then a few diseases can be considered more likely. It is important to know that IVDD is rare at this region. The diseases that are more commonly seen between T2-T10 are spinal neoplasia, discospondylitis, and hemivertebra.

L4-S3 spinal cord segments (lumbosacral enlargement) / L4 - L5 vertebrae (dogs)

This is a small spinal cord region, and most diseases affecting the T3-L3 spinal cord region can also affect this region (e.g. IVDD, trauma, neoplasia). A disease that frequently affects this specific region is fibrocartilaginous embolic myelopathy.

L6-L7 vertebrae and sacrum in dogs

Problems affecting the caudal lumbar region are very common in large breed dogs. Spinal diseases affecting this region can appear similar to musculoskeletal disorders, as lameness may be the only clinical sign. It is very important to know that degenerative lumbosacral disease (cauda equina syndrome) in dogs will NOT cause paralysis (paraplegia) as the femoral nerve emerges before the L7-S1 vertebral region. The primary differential diagnoses for diseases affecting this region are degenerative lumbosacral stenosis, discospondylitis, neoplasia and extradural synovial cysts. Spinal pain is often a consistent feature of diseases affecting the caudal lumbar/lumbosacral spine.

A step-wise process facilitates the diagnostic approach to patients with spinal disorders. The VITAMIN-D system can be used to select the primary pathophysiologic mechanisms causing the patient's clinical signs. Using the acronym VITAMIN-D, the clinician may be able to go from a large list of diagnostic possibilities, to a shorter list of diagnostic probabilities.

Top 4 Chronic Spinal Diseases Main Differentials for Chronic Progressive Paresis

Ronaldo C. da Costa, DMV, MSc, PhD, DACVIM (Neurology)
Columbus, OH

Older dogs frequently have orthopedic diseases that affect their mobility. When faced with these cases the clinician must often decide whether the patient has neurologic disease in addition to musculoskeletal conditions. Clinical signs consistent with neurologic disease are proprioceptive ataxia, weakness (paresis), spinal pain, and delayed postural reactions (proprioceptive deficits). The main neurologic differentials causing pelvic limb paresis in older dogs are degenerative lumbosacral stenosis, intervertebral disc protrusion, degenerative myelopathy and spinal neoplasia. The main features of these diseases are summarized below.

DEGENERATIVE LUMBOSACRAL STENOSIS

This is a common disease syndrome that typically affects adult (usually middle-aged to older), large-breed dogs. Clinical signs are variable, but lumbosacral pain is an early and consistent finding. Pain may be manifested in a number of ways, such as reluctance to rise or sit, and unilateral or bilateral pelvic limb lameness. Pain and lameness may be acute or chronic and may be persistent or episodic. These clinical signs may be misinterpreted as being due to orthopedic disease, most notably hip dysplasia. If untreated, clinical signs of dysfunction may progress to proprioceptive loss in the pelvic limbs, voluntary motor weakness (in the distribution area of the sciatic nerve), and urinary/fecal incontinence, usually in that order. Importantly, paraplegia (complete loss of motor function in the pelvic limbs) is not seen with lumbosacral disease in dogs because the femoral nerve (the main extensor nerve of the pelvic limbs) emerges before the lumbosacral region.

Diagnosis of degenerative lumbosacral stenosis is based upon signalment, historical and clinical findings, and results of diagnostic imaging of the lumbosacral region. The definitive diagnosis of degenerative lumbosacral stenosis is made by imaging of the lumbosacral area and demonstrating compression of the cauda equina. One must be careful with the interpretation of imaging findings because lumbosacral compression is also frequently seen in asymptomatic dogs, thus clinical and imaging findings must be correlated for a diagnosis of lumbosacral stenosis.

Treatment of the patient with degenerative lumbosacral stenosis may be conservative or surgical. Treatment decisions are based primarily on severity of clinical signs, age of the patient and concurrent diseases; as mentioned previously, there appears to be no clear correlation between extent of cauda equina compression evident on imaging and disease severity or postoperative outcome. Conservative therapy consists of enforced rest initially for a few weeks. Additionally, anti-inflammatory medication (either nonsteroidal drugs or prednisone, not both), analgesics (such as gabapentin), and body weight reduction are recommended. In patients with neurologic deficits, or patients for whom pain is refractory to conservative management, surgery is chosen as the preferred mode of therapy. Most dogs are treated with decompressive surgery (dorsal laminectomy), though other techniques must be considered based on imaging findings (distraction, stabilization or lateral foraminotomy).

INTERVERTEBRAL DISC PROTRUSION

There are two basic types of disk degeneration, referred to as chondroid and fibroid degeneration. These two types of degeneration typically cause two distinct types of disk disease. The one that causes progressive, chronic signs is the fibroid degeneration that involves a progressive thickening of the dorsal annulus fibrosus, which protrudes dorsally into the vertebral canal. This type of disk disease is called Hansen Type II, (Type II) disk protrusion. Hansen Type II protrusions typically occur in dogs five years of age and older. It typically causes progressive signs of paraparesis, often with some degree of thoracolumbar spinal pain detected on palpation.

Diagnosis of disk disease is based upon signalment, history, clinical signs, and results of diagnostic tests, primarily imaging of the vertebral column. Because these dogs can have multiple spinal abnormalities, MRI is the preferred imaging modality.

Type II disk disease is typically managed medically with restricted activity (but not cage confinement) and anti-inflammatory drugs. Most patients seem to improve and stabilize. When surgical intervention is indicated it requires decompression of the spinal cord and removal of disk material from the vertebral canal. This can be accomplished via hemilaminectomy, pediculectomy or corpectomy.

DEGENERATIVE MYELOPATHY

This is a degenerative disease that affects the thoracolumbar spinal cord of medium- to large-breed dogs over 5 years of age. Among the large breed dogs, the German shepherd dog and Boxers are the most commonly affected breeds. Several other breeds were reported to have histologically confirmed degenerative myelopathy (DM). A mutation in the superoxide dismutase 1 (*SOD1*) gene has been shown in some affected dogs and it is a risk factor for the development of clinical disease.

The clinical picture typically consists of a slowly progressive, nonpainful T3—L3 myelopathy in a middle-aged to older large-breed dog. Age of onset of neurologic signs is usually 5 years or older with a mean age of 9 years in large dog breeds with DM. Most dogs are at least 8 years of age at onset of clinical signs.

A tentative diagnosis is based upon signalment, clinical features, and exclusion of other spinal cord disorders. Spinal imaging (MRI, CT, myelogram) is typically normal, but some dogs may have concurrent mild Type II disk lesions that are probably clinically insignificant. A DNA test based on the *SOD1* mutation is commercially available. The dogs homozygous for the mutation are *at-risk* for developing DM and will contribute one chromosome with the mutant allele to all of their offspring. The heterozygotes are DM carriers that are unlikely to develop clinical DM but could pass on a chromosome with the mutant allele to half of their offspring.

Treatment options are limited and no medical therapy has been shown to alter the survival of these dogs. Physical therapy appears to be the only modality that alters their survival. In one retrospective study of dogs with suspected degenerative myelopathy, intensive daily controlled physiotherapy, was shown to significantly improve mean survival time (255 days), compared with dogs receiving moderate physiotherapy (130 days) or no physiotherapy (55 days).

SPINAL NEOPLASIA

There are a large number of tumors that can affect the spinal cord of dogs. Tumors can be conceptually divided into primary and secondary tumors. Primary tumors include those neoplasms that arise from spinal cord parenchyma (e.g., neurons, glial cells), or associated meningeal/ependymal tissue. Secondary tumors include primary or metastatic vertebral neoplasms, malignant nerve sheath tumors, and metastases to the extradural space or the cord parenchyma (intramedullary metastases). Primary tumors are more common than metastatic tumors. It is often clinically useful to classify spinal cord neoplasms based upon the relationship between the tumor and the meninges. Spinal cord tumors are typically classified as extradural, intradural/ extramedullary, or intramedullary.

Large-breed dogs appear to be predisposed to developing spinal neoplasia, in comparison with smaller breeds. Spinal tumors classically cause progressive signs of a myelopathy, but acute or subacute development of spinal cord dysfunction often occurs, especially with intramedullary neoplasms.

A tentative diagnosis of spinal neoplasia is typically based upon signalment, history, clinical signs, and results of spinal imaging.

Therapy for dogs and cats with spinal tumors can be divided into supportive and definitive treatments. Supportive therapies are directed against secondary sequelae of the spinal tumor (e.g., cord edema, pain), whereas definitive therapies are aimed at elimination of neoplastic tissue. Supportive therapy consists of anti-inflammatory doses of glucocorticoids (e.g., prednisone, 0.5 mg/kg, PO, q 12 hr), which can be increased or decreased as needed, with or without additional pain-relieving drugs (e.g., narcotics). Definitive therapy consists primarily of surgery and megavoltage radiation therapy. Chemotherapy is indicated for lymphosarcoma and myeloma. The long-term prognosis for most cases of vertebral neoplasia in dogs is guarded to poor. The type of spinal tumor with the longest survival is myeloma. Early diagnosis and multimodal therapy may yield long survival times.

Clinical Trials on Canine Degenerative Myelopathy

Joan R. Coates, DVM, MS, DACVIM (Neurology)
Columbia, MO

INTRODUCTION

Canine degenerative myelopathy (DM) was first described by Averill in 1973 as an insidious, progressive, general proprioceptive (GP) ataxia and upper motor neuron (UMN) spastic paresis of the pelvic limbs ultimately leading to paraplegia and necessitating euthanasia.[1] After discovery of mutation in the superoxide dismutase 1 gene (*SOD1*), DM is being now recognized as a naturally-occurring progressive adult-onset neurodegenerative disease that has many similarities to human amyotrophic lateral sclerosis (ALS – Lou Gehrig's disease) and may serve as an important, novel disease model for therapy development. A spontaneous canine disease model offers a ready clinical population where therapies can be evaluated in a setting closely mimicking human clinical trials. This approach has proven successful in developing cancer chemotherapies in canine patients which have then been applied to humans.[2] We propose that canine DM will serve as a potentially useful intermediate sized model of ALS, which could be valuable for temporal studies of disease progression and evaluation of ALS and DM targeted new therapeutic/diagnostic regimes.

CANINE DEGENERATIVE MYELOPATHY AS A DISEASE MODEL OF AMYOTROPHIC LATERAL SCLEROSIS

Degenerative myelopathy affects older, adult dogs. Ages at onset range between 8 and 14 years. Mean age at onset is 9 years in large dogs.[1] In the Pembroke Welsh Corgis reported mean age of onset is 11 years; presumably due to breed size differences.[3] Although first described in German Shepherd dogs, DM is now being recognized as a common neurologic problem in many pure breeds and mixed breed dogs.[4] Disease onset and clinical progression are homogeneous across all breeds, representing degeneration of the sensory and upper and lower motor neurons. Clinical presentation has been categorized into four clinical disease stages.[5] Clinical signs emerge as an asymmetric spastic paraparesis and general proprioceptive ataxia (Stage I), progressing to non-ambulatory paraparesis/paraplegia with pelvic limb lower motor neuron signs within 1 year from disease onset (Stage II). Neurologic deficits progress to include thoracic limb weakness (Stage III) followed by flaccid tetraplegia, generalized muscle atrophy and brainstem dysfunction (Stage IV) (Coates 2007; Coates and Wininger 2010; Kanazono 2013; Shelton 2012; Ogawa 2014).[3,5-7] The pattern and rate of clinical progression of canine DM is relatively homogeneous in comparison to the phenotypic heterogeneity for onset of signs, pattern of spread and rate of disease progression in human ALS. We have documented the time of disease through the various diseases stages of DM-affected Boxers, Pembroke Welsh Corgis (corgis) and other breeds.[8] Preliminary data have shown that the median period for disease progression from Stage 1 (ambulatory, onset of signs) to 2 (nonambulatory) is 10 months (95% CI, 9 to 12 months). There was no significant difference in time for transition from Stage 1 to 2 amongst the different breeds. Thus, the homogeneous disease progression of DM will facilitate longitudinal measures and detection of therapeutic efficacy.

Canine DM is definitively diagnosed by histopathology of the spinal cord and nerves. Axonal degeneration of the upper and lower motor neurons is one of the pathologic hallmarks shared by ALS and DM. The pathologic features of DM include axonal degeneration with secondary demyelination and astroglial proliferation (sclerosis) in all spinal cord funiculi, but consistently most severe in the dorsal portion of the lateral funiculus and in the dorsal columns of the middle to lower thoracic region.[1,3,6,9,10] In addition, cytoplasmic aggregates that bind anti-SOD1 antibodies are usually present in intracellular neurons and astrocytes in spinal cords of DM-affected dogs.[4,6,12] Histopathologic studies of dogs in the late disease stage with LMN signs have documented denervation atrophy in muscle, nerve fiber loss with axonal degeneration and secondary myelin loss in myelinated fibers of peripheral nerves.[7]

Because owners of companion dogs with DM have their pets euthanized at different stages of disease progression, by examining tissues from these dogs collected at the time of euthanasia, it is possible to study the disease at all stages of progression and thereby gain a better understanding of the chronological relationships between the pathologies that develop in the spinal cord, peripheral nerves, and muscles. Studies we conducted on the thoracic intercostal muscles of dogs with DM and the associated thoracic motor and sensory neurons indicate that there are significant atrophic changes in these muscles at stages of the disease in which there is no apparent degeneration of the associated motor neurons.[12,13] Neuronal cell body degeneration or loss in the ventral horn of the spinal cord is not a prominent histopathologic finding until terminal disease.[14] Significant sensory neuron degeneration also preceded any evidence of motor neuron pathology.[13] Thus, the distribution of lesions and clinical disease progression in DM are similar to that reported for the UMN-onset ALS[15]; whereby, UMN signs in DM-affected dogs later progress to LMN signs.[5,6] The clinical spectrum of DM has now been broadened to involve both the UMN and LMN systems and is considered a multi-system disease involving both central and peripheral sensory and motor axons.[5]

Awano *et al.* identified a *c.118G>A* transition in Exon 2 of the *SOD1* gene that predicted an E40K missense mutation, which underlies most cases of canine DM.[6] Not all *SOD1:c.118A* homozygotes develop clinical signs. Initially DM appeared to be an autosomal recessive disease with incomplete penetrance; whereas most human *SOD1* mutations in human ALS are autosomal dominant. Thus, homozygosity for the E40K mutation in SOD1 is a major risk factor for canine DM. More recently, DM has also been histopathologically confirmed in few heterozygous dogs.[4] The occurrence of DM in a heterozygote seems plausible since most human *SOD1* mutations cause dominant ALS. Nonetheless, the age at onset for many dogs has exceeded the mean life expectancy of dogs indicating that DM has an age-related, incomplete penetrant mode of inheritance.

We have identified a second *SOD1* missense mutation (*c.52 A> T*) that was homozygous in an affected Bernese Mountain dog.[16] This mutant allele *c.52T* allele appears to be restricted to the Bernese Mountain Dog breed where it is less common than the *c.118A* allele.[4] We have identified a few *c.52T + c.118A* compound heterozygous Bernese Mountain dogs with DM.[4] This finding serves as a reminder that direct DNA tests indicate the presence or absence of disease-causing alleles but cannot be used to rule-out a diagnosis because other sequence variants in the same gene or in a different gene might produce a similar disease phenotype.

The native SOD1 protein catalyzes reduction of superoxide to less deleterious forms of hydrogen peroxide and water. Mutations in *SOD1* alter the conformation of the SOD1 protein and underlie most cases of canine DM and 1-2% of human ALS. Crisp et al., have shown that at the biochemical level, canine and human SOD1 are remarkably similar: both aggregate readily when mutated, both display decreased protein stability and both contain mutants that are functionally indistinct from wild-type protein.[17] Mutant canine SOD1 remains enzymatically active, which supports the hypothesis of a toxic gain-of-function mutation.[17] However, it remains to be elucidated whether or not this mechanism of toxicity is sufficient for the multisystem degeneration observed in DM.

Based on similarities to SOD1-associated human ALS, canine DM can offer some advantages as compared to existing SOD1 models; albeit not a perfect disease model of human ALS (**Table 1**). Rodent (mouse and rat) mutants have been established as useful ALS models in understanding disease mechanisms, as similar CNS pathology manifestations and observables are found. As a naturally-occurring disease in companion dogs, most importantly canine DM differs from many of the transgenic rodent models that require human SOD1 (hSOD1) overexpression to non-physiological levels. Canine DM serves to represent another disease model for therapeutic study on ALS because (i) a naturally-occurring *SOD1* mutation (E40K) is causative for canine DM, a common disease of older aged companion dogs, (ii) dogs have a relatively large spinal cord/brain, (iii) DM-affected dogs have a homogeneous pattern of disease progression, and (iv) histopathologic findings and functional deficits in DM share similarities to some forms of human ALS.

Table 1: Characteristics of Canine Degenerative Myelopathy and *SOD1*-associated Familial ALS

	Canine DM	**SOD1-associated Human ALS**
Clinical Onset	Homogeneous UMN onset – pelvic limbs; asymmetric	Heterogeneous; degrees of UMN and LMN signs Onset – *limb (most common)*, bulbar, UMN, frontotemporal degeneration; asymmetric
Disease Progression	Homogeneous spread and rate from pelvic to thoracic limbs and eventually bulbar; UMN and LMN involvement Proprioceptive involvement first	Variable rate with spread from site of onset ; UMN and LMN involvement No proprioceptive involvement most forms
Pathology	*Sensory neuron pathology predominates, terminal disease* motor neuron *soma loss;* axonal degeneration/loss in CNS (dorsal portion of lateral funiculus and dorsal funiculus) and PNS; NMJ degeneration does not precede motor neuron soma loss; astrogliosis; SOD1 intracellular neuronal and astrocytic aggregates; neurofilament accumulation in neurons	Motor neuron loss predominates, axonal degeneration/loss in CNS (dorsal columns) and PNS, astrogliosis; SOD1 intracellular neuronal and astrocytic aggregates; neurofilament accumulation in neurons
Inheritance	Most autosomal recessive	90% sporadic; 10% inherited – most autosomal dominant
SOD1 Genetics	Inherited with *variable* penetrance; 2 SOD1 mutations: p.E40K and p.T18S	>160 SOD1 mutations; 12% of familial ALS, 2% of sporadic ALS and 2% of all ALS
Biochemistry	Detergent-insoluble SOD1 species (enzyme active) in spinal cord	Detergent-insoluble SOD1 species in spinal cord

As we learn more about the clinical and neuropathological spectrums of DM, we are exploring use of clinical biomarkers: diagnostic, prognostic, predictive and pharmacodynamic. A diagnostic marker is used to categorize patients or stage disease. Prognostic markers are baseline characteristics that provide information about the natural disease history. A predictive marker characterizes patients according to their likelihood of treatment response. Pharmacodynamic measures indicate treatment effectiveness. When translating therapies for ALS from animals to humans or vice versa, sensitive and objective disease markers shared across species are necessary to diagnose and monitor disease progression so that therapeutic efficacy can be accurately evaluated.[18-20] Measures of UMN (i.e. brain and spinal cord imaging, positron emission tomography) and LMN (i.e. electrodiagnostic studies, motor unit number estimation, electrical impedance myography) loss and fluid markers in CSF to follow DM progression are under investigation.

THERAPEUTIC STRATEGIES

Target engagement, cerebrospinal fluid penetration, and delivery methods are areas that are undergoing transformation in treatment of neurodegenerative diseases. In ALS, multiple methods are currently being studied for their ability to decrease levels of unwanted proteins, including immunization strategies, small molecules, gene therapy, RNA interference, and antisense oligonucleotides (ASO).[21,22] Nucleic acid-based therapies offer the promise of modifying or arresting the course of neurodegenerative disease in which down-regulation of a protein is a treatment objective. RNA inference is triggered by short, double-stranded RNA, while ASOs are short, single stranded DNA-like molecules. Recognition of RNA as a regulator of gene expression came with the discovery of interfering RNAs.[22] RNA interference is a collection of small RNA directed mechanisms that result in sequence specific inhibition of gene expression. Both RNA interference and ASOs are posttranscriptional gene down-regulation approaches that destroy messenger RNA (mRNA) in a sequence specific fashion, leads to lower mRNA levels and results in decreased protein production. The use of ASOs and interfering RNA molecules to lower concentrations of mutant mRNA slowed disease progression and increased survival in the *SOD1* transgenic rodent models.[23-27] Nucleic acid-based therapies do not affect a protein that is already expressed; instead they prevent expression of the protein. Thus in ALS, such treatments would reduce synthesis rather than remove aggregated SOD1 protein.

The blood-brain barrier constitutes a challenge for delivering exogenous molecules to the CNS. Parenterally administered ASOs or interfering RNAs do not cross the blood-brain barrier; thus reduced SOD1 expression within neurons and surrounding non-neuronal cells requires direct administration into the CSF via intraventricular or intrathecal (IT) routes. To provide continuous down-regulation of mutant mRNA, these RNA interference compounds would need to be administered directly into the CNS through an intrathecal pump from a reservoir or as a bolus injection. At the University of Missouri, we are investigating IT administration of ASOs and an adeno-associated viral (AAV) gene therapy targeting SOD1.

Antisense Oligonucleotide Therapy

In ASO therapy, expression of a mutant gene is down-regulated by the targeted ASO, reducing the total amount of mutant protein generated. ASOs are 13-25 nucleotides long and are specifically designed to hybridize to corresponding RNA (complementary mRNA) by Watson-Crick binding. The specificity of the hybridization makes antisense treatment an attractive strategy to selectively modulate the expression of genes in pathogenesis of disease.[28,29] The target of the ASO is mRNA; thus, modifies gene expression at the translational step. The mRNA function is inhibited in several ways, including inhibiting splicing, transport and translation of protein. The most important mechanism seems to be the utilization of endogenous RNase H1 enzymes by the ASOs. The RNase H1 recognizes the mRNA-ASO duplex and cleaves the mRNA strand, leaving the ASOs intact to bind to other target RNA. A Phase I human trial of safety, tolerability, and pharmacokinetics of ASOs administered intrathecally to patients with SOD1 familial ALS has been completed (NCT01041222; www.clinicaltrials.gov).[30] One of the major benefits of ASOs and stabilized interfering RNAs is that their doses can be adjusted or treatment can be stopped.

Adeno-associated Viral Gene Therapy

Adeno-associated viral (AAV) vectors are a platform of great potential for therapeutic gene delivery. One of the major challenges of AAV gene therapy is delivering the transgene of interest to target cells at levels that result in expression that is both safe and effective. For diseases of the CNS such as ALS and DM, it is important to identify a dosing paradigm that provides a relatively homogenous distribution of gene transfer along the rostral-caudal axis of the CNS and is translatable. Intrathecal administration is a delivery approach that has shown promise for providing such distribution. AAV dosing via the IT route has been reported in several species including dogs to provide greater CNS distribution, less exposure to peripheral organs and tissues, and reduced impact of immune responses than systemic dosing.[31]

In summary, clinical trial research on canine DM as a disease model for ALS may offer a translational approach based on its homogeneous phenotype, similarities to human ALS and potential share disease measures. We aim to assess the safety, tolerability and pharmacodynamics of these proposed therapies.

REFERENCES

[1]Averill DR. J Vet Med Assoc 1973;162(12):1045. [2]Griffiths IR, et al. J Sm Anim Pract 1975;39(8):1309. [3]Coates JR, et al. J Vet Int Med 2007;21:1323. [4]Zeng R, et al. J Vet Intern Med 2014;28:515. [5]Coates JR et al. Vet Clin N Am Sm Anim Pract 2010;40(5):929. [6]Awano T, et al. Proceed Nat Acad Sci USA 2009;106(8):2794. [7]Shelton GD, et al. J Neurol Sci 2012;318:55. [8]March PA, et al. Vet Pathol 2009;46:241. [9]Matthews NS, et al. J Am Vet Med Assoc 1985;186(11):1213. [10]Wininger FA, et al. J Vet Int Med 2011;25(5):1166. [11]Johnston PEJ, et al. Vet Rec 2000;146(22):629. [12]Morgan BR, et al. J Neurosci Res 2013;91:1639. [13]Morgan BR, et al. J Neurosci Res 2014;92:531. [14]Ogawa M, et al. Vet Pathol 2014;51(3):591. [15]Gordon PH, et al. Neurology 2009;72(22):1948. [16]Polizopoulou ZS, et al. Acta Vet Hungarica 2008;56:293. [17]Kathmann I, et al. J Vet Int Med 2006;20:927. [18]Turner MR, et al. Lancet Neurol 2009;8:94. [19]Paoloni M, et al. Nat Rev Cancer 2008;8:147-156. [31]Katz ML, et al. Sci Transl Med 2015;7(313):1

Update on Thoracolumbar (Hansen Type II) Intervertebral Disk Protrusions in Dogs

Steven De Decker, DVM, PhD, DECVN, MvetMed
London, Hatfield, United Kingdom

INTRODUCTION

Thoracolumbar intervertebral disk disease is the most common spinal disease in dogs. Two types of degenerative intervertebral disk herniation have traditionally been recognized: intervertebral disk extrusion or Hansen Type-I, and intervertebral disk protrusion or Hansen Type-II disk herniation. Intervertebral disk extrusion is characterized by sudden herniation of degenerated and calcified nucleus pulposus through a fully ruptured anulus fibrosus, while intervertebral disk protrusion is characterized by a focal and more gradual extension of the anulus fibrosus and dorsal longitudinal ligament into the vertebral canal.[1] Intervertebral disk extrusions and protrusions are however associated with different clinical characteristics. Intervertebral disk extrusions occur typically in chondrodystrophic dogs, can occur at a young age and are typically associated with an acute onset of severe neurological signs. Intervertebral disk protrusions occur typically in large, non-chondrodystrophic dogs, affected dogs are generally older and can present with a more protracted and insidious clinical history.[1] Although intervertebral disk disease is the most common and best characterized spinal disease in dogs, only a few studies have focused specifically on the clinical characteristics of thoracolumbar intervertebral disk protrusions. In contrast, a multitude of studies have focused on intervertebral disk extrusions, resulting in a thorough characterization of its diagnosis, treatment and prognosis.

CLINICAL PRESENTATION

As illustrated above, thoracolumbar intervertebral disk protrusions and extrusions are characterized by different clinical characteristics. Although thoracolumbar intervertebral disk protrusions typically occur in non-chondrodystrophic, large breed dogs, such as the German Shepherd dog, this type of dog is equally commonly affected by intervertebral disk extrusions.[2,3] Compared to dogs with intervertebral disk extrusions, dogs with protrusions are typically older, have a more chronic and insidious onset of clinical signs, and affected dogs most commonly present with relative mild clinical signs, such as ambulatory paraparesis and spinal hyperesthesia.[2,3] A recent study suggested that the duration of clinical signs before presentation should be considered the most reliable clinical variable to differentiate between large breed dogs with thoracolumbar extrusions and protrusions. A duration of clinical signs of 21 days was associated with the highest combined sensitivity and specificity to differentiate between both types of intervertebral disk herniation.[3]

DIAGNOSIS

Although intervertebral disk herniations can be diagnosed by a variety of imaging modalities, magnetic resonance imaging (MRI) is currently considered the imaging modality of choice. Although several studies have reported MRI findings in dogs with thoracolumbar intervertebral disk herniations, little is currently known about specific MRI abnormalities that can be used to differentiate between thoracolumbar intervertebral disk extrusions and protrusions. A recent study evaluated a large number of, previously published, MRI variables as potential predictors for a diagnosis of thoracolumbar intervertebral disk extrusion or protrusion.[3] Four MRI variables were finally suggested to be independently associated with a diagnosis of a protrusion or extrusion; (1) midline instead of lateralized intervertebral disk herniation, and (2) partial instead of complete intervertebral disk degeneration were associated with a diagnosis of intervertebral disk protrusion, while the presence of a (3) single instead of multiple intervertebral disk herniations and (4) dispersed intervertebral disk material beyond the borders of the intervertebral disk space were associated with a diagnosis of intervertebral disk extrusion.

A further study[4] evaluated if using these four MRI variables as 'diagnostic guidelines' would indeed result in a more accurate diagnosis of intervertebral disk extrusion or protrusion. A large number of randomized MRI studies were presented on two separate occasions to 6 blinded observers, which were divided in different experience categories. During the first round of assessments, the observers had to make a presumptive diagnosis of thoracolumbar intervertebral disk extrusion or protrusion without any guidelines. During the second round of assessments they were asked again to make a presumptive diagnosis of extrusion or protrusion, but they were now provided with the proposed imaging guidelines. Additionally, they were asked to record the presence or absence of each of the four suggested MRI variables. Diagnostic accuracy significantly improved from 70.8% to 79.6% and inter-observer agreement for making a diagnosis of intervertebral disk extrusion or intervertebral disk protrusion improved significantly from fair to moderate after using the proposed guidelines. Diagnostic accuracy was significantly influenced by degree of observer experience, with the least experienced observers always obtaining the lowest scores. Intra-observer agreement for the assessed variables ranged from fair to excellent and inter-observer agreement ranged from fair to moderate. The results of this study suggested that although the proposed imaging guidelines can aid in differentiating thoracolumbar intervertebral disk extrusions from protrusions, they cannot replace thorough clinical training and experience. Although the diagnostic accuracy and reliability improved after using the proposed guidelines, a proportion of cases were still diagnosed incorrectly. This suggests that MRI in isolation cannot be completely relied on to differentiate between thoracolumbar extrusions and protrusions and confirms that imaging findings should always be interpreted in light of a thorough clinical history and results of a complete clinical examination

TREATMENT

Apart from the above mentioned differences in clinical presentation and MRI findings, intervertebral disk extrusions and protrusions are also associated with different suggested treatment options, and possibly also a different prognosis.[2] In stark contrast to the situation of dogs with extrusions, little is known about general treatment recommendations, results of medical treatment, and comparison of different surgical techniques for dogs with intervertebral disk protrusions.

Medical management of dogs with thoracolumbar intervertebral disk protrusions typically consists of restricted exercise in combination with appropriate analgesia, anti-inflammatory drugs, physiotherapy and hydrotherapy. Results of medical management for this type of intervertebral disk herniation have however only been reported rarely and it is unknown which factors currently influence the decision on whether medical or surgical treatment is pursued. A recent study[5] compared the clinical presentation of dogs treated medically or surgically for thoracolumbar intervertebral disk protrusions and evaluated outcome after medical management. Not surprisingly, dogs treated medically and surgically had differences in their clinical presentation and MRI findings. More specifically, medically managed dogs had significantly more often concurrent medical disorders and had less severe spinal cord compression on MRI, while surgically treated dogs had significantly more often neurological deficits, were more commonly non-ambulatory on presentation, and had a lower (more severely affected) neurological grade. Because dogs with medical and surgical management presented with differences in their clinical presentation, it was unfortunately not possible to directly compare the outcome after medical and surgical treatment. Despite having a relative mild clinical phenotype, medical treatment of dogs with thoracolumbar intervertebral disk protrusions was associated with only a guarded to fair prognosis; 40.7% of dogs initially improved with medical treatment, 25.9% remained stable, and 9 progressively deteriorated despite treatment. Recurrence of clinical signs was however common with 55.6% of dogs with an initial successful response, demonstrating a recurrence or acute deterioration of clinical signs later in life. Finally, only 29.6% were therefore considered to have a successful outcome after medical management for thoracolumbar intervertebral disk protrusion.

Thoracolumbar intervertebral disk protrusions have traditionally been managed by a hemilaminectomy followed by surgical excision, or anulectomy, of the protruding intervertebral disk. A common and important complication associated with surgical decompression of these, often chronic, intervertebral disk herniations, is severe neurological deterioration immediately after surgery.[2] Although the exact pathophysiology of this early postoperative neurological deterioration is currently unknown, suggested reasons include reperfusion injuries resulting from oxidative stress and direct manipulation of the chronically compressed spinal cord. More recently, several alternative surgical techniques have been described, including hemilaminectomy with or without additional vertebral stabilization[6,7], a lateral corpectomy[8], or partial lateral corpectomy[9,10]. Although these surgical procedures have been associated with satisfying outcomes, it is currently unclear if these alternative surgical approaches are also associated with a lower risk of early postoperative neurological deterioration.

In the author's institution, thoracolumbar intervertebral disk protrusions are surgically treated by a hemilaminectomy in combination with a partial discectomy. During the latter procedure, the affected intervertebral disk is approached laterally. The dorsal part of the intervertebral disk together with a part of the adjacent vertebral endplates is subsequently removed by a pneumatic drill or surgical aspirator. The most dorsal aspect of the anulus fibrosus and dorsal longitudinal ligament is however left intact, thereby initially not entering the vertebral canal. Finally, the most dorsal aspect of the intervertebral disk is removed by manual traction, thereby achieving spinal cord decompression. Preliminary results of a study, comparing hemilaminectomy with anulectomy and hemilaminectomy with partial discectomy, suggest that the latter technique is associated with reduced risk of early postoperative neurological deterioration, decreased time to neurological improvement, and decreased time until ambulation after surgery. Although no differences were seen in extent of neurological recovery and survival times, dogs undergoing a hemilaminectomy in combination with a partial discectomy demonstrated less often recurrence of clinical signs compared to dogs undergoing a hemilaminectomy in combination with anulectomy.

REFERENCES

[1]Smolders L, et al. Vet J 2013;195:292. [2]Macias C, et al. J Small Anim Pract 2002:43:439. [3]Gomes SA, et al. Vet Radiol Ultrasound 2016;In Press. [4]De Decker S, et al. Proc Annual ECVN-ESVN symposium 2015. [5]Crawford A and De Decker S Proc BSAVA symposium 2016. [6]Downes CJ, et al. J Small Anim Pract 2009;50:525. [7]McKee WM and Downes CJ. J Small Anim Pract 2008;49:536. [8]Moissonnier, et al. Vet Surg 2004;33:620. [9]Flegel T, et al. Vet Surg 2011;40:14. [10]Ferrand FX, et al. Ir Vet J 2015;68:27.

Canine Focal and Absence Epileptic Syndromes

Fiona M.K. James, DVM, MSc, DVSc, DACVIM (Neurology)
Guelph, Ontario, Canada

INTRODUCTION

"The history of classification has rested largely upon astute observations and expert opinions."[1]

As technology evolves, ongoing attempts in human and veterinary medicine to classify and define epilepsies are resulting in much discussion about the appropriate terminology. The most recent, 2010, attempt by the International League Against Epilepsy (ILAE) is still under debate.[2] For veterinary epileptology, the incentives to align terminology with that of the ILAE are the benefits of developing a naturally occurring translational model with associated improvements in understanding epileptogenesis, diagnosis, and therapeutics aiding both humans and small animals.[3] As the the International Veterinary Epilepsy Task Force (IVETF) acknowledges, however, there are significant differences between the current diagnostic approaches to epilepsy in humans and small animal patients that curtail how closely comparisons might be made between syndromes and species.[4] And while the theoretical definitions may be similar, the reality of veterinary medicine is that it is hard to distinguish focal from generalized absence seizures or other transient paroxysmal events.[4,5] Furthermore, despite thorough descriptions of hereditary breed-specific epilepsies in dogs, few genes have been identified.[6-9] Better clarification of epileptic phenotypes might be helpful in this genetic hunt. Better stratification of epileptic phenotypes might also improve treatment of, and prognostication for, canine epileptics, currently challenging as approximately one third of patients respond to the first line antiepileptic drug (AED).[11]

In human medicine, electroclinical syndromes form one of the bases of epilepsy classification.[1] Distinctly recognizable clinical complexes, they provide a focus for investigation, therapy, and prognosis. These specific syndromes are defined by the reliable identification of a cluster of specific electroclinical characteristics, including age of onset, electroencephalographic (EEG) findings and seizure types.

Very few such syndromes have been defined in veterinary medicine. The paucity is likely because veterinary EEG has been considered impractical due to its technical challenge and lack of standardization despite its long history of use.[4,12-15] Nonstandard factors affecting veterinary EEG recording techniques include patient restraint (chemical versus physical if at all), instrumentation, electrode type, and review methods, all of which might contribute to the low frequency of diagnostic success reported between 25-29%.[15-17] The lack of standardization also confounds the comparison of cases between clinics in multicenter studies. Concordance with EEG standards in human medicine might improve comparative epileptology with potential benefits to both species.

ADVANCES IN ELECTROENCEPHALOGRAPHY

Technological advances are improving veterinary scalp EEG at the machine, instrumentation, and technique levels. Tiny (<25-gauge) subdermal wire electrodes are functionally comparable to the more commonly used steel needle electrodes and gold cup electrodes.[16] They are also magnetic resonance imaging (MRI) and computed tomography (CT) compatible, paving the way for future simultaneous EEG and MRI studies.[16,18,19] These electrodes are well tolerated over extended periods of time with minimal movement artifact, opening the door to continuous scalp recordings that in turn reduce the need for more invasive implanted calvarial or cortical EEG.[16,18]

Use of video-synchronized EEG machines has been reported since 2008.[20] Video-synchronized EEG with live telemetry now allows awake ambulatory recordings in dogs with diagnostic success as high as 45%, similar to yields reported in human pediatric EEG.[21-23] With these systems and techniques, identification of hitherto rarely reported focal and absence epilepsy syndromes is accelerating.[24,25]

CANINE ELECTROCLINICAL SYNDROMES

Table 1 summarizes the known canine focal and absence electroclinical syndromes. Three of the descriptions included confirmed ictal EEG patterns with accompanying video-recorded semiology – the Chihuahua, Rhodesian Ridgeback, and Episodic Head Tremor Syndrome (EHTS).[22,24,25] Recordings in these three syndromes were obtained from awake dogs using variations of an ambulatory protocol.[20]

Table 1. Summary of canine focal and absence electroclinical syndromes.

Breed/Syndrome	Age of Onset	EEG Findings	Seizure Type	Reference
Belgian Shepherd	Mean 3.3 years	Interictal spikes & spike-waves, multiple foci	**Focal** onset with secondary generalization	8
Cavalier King Charles	Not given	Interictal generalized spike-waves, ictal rhythmic sharp waves	**Focal** onset (majority) with/out secondary generalization	29
Chihuahua	4 months	**Ictal** generalized 4 Hz spike-waves	Remitting generalized **absence** with myoclonic jerks	22
Finnish Spitz	Median 3 years	Interictal focal or generalized spikes, polyspikes, spike-waves	**Focal** onset with secondary generalization	10, 27, 28
Lagotto Romagnolo	Mean 6.3 weeks	Interictal sharp waves & spikes	Remitting **focal or focal** onset with secondary generalization	26
Rhodesian Ridgeback	Median 6 months	**Ictal** generalized 4 Hz spike-waves	Generalized **absence** with photosensitivity and myoclonus	25
Episodic Head Tremor Syndrome (Head-bobbing)	Mean 29 months	**Ictal** right-side predominant 6 Hz spike-wave	**Focal** onset with secondary generalization	24, 32

IMPLICATIONS

Spike-wave ictal patterns in dogs appear more common than historically reported. Generalized spike-wave (GSW) patterns are the hallmark of the absence seizures seen in human idiopathic generalized epilepsy.[33] This parallel should be of interest to comparative epilepsy research given these are naturally occurring epilepsies in dogs, not induced. The possible causal link between the level of arousal/attention and the occurrence of GSWs is something to be explored in the burgeoning investigation of behavioral comorbidities of epilepsy in canines.[34,35]

Focal onset and generalized absence epilepsies are likely under-recognized in canine epilepsy. Of particular interest are the similar EEG and semiology findings across multiple breeds in canine Episodic Head Tremor Syndrome (EHTS).[24,32] This syndrome was postulated to be a movement disorder, particularly because interruptions may disrupt the tremor.[24,32] However, three of four dogs whose events were recorded with video-EEG demonstrated similar right-side predominant 6 Hz spike-wave epileptogenic patterns.[24] This and the other syndromes reviewed herein both question the reliability of astute observation and expert opinion and demonstrate that video-EEG is necessary for the most confident confirmation of epileptic events and the diagnosis of idiopathic epilepsy.[15,22,24,25] Further discussion will be required to achieve consensus on whether seizures should be classified as "focal" versus "generalized absence" based on the description or observation of semiology or should incorporate the recording of EEG abnormalities and the simultaneous physical manifestation of the seizure.[4,5,15,29]

A final topic to consider is the current nonspecific therapeutic approach to canine epilepsy. An example of this therapeutic challenge is head nodding in Cavalier King Charles Spaniels. Debated to be a movement disorder, it is observed in older dogs and does not appear to be responsive to routine AED therapy.[30,31] Similarly, one of the reasons that EHTS was considered to be a movement disorder was because therapy with phenobarbital, bromide, or diazepam did not appear to affect the paroxysms.[32] If these conditions are shown to be epileptic, the burning question is: are we using the right drugs?[30]

REFERENCES

[1]Berg AT, et al. Epilepsia 2010;51(4):676. [2]Berg AT. Epilepsia 2012;53 Suppl 2:1. [3]Potschka H, et al. Epilepsia 2013;54(4):571. [4]Berendt M, et al. BMC Vet Res 2015;11(1):182. [5]Packer RMA, et al. BMC Vet Res 2015;11(1):39. [6]Lohi H, et al. Science 2005;307(5706):81. [7]Seppala EH, et al. PLoS Genet 2011;7(7):e1002194. [8]Seppala EH, et al. PLoS ONE 2012;7(3):e33549. [9]Koskinen LLE, et al. BMC Genomics 2015;16(1):465. [10]Viitmaa R, et al. J Vet Intern Med 2006;20(2):305. [11]Packer RMA, et al. Vet Rec 2015;176(8):203. [12]Steiss JE. Vet Res Commun 1988;12(4-5):281. [13]Redding RW, Knecht CE. Atlas of electroencephalography in the dog and cat. Praeger; 1984. [14]Fox MW. J Small Anim Pract 1967;8(2):71. [15]De Risio L, et al. BMC Vet Res 2015;11(1):148. [16]James FMK, et al. Am J Vet Res 2011;72(3):384. [17]Brauer C, et al. Vet J 2012;193(1):185. [18]Ives JR. J Clin

Neurophysiol 2005;22(2):119. [19]Rusbridge C, et al. BMC Vet Res 2015;11(1):194. [20]Williams DC, et al. J Vet Intern Med 2008;22(3):630. [21]James FMK, et al. J Vet Intern Med 2014;28(4):1358. [22]Poma R, et al. Epileptic Disord 2010;12(2):138. [23]Watemberg N, et al. Epilepsia 2005;46(5):716. [24]James FMK, et al. J Vet Intern Med 2015;30(1):444. [25]Wielaender F, et al. J Vet Intern Med 2015;30(1):444. [26]Jokinen TS, et al. J Vet Intern Med 2007;21(3):464. [27]Jeserevics J, et al. J Vet Intern Med 2007;21(6):1299. [28]Viitmaa R, et al. J Am Vet Med Assoc 2013;243(7):1001–9. [29]Driver CJ, et al. Vet J 2013;195(2):235. [30]Hülsmeyer V-I, et al. BMC Vet Res 2015;11:175. [31]Rusbridge C. J Small Anim Pract 2005;46(6):265–72. [32]Shell LG, et al. Vet Med Int. 2015;2015:165463. [33]Vaudano AE, et al. PLoS ONE 2009;4(8):e6475. [34]Packer RMA, Volk HA. Vet Rec 2015;177(12):306. [35]Jokinen TS, et al. J Vet Intern Med 2015;29(4):1081.

Canine Post-Traumatic Seizures, Advanced Imaging Prognostic Factors and Future Perspective

Elsa Beltran, Ldo Vet, DECVN, MRCVS, European and RCVS Specialist in Veterinary Neurology
London, United Kingdom

Traumatic brain injury (TBI) is defined as physiological disruption or structural injury of the brain induced by an external force. TBI is clinically indicated by the acute onset of at least one of the following: a period of loss or decreased consciousness, alteration in mental status, neurological deficits and/or the presence of an intracranial lesion. Not all the patients exposed to a head trauma will suffer from TBI, which is an important point to consider when evaluating prognosis.[1,2] TBI in dogs can be a life-threating condition and the mortality rate has been reported up to 24%.[3,4] A single-center retrospective study showed that up to 25% of dogs with severe blunt trauma suffer from TBI.[4] Dogs with TBI are assessed clinically using the modified Glasgow coma scale (MGCS), which evaluates motor activity, brainstem reflexes and level of consciousness.[3] The MGCS total score ranges from 3 to 18 (Table 1), with the lowest score representing the worst neurological status and the lowest probability of survival within the first 48 hours after head trauma.[3]

The TBI in dogs has been associated with significant risk of developing post-traumatic epilepsy (PTE) with an incidence up to 10 %.[2,5] In other words, dogs that have suffered TBI have a 5- to 18- fold increased risk of developing epilepsy compared to the general hospital population and based on the expected incidence of canine epilepsy. TBI is still one of the leading cause of acquired epilepsy in humans.[6] The onset of seizures after TBI has been categorized into immediate (<24 hours after TBI), early (within a week after TBI) and late (more than a week after TBI) onset.[2]

Seizures in particular are a significant source of morbidity and mortality early in the course of TBI. In human medicine, there is level II evidence (evidence obtained from well-designed cohort or case-control analytic studies) to recommend the use of prophylactic antiepileptic medication for seven to fourteen days post-TBI as it has shown a significant reduction in early post-traumatic seizures (seizures that occur within seven days after TBI). The prevention and clinical management of canine PTE are far from having standardized protocols. Regardless of prophylactic treatment recommendations, it is important to treat seizures that do occur following TBI. Benzodiazepines (diazepam, midazolam) can be given intravenously to treat seizures. Additionally, phenobarbital may also be given intravenously or intramuscularly and continued parenterally following a loading dose if necessary. Recently, the use of intravenous levetiracetam has been described for emergency seizure treatment. Levetiracetam is an attractive option because it works quickly, lacks of hepatic metabolism by CP450 enzymes, lacks of significant sedative effects, which could alter neurological assessment, and may be effective for up to 8 hours.[7] Moreover, preclinical studies using animal models of TBI have shown efficacy of levetiracetam as a neuroprotectant and also improvements in motor and memory performances.[8,9] The risk of developing PTE increases with the severity of TBI and further research is needed to investigate if these patients could benefit from a 14-day course of levetiracetam (starting within 8 hours after the TBI) as prophylactic antiepileptic treatment and possibly also as a neuroprotectant.[10,11]

On the other hand, the role of advanced imaging in predicting outcome is still emerging. There is a growing recognition that the prediction of outcome in humans after TBI based only on the Glasgow coma scale is very limited.[12] There are recent studies focusing on MRI findings as predictors of clinical outcome due to its high sensitivity for the detection of non-hemorrhagic contusions, brainstem lesions and diffuse axonal injuries.[13-15] A recent study in dogs with TBI supported the clinical relevance of the use of advanced imaging (MRI) due to their significant correlation with outcome.[2] This study used a grading system modified from human medicine, where the severity and location of the lesions in MRI were taken into account (Table 2). Brainstem lesions were associated with poor outcome with bilateral injuries having the poorest prognosis (Grade V and VI).[2] The risk of developing seizures increased with the severity of the TBI. Significant association has been found between early MRI findings and prognosis in dogs with TBI. The presence of post-traumatic seizures in dogs has been associated with higher size of intraparenchymal lesions (>25%), skull fractures and herniation through a fracture site (Fig.1). Multicenter and prospective studies, including advanced MRI techniques (diffusion tensor imaging, MR spectroscopic imaging), may provide insight in understanding canine TBI and imaging implications for treatment and functional outcome.

Modified Glasgow Coma Scale	Score
Motor activity	
Normal gait, normal spinal reflexes	6
Hemiparesis, tetraparesis, or decerebrate activity	5
Recumbent, intermittent extensor rigidity	4
Recumbent, constant extensor rigidity	3
Recumbent, constant extensor rigidity with opisthotonus	2
Recumbent, hypotonia of muscles, depressed or absent spinal reflexes	1
Brainstem reflexes	
Normal pupillary light reflexes and vestibulo-ocular reflex	6
Slow pupillary light reflexes and normal to reduced vestibulo-ocular reflex	5
Bilateral unresponsive miosis with normal to reduced vestibulo-ocular reflex	4
Pinpoint pupils with reduced to absent vestibulo-ocular reflex	3
Unilateral, unresponsive mydriasis with reduced to absent vestibulo-ocular reflex	2
Bilateral, unresponsive mydriasis with reduced to absent vestibulo-ocular reflex	1
Level of consciousness	
Occasional periods of alertness and responsive to environment	6
Depression or delirium, capable of responding but response may be inappropriate	5
Semicomatose, responsive to visual stimuli	4
Semicomatose, responsive to auditory stimuli	3
Semicomatose, responsive only to repeated noxious stimuli	2
Comatose, unresponsive to repeated noxious stimuli	1

Table 1. Scoring system of the current Modified Glasgow Coma Scale[3]

However, one of the limitations of using MRI is that usually it is performed in referral centers and requires patients that are hemodynamically stable.

	Modified MRI Brain Grading System in dogs with TBI
Grade	**MRI findings**
I	Normal brain parenchyma
II	Lesions only affecting the hemisphere and/or cerebellar parenchyma without midline shift
III	Lesions only affecting the hemisphere and/or cerebellar parenchyma and causing midline shift
IV	Lesions affecting corpus callosum, thalamus or basal nuclei with or without any of the foregoing lesions of lesser grades
V	Unilateral lesions in the brainstem with or without any of the foregoing lesions of lesser grades
VI	Bilateral lesions affecting the brainstem with or without any of the foregoing lesions of lesser grades

Table 2. Severity and location of lesions-modified MRI grading system[2,16]

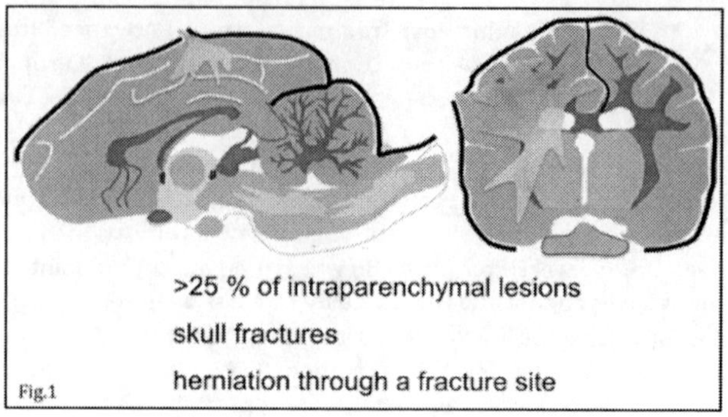

Fig.1

Moreover, it is also important to take into consideration that dogs with a history of TBI can develop endocrine abnormalities indicative of hypothalamic-anterior pituitary dysfunction (severe/partial growth hormone deficiency, ACTH/cortisol and TSH/T_4). [17] If these deficiencies are not recognized, they could contribute to increased morbidity and mortality in patients after TBI.

Thinking about future perspective, it is known that seizure prophylaxis can potentially prevent early onset of seizures in humans, however there are not available treatment that effectively prevent late-onset of seizures.[18] Current studies are focused on learning about the progression of neural injury over time and assessing how this neural injury could contribute to late onset of seizures development. The aim of understanding this process is to design appropriate treatments with molecular targets in order to prevent late-onset of seizures after TBI. The main mechanisms of injury that contribute to seizures include: glutamate toxicity, neuroinflammation and tau pathology.[19–21] During the acute phase, glutamate excitotoxicity contributes to the early onset of seizures and over time secondary injury cascades (neuroinflammation) facilitates the transition towards PTE. Late onset of epilepsy has been associated with tau pathology. Neuronal injury after TBI contributes (over months to years) to tau aggregation interacting with zinc to generate further free radical damage, and therefore reducing the threshold for seizures.[21–23] Novel approaches and treatment options currently investigate in pre-clinical settings appears promising, however the cornerstone is the overall understanding of the pathophysiology leading to PTE, where the knowledge gap is still present.

Randomized and multicentric controlled double-blinded trials are needed to investigate whether treatment (antiepileptic/antiepileptogenic) can alter outcome in dogs with TBI and make a useful contribution to the evaluation of treatment for posttraumatic seizures in Veterinary Medicine.

REFERENCES

1. West TA, Marion DW.J Neurotrauma [Internet]. 2014;31(2):159–68.
2. Beltran E, et al. J Vet Intern Med 2014;28(4):1256–62.
3. Platt SR, et al. J Vet Intern Med. 2001;15(6):581–4.
4. Simpson SA, et al. J Vet Emerg Crit Care. 2009;19(6):588–602.
5. Steinmetz S, et al. Epilepsia. 2013;54(4):580–8.
6. Agrawal A, et al. Clinical Neurology and Neurosurgery. 2006. p. 433–9.
7. Patterson E (Ned) Veterinary Clinics of North America - Small Animal Practice. 2014. p. 1103–12.
8. Kochanek P, et al. J Neurotrauma. 2015;(305):1–42.
9. Browning M , et al. J Neurotrauma [Internet]. 2015;
10. Klein P, et al. Arch Neurol. 2012;69(10):1290–5.
11. Pearl PL, et al. Epilepsia. 2013;54(9).
12. Zuercher M, et al. Brain Inj. 2009;23(5):371–84.
13. Yuh EL, et al. Ann Neurol. 2013;73(2):224–35.
14. Chew BG, et al. J Neurosurg. 2012;117(4):722–8.
15. Lee S-Y, et al. J Clin Neurol. 2012;8(3):224–9.
16. Firsching R, et al. Acta Neurochir (Wien). 2001;143(3):263–71.
17. Murtagh K, et al. Vet Record. 2015; 176:20.
17. Gabriel WM, Rowe AS. Ann Pharmacother. 2014;48(11):1440–4.
18. Hameed MQ, et al. Neuroreport [Internet]. 2014;25:532–6.
19. Diamond ML, R et al. Epilepsia. 2014;55(7):1109–19.
20. Zheng P, et al. Molecular Neurobiology. 2014. p. 1532–9.
21. Szewczyk B. Frontiers in Aging Neuroscience. 2013.
22. Lucke-Wold BP, et al. Seizure. 2015. p. 13–23.

Seizures and Other Involuntary Movements in Cats

Katia Marioni-Henry, DVM, PhD, DACVIM (Neurology), ECVN, MRCVS
Edinburgh, United Kingdom

Recent advances in feline neurology have generated new approaches for the investigation and treatment of seizures and other types of feline involuntary movements.

A peculiar form of feline epilepsy characterized by complex partial seizures with orofacial involvement and hippocampal lesions has been recently associated with voltage-gated potassium channel (VGKC) complex antibody found in cases of limbic encephalitis in people.[1, 2] This type of feline epilepsy was first described fifteen years ago, when it was characterized by post-mortem findings of necrosis of the hippocampus and piriform lobe.[3] Initially, only cats from central Europe appeared to be affected and the prognosis for recovery was considered poor. The presence of clusters of cases from the same area or household led to hypothesize a possible toxic cause.[3] However, in recent studies cats diagnosed with hippocampal necrosis were successfully treated[1] and cases of feline hippocampal necrosis were reported also in United Kingdom[4], United States[5], Australia[6] and Finland[7]. The presence of clusters of cases from the same area has yet to be explained. In human patients, limbic encephalitis are usually autoimmune conditions associated with distant tumours, however, the VGKC- complex antibodies have limited specificity and may be found in non-autoimmune encephalitis, including Creutzfeldt–Jakob disease (CJD).[8]

Increased awareness of feline hippocampal necrosis and the introduction of novel terminology and diagnostic tools such MRI, have influenced our knowledge of feline epilepsy.[9, 10] In a recent retrospective study on feline epilepsy including 81 cats with MRI or post-mortem examination, epilepsy of unknown cause was diagnosed in 22% of cats with recurrent seizures, compared to 38% and 66% previously reported by authors who called this entity primary or suspected idiopathic epilepsy combining genetic epilepsy and epilepsy of unknown cause.[9, 11, 12] Cats with epilepsy of unknown cause had a significantly longer survival rate compared with cats with epilepsy secondary to structural brain lesions or acute symptomatic seizures with 73% of the cats alive after 1 year and 44% of the cats achieving remission with phenobarbital treatment.[9]

Other conditions that will be discussed in this video-based presentation are permetrin toxicosis[13], REM sleep behaviour disorder[14] and feline hyperaesthesia syndrome[15]. Particular emphasis will be given to clinical features that help differentiating these conditions.

REFERENCES

1. Pakozdy A, et al. J Feline Med Surg 2011;13: 687.
2. Pakozdy A, et al. J Vet Intern Med 2013;27: 212.
3. Fatzer R, J Vet Intern Med 2000:14: 100.
4. Marioni-Henry K, et al. Vet Rec 2012;170: 471.
5. Gelberg HB, et al. Vet Pathol 2013;50: 1156.
6. Adagra C, et al. Case Rep Vet Med 2014.
7. Fors et al. Acta Vet Scand 2015;57:41
8. Leypoldt F, at al. Ann. N.Y. Acad. Sci. 2015;1338: 94.
9. Wahle AM, et al. J Vet Intern Med 2014;28: 182.
10. Pakozdy A, et al. J Vet Intern Med 2014;29: 485.
11. Pakozdy A, et al. J Feline Med Surg 2010;12: 910.
12. Schwartz-Porsche D, et al. Probl Vet Med 1989;1: 628.
13. Peacock RE, J Vet Emerg Crit Care 2015;25(5): 597.
14. Hendricks JC, et al. Brain Res. 1982;239(1): 81.
15. Tuttle J. J Am Vet Med Assoc. 1980;176(1): 47.

Diagnosing Seizure-like Episodes: Startling Results of EEG Video Case Studies

William Bush, VMD, DACVIM (Neurology)
Rockville, MD

Introduction

The International League Against Epilepsy (ILAE) defines seizure as the transient occurrence of signs, symptoms or both due to abnormal excessive or synchronous neuronal activity in brain. Although this seems straight forward, there are many events that cause a transient occurrence of signs, symptoms or both that are NOT from abnormal excessive or synchronous neuronal activity in the brain. These events are seizure-like and masquerade as a true epileptic seizure (see Table 1). Without confirming with Electroencephalography (EEG) these events could be falsely classified as seizure. Conversely, abnormal excessive or synchronous neuronal activity in the brain does not always have a detectable transient occurrence of signs or symptoms despite having electrical, non-convulsive seizure. Thus using observation alone, without EEG, could lead to transient events being characterized incorrectly. The goal of this talk is to discuss the application of EEG in a clinical, private practice setting to discern seizure-like episodes from true epileptic seizure. Video-EEG case studies will be presented with goal of improving attendees ability to discern seizure from seizure-like episodes.

Role of EEG

In human medicine EEG is used to discriminate movement disorder, psychogenic events, and metabolic disease from true epileptic seizure. EEG is also used to detect subtle or non-convulsive seizure and then evaluate the effectiveness of anti-epileptic drug (AED) therapy. We have EEG in 5 of our clinics as well as an intramural EEG certification program for our technical staff. The EEG is read with the assistance of an extramural physician expert in the interpretation of both human and animal EEG. We perform EEG for the exact same reasons it is performed in human medicine.

Electrical and Observed Phenomenon During Seizure

The first step in seizure generation is that a group of neurons synchronizes and depolarizes / repolarizes autonomously and spreads within that hemisphere of the brain due to failure of spatial containment. This hypersynchronous electrical activity then crosses to the opposite hemisphere capturing the entire brain before eventually being contained, usually within 2 minutes. During the seizure there is excess glutamate release that can lead to temporary neuronal dysfunction and/or neuronal necrosis.
In the **pre-ictal state**, as the focus is developing and spreading the patient may experience abnormal visual, auditory, physical, or autonomic nervous system abnormalities manifested as staring off into space, searching a room, restlessness, clingy behavior, fly biting, circling, odd vocalization, a limb becoming stiff or rhythmically moving, elevated heart rate, dilated pupils, salivation, vomiting.

In the **ictus or seizure**, the focus has captured both cerebral hemispheres and the patient may experience loss of consciousness, collapses, rigidly extend the neck and all 4 limbs. The hypersynchronous or rhythmic nature of the electrical focus can be noted as paddling of all 4 limbs. A failure to control and regulate the breathing can manifest as apnea and paradoxical breathing where the diaphragm and intercostal muscles are not working together. Perturbations in the autonomic nervous system can lead to bradycardia or tachycardia, profuse salivation, urination, defecation, miosis or mydriasis, and piloerection. The **post-ictal period** results from excitotoxicity and typically lasts 30-60 minutes where the patient emerges from being confused, blind, weak, and side-stepping. If the seizure is not recognized (non-convulsive) or difficult to treat neuronal necrosis and death can result from seizure.

Practical Criteria for Distinguishing Seizure

In a recent paper the inclusion criteria for seizure was when 3 of the 4 of the following were noted.

1) Salivation, urination or defecation
2) Tonic or tonic-clonic posture or movements or rhythmic contractions of facial or appendicular muscles
3) Decreased responsiveness intra-ictally
4) Postictal phase in which abnormal behavior or mental state was noted

Observing 3 of the 4 criteria to make a diagnosis of seizure is an effective, practical strategy, however there are many documented seizure-like events (Table 1) that would be falsely categorized as seizure. This could lead to a delayed diagnosis and treatment of the underlying cause and the needless and potentially harmful application of AED. Many video-EEG case examples will be presented of cases that could have potentially been misclassified as seizure.

Electrical Seizure (ES) and Electrical Status Epilepticus (ESE)

As mentioned above a seizure is fundamentally from abnormal excessive or synchronous neuronal activity in the brain. Using EEG, electrographic seizures are defined as ictal discharges consisting of a rhythmic pattern with definitive evolution in frequency, amplitude and/or morphology persisting for at least 10 seconds. When ictal discharges are present for more than 10-15 minutes then patient is suffering from electrical status epilepticus.

Convulsive vs. Non-Convulsive Seizure

Electrical seizure can occasionally manifest as convulsions (generalized tonic-clonic) with patient flailing on its side, paddling all 4 limbs or holding the limbs, head and neck in rigid extension. A non-convulsive seizure (NCS) is defined as a seizure where there is no overt convulsive movements. Another term used for non-convulsive seizure is complex partial seizure where there is only an acute alteration in consciousness. NCS is more common than convulsive seizure in people and cats, and potentially dogs as well. When an electrical seizure lasts for more than 30 minutes it is almost always non-convulsive and termed non-convulsive status epilepticus (NCSE). Using observation alone to detect seizure, non-convulsive seizure could be missed or falsely negative, AED would not be applied and the patient would be at higher risk for mortality and permanent disability or poor seizure control. Many video-case examples will be presented of non-convulsive seizure.

What is the Incidence of ES and ESE?

We have conducted a retrospective study of 94 dogs and 15 cats who had an EEG for suspected seizure with the goal of identifying the incidence, mortality, clinical characteristics and risk factors for ES and ESE. We found 21/104 (20%) of patients had ES and 12/104 (12%) met the definition of ESE. Non-convulsive seizure was noted in 18/21 (86%) of the ES cases and manifested as coma, twitching of the ears or facial muscles, pupil dilation, paroxysmal elevations of temperature or respiratory rate. The incidence of electrical seizure, non-convulsive seizure and non-convulsive status epilepticus in this study mirrored findings in pediatric and adult studies of hospitalized human patients.

Are Mortality Rates Different Between Patients With and Without ES and ESE?

10/21 (48%) of the patients with ES died during their hospital admission while only 16/83 (16%) of patient without ES died in hospital. This statistically significant difference suggests that the presence of ES has an influence on mortality. In human studies, statistical analysis shows that ES is an independent risk factor for mortality. It is likely that elaboration of glutamate is leading to neuronal necrosis, elevated intracranial pressure and death.

What are Risk Factors for ES and ESE?

EEG is not readily available to in any setting in veterinary medicine, therefore it would be useful to know the risk factors for a diagnosis or ES or ESE. In our study we examined many variables including diagnosis, MRI and CSF findings, age, time since seizure and other historical factors using bivariable analysis and multivariable logistical regression. Cluster seizure, younger age and time since last seizure were predictors of ES. ES patients had a mean age of 3.9 years compared to 7.1 years for non-ES patients. History of cluster seizure was seen in 27% of the ES cases and only 10% of the non-ES cases, Observation of a clinical seizure within 8 hours of the EEG also placed patient at significantly higher risk for ES. Only a history of cluster seizure was a risk factor for ESE. Importantly there was no clinical characteristic of the event itself that could distinguish an ES from a non-ES, in other words without EEG we could not distinguish seizure from non-seizure by using characteristics like twitching or altered mentation.

What About Convulsive Status Epileptics Patients that are Still Twitching?

In recent human study of 164 patients presenting for convulsive status epilepticus (defined as a seizure lasting more than 5 minutes, 2 seizure without becoming normal between or seizing at presentation) patients were treated with a standard protocol for status epilepticus. Once the convulsive seizure resolved, patients had an EEG and about 50% were found to still have electrical seizure and 14% were in electrical status epilepticus. The mortality rate in the electrical seizure group was 50% and independent of age and underlying diagnosis. This paper also notes that once the EEG was used as the end point for treatment and not observation, outcomes were thought to improve. In the only current veterinary study of patients presenting for status epilepticus and then having an EEG, all 10 patients had electrical seizure despite resolution of clinical seizure.

This begs the question for veterinary medicine of how and when we should treat without the benefit of EEG. A general recommendation can be to treat a convulsive seizure patient with AED until there are no signs of twitching or autonomic changes. It is advised to give Leveteriacetam 60 mg/kg intravenously and then 10 mg/kg boluses of Phenobarbital every 20 minutes provided the systolic blood pressure is greater or equal to 90 mmHG. In the authors experience a total dose of 60 mg/kg can safely be administered to both dogs and cats.

Conclusion

When a patient presents for seizure the veterinarian must ask themselves if the event was a seizure. Falsely identifying events as seizure or failing to recognize non-convulsive seizure will lead to worse outcomes. Although EEG is needed to allow best diagnosis of seizure-like events, a thorough history and examination, considering the list of seizure-like diseases, and being cognizant of varied presentation for non-convulsive seizure can assist in better diagnosis and treatment of seizure-like diseases.

Table 1. Disease Processes with Seizure-like Appearance

▪ Atlantoaxial subluxation
▪ Breed and drug induced dyskinesia / movement disorders
▪ Cataplexy, narcolepsy, rapid eye movement (REM) sleep disorders
▪ Cervical muscle spasm
▪ Chiari-like malformation / syringomyelia
▪ Encephalitis
▪ Exercise induced collapse
▪ Extreme agitation / psychogenic seizure
▪ Feline hyperesthesia syndrome
▪ Head bobbing / Tremor syndromes
▪ Intermittent decerebrate/decerebellate rigidity
▪ Jaw chomping / fly biting / lip smacking
▪ Metabolic encephalopathy
▪ Myoclonus
▪ Neuromuscular disease
▪ Psychogenic Non-epileptic Spells / Panic Attacks
▪ Syncope
▪ Toxicity

SELECTED REFERENCES

Bush W, et al. Diagnosis REM sleep disorder with EEG and treatment with antidepressant, JAAHA, 2004

Cuff DE, Bush WW, Williams DC, Stecker MM. Use of continuous electroencephalography for the diagnosis and monitoring of treatment in nonconvulsive status epilepticus. JAVMA 15;244(6):708-14.

DeLorenzo RJ, et al. Persistent NCSE after control of convulsive SE. Epilepsia 39 (8): 833-840, 1998

Granum L, et al. Clinical characteristics and outcome of cases diagnosed with electrical seizure in 73 dogs and 13 cats. Proceedings Abstract, 2015

Muñana KR,et al. Evaluation of levetiracetam as adjunctive treatment for refractory canine epilepsy: a randomized, placebo-controlled, crossover trial. *J Vet Intern Med.* 2012;26(2):341-8

Raith K, et al. Continuous EEG monitoring of status epilepticus in dogs and cats: 10 patients (2004-2005). JVECC 20 (4) 2010, pp. 446-455

Schreiber JM, et el. Continuous Video EEG for Patients with Acute Encephalopathy in PICU. Neurocrit Care (2012) 17: 31-38

Trinka E, et al. A definition and classification of status epilepticus – report of the ILAE task force on classification of status epilepticus. Epilepsia; 2015 (10); 1515-1523

Roundtable: Diagnosis and Treatment for Lumbosacral Stenosis

Nick Jeffery, BVSc, PhD, MSc, DECVS, DECVN, FRCVS
Ames, IA

INTRODUCTION

Degenerative lumbosacral stenosis (DLSS) in dogs has been a well-recognized syndrome since the first descriptions appeared in the late 1970s. Despite this, there remains considerable variation in case definition and optimal methods of diagnosis and treatment, even for specific sub-classifications of this syndrome, have not been determined. Although there has been tremendous progress in imaging of DLSS, mainly because of the availability of cross sectional modalities, it is arguable whether management of this condition has improved during the past 30 years.

DIFFICULTIES IN PHYSICAL DIAGNOSIS

DLSS presents problems in initial diagnosis because the syndrome is not associated with invariable clinical signs, traditional methods of neurological diagnosis are not often helpful and the main clinical signs can readily be associated with many other lesions affecting the hindquarters of dogs.

Although some cases of DLSS will show neurologic deficits (in which case diagnosis is straightforward) most show ill-defined signs of dysfunction around the hindquarters, such as apparent reluctance to rise from recumbency or jump, and may also exhibit signs of pain on hip extension or ventrally-directed pressure on the lumbosacral junction. These clinical signs can of course be associated with a great variety of orthopedic and neurologic disease affecting that region of the body. To overcome this problem some clinicians rely on highly individual methods of examination (e.g. deep palpation of the sciatic nerve) to indicate DLSS.

DIFFICULTIES IN IMAGING DIAGNOSIS

When DLSS was first recognized there were many difficulties in confirming the diagnosis through imaging. Although plain radiographs are reasonably accurate in ruling out destructive bone lesions as a differential diagnosis they are not adequate to positively diagnose herniated discs or impingement of the ligamentous structures on the L7 nerves or cauda equina, which constitute the majority of cases of DLSS. It was hoped that access to cross sectional imaging using CT and MR would solve these problems but that has not entirely been the case. Although severe entrapment of nerves by soft tissue or bone structures can sometimes be unequivocally diagnosed using these modalities, in many potential DLSS cases there remains doubt in diagnosis even after MR or CT. Unfortunately, it has been shown that many abnormalities seen on cross sectional images in affected dogs are also apparent in normal dogs.

DIFFICULTIES IN ELECTROPHYSIOLOGICAL DIAGNOSIS

The characteristic lesion of DLSS implies compression of L7 and, or, nerves of the cauda equina. It would be expected that, in addition to pain, there might also be associated neurologic dysfunction, detectable, for instance, by using electromyography. While this is sometimes true, many cases do not show obvious EMG abnormalities. Although there is some evidence for increased F wave latency in some affected cases, reference intervals for these values are, unfortunately, not well defined and rely largely on measurements made many decades ago.

DIFFICULTIES IN DECISION-MAKING

The multitude of questions that cannot be answered definitively about diagnosing this condition highlight the risk of biased diagnosis; for instance a case one clinician may suspect to be DLSS may be diagnosed by another as iliopsoas injury. Furthermore, even if clinicians agree about a diagnosis of DLSS they may disagree as to the exact cause of the dog's clinical signs; for instance whether the pain arises from synovial joint inflammation and distention, proliferation of peri-foraminal soft tissue or simple disc protrusion. Although this may seem like a sorry state of affairs it is worth noting that physicians dealing with human low back pain have more or less the same problem even without the inherent difficulties in dealing with a painful condition in non-verbal patients.

Nevertheless, difficulties in establishing an objective diagnosis of DLSS imply the risk that clinician prejudice, which is inherent in all of us, may go a long way towards establishing a 'diagnosis' of DLSS and its specific sub-type. It is probable that individual clinicians that 'look for' DLSS (especially from the point-of-view of selection of surgical candidates) will diagnose many more than those who are reluctant to make definitive diagnoses (and might favor conservative interventions).

TREATMENT PLANNING

Overall there are three main therapies for DLSS: conservative, decompressive surgery to excise tissue that is impinging on nerves and fixation-fusion procedures. Conservative therapy can include such simple interventions as restriction to leash exercise, ranging through oral NSAIDs to include injection of corticosteroid into the epidural space or intervertebral foramen. Decompressive surgery is primarily a decision between laminectomy (+/- disc or annulus excision) and foraminotomy and the decision-making here might focus on the structure that is thought to be compressed – is it the L7 nerve(s) or the body of the cauda equina? Fixation-fusion procedures can be used in cases in which a dynamic component to compression is suspected or cases in which a bilateral foraminotomy might be thought useful (because one fixation surgery might be considered more straightforward than a bilateral decompressive procedure). Each of these therapies has its advantages and disadvantages and so a specific intervention can be selected for individualized treatment of each dog, although some also have (sometimes very vocal) advocates who suggest that specific procedures are applicable to a wide spectrum of DLSS cases.

Designing a prescription for treatment of any condition should usually be straightforward once the diagnosis has been clearly established. The main difficulties in planning treatment for DLSS are that the diagnosis is often equivocal and that conservative therapy, for both DLSS itself and for many of its differential diagnoses, is often efficacious. Such a position means that surgical interventions may appear to be effective even when they have made no difference to the outcome (because the dog would have spontaneously recovered anyway). Again, it is worth stressing that veterinarians are not the only clinicians that have difficulties in recommending therapy for back pain; the argument about whether surgery or conservative therapy should be preferred for human patients with low back pain has been raging for more than 20 years.

Of course, the ideal method to resolve these difficulties in treatment of DLSS in veterinary patients would be to run clinical trials to compare outcomes. There is no reason why this should not be done, but some clinicians are reluctant to submit specific patients to randomized therapies that they believe may not be appropriate (despite the lack of objective evidence to make these decisions). Careful categorization of inclusion and exclusion criteria could overcome this problem. I would suggest that the main questions to be addressed would be choices between: i) conservative therapy versus laminectomy for dogs with back pain alone; ii) foraminotomy versus laminectomy versus fixation-fusion for dog with neurologic deficits including pelvic limb lameness. Agreement about suitable outcome measures would also be required; these could be very simple (such as owner assessment of dog quality of life and ability to complete everyday tasks) or could rely on more specialized measurements, such as force plate. There are merits in both approaches, but each has important implications regarding required study sample size.

REFERENCES

1. Denny, H.R., et al. J Small Anim Pract 1982;23:425.
2. Mayhew PD, et al. JAAHA 2002;38:555.
3. Deyo RA, et al. NEJM 2001; 344: 363.
4. Jeffery ND, et al. Vet J 2014;201:9.

Diagnosis and Treatment of Lumbosacral Stenosis
Rodney S. Bagley DVM, DACVIM (Neurology and Small Animal Internal Medicine)
Ames, IA

At the ACVIM forum in 2006 a panel discussion was presented on lumbosacral (LS) disease. While there has been clinical scientific information published in the last 10 years, the convening of this panel suggests that a number of clinical questions remain unanswered.

- What is the spectrum of pathophysiologic abnormalities that occur under the clinical label of 'lumbosacral disease'?
- How are each of the pathophysiologic components definitively diagnosed?
- What is the role of the various imaging modalities in supporting/confirming the diagnosis?
- What is the contribution of foraminal size alterations to the clinical signs and progression?
- What is the contribution of instability to the clinical signs and progression?
- What is the contribution of vascular alterations to the clinical signs and progression?
- What is the most effective treatment?
- What are the advantages and disadvantages of the various surgical treatments?
- What is the natural course of the disease process without treatment?

This is a partial list of important clinical questions that remain without consensus. As with other diseases, it is important to understand the pathophysiologic cause (s) of the lumbosacral abnormality prior to performing any surgical treatment so that the most appropriate surgery can be chosen. For the typical cases of degenerative lumbosacral disease, there is often some abnormality of the lumbosacral joint complex including the L7 and sacral vertebrae, the intervertebral disk elements, the articular facet joints, and associated ligaments. Pathology in this combination of structures results in compression of the nerves of the cauda equine in the lumbosacral area. Alterations in other structures such as the sacroiliac joints and the various muscles, tendons, and ligaments are less well elucidated. From a surgical standpoint, how best to treat lumbosacral disease remains debatable. Some would argue that decompression alone via laminectomy is often helpful, however not always successful, and failure to improve or recurrence of clinical signs may be attributable to instability. The extent of surgical decompression necessary is often variable ranging from simple laminectomy to laminectomy with partial diskectomy, foraminotomy, and/or facetectomy. Many describe performing a 'foramenotomy' however, what the physical outcomes of this procedure are are commonly undocumented with post-operative imaging. "Diskectomy" is sometime performed, however, the terminology is misleading as this procedure is more appropriately defined as partial given that the entire intervertebral disk components are not removed during the procedure. With regards to outcomes, some dogs with lumbosacral compression do not respond or have recurrence of clinical signs after laminectomy and diskectomy. Dogs treated with laminectomy and partial diskectomy alone may have recurrence of clinical signs.

A significant amount of the debate in the last number of years has centered on the contribution of instability to the pathophysiologic sequelae. There is clinical and radiological evidence that there is instability in instances of LS disease. Dynamic imaging studies of the LS joint tend to show that the LS joint is movable and that LS compression is worse in most instances when the joint is in extension as compared to flexion. Physiologic information in human beings show that epidural pressures fluctuate with movement of the joint during walking. There are instances of animals with acute onset of signs of LS disease that resolved quickly (within 24 hours) suggesting that dynamic changes in the nerves of this area were responsible for the clinical decline. Additionally, during surgery at the LS joint, as has been reported previously, there is often grossly visible collapse of the LS joint with the L7 articular facet joints being caudally displaced relative to the sacrum. When these articulations are distracted during surgery with a lamina spreader or similar instrument, motion in this area is obvious.

The answers to these important aspects of LS disease will allow the profession to move forward with a better understanding of this complex disease process. Concurrently, the hope would be that this understanding will result in better judgments regarding acceptable imaging criteria and surgical outcomes for out patients.

PUBMED Search January 28, 2016 10:50 am CST – Relative articles

Pedicle screw-rod fixation: a feasible treatment for dogs with severe degenerative lumbosacral stenosis. Tellegen AR, Willems N, Tryfonidou MA, Meij BP. BMC Vet Res. 2015 Dec 7;11(1):299. doi: 10.1186/s12917-015-0614-3.

Degenerative lumbosacral stenosis in dogs: will we see progress in the next 30 years? da Costa RC. Vet J. 2014 Nov;202(2):201-2. doi: 10.1016/j.tvjl.2014.09.011. Epub 2014 Sep 16.

Lumbosacral foraminal ratios and areas using MRI in medium-sized dogs. Reynolds D, Tucker RL, Fitzpatrick N. Vet Comp Orthop Traumatol. 2014;27(5):333-8. doi: 10.3415/VCOT-13-05-0085. Epub 2014 Aug 27.

Clinical signs and outcome of dogs treated medically for degenerative lumbosacral stenosis: 98 cases (2004-2012). De Decker S, Wawrzenski LA, Volk HA. J Am Vet Med Assoc. 2014 Aug 15;245(4):408-13. doi: 10.2460/javma.245.4.408.

What progress has been made in the understanding and treatment of degenerative lumbosacral stenosis in dogs during the past 30 years? Jeffery ND, Barker A, Harcourt-Brown T. Vet J. 2014 Jul;201(1):9-14. doi: 10.1016/j.tvjl.2014.04.018. Epub 2014 May 5. Review.

Transarticular fixation with cortical screws combined with dorsal laminectomy and partial discectomy as surgical treatment of degenerative lumbosacral stenosis in 17 dogs: clinical and computed tomography follow-up. Golini L, Kircher PR, Lewis FI, Steffen F. Vet Surg. 2014 May;43(4):405-13. doi: 10.1111/j.1532-950X.2014.12117.x. Epub 2014 Mar 11.

Pedicle screw-rod fixation of the canine lumbosacral junction. Smolders LA, Voorhout G, van de Ven R, Bergknut N, Grinwis GC, Hazewinkel HA, Meij BP. Vet Surg. 2012 Aug;41(6):720-32. doi: 10.1111/j.1532-950X.2012.00989.x. Epub 2012 Jul 23.

Transarticular facet screw stabilization and dorsal laminectomy in 26 dogs with degenerative lumbosacral stenosis with instability. Hankin EJ, Jerram RM, Walker AM, King MD, Warman CG. Vet Surg. 2012 Jul;41(5):611-9. doi: 10.1111/j.1532-950X.2012.01002.x. Epub 2012 May 18.

Degenerative lumbosacral stenosis in dogs. Meij BP, Bergknut N. Vet Clin North Am Small Anim Pract. 2010 Sep;40(5):983-1009. doi: 10.1016/j.cvsm.2010.05.006. Review.

Degenerative lumbosacral stenosis in working dogs: current concepts and review. Worth AJ, Thompson DJ, Hartman AC. N Z Vet J. 2009 Dec;57(6):319-30. doi: 10.1080/00480169.2009.64719. Review.

Lumbosacral degenerative stenosis in the dog. The results of epidural infiltration with methylprednisolone acetate: a retrospective study. Janssens L, Beosier Y, Daems R. Vet Comp Orthop Traumatol. 2009;22(6):486-91. doi: 10.3415/VCOT-08-07-0055. Epub 2009 Oct 29.

Review and retrospective analysis of degenerative lumbosacral stenosis in 156 dogs treated by dorsal laminectomy. Suwankong N, Meij BP, Voorhout G, de Boer AH, Hazewinkel HA. Vet Comp Orthop Traumatol. 2008;21(3):285-93.

Results of 2015 Neurology Diplomate Survey: Terminology

Rodney S. Bagley DVM, DACVIM (Neurology and Small Animal Internal Medicine)
Ames, IA

In an effort to determine if clinical neurologist agree on medical terminology used to describe clinical assessments, a study was performed at the ACVIM forum in 2015 by the authors. The following, were goals of this session:

1. Identify terminology associated with dog gait assessment where there is not universal agreement on operational usage
2. Determine the range of current operational definitions
3. Establish a plan for resolution of any discrepancies among board-certified neurologists

Specific Aims

- To collect survey data on the operational literal definition and associated clinical connotation of medical terminology used to describe neurologic signs by expert clinicians (board certified veterinary neurologists)
- To determine from a range of options the most consistent terminology used to describe clinical signs in dogs

Background

Gait is commonly defined as a regularly repeating series of leg movements during walking or running. The nervous system controls the actions of the muscles associated connective tissue important for this function. Bones, joints, and ligaments provide a mechanical support system for movements. Disease of any or all of these components can result in a gait abnormality. Normal walking is produced by the recruitment of stepping reflexes, which alternate between the extensor and flexor muscles.

All board-certified neurologists are human beings. Therefore, they are susceptible to the vast array of cognitive biases present in human beings. These cognitive biases, however, can result in inaccurate medical descriptions of patient abnormality. Medical description inaccuracies have a number of negative consequences in both clinical practice and clinical science. Some of the most common sources of potential for mislabeling are associated with terms such as "ataxia", "hypermetria", "paresis", and "lameness".

Ataxia

Ataxia - Function: *noun*: an inability to coordinate voluntary muscular movements that is symptomatic of some nervous disorders
http://www.merriam-webster.com/medlineplus/ataxia
Definition of *ATAXIA*: an inability to coordinate voluntary muscular movements that is symptomatic of some central nervous system disorders and injuries and not due to muscle weakness —called also *incoordination*
http://www.merriam-webster.com/dictionary/ataxia

Noun -ataxia (***countable and uncountable, plural*** ataxias) - (pathology) Lack of coordination while performing voluntary movements, which may appear to be clumsiness, inaccuracy, or instability.
http://en.wiktionary.org/wiki/ataxia

Ataxia is a failure of muscular coordination or an irregularity of muscle action. In some situations, it is defined as lack of an axis, implying an inability to travel along a straight line. Ataxia is usually associated with a "falling" or "swaying" of the body from side to side. It is also commonly present in association with other forms of abnormal limb movements such as dysmetria.

Ataxia results from a variety of anatomic lesions within the nervous system, most commonly of the cerebellum, vestibular system, and spinal cord sensory pathways. Ataxia, without motor dysfunction, usually implies cerebellar or cerebellar pathway disease. Sensory ataxia due to loss of joint position sense is often made worse by blindfolding the animal, effectively preventing visual compensatory mechanisms.

Hypermetria

Function: *noun*: a condition of cerebellar dysfunction in which voluntary muscular movements tend to result in the movement of bodily parts (as the arm and hand) beyond the intended goal
http://www.merriam-webster.com/medlineplus/Hypermetria

Definition of *HYPERMETRIA*: a condition of cerebellar dysfunction in which voluntary muscular movements tend to result in the movement of bodily parts (as the arm and hand) beyond the intended goal—compare hypometria
http://www.merriam-webster.com/medical/hypermetria
Noun- hypermetria

1. dysmetria in which the patient tends to overshoot the intended target

http://en.wiktionary.org/wiki/hypermetria

Dysmetria is improper estimation of distance during muscular activity. Dysmetria includes both *hypo-* and *hypermetria.* With hypermetria, voluntary muscular movement overreaches the intended goal and is essentially a failure to limit the degree of flexion. With hypometria, voluntary movement falls short of the intended goal. Hypermetria is more commonly recognized than hypometria. Both of these abnormalities are most often associated with lesions of the cerebellum or cerebellar pathways.

Paresis

Function: *noun*
Inflected Form(s): *plural* **pa·re·ses** \-ˌsēz\
1: slight or partial paralysis **2:** general paresis
http://www.merriam-webster.com/medlineplus/paresis

Definition of *PARESIS*-
: slight or partial paralysis
: general paresis
http://www.merriam-webster.com/medical/paresis

Noun -paresis (***plural*** pareses) – 1) A paralysis which is incomplete or which occurs in isolated areas, 2) Inflammation of the brain as a cause of dementia or paralysis.
http://en.wiktionary.org/wiki/paresis

Paresis is derived from the Greek word for relaxation and implies weakness of motor function without complete paralysis (some voluntary motion remains). *Paralysis* implies loss of motor function from either a neurologic or muscular abnormality. The term "paresis," however, has different connotations including limb dysfunction from various causes. In many clinical descriptions, paresis and conscious proprioceptive abnormalities are used interchangeably, with the term "paresis" implying a conscious proprioceptive abnormality.

Lame (ness)
Function: *adjective*
Inflected Form(s): **lam·er; lam·est**
: having a body part and especially a limb so disabled as to impair freedom of movement : physically disabled
—**lame·ly** *adverb*
—**lame·ness** *noun*
http://www.merriam-webster.com/medlineplus/lameness

Definition of *LAME*
: having a body part and especially a limb so disabled as to impair freedom of movement : physically disabled
http://www.merriam-webster.com/medical/lameness

Noun- lameness (***countable and uncountable, plural*** lamenesses) -A impediment to walking due to the feet or legs.
http://en.wiktionary.org/wiki/lameness

The aim of this session is two-fold. First the current state of medical descriptions will be discussed, followed by audience feedback opportunities. The outcome is to record audience perspectives on currently used medical terminology often used to describe abnormalities of gait in dogs.

Materials and Methods (intended)

This study was assessed by the institution review board of Iowa State University as exempt. All participation was voluntary. Appropriate study explanations and consent was obtained prior to participation. ACVIM Board certified veterinary neurologists attending the premiere international scientific meeting of this specialty group were invited to participate in this study as well as other audience members who were not ACVIM board certified neurologists.

The study was performed through audience participation during a 50 minute session of the American College of Veterinary Internal Medicine Meeting between June 3, 2015 in Indianapolis Indiana, USA.

Data was collected in hard copy provided to each of the participants. The study was performed immediately at the beginning of the session without any other topic introduction.

The survey instrument consisted of 4 components. In component 1, participants were asked to match in a multiple choice answer format the following words with a standard medical dictionary definition:

- Ataxia
- Paresis
- Hypermetria
- Lameness

In component 2, 10, 30 second video clips of dogs at gait that were either normal or had spontaneously occurring gait abnormalities. Participants were asked to determine their assessment using the aforementioned subsequently medical terminology they would enter into the patient medical record to describe the their assessment from a list of appropriate terminology.

In component 3, participants were provided a free text opportunity to record their operational clinical definitions for these abnormalities.

In component 4, participants were asked to provide demographic information on diplomate status and years of clinical practice experience.

In this session the **results** of this assessment will be shared with the audience and an action plan for future assessments will be formulated. There were 71 participants that were board-certified neurologist and 35 other participants. The data presented will be from this cohort of individual participants.

REFERENCES

Bagley, R. S. 2005. Fundamentals of Clinical Neurology, Blackwell Publishing.
Brown, C. M. 1986. Dog Locomotion and Gait Analysis. Hoflin Publishing.
Dewey C.W. 2008. A Practical Guide to Canine and Feline Neurology, Wiley Blackwell.

AUTHOR INDEX

AUTHOR INDEX

AUTHOR INDEX

AUTHOR INDEX

AUTHOR INDEX

American College of Veterinary Internal Medicine (ACVIM)
1997 Wadsworth Blvd.
Lakewood, CO 80214-5293
www.acvim.org

ISBN 978-1-5108-2540-6
9 781510 825406

CURRAN ASSOCIATES INC.
proceedings.com

American College of Veterinary Internal Medicine (ACVIM) Forum 2016

Denver, Colorado, USA
8 - 11 June 2016

Volume 2 of 2

ACVIM
American College of Veterinary Internal Medicine

American College of Veterinary Internal Medicine (ACVIM) Forum 2016

Denver, Colorado, USA
8 - 11 June 2016

Volume 2 of 2

ISBN: 978-1-5108-2540-6

Printed from e-media with permission by:
Curran Associates, Inc.
57 Morehouse Lane
Red Hook, NY 12571

CURRAN ASSOCIATES INC.
proceedings
.com

Some format issues inherent in the e-media version may also appear in this print version.

Printed by Curran Associates, Inc. (2016)

Address correspondence to:
AMERICAN COLLEGE OF VETERINARY INTERNAL MEDICINE (ACVIM)
1997 WADSWORTH BLVD.
LAKEWOOD, CO 80214-5293
www.ACVIM.org

Additional copies of this publication are available from:

Curran Associates, Inc.
57 Morehouse Lane
Red Hook, NY 12571 USA
Phone: 845-758-0400
Fax: 845-758-2633
Email: curran@proceedings.com
Web: www.proceedings.com

TABLE OF CONTENTS

VOLUME 1

CARDIOLOGY

ONCOLOGY

NEUROLOGY

SMALL ANIMAL-ENDOCRINOLOGY

SMALL ANIMAL-GASTROENTEROLOGY

SMALL ANIMAL-HEMATOLOGY

SMALL ANIMAL-HEPATOLOGY

SMALL ANIMAL-INFECTIOUS DISEASE

SMALL ANIMAL-NUTRITION/METABOLISM

SMALL ANIMAL-NEPHROLOGY/UROLOGY

SMALL ANIMAL-OTHER

SMALL ANIMAL-PHARMACOLOGY

SMALL ANIMAL-RESPIRATORY

EQUINE

FOOD ANIMAL

AMERICAN ASSOCIATION OF VETERINARY NUTRITION/AMERICAN COLLEGE OF VETERINARY NUTRITION

AMERICAN ASSOCIATION OF VETERINARY NUTRITION/AMERICAN COLLEGE OF VETERINARY NUTRITION FOOD ANIMAL

AMERICAN COLLEGE OF VETERINARY CLINICAL PHARMACOLOGY/AMERICAN ACADEMY OF VETERINARY PHARMACOLOGY AND THERAPEUTICS

AMERICAN SOCIETY FOR VETERINARY CLINICAL PATHOLOGY

AMERICAN SOCIETY OF VETERINARY NEPHROLOGY & UROLOGY

ASSOCIATION OF VETERINARY HEMATOLOGY AND TRANSFUSION MEDICINE

BUSINESS MANAGEMENT

COMPANION ANIMAL PARASITE COUNCIL

COMPARATIVE GASTROENTEROLOGY SOCIETY

INTERNATIONAL SOCIETY FOR COMPANION ANIMAL INFECTIOUS DISEASES

LIVER STUDY GROUP

ONE MEDICINE

SOCIETY FOR COMPARATIVE ENDOCRINOLOGY

VETERINARY COMPARATIVE RESPIRATORY SOCIETY

VETERINARY EMERGENCY AND CRITICAL CARE SOCIETY

CARDIOLOGY

NEUROLOGY

VOLUME 2

ONCOLOGY

SMALL ANIMAL INTERNAL MEDICINE

EQUINE

FOOD ANIMAL

LARGE ANIMAL INTERNAL MEDICINE

TECHNICIAN PROGRAM

ADDITIONAL PAPER

Exploiting Proteases and their Inhibitors as Targets for the Treatment of Canine Intracranial Tumors

John H. Rossmeisl, Jr. DVM, MS, DACVIM
Blacksburg, VA

INTRODUCTION

The post-genomic era has resulted in a rapid shift in the understanding of protease function. They are no longer considered blunt instruments limited to protein catabolism via hydrolysis. Their actions extend far beyond these nonspecific degradative processes, as proteases catalyze highly specific reactions producing new protein products that are crucial to the control of numerous biological functions.[1] Proteases participate in cell proliferation and differentiation, tissue remodeling, DNA replication and transcription, coagulation, immunity, angiogenesis, neurogenesis, autophagy, necrosis, and apoptosis. Given their important and ubiquitous roles in the cellular life cycle, proteolytic alterations are fundamental to the pathogenesis of conditions such as neuroinflammation, neurodegenerative disorders, and cancer.[1]

PROTEASES- CLASSIFICATION, FUNCTION AND ROLES IN CANCER

Proteases, which can also be referred to as peptidases or proteinases, are found in all prokaryotic and eukaryotic organisms, as well as viruses. Traditionally, the common denominator among proteolytic enzymes was their initiation of protein catabolism by hydrolysis of peptide bonds that link amino acids into a polypetide chain.[1] Proteases have also been classified into soluble and intramembranous forms. Intramembrane proteases (IMP) are themselves integral membrane proteins of the lipid bilayer of cells whose function is to cleave the transmembrane domains of other proteins. However, the classification of proteases has undergone significant expansion, editing, and revision in the last decade. Currently, seven groups of proteases have been described (**Table 1**), including the asparagine peptide lysases, whose catalytic action does not rely on hydrolysis.[2] A current, searchable database describing the evolutionary classification of protease superfamilies can be found at the MEROPS website (https://merops.sanger.ac.uk/).[3] MEROPS is a hierarchal system that classifies proteases into families and clans based on statistically significant similarities in structure, mechanism and catalytic residue order, and families may contain dozens to hundreds of interrelated proteases.[3]

Table 1- Classification of Proteases and Their Inhibitors

Class (Active Site)	**Subtypes**	**Examples and Comments**	**Inhibitors**
Aspargine peptide lysases		Picornavirus capsid protein	Unknown
Aspartic	Soluble	Pepsin, cathepsin, renin, HIV protease	Pepstatin, Ritonavir, Indanivir, Zankiren
	Intramembrane	Presenilins, signal peptide peptidase (SPP)	
Cysteine	Soluble	Bromelain, calpain, caspases, papain	Cystatins
	Intramembrane		
Glutamic acid		Primarily found in pathogenic fungi	
Metalloproteases	Soluble	Stromelysins, angiotensin converting enzyme	TIMPs, Doxycycline, Marimastat
	Intramembrane	Site-2 protease (S2P) family	
Serine/Threonine	Soluble	Trypsin, chymotrypsin, elastase, clotting enzymes, proteasome	Serpins, Bortezomib
	Intramembrane	Rhomboid family	

Nearly 60 years of research has implicated aspartic, cysteine, metalloproteinase, serine, and threonine proteases in numerous fundamental oncogenic events. Early studies focused on the roles of proteases in the late stages of tumor progression, demonstrating the local proteolytic mechanisms underpinning tumor invasion via attachment of tumor cells to the basement membrane (BM), degradation of the extracellular matrix (ECM), and subsequent migration of tumor cells through the proteolytically modified microenvironment.[4] Proteases are also critical for the step-wise events that define metastasis, including intravasation, arrest, extravasation, distant invasion, and angiogenesis. More recent evidence has emerged demonstrating the involvement of proteases in tumor growth both at primary and metastatic sites, as several classes of proteases have been shown to be tumor growth promoters, acting via regulation of gene and tumor growth factor expression.[5] It has been documented in numerous human and several canine cancers that increased protease expression and activity, altered protease localization, and imbalances between proteases and their inhibitors are associated with tumor progression.[4-7] This review will focus on members of serine protease and matrix metalloprotease (MMP) families, as the pathophysiology roles of these peptidases have been well described in human brain tumors.

A characteristic feature of high-grade meningiomas and malignant gliomas is their tendency to locally or diffusely infiltrate the neural parenchyma, which often precludes complete surgical resection. Proteases play a decisive role in this invasive process, by degradation of brain ECM, cellular adhesion molecules, or by influencing the activity of growth and chemotactic factors. MMP, which are also designated matrixins, are a family of calcium dependent endopeptidases that can collectively cleave the majority, if not all, of the components of the extracellular matrix (ECM) and basement membrane (BM), as well as act on a number of biologically active molecules.[8] MMP differ from other proteases in that they are dependent on zinc ions for their activity. The majority of MMPs are synthesized as inactive zymogens. They are kept inactive by an interaction between a cysteine-sulphydryl group in the propeptide domain and the zinc ion bound to the catalytic domain. Their activation requires proteolytic removal of the propeptide domain, which under physiological conditions is tightly regulated by tissue inhibitors of metalloproteases (TIMP). On the basis their specificity for ECM components, MMPs are divided into collagenases, gelatinases, stromelysins, matrilysins, and membrane-type MMPs (MT-MMPs). In humans, MMP2, MMP9, and MT1-MMP have been documented to participate in glioma and meningioma cell invasion and motility, and expression of these proteases has been associated with tumor grade, neuroinvasive phenotype, VEGF and EGFR expression, intratumoral cyst formation, and extent of peritumoral edema.[9,10]

Expression of multiple components of the plasminogen-plasmin system within tumors has been shown to have prognostic significance in humans, including in bone, brain, breast, colon, lung, ovary, prostate, and renal cancers.[11] Uroplasminogen activator (uPA) is a serine protease that catalyzes the conversion of inactive plasminogen into plasmin, a broadly-acting enzyme able to degrade a variety of extracellular matrix proteins and activate MMPs, growth factors, and pro-uPA. uPA is involved in the regulation of coagulation, angiogenesis, and tumor growth.[11] uPA binding to its specific receptor (uPAR) directs plasmin activity to the migrating tumor cell surface. However, the uPAR receptor is a multifunctional protein, capable of interactions with a wide variety of other membrane receptors and external ligands, and can function as an adhesion-modulating and intracellular signaling receptor. uPA and uPAR overexpression has been documented in multiple types of human brain tumors, and uPA/uPAR dysregulation has been implicated in brain tumor gene expression, cell proliferation, adhesion, migration, and invasion.[12]

Given the decisive role of proteases in the pathogenesis of brain tumors and other cancers, inhibition of the proteasome is a rapidly evolving field in oncology. Combination protocols with conventional cytotoxic agents, targeted biologics and selective proteasome inhibitors, such as bortezomib, have resulted in high remission rates and durable responses in some hematological malignancies.[13] These successes in multiple myeloma have led to the development of multiple novel proteasome inhibitors, as well as a resurgence of the investigation of anti-cancer properties and indications for older generation protease inhibitors. Proteasome inhibition has several potential clinical benefits for the treatment of cancer, including the ability to overcome conventional chemoresistance, direct induction cell cycle arrest and tumor apoptosis, and targeting the tumor microenvironment.[13] Despite extensive study of the mechanisms of action of the individual components of the proteasome in tumor cells *in vitro* and in animal models, and demonstration of the ability to achieve tumor control in experimental models of brain cancer by interrupting putative proteolytic pathways involved in tumor progression, modulation of the proteasome thus far has not been able to be translated into any clinical benefit in humans with brain tumors.

PROTEASE EXPRESSION AND ACTIVITY IN CANINE BRAIN TUMORS

The expression and activity of several proteases have been described in canine intracranial tumors. In an immuohistochemical study examining Grade I meningiomas, immunoreactivity to MMP-9 was demonstrated in 22/22 cases, immunoreactivity to latent proMMP-2 in 21/22 cases, and immunoreactivity to latent/active proMMP-2 in 9/22 cases.[14] Mariani and colleagues demonstrated that dogs with intracranial tumors were significantly more likely than control dogs to have detectable MMP-9 in cerebrospinal fluid (CSF), and documented latent MMP-9 activity in the CSF of dogs with intracranial meningiomas, gliomas, pituitary tumors, choroid plexus tumors, and lymphoma.[15] The presence of a detectable MMP-9 concentration was also associated with higher CSF cell counts and protein concentrations. Similarly, an earlier study demonstrated that CSF leukocytes were the source of elevated concentrations of CSF MMP-9 observed in dogs with intracranial tumors.[16] Indirect evidence of involvement of several serine proteases involved in the coagulation cascade in the pathogenesis of canine brain tumors has also been reported. In one study fibrin/fibrinogen and D-dimer immunoreactivity was documented in 100% (29/29) of canine intracranial meningiomas.[17] The same study found a significant difference between the semi-quantitative fibrin/fibrinogen immunohistochemical score among the histologic grades of meningiomas. Intratumoral fibrin/fibrinogen immunoreactivity was universally present in 15 oligodendroglial, 6 astrocytic, and 3 mixed glial tumors examined in another investigation.[18] Collectively, these studies demonstrate that extravascular expression of serine proteases involved in the coagulation cascade occurs in canine brain tumors, and provide justification of the development of therapies designed to modulate these pathways.

We have characterized the expression of uPA/uPAR and their inhibitors in canine intracranial neoplasms as the initial phase of development and validation of uPAR-targeted brain tumor therapeutics. Using immunohistochemistry, ELISA, immunoblotting, casein zymography, and RT-PCR, we demonstrated the expression and activity of the urokinase plasminogen activator system in multiple tumoral microenvironmental niches, including neoplastic cells, stroma, and vasculature of canine meningiomas, gliomas, and choroid plexus tumors. Relative to normal brain tissues, uPA activity and expression was greater in all tumor types examined, and correlated with tumor grade in gliomas. We found no differences in concentrations of PAI-1 between normal and neoplastic brain tissues.

PROTEASE TARGETED THERAPIES FOR CANINE BRAIN TUMORS

Targeted Oncolytic Viruses

Newcastle Disease Virus (NDV), an avian paramyxovirus, has a long and established history as an oncolytic agent, being capable of selectively killing cancer cells through the induction of apoptosis.[19] NDV is particularly attractive as an agent for the treatment of brain tumors, as it is capable of crossing the blood-brain barrier and exerting biological activity in the brain when administered systemically, and vectors have been developed for use as vaccines and safely administered to dogs and cats.[20] Several human Phase I/II clinical trials using naturally occurring strains of NDV such as 73-T, MTH68, PV701 and HUJ have been successfully conducted. Recently, the safety and anti-tumor activities of the NDV mutant (NDV-HUJ) administered IV were studied in a phase I/II clinical trial in humans with recurrent malignant glioma. Each patient received eight cycles of NDV-HUJ IV daily for the study duration. This regime was well tolerated, with no significant toxicities observed, but no maximum tolerated dose was reached. A single patient had a complete tumor response and three patients were deemed to have 'long-term' survival.[21] It is evident from these clinical studies that optimization of viral dose and delivery route without compromising the safety margin is imperative. Systemic administration of NDV requires the use of multiple doses in large quantities.

The laboratory of my collaborator, Dr. Kumaran Subbiah, focuses on the specific use NDV as an anticancer agent. The unique requirement that NDV fusion (F) protein be activated on the target cell surface provides an opportunity to re-engineer this protein to make its activation dependent on tumor cell surface proteases. Based on the known overexpression of MMPs and uPAR in many canine brain tumors, we have developed several protease targeted recombinant NDV strains that have been demonstrated to be well tolerated following systemic administration and have potent *in vitro* and *in vivo* antitumor activity against a range of human and animal cancers, including brain tumors. We have also recently identified the reason for the low therapeutic index of systemically administered NDV in human clinical trials. We found that NDV grown in chicken eggs is highly susceptible to the neutralizing action of mammalian complement.[22] To date, most human clinical trials used egg-grown NDV. Further, we have shown that amplifying NDV in cells of homologous species allows the virus to incorporate species-specific complement inhibitors on the surface and enable the virus to evade complement.[22]

We are currently conducting a Phase I clinical trial in dogs to evaluate the safety, tolerability, and preliminary efficacy of parenterally administered recombinant LaSota-uPA (rLAS-uPA) NDV strain in dogs with intracranial meningiomas. The trial follows a conventional 3+3 dose-escalation design with predetermined dosing cohorts. The safety and tolerability of rLAS-uPA is determined every 14 days for six weeks through assessment of toxicity by standard Veterinary Cooperative Oncology Group Common Terminology Criteria for Adverse Events (VCOG-CTCAE) metrics. Antitumor activity and local treatment toxicity within the brain will be determined by comparing serial MRI scans of the brain. We are also evaluating host immunological endpoints. Preliminary results of this clinical trial will be discussed during this session.

New Tricks for Old Drugs? Pharmacologic Protease Inhibition for Brain Tumors

Several classes of protease inhibitors (PI) have been shown to have demonstrable anticancer activity in human cancer cells.[23] Nelfinavir, a potent inhibitor of human HIV-1 protease, and the angiotensin converting enzyme inhibitor captopril have both been shown to have anti-neoplastic properties, and have known pharmacokinetics and favorable adverse effect profiles in the dog. We will present data demonstrating that these PIs can inhibit proliferation and cause death of human and canine malignant glioma cell lines *in vitro*, which is mediated at least in part by inhibition of MMP-2 and MMP-9.

REFERENCES

[1]Neurath H, et al. Proc Natl Acad Sci USA 1976; 73:3825. [2]Rawlings ND, et al. J Biol Chem 2011;286:38321. [3]Rawlings ND, et al. Nucleic Acid Res 2014;42:D503. [4]Liotta LA, et al. Lab Invest 1983;49:636. [5]Berchem GJ, et al. Oncogene 2002; 51:5951. [6]Golshahi H, et al. Bull Vet Institut Pulawy 2013; 57:587. [7]Faleiro MBR, et al. Ci Anim Bra Golania 2013;14:237. [8]Visse R, et al. Circ Res 2003;92:827. [9]Wang M, et al. Brain Tumor Pathol 2003;20:65.[10]Nordqvist AC, et al. J Neurosurg 2001;95:839.[11]Gonias SL, et al. Front Pharmacol 2015;6:154. [12]Rao JS. Nat Rev Cancer 2003; 3:489.[13] Montagut C, et al. Clin Transl Oncol 2006 8:313. [14] Beltran E, et al. Vet Pathol 2013; 50:1091.[15] Mariani CL, et al. Am J Vet Res 2013;74:122. [16] Turba ME, et al. J Neurooncol 2007: 81:123. [17] Font C, et al. Vet J 2015; 3:426. [18] de la Fuente C, et al. Vet J 2014; 3:387. [19]Cassel WA, et al. Cancer 1965; 18: 863. [20]Ge J, et al. J Virol 2011; 85:8421. [21]Freeman AI, et al. Mol Ther 2006; 13: 221. [22]Biswas M, et al. J Virol 2012; 86(23): 12708. [23]Arodola OA, et al. Drug Des Devel Ther 2015; 13:6055. [24]Shetty BV, et al. Antimicrob Agents Chemother 1996; 40:110.

MRI Principles and Pulse Sequence Selection for Neuroimaging

Derek D. Cissell, VMD, PhD, DACVR
Davis, CA

PRINCIPLES OF MAGNETIC RESONANCE IMAGING

Magnetic resonance imaging (MRI) is based on the phenomenon of nuclear magnetic resonance associated with certain atoms, including hydrogen protons (^{1}H), carbon-13 (^{13}C), and sodium (^{23}Na). These atoms possess unbalanced electrical charge associated with the distribution of protons and neutrons comprising their nuclei. Additionally, all nuclei exhibit intrinsic angular momentum or "spin." The combination of unbalanced electrical charge and spin causes these atoms to exhibit magnetic dipoles and act like tiny spinning bar magnets. Hydrogen protons, or simply "protons," are the most abundant atoms with magnetic dipoles in biologic tissues and are the source of signal in MRI. In the absence of an external magnetic field, there is no order among the orientations of protons and nothing interesting happens from an electromagnetic perspective. Placing a group of protons into a magnetic field, such as the bore of an MRI system, causes the protons to align parallel ("spin up") or anti-parallel ("spin down") to the magnetic field. After a few seconds, the group of protons will reach an equilibrium state in which a slight excess of protons will occupy the lower energy, spin up state. The slight excess of protons in the spin up state will generate a net magnetization parallel to the magnetic field, defined as **longitudinal magnetization**. Interestingly, protons do not perfectly align with the magnetic field, but orient themselves at an angle to it. This imperfect alignment coupled with spin causes the protons to "precess," or wobble, around the axis of the magnetic field. The frequency of precession, known as the **Larmor frequency**, will be specific to a given element and proportional to the magnetic field strength. At equilibrium, the position or "phase" of each proton in its precessional orbit will be random. Among a large group of protons, randomly distributed phases will cancel each other out such that there is no net magnetization perpendicular to the magnetic field. Therefore, at equilibrium in a magnetic field, a group of protons will exhibit a positive longitudinal magnetization and zero **transverse magnetization.**

To generate an image, it is necessary to disrupt the equilibrium state by briefly applying a secondary magnetic field that oscillates at the Larmor frequency. This secondary magnetic field is referred to as the **radiofrequency (RF) pulse**. Resonance between the RF pulse and proton precession stimulates some protons to flip from the spin up state to the spin down state. Additionally, the RF pulse induces protons to align the position of their precessional orbits, a condition described as being "in phase." RF pulse application reduces the net longitudinal magnetization and generates transverse magnetization in the population of protons. Transverse magnetization is the signal detected by the MRI system for creating an image. Immediately following an RF pulse, the protons begin to return to the equilibrium state by two independent processes: (1) recovery of longitudinal magnetization and (2) decay of transverse magnetization. Longitudinal recovery, also known as **T1 relaxation** or spin-lattice relaxation, occurs as excited protons in the spin down state slowly return to the lower energy, spin up state. Transverse decay (aka **T2 relaxation** or spin-spin relaxation) is the relatively rapid process by which protons lose their phase coherence and return to a random distribution of precessional orbits. The rates at which longitudinal recovery and transverse decay take place are determined by tissue properties defined as T1 and T2 relaxation times, respectively. A tissue's T1 and T2 relaxation times and its proton density are the major intrinsic properties that contribute to its appearance in an MR image.

Anatomy of a pulse sequence

A MRI pulse sequence consists of carefully coordinated RF pulses, gradient magnetic fields, and signal detection to achieve (1) excitation, (2) spatial encoding, and (3) refocusing (Figure 1). In addition to intrinsic tissue properties, the timing of specific events in a pulse sequence has a major influence on the appearance of the final image. Innumerable pulse sequences have been developed for specific applications and increasingly exotic pulse sequences are constantly being added to the MRI lexicon, but all pulse sequences are based on the same essential principles.

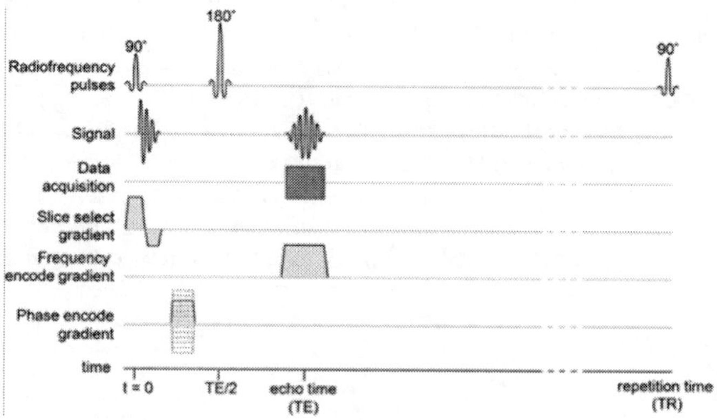

Figure 1. Schematic of events occurring during a spin echo pulse sequence.

Excitation is the first event in most pulse sequences and occurs via application of an RF pulse. The amount of time between sequential excitation pulses is defined as the pulse sequence **repetition time (TR)**. A 90° RF pulse flips exactly enough protons so that equal numbers will inhabit the spin up and spin down states, resulting in zero net longitudinal magnetization at the end of the pulse. Phase coherence following a 90° RF pulse will also result in the maximum possible transverse magnetization. Considering a large group of protons, application of a 90° RF pulse can be thought of as tipping the net magnetization from the longitudinal axis to the transverse plane. A "tip angle" of 90° is commonly used for excitation RF pulses.

Spatial encoding is required for image formation. The transverse magnetization that forms after an excitation RF pulse results in signal emanating from excited protons in a tissue, but this initial signal does not contain spatial information and cannot be used for image formation via basic MRI pulse sequences. Spatial encoding involves applying a series of three gradient coils that add to or subtract from the main magnetic field in a linear, spatially dependent fashion along the x-, y-, and z-axes of the MRI system. Because a proton's Larmor frequency depends on the magnetic field strength it experiences, a spatially varying magnetic field gradient (created by a gradient coil) will cause protons in a tissue to precess with varying frequencies depending on their location in space. Three steps are employed for spatial encoding during conventional MRI pulse sequences: (1) slice selection, (2) phase encoding, and (3) frequency encoding. Slice selection occurs by turning on a gradient coil during the excitation RF pulse to selectively excite a thin "slice" of protons in a patient. The strength of the magnetic field gradient and frequencies of the RF pulse will determine the location and thickness of the excited slice. Spatial information within an excited slice is imprinted on protons by applying phase encode and frequency encode gradients. A phase encode gradient is applied between excitation and data acquisition to briefly speed up precession and advance the phase of some protons while slowing precession and retarding the phase of other protons depending on their location along one axis within the slice. Image acquisition typically requires many phase encode steps and only a single or few phase encode steps are usually performed per TR interval. Thus, the need for phase encoding is often the rate-limiting step in MR image acquisition. Lastly, frequency encoding occurs by applying a gradient coil at the time of data acquisition to vary the frequency of signal emanating from protons along the final remaining axis within the slice. Thus, at the time of data acquisition, the recorded signal is encoded with spatial information pertaining to the three-dimensional location of protons within a patient.

Refocusing is necessary because the initial signal that forms immediately after an excitation RF pulse rapidly decays and is not be detectable by the time spatial encoding has taken place. Refocusing reverses the dephasing effects of gradient coil application and some magnetic field inhomogeneity to generate an "echo" of the original signal. The time between excitation and the center of echo formation is defined as the pulse sequence **echo time (TE)**. Refocusing can be elicited by one of two means: (1) a 180° RF pulse is used to generate an echo for **spin echo** imaging and (2) reversing the application of gradient coils generates an echo for **gradient echo** imaging. Many clinically used pulse sequence generate multiple echoes per excitation to reduce image acquisition time and/or alter the image contrast. **Fast spin echo (FSE)** imaging (aka, turbo spin echo or rapid acquisition with refocused echoes) generates multiple echoes by applying multiple 180° refocusing RF pulses at regular intervals after excitation. Because signal is derived from multiple echoes, FSE pulse sequences will produce images with contrast that differs somewhat from single echo images. Multi-echo imaging can also be performed using gradient echo refocusing, but the number of useful echoes that can be acquired by gradient echo imaging is limited relative to FSE imaging.

PULSE SEQUENCES FOR NEUROIMAGING

The appearance of a MR image depends on intrinsic tissue properties as well as extrinsic factors, including the type of pulse sequence and certain variable pulse sequence settings such as the TR and TE. Intrinsic tissue properties that contribute to a tissue's signal intensity (i.e., how bright or dark the tissue appears in an image) include a tissue's T1 relaxation time, T2 relaxation time, and proton density. Movement of water via flow or diffusion can also impact the appearance of a MR image. Pulse sequences have been developed to exploit specific intrinsic tissue properties and biological processes (e.g., blood flow) for generating contrast in an image.

Basic pulse sequences

The major MRI pulse sequences used for general applications include T1-weighted (T1W), T2-weighted (T2W), or proton density-weighted (PDW) sequences. A T1W pulse sequence employs a relatively short TR and short TE to generate images with contrast between tissues with different T1 relaxation times. A tissue with a short T1 relaxation time, such as fat, will rapidly recover its longitudinal magnetization between sequential excitation RF pulses and develop a large transverse magnetization following each excitation pulse. The short pulse sequence TE will minimize transverse decay of the signal and lead to a bright appearance of fat in a T1W image. Conversely, a tissue with a long T1 relaxation time, such as cerebrospinal fluid (CSF), will not completely recover its longitudinal magnetization becoming partially saturated by sequential excitation pulses. Partial saturation makes less longitudinal magnetization available to be tipped into the transverse plane during excitation, reducing the amplitude of the transverse magnetization and causing a dark appearance of CSF on a T1W image. The short TE of a T1W pulse sequence minimizes the effect of T2 relaxation times on contrast in the image. Although T1W images are somewhat insensitive to detecting many pathologic processes associated with neuroimaging, T1W images are highly sensitive to the presence of paramagnetic materials and other materials with short T1 relaxation times. Paramagnetic materials shorten the T1 relaxation times of other tissues in their presence. Examples include gadolinium, methemoglobin, and melanin, all of which will appear bright on a T1W image. Gadolinium is not present in the body naturally, but is used for intravenous MRI contrast media to increase the conspicuity of blood vessels and some pathologic processes. Other examples of tissues that appear hyperintense (i.e., bright) on T1W images include proteinaceous fluids, some mineral, fat, cholesterol, and slow blood flow. Low-protein fluids and tissues with high free water content will appear relatively hypointense (i.e., dark) on T1W images.

A T2W pulse sequence uses a long TR and relatively long TE. A long TR allows most or all tissues to completely recover their longitudinal magnetization between excitation RF pulses, thus minimizing the contribution of differing T1 relaxation times on image contrast. The relatively long TE of a T2W pulse sequence allows tissues with different T2 relaxation times to experience different degrees of transverse decay between excitation and echo formation. A tissue with a long T2 relaxation time, such as CSF, will slowly lose transverse magnetization and appear bright in a T2W image. A tissue with a short T2 relaxation time, such as muscle, will rapidly lose transverse magnetization and appear dark in a T2W image. T2W images are sensitive to changes in tissue water content and are often considered to be "pathology weighted." Increased water content in tissues due to edema, infection, or neoplasia all result in hyperintensity on T2W images. Collagenous tissues tend to have short T2 relaxation times and appear hypointense on T2W images. Clinical T2W images are nearly always obtained using FSE pulse sequences to reduce image acquisition time. While most expected tissue contrast is preserved on T2W FSE images, fat will appear bright despite its relatively short T2 time due to the contribution of echoes with relatively short TE to the final image.

PDW images are obtained using pulse sequences with long TR and short TE. The long TR allows most tissues to recover their longitudinal magnetization between excitation RF pulses and the short TE minimizes transverse decay prior to image data acquisition. Thus, images obtained using PDW pulse sequences are minimally affected by tissue T1 or T2 relaxation times and contrast is generated primarily by differences in tissue proton densities. PDW images provide excellent signal:noise ratio, but are generally insensitive to detecting pathology. PDW images can be useful for evaluating tissues with short T2 relaxation times that would appear dark on a T2W image. It is important to note that some intrinsic tissue properties can have a disproportionate affect on signal intensity in an image. For example, a tissue with low proton density or very short T2 relaxation time will tend to appear dark on all images, regardless of the intended image weighting.

Tissue nulling and inversion recovery

For some applications, it may be desirable to null the signal from a specific tissue. There are several ways to achieve tissue nulling such as inversion recovery, frequency-selective saturation, and chemical shift-based techniques such as iterative decomposition of water and fat with echo asymmetry and the least-squares estimation (IDEAL)[1,2].

Commonly used inversion recovery pulse sequences include short tau inversion recovery (STIR) and fluid attenuating inversion recovery (FLAIR), which null signal from fat and low-protein fluid, respectively. Inversion recovery pulse sequences begin with a 180° inversion RF pulse followed by a delay, known as the inversion time (TI), before the excitation RF pulse. The remaining events of the pulse sequence, including spatial encoding, refocusing, and data acquisition, proceed as for other spin echo and FSE pulse sequences. The purpose of the initial inversion pulse is to induce negative longitudinal magnetization. During the TI, longitudinal magnetization recovers in tissues according to their T1 relaxation times. If a tissue passes through the point of zero net longitudinal magnetization at the time of the excitation RF pulse, there will be no magnetization to tip into the transverse plane and the tissue will not contribute signal to the image. A STIR pulse sequence has a relatively short TI, such that fat will have zero longitudinal magnetization at the time of excitation and appear black in the image. STIR pulse sequences are typically designed with T2-weighting and are useful for detecting high-fluid lesions in regions where the presence of bright fat signal may mask the lesion on T2 FSE images. STIR images are particularly useful for evaluating lesions in trabecular bone, the epidural space, and retrobulbar space [3,4]. FLAIR images are obtained using pulse sequences with long TI in order to null the signal from low-protein fluids, such as CSF. Proteinaceous fluids, edema, and cellular lesions will not be suppressed and appear hyperintense on FLAIR images, making this technique especially useful for detecting periventricular or subarachnoid lesions and for differentiating normal CSF or low-protein fluid from exudate associated with inflammatory disease processes, e.g., abscesses or empyema[5,6].

Frequency selective saturation relies on the slightly different Larmor frequencies of protons in water vs. protons in fat. In a 1.5 T MRI system, protons will precess at a frequency ~210 Hz less than water protons. Applying a saturation RF pulse at the specific Larmor frequency of fat protons, it is possible to null the signal from fat such that fat will appear dark or black in the image. Chemical saturation can be performed as part of most conventional pulse sequences to acquire T1W, T2W, or PDW (or other) images with fat saturation. One of the major applications of chemical fat saturation is during acquisition of post-contrast T1W images to detect contrast enhancement in tissues with a high fat content, such as the retrobulbar or epidural spaces. A disadvantage of chemical saturation is that inhomogeneity of the primary magnetic field or RF pulse will result in incomplete or heterogeneous fat saturation. Inversion recovery pulse sequences typically achieve more uniform tissue nulling than chemical fat saturation.

The chemical shift-based concept of tissue nulling is also based on the different Larmor frequencies of protons in water and fat. Due to their different frequencies of precession, peaks and troughs in the signal waves originating from fat and water will occur at slightly different times. As the signal from both tissues develops, there will be specific moments when the signal for fat and water peak at the same time ("in phase") and other moments when signal from water peaks and fat signal reaches a trough ("out of phase"). IDEAL MRI acquires data with multiple TE's to capture the in-phase and out-of-phase signals to generate separate images depicting only water signal or only fat signal. IDEAL MRI of the spine has been reported to achieve improved fat nulling and image quality compared to frequency-selective fat suppression[2].

T2*, DWI, and other pulse sequences

Other pulse sequences commonly used for neuroimaging include T2*-weighted (T2*W), diffusion weighted imaging (DWI), and rapid acquisition pulse sequences. T2*W images are produced using a gradient echo pulse sequence with relatively long TE. Gradient echo imaging incompletely reverses the effects of magnetic field inhomogeneity compared to spin echo imaging, resulting in more rapid decay of transverse magnetization according to a rate constant known as the **T2* relaxation time.** This behavior of gradient echo pulse sequences makes T2*W images susceptible to subtle fluctuations in magnetic field homogeneity. T2*W images are very sensitive to the presence of magnetically heterogeneous materials and their primary clinical application is for detection of hemorrhage. As hemoglobin in a hematoma progresses from oxyhemoglobin to deoxyhemoglobin, methemoglobin, and hemosiderin / ferritin, the local magnetic field becomes disrupted, causing rapid T2* decay and a dark appearance of acute and subacute hemorrhage on T2*W images. Chronic hematomas may eventually develop a hyperintense center surrounded by a dark rim on T2*W images[7].

DWI takes advantage of the exquisite sensitivity of MRI to movement of water. DWI is based on pulse sequences with added gradient pulses that intentionally dephase then rephase signal in a spatially dependent manner. Protons associated with water that move due to diffusion between the dephasing and rephrasing gradient pulses will not be appropriately rephased. Thus, freely diffusing water molecules will contribute less signal to the image. The strength of diffusion weighting in a diffusion-weighted image is denoted by its b-value, which depends on the magnitude and time between the dephasing and rephasing gradients. DWI pulse sequences typically produce two or more images including a baseline T2W image with minimal diffusion weighting ($b \approx 0$ s/mm^2) and a strongly diffusion weighted image (e.g., $b = 1000$ s/mm^2). Comparison of the baseline and diffusion weighted images can yield a subjective assessment of diffusion restriction with freely diffusing fluid (e.g., normal CSF) appearing dark and restricted diffusion appearing bright in the diffusion weighted image. Apparent diffusion coefficients (ADC) for each pixel in the image can be calculated to create an ADC map for quantitative analysis of diffusion restriction. Although DWI is often used to diagnose the presence of vascular infarcts that cause diffusion restriction, other disease processes also cause diffusion restriction in human and veterinary patients[8]; however, in human neuroimaging, the degree and distribution of diffusion restriction in conjunction with other MRI findings help inform differential diagnoses and response to therapy[9].

Rapid acquisition pulse sequences typically rely on acquiring many phase encode steps in a single TR interval and/or acquiring less image data than conventional pulse sequences. Echo planar imaging (EPI) is an extreme form of multiple echo acquisition in which up to 128 echoes are refocused following a single excitation pulse and each echo completes a different phase encode step. Using EPI, it is possible to complete all phase encode steps in a single TR interval and acquire an image in only a few seconds. EPI pulse sequences usually result in heavy T2-weighting due to the long effective TE that results from acquiring many echoes after a single excitation. It should be noted that EPI forms the basis of many clinically used DWI pulse sequences, allowing for rapid acquisition of diffusion-weighted images and ADC maps. Another approach to rapid image acquisition is to reduce the amount of information acquired for image construction. For example, partial k-space filling pulse sequences rely on the principle that conventional pulse sequences acquire more data than is necessary to generate an image. Partial k-space filling pulse sequences acquire less data than conventional pulse sequences to reduce image acquisition time, but sacrifice signal:noise ratio. A number of different strategies for partial k-space filling pulse sequences have been developed with various effects on image contrast, signal:noise ratio, and contrast. The fastest available pulse sequences typically combine EPI and partial k-space filling to acquire images very rapidly. Although spatial resolution and signal:noise ratio are often sacrificed for rapid image acquisition, pulse sequences are emerging that preserve contrast and signal:noise ratio with short image acquisition times.

Many other pulse sequences and variations of pulse sequences are available for MRI in neuroimaging applications. In particular, exciting developments have been made using diffusion tensor imaging, perfusion weighted imaging, magnetic resonance angiography, and magnetic resonance spectroscopy, which have applications for surgical planning, guiding therapy, and differentiating some disease processes[9]. Additionally, researchers and MRI manufacturers are continually developing new pulse sequences for specific applications. Understanding how basic and novel pulse sequences work is essential for appropriate application and accurate interpretation of their images.

REFERENCES

1. Costa DN, et al. *AJR Am J Roentgenol* 2008;190(4):1076-1084.
2. Ren AJ, et al. *Korean J Radiol* 2012;13(1):44-52.
3. Colosimo C, et al. *Neuroradiology* 2006;48 Suppl 1:18-33.
4. Lee AG, et al. *Am J Ophthalmol* 2004;138(5):852-862.
5. Castillo M, et al. *Semin Ultrasound CT MR* 2000;21(6):417-427.
6. Maeda M, et al. *Eur Radiol* 2003;13 Suppl 4:L192-201.
7. Kidwell CS, et al. *Lancet Neurol* 2008;7(3):256-267.
8. Sutherland-Smith J, et al. *Vet Radiol Ultrasound* 2011;52(2):142-148.
9. Mabray MC, et al. *Brain Tumor Res Treat* 2015;3(1):8-23.

SUGGESTED READING

1. Westbrook, C, et al. *MRI in practice.* 2005; Malden, MA: Blackwell Pub.
2. Bushberg, JT, et al. *The essential medical physics of medical imaging.* 2nd Ed. 2002; Philadelphia: Lippincott, Williams, & Wilkins.

GENE EXPRESSION ANALYSIS IN CANINE TUMORS

Gerald Post[1, 2], Barbara Davis[2], David Duchemin[1, 2], J. Carl Barrett[2]

[1]*The Veterinary Cancer Center, Norwalk, CT, USA,* [2]*Innogenics, Harvard, MA, USA*

Gene expression and mutation analysis has become exceedingly important in human medicine, particularly in oncology. Unfortunately genomic testing in companion animals has been impeded by a number of challenges including availability of technology platforms, tissue acquisition requirements, and potential costs. To meet these challenges, we developed an automated, robust, reproducible, and cost-effective canine-specific RNA-based quantitative nuclease protection gene (qNPA) signature assay that uses small amounts of formalin-fixed paraffin embedded (FFPE) samples prepared on glass slides.

Using the HTG Edge system, we identified relevant housekeeping genes, developed our proprietary gene signature assay and performed biological validation. Final implementation was performed using 30 FFPE samples spanning carcinoma, sarcoma, melanoma, and lymphoma.

Our assay reliably distinguished different types of lymphomas, carcinomas and sarcomas. In lymphomas, myc and bcl2 ratios were found to correlate with response to chemotherapy. Activated tyrosine kinase pathways were significantly expressed in a variety of tumors, potentially affecting clinical decision-making and success of targeted therapeutic interventions. Estrogen and progesterone receptors were identified among mammary carcinomas. Cox-2 expressions was identified in subsets of multiple samples including transitional cell and anal sac carcinomas.

Introducing an affordably priced, clinically relevant, molecular approach to companion animal diagnostics and therapeutics has substantial advantages over current practices, particularly reliance on IHC or need for additional tissue collections for PCR or flow cytometry. Because this technology works reliably on FFPE specimens, it can help expedite the broad introduction of personalized cancer therapy by providing comprehensive, sensitive and accurate cancer genome analysis in 'real-time'.

INVESTIGATING MULTIPLE IN VITRO PATHWAYS AFFECTED BY METFORMIN IN FELINE CANCER CELLS

Jackie Wypij, Holly Pondenis, Zachary Cowden

University of Illinois at Urbana-Champaign, Urbana, IL, USA

Metformin (N', N'-dimethylbiguanide) is an oral anti-hyperglycemic agent associated with a decreased risk of cancer in human diabetic patients. In addition to its chemopreventative effects metformin also appears to have anti-neoplastic effects in vitro and in tumor models. While metformin has potential in feline cancer patients there appears to be a risk of developing hyperlactatemia which is a rare adverse effect in humans treated with metformin. The exact mechanisms by which metformin exerts its anti-neoplastic effects remains under investigation and multiple mechanisms have been proposed. The purpose of this study was to evaluate effects of metformin in feline cancer cells on 1) secretion of interleukins IL-6 and IL-10 and Stat-3 phosphorylation, 2) secretion of growth factors VEGF and IGF-1, 3) secretion of lactate and alteration of pH, and 4) modulation of AMPK phosphorylation.

Commercial assays were used for quantification of interleukins, growth factors, lactate, and Stat-3 phosphorylation. Western blot was used to evaluation AMPK protein expression. Multiple feline cancer cell lines were evaluated (SCCF1, CAT-MT, K12, FIB).

Metformin has modest effect on interleukin and growth factor secretion. Metformin increases lactate production with concomitant decreased pH which can be reversed by lactate pathway inhibition. Feline cancer cell lines express AMPK1/2 protein with modulation by metformin.

Metformin affects multiple pathways in feline cancer cell lines that may contribute both to its anti-neoplastic effect and adverse side effects. Further investigation of the molecular mechanisms of metformin may aid in guiding optimal therapeutic use and monitoring.

USE OF THE NOVEL OXYGEN CARRIER PROTEIN, ZOX, IN DOGS WITH INTRACRANIAL MASSES

Michael Kent[1], Philberta Leung[2], Peter Dickinson[1], Beverly Sturges[1], Sarah Ng[2], Catherine Bedard[2], Johnathan Winger[2], Teri Guerrero[2], Stephen Cary[2], Ana Krtolica[2]

[1]*University of California, Davis, CA, USA,* [2]*Omniox, Inc, San Carlos, CA, USA*

Tumor hypoxia promotes tumor invasiveness as well as therapeutic resistance and recurrence. The cytotoxic effect of radiation therapy (RT) is severely blunted by tumor hypoxia since ionizing radiation requires molecular oxygen to damage DNA permanently. To increase tumor oxygenation, Omniox has engineered a novel PEGylated oxygen delivery protein, ZOX, which is tuned to release oxygen specifically in hypoxic tissue and has been shown to enhance RT efficacy in rodent tumor models. Patient dogs with intracranial masses underwent 1) resection surgery (n=6), 2) surgery with RT (20x2.5Gy; n=5), or 3) stereotactic surgery (SRS; 3x8Gy; n=4), with ZOX administered pre-surgery or in combination with RT or SRS (75 – 100 mg/kg, IV). Pharmacokinetic data show that ZOX had a mean±SD plasma half-life of 35.4±6.3 hours (single dose, 75 mg/kg). Besides an acute and transient decrease in WBC, there were no significant changes in CBC, blood chemistry, or coagulation. Importantly, generation of anti-drug antibodies was not observed in any dogs. Immunohistochemical and ELISA analysis of resected tumors shows that ZOX penetrates into tumor tissue and that there is a strong correlation between ZOX accumulation and a reduction in hypoxia levels. Dogs were re-imaged at 3 and 6 months after treatment. To date, 2 surgery+RT dogs have reached 6 months following RT without recurrence. Among the 4 SRS dogs, all have reached 3 months with a decrease in tumor volume of 72.4±25.6%, and 2 dogs have reached 6 months with 1 showing a complete response and the other a 66.9% decrease in tumor volume. Thus, ZOX can be safely administered to dogs with intracranial masses and shows promise in decreasing tumor hypoxia for the enhancement of RT efficacy.

ADVERSE EVENT REPORTING IN COMPANION ANIMAL CLINICAL TRIALS EVALUATING CANCER THERAPY: A SYSTEMATIC REVIEW

Michelle Giuffrida

University of California - Davis, Davis, CA, USA

Aim: to systematically evaluate methods used to define, ascertain, and report adverse events (AEs) in companion animal clinical trials, and to identify study characteristics associated with quality of AE reporting.

English-language articles in peer-reviewed journals (2008-2014) of cancer clinical trials in dogs and cats were identified via systematic online database search. Article characteristics were abstracted and summarized. AE data were collected and summarized using a 14-item reporting checklist adapted from the CONSORT harms reporting extension. Study characteristics associated with AE checklist score were assessed with multiple linear regression.

168 articles involving 6,132 animals were included. Standardized terminology was used to describe AEs among 95/115 (83%) trials that included chemotherapy, and 16/53 (30%) trials that did not include chemotherapy ($p<0.001$). Median AE reporting checklist score was 5 out of 14 (range 0-12). Poorly reported items included methods and time frame of AE ascertainment, handling of recurrent events, AE data analysis, and reasons for treatment discontinuations and deaths. Articles rarely discussed limitations of AE data collection or analysis, or generalizability of harms results. Active surveillance for both laboratory and clinical AEs was reported in 16% of trials. Industry funding, single-arm design, and use of chemotherapy were associated with higher quality AE reporting.

Reporting of AEs in veterinary oncology trials is selective and heterogeneous. Harms associated with cancer treatments could be underestimated due to suboptimal collection and reporting of AE data. These findings support the adoption of a higher standard for AE surveillance and reporting in veterinary clinical research.

DEVELOPMENT AND VALIDATION OF THE CANINE OWNER-REPORTED QUALITY OF LIFE SCALE IN DOGS WITH CANCER

Michelle Giuffrida[1], Dorothy Cimino Brown[1], Susan Ellenberg[2], John Farrar[2]

[1]University of Pennsylvania School of Veterinary Medicine, Philadelphia, USA, [2]University of Pennsylvania Perelman School of Medicine, Philadelphia, USA

Aim: To develop and psychometrically test an owner-reported questionnaire instrument designed to standardize measurement of general quality of life (QOL) in dogs with cancer.

Standard methods for development and testing of instruments intended to measure subjective states were used. Key-informant interviews with dog owners and veterinary experts were used to generate, select, scale, and pre-test items for content, meaning, and readability. A prospective field trial was conducted involving 90 dog owners who completed the questionnaire twice in a 2-week period. Response items were evaluated using exploratory factor analysis, Cronbach's α, and convergence with a global QOL item. Preliminary tests of stability and responsiveness were performed.

The Canine Owner-Reported Quality of life (CORQ) scale is a 4-factor, 17-item questionnaire containing observable items owners use to evaluate their dogs' QOL. Several items pertaining to physical disturbances, such as vomiting and diarrhea, performed poorly and were omitted. The remaining items loaded into 4 factors, designated Vitality, Companionship, Pain, and Mobility based on the items they contained. The CORQ and its factors had high internal consistency (Cronbach's α=0.68-0.90) and moderate to strong correlations (r=0.49-0.71) with global QOL as measured on a separate visual analog scale. Preliminary testing indicated good test-retest reliability and responsiveness to improvements in overall QOL.

The CORQ is a valid, reliable owner-reported questionnaire that standardizes assessment of factors thought to represent QOL among dogs with cancer, and shows promise as a clinical trials outcome measure for quantifying QOL changes that occur in response to cancer treatment, progression, and cure.

MYCOBACTERIUM CELL WALL FRACTION (MCWF) AS AN AID IN THE TREATMENT OF CHEMOTHERAPY- INDUCED NEUTROPENIA IN DOGS

Juan Mangieri[1], Lucas Rodriguez[2], Aleksandar Masic[3]

[1]*Veterinaria CUCCARO, San Martin, Mendoza 5570, Argentina,* [2]*Estima Hospital Veterinário, Centro – Taubaté, Brazil,* [3]*NovaVive Inc, Belleville, Ontario, Canada*

Chemotherapy-induced neutropenia in clinical oncology may have negative implications as it increases the risk of infection, interrupts chemotherapy protocols and requires treatment dose reduction. Currently, there is no veterinary product available for use in the prevention of chemotherapy-induced neutropenia. For many years immune response modifiers have been used for cancer treatment either alone or in conjunction with standard treatment protocols including surgery, radiation or chemotherapy. Mycobacterium Cell Wall Fraction (MCWF) is a non-specific immunomodulator derived from a non-pathogenic mycobacterium, *Mycobacterium phlei* which has been used for treatment of mammary tumors, osteosarcoma, transitional cell carcinoma, hemangiosarcoma and transitional venereal tumors in dogs. In addition, MCWF is currently used in Phase III clinical studies in humans for patients with bladder cancer who failed standard treatment protocols (BCG). Main purpose of our research was to demonstrate MCWF immunomodulatory potential in ameliorating the chemotherapy-induced neutropenia in dogs.

Efficacy of three different doses of MCWF as an aid in the treatment of neutropenia induced by chemotherapeutic agent vinblastine in healthy dogs was evaluated. Study design included four experimental groups with ten dogs per group (1-8 years of age). Dogs were required to be healthy to proceed to vinblastine treatment for induction of neutropenia (defined as Day 0), and further, were required to demonstrate neutropenia within 7 days of vinblastine treatment, to proceed to inclusion in the study, and treatment with MCWF. Neutropenia was defined as less than 2,000 neutrophils per µL of blood for the purpose of this study. All dogs received 3 mg/m^2 of vinblastine on day 0, by the intravenous route (IV), to induce neutropenia, then, once neutropenia was attained, a dog was randomly allocated to one of the 4 treatment groups (100, 200 or 500 µg/kg MCWF IV, or no treatment). MCWF treatment was administered by IV route as a slow bolus injection (3-5 min.) within 24 hours. CBC counts were measured on Day 0, Day 2 and daily until the end of the study. The end of the study was defined as neutrophil count ≥ 2000/µL for two consecutive days (48 hours) or day 10, whichever came first. CBC counts included monitoring of packed cell volume (PVC), red blood cells (RBC), thrombocytes, total white blood cells (WBC), and differential WBC (neutrophils, lymphocytes, monocytes, eosinophils and basophils). Efficacy of MCWF as a treatment for neutropenia was determined by comparing the duration of neutropenia in the control group versus the three different MCWF dose groups. Duration is defined as the number of days from MCWF treatment administration (24 hours after neutropenia was detected) until neutropenia was resolved (neutrophil count ≥ 2000 cells/ µL).

A clear dose-response was apparent in the time to recovery from neutropenia; mean duration of neutropenia was 4.2, 2.5, 1.8, and 1.2 days in the control, 100, 200 and 500µg/kg MCWF-treated groups respectively (Table 1.). The difference in the mean duration of neutropenia between the control and 100 µg/kg MCWF-treated groups was statistically-significant at $p<0.05$, the 200 and 500 µg/kg MCWF-treated groups also had a mean duration of neutropenia significantly shorter than the control group, with $p<0.01$ and $p<0.0001$ respectively (Figure 1). The differences in mean recovery time between the 100 and 200 µg/kg dose groups ($p<0.05$), and the 200 and 500 µg/kg dose groups ($p<0.05$) were also statistically-significant. Dogs were observed for safety of administration of MCWF, for 30 minutes immediately after administration, and again at 12 hours after administration. Adverse events such as dizziness, lethargy, and mild transient hyperthermia were observed in some dogs immediately after MCWF administration but resolved within 12-24 hours and did not require additional treatment. Vomiting and diarrhea were observed to a lesser extent as expected. All GIT adverse events were mild and lasted for 2 to 3 days following vinblastine treatment (strictly chemotherapy associated).

Administration of MCWF demonstrated efficacy in ameliorating chemotherapy-induced neutropenia in healthy dogs. These findings could have significant importance from a clinical standpoint and could support the use of MCWF in conjunction with standard chemotherapy protocols. Additional studies are underway to demonstrate the efficacy of MCWF in preventing chemotherapy-induced neutropenia following concurrent administration in dogs.

Table .1. Median and mean neutropenia grades by treatment group

Treatment group	Neutropenia grade (median value)	Median duration (days)	Neutropenia grade (mean value)	Mean duration (days)
Control	4	4	3.7	4.2
100 µg/kg	3.5	2	3.3	2.5
200 µg/kg	3	2	3.2	1.8
500 µg/kg	4	1	3.6	1.2

Severity grading system:	1	1,500 – 1,999 neutrophils/µL	3	500 – 999 neutrophils/µL
	2	1,000 – 1,499 neutrophils/µL	4	< 500 neutrophils/µL

Figure1. Time-course of neutropenia development and resolution following vinblastine administration.

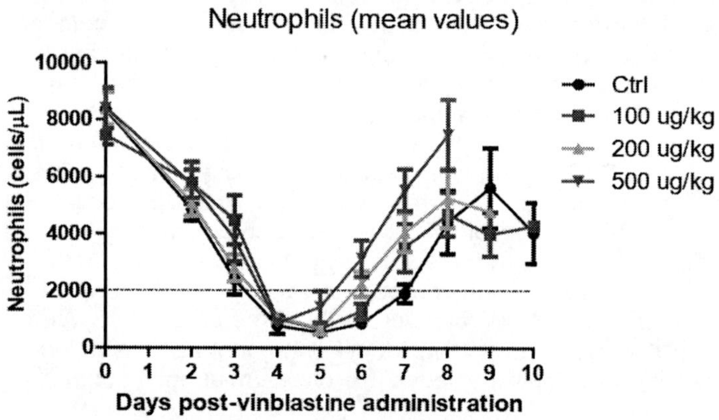

STAT3 PATHWAY IS UPREGULATED IN CANINE DIFFUSE LARGE B CELL LYMPHOMAS AND IS ASSOCIATED WITH A POOR PROGNOSIS

Xuan Pan, Paulo Jark, Courtney Hong, Hannah Ruetten, Nathan Bollig, Marie Pinkerton

University of Wisconsin, Madison, USA

Lymphoma comprises 83% of all canine hematopoietic malignancies and approximately 24% of all canine cancer. The most common form of canine lymphoma is the diffuse large B cell lymphoma (DLBCL). Due to the poor prognosis with the current standard of care, there is a substantial need for further understanding the molecular pathway of canine lymphomagenesis, and for identifying novel prognostic and therapeutic targets. The Janus Kinase (JAK) and signal transducer and activator of transcription (STAT) pathways play important roles in the pathogenesis of many hematologic malignancies. JAK2-mediated activation of STAT3 promotes growth and survival of a variety of human lymphomas. Therefore, there is a great demand for veterinary oncologists to understand the JAK2-STAT3 pathway in pathogenesis, prognosis and therapeutic treatment of canine DLBCLs. Our study aims to assess **1)** whether the STAT3 pathway is deregulated in canine DLBCLs, **2)** whether the deregulation of STAT3 pathway is associated with a clinical prognosis, and **3)** the therapeutic potential of using JAK1/2 inhibitors to treat canine DLBCLs.

To assess whether the STAT3 pathway is deregulated, total STAT3 and phosphorylated STAT3 (pSTAT3) protein expressions were evaluated by immunohistochemistry. Histopathology samples of forty-two canine DLBCL lymph nodes, ten reactive and ten normal lymph nodes were assessed. The mean percentage of positive cells per sample was calculated by averaging the percentage of immunolabelled cell/total cell in each of five representative fields. The percentage of both STAT3 and pSTAT3 immunolabelled cells were significantly increased in canine DLBCLs (STAT3=76%, pSTAT3=15%) compared with the normal lymph node (STAT3=49%, pSTAT3=3.9%), ($p<0.05$). STAT3 or pSTAT3 expression was not significantly different when comparing DLBCLs to reactive lymph nodes, or reactive lymph nodes to normal lymph nodes.

To evaluate the impact of STAT3 or pSTAT3 activation on overall clinical prognosis in canine DLBCL patients, we divided forty-two canine DLBCL patients into four cohorts of low total STAT3, high total STAT3, low pSTAT3 or high pSTAT3, based on the percentage of STAT3 or pSTAT3 immunolabelled cells. High total STAT3 group had decreased overall median survival time (MST) compared with the low total STAT3 expressing cohort (143 d versus 283 d, $P<0.05$). MST was not significantly different between high pSTAT3 and low pSTAT3 cohorts.

To explore the therapeutic potential of inhibiting JAK2-STAT3 pathway in canine DLBCLs, we treated canine DLBCL cell line CLBL-1 and canine normal kidney cell line MDCK with novel JAK1/2 inhibitors AZD1480 and CYT387 respectively. There was a significant reduction of viable CLBL-1cells while increasing the concentrations of JAK1/2 inhibitors. To explore the underlying mechanism of JAK inhibitor mediated inhibition of lymphoma cell growth, we performed an apoptosis assay to determine if the reduction in cell number occurred as a consequence of increased apoptosis. CLBL-1 and MDCK cells were stained with Annexin V and SYTOX Red dead cell stain and were subjected to FACS analysis at 96 hours post AZD1480 or CYT387 treatments. While the percentage of healthy CLBL-1 cells decreased with increasing concentration of AZD1480 and CYT387, the percentage of early apoptotic cells was increasing. There was no significant change of viable cells or early apoptotic cells in AZD1480 or CYT387 treated MDCK cells. Therefore, JAK1/2 inhibitors AZD1480 and CYT387 can induce apoptosis in canine DLBCLs in a dose dependent and lymphoma tissue-specific manners.

We conclude that STAT3 pathway is up-regulated in canine DLBCLs. A group of DLBCL patients expressing high total STAT3 is associated with an overall poor clinical prognosis. JAK1/2 inhibitors can inhibit DLBCL cell growth by inducing apoptosis in a lymphoma-specific and dose-dependent manner. This is the first study to evaluate the JAK2/STAT3 pathway in canine DLBCLs. The knowledge of STAT3 activity in canine DLBCLs enables us to understand and explore the therapeutic potential of JAK inhibitors in future.

ALTERNATING RABACFOSADINE/DOXORUBICIN: EFFICACY AND TOLERABILITY IN NAÏVE CANINE MULTICENTRIC LYMPHOMA

Douglas Thamm[1], David Vail[2], Gerald Post[3], Timothy Fan[4], Brenda Phillips[5], Robyn Elmslie[6], Sandra Bechtel[7], Mary Kay Klein[8]

[1]Colorado State University, Fort Collins, CO, USA, [2]University of Wisconsin - Madison, Madison, WI, USA, [3]The Veterinary Cancer Center, Norwalk, CT, USA, [4]University of Illinois, Urbana, IL, USA, [5]Veterinary Specialty Hospital of San Diego, San Diego, CA, USA, [6]Veterinary Referral Center of Colorado, englewood, CO, USA, [7]University of Missouri, Columbia, MO, USA, [8]Southern Arizona Veterinary Specialty & Emergency Center, Tucson, AZ, USA

Lymphoma is one of the most common canine cancers. While capable of inducing remission in most naïve canine lymphomas, CHOP-based standard-of-care therapies can be cumbersome, owing to the number and frequency of treatments required (typically 12-16 visits over 15-25 weeks). Single-agent doxorubicin (DOX) is an alternative chosen by some for issues of convenience and/or cost, although it is considered inferior to CHOP in terms of efficacy. Rabacfosadine (TANOVEA™/VDC-1101/GS-9219), a novel double prodrug of the acyclic nucleotide phosphonate 9-(2-phosphonylmethoxyethyl)-guanine (PMEG), preferentially targets lymphoma cells with reduced systemic toxicity compared to PMEG. Single-agent rabacfosadine has been administered on a variety of dosing schedules to dogs with lymphoma. Objective response rates of 60-100% have been reported in previous studies, with improved outcomes in naïve patients. Given the convenience of an every-21-day administration schedule and non-overlapping mechanism of action/resistance with DOX, we sought to evaluate the efficacy and tolerability of treatment with alternating doses of rabacfosadine and DOX in dogs with naïve multicentric lymphoma.

Dogs with cytologically or histologically confirmed and immunophenotyped lymphoma were treated with alternating doses of rabacfosadine (1.0 mg/kg free base as a 30-minute IV infusion weeks 0, 6, 12) and DOX (30 mg/m^2 weeks 3, 9, 15). Dogs experiencing a complete response (CR) received up to six total treatments, followed by monthly rechecks. Complete clinicopathological assessment and clinical assessment of remission and adverse effects (AEs) were performed every 21 days. Response was assessed according to published VCOG criteria and AEs according to the VCOG-CTCAE v1.1.

54 dogs were prospectively enrolled. 51 were evaluable for response assessment and 52 were evaluable for progression free interval (PFI) and AE assessment. 44 dogs had B cell lymphoma and 8 had T cell lymphoma. 40 were considered substage A and 12 substage B. The overall response rate (ORR) was 81% (62% CR, 19% PR). The overall median PFI was 200 days (218 days for dogs experiencing CR and 148 days for PR). Both ORR and PFI were significantly higher in dogs with B cell lymphoma than in T cell (91% vs 25% and 215 vs 43 days). When compared with historical data, the observed overall PFI is superior to that reported with single-agent DOX (81-169 days) and comparable to that reported with standard 15-25 week CHOP protocols (140-219 days).

The majority of AEs were mild and self-limiting: gastrointestinal and hematologic AEs were most common. Grade 3 AEs included neutropenia (8), weight loss (6), hyporexia (4), diarrhea (4), liver enzyme elevations (3), thrombocytopenia (1), and dermatopathy (1). 5 dogs experienced grade 4 hematologic toxicity, and 1 developed transient grade 4 ALT elevation. 12 dogs experienced grade 1/2 dermatologic AEs, and 2 dogs developed grade 5 pulmonary fibrosis, several months after treatment discontinuation and while in CR. Modified ECOG performance status was generally maintained throughout the study. Rabacfosadine dose reductions were performed in 8 dogs and DOX dose reductions occurred in 6 dogs. No dose delays were required.

In conclusion, an alternating every-three-week rabacfosadine/DOX regimen was generally well tolerated and appears to result in outcomes comparable to those observed after standard CHOP-based protocols, with substantially fewer treatment visits (6 vs. 12-16). Adverse events were generally mild and self-limiting. Further studies are warranted to explore the impact of re-treatment as well as other rabacfosadine combinations.

PRELIMINARY INVESTIGATION OF THE INSULIN-LIKE GROWTH FACTOR 1 AXIS IN DOGS WITH MULTICENTRIC LYMPHOMA

Cailin Heinze[1], Bobbi McQuown[2], Kristine Burgess[1]

[1]Cummings School of Veterinary Medicine at Tufts University, North Grafton, MA, USA, [2]VCA Veterinary Referral Associates, Gaithersburg, MD, USA

Insulin-like growth factor 1 (IGF-1) is a peptide hormone related to insulin. This hormone has effects on glucose metabolism by binding the insulin receptor and is also involved in regulation of somatic growth, cell proliferation, and apoptosis through the IGF-1 receptor. Studies in humans and animals have suggested higher blood and tissue concentrations of IGF-1 are associated with increased cancer risk and worse cancer prognosis. The aim of this study was to compare concentrations of serum IGF-1, insulin, glucose, ketones, and lactate concentrations in dogs with lymphoma compared with age, sex, and weight-matched controls. Dogs with naïve lymphoma ($n = 16$) were identified and matched with controls ($n = 16$) on the basis of age, sex, and weight. IGF-1, insulin, glucose, ketones, and lactate concentrations were measured in fasted blood samples. Variables were compared between groups and analysed for correlations between the individual variables. Insulin, IGF-1 and glucose were not different between cases and controls. However, lactate and ketones were higher in cases than in controls ($P = 0.007$, $P < 0.001$) and these variables were correlated with each other ($r = 0.450$, $P = 0.011$). Insulin and IGF-1 ($r = 0.457$, $P = 0.009$) were also correlated across both groups of dogs. This study did not demonstrate increased IGF-1 and insulin in this small group of dogs with lymphoma compared to the controls. However, ketones were higher in the cases and correlated with lactate, which suggests that they could also be a useful biomarker of metabolism in dogs with lymphoma.

ELECTROCHEMOTHERAPY WITH INTRAVENOUS BLEOMYCIN FOR TREATMENT OF FELINE SQUAMOUS CELL CARCINOMA: EXPERIENCE ON 12 CATS

Alessio Pierini[1], Ron Lowe[2], Valentina Granziera[2], Veronica Marchetti[1], George Lubas[1]

[1]*Department of Veterinary Sciences, University of Pisa, Pisa, Italy,* [2]*Ashleigh Veterinary Clinic Limited, Knaresborough, UK*

The aim of this study was to evaluate the efficacy and safety of neoadjuvant electrochemotherapy (ECT) with intravenous bleomycin for the treatment of skin squamous cell carcinoma (SCC) of the head in cats.

Twelve client-owned cats with histological diagnosis of SCC of the head were enrolled. The owners elected to carry out ECT as an alternative treatment to surgery. All cats were staged by physical examination, fine-needle aspiration of mandibular lymph node (if palpable) and three-view thoracic X-rays. Complete blood cell count, serum biochemistry profile and urinalysis also were performed as the patients underwent general anesthesia as well. Cats with lymph node involvement or radiographic features of pulmonary metastases were excluded from this study. Clinical T stage was assigned on the basis of WHO criteria. The longest tumor diameter based on physical examination was used for clinical staging and follow-up.

For ECT treatment, cats were premedicated with medetomidine and butorphanol. Once sedated, an IV catheter was placed in a peripheral vein, cats were preoxygenated via a face mask, and anesthesia was induced with alfaxalone administered IV. Then cats were intubated and anesthesia was maintained with oxygen, nitrous oxide and isoflurane. Bleomycin was diluted in three milliliters of saline solution and administered IV at a dosage of 15-20 mg/m2.

The electroporation was performed using a Cytopulse PA4000 (Cyto Pulse Sciences, Inc) in nine cats or a Cytopulse Oncovet in two cats with Gehl needle electrodes of 1 cm length from the same manufacturer. The pulse pattern employed was 8 monophasic square pulses of 100 microseconds each at a frequency of 1 Hz (interpulse interval 0.9999 sec) with the PA4000 or 5 kHz (interpulse interval 0.01 seconds) with the Oncovet. The latter equipment allowed more rapid application of the therapy because of the higher pulse frequency and the improved user interface. The Gehl electrode pattern consists of two parallel 1-1.5 cm rows of 6 needles, the rows being 6 mm apart (Gehl and others 1999). The pulse amplitude was 600-720 volts (1000-1200 volts/cm). This pulse pattern was established by Gehl and others (1999). In one case (periocular SCC) a Cliniporator (IGEA S.p.a.) with 8-needle electrodes and 20mm length needles were used. The voltage setting for the Cliniporator was pre-set by the manufacturer at 1000 volts/cm with current varying between applications depending on tissue conductivity (safety limit of 20A).

Patients were assessed every 3-10 days for the initial 4 weeks after treatment. Tumor response was assessed every 6-8 weeks following the response evaluation criteria for solid tumours estabilished by veterinary cooperative oncology group consensus document by physical examination, fine-needle aspiration of mandibular lymph node if palpable and 3-views thoracic X-rays. Tumor measurements were made until remission or until relapse occurred. A complete remission (CR) was defined as total reduction of the tumor. A partial remission (PR) was defined as ≥30% reduction in tumor diameter. Stable disease (SD) was defined as <30% reduction in tumor diameter or <20% increase in tumor diameter, and progressive disease (PD) was defined as ≥20% increase in tumor diameter. Presence of new lesions near to the primary tumor or presences of metastasis were considered as PD. A minimum duration of two to three weeks was required for a response to qualify as positive.

Progression-free survival (PFS), median survival time (MST) and overall response rate (ORR) were calculated. PFS was defined as time from ECT until tumor progression, recurrence or death. ST was defined as time from ECT until death. ORR was defined as proportion of cats that reached CR or PR.

The toxicity treatment (especially locally) was assessed by interviewing the owner and carrying out a physical examination up to 7-14 days after ECT. Only early adverse effects were evaluated and a subjective scoring system from zero (no adverse effect) to 4 (toxicity-related death) was used. For each cat, the highest toxicity score was used for analysis.

Eleven of twelve SCC were localized on the nasal planum. One was localized on the periocular region. Ten were neutered males and 2 were spayed females. Ages ranged from 7 to 16 years old (median, 11.4). SCCs were classified with WHO T-stage as T1 (#7), T2 (#3), T3 and T4 (one each).

Sixteen ECT treatments were done. The cat with periocular SCC had four treatments and one cat with nasal planum SCC had two treatments. Eleven of 12 cats were eligible for survival analysis. A cat with nasal planum T1 SCC due to treatment toxicity (toxicity score = 4) at 9 days after ECT was euthanized.

For survivor cats, ORR was 100%. There were 8 (73%) CR and 3 (27%) PR. Six (100%) of 6 cats with T1 SCC had CR and two cats (66%) of 3 T2 SCC had CR. One cat with T2 and two cats with T3 and T4 SCC had PR. All cats were dead at the end of the study. Three cats developed PD, with a median PFS of 253 days (range, 97–468 days). All of them had PR. One cat with a nasal planumT3 SCC developed recurrence after 97 days after the first ECT treatment and was treated once again. This cat was euthanized for PD 231 days after the first ECT. One cat had a periocular T2 SCC developed the first recurrence 468 days after ECT. Overall, this cat was treated for a total of four times for new recurrences and was euthanized 730 days after the first ECT. The third cat had a nasal planum T4 SCC and was euthanized 194 days after ECT. The other eight cats were died for unrelated causes. No cats developed metastasis. Overall MST was 452 days (range, 194–2973 days). MST for cats that achieved CR was 452 days (range, 194–2973 days) and 231 days (range, 194–730 days) for cats that achieved PR. MST was calculated for any T stage and was 397 days for T1 (range, 251–2973), 730 days for T2 (range, 599–1731), 231 days for T3 and 194 days for T4.

All twelve cats were evaluated for toxicity. All cats developed early effects. Toxicity was classified as grade 1 (#8), grade 2 (#1), grade 3 (#2) and grade 4 (1). Toxicity was mild in almost all T1 SCC (6 grade 1, 1 grade 4).

Several therapeutical strategies are described to treat feline SCC of the head such as surgery, external-beam radiation therapy, strontium-90 plesiotherapy, cryosurgery, photodynamic therapy, laser application, hyperthermia, thermocautery with curettage and electrochemotherapy. However, only superficial cancers can be managed effectively with almost the treatments described above.

Nowadays, ECT has become a reliable treatment for cutaneous SCC of the head in cats. Recently, two papers evaluated the efficacy of ECT delivered five-minute after intravenous bleomycin in feline cutaneous SCC at varies WHO T stage. Tozon et al. (2014) observed an ORR of 87.5% for sixteen superficial SCC in cats. Spugnini et al. (2015) reported the capability of ECT to improve bleomycin efficacy for feline SCC. Twenty-six cats were treated with intravenous bleomycin coupled with ECT and 21 cats were treated with bleomycin alone. ORR in the ECT treated group was 88.5% versus 33% for control group. Median time to progression in the ECT treated animals was 30.5 months, whereas in controls it was 3.9 months. In the present study an overall response rate of 100% and a MST of 452 days (about 15 months) were observed. Many reports have indicated that aggressive local tumor control offers the best chance to obtaining clinical tumor response and long survival times. However, tumor control achieved by aggressive surgical excision or external beam radiation therapy is dependent on reaching adequate surgical margins or the degree of response to radiation therapy. In a previously study, cats with SCC treated with ECT that did not reach CR were more likely to be in an advanced WHO T stage. In the present cohort of cats PR was reached in T3 and T4 SCC and in 4-cm T2 SCC. Authors speculated that more advanced T stage SCCs can benefit of more ECT treatments until CR is reached or can eligible to have a multimodal therapy.

Moreover, aggressive surgical excision can be limited by anatomical localization and cosmetic concerns for owners. On the other hand, radiation therapy is limited by early and late effects on the eyes in periocular SCC and by financial concerns for owners. In the present study early effects were considered well tolerated in nine of twelve cats. Two cats with severe early effects had a T4 SCC and a 4-cm periocular SCC. Severe effects can occur more frequently in cats with bulky or infiltrative tumor. Discharge instructions for owners advised application of an Elizabethan collar on cats. Nevertheless, the present study did not review symptomatic therapeutic modalities that authors used in management of pain and/or scratching. Non-steroidal antinflammatory drug such as meloxicam was previously reported for successfully early effect management in ECT treated cats. Furthermore in the present study one cat with a T1 SCC developed an ECT toxicity-related death. The major limit to interpret toxicity-related death is that cat was euthanized due to owner's decision and authors were not able to confirm whether the owner followed discharge instructions for scratching control.

This study has some limitations. First, the small number of enrolled cats can influence ORR, MST and toxicity evaluations. Second, the subjective evaluation of toxicity makes interpretation of these data difficult. Finally, ORR and MST can be influenced by the use of three different electroporation machines and a non-standard dose of intravenous bleomycin.

In conclusion, intravenous bleomycin coupled with ECT is well tolerated for cutaneous SCC in cats. The results of this study suggest that ECT should be considered as an alternative treatment option, especially in superficial SCC, and when owners do not accept other treatment approaches due to cosmetic or financial concerns.

Gehl J, Sorensen TH, Nielsen K, Raskmark P, Nielsen SL, Skovsgaard T, et al. In vivo electroporation of skeletal muscle: threshold, efficacy and relation to electric field distribution. Biochimica et Biophysica Acta 1999; 1428: 233–240.

Spugnini EP, Pizzuto M, Filipponi M, Romani L, Vincenzi B, Menicagli F, et al. Electroporation Enhances Bleomycin Efficacy in Cats with Periocular Carcinoma and Advanced Squamous Cell Carcinoma of the Head. J Vet Intern Med 2015;29(5):1368–75.

Tozon N, Pavlin D, Sersa G, Dolinsek T, Cemazar M. Electrochemotherapy with intravenous bleomycin injection: an observational study in superficial squamous cell carcinoma in cats. J Feline Med Surg. 2014 Apr;16(4):291-9.

NOVEL ONCOLYTIC MARABA VIRUS FOR THE ADJUVANT TREATMENT OF FELINE MAMMARY CARCINOMA

J. Paul Woods[1], Byram Bridle[2], Michelle Oblak[1], Robert Foster[2], Victoria Sabine[1], Jeff Hummel[3], Brian Lichty[3]

[1]*Clinical Studies, Ontario Veterinary College, University of Guelph,, Guelph, ON, Canada,* [2]*Pathobiology, Ontario Veterinary College, University of Guelph, Guelph, ON, Canada,* [3]*McMaster Immunology Research Centre, McMaster University, Hamilton, ON, Canada*

Feline mammary carcinomas (FMC) are highly metastatic tumours frequently (>80%) displaying metastasis to the regional lymph nodes, lungs, pleura, and liver. Historically, the treatment of FMC has been surgery; however, at the time of diagnosis, FMC is often in an advanced stage due to its rapid progression. Following surgery, the median disease free survival reported is less than a year with a 2-year survival rate of less than 20%. Adjuvant chemotherapy (e.g. doxorubicin) has not shown benefit in delaying relapse or prolonging life with FMC.

Oncolytic viruses are viruses whose replication is restricted to tumour cells leading to cytolysis. A growing number of viruses are being developed as oncolytic viruses and various strategies have been utilized to target tumour cells. Our group developed a biotherapeutic that combines two emerging cancer treatment modalities: tumour vaccination and oncolytic viruses. We recently determined that oncolytic viruses expressing tumour antigen transgenes are an excellent means to boost and enhance tumour vaccination.

Therefore, we are conducting a clinical trial combining the two emerging cancer treatment modalities, tumour vaccination and oncolytic viruses, to treat FMC following surgery (i.e. adjuvant therapy in addition to surgical excision). Our goal is to test our heterologous prime:boost strategy targeting tumour antigens relevant to FMC. We hypothesize that induction of anti-tumoural immunity and direct oncolysis of metastatic (often occult) disease using our oncolytic vaccine booster will delay or prevent post-surgical relapse and extend survival in cats with mammary carcinoma.

Cats with FMC (histologically confirmed) were staged (CBC, biochemical profile, thoracic radiographs, abdominal ultrasound or CT). Cats were vaccinated with an adenovirus expressing a cancer gene, then 2 weeks later the tumour was excised (radical mastectomy), and then 4 weeks postoperatively the cats received a booster with an oncolytic virus (Maraba virus) also expressing the cancer gene. Cats were followed up with physical exams and thoracic radiographs.

Twelve female (spayed) cats have entered the study with a mean age of 12.5 years (range 7.5-16.8), median weight of 4.4 kg (range 3.25-5.89), consisting of 7 DSH, 4 DLH, 1 Turkish Van. The FMC consisted of 7 stage 1, 1 stage 2, and 4 stage 3. During oncolytic virus infusion 1 cat died of putative anaphylaxis (severe acute diffuse pulmonary hemorrhage and edema on post mortem). Otherwise toxicity of the adenovirus vaccine, surgery and marabavirus infusion was manageable. Following the anaphylactic death, pre-infusion skin testing of the marabavirus was introduced for the 5 subsequent cats and all 5 cats were negative. Six cats are dead with 5 dying of disease and 1 dying due to putative anaphylaxis during the oncolytic virus infusion treatment. Six cats are still alive (ranging from 92 - 568 days). The median overall survival is 240 days.

Except for the one catastrophic anaphylactic reaction, the treatment is well tolerated. Employing a novel oncolytic virus to implement the heterologous prime:boost strategy to target tumour antigens may result in a new modality for the adjuvant treatment of FMC following surgical excision.

OSTEOSARCOMA-DERIVED EXOSOMES IMPAIR CD4+ AND CD8+ T-CELL PROLIFERATION AND INDUCE T-REGULATORY CELL EXPANSION

Shay Bracha[1], Liping Yang[3], Claudia Maier[3], Cheri Goodall[1], Ryan Troyer[2]

[1]Oregon State University-Department of Clinical Sciences, Corvallis, Oregon, USA, [2]Oregon State University-Department of Biomedical Sciences, Corvallis, Oregon, USA, [3]Oregon State University-Department of Chemistry, Corvallis, Oregon, USA

Exosomes are microvesicles secreted by cells that function in cell-to-cell communication through cellular uptake of exosomal protein and RNA cargo, as well as interaction with surface-expressed proteins. Exosomes are crucial for tumor progression, angiogenesis and immune evasion. Although the role of tumor exosomes has been described in numerous studies, the effect of osteosarcoma exosomes on T-cells is lacking. We hypothesize that osteosarcoma-derived exosomes carry a vastly different cargo in comparison to exosomes from healthy osteoblasts, and include immunosuppressive proteins which directly impact healthy T-cells by reducing the proliferation and activation of CD4+ and CD8+ cells and enhancing the expansion of T-regulatory cells. Exosomes were isolated from the media of osteosarcoma (HmPos) and healthy canine osteoblast (CnOb) cells (Total Exosome Isolation, Invitrogen). Exosome size and concentration was validated by NanoCyte single molecule imaging. Exosomal proteins were digested and subjected to mass spectrometry (LC-MS/MS), followed by data analysis using MASCOT algorithm and Scaffold software. Additionally, intact exosomes were co-incubated with lymphocytes from a healthy dog for 3 and 5 days; controls included unexposed lymphocytes and lymphocytes exposed to an equal concentration of exosomes from normal osteoblasts (CnOb). Flow cytometry analyses were performed to measure expression of T cell markers (CD4, CD8, CD25, and Foxp3), cell viability, and cell proliferation. HmPos and CnOb exosomes shared 55 proteins, while 43 and 62 proteins were unique, respectively. HmPos exosomes included proteins which are known to induce immune modulation and evasion (Tenascin, alpha-fetoprotein). T-cells incubated with malignant exosomes exhibited a profound reduction in proliferation and activation of CD4+ and CD8+ T cells at days 3 and 5 ($p < 0.001$). Expansion of the T-regulatory population in the malignant exosome-treated group was significant at day 5 ($p < 0.001$) in comparison to normal osteoblasts. Our study is the first to identify and characterize the exosomal cargo in canine osteosarcoma and healthy osteoblasts and to investigate their impact on healthy T-cells.

New Therapeutics Impacting Oncology Practice

Chad M. Johannes, DVM, DACVIM (SAIM, Oncology)
Ames, IA

IMMUNO-ONCOLOGY DEVELOPMENTS

The area of immuno-oncology is rapidly expanding in human medicine and these advances are having significant impact on treatment options for many types of cancer. The most active areas of development within this realm include monoclonal antibodies (MAb) and therapeutic anti-cancer vaccines. In recent years, immuno-oncology therapeutics tailored for canine and feline oncology patients have gained attention. Although many of these therapeutics are in the earlier stages of development, they hold strong potential and are beginning to become clinically available for patient care.

Anti-Cancer Monoclonal Antibodies

1. **Monoclonal Antibody for Aid in Treatment of Canine B-cell Lymphoma (Aratana Therapeutics, Inc.)**
   This caninized MAb (AT-004) was fully licensed by the USDA in January 2015 as an aid in the treatment of canine B-cell lymphoma. Initial clinical data presented in abstract form showed signs of biologic activity[1-3] however no peer-reviewed data is available to date. Additional clinical studies are reportedly underway to further evaluate the best use and timing of this MAb in conjunction with cytotoxic chemotherapy, however it is not widely commercially available at this time. Aratana has stated that scientific studies suggest that AT-004 is not as specific to the CD20 target as expected.[4]
2. **Monoclonal Antibody for Aid in Treatment of Canine T-cell Lymphoma (Aratana Therapeutics, Inc.)**
   This caninized MAb (AT-005) received conditional licensure by the USDA in January 2014 as an aid in the treatment of canine T-cell lymphoma. The stated target of this MAb, CD52, has not yet been described in canine lymphoma cells. Aratana is currently conducting two prospective studies to further evaluate AT-005 in conjunction with cytotoxic chemotherapy (CHOP-based protocol and lomustine) for treatment of canine T-cell lymphoma. Unfortunately, the MAb does not seem to be adding significant progression free survival in these two studies. Aratana has also stated that scientific studies suggest that AT-005 is not as specific to the CD52 target as expected.[4] AT-005 remains available to many veterinary oncologists however no peer-reviewed clinical evidence of efficacy has been published at this time.
3. **Anti-CD20 Monoclonal Antibody for Canine Lymphoma (Elanco)**
   The generation and characterization of a rituximab-like anti-CD20 antibody intended as a candidate treatment for canine B-cell lymphoma has been described.[5] Monoclonal antibody 1E4 binds to approximately the same location in the extracellular domain of CD20 as rituximab, and 1E4-based chimeric antibodies co-stain canine B cells in flow cytometric analysis of canine leukocytes using an anti-canine CD21 antibody. Chimeric monoclonal antibodies were assembled. Both 1E4-cIgGB and 1E4-cIgGC were observed to significantly deplete B-cell levels in healthy beagle dogs. The *in vivo* half-life of 1E4-cIgGB in a healthy dog was ~14 days. The antibody 1E4-cIgGB has been selected for further development as an agent for the treatment of canine B-cell lymphoma.[5]

Therapeutic Anti-Cancer Vaccines

1. **Canine Lymphoma Vaccine, DNA (Merial)**
   This xenogeneic murine CD20 DNA therapeutic vaccine for use in dogs with B-cell lymphoma was conditionally licensed by the USDA in September 2015. Utilizing a similar technology platform as Oncept® for canine melanoma, the anti-CD20 vaccine is administered using a needle-free transdermal device. A prospective, multi-center clinical trial in dogs with B-cell lymphoma receiving CHOP-based chemotherapy followed by 4 doses of anti-CD20 vaccine has been described but results have not yet been published.[6]
2. **Canine Osteosarcoma Vaccine (Aratana Therapeutics, Inc.)**
   A recombinant HER2/neu expressing Listeria therapeutic vaccine (AT-014) is being studied as an aid in the treatment of canine osteosarcoma. In a study conducted at the University of Pennsylvania, Dr. Nicola Mason administered AT-014 to 18 dogs with appendicular osteosarcoma following amputation and chemotherapy (4 doses of carboplatin). The median survival time (MST) of 11 historical control dogs was 316 days; the MST for the treated group had not yet been reached ($p<0.0001$). However, eleven of the 18 treated dogs had surpassed the MST of the control group. Adverse events were mild to moderate and primarily consisted of fever, lethargy, and nausea/vomiting.[7]
3. **Feline IL-2 Immunomodulator (Merial)**
   Feline IL-2 Immunomodulator is a recombinant canarypox virus (ALVAC) expressing feline interleukin-2 (IL-2). This therapeutic was approved by EMA in 2013 (Oncept IL-2). Conditional license was granted by USDA in March 2015 (indicated to delay postsurgical recurrence of feline fibrosarcoma in adult cats with stage 1 disease).[8] A recent report outlines adjuvant treatment of feline injection-site sarcomas with this immunotherapy in complement to surgery and brachytherapy.[9]
4. **Telomerase Vaccine (Takis)**[10]
   A DNA telomerase therapeutic vaccine has been described. As observed in humans, telomerase reverse transcriptase (TERT) activity is largely confined to tumor tissues and absent in the majority of normal dog tissues. Therefore, dog TERT (dTERT) may constitute a valid target for translational cancer immunotherapy. The ability of adenovirus serotype 6 (Ad6) and DNA

electroporation to induce immune responses against dTERT in dogs affected by lymphoma was evaluated in combination with standard chemotherapy [cyclophosphamide, vincristine, prednisone (COP)].[11] Current studies are evaluating this immunotherapeutic in treatment of several other types of cancer.[12]

5. **ImmuneFx™ Cancer Vaccine (Morphogenesis Inc)**
   A multi-indication cancer vaccine, IFx-VET, is available. This product reportedly combines cell and gene therapy to promote cross-presentation and priming of the dog's immune system. It can be delivered as direct DNA or autologous vaccine.[13] Clinical benefit of this therapeutic is not well defined as no peer-reviewed data is available to date. As an autologous vaccine, this therapeutic is not currently approved or licensed by a regulatory agency.

SMALL MOLECULE INHIBITOR DEVELOPMENTS

1. **Verdinexor/KPT-335 for Canine Lymphoma (Karyopharm Therapeutics)**
   KPT-335 is a novel orally bioavailable XPO1 inhibitor. Exportin 1 (XPO1) is the sole nuclear exporter of several major tumor suppressor and growth regulatory proteins (TSPs and GRPs, including p53, p75, Rb, p21, p27, STAT3, FOXO and IkB among others). Expression of XPO1 is known to be upregulated in a variety of both hematologic malignancies and solid tumors and this correlates with a poor prognosis, indicating that changes in nuclear-cytoplasmic trafficking resulting in aberrant localization of key proteins can contribute to tumorigenesis and potentially resistance to therapy.[14] A Phase I clinical trial of KPT-335 was performed in 17 dogs with non-Hodgkin lymphoma (NHL, naive or relapsed), mast cell tumor or osteosarcoma. The maximum tolerated dose was 1.75 mg/kg given orally twice/week. Clinical benefit including partial response (PR, n = 2) and stable disease (SD, n = 7) was observed in 9/14 dogs with NHL with a median time to progression (TTP) for responders of 66 days (range 35–256 days). Toxicities were primarily gastrointestinal and were manageable; hepatotoxicity, anorexia and weight loss were the dose limiting toxicities.[14] A Phase 2b clinical trial of Verdinexor in dogs with lymphoma has been conducted. Verdinexor has received a Minor Use / Minor Species, or MUMS, designation from the Center for Veterinary Medicine (CVM) of the FDA for the treatment of lymphomas in dogs.[15]

2. **RV-1001 for Canine Lymphoma (Rhizen Pharmaceuticals)**
   PI3 kinase (PI3K) is a protein involved in cell signaling. It has been shown that for lymphomas and leukemias, PI3K isoforms γ and δ are particularly important in maintaining tumor growth. RV-1001 is an orally bioavailable inhibitor of PI3K family members, having a strong binding affinity towards PI3Kδ. A phase I clinical trial performed in dogs with naïve or relapsed B or T-cell lymphoma was conducted with oral dosing beginning at 10mg/kg PO q24h and escalating to 25mg/kg PO q24h. Grade 3 and 4 hepatotoxicity was noted in all dosing groups, prompting a change in dosing to once daily on Monday through Friday rather than 7 days/week. This Monday-Friday dosing regimen markedly reduced toxicity while maintaining response to therapy.[16] Nine dogs have been entered into the Phase I clinical trial beginning at a dose of 10 mg/kg q24h using a standard 3x3 design. One dog with T cell NHL has had a durable complete response to therapy (10 mg/kg), and 3 additional dogs have had partial responses (15 mg/kg [n=2] and 25 mg/kg [n=1]) for an objective response rate of 44%.[17]

3. **Toceranib phosphate/PALLADIA® (Zoetis)**
   Although first approved by the FDA in 2009, there are a number of recent publications to review that help to further define the best clinical use of this powerful therapeutic.
   A. **Pulse-dosed Palladia + lomustine for nonresectable canine mast cell tumor (MCT)**
      Lomustine at a dosage of 50 mg/m^2 once every 3 weeks combined with pulse-dosed Palladia was well-tolerated. The objective response rate of 46% for this protocol is comparable to what previously has been reported for single-agent protocols, but considerably higher than that reported with single-agent lomustine. Combining pulse-administered Palladia with lomustine may be a reasonable treatment option for dogs with unresectable or metastatic MCT.[18]
   B. **Maintenance Palladia following doxorubicin-based chemotherapy for canine splenic hemangiosarcoma (HSA)**
      Results of this study indicate that use of Palladia following doxorubicin chemotherapy does not improve either DFI or ST in dogs with stage I or II HSA compared to previously published reports of adjuvant chemotherapy for splenic HSA.[19]
   C. **Impact of Palladia/piroxicam/cyclophosphamide maintenance therapy on canine osteosarcoma (OSA)**
      The addition of Palladia to metronomic piroxicam/cyclophosphamide therapy following amputation and carboplatin chemotherapy did not improve median DFI, OS or the 1-year survival rate in dogs with OSA.[20]
   D. **Evaluation of Palladia administered to dogs with solid tumors at doses below the maximum tolerated dose**
      Palladia administered at doses between 2.4-2.9 mg/kg PO q48h resulted in an average 6–8 hr plasma concentration ranging from 100–120 ng/ml, well above the 40 ng/ml concentration associated with target inhibition. The lower doses of Palladia used in this study were associated with a substantially reduced adverse event profile compared to the established label dose of 3.25 mg/kg q48h.[21]
   E. **Continuous Palladia dosing in combination with cytotoxic chemotherapeutics**
      Studies evaluating the use of continuous Palladia dosing in combination with vinblastine,[22] CCNU,[23] and doxorubicin[24] have been described. These reports provide initial information on how to potentially utilize Palladia concurrently with other anti-cancer agents.

CYTOTOXIC CHEMOTHERAPEUTIC DEVELOPMENTS

1. **Paccal Vet®-CA1 (Oasmia Pharmaceutical)**
   The active ingredient in Paccal Vet-CA1 is paclitaxel, an antimicrotubule agent that works by stabilizing microtubules. Paccal Vet-CA1 has conditional approval and is labeled to treat: 1) Nonresectable stage III, IV, or V mammary carcinoma in dogs that have not received previous chemotherapy or radiotherapy; and 2) Resectable and nonresectable squamous cell carcinoma in dogs that have not received previous chemotherapy or radiotherapy. Federal law prohibits extra-label use of conditionally approved drugs meaning that Paccal Vet-CA1 can only be utilized for the two labeled indiations.[25]

2. **Doxophos® Vet (Oasmia Pharmaceutical)**
   Doxophos Vet is a patented formulation of doxorubicin which is being developed for treatment of canine lymphoma. Oasmia is reportedly conducting a Phase I study of Doxophos Vet which will include approximately 15 dogs. The FDA has granted MUMS Designation for Doxophos Vet for the indication of canine lymphoma.[26]

3. **Rabacfosadine/VDC-1101/Tanovea™ (VetDC)**
   Tanovea is a prodrug of the nucleotide analogue 9-(2-phosphonylmethoxyethyl) guanine (PMEG). A phase I/II trial was conducted in 38 dogs with non-Hodgkin lymphoma (NHL) using different dose schedules. Thirty (79%) dogs receiving Tanovea monotherapy achieved clinical remission [23 (61%) CR and 7 (18%) PR] and responses were noted in all eight treatment cohorts. Dose-limiting toxicities were generally manageable and reversible and included dermatopathy, neutropenia, and gastrointestinal signs. Two dogs developed signs of pulmonary dysfunction after they had been withdrawn from study after completing all Tanovea treatments.[27] In June 2013, the FDA granted Tanovea MUMS designation for use in canine lymphoma, allowing VetDC to move toward filing for full regulatory approval. Tanovea is also currently being evaluated for use in cats with lymphoma.[28]

ONCOLOGY SUPPORTIVE THERAPEUTIC DEVELOPMENTS

1. **Canalevia™ (Jaguar Animal Health)**
   Canalevia is a canine-specific formulation of crofelemer, an active pharmaceutical ingredient isolated and purified from the *Croton lechleri* tree, that contains anti-secretory properties. As an oral, enteric-coated, twice daily formulation of crofelemer, Canalevia is being developed for the treatment of chemotherapy-induced diarrhea, or CID, in dogs. The product is not absorbed systemically at the therapeutic dose, but acts locally in the gastrointestinal tract. Jaguar is pursuing a Minor Use/Minor Species (MUMS) approval from the CVM for Canalevia for CID in dogs. If conditional approval is received, Jaguar expects to launch Canalevia for CID in dogs in 2016.[29,30]

2. **Capromorelin/AT-002 (Aratana Therapeutics, Inc.)**
   Capromorelin is an orally active small molecule that mimics the action of ghrelin which causes growth hormone secretion and appetite stimulation. Capromorelin has been shown to cause increased food intake and weight gain in both laboratory and client-owned dogs.[31,32] Additionally, capromorelin has been demonstrated to cause increases in IGF-1 and increased food intake and body weight in cats.[33] Based on current timelines, Aratana anticipates FDA approval and commercialization of capromorelin for treatment of inappetence in dogs in 2016.[34] This therapeutic has significant potential to positively impact the treatment of inappetence related to chemotherapy and/or underlying cancer in our canine and feline patients.

3. **Grapiprant/AT-001 (Aratana Therapeutics, Inc.)**
   Grapiprant is a selective antagonist of the EP4 receptor, one of four receptors that mediates the action of prostaglandin E_2 (PGE_2). This receptor is believed to play a key role in mediating pain associated with osteoarthritis. Long-term safety in laboratory dogs has been demonstrated.[35] FDA approval and commencement of commercialization is anticipated in 2016.[36] This may provide an alternate pain management therapeutic for cancer patients experiencing chronic pain.

REFERENCES

[1]Bulman-Fleming J, et al. Abstract, 2014 Veterinary Cancer Society; St. Louis, MO. [2]Ogilvie GK, et al. Abstract, 2014 Veterinary Cancer Society; St. Louis, MO. [3]Rodriguez CR, et al. Abstract, 2015 ACVIM Forum; Indianapolis, IN. [4]http://aratana.investorroom.com/2015-09-24-Aratana-Therapeutics-Provides-Product-Updates. [5]Rue SM, et al. Vet Immunol Immunopathol 2015;164(3-4):148-159. [6]http://c.ymcdn.com/sites/www.nysvms.org/resource/collection/824A08A0-F7AF-4D95-A850-C088F60C9C8E/NYSMVS_SeptOct_proof_06.pdf. [7]Mason N, et al. Abstract, 2014 ACVIM Forum; Nashville, TN. [8]http://todaysveterinarypractice.epubxp.com/i/504690-may-jun-2015/18. [9]Jas D, et al. Trials in Vaccinology 2015;4:1–8. [10]http://www.takisbiotech.it/index.php/news/30-takis-veterinary-cancer. [11]Peruzzi D, et al. Mol Ther 2010;18(8):1559-67. [12] http://www.vetcancertrials.org/studies/vschv-telomerase-vaccine-for-canine-cancer. [13]http://www.morphogenesis-inc.com/veterinary-oncology/veterinarians/. [14]London CA, et al. PLOS One 2014;9(2):e87585. [15]http://karyopharm.com/drug-candidate/oral_verdinexor_kpt-335/. [16]https://iom.nationalacademies.org/~/media/Files/Activity%20Files/Disease/NCPF/2015-JUN-8/London.pdf. [17]London CA, et al. Abstract, 2015 American Association for Cancer Research; Philadelphia, PA. [18]Burton JH, et al. J Vet Intern Med 2015;29:1098–1104. [19]Gardner HL, et al. BMC Veterinary Research 2015;11:131. [20]London CA, et al. PLOS One

2015:DOI:10.1371/journal.pone.0124889. [21]Bernabe LF, et al. BMC Veterinary Research 2013;9:190. [22]Robat C, et al. Vet Comp Oncol 2012;10(3):174-183. [23]Pan X, et al. Vet Comp Oncol 2014. [24]Pellin M, et al. Abstract, 2015 Veterinary Cancer Society; Vienna, VA. [25]http://www.fda.gov/AnimalVeterinary/ResourcesforYou/ucm402476.htm [23] http://www.oasmia.com/news.asp?c_id=373. [26]http://www.oasmia.com/pages.asp?c_id=313 [27]Vail DM, et al. Clin Cancer Res 2009;15(10):3503-10. [28]http://vet-dc.com/products/tanovea-for-lymphoma/. [29] http://jaguaranimalhealth.com/animal-health/. [30] http://www.businesswire.com/news/home/20150701005278/en/Jaguar-Completes-Study-Dogs-Chemotherapy-Induced-Diarrhea-Commercial#.Vc4B8ZNViko. [31] Zollers B, et al. Abstract, 2014 American College of Veterinary Internal Medicine; Nashville, TN. [32]Zollers B, et al. Abstract, 2014 ACVIM Forum; Nashville, TN. [33]Zollers B, et al. Abstract, 2015 ACVIM Forum; Indianapolis, IN.[34]http://www.aratana.com/aratana-announces-positive-results-in-a-pivotal-field-effectiveness-study/. [35]Rausch-Derra L, et al. Abstract, 2014 ACVIM Forum; Nashville, TN. [36]http://www.aratana.com/aratana-reports-positive-results-from-its-pivotal-field-study-of-at-001-grapiprant/.

The Delivery of Precision Medicine in Veterinary Oncology

C. Khanna[1], DVM, PhD, DACVIM (Oncology), DACVP (Hon)[1], and William Hendricks, PhD[2],
[1]Washington DC, Leesbrurg,VA , and [2], Phoenix, AZ
Phoenix, AZ

THE GENETIC BASIS OF CANCER

Cancer is a genetic disease that arises as a consequence of the stepwise accumulation of disruptive mutations in genes that regulate cell life and death. Clonal expansion of cell populations bearing cancer gene mutations fuels the formation of malignant tumors capable of invasion and metastasis. This genetic model of cancer emerged in the latter 20th century as a product of advances in genetics, evolution, and cancer medicine. Small targeted gene sequencing studies conducted in colorectal cancer samples revealed that inactivating mutations in tumor suppressor genes (genes whose normal function serves to prevent aberrant cell growth) such as *APC, TP53,* and/or *PTEN* could drive malignant transformation of normal epithelium into an adenoma. Subsequent activating mutations in oncogenes (genes whose normal function is to accelerate cell growth) such as *KRAS, BRAF,* and/or *PIK3CA* coupled with additional cancer gene mutations were then shown to confer increasingly malignant phenotypes in the course of adenocarcinoma development and metastatic spread. Through this mutational process, cancers acquire specific key properties including self-sufficient growth signaling, resistance to growth inhibitory signaling, invasion and metastasis, unlimited replication potential, angiogenic signaling, immune modulation, DNA instability, metabolic dysregulation, and immune evasion. The genetic model and its downstream phenotypes have since been validated in many cancer types and provided a framework for our growing genomic understanding of cancer.

Human cancer genome landscapes

The 21st century genomics revolution is now expanding the genetic model of human cancer through exhaustive annotation of cancer-driving mutations and identification of profound genomic diversity existing not only between tumor types across patients, but also even within tumor cell populations in individuals. The first large-scale cancer mutation screens used Sanger sequencing of individual genes or gene families such as BRAF or protein kinases in large cohorts of varying cancer types. Frequent activating BRAF mutations were thereby discovered mutations in human melanomas and nevi as were ERBB2 mutations in breast cancer. In 2003, the first cancer exomes (the complement of all protein-coding genes in the genome) were sequenced in 140 acute myeloid leukemias (AML). Soon after, in 2006, Sanger-based exome sequencing of 13,023 genes in 11 breast and 11 colorectal cancers were published. This study revealed the first sign of cancer's truly overwhelming complexity –189 frequently-mutated genes were identified with an average of 90 mutant genes (11 putative cancer genes) per tumor. Follow-up analysis of an additional 7,000 genes in the same tumors confirmed that a small number of commonly mutated genes in the genomic landscape (the "mountains") existed among a large number of less frequently mutated genes ("hills"). In 2008, four Sanger sequencing based exome projects additionally characterized over 20,000 genes in glioblastoma multiforme and pancreatic cancer with similar results, although these studies also uncovered *IDH1* and *PIK3R1* as new recurrently mutated genes. In these and subsequent genomic analyses, it was found that the high-frequency mountains in the broad cancer genomic landscape had for the most part already been identified. Identification and characterization of the many hills has since been the focus of much genomic and functional work.

After the introduction in 2006 of widely available massively parallel sequencing technology, also know as next generation sequencing (NGS), the number and scope of cancer genome projects has accelerated at a dizzying pace. Whole cancer genomes sequenced via NGS were first described in 2008 (two lung cancers) and were followed by descriptions of single AML cases in 2008 and 2009. In 2009, analysis of paired primary and metastatic tumors from the same patient was conducted in breast cancer, identifying important similarities and differences. Comprehensive assessment of genome-wide mutations in melanoma and small cell lung cancer in 2010 uncovered enormous mutation burdens in these tumor types in addition to mutation signatures linked to carcinogens such as ultraviolet radiation and cigarette smoke. As of today, large genomic analyses having been completed in dozens of cancer types and are increasingly incorporating larger cohort sizes, aspects of tumor heterogeneity, and assessment of cancer genomes in the context of key clinical parameters such as treatment response.

The COSMIC (Catalogue Of Somatic Mutations In Cancer) database, which actively collects all published cancer sequencing data, contains data on nearly 1.2 million samples containing 3.7 million mutations (see http://cancer.sanger.ac.uk/cosmic, Release 75). Two large multi-institutional and international consortia – The Cancer Genome Atlas (TCGA) and the International Cancer Genomics Consortium (ICGC) – have additionally overseen genomic analysis of more than 30,000 tumors across nearly 130 studies (see http://cancergenome.nih.gov and http://icgc.org). Despite the extraordinary depth of complexity and diversity in cancer genomic data, it is thought that most mutations target genes that are members of a relatively small group of approximately 12 core signaling pathways. These data suggest an avenue for treating tumors based on aberrant pathways rather than individual aberrant genes. Human cancer genomic data and genomic tools are now extending into the cancer clinic where they are driving diagnostic, prognostic, and treatment innovation. This transfer of genomic knowledge and tools to the clinic is at the heart of the Precision Medicine (PMed) paradigm.

The genetic basis of canine cancer

Dogs can develop many of the same cancers as humans at similar frequencies and with a similar enrichment for disease incidence late in life. Many studies have suggested that canine and human cancers share many features of their biology and clinical presentation. Although canine cancer epidemiology is notoriously incomplete, notable differences do exist in distribution of the most prevalent histologies across species. For example, sarcomas are more common in dog whereas carcinomas are more common in human. Some of these differences may be accounted for by differences in diet for cancers such as colorectal as well as spaying and neutering practices in the United States for breast and prostate cancer. However, dogs do share human environments and thus different patterns of incidence may also be due instead to genetics. Due to selective breeding practices and the unique genetic history of the dog, with over 350 isolated breeds, a number of breeds are predisposed to specific cancer subtypes. Generally, however, it is clear that cancer is fundamentally a genetic disease across species.

Canine cancer genome landscapes

Canine genetics has advanced rapidly over the course of the past 20 years. The canine chromosomes were first identified in 1999 with increasingly detailed genetic maps generated across the 90s and early 2000s. The canine genome sequence followed shortly on the heels of the human. Whereas the first high quality human genome sequence was completed in 2001 at an estimated cost of $3 billion, the first such canine genome sequence was completed in 2005 for about $30 million. The dog genome is, of course, unique despite similarities with human. Around 94% of the canine genome is arranged similarly to that of the human with about 5% of the actual genome sequence under purifying selection (i.e. highly conserved) between human and canine, a large fraction considering that only about 2% of each genome contains protein-coding genes. Canines bear 39 chromosome pairs consisting of 2.4 billion base pairs compared to 3.2 billion base pairs in 23 human chromosomes. Both canine and human genomes are predicted to contain roughly 20,000 protein-coding genes. However, only about 2,000 transcripts have been well annotated in the dog genome whereas this number approaches 50,000 in the human. Overall, the canine genome is actually more similar to the human than is the mouse, a striking picture given that mice and humans share a more recent ancestor than do dogs and humans.

A first canine oncogenic mutation – *KIT* duplication in malignant mast cell tumors – was discovered in 1999. Additional targeted analyses subsequently revealed alterations in specific canine cancer types and/or breeds such as *MET* mutations in Rottweilers and putative *BRCA1/2* pathogenic variants in canine mammary cancer. Coupled with the high quality canine genome sequence, genome-wide association studies (GWAS) of breed stereotypes such as that published across 148 breeds in 2008 have enabled deeper analysis of germline factors influencing cancer development in the dog. Comprehensive cancer GWAS studies have been published for a number of nonsyndromic canine cancers including osteosarcoma, histiocytic sarcoma, and squamous cell carcinoma of the digit. This substantial body of work has pointed to key genetic loci and cancer genes that are implicated in these tumor types. Much of our knowledge of somatic canine cancer genomics (i.e. mutations arising sporadically in somatic tissues or tumors), however, has come from molecular cytogenetic studies that have identified genome-wide somatic copy number alterations in cancers such as lymphoma and osteosarcoma from 2009 onwards. Relatively few studies of canine cancer genomes incorporating NGS platforms have yet been published. Contrasted to the tens of thousands of sequenced human tumor genomes, fewer than 250 canine tumors have yet had their genomes deeply characterized. Only about 10 such studies have thus far been published. One of the first such studies undertook analysis of 25 canine mammary cancers in 2013 (5 whole genomes). Additional studies include: soft tissue sarcoma (1 study – 7 exomes), lymphoma (3 studies – 110 exomes, 36 transcriptomes), mammary carcinoma (1 additional study – 3 whole genomes, 4 exomes, and 10 transcriptomes), transitional cell carcinoma (1 study – 4 transcriptomes), hemangiosarcoma (2 studies – 69 transcriptomes), and canine transmissible venereal tumors (1 study – 2 whole genomes). These projects are only just beginning to shed light on the genomic landscapes of canine cancers. They have revealed that many of the mutations driving canine cancer are similar to those seen in humans. In some case, such as non-Hodgkin lymphoma, discovery in the dog (e.g. identification of frequent *TRAF3* mutations) has even informed analysis and discovery in the human. Canine studies have also pointed to novel findings such as candidate cancer genes not previously identified in human cancers or cancer genes occurring in cancer subtypes in the dog, but not the human. For example, the oncogene BRAF has been found to be mutated a high frequency in canine transitional cell carcinoma whereas this mutation is uncommon in the human cancer. We have only begun to chart the rough outline of canine cancer genome landscapes, but what we have thus far mapped out has already proven to be of enormous value for both canine and human.

CANCER GENOMICS PLATFORMS: A HISTORICAL PERSPECTIVE

Molecular cytogenetics: Identification of chromosomal alterations in cancer

Mid-20th century advances in cytogenetics such as chromosome staining were the first to enable empirical observation of the chromosomal complement of a cell. This technology enabled identification of the "Philadelphia chromosome" consisting of the BCR-ABL1 translocation in chronic myelogenous leukemia, the first specific genetic change found to be characteristic of a cancer type. This discovery inspired subsequent assessment of such structural alterations across multiple cancer types such as sarcomas, leukemias, and lymphomas. New approaches to chromosomal staining soon afterwards allowed progressively detailed views of chromosome aberrations in cancer cells. These included fluorescence in situ hybridization (FISH, using fluorescent DNA probes that hybridize to specific chromosomal regions), spectral karyotyping (genome-wide chromosome staining)(29), multiplex FISH (M-FISH), chromosome microdissection, and comparative genomic hybridization (CGH).

A paradigm-shifting idea then emerged that impacted early high-throughput genomics and ultimately shaped the emergence of NGS technology. This shift was based on printing nucleic acids on slides (microarrays), an innovation that dramatically increased the number probes that could be assayed in a single experiment thereby reducing cost and increasing throughput. This paradigm was applied to molecular cytogenetics (array CGH or "aCGH" which is the spotting of DNA capture probes onto slides in intervals), characterization of single nucleotide polymorphisms or "SNPs" (small DNA probes that cover millions of SNPs genome-wide), and gene expression analysis (discussed later).

Three broad classes of mutation can be considered in cancer genomics: structural variation (SV), copy number variation (CNV), and single nucleotide variation (SNV). SVs are large-scale rearrangements of chromosomes. CNVs are amplifications or deletions of chromosome segments. SNVs are individual mutated bases or small stretches of DNA that are added or deleted from the genome. Both SVs such as the *BCR-ABL1* translocation and CNVs such as *KIT* amplification can be characterized using the above cytogenetic tools. Translocations are very common and are pathognomonic in many cancers, particularly liquid cancers and sarcomas. These mutations consist of the aberrantly positioned DNA segments occurring either in the wrong place in their original chromosome (intrachromosomal translocation) or different chromosomes (interchromosomal translocation). They are the result of defects or accidents in DNA repair and recombination processes. Translocations can activate oncogenes (i.e. constitutive activation of the ABL tyrosine kinase via fusion with BCR) or inactivate tumor suppressor genes (such as translocations that interrupt TP53 tumor suppressor function). CNVs that involve gains or losses of entire chromosomes are some of the most common alterations seen in cancers (i.e. aneuploidy), but CNVs can also occur within smaller chromosomal regions. Defective chromosomal segregation typically leads to these phenomena. Overall, CNVs tend to alter the dosage of oncogenes or tumor suppressor genes such as amplification of *ERBB2* in breast cancers or deletion of the TP53 tumor suppressor in diverse cancers.

Molecular genetics: Identification of sequence alterations in cancer

Just as molecular cytogenetic techniques described above enabled sensitive and specific detection of SVs, advances in molecular genetics in the late 1970s enabled detection of SNVs. The combined developments of molecular cloning, the polymerase chain reaction (PCR), the Sanger dideoxynucleotide chain termination sequencing method, and agarose gel electrophoresis powered early gene sequencing studies. These methods were then progressively refined (through use of fluorescently labeled nucleotides, capillary electrophoresis, paired-end sequencing, shotgun sequencing, and laboratory automation) to the point that entire genomes could be sequenced with substantial effort and dedicated resources such as those committed to the Human Genome Project at the end of the 20th century and those applied to the first sequenced cancer genomes in 2003-2005.

Gene expression microarrays: Identification of cancer gene expression signatures

The application of microarray technology to genomics was first deployed in gene expression studies even prior to its use for molecular cytogenetics. Gene expression microarray probes complementary to target RNAs are created from complementary DNA (formed via reverse transcription of RNA) or synthesized small nucleotides and attached at high density to glass slides to enable high-throughput measurement of large numbers of transcripts. Studies implementing these arrays provided some of the first genome-wide views of aberrant genes and signaling pathways in cancer. Gene expression signatures have been used to classify cancers according biological or clinical phenotypes, at times revealing unique features not discernable through standard histopathology. Although advances in RNA-based NGS have now replaced microarrays in terms of sheer power, gene expression profiling continues to play an important role in biological studies and in the clinic. Commercially-available clinical tests based on microarray studies include OncoType DX and MammaPrint for breast cancer.

Massively parallel sequencing: Charting cancer genome landscapes

The extraordinary depth of cancer genomic data now available has been made possible by the NGS revolution that followed the completion of the human genome. By the genome's completion, the cost of a single genome sequenced using Sanger technology had dropped to the $1 million mark. This cost was still prohibitive for most genome studies, however. Thus, the National Human Genome Research Institute created the $70 million sequencing technology program to stimulate the development of the $1,000 genome. This quest for the $1,000 genome inspired a complete overhaul of DNA sequencing methods. The Sanger sequencing workhorse had to be replaced with technology capable of eliminating the requirement for molecular cloning steps, individual PCR reactions, and small sequencing regions. Thus, new methods that generated sequencing templates without cloning, that massively multiplexed the PCR reactions, and that immobilized full sequencing libraries in a single reaction chamber enabled massively parallel sequencing of millions of templates at once. These innovations formed the backbone of the new approach to NGS that has dramatically enhanced throughput and reduced cost. It is now standard to sequence 500 billion DNA bases, equivalent to 156 human genomes, in the course of a single 6-day experiment.

NGS instruments and analysis platforms

Many NGS platforms are now available with each having unique strengths and applications. Most platforms are capable of performing diverse genomic analyses when coupled with appropriate informatics tools. These analyses include DNA sequencing, copy number, and structural analysis in addition to RNA sequencing and epigenomic sequencing. Major NGS platform manufacturers currently include Illumina, Life Technologies, Pacific Biosciences, and Roche. Platforms vary in their approach to sequencing template preparation, sequencing chemistry, maximum sequencing read length, bases generated per sequencing run, accuracy, sequencing run time, and their application. For example, one of the most commonly employed platforms is many genomics research laboratories is the Illumina HiSeq 2500 which is based on solid-phase template preparation, uses reversible terminator chemistry, generates reads of 125 nucleotides up to 500 billion bases per run at >99% accuracy and has a 6-day run time. This instrument can be used for whole genome, exome, RNA, targeted, epigenomic, metagenomics, and de novo sequencing. Newer "benchtop" instruments have also been developed with reduced cost, smaller output and shorter run times in order to increase accessibility and enable clinical studies.

NGS approaches are also shaped by the nucleic acid preparation approach (or the type of "library" prepared for sequencing on the instruments described above). These approaches can be either whole genome or base on candidate gene selection. To facilitate more rapid, affordable, and focused assessment of genomic regions, a number of target enrichment strategies have also been developed. While the cost of whole genome analysis has dropped dramatically, it still often remains costly and unnecessary to sequence entire genomes. One of the most common platforms enabling candidate gene analysis is hybridization-based capture with oligonucleotides complementary to target DNAs or RNAs in solution or on microarrays in addition to highly multiplexed PCR or microdroplet PCR. This approach is frequently brought to bear on assessing the mere 1% of the human genome containing coding regions (all exons = the exome). Commercially available candidate gene platforms therefore include human exome kits, other targeted capture kits (e.g. cancer gene panels), and even platforms that facilitate custom capture designs. Overall, the capability for a candidate-gene approach on benchtop sequencers is powering affordable genomic analysis that is both driving discovery in the research setting and is also enabling clinical laboratory testing.

Genomic data analysis and visualization

NGS data is immensely complex and therefore requires substantial informatic resources to enable analysis including high performance computing, data storage, and integrated analysis pipelines incorporating diverse data tools. Most NGS technology is based on generation of raw sequence data consisting of short "reads" of several hundred nucleotides that are generated as text files. An average run of an NGS instrument typically sequences around 500 billion bases equivalent to about 4 billion short reads. Once these reads have passed quality control filters, they must then be mapped (aligned) to a reference genome available from a genome database – GRCh38 from the Genome Reference Consortium in the case of the human genome. A number of alignment software tools is available. The computational power and run time required to map these reads depends of course on the size of the sequenced genome. Downstream informatic analysis is then performed on aligned data from tumors (and, often, also from matched germline DNA) to identify alterations present in tumors that are not found in the patient's normal DNA. This process is referred to as "variant detection" and incorporates multiple algorithms for identifying mutation types (e.g. SNVs, CNVs, and SVs) and annotating the mutations according to their function.

Canine-specific genomic tools and resources

Although the core elements of NGS benchtop and informatic technology are applicable across species, a number of species-specific features are needed to enable canine genomics. First and foremost, the reference genome (current version CanFam3.1) has been critical for enabling mapping of canine sequencing reads to their appropriate locations. Next, the species-specific annotation of genomic features based on computational prediction as well as experimental validation is critical for enabling interpretation of mutations identified in tumors. Along similar lines, knowledge of the natural variation and single nucleotide polymorphisms (SNPs) occurring between individuals is critical for interpretation of mutations as simply normal variation or as potentially pathogenic mutations. Finally, the reagents necessary for candidate gene analysis are typically species-specific insofar as they are based on targeted capture using sequence-dependent capture probes. For example, descriptions of only three canine exome-capture kits based on the latest canine genome build have thus far been published whereas numerous such kits exist for human genomes.

PRECISION MEDICINE FOR HUMAN AND CANINE ONCOLOGY

PMed is ushering in a new era in cancer therapy in which clinical and translational value are applied to advances in the genomic analysis of cancer. The discoveries and tools described above have provided new opportunities to tailor cancer therapy to the individual molecular characteristics of a specific cancer in a specific patient to guide diagnosis, prognosis, and treatment selection. In many cases, these genetic alterations can be matched to specific therapeutic agents as a means to uniquely improve outcomes for patients. This novel therapeutic approach is now available to patients through The Oncology Service. Tumor samples and matched germline samples (from peripheral blood or cheek swabs) are collected, preserved, and then analyzed for genetic alterations in a core set of cancer genes which are ultimately matched to an individualized therapeutic recommendation.

How Precision Medicine differs from the current practice of oncology

The use of patient-specific information as a means to deliver precision medicine is not new to the treatment of cancer patients. In general through the history of cancer medicine, treatments have been administered in a patient-specific and personalized manner. Even in the modern era, it is the use of molecular data to guide the therapy of specific individuals with cancer is not entirely novel. The use of specific immunohistochemical or cytogenetic markers to guide diagnosis and prognosis has been a critical and routine practice in pathology laboratories for many decades. Further, in some cancers, molecular markers have been used to guide treatment selection as well. For example, in human breast cancer, it has been long-standing practice to define the expression of hormone receptors as a means to deliver specific therapeutics that alter downstream signaling pathways. More recently in veterinary medicine our understanding of the clinical and biological importance of c-kit mutations has not only directed distinct therapeutic plans for patients with and without such mutations, but knowledge of these mutations has been the basis in many cases to select specific small molecule inhibitors of these altered receptors.

Human cancer Precision Medicine

The potential success of Precision Medicine has received widespread recognition in the human field:

The path to personalized medicine. N Engl J Med. 2010 Jul 22;363(4):301-4.

Cancer medicine is ultimately aimed at recommendation of the most effective treatment for an individual patient based on supporting scientific evidence. The use of a PMed approach to cancer diagnosis and therapy may deliver a variety of outcomes related to this goal:

1. Improved outcomes for patients with a specific cancer.
2. Improved outcomes for patients with cancers that are not effectively managed with conventional treatment approaches.
3. Data that would support the grouping of patients in clinical trials that is agnostic to histology (i.e. "basket study designs").
4. Data that provides an accelerated rationale for a conventional clinical trial of a specific therapy in a specific cancer (i.e. "accelerated reverse directional drug development").

Although genomic data is expected to dramatically impact the above outcomes, genomic data have been slow to enter the clinic given the complexity of their implementation. One of the first pilot studies incorporating molecular profiling to guide therapy in advanced cancers was published by our collaborators in 2010:

Von Hoff DD, Stephenson JJ, Rosen P, Loesch DM, Borad MJ, Anthony S, et al. Pilot study using molecular profiling of patients' tumors to find potential targets and select treatments for their refractory cancers. Journal of clinical oncology. 2010;28(33):4877-83.

This study faced considerable challenges, but found that 27% of 68 patients treated according to molecular profiling recommendations experienced a longer progression-free survival than during the most recent treatment on which they had progressed. Although perhaps intuitively beneficial to incorporate precise target identification into patient treatment, this approach still faces significant hurdles and it remains to be proven through prospective trials that treatment based on PMed outperforms physician's choice of treatment. Now, multiple clinical trials incorporating genomics-guided therapy selection are underway to test this very hypothesis with some early results published in recent years. Trial designs include adaptive, basket, and umbrella trials. Adaptive trials such as I-SPY 2 (Investigation of Serial Studies to Predict Your therapeutic response with imaging and molecular analysis) modify treatment arms based on patient response. Basket trials such as the NCI MATCH (Molecular Analysis for Therapy Choice) trial assign patients with diverse cancers to treatment baskets based on mutation status. Umbrella trials incorporate study of multiple drugs in a single disease based on mutation signatures. One such trial - the Stand Up To Cancer and Melanoma Research Alliance Dream Team Clinical Trial is assessing molecularly-guided therapy in non-V600 BRAF metastatic melanoma. This is now an ongoing randomized treatment study. The non-treatment pilot study is described here:

LoRusso PM, Boerner SA, Pilat MJ, Forman KM, Zuccaro CY, Kiefer JA, et al. Pilot Trial of Selecting Molecularly Guided Therapy for Patients with Non–V600 BRAF-Mutant Metastatic Melanoma: Experience of the SU2C/MRA Melanoma Dream Team. Molecular cancer therapeutics. 2015;14(8):1962-71.

Finally, beyond informing targeted treatment, cancer genomic studies in clinical contexts have also identified diagnostic and predictive biomarkers that have been rapidly commercialized.

Returning clinically relevant and actionable information based on genomic analysis within a window that enables effective treatment selection is a substantial challenge. Notable hurdles include those associated with patient consent, tumor biopsy, sample preservation and transport, nucleic acid extraction and quality control, genomic sequencing infrastructure and platform, data analysis and integration, generation of digestible genomic reports for physicians and scientists alike, and conduct of tumor board review to provide a treatment recommendation. These processes are not yet sufficiently streamlined to enable implementation beyond major academic research centers. Additional hurdles to the implementation of Precision Medicine in the clinic include:

1. <u>Matching biological targets to drugs: A hierarchy of molecular events</u>. Any analysis of cancer at a molecular level reveals a variety of alterations when compared to normal tissues including those of DNA, RNA, and protein. However, not all such alterations provide clinically relevant or actionable information. For example, many gene expression (RNA) signatures have been hypothesized to predict treatment response, however few such alterations have been validated in clinical studies. As such, the simple expression of a cancer gene or its associated protein is not necessarily sufficient to be considered a highly

valuable target in the biology or therapy of that cancer. Conversely, mutations in DNA that comprise the fundamental reducible phenomena behind oncogene activation or tumor suppressor inactivation are often considered the gold standard for predictive capacity.

2. Discrimination of driver versus passenger mutations. Even though cancer is a fundamentally genetic disease driven by mutations in cancer genes, not all mutations contribute to cancer fitness (i.e. not all mutations are cancer "drivers"). Many mutations are "passenger" mutations. They occur as collateral damage of normal replicative processes and/or mutational processes in cancer cells, but do not directly contribute to the growth phenotype. It is not always straightforward to identify the functional impact of a specific mutation, particularly if that mutation has not been previously extensively characterized. Therefore driver/passenger discrimination remains a key challenge for individualized genomic analysis.
3. Context may influence the role of a driver. In keeping with the above observation, driver mutations may have variable clinical relevance depending on the presence or absence of other mutations. For example, activating mutations in the *BRAF* oncogene are common in both melanoma and colorectal cancer. In melanoma, these mutations confer sensitivity to selective BRAF inhibitors. However, colorectal cancer has proven to be resistant to such drugs due to the presence of concomitant activation of genes upstream of BRAF. Mutation status of single genes has not always proven to be predictive across cancer types in diverse mutational backgrounds.
4. Drug matching rules, evidence, report generation, and molecular tumor boards. The above considerations all factor into the daunting task of cataloguing rules for matching drugs to specific aberrations based on various levels of evidence from preclinical and clinical literature. Evidence levels range from direct evidence of a positive predictive biomarker in prospective randomized clinical trials to laboratory experiments assessing molecular alterations and drug sensitivity in cell lines or animal models. An actively curated, literature-linked drug rule database must be created, maintained, and intersected with genomic reports. Ultimately, these reports must be automated, precise, and comprehensive enough to enable an expert team of clinicians and scientists to interpret, discuss, and reach consensus on a treatment recommendation. The molecular tumor board thereby necessitates a high degree of art, interpretation, and expertise.

An ever-growing array of tests designed to inform diagnostic and treatment decisions in the clinic is available from more than 100 academic and 50 commercial laboratories. These tests range in scope from single genes to gene panels, exomes, and even whole genomes. In everyday clinical practice, cost is still prohibitive and the expertise and infrastructure that are required to bring them to bear on patient care are also largely lacking. Finally, more comprehensive data showing improvements in genomics-correlated clinical outcomes is needed to support the use of these tests and of off-label drug use. Commercial sequencing laboratories are also faced with the need for approval from the FDA and Medicare in order to guarantee that insurance companies will cover the tests. Notably, procedural terminology codes and an initial group of 21 tests have been added to the Centers for Medicare and Medicaid Services fee schedule. A final challenge faced by these tests is the risk of incidental secondary findings of possible pathogenic germline variants. Determining the best approach to manage such findings is an area of active controversy.

THE PATH TO PMED FOR CANINE CANCER PATIENTS

Although much work remains to be done to chart the genomic landscapes of canine cancers, the precision medicine approach is nonetheless poised to make a dramatic impact on the care of canine cancer patients. Indeed, we have published on the clinical feasibility of this approach in dogs:

> *Paoloni M, et al. Prospective molecular profiling of canine cancers provides a clinically relevant comparative model for evaluating personalized medicine (PMed) trials. PLoSOne. 2014 Mar 17;9(3)e90028.*

Precision medicine now represents a cutting edge opportunity in veterinary cancer care. In fact, not only would this approach make great headway in the care of canine cancer patients, but given the unique aspects of naturally occurring cancer in pet dogs as a comparative cancer model for humans, the veterinary profession on the whole may be able to provide key data validating this model and refining its implementation for human medicine.

Emerging opportunities for development of canine cancer Precision Medicine

As noted above our experience and publications to date have demonstrated the feasibility of deriving a molecular analysis of a canine patient tumor and, in a clinically relevant time frame, to deliver a patient-specific drug matching report. Beyond this feasibility, there is now an opportunity to test the clinical advantage of PMed approaches to cancer management. For a variety of reasons, the canine endothelial malignancy, hemangiosarcoma, will be the first disease entity in which we ask whether a specific PMed approach can improve outcomes in this currently devastating disease. In addition there is a similar opportunity to better understand the values and shortcomings of a PMed approach to cancer management in a prospective trial that would include a series of cases, so called "n=1" studies. We expect that this entry into a personalized and molecularly guided approach to cancer diagnosis and therapy will be considered historically as another important transition from organ to cell, and now to molecular based medicine

Electroporation in Oncology: Clinical Applications and Future Perspectives

Enrico P. Spugnini, DVM, PhD, DACVIM (Oncology), DECVIM-CA (Oncology)
Rome, Italy

INTRODUCTION

Electroporation is a technique that aims at ablating tumor cells through the direct application of current (irreversible electroporation or IRE) or at increasing the delivery of drugs through the cell membrane by applying proper permeabilizing pulses. Chemotherapy drugs are effective provided that they reach their site of action. The primary obstacle for chemotherapy agents is the cytoplasmic membrane, especially for lipophobic agents such as bleomycin. Electrochemotherapy (ECT) couples the administration of an anticancer agent (usually bleomycin or cisplatin), to the delivery of electric pulses having appropriate waveforms. The application of permeabilizing pulses leads to perturbation of the cell membrane, thus resulting in increased uptake of chemotherapeutic drugs, ultimately leading to cell apoptotic death.[3] Clinical trials evaluating humans treated with ECT for cutaneous neoplasms were first reported in the early 90s, and have been followed by many trials that confirm its potential for the treatment or palliation of dermatological tumors.[3-5] Recently, a standard operating procedure has been proposed.[6]

ELECTROCHEMOTHERAPY EQUIPMENTS

Different waveforms have been adopted by investigators: exponentially decaying, square, rectangular or biphasic. In general the number of pulses delivered is set in 8 single pulses applied per cm of tumor area at a voltage of 1300 V/cm (800 V/cm for intraoperative use), with a duration of 100 μs and a frequency of 1 Hz. Treatments are repeated until the whole tumor area is covered. We recently developed a shortened protocol with decreased morbidity and maintained efficacy.[7]

RESULTS IN VETERINARY MEDICINE USING SQUARE PULSES BASED ECT

The first clinical trial in veterinary patients involved the treatment of feline sarcomas recurring after radiation therapy using bleomycin and Interleukin 2 producing cells coupled with square pulses.[8] The authors did not report any clinical response but claimed prolonged control for cats undergoing ECT compared to untreated patients. More recently, CDDP or bleomycin based ECT has been reported to treat cutanous neoplasms in dogs and cats as well as sarcoids in horses.[9-11]

EVALUATION OF BIPHASIC PULSES BASED ECT IN PETS

Experimental design

Our group investigated in the past 12 years the feasibility of electrochemotherapy in companion animals affected by different neoplasms. We adopted trains of biphasic pulses delivered as a burst instead of single impulses. The electric waveform were generated by an equipment developed by our group. This device is built up by a toroidal core transformer generating a roughly rectangular pulse which is split in two halves that are sequentially driven to obtain a biphasic pulse. Pulses are not singularly produced but are created in bursts. The equipment allows choosing electric field intensity, number of pulses and pulse duration. The standard wave train used for this experiment was set to 8 pulses of 50 + 50 μs (biphasic pulse), and the electric field intensity was 1250 V/cm. The pulse repetition frequency is 1 Hz while the frequency of burst repetition is 1 kHz, resulting in a total burst duration of 7.1 ms. Bleomycin was locally injected at the tumor site and in the tumor's surrounding tissues. Five minutes after the injection, trains of biphasic pulses were administered through caliper electrodes as above described. Adherence of the caliper electrodes to the lesions was maximized using an electrophoresis gel. The treatment was repeated after 1 week.[12] More recently, a shortened protocol involving 8 pulses with 10μs interpulse interval and a total duration of 3.1ms has been devised.[7] The adoptions of a dedicated software for the impulse stabilization and optimization resulted in improved control and minimal side effects.[7]

RESULTS

In the first trial, ECT was used to directly attack neoplasms.[13] Overall response rate was 80%, with 40% long lasting remissions (in excess of 1 year). This phase I/II study led to the development of several electrodes specific for the different body districts.[14] On the basis of the preliminary studies, cohorts of dogs and cats affected by melanoma, soft tissue sarcoma mast cell tumor and nasal squamous cell carcinoma were enrolled in phase II studies.

Melanoma Ten patients with malignant melanoma (MM) of the oral cavity were enrolled in a phase II study. The overall response rate was 80% (median time to recurrence 6 months), with a 50% of patients in remission after 1 year.[15] Of interest, most of the long term responders experienced a vitiligo like discoloration at the treatment site, potentially suggestive of recruitment of the immune system after uncovering of deep antigens.

Soft tissue sarcoma A total of 72 cats affected by soft tissue sarcoma were assigned to three different groups: 1) surgery alone, 2) surgery coupled with intraoperative ECT, 3) surgery coupled with post-operative ECT.[16] The median control times were respectively 4, 12 and 19 months. In this study were identified some prognostic factors such as previous treatment and tumor size. A cat with a recurring hemangiopericytoma was also referred at the time of the study, and is still in remission 2.5 years after the completion of the ECT treatment.[17] The only side effects were tumor dehiscence in 2 patients and focal inflammation in 2 patients. Systemic side effects were not reported among the enrolled patients.

In another study, adjuvant ECT was adopted to increase local control in a group of 22 dogs, obtaining a mean time to recurrence of 730 days, with 50% of the dogs still desease free at the time of writing.[18] In one dog with a large, unresectable sarcoma, ECT has been used to decrease tumor size and allow surgical excision.[19] More recently, ECT has been exploited to reintroduce potentially toxic drugs in the oncology protocols. Specifically, a cohort of 64 cats with incompletely excised soft

tissue sarcoma have been enrolled in a study involving the use of cisplatin as ECT agent and matched with a control cohort of 14 cats treated with surgery. All the cats tolerated the treated very well, despite the high reported toxicity of the drug when administered systemically. In terms of tumor control mean time to recurrence was 666 days and 180 days, respectively.[20]
A potentially novel application of ECT as suggested by a recent case report, is in the treatment of aponeurotic fibromatosis, this rare neoplasms, albeit benign, poses a significant challenge for its local invasiveness and tumor location.[21]

Mast cell tumor Adjuvant bleomycin-based ECT was used in 28 dogs with incompletely excised mast cell tumors. Overall response rate was 85%. The median time to recurrence was not reached. Estimated time to recurrence was 52.76±6.5 months.[22] One dog had wound dehiscence. In 2009 Kodre et coll. compared square pulses based ECT to surgery alone in the treatment of mast cell tumors, using intralesional cisplatin, and claiming a longer control with ECT.[23] Another study, enrolling 37 dogs, evaluated the efficacy of cisplatin for the treatment of incompletely excised canine mast cell tumors obtaining a mean disease-free interval of 1218 days with 78% of the patients experiencing local control at different times.[24]

Squamous cell carcinoma a small cohort of 9 cats with sun induced squamous cell carcinoma of the nasal planum were treated with intralesional bleomycin coupled with ECT with a rate of complete responses of 77%.[25] A more recent article describes the use of square pulses based ECT for superficial squamous cell carcinoma in cats, reporting a 81% of complete responders.[26] A recently published article reports an high percentage of responders in cats with advanced head and neck carcinoma using biphasic pulses with minimal side effects.[27]

CURRENT PROTOCOLS

Electrochemotherapy is mostly used as a adjuvant therapy, locally injecting the chemotherapy drugs bleomycin or cisplatin and then following with the application of permeabilizing pulses. For larger tumors drugs can be administered locally or systemically, mostly on the basis of tumor volume, firmness or resistance to insertional electrodes.[12] Caution should be exerted when using electrochemotherapy in patients previously treated with radiation therapy due to the risk of radiation recall.[28]

ELECTROCHEMOTHERAPY IN OTHER SPECIES

ECT has been used successfully in horse to treat sarcoids using intralesional bleomycin and to treat melanomas using cisplatin as chemotherapy agent, either directly or combined with surgery. The treatment has been well tolerated and resulted in high percentage and durable responses.[29-31]
ECT has been used to treat tumors in exotic animal as well. Our group described the usefulness of adjuvant ECT in the treatment of advanced cutaneous tumors in rabbits and in a turtle.[7,32]

ELECTROGENETHERAPY

The first report of electrochemogene therapy in veterinary cancer patients has been presented by Draghia and colleagues in 2002. The authors investigated the possibility to palliate several cancer induced complications such as cachexia, anemia, anorexia, and decreased activity levels that many veterinary patients experienced by delivering a plasmid encoding for insulin-like growth factor-I.[33] More recently, electrochemogene therapy using plasmids encoding for IL-12 has been successfully investigated in small cohorts of dogs with a variety of spontaneous neoplasms.[34] Finally a recent prospective study on canine mast cell tumor recruited eight dogs for a total of 11 lesions that received from one to four sessions of electrochemogene therapy with IL-12.[35] The treatment was well tolerated but the tumor response was extremely variable.

IRREVERSIBLE ELECTROPORATION

This technique aims at inducing focal tumor destruction by using direct tumor ablation through the delivery of electric pulses at higher voltage than RE (used in ECT), alone or combined with chemotherapy. Two reports have been published so far describing the successful treatment of a canine glioma patient and of a canine patient with a large histiocytic sarcoma.[36,37] These two reports are very promising and warrant further investigation to confirm the efficacy of this approach.

CONCLUSION

Electroporation in its different applications (irreversible and reversible) is a safe and efficacious approach to veterinary. Its low cost and ease of administration makes it a valuable addition to the currently available oncological therapies. One advantage of this technique is the possibility of repeated treatments in case of local recurrence and the recruitment of the patients' immune system, as suggested by the low incidence of metastatic disease among most treated patients and by the tumor selection observed among some pets with recurring disease.

FUTURE DIRECTIONS

Ongoing studies are evaluating the efficacy of ECT when other drugs than bleomycin are used, in particular anthracyclines and their analogues (mitoxantrone) and platinum compounds such as CDDP and carboplatin. Studies are being performed to characterize the membrane alterations induced by trains of biphasic pulses in order to improve the efficacy of the technique.[38] Novel electrochemotherapy equipments adopting different electrical parameters are currently being developed.[39]

REFERENCES

[1] Keet, Gehl and Lee (Eds) Clinical Aspects of Electroporation, Springer Ed, 2011. [2] Gothelf et al. **Cancer Treat Rev** 2003; 29: 371. [3] Belehradek et a. **Cancer** 1993; 72: 3694. [4] Daskalov et al.**IEEE Eng Med Biol** 1999 ; 18 : 62. [5] Gehl et al. **Melanoma Res** 2000; 10: 585. [6] Mir et al. **Eur J Cancer** 2006 ; 4(S) : 14. [7] Spugnini et al. **J Cell Physiol.** 2014;229:1177-81. [8].Mir et al. **Br J Cancer** 1997 ; 76 : 16. [9] Tozon et al. **Anticancer Res** 2001; 21: 2483. [10] Rols et al. **Bioelectrochem** 2002 ; 55 : 101. [11] Tozon et al. **Anticancer**

Res 2005 ; 25 : 839. [12] Spugnini et al. **Methods Mol Biol.** 2014;1121:247-56. [13] Spugnini et al. **J Exp Clin Cancer Res** 2003 ; 22 : 571. [14] Spugnini et al. **J Exp Clin Cancer Res** 2005; 24 : 245. [15] Spugnini et al. **Melanoma Res** 2006; 16 : 23. [16] Spugnini et al. **Cancer Chemoter Pharmacol**; 2007; 59 : 375. [17] Spugnini et al. **Vet Rec** 2006; 159 : 598. [18] Spugnini et al. **In Vivo** 2007; 21: 819. [19] Spugnini et al. **Vet Rec** 2008; 162: 186. [20] Spugnini et al. **J Transl Med** 2011; 9; 152. [21] Spugnini et al. **Can Vet** J. 2013; 54:606-9 [22] Spugnini et al. **Anticancer Res** 2006; 26: 4585. [23] Kodre et al. **In Vivo**. 2009; 23:55-62. [24] Spugnini et al. **J Vet Intern Med** 2011; 25: 407. [25] Spugnini et al. **Vet J** 2009; 179: 117. [26] Tozon et al. **J Feline Med Surg.** 2014;16:291-9. [27] Spugnini et al. **J Vet Intern Med** 2015;29:1368-75. [28] Spugnini et al. **In Vivo** 2008; 22: 751-53. [29] Tamzali et al. **Equine Vet J** 2012; 44:214. [30] Spugnini et al. **J Equine Vet Sci** 2011; 31: 430. [31] Scacco L et al. **J Equine Vet Sci** . 2013; 33: 637-639. [32] Lanza et al. **J Am Vet Med Assoc** 2015;246:455-7. [33] Draghia et al. **Mol Ther** 6:830–836. [34] Reed et al . **Cancer Gene Ther** 17:457–464. [35] Pavlin et al. **Radiol Oncol** 45:31–39. [36] Garcia et al. **Technol Cancer Res Treat.** 2011;10:73-83. [37] Neal et al. **J Clin Oncol.** 2011; 29):e372-7. [38] Spugnini et al. **Microsc Res Tech** 2007; 70: 1041. [39] Spugnini et al. **Curr Cancer Drug Targets** 2015;16:43-52.

Electrochemotherapy in Private Practice

Chelsea Tripp, DVM, MS, DACVIM (Oncology)
Seattle, WA

Several novel tumor-targeting and drug-delivery approaches in cancer treatment are currently undergoing intensive investigation, among them physical approaches such as tissue electroporation. Since the early 1980s, electroporation has proved to be effective in facilitating the transport of various molecules across the plasma membrane. By using short, intense electric pulses, the plasma membrane becomes permeable to molecules such as bleomycin and cisplatin that are otherwise deprived of membrane transport mechanisms. The most appropriate electric pulses are those of a duration and amplitude suitable to preserving the viability of cells, allowing for resealing of the cell membrane a few minutes after the application of the pulses.

These electroporation-based cancer approaches are currently undergoing investigation as well in the field of drug delivery and gene therapy. The first biomedical application of this method of treating cancer came in the form of electrochemotherapy, which, since its beginnings in the late 1980s, has evolved into a clinically verified treatment approach for cutaneous and subcutaneous tumor nodules in both humans and our veterinary patients.

What is electrochemotherapy?

Electrochemotherapy is an anti-cancer treatment that couples local administration of chemotherapy drugs to the delivery of pulses of a specific waveform. The application of these pulses allow for alterations in the cell membrane through an electrically mediated reorganization of the plasma membrane of cells such that there is increased uptake of the chemotherapy drugs. Ultimately this leads to cancer apoptotic death.

What is the difference between electropulsation and electrochemotherapy?

- **Electropulsation** – *the direct delivery of electric pulses to cells*
- **Electropermeabilization/Electroporation/Electrochemotherapy** – *using exogenous chemotherapy to enter cells after electropulsation*

Electrochemotherapy uses a substance (in this case a chemotherapy agent) in conjunction with the pulses to help facilitate higher delivery of the drug to the cytosol of the cell. Traditional chemotherapy is widely used for a broad range of cancers. In many instances, the response rates are low. This is because the drug cannot reach its intracellular site of action. The cell membrane is often the barrier that reduces the effectiveness by restricting drug access. Electric pulses can be utilized to cause temporary and reversible permeability of cell membranes to augment the delivery of chemotherapy. The two drugs of choice are Bleomycin and Cisplatin; as they both are relatively large molecules and poorly permeable on their own. Electroporation does not enhance the therapeutic potential of lipophilic agents. Gehl, et al observed that, for the lipophilic anticancer drugs daunorubicin, doxorubicin, etoposide and paclitaxel, there was no increase in cytotoxicity when combined with electropermeabilization. Investigators are looking at carboplatin, 5-FU and using concomitant peri-lesional administration of cytokines, such as IL-2.

Are there other effects to the microenvironment with electrochemotherapy?

It has been further demonstrated that endothelial cells are directly sensitive to bleomycin and cisplatin delivered by electropulsation. It is also thought that there is prolonged drug entrapment within the tumor because of a transient induced but reversible reduction of tumor blood flow and a vascular disruption effect that also enhances the efficacy of ECT. This vascular constriction is advantageous for electrochemotherapy because drugs present in the tumor tissue are retained at the time when the cells in the tissue are permeabilized, thus enhancing the potential for intracellular traffic. This phenomenon has been termed the "vascular lock." It is suggested that the effect of the vascular lock is more prolonged in tumors than in normal tissue since they usually have a more chaotic irregular microvasculature, with fenestrated capillaries such that no continuous barrier exists between the vascular compartment and the interstitial fluid

What happens to the tissue histologically with electrochemotherapy?

Histologic evaluation of the tissues after treatment reveals a large detectable inflammatory reaction around and inside the treated regions. Macrophages and lymphocytes are usually present in large numbers as well as eosinophilic infiltration is possible. Appearance of the tumor cells changes into enlarged cells with pyknosis and cytoplasmic vacuolization, indicative of massive necrosis.

Electrochemotherapy for humans

In humans, electrochemotherapy is being used in approximately 40 cancer centers in Europe and the United States. Current clinical use of electrochemotherapy is focused on palliation of progressive disease; however, it can also be used as cytoreductive treatment before surgical resection in an organ-sparing attempt. It has been used in such a setting before a sphincter-sparing resection of anal melanoma and in digital chondrosarcoma, rescuing the finger from amputation. Because of its vascular disruption effect, electrochemotherapy is also effective in the treatment of bleeding metastases. Furthermore, it could be used to treat basal cell carcinoma of the face with curative intent. Its beneficial anti-tumor effects have been proved, giving better cosmetic results than excisional surgery does.

Electrochemotherapy for our veterinary patients

Electrochemotherapy, in veterinary oncology (dogs, cats, exotics and horses), has been dominated by centers in Brazil, France, Italy, Ireland, Slovenia and the UK. We are seeing growth in the application within the United States & Canada with facilities that have trained specialists.

How is electrochemotherapy performed?

The process does require heavy sedation but is overall well tolerated and many tumors can be treated in 1-3 treatment sessions. Larger tumors may require more sessions and more time to regress.

There are two types of probes that are typically used to administer the pulsation for electrochemotherapy; planar plate electrodes and parallel needle based electrodes. These probes offer different applications and abilities to treat gross disease or deeper tissue segments. The electrical field generated by these probes will differ somewhat and there is variability within tumor characteristics that will further alter the electrical field. Electrode systems have recently been developed for intraluminal delivery by endoscopes, laparoscopic systems, or intravascular catheters.

What tumor types should consider electrochemotherapy?

The following are examples of tumor types which are amenable to electrochemotherapy. In general, tumors need to be superficial or easy to access the tumor bed to be successful with treatment:

- Fibrosarcoma in the mouth or eyelid.
- Melanoma in the mouth, eyelid or paw
- Squamous cell carcinoma in mouth, eyelid, nasal planum or paw.
- Soft tissue sarcomas post debulking surgery where aggressive surgery would necessitate amputation
- Feline vaccine-associated fibrosarcoma after surgery
- Localized cutaneous lymphoma in dogs or cats
- Plasmacytic tumors of the footpad, lip or skin
- Low to intermediate grade mast cell tumors on distal limb, trunk and face.
- Perianal & rectal tumors
- Operation site margins where there is a significant risk of remaining tumor tissue
- Sarcoids and squamous cell carcinoma in horses
- Superficial tumors on exotics (small mammals, birds & zoo animals)
- Some internal tumors – tracheal and esophageal (on a case by case and location basis)

Chemotherapy Safety and Administration

Kim Albin, LVT, VTS (oncology)
St. Johns, Michigan

Exposure

Let's talk about the drugs that oncology workers have to handle. On the whole, they have characteristics that are not very pleasing to the "healthy, non-cancerous" person. These drugs are not good for the people who take them to fight the cancer in their bodies let alone for the people who administer them. As health care providers we need to know why we get exposed, how we prevent it, what exposure looks like and what to do if we are exposed because you will be.

Cytotoxic drugs are those designed to hit rapidly dividing cells. These include the ones you "know" are dangerous and those you think aren't like metronomics (low dose frequent administration). If it targets rapidly dividing cells in any way they are bad for you. Oral cytotoxic drugs should be considered just as dangerous as the injectable, topical as well.

So, why do we get exposed? There are a few reasons; cost, time, indolence. Yep, you read it correctly. It is expensive to properly protect yourself, it takes time to properly protect yourself and quite frankly some people just don't care about themselves or you. Pretty harsh it's true but true none the less. Cost, time and indolence will be covered in the next lecture.

Exposure happens in 3 main ways. You can inhale, ingest or absorb the cytotoxic drug. Inhaling is pretty self-explanatory, you breathe it in. Think about how you administer the drug. If you use an open system and don't wear proper personal protection equipment (PPE) and the drug gets disconnected and sprays guess what, you are breathing in cytotoxic drug. What about when you get a shipment of drugs and you open the box and one of the liquid bottles has broken within that box. You are breathing in drug. How about when you reconstitute a drug on the counter and you have pressure within the bottle that "escapes" through the top after you remove the needle? That's right, inhalation exposure.

You can ingest by eating it. Eating it? Well, no one actually drinks Adriamycin just because it looks like cool-aid right? True but many people handle these drugs without gloves on and then they touch their key boards, then you touch the key board and then you eat. Did you wash your hands even? Probably not and now you are eating chemotherapy. What about that drug on the counter you made and it splashed out the top well the aerosolizing molecules have landed in the coffee pot and brownies on the counter next to you. Yummy isn't it.

Absorption is pretty easy to think about. If you are not using a needleless system you can imagine how many times you actually inject yourself with drug. How many times do you poke yourself with diphenhydramine or a vaccine or before you draw blood let alone in the chemo hood or on the counter top. If you have a cut on your hand and you are grabbing the phone that the non PPE wearing administrator touched guess what....absorption. You don't wear goggles during administration because they are cumbersome and you are indolent and then you get splashed, here we are again at absorption. Let's not talk about if you chew gum or eat while preparing or administering these drugs. Or if you wear contacts and get some unseen aerosolized drug within your eye because that drug is now caught and sitting on your eye for hours.

Side effects

Remember these drugs target rapidly dividing cells and your rapidly dividing cells will let you know if you have been exposed. Some early side effects from exposure you may see include but are not limited to dermatitis, hyperpigmentation of exposed skin, lesions in the nasal and oral mucosa, burning sensation in the eyes, dizziness, headaches and nausea. Good times right? Long term effects include thinning of hair, reproductive issues, chromosomal abnormalities and carcinogenicity (cancer). Did I mention yet that these changes are irreversible? Well, they are irreversible. Your DNA has been changed and is now replicating in the "damaged" way if you are seeing these things.

Examples of the long term effects, that can be found on the Center for Disease Control's (CDC) website, include miscarriages, difficulty getting pregnant, birth defects, high incidence of leukemias, active drugs in urine (specifically Cytoxan) years after exposure, and RBC changes just to name a few. Oh yeah, and you get cancer. Many times the cancer you get is not easily treatable because you have been exposed to most of the drugs that treat the cancer you have and thusly your cells already recognize it and have made a defense against it, in other words the treatment won't treat. Many studies have been done and are easy to find at www.cdc.gov/niosh

Exposure Happens

Exposure can happen in many ways. Think about drug deliveries. Do you wear PPE every time you open the boxes? You should, who knows how they were packed and what is broken. You should also open the boxes in your chemotherapy administration room with no one else around. If possible, open in your chemotherapy hood. Every delivery is a Pandora's box just waiting to expose you. There are no federal mandates on how cytotoxic drugs have to be handled, boxed and shipped, none. Knowing that, protect yourself.

Let's talk about your chemo hood and administration room. Is the room a negative pressure room and if so do you keep your doors closed? If you don't, it really isn't a negative pressure room is it. Do you have a Class II Biosafety Cabinet that vents outside or inside? Inside doesn't make much sense but you would be surprised how many do vent inside. Do you inspect your cabinet, or hood, every 6 months as recommended by the CDC and change your filters? Do you have a charcoal filter if you use mustargen because charcoal filters are the only filters that actually filter mustargen-true story. Do you start your chemo hood 30 minutes prior to using it and keep the front window below the company set line? Some people put it above the line because it is more comfortable to work that way. Well it is also not catching the particles either and they are venting into the room. Inhalation, absorption, take your pick.

Do you always use PPE? For every single time you may be in contact with these drugs, do you, every time.

Administration

You should always place a catheter into the animal and always used a closed system for administration. There are quite a few options, Equashield, Phaseal just to name a few. Never cut corners because it is a small dose to administer and use a butterfly catheter, you are asking to get sprayed with drug when the animal pulls back. Wear all PPE, always and never ever reuse. One and done for all parts of the PPE. Keep a pee pad or chemotherapy administration pad underneath the administration site during said administration. If there is a leak, the pad will catch it and not the floor that gets walked on all day long and then spread throughout the clinic. Always dispose of anything used during reconstitution and administration in approved chemotherapy waste buckets. Everything including the PPE goes in here.

Clean Up

Animals vomit, pee, defecate and drool all day at the clinic and even after they have had their treatment. There is active drug in all of these items. For this reason, every cage should be labeled "this animal has been treated with _______, it is excreted in the ________, please handle accordingly". Treat as such when cleaning up after them. They are already sitting in whatever they have done so put on PPE. Either 1 person or possibly 2 but no more than that to handle the issue, too many cooks in the kitchen so to speak. If the animal is small, wrap it in a pee pad and take it to wash with detergent. That pee pad has to go into the chemo waste. The cage or run should never be sprayed with anything-think aerosolization- always use wipes or paper towels with cleaner on them. Cleaning should be detergent (round molecule that grabs some chemo) then water, 10% bleach (which kills some chemo) then water and left to air dry. Any items in the cage should either be disposed of in the chemo waste (yes towels and blankets) except stainless steel bowls. Bowls can be cleaned the same way as the cage, detergent water, bleach water and left to air dry. Remember the floor, things drip. Again, all waste into the chemo garbage bin. The reason you have to throw away laundry, think about putting this chemotherapy soaked blanket into the laundry bin. It sits there for days, then is taken down to laundry dripping the whole way and contaminating the floor. At arrival it is picked up by ungloved, non PPE wearing people and put with all the laundry that has neither detergent nor any bleach in the cleaning material and now all the laundry is contaminated because nothing is killed.

Other Considerations

Some people say to me, but everything can kill me, everything has potential to hurt me or change my DNA. I say this, yes it's true almost everything can do damage. The food we eat, the things we drink, the items we drink out of, the sun, the sunscreen, the tobacco and even the air we breathe can change your DNA. This is why many cancers occur. You know sun exposure gives you skin cancer so you wear sunscreen. You know smoking is bad so you don't smoke. You know the plastic bottles are bad for you so you drink out of glass or stainless steel. You know microwaving in plastic allows the plastic particles to leech into your food so you don't microwave in plastic anymore. Why then wouldn't you take steps to protect yourself at your work if you know something there is going to be bad for you? Consider you may also carry this home to your family. If there are cytotoxic drugs on your shoes or on your clothes or in your hair. Do you change as soon as you get home and shower or do you hug your kids, kiss your partner and then make dinner? The CDC says the 93% of Americans have measurable bisphenols (BPA) in their blood and fyi bisphenols are bad for you. BPA's are bad and change your DNA and you work in oncology and are hammering your DNA with yet another item that is bad for you at some point in time your DNA is going to replicate wrong and that leads to.....cancer.

If you work in oncology, go to your gynecologist and let them know. If they don't know how to monitor your blood and hormone levels contact an oncologist or contact the oncology nurses association (ONS). The ONS will be able to point you to a physician in your area that can monitor you and you should be monitored every 6 months according to the CDC. Men, you aren't out of the woods. Guess what your sperm are, rapidly dividing cells. You need to be checked regularly as well.

If you are pregnant, thinking about getting pregnant, breast feeding, thinking about breastfeeding, or thinking about getting a woman pregnant you should cycle out of oncology for your safety and your fetus's safety because what is a fetus but a giant conglomerate of rapidly dividing cells.

You can be safe, you just need to think.

CHEMOTHERAPY SAFETY: REQUIRED PPE

The Center for Disease Control (CDC) says this is the minimum required personal protection equipment (PPE) for all people handling cytotoxic drugs: Non permeable gown with closed cuffs, 2 sets of chemotherapy approved powder free gloves-one set under the gown and one set over the gown completely covering the cuffs, N95 mask preventing 95% of the particles in the air from being inhaled and eye/face shield. They also recommend booties. All of these items are a onetime use. They are not to be hung and reused nor are they to be washed in any way. Booties are recommended so you don't track contaminate through the clinic or to your home.

These PPE should be worn any time you are exposing yourself to chemotherapy. This includes unpacking boxes, stocking, preparation, travel of drug through the clinic, administration, clean-up of administration area, and cleaning up cages and soiled pets. Always, and every time.

I know what you are thinking, you hang your gowns don't you or stuff them in a cupboard. You reuse your mask. You grab gloves from anywhere if the chemo glove box is empty. You have animals to treat in a short amount of time. It doesn't take long to protect you and the consequences are dire.

Gowns are self-explanatory; they are really just plastic bags with sleeves. They keep fluid from touching you thusly keeping your clothing from being contaminated and your body from absorbing the drug.

Chemotherapy gloves, 2 sets always. Powder free, if they are rated for chemo they are powder free anyway, due to the fact that powder can irritate and cause micro-abrasions thusly increasing your absorption if there is a breech and you sweat in powder. How can you know if that is sweat or did you accidentally poke yourself? You can wear a short set under the gown but a longer set should be worn above the gown to ensure there is a good seal and you are not exposed to any aerosol contaminate.

An N95 mask protects you not the pet. A surgical mask protects the patient and not the wearer so a surgical mask lets all particles into your airway. This is not protecting but infecting you. An N95 mask needs to be fitted to each individual person for a good seal and should be rechecked every year. There are dozens of N95 masks, you can buy them for painting, but the one Home Depot carries may not be the one that actually stops 95% for your face. Remember, your face changes as you age; you gain or lose weight, pregnancy etc. Yearly fit testing ensures you are protected.

Face shields should encompass from the hairline down to past the chin from top to bottom and side to side outside the ear to outside the opposite ear. They should be a clear plastic shield and disposed of every time.

Booties are recommended, and disposed of every time. Tracking contaminate through the hospital is fairly common and should be thought about. Also, who wants to take it home anything bad for your family?

This is expensive and cost is quite a problem in the veterinary world. An easy way of getting around this is performing all of your administrations. Wait for all the blood work to return and then do everyone's chemo in a row. Obviously be safe about it, change your outer gloves when needed but do as many as you can in a row and that instantly decreases the amount of PPE used in a day.

Administration Room

A negative pressure room with doors closed should be used for administration. Doors left open do negate the negative pressure aspect and allow any aerosol particles to escape into the clinic. If no negative pressure is available, keep doors closed and during administration only those needed within the room until completely done with said administration, no exceptions.

Your chemotherapy hood should be a class II Biosafety cabinet that vents outside. UV light will destroy many of the particles. Many cabinets vent into the room, please ensure it isn't. Also, these should be checked every 6 months per the CDC. They should have a filter changed every 6 months and if you use mustargen ensure your filter is charcoal. Charcoal is the only filter that catches the active mustard gas. The cabinet should be turned on 30 minutes prior to use and cleaned daily with detergent, water, 10% bleach and water. No sprays only wiped with gauze or the equivalent and then placed in chemotherapy waste bins. Limit the amount of products in the cabinet. Always keep the front glass below or at the line provided by the company. Anything above the line and you are not catching all of the contaminates within. As always, wear the above PPE even when in the chemotherapy cabinet. Nothing works 100% of the time.

Closed Systems

It is recommended by the CDC that a closed system be used from beginning preparation through to administration. At one time Phaseal was the only option, but now there are quite a few more. Closed systems include: ChemoLock, Equashield, ChemoClave, CareFusion and Phaseal. You can contact these companies and ask for samples to try and find the one that works best for you. They are needless which decreases the amount of needle accidents which decreases exposure and sealed. Of course, nothing is 100% so always use PPE and a Biosafety cabinet. At first you will be clumsy and unfamiliar but you will get the hang of it. And yes, these cost a bit extra but your safety is important. If you discuss the safety concerns with your clients, 9/10 are more than happy to pay an extra ~$10/administration for your safety.

Compliance

Compliance is hard. Corners like to be cut whether it is due to cost, time or indolence. Cost can be trickled down to the clients and cut slightly by thinking about how/when you administer. Time should never have its corners cut for so many reasons; your safety is just one of those reasons. Indolence is that employee who never follows the rules, who doesn't wash their hands etc. This is not the person you want in the oncology ward. Limit the amount of people who can handle and administer the cytotoxic drugs, give yearly updates on safety protocols, have repeat training sessions. These things are important and should be performed.

Monitoring your exposure

If you follow all the rules and do all that is recommended you still need to check yourself. Discuss with the Oncology Nursing Society (ONS) about who in your area is a physician that can deal with an oncology technician's needs. You should keep a diary of every day and if you were exposed and the drugs you used. If you are diagnosed with a cancer in the future you will know what you have already been exposed to and help your treatment. Take that diary into the doctor every time you visit. Check your hormone levels every year. Cytotoxic drugs hit those rapidly dividing reproductive cells and change the DNA. Knowing if your hormones are normal can help keep problems at bay in the future. CBC with reticulocyte count is important. Again, rapidly dividing cells, changes, etc. WBC, see last sentence. Finally urinalysis, Cytoxan has been found in almost all of the nurses who have worked in oncology. All is a really big word and one that you don't want to be a part of. Knowledge is power and you should have the power over your safety.

New Laws

There are some states that have implemented regulations about chemotherapy handling and the PPE needed. The CDC has been diligent in preparing guidelines for the states governments and Veterinary Oncologists have been helping set these laws. The problem is that these same Oncologists are not following the rules they set down. Wearing chemo gowns while holding the flowers the technicians received for technician's week is not following the rules. Eating birthday cake within the administration room is not following the rules. These things happen, they aren't on purpose, and need to be thought about.

You can be safe, work in oncology and have a rewarding career. Remember though, if you are pregnant, thinking about getting pregnant, breastfeeding, thinking about getting a woman pregnant or just overtly concerned you should not work in oncology.

RESOURCES

Oncology Nursing Society (ONS) www.ons.org
Center for Disease Control (CDC) www.cdc.gov/niosh
American Society of Health-Systems Pharmacists Guidelines (ASHP) www.asph.org

Cancer Virotherapy

Sanjay Kapil, DVM, PhD, DACVM
Stillwater, OK

Neoplastic diseases are one of the most common causes of death among animals. The common approaches for treatment include surgery, radiation, chemotherapy, hyperthermia, photodynamic and palliative therapy.

Oncolytic Virotherapy is a novel approach to cancer treatment in humans and animals. It uses oncotropic and oncolytic viruses for their ability to find and destroy malignant cells. Oncolytic virotherapy is safe and does not cause harm to normal non-cancerous tissues. Virotherapy is gentle, improves survival of the cancer patient and quality of life. Viral oncotherapy has genuine potential but full potential has not been realized. It is an area of intense investigation and development for human tumors.

Viruses are a useful tool for oncotherapy because of specific cellular tropism. Rabies virus infects neurons, retroviruses target helper T lymphocytes and many viruses naturally target tumor cells. Oncolytic viruses kill tumor cells by virus-mediated cytotoxicity. Viruses have natural preference for tumor cells. A variety of viruses have been experimentally tried for tumor therapy. Examples include H-1 parvovirus, New Castle disease virus, mumps virus, and Maloney leukemia virus.

The International Virotherapy Center (IVC) is located in Riga, Latvia. Biotechnology is used to engineer viruses as therapeutic agents against cancers. Oncolytic viruses are designer viruses that target and lyze only cancer cells. Oncolytic designer viruses are replication competent. It now proven based on preclinical feasibility that a single-shot virotherapy cure is possible in near future.

In 1950's doctors noticed that cancer patients who either vaccinated or suffered from a non-related virus infection spontaneously showed signs of improvements from some cancers. The underlying mechanism was believed to be induction of interferon and tumor necrosis factor. Treatment of human tumors using mumps virus was tried in 1970's at Osaka University, Japan. An attenuated strain of mumps virus was used with some success. Tumor regression was reported in 37 out of 90 patients treated. Interest in viral oncotherapy has been recently revived.

Recently, proof of concept was recently published using thymidine kinase negative human simplex virus. This mutant was found to be attenuated in neurovirulence but active against murine glioblastoma. Clinical tolerance was found to be excellent even at highest feasible doses. One potential risk of oncolytic viruses is that they can spread from treated patient and mutate to regain pathogenic potential. However, so far transmission has not been observed. Recently, phase II trial of intratumoral HSV was published for malignant melanoma (Russell et al., 2014).

There are a variety of surface receptors that are exclusively expressed by tumor cells. These molecules also serve as receptors for virus entry. The examples include EGF receptor, Her 2-neu, folate receptor, prostate specific membrane antigen and CD 20.

Tumor cells lack apoptotic and antimetabolic responses. Attenuated viruses with defective immune combat proteins retain the ability to propagate in tumor cells. To engineer the viruses for tumor therapy the virus promoter is replaced with tumor specific promoter. Further, the virus genes that are dispensable in tumor cells are also deleted.

Adenoviruses have been tried for oncolytic therapy but the efficacy is hampered by wide spread presence of neutralizing antibodies against the virus. The neutralization of adenovirus also hampers the bioavailability of the oncolytic virus. Further, tumor targeting is also hampered by ubiquitous nature of adenovirus receptors. However, these limitations can be overcome by application of less common adenovirus against which neutralizing antibodies will be rare in host populations.

Reoviruses are not associated with any serious clinical diseases. Reoviruses exhibit preference for tumor cells. Thus, reoviruses offer potential for oncolytic therapy. A clinical trial with Reolysin is in phase III against multiple myeloma in humans.

Canine distemper virus live attenuated vaccine strains have been tried for cancer therapy. CDV can also be selectively and efficiently enhanced by engineering against canine tumors.

Imlytic is a oncolytic HSV-1 attenuated strain (JS-1) manufactured by BioVex. It has received FDA approval as an injectable formulation for cancer therapy. TVEC has dual mechanism of action. It directly attacks tumor cells. It can directly injected in tumors. It acts also by helping immune system recognize and destroy tumor cells. Inside the tumor cells, the virus releases GM-CSF. GMCSF also attracts dendritic cells to tumor cells. Finally, cytotoxic cells kill cancer cells. TVEC is a suitable tumor therapy. Currently, TVEC is in phase III clinical trials against metastatic melanomas. It is well tolerated and shows about 17% of patients show complete recovery.

Oncovex is another product that encodes for GMCSF. It is directly injected in tumors. It shows response against metastatic melanomas. In a trial 8/50 patients showed complete remission of injected cancer lesions. It also amplifies antitumor immunity.

JX594 is an oncolytic vaccinia virus that is engineered to express GM-CSF. It is administered intratumor. It has been tried in non resectable hepatocellular carcinomas. Response was observed in 3/10 patients.

One of the lessons learned from these successes in treatment is that very high concentration of virus such as JX594 is needed to cause tumor regression. About nine logs of oncolytic virus were needed to reach tumor in sufficient concentration. It is currently difficult to manufacture oncolytic viruses to reach that high concentration.

Current models to validate these oncolytic viruses have limitations. The models commonly used include cultured cells and xenograft animal models. Application of reporter transgene allows non invasive bio delivery to the tumor. Routes for systemic delivery to metastatic cancers are required. Oncolytic viruses can be coated with PEG and HPMA to prevent clearance.

Other approaches to enhance virus oncolytic activity include treatment with cyclophoshamide, increasing the permeability of tumor blood vessels, evading neutralization of oncolytic viruses and proper delivery of oncolytic viruses in high concentration to tumors.

REFERENCES

Russell et al., 2012, Oncolytic virotherapy, 30, 658-670.

Feline Injection Site Sarcomas: An Evidence-Based Approach

Annette L. Litster, BVSc, PhD, FACVSc (Feline Medicine), MMedSci (Clinical Epidemiology)
Chicago, IL

INJECTION SITE SARCOMAS - HISTORY

1970's: MLV (modified live virus) without adjuvants in use
1985: New Al-adjuvanted FeLV vaccine released
1985: MLV rabies vaccine replaced by killed adjuvanted rabies vaccine because of clinical signs of rabies in 1:500,000 cats
1987: Rabies vaccination mandatory for cats
1991: Hendrick and Goldschmidt: Do injection site reactions induce fibrosarcomas in cats?[1]
1996: Vaccine-Associated Feline Sarcoma Task Force
2002: Gobar and Kass: Incidence of vaccine site-associated sarcomas is low and decreasing[2]
2003: Kass et al.: Vaccine brand, type, practices not associated with increased or decreased risk; long acting injections implicated[3]
2003: Injection site sarcomas can also occur in ferrets and dogs

UNIQUE SUSCEPTIBILITY TO INJECTION SITE SARCOMAS IN THE CAT?

Cats have increased susceptibility to oxidative stress for a number of reasons, including:
(1) Cats have 8 weak sulfhydryl groups on their hemoglobin compared to only 2 for other species;
(2) They have a non-sinusoidal spleen, which reduces their ability to remove Heinz bodies (denatured precipitated hemoglobin on the erythrocyte surface);
(3) They have increased lipid peroxidation of the erythrocyte membrane;
(4) Feline hemoglobin easily converts from tetramers to dimers; and
(5) Cats have a limited concentration of enzymes that catalyze glucuronide conjugation, thereby causing increased exposure of feline erythrocytes to metabolites and increased risk of oxidative injury.

INJECTION SITE SARCOMAS – INCIDENCE STUDIES

The true incidence of feline injection site sarcomas is unknown, but published estimates are based on data from retrospective epidemiologic studies, surveys of biopsy specimens submitted to diagnostic labs, historical records and veterinarian/cat owner recollections linking tumor sites with injection sites. The denominator in the rate equation i.e. (# of sarcomas ÷ # cats at risk) per unit time, is based on current estimates of the US cat population and the average number of annual visits to veterinarians. Additionally, since injection site sarcomas usually develop from 3 months to over 3 years after an injection is administered, data collection is confounded by memory bias and unreported/unknown factors that might have been relevant to tumor development over the period. Because of the uncertainty with which much of this data is collected, current estimates vary from 1-10 cases/10,000 medicalized cats.

One large Canadian epidemiologic study reporting prevalence data showed that despite the widespread use of strategies designed to reduce the risk of injection site sarcomas, such as recommendations to administer vaccines in the lower limbs (1996) and the introduction of recombinant feline vaccines (2000), the prevalence of injection site sarcomas has not reduced over the period 1992-2010[4]. Similar annual statistics from the UK[5, 6] demonstrate that injection site sarcomas are reported in association with inactivated (adjuvanted) vaccines, recombinant vaccines, and non-vaccine injections every year.

SEARCHING FOR A CAUSE AND EFFECT RELATIONSHIP

The published literature on injection site sarcomas mainly consists of case control studies. Because of the study design, this type of study cannot show a cause-and-effect relationship between any particular risk factor and injection-site sarcoma development; it can only show statistically significant 'associations' rather than demonstrate that any single risk factor *caused* a sarcoma to develop. To prove cause and effect, a prospective study must be performed, but with rare disease such as injection site sarcomas, this becomes essentially impossible, since group sizes of hundreds of thousands of cats of cats would be required for statistically verifiable comparisons to be made.

WHICH INJECTIONS?

Vaccines

Many kinds of injections have been associated with injection-site sarcomas. Although vaccines remain the predominant injection type associated with injection-site sarcomas, even recombinant vaccines have been associated with injection-site sarcoma development. There is no published evidence to date to show that recombinant vaccines carry less risk of injection-site sarcoma development than modified live vaccines. One paper demonstrated an association only between adjuvanted (killed) vaccines and injection-site sarcoma development, in the rear limb region only and for rabies and FeLV vaccines only (not FVRCP vaccines)[7]. In this study, cats with injection-site sarcomas were less likely to have received a recombinant vaccine than a killed vaccine. The use of modified live vaccines of any kind (rabies, FeLV, or FVRCP) and at either site (interscapular or rear limb) did not carry any increased risk of injection-site sarcoma development when compared to recombinant vaccines in this study. Interestingly, for the interscapular region only, the use of depot glucocorticoids injections was associated with higher risk of injection-site sarcoma development in this study.

Regarding adjuvants, both adjuvanted and non-adjuvanted vaccines have been reported in association with sarcomas. With respect to adjuvants, three large epidemiologic studies failed to find evidence of increased risk in aluminum containing vaccines compared to those without aluminum. One study reported that adjuvanted (killed) FeLV and rabies vaccines carried increased risk for injection-site sarcoma development, but only in the rear limb region. There is no published evidence to date to show that recombinant vaccines are any safer than modified live vaccines. Multiple injections with vaccines at a single site have been associated with increased risk compared to no vaccine administration as follows: 1 injection increases risk by 50%; 2 injections at the same site increase risk by 127%; and 3 injections at the same site increase risk by 175%. The administration of cold vaccines was associated with mild increased risk in another study. One large epidemiologic study was unable to demonstrate any increased risk in association with specific vaccine brands or manufacturers, vaccine administration factors (e.g. syringe reuse), history of trauma at the tumor site, or concurrent viral infection[8].

Non-Vaccines

Other injection types with reported associations with injection site sarcomas include long-acting antibiotics or steroids, lufenuron and meloxicam. Additionally, non-absorbable suture material, microchip implants (cats and 1 dog), subcutaneous fluid port placed long-term, retained surgical sponge material after spay (1 cat, 1 dog) and surgical implants in laboratory animals, including polymers, cellophane, glass and plastics have also been implicated.

COMPARISON OF FELINE SARCOMAS AT VACCINATION SITES AND NON-VACCINATION SITES

Cats with vaccination site sarcomas were younger (median 8 years vs. 11years old for non-vaccination site cases) in one study[9]. The same study showed a declining risk of sarcoma development at a vaccination site with increasing age. Another study reported local recurrence in up to 45% cases and metastasis in 0-28% cases[8]. Regarding the histological characteristics of injection site sarcomas, the same two studies reported that their location was mainly subcutaneous location and that there was often an inflammatory infiltrate consisting of lymphocytes and macrophages at the periphery of the tumor. In some cases, macrophages contained blue-gray/brown foreign material, which was hypothesized to be the remnants of aluminum adjuvants, although this has never been confirmed. The characteristics of aggressive tumors included increased mitotic activity, pleomorphism and tumor-associated necrosis.

Table 1. Suspected Adverse Reactions, UK[5,6]

Year	Total (n)	Vaccine related * (n)	Feline Injection Site Sarcomas (n)	Inactivated/ Recombinant vaccine/ Non-vaccine (n)	Inactivated/ Recombinant vaccine/ Non-vaccine (%)
2005	694	155	34	9/ 1	26/ 3
2006	802	164	39	6/ 1	15/ 3
2007	857	191	59	10/ 4	17/ 7
2008	926	230	42	8/ 5/ 4	19/ 12/ 10
2009	850	179	40	10/ 4/ 4	25/ 10/ 10
2010	882	223	53	11/ 9/ 1	21/ 17/ 2
2011	962	239	34	3/ 2/ 2	9/ 6/ 6
2012	1027	295	26	4/ 3/ 0	15/ 12/ 0

*Modified live/inactivated/mixed/vector vaccine/unidentified vaccine

CLINICAL MANAGEMENT OF INJECTION SITE SARCOMAS

Cat owners should monitor all injection sites and incisional, wedge, or multiple tru-cut biopsy specimens should be submitted for histopathology if the 3-2-1 rule applies i.e.

- Mass is present **3** months post-injection
- Mass is ≥ **2**cm in diameter
- Mass increasing in size after **1** month

Report all injection site sarcomas to the manufacturer and United States Pharmacopeia[10]

Tumor factors - The location, shape and size of the tumor should be recorded.

Cat factors – Record full details of signalment, history, clinical signs, treatment etc.

Injection factors – For all injections administered, record details of the product name, manufacturer, serial/lot #, expiration date,

administration date, vaccine type, location of injection, person administering vaccine. Unnecessary injections and previous injection sites should be avoided.
Vaccine factors - Injectable vaccines should be administered SC not IM, to facilitate the detection of a mass. Multiple injectable vaccinations at single sites should be avoided, as should unnecessary vaccinations and previous vaccination sites.
Administer vaccines within 30 minutes of reconstitution and keep refrigerated at all times throughout transport & storage (2-7 °C – in the middle of the refrigerator). Follow the manufacturers' instructions regarding reconstitution, administer 1 vaccine/single-use syringe and needle, dissolve completely and administer immediately. If spillage occurs, clean fur with alcohol swabs and surfaces with a 1:32 bleach solution.
Coordinated Management – For all cases, baseline laboratory data including CBC, serum biochemistry and FIV/FeLV status should be obtained. Chest radiographs (3 views) and abdominal ultrasound should be performed to check for metastases; imaging with CT/MRI prior to biopsy is optimal.

THERAPY OF INJECTION SITE SARCOMAS

Referral to a specialist surgeon is recommended, since aggressive surgical resection tumor removal with 5cm margins and 2 muscle planes is required to minimize the chance of tumor recurrence. In some cases, small sarcomas (<3cm) can be controlled with surgery alone, but a combination therapeutic approach, including surgery, radiation therapy and chemotherapy is thought to provide the best outcome. However, pre-operative radiation therapy can be associated with wound healing complications and post-operative radiation therapy has been associated with increased adverse effects in some studies.

PROGNOSIS

The published prognosis for injection site sarcomas varies according to treatment regimen and study design; medial survival times from 9 months to >2.7 years have been reported[8].

INJECTION SITE SARCOMAS - TAKE-HOME MESSAGES

- Many types of injections have been associated with sarcomas, especially long-acting injections.
- Adjuvanted (killed) FeLV and rabies vaccines carry increased risk for injection-site sarcoma development, in the rear limb region only, in one study.
- No vaccines are risk-free.
- There is no evidence of reduced prevalence of injection site sarcomas in response to changes in vaccine formulation or recommended changes in feline vaccination protocols.
- Veterinarians should document injection sites and details of injections administered.
- Cat owners should monitor injection sites.
- Combination therapy helps provide the best outcome.

REFERENCES

1. Hendrick MJ and Goldschmidt MH. J Am Vet Med Assoc. 1991;199(8):968.
2. Gobar GM, and Kass PH. J Am Vet Med Assoc. 2002;220(10):1477-82.
3. Kass PH, et al. J Am Vet Med Assoc. 2003;223(9):1283-92.
4. Wilcock B et al., Can Vet J. 2012; 53:430-434.
5. Dyer F et al., Veterinary Record (2006-2012)
6. Davis G et al., Veterinary Record (2013)
7. Srivastav A, et al. J Am Vet Med Assoc. 2012;241(5):595-602
8. Martano et al., 2011 The Vet J. 2011;188:136-141.
9. Doddy et al. J Comp Path 1996;114:165.
10. https://www.avma.org/about/alliedorganizations/pages/tfguidelines99.aspx (Last accessed February 3 2016).

Current Options for Diagnosis and Treatment of Primary and Metastatic Lung Cancer in Dogs and Cats

Kim A. Selting, DVM, MS, DACVIM (Oncology), DACVR (Radiation Oncology)
Columbia, MO

INTRODUCTION

Primary lung cancer is uncommon in dogs. However, with the expansion of high quality geriatric care of companion animals, identification may occur prior to clinical signs. Though second-hand smoke has been weakly associated with primary lung cancer in dogs, the cause and course of this cancer may be different than its human counterpart. In addition, the life-limiting complication of solid tumors such as carcinomas and sarcomas in dogs and cats is often pulmonary metastasis. Metastasis to the central nervous system can occur, and paraneoplastic syndromes are common with primary lung tumors.

Lung parenchyma is uniquely hospitable to metastatic cancer cells. Since sarcomas typically spread hematogenously, mechanical factors were thought to be the predominant reason for the pattern of metastasis, with the lungs receiving the entire blood volume, and other organs receiving only a fraction. This concept is also supported by the relative volume of blood delivered to the axial skeleton and certain organs, paralleling other frequent locations for tumor metastasis. When tumor cells migrate and enter the circulation, they often undergo a process of epithelial to mesenchymal transition during which carcinoma cells can acquire or turn on characteristics of mesenchymal cells such as motility traits that facilitate metastasis. These cells follow venous flow and encounter increasingly larger vessels until they reach the heart and lungs, with the lungs constituting the next capillary bed the abnormally large cells encounter, creating a mechanical barrier to passage. This concept is logical but is likely less important than the compatibility of the metastatic cell with its niche in the lungs. Paget's "seed and soil" hypothesis states that there are qualities of the "seed" or cancer cell that are compatible with qualities of the "soil" or distant organ to explain the predilection for lung tissue (or other unique sites of metastasis such as certain tumors that tend to metastasize to bone). Chemokine receptors that are rich in the lungs can bind with ligands expressed on the surface of cancer cells.

Advanced techniques in cross-sectional imaging are becoming increasingly more accessible and computed tomography machines have evolved to have multiple slices for acquisition of images, making total scan time significantly shorter than it used to be. This can allow for improved images of mobile organs such as the lungs. Certain anticancer drugs seem to have advantages over others when treating primary or metastatic lung cancer.

IMPROVED DIAGNOSIS

Traditional staging of cancer involves three-view thoracic radiographs. Metastatic pulmonary neoplasia can be overlooked with fewer views because of superimposition of the heart relative to the increased perfusion of a dependent lung and increased aeration of a superior lung. CT scans have been used to further define pulmonary neoplasia and can be useful in assessing regional lymph nodes. Enlarged lymph nodes on thoracic CT in combination with primary pulmonary carcinoma warrant a poorer prognosis.

With screening for pulmonary metastatic disease, several reports confirm that CT is more sensitive than radiographs. The obvious disadvantages of thoracic CT include cost and the need for anesthesia. Though multislice CT scanners allow rapid acquisition of images, and may be done under sedation, the need for well ventilated lungs when imaging them for metastatic disease obligates the use of general inhalant anesthesia, airway control, and usually the use of a ventilator. In one study of 33 dogs, 4 dogs (all large or giant breed) had no evidence of pulmonary metastatic disease on radiographs but nodules were detected on CT scan. While granulomas or other non-neoplastic causes of pulmonary nodules are possible, the presence of lung nodules in a dog with osteosarcoma most likely warrants a poorer prognosis. As our understanding of the implications of these findings matures, we will be able to determine which dogs are unlikely to benefit from standard cytotoxic chemotherapy, and should be instead treated with novel receptor tyrosine kinase inhibitors (RTKIs) or other tailored therapy. Pulmonary nodules can be seen on CT that are smaller and more numerous than those seen on radiographs. Primary lung tumors are usually solitary, bronchocentric, and have heterogenous contrast enhancement.

Enlarged trachobronchial lymph nodes warrant a poorer prognosis when discovered in association with primary lung cancer in dogs and cats. In one study of 14 dogs, none had evidence of lymphadenomegaly on radiographs, but 5 had enlarged lymph nodes on thoracic CT and all those dogs plus one without CT evidence of lymphadenomegaly all had histologic confirmation of metastatic disease, which was associated with a poorer prognosis. In dogs, the prognosis for lung tumors of low stage versus advanced stage can be as discrepant as ~2 months versus ~2 years.

When considering other advanced diagnostic and imaging techniques, MRI is rarely used to image the lungs, and nuclear scintigraphy is affected by the constant motion produced by breathing. Reports exist, however, of the use of PET and other imaging techniques that may provide more information about location and behavior of tumors.

Biomarkers may hold promise for early detection of cancer and of metastasis. Exhaled breath condensate has been investigated and further work should be performed to determine whether any proteins or other substances can reflect metastatic disease to the lungs. Thymidine kinase 1 (TK1) was assayed in exhaled breath and only small concentrations were noted as an anecdotal observation. However, TK1 in the serum should be investigated since this marker of proliferation is increased with many different kinds of cancer.

IMPROVED THERAPY

Among cytotoxic chemotherapy drugs, vinorelbine (a vinca alkaloid) is unique in its natural accumulation in lung tissue. Vinorelbine reaches concentrations 300-fold greater than plasma, and almost 14 times greater than vincristine in the lungs. In a phase II trial, 2 dogs with bronchogenic carcinoma experienced a partial response with vinorelbine treatment. However, despite this and additional anecdotal success with the use of vinorelbine for both primary and metastatic lung cancer, a large multi-institutional retrospective study found that cisplatin and gemcitabine were superior to vinorelbine and doxorubicin. Given the retrospective study design, it is possible that a global bias affected these results such that dogs with higher grade tumors, or lymph node involvement, or other measures of aggressive behavior were more likely to be treated with chemotherapy. If this is the case, an argument could be made that chemotherapy may improve prognosis such that aggressive tumors fare similarly to less aggressive tumors. A smaller retrospective study also found no benefit to adjuvant chemotherapy following surgical removal of primary lung tumors.

Inhalant chemotherapy using gemcitabine, doxorubicin or cisplatin has been explored. The advantage to this approach is direct delivery of drug to the site of neoplasia such that lower doses can be used, thus sparing the rest of the body from drug exposure. However, delivery of aerosolized chemotherapy results in significant exposure to the environment and nearby personnel. Though changes occur at the cellular level following inhalant chemotherapy, this has not yet translated into clinical benefit. Inhalant IL-2, however, has the advantage of being a cytokine and not a cytotoxic or mutagenic agent. Inhalant liposomal IL-2 was reported to cause complete regression of metastatic osteosarcoma in 2 of 4 treated dogs. This formulation was unfortunately not available for additional studies, though the commercially available formulation of IL-2 (not liposomal) has been anecdotally used to treat primary and metastatic lung cancer with some (estimated ~20%) responses and minimal toxicity. Owners are instructed to administer the IL-aerosol at home. However, the cytokine is very expensive.

Targeted therapies offer a new approach to cancer therapy by exploiting overexpressed or overfunctional molecules in or on cancer cells. Receptor tyrosine kinase inhibitors (RTKIs) are typically administered orally and are well tolerated, with a broad spectrum of activity. Palladia® (toceranib) was designed to inhibit members of the split kinase family, including c-kit (important in mast cell proliferation) and VEGF-R (important in angiogenesis for most tumors). In early trials of toceranib in dogs, 2 dogs with metastatic mammary cancer had more than 30% shrinkage of the lung tumors, and another 2 dogs (out of a total of 5 dogs) experienced stable disease with no growth of existing metastasis and no new lesions for greater than 10 weeks. The importance of VEGF signaling in metastasis for most tumors suggests possible efficacy in a wide variety of tumor types. Toceranib has shown efficacy in dogs with pulmonary metastasis from osteo- and soft tissue sarcomas.

CONCLUSION

Though companion cats and dogs may be euthanized due to complications of local tumors and the impact on their quality of life, metastatic disease to the lungs is a common life-threatening sequella to cancer. New and effective treatments are needed.

SELECTED REFERENCES

1. Armbrust LJ et al. J Am Vet Med Assoc. 2012 May 1;240(9):1088-94.
2. Khanna C at al. Clin Cancer Res 2 (1996) 721-734.
3. Hershey AE et al. Clin Cancer Res 5 (1999) 2653-2659.
4. Selting KA et al. J Aerosol Med Pulm Drug Deliv (2008) 21(3):255-268.
5. Schiller et. al., NEJM 346:92-8, 2002
6. Poirier VJ et al. J Vet Intern Med. 2004 Jul-Aug;18(4):536-9.
7. London CA et al. Clin Cancer Res. 2003 Jul;9(7):2755-68.
8. Paoloni M. J Am Vet Med Assoc. 2006 Jun 1;228(11):1718-22.

A Glimpse into the Tumor Microenvironment: Friend or Foe?

Michelle Goulart, DVM, PhD
London, UK

INTRODUCTION

Tumor progression has long been recognized to involve a crosstalk between different cell types within the tumor mass and its supporting tissue. Established tumors are complex masses composed of two interconnected compartments: parenchyma and stroma. The tumor parenchyma comprises the malignant cells themselves, whereas the stroma is a very heterogeneous microenvironment including a variety of non-malignant cells and the connective tissue. The connective tissue contains the extracellular matrix and adhesion molecules that provide the structural support of the tumor microenvironment, while the diverse stromal cell types consist of fibroblasts, immune cells, as well as cells that participate in the construction of the tumor-associated vasculature, specifically endothelial cells, pericytes and smooth muscle cells. These stromal cells are recruited within the mass and are vital for tumor functional activity, maintenance and growth.[1]

The tumor microenvironment (TME) plays a critical role in modulating tumor cell behavior. Interactions between malignant cells and stroma are highly complex and involve continuous exchange of signals that determines the fate of tumor progression, invasion and metastasis.[2,3] Tumor cells secrete multiple soluble factors that promote the generation and recruitment of specific immune suppressive population. Tumor tolerization and immunoevasion represent important dynamics in tumor progression and occur in many patients with advanced-stage cancer.[4]

CROSSTALK BETWEEN NEOPLASTIC AND STROMAL CELLS

Neoplastic cells modify their adjacent stroma to form a supportive environment for tumor progression by secreting tumor-derived growth factors, cytokines, chemokines and extracellular matrix proteases. These factors act in a paracrine manner, inducing activation and proliferation of surrounding stromal cells, angiogenesis and an inflammatory response. Platelet-derived growth factor (PDGF) and transforming growth factor-β (TGFβ) stimulate fibroblast activation and proliferation. Activated fibroblasts release matrix metalloproteinases that degrade and remodel the tumor extracellular matrix (ECM) components, as well as fibroblast growth factor (bFGF) that induces proliferation of neighboring epithelial cells, contributing to tumor growth.

It is well established that tumors have their own blood supply. The tumor vasculature is an essential source of nutrients and oxygen and constitutes an important component of the tumor microenvironment. Tumors must induce neovascularization, in order to support growth beyond minimal size, through unbalanced secretion of vascular endothelial growth factor (VEGF) resulting in the formation of highly abnormal blood vessels.[1,5] The tumor blood vessels are unevenly distributed, leaky and unorganized, leading to hypoxia within the tumor microenvironment. Hypoxic tumor cells are more resistant to irradiation and chemotherapy, and the hypoxic microenvironment contributes to the generation of immunosuppression. Patients with hypoxic tumors have lower overall survival and decreased response rates to various therapies.[6]

Cancer inflammatory processes contribute to the development of malignancy. Overexpression of inflammatory cytokines and chemokines by neoplastic cells promote expansion and recruitment of immune cells, such as lymphocytes, monocytes (macrophages) and neutrophils into the TME. Immune cells migrate into the TME via the tumor vasculature and can positively or negatively modulate tumor progression.[1,7,8]

THE COMPLEX RELATIONSHIP BETWEEN TUMORS AND THE IMMUNE SYSTEM

A bidirectional interaction between the tumor and immune cells orchestrate the local tumor immunosuppressive network. The cancer immunoediting hypothesis recognizes that, while the immune system can protect the host against tumor development, it may also serve to promote tumor growth and metastasis. According to this hypothesis, cancer development can be divided into three phases: elimination, equilibrium and escape. At early stages, immune cell infiltration favors elimination of malignant cells and limits cancer progression by a process called cancer immunosurveillance. During ongoing neoplastic development, tumor cells that have survived the elimination phase enter into a dynamic balance with the immune system (known as 'equilibrium'). These interactions result in the selection of tumor cell variants that are less immunogenic and therefore, more resistant to destruction. In this way, despite the presence of immune surveillance, tumors that avoid immune-mediated destruction become clinically detectable (escape phase).[9-11]

IMMUNOREGULATORY COMPONENTS OF THE TUMOR MICROENVIRONMENT

Immune evasion and the initiation of metastasis are critical steps in malignant progression of tumors. As tumors progress, they employ multiple immunosuppressive strategies to attenuate the efficacy of immune responses and protect themselves from recognition and destruction.

Tumors are capable of avoiding recognition and cell death through downregulation or complete loss of antigen expression and/or major histocompatibility complex I (MHC class I) molecules, thus rendering the tumor antigens invisible to immune response.[3,12,13] Induction of apoptosis via activation of cell-surface death receptors is one way the immune system can eliminate neoplastic cells; however neoplastic cells can acquire apoptotic resistance by downregulation or inactivation of death receptors (CD95/FAS or TRAIL) or through expression of transmembrane or soluble non-functional decoy receptors.[12-14] Tumors can also avoid killing by inducing T cell anergy and apoptosis via upregulation of cytotoxic T lymphocyte antigen 4 (CTLA-4) and programmed cell death protein-1 (PD-1) ligand. CTLA-4 and PD-1 are inhibitory T cell co-receptors that play a major role in maintaining peripheral T-cell tolerance, and interaction of these receptors with ligands expressed by tumor tissues results in inhibition of T cell activation.[3,12,15]

Tumor immune evasion can also be induced by elaboration and secretion of immunosuppressive cytokines and soluble factors including TGF-β, IL-10, and prostaglandin E2 (PGE2).[12,14] Another important mechanism by which tumors evade immune-mediated elimination is by modulating the recruitment and expansion of immunosuppressive cell populations. Recruitment of regulatory cells such as tumor-associated macrophages (TAM), regulatory T cells (Tregs) and myeloid-derived suppressor cells (MDSC) into the TME induces a tolerogenic microenvironment that not only overpowers anti-tumor immunity, but also serves to promote tumor-dependent angiogenesis, invasion and metastasis.[2,3,8,12,16,17]

Tumor-associated macrophages (TAM) macrophages are cells of the innate immune system and divided into two main phenotypes: M1 and M2 macrophages. M1 macrophages play an immunostimulating role by secretion of pro-inflammatory cytokines, such as IL-12, and reveal phagocytic properties. In contrast, M2 macrophages possess a suppressive function and are the main constituents of TAM population. TAM promote tumor development via immune and non-immune mechanisms.[18] Non-immune mechanisms of tumor progression are associated with stimulation of angiogenesis, tumor metastasis and invasion whereas immune regulatory mechanisms can take place through the production of IL-10 and TGF-β and recruitment of Tregs.[19,20]

Regulatory T cells (Tregs) are a subset of T lymphocytes with immunosuppressive function that contribute to the prevention of autoimmune disorders via regulation or cytolysis of self-reactive lymphocytes. Mechanisms by which these cells exert suppression of T cell activity include secretion of the immunosuppressive cytokines TGF-β and IL-10, direct cell contact through binding of inhibitory co-receptor CTLA-4 on regulatory T cells to cell surface receptors on effector T cells, modulation of antigen-presenting cells (APCs) maturation, or by direct killing of effector cells via granzyme and perforin pathway.[21-24]

A number of studies have confirmed that immunosuppression mediated by Tregs is one of the key mechanisms of tumor immune escape.[4,25] In humans increased frequencies of Tregs were detected in the circulation and within neoplastic tissues of patients [12,26-28] and several studies demonstrate that increased frequencies of Tregs are predictive of decreased survival time and poorer response to treatment. [21,22,25,29] Accumulation of Tregs in the peripheral blood and neoplastic tissue itself has also been identified in dogs with a variety of cancers, such as mammary carcinoma, melanoma, osteosarcoma and B cell lymphoma, and results of these studies provide evidence suggesting that Tregs have a negative prognostic impact in canine cancer patients.[21,30-32] Conflicting studies have also reported that increased frequency of Tregs was associated with improved prognosis in some tumor types, such as hematological malignancies, and the positive impact of these cells is likely due to their ability to control persistent inflammation that leads to proliferative neoplasia, or kill neoplastic T and B cells.[32]

Myeloid-derived suppressor cells (MDSC) have become the focus of intense study in recent years. MDSC comprise a heterogeneous population of myeloid cells in various stages of development that differentiate into mature dendritic cells, granulocytes, or monocytes. Upon activation by tumor-derived factors, MDSC acquire immunosuppressive properties, migrate and accumulate in the circulation and neoplastic tissues of cancer patients.

This population of cells plays a crucial role in the regulation of anti-tumor immunity due its extraordinary ability to suppress T cell responses, as well as to contribute to angiogenesis, cell invasion and metastasis.[33-38] The functional importance of MDSC's suppressive effects on the immune responses has been recently appreciated, and now this distinct cell population is widely accepted as a major contributor to immune suppression in cancer patients and tumor-bearing mice.[39-43]
Expansion of MDSCs has been detected in practically all tumor models and human cancers. Increased frequencies of circulating MDSCs significantly correlate with clinical stage, metastatic burden and poor prognosis, and elimination of these cells dramatically improves immune responses. [44-46] In veterinary oncology, the study of MDSC is still in its infancy, but identification of this population in tumor-bearing dogs has already been documented. As in humans, higher percentages of MDSC were identified in the peripheral blood of canine patients with a variety of histotypes compared to healthy controls, and the increase correlated with clinical stage and metastatic disease.[47,48]

Multiple tumor-derived factors are involved in myelopoiesis and MDSC accumulation, including granulocyte-macrophage colony-stimulating factor (GM-CSF), granulocyte colony-stimulating factor (G-CSF), macrophage colony-stimulating factor (M-CSF), stem cell factor (SCF), TGF-β, VEGF, IL-4, IL-10 and IL-3.18,34,49 Inflammatory mediators produced during cancer-associated inflammation not only induce expansion but also augment the suppressive function of these cells. Cyclooxygenase 2 (COX-2), an enzyme overexpressed in many cancers, and its downstream production of the inflammatory mediator PGE2 are key factors in MDSC accumulation and suppressive activity.[4,18,50,51] The pro-inflammatory cytokine interleukin-1β (IL-1β), which plays a critical role in the induction and maintenance of an inflammatory TME, IL-6 and S100A8/A9 proteins, have also been implicated in MDSC expansion and suppressive function.[34,52] Major mechanisms of T cell suppression include upregulation of the enzymes arginase-1 (ARG-1) and inducible nitric oxide synthase (iNOS2), and reactive oxygen species (ROS), which are implicated in T cell dysfunction and apoptosis, as well as secretion of the immunosuppressive cytokines TGF-β and IL-10 that contribute to the induction of Tregs.[36,37,42]

Tumor-infiltrating lymphocytes (TILs) Cell mediated immune responses, especially cytotoxic T cells (CTLs), play a crucial role in killing neoplastic cells. TILs comprise a population of tumor-specific CTLs that have been activated during tumor development, via effective tumor-antigen presentation by APCs, and migrated into the TME in an attempt control tumor growth.[2,6] Infiltration of immune effectors into the TME is well documented in most if not all cancers, and tumor-specific T cell responses can promote clinical tumor regression and positively correlates with disease outcomes.[4,53,54] Despite their potential to destroy malignant cells, clinical trials exploiting immunotherapeutic approaches have demonstrated that T cell infiltration leads to only a transient tumor regression in established disease, and that the ability of TILs to elimate tumor tissues have not translated into clinical successes for cancer patients. Even when positive responses were achieved, there is usually relapse of the disease due to the immunoregulatory mechanisms that lead to T cell dysfunction and apoptosis.[6,13,16]

REPROGRAMMING THE TUMOR MICROENVIRONMENT TO PROMOTE TUMOR REGRESSION

The dynamic interplay between tumor cells and components of the immune system within the TME yields a multitude of molecules that can be therapeutically targeted. Therefore, understanding the networks between components of the TME and the immunoregulatory processes involved in such interactions is critical to improve the efficacy and safety profile of new cancer treatments.

Targeted therapies are capable of remodeling the TME in many ways. Monoclonal antibodies induce tumor regression via blockade of oncogenic signaling or by stimulating apoptotic signals on tumor cells. This treatment modality includes monoclonal antibodies that target tyrosine kinase receptors overexpressed in malignant tumors, such as HER2, EGFR, and VEGF. Monoclonal antibodies that block the checkpoint molecules PD-1 and CTLA-4 are also being investigated in clinical trials and have demonstrated survival benefits in a various malignancies.[2,15,53] A number of small-molecule kinase inhibitors have already been approved by the FDA and represent first-line treatment for human patients harboring tyrosine kinase receptor mutations, and more recently in dogs.[53,55,56] Strategies to inhibit MDSC and TAM suppressive function include blocking the immunosuppressive enzymes ARG-1, iNOS-2, ROS and COX-2 inhibitors [57,58] and by reversing the hypoxia status in the tumor microenvironment.[59,60] Therapeutic approaches that enhance anti-tumor immunity by inducing the expansion of antigen-specific T cells in the TME include cancer vaccines, potent immune-adjuvants, immunostimulatory cytokines, and adoptive transfer of T cells.

CONCLUSION

Owing to the complexity of the TME there are a wide variety of mechanisms that tumors employ in order to evade immune attack and spontaneously grow, and the use of integrated therapeutic approaches to target specific components within the TME, such as immunotherapy or targeted therapy, could significantly improve the anti-tumor immune response and increase survival of cancer patients.

Naturally occurring canine malignances share a wide variety of biological and clinical features observed in human cancers, and since canine tumors evolve over long periods of time interacting dynamically with an immunocompetent host, it is possible that canine tumors recapitulate mechanisms of resistance observed in human disease.[61] As tumors in dogs progress relatively faster than the same disease in humans, cancer studies in dogs may offer new insights into why some immunotherapeutic strategies fail to induce an antitumor response and elicit survival benefit, as well as allow the development of new treatments that may benefit pet dogs as well as human cancer patients.

REFERENCES

[1]Dvorak HF, et. al J Surg Oncol 2011;103:468-474. [2]Gajewski TF, et. al Nat Immunol 2013;14:1014-1022. [3]Rabinovich GA, et. al Annu Rev Immunol 2007;25:267-296.[4] Zou W. Nat Rev Cancer 2005;5:263-274. [5]Schiavoni G, et. al Front Oncol 2013;3:90. [6]Gross S, et. al Immunol Lett 2008;116:7-14. [7]Elinav E, et. al Nat Rev Cancer 2013;13:759-771. [8]Ostrand-Rosenberg S. Cell Immunol 2016;299:23-29. [9]Dunn GP, et. al Immunity 2004;21:137-148. [10]Swann JB, et. al J Clin Invest 2007;117:1137-1146.[11]Dunn GP, et. al Annu Rev Immunol 2004;22:329-360. [12]Topfer K, et. al J Biomed Biotechnol 2011;2011:918471. [13]Smyth MJ, et. al Nat Immunol 2001;2:293-299. [14]Croci DO, et. al Cancer Immunol Immunother 2007;56:1687-1700. [15]Peggs KS, et. al Clin Exp Immunol 2009;157:9-19. [16]Stewart TJ, et. al Oncogene 2008;27:5894-5903. [17]Murdoch C, et. al Nat Rev Cancer 2008;8:618-631. [18]Gabrilovich DI, et. al Nat Rev Immunol 2012;12:253-268. [19]Bingle L, et. al J Pathol 2002;196:254-265. [20]Siveen KS, et. al Immunol Lett 2009;123:97-102. [21]Kim JH, et. al Vet J 2012;193:222-227. [22]Sakaguchi S, et. al Cell 2008;133:775-787. [23]von Boehmer H. Nat Immunol 2005;6:338-344. [24]Garden OA, et. al International immunopharm 2011;11:576-588. [25]Curiel TJ. Curr Opin Immunol 2008;20:241-246. [26]Wolf AM, et. al Clin Cancer Res 2003;9:606-612. [27]Ormandy LA, et. al Cancer Res 2005;65:2457-2464. [28]Viguier M, et. al J Immunol 2004;173:1444-1453. [29]Lindau D, et. al Immunology 2013;138:105-115. [30]Tominaga M, et. al J Vet Diagn Invest 2010;22:438-441. [31]Biller BJ, et. al J Vet Intern Med 2010;24:1118-1123. [32]Pinheiro D, et. al PLoS One 2014;9:e105027. [33]Ye XZ, et. al J Genet Genomics 2010;37:423-430. [34]Ostrand-Rosenberg S, et. al J Immunol 2009;182:4499-4506. [35]Fujimura T, et. al J Dermatol Sci 2010;59:1-6. [36]Nagaraj S, et. al Cancer Res 2008;68:2561-2563. [37]Serafini P, et. al Semin Cancer Biol 2006;16:53-65. [38]Serafini P. Immunol Res 2013;57:172-184. [39]Meirow Y, et. al Front Immunol 2015;6:523. [40]Gabrilovich DI, et. al Nat Rev Immunol 2009;9:162-174. [41]Greten TF, et. al International immunopharm 2011;11:802-807. [42]Marigo I, et. al Immunol Rev 2008;222:162-179. [43]Serafini P, et. al Cancer Immunol Immunother 2004;53:64-72. [44]Ko JS, et. al Clin Cancer Res 2009;15:2148-2157. [45]Diaz-Montero CM, et. al Cancer Immunol Immunother 2009;58:49-59. [46]Rodriguez PC, et. al Cancer Res 2009;69:1553-1560. [47]Sherger M, et. al BMC Vet Res 2012;8:209. [48]Goulart MR, et. al PLoS One 2012;7:e33274. [49]Condamine T, et. al Trends Immunol 2011;32:19-25. [50]Serafini P. J Leukoc Biol 2010;88:827-829. [51]Sinha P, et. al Cancer Res2007;67:4507-4513. [52]Bunt SK, et. al J Immunol 2006;176:284-290. [53]Xu MM, et. al Trends in Immunology. [54]Rosenberg SA, et. al Curr Opin Immunol 2009;21:233-240. [55]London CA, et. al Clin Cancer Res 2009;15:3856-3865. [56]London C, et. al Vet Comp Oncol 2012;10:194-205. [57]Waldron TJ, et. al Oncoimmunology 2013;2:e24117. [58]Serafini P, et. al J Exp Med 2006;203:2691-2702. [59]Allavena P, et. al Clin Exp Immunol 2012;167:195-205. [60]Doedens AL, et. al Cancer Res 2010;70:7465-7475. [61]Khanna C, et. al Nature Biotechnology 2006;24:1065-1066.

Utilizing Inflammation to Overcome Immune Tolerance in Osteosarcoma

Jonathan E. Fogle, DVM, PhD, DACVIM (Small Animal Internal Medicine)
Raleigh, NC

INTRODUCTION

Despite attempts with various permutations of adjunctive chemotherapeutics, the median survival for dogs with osteosarcoma (OSA) is still 10-12 months, a survival that has not appreciably improved over the last 20 years.[3-6] Metastatic disease continues to be the primary cause of death in ~8000 dogs annually after primary tumor removal and adjuvant chemotherapy.[2] However, one of the most exciting discoveries in canine OSA, is that infection associated with limb-salvage surgery was observed to dramatically improve survival in dogs.[1] Lascelles et al. found that infected dogs were half as likely to die, half as likely to develop metastasis, and these survival effects were due to a delay in metastasis rather than control of local tumor recurrence.[1] Murine studies have confirmed these observations and suggest that the anti-tumor effects may be modulated via monocytes and macrophages.[7] The survival benefit afforded by surgical infection was reaffirmed in a recent study.[8] Subsequent to the observations in dogs, similar findings were described in humans.[9] Collectively, these findings suggest that OSA promotes a tolerogenic immune-microenvironment, which is disrupted by the presence of infection, leading to a vigorous anti-tumor immune response. Infection that confers a survival advantage comes from a variety of Gram-positive and Gram-negative organisms, suggesting induction of a common endogenous anti-tumor pathway.

MONOCYTE DYSFUNCTION IN CANINE OSA

We assessed the expression of monocyte surface receptors using flow cytometry. Peripheral blood monocytes from OSA dogs exhibited significantly decreased expression of CD62L, CCR2, CCR7, CD43, CX3CR1, and CXCR2 when compared to healthy control dogs. This pronounced down-regulation of monocyte adhesion and chemokine receptors of OSA dogs suggests an altered ability of monocytes to migrate to the primary tumor or metastatic lesions. Based upon the flow cytometry results, we hypothesized that OSA inhibits monocyte chemotaxis as an immunosuppressive strategy. Using an in vitro migration assay, we demonstrated decreased monocyte chemotaxis in OSA dogs compared to healthy controls. PGE2 ELISA and a multiplex assay for TNFα revealed increased PGE2 levels and increased TNFα levels respectively in supernatants of cultured monocytes stimulated with LPS of untreated OSA dogs compared to normal controls. The higher PGE2 levels that we observed in monocyte culture supernatants of OSA dogs is consistent with our hypothesis that OSA suppresses monocyte chemotaxis and the host immune response. PGE2 is a main metabolite resulting from the conversion of arachidonic acid by cyclooxygenases (COX-1 and COX-2), and elevated levels of COX-2 and PGE2 have been observed in various tumors.[9] PGE2 secretion from human carcinoma cell lines inhibits monocyte chemotaxis by down-regulating cell surface expression of chemokine and adhesion receptors CCR5 and Mac-1 respectively.[10] In canine OSA, PGE2 is elevated in both canine OSA cell lines and in naturally occurring tumors.[11] A recent study in human OSA showed inhibition of OSA cells by decreasing PGE2 levels via microRNA modulation, demonstrating the role of PGE2 and microRNAs in tumor promotion.[12] Our findings of elevated PGE2 in OSA dogs are in accordance with these previous reports, and emphasize the ability of PGE2 to induce immunopathology in OSA.

MICRORNA DYSREGULATION IN OSA

MicroRNAs, non-coding single-stranded RNAs, have been implicated in the oncogenesis of human OSA.[12] Identifying microRNA dysregulation in OSA informs development of miRNAs as therapeutic targets. Preliminary studies comparing monocyte microRNA suggest expression differs between healthy and untreated OSA dogs. MicroRNAs with differential expression in our study have parallel dysregulation in human OSA, and microRNAs such as miR-125b, miR-101, let-7a, let-7c, let-7f, let-7g, miR-146a help regulate the inflammatory responses and polarization of monocytes and macrophages. Overexpression of microRNAs such as miR-146a and let-7c skew monocytes/macrophages towards an anti-inflammatory, and hence, pro-tumorigenic response. We postulate that canine OSA modulates microRNAs in monocytes to suppress an effective anti-tumor immune response. Further macrophage functional and phenotypic studies are underway in our laboratory to more completely define the role of microRNAs in tumor-induced immunosuppression.

MURINE OSTEOMYELITIS AND NANOPARTICLE STUDIES OSA

Based upon our dog studies described above we used a murine osteomyelitis model to study the effects of bacterial infection upon monocyte function in mice with OSA. Additionally, we asked if the induction of inflammation using metallic nanoparticles and heat would promote immune activation in a similar manner. Using a similar model to that reported by Sottnik et al[7], our initial studies indicate that infection may restore monocyte chemotaxis. Previous nanoparticle studies suggest that heating induces heat shock proteins (HSPs) and more efficient processing of tumor antigens.[13,14] Although our studies are currently ongoing, preliminary data is consistent with the promotion of an inflammatory tumor microenvironment and mobilization of anti-tumor immune responses.

CONCLUSIONS

These results suggest monocyte chemokine receptors are down-regulated in OSA dogs compared to healthy dogs, and monocytes from OSA dogs exhibit decreased chemotaxis and secrete higher levels of PGE2 and TNFα. Preliminary findings using a murine osteomyelitis / OSA model, indicate that osteomyelitis may restore monocyte chemotaxis and nanoparticle therapy may replicate this inflammatory microenvironment. Collectively, these findings suggest that canine OSA evades the immune system by suppressing chemokine receptor expression and increasing PGE2 secretion in monocytes, thus inhibiting their migration and function, and inducing microRNAs that promote a tolerogenic macrophage phenotype. Understanding the mechanisms by which OSA causes dysregulation of the immune response will provide insight into developing novel immunotherapeutics to reverse tumor-induced immunopathology and reduce metastatic disease in OSA.

REFERENCES

[1.] Lascelles BD, Dernell WS, Correa MT, et al. Ann Surg Oncol 2005;12:1073-1083. [2.] Withrow SJ, Powers BE, Straw RC, et al. Clin Orthop Relat Res 1991:159-168. [3.] Selmic LE, Burton JH, Thamm DH, et al. J Vet Intern Med 2014;28:554-563. [4.] Phillips B, Powers BE, Dernell WS, et al. J Am Anim Hosp Assoc 2009;45:33-38. [5.] Bacon NJ, Ehrhart NP, Dernell WS, et al. J Am Vet Med Assoc 2008;232:1504-1510. [6.] Alvarez FJ, Kisseberth W, Hosoya K, et al. J Am Anim Hosp Assoc 2014;50:12-18. [7.] Sottnik JL, U'Ren LW, Thamm DH, et al. Cancer Immunol Immunother 2010;59:367-378. [8.] Culp WT, Olea-Popelka F, Sefton J, et al. J Am Vet Med Assoc 2014;245:1141-1146.[9.] Eberhart CE, Coffey RJ, Radhika A, et al. Gastroenterology 1994;107:1183-1188. [10.] Zeidler R, Csanady M, Gires O, et al. FASEB journal 2000;14:661-668. [11.] Xu H, Mei Q, Shi L, et al. Cell biochemistry and biophysics 2014;69:163-168. [12.] Mohammed SI, Coffman K, Glickman NW, et al. Prostaglandins Leukot Essent Fatty Acids 2001;64:1-4. [13.] Ito A, Tanaka K, Kondo K, et al. Cancer Sci 2003;94:308-313. [14.] Chen T, Guo J, Han C, et al. J Immunol 2009;182:1449-1459.

Demystifying the ABCs of SRS, SRT, IMRT and IGRT

Michelle M. Turek DVM, DACVIM (Oncology), DACVR (Radiation Oncology)
Madison, WI

WHAT IS RADIOTHERAPY?

Radiotherapy is the treatment of neoplasia using ionizing radiation. Ionizing radiation is travelling energy that is energetic enough to result in the ejection of an electron from an atom, creating a positive ion (atom minus electron). Ions created in tissue are highly reactive and cause structural alterations in critical intracellular macromolecules with lethal effects on cells. DNA is believed to be the primary target for this radiation damage leading to cell death. Lesions in DNA prevent normal DNA replication. When a cell whose DNA has been significantly damaged by radiation enters the mitotic phase of the cell cycle, it dies. In other words, radiation injury shows itself when irradiated cells attempt to divide. This point is critical to understanding the effects of radiation in both tumors and normal tissues.

Theoretically, there is a dose of radiation that would sterilize any tumor. The most important factor limiting the dose that can be safely administered, and thus the 'curability' of a tumor, is normal tissue tolerance. The goal of radiotherapy is to deliver the maximum radiation dose to the tumor while keeping the dose to surrounding normal tissues below their tolerance level. It can be said that radiotherapy is the 'art' of sparing normal tissue.

RADIATION SIDE EFFECTS, HOW TO TREAT THEM AND HOW TO AVOID THEM

Since lethally irradiated cells die at the time of division, the response of normal tissue depends on the proliferation rate of the cells that compose it. In general, acute (or early) and chronic (or late) toxicities affect rapidly and slowly renewing tissues, respectively. Rapidly proliferating tissues are epithelial tissues like skin, mucosa, kidney, liver, lung, cornea, conjunctiva and the lining of visceral organs such as intestines and bladder. Slowly proliferating tissues are mesenchymal tissues such as bone, neuronal tissue, muscle, and other 'connective' tissues. Since toxicity is dependent upon the turnover rate of cells, acute effects develop and progress during the course of treatment with resolution within 1- to 3 weeks after therapy. The most commonly affected tissues include skin, mucous membranes of the oral and nasal cavities, eyes (cornea, conjunctiva, tear-producing glands), and the lower gastrointestinal tract. Acute side effects are uncomfortable and often require supportive care including antibiotics, pain medication and prevention of self-trauma, but they are self-limiting because of rapid cellular renewal. On the other hand, late radiation effects involve slowly or non-proliferating tissues such as bone, eyes (lens and retina), nervous tissue, muscle and connective tissue. Lethally irradiated late-responding tissues are able to maintain full or complete function until they are stimulated to divide and will, therefore, not express radiation injury until months to years after irradiation. Since cellular renewal is slow or absent in these tissues, late radiation toxicity is serious, irreversible and is often life threatening when a critical organ is affected. Hence, the primary dose-limiting factor in radiotherapy is the tolerance of the late-responding tissues in the radiation field. Late effects are more likely to occur when radiation is delivered in high doses per fraction (i.e. per treatment). This is why conventionally fractionated definitive protocols are delivered over a period of weeks; the high total dose that is needed to sterilize a tumor is split into small daily fractions (low-dose/fraction) to minimize the risk of the late radiation effects. The risk of acute effects, on the other hand, is higher with daily irradiation when the total dose is high enough to result in tissue damage. The advent of highly conformal radiotherapy equipment, specifically image-guided radiotherapy (IGRT), has allowed for avoidance of critical late-responding tissues. This has created a paradigm shift in radiotherapy as discussed below.

EXTERNAL BEAM RADIATION DELIVERY DEVICES

In veterinary medicine, external beam radiation is the most commonly used technique to deliver radiation to tumors. A linear accelerator is a high-energy x-ray machine. It produces photons, which are "bundles" of high-energy electromagnetic radiation. It functions like a diagnostic x-ray unit in that x-rays are produced by striking an accelerated beam of electrons against a metal target. The interaction of the electrons and the atoms of the metal target results in the production of x-rays that can be collimated and directed towards the patient. X-rays generated by a linear accelerator are high energy (an order of magnitude stronger than x-rays produced by a diagnostic x-ray machine) and penetrate deeply into the body, affecting all tissues in the path of the beam. Deep tissue penetration is needed for treatment of deep-seeded tumors such as brain tumors, nasosinal tumors, and intra-abdominal or intra-thoracic tumors. High energy x-rays produce more biological (DNA) damage than those in the diagnostic x-ray range.

THE BENEFIT OF USING THE ELECTRON BEAM FROM A LINEAR ACCELERATOR

Another way to use a linear accelerator is to utilize the accelerated electrons themselves for treatment of superficial tumors. Unlike high-energy photons that have great tissue penetrability, high-energy electrons deposit radiation dose at the surface of the patient. This allows for adequate dose delivery to superficial tumor beds while sparing underlying normal tissues. This is a great advantage in veterinary medicine because many of the tumors that we treat with radiation are superficially located, including cutaneous mast cell tumors, subcutaneous soft tissue sarcomas, and vaccine-associated sarcomas of cats.

BENEFITS OF ADVANCED TECHNOLOGY IN CONFORMAL RADIATION DELIVERY

As mentioned above, the goal of radiation therapy is to deliver the maximum radiation dose to the tumor while keeping the dose to surrounding normal tissues below their tolerance level. In theory, this should allow us to increase the dose to the tumor while sparing the healthy tissue. As new technologies become available in veterinary medicine, treatment planning and treatment delivery are becoming more sophisticated.

1) Computerized treatment planning using computed tomographic images of the patient in treatment position can greatly enhance normal tissue sparing while ensuring appropriate dose distribution within the tumor. This is especially useful for tumor beds in close proximity to critical normal structures such as the eyes, CNS, heart, lungs, liver, gastrointestinal tract and kidneys.
2) As mentioned above, electron beam therapy allows treatment of superficial tumors while sparing underlying normal tissues.
3) Conformal radiation technologies, including multi-leaf collimators and intensity-modulated radiation therapy (IMRT) software, are available at select veterinary centers. IMRT is a new technology that uses multiple radiation beams with non-uniform beam intensity. With IMRT, larger volumes of normal tissue are exposed to more modest doses of radiation, while a higher total dose is delivered to the volume occupied by the tumor. The result is conformal distribution of the radiation dose to the tumor while sparing sensitive normal tissues.

Patient positioning is an important consideration in radiotherapy, especially when radiation delivery is conformal to the tumor. In order to ensure that tumor localization is accurately repeatable on a day-to-day basis, patient positioning must be easily reproducible. Numerous aids exist to help localize the treatment field, including immobilization devices and dental molds. Ultimately, the only way to verify patient set-up prior to therapy is by documentation using imaging. Portal imaging is the acquisition of images using a radiotherapy beam. Traditionally, images are created using double-exposure 2D techniques, which yield visualization of the treatment field and surrounding anatomy. Newer, more advanced technologies have built-in CT scanners (cone bean CT or helical CT) that are used to image the patient in three dimensions prior to treatment. This ensures accurate positioning for conformal radiotherapy so that the tumor is treated appropriately and critical organs are spared. Image-guided radiotherapy (IGRT) refers to IMRT with on-board 3D imaging, which allows accurate patient positioning for conformal radiation delivery.

WHEN TO OFFER RADIOTHERAPY?

Like surgery, radiotherapy addresses local disease. There will be no systemic effect directly associated with radiation therapy. All the benefits of radiotherapy and any side effects develop in the areas exposed to the radiation beam.

Radiation therapy can be used for two distinct purposes, namely palliation of clinical signs related to tumors and definitive tumor control. Palliative radiation protocols are designed to relieve tumor-related pain and gross organ dysfunction without causing side effects. The goal is not to optimize tumor control, nor to extend patient longevity (i.e. cure), but rather to reduce pain associated with the tumor so that the pet may be more comfortable as it lives with the disease. Side effects would defeat the purpose of palliation, so protocols are designed to minimize acute radiation effects. Palliative radiation treatments are most commonly delivered on a weekly basis using a small number of large-dose fractions (coarse fraction radiation) or given daily for just a few treatments. Due to the large dose per fraction, the risk for late radiation effects is high if late-responding tissues are irradiated. This is considered acceptable since these patients, if selected properly, are not expected to survive long enough for late effects to develop. The palliative approach is indicated for any tumor causing unpleasant clinical signs where the pet-owner wishes only to temporarily alleviate symptoms, or when the presence of advanced disease (distant metastasis) makes a more aggressive treatment approach unfeasible. Palliative radiation provides a less time-intensive and less expensive treatment option and it affords pets an improved quality of life as they live with the disease .The tumor most commonly treated in the palliative setting is osteosarcoma. A response rate of 70% to 90% is reported with an average duration of pain alleviation between 2 and 4 months.

In contrast, definitive therapy (also called curative-intent or full-course radiation) is designed to sterilize as many tumor cells as possible, to achieve long-term tumor control and to extended patient survival. In order to achieve tumor sterilization, high doses of radiation must be delivered to the tumor to kill the malignant cells. To prevent late effects, the dose per fraction is low. So, the high total dose is divided over weeks of daily treatments. Since patients are being treated daily, acute radiation effects are common. This type of treatment is associated with higher cost and time commitment on the part of the pet-owner. It is used primarily for solid tumors with low rates of distant metastasis. Any single site tumor is potentially treatable and regional lymph nodes may be included as well if indicated by the biological behavior of the tumor. Definitive therapy has been described for the following tumors: nasosinal tumors, CNS tumors, soft tissue sarcoma, cutaneous mast cell tumors, cutaneous squamous cell carcinoma, multiple oral tumors (melanoma, squamous cell carcinoma, sarcoma), feline injection site sarcoma, thyroid, anal sac adenocarcinoma, etc. Probability of tumor control, overall clinical outcome, and gross tumor response rate vary amongst the different tumor types.

THE NEW FRONTIER OF CONFORMAL THERAPY

The development of image-guided and highly conformal therapies has opened the door for a new treatment paradigm. Theoretically, it may be possible to deliver higher doses (even in the 15-30 Gy range) per fraction. If this can be done safely by effectively sparing (avoiding) normal tissues, then therapy can be delivered in fewer treatments and tumor kill may be improved. In veterinary medicine, this treatment approach is called stereotactic radiosurgery (when the total dose is delivered in a single fraction), or stereotactic body radiotherapy (when 3-5 fractions are used). This strategy is being investigated at veterinary institutions equipped with image-guided radiotherapy systems. Awareness of these technologies as well as their applications and limitations will allow the practitioner to make thoughtful recommendations for his/her patient and to have nuanced discussions with the oncology team.

The evolution of radiotherapy techniques has allowed for greater conformal dose delivery to tumors. As little as 10 years ago, many radiotherapy facilities delivered radiation by simple, non-graphic manual (point and shoot) techniques that were limited in accuracy and precision. Tumoricidal doses could not be delivered to tumors without causing undesirable side effects, and tumor control was not optimized.

The advent of 3D imaging by computer tomography (CT) and new computerized radiation treatment planning systems (TPS, essentially radiation calculators) allowed for visualization of radiation dose distribution in the patient during the treatment planning process. The radiation oncologist can now manipulate the angle of the radiation beam, block critical tissues from radiation and add other radiation beam modifying devices to optimize the distribution of radiation in the tumor and in the normal tissues. Each time a change is made in how the beam is delivered, the TPS re-calculates the dose distribution. Once the radiation oncologist is satisfied with the dose distribution, the treatment plan is approved and treatment can begin. This process takes hours to days and is called forward planning because the radiation oncologist sets a delivery technique, calculates the dose distribution using the TPS, then modifies the technique and recalculates and so on until an optimal dose distribution is achieved. The graphic visualization of radiation dose distributions and the ability to modify them has dramatically improved the sophistication of radiotherapy.

The era of conformal dose delivery began with the development of the multileaf collimator (MLC). The MLC allows for quick and precise blocking of critical tissues to achieve even better dose distributions and steeper dose gradients around the tumor. This in combination with even more sophisticated computer algorithms that can develop optimal treatment plans on their own through thousands of iterations of plan modifications (inverse planning) allow for even greater dose conformity. These new TPSs determine the optimal way to configure the MLC and the gantry position to deliver an optimized dose distribution that spares normal tissue. This is called intensity-modulated radiotherapy, or IMRT, because the computer-controlled movement of the MLC leaves results in modulation of the beam intensity.

IMRT techniques have made highly conformal dose distributions possible, so accurate patient positioning on the radiotherapy table has become vital to effective treatment delivery. If the patient's tumor is not positioned exactly in the area in space where the linear accelerator is programmed to deliver the radiation, then part of the tumor will be missed, compromising the efficacy of radiotherapy. That brought on the era of IGRT or image-guided radiotherapy, which, in this context, refers to the use of advanced imaging techniques to ensure almost perfect patient positioning each day prior to treatment delivery. CT scans are now built into linear accelerators such that a patient can be imaged by CT daily to verify positioning of the tumor prior to radiation delivery, and the patient position can be corrected as needed. CT scans exist on standard gantry-based linear accelerators (cone beam CT) and on tomotherapy units (helical CT). IGRT is currently the state-of-the-art in veterinary medicine and an increasing number of radiotherapy facilities in the US, Canada and Europe are equipped with this technology. IMRT without regular (usually daily) patient position verification can result in geographic misses of the tumor target.

The next opportunity for improved clinical outcomes using IMRT/IGRT is the delivery of higher doses of radiation (for greater tumor cell kill) in fewer treatments (for greater patient convenience). This approach is what is referred to in veterinary medicine as stereotactic radiotherapy. Stereotactic radiosurgery (SRS) refers to a single ablative dose of radiation, usually 20-30 Gy, delivered to precisely delineated and localized tumors. Stereotactic radiotherapy (SRT) refers to delivery of the ablative dose in 3-5 treatments.

SRS and SRT are relatively new in veterinary medicine. The safety and efficacy of these techniques are being evaluated in clinical trials. SRS is used in people for small intra-cranial lesions such as brain metastases, smaller-sized primary brain or pituitary tumors, and vascular malformations. The small size of the targets, the absence of normal tissue invasion and the complete lack of any motion within the calvarium make SRS feasible. For focal lesions outside of the cranium that are usually larger, less well defined and more susceptible to difficult localization (e.g. breathing motion), the highly-conformal dose delivery needed for SRS is not technically feasible. Thus, the total dose must be split into fractions (e.g. 8 Gy X 3 or 10 Gy X 3) to avoid severe normal tissue complications. This approach is called SRT. In veterinary patients, tumors are rarely small enough to allow for effective SRS, so SRT with daily or every-other-day fraction delivery is used most commonly. It is not clear if fractionation in SRT imparts the same biological, anti-tumor benefits as single ablative-dose SRS. Other benefits of SRS/SRT include fewer anesthetic episodes for the patient, greater convenience for the owner, and minimal exposure of surrounding normal. SRS and SRT have not been validated as effective definitive approaches to tumor control (i.e. curative intent), and safety profiles with respect to normal tissue side effects also have yet to be established in veterinary species. Well-designed studies are needed to investigate SRS/SRT as an alternative to conventionally fractionated protocols.

DEFINITIONS

3D radiotherapy: 3D RT involves the use of patient specific 3D images in the treatment planning process. CT (or less commonly MRI) images are used to define the tumor target and the adjacent normal tissues. Radiation plans are developed on a computer to deliver a radiation dose distribution within the patient where the highest doses are within the tumor while significantly reducing the amount of radiation dose to the neighboring healthy tissues. 3D RT allows for treatment of tumors that in the past might have been considered too close to critical structures for radiation therapy.

IMRT: Intensity modulated radiotherapy is 3D RT with the use of a multileaf collimator and sophisticated computer algorithms that, through thousands of computer-driven iterations of possible plans, create an optimized treatment plan that conforms dose delivery to the tumor and spares critical normal tissues.

IGRT: Image-guided radiotherapy is IMRT with on-board imaging that allows for daily verification of patient position to ensure accurate delivery of the radiation to the tumor. This minimizes the risk of geographic misses.

SRS and SRT: Stereotactic radiosurgery (SRS) refers to delivery of radiation with such conformity of dose to the tumor target that there is no significant radiation dose outside the target. This requires IGRT and precisely delineated and localized tumor targets. Very high, ablative doses of 20-30 Gy can be delivered in a single treatment. Stereotactic radiotherapy (SRT) is fractionated radiosurgery where the total dose of radiation is delivered over 3-5 treatments using the same precise delivery and localization techniques. Fractionation of the dose is required due to limitations related to large tumor size or location. These limitations result in dose conformity that is not as high as with SRS.

WHAT YOU NEED TO KNOW WHEN CONSIDERING OPTIONS FOR YOUR PATIENTS

First, not all cases require conformal radiotherapy. Small superficial tumor targets on the limb, such as an incompletely excised soft tissue sarcoma on the antebrachium, will be adequately treated with a simple, non-computerized treatment plan. Advanced imaging for treatment planning may not be needed in a case like this and sophisticated IMRT or IGRT-capable treatment units may be overkill. Patients that will benefit from conformal therapy are those with tumors of irregular shape or those that are located adjacent to critical normal structures that would suffer consequential side effects if irradiated. For example, most nasal tumors extend from the nasal planum to the cribriform plate and into the sinuses over the brain. The eyes and brain are critical structures that should be spared from irradiation. In these cases, IMRT/IGRT results in more effective therapy (because dose can be adequately delivered to the entire tumor) with significantly less morbidly (because conformal techniques allow for sparing of critical normal tissues) than a non-conformal approach. General practitioners should seek out referral facilities with IMRT/IGRT capabilities for patients that would benefit from them. While not all pet owners will be interested in pursuing state-of-the-art radiotherapy, the pet-owner needs to be able to make an educated decision about where and how to have their companion animal treated. Radiotherapy facilities vary in their ability to offer optimized treatments and the practitioner should have an understanding of these limitations so that he/she can help clients made informed choices.

REFERENCES AND ADDITIONAL READING

Gordon I et al, Cancer Therapy 2008;6:167. LaRue S et al, Radiation Therapy, In: Withrow & MacEwen's Small Animal Clinical Oncology, 5th Ed, 2013: 180-197. Lawrence J et al, Vet Clin Small Anim 2007;37:1151-1165. McEntee M et al, J Am Anim Hosp Assoc 2006;42:94-109.

Update on Diagnostic and Therapeutic Applications for Radioisotopes in Veterinary Oncology

Kim A. Selting, DVM, MS, DACVIM (Oncology), DACVR (Radiation Oncology)
Columbia, MO

INTRODUCTION

While some diagnostic and therapeutic applications of radioisotopes have become germane to routine specialty practice, such as technetium-based scintigraphy and radioiodine therapy, a rapidly expanding armamentarium of isotopes will change veterinary cancer care. Radioisotopes such as ^{18}F-fluorodeoxy glucose (^{18}F-FDG) can exploit the metabolism of the cancer cell, allowing improved and minimally invasive staging, as well as the ability to locate an elusive pathologic process. Additionally, by using inherent and manufactured targeting characteristics, and by selecting the desired energy for a biologic effect, radioisotopes can be used to target the cancer cell while sparing normal tissue. In veterinary medicine, facilities that work with radioactive diagnostic and therapeutic isotopes are sparse, and the goal of this talk is to inform the listener of the appropriate applications for nuclear medicine for veterinary patients. This discussion will help identify cases for which referral to a facility with nuclear medicine will be beneficial.

Radiation therapy causes single and double strand breaks in DNA. Double strand breaks are often lethal to a cell, and an accumulation of single strand breaks can have the same effect. Repeated treatments allow DNA damage to accumulate within a cell. The effects of radiation on both tumor and normal tissue can be seen at various intervals depending on when the cell attempts to complete the cell cycle. A cancer cell can lie dormant for an extended period of time. When the cell eventually attempts mitosis, the damaged cell undergoes a mitotic catastrophe and death. Therefore, tumors can shrink slowly over weeks to months, or sometimes simply stop growing. Lymphocytes are unusual in that they can undergo an intermitotic death.

External beam radiation therapy has been the mainstay of radiation therapy for many years. This entails fractionated treatments over the course of approximately 3-6 weeks. The use of cross sectional imaging (CT scans) for computer-guided planning allows relative sparing of normal tissue and targeting of tumor tissue. Brachytherapy involves implanting a radioactive source into the tumor, and the energy that is emitted is highly concentrated around the point sources, resulting in sometimes complete sparing of the surrounding normal tissue. However, proper orientation of sources can be challenging. Brachytherapy can also be accomplished using targeted pharmaceuticals such as radioiodine and samarium. The University of Missouri houses the largest research reactor on a university campus in North America, and the active radiopharmaceutical program offers unique opportunities for research and clinical treatment of animals with cancer.

NUCLEAR MEDICINE

Diagnostic

Radioisotopes will undergo decay (emit energy) until they are energetically stable. Scintigraphy using Tc-99m-MDP identifies osteoblastic activity and is a common application of diagnostic nuclear medicine for assessing bone pathology in dogs, cats, and horses. Technetium-based scintigraphy has been in common use in veterinary medicine for many years, but positron emission tomography (PET) scanning is a more recent addition to our repertoire, and is available in a small number of centers.

Positron emission tomography (PET) scanning can assess the function of a population of cells. When glucose is labeled with a radioisotope of fluorine, F-18 labeled FDG is produced and will detect any population of cells in the body with an unusually high metabolism of glucose, such as cancer cells. The positron emitted by the F-18 travels a very short distance then undergoes an annihilation reaction with its counterpart, an electron, thus resulting in emission of two gamma rays traveling in opposite directions. These are detected by the PET scanner which can then display the origin. In veterinary medicine, PET has been used to image various tumors including sarcomas, head and neck tumors (nasal and oral), mast cell tumors, and prostatic carcinomas. PET scans are most often performed using ^{18}F-FDG and a growing body of literature has characterized uptake in many canine tumors and normal tissues.

Other agents are being explored to image hypoxia [such as ^{64}Cu-diacetyl-bis(N4-methylthiosemicarbazone) (^{64}Cu-ATSM)] and proliferation [such as ^{18}F-fluorine-labeled thymidine (^{18}F-FLT)]. Staging of canine lymphoma using ^{18}F-FDG is currently being investigated at the University of Missouri, as is the use of ^{18}F-FDG based PET to determine optimal area of soft tissue sarcoma for the most accurate histologic prediction of tumor behavior. PET can also be used to elucidate cases with fever of unknown origin. Inflammatory tissue can also be PET-avid, and standard uptake values (SUV) are used to quantify the amount of activity relative to the injected dose of radioisotope to better distinguish between cancer and non-cancer causes of uptake and proliferation. Eventually, it is likely that PET will be used to assess response to cancer therapy as is done in humans. This can offer prognostic information as well as allow tailored therapy. For example, a mass that has been treated with radiation but did not regress may be inactive on PET scan, suggesting a dormant tumor, whereas PET avidity would suggest that additional therapy is needed immediately. PET has been anecdotally used to distinguish between late effects of radiation (osteopenia) and radiation-induced osteosarcoma 3 years after radiation therapy for soft tissue sarcoma.

Technetium (^{99m}Tc) is used commonly for nuclear scintigraphy. In cats, dogs, and horses ^{99m}Tc-Methylenebis(phosphonic acid), also called ^{99m}Tc-medronic acid (^{99m}Tc-MDP) is used on an almost daily basis for skeletal imaging. The 140 keV gamma provides very good resolution of images. The 6 hour half life allows timely and efficient management of patients and radiation safety. Scintigraphy is very sensitive and as such is an excellent screening tool for skeletal pathology (such as primary and metastatic neoplasia). Increased uptake in potential secondary bone lesions in dogs with primary osteosarcoma has been variable but is estimated to be 8%. An additional 10% of dogs can show non-specific uptake due to possible osteoarthritis and other non-neoplastic conditions, and increased intensity of uptake in dogs with osteosarcoma has also been correlated with earlier time to metastasis.

In addition to skeletal imaging, thyroid tissue will uptake sodium pertechnetate via the sodium iodide symporter, thus allowing scintigraphic imaging of benign and malignant thyroid masses. This can predict uptake of radioactive iodine (^{131}I) for therapy, regardless of thyroid function. Technetium has additional uses and has been conjugated to sestamibi (methoxy-isobutyl-isonitrile) has been correlated to staging of lymphoma in a small number of dogs and a single cat. As a substrate for the membrane pump p-glycoprotein (ABCB1), sestamibi has been used to characterize brain and biliary drug transport in dogs with mutations in this gene (ABCB1-1Delta). For accurate chemotherapy dosing, ^{99m}Tc-labeled diethylenetriaminepentaacetic acid has been used to determine glomerular filtration rate in cats treated with carboplatin chemotherapy.

Though rarely reported, radiolabeled metaiodobenzylguanidine (MIBG, a catecholamine precursor) could be used to detect pheochromocytoma. In addition, Indium-111 labeled octreotide ([^{111}In-DTPA-D-Phe1]-octreotide) has been used to identify insulinomas, including metastatic lesions, in 5 of 6 dogs.

The concept of theranostics exploits radioisotopes that have a gamma ray that is of appropriate energy for imaging such that diagnostic nuclear medicine can be employed in the evaluation of the case, and the same radioisotope has a second form of decay, such as a high energy beta particle that provides therapeutic benefit by damaging the targeted cells. In this way, confirmation of targeting can be accomplished. Diffuse B-cell lymphoma in dogs has been investigated for this approach by targeting BCL2 expression with the radiolabeled peptide nucleic acid-peptide conjugate – ^{111}In-DOTA-anti-BCL2-PNA-Tyr^3-octreotate.

Therapeutic

Therapeutic applications of nuclear medicine include the well-known use of radioactive iodine for the treatment of hyperfunctional thyroid disorders. In addition, dogs with non-functional malignant thyroid tumors have also been successfully treated. The ability to handle the amount of radioiodine to treat dogs and cats with malignant thyroid cancer is limited to only a few centers in the United States. Dogs with thyroid carcinoma treated with radioiodine have median survival times of 2 ½ to 3 years with locoregional disease even if nonresectable, and 1 year with metastatic disease.

Samarium-153-EDTMP (trade name Quadramet™) was developed at the University of Missouri and has been used to treat both primary and metastatic bone tumors. Samarium-153 has a decay half-life of 46 hours, primarily by beta emission. The energy of 0.8 MeV provides a maximum range in bone of 1.5 mm and in soft tissue of 2.8 mm. Using the same premise as the ^{99m}Tc-MDP, Sm-153-EDTMP is injected intravenously and targets areas of high bone turnover; diagnostic bone scans are used to determine optimal candidates for this treatment. The transient myelosuppression that ensues typically occurs 2-4 weeks after treatment, and can be associated with a transient anemia and increased ALP. This has been used primarily for palliation of bone pain. A new formulation to deliver radioactive samarium to sites of bone pathology is being developed at the University of Missouri. The chelant to radioisotope ratio is superior when DOTMP is used in place of EDTMP, and a clinical trial in dogs with osteosarcoma has recently been completed. The new conjugate appears to be less myelosuppressive which may facilitate concurrent or sooner use of chemotherapy.

Samarium has been used to treat nonresectable tumors of the skull (osteosarcoma and multilobular osteochondrosarcoma) with a median survival of 144 days and a 20% clinical response rate. When used to treat osteosarcoma, the intent is primarily palliative to decrease pain, with over 60% of dogs exhibiting less pain 2 weeks after injection. Samarium has also been used for synovectomy in horses, and intra-arterially to osteosarcoma in dogs.

The use of physiologic processes to target radiation to tumor cells aims to deliver radiation to each cell. However, traditional brachytherapy involves implanting physical sources of radiation (such as seeds, ribbons, or needles) directly into the tumor tissue. Sources include radioactive iridium and cesium among others. At the University of Missouri, the injection of radioisotopes suspended in a liquid or colloid has more recently been investigated. This may result in more homogenous dose distribution as the liquid diffuses throughout the tumor tissue. However, determination of the resulting dosimetry is challenging. This technique has been used with bone and soft tissue tumors at the University of Missouri, with radioisotopes such as gold-198 and yttrium-90. Prostate cancer has been treated in men with brachtherapy for many years. This approach has been mimicked in dogs, and we have explored the use of liquid brachytherapy with radioactive gold nanoparticles injected directly into prostatic tumors in dogs.

Horses develop sarcoids, squamous cell carcinoma, and other locally invasive tumors that can benefit from radiation therapy. However, due to their size external beam radiation therapy can be challenging or impossible, and require repeated anesthetic events each of which could result in a difficult recovery and fractured leg. Yttrium-90 in a colloid suspension has been used at our institution to treat a series of horses with sarcoids and squamous cell carcinomas, among other miscellaneous tumors. Unlikely many therapies for these tumors, liquid brachytherapy seems to elicit minimal to no local irritation, and results have been both cosmetic and durable.

In human cancer therapy, radiolabeled monoclonal antibodies using radioisotopes such as yttrium have been used for various cancers including lymphoma. This approach to therapy for lymphoma and other cancers, will certainly be investigated in veterinary medicine when adequate monoclonal antibodies are identified.

RADIATION SAFETY

Radioactive substances naturally exist in our bodies and in our environment (such as radon and potassium-40). Naturally occurring isotopes are a source of radiation exposure, however as caregivers we aim to minimize exposure to both personnel and pet owners. Time, distance and shielding are the most important aspects of keeping radiation exposure as low as is reasonably achievable (ALARA). Radiation exposure decreases exponentially with distance (law of inverse square) so even minor increases in distance from the treated patient can significantly decrease exposure. After diagnostic or therapeutic procedures with radioisotopes, dogs and cats are confined to specially designed facilities for some length of time depending on local, regional, and national regulations. Pets are discharged to their owners with additional instructions to limit close exposure for some period of time based upon the decay half life of the radioisotope. When a minimum of 10 half-lives have been surpassed, the amount of radioactivity remaining in the body is considered to be equivalent to or below background radiation.

Specialized training, facilities, and licensing are needed for the use of radioisotopes. With proper patient identification, animals can be referred to appropriate facilities for advanced care and more treatment options.

REFERENCES

1. Lattimer JC et al. Clinical and clinicopathologic response of canine bone tumor patients to treatment with samarium-153-EDTMP. J Nucl Med. August 1990;31(8):1316-25.
2. Vancil JM et al. Use of samarium Sm 153 lexidronam for the treatment of dogs with primary tumors of the skull: 20 cases (1986-2006). J Am Vet Med Assoc. June 2012;240(11):1310-5.
3. Worth AJ et al. Radioiodide (131I) therapy for the treatment of canine thyroid carcinoma. Aust Vet J. April 2005;83(4):208-14.
4. Turrel JM et al. Sodium iodide I 131 treatment of dogs with nonresectable thyroid tumors: 39 cases (1990-2003). J Am Vet Med Assoc. August 2006;229(4):542-8.
5. Lamb CR et al. Preoperative measurement of canine primary bone tumors, using radiography and bone scintigraphy. J Am Vet Med Assoc 1990;196(9):1474-8.
6. Statham-Ringen KA et al. Evaluation of a B-cell leukemia-lymphoma 2-specific radiolabeled peptide nucleic acid-peptide conjugate for scintigraphic detection of neoplastic lymphocytes in dogs with B-cell lymphoma. Am J Vet Res 2012;73(5):681-8
7. Jankowski MK et al. Nuclear scanning with 99mTc-HDP for the initial evaluation of osseous metastasis in canine osteosarcoma. Vet Comp Oncol 2003;1(3):152-8
8. Robben JH et al. In vitro and in vivo detection of functional somatostatin receptors in canine insulinomas. J Nucl Med 1997;38(7):1036-42

Oncologic Pathology – Interpreting the Data

Sandra M. Nguyen, BVSc (Hons I), DACVIM (Oncology)
Sydney, Australia

INTRODUCTION

Pathology is an essential part of veterinary practice, and this lecture aims to present the important pathologic data from the perspective of oncology, in both diagnosis and treatment. This is will help you interpret the data specifically in the light of your cancer patient.

HEMATOLOGY

There are some aspects to the cancer patient's hemogram that we specifically recognize or look for.

Below is the typical hemogram of a canine patient that has been diagnosed with hemangiosarcoma. Note the regenerative anemia, thrombocytopenia and red cell morphology abnormalities. Lymphoma patients can present similarly, although usually their anemia is non-regenerative. An inflammatory leukogram in non-specific and does not always indicate infection.

Whole Blood - EDTA

	Result		Ref Range
Hematocrit	23	L	42-57
Hemoglobin	7.1	L	14.6-19.7
RBC (mill/uL)	3.8	L	6.1-8.5
MCV (fL)	63		63-74
MCH (pg)	20	L	21-26
MCHC (g/dL)	31	L	32-37
RDW (%)	17.5	H	11.3-14.0
Reticulocyte (%)	3.0	H	0.2-1.1
Absolute reticulocyte (%)	105	H	10.1-75.9
nRBC (/100 WBC)	5		0-1
WBC (thou/uL)	42.6	H	6.2-14.4
Segmented Neutrophils (thou/uL)	37.5	H	3.4-9.7
Band Neutrophils (thou/uL)	0		0-0.1
Lymphocytes (thou/uL)	0	L	1.2-4.7
Monocytes (thou/uL)	4.7	H	0.1-1
Eosinophils (thou/uL)	0.4		0.1-2
Basophils (thou/uL)	0		0-0.1
Platelet count (thou/uL)	105	L	179-483
MPV (fL)	13.8	H	8.4-13.2
Platelet Smear estimation	Low		
Total Protein - Ref (g/dL)	7.3		5.9-7.8
Plasma Appearance	Clear		
RBC Morphology	Anisocytosis-Moderate		
	Schistocytes-Many		
	Target cells - few		
WBC Exam	No Abnormalities		

When assessing a chemotherapy patient at the nadir (typically 7 days after chemotherapy administration) the main cell counts that are important on a hemogram are the neutrophils, haematocrit and the platelet count. Some veterinarians are concerned that the lymphocyte count is often low; this cell line being low is not clinically significant to an oncologist, and is often encouraging, for when we are treating lymphoma we are hoping to target the lymphocytes!

The following is a hemogram of a canine patient that was treated with CCNU (lomustine) 7 days ago. Note the marked leukopaenia due to neutropenia and the thrombocytopaenia. Typically the neutropenia will resolve in 48-72 hours. A result such as this would prompt a future dose reduction of the drug by 25%. CCNU can cause a cumulative, non-reversible thromobocytopaenia, so careful monitoring of this cell line is warranted, particularly in the situation where the patient is expected to have a reasonable life span (and the drug is not being used in the palliative setting).

	Result		Ref Range
Hematocrit	42		42-57
Hemoglobin	16		14.6-19.7
RBC (mill/uL)	7		6.1-8.5
MCV (fL)	63		63-74
MCH (pg)	25		21-26
MCHC (g/dL)	36		32-37
RDW (%)	13		11.3-14.0
Reticulocyte (%)	1		0.2-1.1
Absolute reticulocyte (%)	14		10.1-75.9
nRBC (/100 WBC)	0		0-1
WBC (thou/uL)	2.0	L	6.2-14.4
Segmented Neutrophils (thou/uL)	0.49	L	3.4-9.7
Band Neutrophils (thou/uL)	0		0-0.1
Lymphocytes (thou/uL)	0.7	L	1.2-4.7
Monocytes (thou/uL)	0.3		0.1-1
Eosinophils (thou/uL)	0.4		0.1-2
Basophils (thou/uL)	0		0-0.1
Platelet count (thou/uL)	120	L	179-483
MPV (fL)	10.3		8.4-13.2
Total Protein - Ref (g/dL)	7.3		5.9-7.8
Plasma Appearance	Clear		
RBC Morphology	No abnormalities		
WBC Exam	No abnormalities		

BIOCHEMISTRY

In terms of biochemical abnormalities caused by cancer, the most common paraneoplastic abnormality is hypercalcaemia, most typically seen with T cell lymphoma and anal sac adenocarcinoma. Hypercalcaemia can be considered an oncologic emergency and prompt treatment is necessary.

Hypoglycaemia is associated with insulinomas and some tumours can produce insulin-like growth factor and also cause hypoglycaemia (e.g. lieomyosarcoma).

Concurrent diseases such as hepatopathies and renal disease are screened for in our cancer patients, as this can affect treatment and outcome. A sound understanding of the chemotherapy drugs you are using will enable you to monitor the biochemical abnormalities that can either be caused by the chemotherapy agent, or affect the metabolism, and therefore the toxicity, of the chemotherapy drug you are using.

For example CCNU (lomustine) is associated with hepatoxicity, which can occur after a single treatment, but is typically associated with a few treatments. Doxorubicin can exacerbate renal insufficiency in cats, so a pre-treatment creatinine and USG is recommended. Tyosine kinase inhibitors can also cause hepatotoxicity in the dog.

An example of where the toxicity of the drug is affected is carboplatin. Carboplatin is excreted via the kidneys, so while it doesn't cause nephrotoxic damage per se, renal insufficiency will decrease the rate at which the drug is eliminated, allowing the myelosuppressive effects to be greater than in a patient with normal renal function. For this reason if a dog has an elevated creatinine and low USG we would typically reduce the dose of carboplatin by 25% from the out set.

CYTOLOGY

Cytology is an excellent tool in practice as it is rapid and cheap. While we are looking at the individual cell rather than the cell within the architectural structure of the tissue (which is what you are doing with histopathology), it requires little equipment and you can have an idea of what diagnostic and treatment path to take in the time of the consultation.

For bone marrow and lymph nodes, cytology can be of greater benefit in the clinical setting than histopathology.

After the aspirate is taken, you should check it for diagnostic quality before sending the patient home and the sample to laboratory. User-error causes for a non-diagnostic sample include missing the lesion, spreading the sample inadequately on the slide, and breaking the cells when smearing. Tumour/mass causes of non-diagnostic samples include not exfoliating well, and being highly vascular (resulting in hemodilution). Aspirating necrotic tissue within the lesion will also result in a non-diagnostic sample.

When examining the slide, look at the degree of cellularity, and the main cell type, and consider whether or not you feel that it fits the clinical picture. For example, the tumour may be surrounded by a large amount of inflammation and you may not be aspirating the cause of inflammation. If this is what you suspect, take more samples and check your slides again.

When looking at interpresting the aspirate of the mass, first consider whether you think the sample is inflammatory or neoplastic. Once you have established you think that the mass is neoplastic, you may be able to further characterize the sample into the broad categories of neoplasia – round cell, mesechymal or epithelial.

There are several well published criteria of malignancy of tumours on cytology (1), and these may or may not be present for all cancer types. Criteria of malignancy include anisokaryosis, anisocytosis, clumped and stippled chromatin, nucleoli, abnormal mitoses and increased vacuolation. Cancers which rarely display cyologic criteria of malignancy include anal sac adenocarcinomas, thyroid carcinomas and mammary carcinomas in the dog.

HISTOPATHOLOGY

Particularly where the histopathology doesn't fit the clinical picture, a second opinion is often valuable as the diagnosis or grade of the tumour may be different, which will influence your treatment recommendations. Major disagreements between the first and second opinion have been reported in up to 37% of cases (2) and so a good relationship with your pathology practice is helpful when requesting a review.

When interpreting the histopathology report for a cancer patient, be sure to pay attention to the description as well at bottom line diagnosis, as often the description will contain clues as to the biologic behavior of the tumour. An example is melanoma and mast cell tumors, where the mitotic index is important for prognosticating. If a mitotic index is not reported in these cases, you should request one.

Be careful when using the pathology company treatment macro's in advising your client – these can be incorrect in the clinical setting and can also lead to under- or over- treating of a patient. A personal example is where a patient presented to me with diffuse pulmonary metastatic disease, but had been declared 'cured' of a mammary osteosarcoma 3 months previously after surgical excision. The pathologic recommendation was "The metastatic potential is low considering this is a well differentiated bony mass", yet the literature we have on mammary osteosarcomas (while not extensive) shows that the metastatic rate of this disease is around 60% (3) and we do not have enough information to say whether or not the biologic behavior of this disease is related to cellular differentiation. Your local friendly oncologist will likely be more than happy to offer their opinion of the direction that treatment should take for your patient based on the pathology report.

REFERENCES

(1) Latimer, Duncan and Prasse's Veterinary Clinical Pathology 5th ed. (2) Regan et. al. Vet Comp Oncol. 2015 Jun;13(2):106-16. (3) Langenbach A et. al. J Am Anim Hosp Assoc. 1998 Mar-Apr;34(2):113-20.

Osteosarcoma – Living the Legless Life

Sandra M. Nguyen, BVSc (Hons I), D ACVIM (Oncology)
Sydney, Australia

INTRODUCTION

Osteosarcoma is the most common bone tumour in dogs and accounts for 85% of all bone in that species (1). It more commonly affects the appendicular skeleton in large breed dogs, but this cancer can also affect the axial skeleton, and small dogs.

In the appendicular skeleton the metaphyseal regions of long bones are affected with the distal radius, proximal humerus and distal femur the most common primary tumour locations, and the proximal and distal tibia equally affected when this long bone is involved. The forelimb is more commonly affected than the hindlimb. This has led to the phrase 'away from the elbow and towards the knee' as a means of remembering the locations to be concerned about.

Osteosarcoma is highly metastatic. With amputation of an appendicular lesion alone, the 1-year survival time is only 10%, with death usually because of pulmonary metastasis. At the time of diagnosis, dogs with radiographically appreciable pulmonary metastasis is about 10% (1).

ETIOLOGY

Large breed dogs, such as Rottweilers, Irish wolfhounds and New Foundlands, are more commonly affected. Other breeds with a predisposition to osteosarcoma include Greyhounds, Great Danes and St. Bernards. This then points towards a genetic etiology for this cancer type, or some reason where size would lead to the development of the cancer.

In Scottish deerhounds, about 15% of this breed die from osteosarcoma. It has been found that nearly 70% of risk in this breed is directly attributable to heritable traits (2). Thirty three inherited risk loci have been identified in greyhounds, Rottweilers, and Irish wolfhounds which accounts for 55 to 85% of phenotype variance in each breed (3).

Host factors for development of osteosarcoma include large breed, and it has been postulated that it is the microfractures caused by rapid growth in the first year predispose these bones to osteosarcoma, which has led to the development of 'large breed' dog food, allowing for the appropriate development of the long bones. Early neutering has also been associated with osteosarcoma, along with sites of fractures, both with and without implants, and as a late complication of external beam radiation (1).

DIAGNOSIS

The patient usually presents to the veterinarian with lameness, pain and swelling. The lameness may be acute or chronic, as well as acute on chronic. Survey radiographs are usually adequate to appreciate an aggressive bone lesion. Very occasionally the lesion can be subtle and so advanced imaging, such as computed tomography (CT) and magnetic resonance imaging (MRI) may be required if the survey radiographs are not diagnostic. Radiographs of the distal radius in a Great Dane with confirmed osteosarcoma are below.

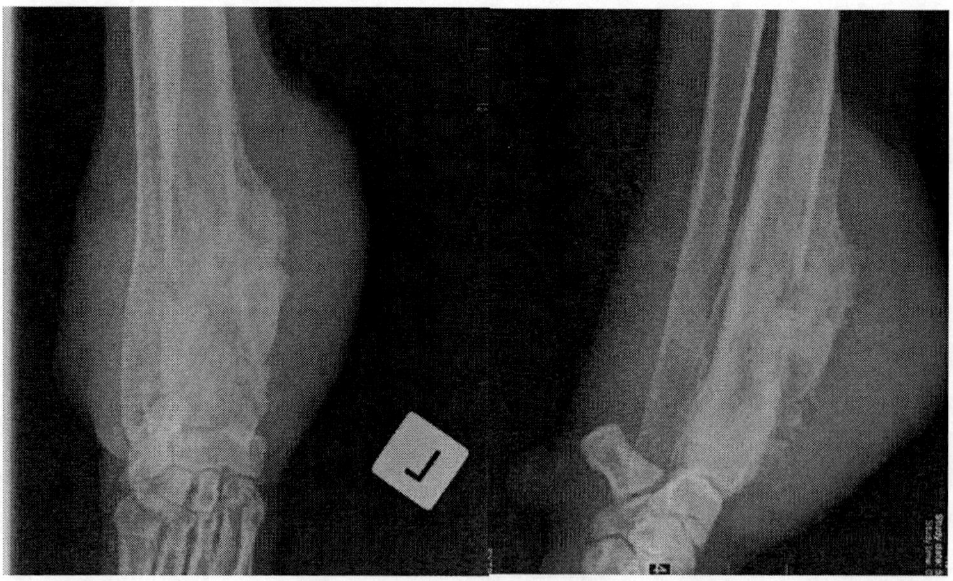

Due to the highly metastatic nature of this cancer, complete staging is recommended. CT is more sensitive than radiographs to assess for metastatic pulmonary disease (4), but survival times quoted are from publications where screening is done with radiography. While abdominal ultrasound usually does not reveal abdominal metastasis, about 5% of patients have concurrent neoplasia (5), which can change treatment plans. Nuclear scintigraphy is sensitive when screening for bony metastatasis, but not very specific.

If a biopsy for definitive diagnosis is to be undertaken, it must be taken from the center of the lesion in order to increase the chance of a diagnostic yield. Personally I have also had good success with aspirates from the center of the lesion, and if either (biopsy or fine needle aspirate) is to be undertaken the client needs to be counseled regarding the chance for pathologic fracture.

PROGNOSIS

Histologic grade and mitotic index are tumor associated factors that have been shown to influence survival, with high grade tumours having a worse prognosis. Patient factors include age, with young dogs having a shorter survival time than older dogs, serum ALP; those with high serum ALP have a poorer outcome, and those with a lymphocytosis and monocytosis at the time of diagnosis also having shorter disease free intervals (1). In another study of dogs with osteosarcoma living greater than one year after diagnosis, 48% of those dogs had an elevated serum ALP at the time of diagnosis (6).

From a clinical perspective, it is difficult to know what to do with some of this data, as I still recommend definitive treatment in dogs with an elevated ALP and monocytosis. When I am recommending amputation I rarely have a grade of tumour. Most studies give us the median survival times and disease free intervals for a group of dogs with osteosarcoma, and at this point we still cannot predict with certainty what will happen in the individual. In terms of the conversation I have with the owner, I am inclined to ask whether or not they want to know what haematologic factors for survival their dog has and whether it will help or hinder them in their treatment making decisions.

TREATMENT

The treatment of choice remains amputation and chemotherapy (platinum based protocol) resulting in a median survival of 10-12 months. In an older study of 107 dogs 28% were alive at 2 years and 16% alive at 3 years (7). In the more recent study of 90 dogs that lived to greater than one year post diagnosis, only 19 (21%) survived to 2 years. (6).

Amputation can be a confronting decision for owners, and the anthropomorphic tendencies both owners and some veterinarians have can be a significant psychological barrier to overcome. Thankfully, I have never had a patient not rise from an amputation, and some of those patients have had significant concurrent orthopaedic disease. The one obese patient that did not rise within 24 hours of amputation was discovered to have a mis-placed nerve block, which resolved with time.

In a survey of owners after amputation, 91% perceived no change in their dog's attitude after amputation with 73% reporting no change in the dog's recreational activities. Only 6% said they would not do it again, with most saying that the recovery was much better than they had anticipated (8).

Metastatic disease

When dogs develop metastatic disease there are additional therapies now available. In a study looking at non-mast cell tumor response to tocerinib, dogs with metastatic osteosarcoma were enrolled and 48% of them experienced clinical benefit some of which also received metronomic cyclophosphamide and a non steroidal anti inflammatory (9). I have personally had 3 patients now live more than 12 months with pulmonary metastatic disease from osteosarcoma, which a few years ago would have been unheard of. Unfortunately adding this therapy after carboplatin chemotherapy has not resulted in an increase in survival time (10).

COMPARATIVE ASPECTS

Osteosarcoma is the most common form of bone cancer in children, and because of this the dog has and is used as a spontaneous model for cancer. The intact immune system of the dog, along with the similar histologic and molecular aspects of the neoplasia make it an effect large animal model for drug development (11).

REFERENCES

(1)Withrow et.al Small Animal Clinical Oncology 5th ed. (2). Phillips JC1 Genomics. 2007 Sep;90 (3):354-63.3. Karlsson EK, Genome Biol. 2013 Dec 12;14(12):R132 (4) Eberle et al Vet Comp Oncol. 2011 Jun;9(2):131-40 (5) Wallace et al J Am Anim Hosp Assoc. 2013 Jul-Aug;49(4):243-5 (6) Culp WT, et al J Am Vet Med Assoc. 2014 Nov 15;245(10):1141-6. (7) Berg J et al: Effect of timing of postoperative chemotherapy on survival of dogs with osteosarcoma. Cancer 1997;79:1343. (8) Dickerson VM J Am Vet Med Assoc. 2015 Oct 1;247(7):786-92. (9) London C, Vet Comp Oncol. 2012 Sep;10(3):194-205
(10) London et. Al PLoS One. 2015 Apr 29;10(4):e0124889. (11) Fenger ILAR J. 2014;55(1):69-85.

Companion Animals in Comparative Oncology: One Medicine in Action

Oliver A. Garden, BSc, BVetMed, PhD, FHEA, FRSB, FRSA, MRCVS, DACVIM, DECVIM-CA
London, UK

INTRODUCTION

Cancer is a major cause of morbidity and mortality in companion animals [1-10]. Epidemiological studies in dogs, for which most data are available, suggest that cancer kills 40 to 50% of individuals over the age of 10 years [1,2,4,8]. Cancer in dogs resembles cancer in humans in various ways, including its long latency, clinical presentation and metastatic potential; its pathobiological characteristics, including tumor cell heterogeneity and suppressive microenvironment; its genomic instability and pharmacogenomic signatures, including chemoresistance; and its multifactorial nature, including both genetic and environmental risk factors. Increasing recognition of the inability of murine models to recapitulate all of these characteristics of human tumors and the frequent failure of such models to inform Phase II and III clinical trials has brought to light the huge potential value of spontaneous canine and, to a lesser extent, feline cancer in the drug discovery and validation pipeline, for the mutual benefit of all parties [11-16]. The term 'Comparative Oncology' – the study of naturally occurring cancers in animals as models for human disease – has been borne. Although apparently a recent phenomenon, the wider concept of One Medicine is by no means new: such luminaries as Hippocrates, Plato, Virchow and Osler all espoused the virtues of comparative medicine, which is once again gaining traction in biomedical research today [17,18].

THE SHORTCOMINGS OF MURINE CANCER MODELS

A mere 11% of anti-neoplastic therapies that demonstrate efficacy in murine models are ever approved for human use, representing a very disappointing and expensive attrition of candidate molecules [19]. Apart from the multiple failures of murine models to recapitulate the complexities of human cancer, murine bone marrow is generally less sensitive to chemotherapy than human bone marrow, precluding the use of mice in the generation of safety data. Furthermore, the murine tumor microenvironment is often different in a number of ways from that of human cancer, reflecting the xenotransplantation of 'doses' of human tumor cell lines or patient-derived biopsies into the subcutaneous tissue of immunodeficient recipient mice, where the neoplastic cells are generally more responsive to therapeutic agents [19,20]. So while murine work is often still the first province of preclinical research and remains an important initial 'port of call', its shortcomings and the need for additional animal models are being increasingly recognized.

THE CURRENT DRUG DEVELOPMENT PIPELINE

The development of new cancer therapies is protracted and expensive: up to 16 years and a staggering $1.8 billion may be required to bring a new therapeutic agent from target validation to the marketplace [21]. Attrition rates for therapies in oncology are also significantly higher than those in other therapeutic areas: approximately 59% of anti-neoplastic drugs entering Phase III clinical trials fail, more often as a consequence of therapeutic inefficacy than toxicity [22,23]. In the conventional drug development pathway, Phase I human clinical trials are used to assess toxicity and dose; Phase II trials explore anti-tumor activity; and Phase III trials compare outcomes of patients treated with the new agent *versus* standard of care. Therapies demonstrating better outcomes than the standard of care emerge successfully from this pipeline, but its reliance on preclinical murine models that poorly predict efficacy and toxicity, and its largely linear, non-iterative trajectory, markedly limit its success. A superior paradigm is clearly needed. An integrated model, in which comparative studies are undertaken immediately before or after Phase I human clinical trials to help predict pharmacokinetic (PK) and pharmacodynamic (PD) properties, as well as dosing regimens, is likely to decrease attrition and cost considerably [24]. To this end, the National Cancer Institute (NCI) established the Comparative Oncology Trials Consortium (COTC) in 2003, to provide the infrastructure to integrate data derived from clinical trials in cancer-bearing pet patients into the development pathway of new drugs, devices and imaging modalities for human cancers [25]. This bold initiative has already borne fruit, as the exemplars in the next section will showcase, but much more can yet be done.

COMPARATIVE ONCOLOGY AND ONE MEDICINE

Cancer in dogs: pathobiology, genomics and pharmacogenomics

Of the companion animals, the dog is the species in which the discipline of comparative oncology has shown most growth [3,12,26]. Cancer in dogs shares clinical and pathological similarities to cancer in humans, and common diagnostic tests are employed in the oncological workup of these two species. However, the dog also affords a rich resource for the genetic dissection of cancer by virtue of the distinctive breed structure of the species, characterised by striking heterogeneity between, but minimal heterogeneity within, breeds, coupled with the known predisposition of particular breeds to certain cancers [27] – for example, Scottish Terriers to transitional cell carcinoma, Bernese Mountain Dogs to histiocytic sarcoma, Golden Retrievers to lymphoma, and Boxers to glioblastoma. The shorter lifespan of dogs and the generally faster progression of cancer in this species also allows more timely assessment of clinical outcomes.

Mapping mutations driving neoplastic behavior has been facilitated by the release of CanFam 3.1, the most recent version of the canine genome assembly that includes 85Mb of new sequence data to increase coverage from 99.2% to 99.6% of the genome [28]. Behind the murine and human genomes, the canine is the third most detailed mammalian genome assembled to date. Efforts to map canine oncogenic mutations have also benefited from the creation of the Canine Comparative Oncology and Genomics Consortium (CCOGC; http://ccogc.net/), incorporated as a non-profit organisation in 2007, which has established a large biospecimen repository containing 60,000 samples from approximately 2,000 patients. The value of combining genome-wide association studies (GWAS) with the genetic analysis of tumor biopsies was demonstrated in recent work that identified 33 different loci associated with the development of osteosarcoma (OSA) in Greyhounds, Irish Wolfhounds and Rottweilers [29]. While there was no overlap in loci between breeds owing to their individual genetic homogeneity, the genes all impacted signaling pathways for osteoblast differentiation and proliferation, skeletal growth, and epithelial to mesenchymal trans-differentiation [29]. This work complemented an earlier study that disclosed two genes uniformly expressed in dogs with OSA – *il8* and *slc1a3* – that were associated with a poor outcome in human pediatric OSA patients, vindicating the value of such cross-species approaches [30].

The ultimate goal of cancer therapy will be to target specific treatments to specific patients according to their 'oncogenomic' signatures [31,32]. This so-called precision or personalized medicine approach has been championed by the NCI and the Translational Genomics Research Institute (TGEN), involving deep exome sequencing to identify low-frequency mutations in heterogeneous tumor samples, alongside RNA sequencing to confirm mutations, identify oncogenic fusions, and examine the differential expression of genes. A proof-of-principle trial examining the mRNA transcriptome of 24 canine tumor biopsy samples of various histotypes by means of Affymetrix gene expression profiling revealed the feasibility of this approach in dogs, with a mean turnaround from receipt of biopsies to the generation of a personalized drug report of 4.8 days, considered to be sufficiently fast to allow the dog to model advancements in human personalized medicine [33].

Environmental carcinogens: dogs and cats as sentinels

One of the obvious benefits of studying cancer in dogs and cats is their exposure to the same environmental xenobiotics as humans, since they usually live in their owners' homes. They may thus act as sentinels for carcinogens in the environment [34,35], allowing the dissection of the relative contribution of genes and environment to the pathogenesis of cancer. For example, significant associations between exposure to tobacco smoke and sinonasal cancer in dolicocephalic dogs, between exposure to pesticides and lymphoma, urothelial transitional cell carcinoma and mammary carcinoma, and between exposure to asbestos and mesothelioma have all been established in dogs [34,36-42]. Similar studies in cats have also yielded evidence that hyperthyroidism, caused by a thyroid gland adenoma in a proportion of cases, is associated with the consumption of both canned food, thought to be related to the chronic ingestion of bisphenol A, a building block for the epoxy resins used to line metallic cans [43,44], and polybrominated diphenyl ethers (PBDEs), used as flame retardants in a variety of household consumer products and ingested in dust as a consequence of cats' grooming behavior [44]. Recent studies have also suggested a link between the ingestion of organohalogenated xenobiotics in dust and feline somatotrophinomas, which cause hypersomatotropism and the clinical entity acromegaly [45].

Veterinary oncology and cancer clinical trials

Comparative oncology is gaining traction and credibility in the medical community, with increasing recognition of the mutual benefit to both human and canine health gained from integrative human and veterinary basic research and clinical trials. A number of exemplars of such studies are considered in this section.

With the creation of the COTC in 2003 came the infrastructure to undertake multi-center, nationwide clinical trials in tumor-bearing dogs using cancer drugs under development for human patients. The first published study by the COTC vindicated the huge potential of this consortium, examining the safety and efficacy of adeno-associated bacteriophage delivery of tumor necrosis factor-α to αV integrins expressed by tumor endothelium, thus providing valuable data to inform the design of first-in-man studies [25]. Work on canine OSA has defined genomic targets associated with the progression of disease, emphasising the merit of focusing drug development on metastatic pathways, rather than solely on regression of measurable lesions, to improve the stagnant long-term outcomes for patients with OSA [29,30]. Multicenter clinical trials organized by the COTC and the high prevalence of OSA in dogs would allow the timely accrual of data comparing and prioritizing agents for evaluation in human patients [46].

Clinical trials of novel cancer drugs in dogs also allow proof of target and the assessment of toxicity in a manner that is often not possible in murine models. For example, the tyrosine kinase inhibitor (TKI) toceranib, administered to dogs, is similar in structure to the drug sunitinib, administered to humans. The two TKIs elicit some toxicities that are similar in the two species, such as weakness, vomiting and neutropenia [47,48]. Lethargy or weakness can readily be assessed in dogs, but much less so in mice; graded and standardized toxicity assessments are possible in dogs and indeed also cats. Longitudinal tissue sampling is also possible in dogs to assess PK/PD correlations. For example, studies of the autophagy inhibitor hydroxychloroquine in dogs [49] informed the design of subsequent Phase I human clinical trials [50], revealing that tumor tissue concentrations of the drug were approximately 100 times greater than plasma concentrations: the latter were therefore a poor measure of drug exposure of the tumor.

Further cogent examples of canine clinical studies informing human clinical trials populate the literature. The Bruton tyrosine kinase inhibitor ibrutinib blocked B cell activation in preclinical research and showed efficacy in B cell lymphoma in dogs, from which paired tumor biopsies and PBMC samples revealed that the latter could act as a surrogate endpoint for human clinical trials [51]. Effective doses were also defined in dogs, leading ultimately to lower doses than initially envisioned being administered to human patients [52]. Recent Phase I and Phase II clinical trials of the exportin-1 (XPO1) protein inhibitor KPT-335 in dogs with lymphoma were helpful in defining the toxicity profile and dosing regimen prior to human clinical trials of the same molecule [53]. Similarly, a Phase I/II clinical trial of the prodrug GS-9291 in canine multicentric lymphoma, a model of non-Hodgkin lymphoma (NHL), defined PK parameters, toxicity, and antitumor activity, providing proof of concept and refining the dosing schedule for a Phase I human clinical trial [54].

The value of canine clinical studies in extending preliminary findings in murine models of cancer was demonstrated by work on the signal transducer and activator of transcription 3 (STAT3) LY5, discovered by site-directed fragment-based drug design *in silico*, which showed good oral bioavailability in dogs [19] after initial murine work demonstrated anti-tumor activity [55]. Future preclinical work on healthy dogs and clinical trials in canine cancer patients will define the maximum tolerated dose, PK/PD relationships, adverse event profile, and optimal dosing regimen of this small molecule inhibitor. Exciting work on the procaspase-3 activator 'procaspase-activating compound-1' (PAC-1) in various murine models of cancer suggested its potential as an adjunctive treatment [56]. The next step in its assessment as a treatment involved its administration to dogs with cancer in clinical trials to investigate PK, toxicity and combinations with conventional drugs [57]. In particular, its penetration of the blood-brain barrier suggested that the molecule would have promise as an agent to treat cancer of the central nervous system and clinical trials in dogs with glioma are planned alongside ongoing Phase I clinical trials of human patients with glioblastoma, informed by the preliminary canine work [19]. Novel derivative molecules based on PAC-1 are also showing promise in murine models [58].

Canine clinical studies have also advanced oncology in other, unanticipated ways. Currently the only licensed DNA cancer vaccine in any species is a xenogeneic product based on human tyrosinase, Oncept® (Merial), used in the treatment of stage II or III oral melanoma in dogs after control of local disease [59,60]. This pivotal advance in the field has encouraged the development and assessment of several xenogeneic DNA vaccines in Phase I/II clinical trials in human cancer patients [61], complementing a number of exciting advances in the immunotherapy of cancer in general in both veterinary and human patients [62].

Integrating veterinary clinical trials into drug development

Despite the obvious benefits to be gained from comparative oncology, as shown by the preceding exemplars, pharmaceutical companies generally remain hesitant to commit resources to canine clinical trials because of the intense competition for resources between the various teams involved in developing a therapeutic agent, a lingering perception that they add little to the overall development pipeline, and a fear that new safety signals will be generated that could derail the drug development process. But things are turning around, albeit slowly. One company bucking the trend is Genentech (www.gene.com/), which has introduced clinical trials in pet patients into their overall work-flow, championed by a cross-functional team working in collaboration with molecule teams [19].

RESEARCH NEEDS AND CONCLUSIONS

A recent workshop on translational cancer research, focusing on the contribution of clinical studies on pets – chiefly dogs – with cancer to the development pipeline of human cancer therapies [19], concluded that the principal research needs in the community were a better characterization of the canine genome to facilitate the molecular comparison of canine and human cancers; generation of a larger armamentarium of canine-specific assays and reagents to allow canine clinical studies to recapitulate fully the complexities of human clinical studies, including a diversity of biomarkers; and a drive to increase awareness of the merit of canine clinical trials amongst the human medical community and pharmaceutical industry, who were often oblivious to their value. While there are still many hurdles to overcome before clinical studies in pet animals, particularly dogs, become a routine part of drug development by Big Pharma, much has been achieved in the past decade and the tide is turning, slowly but surely. The next decade is poised to deliver a number of breakthroughs in cancer therapy, predicated on the exciting groundwork undertaken in the past five years or so: let us hope that man's best friend will be part of this journey, not only in enabling more rapid advancement of therapies, but also in saving millions of dollars along the way.

REFERENCES

[1]Albert RE, et al Mechan Ageing Develop 1994;74:149; [2]Bonnett BN, et al Acta Vet Scand 2005;46:105; [3]Di Cerbo A, et al J Cancer Res Clin Oncol 2014;140:1807; [4]Egenvall A, et al Acta Vet Scand 2005;46:121; [5]Landes C, et al Tierarz Prax 1984;12:369; [6]Misdorp W Vet Q 1996;18:32; [7]Moore GE, et al J Am Vet Med Assoc 2001;219:209; [8]Vascellari M, et al BMC Vet Res 2009;5:39; [9]Miller MA, et al Vet Pathol 2015;Oct 12.pii:0300985815608674; [10]Adams VJ, et al J Small Anim Prac 2010;51:512; [11]Alvarez CE ILAR J 2014;55:16; [12]Fürdös I, et al J Trans Med 2015;13:265; [13]Ito D, et al Vet Immunol Immunopathol 2014;159;192; [14]Rankin KS, et al Pediatr Blood Cancer 2012;58:327; [15]Riccardo F, et al Cancer Immunol Immunother 2015;64:137; [16]Richards KL and Suter SE Immunol Rev 2015;263:173; [17]Bradley OC Proc R Soc Med 1927;21:129; [18]Teigen PM Can Vet J 1984;25:400; [19]National Acad Sci The Role of Clinical Studies for Pets with Naturally Occurring Tumors in Translational Cancer Research: Workshop Summary (Washington (DC), 2015; [20]Holden SA, et al Cancer Chemother Pharmacol

1997;40:87; [21]Paul SM, et al Nat Rev Drug Discov 2010;9:203; [22]Kola I and Landis J Nat Rev Drug Discov 2004;3:711; [23]Hay M, et al Nat Biotechnol 2014;32:40; [24]Paoloni M and Khanna C Nat Rev Cancer 2008;8:147; [25]Paoloni MC, et al PLoS ONE 2009;4:e4972; [26]Gordon I, et al PLoS Med 2009; 6:e1000161; [27]Dobson JM ISRN Vet Sci 2013;941275; [28]Hoeppner MP, et al PLoS ONE 2014;9:e91172; [29]Karlsson EK, et al Genome Biol 2013;14:R132; [30]Paoloni M, et al BMC Genomics 2009;10:625; [31]Hu X and Zhang Z Trends Genet 2016;32:127; [32]Wurz GT, et al Ther Adv Med Oncol 2016; 8:4; [33]Paoloni M, et al PLoS ONE 2014;9:e90028; [34]Reif JS Public Health Rep 2011;126(S1):50; [35]van der Schalie WH, et al Environ Health Perspect 1999;107:309; [36]Andrade FH, et al Diag Pathol 2010;5:45; [37]Bukowski JA, et al J Toxicol Env Health 1998;54:579; [38]Gavazza A, et al J Vet Intern Med 2001;15:190; [39]Hayes HM, et al Am J Epidemiol 1981;114:229; [40]Pastor M, et al J Vet Intern Med 2009;23:301; [41]Severe S, et al SpringerPlus 2015;4:27; [42]Glickman LT, et al Environ Res 1983;32:305; [43]Edinboro CH, et al J Amer Vet Med Assoc 2004;224:879; [44]Peterson M J Fel Med Surg 2012;14:804; [45]Dirtu AC, et al Environ Int 2013;57:60; [46]Khanna C, et al Clin Cancer Res 2014;20:4200; [47]Faivre S, et al J Clin Oncol 2006;24:25; [48]London CA, et al Clin Cancer Res 2003;9:2755; [49]Barnard RA, et al Autophagy 2014;10:1415; [50]Rangwala R, et al Autophagy 2014;10:1369; [51]Honigberg LA, et al Proc Natl Acad Sci USA 2010;107:13075; [52]Burger JA Curr Hematol Malign Rep 2014;9:44; [53]London CA, et al PLoS ONE 2014;9:e87585; [54]Vail DM, et al Clin Cancer Res 2009;15:3503; [55]Yu W, et al J Med Chem 2013;56:4402; [56]Peterson QP, et al J Med Chem 2009;52:5721; [57]Peterson QP, et al Cancer Res 2010;70:7232; [58]Wang F, et al Mol Oncol 2014;8:1640; [59]Bergman PJ, et al Vaccine 2006;24:4582; [60]Liao JC, et al Cancer Immun 2006;6:8; [61]Yang B, et al Hum Vaccin Immunother 2014;10:3153; [62]Regan D, et al Vet J 2016;207:20.

From Theory to Therapeutic: Drug Development Basics for the Clinical Scientist

Linda Rhodes, VMD, PhD
Durham, NH

INTRODUCTION

There is no course to teach veterinarians about what is required to get an approval for a veterinary drug. Those of us in industry learn on the job. Why should clinical veterinarians care about drug development? There are two main reasons: first, to help you access the best information on drugs that are available for your use, and second, to give you help you understand how you can incorporate clinical studies into your practice, both for the excitement of participating in research, and also, if you are a private practitioner, to generate revenue.

REGULATORY APPROVAL

This talk will focus on Food and Drug Administration (FDA) Center for Veterinary Medicine approval of drugs for companion animals. As you may know, in the US, other regulatory bodies are involved in approval of drugs used in veterinary practice – the Environmental Protection Agency (EPA) regulates pesticides that are used topically for fleas and ticks, and the Center for Biologics at the United States Department of Agriculture (USDA) regulates vaccines and some monoclonal antibodies. A drug approved by the FDA is given an NADA (New Animal Drug Approval) number. This number is on the label and the packaging and assures you of FDA approval. Prior to FDA approval, studies can be conducted in client owned animals if a Sponsor (the company working on getting the drug approved) has an INAD – an Investigational New Animal Drug number.

In order to achieve approval of a drug in the US, the Sponsor has to complete extensive research and development. The FDA (and most regulatory agencies world-wide) require the following:

- Manufacturing – companies have to show that their drug can be manufactured in a reproducible way (batch after batch); they must define the stability and formulation, develop and validate analytical methods to measure the specifications of each batch, and demonstrate the quality of the manufacturing process of both the active pharmaceutical ingredient and the formulated product, under Good Manufacturing Practices (GMP).
- Safety – safety studies at exaggerated doses need to be conducted in the 'target' species – the species in which the drug will be used, and safety is also evaluated in field studies.
- Effectiveness – determination of the dose and demonstration that the drug does what the label claims.

Sponsors invest a great deal of time and effort to generate the data required to achieve a drug approval. Developing a new drug for cats or dogs takes about 5-7 years and costs about $10-$12M dollars.

Let's focus on how effectiveness is demonstrated – that is the part of the drug development process in which you, as veterinary clinical scientists, will be most involved. I will then touch briefly on safety, and finish with some hints on good places to go to review data on approved drugs and get more information.

LABEL CLAIM

The first step in determining the best way to show that a drug is effective is to carefully define the label claim. The claim, or indication, must clearly define what the drug can do. Some examples of claims are:

- "For the control of pain and inflammation associated with osteoarthritis in dogs"
- "For the prevention of acute vomiting in dogs"
- "For appetite stimulation in dogs"
- "For the treatment of Patnaik grade II or III, recurrent, cutaneous mast cell tumors with or without regional lymph node involvement in dogs".

Once the claim is clearly defined, the Sponsor must work with the FDA to propose how the drug will be proved to be effective for that claim, defining endpoints based on what needs to be demonstrated. Notice that the claim for the cancer drug is very specific, narrowing the patient population in which the drug was evaluated.

As our profession is working toward making decisions for our patients using evidence based medicine, it is important to know what evidence is required from Sponsors to demonstrate the effectiveness of new drugs in development. Some of the largest veterinary clinical studies conducted are those completed for drug approval. They use strict criteria for clinical trial design – they must be prospective, multi-site, randomized, masked (blinded), and have placebo controls. Sponsors usually start by studying new drugs in pharmacokinetic studies in laboratory animals to select a dose, and then in pilot clinical studies in patients, where efficacy is evaluated. Understanding safety in the pet population is an important component in the development process. Once pilot studies are complete, Sponsors then proceed to conduct what is called a "pivotal" effectiveness study.

PIVOTAL EFFECTIVENESS STUDIES

Pivotal effectiveness studies are required for drug approval, and generally consist of a minimum of 100 animals treated with the test drug, and either 50 or 100 placebo treated animals. The FDA is strict about the design of pivotal studies; they must support the proposed label claim, they must be powered to assure the predictability of the results, and the patient population in which the study is conducted must be representative of the target population for the drug. In addition, prior to conducting the study, the

Sponsor submits the protocol to the FDA for review and concurrence. If the FDA concurs with the protocol that means that they have approved the study design, and if the Sponsor conducts the study according to the protocol, the FDA will consider those data as adequate for evaluation of the drug's effectiveness. Also, the FDA is careful about making sure that the data on the safety of the drug are evaluated in these studies, as discussed below.

The protocols are very detailed – usually about 20-30 pages in length, and they narrowly constrain the study so that at each site, the same procedures are followed. Inclusion and exclusion criteria are carefully defined to help minimize variability. For example, in a study evaluating dogs with osteoarthritis for pain, other treatments that may impact the pain evaluation, such as glucosamine, are not allowed. Sponsors are careful not to be too restrictive in what dogs are allowed in the study, because each restriction might end up as a restriction on the label. For example, it might seem to make sense to exclude dogs that are on thyroid medication, or medication for congestive heart failure, from a study of osteoarthritis, because of a concern that other medications might interfere with evaluation of the treatment effect of the test drug. However, Sponsors generally use language such as "Dogs with well controlled medical conditions that are on stable medications for more than 30 days may be enrolled." This is because in the real world, these dogs would likely be considered for treatment with a new osteoarthritis drug. The FDA wants to include these dogs are representative of the patient population in which the drug will eventually be used.

Pivotal effectiveness studies are conducted according to Good Clinical Practices (GCP). These are laid out in a Guidance Document[1] and include a long list of what needs to happen for a trial to qualify as being GCP. The core of GCP is to assure conduct of the study according to the protocol and accurate documentation of the study, as well as to guard against fraud. All study activities are carefully recorded, and details of who did what and when must be clear. If changes were made in the protocol, or if deviations from the protocol occur, these must all be recorded so that the Sponsor and the FDA can follow exactly what occurred in the study. There are multiple layers of trained people to help Investigators and to assure the quality of the data.

PIVOTAL SAFETY STUDIES

Safety of a new drug is evaluated extensively. There are many components to this; this talk will focus on two: the pivotal target animal safety study and the field safety evaluation. The FDA requires Sponsors to conduct a study in laboratory animals of the target species (the species in which the drug will be used) where groups of males and females are exposed to multiples of the proposed dose (generally up to 5X) and which are conducted over a period of time that is approximately 3X the proposed length of dosing. For example, if a new drug has a dose of 1 mg/kg and is intended to be given for 1 week, the FDA might require the Sponsor to do a study where the dose groups are 0 (placebo), 1, 3 and 5 mg/kg for three weeks. For any drug intended to be given chronically, a 6 month laboratory safety study is required. During these studies animals are observed, blood and urine are collected for clinical pathology, often ECG and ophthalmic exams are completed, and then all animals are euthanized, necropsied and have an extensive list of tissues evaluated for any histopathological changes.

In addition to this laboratory study, safety data are collected in more real world settings. Adverse events (AEs) are collected from animals in the pivotal effectiveness study. These safety observations are important in that they are made in animals that will be the actual patient population when the drug is approved. And what is an adverse event? It is defined as any observation that is unfavorable and unintended and occurs after the use of study drug (either active drug or placebo). The FDA pays very close attention to these adverse events, and any serious adverse event (one that causes death or is life threatening) must be reported to the FDA within a few days of its occurrence. Investigators are blinded to whether the animal is being treated with active drug or placebo, and are asked to describe the likelihood of the AE being caused by the treatment. In most cases Sponsors want these AEs to be fully worked up, so that they can see how they resolve, and in most cases, Sponsors pay for the necessary diagnostics and treatment.

All the safety information is submitted to the FDA for review, and the results are reflected on the drug label and insert, where safety findings are given for the information of the prescribing veterinarian.

SHOULD YOU BE A CLINICAL INVESTIGATOR?

All this sounds like a lot of work and it is. So why would a veterinary clinical scientist be interested in conducting these clinical studies? Being in Investigator allows you and your patients to participate in cutting edge science, testing some of the most exciting new veterinary drugs. In addition, it allows your patients to benefit by having access to new drugs otherwise unavailable, as well as having much of their diagnosis and treatment costs during the study covered by the Sponsor. In addition, Investigators usually get paid on a per case basis, and depending on how many cases you enroll, this could be a significant supplement to your practice income.

On the other side of the coin, participation takes a lot of time and effort. Investigators must be proficient in the protocol, and they and any staff involved must participate in detailed training sessions. Time is required to explain the trial to clients whose pets qualify for enrollment, and data collection and other study activities also take considerable time and effort. Much paper used to be involved, and still is in some studies, but many studies now use electronic data capture, making the process a bit easier. Some clients might be averse to participation because they are concerned their pet will be assigned to the placebo group. Finally, once the study is complete, the Investigator is subject to FDA audit – about 10% of the clinics involved in a study are audited by the FDA. Although Sponsors help with this process, it is stressful and does take time.

Some Investigators participate in many clinical trials and have made this activity part of their business. One very effective way of doing this is to have a dedicated veterinary technician serve as study coordinator for several studies at a time; they can support the Investigator, make sure the study runs smoothly, explain the protocol to clients, and handle much of the study logistics and paper work. It is not unusual for Investigators with study coordinators to be participating in studies with several Sponsors at a time

In the past 10 years, clinical trial groups have been developed at veterinary colleges to run clinical studies. However, if you look at the records of who conducts the pivotal effectiveness studies for FDA approval of new animal drugs, generally it is Investigators in private practice. Oncology may be the exception here because of the need for a large number of animals in a specific patient population which may only be seen in the University setting or a very busy oncology referral practice. If you, as a veterinary clinical scientist would like to get involved, or get more information on how you might participate, a good source of information are the several contract research organizations (CROs) that conduct these clinical trials (AlcheraBio, Animal Clinical Investigations, etc.). But don't agree to participate if you don't see the specific patient population called for in the protocol – nothing is more frustrating to both Investigator and Sponsor than to spend the time and money to contract with an Investigator, train the staff, ship the drug, and then have no or very limited enrollment.

MORE INFORMATION

Busy veterinarians often don't take the time to carefully read the product label and insert of veterinary drugs, but if you do, you will find it is a wealth of information. This is particularly important for newly approved drugs. The Sponsor proposes 'label language' but the FDA has the final say on what exactly is on the label; each word is carefully chosen. That language represents the best effort to provide the veterinarian with all the information needed to use the drug; it distills many years of data into important highlights.

One of the best kept secrets of pivotal trial conduct and design is that there is a section of the FDA Center for Veterinary Medicine's website[2] that has fairly detailed descriptions of the pivotal clinical studies that were done to approve each new veterinary drug. These are called Freedom of Information Summaries (FOIs). Not only do they summarize the pivotal safety and effectiveness information the FDA evaluated to decide to approve the drug, but they also list the clinical investigators that participated in the clinical studies. These veterinarians could be an excellent source of real world information regarding participation in industry-sponsored clinical trials.

There are very few publications that describe the drug approval process by the FDA and other regulatory agencies world-wide. Some resources are given in the reference list below.[3,4]

CONCLUSION

As veterinarians, we have many products available for our patients, but few have been through full FDA approval for the specific indication for which we use them. Walking the aisles of the national veterinary meetings, we are swamped with products that are used for "support". Why do we see language such as "supports the immune system" or "supports liver function" or "supports a health GI tract"? Because this language is not a drug claim, and the FDA will not require that these products submit data for FDA approval, since they are not making drug claims. When in doubt, look for the NADA number – if a product has this, you know the manufacturer has provided sufficient data on safety, effectiveness and manufacturing, for FDA approval. If you see this number, you will know that the drug is manufactured to the highest quality standard, its safety in the target species has been extensively evaluated, at least one large prospective, multi-site, randomized masked, placebo controlled study has been done to demonstrate effectiveness, and best of all, the key safety and effectiveness data are available for you on the Center for Veterinary Medicine of the FDA's website. I hope you will consider being part of a new drug clinical trial and experience the excitement of seeing the most innovative new treatments being tested in your practice.

REFERENCES

[1]Guidance for Industry 85, Good Clinical Practice, VICH GL9, 2001,[2]http://www.fda.gov/AnimalVeterinary/default.htm, [3]Rhodes L. J Fel Med Surg 2015: 17:783, [4]http://www.fda.gov/AnimalVeterinary/DevelopmentApprovalProcess/default.htm.

Chemotherapy for Lymphoma: Have We Become Complacent?

Carlos H. de M Souza, DVM, MS, DACVIM (Oncology), DACVS
Gainesville, Florida

The objective of cancer treatment is to reduce the tumor cell population to zero. Models of transplanted tumors in mice have validated the hypothesis of fractional cell kill. This hypothesis states that at a given concentration and time period, a chemotherapy drug will kill a constant fraction of the cell population, independent of the absolute number of cells. Tumor regrowth will occur in between treatment cycles. The results of chemotherapy treatment are directly related to the dose of drug administered and the number of cycle repetitions. This fractional cell kill hypothesis applies well to fast growing tumors such as lymphoma and leukemia.

Although cancer progress in a clonal fraction, DNA instability leads to individual cell mutations with heterogeneity being the end result. The concept that in a given cell population multiple different clones exist led to the development of multi-drug chemotherapy protocols. The idea is that multiple drugs, with different mechanisms of action and different mechanisms of resistance will overcome the inherited resistance of most tumors. The goal then becomes to give the higher number of drugs, at the highest concentration without severe side effects to systems such as bone marrow and gastrointestinal tract. Multi-drug chemotherapy protocols used for the treatment of lymphoma and leukemias in humans apply this concept, which is known as dose-intense protocols. Dose intensity is defined as the amount of drug delivered per unit of time, expressed in milligrams per M2/week.

In people, dose-intense protocols have shown to improve cure rates even when less intense protocols are associated with the same complete remission rate.

This concept has long been forgotten in veterinary medicine where zero side effects are the main goal. A recent veterinary study showed that patients with lymphoma that experienced treatment delays due to bone marrow toxicity lived significantly longer, when compared to patients that did not experience treatment delays.

This presentation will discuss the current literature and offer a new approach of a dose-intense CHOP base lymphoma and leukemia protocol. Such protocols are expected to increase the complete response rate, the cure rate, and with minimal severe side effects.

*** Formatting Issue - Best Available Paper/Graphic ***

Just Because We Can Doesn't Mean We Should

Carolyn J. Henry, DVM, MS, DACVIM (Oncology)
Columbia, MO

INTRODUCTION

From early in our veterinary medical education, we are taught the axiom, "above all, do no harm". This is regarded as a guiding principle of our profession, yet this very principle is often disregarded when it comes to decision making in the day-to-day practice of veterinary clinical oncology. One need look no further than the latest online case discussion to find a flurry of anecdotes offered up as suggestions for therapy. Posts often are made with the disclaimer, "for what it's worth" (FWIW) or "my two cents' worth" (MTCW), suggesting that one should not place too much stock in the information. And yet we see examples every day where treatment decisions are made on the basis of wanting to do SOMETHING, as opposed to nothing---even if that something is, at best, unproven or, at worst, harmful. This phenomenon occurs in human medicine, as well. One poignant example is treatment of "the death rattle" in terminal human patients. Anticholinergic medications are often administered in order to decrease the excessive respiratory secretions associated with the end of life in human patients. However, systematic review of clinical outcomes associated with such intervention has demonstrated no clinical benefit to the administration of antimuscarinic drugs.(1,2) So, why does the practice persist? Published surveys and focus group discussions point to concerns for caretaker anxiety related to hearing the excessive respiratory noise and, again, the desire to "do something". (2,3,4)

It is time that we, as veterinarians, have a dialogue about the cost of "doing something" that is unproven, anecdotal, sub-standard-of-care, or even known to be ineffective. All treatment decisions come at a cost, whether it be time, money, emotion, quality of life, or some combination of these currencies. Therefore, we must serve as better stewards of our clinical knowledge and of the available credible literature to help guide our clients through the decision-making process. A discussion of the levels of evidence applicable to clinical practice decisions and examples of where veterinary oncology patients are routinely treated with products and methods that fall woefully short of these standards is necessary and will serve as the basis for this session.

EVIDENCE-BASED MEDICINE

Evidence-based medicine (EBM) is considered the appropriate means by which to practice veterinary medicine today and has been integrated into the modern veterinary curriculum. By definition, EBM is "the conscientious, explicit, and judicious use of the current best evidence in making decisions about the care of individual patients"(2,5) The general hierarchy of evidence is depicted in the pyramid below, with the highest level of evidence being meta-analysis of multiple randomized, prospective controlled, clinical trials.

Levels of Evidence in Veterinary Oncology Literature

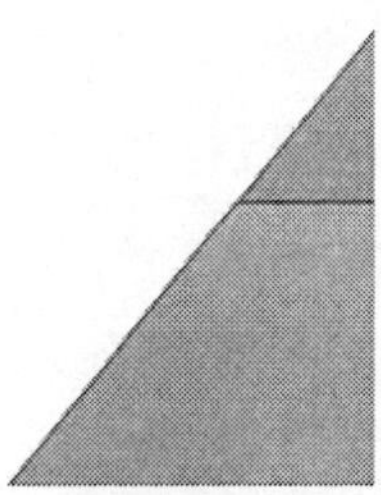

It has been suggested that there are contexts within which the principles of EBM are not entirely appropriate. In his 2008 article, "Evidence-based medicine beyond the bedside: keeping an eye on context", Dr. Jon Tilburt of the Mayo Clinic asserted that additional factors such as clinical circumstances and patient preferences should be taken into consideration for the most appropriate application of EBM to daily practice.(6) In fact, the authors of the aforementioned definition of EBM later updated their definition to include the "integration of best research evidence with clinical expertise and patient values".(7) This latter definition is more in keeping with the principles applied to veterinary medicine, yet does not entirely address three important components of decision-making that we face as veterinary oncologists; namely, product and treatment availability, client financial constraints, and the option of euthanasia.

PRODUCT AND TREATMENT AVAILABILITY

The decision to apply evidence vs. anecdote to clinical decision-making is often influenced by the inaccessibility of some treatment modalities to veterinary practitioners. The paucity of drugs approved for use in veterinary cancer patients has made off-label use of human oncology drugs a common practice. Dosages and indications are sometimes extrapolated from human oncology literature, without prospective evaluation of safety or efficacy in companion animal species, much less veterinary patients with cancer. However, lack of drug approval does not negate the need for clinical practice to be evidence-based. Anecdotal reports of drug efficacy in a handful of patients is often the basis for justification of therapy decisions. This is despite our fundamental understanding of the need to objectively evaluate clinical responses in light of potential biases in patient selection or outcome assessment. The practice also extends to other cancer treatment interventions including surgery, radiation therapy, and complementary or alternative medical care.

THE IMPACT OF FINANCIAL RESOURCES

It is reported that only 5.7% of the dog-owning public and 2.6% of cat owners in the US have pet insurance.(8) Accordingly, treatment decisions are considered in the context of financial obligation, with the cheaper alternative often selected over data-proven, effective therapy. This is a seemingly inescapable reality of current veterinary practice. Provided we have not made incorrect financial assumptions about our clients, we are obligated to advise them as to the best way to achieve their treatment goals within their financial constraints. An increase in number of available funded clinical trials has provided a less expensive option for the [fully-informed, consenting] client needing to balance treatment costs with financial limitations. A searchable database of oncology clinical trials in the US can be found at www.VetCancerTrials.org

At the other end of the spectrum is the 'do-everything' client without financial constraints. In this situation, it is important to advocate for therapy that is truly in the best interest of our patient. Advanced care including tumor vaccines, radiation therapy, multimodality therapy, and bone marrow transplantation have become available at a higher price tag for selected companion animal cases. Where financially plausible, these options offer the chance for a cure and, as such, may be reasonable pursuits. However, it is important that the client with greater financial resources is counseled regarding potential side effects and has accurate expectations for success.

PALLIATIVE CARE, END-OF-LIFE CARE, AND EUTHANASIA

At the center of the EBM debate in human medicine is the field of palliative care. In many regards, the same issues apply to veterinary oncology. Whereas the veterinary oath recognizes relief of suffering as a tenant of good veterinary practice, this is not part of the Hippocratic Oath for physicians.(10) The World Health Organization has addressed this in their definition of palliative care as "an approach that improve the quality of life of patients and their families facing the problem associated with life-threatening illness, through the prevention and relief of suffering by means of early identification and impeccable assessment and treatment of pain...".(9) The WHO definition falls noticeably short of advocating for euthanasia as a means of preventing and relieving suffering. The Hippocratic Oath, in fact, speaks against it: "I will give no deadly medicine to anyone if asked, nor suggest any such counsel". (10)

ANECDOTE VS. EBM IN PRACTICE

The practice of applying solid EBM to clinical decision making is predicated on the existence of high-quality clinical research and literature to inform our choices. As has been noted by others, this prerequisite is not always available to those in the veterinary profession.(11) Furthermore, even when such evidence is available, there is a certain reassurance that comes from hearing the anecdotal experience of our colleagues. This phenomenon has been reported for physicians, as well. Hay, et al. conducted a survey of 29 individual physicians and three focus groups (n = 10) to explore physicians' use of multiple sources of information in clinical decision-making.(12) They reported the following:

> *"We found that physicians tend to favour experience, either theirs or that of trusted colleagues, in making clinical decisions, referring to the EBM literature (largely through electronic clinical resources) either for general information about a condition or to double check that a therapy does not have a published negative outcome. Decisions are made with some reference to EBM, but experience weighs more heavily in clinical decision making about therapeutics. Experience certainly is built up with reference to EBM, but the learning of traditions of practice through apprenticeships and learning from one's particular cases seem to be at the heart of clinical practice."*

Thus, it would appear that the reliance upon our own clinical experience and that of our colleagues in clinical decision-making is not unique to the veterinary oncology community. That is not to say, however, that we should endorse such practice. In order to understand the impact of this approach, we will consider examples from the board-certified oncology veterinary listserve community and the first-opinion practitioners who rely upon their recommendations. Specifically, we will explore the following topics as time permits:

- metronomic therapy
- tumor vaccines for melanoma and lymphoma
- complementary and alternative medicine
- chemotherapy protocols that have not been proven efficacious

CONCLUSIONS

One of the perks of practicing veterinary medicine is the flexibility to individualize treatment plans beyond the confines of third party payer or institutionally-defined standard-of-care practice. Along with this flexibility must come a commitment to provide care that is in keeping with the principles of the veterinary oath, medical ethics, and our professional obligation to be responsible stewards of our medical knowledge and expertise.

REFERENCES

1) Wee B and Hillier R. Cochrane Database Syst Rev 2008:CD005177.
2) Visser C, et al. Cancer Biol Med 2015;12:193.
3) Wee B, et al. Palliat Med 2008;22:173.
4) Hirsch CA, et al. Palliat Med 2013;27:732.
5) Sackett DL, et al. BMJ 1996;312:71.
6) Tilburt.J. Eval Clin Pract. 2008 Oct;14(5):721.
7) Sackett, D. L., Straus, S. E., Richardson, W. S., Rosenberg, W. & Haynes, R. B. (2000) Evidence-Based Medicine: How to Practice and Teach EBM, 2nd edn. Edinburgh: Churchill Livingstone.
8) US Pet Ownership and Demographics Sourcebook. AVMA, 2012.
9) World Health Organisation. WHO Definition of Palliative Care 2015.Available online: http://www.who.int/cancer/palliative/definition/en/
10) Hippocratic oath. (2016). In Encyclopædia Britannica. Retrieved from http://www.britannica.com/topic/Hippocratic-oath
11) Sahora A and Khanna C. J Vet Intern Med. 2010 Jan-Feb;24(1):51.
12) Hay MC, et al. J Eval Clin Pract. 2008 Oct;14(5):707.

Fundamentals and Clinical Applications of Chemotherapy

Michael A. Kiselow, DVM, DACVIM (Oncology)
Campbell, CA

INTRODUCTION

The primary goal of this lecture is to understand the fundamental principles of anti-neoplastic, chemotherapeutic agents. This will include defining basic, essential terminology, as well describing theoretical concepts of how cancer cells are affected by chemotherapy. This will be followed by a review of the most-commonly used chemotherapeutic medications and their specific applications in veterinary medicine. Finally, basic practices of managing patients undergoing chemotherapy treatments will be outlined in a clinical context. These concepts will focus primarily on conventional, cytotoxic, anti-neoplastic agents. However the new-generation, targeted-therapeutic category of anti-cancer medications will also be addressed.

BACKGROUND

The earliest innovations in anti-cancer therapy were achieved with the study of the effects of mustard-based gas used during both World War I and World War II. The most relevant observations were in regard to the consequence of bone marrow - most specifically lymphoid - hypoplasia. Clinically, this category of chemicals was subsequently applied in human patients with Hodgkin's lymphoma, which resulted in dramatic, albeit transient, regression of disease and physical symptoms. Concurrently, unrelated research conducted by Dr. Sidney Farber revealed that folic acid promoted proliferation of leukemia cells. This led to the development of analogue chemicals that inhibit folic acid metabolism, thereby inducing remission in patients (primarily children) with acute lymphoblastic leukemia.

In the generation following these initial discoveries, a revolution of effort was invested into the development of additional anti-neoplastic agents. This involved both the investigation of compounds derived from 'natural' (mostly plant-based) sources, as well as the synthesis and modification of man-made chemicals. Coupled with rapidly-evolving technologic innovations that facilitated greater understanding of cell biology and cycle kinetics, these scientific advancements resulted in the production of several compounds that exhibited clinically significant effects for cancer patients. Initially, these agents were used individually so that their independent efficacies could be assessed. However as remissions were typically temporary, the concept of combination chemotherapy protocols was subsequently developed in effort to enhance disease control. Indeed, improved outcomes were achieved in several instances, and multi-drug regimens are now considered standard of care for most cancers.

In these early stages of chemotherapy development, most agents exerted relatively nonspecific biologic mechanisms of action, usually damaging DNA and/or inhibiting enzymatic pathways and other proteins. In contrast, with more-recent advancements in the fields of molecular biology and biotechnology, the current direction of drug research has focused on 'targeted' therapies. These latter medications are created with particular emphasis on addressing unique genetic mutations that are often highly specific to individual cancer types. In addition, an increasing understanding of immunology has fostered the development of agents intended to both enhance tumor cell killing by the immune system, and to improve surveillance by reversing the specific immune-inhibitory effects induced by cancer.

DEFINITIONS

Chemotherapy can be applied in several clinical contexts, depending on the type of cancer being treated, as well as the needs and limitations of the individual patient. These applications include the following:

Primary – Refers to chemotherapy used as initial treatment, prior to or in lieu of, surgery and radiation therapy.

Neoadjuvant – Chemotherapy applied in effort to cytoreduce a tumor burden in effort to improve the success of subsequent surgery or radiation therapy.

Induction – Chemotherapy used as the sole treatment for cancers for which there are no other viable therapies (i.e., patients with advanced, metastatic disease).

Adjuvant – Refers to chemotherapy used following surgery and/or radiation therapy. In this context, the disease burden is substantially reduced – ideally to a microscopic level – such that chemotherapy is used in effort to prevent local recurrence of the primary tumor and to prevent metastasis.

The dosing strategies of most cytotoxic chemotherapy agents focus on the concepts of cell growth and kinetics. In general terms, chemotherapy causes damage to vital cell machinery involved in metabolism and reproduction. If the damage is irreparable, affected cells die by either necrosis or, more commonly, apoptosis during subsequent attempts to undergo mitotic division. As such, rapidly growing cells are typically most vulnerable to these effects. In comparison to most normal tissue types, cancer cells generally reproduce more-quickly and, theoretically, should be preferentially susceptible. However there are select, normal tissue types which also proliferate rapidly and are therefore also easily impacted by chemotherapy. These include myeloid progenitor cells in the bone marrow, gastrointestinal epithelium, oral mucosa, skin, and hair follicles. Clinically, this is manifested as damage to these tissues and physical, adverse effects.

On a fundamental level, the relative effects of chemotherapy on different tissue types is studied when compared to a cancer that exhibits a very high **growth fraction** – this refers to the proportion of cancer cells actively dividing within a tumor:

Growth fraction = # cells actively dividing / total cell population

Cell Kill Theory – A mathematical concept that suggests that for a cancer with 100% growth fraction (logarithmic growth rate), each dose of chemotherapy kills a certain proportion of cells within a tumor, not an absolute number.

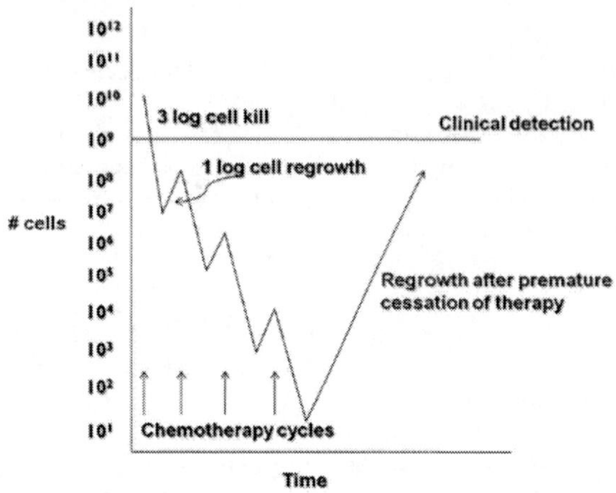

Figure 1. Graphic depiction of cell survival based on Cell Kill Theory

Goldie-Coldman Hypothesis – Concept that suggests that a tumor undergoes its most-rapid growth prior to becoming clinically detectable (approximately 10^9 cells). In addition, most genetic mutations resulting in heterogeneity and inherent resistance to chemotherapy occur earlier in development ($10^4 - 10^6$ cells).

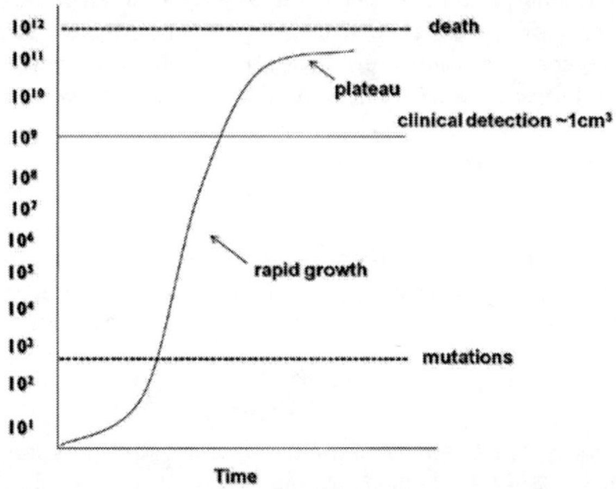

Figure 2. Graphic depiction of Goldie-Coldman Hypothesis

CONCEPTS

Dose Limiting Toxicity (DLT) – The primary adverse effect of a chemotherapy agent that restricts the milligram dosage that can be safely administered.

Maximum Tolerated Dose(MTD) – The highest recommended dosage of a chemotherapy agent based on its known toxicities. *Primarily applicable to conventional, cytotoxic medications.

Dose Intensity (DI) – The amount of a chemotherapy agent administered over a specified interval of time (i.e., mg per week) when used at its MTD.

Summation Dose Intensity (SDI) – The numerical, calculated summary of more than one, co-administered administered chemotherapy agents. This is best-achieved if medications in a protocol have unrelated DLTs, or otherwise requires a decrease in the dose of one or both medications in order to prevent unacceptable toxicities. This calculation therefore provides theoretical guidance regarding whether the combination will act synergistically against a tumor, in order to justify the necessary dose reductions.

Area Under Curve (AUC) – Relationship between drug concentrations in the plasma over time, which is used to assess total drug exposure.

AUC = Dose/Clearance

Dose Response Curve: Graphical representation illustrating the relationship between effects of chemotherapy on both target (tumor) tissues and normal, healthy tissues.

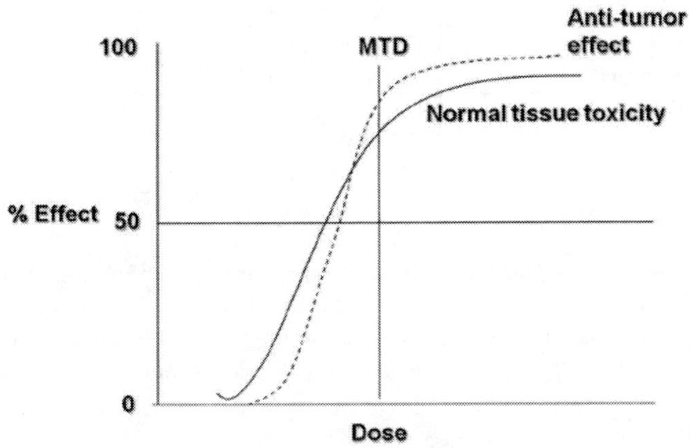

Figure 3. Dose Response Curve

By adjusting chemotherapy protocols (dose adjustments, altered route of administration, concurrent protective medications), the relationship between these curves can be augmented in effort to reduce the risk of toxicity.

CLINICAL MANAGEMENT

Dose calculation

Most chemotherapy medications are dosed based on body surface area (BSA) rather than on patient mass. Body surface area is theoretically better-representative of drug metabolism because it correlates more closely to metabolic rate and is less-influenced by adipose than weight.

$$BSA = \frac{wt\,(g)^{2/3} \times K}{10{,}000}$$

K = body shape constant: Cat = 10; Dog = 10.1; Horse = 10.5

However BSA is an imperfect guideline, as it fails to account for several other factors that impact absorption and clearance of drugs within individuals: hepatic/renal function, age, extreme obesity, concurrent diseases. As such, there may still be great variability in systemic drug concentrations that are achieved when all patients' doses are calculated using an identical formula. This may result in either underdosing - wherein inadequate blood concentrations of drugs are achieved, leading to a subtherapeutic effect, or overdosing - resulting in excessive patient toxicity. Indeed in veterinary medicine, this challenge has been well-documented in canine patients receiving Doxorubicin, wherein smaller dogs (<10kg) experience a higher incidence of adverse effects when treated based on BSA compared to larger dogs. Similarly, the BSA-based dosages of platinum drugs are calculated for dogs using a sliding scale related to body weight.

Variability has been documented in cats receiving Carboplatin using standard BSA dosing due to differences in renal clearance. Subsequent studies evaluating drug dosing based on individual patient glomerular filtration rates yielded less inconsistency in blood concentrations. Despite these known shortcomings, BSA remains the most-commonly used calculation for the dosing of most chemotherapy agents in veterinary medicine.

Monitoring

Management of patients undergoing chemotherapy involves evaluation of both tolerance and efficacy. Toxicity is assessed via regularly-scheduled physical exams and laboratory testing. As outlined below, the most common, clinically-relevant hematologic toxicities include neutropenia and thrombocytopenia, as well as damage to individual organ function. Secondary medications (anti-emetics, antibiotics) can be combined with chemotherapy in effort to assuage expected toxicities. However if unacceptable adverse effects are observed, the dose of the offending chemotherapy agent can be reduced (usually 10-25%) to prevent recurrence with subsequent treatments. Established toxicity criteria may be referenced to determine the relative severity ('grade') of side effects in order to objectively guide the decision to reduce the dose of chemotherapy. This is an important concept, as reducing chemotherapy dosages in order to prevent toxicity can also significantly impact anti-tumor activity. Therefore, dose reduction should be a last resort after other measures fail to mitigate negative effects.

Effectiveness of chemotherapy should also be monitored via objective criteria by recording measurements of known target lesions. This helps to guide decisions regarding continuing ongoing chemotherapy regimens vs. transitioning to secondary rescue protocols. Two classification schemes for responsiveness to chemotherapy have been adapted from human oncology practices.

	WHO	RECIST
Measurement	Total tumor volume	Longest 1 diameter
Complete Remission	100% reduction	100% reduction
Partial Remission	>50%, <100% reduction	>30%, <100% reduction
Stable Disease	<50% reduction, <25% growth, No new lesions	<30% reduction, <20% growth, No new lesions
Progressive Disease	≥25% growth new lesions	≥20% growth new lesions

Figure 4. World Health Organization and Response Evaluation Criteria in Solid Tumors

Along with these conventional remission criteria, the recent advent of targeted treatments (small molecule inhibitors, immunotherapies) has led to a revised definition of clinical response. Specifically, whereas stable disease has historically been considered an inadequate response to cytotoxic chemotherapy, it is now considered a reasonable goal in the context of chronic disease management with newer therapies. This is reflected in the incorporation of complete remission, partial remission and stable disease into the single category of 'biologic response.'

CHEMOTHERAPY AGENTS

Chemotherapy medications can be categorized based on mechanism of action (MOA) and phase of cell cycle during which they exert anti-neoplastic effects (if phase-specific). It is important to understand the specific biologic mechanisms by which each agent induces cells death, as well as the resistance strategies that cancer cells develop against cell killing. Each major category of cytotoxic chemotherapy agents is outlined below, including tables that summarize clinically relevant information specific to veterinary medicine.

Anti-mitotics (M-phase specific)

MOA: Bind tubulin protein of mitotic spindle, hindering polymerization/depolymerization, preventing progression through metaphase.

Resistance: MDR1/MRP1 expression, altered tubulin structure, upregulation of anti-apoptotic growth factors (IGF-1, survivin, p21).

Vinca alkaloids: Vincristine, Vinblastine, Vinorelbine

Taxanes: Docetaxel, Paclitaxel

Agent	Indications	Toxicities	Vesicant	Dose
Vincristine	Lymphoma, Hemangiosarcoma	Myelosuppression, GI disturbance	Yes	Dog: 0.7m/m^2 IV Cat: 0.5m/m^2 IV
Vinblastine	Mast cell tumor, Lymphoma, Transitional cell carcinoma	Myelosuppression, GI disturbance	Yes	Dog: 2-3.5mg/m^2 IV q 1-2wk
Vinorelbine	Lung tumors	Myelosuppression, GI disturbance	Yes	Dog: 10-15mg/m^2 IV q 1-2wk Cat: 11.5mg/m^2 IV q 1-2 wk
Docetaxel	Carcinomas	Myelosuppression, GI disturbance	Yes	Dog: 1.625mg/kg IV q 2 wk Cat: 1.75mg/kg IV q 2wk
Paclitaxel	Mammary carcinoma, Squamous cell carcinoma	Myelosuppression, GI disturbance	Yes	Dog: 150mg/m^2 IV q 3 wk

Topoisomerase Inhibitors (S-phase specific)

MOA: Bind + stabilize TP I or II, preventing DNA replication.

Anthracyclines/Anthracenediones

MOA: Inhibition of Topoisomerase II, DNA intercalation, free radical formation/oxidation.

Resistance: MDR1/MRP expression, altered TPII activity, p53 mutations, bcl-2 expression, DNA repair, mutations in mismatch repair (increased tolerance for DNA damage).

Agent	Indications	Toxicities	Vesicant	Dose
Doxorubicin	Lymphoma, Sarcomas, Carcinomas	Myelosuppression, GI disturbance, Myocardial damage, nephrotoxicity (cats)	Yes – severe	Dog: 30mg/m^2 large; 1mg/kg small IV q 2-3wk Cat: 25mg/m^2 or 1mg/kg IV q 3wk
Idarubicin	Efficacy demonstrated against Lymphoma in dogs; uncertain for cats	Myelosuppression, GI disturbance (cat)	No	Dog: 22mg/m^2 PO q 3wk Cat: 2mg/cat/d x 3d, q 3wk
Epirubicin	Hemangiosarcoma	Myelosuppression, GI disturbance, Myocardial damage	Yes	Dog: 30mg/m^2 IV q 3wk
Mitoxantrone	Transitional cell carcinoma, Lymphoma, Sarcomas	Myelosuppression, GI disturbance	Yes	5-6mg/m^2 IV q 3wk

Alkylating Agents

MOA: Covalent binding of alkyl group to varying DNA targets (bases). They are not cell cycle phase-specific. They may contain either one or two reactive groups (mono or bi-functional). Bifunctional agents form crosslinks between molecules (inter-strand), and are the most clinically effective.

Resistance: Reduced cellular uptake, increased glutathione (GSH) expression, enhanced crosslink repair (guanine O^6-alkyl transferase), nucleotide excision repair, homologous recombination, decreased/mutated p53 expression.

Nitrogen Mustards: Bifunctional; Alkylate the N-7 position of guanine

Agent	Indications	Toxicities	Vesicant	Dose
Mechlorethamine	Lymphoma	Myelosuppression, GI disturbance	Yes – severe	3mg/m^2 IV q 1 wk
Cyclophosphamide	Lymphoma, Sarcomas, Carcinomas	Myelosuppression, GI disturbance	No	200-250mg/m^2 IV, PO q 2-3wk
Ifosfamide	Lymphoma, Sarcomas	Myelosuppression, Hemorrhagic cystitis	No	Dog: 275-350mg/m^2 IV q 3wk Cat: 900mg/m^2 IV q 3wk
Chlorambucil	Lymphoma, Chronic lymphocytic leukemia, Metronomic protocols	Myelosuppression, GI disturbance	No	6-8mg/m^2 PO q 2-3d; other regimens also
Melphalan	Multiple myeloma, Anal sac adenocarcinoma	Myelosuppression	No	0.1mg/kg PO SID, then EOD

Nitrosureas/Nonclassical: Monofunctional, highly lipophilic; Alkylate the O-6 position of guanine

Agent	Indications	Toxicities	Vesicant	Dose
Lomustine	Lymphoma, Histiocytic sarcoma	Myelosuppression, Hepatic damage	No	Dog: 90mg/m^2 PO q 3wk Cat: 60mg/m^2 PO q 3-6wk
Carmustine	Brain tumors	Myelosuppression, GI disturbance nephrotoxicity	No	Dog: 50mg/m^2 IV q 6wk
Streptozotocin	Insulinoma	Myelosuppression, GI disturbance nephrotoxicity	No	Dog: 500mg/m^2 IV q 3wk
Procarbazine	Lymphoma	Myelosuppression, GI disturbance	No	50mg/m^2 PO SID
Dacarbazine	Lymphoma, Hemangiosarcoma	Myelosuppression, GI disturbance	Yes	Dog: 1000mg/m^2 IV q 3wk

Antimetabolites (S-phase specific)

L-asparaginase:

MOA: Catalyzes hydrolysis of asparagine (essential amino acid) to aspartic acid and ammonia, resulting in inhibition of protein synthesis.

Resistance: Increased expression of asparagine synthetase.

Gemcitabine

MOA: Inhibition of DNA polymerase, inhibition of ribonucleotide reductase.

Resistance: Reduced deoxycytidine kinase expression (enzyme that converts/activates Gemcitabine), increased ribonucleotide reductase expression, increased heat shock protein expression, inhibition of nucleoside transporters (decreased drug uptake).

Cytosine Arabinoside (ara-C)

MOA: Competitive inhibition of DNA polymerization α and β.

Resistance: Reduced deoxycytidine kinase expression (enzyme that converts/activates) parent compound.

5-Fluorouracil

MOA: Inhibition of Thymidylate Synthase, incorporation into RNA (interferes with RNA synthesis), uracil incorporation into DNA.

Resistance: Reduced expression of activating enzymes.

Hydroxyurea

MOA: Inhibition of ribonucleotide reductase (RR), rate limiting enzyme for synthesis of deoxyribonucleotide triphosphates.

Resistance: Increased RR expression.

Methotrexate

MOA: Inhibition of dihydrofolate reductase (DHFR - essential enzyme for folate metabolism).

Resistance: Mutated transmembrane carrier proteins, altered intracellular transformation, upregulation of DHFR.

Agent	Indications	Toxicities	Vesicant	Dose
L-asparaginase	Lymphoma, Lymphoid leukemia	Anaphylaxis, pancreatitis	No	Dog: 400IU/kg SQ, IM Cat: 10,000IU/m^2 SQ, IM
Gemcitabine	Carcinomas	GI disturbance	No	Dog: 300-675mg/m^2 IV q 2wk
Cytosine Arabinoside	Lymphoma, Acute leukemia	Myelosuppression, GI disturbance	No	100-400mg/m^2 IV, SQ q 1wk
5-Fluorouracil	Carcinomas	Myelosuppression, GI disturbance, Seizures	No	Dog: 150mg/m^2 IV q 1 wk *Toxic to cats (neurotoxic)
Hydroxyurea	Polycythemia, Chronic myeloid leukemia, Mast cell tumor	Myelosuppression, GI disturbance, Sloughing of toenails	No	Dog: 50mg/kg PO SID Cat: 10mg/kg PO SID
Methotrexate	Lymphoma	GI disturbance	No	0.6-0.8mg/kg PO, IV

Platinum Agents

MOA: Aquation reaction replaces Cl (or other leaving groups) with –OH groups, which create a more-active compound and prevents drug from diffusing out of cell. Covalent binding to purine DNA bases (especially N7 of guanine) occurs, resulting in intrastrand crosslinks. This can lead to both apoptotic and non-apoptotic cell death.

Resistance: Upregulation of efflux transporters, increased production of glutathione + metallothionein (deactivation), nucleotide excision repair, mutations in mismatch repair (increased tolerance for DNA damage).

Agent	Indications	Toxicities	Vesicant	Dose
Cisplatin	Osteosarcoma, Squamous cell carcinoma, Transitional cell carcinoma, Sarcoids (horses)	Myelosuppression, GI disturbance, Nephrotoxicity	No	Dog: 50-70m/m^2 IV q 3wk *Toxic to cats (pulmonary edema)
Carboplatin	Osteosarcoma, Fibrosarcoma, Carcinomas	Myelosuppression, GI disturbance	No	Dog: 300mg/m^2 IV q wk Cat: 200-260mg/m^2 IV q 3-4 wk
Satraplatin	**	Myelosuppression (neutrophils, platelets), GI disturbance	No	Dog: 35mg/m^2 PO SID x 5d
Lobaplatin	Osteosarcoma	Myelosuppression, GI disturbance	**	Dog: 35mg/m^2 IV q 3wk

TARGETED THERAPIES

The primary concept of targeted therapies is to identify individual genetic mutations that are involved in the malignant phenotype of specific cancers. These often affect vital cell processes, such as angiogenesis, proliferation, and avoidance of normal cell cycle checkpoints. If these unique molecular targets – which should be specific to neoplastic cells – can be addressed with complementary agents, the biologic effects should theoretically only impact the malignant population, while sparing normal host tissues. This is in stark contrast to conventional cytotoxic chemotherapeutics, the cellular targets of which are less-specific. However, while many therapeutics have been developed with this concept in mind, clinical experience has revealed that these agents readily interact with multiple cell types, resulting in adverse effects which still limit applicability. Research has revealed this cross-over effect to be the result of binding to multiple receptors (other than the primary molecular target) by the agents.

Through pharmacodynamic studies, it has been demonstrated that many targeted therapies do not need to be administered at their maximum tolerated doses in order to achieve complete saturation/binding of the cellular target and to exert a clinically relevant effect. This concept exemplifies another important difference from conventional chemotherapy agents, which should always be used at the MTD.

In veterinary medicine, there is currently only a single FDA-approved small molecule inhibitor – Palladia (Toceranib phosphate) – which is licensed for the treatment of recurrent, grade II or grade III cutaneous mast cell tumors in dogs. Toceranib exhibits inhibitory effects against the receptor tyrosine kinases Kit, VEGFR2, and PDGFRβ. As mutations in these receptors are common, subsequent studies of off-label usage of Palladia has also demonstrated clinically relevant effects against a variety of other tumor types (especially neuroendocrine carcinomas).

IMMUNOTHERAPY

As evasion of normal immune surveillance is a critical mechanism for survival of malignant cells, much research and effort have been applied to understanding the molecular mechanisms responsible for these mutations. This has allowed for the development of new therapies intended to overcome immune incompetency, thereby employing the host's immune system for combating cancer.

In veterinary medicine, there is currently a single USDA-approved vaccine therapy – Oncept – which is licensed for the treatment of stage II or stage III oral melanoma in dogs. The vaccine utilizes xenogeneic plasmids encoding human tyrosinase (huTyr), which is the rate-limiting enzyme in the production of melanin. Research has demonstrated that dogs treated with the Oncept vaccine develop anti-huTyr antibodies which cross react with canine tyrosinase. Subsequent clinical studies revealed survival benefits in dogs with oral melanoma following conventional therapy (surgery, radiation therapy) to control the primary tumor.

REFERENCES

Chabner, BA, et al. Cancer Chemotherapy & Biotherapy – Principles and Practice, Fourth Edition. Lippincott Williams & Wilkins; Philadelphia: 2006.
DeVita VT, et al. DeVita, Hellman, and Rosenberg's Cancer, Principles and Practice of Oncology. Lippincott Williams & Wilkins; Philadelphia: 2008.
Tannock IF, et al, The Basic Science of Oncology. The McGraw-Hill Companies, Inc; USA: 2005.
Withrow SJ, et al. Withrow & MacEwen's Small Animal Clinical Oncology, Fifth Edition. Saunders; St. Louis: 2013.

OVERWEIGHT/OBESITY IN GOLDEN RETRIEVERS AS A FUNCTION OF NEUTER, AGE, ACTIVITY LEVEL, AND U.S. REGION

Missy Simpson[1], Kelly Diehl[1], Erin Searfoss[1], Sharon Albright[1], John Bauer[2], Debbie Davenport[1], Rog Page[3]

[1]*Morris Animal Foundation, Denver CO, USA,* [2]*Texas A&M University, College Station TX, USA,* [3]*Colorado State University, Fort Collins CO, USA*

Overweight and obesity (O/O) are implicated as contributing factors to the morbidity of multiple disease processes in dogs including osteoarthritis, neoplasia, and endocrine derangements. In spite of the importance of O/O, very few studies exist that systematically investigate exposures thought to be important in the development of O/O, and those that are published present conflicting information. *The purpose of this study was to examine associations between O/O and age at neuter, region of residence, owner reported physical activity, and age at clinical visit.*

This study was conducted using data collected at the first clinical visit in the Golden Retriever Lifetime Study (GRLS), a prospective cohort of 3,043 privately-owned golden retriever dogs that are followed over their lifetime for four major cancer outcomes. O/O was measured by veterinarians using a nine level body condition scoring system; a BCS of 6 or higher was classified as O/O (n = 868, 29%). Age at neuter was divided into four categories: ≤ 6 months (n = 375, 12%), 6 months to one year (n = 666, 22%), > one year (n = 417, 14%), and intact (n = 1586, 52%). Physical activity had three levels: little or none (n = 29, 1%), moderate (n = 1470, 48%), and very active (n = 1552, 51%). The regions of residence are: Northeast (n = 638, 21%), Midwest (n = 723, 24%), Mountain (n = 414, 14%), South (n = 842, 28%), and Pacific (n = 430, 14%). Age at clinical visit was analyzed continuously. Logistic regression was used to estimate unadjusted odds ratios (OR) and confidence limits (CL) for O/O and the potential correlates listed above. The adjusted model included age at neuter, region of residence, physical activity, and age at clinical visit.

The mean age at clinical visit was 1.2 years (± 0.5 years). Compared to intact animals, age at neuter was associated with O/O in unadjusted analysis (≤6 mos. OR: 2.7, 95% CL: 2.1-3.4; 6mos.-1yr OR: 2.2, 95% CL: 1.8-2.7; ≥1yr OR: 1.8, 95% CL: 1.5-2.3). Unadjusted older age at clinical visit was associated with O/O (OR: 1.7, 95% CL: 1.4-1.9). Compared to dogs with little or no activity, the highest level physical activity was associated with decreased odds of O/O (OR: 0.4, 95% CL: 0.2-0.9), moderately active dogs did not differ from those with little or no activity (OR: 0.5, 95% CL: 0.2-1.1). Region of residence was not associated with O/O. In the fully adjusted model, very active was inversely associated with O/O (OR: 0.4, 95% CL: 0.2-0.9), older age at clinical visit, and being neutered before one year of age were also independently associated with O/O (age at clinical visit OR: 1.5, 95% CL: 1.3-1.7; neuter ≤6 mos. OR: 2.5, 95% CL: 2.0-3.2; neuter 6mos.-1yr OR: 2.1, 95% CL: 1.7-2.5). The full model accounted for 4.7% of the variability in O/O.

These data lend systematic evidence that earlier age at neuter is a risk factor for O/O independent of age at the time BCS was measured and physical activity. While moderate physical activity was not significantly different than dogs with little or no activity, the OR suggests that an inverse trend exists between physical activity and O/O. Nutritional data on this cohort of dogs was not available at the time this abstract was submitted. Because only 4.7% of the variability in O/O was explained by our statistical model, additional exploration of the correlates is needed, including nutrition and genetics. Further studies will examine the impact O/O has on important health events. These data were collected on the largest cohort to date of dogs with both veterinary and owner-reported data available.

EVALUATION OF BODY WEIGHT, BODY CONDITION, AND MUSCLE CONDITION IN CATS WITH HYPERTHYROIDISM

Mark Peterson[1], Carol Castellano[1], Mark Rishniw[2]

[1]*Animal Endocrine Clinic, New York, New York, USA,* [2]*Cornell University, Ithaca, New York, USA*

Hyperthyroidism is a catabolic state associated with increased energy expenditure, increased lipolysis, and increased protein turnover. In human hyperthyroid patients, weight loss is predominantly due to the decrease in lean body mass (primarily muscle) with the loss of fat being less important (at least initially). After treatment, recovery of lost muscle mass generally occurs before fat deposits are completely restored. In hyperthyroid cats, the contribution of muscle wasting (sarcopenia) to the overall loss of body weight is not known. However, many cats are currently being diagnosed at an early or mild stage of hyperthyroid disease and display minimal weight loss or other clinical signs. If these cats are similar to humans, loss of muscle mass might still be expected, but again, this has not been reported.

Our aim in this study was to investigate the body weight, body condition score (BCS), and muscle condition score (MCS) of a large population of hyperthyroid cats, both before and after successful treatment with radioiodine. We studied 436 hyperthyroid cats, of which 60 had a follow-up examination 3-12 months after successful treatment. For inclusion, all cats had a detailed history (including diet fed) and complete physical examination. Diagnosis was confirmed by use of a complete serum thyroid panel (T_4, T_3, fT_4, TSH) and results of thyroid scintigraphy. Body weight, BCS (5-point scale) and MCS (4 point scale) were measured (J Small Anim Pract 2011; 52: 385) in all cats by a single investigator (MEP).

Before treatment, body weight in the 436 cats ranged from 1.9 to 8.8 kg (median, 4.36 kg); this hyperthyroid weight was significantly less (*P* <0.0001) than the premorbid weight (5.45 kg) recorded 1-2 years prior to diagnosis. The median pre-treatment BSC was ideal (3/5); of the 436 cats, 72 (16.5%) were overweight, 210 (48.2%) were ideal, and 154 (35.3%) were underweight. The median pre-treatment MCS was consistent with mild muscle loss (2/3); of the 436 cats, 94 (21.6%) had a normal MCS, whereas mild, moderate, and severe muscle loss was recorded in 173 (39.7%) 128 (29.4%), and 41 (9.4%), respectively.

Weak but significant (*P* <0.05) correlations were found between serum T4 and both BCS (r=0.16) and MCS (r =0.15). When the 436 cats were subdivided into 3 groups based on the magnitude of the serum T4 value (<7 µg/dl; 7-12 µg/dl; >12 µg/dl), the prevalence of both low BSC and moderate to severe muscle wasting increased with worsened severity of disease (*P*<0.05). Significant (*P*<0.0001) correlations were also found between age and both BCS (r=0.25) and MCS (r=0.23). When the 436 cats were subdivided into 3 groups based on their life stage (mature, senior, or geriatric), prevalence of low BSC and moderate to severe muscle wasting was significantly higher (*P* <0.005) in geriatric cats than the younger groups. There was no difference in median serum T4 concentrations between cats of these 3 age groups. Finally, no relationship between the diet fed (dry, canned, or both) and BCS or MSC could be identified.

Median body weight, BCS, and MCS all increased significantly (*P*<0.001) in the 60 cats re-evaluated after treatment. Of these, 21 (35%) were originally underweight: after treatment, 18 had an increase in BCS, 3 were stable, and 6 remained underweight. Of the 60 cats, 47 (78.3%) originally had low MSCs; after treatment, MCSs normalized in 16, improved but remained low in 20, and showed no change in 11.

In conclusion, most hyperthyroid cats lose body weight but maintain an ideal body condition or are overweight, with only about a third being underweight. However, loss of muscle mass (sarcopenia) is very common (over 75%). Severity of hyperthyroidism and older age both contribute to lowered BCSs and MCSs. Successful treatment leads to weight gain and increase of BSC in most cats, many times into the overweight range. Improvement in low MCSs are seen in most cats after treatment, but many cats fail to regain normal MCSs. Further studies are needed to investigate the role of nutrition (especially protein intake) in the pathogenesis of the low MCS observed in cats with hyperthyroidism.

RESPONSE TO SILDENAFIL CITRATE IN DOGS WITH PULMONARY HYPERTENSION AND PDE5A:E90K POLYMORPHISM

Joshua Stern, Yu Ueda, Lance Visser, Lynelle Johnson

Department of Medicine & Epidemiology, University of California Davis, Davis, CA, USA

In apparently healthy dogs, a polymorphism in the phosphodiesterase 5A gene (PDE5A:E90K) has been shown to alter basal levels of plasma cyclic guanosine monophosphate. Similar functional polymorphisms in humans have been demonstrated to impact response to therapy with PDE5 inhibitory drugs such as sildenafil citrate. This study aims to investigate the impact of PDE5A:E90K genotype on clinical response to sildenafil in dogs with moderate to severe pulmonary hypertension.

A prospective trial was performed in dogs with moderate to severe pulmonary hypertension associated with varied causes. All dogs underwent an echocardiogram with estimation of pulmonary pressures, and a quality of life questionnaire was completed before and after sildenafil (4mg/kg/day) for 30 days. Select echocardiographic measurements and quality of life scores were compared before and after sildenafil therapy overall and between each PDE5A:E90K genotype group (wild-type=WT, heterozygous=HET, homozygous=HO).

27 cases were enrolled, and 19 cases have completed the study. The genotype distribution was 5 WT, 8 HET, and 14 HO. Echocardiographically estimated pulmonary pressures were significantly reduced after sildenafil therapy (pre-sildenafil TR gradient 4.6 ± 0.6 m/sec; post-sildenafil 4.0 ± 0.7 m/sec; P=0.001), which was also reflected in improved quality of life scores (pre-sildenafil 36 +/- 20; post-sildenafil 14 +/- 10; P=0.0002). Non-responders were identified within each genotype group (2/5 WT; 1/5 HET, 1/9 HO).

The WT genotype of PDE5A:E90K is uncommon and thus far, was more frequently identified in dogs that failed to respond to sildenafil. Further investigation into the role of this polymorphism in diseased states is warranted.

RESPONSE TO SILDENAFIL DIFFERS IN DOGS WITH PULMONARY HYPERTENSION ASSOCIATED WITH CARDIAC AND RESPIRATORY ETIOLOGIES

Lynelle Johnson, Joshua Stern

University of California, Davis, Davis, CA, USA

Pulmonary hypertension (PH) can develop with many common cardiopulmonary diseases. This study aimed to evaluate the effect of the underlying etiology of PH on clinical signs, disease severity, and response to therapy.

This prospective trial used Doppler echocardiography to identify dogs with moderate to severe PH. Additional diagnostic tests were performed at the discretion of the attending clinician and with owner consent to detect underlying diseases. A quality of life questionnaire was completed, and sildenafil was prescribed at 4mg/kg/day for 35 days, along with other medications deemed necessary for management of the primary disease. After 30 days, dogs returned to the hospital for repeat echocardiogram and a second quality of life survey. Select echocardiographic measurements and quality of life scores were compared before and after sildenafil therapy.

Twenty-seven dogs were enrolled. Acceptable diagnostic testing and follow-up were available in 22 cases, 8 of which had PH of cardiac origin and 14 with PH of respiratory origin. Baseline parameters did not differ between groups, with the exception of greater left atrial size in dogs with cardiac-related PH. Following 1 month of therapy, there was a trend towards lesser improvement in quality of life score in dogs with underlying respiratory versus cardiac disease, ($p=0.057$) despite a greater decrease in pulmonary pressure in dogs with respiratory disease compared to those with cardiac disease ($p=0.0037$).

It is unclear why pulmonary hypertension develops in only a sub-population of dogs with common cardiorespiratory diseases. Further research is needed to predict optimal response to therapy.

A DYSBIOSIS INDEX TO ASSESS MICROBIAL CHANGES IN FECAL SAMPLES OF DOGS WITH CHRONIC ENTEROPATHY

Mustafa AlShawaqfeh[1], Bilal Wajid[2], Melissa Guard[1], Yasushi Minamoto[1], Jonathan Lidbury[1], Jörg Steiner[1], Erchin Serpedin[2], Jan Suchodolski[1]

[1]*Gastrointestinal Laboratory, Texas A&M University, College Station, TX, USA,* [2]*Department of Electrical and Computer Engineering, Texas A&M University, College Station, TX, USA*

The abnormal interaction between the gastrointestinal microbiome and the immune system is thought to be a major contributing factor to the pathogenesis of chronic enteropathies (CE) in dogs. Recent 16S rRNA gene sequencing studies have described alterations in the gut microbiota of dogs with CE compared to healthy dogs and have identified various bacterial groups that are believed to be important for intestinal health. Disadvantages of taking a sequencing based approach are the high cost, as well the turnaround time to receive sequencing results. Therefore, the aim of this study was to develop a PCR panel, which allows for rapid diagnosis of fecal dysbiosis in dogs with CE. Furthermore, a mathematical algorithm was developed to report these changes of the microbiota as a single numerical value (dysbiosis index).

Fecal DNA from 285 healthy dogs and 172 dogs with CE was initially analyzed for 13 bacterial groups using quantitative PCR (qPCR) assays. These bacterial groups evaluated had previously been identified as being significantly altered in fecal samples from dogs with chronic GI diseases using 16S rRNA gene sequencing studies. Various mathematical algorithms were tested to determine the one with the best diagnostic utility to separate healthy dogs from dogs with CE based on a combination of bacterial groups. The final dysbiosis index model was built using a nearest prototype (centroid) classifier. Initially, the model was trained with samples from 43 healthy dogs and 64 dogs with CE to identify the prototypes of the healthy and diseased classes. Then, the model was tested with an independent set consisting of 242 healthy dogs and 108 diseased dogs to assess its clinical diagnostic performance and its ability to differentiate the deviation in gut microbial communities. Receiver-operating characteristic (ROC) curve analysis was performed for sensitivity and specificity calculations.

The optimal PCR panel consisted of 6 bacterial groups: *Faecalibacterium*, *Turicibacter*, *E. coli*, *Streptococcus*, *Blautia*, and *Fusobacterium*. The degree of dysbiosis was quantified in a single numeric value, called the Dysbiosis Index (DI), that measures the closeness (in the $l_2 - norm$) of the test sample to the mean (prototype) of both disease classes (healthy and diseased). A negative DI indicates normobiosis, whereas a positive DI indicates dysbiosis. Furthermore, an increasing DI indicates stronger deviation of the sample from normobiosis. ROC curves indicated an area under the curve (95% CI) of 0.95 (0.91 – 0.98). A DI of 0 showed 84% sensitivity (95% CI: 73% - 92%) and 86% specificity (72% - 94%) for the differentiation of dysbiosis and normobiosis.

In conclusion, a qPCR based dysbiosis index was developed that allows rapid assessment of fecal dysbiosis present in dogs with CE.

UTILITY OF PARR ANALYSIS FOR IMPROVED DETECTION OF LYMPHOMA IN FELINE ENDOSCOPIC DUODENAL BIOPSIES

John Peauroi[1], Stanley Marks[2], Philip Kass[2], Catherine Tran[1], Elise Peauroi[1], Peter Moore[2]

[1]*VDx - Veterinary Diagnostics, Davis, CA, USA,* [2]*University of California, Davis, Davis, CA, USA*

The purpose of this study was to assess the sensitivity and specificity of PARR (PCR for Antigen Receptor Rearrangement) analysis performed on feline endoscopic duodenal biopsies for detection of intestinal lymphoma, and to compare the sensitivity and specificity of PARR analysis to routine histologic evaluation.

A retrospective analysis of feline endoscopic GI biopsies submitted to a commercial diagnostic laboratory was undertaken by identifying 190 sequential cases of feline endoscopic duodenal biopsy samples from the laboratory database over a five-month period. A survey was mailed to the submitting veterinarian to determine clinical progression following biopsy. Of 83 follow-up surveys received, information on clinical behavior was available for 75 cases. Median time from biopsy to survey completion was 3.7 years and median age of the cats was 11.9 years. Analysis of the formalin-fixed paraffin-embedded duodenal biopsy samples for the 75 cases was performed by PARR to assess for clonality in TRG (T cell receptor gamma).

PARR analysis found 51/75 cases (68%) with clonal rearrangements for TRG, 16 cases (21%) with polyclonal rearrangements, and 8 cases (11%) exhibiting minor clones (equivocal results). Histologic evaluation found 25/75 cases (33%) with histologically confirmed small cell lymphoma, 10 additional cases (13%) with findings suggestive of, but not conclusive (equivocal) for small cell lymphoma, and 40 cases (53%) with findings neither consistent with nor suggestive of lymphoma (ie normal or inflammatory changes).

There was a statistically significant association between histologic diagnosis and clinical behavior ($P<0.0001$), PARR status and clinical behavior ($P<0.0001$) and histologic diagnosis and PARR status ($P=0.0078$).

To determine the sensitivity of PARR analysis for the diagnosis of GI lymphoma, the frequency of clonality for TRG was analyzed on the subset of cases with histologically confirmed lymphoma (n=25). PARR analysis on this subset found 22 of 25 cases clonal for TRG, for an overall sensitivity of 88% (95% CI 69-97%). One case had polyclonal results, and 2 cases had equivocal results (minor clones). Specificity was determined to be 93% (95% CI 82-99%), based on 43 of 46 cases in which clonality was demonstrated exhibiting either histologically confirmed lymphoma, or clinically progressive and aggressive GI signs. Three of 19 cases (16%) with indolent disease and without histologically evident lymphoma had clonal results, representing either cases of histologically undiagnosed lymphoma controlled by treatment, cases of early emerging lymphoma, or potential false positive results.

The sensitivity of histologic evaluation for the diagnosis of GI lymphoma was determined by assessing the frequency of histologically diagnosed lymphoma among the subset of 36 cases exhibiting progressive and aggressive GI signs over the follow-up period. Lymphoma was diagnosed histologically in 16 of these cases, indicating a sensitivity of 44% (95% CI 28-62%). An additional 5 cases (14%) had findings suggestive of, but not definitively diagnostic for, lymphoma, while 15 cases (42%) had histologic findings consistent with normal or inflammatory disease, suggesting a high incidence of false negative diagnosis by histology alone. The specificity of a histologic diagnosis of lymphoma could not be calculated as it was impossible to determine the true disease state of the subset of cases (9 of 25) in which lymphoma was diagnosed histologically and which exhibited indolent clinical behavior.

Of 51 cases where clonality was demonstrated, only 22 (43%) had a histologically definitive diagnosis of lymphoma, while 9 of 10 cases (90%) where lymphoma was suspected but not confirmed histologically were clonal and 20 of 40 cases (50%) where lymphoma was neither identified nor suspected histologically were clonal.

Results of this study show that histologic evaluation of endoscopically derived duodenal biopsy specimens has a low sensitivity for the diagnosis of small cell GI lymphoma and that PARR analysis for TRG performed on these specimens has a markedly superior sensitivity to histology alone, greatly increasing the detection of feline intestinal lymphoma. Furthermore, clonal rearrangements are highly specific as a predictor of clinically aggressive and progressive GI disease. PARR analysis of formalin-fixed paraffin-embedded feline endoscopic duodenal biopsy samples, when used as an adjunct to histologic evaluation, dramatically improves the sensitivity of detection of lymphoma and is able to identify cats likely to exhibit progressive and aggressive GI signs (ie GI lymphoma) among cats without histologic evidence of neoplasia.

THREE DIVERSE MUTATIONS UNDERLYING CANINE XANTHINE UROLITHIASIS

Eva Furrow, Nicole Tate, Katie Minor, James Mickelson, Kasey Peterson, Jody Lulich

University of Minnesota, College of Veterinary Medicine, Saint Paul, MN, USA

Hereditary xanthinuria in people is an autosomal recessive disease caused by mutations in xanthine dehydrogenase (*XDH*) or molybdenum cofactor sulfurase (*MOCOS*). There are rare reports of hereditary xanthinuria in dogs, but genetic investigations have not previously been described. The objective of this study was to identify mutations underlying risk for canine xanthine urolithiasis by sequencing *XDH* and *MOCOS* in affected dogs.

Five dogs with primary xanthine urolithiasis (i.e. no history of xanthine dehydrogenase inhibitor therapy) were recruited. This group included 2 Cavalier King Charles Spaniels (CKCS), 2 Toy Manchester Terriers (TMT), and a mixed-breed dog. Three homozygous mutations were identified. The CKCSs had a mutation resulting in a premature stop codon in *MOCOS*, the TMTs had a splice site mutation in *MOCOS*, and the mixed breed dog had a splice site mutation in *XDH*. cDNA sequencing confirmed exon skipping for the splice site mutations.

Mutation-specific assays were developed to genotype 108 CKCSs and 50 TMTs without a history of urolithiasis. Of the 108 CKCS, 105 were clear, 3 were carriers, and none were homozygous for the CKCS mutation. Of the 49 TMTs, 37 were clear, 10 were carriers, and 2 were homozygous for the TMT mutation. Urine was analyzed from the 2 homozygous TMTs and revealed xanthinuria.

In conclusion, diverse mutations underlie canine hereditary xanthinuria. Genetic testing can help inform breeders and identify dogs that may benefit from preventative therapies. Future studies are needed to determine mutation frequencies in other breeds and whether clinical outcome differs between the mutations.

URINARY MICROBIOTA IN HEALTHY DOGS

Erin Burton, Leah Cohn, Carol Reinero, Hans Rindt, Aaron Ericsson

University of Missouri, Columbia, MO, USA

Areas of the body previously believed to be sterile during health actually possess a microbiota that may protect from pathogenic infection. The urinary microbiota of healthy dogs has not been previously described. As urinary tract infections (UTI) are common in dogs, an understanding of "normal" urinary microbiota may prove helpful in prevention or treatment of UTI. We sought to describe the urinary microbiota of healthy dogs, with comparison to reproductive and rectal microbiota.

Dogs >15 kg and 1 to 10 years of age with no signs of urinary disease and that had not received antibiotics, probiotics, or corticosteroids within the prior 30 days were sampled with informed owner consent. Cystocentesis urine samples, rectal, and vaginal or preputial swabs were collected. Aliquots of urine were used for routine urinalysis and urine culture, and dogs were excluded from further evaluation if pyuria, bacturia, or bacterial growth were detected. DNA was extracted from urine and swab samples and subject to 16S rRNA sequencing. Assembly, binning, and annotation of DNA sequences was performed at the MU Informatics Research Core Facility. Taxonomy was assigned to operational taxonomic units (OTUs) using BLAST against the Greengenes database.

After excluding 4 dogs with cultivable bacteria, samples from 10 male (M; 1 intact) and 10 female (F) spayed dogs remained. All samples provided adequate genetic material for analysis. Four taxa (*Pseudomonas* sp, *Acinetobacter* sp, *Sphingobium* sp and *Sphingomonas* sp, and *Bradyrhizobiaceae* sp) dominated the urinary microbiota in all dogs of both sexes. These taxa were also detected in the genital swabs of both sexes, while the rectal microbiota differed substantially from the other sample sites (figure). Principal component (PC) analysis of PC1 through PC3 showed overlap of urinary and genital microbiota and a clear separation of rectal swabs from the other sample sites along PC1, which explained 44.94% variation. Surprisingly, the urinary microbiota (mean # OTU 92.6 F, 90.2M) was significantly richer than the genital (67.8 F, 66.6M) or rectal swabs (68.3F, 71.2M) ($p = 0.001$, 2-way ANOVA), with no difference between sexes at any sample site.

Healthy dogs have a rich urinary microbiota, with little difference between sexes but marked differences from the rectal microbiota.

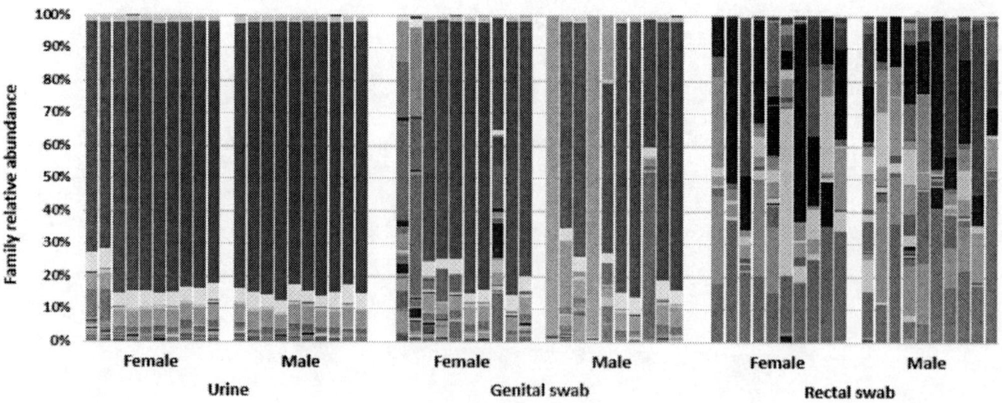

VITAMIN D METABOLITES, PARATHYROID HORMONE AND FIBROBLAST GROWTH FACTOR-23 IN CANINE CHRONIC KIDNEY DISEASE

Valerie Parker, Laura Harjes, Katarzyna Dembek, Gregory Young, Dennis Chew, Ramiro Toribio

The Ohio State University, Columbus, OH, USA

In people with chronic kidney disease (CKD), hypovitaminosis D is associated with disease progression and development of renal secondary hyperparathyroidism (RHPT). Little is known about vitamin D metabolism in dogs with CKD. The purpose of this study was to evaluate vitamin D metabolites in dogs with different International Renal Interest Society (IRIS) stages of CKD and to determine their association with factors that contribute to renal vitamin D activation and development of RHPT [parathyroid hormone (PTH) and fibroblast growth factor-23 (FGF-23)].

Thirty-seven CKD dogs and 10 healthy dogs were included. Vitamin D metabolites, including 25-hydroxyvitamin D [25(OH)D], 1,25-dihydroxyvitamin D [1,25(OH)$_2$D], and 24,25-dihydroxyvitamin D [24,25(OH)$_2$D] were measured. All vitamin D metabolites were decreased in CKD dogs compared to healthy dogs, reaching statistical significance at stage 3 CKD ($p<0.05$), and were negatively correlated with creatinine concentrations ($p<0.05$). Values are presented as medians (ranges). In healthy dogs and dogs with IRIS stages 3 and 4 CKD, serum 25(OH)D concentrations were 75.1 ng/ml (50.4-97.9), 42.7 (3.5-95.8), and 25.0 (5.7-52.5), respectively. Serum 1,25(OH)$_2$D concentrations were 209.6 pg/ml (168.9-428.0), 104.8 (29.2-228.7), and 64.7 (19.0-91.1). Serum 24,25(OH)$_2$D concentrations were 38.7 ng/ml (24.0-89.5), 10.3 (0.3-42.1), and 7.0 (2.0-16.4). In CKD dogs, serum PTH and plasma FGF-23 concentrations were 2.5 pmol/L (0.8-229.0) and 467 pg/ml (122-41,265), respectively. All vitamin D metabolites were negatively correlated with PTH ($p<0.01$) and FGF-23 ($p<0.01$) concentrations.

Vitamin D metabolism dysregulation in canine CKD is best characterized by comprehensive vitamin D metabolite measurements. Therapeutic strategies to increase vitamin D metabolite concentrations should be considered.

ASSOCIATION BETWEEN VITAMIN D METABOLITES AND PROTEINURIA IN DOGS

Valerie Parker, Chen Gilor, Adam Rudinsky, Gregory Young, Dennis Chew, Ramiro Toribio

The Ohio State University, Columbus, OH, USA

Proteinuric kidney disease causes significant morbidity and mortality in dogs. In people with proteinuria, hypovitaminosis D is an independent predictor of disease progression and death, and vitamin D supplementation improves proteinuria and survival. There is limited data regarding vitamin D status in dogs with proteinuric kidney disease.

Forty dogs with proteinuria of renal origin were included [urine protein:creatinine (UPC) median was 2.1; range, 0.2-27.5]. Dogs were excluded if extra-renal causes of proteinuria or concurrent diseases that would affect vitamin D metabolism were documented. Vitamin D metabolites (25-hydroxyvitamin D [25(OH)D], 1,25-dihydroxyvitamin D [$1,25(OH)_2D$], and 24,25-dihydroxyvitamin D [$24,25(OH)_2D$]) were measured. Spearman correlations and multivariate analysis were performed. Values are reported as medians (ranges).

In this cohort, blood pressure was 155 mmHg (120-240), serum creatinine concentration was 1.4 mg/dl (0.5-12.9) and serum albumin concentration was 3.4 g/dl (0.9-4.3). Serum 25(OH)D concentrations [39.9 ng/ml (2.5-95.8)] were lower than previously reported for healthy dogs [75.1 ng/ml (50.4-97.9)]. Serum $1,25(OH)_2D$ and $24,25(OH)_2D$ concentrations were lower than reported in healthy dogs. Serum 25(OH)D concentrations were negatively correlated with UPC ($rs= -0.61$; $p<0.001$) and positively correlated with albumin ($rs=0.55$; $p=0.0003$). After adjusting for creatinine and albumin in a multivariable regression model, UPC remained a significant predictor of 25(OH)D concentrations ($p=0.0011$)

Hypovitaminosis D was associated with proteinuria, although the exact mechanism for this relationship is unclear. Given the known effects of vitamin D on the renin-angiotensin-aldosterone system and glomerular function, further consideration should be given to vitamin D metabolites as potential markers of renal disease and therapeutic intervention.

SHORT TELOMERES ARE ASSOCIATED WITH FELINE CHRONIC KIDNEY DISEASE AND HYPERTENSION

Jessica Quimby[1], Andrea Herndon[1], David Maranon[1], Sabina Ligas[1], Katharine Lunn[3], Jonathan Elliott[2], Jack Lawson[2], Rachel Cianciolo[4], Ann Hess[1], Susan Bailey[1]

[1]Colorado State University, Fort Collins, CO, USA, [2]Royal Veterinary College, London, England, UK, [3]North Carolina State University, Raleigh, NC, USA, [4]The Ohio State University, Columbus, OH, USA

Telomeres are protective structures at the ends of chromosomes, maintenance of which has important implications for aging and age-related pathologies. The purpose of the current study was to explore the relationship between telomere length and age, chronic kidney disease (CKD), hypertension and proteinuria in cats.

Telomere length was assessed in kidney tissue from 62 cats with CKD substaged for hypertension and proteinuria (IRIS Stage 2:16 cats, IRIS Stage 3: 20 cats, IRIS Stage 4: 26 cats), and 40 non-CKD cats (0 – 15+ years) utilizing telomere fluorescence in situ hybridization combined with immunohistochemistry to specifically identify proximal tubular epithelial cells.

Statistical analysis of mean telomere fluorescence intensity revealed no significant differences in telomere length, or in the percentage of short telomeres, with age. However, telomeres were significantly shortened in CKD cats as compared to normal cats, whether young (<10 years; p=0.006) or geriatric (>10 years; p=0.04); the percent short telomeres was also significantly higher in CKD cats compared to normal young cats (p=0.04). Within CKD cats themselves, there were no significant differences in mean telomere length or percent short telomeres with IRIS stage or proteinuria. Interestingly, CKD cats with hypertension displayed significantly shorter telomeres (p=0.01), as well as a trend towards higher percentages of short telomeres (p=0.06), compared to CKD cats without hypertension.

These results support our previous observations and further, suggest that telomere shortening likely occurs as a consequence of CKD, rather than being directly involved in the pathogenesis of progressive CKD.

REGENERATIVE MEDICINE APPROACH TO THE TREATMENT OF URINARY INCONTINENCE IN FEMALE DOGS

Shelly Vaden[1], Kyle Mathews[1], Koudy Williams[2], Hannah Reynolds[1], Katherine Gleason[1], Tonya Harris[1], Kristin Manning[1], Jorge Piedrahita[1]

[1]North Carolina State University, Raleigh, NC, USA, [2]Wake Forest Institute for Regenerative Medicine, Winston-Salem, NC, USA

Urinary incontinence (UI) affects more than 20% of spayed female dogs, the majority of which have dysfunction of the urethral sphincter. Those dogs that fail to respond to medical management (i.e., phenylpropanolamine, estrogen) for UI can be very difficult to treat, requiring repeat urethral injections of bulking agents or surgical implantation of a urethral occluding device. Recent studies have assessed the feasibility of using muscle-derived progenitor cell (MPC) injections for the management of UI in women. We previously demonstrated dogs with induced urethral sphincter mechanism incontinence have functional recovery after injection of autologous MPCs. The purpose of this study was to determine if urethral injection of autologous MPCs in dogs with naturally occurring UI leads to restoration of urinary continence.

Dogs were included in the study if they had UI that was presumed to be due to urethral sphincter mechanism incompetence and unresponsive to traditional medical management. Dogs were excluded if they had an anatomic abnormality causing their UI (e.g., ectopic ureter) or unresolved urinary tract infection. UI was quantified prior to and 3, 6, 12 and 24 months following injection by use of an incontinence questionnaire (1=never continent, 5=always continent). In addition, maximal urethral pressure (MUP, average of 3 recordings) was measured prior to and at 6, 12 and 24 months after injection. Approximately 14 days prior to injection, a muscle biopsy was obtained from the left triceps of the anesthetized patient. Myoblasts were isolated from biopsy by mincing followed by enzymatic digestion with collagenase type I, then plated on matrigel-coated plates for initial expansion in myogenic media. Cells were passaged 2-3 times onto non-coated, tissue-culture treated plates until at least 100 million cells were present. After harvesting with trypsin and myogenic media, cells were counted and resuspended in phenol-free DMEM. For injection, 100 million cells were put into an additive-free serum collection tube and resuspended in 0.6 mL phenol-free DMEM. Cells were combined with 0.4 mL of the patient's plasma and then injected via a surgical procedure into the urethral submucosal layers just distal to the trigone.

7 spayed female dogs (mean age 4.9 years) have been injected with MPCs; 2, 3 and 1 are past the 6, 12 and 24 month post-injection time point, respectively. Median continence scores were as follows: baseline, 3; 3 months post, 4.5; 6 months post, 5; 12 months post, 5. 2 dogs no longer require additional medical management to maintain continence while the remaining dogs require either phenylpropanolamine alone (2 dogs) or in combination with estrogen (1 dog). One dog is deceased and had not improved at the time of her death. Of the 5 dogs that had MUP measured at baseline and 6 months post injection, 3 dogs had an increase, 1 had no change and 1 had a decrease. These preliminary results suggest that urethral injection of autologous MPCs can lead to improvement or resolution of UI in female dogs with presumed urethral sphincter mechanism incompetence.

DEVELOPMENT OF BIOMARKER ASSAYS FOR THE PHARMACODYNAMIC EVALUATION OF MYCOPHENOLATE MOFETIL IN THE DOG

Andrew Mackin, Charlee Mulligan, Lakshmi Narayanan, Kari Lunsford, Claire Fellman, John Thomason, Ben Lee, Todd Archer

Mississippi State University College of Veterinary Medicine, Starkville, MS, USA

Mycophenolate mofetil (MMF) is a pro-drug for mycophenolic acid (MPA), which inhibits lymphocyte purine synthesis. MMF is considered to be a potent immunosuppressive agent in human medicine, associated with inhibition of B and T cell proliferation, and of T cell function. The immunosuppressive effects of MMF in dogs have not been well established. Our study was designed to develop and evaluate a panel of potential biomarkers that could be utilized as pharmacodynamic assays for measuring the effects of MMF on the canine immune system.

Biomarker assays evaluated were T and B cell lymphocyte proliferation (LP), T cell ATP production, T cell expression of the CD25 surface marker and CD4+CD25+Foxp3+ regulatory T cell (T-reg) numbers. Newly developed assays were evaluated in an *in vitro* study in samples from normal dogs, with PBMC incubated in different MPA concentrations, as well as the DMSO carrier and a no drug/carrier control. The new assays, as well as previously established assays, were then used in an *in vivo* study in 7 normal dogs before and on the 7th day of oral MMF at 10 mg/kg PO BID. For LP, PBMCs were activated with pokeweed mitogen (B cell), or concanavalin A (T cell), and proliferation was evaluated by measuring incorporation of tritiated thymidine. For T cell ATP production, cells were activated with concanavalin A and, after 1 day of incubation post-activation, incubated with anti-dog CD4, and then anti-dog immunoglobulin magnetic beads were used to extract CD4+ T cells. Extracted cells were lysed, and ATP was measured by luminescence. T cell CD25 expression and T-reg numbers were measured using flow cytometric methods previously developed and published by our laboratory.

In vitro, there was a significant decrease in T cell proliferation and T cell ATP following exposure to MPA. LP (T cell only) after exposure of PBMCs to high MPA concentrations (5, 10, 20, 40, 80, 160, 320 & 640 µg/mL) revealed marked suppression at all MPA levels (control 133,773, DMSO 48,722, compared to MPA 145-260 [average for all concentrations 200] mean counts/minute). T cell ATP after exposure of PBMCs to lower levels of MPA revealed a concentration-dependent decrease: control 1,819, DMSO 1,480, and MPA 1,229, 1,048, 1,365, 947, 741, 613, 936 & 846 mean fluorescent units at 10, 20, 40, 80, 160, 320, 640 & 1,280 ng/mL respectively.

Oral MMF had no significant effect on B or T cell LP, T cell ATP, T cell expression of CD25, and T-reg numbers. For B cell proliferation, pre-treatment LP (mean counts/minute ± SEM) was 25,926 ± 3,859, and post-treatment was 30,654 ± 5,499. For T cell proliferation, pre-treatment LP was 57,930 ± 5,760, and post-treatment was 68,360 ± 9,182. For T cell ATP production, pre-treatment ATP (mean fluorescent units ± SEM) was 374 ± 57, and post-treatment ATP was 1,033 ± 191. For flow cytometry, CD25 single positive T cells were 0.32 ± 0.21% gated before and 2.96 ± 1.83% after treatment, and T-regs were 0.06 ± 0.02% before and 0.02 ± 0.01% after.

The results of our study indicate that, although *in vitro* exposure to MPA has a significant suppressive effect on measures of canine lymphocyte function, oral administration of MMF to dogs at the low end of the standard dose range has no detectable suppressive effect on the immune system. Further studies at higher drug doses, or utilizing more sensitive assays, may be needed to determine if MMF is truly likely to be an effective immunosuppressive agent in dogs.

EFFECTS OF IMMUNOSUPPRESSIVE DRUG THERAPY ON CANINE ACTIVATED WHOLE BLOOD EXPRESSION OF INTERLEUKIN-2 AND INTERFERON-γ

Claire Fellman, Lakshmi Narayanan, Caitlin Riggs, Charlee Mulligan, Kari Lunsford, John Thomason, Andrew Mackin, Todd Archer

Mississippi State University, Mississippi State, MS, USA

Immunosuppressive drugs work through a variety of mechanisms to reduce adaptive and innate immune responses. T-cells are an important mediator of inflammatory and autoimmune diseases. Previous investigations by our laboratory have evaluated expression of IL-2 and IFN-γ after *in vitro* and *in vivo* exposure to cyclosporine, and have found that these T cell-associated cytokines are valuable markers of cyclosporine's effect in dogs. Since other immunosuppressive medications are frequently co-administered with cyclosporine but have other mechanisms of action, it is important to determine what effect they may have on T-cell cytokine production.

This study evaluated IL-2 and IFN-γ expression in 3 dogs treated with 5 oral immunosuppressive drugs at standard doses using a crossover study design. Cyclosporine (10 mg/kg every 12 hrs), mycophenolate mofetil (10 mg/kg every 12 hrs), prednisone (2 mg/kg every 24 hrs), azathioprine (2 mg/kg every 24 hrs), and leflunomide (4 mg/kg every 24 hrs) were given orally for 7 days, with a minimum of a 3 week washout period between each drug. Whole blood gene expression of IL-2 and IFN-γ was evaluated before treatment and again on Day 7 of drug dosing using quantitative reverse transcription polymerase chain reaction (qRT-PCR). Samples were activated for 5 hours with phorbol 12-myristate 13-acetate and ionomycin, RNA was extracted, and relative gene quantification was performed comparing gene expression after drug administration to pre-treatment levels.

Suppression of both IL-2 and IFN-γ was most pronounced for cyclosporine, followed by much milder suppression for prednisone and azathioprine. Leflunomide effects were less consistent, with 2 dogs showing decreased and 1 dog increased cytokine expression. After mycophenolate mofetil, all 3 dogs had increased cytokine gene expression relative to pre-treatment levels.

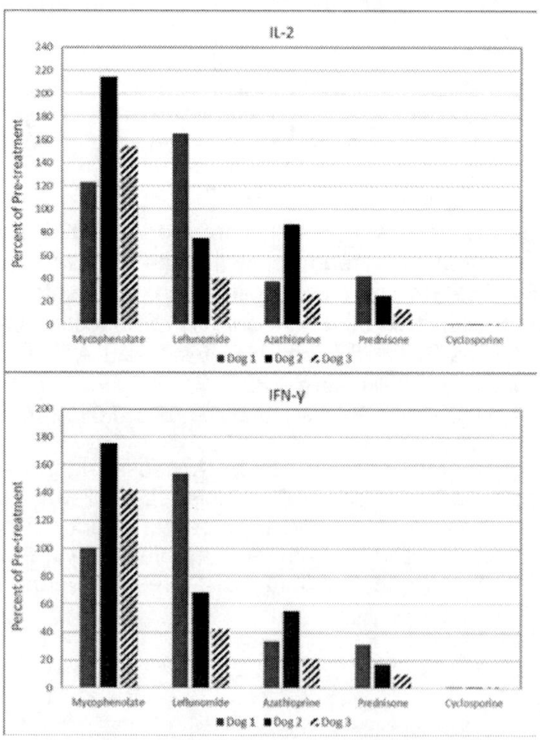

Our results suggest that the effects of immunosuppressive drugs on gene expression of nuclear factor of activated T cell-regulated cytokines such as IL-2 and IFN-γ are highly drug specific, and that the effects of other non-calcineurin inhibitor immunosuppressive agents are relatively

mild compared to cyclosporine. Further investigation is warranted with co-administered drugs and lower doses of cyclosporine to refine intepretation of biomarker assays in clinic patients.

SUBCUTANEOUS URETERAL BYPASS DEVICE PLACEMENT FOR BENIGN URETERAL OBSTRUCTION IN CATS: 137 CATS (174 URETERS)

Allyson Berent[1], Chick Weisse[1], Demetrius Bagley[2]

[1]*The Animal Medical Center, New York, NY, USA,* [2]*Thomas Jefferson University, Philadelphia, PA, USA*

SUB device placement has been used as an alternative to traditional ureteral surgery, however short and long-term outcomes have not been described.

The objective is to evaluate pre-operative, peri-operative(< 7 days), short(7-30 days) and long-term(> 30 days) parameters in cats treated with a SUB for benign ureteral obstruction(s). The hypothesis is that SUBs associated with a favorable outcome compared with alternatives.

The SUB was placed using fluoroscopic- and surgical-assistance creating an artificial ureter for ureteral diversion. Medical records were reviewed for pre-, intra-, and post-operative data.

174 SUBs were placed in 137 cats (20% bilateral). The cause of obstruction was: ureterolithiasis(67%), stricture(13%), both(20%), or pyonephrosis(0.5%). Placement was successful in all cases. 96.4% of cats were azotemic at presentation. The median creatinine at presentation and 3 month post-SUB was 6.6 and 2.6mg/dL, respectively. The median renal pelvis diameter pre- and post-procedure was 9.1mm and 1.5mm, respectively. Peri-operative complications included device leakage(3.4%), kinking(5%), and occlusion with blood clot(s)(7.5%). 93.7% survived to discharge. Catheter mineralization occurred in 25% of systems (median 364 days), with only13% required exchange due to re-obstruction. Dysuria was reported in 23% of pre-SUB and 8.2% at any time thereafter. The overall MST was 827 days (range,1-2397) and 65% eventually died of renal-related disease. Limitations are related to the retrospective nature of this study.

The SUB device can be considered a treatment option for feline ureteral obstruction(s), regardless of cause, location, or stone number. The prognosis after SUB placement could be considered good to excellent.

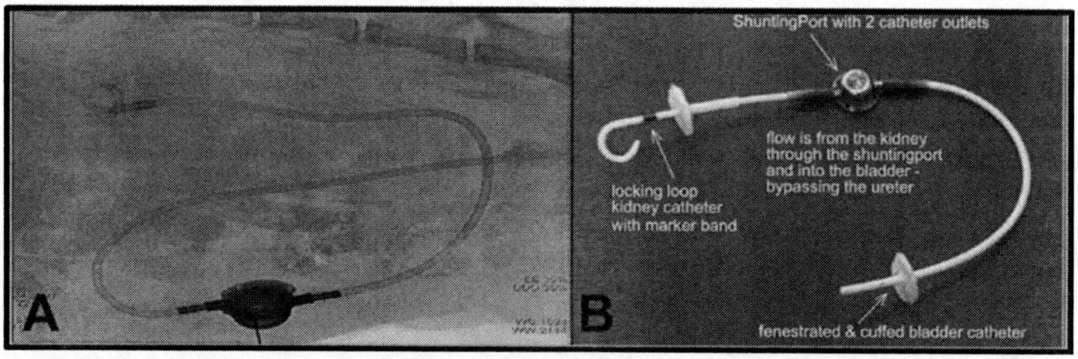

ALICAM AND GASTROINTESTINAL DISEASE IN DOGS

PART A: UTILITY OF ALICAM IN THE IDENTIFICATION AND LOCALIZATION OF GASTROINTESTINAL BLEEDING IN DOGS

Jill Pomrantz[1], Jeffrey Solomon[1], Jonathan Lidbury[2]

[1]Infiniti Medical, LCC, Menlo Park, CA, USA, [2]Texas A&M University, College Station, TX, USA

Identifying the presence and source of gastrointestinal bleeding (GIB) in dogs can be problematic, but it is essential in selecting an appropriate therapy. This retrospective study assessed the ability of ALICAM, an ingestible capsule imaging system, to identify the presence of GIB, determine its location and characterize underlying lesions.

Twelve dogs were administered ALICAM for overt GIB, unexplained anemia and/or microcytosis. A board-certified internist evaluated the studies for the presence of intraluminal blood and the ability of the study to localize the source of bleeding and characterize underlying lesions.

Eleven of 12 (92%) dogs were anemic (5 regenerative, 6 nonregenerative). Three of 12 (25%) dogs had a microcytosis. Eight of 12 (67%) dogs had overt GIB (hematemesis, melena and/or hematochezia). Eight of 12 (67%) dogs had received glucocorticoids or nonsteroidal anti-inflammatories in the preceding month. The gastrointestinal tract was sonographically normal in 6 of 6 (100%) dogs. ALICAM identified intraluminal blood in 12/12 (100%) examinations. The study identified the location of bleeding in 12/12 (100%) dogs. Three dogs had multiple sources (gastric polyp with gastric erosions, small intestinal nodules with colonic erosions, gastric erosions with colonic mass). One dog had a solitary jejunal mass. Six dogs had gastric or small intestinal erosions or ulcers. Underlying lesions were not visualized in two cases, likely due to overlying hematomas.

ALICAM is useful in detecting GIB in dogs and identifying the location and type of lesion responsible. Its role in the work up of anemia and the treatment of GIB warrants further investigation.

ALICAM AND GASTROINTESTINAL DISEASE IN DOGS

PART B: VISUALIZATION OF THE DUODENAL PAPILLA USING ALICAM COMPARED TO CONVENTIONAL DUODENOSCOPY IN DOGS

Jill Pomrantz[1], Jeffrey Solomon[1], Jonathan Lidbury[2]

[1]*Infiniti Medical, LLC, Menlo Park, CA, USA,* [2]*Texas A&M University, College Station, TX, USA*

Anecdotally, the duodenal papilla (DP) is difficult to identify with conventional, front-view duodenoscopy. A side-view endoscope can improve detection, but is not routinely available. ALICAM is an ingestible capsule imaging system that utilizes 4 side-view cameras to record intraluminal images. The objective of this study was to calculate the rate of DP visualization in dogs using ALICAM and compare that to conventional duodenoscopy.

The rate of DP visualization (major or minor DP) was retrospectively assessed in 66 ALICAM studies consisting of 15 healthy and 51 symptomatic dogs evaluated for gastrointestinal disease (Group 1) and prospectively in 12 dogs undergoing gastroduodenoscopy for gastrointestinal disease (Group 2). ALICAM interpretations and endoscopic procedures were performed by board-certified internists.

Group 2 was slightly older than Group 1 (mean±SE years; 9.42±1.07 vs. 7.14±0.50, p=0.07) and weighed significantly less (mean±SE kg; 13.90±3.63 vs. 25.17±1.40, p=0.003), however, neither age (p=0.44) nor weight (p=0.94) was independently associated with DP visualization. The DP visualization rate was significantly greater with ALICAM (45/66; 68.2%) compared to endoscopy (4/12; 33.3%). Dogs that had ALICAM were 4.3 times more likely to have their DP visualized (p=0.024). Healthy dogs had a higher DP visualization rate (13/15; 86.7%) compared to symptomatic dogs (36/63; 57.1%, odds ratio [OR]=4.9, p=0.024). In symptomatic dogs, the DP visualization rate remained greater with ALICAM (32/51; 62.7%) compared to endoscopy (4/12; 33.3%) and approached significance (OR=3.4, p=0.06).

Overall, ALICAM detected the DP at a higher rate than endoscopy. The clinical implications of this are not known, but warrant future studies.

ALICAM AND GASTROINTESTINAL DISEASE IN DOGS

PART C: ALICAM FINDINGS IN DOGS WITH GASTROINTESTINAL SIGNS AND A NORMAL GASTROINTESTINAL TRACT ON ULTRASOUND

Jill Pomrantz, Jeffrey Solomon

Infiniti Medical, LLC, Menlo Park, CA, USA

The purpose of this study was to evaluate the ability of ALICAM, an ingestible capsule imaging system, to detect abnormalities in dogs with gastrointestinal signs and a normal ultrasound.

Fourteen dogs with gastrointestinal signs and a normal gastrointestinal tract on ultrasound were given ALICAM. Board-certified internists interpreted ALICAM studies and radiologists performed or interpreted ultrasounds.

The median age was 7.9 years (0.875-10.9) and median weight was 30 kg (6.2-40). Presenting signs included a combination of vomiting (7/14), hematemesis (2/14), diarrhea (7/14), melena (2/14), anorexia (4/14), weight loss (6/14), anemia (3/14) and microcytosis (1/14). Signs were present from 24 hours to 2 years. ALICAM identified abnormalities that could localize and/or explain presenting signs in 13/14 dogs. Findings included diffusely dilated lacteals (2/14) in dogs with weight loss and diarrhea, bleeding jejunal mass (1/14) in a dog with anemia and weight loss, large gastric hematoma (1/14) in a dog with anemia, hematemesis and melena, findings consistent with inflammatory bowel disease (6/14), including moderately to severely irregular gastric and small intestinal mucosa with erosions in dogs with weight loss, vomiting and/or diarrhea, gastric erosions (2/14) in dogs with anemia and melena or microcytosis and markedly delayed gastric emptying with erosions (1/14) in a dog with subacute anorexia. Delayed gastric emptying was also seen in 4 other previously aforementioned dogs with either vomiting or moderate to severe gastric mucosal abnormalities. One study was inconclusive.

The results show that ALICAM can identify abnormalities undetected by ultrasound and may facilitate the diagnosis of gastrointestinal disease.

NEUTROPHIL GELATINASE-ASSOCIATED LIPOCALIN URINARY CONCENTRATION IN HEALTHY AND ILL DOGS – NEW PROPOSAL FOR THE INTERPRETATION

Marileda Carvalho, Marcelo Alves, Amanda de Vasconcellos

Universidade Estadual Paulista, Jaboticabal, São Paulo, Brazil

The neutrophil gelatinase-associated lipocalin (NGAL), a biomarker of kidney disease, has been considered an important tool, mainly for early diagnoses of acute kidney injury (AKI). A prospective study was performed in dogs referred to a Veterinary Nephrology-Urology Service, because of urinary tract sign complaint or checkup request (pre-anesthesia or urinary health status certification). The aim of the study was to evaluate the urine NGAL (uNGAL) as a tool for a further comprehensive understanding of chronic kidney disease (CKD). The study was approved by the Local Ethics Committee and informed owner's consent was obtained. The patient's inclusion criteria had no restriction related to gender, age, breed or possible extra urinary comorbidity. The samples for uNGAL determination and overall patient's evaluation were obtained at the consultation day, prior to any medication, except when rehydration was required. In addition to routine required lab work, fractions of the urine supernatants were storage at -80ºC for later NGAL analysis (Dog NGAL ELISA Kit 043, Bioporto). Any complementary evaluations were made within 72 hours from admission. As referred patients, those with extra-urinary comorbidities, already had documented diagnoses. Out of the 102 patients enrolled in the study, 31 were excluded. The reasons for exclusion were cystitis with no kidney disease, clinically stablished AKI, inconclusive diagnoses or CKD risk. From the remaining 71 dogs, 17 (10 males, 7 females) were healthy, and 54 (24 males, 30 female) had kidney disease (IRIS CKD stage: 6 CKD 1; 11 CKD 2; 26 CKD 3; 11 CKD 4). The results of healthy dogs were included as control data. For the purpose of experiment, the values of uNGAL (pg/mL) were mathematically transformed (1) by being divided per creatinine (Cr) concentration of the same urine specimen (uNGAL/uCr), as it is in current usage, and (2) by being multiplied by the urine concentration factor (UCF), obtained, for each patient, by the urineCr-to-serumCr ratio, (uNGAL.UCF), as a new proposal. The results are the following.

parameter	dogs	Median	Lower CI	Upper CI	Actual CL
uNGAL/uCr (pg/mg)	healthy	3.71	2.05	9.15	95.10%
	CKD	461	323	692	95.98%
uNGAL.UCF (pg/mL)	healthy	1800	921.6	5282	95.10%
	CKD	7149	4460	11575	95.98%

CI: confidence limit; CL: confidence level.

Plotting uNGAL.UCF values and respective serum Cr levels shows a profile particularly related to the clinical condition of each patient. It was possible to connect the highest uNGAL.UCF values with the renal disease severity or ongoing comorbidities known to be harmful to the kidney (CKD 1 and 2, mainly). Additionally, the lowest values were found to be related to advanced renal failure (CKD 4 and 76% of the CKD 3 patients), highlighting the loss of nephrons. However, for the uNGAL/uCr, an oscillatory pattern, apparently not related to the CKD stage or comorbidities, is observed. No significant correlation was found between uNGAL.UCF and serum NGAL, except for the, not expressive, CKD 3 result (r^2 0.69; $p < .0001$). Save for the need of corroborating studies with similar dog population, our results indicate the uNGAL.UCF as a better tool to clarify the way we see the renal disease processes in practice, and as an additional method for CKD physiopathology investigation.

EFFECTS OF IMMUNOSUPPRESSIVE AGENTS ON THE HEMOSTATIC SYSTEM IN DOGS

John Thomason, Todd Archer, Robert Wills, Andrew Mackin

Mississippi State University College of Veterinary Medicine, Mississippi State, Mississippi, USA

Immune-mediated hemolytic anemia (IMHA) is a common condition in dogs, and is associated with a high mortality rate, with death often due to thromboembolic complications such as pulmonary thromboembolism. Cyclosporine has become popular for the treatment of IMHA, however, at immunosuppressive doses in normal dogs, cyclosporine has been shown to cause a marked increase in the synthesis of thromboxane A_2, a potent vasoconstrictor and platelet activator. Fortunately, there are other immunosuppressive agents that can be used instead of cyclosporine in dogs. However, the effects of these immunosuppressive agents on the hemostatic system have not been evaluated, and it is therefore uncertain whether these drugs might actually be potentially safer than cyclosporine in hypercoagulable dogs. The objective of this study was to determine the effects of a range of immunosuppressive agents on a comprehensive panel that evaluates primary and secondary hemostasis in dogs.

In a cross-over study design, eight healthy adult dogs were given either prednisone (2 mg/kg SID, PO), azathioprine (2 mg/kg SID, PO), cyclosporine (10 mg/kg BID, PO), mycophenolate mofetil (10 mg/kg, BID, PO), or leflunomide (4 mg/kg, SID, PO) for 7 days, with a washout period of at least three weeks between each cross-over. Platelet function and number were evaluated both prior to (Day 0) and during (Day 7) therapy using turbidimetric aggregometry (collagen and ADP as agonists), PFA-100® closure time (collagen/epinephrine cartridge), and platelet count. Clot formation was evaluated via viscoelastometry, prothrombin time (PT) and activated partial thromboplastin time (PTT). Urinary levels of 11-dehydro-thromboxane B_2 (11-$dTXB_2$, a stable thromboxane A_2 metabolite) and 6-keto-prostaglandin $F_{1\alpha}$ (6-keto-$PGF_{1\alpha}$, a stable prostacyclin metabolite) were measured via ELISA and normalized to urine creatinine concentration.

On Day 7 of leflunomide administration, when compared to Day 0, there was a significant ($p=0.0015$) decrease in the maximal amplitude on aggregometry with ADP and a significant ($p=0.0118$) increase in the PFA-100® closure times. For the remaining immunosuppressive agents, there were no change in platelet count or any measure of platelet function. With all medications, there were no significant changes for all viscoelastometry parameters and PTT; however, compared to pretreatment values, on Day 7 of prednisone administration there was a significant ($p<0.0001$) decrease in PT. Although urinary 11-$dTXB_2$ levels on Day 7 increased by an average of 115% compared to Day 0 during cyclosporine administration (a result comparable to previous results of previous studies in our laboratory), this increase was not statistically significant. During cyclosporine administration, there was a significant ($p<0.0001$) increase in urinary 6-keto-$PGF_{1\alpha}$ levels on Day 7. For both urinary 11-$dTXB_2$ and 6-keto-$PGF_{1\alpha}$, there were no significant changes for any of the other medications.

Our study suggests that most of the commonly used immunosuppressive agents do not significantly influence platelet function, platelet number or coagulation, expect for leflunomide, which appears to decrease platelet function in dogs. Our study also suggests that, during cyclosporine therapy, prostacyclin synthesis increases. Since prostacyclin causes vasodilation and inhibition of platelet function, increased prostacyclin production potentially counteracts the prothrombotic properties of thromboxane A_2 in dogs receiving cyclosporine. Further investigation of the effects of immunosuppressive therapy on a comprehensive panel of hemostatic parameters is warranted in patients at risk for thromboembolism, such as dogs with IMHA.

PREVALENCE OF GENETIC DISEASE VARIANTS IN 100,000 PUREBRED AND MIXED BREED DOGS

Jonas Donner[1], Heidi Andersson[1], Stephen Davison[2], Bala Ganesan[3], Cindy Cole[3], Hannes Lohi[4]

[1]*Genoscoper Ltd, Helsinki, Finland,* [2]*Waltham, Leicestershire, UK,* [3]*Mars Veterinary, Portland, Oregon, USA,* [4]*University of Helsinki, Helsinki, Finland*

The annotation of the canine genome and subsequent development of efficient genomic tools over the past ten years has greatly facilitated gene discoveries in many genetic conditions across purebred dogs. The list of known mutations that an individual dog can potentially be tested for has grown from tens to hundreds, which has important veterinary diagnostic and breeding implications. The rapid growth of the number of genetic tests available has fostered a conceptual change in DNA-based diagnostics, transforming the field from one-by-one single gene testing to more comprehensive screening based on gene panels. Panel testing in large numbers of dogs, of diverse breed background and representing different geographic populations, provides a unique view to the epidemiology of risk variants and their relevance for veterinary care, breeding programs and animal welfare. In particular, panel tests provide a uniquely efficient platform for exploring the poorly understood risk variant prevalence in mixed breed dogs, which is particularly relevant given that roughly half of the dogs in US are mixed breeds.

We have recently developed a comprehensive multiplex panel assay, enabling simultaneous testing for approximately 150 known canine disease variants across a wide variety of conditions. We report here on the information gained by the testing of approximately 100,000 dogs (90,000 mixed breed and 10,000 purebred dogs), representing both the US and Europe, and forming the largest cohort for a canine disease frequency investigation to date.

We found a high and equal overall prevalence of disease risk variants in both purebred and mixed breed dogs. Approximately every third dog tested carried at least one risk variant (four at most). Around 70% of the mutations screened for were observed at least once in the cohort, indicating that the remainder of the mutations are extremely rare, limited to specific dog populations, or have been eradicated through selective breeding. Some of the most common mutations in mixed breeds and purebreds alike were the variants associated with degenerative myelopathy (DM), hyperuricosuria (HUU), exercise-induced collapse (EIC), multidrug resistance 1 (MDR1) and primary lens luxation (PLL). The disease variants found in mixed breed dogs, to some extent, reflects purebred ancestry known to be prevalent in US mixed breeds. In purebreds, we found several new genetically affected breeds for a number of conditions, further breaking down patterns of perceived breed specificity for certain disorders.

This study demonstrates the power and efficiency of panel screening for inherited disorders in dogs, which has significant implications for veterinary medicine. The prevalence of risk variants in the general dog population is high, and mixed breeds are equally susceptible to common canine inherited conditions. This emphasizes the need for genetic testing as a tool to improve the diagnostics, care and ultimately the welfare of all dogs. Further studies are needed to comprehensively understand the clinical significance of the disease variants in mixed breeds, and in new affected purebreds.

AMINOACIDURIA MAY EXPLAIN HYPOAMINOACIDEMIA IN CANINE HEPATOCUTANEOUS SYNDROME (n=20)

John Loftus, Sharon Center

Cornell University, Ithaca, NY, USA

Hepatocutaneous syndrome (HCS) is a rare canine disorder characterized by Superficial Necrolytic Dermatitis (SND) and a distinctive hepatopathy. A unifying feature of HCS is hypoaminoacidemia, thought to provoke the unique cutaneous and hepatic lesions; yet, the pathomechanism remains enigmatic. Disorders in human beings with similar skin and hepatic lesions include Hartnup's Disease (congenital tryptophan deficiency due to a dysfunctional neutral amino acid [AA] transporter), Lysinuric Protein Intolerance, and Prolidase Deficiency, each associated with some degree of aminoaciduria or imidodipeptiduria. Characterization of urine amino acids (UAA) vis-à-vis plasma AA (PAA) was undertaken in a cohort of canine HCS patients (n=20) to explore a possibility of pathologic aminoaciduria as a cause of hypoaminoacidemia. All dogs had characteristic HCS hepatic features on abdominal ultrasonography (Swiss Cheese pattern); 18 with characteristic biopsies of skin and/or liver (n=4 skin only, n=12 liver only, n=2 skin & liver biopsies) and two dogs with cutaneous lesions and health status examined by a board certified veterinary dermatologist. Additional diagnostics also supported HCS. Heparinized plasma and urine AA profiles were performed by an established veterinary laboratory on samples acquired within 6-hrs of each other. Urine creatinine (Cr) concentration was used to normalize UAA concentrations. PAA were compared to mean reference values and UAA:Cr concentrations were compared to maximum UAA:Cr in aged matched controls.

All patients had a generalized hypoaminoacidemia; numerous AA were <50% of control means. The most consistently severe hypoaminoacidemias (median measured:reference mean ratios; Figure panel A) were seen with glutamine (0.3x), proline (0.2x), cysteine (< 0.1x), and hydroxyproline (< 0.1x). A subset of AA were increased relative to mean values, including amino L-alpha-amino-n butyric acid (3.5x),

3-methylhistidine (1.8x), glutamic acid (1.7x), and phenylalanine (1.7x). All HCS dogs had marked lysinuria. UAA:Cr (Figure-panel B) with mean measured:maximum ratios >1.0 included lysine (5.0x), 1-methylhistidine (2.1x) and proline (1.8x). PAA reductions prominently involved AA associated with the urea cycle and synthesis of glutathione or collagen. Lysinuria and prolinuria implicate defective renal amino acid transporters, either acquired with disease or associated with HCS pathogenesis. Further investigations are needed to determine: i) pathogenesis of the selected aminoaciduria, ii) role of aminoaciduria in pathophysiology of HCS, iii) if aminoaciduria has diagnostic or prognostic value, and iv) if tailored nutritional supplements can improve HCS management compared to dismal response of general amino acid supplementation.

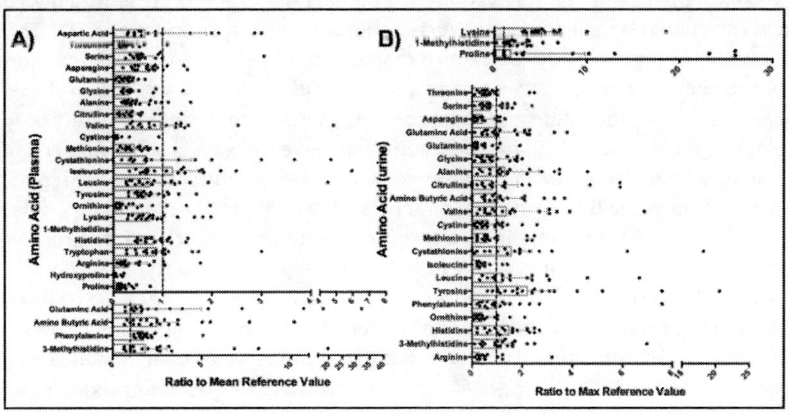

Renal Markers in Acute Kidney Injury

Larry D. Cowgill, DVM, PhD, DACVIM
Davis, CA

INTRODUCTION

Acute kidney disease (AKI) represents a spectrum of disease associated with a sudden onset of renal parenchymal injury and proportionate immediate losses of renal excretory functions including: glomerular filtration, tubular reabsorption, tubular secretion, and urine concentration/dilution. Over time, additional metabolic and endocrine functions of the kidney may become dysregulated and manifested clinically. Initially, the renal injury may be mild and go unnoticed, but more characteristically in veterinary medicine, patients are presented primarily at advanced stages of the spectrum well established into the maintenance phase of AKI. For animals at more established stages of AKI, the opportunity for recovery of renal function is less predictable and may be preceded by death from the consequences of the renal failure before function is reestablished.

Of foremost importance is the potential to recognize the initiation and induction of AKI at its very inception when prophylaxis and therapeutic intervention are more likely beneficial. Unfortunately, the diagnosis of kidney injury at this stage using routine clinical parameters can be a diagnostic challenge. Strategic biomarkers that predict the presence of kidney injury before overt clinical abnormalities are present and before development of azotemia would greatly facilitate the detection of kidney injury when it might be most responsive to medical management and when prevention of progression may be possible. The overall mortality rate for human beings and animals with AKI is approximately 50%, but outcome is highly dependent on the underlying etiology and available treatment options. This high mortality rate has remained almost unchanged, despite advances in diagnostic tests and available treatments including renal replacement therapy. One of the suggested reasons for the high mortality is the late recognition of AKI using traditional clinical diagnostic tests including serum creatinine concentration and urinalysis, leaving a narrow window of opportunity for therapeutic or preventive intervention before the development of kidney failure. The need for sensitive and specific markers for early identification and stratification of AKI has generated growing interest and active search. Appropriate and sensitive biomarkers offer unprecedented utility for diagnosing, classifying, and grading the severity of AKI.

Ideally, clinicians could monitor specific biomarkers individually or in combination to maximize sensitivity and specificity of the diagnosis and forecast precisely the phase and extent of morphologic injury. Of most importance would be predictors to indicate the onset and duration of the earliest phases of kidney injury before traditional markers of kidney function become abnormal. The hunt for such biomarkers is ongoing, and several tubular-specific candidates currently are nearing commercial availability.

CONVENTIONAL MARKERS OF AKI

Until novel biomarkers are validated further and established clinically, it is necessary to rely on conventional predictors of kidney function or dysfunction to identify early AKI. Recently, IRIS has established a classification scheme to grade the severity of AKI in dogs and cat. Unlike the IRIS staging for CKD, the IRIS AKI Grade represents a discrete moment in the course of the disease and is predicted to change as the condition worsens, improves, or transitions to CKD. The IRIS AKI grading scheme is based on serum creatinine and urine formation as the traditional markers of injury. IRIS AKI grading is intended to facilitate classification, functional stratification, and therapeutic decision making for AKI but requires close scrutiny of these clinical parameters.

Changes in urine production and urine concentration are perhaps the oldest predictor of acute kidney dysfunction and are reliable harbingers of early kidney injury. They suffer, however, both a lack sensitivity and specificity and may not discriminate acute change from an established baseline in the absence of serial assessment or knowledge of the historical baseline. A measured oliguria of <1 ml/kg/hr over a 6 hour interval or anuria have been established as IRIS AKI criteria to define or diagnose an AKI. Unfortunately, some acute kidney injuries are associated with non-oliguric states which would mask the early stages of the disease. In addition, quantitative urine production may be difficult (especially in cats) or inconvenient to access in the clinical setting. Changes in urine specific gravity can be a sensitive predictor of kidney injury if a baseline value is known, therapeutic interventions (i.e., fluids, diuretics) do not interfere with urine concentration, and changes are assessed serially over time. In many circumstances the change in urine specific gravity will precede changes in other traditional markers like serum creatinine. As concentrating function is reestablished, urine specific gravity can also correlate or document renal regeneration and repair.

Progressive (and often subtle) azotemia is the hallmark of AKI, but the progressive component is observed inconsistently in animals because of variations in the stage and chronicity of disease at presentation. Serum creatinine increases proportionally with the severity of the renal injury at presentation. Oppositely, reductions in lean body mass, creatinine generation, and overhydration can lower the serum creatinine concentration and overestimate monitored renal function. There is a curvilinear and inverse relationship between serum creatinine and GFR. Incremental changes in monitored serum creatinine will be small in IRIS AKI Grades I and II despite significant changes (increases or decreases) in GFR. Notwithstanding the insensitivity and limitations of serum creatinine as a biomarker of kidney injury, it currently stands as the best, most time-tested and most familiar clinical marker of AKI. Serum creatinine has been designated as the surrogate marker of renal function for the IRIS grading of AKI, and changes in serum creatinine are used to predict improving or worsening stages of the condition. With **careful interpretation**, sequential monitoring of serum creatinine often will provide sufficient clinical information to direct re-grading of the AKI, adjust ongoing therapy, modify pharmaceutical doses, score the severity of the disease, and facilitate prognosis and outcomes predictions. In the absence of renal replacement therapy, the magnitude of serum creatinine clearly correlates with the degree of kidney dysfunction and outcome. However, with the establishment of renal replacement therapy to extend the window for kidney repair, presenting serum creatinine concentration is not a predictor of outcome. Serum SDMA is a recently available GFR marker with complementary diagnostic utility to serum creatinine in animals with chronic kidney disease (CKD). Under a variety of clinical circumstances, SDMA may predict the progressive development of CKD with greater sensitivity than serum creatinine. In limited observations in experimental AKI, SDMA changes identically and proportionally with serum creatinine and predicts the onset and progression with equal sensitivity.

Generally, renal function can be monitored effectively with routine clinical and laboratory parameters discussed above. However, estimation of glomerular filtration rate (GFR) is regarded the most important predictor of renal function. It can be measured from either the urinary clearance or the plasma clearance of appropriate endogenous or exogenous markers of filtration. Urinary clearance requires accurate collection of timed urine samples and appropriately timed blood collections. Ordinarily, measurement of GFR is beyond the scope and utility of routine clinical practice, but many patients in the ICU have indwelling urinary catheters to monitor urine production and are readily poised for GFR measurement. The urinary clearance of endogenous creatinine is the most convenient and provides clinically relevant estimates of GFR. Procedurally, a well-mixed aliquot from an accurate, timed (30 minutes to several hours) urine collection from the catheter or closed collection system and a serum sample collected at the midpoint of the urine collection are submitted for creatinine determinations. The clearance of creatinine (as an estimate of GFR) is calculated from the equation, $C_{creat} = (U_{creat} \cdot V) \div P_{creat}$ where,

C_{creat}, clearance of creatinine (ml/min)
U_{creat}, Urine creatinine concentration (mg/dl)
V, urine flow (ml/min)
P_{creat}, plasma creatinine concentration (mg/dl)

If the laboratory analyses are extended to include electrolytes and minerals on the urine and plasma samples, the clearance and fractional excretion of sodium, potassium, chloride, calcium, and phosphate can be determined concurrently with the GFR estimates. The fractional excretion (FE) is calculated as the $(C_x \div C_{creat}) \times 100$, where C_x, the urinary clearance of solute, x. The FE_{Na} has been shown to correlate with improvement of renal function in some etiologies of AKI in dogs.

NOVEL (NONTRADITIONAL) BIOMARKERS FOR AKI

As mentioned, there has been growing interest and research in human nephrology with the goal to identify sensitive and specific biomarkers that would predict the early onset of AKI. Similar efforts are emerging in veterinary medicine and show great promise. An AKI biomarker should be a urine and/or plasma constituent that can be assessed routinely and serve as an indicator of kidney function or dysfunction or response to injury. The marker must reflect kidney specific events and reflect early phases of the pathogenesis or repair process. An ideal marker should reflect the extent of injury and changes in the pathology or therapeutic intervention over time. For biomarkers to be useful, clinicians need to test patients with risk for kidney injury before identified by conventional clinical parameters or overt kidney failure is evident. Consequently, the establishment of AKI biomarkers will require a proactive change in the practice patterns related to kidney injury. Clinicians will by necessity need to recognize and anticipate the potential risks and clinical circumstances that might predispose to kidney injury including comorbid diseases (e.g., heart failure, pancreatitis, heat stroke, vomiting), medications (e.g., ACEI, diuretics, antimicrobics), and diagnostic and therapeutic procedures (e.g., anesthesia, surgery, dentistry, contrast administration).

Novel "AKI biomarkers" historically have been discovered and exploited for their application in the early detection of acute kidney injury. However, this restricted view will constrain broader and potentially more clinically relevant application of these same markers and potentially blur their specificity for AKI as currently defined. Novel, kidney-specific biomarkers that localize to functional renal tubular epithelium (or other kidney-specific loci) and respond to diverse stresses or pathophysiologic disruption of normal cellular function have potential to signal the early, specific, and sensitive detection of kidney injury and are perhaps better termed "active injury biomarkers". These "active injury biomarkers" could expose ongoing or progressive kidney diseases in advance of conventional diagnostic methods that document consequent alterations to glomerular filtration or substantive loss of functional parenchyma over time. A static kidney disease (e.g., CKD) is a state of long-standing, sustained, and irreversible loss of variable degrees of kidney parenchyma and function associated with the absence of detectable ongoing kidney injury and stable biochemical markers of kidney function including serum creatinine and/or SDMA concentrations over a prolonged but variable period of time. Progressive CKD is a state of long-standing, sustained, and irreversible loss of variable degrees of kidney parenchyma and function associated with progressive worsening of biochemical markers of kidney function including serum creatinine and/or SDMA concentrations over a prolonged but variable period of time. Progressive CKD implies there an "active", albeit subtle or occult, ongoing loss of structural kidney parenchyma or kidney function secondary to persistent or ongoing pathologic processes that promote cumulative injury and damage to the kidney. These changes ultimately are detectable by conventional biochemical markers of kidney function. Evidence of serum creatinine and/or SDMA trending up (as measured by these biochemical markers over time) is the current standard for identifying progressive CKD. Detection of biomarkers associated with "active kidney injury" in animals with CKD has the potential to predict or identify those patients whose underlying kidney disease is ongoing and likely to progress in advance of biochemical markers whose short-term values reflect relatively static kidney function.

Many candidate serum and urinary biomarkers have been assessed in human medicine, and many of the promising markers are now being evaluated and validated in animals. Some of the most promising include retinol binding protein (RBP), cystatin-C, kidney injury molecule-1 (KIM-1), neutrophil gelatinase-associated lipocalin (NGAL), and interleukin-18. Retinal binding protein has been evaluated in dogs associated with a variety AKI models and preceded routine clinical predictors of AKI. In our preliminary investigations, there was no detectible KIM-1 for any of the available human, primate, or rodent assays for KIM-1 in dogs with experimental AKI.

Neutrophil gelatinase-associated lipocalin (NGAL) is a 25-kDa protein initially identified bound to gelatinase in specific granules of the neutrophil. Subsequently, NGAL has been demonstrated to be expressed by a variety of epithelia and specifically up-regulated more than tenfold within the first few hours after ischemic, obstructive, and toxic renal injuries in experimental models of AKI. Recent studies in dogs with experimental nephrotoxic injury and in dogs with naturally acquired AKI, NGAL demonstrated sensitivities and specificities approaching 95% or greater for AKI. There is a commercially available assay for canine NGAL (Abbott Laboratories, Abbott Park, IL) that has been validated for AKI. Urinary NGAL is currently one of the most promising and available of these novel biomarkers to date.

Other promising biomarkers including serum inosine, and urinary clusterin, cystatin B and C are under active evaluation in a variety of dog and cat models of AKI, active kidney injury, and progressive CKD. The promise for kidney-specific biomarkers is the potential to document subclinical kidney injury within hours of its initiation, predict the pathophysiologic phase of the current injury, forecast the onset of renal repair and regeneration, and potentially predict the outcome of the patient. For future diagnostics, individual markers or a panel of markers will provide the potential to recognize subtle and often subclinical kidney disease that otherwise remains undiscovered until more substantial injury and dysfunction triggers alterations in conventional diagnostic methods. Even more importantly, "active injury biomarkers" have potential to establish a new understanding of our traditional views of CKD including early identification and possible mediators of its progression which have not advanced in many decades. Novel, kidney-specific biomarkers will likely establish a completely new and sophisticated paradigm in our approach to the understanding, diagnostic evaluation, and treatment of kidney disease in dogs and cats.

SELECTED READING

1. J. De Loor, S. Daminet, P. Smets, B. Maddens, and E. Meyer. Urinary Biomarkers for Acute Kidney Injury in Dogs. J Vet Intern Med 2013;27:998-1010.
2. Maddens B, Heiene R, Smets P, et al. Evaluation of kidney injury in dogs with pyometra based on proteinuria, renal histomorphology, and urinary biomarkers. J Vet Intern Med 2011;25:1075-1083.
3. G. Segev, C. Palm, B. LeRoy, L.D. Cowgill, and J.L. Westropp. Evaluation of Neutrophil Gelatinase-Associated Lipocalin as a Marker of Kidney Injury in Dogs. J Vet Intern Med 2013;27:1367-1367.
4. C.A. Palm, G. Segev, L.D. Cowgill, B.E. LeRoy, K.L. Kowalkowski, K. Kanakubo and J.L. Westropp, Urinary Neutrophil Gelatinase-associated Lipocalin as a Marker for Identification of Acute Kidney Injury and Recovery in Dogs with Gentamicin-induced Nephrotoxicity. J Vet Intern Med 2013;30:200-205.

Renal Markers LUT: Biomarkers of Lower Urinary Tract Disease

John M. Kruger, DVM, PhD, DACVIM; Maciej Parys, DVM; Peter Schall, BS; Taylor Aiello, BS; Lorraine M. Sordillo, PhD; Vilma Yuzbasiyan-Gurkan, PhD
East Lansing, Michigan

INTRODUCTION

Lower urinary tract disease (LUTD) biomarkers can be any characteristic that is objectively measured and evaluated as an indicator of normal biological processes, pathological processes, or response to therapy. Ideally, biomarkers should be sensitive, specific, and early predictors of disease; provide insight into the cause, location, and degree of injury; and predict and assess patient responses to therapeutic interventions. Over the past several decades, a large number of putative biomarkers have been identified for detection and characterization of LUTDs in people, dogs, and cats (**Tables 1-4**). Yet, despite extensive investigation, development and validation of clinically useful biomarkers for urinary tract infection, urolithiasis, neoplasia, and idiopathic cystopathies has been challenging. However, recent advances in molecular techniques, especially metabolomic, proteinomic, lipidomic, genomic, and transcriptomic profiling, have enabled identification of a new generation of compound molecular signatures that may provide more reliable biomarker assays for noninvasive diagnosis and monitoring of patients with various LUTDs.[65,71]

BLADDER NEOPLASIA

Bladder cancers comprise approximately 2% of all naturally occurring cancers in dogs and are uncommon in cats.[33] Diagnosis is currently based on histological or cytologic exam of tissue biopsies collected by surgery, cystoscopy, or catheter biopsy. A large number of potential biomarkers have been developed and evaluated for detection of bladder cancer in people; fewer have been evaluated in dogs and none have been reported in cats (**Table 1**). Unfortunately none of these tests to date have achieved the combined sensitivity and specificity to replace established clinical evaluations.[14,33,65] However, it is anticipated that rapid developments in "omic" profiling will ultimately lead to more reliable biomarkers for detection and management of TCC.[14,33]

Table 1. Potential Biomarkers for Urinary Bladder Neoplasia

Marker	Sample	Method	Species Investigated			References
			Human	Dog	Cat	
Bladder Tumor Antigen (BTA)	U	Rapid latex agglutination	X	X	NR	5,8,14, 27, 34
BRAF Gene Mutations	U	RNA/DNA sequencing, RFLP	X	X	NR	16, 48
Calgranulins (S100 A/8/A9, A12)	U	RIA	X	X	NR	26
Cytokeratins (8,18,19,20)	U	ELISA, qPCR	X	NR	NR	34
Hyaluronic acid/Hyaluronidase	U	ELISA	X	NR	NR	34
Lipidomic Profile	T	DESI-MS	X	X	NR	18, 33
Metabolomic/Proteinomic Profiles	U	CE-MS; LC-MS/MS; NMR	X	X	NR	9,65
MicroRNAs	T/C	qPCR, RNA sequencing	X	X	NR	68
Multiantibody immunofluorescence	U	IFA	X	NR	NR	14,34
Multiplex gene expression	U	qPCR	X	NR	NR	14, 28
NMP22	U	ELISA	X	NR	NR	14, 34
Survivin	U, T/C	IHC; qPCR	X	X	NR	33,34,39
Telomerase	U	TRAP	X	X	NR	33, 47

URINARY TRACT INFECTION

Urinary tract infection (UTI) encompasses a variety of clinical entities whose common denominator is microbial invasion of any or all of its components. Establishing a diagnosis of bacterial UTI, localizing the site(s) of infection, and characterizing host responses are based on clinical signs, physical examination findings, and results of diagnostic imaging, urinalysis, and quantitative urine culture. However, establishing a diagnosis of concurrent pyelonephritis in patients with UTI and predominately lower urinary tract signs can be difficult and may require more invasive evaluations. In addition, dogs and cats may develop *asymptomatic bacteriuria* (ASB), which is detection of significant bacteriuria unaccompanied by signs referable to the urinary system.[52] Although treatment of ASB is generally not recommended,[70] optimal clinical management strategies for ASB in dogs and cats have yet to be determined. Development of rational therapeutic approaches to ASB would be facilitated by more objective screening strategies that allow identification of patients in whom antimicrobial therapy would be beneficial.[43,62] Urinary biomarkers have the potential of being a rapid means of diagnosing UTI, localizing to upper and/or lower urinary tracts, and identifying patients with ASB who warrant therapy. A number of potential biomarkers have been evaluated for detection and localization of UTI in people; however, few have been evaluated in dogs and cats (**Table 2**).

Table 2. Potential Biomarkers for Urinary Tract Infection

Marker	Sample	Method	Species Investigated			References
			Human	Dog	Cat	
Elastase-alpha1-proteinase inhibitor	U, P	ELISA	X	NR	NR	22,51
Heparin Binding Protein	U	ELISA	X	NR	NR	35
Interleukins: IL-6, IL-8	U, P	ELISA, BB-CLIA	X	NR	NR	6,35,51,56,59
Lactoferrin	U	ELISA, Electrochemical biosensor	X	NR	NR	2,49,53
Multiplex biosensor bacterial 16SrRNA	U	Electrochemical biosensor	X	NR	NR	46,49,53
Procalcitonin	S	BB-CLIA	X	X	NR	41,51,72

UROLITHIASIS

Central processes involved in urolith formation include supersaturation, crystal nucleation, aggregation, growth, and retention. While supersaturation of urine with calculogenic substances is a fundamental prerequisite, stone formation is also influenced by organic and inorganic urinary molecules that can inhibit or promote crystal nucleation and aggregation, or form soluble complexes with specific crystalline lattice ions.[1,21] Urine of dogs and cats is often supersaturated with respect to various calculogenic substances; however, only a small fraction of these animals form stones.[21] Although conventional markers of urine saturation (e.g. crystalluria, urine electrolyte concentrations, relative supersaturation, etc.) are important indicators of the overall propensity (risk) of crystal formation, they are not necessarily predictive of the risk of stone formation in an individual patient. A number of potential biomarkers have been evaluated in people for their ability to further discriminate between stone formers and non-stone formers; however, few been evaluated in dogs and cats (**Table 3).**

Table 3. Potential Biomarkers for Urolithiasis

Marker	Sample	Method	Species Investigated			Reference s
			Human	Dog	Cat	
Calgranulin	U	RIA	X	NR	NR	1,3
Genomic profile	T/C	DNA sequencing, GWAS, Single locus-analysis	X	NR	NR	67,69
Proteinomic Profile	U,SM,T/C	CE-MS; LC-MS/MS; MALDI-TOF-MS, SELDI-MS, Q-TOF-MS	X	X	NR	3,21,30,75
Nephrocalcin	U	DEAEC- C, ELISA	X	X	NR	1,12,13
OPN Polymorphism Genotyping	T/C	PCR-RFLP, PCR SSCP	X	NR	NR	42,63
Osteopontin	U, S	ELISA	X	NR	NR	1,42,63
Tamm-Horsfall Protein (Uromodulin)	U	ELISA, RIA, BB-CLIA, SDS-PAGE, LC-MS/MS	X	X	NR	1,12,17
Urinary Prothrombin Fragment 1	U,SM,T/C	ELISA, IFA	X	NR	NR	1
Urinary Trefoil Factor 1	U	ELISA, Q-TOF-MS	X	NR	NR	1

FELINE IDIOPATHIC CYSTITIS

Feline idiopathic cystitis (FIC) is the most common urinary disorder of young to middle-aged cats. The clinical and morphologic features of FIC are strikingly similar to those of an idiopathic cystopathy of people called interstitial cystitis/painful bladder syndrome (IC/PBS). Our observations and those of others suggest that FIC is associated with different clinical and pathologic phenotypes which include ulcerative inflammatory and nonulcerative noninflammatory forms. These phenotypes appear to be similar to nonulcerative (Type I) and ulcerative (Type II; classic) forms encountered in people with IC/PBS. As of yet, there are no consistent specific diagnostic markers that establish a diagnosis of FIC or IC/PBS, or differentiate between various disease phenotypes. Over the past several decades a large number of potential biomarkers have been identified and evaluated in people; however, relatively few have been evaluated in cats with FIC (**Table 4**). We have recently been investigating potential urine, serum, and tissue biomarkers of FIC using genomic, transcriptomic, proteinomic, and lipidomic analyses. Results of RNA sequencing pilot studies have identified over 70 genes that are differentially expressed in cats with ulcerative/inflammatory and nonulcerative forms of FIC.[54] Tissue lipid profiles and relative gene expression studies have revealed increased expression of COX-2 (*PTGS2*) 12-LOX (*ALOX12B*) and PGDH (*HPGD*) and decreased tissue concentrations of anti-inflammatory/pro-resolving oxylipid mediators in affected cats compared to controls.[38] We have also observed significantly increased serum concentrations of cytokines IL-12(p40), IL-18, and SDF1 in affected cats compared to controls.[54] Additional studies are necessary to establish the diagnostic utility of these potential biomarkers for diagnosis and characterization of FIC.

Table 4. Potential Biomarkers for Idiopathic Cystitis

Marker	Sample	Method	Species Investigated			References
			Human	Dog	Cat	
Antiproliferative Factor (APF)	U	Bioassay	X	NR	NR	20,31,36,74
Chemokines; CXCL-1, -10	U	BB-CLIA	X	NR	NR	64
Epidermal Growth factor (EGF)	U	ELISA	X	NR	NR	20,31,74
Genomic Profile	T/C	DNA sequencing	X	NR	NR	61,71
GP51	U	ELISA	X	NR	X	10,55
Interleukins: IL-1β, -2, -6, -8, -12, -18	U, S	ELISA, BB-CLIA	X	NR	X	15,19,20,29,36,44, 54,64
Lipidomic Profile	T	LC-MS/MS	ND	NR	X	37,38
Metabolomic Profile	U, S	IRMS, LC-MS/MS, NMR, Q-TOF-MS	X	NR	X	23,25,57,66,71
MicroRNAs	T	qPCR	X	NR	X	50,54,58,
Nerve Growth Factor (NGF)	U, S, T/C	ELISA, BB-CLIA	X	NR	NR	4,32,36,44,64
Prostaglandin E2 (PGE2)	U, T/C	ELISA, LC-MS	X	NR	X	4,19,32,38
Proteinomic Profile	U, S	LC-MS/MS, MALDI-TOF-MS NMR, Protein array	X	NR	X	11,25,40,66,71
Stromal Cell-Derived factor-1 (SDF-1)	U, S	BB-CLIA	X	NR	X	15,54
Tumor Necrosis Factor alpha (TNFα)	U,S,T/C	ELISA, BB-CLIA	X	NR	X	29,44,54
Transcriptomic Profile/Gene Expression	T/C,US, PBMC	RNA sequencing, Gene array, qPCR	X	NR	X	7,24,45,38,54,71
Urinary Trefoil Factor 2	U	MALDI-TOF-MS	NR	NR	X	40

ABBREVIATIONS AND ACRONYMS

BB-CLIA= bead-based chemiluminescent immunoassay; CE-MS= capillary electrophoresis mass spectroscopy; DEAEC-C= Diethylaminoethyl cellulose ion exchange chromatography; DESI-MS= desorption electrospray ionization mass spectroscopy; ELISA= enzyme-linked immunosorbent assay ; GWAS= genome-wide association study; IRMS= infrared microspectroscopy; LC-MS/MS= liquid chromatography tandem mass spectroscopy; MALDI-TOF-MS= Matrix-assisted laser desorption ionization time-of-flight mass spectroscopy ; NR= not reported; PBMC= peripheral blood mononuclear cells; P=plasma; qPCR=quantitative real-time polymerase chain reaction; Q-TOF-MS=quadrupole time-of-flight mass spectroscopy; RFLP= restriction fragment length polymorphism; RIA= radioimmunoassay; SELDI-MS= Surface-enhanced laser desorption/ionization mass spectroscopy; S=serum; SM= stone matrix; T/C= tissue or cells; TRAP= telemetric repeat amplification protocol; U= urine

REFERENCES

[1]Aggarwal KP, et al. *BioMed Res Int* 2013;292953.[2]Arao S, et al. *J Clin Microbiol* 1999;37:553.[3]Bergsland KJ, et al. *Am J Physiol Renal Physiol* 2006;291:F530. [4]Bhide AA, et al. *Int Urogynecol J* 2013;24:1065.[5]Billet JPHG, et al. *Am J Vet Res* 2002;63:370. [6]Bitsori M, et al. Pediatr Nephrol 2001;26:2003.[7]Blalock EM, et al. *J Urol* 2012;187:725. [8]Borjesson DL, et al. *Vet Clin Path* 1999;28:33.[9]Bracha S, et al. *Anal Biochem* 2014:455:41.[10]Byrne DS, et al. *J Urol* 1999;161:1786.[11]Canter MP, et al. *Am J Obstet Gynecol* 2008;198:553.e1.[12]Carvalho M, et al. *Urology* 2003;62:566.[13]Carvalho M, et al. *Int Braz J Urol* 2006;32:342.[14]Chou R, et al. *Ann Intern Med* 2015;163:922.[15]Corcoran AT, et al. *World J Urol* 2013:31:241.[16]Decker B, et al. *Mol Cancer Res* 2015;13:993.[17]Devuyst O, et al. *Nephro Dial Transplant* 2005; 20:1290. [18]Dill AL, et al. *Anal Chem* 2009;81:8758.[19]Erickson DR, et al. *Urology* 2001; 57(Suppl 6A):15.[20]Erickson DR, et al. *J Urol* 2002;167:2461.[21]Forterre S, et al. *J Anim Physiol Anim Nutr* 2006;90:192.[22]Fretzayas A, et al. *Pediatrics* 2000;105:e28.[23]Fukui Y, et al. *J Chromatogr B* 2009;877:3806.[24]Gamper M, et al. *BMC Genomics* 2009;10:199. [25]Goo YA, et al. *Int Braz J Urol* 2010;36:464.[26]Heilmann RM, et al. *J Vet Diag Invest* 2014;26:104.[27]Henry CJ, et al. *Am J Vet Res* 2003; 64:1017.[28]Holyoake A, et al. *Clin Cancer Res* 2008;14:742.[29]Jiang YH, et al. *PLoS ONE* 2013;8:e76779.[30]Kaneko K, et al. *Int J Urol* 2013;19:765. [31]Keay SK, et al. *Urology* 2001;57(Suppl 6A):9.[32]Kim JC, et al. *J Urol* 2006;175:1773. [33]Knapp DW, et al. *ILAR* 2014;55:100.[34]Konety BR. *Urol Oncol-Semin Ori* 2006;24:326. [35]Kjölvmark C, et al. *OFID* 2014;1:ofu004. [36]Kou HC, *Int J Urol* 2014;21:34. [37]Kruger JM et al. *J Vet Intern Med* 2013;27:740.[38]Kruger JM, et al. *Proc ACVIM Forum* 2015. [39]Ku JH, et al. *BJU Int* 2012;110:630.[40]Lemberger SIK, et al. *BJU Int* 2011;107:670. [41]Leroy S, et al. *Pediatrics* 2013;131:870.[42]Li X, et al. *BioMed Res Int* 2015;315043.[43]Litster A, et al. *Vet Microbiol* 2009;136:130.[44]Liu HT, et al. *Urol Sci* 2015;26:225.[45]Logadottir Y, et al. *J Urol* 2014;192:1564.[46]Mach KE, et al. *J Urol*2009;182:2735.[47]McCleary AL, et al. *Am J Vet Res* 2010;71:1468.[48]Mochizuki H, et al. *PLoS ONE* 2015;10:e0129534.[49]Mohan R, et al. *PLos ONE*;201:e26846. [50]Monastyrskaya K, et al. *Am J Pathol* 2013;182:431. [51]Nanda N, et al. *Biomarker Insight* 2009;4:111.[52]Olin SJ, et al. *Vet Clin Small Anim* 2015;45:721.[53]Pan Y, et al. *Biosens Bioelectron* 2010;26:649. [54]Parys M, et al. *Unpublished Data* 2016[55]Press SM, et al. *J Urol* 1995;153:288A. [56]Rodhe N, et al. *Scand J Prim Health*

2009;27:74. [57]Rubio-Diaz DE, et al. *Analyst* 2009;134:1133.[58]Sanchez-Freire V, et al. *Am J Pathol* 2010; 176:288.[59]Sheu JN, et al. *Cytokine* 2006;36:276.[60]St John A, et al. *Am J Clin Path* 2006;126:428.[61]Sugaya K, et al. *J Urol* 2002;168:2668.[62]Trautner BW, et al. Infect Dis Clin North Am 2014;28:15.[63]Tugcu V, et al. *Ren Fail* 2013;35:825. [64]Tyagi P, et al. *J Urol* 2012;187:2243[65]Urquidi V, et al. *Eur Med J Urol* 2013;1:70.[66]Van QN, et al. *Dis Markers* 2003-2004;19:169.[67]Vezzoli G, et al. *Kidney Int* 2011;80:587. [68]Vinall RL, et al. *Am J Vet Res* 2012;73:1626.[69]Wang ET, et al. *J Pharmacogenom Pharmacoproteomics* 2013;4:1000e136. [70]Weese JS, et al. *Vet Med Int* 2011;263768.[71]You S, et al. *Int Neurourol J* 2012;16:159.[72]Xu RY, et al. *BMC Urol* 2014;14:45[73]Zhang CO, et al. *BMC Urology* 2005;5:7.[74]Zhang J, et al. *Biochim Biophys Acta* 2012;1822:1807.[75]Zhu W, et al. *BioMed Res Int* 2014;415651.

Biomarkers for Chronic Kidney Disease
Rosanne Jepson, BVSc, MVetMed, PhD, DACVIM, DECVIM
London, UK

INTRODUCTION: UTILITY OF BIOMARKERS IN CKD

A biomarker is an indicator of a biological state, which may be a normal biological process, pathological process or a response to therapeutic intervention. An ideal biomarker will be objectively measured, demonstrating a high sensitivity and specificity for the disease process or response of interest. Chronic kidney disease (CKD) is a common condition in feline and canine medicine. Our current diagnosis of CKD is based on a combination of historical and physical examination findings, surrogate markers of GFR (e.g. plasma creatinine/urea) in addition to urinalysis, diagnostic imaging and assessment of proteinuria.

However, it is widely accepted that, creatinine, particularly in the early stages of CKD, is an insensitive marker of GFR. Patients can therefore have chronic intrinsic renal damage prior to the onset of azotemia or any demonstrable clinical signs relating to CKD; a concept that is reflected by the International Renal Interest Society (IRIS) staging system (stages 1 and early stage 2). Biomarkers are sought which predict patients at risk of developing CKD and that give us information about change or decline in renal function before the onset of overt azotemia. Such biomarkers would facilitate early identification of patients with CKD that may allow interventions that improve quality of life, slow disease progression and improve outcome.

Due to the intrinsic functional and structural response of the kidney to loss of nephrons, CKD should always be considered a progressive condition. In the azotemic stages of CKD, plasma creatinine performs well as an indicator of change in GFR. However, given that renal damage should be considered an irreversible process, by the time an increase in plasma creatinine has been documented, this already implies further decline in renal function. Biomarkers that indicate incipient risk of progression or which inform clinicians about the presence of risk factors for progression of disease would therefore be of interest in order to optimise therapy and limit progression of CKD.

When patients diagnosed with CKD are started on medical management, biomarkers can be used to determine whether management strategies or pharmacological interventions are having their desired effect. Where an evidence base exists, biomarkers may be translated into targets for therapeutic intervention with the goal of optimising treatment. Perhaps the commonest examples of this currently are dietary intervention with subsequent monitoring and IRIS targets for plasma phosphorus concentration as a biomarker of mineral and bone disorder, monitoring of urine protein to creatinine (UPC) ratio when instituting anti-proteinuric therapy and monitoring of blood pressure when instituting antihypertensive therapy to prevent further target organ injury. However, as we gain a greater understanding of the underlying pathophysiological mechanisms, it may be that there are more sensitive biomarkers that inform us about the effects of therapeutic interventions than those traditionally measured.

In human medicine, use of biomarkers in CKD extends beyond the kidney and considers other body systems that may be influenced in CKD, in particular the cardiovascular system (CVS).[1,2] The importance of biomarkers relating to the CVS stems from the significant cardiovascular morbidity and mortality risks that are identified in human patients with CKD. Current information regarding cardiorenal syndrome in dogs and cats is limited but future expansion of knowledge and evaluation of biomarkers pertaining to the cardiovascular system in canine and feline patients with CKD is likely to emerge in coming years.[3]

Finally biomarkers may be of use in the design of clinical trials. A common end point that is used to determine the effect of risk factors or therapeutic intervention is survival and yet studies that rely on the outcome of survival are both expensive and time consuming to complete. Biomarkers may therefore provide surrogate end points for clinical trials.[4]

THE SEARCH FOR BIOMARKERS

Discovery of truly novel biomarkers is a costly venture for which a number of different methodologies can be used depending on whether biomarker discovery is based on prior pathophysiological knowledge or de-novo discovery and what samples are available for evaluation e.g. blood, urine or kidney tissue.[5] Currently utilised technologies include proteomics[6], metabolomics[7,8], lipidomics[9], genomics,[10,11] epigenomics[12-14] and transcriptomics (expression profiling of mRNA[15] or microRNA[16]). These methodologies focus on enhancing our understanding of genetic variation, genetic expression and pathophysiological mechanisms which underpin the development and progression of CKD and may therefore lead to novel biomarker discovery. However, even after such studies are performed there is considerable work required in order to ensure validity of any potential biomarker and to establish whether either alone or in conjunction with a panel of markers the use of any novel biomarker will translate into a quick, simple, sensitive, specific, cost-effective and informative test for clinical practice.

Only a few novel biomarker discovery studies have been performed in relation to CKD in veterinary medicine. These have provided preliminary investigation of the urinary proteome and biomarkers that are species specific (e.g. cauxin).[17-21] A number of methodologies have been used including electrophoresis based techniques through to more sensitive protein array and mass spectrometer based technologies.[17-19,22] However, these discovery techniques are not without limitations including the limited protein concentration of urine, effects of handling and storage of urine, effects of day to day variability, diet or other disease processes on the protein profile of urine, repeatability, limited throughput of samples and financial and time related expense.[23] Therefore overall, the search for biomarkers for CKD in veterinary medicine has largely been extrapolated from studies that have been performed in human medicine.

BIOMARKERS FOR CKD

Biomarkers for CKD can be considered in terms of whether they inform regarding renal function (i.e. GFR), tubular or glomerular injury, are associated with the pathophysiological mechanisms of CKD progression (inflammation, oxidative stress, fibrosis) or whether they are informative in terms of metabolic disturbances that occur during CKD.

Surrogate markers of GFR

Creatinine and urea are commonly used and readily available surrogate markers for GFR. However, both can be affected by factors extraneous to renal function, which renders them less than ideal as markers for the early identification of CKD. In recent years two alternative markers of GFR have received considerable attention, cystatin C and symmetric dimethylarginine (SDMA), the latter of which has reached commercial availability both in the USA and Europe.

Cystatin-C: Cystatin-C is a cysteine protease inhibitor that is produced by all nucleated cells at a constant level. It has a molecular weight of ~13 Kilodaltans (KDa) and is freely filtered at the glomerulus with no tubular secretion. Cystatin-C is reabsorbed in the tubules but is subsequently broken down so that reabsorption should not impact on plasma concentrations.[24] However, concern has been raised regarding the effect that hyperthyroidism, cortisol and some malignancies may have on cystatin-C concentrations and hence it's utility as a marker of GFR. Studies investigating cystatin-C in dogs suggest that it correlates well as a marker of GFR with some studies indicating that it may be more sensitive than creatinine in the early stages of CKD.[25-27] Numerous studies have recently evaluated cystatin-C in cats but unfortunately these studies indicate that it offers little discriminatory power between cats of different stages of CKD.[28-33]

Symmetric dimethylarginine (SDMA): SDMA is the stereoisomer of asymmetric dimethylarginine (ADMA) and is produced from intra-nuclear methylation of L-arginine.[34,35] Whilst ADMA is degraded, SDMA is almost entirely excreted via the renal route.[35] In human medicine, strong correlations exist between GFR, creatinine and SDMA.[36] Plasma SDMA has been demonstrated to correlate well with plasma creatinine concentration.[37,38] In azotemic and non-azotemic cats, a linear relationship has been shown between the reciprocal of SDMA (1/SDMA) and GFR assessed by plasma iohexol clearance performing in a similar manner to the reciprocal of creatinine (1/creatinine).[39] Studies suggest that SDMA concentrations are less influenced by body muscle mass than plasma creatinine which may be of particular benefit in aging sarcopenic patients.[38,40,41] Preliminary studies suggest that SDMA may be a more sensitive marker of early decline in renal function in cats.[42] However, further work is still required to determine the effect of concurrent disease on this novel marker particularly in relation to hyperthyroidism.

Direct GFR measurement, limited sampling protocols, novel methods for GFR assessment and eGFR: Gold standard evaluation of renal function has long been considered to be direct measurement of GFR.[43,44] In clinical practice we are frequently put off direct GFR measurements due to the time required to perform this diagnostic test and requirement for hospitalisation. However, numerous studies have been published in recent years focussing on limited sampling protocols[45-47] and on novel technologies (e.g. computed tomography or transcutaneous markers) [48-53], which may in the future promote the more frequent use of GFR measurement in clinical practice for those patients with early non-azotemic CKD where current markers may be insensitive. A further interesting concept, which warrants further evaluation in veterinary medicine, is the use of estimated GFR calculations (eGFR) which have the potential to improve sensitivity of our commonly used serum or plasma based markers.

Other potential markers of GFR yet to be explored: B-trace protein a lipocalin glycoprotein

Markers of tubulointerstitial injury

Markers of tubulointerstitial injury reflect either change in function of the tubular cells, such as altered reabsorption of low molecular weight proteins, or provide evidence of direct cellular damage, such as release of enzymes from the brush border to the urine or epithelial cell markers.

Proteinuria: Assessment of proteinuria has become part of the routine for assessment of renal function and is a fundamental part of IRIS staging. Proteinuria as assessed by UPC gives a global assessment of the quantity of protein being excreted by the kidney and, when renal in origin, can imply a change in renal hemodynamics, tubular handling of protein or when marked (UPC >2.0) is more likely to imply glomerular disease. In feline CKD UPC has been associated with survival and identified as a predictor of both the development of azotemia and having a more progressive phenotype of CKD.[54-56] A similar association with progression of CKD has been identified in dogs.[57] Although quantifying individual proteins such as albumin (urine albumin:creatinine, UAC) have been explored, there is yet to be convincing evidence that assessing UAC out-performs UPC.

Low molecular weight proteins: Low molecular weight (LMW) proteins should be freely filtered by the glomerulus and reabsorbed by megalin and cubulin mediated endocytosis in the proximal tubules. The presence of tubulointerstitial damage may therefore impede reabsorption resulting in increased concentrations of urinary LMW proteins. Specific proteins that have undergone investigation in veterinary medicine include direct quantification of retinol binding protein (RBP), β_2-microglobulin and cystatin-C.[54,58,59 60-63] However, to date these LMW proteins have not been documented to out-perform evaluation of UPC in the prediction of renal damage. High molecular weight (HMW) proteins such as IgG (160KDa) have also been investigated in relation to glomerular disease but again, whether assessment of IgG provides additional benefit to evaluation of UPC remains to be determined.[60,61,64]

Urinary enzymes: Enzymes located either within or on the brush border of tubular cells can be released into urine during cellular damage and may therefore represent biomarkers of tubular injury. Knowledge of the location of tubular enzymes and their detection in urine could therefore give additional information regarding the region of kidney damage. Urine enzymes that have been investigated include N-acetyl-β-D-glucosaminidase (NAG), γ-glutamyl-transferase (GGT) and the feline specific enzyme, cauxin. One of the primary advantages in veterinary medicine of assessing enzymuria is the potential to quantify enzymatic activity rather than the requirement for cross reactivity or species specific antibodies for the detection of proteins.

Due to its superior stability with storage, NAG index (NAG activity as a ratio to urine creatinine) has received the most attention in association with CKD. In dogs with X-linked hereditary nephritis (XLHN) although urinary biomarkers including NAG index increased prior to the onset of azotemia this finding did not precede onset of proteinuria as indicated by UPC.[60] In cats, no association was found between NAG index and plasma creatinine concentration although there was a moderate positive association with proteinuria.[65] In a prospective study evaluating predictors of the development of azotemia in cats, although NAG index was significantly higher in those cats that went on to develop azotemia compared to those that remained non-azotemic, it did not provide additional information compared to assessment of UP/C.[54]

Cauxin is a 70KDa feline specific carboxylase enzyme that has been located to the proximal tubule. Cauxin has a biologically important role in the production of the pheromone, felinine. Studies have investigated urine cauxin to creatinine ratio as a potential marker of proximal tubular damage in cats > 9 years. However, no association was found between urine cauxin: creatinine and plasma creatinine concentration although a weak positive association was found with proteinuria. Similar to NAG, although urine cauxin to creatinine ratios were significantly higher in geriatric cats that developed azotemia within a 12 month period there was substantial overlap compared to cats that remained non-azotemic, questioning any future role of cauxin as a predictor of the development of azotemic CKD.[20]

Neutrophil gelatinase associated lipocalin (NGAL): Neutrophil gelatinase-associated lipocalin (NGAL) is a 25 KDa protein expressed not only by neutrophils but also by epithelial cells and specifically the proximal tubular cells within the kidney. In human medicine, NGAL may exist as a monomer, dimer or complexed to matrix-metalloproteinase-9 (MMP-9) and a recent study has identified these different forms of NGAL exist in canine urine.[66] A canine specific NGAL ELISA has been developed and validated.[60] Studies have evaluated NGAL both in urine where it may act as a marker of tubular damage and also in serum/plasma where it may be indicative of renal function. Studies indicate that urine NGAL:creatinine ratio may be increased in dogs with CKD although there is overlap compared to dogs with AKI.[67,68] Caution must be taken, to avoid assessment in dogs where urine NGAL may be falsely increased due to the presence of a urinary tract infection.[69] Serum/plasma NGAL concentration have been reported to be higher in dogs with CKD than controls, to increase with advancing stage of CKD but are typically lower than in dogs with AKI.[67,70] To date NGAL quantification in cats has not been reported in the peer reviewed veterinary literature.

Other markers of tubulointerstitial injury yet to be explored: Other markers of tubulointerstitial injury that have been evaluated in human medicine but which remain to be comprehensively evaluated in veterinary medicine include kidney injury molecule 1 (KIM1)[71], clusterin and liver-type fatty acid binding protein (L-FABP). KIM-1 has been identified in a single publication in urine and kidney tissue from cats with acute kidney injury.[72]

Inflammation, fibrosis, hypoxia and oxidative stress

Urinary markers of inflammation (C-reactive protein, IL-18, Tenascin, TIMP-1, PTX3, sTNFrii), fibrosis (TGF-β) and oxidative stress (Ox-LDL, TBARS, AGE, plasma and urinary F2-isoprostanes, protein reduced thiols, protein carbonyls and antioxidant enzyme activities (e.g. superoxide dismutase, glutathione peroxidase, catalase)) have been evaluated in experimental and human studies of kidney disease. Such markers are of interest because they may provide more direct evidence of pathophysiological mechanisms that are taking place within the kidney and contributing to disease progression. To date, preliminary studies have begun to explore oxidative stress,[73,74] urinary cytokine markers [75] and TGF-β[76] in cats with CKD although conclusive evidence of their role as biomarkers is lacking.

Metabolic disorders

Whilst not directly considered a biomarker in the sense of informing about GFR, biomarkers that inform about metabolic derangements occurring as a consequence of reduced renal function can be of considerable clinical benefit. Perhaps the marker that has received the most attention as an early indicator of reduced renal function and the secondary consequence of mineral and bone disorder is fibroblast growth factor 23 (FGF23).[77] In particular, studies support that FGF23 is elevated in advance of the development of azotemic CKD in cats and that FGF23 may indicate alteration in phosphorus homeostasis in advance of changes that occur to either parathyroid hormone or plasma phosphorus concentrations.[78-80] In addition, FGF23 concentrations have been shown to decrease in patients where a renal diet is introduced even if their phosphorus concentration was initially within IRIS target and did not significantly change with dietary modification.[81] Further work is required to determine whether this modulation in FGF23 translates to survival or outcome benefit. Like many other markers however, currently FGF23 quantification is available only on a research basis.

This is an exciting time in biomarker discovery and development. It is perhaps disappointing that looking at the past decade of work there has been really only one biomarker that has gone from discovery to becoming available as a clinically usable diagnostic test. However, the situation is not greatly different in human medicine but the process of biomarker discovery is proving informative in terms of improving our understanding of the disease process of CKD.

REFERENCES

1. Fassett RG, Venuthurupalli SK, Gobe GC, et al. Kidney Int 2011;80:806-821. 2. D'Marco L, Bellasi A, Raggi P. Disease Markers 2015;2015:586569. 3. Pouchelon JL, Atkins CE, Bussadori C, et al. The Journal of Small Animal Practice 2015;56:537-552. 4. Hartung E. Pediatric Nephrology 2015:1-11. 5. Cisek K, Krochmal M, Klein J, et al. Nephrology Dialysis Transplantation 2015. 6. Mischak H, Delles C, Vlahou A, et al. Nat Rev Nephrol 2015;11:221-232. 7. Barrios C, Spector TD, Menni C. Archives of Biochemistry and Biophysics 2016;589:81-92. 8. Breit M, Weinberger KM. Archives of Biochemistry and Biophysics 2016;589:62-

80. 9. Zhao Y-Y, Vaziri ND, Lin R-C. In: Gregory SM, ed. Advances in Clinical ChemistryElsevier; 2015:153-175. 10. Keller BJ, Martini S, Sedor JR, et al. Kidney Int 2012;81:14-21. 11. Boger C, Heid I. Kidney & blood pressure research 2011;34:225 - 234.12. Ko Y-A, Susztak K. Seminars in nephrology 2013;33:10.1016/j.semnephrol.2013.1005.1007 13. Wing MR, Ramezani A, Gill HS, et al. Seminars in nephrology 2013;33:10.1016/j.semnephrol.2013.1005.1008. 14. Dwivedi RS, Herman JG, McCaffrey T, et al. Kidney international 2011;79:10.1038/ki.2010.1335. 15. Rudnicki M, Perco P, D'haene B, et al. European Journal of Clinical Investigation 2015:n/a-n/a. 16. Trionfini P, Benigni A, Remuzzi G. Nat Rev Nephrol 2015;11:23-33. 17. Nabity MB, Lees GE, Dangott LJ, et al. Veterinary clinical pathology / American Society for Veterinary Clinical Pathology 2011;40:222-236. 18. Forterre F, Raila J, Schweigert FJ. J Vet Diag Invest 2004;16:271-277. 19. Jepson RE, Coulton GR, Cowan ML, et al. American Journal of Veterinary Research 2013;74:333-342. 20. Jepson RE, Syme HM, Markwell P, et al. American Journal of Veterinary Research 2010;71:982-987. 21. Nabity MB, Lees GE, Dangott LJ, et al. Veterinary Clinical Pathology 2011;40:222-236. 22. Ferlizza E, Campos A, Neagu A, et al. The Veterinary Journal 2015;204:73-81. 23. Decramer S, de Peredo AG, Breuil B, et al. Molecular & Cellular Proteomics 2008;7:1850-1862. 24. Laterza OF, Price CP, Scott MG. Clinical Chemistry 2002;48:699-707. 25. Almy FS, Christopher MM, King DP, et al. Journal of Veterinary Internal Medicine 2002;16:45-51. 26. Wehner A, Hartmann K, Hirschberger J. Journal of the American Animal Hospital Association 2008;44:131-138. 27. Miyagawa Y, Takemura N, Hirose H. Journal of Veterinary Medical Science 2009;71:1169-1176. 28. Ghys L, Paepe D, Lefebvre H, et al. Proceedings of 25th ECVIM-CA congress, Lisboa, Portugal 2015. 29. Ghys LFE, Paepe D, Duchateau L, et al The Veterinary Journal 2015;204:168-173. 30. Paepe D, Ghys LF, Smets P, et al. J Feline Med Surg 2014. 31. Ghys LFE, Meyer E, Paepe D, et al. Veterinary Clinical Pathology 2014;43:226-234. 32. Ghys L, Paepe D, Smets P, et al. Journal of Veterinary Internal Medicine 2014;28:1152-1164. 33. Ghys LF, Paepe D, Taffin ER, et al. Journal of Feline Medicine and Surgery 2015. 34. Bode-Böger SM, Scalera F, Kielstein JT, et al. Journal of the American Society of Nephrology 2006;17:1128-1134. 35. Kielstein JT, Tsikas D, Fliser D. Eur J Clin Pharmacol 2006;62:39-44. 36. Fliser D. Eur J Clin Invest 2005;35:71-79. 37. Jepson RE, Syme HM, Vallance C, et al. J Vet Intern Med 2008;22:317-324. 38. Pedersen LG, Tarnow I, Olsen LH, et al. Research in Veterinary Science 2006;80:336-342. 39. Braff J, Obare E, Yerramilli M, et al. Journal of Veterinary Internal Medicine 2014;28:1699-1701. 40. Hall JA, Yerramilli M, Obare E, et al. Journal of Veterinary Internal Medicine 2015;29:808-814. 41. Hall JA, Yerramilli M, Obare E, et al. The Veterinary Journal 2014;202:588-596. 42. Hall JA, Yerramilli M, Obare E, et al. Journal of Veterinary Internal Medicine 2014;28:1676-1683. 43. Finch N. J Feline Med Surg 2014;16:736-748. 44. Von Hendy-Willson VE, Pressler BM. Vet J 2011;188:156-165. 45. Finch NC, Heiene R, Elliott J, et al. J Vet Intern Med 2013;27:782-790. 46. Paepe D, Lefebvre HP, Concordet D, et al. J Feline Med Surg 2014. 47. Finch NC, Syme HM, Elliott J, et al. Am J Vet Res 2011;72:1652-1659. 48. Granger LA, Armbrust LJ, Rankin DC, et al. Veterinary radiology & ultrasound 2012;53:181-188. 49. Schmidt DM, Scrivani PV, Dykes NL, et al. Am J Vet Res 2012;73:463-469. 50. Steinbach SML, Sturgess CP, Dunning MD, et al. Research in Veterinary Science 2015;100:263-264. 51. Steinbach S, Krolop N, Strommer S, et al. PLoS One 2014;9:e111734. 52. Alexander K, Authier S, del Castillo JR, et al. Contrast media & molecular imaging 2010;5:133-139. 53. Wang E, Meier DJ, Sandoval RM, et al. Kidney Int 2012;81:112-117. 54. Jepson RE, Brodbelt D, Vallance C, et al. J Vet Intern Med 2009;23:806-813. 55. Syme HM, Markwell PJ, Pfeiffer D, et al. J Vet Intern Med 2006;20:528-535. 56. Chakrabarti S, Syme HM, Elliott J. Journal of Veterinary Internal Medicine 2012;26:275-281. 57. Jacob F, Polzin DJ, Osborne CA, et al. J Am Vet Med Assoc 2005;226:393-400. 58. van Hoek I, Lefebvre HP, Peremans K, et al. Domest Anim Endocrinol 2009;36:45-56. 59. van Hoek I, Daminet S, Notebaert S, et al. J Immunol Methods 2008;329:208-213. 60. Nabity MB, Lees GE, Cianciolo R, et al. Journal of Veterinary Internal Medicine 2012;26:282-293. 61. Maddens B, Heiene R, Smets P, et al. Journal of Veterinary Internal Medicine 2011;25:1075-1083. 62. Raila J, Brunnberg L, Schweigert FJ, et al. American Journal of Veterinary Research 2010;71:1387-1394. 63. Williams TL, Archer J. Journal of Small Animal Practice 2015:n/a-n/a. 64. Smets PMY, Meyer E, Maddens BEJ, et al. Journal of Veterinary Internal Medicine 2010;24:65-72. 65. Jepson RE, Vallance C, Syme HM, et al. Am J Vet Res 2010;71:241-247. 66. Hsu W, Chiou H, Tung K, et al. Veterinary Research 2014;10:202. 67. Steinbach S, Weis J, Schweighauser A, et al. Journal of Veterinary Internal Medicine 2014;28:264-269. 68. Segev G, Palm C, LeRoy B, et al. Journal of Veterinary Internal Medicine 2013;27:1362-1367. 69. Daure E, Belanger MC, Beauchamp G, et al. Research in Veterinary Science 2013;95:1181-1185. 70. Ahn H-J, Hyun C. Veterinary Record 2013;173:452. 71. Lim AI, Tang SCW, Lai KN, et al. Journal of Cellular Physiology 2013;228:917-924. 72. Bland SK, Côté O, Clark ME, et al. Journal of Veterinary Internal Medicine 2014;28:1454-1464. 73. Krofič Žel M, Tozon N, Nemec Svete A. Journal of Veterinary Internal Medicine 2014;28:130-136. 74. Keegan RF, Webb CB. Journal of Veterinary Internal Medicine 2010;24:514-519. 75. Habenicht LM, Webb TL, Clauss LA, et al. J Feline Med Surg 2013;15:99-104. 76. Arata S, Ohmi A, Mizukoshi F, et al. J Vet Med Sci 2005;67:1253-1255. 77. Wolf M. Kidney Int 2012;82:737-747. 78. Geddes RF, Finch NC, Elliott J, et al. Journal of Veterinary Internal Medicine 2013;27:234-241. 79. Finch NC, Geddes RF, Syme HM, et al. Journal of Veterinary Internal Medicine 2013;27:227-233. 80. Geddes RF, Elliott J, Syme HM. Journal of Veterinary Internal Medicine 2015;29:1494-1501. 81. Geddes RF, Elliott J, Syme HM. Journal of Veterinary Internal Medicine 2013;27:1354-1361.

Infection Control in Veterinary and Human Hospitals

Jason W. Stull, VMD, MPVM, PhD, DACVPM
Columbus, OH

INTRODUCTION

Hospital-associated infections (HAIs) are infections acquired by patients during hospitalization and are an inherent risk in both human and veterinary medicine. In human hospitals, HAIs are a well-recognized contributor to illness and death, with an estimated 5% of US patients developing a HAI, and tens of thousands dying each year.[1] Substantial direct and indirect costs are estimated to result from HAIs (eg., in the US $28-45 billion in direct costs annually).[2] While there are limited data for these estimates in veterinary medicine, similar HAI risks have been reported, such as HAIs in 16% of ICU patients.[3]

Given these health risks and associated costs, the field of infection control has become well-established in human medicine to limit HAIs. In comparison, veterinary medicine (notably small animal medicine) has lagged behind human medicine in infection control recommendations and uptake. Due to limited applied research in veterinary medicine related to infection control, most veterinary recommendations are modeled from human hospitals. Despite the numerous similarities between infection control goals and challenges in the veterinary and human healthcare environments, there are also considerable differences. A considered comparison between human and veterinary infection control practices and structure is important as we evaluate the infection control needs and future directions in veterinary medicine.

PATHOGENS AND TYPES OF HOSPITAL-ASSOCIATED INFECTIONS

The human and veterinary fields share many of the same pathogen concerns and types of HAIs. In human medicine, urinary tract infections (UTIs), pneumonia, surgical site infections (SSIs), and blood-stream infections (BSIs) account for approximately 80% of all HAIs.[4] In veterinary medicine, these sites (along with infectious diarrhea) are likely to be the most common presentations for HAIs.[5,6] Where data are available, similar pathogens appear to be the main culprits for HAIs in both veterinary and human medicine.[5,7] Bacteria are responsible for approximately 90% of human HAIs, with commonly identified groups including *Staphylococcus aureus* (16%), *Enterococcus* spp. (14%), *Escherichia coli* (12%), coagulase-negative staphylococci (11%), *Candida* spp. (10%), *Klebsiella* spp. (8%), *Pseudomonas aeruginosa* (8%), *Enterobacter* spp. (5%), and *Acinetobacter baumannii* (2%).[7] Data for veterinary medicine are less clear. Based on reported HA-outbreaks in veterinary medicine, several important pathogens responsible for HAIs are identifiable: *S. aureus* and *Staphylococcus pseudintermedius*, enterococci, *Salmonella* spp., *Acinetobacter* spp., *Pseudomonas* spp., and *E. coli.*[5]

KEY PREVENTION PRACTICES

In human and veterinary medicine, several areas have been identified as key to successful reduction of HAIs. These areas include appropriate timing and use of hand hygiene, contact precautions (personal protective equipment (PPE) such as gloves and gowns), environmental disinfection, antimicrobial stewardship, patient procedures (e.g. use of isolation), and rapid identification of infectious patients. Despite the perceived importance of these areas by infection control practitioners there remains poor use of these practices for both medical fields. For instance, appropriate hand hygiene occurs in fewer than 40% of the opportunities for which it is indicated in human and veterinary healthcare workers,[8,9] while PPE is either not used or inappropriately used resulting in frequent contamination of the worker.[10,11]

HOSPITAL DEMOGRAPHICS AND FINANCIAL MODELS

Hospital demographics in human and veterinary medicine are markedly different. For instance, in the United States there are approximately 6,000 human hospitals, with close to one million staffed beds and 37 million annual admissions. On average, each hospital has 60 employed physicians. In 2006, the average length of stay was 4.8 days.[12] As a comparison, there are approximately 30,000 veterinary practices in the United States with on-average two veterinarians per practice.[13] Other demographic information for the veterinary sector is largely unknown, such as average length of patient stay, but anecdotally is likely much less than 4 days. Given these differences it would be expected that veterinary facilities overall have limited space, funds and other resources that can be dedicated to infection control. For instance, 65% (n = 66) of the small animal veterinary practices in one study reported they did not have a dedicated isolation area.[14] Additionally, the ability to detect the relatively infrequent HAIs with a surveillance system, irrespective of its sensitivity and specificity, may also be difficult with the lower number of in-patients per veterinary facility. At the same time, with smaller-sized practices and shorter patient hospital stays, veterinary facilities may also have a lower risk for HAIs as compared to human hospitals. They may be better equipped to rapidly develop and implement infection control practices if interested, and these practices may be more likely to be effective.

Access-to-care has become an important area for veterinary and human medicine alike. Significant efforts have been expended attempting to "encourage" those in-need to seek care. Although varying largely by country and healthcare model, human healthcare is more accessible than veterinary through the use of national programs and assistance. A direct pay-for service remains in place for veterinary practices, whereby only those able to afford care generally seek it. This may result in a potentially on-average higher socio-economic status clientele for the veterinary as compared to human healthcare field. The relevance of this difference related to infection control is difficult to forecast, but might result in different barriers to effective infection control practices.

NATURE OF PATIENT CONTACT

For the average patient encounter, the human and veterinary settings are likely to differ. Veterinary care is dependent on close contact with the patient for most encounters, through the need of restraint and similarly close components of the physical examination and diagnostic process. In general, this level of direct patient contact is expected to be less common in human healthcare. An important exception may be intensive care where close and frequent patient contact is expected for both settings. Interestingly, it is this setting where most HAIs are identified in veterinary and human medicine.[3,15]

The ability to control patient contact it also likely to differ between these two settings. In human healthcare, restricting patient movement can be challenging and effective cleaning and disinfection of the large, patient room extremely difficult. This is evidenced by an increased risk for patients housed in a room previously occupied by a patient with methicillin-resistant *S. aureus* (MRSA) or *Clostridium difficile*.[16] Theoretically, controlling patient contact should be more achievable in the veterinary setting, where healthcare personnel "decide" what level of contact is acceptable and housing is relatively contained and surfaces easy to clean and disinfect. Despite these advantages in veterinary medicine, cage/run cleaning and disinfection is generally poor (26% of surfaces adequately cleaned in animal housing wards)[17] and HAI outbreaks have resulted from dog-dog contact in practices.[18]

DECISION-MAKING AND INFRASTRUCTURE

Dependent on country, some infection control-related decisions in human healthcare are determined by groups outside of the specific facility (e.g., insurance carriers, regulatory bodies). For instance, there are specific guidelines and standards that are required when managing patients with tuberculosis in human hospitals.[19] Outside of foreign animal and national animal disease programs, such oversight is rare in veterinary medicine. Instead, decisions are made by the practitioner and/or practice owner (ideally with direction from an infection control plan). Since relatively few veterinary practices have infection control plans (<30%),[10] these decisions may not be made with infection control in-mind and client financial concerns may play a larger role. Thus, the veterinary field may have more control at the local level (for good or bad), but need strong protocols to offset potential barriers (e.g., reluctance from clientele). Additionally, perhaps on a positive note, the lack of support and regulatory guidance forces an individualized risk-based approach to infection control by veterinary facilities. Unfortunately expertise in this area is needed for this individualized approach to be useful.

Based on the demographics previously described, it is not surprising that infection control-dedicated resources and infrastructure will be markedly different between the healthcare sectors. In many jurisdictions infection control-dedicated staff, such as infection control practitioners with established minimum standards, is required at human healthcare facilities. A similar requirement does not extend to the veterinary sector, and in fact most practices are unlikely to have staff whose role specifically includes infection control practices. Further, training specific for veterinary infection control is limited, while this training is well-established in human healthcare. Finally, the use of computerized record keeping is markedly different between the groups, with many veterinary practices either not having electronic records, or using a system that does not lend itself to infection control needs. Most recently, the importance of electronic records has been emphasized in recommended strategies to achieve compliance with antimicrobial stewardship programs.[20] These strategies include the use of post-prescription review and computer-based decision support, both of which are cumbersome and unlikely to be successful without electronic record systems that can meet these specific needs.

AVAILABLE RESOURCES

Clinical laboratories are frequently an important supporting tool in infection control. Rapid turn-around of test results can assist in prompt isolation, PPE use and infection control-related treatment decisions. Additionally, laboratories can provide critical assistance in antimicrobial stewardship decisions. Laboratories are general well-integrated into the human healthcare field, however outside of academia and large referral practices, this is not the case in veterinary medicine. This difference is an important weakness for veterinary medicine. Although it is possible that depending less on immediate diagnostic results and moving toward a syndromic approach may be overall beneficial to achieve infection control goals (e.g., novel pathogens, insensitive tests).

Written and on-line guidance on the topic of infection control is markedly different between the fields. A PubMed search for infection control/biosecurity articles displayed over 194,000 hits when restricted to humans as comparted to 73,000 for all other animals. In reality, this comparison does not do justice to the limited guidance for infection control for veterinary practitioners. Until recently, a comprehensive collection of targeted articles on small animal infection control was not available.

SUPPORTING/REGULATORY AGENCIES

Surveillance programs allow for early recognition of HAI transmission and outbreaks. They also provide 'benchmarking' data that can be used to more accurately and rapidly identify changes in HAI rates. In human medicine, regulatory and supporting agencies (e.g., national and regional public health programs) generally have well-established surveillance programs. These include specific, well-defined case definitions as well as on-going active, passive, and syndromic surveillance efforts. Examples include mandatory MRSA screening for defined high risk groups, reporting of HAIs for pre-determined pathogens (e.g., HA-MRSA bacteremia) and standard case definitions for identifying and classifying surgical site infections.[21] Without a regulatory body or wider organizational system, case definitions and surveillance programs are generally not available in veterinary medicine. Several independent regional efforts are underway, but without an overseeing body they are unlikely to become commonplace.

EVIDENCE-BASED, OUTCOME ORIENTED RECOMMENDATIONS

Recently, there has been a push toward evidence-based practice (EBP) as a problem-solving approach to health care delivery. This same principle extends into the infection control field, where best evidence from well-designed studies are integrated with an infection control practitioner's expertise for the setting to yield practical solutions with proven results. In human medicine there are many developed EBPs (e.g., MRSA decolonization or routine chlorhexidine bathing protocols).[22] To-date, there are few such practice guidelines in veterinary medicine. Given the varying interaction between the pathogen-host-environment in the veterinary as compared to human field, blindly incorporating such human EBPs is likely to be problematic.

IMPLICATIONS OF HOSPITAL-ASSOCIATED INFECTIONS

Regardless of setting, HAIs can have profound effects on an individual healthcare facility. Between 2003 and 2008, 82% of veterinary teaching hospitals in North America and Europe reported at least one HAI outbreak. Many of these outbreaks required restricted patient admissions or closure of the hospital or section.[6] Similar outcomes have occurred in human healthcare facilities.[23] In addition to directly reducing revenue through section or hospital closure, indirect effects due to damaged facility and healthcare staff reputation may occur. In addition, in the human healthcare field governmental clawbacks (i.e., the recovery of money already disbursed), on payments and lawsuits may further damage the healthcare facility. In veterinary medicine, the direct payment model negates clawbacks, however there may be internal (practice) or external (client) pressure to discount bills. Lawsuits do occur in relation to infection control in veterinary medicine, but are relatively infrequent at present.[24]

CONCLUSIONS

Hospital-associated infections have been reported in human and veterinary medicine. In veterinary medicine their frequency is likely to increase with the rise in intensive care practices, prolonged hospitalizations, and use of invasive devices. Given a number of important differences between the human and veterinary fields relative to demographics, infection control resources, internal/external support, decision-making, and infrastructure the most useful approaches to address individual facility infection control needs are likely different. Further attention to these differences with additional targeted efforts toward the development of veterinary-specific evidence-based recommendations is needed.

REFERNCES

[1]Klevens R. M., et al. Public Health Rep 2007;122:160-166. [2]Scott R. D. Division of Healthcare Quality Promotion, Centers for Disease Control and Prevention 2009. [3]Ruple-Czerniak A., et al. J Vet Intern Med 2013;27:1392-1399. [4]Horran T. C., et al. Am J Infect Control 2008; 36:309-332. [5]Stull, J. W., et al. Vet Clin North Am Small Anim Pract. 2015;45: 217-233. [6]Benedict K. M., et al. J Am Vet Med Assoc 2008;233:767-773. [7]Sievert D. M., et al. Infect Control Hosp Epidemiol 2013; 34:1-14. [8]Erasmus, V., et al. Infect Control Hosp Epidemiol 2010;31: 283-294. [9]Anderson, M. E. C., et al. BMC Vet Res 2014;10: 1. [10]Wright, J. G., et al. J Am Vet Med Assoc 2008;232:1863-1872. [11]Tomas, M. E., et al. JAMA 2015;175: 1904-1910. [12]Nettleman, M. D. et al. Principles of Healthcare Epidemiology 2012. [13]AVMA. Veterinary Market Statistics 2005. [14]Murphy, C. P., et al. Zoonoses Public Health 2010;57:429–438. [15]National Center Health Statistics 2014. [16]Donskey, C. J. Am J Infect Control 2013; 41:S12-9. [17]Weese, J. S. et al. Vet Rec 2012;171. [18]Weese, J. S., et al. Can Vet J 2013;54:79-82. [19]Centers for Disease Control and Prevention. MMWR 2005;54: 2-107. [20]Guardabassi, L., et al. Vet Clin North Am Small Anim Pract. 2015;45: 361-376. [21]Centers for Disease Control and Prevention. Procedure-associated events: Surgical site infections 2013. [22]Liu, C., et al. Clin Infect Dis 2011;52: e18-e55. [23]Rhinehart, E., et al. Am J Infect Control 2012;40: 2-8. [24]Marsh, A. E., et al. Vet Clin North Am Small Anim Pract. 2015;45: 393-408.

Emerging and Endemic MDR Pathogens in the Veterinary Hospital

J. Scott Weese, DVM, DVSc, DACVIM
Guelph, Ont, Canada

INTRODUCTION

Ready access to antimicrobials has revolutionized veterinary medicine, allowing for advances in care that would otherwise have been impossible. Unfortunately, the parallel emergence and dissemination of antimicrobial resistance threatens these gains. The ability of antimicrobials to resist the effects of antibiotics existed before the advent of medical antimicrobials, yet use (prudent use and misuse) of antimicrobials almost inevitably leads to some pressure for resistance. Resistance of any pathogen to any antimicrobial can be a clinically important event, but certain bacterium-resistance combinations are of particular concern because of the potential for broad impacts on animal (and sometimes human) health.

STAPHYLOCOCCI

From the first emergence of penicillin resistance shortly after the discovery of penicillin to identification of virtually pan resistant strain in humans, staphylococci have been notoriously adept at becoming resistant to antimicrobials. The emergence of methicillin resistant (MR) staphylococci, particularly *S. pseudintermedius* (MRSP) and *S. aureus* (MRSA) as important veterinary pathogens in the 1990s and 2000s was a watershed period in veterinary medicine, necessitating changes in approaches to common infectious diseases. MR staphylococci are of concern for many reasons, including inherent resistance to many commonly used drugs (i.e. beta-lactams), commonness of acquisition of resistance to other antimicrobials, commonness of infections (particularly skin and soft tissue infections) and the potential for zoonotic transmission.[1-5] There has been widespread dissemination of MRSA and MRSP in animals, both in clinically inapparent carriers and animals with infections (predominantly skin and soft tissue infections) and these can be considered endemic (if not hyperendemic) in most regions rather than emerging issues. However, continued acquisition of resistance genes can pose new challenges, particularly with MRSP. As a highly clonal organism, a limited number of MRSP genotypes are found internationally[6] and MRSP has shown a concerning propensity to rapidly acquire resistance. Interestingly, and rather inexplicably, MRSP isolates from animals tend to be much more resistant than MRSA in humans, despite the fact that MRSA emerged and disseminated decades before MRSP. MRSP isolates are commonly susceptible to a narrow list of drugs such as amikacin and chloramphenicol. As resistance to these is now encountered in some regions, the highly rapid and clonal spread of MRSP suggests the potential for similarly rapid expansion of new resistance genes in the predominant clones. Continued emergence and dissemination of resistance in MRSP may pose even greater challenges in the limited range of effective drugs further decreases.

ENTEROCOCCI

This genus includes a variety of species of varying clinical relevance. Most infections are caused by two species, *E. faecium* and *E. faecalis,*[7,8] both of which can be associated with multidrug resistance. Enterococci are inherently resistant to a variety of drug classes (e.g. penicillin, clindamycin, trimethoprin), some drugs classes have poor efficacy in vivo (e.g. cephalosporins) and acquisition of resistance genes is not uncommon. Therefore, they are at best potentially susceptible to only a limit range of 'routine' veterinary drugs, and multidrug resistance is not uncommon. Enterococci resistant to all routinely tested drugs are not unusual and pose challenges.

Despite the potential for highly drug resistant strains, enterococci tend to be of limited virulence and infections are most common in highly compromised hosts. Enterococci may be isolated as causes of disease but are also often found as contaminants, colonizers or of uncertain clinical significance, complicating treatment decisions. Multidrug resistant enterococci can be considered endemic pathogens. They are, and have been, widespread in animals as commensals and opportunists for decades. As opposed to staphylococci and Enterobacteriaceae, changes in resistance trends do not appear to have been profound in enterococci in recent years. Certainly, highly antimicrobial resistant isolates are not uncommon,[7-10] but this has been the case for some time and it is unconvincing that there has been a remarkable change in the epidemiology of either disease or antimicrobial resistance in this genus. Perhaps the main change that has occurred over the past decade has been increasing recognition that enterococci may not need to be treated in many cases. As bacteria of limited virulence, there is increasing scrutiny being applied to treatment decisions in humans,[11] and animals (personal observation). In many situations, isolation of enterococci (whether multidrug resistant or not) may not be of clinical significance. This is a particularly important aspect to consider with multidrug resistant isolates where drugs such as linezolid or vancomycin may be the only effective options based on *in vitro* testing, but where treatment of any sort may not be required.

Vancomycin resistant enterococci (VRE) are important pathogens in humans.[12] Their virulence still tends to be limited and disease occurs most commonly in compromised hosts, but transmission of VRE within healthcare facilities can be problematic. VRE was a concern in some European countries associated with use of the related drug avoparcin in food animals, with increasing VRE rates in food and companion animals.[13,14] However, since this drug was banned in the EU in 1997, reported VRE rates from companion animal isolates tend to be low.[13,15,16] Despite the importance of VRE in human hospitals, reports of VRE infections in companion animals are rare and difficulties in treating enterococci in companion animals are more related to resistance to typical veterinary drugs than focused on resistance to the rarely used drug vancomycin.

ENTEROBACTERIACEAE

Enterobacteriaceae is a family that includes various animal and human pathogens, including *E. coli, Enterobacter, Proteus* and *Klebsiella*, among other relevant genera. This family therefore encompasses a wide range of opportunistic pathogens that are commonly found as part of the commensal microbiota. The large gastrointestinal population of these bacteria creates ample potential for exposure to antimicrobials, intra- and inter-species transmission of plasmid-mediated resistance and creation of a large pool of resistance. Multidrug resistant Enterobacteriaceae have emerged as an important problem, particularly in recent years. While resistance to various antimicrobials has been widely reported in Enterobacteriaceae, the emergence and widespread dissemination of some multidrug resistance genes has greatly heightened concerns. Multidrug resistant Enterobactericeae might represent the most dramatic and import shift in resistance patterns in companion animals over the past few years. A few different types of resistance, mediated by a wide (and sometime confusing) collection of resistance genes, can be involved.

Extended spectrum beta-lactamases (ESBLs)

Production of beta-lactamases, enzymes that are able to hydrolyze beta-lactam antimicrobials, is widespread in Enterobacteriaceae (and other bacteria) and is well known. These are typically easily managed through use of potentiated penicillins (e.g amoxicillin/clavulanic acid), cephalosporins or other drug classes. However, extended spectrum beta-lactamases (ESBLs) are enzymes that confer a broader degree of resistance. ESBLs are effective against extended spectrum cephalosporins (e.g. cefpodoxime, cefovecin, ceftiofur, cefotaxime, ceftazidime) and monobactams (e.g. aztreonam) but are ineffective against cephamycins (e.g. cefoxitin) and carbapenems (e.g. meropenem).[17] They are inhibited by beta-lactamase inhibitors such as clavulanic acid and can be mediated by a variety of genes, such as $bla_{CTX\text{-}M}$, bla_{TEM} and bla_{SHV} that are typically encoded on mobile genetic elements and can be transmitted readily between various bacteria.[18-23] ESBL-production can be found in a range of Gram negatives, but is most commonly reported in *Klebsiella, E. coli* and *Enterobacter*. ESBL-producing bacteria disseminated widely in humans in a highly diverse population of Enterobacteriaceae, and are significant causes of disease, particularly in hospitals and longterm care facilities. While lagging behind humans, ESBLs are now being identified in small animal patients.[18,19,22-28] This may be a reflection and consequence of the emergence of ESBLs in humans, as bacteria carrying ESBLs that have been identified in dogs can be the same strains as are found in people,[28-30] a result of exposure through food and food animals (including the environment) or through development of resistance within companion animals. The similarity of some ESBL genes and ESBL-producing bacteria found in humans and companion animals human-animal transmission may be an important route of infection of veterinary patients, but also indicates that there must be concern about the potential for zoonotic transmission from infected or colonized animals, although the true role of animals in human infections is not known.[19,26,30-32]

As is the case with most multidrug resistant bacteria, ESBL-producing bacteria cause the same types of infections as their susceptible counterparts. As such infections such as urinary tract infection are most common. A complicating factor with ESBLs is their ability to be harboured by bacteria that are common components of the intestinal microbiota, and a small percentage of healthy dogs and cats are shedding ESBL-producing bacteria at any time.[18,25,32] This can lead to accumulation of this resistance gene in the population, with healthy carriers acting as a source of infection of themselves or others. Exposure to antimicrobials, particularly 3rd generation cephalosporins, is an important risk factor for multidrug-resistant *E. coli* shedding,[25] and the potential for widespread transmission in veterinary hospitals and the community is of concern.

Antimicrobial options for treatment depend on susceptibility to other antimicrobials as ESBL-producing bacteria often possess other resistance determinants.[33] Cefoxitin can be an effective option in some situations; however, response to this antimicrobial can be unpredictable. Dosing regimens (route and frequency) are also problematic for outpatients. Beta lactam/beta-lactamase inhibitor (e.g. beta-lactam with clavulanic acid, sulbactam and tazobactam) combinations may be effective, but that depends on the amount of ESBL production, as high production of ESBLs (especially when more than one gene is involved) can overwhelm the beta-lactamase inhibitor. Amikacin and carbapenems are often the main viable choices, both of which are typically effective but often problematic to use. There is some evidence in humans that cefepime might be effective against ESBLS,[34] but there is little veterinary information.

AmpC beta-lactamases

Similar to ESBLs, AmpC (CMY-2) beta-lactamases confer resistance to penicillins and cephalosporins, but confer a broader degree of resistance than ESBLs, since they are active against cephamycins such as cefoxitin. They are also resistant to inhibition by beta-lactamase inhibitors. Carbapenems or aminoglycosides tend to be the main options for treatment of veterinary patients.

Exposure to raw meat has been associated with increased risk of fecal shedding of multidrug resistant *E. coli*,[25,35] something that is probably more attributable to AmpC-producing *E. coli* than ESBLs, because of the prevalence of AmpC producing bacteria in food animals and food. Otherwise, the epidemiology is still poorly described.

Carbapenemase producing *Enterobacteriaceae* (CPE)

In humans, CPE are a significant concern. Carbapenemases are a diverse group of enzymes with variable efficacy against carbepenems (e.g. meropenem) as well as most other beta-lactams. They confer resistance to penicillins, carbapenems, and, depending on the gene that is involved, potentially all cephalosporins, and therefore have a broad range of resistance that is often complemented with acquired resistance to various other drug classes. The effect of beta-lactamase inhibitors is variable and often weak. CPE are of particular concern because carbapenems represent one of the best and last lines of treatment for infections caused by some multidrug resistant bacteria, such as ESBL-producers. There are only rarely reports of CPE in animals;[30,36] however, anecdotal information suggests that infections are occurring in some regions, particularly in *E coli* from urinary tract infections. Given the perceived rarity of use of carbapenems in veterinary medicine, this might be a reflection spillover of CPE from humans into domestic animals, but that has yet to be confirmed.

As with other resistance mechanisms described above, CPE can become part of the commensal gut microbiota, something that could result in longterm silent shedding that could be difficult to eliminate. The rarity use of carbapenems and uncommon presence of CPE in people in the community should help limit the emergence of this pathogen in animals; however, use of any beta-lactam antimicrobial could select for CPE if an animal is exposed to the bacterium, so diligence is required. Amikacin would likely be a last (or only) line of treatment of companion animals.

New Delhi metallo-beta-lactamase-1 (NDM-1)

This enzyme was first identified in *Klebsiella pneumonia* in 2008 and its origin traced to India, where was subsequently identified to be widespread in various Enterobacteriaceae in humans, animals and the environment. Its emergence has caused much concern because of the high degree of resistance that is present, with some isolates being resistant to all available antimicrobials or susceptible only to a limited range of rather undesirable drugs like colistin.

There has been very little study of NDM-1 producing bacteria in animals; however, they have been identified in *E. coli* from urine samples from dogs and a cat.[37] As with other resistance genes, the potential for carriage of NDM-1 genes in components of the commensal microbiota is of profound concern, along with the potential for transmission of plasmids between bacteria (including zoonotic pathogens). Other bacterial species may carry this mobile gene as well, including other Enterobacteriaceae, *Acinetobacter* and *Salmonella*.[37-40] Currently, NDM-1 producing bacteria are rare in people outside of hospitals, and the risks to animals are probably greatest in those that have direct contact with infected or colonized people or that have exposure to wildlife in areas where this gene is found in the environment (e.g. India).[41] Current risks are probably limited and will depend on the continuing epidemiology in humans. Should NDM-1 producing bacteria cause SSIs in animals, there might be no viable treatment options.

mcr-1

The detection of this resistance gene recently raised much concern because it confers resistance to colistin, a toxic drug but one that is of last resort for some people with multidrug resistant Enterobacteriaceae infections. As another plasmid mediated (and therefore potentially transferrable) resistance gene that can be found in the commensal microbiota, much concern has been expressed about its emergence. Heightening this concern was the finding of *mcr-1* in food animals and food, indicating a potential for broader community exposure, not just exposure of people in hospitals. Currently, the relevance to companion animals is probably greatest from the scrutiny this finding could have on antimicrobial use in veterinary medicine, given the extreme rarity of use of colistin as a therapeutic agent in companion animals and the likely rarity of potential exposure. Yet, as with all of these resistant Enterobacteriaceae, the potential for *mcr-1* possessing bacteria to reside in a poor of healthy individuals in the community, often in bacteria that have various other resistance genes, cannot be ignored.

Fluoroquinolone resistance

Fluoroquinolones can be excellent drugs against Gram negative pathogens but resistance can emerge through various mechanisms, including mutation (e.g. mutations DNA gyrase or topoisomerase IV genes) or acquisition of genes (e.g. those that decrease drug uptake or mediate efflux pumps). Single step mutations may have variable effects on bacterial MICs, ranging from subtle changes in MIC with continued susceptibility to resistance. Accumulation of multiple mutations is most often associated with clinical resistance. Resistance to one fluoroquinolone typically means that resistance to all fluoroquinolones is present; however, there may be some variability between drugs, with MICs often being lower to newer fluoroquinolones such as pradofloxacin.[42] Fluoroquinolones are excellent options when the bacterium is susceptible and can be considered when there is intermediate susceptibility but high drug levels can be achieved (less relevant for surgical site infections) and where other options are less desirable.

OTHERS

Antimicrobial resistance is not restricted to the above-described groups. Some, such as *Serratia, Acinetobacter* and *Pseudomonas* are commonly multidrug resistant, while resistance is uncommon in others such as streptococci. Regardless, virtually any clinically relevant bacterium can be resistant to antimicrobials, and on the individual patient level, that resistance may range from innocuous to life-threatening, largely based on the rapidity of identification of resistance and starting appropriate therapy.

REFERENCES

[1] Weese JS, et al. **Vet Microbiol** 2006;115:148-155.

[2] Manian FA. **Clin Infect Dis** 2003;36:e26-28.

[3] van Duijkeren E, et al. **J Clin Microbiol** 2005;43:6209-6211.

[4] Pottumarthy S, et al. **J Clin Microbiol** 2004;42:5881-5884.

[5] Stegmann R, et al. **J Antimicrob Chemother** 2010;65:2047-2048.
[6] Perreten V, et al. **J Antimicrob Chemother** 2010;65:1145-1154.
[7] Iseppi R, et al. **New Microbiol** 2015;38:369-378.
[8] Kukanich KS, et al. **J Am Anim Hosp Assoc** 2015;51:148-154.
[9] Kataoka Y, et al. **J Vet Med Sci** 2014.
[10] Wong C, et al. **J Vet Int Med** 2015.
[11] Lin E, et al. **Archives of internal medicine** 2012;172:33-38.
[12] Simor AE, et al. **Infect Control Hosp Epidemiol** 2013;34:687-693.
[13] López M, et al. **Zoonoses and Public Health** 2013;60:160-167.
[14] Kruse H, et al. **Microbial drug resistance (Larchmont, NY)** 1999;5:135-139.
[15] Gozalan A, et al. **Journal of medical microbiology** 2015;64:759-766.
[16] Wagenvoort JH, et al. **J Antimicrob Chemother** 2003;52:532.
[17] Rubin JE, et al. **Veterinary Microbiology** 2014;170:10-18.
[18] Baede VO, et al. **Antimicrobial Agents and Chemotherapy** 2015;59:3117-3124.
[19] Bogaerts P, et al. **Microbial drug resistance (Larchmont, NY)** 2015.
[20] Timofte D, et al. **Antimicrob Agents Chemother** 2014;58:789-794.
[21] Ewers C, et al. **Journal of Antimicrobial Chemotherapy** 2014;69:2676-2680.
[22] Albrechtova K, et al. **Microbial drug resistance (Larchmont, NY)** 2014;20:372-375.
[23] Falgenhauer L, et al. **Berliner und Münchener tierärztliche Wochenschrift** 2014;127:458-463.
[24] Dierikx CM, et al. **J Antimicrob Chemother** 2012;67:1368-1374.
[25] Schmidt VM, et al. **Preventive Veterinary Medicine** 2015;119:31-40.
[26] Schaufler K, et al. **Infection ecology & epidemiology** 2015;5:25334.
[27] Wagner S, et al. **Vet Microbiol** 2014;169:171-178.
[28] Ewers C, et al. **Journal of Antimicrobial Chemotherapy** 2014;69:1224-1230.
[29] Ewers C, et al. **Journal of Antimicrobial Chemotherapy** 2010;65:651-660.
[30] Schmiedel J, et al. **BMC Microbiology** 2014;14:187.
[31] Rocha-Gracia RC, et al. **Veterinary journal (London, England : 1997)** 2014.
[32] Wu G, et al. **PLoS ONE** 2013;8:e75392.
[33] Donati V, et al. **PLoS ONE** 2014;9:e90564.
[34] Tamma PD, et al. **Clin Infect Dis** 2013;57:781-788.
[35] Lefebvre SL, et al. **Zoonoses and Public Health** 2008;55:470-480.
[36] Stolle I, et al. **J Antimicrob Chemother** 2013;68:2802-2808.
[37] Shaheen BW, et al. **Antimicrob Agents Chemother** 2013;57:2902-2903.
[38] Barantsevich EP, et al. **J Antimicrob Chemother** 2013.
[39] Borgia S, et al. **Clin Infect Dis** 2012;55:e109-117.
[40] Wang Y, et al. **PLoS ONE** 2012;7:e37152.
[41] Walsh TR, et al. **The Lancet infectious diseases** 2011;11:355-362.
[42] Liu X, et al. **World J Microbiol** 2013;29:347-354.

Creating an infection control culture in veterinary practice

Brandy A. Burgess, DVM, MSc, PhD, DACVIM, DACVPM
Blacksburg, VA

Veterinary hospitals, by their very nature, create a high risk environment for the transmission of infections agents – bringing together animals from many different farms with varying levels of compromise. Outbreaks of healthcare-associated infections commonly occur in veterinary hospitals with 82% of American Veterinary Medical Association (AVMA) accredited veterinary teaching hospitals (VTHs) reporting such events within the previous 5 years [1]. In addition, 50% reported the occurrence of zoonotic infections during the previous 2 years [1]. In order to practice high-quality patient care, infection control should be integrated into the patient management plan to protect the people, the patients, and the hospital.

While veterinarians are in a profession recognized to increase the lifetime risk for contracting a zoonotic infection, there is a general disregard for utilizing those practices known to be effective prevention measures. Among AVMA member veterinarians, only 55.2% of small animal practitioners and 28.1% of large animal and equine practitioners reported always washing their hands prior to eating, drinking, or smoking at work; and only 48.4% of small animal practitioners and 18.2% of large animal and equine practitioners reported always washing their hands between patient contacts [2]. Despite veterinarians expressing concern about risks associated with gastrointestinal parasites and gastrointestinal bacteria, 71.2% and 70.9%, respectively, of small animal practitioners, and 35.7% and 39.8%, respectively, of large animal practitioners, reported not using appropriate personal protective equipment (PPE) when evaluating an animal with gastrointestinal signs [2].

Clearly, the biggest challenge in veterinary infection control is simply to do it. Why don't practitioners use best practices with respect to infection control in the practice of veterinary medicine? While there is a paucity of information in the literature specific to veterinarians, we can glean something from evaluations of this issue in human medicine. Consider the barriers which may affect effective implementation of practice guidelines – such as knowledge, attitudes, and behavior. In a systematic literature review, there were multiple factors contributing to each of these barriers including intrinsic factors such as a general lack of familiarity or of awareness with published practice guidelines, a lack of agreement with specific guidelines or with guidelines in general, a belief in a lack of expected outcome, a lack of self-efficacy, and the habit or routine of previous practices; and extrinsic factors such as guideline characteristics, patient influences, and environmental factors (e.g., time and available resources) [3].

While some of these barriers may be relatively simple to overcome (i.e., time and available resources), what it really comes down to is social cognitive behavior – specifically knowledge, motivation, intention, perception of a threat, outcome expectancy, perceived behavioral control, and social pressure – and understanding what motivates a particular behavior [4]. Case in point – Ignaz Semmelweis, (1818-1865), the father of hand hygiene, deduced in 1846 that patients were dying due to "pathogenic causes" associated with deficient hand hygiene practices by physicians [5]. While the prevention strategy he implemented was epidemiologically sound, it was not well received by colleagues and in fact he was relieved of his hospital duties [4]. So why the failure? Not only did this event occur before the germ theory even existed, but his observations were also not reported in a timely fashion, thus there was a general lack of knowledge leading to disbelief by his colleagues. Rather than educating his colleagues on the threat and expected outcome, he chose a more autocratic approach to intervention implementation. Had he approached this in a different manner – taking into account social behavior theory – results may have differed.

An infection control culture is part-in-parcel with creating a workplace safety culture through the integration of behavior based safety programs and the organization's safety culture [6]. Behavior-based safety programs use a bottom-up approach that rely on continuous processes which are data-driven and setting-specific. These programs define critical behaviors, set performance goals, monitor performance, provide feedback, and strive to reinforce the desired behavior. Organization safety culture uses a top-down approach that is a more subjective, self-sustaining process that typically uses surveys to identify and assess personnel assumptions and beliefs. Findings are then used to stimulate organizational change through participation of personnel including leadership. A strong safety culture should have 3 components; 1) norms and rules which explicitly define risks and provide guidelines; 2) individual and group attitudes that place safety as highly important and that motivates personnel to act in this regard; and 3) use reflexivity to continuously understand risk and its mitigation [7].

In practice, fostering an infection control culture can be very difficult – everything we ask personnel to do is inconvenient. It takes time to wash your hands before and after every patient contact, it takes time to don and doff PPE. So how do veterinary practices develop a strong safety culture of infection control and worker safety when it comes down to influencing behaviors and attitudes? The Ecological Model of Behavioral Change suggests that behavior is both affected by and affects several levels of influence including intra-personal factors (e.g., knowledge, attitudes, beliefs), inter-personal factors (e.g., social identity, support network, peer group pressure), community factors (e.g., social networks and norms), and institutional factors (e.g., administrative support) [4]. When applying this to organizational culture – consider that this is a multidimensional concept with core assumptions being at its deepest level, followed by beliefs and values, and finally norms at its most superficial level [7]. As such, it will take a multifaceted approach to implement change. For example, if a facility wants to embark on a hand hygiene campaign it must first gain personnel involvement and ownership of the issue with a commitment to a solution. Individuals within the organization may champion the cause, providing verbal instruction and training on proper procedures based on the best evidence currently available, in addition to demonstrating best-practices in their daily practice. Management must be committed to the cause, providing resources such as hand hygiene stations and financial support for the effort. In addition, the facility may choose to endorse the program through the development of slogans and posters to promote effective hand hygiene practices. In this way the "culture" of the facility can gradually change to one where the expected norm is one of excellent hand hygiene practices where personnel ensure effective implementation through social influences, alteration of individual beliefs and intentions, and ultimately, permanent changes in behavior.

REFERENCES

[1] Benedict, K.M., Morley, P.S. and Van Metre, D.C. (2008) Characteristics of biosecurity and infection control programs at veterinary teaching hospitals. *J Am Vet Med Assoc* **233**, 767-773.

[2] Wright, J.G., Jung, S., Holman, R.C., Marano, N.N. and McQuiston, J.H. (2008) Infection control practices and zoonotic disease risks among veterinarians in the United States. *J Am Vet Med Assoc* **232**, 1863-1872.

[3] Cabana, M.D., Rand, C.S., Powe, N.R., Wu, A.W., Wilson, M.H., Abboud, P.A. and Rubin, H.R. (1999) Why don't physicians follow clinical practice guidelines? A framework for improvement. *JAMA* **282**, 1458-1465.

[4] Pittet, D. (2004) The Lowbury lecture: behaviour in infection control. *J Hosp Infect* **58**, 1-13.

[5] Cork, D.P., Maxwell, P.J.t. and Yeo, C.J. (2011) Remembering Semmelweis: hand hygiene and its importance on today's clinical practice. *Am Surg* **77**, 123-125.

[6] DeJoy, D.M. (2005) Behavior change versus culture change: Divergent approaches to managing workplace safety. *Safety Science* **43**, 105-129.

[7] Clarke, S. (2000) Safety culture: under-specified and overrated? *International Journal of Management Reviews* **2**, 65-90.

Creating a Culture of Infection Control within the Private Practice Setting

Dennis Spann, DVM, DACVIM, (Small Animal Internal Medicine)
Roseville, CA

INTRODUCTION

From the beginning, bacteria have had an identity crisis. The identity crisis however, was ours, not theirs. It has always been due to our inability to recognize and identify bacterial pathogens readily. The gross effects of infectious agents are apparent enough, but we are frequently complacent about the threat of infectious agents because they are invisible in their early stages and invisible as we spread them amongst patients and through the environment. This invisibility leads to complacency. Our complacency leads to lax behavior and provides the opportunity for bacterial pathogens to spread in our hospital environments. Healthcare acquired infections (HAI's) have emerged as a major health care risk in human and veterinary medicine. Extensive antibiotic resistance is an increasingly prevalent aspect of HAI's. Multiple governmental bodies and the popular press have sounded the alarm to increase awareness and mobilize health care professionals in human and veterinary medicine to adapt strategies that minimize the development of antibiotic resistance and the development of HAIs. The frequent association of antibiotic resistance with HAIs challenges us to be a part of the solution rather than a part of the problem. Development of an infection, particularly a resistant infection in the hospital is not always preventable, but it risks undermining the trust and confidence that is so essential in medical practice. In addition, hospital acquired infections raise issues of added financial burden, financial responsibility and sometimes can result in premature euthanasia. Increasingly we have limited antibiotic choices available, so preventing infections becomes a major tactic to preserve the usefulness of the antibiotics that remain. The magnitude of the problem is uncertain in veterinary medicine, but it is clearly significant and on the rise.

On the human side, the World Health Organization has called bacterial resistance a serious enough problem to" reverse the advances of modern human medicine". Multipronged strategies need to be employed to prevent occurrence and spread of HAIs. One of the key elements to development of infection control cultures is to raise the awareness of the medical staff and reception personelle. There are several key components of an infection control program: infection surveillance (passive or active depending on facility size and resources), minimization of patient and staff exposure, environmental contamination control, patient risk assessment, education, and antibiotic stewardship. The key to successful infection control is staff buy- in, and the key to staff buy in is education and organization. Where larger universities have advantages of certain advantages forming communities, smaller individualized protocols can readily be adapted, instituted and adjusted in the smaller private practice setting.

ESTABLISHMENT OF AN INFECTION CONTROL POLICY.

The first step to establishing an infection control policy is to create an infection control committee and appoint an infection control officer. The committee will develop a written policy and the infection control officer will be the go-to person for case management and troubleshooting. This person may also take the lead in coordinating the development of an infection control policy. This policy should be a written document that is readily available and disseminated to all staff. It will be most useful if it is practical, terse and readable. A bulleted reference version can also help make the policy more accessible to post and implement. The protocol should address specific procedures starting with handling patient inquires on the phone through in- hospital management, discharge, case transfer and possibly autopsy.

Surveillance is more frequently passive in small hospital settings because active surveillance is more costly. Surveillance can be mixed if certain key areas are actively surveyed and the cost of doing so is incorporated into hospitalization costs. Passive surveillance may involve keeping a categorized binder tracking cultures by organ system or organism. This is called an antibiogram. It is most useful if you can capture the information from many clinics in area. This can be done in a binder and more readily in a computer spreadsheet. Collaborating with other clinics, or better, with reference labs could provide antibiograms which improve outcomes during the empirical part of antibiotic selection.

Important components of infection control are outlined below:

Patient factors must also be considered. Identify at risk patients. These are critically ill patients, patients that have gastroenteritis, immune compromised patients (from immune suppressants, chemotherapy or sepsis), and patients with difficult to heal wounds or wounds that produce copious discharge. Animals with coughing and or oculonasal discharge. Animals with these problems should have hygiene practiced aggressively to prevent infection and to prevent spread of that infection should it occur. Immobile, debilitated, heavily instrumented animals are all at risk for developing and potentially spreading ordinary and resistant infections. Known infectious animals should be isolated to prevent exposure to other patients. Suspected infectious patients should be separated from healthy and at-risk patients.

Tracking infections is an important aspect of infection control. Records of patients' infections and particularly records of resistant infections should be maintained. Ideally, the hospital location of every patient should be recorded. A binder of positive cultures can be maintained and resistance patterns monitored. Maintaining and reviewing records avoids conjecture and vague anecdotal assessments. One of the significant limitations in veterinary medicine is that lack of centralized collection of data from multiple hospitals. This handicaps the recognition of evolving resistance patterns in our patients and hospitals. Reference labs could serve as a valuable resource for recognition of local and regional resistance patterns. Knowledge of local susceptibility patterns would allow for educated empirical choices of antibiotics while waiting for culture results. A collaborative effort between academia and reference labs could be very productive in this effort.

Hand hygiene is most important aspects of infection control. The act of disinfecting hands with soap and water or utilizing an antiseptic sanitizer decreases the bacterial load on health-care workers' hands. Pathogens can survive an hour or more on clinicians' hands and the degree of contamination is proportional to the time spent with clinical cases without effective washing. Despite this knowledge, clinicians routinely have compliance less than 50 % in observational studies with reports as low as 17 %. Washing should last 20 seconds and contact all surfaces and up the forearms. The time to sing "happy birthday to you" twice approximates 20 seconds. Washing should occur before an after contact with a patient, before and after contact with the patients environment (cages, blankets, etc.), bodily fluids, and before putting on or taking off gloves. Wearing gloves does not replace hand washing, and wearing the same pair of gloves through multiple cases is not helpful.

Personal protective equipment is the second component to prevent us from becoming fomites. Street clothes should be covered by smocks or lab coats in the clinic. Scrubs and lab coats shouldn't be worn outside of the clinic. Lab coats, gowns, scrubs and smocks should be laundered frequently (daily would be best), and changed with first gross contamination or after handling any patient with know or suspected infectious disease. Gloves should be worn when handling bodily fluids, when performing laundry or when cleaning environmental surfaces. Face protection should be used if splashes or other aerosols are expected particularly during dentistry. Foot covers should be worn when in isolation or when cleaning grossly contaminated laundry or working with patients with infections. Many newer recommendations are suggesting clinicians should be bare below the elbows, free of jewelry and watches, and ties. Stethoscopes and other shared equipment should be cleaned and disinfected between patients.

Environmental risk reduction includes assuring high levels of patient cleanliness, identification of patients at risk of soiling themselves and the environment, removal of organic debris and sterilization from the environment (cages, floors, exam tables, sinks, stethoscopes, thermometers, clippers and other commonly used equipment), and appropriate use of patient spacing or isolation. Soiling with any bodily discharges should be rapidly cleaned and disinfected. Linens (towels, blankets, etc.) should be changed at least daily. Cleaning and disinfection need to be separate steps. Cleaning removes gross organic material mechanically usually with soap and water, while disinfection helps remove microscopic pathogens with disinfectant solutions Mop buckets should be changed frequently through the day.

A review of disinfectants is beyond the scope of this talk, but the choice of disinfectant depends on the surface, degree of penetration required, the pathogens being destroyed and toxicity considerations. Choices often involve alcohols, biguanides, halides, peroxides, phenolics and quaternary ammoniums. Care should be given to assure hygienic handling of all bodily fluids and discharge. These biological wastes should be treated and disposed as such. Bandages should be kept clean and dry and free of strike though. Areas in the hospital and lab that are contacted by biologics should be cleaned and disinfected. Waste should be categorized as routine versus biological waste. Biomedical waste includes: Sharps, blood or bodily fluid soaked material including bandages, tissues and dead animals. Elimination areas should be separated for patients shedding infectious organisms. Ideally a cleanable surface should be chosen for elimination for those patients. Clients should generally be restricted from visiting animals with infectious diseases. Raw food should never be fed to hospitalized patients.

Surgical patients are at risk for hospital-acquired infection. First, surgery creates a breach in the skin, which is the natural barrier for infectious agents. Surgery creates some degree of vascular and lymphatic compromise and tissue damage. Environmental cleanliness, surgical scrub and gown cleanliness and hand hygiene are paramount. Sterilization of equipment is an obvious necessity and indicators of sterilization are essential. Anesthetic equipment should be clean after each procedure since hands touch patients and then the equipment. Endotracheal tubes should be meticulously clean and ideally sterilized. Preoperative clipping should be as close to the surgical time as possible. Dirty patients may require a bath prior to surgery. Different clipper blades should be used for contaminated and clean procedures and both should be regularly cleaned and replaced. Perioperative antibiotics when indicated should be given so that they are at therapeutic levels at the time of first incision. Post operatively; incisions should not be touched with bare hands to avoid introducing bacteria. Other important possible sources of infection could be failure to change disinfection scrub pots, or inappropriate cleaning of multidose vials.

Wounds are important avenues of infection. When handling wounds, sterile gloves should be used. Glove changes should occur as necessary as the wound prep becomes cleaner. Bandages must be carefully monitored as specific timed patient orders and any strike-through warrants a bandage change. Soiled bandages should be treated as infectious material. Personnel protection is needed if the wound has known multiresistant bacteria and these patients should have some form of isolation.

Antibiotic stewardship is integral to the preventing the increase of antibiotic resistance and hospital acquired infections. Antibiotics should be delayed long enough to obtain appropriate cultures, and then administered aggressively. Antibiotic choice should be based on culture results where available and de-escalation should occur as indicated once cultures are final. Infection control committees should have a decision process and create a criterion for use of higher tier antibiotics. Where empirical choices are required, antibiotics should be chosen based on the likely organisms for the organ system affected. Antibiotics should be chosen on pharmacologically sound principles. They should be able to reach therapeutic levels in the affected organ. They should be used for an appropriate length of time for the disease process. Owners should be educated about the importance of antibiotic compliance to avoid treatment failure and contributing to antibiotic resistance.

Summary

Prevention and limitation of hospital acquired infections and prevention or antibiotic resistance requires an organized multifaceted approach. Veterinary hospitals of every size should have formalized infection control policies. We currently lack some of the centralized information collection that we need to track infection and resistance patterns. Despite this limitation, institution of well organized infection control policies can provide guidelines for optimal management of our routine and at-risk patients. Improved infectious disease control is going to require the development of centralized information about infection patterns and resistance. These will require greater efforts individually to obtain appropriate cultures, coupled with academic and industry collaboration to develop a clearing house for larger patterns of infection and resistance than can be appreciated at by individual clinicians or practices.

REFERENCES

Infection Prevention and Control Best Practices for Small Animal Veterinary Clinics – Canadian Committee on Antibiotic Resistance. Healthcare Personnel Attire, Recommendations to help prevent healthcare-associated infections transmitted through clothing. Society for Healthcare Epidemiology of AmericaStudy contact: Gonzalo Bearman MD, MPH. Ther Adv Drug Saf 2014, Vol 5(6) 229 – 241. B. Spellberg et al. Trends in Antimicrobial Drug Development: Implications for the Future Clin Infect Dis. (2004) 38 (9): 1279-1286. Chang-Ro Lee et al. Strategies to Minimize Antibiotic Resistance Int. J. Environ. Res. Public Health 2013, 10, 4274-4305, Sabino et al. Can Vet J 2006;47:779–782. Jumaa et al. Hand hygiene: Simple and complex. International Journal of Infectious Diseases (2005) 9, 3—14

Cachexia and Sarcopenia in Small Animals

Lisa M. Freeman, DVM, PhD, DACVN
North Grafton, MA

Cachexia and sarcopenia are two important syndromes that occur in chronic disease and aging, respectively. Cachexia has been recognized in human patients for over 2000 years and is becoming increasingly acknowledged as a common and negative finding that is associated with increased morbidity and mortality. As a result, efforts are being made to combat this syndrome and to improve outcome for the wide variety of disease with which it is associated. Sarcopenia is similar to cachexia in that it is characterized by muscle loss, but sarcopenia is a syndrome seen during aging in the absence of disease. Both of these syndromes are becoming increasingly important in human and veterinary medicine because of their high prevalence and deleterious effects, and a better understanding of these syndromes is critical to optimize patient care. This high interest is spurring the development of new drugs, diets, and other treatments to combat these syndromes in people. Our veterinary patients will benefit from this research.

Cachexia

Cachexia, a loss of lean body mass, has been described since the time of Hippocrates. It is estimated to be present in more than 5 million people in the United States. Cachexia occurs in a variety of acute and chronic diseases in people, including cancer (cancer cachexia), heart failure (cardiac cachexia), chronic kidney disease (renal cachexia), chronic obstructive pulmonary disease, and a variety of acute illnesses and injuries seen in the intensive care unit. The syndrome of cachexia also appears to be common in companion animals with these conditions. A review of cachexia and sarcopenia in dogs and cats is available.[1]

The weight loss that occurs in cachexia is unlike that seen in a healthy dog, cat, or person that loses weight. In a healthy animal that is receiving insufficient calories to meet requirements, metabolic adaptations allow fat to be used as the primary fuel source, thus preserving lean body mass. Conversely, the primary fuel source in animal with chronic disease is amino acids from muscle so that these animals quickly catabolize muscle and lean body mass. Therefore, loss of lean body mass is a hallmark of cachexia. Fat and bone also are lost to lesser degrees although in advanced cases of cachexia, depletion of all compartments is apparent. The loss of lean body mass has direct and deleterious effects on strength, immune function, and survival. In people, cachexia is an independent predictor of survival.

One of the most pressing needs in the area of cardiac cachexia research is the need for a definition and a clinically relevant way to diagnose this syndrome. Current definitions in people rely primarily on loss of body weight (typically 5% over a period of 6-12 months). Total weight loss is an insensitive measure of muscle loss, so using weight loss as a diagnostic criterion reduces the ability to identify cachexia until its more advanced stages and a resulting underdiagnosis of the prevalence of this syndrome. In addition, there are certain types of cachexia (e.g., cardiac cachexia with fluid accumulation) in which the weight loss is masked by accumulation of fat or water. Therefore, waiting until weight loss occurs often prevents an early diagnosis and misses the underlying issue of inflammation and lean body mass losses. The critical role of inflammation in cachexia, as well as the clinical importance of anorexia, makes inclusion of these factors important for patient assessment.

In addition to the insensitivity of weight loss as the criterion for cachexia, it is not only the quantitative loss of muscle that results in deleterious effects. There also are qualitative changes in muscle function. Skeletal muscles in heart failure have increased collagen content, altered mitochondrial function, and a shift in fibers from type I (oxidative) to type IIb (glycolytic) fibers. This shift may further predispose muscle fibers to atrophy as part of the mechanism for cachexia as glycolytic fibers are more susceptible to atrophy. Another reason for using factors other than total weight loss for a diagnosis of cachexia is that cachexia is a process – i.e., a loss of lean body mass - and not necessarily an end-stage syndrome. Lean body mass loss occurs before significant weight loss can be detected. If a loss of lean body mass were used as a criterion for the diagnosis of cachexia, the prevalence of this syndrome in chronic diseases would be even higher. Severe weight loss in a patient with advanced heart failure or cancer represents a classical picture of cachexia and offers little diagnostic dilemma. However, identification of cachexia is more difficult when it is more subtle. Therefore, it is important to develop clinically applicable techniques to identify lean body mass loss at an early stage in which treatments are much more likely to be successful. The challenge is that to detect cachexia at an earlier stage requires evaluation techniques that currently are not clinically applicable. Even some of the research tools that are used, such as dual-energy x-ray absorptiometry (DEXA), have limitations for measurement of lean body mass due to inherent assumptions (e.g., constant hydration of lean tissues, which is not the case in conditions associated with fluid accumulation or dehydration). Clinically-applicable methods of assessing muscle include the muscle condition score (MCS) which should be evaluated in every patient at every visit, in addition to body weight, body condition score (BCS), and diet history.[2]

Cardiac Cachexia

One of the forms of cachexia that has been studied in dogs and cats is cardiac cachexia. In one study of dogs with dilated cardiomyopathy (DCM), over 50% of patients had some degree of cachexia.[3] The loss of lean body mass in cardiac cachexia is a multifactorial process caused by the anorexia, increased energy requirements, and metabolic alterations. The anorexia may be secondary to the fatigue or dyspnea or may be due to medication toxicity or an unpalatable diet. Anorexia is present in 34-84% of dogs and cats with heart disease.[4-6] However, it is inflammatory cytokines, such as tumor necrosis factor (TNF) and interleukin-1 (IL-1), that are primary mediators of cachexia.[3,7] These inflammatory cytokines are known to directly cause anorexia, to increase energy requirements, and to increase the catabolism of lean body mass. Of particular pertinence to heart disease, TNF and IL-1 also cause cardiac myocyte hypertrophy and fibrosis and have negative inotropic effects.

The deleterious effects of cachexia have been emerging and recent studies have emphasized the role of body weight and body composition in heart failure. While obesity is a risk factor for development of heart disease in people, obesity may actually be associated with a protective effect once heart failure is present - this is known as the obesity paradox. Given the adverse effects associated with cachexia, the association between obesity and improved survival in heart failure appears to be due to a *lack* of cachexia, rather than to the obesity per se. This is due to the increased reserve of lean body mass in overweight and obese people. The obesity paradox also has been demonstrated in dogs and cats with heart failure.[8,9] These data emphasize the importance of avoiding weight (and muscle) loss in the patient with heart failure.

Cardiac cachexia is typically recognized only after congestive heart failure (CHF) has developed. Loss of lean body mass is most readily evident in the epaxial, gluteal, scapular, or temporal muscles. Nutritional considerations for cardiac cachexia should include management of anorexia, if present and nutritional modulation of cytokine production also may be helpful. One method of decreasing the production and effects of cytokines is with n-3 polyunsaturated fatty acid supplementation. Supplementation of fish oil, which is high in n-3 fatty acids can decrease cytokine production in dogs with CHF and improve cachexia.[3] In some, but not all dogs with CHF-induced anorexia, fish oil supplementation can improve food intake.[3] In addition, reduction of cytokines has been correlated with survival in dogs with CHF.[3]

Cancer Cachexia

In people, cancer is one of the most common diseases in which cachexia is present. In people with cancer, it is estimated that over 50% lose weight unintentionally, although the prevalence depends on the type of cancer. In dogs, only 4% had a low body condition score (≤3/9) at the time of diagnosis and 55% had a body condition score ≥6/9.[10] However, 35% of dogs had mild to severe muscle wasting. In addition, 69% of dogs for which pre-diagnosis body weights were available had experienced some weight loss (31% had <5% weight loss, 14% had lost 5-10%, and 23% had a weight loss of >10%.)[10] A study in cats with cancer showed muscle loss in 91% of patients.[11] In addition, cats that were below optimal body condition had a significantly shorter survival time compared to those with a body condition score of five or greater. Therefore, specific attention to muscle mass, as well as changes in body weight, is important in order to detect cancer cachexia.

Sarcopenia

The human population is aging; it is estimated that by 2030, >20% of the American population will be over 65 years of age. While 2/3 of older Americans are overweight or obese,[12] there is significant loss of lean body mass that occurs with aging. This loss of lean body mass, termed sarcopenia, occurs in the absence of disease (this is key to the definition of sarcopenia because lean body mass loss associated with disease would be cachexia). The current definition of sarcopenia is lean body mass less than two standard deviations below that of young adults. Sarcopenia actually begins early in life, around 30 years of age, and from 20 to 80 years of age, there is a 30% reduction in muscle mass.[13] In sarcopenia, the loss of lean body mass often is accompanied by an increase in fat mass so the total weight may not change (or may even increase), thus masking the sarcopenia. Like cachexia, sarcopenia has important clinical implications because it is associated with increased mortality.[13] The mechanism of sarcopenia appear to be multifactorial and involve physical inactivity, elevated cytokine production, decreased concentrations of growth hormone and testosterone, changes in type II muscle fibers, insulin resistance, and decreased protein synthesis.[14]

As in people, dogs and cats also are aging with between 1/3 and 1/2 of pet cats and dogs at least 7 years of age. Few studies investigating sarcopenia have been conducted in dogs and cats but available information suggests that dogs and cats also lose lean body mass during aging.[15] This is an important area for future research.

New Interventions

Because of the important implications of cachexia and sarcopenia on morbidity and mortality in people, there is extensive research into the prevention, diagnosis, and treatment of this syndrome. Nutritional modification, hormone replacement, and exercise (both resistance and aerobic) are being actively studied to combat sarcopenia and to enhance healthy aging. For cachexia (and when sarcopenia is not prevented), there are exciting opportunities for new and effective targets to reduce energy requirements, enhance energy intake, improve nutrient absorption, and modify metabolic alterations to prevent and even reverse the effects of cachexia and sarcopenia as the mechanisms of these syndromes are becoming better understood. Myostatin antagonists, for example, may have beneficial effects in animals with muscle loss due to heart failure and other diseases.[16] Blocking single pathways may not be completely effective; treatments that have multiple effects may prove to be the most valuable. For example, exercise training increases muscle mass but also can modify some of the alterations seen in cachexia, such as the reduced GLUT4 expression and increased myostatin concentrations. Ghrelin improves food intake, gastrointestinal motility, and also increases growth hormone so may have beneficial direct effects on heart.[1] These multiple effects can be exploited to address the multiple pathways affected in this syndrome.

Practical Implications

One of the keys to the practical management of cachexia and sarcopenia in dogs and cats is recognizing it in its earliest stages, and to achieve this, BCS and MCS must be consistently assessed. The goal for BCS in a healthy young to middle-aged dog or cat should be 4 to 5 on a 9-point BCS scale. However, in certain diseases (e.g., heart failure, chronic kidney disease), a slightly higher BCS may be desirable (i.e., a BCS of 6 to 7/9 but further research is required to make specific recommendations. Severe obesity should be avoided at all times.

The MCS differs from the BCS in that it specifically evaluates muscle mass. Evaluation of muscle mass includes visual examination and palpation of the head, scapulae, thoracic and lumbar vertebrae, and pelvic bones. BCS and MCS are not directly related as an animal can be obese but still have significant muscle loss (or conversely, be very thin but have a normal MCS). Palpation is required for accurately assessing BCS and MCS, especially in animals with medium to long hair coats. Consistently evaluating MCS in all patients will help to identify muscle loss at an early stage (i.e., mild) in aging or ill animals, rather than waiting until muscle loss is moderate or severe, when it may be more difficult to successfully intervene. Muscle condition scoring systems have been developed for both dogs and cats.[17,18]

For animals with chronic diseases in which weight and/or muscle loss is a component (e.g., renal, heart or hepatic failure; cancer; respiratory disease), a variety of treatments can be employed, including careful attention to nutrition. This is particularly important because anorexia is one of the most common contributing causes to an owner's decision to euthanize their pet.[4] Dietary modification, assisted feeding, or feeding strategies often are beneficial in improving food intake and quality of life for these patients. Pharmacologic and dietary management will be discussed further.

Conclusion

Sarcopenia and cachexia are becoming important in veterinary practice because of their high prevalence and deleterious effects, and a better understanding of these syndromes is critical to optimize patient care. New drugs, diets, and other treatments to specifically target sarcopenia and cachexia are being developed and are likely to benefit canine and feline, as well as human patients.

REFERENCES

1. Freeman LM. Cachexia and sarcopenia: Emerging syndromes of importance in dogs and cats. J Vet Intern Med 2012;26:3-17.
2. WSAVA Nutritional Assessment Guidelines Taskforce: Freeman L, Becvarova I, Cave N, MacKay C, Nguyen P, Rama B, Takashima G, Tiffin R, Van Beukelen P, Yathiraj S. 2011 Nutritional Assessment Guidelines. J Small Anim Pract 2011; 52: 385-396.
3. Freeman LM, Rush JE, Kehayias JJ, Ross JN, Meydani SN, Brown DJ, Dolnikowski GG, Marmor BN, White ME, Dinarello CA, Roubenoff R. Nutritional alterations and the effect of fish oil supplementation in dogs with heart failure. J Vet Int Med, 12:440-448, 1998.
4. Mallery KF, Freeman LM, Harpster NK, Rush JE. Factors contributing to the euthanasia decision in dogs with congestive heart failure. J Am Vet Med Assoc, 214:1201-1204, 1999.
5. Freeman LM, Rush JE, Cahalane AK, Kaplan PM, Markwell PJ. Dietary patterns in dogs with cardiac disease. J Am Vet Med Assoc 2003; 223: 1301-1305.
6. Torin DS, Freeman LM, Rush JE. Dietary patterns of cats with cardiac disease. J Am Vet Med Assoc 2007; 230: 862-867.
7. Meurs KM, Fox PR, Miller MW, Kapadia S, Mann DL. Plasma concentrations of tumor necrosis factor-alpha in cats with congestive heart failure. Am J Vet Res 2002; 63:640-642.
8. Slupe JL, Freeman LM, Rush JE. The relationship between body weight, body condition, and survival in dogs with heart failure. J Vet Intern Med 2008; 22: 561-565.
9. Finn E, Freeman LM, Rush JE, Lee Y. The relationship between body weight, body condition, and survival in cats with heart failure. J Vet Intern Med 2010; 24: 1369-1374
10. Michel KE, Sorenmo K, Shofer FS. Evaluation of body condition score and weight loss in dogs presented to a veterinary oncology service. J Vet Intern Med 2004; 18: 692-695.
11. Baez JL, Michel KE, Sorenmo K, Shofer FS. A prospective investigation of the prevalence and prognostic significance of weight loss and changes in body condition in feline cancer patients.
12. Biggs ML, Mukamal KJ, Luchsinger JA, et al. Association between adiposity in midlife and older age and risk of diabetes in older adults. J Am Med Assoc 2010; 303(24): 2504-2512.
13. Lang T, Streeper T, Cawthon P, et al. Sarcopenia: etiology, clinical consequences, intervention, and assessment. Osteoporos Int 2009;21: 543-559.
14. Evans WJ. Skeletal muscle loss: Cachexa, sarcopenia, and inactivity. Am J Clin Nutr 2010; 91: 1123-1127.
15. Hutchinson D, Sutherland-Smith J, Watson AL, Freeman LM. Assessment of methods of evaluating sarcopenia in old dogs. Am J Vet Res 2012;73:1794-1800.
16. Freeman LM, Rush JE, Cunningham SM, Yang VK, Bulmer BJ. Use of a myostatin antagonist to combat cardiac cachexia in dogs with congestive heart failure. J Vet Cardiol 2015; 17: 210-215.
17. World Small Animal Veterinary Association Global Nutrition Committee. Muscle condition score chart, body condition score chart, diet history form, and other nutritional tools. http://www.wsava.org/nutrition-toolkit. Accessed January 31, 2016.
18. Michel KE, Anderson W, Cupp C, Laflamme DP. Correlation of a feline muscle mass score with body composition determined by dual-energy x-ray absorptiometry. Br J Nutr 2011;160:S57-S59.

Abolishing Anorexia: The Physiology of Appetite Regulation

Audrey K Cook, BVM&S, MRCVS, DACVIM (SAIM), DECVIM-CA, DABVP (Feline)
College Station, TX

INTRODUCTION

The regulation of appetite is a complex process, involving integrated endocrine and neurological responses from the brain, pituitary, gastrointestinal (GI) tract and adipose tissue. Adequate nutritional intake is essential for survival, and disease-associated hyporexia markedly impacts quality of life and outcome for companion animal patients. The goal of this lecture is to review the physiology of appetite regulation and set the stage for a better understanding of pharmacological interventions in anorexic patients.

CENTRAL INFLUENCES

The hypothalamus plays a key role in food intake, with two distinct neuronal subpopulations driving food seeking or promoting satiety.[1] Within the ventrolateral arcuate nucleus are the anorexigenic neurons; these are pro-opiomelanocortin (POMC) neurons, and produce cocaine and amphetamine-related transcript (CART) and α-melanocyte stimulating hormone (α-MSH). These bind to melanocortin receptors and activate the satiety center in the hypothalamus and increase energy expenditure. Within the ventromedial arcuate nucleus are the orexigenic neurons; these release neuropeptide Y (NPY) and agouti-related peptide (AgRP) which activate the hunger center in the hypothalamus. They also reduce energy expenditure, which is an appropriate response perceived food scarcity. Both groups also release their mediators within other areas of the brain, brainstem and spinal cord.

PERIPHERAL NERVOUS SYSTEM

The vagus nerve, via the general visceral axons, controls digestive and propulsive events in the GI tract. Activation of afferent fibers by gastric distention inhibits appetite and diminishes food seeking behavior. The vagus also contains receptors for various hormones and can therefore rapidly relay signals to the brain regarding digestive events, fullness, etc. Transection of all gut sensory fibers results in increased meal size and meal duration.[2,3]

ENDOCRINE EVENTS

The impact of cortisol, thyroxine and insulin on metabolism and food intake are well understood. These hormones modulate energy homeostasis and have long term influences on appetite. However, numerous hormones secreted by the GI tract, pancreas, and adipose tissue directly influence food intake, often on a short term basis.

Cholecystokinin

Cholecystokinin (CCK) is released by cells in the proximal small intestine, in response to the arrival of chyme from the stomach. Proteins and fat are particularly potent secretogues, with plasma levels rising within 15 minutes of the start of a meal. In addition to triggering pancreatic acinar secretion and gallbladder contraction, CCK suppresses food intake via activation of anorexigenic neurons within the hypothalamus. In fact, CCK was the first GI hormone shown to modulate food intake. It also mediates satiety via vagal afferent fibers innervating the GI tract and plays a key role in the cessation of ingestion.[2-4]

The incretins

The incretins (glucose-dependent insulinotrophic peptide [GIP]; glucagon-like peptide 1 [GLP-1]) are also released by specialized cells in the small intestine and suppress the further intake of food. GLP-1 may act synergistically with CCK to potentiate its satiety signaling effect. GLP-1 analogues are routinely used in people for the management of type 2 diabetes mellitus, and have been shown to delay gastric emptying and abate feelings of hunger. Meal size is also reduced, suggesting impacts on satiation.[5,6]

Ghrelin: 'the hunger hormone'

Ghrelin is essential for survival as it plays a pivotal role in the generation of appetite. It is a 28 amino acid protein, produced primarily by X/A cells of the gastric mucosa (>75%), but also from the small and large intestine, pancreas and hypothalamus. There are two forms: acyl (10-30%) and des-acyl (70-90%) ghrelin. The acyl form has a half-life of about 30 minutes and is able to attach to the ghrelin receptor. Plasma ghrelin levels rise before eating. Secretion is quickly suppressed by the arrival of food within the gastric lumen, although it has been suggested that absorbed nutrients within local vessels may also inhibit ghrelin release.[4]

Ghrelin is a relative endocrine newcomer, as it was first reported in 1999.[7] Interestingly, its receptor (the growth hormone secretagogue receptor [GHS-R]) was discovered first and the subsequent hunt for its endogenous ligand led to the discovery of ghrelin. Receptors for ghrelin have been identified in numerous peripheral tissues (pituitary, stomach, intestine, pancreas, adipose tissue, adrenal gland, thyroid gland, myocardium, spleen, immune cells, vagal afferents) and within the brain (arcuate nucleus, ventromedial nucleus, hippocampus, and other sites). Ghrelin directly activates the orexigenic neurons in the hypothalamus, and indirectly inhibits the anorexigenic neurons via gabaminergic inputs. These effects trigger the sensation of hunger and the search for food. In addition, ghrelin impacts homeostasis by slowing down energy expenditure and supporting the conservation of resources. It also activates mesolimbic cholinergic-dopaminergic reward systems in the brain, and is therefore linked to perceptions of pleasure.

Activation of GHS-Rs within the hypothalamus results in the secretion of GH releasing hormone (GHRH) and the subsequent secretion of GH by the somatotrophs of the pituitary.[8] Ghrelin also directly stimulates GH release from the anterior pituitary. GH and its product, insulin-like growth factor 1 (IGF-1), promote weight gain and an increase in muscle mass.

In addition to its appetite promoting and anabolic effects, ghrelin is also an immunomodulator, with an overall anti-inflammatory action. It increases the synthesis of anti-inflammatory interleukin (IL) 10, whilst decreasing the production of various pro-inflammatory cytokines (e.g., IL-1β, IL-6 and TNF-α). It also inhibits the activation of nuclear factor-κB, a substance which stimulates proinflammatory cytokine synthesis and protein degradation.

Ghrelin also impacts GI motility and gastric acid secretion, likely via the vagus. Much like its close family member, motilin, ghrelin induces hunger contractions in the fasted state and speeds gastric emptying post-prandially.[9] In patients with idiopathic gastroparesis, ghrelin has been shown to hasten the exit of both fluids and semisolid material from the stomach.[10]

Leptin: 'the satiety hormone'

Leptin is secreted by adipocytes, primarily those in white adipose tissue, and circulating levels are exponentially proportional to fat mass.[2] This hormone is primarily involved in the long-term regulation of food intake and energy balance; low levels act as a starvation signal to promote food seeking behaviors and the conservation of resources whereas adequate leptin levels communicate repletion of fat stores to the brain and facilitate non-food seeking activities. Leptin is moved across the blood brain barrier by a saturatable transport system, and exerts is anorectic effect in the arcuate nucleus. Interestingly, receptors are present on both the anorexigenic and orexigenic nuclei, which are stimulated or inhibited respectively. Leptin receptors have also been identified on the vagus, and may play a role in satiety (i.e., the cessation of food ingestion), possibly in synergy with CCK.[3] Ironically, leptin levels are persistently elevated in obese individuals but no longer inhibit food intake. This phenomenon, known as leptin resistance, may be due to changes in the transportation across the blood brain barrier or the reduced receptor signal transduction.

PATHOLOGICAL SUPPRESSION OF APPETITE AND THE CACHECTIC SYNDROMES

Numerous conditions in companion animals are associated with anorexia and weight loss. Transient poor nutritional intake is variably well tolerated in dogs (less so in cats) but may still impact patient outcomes. Sustained hyporexia however has substantial impact on morbidity and mortality, and is associated with poor healing, vulnerability to infection, ability to ambulate, etc. It is also important to recognize that metabolic changes associated with chronic illnesses may result in the loss of muscle mass despite apparently appropriate caloric intake. This process is referred to a cachexia: 'a complex metabolic syndrome associated with underlying illness and characterized by loss of muscle with or without loss of fat'.[11] This is different to malnutrition, defined the consequences of insufficient food intake or an improper diet, and associated with hunger and reduced energy expenditure.[12]

Cancer

Cachexia is commonly associated with neoplastic disease and influences owner decisions about therapy and quality of life. Inflammation-related anorexia likely contributes to muscle loss, but recent studies indicate that increased brown fat thermogenesis and higher resting energy expenditure play a central role.[13] It is likely that numerous factors are involved, including tumor-derived parathyroid hormone-related protein (PTHrp) and cytokines including IL-4, IL-15, and tumor necrosis factor (TNF)-α.[14,15] Some of the anorexigenic effects of cytokines include changes in GI motility and alterations of efferent signals that regulate satiety.

Chronic kidney disease (CKD)

In human medicine, cachexia or protein-energy wasting (PEW) is strongly associated with CKD. Inflammatory cytokines likely play a key role, but changes in appetite regulating hormones are also involved.[11] Leptin is degraded by the kidney, and levels are inversely correlated with glomerular filtration rate (GFR) in children.[16] The kidney also degrades ghrelin, and as expected, total levels are substantially elevated in CKD patients. However, the acyl/total ghrelin ratio is decreased, reflecting a preponderance of the des-acyl form. Although a receptor for this has not been identified, des-acyl ghrelin appears to induce a negative energy balance and can be regarded as antagonistic to the acyl form.[16,17]

SUMMARY

The regulation of appetite and the impact of illness upon energy homeostasis is complex, but a better understanding may pave the way for successful pharmacological manipulations of these complex neuroendocrine systems and improved patient outcomes.

REFERENCES

[1]Meister B. Physiol Behav 2007;92:263.

[2]Suzuki K, et al. Endocr J 2010;54:359.
[3]Owyang C, et al. J Neurogastroent Motil 2011;17:338.
[4]Stengel A, et al. Curr Protein Pept Sci 2011;12:293.
[5]Horowitz M, et al. Diabetes Res Clin Pr 2012;97:258.
[6]Naslund E, et al. Int J Obesity 1999;23:304.
[7]Kojima M, Nature 1999;402:656.
[8]Broglio F, et al. J Clin Endocr Metab 2001;86:5083.
[9]De Smet B, et al. Pharmacol Therapeut 2009;123:207.
[10]Tack J, et al. Aliment Pharm Ther 2005;22:847.
[11]Evans WJ, et al. Clin Nutr 2008;27:793.
[12]Mak RH, et al. Pediatr Nephrol 2011;DOI 10.1007/s00467-011-1765-5.
[13]Fearon K, et al. Nat Rev Clin Oncol 2012;10:90.
[14]Kir S, et al. Nature 2014;513:100.
[15]Argiles JM, et al. Nat Rev Cancer 2014;14:754.
[16]Suzuki H, et al. Pediatr Nephrol 2031;28:521.
[17]Gunta S, et al. Pediatric Neprhol 2013;28:611.

Abolishing Anorexia: The Use of Ghrelin Receptor Agonists

Julie Allen, BVMS, MS, MRCVS, DACVIM (SAIM)
Raleigh, NC

INTRODUCTION

A good appetite is essential for adequate food intake and the maintenance of energy homeostasis. A lack of appetite can lead to serious metabolic issues, decreased quality of life and even death. Inappetence is defined as a reduced or lack of appetite or desire to eat. It can be the first sign an animal displays to indicate it is sick; in some situations, it may be the only sign of illness. Almost every disease can manifest with, or develop, decreased appetite (hyporexia), complete lack of appetite (anorexia), or changes in appetite (dysrexia). It can also be a side effect of the treatment of some diseases, e.g., certain antimicrobials and chemotherapeutics. As a result, inappetence is exceedingly common in veterinary medicine. Prolonged inappetence – if left untreated – can lead to inadequate nutrition, and this can sometimes be more detrimental to the patient than the underlying primary disease. The problem is not limited to veterinary medicine either – anorexia and cachexia are significant causes of morbidity and mortality in humans with chronic diseases, such as AIDS or cancer, and there are ongoing efforts to find effective solutions as the current therapeutic options are inadequate.

Inappetence is a serious unmet medical need in dogs and cats and is caused by numerous underlying conditions. Today there are limited therapeutic solutions, most of which are not FDA approved for dogs or cats. Capromorelin, a ghrelin agonist, is being evaluated for its specific effects on appetite stimulation. Capromorelin has the potential to be the first FDA-approved veterinary product for appetite stimulation and will prove to be an important tool in the medical management of conditions where appetite is reduced.

GHRELIN

Ghrelin is a 28 amino acid protein hormone that is produced mainly in the stomach, although smaller amounts are also produced in the intestine, pancreas, and hypothalamus. The active form accounts for only 10% to 30% of circulating ghrelin and is acylated by an enzyme called ghrelin O-acyltransferase (GOAT). Ghrelin is integral to appetite regulation and the maintenance of homeostasis and energy metabolism. Ghrelin also binds to the growth hormone secretagogue receptor causing increased growth hormone (GH) release and subsequent increased production of insulin-like growth factor (IGF-1) in the liver. Increased levels of GH may lead to increased lean muscle mass. This effect is regulated by negative feedback; as IGF-1 rises, it suppresses subsequent increases in GH release thereby preventing overstimulation. Ghrelin has anti-inflammatory properties – decreasing production of inflammatory cytokines from various cells, inhibiting NF-κB (a potent proinflammatory pathway) and increasing the production of anti-inflammatory cytokines, such as interleukin-10. Ghrelin also increases the rate of gastric emptying and intestinal transit.[3]

CAUSES OF INAPPETENCE

The causes of inappetence are numerous and can be divided into pseudoinappetence and true inappetence. Pseudoinappetence can be due to physical problems affecting eating, e.g., pain. Some animals may choose not to eat due to poor diet palatability or food aversions. Primary true inappetence can be due to lesions affecting the hypothalamus or psychological/behavioral causes. Secondary causes are diverse and more common. Systemic disease causes inappetence in a multitude of ways. In renal failure, uremic toxins act on the chemoreceptor trigger zone to cause nausea; and hypergastrinemia can lead to gastrointestinal erosions/ulcerations. Hypokalemia, metabolic acidosis, and activation of the renin-angiotensin-aldosterone system can lead to decreased appetite too. The underlying cause of chronic renal disease is often inflammatory in nature, and increased levels of cytokines often exist – some of these mediators can elicit effects mimicking leptin and suppressing orexigenic ghrelin.[3] A scientific report by Freeman et al at last year's ACVIM Forum, highlighted the significant weight loss seen in cats with CKD. 569 cats with various IRIS stages of kidney disease were evaluated and it was found that cats lost a median of 8.9% of their body weight in the 12 months preceding their diagnosis of CKD and weight loss even began as early as 3 years prior to diagnosis. This weight loss accelerated after the cats were diagnosed with CKD and it was noted that cats that weighed less than the median body weight of 4.2 kg at diagnosis had significantly shorter survival times compared to those that weighed more than 4.2 kg at diagnosis.[4]

A decreased appetite is also common in human cancer patients – surprisingly there is minimal data on the prevalence in veterinary cancer patients. A 2004 study by Michel et al revealed that 37% of veterinary oncology patients had lost greater than 5% of their body weight (one of the criteria of cachexia in people) at diagnosis.[5] Additionally an abstract presented at VCS last year found that 23% of dogs receiving chemotherapy developed inappetence and of those 9/10 lost weight with 4/9 dogs also losing greater than 5% of their body weight. [6] Inappetence and weight loss may be much more prevalent in our veterinary cancer patients than we realize. Cancer patients may not eat for a variety of reasons, e.g. chemotherapy, increased levels of inflammatory mediators, hormonal disruptions. When considering the multifactorial nature of inappetence, it is not surprising that most of our current treatment options are ineffective.

CONSEQUENCES OF INAPPETENCE

When animals do not eat, they experience weight loss, muscle wasting, and poor healing. Owners are distressed by these effects, often deem their pet's quality of life to be poor as a result, and may elect euthanasia. Weight loss is an obvious result of a decreased appetite; and, when nutrition is inadequate, cats in particular actually lose muscle as well. Inappetent cats are also predisposed to hepatic lipidosis, especially cats that are initially overweight. Decreased appetite also upsets the GI tract flora, and increased permeability can result, as the enterocytes become hypoplastic and hypofunctional. Decreased appetite and weight loss have also been negatively correlated with survival.

MANAGEMENT OF INAPPETENCE

The focus when faced with an inappetent patient should always be to identify and address the underlying cause. Some factors that contribute to inappetence, e.g., nausea, can be easily resolved. If the patient is stable, tempting with different foods can be tried; but, if a patient is anorexic for more than three to five days, a feeding tube is recommended. Short-term use of an appetite stimulant can be helpful in a variety of situations, e.g., while investigating the cause of the anorexia. In some situations, the underlying reason for the inappetence cannot be corrected and, in these patients, long-term support of their nutritional intake is needed in order to maintain their quality of life.[7] There are no FDA-approved veterinary products for inappetence, and none of the drugs currently used were developed for the sole purpose of appetite stimulation.

GHRELIN AGONISTS

Due to the effects described above, it would seem reasonable to use ghrelin as a potential therapeutic agent, but ghrelin's short half-life and poor oral bioavailability have precluded this, leading to the development of the small-molecule, orally active ghrelin agonists. These drugs (also known as growth hormone secretagogues) are being evaluated in a variety of human conditions, including anorexia and cachexia due to chronic diseases, such as renal failure, AIDS, and cancer.[3] Lung cancer is the most common cancer affecting people globally with non-small-cell lung cancer (NSCLC) accounting for 85% of all cases. There is a high incidence of cancer anorexia-cachexia syndrome in these patients. Two phase III clinical trials using the ghrelin agonist, anamorelin, were performed in advanced NSCLC patients; and, for the first time, consistent improvements in appetite and body weight (although not muscle strength) were seen. Additionally, patients who were able to stabilize or increase lean body mass had a significant improvement in overall survival ($p<0.001$) vs. patients who lost lean body mass. These results, although preliminary, demonstrate how maintaining or improving body condition in patients with cachexia can enable them to survive longer.[8] Additionally the prokinetic activity of this drug class is being investigated in diabetics with gastroparesis and patients with constipation following spinal cord injury. So what about veterinary medicine?

CAPROMORELIN, A GHRELIN AGONIST: A NEW APPROACH TO MANAGEMENT OF APPETITE AND WEIGHT LOSS

Capromorelin is a small-molecule ghrelin agonist; that is, it mimics the effect of ghrelin, but it has a longer serum half-life and, therefore, prolonged effects. Capromorelin has been formulated into a flavored liquid formulation for ease of administration to dogs and cats. Recently this compound has been tested for its effects on body weight and appetite in both dogs and cats. It is the first drug ever developed specifically for appetite stimulation in veterinary medicine. Several abstracts have been presented showing data on its safety and effectiveness in both laboratory dogs and cats and in canine patients presented at veterinary clinics for lack of appetite; but, as the drug is not yet FDA approved, claims cannot yet be made that it is safe and effective.

A laboratory study in Beagle dogs showed that when capromorelin was given once daily orally for one week, all the treated dogs had increased food consumption and all gained weight.[9] GH and IGF-1 levels were also increased, and the increase in GH was moderated due to the intact negative feedback loop of IGF-1. A second placebo-controlled laboratory study in 24 Beagle dogs has been conducted. Dogs who received capromorelin at 3 mg/kg once daily demonstrated a statistically significant increase in food consumption from day 0 to day 3 compared to placebo ($p<0.001$; 61% increase in capromorelin-treated dogs vs. reduction of 11% in placebo-treated dogs). Percent change in body weight from day 0 to day 3 was also statistically significant compared to placebo ($p<0.001$) with capromorelin-treated dogs demonstrating weight gain of approximately 6% vs placebo-treated dogs whose weight was stable.

The drug was also evaluated in a multisite, randomized, placebo-controlled, masked field study and again, treated dogs, regardless of the etiology of the inappetence, were observed to have an increased appetite and weight gain with minimal adverse effects.[10] The pivotal field study (conducted to meet FDA requirements to evaluate effectiveness and safety) enrolled more than 200 client-owned dogs with inappetence from a variety of causes. Dogs were randomized to either a group treated with capromorelin at 3 mg/kg or a placebo treatment once daily for 4 ± 1 days. Dogs were assessed for appetite on day 4 using an owner appetite assessment instrument. Clinical success rates were approximately 70% for the once-daily capromorelin group vs approximately 45% for the placebo group ($p<0.05$).

Capromorelin has also been evaluated in laboratory cats, as well as client owned cats, with chronic kidney disease. In laboratory cats, capromorelin demonstrated positive effects on food intake, weight gain, GH, and IGF.[11] A multisite, placebo-controlled, pilot field study of approximately 40 cats with chronic kidney disease (CKD) measured weight gain/loss in cats with documented weight loss. Cats were treated with capromorelin once daily and compared to cats that received placebo treatment. Both groups were treated for 90 ± 3 days. The capromorelin-treated cats had statistically significant increases in body weight compared to placebo after 90 days ($p<0.01$). The drug was well tolerated. Safety studies in laboratory cats treated for 90 days at three times the proposed clinical dose showed the drug treatment caused no adverse effects, and this was confirmed in the clinical study where no specific drug-related adverse events were seen.[12] Based on the afore mentioned study, restoring body weight in CKD cats could have a significant effect on cats that have lost weight due to their disease.

CONCLUSION

Inappetence is a common complaint in dogs and cats, and its effects are far-reaching. The complicated control of appetite has limited our therapeutic options in the past, and we have been limited to using drugs that stimulate appetite as a side effect. Capromorelin, on the other hand, is being specifically developed for its effects on appetite and weight maintenance/gain. Because its mechanism of action mimics an endogenous hormone that regulates appetite, it represents a breakthrough treatment. Capromorelin is anticipated to be the first FDA-approved veterinary product labeled for stimulation of appetite in dogs and weight management in cats with chronic kidney disease and will prove to be an important tool in the medical management of conditions in dogs and cats where appetite is reduced.

References

1. 1.Hainerova IA, et al. *J Pediatr Gastroenterol Nutr*. 2010; 51(3):S123-S124.
2. 2.Perry B, et al. *Nutrition and Diabetes*. 2012; 2, 1-7.
3. 3.Smith RG, et al. *Ghrelin in Health and Disease. Contemporary Endocrinology*, 2012.
4. 4.Freeman et al. *Journal of Veterinary Internal Medicine*. 2015; 29: 1257–1283.
5. 5.Agnew W, et al. *J Feline Med Surg*. 2014; 16[9]:749-756.
6. 6.Michel KE et al. *Journal of Veterinary Internal Medicine*. 2004; 18: 692–695.
7. 7.Rhodes L, et al. *VCS Conference Proceedings*. 2015.
8. 8.Garcia JM, et al. *Lancet Oncol*. 2015; 1(1):108-116.
9. 9.Rhodes L, et al. *Journal of Veterinary Internal Medicine*. 2014; 28: 1032.
10. 10.Zollers B, et al. *Journal of Veterinary Internal Medicine*. 2014; 28: 1034.
11. 11.Zollers B, et al. *Journal of Veterinary Internal Medicine*. 2015; 29: 1219.
12. 12.Zollers B, et al. *Journal of Veterinary Internal Medicine*. 2015; 29: 1219.

Feeding the Finicky Feline

Jessica Quimby, DVM, PhD, DACVIM (Small Animal Internal Medicine)
Fort Collins, CO

IMPORTANCE OF NUTRITION

Anorexia is one of the most common clinical signs of disease in cats presenting to the veterinary hospital. Unfortunately it is non-specific and a thorough medical workup is required to determine the underlying cause. Nutritional support is critical during the period of time that the primary problem is being sought and addressed. Management of acute and chronic dysrexia should not be overlooked in the management of feline patients. Assessment of nutrition should be performed in every patient and assisted intervention should be instituted in patients that have had inadequate food intake for longer than 3-5 days (including the time before presentation to the veterinary clinic).[1] Adequate caloric support is crucial for ill or injured patients. Poor body condition is associated with decreased prognosis in several species and has a negative effect on immune function, wound healing and strength.[2] Additionally, poor appetite is perceived by owners as a significant quality of life concern and anorexia in companion animals can cause emotional distress to owners.[3]

Specially formulated diets have been shown to benefit patients affected by several disease processes, but are irrelevant if rejected. Therefore a key therapeutic target for these patients should be the maintenance of appetite and food intake. Common chronic disorders that may be nutritionally managed include: Dental disease, diabetes mellitus, hyperlipidemia, obesity, gastrointestinal disorders, pancreatitis, liver disorders, kidney disease, dermatopathies, urolithiasis, osteoarthritis, cognitive disorders, and hyperthyroidism.[4]

ASSESSMENT OF NUTRITION: THE 5TH VITAL SIGN

A nutritional assessment including body weight, body condition score, muscle condition score, adequacy of caloric intake (including open ended questions about how the pet is eating), and a complete dietary history (including pet food, treats, supplements and items used to give medications) should be obtained. A nutritional plan should be performed for every patient for both acute and chronic illness. Awareness of these parameters and tools for assessment have been made available due to a global nutritional initiative by WSAVA.[5] http://www.wsava.org/nutrition-toolkit.

THE UNIQUE FELINE

Cats have a higher requirement for protein and amino acids than other species. When nutrition is inadequate, energy is derived from mobilization of amino acids from muscles stores as opposed to fat.[2] Elderly cats are also unique in comparison to other species as they have stable to increased level of metabolism, as opposed to decreased metabolism.[6] A reduced ability to digest protein and fat has been documented in elderly cats. These combined factors make the utilization of high quality, easily digestible food product critical for these patients.[7]

WHAT IF THEY WON'T EAT IT?

Environmental factors and food choices

As cats are particularly sensitive to environmental factors, suggestions for appetite enhancement include: quiet environment with exclusive access without interference from bothersome household members, novel food type in cases of possible aversion, similar food type in case of food exclusivity, warming the food particularly if olfaction is an issue, alternatively chilling the food if aromas appear to result in nausea, social interaction while eating, feeding small frequent meals as premature satiety is associated with many disease states. Cats are also particularly sensitive to changes in environmental factors, such as the timing and location of feeding, as well as the food type; smell and "mouth feel" may also play into their willingness to eat.[8]

Medical management of underlying disease

Whenever possible it is obviously ideal to address the underlying disease condition. However in some chronic diseases, i.e. kidney disease, this is not possible and therefore medical management of complications of chronic disease are ideal. Dehydration and anemia can both potentially play a role in inappetence and therefore should be addressed when applicable.

Pain management

For both acute and chronic medical conditions, appropriate management of pain is critical to improving appetite. Opioids are perhaps most commonly utilized. Buprenorphine has a significantly longer half-life and may be useful for in home palliation of chronic pain. Butorphanol may also have some visceral analgesic properties. Opioids should be used with caution in patients with ileus. Maropitant may also have some visceral analgesic properties.[9]

Learned food aversion

Care should be taken to select the appropriate patients for appetite enhancement as learned food aversion is thought by most to be prevalent in cats.[8] Learned food aversion occurs when the patient associates nausea, pain or other physical manifestations of disease with the act of eating or the sight or scent of food. Even after the underlying illness is resolved, this aversion may remain. Therefore it is critical that cats that are overtly nauseous – drooling, gagging, turning away from food - particularly in hospital or in acute illness are not forced to eat lest food aversion be created.[8]

If cats are too nauseous or critical to even consider oral feeding, or have not responded to appetite encouragement after 3-5 days, placement of an enteral feeding tube should be considered. Nasoesphageal, esophageal or gastrotomy tube can be chosen depending on the type and duration of feeding desired.[2] Esophageal feeding tubes can be a valuable tool for long-term management of CKD patients as food, medications and water can be easily given without stressing the patient. Parenteral feeding should be considered in cats that cannot tolerate enteral feeding. Additionally many clinicians feel that prescription diets (i.e. renal) should not be fed in hospital during a crisis lest an aversion be created to the diet desired for long term management. The best candidates for pharmacological enhancement of appetite are cats leaving the hospital with their acute crisis resolved, and cats with chronic disease in the home environment.

ANTI-NAUSEA THERAPY

Cats with CKD likely suffer from nausea, vomiting and inappetence as a result of uremia – a buildup of toxins in blood– that affects the chemoreceptor trigger zone in brain. In addition, uremia may have effects on the intestinal tract that lead to further unwillingness to eat. Several anti-nausea therapies have become recently available. These include maropitant, ondansetron and dolasetron. These drugs work at nausea center in brain as well as in gut and can be given as an injection. Ondansetron has been documented to be helpful in human patients suffering from uremia.[10] However recent pharmacokinetic studies in cats have demonstrated that oral bioavailability of ondansetron is poor in cats (~35%) and the half -life is very short (approximately 1 hour) making it a q 8hr medication.[11] Subcutaneous ondansetron has a slightly longer half-life of 3 hours. In addition to its appetite-stimulating properties, mirtazapine demonstrates anti-nausea properties as it acts at the 5HT3 receptor similarly to ondansetron and dolasetron.[12,13] Limiting gastric acidity with the use of H_2 blockers or proton pump inhibitors such as famotidine or omeprazole respectively anecdotally appears to palliate inappetence in some patients, however as previously mentioned both the degree of hyperacidity present in CKD and the efficacy of these medications for management of cats with CKD remains unproven. Recent studies of the effect of omeprazole on the gastric pH in cats indicates that it is superior to famotidine in its ability to inhibit acid production and twice daily dosing is necessary.[14]

Maropitant is commonly used for acute vomiting. A pharmacokinetic and toxicity study in cats indicated that longer-term usage appears safe.[15] A recent study assessed the efficacy of maropitant for management of chronic vomiting and inappetence associated with feline CKD.[16] When given daily for two weeks, maropitant was demonstrated to palliate vomiting associated with chronic kidney disease, however did not appear to significantly improve appetite or result in weight gain in cats with Stage II and III CKD within the timeframe of the study.

APPETITE STIMULANT THERAPY

Cyproheptadine has been used for some time as an appetite stimulant and has anecdotal efficacy in many patients, however its efficacy has never been scientifically evaluated. Twice daily administration is necessary in many cases and this can prove a challenge for owners, particularly long term. Sedation is a common side effect. Mirtazapine has become more commonly used and recent exploration of its pharmacodynamics and pharmacokinetics have provided information for more effective use in cats.[17,18] Pharmacodynamic studies have illustrated that it can be a potent appetite stimulant, but higher doses are more commonly associated with side effects (hyperexcitability, vocalization, tremors). Smaller, more frequent doses are recommended. The half-life is short enough that it could be administered daily in normal cats. Renal disease delays clearance and in these patients, every other day administration is recommended.[17] Owners should be aware mirtazapine and cyproheptadine cannot not be administered concurrently; cyproheptadine is in fact used as an antidote for serotonin effects of mirtazapine overdose.

A recent placebo-controlled, double-masked crossover clinical trial was performed to evaluate the effects of mirtazapine on body weight, appetite, and vomiting in cats with CKD.[13] Mirtazapine is an effective appetite stimulant in cats with CKD and resulted in significantly increased appetite and weight. Mirtazapine also appears to have anti-emetic properties and resulted in significantly decreased vomiting in cats with CKD. This drug could be a useful adjunct to the nutritional management of cats with CKD.

Table 1: Commonly used appetite stimulant and anti-nausea medications

Drug	Receptor	Location of action	Dosage
Butorphanol	ENK	Cerebral cortex and CRTZ	0.2 – 0.4 mg/kg SQ, IV q 6-8 hrs
Buprenorphine	*mu*	Various	0.005 – 0.01 mg/kg SQ, buccal mucosa
Metoclopramide	D2 (weak)	CRTZ (weak)	0.2 – 0.4 mg/kg SQ, IV q 6-8hrs
Ondansetron	5HT3	CRTZ and GI afferent	1.0 mg/kg IV, SQ, PO q 8 hrs
Dolasetron	5HT3	CRTZ and GI afferent	1.0 mg/kg IV, SQ, PO q 24hrs
Maropitant	NK-1	Emetic center, CRTZ, GI	1 mg/kg SQ, PO q 24hrs
Mirtazapine	5HT3	CRTZ and GI afferent	1.87mg q 24 hrs in normal cats, And q 48 hrs in kidney disease.
Cyproheptadine	H1	Various	2-4 mg per cat q 12-24 hrs

REFERENCES

[1]Freeman LM J Vet Emerg Crit Care 2015; 25: 4-5. [2]Chan DL. J Feline Med Surg 2009; 11: 925-33. [3]Reynolds CA, et al. J Vet Intern Med 2010; 24: 1421-6. [4]Chandler M Vet Clin North Am Small Anim Pract 2005; 35: 713-42. [5]Freeman L, et al. Compend Contin Educ Vet 2011; 33: E1-9. [6]Laflamme DP. Vet Clin North Am Small Anim Pract 2005; 35: 713-42. [7]Sparkes AH. Top Companion Anim Med 2011; 26: 37-42. [8]Michel KE. J Feline Med Surg 2001; 3: 3-8. [9]Boscan P, et al. Am J Vet Res 2011; 72: 1576-9. [10]Ljutic D, et al. Kidney Blood Press Res 2002; 25: 61-4. [11]Quimby JM, et al. J Vet Pharmacol Ther 2014; 37:348-353. [12]Riechelmann RP, et al. Am J Hosp Palliat Care 2010; 27: 106-10. [13]Quimby JM, et al. Vet J 2013; 197: 651-5. [14]Parkinson S, et al. J Vet Intern Med 2015; 29: 104-12. [15]Hickman MA, et al. J Vet Pharmacol Ther 2008; 31: 220-9. [16]Quimby JM, et al. J Feline Med Surg 2014 [17]Quimby JM, et al. J Vet Intern Med 2011; 25: 985-9. [18]Quimby JM, et al. J Vet Pharmacol Ther 2011; 34: 388-96.

Individualized Medicine: The Canine Patient

Katrina L. Mealey DVM, PhD, DACVIM, DACVCP
Pullman, WA

INTRODUCTION

Most of us, in the course of veterinary practice, have observed extreme variability among patients with the same disease with respect to response to drug treatment (drug efficacy and/or toxicity). Why does the identical chemotherapy protocol induce long-term remission in some lymphoma patients but not others? Why does that same chemotherapy protocol cause severe myelosuppression or vomiting in some dogs but not others? More importantly, can patient care be improved by being able to customize drug treatment for each canine patient to optimize drug efficacy and prevent adverse drug reactions?

The concept of "individualized medicine" (also called personalized medicine or precision medicine) goes beyond the one-on-one relationship that a practitioner has with a patient. Rather, the intent is to use biomarkers (genomic, proteomic, metabolomic, etc.) to improve both the characterization of a disease as well as identify specific, effective treatments. Currently, the approach to treating most canine patients with a given disease (e.g., epilepsy) is to utilize an algorithm based primarily on trial and error. Because epilepsy, for example, is not the same in every dog, it is unreasonable to expect that individual animals will have the same response to a particular antiepileptic drug. Incorporation of individualized medicine has the potential to identify the specific cause of the disease, and therefore enable selection of the most effective treatment immediately (no algorithm, no trial and error). For example, does the epileptic patient have a loss-of-function mutation in a GABA receptor or a gain-of-function mutation in a potassium channel? These represent very different potential causes of epilepsy. It is logical that optimal treatment would differ depending on the precise molecular defect responsible for causing the disease in a given patient. Devoid of the information provided by an individualized medicine approach, the epileptic patient could be treated with half a dozen different drugs before an effective therapeutic regimen is identified. Equally important, the concept of individualized medicine includes the ability to predict the safety as well as the efficacy of a drug for each patient. Genomic or other biomarkers from a patient can be used to determine if a particular drug should be avoided in that patient or if the dose should be altered in order to prevent a severe adverse reaction.

Physicians have access to resources that summarize information on hundreds of disease-gene and gene-drug combinations that can be used to optimize drug therapy. No such resource exists in veterinary medicine but is currently being organized by the Program in Individualized Medicine (PrIMe) at Washington State University. The goal of this presentation is to summarize the state of individualized medicine in dogs and encourage coordinated efforts to generate new gene-drug associations, critically evaluate the clinical impact of those associations and determine genotype-based dosing guidelines. Examples of some of the most well-documented gene-drug associations are presented below.

Acepromazine

Acepromazine is a phenothiazine used in dogs primarily as a preanesthetic agent. One gene-drug interaction has been reported for acepromazine. Specifically, dogs with the ABCB1-1Δ allele (more widely known as the MDR1 mutation) experience more profound sedation than dogs with the ABCB1 wildtype allele.[1] ABCB1 encodes P-glycoprotein (P-gp), a drug transporter that minimizes gastrointestinal absorption, enhances biliary and renal drug excretion and is an integral component of many tissue barriers (e.g., blood brain barrier). Because acepromazine is a substrate for P-gp, penetration into the brain is normally limited while biliary and or renal excretion is enhanced. Defective P-gp function (as occurs in dogs with the ABCB1-1Δ allele), alters disposition of substrate drugs. Specifically, defective P-gp allows greater brain penetration of acepromazine resulting in more profound sedation, slower clearance and longer duration of drug action. Anecdotal dose guidelines suggest decreasing the dose by 25% or 50% in patients with one or two ABCB1-1Δ alleles, respectively. Both purebred and mixed herding breed dogs have a higher incidence of the ABCB1-1Δ (MDR1) mutation than other breeds.[2]

Atenolol

Atenolol is a selective Δ_1 adrenergic receptor antagonist used to treat cardiovascular diseases in dogs. One gene-drug interaction has been reported for atenolol in dogs. Specifically, Meurs, Stern, et al. describe two deletion mutations in the canine adrenoreceptor beta-1 (ADRB1) gene.[3] Dogs with the ADRB1 deletions had lower basal heart rates, lower heart rates after treatment with atenolol, and a significantly smaller reduction in heart rate compared to control dogs after treatment with atenolol. The two ADBR1 deletions are predicted to alter pharmacodynamics of atenolol because they alter structure of the Δ_1 adrenergic receptor, likely interfering either with drug binding and/or G-protein-mediated cell signaling. No dosing guidelines have been suggested, however the researchers suggest that dogs with heart failure that have the two deletions might be expected to have less of a response to treatment with beta adrenergic receptor antagonists than dogs without the polymorphisms. Although it has not been investigated, one would presume that the ADBR1 would alter the efficacy of other Δ_1 adrenergic receptor antagonists in a similar manner. No breed predilection has been reported.

Butorphanol

Butorphanol is a partial opiate agonist/antagonist FDA-approved as an antitussive for dogs. One gene-drug interaction has been reported for butorphanol in dogs based on observations of a colony of Collies.[11] Specifically, Collies that were homozygous for the ABCB1-1Δ mutation appeared to experience more profound and prolonged sedation compared to Collies with the wildtype ABCB1 allele. Prospective blinded studies have not been performed. Anecdotal dose guidelines suggest decreasing the dose by 25% or 50% in patients with one or two ABCB1-1Δ alleles, respectively. Both purebred and mixed herding breed dogs have a higher incidence of the ABCB1-1Δ (MDR1) mutation than other breeds.

Doxorubicin

Doxorubicin is an anticancer drug in the 'antibiotic' class. One gene-drug interaction has been observed by the author during the initial phase of a prospective study to determine whether dogs with the ABCB1-1Δ mutation were more susceptible to doxorubicin-induced hematologic toxicity than ABCB1 wildtype dogs. The study was abandoned because the first two dogs homozygous for the ABCB1-1Δ allele that received doxorubicin (30 mg/m^2) died from overwhelming neutropenic sepsis. For drugs like doxorubicin that are P-gp substrates, clearance is highly dependent on biliary excretion. Defective P-gp function (as occurs in dogs with the ABCB1-1Δ allele), essentially prevents biliary excretion resulting in greater overall exposure to doxorubicin. Anecdotal dose guidelines suggest that decreasing the dose by 25% or 50% in patients with one or two ABCB1-1Δ alleles, respectively minimizes myelosuppressive effects. Both purebred and mixed herding breed dogs have a higher incidence of the ABCB1-1Δ (MDR1) mutation than other breeds.

Halothane

Halothane is an inhalant anesthetic agent that has been associated with malignant hyperthermia in several species including dogs. Other volatile anesthetic agents and some neuromuscular blockers have also been implicated. One gene-drug interaction, a single nucleotide polymorphism (SNP) in the ryanodine receptor (RYR1) gene, has been reported in a colony of dogs susceptible to malignant hyperthermia.[4] RYR1 encodes a calcium channel of the sarcoplasmic reticulum. Based on the pharmacogenetics of malignant hyperthermia in other species, it is likely that different mutations in the RYR1 gene or other genes may also be responsible for malignant hyperthermia in other dogs. This is the subject of PrIMe research being conducted by Dr. Michael Court; he is seeking DNA samples from patients that have exhibited malignant hyperthermia (Michael.court@vetmed.wsu.edu).

Loperamide

Loperamide is an opioid that is generally devoid of CNS activity because it is excluded from the brain by P-gp. It is an over-the-counter human product labeled for symptomatic treatment of diarrhea. One gene-drug interaction has been described for loperamide.[5] Loperamide causes marked CNS depression in dogs homozygous for the ABCB1-1Δ allele (MDR1 mutation) whereas dogs with the wildtype ABCB1 allele experience no CNS depression. Because loperamide is a substrate for P-gp, penetration into the brain is normally limited. Defective P-gp function (as occurs in dogs with the ABCB1-1Δ allele) allows greater brain penetration of loperamide resulting in typical opioid-induced CNS depression. Opioid antagonists can reverse the CNS depression. Alternative antidiarrheal drugs should be used in dogs with ABCB1-1Δ. Both purebred and mixed herding breed dogs have a higher incidence of the ABCB1-1Δ (MDR1) mutation than other breeds.

Macrocyclic lactones

Macrocyclic lactones include a number of drugs that act as parasiticides (ivermectin, moxidectin, milbemycin, selamectin, etc.). One gene-drug interaction has been reported for macrocyclic lactones. Specifically, dogs with the ABCB1-1Δ allele (more widely known as the MDR1 mutation) are highly susceptible to CNS toxicity associated with macrocyclic lactones than dogs with the ABCB1 wildtype allele.[6] ABCB1 encodes P-gp, a drug transporter that minimizes gastrointestinal absorption, enhances biliary and renal drug excretion and is an integral component of many tissue barriers (e.g., blood brain barrier, blood placental barrier). Because macrocyclic lactones are substrates for P-gp, penetration into the brain is normally limited. Defective P-gp function (as occurs in dogs with the ABCB1-1Δ allele) allows greater brain penetration of macrocyclic lactones resulting in central nervous system toxicity (usually CNS depression that can progress to coma and death) and slower clearance, resulting in extended (weeks) CNS depression depending on the dose. A dog that is homozygous for ABCB1-1Δ will experience mild-to-moderate CNS toxicity after a single oral or subcutaneous administration of ivermectin at a dose of ~100 micrograms per kilogram. Repeated or higher doses will cause severe CNS toxicity. Label doses of macrocyclic lactones in FDA approved heartworm prevention products are safe in dogs with ABCB1-1Δ. Dogs with herding breed ancestry have a higher incidence of the ABCB1-1Δ (MDR1) mutation than other breeds.

Phosphodiesterase 5A inhibitors (Sildenafil)

Vasodilators such as sildenafil act by inhibiting cyclic guanosine monophosphate (cGMP)-specific phosphodiesterase (PDE5A). One potential drug-gene interaction has been reported.[7] Dogs with a nonsynonymous SNP in PDE5A (PDE5A: E90K) have significantly lower circulating cGMP concentrations than dogs with the wildtype allele. Whether or not these dogs have altered responses to treatment with phosphodiesterase 5A inhibitors has not yet been investigated, however, beneficial effects of PDE5A inhibitors such as sildenafil are reported to be variable on an individual dog basis. No breed predilection has been reported.

Propofol

Propofol is a short-acting anesthetic agent FDA-approved for use in dogs. Some greyhounds and other sighthound breeds have experienced severely delayed recovery compared to other breeds. Although a specific gene-drug interaction has not yet been reported, a relationship between the slow recovery phenotype and a specific drug metabolizing enzyme, cytochrome 2B11 (CYP2B11) has been documented by Dr. Michael Court. The molecular genetic basis for this breed-dependent difference in drug metabolism is the subject of ongoing research by Dr. Court (Michael.court@vetmed.wsu.edu); he is seeking DNA samples from patients that have exhibited this phenotype.

Vinca alkaloids (Vincristine, Vinblastine, Vinorelbine)

Vinca alkaloids are microtubule inhibitors used to treat various types of cancer. One gene-drug interaction has been documented.[8] Dogs that carry one or two ABCB1-1Δ alleles are significantly more likely to develop vincristine-induced neutropenia and thrombocytopenia than dogs homozygous for the ABCB1 wildtype allele. A case report also describes vincristine-induced CNS toxicity in a dog homozygous for ABCB1-1Δ. Defective P-gp function (as occurs in dogs with the ABCB1-1Δ allele) profoundly delays biliary excretion resulting in greater overall exposure to vinca alkaloids and thus predisposes to myelosuppression. Enhanced susceptibility to CNS toxicity is due to defective blood brain barrier function of the variant P-gp. Anecdotal dose guidelines suggest decreasing the dose by 25% or 50% in patients with one or two ABCB1-1Δ alleles, respectively to minimize myelosuppressive effects. Both purebred and mixed herding breed dogs have a higher incidence of the ABCB1-1Δ (MDR1) mutation than other breeds.

Reported gene polymorphisms with unknown phenotype

Described briefly below are polymorphisms that have been identified but a specific phenotype (i.e., susceptibility to drug toxicity, lack of drug efficacy) has not yet been identified. If a patient is identified that experiences either unusual susceptibility to drug toxicity or resistance to pharmacological effects when treated with a drug that interacts with the gene products described below, veterinarians are encouraged to contact the author (kmealey@vetmed.wsu.edu) or the Program in Individualized Medicine at Washington State. DNA samples from patients like these are extremely helpful in identifying the molecular genetic basis for many unexpected drug responses.

Angiotensin converting enzyme

Meurs et al. describe several SNPs in canine angiotensin converting enzyme.[9] Whether or not any of these SNPs play a role in sensitivity or resistance to angiotensin converting enzyme (ACE) inhibitors is unknown.

Glucocorticoid receptor α

Costa and Mealey et al. describe six polymorphisms in the canine glucocorticoid receptor (NR3C1α) gene.[12] Two of those polymorphisms were nonsynonymous and were predicted by in silico analysis to be "probably damaging" to structure and function of the resulting protein. It was suggested that NR3C1 polymorphisms may contribute to individual variation in response to corticosteroid therapy in dogs. The author is seeking DNA samples from patients that have exhibited extreme sensitivity or resistance to corticosteroids (kmealey@vetmed.wsu.edu).

REFERENCES

1. Deshpande et al. Effect of canine ABCB1-1Δ on sedation JVIM 2016 epub.
2. Mealey et al. ABCB1-1Δ breed distribution JAVMA 2008,233(6):921-4.
3. Meurs et al. Canine double deletion Δ1 adrenoreceptor polymorphisms Pharmacogenet Genomics. 2015,25(9):427-31.
4. Roberts et al. Canine malignant hyperthermia gene mutation Anesthesiology 2001,95:716-25.
5. Mealey et al. P-gp contributes to blood brain barrier Drug Metab Disp 2008,36(6)1073-9.
6. Mealey et al. Ivermectin sensitivity in collies Pharmacogen 2001,11(8):727-33.
7. Stern et al. PDE5A:E90K polymorphism in canine PDE5A gene. JVIM 2014,28(1):78-83.
8. Mealey et al. ABCB1 polymorphism predicts hematologic toxicity JVIM 2008,22(4):996-1000.
9. Meurs et al. Polymorphisms in RAAS genes. An Gen 2015,46:226.
10. Costa et al. Polymorphisms in canine NR3C1. JVPT 2015 epub.
11. Mealey KL. Adverse drug reactions in herding-breed dogs: The role of P-gp.Comp Cont Ed Prac Vet 2006, 28:23-33.
12. Costa A et al. Polymorphisms in the canine glucocorticoid receptor alpha gene (NR3C1α). J Vet Pharmacol Ther 2016, 39(1):16-21.

Individualized Medicine: The Feline Patient

Katrina L. Mealey DVM, PhD, DACVIM, DACVCP
Pullman, WA

INTRODUCTION

Most of us, in the course of veterinary school or practice, have heard the phrase "cats are not small dogs" with regard to their ability to metabolize drugs. A number of feline-specific "defects" have been identified in drug metabolizing enzymes in cats compared to other species, particularly dogs and people. Many of these differences are thought to result from their status as obligate carnivores compared to the omnivorous status of dogs and people, meaning that these are not actually defects in animals that would not naturally encounter xenobiotics found in plants. They only appear to be defects when cats encounter these xenobiotics in the course of medical treatment. This discussion will summarize some of these pan-species differences that characterize the cat as "sensitive" to a number of drugs.

Most of us have also observed inter-individual variability in drug response (drug efficacy and/or toxicity) among different cats. Why does the identical chemotherapy protocol induce long-term remission in some lymphoma patients but not others? Why does that same chemotherapy protocol cause severe myelosuppression or vomiting in some cats but not others? More importantly, can patient care be improved by being able to customize drug treatment for each feline patient to optimize drug efficacy and prevent adverse drug reactions? The concept of "individualized medicine" (also called personalized medicine or precision medicine) goes beyond the one-on-one relationship that a practitioner has with a patient. Rather, the intent is to use biomarkers (genomic, proteomic, metabolomic, etc.) to identify patients that are most likely to respond to a specific drug as well as patients that are likely to experience an adverse drug reaction (so that an alternate drug or dose can be used in that patient. Currently, the approach to drug therapy in feline patients is based on trial and error.

Physicians have access to resources that summarize information on hundreds of disease-gene and gene-drug combinations that can be used to optimize drug therapy. No such resource exists in veterinary medicine but is currently being organized by the Program in Individualized Medicine (PrIMe) at Washington State University. The goal of this presentation is to summarize the state of individualized medicine in dogs and cats and to encourage coordinated efforts to generate new gene-drug associations, critically evaluate the clinical impact of those associations and determine genotype-based dosing guidelines. Examples of some of the most well-documented gene-drug associations in cats are presented below.

DRUG DISPOSITION IDIOSYNCRASIES OF CATS COMPARED TO OTHER SPECIES

Glucuronidation

Cats are sometimes too broadly characterized as being poor metabolizers or slow metabolizers of all drugs compared to other species. This is not the case and is well described in an article by Court.[1] In that article, Court demonstrates that for cats, drug elimination that is independent of conjugation is often greater than or equal to that in dogs or humans (i.e., oxidation reactions, or excreted via the bile or urine as parent drug). Furthermore, not all conjugated drugs are cleared more slowly in cats. Acetylsalicylic acid, propofol, acetaminophen and carprofen are drugs that are primarily eliminated by conjugation and are slowly eliminated in cats compared to people or dogs. However, chloramphenicol, ketoprofen, morphine and lorazepam are also primarily eliminated by conjugation but these drugs are eliminated by cats at an equivalent or faster rate than by people or dogs. The mechanisms for these species differences in conjugation-dependent elimination rates, particularly as they relate to predicting drug dose regimens in cats based or human or canine pharmacokinetic data, have yet to be determined. A brief summary of what is known about conjugation in cats follows.

Many, *but not all* conjugation reactions that depend on glucuronidation are deficient in cats compared to other species. Cats appear to lack the ability to glucuronidate simple planar phenolic compounds, which tend to be metabolized by several uridine diphosphate glucuronosyltransferase (UGT) isoforms in other species, particularly UGT1A6 and UGT1A9. Both UGT1A6 and UGT1A9 are deficient in cats. Cats express only 2 UGT1A1 isoforms (UGT1A1 and UGT1A2) whereas dogs and humans express 10 and 9 isoforms, respectively. UGT1A is essential for bilirubin glucuronidation. Little is known about the function of UGT1A2. Since both UGT1A6 and UGT1A9 are involved in glucuronidation of acetaminophen (the major metabolic pathway in dogs and humans), cats must use alternate metabolic pathways to clear the drug. This deficiency in glucuronidation is likely the key reason that cats are exquisitely sensitive to acetaminophen toxicity. Whether or not glucuronidation deficiency is the mechanism responsible for slower elimination of propofol or carprofen in cats compared to dogs and people has not been elucidated. Other drugs (ibuprofen, lorazepam, salicylate, telmisartan) are glucuronidated with similar efficiency in cats compared with other mammalian species. Thus, the fact that a drug is eliminated via glucuronidation in dogs and humans does not preclude its use in cats. Nor does it necessarily mean that a reduced dose or dosing frequency should be employed in cats. Cats may use alternate conjugation pathways to achieve similar levels of total drug clearance compared to other species. Some drugs that are eliminated primarily via phenolic glucuronidation in canine and human liver microsomes, are eliminated primarily via phenolic glucosidation in feline liver microsomes such that total conjugation rates among these species are similar despite large differences in the type of conjugation reaction (i.e., glucuronidation versus glucosidation). Conversely, cats may be deficient in other conjugation pathways such as glycination or sulfation. In fact, there is evidence to suggest that slow clearance of aspirin in cats is due to deficient conjugation of salicylate with glycine rather than glucuronic acid. Clearly further investigation is needed to understand conjugation differences in cats compared to other species to improve our ability to predict safe and effective doses of drugs currently approved for dogs or people.[2]

Thiopurine Methyltransferase (TPMT)

Cats are much more susceptible than dogs or people to the myelosuppressive effects of azathioprine.[9] Deficient TMPT activity has been demonstrated and several polymorphisms in the coding region of the feline TMPT gene have been identified.[3]

ABCG2 Transporter

Transport function of the feline ABCG2 drug transporter is deficient in cats compared to the canine and human ortholog.[3] The most important clinical consequence of this transport defect identified to date is fluoroquinolone-induced retinal degeneration. Despite the widespread use of fluoroquinolones in dogs and humans, retinal degeneration and blindness does not seem to be a problem, likely because ABCG2 transport function prevents fluoroquinolone-induced retinal damage. Most fluoroquinolones generate reactive oxygen species (ROS) when exposed to light. In humans, these ROS can be generated in tissues exposed to light. Indeed, cutaneous photosensitivity is one of the most common adverse reactions reported after fluoroquinolone use in people. Because the skin of most dogs and cats are covered (shaded) by fur, cutaneous photosensitivity is not a problem. However, the retina is also exposed to light and, if adequate concentrations of fluoroquinolones are present in retinal tissue, ROS that are generated may attack cellular lipid membranes causing tissue damage. In other species, ABCG2 present at the blood-retina barrier prevents accumulation of fluoroquinolones in retinal tissue. Because ABCG2 function is defective in cats, fluoroquinolones are able to accumulate in retinal tissue and, when exposed to light, generate ROS that damage the retina. Whether or not this deficiency in ABCG2 transport function causes increased susceptibility to toxicity of other substrate drugs has not yet been elucidated.

N-acetyltransferase 2 (NAT2)

N-acetylation mediated by NAT1 and NAT2 is an important metabolic pathway for clearance of a variety of drug used in human patients (isoniazid, procainamide, hydralazine, dapsone and some sulfonamide antibiotics. Cats lack NAT2 which results in low acetylation of several sulfonamide antibiotics and isoniazid.[5] The author is not aware of any studies that have assessed acetylation of procainamide, hydralazine or dapsone.

GENETIC VARIATION AMONG CATS

Angiotensin converting enzyme

Meurs et al. describe several single nucleotide polymorphisms (SNPs) and two deletion mutations in feline angiotensin converting enzyme in a small (n=15) sample of cats.[6] Whether or not any of these SNPs play a role in sensitivity or resistance to angiotensin converting enzyme (ACE) inhibitors (e.g., enalapril, benazepril, lisinopril) is unknown.

Angiotensin II receptor

Meurs et al. also describe one SNP in the feline angiotensin II receptor (type 2) receptor in the same sample of cats.[6] Again, whether or not this SNP alters the pharmacological response to angiotensin II receptor antagonists (e.g., telmisartin) in affected cats is unknown.

ABCB1

The ABCB1 gene encodes P-glycoprotein (P-gp) a drug transporter that substantially affects the disposition of substrate drugs. P-gp is expressed by many tissues including the enterocytes of the intestines, biliary canalicular cells and brain capillary endothelial cells. At these strategic locations, P-gp functions to actively extrude substrate drugs thereby limiting their oral absorption, enhancing their biliary excretion, and limiting their penetration into the central nervous system. A genetic defect of ABCB1 (ABCB1-1delta or "MDR1 mutation") has been well characterized in dogs.[10] Recently, a 2 base pair deletion mutation (*ABCB1*1930_1931del TC) was identified in a cat that experienced ivermectin toxicity.[7] Screening of a nonbiased cat population (n=100) yielded 4 additional cats with *ABCB1*1930_1931del TC. Studies to determine the phenotype of affected cats receiving P-gp substrate drugs are ongoing. The fact that the mutation creates a stop codon interrupting ~50% of protein synthesis suggests that affected cats would have a similar phenotype to dogs with the well-characterized ABCB1-1delta mutation.

CURRENTLY UNEXPLAINED DRUG SENSITIVITIES REPORTED IN CATS

APPARENT SPECIES-WIDE SENSITIVITIES

5 Fluorouracil (5FU)

It has been recommended that cats should not be treated with 5FU because of extreme susceptibility to neurotoxicity.[11] Neurotoxicity from 5FU has been reported in dogs and people, but the literature suggests that cats are more susceptible..[12] Mutations in the dihydropyrimidine dehydrogenase (DPYD) gene in people have been associated with predisposition to 5FU toxicity in people.[13] DPYD is the enzyme responsible for metabolizing 5FU to nontoxic metabolites. One group of investigators indirectly assessed DPYD activity in cats and determined that 23% of cats studied had a low uracil:dihydrouracil ratio which is suggestive of low

DPYD activity[8]. No genotype-phenotype studies have been performed. The mechanism responsible for the apparent sensitivity of cats to 5FU induced neurotoxicity is unknown.

Cisplatin

Treatment of cats with cisplatin is listed as contraindicated due to fulminant pulmonary edema (even after one treatment).[14] The author is unaware of any investigations on the cause of this unusual feline-specific drug sensitivity.

INDIVIDUAL ANIMAL SENSITIVITIES

Methimazole

Idiosyncratic toxicities attributed to methimazole use in cats include hepatotoxicity, blood dyscrasias and facial pruritus and excoriation.[15] Pruritus and facial excoriation have also been described for phenobarbital and aldosterone in a fraction of cats receiving these drugs.

Bromide

Adverse effects involving the respiratory tract limit the usefulness of bromide in cats.[16] Coughing that is thought to be due to allergic bronchial disease was reported in 38% of cats in one study. Respiratory compromise can be life-threatening.

Diazepam

Oral diazepam has been associated with acute hepatic necrosis in a small number of cats.[17]

REFERENCES

1. Court MH Feline drug metabolism and disposition. Vet Clin N Am Sm An Pract 2013, 43(5):1039-54.
2. Van Beusekom et al. Comparing glucuronidation capacity of feline liver. JVPT 2014,37(1):18-24.
3. Salavaggione et al. Cat RBC TMPT pharmacogenetics J Pharm Exp Ther 2004;308(2):617-26.
4. Ramirez et al. Molecular genetic basis of FQ retinal toxicity in cats. Pharmacogen Gen 2011,21(2):66-75.
5. Trepanier LA et al. Deficiency of N-acetylation in felids caused by the presence of a single NAT1-like gene. Pharmacogen 1998,8(2):169-79.
6. Meurs et al. Polymorphisms in canine and feline RAAS genes. An Gen 2015,46:226.
7. Mealey et al. Identification of a nonsense mutation in feline ABCB1 J Vet Pharmacol Ther 2015, 38(5):429-33.
8. Saba et al. Assessment of DPYD activity in cats. Vet comp Oncol 2013,11(4):265-71.
9. White SD, et al. Thiopruine methyltransferase in red blood cells of dogs, cats, and horses. J Vet Intern Med 2000, 14(5):499-502.
10. Mealey et al. P-gp contributes to blood brain barrier Drug Metab Disp 2008,36(6)1073-9.
11. Saba CF et al. Indirect assessment of dihydropyrimidine dehydrogenase activity in cats. Vet Comp Oncol 2013,11(4):265-71.
12. Plumb DC. Fluorouracil (5-FU).Plumb's Veterinary Drug Handbook 2015, 461-462.
13. Mattison LK, et al. Implications of dihydropyrimidine dehydrogenase on 5-fluorouracil pharmacogenetics and pharmacogenomics. Pharmacogenomics 2002, 2(4):485-92.
14. Knapp DW et al. Cisplatin toxicity in cats. J Vet Intern Med 1987, 1(1):29-35.
15. Trepanier LA et al. Efficacy and safety of once versus twice daily administration of methimazole in cats with hyperthyroidism. J Am Vet Med Assoc 2003, 222(7):954-8.
16. Boothe DM et al. Disposition and clinical use of bromide in cats. J Am Vet Med Assoc. 2002, 221(8):1131-5.
17. Center SA, et al. Fulminant hepatic failure associated with oral administration of diazepam in 11 cats. J Am Vet Med Assoc 1996, 209(3):618-25.

Mechanical Ventilation – Set-Up and Strategies

James Barr, DVM, DACVECC
Medora Pashmakova, DVM, DACVECC
College Station, TX

INTRODUCTION

Intermittent positive pressure ventilation (PPV) through mechanical ventilation is a fundamental aspect of respiratory support in the patient with respiratory failure. Long-term PPV is life-saving. It also requires an understanding of equipment, patient pathology, and ventilator-patient interactions. This manuscript describes how to initiate PPV, strategies for specific respiratory pathologies, and the fundamentals of ventilator settings. The accompanying manuscript describes targets, complications, and weaning strategies once a patient is on PPV.

PPV – INDICATIONS

Mechanical ventilation is indicated for patients that meet the following criteria:[1]

1. Hypoxemia ($PaO_2 < 60$ mmHg or $SpO_2 < 90\%$) despite supplemental O_2 administration
2. Hypercapnia ($PaCO_2 > 60$ mmHg or $PvCO_2 > 60$ mmHg)
3. Excessive work of breathing
4. Severe circulatory shock

Hypoxemia

Hypoxemia is defined as a PaO_2<80 mmHg and severe hypoxemia is defined as a PaO_2<60 mmHg, at which stage intubation and PPV are indicated. This is approximately equivalent to a SpO_2 of 90%, however, it is well recognized that pulse oximetry may be inaccurate in many instances and for this reason, arterial blood gas analysis is the gold standard for determining hypoxemia. Hypoxemia can result from 1 of 3 general causes: 1) Hypoventilation, 2) Low inspired FiO_2 (rarely a problem in the spontaneously breathing patient), or 3) Venous admixture (blood that returns to the left side of the heart without undergoing gas exchange, as with anatomic shunts, V/Q mismatch, or significant diffusion barriers).

Hypoventilation

Hypoventilation is defined by hypercapnia. It is crucial to understand that a patient's respiratory pattern or effort is **not** an indicator of ventilatory status. Patients that pant do not necessarily hyperventilate. One may use venous blood gas analysis rather than arterial blood gas analysis to determine hypoventilation since the arterio-venous CO_2 gradient is small enough to where a venous analysis is adequate in most cases. Importantly, however, all blood gas samples, whether venous or arterial, that are used to determine oxygenation or ventilation status must be analyzed immediately and not allowed to equilibrate with room air, even the air within a blood tube. This will lead to an inappropriate representiaton of what is happening in the patient. Causes of hypoventilation affect patient's respiratory rate or tidal volume (TV) and usually include brain disease, cervical spinal cord disease, peripheral neuropathies, myopathies, NMJ diseases, and metabolic disorders. General anesthesia with inadequate ventilation is the most common cause of hypoventilation in small animals.

Excessive work of breathing

Regardless of equipment or blood gas indicators of oxygenation and ventilation, respiratory distress resulting in exhaustion (orthopnea or eyes closed during breathing) require PPV to prevent imminent death.[1]

Severe circulatory shock

The goal of PPV in the patient with circulatory shock is to reduce oxygen consumption by relieving the workload on respiratory muscles. Additionally, these patients may benefit from airway support if severely obtunded or anesthetized for management of their primary disease process.

PPV – OUTCOMES

PPV requires significant owner financial and emotional investment as well as extensive hospital resources. Communicating the likely success of PPV therapy early in the process is an important part of managing expectations. In general, PPV for hypoventilation has a better outcome than PPV for hypoxemia. Disease processes that have a fair prognosis for weaning include congestive heart failure[2], hypoventilation secondary to neurologic disease[3-5], and toxicities (specifically baclofen and ivermectin affecting the CNS)[6,7]. Outcomes are significantly lower (20-30% successful weaning) in patients where hypoxemia is the primary, or compounded, reason for mechanical ventilation. Specifically dogs with LMN disease that also have aspiration pneumonia, acute respiratory distress syndrome, or post cardiopulmonary arrest, have a poor prognosis.[1,3,5] In comparison to all dogs hospitalized in a university ICU, brachycephalic dogs were more likely to receive PPV than non-brachycephalic dogs, with aspiration pneumonia as the primary cause, however, survival was unrelated to brachycephalic conformation and was similar to previous reports (~47% weaning).[8]

INITIATION

Once the decision to employ PPV has been made, the following sequence of events must take place:

1. Induce anesthesia with agents you are comfortable using: traditionally propofol or ketamine/diazepam are excellent choices.
2. Intubate the patient with sterile equipment and manually ventilate (ambubag) if mechanical ventilator is not readily available or requires time to set up.
3. Obtain lower airway samples for in-house Wright and Gram stain and official cytology and microbiology, especially in cases where infectious lower airway disease is the cause for hypoxemia. The authors do not perform this routinely in cases where hypoventilation is the only cause and lung function is deemed normal.
4. Attach in-line capnograph and provide 100% FiO_2 while determining initial ventilator settings.
5. Total intravenous anesthesia (TIVA) is recommended for the duration of PPV. A table of recommended drugs and dosages is included (Table 1).

Ventilator modes

There are a few basic terms with which the operator should be familiar in choosing the ventilation mode for a patient. First is the understanding of how the ventilator delivers a breath - **pressure** versus **volume** control (PC and VC). PC means the ventilator delivers a breath up to a pre-set airway pressure. The amount of breath delivered can vary based on the lung compliance. A VC breath delivers a pre-set tidal volume and the airway pressure depends on the lung compliance. Incompliant lungs, such as with ARDS, can result in high airway pressures in volume control despite delivering a low TV. Traditional anesthetic ventilators are VC machines, while modern long-term mechanical ventilators can deliver either. The authors use PC in essentially all patients, as this is a more lung-protective strategy.

Next is the understanding of when the ventilator delivers a breath. **Assist-control** (AC) is the type of PPV that provides maximum support and is used for patients with minimal to no respiratory drive, such as cervical disc patients. With AC, the operator sets a mandatory number of breaths per minute that are delivered in a regular pattern. The patient can trigger (initiate) some own breaths, but every breath is a full ventilator breath, either PC or VC. Breaths triggered by the ventilator are controlled breaths, those triggered by the patient are assisted breaths.

Synchronized intermittent mandatory ventilation (SIMV) is the most commonly used mode in patients with lung pathology but a normal ventilator drive or cervical patients who are able to generate some breaths without fatiguing. With SIMV, the patient can breathe spontaneously around the pre-set mandatory breaths. Modern machines try to synchronize with the patient in order to improve patient-vent interactions.

Continuous positive airway pressure (CPAP) is a weaning mode or a type of support for patients who are entirely able to ventilate for themselves but need some constant baseline airway pressure to prevent lung or airway collapse. The ventilator delivers no breaths. This mode works for patients that have a strong respiratory drive and minimal lung dysfunction.

Finally, **pressure support (PS)** ventilation allows the patient to generate every breath, but augments the breath with added pressure in order to maximize (reward) the effort. PS may be used on its own or to support the breaths in SIMV or CPAP.

Ventilator settings and initial selections

The following settings are present on most ventilators and must be determined at the time of initiation. Table 1 includes recommended initial settings for patients with normal lungs and patients with lung pathology.

1. **Trigger** – this is the variable that determines what starts a breath. In patients without any respiratory drive, the trigger is time. When a certain number of seconds pass, the ventilator delivers a breath. In patients with respiratory drive, the trigger can be a change in airway pressure, so that when the patient attempts a breath, the ventilator recognizes the trigger. Overtriggering can happen if the setting is too sensitive and there is patient movement, leaning over the patient, or coupage.
2. **Inspiratory:expiratory ratio** – the ratio of time spent on inspiration to time spent on expiration. It is most comfortable to set this to 1:2. This may need to be adjusted indirectly by adjusting the inspiratory time.
3. **Respiratory rate** – the number of breaths per minute. Commonly set at 10-20 initially.
4. **Inspiratory time** – the time to deliver the inspiratory part of the breath, commonly set at 1 second.
5. **Tidal volume** – 10-15 ml/kg though should be decreased in animals with significant lung pathology, such as ARDS, where lung protective strategies are used. In those patients, TV of 6 mL/kg may be more appropriate. Because patient pathology and conformation can impact TV, observing chest excursions is a reasonable subjective way to determine if TV is adequate. Caution should be exercised to not overdistend the lungs as this is a major cause of ventilator induced lung injury (VILI).
6. **Positive end expiratory pressure** – PEEP maintains positive pressure in the airways during exhalation and holds the lung with some inflation at all times. This is particularly helpful in patients with collapsed alveoli. Because it decreases the amount of recurrent alveolar inflation and collapse, PEEP can be helpful to prevent VILI and is an important strategy for ARDS ventilation. Healthy lungs should have a setting of 1-4cm H20 PEEP, which can easily double or triple in unhealthy lungs.

7. **Peak inspiratory pressure** – Healthy anesthetized patients do not need more than 10-15 cm H20 of PIP, however, patients with stiff incompliant lungs may need PIP of 25-30 cm H20 or more.

MAINTENANCE OF ANESTHESIA

Long-term ventilation requires the use of total intravenous anesthesia (TIVA) as inhalant anesthesia is not a viable option in this circumstance. A combination of drugs is used, being mindful of each drug's cardiovascular effect, duration of action/accumulation over prolonged time, and cost. It is recognized in cats that recovery after prolonged anesthesia for PPV is long and can, in itself, require ventilator support for an additional day.[9] In addition, propofol can cause Heinz body formation in cats and lipemia with prolonged use in both species, so familiarity and ability to use multi-drug protocols is imperative. The authors use a neuroleptanalgesia (opioid + sedative) in addition to propofol most commonly. Some options are: fentanyl-dexmedetomidine-propofol, fentanyl-ketamine-propofol. The use of benzodiazepenes can delay the ability to wean as these drugs accumulate and cause prolonged recoveries. It is reasonable to use them early in the course of PPV and if the patient shows signs of improvement, discontinue in preparation for weaning. Table 2 includes some commonly recommended doses for TIVA.

Table 1. Suggested initial settings for patients with normal lung function and patients with lung pathology

Patient with normal lungs		Patient with lung pathology	
Fraction of inspired O_2	100%, decrease to lowest necessary	Fraction of inspired O_2	100%, decrease to lowest necessary
Tidal volume	10-15 mL/kg	Tidal volume	6 mL/kg
Resp rate	10-20 bpm	Resp rate	20-30 bpm
Inspiratory pressure above PEEP	8-12 cm H20	Inspiratory pressure above PEEP	10-15 cm H20
PEEP	1-4 cm H20	PEEP	5-10 cm H20
Inspiratory time and I:E	1s ; 1:2	Inspiratory time and I:E	1s ; 1:2
Trigger	-1 to -2 cm H20	Trigger	-1 to -2 cm H20

Table 2. Suggested drugs to maintain TIVA.

Drug*	Loading dose	CRI
Propofol**	2-4mg/kg (induction)	0.1–0.5 mg/kg/min
Alfaxalone	2-3mg/kg (induction)	0.1-0.2 mg/kg/min
Fentanyl**	2–5 mg/kg	1–7 ug/kg/hr
Morphine	0.1–0.2 mg/kg	0.2 mg/kg/hr
Midazolam**	0.2–0.4 mg/kg	0.1–0.5 mg/kg/hr
Diazepam	0.5 mg/kg	0.1–1.0 mg/kg/hr
Dexmedetomidine	2-5ug/kg	0.5 to 1.0 ug/kg/hr
Ketamine	0.5 mg/kg	2-20 ug/kg/min
Lidocaine	1–2 mg/kg	3 mg/kg/hr

*Suggest using only one drug per class. Each additional drug class used together will reduce your need for a particular anesthetic agent or increase the depth of anesthesia.
**A combination of propofol, fentanyl and midazolam is the most commonly used protocol in the author's institution.

REFERENCES

[1]Hopper K, et al. VCNA 2013;43:955-969. [2]Edwards TH, et al. JVECC 2014;24(5):586-593. [3]Hopper K, et al. JAVMA 2007;230:64-75. [4]Webster RA, et al. Aust Vet J 2013;91:233-239. [5]Lee JA, et al. JAVMA 2005;226:924-931. [6]Wright HM, et al. JVECC 2011;21(6):666-672. [7]Khorzad R, et al. JAVMA 2012;241:1059-1064. [8]Hoareau GL, et al. JVECC 2011;21(3):226-235. [9]Boudreau AE, et al. JVECC 2012;22(2):239-252.

Mechanical Ventilation – Management and Weaning

Medora Pashmakova, DVM, DACVECC
James Barr, DVM, DACVECC
College Station, TX

INTRODUCTION

The mechanically ventilated patient is entirely reliant on clinicians and nursing staff to monitor, correct, and maintain ventilator controls and physiologic homeostasis. Positive pressure ventilation (PPV) has very specific impacts on hemodynamics that are in contrast to spontaneous ventilation (a negative pressure process). The mechanically ventilated patient is also predisposed to significant complications. Finally, if the patient's disease process improves, an understanding of when and how to wean is imperative. This manuscript will discuss possible complications, patient care issues, and weaning strategies for the mechanically ventilated patient.

IMPORTANT TOOLS

There are several key tools that are imperative for optimal PPV and optimizing patient outcomes. The author strongly recommends that practices that perform PPV are equipped with a **blood gas analyzer** and **capnograph**. Many different models of mechanical ventilators exist, however, investing in one with graphical waveform analysis can dramatically increase the clinician's understanding of patient pathophysiology and patient-ventilator interactions. A great review of ventilator waveforms[1] and a user-friendly **ventilation textbook[2]** are both available and are helpful to troubleshoot specific issues.

MANAGEMENT

Targets

Once a patient is started on PPV, the goals are to use the least aggressive settings for the shortest period of time before a weaning process is initiated once patient pathology allows. Intubation and prolonged anesthesia are profoundly disruptive to the host local and systemic immune defenses. Aggressive ventilator settings can additionally lead to baro- and volu- trauma (components of ventilator induced lung injury). Further, using high inspired oxygen (FiO_2) concentrations on its own will promote lung injury.

- **PaO_2** – A PaO_2 of 80-120 mmHg is normoxemic and corresponds to approximately 93-100% SpO_2 where an arterial blood gas (ABG) is not available. One of the first targets is to use the lowest FiO_2 setting necessary to maintain normoxemia. Supraphysiologic PaO_2 is unnecessary and indicates that settings can be decreased. Increasing the positive end expiratory pressure (PEEP) can allow a decrease in FiO_2 while maintaining O_2 saturation.
- **$PaCO_2$** – A $PaCO_2$ of 35-45 mmHg is normocapnic. $PaCO_2$ depends on the minute ventilation, which is equal to respiratory rate (RR) x tidal volume (TV). Increasing either RR or TV results in increased minute ventilation and resultant decrease in PCO_2. If ABG is not available, a venous blood gas (VBG) can also be helpful to guide $PaCO_2$ analysis. VBG **cannot** be used for PaO_2 analysis. Once initial setting have been determined on the ventilator, it is helpful to obtain an initial ABG (or at least VBG) and determine PaO_2 and $PaCO_2$. At the time that the blood gas is obtained, also note the $EtCO_2$ capnograph reading. This will determine the $PaCO_2$ – $EtCO_2$ gradient, which can "calibrate" the patient to the capnograph and allow non-invasive ventilation monitoring for some period of time. Because this gradient can change, it is recommended that a blood gas is performed q6h to assess progress in PaO_2 as well as determine the gradient each time.
- **Optimizing targets in ARDS** – in most patients, these targets can be achieved with $FiO_2 < 60\%$ and moderate ventilator settings. If these are not enough and significant lung pathology exists, such as ARDS, specific strategies can be employed. It is recognized that using a low TV high PEEP strategy reduces mortality in human ARDS patients.[2,3] ARDS patients also benefit from a balance of ventilator settings and oxygenation/ventilation targets called "permissive hypercapnia". This means that aggressive ventilator settings can be more damaging than allowing some hypoxemia and hypercapnia. Current ARDS ventilation recommendations are a PaO_2 of 55-80 mmHg (SpO2 85-90%) and allowing $PCO_2 > 50$ mmHg as long as pH remains >7.2 (another reason for blood gas analysis).
- **Brachycephalic dogs** – because of the frequency with which brachycephalics need PPV, it is interesting to note that they may have normally lower PaO_2 and higher $PaCO_2$ then their non-brachycephalic counterparts.[4]

COMPLICATIONS

Ventilator Associated Pneumonia (VAP)

One of the most significant morbidities incurred by ventilated patients is ventilator associated pneumonia (VAP). VAP is defined as new onset pneumonia that starts 48 hours or later after intubation, with reported mortality rates of 20-70% in people.[5] Because aspiration of oropharyngeal secretions is the major mechanism of VAP, much research has gone into oral decontamination strategies. The authors emphasize the following veterinary applicable recommendations for preventing and addressing VAP:

1. Acute change in lung function based on oxygenation and compliance changes **must** be investigated with radiography and repeat lower airway sampling (cytology, Gram stain, microbiology).

2. Disconnecting the patient from the circuit should happen as infrequently as possible. Mandatory changes of circuits and tubes may predispose to, not prevent, VAP.[5]
3. Oral care with a chlorhexidine based oral disinfectant/lubricant (C.E.T. Rinse, Virbac) should be performed q6h. This recommendation is in agreement with the 2012 Surviving Sepsis Guidelines.[6]
4. The pharynx should be packed off with soft gauze and the patient must be orally decontaminated and **suctioned** of all secretions (subglottic and directly around the endotracheal tube) **before** any deflation of the cuff or extubation.

A recent retrospective study revealed that dogs and cats with respiratory failure requiring PPV have a different population of organisms (Gram negative enteric) and resistance patterns (highly resistant to standard antimicrobials) in comparison with respiratory disease of milder form. Although this information depends on the antibiogram of a particular facility, it is noteworthy that amikacin and imipenem were the antibiotics most likely to successfully treat patients intubated and ventilated for pneumonia.[7]

Ventilator induced lung injury (VILI)

VILI is an umbrella term describing injury to the lungs at the microscopic level. It is notoriously difficult to recognize, however, because clinical and bloodwork findings can overlap with the patient's primary pathology (for example, ARDS). It is reasonable to assume that acute lung injury (ALI) and ARDS may be partially a result of ventilation, rather than natural progression.[2] Barotrauma is a type of VILI that occurs when high pressure causes alveolar rupture. This can result in air dissecting the mediastinum, subcutaneous space, or cause a pneumothorax. Using very high tidal volumes can also result in a type of injury called volutrauma. Atelectrauma described the concept of injuring the lung from repeated opening and closing of alveolar units at low lung volumes. Biotrauma describes the inflammatory response of the lungs to mechanical stress and cell membrane disruption of the lungs. This is made worse by high inspired oxygen fractions, which in itself is inflammatory from the generation of reactive oxygen species. Detailed explanations of these concepts is available[2], however, for the purposes of this manuscript, the fundamental point is that ventilation itself inherently causes lung injury, which may ultimately be the cause for patient demise.[2]

Extrapulmonary considerations

PPV exerts positive intrathoracic pressure with each delivered breath. This is in contrast to negative pressure ventilation in a spontaneously breathing patient. Positive pressure alters hemodynamics of the heart and great vessels which in turn affects the GI, renal, and neurologic systems. It has been recognized that PPV reduces cardiac output because the increased intrathoracic pressure is transmitted to the great vessels, resulting in decreased ability of blood to drain into the heart (and therefore decreased preload). This effect is enhanced when PEEP is applied, which maintains constant positive pressure upon the cardiovascular structures in the thorax.

These effects can be minimized by reducing ventilator settings, specifically lower airway pressures, lower PEEP, and shorter inspiratory:expiratory (I:E) ratios. This is easier said than done, unfortunately, because the patient's pathology may require exactly the opposite. For example, extrapulmonary effects are particularly difficult to balance with the ventilation needs of an ARDS patient.

What are the systemic consequences of the extrapulmonary effects of PPV? Because of a perceived decrease in cardiac output, patients will upregulate fluid-retention hormonal systems (RAAS, ADH) and start to retain fluid, decrease urine output, gain weight, and become interstitially overhydrated. The authors recognize this clinically by 48 hours of PPV. Additionally, PPV can increase intracranial pressure (by increased central venous pressures) and cause a decrease in cerebral perfusion pressure. This can be observed clinically by jugular venous distention. In the same way as intracranial pressure, PPV can cause an increase in abdominal pressure and therefore splanchnic resistance. This can contribute to GI mucosal ischemia, ileus, decreased nutrient absorption, and GI ulceration, which is already recognized commonly in critical illness.[8] The impetus to start prophylactic antacid therapy must be balanced against the change in oropharyngeal bacterial composition documented in dogs, as this is the main nidus for aspiration pneumonia.[9]

TROUBLESHOOTING

Several time-sensitive issues can arise in the mechanically ventilated patient that require quick assessment and resolution. These include hypoxemia, hypercapnia, patient-ventilator dyssynchrony, increased airway resistance, barotrauma, and air leak.

1. **Hypoxemia** – acute hypoxemia indicates a dramatic and potentially life-threatening change in lung function. Specifically, this may be a life-threatening tension pneumothorax. The sequence of actions that should take place is 1- increase FiO2 to 100%, 2 – auscult thorax, 3- evaluate ventilator function and circuit. A thoracocentesis should be performed if any concern exists. The authors often have a chest tube prepared near the patient at all times to handle such an emergency. A more gradual hypoxemia usually indicates a new onset of progressive lung disease, such as VAP, VILI, or aspiration pneumonia.
2. **Hypercapnia** – this may develop due to patient or ventilator factors. The most important patient factor to rule out is a pneumothorax, a significant PTE, or an endotracheal tube kink or obstruction. Ventilator factors include increased apparatus dead space, incorrect circuit assembly/leaks, or inadequate ventilator settings such as a low TV, low RR, or both. Insufficient expiratory time can also cause hypercapnia.
3. **Patient-ventilator dyssynchrony** is the situation where there is conflict between the patient's respiratory needs/efforts and what the machine is delivering. It is also recognized as "bucking" the vent. This results in ineffective ventilation,

patient discomfort, and can lead to hypoxemia and hypercapnia. Systematically assess if the patient is air hungry because of hypoxemia or hypercapnia, or has developed a pneumothorax. Hyperthermia can cause animals to pant, which can result in dyssynchrony. Inadequate ventilator settings (evaluate if the patient's inspiratory and expiratory efforts are lined up with the machine) and lastly, depth of anesthesia, must also be assessed. Caution should be exercised to not automatically increased depth of anesthesia of there is a physiologic or equipment issue that must first be addressed.

PATIENT CARE

Personnel

At the author's institution, a clinician and a technician are dedicated to the mechanically ventilated patient at all times. Whenever possible, these should be in a shift rotation where there is continuity of care or seamless communication between all members of the patient care team. The technician assigned to the ventilator patient should be one of the most highly skilled available.

Instrumentation

When PPV is initiated, it is recommended for the patient to be instrumented for with an arterial catheter (if possible this gives access for direct blood pressure and ABG analysis), central venous access (to the level of the right atrium for CVP measurement, in addition to readily available sampling and infusion routes), urinary catheter, and nasogastric tube (for suctioning and nutritional support).

Monitoring

Electrocardiogram, continuous blood pressure (invasive or non-invasive), capnography, and pulse oximetry are monitored continuously. Where possible, continuous CVP measurement is also desirable. As necessary, thoracic imaging (radiographs, echo) to assess lung pathology and volume status and cardiac contractility may be beneficial if those issues arise.

Drug therapies

The authors prefer to keep this simple until it requires becoming complicated. Antibiotics should be used carefully and only with pending microbial cultures in patients where pneumonia is the concern. Antibiotics should **not** be prophylactically started in patients ventilated for non-infectious reasons (smoke inhalation, ARDS, lung hemorrhage, CHF, etc). Prokinetics and anti-emetics are routinely employed. Antacids may be used if a significant suspicion for ulceration exists, however, this can dramatically change the propensity for microbial aspiration into the lungs. The authors do not routinely use antacids.

Nursing care

Every 4-6 hours, the patient is rotated, undergoes passive range of motion of all limbs, the eyes are lubricated, the mouth is suctioned gently and if tracheal tube care is required, the mouth should be decontaminated first. The patient should be weighed at least q24h, if not q12h in patients where fluid balance is particularly challenging. The nasogastric tube does not have to be suction unless gastric stasis is an issue. Ideally, enteral nutrition is provided as early as possible.

WEANING

The process of transitioning the work of breathing from the mechanical ventilator back to the patient is called weaning from the ventilator.[10] The goal of weaning is to maintain patient safety by meeting their ventilatory needs, measuring their capabilities and ensuring that the patients will be able to maintain themselves without ventilatory support. An absolute requirement before this can happen is that the disease that necessitated ventilatory support must begin to resolve. How a patient is challenged to wean is largely dependent on their underlying pathology. Patients that were hypoventilating (i.e. cervical IVDD patients) must be able to generate adequate negative airway pressure with their respiratory muscles. Patients that have lung pathology and hypoxia must have the ability to oxygenate adequately with non-invasive oxygen supplementation. Figure 1 is a simplified algorithm for weaning.

Ensuring the patient's ability to take on the work of ventilation is performed by sequentially removing support and increasing the difficulty of triggering breaths while making sure that the patient is able to maintain normal $PaCO_2$. Decreasing ventilatory support may be as simple as decreasing the number of mandatory breaths per minute while still supporting the breaths that the patient triggers. The way to measure if a patient is able to generate negative pressure is by decreasing the ventilator's sensitivity to trigger a breath (i.e. making it more difficult to trigger a breath). If a patient is able to trigger a breath with a high patient trigger setting (>10cm H_20), then they are more likely to be able to breathe on their own. Weaning is typically done with the ventilator set in the synchronized intermittent mandatory ventilation (SIMV) setting.

A patient with lung disease is challenged by decreasing the inspired oxygen percentage and monitoring their ability to oxygenate. While not ideal, this can be done as simply as evaluating the pulse oximetry reading while decreasing the FiO_2. At the author's institution, ABG measurements are used to guide weaning. A useful measure of lung function is determining PaO_2 to FiO_2 ratios to determine the ability of the patient to oxygenate adequately. If a patient has a normal or high PaO_2 on 40% oxygen, then they will likely be able to oxygenate better when they are off of the ventilator. A PaO_2:FiO_2 of 250 equates to a patient that has a normal PaO_2 of 100mmHg on 40% oxygen and usually means that their lung function will be good enough to be able to have normal oxygenation on supplemental oxygen. 40% oxygen is well within the range of supplemental oxygen that can be administered non-invasively.

If the patient is able to generate breaths on their own, maintain themselves on few mandatory breaths and oxygenate appropriately, then they are either switched to continuous positive airway pressure (CPAP) or ventilation is discontinued while keeping the patient anesthetized. During weaning the patient is monitored to ensure safety. It is important to note that the patients must be at a plane of anesthesia that ensures that they are able to spontaneously ventilate. A patient may only be extubated when they have regained control of their airway and a gag reflex. A review of ventilator weaning and an algorithm for weaning can be found in this excellent review article.[10]

REFERENCES

[1]Corona TM, et al. JVECC 2011;21(5):496-514. [2]Pilbeam's Mechanical Ventilation. Saunders; 2009. [3]ARDSNet. New Engl J Med 2000;342:1301-8. [4]Hoareau G, et al. JVIM 2012;26:897-904. [5]Kollef MH. Crit Care Med 2004;32:1396-1405.[6] Dellinger RP, et al. Crit Care Med 2013;41(2)580-637.[7] Epstein SE, et al. JVECC 2010;20(6):587-594. [8]Monnig AA, et al. JVECC 2011;21(5)484-495.[9] Sullivan LA, et al. JVECC 2015;00(0):1-5.[10] Mellema MS, Haskins SC. Clin Tech in Sm An Practice, 2000; 15(3): 157-164.

Figure 1. Weaning algorithm.
SIMV: Synchronized Intermittent Mechanical Ventilation

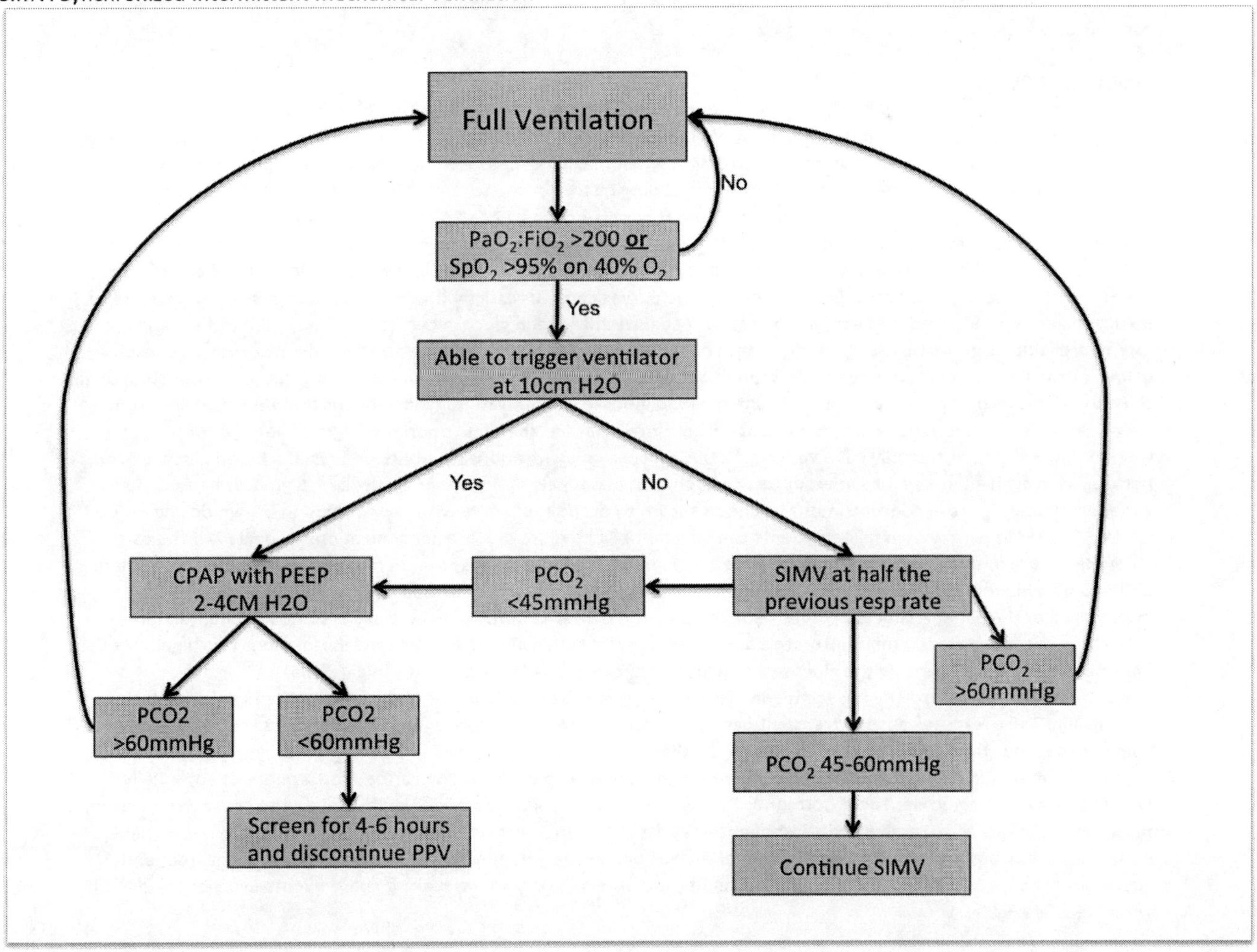

How to Get the Most Value Out of Serial Blood Glucose Curves

Linda Fleeman, BVSc, PhD, MANZCVS; Ann Thompson, BSc, BVSc, MS, GCHEd, MANZCVS, DACVIM; Patty Lathan, VMD, MS, DACVIM
Melbourne, Victoria, Australia

INTRODUCTION

Blood glucose curves can be very useful in the assessment of diabetes, although they also have limitations and there is considerable day-to-day variation in results, especially in patients with poor glycemic control. Therefore they must always be interpreted with great caution in light of other clinical information, including history, physical examination, and review of owner's records.

The primary aims of therapy in diabetic animals are to achieve resolution of clinical signs, avoid insulin-induced hypoglycemia, and prevent other complications such as diabetic ketoacidosis. An additional goal in recently diagnosed cats is to optimize the chance of diabetic remission. However, it is also very important that the treatment and monitoring regimen fits well into the owners' daily routines because euthanasia can occur due to either unmet owner expectations or the negative impact of disease management on their lives.[1]

A common error when managing diabetes is to assume that the condition can be 'stabilized' and that measurements taken on one day are representative of diabetic control on other days. A more appropriate approach is to establish an on-going monitoring regimen that will detect the changing trends in the patient's response to treatment. Improved activity, resolution of polydipsia and polyuria, and weight gain are all indicators of improved glycemic control. When this occurs, the animal's insulin requirement will usually decrease substantially in dogs and cats, and remission might occur in cats. This may happen within days, weeks, or even months of the last insulin dose adjustment.

Successful monitoring of diabetics must aim to recognize when the clinical signs are resolving and then identify when the insulin dose needs to be decreased to avoid hypoglycemia. Performing serial blood glucose curves in patients with good clinical control provides a method of determining if the current insulin therapy is likely to cause blood glucose concentrations to go too low.

BLOOD GLUCOSE CURVES

The principle reason for performing a glucose curve is to find the lowest blood glucose concentration (nadir) resulting from the current insulin therapy. Multiple sampling is required because the timing of the nadir is typically very variable. A glucose curve can thus provide information when making decisions about changing insulin doses that cannot be obtained from any other method of monitoring including history or fructosamine levels. The goal of the glucose curve is to track the changes in blood glucose concentration that occur over a 12 to 24 hour period after insulin administration. The ideal curve reflects what happens to the patient's blood glucose in the home environment. This information is used to assess how the blood glucose concentrations for a given dose of insulin and a set amount of food are likely influencing the animal's clinical signs at home. Serial blood glucose curves can thus assist with decision making when adjustment of the insulin dose is indicated. It is very important to recognize that blood glucose curves have many limitations and are often unreliable diagnostic tools. Multiple factors can influence blood glucose concentrations on any given day resulting in lower or more commonly higher results than might occur on a 'typical' day. The diabetic state causes dysfunction of all the normal physiological mechanisms that control blood glucose concentration, which means that diabetics are prone to widely variable and unpredictable blood glucose. Therefore a patient that has experienced good diabetic control for a week or more can unexpectedly have very high blood glucose concentrations for hours or days. If a blood glucose curve is performed at such a time, results may suggest poor control in animals that are well controlled. Significant day to day variation in blood glucose concentrations has been shown to occur in patients with no variation in insulin dosing or food intake.[2,3] Stress in particular in feline patients can substantially increase glucose concentrations, especially if the cat struggles before or during blood sampling or has a concurrent illness.[4] For this reason, minimizing stress for any patient undergoing a glucose curve is essential. It is important that all sampling is performed with minimal handling and disturbance of the patient. Glucose curves performed by the owner at home may help avoid some hospital related stressors.[5] These factors can make interpretation of a curve difficult, especially if assessed in isolation. This highlights the importance of always assessing the glucose curve in light of the animal's history and clinical signs.

The Somogyi effect (hypoglycemia-induced hyperglycemia) may occur when diabetic patients are overdosed with insulin. The overdose causes a transient period of hypoglycemia. Counter-regulatory hormones including glucagon, epinephrine, and cortisol are released to prevent further lowering of blood glucose. This may result in a period of hypoglycemia that lasts only minutes and so prevents the animal experiencing the serious consequences potentially caused by low blood glucose. These hormones thus cause life-saving insulin resistance; however, the counter-regulatory response is sometimes excessive and subsequently results in hyperglycemia that may persist for hours to days. These patients show symptoms of poor diabetic control such as polydipsia and polyuria. The hypoglycemic event frequently goes unnoticed by owners; it may also occur at night (nocturnal hypoglycemia) making these events even less likely to be witnessed. See Figure 1.

Information required to improve the usefulness of blood glucose curves

Glucose curves in isolation are of limited use but their clinical value can be greatly improved when combined with a detailed history. Owners of diabetic animals should be encouraged to keep a daily diary or spreadsheet (an example is available at http://www.animaldiabetesaustralia.com.au/) of their pet's progress at home and to provide this for review at each reassessment. The following basic information should be recorded for all diabetics:

1. Insulin type, dose, and time of administration
2. Meals including type and amount of food fed. The amount consumed by the animal should also be recorded.
3. Appetite, general demeanor and behavior.
4. Water intake. This should ideally be measured every day when the animal is receiving insulin. For diabetic cats that no longer require insulin, it should be measured once a week. The key is to determine how many mL of water the animal drinks over a known number of hours. If the water bowl is shared with one or more other pets, then the volume drunk by all should be measured. Owners of small dogs and cats will need a measuring jug with 10 mL increments.
5. Results of dipstick testing of urine glucose and ketones. This ideally requires collection of a sample of urine every day in cats and at least once a week in dogs.
6. Body weight. A diabetic animal's body weight should be recorded once each week. It is important to use the same scales each time. Scales designed for weighing adult humans are not suitable for cats and small dogs. Human baby scales are a good option.

Monitoring daily water intake provides a very useful guide to the current level of glycemic control. Average blood glucose concentration correlates positively with 24-hour water intake.[6] It is helpful to educate clients on the 'normal' water intake of dogs and cats because most owners are not familiar with this. Normal young adult cats eating a 100% diet of dry cat food drink 80 to 150 mL daily (20 to 30 mL/kg/day), while they drink little or nothing when eating a diet of canned food. If a diabetic cat drinks more than 40 mL/kg/day (typically >150-250 mL/day, depending on the cat's size and diet) or is lethargic or losing weight, then adjustment of the insulin dose is probably required. Healthy dogs typically drink 60-75 mL/kg/day if fed 100% dry food and 5-40 mL/kg/day if fed wet food only. Diabetic dogs with a water intake greater than 75 mL/kg/day may warrant investigation.

Most animals will readily cooperate with urine glucose testing. A practical method for many owners is to take the dipsticks when walking their dog on a lead and directly test the urine stream. Alternatively, dog's urine can be easily collected in a clean container or using a clean soup ladle that is kept specifically for this use. Cats that use a litter tray will often urinate immediately if their owner changes the litter while the cat is in the room. Cats that do not use a litter tray will typically allow urine sampling if their owner accompanies them when they are let out into the garden in the morning. A dipstick can be applied to urine-soaked litter or soil while it is still wet. If the urine patch has dried out, tap water can be added to allow application of a dipstick. This will dilute the sample but still allow determination of positive versus negative urine glucose. Some owners prefer to use non-absorbent litter or a litter tray with a sieve-like bottom that allows urine to collect in a lower tray. It is necessary to counsel owners that no importance can be attributed to the amount of glucose recorded on the dipstick – that is, there is no important difference between 1+ and 4+ results; it is simply a question of positive or negative glucose. Persistent negative glycosuria generally identifies periods when the exogenous insulin dose should be decreased.

A comprehensive assessment form can help make compilation of the required clinical information more efficient. Both body weight and body condition should always be recorded. The dose of insulin and time given should be confirmed. In addition to appraisal of the owner's insulin dosing technique, compliance of both the owner and the patient with the feeding recommendations must be routinely evaluated. The owner should be asked about activity level, water intake, appetite, and any other concerns. Information that suggests good control includes normal water intake, stable body weight, normal appetite and activity levels. They should also be specifically asked about possible hypoglycemic events. If owners have done any additional blood glucose measurements at home such as spot checks or pre-insulin testing, these should also be reviewed. The lowest blood glucose concentration obtained during the previous 2 weeks is generally of most interest, along with the date, time, and the preceding insulin dose.

Tips for performing blood glucose curves in the clinic

Glucose curves should commence as soon as possible after the morning meal and insulin dose, which should be given at the usual time. Animals are more likely to eat if fed by their owner at home. Insulin may also be administered at home; however if given at the clinic this allows appraisal of the owner's insulin dosing technique. If a dog does not eat, blood glucose results will likely be lower than typically occur after a meal. However glucose curves can continue if a cat does not eat all of its usual meal because the low carbohydrate diets that are typically recommended for diabetic cats do not seem to influence blood glucose concentrations.[7] However, it is preferable to follow the cat's usual daily routine as closely as possible.

Once admitted, the patient should be housed in a quiet place in the hospital. For cats or timid dogs, if the clinic contains loud barking dogs consider housing the animal in an unused consultation room or other area to reduce anxiety in the patient. Ideally no other testing or procedures should be performed on the patient on the day of the glucose curve. The curve should commence immediately after admission to hospital. Rather than shaving and restraining the dog or cat to take blood from veins, a minimally invasive technique should be used to sample from the ears, paw pads, or buccal mucosa to reduce the stress associated with blood draws. Blood should be sampled every 2-3 hours ideally for 12 hours. It should be noted that if the patient is receiving a longer acting insulin preparation, the peak effect might be 12 hours after administration hence the lowest blood glucose concentration may occur close to the time of insulin dosing.

Tips for owners performing blood glucose curves at home

Blood glucose monitoring at home will particularly appeal to those owners who desire more autonomous control over their pet's diabetes. Owners who choose this method of monitoring sometimes need to be advised against overzealous blood glucose measurement and interpreting the results themselves. Others might be unwilling to add home blood glucose monitoring because they feel it is an added source of stress. Key advantages of performing glucose curves at home are that they are less costly for the owner and, if confusing results are obtained on one day, it is often more practical to repeat a curve at home than for the pet to return to the clinic. It can also be a great advantage for an owner to have the ability to test blood glucose concentration whenever a diabetic pet shows signs consistent with hypoglycemia. The same protocol should be followed for serial blood glucose curves at home as for those performed in the hospital. Spot checks are useful whenever the owner is concerned about the well-being of the animal but are not a substitute for a full blood glucose curve. A practical approach is to use knowledge of the animal's clinical signs to guide the timing of home-generated blood glucose curves. For example, if there is marked variability of 24-hour water intake and/or urine glucose results, owners can be advised to perform a glucose curve on a day when the cat has negative glycosuria or does not drink much water. This approach would increase the chance of detecting hypoglycemia.

CONTINUOUS GLUCOSE MONITORS

Continuous glucose monitoring systems (CGMS) provide several advantages compared with traditional methods of serial blood glucose monitoring, with the major one being that glucose measurements are obtained every 3-5 minutes allowing clinicians to base treatment decisions on much more detailed glycemic data. Importantly, CGMS can facilitate detection of brief periods of hypoglycemia during both night and day. Unfortunately, a recent label change in North America prohibits use of CGMS devices in non-human patients. Similar restrictions might be introduced elsewhere.

CONCLUSION

Compliance is markedly improved if there is close rapport between the owner and the veterinary team and appropriate individualization of the pets monitoring regimen. No matter how blood glucose samples are collected they only provide a small piece of the puzzle of good diabetes management. Our role includes obtaining relevant patient information at each reassessment, minimizing stress when taking blood in hospital, and supporting the client when learning injection and blood sampling techniques. Knowing what information to collect and correct patient handling will minimize over-interpretation of glucose curves and inappropriate insulin dose changes.

Figure 1: Serial blood glucose concentration curves from a 12 year old, spayed female, Burmese cat receiving subcutaneous detemir insulin (Levemir® Flexpen®) every 12 hours at 7.00 am and 7.00 pm.

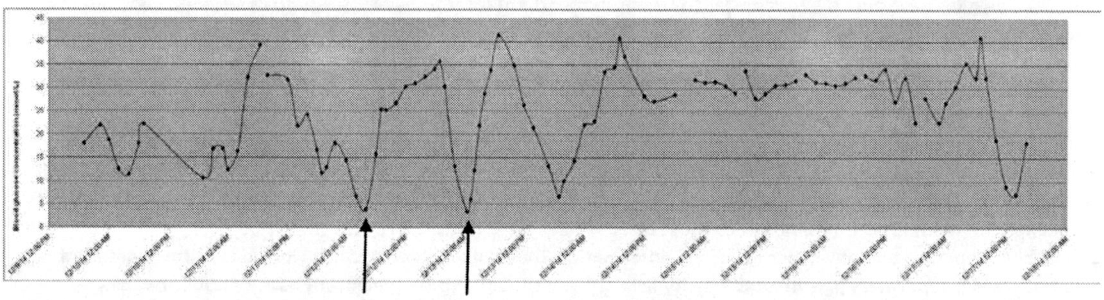

(a) Around the clock serial blood glucose monitoring over 8 days showing nocturnal hypoglycemia-induced hyperglycemia (Somogyi effect) on days 3 and 4. Both hypoglycemic events (indicated by arrows) occurred soon after midnight and were followed by a rapid increase of blood glucose >630 mg/dL (>35 mmol/L). The insulin dose was then decreased by 50% on day 4. There was immediate resolution of the hypoglycemia and improved glycemic control within a few days.

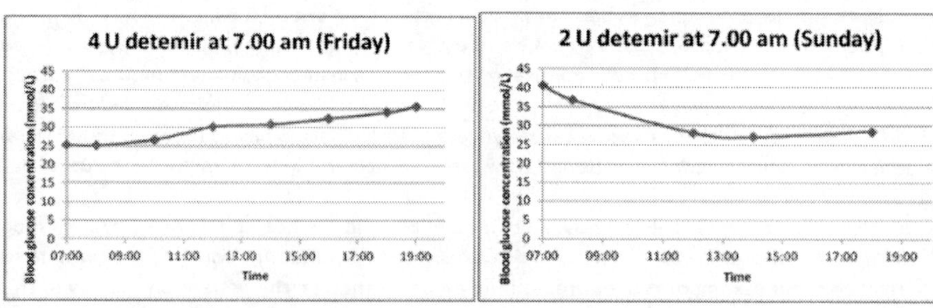

(b) Daytime glucose curves for days 2 and 4 showing hyperglycemia consistently >450 mg/dL (>25 mmol/L). If nocturnal hypoglycemia had not been identified, the insulin dose may have been increased resulting in worsening clinical signs and risk of severe hypoglycemia. This portrays a major disadvantage when daytime curves are the only information gathered.

REFERENCES

[1]Sparks AH et al, J Fel Med Surg 2015;17:235, [2]Alt N et al, J Am Vet Med Assoc 2007;230:1011, [3]Fleeman LM, Rand JS, J Am Vet Med Assoc 2003;222:317, [4]Rand JS et al, J Vet Intern Med 2002;16:123, [5]Reusch CE et al, J Fel Med Surg 2006;8:119, [6]Martin GJ, Rand JS, Vet Rec 2007;161:52, [7]Martin GJ, Rand JS, J Fel Med Surg 1999;1:241.

Advances in Diabetes Therapy: GLP-1 Based Drugs in Veterinary Medicine

Chen Gilor, DVM, PhD, DACVIM
Columbus, OH

INTRODUCTION

Incretin-based therapies are revolutionizing the field of diabetes therapy by replacing insulin therapy with safer and more convenient, long-acting drugs. Incretin hormones (GLP-1 and GIP) are secreted from the intestinal tract in response to the presence of food in the intestinal lumen. GLP-1 augments insulin secretion and suppresses glucagon secretion during hyperglycemia in a glucose-dependent manner. Clinical data have revealed that incretin-based drugs are as effective as insulin in improving glycemic control while reducing body weight (GLP-1 analogs, specifically) in patients with type 2 diabetes. Furthermore, incidence of hypoglycemia is relatively low with these drugs because of their glucose-dependent mechanism of action. Another significant advantage of these drugs is their duration of action. While insulin injections are administered at least once daily, long-acting GLP-1 analogs have been developed as once-a-week injections and could potentially be administered even less frequently than that in diabetic cats. This review considers the physiology of incretin hormones, and the pharmacology and use of GLP-1 analogs in people. We will also review up-to-date research data on GLP-1 analogs in veterinary medicine.

INCRETIN PHYSIOLOGY

Incretin-secreting cells (K and L cells) are interspersed in the epithelium of the gut. Their physiological role is to sense the type and quantity of digested nutrients in the gut. They then secrete incretin hormones as preparatory signals to other remote organs (e.g. brain, pancreas etc.). In the pancreas, the incretin signal is translated to increased sensitivity to the stimulatory effect of glucose. This effect is responsible for the observed difference in insulin secretion between oral and intravenous glucose and is defined as the "incretin effect"; oral glucose leads to much greater insulin secretion compared to intravenous glucose even when blood glucose concentrations are equal.[1,2] The incretin effect is thought to be exclusively mediated by 2 peptide hormones: Glucagon-like peptide-1 (GLP-1) which is secreted from L cells, and glucose-dependent insulinotropic polypeptide (GIP) which is secreted from K cells. Incretin hormones also promote expansion of pancreatic beta-cell mass: They promote differentiation of pancreatic ductal cells into beta cells and increase beta cell proliferation. Importantly, incretin hormones protect beta cells from apoptosis induced by various cytotoxic agents or dexamethasone. They also prevent apoptosis secondary to glucotoxicity and lipotoxicity.[3]

Incretin hormones affect glucagon secretion (GLP-1 inhibits, GIP stimulates), slow the rate of gastric emptying (GLP-1), increase satiety (GLP-1), and increase the sensitivity of adipose tissue to insulin (GIP). Thus, they play a major role in glucose homeostasis.[3,4] Active GIP and active GLP-1 are degraded by the enzymes dipeptidyl peptidase-4 (DPP-4, also known as CD26) and neutral endopeptidase 24.11 (NEP-24.11) into inactive forms, thereby modifying or inhibiting their activity.[5] DPP-4 and NEP are ubiquitous in tissues, and DPP-4 is also present in a soluble form in the blood. Consequently, when injected intravenously, active GLP-1and GIP are quickly degraded into inactive forms. When injected intravenously, the half-life of GLP-1 is only about 1-2 minutes and the half-life of GIP is only about 5 minutes. The inactive GLP-1 and GIP are quickly cleared by the kidneys.

INCRETIN HORMONES IN DIABETES AND OBESITY

The incretin effect is severely reduced or absent in people with type 2 DM, contributing to glucose intolerance and post-prandial hyperglycemia. In diabetics, the secretion of GIP is normal or slightly reduced but its effect on the pancreas is markedly decreased. In contrast, GLP-1 retains its insulinotropic effects in type 2 DM (at least in supraphysiologic concentrations), but its secretion is decreased.[4] Whether a blunted incretin response is a primary process in type 2 DM or a secondary process caused by diabetes, is debatable. There is some evidence that obesity contributes to attenuation of the incretin effect independently of diabetes. In a recent study, obesity was associated with decreased incretin effect including decreased beta-cell response to GLP-1 but also decreased GLP-1 secretion.[5] In some studies, GLP-1 secretion was normal in obese people but GIP secretion was increased during fasting and early after a meal.[6] Insulin resistance in itself, regardless of obesity or diabetes, has been associated with abnormal secretion of GIP and GLP-1.[7] Increased insulin concentrations in the insulin-resistant subjects might down-regulate incretin secretion.

Traditional treatments for diabetes are either just partially effective (diet change, oral drugs) or associated with significant side effects (insulin therapy). Owner compliance is a major problem because twice-a-day injections are required with most commonly used insulin preparations. Weight gain may indicate a good response to therapy initially but can eventually become a problem. Hypoglycemia is a common complication of insulin therapy and can be life-threatening. These side effects have been significantly reduced in human medicine with the introduction of incretin-based treatments. GLP-1 stimulates glucose-dependent insulin secretion. Its effect is blunted in euglycemia and therefore they are less likely to cause hypoglycemia. GLP-1 also promotes weight loss. GLP-1 analogs are not only effective in controlling blood glucose in diabetics but they are also associated with weight loss in diabetic and non-diabetic obese people. One major advantage of incretin-based therapies is their potential to reverse the course of the disease. Incretin-based treatments in diabetic people correct not only hyperglycemia but also all the major markers of beta-cell dysfunction.

The very short half-life of GLP-1 limits its clinical use. There are numerous treatment strategies that overcome this problem and take advantage of incretin-hormone physiology. Novel GLP-1 analogs have a prolonged half-life and are used as SQ injections at intervals of up to one week.

INCRETIN-BASED THERAPY STRATEGIES

There are several treatment modalities that overcome the problems associated with direct administration of incretin hormones. Long-acting, DPP-4-resistant, synthetic GLP-1 mimetic peptides are most commonly used. Available formulations are all for SQ injections but other delivery systems are being developed (including enteral, pulmonary, or sublingual). Also under development are non-peptidic GLP-1 receptor agonists.[8] Oral DPP-4 inhibitors that prolong the half-life of incretin hormones are also commercially available.[3] Non-peptidic drugs that activate free fatty acid receptors in L cells and stimulate GLP-1 secretion are being developed (GRP120 and GRP119 agonists).[9, 10,11] Diet manipulation should also be investigated.

Drugs that are commercially available or that are currently in advanced clinical trials are described below.

GLP-1 MIMETICS AND ANALOGS

Exenatide

The peptide exendin-4 was first isolated from the poisonous venom of the Gila Monster (*Heloderma suspectum*). Exendin-4 is a 39-amino acid peptide that shares only a 53% sequence homology with GLP-1 but its affinity for the GLP-1 receptor is 1000 times greater than the affinity of GLP-1. Unlike GLP-1, exendin-4 is not a substrate for DPP-4 and NEP.[4] Exenatide is a synthetic exendin-4. Resistant to degradation, exenatide is eliminated by the kidneys and has a half-life of 3-4 hours in people. Its biological effect lasts about 8 hours after subcutaneous injection and it can be detected in the plasma for up to 15 hours.[12] Multiple studies, both *in vitro* and *in vivo*, have shown that, in general, exendin-4 has the same physiologic effects as GLP-1 in the pancreas, GI tract, and brain.[3] Exenatide is associated with improvement in some of the earliest and most fundamental abnormalities of type 2 diabetes: diminished "first-phase insulin response" and proinsulin/insulin ratio. Acute administration of exenatide in type 2 diabetic patients corrects the abnormal insulin secretion pattern after an IV glucose bolus (first phase and second phase insulin responses) and restores the ability of beta cells to respond to rapid changes in blood glucose concentrations.[3] Exenatide also improves proinsulin/insulin ratio after 30 weeks of treatment.[13]

Exenatide has been shown to be as effective as insulin glargine in the treatment of DM but with less side effects (e.g. hypoglycemia and weight gain).[14] In a 2-year follow-up of patients receiving exenatide, patients achieved sustained and significant reductions in glycosylated hemoglobin, accompanied by significant weight loss (instead of weight gain commonly seen in diabetics receiving insulin) and improvement in serum liver enzyme activity and blood pressure. Most importantly, treatment with exenatide improved beta cell function as measured by homeostasis model assessment of beta cell function (HOMA-B).[15]

Exenatide has minimal side effects in people. It is mostly associated with nausea and less frequently with vomiting. Infrequently, it might cause hypoglycemia. Severe hypoglycemia (requiring assistance) was reported rarely (only 5 of 2781 patients) and only in patients who also received sulfonylurea drugs. Antibodies to exenatide developed in 67% of patients but this did not affect outcome and was not associated with side effects.[16]

When first discovered, exendin-4 was shown to potentiate amylase release from rat acinar pancreatic cells in response to other hormones such as cholecystokinin. This was shown *ex vivo* and in high doses. A possible association between GLP-1 analogs and pancreatitis as well as medullary thyroid cancer has been suggested and lead to issuing of a warning by FDA. However, multiple studies have refuted these concerns and it is widely accepted that the proven enormous benefits of GLP-1 analogs outweigh its hypothetical risks.[17,18]

Exenatide is commercially available in the USA under the trade name Byetta®.

Exenatide in cats

In healthy cats, exenatide was quickly absorbed after a SQ injection and caused glucose-dependent insulin secretion. Increased glucose tolerance, however, was not observed after a single SQ injection.[19,20] At a dose of 1.0 mcg/kg SQ (about 10 times the dose that is used in diabetic people), exenatide injection did not cause any side effects in healthy cats, except for hypoglycemia in 1 out of 9 cats.[19] Exenatide has led to significant weight loss in healthy cats of 7.0 ± 4.9% (from 4.78 ± 1.5 kg to 4.48 ± 1.5 kg) with a dose of 1.0 mcg/kg SQ BID for 28 days.[20]

Exenatide Extended-Release

A long-acting sustained-release formulation of exenatide (Bydureon®) has recently been approved by the FDA as the first once-weekly subcutaneous injection for treatment of type 2 diabetes people. It consists of injectable microspheres of exenatide and poly(D,L-lactic-co-glycolic acid), a common biodegradable medical polymer with established use in absorbable sutures and extended-release pharmaceuticals, that allows gradual drug delivery at controlled rates. In people, Exenatide plasma concentrations are sustained at an effective concentration (50 pg/mL) for longer than 60 days after a single injection at doses of 5mg, 7mg or 10mg.[21]

It has been shown in a recent clinical study to be more effective than once-a-day insulin glargine in achieving glycemic control with decreased risk of hypoglycemia and with reduction (instead of gain) in body weight.[22] This extended release formulation was also more effective than regular exenatide (Byetta) in achieving glycemic control with no increased risk of hypoglycemia, decreased side effects like nausea, and with similar reductions in body weight.[23]

Exenatide Extended-Release in cats

We have studied Exenatide Extended-Release in healthy cats at a dose of 0.13 mg/kg. Three weeks after a single subcutaneous injection fasting BG was decreased and during a hyperglycemic clamp at that time glucose tolerance improved, insulin concentrations increased and glucagon concentrations decreased.[24] Recently, exenatide-ER was assessed in a group of newly diagnosed diabetic cats treated with insulin glargine and fed a diabetic diet. Cats in this study were treated with once-weekly injection of placebo or exenatide-ER at a dose of 0.2 mg/kg.[25] Despite using what seems in retrospect like a very high dose, this study found only a trend towards a small effect of exenatide-ER on remission rates and improved glycemic control. At first glance these are disappointing results because they suggest lack of efficacy in diabetic cats. However, it is possible that a more obvious positive effect would have been seen if the target population was more similar to the target population used in exenatide studies in people, i.e. non-insulin dependent type 2 diabetics (relatively early in the course of the disease). No side effects were observed in cats in the two studies described above.

Liraglutide

Liraglutide is a GLP-1 synthetic analog with 2 amino acid substitutions and a fatty acid acyl group that enables noncovalent binding to albumin, thereby extending the pharmacokinetic profile of the GLP-1 molecule. Liraglutide exhibits a prolonged pharmacokinetic profile after a single injection, and exhibits all of the actions of native GLP-1.[4] Liraglutide (once-a-day) was recently compared to exenatide (twice-a-day). Liraglutide provided significantly greater improvements in glycemic control than did exenatide and was generally better tolerated.[26] Liraglutide once-daily was also more effective than Exenatide-ER once-weekly in achieving glycemic controlling but nausea, vomiting and diarrhea occurred less frequently with Exenatide-ER. [27] Liraglutide has been used successfully to treat obesity in non-diabetic patients.[28]

Liraglutide is commercially available in the USA and Europe under the trade name Victoza®.

Liraglutide in cats

We have studied liraglutide in healthy cats at a dose of 0.6 mg/cat once daily for 7 days. Liraglutide caused significant weight loss in all cats at day (9% ± 3). Appetite was subjectively decreased in all cats and one cat was withdrawn on day 4 because of 48 hours of anorexia. During a hyperglycemic clamp, liraglutide was associated with a trend towards improved glucose tolerance, higher insulin concentrations and lower glucagon concentrations. Fasting glucose concentrations were not affected.[28]

Albiglutide

Albiglutide is a recombinant GLP-1-albumin fusion protein that exhibits a reduced affinity for the GLP-1R, but displays a broad spectrum of GLP-1R-dependent actions in preclinical studies, including inhibition of food intake and gastric emptying and reduction of glycemia excursion after meal ingestion. Importantly, it has the potential to be used as a very long-acting drug: It has been investigated in people for use as a weekly, biweekly and monthly injection.[30] Albiglutide was approved by FDA in 2014.

DPP-4 INHIBITORS

DPP-4 inhibitors (e.g. sitagliptin, vildagliptin) are administered orally. In people, vildagliptin and sitagliptin are well-tolerated and not associated with hypoglycemia when used alone. Both agents increase plasma concentrations of GLP-1 and GIP after meal ingestion, enhance glucose-stimulated insulin secretion and reduce ratios of proinsulin:insulin, consistent with an improvement in beta cell function. They are, however, less potent than other oral hypoglycemic drugs.[3] In contrast to GLP-1 analogs, DPP-4 inhibitors are not associated with nausea or vomiting in people, but they are associated with weight gain. DPP-4 inhibitors are also associated with increased risk of nasopharyngitis, urinary tract infections and headaches.[16] Increased risk of infections might be related to the action of DPP-4 in T-cells as a co-stimulatory molecule (CD26).

A DPP-4 inhibitor has been used experimentally in cats and was effective in enhancing insulin secretion and inhibiting glucagon secretion after an intravenous glucose challenge. Glucagon inhibition was also observed after a meal challenge.[31] In that study, the drug was administered subcutaneously and not orally.

REFERENCES

1. McIntyre N, et al. J Clin Endocrinol Metab. 1965;25:1317-24.
2. Nauck MA, et al. J Clin Endocrinol Metab 1986;63:492-498
3. Kim W, Egan JM. Pharmacol Rev 2008;60:470-512
4. Holst JJ, et al. Mol Cell Endocrinol. 2009;297:127-36.
5. Muscelli E, et al. Diabetes 2008;57:1340-8.
6. Vilsboll T, et al. J Clin Endocrinol Metab. 2005;88:2706-13.
7. Rask E, et al. Diabetes Care. 2001;24:1640-5.
8. Sloop KW, et al. Diabetes, 2010 Dec;59(12):3099-107

9. Adachi T, et al. Biol Pharm Bull. 2008;31:1021-3.
10. Morishita M, et al. J Control Release. 2008;132:99-104.
11. Overton HA, et al. Br J Pharmacol. 2008;153:S76-81.
12. Kolterman OG, et al. Am J Health Syst Pharm. 2005;62:173-81.
13. DeFronzo RA, et al. Diabetes Care. 2005;28:1092-100.
14. Glass LC, et al. Curr Med Res Opin. 2008;24:639-44.
15. Buse JB, et al. Clin Ther. 2007;29:139-53.
16. Amori RE, et al. J Am Med Assoc. 2007;298:194-206.
17. Alves C, et al. Diabetes Res Clin Pract. 2012 Nov;98(2):271-84.
18. Chiu WY, et al. Exp Diabetes Res. 2012;2012:924168.
19. Gilor C, et al. Domest Anim Endocrinol 2011 Jul;41(1):42-9
20. Seyfert TM, et al. Intern J Appl Res Vet Med 2012 ;10(2) :147-156
21. Fineman MS, et al. Clin Pharmacokinet. 2011 Jan;50(1):65-74.
22. Diamant M, et al. Diabetes Care. 2012 Apr;35(4):683-9.
23. Buse JB, et al. Diabetes Care. 2010 Jun;33(6):1255-61
24. Rudinsky AJ, et al. Domest Anim Endocrinol, 2015; 51:78-85.
25. Riederer AR et al. J Vet Intern Med 2015; 29(1): 448
26. Buse JB, et al. Lancet. 2009 Jul 4;374(9683):39-47.
27. Buse JB, et al. Lancet. 2013 Jan 12;381(9861):117-24.
28. Astrup A, et al. Lancet. 2009;374:1606-16.
29. Hall MJ, et al. Domest Anim Endocrinol, 2015 51: 114-21
30. Rosenstock J, et al. Diabetes Care. 2009 Oct;32(10):1880-6.
31. Furrer D, et al. Vet J. 2010;183:355-7.

SDMA and BAIB as Biomarkers for CKD Associated with Kidney Stones in Cats
Jean A. Hall, DVM, PhD, DACVIM (Small Animal Internal Medicine)
Corvallis, OR

INTRODUCTION

Beginning in the mid-1980s, a noticeable increase in the frequency of calcium oxalate uroliths occurred in cats concomitant with a decrease in the frequency of struvite uroliths.[1] In 2002, approximately 55% of the 95,000 feline uroliths submitted to the Minnesota Urolith Center were composed of calcium oxalate.[1] More recently (2007), the frequency of feline calcium oxalate uroliths declined to 41%, perhaps associated with reformulation of adult maintenance diets to minimize risk factors for calcium oxalate crystalluria.[1] The increase in calcium oxalate urolithiasis in cats has been associated with a parallel increase in the occurrence of these uroliths found in the upper urinary tract (kidneys and ureters).[1]

Calcium oxalate uroliths tend to develop in mid-to-older aged cats.[2] Factors associated with increased risk of calcium oxalate urolithiasis are feeding urine-acidifying diets, feeding a single brand of cat food, being kept indoors only, and being of the Persian breed.[2] It has been reported that calcium oxalate urolith formation occurs when urine is supersaturated with calcium and oxalate,[3] although all healthy cats have urine that is supersaturated with calcium and oxalate. In one study, urine relative supersaturation with calcium oxalate was higher in healthy cats fed an acidifying diet, supporting a pH effect on urinary saturation with calcium oxalate, however, there were no significant differences in urinary excretion of calcium, oxalate, or citrate.[4] Thus, it is likely that other unknown risk factors, perhaps dietary or perhaps individual, are operative.

Under normal conditions, concentrations of oxalate in blood and urine depend on the content of oxalic acid in foods and on metabolic conversion of endogenous oxalate precursors.[5] The major precursor of oxalate under normal circumstances appears to be the amino acid hydroxyproline from collagen.[5] Hydroxyproline is one of the most abundant amino acids in collagen and every day the body turns over several grams of collagen. Glyoxylate is the two-carbon end-product of hydroxyproline catabolism (pyruvate is the other product). The conversion of glyoxylate to oxalate is catalyzed by lactate dehydrogenase. Normally, most of the glyoxylate resulting from hydroxyproline catabolism is diverted to glycine/pyruvate synthesis by the reaction catalyzed by alanine-glyoxylate aminotransferase 2 (AGXT2), rather than converted from glyoxylate to oxalate by lactate dehydrogenase. Note that the key intermediate in calcium oxalate urolithiasis is glyoxylate, which is the focal point in pathways leading to and away from oxalate.

We have previously reported that serum concentrations of symmetric dimethylarginine (SDMA) were elevated in three azotemic and two nonazotemic cats with calcium oxalate nephrolithiasis out of 21 cats with chronic kidney disease (CKD).[6] We have shown that serum SDMA correlates with glomerular filtration rate (GFR) in cats.[6,7] Furthermore, serum SDMA concentrations are not affected by lean body mass in cats.[7] Thus, cats with calcium oxalate nephrolithiasis may have IRIS stage 1 CKD with increased SDMA and nonazotemia, or more advanced IRIS stages of CKD with azotemia.

Symmetric dimethylarginine is produced when proteins containing methylated arginines are degraded. Free methylarginines are released into the cytosol following proteolysis, and then enter the blood circulation. Symmetric dimethylarginine is excreted primarily (≥ 90%) by renal clearance.[8,9] Because serum SDMA is eliminated by the kidneys, plasma concentrations are affected by changes in GFR. There are 3 main species of methylated arginines: monomethylarginine (MMA), asymmetric dimethylarginine (ADMA), and SDMA.[10] The dimethylarginine dimethylaminohydrolases ($DDAH_1$ and $DDAH_2$) are involved in enzymatic degradation of ADMA, but not SDMA.[8] ADMA is also degraded to a lesser extent by AGXT2.[11]

AGXT2 is also the key enzyme responsible for the degradation of endogenously formed β-aminoisobutyrate (BAIB). BAIB is formed in the cytosol from thymine degradation. BAIB is then transported into mitochondria and further metabolized to malonate semialdehyde by AGXT2.[12] AGXT2 has a higher affinity for BAIB as a substrate compared with dimethylarginine. If AGXT2 is the main enzyme involved in degrading BAIB and possesses high affinity for BAIB as a substrate, then increased BAIB concentrations could serve as an indicator of overall AGXT2 enzyme activity.[13,14] Based on an mRNA analysis across tissues, AGXT2 was found to be highly expressed in the kidney (Hill's Pet Nutrition, Inc., unpublished results).[13] Thus, an accumulation of BAIB could accompany several altered physiological functions in the kidney, including altered detoxification of glyoxylate and increased concentrations of oxalate in blood and urine.

CHARACTERIZATION OF 43 CATS WITH KIDNEY STONES BY RENAL BIOMARKERS FOR CKD

The purpose of this retrospective study was to report on the utility of measuring serum SDMA and BAIB concentrations, along with serum Cr concentrations, in cats with kidney stones. Cats with kidney stones came from a colony of over 400 DSH cats. Cats in the colony ranged in age from 1 to 19 years, with approximately 25% of cats >10 years. All cats had been used historically for palatability studies. Many different types of foods of varying nutrient compositions from multiple manufacturers were fed, including dry and canned cat foods. All foods met the requirements established by the Association of American Feed Control Officials (AAFCO) for complete and balanced pet foods for adult cats.

Cats with kidney stones were diagnosed by radiographic imaging (antemortem) or by postmortem necropsy. All cats with renal stones diagnosed from August 2010 to December 2015 were included (n=43). Cats with kidney stones ranged in age from 4.7 to 18.1 years, with mean age of 11.7 years at the time of diagnosis. Approximately one-third of kidney stones were diagnosed by radiographic imaging antemorten, and two-thirds were discovered at necropsy. Approximately 90% were calcium oxalate monohydrate. Over 90% of these cats had increased serum SDMA concentration at some point prior to death, whereas less than

25% were azotemic prior to death. The median (range) in months that SDMA concentrations were shown to increase above the normal reference interval before serum creatinine concentrations increased above the normal reference interval was 26.5 months (0.0 to 60.0 months).

If SDMA concentrations are elevated in mid-to-older aged cats along with normal serum creatinine concentrations, further imaging studies are warranted to rule in/out kidney stones.

REFERENCES

1. Osborne CA, Lulich JP, Kruger JM, et al. Analysis of 451,891 canine uroliths, feline uroliths, and feline urethral plugs from 1981 to 2007: Perspectives from the Minnesota Urolith Center. Vet Clin North Am Small Anim Pract 2009;39:183-197.

2. Kirk CA, Ling GV, Franti CE, et al. Evaluation of factors associated with development of calcium oxalate urolithiasis in cats. J Am Vet Med Assoc 1995;207:1429-1434.

3. Bartges JW, Osborne CA, Lulich JP, et al. Methods for evaluating treatment of uroliths. Vet Clin North Am Small Anim Pract 1999;29:45-57.

4. Bartges JW, Kirk CA, Cox SK, et al. Influence of acidifying or alkalinizing diets on bone mineral density and urine relative supersaturation with calcium oxalate and struvite in healthy cats. Am J Vet Res 2013;74:1347-1352.

5. Glew RH, Sun Y, Horowitz BL, et al. Nephropathy in dietary hyperoxaluria: A potentially preventable acute or chronic kidney disease. World J Nephrol 2014;3:122-142.

6. Hall JA, Yerramilli M, Obare E, et al. Comparison of serum concentrations of symmetric dimethylarginine and creatinine as kidney function biomarkers in cats with chronic kidney disease. J Vet Intern Med 2014;28:1676-1683.

7. Hall JA, Yerramilli M, Obare E, et al. Comparison of serum concentrations of symmetric dimethylarginine and creatinine as kidney function biomarkers in healthy geriatric cats fed reduced protein foods enriched with fish oil, L-carnitine, and medium-chain triglycerides. Vet J 2014;202:588-596.

8. Schwedhelm E, Boger RH. The role of asymmetric and symmetric dimethylarginines in renal disease. Nat Rev Nephrol 2011;7:275-285.

9. Kielstein JT, Boger RH, Bode-Boger SM, et al. Marked increase of asymmetric dimethylarginine in patients with incipient primary chronic renal disease. J Am Soc Nephrol 2002;13:170-176.

10. Bedford MT, Richard S. Arginine methylation an emerging regulator of protein function. Mol Cell 2005;18:263-272.

11. Martens-Lobenhoffer J, Rodionov RN, Bode-Böger SM. Probing AGXT2 enzyme activity in mouse tissue by applying stable isotope-labeled asymmetric dimethyl arginine as substrate. J Mass Spectrom 2012;47(12):1594-600.

12. Kontani Y, Kaneko M, Kikugawa M, et al. Identity of D-3-aminoisobutyrate-pyruvate aminotransferase with alanine-glyoxylate aminotransferase 2. Biochim Biophys Acta 1993;1156:161-166.

13. Yerramilli M. Prognostic value of SDMA to creatinine ratio in dogs and cats with chronic kidney disease. 2015 ACVIM Forum Research Reports Program. J Vet Intern Med 2015;29(4):1257–1283.

14. Kittel A, Muller F, Konig J, et al. Alanine-glyoxylate aminotransferase 2 (AGXT2) polymorphisms have considerable impact on methylarginine and beta-aminoisobutyrate metabolism in healthy volunteers. PLoS One 2014;9:e88544.

Novel Options for Refractory Incontinence: Stem Cells to Silicone

Christopher A. Adin, DVM, DACVS
Shelly L. Vaden, DVM, DACVIM (Small Animal Internal Medicine)
Raleigh, NC

INTRODUCTION

It is estimated that 20% of dogs in the United States will develop urinary incontinence during their lifetime. Urethral sphincter mechanism incompetence (USMI) is the most common cause of incontinence and is caused by loss of urethral tone and insufficient sphincter pressure. USMI is multifactorial with breed, neutering, caudal bladder position, shortened urethral length and obesity all being possible contributing factors. Other potential causes include ectopic ureters, paradoxical incontinence, detrusor hyperactivity, vestibulovaginal abnormalities and neurologic disorders. Up to 50% of dogs suffer from continued urine leakage after ureteral reconstruction for ectopic ureter, often due to urethral sphincter dysfunction. Signalment, historical information, pattern of incontinence and response to medical therapy can lead to an accurate diagnosis in most dogs with urinary incontinence. However, advanced imaging studies or cystourethroscopy or both may be needed to render a diagnosis in dogs that fail traditional medical management.

Reproductive hormones or alpha agonists or a combination of these drugs can be used to treat dogs with USMI as well as some with ectopic ureters and work through increasing the strength of urethral contraction. The initial response rate may be as high as 85-90% but up to 27% of the dogs either have no initial response or eventually become refractory to treatment, requiring other forms of management.[1] Furthermore, these drugs are associated with a number of potential side effects and must be administered for the remainder of the animal's life.[2-6] In addition, some owners will seek out other management options because of medication side effects or the unwillingness to administer daily medications for the life of their pet. The purpose of this presentation is to discuss novel options for dogs with urinary incontinence.

URETHRAL BULKING AGENTS

Dogs with refractory USMI may have improved urethral closure and continence following the submucosal injection of bulking agents in the proximal urethra during cystourethroscopy.[7-8] Submucosal bulking creates a central filling volume that in turn increases the length of the muscle fibers and closure power of the urethral sphincter thereby increasing resistance to urine flow. Submucosal bulking is performed during cystourethroscopy and requires specialized injection equipment. Injections are made at the level of the proximal urethra, just distal to the trigone. Generally the bulking agent is injected in 3-4 sites around the circumference of the urethra until the deposits close the urethral lumen, as seen through the cystoscope. Mild complications (e.g., dysuria) occur in up to 15% of dogs in the immediate post-procedural period. Success is usually apparent within 1-3 days of the procedure.

Although several substances have been used for urethral bulking, collagen is the one that has been used most commonly in veterinary medicine. Bovine collagen used for urethral bulking is purified and cross-linked with glutaraldehyde for injection (Regain, Avalon Medical, Stillwater, MN). Submucosal collagen injections are effective at restoring continence in approximately 68% of dogs, with an additional 15% achieving continence after adding standard medical therapy. Approximately 10% of dogs will demonstrate improvement with collagen injections and standard medical management but will not be completely continent and the remaining 10% will not have any demonstrable improvement after injection. Continence obtained following collagen injections is not permanent, with most dogs requiring subsequent injections within an average of 17 months.[7] Polydimethylsiloxane (Macroplastigue, Urolplasty, Inc., Minnetonka, MN) may be a safe and effective alternative to collagen for use as a urethral bulking agent in dogs.

REGENERATIVE MEDICINE APPROACH

The normal urethra consists of layers of striated muscle, smooth muscle, connective tissue, submucosal vascular plexus and epithelium. The striated muscle layer appears to contribute the most to maintenance of urethral tone and continence.[9] It is logical to assume that conditions where there is reduced striated muscle would be associated with urinary incontinence. Stress urinary incontinence (SUI), the most common form of urinary incontinence in women, is characterized by urinary leakage when increased abdominal pressure causes intravesical pressure to exceed the urethral closure pressure.[10] Men can also develop SUI, primarily after prostate surgery. People with SUI have reduced urethral striated muscle, the degree of which corresponds to the decrease in urethral closure pressure. Smooth muscle is also reduced but plays a less important role in maintaining urethral tone.

Certain stem cells differentiate into striated and smooth muscle cells. When injected into the urethra, these stem cells have the potential to restore the musculature. In addition, certain stem cells might secrete angiogenic or musculotrophic growth factors. In rats where conditions resembling SUI have been induced by sphincter injury, stem cell injections were associated with improvement of urethral function as well as histologic appearance of the musculature. Adipose-derived stem cells are able to produce collagen and elastin, and have been shown to improve urethral connective tissue in rats.[9]

Injection of muscle-derived stem cells, obtained from skeletal muscle biopsies, have been used for the regenerative repair of deficient sphincter muscle in women with SUI who failed traditional medical therapy. Improved continence rates of up to 89% were reported 2 years following injection in women. The onset of improvement was between 3 and 8 months and continued for a median of 10 months suggesting regeneration of sphincter function rather than a response to urethral bulking.[10]

A regenerative medical approach for urethral incontinence was studied in dogs that had induced urinary sphincter insufficiency following urethral damage. Autologous muscle progenitor cells were isolated from skeletal muscle biopsies, expanded in culture and injected into the damaged sphincter. Six months following injection, dogs had urethral pressures that approximated 80% of normal compared with control dogs that remained at around 20% of normal. These preliminary results suggested that autologous derived muscle progenitor cells could be used to restore damaged or weakened urethral muscle and sphincter function.[11] Ongoing clinical trials suggest that urethral injection of autologous muscle progenitor cells can also lead to improvement or resolution of urinary incontinence in female dogs presumed to be secondary to naturally acquired urethral sphincter mechanism incompetence.

ARTIFICIAL URETHRAL SPHINCTER (HYDRAULIC OCCLUDER)

The canine artificial urethral sphincter (AUS) can be used to manage incontinence in males and females that is a result of unresponsive USMI or failed ectopic ureter repair. The device (Figure 1) was initially created by combining an inflatable silicone vascular occluder (DOCXS, Ukiah California) with a titanium subcutaneous injection port (Le Grande Vascular Access Port, Access Technologies) and then modified by widening the cuff to minimize urethral trauma during compression. We also changed the entry point of the actuating (inflation) tubing, making a side entry that would be parallel to the urethra and would allow easier implantation. Inflation of the occluder is performed by injecting fluid into a titanium vascular access port, which is placed in the subcutaneous tissues along the ventral abdomen. The occluders are composed of medical grade silicone and have been used in physiology research for many years. An in vitro experiment in our lab demonstrated reliable maintenance of occlusion during a 5-month period of immersion in simulated body fluid when occluders were infused with either isotonic saline or sodium hyaluronate.[12] Following the in vitro work, maximal urethral closure pressure (MUCP) and cystourethral leak point pressure (CLPP) were measured before and after application of the HUS to 6 female dog cadavers.[13] MUCP and CLPP were directly proportional to the volume injected into the hydraulic occluder and equaled or exceeded values reported in live, continent dogs after 50% occlusion of the HUS.[13] The data obtained in this cadaver model suggested that urinary continence may be achieved after application of the HUS to spayed female dogs with USMI. Although adjustable, this system is not intended to be patient-controlled. Rather, the HUS would provide a low level, static increase in urethral resistance, similar to that achieved through submucosal collagen injection, but of a more durable nature. The amounts of saline that were injected in the cadaver study were later modified to smaller volumes (0.1 to 0.2 mLs) based on observations in live animals with USMI. A pilot study was funded by the American Kennel Club to test application in dogs with refractory USMI and long-term results (2 years) in 4 dogs were reported.[14]

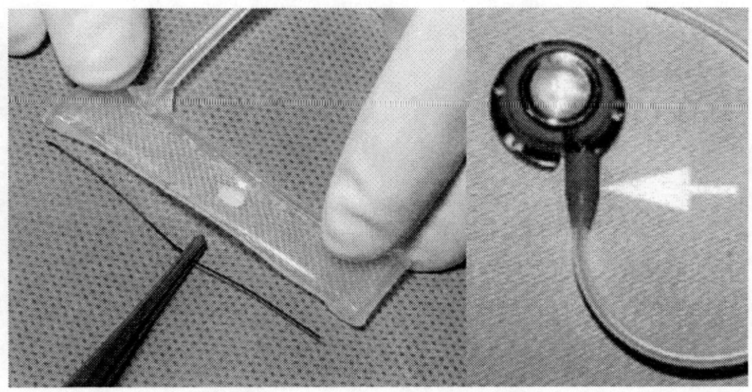

A B

Figure 1 (A and B): A, An artificial urethral sphincter with side entry tubing is sized by comparing to a piece of suture that was cut to approximate urethral circumference. The sphincter is closed around the urethra by placing a strand of non-absorbable suture material through the suture eyelets in both ends of the cuff. The cuff may then be inflated by injecting fluid into the subcutaneous port, causing mild, static occlusion of the urethra. **B,** The actuating tubing is attached to a titanium subcutaneous injection port using a "boot" (white arrow) supplied by the manufacturer of the port. (modified from Reeves, et al, *Vet Surg*, 2012)

We have now implanted this device in over 30 animals, including 5 male dogs and two female cats. Outcome for the majority of these animals (27 dogs) was presented in a recent manuscript.[15] Using a continence scale (1 representing constant leakage and 10 representing complete continence), animals improved from a median of 2 preoperatively to a median of 9 at last follow up. The advantage of the technique is that the device is permanent and adjustable, so injections can be performed with minimal expense or invasiveness to increase urethral occlusion if incontinence recurs. As with all procedures that involve implantation of a permanent device, there are several potential complications that may be recognized. In this study, two of 27 dogs developed progressive signs of urethral obstruction that required recommendation of device removal. Gross examination suggested that fibrosis around the device may have contributed to excessive urethral compression, although both dogs could be catheterized easily and obstruction was incomplete. A small number of dogs (2) failed to achieve adequate continence with the device, although this was associated with some compliance problems and difficulty with return for recheck exams or injections in both cases. Overall, the device appears to be approximately 90% effective, but clients should be warned about the potential need for adjustment or removal of the device in some animals.

REFERENCES

[1]Arnold S, et al. Schwez Arch Tierheilkd 1989;131:259. [2]Richter KP, et al. JAVMA 1985;187:605. [3]Mandigers RJ, et al. Vet Rec 2002;149:764. [4]Plum DC. *Veterinary Drug Handbook*. White Bear Lake, MN Pharma Vet Publishing, 2nd Edition, 1995, p 209, 547. [5]Moreau PM, et al. Kirk RW (ed): *Current Veterinary Therapy, X*. Small Animal Practice. Philadelphia, PA, WB Saunders Co, 1992, p 875. [6]Gookin JL, et al. Compend Contin Educ Pract Vet 1996;18:525. [7]Barth A, et al. JAVMA 2005; 226:73. [8]Byron J, et al. JVIM 2011;25:980. [9]Lin, CS, et al. Stem Cells Dev 2012. [10]Carr L, et al. Int Urogynecol J Pelvic Floor Dysfunct 2008;19:881. [11]Eberli D, et al. Cell Transplant, 2012. [12]Sereda C, et al. ACVS Abstract, 2005. [13]Adin CA, et al. AJVR 2004;65:283. [14]Rose S, et al. Vet Surg 2009;38:747. [15]Reeves L, et al. Vet Surg 2013;42:12.

Characterization of Bronchointerstitial Disease in Cats: Spontaneous Disease & Exp Models

A. Ray Dillon, DVM, MS, MBA; DACVIM (SAIM)
Auburn, AL

INTRODUCTION

A clear understanding of the nature of feline lower respiratory disease has been lacking. Based on response to therapy, the general approach to a chronic coughing cat with or without intermittent dyspnea has been to rule out obvious infectious agents and treat with a combination of corticosteroids via different routes with bronchial dilating agents. Thoracic radiographs demonstrating a "bronchial pattern" and a response to therapy resulted in a tentative diagnosis of feline asthma. On examination of bronchial alveolar lavage fluid, an increase in percent eosinophils tended to reinforce the tentative diagnosis. The gradual increase in severity or frequency of clinical signs resulted in a confirmation of the need for long term management and often maintenance medications.

SPONTANEOUS DISEASE

A prospective study of first time coughing/dyspneic client owned cats over 6 months of age was conducted in 14 practices in 212 cats over a 2 year time. At presentation, 45 days and 60-90 days after initial presentation thoracic radiographs, CBC, heartworm antigen and antibody were collected. Practitioners treated clients as per their normal approach with most all receiving corticosteroid medications resulting in abatement of signs. All data analysis and radiographs interpretations were blinded. The original intent to identify the incidence of feline *D. immitis* infection was supported by the finding that approximately 40% were antibody positive at one of the 3 observation point. However, a pattern emerged of cats which were presented for coughing but the antibody titer often became negative even as lung lesions persisted for 90 days. Likewise, of 41% of cats with normal lungs on initial presentation, 43% demonstrated bronchointerstitial changes at later examinations even though asymptomatic. Of cats with lung bronchial scores > 1 (0-3 scale) on presentation, at 45 or 60-90 days later the disease was unchanged (54%), improved (16%), or became more severe (30%). Clinical outcomes were unrelated to radiographs, CBC, or heartworm antibody/antigen test results. Thus, these results lead to the consideration of a transient immature heartworm infection with lung disease (later defined as HARD) which does not develop into an adult heartworm.

This consideration was further bolstered by histopathologic necropsy studies from shelter cats where a high incidence of bronchial disease and pulmonary arterial changes were noted without adult heartworms at necropsy and a negative antibody titer. The high incidence of pulmonary arterial disease was assumed to be associated with prior heartworm infections.

To investigate this further, 120 clinically normal shelter cats were studied using thoracic radiographs, HW serology, fecal examination, and necropsy and lung histopathology at fixed infusion pressure. Included in the radiographic data was the finding that 40 % of the cats had bronchial grades of > 2 (scale 0 3) and 95% had grades of > 1. Prominent caudal pulmonary arteries (>2, scale 0-3) were noted in 50%. Adult heartworms were recovered in 3 cats, 8 were antigen positive, and 29 antibody positive. Greater than 20 % of the pulmonary arteries were abnormal (histologic thickened wall/lumen ratios) in 47% of cats. The correlation of radiographic bronchial scores of > 2 was most consistently associated with alveolar and interstitial proliferation with or without airway lesions (bronchioles or bronchi). The high incidence of abnormal radiographic lung patterns and interstitial lesions with mixed bronchial and pulmonary arterial changes was not consistent with *D. immitis* being the only etiology.

Thus many clinically normal cats had radiographically abnormal "bronchial patterns" which were not directly related to histologic evident of airway disease but instead more closely correlated with interstitial disease. The interstitial densities added to the radiographic peribronchial contrast around unaffected air filled bronchioles. The result was "railroad tracks" and "donuts". This concept was later reinforced by CT studies in experimental models.

Confusion of these and other studies and the lack of correlation of spontaneous disease with radiographic patterns, histologic lesions, and other diagnostic modalities opened the consideration of different causes of feline lung disease. The use of only radiographs, BALF, and serology to accurately diagnosis and treat active lung disease is improbable.

EXPERIMENTAL MODELS

For the purpose of this presentation, the extensive evidence and modeling of induced feline bronchial reactivity via hypersensitization will not be discussed. However, the recent studies of parasitic induced lung disease provide direct clinical relevance. Evaluated in total, a different picture is emerging relative to feline lung disease. Consideration of these studies indicate that an eosinophilic BALF has multiple clinically relevant causes associated with common parasites.

Eosinophilic BALF cytology of > 16% of nucleated cells was consistently observed:

1. Adult *D. immitis*
2. Immature *D. immitis* w/ no adult development (HARD)
3. D. *immitis* when preventative (selamectin) initiated 1 mo post L3 infection

4. *D. immitis* even when preventative (selamectin) initiated 1 mo prior to L3 infection (transient)
5. *T cati* infection per os which developed intestinal worms (both kittens & adults)
6. *T cati* infection per os which never developed intestinal worms (both kittens & adults)
7. *T cati* infection per os even when moxidectin initiated 2 month prior to infection
8. *Aeolurostrongylus* infection

It should be noted that all in the above list (except 3 & 4) are associated with a typical bronchial-interstitial lung pattern on thoracic radiographs, with the notable exception of *D. immitis* on preventative medications. Cats placed on preventative medications one month after infection may become antibody positive, but cats placed on prevention at least one month before infection will remain feline HW antibody negative.

***D. immitis* infection L3**

Various studies have long described significant lung parenchymal and bronchial changes associated with feline heartworm disease. The concept that it is only a pulmonary arterial disease is misguided. The evidence suggested in the field study of 212 cats suggested a transient heartworm infection which differs from the traditional thinking of feline heartworm disease being like canine heartworm disease where the adult is the primary clinical concern. The subsequent series of studies demonstrated that immature adults, which die and never develop, induce pulmonary parenchymal and interstitial disease (HARD) and a transient positive antibody titer. The high mortality of immature adults (2 ½- 4 months post infection) also accounts for lung disease in cats even if a couple of the immature adults survive to fully mature adults and live 3-4 years. Further, interstitial and pulmonary lesions were induced with only ultra-filtered heartworm homogenate over 14 days in the absence of significant bronchial lesions. The character of the parenchymal reaction was myofibrocyte proliferation in the alveolar struts and interstitial parenchyma. Consideration for the role of pulmonary intravascular macrophages (PIMs) to potentially capture and initiate this myofibrocyte proliferation has not been investigated. The interaction of the parasite and host is complex as it has been demonstrated that live adult *D. immitis* infections result in the down regulation of PIM. Cats with adult *D. immitis* and bronchial disease had blunting of the *in vitro* bronchial reactivity response of cats; and moreover blunting was also noted in cats administered only the filtered homogenate, in the absence of bronchial disease.

The bronchial and interstitial disease of HARDs and adult heartworm infection is histologically often unrelated to the location of pulmonary arteries. Homogenate induced interstitial disease enforces the concept that the interstitial and bronchial disease is not an extension of the pulmonary arterial lesion, may be related to soluble products of heartworms, and may or may not be related to PIM activation. The result of *D. immitis* infection is not only a transient eosinophilic dominated airway reaction, limited ability to bronchodilate or constrict. The interstitial, peribronchial, and pulmonary arterial myocyte proliferation was noted even in lobes not associated with the physical presence of an adult worm. The significant pulmonary myocyte proliferation is not associated with cor pulmonale in cats.

Feline heartworm disease ultimately results in lung parenchymal myofibrocyte proliferation, unresponsive diseased airways, but also a restrictive lung disease. CT of the thorax of cats and 3D reconstructions have demonstrated not only a decrease in total air capacity but also a shift in the densities towards more dense tissues. The analysis of the density of Hounsfield Units can only be calculated with a fixed controlled inflation positive pressure, as the use of estimated tidal volumes based on body weight create nonphysiologic tissue pressures in restrictive lung disease.

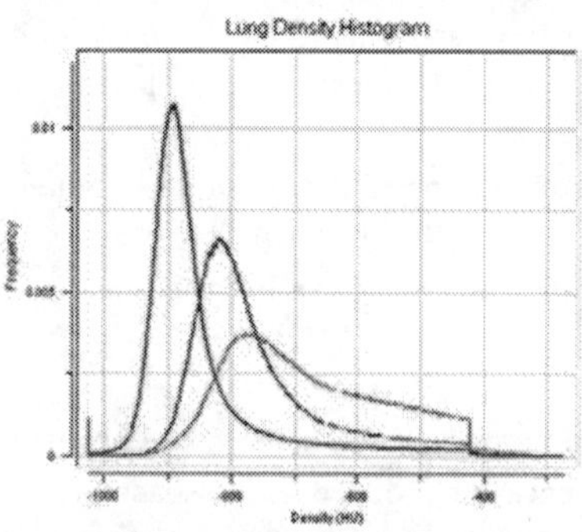

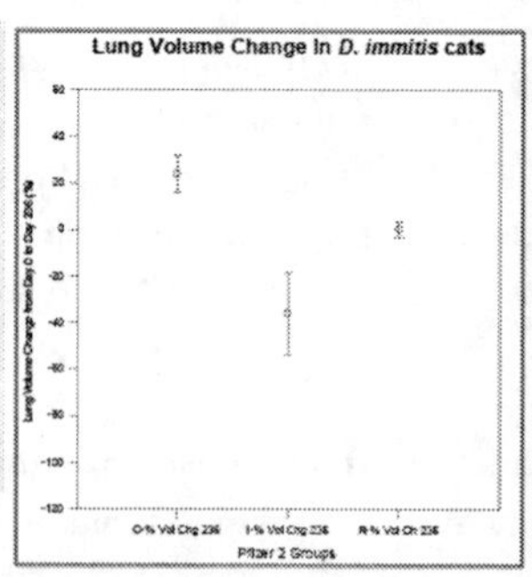

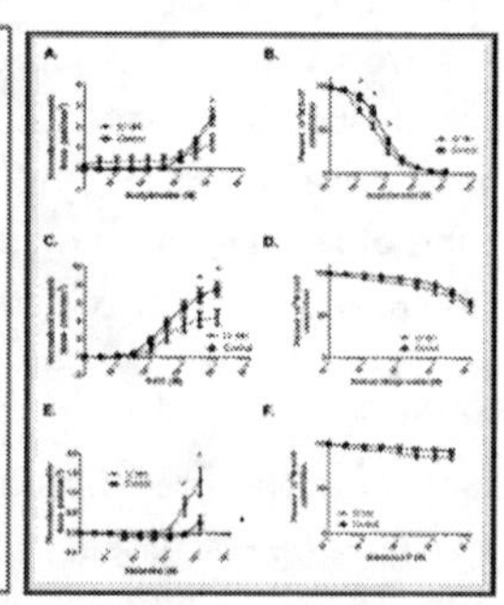

The graph (left above) is total lung CT 3D reconstructed histogram of a *D. immitis* infected cat at baseline (red), 4 month pi (blue), and 8 months pi (green). Demonstrates the shift in percent of lung from the air density (-900 HU) towards more of the interstitial densities (-600 to -700 range). The middle graph demonstrated the mean of the cats in each group where baseline lung volume is compared to 8 months post infection. Control cats (C= first bar) had over 20% increase in total lung over time, *D. immitis* infected cats (I=middle) had about 30% decrease in lung volume, and selamectin pretreated cats (R=right) were unchanged. The *in*

vitro response (right figure) of isolated bronchial rings demonstrated blunted relaxation and constriction even in cats only administered homogenate.

T. cati infection per os

Lung disease and pulmonary artery hypertrophy in cats has been induced by *T. cati* but without careful clinical examination. Recent studies are reason for serious concern to practicing veterinarians. Approximately 25% of shelter cats from all regions of U.S. are shedding *T. cati* eggs in the National Parasite Survey. *T. cati* eggs are viable for years and are found in high concentration in the soil in most of the world. The orally ingested eggs from soil or grooming develop quickly and the L3 penetrate the GI, migrate through the liver, then the lungs, coughed up and swallowed to become adult worms. This migration does not occur when infections is via paratenic hosts.

In a series of *T. cati* infected kitten and adult cats, lung lesions were demonstrated at the first CT time point of 10 days after a single oral dose. The lung injury became more severe over time and repeat infections and was independent of whether GI adults ever developed or not. Thus a fecal was not a valid test for ruling in or out *T cati* induced lung injury. Cats with *T cati* lung injury can be negative on repeated fecal examinations. Adult cats and kittens developed the same severity of lung injury and of more concern was the finding that even pretreatment with moxidectin/imidacloprid for 2 months before infections did not prevent the same degree of lung injury. Thus cats (adults and kittens) infected with oral *T. cati* eggs will develop lung injury, even if on preventative medications. Preventative medications are labeled to prevent the adult stages of the roundworm infection, not the migratory stages. The rapid transient of the L3 from the stomach through the liver and lung would appear to be completed by 7 days in cats. The ability of medications to abbreviate this migration prior to reaching the lungs is being investigated but offers no current practical solution. Thus, outdoor cats can be expected to develop some degree of pulmonary parenchymal disease over the course of their routine exposure.

The nature of the lung injury from *T. cati* is difficult to distinguish from *D. immitis* infection. On radiographs, both develop the typical "bronchial" pattern which, based on CT and histopathologic correlation is the involvement of the interstitial tissue. On radiographs, the enlargement of the caudal pulmonary artery is just as prominent in *T. cati* infections as it is in *D. immitis* infections. Histologically, the myocyte proliferation was marked in the alveolar interstitial and the pulmonary arterial, but the airway proliferation is less severe than in *D. immitis* infected cats. It should be noted that *Aelurostrongylus* infection in cats creates similar radiograph patterns but pulmonary arterial enlargement is not prominent and lymph nodes may be enlarged. The *in vitro* bronchial ring reactivity was normal in all infected cats even with bronchial disease, as compared to the blunted response in *D. immitis* cats with adults or even just the homogenate.

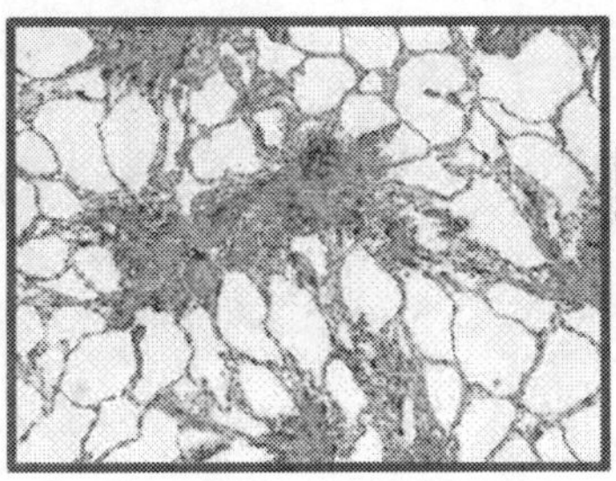

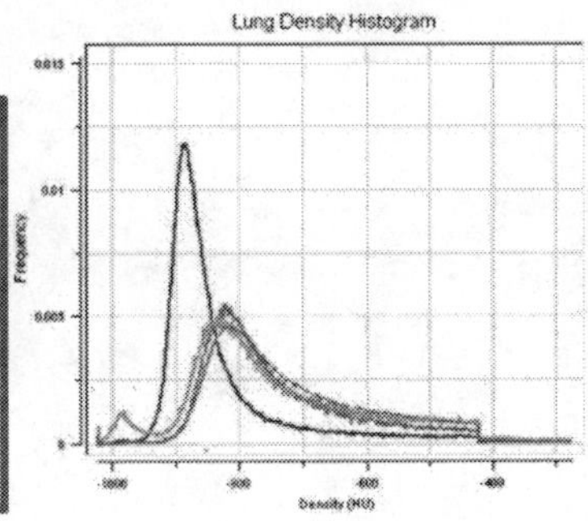

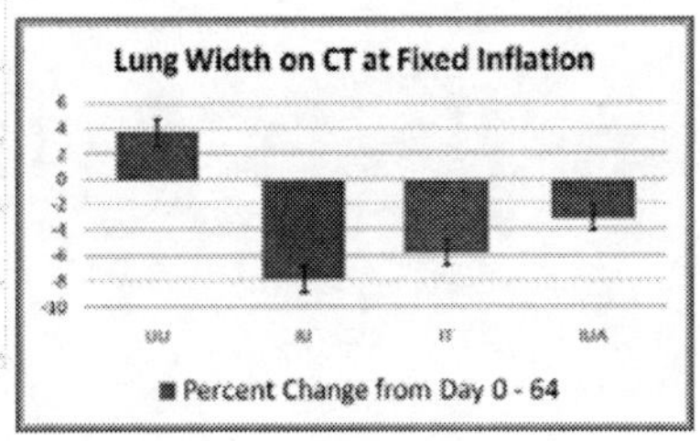

The nature of parenchymal myofibrocyte (SMA staining fibrocytes, left figure) proliferation can be severe and be "honey combed" on CT. Sequential (over time) 3D CT lung density histogram (middle figure) demonstrates a shift away from air (-845 to -900 at baseline, red) towards tissue densities (gold, blue, green). The result is a restrictive lung disease and decrease in the total lung volume (right figure) where endpoint volume is compared to baseline. All infected groups (IT=treated kittens, IU=untreated kittens, IUA=adults) compared to normal controls (UU=uninfected kittens). All infected groups demonstrated abnormal CT, bronchial pattern on radiographs, and eosinophilic BALF. The preventative provided no significant protection against the lung injury.

CLINICAL IMPLICATIONS

Based on these data, transient unsuccessful infections from *D. immitis* can result in HARDs and oral *T. cati* infection in adult or kittens, even on appropriately prescribed and administered preventatives, will result in radiographic evident bronchial patterns and eosinophilic BALF. *Aeolurostrongylus* infection can also mimic this presentation. Common to these infections is the myofibrocyte interstitial proliferation resulting in restrictive lung disease. More confusing is the finding that both *D. immitis* and *T cati* can induce radiographically evident enlarged pulmonary arteries. Although after evaluating many CTs of these experimentally infected cats and clinical experience, there is a subtle different pattern of increased interstitial densities between the *D. immitis, T. cati, and A abstrusus,* but this is not a consistent nor uniformly distributed through all lung and the patterns vary with the stage of the infection. The heartworm antibody and fecal examination offer no absolutes on ruling in or ruling out prior disease. Although the lung disease from *D. immitis* can be prevented by initiating heartworm preventatives before possible exposure to infective mosquitoes, there is currently no practical preventive medication which is believed to abbreviate the lung migration of *T cati* larvae. Adult cats and kittens on or off preventatives would be expected to give additive insults to the lung with repeated exposure. Although there are limited long term studies, it would appear that the myofibrocyte proliferation of interstitium and pulmonary arteries is long term and may not be reversible.

Given the wide distribution of *D. immitis* dogs and mosquito vectors and ubiquitous presence of *T. cati* eggs in the environment, the clinician can assume that many client cats have been infected, subclinically, with these parasites and have subsequent radiographic "bronchial" patterns which become the background on which compounding disease or infection induces clinical signs. When presented for clinical disease, a clinician must consider that the thoracic radiographs may represent disease which has been present for months or years and may or may not be related to present clinical signs.

REFERENCES

McCall JW et al, AmHWSocProc 1992:71. Dillon AR, VetClinNA 1984;(14):1185. Holmes RA et al, AmHWSym 1992;81. Brawner, B et al Vet Therap: Res Applied Vet 2000;(1,2):81. Robertson, C et al Vet Therap: Res Applied Vet Med 2000;(1,2):88. Dillon, AR et al Vet Therap: Res Applied Vet Med 2000;(1,3):176. Browne LE AmJVetRes; 2005;(66);1544. Dillon AR et al, Vet Par 2008;(158):171. Blagburn BL et al, Vet Med; 2007; (102);7. Wooldridge A, AmJVetRes; 2012;(73,3):439. Dillon AR et al. VetPar 2014;(206,1);14. Venco L et al, VetPar 2008;(158,3);232. Dillon AR et al, VetPar 2013(193:4);413. Dillon AR et al, Vet Par 2014;(206,1):24. Dennler M et al VetRadUlta 2013;(54,5):459.

Changing Paradigms in Interventional Radiology: Observations on Tracheobronchial Collapse and Stenting

Chick Weisse, VMD, DACVS
New York, NY

Intraluminal stenting is a palliative, minimally invasive therapy used for restoration of an obstructed or narrowed lumen. Interventional radiologists have employed these techniques in virtually every accessible lumen of the human body and, if recent history is any indication, it appears it is now the veterinarians' chance to do the same in our animals. Following even a preliminary review of the human interventional radiology (IR) literature, it does not take a tremendous amount of imagination to conclude our patients can likely benefit from similar procedures as there is such abundant overlap of the disease processes. The respiratory system has the potential to be particularly well suited to these types of interventions as these patients are often severely affected and fragile, comorbidities are not uncommon, and the risk of associated surgical complications from traditional therapies can be relatively high.

Veterinary IR has evolved to encompass many different body systems, each with their own successes and failures. However, even with the advancements in intraluminal tracheal stenting devices and technical modifications, a number of these patients continue to be plagued by complications including intractable coughing, infection, tissue ingrowth, stent fracture, and others. Although these procedures are still not providing the ideal outcomes we might hope for at this time, veterinarians performing this procedure with even some regularity will generally agree that dramatic improvements can be achieved in certain cases that have failed aggressive medical management and surely would have not tolerated a surgical intervention. Clearly there is a role for tracheal stenting in our veterinary patients, and this may expand to bronchial stenting for similar collapse/compression of these structures in certain circumstances.

STENTS, TECHNIQUE, AND OUTCOMES

Over the last decade, tracheal stents have become increasingly available to veterinarians, and their design and delivery systems have evolved accordingly. We have come a long way from the originally described metallic balloon-expandable stents to shape-memory self-expanding metal (and even tapered) stents. Unfortunately, currently manufactured metallic stents remain a long way away from mimicking the complexity of the living trachea. Our human medicine colleagues have described complications including "obstructive granulation tissue, stenosis at the ends of the stent, migration of the stent, mucus plugging, infection, and stent fracture"[1]; very similar to some of the major complications identified in our veterinary patients. However, we must interpret this information with an understanding of the different indications and circumstances in humans compared with our patients. Much of the tracheomalacia (TM) treated in humans is temporary so removable stents (silicone) are preferred. In addition, even for permanent indications, stent exchanges and "toilet bronchoscopy" procedures to suction, remove, and reculture infectious mucus accumulations are not uncommon in people with silicone stents; perhaps a limiting factor for routine use silicone stents in our patients due to cost and owner compliance.

Fortunately, it seems self-expanding metallic tracheal stents have better outcomes in dogs compared to those routinely reported in humans for benign diseases. Our problem has been determining which of the canine patients are mostly likely to respond favorably to this treatment, as the current outcomes are widely variable and difficult to predict. Perhaps the most important advancements in my own stenting experiences have been in recognizing the variability of the "tracheal collapse syndrome" (weakened dorsal membranes versus tracheal malformations[2], for instance) and in refining our stent placement techniques. The technical advancements include improved medical management and recognition of comorbidities, appropriate fluoroscopic technique and tracheal measurements, proper stent diameter oversizing and preventing overlong stents, pre- and post-stent tracheoscopy, and vigilant client communication with regular scheduled follow-up examinations.

Our approach to traditional flattened tracheal lumens (flaccid dorsal membranes) is different from the "W"-shaped cartilage rings seen in tracheal malformation cases. More involved upper airway examinations identifying comorbidities such as overlong soft palates or epiglottic retroversion have led to concurrent airway surgeries performed more routinely in our patients. Performing careful patient positioning and both positive- and negative-pressure ventilation images with esophageal marker catheter placement confirms precise collapse location and minimizes error in radiographic magnification, one of the most common mistakes made during the stenting procedure. Pre- and post-stent tracheoscopy confirms the variety of collapse as well as appropriate stent-to-tracheal wall contact and positioning during the procedure so inappropriately positioned stents can be removed or repositioned while it is still possible.

Although tracheal stenting is one of the first IR techniques veterinarians attempt, it is a demanding technical procedure with little room for error, and advanced training is highly recommended to avoid some of the more common mistakes.

COMPLICATIONS

Fortunately there are some complications we can control; Those include mistakes in case selection, stent choice and technical decisions, inappropriate monitoring, late detection of complications, and poorly executed medical management. Unfortunately there remain some complications we currently cannot easily control; These include granulation or inflammatory tissue ingrowth, infection, stent fatigue and fracture, and disease progression. The combination of these problems contributes to an approximately

40% need for restenting procedures (personal unpublished data). As stent technology and designs improve in the future, some of these problems may become less of a concern, however in the meantime, appropriate client counseling, early detection, and aggressive medical management or intervention may reduce the consequences of these complications.

Tissue ingrowth into tracheal stents has been reported in both human and veterinary patients. In animals, any growth within the stent was initially considered "granulation tissue" due to human reports and appearance. More recently, we and others have appreciated tissue regression following high dose corticosteroid administration. We know granulation tissue does not regress with corticosteroids leading us to conclude this is often inflammatory tissue/granulomas rather than granulation tissue. This tissue typically develops at the cranial aspect of the stent but can develop elsewhere. Over the years we have recognized this tissue can respond to high dose steroids when administered early, however when treatment is delayed responses are more uncommon. Rather than granulation tissue "versus" inflammatory tissue developing independently, it appears that the longer inflammatory tissue is present without treatment the more this tissue will become organized and infiltrated with fibrous scar tissue; This what a number of our biopsies have suggested (currently unpublished data). This highlights the importance of regular monitoring and early detection of tissue ingrowth (~20% tissue ingrowth rate currnetly). Further we have recognized that tissue growth adjacent to or within the stent can often be unrecognized due to forelimb positioning in a lateral radiographic projection. For this reason we have instituted mandatory 3-view chest and neck radiographs for all tracheal stent cases with the forelimbs pulled cranially in one lateral and caudally in another. This position permits complete evaluation of the cervical and intrathoracic trachea without interference of the forelimb soft tissue and boney structures.

Infection has not been considered a major component of the tracheal collapse syndrome previously but pre-stent cultures are positive in ~80% of cases and once a tracheal stent has been placed, it is clear that the trachea remains susceptible to infections. This is exacerbated when there are persistent gaps between the stent and tracheal wall, permitting mucus accumulation, most commonly seen in the tracheal malformation cases we recently reported.[2] Infections alone can be problematic, however the greater concern may perhaps be the increased risk of tissue ingrowth that occurs when tracheal stents become infected. This is well documented in humans with tracheal stents. It had been unclear whether the tissue ingrowth led to infection or the infection led to tissue ingrowth. More recent data suggests that both infection and stent micromotion independently contribute to the development of tissue ingrowth into the stent.[4] Over the years we have been more aggressive about early intervention of cases with persistent clinical signs; early tracheoscopy and culture permit rapid, appropriate antibiotic therapy. Our impression has been diffuse stent infiltration with tissue is most often associated with infection and appropriate antibiotic therapy can often result in regression. Hesitation to interrogate the problem can lead to inappropriate antibiotic selection, tissue progression and organization/fibrosis, and ultimately require additional stent placement during respiratory crises.

Stent fracture will continue to remain a concern when metallic stents are used (~20% major fracture rate). There is no perfect stent; The unforgiving environment and extreme forces encountered are more than we can currently expect a stent to sustain for the many years we are anticipating it to be in place. Super-elastic titanium alloys, special coatings, improved designs, and other technological advances will continue to improve the longevity of these stents. Currently, the most important aspect of tracheal stent fracture in our practice is early detection and restenting prior to patient decompensation. Good outcomes can still be achieved when stent fractures are identified and restented quickly.

Although previous literature cited stent migration and substantial stent shortening as common complications, our experience has not supported these findings (this information is being presented at this conference). When appropriately sized diameter stents are used, migration is almost negligible (as demonstrated by changing distance from stent to cricoid and/or carina) and our stent shortening rates are approximately 11% over the long-term compared to the approximately 30% previously reported (**Figure 1**).[3] The stent type and diameter used and/or the placement technique likely explain these differences.

FIGURE 1: Percent change in tracheal dimensions ("Shortening") and change in distance from cricoid and carina ("Migration") for each category and total ultimate percent change [Raske M et al., ACVS 2014]

Group	Follow up time	% length achieved by stent (# of dogs)	% change in stent length over time (# of dogs)	Change (mm) in distance from cricoid (# of dogs)	Change (mm) in distance from carina (# of dogs)
1	**Immediate post placement**	**126.74% (n=53)**	**N/A**	**N/A**	**N/A**
2	≥ 1 day & < 90 days	104.20% (n=43)	-6.38% (n=43)	0.02mm (n=23)	-0.40mm (n=41)
3	>90 days & < 180 days	108.30% (n=8)	-12.61% (n=8)	-2.15mm (n=5)	-3.58mm (n=8)
4	>180 days & < 365 days	118.82% (n=8)	-10.96% (n=8)	0.34mm (n=2)	-0.44mm (n=8)
5	>365 days & < 730 days	108.30% (n=11)	-12.58% (n=11)	-0.96mm (n=4)	-0.17mm (n=10)
6	>730 days	107.82% (n=12)	-12.39% (n=12)	-3.76mm (n=6)	-0.50mm (n=11)
TOTAL (Mean)	**Mean 440 days**	**113.13% (n=49)**	**-10.77% (n=49)**	**0.8mm (n=28)**	**1.6mm (n=47)**

CONSIDERATIONS ON BRONCHOMALACIA

Bronchial stenting may be the natural progression of veterinary respiratory endoluminal therapy. The presence of left mainstem bronchomalacia and bronchial collapse has been well reported in the veterinary literature, commonly associated with left atrial enlargement. It has also been reported as an isolated, incidental finding[5, 6] and in dogs with tracheal collapse[7] with or without concurrent heart disease. Interestingly, brachycephalic breeds have also been shown to have unilateral bronchial collapse (84% left-sided) as well – even without the presence of left atrial enlargement[8]; This has been explained due to chronic abnormal airway pressures leading to progressive weakening and subsequent collapse of the left mainstem bronchus. The unilateral nature remains unclear but theories have included different airway pressures encountered along this longer left bronchial segment when compared with the relatively shorter right mainstem bronchus. This left mainstem bronchomalacia has even been cited as a possible cause for left cranial lung lobe torsions in Pug dogs due to the weakened cartilage integrity. The wide variety of dogs experiencing mostly left bronchial collapse, with or without heart disease, with or without upper airway collapse, and with or without clinical signs suggest a variety of possible explanations, likely important to understand more clearly if we hope to predict outcomes following invasive, or even noninvasive, interventions.

Comparatively, bronchomalacia in humans is typically bilateral and when unilateral, the left and right sides seem generally equally affected for the most part. Anatomical differences between humans, animals, and even different shaped animals (narrow, deep-chested versus barrel-chested breeds) inspired a more detailed investigation into the thoracic anatomy of different breeds. Some examples of this investigation are included in **Figure 2** demonstrating the possible "compressive, sandwiching" nature of the left mainstem bronchus between the left atrium and aorta, particularly apparent in barrel-chested dogs (not uncommon conformations of brachycephalic breeds and some of the at-risk collapsing trachea breeds including barrel-chested Yorkshire terriers). A static compression of the bronchus could lead to cartilaginous weakness and further collapse over time, progressing from "static" to "dynamic" compression. This phenomenon could explain why some bronchoscopically-identified bronchial compression is "static" or "fixed" (as might be identified with compression between two structures) versus "dynamic" (as might be expected with malacia of the cartilage during dynamic airway pressures). In fact the bronchial collapse described in the brachycephalic breeds was described as "fixed" in 87.5% of the dogs.[8] The location of the pulmonary arteries may also contribute to airway compression and could even help explain the narrow pedicle of the left cranial lung lobe in Pug dogs and possible predilections to bronchomalacia, regional atelectasis, torsion, or combination. The more typical compression of the left mainstem bronchus may even be another possible explanation for the more commonly located right-sided (46%) bronchial versus left-sided (29%) bronchial airway foreign bodies in dogs.[9] Figure 1 demonstrates just some of the wide variability of cardiovascular and airway structures positioned in different breeds, as well as individual dogs within a breed. If "bronchial compression" is confirmed in certain breeds and/or thoracic conformations, this might help determine which patients may benefit from bronchial stenting; It may also help determine which patients might not.

This information is preliminarily under investigation at our institution but further characterization of anatomical differences between breeds, anatomical conformations of vascular and respiratory structures, and even gestational growth patterns leading to cartilage weakness and future collapse could provide useful information for our patients in the future. Much work needs to be done to determine which patients may benefit from bronchial stenting. Personally I feel there will be a subset of patients that will benefit and others that will not; How we will categorize these patients remains to be determined and will likely take the work of many researchers around the world for years to come. Interestingly, our recent research has not found a negative impact of pre-stent bronchial collapse in our patients survival.

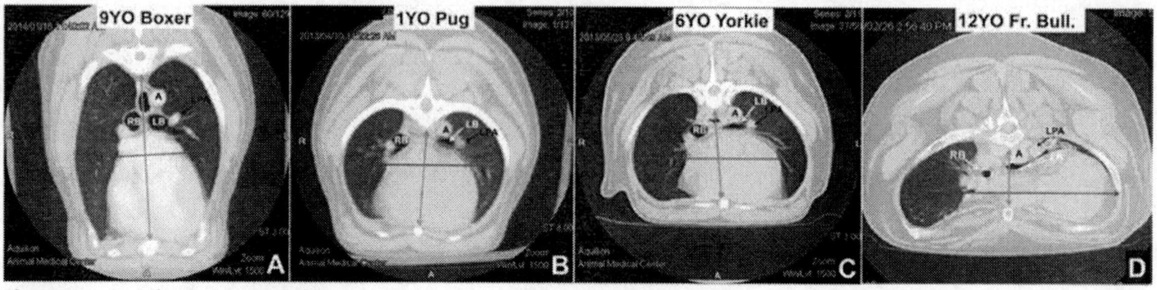

Figure 2: Serial, static, axial thoracic CT images in 4 dogs without cardiac or pulmonary disease. Note the variable thoracic width (red line) and depth (blue line) conformations and corresponding cardiovascular structure locations (A-Aorta, RB-right bronchus, LB-left bronchus, LPA, left pulmonary artery). A. Deep-chested Boxer with considerable space between thoracic vertebral body, aorta, and left bronchus. B. Intermediate barrel-chest conformation Pug with left bronchus compressed between aorta and heart. Note the lack of available space between the thoracic vertebral body, extremely dorsally located aorta, and the left bronchus. The right bronchus is free of compression from any cardiovascular structures. C. Intermediate barrel-chest conformation Yorkshire terrier with left bronchial compression between aorta and heart. The right bronchus is free of compression from surrounding structures. D. Severely dorsoventral flattened thoracic conformation in a French Bulldog with a retrobulbar sarcoma and no respiratory signs having thoracic CT for metastasis screening. Note considerable left bronchus compression between aorta and the left-sided deviated heart with additional compression from the left pulmonary artery.

REFERENCES

1. FDA medical safety warning: "Metallic Tracheal Stents in Patients with Benign Airway Disorders"(2005);http://www.fda.gov/MedicalDevices/Safety/AlertsandNotices/PublicHealthNotifications/ucm062115.htm
2. Branter E., Weisse C., Berent, A. Tracheal malformations in Yorkshire Terriers (2012) ACVIM Forum, New Orleans, Louisiana
3. Moritz, A., Schneider, M. & Bauer N. (2004) Management of advanced tracheal collapse in dogs using intraluminal self-expanding biliary wallstents. *Journal of Veterinary Internal Medicine* 18, 31-42
4. Ost, D.E., Shah, A.M., Lei, X., et al. (2012) Respiratory infections increase the risk of granulation tissue formation following airway stenting in patients with malignant airway obstruction. *Chest* 141(6), 1473-81
5. Clercx, C., Peeters, D., Snaps, F., et al. (2000) Eosinophilic bronchopneumopathy in dogs. *Journal of Veterinary Internal Medicine* 14, 282-291
6. Norris, C.R., Griffey, S.M., Samii, V.F., et al. (2001) Comparison of results of thoracic radiography, cytologic evaluation of bronchoalveolar lavage fluid, and histologic evaluation of lung specimens in dogs with respiratory tract disease: 16 cases (1996-2000). *Journal of the American Medical Veterinary Association* 218, 1456-1461
7. Johnson, L.R., and Pollard, R.E. (2010) Tracheal collapse and bronchomalacia in dogs: 58 cases (7/2001-1/2008). *Journal of Veterinary Internal Medicine* 24, 298-305
8. De Lorenzi, D., Bertoncello, D., and Drigo, M. (2009) Bronchial abnormalities found in a consecutive series of 40 brachycephalic dogs. *Journal of the American Medical Veterinary Association* 235, 835-840.
9. Tenwolde A.C., Johnson, L.R., Hunt, G.B., et al. (2010) The Role of Bronchoscopy in Foreign Body Removal in Dogs and Cats: 37 Cases (2000–2008) *Journal of Veterinary Internal Medicine* 24, 1063-1068

Chronic Hypertonicity and its Possible Consequences for Diabetic Dogs and Cats

Thomas Schermerhorn, VMD, DACVIM(SAIM)
Manhattan, KS

INTRODUCTION

Serum tonicity is related to but distinct from the serum total osmolality. Serum total osmolality represents the total number particles with osmotic activity in serum but does not distinguish ineffective osmoles, substances that freely permeate biologic membranes (and thus equilibrate with serum), from effective osmoles, substances that do not freely permeate membranes (1). Serum tonicity (also known as serum effective osmolality) represents the portion of serum osmoles that are not freely permeable and thus can influence the movement of water across cell membranes (2). Serum tonicity is a better clinical measure than total osmolality for estimating the risk for osmotic complications in humans (3).

TONICITY MEASUREMENT

Total serum osmolality can be measured using a commercial osmometer (eg, freezing point depression or vapor pressure) or calculated using a standard formula. Serum tonicity cannot be directly measured by commonly available techniques but can be calculated using the serum concentrations of sodium, potassium, and glucose. Calculation of total osmolality (OsM_C) (Eq. 1) and effective osmolality (tonicity; OsM_E) (Eq. 2) using standard clinical formulas. Note that equations use concentrations in mEq/L or mg/dl (4).

$$\text{Eq 1. } OsM_C = + 2(Na + K) + BUN/2.8 + glucose/18$$
$$\text{Eq 2. } OsM_E = + 2(Na + K) + glucose/18$$

Measurement of tonicity is typically carried out in vitro and usually involves detection of volume changes in intact cells or cell preparations. RBCs have been frequently employed in experimental methods to assay tonicity, despite observations that the RBC does not behave as a perfect osmometer system (5). The tonic effects of major serum components and select exogenous components (such as mannitol) are established (1) but possible roles for other effective osmoles remain to be completely elucidated. Canine RBCs incubated for 24-hr in a hypertonic solution containing 11.1 mM (200 mg/dl) glucose show an increase in volume when cells were placed in a solution isotonic to normal plasma (6). This volume response may be modulated by ketones but the mechanism is unclear (6). Recently, we presented evidence that ketones are not likely to act as effective osmoles in vivo but the role of lactate, another organic acid, was less clear (7).

The mean cell volume difference (dMCV) is a red blood cell parameter that correlates with serum osmolarity and has recently been shown to be a reasonably reliable clinical indicator of serum hypertonicity in dogs (8,9). In a study evaluating dogs with various non-diabetic illnesses, a dMCV >3 fl was a reasonable predictor (76% sensitivity, 71% specificity) of hypertonicity (serum tonicity ≥320 mOsm/L) (8) and dMCV increased in normal dogs after mild hypertonicity was induced by water deprivation (9).

FUNCTIONAL CONSEQUENCES OF HYPERTONICITY

Serum hypertonicity describes an increase in the concentration of serum solutes (effective osmoles) that promote water loss of cells via osmosis. Tonicity of serum is viewed relative to the prevailing intracellular tonicity. Acute hypertonicity results when the serum effective osmole concentration exceeds that of the intracellular compartment. The net effect is a reduction in intracellular water, moving into the extracellular compartment, until osmotic equilibrium is restored. Chronic hypertonicity is associated with osmotic equilibrium but is characterized by concentrations of extra- and intracellular effective osmoles that are elevated above normal. Hypertonicity has been linked to cellular dysfunction, including impaired protein metabolism (10), altered ion channel activity (11) and DNA damage (12). Hypertonicity can result as a direct consequence of a defined pathological process, such as hypernatremia (13) or hyperglycemia (1), or can develop insidiously. In the latter circumstance, chronic hypertonicity develops in some individuals because of subtle alterations associated with compensatory physiologic or pathophysiologic mechanisms, such as endocrine alterations (disturbed vasopressin or glucocorticoid secretion), presence of exogenous solutes that contribute to tonicity, altered volume status, and non-osmotic factors (such as stress) (14).

Population-based human studies have demonstrated that hypertonicity is prevalent among geriatrics and individuals at risk for obesity and diabetes (15). Hypertonicity increases extracellular fluid (ECF) at the expense of intracellular fluid (ICF) and has been proposed to mediate the increased ECF/ICF ratio that characterizes obesity (16). An epidemiologic study that analyzed diabetes risk in geriatric (≥70 yrs old), non-diabetic people demonstrated that hypertonicity (plasma tonicity ≥ 300 mOsm/L) was an independent risk factor for diabetes development (Odds Ratio = 2.0) (17). Furthermore, individuals with hypertonicity and mild hyperglycemia (glucose ≥125 mg/dl) were at greater risk for and developed diabetes at a faster rate than individuals with hyperglycemia alone (17).

PREVALENCE OF HYPERTONICITY

In a large cross-sectional U.S. study, the prevalence of hypertonicity in human adults was estimated at 20% with an additional 40% of the population affected with borderline hypertonicity. (15). Advanced age, obesity, race, and diabetic status may exacerbate the risks of chronic hypertonicity (15, 16). Plasma hypertonicity was also reported as an independent predictor for the onset of disability within a 4-year period and mortality within an 8-year period and may be an early indicator of frailty, a condition common among elderly humans that is characterized by decreased functional reserve and increased systemic morbidity (18, 19).

Hypertonicity is prevalent (as high as 78%) among insulin resistant and diabetic humans (15). In pre-diabetic humans, hypertonicity *per se* – independent of glucose - has been shown to be a risk factor for progression to diabetes mellitus (17). Hypertonicity is equally common in diabetic dogs and cats. In diabetic dogs and cats, the prevalence of effective hyperosmolality was reported as 81% in one study (20). Unpublished data from another population of diabetic dogs and cats shows similarly high prevalence rates of moderate to severe hypertonicity in treated diabetics. In diabetic dogs and cats, hypertonicity is primarily because of hyperglycemia with up to 5% of total osmoles contributed by glucose. Interestingly, in diabetic animals, sodium (Na) concentrations are more closely associated with osmolality (tonicity) than glucose, emphasizing the importance of sodium in mediating tonicity in diabetic patients (2, 20). The correlations between sodium and tonicity and glucose and tonicity tend to be higher in acute diabetes (newly diagnosed or treated for a short time) than in dogs and cats with chronic diabetes (unpublished observations).

CONSEQUENCES OF HYPERTONICITY IN DIABETES

Serum hypertonicity reflects a condition characterized by an increase in serum effective osmoles and is a feature of acute and chronic diabetes in dogs and cats. Acute serum hypertonicity results in an osmotic gradient that favors water movement from the intracellular to the extracellular compartment and may lead to cellular dehydration in severe hypertonicity. Chronic serum hypertonicity results from persistent elevations in serum tonicity and is less likely to cause rapid and marked water shifts. The osmotic consequences of chronic serum hypertonicity are blunted by production of intracellular osmoles (idiogenic osmoles) that oppose the osmotic gradient that favors water movement into the extracellular fluid compartment.

Hypertonicity is both a mediator and a consequence of pathophysiology in diabetes. It is important to note that the hypertonic state exerts detrimental effects that are independent of the elevated osmole(s) (21). In laboratory studies, some of the glucose-induced changes in osteoblast function and gene expression are responses to tonicity and can be elicited by effective osmoles other than glucose (21). Furthermore, the neuronal response to hypertonicity and osmotic stress is similar regardless of the osmole used to induce the response. In chronic hypertonicity, 50% of the increase in intracellular osmoles in brain is because of the production of idiogenic osmoles, which are the same regardless of whether extracellular tonicity is increased by glucose, sodium or another effective osmole. Thus, the presence of serum hypertonicity may be associated with myriad adverse effects, many of which may be currently unknown or underappreciated, that are independent of the glucose concentration.

Hypertonicity impairs immune function, which may contribute to secondary complications of diabetes such as increased susceptibility to infection (22). Serum hypertonicity increases hepatic glucose production and blood glucose concentration in non-diabetic men and may contribute to hyperglycemia in diabetics through similar mechanisms (23). The pathophysiology of acute complications of diabetes, such as cerebral edema, described in humans and veterinary patients is multifactorial but hypertonicity appears to play a role (24). Serum hypertonicity correlates with progressive neurological impairment in humans with diabetic ketoacidosis or non-ketotic hyperosmolar diabetes (3) and is likely to have similar correlation in diabetic dogs and cats (20). Most diabetic dogs and cats have hypertonicity when presented for treatment (20). A study in cats with DKA undergoing treatment found that hypertonicity was present at presentation and persisted for at least 72 hours during treatment (2). In diabetic dogs and cats, serum tonicity correlates best with serum sodium but severe hyperglycemia contributes substantially in animals with marked hyperglycemia (2,20).

In addition to acute effects, the chronic hyperglycemia that accompanies diabetes, which may persist despite insulin therapy, produces a chronic hypertonic state. As in acute diabetes, sodium and glucose remain the major contributors to serum tonicity in dogs and cats with chronic diabetes. It can be hypothesized that diabetic dogs and cats are particularly at risk for chronic hypertonicity caused by hyperglycemia because glycemic control is typically much more "loose" in veterinary patients compared with humans and because target glucose concentrations for veterinary patients are generally higher than the species reference range for serum glucose. For example, the upper level of the target glycemic range in diabetic dogs and cats is generally about 250 mg/dl (`14 mM). Persistent hyperglycemia near or above the target range will add a minimum of 13-15 mOSM to serum. The additional osmoles added by glucose in insulin-treated diabetics may be offset by decreased serum sodium; this effect is particularly pronounced in dogs. Even though serum sodium concentrations tend to be lower during diabetes, which blunts the glucose effect on tonicity (2), mean serum tonicity in treated diabetic dogs (315 mOSM) and diabetic cats (330 mOsm) were found to be elevated above that of healthy dogs and cats (unpublished observations).

CONCLUSION

Diabetes is associated with a high prevalence of hypertonicity in dogs and cats. Hypertonicity is a feature of untreated and complicated forms of diabetes and persists in insulin-treated animals with chronic diabetes. The spectrum of possible complications arising from chronic hypertonicity is not yet defined in diabetic dogs and cats and the role of hypertonicity in the development and propagation of diabetic complications deserves closer study.

REFERENCES

1. Lord RC. Osmosis, osmometry, and osmoregulation. Postgrad Med J. 1999 Feb;75(880):67-73.
2. Kotas S, et al. Changes in serum glucose, sodium and tonicity in cats treated for diabetic ketoacidosis. *J. Vet. Emer. Crit. Care* 18(5):488-495, 2008.
3. Siperstein MD. Diabetic ketoacidosis and hyperosmolar coma. Endo. Metabol. Clin. North Am. 1995; 21:415-432.
4. Wellman ML, et al. Applied physiology of body fluids in dogs and cats. In: DiBartola, S.P. (Ed.), Fluid, Electrolyte, and Acid-Base Disorders in Small Animal Practice. Elsevier-Saunders, St. Louis, MO, 2012, pp. 2–25.
5. Wong P. The behavior of the human erythrocyte as an imperfect osmometer: a hypothesis. J Theor Biol. 2006 Jan 7;238(1):167-71.
6. Comazzi S, et al. Erythrocyte changes in canine diabetes mellitus: in vitro effects of hyperglycaemia and ketoacidosis. Comparative Clinical Pathology 2004 April;12(4): 199-205.
7. Reinhart JM, et al. Determination of tonicity effects of ketoacids and lactate by use of two canine red blood cell assays. Am J Vet Res. 2015 Jan;76(1):77-83.
8. Reinhart JM, et al. In vitro increase of mean corpuscular volume difference (dMCV) as a marker for serum hypertonicity in dogs. Res Vet Sci. 2014 Jun;96(3):442-6.
9. Reinhart JM, et al. Evaluation of mean corpuscular volume difference as a marker for serum hypertonicity during water deprivation in dogs. Am J Vet Res. 2015 Feb;76(2):170-3.
10. Garner MM, Burg MB. Macromolecular crowding and confinement in cells exposed to hypertonicity. Am J Physiol. 1994 Apr;266(4 Pt 1):C877-92.
11. Lang F, et al. Functional significance of cell volume regulatory mechanisms. Physiol Rev. 1998 Jan;78(1):247-306.
12. Alfieri RR, Petronini PG. Hyperosmotic stress response: comparison with other cellular stresses. Pflugers Arch. 2007 May;454(2):173-85. Epub 2007 Jan 6.
13. Marks SL, Taboada J. Hypernatremia and hypertonic syndromes. Vet Clin North Am Small Anim Pract. 1998 May;28(3):533-43.
14. Schliess F, et al. Osmosensing and signaling in the regulation of mammalian cell function. FEBS J. 2007 Nov;274(22):5799-803. Epub 2007 Oct 17.
15. Stookey JD. High prevalence of plasma hypertonicity among community-dwelling older adults: results from NHANES III. J Am Diet Assoc. 2005 Aug;105(8):1231-9.
16. Stookey JD, et al. The altered fluid distribution in obesity may reflect plasma hypertonicity. Eur J Clin Nutr. 2007 Feb;61(2):190-9. Epub 2006 Oct 4.
17. Stookey JD, et al. Hypertonic hyperglycemia progresses to diabetes faster than normotonic hyperglycemia. Eur J Epidemiol. 2004;19(10):935-44.
18. Clegg A, et al. 2013. Frailty in elderly people. Lancet 381, 752–762.
19. Stookey JD, et al. Plasma hypertonicity: another marker of frailty? J Am Geriatr Soc. 2004 Aug;52(8):1313-20.
20. Schermerhorn T, Barr SC. Relationships between glucose, sodium, and effective osmolality in diabetic dogs and cats. *J Vet Emer Crit Care*. 16(1):19-24, 2006.
21. Botolin S, McCabe LR. Chronic hyperglycemia modulates osteoblast gene expression through osmotic and non-osmotic pathways. J Cell Biochem. 2006 Oct 1;99(2):411-24.
22. Lang KS, et al. Inhibition of interferon-gamma expression by osmotic shrinkage of peripheral blood lymphocytes. Am J Physiol Cell Physiol. 2003 Jan;284(1):C200-8.
23. Keller U, et al. Effects of changes in hydration on protein, glucose and lipid metabolism in man: impact on health. Eur J Clin Nutr. 2003 Dec;57 Suppl 2:S69-74.
24. Dunger DB, et al. ESPE/LWPES consensus statement on diabetic ketoacidosis in children and adolescents. Arch. Dis. Child. 2004; 89(2):188-94.

Advances in Insulin Therapy: Rapid-Acting Insulin Analogs in Cats

Jon M. Fletcher, DVM, DACVIM
Baton Rouge, LA

INTRODUCTION

A number of major advancements have been made in the formulation of insulin since the United States Food and Drug Administration first approved synthetic insulin in 1982. Recombinant DNA technology has made it possible to modify the amino acid structure of insulin so as to yield analogs of human insulin. With the development of insulin analogs came a new category of insulin with a more rapid onset of action than regular insulin (i.e., rapid-acting analogs) as well as additional long-acting formulations (e.g., insulin glargine and insulin detemir). The time-action profiles of these analogs better mimic the normal physiologic secretion (i.e. bolus phase and basal phase) of endogenous insulin by the pancreas than do crystalline insulins. By combining mealtime administration of a rapid-acting insulin analog that mimics prandial insulin secretion with a long-acting analog that mimics basal insulin secretion, it is possible to more closely mimic physiological secretion of insulin. In people, there is less variation in absorption of insulin analogs and the action profiles are more predictable, which decreases the risk of hypoglycemic events.[1]

RAPID-ACTING ANALOGS

The rapid-acting insulin analogs include insulin lispro (Humalog®, Lilly), insulin aspart (Novolog®, Novo Nordisk) and insulin glulisine (Apidra®, Sanofi Aventis). Insulin lispro was created by reversing the amino acids at B28-B29 from proline-lysine to lysine-proline. Insulin aspart differs from human insulin in that proline at B28 is replaced by aspartic acid. Insulin glulisine was produced by replacing asparagine at position B3 with lysine and lysine at B29 with glutamic acid.[2] Each of these unique alterations in the amino acid structure of the insulin B chain decreases the tendency of the molecules to form hexamers, which means they remain monomeric and dimeric in solution.[3] The amino acid changes do not interfere with receptor binding, circulating half-life, or immunogenicity. These monomers and dimers diffuse rapidly and more consistently from the subcutaneous space into the circulation.[2,4,5] This is in contrast to regular insulin, which has a much higher tendency to associate and form hexamers. Hexamer dissociation is the rate-limiting step in the absorption of regular insulin and results in a slower onset and longer duration of action. It also increases the potential for variability in absorption, which may lead to greater variability in the blood glucose concentration and risk of hypoglycemia.[1,3]

Pharmacodynamics (PD) and pharmacokinetics (PK) of the rapid-acting insulin analogs have been evaluated in healthy subjects. The onset of action of the rapid-acting analogs in people ranges from 5 to 20 minutes with peak action occurring 30 to 90 minutes after subcutaneous (SC) injection.[6-8] This allows human diabetics to eat within 20 minutes of insulin administration rather than waiting as long as 60 minutes after injection of regular insulin before eating. With a very rapid onset of action and 2 to 6 hour duration of effect, these analogs more closely mimic pancreatic insulin secretion in response to a meal. Patient convenience, improved predictability, and decreased frequency of hypoglycemic episodes has led many physicians to replace meal-time regular insulin with a rapid-acting analog.[1,9] In addition to being used in the long-term management of diabetes, the consistent and predictable absorption following SC injection has led researchers to investigate the use of SC rapid-acting analogs in the treatment of diabetic ketoacidosis (DKA).[10-12]

RAPID-ACTING ANALOGS IN DOGS

The published data on the use of rapid-acting insulin analogs in veterinary medicine is limited to dogs. Some of the early pharmacokinetic work with insulin aspart was performed in research beagles. In these dogs, insulin aspart tended be more rapidly absorbed than regular insulin when administered by SC injection at a therapeutic dose (0.25 U/kg), although no difference was observed in the maximum plasma insulin concentration achieved.[13] Insulin lispro has been evaluated as prandial or mealtime insulin in six dogs with well-regulated diabetes that were receiving NPH insulin.[14] The addition of SC insulin lispro (0.1 U/kg) at mealtime was well tolerated and may be useful in reducing postprandial hyperglycemia. Although this approach may prove beneficial in dogs that have unacceptable glycemic control related to postprandial hyperglycemia, this is an unlikely use of rapid-acting analogs in cats given the glycemic control that can be achieved in the majority of cats with twice daily administration of a long-acting insulin formulation.

INSULIN ASPART IN CATS

We have studied the PD and PK of insulin aspart in healthy cats following intramuscular (IM) and SC administration. Subcutaneous administration resulted in rapid absorption and the duration of action was approximately 2.5 hours. Peak plasma insulin aspart concentrations were higher following SC administration and occurred 15-30 minutes after injection. The time-action profile and total metabolic effect of insulin aspart was similar for IM and SC routes of administration. The results in cats following SC administration are similar to what has been described in dogs and people.[6,7,13]

RAPID-ACTING ANALOGS IN DIABETIC KETOACIDOSIS

Despite substantial advances in insulin therapy and the management of diabetes during the last 35 years, minimal changes have been made to the protocols that are recommended for the management of human DKA during this time.[15] The standard treatment protocols continue to involve the administration of low-dose regular insulin by continuous rate infusion (CRI) or by frequent SC or IM injections.[16-18] Although there is evidence that low-dose insulin therapy is effective regardless of the route of administration[17,19-22], most experts continue to recommend intravenous insulin CRI because of the delayed onset of action and prolonged half-life of IM and SC regular insulin.[18,23-27] More recently, SC administration of rapid-acting insulin analogs has been compared to intravenous regular insulin. Insulin lispro, insulin aspart, and insulin glulisine were found to be as effective as regular insulin CRI in the treatment of people with mild to moderate DKA. The SC rapid-acting analog protocols were associated with fewer intravenous catheter complications, required fewer resources, and had reduced cost when compared to an intravenous CRI protocol.[10-12]

Veterinary Medicine

The three protocols for the treatment of DKA that have been described in veterinary medicine include intravenous CRI, hourly IM insulin, and IM insulin administered every 4 to 6 hours.[28-30] Many clinicians consider intravenous CRI the standard of care although the ideal route of administration remains a matter of debate. Insulin lispro and insulin aspart administered as an intravenous CRI have been successfully used to treat DKA in dogs.[31,32] The studies found that these rapid-acting analogs were a safe and effective alternative to regular insulin although a clinically significant benefit was not identified.

To the author's knowledge, subcutaneous administration of rapid-acting insulin analogs for the treatment of DKA in dogs and cats has not yet been investigated. This treatment may provide an alternative to CRI and IM regular insulin protocols in cats and dogs, and may have advantages when compared to traditional protocols. While it cannot be assumed that the PD and PK following SC injection will be the same in cats with DKA because of dehydration and perfusion abnormalities, our results in healthy cats combined with the clinical data obtained in people suggests that a SC insulin aspart protocol could be an effective treatment for cats with DKA. This type of intermittent treatment protocol could be a better option for intermediate care wards or veterinary clinics that do not have an intensive care unit or access to numerous intravenous fluid pumps. The ability to use rapid-acting analogs to treat dogs and cats with DKA may be of greater importance in the future if regular insulin becomes unavailable due to a decreasing demand from the human medical field.

REFERENCES

1. Brunton S, et al. *J Fam Practice*. 2012;61:S6.
2. Hirsch I. *New Engl J Med.* 2005;352:174.
3. Holleman F, et al. *New Engl J Med.* 1997;337:176.
4. Howey D, et al. *Diabetes*. 1994;43:396.
5. Kang S, et al. *Diabetes Care*. 1991;14:942.
6. Home P, et al. *Eur J Clin Pharmacol*. 1999;55:199.
7. Mudaliar S, et al. *Diabetes Care*. 1999;22:1501.
8. Becker R, et al. *Exp Clin Endocrinol Diabetes*. 2005;113:292.
9. Rodbard HW, et al. *Endocr* Pract. 2009;15:540.
10. Umpierrez G, et al. *Diabetes Care*. 2004;27:1873.
11. Umpierrez G, et al. *Am J Med.* 2004;117:291.
12. Umpierrez G, et al. *Diabetes Care*. 2009;32:1164.
13. Plum A, et al. *Drug Metab Dispos*. 2000;28:155.
14. Bertalan A, et al. *J Vet Intern Med.* 2014;28:976 (abstr)
15. Kitabchi A, et al. *J Clin Endocrinol Metab.* 2008;93:1541
16. Kitabchi A, et al. *Diabetes Care*. 2001;24:131
17. Alberti K, et al. *Arch Intern Med*. 1977;137:1367.
18. Lebovitz H. *Lancet*. 1995;345:767.
19. Fisher N, et al. *New Engl J Med*. 1977;297:238.
20. Alberti K, et al. *Lancet*. 1973;2:515.
21. Sacks H, et al. *Ann Intern Med.* 1979;90:36.
22. Piters K, et al. *Diabetologia*. 1977;13:317.
23. White N. *Endocrinol Metab Clin North Am*. 2000;29:657.
24. Freire A, et al. *J Crit Care*. 2002;17:207.
25. Soler N, et al. *Lancet*. 1973;1:951.
26. Edwards G, et al. *J Pediatr*. 1977;91:701.
27. Ellemann K, et al. *Diabetes Care*. 1984;7:528.
28. Macintire D. *JAVMA*. 1993;202:1266
29. Broussard J, et al. *Kirk's Current Vet Therapy XII*. 1995:393
30. Chastain C, et al. *JAVMA*. 1981;178:561.
31. Sears K, et al. *J Vet Emerg Crit Care*. 2012;22:211.
32. Walsh E, et al. *J Vet Emerg Crit Care*. 2016;26:101.

Canine Pituitary Dependent Hyperadrenocorticism: Other Treatment Options

Katharine F. Lunn, BVMS, MS, PhD, MRCVS, DACVIM
Raleigh, NC

INTRODUCTION

Hyperadrenocorticism (HAC; also termed hypercortisolemia or canine Cushing's syndrome) is a common endocrine disease of middle-aged and older dogs. This clinical syndrome results from chronic exposure to excessive blood levels of glucocorticoids. Naturally occurring HAC may be pituitary-dependent (pituitary-dependent hyperadrenocorticism; PDH), or may result from excessive glucocorticoid secretion from an adrenocortical tumor (adrenal tumor hyperadrenocorticism; ATH). The clinical signs in dogs with HAC are the result of the gluconeogenic, catabolic, immunosuppressive, and anti-inflammatory effects of excessive circulating glucocorticoids. These signs include polyuria, polydipsia, polyphagia, abdominal enlargement, lethargy, panting, exercise intolerance, muscle weakness, alopecia, calcinosis cutis, thinning of the skin, poor wound healing, muscle wasting, decreased bone density and reproductive abnormalities. Dogs with HAC are also predisposed to diabetes mellitus,[1,2] and are more susceptible to infection, particularly urinary tract infections.[3] More serious disorders associated with canine HAC include hypertension and proteinuria.[4] Although uncommon, pulmonary thromboembolism is another potentially life-threatening complication of HAC.[1,5]

Pituitary-dependent HAC is the most common form of spontaneous canine Cushing's syndrome, accounting for 80-85% of naturally occurring cases. This disorder is a consequence of autonomous synthesis and secretion of adrenocorticotropic hormone (ACTH) from the pituitary gland, because of the presence of a pituitary tumor. Chronic excessive ACTH secretion from the pituitary tumor leads to bilateral adrenal cortical hyperplasia and hypercortisolemia.[6] Although almost all cases of PDH are associated with a pituitary tumor, patients do not often show clinical signs directly referable to the local effects of the tumor. The majority of cases are presented for veterinary care because of the typical clinical signs of HAC. Pituitary tumors may be detected by brain imaging,[7-9] but the diagnosis of PDH is usually based upon typical clinical signs and clinicopathological changes of hypercortisolemia, and results of endocrine testing.

The most commonly used screening tests for HAC are the low dose dexamethasone suppression test (LDDST) and the ACTH stimulation test. For patients with typical clinical signs of HAC and positive results on a screening test, further testing is often necessary to differentiate between pituitary and adrenal-dependent disease. Differentiation tests that are commonly used include the high dose dexamethasone suppression test (HDDST) and the measurement of endogenous ACTH levels. The results of imaging studies, including ultrasonography,[10] CT,[7,8] or MRI,[7-9] may also assist in differentiating between PDH and ATH.

Although PDH is a disease of the pituitary gland, the majority of dogs with this disorder are treated by medical therapies that address adrenocortical hyperplasia and hyperfunction. The most commonly used medications are mitotane[11] and trilostane.[12,13] Trilostane is a competitive inhibitor of adrenocortical steroid synthesis that is increasingly used for the management of canine PDH. Both mitotane and trilostane are widely and successfully used in the management of PDH, and both have their own advantages and disadvantages.[11-14] The purpose of this review is to discuss other options for management of PDH in dogs.

DIRECT TREATMENT OF THE PITUITARY LESION

As PDH results from the presence of a pituitary tumor, direct removal of the causative lesion is a logical approach to the management of this disorder. This can be achieved by surgical hypophysectomy, or by radiation therapy.

Surgery

Hypophysectomy is the treatment of choice for PDH in humans, and has been successful in dogs.[15-18] One prospective study of 150 dogs that underwent transsphenoidal hypophysectomy for PDH gave 1-, 2-, 3-, and 4-year estimated survival rates of 84%, 76%, 72%, and 68% respectively. Twelve dogs died post-operatively and 127 went into remission with 32 of those dogs later experiencing a recurrence of disease. Complications included central diabetes insipidus in 53% of dogs undergoing remission, and incomplete hypophysectomy in 9 dogs. The overall success rate of transsphenoidal hypophysectomy was determined to be 65% in this study.[17] Until recently, most reports of this procedure originated in Europe; however, this treatment option is now available in the USA at Washington State University. The clinician leading the team offering this procedure recently published the outcome of transsphenoidal surgery, using a high definition video telescope with a transoral approach, for dogs with PDH.[19] A total of 26 dogs was treated, with 19% mortality overall. It is important to note that mortality was 50% in the first 10 cases, and 0% in the last 16, indicating the importance of experience in the outcome of these cases. For dogs that survived to discharge, there was clinical and laboratory evidence of remission of PDH at 1 month post-surgery. All of the dogs required permanent therapy with thyroid hormone replacement, all but one required permanent therapy with glucocorticoid replacement, and DDAVP therapy was discontinued by 30 days in all but 2 dogs.[19]

Radiation Therapy

There are several reports of the successful use of radiation in the treatment of canine PDH.[20-25] Dow and colleagues treated six dogs with functional pituitary macrotumors with 40 Gy given in 10 equal fractions. Median survival was reported to be 743 days, neurological signs resolved in all dogs, and ACTH levels remained high for at least one year after therapy.[22] Goossens et al. used cobalt 60 radiotherapy to treat six dogs with PDH because of a pituitary tumor that was detectable on MRI. The radiation was given in 11 fractions for a total dose of 44 Gy, and tumor size was significantly reduced in all cases. However clinical signs of PDH were only adequately controlled in one dog.[23] Théon and Feldman evaluated the effects of megavoltage irradiation on pituitary tumors in 24 dogs with neurological signs. Ten dogs experienced complete remission of neurological signs, and another 10 achieved partial remission. As in previous studies, these authors noted a correlation between relative tumor size and the severity of neurological signs in dogs with pituitary tumors. They also noted a correlation between tumor size and remission of neurological signs after pituitary irradiation, suggesting that early treatment of these tumors should improve prognosis. The dogs in this study received a total of up to 48 Gy of radiation given in equal fractions of 4 Gy, and the authors noted that therapy was more effective in delaying tumor growth than in controlling ACTH secretion.[21] A recent retrospective study of radiation therapy for the treatment of pituitary masses demonstrated significantly improved survival and control of neurological signs in 19 dogs that received radiation therapy, compared to 27 untreated control dogs with pituitary masses. Mean survival time in the treated group was 1,405 days, compared to 551 days in the nonirradiated group. The 1-, 2-, and 3-year estimated survival rates were 93, 87, and 55%, and 45, 32, and 25% for the irradiated and nonirradiated dogs respectively. Treated dogs with smaller tumors lived longer than those with larger tumors, again suggesting that early diagnosis and treatment of pituitary tumors are beneficial. The dogs in this study were irradiated with 48 Gy in 3 Gy daily fractions of radiation, using a linear accelerator. Five of 14 dogs with PDH in this study were reported to show resolution of clinical signs of hyperadrenocorticism together with at least one normal ACTH stimulation test result after completion of radiation therapy.[20]

Most of the reports that document the use of radiation therapy for the treatment of pituitary tumors in dogs provide little detailed information regarding the progress of the clinical syndrome of HC in these patients.[20-24] In some cases the reported studies were retrospective and therefore the investigators were only able to access pre-existing medical data regarding the progress of endocrine disease in these dogs.[20,24] In the prospective studies of radiation therapy in dogs with PDH, the follow-up assessments of control of HC were not standardized, and were performed at highly variable time points.[21,22] Thus while radiation therapy appears effective in controlling neurological signs and increasing survival,[20] it is difficult to predict the endocrinological outcome of radiation therapy for dogs with PDH.

Stereotactic Radiation Therapy

A significant disadvantage of conventional pituitary irradiation, as performed in the studies mentioned above, is that it requires up to 16 fractions of radiation, given daily or on alternate days, with associated long periods of hospitalization. It also appears that this therapy does not consistently remove the need for medical treatment in dogs with PDH. Furthermore, conventional radiation therapy is not well targeted and requires the inclusion of more normal brain in the radiation field to insure treatment of the tumor, thus limiting radiation dose. While this technique can resolve neurological signs by shrinking the tumor, it does not ablate tumor cells and residual cells can continue to produce hormones. Stereotactic radiation therapy (SRT) is an exciting newer technology that uses a state-of-the-art linear accelerator to deliver radiation to a specific area while minimizing the dose to normal surrounding tissues. This technique is a standard treatment for pituitary tumors in human patients,[26-30] and now SRT capabilities are becoming more widely available to veterinary patients. The technical advances provided by SRT system allow the use of higher doses per fraction of radiation that cannot be used with conventional radiation therapy equipment. In addition to biological benefits for tumors such as pituitary adenomas,[28] this also has the added benefit of decreasing the number of anesthetic episodes for veterinary patients. Acute effects are also minimal with SRT, ensuring that the patient remains comfortable during and after treatments.

This author participated in a study of SRT for the treatment of canine PDH, with the specific goal of evaluating control of endocrine disease in these patients.[31] Twelve dogs with PDH were enrolled in the study and received three consecutive fractions of stereotactic radiation therapy directed at the pituitary gland. Adrenal and thyroid function was assessed at several time points after SRT. Three patients were euthanized prior to completion of the study: two because of neurological disease and one because of progression of clinical signs, each at 6 months post-SRT. Pre-ACTH cortisol concentrations decreased significantly over time, but post-ACTH cortisol and endogenous ACTH concentrations did not change significantly. Six owners reported improvement in their dogs' clinical signs throughout the course of the study; of these dogs, four had post-cortisol levels within normal limits at the time the last sample was collected. Three dogs required thyroid supplementation because of hypothyroidism. Interestingly, many of the dogs that showed good control of their PDH had been previously treated with medical therapy. Although this preliminary study showed that SRT was effective in treating a subset of patients with PDH, it was not uniformly effective in all patients, at least for the time period studied.

OTHER MEDICAL TREATMENT OPTIONS

Several other options for medical therapy of PDH have been explored, but none have demonstrated superior efficacy to mitotane or trilostane.

Ketoconazole

This acts by inhibiting adrenocortical enzymes responsible for steroid synthesis. This medication has fallen out of favor as it is less efficacious than both mitotane and trilostane, and offers no significant benefit compared to either of these medications. Ketoconazole could be a reasonable alternative therapy in situations where neither mitotane nor trilostane were available.

Monitoring is with ACTH stimulation testing. A low starting dose is recommended (5 mg/kg q 12hr) with evaluation for side effects including gastrointestinal signs and hepatotoxicity.

L-Deprenyl

This acts by inhibiting degradation of dopamine, thereby potentially inhibiting ACTH secretion from the intermediate lobe of the pituitary gland; unfortunately only 20% of canine PDH cases arise from disease in this area. Additional disadvantages are documented poor efficacy,[32] expense, and inability to monitor response with ACTH stimulation testing.

Cabergoline

This medication also acts to decrease pituitary ACTH secretion through increasing dopaminergic tone. In the only published study of cabergoline use in dogs with PDH, 17 of 40 dogs showed a favorable clinical response. Decreases in ACTH concentrations, urine cortisol:creatinine ratios, and pituitary tumor size were also reported.[33] Further studies are needed before this is likely to become a widely adopted treatment for canine PDH.

Bromocriptine

Bromocriptine is a dopamine agonist and also acts to decrease plasma ACTH concentrations. Because of lack of demonstrated efficacy it is not recommended for treatment of canine PDH.

Cyproheptadine

Use of this antiserotoninergic medication arose from the hypothesis that excessive ACTH secretion could result from excessive serotoninergic stimulation of the pituitary gland. This drug has been shown to be ineffective in clinical cases.[34]

Aminoglutethimide and Metyrapone

These medications are enzyme inhibitors that non-selectively block adrenal steroid synthesis. They have been used in combination therapy in humans. One report of the use of aminoglutethimide in dogs demonstrated poor efficacy of this treatment with several adverse effects.[35] There are no reports of the use of metyrapone for the treatment of canine PDH.

Retinoic Acid

This medication may inhibit pituitary tumor development, decrease ACTH production, and inhibit cell proliferation. One study showed promise in terms of decrease in size of pituitary tumors, and subjective improvement in clinical signs.[36] Unfortunately experience with this medication is very limited, particularly when contrasted with proven therapies such as mitotane and trilostane. Availability and cost of the appropriate formulation of retinoic acid are also significant concerns.

Melatonin

Melatonin has been suggested as a therapy for canine PDH. Melatonin in combination with lignan has been shown to decrease steroid hormone secretion in human adrenocortical carcinoma cells in vitro.[37] Oral melatonin has also been demonstrated to affect serum sex hormone levels in male and female dogs,[38] and oral melatonin has been reported to decrease clinical signs in ferrets with adrenocortical disease.[39] Despite these findings, and despite ongoing anecdotal reports of the use of melatonin in canine PDH, there are no published data to support this recommendation.

REFERENCES

[1]Nichols R. Vet Clin Small Anim 1997;27:309. [2]Peterson ME, et al. Res Vet Sci 1984;36:177. [3]Forrester SD, et al. J Vet Intern Med 1999;13:557. [4]Ortega TM, et al. J Am Vet Med Assoc 1996;209:1724. [5]Johnson LR, et al. J Vet Intern Med 1999;13:338. [6]Kemppainen RJ, et al. Vet Clin Small Anim 1997;27:173. [7]Wood FD, et al. J Am Vet Med Assoc 2007;231:1081. [8]Moore SA, et al. Compend Contin Educ Pract Vet 2008;30:33. [9]Bertoy EH, et al. J Am Vet Med Assoc 1996;208:1268. [10]Gould SM, et al. J Small Anim Pract 2001;42:113. [11]Kintzer PP, et al. J Vet Intern Med 1991;5:182. [12]Neiger R, et al. Vet Rec 2002;150:799. [13]Vaughan MA, et al. J Am Vet Med Assoc 2008;232:1321. [14]Barker EN, et al. J Vet Intern Med 2005;19:810. [15]Meij B, et al. Mol Cell Endocrinol 2002;197:89. [16]Meij BP, et al. Vet Surg 1998;27:246. [17]Hanson JM, et al. J Vet Intern Med 2005;19:687. [18]Hara Y, et al. J Vet Med Sci 2010;72(4):397. [19]Mamelak AN, et al. Vet Surg 2014;43:369. [20]Kent MS, et al. J Vet Intern Med 2007;21:1027. [21]Theon AP, et al. J Am Vet Med Assoc 1998;213:225. [22]Dow SW, et al. J Small Anim Pract 1990;31:287. [23]Goossens MM, et al. J Am Vet Med Assoc 1998;213:374. [24]De Fornel P, et al. Can Vet J 2007;48:481. [25]Marcinowska A, et al. Vet Radiol Ultrasound 2015;56(5):554. [26]Hentschel SJ, et al. Neurosurg Focus 2004;16:E5. [27]Laws ER, et al. J Neuro-Onc 2004;69:257. [28]Prasad D. Neurosurg Clin N Am 2006;17:129. [29]Voges J, et al. Cancer 2006;107:1355. [30]Kong D-S, et al. Cancer 2007;110:854. [31]Timian J, et al. J Vet Intern Med 2013;27:687. [32]Braddock JA, et al. Aust Vet J 2004;82(5):272. [33]Castillo VA, et al. Res Vet Sci 2008;85(1):26. [34]Stolp R, et al. J Endocr 1984;101:311. [35]Perez, AM et al. J Small Anim Pract 2002;43(3):104. [36]Castillo V, et al. Endocrinology 2006:147:4438. [37]Fecteau KA, et al. Am J Vet Res 2011;72(5):675. [38]Ashley PF et al. J Am Vet Med Assoc 1999;215(8):1111. [39]Ramer JC et al. J Am Vet Med Assoc 2006;229(11):1743.

Clinical Approach to Adrenal Incidentalomas

Ellen N. Behrend, VMD, PhD, DACVIM (Small Animal Internal Medicine)
Auburn, AL

Ultrasonography has become a routine diagnostic tool for evaluation of the abdominal cavity. One consequence is the unexpected finding of a seemingly incidental adrenal mass. It should be remembered that an adrenal mass is not necessarily a primary adrenal tumor (AT), and not all AT are cortisol-secreting. In one study, 27% of canine and 60% of feline AT were metastatic lesions.[1] A primary AT can be benign or malignant and may or may not be functional, i.e. secreting a hormone; myelolipomas or lipomas can occur in the adrenal cortex. Functional masses can arise from the cortex and secrete steroid hormones or arise from the medulla, i.e. a pheochromocytoma. Of primary AT, approximately 75% are adrenocortical with the remainder of neuroendocrine origin.[2] Non-tumorous possibilities for masses also exist, including nodular hyperplasia, cyst, abscess, hematoma, and granuloma.

INCIDENTALOMA

An incidental adrenal mass (IAGM), i.e. an "incidentaloma", is defined as a focal enlargement of the adrenal gland in patients without prior evidence of adrenal gland disease. An AT should be suspected when there is loss of the typical shape of an adrenal gland regardless of size, asymmetry in shape and size between the affected and contralateral adrenal glands is present, or the mass has infiltrated the phrenicoabdominal vein, vena cava, or surrounding soft tissues. The incidence of IAGM is estimated at 4% in dogs overall, but with a higher incidence in older and larger dogs.[3] Only 17% of dogs with IAGM were < 9 years of age and the median weight of dogs with an IAGM was 21 kg. Incidence of adrenal incidentalomas in cats is unknown.

As the first step, abdominal ultrasound should always be repeated to ensure the mass is a consistent finding. Once an IAGM is confirmed, a number of differentials, as discussed above, need to be considered. Many factors determine the aggressiveness of the diagnostic and therapeutic approach, including the severity of concurrent problems, the original reason for performing abdominal ultrasound, patient age, the likelihood the mass is hormonally active and/or malignant, the size and invasiveness of the mass, and the client's desires.

The history, physical examination, and results of routine blood and urine tests should be assessed for evidence of hypercortisolism, hyperaldosteronism, elevated sex hormone concentrations or of a pheochromocytoma. Appropriate tests should be performed as indicated to confirm the diagnosis.

The chance of an IAGM being malignant may be between 14 and 30%.[3] Unfortunately, it can be difficult to determine whether an IAGM is neoplastic and malignant or benign without histopathology. Although cytology has 90-100% accuracy for differentiating cortical from medullary origin, it is not reliable for distinguishing benign from malignant.[4] Imaging may be helpful but must be interpreted carefully. Characteristics that do not help differentiate adrenal adenoma and carcinoma are mineralization or ultrasonographic echogenicity. Both adenomas and carcinomas can contain mineral densities or appear as a mass cranial to the kidney While diffuse, ill-defined mineralization usually is associated with adrenal neoplasia, discrete, well-marginated mineralization develops in normal animals and may be a dystrophic change.[5] On the other hand, possible metastases may be identified by thoracic radiography. Certain ultrasonographic findings are highly suggestive of malignancy. Suspect metastases to abdominal organs, especially the liver, can be visualized and confirmed by ultrasound-guided biopsy. If the maximum diameter of the AT is >2 cm, chance of malignancy[3] and growth[6] may be high. Evidence of invasion into surrounding tissues and/or the vena cava is suggestive of a carcinoma; however, it can be missed.[7] Presence of tortuous vessels and heterogeneity of contrast enhancement as judged by contrast-enhanced ultrasonography may also be markers of malignancy.[8]

If the IAGM is suspected to be malignant, adrenalectomy is the preferred treatment, but it may not be indicated if the mass is small, hormonally inactive, not infiltrating surrounding structures and believed likely to be benign. If no suggestion of malignancy is found, clinical signs or findings on physical examination and routine blood and urine tests do not support a functional AT and the tumor is <2 cm in diameter, a conservative approach should be adopted. Monthly monitoring with ultrasound to determine the rate of growth of the mass and changes in the appearance of the adrenal gland is recommended. Growth of IAGM is unpredictable. Of 7 dogs with an IAGM, 3 lesions were not found when ultrasound was performed again >4 months later. No growth occurred after >6 months in 2 dogs. In 2 dogs with initially larger tumors, growth occurred. In one, the mass grew from 1.6 to 2.5-cm diameter and invaded the vena cava within 10 months; in the other, the mass grew from 2.5 to 3.1-cm diameter within 8 months.[3] In 9 dogs with non-cortisol-secreting AT followed for 12 months, no change was seen in 7. The two tumors that did grow were originally 2.0 and 2.5 cm in length.[6]

If the IAGM has not increased in size after 3 months, the time interval between ultrasound evaluations can be increased. However, if the IAGM is enlarging, changing in appearance, or compressing or infiltrating surrounding blood vessels or soft tissues, or if clinical signs affiliated with excess hormone secretion develop, adrenalectomy may be warranted. For non-cortisol-secreting AT, median survival without surgery in 14 dogs was 29.8 ± 8.9 months (range 1.0 - 96.0 months). Larger tumor size was associated with shorter survival.[6]

HYPERADRENOCORTICISM

Given the vast amounts written on hyperadrenocorticism (HAC), it will not be considered in depth here and the reader is referred elsewhere.[9] For HAC, the classic clinical signs in dogs are polyuria/polydipsia, polyphagia, panting and a pot-belly appearance. The tumor itself can cause issues, such as rear limb edema because of vena caval invasion, but these are rare. On routine laboratory testing, the vast majority have an elevated ALP activity; hypertension and proteinuria are also quite common. If HAC is suspected, a low-dose dexamethasone suppression test is the screening test of choice. The sensitivity of the ACTH stimulation test for diagnosing HAC secondary to an adrenal tumor is approximately 60%.

HYPERALDOSTERONISM

Primary hyperaldosteronism (PHA) because of an AT, i.e. aldosteronomas, are being increasingly recognized in cats. Given the rarity in dogs, the discussion will focus on what is known in cats. However, information on clinical signs and diagnosis likely apply to dogs. Feline tumoral PHA is usually caused by a unilateral adrenocortical adenoma or carcinoma, although bilateral adrenal adenomas have been identified.[10] Malignant tumors are more common than benign.[11] No breed predisposition for feline aldosteronoma is apparent. Median age at diagnosis is approximately 13 years and most cats are >10 years. No gender predisposition has been noted.[12]

As aldosterone is responsible for increasing potassium excretion and blood pressure, clinical signs of hyperaldosteronism relate mainly to hypokalemia or hypertension. The most common clinical sign in affected cats is persistent, progressive weakness, i.e. "hypokalemic polymyopathy"[11], which typically occurs when serum potassium concentrations is <3 mg/dL. Cervical ventroflexion, hindlimb weakness (sometimes plantigrade stance), difficulty jumping, listlessness, and ataxia are the most common owner complaints. Episodic signs or an acute onset can be reported.[13] Respiratory failure secondary to weakness of the respiratory muscles can occur rarely. Owners may note acute blindness and/or a sudden change in eye color because of hypertension, usually induced by intraocular hemorrhage or retinal detachments. On occasion, hypertension may cause neurological signs such as seizures, ataxia and behavioral changes.[12] Tumors secreting both mineralocorticoids and glucocorticoids occur uncommonly, so clinical signs of and clinicopathologic changes consistent with glucocorticoid excess may also be present.

The single abnormality typical for PHA on routine laboratory testing is hypokalemia. Although hypernatremia can occur, it is uncommon. Creatine kinase (CK) is usually elevated because of the myopathy, but the degree is variable. Metabolic alkalosis is common. An aldosteronoma should be strongly considered in a cat with unexplained hypertension, especially if it is refractive to therapy. Approximately 85% of affected cats are persistently hypertensive.[14]

Radiology is often not helpful, but a large tumor may be visible. Adrenal calcification is present in approximately 33% of older, healthy cats, and should not be over-interpreted. Pulmonary metastases are uncommon. In all reported cases of aldosteronoma in cats, an AT has been identified by ultrasound; however, in a few cats with bilateral AT, only one tumor was visualized.[10,15] The contralateral adrenal gland needs to be evaluated; a decreased size suggests the AT is secreting a hormone with glucocorticoid or progestogen activity[16] and bilateral masses may occur.[10] Signs of malignancy should be sought (see above). However, ultrasound has not correctly identified vascular invasion in a high proportion of cats with aldosteronoma.[15,17]

If a mass in the area of the adrenal gland is identified, cytology can be helpful in determining origin.[18] While presence of an AT in a cat with persistent hypokalemia and/or hypertension is highly suspicious for aldosteronoma, confirmation of the diagnosis relies on demonstration of an elevated basal aldosterone concentration. The assay is relatively easily available, and no special sample handling is required. To date, the highest aldosterone concentrations reported have occurred with AT. However, as suspicion for PHA is becoming stronger because of an increasing prevalence, testing will likely occur earlier in the course of disease,when aldosterone concentrations likely can be in the upper end of the reference range. Furthermore, the ranges of aldosterone concentrations reported for primary and secondary hyperaldosteronism overlap to a large extent. Secondary hyperaldosteronism is the result of a condition, e.g. heart failure or chronic kidney disease, which stimulates the renin-angiotensin-aldosterone system. Distinguishing primary and secondary hyperaldosteronism can be quite difficult. Concentrations > 1000 pmol/L have been reported with secondary PHA, especially in cats with chronic renal failure. The ideal would be to measure plasma renin activity (PRA). Unfortunately, a PRA assay is not commercially available in the US.

Desoxycorticosterone-secreting tumors have been reported rarely.[19,20] Clinical findings are the same as with primary hyperaldosteronism but aldosterone concentrations are low. Because a desoxycorticosterone assay is not commercially available, a diagnosis is presumptive based on the constellation of diagnostic test results.

PHEOCHROMOCYTOMA

Pheochromocytomas are uncommon in dogs and rare in cats. They occur in middle-aged to older patients with no known sex or breed predisposition. For pheochromocytomas, the most common clinical signs are weakness, lethargy, tachypnea or panting and collapse. The severity of clinical signs may correlate with the size of the tumor. They often occur intermittently because of episodic catecholamine secretion. They may, however, also present as acute events, e.g. associated with tumor rupture. Other clinical signs and physical examinations include anorexia, weight loss, cardiac arrhythmias, tachycardia, polyuria/polydipsia, vomiting and abdominal pain.

Diagnosis of a pheochromocytoma can be a challenge. No consistent changes in the CBC, serum biochemistry panel and urinalysis exist which support the suspicion of a pheochromocytoma. Hypertension is present in more than 50% of dogs with a pheochromocytoma. As it can be episodic, repeat blood pressure measurements should be performed. The higher the systolic blood pressure the more likely a pheochromocytoma is present, e.g. systolic blood pressures > 300 mm Hg have so far only been seen in dogs with pheochromocytoma. A retinal examination may also provide proof of the presence of chronic hypertension. No definitive, validated biochemical test for diagnosis of a pheochromocytoma exists in veterinary medicine. Research into measurement of urinary metanephrine concentrations is promising.

SEX HORMONE-SECRETING TUMORS

Adrenocortical tumors can synthesize and secrete sex hormones. A small number of cats and dogs with sex hormone-secreting AT have been described. The cats ranged in age from 7-15 years. A 14-year-old, female spayed domestic shorthair cat with bilateral adrenal enlargement and excessive estradiol and testosterone production has been reported. Excess production of a progestin, e.g. progesterone or 17-hydroxy-progesterone, will cause signs typical of cortisol excess, such as a poor haircoat, dermal atrophy and cutaneous fragility. Diabetes was present in 3 of 4 cats with a progesterone-secreting tumor, and was poorly regulated in 2. Excess androgen production in cats can cause a change in the urine odor, urine spraying behavior, aggression and development of penile spines. Hyperestrogenemia may cause cyclic estrus behavior and vulvar enlargement. No dogs with clinical signs of hyperestrogenemia or hyperandrogenemia because of an AT have been reported. Tumor type was known in 5 cats, and only one was an adenoma.[16,21-25]

Suspicion of a sex-hormone secreting tumor will arise with clinical signs. No result of a routine test, e.g. complete blood count or serum biochemistry profile, will aid in the diagnosis. One of the main indications for sex hormone measurement is the finding of cortisol concentrations, including basal, below the reference range in a dog or cat tested for hyperadrenocorticism with an ACTH stimulation test or low-dose dexamethasone suppression test. If administration of exogenous glucocorticoids of any form or of medications that alter cortisol synthesis (e.g. ketoconazole) is ruled out, a sex hormone-secreting AT may be present. Finding of an AT on ultrasound further increases suspicion. Signs of malignancy should be sought and thoracic radiographs taken to assess for metastatic disease. Another indication would be the presence of sex hormone-associated changes and behaviors, e.g. urine spraying, mounting, vulvar enlargement, etc.

To document sex hormone elevations, an ACTH stimulation test should be performed as for typical hyperadrenocorticism. However, rather than cortisol, sex hormones are measured in samples drawn before and one hour after ACTH administration. In dogs at least, sex hormone concentrations should be interpreted cautiously. For estradiol, a wide range of variability exists within and between dogs; random, basal estradiol concentrations in individual dogs often exceed the reference range. In 6 dogs with a pheochromocytoma or a non-functional AT, androstenedione, progesterone, 17OHP, testosterone and/or estradiol concentrations were elevated.[26]

REFERENCES

[1]Labelle P, De Cock HEV. Vet Pathol 2005;42:52-58.; [2]Capen CC. Endocrine glands. In: Maxie MG, ed. Jubb, Kennedy and Palmer's Pathology of Domestic Animals. Philadelphia: Saunders; 2007:325-428.; [3]. Cook AK, Spaulding KA, Edwards JF. J Am Vet Med Assoc 2014;244:1181-1185.; [4]Bertazzolo W, Didier M, Gelain ME, et al. Vet Clin Pathol 2014;43:453-459.; [5]Widmer WR, Guptill L. J Am Vet Med Assoc 1995;206:1857-1864.; [6]Arenas C, Perez-Alenza D, Melian C. Vet Record 2013;173:501-506.; [7]Besso JG, Pennick DG, Gliatto JM. Vet Radiol Ultrasound 1997;38:448-455.; [8]Pey P, Rossi F, Vignoli M, et al. Am J Vet Res 2014;75:886-892.; [9]Behrend EN. Canine hyperadrenocorticism. In: Feldman ECN, R.W.; Reusch, C.E.; Scott-Moncrieff, J.C.R., ed. Canine and Feline Endocrinology, 4 ed. St. Louis, MO: Elsevier; 2015:377-451.; [10]Ash RA, Harvey AM, Tasker S. J Feline Med Surg 2004;7:173-182.; [11]Refsal KR, Harvey AM. Feline primary hyperaldosteronism. In: Bonagura JD, Twedt DC, eds. Kirk's Current Veterinary Therapy XV. St. Louis: Elsevier; 2014:238-242.; [12]Harvey AM, Refsal KR. Primary hyperaldosteronism. In: Rand J, ed. Clinical Endocrinology of Companion Animals. Ames, IA: Wiley-Blackwell; 2013:115-127.; [13]Javadi S, Djajadiningrat-Laanen SC, Kooistra HS, et al. Domest Anim Endocrinol 2005;28:85-104.; [14]Djajadiningrat-Laanen S, Galac S, Kooistra H. J Feline Med Surg 2011;13:641-650.; [15]Combes A, Pey P, Paepe D, et al. J Feline Med Surg 2013;15:445-457.; [16]Briscoe K, Barrs VR, Foster DF, et al. J Feline Med Surg 2009;11:758-762.; [17]Lo AJ, Holt DE, Brown DC, et al. J Vet Intern Med 2014;28:137-143.; [18]Renschler JS, Dean GA. Vet Clin Pathol 2008;38:1-4.; [19]Davies DR, Foster SF, Hopper B, et al. Aust Vet J 2008;86:139-146.; [20]Reine NJ, Hohenhaus AE, Peterson ME, et al. J Vet Intern Med 1999;13:386-390.; [21]DeClue AE, Breshears LA, Pardo ID, et al. J Vet Intern Med 2005;19:355-358.; [22]Boord M, Griffin C. J Am Vet Med Assoc 1999;214:666-669.; [23]Rossmeisl JH, Scott-Moncrieff JC, Siems J, et al. J Am Anim Hosp Assoc 2000;36 512-517.; [24]Millard RP, Pickens EH, Wells KL. J Am Vet Med Assoc 2009;234:505-508.; [25]Meler EN, Scott-Moncrieff JC, Peter AT, et al. J Feline Med Surg 2011;13:473-478.; [26]Hill KE, Scott-Moncrieff JCR, Koshko M, et al. J Am Vet Med Assoc 2005;226:556-561

Epigenetics nd MicroRNAs Modulate T-Cell Dysfuntion in the FIV Model

Jonathan E. Fogle, DVM, PhD, DACVIM (Small Animal Internal Medicine)
Raleigh, NC

INTRODUCTION

For most individuals infected with HIV, virus escape from a robust CD8+ cytotoxic T lymphocyte (CTL) response within the early, acute stage of infection contributes to a cascade of events including alterations in the viral set point, alterations in lymphocyte homeostasis, and the initiation of a cycle of persistent immune activation combined with immune dysfunction. [1-4] Other acute viral infection studies clearly demonstrate that CD8+ lymphocytes acquire specific epigenetic programs, including histone acetylation and DNA demethylation (euchromatin formation), as these cells become activated and differentiate into effector / memory (e/m) populations. Disruption of this process leads to long-term CD8+ T cell dysfunction, persistent viremia, chronic immune activation, and T cell exhaustion.[5-7] Induction of CD8+ T cell dysfunction is mediated by a number of factors including lentivirus-activated CD4+CD25+ T regulatory (Treg) cells.[8] Using the feline immunodeficiency virus (FIV) model, we have clearly demonstrated that activated Treg cells suppress IL2, TNFα, and IFNγ in activated CD8+ targets.[9-13]

THE ROLE OF EPIGENETICS IN CD8+ PHENOTYPIC AND FUNCTIONAL DEVELOPMENT

Previous investigations in murine models have clearly demonstrated that the generation of phenotypically mature, fully functional antiviral CD8+ T cell e/m lymphocytes is dependent upon DNA demethylation at the IL2 and IFNγ promoter regions and histone acetylation at the IFNγ promoter.[7,13] Disruption of this process early during the course of infection sets the stage for chronic CTL dysfunction, persistent viremia, immune activation and a step-wise loss of IL2, TNFα, and IFNγ production in virus-specific CD8+ T cells.[14-16] Likewise, for HIV-infected individuals, other than elite controllers (ECs), virus-specific CD8+ cytotoxic T lymphocyte (CTL) activation is followed by the induction of CD8+ T cell dysfunction, which is characterized by phenotypic activation, persistent viremia, upregulation of exhaustion markers, and altered e/m differentiation.[1-4] For simplicity, the various CD8+ subsets (central memory, effector / memory, terminal effector) are collectively referred to here as e/m populations.

TREG CELLS AND TGFβ

Treg cells can suppress other T cell responses by production of immunosuppressive cytokines or through contact-mediated inhibition. Different murine models have demonstrated that Treg cell suppression of C8+ T cell responses can occur through TGFβ / TGFβ receptor II (TGFβRII) signaling.[24] Consistent with murine models of Treg-mediated suppression, using the FIV lentiviral model, we have clearly demonstrated that activated Treg cells display a novel membrane-associated TGFβ (mTGFβ), CD8+ T cells upregulate TGFβRII, and that Treg-mediated suppression is TGFβ dependent.[12,30] Recent observations indicate that infection with a recombinant SIV/HIV virus (R5-SHIV) in macaques leads to accumulation of Treg cells in LNs and Treg cell depletion restores antiviral CD8+ T cell responses.[18] Likewise, a recent report has demonstrated a correlation between Treg cell numbers, CD8+ suppression and poor virus control in HIV patients.[19] Miles et al. recently described accumulation of Treg cells and TGFβ-dependent suppression of follicular Th cells in LN follicles and in a previous investigation, the same group clearly demonstrated lower numbers of SIV-specific CTLs within the follicles was associated with increased SIV RNA.[21-22] Collectively, these studies suggest TGFβ signaling is dysregulated in lentiviral infection and that Treg cells are integral to inducing TGFβ signaling and suppression of targeted lymphocytes.

FOXP3 AND LYMPHOCYTE DYSFUNCTION

In activated CD4+ T cells, TGFβ / TGFβRII signaling can induce expression of the forkhead transcription factor Foxp3. Foxp3 is one of the defining factors for CD4+CD25+ Treg cell phenotype and function and in CD4+ T cells can bind to IL2, TNFα, and IFNγ promoter regions to actively suppress transcription.[23] In a series of experiments, we (and others) have demonstrated increased Foxp3 expression in activated CD8+ lymphocytes following interaction with lentivirus-activated Treg cells.[8] Although others have demonstrated that these CD8+Foxp3+ are indeed suppressor cells, our laboratory was unable to document that CD8+ lymphocytes exhibited suppressor function following Treg coculture, despite the induction of CD8+ Foxp3 expression. As part of the same series of experiments in our laboratory, we were able to demonstrate that Treg cells inhibit IL2 and IFNγ mRNA expression and induce G1 cell cycle arrest in CD8+ T cell targets during both acute and chronic FIV infection. Because Treg cells can suppress genes associated with CD8+ lymphocyte function and induce dysfunction in CD8+ lymphocyte targets during acute and chronic lentiviral infection; we asked if Treg-induced Foxp3 expression in CD8+ T cells plays a role in CD8+ T cell dysfunction. We have recently reported that Foxp3 is increased in CD8+ lymphocytes following interaction with lentivirus-activated Treg cells and binds the IL2 promoter, which can be blocked by inhibiting DNA demethylation.[9] Further, our recent preliminary data indicate that Treg-induced Foxp3 occupies the IL2, TNFα, and IFNγ promoter regions in virus-specific CD8+ T cells. Collectively, these novel findings suggest that Foxp3 likely plays a role in CD8+ T cell gene suppression, the induction of dysfunction, and poor e/m development.

MICRORNAs STABILIZE FOXP3 EXPRESSION IN LENTIVIRUS-ACTIVATED TREGS

Treg cells are characterized by constitutive expression of surface CD25 and stable expression of the repressive transcription factor FoxP3, which is required for their development and function. In the context of appropriate immune responses, Treg cells are essential for maintaining peripheral tolerance, preventing autoimmune diseases, and limiting chronic inflammatory diseases.

Infection with lentiviruses such as HIV and FIV pathologically activates Treg cells leading to suppression of other T cell subsets, including CD4+ T helper (Th) cells and CD8+ CTLs (as described above). MicroRNAs (miRNA) are small, non-coding RNAs typically about 20-25 nucleotides in length and play important gene-regulatory roles in both plants and animals. Since first being identified *in C. elegans* in 1993, more than 700 miRNAs have been identified in mammalian cells and play important roles in development, cellular differentiation and homeostasis, adaptation to the environment, oncogenesis, and host cell interactions with pathogens. miRNAs usually interact with the 3-prime untranslated region (3'UTR) of target messenger RNAs (mRNAs) leading to mRNA degradation and/or translational repression (i.e. suppressing gene expression). Traditionally, miRNAs have been ascribed the role of gene suppression; however, recent evidence suggests miRNAs may have the ability to promote the expression of some genes. We hypothesized that stable Foxp3 expression in CD4+CD25+ Treg cells from FIV infected cats is mediated by miRNA stabilization of Foxp3 transcription. Th cells and Treg cells from FIV infected were examined for differential miRNA expression. Three miRNAs reported to positively regulate Foxp3 and two miRNAs reported to negatively regulate Foxp3 in other species were identified in our study. Using a luciferase reporter assay, we identified miRNA binding to the Foxp3 3'UTR. Feline Mya-1 cells (T cell line) were transfected with the identified miRNA sequences and Foxp3 mRNA and protein was assessed by reverse transcription quantitative (RT-qPCR) and western blot, respectively. Our results suggest that Foxp3 mRNA remains unchanged following transfection of Mya-1 cells with miRNAs reported to stabilize Foxp3; however, surprisingly, we have consistently demonstrated increased Foxp3 protein following transfection. Studies are underway to clarify the molecular mechanism(s) responsible for this observation.

CONCLUSIONS

In summary, our data suggest that specific miRNAs can modulate Foxp3 expression in lentivirus-activated Treg cells. Thus, stable Foxp3 expression within the Treg cell may be a key event leading to a series of immunosuppressive events including suppression of IL2, TNFα, and IFNγ in activated CD8+ targets. Further, we have shown that Treg cells induce the expression Foxp3 in CD8+ targets following coculture, and that Foxp3 binds the CD8+ T cell IL2, TNFα and IFNγ promoters. These data suggest early epigenetic changes, while essential for CD8+ T cell antiviral function, render the CTL highly permissive to Treg-induced suppression. In essence, the "open" euchromatin formation at the "active" areas of CD8+ T cell transcription permits access of repressive transcription factors such as Foxp3. These findings help, in part, explain the unusual finding chronic T cell immune activation combined with T cell immune dysfunction that is a hallmark of lentiviral infection.

REFERENCES

[1.]Tompkins MB, Tompkins WA. Vet Immunol Immunopathol 2008.[2.]Trautmann L, Mbitikon-Kobo FM, Goulet JP, Peretz Y, Shi Y, Van Grevenynghe J, et al Blood 2012.[3.]Stephenson KE, Li H, Walker BD, Michael NL, Barouch DH. J Virol 2012,86:9583-9589.[4.]Kaul R, MacDonald KS, Nagelkerke NJ, Kimani J, Fowke K, Ball TB, et al. AIDS 2010,24:1449-1454.[5.]Agarwal P, Raghavan A, Nandiwada SL, Curtsinger JM, Bohjanen PR, Mueller DL, et al. J Immunol 2009,183:1695-1704.[6.]Denton AE, Russ BE, Doherty PC, Rao S, Turner SJ. Proc Natl Acad Sci U S A 2011,108:15306-15311.[7.]Northrop JK, Thomas RM, Wells AD, Shen H. J Immunol 2006,177:1062-1069.[8.]Fogle JE, Tompkins WA, Tompkins MB. Retrovirology 2010,7:97.[9.]Miller MM, Akaronu N, Thompson EM, Hood SF, Fogle JE. J Immunol 2015,194:990-998.[10.]Walker-Sperling VE, Cohen VJ, Tarwater PM, Blankson JN. J Virol 2015,89:9631-9638.[11.]Shan L, Deng K, Shroff NS, Durand CM, Rabi SA, Yang HC, et al. Immunity 2012,36:491-501.[12.]Fogle JE, Mexas AM, Tompkins WA, Tompkins MB. AIDS Res Hum Retroviruses 2010,26:201-216.[13.]Northrop JK, Wells AD, Shen H. J Immunol 2008,181:865-868.[14.]Brooks DG, Trifilo MJ, Edelmann KH, Teyton L, McGavern DB, Oldstone MB. Nat Med 2006,12:1301-1309.[15.]Sevilla N, Kunz S, Holz A, Lewicki H, Homann D, Yamada H, et al. J Exp Med 2000,192:1249-1260.[16.]Wherry EJ, Blattman JN, Murali-Krishna K, van der Most R, Ahmed R. J Virol 2003,77:4911-4927.[17.]Fuller MJ, Khanolkar A, Tebo AE, Zajac AJ. J Immunol 2004,172:4204-4214.[18.]Santosuosso M, Righi E, Hill ED, Leblanc PR, Kodish B, Mylvaganam HN, et al. PLoS One 2011,6:e18465.[19.]Shaw JM, Hunt PW, Critchfield JW, McConnell DH, Garcia JC, Pollard RB, et al. AIDS Res Hum Retroviruses 2013,29:172-177.[20.]Miles B, Miller SM, Folkvord JM, Kimball A, Chamanian M, Meditz AL, et al. Nat Commun 2015,6:8608.[21.]Connick E, Folkvord JM, Lind KT, Rakasz EG, Miles B, Wilson NA, et al. J Immunol 2014,193:5613-5625.[22.]Chen ML, Pittet MJ, Gorelik L, Flavell RA, Weissleder R, von Boehmer H, et al. Proc Natl Acad Sci U S A 2005,102:419-424.[23.]Beavis PA, Gregory B, Green P, Cribbs AP, Kennedy A, Amjadi P, et al. Proc Natl Acad Sci U S A 2011,108:16717-16722.[24.]Chen C, Rowell EA, Thomas RM, Hancock WW, Wells AD. J Biol Chem 2006,281:36828-36834.

Diseases Caused by Canine Herpes Virus

Sanjay Kapil, DVM, PhD, DACVM
Stillwater, OK

Canine herpes virus (Canid herpes virus, CaHV-1) is a member of Family Herpesviridae, subfamily Alphaherpesvirinae. CaHV-1 has world-wide distribution. CaHV-1 serological incidence varies in different countries of the world (40-93%). In Scandinavia, about 86% seropositivity has been reported in unvaccinated bitches. CaHV-1 host range is canids (dogs, wolves and coyote). CaHV-1 does not infect humans. CaHV-1 is monotypic and, is genetically closely related to feline herpes virus. CaHV-1 isolates from different parts of the world cannot be distinguished genetically because of high homology. Because CaHV-1 is poorly immunogenic, the diagnosis can be missed based on serology alone. Latently infected dogs can convert to seronegative status.

CaHV-1 has an incubation period of about 4-6 days. The virus is transmitted by various routes such as oral, aerosol transmission through respiratory route, and *in utero* route leading to high puppy mortality shortly after birth. CaHV-1 is labile. Thus, successful transmission of the virus requires direct contact with nasal and genital secretions. Infected bitches remain symptomatic and puppies become anorectic (1).

CaHV-1 produces latency in trigeminal and lumbosacral ganglia. Abortion and still birth puppies may also be associated with CaHV-1 reactivation. No CaHV-1 reactivation occurs from PBL. CaHV-1 is latent in dogs, worldwide.

The predisposing factors in puppies include exposure to cold weather leading to hypothermia that allows higher replication of the virus.

Clinical signs in neonatal puppies include vocalization, anorexia, loss of sucking reflex, abdominal pain, dyspnea and hypothermia. Incubation Period is about 7-10 days. Sometimes total litter mortality can happen. Older Puppies remain asymptomatic and mount a febrile response.

Reproductive disorders include lesions on penis in male dogs and vaginal hyperemia and lymph follicular lesions in females. In pregnant females, CaHV-1 causes mid-gestation abortions and stillborn puppies.

Respiratory diseases due to CaHV-1 include mild rhinitis, pharyngitis, and pneumonia. CaHV-1 infections are detected about 3-4 weeks after introduction of dogs in the kennel. Underlying factors such as immunosuppressant drugs during cancer therapy can exacerbate CaHV-1 induced tracheobronchitis in mature dogs (2). I have diagnosed few cases of pneumonia in adult dogs from South-East USA with wide spread infection of lungs confirmed by CaHV-1 fluorescent antibody test. There are recent reports of CaHV-1 associated acute respiratory disease in 3 breeds of dogs. The lesions included hemorrhagic rhinitis and tracheaitis. No other CIRDC pathogens were detected in this case report of CaHV-1 associated lung disease (3).

Ocular disease due to CaHV-1 has been extensively studied recently by Dr. Ledbetter at Cornell University, NY. The disease symptoms include conjunctivitis, blepharitis, keratitis, and recurrent eye infections. In newborn puppies, eye infections are bilateral associated with blindness, cataracts, optic nerve atrophy, and retinal degeneration. Dendritic conjunctival ulceration is strongly suggestive of CaHV-1 infection of eye. Spontaneous reactivations of CaHV-1 associated ocular infections are not common.

In adult dogs clinical cases of systemic CaHV-1 have been reported. Lesions include multifocal hepatic necrosis and intranuclear inclusions in liver, adrenal and small intestines. Role of CaHV-1 in neurologic disease has been documented.

In other canids, such as Iberian wolves, CaHV-1 infections have been recently reported.

CaHV-1 lesions include multifocal areas of necrosis and hemorrhage in lungs, liver, brain, and kidneys. Lesions in very young puppies include red spots on kidneys. CaHV-1 causes necrotizing vasculitis. Lymph nodes are swollen and spleen is enlarged.

Specimens of choice include fresh frozen kidneys, lungs and liver. For kennel diagnosis, CaHV-1-SN is a test of choice. Specimens should be submitted overnight to a diagnostic Laboratory for testing by direct fluorescent antibody test.

CaHV-1 is transmitted to newborn puppies in the birth canal. A breeding kennel can check the potential shedding of CaHV-1 just before whelping. If the virus is detected, caesarian section can be used to deliver the puppies to prevent transmission of the virus.

Role of CaHV-1 in canine diseases is expanding because of the availability of more sensitive tools such as CaHV-1 PCR on formalin-fixed tissue sections. If a kennel is experiencing fertility problems, testing for canine herpesvirus by SN test and canine Brucella card test are recommended. Serology indicates that dog has been exposed but may or may not be shedding the virus.

Antiviral therapy for CaHV-1 ocular disease includes Cidofovir ocular drops 0.5%, one drop twice daily. Administration of immune plasma intra-abdominally has been reported to provide some protection. Rearing the puppies in increased temperature (95F) and 50% relative humidity incubator reduces losses (4).

A vaccine for CaHV-1 is available in Europe but not in the USA. A titer of 1:4 on CaHV-1-SN is protective. Vaccination is performed towards the end of pregnancy and boosting at whelping. Vaccine has been reported to be effective and safe in pregnant bitches. CaHV-1 is susceptible to most disinfectants used for cleaning the contaminated surfaces.

REFERENCES

1.Evermann et al., 2011 Vet Clin Small Anim 41:1097-1120.
2. Kawakami et al., 2010, JCM, 48:1176-1181.
3. Kumar et al., 2015, Vet Pathol 52:681-687.
4. Ledbetter et al., 2015, J Ocul Pharmacol Ther 31: 642-649.

Current Status of Canine Parvovirus-2 Genetic Variants in the USA (2015)

Sanjay Kapil
Stillwater, OK

Canine parvovirus (CPV-2) is the most clinically significant virus pathogen of dogs. In spite of the availability of effective commercial vaccines, CPV-2 continues to be a problem in the USA and other countries, worldwide. Properties of parvovirus make them an ideal pathogen of dogs. All parvoviruses have a single-stranded DNA genome that allows a very high rate of mutation. Parvoviruses are non-enveloped and highly stable in environment allowing high levels of contamination and infectivity for long time. CPV-2 can multiple productively in the rapidly dividing cells of the intestinal crypts and also in the basal layer of the tongue epithelium.

CPV-2 is a disease of canidae, such as dogs, cats, and wild carnivores. The virus is highly contagious and survives in environment of dogs and on the dog hair coat from fecal contamination. The most common age range is 6 weeks to 6 months. Incubation period is about 1-2 weeks. The risk of exposure to CPV-2 is proportional to the potential contact with canine hosts. Most common areas of exposure include social venues such as doggy parks, contaminated surfaces, and wild life infected with CPV-2. A CPV-2-like virus has been detected in raccoons but its ability to infect dogs is not known.

Canine parvovirus-2 emerged as a new pathogen of dogs in 1978. In 1979, a new genetic variant designated CPV-2a was detected. In 1984, another genetic variant CPV-2b was detected. First case of a third genetic variant CPV-2c was first detected in the USA in 2006.

CPV-2 genotyping is a PCR-based procedure that classifies the CPV-2 isolates based on the amino acid at 426 position of the viral protein-2 (VP-2). If the codon at 426 is AAT (asparagine) in CPV-2 and CPV-2a; codon is GAT encoding aspartic acid in CPV-2b; and GAA encoding glutamic acid in CPV-2c genotype. I have observed another significant change in position 440 among CPV-2c isolates over the last 10 years (2006-2015). Codon at position 440 can be ACA or GCA. Amino acids 426 and 440 are both present on the top of the VP-2 protrusion of the CPV-2 capsid. Older dogs affected by CPV-2 tend to have GCA at position 440 of the VP-2.

After contacting CPV-2 present in contaminated feces, susceptible dogs begin multiplying and excreting the virus 3-4 days after exposure. CPV-2 excretion begins before clinical signs appear. All breeds of dogs appear to be susceptible to CPV-2c. However, smaller breeds of dogs were more commonly involved in my study. Peak of CPV-2 infections occur at 2 months of age because the maternal immunity is lost and then puppies become more susceptible to CPV-2 infections.

I began surveillance for genotypes of CPV-2 in 2006 after the detection of CPV-2c in the USA. CPV-2c has gradually become more predominant in the USA over the time span of 10 years as will be expected for newly introduced variant in the dog populations in the USA. CPV-2c is fitter to bind to the canine transferrin receptor based on in vitro studies. Based on reports from USA, Europe, Asia and South America, CPV-2c has been detected as a cause of diarrhea and vaccine failures in many parts of the world.

Clinical signs of CPV-2c are similar to all other CPV-2 cases. Intestines are the main target of the infection leading to enteritis and vomiting. However, veterinarians have reported that CPV-2c clinical signs can be more severe and the response to symptomatic therapy is less successful. The feces are normally yellow-gray and may not contain blood.

To confirm the clinical diagnosis, testing can be performed animal side using CPV-2 Lateral Flow immunoassays that are commercially available. These field tests have good specificity and limited sensitivity. If the testing is negative, the animal specimens can be sent to a diagnostic laboratory that offers direct FAT on small intestines and fresh tongue. In most cases (99%) the results match between intestines and tongue. The failure of diagnostic tests for CPV-2 correlates with emergence of newer CPV-2c genotypes because the SNAP test uses a monoclonal that may not capture the newer variants of CPV-2c. An inexpensive slide agglutination test has been reported but is not widely used.

CPV-2c is widely distributed in the USA. However, CPV-2c has not been detected in the cats in the USA. It is also possible to have both CPV-2c and CPV-2b detected in the same kennel and same litter of puppies.

Commercial vaccines available for canine parvovirus in the USA are either genetically CPV-2 or CPV-2b. Currently, no CPV-2c vaccine is available. The success of puppy vaccination for CPV-2 depends on various factors such as antibody titer of the bitch; amount of colostral antibody transfer in the first 24 hours of suckling when the gut is open to antibody absorption; age of last puppy vaccination; and the time when the puppy encounters the virulent CPV-2. Although commercial vaccines provide good cross-protection against CPV-2c, vaccine failures have been reported in isolates cases mainly due to CPV-2c. The current AAHA guidelines for puppy vaccination recommend that puppies receive a minimum of 3 vaccines at the intervals of 3-weeks. A final dose of CPV-2 vaccination at 15 weeks is recommended. Dogs should receive CPV-2 vaccines according to label guidelines written on the vaccines used. No vaccine protocol is perfect because maternal antibody interferes with the uptake of the CPV-2 vaccines. In spite of detail examination, I do not have any evidence that current commercial vaccines can cause CPV-2 associated clinical disease because most modified live vaccines for CPV-2 are attenuated and contain a small amount of attenuated CPV-2 that fails to cause lesions and significant clinical disease. It is critical that if dogs develop bloody diarrhea after vaccination that other potential pathogens of hemorrhagic enteritis are ruled out, such as Clostridium sp. by culture and PCR on the feces.

In an outbreak situation, the sick dogs should be isolated and bio security measures should be followed. CPV-2 is a very resistant virus and house hold bleach can be used as a disinfectant (1:10 dilution) and a minimum exposure of 10 minutes is recommended to disinfect the contaminated and cleaned surfaces.

REFERENCES

Kapil et al., 2007. J Clin Microbiol, 45:4044-4047.

Recent Advances in Antiviral Therapy for Dogs and Cats

Sanjay Kapil
Stillwater, OK

Virus infections are common and clinically significant in companion animals. These virus outbreaks cause huge demands for specific antiviral therapy. Most common and clinically significant viral infections in dogs include canine parvovirus and canine distemper virus. Most clinically significant viruses of cats include feline parvovirus, feline retroviruses and feline calicivirus. Despite the huge need for specific antivirals for companion animals, only few have been investigated and fewer have been approved in the USA and EU.

Currently, there are limitations in the discovery and development of antivirals because of huge cost of discovery and price of using them clinically in hospital settings. Most of these antivirals have been found to be useful only for single virus type and species of animals. Broad spectrum antivirals are not available. The discovery of new antivirals is hampered by need for rapid diagnostic tests. One of limitations of antivirals is potential side effects of these drugs. Because viruses are obligate intracellular pathogens, development of antivirals can be challenging.

The only approved antiviral for cats is feline omega interferon in UK. The only approved antiviral for cats in the USA is polyprenyl immunostimulant.

There are about 25 families of animal viruses with DNA and RNA genomes. All these viruses have unique mechanisms of replication. Due to this huge diversity of replication mechanisms, specific antivirals can be designed to target specific viruses.

DNA Viruses

Single Stranded DNA

Canine Parvovirus: Recombinant canine granulocyte colony stimulating factor has been reported to be clinically effective for CPV-2 induced neutropaenia. The treatment reduced the duration of hospitalization after CPV-2 infection. Oseltamivir treatment of dogs with CPV-2 had reduced WBC drops compared to untreated dogs with CPV-2 infections. No major side effects of Oseltamivir were observed.

Feline panleukopaenia is caused by a parvovirus. Feline omega interferon shows clinical efficacy against parvoviruses.

Double Stranded DNA

For ocular infections with canine herpes virus, one drop of 0.5% Cidofovir in both eyes, twice daily for 14 days significantly reduced the ocular CHV-1 infection. However, ocular toxicity was observed.

For ocular infections with feline herpes virus, famcicolvir therapy at a dose of 62.5 to 125 mg three times daily has been reported to be clinically effective. Famciclovir was well tolerated for duration of 4 months treatment. Oral therapy was found to be more convenient than topical therapy.

RNA Viruses

Canine distemper virus infected dogs can be treated with ribavarin. Ribavirin has been found to be effective in vitro. Ribavirin also inhibits the replication of human measles virus that is closely related to CDV.

Feline calicivirus causes significant morbidity and mortality in cats. In viro lithium chloride inhibits FCV replication. In vivo, mefloquine combination therapy with feline interferon has been found to have additive protective effects against FCV.

For feline leukemia virus, azidothymidine (AZT) at the dose of 30-60 mg per kilogram was used. Efficacy of AZT against FeLV is unclear because in addition to toxicity, clinical efficacy of AZT against FeLV was unclear.

Feline omega interferon has shown clinical efficacy against feline retroviruses.

Rabies virus is an important zoonosis. Ribavirin and alpha interferon were found to lack efficacy against rabies virus infections.

Approved Antivirals

Recombinant Feline Omega Interferon

Recombinant feline omega interferon is manufactured by Virbac and sold as Vibragen. It enhances antiviral immunity. It is licensed in UK against feline retroviruses (FeLV and FIV). It is given to cats more than 9-weeks of age. Injections are given at 0, 14 and 60 days. Each treatment cycle is 1MU/kg once daily for 5 days. It has been demonstrated to show clinical improvement against feline retroviruses.

Vibragen Omega has used at a concentration of 10MU/vial against CPV-2 infections in dogs. It was tested in dogs of one month of age infected with CPV-2. The mortality was reduced 4.4 to 6.4 times after Vibragen omega therapy. Suspension was injected SQ once daily for 5 consecutive days.

Side effects of Vibragen include increased temperature about 3-6 hours after injection, transitory vomiting, soft feces to diarrhea in cats, fatigue. Other side effects include slight decrease in WBC, platelets, erythrocytes and rise of alanine amino traneferase. Cats return to normal in about 1 week after last injection of Virbragen. Vaccination is contraindicated during Vibragen therapy.

Polyprenyl Immunostimulant

Polyprenyl immunostimulant is manufactured by Sass and Sass, TN, USA. Polyprenyl immunostimulant is an oral drug shown to have clinical efficacy against feline herpes virus and dry form of feline infectious peritonitis. It is a plant-based drug that promotes increased cytokine production. It was found to be safe and well tolerated in cats. PI has been given conditional license for the treatment of feline herpesvirus. PI has been used off-label in dry form of FIP. PI was not found to be effective in wet form of FIP. Drug is expensive but shows clinical efficacy against dry form of FIP.

REFERENCES

Caddy, 2014, Efficacy of veterinary antivirals in small animals, Vet times.

Why HGE is an Outdated Name: New Insights in the Pathogenesis of Acute Hemorrhagic Diarrhea Syndrome

Stefan Unterer, DVM, DECVIM
Munich, Germany

INTRODUCTION

Acute hemorrhagic diarrhea (AHD) in dogs can have numerous potential causes. However, in many cases no underlying cause for the AHD can be found and a tentative diagnosis of "hemorrhagic gastroenteritis" (HGE) is made. The term HGE was established in 1970 and defines a syndrome characterised by acute-onset vomiting, anorexia, and lethargy, progressing to hematemesis and severe, hemorrhagic diarrhea. Additionally, a significant increase in PCV is considered characteristic for this disorder. Allergic, hereditary, autoimmune, and infectious disorders have been proposed as causes. However, the exact pathogenesis of this syndrome remained unknown in the last 40 years. Recently, the term HGE was replaced by "acute hemorrhagic diarrhea syndrome" (AHDS). The aim of this lecture is to give an update of important aspects on the pathogenesis of AHDS in dogs.

SIGNALEMENT

AHDS occurs more commonly in young to middle-aged and small breed dogs. A recent study showed that body weight in dogs is strongly and inversely related to excitability and hyperactivity, which can be associated with stress. It is known that stress influences gastric motility and promotes intestinal bacterial overgrowth, which might lead to gastrointestinal symptoms. Therefore, increased excitability and its effect on the gastrointestinal tract might represent one explanation for the age and breed predisposition of dogs with AHDS.

LABORATORY CHANGES

A significant increase in PCV due to intensive fluid loss into the gut is frequently observed in dogs with AHDS, and some authors used a PCV ≥ 60% as inclusion criterion for this syndrome. However, a recently published study showed that the PCV exceeds 60% in only 31% of cases. In contrast, the PCV was within the reference range in 51.9% of dogs with AHDS. Therefore, a PCV < 60% cannot be used as a criterion to exclude AHDS. A stress leukogram can often be observed, but in almost half of the dogs, a left shift is present. This is most likely due to mucosal destruction and associated defence mechanisms and cytokine production. Other laboratory abnormalities (e. g. ALT elevation, metabolic acidosis) primarily reflect hypovolemia and subsequent hypoperfusion of organs.

HISTOLOGIC CHANGES OF THE GASTROINTESTINAL TRACT

Biopsies of stomach, duodenum, ileum, and colon taken *intra vitam* (ie, while alive) from 10 dogs with AHDS demonstrated acute intestinal mucosal necrosis (10/10) and intestinal neutrophilic infiltration (10/10) as characteristic histologic findings. As most dogs with AHDS vomit, which is hemorrhagic in 50% of cases, it was assumed that the stomach is involved in the disease process. Therefore, it was very interesting and unexpected that no gastric lesions could be detected macroscopically and histologically in dogs, in which an endoscopy of the gastrointestinal tract during the acute phase of AHDS was performed. Since mucosal lesions were restricted to the intestines, the old term "HGE," which still is used commonly in the current literature and implies involvement of the stomach, is inadequate.

A striking feature observed in the histologic sections of the intestinal biopsies of most dogs with AHDS was a dense layer of large rod-shaped bacteria adherent to necrotic villous tips. These bacteria were identified as *C. perfringens* by immunohistochemical staining and culture.

PATHOGENIC ROLE OF *C. PERFRINGENS* AND ITS TOXINS

A primary pathogenic role of *C. perfringens* in dogs with AHDS was suspected and supported by the facts that no other explanation for the necrosis of the superficial intestinal epithelium could be identified, and that the presence of bacteria and epithelial lesions was closely associated. In addition, similar lesions as detected in the intestine of dogs with AHDS are commonly observed in different species with *C. perfringens* infections. *C. perfringens* is a Gram-positive, anaerobic bacterium capable of producing toxins that cause different forms of enteritis and enterotoxaemia in various hosts. Influence of *C. perfringens* enterotoxin (CPE) could potentially explain the extensive fluid loss and intestinal necrotic lesions characteristically observed in dogs with AHDS and was suggested as toxin possibly responsible for diarrhea in dogs. Therefore, fecal samples of 54 dogs with AHDS were tested by qualitative CPE ELISA. CPE was detected significantly more often in the feces of dogs with AHDS than in the feces of dogs in the control group. However, the prevalence of CPE in the feces of dogs with AHDS was only 24%, and there was no difference in severity of clinical signs, duration of hospitalisation, or outcome between CPE-positive and -negative dogs with AHDS. Thus, the results of this study suggest that CPE does not play a significant role in dogs with AHDS.

Recently, a novel toxin in type A *C. perfringens* strains isolated from a dog with fatal hemorrhagic enteritis was identified. It was speculated that this pore-forming toxin (*netF*), which is related to the leukocidin/hemolysin superfamily, was responsible for the necrotic enterocolitis in this dog. To confirm this hypothesis, *C. perfringes* strains isolated from the duodenum of six dogs with acute hemorrhagic diarrhea and of two dogs, in which endoscopy was performed for unrelated disease, were genotyped. Additionally, PCR was performed for the pore forming toxins *netE* and *netF*. In all dogs, clostridial strains were categorized as *C. perfringens* type A. However, only the clostridial strains of dogs with AHDS and not those of the two control dogs encoded for pore forming toxins.

Although the number of evaluated dogs with AHDS and control dogs is low, it is currently believed that intestinal lesions of dogs with AHDS are caused by pore forming toxins produced by type A *C. perfringens* strains. Further studies are needed to confirm this hypothesis and to identify the cause of clostridial overgrowth and toxin release.

SUMMARY

In dogs with AHDS, mucosal lesions are restricted to the large and small intestines. Therefore, the term "hemorrhagic gastroenteritis" ("HGE") is misleading and should be replaced by "acute hemorrhagic diarrhea syndrome" ("AHDS"). Several studies demonstrated an apparent association between *C. perfringens* and the occurrence of acute hemorrhagic diarrhea. There is strong evidence that clostridial toxins are responsible for the intestinal epithelial necrosis in dogs with AHDS. Pore-forming toxins are capable to cause such lesions and every *C. perfringens* strain isolated from dogs with AHDS encoded for these toxins, which are designated as *netE* and *netF*.

REFERENCES

1. Burrows C. Canine Hemorrhagic Gastroenteritis. J Am Anim Hosp Assoc 1977;13:451-458.
2. Busch K, Suchodolski JS, Kuhner KA, et al. Clostridium perfringens enterotoxin and Clostridium difficile toxin A/B do not play a role in acute haemorrhagic diarrhoea syndrome in dogs. Vet Rec. England, 2015;253.
3. Cave NJ, Marks SL, Kass PH, et al. Evaluation of a routine diagnostic fecal panel for dogs with diarrhea. J Am Vet Med Assoc 2002;221:52-59.
4. Goldstein MR, Kruth SA, Bersenas AM, et al. Detection and characterization of Clostridium perfringens in the feces of healthy and diarrheic dogs. Can J Vet Res 2012;76:161-165.
5. Spielman B, Garvey M. Hemorrhagic Gastroenteritis in 15 Dogs. J Am Anim Hosp Assoc 1993;29:341-344.
6. Suchodolski JS, Markel ME, Garcia-Mazcorro JF, et al. The Fecal Microbiome in Dogs with Acute Diarrhea and Idiopathic Inflammatory Bowel Disease. PLoS ONE 2012;7:e51907.
7. Unterer S, Busch K, Leipig M, et al. Endoscopically visualized lesions, histologic findings, and bacterial invasion in the gastrointestinal mucosa of dogs with acute hemorrhagic diarrhea syndrome. J Vet Intern Med 2014;28:52-58.
8. Unterer S, Lechner E, Mueller RS, et al. Prospective study of bacteraemia in acute haemorrhagic diarrhoea syndrome in dogs. Vet Rec. England, 2015;309.
9. Unterer S, Strohmeyer K, Kruse BD, et al. Treatment of aseptic dogs with hemorrhagic gastroenteritis with amoxicillin/clavulanic acid: a prospective blinded study. J Vet Intern Med 2011;25:973-979.
10. Mehdizadeh Gohari I, Parreira VR, Nowell VJ, et al. A novel pore-forming toxin in type A Clostridium perfringens is associated with both fatal canine hemorrhagic gastroenteritis and fatal foal necrotizing enterocolitis. PLoS One 2015 Apr 8;10(4):e0122684.
11. Guard B, Barr J, Steiner J, Gaschen F et al. Effects Of A 2-Week Treatment With Metronidazole On The Fecal Metagenome Of Healthy Dogs. Journal of Veterinary Emergency and Critical Care 2015, 25: 1-32.

Current Concepts in the Treatment of Acute Hemorrhagic Diarrhea Syndrome: A Shift Away from Antibiotics

Stefan Unterer, DVM, DECVIM
Munich, Germany

INTRODUCTION

Antibiotics have been generally recommended to treat acute hemorrhagic diarrhea (AHD) in dogs. Hemorrhagic diarrhea obviously reflects destruction of the protective intestinal epithelial barrier, and loss of an intact intestinal mucosa can result in bacterial translocation with bacteremia and potentially sepsis. Without any doubt, in patients with hemorrhagic diarrhea due to canine parvovirus infection, antibiotics are-life saving, because the combination of intestinal damage and neutropenia commonly results in sepsis with Gram-negative bacteria from the intestinal microbiota.

INDICATIONS FOR ANTIBIOTICS IN DOGS WITH ACUTE HEMORRHAGIC DIARRHEA SYNDROME

In dogs with acute hemorrhagic diarrhea syndrome (AHDS), necrosis of the intestinal mucosal surfaces is present. In addition, intestinal bacterial overgrowth with *Clostridium perfringens* strains was demonstrated in dogs with AHDS. Although these are risk factors for bacterial translocation, the actual incidence of bacteremia in dogs with AHDS was unknown. A prospective study including 87 dogs with AHDS and 21 healthy dogs showed for the first time that the incidence of bacteremia in dogs with AHDS is low and not different from those of healthy control dogs. Moreover, in dogs with AHDS bacteremia neither influenced the clinical course nor the survival. These findings led to the assessment that dogs with a severe destruction of the intestinal barrier still can effectively clear bacteria entering the portal blood as long as the liver and immune system are functioning as a cellular back-up system. Therefore, antimicrobial administration as a routine treatment in dogs with AHDS has to be questioned.

In a prospective placebo-controlled treatment study, it could be confirmed that in dogs with aseptic AHDS, antibiotics do not change outcome or time to recovery. In this study, no obvious adverse effects due to the administration of amoxicillin/clavulanic acid were observed. This could lead to the question – why should antibiotics not generally be used in every patient with hemorrhagic diarrhea as "prophylactic" treatment?

CONTRAINDICATIONS FOR ANTIBIOTICS IN DOGS WITH ACUTE HEMORRHAGIC DIARRHEA SYNDROME

One reason not to use antibiotics is the fact, that inappropriate usage of antibiotics can cause disruption of the protective intestinal flora. Some of the enteropathogenic bacteria can only colonize the gut if the normal intestinal microbiota is disturbed or absent.

There is strong evidence that clostridial toxins are associated with AHDS in dogs. Toxin production can be stimulated by antibiotic treatment. This is the reason why gastrointestinal infections in humans with Shiga toxin-producing *E. coli* are not treated with antimicrobials. Therefore, since injudicious administration of antimicrobials could be more harmful than beneficial in individual patients, antibiotic treatment is only recommended in dogs with suspected or confirmed bacterial-associated diarrhea manifesting with systemic signs of illness.

Another reason to restrict unnecessary antibiotic treatment is that resistance to antimicrobials is increasing among bacteria isolated from pets, and inappropriate usage of prophylactic antibiotics, especially b-lactam antibiotics, can lead to multi-resistant bacteria, including methicillin-resistant *Staphylococcus aureus*, which affects survival in human patients. Although side effects, such as nausea, vomiting, diarrhea, abdominal pain, and headaches are commonly experienced by human patients with antibiotic treatment, it is believed by many clinicians that dogs tolerate antibiotics very well. However, first studies performed in healthy pet dogs showed that 9/16 dogs developed loose stools when receiving metronidazole for 2 weeks.

CRITERIA FOR SYSTEMIC BACTERIAL INFECTION IN DOGS WITH ACUTE HEMORRHAGIC DIARRHEA

Ruling out an enteral infection with pathogenic bacteria is challenging, and bacterial translocation is a potentially life-threatening complication. Especially in significantly dehydrated dogs with AHDS, it is difficult to assess if sepsis is present. Proposed criteria for sepsis, such as hypothermia, tachycardia, and tachypnea, are common physical examination findings in patients in hypovolemic shock and not specific indicators for bacteremia. The following parameters should be used as indication for antibiotic treatment in dogs with AHDS:

• rectal temperature	> 39.5°C
• tachycardia after rehydration and pain management	> 120/min
• WBC	< 4 oder > 25 x 10^9/l
• banded neutrophils	> 1.5 x 10^9/l

Beside these criteria, evaluation of the clinical course and response to fluid therapy is very important to detect those few patients that require antibiotic treatment. With adequate fluid therapy (crystalloids; fluid amount depended on dehydration, maintenance demands, and ongoing losses) a rapid improvement of clinical signs characteristically can be observed during the first 48 hours without antibiotic treatment **(Table 1)**. Dogs, that are not adequately responding to fluid therapy after 4 hours (i. e. alert mental status, heart and respiratory rate dropping into normal range, adequate production of urine) should be categorized as "complicated AHDS" patients and prompt the clinician to look for other underlying causes (i. e. acute pancreatitis, viral enteritis). In patients with suspected bacterial infection with an enteropathogen or bacterial translocation from the intestinal tract resulting in sepsis, bacterial cultures from blood, aspirated material from enlarged abdominal lymph nodes, bile or even feces might help to isolate the responsible pathogen and to select an appropriate antibiotic. Until sensitivity testing is available, suspected septic patients should be treated with intravenously administered antibiotics covering primarily Gram-negative and anaerobic bacteria (e. g. fluoroqinolones and metronidazole).

SUMMARY

In summary, most dogs with acute hemorrhagic diarrhea are significantly dehydrated and show a rapid clinical improvement with adequate fluid therapy. Septic events are uncommon, and prognosis is good even in patients without antibiotic treatment. Close patient monitoring is essential to identify those dogs, which might require antibiotic treatment.

Table 1. Characteristic clinical course in dogs with AHDS

Day	Median	Range
AHD index day 1	12.5	3.0–16.0
AHD index day 2	3.0	0.0–12.0
AHD index day 3	2.0	0.0–12.0
AHD index day 4	2.0	0.0–11.0
AHD index day 5	2.0	0.0–7.0
AHD index day 6	1.0	0.0–5.0
AHD index day 7	1.0	0.0–5.0

Acute hemorrhagic diarrhea (AHD) index is calculated as the sum of the scoring of the parameters attitude, appetite, vomiting, stool consistency, stool frequency, and dehydration. Each parameter was scored (0 = normal, 1 = mild, 2 = moderate, 3 = severe) in 30 patients on any individual treatment day, and the sum of scores (max. 18) yielded a total cumulative score.

REFERENCES

1. Burrows C. Canine Hemorrhagic Gastroenteritis. J Am Anim Hosp Assoc 1977;13:451-458.
2. Busch K, Suchodolski JS, Kuhner KA, et al. Clostridium perfringens enterotoxin and Clostridium difficile toxin A/B do not play a role in acute haemorrhagic diarrhoea syndrome in dogs. Vet Rec. England, 2015;253.
3. Cave NJ, Marks SL, Kass PH, et al. Evaluation of a routine diagnostic fecal panel for dogs with diarrhea. J Am Vet Med Assoc 2002;221:52-59.
4. Goldstein MR, Kruth SA, Bersenas AM, et al. Detection and characterization of Clostridium perfringens in the feces of healthy and diarrheic dogs. Can J Vet Res 2012;76:161-165.
5. Spielman B, Garvey M. Hemorrhagic Gastroenteritis in 15 Dogs. J Am Anim Hosp Assoc 1993;29:341-344.
6. Suchodolski JS, Markel ME, Garcia-Mazcorro JF, et al. The Fecal Microbiome in Dogs with Acute Diarrhea and Idiopathic Inflammatory Bowel Disease. PLoS ONE 2012;7:e51907.
7. Unterer S, Busch K, Leipig M, et al. Endoscopically visualized lesions, histologic findings, and bacterial invasion in the gastrointestinal mucosa of dogs with acute hemorrhagic diarrhea syndrome. J Vet Intern Med 2014;28:52-58.
8. Unterer S, Lechner E, Mueller RS, et al. Prospective study of bacteraemia in acute haemorrhagic diarrhoea syndrome in dogs. Vet Rec. England, 2015;309.
9. Unterer S, Strohmeyer K, Kruse BD, et al. Treatment of aseptic dogs with hemorrhagic gastroenteritis with amoxicillin/clavulanic acid: a prospective blinded study. J Vet Intern Med 2011;25:973-979.
10. Mehdizadeh Gohari I, Parreira VR, Nowell VJ, et al. A novel pore-forming toxin in type A Clostridium perfringens is associated with both fatal canine hemorrhagic gastroenteritis and fatal foal necrotizing enterocolitis. PLoS One 2015 Apr 8;10(4):e0122684.
11. Guard B, Barr J, Steiner J, Gaschen F et al. Effects Of A 2-Week Treatment With Metronidazole On The Fecal Metagenome Of Healthy Dogs. Journal of Veterinary Emergency and Critical Care 2015, 25: 1-32.

How to Modulate the GI Microbiome: Effect of Diet, Antibiotics and Probiotics

Jan S. Suchodolski, DMV, PhD, AGAF, DACVM (Immunology)
College Station, TX

INTRODUCTION

The intestinal microbiota is the collection of all live microbes in the GI tract. It is estimated that the intestine harbours several thousand bacterial phylotypes and a total microbial load between 10^{12}-10^{14} cells, about 10 times the number of cells composing the host body, and 100 times the number of host genes. It is obvious that this tremendous bacterial mass has a crucial impact on host health. The resident microbiota provides nutritional substrates to the host (e.g., short chain fatty acids), stimulates mucosal growth, and protects the host from invasion by pathogens (through competition for oxygen, nutrients and for epithelial adhesion sites). For example, nutrient sources such as complex carbohydrates (e.g., starch, cellulose, pectin) are fermented by bacteria, resulting in the production of short chain fatty acids (SCFA). These act as energy sources for the host, regulate intestinal motility, and are important growth factors for epithelial cells. SCFA also have direct anti-inflammatory properties through expansion of immunoregulatory lymphocytes. Other bacterially-derived metabolites such as indole, a byproduct of tryptophan degradation, and secondary bile acids are also immuno-modulatory, thereby maintaining immune homeostasis and strengthening intestinal barrier function. These beneficial effects of the gut microbiota reach beyond the GI tract.

The intestinal microbiota also plays a role in various GI disorders. Invading pathogens may cause acute or chronic diarrhea. Furthermore, the resident microbiota has been implicated in chronic disorders such as inflammatory bowel disease (IBD) in dogs. Historically, *Lactobacillus* and *Bifidobacterium* spp. were considered the most important beneficial bacterial groups in the intestine of dogs and cats. However, recent studies have shown that members of Clostridium clusters XIVa and IV (i.e., *Dorea* spp., Lachnospiraceae, *Ruminococcus* spp., *Faecalibacterium* spp., and *Roseburia* spp.) are more consistently depleted in humans and dogs with IBD and acute colitis, suggesting that these groups are highly important in maintaining intestinal homeostasis. A therapeutic modulation of the intestinal ecosystem is a reasonable approach in animals with GI disease. Changes in the intestinal ecosystem may results in a reduced load of stimulating antigens or in the creation of an environment within the GI tract that allows a more effective utilization of ingesta. However, the exact effects of the various treatments such as antibiotics, probiotics, and diets on the intestinal microbiota are still largely unknown. Use of metagenomics and metabolomics tools will help to elucidate these effects and may aid in the decision making process as to which treatment approaches are best suited for individual patients.

DYNAMICS OF THE INTESTINAL MICROBIOTA

Host genetics and the environment are the most important factors influencing the overall composition of the intestinal ecosystem. Within hours after birth, the sterile intestine becomes colonized by bacteria present in the birth canal, milk, and surrounding environment. The qualitative composition is quite stable over long periods of time, but the proportions of individual bacterial groups can show substantial inter-day variation.

While there are specific bacterial genera that constitute the basic intestinal microbiome, each dog and cat possess a very unique microbial profile on a species and strain level, with only a minor overlap of bacterial species between individuals. Therefore, changes in the intestinal microbiota in response to treatment are usually very individualized. Environmental factors such as antibiotic administration and dietary changes in dogs and cats cause changes in microbial groups, but the microbiota is typically resilient to these changes and returns to its pre-treatment state (with the exception of antibiotic treatment which may lead to prolonged depression of specific bacterial groups for several months). Long term treatment strategies for modulating the microbiota are typically necessary. New functional studies (based on shotgun sequencing of genomic DNA) suggest the presence of a "core microbiome" in the intestine of dog and cats, because despite obvious differences in bacterial phylotypes between individuals, the microbiome of each individual appears to have similar gene content and therefore similar functions.[1,2] Furthermore, a functional redundancy exists in the GI tract. Various members of the intestinal ecosystem perform similar functions, and if one bacterial group is displaced because of for example antibiotic therapy, other members of the ecosystem take over the function, thus maintain a stable ecosystem function. Therefore, correlation of treatment with microbiota shifts and changes in metabolic function remains challenging.

MODULATING THE INTESTINAL MICROBIOTA

Due to vast inter-individual differences in the intestinal microbiota it remains challenging to define a normal and balanced intestinal ecosystem. Dietary change, treatment with antibiotics or administration of pre- and/or probiotics are commonly associated with changes in the microbial composition. However, these changes are largely individualized for each animal. These changes are rarely associated with one specific bacterial phylotype that is consistently altered in every individual. The vast majority of microbial groups within the ecosystem remain typically uninfluenced with diet or nutraceuticals.

Antibiotic administration, such as tylosin or metronidazole has proven to be useful in treatment of GI disease. Tylosin is beneficial in dogs with chronic colitis or presumed small intestinal dysbiosis.[3] Enrofloxacin ameliorates clinical signs and improves histologic lesions in dogs with granulomatous colitis.[4] The effect of antibiotics is most likely a combination of decreases in total bacterial load, changes in specific bacterial groups and direct immunomodulation. For example, metronidazole increases *Bifidobacterium* populations and reduces the colonic oxidative damage to proteins.[5] Tylosin has been shown to increase

Enterococcus populations and this has been correlated with clinical outcome. However, also a direct anti-inflammatory effect of tylosin has been proposed. This anti-inflammatory effect has been speculated to be due to the modulation of cyclooxygenase-2, nitric oxidase synthase, and several cytokines.[6]

Antibiotic usage, however, has a very pronounced effect on the intestinal microbiome and may disrupt the microbial ecosystem for prolonged periods of time (weeks to months). In one study evaluating the fecal microbiota of healthy humans, approximately 30% of all bacterial taxa were affected, some of them for up to six months.[7] Similarly, administration of tylosin for 14 days led to significant modifications in the jejunal microbiota of dogs, with some bacterial groups depressed for more than then study period of 14 days.[8] Highly individualized responses for some bacterial groups were observed in each animal. Furthermore, recent controlled studies are appreciating that antibiotics cause severe disruptions in major metabolic pathways, including bile acid metabolism and increase oxidative stress. Furthermore, these metabolic changes are associated with clinical signs of GI disease in a subset of animals. For example, 14 day administration of metronidazole to healthy pet dogs led to development of diarrhea in 9/16 dogs while on metronidazole, which resolved after the end of administration.[9] Similarly, 13/18 dogs that received intramuscular injections of lincomycin developed diarrhea.[10] Also, in a recent study, 34 cats were receiving amoxicillin-clavulanate twice daily for 7 days, and the majority of cats developed signs of GI disease with worsening of fecal scores.[11] Furthermore, new epidemiological data from human medicine would suggest that antibiotic induced dysbiosis in early childhood is clinically relevant, as it is one of the major risk factors for the development of chronic diseases such as allergies and IBD.[12] Therefore, antibiotics usage and its potential side effects requires careful study in dogs and cats.

Probiotics are currently defined as live microorganisms, which when administered in adequate amounts confer a health benefit on the host. Probiotic strains interact with the host on various levels. Colonization with probiotic bacteria modulates gene expression within the host and also affects the immune system. Within the gut lumen, probiotic strains have also direct interactions with commensal bacteria. Probiotics can help breakdown complex nutrients in the diet, thereby releasing metabolites that are taken up by other bacteria, thus providing nutrients for their growth. Probiotics can also function by acting as growth inhibitors for other bacteria. This can be accomplished through competition for specific nutrients and mucosal adhesion sites, production of acids and antimicrobials, and stimulation of intestinal immune response. The impact of probiotics on the composition of the gut microbiota was thought to be one of the major mechanisms of their beneficial action. Various studies have evaluated the composition of the gut microbiota with various techniques, including traditional cultivation methods, molecular based tools based on PCR of FISH for specific bacterial groups, and most recently high-throughput sequencing of 16S rRNA genes and metagenomics. The observed changes in the microbiota are depending on the methodology used and the evaluated localization within the gut. Generally, studies using metagenomics approaches have illustrated that probiotics can induce some microbiota changes in the large intestine, but these changes typically are minor and transient, persisting only a few days after the end of administration. Probiotics can lead to a transient increase in the administered target species, but this has usually an insignificant impact on the entire intestinal microbiome. In one study, Lactobacillales increased from 1% to 2.5% of total bacteria after administration of a multi-species probiotic containing several different *Lactobacillus* spp.[13] Furthermore, probiotics are typically eliminated from the intestine within a few days after end of administration. Therefore, administration of high doses over prolonged periods of time is usually required to maintain viable counts of probiotic species. The ability to detect an administered probiotic in feces of healthy dogs is dose dependent.[14] After administration of 5×10^{11} cfu, the strain was recovered in 4/4 dogs but only in 4/8 dogs when administered at commonly used dosages of 10^8 cfu.[15] However, studies that have looked a mucosa-adherent microbiota have identified major changes induced to probiotic administration. For example, after administration of VSL#3 to mice, major changes in ileal microbiota were noticed using FISH analysis.[14]

Diet is considered one of the most important modulators of the intestinal microbiota. However, we are still in the process of understanding which components of the diet are able to induce shifts in the microbiota. It is also challenging, as some dietary components are able to induce phylogenetic changes that are readily detectable in feces while other dietary components may induce functional changes that are difficult to assess with standard methods. Pronounced increases in macronutrients such as protein and fat content are able to induce shifts in the composition of the microbiota. For example, two different diets differing in protein content 34% vs. 53% of dry matter led to almost complete altered microbiota structure.[16] Similarly, diets differing in fat content (11% vs. 42% of dry matter) also led to complete change in microbiota structure as evidenced by next generation sequencing.[17] Also, dogs fed a BARF diet had a complete different microbiota structure compared to dogs fed a commercial kibble diet, most likely due to significant differences in protein, fat, and fiber content (unpublished data). Interestingly, a hydrolysed protein diet did not alter the fecal microbiota of healthy pet dogs, even after feeding for 6 weeks.[9] This was most likely due to the fact that the hydrolysed soy based diet did not differ significantly in macronutrient content from the original diets. Similarly, fibre in form of prebiotics, added to diets to enhance the growth of endogenous microorganisms, has typically only a minor impact on the entire ecosystem, as the administered prebiotics fulfil only some of the nutrient requirements for their target bacteria. However, some functional alterations, for example increases in butyrate may be observed.

CONCLUSIONS

It remains difficult to provide a universal algorithm for the modulation of the microbiota. Every animal may require an individualized approach consisting of combinations of dietary therapy, antibiotics and/or nutraceuticals. Furthermore, changes in microbiota due to diet and other modulators may escape when only fecal samples are assessed. Therefore, comprehensive studies evaluating the microbiota and the metabolome, as well as the host response via transcriptome will be needed to better understand what components of antibiotics, probiotics, or diets have the most important impact on the host.

REFERENCES

1. Swanson KS, Dowd SE, Suchodolski JS, et al. Phylogenetic and gene-centric metagenomics of the canine intestinal microbiome reveals similarities with humans and mice. *ISME J* 2011 5(4):639-49.
2. Barry KA, Middelbos IS, Vester Boler BM, et al. Effects of dietary fiber on the feline gastrointestinal metagenome. *J Proteome Res* 2012;11:5924-5933.
3. Kilpinen S, Spillman T, Syrja P, et al. Effect of tylosin on dogs with suspected tylosin-responsive diarrhea: a placebo-controlled, randomized, double-blinded, prospective clinical trial. *Acta Veterinaria Scandinavica* 2011;53 1-10.
4. Mansfield CS, James FE, Craven M, et al. Remission of Histiocytic Ulcerative Colitis in Boxer Dogs Correlates with Eradication of Invasive Intramucosal Escherichia coli. *J Vet Intern Med* 009 23(5):964-9
5. Vasquez N, Suau A, Magne F, et al. Differential effects of Bifidobacterium pseudolongum strain Patronus and metronidazole in the rat gut. *Appl Environ Microbiol* 2009;75:381-386.
6. Cao XY, Dong M, Shen JZ, et al. Tilmicosin and tylosin have anti-inflammatory properties via modulation of COX-2 and iNOS gene expression and production of cytokines in LPS-induced macrophages and monocytes. *Int J Antimicrob Agents 2006 May;27(5):431-8.*
7. Dethlefsen L, Eckburg PB, Bik EM, et al. Assembly of the human intestinal microbiota. *Trends Ecol Evol* 2006;21:517-523.
8. Suchodolski JS, Dowd SE, Westermarck E, et al. The effect of the macrolide antibiotic tylosin on microbial diversity in the canine small intestine as demonstrated by massive parallel 16S rRNA gene sequencing. *BMC Microbiol* 2009;9:210.
9. Olson E HJ, Waddle M, Steiner JM, Suchodolski JS, Gaschen F. Evaluation of the effects of a 2-week treatment with metronidazole on the fecal microbiome of healthy dogs. *J Vet Intern Med* 2015;29:1184-1185.
10. Aktas M BM, Ozkanlar Y. Efficacy of Saccharomyces boulardii as a probiotic in dogs with lincomycin induced diarrhoea. *Bull Vet Inst Pulawy* 2007;51:365-369.
11. Torres-Henderson C SS, Caress A, Olea-Popelka F, Lappin M. Effect of Enterococcus Faecium Strain SF68 on Gastrointestinal Clinical Signs of cats administered Amoxicillin-Calvulanate. *J Vet Intern Med J Vet Intern Med* 2015;29:1183
12. Ungaro R, Bernstein CN, Gearry R, et al. Antibiotics associated with increased risk of new-onset Crohn's disease but not ulcerative colitis: a meta-analysis. *Am J Gastroenterol* 2014;109:1728-1738.
13. Garcia-Mazcorro JF, Lanerie DJ, Dowd SE, et al. Effect of a multi-species synbiotic formulation on fecal bacterial microbiota of healthy cats and dogs as evaluated by pyrosequencing. *FEMS Microbiol Ecol* 2011;78:542-554.
14. Mar JS, Nagalingam NA, Song Y, et al. Amelioration of DSS-induced murine colitis by VSL#3 supplementation is primarily associated with changes in ileal microbiota composition. *Gut Microbes* 2014;5:494-503.
15. Weese JS, Anderson ME. Preliminary evaluation of Lactobacillus rhamnosus strain GG, a potential probiotic in dogs. *Can Vet J* 2002;43:771-774.
16. Hooda S, Vester Boler BM, Kerr KR, et al. The gut microbiome of kittens is affected by dietary protein:carbohydrate ratio and associated with blood metabolite and hormone concentrations. *Br J Nutr* 2013;109:1637-1646.
17. Bermingham EN, Young W, Kittelmann S, et al. Dietary format alters fecal bacterial populations in the domestic cat (Felis catus). *Microbiologyopen* 2013;2:173-181.

Cytokines and Inflammatory Mediators in Canine IBD

Karin Allenspach, Dr.med.vet. FVH, DECVIM-CA, PhD, FHEA
London, UK

PATHOGENESIS OF CANINE IBD: THE INTERPLAY OF MUCOSAL IMMUNE SYSTEM WITH MICROBIOTA

Recent research in mouse models and human beings with Inflammatory Bowel Disease (IBD) has shown that the basis of the inflammation in the gut is triggered by the microbiome interacting with the gut immune system, which leads to the increase of inflammatory mediators in affected areas. In the dog, several studies have focused on trying to identify which factors lead to the inflammation seen in the intestine, and which mediators are particularly important. The general hypothesis states that the gut epithelium detects the presence of both commensal and pathogenic bacteria via Pattern Recognition Receptors (PRRs) [1]. This group contains Toll-like receptors (TLRs) and Nucleotide Binding Oligomerisation Domain-like receptors (NODs) [2]. These receptors recognize specific molecules termed Pathogen-Associated Molecular Patterns (PAMPs), which are conserved molecules found on bacteria and other infectious agents. Recognition of PAMPs by TLRs is a crucial part of the innate immune response to invading bacteria in the gut and initiates a complex intracellular signaling pathway culminating in the activation of the transcription factor NF-κB [3]. This results in the transcription and secretion of a variety of pro- and anti-inflammatory cytokines and chemokines from the cell bearing the TLR. If successful, immune cells activated by this response clear offending pathogens. Despite the fact that commensal and pathogenic bacteria share PAMPs, the immune system remains unresponsive towards commensal organisms present in the intestinal lumen in the normal case scenario [4]. The inflammatory response which is normally only seen as a reaction towards pathogenic bacteria breaching the intestinal barrier is similar to the response seen in the mucosa of dogs affected with IBD. However, the response in IBD occurs in the absence of pathogens. It is believed that the innate immune system reacts to normal commensals in the intestinal lumen as if they were pathogens. Several areas of these pathways have been investigated in dogs with IBD and are discussed in the paragraphs below.

EVIDENCE OF INNATE IMMUNITY HYPER-RESPONSIVENESS IN CANINE IBD

TLRs have been shown to be upregulated in the intestine of human beings with Crohn's disease and ulcerative colitis. This may be either a consequence of the ongoing stimulation of TLRs by an altered microbiota or it may be a causal factor contributing to the pathogenesis of disease. Most human studies show that the mRNA and protein expression of TLR2 as well as TLR4 are increased in the intestine of people with active IBD [5, 6]. In a recent clinical study in dogs, it was shown that dogs of any breed with clinically severe, active IBD express higher levels of TLR2 receptors in the duodenum compared to healthy dogs when measured by real-time PCR in whole endoscopic biopsies [7]. In addition, TLR2 expression was correlated with the clinical severity of IBD using the Canine Chronic Enteropathy Clinical Activity Index (CCECAI). However, TLR4 expression levels were similar to those in healthy canine intestine. Other studies have found that only a subgroup of dogs with IBD (the ones responding only to steroid administration) showed an increased expression of TLR2, TLR4 and TLR9 compared to healthy intestines when expression was measured by real-time PCR [8]. In further studies looking at German Shepherd dogs with IBD, it was found that TLR4 expression was 60-fold higher in the duodenum, ileum and colon of dogs with IBD compared to samples from healthy dogs, however, TLR2 and TLR9 were similarly expressed as in healthy dogs [9]. These data show that it is important to look at similar phenotypes of dogs when choosing cases for such studies, as the results will vary depending on the severity of disease, the treatment response in the cases and the specific breed of dogs. In addition, care must be taken to compare studies using real-time PCR as the standardization method used depends on the reference genes, which need to be carefully chosen for each study. TLR2 has recently been shown to be overexpressed in the diseased intestine in mouse models of IBD[10]. TLR2 in this context is implicated in the homeostasis and repair of intestinal tissue after injury. It is therefore possible that the high expression of TLR2 found in dogs with IBD in the studies mentioned above could be a marker of intestinal inflammation and its physiological action is to downregulate ongoing inflammation.

TLR5 expression was seen to be consistently downregulated in the intestine of German Shepherd Dogs with IBD as compared to healthy dogs [9]. In mice and human beings, TLR5 is highly expressed in the healthy small intestine, with CD11c+ dendritic cells (DCs) in the lamina propria mucosae expressing most TLR5 [11]. It is believed that this tolerogenic phenotype of DCs induces T regulatory cells and stimulates the production of anti-inflammatory cytokines such as IL-10 in response to flagellin [12]. In contrast, in intestinal inflammation characterized by the upregulation of Th1– and Th17 cytokines, CD11c- DCs express low levels of TLR5 but instead high levels of TLR4 [12]. In this context, TLR4 is thought to be upregulated to compensate for the low TLR5 expression. It could be speculated that the differentially low expression of TLR5 and very high expression of TLR4 seen in the intestine in the GSD of our study indicates a similar compensatory role of TLR4, as gram-negative flagellated bacteria can also be recognized through binding of LPS by TLR4.

CYTOKINE PROFILE IN THE INTESTINES OF DOGS WITH IBD

For years, many research groups have tried to identify a specific pattern of pro-inflammatory cytokines in the intestines of dogs with IBD, without convincing evidence for either Th1, Th2, or Th17 bias. A meta-analysis of available studies looking at cytokine expression in the intestine of dogs with IBD was performed. Results showed that the intestines of healthy dogs and those with IBD express numerous cytokines and that a pro-inflammatory expression profile is not a feature of small or large-intestinal IBD[13]. The meta-analysis data included 158 dogs characterized as healthy (n = 45), diarrheic nonIBD dogs (n = 6), nonresponders (n = 2), small-intestinal IBD (n = 41), colonic IBD (n = 25), and chronic enteropathy (n = 39). German shepherd dogs were overrepresented. Healthy dogs showed mRNA expression for most cytokines including IL2, IL4, IL5, IL10, IL12, IFNgamma, TNFalpha, and TGFbeta. Only IL12 mRNA expression was increased consistently in small-intestinal IBD, whereas IBD colitis lacked consistent patterns of expression. Current evidence therefore suggests that dogs with IBD fail to express a predominant Th1- or Th2 cytokine bias in inflamed mucosa. The aim of another recent study was to assess Th17 gene expression of signature cytokines in duodenal tissues from 18 German shepherd dogs with IBD (group 1), 33 dogs of other breeds with IBD (group 2) and 15 control dogs (group 3)[14]. Relative quantification of IL-17A, IL-22, IL-10, IFNγ and TGFβ was performed. Expression of IL-17A was significantly lower in groups 1 and 2 compared to group 3 ($p=0.014$), but no difference in the expression of IL-22 ($p=0.839$), IFNγ ($p=0.359$), IL-10 ($p=0.085$) or TGFβ ($p=0.551$) across groups was detected. Thus, it was concluded that no clear evidence for the involvement of Th-17 signature cytokines in canine IBD at the mRNA level could be shown.

EVIDENCE OF INFLAMMASOME ACTIVATION IN DOGS WITH IBD

Inflammasomes are intracellular multi-protein complexes that coordinate the maturation of Interleukin (IL)-1β and IL-18 in response to pathogens and metabolic danger signals[15]. In human beings with IBD, both of these cytokines are implicately involved in the pathogenesis of the disease, with numerous studies showing elevated levels in inflamed mucosa and decreased levels after successful treatment[16]. Recent research has shown imbalances in the mRNA expression of several signaling proteins of the inflammasome pathway in dogs with chronic enteropathies on the mRNA level, therefore suggesting that IL-1beta could play a similarly important role in canine IBD[17]. Furthermore, protein expression was assessed in another recent study. IL-1β expression in the duodenal mucosa of dogs with food-responsive chronic enteropathies (FRD) was assessed before and after treatment in dogs and in healthy dogs using immunohistochemistry[17]. Nine dogs with FRD were included in the study and treated with hydrolysed diet for 6 weeks. Biopsies from 5 of these dogs were available after treatment. Clinical scores were obtained before and after treatment. Duodenal biopsies from 11 healthy Beagles were used as controls. Immunohistochemistry was performed using a polyclonal rabbit anti-human primary antibody, and cells were counted in 10 randomly selected villous regions and in 10 regions of the sub-villous lamina propria (LP) and expressed as positive cells per high power field (hpf). The total count of IL-1β producing cells was increased in FRD dogs compared to controls, and decreased after treatment with elimination diet. Interestingly, this was not reflected by a quantitative change of IL-1β protein producing cells in the villi but was mostly caused by an increase in positive cells in the LP, which did not entirely return to baseline values after treatment. The decrease in IL-1β positive cell in the LP correlated with the clinical activity index. Although only a pilot study, these results indicate that IL-1beta expression in intestinal biopsies from dogs with IBD should be investigated further in other groups of dogs with IBD, such as more severely affected dogs and/or dogs that need immunosuppressive treatment to keep their clinical signs under control.

RAGE/S100A12 AS INFLAMMATORY MEDIATORS IN CANINE IBD

The receptor for advanced glycation end products (RAGE) and the S100A12 protein family combine some characteristics of potentially ideal biomarker (i.e., organ- specificity of expression and secretion, disease-specificity, measurability without temporal delay, minimal or non-invasiveness, and stability in specimens under clinical conditions). S100A12 has cytokine-like effects functioning as an endogenous danger-signaling molecule that is associated with neutrophil and macrophage 82 activation, and plays an important role in acute and chronic inflammatory responses. S100A12 also plays a role in the host defense against microorganisms and parasites and has chemotactic properties. After its release from neutrophils/ monocytes due to cell damage, infection, or inflammation, S100A12 functions as a ligand for the receptor for advanced glycation end products, a pattern recognition receptor, and activates the nuclear factor kappa B signal transduction pathway. This pathway has been proposed to be involved in the amplification and perpetuation of the traditional inflammatory response and to be linked to the pathogenesis of inflammatory disorders in humans. In human medicine, S100A12 has been shown to be a sensitive and specific marker of localized inflammatory processes, such as gastrointestinal inflammation and to be increased in plasma and serum in patients with various inflammatory disorders.

S100A12 also has therefore been investigated for its potential utility as a biomarker of gastrointestinal inflammation in dogs[18, 19]. Fecal S100A12 concentrations have been shown to be increased in dogs with IBD, and correlate with the severity of clinical signs, endoscopic lesions, and colonic inflammation[20]. A preliminary study also showed some evidence for Fecal S100A12 to be useful as a marker of negative outcome in dogs with CE. While these fecal markers have been shown to be sensitive to detect gastrointestinal inflammation in people, they are also fairly non-specific and are elevated in many different GI inflammatory conditions. The specificity of these markers have yet to be investigated further for use in canine IBD, and large scale studies are needed to investigate further clinical usefulness.

REFERENCES

1.Sandor, F. and M. Buc, *Toll-like receptors. III. Biological significance and impact for human medicine.* Folia Biol (Praha), 2005. **51**(6): p. 198-203. 2.Cario, E., *Bacterial interactions with cells of the intestinal mucosa: Toll-like receptors and NOD2.* Gut, 2005. **54**(8): p. 1182-93. 3. Magalhaes, J.G., I. Tattoli, and S.E. Girardin, *The intestinal epithelial barrier: how to distinguish between the microbial flora and pathogens.* Semin Immunol, 2007. **19**(2): p. 106-15. 4. Sirard, J.C., M. Bayardo, and A. Didierlaurent, *Pathogen-specific TLR signaling in mucosa: mutual contribution of microbial TLR agonists and virulence factors.* Eur J Immunol, 2006. **36**(2): p. 260-3. 5.Cario, E. and D.K. Podolsky, *Differential alteration in intestinal epithelial cell expression of toll-like receptor 3 (TLR3) and TLR4 in inflammatory bowel disease.* Infect.Immun., 2000. **68**(12): p. 7010-7017. 6. Szebeni, B., et al., *Increased expression of Toll-like receptor (TLR) 2 and TLR4 in the colonic mucosa of children with inflammatory bowel disease.* Clin Exp Immunol, 2008. **151**(1): p. 34-41. 7. McMahon, L.A., et al., *Expression of Toll-like receptor 2 in duodenal biopsies from dogs with inflammatory bowel disease is associated with severity of disease.* Vet Immunol Immunopathol, 2009. 8. Burgener, I.A., et al., *Upregulation of toll-like receptors in chronic enteropathies in dogs.* J Vet Intern Med, 2008. **22**(3): p. 553-60. 9. K Allenspach, J.S., FM McNeill, A House, A Hendricks, JM Steiner, D Werling, *Characterization of the bacterial microflora and the innate immune response in German Shepherd Dogs with chronic enteropathies.* Abstract, ACVIM Forum San Antonio, TX, 2008. 10.Gibs on, D.L., et al., *Toll-like receptor 2 plays a critical role in maintaining mucosal integrity during Citrobacter rodentium-induced colitis.* Cell Microbiol, 2008. **10**(2): p. 388-403. 11. Uematsu, S., et al., *Regulation of humoral and cellular gut immunity by lamina propria dendritic cells expressing Toll-like receptor 5.*

Nat Immunol, 2008. **9**(7): p. 769-76. 12. Uematsu, S. and S. Akira, *Immune responses of TLR5(+) lamina propria dendritic cells in enterobacterial infection.* J Gastroenterol, 2009. **44**(8): p. 803-11. 13. Jergens, A.E., et al., *Intestinal cytokine mRNA expression in canine inflammatory bowel disease: a meta-analysis with critical appraisal.* Comp Med, 2009. **59**(2): p. 153-62. 14. Schmitz, S., et al., *Gene expression of selected signature cytokines of T cell subsets in duodenal tissues of dogs with and without inflammatory bowel disease.* Vet Immunol Immunopathol, 2012. **146**(1): p. 87-91. 15. Hirota, S.A., et al., *NLRP3 inflammasome plays a key role in the regulation of intestinal homeostasis.* Inflamm Bowel Dis. **17**(6): p. 1359-72. 16. Zaki, M.H., M. Lamkanfi, and T.D. Kanneganti, *The Nlrp3 inflammasome: contributions to intestinal homeostasis.* Trends Immunol. **32**(4): p. 171-9. 17. Schmitz, S., D. Werling, and K. Allenspach, *Effects of ex-vivo and in-vivo treatment with probiotics on the inflammasome in dogs with chronic enteropathy.* PLoS One, 2015. **10**(3): p. e0120779. 18. Equilino, M., et al., *Evaluation of serum biochemical marker concentrations and survival time in dogs with protein-losing enteropathy.* J Am Vet Med Assoc, 2015. **246**(1): p. 91-9. 19. Heilmann, R.M., et al., *Serum calprotectin concentrations in dogs with idiopathic inflammatory bowel disease.* Am J Vet Res, 2012. **73**(12): p. 1900-7. 20. Heilmann, R.M., et al., *Development and analytical validation of a radioimmunoassay for the measurement of alpha1-proteinase inhibitor concentrations in feces from healthy puppies and adult dogs.* J Vet Diagn Invest, 2011. **23**(3): p. 476-85.

ANCA and Other Autoimmune Antibodies in Dogs: When and How to Use Them

Karin Allenspach, DMV, DECVIM-CA
London, UK

INTRODUCTION

Anti-neutrophil antibodies (ANCA) have been used in human medicine for decades to help in the diagnosis of various autoimmune disorders, such as various presentations of systemic vasculitis disorders, rheumatoid arthritis, and inflammatory bowel disease (IBD). Over the last 10 years, reports describing the use of ANCA diagnostics are accumulating in the veterinary literature as well. The following paragraphs will discuss the use of these diagnostic assays in human medicine and in dogs, explain the assays available to detect them and give some indications for their use.

WHAT ARE ANCA?

ANCA are auto-antibodies that can be detected in the serum of people and dogs that are directed against neutrophils. The main target antigens of ANCA in people are proteinase-3 and (PR-3) and myeloperoxidase (MPO) located on primary granules of neutrophils and lysosomes of monocytes[1]. When ANCA bind to their target antigens, they activate neutrophils and induce the release of granule enzymes, pro-inflammatory cytokines, and the generation of a respiratory burst, leading to inflammation in one or several target organs and producing the typical phenotype of disease.

IN WHICH DISEASES ARE ANCA FOUND?

Human beings

In people, ANCA are the diagnostic markers of choice for single organ or systemic vasculitis cases[1-3]. This is a heterogeneous group of diseases characterized by inflammation and necrosis of the vessel wall. The diagnosis in these cases can be challenging, as patients frequently present with multi-organ signs. Definite diagnosis depends on tissue biopsies and angiography, which are invasive tests and therefore not always risk-free in these patients. Therefore, serologic tests have always been combined with more invasive tests to help in the diagnostic workup of such cases. There is now consensus that some small-vessel necrotizing vasculitides such as granulomatosis with polyangitis (GPA) (formerly called Weger's granulomatosis), microscopic polyangitis (MPA), eosinophilic granulomatosis with polyangitis (EGPA, formerly called Churg-Strauss syndrome), and renal limited vasculitis (RLV), are all variably associated with positive serum tests for ANCA, and are therefore recommended to rule in or out various disease syndromes. Other auto-immune disorders in people that can present with positive ANCA tests include pulmonary fibrosis, ocular vasculitides, rheumatoid arthritis, Behcet's disease (a chronic multisystem inflammatory disorder characterized mainly by recurrent oral ulcers, ocular involvement, genital ulcers, and skin lesions), autoimmune hepatitis, primary sclerosing cholangitis and Systemic Lupus Erythematosus. In Inflammatory Bowel Disease (Crohn's Disease =CD and Ulcerative Colitis =UC), ANCA have been used for decades to help diagnose the disease[4,5]. Similar to the vasculitis cases, these patients can present a challenge to work up, as the ultimate diagnosis is made by endoscopically obtained intestinal biopsies. ANCA have been detected in 70-90% of people with UC, and are a recognized standard serum biomarker in the diagnostic investigation of suspected cases[6]. Cases positive for ANCA have also been shown to have a higher rate of severe and complicated disease, often needing surgery to control clinical signs. Specificity of these assays is higher in any of the above mentioned diseases if the pre-test probablility is high. Many other disease, including infections, tumors, and drug-induced vasculitis, have been associated with a higher frequency of positive samples for ANCA, and therefore, excluding other causes and tissue diagnosis remains the gold standard of diagnostics.

Dogs

ANCA in dogs have been reported in various diseases. The first publication investigated the occurrence of ANCA by indirect immunofluorescence technique (IIF) in dogs with idiopathic inflammatory bowel disease[7]. In this study, sensitivity and specificity of perinuclear ANCA (pANCA) for canine IBD were 50% and 95%, respectively. This is similar to what has been reported in UC in people for pANCA positivity, and was therefore encouraging. Further reports have substantiated the utility of this serological marker for the diagnosis of canine IBD[8]. Furthermore, in a long-term follow-up study of dogs with IBD, it was shown that pANCA was significantly more likely to be seen in dogs with food-responsive disease than compared to a group of dogs with steroid-responsive disease[9]. pANCA have also been detected in Soft Coated Wheaten Terriers (SCWT) with protein losing enteropathy and protein losing nephropathy[10]. In this group, it was shown that pANCA could be detected in the serum of dogs up to 2 years before they developed hypoalbuminemia and several years before they developed clinical signs[10]. It has therefore been suggested that pANCA could be a useful early marker of disease in this breed. There are also some case reports in the literature which describe positive ANCA in the serum of dogs with immune-mediated neutropenia and thrombocytopenia[11] as well as a case of possible Wegener's granulomatosis in a dog[12]. In addition, ANCA have been detected in up to 50% of dogs with IMHA[13], which is a similar sensitivity as was found in the original studies in dogs with IBD. However, as in human medicine, ANCA can be detected in dogs with neoplasia (17% of dogs with intestinal lymphoma in one report[14]), in dogs infected with various vector-borne diseases such as Leishmania, Ricksettsia and Bartonella (48% of dogs[13]), and anecdotally in dogs with drug-associated systemic vasculitis.

ASSAYS USED TO DETECT ANCA IN PEOPLE AND IN DOGS (AND PITFALLS OF ASSAYS CURRENTLY USED)

Currently, the main assay for detection of ANCA in human sera are the indirect immunofluorescence (IIF) on ethanol-fixed human neutrophils and enzyme-linked immunosorbent assay (ELISA) for antigen specificity. In IIF, the granulocyte slides are incubated with the patient serum first and subsequently with a FITC-labeled secondary antibody which can then be detected under fluorescent light. The main patterns observed in IIF are the cytoplasmic (cANCA) pattern that displays a diffuse granular cytoplasmic staining and the perinuclear (pANCA) pattern that shows mainly perinuclear fluorescence[1]. C-ANCA is mostly associated with anti-PR3 antibodies, whereas pANCA is associated with anti-MPO antibodies. Another ANCA pattern described is the atypical ANCA (A-anca), which can show diffuse flat cytoplasmic staining in combination with nuclear or perinuclear staining. A-Anca are associated with antibodies against other target antigens in cytoplasmic granules of neutrophils, such as lactoferrin, lysozyme, azurocidin, elastase, cathepsin G and bactericidal/permeability increasing enzyme (BPI). In human medicine, different immunoassays have been developed for the detection of specific antigens found in ANCA. The newer assays are second- or third generation ELISAs with improved sensitivity to previously available assays. These ELISAs are also referred as capture ELISAs which use monoclonal antibodies bound on the ELISA plate to capture and immobilize the antigens. With IIF as a screening test and follow-up with one of these newer ELISAs, a specificity for antigen detection of up to 100% can be reached[1]. In dogs, the assays used have been IIF methods adapted from human medicine. The first assay described used ethanol-fixed purified canine granulocytes[7]. This assay has since then been used for various publications for detection of ANCA in dogs. The assay has the advantage of being canine-specific, the pitfalls being like in any other IIF test that interpretation can be difficult and only improves with experienced readers. Another downside of this assay is its inherent variability between assays, as the coating of slides with canine granulocytes is not performed in an automated system, therefore inter-assay variation is as big concern in the interpretation of the assay. A system of standardized, automated and miniaturized assay production as is available in human medicine would therefore address these issues and make larger scale evaluations of ANCA across different diseases in dogs easier to undertake. To this end, a human IIF assay has been trialed for the detection of ANCA in dogs, and has shown very good inter-observer agreement for (kappa 0.75)[14]. The antigens for canine ANCA have so far been unknown, however with the use of this human assay it was possible to identify identify either MPO or PR-3 as antigens on the biochips. These early results indicate that the use of an automated system to produce the slides for the IIF assays for ANCA in dogs would be a significant step forward, however, it also seem that there is a loss of sensitivity when using the human assay instead of the dog assay. Once the possible antigens in canine ANCA are identified, this can lead the way t the development of ELISA assays for dog ANCA.

INDICATIONS FOR USE OF ANCA IN DIAGNOSTIC INVESTIGATIONS

Most documented studies so far pertain to the clinical utility of pANCA in dogs with IBD, and particularly, in the subgroup of dogs that respond to elimination diet (food-responsive diarrhea or FRD). It is therefore feasible to include measurement of pANCA in the workup of dogs with suspected IBD. In addition, perinuclear pattern of ANCA have been shown to be prevalent in many immune-mediated diseases in dogs, such as IBD, IMHA and systemic vasculitides. However, larger scale studies have so far been hampered by the cumbersome dog granulocyte assay used for IIF. It is likely that other immune-mediated diseases may show positive ANCA seroreactivity as well, such as immune-mediated thrombocytopenia, immune-mediated neutropenia, and some cases of aplastic anemia in dogs. Once more data are available, the utility of ANCA can be assessed in clinical cases. Further studies are needed to investigate possible ANCA reactivity in a number of immune-mediated diseases in dogs. The development of an automated system to produce granulocyte-fixed slides for IIF would greatly facilitate further studies.

REFERENCES

[1.] Kallenberg CG. P Clinical and experimental rheumatology 2015;33:S11-14. [2.] Miloslavsky EM, Specks U, Merkel PA, et al. Arthritis & rheumatology 2015;67:1629-1636. [3.] Grayson PC, Carmona-Rivera C, Xu L, et al. Arthritis & rheumatology 2015;67:1922-1932. [4.] Dubinsky MC, Ofman JJ, Urman M, et al. AmJGastroenterol 2001;96:758-765. [5.] Peeters M, Joossens S, Vermeire S, et al. AmJGastroenterol 2001;96:730-734. [6.] Yu JE, De Ravin SS, Uzel G, et al. Clin Immunol 138:14-22. [7.] Allenspach K, Luckschander N, Styner M, et al. Am J Vet Res 2004;65:1279-1283. [8.] Mancho C, Sainz A, Garcia-Sancho M, et al. J Vet Diagn Invest 22:553-558. [9.] Luckschander N, Allenspach K, Hall J, et al. J Vet Intern Med 2006;20:221-227. [10.] Allenspach K, Lomas B, Wieland B, et al. Am J Vet Res 2008;69:1301-1304. [11.] Vargo CL, Taylor SM, Haines DM. Can Vet J 2007;48:1159-1163. [12.] Bohm M, Basson S. Journal of the South African Veterinary Association 2015;86:1187. [13.] Karagianni AE, Solano-Gallego L, Breitschwerdt EB, et al. Am J Vet Res 2012;73:1403-1409. [14.] Mancho C, Sainz A, Garcia-Sancho M, et al. m J Vet Res 2011;72:1333-1337. [14] Abstract, ACVIM Forum Denver, 2016.

IBD: Diagnosis and Management
Marnin A Forman, DVM, DACVIM (SAIM)
Stamford, CT

INTRODUCTION

Inflammatory bowel disease (IBD) is a collective term for a group of chronic enteropathies that are common in cats and dogs. IBD is a complex interaction among the intestinal microbiome, dietary constituents, the immune system, host genetics and environmental triggers of intestinal inflammation. This talk will focus on techniques to optimize the diagnosis of IBD in cats and dogs, include advances in ultrasound interpretation, endoscopic utilization and considerations for additional testing following standard light microscopy. Global and specific therapeutic approaches to IBD will be discussed with a focus on management options for steroid-refractory IBD.

IBD DIAGNOSIS

The accurate diagnosis of IBD involves a systematic integration of signalment, medical and dietary history, physical findings, clinicopathologic testing, diagnostic imaging, and intestinal histopathology. This evaluation permits exclusion of non-IBD conditions causing similar clinical signs including gastrointestinal infections (endoparasites and enteric pathogens), exocrine pancreatic insufficiency, hypoadrenocorticism, hepatopathies, gastrointestinal obstructions and cancers.

During the initial diagnostic evaluation, often prior to obtaining a definitive diagnosis, therapeutic and prognostic information can be obtained. The clinical severity of the disease process can be estimated by calculating the canine IBD activity index (CIBDAI) and/or measuring serum C-reactive protein (CRP) levels. These parameters are also helpful as a baseline when evaluating response to treatment. Detecting hypoalbuminemia, low serum cobalamin concentration (<200 ng/L) and/or elevated pancreatic lipase immunoreactivity (PLI) concentrations has been associated with a poor prognosis. In patients with hypoalbuminemia secondary to a protein loosing enteropathy (PLE), screening hemostatic function (in particular thromboelastography) is recommended to detect and if abnormal, guide treatment of hypo or hypercoagulability. Fecalalpha1-proteinase inhibitor (α1-PI) is a valuable test in dogs with suspected PLE, in particular prior to overt clinical signs (i.e. vomiting, diarrhea, weight loss) or in dogs with intestinal disease and concurrent renal or hepatic disease.

Abdominal ultrasound is an important tool during the diagnostic evaluation of cats and dogs suspected to having IBD. It provides a rapid, non-invasive method to screen for gastrointestinal obstructions and masses and concurrent conditions, including pancreatitis, gall bladder or hepatic disorders. During evaluation of the gastrointestinal tract the wall thickness, appearance of wall layers including layer distinction, peristaltic activity, luminal contents and diameter, adjacent lymph nodes, peri-intestinal echogenicity and presence of free fluid are evaluated. The degree and distribution (focal, multifocal, or diffuse) of disease are described. Five distinct layers are visible in the wall of the stomach and intestine (luminal to peripheral - bright hyperechoic mucosal-luminal interface, hypoechoic mucosal layer, thin hyperechoic submucosa, thin hypoechoic muscularis layer, outermost hyperechoic serosa). Normal peristaltic activity for the stomach and small intestines is 3-5 contractions per minute.

Ultrasound abnormalities reported with IBD include diffuse or multifocal, mild to moderate, transmural thickening of the intestinal wall with preserved wall layering and no to mild mesenteric lymphadenopathy; however, intestinal wall thickness, wall layer definition or mesenteric lymph node size have not been reported to be specific or sensitive for diagnosis of IBD. Increased prominence of the muscularis layer (equal or greater in thickness to the mucosal layer) has been reported with IBD, in particular in cats. Mucosal echogenicity has been suggested to be a more sensitive indicator of IBD. The presence of hyperechoic striations, but not hyperechoic speckles, in the jejunum had a sensitivity of 75% and a specificity of 96% for dogs with protein-losing enteropathy. A complete loss of wall layer definition and wall thickness >1.5cm are considered more specific indicators of gastrointestinal neoplasia. Definitive differentiation of IBD, cancer, or other chronic enteropathies is not possible with ultrasound.

Intestinal biopsies, acquired endoscopically or surgically, are indicated in patients with suspected IBD that have failed empirical therapy to confirm the diagnosis, classify the type of IBD and to eliminate non-IBD disorders, in particular neoplastic or infectious enteropathies or lymphangiectasia. Some authors prefer diagnostic endoscopy to procure samples in patients without an indication for surgery (i.e. gastrointestinal obstructions, intestinal masses, perforation and/or indication for hepatic/pancreatic biopsies). In some, but not all, studies the endoscopic appearance of the small intestines provide better prognostic information than the histopathology. Suggested indications for surgical biopsies is apparent involvement of the submucosa or muscularis, often inferred from ultrasound studies, concern for intestinal leiomyositis and/or prior endoscopic biopsy findings that do not correlated with the clinical picture.

If ileal involvement is suspected based on low cobalamin levels and/or ultrasonographic findings, transcolonic ileoscopy, in addition to standard upper GI tract endoscopic examination, should be performed. Independent of localization of the small intestinal disease, ileal biopsies should be considered to screen for small cell lymphoma (LSA). In one retrospective study, 44% of cats were diagnosed with LSA based solely on ileal histopathology without involvement of the duodenum. The quality and number of endoscopic biopsies and tissue processing at the histopathology laboratory significantly impacts identifying certain intestinal lesions and potentially a diagnosis of IBD. Although requiring future research, a reasonable goal for the number and quality of endoscopic biopsies in dogs are 13 marginal / 7 adequate gastric and 10-15 marginal / 6 adequate duodenal samples and for cats are 6 marginal / adequate gastric or duodenal samples.

The emphasis of gastrointestinal histopathologic evaluation in patients with IBD is a semiquantitative estimate of the type and degree of cellular accumulation in the lamina propria. The type of inflammation is based on the primary cell type – eosinophilic, neutrophilic, lymphoplasmacytic and granulomatous and the extent of the inflammation ranges from focal to diffuse and is subjectively categorized as normal, mild, moderate, or severe. Mucosal architecture findings (villous morphology, lymphatic dilatation, goblet cell mucus content, and crypt lesions) correlate with clinical severity and proinflammatory cytokines and may prove to be a more accurate categorization for IBD. For example, in cats with GI disease, villus atrophy and fusion in the duodenal mucosa correlated with the severity of clinical signs and in dogs, loss of mucus and goblet cells correlated with the severity of lymphoplasmacytic colitis.

To more accurately define certain abnormalities noted on standard light microscopy, additional studies should be considered. Neutrophilic and/or granulomatous inflammation raises the possibility of an infectious process, and cultures (mucosal biopsies, intestinal lymph nodes), cytochemical 'special' stains (Gomori methenamine silver, periodic acid–Schiff, Gram, and modified Steiner stains), and/or fluorescent in situ hybridization with a probe directed against eubacterial 16S ribosomal RNA (FISH) are indicated. Imaging of the chest and retinal examination can be considered to screen for systemic infections. Differentiation of severe IBD and intestinal lymphoma (enteropathy-associated T-cell lymphoma) can be challenging based in histopathology alone, especially when only endoscopic biopsies are available. To improve the accuracy of the diagnosis, a stepwise diagnostic approach has been suggested. Histopathology is the first step, followed by immunohistochemistry (CD3, CD20, and Ki-67) and polymerase chain reaction (PCR) for antigen receptor rearrangement (T-cell clonality).

IBD MANAGEMENT

The global therapeutic approach to biopsied confirmed IBD (enteritis) is directed at counteracting inflammation and dysbiosis and correcting nutritional deficiencies (i.e. cobalamin and/or folate deficiency). Empirical anthelmintic therapy (i.e. fenbendazole 50 mg/kg PO q 24hrs for 5 days) for Giardia and antibiotic therapy (i.e. tylosin 10 - 15 mg/kg PO q8hrs, oxytetracycline 20 mg/kg PO q8hrs, or metronidazole 10 mg/kg PO q12hrs for 14-28 days) are often administered, if not already initiated. In patients with a positive response to antibiotics (antibiotic-responsive enteropathy), transition to a probiotic (i.e. VSL #3) can be considered. In dogs with IBD, combination therapy including VSL#3 demonstrated a protective effect with a significant decrease in clinical and histological scores and CD3+ T-cell infiltration. Cobalamin deficiency has traditionally been treated with parentally administered Vitamin B12; however, oral therapy (cyanocobalamin 1mg tablets, dogs with a body weight of 1–10 kg ¼ tablet, >10–20 kg ½ tablet, and >20 kg 1 tablet PO q 24h) has been shown to be effective.

Granulomatous or neutrophilic IBD

The initial focus is screening (see above) for underlying infectious causes (i.e. Bacterial (E. coli (GC in boxers), Streptococcus, Campylobacter, Yersinia, and Mycobacteria; fungal (i.e., Histoplasma) or algal (i.e., Prototheca) infections). Signalment, historical findings (i.e. raw food diet) or travel history assists in prioritizing pathogen screening. Therapy is directed at the underlying infectious agent. Immunosuppression should not be considered until infectious agents have been excluded. The prognosis for idiopathic granulomatous or neutrophilic IBD is considered poor.

Lymphoplasmacytic IBD

The mainstay of therapy remains a combination of corticosteroids, antibiotics and dietary therapy. Multiple studies have evaluated the management of IBD; however, many of these studies have evaluated multiple parameters (different immunosuppressive therapies and/or diets) or mixed IBD types (i.e. lymphoplasmacytic or eosinophilic or presumptive IBD). Greater than 80% of dogs with IBD have been reported to go into remission, based on a >75% decrease in CIBDAI score, with prednisone (1 mg/kg PO q 12hr) alone and prednisone with metronidazole (10 mg/kg PO q 12hr). However, a different retrospective study including dogs treated with prednisolone, sulphasalazine, metronidazole ± tylosin, suggested a much lower rate of remission (26% complete, 50% intermittent signs, 4% uncontrolled, 13% euthanized (GI), 7% euthanized [non-GI]).

A sequential therapeutic approach has been suggested, adjusted based on clinical disease activity, histopathologic severity and presence of hypoalbuminemia. In patients with mild to moderate clinical disease activity and intestinal histopathology and serum albumin levels >2 g/L, a dietary and antibiotic trial is initiated first and if an incomplete response, followed by immunosuppression with glucocorticoids. If a poor response, escalating immunosuppression is then considered. In patients with moderate to severe clinical disease activity and intestinal histopathology (atrophy, fusion) and serum albumin levels <2 g/L, immunosuppression is started concurrently.

Considering the dietary studies as a whole, a positive response to diet is reported in approximately 50% of dogs with chronic GI signs and IBD. The descriptive terms food responsive or dietary intolerant have been suggested as a more appropriate term than food allergy or intolerance since many patients do not relapse on rechallenge with the original diet. The focuses of dietary therapies are antigenic modification (antigen-restricted /novel protein source or protein hydrolysate) or optimize assimilation (highly digestible, fat restricted and/or restricted fiber).

In patients with lymphoplasmacytic IBD and a poor response to immunosuppression with prednisone, it is important to reappraise all aspects of the diagnosis and therapy. Screening for concurrent or newly development conditions to explain the clinical signs should be considered and treated if detected. Prior to implementing escalation of immunosuppression, switching from oral prednisone to injectable dexamethasone can be considered. Budesonide (3-7 kg: 1 mg PO q24h, 7.1-15 kg: 2 mg PO q24h, 15.1-30 kg: 3 mg PO q24h, >30 kg: 5 mg PO q24h) has been suggested to reduce the side effects of prednisone and has reported to have a similar remission rate of 78%, based on >75% decrease in CIBDAI score; however, the frequency adverse effects was similar to prednisone. In a small study of dogs with steroid-refractory IBD (n=14), cyclosporine therapy (5 mg/kg PO q24h) was reported to be effective therapy. Improvement in clinical activity score, infiltrating lymphocytes and T cells was noted in 86% of the dogs. Masitinib, Azathioprine and Mycophenolate have all been suggested in the management of these patients. Elemental diets and partial parenteral nutrition may be indicated in dogs with prednisone resistant severe protein-losing enteropathy.

Eosinophilic IBD

An immunologic reaction to diet or parasites is speculated to result in eosinophilic IBD. Repeated zinc-sulfate fecal centrifugation evaluation, potentially with PCR pathogen screening, should be considered to screen for parasites. A prophylactic anthelmintic therapy (i.e. fenbendazole 50 mg/kg PO q 24hrs for 5 days) is warranted. Some patients respond to diet therapy (antigen-restricted or protein hydrolysate diets) and those failing diet therapy often improve with immunosuppressive therapy (prednisone / prednisolone 1 mg/kg PO q12 hrs or 2mg/kg PO q 24hrs).

References

1. Simpson, K.W. and A.E. Jergens, Pitfalls and progress in the diagnosis and management of canine inflammatory bowel disease. Vet Clin North Am Small Anim Pract, 2011. 41(2): p. 381-98.
2. Gaschen, L., et al., Comparison of ultrasonographic findings with clinical activity index (CIBDAI) and diagnosis in dogs with chronic enteropathies. Vet Radiol Ultrasound, 2008. 49(1): p. 56-64.
3. Larson, M.M. and D.S. Biller, Ultrasound of the gastrointestinal tract. Vet Clin North Am Small Anim Pract, 2009. 39(4): p. 747-59.
4. Gaschen, L., Ultrasonography of small intestinal inflammatory and neoplastic diseases in dogs and cats. Vet Clin North Am Small Anim Pract, 2011. 41(2): p. 329-44.
5. Jergens, A.E., et al., Comparison of oral prednisone and prednisone combined with metronidazole for induction therapy of canine inflammatory bowel disease: a randomized-controlled trial. J Vet Intern Med, 2010. 24(2): p. 269-77.
6. Craven, M., et al., Canine inflammatory bowel disease: retrospective analysis of diagnosis and outcome in 80 cases (1995-2002). J Small Anim Pract, 2004. 45(7): p. 336-42.
7. Dye, T.L., et al., Randomized, controlled trial of budesonide and prednisone for the treatment of idiopathic inflammatory bowel disease in dogs. J Vet Intern Med, 2013. 27(6): p. 1385-91.
8. Allenspach, K., et al., Pharmacokinetics and clinical efficacy of cyclosporine treatment of dogs with steroid-refractory inflammatory bowel disease. J Vet Intern Med, 2006. 20(2): p. 239-44.
9. Jergens, A.E., et al., A scoring index for disease activity in canine inflammatory bowel disease. J Vet Intern Med, 2003. 17(3): p. 291-7.
10. Willard, M. and J. Mansell, Correlating clinical activity and histopathologic assessment of gastrointestinal lesion severity: current challenges. Vet Clin North Am Small Anim Pract, 2011. 41(2): p. 457-63.
11. Andrews, C., et al., Cross Lineage Rearrangement in Feline Enteropathy-Associated T-cell Lymphoma. Vet Pathol, 2015.
12. Carrasco, V., et al., Distinguishing Intestinal Lymphoma From Inflammatory Bowel Disease in Canine Duodenal Endoscopic Biopsy Samples. Vet Pathol, 2015. 52(4): p. 668-75.
13. Willard, M.D., et al., Effect of sample quality on the sensitivity of endoscopic biopsy for detecting gastric and duodenal lesions in dogs and cats. J Vet Intern Med, 2008. 22(5): p. 1084-9.
14. Willard, M.D., et al., Effect of tissue processing on assessment of endoscopic intestinal biopsies in dogs and cats. J Vet Intern Med, 2010. 24(1): p. 84-9.
15. Allenspach, K., Diagnosis of Small Intestinal Disorders in Dogs and Cats. Clin Lab Med, 2015. 35(3): p. 521-34.
16. Berghoff, N. and J.M. Steiner, Laboratory tests for the diagnosis and management of chronic canine and feline enteropathies. Vet Clin North Am Small Anim Pract, 2011. 41(2): p. 311-28.
17. Rossi, G., et al., Comparison of microbiological, histological, and immunomodulatory parameters in response to treatment with either combination therapy with prednisone and metronidazole or probiotic VSL#3 strains in dogs with idiopathic inflammatory bowel disease. PLoS One, 2014. 9(4): p. e94699.
18. Scott, K.D., et al., Utility of endoscopic biopsies of the duodenum and ileum for diagnosis of inflammatory bowel disease and small cell lymphoma in cats. J Vet Intern Med, 2011. 25(6): p. 1253-7.
19. Zacuto, A.C., et al., Intestinal Leiomyositis: A Cause of Chronic Intestinal Pseudo-Obstruction in 6 Dogs. J Vet Intern Med, 2016. 30(1): p. 132-40.

Inflammatory Bowel Disease: Complications and Comorbidities
Marnin A. Forman, DACVIM (SAIM)
Stamford, CT

INTRODUCTION

Subgroups of dogs and cats, diagnosed correctly with inflammatory bowel disease (IBD), will fail to respond to standard therapies. Screening for established complications and comorbidities is useful in predicting response to therapy, modifying dietary and medical therapies and predicting outcome. Although similarities have been noted between human, canine and feline IBD, complications noted in humans, including intestinal fistulae, strictures, intramural abscessation, intestinal obstruction and intestinal perforation, are rare in dogs and cats. Ensuring the accuracy of the IBD diagnosis is paramount; however will not be reviewed in this talk.

HYPOALBUMINEMIA / PROTEIN LOSING ENTEROPATHY

Hypoalbuminemia (<20 g/L) has been established as a negative prognostic indicator and higher likelihood of refractoriness to treatment in dogs and cats with IBD. Hypoalbuminemia secondary to a protein losing enteropathy (PLE) may be due to a reduction in intestinal surface area, such as with villus atrophy or fibrosis, hemorrhage or exudation of protein into the gastrointestinal tract and/or increased intestinal permeability. The loss of protein in PLE is independent of the molecular weight, therefore panhypoproteinemia would be expected; however, there are cases in which only hypoalbuminemia is present. The presence and magnitude of PLE is associated with IBD more often in dogs than in cats. Lymphangiectasia can occur secondary to IBD and can result in intestinal protein loss through ruptured lacteals. However, PLE can occur in patients with IBD in the absence of lymphangiectasia, likely due to increased intestinal permeability. In rodent models heparin sulfate deficiency has been demonstrated to induce intestinal protein loss and inflammation. This may be a future therapeutic direction by administering a nonanticoagulant heparin derivative. Albumin is responsible for approximately 70% of plasma oncotic pressure; therefore hypoalbuminemia can result in decreased oncotic pressure and the formation of pleural and/or peritoneal effusion and peripheral pitting edema.

Primary therapy for hypoalbuminemia / PLE is the treatment of the underlying intestinal pathology with medical and dietary therapy. Patients with hypoalbuminemia should be screened for pleural effusions, prior to anesthesia or when displaying tachypnea or labored breathing. Therapeutic thoracocentesis is indicated in patients with moderate to severe pleural effusions. Abdominocentesis is typically reserved for patients with moderate to severe volumes of ascites associated with apparent discomfort and inappetence. Hypoalbuminemia can contribute to other life-threatening complications including hypotension, hypovolemia and pulmonary edema. Colloid therapy has been recommended to restore colloid osmotic pressure [if documented or decreased oncotic pressure is suspected or when the total protein is <3.5 mg/L (or albumin is <1.5 g/L)] and blood volume; however, the choice of colloid therapy has been debated in human and veterinary medicine. Synthetic colloids [hydroxyethyl starch (dogs up to 20 mL/kg/24 hr, divide into 5 mL/kg boluses to effect; cats 10–20 mL/kg/24 hr, 2.5–3 mL/kg boluses to effect)] are typically the first line of therapy. Although synthetic colloids increase osmotic pressure and help to maintain fluid in the intravascular space, they do not appreciably change total solids as measured by refractometry. A large, retrospective study demonstrated it appeared safe to administer 5% human serum albumin (2 mL/kg/hr for 10 hours/day (total volume 20 mL/kg/d) until albumin reached 20 g/L) to critically ill hypoalbuminemic cats and dogs. In this study, a majority of the patients (75% dogs, 72% cats) survived to discharge and no signs of severe hypersensitivity reactions such as anaphylaxis, angioedema and urticaria were detected. Human serum albumin is costly and the therapeutic benefit in dogs and cats with PLE has not been determined. The author rarely uses human albumin in dogs or cats with PLE. Plasma transfusions are not typically effective is raising albumin levels due to the relatively low albumin levels in plasma. Rarely total parental nutrition is considered in the management of severe PLE patients that are intolerant of enteral nutrition.

THROMBOEMBOLISM / HYPERCOAGULABILITY

One of the more potentially severe complications of PLE is hypercoagulability and thromboembolic complications, both venous and arterial. Dogs with PLE secondary to IBD have been demonstrated to have an appreciable prevalence of hypercoagulability. Thromboembolism (TE) was documented in 12-18% of Wheaten terriers, suspected or confirmed as a cause of death in Yorkshire terriers, and clinically apparent in 7.5% of all dogs in a report compiling several case series of PLE. It is likely that these numbers underestimate the prevalence of TE; however, due to the difficulty in confirming these events antemortally and postmortally. Cats with PLE have been reported to have pulmonary TE (PTE).

The mechanism of hypercoagulability is multifactorial, and includes antithrombin deficiency, hypoalbuminemia-related platelet hyperaggregability and hyperfibrinogenemia. Hypercoagulability is further exacerbated by corticosteroid therapy. Laboratory confirmation of hypercoagulability is difficult in practice. The routine coagulogram does not identify hypercoagulability. (Neither a decreased PT or aPTT nor thrombocytosis is correlated.) Documentation of antithrombin (AT) deficiency may be helpful, but normal AT does not exclude hypercoagulability. Elevated D-dimer concentrations are indicative of active coagulation and fibrinolysis but do not correlate with TE risk, and normal D-dimers do not exclude it. Thromboelastography (TEG) has been shown to be a useful tool to identify hypercoagulability in small animals and in humans. In a small prospective study of 15 dogs with PLE, all dogs were hypercoagulable [decreased reaction time (R), decreased clotting time (K), increased angle (α), increased maximal amplitude (MA)] compared to 30 normal dogs. In the author's practice, TEG is routinely used for the assessment of patients considered at risk for TE and those suspected to have TE complications. A platelet to albumin (expressed in g/dL) ratio above 240,000 has also been proposed as a marker for hypercoagulability in dogs. In one abstract, it identified all dogs with PLE and increased thrombin-antithrombin complexes.

Protocols for the treatment of TE and for primary thromboprophylaxis in dogs and cats have not been definitively established. Anticoagulants are the mainstay of management for confirmed or suspected venous TE, to prevent thrombus extension and rethrombosis (secondary thromboprophylaxis). Unfractionated heparin (UFH) therapy is best achieved using a heparin nomogram. UFH is initiated with bolus injection (80-100 U/kg IV), followed by a continuous rate infusion of 18-20 u/kg/hr. Because of the unpredictability of UFH bioavailability and response, monitoring and titration are essential. An aPTT is monitored every 6 hours, and the infusion rate titrated to maintain an aPTT of 1.5 – 2.0 X pre-treatment. A nomogram has not been established for cats, and aPTT monitoring in this species appears to be unreliable. In humans, LMWHs have shown to be of at least equivalent efficacy, but do not require monitoring in most cases. No evidence exists in small animals comparing UFH and LMWH regimens for TE therapy. The LMWH, Dalteparin can be administered in dogs at 150 U/Kg SC q8h and in cats at 180 U/Kg SC q6h. Enoxaparin at 0.8 mg/kg SC q6h has been demonstrated to maintain target levels of anti-Xa in healthy greyhounds. In cats, an enoxaparin dose of 1.25mg/kg SC q 6h has been recommended. The frequency of dosing is based on pharmacokinetic studies in these species. More recent venous stasis models, however, imply that adequate anticoagulation does not require maintenance of target anti-Xa levels and, in humans, the dose is based on the achievement only of a peak anti-Xa in the target range. For these reasons, and anecdotal experience, some veterinary clinicians recommend enoxaparin at 1.0 mg/kg q8-12h. Because LMWH regimens are not well validated in small animals, the author recommends measurement of anti-Xa activity at peak (2-3 hours in the cat, 3-4 hours in the dog) and dose adjustment as needed to achieve therapeutic target range (0.5 – 1.0 Units/ml).

Intermittent, subcutaneous UFH is not considered ideal for secondary thromboprophylaxis due to the difficulty in achieving a therapeutic effect in a timely fashion. If it is to be used, a dose of 250-300 u/kg SC q6h is recommended. Dose titration based on frequent aPTT monitoring is essential due to the huge variability and unpredictability of response. If AT is decreased, administration of fresh frozen plasma (FFP, 6-10 ml/kg) may be of benefit. The addition of an antiplatelet drug to the therapeutic protocol has merit – for arterial thromboprophylaxis as well as a potential synergistic effect with the anticoagulant for venous thromboprophylaxis. This can be achieved with low-dose aspirin (dogs 0.5-1.0 mg/kg PO q12h, cats 5 mg/cat PO q72h). Clopidogrel bisulfate (dogs 2 mg/kg q24h, cats 18.75mg PO q24h) is an alternative option for patients that have failed, or do not tolerate, aspirin therapy. It should be noted that the onset of effect is delayed, 5-7 days.

Primary thromboprophylaxis to mitigate thrombotic risk in hypercoagulable patients with PLE will likely improve outcome; however, this has not been critically evaluated. Risk stratification in these patients is subjective. Hypoalbuminemia is not predictive. A persistent hypercoagulable state has been documented in dogs with PLE despite a significant improvement in albumin concentration, likely due, in part, to corticosteroid therapy. Risk is increased in the presence of a concomitant hypercoagulable state or thrombophilic scenario (e.g., hypovolemia, catheterization). The choice of antithrombotic drugs for primary thromboprophylaxis is based on several parameters including anticipated duration of therapy, route of administration, and cost. Antiplatelet drugs (aspirin, clopidogrel) are indicated for arterial thromboprophylaxis. For venous thromboprophylaxis, anticoagulants (UFH, LMWH, or warfarin) should be prioritized; antiplatelet drugs can be considered as adjunctive agents. Since PLE-related TE can be arterial or venous, the combined use of anticoagulant and antiplatelet drug is prudent.

HYPOCOBALAMINEMIA

Low serum cobalamin concentrations (<200 ng/L) has a wide-ranging reported prevalence (6-73%) and has been associated with refractoriness to treatment in cats and dogs with chronic enteropathies. In dogs with PLE it has been correlated with the serum albumin levels. Hypocobalaminemia can be detected in combination with hyperfolatemia. Absorption of cobalamin in cats and dogs is a complex, multistep process mediated by carrier proteins and intrinsic factor (IF). IF is produced by the stomach and pancreas in dogs and by the pancreas, but not the stomach, in the cat. The cobalamin-IF complex is absorbed in the ileum by receptors; however, in humans a small amount (1%) of unbounded cobalamin is absorbed along the entire intestine. Suggested causes of hypocobalaminemia in patients with chronic enteropathies include damage to the ileal mucosal receptors for binding cobalamin–IF complexes or decreased amounts of cobalamin available for absorption due to bacterial competition of nutrients. Hypocobalaminemia can cause inappetence, weight loss, villous atrophy, malabsorption of other vitamins and nutrients, central and peripheral neuropathies and immunodeficiency.

Cobalamin supplementation has traditionally been with a repeated, parentally administered, Vitamin B12 injection protocol (500 to 1500 mg/dog depending on the size; 250 mg/cat). This protocol was based on the pathophysiologic justification that oral Vitamin B12 would be poorly absorbed and would not normalize hypocobalaminemia. Several studies in humans and a Cochrane review suggested oral cobalamin might be as effective as parenteral administration in various conditions, including gastrointestinal disorders. A retrospective study in dogs with chronic enteropathies and hypocobalaminemia (cobalamin concentration of ≤270 ng/L used) demonstrated oral cobalamin supplementation can restore normocobalaminemia. Cyanocobalamin 1mg tablets [dogs with a body weight of 1–10 kg ¼ tablet, >10–20 kg ½ tablet, and >20 kg 1 tablet PO q 24h] were shown to be effective. Daily, oral cobalamin supplementation has been suggested to be likely cheaper, simpler, and pain-free alternative to parental cobalamin injections. Cobalamin supplementation can be started while pending Vitamin B12 levels and based on a study suggesting cobalamin deficiency on a cellular level [increased methylmalonic acid (MMA)] with Vitamin B12 in the lower end of the reference interval, therapy can be considered even with low normal measurements. Despite therapy with parental vitamin B12 supplementation for 6 weeks, a prior study suggested hypocobalaminemia remains a negative prognostic indictor and associated with an increased risk of euthanasia.

ASSORTED COMPLICATIONS AND COMORBIDITIES

Iron deficiency may result from chronic blood loss in patients with erosions or ulceration secondary to IBD. Iron deficiency anemia is typically regenerative early in the disease process, but ~50% of patients with iron deficiency anemia will eventually become non-regenerative. Patients with iron deficiency anemia secondary to IBD do not always exhibit clinical signs of gastrointestinal disease such as vomiting, diarrhea, or anorexia, and they do not always have overt evidence of gastrointestinal bleeding. Iron administration via iron dextrans injections (dogs 10-20 mg/kg IM once, cats 50mg/cat IM once) and followed by oral ferrous sulfate (dogs 100-300 mg/dog PO daily, cats 50-100 mg/cat PO daily) are appropriate treatments for iron deficiency. Some patients, in particular cats, will not tolerate oral iron products. Ulcer therapies should also be provided if suspected or confirmed.

Dogs with IBD, and particularly those with PLE, have been shown to have lower concentrations of 25 hydroxyvitamin D (25(OH)D) than normal dogs. Additionally, dogs with poor outcomes secondary to IBD have been shown to have lower Vitamin D levels compared to survivors. Mechanisms by which IBD may cause hypovitaminosis D include malabsorption of vitamin D, loss of vitamin D bound to albumin, or decreased intake of vitamin D due to anorexia. Conversely, it is possible that hypovitaminosis D may contribute to the development of IBD via several mechanisms. Vitamin D has been shown to modulate pro-inflammatory responses, suggesting that a lack of vitamin D may incite a pro-inflammatory state. Additionally, Vitamin D has been shown to enhance tight junctions, therefore a lack of vitamin D may result in increased gastrointestinal epithelial permeability, resulting in increased antigen exposure to the mucosal immune system.

An association between feline and canine IBD and pancreatitis has been suggested based on limited histopathologic studies, involving small numbers of patients, and larger, but still relatively limited, serologic studies utilizing pancreatic lipase immunoreactivity (PLI) concentrations. It is unknown if there is a common underlying cause for the apparent association; however, there does appear to be a clinical significance of concurrent IBD and pancreatitis (based on elevated PLI concentrations). Dogs with IBD and increased serum cPLI concentrations have worse outcome and increased risk of euthanasia compared to dogs with IBD and normal cPLI concentrations. Increased fPLI concentrations did not significantly affect survival outcome in cats with IBD; however, hypoalbuminemia and hypocobalaminemia was observed more frequently with serum fPLI concentration of ≥12.0 µg/l.

References

1. Allenspach K, Wieland B, Grone A, et al. Chronic enteropathies in dogs: evaluation of risk factors for negative outcome. J Vet Intern Med 2007;21:700-708.
2. Bailey S, Benigni L, Eastwood J, et al. Comparisons between cats with normal and increased fPLI concentrations in cats diagnosed with inflammatory bowel disease. J Small Anim Pract 2010;51:484-489.
3. Craven M, Simpson JW, Ridyard AE, et al. Canine inflammatory bowel disease: retrospective analysis of diagnosis and outcome in 80 cases (1995-2002). J Small Anim Pract 2004;45:336-342.
4. Dossin O, Lavoue R. Protein-losing enteropathies in dogs. Vet Clin North Am Small Anim Pract 2011;41:399-418.
5. Gow AG, Else R, Evans H, et al. Hypovitaminosis D in dogs with inflammatory bowel disease and hypoalbuminaemia. J Small Anim Pract 2011;52:411-418.
6. Kathrani A, Steiner JM, Suchodolski J, et al. Elevated canine pancreatic lipase immunoreactivity concentration in dogs with inflammatory bowel disease is associated with a negative outcome. J Small Anim Pract 2009;50:126-132.

7. Vigano F, Perissinotto L, Bosco VR. Administration of 5% human serum albumin in critically ill small animal patients with hypoalbuminemia: 418 dogs and 170 cats (1994-2008). J Vet Emerg Crit Care (San Antonio) 2010;20:237-243.
8. Lunsford KV, Mackin AJ, Langston VC, et al. Pharmacokinetics of subcutaneous low molecular weight heparin (enoxaparin) in dogs. J Am Anim Hosp Assoc 2009;45:261-267.
9. Kerl ME, Harnett BH. Unfractionated and low-molecular-weight heparin for hypercoagulability in dogs and cats. Veterinary Medicine 2007.
10. Ristic JM, Stidworthy MF. Two cases of severe iron-deficiency anaemia due to inflammatory bowel disease in the dog. J Small Anim Pract 2002;43:80-83.
11. Simmerson SM, Armstrong PJ, Wunschmann A, et al. Clinical features, intestinal histopathology, and outcome in protein-losing enteropathy in Yorkshire Terrier dogs. J Vet Intern Med 2014;28:331-337.
12. Titmarsh H, Gow AG, Kilpatrick S, et al. Association of Vitamin D Status and Clinical Outcome in Dogs with a Chronic Enteropathy. J Vet Intern Med 2015;29:1473-1478.
13. Goodwin LV, Goggs R, Chan DL, et al. Hypercoagulability in dogs with protein-losing enteropathy. J Vet Intern Med 2011;25:273-277.
14. Toresson L, Steiner JM, Suchodolski JS, et al. Oral Cobalamin Supplementation in Dogs with Chronic Enteropathies and Hypocobalaminemia. J Vet Intern Med 2016;30:101-107.

Inflammatory Bowel Disease: Nutrition in the Pathogenesis & Management

Joe Bartges, DVM, PhD, DACVIM, DACVN
Stamford, CT

INTRODUCTION

Canine and feline IBD is a heterogeneous group of chronic enteropathies characterized by persistent or recurrent gastrointestinal signs with inflammatory infiltrate.[1] The infiltration is most often lymphoplasmacytic, but may include eosinophilic and neutrophilic and may be associated with crypt abscessation and/or lacteal dilation with protein losing enteropathy (PLE). Patients with IBD may respond to diet (food responsive), antimicrobial agents, and/or immunosuppressive agents. Clinical signs include vomiting, diarrhea, and/or changes in appetite and often patients have variable degrees of weight loss. Inflammation may involve the stomach, small intestine, and or large intestine; however, PLE usually involves only the small intestine.[2] The cause(s) of IBD is/are unknown in dogs and cats but likely involves interactions between the GI microbiota and dysregulated responses in a genetically susceptible individual.[3] The genetic basis of IBD in certain breeds of dogs is recognized. The roles of dysbiosis and abnormal innate immunity have also been recognized. Dysbiosis is an important component of IBD, as evidenced by response to metronidazole, tylosin, or quinolones.[4, 5]

How is IBD diagnosed?

IBD is diagnosed based on historical information and clinical examination findings in addition to supporting laboratory evaluation (CBC, biochemistry, B12, folate, TLI, PLI, etc.), imaging (radiography, ultrasonography), and biopsy (endoscopic or surgical). There is an association between IBD and cutaneous and other intra-abdominal organ inflammation. The role of biopsy is important as it not only provides a histologic diagnosis, but it can be submitted for fluorescent in situ hybridization for examination of enteroinvasive pathogens.[6, 7] Additionally, there are clinical activity indices for dogs and cats with chronic GI disease.[1] Dietary modification is an important component in management of dogs and cats with IBD. There are diagnostic tests available for determining serum allergen-specific IgE and salivary allergen-specific IgA. Unfortunately, these tests have no validation as to their usefulness in diagnosing a dog or cat with a food allergy. Serum allergen-specific testing may be sensitive but lack specificity.[8, 9] Additionally, serum allergen-specific IgG and IgE levels do not change with feeding an elimination diet despite resolution of clinical signs.[10] Conventionally, a food elimination trial with challenge can be undertaken to evaluate for a food allergy; however, response to diet change does not confirm food allergy, only that the patient has a food-responsive disorder. The most useful diagnostic test of dietary sensitivity is feeding an elimination diet followed by challenge with a test meal. Elimination diets must be individualized based on previous dietary exposure. A detailed study of the animal's diet will allow identification of foods that have not been fed before and that could be used to formulate a nutritionally balanced elimination diet. If it is not possible to formulate a suitable elimination diet, then a restricted diet may be used that contains only one or two potential allergens, preferably ones that have not been eaten in the preceding month. Many homemade diets used as elimination diets are not complete and balanced (e.g. cottage cheese and rice, or chicken and rice). Supplementation with vitamins and minerals is encouraged, but avoid use of supplements that contain potentially offending foodstuffs.

NUTRITION IN PATHOGENESIS OF IBD

Is dietary protein source a cause of IBD?

Throughout life, animals are exposed to a variety of potential dietary allergens; however, after a variable period of time, some may develop an immune response against a particular foodstuff that activates one or more immunopathogenic pathways. After development of this response, subsequent ingestion of this foodstuff results in clinical signs. These dietary antigens do not normally cause problems because the intestinal mucosa forms a barrier that limits absorption of macromolecules. There is evidence that antigens are absorbed through both normal and abnormal gut as allergen specific food antibodies are demonstrable in healthy individuals. Upon initial presentation of the antigen to the gut mucosa, there is an immune response involving IgA. This reduces the amount of antigenic material absorbed. Immune complexes of antigen and IgA antibody are transported across hepatocytes, into bile, and re-circulated to the intestine. This local IgA response may be followed by a transitory systemic immune response, but immunologic tolerance follows. Thus, there is an apparent paradox of a vigorous local immune response followed by a systemic tolerance. Absorption of macromolecules can be altered by local immunity. Decreased uptake has been demonstrated experimentally following oral or parenteral immunization in rats and increased absorption occurs in IgA-deficient human beings. Absorption is also enhanced by gut mucosal vasodilatation. In this case, the patient becomes caught in an immunological vicious circle because local hypersensitivity reactions favor access of allergens that in turn heightens the antibody response. Factors that lead to development of hypersensitivity to ingested antigens are speculative. Those most frequently implicated are heat- and acid-stable glycoproteins with molecular weights of 18-30 kD. Hypersensitivity reactions involved in food allergies have been shown to involve types I, III, and IV reactions. Some studies indicate that IgE is implicated in some instances and the reactions involved include both classic, immediate Type I reaction and late-phase IgE-mediated reactions. What initiates the original immunologic reaction is not clear. If clinical or subclinical gastrointestinal disease occurs altering mucosal integrity, absorption of antigenic proteins may occur which may initiate the processes. Additionally, dysbiosis is being recognized as a common finding in patients with IBD.

What are advanced glycated end-products (AGE's) and do they have a role in IBD?

Many pets eat processed foods. Temperature and residence time during processing are known to influence Maillard reactions in which a reducing sugar binds to a free reactive amino group of an amino acid blocking its reactive site. In intact proteins, the ε-amino group of lysine is the most abundant free amino group. The reaction results in Amadori compound leading eventually to decreased lysine bioavailability and Maillard reaction products (MRP's) that are also advanced glycation end-products (ABE's). During formation, AGE's can covalently cross-link tissue protein and modify structural and functional properties. Tissue AGE's are associated with various diseases such as diabetes mellitus, cataract formation, osteoarthritis, and pro-inflammatory responses. Dietary MRP's can be absorbed and contribute to the body's AGE pool and possibly to associated diseases.[11] While no studies have evaluated the role, if any, of MRP's with animal diseases, one study showed that dogs may ingest up to 122 times more MRP when consuming an average extruded dog food than the calculated average for adult human beings and up to 36 times more MRP compared to human infants. Likewise, cats consuming an average extruded cat food may ingest up to 38 times more MPR compared to adult human beings and 11 times more MRP compared to human infants. Those fed canned pelleted diets are more likely to ingest higher amount of MRP compared to extruded diet.[11]

NUTRITION IN MANAGEMENT OF IBD

Treatment of canine and feline IBD is empirical and consists of dietary and pharmacological therapy. There are several options for dietary trials: (1) global modification: switch to a different diet or a different manufacturer, (2) optimize assimilation: highly digestible (usually rice based), fat restricted (< 15% dry matter basis), easy-to digest fats, restricted fiber, (3) antigen modification: antigen-restricted novel protein source, protein hydrolysate, (4) immunomodulation: altered fat composition (e.g. omega 3 fatty acids), prebiotics (e.g. inulin, fructooligosaccharides, etc.), (5) homemade complete and balanced diets. [12] The rationale for dietary therapy of IBD is that restricting exposure to antigens will mitigate clinical signs and prevent recurrence.

Does switching to a different diet work?

There are no specific studies that have evaluated changing from one commercial diet to another commercial diet without utilizing a novel protein or hydrolyzed protein source. Response to a non-therapeutic diet that has a different nutrient composition than the diet would be presumed to be associated with removing the offending dietary component responsible for clinical signs. Food or food ingredients reported to cause adverse reactions in 330 dogs included in descending order: beef, dairy, chicken, wheat, chicken egg, soy, lamb, pork, fish, corn, turkey, rice, and duck; while in 56 cats they included: beef, dairy, fish, chicken, corn/corn gluten, lamb, wheat, and chicken egg.[13] Unfortunately, the references for these data are from 2007 or before. However, choosing a diet based on previous exposure and this information could allow for selection of a commercial non-therapeutic diet. Recommendations of choosing diets for elimination trial is concerning as there have been publications documenting protein sources analyzed using species specific DNA being present in pet foods and therapeutics that were not specifically listed including 3 studies where 14 of 17 samples[14], 20 of 52 samples[15], and 4 of 4 samples[16] were mislabeled, and one study of therapeutics where 4 of 7 products were mislabeled.[17] Additionally soy may be present in diets that are claimed to be soy free and in one study 3 of 4 such over the counter diets contained appreciable amounts of soy protein.[18]

What are easily assimilated diets?

Easily assimilated diets refer to easily digestible diets that are moderate in protein, lower in fat, low in fiber, and moderate in carbohydrates. There are no studies on the amount of dietary protein in management of IBD in dogs and cats. While providing a higher protein diet would seem logical, a study of human beings with ulcerative colitis demonstrated that feeding a higher protein diet was associated with 3-fold risk of relapse when compared with patients fed a lower protein diet. One possible explanation is related to increased amounts of some deleterious amino acid derived metabolites produced from undigested protein in the large intestine lumen by microbiota.[19] In one study of cats with IBD, cobalamin deficiency was associated with decreased fat digestibility.[20] Dietary fat restriction has been shown to be beneficial in dogs with PLE.[21] In human beings with ulcerative colitis, increased dietary total fat, animal fat, omega-6 fatty acids, and omega-6-to-omega-3 fatty acid ratios were associated with increased incidence.[22] Carbohydrate maldigestion is a feature of feline IBD; however, it has not been shown to contribute to clinical signs.[23] Gluten likely does not play a role in most patients with IBD except for Irish Setters with known gluten sensitivity. In addition to macronutrients, micronutrients such as vitamin and mineral deficiency have been reported including vitamins B12 and D and calcium.[24-27]

Are novel protein diets beneficial in treating IBD?

Feeding a novel protein diet assumes several things: (1) a knowledge of dietary protein exposure, (2) clinical signs are due to the dietary protein serving as an allergen, and (3) the novel protein diet chosen contains only the protein listed. What constitutes an ideal novel protein is unknown and likely variable depending on the patient's previous exposure. There are no published studies that have compared response of dogs or cats with IBD to an intact novel protein diet with a hydrolyzed protein diet. Response to controlled dietary change using intact proteins is good ranging from 50-100%. Dogs with food responsive IBD tend to be younger in age and have less severe clinical signs than in dogs requiring immunosuppressive therapy.[2, 28] In one study, dogs with food responsive IBD responded to an elimination diet of intact protein and response occurred in 10-14 days.[2] The diet was fed for 14 weeks and 31 of 39 dogs were able to be transitioned to their original diet while 8 required continued dietary modification. A similar result was found in two other studies.[29, 30] In a study of 55 cats with chronic GI signs, 27 were diagnosed as food responsive or food allergic based on response to a diet utilizing intact proteins and 11 of these cats could be converted to their original diet.[8] In one study, approximately 50% of cats with lymphocytic/plasmacytic gastroenteritis responded to a change in intact protein diets.[31] In another retrospective study, 100% of cats responded to dietary change using a different intact protein with or without additional pharmacologic therapy.[32]

Are protein hydrolysate diets beneficial in treating IBD?

The primary aim of hydrolyzing proteins for specialized diets is to sufficiently disrupt the protein structure within the diet to remove any existing allergens and allergenic epitopes and thereby prevent immune recognition by patients already sensitized to the intact protein. Because proteins with molecular weights over 18,000 Daltons are incriminated as being antigenic, modification of proteins to compounds having lower molecular weight may be of benefit. Reducing the average weight of the protein molecule to less than 18,000 Daltons can result in a protein that may be truly hypoallergenic.[33] There are a few studies demonstrating 50-100% efficacy of hydrolyzed protein diets in managing IBD in dogs and cats;[34-37] however, some patients may have increased cutaneous signs.[38] Available commercial hydrolyzed protein diets are complete and balanced for adult dogs or cats. Consumption of protein hydrolysates may result in a more rapid uptake of amino acids when compared with whole proteins or amino acids in mixtures.[19] Soy hydrolysate diets have been shown to be tolerated in dogs with known soy hypersensitivity.[39]

Are homemade diets beneficial in treating IBD?

Homemade diets allow control over ingredients and truly novel ingredient diets may be formulated. Because some dietary ingredients may become antigenic due to processing, feeding a whole ingredient or raw food diet means there is no processing. Homemade diets are typically more digestible and smaller quantities may be fed. In one survey, 90% of homemade elimination diets prescribed by 116 veterinarians in North America were not nutritionally adequate for adult dog or cat maintenance. Few of the recipes available in books, magazines, and on-line have been tested to document the nutritional adequacy of the diet. There are common nutrient problems in many homemade foods. Many formulations are deficient in calories, calcium, vitamins, and micro-minerals.[40-42] Nutritionally complete and balanced diets can be formulated by board-certified veterinary nutritionists and a listing can be found at the American College of Veterinary Nutrition website: http://www.acvn.org.

Are there other nutritional modifications that may be useful with IBD?

Probiotics are live microbial feed supplements that beneficially affect the host by improving the intestinal microbial balance and so correct dysbiosis that occurs with IBD. Prebiotics are indigestible fibrous products that selectively promote the growth or activity of commensal microorganisms, also improving the well-being of the host. Prebiotics are carbohydrates that are not digestible in monogastric animals but are selectively fermented by commensal bacteria (e.g. Bifidobacteria) including inulin, fructooligosaccharides, mannanoligosacharides and resistant starch. Their fermentation produces short-chain fatty acids that have anti-inflammatory properties, promotes colonic epithelial integrity and healing, reduces neoplasia, and more. Their use has been minimally evaluated in dogs and cats but have shown benefit in dogs with large bowel diarrhea[43] and in altering small intestinal bacterial population.[44, 45] Probiotics have been shown to be beneficial in dogs with IBD.[46, 47] Omega-3 fatty acids modulate the immune response and exert an anti-inflammatory effect. Their use is documented in human beings with IBD; however, there are no studies in dogs and cats evaluating their effect.[48]

REFERENCES

[1] Jergens AE, et al. Front Biosci (Elite Ed). 2012;4(1404. [2] Allenspach K, et al. J Vet Intern Med. 2007;21(4):700. [3] Sartor RB Nat Clin Pract Gastroenterol Hepatol. 2006;3(7):390. [4] Nitzan O, et al. World J Gastroenterol. 2016;22(3):1078. [5] Westermarck E, et al. J Vet Intern Med. 2005;19(2):177. [6] Simpson KW, et al. Infect Immun. 2006;74(8):4778. [7] Twedt DC, et al. J Feline Med Surg. 2014;16(2):109. [8] Guilford WG, et al. J Vet Intern Med. 2001;15(1):7. [9] Bethlehem S, et al. Vet Immunol Immunopathol. 2012;145(3-4):582. [10] Zimmer A, et al. Vet Immunol Immunopathol. 2011;144(3-4):442. [11] van Rooijen C, et al. J Agric Food Chem. 2014;62(35):8883. [12] Simpson KW, et al. Vet Clin North Am Small Anim Pract. 2011;41(2):381. [13] Roudebush P Vet Dermatol. 2013;24(2):293. [14] Maine IR, et al. Acta Vet Scand. 2015;57(7. [15] Okuma TA, et al. Food Control. 2015;50(9. [16] Raditic DM, et al. J Anim Physiol Anim Nutr (Berl). 2011;95(1):90. [17] Parr JM, et al. J Am Anim Hosp Assoc. 2014;50(5):298. [18] Willis-Mahn C, et al. J Am Anim Hosp Assoc. 2014;50(6):383. [19] Lan A, et al. Inflamm Bowel Dis. 2015;21(1):198. [20] Salas A, et al. J Nutr Sci. 2014;3(e62. [21] Okanishi H, et al. J Vet Intern Med. 2014;28(3):809. [22] Shah ND, et al. Nutr Clin Pract. 2015;30(4):462. [23] Ugarte C, et al. J Nutr. 2004;134(8 Suppl):2068S. [24] Lalor S, et al. J Vet Intern Med. 2014;28(2):351. [25] Berghoff N, et al. Am J Vet Res. 2013;74(1):84. [26] Gow AG, et al. J Small Anim Pract. 2011;52(8):411. [27] Whitehead J, et al. J Am Anim Hosp Assoc. 2015;51(6):380. [28] Munster M, et al. Berl Munch Tierarztl Wochenschr. 2006;119(11-12):493. [29] Luckschander N, et al. J Vet Intern Med. 2006;20(2):221. [30] Schreiner NM, et al. J Vet Intern Med. 2008;22(5):1079. [31] Dennis JS, et al. J Am Vet Med Assoc. 1992;200(11):1712. [32] Jergens AE, et al. J Vet Intern Med. 2010;24(5):1027. [33] Cave NJ Vet Clin North Am Small Anim Pract. 2006;36(6):1251. [34] Mandigers PJ, et al. Tijdschr Diergeneeskd. 2010;135(18):668. [35] Mandigers PJ, et al. J Vet Intern Med. 2010;24(6):1350. [36] Marks SL, et al. Vet Ther. 2002;3(2):109. [37] Biourge VC,

et al. J Nutr. 2004;134(8 Suppl):2062S. [38] Olivry T, et al. Vet Dermatol. 2010;21(1):32. [39] Puigdemont A, et al. Am J Vet Res. 2006;67(3):484. [40] Freeman LM, et al. J Am Vet Med Assoc. 2001;218(5):705. [41] Larsen JA, et al. J Am Vet Med Assoc. 2012;240(5):532. [42] Remillard RL Top Companion Anim Med. 2008;23(3):137. [43] Leib MS J Vet Intern Med. 2000;14(1):27. [44] Willard MD, et al. Am J Vet Res. 1994;55(5):654. [45] Zentek J, et al. J Nutr. 2002;132(6 Suppl 2):1682S. [46] Sauter SN, et al. J Anim Physiol Anim Nutr (Berl). 2006;90(7-8):269. [47] Rossi G, et al. PLoS One. 2014;9(4):e94699. [48] Parian AM, et al. Nutr Clin Pract. 2015;30(4):551.

Inflammatory Bowel Disease: Complementary and Integrative Therapies

Donna M. Raditic, DVM, DACVN, CVA
Stamford, CT

INTRODUCTION

In 2014 the National Center for Complementary and Integrative Care (NCCIH) stated that complementary approaches are no longer considered an 'alternative' to conventional medical care, but are used in combination to provide what is defined as integrative health care. Complementary approaches include herbs, dietary supplements, acupuncture and other modalities that are rational and supported by evidence to alleviate physical symptoms, improve quality of life (QOL) and prevent disease. There is consumer demand for the integrative health care and we are challenged to understand these nontraditional therapies. The NCCIH states that despite widespread use of integrative health care, scientific evidence for many complementary approaches is still needed.[1]

A survey of adults attending an outpatient GI clinic reported the prevalence of complementary medicine (CM) use was 44%, more often in females and those dissatisfied with conventional treatments. Among those using CM, 54% reported taking a supplement or dietary modification, citing "they wish to feel better" (68%) as the most common reason for CM use. In this study 30% of patients stated they did not discuss CM with their physicians citing the primary reason was the physician did not ask (82%). It was concluded that a discussion of CM should be part of a GI office visit and gastroenterologists should have a greater awareness of CM to identify potential side effects and promote a more holistic, patient-centered approach to health.[2] Another study surveyed the American Gastroenterological Association, reporting that although gastroenterologists lacked formal training in CM (65%) they were receptive to its use and most felt it had a role in the treatment of IBD without necessarily compromising conventional therapy.[3]

Considering the complex bond between owners and pets, it would be logical that owners of pets with GI disease will apply complementary approaches to the treatment of their pets. This would be especially applicable in patients where conventional treatment is not eliminating all symptoms such as with IBD. Herbs and dietary supplements (HDS) are the most accessible form of complementary medicine. In 2014, pet owners spent approximately $191 million for dietary supplements for their companion animals. Owners reported wanting not only supplements for joints, but supplements for digestive issues. The evidence base for the use of relevant HDS for IBD was assessed via literature searches for dog/cat, human or in vitro cell lines. No meta- analysis and only a few randomized controlled clinical trials (RCCT) using HDS in canine IBD were found. There is a paucity of research in feline IBD.

TRANSLATIONAL RESEARCH

Human idiopathic IBD may be similar to canine IBD. Numerous studies of IBD in the veterinary literature compare and contrast results to human IBD including gene expression, helper T lymphocyte (Th) profiles, histopathology, biological markers, and response to treatments. Studies using high-throughput DNA sequencing techniques have determined fecal microbial phylogeny (e.g. predominance of Firmicutes, Bacteroidetes, Proteobacteria, and Actinobacteria) and their functional capacity (e.g. related to carbohydrate, protein, DNA, vitamin metabolism, virulence factors) in the canine and feline gut. These findings are overall very similar to what has been found in the human gut. Canine IBD represents a potential model to study human IBD and parallel studies could lead to more information in man while providing better therapeutics in canine IBD.[4-12] Acknowledging the translational and clinical science role of this disorder, the evidence base for the use of relevant HDS for human IBD was assessed. Over a hundred randomized controlled clinical trials (RCCT) using HDS in human IBD were found.

THE INTESTINAL MICROBIOTA

IBD is an idiopathic cause of chronic gastrointestinal disease in dogs, cats and humans. Three main factors are considered fundamental to the pathogenesis of idiopathic IBD: the interactions between the mucosal immune system, host genetic susceptibility, and environmental factors that include nutrition and the intestinal microbiota. As noted in human idiopathic IBD there is an 'intestinal dysbiosis' with a decrease in bacterial diversity; this same phenomenon has been reported in companion animal IBD.

Human idiopathic IBD is mainly defined as either ulcerative colitis (UC) or Crohn's disease (CD). The gut microbiota shows little temporal change in healthy individuals, whereas the microbiota of the IBD patient is unstable differing between active and quiescent stages. A longitudinal human study showed the microbiota in IBD patients remained unstable even in patients with UC in remission. Before relapses normal anaerobes are decreased and the intestinal microbiota is reduced. It is controversial whether dysbiosis is a cause or consequence of intestinal inflammation in IBD, but currently the literature favors the former. Studies in twins, affected and unaffected relatives and studies of NOD2 and ATG18L, two major CD susceptibility genes, suggest dysbiosis is caused by genetics and environmental factors rather than a consequence of chronic inflammation. In canine IBD, affected dogs have a lower abundance of Bacteroidetes and Firmicutes and higher abundance of Proteobacteria when compared to healthy dogs. The role of the intestinal microbiota in the pathogenesis of idiopathic IBD is supported by both human and animal studies and direct modulation of the intestinal microbiota may potentially be a treatment.[11-20]

DIETARY SUPPLEMENTS IN MANAGEMENT OF INFLAMMATORY BOWEL DISEASE

Probiotics

The most common HDS approach noted in PubMed searches was the use of probiotics. Probiotics are defined as live microorganisms that, when administered in adequate amounts confer a health benefit(s) to the host. In the human literature, two meta-analyses of IBD found some benefit from probiotics in inducing and maintaining remission in UC, especially when combined with conventional therapies. Meta-analysis in patients with CD did not find a significant benefit of probiotics in preventing post-operative recurrences, decreasing relapses or maintaining remission.[11-17]

Probiotic studies in the veterinary literature have looked at probiotics in acute diarrhea or chronic enteropathies such as canine food responsive diarrhea with only one study, Rossi et al, evaluating probiotic in 20 dogs with idiopathic IBD.[17-20] This randomized 90 day open label study compared the effects of the probiotic SIVOY (VSL Pharmaceuticals, Inc. Gaithersburg, MD, USA), a high- dose multi-strain probiotic product containing four strains of viable lyophilized bacteria to combination therapy with prednisone and metronidazole in canine IBD. The probiotic is a pet product using bacteria strains from VSL#3 which has shown some efficacy in humans for the prevention, treatment and maintenance of remission of pouchitis and ulcerative colitis in adults and children. The probiotic group (D-VSL#3; n=10) received 112-225 billion bacteria per 10 kg; the combination drug therapy (D-CT; n=10) received a combination of 20 mg/kg q 12h metronidazole and 1mg/kg/day prednisone for 60 days.

Tissue sampling for histological scores, CIBDAI and TGF-β+, FoxP3+, and CD3+ T-cell expression were determined after enrollment (time point T0) and 30 days after completing 60 days of treatment(T1) in both groups. The study reported a decrease in duodenal histological scores, CIBDAI and infiltration with mucosal CD3+ in both treatment groups with no significant effect between treatments. TGF-β+ increased in both groups, but the magnitude of this increase was greater in the D-VSL#3 dogs. FoxP3+ increased in the D-VSL#3 group after treatment, but not in the D-CT group. Microbiota analysis showed that at T0 all dogs showed a statistically significant decreased abundance of *Faecalibacterium spp.* and *Turicibacter spp.* when compared to 10 healthy dogs (one time only). *Faecalibacterium spp.* increased significantly in the D-VSL#3 dogs as well as with a trend to increase in the D-CT. This bacterium has been reported to increase with conventional therapies in IBD dogs and could be an indicator of normalization of dysbiosis.

The study suggested that the enhancement of the regulator T-cell markers (FoxP3+ and TGF-β+) along with normalization of dysbiosis might indicate a protective effect of the probiotic as compared to combination therapy. It was concluded probiotics induced differential anti-inflammatory responses when compared to conventional combination treatment. This was not a superiority study as there was no placebo group, but findings complement other studies of how probiotics may affect the mucosal immune system. Other studies of probiotics have demonstrated increases in anti-inflammatory cytokines IL-10 as well as TGF-β+ by modulation of NF-κB pathway and influencing the cross talk between natural killer cells and dendritic cells. Although study limitations exist, this study suggests a role for probiotics in canine IBD.

Prebiotics & Symbiotics

Prebiotics are defined as selectively fermentable ingredients that result in specific changes in the composition and/or activity of the GI microbiota, thus conferring benefit(s) upon host health. When prebiotics reach the colon they are rapidly fermented to short chain fatty acids (SCFA) acetate, propionic, and butyrate by resident bacteria. These SCFA especially butyrate activate nuclear peroxisome-proliferator- activated receptor γ which antagonizes inflammatory pathways. Studies using fructo-oligosaccharide (FOS) and oligofructose-enriched inulin in CD patients did not show any significant benefits. One small study in CD and prebiotic reported a decrease in activity scores and increased *Bifidobacteria* species. A symbiotic, defined as the combination of a probiotic and prebiotic, was shown to improve symptoms, pro-inflammatory cytokines and increased *Bifidobacteria* content in active CD patients. The role of prebiotics and IBD in veterinary medicine has been minimally investigated.[11-17]

Fecal Microbiota Transplantation

Fecal microbiota transplantation (FMT) has emerged as a treatment for severe, recurrent human *Clostridium difficile* infection (CDI). It is not a new therapy, first reported in 4th century China and in the 17th century in veterinary medicine. Tranplantantion has been carried out both orally and rectally, and has been termed transfaunation. Systematic review and meta- analysis of FMT in IBD determined more RCT's are needed. Weese et al. reported using stool transplantation enemas in a dog with eosinophilic IBD and a cat with refractory GI symptoms.[21-24]

Nutrient: Vitamin D

Studies in the human literature have investigated the role of specific nutrients and IBD including omega three fatty acids, fat-soluble vitamins (A, D, E, K), ascorbic acid, B12, folate, iron, and zinc. Prevalence of vitamin A and D deficiency has been reported as 26-93% (UC), 11-50% (CD) and 35% (UC), 75% (CD) respectively. Iron deficiency has been reported to have a prevalence of 81% in UC and 39% in CD.[25] As most dogs and cats are fed complete and balanced diets concerns for overt nutrient deficiencies in disease states has not been intensely investigated, including in canine and feline IBD.

Studies of the role of vitamin D in human idiopathic IBD including prevalence, optimal levels, prevention, quality of life and genomic technologies are abundant and would suggest a need for further investigation in vitamin D levels in canine and feline IBD.[25-30] A recent study evaluated vitamin D status in hospitalized cats. It was reported that the odds ratio of mortality within 30 days was 8.27 (95% confidence interval 2.54-31.52) for cats with a serum vitamin D in the lower tertile.[30]

Dietary levels of vitamin D in pet diets are adequate to prevent nutritional rickets however the amounts that correspond to adequate serum levels for cellular health have not been established. Sharp et al. has defined sufficient serum vitamin D levels to be greater than 100 ng/ml and evaluated serum vitamin D levels in dogs fed commercial dog food, homemade diets or combination. Serum concentrations varied in all diets, ranging from 9.5 to 249.2 ng/ml. It was concluded that serum vitamin D concentrations in dogs vary widely, likely as a consequence of dietary intake.[31]

Gow et al. reported dogs with IBD and hypoalbuminemia frequently have elevated ionized calcium, high PTH and low serum vitamin D concentrations. It could not be determined whether low serum vitamin D was due to low dietary intake from inappetence or increase loss of vitamin D through the GI tract. The authors favored the latter explanation, but concluded further studies were indicated to establish the pathogenesis of this IBD complication.[32] A recent study evaluated the relationship between vitamin D and markers of systemic and gastrointestinal inflammation in a cohort of dogs with CE. This study concluded serum vitamin D concentrations were inversely associated with systemic markers of inflammation in dogs with CE and with severity of inflammatory changes seen in histopathological samples. The authors reported that although hypovitaminosis D can be seen with intestinal disease, newer information to identify its contribution to the initiation and perpetuation of intestinal inflammation needs to be determined.[33] The role of vitamin A status in human IBD and intestinal inflammation has been similarly questioned. It is known that retinoic acid induces FoxP3+ T cells, which regulate intestinal inflammation. Low levels of carotene, a vitamin A precursor, and vitamin A (as serum retinol) have been reported in CD.[25, 34]

These findings suggest that, despite the concept of complete and balanced commercial pet diets, there is a need to assess vitamin D status in canine and feline IBD where inappetence, vomiting, diarrhea, and malabsorption together with diet variation may affect serum vitamin D levels, similar to human IBD. The diet study also noted higher serum vitamin D in dogs given salmon oil. As vitamin D is fat soluble and absorption is enhanced with fat intake, IBD patients with severe disease and low fat diet may be more at risk of hypovitaminosis D. Studies in the role of vitamin D and canine and feline IBD suggest vitamin D status should be assessed, treated and monitored. It should be noted that parental vitamin A and D solutions can be easily administered orally or subcutaneously, and have been used safely for other companion animal disease states.

Nutrient: Iron

In the human IBD literature, it is noteworthy that anemia is considered the most common systemic complication and extra intestinal manifestation of IBD. Consensus statements recommend prevention of anemia and maintenance of iron stores in IBD patients as the impact of anemia on QOL, ability to work, hospitalization and health care costs is substantial. The major forms of anemia of IBD are defined as iron deficiency anemia (IDA), anemia of chronic disease (ACD) and anemia of mixed origin. Numerous studies evaluating diagnosis, prevention and treatment of IBD associated anemia abound in the human literature. The concept of our pets' daily consumption of a complete and balanced diet may have slowed translational research and investigation into the role of IDA in canine and feline IBD. As the diagnostics of anemia and iron metabolism in dogs and cats are being reevaluated, it would be timely to reassess the role of anemia in veterinary IBD patients.[35-43] Parental iron has been used in suspect cases of IDA in canine IBD with positive outcomes.

HERBS AND THE MANAGEMENT OF INFLAMMATORY BOWEL DISEASE

A review of complementary therapies compared to conventional treatments assessed the level of evidence based on Strength of Recommendation Taxonomy (SORT) as follows: Grade A based on consistent, good-quality oriented evidence such as systematic reviews or meta-analysis or high quality patient orientated RCCT, Grade B is based on inconsistent or limited-quality patient-oriented evidence, and Grade C is based on consensus, usual practice, opinion, or disease-oriented evidence. This comparison also assessed level of risk of each therapy. Curcumin and Boswellia were given a SORT rating of B, with level of risk rated low for Boswellia. Corticosteroid, aminosalicylates, and Infliximab had a SORT of A, but level of risk was high. Similar results are noted in other systematic reviews of studies of the use of curcumin and Boswellia in IBD.[11,44-46]

REFERENCES

[1]Raditic D Vet Clinics Sm Anim 2015;45:857.[2]Hung A, et al. Dig Dis Sci 2015;60:1883.[3]Gallinger ZR, et al. J Compl Integr Med 2014;11(4):297.[4]Cerquetella M, et al, World J Gastroenterol 2010;16(9):1050.[5]Junginger J, et al. Innate Imm 2014;20(2):145.[6]Wilke VL, et al. AJVR 2012;73(8):1219.[7]Heilimann RM, et al. J Vet Immuno Immunopath 2015;168:131. [8]Heilimann RM, et al. J Vet Immuno Immunopath 2014;158:156.[9]Jergens AE, et al. Frontiers in Biosci 2012;E4:1404.[10]Minamoto Y, et al. Gut Microbes 2015;6(1):33.[11]Parian AM, et al. Nutr Clin Pract 2015;30(4):551.[12]Matsuoka K, et al. Semin Immunopathol 2015;37:47.[13]Verbeke KA, et al. Proc Nutr Society 2014;73:490.[14]Mardini HE, et al. Inflamm Bowel Dis 2014;20:1562.[15]Steed H, et al, Ailment Pharmacol Ther 2010;32:872.[16]Cammarota G, et al. Pharm & Therap 2015;149:191.[17]Honneffer, et a.l World J Gastroenterol 2014;20(44):16489.[18]Schmitz A, et al. JVMA 2015;29(2):533.[19]Sauter SN, et al. J Anim Physiol Anim Nutr 2006;90(7-8):269.[20]Rossi G, et a.l PLoS One 2014;9(4):e94699.[21]Colman, et al. J Crohn's

Colitis 2014;8(2):1569.[22]Brandt, et al, Gastrointest Endoscopy 2013;78(2):240.[23]Wang Z, et al. World J Gastroenterol 2014;20(40):14805.[24]Weese et al ACVIM 2013.[25]Massironi S, et al. Clinic Nutr 2013;32:904.[26]Wasko, et al. Adv Clin Exp Med 2012;21(1):5.[27]Reich KM, et a. World J Gastroenterol 2014;20(17):4934.[28]Raftery T, et al. Proc Nutr Soc 2014;74:56.[29]Ferguson LR, et al. Mol Nutr Res 2016;60:119.[30]Titmarsh HF, et al. PLoS One 2015;10(15):e0125997.[31]Sharp, et al. BMC Res Notes 2015;8:442.[32]Gow AG et al. J of Sm An Pract 2011;52:411.[33]Titmarsh HF, et al. PLoS One 2015;10(9):e0137377.[34]Kang, et al. Gastroenterol 2009;137(4):1391.[35]Dignass AU, et al. J of Crohn's Colitis 2015:211. [36]Neilsen O,H et a. Medicine 2015;94(23):1.[37]Guagnozzi, et al. World J Gastroenterol 2014;20(3):3542.[38]Ridgeway J et al JAAHA 2001;37(1):65.[39]Ristic JE et al J Sm An Pract 2002;43(2):80.[40]Craven M et al J Sm An Pract 2004;45(7):336.[41]Radokovich LB, et al. Vet Clin Path 2015;44:511.[42]Grimes CN, et al. Vet Path 2015;52(2):298.[43]Kogika MM, et al. Vet Clin Path 2015;44:70.[44]Ng SY, et al. Ailment Pharmacol Ther 2013;38:854.[45]Langhorst J, et al. J of Crohn's and Colitis 2015:86.[46]Triantafyllidi A, et al. Ann Gastroenterol 2015:28(2):210.

IMHA 101: An Update on Pathophysiology and Diagnosis

Oliver A. Garden, BSc, BVetMed, PhD, FHEA, FRSB, FRSA, MRCVS, DACVIM, DECVIM-CA
London, UK

INTRODUCTION

Despite numerous studies on the pathobiology, diagnosis and treatment of canine immune-mediated hemolytic anemia (IMHA) since its first description in 1957, it continues to be a disease that claims the lives of up to 70% of affected dogs[1-3]. It is a prominent problem in a number of popular breeds, including Cocker and Springer Spaniels, Old English Sheepdogs, Bichon Frise, Bearded and Rough-coated Collies, Poodles, and Flat-coated Retrievers[4]. Though sparse in comparison, knowledge of feline IMHA is growing[5]. Primary disease is being increasingly recognized in cats[5] and factors impacting prognosis have been characterized in both species[3,5-8]. While there is still much to be learned about the pathophysiology of this disease, the prominent inciting red blood cell (RBC) autoantigens and principal cellular players implicated in RBC lysis are well recognized. This presentation will summarize what is known about the pathophysiology of IMHA in dogs and cats, making reference to relevant murine and human literature, and consider both well-established and more controversial diagnostic tests for the disease.

PATHOMECHANISMS OF HEMOLYSIS: IMMUNOBIOLOGY AND TRIGGER FACTORS

Canine IMHA resembles human warm autoimmune haemolytic anemia (wAIHA), for which various pathomechanisms have been characterized[9,10]. RBC destruction occurs following the binding of autoantibodies to red cell surface epitopes, with or without complement activation. Opsonized RBCs may be phagocytosed by activated mononuclear cells expressing Fc receptors, commonly yielding spherocytes in the process if phagocytosis is incomplete; mononuclear ectoenzymes may also cause micro-perforations of the RBC surface, facilitating the formation of spherocytes[11]. RBCs are commonly opsonized by immunoglobulin (Ig) G – isotype IgG1 and less frequently IgG3 – and phagocytosis of IgG-coated RBCs, a route to extravascular hemolysis, occurs predominantly in the spleen; IgM is much less commonly detected on the surface of RBCs in wAIHA, though this may represent their low affinity and artefactual elution from the RBC surface, while IgA is most often found in association with IgG in up to 20% of patients[9,10]. The amount of antigen-antibody complex on RBCs may be sufficient to activate the complement factor C1 by the classical complement pathway, leading to the synthesis of C3b[10]. Certain isotypes of IgG (e.g. IgG3 and to a lesser extent IgG1) activate complement efficiently, as does IgM; other isotypes of IgG (e.g. IgG4) and IgA are thought not to activate complement. Phagocytosis of C3b-opsonized RBCs by hepatic Kupffer cells represents another pathomechanism of extravascular hemolysis in wAIHA, while intravascular hemolysis mediated by the terminal complement pathway yielding the membrane attack complex is less prominent in this disease[12]. Antibody-dependent cellular cytotoxicity may also be induced by surface autoantibodies of the IgG class upon ligation of Fc receptors expressed by NK cells and CTLs; this process occurs predominantly in the spleen. Some autoantibodies may induce eryptosis (apoptosis of RBCs) and rare forms of exclusively IgA-mediated wAIHA are thought to involve membrane exchange (trogocytosis) between opsonized RBCs and mononuclear cells in the spleen, leading to Fc-independent hemagglutination and sequestration *in situ*[13].

The predominant autoantigens in wAIHA are Rhesus peptides, Band 3 protein and Glycophorin A[14]. While much is known about the distal pathomechanisms in wAIHA, less is known about the proximal triggers to loss of self-tolerance[15,16]. Why does the immune system inappropriately target RBCs? The answer is unlikely to be simple and involves a complex interplay of genetic and environmental factors, the latter of which may include exposure to drugs, vaccines, xenobiotics, infectious agents, or neoplasia[15,16]. There is increasing interest in the way in which infectious agents may trigger immune-mediated disease by means of molecular mimicry between pathogenic and auto-antigens; similar mechanisms are likely to play a role in para-neoplastic or drug-induced wAIHA, the latter of which may involve the creation of neo-epitopes by the association of small drug molecules acting as haptens with host tissues, eliciting autoaggressive responses to the hapten-carrier complex. Much research has been undertaken on the genetics and epigenetics of autoimmune disease in human patients, including wAIHA, but the mechanistic ramifications of these genetic signatures often remain incompletely characterized[17,18]. Decreased frequency and function of peripheral regulatory T cells (Tregs) have been documented in murine models[19,20] and human patients[21,22], leading to the prevailing view that defective peripheral regulation of the adaptive immune response is a pre-requisite for clinical disease. This reconciles with models of the pathogenesis of a number of human organ-specific and polysystemic autoimmune diseases, in which similar observations have been made[23]. However, Tregs operate within a complex network of interacting lymphoid and myeloid regulatory cells and the situation is likely to be much more complex than these isolated observations would suggest.

Similar pathomechanisms to those of wAIHA are assumed to be responsible for IMHA of dogs and cats, in which hallmarks of extravascular hemolysis (spherocytosis) or intravascular hemolysis (ghost cells), or both, are recognized in individual patients[2,5,24]. Surface antibodies of IgG and IgM class, and C3, have been documented on canine[25-27] and feline[28] RBCs in IMHA, but there is little consensus on their respective prevalence or pathogenic significance. The major canine RBC autoantigens are thought to be the glycophorins[29,30], though recent work has also implicated additional membrane antigens[31]. Antibodies against Band 3 are found in both canine pIMHA patients and healthy, non-anemic dogs[31], underlining the multifactorial nature of the disease, in which the presence of anti-RBC antibodies alone is insufficient to elicit unregulated, pathological hemolysis. Indeed, the role of autoantibodies in the removal of senescent RBCs in health is well recognized[32]. The specific autoantigens implicated in feline IMHA have not been investigated to date.

Reconciling with the known breed disposition to IMHA in dogs and the immune-mediated pathogenesis of erythrolysis, specific MHC (dog leukocyte antigen) class II haplotypes were associated with IMHA within individual breeds[33]. The association of different MHC haplotypes with IMHA in different canine breeds accords with the observation that different human ethnic groups can have different human leukocyte antigen associations with the same immune-mediated disease. Little is known about the genetics of IMHA in cats, the majority of cases occurring in the relatively outbred domestic short-haired breed[5,28,34].

PATHOMECHANISMS OF HYPERCOAGULABILITY

Pulmonary thromboembolism (PTE)[35] is a significant cause of mortality in IMHA[35-40], prompting consideration of Virchow's triad of contributing factors – vascular stasis, hypercoagulability and endothelial damage[41]. Stasis alone does not appear to be sufficient to induce coagulation unless it is combined with endothelial damage and, or hypercoagulability[42]. The pathogenesis of thromboembolism in IMHA is thought to be multi-factorial[43,44], including:

- Hypercoagulability – potential causes include (i) disseminated intravascular coagulation (DIC)[36,38] secondary to tissue factor activation by anti-red cell antibodies, lyzed RBCs and inflammatory mediators[45]; (ii) platelet activation[46,47]; (iii) corticosteroid administration[48]; and (iv) loss of antithrombin (AT) and hypoalbuminemia – with increased platelet aggregability – if there is concurrent glomerulonephritis (GN) and protein-losing nephropathy (PLN)[49-51]. Current evidence suggests that antiphospholipid antibodies do not play a major role in thromboembolism of canine IMHA[38], in contrast to their significance in human IMHA and SLE; however, a lupus-like 'anticoagulant', which prolongs the activated partial thromboplastin time (APTT) but paradoxically predisposes to thromboembolism *in vivo*, has been documented in the dog[52].
- Endothelial injury mediated by inflammatory mediators and hypoxia[43] – in contrast to human IMHA, anti-endothelial antibodies are unlikely to play a major role in the canine disease[53].
- Vascular stasis – thromboemboli tend to inhibit blood flow, exacerbating the ongoing process[43]; furthermore, vasodilatory responses may be blunted by the binding of nitric oxide by even trace amounts of free hemoglobin[54].

Recent studies have also implicated procoagulant microparticles in the thromboembolism of canine IMHA[55], and the presence of neutrophil extracellular traps – pro-thrombotic webs of DNA and protein extruded from activated neutrophils – in dogs with IMHA[56] raises the intriguing possibility that they also contribute to the thromboembolic complications of this disease. Thromboembolism appears to be an uncommon sequela of feline primary IMHA[5,34,57-59]. Little is known about thromboembolism in IMHA of cats, but similar pathomechanisms to those of the dog are assumed to play a role.

A RATIONAL DIAGNOSTIC APPROACH

How do we diagnose IMHA: what's in a name?

The diagnosis of human wAIHA is essentially based on two criteria, namely (i) serological evidence of an autoantibody by means of a positive direct antiglobulin (DAT; Coombs') test coupled with (ii) clinical and/or laboratory evidence of hemolysis[60]. Demonstration of a positive Coombs' test result is thus a key diagnostic criterion for wAIHA, though 3-10% of such patients have 'Coombs' test-negative wAIHA', in which the standard DAT yields a negative result but alternative tests, such as the complement-fixation antibody-consumption test or flow cytometry, confirm opsonisation of RBCs with antibody[61]. Potential reasons for a negative Coombs' test result in these cases include IgG sensitization below the threshold of detection of the commercial antiglobulin reagent; low affinity IgG, removed by preparatory washes not conducted at 4°C or low ionic strength; and RBC sensitization by IgA – or rarely monomeric IgM – alone, not accompanied by complement fixation and thus undetectable by the commercial reagent containing anti IgG and anti-C3[61].

Criteria for the diagnosis of IMHA in dogs are less clearly defined, not least because there is no single 'gold standard' diagnostic test in this species. Most clinicians make a diagnosis on the basis of (i) evidence of hemolysis, either of direct nature such as hemoglobinuria and/or the presence of ghost cells on a blood smear, or of indirect nature such as the absence of any plausible source of blood loss in a patient with moderate to marked anemia (i.e. not anemia of chronic disease); and (ii) at least one hallmark of immune-mediated pathogenesis, including the observation of in-saline agglutination, spherocytosis, a positive Coombs' test result, or splenic and/or bone marrow erythrophagocytosis by mononuclear cells[2,62,63]. Further retrospective support may come in the form of an encouraging response to immunosuppressive therapy. Congenital enzymopathies should always be considered in the differential diagnosis of regenerative anemia, since the clinicopathological manifestations may be deceptively similar to IMHA in some cases.[64-70]

Criteria for the diagnosis of feline IMHA are also less well defined than those for human wAIHA, usually including evidence of hemolysis coupled with one or more hallmarks of immune-mediated pathogenesis, such as in-saline agglutination and/or splenic or bone marrow erythrophagocytosis[5]. In both dogs and cats, as in human patients, appropriate diagnostic tests are undertaken to explore underlying trigger factors and some centers will submit tests to assess platelet function and coagulation status prior to anti-platelet or anti-coagulant therapy (assuming that thrombocytopenia is not present). However, each one of the diagnostic criteria in dogs and cats is open to debate and controversy, as discussed in the following sections.

In-saline agglutination: hallmark of IMHA #1?

Autoagglutination, either observed on a blood smear or following the admixture of a drop of EDTA blood with two to four drops of 0.9% saline solution on a slide, is commonly cited as a hallmark of an immune-mediated process. Positive in-saline agglutination is thought to be the consequence of IgM or large quantities of IgG autoantibody coating the RBCs[24,71] and has been observed in 35% to 78% of canine cases[36,63,72,73]. However, one study noted that only 4/42 (9.5%) of dogs demonstrated persistent autoagglutination following the washing of patient RBCs in 0.9% saline[74], reconciling with the findings of a more recent paper, which demonstrated that not a single one of the 20 autoagglutinating dogs – of a total of 46 anemic canine patients – showed persistence of agglutination after washing[75]. Moreover, Coombs' tests were performed in every one of these 20 dogs after washing the RBCs, yielding positive titers in 16 and negative titers in 4: in other words, there was no evidence of RBC opsonization in 4/20 (20%) of the autoagglutinating anemic dogs, raising the possibility that autoagglutination may represent a non-specific finding unrelated to IMHA in some cases[75]. Nevertheless, both autoagglutination without prior washing[63] and persistent autoagglutination with washing[74] have been associated with severe disease and higher rates of mortality in canine patients.

Data in the cat are more sparse, one report documenting autoagglutination in 10% of cases[76] and a second describing macroscopic agglutination in 100% cases, but persistent agglutination post-washing in only 21%[34]. The most recent and largest study of feline IMHA documented positive in-saline agglutination in 70/107 (65.4%) of cases, but RBCs were not washed prior to assessment in any of these cases[5]. On balance, the available literature suggests that a positive in-saline agglutination test result in an anemic dog or cat is strongly supportive of IMHA, but is neither pathognomonic for the disease nor precludes the performance of a Coombs' test if the RBCs are washed.

Spherocytosis: hallmark of IMHA #2?

Spherocytes, RBCs that have suffered damage to the plasma membrane or partial phagocytosis – with decreased surface area to volume ratio, assumption of a spherical shape and loss of central pallor – are often associated with immune-mediated disease, being observed in up to ~90% dogs with IMHA[77]. However, alternative differential diagnoses for spherocytosis may include severe hypophosphatemia[78,79], microangiopathic injury referable to disseminated intravascular coagulation, vasculitis and other causes such as splenic torsion[80], Heinz body haemolytic anemia[81], congenital red cell enzymopathies such as pyruvate kinase deficiency[66], zinc toxicosis[82] and coral snake envenomation[83]. Since feline erythrocytes lack obvious central pallor, spherocytes are much more difficult to discern and are generally not used as a diagnostic criterion in this species [81].

The Coombs' test: something old, something new?

The DAT or direct Coombs' test detects antibodies associated with the surface of RBCs and has been recommended as a diagnostic test for IMHA in both dogs and cats. The primary reagent is polyvalent rabbit anti-dog or anti-cat IgG, IgM and C3 antiserum, but a full test also uses a panel of monospecific antisera[72,76,84-90]. A recent study suggested that the sensitivity of the test could be optimised by inclusion of the incubation step at 4°C, which identified 10 of 65 canine cases that yielded negative results at 37°C[27]. A positive response is detected by agglutination of patient RBCs and an end-point dilution of the antiserum reagent is determined. Results are either reported as a reciprocal titer or a 'positive / negative' result, depending on the house style of the clinical pathology laboratory: while some studies have suggested that the magnitude of the titer may have diagnostic significance, becoming negative with remission[89,91], others have been unable to show such a correlation and indeed have suggested that in a minority of patients a positive titer may persist during remission[72,85].

The sensitivity of the direct Coombs' test in the dog has varied from 33% to 96% in different studies[36,63,73,88,91-95], though as a rule of thumb is generally thought to be positive in approximately two thirds of cases as traditionally performed; its specificity – when examined – has generally exceeded 90%[84,86,88], though one study documented weak positive titers in 19 of 38 dogs that did not meet additional criteria for IMHA[72]. Surface IgG – with or without IgM and, or C3 – is the most common finding in Coombs' tests employing monospecific antisera, accounting for 42 to 95% of positive results; IgM – with or without IgG and, or C3 – is less commonly observed, being present in 1.5 to 52% of cases; while C3 – with or without IgG and, or IgM – accounts for 5.3 to 87% of positive results[27,74,84,88,92,94]. One study documented a high prevalence of C3 alone in minimally anemic dogs, often with underlying disorders[84], while a more recent study suggested that dogs with specific reactivity at 4°C – rather than 37°C – were more likely to be suffering from secondary IMHA[27].

Data in the cat are sparse, reflecting the paucity of documented cases: a recent study suggested that IgG autoantibodies are more common than those of IgM isotype; furthermore, none of the cats in this series demonstrated Igs against C3[34]. However, an earlier study suggested that the direct Coombs' test lacked specificity in the cat, 8 of 20 non-anaemic, control cats yielding weak positive titers[76]. The predominant cause of false-negative results is thought to be the limited analytical sensitivity of the assay, which is unable to detect low surface densities of Igs or C3 on RBCs, which may nevertheless be of pathogenic significance[88,96-98]. This short-coming has prompted the development of the direct enzyme-linked antiglobulin test (DELAT), which is based on the elaboration of a color product by an enzyme conjugated to the antiserum reagent in the style of an ELISA test[96-98]. Though the DELAT has performed well in a number of studies, it is not widely available in practice. Still greater sensitivity has been achieved by the development of flow cytometric assays[93,99], though availability is limited to those diagnostic laboratories offering flow cytometry as a specific expertise. Recent studies have assessed the performance of variants of the conventional microtiter DAT, including gel microcolumn, capillary microtube and immunochromatographic strip techniques[75,100]. The gel and capillary methods corresponded well to the microtiter DAT, but the immunochromatographic strip method frequently yielded weak positive band intensities, which made interpretation of a 'positive' result challenging in some cases[75]. The gel method, though encouraging, is no longer commercially available.

The mediocre performance of the Coombs' test reported in much of the established literature has recently been challenged, in particular the impact of therapy and prior blood transfusion[75]. The results of Coombs' tests in anemic dogs were not influenced by immunosuppressive therapy for days to weeks after its inception in one study, becoming negative only when remission of the anemia occurred[75]. The impact of blood product administration was also assessed by the same authors, since the most common reported cause of false positive Coombs' test results is prior blood transfusion. Incompatible RBCs are thought to react with pre-existing alloantibodies or stimulate the production of new antibodies: both are assumed to be detectable in the Coombs' test, even in the absence of clinical evidence of transfusion incompatibility[85]. However, 11 anemic dogs not suspected of having IMHA, transfused one to 21 days before Coombs' testing took place, all yielded negative results; another 10 dogs suspected of having IMHA, transfused one day to five years before testing, were also DAT-negative[75]. Positive Coombs' test results have occasionally been reported in association with overt infection or neoplasia, even in the absence of anemia, thought to be of questionable pathogenic significance[84]; if there is attendant anemia, the positive Coombs' test results may signal secondary IMHA[27,84]. Finally, the low-level auto-Ig involved in the clearance of senescent RBCs in health[32,101] is thought to be below the sensitivity of the Coombs' test and is thus not a cause of false positive results.

Primary *versus* secondary disease: needles in haystacks?

A plethora of iatrogenic, xenobiotic, infectious and neoplastic trigger factors have been associated with immune-mediated disease in human and veterinary patients. Once a diagnosis of IMHA has been made, appropriate diagnostic tests to explore potential trigger factors should be considered. There is currently no consensus on the minimum or preferred suite of diagnostic tests to rule out known or suspected trigger factors, which will be predicated on a number of variables, such as owner wishes and resources, geographical locale, and access to advanced diagnostic techniques, but most clinicians would undertake imaging of the thorax and abdomen to hunt for occult septic or neoplastic lesions, and tests for vector-borne and other infectious agents according to the probability of exposure – in addition to a complete blood cell count with blood smear, serum biochemical profile and routine urinalysis. Additional tests, such as urine and/or blood culture, fecal flotation, Baermann assay or Knott test, echocardioagraphy, arthrocentesis, limb or spinal imaging, cerebellomedullary cisternal puncture, antinuclear antibody assay, and biopsies, would be considered if historical, physical examination and initial diagnostic test findings prompt them, especially in the context of polysystemic disease.

Successful therapy of secondary immune-mediated disease relies in part on addressing the underlying cause or trigger factor. In some cases, disclosure of such factors may have a major impact on therapy, for example if a septic focus is discovered. With increasing recognition of the ever-expanding spectrum of infectious agents and the impact of the microbiome on global immune function, the concept of primary disease is controversial and this author prefers use of the term 'cryptogenic' in recognition of the fact that some trigger factors may remain occult in a proportion of apparently idiopathic cases despite all of our best diagnostic efforts.

Tests of coagulation status: what do they add?

Hypercoagulability presents a challenge to measure in clinical practice, but the recent validation of thromboelastography (TEG) as a technique to assess global clotting function in dogs *in vitro* goes some way towards addressing this deficit[47,102-108]. However, samples need to be run within one to two hours of collection and the availability of TEG machines is currently limited; nevertheless, when available, this methodology yields information on both coagulation and fibrinolysis[35,109]. Clinical utility of TEG in cats is still in its infancy, though reference ranges for TEG parameters in cats have been defined[110,111]. Complementing TEG, the technique of platelet aggregometry is available in some centers to assess platelet function[112-114]. However, the correlation between both TEG and platelet aggregometry data with thromboembolism *in vivo* remains unclear, reflecting the many and varied factors impacting coagulation that cannot be measured *in vitro*. Perhaps one of the most informative applications of these tests is in the monitoring of patients on anti-platelet or anti-coagulant therapy, refining and complementing the information gained from more conventional tests of coagulation.

RESEARCH NEEDS AND CONCLUSIONS

Despite significant advances in our understanding of the clinical behavior of canine and feline IMHA, treatment of the disease has not progressed significantly in the past 50 years, still relying on non-specific immune suppression that is associated with significant morbidity[62,115]. There is thus an urgent unmet need to advance our understanding of the pathogenesis of IMHA in order to develop more targeted therapies in the future. Such work also promises to disclose novel biomarkers of disease that may facilitate an assessment of the impact of treatment on underlying pathophysiology, thus allowing more confident prediction of patient prognosis. Allied to such fundamental research efforts is a need to provide consensus on potential trigger factors for the disease in dogs and cats, and thus what diagnostic tests should be undertaken in order to diagnose 'primary IMHA'. The issue of vaccination as a trigger factor also remains controversial: should we be re-vaccinating dogs or cats whose immune-mediated disease occurred within three weeks of previous vaccination? These and other burning clinical questions relating to IMHA are being addressed by the *Hematology Task Force* of the Clinical Immunology Special Interest Group, who will be reporting their preliminary findings alongside a number of other task forces in the Interactive Session starting at 08:00 on 10th June here at the Forum, entitled: *'Veterinary Clinical Immunology: A New Sub-Specialty of Internal Medicine?'*

REFERENCES

[1]Lewis RM, et al Annals New York Acad Sci 1975;124:178; [2]Piek CJ Vet Q 2011;31:129; [3]Swann JW and Skelly BJ J Vet Intern Med 2015;29:7; [4]Miller SA, et al J Am Vet Med Assoc 2004;224:232; [5]Swann JW, et al J Vet Intern Med 2016;30:147; [6]Goggs R, et al J Vet Intern Med 2015;29:1603; [7]Balch A and Mackin A Comp Contin Educ Vet 2007;29:230; [8]Ishihara M, et al J Vet Med Sci 2010;72:465; [9]Barcellini W Transfus Med Hemother 2015;42:287; [10]Berentsen S and Sundic T BioMed Res Intern 2015;363278; [11]LoBuglio AF, et al Science 1967;158:1582; [12]Kurlander RJ, et al J Clin Invest 1978;61:1309; [13]Chadebech P, et al Blood 2010;116:4141; [14]Leddy JP, et al J Clin Invest 1993;91:1672; [15]Barros MM, et al Transfus Med Rev 2010;24:195; [16]Michel M Presse Med 2014;43:e97; [17]Aslani S, et al Autoimmunity 2016;doi 10.3109/08916934.2015.1134511; [18]Seldin MF J Autoimmun 2015;64:1; [19]Mqadmi A, et al Blood 2005;105:3746; [20]Russell PJ, et al Clin Exp Immunol 1981;45:496; [21]Ahmad E, et al J Investig Allergol Clin Immunol 2011;21:297; [22]Soyano A, et al Clin Immunol Immunopathol 1982;23:70; [23]Miyara M, et al Autoimmunity Rev 2011;10:744; [24]Balch A and Mackin A Compen Contin Educ Vet 2007;29:217; [25]Barker RN, et al Res Vet Sci 1993;54:170; [26]Quigley KA, et al J Vet Diag Investig 2001;13:297; [27]Warman SM, et al J Small Anim Pract 2008;49:525; [28]Tasker S, et al J Small Anim Pract 2010;51:192; [29]Barker RN and Elson CJ Vet Immunol Immunopathol 1995;47:225; [30]Corato A, et al Vet Immunol Immunopathol 1997;59:191; [31]Tan E, et al Vet Clin Pathol 2012;41:45; [32]Berkman P, et al Autoimmunity 2002;35:415; [33]Kennedy LJ, et al Tissue Antigens 2006;68:502; [34]Kohn B, et al J Vet

Intern Med 2006;20:159; [35]Goggs R, et al J Vet Emerg Crit Care 2009;19:30; [36]Carr AP, et al J Vet Intern Med 2002;16:504; [37]Johnson LR, et al J Vet Intern Med 1999;13:338; [38]Scott-Moncrieff JC, et al J Am Anim Hosp Assoc 2001;37:220; [39]LaRue MJ and Murtaugh RJ J Am Vet Med Assoc 1990;197:1368; [40]Mischke VR Dtsch Tierarztl Wschr 1998;105:13; [41]Virchow R Gessammelte Abhandlungen zur Wissenschaftlichin Medicin (Meidinger Sohnu Comp, Frankfurt, Germany) 1856:148; [42]Thomas DP, et al Br J Haematol 1985;59:449; [43]Johnson LR J Vet Intern Med 2002;16:501; [44]Kidd L and Mackman N J Vet Emerg Crit Care 2013;23:3; [45]Rabiner SF and Friedman LH Br J Haematol 1968;14:105; [46]Weiss DJ and Brazzell JL J Vet Intern Med 2006;20:682; [47]Hamzianpour N and Chan DL J Vet Emerg Crit Care 2015; doi 10.1111/vec.12315; [48]Larue MJ and Murtaugh RJ J Am Vet Med Assoc 1990;197:1368; [49]Green RA, et al J Am Vet Med Assoc 1985;186:485; [50]Rasedee A , et al Acta Vet Scand 1986;27:369; [51]Rasedee A and Feldman BF Vet Res Commun 1985;9:199; [52]Stone MS, et al J Vet Intern Med 1994;8:57; [53]Wells R, et al J Vet Intern Med 2009;23:295; [54]Houston DS, et al Blood 1990;76:953; [55]Kidd L, et al J Vet Intern Med 2015;29:908; [56]Jeffery U, et al 2015;168:262; [57]Schermerhorn T, et al 2004;18:533; [58]Norris CR, et al J Am Vet Med Assoc 1999;215:1650; [59]Husbands BD, et al J Vet Intern Med 2002;16:350; [60]Naik R Hematol Oncol Clin North Am 2015;29:445; [61]Segel GB and Lichtman MA Blood Cells Mol Dis 2014;52:152; [62]Swann JW and Skelly BJ J Small Anim Pract 2011;52:353; [63]Weinkle TK, et al J Am Vet Med Assoc 2005;226:1869; [64]Searcy GP, et al Am J Pathol 1979;94:689; [65]Searcy GP, et al Can J Comp Med 1971;35:67; [66]Giger U and Noble NA J Am Vet Med Assoc 1991;198:1755; [67]Schaer M, et al J Am Anim Hosp Assoc 1992;28:233; [68]Skibild E, et al J Small Anim Pract 2001;42:298; [69]Giger U, et al J Am Vet Med Assoc 1992;201:1569; [70]Kohn B and Fumi C J Fel Med Surg 2008;10:145; [71]Sharp C and Kerl ME Stand Care: Emerg Crit Care Med 2008;10.10:1; [72]Jacobs RM, et al J Am Vet Med Assoc 1984;185:66; [73]Reimer ME, et al J Am Anim Hosp Assoc 1999;35:384; [74]Klag AR, et al J Am Vet Med Assoc 1993;202:783; [75]Caviezel LL, et al J Vet Intern Med 2014;28:583; [76]Dunn JK, et al Can J Comp Med 1984;48:349; [77]Sharp C and Kerl ME Stand Care: Emerg Crit Care Med 2008;10.10:1; [78]Willard MD, et al J Am Vet Med Assoc 1987;190:1007; [79]Adams LG, et al J Vet Intern Med 1993;7:266; [80]Rebar AH, et al Vet Pathol 1981;18:415; [81]Allison RW and Meinkoth JH Vet Clin N Amer Small Anim Pract 2007;37:245; [82]Latimer KS, et al J Am Vet Med Assoc 1989;195:77; [83]Marks SL, et al J Am Anim Hosp Assoc 1990;26:629; [84]Slappendel RJ Vet Immunol Immunopathol 1979;1:49; [85]Werner LL Contin Educ Pract Vet 1980;II:96; [86]Jones DR, et al Res Vet Sci 1990;48:53; [87]McCullough S Vet Clin N Amer Small Anim Pract 2003;33:1295; [88]Overmann JA, et al Vet Clin Pathol 2007;36:179; [89]Day MJ Fel Pract 1996;24:7; [90]Day MJ Fel Pract 1996;24:14; [91]Mason N, et al J Vet Intern Med 2003;17, 206; [92]Jackson ML and Kruth SA Can Vet J 1985;26:245; [93]Quigley KA, et al J Vet Diag Invest 2001;13:297; [94]Piek CJ, et al J Vet Intern Med 2008;22:366; [95]Burgess K, et al J Vet Intern Med 2000;14:456; [96]Jones DR, et al Vet Immunol Immunopathol 1987;16:11; [97]Jones DR, et al Am J Vet Res 1992;53:457; [98]Barker RN, et al Vet Immunol Immunopathol 1992;34:1; [99]Morley P, et al J Vet Intern Med 2008;22:886; [100]Piek CJ, et al Acta Vet Scand 2012;54:10; [101]Kay MM Adv Exp Med Biol 1994;347:161; [102]Jessen LR, et al Vet Clin Pathol 2008;37:363; [103]Wiinberg B, et al J Vet Intern Med 2008;22:357; [104]Kristensen AT, et al J Vet Intern Med 2008;22:140; [105]Wiinberg B, et al Vet J 2007;174:62; [106]Wiinberg B, et al Vet Clin Pathol 2005;34:389; [107]Vilar P, et al J Vet Intern Med 2008;22:774; [108]Vilar P, et al J Vet Intern Med 2008;22:374; [109]Donahue SM and Otto CM J Vet Emerg Crit Care 2005;15:9; [110]Alwood AJ, et al J Vet Emerg Crit Care 2004;14:S1; [111]Bjornvad CR, et al J Vet Intern Med 2008;22:739; [112]Marschner CB, et al J Vet Emerg Crit Care 2012;22:107; [113]Magee AN, et al Anm J Vet Res 2014;75:309; [114]Defontis M, et al Exp Toxicol Pathol 2013;65:637; [115]Swann JW and Skelly BJ J Vet Intern Med 2013;27:1

Immunosuppressive Therapy in Dogs and Cats: Rational Use of Established Drugs

Andrew Mackin, BSc, BVMS, MVS, DVSc, FANZCVSc, DACVIM (SAIM)
Starkville, MS

A number of established immunosuppressive agents have been used in small animal medicine for many decades. Some have justifiably fallen out of favor whereas, for others, new and promising uses have been described in the recent veterinary literature. This lecture will discuss some 'old favorites': cyclophosphamide, chlorambucil, azathioprine, danazol and vincristine.

CYCLOPHOSPHAMIDE

Cyclophosphamide, a cell-cycle nonspecific nitrogen mustard derivative alkylating agent, was approved by the FDA over 50 years ago, and has since become well-established in human medicine as an antineoplastic drug and immunosuppressive agent. Cyclophosphamide has persisted as a core drug used in many small animal cancer chemotherapeutic protocols. In contrast, after years as one of the most commonly immunosuppressive drugs utilized to treat immune-mediated diseases in cats and dogs, the use of cyclophosphamide as an immunosuppressive agent in small animal patients has in the past two decades essentially faded away.

Cyclophosphamide is a prodrug that is metabolized by the hepatic cytochrome P450 enzyme system to eventually form active metabolites such as 4-hydroxycyclophosphamide, 4-hydroperoxycyclophosphamide, and aldophosphamide. These metabolites can enter cell cytoplasm, where they are ultimately metabolized to phosphoramide mustard and acrolein. Phosphoramide mustard is an alkylating agent that replaces a hydrogen atom with an alkyl group on the guanine base of DNA, which interferes with nuclear DNA replication and cytoplasmic RNA transcription by forming crosslinks. Cyclophosphamide has long been reported to be a potent immunosuppressive that inhibits humoral and cell-mediated immunity, including inhibition of primary and secondary immune responses, reduction of antigen trapping in lymph nodes, and inhibition of local inflammatory responses, although in a number of experimental studies in dogs, cyclophosphamide often appears to be mildly immunosuppressive compared to other drugs[1-3].

Cyclophosphamide, like most alkylating agents, can cause including GI side effects, myelosuppression, and hair loss. GI signs are relatively common, and include nausea, anorexia, vomiting and diarrhea. Myelosuppression appears to be dose dependent, and is associated with both the use of high drug doses and the use of lower doses over a sustained period of time. Myelosuppression is typically reversible with drug dose reduction or discontinuance, but can occasionally persist for weeks or even months, particularly after chronic cyclophosphamide therapy. Alopecia is most common is susceptible breeds such as poodles and old English sheepdogs. A major side effect of cyclophosphamide is sterile hemorrhagic cystitis[4,5]. Cystitis is mediated by urinary excretion of acrolein. Cystitis is often debilitating to the patient, and will not resolve until the drug is discontinued. Unfortunately, cystitis can sometimes persist for days or even weeks after drug discontinuation. The chronic local bladder inflammatory effects of cyclophosphamide have also been reported to predispose to the development of irreversible bladder wall fibrosis and transitional cell carcinoma[6].

Over the years, a number of immune-mediated and inflammatory diseases in dogs and cats have been treated with cyclophosphamide, including immune-mediated hemolytic anemia (IMHA), immune-mediated thrombocytopenia (IMT), megakaryocyte hypoplasia, pure red cell aplasia, systemic lupus erythematosus, immune-mediated polyarthritis, inflammatory bowel disease (IBD), glomerulonephritis, non-infectious inflammatory meningoencephalitis, immune-mediated vasculitis and pemphigus. Cyclophosphamide was long considered to be a 'big gun' for dogs with severe IMHA. However, nearly two decades ago, several case studies reported that cyclophosphamide was not better than steroids alone for the treatment of IMHA and, at worst, associated with a higher mortality rate[7-10]. Given the limitations of retrospective studies, including potential 'case selection bias' (dogs with severe IMHA may have been given cyclophosphamide because it was perceived to be more potent), it is hard to know with certainty whether cyclophosphamide actually worsens prognosis in dogs with IMHA. Nevertheless, there is no doubt that, since publication of these papers, the use of cyclophosphamide to treat conditions such as canine IMHA has markedly decreased.

Cyclophosphamide has historically been relatively cheap, but prices have increased markedly recently. Cyclophosphamide is difficult to dose accurately, and even more difficult to taper, especially in smaller patients. Cyclophosphamide tablets are composed of an active inner core surrounded by an outer shell. Because of uneven drug distribution through the tablet, cyclophosphamide tablets cannot be split without a risk of dosing inaccuracy. Without drug compounding, cyclophosphamide doses must therefore be in multiples of 25 or 50. Published immunosuppressive doses for cyclophosphamide in dogs include 50 mg/m^2 or 1.5 to 2.5 mg/kg every second day or daily on a '4 days on, 3 days off' weekly protocol. In cats and small dogs, similar total weekly doses can be used, but 'pulsed' at infrequent dosing intervals that ensure that the total weekly dose is equivalent to 7× the calculated daily dose. Complete blood counts must be performed regularly throughout the course of treatment. Cyclophosphamide is available in an IV form as well as an oral form, and a recent study in dogs confirmed that equivalent oral and IV doses of cyclophosphamide achieved comparable blood levels of the active metabolite 4-hydroxycyclophosphamide[11]. Intravenous cyclophosphamide may therefore be a viable treatment option in vomiting animals that are unable to tolerate oral immunosuppressive agents.

CHLORAMBUCIL

Chlorambucil is a nitrogen mustard derivative cell-cycle nonspecific alkylating agent that has, for many decades, been used in both human and veterinary medicine as an antineoplastic agent. Antineoplastic cytotoxicity is derived from cross-linkage of DNA and RNA by insertion of alkyl radicals on the purine base, guanine. Chlorambucil also has immunosuppressive properties, and has occasionally been used in human medicine to treat conditions such as glomerulonephritis. More than 30 years ago, some clinicians began suggesting the use of chlorambucil as an immunosuppressive agent for our small animal patients. Since then, the use of chlorambucil for the treatment of a number of feline dermatoses, such as pemphigus and eosinophilic granuloma complex, and for IMT and refractory IBD, has become well established[12-14], perhaps because of a paucity of viable alternative medications that could be accurately dosed with safety in cats. The use of chlorambucil as immunosuppressive agent in dogs has been slower to evolve, but its use has been described for the treatment of pemphigus, glomerulonephritis and, most recently, IBD[15].

Chlorambucil is metabolized predominantly in the liver, primarily to the active metabolite phenylacetic acid mustard. Compared to other alkylating agents, chlorambucil is relatively well tolerated, but does occasionally cause GI side effects such as vomiting and diarrhea, and/or myelosuppression. Alopecia and poor hair growth are sometimes reported in susceptible dog breeds, such as poodles. Chlorambucil-associated neurologic signs (including myoclonus, twitches and seizures) have been reported in cats[16].

Chlorambucil is available as a coated 2 mg tablet that cannot feasibly be divided, and dosing recommendations in smaller patients are therefore typically provided in multiples of two, and/or 'pulsed' at infrequent dosing intervals in order to avoid overdose. Chlorambucil is almost always given in combination with a glucocorticoid. In dogs, recommended starting oral immunosuppressive chlorambucil doses range from 0.1 to 0.2 mg/kg (or, alternatively, 4-6 mg/m^2) every 1-2 days. In cats (and small dogs), a starting oral chlorambucil dose of 2 mg every 2nd day, tapered to every 3rd or 4th day, is my preferred dosing immunosuppressive regime. Lower daily doses of chlorambucil, comparable to dog dosing regimes, can also be used in cats if the drug is compounded, but the effects of compounding on drug efficacy are not known. Complete blood counts must be monitored regularly and, since myelosuppression tends to be dose-dependent rather than idiosyncratic, doses can be tapered 'to effect' rather than discontinued completely.

Chlorambucil has until recently been moderately priced. Unfortunately, the patent on the only available product, Leukeran®, recently expired, leading to a change in drug company ownership, and the US price of chlorambucil has quadrupled as a result, to over $20 for a 2 mg tablet. There are currently no other US generic alternatives, apart from compounded products.

AZATHIOPRINE

Azathioprine has been used as an immunosuppressive agent in dogs for over 50 years. The drug was initially primarily used in studies that utilized dogs as a model for investigations of immunosuppression and organ transplantation, and rapidly became the immunosuppressive drug of choice to prevent organ rejection in early human transplant medicine (although the modern era of transplantation, with low organ rejection rates, really only commenced when prednisolone was added to azathioprine regimes). Within a few years, azathioprine was also being used to treat naturally occurring diseases in dogs. Despite almost half a century of clinical and research experience on the use of azathioprine in dogs, however, there have been remarkably few studies that actually elucidate the precise effects that azathioprine has on the canine immune system[17-19]. Azathioprine is a prodrug for the active metabolite 6-mercaptopurine, and the primary mechanism of action was long believed to be inhibition of the synthesis of the purines adenine and guanine by blockage of enzymes such as amidophosphoribosyltranserase, with resultant production of nonfunctional nucleic acid strands. Disruption of purine synthesis inhibits DNA and RNA synthesis, thereby inhibiting the proliferation of fast-growing cells such as lymphocytes. In the past few decades, however, multiple other mechanisms of action mediated by various azathioprine metabolites have been proposed, including blockage of T cell activation and stimulation of T cell apoptosis. Azathioprine has long been reported to be more effective against T cell function than B cell function, although strong evidence supporting this is lacking[20].

One of the key enzymes involved in azathioprine metabolism is thiopurine methyltransferase (TPMT). Individual people inherit a marked deficiency in the TPMT enzyme that renders them highly susceptible to azathioprine toxicity. Interestingly, cats also have a marked deficiency in TPMT activity[21], which may explain why azathioprine causes marked myelosuppression in cats at standard canine doses[22]. Although the use of azathioprine at a very reduced dose rates has previously been published in cats, given the narrow margin for safety it is probably wisest to recommend that azathioprine never be used in cats at any dose. Although TPMT expression in dogs is widely variable, severe deficiencies in enzyme activity of the magnitude seen in cats and some people have not been widely reported, and TPMT deficiency does not appear to be related to the severe drug toxicities sometimes seen in dogs[23,24].

The standard azathioprine starting dose in dogs is 2 mg/kg orally once daily. This dose is usually well tolerated and, although GI side effects such as nausea, anorexia, vomiting and diarrhea are occasionally reported, they are typically mild and self-limiting. Chronic azathioprine usage can occasionally cause anemia in human patients, and a comparable mild to moderate sub-clinical anemia is also occasionally anecdotally reported in dogs. Hepatotoxicity, as characterized by a significant elevation in liver enzymes (particularly ALT), typical with minimal or no associated clinical signs, has been reported within a few weeks of commencing therapy in approximately 15% of dogs receiving azathioprine, but appears to be readily reversible with drug discontinuation[25]. Uncommonly, azathioprine can also cause profound myelosuppression in dogs[26-28]. Myelosuppression appears to be a non-dose-dependent idiosyncratic reaction, and are typically reversible if the problem is recognized early and azathioprine is discontinued. Complete blood counts and serum biochemistry (especially ALT) should therefore be monitored regularly during azathioprine therapy. Individual case reports have reported pancreatitis in dogs receiving azathioprine, but cause and effect has not been established[28-29].

Azathioprine has become well established as an 'add on' agent for the treatment of many different immune-mediated and inflammatory conditions when steroids alone are ineffective or poorly tolerated, including IMHA, IMT, IBD, chronic hepatitis, glomerulonephritis, immune-mediated polyarthritis, myasthenia gravis, non-infectious meningoencephalitis, immune-mediated skin diseases, and anal furunculosis. Despite decades of azathioprine usage, evidence supporting efficacy for many of these diseases is remarkably limited. One perceived 'limitation' of azathioprine compared to other immunosuppressive agents, that it can take many weeks or even months to exert its effects, is based on limited and dated data derived predominantly in humans. In my experience, azathioprine in a clinical setting exerts its immunosuppressive effects in dogs about as rapidly as most other comparable agents.

Compared to most other immunosuppressive agents, azathioprine is relatively inexpensive, which is an important consideration with long-term immunosuppressive therapy, especially in large dogs. While the proprietary product (Imuran® or Azasan®) typically still costs over $8 per 50 mg tablet, the generic equivalent can be obtained for less than $1 a tablet. The smallest tablet size is 50 mg (although tablet scoring permits a 25 mg dose), which can present dosing problems in small (under 20 lb) dogs.

DANAZOL

Danazol, a synthetic androgen with weak ('impeded') androgenic effects, has in the past been suggested for the treatment of canine IMHA and IMT, in combination with glucocorticoids[30]. Danazol is derived from the synthetic steroid ethisterone, a modified progestogen. Danazol's most important mechanism of action is probably to reduce macrophage Fc receptor/antibody binding affinity. Danazol also competes with glucocorticoids for combination with steroid-binding globulin, consequently increasing the availability of active unbound steroid. Danazol therefore enables significant steroid dose reduction. Danazol may also reduce the binding of antibody and complement to cell surfaces. Side effects are uncommon, and include hepatotoxicity and masculinization of female dogs. However, although some dogs with refractory IMHA and IMT may benefit from danazol[31-33], the drug fell out of favor a few decades ago, probably because it was very expensive at the time, and response to therapy was sluggish and highly unpredictable. Reported oral danazol doses in dogs with IMHA or IMT range from 5 to 15 mg/kg daily, either given as a single dose or 2-3 divided doses. Danazol comes in 50 mg, 100 mg, and 200 mg capsules, and currently costs around $3 for a 50 mg capsule.

VINCRISTINE

The vinca alkaloids are biologically-active dimeric alkaloids derived from the Madagascar periwinkle plant, *Catharanthus roseus*. Vincristine, a naturally-occurring vinca alkaloid, were characterized phytochemically more than 50 years ago. The biological effects of vincristine have traditionally been attributed to drug-induced disruption of intracellular microtubules. Microtubules are composed predominantly of complex helical polymers of the structural protein tubulin. Vincristine binds directly to tubulin, causing both inhibition of microtubule synthesis and disruption of intact microtubules. Microtubular structures susceptible to the effects of vinca-tubulin binding include the mitotic spindle in dividing cells, the neurotubules in neurons, and the cytoskeletal microtubules in platelets. Vincristine may also exert biological effects that are independent of disruption of intracellular microtubules, such as inhibition of RNA, DNA and protein synthesis, and modification of prostaglandin production.

Vincristine is a cell-cycle-specific cytotoxic agent. Vincristine disrupts microtubules within the mitotic spindle of dividing cells, thereby arresting chromosomal separation in metaphase. Vincristine at standard doses is minimally myelotoxic in most dogs and cats, and is therefore used in in veterinary cancer chemotherapy in combination with more myelosuppressive agents. The degree of immunosuppression induced by vincristine at therapeutic doses is minimal compared to other immunosuppressive agents, and vincristine is therefore not used for the treatment of most immune-mediated diseases. The one exception is IMT, where vincristine has become a mainstay of treatment.

Vincristine is usually administered IV, and inadvertent SC or IM administration causes severe local tissue irritation and necrosis. Platelets, because of their intracellular tubulin, demonstrate a remarkable ability to concentrate vincristine from plasma, and are therefore the principal circulating cellular carriers of the drug. During early clinical trials in human cancer patients, it was observed that vincristine typically caused significant but transient increases in circulating platelet numbers. A similar phenomenon has since been reported in dogs, both in research animals and in cancer patients[34-35], and appears to be due to increased megakaryocytopoiesis and thrombopoiesis, although the precise mechanisms of thrombocytosis are still uncertain. Vincristine, typically in combination with prednisone, has been reported to similarly facilitate remission of thrombocytopenia in many canine patients with IMT. Original case reports demonstrating a rapid response to vincristine in dogs with IMT[36-39] have been supported, decades later, by evidence obtained from prospective studies[40-41]. Circulating platelet numbers increase markedly within 3-5 days of vincristine administration, and the addition of vincristine to standard therapy in dogs with IMT appears to shorten hospitalization time by several days. Most authors currently recommend an intravenous vincristine bolus dose of 0.02 mg/kg for the treatment of canine IMT. Vincristine has been used in cats with IMT, although evidence of clinical efficacy is lacking. One significant advantage of vincristine compared to other therapeutic options for IMT is that vincristine is inexpensive (a 1 ml vial of 1mg/ml vincristine sulfate costs around $30).

The pathogenesis of vincristine-induced platelet responses in IMT patients is uncertain. Studies in people suggest that the main therapeutic effect of vincristine in IMT patients is not increased thrombopoiesis. Post-treatment average platelet life-spans are significantly prolonged in patients that respond to vincristine, suggesting that remission is due to reduced platelet destruction rather than increased platelet production. Since platelets are the major circulating cellular carriers of vincristine, researchers have speculated that antibody-coated platelets selectively deliver vincristine to those phagocytes within the mononuclear phagocytic system that are actively involved in platelet destruction. This proposed mechanism explains why, despite being an ineffective immunosuppressive agent for the treatment of most conditions, vincristine can still be very effective for the treatment of IMT.

Neurotoxicity, although uncommon, is the most frequent significant side-effect associated with therapeutic doses of vincristine in dogs and cats[43]. Reversible vincristine-induced neurotoxicity in the dog has been reported with chronic cancer chemotherapy, but is not likely to be an issue with the single doses used to treat IMT[44]. Other side effects such as GI disorders and alopecia, occur less frequently, and are typically mild and temporary. Vincristine at doses used for IMT typically causes minimal myelosuppression in dogs, although dogs with the ABCB1-1Δ (MDR1) gene mutation and some Border Collies have been reported to be more susceptible than other dog breeds to myelosuppression at antineoplastic vincristine doses[45-46]. An unusual transient pulmonary toxicity has been reported in a cat receiving chemotherapeutic doses of vincristine[47].

REFERENCES

[1]Legrand J, et.al. **J Immunotoxic** 2013;10:90. [2]Ogilvie G, et.al. **Vet Imm Immunopath** 1988;18:119. [3]Putnam C, et.al. **Clin Exp Imm** 1975;22:323. [4]Charney S, et.al. **JAVMA** 2003;222:1388. [5]Crow S, et.al. **JAVMA** 1977;171:259. [6]Macy DW, et.al. **JAAHA** 1983;19:965. [7]Burgess K, et.al. **JVIM** 2000;14:456. [8]Grundy S, et.al. **JAVMA** 2001;218:543. [9]Mason N, et.al. **JVIM** 2003;17:206. [10]Reimer M, et.al. **JAAHA** 1999;35:384. [11]Warry E, et.al. **JVIM** 2011;25:903. [12]Irwin K, et.al. **Vet Derm** 2012;23:403. [13]Preziosi D, et.al. **Vet Derm** 2003;14:313. [14]Wondratschek C, et.al. **JAAHA** 2010;46:12. [15]Dandrieux J, et.al. **JAVMA** 2013;242:1705. [16]Benitah N, et.al. **JAAHA** 2003;39:283. [17]Harkin K, et.al. **JAAHA** 2007;43:21. [18]Ogilvie G, et.al. **Vet Imm Immunopath** 1988;18:119. [19]Rinkardt N, et.al. **Can J Vet Res** 1999;63:18. [20]Dowling P. **Can Vet J** 1995;36:781. [21]Foster A, et.al. **JVIM** 2000;14:552. [22]Beale K, et.al. **AJVR** 1992;53:1236. [23]Kidd L, et.al. **JVIM** 2004;18:214. [24]Rodriguez D, et.al. **JVIM** 2004;18:339. [25]Wallisch K, et.al. **JVIM** 2015;29:513.
[26]Rinkardt N, et.al. **Can Vet J** 1996;37:612. [27]Favrot C, et.al. **Vet Rec** 2007;160:520. [28]Houston D, et.al. **Can Vet J** 1991;32:496. [29]Moriello K, et.al. **JAVMA** 1987;191:695. [30]Miller E. **Sem Vet Med Surg** 1997;12:167. [31]Bloom J, et.al. **JAVMA** 1989;194:76. [32]Holloway S, et.al. **JAVMA** 1990;197:1045. [33]Roseler B, et.al. **J Vet Allerg Clin Imm** 1996;4:53. [34]Mackin A, et.al. **AJVR** 1995;56:100. [35]Neel J, et.al. **Vet Clin Path** 2012;41:216. [36]Greene C, et.al. **JAVMA** 1982;180:140. [37]Jans H, et.al. **JVIM** 1990;4:4. [38]Williams D, et.al. **JAVMA** 1984;185:660. [39]Wolf A. **Southwest Vet** 1983;35:209. [40]Balog K, et.al. **JVIM** 2013;27:536. [41]Rozanski E, et.al. **JAVMA** 2002;220:477. [42]Tasker S, et.al. **JSAP** 1999;40:127. [43]Cho E, et.al. **Arch Toxicol** 1983;52:83-90. [44]Hamilton T, et.al. **JAVMA** 1991;198:635. [45]Lind D, et.al. **AJVR** 2013;74:257. [46]Mealey K, et.al. **JVIM** 2008;22:996. [47]Polton G, et.al. **Vet J** 2008;177:130.

Immunosuppressive Therapy in Dogs and Cats: Recent Drug Development

Andrew Mackin, BSc, BVMS, MVS, DVSc, FANZCVSc, DACVIM (SAIM)
Starkville, MS

Several potent immunosuppressive drugs developed over the past few decades in human medicine have recently made the leap to our small animal patients, and our use of them is growing. This lecture will discuss cyclosporine, leflunomide and mycophenolate.

CYCLOSPORINE

Cyclosporine is a potent immunosuppressive drug indicated for the treatment of inflammatory and immune-mediated diseases, and for organ transplantation. Cyclosporins are cyclic polypeptide macrolides originally derived from the soil fungus *Beauveria nivea*. Cyclosporine A is the molecule developed for commercial use. The use of cyclosporine to prevent rejection of renal allografts in people was first described in the 1970s, and cyclosporine rapidly became the "go to" drug (in combination with prednisolone) to prevent transplant rejection, although recently tacrolimus has superseded cyclosporine as the immunosuppressive drug of choice in transplantation medicine. Unfortunately, tacrolimus has long been considered too toxic (hepatotoxicity and nephrotoxicity) for use as a systemic immunosuppressive agent in dogs, although a number of individual case studies originating from Japan report that, at very low doses, tacrolimus appears to be well-tolerated, and may have beneficial clinical effects[1-4]. However, it is yet to be demonstrated that very low doses of tacrolimus actually have any detectable effects on the canine immune system. In veterinary medicine, oral cyclosporine received FDA approval in 2003 for treatment of canine atopy, and was recently also approved for allergic skin disease in cats. Cyclosporine has been used in an extra-label fashion for many years for renal transplantation in dogs and cats[5,6], and for a range of inflammatory and immune-mediated conditions[7].

Cyclosporine's primary immunosuppressive mechanism of action is inhibition of T cell function. Cyclosporine acts to inhibit calcineurin, an intracellular protein phosphatase that activates gene transcription factors. Tacrolimus is also a calcineurin inhibitor. Activation of T cells results in activation of calcineurin, which dephosphorylates inactive nuclear factor (NFAT). NFAT translocates into the nucleus, where it upregulates transcription of genes coding for several important cytokines, including IL-2, IL-4, TNF-α, and INF-γ. Calcineurin inhibitors, including cyclosporine, act by binding to intracellular cyclophilins, which are proteins that facilitate protein folding. Binding of cyclosporine to cyclophilin A creates a complex with high affinity for calcineurin. Through inhibition of calcineurin, cyclosporine specifically inhibits T cell function and thus, cell-mediated immunity, but has little immediate impact on humoral immunity. Decreased IL-2 expression in CD4+ Th1 cells associated with cyclosporine therapy leads to inhibition of proliferation and activation of both T-helper and T-cytotoxic lymphocytes, and blunting of the immune response. Cyclosporine has also been shown to have many other local anti-inflammatory and immunosuppressive effects, especially in the skin.

Cyclosporine is a large lipophilic molecule which must be solubilized prior to intestinal absorption. Commercial cyclosporine is available as two very different types of oral formulations. Cyclosporine was initially approved in humans as a vegetable-oil based preparation (Sandimmune®), but poor oral bioavailability caused marked variability in blood drug concentrations. A more recent formulation, an ultramicronized preparation approved in 1996 (Neoral®), forms a microemulsion upon contact with aqueous fluids, resulting in more predictable absorption. Oral bioavailability of the microemulsion is improved by up to 50% compared to the oil-based preparation. Because of the marked variability in bioavailability of the non-ultramicronized preparation, it is not recommended for oral use in small animals. Cyclosporine has a high binding affinity for red blood cells and plasma lipoproteins. Because up to 50% of the drug in blood is located in red cells, whole blood is recommended for therapeutic drug monitoring (TDM). Once in the circulation, cyclosporine distributes widely, accumulating in skin, liver, kidneys and fat, resulting in a large volume of distribution. Tissue levels exceed levels in serum by a factor of 3 to 14. Peak blood concentrations generally occur approximately 2 hours after oral administration. Blood concentrations then rapidly decrease over the remainder of the dosing interval. Extensive metabolism by the hepatic cytochrome P-450 system yields many different metabolites, some of which may retain therapeutic efficacy. Drugs that inhibit P-450 enzymes have been given concurrently with cyclosporine in order to decrease the dose needed to maintain adequate blood drug concentrations. Ketoconazole, in particular, has been used to decrease in oral cyclosporine dosages in dogs by as much as 75%[8,9].

The complexities of cyclosporine disposition in normal animals contribute to markedly variable blood drug concentrations both between patients and even within the same patient. Therapeutic management may therefore be facilitated by monitoring blood cyclosporine concentrations. Currently available methods for TDM include HPLC and antibody-based methods such as a specific monoclonal RIA. HPLC has the advantage that the parent drug can be discriminated from metabolites. Since antibody-based methods typically measure metabolites as well as the parent drug, blood cyclosporine concentrations will therefore be higher by a factor of 1.5 to 1.7 compared to the same sample analyzed by HPLC. HPLC is labor intensive, and so antibody-based methods are often employed in clinical situations, with the laboratory performing the assay typically providing recommendations regarding ideal target drug concentrations. In human medicine, trough blood levels were the initial basis for adjustment of drug dosages. Many studies in people have since suggested that area under the plasma drug concentration time curve or 2 hour peak drug concentrations are preferred. With peak cyclosporine concentrations requiring only a single sample, adjusting drug doses to attain target peak drug levels has become the single best blood concentration measurement for use during human organ transplantation.

In veterinary medicine, measurement of trough cyclosporine concentrations also prevailed for many years based on initial work done in canine and feline renal transplant studies. Recommendations from laboratories offering TDM have often involved measurement of both peak and trough cyclosporine blood levels, although target peak concentrations have not been well established in small animals.

Pharmacodynamic assays investigate a drug's effect on target cells. Many pharmacodynamic markers of the immunosuppressive effects of cyclosporine have been studied in human medicine, including lymphocyte proliferation, calcineurin enzyme activity, lymphocyte surface antigen expression, and intracellular cytokine quantification. Through pharmacodynamic monitoring, human studies have shown individually distinct degrees of calcineurin inhibitor sensitivity in patients. Pharmacodynamic monitoring shows great promise for optimizing cyclosporine therapy and delivering individualized therapy. At MSU, we recently measured activated T cell expression of IL-2, IL-4, and IFN-γ via flow cytometry in dogs given two different oral cyclosporine dosages[10,11]. The dogs were first administered a high dose (10 mg/kg orally twice daily), with doses adjusted upwards to attain a target trough drug concentration greater than 600 ng/mL (HPLC), a protocol known to be sufficiently immunosuppressive for organ transplantation. With this high dose, activated T cell expression of IL-2 and IFN-γ was significantly suppressed. The dogs were then administered the dose of cyclosporine used to treat canine atopy (5 mg/kg orally once a day), a dose which has been considered to be low enough to avoid predisposing to immunosuppression and infection. Even with this low dose, however, expression of IFN-γ and IL-2 was still markedly suppressed in some dogs. Subsequent studies determined that activated T cell mRNA IL-2 and IFN-γ expression using a quantitative reverse transcriptase polymerase chain reaction (qRT-PCR) assay was comparable to flow cytometry results[12,13]. One advantage of the qRT-PCR assay is that it can be performed on blood samples mailed in by practitioners. Cyclosporine has been shown to have much the same effect on T cell cytokine production in cats as it does in dogs[14,15].

Cyclosporine is FDA-approved for the treatment of canine atopic dermatitis and feline allergic skin disease, and has also been used to prevent transplant rejection and to treat sebaceous adenitis, pemphigus foliaceus, anal furunculosis, feline stomatitis, inflammatory bowel disease (IBD), myasthenia gravis, non-infectious inflammatory meningoencephalitis, pure red cell aplasia, immune-mediated hemolytic anemia (IMHA), immune-mediated thrombocytopenia (IMT), and immune-mediated polyarthritis in dogs and cats. Our recent pharmacodynamic research has confirmed that canine responses to the drug are comparable to the response profile that is well recognized in people: individual responses are extremely variable from dog-to-dog, both in dogs receiving the same standard oral dose, and in dogs with oral doses adjusted to attain comparable blood levels. Given this high degree of variability of individual responsiveness to cyclosporine in dogs, dosing protocols should be tailored to allow for this patient-to-patient variability. In my opinion, recommended dosing protocols in dogs with chronic, non-life-threatening inflammatory skin and gastrointestinal diseases should be quite different from the protocols used in dogs with more acute and life-threatening immune-mediated diseases.

In chronic diseases that are typically not life-threatening, such as skin conditions, anal furunculosis, mild IBD and, arguably, chronic hepatitis, cyclosporine is often effective at a standard, relatively low starting dose. Cyclosporine is typically given long term, with doses adjusted upwards if needed 'to effect', based predominantly on clinical signs. Typically, however, starting doses do not need to be increased and, in the long-term, the drug is tapered to the lowest effective dose needed to maintain remission. Recommended starting cyclosporine doses in dogs are 5 mg/kg once daily for most skin diseases and IBD[16,17], and 5 mg/kg once to twice daily for anal furunculosis[18]. In cats with skin conditions such as allergic skin disease, eosinophilic granuloma complex and pemphigus foliaceus, a starting cyclosporine dose of around 5 mg/kg daily is recommended[19]. Blood concentrations are usually not necessary for treatment of these conditions, as remission of disease is the main criterion used to decide whether drug therapy is adequate. In fact, for many of these conditions, cyclosporine blood levels have been shown to have minimal correlation with remission, perhaps because the drug is selectively concentrated in tissues such as the skin. Even at standard low FDA-approved doses, some dogs can still develop significant suppression of T cell biomarkers of immunosuppression despite very low trough cyclosporine concentrations. This could explain the phenomenon reported by some dermatologists, that individual dogs treated for atopy can develop severe secondary skin and systemic infections[20].

In dogs suffering from more life-threatening diseases such as severe IMHA and IMT, in contrast, cyclosporine must be targeted to attain effective immunosuppression as rapidly as possible. In these patients, starting cyclosporine at a low dose and adjusting doses upwards 'to effect' is not recommended. Attaining effective oral doses as rapidly as possible is essential for ensuring adequate immunosuppression whilst avoiding overdosage. Currently recommended starting cyclosporine doses for life-threatening diseases range from 5 mg/kg to 10 mg/kg twice daily. Subsequent measurement of blood drug levels and/or assessment of activated T cell mRNA IL-2 and IFN-γ expression using qRT-PCR within a week of starting therapy, with dose adjustments as needed, are the best methods that are currently available to assess adequacy of therapy, and are strongly recommended in dogs with severe disease.

Side effects are uncommon with cyclosporine, with the exception of GI signs such as vomiting, diarrhea, anorexia and nausea. Giving the drug frozen and/or with food can reduce GI side effects, although such measures may also alter drug absorption profiles. Uncommonly, cyclosporine can cause an idiosyncratic hepatotoxicity, which does not seem to be dose dependent. Gingival hyperplasia and hypertrichosis have also occasionally been reported with cyclosporine. Chronic cyclosporine therapy may also predispose to neoplasia such as lymphoma. One advantage of cyclosporine is that it is not myelosuppressive. Experimentally, cyclosporine has been shown to increase some markers of platelet activation in dogs, which may be a concern in patients with

IMHA, where hypercoagulability and resultant pulmonary thromboembolism can be a major contributor to patient mortality[21]. However, to date, it has not been demonstrated whether this phenomenon is clinically relevant in IMHA patients with naturally occurring disease.

Cyclosporine is an expensive drug, and clinicians are therefore tempted to explore cheaper forms of the drug. There are many approved human generic microemulsion preparations similar to Neoral®, and these have been shown to have therapeutic equivalency in people. Studies investigating the pharmacokinetics of these preparations in dogs have not been performed, and it is not safe to assume that a generic formulation is therapeutically equivalent to the approved canine product (Atopica®). Neoral® currently costs around $2 for a 25 mg capsule and $6 for a 100 mg capsule, while the generic equivalents cost around $2 and $4 for the 25 mg and 100 mg capsules accordingly. The approved veterinary product tends to be priced comparably to the human proprietary products, but has the advantage of being FDA-approved and available in capsule sizes that are convenient for dosing accuracy (10 mg, 25 mg, 50 mg and 100 mg), as well as a 100 mg/ml oral suspension. In my opinion, the compounded cyclosporine that is offeredat significant cost reductions by compounding pharmacies should not be used for systemic immunosuppression, because adequate drug absorption is likely to be highly unreliable. Unfortunately, transdermal cyclosporine has been shown to be inadequately absorbed in cats[22].

LEFLUNOMIDE

Leflunomide is an isoxazol derivative immunosuppressive drug that was developed within the past two decades for treatment of rheumatoid arthritis and prevention of transplant rejection. Leflunomide is a prodrug for its primary active malononitriloamide metabolite, A77 1726 (teriflunomide). Malononitriloamides reversibly inhibit the mitochondrial enzyme dihydroorotate dehydrogenase, a key enzyme in pyrimidine synthesis, with resultant inhibition of the pyrimidine ribonucleotide uridine monophosphate (rUMP), and decreased DNA and RNA synthesis and G1 cell cycle arrest. Leflunomide inhibits B and T cell function, suppresses antibody production and has anti-inflammatory effects, possibly via inhibition of *de novo* pyrimidine biosynthesis and cytokine-associated and IL-2-stimulated tyrosine kinase activity.

Prior to commercial development, leflunomide was made available to the transplant research group at UC-Davis. This group described treatment of small number of canine patients with refractory inflammatory and immune-mediated diseases such as IMHA, IMT, non-infectious inflammatory meningoencephalitis, systemic histiocytosis, immune-mediated polymyositis, immune-mediated polyarthritis, and pemphigus foliaceous with leflunomide, typically with promising success rates[23]. Unfortunately, when these results were reported in the late 1990s, the drug was no longer commercially available. When leflunomide did become available as the proprietary product Arava®, drug cost limited its use in small animal clinic patients. Only recently did an affordable generic equivalent become available and, as a result, preliminary reports of leflunonide use in small animal patients are beginning to surface. There are still currently very few published reports discussing the use of leflunomide in dogs and cats[24,25]. Recently, a case series describing the use of leflunomide in 14 dogs with immune-mediated polyarthritis reported a high response rate with few side effects[26]. Another recent case series reported treatment of refractory inflammatory colorectal polyps in 15 Miniature Dachschunds, again with a high response rate, although a number of side effects, including reversible myelosuppression and hepatopathy, were noted in several dogs[27].

Leflunomide appears to be very well tolerated in dogs although, if the drug attains more common usage, it is likely that less frequent but more serious side effects will be recognized. The most common side effect seen with leflunomide in dogs is occasional inappetance, lethargy and vomiting. Serious side effects occasionally reported in people include myelosuppression, cutaneous drug reactions and hepatotoxicity, and similar issues are beginning to appear in the veterinary literature[27]. In humans, traces of teriflunomide can persist for many months after drug discontinuation, and in the instance of drug reactions, cholestyramine or activated charcoal are needed to rapidly reduce drug levels. In dogs, the terminal half-life of teriflunomide is much shorter than in humans, so the potential for persistent side effects is probably significantly less[28]. Complete blood counts and serum biochemistry (especially ALT) should be regularly monitored in small animal patients on leflunomide.

The initial recommended starting oral dose for leflunomide in dogs is 2-4 mg/kg daily, with doses adjusted to attain a plasma trough teriflunomide levels of 20 µg/ml within a few weeks of commencing therapy. Measurement of leflunomide levels is available through Auburn University. For cats with immune-mediated polyarthritis, a leflunomide dose of 10 mg (total dose) orally, once daily, combined with methotrexate, has been suggested, with dose reductions to effect[29]. Leflunomide comes in tablet sizes (10 and 20 mg) that are convenient for dosing our smaller patients. Proprietary leflunomide costs about $30 to $40 for a 10 mg or 20 mg tablet, while the generic equivalent can sometimes be found for around $1 for both 10 and 20 mg tablet, although availability is inconsistent. Leflunomide generics have an 'AB' rating by the FDA, meaning that the generic is 'equivalent' to Arava®. However, since 'equivalence' is often determined by pharmacokinetic data in healthy individuals, an AB rating does not guarantee identical performance in clinical patients.

MYCOPHENOLATE

Mycophenolate is the synthesized prodrug form of mycophenolic acid, a selective inhibitor of inosine monophosphate dehydrogenase, an enzyme that controls the rate of synthesis of guanine monophosphate in the *de novo* pathway of purine synthesis. Mycophenolate is a fermentation product derived from *Penicillium* fungi. Mycophenolic acid inhibits B and T cell proliferation, and decreases antibody production. Mycophenolate is primarily used in human medicine for prevention of rejection of transplanted organs, although it also used to treat immune-mediated diseases such as systemic lupus erythematosus, IMHA, IMT and pemphigus vulgaris.

The original proprietary mycophenolate mofetil, CellCept®, and the related mycophenolate sodium, Myfortic®, were expensive, and as a result the products only achieved limited usage in small animal medicine. However, recently, the availability of cheaper generic alternatives has led to dramatically increased usage of mycophenolate in small animal patients. A 250 mg CellCept® capsule currently costs around $10, whereas the equivalent generic 250 mg capsule costs less than 50c. An oral suspension version of mycophenolate mofetil (200 mg/ml) is available for more convenient dosing in smaller patients. The successful use of mycophenolate mofetil in a small animal patient with naturally-occuring disease was first described in a dog with myasthenia gravis[30]. Much of the subsequent anecdotal usage of mycophenolate for a variety of different immune-mediated diseases used dosing similar to that reported in this original paper[31-33]. Mycophenolate mofetil is also available in an injectable form, and its IV use has been described during the successful stabilization of 3 dogs with myasthenia that could not tolerate oral medications[31]. Ironically, a recent retrospective study of 15 dogs with myasthenia gravis treated with mycophenolate mofetil reported that the drug was ineffective at attaining clinical remission[34]. Another recent retrospective study of 30 dogs with IMHA receiving oral mycophenolate mofetil reported that "the combination of glucocorticoids and mycophenolate mofetil has similar efficacy to alternate immunosuppressive protocols", although the study was underpowered for truly proving equivalency and, in fact, for some outcome categories, dogs receiving mycophenolate appeared to do worse than dogs receiving other therapy[35]. Some individual patients, however, do appear to respond favorably to mycophenolate mofetil, as was demonstrated in several recent small case series in dogs with IMHA and IMT[36,37].

The correct starting dose of mycophenolate mofetil in dogs, in my opinion, has yet to be determined. In clinical case reports, total daily doses range from about 15 mg/kg to 45 mg/kg, with the drug given either once daily or, more commonly, split into twice or three times daily doses. Consistently throughout recent case reports, the higher end of the dose range can cause diarrhea (sometimes severe) in many dogs. Generally, a total daily dose of 20 mg/kg (most typically split into 10 mg/kg twice daily) is reasonably tolerated by most dogs, although individual dogs can tolerate much higher doses. Unfortunately, to date, the relatively reliably tolerated dose rate of 10 mg/kg twice daily has not been established to have consistent measurable immunosuppressive effects in dogs and, in fact, preliminary T and B cell function studies in our laboratory suggest that this dose may be too low to be effective in many dogs. Mycophenolate mofetil has variable oral bioavailability in dogs, so variability in response to therapy should also be expected[38]. Mycophenolic acid also has extensive and variable protein binding in dogs, which can have a significant and widely variable impact on the amount of free drug available to actually exert immunosuppressive effects (personal communication, Dr. Nicolas Villarino, Drug Pharmacokinetics and Pharmacodynamics Laboratory, Washington State University). Mycophenolate has not been used widely enough to establish the frequency of serious side effects in dogs but, in people, GI signs and, less commonly, marked myelosuppression and a rare and fatal neurologic disease (progressive multifocal leukoencephalopathy) have been reported. Complete blood counts should therefore probably be regularly monitored in dogs receiving mycophenolate. In humans, GI side effects can be reduced by replacing mycophenolate mofetil with mycophenolate sodium but, in dogs, this substitution does not appear to reduce the incidence of diarrhea[39]. Mycophenolic acid in humans is primarily excreted conjugated to glucuronide and, since cats lack the glucuronyl transferases needed for glucuronidation, it has been suggested that the drug should used with caution in this species (at least until more extensive drug metabolite studies have been completed), although the use of mycophenolate mofetil has been described at a dose rate of 10 mg/kg twice daily, with no obvious side effects, in two cats with IMHA[40].

REFERENCES

[1]Kano R, et. al. **Vet Rec** 2004;154:536. [2]Kano R, et.al. **J Vet Med Sc** 2006;68:95. [3]Yuki M, et.al. **J Jpn Vet Med Assoc** 2004;57:721. [4]Yuki M, et.al. **Open J Vet Med** 2012;2:158. [5]Bernsteen L, et.al. **Clinic Tech Small Anim Pract** 2000;15:40. [6]Bernsteen L, et. al. **Vet Surg** 2003;32:213. [7]Archer T, et.al. **JVIM** 2014;28:1. [8]Dahlinger J, et.al. **Vet Surg** 1998;27:64. [9]Patricelli AJ, et.al. **JAVMA** 2002;220:1009. [10]Archer TM, et.al. **JVIM** 2011;25:1391. [11]Fellman CL, et.al. **Vet Immunol Immunopathol** 2011;140:175. [12]Riggs C, et.al. **Vet Immunol Immunopathol** 2013;156:229. [13]Fellman C, et.al. **J Vet Pharm Ther** 2015 (Doi: 10.1111/jvp.12280). [14]Aronson L, et.al. **AJVR** 2011;72:541. [15]Kuga K, et.al. **J Vet Med Sc** 2008;70:1011. [16]Olivry T, et.al. **JAVMA** 2002;221:370. [17]Allenspach K, et.al. **JVIM** 2006;20:239. [18]Griffiths LG, et.al. **JSAP** 1999;40:569. [19]Wisselink M, et.al. **Vet J** 2009;180:55. [20]Siak M, et.al. **Vet Derm** 2013;24:453. [21]Thomason J, et.al. **JVIM** 2012;26:1389. [22]Miller R, et.al. **JAAHA** 2014;50:36. [23]Gregory C, et.al. **Transpl Proc** 1998;30:4143. [24]Bianco D, et.al. **JAAHA** 2009;45:147. [25]Ohno K, et.al. **J Jap Vet Med Associ** 2003;56:669. [26]Colopy S, et.al. **JAVMA** 2010;236:312. [27]Fukushima K, et.al. **J Vet Med Sc** 2015 (on-line advanced publication). [28]Singer L, et.al. **J Vet Pharm Ther** 2011;34:609. [29]Hanna F. **VCOT.** 2005;18:94. [30]Dewey D, et.al. **JVECC** 2000;10:177. [31]Abelson A, et.al. **JVECC** 2009;19:369. [32]Ginel P, et.al. **J South Afr Vet Assoc** 2010;81:253. [33]Yuki M, et.al. **Austral Vet J** 2007;85:495. [34]Dewey C, et.al. **JAVMA** 2010;236:664. [35]Wang A, et.al. **JSAP** 2013;54:399. [36]Yau V, et.al. **JSAP** 2014;55:330. [37]West L, et.al. **J Vet Emerg Crit Care** 2014;24:226. [38]Lupu M, et.al. **J Am Soc Blood Marrow Transpl** 2006;12:1352. [29]Bacek L, et.al. **JVECC** 2011;21:45. [39]Chanda SM, et.al. **Transpl Proc** 2002;34:3387. [40]Bacek L, et.al. **J Vet Emerg Crit Care** 2011;21:45.

Immunosuppressive Therapy and Hypercoagulability

John M. Thomason, DVM, MS, DACVIM
Starkville, MS

INTRODUCTION

Immune-mediated hemolytic anemia (IMHA) results from the antibody-mediated destruction of circulating red blood cells.[1] In dogs, IMHA is the most frequent cause of hemolytic anemia, and the most common form of immune-mediated disease. Unfortunately, despite our best efforts, mortality rates for canine IMHA remain high, ranging from 50 to 70%.[2] One of the most common cause of death in dogs with IMHA is thromboembolic disease, particularly pulmonary thromboembolism (PTE). The incidence of PTE in IMHA patients is reported to be over 50%, and may even be as high as 80%.[3] Glucocorticoids are the most commonly used immunosuppressive agents for the treatment of IMHA in dogs. However, glucocorticoids are commonly associated with side effects that are undesirable to the patient and owner. Although glucocorticoids are the cornerstone of IMHA therapy, additional immunosuppressive agents such as cyclosporine, azathioprine, mycophenolate mofetil and leflunomide are often concurrently utilized to reduce the undesirable side effects of glucocorticoid and further suppress the immune system.

THROMBOEMBOLIC DISEASE OVERVIEW

Normal platelet function is necessary for the prevention of hemorrhage, but excessive platelet activation can lead to the development of thromboembolism. There are many conditions in veterinary medicine, such as IMHA, protein-losing nephropathies, hyperadrenocorticism, and disseminated intravascular coagulation (DIC), which are associated with an increased risk of thromboembolic complications.

Arterial Thrombosis

Thrombosis and thromboembolism have long been recognized as major contributors to the morbidity and mortality of many disease states in human medicine. In humans, arterial thromboembolism (ATE) involving the coronary circulation is the most common cause of myocardial infarction and the thrombotic occlusion of cerebral arteries is a leading cause of stroke. Cardiogenic embolism, a leading cause of ischemic stroke in human patients, typically results from thrombus formation secondary to blood stasis within the heart. Fragments may then break off from these thrombi leading to peripheral embolization. Cardiogenic embolism is a frequent form of ATE in veterinary patients. In cats, ATE most frequently results when emboli break off thrombi that have formed in the left atrium or ventricle secondary to cardiomyopathy. In dogs, ATE has been reported associated with gastric dilation and volvulus with thrombus formation in the left ventricle. Canine ATE has been associated with neoplasia and hyperadrenocorticism, however the specific etiology and origin of these thrombi are not known with certainty.

Venous Thrombosis

In contrast to arterial thrombi, in which platelet aggregates often predominate, the thrombi that form in the venous circulation are composed primarily of fibrin and red blood cells. The pathophysiologic processes that underlie the formation of venous thrombi cause pro-coagulant stimuli to overwhelm the counter-regulatory mechanisms that normally ensure hemostatic balance. The stimuli leading to thrombus formation often involve inappropriate or excessive activation of the coagulation system, or a condition of hypercoagulability. The potential for pathologic thrombus formation is heightened when this situation occurs in the face of other abnormalities such as vascular injury or blood stasis.

In people, venous thrombi most frequently develop in the veins of the calf under low-flow conditions. When venous stasis occurs, local hypoxemia develops and the expression of tissue factor is induced. This does not tend to be the case in veterinary medicine, where venous thrombosis occurs when hypercoagulable conditions combine with localized alterations in blood flow and disruptions in endothelial integrity. Thrombi then form at sites of local pathology, or throughout the body as in the case of DIC where endothelial alterations occur system wide. Tissue damage, surgical procedures and medical conditions all may generate inflammatory cytokines that can contribute to a pro-coagulant state. These cytokines activate endothelial cells, causing them to express adhesion molecules to bind circulating leukocytes. When monocytes bind, they synthesize tissue factor and express receptors for factor X and fibrinogen, thus promoting coagulation on their surfaces.[4]

CYCLOSPORINE

Cyclosporine has become popular for the treatment of IMHA partly to the perception that cyclosporine has a more favorable adverse effect profile when compared to other immunosuppressive agents.[5] However, recent studies in humans have shown that cyclosporine can alter the platelet plasma membrane, increase the thrombogenic properties of platelets, and accelerate thrombin generation.[6-9] Additionally, in humans, cyclosporine also has been shown to increase the synthesis of thromboxane A_2.[8] Synthesized by activated platelets, thromboxane A_2 triggers vasoconstriction, platelet activation, and enhances platelet aggregation, all of which will decrease blood flow, increase blood stasis, and promote coagulation which could predispose patients to thromboembolism.[10] When compared to renal transplant recipients treated with azathioprine, human patients receiving cyclosporine have increased thromboxane A_2 production and increased rates of thrombus formation within the renal parenchyma.[8]

In dogs, similar to humans, the administration of cyclosporine is associated with a significant increase in thromboxane synthesis, especially at immunosuppressive cyclosporine dosages and at times of peak blood drug concentrations.[11] This increase in thromboxane synthesis suggests that standard immunosuppressive dosages of cyclosporine might enhance platelet activation in dogs, potentially predisposing hypercoagulable patients to thromboembolic complications.

In an attempt to prevent thromboembolism, anticoagulant medications are routinely used prophylactically in canine IMHA patients. Many of these drugs can only be given by injection and/or are cost prohibitive, leaving oral low-dose aspirin as the most affordable prophylactic therapy. In fact, the use of low-dose aspirin has been shown to improve survival rates in dogs diagnosed with IMHA.[12] Aspirin is a cyclooxygenase (COX) inhibitor that irreversibly inhibits platelet function. The COX enzyme is essential for the conversion of arachidonic acid to several biologically active prostaglandins, including thromboxane A_2.[13] By inhibiting COX, aspirin reduces platelet synthesis of thromboxane A_2, thereby inhibiting platelet function.[14] Since platelets do not contain a nucleus, and are unable to synthesize new and functioning COX enzymes, aspirin-induced platelet dysfunction will persist for the entire lifespan of the platelet, and function will not return until the drug is discontinued and new platelets, not exposed to aspirin, are released from the bone marrow. Since, cyclosporine and low-dose aspirin are often used concurrently in canine IMHA patients, it is possible that low-dose aspirin could counteract cyclosporine-associated thromboxane synthesis. Alternatively, when used concurrently, cyclosporine associated thromboxane synthesis could counteract the anti-platelet effects of aspirin. Recently, in a randomized, cross-over study, which utilized turbidimetric aggregometry and the PFA-100® platelet function analyzer, there was no difference in platelet function between dogs treated with low-dose aspirin (1 mg/kg, PO q24h) and dogs treated with low-dose aspirin (1 mg/kg, PO q24h) and cyclosporine (10 mg/kg, PO q12h). Additionally, there was a significant difference in thromboxane synthesis between a group of dogs treated with cyclosporine and cyclosporine with low-dose aspirin. Additionally, there was no difference in thromboxane synthesis between dogs treated with low-dose aspirin and cyclosporine and low-dose aspirin. Therefore, this study suggests that low-dose aspirin inhibits cyclosporine-induced thromboxane synthesis, and the concurrent use of these medications does not alter the anti-platelet effects of aspirin.[15]

AZATHIOPRINE, MYCOPHENOLATE MOFETIL AND LEFLUNOMIDE

Fortunately, there are other immunosuppressive agents than can be used instead of cyclosporine in dogs with IMHA, including azathioprine, mycophenolate mofetil and leflunomide. The effects of these immunosuppressive agents on the hemostatic system have not been evaluated as comprehensively as the effects of cyclosporine, and it is unknown whether these drugs might be safer than cyclosporine in hypercoagulable dogs. Because immunosuppressive therapies are commonly used in hypercoagulable IMHA patients, it is important to characterize the potential for these common immunosuppressive agents to increase the risk of thromboembolism in dogs, beyond what is only now known for cyclosporine.

ANTI-COAGULANT AND ANTI-PLATELET THERAPY

Unfortunately, mortality rates for canine IMHA remain high, with thromboembolic disease being one of the most common causes of death, making treatment very difficult and challenging. Therefore, minimizing the risk of thromboembolism, such as appropriate selection of immunosuppressive therapies, will improve survival. However, removal of risk factors are only one aspect of IMHA therapy. Early and aggressive thrombolytic treatment may help dissolve the emboli, however, this treatment modality can be associated with uncontrollable hemorrhage. To prevent thrombus formation, prophylactic medications are often administered to inhibit either platelet function or the clotting cascade. Anticoagulant therapy with unfractionated or low-molecular weight heparin or warfarin can be administered to prevent additional thrombus and emboli. Many of these preventative medications can only be given by injection and/or are cost prohibitive for long term administration, leaving oral anti-platelet therapy, low dose aspirin and ADP-receptor antagonists, as the most affordable option for prophylactic therapy.

REFERENCES

1. Archer T, et al. *Today's Veterinary Practice* 2014:4146. 2. Reimer ME, et al. *J Am Anim Hosp Assoc* 1999;35:384. 3. Miller SA, et al. *J Am Vet Med Assoc* 2004;224(2):232. 4. Weitz J, et al. *Chest* 2004;126:265S. 5. Kahan BD. *Transplant Proc* 2004;36(Suppl 2S):378S. 6. Tomasiak M, et al. *Nephrol Dial Transplant* 2007;22:1750. 7. Grace A, et al. *Kidney Int* 1987;32(6):889. 8. Averna M, et al. *Kidney Int* 2001;59:1574. 9. Sahin G, et al. *Ren Fail* 2009;31(2):111. 10. Oskarsson HJ, et al. *Hypertension* 1997;29:1314. 11. Thomason J, et al. *J Vet Intern Med* 2012;26:1389. 12. Weinkle T, et al. *J Am Vet Med Assoc* 2005;226:1869. 13. Ruan C, et al. *Life Sci* 2011;88:24. 14. Patrono C, et al. *Chest* 2001;119:39s. 15. Thomason J, et al. *J Vet Intern Med* 2014;28(4):1364.

Interventional Endoscopy: Just Because We Can Does That Mean We Should?

Allyson Berent, DVM, DACVIM
New York, NY

Introduction

Interventional Endoscopy (IE) involves the use of endoscopic equipment with other contemporary imaging modalities, such as fluoroscopy and/or ultrasound, to perform diagnostic and therapeutic procedures in virtually any part of the body accessed endoscopically (gastrointestinal, biliary, respiratory, urinary tract, etc).

Currently, an expanding investigation of the use of some novel techniques in veterinary medicine has been undertaken. The combination of endoscopy and fluoroscopy allows for one to visualize and gain access into small orifices that would otherwise require more invasive surgical techniques. A good example of this is the placement of a biliary stent into the common bile duct via the major duodenal papilla with endoscopic and fluoroscopic guidance. Many of these interventional procedures are considered the standard-of-care in human medicine, and are currently being more intensely investigated in veterinary medicine.. The use of these techniques in our veterinary patients is expanding as these modalities are becoming more widely available.

The invasiveness and morbidity associated with some traditional surgical techniques (e.g. ureterotomy for ureteral obstructions or strictures, biliary diversion surgery, nasopharyngeal surgery for nasopharyngeal stenosis) makes the use of alternatives using IE appealing. The advantages of such procedures are not only their minimally invasive nature but the lower morbidity, shorter hospital stays, and most importantly, the lack of alternative options some our patients have. The limitations are that these procedures are technically challenging and require specialized equipment and extra training.

This talk is a brief overview of some of the minimally invasive interventional endoscopic procedures actively being performed in veterinary medicine as well as some promising future applications currently under investigation. This talk will focus on the current evidence behind some of the procedures that have been actively investigated in veterinary medicine in the area of IE and a discussion of the risk and benefits.

RESPIRATORY INTERVENTIONS

Nasopharyngeal stenosis

Nasopharyngeal stenosis (NPS) is a pathologic narrowing within the nasopharynx caudal to the choanae resulting in a variable degree of inspiratory stertor. This can occur as a congenital anomaly or be secondary to an inflammatory condition (e.g. aspiration rhinitis), surgery, trauma, or space-occupying lesion. Traditional therapy involves surgery or serial balloon dilatation procedures. Balloon dilatation is minimally invasive and utilizes interventional techniques via fluoroscopy and endoscopy, but can result in re-stricture in a few days to weeks. We have found that stenting of the nasopharyngeal region allows for a more permanent fix with fewer procedures. Results in both dogs and cats have been extremely promising.

Under both fluoroscopic and rhinoscopic guidance a hydrophilic Weasel® (Infiniti Medical, LLC) guidewire is advancedcaudally from the nares into the ventral nasal meatus, through the stenotic opening, and down the esophagus. This is viewed inside the nasopharynx with retroflex rhinoscopy, and from the outside with fluoroscopy. Once the stenotic lesion is identified a percutaneous transluminal angioplasty (PTA) balloon, preloaded with a metallic stent (balloon expanded metallic stent—BEMS) is advanced over the guidewire and centered over the stenotic lesion. Using both fluoroscopic and endoscopic viewing the balloon is inflated (with a 50:50 mixture of contrast and saline) and the waist of the stenosis is subsequently broken with the balloon. As the balloon expands, the stent deploys. Once the stenosis is open, the balloon is deflated and removed over the wire, and the stent is left in place. More recently, the authors are infusing 2-5 ml of topical mitomycin C (0.1%) onto the stenotic lesion after the stent is deployed to prevent aggressive fibrosis. The stent will re-epithelialize in a few days to weeks (approximately two to - six weeks). The size (length and width) of the stent and balloon are typically chosen based on Computed Tomography (CT), which is done prior to the procedure. The patients usually go home the same day as the procedure with anti-fibrotic doses of gluococorticoids (prednisone 0.5 mg/kg BID) tapered over six to eight weeks, two weeks of antibiotics and tramadol as needed for any discomfort. The author's recommend feeding moist food for the first two months as well.

Tracheal stricture balloon dilatation

For tracheal strictures, balloon dilatation prior to considering stent placement is often ideal. In order to avoid tearing the trachea longitudinally, a hot-knife can be used to weaken specific locations at the stricture site and pre-determine where the stricture should tear, preventing a longitudinal tear in the normal tracheal tissue.

GASTROINTESTINAL INTERVENTIONS

Esophageal balloon dilatation and esophageal stenting

Esophageal strictures are frustrating to treat for both veterinarians and physicians. Patients classically present with regurgitation. Strictures in the esophagus can be secondary to reflex esophagitis (common post-anesthesia), caustic substance ingestion, medications sitting on the esophageal mucosa for lengths of time (e.g. doxycycline tablets in cats), esophageal foreign bodies, etc. Balloon dilatation or bougienage procedures, using endoscopic guidance, are currently the treatments of choice in veterinary medicine. Regardless of the intervention chosen, many of these strictures recur in our feline and canine patients and present as a monetary and clinical dilemma for veterinarians and clients. In human medicine, fluoroscopy, in conjunction with endoscopy, is used for dilatation of esophageal strictures because they allow for better visualization and assure that the waist of the stricture is not just stretched, but completely broken. With a similar theory to the other stricutres, esophageal strictures would ideally be balloon dilated, with a stent left in place to keep the stenotic lesion open for the time it would normally take the tissue to re-stenose. The biggest concern about doing this in the esophagus is that this area is very motile (vs. the nasopharynx) and food will need to pass through the area. The risk of the stent migrating into the stomach, or proliferative tissue growth around the ends of the stent material, makes permanent stenting for benign disease less than ideal. In order to circumvent these concerns pliable stents with a shape that would hold up again peristalsis (dumb-bell and self expanding) have been tried. Knowing that the stenotic tissue will heal over 14 or more days, having a stent that can be removed or resorbed (polylactic acid or PDS stents) in a few months are being investigated. These various types of stents have been studied in humans for some time and a small handful of clinical veterinary cases. The technique holds a more clear indication for non-resectable esophageal malignancies, and this has been performed with success.

Using both fluoroscopic and endoscopic guidance, the stent is placed over a guidewire and centered over the esophageal stenosis or tumor inside its delivery system. Once deployed the stent is expanded across the stenosis. For benign lesions it is recommended that the stent be tacked with a suture (either endoscopically or manually placed suture) to prevent stent migration into the stomach. The preliminary results thus far have been variable. In the future this may be a consideration at the time of 2nd or 3rd balloon dilatation to avoid serial anesthetic procedures and high costs but the ideal stent needs to be investigated before this is routinely recommended.

A newer device has become available which is a removalable balloon esophagostomy tube (BE tube) that is made for balloon dilation of an esophageal stricture to be done twice daily at home for 6 weeks. To date this device has very promising results, with little to no recurrence after placement.

Esophageal-jejunal feeding tubes

Enteral methods of feeding are preferred over parenteral nutrition in humans due to the benefits on gut mucosal integrity, barrier function, and lower complication rates. Jejunal feeding in small animal patients is controversial. In animals that are intolerant of gastric feedings(e.g. intractable vomiting), those where pancreatic exocrine duct by-pass is desired, or animals that are unconscious and regurgitation or reflux is a concern (e.g. ventilated animals), feeding directly into the jejunum is often considered. Classically, this has been done via surgical or laparoscopic techniques with a high complication and orad dislodgement rates. More recently, PEG-Jejunal feeding tube placement using endoscopy has been investigated and successful.

Due to the ease of placing a nasal feeding tube or an esophagostomy feeding tube, tubes have been able to be placed into the jejunum from the nares (NJ) or esophagus (EJ) with fluoroscopy +/- endoscopy. This eliminates the complications associated with enterotomy or gastrotomy such as septic peritonitis or unnecessary gastric or jejunal orificies). Fluoroscopy is used with NJ and EJ tube placement to visualize the guidewire and catheter placement into the duodenum and jejunum. If an upper GI endoscopic procedure is being performed at the same time, wire placement across the pylorus can be done through the endoscope. This technique is fast, effective and fairly inexpensive when compared to surgical placement or the intensive care monitoring of TPN. NJ and EJ tubes have been left in place for over two weeks, and once the J-feedings are unnecessary an E-tube remains for convenience.

Colonic stenting

Colonic obstructions are rare in small animal patients. They can be due to neoplastic lesions, strictures, or granulomatous lesions. In humans, colonic stents have been available for over a decade and are most commonly placed in people with neoplasia who are a prohibitive surgical risk or those in which resection holds little chance of surgical cure. They have been used as a mechanism to help de-obstipate patients prior to resection and anastomosis. In humans, colonic stents can either be placed through the endoscope using direct visualization for deployment, or they can be placed over a guidewire under fluoroscopy. Endoscopic placement is preferred in more proximal tumors, for precise stricture localization, and to guide the stent across acute angulations in the colon. In humans, clinical success is seen in up to 95% of patients. At the University of Pennsylvania, four colonic stents have been placed in cats; three for tumors and one for a stricture. In all cases colonoscopy was done to visualize the lesion and help localize the lesion fluoroscopically. A guidewire was then advanced through the stenotic lesion. Under fluoroscopic guidance a self-expanding metallic stent (SEMS) was placed across the stenotic lesion or tumor and the stent was deployed in that position. Patency was re-established immediately in all cases and subsequent de-obstipation was achieved. All cats were fecally continent, and no stent migrations were seen. The stent was visualized to be incorporated into the colonic mucosa within four days in one cat that was re-scoped.

BILIARY INTERVENTIONS

Extrahepatic biliary obstructions present a therapeutic dilemma as they induce life-threatening metabolic derangements e associated with high morbidity and mortality. Surgical treatment is often indicated, but biliary diversion surgery has mortality rates ranging from 25-70% in dogs and over 75% in cats. If the metabolic derangements can be relieved by a fast and effective decompressive procedure, future surgical interventions for more definitive decompression may be safer for the patient. We will discuss two options that can be performed in veterinary patients 1) endoscopic drainage through the common bile duct (ERCP) and 2) laparoscopic assisted biliary drainage via cholecystostomy tube placement.

Endoscopic retrograde cholangiopancreatography (ERCP) and biliary stent placement

Endoscopic retrograde cholangiopancreatography is an IE technique used for the diagnosis, and potential treatment, of biliary tract disease, pancreatitis, or pancreatic obstructive lesions. To date, biliary stents have been successfully placed in a small handful of normal purpose-bred dogs, and clinical investigation is underway. Using a side-view duodenoscope the major duodenal papilla is visualized and cannulated with a sphinctertome catheter. Once a retrograde cholangiogram and pancreatogram are performed, a guidewire is advanced into the common bile duct under fluoroscopic guidance. Then, through the endoscope, a polyurethrane stent is advanced over the wire. With fluoroscopic and endoscopic guidance, the stent is advanced into the common bile duct, transverses the major duodenal papilla and exits into the duodenum. This can be left in place until the obstructive lesion resolves (e.g. pancreatitis), or a permanent metallic stent can be used (e.g. neoplasia). This bypasses the need for diversion biliary surgery for EHBDO.

Laparoscopic cholecystostomy tubes

Laparoscopic placement of cholecystotomy tubes has been reported in a small number of clinical veterinary patients. They can be placed with very short anesthesia times. An 8 or 10 french locking-loop pigtail catheter can be advanced using a right paracostal approach laparoscopic visualization. With transhepatic penetration of the catheter, the gallbladder is accessed. Once the trocar is in the lumen of the gallbladder, as visualized via laparoscopy, the trocar and stylette are slowly removed and the locking loop mechanism is set in place. The catheter can drain the gallbladder and can be sutured securely to the body wall. This can remain in place until the patient is a better anesthetic candidate for surgery or for four to sixweeks while a more benign lesion resolves (e.g. pancreatitis) and a seal forms. In the clinical reports, most patients needed to have the rent in the gallbladder closed surgically once stabilized, as a seal did not form very quickly.

URINARY INTERVENTIONS

Ureteral stenting

Ureteral stenting is performed for a variety of disorders to divert urine from the renal pelvis into the urinary bladder. This technique can be useful in patients with ureteral obstruction due to ureterolithiasis, ureteral or trigonal obstructive neoplasia, ureteral spasm, ureteritis, and ureteral stenosis/stricture.. Stenting may also be beneficial following ureteroscopy, percutaneous nephrolithotomy, ureteral stone retrieval (basket retrieval or via laser lithotripsy), for post-operative ureteral anastomosis, and ureteral tears. In addition, the presence of the ureteral stent may result in subsequent passive ureteral dilation to permit passage of previously obstructive ureteroliths or allow passage of the flexible ureteroscope for appropriate ureteral intervention. This technique has been performed in the authors practice in over 120 patients (dogs and cats) for various causes.

Cystoscopic-guided laser ablation (CLA) of ectopic ureters

Ectopic ureters are a common congenital anatomic deformity in dogs with the ureteral orifice being positioned distal to the bladder trigone within the ureter, vagina, vestibule or uterus. Over 95% of ectopic ureters in dogs transverse intramurally and are candidates for this minimally invasive IE procedure. Endoscopic repair of ectopic ureters has been performed in over 70 dogs (males and females) in the United States. This is done with the use of fluoroscopy, cystoscopy and a diode or holmium: (yttrium, aluminum, garnet) laser. To avoid more than one anesthetic procedure, this procedure is performed on an outpatient basis at the time of cystoscopic ectopic ureter diagnosis . This will be discussed in more detail in the incontinence section.

Ureteroscopy

Ureteroscopy is limited to dogs larger than 20 kg and performed most commonnly for evaluation of idiopathic renal hematuria. This procedure is difficult to perform in dogs through a normal ureteral orifice, because the ureter in a normal dog is less than 2 mm and the smallest ureteroscope is approximately 2.5 mm. Ureteral access is obtained via cystoscopy, with a guidewire being advanced up the ureter into the renal pelvis and the endoscope then advanced over the guidewire under fluoroscopic and endoscopic guidance. As mentioned previously, ureteroscopy is ideal for the evaluation of the ureteral and renal pelvic mucosa in cases of idiopathic renal hematuria, but is also used for ureteral stone disease and laser lithotripsy if ESWL is not effective, or for evaluation of ureteral obstruction and hydroureter when other imaging modalities are not diagnostic (e.g. ureteral stricture/stenosis or neoplasia).

Idiopathic renal hematuria is a rare condition in which a focal area of bleeding in the upper urinary tract results in hematuria, iron deficient anemia (chronically) and possible clot formation, or calculi due to blood clots. The result is ureteral colic or signs of lower urinary tract disease. In people, the presence of a hemangioma or vascular malformations have been visualized ureteroscopically, which is cauterized through the working channel of a ureteroscope. This has also been performed in a small number of dogs to date.

Laser lithotripsy

Laser lithotripsy is an innovative technique involving the intracorporeal fragmentation of uroliths, which is assisted with a rigid or flexible cystoscope or ureteroscope. The holmium:YAG laser combines tissue cutting and coagulation properties with the ability to fragment stones upon contact. The laser energy is focused on the urolith surface, directed via cystoscopy. This process is useful for ureteral, cystic and urethral calculi. All stone types are able to be fragmented using laser lithotripsy.

Other urologic applications for laser include incision of urethral and ureteral strictures; ablation of superficial transitional cell carcinoma/prostatic adenocarcinoma within the urethral lumen (using a diode laser), and laser ablation of urinary polyps. Bladder polyps are common findings in dogs and can be associated with chronic, recurrent urinary tract infections, cystolith formation, and are often misinterpreted for cystic neoplasia. Using cystoscopy and baskets or laser lithotripsy the polyps can be removed without surgical intervention by cauterizing the stalk.

Urethral stents

Malignant obstructions of the urethra can cause severe discomfort, dysuria and life-threatening azotemia. Greater than 80% of animals with transitional cell carcinoma (TCC) of the urethra, and/or prostatic carcinoma experience dysuria and approximately 10% developing complete urinary tract obstruction. Chemotherapy and radiation therapy has been successful in slowing tumor growth but complete cure is uncommon. When signs of obstruction occur, more aggressive therapy is indicated. Placement of cystostomy tubes, transurethral resections, urethral lasering procedures, and surgical diversionary procedures have been described but few hold long-term promise, most are very expensive, and many are associated with undesirable outcomes such as the need for manual urine drainage, associated morbidity, frequent urination, and chronic infection. Placement of self-expanding metallic stents using fluoroscopic guidance through a transurethral approach can be a fast, reliable, and safe alternative to establish urethral patency in both males and females with a greater than 90% good to excellent palliative outcome. Most procedures can be done on an out-patient basis. Urethral stenting has also been useful in patients with benign urethral strictures, granulomatous/proliferative urethritis or reflex dysynnergia, when traditional therapies have failed.

Transurethral submucosal collagen implantation

Transurethral submucosal collagen implantation is performed with urethroscopic guidance for urethral sphincter mechanism incontinence (USMI) at many institutions. This procedure is indicated if medical management for USMI has failed, is contraindicated, or not tolerated. Overall success of the procedure is good (>80%), though the average maintenance of continence rates following this procedure is reported to be 68% at 17 months, with re-injections potentially being necessary thereafter. Different materials have been tried with variable success.

CONCLUSION

In conclusion, IE provides new, safer alternatives to the treatment of certain conditions that are traditionally very difficult to manage or have required more invasive techniques for disease management.

REFERENCES

Available upon request.

Interventional Radiology: Lessons Learned Over 15 Years of Discovery

Chick Weisse, VMD, DACVS
New York, NY

Interventional Radiology (IR) involves the use of contemporary imaging modalities to gain access to different structures in order to deliver materials for therapeutic purposes. Over the last decade, IR has become a relatively well-known entity in veterinary medicine. The arrival of these techniques into the "mainstream" of veterinary medicine can be demonstrated by the growing number of journal articles published using these techniques as well as "reported sightings" on recent medicine and surgery board certification examinations.

The development of these techniques for peripheral use, rather than the historic cardiology applications, began slowly int the veterinary community. Initial lectures and publications progressed from case reports of what we could do, to reports on what we had done, and then to case series and larger retrospective studies currently scattered throughout the veterinary literature and conferences/symposia. Still, few training opportunities and high initial capital costs have limited the bulk of these procedures to major institutions and large referral centers. The subject of this lecture will differ from previous IR lectures given at the ACVIM symposium. Rather than demonstrating cases performed, or describing certain techniques, the following lecture will discuss the "lessons learned" about veterinary medicine through the use of these procedures. A number of different body systems and pathological processes will be discussed, with the insight the author has gained over the years through IR techniques.

TRACHEAL COLLAPSE

Performing tracheal stenting, a procedure initially offered in few locations, provided our service with a large number of cases quickly. There is no better way to advance ones knowledge about a condition than to see large numbers of patients quickly. Some of the conventional wisdom was challenged below:

1) Canine tracheal collapse syndrome (CTCS) is associated with either chondromalacia or a weakened dorsal tracheal membrane: This was rejected in our patients as 55% had tracheal "malformations"
2) Concurrent upper airway comorbidities are common including laryngeal paralysis or collapse: This was confirmed, however laryngeal paralysis was never identified, rather long soft palates (23%) and epiglottic retroversion (10%) were more commonly encountered.
3) The tracheal stenting procedure is associated with low morbidity and low mortality: Confirmed to be very low (6% mortality rate).
4) Bronchial collapse is a common comorbidity and often carries a worse prognosis: Stenting these patients should be reconsidered: Theory that tracheal stenting should NOT be performed in these patients is REJECTED, we will stent these patients.

VASCULAR ANOMALIES

Like the patients above, receiving a "relatively" large population of intrahepatic portosystemic shunt (IHPSS) and other vascular anomaly dogs and cats permitted closer evaluation of these patients. Some previously unknown insights reported below:

1) Gastrointestinal ulceration is a common pre- and post-operative problem in these dogs. Often GI biopsies are associated with varying degrees of inflammatory bowel disease. The risk of GI bleeding remains throughout these patients' lives and preoperative proton pump inhibitors reduces mortality.
2) Multiple IHPSS are more common than previously reported with 10% of dogs having more than 1 IHPSS. If there are greater than 3 IHPSS, these patients will often demonstrate elevated portal:CVP gradients.
3) Pre-operative bloodwork or shunt type is NOT associated with outcome. Smaller patients had improved outcomes as did patients with shunts located anatomically more caudal.
4) Acquired intrahepatic collaterals can develop following post-hepatic shunt attenuation which limits resolution of bloodwork changes but also reduces risks of the development of portal hypertension.
5) Some patients with "acquired multiple EHPSS" can benefit from shunt attenuation.
6) Hepatic arteriovenous malformation (HAVM) patients are NOT AVFs! They also have poorly developed portal perfusion and IR techniques may provide a safer, alternative treatment option.

THROMBOSIS/THROMBECTOMY/THROMBOLYSIS

These are some of the most difficult and frustrating cases to manage as time is critical and the need for anticoagulation limits our ability to safely perform surgery. IR techniques are well suited to these cases. Some lessons below:

1) IR techniques are safe and feasible in a wide range of patient sizes using thrombectomy devices or stents.
2) Local perfusion of TPA can be safe and more effective than systemic administration.
3) Delayed reperfusion of limbs may be safer and advantageous.
4) Evaluation of distal limb run-off is extremely important, and may predict outcome, as is immediate post-op physical therapy and rehabilitation.

5) Dogs with aortic thrombi (NOT cats) have a good prognosis given the chance!!
6) Don't mistake aortoiliac occlusive disease with saddle thrombus!

"NON-RESECTABLE" CANCER

Perhaps one of the greatest lessons learned from IR techniques is the ability to not simply replace more invasive surgical techniques but rather provide therapeutic options for veterinary patients with few options. Some of these lessons are listed below:

1) Palliative stenting for malignant obstructions can be safely performed in many lumens of the body. Prolonged survival times can be achieved when these therapies are followed with adjuvant systemic therapies.
2) Local delivery of chemotherapy can provide improved tumor responses and reduced systemic toxicities.
3) Hybrid therapies involving IR techniques and surgery can be combined to offer improved options for more aggressive tumors such as bladder and prostate cancer.

The use of IR therapies throughout the body over the last 15 years has taught this author much of what he may never have known without these techniques. The expansion of these techniques throughout the veterinary profession can offer others unique opportunities to further elucidate a variety of conditions in our patients.

Complications: Lessons in Humility

Allyson Berent, DVM, DACVIM
New York, NY

Introduction

This lecture will focus on some of the most basic to some of the most severe complications seen in our practice in dealing with interventional cases. The focus will be in endourology and this lecture will be given by Dr. Allyson Berent (internist/interventionalist), Dr. Chick Weisse (surgeon/interventionalist) and Dr. Kelly Healy (human urologist). We will discuss our worst cases, biggest mistakes, ultimate solutions and biggest lessons learned from these complications.

This lecture will be informal and we will bring some of our personal cases, which in hindsight we may have done things a little differently. The goal of this lecture is to show how you learn from mistakes, and must think carefully about decisions being made during some of the most complicated, and straightforward, cases. We welcome the audience to participate and bring some of their own case examples along so we can all discuss them and share in the humility.

What Would the "Human Doctor" Do Now?

Allyson Berent, DVM, DACVIM
New York, NY

Introduction

This lecture will focus on controversial cases in veterinary medicine, mostly involving interventional radiology and endourology. The veterinarians will present some of their cases that were controversial in regards to treatment approaches, and discuss our decision making process while the human urologist tells the audience what would have happened in human medicine. These cases will not be shared with the MD prior to the presentation so that honest criticism and comparisons can be made. This discussion will highlight ethical dilemmas in decision making for patient quality of life, longevity, owners expectations, and the opinions of colleagues.

We will encourage audience participation and questions about their own patients to the panel for comments, recommendations, and discussion.

Understanding Sepsis: Is It Time For a Paradigm Shift?

Amy E. DeClue, DVM, MS, DACVIM (SAIM)
Columbia, MO

Sepsis is a syndrome of immune dysregulation and subsequent globalized tissue damage secondary to infection. For years sepsis has been viewed as a pro-inflammatory phenotype, yet clinical trials aimed at blunting this pro-inflammatory shift have been largely unsuccessful. One of the hypotheses for why many clinical trials aimed at restoring immune homeostasis have failed is that all patients have been treated as if they have a hyper-inflammatory response when in fact they may not. For years, clinicians have recognized that some patients with sepsis develop additional infections during treatment, often with atypical pathogens that are generally considered opportunistic. This suggested that sepsis could lead to not just a pro-inflammatory phenotype, but also to immunosuppression. The idea that sepsis could include an immunosuppressive phase seemed logical considering that administration of immunosuppressive therapy had limited clinical utility. The realization that sepsis is not simply inflammation but a complex, variable syndrome that encompasses aspects of both exuberant inflammation and immunosuppression has revolutionized the classical paradigm of sepsis.

Classically sepsis has been considered an imbalance between hyper-inflammatory and hypo-inflammatory responses. The hyper-inflammatory response has largely been blamed for the morbidity and mortality associated with sepsis. The hyper-inflammatory response (i.e., SIRS) is characterized by activation of the innate immune system and production of a cytokine storm. This leads to cellular recruitment, tissue damage and many of the classical manifestations of sepsis. However, animals with sepsis can also develop a hypo-inflammatory response which is characterized by immunoparalysis or lack of effective immune response to infection. During sepsis, it is likely that there are dynamic, oscillating shifts in the type of immune response to infection. Therefore, sepsis is a complex, variable and prolonged host response that is triggered by infection with both pro- and anti-inflammatory responses which could result in inflammation and tissue damage, immunosuppression and infection, recovery or a combination of these outcomes.

The manifestations of sepsis vary based on both pathogen and host factors. Pathogen factors include the type of pathogen, load, virulence and site of inoculation while the host factors include genetic factors, anatomic regions of infection, and co-morbid diseases. The innate immune system is predominately responsible for the initial manifestations of sepsis and is primarily activated by danger associated molecular patterns (DAMPs). Danger associated molecular patters include highly conserved molecules found in or on pathogens. These molecules are referred to as pathogen associated molecular patterns (PAMPs). The innate immune system recognizes PAMPs using pattern-recognition receptors including toll-like receptors, C-type lectin receptors, retinoic acid inducible gene 1-like receptors, and nucleotide-binding oligomerization domain-like receptors. A second class of DAMPs involved in sepsis come from the host. Tissue damage leads to release of damage associated molecular patterns or alarmins such as high-mobility group protein B1, S100 proteins, RNA, DNA and histones.

In the initial phases of in infection, PAMPs (e.g. endotoxin from gram negative bacteria; exotoxins, peptidoglycans and superantigens from gram positive bacteria; and fungal cell wall material) induce systemic inflammation initially through activation of local innate immune cells. For example, during Gram negative sepsis, lipopolysaccharide (LPS), the glycolipid component of the cell wall of Gram negative bacteria, is released. The lipid A portion of LPS binds to LPS binding protein. LPS is recognized via macrophage cell surface receptors like CD14. The main function of CD14 is to assist in toll like receptor (TLR)-4 and MD-2 signaling. Once LPS binds to these cell surface receptors, the macrophage becomes activated. Activation of inflammatory cells results in the production of multiple inflammatory mediators that have been implicated in the induction and maintenance of sepsis. There are many inflammatory mediators involved with sepsis and SIRS. TNF-α, IL-6, nitric oxide and leukotrienes are examples of important mediators contributing to the pathology of sepsis in dogs and cats. Production of these inflammatory mediators results in activation of vascular endothelial cells and upregulation of adhesion molecules. These inflammatory messages enter into the local milieu and then into the systemic circulation. Neutrophils, monocytes and lymphocytes travel to the site of infection based on these inflammatory mediator signals and secrete a second wave of inflammatory mediators and provide killing mechanisms for microorganism clearance. Neutrophils also release neutrophil extracellular traps that consist of filamentous DNA containing histones and granular proteins. Neutrophil extracellular traps have beneficial effects such as enhanced bacterial killing but they also interact with the coagulation system and may promote thrombosis. The transition from a localized infection/ inflammatory focus to sepsis is dependent on the systemic circulation of inflammatory messages and activation of immune cells distant to the initial source of infection.

The adaptive immune system is involved in the pathogenesis of sepsis. Antigen presenting cells that have ingested a pathogen activate the adaptive immune response. Subsequently, naïve T cells proliferate and effector cells are generated. Early in the course of sepsis, naïve T cells differentiated into Th1 cells and Th17 cells which produce additional cytokines such as TNF, IL-2, IL-12, IFN-γ, and leukotrienes or IL-17A, respectively, which typically drives a pro-inflammatory phenotype. Conversely, an anti-inflammatory phenotype is observed when there is a Th2 response, in part, due to production of cytokines such as IL-4, IL-5, IL-9, IL-10 and IL-13. The balance between the Th1/Th17 and Th2 response contributes to the clinical manifestations of a pro-inflammatory shift, anti-inflammatory shift or recovery.

The compensatory anti-inflammatory response syndrome (CARS) is characterized by varying immunoparalysis or relative lack of immune response to infection. Immunoparalysis is the result of a loss of lymphocytes, dendritic cells, endothelial cells, GI epithelial cells and thymocytes. The predominant mechanism for this loss is apoptosis. During the hyper-inflammatory response, activation of inflammatory cells leads to nuclear translocation of NF-kappa-B and IL-1β activation both of which induce apoptosis of adaptive immune cells. People with sepsis have depletion of CD4+ and CD8+ T cells, B cells and dendritic cells. The anti-inflammatory cytokine IL-10 is produced in abundance leading to suppressed CD8+ T cell function and reduced monocytic production of pro-inflammatory cytokines. A reduction in cytokine production by and expression of inhibitory receptors on T cells is also observed during immunoparalysis. There is an increase in T regulatory cells and myeloid-derived suppressor cells which contributes to immune cell quiescence. The neuroinflammatory reflex leads to norepinephrine release in the spleen and production of acetylcholine by CD4+ T cells which in turn targets alpha-7 cholinergic receptors on macrophages suppressing cytokine production. There is a reduction in HLA-DR expression by myeloid cells which results in a reduction of antigen presentation. Clinically, immunoparalysis leads to an inability to fight an eliminate infection and the development of new infections from opportunistic or otherwise normally weakly virulent organisms. Reactivation of latent viruses is another possible sequela.

When we evaluate the physiologic outcomes from these immunologic changes the resultant pathology is vast and variable. Some animals will suffer tissue damage secondary to a hyperactive immune system while other animals will develop secondary infections. In some cases, animals will cycle from a pro to an anti-inflammatory response or could develop features of each. The fortunate ones go on to have resolution of immunodysfunction and recovery. As we learn more about the immune response during sepsis, it has become clear that sepsis is a complex and dynamic syndrome of immune dysregulation and subsequent globalized tissue damage secondary to infection. Because of the widely varied and dynamic nature of the immune dysregulation, we need to reframe our clinical and research perspectives about the pathophysiology of sepsis. There is a major need to develop mechanisms to identify the type of immune system dysregulation so that we might develop targeted therapies. However, it is important to note that much of what we "know" about sepsis comes from animal models, often murine models, and people. Therefore, it is unclear how this information applies to the dog and cat population. As small animal clinicians, we must being the long process of teasing out the pathophysiology of sepsis in dogs and cats in our ICUs and how to better identify and describe immune dysfunction related to sepsis.

REFERENCES

Boomer JS, et al. **Virulence** 2014;5(1):45. Chang K, et al. **Critical care (London, England)** 2014;18(1):R3. Hotchkiss RS, et al. **The Lancet infectious diseases** 2013;13(3):260. Hotchkiss RS, et al. **Nature reviews Immunology** 2013;13(12):862. Hotchkiss RS, et al. **Nature medicine** 2009;15(5):496. Hutchins NA, et al. **Trends in molecular medicine** 2014. LeMay D, et al. **Am J Physiol** 1990;259(1 Pt 2):126. McDunn JE, et al. **Critical care (London, England)** 2009;13(2):127. Muenzer JT, et al. **Infection and immunity** 2010;78(4):1582. Miyamoto T, et al. **Jpn J Vet Res** 1996;44:107. Senturk S. **J Vet Pharmacol Ther** 2005;28:57. Schwulst SJ, et al. **Journal of the American College of Surgeons** 2008;206(2):335. Schutzer K, et al. **Circ Shock** 1989;29:13. Schutzer K, et al. **Am Rev Resp Disease** 1993;147:1380. Schutzer K, et al. **Europ Resp J** 1994;7:1131. Schwulst SJ, et al. **Journal of the American College of Surgeons** 2008;206(2):335. Sherwood ER, et al. **Critical care (London, England)** 2013;17(6):1022. Unsinger J, et al. **Journal of leukocyte biology** 2009;86(2):219 Venet F, et al. **Current opinion in immunology** 2013;25(4):477.

Current Concepts and Future Directions in Sepsis Management

Amy E. DeClue, DVM, MS, DACVIM (SAIM)
Columbia, MO

CURRENT CONCEPTS

Our current philosophy of sepsis management focuses on three key aspects: 1.) initial resuscitation, 2.) alleviating the inciting cause and 3.) aggressive supportive care. The treatment of sepsis can be roughly organized into initial resuscitation (i.e., first 1-6 hours) and long-term management (i.e., > 6 hours to discharge from hospital) phases. Initial resuscitation includes not only restoration of hemodynamic homeostasis, but also alleviating the inciting cause through administration of antimicrobial drugs and/or debridement. This resuscitation period is based on the idea of tailoring treatment to specific physiologic parameters thought to be associated with restoration of homeostasis and improved outcome.

INITIAL RESUSCITATION

Restoration of hemodynamic stability

Initial hemodynamic resuscitation of animals with sepsis is important for restoration of homeostasis. While there are no generally accepted, standardized initial resuscitation protocols for animals, guidelines have been established for people. These guidelines are outlined in the Surviving Sepsis Campaign: International Guidelines for Management of Severe Sepsis and Septic Shock. These guidelines were initially established in 2004 and then updated in 2008 and 2013. The Surviving Sepsis Campaign guidelines were developed by a group of physicians based on an extensive literature review. Recommendations were graded based on the quality of scientific evidence available. The authors recommended protocolized, quantitative resuscitation of patients with sepsis-induced tissue hypoperfusion. This approach is also termed early goal-directed therapy. The purpose of early goal-directed therapy in people is to achieve each of the following during the first 6 hours of intervention:

- CVP 8-12 mmHg
- MAP ≥65 mmHg
- Urine output ≥ 0.5 ml/kg/h
- Superior vena cava oxygenation saturation of 70% or mixed venous oxygen saturation of 65%
- Additionally, resuscitation should be targeted to normalize lactate in patients with increased blood lactate concentrations.

While these goals have not been specifically tested in randomized, multicenter trials in animals, the general concept behind the goals could have value in veterinary patients. Resuscitation goals should be tailored to the unique physiologic parameters of the species.

There is a continuum between early goal directed therapy and later management. While it is important to consider early resuscitation goals, maintenance of these goals during continued care outside of the initial few hours of management is important. After initial resuscitation, it is not uncommon for animals to have recurrence of hemodynamic derangements. Achieving hemodynamic support goals can be accomplished using boluses of isotonic crystalloids or colloids. Then, ongoing fluid therapy should be tailored to the needs of the patient to provide for maintenance requirements, correct interstitial hydration deficits, replace ongoing losses and continue to correct any hemodynamic derangements. Patients with sepsis will have a propensity to develop interstitial edema, including subcutaneous and pulmonary edema, because of increased vascular permeability and decreased blood colloid osmotic pressure. Administration of colloid fluids (e.g., hydroxyethyl starch) may help prevent interstitial edema and should be considered for volume resuscitation and during maintenance of sepsis patients, in addition to crystalloid fluid support. However, there is evidence in people that hydroxyethyl starch may increase the risk for acute kidney injury during sepsis. Conversely, in goats with experimental endotoxemia, hydroxyethyl starch did not alter creatinine clearance or ultrastructural tubular integrity indicating that there may be species variation in response to colloids. No evidence supporting or refuting this phenomenon is available in dogs and cats with sepsis at this time. A possible alterative option is 5% lyophilized canine-specific albumin. In one study of dogs with abdominal source sepsis, administration of canine-specific albumin resulted in increased plasma albumin, COP and blood pressure. While further study is indicated, canine-specific albumin might be a good alternative to hydroxyethyl starch in dogs with sepsis.

Despite aggressive volume resuscitation, some patients will require additional support to maintain normal blood pressure and tissue perfusion. The safest and most efficacious treatment for septic shock has not been determined for dogs or cats. Prior to administration of any sympathomimetic drug, care should be taken to ensure that the patient is not hypovolemic. Since the goal of treating septic shock is to maintain tissue perfusion, medications that cause vasoconstriction should be used only if absolutely necessary. Positive inotropic drugs (e.g., dobutamine) may be a good initial management choice for septic shock since they help combat decreased cardiac output caused by myocardial dysfunction without inducing peripheral vasoconstriction. However, care should be taken to avoid positive inotropic drugs in cats with evidence of HCM. If volume resuscitation and positive inotropic support has failed to restore adequate blood pressure, a vasopressor agent (e.g., dopamine, norepinephrine, epinephrine or vasopressin) could be added. Vasopressors should always be administered as a CRI and ideally, direct blood pressure evaluation should be used for monitoring. Although there are no clinical trials evaluating these drugs, epinephrine is generally considered the least ideal vasopressor. Critical illness related adrenal cortical insufficiency has been documented in dogs and cats with sepsis and refractory hypotension and should be considered in any animal requiring vasopressor therapy during sepsis. Judicious use of corticosteroids may address therapeutic failure of vasopressor therapy in these patients.

Alleviate the cause

An important aspect of treating bacterial sepsis centers on the identification and eradication of the inciting cause. Although vigorous efforts should be made to identify the cause of sepsis, early antimicrobial treatment is critical for survival and should not be withheld pending culture results. In humans, early antibiotic administration has been shown to decrease mortality. Broad spectrum, bactericidal antimicrobial agents administered IV should be instituted as quickly as possible in patients with suspected sepsis, ideally within the first 30-60 minutes. Antibiotic therapy should be selected based on the most likely type of organism given the site of infection. Depending upon the source of infection, surgical debridement may be necessary for infection control. Once culture specific antibiotic sensitivity is determined, the antibiotic with the narrowest spectrum of activity should be chosen and administered until there is complete clinical resolution.

LONG TERM MANAGEMENT

Supportive care

GI/Nutrition- Bacterial translocation from the GI tract can contribute to systemic inflammation during sepsis. Early placement of a feeding tube and initiation of enteral or, in patients with vomiting, parenteral nutrition will help maintain GI barrier function. Additionally, medication aimed at gut health (e.g., omeprazole, famotidine, sucralfate, probiotics) could be considered. Hyperglycemia can be a complication of nutritional therapy, especially parenteral nutrition. Hyperglycemia has been associated with increased inflammation and a poorer prognosis in people with sepsis and SIRS. Although the importance of glucose homeostasis in small animals is unknown, iatrogenic hyperglycemia should be avoided.

Cellular oxygenation- Oxygenation, acid-base status, PCV or hemoglobin concentration and organ function should be closely monitored. Along with maintaining good tissue perfusion, maximizing cellular oxygenation will help maintain tissue viability and avoid multiple organ dysfunction. This may be accomplished by ensuring good oxygen carrying capacity, providing supplemental oxygen when indicated and preventing microthrombi formation. For the majority of patients, acid-base abnormalities are related to lactic acidosis secondary to poor tissue perfusion. Typically these abnormalities will resolve with resolution of hypotension and normalization of perfusion. Therefore, bicarbonate administration is rarely needed and in fact may be contraindicated.

Nursing care/Pain management- Care should be taken to ensure adequate patient comfort including management of pain, careful catheter maintenance, bladder/bowel care and frequent patient turning/movement to prevent decubital ulcers. Almost all patients with sepsis will require analgesic administration for pain management. Although many analgesics like ketamine and buprenorphine may offer specific anti-inflammatory advantages during endotoxemia, some may be detrimental. Morphine, for instance, augments the inflammatory response to endotoxin, has a detrimental effect on mean arterial pressure and increases mortality in endotoxemic rats. While it is not known if morphine is detrimental during sepsis in dogs and cats its routine use in septic patients should be avoided where possible.

NEWER CONCEPTS IN THE TREATMENT OF SEPSIS

Appropriate resuscitation fluid selection

For many years electrolyte solutions used for intravascular volume expansion have been generally considered equivalent in the context of restoring hemodynamic stability. Recent evidence in people and animal models of sepsis suggest that the composition of the electrolyte solution used for resuscitation impacts physiology and clinical outcome. Rapid administration of large volumes of 0.9% NaCl results in hyperchloremia, decreased blood pH and reduced base excess compared to administration of a similar volume of plasmalyte. The use of plasmalyte, compared to 0.9% NaCl, has the advantages of improving acid base status and reducing the incidence of hyperchloremia in people. In people with sepsis, hyperchloridemia is associated with higher mortality rates. This is because chloride increases production of cytokines, induces hypotension, promotes shock and increases the risk for acute kidney injury. Hyperchloremic acidosis causes activation of the immune system through induction of NF-kappa B and production of TNF, IL-6 and IL-10. Increased risk of acute kidney injury is thought to be secondary to alterations in renal blood flow. In a study of healthy people, administration of 0.9% NaCl reduced renal blood flow by 40%, decreased urine production and reduced Na excretion. In comparison, plasmalyte does not promote cytokine production, induce hypotension, promote shock or increase the risk of kidney injury. Plasmalyte also produces significantly greater diuresis than 0.9% saline. For these reasons, chloride rich solutions like 0.9% saline should be avoided for volume resuscitation during sepsis.

Optimizing pathogen clearance

Shock is a key contributor to morbidity and mortality during sepsis. The concept of irreversible shock was first introduced by Wigger et al. in 1950. Wigger suggested that shock can only be tolerated for a defined period of time and that if not reversed, and by extension the cause of shock is not reversed, the animal will die. This led to many of the current philosophies in emergency medicine such as early goal directed therapy and the concept of the "golden hour". While there is great emphasis placed on reversing shock and establishing hemodyanmic homeostasis, classically there has been less focus placed on alleviating the inciting cause in a rapid fashion. However, since the pathogen is the driving force for immunostimulation which leads to septic shock, eradication of the pathogen should reduce further immune system stimulation. In recent years, the concept of pathogen clearance optimization has come into favor. Optimizing pathogen clearance involves early administration of appropriate antimicrobials and source control when indicated. Indeed, rapid initiation of appropriate antimicrobial therapy reduces mortality in people with sepsis. Since identification of infection may be challenging in some animals, early administration of antimicrobials should be considered any time sepsis is suspected. Factors to consider when developing protocols for optimizing pathogen clearance in the dog and cat are noted in table 1.

Table 1: Factors to consider when optimizing pathogen clearance in dogs and cats (adapted from Kumar, 2014).

Goals	Considerations
Early administration	Administration within 30 min-1 hour of presentation or identification of sepsis
	Maximize the bioavailability (IV)
	Maximize the speed at which the antimicrobial becomes bioavailable
	Consider loading doses for some antimicrobials
Maximizing antimicrobial potency	Use -cidal antimicrobials when possible
	Utilize pharmacokinetics to maximize potency; time vs concentration dependency
	Consider combining multiple antimicrobials with different mechanisms of action
	Provide broad spectrum coverage
	Consider antibiotic penetration into the infection microenvironment and anatomic location; community vs hospital acquired infection
Pathogen removal/reduction in pathogen exposure	Source control via surgical debridement

Modulating the immune system

Sepsis is a complex, dynamic syndrome characterized by immune system dysregulation. Both the innate and adaptive immune systems are activated and potentially become dysregulated during sepsis. Dysregulation includes the development of a pro-inflammatory, hyperactive immune response or immunosuppression of the innate and/or adaptive immune systems. Characteristics of sepsis-induced immune dysregulation include persistent inflammation, catabolism, inappropriate cytokine production, reduced phagocytosis, contracted antigen presentation, cellular immortality, T cell anergy, lymphocyte apoptosis, reduced T cell cytotoxicity, inappropriate T cell polarization and increased Treg suppressor function. The outcomes of these immune changes vary and include rapid, severe tissue damage, organ failure and death; immunoparalysis and secondary opportunistic infections; long-term tissue damage and blunted regenerative responses, and recovery. As we have learned more about variation in the immunophenotype associated with various forms of sepsis, new targets for immunotherapy have been developed. What once was a quest to simply suppress an overzealous, pro-inflammatory response has now changed to a focus on restoring immune homeostasis. It is likely that in the future we will use immunophenotyping to determine the nature of the immune dysregulation in an individual patient and then target those specific pathways with either an immunostimulant or an immunosuppressant. Several possible immunomodulating therapies aimed at restoring immune system regulation have been proposed including TLR4 antagonists, C5aR antagonists, caspase inhibitors, C-GSF, GM-CSF, IFN-γ, Anti-PD1/PD-L1, IL-3, IL-7, IL-15, IL-17, and statins.

Blood purification for removal of DAMPs and inflammatory mediators

Blood purification has been used target and remove danger associated molecular patterns and other mediators of sepsis from the systemic circulation. The goal of these therapies is to interrupt the signaling that leads to immune system dysregulation and tissue injury. There are several methods being investigated including high volume hemofiltration, high absorption hemofiltration, high cut-off membrane hemofiltration and coupled plasma filtration adsorption. Some of these methods are selective in the sense that they utilize techniques that specifically remove a substance of interest while others are non-selective in nature. Hemoperfusion cartridges with fibers coated in polymyxin-B, a polypeptide antibiotic that binds to endotoxins, is a selective method used to remove circulating endotoxins with a low incidence of adverse events. Non-selective methods include coupled plasma filtration and absorption which reduces circulating cytokines such as TNF, IL-1, IL-6, CXCL-8, and IL-10. These methods have shown promise for the management of sepsis in people, however evidence of efficacy is limited at this time.

REFERENCES

Aziz et. al. J Leukocyte Bio. 2013;93:329-342. Conti-Patara et. al J Vet Emerg Crit Care 2012;22(4):409-18. Craft et. al. J Vet Emerg Crit Care. 2012; 22(6):632-639. Delano et. al. J Clin Investigation 2016; 126(1):23-31. Dellinger et. al Crit Care Med 2012; 41(2):580-637. Esteban et. al. Mediator Inflamm. 2013; Epub. Ertmer et. al Crit Care Med 2011; 39(4):783-92. Hassan et. al. EXCLI J. 2013; 12: 681–692. Hutchin et. al. Trends Mol Med. 2014;20(4):225-33. Kellum et. al Chest 2006 Oct;130(4):962-7. Kellum et. al Chest 2004 Jan;125(1):243-8. Kumar. Virulence. 2014;5(1): 80–97. Laszlo et. al. J Immuno Res. 2015; 2015: 510436. Mutter et. al Cochrane Collaboration 2013; 1-27. Nierhaus et. al. Intensive Care Med. 2003;29:646–651. Shukla et. al. Br J Pharmacol. 2014;171(22):5011-31. Silverstein et. al J Vet Emerg Crit Care 2012; 22(6);666-73. Young et. al Ann Surg 2014;259(2):255-62.

Vasopressor Use in Small Animal Critical Care Medicine

Medora Pashmakova, DVM, DACVECC
College Station, TX

INTRODUCTION

Systemic hypotension is a state of low arterial blood pressure such that delivery of oxygen to the tissues is inadequate. A state of inadequate oxygen delivery to, or oxygen utilization by, the tissues resulting in cellular dysfunction is the definition of shock. Understanding the principles of hypotension and the therapeutic options we have to augment vascular tone are critical to our ability to treat critically ill patients, optimize organ perfusion, and improve patient outcomes.

OXYGEN DELIVERY

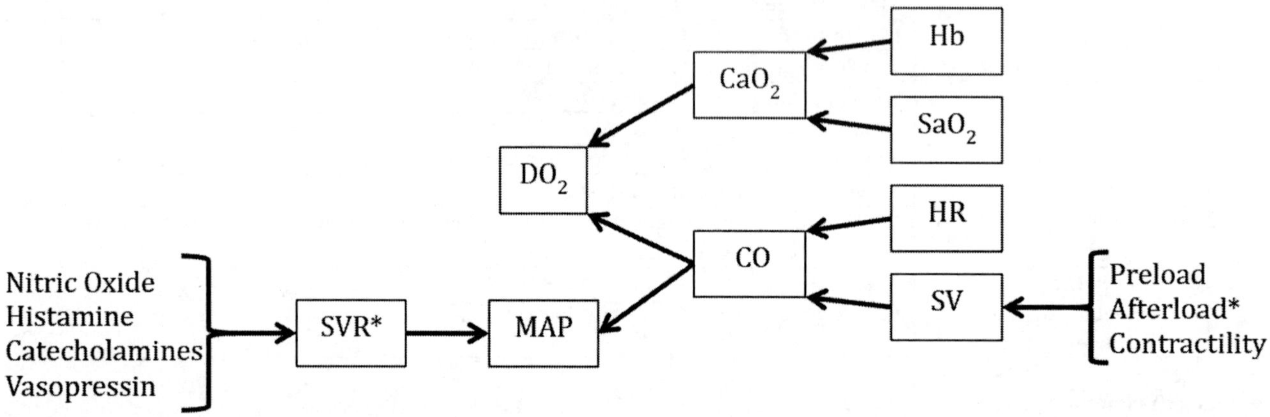

Figure 1. Determinants of oxygen delivery.

"Tree of Life"

Optimization of oxygen delivery (DO_2) depends on 2 determinants: arterial oxygen content (CaO_2) and cardiac output (CO). The former relies on adequate concentrations of hemoglobin (Hb) and oxygen saturation (SaO_2). The relationship between oxygen tension and oxygen saturation of hemoglobin is described by the oxy-hemoglobin dissociation curve. Cardiac output is the result of heart rate (HR) multiplied by stroke volume (SV), which is dependent on preload, afterload, and cardiac contractility. Cardiac output and systemic vascular resistance (SVR) are the main determinants of mean arterial pressure (MAP). SVR responds to both locally produced (nitric oxide, histamine) and systemically circulated (catecholamines, vasopressin, angiontensin II) modulators.

This algorithm (Figure 1) is an effective way to approach the patient in shock (i.e. lack of DO_2) as most major determinants can be isolated and optimized using a systematic process. The majority of patients that present in shock to emergency services suffer from hypovolemia (i.e. inadequate preload) secondary to excessive fluid loss or hemorrhage. The patient with dilated cardiomyopathy will present in shock due to inadequate contractility. The clinical anemic patient is also a patient in shock due to the lack of oxygen carrying capacity, i.e. hemoglobin, of the blood. A patient may have normovolemic anemia (such as with IMHA) or hypovolemic anemia (hemorrhagic shock). Distinguishing the etiology of shock in any particular patient is instrumental to choosing the most appropriate intervention.

In rare but notable instances, adequate DO_2 is delivered to the tissues, but tissue utilization (VO_2) is impaired. This is most significant in patients with mitochondrial dysfunction secondary to sepsis. Often these patients also have inadequate vascular tone, vascular leak syndromes, and present a more complex challenge to optimizing oxygen availability than simply optimizing vascular volume.

VASCULAR TONE AND VASOPLEGIA

Vascular tone is the sum total of vasoconstricting and vasodilating influences on a blood vessel. Both *extrinsic* and *intrinsic* factors contribute to vascular tone. Extrinsic factors in general are neurohumoral influences such as catecholamines secreted by sympathetic nerves and the production of angiotensin II. Intrinsic factors include those that regulate local blood flow, such as endothelial production of nitric oxide and endothelin, and local mediators such as bradykinin and histamine. Importantly, states of hypoxia result in a general decrease in vascular tone (shock begets more shock).[1]

The vascular endothelium is lined by smooth muscle cells containing receptors for angiotensin II, norepinephrine, and endothelin (the body's most potent endogenous vasoconstrictors). These substances activate smooth muscle contraction via the IP3 and DAG (inositol triphosphate and diacylglycerol) second messenger systems that ultimately result in increasing the cytosolic concentration of Ca^{2+}.[2] In order for vascular smooth muscle cells to function adequately, membrane potential gradients must be maintained. Conditions resulting in alterations of this gradient result in inadequate ability for cells to depolarize and contract. K_{ATP}

channels located on vascular smooth muscle cells are sensitive to decreases in cellular ATP, which follows inadequate aerobic cellular metabolism. Activation of K_{ATP} channels results in an efflux of K^+, hyperpolarization of the cell interior, and prevention of Ca^{2+} entry into cells. Activation of K_{ATP} channels can also take place in states of intracellular acidosis, lactate accumulation, hypoxia, or nitric oxide (NO) build-up.

Nitric oxide is a potent vasodilatory molecule that is produced locally from the amino acid arginine and the upregulation of the enzyme inducible nitric oxide synthase (iNOS). Amplification of NO production occurs in the presence of inflammatory cytokines, bacterial endotoxins, sepsis, and late hemorrhagic shock. The overproduction of reactive NO substances has been implicated in the downregulation of adrenoreceptors and contribution to the development of septic shock.[3]

The third mechanism that propagates vasoplegia is the development of vasopressin (AVP) deficiency, which can play an important role in sepsis, prolonged hemorrhagic shock, and cardiac arrest. Vasopressin is produced in the hypothalamus and stored in the posterior pituitary for secretion in response to 1) increased plasma osmolality, 2) decreased blood pressure, or 3) decreased effective circulating volume. This happens early in shock, however, circulating levels of AVP are markedly decreased in the later stages of shock, presumably secondary to neurohypophyseal depletion. Vasopressin modulates vascular tone via the V_1 receptor. This is mediated by a G protein-coupled mechanism and the IP3-DAG second messenger system.[4] Vasopressin also inactivates the K_{ATP} channels. Both of these mechanisms result in an ultimate increase in intracellular Ca^{2+} availability.

The vasoplegic patient (such as one with septic shock or anaphylaxis) requires vascular smooth muscle augmentation in order to restore vascular tone, improve SVR, and optimize DO_2 by restoring MAP. Without intervention, these patients embark on a vicious cycle of systemic hypoxia propagating further vasodilatory mechanisms that eventually terminate in death.

WHAT IS THE EVIDENCE?

Pressor therapy remains a topic of great debate in human medicine. Large-scale studies and consensus statements on goal directed therapy and management of sepsis have focused on bundles of care and algorithms to determine when pressor therapy should be initiated.[5-7] The debate continues largely due to lack of consistent evidence in improved outcomes with protocol based goal directed therapies, of which normotension is a primary endpoint.[8,9] The most recent specific evaluation of the Surviving Sepsis Campaign Guidelines' impact on patient care documented a substantial increase in participating institutions' implementation of the guidelines concurrent with a decrease in overall hospital mortality (though cause and effect may not be easy to establish).[10] Because of the recent and wide-scale implementation of these guidelines, more time and data interpretation is needed to determine if the medical field is finally making strides towards improved survival.

Veterinary medicine offers a dearth of information on how hypotension impacts patient outcomes, the utility of vasopressors in our species, and even how to best obtain blood pressure measurements in dogs versus cats, and anesthetized versus awake patients. A discussion on blood pressure monitoring is beyond the scope of this manuscript, however, the reader is encouraged to understand the benefits and pitfalls of indirect (oscillometric vs. Doppler) vs. direct blood pressure methods. There is unanimous agreement that patients on pressor therapy should, whenever possible, have direct arterial blood pressure measured. The author recommends Doppler BP in all dogs and cats < 10 kg with a reliable cuff width measuring 40% of leg circumference.

A recent review article highlights the debate and various vasopressor practices in veterinary medicine.[11] A retrospective study of dogs with septic peritonitis identified that perioperative hypotension was one of the main factors associated with non-survival.[12] This study also identified that dogs on 2 or more pressors had a significantly higher likelihood of death compared to dogs on 0-1 pressors. A case series of 5 dogs in 2007 further established that vasopressin in conjunction with catecholamine pressors improved MAP in patients that were otherwise in refractory hypotension[13] and is currently recommended in vasodilatory shock unresponsive to catecholamines[2]. Vasopressors are also recommended in the management of veterinary acute kidney injury (AKI) associated with severe sepsis to optimize kidney perfusion.[14]

VASOPRESSORS – WHEN, WHAT, AND HOW MUCH

Pressor therapy is indicated in the vasodilated volume-replete patient with ongoing hypotension. Pressor therapy augments systemic vascular resistance, which is a component of afterload (both are marked with an * on the Tree of Life above). It is critical for the clinician to understand the cardiovascular impact of pressor therapy, and the potential complications. In addition to volume, cardiac contractility must be normal when pressor therapy (resulting in an increased afterload) is initiated. In patients where cardiac contractility is the cause for hypotension, inotropes (β-selectivity) are indicated before pressor therapy in order to prevent even more substantial cardiovascular failure. One such instance is transient myocardial stunning following cardio-pulmonary arrest and return to spontaneous circulation. These patients often require positive inotropy to sustain normotension, rather than vasopressors. In the instance where both cardiac contractility and vascular tone must be addressed, both α and β acting agonists must be used.

Pressor therapy is an integral part of early goal-directed therapy. The veterinary literature has extrapolated numerical targets for optimal resuscitation from the human literature, recommending a MAP>65 mmHg with pressor therapy if fluid therapy alone is insufficient.[15] Because of the logistical difficulty in instrumenting and maintaining an arterial catheter in companion animals, direct arterial pressures are often unavailable. This should not preclude the use of vasopressors. The use of trends in Doppler systolic pressures to guide pressor therapy is often adequate. In this instance, the author recommends targeting Doppler systolic pressures of 90-100 mmHg in previously normotensive patients.

Target blood pressures may be higher in patients that are chronically hypertensive (where a systolic of 120 mmHg may be relatively "hypotensive" for the established autoregulatory mechanisms). In sepsis, a high output vasodilatory shock can develop where tachycardia is sustained and the patient maintains a "normal" blood pressure. In the author's opinion, these patients may benefit from pressor therapy to augment vascular tone and reduce the cardiac demand (if the heart doesn't have to beat fast, it shouldn't beat fast), in effect treating 2 problems with 1 specific therapy. Assessment of capillary refill time (rapid or undetectable from vasodilation) and bounding pulses in a tachycardic septic patient may distinguish these patients as needing pressor therapy before overt hypotension is documented.

Catecholamines

Alpha agonists function to increase vascular tone, but due to their effect on increasing afterload, may result in a decrease in cardiac output and regional blood flow, such as in the cutaneous, splanchnic, and renal capillary beds.[11] Alpha receptors are located on post-synaptic regions of sympathetic nerves on smooth muscle cells and their stimulation causes vasoconstriction. Beta receptors are primarily located in the myocardium, though β-2 receptors are also located on bronchial and peripheral smooth muscle, resulting in broncho- and vaso-dilation, respectively, while β-1 agonism results in positive inotropy and chronotropy. A significant adverse consequence of using β agonists is precipitation of dysrhythmias as well as some metabolic alterations (lipolysis, glycogenolysis, insulin secretion).[11]

Dopamine receptors (D1-5) are located in the smooth muscle of cardiovascular, cerebral, coronary, and splanchnic circulations. Stimulation of dopamine receptors results in inhibition of norepinephrine release from the nerve terminals, resulting in local vasodilation. When dopamine is used as a pressor, the lowest dosage is thought to stimulate primarily dopamine receptors (and result in vasodilation), while the middle and higher range is considered β- and α- stimulating, respectively, and exert the desired pressor effects. If the reader is interested in a more thorough review of catecholamine receptors and distribution, one is available.[16]

A recent survey of Diplomates of the ACVECC revealed that catecholamines are the most commonly used first-line pressor.[17] Large-scale studies in human medicine demonstrate that when comparing **norepinephrine versus dopamine**, patients receiving norepinephrine had a survival benefit in some studies, and fewer dysrhythmias in many studies. These findings suggest that overall norepinephrine may be preferable as a first-line pressor, although no such comparisons have been evaluated in veterinary medicine.

When comparing **norepinephrine versus epinephrine**, human studies demonstrate a higher propensity for tachycardia and hyperlactatemia in patients receiving epinephrine, which can be explained by the stronger affinity of that drug for β receptors. The evidence is less clear about survival outcomes and at this stage; further evaluation is needed to determine if 1 drug is superior. The current Surviving Sepsis Guidelines recommend that vasopressors are initiated within 6 hours of admission in non-responders to fluid resuscitation. The first choice vasopressors is norepinephrine (grade 1B recommendation), followed by epinephrine (added or potentially substituted) if norepinephrine is not enough (grade 2B recommendation).[7]

Vasopressin

In situations where a catecholamine pressor is insufficient, cases in which vasopressin depletion (prolonged shock) is suspected, or if adrenergic receptors are unresponsive due to hypothermia or acidosis, vasopressin is the second-line pressor used in human and veterinary medicine. The VASST (Vasopressin and Septic Shock Trial) evaluating vasopressin vs. norepinephrine in patients with septic shock found no 28-day mortality benefit in patients with more severe septic shock, however, in those with less severe septic shock, a significant mortality benefit was identified with the use of vasopressin.[18] It is noteworthy also that in patients on norepinephrine where vasopressin was also added on, the concurrent use of physiologic corticosteroids and vasopressin was associated with less organ dysfunction and death compared to corticosteroids and norepinephrine alone.[19] The Surviving Sepsis Guidelines recommend that vasopressin is added to norepinephrine with intent of either raising MAP or decreasing norepinephrine dosage, however, vasopressin is **not** recommended as a single first-line pressor. Lastly, dopamine may be chosen as an alternative to vasopressin in select patients (those with relative bradycardia or at low risk for tachyarrhythmias).[7]

Practical pressor use

Because of the mixed receptor profile of most clinically useful pressors, a table is included to help the clinician select the most appropriate drug for each individual case (Table 1). Predictably, no large studies exist in veterinary medicine to guide our therapy, however, several prospective trials are in progress. In the meantime, the author's practice is to use norepinephrine as a first-line pressor in all cats and most dogs, followed by vasopressin. Titrations of dosage should occur every 10-15 minutes until the desired pressure is achieved. It is imperative that pressors are never discontinued abruptly, not even for a short time. Once hemodynamic stability is achieved, the time to initiate weaning of pressors should not be rushed. It is not uncommon for critically ill patients to be pressor dependent for several days. When underlying disease process is improving, nutrition is addressed, and organ perfusion (based on bloodwork parameters, urine output, mentation) is achieved, a slow and gradual tapering of pressors in the reverse order in which they were increased is performed. Ideally, 6-8 hours of maintained normotension should be achieved before the next "step down". The exception to this is if a patient becomes hypertensive or tachyarrhythmic and requires a faster discontinuation. Typically, weaning takes 24 hours in the optimal scenario. If at any point hypotension recurs, a reinstitution of the previous dosage must occur.

Table 1. Receptor activity, physiologic responses, and dosages for most commonly used pressors in small animal medicine.

Pressor	β_1	β_2	α	Contractility	Heart rate	Vascular tone	Blood pressure	Dosage
Dobutamine	++	+	+	↑↑	↑	↓	Varies	5-20 mcg/kg/min
Dopamine	++	+	++	↑↑	↑↑	↑↑	↑↑	5-20 mcg/kg/min
Epinephrine	+++	+++	+++	↑↑↑	↑↑↑	↑↑↑	↑↑↑	0.05-1 mcg/kg/min
Norepinephrine	+	0	+++	↑	Variable	↑↑↑	↑↑↑	0.1-2 mcg/kg/min
Vasopressin	0	0	0	0	↓	↑↑	↑↑	0.5-5 mU/kg/min

REFERENCES

[1]Klabunde RE. www.cvphysiology.com 2013. [2]Scroggin RD, et al. JVECC 2009;19(2):145-157. [3]Takakura K, et al. Anesthesiology 2003;98(4)928-934. [4]Silverstein DC. Small Animal Critical Care Medicine 2nd ed, Ch.158. [5]Havel C, et al. Cochrane Database 2011;(5):CD003709. [6]Rivers EP, et al. New Engl J Med 2001;345(19)1368-1377. [7]Dellinger RP, et al. Crit Care Med 2013;41(2)580-637. [8]ProCESS Investigators. New Engl J Med 2014;370(18)1683-1693. [9]Asfar P, et al. New Engl J Med 2014;370(17):1583-1593. [10]Mitchell LM, et al. Crit Care Med 2010;38(2):367-374. [11]Silverstein DC. JVECC 2015;25(1):48-54. [12]Bentley AM, et al. JVECC 2007;17(4)391-398. [13]Silverstein DC, et al. JVECC 2007;17(4)399-408. [14]Keir I, et al. JVECC 2015;25(2)200-209. [15]Butler AL. VCNA 2011;41:817-838. [16]Long KM, et al. JVECC 2008;18(1):2-25. [17]Silverstein DC, Rishniw M, University of Pennsylvania, VIN survey, 2014. [18]Russell JA, et al. NEJM 2008;258(9)877-887. [19]Russell JA, et al. CCM 2009;37:811-818.

Fluid Therapy in Critical Illness

Geoff Heffner, DVM, DACVECC
Fort Collins, CO

INTRODUCTION

Both insufficient and excessive fluid resuscitation has been associated with increased patient morbidity and mortality. Identifying patients in need of fluid resuscitation or who have achieved sufficient volume expansion remains challenging, however. Complicating the matter, there is not yet a consensus on the most appropriate combination of biomarkers and endpoint parameters by which to determine if a patient requires additional fluid. Finally, there has been renewed debate about what type of fluid should be prescribed. As such, fluid therapy in critically ill patients remains an evolving topic. This lecture will review some common resuscitation endpoints, provide updates on early goal directed therapy, and discuss trends in types of fluid therapy including hypertonic fluid therapy, human albumin, synthetic colloid fluids, and plasma transfusions.

Cardiopulmonary monitoring

Shock has been associated with a countless number of physical examination, biochemical, and cardiovascular derangements. Although often cited and easily attainable, physical exam findings are insufficient markers of adequate tissue perfusion [1]. This led to the introduction of more invasive methods of characterizing a patient's cardiovascular status, including wide spread adoption of the pulmonary artery catheter (PAC). The PAC provides hemodynamic information not supplied by clinical diagnosis alone such as cardiac output and left ventricular filling pressure. Studies have shown that incorrect interpretation of the data is common, and even when correctly interpreted, information derived from the PAC has not been shown to improve patient outcome [2-4]. Additionally, use of a PAC has been associated with complications and increased patient morbidity, resulting in a dramatic decline in clinical utilization. While venous pressure assessment may still be interpreted for determining patient volume status, the sole use of central venous pressure (CVP) is waning in the clinical decision making to administer fluids, vasopressors, or other interventions.

Instead, there has been growing interest in adding parameters to predict whether a patient's cardiac index will increase in response to a fluid challenge. Examples of studies of dynamic indexes in human medicine have included passive leg raise, arterial pulse contour analysis, and bioimpedence, and bedside ultrasonography [5-10]. Clinicians evaluate abbreviated ultrasound parameters such as ejection fraction, stroke volume variation, and caudal vena cava dimension to characterize cardiovascular status. For instance, >50% inferior vena cava collapse during inspiration is thought to be fluid responsive, but there is controversy in absolute numbers. This continues to be an active area of research in both human and veterinary medicine [11, 12].

Endpoints of fluid resuscitation

The cardiopulmonary effects of hypovolemic shock have been well described in veterinary medicine [1]. Early and ongoing research utilizes many of these parameters to identify deficits in cardiopulmonary function, and then assess any clinical benefit when correcting the identified abnormality. Studies enhancing cardiovascular performance have not uniformly shown survival advantage, however. One explanation is that there are various forms of shock. Hypovolemic shock has been associated with a decrease in cardiac output, while the mal-distributive shock caused by sepsis is a condition of increased cardiac output when appropriately restoring intravascular volume. Complicating the matter, a single patient may have multiple simultaneous forms of shock. Therefore, it is perhaps of no surprise that studies titrating dobutamine to a pre-determined cardiac output have actually been associated with an increased risk of death [13]. Other studies examining CVP or mean arterial blood pressure (MAP) suggest that while there may be pre-requisite minimum values for tissue perfusion, achieving or exceeding these parameters does not ensure sufficient oxygen tissue delivery [14]. These types of parameters have become known as macrocirculatory endpoints, and correction of these parameters has not consistently resulted in adequate tissue perfusion. The reasons for this are multifactorial, including the complex relationship between volume, pressure and flow, physiologic compensatory mechanisms, and endocrine, humoral, and autonomic mediators of regional tissue perfusion [15].

New paradigms in endpoint resuscitation pivot around a central, basic definition of shock: a disease process in which oxygen consumption exceeds oxygen delivery. Clinical reliance on blood pressures or cardiac output may overemphasize systemic hemodynamics while underappreciating the significance of tissue perfusion. Canine studies evaluating the effect of hypovolemic shock show that as oxygen delivery declines, tissue oxygen extraction increases [1]. Tissue oxygen extraction is measured by sampling venous blood from the tissue of interest, or by assessing global oxygen extraction from either the cranial vena cava (central venous oxygen saturation; $ScvO_2$) or pulmonary artery (mixed venous oxygen saturation; SvO_2). A decrease in $ScvO_2$ has been utilized not only as a marker of hypoperfusion in sepsis, but also an increase in SvO_2 has been suggested as a therapeutic endpoint for prescribed therapies [14]. Although utilized as a marker for many years in veterinary medicine, there has been renewed interest in lactate as a specific endpoint of fluid resuscitation. Studies have described titrating IV fluids to achieve a 10% improvement in lactate [16].

Identifying persistent tissue hypoperfusion remains an extremely difficult task. Despite a multitude of published normal and abnormal cardiovascular parameters, additional studies are required to determine whether a combination of parameters titrated to specific, as of yet unknown endpoints can reduce morbidity and mortality in critically ill patients.

Hypotensive resuscitation

Any topic of resuscitation should include a brief discussion about whether resuscitation should be pursued at all. Hypotensive resuscitation is defined as restricting initial fluid resuscitation in patients with severe, uncontrolled hemorrhage until attaining surgical hemostasis [17]. Patients in which hypotensive resuscitation may be considered include those with severe abdominal trauma. These protocols target endpoint parameters traditionally considered lower than necessary for tissue perfusion, with the goal of avoiding complications associated with fluid therapy. Studies suggest that aggressive fluid therapy has been associated with intravascular volume overload, increased risk of SIRS leading to MODS, increased risk of acute lung injury, increased requirement of mechanical ventilation, dilutional coagulopathy, disruption of thrombus formation, increased bleeding, increased blood transfusion requirements, and decreased survival [17-19]. A landmark study published in 1994 by Bickell et al, found a significant survival advantage in human patients with penetrating abdominal injury, when fluid therapy was delayed approximately 30 minutes, during which time definitive surgical hemostasis was achieved [20]. Additional studies have since found inverse correlations between the volume of fluid resuscitation and incidence of cardiopulmonary complications, acute respiratory distress syndrome, bacteremia, and surgical site infections [17,19].

Studies evaluating hypotensive resuscitation in clinical veterinary models are lacking, and several factors need to be considered prior to clinical application. In the Bickell study, patients in the fluid restricted group did not receive fluids for approximately the first 30 minutes following the time of trauma, during which time they were undergoing surgical hemostasis. In addition, there has been a shift in human medicine towards multiple staged surgical procedures to control hemostasis allowing for correction of metabolic abnormalities, rather than long definitive operations. Unlike human medicine, veterinary patients may not receive care within minutes of the trauma incident, do not receive immediate pre-hospital medical care, and do not often have access to surgical hemostasis within 30 minutes. Finally, exsanguination remains the second most common cause of death in human trauma patients and hypotensive resuscitation may be implemented under the guidance of additional cardiopulmonary monitoring systems not routinely applied to veterinary patients [21].

Permissive hypotension

Some clinicians may inadvertently mistake permissive hypotension with hypotensive resuscitation. Permissive hypotension is the restriction of additional fluid therapy and/or vasopressor use in hypotensive patients that do not have evidence of tissue hypoperfusion. Observational data in human medicine suggest a subset of hypotensive patients with septic shock maintain normal lactate levels. These patients exhibit lower severity scores with a significantly reduced mortality [22]. Clinical data evaluating this concept is sparse in the veterinary literature. A 2015 study of cats presenting to the emergency department published by Reinke et al, identified 15/43 (35%) of hypotensive cats (SAP < 90 mmHg) had a normal lactate (< 2.5 mmol/L) [23]. This suggests there could be a similar population of hypotensive veterinary patients who do not have severely reduce tissue perfusion. Alternatively, this could support discussion about whether hyperlactatemia should be a criterion for septic shock. A 2008 study by Silverstein et al, evaluated the relationship between blood pressure and survival in critically ill cats. Unsurprisingly they found an increased mortality in the 39/83 of hypotensive cats [24]. Within the hypotensive cat population, survival increased if blood pressure increased more than 20 mmHg. Interestingly, in hypotensive cats that received catecholamines there was no significant increase in blood pressure as compared to cats not receiving catecholamines [23]. Additionally, hypotensive dogs without hyperlactatemia had a better prognosis and chance of surviving to hospital discharge than hypotensive dogs with hyperlactatemia [25]. These combined observations could invite debate about how aggressively clinicians choose to treat hypotension in patients with a normal lactate.

Early Goal Directed Therapy

Early goal directed therapy (EGDT) is the practice of implementing sequential therapies, to achieve specific endpoints in a time-specific resuscitation protocol. In 2001, a landmark publication in human medicine, investigating EGDT in severe sepsis evaluated CVP, MAP, urine output (UOP), and $ScvO_2$ within the first six hours of hospital presentation [14]. The study protocol resulted in a significant decline in patient mortality from 46% to 30%. Robust debate ensued about whether all endpoints were required, which were attributed to the survival advantage, and if the selected target was appropriate. A veterinary observational study identified a decreased $ScvO_2$ in 37.8% of patients despite achieving normal conventional perfusion parameters, leading the authors to conclude that occult shock may exist in a high percentage of canine patients following standard fluid therapy [26]. Additional studies targeting lactate clearance in humans can result in similar survival advantage as compared to protocols designed to normalize $ScvO_2$. In contrast, three recent studies show that EGDT may not result in reduced mortality [27-29]. Alternate interpretation of these studies could illustrate the true impact EGDT has had on usual care provided over the past 15 years.

Which fluid to choose

Once the clinician has decided which patients require fluid therapy, the task is then choosing a fluid type. This requires comprehensive understanding of the biological fluid compartments, physiologic barriers dividing these compartments, and principles of fluid and solute movement across these barriers. Conventionally the principles of fluid therapy have been based on effective serum osmolality and Starling's principle. Effective serum osmolality can be calculated by the equation $2(Na^+) + Glu/18$

[30]. Starling's principle states that hydrostatic pressure favors movement of fluid and solute from the vascular compartment into the interstitium, whereas oncotic pressure favors movement in the opposite direction as defined by the equation $K_f([P_c-P_i]-\sigma[\pi_c-\pi_i])$. More recent observations that IV fluid therapy does not result in the expected extracelluar volume distribution has intensified interest in the physiologic role of the endothelial glycocalyx, endothelial basement membrane and the extracellular matrix. This research has led to a revision of Starling's equation and helps explain the limitations of colloidal therapy in the treatment of tissue edema [15].

Synthetic colloidal debate

Recent research in human medicine has shown a number of adverse effects to synthetic colloid fluid administration including coagulopathy, increased incidence of acute kidney injury, and increased requirement for renal replacement therapy [31-34]. These have been attributed to intracellular accumulation of osmotically active particles. These result in intracellular swelling, dysfunction of renal tubule transport mechanisms, and increased intra-capsular pressure with resultant decreases in renal blood flow. Additional data suggests that these particles may also be immunogenic, resulting in an inflammatory response within the renal interstitium, tubular plugging, and osmotic nephrosis.

Over the past 50 years, 24 multicenter randomized controlled trials have been characterized as landmark studies, based on their survival impact in human medicine. Recently, researchers analyzed the impact various interventions had based on recommendations from these 24 studies. They found 15 non-surgical interventions were found to impact mortality in critically ill human patients (7 improved mortality/8 negatively impacted mortality). The use of hetastarch was identified as one of the 8 treatments that negatively impacted mortality [35].

Based on the available cumulative information the use of hetastarch is no longer recommended in septic human patients. In fact, hetastarch has been withdrawn from the European market and the FDA advises against it's use in the critically ill, patients with renal dysfunction, and cardiopulmonary bypass patients. Large, comprehensive studies in veterinary medicine are lacking. There are several active research projects evaluating the use of colloids in veterinary septic patients, however until reviewing this data, caution may be advised in selecting synthetic colloid solutions.

Exogenous albumin

In health, albumin accounts for roughly 70-80% of the plasma colloid oncotic pressure with smaller contributions by fibrinogen, globulins and other plasma proteins [36, 37]. Albumin also plays an important role in acid-base regulation, intravascular molecular transport, immune response, and coagulation. Albumin is considered a negative acute phase protein, and hypoalbuminemia has been associated with critical illness due to decreased nutritional intake, blood loss, gastrointestinal loss, and reduced hepatic production [36,37]. In addition, hypoalbuminemia has been a poor prognostic marker although exogenous administration / correction has not been shown to be beneficial in human medicine. Human albumin has been shown to increase serum albumin and COP when administered to critically ill dogs [38, 39]. When administered to healthy dogs, human albumin has resulted in significant patient morbidity and mortality, however [40-42]. This may illustrate both the significant immunogenic potential that administering foreign protein may have and and the possible immunosuppression associated with critical illness. The use of human albumin is no longer currently recommended in dogs or cats and species specific albumin is not commercially available. Some clinicians have prescribed fresh frozen plasma in hypoalbuminemic patients. The albumin content in stored plasma is low however, requiring approximately 45 mL/kg of body weight to increase serum albumin 1 g/dL [37]. Based on conventional beliefs, administration of plasma is ineffective at significantly raising COP. Albumin has been shown to bind to the endothelial glycocalyx, however. With renewed perspective based on the modified Starling equation, future investigations are warranted to determine if administering exogenous albumin alters the function of the endothelial glycocalx or endothelial basement membrane in critical illness.

Hypertonic solutions

Hypertonic fluids have been advocated in a number of disease conditions such as those patients with elevated intracranial pressure and hyperosmolar syndromes. Elevated intracranial pressure may be the result of traumatic brain injury, intracranial hemorrhage, recent craniotomy, or intracranial tumors. While mannitol remains the most commonly administered hyperosmolar agent in patients with elevated intracranial pressure, there is some evidence suggesting that hypertonic saline could be a more effective agent [43]. Currently however, there is no consensus as to appropriate indications for use, the best concentration, and the best method of delivery. The available data are limited by low patient numbers, limited randomized control studies, and inconsistent methods between human studies.

Patients with hyperosmolar syndromes such as hyperglycemic hyperosmolar syndrome (HHS) will typically present dehydrated and hypovolemic. Rapid correction of their intravascular volume is required, however the initial fluid of choice should be carefully chosen to avoid rapid shifts in patient serum osmolality inducing unwanted fluid shifts. A study of feline HHS found a median serum osmolarity of 384 mOsm/kg, whereas the osmolarity of 0.9% NaCl is 308 mOsm/kg [44]. Therefore, a clinician may choose to formulate a hypertonic resuscitation fluid, making it isotonic to the hyperosmolar patient in order to avoid rapid movement of fluid into the central nervous system possibly resulting in cerebral edema. While studies have shown that abnormal neurologic status is a poor prognostic indicator in HHS, additional studies comparing the impact of fluid osmolarity to patient osmolarity are needed.

REFERENCES

1. Haskins SC et al. J Vet Emerg Crit Care 2005;15(2):100.
2. Yu T et al. Crit Care Med 2003;31(12):2734.
3. Afessa B et al. Crit Care Med 2001;29(6):1145.
4. Rajaram SS et al. Cochrane Database Syst Rev 2013;2:CD003408.
5. Mahjoub Y et al. Br J Anaest 2014:112(4):681.
6. Feissel M et al. Intensive Care Med 2004:30(9):1834.
7. Stawicki SP et al. J Am Coll Surg 2009;209(1):55.
8. Duus N et al. J Crit Care 2015; 30(1):217.
9. Monnet et al. Crit Care Med 2006;34(5):1402.
10. Picolli A et al. Crit Care Med 2000;28(1):132.
11. Kircher BJ, Himelman RB, Schiller NB. Am J Cardiol 1990;66(4):493.
12. Ommen SR, Nishimura RA, Hurrell DG, Klarich KW. Mayo Clin Proc 2000;75(1):24.
13. Inmaculada A et al. Chest 1999;115:453.
14. Rivers E, et al. NEJM 2001;345:1368.
15. Woodcock TE and Woodcock TM. Br J Anaesth 2012;108(3):384.
16. Jones et. al. JAMA 2010;303(8):739.
17. Revell M, et al. J Trauma 2003;54:S63.
18. Marrison CA et al. J Trauma 2011; 70(3);652.
19. Mazzaferro EM. Vet Clin Small Anim 2008:38:607.
20. Bickell WH, et al N Engl J Med 1994;331:1105.
21. Smith JB et al. Curr Anesthesiol Rep 2014; 4(3):209.
22. Casserly B et al. Crit Care Med 2015; 43:567.
23. Reinke et al. J Am Vet Med Assoc 2015;247(1):79.
24. Silverstein DC, et al J Am Vet Med Assoc 2008;232:893.
25. Ateca et al. J Am Vet Med Assoc 2015;246(1):100.
26. Young BC, et al JVECC 2014;24(2)154.
27. Peake SL, et al. N Engl J Med 2014; 371(16):1496.
28. Yealy DM, et al. N Engl J Med 2014; 370(18):1683.
29. Mouncey, et al. N Engl J Med 2015; 372(14): 1301.
30. Guyton AC, Hall JE. Textbook of Medical Physiology. 11th ed. Saunders 2001.
31. Glover PA, et al. J Vet Emerg Crit Care 2014; 24(6):642.
32. Wierenga JR et al. Am J Vet Res 2007;68:605.
33. Hembold KA et al. J Vet Emerg Crit Care 2014; 24(4):381
34. Bayer O et al. Crit Care Med 2011;39:1335.
35. Landoni G et al. Crit Care Med 2015;43(8):1559.
36. Mazzaferro EM et al. J Vet Emerg Crit Care 2002;12(2):113
37. Mazzaferro EM et al. J Vet Emerg Crit Care 2002;12(2):113.
38. Trow AV et al. J Am Vet Med Assoc 2008;233:607.
39. Mathews KA, Barry M. J Vet Emerg Crit Care 2005;15(2):110.
40. Martin LG, et al. J Am Vet Med Assoc 2008;232:1004.
41. Cohn LA, et al. Am J Vet Res 2007;68:657.
42. Francis AH, et al. J Am Vet Med Assoc 2007;230:873.
43. Kamel H et al. Crit Care Med 2011;39(3):554.
44. Koenig A, et al. J Vet Emerg Crit Care 2004;14(1):30.

Gastrointestinal Tract Function in Critical Illness

Tracy L. Hill, DVM, DACVIM, PhD, MRCVS, DECVIM-CA
Edinburgh, United Kingdom

INTRODUCTION

Gastrointestinal (GI) disturbances that occur with critical illness fall into one of two broad categories of dysfunction: dysmotility and barrier dysfunction. There is a much broader understanding of both of these complications in critically ill people as compared to the small animal population. In this discussion, the mechanisms of both types of gastrointestinal dysfunction will be explored. We will discuss the prevalence and pathophysiology of gastrointestinal dysfunction in people and veterinary patients along with appropriate treatment protocols. Because of the paucity of data regarding gastrointestinal function in dogs and cats, there will be a focus on treatment of gut dysfunction in critically ill people.

PREVALENCE AND DEFINITION OF GUT DYSFUNCTION WITH CRITICAL ILLNESS

Various clinical signs indicate the development of gastrointestinal tract dysfunction: absent/abnormal bowel sounds, abdominal distension, vomiting, diarrhea, constipation, increased gastric residual volume, GI haemorrhage, as well as discontinuation of enteral feedings. In people, clinical signs of GI dysfunction occur in approximately 60% of critically ill patients.[1,2] Higher age, body mass index, Acute Physiology and Chronic Health Evaluation (APACHE) II scores and C reactive protein are associated with a greater number of GI signs. The development of one or more signs of gut dysfunction are consistently associated with poorer outcomes.[3]

In 2012, the European Society of Intensive Care Medicine proposed a definition of acute gastrointestinal injury (AGI) to improve preventative and therapeutic measures.[4] AGI is graded on a scale I to IV.

Grade	Definition	Consistent clinical signs
I	Mild, self-limiting, partial impairment of GI function (often post-operative)	Nausea, vomiting, decreased bowel motility
II	GI tract unable to meet nutrient and fluid requirements, no systemic signs	Reflux, ileus, mild intra-abdominal hypertension, hematemesis or melena
III	Persistent loss of GI tract function in spite of interventions, possible worsening of systemic signs	Persistent feed intolerance, ileus, moderate intra-abdominal hypertension
IV	Life-threating GI failure, worsening of MODS and shock	Bowel ischemia/necrosis, GI haemorrhage with shock

AGI is also defined as either primary or secondary. Primary AGI develops when the underlying disease has a direct effect on the gastrointestinal tract (such as abdominal surgery). Secondary AGI defines GI dysfunction in cases of critical illness distant to the GI tract (such as pneumonia or cardiac disease). The definitions have guided the development of an algorithm describing appropriate choice of enteral or parenteral nutrition, additional management and supplemental monitoring based on AGI grade.

In dogs and cats, the prevalence of gut dysfunction is largely unknown and there is no similar definition of AGI. Enteral feeding failure and the development of gastrointestinal signs are infrequently described in the veterinary literature though similar GI signs are observed in critically ill veterinary patients. Because the gastrointestinal tract exhibits hypoperfusion with shock in dogs as it is in people, it would be expected that critical illness would result in similar AGI.[5] In cats, the gut is not considered to be a shock organ, though AGI may occur in cats, albeit with less frequency than in dogs or humans.[6]

CLINICAL SIGNIFICANCE OF GUT DYSFUNCTION

The development of gastrointestinal tract dysfunction in critically ill people is associated with a poorer outcome.[3] Patients with absent or abnormal bowel sounds, bowel distension, or GI haemorrhage are less likely to survive hospitalisation. People with two or more GI signs have prolonged ICU stays. Mortality rates increase with higher AGI grades (Grade I: 95% survival, Grade IV: 10% survival) and is nearly twice as high in patients with secondary versus primary AGI (43% vs 24%).[7]

Gut dysfunction would be expected to similarly affect outcome in veterinary patients. Gastrointestinal hemorrhage in canine ICU patients was associated with a higher mortality rate.[8] The effect of other manifestations of gut dysfunction on survival in dogs and cats is unknown.

ASSESSMENT OF GUT FUNCTION

Gut function in people with critical illness is measured by a variety of methods. Assessments of barrier dysfunction can be made either by direct examination of the mucosa for signs of ulceration or erosion or by measurements of movements of ions or other molecules across the gastrointestinal epithelium. Endoscopic examination of the gastrointestinal mucosa can demonstrate lesions of stress related mucosal disease in the gastric mucosa. Impedance spectroscopy assesses gastric permeability *in vivo* by measurement of electrical resistance. Intestinal permeability is commonly assessed by gastrointestinal absorption of various orally administered sugars, most commonly lactulose and rhamnose. All of these techniques are theoretically possible in veterinary patients. Lactulose: rhamnose or 5-sugar permeability testing, iohexol absorption and ^{51}Cr-EDTA absorption tests have been described in dogs with various disease state.[9-11] Mannitol and ^{51}Cr-EDTA absorption has been performed in cats as well.[12]

Assessment of gastrointestinal motility in people consists primarily of measurement of gastric residual volume, paracetamol (acetaminophen) absorption test, carbon isotope breath testing, ultrasound calculation of gastric emptying, and gastric impedance monitoring. Nuclear scintigraphy is considered the gold standard, though it does suffer from individual variability and requires the use of a radioactive isotope.[13] Gastric residual volume measurement is the technique most frequently applied in people, though it may be significantly affected by endogenous gastric and salivary secretions. It also poorly correlates with the incidence of regurgitation or aspiration and is not standardized.[14]

In dogs, gastric motility has been assessed by ultrasound calculation of gastric emptying.[15] Capsule endoscopy can also establish gastrointestinal motility.[16,17] 13C-sodium-acetate breath test has been validated in dogs, but is not readily available for use in most ICUs as it requires specialized equipment.[18] Nuclear scinitigraphy is the gold standard in veterinary medicine as with people, but its clinical utility is likewise limited due to the necessity of administration of a radioisotope.

MANIFESTATIONS OF ACUTE GASTROINTESTINAL INJURY

Dysmotility

Clinical signs of dysmotility include: gastroesophageal reflux (with or without aspiration pneumonia), absent/altered bowel sounds, diarrhea, constipation, and delayed gastric emptying, and failure of enteral feeding. Critically ill patients often receive opioids as part of therapy, which have a direct effect on GI motility. Additional treatments (pressor support, non-steroidal anti-inflammatory drugs (NSAID), over-resuscitation with intravenous fluids) may also alter effective motility patterns. Surgical manipulation of the intestines during abdominal surgery may also alter intestinal motility by stimulation of corticotrophin releasing factor pathways.[19] Postoperative ileus occurs in up to 90% of people subsequent to abdominal surgery.[20]

Gastric emptying times are prolonged in dogs following elective abdominal surgery.[21] Gastrointestinal motility in dogs and cats with critical illness remains otherwise uncharacterized.

Barrier dysfunction

Clinical signs of barrier dysfunction include: stress related mucosal disease (SRMD), GI hemorrhage, bacterial translocation, sepsis, and in severe cases the development of MODS. Alterations in barrier function are likely multi-factorial. First, many conditions resulting in critical illness will also result in relative splanchnic hypoperfusion. Decreased submucosal blood flow to the gut results in ischemia and hypoxia-induced cell injury.[19] Gastric ischemia, as indicated by impedance spectroscopy, predicts persistent shock, mortality, and longer hospitalisation time in people undergoing elective cardiovascular surgery.[22] Cytokines produced in systemic inflammatory response syndromes can also directly induce cytotoxic injury to gastrointestinal epithelial cells. Corticotrophin-releasing factor is elevated in trauma patients undergoing surgery compared to those undergoing elective small bowel resection although both groups of patients had increased intestinal permeability.[19]

Male dogs with a history of blunt force trauma had increased lactulose: rhamnose ratios compared to control dogs 48-hours after injury, which provides evidence of gastrointestinal barrier dysfunction in small animal veterinary patients.[11] Alaskan sled dogs, who when racing may have similar relative hypoperfusion to the gut as with critical illness, have both gastric lesions and increased intestinal permeability.[10] In the single available study that has examined SRMD in dogs and cats, 19% of dogs developed melena, hematochezia, or hematemesis during ICU hospitalisation; no cats were identified with signs of SRMD.[8] In this study, no direct observance was made of the gastric mucosa, so the actual prevalence of SRMD lesions is unknown. Mortality was higher in dogs with evidence of GI hemorrhage.

TREATMENT AND PREVENTION

Gut dysfunction, or AGI, is well recognized in critically ill people and is associated with poorer outcomes. Treatment and prevention of these complications, then, is of paramount important to maximizing success. The ESICM Working Group has also developed a treatment algorithm for management of AGI in people.[3]

AGI Grade	Management
I	Appropriate fluid therapy, early enteral feeding, avoid drugs that impair GI motility
II	Prokinetic therapy, continue enteral feeding (small amounts frequently if necessary), post-pyloric feeding if gastroparesis
III	Discontinue drugs affecting motility, use small amounts of enteral feeding, post-pyloric if necessary, avoid parenteral nutrition in the first 7 days due to increased risk of hospital acquired infection
IV	Laparotomy or other emergency intervention, no known conservative treatment is effective

Prevention

Optimizing peri-operative management

Several strategies have proven efficacy in the prevention of AGI in people. First, minimal handling of the gut and use of laparoscopic procedures decreases alterations in barrier function and dysmotility induced by physical trauma of surgery.[23] The use of epidural analgesia reduces time to recover of GI function by reducing requirements for systemic opioid induced analgesia which affects GI motility.[20] The concomitant use of other analgesic agents, including acetominophen, NSAIDs, and ketamine can reduce the need for motility-altering opioids. Perfusion to the gut should be maintained with the use of appropriate fluid therapy and maximization of oxygen delivery. This may include crystalloid or colloid fluid therapy, pressor support, and/or blood transfusions where appropriate. Early mobilisation of patients after surgery is usually recommended to decrease ileus.

There is less information regarding the prevention of GI signs in veterinary patients. In horses, intravenous lidocaine leads to early bowel movement (decreased post-op ileus) and enhanced short-term survival. The use of a bupivacaine epidural had a similar effect.[24] In dogs with septic peritonitis, lidocaine continuous rate infusions resulted in improved 48-hour survival. In this study, the presence of GI signs is not described, nor is the dose of concurrent opioid administered, so it is unclear whether the lidocaine improved survival in these dogs in part by decreasing AGI or by a mechanism unrelated to the GI tract.[25] In a sepsis rat model, ketamine given during anaesthesia diminished deleterious effects of LPS, decreasing gastric luminal fluid accumulation and attenuating COX-2 stimulation.[26] The use of these peri-operative management strategies has not been examined in cats.

Overall, it is prudent to undertake steps to reduce the need for opioid-based analgesia and maintain gut perfusion in both the intraoperative and post-operative period. Epidural analgesia should be considered strongly for any abdominal surgery, particularly one involving the gastrointestinal tract. Continuous rate infusions of lidocaine or ketamine both intra- and post-operatively decrease the doses required of opioids and may have independent mechanisms of AGI prevention. A strategy of balanced analgesia, including the use of other agents such as acetaminophen (in dogs) and NSAIDs (when appropriate), should likewise reduce dependence on opioids. Gut perfusion should be maintained with judicious fluid therapy and oxygen-carrying support when indicated.

Enteral nutrition

Early enteral nutrition (EEN) is a key strategy in reducing the development of AGI. In people, EEN reduces mortality, infectious complications, and length of hospitalisation with a variety of critical illnesses.[27] In dogs, early nutritional supplementation has also been associated with decreased hospitalisation time.[28] In this study, however, early nutrition included both enteral and parenteral so it is not possible to determine the specific effect of EEN in dogs. Gastrointestinal motility, measured in a dog model of colorectal cancer surgery, had faster return to normal in dogs receiving enteral nutrition versus parenteral nutrition.[29] Dogs with acute pancreatitis had fewer vomiting and regurgitation episodes when fed enterally versus parenterally.[30] The robust evidence to support the use of EEN in people along with early studies to suggest its benefit in dogs warrants the use of EEN to prevent AGI. Early enteral nutrition has not been examined in cats.

Treatment

Acid suppression/treatment

Acid suppressants are used primarily to prevent complications of SRMD. The overall prevalence of SRMD is often quoted to be up to 100%, though these findings are based on two studies from 1971 and 1985. In the first, candidates were selected for mucosal examination that were "prime candidates to develop gastric bleeding" and in this patient population, the presence of SRMD was 100%.[31] In the second, 14 of 39 patients had ulcers or endoscopic bleeding, and an additional 15 had erosions or petechiation, leading to a maximal prevalence of 74%.[32] The incidence of clinically significant hemorrhage due to SRMD was as high as 25% a few decades ago and is associated with increased mortality. As a result, people are almost always treated with some form of ulcer prophylaxis and now the rate of significant GI hemorrhage has been reduced to 1-4%.

Two-thirds of prophylaxis against SRMD is with the use of proton pump inhibitors (PPI). Though effective at suppressing acid production, this class of drug increases the risk of nosocomial pneumonia and *Clostridium difficile*-associated diarrhea, which carry an odds ratio for mortality of 1.8 and 1.6, respectively.[33] Proton pump inhibitors are also problematic because of their effect on leukocyte function and on P450 enzyme inhibition. In comparative studies, PPIs did not provide an advantage over histamine-2 receptor antagonists or sucralfate.[34] Sucralfate was protective against the development of nosocomial pneumonia.[33] In the author's unpublished work, sucralfate was protective against and accelerated recovery from acid injury in a canine *ex vivo* injury model.

In dogs and cats, as in people, the most appropriate drug for treatment or prevention of SRMD in unknown. In both dogs and cats, proton pump inhibitors provide the most effective acid suppression.[35,36] The importance of SRMD as a manifestation of AGI in dogs and cats is unknown so the widespread use of acid suppressants to prevent SRMD without signs of gastrointestinal haemorrhage cannot be recommended.

Motility agents

Although the use of prokinetic agents is widely recommended for treatment of ileus in people, there is little high-quality evidence that metoclopramide, cisapride, ranitidine, and erythromycin have a significant effect.[23] The effect of these drugs on dysmotility in critically ill dogs and cats is unknown; evidence of the efficacy of these drugs is largely limited to studies using dogs as models for human drug development. Therefore, prokinetics as treatment for GI dysmotility in dogs and cats with critical illness requires further investigation for its use to be justified.

Novel agents

There are a number of newer agents that are undergoing various stages of drug development for treatment of AGI in people. Many of these use dogs during some stage of efficacy testing, so there is at least support for their eventual use in veterinary patients. These include: mast cell stabilisers, peripherally acting u-opioid receptor antagonists, and acetylcholinesterase inhibitors, among others. Further work would be needed before these newer agents could be recommended for veterinary patients.

REFERENCES

[1]Reintam A, et al. Acta Anaesthesiol Scand 2009;53:318-324. [2]Montejo J, et al. Crit Care Med 2002;30(4):796-800. [3]Reintam A, et al. Crit Care 2008;12:R90. [4]Reintam A, et al. Intensive Care Med 2012;38:384-394. [5]Lagoa CE, et al. Crit Care 2004;8:R221-R228. [6]Brady CA, et al. JAVMA 2000;217:531-5. [7]Chen H, et al. Therapeut Clin Risk Management 2015;11:1523-1529. [8]Swann J, et al. JVECC 2015;00(0):1-9.[9]Frias R, et al. Vet J 2012;1:123-125. [10]Davis M, et al. JVIM 2005;19:34-39. [11]Streeter E, et al. JVIM 2002;16:669-673. [12]Bijlsma P, et al. Scand J Gastroenterol 2002;37(7):799-806. [13]Moreira T, et al. Nutr Clin Prac 2009;24(2):261-273.[14]McClave S and Snider H. J Parenteral Enteral Nutr 2002;26(2):S43-S50. [15]Tsukamoto A, et al. J Vet Med Sci 2011;73(9):1133-8. [16]Boscan P, et al. Vet Anaesth Analgesia 2014;41:73-81. [17]Lidbury J, et al. Vet Med International 2012;1-8. [18]Schmitz S et al. Tierarztl Prax Ausg K Kleintiere Heimtiere 2010;38(4)211-6. [19]Hill L, et al. Nutr 2013;29:650-4. [20]Liu Anesth Analg 2007;104:689-702. [21]Schmitz S, et al. Tierarztl Prax Ausg K Kleintiere Heimtiere 2013;41:87-92. [22]Beltran N, et al. Physiol Measurement 2006;27:265-77. [23]ythen M. Anesth Analg 2005;100:196-204. [24]Wallstrom A and Hollman G. J Clin Nurs 2013;23:24-44. [25]Bellini L and Seymour C. JAVMA 2016;248:422-9. [26]Suliburk J, et al. Eur Surg Res 2008;40:184-9. [27]Heighes PT, et al. Anaesth Intestive Care 2010;38:167-174. [28]Liu D, et al. JVECCS 2012;22(4):453-9. [29]Kawasaki N, et al. Sug Today 2009;39:225-30. [30]Mansfield C, et al. JVIM 2011;25:419-25. [31]Lucas C, et al. Arch Surg 1971;102:266-73. [32]Peura D and Johnson L. Ann Int Med 1985;102(2):173-7.[33]Buendgens L, et al. J Crit Care 2014;29:696.e11-e15. [34]Krag M, et al. Acta Anaesth Scand 2015;59:576-85.[35]Parkinson S, et al. JVIM 2015;29(1):104-112. [36]Tolbert MK, et al. JVIM 2011;25(1):47-54.

Improving the Residency Training Progam

Robert J. Washabau, VMD, PhD, DACVIM
St.Paul, MN

The ACVIM Improved Oversight of Credentials and Residency Training (IOCART) Taskforce has expressed concern about the growing number of residency-trained individuals who have failed to achieve board certification. The Taskforce has identified the certification exam and the publication requirement as significant barriers to board-certification. Post-Forum listserve discussion has focused primarily on the merits and demerits of the publication requirement. This session will not take a position on the validity of the publication requirement, but will instead offer alternatives to the requirement. The presentation will outline (1) the implementation of a physiology course designed to improve Resident understanding of core physiologic concepts, and (2) the use of critical scientific reading concepts to improve comprehension of scientific manuscripts with basic, clinical, or translational significance.

PHYSIOLOGY REDUX

In 2004, a three-year cycle of pathophysiology lectures was developed for Internal Medicine Residents-in-training at the University of Minnesota. Our purpose was the (re)education of Internal Medicine Residents in the biology, physiology, pathology, and medicine of the disciplines relevant to Internal Medicine thereby improving their core scientific knowledge. The specific goals of the course were:

(1) Understand anatomic strategies adopted by different animal species to achieve same/similar function.
(2) Understand important physiologic processes used by animals to maintain homeostasis.
(3) Understand neural, endocrine, and paracrine regulation of organ systems.
(4) Understand intermediary metabolism as it relates to development, growth, and maintenance.
(5) Understand processes used by different organ systems to accomplish the same biologic effect.
(6) Understand host defense mechanisms that prevent or modulate inflammation of infection.
(7) Understand the application of principles of physiology in the diagnosis and treatment of disease.
(8) Understand animal physiology in terms of ecosystems and population biology.

Lectures were presented on consecutive Monday mornings from 8:00-9:00 am over the three-year period of the residency. Experts in their disciplines were asked to present core physiology concepts integrated with clinical correlates where appropriate. Disciplines that were covered included: anesthesia, cardiology, clinical pathology, dermatology, endocrinology, gastroenterology, hepatology, immunology, oncology, nephrology, neurology, nutrition, respiratory, and theriogenology. As an example, the gastroenterology section was organized in the following way:

G.I. Physiology – Motility – Oral/Esophageal
G.I. Physiology – Motility - Stomach
G.I. Physiology – Motility – Intestine/Colon
G.I. Physiology – Secretion – Salivary/Gastric
G.I. Physiology – Secretion – Intestine
G.I. Physiology – Secretion – Pancreas
G.I. Physiology – Secretion – Biliary
G.I. Physiology – Digestion - Hydrolytic
G.I. Physiology – Digestion - Fermentative
G.I. Physiology – Absorption - Intestine
G.I. Physiology – Absorption – Colon
G.I. Physiology – Blood Flow
G.I. Physiology – Liver Metabolism I
G.I. Physiology – Liver Metabolism II
G.I. Physiology – Integration of Function

There are many intrinsic and extrinsic factors that influence a trainee's ability to take and pass the ACVIM certification exam. Internal Medicine faculty at the University of Minnesota believe that the institution of the physiology course has had several beneficial effects in our training program, including better basic science knowledge, improved clinical application of scientific knowledge, better internists completing the program, and higher completion rates in the certification process.

CRITICAL SCIENTIFIC READING

We have used our journal club to advance the concept of critical scientific reading.

Critical scientific reading is a technique for discovering information and ideas within a manuscript. Critical reading refers to careful, active, reflective, and analytic reading.

Critical scientific thinking is a technique for evaluating information and ideas, for deciding what to accept and believe. Critical thinking involves reflecting on the validity of what you have read in light of our prior knowledge and understanding of the world.

This strategy was originally developed as a core course for 1st year veterinary students at the University of Minnesota, elements of which have now been adopted into our weekly Journal Club. This is a strategy to improve comprehension of scientific manuscripts with basic, clinical, or translational significance. The specific objectives are:
(1) Stimulate critical reading of scientific manuscripts with basic, clinical, or translational significance.
(2) Reinforce the use of the scientific method in the evaluation of biologic and clinical problems.
(3) Encourage residents to extrapolate and integrate concepts across disciplines.
(4) Critique the use of statistical methods in the analysis of biological data.
(5) Evaluate scientific manuscripts for the ethical use of animals.

DIFFERENTIATION OF JOURNAL CLUB vs. JOURNAL REVIEW

Internal Medicine Residents at the University of Minnesota hold a weekly meeting at which they review papers published in a wide variety of disciplines covering a 5-year publication period. While useful for staying current in the scientific literature, the papers are often read in a superficial, non-critical way, hence the need to differentiate Journal Review from Journal Club.

Journal Club

The Journal Club is a regular meeting of the academic principals to appraise the validity and merit of scientific papers in common areas of interest (Veterinary Internal Medicine broadly defined). A major aim of Journal Club is to develop and sustain critical reading skills. Journal Club participants generally present one scientific paper for in-depth discussion by the Group. Papers are critiqued for validity of experimental design, statistical analysis, results, contributions to the literature, and merit of the study conclusions. Critical reading is emphasized in this format (see Appendix). Review papers are not acceptable in this format.

Journal Review

In contrast to the Journal Club, Journal Review is a more relaxed review of multiple papers with the specific aim of staying current in one's discipline and/or preparing for the Internal Medicine certification exam. Papers should still be read and presented critically, but the overall goal is to assimilate the contemporary literature in the area of interest (Veterinary Internal Medicine).

All three components – physiology redux, critical scientific reading, journal club – dramatically improve a residency training program, and could be considered in a new model for the ACVIM.

REFERENCES

Larson RL, White BJ. Importance of the role of the scientific literature in clinical decision making. J. Am. Vet. Med. Assoc. 2015: 247: 58-64.
Yudkin B. Critical Reading: Making Sense of Research Papers in Life Sciences and Medicine. P Routledge Press, London, 2006, 129 pages.

APPENDEX A – CRITIQUE OF A MEDICAL REPORT

I. **Object or Hypothesis**
A. What are the objectives of the study or the questions to be answered?
B. What is the population to which the investigators intend to refer their findings?

II. **Design of the Investigation**
A. Was the study an experiment, planned observations, or an analysis of records?
B. How was the sample selected? Are there possible sources of selection error which would make the sample atypical or non-representative? If so, what provision was made to deal with this bias?
C. What is the nature of the control group(s) or standard of comparison?

III. **Observations**
A. Are there clear definitions of the terms used, including diagnostic criteria, measurements made, and criteria of outcome?
B. Was the method of classification or of measurement consistent for all the subjects and relevant to the objectives of the investigation? Are there possible biases in measurement and, if so, what provisions were made to deal with them?
C. Are the observations reliable and reproducible?

IV. **Presentation of Findings**
A. Are the findings presented clearly, objectively, and in sufficient detail to enable the reader to judge them for him/herself?
B. Are the findings internally consistent, i.e., do the numbers add up properly, can the different tables be reconciled, etc.?

V. **Analysis**

A. Are the data worthy of statistical analysis? If so, are the methods of statistical analysis appropriate to the source and nature of the data and is the analysis correctly performed and interpreted?

B. Is there sufficient analysis to determine whether significant differences may in fact be due to lack of comparability of the groups in sex or age distribution, in clinical characteristics, or in other relevant variables?

VI. <u>Conclusions</u>

A. Which conclusions are justified by the findings? Which are not? Are the conclusions relevant to the questions posed by the investigators?

B. What was good about the study? What was bad about the study? What additional studies should be performed to support the conclusions?

Alternative Residency Training Programs: Competency-Based Resident Education

Jean-Sébastien Palerme, DVM, DACVIM
Ames, IA

INTRODUCTION

Residency training, both in human and veterinary medicine, has historically been centered on a time-based model of preceptorship. This model, developed by Halsted in the 19th century, assumes that a pupil becomes competent in his area of study after a specific period of mentor-assisted training. Success of this model is based on the assumption that competence is forged by repetitive exposure to experiences (patients) and chances to perfect one's abilities.[1] The success of this model has recently been put into question in human medicine following the imposition of work hour restrictions for residents. In the United States, Canada and Europe, work hours have been restricted by 20-50% as a measure to decrease resident fatigue.[2] With this decrease in hours on clinics, there is less time for procedural practice and teaching. Consequently, some concern has been raised that resident training may be suboptimal.[3]

COMPETENCY-BASED FRAMEWORK

In an effort to improve the current training model, some medical schools and organizations have developed integrative competency-based frameworks to guide the teaching and evaluation of medical specialists. The most influential of these, the Canadian Medical Education Directions for Specialists (CanMEDS), proposed 7 domains of competency that residents should be proficient in before being granted their title. These include medical expert, communicator, collaborator, manager, health advocate, scholar and professional.[4] Similar domains of expertise have been introduced in the United States by the American Accreditation Council for Graduate Medical Education (AACGME) and have been incorporated into training curricula. This training model defines successful training as attainment of competence (as judged by mentors) instead of as having fulfilled a specific time requirement. While proponents of this model emphasize its learner centeredness and transparency, there are concerns that this approach to graduate education might result in a more reductive training that is devoid of the variety and richness of the traditional time-based model. Theoretically, the competency-based program does allow for trainees to finish their training earlier or later, as dictated by their level of expertise. However, with very few exceptions, residency training remains time-based.

The limited studies evaluating the efficacy of competency-based education suggest that it is promising. Students from two American schools that implemented competency-based curricula have outperformed students trained in traditional programs on standardized testing. In the field of graduate training, a study compared the orthopedic skills of residents trained with traditional clinical rotations to residents that participated in an intensive 30 day competency-based "boot camp" to learn basic surgical skills.[5] While there were no differences in their baseline skill, residents that went through the "boot camp" had higher scores than traditional residents when surgical skills were assessed after the training. In addition, subsequent evaluations found that this difference persisted for at least 6 months after the "boot camp" and junior residents that had competency-based training had technical skills similar to senior residents.[6]

COMPETENCY-BASED EDUCATION IN VETERINARY MEDICINE

Though competency-based education has not yet been adopted in veterinary medicine, some domains of expertise have been proposed.[7] Based on a review of the opinions of "stakeholders" in veterinary medicine, Bok et al proposed veterinary expertise, communication, collaboration, entrepreneurship, scholarship and personal development as basic domains for veterinary competency.[7] As with human medicine, competency-based education would initially be easiest to apply to improvement of technical skills and knowledge base (veterinary expertise and scholarship). The use of intensive training courses early on the course of the residency would likely pay dividends later on by increasing not only knowledge and technical skills but also the clinician's confidence. Competency-based curricula could be an innovative framework to consider as individual residency training programs or larger regulatory bodies seek to restructure or improve veterinary residency training,

REFERENCES

1. Ericsson et al. Psychol Rev. 1993; 100: 363. 2. Winslow et al. J Am Coll Surg 2004; 198: 111. 3. Bell et al. Ann Surg. 2009 249: 719. 4. Frank et al. Med Teach 2007; 29: 642. 5. Sonnadara et al. Surgery 2011; 149: 745. 6. Sonnadara et al. Surgery 2012; 151: 803. 7. Bok et al. JVME 2011; 38:262.

Career Opportunities and Considerations Post-Residency: Things You May Not Know

Annie Corrigan, DVM, MS, DACIM (SAIM) and Sandra (Axiak) Bechtel, DVM, DACVIM (Oncology)
Grenada, West Indies and Columbia, MO

INTRODUCTION

Perhaps one of the most difficult decisions to make post-residency is your career path. Should you choose academia or private practice? What are the benefits of each career choice? What information should I know about each path prior to making that decision? Once I've decided where to go, what are some tips that can make negotiation more clear? What is expected of me once I begin my new job? The goal of this presentation is to provide an open discussion forum on available career paths post-residency.

CURRICULUM VITAE

Perhaps the best advice received during and upon finishing our residencies was to update the CV at least once monthly. Record everything you do – even those things that seem small (such as reviewing abstracts or journal articles, moderating talks, etc). You would be surprised at how much you actually accomplish! Most positions also involve yearly reviews, which is another stimulus to update your CV and make sure it is complete and accurate. Don't forget to list every research abstract with its funding source and resultant manuscript, and to note research mentees. Also to make sure you list any courses you have taught in, including clinical rotations, and any research abstracts, continuing education lectures, and other presentations.

ACADEMIA

Full-time equivalent (FTE)

FTE indicates workload with 1.0 equaling full time. In academia, it's used as a metric for measuring contribution to different areas of your job, with the big 3 being teaching, research, and service. An example standard position is 0.4 (40%) teaching, 40% research, 20% service. This differs for the practicing veterinarians because of clinical responsibilities, which can be considered teaching or service. For example, of the 52 weeks of the year, 20.8 should be spent on research, 20.8 weeks teaching, and 10.4 for service in the 40/40/20 situation (this is not accounting for vacation – which is important!). FTEs are different at every university and for every faculty member, depending on the position negotiated. Make sure your position is clearly defined, and is one you are comfortable with (especially research commitment).

Tenure track versus clinical track

Many universities will offer both tenure track and clinical track positions, and it can be difficult to decide which the best route is. The definition of each and the availability for promotion for clinical track positions are variable, so make sure to carefully read the job description before interviewing. Make a list of questions in regards to position expectations to ask during your interview. Utilize any online resources available at that university and contact faculty members in each of the tracks to determine which a better fit is for you. Make sure the amount of research expected of you is well defined regardless of position type (grant funding, first author articles, etc.).

If you elect a tenure track position, start reading and building your tenure dossier. Every university has a different format for the dossier, and it needs to be written to explain what we do, and how your position differs from the typical university faculty member due to clinical services. Start early, and update often.

Teaching

Teaching is a large portion of time for most ACVIM diplomates in academia, be it standard didactic lectures, hands-on labs, or in-clinic teaching of veterinary students. Make sure to clarify the definition of clinical teaching and expected weeks on and off clinics. Diversifying your teaching portfolio is also important regardless of tenure-track versus clinical track. For example, teaching didactic lectures to veterinary students, graduate students, and even undergraduate students are an integral part of your teaching portfolio. Student evaluations and peer evaluations will not only provide constructive feedback for improvement, but will also be included in your tenure dossier or in your CV when applying for promotion.

Research

The percentage of time you negotiate for research will depend on the needs of the service and your research experience. If you don't feel you have a strong research background, discuss this prior to negotiating percentages as this will be important for promotion and/or tenure. Programs are available for new clinician scientists that will discuss grant writing, how to start a project, and molecular biology skills. Participation in these programs can be negotiated into your contract. Furthermore, a start-up package may be available to assist in setting up or adding to a laboratory. Get details on what a lab has, and what you might need to start your research program when negotiating this package. Also research on-campus funding opportunities and grant writing seminars that may aid your research. Discuss potential research collaborations with established researchers to help get your program started.

Service

Service includes committee appointments in your department, college, university, and to your profession. This also can include outreach (speaking at school career days and to general public) and continuing education. Discuss your service commitments and know what the expectations are shortly after starting, to prevent over-commitment.

PRIVATE PRACTICE

Private practice offers a high caseload with the opportunity to focus on clinical care and client communication. If you elect this route, make sure you have interviewed at multiple practices to pick the best fit for you. Investigate health and life/ disability insurance that is offered, in addition to expectations for licensing fees and continuing education (vacation time or is time off for CE in addition to vacation time? Stipend for CE?). Also investigate licensing requirements in each state you interview in, as the requirements can be variable and sometimes time consuming/ difficult to navigate. Don't forget about your retirement! Now is the time to start saving, and some practices will have matching programs for retirement contribution programs. Meet with current doctors and support staff in the practice, and ensure that the practice can provide adequate support staff for an additional specialist. Also make sure you have other specialists important for your practice readily available for consultation (such as surgeons, radiologists, etc.) in the practice itself or close by. In your contract, make sure to define your working hours, salary/ bonus system, holidays, vacation time, etc.

CONCLUSION

Residency completion and the search for your permanent job is an exciting time in your career! Interview and consider more than one job opportunity, so you can explore different options. And most of all, enjoy the process!

Case Series Versus Population Cohort Studies

Jan M. Sargeant[1], DVM, MSc, PhD; Jonah N. Cullen[2], BS, MS; Annette M. O'Connor[2], BVSc, MVSc, DVSc, FANZCVS; Kelly Makielski[2], DVM
[1]Guelph, Ontario, Canada; [2]Ames, Iowa

STUDY DESIGNS FOR CLINICAL RESEARCH

Clinical research in the veterinary sciences provides knowledge that is directly applicable to veterinarians to aid in the diagnosis, treatment, or prevention of disease, or to understand the clinical course or prognosis of diseases. Clinical research encompasses a variety of study designs, each best suited to addressing specific types of research questions, and each with its own strengths and limitations. Evidence pyramids are one way of illustrating the relative level of evidence provided by study designs when evaluating the efficacy of treatments or other interventions under real-world conditions. An example based on common study designs for interventions is shown in Figure 1 (modified from [1]).

HIGHEST EVIDENCE Systematic review / meta-analysis of randomized controlled trials
Randomized controlled trials
Non-randomized trials and deliberate disease induction trials
Incidence Observational studies: Cohort studies, incidence case controls
Prevalence observational studies: Prevalence case-control, cross sectional
LOWEST EVIDENCE Case series / case reports

Figure 1: Evidence-pyramid for evaluating the efficacy of interventions under real-world conditions.

There are a number of key concepts included in this evidence pyramid; the evidentiary importance of synthesized results (systematic review / meta-analysis), the distinction between experimental and observational designs, the distinction between incidence and prevalence studies, and the distinction between descriptive studies and hypothesis-testing studies. Here, we focus on the distinction between experimental and observational studies and the distinction between descriptive studies and hypothesis-testing studies.

Experimental studies are studies where the investigator controls the allocation of study subjects to treatment group. Randomized controlled trials have high evidentiary value for evaluating the efficacy of interventions due to the use of a formal random process for treatment allocation, which minimizes the potential for bias (assuming that the sample size is large). However, clinical trials are not ethical or feasible for every type of intervention, and there are a number of clinically relevant research questions for which clinical trials are not the most appropriate design. An example of an important research question for which a clinical trial would not be suitable would be "what is the difference in survival time between dogs with differing stages of cancer at the time of diagnoses". Clearly, it would not be possible for the investigator to allocate dogs to different stages of cancer to explore this research question! Thus, the evidence pyramid, as shown in the figure, is best used when considering preventive and therapeutic research questions, rather than for considering clinical research in general.

Although not suitable for a clinical trial, the question on the association between stage of disease and survival could be answered using an observational study design. In observational studies, the researcher does not control allocation of individuals to intervention (or exposure) groups, but rather observes the differences that occur naturally. There are a number of different types of observational studies, distinguished from one another by the method of selecting the study population and the use of incident versus prevalent cases.

The final important distinction of relevance to this paper, is descriptive versus hypothesis-testing studies. Hypothesis-testing studies may be either experimental or observational. What differentiates hypothesis-testing studies from descriptive studies is that they compare an outcome between two or more exposure groups (where "exposure" could be an intervention, risk factor, or other independent variable). Specifically, hypothesis tests evaluate whether the difference between the groups being compared is larger than would be expected based on sampling error (or "chance" in selecting the study population from the larger source population from which it was drawn). Inherent in this definition is that a hypothesis-testing study must have at least two levels of the exposure variable and at least two levels of the outcome (often diseased or not diseased in clinical research). Descriptive observational studies, by contrast, do not include a comparison group and therefore cannot be used for testing a hypothesis. Descriptive studies are used to describe a phenomenon or to estimate frequency or occurrence of a parameter of interest. Examples of descriptive observational studies include case reports and case series.

CASE SERIES AND CASE REPORTS

Case series and case reports are common study designs in the veterinary clinical literature. A case report describes a condition in a single animal (and will not be discussed further here) whereas a case series describes a condition in two or more animals. Case series are a descriptive observational design. The purpose is generally to describe the presenting history, clinical presentation and progression, diagnosis, treatment, and/or prognosis for a condition of interest. Often, the condition of interest represents a rare or novel disease, or an unusual clinical presentation of a more familiar disease. The value of case series is that the veterinary community can be alerted to a new or unusual disease condition, and they also may provide ideas for novel diagnostic or therapeutic approaches, or risk factors for a disease condition. However, case series are a descriptive study design; there is no comparison group (e.g., neither treated and non -treated animals nor diseased and non-diseased animals). Therefore, the case series design may provide ideas for possible treatments or risk factors, but the design is incapable of allowing hypothesis testing, and suggestions as to risk factors or appropriateness of treatments based on cases series should be viewed with extreme caution. For this reason, case series rank very low as an evidence source for identifying risk factors, evaluating efficacy of interventions, or assessing the accuracy and usefulness of diagnostic approaches.

COHORT AND POPULATION COHORT STUDIES

Cohort studies are a hypothesis testing observational study design. In the "classic" cohort study, subjects are selected based on their exposure status and followed over time to compare the incidence of disease (or any other outcome of interest) between two or more exposure groups. An example of this type of cohort study would be where individuals exposed to a contaminant in feed and individuals not exposed to the contaminant are identified and followed over time to compare the incidence of a specific disease (or other outcome) between those exposed and those unexposed. This is an excellent study design for investigating the effects of rare exposures and to evaluate multiple possible outcomes from an exposure. Like all observational studies, confounding is an issue with cohort studies. However, because the outcome is measured using incident cases and the exposure clearly precedes the outcome, cohort studies are considered to provide the strongest evidentiary value of the hypothesis-testing observational studies for investigating exposures, risk factors, or interventions.

A variation on the "classic" cohort study is the population cohort. In this design, a "cohort" with some distinguishing characteristic is selected for inclusion as the study population. Individuals within the cohort (who do not have the outcome of interest) are characterized based on their exposure status, and then followed over time to compare the incidence of the outcome between exposure groups. A well-known example of a population cohort is the Framingham heart study (www.framinghamheartstudy.org), which began in 1948 with a cohort of approximately 5,000 residents of Framingham, Massachusetts and currently is on its third generation of participants. Individuals in the Framingham cohort are categorized based on a number of exposures and risk factors when they enter the study, and are followed over time to identify risk factors for cardiovascular disease. Many of the risk factors that are studied in population cohort studies are factors that are not amendable to experimental study, such as gender, age, and other risk factors for the outcome. An example of a population cohort in veterinary medicine might be a study where a cohort of animals with a specific condition of interest – such as lymphoma – are selected as the study group, either prospectively or retrospectively. The risk factor of interest in this example might be the stage of tumor at the time of diagnoses and the outcome might be death (which is an incident measure). The research hypothesis being tested would be that animals with more severe stages of the tumor at diagnosis have lower survival times. The population cohort would be the best study design available to test this hypothesis.

EVALUATION OF CASE SERIES IN THE PUBLISHED VETERINARY LITERATURE

Clearly, case series and population cohort studies address very different questions and, for evaluating risk factors and interventions, provide very different levels of evidence. However, a recent study reported that the term "case series" was used inconsistently – and often incorrectly - in the human healthcare literature[2]. We were interesting in evaluating the use of the word "case series" in the published veterinary literature, in particular to determine whether some of these studies were actually population cohort studies. We conducted a literature search to identify studies described by the authors as "case series" (in the title or abstract) involving dogs or cats, and published after 2010. From the citations identified by the search, we randomly selected 100 articles for inclusion in the study. We determined that a substantive proportion of case studies without a comparison group included qualitative inferences about the efficacy of interventions or the significance of risk factors. Additionally a relatively large proportion of the studies identified by the authors as case series were actually population cohort studies, or included a population cohort component. Full results of this study will be included in the presentation.

REFERENCES

[1] Sargeant JM, et al. Zoon. Public Health. 2014;61(S10):10-17. [2] Dekkers OM, et al. Ann Intern Med. 2012;156:37-40.

Understanding the Case Control Study Design

Jonah Cullen BS, MS[1]; Annette O'Connor, BVSC, MVSC, DVSC, FANZCVS[1]; Kelly Makielski, DMV[1]; Jan Sargeant DVM, PhD[2]
[1]Ames, IA, [2]Guelph, Ontario

INTRODUCTION

Study designs in disease research are often broadly divided into those that are considered experimental designs and those considered observational designs. Of the observational designs, cohort, case-control, and cross sectional designs are often called the "big three" These designs are frequently used to assess exposure and outcome associations. Often the outcome of interest is a disease, although other outcomes of interest are possible such as performance.

The case-control design is often described as one of the most efficient choices for a research question regarding an exposure and outcome association owing to the inexpensive nature of the design in both assessment of exposure and the lack of required follow-up time. These benefits are even more pronounced if the disease of interest is rare. For rare outcomes, population-based cohort studies would likely require enrollment of a large number of participants to ensure enough incident cases of disease occur for the conduct of meaningful statistical analyses. For these reasons, the case-control design is often conducted in the early stages of investigating an exposure and outcome association, i.e. in the hypothesis generating stage. Further, clinicians might favor the case-control design as it is often associated with the use of preexisting records and therefore shortened data collection periods. It should be noted that the use of the retrospective design as a synonym for case control design is incorrect and should be avoided. However, while case control studies are comparatively inexpensive and efficient, the case-control design can be deceptively complex to conduct, requiring close attention to detail, suitable case and control selection, and correct interpretation of the effect measure, the odds ratio.

NATURE OF THE OUTCOME IN CASE CONTROL STUDIES

Case-control designs are characterized by the selection approach of study participants into the study based on the outcome rather than exposure. Although often the outcome of interest will be a disease status, this is not a requirement of the case-control design, and a common source of confusion.

The use of the terms case and control as synonyms for diseased and non-diseased respectively is a source of confusion in the design of observational studies. For example, a cohort study might be conducted to determine if dogs with diabetes are more likely to develop tumors compared to dogs without diabetes. In this example both the exposure and the outcome contain diseased and not diseased animals. In a case control study looking at risk factors for diabetes, the outcome might be dogs with diabetes (cases) and dogs without diabetes (controls) and the exposure might be history of arthritis or not. In this situation, again both the exposure and the outcome are disease categories. The best approach to thinking about outcomes and exposures in a study design is to ask what causal inference are we seeking to make. That is, in the cohort study we are asking does diabetes (exposure) cause tumors (outcome) and in the case control study we are asking does arthritis (exposure) causes diabetes (outcome).

Subjects included in the study with the outcome of interest are considered the case population (cases) and subjects from the same source population without the defined outcome are considered the control population (controls). Hereafter we will adopt that same usage.

CLASSIFICATION OF CASES: INCIDENT CASES & PREVALENT CASES

One of the first considerations when designing a case-control study is the nature of the outcome, which may be incident or prevalent. Prevalent cases are those already in the population (i.e. not newly detected and diagnosed). Case-control studies using prevalent cases may be preferred when it is difficult to detect and measure incident cases of the disease of interest. These designs based on prevalent cases assess the population burden of disease at a time and are useful for non-fatal, chronic diseases (i.e. idiopathic inflammatory bowel disease or hypothyroidism). However prevalence-based case-control designs based on prevalent cases are not optimal for assessing disease causation due to differences in disease duration or age-specific disease incidence between different groups. An alternative to selecting prevalent cases for the case control design is to select incident cases, which represent newly diagnosed patients. Incident cases in a case-control design are more advantageous for determining disease causation. These cases are often identified in hospital settings at the first presentation of a disease or defined outcome of interest.

TIME OF CONTROL SELECTION

A second consideration in designing a case-control study is the selection timing of controls (subjects without the outcome of interest). Ideally, controls should be selected from the same source population as the cases in order that the exposure distribution of the controls is the same as that of the cases. Traditionally, case-control studies are taught with the controls being sampled at the end of follow-up (cumulative sampling or exclusive sampling) and that when the disease of interest is rare, the odds ratio will estimate the risk ratio. Alternatively, controls may be randomly sampled from the entire source population at the beginning of follow-up (case-cohort sampling or inclusive sampling). Finally controls may be sampled concurrently as cases are identified meaning cases and controls are matched on time (density sampling or concurrent sampling). The rare disease assumption is not necessary for either case-cohort or density sampling for the odds ratio calculation to estimate the risk or rate ratio, respectively.

CONDUCT OF CASE CONTROLS IN VETERINARY MEDICINE

Recently Knol *et al.* conducted a survey of case-control studies in journals and textbooks considering human medicine with the objective of assessing the relevant study design elements, underlying assumptions, and interpretation of the outcome. The findings suggest that many authors misidentified the type of case control design used, and consequently, the inference obtained from studies was often incorrect. For example a number of authors did not acknowledge or recognize the use of an incidence density cases control design and the potential to estimate the risk ratio. Given the preference to estimate the risk or rate ratio these findings were viewed as lost opportunity and evidence that the case control design is perhaps poorly understood in human health. We have conducted a similar study for veterinary science and will present the results at the conference.

REFERENCES

von Elm, E., et al., Epidemiology, 2007, 18 (6):
Schulz, K. F., et al., Lancet, 2002, 359 (9304):
Pearce, N., Environ Health Perspect, 2004, 112 (10):
Pearce, N., Int. J. Epidemiol., 2012, 41 (2):
Rodrigues, L., et al., Int J Epidemiol, 1990, 19 (1):
Prentice, R. L., Biometrika, 1986, 73 (1):
Greenland, S., et al., Am J Epidemiol, 1982, 116 (3):
Knol, M. J., et al., Am. J. Epidemiol., 2008, 168 (9):

Humor as a Teaching Tool

Shane D. Lyon, DVM, DACVIM (SAIM)
Stillwater, OK

INTRODUCTION

Humor, as it relates to education, has been evaluated by psychologists, communication researchers and educational psychologists for nearly half a century. The topic of educational humor is complicated taking into account instructor characteristics, student characteristics, type of humor, context, delivery/timing, and even frequency of utilization. When used appropriately, humor can result in increased learner attention, motivation, immediacy, learning and improved instructor and course evaluations.

POTENTIAL BENEFITS OF HUMOROUS INSTRUCTION

Benefits for the student

Several aspects of how humorous instruction may benefit students have been studied. Instructors who effectively utilize humor have been demonstrated to enhance students' cognitive learning, increase student effort and participation, reduce tension, stress and boredom, as well as increase out of class communication with the instructor.[1,2,3,4]

Benefits for the instructor

There are several potential benefits for the humorous instructor. Humorous instructors are seen as more approachable (immediate), compassionate, entertaining, stimulating and enjoyable.[3] This may lead to increased instructor and course evaluations, honors and awards which can enhance instructor dossier and ultimately impact reappointment, promotion or tenure. Among other traits, the use of humor is a commonplace among award winning instructors.[5] When commonly utilized words were analyzed in student nominations for outstanding educators, "humorous" was the fourth most common finding.[6] Lastly, humor can create a positive learning environment which can increase student attendance and communication.[2,7]

GUIDELINES FOR EFFECTIVE UTILIZATION OF HUMOR IN THE CLASSROOM

How do you effectively utilize humor in the classroom? To be interpreted as funny, a statement must meet three criteria. It must be recognized as a humorous attempt, the recipient must process and understand the humorous attempt, and lastly the recipient must find it humorous.[8] Simply put, the student must recognize that the instructor is attempting a joke, the student must understand the joke, and most importantly the student must find the joke funny.

Quantitate and qualitative data have been collected as it relates to humorous instruction. More experienced instructors tend to utilize humor more frequently than novice instructors. Randomly selected recordings of college professors found humorous messages to occur at a rate of 3.34 per 50 minute lecture, or once every 15 minutes. Inexperience lecturers utilized humor at a rate of 1.6 per 50 minute lecture compared to experienced instructors at 6.5 per 50 minute lecture. Award winning instructors utilized humor more than the average at a rate of 7.44 times per 50 minute lecture.[7]

Scholars debate on the classification of humor. The number of categories range from 3 to 13 with up to 26 sub-categories depending on classification system. Basic categorizations include content-related, content-unrelated, self-disparaging, other-disparaging, and offensive. Sub-categories may include jokes, funny stories, puns, riddles, physical humor, humorous images, accidental humor, etc..[7]

To increase effectiveness, the humorous content should be related to the subject matter. Utilization of content humor has been shown to increase recall of information among learners.[9,10,11] In order for humor to be effective, the student(s) must find the instructor "funny" and "get the joke". In 2010, Wazner et al. collected data from students regarding their perception of instructors' humor and how they felt it related to learning. Positive correlations were found between students' reports of content humor and learning.[8] This supports previous analysis that indicated 74% of college students appreciated instructors' constructive use of humor and 80% of those respondents said humor "often" or "always" aided in learning.[12]

Another component of effective utilization of humor is related to appropriateness of humor. Wazner et al. in 2006 evaluated appropriate and inappropriate uses of humor by teachers. Not surprisingly, offensive humor (jokes or comments based on race, religion, sex, sexual orientation, culture, and political affiliation) were determined to always be an inappropriate use of humor. The appropriateness of the subject was also evaluated in this study. Again, not surprisingly, student-directed or student targeted humor was also found to be inappropriate by students. This would include not only isolating an individual student, but comments referring to the class as a collective. In fact, students rated this as the second most offensive use of humor behind sexual humor.[13] Another area for concern is self-disparaging humor. It is thought that self-disparaging humor may result in decreased instructor credibility which could negatively impact learning. This theory has not been supported. Interestingly, Wazner et al. found a weak positive correlation between self-disparaging humor and student perception of learning.[8] Additional studies are needed to further investigate this specific aspect of educational humor.

SUMMARY

In conclusion, humor has been demonstrated to be an effective educational tool. When used appropriately, it can increase student attention, motivation, and learning and also offers benefits to the instructor, beyond making the class more enjoyable to teach. To be utilized effectively, humorous statements should be related to course content. Care should be taken to avoid subjects which may be considered offensive to some and students should never be the target of humor.

REFERENCES

[1]Berk RA, et al. Humor 2009;11(4):383, [2]Garner RL. College Teaching 2006;54(1):177, [3]Goodboy AK, et al. Comm Quart 2015;63(1):44, [4]Martin RA. The Psych of Humor 2007, [5]Kelly N, et al. ERIC, ED 230080, [6]Lowman J. College Teaching 1994;42(4);137, [7]Banas JA, et al. Comm Ed 2011;60(1):115, [8]Wazner MB, et al. Comm Ed 2010;59(1):1, [9]Kaplan RM, et al. J of Ed Psych 1977;69(1):61, [10]Wazner MB, et al. Comm Ed 1999;48:48, [11]Ziv A. J of Exp Ed 1988;57:5, [12]Torok SE, et al. College Teaching 2004;52:14, [13]Wazner MB, et al. Comm Ed 2006;55(2):178

Self-Directed Learning

Audrey K Cook, BVM&S, MRCVS, DACVIM (SAIM), DECVIM-CA, DABVP (Feline)
College Station, TX

INTRODUCTION

In 2014, Kaufman and Mann described the ideal veterinary graduate as a self-directed, life-long learner, aware of his/her own limitations, equipped with a body of knowledge that can be applied in the abstract and strengthened through regular reflection on action.[1] As clinicians in a teaching environment, most of us devote substantial energy to the maintenance and improvement of our own skill sets, both medical/surgical and pedagogical. This may be deliberate (e.g., researching a novel therapy) or casual (e.g., looking up a fact during a conversation with a colleague), but this behavior pervades our daily activities. We are by nature self-directed learners, but how do we pass this on to our students? In essence, how do we "teach" them to learn for themselves?

DEFINING THE PROCESS OF SELF-DIRECTED LEARNING (SDL)

SDL is not simply independent study or a willingness to work without direct supervision; instead it involves learners as active participants, with clear responsibilities for their own learning.[2] As described by Knowles[3], effective SDL encompasses a series of deliberate activities: 1) identification of a need or knowledge deficit; 2) formulation of specific goals; 3) identification of resources; 4) implementation of appropriate activities; and 5) evaluation of outcomes. From a pedagogical standpoint, SDL supports one of the tenets of constructivism, namely that the learner is central to the creation of meaning. It is the learner and not the teacher who determines the direction and quality of the learning.[4]

SDL MODELS AND METHODS

Numerous educational models include some elements of SDL and are often labelled as such, even if all the classic components are not included. These include problem-based learning (PBL), guided discovery learning, task based learning and reflective small group learning. Of these, PBL is probably the most well-known, although there is substantial heterogeneity in the definition of a PBL curriculum. When McMaster University pioneered the concept in 1969, the intent was to make medical education more interesting and relevant. There was no evidence at that time to suggest that the PBL-trained learner would become a "better" doctor. Forty years later, this question is still under debate... knowledge *acquisition* may be weaker but clinical *performance* may be improved.5 In a large meta-analysis conducted in 2010, SDL (all forms) was moderately more effective with respect to the acquisition of knowledge for advanced learners, and was reported to be as effective with respect to skills and attitudes when compared to traditional educational models.[6]

Even within a more traditional curricular model, SDL can be incorporated into routine teaching/learning activities. An institution does not need to adopt a global approach; instead individual instructors can include elements of SDL into their interactions with students. This may need to be incremental, as students accustomed to a teacher-centric culture may be unsettled by a change in methods. More than two decades ago, Gerald Grow proposed the staged SDL model in which the learner progresses from dependent (stage 1), through interested (stage 2) and involved (stage 3), to self-directed (stage 4).[7] This is evolution is mirrored by the teacher, who shifts from authority (stage 1), through motivator (stage 2) and facilitator (stage 3) to consultant (stage 4). This model reflects the idea that just as dependency and helplessness can be learned, so can self-direction. The teacher is responsible for adjusting to meet the needs of the students while empowering them towards greater autonomy.

THE POSITIVES OF SDL

Effective clinical reasoning requires more than simple factual recall; instead students must apply, analyze and interpret information. These higher cognitive domains are fostered by active learning and discovery; by students identifying and answering their own questions.[8] Traditional teaching methods tend to encourage memorization and (unintentionally) foster superficial learning instead of deep understanding. Facts may be recalled, but meaning may be lost. SDL lets students acquire knowledge in a way that can be mobilized to solve similar problems, and new information can be anchored into existing frameworks.

Active learning processes also improve retention. Information that is discovered by the learner has relevance, which helps place it where it can be used in the future. When a concept is known and understood, it is more easily recalled when needed. Appropriate contextualization also improves retention as information is incorporated into an appropriate paradigm. Learning that has been earned is more likely to be retained.

Another positive aspect of SDL is a reduction in the student's dependency on the teacher. Most traditional educational approaches foster dependency; students feel that information must be provided by a perceived expert for it to be real, valid or necessary.[8] Standard assessment methods support this viewpoint, as there is a reluctance to test beyond lecture content and students can achieve a perfect score (regarded as proof of competency) without the need for independent discovery. There is also a tendency for students to assume that material presented in the formal lecture setting is complete and sufficient; students may be unaware of their own limited or superficial grasp of a topic. In essence, they don't know what they don't know. SDL creates an awareness of one's own limitations and the vast body of knowledge that remains to be explored. More than ever before, new veterinary graduates need the ability to assess their own skills and improve those as necessary. Most of their learning will occur post-graduation and will have to be self-directed.

FOSTERING SDL IN THE CLINICAL SETTING

Traditional 4VM clinical rotations are an ideal setting for SDL as small group and one-on-one interactions between students and faculty are the norm. However, setting the stage for SDL in the clinic requires more than simply asking the students to look things up rather than answering their questions. Instead, we need a culture in which faculty and residents model reflective behavior and provide support as students begin to take control of their own learning.[9] As students enter the clinics, most begin to realize the limitations of their knowledge and start to reconcile the differences between classroom learning (in which material is presented and then recall is tested) and clinical learning (which is exquisitely problem oriented and carries substantial uncertainty). The occasional student may enter the clinical year with strong SDL skills, but this is unusual. Even within a PBL curriculum, individuals may still be somewhat teacher-dependent, with limited ability to identify and address their own learning needs. The time is ripe for a shift in learning behaviors and good clinical teachers can facilitate this change.

First, we need to raise awareness of SDL and deliberately model the process. As teachers, we need to share our own uncertainties and involve students in our own efforts for improvement. This is often referred to as 'reflection on action' and is the hallmark of a self-aware professional. We need to talk openly about how we make decisions and admit when we could have done better. Modeling our own efforts at SDL empowers students to solve their own learning needs. This demystifies our status as apparent experts; we no longer have ownership of knowledge and demonstrate that it is accessible to all. This also encourages students to move from a dependent state to a more autonomous one. If we are seen to find ways to answer our own questions, they are empowered to do the same.

Second, we need to build motivation. This plays a key role in SDL as the process requires effort and is not necessarily immediately rewarding. Most of us would struggle if asked to explain what drives us forward as professionals, and the same is true for our students. Motivation is a complex mixture of intrinsic and extrinsic components, and can wax and wane during training. Within an individual, sources of motivation can shift depending on inherent interests and aptitudes, and is strongly influenced by the relationship with the teacher.[10] Although intrinsic motivation is likely the most enduring form, extrinsic influences can be significant and can set the stage for greater efforts. Good teachers can provide substantial motivation through creating a sense of inclusion and by strengthening the student's sense of professional identity. For the 4VM students, the end is in sight, and we can boost motivation by creating a collegial, team-like environment. They want to be good veterinarians and we can support that outcome by treating them as such. It is also helpful to praise effort rather than result; students are more likely to continue to invest time and energy on challenging tasks if their labors-rather than the specific outcomes of those labors-are acknowledged.[11] The process of SDL must be celebrated so that it brings its own rewards.

Third, we need to set the scene for SDL on the clinic floor and provide opportunities for students to take control of their own learning. These efforts can be simple; if the group is stumped by a question or topic in rounds, identify it as a learning issue and ask one of the team to share a solution or new information later in the day. Providing careful feedback is also important, as students need help in evaluating their own outcomes. Discuss about the resources used and challenge students to become more critical. When students present cases to the group, ask them for one 'pearl', i.e., a valuable piece of learning or insight they will use in the future. Take a few minutes to talk about how this learning took place and the gap that it fills. Try and integrate this with other information and contextualize it for future use. Discrepant questioning is another technique for triggering SDL. If a student writes or says something factually incorrect, our tendency as teachers is to provide the 'right' answer. However, for a student with an incomplete understanding, this simply creates two disparate pieces of information which are both retained and often given equal merit. Instead of correcting the mistake, it is more helpful to explore it with the student. Take the idea further until the student becomes aware of its failings. At that point, highlight the gap in the student's understanding and talk about ways this can be resolved.

SUMMARY

Supporting SDL in a clinical context involves modeling appropriate behaviors, and setting the environment for students to identify their own learning needs and ways to address them. Curricular models that formalize SDL may have some educational advantages, but SDL can also be incorporated by clinical faculty in to the routines and rhythms of senior student rotations. Successful practitioners engage in constant self-education, and we have a responsibility to start that process prior to graduation.

REFERENCES

[1]Kaufman DM, et al. in Swanwick T (ed) Understanding medical education: Evidence, theory and practice 2nd ed; Oxford: Wiley Blackwell 2014;7. [2]Spencer JA, et al. BMJ 1999;318:1280. [3]Knowles M. Self-Directed Learning: A Guide for Learners and Teachers. New York, NY: Associated Press 1975;18.[4]Biggs J. Higher Education 1996;32:347. [5]Neville AJ. Med Princ Pract 2009;18:1. [6]Murad MH, et al. Med Educ 2010;44:1057. [7]Grow GO. Adult Ed Quart 1991;41:125. 8Dale VHM, et al. J Vet Med Educ 2015;http://dx.doi.org/10.3138/jvme.35.4.581 [9]Sandars J, et al. Med Teach 2011;33:875. [10]Parkinson TJ, et al. J Vet Med Ed 2006;32:253. [11]MacLellan E. Active Learn Higher Ed, 2005;6:194.

ACVIM Candidate Function: GI Endoscopy: 100 Years of Experience in 75 Minutes

A. Jergens, DVM, PhD, DACVIM; S. Marks, DVM, PhD, DACVIM; D. Twedt, DVM, MS, DACVIM; M. Willard DVM, MS, DACVIM
College Station, TX

ENDOSCOPIC BIOPSY OF THE SMALL INTESTINES

Indications: Patients with small intestinal disorders including chronic small bowel diarrhea, weight loss, alterations in appetite (often anorexia), vomiting, and melena. Specific small intestinal diseases diagnosed via enteroscopy include inflammatory bowel disease (IBD), intestinal neoplasia, duodenal ulcer/erosions, lymphangiectasia, and GI histoplasmosis. Enteroscopy is particularly useful in obtaining mucosal biopsies in protein-losing enteropathy patients where patient debilitation and hypoalbuminemia may increase the risk for complications.

Patient preparation: Withhold food for 12-18 hours for upper small bowel examination. Retrograde ileoscopy will require more extensive patient preparation (see colonoscopy).

Instrumentation: Flexible endoscopy is required to traverse the pylorus and to visualize the descending duodenum. The jejunum is sometimes accessible to endoscopic evaluation in small dogs and most cats. Serrated cup forceps (without bayonet) are often useful for small bowel biopsy procedures but personal preference will influence selection of forceps type. Larger biopsy forceps (2.8 mm or >) procure the best diagnostic specimens.

Abnormal findings: Alterations in mucosal texture (increased granularity), friability, and hyperemia are commonly observed. Increased granularity, friability, and erosions are often associated with mucosal inflammation seen with IBD and intestinal neoplasia. Ulcers (uncommon) and erosions (more common) are characteristic of inflammatory and neoplastic lesions. A milky-white mucosal appearance (focal or diffuse striations or "spots") from distended villi or milky exudate within the lumen from ruptured lymphatics may be seen in dogs with intestinal lymphangiectasia.

Biopsy recommendations and technique: Duodenal/jejunal biopsies should always be obtained regardless of mucosal appearance. Duodenal tissue is normally quite friable and good technique is essential. Serrated cup (non-bayonet) biopsy forceps should be used to obtain 10-15 good quality specimens. Focal mucosal lesions (e.g., erosions or granular areas) are biopsied directly since these abnormalities often correlate to histopathologic lesions. I generally use two biopsy techniques when obtaining small intestinal samples. The best biopsies are obtained by directing the instrument perpendicular (90 degrees) to the mucosal surface at flexure locations, if possible. I choose flexure sites since these locations facilitate targeted, deep mucosal tissue purchases while reducing the tendency of the forceps cups to slide along the superficial mucosa. Tissue biopsies procured at non-flexured sites will require directing the endoscope tip (and biopsy forceps) obliquely towards the mucosa. The use of high quality, sharp forceps with large size cups will maximize sample size regardless of the biopsy technique utilized.

Alternatively, the patient may be large necessitating the use of a 'long-distance' biopsy technique. The biopsy forceps is passed as far as possible down the lumen until resistance is met. Usually the forceps cups are beyond the operator's view distally. Once resistance is encountered, the forceps are retracted slightly (to open up the jaws) then re-advanced firmly against the mucosa where a tissue specimen is obtained. The disadvantages of this procedure are its 'blinded' approach and the tendency to biopsy repeatedly in the same site. Mucosal perforation is a potential but extremely unlikely complication of this procedure. Nevertheless, this biopsy technique is often used and is quite successful. Due to their friability, special care is advised when removing small intestinal specimens from biopsy forceps. Gentle teasing from the forceps with a 25-gauge needle is recommended.

ENDOSCOPIC BIOPSY OF THE LARGE INTESTINES

Indications: Clinical signs of chronic colonic disease including large bowel diarrhea (tenesmus, dyschezia, hematochezia, or mucoid feces). Colonoscopy is particularly useful in the diagnosis of IBD (lymphocytic-plasmacytic colitis), colonic neoplasia, and rectal masses in both dogs and cats.

Patient preparation: Withhold food for 18-24 hours. While longer periods (24-36 hours) of food denial may result in less retained feces, this time period is often impractical for many clients. In dogs, I prefer to administer two doses of a colonic electrolyte lavage solution (GoLYTELY, 20 ml/kg/dose given 4-6 hours apart orally) the afternoon before an AM endoscopy. It has been the author's experience that these lower doses of lavage solution provide excellent mucosal cleansing while higher doses (40-50 ml/kg/dose) of solution may cause vomiting in some patients. The cleansing solution may be administered either by orogastric intubation with a stomach tube or by offering the solution free-choice in a water bowl. A warm-water enema (20ml/kg) can then be administered the morning of the procedure to both dogs and cats. Miralax is an effective colonic cleansing agent in cats.

Instrumentation: Flexible endoscopy allows for visualization of all colonic regions as well as performance of retrograde ileoscopy. Biopsy instruments should include serrated cup forceps or other biopsy forceps based on clinician preference.

Abnormal findings: Similar as for enteroscopy including increased mucosal granularity, friability, and the presence of ulcer/erosions. Loss of submucosal vascularity is a significant finding and may be caused by mucosal edema, accumulation of exudate (blood, mucous, necrotic debris), or infiltration by inflammatory or neoplastic cells. Rectal masses and colonic nematodes (*Tricuris vulpis*) are less common endoscopic observations.

Biopsy recommendations and technique: Colonic biopsies are always obtained regardless of mucosal appearance. Flexible endoscopy is preferable since examination and biopsy of the transverse and ascending colons may also be performed. Focal mass lesions or mucosal lesions of increased granularity and/or erosions are biopsied directly. In the absence of gross mucosal abnormalities, obtain 3-4 biopsy specimens from each colonic region (e.g., ascending, transverse, and descending colon) using serrated cup biopsy forceps.

Retrograde ileoscopy should be performed as part of routine upper and lower GI endoscopic examination in dogs and cats. Furthermore, this procedure is useful to perform in patients where poor quality duodenal biopsies have been obtained. Ileoscopy will necessitate thorough colonic cleansing and advancement of the endoscope tip through the ileocecal sphincter in dogs. Note that the ileal mucosa has an identical appearance to that of the proximal duodenum. Alternatively, 'blind' biopsies may be procured by passing the biopsies forceps through the ileocecal orifice in small dogs and cats. Potential risks of this procedure include repeated sampling of similar areas of ileal mucosa and intestinal perforation (rare). Ileal biopsies may be particularly useful in differentiating severe IBD from lymphosarcoma in cats.

POST-BIOPSY HANDLING/PROCUREMENT CONSIDERATIONS

Biopsy specimen handling and histopathologic processing are critically important considerations. Specimens should be carefully removed from forceps cups with a 25-gauge needle being careful to position the specimen 'serosal' side down onto a biopsy sponge. Note that biopsy artifacts, such as procurement of superficial mucosa and crush artifacts, are inherent in most biopsy procedures regardless of operator experience and technique. Obtaining at least 8 biopsy specimens per area will result in the greatest diagnostic accuracy for detecting most mucosal diseases affecting the small intestine. Similar studies evaluating the optimum number of endoscopic biopsies required for diagnosis of colonic diseases have not been performed. Lastly, utilize an experienced histopathology laboratory staffed by pathologists who are experienced in reading and interpreting endoscopic tissue specimens to minimize misdiagnosis.

ESOPHAGEAL STRICTURES

The most common causes of esophageal strictures in dogs include GER (gastroesophageal reflux) from general anesthesia, chemical injury from swallowed substances, esophageal foreign bodies, esophageal surgery, and intraluminal or extraluminal mass lesions (neoplasia or abscess). Cats are particularly susceptible to doxycycline and clindamycin-associated esophagitis and esophageal stricture. Gastroesophageal reflux associated with general anesthesia has been reported to occur in up to 65% of cases of esophageal stricture, with a median onset of clinical signs occurring 7.5 days post-anesthesia. Damage to the muscularis layer of the esophagus is associated with fibroblastic proliferation and contraction leading to stricture formation.

There are at least 5 non-surgical techniques for treating benign esophageal strictures in dogs and cats; however, success cannot be guaranteed. Management of esophageal strictures can vary from relatively simple to extremely challenging depending on the nature of the stricture(s). The main points to remember are:

1) Your first goal is to restore functionality, not anatomic normalcy,

2) The more trauma you cause when you dilate the stricture, the more likely the stricture is to recur, and

3) There is NO technique that will always work, so don't go making any promises, explicit or implied. There are a lot of nuances to dilating the more challenging esophageal strictures, and experience in performing the procedure is pivotal.

Balloon dilatation: This is the most popular technique because it has been touted as the best (which is debatable) based on the forces that are applied to the stricture that are a radial stretch, in contrast to the longitudinal forces applied with rigid bouginage. There are esophageal balloon catheters that can withstand a relatively high pressure up to 45 psi and are either passed via the biopsy channel of the endoscope or are passed alongside the endoscope. The former option is generally easier and quicker to use, but typically restricts the devices to single use. After the balloon has been fully inflated, it is difficult to compress the balloon enough so that it can be successfully passed through the biopsy channel again. The balloon catheters that are designed to be passed alongside the endoscope can be used multiple times (assuming that one takes care of them), but can be a little more cumbersome to use because they are typically used with a guidewire.

How one proceeds to dilate the stricture depends upon what one sees when the stricture is first visualized. If the stricture is relatively short and the esophageal diameter is not too narrow, one can probably balloon it once and solve the issue (i.e., being aggressive is ok). The longer the stricture and the more narrow the esophageal diameter, the more the potential for problems when it is dilated. If in doubt, it is better to plan on multiple, small dilations versus being overzealous and attempting to dilate the stricture with a single dilation procedure. One can repeat dilations every 2-3 days if needed (for severe strictures) or one can stretch out repeat dilations to about a week, but be aware that a stricture can easily reform and be as bad as before within 7 days. It can take multiple balloon dilation procedures (> 10) in some animals for restoration of esophageal function. As eluded to above, every effort must be made to avoid making a major, deep fissure into the esophagus. While perforation is possible, it tends to be a rare event (unless one is overzealous during dilation). The more pressing issue is the fact that the more damage one causes, the more inflammation is likely, and the more likely the stricture is to reform (and often within days to a week). In addition to avoiding deep fissures, one should also strive to avoid unnecessarily removing mucosa during the dilation procedure. It is easy to remove mucosa if one a) does a lot of suction/aspiration to get a nice, clear picture of the lesion, or b) if one moves the dilated (or is some cases deflated) balloon (or even the endoscope, for that matter) back and forth across the dilated stricture site.

Bouginage: While there has been a lot of press about how much better balloons are than bouginage, the truth is that either will work very well. It all depends upon the training of the operator. The only real difference is that it is easier to perforate the esophagus during bouginage.

There are many different types of bougies. While using a bougie that goes over a guidewire is probably optimal, one can use an olive tip bougie that has no such guide. The two places that are most susceptible to perforation include a) the thoracic inlet, and b) just beside the opening of the stricture. To avoid perforating at the first site, one should depress the head at the occipital protuberance and simultaneously lift the lower jaw as the tip of the mandible so that the esophagus is in as straight a line as possible. To avoid perforating near the stricture, one needs to make sure that the tip of the bougie is in the stricture orifice so that the operator is not forming a blind sac next to the orifice. It is surprisingly easy to perforate an esophagus.

Bouginage is often reserved for those patients whose strictures are so dense that the pressure generated by the balloon is insufficient to dilate the stricture. The procedure is performed under direct visualization with the tip of the endoscope a few cm behind the tip of the bougie).

There are two big "NO's". First, never try to use the endoscope as a bouginage in light of the risk of perforating the esophagus. Second, do not use any bougie to try to find the continuation of the lumen if one is faced with a stricture that has completely occluded the lumen. If one cannot see a lumenal orifice, one should try going through the suspected site of the lumen with the stiff side of a wire guide. Then advance it slowly and see if it goes into the stomach (i.e., you will have to snap a quick radiograph to know).

"Nicking" the stricture: Another technique that has been used to try to avoid severe fractures is to use either an electrocautery polypectomy loop, an electrocautery papillotomy knife, or a laser to make 3 or 4 small "knicks" in the stricture. The objective is to facilitate the stricture breaking at 3 or 4 points (creating 3 or 4 small fractures) instead of one large fracture and fissure. The author finds that using an electrocautery polypectomy snare is preferable to using a knife. The knife "blade" can be surprisingly hard to position. The idea is to nick the stricture at 12 o'clock, 3 o'clock, 6 o'clock, and 9 o'clock. One does not need to make a deep nick; all you need is a nick about the depth of the wire you are using to cut it.

Injecting the stricture with steroids: In an effort to slow down healing (and reformation of the stricture), one can use endoscopic transbronchial or sclerotherapy needles to inject steroids into the stricture site. The catheter keeps the needle sheathed until the needle is extended past the tip after the catheter is exteriorized out of the endoscope. One can use single-use, disposable catheters, or one can get a re-useable sheath and just buy new needles. A 23 to 25-gauge needle is generally recommended.

The stricture site can be injected with steroids in a 4-quadrant pattern before balloon dilation or following balloon dilation, and the choice depends on clinician preference. Correct injection technique typically causes bleb formation at the site. Triamcinolone acetonide (10 mg/mL) or dexamethasone can be injected. In general, one injects approximately 1-2 ml of triamcinolone or dexamethasone per dog.

Painting the stricture with Mitomycin C: Another technique is to "paint" the site where the stricture was broken down with Mitomycin C (NOT mithromycin C, there is a big difference). A 5 mg bottle is reconstituted and soaked up into a gauze sponge. The sponge is endoscopically placed on the exposed tissue at the site where the stricture was broken up and kept in contact for 5 min. The site is then rinsed with 60 ml of water following the removal of the sponge. One needs to take all the precautions associated with handling any other chemotherapeutic drug. There have been recent reports of intralesional injection of Mitomycin C into the stricture site; however, the authors have no experience with this technique.

Stenting: For particularly difficult cases, stents may be placed in the esophagus. The stents are made by several corporations, although Infiniti corporation manufactures a variety of products for veterinary use (http://www.infinitimedical.com/p_stents.html). There are three types of stents: biodegradable, wire-uncovered, and wire-covered. The decision to select one type over another depends on the nature of the stricture and personal preference, although most veterinarians prefer the wire-covered stents. Esophageal stents are expensive, but can be less expensive than multiple ballooning procedures.

The stents can be placed via fluoroscopy or endoscopically. Some clinicians prefer to place the stents without the advent of imaging and rely upon endoscopy to ensure that the stent is placed correctly. If a stent is placed incorrectly, one can use a rigid endoscope and forceps to pull it forward (rather easy to do) or push it further in (much harder to do). Remember that the stent will become shorter when it expands.

The most important aspect of placing the stent is to ensure that it expands the stricture site. The second most important aspect is to ensure that the orad most aspect is in the cervical esophagus. One does this so that a surgeon (who has cut into the cervical esophagus and exposed the esophagus at the level where the stent is) can be guided to place a large, monofilament, non-absorbable suture though the esophagus and the stent, thereby fixing the stent in place. Doing this correctly is critically important. If one does not fix the stent in place, then the stent will invariably migrate (usually orad). If the stent gets too close to the upper esophageal sphincter, the animal can gag intractably. The stent is usually left in place for 2-3 weeks. A rigid endoscope can be used to find the anchoring suture and cut it with rigid biopsy forceps at the time of stent removal. Stents can cause substantial problems and the long-term prognosis in dogs treated with stents is guarded. Many patients have done poorly. Migration is one of the most common problems followed by recurrence of the stricture(s).

ESOPHAGEAL FOREIGN BODIES

Endoscopy is almost always the preferred method of removing esophageal foreign objects, but fluoroscopic and surgical techniques can be effective if the clinician is well trained. Rigid endoscopes (i.e., sigmoidoscopes or proctoscopes) allow much more control of the foreign object and are preferred to flexible endoscopes for removal of these foreign objects. It is especially useful to be able to pull the object into the lumen of a rigid endoscope and then withdraw it and the scope as a unit, thus protecting the esophagus. The main disadvantage of rigid endoscopes is that they are often not long enough in larger dogs.

Finesse is required because brut force can easily lacerate/perforate the esophagus. If a large object or a bone cannot be easily dislodged, do not force it lest you perforate a previously intact esophagus; instead, one can use rigid equipment (bone Rongeurs) to break it up and hopefully dislodge it. If that fails, one can attempt to pass a large Foley catheter behind the foreign object and inflate the balloon aborad to the foreign body; the catheter will distend the esophagus (thus freeing the foreign object) allowing one to pull the object out. If one cannot pull a foreign object out of the esophagus, one can try to push it into the stomach with the Foley catheter in front so that it helps open up the LES. However, do not push bones or other foreign objects into the stomach unless one is certain that it is smooth on the aborad side and will not further damage the mucosa. Finally, be careful to avoid over insufflating the esophagus, lest you rupture a weakened area in the mucosa and/or cause a fatal tension pneumothorax.

Fish hooks (including treble hooks) can typically be successfully removed endoscopically as long as they are fresh water hooks. Salt water hooks tend to have much larger barbs on them, making it difficult/impossible to forcibly tear it out of the mucosa. Fish hooks have usually penetrated the mucosa (and sometimes the muscular tunics) requiring one to use rigid equipment to carefully force the tip of the hook back out of the mucosa. A small hole is left, but there are very seldom any complications. In rare cases, one can use rigid biopsy forceps to cut the mucosa over a hook with a large barb so that it can be removed endoscopically.

After removing the foreign object, retake plain chest radiographs to be sure that a pneumothorax or pneumomediastinum (which would indicate a perforation) is not present. Antibiotics are indicated if there is substantial esophageal mucosal ulceration (and especially if one removes a fish hook which had been used with various baits that can harbor anaerobic bacteria). Depending upon the amount of damage, corticosteroids may be used to try to prevent cicatrix formation; however, it is not clear that they are effective.

If one is dealing with a small perforation and the edges are not necrotic (i.e., they are not due to a bone which has been there for 4 days), one can often place a PEG tube, assign the dog NPO and wait. Many times, keeping the esophagus contracted (i.e., there is no food or water going down the tube) will allow a small perforation without necrotic edges to spontaneously heal. If that does not work, one can always perform a thoracotomy later.

ESOPHAGEAL HEMORRHAGE

Although rare, it is surprising how brisk a hemorrhage can come from the esophagus. Esophageal hemorrhage is primarily associated with esophageal foreign bodies, but may also occur when there is severe esophagitis, endoscopically removing gastrointestinal foreign bodies, dilating esophageal strictures, or endoscopically injecting or cutting the esophagus. If esophageal hemorrhage occurs, often the best option is to get the largest esophageal balloon you can, place it at the site, and inflate it in an attempt to tamponade and stop the hemorrhage.

ESOPHAGEAL BIOPSY

It is extremely challenging to acquire good quality endoscopic biopsies from the canine and feline esophagus. Simply put, if one can obtain a good tissue sample of the canine esophagus with flexible endoscopic biopsy forceps, the esophagus is not normal (unless one happens to get right at the LES, in which case you might get a decent sample). One might be able to more easily acquire a sample of the feline esophagus in the region where there is smooth muscle.

POLYPECTOMY IN DOGS AND CATS

Esophageal polyps are rare in dogs and cats. It is important to note that most benign esophageal polyps are associated with an underlying malignancy. Most GI polyps that are resected endoscopically will be gastric. The main reason to remove them is if they are so large as to be causing gastric outflow obstruction. Duodenal polyps are rarely resected endoscopically.

Endoscopic removal of colonic/rectal polyps is primarily indicated when they are located so far orad that they cannot be everted for surgical removal. If they are large, do not attempt to remove them all at once. Piece-meal removal is indicated. One should never just grab a bunch of tissue and hit the pedal. You must be absolutely sure where the snare is so that you know that you are not grasping so far down the stalk that you will cause perforation.

Performing electrocautery polypectomy in airways poses special concerns. First and foremost, if you are pushing pure oxygen in via an endotracheal tube, there is a real risk of electrocautery igniting the tracheal/bronchial mucosa (or skin, or endotracheal tube, or any cloth material) with catastrophic results (think "flash paper"). Insufflation of room air through the endoscope during the procedure will generally be sufficient to support the patient and allow the procedure to occur safely. Second, be ready to aspirate blood quickly. Tracheal masses can be surprisingly vascular, and if you are working on a small patient, it will not take much blood to occlude the airway and suffocate the patient.

LAPAROSCOPIC ORGAN BIOPSY

Laparoscopy is considered a minimally invasive procedure that can be easily performed by the internist for liver, pancreas or kidney biopsy. There are some inherent benefits of obtaining laparoscopic assisted biopsies over either surgical or ultrasound guided needle biopsies. Those include the ability to visually examine the organ of interest, direct biopsy forceps or needles to an appropriate site and the ability to monitor the biopsy site for excessive bleeding and to take measures to stop any excessive hemorrhage if it were to occur. There is almost never a post procedure infection. The perceived downside of laparoscopic tissue biopsies over ultrasound guided needle biopsy include the instrumentation cost, the learning curve required to perform the procedure, the more invasiveness and the greater procedure time required for laparoscopy. However, with experience tissue biopsy can be performed quickly and often can be performed as an outpatient procedure. I generally will perform laparoscopy for the evaluation and biopsy of the liver, extrahepatic biliary system, pancreas, kidney, lymph nodes and/or spleen. The basic advantage of laparoscopic cup biopsies over ultrasound directed needle biopsies include the sample size, the ability to visualize the liver and extrahepatic biliary system and the ability to select an appropriate biopsy site.

Instrumentation: I recommend using a 5 mm zero degree laparoscope in conjunction with a second cannula site for instrument placement. This is referred to as a two puncture technique. Some recommend using an operating telescope that has a biopsy channel within the telescope so that instruments and scope are passed through a larger single cannula port (one puncture technique). I personally believe that the downside of the operating telescope technique is that one is limited to only tissue biopsy and it becomes is more difficult to explore the abdomen. I generally use a 5X10mm oval cup biopsy forceps for the liver and a 5 mm punch type biopsy forceps for the pancreas.

Liver Biopsy: There are two basic approaches for liver biopsy; midline approach and right lateral approach. I personally like the right lateral approach because I can easily examine the right wing of the pancreas as well as the gallbladder, cystic and common bile duct. The lateral approach limits the complete examination of the left lateral liver lobe. With the midline approach visualization is sometimes hindered by falciform ligament especially in fat animals and it is more difficult to follow the extrahepatic biliary system and to visualize the pancreas. I almost exclusively use a 5X10 mm oval biopsy cup forceps for liver biopsy. Either the edge or surface of a liver lobe can be biopsied with these forceps. Once the liver tissue is grasp the forceps are closed and held for 15-30 seconds and then the sample is gently pulled away from the liver. In fibrotic livers the forceps may need to be twisted first to pull the sample away from the liver. Generally at least 4 liver samples from different lobes should be collected. The size of the sample is adequate for most all liver evaluations. In small dogs and cats I may take a smaller biopsy (half a biopsy cup full approximately 5X5mm sample).

There can be non-specific reactive changes adjacent to the liver capsule and it has been argued by some that cup biopsies only get the surface of the liver. That is true but the biopsy sample size is large and deeper parenchymal changes can be adequately evaluated. Pathologist tends to disregard the nonspecific subcapsular changes or at least note those changes are limited to the subcapsular area. One study compared 2-18g needle biopsies (in my opinion 18g is too small) to a larger laparoscopic or surgical wedge biopsy and found more than half the time the needle biopsy did not reflect the same diagnosis as the larger samples. I would only consider laparoscopically guided needle biopsies if ultrasound showed a deep parenchymal lesion and I was not able to visualize it at laparoscopy. I would use either a 16-14g semi-automatic Tru cut™ type biopsy needle.

I will almost always take one biopsy cup sample for quantitative hepatic copper analysis and also culture a small piece of liver for aerobic and anaerobic organisms. I avoid collecting tissue from regenerative nodules as they tend to contain less copper than surrounding parenchyma. I generally transport my sample for copper analysis in a blood clot tube that is copper free and I avoid tissue contact with the tube lid.

Following liver biopsy the site is examined to assure adequate clotting. Normally only 0.5 to 2 milliliters of blood is lost from a site. I always use a palpation probe to examine the area for active bleeding. The probe can also be used to apply local pressure over the bleeding area if bleeding is felt to be excessive. Although infrequently required, excessive bleeding can usually be managed by placing a small piece of Gel Foam™ into the bleeding area using laparoscopic grasping forceps. Electrocoagulation (biopsy forceps have an attachment for cautery) can also be performed at the bleeding area however I find this rarely necessary.

Abnormal coagulation times are not always a contraindication to liver biopsy. I will frequently perform laparoscopy in cases having mild thrombocytopenia or slightly prolonged PT or APTT times. We also perform BMBT on our cases prior to laparoscopy because I find the cases having the most serious bleeding are often the von Willebrans dogs. Low serum fibrinogen concentrations may be a more reliable prediction of bleeding and we are now currently investigating that and other coagulation parameters prospectively in our clinical cases. Surprisingly the advanced cirrhotic or fibrotic livers often have little bleeding because fibrosis is less vascular and fibrotic areas tend to collapse the biopsy site periphery. It has been consistently shown that abnormal coagulation tests do not reliability predict which patient will have excessive bleeding.

Gallbladder Aspiration: In patients suspected of having gallbladder (GB) disease I generally aspirate the GB for cytology and culture. I believe this is especially important in cats having cholangitis and dogs with evidence of cholestatic liver disease. It is a relatively safe procedure to perform. With a right lateral approach an 18g 6-inch spinal needle is directed into the body of the GB and bile is aspirated. It is important to make sure the needle does not penetrate the diaphragm for fear of a pneumothorax and that the GB is decompressed as much as possible so leakage does not occur after the needle is removed. In the ventral approach the needle can be directed through the quadrate lobe of the liver and into the GB at a point where it is attached to the liver. In doing so when the needle is removed and if there is any leakage of bile it goes directly into the liver and not the abdominal cavity. To date I have not observed serious complications associated with GB aspiration.

Pancreatic biopsy: The pancreas can be evaluated with a right lateral or ventral abdominal approach. It is almost impossible to visualize the left wing of the pancreas in any approach so only the right wing as it sits next to the duodenum can be evaluated and biopsied. Often the diagnosis of acute or chronic pancreatitis can be made based on visual inspection alone. Most of the cases I biopsy are in cats suspected to have chronic pancreatitis or concurrent cholangitis and pancreatitis. Viewing the pancreas in acute pancreatitis is sometimes difficult when there is considerable inflammation, effusion or adhesions around the organ. For the pancreatic biopsies I prefer using a punch type biopsy forceps however the oval cups can also be used but they tend to take much larger bites. The biopsy should be obtained from an edge of the organ away from the pancreatic ducts that traverse the center of the gland. Because the pancreas is held in the omentum pulling back the forceps once closed also tends to pull the pancreas with it. The pancreas will however drop off as the forceps are withdrawn through the cannula. Complications from laparoscopic pancreatic biopsies are rare.

Kidney Biopsy: Kidney biopsies require an assistant and samples are generally obtained using a standard 14-16g semi-automatic Tru Cut™ type biopsy needle. I avoid the automatic gun type biopsy needles for cats and small dogs because of the reported fatal vasogenic shock that can occur after firing. The right kidney is preferred because it is less movable than the left kidney and for that reason a right lateral approach is most often used. The abdominal entry site for the biopsy needle is determined during laparoscopy and is generally located caudal to the kidney. A small skin incision is made at the biopsy needle entry site and the biopsy needle is passed directly through the abdominal wall and advanced toward the kidney. Always make sure the needle entry is caudal to the diaphragm. The biopsy needle is visually directed at an angle to best obtain renal cortex, avoiding the large arcuate arteries at the cortico-medullary junction. One should be also aware of the biopsy needle throw length when taking a biopsy. I will seat the tip of the needle in the kidney capsule and then rapidly fire the semi-automatic biopsy needle. Most all kidney biopsies bleed excessively so I always have a palpation probe in place so the assistant can apply pressure at site as soon as the needle is removed. I generally take 2 biopsies and find that is adequate to provide enough glomeruli for a good evaluation. One study comparing ultrasound needle biopsies vs laparoscopic biopsies found better histologic sample quality when biopsies were obtained laparoscopically. Prior to kidney biopsy make sure the patient is not receiving pressor drugs or mannitol that will accentuate bleeding considerably.

Other Organ Biopsies: I will occasionally biopsy enlarged lymph nodes and or spleen laparoscopically. The technique is similar to the technique I do with liver biopsies using the oval cup forceps. I almost always apply Gel Foam™ to a splenic biopsy as they tend to bleed a fair amount.

REFERENCES

Scott, KD, Zoran DL, Mansell J, et al. Utility of endoscopic biopsies of the duoduenum and ileum for diagnosis of inflammatory bowel disease and small cell lymphoma in cats. J Vet Intern Med 25 (6): 1253-1257, 2011.

Small Animal Endoscopy, 3rd edition, Tams TR and Rawlings CA editors, Elsevier-Mosby, St. Louis, MO, 2011.

WSAVA guidelines for endoscopy reporting: http://www.wsava.org/guidelines/gastrointestinal-guidelines

THE TRANSCRIPTOME OF *SARCOCYSTIS NEURONA* INFECTED HORSES DEMONSTRATES T-CELL MEDIATED IMMUNOPATHOGENESIS WITH CYTOKINE DYSREGULATION

JunJie Liu[1], Lei Zhou[2], Nancy Denslow[4], Lisa Farina[1], Michael Dark[1], Robert MacKay[3], Maureen Long[1]

[1]*University of Florida Department of Infectious Diseases and Pathology, Gainesville, FL, USA,* [2]*University of Florida Department of Molecular and Microbiology, Gainesville, FL, USA,* [3]*University of Florida Department of Large Animal Clinical Sciences, Gainesville, FL, USA,* [4]*University of Florida Department of Physiology, Gainesville, FL, USA*

The primary goal of this research was to contribute new knowledge to the understanding of *Sarcocystis neurona* (SN) infection in horses. The hypothesis investigated was that the CNS of gravely SN affected horses had unique cassettes of differently expressed (DE) genes that define disease pathogenesis. Tissues from 2 groups of humanely euthanized horses (6 each) with either grave or no neurological symptoms were collected under University of Florida IACUC protocols 3984, 4109, and 4559. Sets of CNS tissues from all levels of brain and spinal cord were fixed for pathology and frozen for nucleic acid extraction. Histopathologically, encephalomyelitis and normal CNS was confirmed in affected and unaffected horses, respectively. SN, *Neospora hughesi*, EHV-1, WNV, and EEEV virus PCR was performed to confirm neuropathogen status in all horses. A microarray (Agilent, Santa Clara, CA) was used that targeted most known equine reading frames (EqCAb2 and WNVHorse projects). A total of 36 arrays consisting of 12 SN^+ lesion tissues (LTs), 12 SN^- adjacent tissues (ATs), and 12 location-matched normal tissues (NT) were used. Linear models for microarray data (LIMMA, R/Bioconductor.org) was used for data analysis [5% false discovery rate (FDR); 1.5-fold ($\log_2$) cutoff, $P<0.05$]. A second analysis was performed wherein relative SN load was calculated for each tissue and using Pearson correlation coefficients ($r>0.65$, R) DE was correlated to SN burden. Several platforms were used to perform hierarchical clustering [Bonferroni (FWER) and FDR corrections], gene enrichment (Markov), and GO analyses ($P<0.05$). Comparing LT to AT, 754 DE genes with 618 and 136 up- and downregulated genes were detected. Comparing LT to NT, 1535 DE genes with 835 and 618 up- and downregulated were detected. Positive activation/proliferation of T-cells were emphasized in 13/20 top canonical pathways. T-cell (C-X-C) chemokines CXCL9 (200-fold), -10 and -11 were upregulated. A mixed T_H response was demonstrated with interferon(IFN)-γ (20-fold), interleukin(IL)-2, IL-4, and IL-10 upregulated. Although TNF-α, IL-6, and IL-12 were upregulated, the IL-1 family was downregulated except for IL-1 neutralizing receptor. The only top DE cassette of neurological genes was downregulation of dopaminergic signaling. Parasite burden was positively and negatively associated with 398 and 71 genes. Natural killer(NK)-lysin and IL-1α,β were the most highly associated up- and downregulated genes. The hallmark of gene expression was a T-cell response with immune dysregulation based on cytokine profiles confirmed by rt-PCR. Although the IFN-γ KO mouse and other literature indicate a role for reduced IFN-γ, horses that succumb to grave SN infection had robust responses. The intense T-cell reactivity may indicate that IFN-γ is present, but killing is inhibited. The downregulation of the proinflammatory cytokine IL-1 was surprising given the inflammatory profile and may contribute to susceptibility of horses to SN.

EVALUATION OF TRIAXIAL ACCELEROMETERS FOR DETECTION OF GAIT DEFICITS IN HORSES WITH NEUROLOGIC DISEASE

Yvette Nout-Lomas[1], Birgitte Luining[2], Jeremiah Easley[1]

[1]*Colorado State University, Fort Collins, CO, USA,* [2]*Utrecht University, Utrecht, The Netherlands*

The severity of ataxia in horses is graded on a scale from 0 (normal) to 5 (recumbent, unable to rise). The correct assessment of ataxia in horses is critical to diagnostics, treatment, and evaluation of response to treatment. However, this grading scale is somewhat subjective, has only moderate to poor agreement between observers, and lacks sufficient discriminatory ability.

The objective for this study was to evaluate the use of triaxial accelerometers for gait analysis in horses and specifically to determine whether they would differentiate between sound horses and those with neurologic disease.

For this study we used 7 adult horses. Four horses (1 – 2 years of age) were clinically normal and 3 horses (2 – 10 years of age) were diagnosed with cervical vertebral stenotic myelopathy (CVSM) via myelography. Ataxia was graded at 2 – 3 out of 5 in all 3 horses. Gait analysis included a full neurologic examination and analysis of data recorded by 6 triaxial accelerometers (Gulf Coast Data Concepts X16-4 and X16-1C 3-axis) placed on the lateral metacarpus and metatarsus of all four limbs, on the head, and on the sacrum, Once all accelerometers were placed on the horse's body, data was collected while the horse was walked in a straight line at its own natural pace, and while walking it in a straight line with its head elevated. During data collection any abnormal events (e.g. sudden stop or spook) were recorded for each run, to ensure a suitable run was used for data analysis. For data analysis, five sequential strides from the midpoint of each run were used. The raw data was initially plotted using the XLR8R 2.1 application provided by Gulf Coast Data Concepts. A usable run was then identified based on the data collection notes and the regularity of the plotted data. A five-stride sequence was then selected and exported to Microsoft Excel 2013 for further analysis. In the leg-based accelerometer data it was possible to determine stride characteristics using all three axes. Single steps, the moment of lifting the heel and toe and the moment of touching the toe down were determined. For the sacrum-based accelerometer, the z-axis (vertical movement) and y-axis (mediolateral movement) could be used to determine steps with either the left or right leg. Steps with the right leg corresponded to a positive peak in the y-axis while steps with the left leg corresponded to a negative peak. Raw data from both head and sacrum accelerometers were converted to g-forces.

The Ax and Az range values recorded from the limbs were significantly higher in ataxic horses when compared to normal horses ($p<0.005$). Relative swing phase values were significantly higher in clinical cases compared to normal horses for both thoracic and pelvic limbs during the normal walk ($p<0.0001$) and when the head was elevated ($p<0.0001$). Mediolateral accelerations recorded from the sacrum in ataxic horses were significantly higher when compared to normal animals ($p<0.05$).

In conclusion we were able to differentiate between normal and ataxic horses using data recorded from triaxial accelerometers mounted on the horse's body. Studies are ongoing to determine how vertebral interbody fusion surgery affects the gait in these horses. Future work will require evaluation of these devices to distinguish between different severities of ataxia development of a scale in which objective data such as that derived from accelerometers can be incorporated.

IMMUNOGENICITY AND EFFICACY OF A NOVEL VACCINE AGAINST RHODOCOCCUS EQUI PNEUMONIA IN FOALS

Joana Rocha[1], Noah Cohen[1], Colette Cywes-Bentley[2], Courtney Brake[1], Angela Bordin[1], Gerald Pier[2]

[1]Texas A&M University, College Station, TX, USA, [2]Harvard Medical School, Boston, MA, USA

The purpose of our research is to develop an effective, safe vaccine to prevent rhodococcal pneumonia in foals that targets the surface polysaccharide poly-N-acetyl glucosamine (PNAG). The purpose of this research report will be to present background information about the dPNAG vaccine, the aforementioned results, implications of our results, and our plans for future work with this vaccine. We have demonstrated that PNAG is expressed on the surface of R. equi and on the surface of macrophages infected with R. equi using fluorescence and confocal microscopy. We have demonstrated that a conjugate vaccine based on synthetic motifs of deacetylated PNAG are immunogenic in mares and foals using enzyme-linked immunosorbent assays (ELISAs). Using a randomized, blinded study design, we have further demonstrated protection against intrabronchial challenge at approximately 24 days of age with virulent R. equi of foals (n = 5) born to mares vaccinated with the dPNAG vaccine, whereas 3 of 4 foals born to unvaccinated mares developed moderate to severe clinical signs of rhodococcal pneumonia and thoracic ultrasonographic findings of pulmonary abscessation or consolidation. All pneumonic foals were successfully treated with clarithromycin combined with rifampin. Strong positive correlation was observed between maternal, colostral, and foal serumal anti-dPNAG antibody concentrations. Moreover, these titres also strongly correlated with clinical outcome of foals. Our findings indicate that maternal vaccination can protect foals against R. equi pneumonia and that efficacy is likely mediated by antibodies.

TRACHEOBRONCHOSCOPIC ASSESSMENT OF EXERCISE-INDUCED PULMONARY HEMORRHAGE (EIPH) AND AIRWAY INFLAMMATION IN BARREL RACING HORSES

Renaud Leguillette, Mei Steinmann, Stephanie Bond, Bailee Stanton, Persephone Grecco-Otto

University of Calgary, Calgary, AB, Canada

Poor performance is often suspected to be associated with EIPH in barrel racing horses, however there are no published reports of EIPH for this discipline. The prevalence of EIPH in barrel racing horses is also unknown.

This study was performed to determine the prevalence of EIPH and signs of airway inflammation in barrel racing horses under normal racing conditions in Alberta.

170 barrel racing horses

Tracheobronchoscopic examinations were performed at least 30 minutes post-race. Video recordings were scored off-site independently by two observers for EIPH and tracheal mucus accumulation (TMA). Horses with an EIPH score ≥2 were not assessed for TMA. Inter-observer agreement was calculated using weighted κ statistics. Run times, environmental parameters and clinical information were also recorded for analysis.

77/170 (45.3%) of horses examined showed evidence of EIPH (grade ≥ 1). Inter-observer agreement was 0.94. 140/141 (99.3%) of horses assessed for TMA showed evidence of tracheal mucus accumulation (grade ≥1) with 104/141 (73.8%) having a TMA score ≥2. Inter-observer agreement was 0.73. A weak positive association was found between EIPH scores and average run speed.

The high prevalence of EIPH and its weak positive association with average running speed observed in the sampled population of barrel racing horses indicates that the explosive, anaerobic exercise demanded in this sport induces substantial stress on the lungs. Factors such as environmental dust and frequent travelling may have contributed to the very high prevalence of TMA observed.

MISOPROSTOL IS SUPERIOR TO COMBINED OMEPRAZOLE AND SUCRALFATE FOR HEALING GLANDULAR GASTRIC LESIONS IN HORSES WITH CLINICAL DISEASE

Georgina Varley[1], Mark Bowen[1], Vikki Nicholls[2], Jocelyn Habershon-Butcher[1], Gayle Hallowell[1]

[1]University of Nottingham, Sutton Bonington, UK, [2]University of Liverpool, Neston, UK

It is well recognized that omeprazole monotherapy results in poor healing and improvement rates in horses with gastric glandular disease (EGGD). The purpose of this study was to evaluate whether misoprostol (M) was superior to combined omeprazole and sucralfate (OS) therapy for the treatment of EGGD. Sports horses, primarily Warmbloods with clinically significant glandular lesions were enrolled following gastroscopic examination and were included if re-examinations were performed. Horses received either 5ug/kg misoprostol PO BID or enteric coated omeprazole (4mg/kg PO SID) and sucralfate (10mg/kg PO BID) administered before food. Treatment was allocated based on the clinic. Repeat gastroscopy was performed at 28±5 days. The data was largely not normally distributed (except age) and was analysed using Student's T(age), Chi squared, Friedman and Mann Whitney U test. Data is displayed as mean±SD (age) or median (IQR). Significance was assumed at $p<0.05$.

Forty horses met the inclusion criteria with 22 horses receiving misoprostol and 18 omeprazole and sucralfate. Majority of the lesions were multifocal, raised and haemorrhagic in appearance and subjectively this did not differ between groups. There were no differences between the groups regarding mean age (M 10±3.2 years; OS 12±4.3 years; $p=0.08$), gender ($p=0.21$), median glandular score at enrolment (M and OS; 2(IQR=2-3); $p=0.12$). A reduction in median glandular score was observed over time in both groups ($p<0.0001$; M to 0(0-1) and OS to 2 (1-2)). Improvement, defined as a decrease in score by at least 1 grade, occurred in both groups (M=95% and OS=89%; $p=0.60$). Healing, defined as a return to normal appearance occurred in 73% of horses receiving misoprostol, but only 22% of horses receiving omeprazole and sucralfate ($p=0.03$). Worsening of ulcer grade was not observed in any animal. No side effects were reported in any animal. In conclusion, the results of the study suggest that misoprostol is superior for healing when compared with a combined omeprazole and sucralfate for the treatment of EGGD and thus this drug warrants further large-scale study.

EXPERIMENTAL MODEL OF DUODENITIS PROXIMAL-JEJUNITIS IN HORSES INOCULATED WITH *CLOSTRIDIUM DIFFICILE* TOXINS

Luis Arroyo[1], Marcio Costa[2], Bruce Guest[1], Brandon Plattner[2], Brandon Lillie[2], J. Scott Weese[2]

[1]Department of Clinical Studies/Ontario Veterinary College/University of Guelph, Guelph, Ontario, Canada, [2]Department of Pathobiology/Ontario Veterinary College/University of Guelph, Guelph, Ontario, Canada

Duodenitis proximal-jejunitis (DPJ) is a clinical disease of horses characterized by signs that include depression, colic, ileus and endotoxemia. The cause remains unknown, however, toxigenic strains of *Clostridium difficile* have been isolated from refluxed stomach contents of DPJ cases, suggesting that this organism may play a role in the etiology of this disease. The aim of this study was to determine whether horses inoculated with *C. difficile* toxins develop similar clinical and histopathological changes to those seen in naturally occurring DPJ.

Six healthy mature horses were inoculated with *C. difficile* crude toxins harvested from a toxigenic strain ($A^+B^+CDT^+$) isolated from the reflux of a DPJ case. Horses were fasted overnight and toxins were inoculated into the gastric pylorus region via endoscopy. Horses 1 and 2 were inoculated with 0.5 L and 3 L of toxins, respectively. Horses 3-6 were administered 3 L of 2X concentrated toxins. All horses were closely monitored for up to 48 h or until euthanasia. Physical examination was performed at -24, -1, 1, 2, 3, 4, 5, 6, 8, 10, and 12 h, then every 4 h for a total of 48 hours post-inoculation. Horses were monitored for changes in general demeanor, heart rate, respiratory rate, rectal temperature, hydration status, gastrointestinal sounds and signs of colic/discomfort. The small intestine (SI) was examined by ultrasonography in each horse prior to toxin administration, then at 1, 3, 6, 12, 18 and 24 h post-inoculation. Before inoculation and before euthanasia, blood was collected for complete blood cell count, biochemistry profile and plasma fibrinogen assay and abdominal fluid was collected for cytological analysis and total solids assay/determination. Tissue samples were collected after euthanasia from stomach, duodenum, jejunum, ileum, cecum and ventral colon; all samples were formalin fixed, paraffin embedded and processed for routine histological analysis. All tissue sections were blindly assessed for lesions, and assigned a combined score for epithelial desquamation/necrosis, lamina proprial necrosis, villus blunting, inflammation, submucosal edema and hemorrhage.

Horses 1, 4 and 6 remained clinically normal throughout the observation period and horse 2 showed mild colic signs. Horses 3 and 5 became depressed, tachycardic, tachypneic, pyrexic, and developed mild signs of colic within 6 h of inoculation and were euthanized at 9 and 12 h, respectively. Ultrasonographically, horse 2 had transient distention of SI and horses 3 and 5 had amotile distended loops of SI (□5 cm in diameter) prior to euthanasia. Clinico-pathological changes included leukopenia and hemoconcentration.

Microscopically, lesions were found in all horses. The anatomical sites with highest combined scores in all horses were the duodenum and jejunum. In some sections, the SI mucosa was flattened and/or mildly dysplastic and less frequently neutrophils and/or fibrin was present within the superficial villar lamina propria. In horses 2, 4 and 6 the lesions were mild and the most consistent lesion was vascular congestion of the duodenum; this was also present in other sections of the gut. Horses 3 and 5 had the most severe lesions with vascular congestion, hemorrhage, edema and occasional neutrophils most prominent in the gastric and duodenal mucosa, but also to a lesser degree in the more distal SI.

In summary, the inoculation of healthy horses with a crude mixture of *C. difficile* A/B toxins elicited clinical signs in 2/6 horses, and histopathological lesions in all 6 horses similar to those reported in horses with naturally-occurring DPJ. The clinical signs observed in this subset of inoculated horses are typical though not specific of horses with the naturally occurring disease. Dose dependency or individual animal variation may play a role in response to administration of *C. difficile* toxins. DPJ may be a condition with multiple etiologies that result in the same clinical and pathological findings, and *C. difficile* may be one of the etiologies of this syndrome.

TRANSFORMING GROWTH FACTOR BETA AT THE HOOF LAMELLAR INTERFACE IN EQUINE LAMINITIS

Lori Gutzmann[1], Philip Johnson[1], Rajiv Mohan[2]

[1]University of Missouri College of Veterinary Medicine, Columbia, MO, USA, [2]Harry S. Truman Veterans Administration Hospital, Columbia, MO, USA

Transforming growth factor β1 (TGF-β1) is an important cytokine known to regulate wound healing, in part through effects on fibronectin (FN) and collagen α1, type III (COL3A1). These proteins have been identified in the equine hoof-lamellar interface, where the initiation cascade leading to laminitis occurs. Expression of FN and COL3A1 protein is stimulated by TGF-β1, which also uses these proteins for activation from latent TGF-β1 in an autocrine fashion. We sought to compare mRNA expression and protein levels of TGF-β1, FN and COL3A1, and map their distribution at the hoof-lamellar interface of healthy horses, and in horses with carbohydrate overload-induced (CHO) laminitis to localize and determine whether TGF-β1 is active during early stages of laminitis development.

Frozen tissue samples from twenty-four adult horses in three treatment groups (n=8) were used. Tissues from developmental (DEV) and Obel Grade 1 (OG1) stages of experimental laminitis were compared with sham-treated controls (CON). Hoof lamellar tissue extracts processed from frozen tissues were used to determine relative mRNA expression using real-time quantitative polymerase chain reaction (qPCR). Quantitative immunofluorescence (QIF) microscopy of non-fixed frozen sections was used to map protein distribution and perform a semi-quantitative analysis. QIF was accomplished by imaging a minimum of 10 non-overlapping sections from each animal, followed by processing using an ImageJ macro to quantify relative area, and statistical comparison of groups based on the mean values from each animal.

Quantification of mRNA did not differ between groups for gene expression of TGF-β1, FN, or COL3A1 when measured by qPCR. QIF demonstrated significantly increased TGF-β1 protein expression (mean fold change 1.3, p=0.025) in DEV tissues compared with CON, while FN and COL3A1 remained unchanged (mean fold change range 0.9 to 1.0). TGF-β1 protein was localized to primary and secondary epidermal lamellae (PEL, SEL), while FN and COL3A1 were diffusely present in primary and secondary dermal lamellae (PDL, SDL). COL3A1 showed more intense immunofluorescence in the region of SDL than other areas, and this relative intensity appeared similar between groups.

While gene expression did not change significantly, an increase in TGF-β1 protein expression was shown at the hoof-lamellar interface. TGF-β1 in the equine hoof-lamellar interface was localized to suprabasal cells of the epidermal layers, while FN and COL3A1 were diffusely distributed in the dermis. Localization of TGF-β1 to epithelium with increased protein expression during acute laminitis development suggests a role for TGF-β1 in acute CHO-induced laminitis.

ARE HORSES SUBJECT TO LONG-QT SYNDROME?

Rikke Buhl[1], Julie Liebmann[1], Dagmar Trachsel[1], Dirk Lebelt[2], Diana Stucke[2]

[1]*University of Copenhagen, Taastrup, Denmark,* [2]*Pferdeklinik Havelland, Beetzsee, Germany*

Preliminary data from horses castrated under general anesthesia.

Long-QT Syndrome (LQTS) is characterized by a delayed repolarization of cardiac cells resulting in prolonged action potentials and QT intervals on surface ECG. In humans it is associated with the predisposition to develop severe ventricular tachyarrhythmia, which can progress to sudden death. Altered function of K^+ channels either due to mutations (inherited form of LQTS) or due to interaction with drugs affecting the function of K^+ channels (acquired form of LQTS) are known in humans and dogs. LQTS has not yet been described in horses and the effect of hormones and various drugs on the duration of QT interval has only been scarcely described in this species. However, functional similarities between equine and human K^+ channels allow the hypothesis that LQTS might also be present in horses. The purpose of this study was to assess the effect of withdrawal of androgens on cardiac repolarization by measuring the QT interval before and after castration. Further, as the same medication and anesthetic protocol was applied on each horse; we also studied the effect of drug combinations on the QT- interval.

The study population consisted of 51 stallions of various breeds, mean age 2.3 years, referred to an equine private practice for castration. The peri-operative medication protocol included administration of trimethoprim-sulfadiazine (Sulfa-TMP) (Synutrim® 72 %: 5.5 mg/kg bodyweight (BWT) trimethoprim plus 25 mg/kg BWT sulfadiazine) po for 3 days starting in the morning before castration. For induction of anesthesia horses were sedated with 80 µg/kg BWT romifidine (Sedivet®) iv. Immediately after sedation, all horses received 1.1 mg/kg BWT flunixin meglumine (Flunixin 5%) iv. Anesthesia was inducted with 0.1 mg/kg BWT diazepam (Diazepam-ratiopharm) and 2.2 mg/kg BWT ketamine (Ketamine 10%). The anesthesia lasted about 15 min. For all horses, two ECGs were obtained the day before surgery. Three ECGs were recorded on surgery day; in the morning before any medication, 4 and 8 hours after surgery. Finally three ECGs were obtained on day 1 and 2 after surgery. Seventeen horses had ECG recordings obtained again, 1.5 to 3 years after castration. For all ECGs HR, RR intervals, QT duration from Q to the peak of the T wave (QT_{peak}), and QT duration from Q to end of the T-wave (QT_{end}) were measured manually with on-screen calipers by one observer. The HR corrected QT interval (QT_c) was obtained with the Framingham formula. ECG measurements taken before and until day 2 after castration were compared by one-way ANOVA with Dunnett's post hoc test and QT_c before castration was compared to those obtained after castration by unpaired Students t-test.

The QT_c interval were significantly longer 4 hours after castration in comparison to baseline ($P<0.001$), hereafter it returned to normal values. Also, when ECG intervals of the horses were measured 1.5-3 years after castration, a significant prolongation of the QT_c was found ($P=0.041$).

The increase of the QT interval after castration is in agreement with the findings in humans and dogs where increased QT interval due to decreasing level of androgens is reported. Secondly, profound prolongation of QT interval was found 4 hours after surgery. At that moment several of the administrated drugs could have affected the function of K^+ channels, but it is not possible to identify which drugs or which drug combination could have caused this prolongation. We hypothesized that the combination of an alpha-2 agonist and Sulfa-TMP might be the potential causative agents. Indeed, trimethoprim-sulfonamide combination is known to cause acquired LQTS in humans, and the combination of alpha-2 agonist and Sulfa-TMP has been suggested to cause arrhythmias and death in horses. Further studies focusing on the drugs and their combinations will have to elucidate their electrophysiological effect on K^+ channels in horses.

COMPARISON OF FIBRINOLYSIS IN PERIPARTUM AND NON-PREGNANT MARES USING MODIFIED THROMBOELASTOGRAPHY

Kira Epstein, Kelsey Hart, Steeve Giguere

University of Georgia, Athens, GA, USA

Hypercoagulation and hypofibrinolysis identified in peripartum women is believed to protect against peripartum hemorrhage. Although hypercoagulation has been identified in peripartum mares, little is known about fibrinolysis. Utility of measuring plasma components related to fibrinolysis for identifying changes in fibrinolysis is questionable. Clot lysis assays performed on activated (tissue factor) samples with induced lysis (tissue plasminogen activator) accurately reflect fibrinolytic potential in humans and dogs but have not been evaluated in horses.

This study was performed to characterize and compare the fibrinolytic potential of the healthy non-pregnant mares and peripartum mares using tissue plasminogen activator modified/tissue factor-activated thromboelastography.

Tissue factor-activated thromboelastography modified with tissue plasminogen activator (500 and 650 U/ml) was performed on plasma from 9 pregnant mares at 3-, 2- and 1-month pre-partum and 1-, 7-, and 30-days postpartum and on time-matched samples from 6 non-pregnant mares.

There was evidence of hypofibrinolysis at both tissue plasminogen activator concentrations with significantly increased maximum amplitude (peak clot strength) and clot lysis at 30 minutes post-MA (amplitude at 30 minutes post-MA/MA *100) and decreased lysis at 30 minutes post-MA (percent decrease in area under the curve compared to MA 30 minutes after reaching MA) in pregnant mares compared to non-pregnant controls. These changes were most frequently detected 1-month pre-partum and 1- and 7-days post-partum and had returned to similar to observed in non-pregnant controls by 30-days post-partum.

Further research on hypofibrinolysis in peripartum mares and the use of modified thromboelastography in horses with hemorrhage and being treated with anti-fibrinolytic drugs is warranted.

DIFFERENTIAL PROTEOMIC EXPRESSION OF EQUINE CARDIAC AND LAMELLAR TISSUE DURING INSULIN-INDUCED LAMINITIS

Allison Campolo, Shanell Shoop, Melody De Laat, Stephen Hartson, Veronique Lacombe

[1]*Oklahoma State University, Stillwater, USA,* [2]*Queensland University of Technology, Queensland, Australia*

Although insulin dysregulation is an important risk factor for laminitis, the mechanisms linking equine metabolic syndrome with laminitis remain elusive. To explore the relationship between these 2 conditions, we aimed to identify the molecular processes underlying hyperinsulinemia-induced laminitis. Using mass spectrometry (FDR=0.05), we evaluated the cardiac and lamellar proteomes from horses treated with a prolonged insulin infusion or a balanced electrolyte solution for 48 hours (n=4/group). All hyperinsulinemic horses developed clinical laminitis, while the controls did not. We identified 737 proteins in the lamellae tissue, 24 of which were differentially expressed ($P<0.05$) between the control and laminitic groups. Of the differentially expressed proteins, 8 were upregulated in the laminitic group. A STRING analysis of protein-protein interactions identified that these upregulated proteins were primarily involved in the coagulation and complement cascades, platelet activation and ribosomal function. In contrast, 13 of the differentially expressed proteins were downregulated in the laminitic group compared to the controls, and were primarily involved in focal adhesions and spliceosomes. Interestingly, 3 proteins uniquely expressed in the hyperinsulinemic group were involved in ribosomal function and B-cell-mediated immunity. Although we identified 538 proteins in the heart (mostly involved in metabolic pathways), none were significantly differentially expressed between groups. These data indicate that while hyperinsulinaemia induced, in part, microvascular damage, complement activation, and ribosome dysfunction in the diseased lamellae, a similar effect was not seen in the heart. In brief, this proteomic investigation of an equine model of hyperinsulinemia identified novel proteins, which may lead to the discovery of molecular biomarkers and/or therapeutic targets for endocrinopathic laminitis.

ASSOCIATION BETWEEN HYPERINSULINEMIA AND LAMINITIS SEVERITY AT THE TIME OF PITUITARY *PARS INTERMEDIA* DYSFUNCTION DIAGNOSIS

Elizabeth Tadros[1], Jennifer Fowlie[2], Kent Refsal[1], Judith Marteniuk[2], Harold Schott II[2]

[1]michigan State University College of Veterinary Medicine Diagnostic Center for Population and Animal Health, Lansing, MI, USA, [2]Michigan State University College of Veterinary Medicine Department of Large Animal Clinical Sciences, East Lansing, MI, USA

Hyperinsulinemia is the suspected component of insulin dysregulation having the strongest association with laminitis and occurs variably in equids with pituitary *pars intermedia* dysfunction (PPID). When present with PPID, hyperinsulinemia is a negative prognostic indicator; euthanasia is sometimes due to laminitis. We hypothesized that magnitude of hyperinsulinemia correlates with laminitis severity in PPID-affected equids. Furthermore, we hypothesized that owners can be unaware of mild chronic endocrinopathic laminitis.

Resting serum insulin concentrations, owner-reported laminitis history, and radiographic evidence of laminitis were determined in 38 client-owned horses and ponies with confirmed PPID. Laminitis severity was classified into 5 categories based on degree of distal phalangeal rotation. Non-laminitic equids had no prior history, hoof morphological, or radiographic evidence of laminitis. Animals were also categorized as normoinsulinemic (< 20 µU/mL), mildly hyperinsulinemic (20-50 µU/mL), and severely hyperinsulinemic (> 50 µU/mL). One-way ANOVA, *t*-tests, and Fisher's exact tests were performed.

While owners reported historical laminitis in 37% of animals, 76% were laminitic based on study criteria ($P = 0.014$). Owners reported laminitis more frequently in hyperinsulinemic versus normoinsulinemic animals; recognition increased with severity of hyperinsulinemia ($P = 0.025$). Mean insulin concentrations were higher in equids with moderate to severe radiographic laminitis versus those classified radiographically as normal to mild ($P = 0.031$).

Although radiographic abnormalities were present in most animals at the time of PPID diagnosis, chronic laminitis remained unrecognized by many owners. Owner awareness of laminitis increased with severity of hyperinsulinemia and higher insulin concentrations were detected in association with more severe radiographic changes.

IDENTIFICATION OF A GENETIC LOCUS ASSOCIATED WITH HEIGHT AND FASTING INSULIN IN WELSH PONIES

Molly McCue[1], Elaine Norton[1], Felipe Avila[1], James Mickelson[1], Ray Geor[2], Nichol Schultz[1]

[1]*University of Minnesota, Saint Paul, MN, USA,* [2]*Massey University, Palmerston North, New Zealand*

Pony breeds are thought to be susceptible to equine metabolic syndrome (EMS), a clustering of metabolic derangements, namely hyperinsulinemia, insulin resistance and adiposity. In particular, ponies are reported to have higher fasting insulin concentrations and are more insulin resistant than horses. In humans and mice, several alleles resulting in small skeletal size have also been associated with alterations in insulin and glucose metabolism. We hypothesize that loci affecting height in ponies have pleiotrophic effects on metabolic traits, leading to predisposition to EMS. Genome-wide association studies (GWAS) were performed for height and fasting insulin concentration in Welsh ponies (WP, n=232) and Morgan horses (n=286). A locus on chromosome 6 (ECA6) was significantly associated with height and fasting insulin in the WP, but not in Morgan. The same locus was also identified as a region undergoing positive selection in the WP using an F_{ST}-based statistic (*di*). Two candidate genes were identified in this region. *HMGA2* was previously associated with height in ponies as well as altered adiposity in mice. *IRAK3* has been associated with obesity and metabolic syndrome in humans. Haplotype analysis shows that the association for height and fasting insulin spans both candidate genes. These data suggest that a variant in *HMGA2* has an effect on height and insulin, or an *IRAK3* allele that affects insulin is "hitchhiking" along with the known height allele in *HMGA2*. Identification of functional alleles in *IRAK3* and *HMGA2,* and association between height and fasting insulin will be necessary to differentiate between these possibilities.

THE ASSESSMENT OF BEHAVIOURAL CHANGES DISPLAYED IN HORSES WITH EQUINE GLANDULAR GASTRIC DISEASE

Aleina Binns[1], Gayle Hallowell[1], Teresa Hollands[2], Jocelyn Habershon-Butcher[1], Mark Bowen[1]

[1]University of Nottingham, Loughborough, UK, [2]University of Surrey, Guildford, UK

The pathophysiology of Equine Glandular Gastric Disease (EGGD) is poorly understood. Clinical signs including behavioural changes displayed in horses with EGGD are relatively undefined and difficult to quantify. The objectives of this study were to compare clinically apparent signs and risk factors in horses with EGGD as well as identify horse owner and veterinary surgeon perceptions regarding Equine Gastric Disease (EGD). This Questionnaire based study was approved by the UoN ethics and welfare committee.

36 horses displaying clinical signs associated with EGD underwent gastroscopy. Owners of these horses completed a questionnaire regarding behavioural changes, medical history and husbandry factors. 77 veterinary surgeons and 144 horse owners completed separate questionnaires assessing knowledge surrounding EGD.

28% of horses presented for gastroscopy had clinically relevant EGGD. The only husbandry, exercise or medical factor associated with EGGD was time spent grazing at pasture; 80% of horses with confirmed EGGD had less than or equal to 5 hours grazing daily. Husbandry factors that have previously been linked to EGD that were not found to be associated with the prevalence of EGGD included concentrate feeding, timing of concentrate feeding in relation to exercise, roughage access when stabled and roughage quantity when stabled. Horses with a history of colic, a history of EGD or a recent history of NSAID therapy were not more likely to have EGGD. Of the horses with confirmed EGGD, 90% were geldings. Common clinical signs previously associated with EGD including an unexplained loss of condition, appetite change and a change in coat condition had no association with the occurrence of EGGD. Of the 10 horses with confirmed EGGD, 8 (80%) were showing signs of 'stress' and 7 (70%) had changes in behaviour. Behavioural changes that had no association with the occurrence of EDG include aggressive behaviour, stereotypical behaviour and showing signs of discomfort/pain. Of those presented for gastroscopy, 25 (71%) horses displayed a change in behavior and 18 (84%) were displaying signs of discomfort/pain. Furthermore, horses showing obvious signs of discomfort/pain were more likely to develop behavioural changes. Over 80% of horse owners had heard of EGD. Ownership of a 'competition' or 'leisure' horse had no impact upon horse owner knowledge surrounding EGD, or the source of advice acquired in different clinical scenarios. 67% of 'competition' owners and 60% of 'leisure' owners believed an incorrect cause of EGD to be correct and 86% of competition owners and 89% of leisure owners thought an incorrect clinical sign of EGD to be correct. Of the horses previously diagnosed with EGD, a veterinary surgeon did not diagnose 30% of these cases. 45% and 67% of veterinary surgeons used Omeprazole to treat mild and severe EGGD respectively. Over 60% and 90% of veterinary surgeons identified causes and clinical signs of EGD not supported by evidence. Level of veterinary training had no impact upon understanding surrounding EGD, nor the first diagnostic test of choice in different clinical scenarios. Where veterinary surgeons qualified had no impact upon the use of particular drugs in the treatment of EGD.

Clinical signs and behavioural changes associated with EGGD remain to be elucidated. Knowledge of veterinary surgeons and horse owners surrounding EGD lags behind evidence-based medicine.

PHARMACODYNAMICS OF A LONG-ACTING INJECTABLE FORMULATION OF OMEPRAZOLE IN THE HORSE

Ben Sykes[1], Claire Underwood[1], Krishnah Kathawala[2], Yunmei Song[2], Sanjay Garg[2], Mills Paul[1]

[1]*The University of Queensland, Gatton, QLD, Australia,* [2]*The University of South Australia, Adelaide, SA, Australia*

A recent study demonstrated that the duration of acid suppression achieved following oral administration of omeprazole may be inadequate for healing of equine gastric ulcer syndrome to occur in horses consuming a high roughage diet. The objective of this study was to investigate the pharmacodynamics of a long-acting injectable formulation of omeprazole.

Six adult, Thoroughbred horses fitted with percutaneous gastrotomy (PEG) tubes were used. Intra-gastric pH was measured at two regions for a 23 hour period each day using a data logger fitted to a pH probe that was inserted into the PEG tube. Median intra-day intra-gastric pH and the percentage of time that intra-gastric pH was greater than 4 (%tpH>4) were measured. A baseline measurement was performed on day 0 and a single, 2 g dose of a long-acting formulation of omeprazole was administered intra-muscularly on day 1. Measurement of intra-gastric pH was performed for a further 7 days. Median intra-gastric pH and %tpH>4 for days 1 – 7 were compared to day 0 using a repeated measures ANOVA with Dunnett's multiple comparisons. The study was conducted under an ethics permit from the NSW Department of Primary Industries.

Median intra-day pH was higher than day 0 for days 1-7 at measurement region 1 ($p<0.0001$) and for days 1-5 at measurement region 2 ($p=0.025$). The %tpH>4 was higher than day 0 for days 1-7 at measurement region 1 ($p<0.0001$) and measurement region 2 ($p<0.025$).

The formulation warrants further investigation as a once weekly injectable medication for the treatment of EGUS.

ENANTIOSELECTIVE BRONCHOPULMONARY PHARMACOKINETICS OF SALBUTAMOL IN HORSES

Sharanne Raidal[1], Kate Robson[1], Glenn Jacobson[2]

[1]School of Animal and Veterinary Sciences, Charles Sturt University, Wagga Wagga, NSW, Australia, [2]School of Medicine, University of Tasmania, Hobart, Tasmania, Australia

Salbutamol (albuterol) is a β2-agonist routinely used for bronchodilation in horses, typically administered as a 50:50 racemic mixture. Studies in horses and other species have suggested that the R- enantiomer is responsible for therapeutic effects, and the S- enantiomer may be responsible for airway hyper-responsiveness. Despite widespread usage, there is comparatively little known about the enantioselective pharmacokinetics for this agent in horses. The current study was undertaken to characterise enantiomers of salbutamol in bronchial epithelial lining fluid (ELF), bronchial epithelium, pulmonary tissue and peripheral blood. Twelve healthy horses were administered 1mg (10 x 100 µg actuations) of racemic (*rac-*) salbutamol via a pressurised metered dose inhaler with Aerohippus® equine aerosol chamber. Samples of ELF were collected 2, 5, 10 and 15 min following medication. Endoscopic biopsies of bronchial epithelia were collected approximately 20 to 25 min following treatment, and a pulmonary biopsy was taken at 30 min. Peripheral blood samples were collected at each sampling time. Salbutamol enantiomers were analysed by an enantioselective UPLC-MS/MS assay. Peak concentrations of salbutamol were approximately 400 ng/g ELF for both enantiomers. A reduction of approximately 50% was observed in samples collected over the 2 to 15 min sampling period, consistent with mucociliary clearance and tissue absorption, and no stereoselectivity was observed in the lung. Salbutamol was detected in bronchial epithelium and peripheral lung tissue in equal amounts for both R- and S- forms. In contrast to pulmonary findings, plasma concentrations evidenced enantioselectivity, with the S- enantiomer present at approximately 1.5x the concentration observed for R-. Our findings demonstrated local concentrations of salbumatmol in ELF sufficient to exert biological effect, and have established non-enantioselective uptake into bronchial and pulmonary tissues. Differences in enantiomer concentrations in peripheral blood are likely to reflect enantioselective differences in uptake into peripheral tissues.

IMMUNOPROTEOMIC ANALYSIS OF INHALABLE BARN DUST

Kathleen Ivester, Laurent Couetil

Purdue University, West Lafayette, Indiana, USA

Aeroallergens are implicated in the pathogenesis of both recurrent airway obstruction (RAO) and inflammatory airway disease (IAD), two chronic inflammatory diseases on the spectrum of equine asthma. However, the antigenicity of proteins present in airborne barn dust has not been explored. A technique termed "immunoproteomics," which combines immunodetection methods with mass spectrometry (MS)-based protein analysis, has been used to identify allergens involved in a number of human diseases. This report details the immunoproteomic identification of antigens in inhalable barn dust.

Soluble proteins from inhalable dust collected from the breathing zone of stabled horses were probed with bronchoalveolar lavage fluid (BALF) from healthy horses, horses diagnosed with RAO, and horses diagnosed with IAD. Differentially recognized protein bands were chosen for analysis by liquid chromatography (LC)- tandem MS/MS.

A 60kDa protein recognized by horses with RAO only was identified as the chaperonin GroEL, a protein shown to exert immunomodulatory effects in murine models of asthma. Additionally, a 37kDa protein recognized by all groups with varying intensity was identified as alcohol dehydrogenase (ADH), previously identified as a major fungal allergen in human studies. Protein identification by MS analysis was confirmed by western blot.

This method visually demonstrates the antigenic potential of barn dust and the differential recognition of airborne antigens between disease groups, offering an exciting new tool to examine the interaction between the environment and the pulmonary immune system of the horse.

EFFECTS OF FREE AND CARRIER-BOUND CORTISOL ON EQUINE NEUTROPHIL FUNCTION

Melanie Fratto, Kelsey Hart, Natalie Norton, Michelle Barton, Steeve Giguere, David Hurley

University of Georgia, Athens, GA, USA

Cortisol is a key anti-inflammatory hormone that increases in bacterial sepsis and circulates predominantly bound to cortisol binding globulin (CBG) and albumin. Only unbound cortisol was believed to be biologically active, but recent evidence suggests that CBG-bound cortisol also regulates inflammation. Both healthy and septic foals are CBG-deficient compared to adult horses, which could impair foals' ability to direct and temper inflammatory responses. The objective of this study was to evaluate the effects of free and CBG-bound cortisol on equine neutrophil function. We hypothesized that CBG would enhance cortisol-mediated suppression of neutrophil pro-inflammatory responses. Isolated neutrophils from 8 foals and 6 adult horses were exposed to *Staphylococcus aureus* antigen (SAA) alone and with hydrocortisone (HC), CBG, or both (CBG+HC). Inflammatory cytokine (TNF-α, IL-8) and reactive oxygen species (ROS) production were measured and compared among stimulants and between ages with linear mixed-effects models. CBG and CBG+HC inhibited ROS production in response to SAA in both foal and horse neutrophils, maintaining it at a level comparable to baseline production ($P \leq 0.060$-0.907). TNF-α production was not significantly different among stimulants ($P=0.284$). CBG+HC significantly ($P \leq 0.016$) increased IL-8 production by neutrophils in response to SAA in both foals and adults, although the response of foals was significantly greater that that of adults ($P<0.001$). These findings suggest that CBG directly modulates equine neutrophil responses both with and independently of cortisol, but effects are cytokine- and age-specific.

BONE MARROW TRANSPLANTATION AND EPIGENETIC MODULATION OF HEMATOPOIETIC PRECURSORS IN EQUINE COMMON VARIABLE IMMUNODEFICIENCY

M. Julia Felippe, Rebecca Tallmadge, Cleanth Toledano, Ute Schwab

Cornell University College of Veterinary Medicine, Ithaca, NY, USA

Our laboratory studies common variable immunodeficiency (CVID) in horses, a rare and fatal condition that manifests with late-onset impaired B cell production in the bone marrow (BM) and, consequently, hypogammaglobulinemia. Our studies have shown that the B cell production is halted at the pre-pro B cell developmental transition due to epigenetic aberration of essential lineage-specific transcription factor regulatory regions, with resulting gene silencing of PAX5.

In order to reestablish B lymphopoiesis, we performed hematopoietic stem cell transplantation (HSCT) in a Thoroughbred mare diagnosed with CVID using bone marrow from a major histocompatibility complex (MHC) compatible healthy female full-sibling donor, after non-myeloablative reduced-intensity conditioning treatment of the recipient based on cyclophosphamide and fludarabine. Chimerism (presence of donor genomic DNA) was detected at low level (< 1%) in the recipient, and peripheral blood B cell distribution increased from 1.5% baseline to 4%, but with no changes in serum IgG and IgM concentrations.

In order to overcome challenges and side effects using HSCT, our research group studies therapeutic strategies to correct aberrant epigenetic mechanisms in vitro for subsequent autologous transplantation of B cell precursors. Our overall hypothesis is that epigenetic aberrations can be reversed and B cell lymphopoiesis rescued in BM-derived hematopoietic stem cells of CVID-affected horses using inhibitors of DNA methyltransferase (5-azacytidine) and histone deacetylase (valproic acid).

$CD34^+$ BM cells from healthy control and CVID-affected horses were cultured in the presence of stromal cells and cytokines conducive to B cell differentiation, and characterized using immunocytofluorescence microscopy. The effect of epigenetic modulators was tested in these culture conditions: the presence of 5-azacytidine increased the distribution of $CD34^+$ cells from CVID-affected (20%) and healthy (10%) horses in a dose dependent manner (0.5 µM), whereas the effect of valproic acid was minor (<10%). Current research efforts include optimization of culture conditions of B lymphopoietic precursors of equine CVID patients for subsequent autologous transplantation.

SERUM AND CSF LYME MULTIPLEX RESULTS FOR NEUROLOGIC HORSES WITH AND WITHOUT NEUROBORRELIOSIS

Amy Johnson, Laura Johnstone

University of Pennsylvania, New Bolton Center, Kennett Square, PA, USA

Paired serum and CSF samples from horses with neurologic signs are frequently analyzed using the Lyme Multiplex assay to confirm or refute Lyme neuroborreliosis (LNB) as the cause of disease. The purpose of this retrospective study was to describe Lyme Multiplex results obtained for neurologic horses with postmortem diagnoses.

Seventy-eight horses with paired serum and CSF test results as well as subsequent postmortem nervous system examination were identified. Ten horses had a postmortem diagnosis of LNB, and 68 horses did not. Not all horses with LNB had positive serum or CSF results; only 6/10 had at least 1 positive value. Four of ten had at least 1 positive serum value, 5/10 had at least 1 positive CSF value higher than the corresponding serum value, and 3/10 had at least 1 CSF value 4-fold greater than the corresponding serum value.

Results were similar for the 68 horses that did not have LNB. Forty-three (63%) had at least 1 positive value on serum or CSF. Thirty-seven (54%) had at least 1 positive serum value, 28 (41%) had a positive CSF value higher than the corresponding serum value, and 11 (16%) had at least 1 CSF value 4-fold greater than the corresponding serum value.

Positive Lyme Multiplex values were common in neurologic horses, and did not adequately differentiate horses with LNB from horses with other disorders, regardless of whether absolute values or ratios were used. A better diagnostic technique is needed to diagnose LNB in the living horse.

USE OF ENRICHMENT AND QUANTITATIVE PCR TO IMPROVE DETECTION OF SALMONELLA IN REFERRAL HOSPITALS

Adam Krull, Carly Kanipe, Amanda Kreuder

Iowa State University, Ames, IA, USA

Rapid detection of Salmonella-shedding animals, along with environmental surveillance testing in large animal referral hospitals, is of utmost importance in avoiding disease outbreaks. Testing has traditionally relied on culture with enrichment with an average turnaround time of 5 to 7 days. Use of qPCR can decrease turnaround time, although detection of residual DNA following cleaning has led to false positives. Our studies aimed to determine whether an overnight enrichment followed by qPCR could be as highly sensitive and specific as enrichment culture. Experiments varied the type of enrichment media, the starting concentration of Salmonella, and the effects of disinfectants and residual Salmonella DNA on the assay. Buffered peptone water was found to be a superior enrichment media to tetrathionate broth, effectively amplifying as little as 1 colony-forming unit per 50 mL of starting Salmonella to detectable levels in less than 18 hours. In-vitro testing of commonly used disinfectants on known positive Salmonella environmental samples revealed inadequacies in the ability to completely eliminate live Salmonella. Based on known starting concentrations of bacteria, interpretive criteria were able to be assigned to ranges of CT values to categorize live Salmonella versus dead and demonstrated that by utilizing enrichment, the presence of live bacteria could clearly be determined. Our interpretive criteria were then compared to standard culture techniques and shown to be highly correlated. Results show that the use of qPCR following an overnight enrichment could have sensitivity and specificity on-par with enrichment culture over a much shorter time frame.

Sleep Disorders in Horses

Joseph J. Bertone, DVM, MS, DACVIM (LAIM)
Pomona, CA

INTRODUCTION

Sleep, patterns of sleep, and sleep behavior are, in general, not areas of concern in equine veterinary medicine. However, many practitioners have seen single to multiple cases of sleep disorder and have had few resources to develop diagnostic approaches and protocols. In addition, we have been limited to narcolepsy. That term and the lack of large case number reviews have hampered a closer inspection of these horses that present with partial drowsy collapse.[1,2] In addition, the physostigmine challenge test has been incorrectly applied as a diagnostic tool in equine narcolepsy.[3,4] This test's use in dogs is based on the unique pathophysiology of the disease in that species as compared to that disease in people and as yet an unknown pathophysiology in horses. In dogs the problem is the receptor type 2 gene. And in fact, hypocretin concentrations do not change.[5] In human beings, hypocretin concentrations are diminished to completely absent.[6]

There is a traditional lack of concern with sleep and its impact on performance and clinical outcome in the equine veterinary environment. Owners have commonly sought help outside of practicing veterinarians or university sources because few answers or expertise exist in this issue.[a] That is difficult to understand in light of the need for sleep's essential, sustaining, recuperative physiology. There is no definitive explanation for the necessity or the function of sleep in horses, or for that matter in most other species where the need for sleep time varies widely. Likely, it is a complex of metabolic and adaptive needs developed and dependent on the niche the species fills in evolution.[7,8,9]

Recent studies have suggested that the restorative function of sleep may be the result of increased convective exchange of interstitial and cerebrospinal fluid. These fluxes increase the rate of β-amyloid clearance. Therefore, the need may be to enhance removal of potentially neurotoxic waste products that accumulate in the awake central nervous system.[7] Sleep is commonly defined as a period of immobility in which individuals seem unresponsive to their environment. Sleep now is accepted as a behavior with clear physiologic necessities and when lacking has dysfunctional impacts. As a component of the day, it is estimated that horses require 2 hours of drowsiness, 3 hours of slow wave sleep and less than 1 hour of paradoxical sleep. This yields a need of approximately 4 hours of total sleep time.[9,10] However, a 24 hour sleep cycle is not the horse, or other, herd herbivore model in that herd animals can go for much longer periods without sleep.[8,9]

Sleep can be altered by environmental and physical factors. Horses assume recumbency in a comfortable environment. When moved to an uncomfortable environment, sleep can be adversely affected for several days at a minimum.[9-12] When horses are put outdoors, it may be one or two days before they first lay down.[9] This habituation may be socially facilitated. If one horse is familiar with the new environment, it will lie down and the others will follow.[10,11] The dominant animal is usually the first to lay down.[11,12] Observations of wild ponies indicated that recumbent resting behavior occupies 16% of the day in these animals.[11,12] This behavior reaches a peak between 12:00 AM to 4:00 AM. All ponies were never seen to be lying down simultaneously.[11] However, this contrasts with a study with feral Appaloosa horses where the entire herd was seen recumbent at one time during each day.[11,12]

Normal sleep behavior in horses has been well defined, although continued work is necessary.[9-13]
Horses that frequently partially collapse may be suspected of recumbent sleep deprivation.[1,2,13] Sleep disorders that lead to this behavior continue to been elucidated, but the causes of this behavior have not been reviewed in a large number of cases.

MATERIALS AND METHODS

The author experience and conclusions come from tabulation of four hundred and twenty (420) client internet referred cases from 2000 to 2016 for which the author has consulted. Inclusion criteria was the submission of video and images demonstrating the partial collapse behavior. At no time was or will there be any financial remuneration requested for consultation on these cases. The only requests made of clients and veterinarians in consultation were video recordings of episodes, images of horses and the environment, continued follow and case updates.

Figure 1 represents a static illustration of sleep deprivation behavior needed for entry.

Figure 1. A chronically sleep deprived horse. The image shows him as he becomes diffusely drowsy.[1,2,13] Rather than arousing and lying down (transitional phase), he ends in the position in the right image for a few seconds. Abruptly, the horse rises, wakes and cycles back if not disturbed. This occurs repeatedly. This horse had been seen rolling, but he would not lie down to go through the entire, required sleep cycle. This was a solitary individual whose abnormal behavior ceased on introduction of another horse to the facility.

RESULTS

Categories were developed out of case review and patterns evident. These are based on response to therapy, changes in environment as described in the discussion section.

Table 1 summarizes the cases as categorized.

Table 1. Summary of horses with behavior associated with sleep deprivation. Unclassified cases are those horses where no association could be identified. Classified are those where an association could be found.

	Total	Gelding	Mare	Sex ratio
Total	420	336	84	4
Unclassified	176	148	28	5.3
Classified	244	188	56	3.4
Physical discomfort		18	17	1.1
Late pregnancy			7	0
Monotony		14	12	1.2
Sleep terror		8	8	1
Lyme		4	7	0.6
Aggression displacement		49	1	49
Environmental insecurity		95	4	23.8

Horses were categorized as described in the discussion.

DISCUSSION

It needs to be clear that cases were categorized when owners indicated that response to an advised therapy or changes in management (e.g., additional herd mates, changes in stall configuration, minocycline treatment, pain management, etc.) were incorporated and resolved the behavior. In the case of sleep terror, the decision to categorize horses was based on videography that indicated the behavior.

The following clinical categories were developed out of case review. These are based on response to therapy, changes in environment, or changes in status as described in the results.

1. PAIN OR PHYSICAL DISCOMFORT ASSOCIATED BEHAVIOR: Horses were categorized into this group when these issues were clearly diagnosed. These varied from severe joint disease, polysaccharide storage myopathy, other musculoskeletal disease, gastric ulcers, abdominal adhesions, late pregnancy, etc. Cases that responded to non-specific pain management including non-steroidal anti-inflammatory drug administration, or a response to omeprazole were also included. Heavy late pregnancy mares were included in this group and invariably resolved post-parturition.

2. ENVIRONMENTAL INSECURITY ASSOCIATED BEHAVIOR: Horses were categorized into this group when issues in the environment were identified and changed and cases resolved. These included stall location and/or size changes, stall relocation, loss of other horses to which the patient seemed be behaviorally attached, light bulb wattage changes, blanket issues, window coverings, weather, and many more seemingly incidental environmental issues that when addressed lead to behavior resolution.

3. MONOTONY ASSOCIATED BEHAVIOR. Horses were categorized into this group when they manifest signs only when placed long term in cross-ties or forced to stand quietly for prolonged periods. These horses do not express this behavior at other times. Simple intermittent stimulation was effective in avoiding the behavior.

4. AGGRESSION DISPLACEMENT (AGGRESSIVE GELDING) ASSOCIATED BEHAVIOR: Horses were characterized in this group when they was a history or presentation of excessive and continuous aggression towards single or all horses in a group. These horses responded to the addition of an alpha female or, more rarely, an alpha male when available.

5. LYME DISEASE ASSOCIATED BEHAVIOR: Horses were characterized into this group when horses were positive for Lyme disease[b] and when no other category applied. These horses did not show evidence of joint pain. They also responded to therapy for Lyme disease.

6. SLEEP TERROR ASSOCIATED BEHAVIOR: Horses were characterized into this group when there was evidence of behavior similar to sleep terror behavior in human patients. This is a relatively new finding in the data set as all cases have been identified since January 2012. This may have a higher incidence, since night videography capabilities have become relatively inexpensive.

The nature of internet case referrals and the evolving approach to these cases requires no conclusions be drawn as to breed or geographic predilection, prevalence or incidence information. All are considered by the author to be heavily biased. In addition, technical capabilities of the owners and improving video capability have also evolved through the time of this data set collection and clearly effected the results. This is especially true in the case of sleep terror. This relatively recent find was identified due to submitted night-time video camera files and may explain many uncategorized cases in the past. An evolving protocol may also be an issue. Until recently, a screening for Lyme disease was not requested of cases. Again that may have impacted earlier diagnoses and classification because and accurate test was not available and it was not considered likely that this behavior could be induced via Lyme disease. This category has developed from one serendipitous case who was treated for the disease and responded. New cases have occurred due to the suggestion of adding this to the protocol and again response to therapy. More severe and chronic clinical signs were most likely over-represented in this case group as well. Another possible bias in recent data is that as time has passed, more veterinarians have become aware of this syndrome. More seem to deal with it without seeking consultation.

There were clear disproportionate sex predilections. Pain, monotony and sleep terror were evenly distributed across sex as would be expected. Lyme disease may have been over-represented by mares due to the small number of cases. There was clear sex predilection for geldings to mares in aggression displacement (49:1) and environmental insecurity (23.8:1). Interestingly, no stallions have been presented. This leads one to hypothesize that there may be sex hormone and herd structure issues with these 2 presentations that trend risk to geldings.

Sleep terror associated behavior was diagnosed when there was evidence of behavior similar to sleep terror behavior in human patients. This is a relatively new finding in the data set as all cases have been identified since June 2012. This may be at higher incidence and more elucidation of this syndrome may occur as night videography has becomes more common and inexpensive. Appeasing Pheromone Gel[c] has been applied in 2 recent cases and seems to have resolved the issue. Other recommendations include late night grain feeding, melatonin and diazepam have been attempted with varying success. More work is needed prior to making further recommendations.

An algorithm guidance for dealing with these cases is proposed (see Figure 2). Firstly, one has to be assured that one is dealing with this behavior firsthand, or captured on video. Historically, an important issue is whether or not the horse is seen to roll in the paddock or stall. This is important to assure the physical capability or incapability of the patient to become recumbent. Reluctance to roll should trigger a soft-tissue and orthopedic workup for pain. If no orthopedic issues are evident, then an abdominal evaluation is recommended, as enteroliths, abdominal adhesions, gastric squamous cell carcinoma, gastric ulcers have been implicated. The review also suggests that a night-time video and Lyme analysis may be called for. Monotony horses are those that usually occur in cross ties during braiding, or other long situations where the horse receives other stimuli. In these cases, intermittent stimulation seems all that is necessary. Late pregnancy mares that have difficult rising and acquiring recumbency have a self-limiting problem.

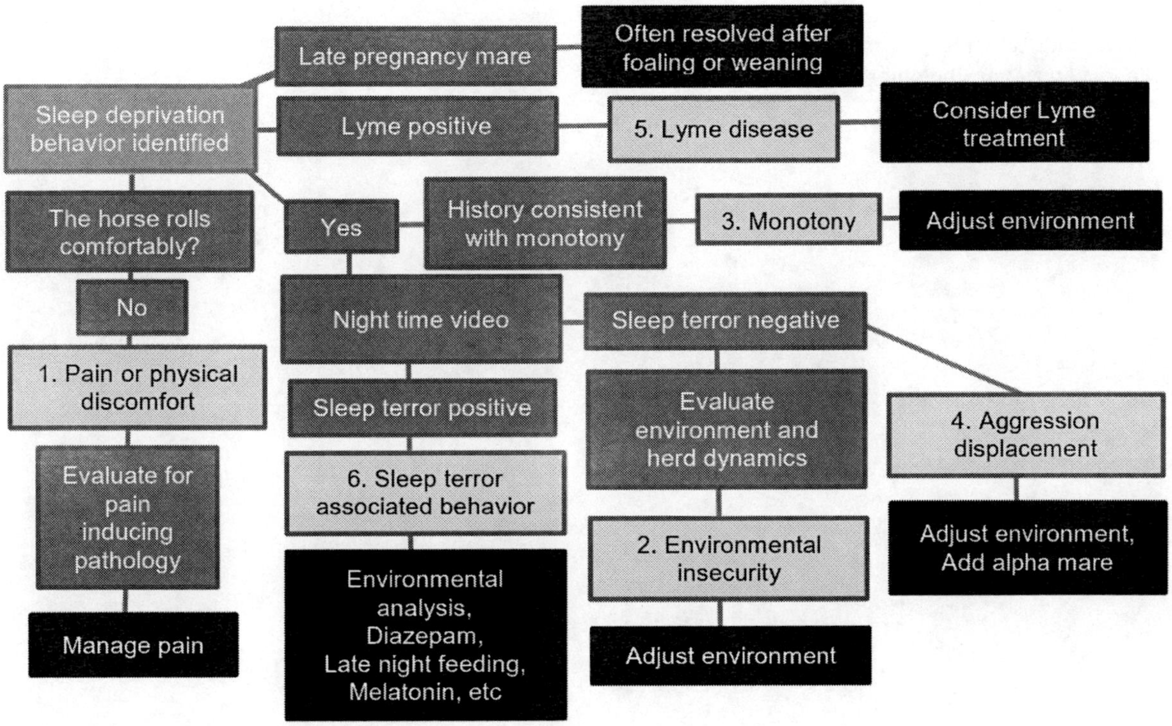

Figure 2. Proposed algorithm for diagnosis and managing of sleep deprivation in horses. Blue boxes are diagnostics recommendations and findings. Yellow boxes are diagnoses.

Once the behavior is documented, then the first line of questions should involve an exact description of when the episodes occur and if the horse is ever seen recumbent. Then follow through to a possible cause.

In conclusion, sleep deprivation is a widely known issue to horse owners and cases continue to occur. This paper seeks to categorize cases advised in internet consultation and seeks to provide field veterinarians with a greater diagnostic armamentarium to deal with these cases locally.

REFERENCES

1. Bertone, JJ. Sleep disorders in horses. In, Martin Furr, Stephen Reed (Eds). Equine Neurology. Wiley-Blackwell, Hoboken, NJ. *In Press.*
2. Aleman M, Williams DC, Holliday T. Sleep and Sleep Disorders in Horses. Proceedings AAEP scientific meeting. 2008;54:180-185.
3. Ludvikova, E., Nishino, S., Sakai, N., & Jahn, P. (2012). Familial narcolepsy in the Lipizzaner horse: a report of three fillies born to the same sire. *32*(2), 99-102.
4. Lacombe, V., & Furr, M. (2008). Differential diagnosis and management of horses with seizures or alterations of consciousness. In J. Mayhew, *Equine neurology* (pp. 91-92). Ames: Blackwell.
5. Hungs, M., Fan, J., Lin, L., Lin, X., Maki, R., & Mignot, E. (2001). Identification and functional analysis of mutations in the hypocretin (orexin) genes of narcoleptic canines. *Genome Res, 11*(4), 531-9.
6. Alakuijala A, Sarkanen T, Partinen M. Hypocretin-1 Levels Associate with Fragmented Sleep in Patients with Narcolepsy Type 1. Sleep. 2016 Feb 1: 481-15
7. Xie L, et al. Sleep drives metabolite clearance from the adult brain. Science 18;342(6156):373-7
8. Meddis R. Anim Behav. On the function of sleep. 1975 Aug;23(3):676-91.
9. Dallaire A. Rest behavior. Vet Clin North Am Equine Pract. 1986 Dec;2(3):591-607.
10. Ruckebusch, Y .: The relevance of drowsiness in the circadian cycle of farm animals. Anim. Behav., 20:637-643, 1972.
11. Keiper, R. R., and Keenan, M. *A.:* Nocturnal activity patterns of feral ponies. J. Mammal., 61(1):116-118, 1980.
12. Stebbins, M. C.: Social organization in free-ranging Appaloosa horses. Unpublished M.S. Thesis. Pocatello, Idaho State University, 1974.
13. Bertone JJ. Excessive drowsiness secondary to recumbent sleep deprivation in two horses. *Vet Clin North Am [Equine Pract]* 2006;22:157–162.

[a] Joseph Bertone: personal communication.
[b] Lyme multiplex. Diagnostic Laboratory. Cornell University
[c] Confidence EQ Equine Appeasing Pheromone Gel. CEVA Animal Health.

Differentiating Between Equine Botulism and Equine Dysautonomia

Neil P.H. Hudson, MA, VetMB, PhD, DEIM, DipVetClinStud, SFHEA, MRCVS
Edinburgh, United Kingdom

INTRODUCTION

Equine botulism and equine dysautonomia (grass sickness) are important diseases with significant morbidity and mortality in the horse.[1,2] The two diseases share some common features in terms of clinical signs and hence are often considered important differential diagnoses for each other.[3,4] In addition, the known etiological agent for botulism, *Clostridium botulinum*, has been postulated as a possible causative agent in the development of grass sickness, the definitive etiology of which at this stage still remains elusive. This presentation will describe the features of both diseases and highlight differentiating factors and areas of commonality both in terms of clinical diagnostic approach and also pathophysiological mechanisms and research.

EQUINE BOTULISM

Botulism is a syndrome of diffuse neuromuscular weakness which affects most mammals. The horse appears to be one of the most susceptible species. The disease is caused by one of the neurotoxins produced by the gram-positive spore-forming anaerobic bacterium *Clostridium botulinum*. The bacterium produces eight different serotypes of neurotoxins (A, B, C1, C2, D, E, F and G). Most cases of equine botulism appear to result from serotypes B and C,[5] although type A [6,7] and possibly type D cases have also been reported. This presentation will concentrate largely on the syndrome seen in adult horses.

Epidemiology and Pathogenesis

Botulism occurs by three routes of infection: 1) ingestion of pre-formed toxin ('forage poisoning'), 2) ingestion of spores leading to toxico-infectious botulism (usually in foals, largely in the USA; 'shaker foal' syndrome) and, more rarely, 3) wound botulism through *C. botulinum* contamination of anaerobic wounds with subsequent bacterial growth and release of toxin. The disease can occur in outbreaks, as has been seen in the USA and Australia, and as sporadic cases as seems to be the case more often in the United Kingdom. The highest risk to the horse exists in contamination of feedstuffs (e.g by rodent carcasses) and spoiled silage and haylage. There is significant geographical variation in serotype prevalence and associated disease. Type B predominates in the mid-Atlantic USA, [5] whereas type A is more common in the western USA.[6] Outbreaks in Australia have been associated with serotype C and serotype D [8,9].

Botulinum toxin acts primarily presynaptically at the peripheral cholinergic neuromuscular junction where it blocks the release of the neurotransmitter acetylcholine (ACh). Such neuromuscular blockade occurs in three stages: rapid irreversible binding of the toxin to receptors on the presynaptic nerve terminal; internalisation involving receptor-mediated endocytosis of toxin; and a final blocking step preventing release of ACh from the vesicle resulting in flaccid paralysis. Nerves with high efferent traffic such as cranial nerves and those supplying the anti-gravity (extensor) muscles are often the earliest and most prominently affected, explaining the resulting anatomical distribution of clinical signs. However, the neurotoxins do not affect the central nervous system or sensory nerves, facilitating differentiation from many other neurological diseases.

Clinical Signs, Diagnosis and Management

The actions of the Botulinum toxin explain the clinical manifestation of botulism as a symmetrical, diffuse neuromuscular weakness and accompanying cranial nerve paresis, progressing to generalised flaccid paralysis. It is not really possible to differentiate between the different serotypes of neurotoxin based on clinical signs. Clinical signs in the horse may include: signs of mild depression, dysphagia, reduced tongue tone and strength, reduced eyelid tone, mydriasis, sluggish pupillary light responses, reduced tail tone, muscle fasciculations, progressive myasthenia, recumbency, secondary complications (e.g. aspiration pneumonia, colic, urinary retention, decubital ulcers), respiratory distress and even death.

The diagnosis of botulism is largely made on characteristic history and clinical signs and through the exclusion of other possible differential diagnoses, such as equine grass sickness, equine motor neuron disease, equine herpes virus-1 myeloencephalopathy, seasonal pasture myopathy, hyperkalaemic periodic paralysis (HYPP) and West Nile Virus. Obtaining a definitive diagnosis is difficult and there are limited laboratory tests available to the equine veterinarian, especially in the United Kingdom (although this is not the case in North America). Laboratory tests include the (less acceptable) mouse bioassay and the recently reported quantitative real-time PCR for the detection of the neurotoxin gene of *C. botulinum* types A, B and C offered by the University of Pennsylvania, USA [10,11]. Needle electromyography (EMG) appears to have been of use in foals but less so in adult animals.

Antitoxin to neutralise circulating toxin is valuable in the treatment of botulism cases but although commercially available in the USA, this is not the case in the UK and many other countries. Antimicrobial drugs may be indicated, for instance for secondary aspiration pneumonia; it is important to avoid drugs that may make the neuromuscular blockade worse e.g. aminoglycosides, tetracyclines or procaine penicillin. Treatment of equine botulism involves significant supportive and nursing care such as feeding support and management of recumbency issues.

Prognosis and prevention

Generally the prognosis is poor and is associated with the severity of clinical signs, rapidity of onset, and the progression, or otherwise, to recumbency. The prognosis for standing horses receiving antitoxin is good (approximately 70% survival has been reported) but is significantly reduced (as low as 10% survival) without the provision of antitoxin. A recent retrospective study of 92 adult botulism cases from the University of Pennsylvania reported an overall survival rate of 48%.[1] In terms of prevention, there is a vaccine against type B toxico-infectious botulism that is commercially available in the USA. Careful monitoring of equine feedstuffs for signs of spoilage etc. is an important managemental step.

EQUINE DYSAUTONOMIA (GRASS SICKNESS)

Equine dysautonomia (grass sickness) is a polyneuronopathy affecting both the central and peripheral nervous systems of horses. As the name suggests, grass sickness almost exclusively affects grazing horses resulting in the development of a characteristic array of clinical signs, most of which can be attributed to neuronal degeneration in the autonomic (ANS) and enteric (ENS) nervous systems[3,12]. Despite extensive research efforts since grass sickness was first reported over a century ago, the precise etiology of this frequently fatal disease remains elusive.

Etiological hypotheses

Since grass sickness was first reported, a vast array of etiological hypotheses have been proposed and addressed experimentally; yet to date, the definitive cause still remains unclear. Historical routes of investigation have included the contributing role of alsike clover (*Trifolium hybridum*) ingestion, cyanogenic glycosides in white clover, *C.perfringens* enterotoxicity, fungi and *C. botulinum* toxico-infection. The results of a study reported by Gilmour *et al.*[13] were confirmatory of the presence of a toxic component within the serum of acute, but not chronic, grass sickness cases, whereby the intra-peritoneal injection of serum into recipient ponies (but not other species) induced the classic neuropathology within autonomic ganglia but not clinical disease. To date, the identity of such a serum derived neurotoxic compound remains unknown.

Most recently, the principal research focus has been the investigation into the role of *C. botulinum*, an area of study which also dominated grass sickness research efforts immediately prior to and during the early 1920s, whereby it was hypothesised that grass sickness resulted from a toxico-infection (not intoxication) with *C.botulinum (previously termed Bacillus botulinus)*. Recently, an increased body of experimental evidence demonstrating an association with *Clostridium botulinum* C1 neurotoxin has been generated, initially via the significantly greater frequency of detection of *C.botulinum* C1 neurotoxin in the ileal contents and/or faeces of grass sickness cases compared with healthy horses. However, a similar difference in detection frequency of *C. perfringens* between grass sickness cases and controls and an increase in the number of other Clostridial species in grass sickness has also been reported, thus potentially reflecting a generalised Clostridial overgrowth in a dysfunctional bowel. In comparison, the identification of a lower antibody status to *C. botulinum* C1 neurotoxin and *C. novyi* (surrogate marker of *C. botulinum* types C and D) in grass sickness cases compared to control horses and the recognition that a low antibody status to these components (and *C. botulinum type C)* constitutes a risk factor for grass sickness was more suggestive of a causal association with this bacteria. Such a causal role for *C. botulinum* was also supported by historical evidence of two reportedly efficacious vaccine trials conducted in 1922 and 1923. However, despite evidence that, unlike other *C. botulinum* neurotoxins, the C1 serotype can induce neurodegeneration in cultured spinal cord cells *in vitro*, there remains a lack of evidence that *C. botulinum* neurotoxin can induce the classic histopathological lesions characteristic of grass sickness *in vivo*. At present, especially upon consideration of the association between grass sickness and *C. botulinum* serological status and the failure to experimentally induce grass sickness in a series of challenge studies (R.S.Pirie, *personal observation)*, further investigation of this hypothesis is focussed on establishing once again whether vaccination against *C. botulinum* type C affords any protection against disease. A UK nationwide vaccine trial commenced in 2014 with a span of 2 to 3 years involving approximately 1100 horses and ponies[14]. The results obtained from this trial may support or refute a role for *C. botulinum* C and D neurotoxins in grass sickness etiopathogenesis.

Epidemiology

Since first reported from 1909 in Eastern Scotland, grass sickness has been reported throughout the UK and many Northern European countries. Grass sickness was recently described in a mule in the USA [15], a country mainly considered to be free of the disease. Interestingly, a clinically and pathologically identical disease known as Mal Seco, also occurs in South America, predominantly Chile. As the name implies, grass sickness has a strong association with grazing with only a few cases reported to have had no exposure to pasture. In addition to grazing, numerous epidemiological studies have been instrumental in identifying a variety of other risk factors, at the horse-, premises-, management- and climate-level. Although a degree of inconsistency exists between studies in relation to certain reported risk factors, those which are consistently reported include age, recent movement to new pasture or premises and time of year, with a peak incidence in the northern hemisphere during the Spring and early summer months. When considered in combination, many of the identified associations support the role of a resilient soil-borne agent which, under certain conditions, results in the production and/or liberation of a putative neurotoxin to which the horse may mount a variable mucosal and/or systemic immune response.

Pathology and pathogenesis

The gross pathological manifestations of grass sickness largely reflect the dysautonomic effects on alimentary function whereby specific gross pathological findings are largely dependent on the severity of intestinal dysfunction present *ante mortem*.

The histological features of grass sickness are almost exclusively restricted to neuronal degeneration, something that differentiates the disease from botulism where there are no neuronal histopathological changes. Typical histopathological findings in grass sickness include extensive chromatolysis with loss of Nissl substance, eccentricity or pyknosis of the nuclei, neuronal swelling and vacuolation, accumulation of intracytoplasmic eosinophilic spheroids and axonal dystrophy. Histological evidence of neuronal pathology is evident in both the peripheral (including enteric) and central nervous systems, although lesions are most evident in the prevertebral and paravertebral ganglia of the autonomic nervous system and enteric neurons (myenteric and submucous plexuses). Additionally, grass sickness cases may have a reduced number of interstitial cells of Cajal (ICC), the pacemakers and mediators of neurotransmission in the gastrointestinal tract, at both the myenteric plexus and circular muscle regions of both ileum and pelvic flexure of the large colon.[16]

Clinical Signs and Diagnosis

The clinical presentation of grass sickness includes varying degrees of colic, ileus, inappetance, tachycardia, weight loss, drooling of saliva, dysphagia, patchy sweating, ptosis, muscle fasciculations and *rhinitis sicca*[3]. Grass sickness can be classified according to the duration and severity of the clinical signs. Acute grass sickness usually results in death in less than 2 days, subacute grass sickness in 2-7 days, and in chronic grass sickness, the duration of the disease is over 7 days. Although this classification scheme is useful, it should be recognised that there is a continuum between these categories with some overlap observed. The prognosis for acute grass sickness and subacute grass sickness is hopeless. Despite some authors' views that the prognosis for chronic grass sickness is equally hopeless, in certain cases recovery is possible with intensive nursing (approximately 40% survival). A presumptive diagnosis of grass sickness can be made upon consideration of the nature and progression of clinical signs, history and epidemiological information and following the systematic elimination of alternative diagnoses. Examination *per rectum* and transabdominal ultrasonography are helpful. Clinical pathological evaluation of blood and peritoneal fluid samples broadly are useful as adjuncts without giving a definitive diagnosis. As previously mentioned, bilateral ptosis is a recognised feature of grass sickness. Confirmation of the presence of smooth muscle paralysis underlying the ptosis, and therefore provision of additional diagnostic information, can be achieved by the topical administration of 0.5 per cent phenylephrine to the cornea which results in temporary reversal of this clinical sign;[17] it is important to note that this test is non-specific and may be positive in diseases such as botulism. Definitive diagnosis can only be confirmed by demonstration of characteristic histopathology in the ENS or autonomic ganglia examined at *post mortem*. *Ante mortem* histopathology of ileal biopsies taken at laparotomy can be diagnostic.

CONCLUSIONS: COMPARING BOTULISM AND GRASS SICKNESS

Clearly, there is some degree of overlap in some of the clinical signs of the two diseases [18], but careful attention should be paid to some helpful historical and clinical differentiating features between the two diseases (**Table 1**), in addition to a school of thought that grass sickness may be in some way caused by the same known causative agent of botulism, namely *C. botulinum*. However the ubiquitous nature of *C. botulinum* and the susceptibility of horses to developing botulism does raise some questions regarding the possible link when one considers the more specific geographical distribution of grass sickness.

Table 1. Historical and clinical differentiating features between botulism and grass sickness. * = if indicated or available.

Differential Diagnosis	Clinical Similarities	Clinical and Historical Differentiating Features		*Ante mortem* Diagnostic Tests in Grass Sickness
		In the Differential diagnosis (grass sickness)	In Botulism	
Equine grass sickness	-Salivation -Dysphagia -Ptosis -Muscle tremors -'Tucked up' abdomen -Narrow-based stance -Colic	-Tongue normal -Access to grass -Rhinitis sicca -Patchy sweating -Tachycardia -Gastrointestinal signs more severe	-Tongue paresis/paralysis -Access to big bale silage -Profound myasthenia -Muscle tremors stop when recumbent -Mydriasis -Reduced tail tone	-Phenylephrine eye drop test -Ileal biopsy at exploratory laparotomy*

In addition, recent histopathological and immunohistochemical studies have shown pathology in the autonomic and enteric neurons in grass sickness but not botulism[19,20]; also there is increased expression of SNARE proteins in the neuronal perikarya in grass sickness but not botulism, suggesting that the former may not be caused by botulinum neurotoxins.[20] Therefore, although there are many similarities between these two diseases, important differences exist that help the clinician and scientist differentiate between the two, although this can often be difficult.

REFERENCES

[1]Johnson AL et al. J Vet Intern Med 2015;29:311. [2]Pirie RS et al. Equine Vet J 2014;46:545. [3]Pirie RS. Clin Tech Equine Pract 2006;5:30. [4]Stratford CH et al. Equine Vet Educ 2014;26:441. [5]Whitlock RH et al. Clin Tech Equine Pract 2006;5:37. [6]Johnson AL et al. J Vet Diagn Invest 2010;22:165. [7]Ostrowski SR et al. J Vet Diagn Invest 2012;24:601. [8]Kelly AR et al. Equine Vet J 1984;16:519. [9]Hutchins R Aust Equine Vet 1994;12:54. [10]Johnson AL et al. Vet J 2012;194:118.. [11]Johnson AL et al. 2014;199;157.[12]Cottrell DF et al. Neurogastroenterol Motil 1999;11:79. [13]Gilmour JS et al. Res Vet Sci 1977;22:1. [14]Ireland J Vet Rec 2014;175:114. [15]Wright A et al. Equine Vet J 2010;42:170. [16]Hudson NPH et al. Autonom Neurosci: Basic and Clinical 2001; 92:37. [17]Hahn CN et al. Vet Rec 2000;147:603. [18]McGorum BC et al. Vet Rec 2000;152:334. [19]Waggett BE et al. J Comp Path 2010;142:284. [20]McGorum BC et al. Equine Vet J 2015;doi:10.1111/evj.12543.

Myofibrillar Myopathy: A Newly Discovered Cause of Exertional Myopathy in Arabian and Warmblood Horses

Stephanie J. Valberg DVM, PhD, DACVIM, ACVSMR and Erica C McKenzie BVMS, PhD, DACVIM, ACVSMR
East Lansing, MI

Research into the causes of exertional myopathies in horses has focused on dysfunction of glycogen metabolism, oxidative metabolism, and excitation contraction coupling [1]. Disturbances in these pathways are characterized by the classic clinical signs of 'tying-up'; painful muscle contractures during exercise and concurrent marked elevations in serum creatine kinase (CK) activity (indicating damage to the muscle cell membrane). Two specific equine genetic mutations have been identified from this research; the glycogen synthase 1 mutation (*GYS1;* G309H) responsible for polysaccharide storage myopathy and the ryanodine receptor mutation (*RYR1;* R2454G) responsible for malignant hyperthermia [2,3]. Genetic tests have greatly facilitated identifying the cause of tying-up in some breeds, yet there are many equine exertional myopathies for which a cause is unknown. Particularly challenging are cases of exercise intolerance that are not typified by concurrent elevations in serum CK activity. Recognition of such subtle primary muscle disorders is difficult given the horse's inability to verbalize symptoms, and a lack of a clinical marker for muscle weakness or pain. Based on the number of Warmblood and Arabian horses with exertional myopathies that lack a primary diagnosis, we have expanded our investigations into causes of exercise intolerance in horses to include pathways not typically associated with exertional rhabdomyolysis, yet likely to impact athletic performance as a result of weakness or muscle pain. One such avenue of exploration involves perturbations in the sarcomere Z-disc and muscle cytoskeleton. Mutations in proteins involved in these structures are increasingly being recognized as a cause of late onset skeletal muscle weakness in humans [4-6].

The sarcomere

Although actin and myosin are the main contractile elements of the sarcomere, numerous other proteins act as scaffolds, control ultrastructure composition, regulate muscle contraction, and transmit tension between sarcomeres to the whole myofibril. The Z-disc defines the lateral borders of the sarcomere and actin filaments extend from the Z-disc into adjacent sarcomeres crosslinked by α-actinin molecules. Z-discs have one of the most complex macromolecular structures in biology composed of at least 200 novel Z-disc proteins and interacting partners [7]. Recent research has shown that Z-discs are not only vital to structural stability but also act as a nodal point for intracellular signaling, with molecular links to metabolic sensors such as AMP kinase, mitochondria, the t-tubular system, the sarcoplasmic reticulum and E3 ubiquitin ligases which are important for protein turnover and autophagy [7,8].

The cytoskeleton

Intermediate filaments such as desmin maintain the precise alignment of Z-discs across the width of the fiber and tether mitochondria to sarcomeres, the peripheral myofibrils to the sarcolemma and the sarcomeres to the nuclear membrane. Desmin plays a critical role in maintaining the integrity of the cell, ensuring force transmission and providing a pathway for signaling. Knocking out desmin in mice does not impair muscle formation, however, it does result in myofibril misalignment, nuclear and mitochondrial malpositioning within the cell, and susceptibility to damage following contraction [9]. Immunohistochemical stains for desmin have been used as a marker for skeletal muscle in veterinary medicine, and because desmin is the first intermediate filament to be produced in developing muscles, desmin stains have also been used as a marker for regenerative muscle fibers.

Myofibrillar myopathies

Myofibrillar myopathies (MFM) represent a group of dystrophic muscle diseases with a similar morphologic phenotype characterized by disintegration of the Z-disk, accumulation of myofibrillar degradation products and ectopic expression of multiple proteins including desmin and alpha beta crystallin [4-7]. In humans, mutations in genes that encode Z-disc associated proteins such as desmin, alpha beta crystallin, myotilin, filamin and Z-disc alternatively spliced PDZ motif (ZASP,) among others, are known to cause MFM [4-7]. The specific genetic basis for many other MFM in humans, diagnosed on the basis of muscle histopathology, remain unknown. The clinical presentation of MFM in humans can be highly variable. In adults, MFM usually present with late onset progressive muscle weakness with or without signs of cardiomyopathy. Serum creatine kinase activity can be normal or up to 7 fold above the normal range [4-7].

Investigation of Arabian and Warmblood horses

The hypothesis for this study was that a subset of horses with exercise intolerance and exertional muscle pain have an underlying defect in the Z-disc and associated structures. This was derived from reviewing hundreds of muscle biopsy submissions from Arabian and Warmblood horses that were submitted to the Neuromuscular Diagnostic Laboratory now located at Michigan State University. Arabian horses were comprised largely of endurance horses that had a history of intermittent elevations in serum CK activity whereas Warmblood horses typically had a history of reluctance to collect, exercise intolerance and undiagnosed subtle lameness without demonstrable elevation in serum CK activity. Immunohistochemical (IHC), immunofluorescent and electron microscopy methods were used to investigate the cytoskeleton and Z-disc. Muscle samples for electron microscopy were obtained from horses that had been resting for at least 48 h before biopsy to avoid any potential effects of exertion on morphology.

Muscle histopathology

In a subset of muscle biopsies from the Arabian and Warmblood horses, mature muscle fibers with internalized myonuclei, aggregates of amylase-sensitive cytoplasmic glycogen and large aggregates of cytoplasmic desmin in scattered mature muscle fibers were identified. The desmin aggregates were found in type 2A and 2X muscle fibers, coinciding with positive staining for the Z-disc protein alpha beta crystallin. The desmin aggregates strongly resembled those described in MFM of humans and dogs. With few exceptions, classic regenerative fibers (small basophilic fibers with prominent nucleoli) were absent in the muscles examined. When present, small regenerative fibers had uniform deep staining for desmin in contrast to the multiple large aggregates of desmin found in mature myofibers of horses with suspected MFM. Results for Arabian horses have been published [10]

Electron microscopy

Most areas of the muscle samples evaluated had a regular confluent alignment of sarcomeres. Z-disc streaming, rupture through the Z-disc where triads of sarcoplasmic reticulum remained intact and accumulation of beta glycogen particles in the ruptured sarcomeres were found in scattered regions of the samples examined. Remnants of Z-disc proteins, abnormal mitochondria, degenerating myeloid bodies and flag like semi-dense extensions of protein streaming for the Z-disc were also identified in samples from some horses. Z-disc degeneration was similar to that described for MFM in humans, however, the amount of ectopic granulofilamentous material often described in human MFM was less than that reported in human MFM [4-7]. Streaming of Z-discs occurs following very strenuous eccentric exercise, however, eccentric exercise was not likely the cause of the Z-disc streaming in the horses studied as they not been exercised prior to obtaining the muscle biopsy. The large accumulation of beta glycogen particles within disrupted myofibrils could have given rise to the cytoplasmic aggregates of PAS-sensitive glycogen seen in histochemical stains. Results for Arabian horses have been published [10]

Clinical presentation

Horses with suspected MFM did not have the classic presentation of late onset progressive muscular weakness reported in human MFM. Rather Warmblood horses showed reluctance to engage and evidence of muscle pain with exercise whereas Arabians had a history of intermittent muscle pain and intermittent elevations of serum CK with endurance riding. The lack of concordance of clinical signs between horses and humans does not rule out a Z-disc disorder in horses. Whereas humans with weakness or subtle muscle pain can elect to be sedentary, horses are forcibly exercised often beginning at 3 to 5 years of age. Thus, exercise intolerance would be more likely to be a feature of an equine myopathy. In addition, there are a host of proteins in the Z disc which could be dysfunctional in a potential MFM of horses, many of which could involve the link between Z-discs, the cytoskeleton and mitochondrial function. The link between Z-disc proteins and performance is further underscored by the finding that a mutation in the gene encoding the Z-disc protein α-actinin-3 expressed in fast-twitch skeletal muscle fibers is linked to elite sprinting capacity in human athletes [11]. Thus it is possible that abnormalities of yet to be identified Z-disc/cytoskeletal proteins could cause exercise intolerance in horses.

Conclusions

Results of this ongoing research suggest that muscle pain and exercise intolerance in a subset of Arabian and Warmblood horses may be related to damage to the Z-disc resulting in proliferation of the cytoskeletal protein desmin. Further studies of Z-disc proteins and regenerative markers as well as gene expression and gene sequencing are underway to determine whether the ectopic accumulation of desmin and other proteins that typify this subset of horses represents a true myofibrillar myopathy.

REFERENCES

1. Valberg SJ. Proceed. *American Assoc Equine Pract* 2012: 85-123.
2. Mccue MM. et al *Genomics* 2008;91(5):458-66.
3. Aleman MH. *Muscle Nerve* 2004;30:356–365.
4. Selcen, D. *Neuromuscul. Disord.* 2011;21: 161-171.
5. Olive, M. et al. *Neuromuscul. Disord.* 2011;21: 533-542.
6. Selcen, D et al. *Brain* 2004:127: 439-451.
7. Knoll R et al. *J Biomed, Biotech* 2011; 1-12.
8. Gregor M et al. *J Cell Sci* 2006:119:1864-1875.
9. Carlsson L, Thornell L-E. *Acta Physiologica Scandinavica* 2001:171(3);341–348.
10. Valberg SJ et al *Equine Vet J* 2015 Aug 1. doi: 10.1111/evj.12493.
11. Yang et al. *Am J Hum Genet* 2003;73;627-631

WALTHAM Collaborative Research into Equine Laminitis and Obesity

Pat Harris, MA, PhD, DECVCN, VetMB, MRCVS
Waltham-on-the Wolds, UK

INTRODUCTION

Laminitis, is not a 'modern' condition; it has been recognised for well over 2000 years. Aristotle even referred to it around 350 BC as 'Barley Disease' presumably because it was associated even then with excessive consumption of grain. However, recently there appears to have been an increased recognition of the importance of pasture-associated laminitis (PAL) and similarly, whilst an obese horse is not a new phenomenon, the number of obese animals especially within the pleasure/leisure arena appears to be increasing and it is becoming a globally recognised important welfare issue.

WALTHAM's journey into equine laminitis and obesity started with the collaborative research that we undertook initially with Prof David Kronfeld and then Prof Ray Geor and their students at the Virginia Tech MARE Centre in the late 1990's and early 2000's. Around the same time we also started collaborating with Prof Jonathan Elliott, Dr Simon Bailey and Dr Nicola Menzies Gow and their team at the Royal Veterinary College, London (RVC). In the mid-2000's WALTHAM established the WALTHAM international Laminitis Research consortium followed by the WALTHAM international Obesity consortium. These consortia bring together world-leading equine veterinary, nutrition and research experts interested in collaborating on the important topic of laminitis, and have resulted in many key publications in these areas.

This talk will provide an overview of some of the collaborative research that WALTHAM has carried out into equine laminitis and obesity, in particular in collaboration with members of the WALTHAM International consortia. It will concentrate on laminitis and highlight our work looking at how to predict which animals may be at increased risk of developing laminitis, and how we may reduce this risk.

OUR COLLABORATIVE RESEARCH INTO PREDICTION OF LAMINITIS RISK

Insulin and glucose responses

With respect to the naturally occurring disease perhaps the most important concept that may enable us to solve the question 'why this individual rather than that one ' is that any vascular, inflammatory etc. perturbations, that may occur as a result of the final triggering factor, may be exacerbated by pre-existing metabolic and/or endocrine imbalances within that individual animal. Insulin resistance or insulin dysregulation may, for example, lower the threshold for the development of laminitis in the face of hindgut disturbances that initiate the cascade of events that ultimately lead to the clinical expression of laminitis. Alternatively the insulin dysregulation itself may initiate the final cascade in certain individuals with underlying pre-existing imbalances/predisposition.

Much of our work in this area therefore has concentrated on the role of insulin, insulin resistance and insulin dysregulation in laminitis risk. This required us initially to modify the frequently sampled intravenous glucose tolerance test for use in the horse as an indicator of the level of insulin sensitivity [1] and led to a series of studies concentrating on static and dynamic tests looking at the role that insulin, and the insulin response to various stimuli, may play in animals that have not had laminitis, or have had several episodes of laminitis or go onto develop laminitis. It also led to a couple of studies evaluating insulin determination and confirmed the variation with different analytical methods and the need for insulin-depleted equine serum (IDS) to be used as a diluent [2,3,]

Initially it was thought that diagnosis of 'insulin resistance' (IR) would help identify laminitis-prone animals and therefore enable appropriate preventive countermeasures to be initiated (e.g. [4]). However, such a diagnosis in equines is difficult, especially as quantitative measurements of insulin sensitivity using a euglycaemic-hyperinsulinaemic clamp or computerised minimal model analysis of the frequently sampled intravenous glucose tolerance test (fsivGTT) are not practical or cost effective for use in the field. Basal insulin concentration is affected by individual, diet, level of pain, stress plus diurnal and seasonal variation, and is unsuitable to estimate insulin sensitivity. Proxy measurements that we investigated[5] of insulin sensitivity (reciprocal of the square root of insulin & quantitative insulin sensitivity check index and insulin secretory response (modified insulin-to-glucose ratio) based on basal glucose and insulin concentrations, suggested early promise as a practical method of evaluating level of insulin resistance and possibly predicting laminitis risk. However, whilst in another study we showed that these could distinguish between the groups of normal (NP) and previously-laminitic (PLP) ponies, none of the proxies could accurately predict predisposition to laminitis in individual animals [6]. Seasonal differences were also apparent, suggesting seasonal changes in insulin sensitivity and insulin secretory response. In addition, this confirmed previous findings that not all laminitis-prone animals are insulin resistant and not all insulin resistant animals are prone to laminitis.

Considering the potential role of insulin dysregulation rather than insulin resistance *per se* led us to evaluate the potential value of dynamic tests looking at blood glucose and insulin concentrations following the feeding of supplementary water soluble carbohydrates. Laminitis-prone ponies, some of which did not have elevated basal plasma insulin concentrations but were shown to be IR using the fsivGTT, demonstrated exaggerated increases in serum insulin concentrations in response to the feeding of inulin, a type of fructan[7]. We therefore undertook a more in-depth study to assess the glycaemic and insulinemic responses of seven normal (NP) and five previously laminitic (PLP), mixed breed, native UK ponies when fed glucose, fructose and inulin (each at 1g kg^{-1} day^{-1} for 3 days) or no supplementary carbohydrate (control) in spring and autumn after a 7 day adaptation to a pasture or hay diet. The results[8] suggested that the measurement of peak insulin 2 hours after feeding a single dose of glucose (1 g kg^{-1} : oral glucose test [OGT]) may be a simple and practical way to aid identification of laminitis-prone ponies before the onset of clinical disease, particularly when ponies are adapted to eating autumn pasture. We subsequently demonstrated that peak serum insulin concentration following the OGT, using a higher dose of 1.5g/kg glucose powder, correlates with insulin dysregulation defined using the fsivGTT [9] and that non-obese animals ponies and an 'easy keeping' horse breed (Andalusians) are more insulin resistant (fsivGTT) with a greater peak insulin post the OGT than Standardbreds .

An even simpler challenge , the oral sugar test (OST) has recently been advocated which currently requires 0.15ml/kg corn syrup to be syringed *per os* and, for example, defines an abnormal result as an insulin concentration >60µIU/ml at 75 minutes (see [10]). In our study [10] comparing the insulin response to OGT (1g/kg BW) and OST in eight ponies and five horses of unknown insulin sensitivity there was only agreement in the identification of insulin dysregulation in 85% of the animals. In this study, animals were not ranked in the same order under the OST and OGT AUC_i and the Tmax was significantly different for horses and ponies. Much more work is therefore required to optimise these tests and to establish which test more accurately diagnoses insulin dysregulation in both horses and ponies. We are currently working on further validation with the RVC. Our work has also shown that some ponies with marked insulin dysregulation do not seem to be prone to laminitis; we are currently working on this important aspect with colleagues at the Queensland University of Technology as well as Melbourne Veterinary School.

It is possible that other dynamic tests may be of value. For example, one of our studies suggested that the insulin response to a low dose dexamethasone test could detect ponies with an abnormal insulin response, which in turn could be linked with increased insulin resistance and an increased risk of laminitis [7]. Again, similar to the seasonal effects on the cortisol response to this test, we found seasonal effects on the insulin responses and the current recommendation would be to carry out the DST in spring to potentially optimise both the diagnosis of PPID and laminitis susceptibility [11]. The challenge, however, remains to identify at the individual rather than the group level and we are currently working on this with the RVC.

Other potential biomarkers identified from our collaborative research

Multiple variables have been evaluated as potential risk factors for laminitis and an association between the occurrence of laminitis and being a pony, time of year, being female , increased age, change of grass, as well as obesity have been demonstrated in some studies. In one of our studies, for example, in a group of inbred ponies with a high incidence of laminitis, it was found that generalized obesity, regional accumulation of neck crest adipose tissue ('cresty neck'), hyperinsulinaemia (>30 IU/ml) and hyperleptinemia (>7.5 ng/ml) were predictors of the development of laminitis when ponies were subsequently exposed to spring pasture [12.13]. Further work is therefore required into this important area and we are currently working on this aspect with the RVC.

Critical evaluation of the previous data relating to plasma TNF-α concentrations and laminitis predisposition and IR suggested that age could be a confounding factor (see [14]) thus bringing into question the role of inflammation in laminitis predisposition. We, therefore, undertook a study [14] to see if there were any differences in the inflammatory profile of a group of laminitis-prone ponies (PL; n = 38 and 42) compared with control animals at pasture in late spring and winter (NL; n = 41 and 39). Body condition score, height, weight and crest height and thickness were measured. Plasma concentrations of TNF-α, serum amyloid A, haptoglobin, insulin, adiponectin, triglyceride, fibrinogen, interleukin-17, interleukin-4 and interferon-γ were determined; only plasma [adiponectin] concentrations were significantly influenced by laminitis predisposition, being lower in PL (median[interquartile range] 2.1[1.4-3.2]ng/ml) compared to NL (3.4[2.6-4.1]ng/ml; p<0.0001). No other cytokines or inflammatory markers were associated with laminitis predisposition. This suggested that predisposition to laminitis may be associated with reduced anti-inflammatory capacity rather than a pro-inflammatory state.

Most studies, including the one reported above, however, have investigated multiple variables as laminitis risk factors but only after disease occurrence, rather than prior to disease occurrence. We have recently, therefore, been undertaking a series of prospective studies. In the first of these [15] ponies >7 years old with no previous laminitis were recruited. Body condition score, height, weight and crest height and thickness were measured and an overnight dexamethasone suppression test performed. Plasma/serum concentrations of adiponectin, leptin, triglyceride, insulin, insulin-like growth factor (IGF)-1, C-reactive protein, von Willebrand's factor, soluble (s) E-selectin and sP-selectin were assayed. Follow-up was obtained annually for 3 years to ascertain laminitis development. In total 446 animals were recruited to the study and the results from the first 2 years of data showed that after 1 and 2 years, 18 (4%) and 30/446 (7%) had developed laminitis. Very interestingly plasma [adiponectin] and basal and post dexamethasone serum [insulin] were significantly ($p \leq 0.05$) associated with development of laminitis after 1 and 2 years. Plasma [IGF-1] was significantly ($p \leq 0.05$) associated with development of laminitis after 2 years. Further data are being evaluated from this study and will hopefully be published in full in 2016.

Role of the microflora

We have shown in some pilot data that 16S *rRNA* t-RFLP profile from bacteria extracted from equine faeces clustered weakly into two groups representing normal versus laminitis-prone ponies, suggesting a difference in gut microflora between the two groups. In a follow up study we found that the bacterial population of ponies responds differently to the addition of inulin to the diet in laminitic and non-laminitis prone animals with a dramatic increase in *streptococci* numbers recovered from the faeces of lamintis prone ponies, but not non-laminitis prone ponies, following the addition of inulin [16]. We are currently working with the University of Aberystwyth to understand in more depth the role of the microflora in laminitis and obesity.

COLLABORATIVE RESEARCH INTO OBESITY

In summary, (non-exhaustive list) over the past 12 years or so we have concentrated on 4 main areas 1) How can we measure, define and monitor equine obesity?[12,17,18,19,20] 2) What is the prevalence of equine obesity?[21,22,23] 3) Why does it occur and why is it important? [24, 25, 26] 4) How is it best to manage the obese horse or an animal prone to obesity [27,28,29,30,31,32,33.34]. This work has led for example, to deuterium oxide dilution becoming the gold standard method in horses to determine % body fat [18] and to the appreciation that horses and ponies with a body condition score of ≥7/9 should be considered obese [19]. We have shown that breed differences exist in insulin sensitivity and insulin responses even in non-obese animals [9] and those animals adapted to grain based diets [35]. Perhaps most surprisingly we have shown that obesity is not always associated with the development of insulin resistance and can, depending on the diet, even result in an increase in insulin sensitivity [36,37]. Pasture intake is a major factor for obesity [22] and we have observed that ponies can ingest up to 41% of their daily DM intake within a 3-h turnout period [38]; up to 1% of body weight in DM within a 3-h period [39] and nearly 5% of their body weight in DM per day [24, 40]. We have evaluated the use of grazing muzzles [39, 41] as well as types of haynets [42] as part of the management strategy for obese animals. We have shown that if required, marked dietary restriction to 1% of body mass in DM of appropriate, good hygienic quality, grass hay can be clinically safe (although significant shavings ingestion can be a potential risk [28]) and is likely to elicit weekly weight losses of ~0.5% of outset BM weekly [27,30]. However, weight loss responses vary widely between individuals and importantly we have shown that some animals are more weight loss resistant than others [30]. In addition, the severity of any energy restriction may influence subsequent weight gain [34]. We have shown that whilst hay soaking can be useful it cannot be relied upon to reduce the level of water soluble carbohydrate (WSC) to less than a specific content level, as the loss can be highly variable [29]. The temperature of the water used to soak and the time of soaking will influence not only the loss of the WSC but also the microbial contamination as well as the extent of any dry matter loss[29,33 43]. This has resulted in the recommendations that hay composition analysis should be undertaken post-soaking for laminitis (as well as PSSM etc.) prone animals. Importantly we have confirmed that owner's may be less able to correctly identify overweight animals by visual appearance alone [44] and we have, therefore, developed the body condition index as a potential additional monitoring aid[45].

ACKNOWLEDGMENTS

We would like to take this opportunity to thank all the excellent researchers with whom WALTHAM and its equine studies team has collaborated with over the years.

REFERENCES

1. Hoffman, R.M., et al. J Anim Sci 2003 81:2333. 2. Tinworth, KD, et al Domestic Animal Endocrinology 2011, 41 81. 3. Borer KE, et al Am J Vet Res 2012;73:1596 . 4. Treiber KH, et al Pferdeheilkunde 2007 23 237 .5. Treiber KH, et al Am J Vet Res 2005 ;66:2114. 6. Borer KE, et Equine Veterinary Journal 2012; 44 (2012) 444–448 . 7. Bailey SR, et al. Am Vet Med Assoc 2007; 231 1365 . 8. Borer, K. E., et al. Journal of animal science 90.9 (2012): 3003. 9. Bamford NJ, et al Domestic Anim Endocrinol 2014 Apr 9;47:101. 10. Smith, S., et al Equine veterinary journal; 2015 On line. 11. Borer-Weir K, et al Equine Vet J. 2013; 45 688 12. Carter RA et al Veterinary Journal 2009a; 179 204. 13. Carter RA et al Equine Vet Journal 2009b; 41 (2) 171 14. Wray H.' et al Equine Veterinary Journal 2013; 45 546 . 15. Menzies-Gow, NJ. Et al ECVEIM proceedings 2015. 16. Newbold CJ, et al. In proceedings of the British Society of Animal Science : Advances in animal Biosciences 2009; Cambridge University Press 16. 17. Dugdale, A. H. A. et al Equine Veterinary Journal 2011a; 43(5): 552. 18. Dugdale, A. H. A., et al Equine Veterinary Journal 2011b; 43(5): 562. 19. Dugdale A, et al The Veterinary Journal 2012; 194 173. 20. Morrison P et al ACVIM proceedings (2014) Nashville no 93064 21. Harker I.J et al J Equine Vet Sci 2011 31 253. 22. Giles SL, et al PeerJ 2014; 2: e299. 23. Giles SL, et al BMC Veterinary Research.2015, 11:13 DOI: 10.1186/s12917-015-0327-7. 24. Dugdale A,. et al Vet J 2011; 190 p 329. 25. Giles SL et al Applied Animal Behaviour Science, 201; 166 (1), pp. 71. 26. Bamford, N. J., et al Equine Veterinary Journal. 2015; doi: 10.1111/evj.12434. 27. Dugdale A,. et al Equine Veterinary Journal 2010; 42 (7) 600. 28. Curtis GC, et al . Br J Nutrition 2011; 106, S178. 29. Longland AC et al Vet Rec 2011 ; 168 p 618. 30. Argo McC et al The Veterinary Journal 2012; 194 179. 31. Tinworth, KD, et al The Veterinary Journal, 2012; 191, 79. 32. Bruynsteen L, et al 2014; 112 1402 . 33. Longland A, et al Veterinary Record 2014; 174 68. 34. Bruynsteen L. et al Vet J 2016 (1) 61. 35. Bamford et al J Animal Science 2015; 93(7) 3377. 36. Bamford, N. J., et al Equine Veterinary Journal 2015 . doi: 10.1111/evj.12434. 37. Bamford NJ, et al the Veterinary Journal in press. 38. Ince J, et al J Equine Vet Sci 2011; 31: 283.39. Longland AC, et al EAAP publication no 132 2012; pp.185 . 40. Longland AC, et al J Equine Veterinary Science 2011; 31 : 275 . 41. Longland AC , et al JEVS 2015; 35 p 430. 42. Ellis A. D., et al Applied Animal Behaviour Science 2015; 171 108. 43.Moore-Colyer MJS, et al PLoS ONE 2014; 9(11): e114079. doi:10.1371/journal.pone. 44 Morrison, Pet al Acta Veterinaria Scandinavica 2015; , 57(Suppl 1), O6. http://doi.org/10.1186/1751-0147-57-S1-O6. 45 Potter SJ et al Journal of equine Veterinary Science 2015; 35 p 410 - 411

A Role for Somatostain in Insulin Dysregulation?

Nicholas Frank, DVM, PhD, DACVIM
North Grafton, MA

INTRODUCTION

Somatostatin (SST) may be the forgotten hormone of the pancreatic Islet cells. Alpha cells secrete glucagon and beta cells release insulin, but delta cells and pancreatic polypeptide cells are rarely mentioned. Delta cells are found adjacent to alpha and beta cells and produce somatostatin, a hormone that inhibits secretion of glucagon and insulin. Somatostatin secreted from delta cells acts in a paracrine fashion and binds to somatostatin receptors (SSTRs) on the cell membranes of alpha and beta cells to inhibit glucagon or insulin secretion, respectively. It can therefore be hypothesized that hyperinsulinemia reflects decreased somatostatin action on beta cells, with insulin secretion increasing as a result. Hyperinsulinemia is detected in horses and ponies and evidence provided by experimental models establishes hyperinsulinemia as a cause of laminitis.[1,2] We must therefore consider a role for somatostatin in the development of hyperinsulinemia in the horse and other domestic animals.

INSULIN DYSREGULATION

The terminology used to describe the endocrine and metabolic problems of horses or ponies with laminitis has changed over the years as more research has been performed and our understanding of equine physiology has expanded. Terms in use at present include *hyperinsulinemia*, which is easily defined as a plasma or serum insulin concentration above reference interval. More recently, the term *postprandial hyperinsulinemia* (or sometimes postprandial dysinsulinemia) has been used to identify horses or ponies with higher than normal plasma or serum insulin concentrations after an oral sugar test (OST) or test meal. *Insulin resistance* (IR) is also detected in equids, but this term should be reserved for horses and ponies with a detectable reduction in insulin action at the tissue level. These actions are evaluated by assessing the decrease in blood glucose concentrations that follows intravenous injection of exogenous insulin, and the combined glucose-insulin test (CGIT) and insulin tolerance test (ITT) are used for this purpose in clinical practice. However, it is very important to note that there exists a population of horses and ponies with postprandial hyperinsulinemia that have negative test results when evaluated for IR. There are at least two potential explanations for this finding: insulin sensitivity is normal in these animals or the diagnostic tests that are currently used to detect IR have low sensitivity.

If we return to the horse or pony that presents with laminitis, it is essential that we recognize hyperinsulinemia as a cause of laminitis. The term *insulin dysregulation* (ID) was therefore introduced in 2014 and this name encompasses postprandial hyperinsulinemia, fasting hyperinsulinemia, and IR, or any of these factors alone or in combination. The primary reason for adopting this term is to encourage assessment of insulin status in horses and ponies so that at-risk animals are identified and appropriately managed before laminitis occurs.

Where then does *equine metabolic syndrome* (EMS) fit into the picture? Insulin dysregulation is a key component of this syndrome, along with increased adiposity (generalized obesity and/or regional adiposity), and dyslipidemia.[3] Metabolic components of EMS also include the easy keeper metabolism of many affected horses and weight loss resistance when caloric intake is restricted.[4] Equine metabolic syndrome has a significantly higher incidence in certain breeds of horse and we now hypothesize that ID is the genetic disorder responsible for this syndrome and other components develop as exacerbating factors are encountered.

Pituitary pars intermedia dysfunction (PPID) is an endocrine disorder associated with aging in horses that appears to occur with greater frequency in horses and ponies with EMS. There are two subpopulations of horses with PPID, those with ID and those with normal insulin status. Our hypothesis that ID is a genetic disorder in Equidae can be extended to PPID if we assume that some horses with the ID genetic trait develop EMS first and then transition into PPID, and others remain subclinical until PPID develops and insulin sensitivity decreases because of glucocorticoid excess. It should not be assumed from this discussion that ID is required for the development of PPID, but it does appear to be important risk factor. Insulin status must therefore be evaluated in all horses and ponies with PPID to assess the risk of laminitis.

In this discussion, we focus on ID and assume that this problem exists in all horses and ponies with EMS, some equids with PPID, and a population of genetically predisposed animals without clinical signs of EMS or PPID that are at risk of developing laminitis if exacerbating factors move them from the subclinical to clinical state.

SOMATOSTATIN

Somatostatin is also known as growth hormone inhibiting hormone (GHIH) and is produced by delta cells found within the pancreas, hypothalamus, central nervous system, peripheral neurons and the gastrointestinal tract. A prohormone (pro-SST) is synthesized and then cleaved by convertases into two active forms consisting of 14 or 28 amino acids (SST-14 and SST-28, respectively).[5] In the Islets of Langerhans, somatostatin secretion is stimulated by glucose, amino acids and glucagon-like peptide-1 (GLP-1), and the hormone acts directly on islet cells to inhibit secretion of insulin, glucagon, and pancreatic polypeptide.[6] Somatostatin is thought to act as a 'buffering hormone' that prevents over-secretion of glucagon and insulin, with insulin secretion decreasing as blood glucose concentrations decrease and glucagon secretion decreasing as hyperglycemia develops.[7] Paracrine signaling occurs through binding of somatostatin to SSTRs on alpha and beta cells, and five different somatostatin receptor (SSTR1-5) subtypes have been cloned.[8,9] Somatostatin receptors are distinguished by their pharmacological specificity for different analogs, tissue distribution, regulation, and intracellular signaling pathways.[10] Immunohistochemistry studies in humans have revealed that

SSTR1, SSTR2 and SSTR5 are expressed on 100%, 44% and 87% of beta cells and on 26%, 89% and 35% of alpha cells, respectively, and inhibition of hormone secretion is mediated mainly through SSTR2 and SSTR5.[11,12] Somatostatin receptor 2 is the dominant receptor for glucagon and insulin secretion in alpha and beta cells in humans, whereas SSTR2 (predominantly expressed in alpha cells) and SSTR5 (predominantly expressed in beta cells) are both important in the rat.[13-15] Activation of SSTRs suppresses hormone secretion by reducing cAMP levels, inhibiting electrical activity, decreasing Ca^{2+} influx through voltage-gated Ca^{2+} channels and directly reducing exocytosis in a Ca^{2+} and cAMP-independent manner.[5] Evidence from rodent models suggests that somatostatin secretion from delta cells is dysregulated in diabetes mellitus, which may contribute to the metabolic disturbances in this disease.[5,16]

SOMATOSTATIN ANALOGS

Somatostatin analogs were developed for the treatment of acromegaly and adult Cushing's disease in humans. They inhibit growth hormone (somatotropin) secretion from the pars distalis and therefore lower insulin-like growth hormone 1 (IGF-1; somatomedin C) concentrations in acromegalic patients. Expression patterns of SSTRs are relevant to the use of somatostatin analogs in human and veterinary medicine. Octreotide (Sandostatin®; Novartis Pharmaceuticals Corporation, East Hanover, NJ) and lanreotide (Somatuline®; Ipsen Biopharmaceuticals, Inc., Basking Ridge, NJ) have strong affinity for SSTR2 receptors, whereas pasireotide (Signifor®; Novartis Pharmaceuticals Corporation, East Hanover, NJ) has highest affinity for SSTR5, but also binds to SSTR1, SSTR2, and SSTR3.[17] Our research group is examining the immunohistochemical staining patterns of islets within pancreas samples collected from horses and cats; and both alpha and beta cells stain positive for SSTR2 in equine and feline Islets of Langerhans. However, preliminary results from a concurrent *in vivo* study of somatostatin analogs in cats indicate that pasireotide, but not octreotide alters glucose and insulin concentrations, and this suggests that SSTR5 is more important for feline beta cells. In contrast, octreotide inhibits insulin secretion more effectively than pasireotide in horses, indicating that SSTR2 is the predominant receptor on equine beta cells. Inter-species differences in SSTR expression warrant further study as we compare glucose and insulin disorders in cats, dogs, horses, and humans, and explore the use of somatostatin analogs as medical treatments for these conditions.

Octreotide is also used for the medical treatment of insulinomas in humans and tumors expressing SSTR2 receptors are more responsive to treatment.[18] An octreotide test is performed in humans to determine whether SSTR2 receptors are present on insulinomas and this procedure involves injection of 100 μg of octreotide subcutaneously with blood samples collected before and 1, 2, 4, 6, and 8 hours after injection. Insulinomas secrete excessive amounts of insulin and induce hypoglycemia, so patients responding to octreotide show suppression of insulin secretion (< 3 μU/mL) and plasma glucose concentrations increase above 3 mmol/L (54 mg/dL) during the test.

Our research group has recently developed an octreotide response test (ORT) to evaluate blood insulin and glucose responses to octreotide in horses with ID. Octreotide is injected intravenously and plasma insulin concentrations immediately decrease. Blood glucose concentrations increase when insulin concentrations decrease and hyperglycemia develops as a result. Suppression of insulin secretion by octreotide is only transient and insulin concentrations subsequently increase as the horse attempts to regain glycemic control. The maximum insulin concentration (C_{max}) reflects the amount of insulin secreted by beta cells to regain glycemic control and the rate of decrease in blood glucose concentrations serves as a measure of insulin sensitivity. Normal horses and those with ID have been compared and higher insulin C_{max} values are detected in affected horses.

Plasma concentrations of connecting (C)-peptide have also been measured to confirm that insulin secretion decreases and then increases after octreotide injection. This peptide connects the A and B chains of the insulin molecule and is released into the blood in equimolar amounts as beta cells secrete insulin. In contrast to insulin, which is estimated to undergo approximately 74% clearance from the portal blood by the liver in healthy horses,[19] C-peptide is not cleared by the liver and plasma concentrations reflect the amount of insulin secreted. Our first project with somatostatin was published in 2010 and demonstrated that the secretion of insulin and C-peptide by beta cells was suppressed when biosynthetic human somatostatin was infused intravenously as a bolus injection (500 μg) followed by continuous rate infusion (500 mg/h).[19] The ORT has been developed from this initial investigation and involves transient suppression of insulin secretion and then assessment of the perturbations in blood glucose and insulin concentrations that follow.

SOMATOSTATIN, INCRETIN HORMONES, AND NEW THEORIES

Recent research provides evidence that the intestine plays an important role in ID and therefore laminitis. Preliminary findings first reported by our research group in 2010 at the ACVIM Forum suggested that horses with EMS experienced higher active GLP-1 concentrations during an oral sugar test (OST) than normal horses. This has been followed by a more comprehensive study of the enteroinsular axis in ponies performed by de Laat and colleagues at the Queensland University of Technology.[20] Ponies with ID were compared to normal ponies and affected animals had significantly higher plasma active GLP-1 concentrations in response to an oral glucose tolerance test and test meal. An equally fascinating finding from the study was the detection of higher glucose absorption from the intestine in ponies with ID, when compared with unaffected controls. These findings suggest that two ponies grazing on the same pasture can absorb glucose differently and vary markedly in their postprandial insulin responses to ingested sugars and amino acids. What role then might somatostatin play in this new theory of ID? In order for postprandial hyperinsulinemia to develop, beta cells must increase the rate of insulin secretion. It can then be hypothesized that somatostatin secretion from delta cells decreases to permit hyperinsulinemia or that SSTRs on beta cells are down regulated. These processes warrant further study in horses so that we can better understand the pathophysiology of ID and explore new approaches of medically managing hyperinsulinemia.

ACKNOWLEDGEMENTS

Drs. Ray Boston and Jim Orsini from the University of Pennsylvania, Drs. Pilar Hermida, Alfredo Sanchez Londoño, and Melissa Restifo from the Cummings School of Veterinary Medicine at Tufts University, and Dr. Cassandra Uricchio from the University of Massachusetts at Amherst are all acknowledged for their contributions to this area of research and ongoing collaborations.

REFERENCES

1. Asplin KE, et al. Vet J 2007;174:530-535.
2. de Laat MA, et al. Equine Vet J 2010;42:129-135.
3. Frank N, et al. J Vet Intern Med 2010;24:467-475.
4. Argo CM, et al. Vet J 2012;194:179-188.
5. Braun M. Vitam Horm 2014;95:165-193.
6. Patel YC. Front Neuroendocrinol 1999;20:157-198.
7. Brereton MF, et al. J Histochem Cytochem 2015;63:575-591.
8. Yamada Y, et al. Proc Natl Acad Sci U S A 1992;89:251-255.
9. Yamada Y, et al. Biochem Biophys Res Commun 1993;195:844-852.
10. Florio T, Schettini G. J Mol Endocrinol 1996;17:89-100.
11. Chen C, et al. Trends Endocrinol Metab 1994;5:227-233.
12. Kumar U, et al. Diabetes 1999;48:77-85.
13. Hunyady B, et al. Endocrinology 1997;138:2632-2635.
14. Mitra SW, et al. Endocrinology 1999;140:3790-3796.
15. Kimura N, et al. Endocr J 2001;48:95-102.
16. Portela-Gomes GM, et al. Pancreas 2010;39:836-842.
17. McKeage K. Drugs 2015;75:1039-1048.
18. Furukawa F, et al. Cancer Lett 2000;159:43-48.
19. Toth F, et al. Equine Vet J 2010;42:149-155.
20. de Laat MA, et al. Am J Physiol Endocrinol Metab 2016;310:E61-72.

Changing Concepts of Hyperglycemia, Hyperlactatemia and Hypertriglyceridemia

Bettina Dunkel, DVM, PhD, DACVIM, DECEIM, DACVECC, FHEA, MRCVS
London, UK

INTRODUCTION

Energy metabolism in healthy individuals is a tightly regulated and intricate process. Under normal conditions, most energy is derived from ingested carbohydrates via aerobic glycolysis. Blood glucose concentrations are maintained within a narrow range with rapid endocrine responses preventing development of hyper- or hypoglycemia. During starvation, energy is generated by mobilization and break down of fat reserves and protein catabolism. Again, tight endocrine control prevents development of hypoglycemia or accumulation of fat metabolites in the circulation or organs. However, systemic illness with secondary endocrine dysregulation or primary endocrine disease can lead to derangements in energy metabolism with potentially severe consequences for short and long term outcome.

HYPO-, HYPER-, AND DYSGLYCEMIA IN CRITICAL ILLNESS

Following reports of better outcomes in critically ill patients in which blood glucose concentrations were maintained in narrow limits, strict blood glucose control was widely advertised[1-3]. Intensive glucose control was recommended in guidelines and adopted by many intensive care units around the world4. In subsequent large-scale trials, most notably the NICE SUGAR trial, benefits of strict glycemic control could not be replicated. In contrast, an increased risk of hypoglycemia and/or increased risk of death was identified[5,6] and tight glucose control remains an intensely debated topic[7]. Thus current recommendations for critically ill human patients state that insulin therapy should be initiated once blood glucose concentrations exceed 180mg/dL aiming to maintain concentrations of 144-180mg/dL using insulin as necessary[4]. In people, common reasons for hyperglycemia comprise previously diagnosed or undiagnosed diabetes mellitus and stress induced hyperglycemia. Stress induced hyperglycemia is defined as increase in blood glucose concentrations above the normal reference range which return to normal limits once illness subsides8. Causes of stress hyperglycemia in critical illness are poorly defined. Inflammatory mediators, such as IL-1 and TNF are thought to decrease insulin secretion relative to the blood glucose concentration and increase the release of hormones favouring development of insulin resistance such as glucagon, cortisol, catecholamines and growth hormone[9]. Several studies have identified a link, but not causality, between admission hyperglycemia and increased morbidity and mortality[10-12]. Interestingly, the relationship between hyperglycemia and mortality is weakened or absent in diabetic patients[9]. Hypoglycemia is commonly defined as blood glucose concentrations of 40–70 mg/dl (moderate) or <40 mg/dl (severe). Risk factors for hypoglycemia in people include co-existing illness and illness severity; however, the most important risk factor is intensive insulin therapy[9]. Hypoglycemia has a deleterious impact on ICU patient survival and the lower the blood glucose concentrations, the stronger the association[6]. An increased incidence of hypoglycemia has several times been associated with tight glycemic control which is one of the reasons why a more lenient approach to blood glucose management has been suggested[13]. Indeed, some investigators have speculated that tight glucose control would undoubtedly be beneficial, were it not for the increased incidence of hypoglycemia. Blood glucose concentrations in critically ill horses have been investigated. Studies in neonatal and adult equine patients have identified an association between hyperglycemia and poor outcome[14-16]. In search for a possible causal relationship, the effects of experimental hyperglycemia alone and hyperglycemia in combination with endotoxemia on coagulation parameters of healthy horses were investigated but no effect of hyperglycemia was identified[17,18]. In neonatal foals, hypoglycemia was also associated with non-survival, sepsis, positive blood cultures and SIRS[14]. Hypoglycemia is exceedingly rare to non-existing in adult horses with acute abdominal disease which probably reflects a different disease severity and the comparatively rarer use of insulin or ambitions to achieve tight glucose control[15,16]. Blood glucose concentrations in equids are also affected by breed and underlying endocrine diseases. Differences in insulin sensitivity and blood glucose concentrations between horses and ponies and also different horse breeds have been identified, independent of obesity[19]. Normal ponies responded with higher blood glucose and higher insulin concentrations to an oral glucose tolerance test compared to large breed horses suggesting a degree of insulin resistance[19]. These changes were exacerbated in obese and laminitic ponies[20]. While primary diabetes mellitus and uncompensated hyperinsulinemia are rare in equids, 58-94% of equids with PPID reportedly suffer from hyperglycemia[21-23]. Differences in blood glucose concentrations have been detected in horses and ponies with gastrointestinal disease with ponies having higher blood glucose concentrations than horses with comparable disease severity[24]. However, whether pre-existing endocrine diseases alter the blood glucose response and association with increased mortality in systemic illness has not yet been evaluated.

In people, the focus has since shifted from hyperglycemia to inclusion of other abnormalities of glucose metabolism, summarized under the term dysglycemia[25]. Dysglycemia comprises several abnormalities of blood glucose concentrations including hyper- and hypoglycemia and increased glucose variability. Glucose variability (GV) can be measured using a variety of indicators and up to 13 different parameters have been used. The most commonly calculated indicator is standard deviation (SD) of the mean blood glucose concentration and, to a lesser degree, the glycemic lability index (GLI)[26]. The SD determines how tightly various glucose values are clustered around the mean while GLI also takes into account frequency of glucose measurement, measured by the formula GLI = ([mg/dl] 2 /h)/week. Increased blood glucose variation has been associated with mortality, irrespective of the actual blood glucose concentration[27]. However, the reason for this association remains obscure and suggested possibilities include greater severity of illness, less observant medical and nursing team and a direct deleterious effects on the patient provoked by changes in the blood glucose concentration[26]. Glucose variability has to the author's knowledge not been assessed in critically ill horses.

LACTATE CONCENTRATIONS IN CRITICAL ILLNESS

Increased lactate concentrations have been associated with decreased survival under various clinical situations in multiple studies. Equally, low or decreasing lactate concentrations correlated with improved chances of survival[28]. However, reasons for hyperlactatemia vary between patients and diseases and whilst hypoperfusion and tissue hypoxia are a common reason for increased lactate concentrations in critically ill patients, other conditions can also contribute. In models of tissue hypoperfusion due to obstruction, normalization of systemic hemodynamics is associated with rapidly normalizing lactate levels but this is not uniformly the case in tissue hypoperfusion in sepsis models[29]. Obstructive shock, unlike septic shock, maintains intact autoregulatory mechanisms that preserve vital organ perfusion and ensure that the depressed microcirculation can be adequately resuscitated with normalized systemic hemodynamics[29]. Disruption of these compensatory autoregulatory mechanisms during sepsis, rather than global hypoperfusion, causes impaired microvascular perfusion. Preservation or loss of autoregulation explains why the different microvascular perfusion variables returned to normal in the obstructive model but not during endotoxemic shock[29]. While changes in the microvasculature during sepsis, such as endothelial dysfunction and heterogeneous flow patterns, still ultimately result in impaired oxygen delivery and tissue hypoxia, hyperlactatemia in critically ill patients is not always the result of anaerobic metabolism due to hypoxia. The effect of sepsis on metabolism, resulting in increased lactate levels in the absence of tissue hypoperfusion, should also be considered[28]. The presence and extent of mitochondrial dysfunction during sepsis is still intensely debated and findings seemingly depend on what species, tissue and investigation methods are used. In people, with severe sepsis and/or septic shock mitochondrial dysfunction appears to be present in blood mononuclear cells and skeletal muscle[30]. Other, in critically ill patients rarely identified causes, include thiamine deficiency and inherited or acquired mitochondrial disorders. In people, increases in lactate concentrations are closely associated with fasting glucose, fasting insulin, insulin resistance and type 2 diabetes[31,32]. Subclinical or clinical insulin resistance in turn is suspected to be associated with a low oxidative capacity due to impaired mitochondrial phosphorylation. Although it is currently unknown whether the low oxidative capacity is a consequence or cause of insulin resistance, recent studies have suggested that a low oxidative capacity may precede insulin resistance and diabetes[33,34].

Blood lactate concentrations measured either on admission3[5-37], during subsequent hospitalization[38] or both[39] have frequently been associated with outcome in critically ill horses and foals. More recently, the area under the lactate concentration-time curve has been investigated as additional prognostic tool in critically ill neonatal foals[40]. In horses of all ages, blood lactate concentration are viewed primarily as an important marker of systemic hypoperfusion and tissue hypoxia in critically ill patients and lactate reduction has been suggested as a possible target for early fluid resuscitation in horses with colitis[38]. However, impaired oxygen utilization secondary to inflammatory responses and/or reduced clearance as in liver or renal failure has also been suggested as possible cause for hyperlactatemia in foals[37]. Other factors, such as breed and disease process, might also influence blood lactate concentrations. Horses with atypical myopathy anecdotally have lactate concentrations that are disproportionately high compared to the apparent cardiovascular status. It has been speculated that in these cases inadequate mitochondrial oxygen utilization in muscles, related to multiple acetyl-CoA dehydrogenase deficiency, and lowered activity of mitochondrial complexes could play a role[41,42]. In one study evaluating animals with gastrointestinal disease, ponies had higher lactate concentrations than horses despite similar cardiovascular compromise43. Surprisingly, unpublished data suggest that normal ponies (n=16; median 0.7mmol/L; range 0.7-1.3mmol/L) have lower blood lactate concentrations than normal horses (n=17; median 0.9mmol/L; range 0.7-1.8mmol/L; p=0.019) implying that lactate metabolism in ponies could respond differently to either cardiovascular compromises or stresses associated with disease differently compared to horses (Dunkel, unpublished data). To the author's knowledge, lactate concentrations in equids with PPID or equine metabolic syndrome (EMS) have not been investigated. It is possible that an underlying lower oxidative capacity could contribute to the observed differences in lactate, glucose and insulin metabolism in ponies when compared to horses[20,43].

THE LINK BETWEEN GLUCOSE AND LACTATE CONCENTRATIONS

Despite the close metabolic link between glucose and lactate, both metabolites are frequently considered independently and have rarely been studied together. Blood glucose and lactate concentrations both are interconnected in the carbohydrate metabolism and each can lead to the generation of the other. Anaerobic glycolysis of glucose generates lactate as an end product while lactate, via generation of pyruvate, can initiate gluconeogenesis or glycogen formation. The validity of blood glucose as an independent marker of mortality has therefor been questioned. In previous studies that concurrently investigated the role of blood lactate and glucose concentrations, glucose was not found to be associated with mortality[44]. Two studies subsequently found that hyperlactatemia significantly influenced the relationship between blood glucose concentrations and mortality, strengthening the suspicion that blood glucose concentration is not an independent predictor of outcome[45-47]. Unpublished data suggests that this is also the case in equids with gastrointestinal disease[24]. In septic human patients, 45% of lactate is converted into glucose either by gluconeogenesis or glycogen production48. It is therefore conceivable that increased lactate concentrations, at least partially, could contribute to increased glucose concentrations observed in critically ill equids. As discussed above, in equids with gastrointestinal disease and human patients with sepsis or septic shock increased blood lactate concentrations are commonly viewed as markers of tissue hypoxia, presumed to be secondary to hypoperfusion[49,50]. At least in critically ill people, this view is currently changing and some investigators suggest that increased aerobic glycolysis secondary to adrenergic stimulation plays a large, if not the largest, role in the development of hyperlactatemia[49,51]. Interestingly, a strong positive association between plasma adrenalin and lactate concentrations has also been identified in horses with colic[52]. The exact source of increased lactate concentrations in critically ill horses deserves further investigations to confirm or refute that the current concept of using lactate largely as a marker of tissue hypoxia is justified. These findings could have significant therapeutic implications as the focus might shift from control of glucose concentrations to achieving a better understanding, and potentially influencing, glucose and lactate metabolism in critical illness[45].

PLASMA TRIGLYCERIDE CONCENTRATIONS

Increases in plasma lipid concentrations in equids were first described in ponies, miniature breeds and donkeys secondary to a negative energy balance, frequently complicated by a concurrent disease process, pregnancy or lactation. Initial reports highlight the challenging and often unsuccessful treatment attempts and associated high mortality. Clinical signs attributed to mild to moderate increases in plasma lipids are non-specific and include depression, anorexia, colic, fever diarrhea and ventral edema[53-55]. Secondary fatty infiltration of liver, kidney and other organs are thought to contribute to morbidity and mortality. It has since been recognized that increases in plasma triglyceride concentrations can also occur in systemically ill large breed horses and in horses with endocrine diseases[56,57]. Anecdotally, horses with atypical myopathy secondary to hypoglycin A toxicity also appear particularly prone to hypertriglyceridemia, presumably due to disruption of normal lipid metabolism by the effects of the toxin. The milky discoloration of the plasma caused by increased concentrations of very low density lipoproteins (VLDLs), long considered a hallmark of the disease, can be absent, even with significant increases in triglyceride concentrations. This highlights the importance of recognizing risk factors and monitoring of triglyceride concentrations in at risk patients. This is further emphasized by the fact that the success of treatment seems to largely depend on early intervention. Some studies have found an association between hypertriglyceridemia and non-survival but this was not confirmed in others[58,59]. However, even if no statistical association was identified, notably more animals with triglyceride concentrations >500mg/dl did not survive. The authors attributed the lack of an association to the fact that most animals were euthanized for untreatable primary disease rather than treatment failure. This could imply that triglyceride levels are largely a marker of disease severity but not in themselves contributing to mortality if treated appropriately[59]. The detrimental effects of hypertriglyceridemia have further come into question with the observation of marked and prolonged increases in plasma triglycerides found in some horses with endocrine disease. With the exception of mild, subclinical hepatopathy, no apparent ill effects were observed in these animals despite triglyceride concentrations ranging from 930-5336mg/dl[57]. It is currently unknown whether the lipid composition in endocrine hypertriglyceridemia is different or whether other, unidentified factors contribute to morbidity and mortality associated with hypertriglyceridemia in critically ill patients. It is possible, that most observed clinical signs and detrimental effects are indeed related to the primary disease process and that plasma triglycerides largely mark severity of the underlying illness. This is in part supported by the clinical observation that early institution of nutritional support often returns triglyceride levels to normal levels without necessarily improving the outcome. Enteral or parenteral nutritional support have been successfully used and often provision of a relatively small percentage of the resting energy requirements (25-50%; 5-10kcal/kg/d) can return triglyceride concentrations to normal[56]. Concurrent use of insulin appears to be more effective than use of a glucose infusion alone in correcting hypertriglyceridemia[59]. However, hyperglycemia and hyperinsulinemia have been associated with development of laminitis. Infusion of 5.6mg/kg/min (equivalent of 27kcal/kg/d) of glucose induced histopathologic, but not clinical, evidence of laminitis in otherwise healthy horses[60] and the risk of laminitis must be kept in mind.

REFERENCES

[1.] Malmberg K, et al. J Am Coll Cardiol 1995;26:57.[2]van den Berghe G, et al. N Engl J Med 2001;345:1359.[3]Van den Berghe G, et al. N Engl J Med 2006;354:449.[4]Niven DJ, et al. JAMA Intern Med 2015.[5]Investigators N-SS, et al. N Engl J Med 2009;360:1283.[6]Investigators N-SS N Engl J Med 2012;367:1108.[7]Finfer S. Virulence 2014;5:200.[8]Sebranek JJ, et al. Br J Anaesth 2013;111 Suppl 1:i18.[9]Deane AM, et al. Diabetes Obes Metab 2013;15:792.[10]de Miguel-Yanes JM, et al. Heart Lung 2014.[11]Kreutziger J, et al. Injury 2015;46:15.[12]Haddadin F, et al. Int J Clin Pract 2014.[13]Keegan MT, et al. . J Diabetes Sci Technol 2009;3:1288.[14]Hollis ARet al. J Vet Intern Med 2008;22:1223.[15]Hassel DM, et al. J Vet Intern Med 2009;23:1261.[16]Hollis AR, et al. J Vet Intern Med 2007;21:1099.[17]McGovern KF, et al. J Vet Emerg Crit Care (San Antonio) 2012;22:550.[18]McGovern KF, et al. J Vet Intern Med 2013;27:347.[19]Bamford NJ, et al. Domest Anim Endocrinol 2014;47:101.[20]Jeffcott LB, et al. Equine Vet J 1986;18:97.[21]Hillyer MH, et al. Equine Vet Educ 1992;4:131.[22]van der Kolk JH, et al Domest Anim Endocrinol 1995;12:35.[23]Durham AE, et al. Equine Vet J 2009;41:924.[24]Dunkel B, J Vet Emerg Crit Care 2015;25:S15.[25]Badawi O, et al. Crit Care Med 2012;40:3180.[26]Eslami S, et al. Intensive Care Med 2011;37:583.[27]Krinsley JS. Crit Care Med 2008;36:3008.[28]Bakker J. Crit Care Med 2015;43:705.[29]van Genderen ME, et al Crit Care Med 2014;42:e96.[30]Fink MP. Pediatr Crit Care Med 2015;16:89.[31]Lovejoy J, et al. Metabolism 1992;41:22.[32]Chen YD et al. Diabete Metab 1993;19:348.[33]Juraschek SP, et al. PLoS One 2013;8:e55113.[34]Juraschek SP, et al. Ann Epidemiol 2013;23:791.[35]Gray SN, et al. Vet Surg 2014.[36]Radcliffe RM, et al. J Vet Emerg Crit Care (San Antonio) 2012;22:313.[37]Borchers A, et al. Equine Vet J Suppl 2012:57.[38]Hashimoto-Hill S, et al. J Vet Intern Med 2011;25:1414.[39]Wotman K, et al. J Vet Intern Med 2009;23:598.[40]Wilkins PA, et al. J Vet Intern Med 2015;29:659.[41]Westermann CM, et al. Neuromuscul Disord 2008;18:355.[42]van Galen G, et al. J Vet Intern Med 2013;27:186.[43]Dunkel B, et al. Equine Vet J 2013;45:666.[44]Freire AX, et al. Chest 2005;128:3109.[45]Green JP, et al. Acad Emerg Med

2012;19:1268.[46]Kaukonen KM, et al. Crit Care Med 2014;42:1379.[47]Cori CF, et al. Annu Rev Biochem 1946;15:193.[48]Barth E, et al. Crit Care Med 2007;35:S508.[49]Garcia-Alvarez M, et al. Crit Care 2014;18:503.[50]Tennent-Brown BS. Compend Contin Educ Vet 2012;34:E1.[51]Garcia-Alvarez M, Lancet Diabetes Endocrinol 2014;2:339.[52]Hinchcliff KW, et al. J Am Vet Med Assoc 2005;227:276.[53]Naylor JM, et al. Am J Vet Res 1980;41:899.[54]Watson TD, et al. Vet Rec 1992;131:481.[55]Moore BR, et al. J Vet Intern Med 1994;8:376.[56]Dunkel B, et al. Equine Vet J 2003;35:590.

Skin Tumors: Standard Treatments and Emerging Therapies

Krista Estell, DVM, DACVIM (Large Animal Internal Medicine)
Davis, CA

INTRODUCTION

Cutaneous neoplasia is extremely common in general and referral equine practice. Skin tumors represent approximately 30% of equine dermatologic lesions, and the application of the appropriate treatment technique is crucial to the management of these conditions.[1] Though there are numerous treatments that have been reported as effective, no single treatment is universally curative for any given neoplasia. Equine oncology is a continuously growing field, and while new therapies are constantly being applied to horses, it is important to use evidence based medicine from well-planned clinical trials and retrospective studies with histologic confirmation of diagnosis, careful documentation of lesion response to therapy, and long-term follow-up to truly determine the efficacy of these emerging therapies.

DIAGNOSIS

Obtaining an accurate diagnosis is the first step in effective treatment of cutaneous neoplasia. It should no longer be considered standard of care to make treatment recommendations based on clinical appearance alone. Biopsy with histopathology and immunohistochemistry if indicated is the most accurate method of diagnosis. The local lymph nodes should be assessed in all horses, and when metastatic disease is suspected complete blood count, serum biochemical analysis, thoracic radiographs and abdominal ultrasound should be performed prior to treatment. If a rare neoplasia is suspected on histopathology, or if the diagnosed neoplasia does not fit the clinical presentation, a second opinion should be pursued to confirm the diagnosis prior to treatment.

THERAPEUTIC OPTIONS

Surgical excision should be considered the treatment of choice for all cutaneous neoplasias if possible. Too often, however, horse owners are advised to take the "wait and see" approach to cutaneous neoplasia and by the time the tumor is presented for surgical removal; it is too large to be removed with clean margins. If tumors are too large or are in an area that inhibits complete surgical excision, there are many other forms of therapy that may be curative. Cryotherapy and laser ablation are commonly used alone or in combination with surgical excision and are often effective, though they have the obvious disadvantage of the inability to examine surgical margins, as well as the destruction of normal tissue in addition to tumor cells.

Chemotherapy

Intratumoral chemotherapy is most effective for the treatment of small to moderately sized tumors (< 10 cm diameter) before or after surgical debulking, and for the treatment of surgical sites for which complete surgical excision was not obtained. Cisplatin, which causes apoptosis of neoplastic cells by crosslinking with purine bases and interfering with DNA repair mechanisms, is the most effective and most commonly used intratumoral chemotherapeutic agent.[2] Cisplatin can either be injected as a sesame oil emulsion or in a biodegradable bead, though injection of cisplatin emulsion more easily allows for concentration of cisplatin at the site where there is likely to be larger tumor burden. [2,3]

Systemic chemotherapy is indicated when neoplasia is severe or disseminated enough that local treatment is no longer feasible. As with all other therapies, success of treatment depends on the type of neoplasia, the overall health of the horse, and the commitment of the horse owner. Systemic chemotherapy is often met with considerable trepidation by both horse owners and veterinarians as most people are familiar, either from direct experience or from the experience of a loved one, with the devastating side effects of systemic chemotherapy. However, it is important to note that the goals of treatment in horses are quite different from humans. The goal of human chemotherapy is cure or remission for the life of the patient with the average life expectancy being approximately 78 years, which is markedly different from the life expectancy of a horse. Different dose regimens of chemotherapeutic agents are therefore used in horses to avoid intolerable side effects and insure quality of life- but at the risk of a lower chance of complete remission. Though multiple drug chemotherapy protocols are potentially more effective as they allow the use of lower doses of drugs while maximizing the potential synergistic or additive effects of several different agents, single dose protocols have also been effective. Doxorubicin is likely the most widely used systemic chemotherapeutic agent with the exception of corticosteroids. Doxorubicin, administered as a series of 6 treatments, has been shown to be a safe and effective treatment against many types of tumors. In a pilot phase II study, doxorubicin was found to have a response rate of 100% of lymphomas and 100% of carcinomas, but was less effective in the treatment of melanomas and sarcoids. [4] Multicentric lymphoma, which is characterized by involvement of both peripheral and internal lymph nodes and often internal organs, has been treated successfully with a multiple drug therapy protocol including cytarabinine, chlorambucil, and high-dose prednisolone in addition to doxorubicin alone. Corticosteroids have been the most commonly used systemic chemotherapeutic agent and can induce remission for years in some cases. Corticosteroids act on lymphocytes directly by increasing lymphocyte apoptosis as well as reducing T cell proliferation through inhibition of IL-2. Though continuous corticosteroid therapy is necessary in some cases, many cases respond to "pulse-dose" treatment, or treatment with high doses over 4 weeks with remission up to several months. If there is any chance of treatment with other chemotherapeutic agents, corticosteroids should be avoided as tumor cell resistance to steroids may confer resistance to other chemotherapeutic drugs. In other words, if systemic chemotherapy is an option, the most effective treatment option should be pursued first, not only after the inevitable failure of corticosteroids.

Immunotherapy

Immunotherapy seeks to activate the horses' own immune system to actively destroy neoplastic cells. The injection, either intratumorally or intravenously, of mycobacterial cell-wall extracts is a common attempt at immunostimulation, but this treatment lacks evidence to support its use. There has been considerable interest in the application of the canine melanoma vaccine (Oncept®; Merial) to horses with melanomas. This vaccine is a xenogenic DNA vaccine that codes for human tyrosinase and was developed for use in dogs in which all gross tumor and local lymph node metastasis has been eliminated. The Oncept vaccine has been shown to elicit the production of antibodies against tyrosinase in horses, though evidence of cell-mediated immunity has not been documented.[5] Intratumoral immunotherapy has also been performed using injection of DNA plasmids containing interleukin (IL) -12 and IL-18 to stimulate the production of interferon-γ and activate cytotoxic T lymphocytes in equine melanoma with some success.[6-8] Recent developments in the human medical field including the development of oncolytic viruses, which selectively kill cancer cells and induce systemic tumor-specific immune responses, are promising, though to the authors' knowledge this has not been attempted in horses.[9] The application of immunotherapy to equine neoplasias is an exciting step in veterinary oncology.

Electroporation

A limiting factor of intratumoral administration of cytotoxic or immunologic agents is systemic absorption rather than tumor cell uptake. Electroporation has been used in conjunction with intratumoral chemotherapy/immunotherapy to increase intracellular uptake of the cytotoxic agent. Electrochemotherapy is the administration of controlled electrical pulses directly to the tumor immediately after intratumoral chemo/immunotherapy to increase the cellular uptake of the injected agent.[10] In addition to altering cell membrane permeability, electrical stimulation causes local, transient vasoconstriction which may further increase local concentration of the injected agent. A drawback of electrochemotherapy is the necessity of repeated deep general anesthesia for treatment.

Radiation therapy

Radiation therapy is considered one of the most effective non-surgical treatments of localized superficial and deep cutaneous tumors. Lack of both expertise and evidence-based medicine in equine radiation oncology, in addition to the logistical difficulties associated with radiation therapy, make this mode of treatment less frequently applied. Radiation can be administered using a linear accelerator which generates accurately defined x and gamma-rays (teletherapy), and can be implanted directly into the tumor (intersititial brachytherapy), or applied to the surface (plesiotherapy). The goal of radiation therapy is to cause tumor cell double-stranded DNA breaks which ultimately results in destruction of the tumor. Though normal tissue is inevitably affected by radiation, healthy tissue is generally less radio-sensitive and is able to heal faster than neoplastic tissue. Different types of cutaneous neoplasia have differing sensitivities to radiation. Lymphoma, for example, has a high radio-sensitivity, though due to the disseminated nature of most cases of equine lymphoma, radiation is not a good treatment option. The size of the tumor is also a limiting factor, as is the response to previous treatment; tumors that have been treated previously with surgery, or multiple chemotherapeutic drugs and have recurred are less likely to respond to radiation. Radiation therapy has been used to successfully treat sarcoids, myxosarcoma, and squamous cell carcinoma in the horse, though documentation in the literature is sparse. Melanocytic tumors are, unfortunately, relatively radio-resistant, though radiation may still be a good treatment option if the melanoma is singular in nature and not near any excessively radio-sensitive tissues.

Topical treatments

Topical therapies have been used extensively due to ease of administration, though misapplication or discontinuation of treatment by laypeople before the neoplastic condition has completely resolved has resulted in some of the more dramatic treatment failures that the author has witnessed, particularly in the case of sarcoids. All of these topical agents are by their nature quite caustic, and treatment failure often occurs secondary to lack of patient compliance rather than poor client diligence. There is also concern that topical therapeutics do not adequately penetrate deep cutaneous tumors. Lack of drug penetration can be circumvented by continued application, though in the case of ulcerative neoplastic lesions it can be difficult to determine when the tumor has resolved. However, when applied carefully and consistently, topical therapeutics can be curative. Imiquimod, 5% fluorouracil, and mitomycin C are examples of topical therapeutics whose use is supported in the literature; other treatments have been met with anecdotal success though documentation is often lacking.[11-13]

REFERENCES

1 Valentine BA Vet Dermatol 2005; 38(6):698.
2 Theon AP, et. al JAVMA 2007; 230:1506
3 Hewes CA, et. al JAVMA 2006; 229:617
4 Theon AP, et. al JVIM 2013; 27:1581
5 Phillips JC, et al. AJVR 2012; 73:272

6 Heinzerling LM, et. al Jour Mol Med 2001; 34: 692
7 Muller J, et. al J Immunother 2011; 34:58
8 Trinchieri G Ann Rev Immuno 1995;13:251
9 Dharmadhikari N, et. al Current Treatment Oncology 2015; 16:326
10 Cemezar M, et. al JVIM 2008; 22:826
11 Nogueira S, et. al Vet Derm 2006;17:259-265
12 Paterson D Vet Rec 1997;141:626-628
13 Malalana F, et. al Vet Rec 2010;167:373-376

Treatment of Equine Neoplasia: What Are the Options, and What is the Evidence?

Imogen Johns, BVSc, DACVIM (Large Animal Internal Medicine)
North Mymms, United Kingdom

INTRODUCTION

Treatment of equine neoplasia has tended to lag behind that of small animals, although therapeutic modalities for cutaneous neoplasia are well described. Neoplastic conditions of the skin are a common, and typically frustrating, problem encountered in both ambulatory and referral practice, with the main tumour types described being sarcoids, squamous cell carcinomas and melanomas. The scientific literature describing therapeutic options for cutaneous neoplasia is dominated by retrospective studies describing either individual modalities or comparing several different treatment options. These studies tend to be plagued by the well-known weaknesses of retrospective studies, with variable inclusion criteria and prior treatment attempts, uncontrolled selection of treatment modalities, variable definition of treatment 'success' and loss of patients to follow up. As a result, it is next to impossible to make definitive comparisons of treatment efficacy between studies, and sometimes even within studies. The purpose of this summary is to describe the treatment options for cutaneous neoplasia, and the evidence base which exists for each treatment. Although a much smaller evidence base exists for treatment of systemic neoplasia, chemotherapeutic options for the management of systemic lymphoma will also be discussed.

OPTIONS FOR TREATMENT OF NEOPLASIA

The goal in the treatment of any neoplastic condition is to destroy tumour cells, preserve normal cells and thus resolve the condition. Failure to kill all the tumour cells will usually result in recurrence, whilst failure to preserve normal cells can result in local or systemic toxicity. Even within one tumor type, there is likely to be no one universally appropriate treatment, and the development of a therapeutic plan will depend on the tumor type, its size and location, the extent and duration of any previous treatment (especially in the case of sarcoids), as well as the available facilities, experience of clinicians and the expectations of the owners. Oncological therapy can be categorized as ablative (including surgery, laser surgery, cryotherapy and hyperthermy); cytotoxic (chemotherapy (intralesional/topical/systemic); radiation therapy); and biological (immunotherapy and vaccines). In many situations, a combination of these approaches is required, and repeat treatment is typical unless surgical resection allows for complete removal of all tumor cells.

TREATMENT OF SARCOIDS

The wide variety of treatment options available for sarcoids is perhaps an indication that no one treatment is 100% reliable, and owners should thus be warned that most treatments will be prolonged, that recurrence may occur, and that new sarcoids can develop in other sites during and subsequent to treatment. Although no treatment has been uniformly successful, incomplete surgical resection can almost guarantee recurrence, often with a more aggressive form of the tumor. Recurrence in 82% of periocular tumors following surgical resection has been reported.[1] Higher success rates of 60-89% have been described following surgical resection of select cases using either conventional or laser surgery.[2-4] A resolution rate (mean follow up 20 months) of 82% (conventional surgery) and 89% (laser surgery) was reported for tumors not located in the periocular region, that were freely moveable and located in areas where 8mm+ margins were possible.[2] A similarly high resolution rate (83%; mean follow up time of 18 months) has recently been reported following removal via laser of 235 sarcoids in 99 horses, with the authors reporting that poorer outcomes were associated with treatment of a verrucous sarcoid, or sarcoids located on the head or neck.[3] A resolution rate of only 60% was reported following laser removal of a variety of sarcoid types in a number of anatomical locations, suggesting that careful case selection is important to maximise the chance of resolution.[4]

Evidence for the effectiveness of cryotherapy is limited, and most authors suggest it to be of limited value, in particular for treatment of periocular sarcoids.[1,5] A resolution rate of 78% (some tumors were surgically debulked prior to treatment) was however reported by Martens[2], and in a prospective randomized clinical trial, a 100% success rate was reported for cryotherapy of 26 tumors in 10 horses.[6] The high success rate (as determined by measurement of tumor size with mean follow up times of 18 months) in this report may be associated with careful tumor selection (only relatively small tumors and no periocular tumors), careful application of the cryotherapy, in particular use of thermocouples to ensure an appropriately low temperature of treated tissue, and repetition of treatment until tumor resolution.

Intralesional chemotherapy, most commonly using cisplatin as an emulsion or in biodegradable beads, is most applicable for small tumors as a stand-alone therapy, or following surgical debulking of larger tumors. In a retrospective case series, 409 tumors in 386 horses were treated with a protocol of 1mg cisplatin/cm^3 of tumor as an emulsion in sesame oil every 2 weeks for 4 treatments.[7] Larger tumors were debulked prior to treatment. An estimated cure rate (no recurrence at 4 years) of 96% was reported, although a much lower success rate (33%) was reported in a different study for treatment of periocular sarcoids, with the authors reporting technical difficulties including leakage of drug following injection.[1] Cisplatin electrochemotherapy, which involves injection of cisplatin intratumorly followed by administration of an electric current to the area to maximise intracellular drug concentrations has recently been reported, with a resolution rate of close to 100% after a mean of 2.8 treatments.[8] The procedure must be performed under general anesthesia as horses do not tolerate the electrical impulse even with profound sedation. Cisplatin beads, either alone in small tumors or following surgical debulking in larger masses may overcome some of the reported difficulties associated with intralesional injection of the drug.[9] A 2 year relapse free rate of 84% in 13 treated sarcoids has been reported, with a mean of 2.7 treatments required for each horse. Other intratumoral treatments include 5 fluorouracil (5 FU) and bleomycin, with limited evidence for efficacy.[5]

Topical treatments are an attractive option for many owners due to their ease of application and (assumed) lower cost. Although many topical medications have been described, only those with at least some evidence of efficacy in the scientific literature will be mentioned. In the United Kingdom, a preparation known as AW4-LUDES ('Liverpool Cream') is formulated for individual patients, and contains the antimetabolite 5 FU, heavy metals and thiouracil.[10] The cream is not suitable for periocular sarcoids, with resolution in only 35% of 159 treated cases but appears to be beneficial in management of sarcoids in other sites.[1,10] Imiquimod ('Aldara') may be beneficial in the treatment of some sarcoids, likely due to stimulation of a local inflammatory response. The cream is applied every other day or every third day for up to 8 months, with resolution of 9/15 treated sarcoids reported.[11] Various preparations containing blood root, a plant based alkaloid with cytotoxic and immune modulating effects are available including XXTERRA, Newmarket Bloodroot Ointment and Animex. In a retrospective questionnaire based survey, complete regression of tumors was reported in 66% of 74 tumors, with a mean follow up of 3.5 years.[12] The anti-viral acyclovir has been investigated for topical treatment of sarcoids, based on the correlation between bovine papillomavirus 1 and 2 infection and sarcoid development. Applying the cream daily for 2 months resulted in complete regression of 68% of tumors, with smaller, occult tumors less than 5mm diameter having the best response.[13] Similarly, topical 5 FU has been used, with a small number of cases (6/9) achieving complete resolution.[1]

A consistently high success rate has been reported for radiation therapy of sarcoids, in particular for periocular lesions.[1,14] Because of the expense, potential for human radiation exposure and lack of available facilities, it is uncommonly utilised.[5] Although teletherapy (external beam radiation) has been used, more commonly, interstitial brachytherapy (primarily with Iridium-192) which involves implantation of radioactive implants in to the tumor, with the horse then being 'radioactive' until implant removal has been utilised. Reported success rates of 74-98% have been described, with slow tumor regression typical of this therapy.[1,5,14] More recently, high dose interstitial brachytherapy using a technique of remote afterloading has been utilised, and should improve operator and patient safety.[5]

Intratumoral BCG has been used extensively and appears to be most beneficial in the therapy of periocular sarcoids.[1] Complete resolution following treatment has been described in 69% of horses (309 cases)[1]; 100% of 31 equids[15]; 59% of 60 horses[16] and 83% of 29 tumors.[6] A series of injections is required, and due to the method of sarcoid destruction, local inflammation is expected. With repeated injections, the risk of anaphylaxis is increased and pre-treatment with NSAIDs and corticosteroids is recommended.[5] 'Vaccination' is an attractive treatment option as many horses have multiple lesions that require treatment. Autologous 'vaccines' using frozen sections of sarcoid harvested from a sarcoid removed from the horse to be 'vaccinated', and then implanted subcutaneously have been described.[17] Using this technique, regression occurred in 69% of 16 treated horses with a median follow up of 10 months. Other authors have suggested that this method has very poor success rate and is not recommended.[5] Immunotherapy using virus-like particles of chimeric BVP 1 has been described in a prospective placebo controlled trial of 17 donkeys with no difference in regression rate between placebo and control.[18]

TREATMENT OF SQUAMOUS CELL CARCINOMA

Mucocutaneous squamous cell carcinoma (SCC) is the second most common cutaneous neoplasia after sarcoids. Common sites include the periorbital region and the external genitalia, especially in older male horses. Treatment approaches typically differ between locations, although resection with clear margins is curative, assuming metastasis has not occurred. Higher grade (poorly differentiated) tumors are more difficult to treat and more likely to metastasize, suggesting histopathological examination is an important part of the therapeutic plan. Tumors of the nictitans are best treated by complete resection, with no recurrence in 24 treated horses (median follow up 41 months).[19] A second report described a recurrence rate of 19% in 37 horses, and although lack of surgical margins and incomplete resection were not statistically associated with a risk of recurrence, the authors commented that the small numbers on the study may explain this finding.[20] In a large retrospective study investigating the utility of adjunctive radiation therapy on ocular and adnexal SCC, the recurrence rate for horses treated with adjunctive radiation therapy was significantly lower than for horses that did not receive radiation (11.9% vs 44.1%).[21] Of the 115 horses originally evaluated however, 41% were lost to follow up, and at least 7 different types of treatment apart from radiation were utilised, making strong conclusions from this study difficult. Intralesional therapy with cisplatin emulsion had a success rate of 88% in 151 tumors, most of which were located in the periorbital region[7], although only 6/10 tumors responded to cisplatin beads.[9] Complete resection of masses at the cornea-scleral junction by superficial keratectomy is typically limited by an inability to obtain complete margins, and as such, adjunctive therapy is recommended. Mitomycin C is a chemotherapeutic antibiotic which has been used both as a stand-alone treatment of ocular SCC and an adjunctive therapy after surgical resection.[5] A one-off intra-operative application of a 0.4mg/ml solution as well as a repeated 7 day treatment cycles (q6-8 hours) of a 0.04mg/ml solution have been described, with the latter appearing to be most efficacious.[22-24] In a small retrospective study, the non-recurrence rate using mitomycin C after surgery was similar to that of surgery followed by CO_2 laser application (82.4% (14/17) vs 85.7% (6/7) respectively, mean follow up 19 vs 47 months).[22] The drug has also been used as a sole therapy in horses with ocular SCC, with resolution in 6/8 horses.[23] Photodynamic therapy (PDT) using topical verteporfin after surgical excision of periorbital SCC in 10 horses resulted in resolution of all tumors with no regrowth (minimum follow up 25 months).[25] This may be an attractive treatment option for small tumors, or following surgical debulking of larger tumors, although the procedure requires specialized equipment, general anaesthetic, and is only suitable for the most superficial tumors (or following surgical debulking) .

The external genitalia of older male horses is the next most common predilection site for SCC, after the periorbital area. Early treatment is likely to be most effective, especially for surgical resection of tumors where margins of 0.5-1cm are suggested. Although a complete discussion of surgical techniques is beyond the scope of this presentation, a high recurrence rate (29.5%) following surgical resection in 53 treated horses has been described, with incomplete margins (usually following a partial phallectomy) and a higher grade of tumor associated with an increased risk of recurrence.[26] Because of the relatively poor outcome associated with genital SCC (perhaps due to the advanced nature of the disease at first presentation), adjunctive therapy is often recommended, although no good evidence exists as to which treatment may be most effective. Although the use of COX inhibitors such as piroxicam has been described, COX expression appears to be variable in equine SCC, and only one report of apparently effective long term control has been published.[27,28] A recent report described metronomic chemotherapy after surgical debulking of an oral mass.[29] Using meloxicam and cyclophosphamide on a continuous basis at doses significantly lower than the conventional maximally tolerated doses allowed for 5 months of 'excellent quality life' in the treated horse. Although not designed to be curative, this approach may allow for palliative care of cases not amenable to other forms of treatment. Various other treatments have been described, all of which have a relatively small evidence base and most of which are only suitable for treatment of small superficial tumours.

TREATMENT OF MELANOMA

Although melanoma are reported to be the third most common skin tumor in horses, and will likely affect more than 80% of gray horses, relatively little evidence exists to describe treatment efficacy. This may be due to the fact that, historically, the tumor has been considered benign and that treatment was not indicated, and if attempted was unrewarding. However, although early tumors may be benign and typically slow to grow, a significant proportion (up to 30%) become malignant and may metastasize. As such, early treatment, if indicated, may result in a more positive outcome in these typically challenging cases. Early small lesions can be successfully resolved with surgical resection.[30] In addition, surgical debulking, either via conventional surgery or laser surgery can be successfully employed prior to adjunctive therapy, in particular intralesional cisplatin emulsion. Théon (2007) reported an 87.5% estimated cure rate for 16 tumors in 13 horses following 4 injections at 2 week intervals in tumors with a maximum diameter of 5cm.[7] A similarly high success rate was found in 13 horses treated with intralesional cisplatin beads.[9]

Because horses are typically affected by more than one melanoma, systemic therapy is an attractive option. Cimetidine has been advocated as a treatment for melanoma with its anti-tumor effects proposed to be due to immunomodulation and/or via antagonism of H_2 receptors on neoplastic cells. In a prospective randomized placebo controlled trial utilizing either 3.5mg/kg PO BID or 7.5mg/kg PO SID for 60 days, no difference between treated and placebo groups in terms of size and number of tumors was identified.[31] Individual case reports have reported some efficacy, although the consensus appears to be that treatment is unlikely to be beneficial in the majority of cases.[5] Specific immunotherapy utilizing 'vaccines' that target tumor associated antigens has been investigated in horses. Two small studies using autogenous vaccines reported tumor regression and subjective improvement.[32,33] Induction of a non-specific immune response via IL-18 and IL-12 encoding plasmid DNA was shown to cause significant tumor regression in a prospective randomized double blind placebo controlled study in 26 horses, suggesting that this method (although not commercially available) may hold promise for the treatment or control of melanoma in horses.[34] A xenogeneic DNA based vaccine licensed for use in dogs has been shown to produce a tyrosine specific anti-tumor response in healthy horses, and anecdotally has been used with success in melanoma bearing horses.[35] To date, the results of a trial in affected horses have not been published.

TREATMENT OF LYMPHOMA

Lymphoma is the most common hematopoietic neoplasm in horses, with most being multicentric and T cell rich B cell tumors. Although treatment of lymphoma has been described in individual case reports, most treatment protocols are extrapolated from those used in dogs and cats.[36-38] Recently, a Phase I trial and a follow up pilot Phase II study have investigated the efficacy and biosafety of doxorubicin (70mg/m^2 every 3 weeks).[39,40] All horses treated for lymphoma showed tumor regression (tumor volume decrease of greater than 50%) with an estimated mean response rate of 30 months. Table 1 lists reported chemotherapeutic protocols for lymphoma.

SUMMARY

Successful treatment or management of neoplasia in the horse is often limited by lack of available 'cost effective' options, limited facilities (especially radiation oncology) and owner perceptions. Recognition of the need for elimination of all neoplastic cells and of the limitations of available options will allow for both appropriate and realistic treatment plans. Advising owners of the need for realistic expectations, and a commitment to what is likely to be a prolonged and repeated treatment course may improve in particular 'success' rates for cutaneous neoplasia. Palliative treatment of systemic or diffuse neoplasia may become a more realistic option in the future as evidence for appropriate regimes increases.

Table 1: Chemotherapeutic drugs used in the management of equine lymphoma

Drug	Dose and route of administration	Frequency	Reference
Cytarabine Cyclophosphamide	200-300mg/m^2 IM or SQ 200mg/m^2 IV	Every week, alternating with cyclophosphamide Every week, alternating with cytarabine	5,37,38
Prednisolone	1mg/kg PO	Q24-48 hours	5,37,38
Vincristine	0.5mg/m^2 IV	Can be added to above protocol if no response	5
Lomustine (CCNU)	65mg/m^2 PO (NGT)	Q30 days (in combination with pred)	36
Doxorubicin	70mg/m^2 IV	Every 3 weeks	39,40

REFERENCES

1. Knottenbelt D, et al. Vet Ophthalmol 2000;3(2-3):169. 2. Martens A, et al. Vet Rec 2001;149(22):665. 3. Compston P, et al. EVJ 2016 epub ahead of print. 4. Carstanjen B, et al. Can Vet J 1997;38(12):773. 5. Knottenbelt D, et al. Clinical Equine Oncology 2015 Elsevier. 6. Klein W, et al. Cancer Immunolo Immunother 1986;21(2):133. 7. Théon A, et al. JAVMA 2007;230(10):1506. 8. Tamzali Y, et al. EVJ 2012;44(2):214. 9. Hewes C, et al. JAVMA 2006;229(10):1617. 10. Knottenbelt D, et al. EVE 1994;6(2):72. 11. Nogueira S, et al. Vet Dermatol 2006;17(4):259. 12. Wilford S, et al. Can Vet J 2014;55(7):683. 13. Stadler S, et al. Vet Rec 2011;168(7):187. 14. Byam-Cook K, et al. Vet Rec 2006;159(11):337. 15. Lavach J, et al. EVJ 1985;17(6):445. 16. Vanselow B, et al. EVJ 1988;20(6):444. 17. Rothacker C, et al. Can Vet J 2015;56(7):709. 18. Mattil-Fritz S, et al. J Gen Virol 2008;89(1):138. 19. Labelle A, et al. EVJ Suppl 2011;40;42. 20. Scherrer N, et al. JAVMA 2014;245(7):812. 21. Mosunic C, et al. JAVMA 2004;225(11):1733. 22. Clode A, et al. Vet Ophthalmol 2012;15(4):254. 23. Malalana F, et al. Vet Rec 2010;167(10);373. 24. Rayner S, et al. Aus Vet J 2006;84:43. 25. Giuliano E, et al. Vet Ophthalmol 2014;supp 1:37. 26. Van den Top J, et al. EVE 2011;23(12)636. 27. Van den Top EVJ 2014 46:618-24. 28. Moore A, et al. EVJ 2003;35(7):715. 29. Tornago R, et al. EVE 2016 epub ahead of print. 30. Rowe E, et al. JAVMA 2004;225(1):94. 31. Laus F, et al. Israel J Vet Med 2010;65(2):48. 32. Jeglum K. CVET 4. 33. McGillivray M, et al. JVIM 2002;12:452. 34. Muller J, et al. J Immunother 2011;34(1);58. 35. Lembcke L, et al. J Eq Vet Sc 2012;32(10):607. 36. Doyle A, et al. Can Vet J 2013;54(12):1137. 37. Saulez M, et al. JAVMA 2004;224(5)733. 38. Finding E, et al. EVE 2014;26(6):303. 39. Théon A, et al. JVIM 2013;27(5):1209. 40. Théon A, et al. JVIM 2013;27(6):1581.

Equine Lymphoproliferative Disease

Angela E Frimberger, VMD, DACVIM (Oncology), MANZCVS
Wauchope, NSW, Australia

Equine lymphoma is common, and the clinical presentation and course of disease are variable. A wide range of clinical presentations, tumor subtypes, and outcomes have been described. Leukemias however have been rarely reported in the published literature. This lecture will review the literature on both diseases and present anecdotal experiences.

LYMPHOMA

Tumor behaviour and important prognostic factors

Most cases of equine lymphoma are rapidly progressive but a subset, usually cutaneous / subcutaneous disease, appears to have an indolent course and survival for months to years with little or no therapy has been reported in some patients.

A large case series of 203 horses with lymphoma was published in 2013. Ages ranged from 2 months to 31 years and the mean age of presentation was between 10 and 11 years. Of 14 lymphoma subtypes identified, T-cell–rich large B-cell lymphomas (TCRLBCL) were the most common (43%) in both the cutaneous and multicentric groups; and of the cutaneous lymphomas 71% were TCRLBCL. A large recent survey of all types of skin tumors from over 3200 horses reported that thoroughbreds showed an increased incidence of cutaneous lymphoma. In 2014, 4 further cases of neurologic involvement in horses with TCRLBCL were reported. There is no evidence for any heritable or familial component to equine lymphoma.

In dogs, the three most important prognostic factors for lymphoma are stage, substage and immunophenotype. Although immunophenotype of lymphoma in equine lymphomas has been reported in several studies, there have not been any studies to determine if immunophenotype is prognostically significant as it is in dogs and possibly in cats. Likewise the impact of clinical stage has not been evaluated in horses, although there is some obvious difference in outcomes for horses with different anatomic variants (e.g. multicentric versus cutaneous). Finally, substage is a powerful prognostic factor in dogs (a: clinically well or b: unwell), with substage a carrying a better prognosis with treatment.

Horses with cutaneous lymphoma can have a waxing and waning and variable clinical course, but unfortunately most horses with multicentric lymphoma are euthanized because of poor condition weeks to months after diagnosis.

Treatment options

For horses with localized disease that can be resected completely, it may be possible to achieve a cure in selected cases.

An alternative local treatment option in some cases can be radiation therapy. The advantage of treating lymphoma with radiation therapy is that very low doses are lymphocytolytic and therefore can cause a long-term response with low total doses, minimizing the chance for side effects. Four horses that had nasal, perineal, maxillary and tarsal lymphoma with no evidence of systemic disease achieved complete remission for more than 2 years with radiotherapy (all were alive between 29 and 144 months after therapy).

Combination chemotherapy with a regimen containing vincristine, cyclophosphamide, doxorubicin, L-asparaginase, and possibly other drugs, (VELCAP-type protocols) represents the current standard of care in small animal oncology, and has been used in some horses with lymphoma. Long-term follow-up is uncommonly reported, and anecdotally most horses with multicentric disease relapse within 6 to 8 months of chemotherapy; however compared to the median survival time of 6 to 8 weeks (5 of 7 lived 1-3 months, and only 2/7 lived > 7 months) for horses with *multicentric* lymphoma treated only with glucocorticoids, this suggests there is a clinical benefit in horses at least with *systemic* lymphoma. In addition there are case reports of long-term responses.

A relatively recent article reported a case of a 12-year-old gelding with lymphoma in multiple cutaneous sites treated successfully with surgery and lomustine and prednisolone chemotherapy. Lomustine has become a mainstay of therapy for canine cutaneous lymphoma and could be included in a combination protocol for horses with lymphoma.

Cyclophosphamide is a mainstay of first line therapy for lymphoma in all species that have been treated, and there is recent evidence to suggest that it may be much better tolerated in horses than previously thought – it was previously given at essentially the "canine dose". An abstract from the 2013 VCS Conference reported cyclophosphamide pharmacokinetics in 2 horses after the published dose of 200 mg/m^2 IV, then after 400 mg/m^2 IV and 600 mg/m^2 IV. In the Results they only reported the PK data, which showed *no drug or metabolite detected at 200* mg/m^2, so quite possibly this dose should be escalated.

Doxorubicin as a single agent has been reported in 12 horses with 37 measurable tumors (6 with lymphoma (2 T-cell-rich B-cell, 2 T-cell and histiocyte-rich B-cell lymphoma, and 2 T-cell lymphoma)), after a dose-finding study in 17 horses with 34 tumors. In the horses with lymphoma treated at the full dose of doxorubicin, the reported response rate was 100%, with complete responses in the 2 T-cell lymphomas for 9 and 29 months, complete responses in 3 of the B-cell lymphomas lasting beyond 11, 18, and 57 months after treatment, and the fourth B-cell horse having a partial response for 4 months. Cardiotoxicity was not seen in any of the horses in these 2 studies. In contrast to this published data, unpublished anecdotal experience with 2 horses treated with doxorubicin according to the same administration protocol found that both tolerated it poorly; with (1) marked hypersensitivity-type reaction with several days' fever, tachycardia and depression despite full supportive care; and (2) no problems on the first infusion but a similar reaction as in the first horse upon the second infusion, which was unresponsive to supportive care and resulted in the horse's death within a week of treatment. Although anecdotal, these data suggest caution in using doxorubicin as first-line treatment for horses with lymphoma; but it may be possible to re-consider it if the disease is unresponsive to other drugs and the risk / benefit ratio is shifting.

An individual horse with lymphoma associated with EHV-5 has been reported to have had a good response to valacyclovir therapy. In homologous human settings, 3 individual cases of primary effusion lymphoma, which is associated with a herpes virus (HHV-8) have been reported to respond well to antiviral therapy. Epstein-Barr virus (also a herpes virus) -related Burkitt's lymphoma has not been treated with antiviral therapy as sole therapy to my knowledge. However, combination of valacyclovir with chemotherapy is under investigation.

Less aggressive / less costly therapy could include corticosteroids alone or lomustine alone. Although well tolerated in the reported horse and in anecdotal experience, lomustine has the potential for hepatotoxicity and nephrotoxicity in humans and dogs. This has not yet been reported in horses, but very few horses have been treated with this drug so until it is better characterized it may be prudent to avoid relying on it as a single agent to minimize the cumulative dose.

If there is a quality of life impact being caused by a particular lesion, radiation therapy could be palliative even in the absence of chemotherapy.

Responsiveness to hormonal manipulation has been reported in some equine lymphomas, and appears to be related to the expression of estrogen and progesterone receptors. This opens the possibility that hormonal manipulation may offer some treatment benefit.

For some horses fertility may be important. Before starting chemotherapy, consideration should be given to banking as much semen as possible. Male fertility during chemotherapy is unpredictable but it is known that alkylating agents have the greatest impact on fertility; unfortunately these are also among the most effect and safest drugs for horses with lymphoma. In humans the return of fertility after finishing chemotherapy is variable and some men remain infertile. Because of temporary genetic damage, humans are recommended to wait 2 years after finishing chemotherapy before fathering – this time frame may be somewhat reduced in horses but in the situation of a commercial stud it is probably safest to wait that time.

So few horses that have been treated with chemotherapy that it is impossible to make any statistically significant or evidence-based assertions that even chemotherapy compared to no chemotherapy translates into prolonged survival for horses with TCRBCL, let alone to compare different chemotherapy approaches against each other. Having said that, my personal bias is that combination chemotherapy would probably be "better" than shorter / less intense chemotherapy. Overall, dogs and people with high grade lymphoma clearly live much longer and with a better quality of life with any chemotherapy than with none. Further, the more complex the chemotherapy protocol, such as 5- or 7–drug combinations compared to 3-drug or single agent protocols; the better the remission rates and survival times. And, as a general rule of chemotherapy biology, combination therapy is always preferred over single-agent therapy because it is the best way to prevent drug resistance.

LYMPHOID LEUKEMIAS

Tumor behaviour and important prognostic factors

In other species, the major divisions of leukemias are lymphoid and nonlymphoid; and among lymphoid leukemias, the major subdivisions are acute lymphoblastic leukemia (ALL) or chronic lymphoid leukemia (CLL), with each potentially having both T-cell and B-cell forms.

In horses, leukaemias on the whole have been rarely reported in the published literature, with only approximately 20 horses reported with primary lymphoid leukemias. For the horses with details available (Table 1), ages ranged from 3 to 23. Most of them are not specifically described as ALL or CLL, but from reading the case descriptions, 13 appear to have been probably CLL and 3 probably ALL, based on clinical history and cell size. The ALLs were all in horses 8 years old and younger (two 3-year olds and an 8-year old). All the CLLs were in horses 9 years old and over. Thirteen of the leukemias were categorized as B- or T-cell: only 4 of these were B-cell and 9 were T-cell. Of the 3 ALLs, two were not immunophenotyped and 1 was B-cell. Of the 12 CLLs, 2 were not immunophenotyped, 9 were T-cell and 3 were B-cell. Of the 3 B-cell CLLs, one was a hairy cell leukemia in a 9-year old Friesian mare; one was a 20-year old horse with a 2-day history of clinical signs, thrombocytopenia and a neoplastic cell count in the peripheral blood of 70,200 of "slightly large" lymphocytes, and was euthanized at diagnosis; and the other was a 20-year old horse with a 2-year history of clinical signs, thrombocytopenia and a neoplastic cell count in the peripheral blood of 67,000 of medium-sized lymphocytes; was treated with prednisolone and was euthanized on clinical progression after 3 ½ months of therapy. These latter 2 horses may have had ALL.

Canine CLL patients are often asymptomatic at presentation, with lymphocytosis ranging from 6 to greater than 200/uL. There is no obvious sex or breed predisposition with the exception of the subtype derived from large granular lymphocytes (LGL-CLL) which is most common in females. On clinical examination, mild lymphadenopathy is noted in up to 80% of affected dogs with CLL. Splenomegaly and hepatomegaly, attributable to cellular infiltration or to extramedullary haematopoiesis, are often found. Unlike in acute leukemias, peripheral cytopenias tend to occur in a relatively small subset of cases and are generally mild, as marrow is rarely replaced.

In other species, CLL is an uncommon myeloproliferative disease characterized by an increased number of circulating lymphocytes with normal morphology. CLL occurs in three distinct stages: an asymptomatic period, which may last for months despite a high number of circulating lymphocytes; a symptomatic stage when therapy is required; and a third stage, malignant transformation (blast crisis) in which more blast cells arise and eventually predominate. Progression from CLL to lymphoma has been reported but is a rare and usually late event (more than a year after starting therapy). In human patients staging is used to decide when to start treatment (see more under Treatment Options).

In contrast, clinical signs are usually acute with ALL. Thrombocytopenia and anemia are frequent in dogs with ALL, and neutropenia may signal severe bone marrow infiltration. Nonspecific signs such as lethargy and anorexia are common; vomiting, diarrhea, abdominal pain, polyuria, polydipsia, and shifting lameness occasionally are seen. Affected animals are usually thin. Between 50% and 60% have lymphadenopathy, which, as for CLL, is often mild compared to the dramatic lymphadenopathy or other organ infiltration seen with lymphoma. Splenomegaly, hepatomegaly, and pale mucous membranes are the most common abnormalities detected on physical examination. Additional infiltration of other organs may occur, making differential diagnosis between advanced lymphoma and ALL difficult. Although uncommon at the time of first presentation, in dogs ALL is reported to have a higher rate of relapse in the central nervous system (CNS) than lymphoma; signs are usually dullness, depression and changes in behaviour. The response to treatment in dogs with ALL is also usually very brisk, in both clinical signs and circulating counts.

One study in dogs (not all dogs were treated) showed that dogs with lymphocytosis of any cause with positive staining for CD-34 in the tumor cells had an extremely poor prognosis. In this study, dogs with B-cell leukemia had a median survival of 129 days if cells were "large" and >1,000 days if "small". Whether this would also be true for horses is unknown.

Table 1. Summary of published equine leukemia cases

	Report	Age (yr)	Breed	Signs / Duration	CBC	Lymph cell size	Count /ul	B or T	CLL or ALL	Treat-ment	Survival time
1	Dascanio 1992	18		2 mo		Small	100	T	CLL	-	2 mo
2	Dascanio 1992	20		2 days	Thrombo-cytopenia	Slightly large	70.2	B	CLL	-	Euth at diagnosis
3	Lester 1993	3		4 wk	Pan-cytopenia	Marrow blasts	~	~	ALL	-	Euth at diagnosis
4	Lester 1993	11		2 wk	Pan-cytopenia	Marrow well-differentiated lymphocytosis	~	~	CLL	-	Euth at diagnosis
5	Lester 1993	3		2 days	Pan-cytopenia	Marrow blasts	~	~	ALL	-	Euth at diagnosis
6	McClure 2001	20		2 years	Thrombo-cytopenia	Medium	67	B	CLL	pred	3 ½ mo
7	McClure 2001	16		7 mo	Thrombo-cytopenia	Small	208	T	CLL	–	45 days
8	Rendle 2007	14		Lethargy	Thrombo-cytopenia	Small	94	T	CLL	Steroids	12 mo
9	Rendle 2007	23		Incidental, lethargy at 6 mo		Normal initially	8.7 - >88	T	CLL	-	5 y
10	Rendle 2007	20		Fever / weight loss	Thrombo-cytopenia	Atypical	63	T	CLL	Pred	11 mo
11	Rendle 2007	10		Rapid	Thrombo-cytopenia	Small atypical	192-> 492	T	CLL	Sup	1 wk
12	Rendle 2007	20		Fever / weight loss	Thrombo-cytopenia	Atypical small and medium	27	T	CLL	Sup	1 wk
13	Rendle 2007	20		Fever / weight loss	Thrombo-cytopenia	Mostly large atypical	77	T	CLL	Dex	3 days
14	Bezdekova 2009	8	Friesian	Few wk	Thrombo-cytopenia	Blasts	450	B	ALL	Steroids	5 days
15	Cian 2013	16		Incidental		Small	58	T	CLL or SCL	_	194 days
16	Achten-Weiler 2015	9	Friesian	Weight loss / Several mo	Anemia, leukocytosis	Atypical, poorly differentiated, hairy	20	~	CLL		Euth at diagnosis

Treatment options

In human patients, those with lymphocytosis, no evidence of lymphadenopathy or hepato-splenomegaly, and without anemia or thrombocytopenia (Stage 0) have a 60% chance of surviving > 20 years. Those with "smoldering" CLL (also have lymph node and/or spleen and liver enlargement, but no anemia or thrombocytopenia and a lymphocyte count less than 30,000 and that have a lymphocyte doubling time of > 1 year) also survive a similar time to patients without CLL. For human patients with anemia, thrombocytopenia or a doubling time of lymphocytes of <6 months; particularly accompanied by clinical signs, treatment is recommended.

There have been no reports of a horse with primary lymphoid leukemia treated with chemotherapy. Combination chemotherapy can be used for horses with lymphoma; and doses are available for horses for some chemotherapeutics, and this type of protocol should be considered for horses with ALL.

In humans and dogs chlorambucil and prednisone are the most active agents for CLL, and this type of protocol could also be adapted for the horse. In a study of 20 dogs treated for 1 to 3 weeks with vincristine, followed by chlorambucil and prednisone, 14 achieved long-term remission for a median survival of 12 months (range 1 month–3 years). Only 3 dogs failed to respond. Some dogs took up to three months to achieve a remission, and this is common, although symptomatic improvement usually occurs quickly and the response may be very rapid in the neutrophil and platelet lines if they are low. The contribution of vincristine to this protocol is uncertain, as similar results have been achieved with chlorambucil and prednisone alone; in other studies, chlorambucil and prednisone, even on intermittent schedules, have caused remissions of 10 months to 3 years. More aggressive therapy for CLL can be considered in cases where there is no response to chlorambucil and prednisone after 3 months of therapy, or if there is clinical progression the face of such therapy; or if clinical evidence of acceleration to ALL ("blast crisis") develops.

Animals with leukemia may be at risk of infection due to immunosuppression from the disease and the therapy. Secondary infection is a common cause of morbidity and mortality in affected patients, and antibiotics should be used as appropriate.

REFERENCES

Gollagher R D et al, Can Vet J 1993;34:371. Kelley L C and Mahaffey E A, Vet Path 1998;35:241. Rhind S M and Dixon P M, Vet Rec 1999;6:554. Henson K L et al, Vet Clin Path 2000;29(2):40. Oikawa M et al, J Vet Med 2003;50:151. Henson F M D et al, Eq Vet Educ 2004;16(6):312. Meyer et al, Vet Path 2006;43:914. Gerard M et al. Vet Radiol Ultrasound 2010;51(1):97. Campbell M, Proc Vet Cancer Soc Ann Meet, 2013:89. Doyle A J e al Can Vet J 2013;54:1137. Durham A C et al, Vet Path 2013;50(1):86. Schnoke A T et al, Vet Ophth 2013;16(1):35. Theon A P et al, J Vet Int Med 2013;27:1209. Theon A P et al, J Vet Int Med 2013;27:1581. Vender Werf K et al, J Vet Int Med 2013;27:387. Lehmbecker a et al, J Comp Path 2014;151:181. Trope G D et al, Vet Ophth 2014;17(2):139. Westerman T L et al, Can Vet J 2014;55:379.

Dascanio J J et al, J Vet Int Med 1992;6(4):225. Lester G D et al, J Vet Int Med 1992;7(6):360. McClure J T et al J Vet Int Med 2001;15:144. Rendle D I et al, Eq Vet J 2007;39(6):552. Bezdekova B et al, Veterinarni Medicina 2009;54(5):249. Cian F et al, Vet Clin Path 2013;42/3:301. Winter J et al, Tierarztl Prax Ausg G Grosstiere Nutztiere 2014;42(5):311. Achten-Weiler M et al, Vet Q. 2015 Nov 25:1

Electroporation in Oncology: Clinical Applications and Future Perspectives in Equine Medicine

Enrico P. Spugnini, DVM, PhD, DACVIM (Oncology), DECVIM-CA (Oncology)
Rome, Italy

INTRODUCTION

Electroporation is a technique that aims at ablating tumor cells through the direct application of current (irreversible electroporation or IRE) or at increasing the delivery of drugs through the cell membrane by applying proper permeabilizing pulses. Chemotherapy drugs are effective provided that they reach their site of action. The primary obstacle for chemotherapy agents is the cytoplasmic membrane, especially for lipophobic agents such as bleomycin. Electrochemotherapy (ECT) couples the administration of an anticancer agent (usually bleomycin or cisplatin), to the delivery of electric pulses having appropriate waveforms. The application of permeabilizing pulses leads to perturbation of the cell membrane, thus resulting in increased uptake of chemotherapeutic drugs, ultimately leading to cell apoptotic death.[3] Clinical trials evaluating humans treated with ECT for cutaneous neoplasms were first reported in the early 90s, and have been followed by many trials that confirm its potential for the treatment or palliation of dermatological tumors.[3-5] Recently, a standard operating procedure has been proposed.[6]

ELECTROCHEMOTHERAPY EQUIPMENTS

Different waveforms have been adopted by investigators: exponentially decaying, square, rectangular or biphasic. In general the number of pulses delivered is set in 8 single pulses applied per cm of tumor area at a voltage of 1300 V/cm (800 V/cm for intraoperative use), with a duration of 100 µs and a frequency of 1 Hz. Treatments are repeated until the whole tumor area is covered. We recently developed a shortened protocol with decreased morbidity and maintained efficacy.[7]

RESULTS IN SMALL ANIMAL ONCOLOGY

ECT has been used to treat a large variety of solid neoplasms in cats and dogs: soft tissue sarcoma, mast cell tumors, squamous cell carcinoma, melanoma, anal sac tumors, perianal tumors, localized lymphoma and Sticker's sarcoma. The treatment has been applied as single modality, in neoadjuvant or adjuvant fashion, depending on the tumor histotype, size and location. In most studies the technique is adopted in an adjuvant modality after incomplete surgical excision. The therapy has been overall well tolerated and the tumor response, albeit variable among the different histotype, was deemed satisfactory in the different studies.[8-27]

CURRENT PROTOCOLS

ECT is mostly used as a adjuvant therapy, locally injecting the chemotherapy drugs bleomycin or cisplatin and then following with the application of permeabilizing pulses. For larger tumors drugs can be administered locally or systemically, mostly on the basis of tumor volume, firmness or resistance to insertional electrodes.[12] Caution should be exerted when using electrochemotherapy in patients previously treated with radiation therapy due to the risk of radiation recall.[26]

ELECTROCHEMOTHERAPY IN HORSES

The very first preliminary investigation on equine tumors adopting ECT has been published by Rols et coll. in 2002.[28] In that preliminary study three horses with a total of 10 relapsing sarcoids were treated with intratumoral injection of cisplatin followed by ECT. All the horses tolerated the treatment without adverse effect and after multiple courses of ECT, all the lesions achieved a complete remission. used successfully in horse to treat sarcoids using intralesional bleomycin and to treat melanomas using cisplatin as chemotherapy agent, either directly or combined with surgery. A confirmatory study was published few years later describing the extremely favorable outcome of 48 horses carrying 194 sarcoids. The treatment was well tolerated and resulted in high percentage and durable responses: the 4 year nonrecurrence rate was 97.9% for animals (47/48) and 99.5% (193/194) for tumours.[29] In the past years two reports have been published describing the successful use of ECT as single modality or combined with surgery to treat melanoma in horses.[30,31]

ELECTROGENETHERAPY

This application of electroporation involves the administration of DNA plasmids to cancer patients in order to increase the response to chemotherapy or to counter paraneoplastic syndromes. The first report of electrochemogene therapy in veterinary cancer patients has been presented by Draghia and colleagues in 2002. The authors investigated the possibility to palliate several cancer induced complications such as cachexia, anemia, anorexia, and decreased activity levels that many veterinary patients experienced by delivering a plasmid encoding for insulin-like growth factor-I.[32] More recently, electrochemogene therapy using plasmids encoding for IL-12 has been successfully investigated in dogs with a variety of spontaneous neoplasms.[33,34] The treatment was well tolerated but the tumor response was extremely variable. Finally, in an abstract presented to an international congress, it has been described the use of electrochemogenetherapy with cisplatin and IL-12 to treat equine sarcoids.[35]

IRREVERSIBLE ELECTROPORATION

This technique aims at inducing focal tumor destruction by using direct tumor ablation through the delivery of electric pulses at higher voltage than RE (used in ECT), alone or combined with chemotherapy. There are scant reports in small animal oncology, however there are ongoing studies on irreversible electroporation for the treatment of equine tumors in several academic institutions.

CLINICAL CASES

Several horses have been treated so far by our group for different neoplasms using ECT with a novel shortened ECT protocol using an equipment developed by our group. A selection of the most interesting and challenging cases will be presented, including sarcoma, squamous cell carcinoma and melanoma. Treatment protocols, outcomes and pitfalls will be discussed.

CONCLUSION

Electroporation in its different applications (irreversible and reversible) is a safe and efficacious approach to veterinary oncology. Its low cost and ease of administration makes it a valuable addition to the currently available oncological therapies. One advantage of this technique is the possibility of repeated treatments in case of local recurrence and the recruitment of the patients' immune system, as suggested by the low incidence of metastatic disease among most treated patients and by the tumor selection observed among some pets with recurring disease. This technique is gaining momentum in equine oncology on the basis of the recent literature and the increased interest of owners to treat malignancies in horses.

FUTURE DIRECTIONS

Novel electrochemotherapy equipments adopting different electrical parameters are currently being developed and tested by our group. Moreover, our group is assessing the potentials for electrogenetherapy using electric pulses to drive chemosensitizing plasmids within tumor cells as well as participating to the transfer of this technique to human patients.

REFERENCES

[1]Keet, Gehl and Lee (Eds) Clinical Aspects of Electroporation, Springer Ed, 2011. [2]Gothelf et al. **Cancer Treat Rev** 2003; 29: 371. [3]Belehradek et a. **Cancer** 1993; 72: 3694. [4]Daskalov et al.**IEEE Eng Med Biol** 1999 ; 18 : 62. [5]Gehl et al. **Melanoma Res** 2000; 10: 585. [6]Mir et al. **Eur J Cancer** 2006 ; 4(S) : 14. [7] Spugnini et al. **J Cell Physiol.** 2014;229:1177-81. [8].Mir et al. **Br J Cancer** 1997 ; 76 : 16. [9]Tozon et al. **Anticancer Res** 2001; 21: 2483. 55 : 101. [10]Tozon et al. **Anticancer Res** 2005 ; 25 : 839. [11] Spugnini et al. **Methods Mol Biol.** 2014;1121:247-56. [12] Spugnini et al. **J Exp Clin Cancer Res** 2003 ; 22 : 571. [13]Spugnini et al. **J Exp Clin Cancer Res** 2005; 24 : 245. [14]Spugnini et al. **Melanoma Res** 2006; 16 : 23. [15]Spugnini et al. **Cancer Chemoter Pharmacol**; 2007; 59 : 375. [16]Spugnini et al. **Vet Rec** 2006; 159 : 598. [17]Spugnini et al. **In Vivo** 2007; 21: 819.[18]Spugnini et al. **Vet Rec** 2008; 162: 186. [19] Spugnini et al. **J Transl Med** 2011; 9; 152. [201]Spugnini et al. **Can Vet J.** 2013; 54:606. [21]Spugnini et al. **Anticancer Res** 2006; 26: 4585. [22] Kodre et al. **In Vivo**. 2009; 23:55. [23]Spugnini et al. **J Vet Intern Med** 2011; 25: 407.[24] Spugnini et al. **Vet J** 2009; 179: 117. [25]Tozon et al. **J Feline Med Surg**. 2014;16:291. [26] Spugnini et al. **In Vivo** 2008; 22: 751. [27] Spugnini et al. **J Vet Intern** Med 2015; 29:1368. [28]Rols et al. **Bioelectrochem** 2002 ; 55 :101. [29]Tamzali et al. **Equine Vet J** 2012; 44:214. [30]Spugnini et al. **J Equine Vet Sci** 2011; 31: 430: [31] Scacco L et al. **J Equine Vet Sci** . 2013; 33: 637-639. [32] Draghia et al. **Mol Ther** 6:830. [33]Reed et al . **Cancer Gene Ther** 17:457. [34] Pavlin et al. **Radiol Oncol** 45:31. [35] Tamzali Y. 11th Mediterranean Conference on Medical and Biological Engineering and Computing 2007 MEDICON 2007, 26-30 June 2007, Ljubljana, Slovenia.

Treadmill Examination of Horses

Warwick Bayly, BVSc, MS, PhD, DACVIM (Large Animal Internal Medicine)
Pullman, WA

INTRODUCTION

Treadmill exercise testing of horses is nearly always undertaken to either assess the animals' performance potentials or as part of a clinical assessment of one or more body systems, or both. Treadmill tests are usually performed in temperature and humidity controlled environments and in close proximity to electrical outlets which facilitates the use of older electricity-dependent equipment and/or newer, more sophisticated instrumentation. Because of the capital investment required to build and equip an appropriate facility, the cost of performing treadmill tests is can be regarded as considerable by some owners and trainers. The availability of treadmills allows the standardization of exercise testing protocols and obviates the need to be concerned about variations weather and footing that make standardization of field tests and comparison of their results difficult. However, equine treadmills are relatively few in number and usually located in university laboratories or other large equine veterinary practices or training centers. Consequently, horses have to be brought to those centers and the inconvenience of doing so plus thus cost of testing probably deters a number of possible clients. This makes it harder to develop a sizable data bank of results and to operate in a commercially viable manner. Despite these potential disadvantages, the range of tests that can be performed and the accuracy of the results of these tests under standardized tests conditions still make treadmill testing of equine athletes preferable to tests conducted in the field.

TREADMILL SPECIFICATIONS

There has been a proliferation of equine treadmill manufacturers in recent years and a wide range of treadmills are now available around the world. However, not all treadmills are the same and certainly not all are designed with the needs of a veterinarian in mind. In order to gain maximum utility in terms of clinical evaluations or assessment of athletic capability, a suitable treadmill should have a running surface that is at least 15' long and 4' wide with a top speed of at least 15.5m/s (35mph, ~56 kph) and an incline that can be raised to at least 10% (6°). This combination of speed and slope enables brief tests (up to 2mins) requiring maximal effort to be conducted on the most capable horses without having to go to speeds that might be regarded as unsafe. An internet search reveals a limited number of manufacturers of such treadmills and the price of all models that meet the above specifications is >$50,000 when new. The author has some familiarity with treadmills from the following manufacturers: Sato, Mustang, HorseGym and Haico. Usually these and other manufacturers do not carry an inventory and only build their models after an order has been placed.

PREPARING THE HORSE

An acclimatization period is usually necessary when first introducing a horse to treadmill exercise. Most horses (>95%) take well to a treadmill when first exposed to it and walk and trot comfortably. Learning to change gait from the trot to a canter or gallop is generally the biggest challenge to be overcome and it might take some encouragement before the horse has the confidence to do so. For horses being exposed to a treadmill for the first time a minimum of 2 preliminary acclimatizing runs are advised,[1] with more than this number frequently being needed. The first such exercise should consist of a minimum of 4 mins walking at about 2.0m/s (anything slower often prevents the animal achieving a normal walking rhythm and speed) followed by 1 min at a trot (~4m/s is ideal). The primary purpose of this initial bout of exercise is to get the horse used to the sensation of the surface moving beneath its feet and to teach it stay towards the front of the treadmill. Horses can usually be exposed to comfortable gallop speeds (~8m/s) on the 2nd run. If needed, subsequent runs will involve galloping at progressively faster speeds once the horse has demonstrated that it will comfortably break into a gallop. Two acclimatization runs/day can be handled easily by the horse. In this author's experience, most horses are ready to be tested at the gallop after 4 acclimatization runs, which means that the exercise testing can be completed in 3 days for horses that have never previously been on a treadmill. Those with previous experience can generally be tested on an outpatient basis. A more detailed description of the acclimatization process has been described elsewhere.[2]

EXERCISE TESTS PROTOCOLS

In the last 30 years there have been a plethora of publications documenting findings obtained from horses completing a wide variety of treadmill protocols. Despite this variety, exercise test protocols fall into 3 categories – incremental, high-speed and submaximal, with the type of test being used depending on the use of the horse and the nature of the assessments being performed. Racehorses can rarely run faster than 15m/s on a 10% incline and if they do, cannot sustain the speed for long (<20sec). Most healthy and fit racehorses reach maximal heart rate (HR) at speeds between 12 and 14m/s. Endurance horses usually cannot gallop at speeds >11m/s.

Because horses exercising on a treadmill have no momentum and are not moving through space and creating a flow of air over their bodies, the evaporative cooling that is essential for thermoregulation in these animals is largely absent. Consequently it is *critically important* that a large fan or industrial blower be placed in front of the horse in order to help regulate increases in core body temperature in all but the easiest of exercise tests.

Incremental

Incremental tests are the most commonly utilized equine treadmill exercise protocols. Following an appropriate warm-up (eg, 4m/s for 2-4 mins) they usually involve short periods of galloping exercise (usually 60-90sec) at a series of ever increasing speeds until the horse can either no longer keep pace with the treadmill; ie, it is fatigued, or there is no increase in either its HR or oxygen consumption (VO_2) associated with the increase in speed. Because of the ability to control the speed and time at each speed, and because they enable a number of samples or measurements to be taken during the course of the exercise with the results being comparable between horses, incremental tests are commonly referred as standardized exercise tests or SETs. There are no "gold standard" incremental tests. Racehorses being evaluated for performance capability usually undergo 1m/s increments in speed until fatiguing. Endurance horses might run for longer at each speed while experiencing increments of <1m/s.[3]

High-Speed

High-speed tests are usually reserved for racehorses and are designed to replicate the type of exercise associated with racing, particularly Thoroughbred sprint races which are frequently completed in <90sec. The purpose of such tests is to assess maximal aerobic and anaerobic capacity and, by extension, a horse's stamina or ability to maintain a high speed. Following an appropriate warm-up (see above) the treadmill is rapidly accelerated to a predetermined speed, most commonly that associated with an oxygen (O_2) demand that is 1,1, 1.15 or 1.2 times the maximal VO_2 (VO_2max) and the horse gallops at this speed until fatigued. As well as being used to evaluate a horse's athletic ability or potential, high-speed exercise tests of this type are sometimes used to assess whether the administration of different agents could enhance performance.

Submaximal

Strictly speaking, any treadmill test protocol that involves exercise at any speed less than that required to reach VO_2max is classified as submaximal. Therefore, incremental tests incorporate exercise at a number of submaximal speeds. However, some tests involve nothing but submaximal work. Those designed to evaluate endurance capacity or stamina in horses other than Thoroughbreds and racing Quarterhorses[4], or in cases where a clinical problem is suspected to be a cause of reduced performance are good examples of situations in which submaximal testing is indicated. The exact protocol depends on the sport discipline and fitness of the horse, and is usually <30mins duration.

EXERCISE TEST ASSESSMENTS

Exercise tests are utilized for assessment of poor or reduced performance or exercise intolerance, or evaluation of athletic ability in horses presumed to be healthy. With respect to clinical evaluations of poorly performing horses, the majority of these tests will be conducted on patients that are engaged in competition. However, they can also be indicated in recreational or pleasure horses that reportedly have become intolerant of even mild exercise. Investigations of all causes of reduced performance except perhaps lameness, lend themselves to treadmill exercise tests.

The Respiratory System

Endoscopic examination of the upper airway during exercise for assessment of suspected abnormal noise and/or poor performance is now almost mandatory in the wake of studies demonstrating that those that appear normal at rest may be associated with obstruction of airflow during exercise, and that suspected or definitely abnormal appearing upper airways at rest may appear to function normally during exercise.[5-7] This is most likely because the airway pressures and airflows differ markedly Not only is exercising endoscopy necessary to make a definitive diagnosis of upper airway abnormality and the rates at which they change are dramatically different during exercise. For the same reasons, the speed and duration of the exercise can also affect whether or not an abnormality is observed, as breathing frequencies, airflows and airway pressures are all somewhat dependent on speed and duration of exercise. Consequently, while incremental SETs are often preferred for exercise endoscopy, high-speed tests might be more relevant to the horse's competitive situation.[2,8] If Standardbreds and some equitation horses such as those used in dressage are to examined, it is best that this be done in their competition tack in order to ensure that the head and neck are "fixed" in the same manner as in competition.

In some ways, the single best test of respiratory system function is the measurement of arterial blood gases. Measuring venous blood gases serves little value. Arterial blood can be sampled from a catheter placed in the transverse facial artery about 2" lateral to the lateral canthus of the eye during exercise. However, interpretation of results can be difficult because galloping horses normally become hypoxemic at exercise speeds with an O_2 demand that is ~65% of VO_2max and higher and hypercapnia develops at intensities > ~80% VO_2max.[9] Further, hypoxemia and hypercapnia tend to become more marked in fitter and more capable athletes when compared to those that have been trained less and/or have a lower VO_2max.[10] Because intense exercise is associated with marked increases in core temperature it is important that blood gases be corrected for temperature changes rather than determined at 37°C. Actual temperatures can be recorded using a thermistor tipped catheter passed into the right atrium or by extrapolating from previously published reports.[11]

The Cardiovascular System

The subject has recently been well reviewed and the reader is referred to that paper for its extensive list of references.[12] HR and rhythm during exercise can be recorded in a number of ways and constitute the mainstays of evaluation of the cardiovascular system in poorly performing horses. In those presented for assessment of athletic ability the focus is generally on the relationship of HR to speed. In the latter situation any of a variety of relatively inexpensive HR meters are used, although the rapid growth in smartphone apps which utilize Bluetooth technology to determine and store HR data may soon make dedicated HR meters obsolete.

There is a strong linear correlation between speed and HR when the horse is galloping, with a definite plateau in HR being observed despite an increase in treadmill speed when maximal HR (HR_{max}) has been reached. For this reason, most evaluations of HR involve incremental tests. HR_{max} is highest among young, untrained horses but varies with breed and sport discipline, and tends to decrease slowly as with age. Various parameters linking HR and treadmill speed have been proposed as a means of facilitating comparisons between horses. The treadmill speed at which the HR reaches 200 bpm (V_{200}) is the most popular although the HR when speed is in the range of 10-12m/s (eg, HR_{10}) can be used. No HR-based parameters associated with treadmill testing are sacrosanct. Individuals can devise whatever makes physiological sense in terms of the type of activity for which the horse is being evaluated. What is essential is that the treadmill slope be constant in all tests. If it is not, comparisons between horses is meaningless.

Treadmill examination of horses presented for poor performance or exercise intolerance frequently involves electrocardiography (ECG) for assessment of cardiac rhythm. A base-apex ECG is usually obtained continuously throughout the treadmill test and immediately afterwards. Reliable and affordable telemetry equipment is now available and generally preferred to hard-wired systems. The placement of electrodes and the quality of their electrical contact is the biggest source of poor ECG recordings during exercise. The entire electrode must be in contact with the skin which should be well wetted and/or have plenty of electrode gel applied. In horses it may help to clip the area of electrode placement.

In clinical cases it can be difficult to determine whether or not a horse's HR response to exercise is abnormal or not. It helps to have a frame of reference (ie, data bank) on which to base such an assessment. The most ideal situation exists when the patient has results from previous tests. Most commonly, HR responses are regarded as abnormal when the HR is unusually high at the submaximal speeds of an incremental test. In such instances parameters like V_{200} are low and others like HR_{10} are high and the speed at which HR_{max} is reached is slower than expected for a healthy horse. Interpretation of the findings is not always straightforward because the results can be influenced by the ambient temperature, the patient's fitness level, level of hydration and the presence of co-existing disease, particularly those that are associated with pain or interfere with O_2 uptake and/or transport. Abnormal HR responses to exercise are normally also characterized by a prolonged HR recovery period. In fit, healthy horses, HR can be expected to drop dramatically in the first 2 mins post-exercise, generally being <112 after this time. Horses with clinical problems frequently have comparatively prolonged periods post-exercise with HRs in the range of 112-130bpm. While it is important to have a good sense of the normal HR response to exercise, the detection of an abnormal response is not necessarily indicative of cardiovascular disease. Rather, such aberrant findings indicate that further clinical investigation is warranted.

Some abnormal HR responses to exercise can be associated with arrhythmias, however, not all arrhythmias are pathologic or easily detectable. With the exception of atrial fibrillation, detection and interpretation of arrhythmias occurring during or immediately after exercise can be difficult. An ECG recorded at a HR >200 can be difficult to decipher even with the assistance of computer software programs. Arrhythmias are more common in the first 2 mins post-exercise than during exercise and are usually not sustained. Whether during or after exercise, premature supraventricular complexes are the most commonly detected arrhythmias, occurring in up to 50% of racehorses,[13] and they are usually regarded as innocuous unless they occur in spates of more than 1 beat or recur with regularity. Ventricular premature complexes are less frequent[13] and more problematic, especially if superimposed on the T wave as this is when the myocardium is most vulnerable to the development of ventricular fibrillation and sudden death.[12, 14] Interpreting the clinical significance of arrhythmias associated with exercise is a real challenge especially as when they might be primary or secondary to other pathological or temporary abnormalities like electrolyte changes and severe academia. There reader is referred to the ACVIM consensus statement for a more comprehensive discussion of this vexing topic.[14] In summary, arrhythmias must be interpreted in terms of other clinical findings if they exist. For instance, if the patient has auscultatory and/or echocardiographic findings compatible with impaired cardiac function and has a history of reduced performance, the arrhythmia might be significant. If, however, it is an incidental finding on a horse being evaluated for performance potential, it is should best be noted as an isolated but clinically insignificant finding.

Metabolism

Many people regard the VO_2max as the single best indicator of whole body aerobic metabolism. It is measured using an incremental exercise test in which speed is increased until there is no gain in VO_2 despite an increase in speed. Various open- or biased-flow techniques have been described and all seem to be appropriate *providing* the diameter of air flow ports is sufficient to ensure that ventilation is not compromised by the mask that must be worn.[15-17] Most of these systems require accurate measurement of the airflow through the system and it is critical that this be calibrated regularly and carefully. Because of the formula used in the calculation is heavily influenced by the flow rate through the system, even slight inaccuracies can have significant effects in determining VO_2max. While a consistent systematic error might not have major impact on the results determined with a given system, it does make comparison of values produced from different laboratories difficult or impossible.

Most systems capable of measuring VO_2 during treadmill exercise are also capable of measuring carbon dioxide production (VCO_2). VCO_2 increases more rapidly than VO_2 as speed increases with the result that the respiratory exchange ratio (R) starts at values of about 0.9 at easy speeds (eg, 4 m/s) and progressively increases values of about 1.2 at the speed equated with VO_2max. This reflects the increased anaerobic contribution to total energy (ATP) production as exercise intensity rises. VCO_2 and R are important factors in calculation of a useful ventilatory variables like alveolar ventilation and alveolar oxygen tension.

Lactate is an end-product of anaerobic glycolysis and its concentrations in plasma and blood are easily assayed. As such it is often measured in order to assess anaerobic metabolism. Plasma values are usually 30-40% higher than in whole blood although the difference depends to some degree on the intensity of the exercise and the hematocrit. While lactate responses to many different types of exercise have been documented in innumerable papers and its determination has been prominently involved in assessments of athletic potential and fitness to race, the interpretation of results is not as clear cut as many people make out.[18, 19] Firstly, lactate is produced in muscles at all intensities. What varies is the rate at which it is produced, both within an individual horse during the course of an exercise test, and between horses. Some degree of anaerobic metabolism is constantly occurring in muscle through simple mass action effects regardless of exercise level. Anaerobic metabolism does not suddenly "switch on" and horses do not "go anaerobic". The relative contribution of anaerobic metabolism to total energy production progressively increases as the ability of muscles to meet the demand for ATP by aerobic means declines.

The relationship between lactate concentration and speed is exponential or biphasic with the rate of lactate accumulation increasing as exercise intensity increases. The point of inflection in this curvilinear relationship varies according to the level of training and ability. Therefore, comparing lactate results between horses and in the same horse is most meaningful when the conditions of the treadmill test are truly standardized. Over the years a number of variables derived from lactate measurements have been derived to facilitate such comparisons. The best known is the speed at which a horse's plasma lactate concentration reaches 4 mmol/l (V_{La4}). This number was based on the observation that the rate of accumulation of lactate often started to increase significantly at or around a plasma concentration of 4 mmol/l. While a lot of exercise physiological significance has been bestowed on this value, it is in fact an arbitrary designation that has been frequently overinterpreted and/or misapplied over the years. For instance, the variable relies on measurements made in plasma. However, a number of people have used whole blood measures as if they were the same as plasma concentrations. Its value is determined with exponential regression analysis of lactate concentrations measured on blood samples collected at different submaximal speeds. The values recorded in these samples can vary according to the duration of each step in the incremental test and the length of the warm-up. Similarly, once the point of inflection in the lactate-speed relationship is reached, the rate at which lactate increases in response to increasing exercise intensity and its peak value can vary greatly. Generally, the lower the lactate concentration at any given speed and the more slowly it increases, the better the endurance of a horse. Conversely, horses engaged in sprint events, and this tends to include all Thoroughbred races of about 2 mins or less, need to be able to anaerobically generate ATP rapidly and to high concentrations. Peak plasma lactate concentrations after racing or incremental treadmill exercise to fatigue often exceed 20mmol/l and sometimes 30 mmol/l. Consequently, the value of V_{La4} is of little relevance in these animals. More appropriate or relevant variables upon which to base assessments of athletic potential in racehorses of all breeds might be the speed at which lactate concentration reaches 10 or 15 mmol/l (V_{La10} or V_{La15}, respectively), or the lactate concentration after exercise of predetermined duration at a speed like 10m/s at a 10% treadmill gradient (La_{10}).

Skeletal Muscle

It is beyond the scope of this presentation to discuss the use of muscle biopsies collected post-exercise. Measurement of the principal muscle enzymes (creatine kinase (CK) and aspartate aminotransferase (AST)) post-treadmill exercise can be undertaken in situations where a clinical or sub-clinical exertional myopathy is suspected. CK concentrations in particular are very labile and interpretation of values can be difficult because there can be a wide range of what is considered normal.[20] some very normal horses can have CK concentrations> 1000 IU/l 2 hrs post-exercise. Others with signs of post-exercise muscle stiffness might have a CK in the range of 800-1000 IU/l, particularly if the blood sample is collected relatively soon after completing the exercise test. As with all treadmill tests, exercise conditions must be standardized. Treadmill tests are not necessary for horses suspected of having a polysaccharide storage myopathy as they usually show signs of the disease with very mild forms of exercise.

TREADMILL INJURIES

Trainers and owners unfamiliar with equine treadmills often ask questions regarding their safety and the prevalence of injuries associated with tests conducted on them. This author's experience drawn from more than 35 years working with horses on treadmills is that injuries of any kind are uncommon and that major injuries are rare. The treadmill surface is much more consistent than any experienced in the field and horses are not engaged in propulsive activity. The most extensive study of this subject indicated that the likelihood of any injury was about 5% and that of serious injury (fracture, tendon injury or serious lameness) was < 1%.[21] Consequently, providing the patient has been appropriately acclimated to the treadmill, its owner or trainer can be confidently assured that undergoing a treadmill test is quite safe.

REFERENCES

1. Rose RJ, et. al Aust Vet J 1990; 67:345.
2. Rose RJ, et. al In, The Athletic Horse, WB Saunders, 1994; p248.
3. Fraipoint A, et. al Vet Rec 2011; 169:154.
4. Munsters CCBM, et. al Vet J 2014; 202:11.
5. Kannegieter NJ, et. al Aust Vet J 1995; 72:101.
6. Parente EJ, et. al, Proc Amer Assoc Eq Practnrs 1995; 41:170.
7. Lane JG, et. al Eq Vet J 2006; 38:401.
8. Parente EJ, Vet Clin Nth Amer Eq Pract, 1996; 12:421.
9. Bayly WM, et. al, J Appl Physiol, 1989; 67:1989.
10. Christley RM, et, al, Amer J Vet Res, 1997; 58:653.
11. Hodgson DR, et.al, J Appl Physiol, 1993; 74:1161.
12. Allen KJ, et. al, Eq Vet Edu, 2016; 28:99.
13. Ryan N, et. al, Eq Vet J, 2005; 37:265.
14. Reef VB, et. al, J Vet Int Med, 2014; 28:749.
15. Bayly WM, et. al, J Appl Physiol, 1987; 63:1210.
16. Rose RJ, et.al, Eq Vet J Suppl, 1990; 9:39.
17. Butler PJ, et. al, J Exp Biol, 1993; 179:159.

18. Gollnick PD, et. al, Med Sci Sports Ex, 1986; 18:334.
19. Bayly WM, et.al, In, Equine Exercise Physiology II, ICEEP Pubs, 1987; p426.
20. Harris PA, et. al, Eq Vet J Suppl, 1990; 9:61.
21. Franklin SH, et. al, Eq Vet J Suppl 2010; 38:70.

Endocrine Problems in the Equine Athlete

Janice Kritchevsky VMD, MS, DACVIM (LAIM)
West Lafayette, IN

INTRODUCTION

The horse has evolved to be the consummate athlete, and the profound physiologic changes that allow horses to exercise at top levels require a complex interplay of several hormonal systems. As with many of the issues in other body systems, the endocrinologic problems of the equine athlete can often be traced back to inappropriate husbandry or medical intervention. Trainers and owners who are striving to get maximal performance may push the limits of training or use extreme feeding practices that can disrupt normal endocrine function. The use of inappropriate or illegal medications may also adversely affect the equine athlete. Because many consider endocrine diseases to be "old horse" problems, they may not be considered as a potential cause of disease when they occur in a sport horse setting. This is particularly true when confronted with the horse that is just "off" in vague and puzzling ways. The goal of this presentation is to describe some of the more common endocrinologic problems that have been identified in equine athletes.

NORMAL ENDOCRINOLOGIC CHANGES DURING EXERCISE

Hyyppa[1] has recently reviewed many of the endocrinologic changes that occur during exercise and recovery in the equine athlete. Probably the most widely appreciated effect is exerted by epinephrine, the classic "fight or flight" hormone. In addition to its well-known cardiovascular and pulmonary effects, epinephrine stimulates lipolysis, glycogenolysis, and inhibits insulin within seconds of its release. Cortisol, which also increases with exercise, has similar actions, and helps to maintain increased blood glucose and lipid concentrations for a longer period of time than would be possible with epinephrine alone.[2] The cortisol increases that are seen with exercise become less pronounced with training and experience, suggesting that there is a psychic component to the stress response observed in horses as there are with human athletes. Insulin increases during the recovery phase of exercise, which facilitates amino acid and glucose entry into muscle cells. The overall effect of insulin is to increase muscle protein synthesis.[3] Circulating fluid volume and blood pressure is maintained through the balanced actions of atrial natriuretic peptide (ANP), antidiuretic hormone (ADH), and the renin-angiotensin system. Aldosterone, whose secretion is stimulated by angiotensin II, causes increased sodium reabsorption at the kidney. This helps maintain circulating fluid concentrations. ADH is released when blood osmolality increases, which is another effect of intense exercise. Like aldosterone, its actions result in increased water retention by the kidney and increased blood pressure. ANP antagonizes the actions of ADH and aldosterone, and preserves renal blood flow and sodium excretion.

ADRENAL INSUFICIENCY

Horses are extremely sensitive to the effects of exogenous corticosteroids. A single intra-articular injection depresses the hypothalamus-pituitary-adrenal (HPA) axis for over 24 hours,[4] and triamcinolone and prednisolone acetate injections inhibit the HPA axis for 14 and 21 days respectively. Prolonged or inappropriate use of corticosteroids or anabolic steroids can lead to adrenal atrophy and then signs of adrenal insufficiency when they are withdrawn. Adrenal insufficiency, has been reported after the excess or prolonged administration of corticosteroids and anabolic steroids in horses.[5] Many times, the presenting signs and laboratory values are not classic, and adrenal insufficiency should be suspected in any ill horse that has a history of corticosteroid or anabolic steroid administration regardless of blood electrolyte concentrations. In dogs, Addison's disease has been termed "the great pretender" because the clinical signs are often vague and manifest as problems in multiple body systems. The described clinical signs in horses are similarly vague. Often the primary signs are GI, including anorexia and diarrhea. Other clinical signs reported include lethargy and lameness.[5] Hyponatremia is a common finding in horses with GI disease, and a low sodium concentration on diagnostic test may not raise an index of suspicion that adrenal insufficiency is present. The diagnosis is made by performing an ACTH response test.[6] Briefly, blood cortisol concentration is measured before and 30 or 60 minutes after I.V. administration of synthetic ACTH. A normal horse will have at least a two times increase in cortisol post-ACTH administration. Documenting high fractional excretion of sodium in the urine despite low or normal blood values is also suggestive of the problem. Treatment includes giving low levels of exogenous corticosteroids until the adrenal recovers.

NUTRITIONAL SECONDARY HYPERPARATHYROIDISM

Parathyroid gland tissue is extremely sensitive to serum ionized calcium concentrations. Parathyroid hormone (PTH) is released in response to decreased ionized calcium. Secondary hyperparathyroidism occurs when PTH is increased due to non-parathyroid disorders that affect blood calcium concentrations. When they occur due to dietary factors, it is termed nutritional secondary hyperparathyroidism (NSH).[7] PTH has multiple actions including increasing calcium resorption and phosphorus excretion via the renal tubules, stimulating the manufacture of Vit. D, and activating the pathways that lead to calcium resorption from the bone. Osteoclastic resorption of outer bone may disrupt tendinous insertions and bone trabeculae supporting articular cartilage, which can result in tearing and microfractures and subsequently lameness.[7] In its most extreme form, nutritional hyperparathyroidism presents with the distinctive clinical sign of fibrous osteodystrophy (Big head), and may result in severe debilitation. Less overt manifestations have been reported in race horses.[8] The clinical signs reported in the racehorse setting include shifting leg lameness and tendon disease. Interestingly, in at least one instance the lameness resolved after diagnostic nerve blocks. The effects of NSH are more commonly seen in growing animals than in those with mature skeletons. The most common clinical signs include transitory shifting leg lameness, leg swelling, stilted gait and tender joints. NSH is a risk when the Ca: P ratio in the diet is less than 1.5:1. The absolute amounts of either mineral is not as important as their ratio. The presence of phytates, calcium binding proteins that are found in high concentrations in bran, may make the functional ratio in a feed less than the absolute amounts. A survey of racehorses in Australia found that 40% of the horses in that country were being fed inadequate calcium compared to the high phosphorus present in their grain diets,[9] and it was hypothesized that this contributed to the lameness and fractures observed in that population. Ideally, the dietary Ca: P ratio is at least 2:1. Due to homeostatic mechanisms, serum calcium and phosphorus will remain in the normal range, and the problem cannot be identified on routine blood work. Slight alterations in serum phosphorus concentrations and mild increases in alkaline phosphate are usually the only alterations found in the blood. Diagnosis is made by measuring urinary fractional excretions for phosphorus. The normal percent clearance of phosphorus is less than 0.5%, a result of 5.0% or higher s suggestive of the dietary imbalance. Nutritional hyperparathyroidism should be suspected in instances of a horse with vague, shifting leg lameness, particularly if mandibular or facial thickening is present. A careful dietary history should be taken in any horse which may have NSH. Grain and grain products are extremely high in phosphorus, so feeding large amounts of concentrate without balancing for electrolytes and minerals is likely to results in poor calcium: phosphorus ratios. Bran contains high amounts of phytate, which binds calcium and prevents its absorption. Treatment consists of balancing the diet and maintaining a 1.5 -2.5 to 1.0 calcium to phosphorus ratio. Once the diet has been corrected, lameness may improve although the skull and mandible changes are usually permanent.

THYROID DISEASE

Thyroid hormones are integral to proper energy use, and they are crucial to maintenance of proper muscle function. When evaluating the changes in thyroid hormone concentrations observed with exercise in horses, it is important to make the distinction between the changes that occur during and after a single episode of high or low-intensity exercise and alterations in the baseline levels of thyroid hormones which occurs over time as a horse is trained and becomes increasingly fit. Both these situations have been examined repeatedly in horses doing a variety of athletic endeavors, and often the studies produce conflicting results. In general, T3 increases during and shortly after intense exercise while it decreases after prolonged endurance exercise. T4 concentrations are generally unaffected while excercising.[10, 11] Fitness is reported to result in lower baseline thyroid levels in many species, including horses, although this has not been found in all instances.[12] In man, endurance-trained athletes metabolize T3 differently than sedentary controls although the mechanism for this change is not known. Both hyper- and hypothyroidism result in decreased athletic performance in human athletes due to decreased cardiac output and muscle weakness. Hypothyroidism has been documented to decrease athletic performance in horses primarily due to decreased cardiac output.[13] The effect of higher than normal thyroid values on equine performance has not been reported.

Despite extremely scant evidence to support its use, thyroid supplement administration is ubiquitous in the stables of equine sports horses. There are potentially two reasons why thyroid supplement is used in such high amounts. It is possible that some owners, trainers and veterinarians use it because it acts as a non-specific "pick-me-up" and horses that are given thyroid hormone appear to have more energy that they did prior to supplementation. It is also possible that low T3 or T4 values were identified on a screening blood test, and the horse is believed to be truly hypothyroid. Although the finding of a T4 concentration below normal laboratory ranges is fairly common when performing screening blood work on horses, a positive diagnosis of hypothyroidism is exceedingly rare. Before interpreting a T3 or T4 value, practitioners should be aware of the large number of factors that can affect thyroid hormone concentrations in normal horses. Blood thyroid concentrations vary with age, condition, weather, diet, and drug administration. NSAID drugs result in decreased concentrations of T3, T4, and free T4.[14] Extremely high thyroid hormone values have been identified in post-race blood samples in several locations, suggesting that over-supplementation is a common occurrence. Despite this there is no regulatory emphasis on detecting supra-physiologic thyroid hormone supplementation in equine athletes currently.

UNEXPLAINED UNDERPERFORMANCE SYNDROME

Allowing for sufficient rest periods is as important as exercise when training for athletic endeavors. Horses require 72 hours to completely replace glycogen stores in muscle, and strenuous exercise sooner can be detrimental. Over training results when there is insufficient rest time scheduled to allow muscles and other homeostatic mechanisms to completely recover and adapt to a bout of excercise. Prolonged imbalance between training and rest can lead to long term dysfunction in many body systems. The constellation of physiologic alterations that occur after prolonged overtraining has been termed the unexplained underperformance syndrome (UUS).[15] It is defined as the accumulation of training and non-training stress which results in long-term decrease in cardiovascular, hematological, immune, and hormonal function. In human athletes, there is a profound psychological component, with sufferers reporting depression and altered affect. Horses suspected of UUS are often reported to be "stale" or "sour" with poor appetite and no desire to perform at previously attained levels. The pathogenesis of UUS is not well understood, but is thought to arise in the hypothalamus. The result is the failure of appropriate increase in cortisol in response to physiologic stress. Normally, blood cortisol increases after exercise in response to the acute stress. Because the HPA axis is intact, cortisol will increase in response to ACTH administration, which can make diagnosis difficult. Horses may transition into UUS gradually, making it difficult to pinpoint the onset of the problem. UUS should be suspected if there is a decrease in a horse's performance without other explanation, particularly if it is accompanied by weight loss, other medical conditions, and a change in temperament. Treatment consists of prolonged rest.

REFERENCES

[1]Hyyppa, S. *Livestock Production Sci* 2005;92:113-121 [2]Nagata, S et. al *Equine Vet J Suppl* 1999;30:570-574 [3]Fluckey, JD et. al *Acta Physiol Scand* 2001;173:379-384 [4]Soma, LR et. al *J Vet Pharmacol Therap* 2012;36:181-191 [5]Dowling, PM et. al *JAVMA* 1993;203:1166-1169 [6]Stewart, AJ, et. al *JAVMA* 2011;239:834-841 [7]Ronen, N et. al *Tydskr S Afr Vet* 1992;63:134-136 [8]Mason, DK et. al *Equine Practice* 1988;10:10-16 [9]Caple, IW et. al *Australian Vet J* 1982;58:132-135 [10]Gonzalez, O et. al *Equine Vet J* 1998;30:72-78 [11]Graves, EA et. al *Equine Vet J Suppl* 2006;39:32-36 [12]Medica, P et. al *Livestock Production Sci* 2011;140:262-267 [13]Foreman, JH *Vet Clin North Am: Equine Prac* 1996;12:537-554 [14]Sojka, JE et. al *Am J Vet Res* 1993;54:52-55 [15]Graaf-Roelfsema, E et. al *Vet Quarterly* 2007;29:82-101

Leptospirosis: An Important Infectious Disease of Horses

Thomas J. Divers, DVM, DACVIM, DACVECC
Ithaca, NY

INTRODUCTION

Leptospirosis is caused by a highly invasive spiral bacterium of the genus *Leptospira*. The infectious agent is capable of infecting both humans and animals, including the horse. There are at least 22 recognized species of *Leptospira* that have been further classified into three major subgroups: pathogenic, intermediately pathogenic and nonpathogenic (saprophytes), comprising >250 serotypes ("serovars") based primarily on the immunological characterization of surface lipopolysaccharide (LPS) .[1] Serovars are sometimes classified as causing host-adapted infection or incidental host infection. Host-adapted strains seldom cause clinical disease in their maintenance host, infection and shedding are prolonged, and the serologic response following infection is relatively low. Conversely, incidental host serovars are more likely to cause clinical disease in a nonmaintenance host, be associated with a marked serologic response following infection, and be shed only briefly by the host.[2]

In North American horses, *Leptospira interrogans* serovar Pomona type kennewicki is the prominent incidental (pathogenic) serovar for horses. In Europe, important equine strains are *Leptospira kirschneri* serovar Grippotyphosa, strains *duster* (Western Europe) and *moskva* (Eastern Europe). In South America, *L. interrogans icterohemorrhagiae* and *copenhageni* are important strains. *L. interrogans* serovar Bratislava is considered by most researchers to be the host-adapted serovar of the horse and although horses may have high serum antibody titers against serovar Bratislava, it is generally considered to be nonpathogenic to the horse.[2] One explanation for the high serum titers to Bratislava is a high cross reactivity with *Leptospira interrogans* serovar Pomona type kennewicki on the microscopic agglutination assay (personal observation). A variety of wildlife including raccoons, striped skunks, opossums, and red and grey foxes have been shown to host serovar Pomona and multiple strains of *L. Pomona* kenniwicki.[3]

PATHOGENESIS

Leptospira infection most often occurs following contact with *Leptospira* contaminated environmental reservoirs (water, soil), urine or reproductive fluids. The bacteria are believed to penetrate mucosal membranes or enter transdermally through wet or abraded skin. Motility of the organism is thought to be important in penetrating mucosal membranes and crossing cell layers. Upon penetration of membranes, the bacteria adhere to extracellular matrix and host cells.[4] Virulent leptospires are resistant to complement and to killing by neutrophils in nonimmune hosts.[5] Bacteremia may follow within 2-20 days after initial mucosal contact and fever may occur during this acute phase, which generally lasts 2-7 days.[6,7] Although *Leptospira*, like other Gram-negative bacteria, contain lipopolysaccharide (LPS) it has low endotoxin activity[8] and fever is not believed to be associated with the endotoxin. Bacteria are disseminated from the blood to organs of trophism. A primary lesion of acute leptospiremia is damage to endothelial cells of small blood vessels, inhibition of coagulation and inflammation within the trophic tissues.[2] Once specific antibodies are produced, complement and antibody rapidly eliminate the organism from the blood and tissues other than the kidney, eye, CNS, and reproductive tract. A second phase of the disease may occur later in some of those privileged sites and may be driven by the host immune response more than virulence of the organism.[9]

CLINICAL SYNDROMES

Pathogenic *Leptospira* infections in the horse appear to have organ trophism for the kidney, eye, female reproductive tract and, less commonly, the lung. Infection may result in placentitis and abortion, neonatal jaundice, acute renal failure or hematuria, and, most commonly, recurrent uveitis.[2,10] Acute, fatal respiratory distress with pulmonary hemorrhage, thrombocytopenia, jaundice in combination with acute renal failure has been reported in five 1- to 3-month old foals in Europe.[11] The serovar involved in the respiratory cases was not confirmed and I am not aware of confirmed similar cases in North America.

REPRODUCTIVE TRACT INFECTION

Leptospira interrogans serovar Pomona abortions may account for approximately 13% of bacterial abortions in mares in endemic regions, although incidence varies considerably between years.[12] The reason for the yearly variation in incidence of abortions is not clear but rainfall may play a role. Serovar Pomona type *kennewicki* is responsible for most of the *Leptospira* abortions in North America, but serovars Grippotyphosa and Hardjo have also been reported.[13] A single genetic variant of *Leptospira interrogans* serovar Pomona type kennewicki was found to be associated with the majority of mare abortions in Kentucky.[3] Most abortions occur after 8 months of gestation, and rarely, a live foal may be born ill in association with fetal leptospirosis.[2] *Leptospira* may be found in the placenta, umbilical cord, kidney, and liver following abortions. Lesions include placentitis (edema and areas of necrosis in the chorion) that does not involve the cervical star. Microscopic lesions include necrosis and calcification of the placenta.[14] Placental disease may occasionally result in the mare developing hydrallantois.[15] Macroscopically, the fetal liver may have yellow discoloration and microscopic disease includes a multifocal necrosis with lymphocytic/plasmacytic infiltration of hepatic portal triads and sometimes giant cell hepatopathy.[14] Tubulonephrosis and interstitial nephritis may be detected in the kidney of the aborted fetus along with lung hemorrhages and myocarditis. Inflammation of the umbilical cord, funisitis, may be recognized by diffuse yellowish discoloration.[16] Although more than one mare on a farm may abort because of *Leptospira* infection, epidemics of abortions are unusual. A substantial proportion of infected mares deliver healthy foals. This may be explained by timing of infection or a mild degree of placental and fetal pathology and efficacy of the acquired immune response in the infected fetus. It is likely the balance between degree of damage in the placenta and fetus and efficacy of the fetal immune response that is critical to the final outcome.[17] The presence of specific agglutinating antibodies in the infected fetuses implies the infection occurred several days prior to the abortion. In support of this is that very high *Leptospira* antibody titers are found at the time of abortion in the infected mares. Mares that abort their fetus will most commonly do so 2-4 weeks after infection.[17] Aborting mares and other recently infected horses are believed to shed *L. interrogans* serovar Pomona in the urine for approximately 2 to 3 months.[18] A small number of horses on a farm with one or more *Leptospira* abortions may develop uveitis weeks later.

ACUTE RENAL FAILURE

Occasionally, *Leptospira* Pomona causes fever and acute renal failure in horses.[19,20] The kidneys become swollen as a result of tubular necrosis and tubulointerstitial nephritis, and urinalysis may reveal hematuria and pyuria without visible bacteria. Fever has been reported to be persistent for 17-25 days if effective antibiotic treatment is not administered; the animal generally remains bright, alert and eating until azotemia develops.[20] On rare occasions, multiple weanling or yearling horses may be affected with fever and acute renal failure following *Leptospira* Pomona infection. Young horses appear to be more likely to have fever and fulminant clinical disease (renal failure, respiratory disease) associated with marked leptospiremia.[2] Horses with persistent leptospiremia and fever will have very high level of *Leptospira* antibody at the time of azotemia. Chronic renal failure caused by *Leptospira* infection has not been documented in the horse.

RECURRENT UVEITIS

The most important clinical disease associated with *L. interrogans* serovar Pomona infection in adult horses in North America and *L. kirschneri* serovar Grippotyphosa in Europe is equine recurrent uveitis (ERU).[21,22] ERU is believed to be an immune-mediated disease frequently involving antibody against certain *Leptospira* antigens, specifically the LruC outer membrane protein, which cross-reacts with tissues of the lens, cornea, and possibly retina.[23] In North America, approximately 50% of ERU is associated with leptospirosis.[24] Live *Leptospira* organisms can be found in the aqueous or vitreous fluid of many horses with *Leptospira* associated recurrent uveitis.[25,22,26] High concentration of antibody against *L interrogans* serovar Pomona in the aqueous humor, compared with serum titers, also suggests persistent local antigenic stimulation.[27] Survival of the organism in the face of high ocular antibody may indicate an absence of cells or molecules (e.g., complement) involved in bacterial clearance, suggesting an ocular immune privilege similar to that of the central nervous system. Recurrent episodes of the disease may be related to a Th1 and Th17 response of autoreactivity following mimicry and intermolecular or intramolecular epitope spreading or both.[28-30]

Genetic factors are likely involved in the disease process, helping to explain why only some horses infected with *Leptospira* develop uveitis. Appaloosas, Warmbloods and some other breeds are thought to be genetically predisposed and specific MHC markers on ECA1 (Appaloosa) or ELA-A9 (Warmbloods) microsatellite are strongly associated with the disease.[31,32] The prevalence of ERU is unknown but reports suggest that 1% or greater horses will develop the disease during their lifetime.[21] *Leptospira*-associated uveitis may cause corneal, anterior chamber, and posterior chamber disease.[25] Therefore, clinical findings may vary from corneal edema or other changes, to clinically quiet retinal lesions observed on fundus examination, to dramatic recurrent and progressive painful uveitis. Chronic disease may cause cataracts, retinal degeneration, or even glaucoma. There is a high frequency of blindness, globe loss and loss of function in horses with ERU.[24]

PULMONARY HEMORRHAGE AND PNEUMONIA IN FOALS

A syndrome of acute respiratory distress, interstitial pneumonia with hemorrhage into alveoli and thrombocytopenia has been described in 1-5 month old foals in association with leptospirosis.[11] The offending serovar was not determined but *Leptospira* antigen was found in tissues (lung was not tested) and tested foals had positive serology for *Leptospira*. This syndrome has not been documented in North America.

DIAGNOSIS

Diagnosis of *Leptospira* abortion is best accomplished by PCR of aborting tissue or fluids or fluorescent antibody testing (FAT) of the placenta, umbilical cord, fetal liver, or fetal kidney. The sensitivity and specificity of the PCR and FAT in these fluid or tissues are nearly 100%.[33] Examination of silver-stained kidney samples in horses with renal disease is not highly accurate because there may be false-negative and false-positive findings, likely a result of non-pathogenic serovars. Polymerase chain reaction (PCR) testing is preferred for evaluation of fluids, such as urine, ocular fluids, fetal fluid and blood. Collection of a urine sample following furosemide administration may improve sensitivity of PCR, darkfield staining, or culture testing. The microscopic agglutination test is the gold standard for serologic detection of *Leptospira* antibodies. Either IgG or IgM will cause agglutination of the live organisms used in the test.[17] ELISAs for detection of antibodies against immunodominant epitopes of outer membrane proteins of pathogenic *Leptospira* have been evaluated in horses and have similar sensitivities and specificities to MAT.[34] A point of care lateral flow ELISA is available for testing dogs (SNAP Lepto Test IDEXX). *Leptospira* LPS, although minimally toxic, is highly antigenic and after acute infection antibody titers to *Leptospira* develop within 7 days and high titers within 2 weeks (unpublished data from author). Therefore, marked increases in serum antibody titers often accompany *Leptospira* abortions or acute renal failure, but serum titers may be low in horses with recurrent uveitis because of the chronic and localized nature of infection. Acute *L. interrogans* serovar Pomona infections often cause marked increases in antibody titers to several serovars(especially icterohemorrhagiae and Bratislava) but these non-infecting serovar titers usually decline more quickly than the titers to the actual infecting serovar. Interestingly, an aborted fetus does not have the cross reactivity or is not as pronounced as what occurs in horses with recent infection. The reason for this difference is not known but could be related to a predominance of IgM in aborted fetal fluids?[17] A combination of serology, culture, and PCR testing of aqueous fluid may be the only way to confirm *Leptospira*-associated uveitis. In ERU, the organism is most commonly found in the vitreous rather than aqueous fluid, which somewhat limits the practical application of ocular fluid PCR testing.

TREATMENT

Systemic administration of antimicrobials is indicated for horses with fever and acute renal failure caused by leptospirosis. Ticarcillin, penicillin, and enrofloxacin have been used successfully to treat horses with acute renal failure.[19,20] Other antimicrobials to which *Leptospira* may be sensitive include ampicillin, cephalosporin, tetracycline, and doxycycline. Trimethoprim-sulfa and chloramphenicol, two commonly prescribed oral antibiotics, are not effective. Attempts to decrease urine shedding in mares following *Leptospira* abortion by administering oxytetracycline, penicillin G, and streptomycin were surprisingly ineffective in one report.[18] Fluid therapy is indicated as supportive treatment of acute renal failure. If polyuria does not ensue soon after fluid therapy is started, furosemide and administration of drugs that may affect intrarenal hemodynamics should be administered (e.g., dopamine - I know there is no proof of efficacy in a large population of humans but it seemingly works in many large animals with ARF in my experience). Treatment for acute renal failure caused by *Leptospira* infection is often successful.

A variety of treatments for ERU, such as corticosteroids and cyclosporine implants have been used in hope of decreasing the inflammatory response, but these have generally provided only temporary relief and most affected horses either become blind or undergo enucleation because of intractable and persistent pain.[35] Active *Leptospira* infections are present in many horses with ERU in some geographic areas and it might seem prudent that an effort should be directed toward treating the possible infection. Unfortunately, antimicrobial treatment of ocular leptospirosis may not be easy because the blood-ocular barrier inhibits movement of antimicrobials from plasma into the eyes. Even with inflammation, some interference from the blood-ocular barrier may persist. In a study of healthy horses, doxycycline could not be detected in the aqueous humor after a five-day treatment with a dosage of 10 mg/kg given every 12 hours.[36] Minocycline administered at 4mg/kg orally twice daily to healthy horses will result in detectable, albeit low, concentrations in aqueous.[37] Enrofloxacin after repeated intravenous dosing with 7.5 mg/kg results in a peak concentrations in aqueous humor of approximately 0.32 µg/mL and minimal inhibitory concentration (MIC) and minimal bactericidal concentration for enrofloxacin is 0.05-0.30 µg/mL against *L. interrogans* serovar Pomona strains.[38,39] Other antimicrobials that may reach therapeutic concentration in the eye and could be effective against *Leptospira* organisms are intravenously administered oxytetracycline or penicillin. Topical antimicrobials may reach adequate concentrations in the aqueous humor but typically diffuse poorly into the vitreous humor. There is no good data to show that antimicrobial use has been successful in eliminating *Leptospira* from the eye. There is some concern the organism develops a biofilm in the eye which may make antimicrobial therapy ineffective.[40] Intravitreal antibiotic injections have anecdotally been reported to control uveitis in horses. Vitrectomy with gentamicin lavage has been used successfully in Europe in treating *Leptospira* associated uveitis.[41]

PREVENTION

Acutely infected horses and mares aborting from *Leptospira* infection should be isolated for 14 to 16 weeks or the urine should be tested by PCR to determine whether the mare is shedding the organism. Limiting exposure to stagnant water and to potential maintenance hosts of the organism may help in the control of leptospirosis[42] This can be very difficult to do in equine husbandry as horse feeds (hay, grass) and water are frequently contaminated by rodents or wildlife that may be shedding *Leptospira* organisms. Vaccination of horses against *L. interrogans* serovar Pomona using off-label vaccines has historically been performed on farms with endemic abortions or a high rate of uveitis. Numerous canine or farm animal–approved *Leptospira* vaccines have been used in horses. Serovar specific vaccination against leptospirosis has been highly successful in many species providing protection against serovars that cause incidental host infection.[43] Although *Leptospira* endotoxin has little toxicity it is very important in the development of immunity as *Leptospira* antibodies are agglutinating and opsonic to the LPS specific serovar[5] Following natural infection, serovar specific antibody along with complement will rapidly eliminate *Leptospira* from all tissues except from the brain, eyes, and kidneys. Proper vaccination and antibody production prevents bacteremia and therefore, should prevent the organism from invading the eye, CNS, kidney or reproductive tract. There has been concern that vaccination of infected horses with a whole-cell vaccine could theoretically cause ERU (although this has only rarely been observed and could not be demonstrated in one study) but *Leptospira* vaccination of horses with uveitis is not recommended.[44] There is currently one vaccine approved for use in horses, Lepto EQ Innovator® (Zoetis). It is a killed *Leptospira interrogans* serovar Pomona type kennewicki, whole cell bacterin. The product is labeled for vaccination of healthy horses 6 months of age or older as an aid in the prevention of leptospirosis caused by *Leptospira interrogans* serovar Pomona. The vaccine has demonstrated safety in foals as young as 3 months of age and in pregnant mares during the second trimester. Efficacy of this product was demonstrated utilizing an intraperitoneal leptospiral challenge model. Vaccinated horses did not develop leptospiremia or leptospiuria as compared to controls. The duration of immunity of this product has not been determined. One limitation to the vaccine is that it would not protect against *Leptospira kirschneri* serovar Grippotyphosa infection. Although *Leptospira interrogans* serovar Pomona type kennewicki is undoubtedly the predominant *Leptospira* equine pathogen in North America, *L.* Grippotyphosa would be second.

ZOONOSIS

Leptospirosis in humans can vary in severity according to the infecting serovar of Leptospira, and the age, health and immunological competence of the patient. It ranges from a mild, influenza-like illness to a severe infection with renal and hepatic failure, pulmonary distress, and death (the classical Weil's disease). I could find no data on zoonotic potential of *Leptospira interrogans* serovar Pomona type kennewicki; it should be considered a potential zoonotic serovar but of unknown importance. All *Leptospira* are easily and quickly (within 1-2 minutes) killed by common disinfectants and drying.

REFERENCES

[1]Patra KP, et al. BMC Microbiol 2015;15:244. [2]Ellis WA. Curr Top Microbiol Immunol 2015;387:99. [3]Timoney JF, et al. Vet Microbiol 2011;150:349. [4]Murray GL. Curr Top Microbiol Immunol 2015;387:139. [5]Adler B, et al. Vet Microbiol 2010;140:287. [6]Yan W, et al. J Vet Inter Med 2010;24:912. [7]Vasconcellos FA, et al. Trans R Soc Trop Med Hyg 2010;104:259. [8]Shimizu T, et al. Microbiol Immunol 1987;31:727. [9]Rajapakse S, et al. Ann Clin Microbiol Antimicrob 2015;14:2. [10]Verma A, et al. Vet Microbiol 2013;167:61. [11]Broux B, et al. J Vet Intern Med 2012;26:684. [12]Equine Dis Quart News 2015;5. [13]Giles RC, et al. J Am Vet Med Assoc 1993;203:1170. [14]Szeredi L, et al. Vet Pathol 2006;43:755. [15]Shanahan LM, et al. J Vet Intern Med 2011;25:158. [16]Sebastian M, et al. Vet Pathol 2005;42:659. [17]Velineni Sm et al. Equine Vet J 2015;Dec 15. [18]Bernard WV, et al. J Am Vet Med Assoc 1993;202:1285. [19]Divers TJ, et al. J Am Vet Med Assoc 1992;201:1391. [20]Frellstedt L, et al. Equien Vet Educ 2009;21:478. [21]Dwyer AE, et al. J Am Vet Med Assoc 1995;207:1327. [22]Brem S, et al. Berl Munch Tierarztl Wochenschr 1999;112:390. [23]Verma A, et al. Clin Vaccine Immunol 2012;19:452. [24]Gerding JC, et al. Equine Vet J 2015;April 18. [25]Polle F, et al. J Equine Vet Sci 2014;34:1300. [26]Faber NA, et al. J Clin Microbiol 2000;38:2731. [27]Halliwell RE, et al. Curr Eye Res 1985;4:1033. [28]Deeg CA. Vet Ophthalmol 2008;11:61. [29]Regan DP, et al. Vet Ophthalmol 2012;15:145. [30]Deeg CA, et al. Invest Ophthalmol Vis Sci 2006;47:652. [31]Fritz KL, et al. Anim Genet 2014;45:392. [32]Deeg CA, et al. Equine Vet J 2004;36:73. [33]Erol E, et al. Equine Vet J 2015;47:171. [34]Ye C, et al. Clin Vaccine Immunol 2014;21:478. [35]Malalana F, et al. Vet J 2015;206:22. [36]Gilmour MA, et al. Vet Ophthalmol 2005;8:331. [37]Schnabel LV, et al. Equine Vet J 2012;44:453. [38]Kim K, et al. J Vet Sci 2006;7:355. [39]Divers TJ, et al. Equine Vet J 2008;40:167. [40]Niedermaier G, et al. Dtsch Tierarztl Wochenschr 2006;113:418. [41]Toemoerdy E, *et al*.. Pferdeheilkunde 2010;26:251. [42]Kinde H, et al. Equine Vet J 1996;28:327. [43]Koizumi N, et al. J Postgrad Med 2005;51:210. [44]Rohrbach BW, et al. Vet Ophthalmol 2005;8:171.

Borrelia burgdorferi Infection in the Horse: What Do We Actually Know?

Thomas J. Divers, DVM, DACVIM, DACVECC
Ithaca, NY

INTRODUCTION

Borrelia burgdorferi sensu stricto is a common infection in adult horses in many areas of the US and the world. The greatest unknown regarding *B. burgdorferi* infection in horses is the incidence of clinical disease! The high seroprevalence found in many areas and the lack of scientific information on both the incidence and clinical signs of Lyme disease have perhaps made Lyme disease the most controversial equine infectious disease. There are many other unknowns for *B. burgdorferi* infections and Lyme disease in horses including: what are the clinical findings, how can Lyme disease be confirmed both ante mortem and post mortem, interpretation of *B. burgdorferi* serologic test in clinical practice, what are the preferred antibiotics for treating *B. burgdorferi* infection, and is there an appropriate use of canine approved vaccines in horses. Although the list of unknowns is considerable, some evidence-based information is available on *B. burgdorferi* infection in horses that may serve as a guide for interpreting serologic test, confirming Lyme disease, selecting antimicrobials for treatment and expected efficacy of vaccination.

PREVALENCE OF INFECTION AND GEOGRAPHIC DISTRIBUTION

Prevalence of *B. burgdorferi* infection is not well documented in horses with the only published prevalence studies suggesting 77%, 45%, 60%, 60%, 15% prevalence in adult horses in the mid –Atlantic,[1] New England,[2] New York,[3] Minnesota,[4] and Northwestern[5] states, respectively. It should be noted that these figures may be biased as some horses were likely tested based upon some suspicion of infection. The six-month incidence of new infection is reported to be 8% in horses in New York State.[6] Those of us who practice in endemic areas know the seroprevalence is high, even if we cannot provide the precise percentage. The geographic distribution of infection (seropositive horses) might be expected to be similar to that published for both human and canine Lyme disease. Although it is unproven, adult ticks are likely responsible for most of the *Borrelia* infections in horses, whereas nymph stages of the tick are responsible for a high percentage of human infections. CDC maps for human Lyme disease (30,000 cases/year) likely underestimate the endemic areas for horses and dogs since horses and dogs are more likely to have attachment of infected ticks than humans. Serologic studies in unvaccinated horses might indicate what CDC human Lyme maps would look like in future years. It is well documented that *Ixodes* ticks are geographically spreading and nearly 50% of all continental United States counties have *Ixodes* ticks, although *B. burgdorferi* infections remain more common in those areas with large populations of white footed mice and other small rodents serving as reservoirs of the organism.[7] Based upon canine and equine serologic surveys for *B. burgdorferi*, a high adult equine seroprevalence (>25%) would be expected in all or most New England and mid-Atlantic states and in some areas of Minnesota and Wisconsin, a moderate seroprevalence 6 to 15% in the surrounding states and into southern Ontario and along the California coast.[1-5,8]

SEROLOGIC TESTING FOR *B. BURGDORFERI*

There are several serologic tests available for *B. burgdorferi*. A list of tests include: the indirect florescent antibody (IFA) test, Western Blot, whole cell kinetic ELISA, point of care C6 ELISA, and a multiplex ELISA measuring antibodies to outer surface proteins A, C and F. Ponies exposed to *B. burgdorferi* infected ticks developed detectable whole cell k-ELISA antibodies 5–6 weeks after exposure, with antibody titers steadily increased for 3–4 months and remaining high until euthanasia 9 months later.[9] In infected ponies that were treated with antibiotics, the ELISA often declined after 4 weeks of antibiotic treatment but unless the ELISA continued to decrease to a negative value, infection was still present.[10] Western Blot testing is thought to be more specific than ELISA testing and in human medicine ELISA screening and WB follow up (two step testing) or WB testing (using both IgG and IgM) alone remain the recommended test for human Lyme disease. WB tests generally detect antibody against several outer surface antigens including those used in the Lyme multiple ELISA. These are several articles suggesting that the C6 ELISA could replace the two step (ELISA/WB) testing in humans.[11,12] The IDEXX C6 point of care ELISA has good sensitivity and excellent specificity for measuring *B. burgdorferi* antibodies in horses.[13,14] The published sensitivity in horses was 68% for experimentally infected ponies but if the samples collected less than 3 months after tick exposure are removed from the equation, the sensitivity increases to 75%.[14] There are no serologic tests, other than the IgM based test, that claim to consistently detect early (<2 months) infections. Available evidence from pony and dog experimental treatment studies suggest that *B. burgdorferi* infected horses that are successfully treated with antibiotics should have a negative C6 ELISA 2-4 months after completion of successful treatment, although this might vary depending upon duration of infection prior to treatment.[10,14] There is no published data on the multiplex ELISA results following experimental infection in horses, but there is good evidence in other species and from retrospective evaluation of field infected horses to indicate a temporal pattern of change in OspC and OspF antibody in association with infection.[3,15] The OspC antibody is believed to increase within 2-5 months after infection and then returns to a negative range regardless of infection status. OspC is, therefore, a good marker for recent (3-5 months) infection, but on its own is not suited to measure the response to treatment. Recognition of early infection is important if treatment is being considered since treatment of early *B. burgdorferi* infection is believed to easier (shorter treatment and higher success) than when treatment is delayed.[16] Horses with more chronic infections generally have elevations in OspF and this antibody titer likely remains high as long as the horse is infected. An unknown question is:

are horses with persistently high OspF or C6 antibody for several months chronically infected or could it be prolonged antibody production without infection? We have no evidence to suggest anything other than chronic infection. OspF antibody, like C6 antibody, might be used as markers of chronic infection and treatment success (negative antibody).[14,16] High levels of OspA antibody are believed to most commonly result from vaccination, but a small number of unvaccinated horses have persistently elevated OspA antibody. The significance of a high OspA antibody in those horses is unproven but some horses with neuroborreliosis and in one confirmed Lyme uveitis case, only an elevation in OspA antibody was present.[17] Comparison of serum and CSF serology can be somewhat useful in diagnosing neuroborreliosis based upon the concept of intrathecal specific antibody production with CNS infection. The multiplex Lyme assay results cannot be viewed exactly like *Sarcocystis neurona* ELISA ratios because for *B. burgdorferi* multiplex antibody measurements the serum is diluted. Generally, CSF results on the multiplex assay should be 3 or more times greater than serum levels to suggest intrathecal antibody production (personal observation). A marked increase in antibody ratio (CSF:Serum) for only 1 of the 3 antigens may be more significant that an uniform increase for all three. It should be noted that in a few confirmed cases of Lyme disease, horses have been seronegative in serum but positive in other fluid (CSF, aqueous humor). A case of equine neuroborreliosis and common variable immunodeficiency has been reported.[18]

CLINICAL DISEASE

There is strong evidence for a very large number of horses being infected with *B. burgdorferi*, but we have absolutely no clue how many actually have clinical disease! The only well documented and published cases of equine Lyme disease in North America have been in horses with uveitis, neuroborreliosis, or pseudolymphoma.[17-20] Clinical signs observed in horses with neuroborreliosis have included: acute to subacute onset of reluctance to lift the neck to proper height, "root signature" signs, muscle wasting of the neck, top line and gluteal muscles, hyperesthesia in these areas, spinal cord ataxia and less commonly signs of peripheral nerve disease, decreased tail and anal tone. Dysphagia and other cranial nerve signs have also been observed (Dr. Amy Johnson, personal communication). In confirmed cases of equine Lyme disease fever has often been absent. Generalized stiffness, lameness, and hyperesthesia are the most commonly incriminated clinical signs of equine Lyme disease in equine practice but causative proof of Lyme disease in these cases is absent.[21] Suspect Lyme cases, which are mostly sport horses, rarely have severe joint effusion or single leg lameness. Tendon sheath inflammation and bursitis have been rarely associated with *B. burgdorferi* infections. The commonly observed clinical response to doxycycline or minocycline in field cases supports the diagnosis of Lyme disease but improvements in clinical signs may be a result of anti-inflammatory properties of those drugs rather than specific treatment of Lyme disease.[22,23] A dual effect (antimicrobial and anti-inflammatory) of tetracyclines in treating Lyme disease in primates has also been reported.[24] Ponies experimentally infected with *B. burgdorferi* consistently had lymphocytic/histiocytic reaction in the skin near the tick bite, regional lymph nodes, and a lymphocytic infiltrate mostly around synovial nerves and blood vessels of all synovia. In some ponies, meninges, nerve roots and peripheral nerves had similar mononuclear inflammation. These experimental finding could support the most common clinical observations associated with Lyme disease in practice but this remains unproven.[9,10,21,25.] With seroprevalence for *B. burgdorferi* very high in many geographic areas and clinical signs often relatively subtle and likely not proven to be due to *B. burgdorferi* infection there is likelihood that equine Lyme disease might be an over diagnosed disease.

DIAGNOSIS

It is clear that positive antibody test, clinical signs and response to treatment do not provide sufficient basis for the diagnosis of Lyme disease. Instead, diagnosis should be based upon a combination of: clinical findings, ruling out other diseases that may cause similar findings, serologic testing, cytology or histopathology results on affected fluids and tissues, antigen testing if possible and response to treatment in some cases. The sensitivity and specificity of antigen or organism detection tests such as immunohistochemistry, silver staining, fluorescent *in situ* hybridization and even PCR are sometimes not well validated in veterinary testing laboratories.

TREATMENTS

Treatment of *B. burgdorferi* infection in horses has for the past 15 years has consisted mostly of orally administered doxycycline or minocycline. These antimicrobial selections seemed reasonable based upon their known success in treating Lyme disease in humans, *in vitro* susceptibility profiles and ease of administration and safety when given to horses for several weeks.[26,27] Upon closer examination, there are at least two very different variables regarding the use of these drugs in horses verse humans for the treatment of *B. burgdorferi*. Firstly, the duration of infection prior to treatment is likely much shorter in humans that in horses making successful treatment more difficult in the horse. Secondly, the absorption of doxycycline and minocycline are 95% and 100% in humans, respectively, while in horses the absorption is approximately 16% to 26%, respectively.[28-30] Peak serum levels following a 200 mg dose of either drug range from 2.3 to 7.5 µg/ml in humans while in horses given 10 mg/kg of doxycycline or 4 mg/kg minocycline peak serum levels are less than 0.75 µg/ml.[28-30] Peak synovial fluid concentrations of the drugs are similar to plasma concentrations in horses although doxycycline trough concentrations in synovial fluid are higher than serum trough levels.[31] Minocycline appears to have substantially better penetration of the aqueous and CSF in horses than does doxycycline although levels are less than peak serum levels.[30] The tetracycline class drugs, which are time dependent antimicrobials, have an MIC_{90} for *B. burgdorferi* of 0.25-0.8 µg/ml. Therefore, in horses at the current dosing recommendations, tissue and body fluid concentrations would often be below the MIC_{90} for *B. burgdorferi*. This pharmacokinetic and MIC data may help explain the transient decrease in serum antibody following doxycycline treatment of four *B. burgdorferi* experimentally infected ponies but elimination of the organism and negative serology occurred in only 1 of the 4.[10] In addition, field infected horses when treated, even long term, with either minocycline or doxycycline often have only modest or no decrease in *B. burgdorferi* serologic titers.[32] Conversely intravenously administered tetracycline (6.6 mg/kg) reaches high peak serum levels and may maintain trough levels above MIC_{90} for *B. burgdorferi*[33] which might explain oxytetracyline's success in eliminating the organism in the experimental pony infection study.[10] Horses that have been infected for many months prior to treatment may not have the same successful response following oxytetracycline treatment as what occurred in the experimental pony study. The potential for nephrotoxicity, thrombophlebitis, and the practical difficulty in daily, long-term administration have prevented oxytetracycline from becoming a standard of care treatment. Other options for treatment in horses include parenteral administration of a beta-lactam antibiotic, several of which have excellent *in vitro* sensitivity against *B. burgdorferi*. Ceftiofur at a daily IM dose of 2.2 mg/kg for 28 days was able to eliminate *B. burgdorferi* from 2 of 4 experimentally infected ponies (all placebo treated ponies remained infected).[10] The crystalline form of the drug is approved to be administered intramuscularly on days 1, 4, and then weekly and will maintain serum levels >0.22 µg/ml throughout the treatment period and for at least 6 days following end of treatment.[34] Ceftiofur MIC_{90} against *B. burgdorferi* is 0.08 µg/ml (unpublished data). It is unknown if tissue and body fluids levels would be maintained above the MIC. Potassium penicillin administered intravenously could be used, but the practical aspects of long term, 3 or 4 times daily treatments would be difficult except in a hospital and treatment expense would be considerable. Penicillin has a reported MIC_{90} of approximately 4 times that of the tetracyclines.[35] *B. burgdorferi* resistance has been demonstrated to trimethoprim sulfas and enrofloxacin,[36,37] two commonly used oral antibiotics in the horse.

Most *in vitro* testing is performed on viable, motile *B. burgdorferi* as they are readily available for testing and assumed to be the *Borrelia* stage most commonly causing disease. A cystic form of *B. burgdorferi* has been associated with chronic Lyme disease in humans (although not proven) and this form of *B. burgdorferi* is known to have a very high resistance pattern to the macrolides, tetracycline and B-lactam antibiotics that are commonly used for treatment of acute infections.[38,39] Although motile *B. burgdorferi* are reported to be resistant to metronidazole *in vitro*, it has moderate *in vitro* sensitivity for the cystic form of the organism.[35,39] Some equine veterinarians have used metronidazole as treatment for chronic *B. burgdorferi* infections when other antibiotics did not provide a suitable clinical response or decrease in serum antibody. There is currently no proof of clinical efficacy in humans and horses for metronidazole in treating acute or chronic *B. burgdorferi* infections.[40]

VACCINATION

There is good evidence in horses, humans and dogs that a high level of OspA antibody prevents the transfer of *B. burgdorferi* from the infected and attached tick (complete protection).[25] Some canine Lyme vaccines, when administered to horses, will result in a marked level of antibody production but the antibody T½ appears short and more than yearly vaccination would likely be requirc to provide protection from infection. A large number of canine vaccines have been administered to horses in the past 5 years and without reported adverse effects. Although there is good evidence of efficacy and reasonable field evidence of safety, importance of the vaccine for equine practice can only be answered when a determination of incidence of clinical disease is established. The value of adding OspC antigen to an OspA antigen vaccine is unknown; there are genomic strain variations in the OspC antigen.[41] FDA equine approved and effective tickicides or tick repellants for horses are sorely needed.

REFERENCES

[1]Burbelo PD, et al. Clin Vaccine Immunol 2011;18:1562. [2]Magnarelli LA, et al. J Am Vet Med Assoc 2000;217:1045. [3]Wagner B, et al. Equine Vet J 2013;45:533. [4]Durrani AZ, et al. J Equine Vet Sci 2011;31:427. [5]Metcalf KB, et al. J Equine Vet Sci 2008;28:587. [6]Wagner B, et al. Prev Med Med 2012:107:275. [7]Eisen RJ, et al. J Med Entomol 2016;Jan 18. [8]Qurollo BA, et al. Infect Ecol Epidemiol 2014;4:24699. [9]Chang YF, et al. Vet Pathol 2000;37:68. [10]Chang YF, et al. Vet Microbiol 2005;107:285. [11]Barth C, et al. Vet Clin Pathol 2014;43:496. [12]Wormser GP, et al. J Clin Microbiol 2013;51:4045. [13]Chandrashekar R, et al. Intern J Appl Res Vet Med 2008;6:145. [14]Johnson AL, et al. J Vet Diagn Invest 2008;20:321. [15]Wagner B, et al. Clin Vaccine Immunol 2012;19:527. [16]Wagner B, et al. BMC Vet Res 2015;11:163. 17Priest HL, Irby NL, Schlafer DH, et al. Diagnosis of *Borrelia*-associated uveitis in two horses. Vet Ophthalmol 2012;15:398-405. [18]James FM, et al. J Am Vet Med Assoc 2010;237:1180. [19]Imai DM, et al. Vet Pathol 2011;48:1151. [20]Sears KP, et al. Vet Dermatol 2012;23:153. [21]DeVilbiss BA, et al. J Equine Vet Sci 2009;29:237. [22]Maher MC, et al. Equine Vet J 2014;46:198. [23]Fortier LA, et al. J Orthop Res 2010;28:522. [24]Bernardino AL, et al. J Infect Dis 2009;199:1379. [25]Chang Y, et al. Vaccine 1999;18:540. [26]Cameron DJ, et al. Expert Rev Anti Infect Ther 2014;12:1103. [27]Hunfeld KP, et al. Int J Antimicrob Agents 2000;15:11. [28]Saivin S, et al. Clin Pharmcokinet 1988;15:355. [29]Bryant JE, et al. Equine Vet J 2000;32:233. [30]Schnabel LV, et al. Equine Vet J 2012;44:453. [31]Schnabel LV, et al. Equine Vet J 2010;42:208. [32]Divers TJ, et al. Acta Vet Hung 2012;60:421. [33]Teske RH, et al. J Am Vet Med Assoc 1973;162:119. [34]Collard WT, et al. J Vet Pharmacol Ther 2011;34:476. [35]Hunfeld KP, et al. Int J Antimicrob Agents 2000;15:11. [36]Baradaran-Dilmaghani R, et al. Infection 1996;24:60. [37]Kim D, et al. J Vet Sci 2006;7:355. [38]Sapi E, et al. Infect Drug Resist 2011;4:97. [39]Brorson O, et al. APMIS 1999;107:566. [40]Lantos PM, et al. Clin Infect Dis 2014;58:663. [41]Jauris-Heipke S, et al. J Clin Microbiol 1995;33:1860.

EHV-1 Latency: The Best Kept Secret

Lutz S. Goehring, DVM, MS, PhD, DACVIM (LAIM)
Munich, Germany

EHV-1 Latency – the best kept secret

Latency is probably the most powerful and ingenious survival mechanism of herpesviruses and it shows this virus' extreme level of host adaptation and co-existence. Latency is a dormant, non-replicative state which enables the virus to retreat in tissues awaiting periodical re-activation to turn into a lytic, replicative-productive state that allows horizontal spread and infection within the herd. Latency, once established, is considered live-long; periods of latency dominate this persistent infection, and different strains and types (e.g. EHV-4) can be found at one latency location. Latency does not prevent re-infection with same type or type-different virus.

Equid Herpesvirus -1 (EHV-1) infections cause respiratory disease and additional secondary complications such as myeloencephalopathy (Equid Herpesvirus-associated myeloencephalopathy or EHM) in the adult horse or late-term abortions in the pregnant mare. Early on, during lytic EHV-1 infection, virus progresses towards latency. As EHV-1 is a rather common pathogen among horses world-wide and first-time infections occur during adolescence, a large proportion of horses is assumed to be latent-infected. Regional EHV-1 latency prevalence among horses ranges between undetectable up to about 90%.

EHV-1 latency has been shown in trigeminal ganglia, lymphoid tissue of the respiratory tract and in PBMC of horses. Other locations are currently explored. The fact that EHV-1 becomes dormant in distinctly different tissues at the same time is unusual for herpesvirinae as there is typically a preference of one tissue over the other, neural or lympho-reticular. This could be of importance when it comes to recrudescence of EHV-1. Recrudescence is the reactivation of virus at its latent location with a return to productive, lytic replication in the respiratory tract, and thresholds for reactivation could vary among the various latency locations. Reactivation and recrudescence of latent virus requires trigger factors. 'Stress' as a trigger factor has been suggested and explored to some degree with equivocal results. 'Other' trigger factors or in combination probably exist. With some certainty it can be said that recrudescence of EHV-1 is an extremely rare event, as large cohorts of horses sampled under or during stressful conditions or events remained virus-(genome) negative on nasal swabs, yet index cases of EHM outbreaks can quite often be linked to a horse that returns from an event or show.

In analogy with human Herpes simplex 1 (HSV-1), it is likely that latent EHV-1 persists at different stages: deep latency vs. latency which is close to reactivation. HSV-1 research suggests that different stages of latency are not determined only by the virus, but by a triad of tissue-derived factors, modulating factors of the immune system in combination with virus-derived factors .

As it is unlikely that we will be able to prevent first-round infection with EHV-1 in the near future resulting in establishing latency in horses, we need to understand what determines stages of latency at various locations. With this knowledge we might be able to establish 'deep-stage latency', making recrudescence less likely, and thus reducing the possibility of future EHM or abortion outbreaks.

Characterization of Pulmonary Disease in Neonatal Foals with Computed Tomography

Kara M. Lascola, DVM, MS, DACVIM (LAIM)
Urbana, IL

INTRODUCTION

The use of diagnostic imaging, such as ultrasound, radiography or computed tomography (CT), plays a critical role in the diagnosis of respiratory disease. This is particularly important for neonatal foals for whom overt clinical signs of respiratory disease may be variable, subtle, or absent.[1-3] In humans CT imaging has become the imaging modality of choice for the diagnosis and characterization of respiratory disease.[4,5] With the increased availability of CT imaging to veterinary patients, its use has become routine in small animal patients and more common in equine patients. In neonatal foals, radiographic imaging of the lungs has been the mainstay for the diagnosis of respiratory disease and reviews exist which characterize radiographic findings in this age group.[1,2] While reports of CT findings of the lung are limited for neonatal foals with pulmonary disease, in settings where both CT and standard radiography are available, CT imaging of the thorax may offer the advantage of more accurate anatomic and morphologic characterization pulmonary disease.[6]

ADVANTAGES AND LIMITATIONS TO CT IMAGING IN SICK FOALS

Advantages

CT imaging has numerous advantages over standard radiography. One of the greatest advantages to current CT technology is the ability to acquire high resolution, 3-dimensional imaging of the whole lung and thorax within seconds.[7] This may be particularly advantageous to critically ill veterinary patients, such as neonatal foals, particularly when short term sedation can replace the need for general anesthesia. Additional advantages to CT imaging include improvements in disease localization and more complete anatomic and morphologic characterization of the entire lung as well as non-lung thoracic structures.[4,6,8,9] In comparison to standard radiography, CT imaging improves characterization of lung consolidation and is less impacted by respiratory motion artifact or superimposition of overlying thoracic structures.[4,6,8,9] Furthermore, standard CT imaging software allows for instant multiplanar reconstruction and manipulation of CT images.

Limitations

Limitations to the routine use of CT in veterinary medicine have included availability, cost, time, the perceived need for general anesthesia, and particularly for large animal species, body size. Of these, availability and cost are the limitations of greatest relevance to neonatal foals as current CT technology has greatly minimized the time required for image acquisition and often the need for general anesthesia. In contrast to standard radiography which is widely available to most general practitioners, CT imaging is currently limited to select referral practices and veterinary teaching hospitals. At this time, the fixed gantry diameter and table weight restrictions found with most CT systems along with the very high exposure settings needed to penetrate the torso of a larger horse, limit CT imaging of the thorax to foals and small ponies.[6]

CT IMAGING CONSIDERATIONS IN FOALS

At the authors' institution approximately 30-35 seconds are required to scan the entire thorax of a neonatal foal. The entire CT procedure, including sedation, transport to and from the stall, positioning of the foal and scanning, is typically completed within 15-20 minutes.[6] These times may vary at different hospitals according to imaging protocol and the proximity of the CT imaging suite.

Sedation

In contrast to humans or small dogs and cats, CT imaging in awake foals is rarely possible. While awake imaging has been performed in one or two critically ill foals at the author's institution, sedation during CT imaging greatly reduces the risk of injury to the animal or damage to the equipment. For standard CT imaging of the thorax in a sick foal, the administration of short term intravenous sedation is generally sufficient and well tolerated.[6,10-12] At the author's institution clinical cases breathe spontaneously and are maintained on intranasal oxygen during image acquisition. In select cases the administration of propofol may be necessary for effective sedation in healthy foals, but should be used with caution in sick foals with respiratory disease due to the drug's dose- and rate-dependent respiratory depressant properties and the potential for apnea. General anesthesia with mechanical ventilation may be warranted in situations where there is a desire to perform imaging under conditions of controlled apnea or for the delivery of alveolar recruitment maneuvers.

Positioning

The most widely reported protocol for positioning of sedated foals undergoing CT imaging is sternal recumbency with the head facing the CT gantry and the fore- and hind limbs extended cranially and caudally.[6,10-12] Positioning in dorsal recumbency has also been described in healthy, anesthetized neonatal foals.[13] Notable differences to image quality, specifically the degree of dependent lung atelectasis, were not appreciated when foals imaged in sternal recumbency were compared to those imaged in dorsal recumbency.[6,13] While imaging in dorsal recumbency may provide an alternative when pathology involving the ventral lung is suspected, the potential impact on ventilation should be considered in patients with respiratory compromise.

Artifacts

Respiratory motion and positional atelectasis represent two of the most common patient-related artifacts that may impede accurate interpretation of CT images and are of particular relevance to veterinary patients. In spontaneously breathing animals minor motion artifacts may be appear on select axial slices as blurring along distinct margins of thoracic structures such as pulmonary vessels or the heart or may present as a generalized interstitial pattern on involved CT slices.[6] Respiratory motion artifacts are typically minor in healthy foals and sick foals imaged to date.[6,10,11] In patients with significant respiratory compromise and associated tachypnea, they may be of greater concern and could potentially mask or mimic true pathology.

In healthy neonatal foals in sternal recumbency, reported findings attributable to positional atelectasis appear most prominent in the dependent lung caudal to the heart and include a diffuse increase in lung attenuation within the parenchyma along with interstitial and patchy-to-coalescing alveolar patterns with occasional air bronchograms in the areas of greatest attenuation.[6,10,11,,13] It is important to remember that these findings are not unique to atelectasis and thus may complicate image interpretation in animals with concurrent pulmonary disease. Similar to other species, the development of atelectasis in anesthetized or sedated foals most likely occurs fairly rapidly and is primarily attributed to compression of the dependent lung by overlying and adjacent lung and non-lung tissues.[14] Atelectasis formation is most likely also influenced by the increased chest wall compliance of foals and the potential respiratory depressant properties of medications administered for sedation or anesthesia.[10,14,15] Investigation of the use of manual alveolar recruitment maneuvers in healthy, anesthetized neonatal foals did not result in elimination of dependent atelectasis.[13]

SUMMARY OF REPORTED FINDINGS IN NEONATAL FOALS WITH PULMONARY DISEASE

While descriptions of CT imaging of the lung are available for healthy neonatal foals,[6,10,11,13] they are limited to a small number of neonatal foals with pulmonary disease.[6,12] Reported characterizations of pulmonary disease in individual foals includes multiple CT lung patterns often varying throughout the lung in severity, extent, and distribution.[6] More unique patterns, such as "crazy paving", have also been identified.[6,12] When available, direct comparison between findings on CT and radiography suggests that in select cases radiographic findings may under-estimate the severity and extent of changes in the lung as detected by CT imaging.

Reported CT findings for two foals with clinical diagnoses of EqARDS[6,16] and for a foal with diffuse bronchopneumonia are available.[6] In foals with EqARDS, specific findings on CT have included diffuse interstitial and patchy-to-coalescing alveolar patterns, emphysematous lung, and interlobular septal thickening. These changes were found bilaterally distributed throughout both the caudoventral and dorsal lung fields.[6] In a foal with bronchopneumonia, diffuse interstitial and patchy alveolar patterns were noted in addition to diffuse bronchial wall thickening and peribronchial cuffing.[6]

In addition to the cases described in the proceedings, additional cases (all neonatal foals (≤ 14 days of age) representing a wider spectrum of clinical disease will be presented.

QCT APPLICATIONS-

In addition to the image reconstruction and manipulation capabilities offered through standard CT imaging software, more specific image analysis software is available that allows for quantitative CT image (qCT) analysis of the lung. In humans qCT analysis has been used extensively for the diagnosis and characterization of pulmonary disease. [4,5,7,17,18] Currently, the use of qCT analysis has been limited to experimental studies involving healthy foals and to a very small number of foals with pulmonary disease[6,10-13] but may hold future potential as an additional diagnostic tool.

REFERENCES

1. Bedenice D et al. J Vet Intern Med, 2003;17:876.
2. Lester GD et al. Vet Clin North Am Equine Pract, 2001;17: 19.
3. Wilkins PA. Vet Clin North Am Equine Pract, 2003;19:19.
4. Gattinoni L et al. Clin Chest Med 2006;27:559.
5. Grenier PA et al. Eur Radiol 2002;12:1022.
6. Lascola KM et al. Vet Clin North Am Equine Pract, 2015;31(3):497.
7. Rouby JJ et al. Crit Care Med 2003;31(suppl 4):285.
8. Alexander K et al. Vet Radiol Ultrasound. 2012;53(3):258.
9. Rubinowitz AN et al. Crit Care Clin. 2007;23(3):539.
10. Lascola KM et al. Am J Vet Res, 2013;74(9):1239.
11. Schliewert E-C et al. Am J Vet Res, 2014;76(1):42.
12. Lascola KM et al. Proceedings of the 33rd Symposium of the Veterinary Comparative Respiratory Society, 2016.

13. Lascola KM et al., Accepted for publication Am J Vet Res, 2016.
14. Magnusson L et al. Br J Anaesth, 2003;91(1):61.
15. Korterba AM et al. Respir Physiol, 1994; 95:21.
16. Wilkins PA et al. Vet Clin North Am Equine Pract, 2015;31(1):137.
17. Newell JD et al. J Thorac Imaging, 2013; 28:266.
18. McEvoy F et al. Vet Radiol Ultrasound, 2009; 50:47.

HIGHLIGHTS OF THE 8TH EUROPEAN COLLEGE OF EQUINE INTERNAL MEDICINE CONGRESS UTRECHT 2015

Catherine McGowan, BVSc, PhD, DEIM, DECEIM
Liverpool, UK

INTRODUCTION

The European College of Equine Internal Medicine was established in 2012 after full European Board of Veterinary Surgeons recognition in 2010. Since 2014, the ECEIM has had College-wide associate status with the ACVIM (LAIM). To foster the relationship between ECEIM and ACVIM, an ECEIM representative has given a talk at ACVIM as well as there being an ACVIM-ECEIM liaison committee, led by Dr Ben Sykes (Diplomate ACVIM-LAIM). This talk, by Professor Catherine McGowan (DipECEIM) will update highlights from the November 2015 Congress in Utrecht with a summary of some of the latest research from Europe, focussing on endocrine and metabolic disease.

ABOUT ECEIM

ECEIM currently (as of November 2015) has 154 active Diplomates in 22 countries (Table 1.). Of these, 99 Diplomates work in Universities whereas 55 work in private practice (50) or other (5). There were currently 30 residents enrolled in standard residency programs and 7 in alternate programs. Residents are supported by an annual 3-5 day summer school and ACVIM residents are also invited.

Table 1. ECEIM Diplomates and countries of practice.

Country	n	Country	n	Country	n
Australia	8	Grenada, West Indies	1	Spain	7
Austria	3	Hungary	3	Sweden	4
Belgium	11	Ireland	7	Switzerland	11
Czech Republic	2	Israel	1	The Netherlands	11
Denmark	3	New Zealand	1	Trinidad/Tobago	1
Finland	2	Norway	3	United Kingdom	38
France	12	South Africa	2	USA	2
Germany	21				

Cornélie Westermann, president of ECEIM, presented ECEIM's strategic plan, with four aims:

1. Outreach and widening participation; Establishment of ECEIM residency programs across Europe, consolidating training programs in countries with training programs and establishing programs in countries without current training programs and/or ECEIM Dips by engaging with ACVIM Dips
2. Development of the ECEIM family: Improving resident support and training packages. Identify the areas of weaknesses and the centers of excellence in order to promote resource sharing between Diplomates and institutions, focusing on both standard and alternative track programs.
3. Increasing visibility of the ECEIM: Visibility to vet community and public.
4. Maintaining a robust and Professional organization within the College: Consolidating all committees and Board procedures to ensure continuity of College initiative and actions regardless of changes in officers and committee membership. Create an ongoing succession planning system for all Committees and Boards.

HEADLINES 2015

For the first time in 2015, ECEIM gave its special achievement awards to 4 of its Diplomates. These were:

1. The University of Hannover represented by Dr Karsten Feige for their provision of the facilities and help with the ECEIM examinations every year and for their encouragement of their team to become/stay active in quite time-consuming college positions.
2. Dr Ben Sykes for his never tiring force in pursuing and publishing the first ECEIM consensus statement *Equine Gastric Ulcer Syndrome in Adult Horses*[1]
3. Dr Josh Slater for being the driving force in strategic planning and spending a lot of energy and time in quality assurance and enhancement, e.g. in exam procedures.
4. Dr Gunther van Loon for both "inventing" the finances of the College but also his commitment overall to the ECEIM.

ECEIM Congress 2016 is November 3rd to 5th in Helsinki, Finland.

SELECTED ABSTRACTS FROM ECEIM ON ENDOCRINE AND METABOLIC DISEASE

Endocrine and metabolic disease was a notable area of strength of the oral presentations and a number of institutions across Europe presented their research. I have selected six abstracts, and a summary of each follows. Full abstracts are available from the ECEIM Proceedings in the reference list.

Prospective Cohort Study Evaluating Risk Factors For The Development Of Pasture-associated Laminitis in the UK Nicola J. Menzies-Gow et al.[2], the Royal Veterinary College, UK

The study aim was to investigate laminitis risk factors prior to disease occurrence. 446 ponies ≥ 7 years old with no previous laminitis were recruited. Body condition score, height, weight and crest height and thickness were measured and an overnight dexamethasone suppression test performed. Plasma/ serum concentrations of adiponectin, leptin, triglyceride, insulin, insulin-like growth factor (IGF)-1, C-reactive protein, von Willebrand's factor (vWF), soluble (s) E-selectin and sP-selectin were assayed. Follow-up was obtained annually for 2 years to ascertain laminitis development. Laminitis risk factors prior to disease occurrence include low plasma [adiponectin] and [IGF-1] and high basal or post dexamethasone serum [insulin]. However, it was not possible to generate cut-off values to accurately predict future laminitis development in an individual animal.

Ultrasonographic Measurements of Localized Fat Accumulation in Normal and High Energy Fed Shetland Pony Mares Over Time Esther W. Siegers et al.[3], Utrecht University, The Netherlands

In this study effects of a diet-induced increase in body weight on body fat accumulation and localization was determined in 13 Shetland pony mares (3-7 years old),randomly divided into a control and a test group. They were fed either a maintenance diet (100% NE) or high fat diet (200% NE) for 24 weeks. BCS were not significantly different at the start of the diet (4.9 ± 1.2 vs 5.1 ± 1.2) between both groups. Twelve weeks after starting the diet, measurements of the BCS, body weight and ultrasonography of adipose tissue on five different locations (retroperitoneal, axillary, withers, ribeye and rump) started on a monthly basis. Diet induced fat accumulation initially occurred both subcutaneously and retroperitoneally, while after 20 weeks further fat accumulation occurred mainly intra-abdominally.

Residual Effects of a High Energy Diet on Blood Pressure in Obese Shetland Ponies Ellen Roelfsema et al.[4], Utrecht University, The Netherlands.

The aim of this study was to determine the residual effects of a 24 week high-energy diet on the blood pressure of obese Shetland pony mares. Four ponies (test group) received a diet rich in fat and carbohydrates containing 200% of their maintenance energy requirements and four ponies (control group) received a diet containing 100% of their maintenance energy requirements. The high energy diet resulted in weight gain of at least 20% and significant hyperinsulinemia. After the end of the 24 week period (t=0) both groups received only ad libitum hay for 15 weeks. Before (t=0) and after (t=1) the hay-only period non-invasive blood pressures were determined. Systolic and diastolic blood pressures were significantly higher in the test group at t=0 and t=1 ($P<0.001$). Diastolic blood pressure was significantly higher at t=1 compared to t=0 in the control group ($P=0.011$). It can be concluded; that a short term decrease of fat and carbohydrate intake does not change the blood pressure in obese hyperinsulinemic Shetland ponies. Ad libitum feeding of hay compared to meal feeding increases blood pressure in lean Shetland ponies.

TRH-Stimulation Test in Healthy Horses with Pain Judith Christine Winter et al.[5], Free University of Berlin, Berlin

The purpose of this study was to investigate, if the TRH-stimulation test is influenced by pain in horses without PPID. Included in the study were 16 horses with an age of one to 15 years without clinical signs of PPID and moderate pain due to different reasons. There were 13 horses with orthopedic problems including five horses with laminitis and three horses with colic. All horses were judged according to a modified pain score and divided into three different groups according to their pain level. Baseline ACTH and cortisol levels were measured and a TRH-stimulation test performed as previously described. Those horses that reached a painfree or mildly painful level were used as their own controls. All horses except one had baseline ACTH and cortisol levels within the reference ranges. There were no significant differences between the paingroups and the control group or in between the three different pain groups regarding baseline ACTH/cortisol levels or ACTH/cortisol levels after stimulation. Therefore in this study pain does not seem to have an influence on the outcome of the TRH stimulation test.

Morphometric Characteristics, Insulin, Leptin and Triglyceride Concentrations in Obese Andalusian Horses Carla N. Aguirre Pascasio et al.[6], Universidad de Murcia, Spain

The objectives of the study were to determine the association of morphometries and scoring systems with insulin, leptin and triglycerides in Andalusian horses. A cross-sectional study was performed with 127 Andalusian horses. Horses were categorized based on BCS as obese if BCS≥7. Morphometric measurements (in cm) including girth, waist, body length, and height at the withers were taken to calculate body ratios (girth:height, girth:length, waist:height and waist:length). Concentrated feed was withheld for at least 12h before sampling. Blood extractions were performed between 06:00 and 12:00am to measure fasting insulin, leptin and triglycerides concentrations. Correlations between variables and mean comparisons were carried out. Thirty-three horses (26%) were obese. BCS, all morphometric ratios, leptin and triglyceride concentrations were significantly higher in obese horses. In the entire sample, BCS was significantly correlated with leptin, triglycerides, and all morphometric ratios, being girth:length the best correlated ratio Leptin was correlated with insulin and all body ratios. Triglycerides were correlated with waist:height and waist:length ratios. Insulin was correlated with girth: height and girth:length. These results show that in contrast with studies carried out in other breeds, BCS in Andalusian Horses, is not associated with insulin levels, and girth:length ratio is the best correlated parameter with it.

Individual Variations of Dynamic Metabolic Responses to Oral Glucose Test and Combined Glucose/Insulin Test in Horses Tobias Warnken et al.[7], University of Veterinary Medicine, Hannover, Germany

We expected that oral glucose test OGT and combined intravenous insulin/glucose test (CGIT) may not be equally suitable to test insulin sensitivity in horses because of their different physiological mode of action. OGT and CGIT were performed in twelve healthy Warmblood horses. Horses were of different sex, aged 15±6.5 years, weighing 567±81kg, body condition score (BCS) 4.8±1.6. OGT was performed with 1g/kg bwt glucose administered via nasa-gastric intubation. CGIT was performed with injection of 150 mg/kg bwt glucose solution and 0.11U/kg bwt insulin. Blood samples were taken for 3 hours in at least 15 minutes intervals and were analyzed for insulin, glucose, triglyceride and non-esterified-fatty-acid concentrations. The time course of glucose and insulin was similar in all 12 horses in the CGITwith initial peak in glucose and insulin at first sampling time-point followed by continuously and similar decrease during sampling period in al112 horses. In contrast, in the OGT four horses clearly responded with an exaggerated insulin response- area under the curve insulin compared to to the remaining 8 horses ($p < 0.0001$). There was no relationship of age, sex, weight or BCS to OGT insulin concentrations. The present results underline that OGT and CGIT mirror different facets of the metabolic response to a glycemic stimulus.

REFERENCES

1. Sykes BW, et al. J Vet Intern Med 2015;29:1288–1299. **2.** Menzies-Gow NJ, et al. J Proc. 8th Congress ECEIM, 2015 p.110. **3.** Siegers EW, et al. Proc. 8th Congress ECEIM, 2015 p.123. **4.** Roelfsema E et al. Proc. 8th Congress ECEIM, 2015 p.124. **5.** Winter JC et al. Proc. 8th Congress ECEIM, 2015 p.125. **6.** Aguirre Pascasio et al. Proc. 8th Congress ECEIM, 2015 p.144. **7.** Warnken et al. Proc. 8th Congress ECEIM, 2015 p.145.

Dietary Management of Equine Metabolic Syndrome

Catherine McGowan, BVSc, PhD, DEIM, DECEIM
Liverpool, UK

INTRODUCTION

Dietary management is the mainstay of prevention and management of equine metabolic syndrome and will remain so until appropriate pharmaceuticals are developed to better manage the disease. However, it has also been an area of confusion, with many veterinarians feeling uncomfortable with nutrition. This talk will focus on the research from the University of Liverpool in management of obesity and the effects of different forage sources including soaked hay in both research and clinical settings.

EQUINE METABOLIC SYNDROME

First described in 1999, as "Laminitis, Hypothyroidism and Obesity: a peripheral cushingoid syndrome in horses"[1] equine metabolic syndrome (EMS) was defined more recently in the 2010 ACVIM consensus statement.[2] It was described as a phenotype of obesity (regional or generalized), insulin resistance (hyperinsulinemia or abnormal insulin and glucose regulation; now known as insulin dysregulation[3]) and a predisposition to laminitis that has developed in the absence of recognized causes such as grain overload, colic, colitis, or retained placenta.[2] This definition was useful in encapsulating what was known about EMS at the time although it is also important to realise the significance of these three factors in producing the clinical phenotype.

Obesity

Obesity represents an important risk or predisposing factor for EMS, especially in highly susceptible breeds like UK native pony breeds where there is an underlying predisposition. However, obesity does not define EMS, and many ponies and horses are obese with normal insulin sensitivity and, conversely, some EMS horses and ponies have no external evidence of obesity.

Insulin resistance

Insulin resistance is central to EMS and the pathogenesis of laminitis has been shown to be closely associated with hyperinsulinemia in experimental models[4,5], field studies[6], and clinical studies[7,8]: Insulin resistance manifest as resting or basal hyperinsulinemia or an abnormal insulin response to oral or IV carbohydrate (CHO) challenge can collectively be termed insulin dysregulation.[3] Some animals with EMS show a relatively greater response to oral CHO challenge than IV, which is of particular concern for these animals as a risk factor for laminitis.[3,9,10] Either way insulin dysregulation is now considered the core defining factor in EMS.

Predisposition to laminitis

The most important disease expression associated with EMS in horses is laminitis. Despite a general acceptance by veterinarians of the existence of subclinical laminitis and corresponding gross changes in the hoof capsule (divergent rings, dropped sole, separation of the white line and cap horn production),histological evidence of this was only recently published from naturally occurring cases.[8]

DIETARY MANAGEMENT

In the consensus statement[2] it was acknowledged that dietary management was central to the management of EMS, yet up until that point few dietary studies had been performed, none specifically on EMS affected animals and there was some confusion about some of the fundamental principles. The contention of ideal body weight and the measurement and change in body condition score (BCS) to monitor weight loss were unestablished. How much to feed as a percentage of body weight (BWT), the composition (e.g. non-structural carbohydrate (NSC) concentration), insulinemic potential and dry matter (DM) content of different forages were not well established. Furthermore, whether to soak hay or not, how long for and the effect of soaking on dry matter and other nutrients had not been established.

Body condition scoring

Ideal BWT (or bodymass) is a concept that lacks either biological or published supporting evidence.[11] Indeed, trying to restrict caloric intake based on ideal BWT could be dangerous. This is particularly the case where the DM content of the forage concerned has been overlooked (see below). Dugdale and colleagues[12] showed that current BCS scoring systems have a non-linear relationship to whole body fat content, especially at BCS ≥ 7/9 using a modified Henneke scoring system. At the low end of the BCS scale, only a small change in body fat is required for a 1 point change in BCS and the result is close to linear; whereas as the high end of the scale, a much larger change in body fat is required for a 1 point change in BCS and the relationship is lost.[12] This also has implications when monitoring weight loss in horses, where BCS will lag behind losses in BWT in animals where BCS was≥ 7/9.[13]

How much to feed as a percentage of BWT

Current recommendations for safe weight loss in horses have been extrapolated from studies in humans and companion animals where a target loss in BWT of approximately 0.5% - 1% per week is aimed for after week one[14]. Initial weight loss in the first week is usually removed from the calculations due to a considerable reduction in gut fill at the onset of caloric restriction.[13] There have now been several research studies from the University of Liverpool looking at caloric restriction using a commercial chaffed preparation[13], hay[15] and soaked hay[11] where animals were fed at between 1% and 1.25% of their current BWT daily on a DM basis in a controlled environment with only 1 hour a day muzzled pasture access. All studies resulted in safe weight loss within the target weight loss rates, without side effects, and the latter two studies where insulin dysregulation was measured showed a significant improvement of their insulin response to IV challenge.[11,15] There was some variability between animals in each study, which has been termed "weight loss resistance".[15] The rate of weight loss clearly varies between horses by over 50% and is important to factor in to any caloric restriction protocol. In studies using owner-managed clinical cases of EMS, we have developed tailored protocols which allow for variation and adjustment of weight loss in horses.[16]

As mentioned above, it is important to factor in the DM% of the forage fed. For example, most hays will be only around 85% DM and fermented hay (haylages) around 60% DM. This clearly makes a large difference to the daily amount fed as fresh weight. With the abovementioned studies, the amount of weight loss on average per week after week one when ponies were fed 1% of their BWT daily as DM of chaff was 0.7%[13]; when fed 1.25% of their BWT daily as DM of hay was 0.44%[15] and when fed 1.25% of their BWT daily as DM of soaked hay was 1.0%[11]of their week 1 BWT. The much greater weekly reduction in BWT in the soaked hay study was further investigated by correcting the DM in the hay soaked for 7 or 16 hours in accordance with the 'insoluble' ADF content of fresh and soaked hays (and validated using a test-soaking protocol).[14] This showed that there was a loss of 14% DM in soaked hay (with little difference between 7 or 16 hours soaking) and this accounted for a 23.5% greater energy restriction implying that soaked hay was more equivalent to feeding 1% BWT as DM than 1.25%. This could have dangerous consequences if target guidelines for weight loss are not well understood e.g. an underestimation of BWT (or an erroneous application of "ideal" BWT) or a poorer than usual batch of hay or indeed the misguided restriction to 1% BWT forage fed daily as fresh weight without consideration of the DM proportion of the original forage, or indeed, the soaked forage. As such we currently only recommend feeding of soaked hay at 1.5% weighbridge measured actual BWT daily (equivalent to 1% BWT DM) if the horse is being monitored regularly under veterinary supervision.[16] Weight loss at 0.5%-1.0% BWT weekly should only be for a limited period of time (usually 3 – 4 months), with clear outcome targets of improved insulin regulation used as a clear guide as to when to relax to the caloric restriction.[16]

Balancing Calorie Restricted Diets

When horses have their intakes restricted in order to elicit weight loss, careful attention to balancing the diet needs to be employed. Ideally the forage should be analyzed, but in reality many feed companies offer vitamin and mineral premix "balancers" that can be fed to ensure adequate vitamin and mineral intakes. Some chaff based forages are marketed as already containing a vitamin and mineral premix that will provide sufficient nutrient balancing, provided the full manufacturer recommended weight of the chaff is fed. When choosing a balancer attention needs to be paid to the ingredients. Many balancers marketed for laminitic horses or as "low calorie" actually also have quite low protein. This is also somewhat misguided as due to the low volume required to be fed for most balancers (around 125 g/pony/day) the caloric difference typically has negligible effect on energy provision, yet can contain insufficient protein to balance a restricted forage diet with borderline protein provision. We would typically recommend a balancer with a protein of around 25%.[16]

When balancing soaked hay, attention needs to be paid to loss of mineral salts in the soaking process.[11,14,17] Unlike earlier studies with limited analysis and varied soak times[18], our studies have shown a consistent reduction in NSC with soaking of grass hays for either 7 or 16 hours (between feeds) in cold water. This included both hays sampled and soaked as part of experimental studies, but also of hays soaked by their owners.[17] The average reduction in NSC was 50%, accompanyied by 21% loss in calcium, 49% loss of phosphorus, 39% loss of magnesium, 66% loss of potassium, 65% loss of sodium, 72% loss of chloride and 24% loss of suphur. There were no significant losses of crude protein, nor any of the elements (Fe,Zn,Cu,Mn,Mo,Co).[17] For mixed grass hays we have analysed from the Northwest of the UK this is important, as average pre-soaking NSC% were 18% and grass species available do not allow for selection of low NSC (≤10%) varieties. However, importantly, the minerals lost are not found in typical vitamin and mineral balancers and additional mineral supplementation needs to be fed to horses on soaked hay diets.

Role of Nutraceuticals

Nutraceuticals play an important role in the dietary management of people with metabolic syndrome, but less work to date has been performed in horses. A combination of chromium and magnesium failed to improve morphometric measurements or insulin regulation in obese laminitic horses.[19] Supplementation with 4 g per day of L-carnitine resulted in a mild reduction in postprandial insulin and glucose concentrations[20] which is promising, but further work needs to be done. The only nutraceuticals shown to have clear effects on improving insulin sensitivity in obese horses are short-chain fructo-oligosaccharides (FOS).[21] Although in another study, the improvement insulin sensitivity from feeding FOS was not able to be differentiated as additional to the improvements seen following 6 weeks of caloric restriction and marked weight loss in ponies fed soaked hay.[11]

It is interesting to note that FOS are fructans which have been fed in large doses to cause laminitis.[22] However, when fed at such large doses FOS are acting like any CHO overload model, indeed the doses required to induce laminitis from FOS are higher than those required to induce laminitis from glucose.[23] Further, the insulinemic response to FOS is much lower than to glucose[24] Despite speculations of extremely high fructans concentrations possibly ingested by grazing horses being presented in a conference abstract,[25] most grazing horses could only ingest very small quantities of fructans per hour. The contribution of fructans to laminitis are simply via inducing an insulinemic response to CHO ingestion (in combination with other components of the pasture's total NSC content) and not as a CHO overload effect. Indeed FOS has not only been shown to be insulin sensitizing[21] but also protective against a grain overload challenge at typical grazing doses[26].

Insulinemic Responses to Different Forages

Insulin dysregulation is central to the definition of EMS and also central to the risk for laminitis, so in order to fully understand the role of different forages in management of EMS, their effect on postprandial hyperinsulinemia needs to be measured. Reduction in postprandial hyperinsulinemia is the immediate and ultimate goal of EMS dietary management. The insulinemic responses to a meal using hays of different NSC (17%, 10% and 4%) were shown in normal horses and horses with PSSM.[27] The 17% NSC hay produced greater insulinemic responses in both normal and PSSM horses than 10% or 4% (which were not different from each other). This study certainly supports aiming to feed hay of NSC% of 10% or lower. In a recent study conducted at our institute, Carslake et al.[9] performed a randomized crossover trial investigating the insulinemic and glycemic responses to three commonly fed forages. Hay, soaked hay and haylage were fed at an equivalent DM weight (0.25% BWT) as a single feed and blood was sampled at regular intervals for 5 hours and analysed for glucose and insulin concentrations. Area under the curve for insulin (AUCi) was greater for haylage compared to hay (p=0.019) and soaked hay (p=0.002), and greater for hay compared to soaked hay (p=0.002). AUC for glucose (AUCg) was lower for soaked hay compared to hay (p=0.002) and haylage (p=0.003). CGIT positive ponies had greater AUCi for all three forages, compared to CGIT negative ponies. The insulinemic response top soaked hay was very small, which is may explain why there was still greater weight loss in horses fed 1.25% BM DM soaked hay than even 1.0% BM DM chaffed feed under the same experimental conditions. Clinically, it also justifies the feeding of soaked hay in an acute laminitis episode in a horse with EMS.

REFERENCES

1. Johnson PJ, et al. In: Proc. 17th ACVIM Forum, 1999, pp. 192-194. **2.** Frank N, et al. J Vet Intern Med. 2010;24(3):467-475. **3.** Frank N, et al. Equine Vet J. 2014;46(1):103-12. **4.** Asplin KL et al. Vet J. 2007;174(3):530-535. **5.** deLaat MA et al. Equine Vet J. 2010;42(2):129-135 **6.** Treiber KH et al. J Am Vet Med Assoc. 2006;228(10):1538-1545 .**7.** Karikoski N et al. 2011 Dom Anim Endocrin **8.** Kariskoski N et al. Vet Pathol. 2015;52(5):945-956. **9.** Carslake HB et al. Proc. 8th Congress ECEIM, 2015 p.121. **10.** Warnken T et al. Proc. 8th Congress ECEIM, 2015 p.145. **11.** McGowan CM et al. Vet J 2013; 196:153-159. **12.** Dugdale AHA et al. Vet J 2012; 194:173-178. **13.** Dugdale AHA et al. Equine Vet J. 2010; 42(7):600-610. **14.** Argo CMCG et al. Vet J; 2015; 206(2):170-177. **15.** Argo CMCG et al Vet J 2012; 194:179–188. **16.** Morgan RA et al. Equine Vet J 2015. **17.** Mack SJ et al. Vet Rec 2014; 174(18):452. **18.** Longland et al. Vet Rec. 2011;168(23):618. **19.** Chameroy KA et al. Equine Vet J. 2011;43(4):494-499. **20.** Van Weyenberg S et al. J Anim Physiol Anim Nutr (Berl). 2009;93(2):203-208. **21.** Respondek F et al. J Anim Sci. 2011;89(1):77-83. **22.** Van Eps A et al. Equine Vet J. 2006;38(3):203-208. **23.** Davie A PhD thesis, 1995. **24.** Borer et al. **25.** Longland et al **26.** Respondek F. et al. J Anim Sci. 2008;86(2):316-323. **27.** Borgia et al. J Anim Physiol Anim Nutr (Berl). 2011;95(6):798-807.

Use of Next-Generation Sequencing Technologies to Study Disease

Carrie J. Finno, DVM, PhD, DACVIM
Davis, CA

INTRODUCTION

DNA sequencing is the process of determining the exact order of the nucleotides in a DNA segment. In the last 30 years, Sanger sequencing has been the most widely used sequencing technology worldwide.[1] Sanger sequencing is a method based on one or more amplicons that sequence relatively small strings of DNA rather than a complete gene or a set of genes in the equine genome. Therefore, it is expensive and time-consuming procedure if used to determine the entire DNA sequence across a given chromosome, let alone the entire genome. The advent of next-generation sequencing (NGS) technologies has enabled equine genetic research to move beyond single gene analysis to the simultaneous investigation of hundreds of genes and entire pathways. With next-generation sequencing approaches, including whole genome sequencing (WGS), RNA-Sequencing (RNA-Seq) and chromatin immunoprecipitation coupled with NGS (ChIP-Seq), complete genomes, transcriptomes and regulatory regions of specific tissues in specific individuals can be determined. Additionally, investigation of the metagenome can provide insight into the complexity of the microbiota of certain tissues and organs in relation to a variety of physiologic conditions.[2]

TYPES OF NEXT-GENERATION SEQUENCING MODALITIES

Whole genome sequencing

Across species, genome-wide association studies (GWAS) have been the most common approach for identifying disease associations across the entire genome. While current equine GWAS microarrays can interrogate approximately 670,000 single nucleotide polymorphism (SNP) markers in the horse, the most comprehensive method for interrogating the approximately 2.7 billion bases of the equine genome is with whole genome sequencing (WGS). With the rapid reduction in the cost of WGS, a single horse can now be sequenced at 12x coverage for approximately $2,000. Although WGS remains cost prohibitive for large sample sets, subsets of individuals can undergo sequencing and candidate regions from GWAS studies investigated for potential functional variants associated with a specific disease. In brief, a library of genomic DNA fragments is generated with adapters attached to one or both ends. Each molecule, with or without amplification, is then sequenced in a high-throughput manner to either obtain short sequences from one end (single-end sequencing) or both ends (pair-sequencing). Following sequencing, the resulting reads are aligned to a reference genome (WGS or RNA-Seq) or reference transcriptome (RNA-Seq). These reads can then be interrogated for variants (SNPs, insertions/deletions, chromosomal rearrangements) that may be associated within a candidate GWAS region and associated with a phenotype of interest.

RNA-Seq

The transcriptome is the complete set of transcripts, or expressed genes, in a cell and their quantity for a specific developmental stage or physiologic condition. Transcriptome sequencing is a major advance in the study of gene expression as it allows for interrogation of the whole transcriptome rather than a predetermined subset of genes. RNA-Seq provides a more precise measurement of levels of transcripts and their isoforms than other methods. Currently, library preparation kits support precise detection of total RNA, mRNA or miRNA profiles. Additionally, stranded protocols allow for the detection of strand orientation using total and mRNA methods. RNA-Seq can identify new isoforms of a particular transcript and has been demonstrated to be highly accurate for quantifying expression levels as compared to quantitative PCR (qPCR).[3]

Chip-Seq

While genomics involves the study of heritable or acquired alterations in the DNA sequence, epigenetics is the study of heritable changes in gene activity caused by mechanisms other than DNA sequence changes. Mechanisms of epigenetic activity include DNA methylation, small RNA-mediated regulation, RNA-protein interactions and histone modifications. Protein-DNA interactions have a significant impact on many biological processes and disease states. These interactions can be interrogated with NGS by combining chromatin immunoprecipitation (ChIP) assays and NGS methods (i.e. ChIP-Seq). ChIP-Seq captures DNA targets for transcription factors or chromatin modifications across the entire genome of any organism, thereby defining transcription factor binding sites and regulatory regions of the genome.

Metagenomics

Across mammalian species, there are an estimated 10x as many microbial cells as host mammalian cells. These microbes carry out an important number of metabolic reactions that are not encoded by the equine genome and necessary for equine health. Metagenomics is the study of microbial communities within a host. Using 16S rRNA and metagenomic sequencing, the complexity of the equine microbiome can be assayed at various body sites. In humans, there is an effort underway termed the Human Microbiome Project, aimed at characterizing the human microbiome at 15-18 body sites.[4]

APPLICATION OF NGS TECHNOLOGIES TO STUDY EQUINE DISEASE

With the increased availability and affordability of NGS tools in horses, ample opportunities exist to apply these molecular tools to a wide array of equine diseases and traits. To investigate genetic diseases in horses, both the sample population and tools chosen warrant consideration. For disease susceptibility traits, accurate phenotyping is essential. Often, selection of affected cases is more straightforward than selection of an appropriate control group. However, the phenotype to be studied should be decided *a priori* and its severity graded, as this becomes useful when selecting cases for sequencing. Whenever possible, a control population should be selected to maximize the chance that the control horses would never manifest the phenotype and with consideration placed on the age of disease onset, environmental risk factors and degree of relatedness between the control and affected populations.

The utility of genetic tools is affected by the sample size available. When designing an RNA-Seq experiment, it is necessary to factor in the fundamental aspects of sound experimental design; replication, randomization and blocking. Biologic replicates are more important than technical replicates in RNA-Seq study design.[5] It is also important that the RNA obtained in all samples is from a standardized site in the tissue of interest. Simultaneously obtaining RNA-Seq and WGS on a relatively large number of individuals (>50) can allow for correlations between genotype and expression levels, including disease susceptibility.[6]

REFERENCES

[1]Sanger F., et al. **PNAS** 1977;74(12):5463-5467.

[2]Human Microbiome Consortium. **Nature** 2012;486(7402):207-214.

[3]Wang, Z., et al. **Nat Rev Genet** 2010;10(1):57-63.

[4]NIH Human Microbiome Project (HMP) Roadmap Project, Accession PRJNA43021 ID: 43021

[5]Auer, P.L., et al. **Genetics** 2010;185:405-416.

[6]Conesa, A., et al. **Genome Biol** 2016;17(13):DOI 10.1186/s13059-016-0881-8.

Use of Molecular Diagnostics for Infectious Equine Pathogens

Nicola Pusterla, DVM, PhD, DACVIM
Davis, CA

INTRODUCTION

Polymerase chain reaction (PCR) has become an increasingly important tool in microbial diagnosis in recent years, mainly because of its rapidity, affordability, high sensitivity and high specificity. These superior characteristics have propelled the field of PCR-based molecular diagnostics into the arena of applied diagnostics for infectious agents. Because the number of published and offered PCR assays is steadily rising, there is a need for critical evaluation, comparison of performance, and eventually also standardization of methods to enable equine practitioners to select the optimal methodology.

QUALTIY CONTROL AND RESULT INTERPRETATION

There are questions worth asking before samples are submitted to a molecular diagnostic laboratory. These questions should cover 3 areas. First, it is worthwhile to obtain information about the nature of the PCR testing platform (traditional PCR versus qPCR). Secondly, questions addressing the quality control and quality assurance system within a particular laboratory should be asked. In particular it is relevant to know whether whole processes are controlled or just single point controls are used and how contamination is avoided and confirmed to be absent within the laboratory. Thirdly, turnaround time, pricing, and the level of guidance provided when interpreting results are additional questions worth asking before samples are submitted.

Biological samples received by diagnostic laboratories are generally processed the same day and PCR results are usually available within 24–72 hours (including shipping) assuming all relevant standards are met. These standards should include internal sample quality controls (to confirm proper collection, storage, shipping and nucleic acid extraction) and other associated quality controls, such as PCR positive and negative controls (to confirm PCR reagents were properly made and are contaminate free), internal positive control (to confirm absence of PCR inhibitors) and negative extraction control (to confirm absence of cross-contamination during the nucleic acid extraction process).

Interpretation of results obtained with molecular assays for infectious diseases requires understanding of the pathogenesis and biology of the target organism(s). Some challenges are unique to molecular tests and are different from considerations in interpreting other microbiological tests. Such differences are related to the distinction of viable versus nonviable organisms and the correlation of nucleic acid detection with presence of disease or disease association.

Interpretation of a negative PCR result requires information about the sensitivity of the PCR test, limit of detection and the nucleic acid extraction efficiency as indicated by the use of quantitative internal sample controls. A false-negative PCR result may be caused by a degraded or unstable sample. Insufficient or inappropriate sample type, inadequate sampling procedures or transport problems are additional sources of false-negative results. Sample-specific internal positive controls targeting endogenous genes help to overcome this problem. In addition, inhibition phenomena originating from sample matrixes such as feces, urine or environmental samples contaminated with soil or surface water have to be controlled with internal positive controls to assess the inhibitory effects on the PCR process.

The factors requiring consideration for the interpretation of PCR positive results include assay specificity and contamination issues. PCR or any other target amplification method is subjected to these considerations. Quantitative PCR (qPCR) using closed-tube detection procedures reduces the risk of PCR product carry-over as a source for false-positive results.

In general, molecular assays do not provide information about the viability of an infectious agent. Exceptions to this are DNA viruses, bacteria, and parasites analyzed for the presence of RNA molecules such as rRNA and transcribed genes in form of messenger RNA instead of their genomic DNA equivalents.

Detection of nucleic acid of a pathogen does not necessarily indicate the organism is the cause of the disease. However, using the quantitative information of a qPCR result may give further insight and provide a means to evaluate disease association. Primary examples are herpesvirus infections (EHV-1 and EHV-4) where the quantitative detection of DNA may indicate presence of lytic or non-replicating virus. Laboratory specific cut-off values in EHV-4 and EHV-1 DNA load allow differentiation between lytic and non-replicating virus. In these particular cases, high viral loads are generally associated with the presence of clinical signs and the presence of viral RNA transcripts indicating virus replication. Therefore, qPCR can provide a means to obtain information about the disease association, a crucial criteria for the equine practitioner to obtain the correct diagnosis.

CLINICAL APPLICATIONS

An array of nucleic acid amplification techniques may be offered by laboratories in order to establish an etiologic diagnosis of infectious diseases. Several published studies have shown the benefit of nucleic acid amplification techniques in comparison to conventional microbiological techniques. Current efforts are aimed at improvement of the diagnostic efficiency of molecular techniques, both for common and less common pathogens.

Parallel testing of multiple infectious agents in highly standardized platforms is a central component of molecular assays; it essentially allows several detections to happen simultaneously on a single sample for both DNA and RNA pathogen targets. This so called panel strategy allows an efficient work-up of complex clinical syndromes with general symptomatology. In such organ-related complex problems with general or unspecific clinical signs, multiple infectious agents can be responsible for a clinical picture. Even though veterinarians tend to make a single pathogen diagnosis, it has become more evident in recent years that several syndromes are caused by co-infections. Panel testing on a large scale will uncover unknown dual or triple infections in animals which can diffuse the clinical picture.

PCR DETECTION OF SELECTED EQUINE PATHOGENS

Respiratory end enteric pathogens are often contagious, and these infections must be diagnosed rapidly in order to prevent a disease outbreak and to institute the appropriate management plan. The short turn-around time and reliability of qPCR makes this technology an ideal tool for the diagnosis of respiratory and enteric pathogens.

Equine herpesvirus (EHV)-1 and EHV-4 are double-stranded DNA α-herpesviruses that affect the equine respiratory tract and can establish life-long latent infection following exposure. The diagnostic sample of choice is a nasal or nasopharyngeal swab which should be taken early in the febrile phase of the disease. Due to the lymphotropism of EHV-1, detection can also be attempted from whole blood. PCR assays used in the diagnostic field are based on the detection of viral genomic DNA and are therefore unable to distinguish between lytic and non-lytic virus. Alternative molecular approaches have recently been established using qPCR to allow discrimination between the different viral states in horses naturally infected with EHV.[1-3] Discrimination between the different viral states is now possible by (i) targeting several genes (e.g. glycoprotein, latency-associated transcripts), (ii) detecting viral genomic DNA and transcriptional activity of the target genes at the messenger RNA (mRNA) level and (iii) using absolute virus quantification. Quantitative thresholds are used in selected human infectious diseases (i.e. HIV, HCV, herpes simplex virus) to determine disease stage and response to antiviral therapy. A similar concept is used diagnostically for EHV-1 or EHV-4 infected horses in order to discriminate between lytic and non-replicating virus, to determine their infectious risk based on viral load in nasal secretions and to monitor their response to treatment.[1-4] Equine herpesvirus 1 myeloencephalopathy (EHM), although a relatively uncommon manifestation of EHV-1 infection, can cause devastating losses on individual farms, boarding stables, veterinary hospitals, and show and racing venues. In general, EHM is supported by historical and clinical findings, the presence of xanthochromia and elevated total protein concentration in CSF and laboratory detection of EHV-1 in blood and/or nasal secretions by qPCR. Because affected horses can shed the virus in nasal secretions, and thus represent a risk of infection for unaffected, in-contact horses, it is imperative to determine the risk of shedding in a suspected horse in order to initiate an appropriate infectious disease control protocol. A recently identified single nucleotide polymorphism (SNP) at position 2254 in the DNA polymerase gene (ORF 30) correlates with neurological disease.[5] This SNP is responsible for a single amino acid residue at position 752 of the DNA polymerase with EHV-1 strains associated with neurological outbreaks involving a D_{752} genotype (also referred as G_{2254}), whereas most non-neurological outbreaks involve a N_{752} genotype (also referred as A_{2254}). However, the genotyping of field isolates needs to be interpreted carefully as 14% to 24% of EHV-1 isolates from horses with EHM do not have this neuropathogenic marker.[5,6] Strain characterization may be important given that the potential of EHM development is greater in horses infected with a neuropathogenic genotype (D_{752}). Furthermore, detection of a neurotropic EHV-1 strain may influence therapy, especially in the use of antiviral drugs such as valacyclovir, used to decrease viremia and prevent the development of neurological sequelae.

Equine herpesvirus (EHV)-2 and EHV-5 are wide spread in horse populations, therefore, the detection of any of these viruses can occur in healthy but also in sick animals. These two viruses are optimally adapted to their host, which means that significant clinical expression of infection is rarely encountered. Experimental studies have shown that EHV-2 and EHV-5 contribute to various forms of clinical disease in young horses, including upper respiratory tract signs, fever, pharyngitis and enlarged lymph nodes.[7] Recent work has shown that different genetic variants of these viruses circulate amongst healthy horses, hypothesizing that viral re-infections or re-activations may be the origin of clinical disease and also explaining the sporadic occurrence of such cases.[8,9] Another characteristic of the γ-herpesviruses is their ability to immunomodulate the immune system potentially increasing the susceptibility of horses to various infectious pathogens. A recently described atypical interstitial pneumonia named equine multinodular pulmonary fibrosis (EMPF) has been linked to the presence of EHV-5 in affected pulmonary tissue.[10] The diagnosis of this disorder is based on clinical signs, radiographic abnormalities of the lungs, lower airway fluid cytology, histopathology of pulmonary tissue and EHV-5 detection by qPCR in bronchoalveolar lavage fluid and/or pulmonary tissue. Due to their high prevalence in the horse population, and in order to avoid dilemmas with the interpretation of PCR results, testing of EHV-2 and EHV-5 in horses with upper respiratory tract disease is not recommended at this time.

Salmonella enterica is an important cause of enteric disease and death in horses. However, *Salmonella* can also be associated with subclinical shedding, representing a source of infection for other animals. In recent years, PCR assays for the detection of *Salmonella* in fecal samples from horses admitted to veterinary hospitals have been evaluated.[11-15] Collectively, these studies have shown an unquestioned higher analytical sensitivity for the detection of *Salmonella* by qPCR when compared to conventional microbiological culture. The higher detection rate of *Salmonella* by PCR has been attributed to the detection of non-viable organisms and of previously undescribed *Salmonella*-like bacterial organisms. The use of novel virulence target genes for the molecular detection of *Salmonella* has considerably improved the performance and accuracy of such assays. More and more veterinary hospitals across North America are switching from conventional microbiological culture to PCR for the testing of *Salmonella* as part of their infectious disease control program. In such instances, PCR is performed on fecal and environmental samples following a 24-hour selective enrichment step. The use of PCR is very cost-effective and has the potential to reduce contamination risks and turnaround time, with results being available within 22-28 hours from sample collection (18-24 hours enrichment time, 4 hours for DNA purification and amplification). Further, the use of absolute quantitation allows assessment of the infectious nature of hospitalized animals and may be an excellent alternative to conventional culture methods for surveillance and research studies. One must keep in mind, however, that a positive PCR result should be confirmed by culture so that serotyping of the *Salmonella* isolate and antimicrobial susceptibility can be determined.

Equine coronavirus (ECoV) has historically been diagnosed via electron microscopy, antigen-capture ELISA or viral isolation from the feces. All of these detection modalities lack sensitivity, especially if viral particles are not present in sufficient numbers. qPCR for the detection of coronaviruses has supplanted many conventional virological assays, mainly due to its short turnaround time, high throughput capability and increased analytical sensitivity and specificity. The detection of ECoV by qPCR in the feces of foals with fever and diarrhea is difficult to interpret because ECoV has also been detected in the feces of healthy foals.[16] Healthy foals have been found to be infected mostly by ECoV in a single infection without any other co-infecting agents, whereas ECoV was found exclusively in association with other co-infecting agents in sick foals.[16] This is in sharp contrast to the detection of ECoV by qPCR in adult horses, where the virus has been shown to be responsible for a self-limiting disease characterized by lethargy, anorexia, fever and less frequently changes in fecal character and colic.[17-19]

CONCLUSIONS

The number of commercially available PCR assays continues to expand and many molecular assays continue to be developed in the research setting. In the meanwhile, efforts should continue to increase understanding of the strengths and limitations of these new assays. Molecular diagnostic tests may enhance diagnostic capabilities, but they should be interpreted within the clinical context and on the basis of individual laboratory performance. Extensive clinical research and strict adherence to guidelines for method validation are necessary to compare new molecular diagnostic techniques with existing methodologies, to validate new technology when comparable conventional techniques are unavailable, and to determine a method's clinical utility. Probe-based qPCR is an established research tool to quantify infectious agents during disease and after vaccination or therapy. Adaptations of these research applications will continue to impact testing in clinical laboratories

REFERENCES

[1]Pusterla N, et al. J Vet Diagn Invest 2005;17:578. [2]Hussey SB, et al. J Vet Diagn Invest 2006;18:335. [3]Allen GP, et al. Equine Vet J 2006;38:252. [4]Pusterla N, et al. Vet Rec 2008;162:728. [5]Nugent J, et al. J Virol 2006;80:4047. [6]Perkins GA, et al. Vet Microbiol 2009;139:375. [7]Fortier G, et al. Vet J 2010;186:148. [8]Bell S, et al. Vet Microbiol 2006;116:249. [9]Back H, et al. J Gen Virol 2015;doi:10.1099/jgv.0.000332. [10]Williams KJ, et al. Vet Pathol 2007;44:849. [11]Amavist P, et al. Vet Microbiol 2001;79:63. [12]Kurowski PB, et al. Am J Vet Res 2002;63:1265. [13]Ward MP, et al. J Vet Diagn Invest 2005;17:118. [14]Pusterla N, et al. Vet J 2010;186:252. [15]Ekiri AB, et al. Vet J 2015;doi:10.1016/j.tvjl.2015.11.011. [16]Slovis NM, et al. Equine Vet J 2014;46:311. [17]Oue Y, et al. Vet Microbiol 2011;150:41. [18]Oue Y, et al. J Vet Med Sci 2013;75:1261. [19]Pusterla N, et al. Vet Microbiol 2013;162:228.

CONTACT ADDRESS

Dr. Nicola Pusterla, Department of Medicine and Epidemiology, School of Veterinary Medicine, University of California, One Shields Avenue, Davis, CA 95616. Phone: 530-752 1039, Email: npusterla@ucdavis.edu.

Neutrophil Extracellular Traps, Activated Endotheluim and Immunothrombosis

Bettina Dunkel, DVM, PhD, DACVIM, DECEIM, DACVECC, FHEA, MRCVS
London, UK

INTRODUCTION

In the early stages of immunology, leukocytes were presumed to play the major part in the host defense against infection. In contrast, platelets and the coagulation system were assumed to mainly regulate hemostasis while the endothelium was largely seen as vascular wall lining with some regulatory functions to local blood flow. As the tight connection between inflammation and coagulation system became evident, new endothelial functions in defense against pathogens and regulation of both, coagulation and inflammation emerged. More recently, the important function of neutrophils in initiating and perpetuating coagulation have also been explored. The newly discovered functions and tight interactions between leukocytes, particularly neutrophils, platelets and the endothelium and new aspects of the coagulation system will be summarized with specific reference to horses.

NEUTROPHILS AS PART OF THE INNATE IMMUNE SYSTEM

Neutrophils as main effector cells of the innate immune system, have received a large amount of attention and some of their main functions such as phagocytosis and generation of reactive oxygen species (ROS), are well described. In 2004, an active form of neutrophil death associated with the release of a web-like structure into the extracellular space was first described[1]. The web consists of chromatin fibres containing positively charged histones, negatively charged DNA and granule-derived antimicrobial proteins including elastase, myeloperoxidase (MPO), proteinases and defensins. The name neutrophil extracellular traps (NETs) has been introduced and the process of NET release is termed NETosis. Neutrophils appear to be capable of sensing microbial size and release NETs preferentially when confronted with larger microbes or clusters of pathogens that might not be amenable to phagocytosis[2]. Apart from gram positive and negative bacteria, protozoans, fungi and viruses can also be targeted in this fashion[3]. Triggers for NET release include the presence of microbes or inflammatory cytokines (including IL-8 and tumour necrosis factor, TNF) growth factors or immune complexes and contact with activated platelets or endothelial cells[3]. Activated neutrophils preparing for NETosis flatten and adhere tightly to the substratum. Intracellularly, ROS generation and release of cytoplasmic granules into the cytoplasm follows. Neutrophil elastase and MPO migrate into the nucleus promoting chromatin decondensation by histone degradation. Following dissolution of the nuclear membrane, the plasma membrane ruptures and the NET is cast into the extracellular space[3]. NETosis can be "suicidal" (followed by neutrophil death) or "vital" (NET casting neutrophil survives), depending on environmental circumstances. Vital NETosis appears to predominately occur in the context of direct interactions between neutrophils and microorganisms[4]. Cell-free DNA (cfDNA) affects coagulation and fibrinolysis in a multitude of ways, a process that is not always beneficial for the host organism. Knockout mice incapable of forming NETs are susceptible to bacterial infection but protected from thrombosis[5].

NET formation has been induced experimentally in equine neutrophils pre-treated with phorbol-myristate-acetate (PMA) and exposed to bacteria commonly found in infections in horses including *E. coli* and *Streptococcus zooepidemicus*. Equine sperm also induces NET formation in neutrophils[6]. NETs containing neutrophil elastase and MPO were also visualized in samples obtained from horses with endometritis[7] and in bronchoalveolar lavage fluid from horses with exacerbated RAO, but not in fluid from horses with RAO in remission or in challenged healthy horses[8]. Taken together, these studies demonstrate that NET formation occurs in natural infection and states of inflammation in horses and is likely to play a similar role as observed in other species.

THE ROLE OF PLATELETS IN INFLAMMATION AND IMMUNITY

In recent years, platelets have been recognized to play a fundamental part in multiple acute and chronic inflammatory processes much extending beyond hemostasis and wound healing. Platelets contribute to inflammatory conditions directly by release and synthesis of pro-inflammatory mediators and indirectly via interaction with leucocytes and endothelial cells9. Platelets contain dense bodies, alpha granules and lysosomes holding a myriad of mediators including adhesion molecules, mitogenic factors and an array of pro-inflammatory enzymes and cytokine[9-11]. Differential release of presumably alpha granule content, has been demonstrated in human platelets. Human platelets are capable of synthesizing new proteins by translating mRNA originating from the parent megakaryocyte in a process termed signal-dependent translation. This is used for example for the production and release of interleukin 1ß (IL-1ß)[12]. Human platelets possess a cytoskeletal framework allowing active cell locomotion and have been observed to role on activated endothelium *in vivo*, similar to leukocytes. They also contain and release matrix metalloproteinases (MMPs), facilitating degradation of extracellular matrix and transmigration through vascular endothelial cells and basement membranes and are capable of gradient-driven chemotaxis in response to bacterial peptides[13] and allergenic stimuli in human asthma patients. Platelets initiate and perpetuate inflammation at sites of vascular injury[14]. Expression of CD62P upon stimulation mediates their adherence to leukocytes and endothelial cells and enhances expression of tissue factor (TF) on monocytes[15], one of the most important activators of the coagulation cascade. Platelet aggregates have been observed to accumulate in inflammatory lesions either concomitantly with[16] or preceding leucocytes[11] and platelets are believed to be important, if not essential, for leukocyte recruitment into an inflamed area and may significantly contribute to inflammatory organ injury in critically ill patients[17]. Rapid platelet sequestration into lungs and liver has been demonstrated in LPS-induced and anaphylactic shock in mice and platelet degranulation in these organs is believed to contribute to the systemic signs observed18. Human platelets not only augment the antimicrobial functions of leukocytes, but are able to inhibit or kill a variety of bacterial, viral, fungal and protozoal species *in vitro* and *in vivo*[19]. Being among the most rapid and abundant inflammatory cells accumulating at sites of endothelial damage or microbial colonialization platelets contribute to the clearance of microorganisms by binding, aggregating and internalizing invaders in phagosome-like vacuoles and by release of potent antimicrobial peptides[19]. Platelets also share structural and functional features with phagocytic cells and release cytotoxic oxygen metabolites including superoxide anions, hydroxyl radicals, hydrogen peroxide, cationic peptides and other potent antimicrobial peptides[20]. LPS amplifies the interaction between platelets, IgG, microorganisms and immune complexes and might contribute to the development of thrombocytopenia accompanying endotoxemia and gram negative infections. Furthermore, activated platelets trigger the release of NETs from neutrophils.

The role of platelets in equine diseases is also increasingly recognized[21]. Equine platelets contain similar organelles, including dense bodies, alpha granules and lysosomes with an apparently similar mediator content, although far fewer have been specifically identified so far[22-24]. Differential release and *de novo* synthesis of mediators has been demonstrated or suspected in equine platelets[22,25]. Equine platelets up-regulate CD62P expression in response to LPS and lipoteichoic acid (LTA) but formation of platelet-neutrophil aggregates was not observed and equine platelets are *in vitro* 1000 times more sensitive to LPS in terms of thromboxane A_2 (TXA_2) production and p38 MAP kinase phosphorylation than equine leukocytes[26]. However, platelet hyperreactivity was not observed *in vivo* following treatment of horses with LPS[27]. Activated equine platelets appear to be able to inhibit bacterial growth *in vitro*, but effects were less pronounced compared to reports from other species and similar effects were seen with plasma alone[26]. Equine platelets produce superoxide when stimulated with strong stimuli and contain functional myeloperoxidase (MPO)[28](Dunkel and Cunningham, unpublished data). In contrast to other species, superoxide release was not observed following exposure to whole killed bacteria or bacterial products[28]. Several other effects of LPS on equine platelets, such as aggregation and 5-HT release, are dependent on the presence of leukocytes. Both processes are mediated by platelet activating factor (PAF), released from activated leukocytes, but not cyclooxygenase products and can be inhibited by PAF-inhibitors, but not aspirin[29,30]. Indirect evidence for platelet activation and degranulation has been identified in horses with strangulating intestinal lesions where increased concentrations of 5-HT were identified in the circulation and, to a lesser degree, peritoneal fluid. The authors speculated that this increase was due to systemic 5-HT release from platelets; a β-TG/PF4 ratio >2 provided further evidence for systemic platelet activation[31]. The serotonin concentrations in PPP were significantly lower in pre- and post-operative samples from surgical SI colic horses compared to controls. However, no association with postoperative ileus or non-survival could be demonstrated at any time point[32].

ENDOTHELIAL CELLS AS ACTIVE PLAYER IN THE IMMUNE SYSTEM

Endothelial cells have not classical been regarded as immune cells but undeniably contribute significantly to the host defence. Endothelial cells are centrally placed to not only detect vascular injury and invading microorganisms but also to interact with circulating leukocytes and platelets. They play an essential part in regulation of vascular permeability and vasomotor tone, coagulation and hemostasis, and movement of leukocytes between the vasculature and extravascular tissues[33]. Endothelial cells express innate immune receptors, including the Toll-like receptors (TLRs) and, via NF-kB and the MAP kinases, can activate intracellular inflammatory pathways[33]. TLR agonists, including LPS and bacterial lipopeptides, directly upregulate microvascular endothelial cell expression of inflammatory mediators. TLR activation also modulates microvascular endothelial cell permeability and the expression of coagulation pathway intermediaries[33]. Many endothelial functions are mediated via the glycocalyx, a 1-3μm thick network of cell-bound and soluble components attached to the luminal side of endothelial cells. Cell-bound components consist of proteoglycans, glycosaminoglycan (GAG) side chains, and sialoproteins. This complex network continues inside the endothelial clefts, where it extends into the extracellular matrix of the basal membrane[34]. Attached to the luminal side are soluble components including albumin, hyaluronic acid molecules, thrombomodulin and serum proteins (superoxide dismutase, antithrombin, cell adhesion molecules, etc)[34]. An intact glycocalyx is essential for normal endothelial cell function. It regulates microvascular tone, endothelial permeability and leukocyte adhesion and extravasation, maintains an oncotic gradient across the endothelial barrier and inhibits intravascular thrombosis[34]. Healthy endothelium in a normal organism supresses intravascular coagulation by generation of activated protein C (APC) on the cell surface, production of tissue-type plasminogen activator to stimulate fibrinolysis and by impeding thrombin formation and platelet adhesion. Glycosaminoglycans present on the vessel wall play a major role in anti-coagulation by supporting antithrombin-mediated inhibition of thrombin[35]. In addition, tissue factor pathway inhibitor (TFPI) is attached to the endothelium via proteoglycans. During systemic inflammation, endothelial cells contribute to the development of a pro-coagulant environment by expression of TF on their surface upon activation. Furthermore, anticoagulant properties are lost due to decreased antithrombin concentrations and decreased availability of TFPI and activated protein C. During sepsis, the protein C system is impaired as a result of several factors, most notably decreased synthesis and increased consumption of protein C and decreased activation of protein C due to reduced expression of thrombomodulin and endothelial protein C receptors on endothelial cells[35]. The interactions between platelets and neutrophils further compromise endothelial function. Platelet-induced NET formation is tightly regulated by endothelial-derived prostacyclin and is selectively controlled by acetylsalicylic acid-induced platelet inhibition[36]. Histone H4, a component of NETs, is cytotoxic towards endothelial and epithelial cells.

Little is currently known about equine endothelial function in health and during systemic inflammation. Clinically relevant concentrations of LPS activate equine digital endothelial cells which leads to increased E-selectin mediated neutrophil adhesion[37]. An upregulation of E-selectin has also been observed in equine pulmonary endothelial cells following exposure to LPS[38]. Responses appear to be maximal within 2h of exposure. Endothelial cell activation was dependent on p38 MAPK phosphorylation. Studies investigating endothelial cells in the context of laminitis have identified endothelial dysfunction (decreased relaxation in response to carbachol) in digital arteries following overnight incubation with 10ng/ml TNF under hypoxic conditions. Short-term incubation for 1.5h had no effect in this study[39] but in another study, incubation with TNF (0.1 or 1 ng/ml) for 10 min decreased endothelial-dependent relaxation of equine digital arteries to acetylcholine and increased the maximal contraction to noradrenaline40. Reported plasma TNF concentrations in normal horses range from 1-14ng/ml[41,42].

THE CONCEPT OF IMMUNOTHROMBOSIS

Immunothrombosisis a recently introduced term describing an innate immune response in which local activation of the coagulation cascade enables recognition, containment and destruction of pathogens[5,43]. Similar to NETosis, the primary aim of the process is confinement and destruction of pathogens. Four major functions of immunothrombosis that support the innate immunity have been suggested: First, ensnaring and immobilizing pathogens within a fibrin network, thereby preventing dissemination. Second, prevention of tissue invasion by formation of microvascular thrombi. The generation of a distinct compartment in which antimicrobial elements can be concentrated to target pathogens and favor their killing is the third function and lastly, recruitment of additional effector cells by deposition of fibrinogen and fibrin[43]. It has been hypothesized that leukocyte adhesion to the endothelium and microvascular thrombosis direct obstruction of vessels draining sites of infection are initially beneficial responses, containing the infection. In contrast, dysregulation and systemic spread of these pathways can lead to diffuse microvascular thrombosis, increased leukocyte activation within multiple organs and capillary leak leading to intravascular hypovolemia and edema. These dysfunctional responses are believed to promote the life-threatening syndromes of septic shock, disseminated intravascular coagulation (DIC) and multi-organ failure[33]. Treatment of inappropriate intravascular coagulation includes systemic use of anticoagulants which inadvertently also affect normal hemostasis, leading to complications with systemic hypocoagulation and bleeding. However, it has recently been suggested that several aspects of immunothrombosis are different from normal hemostatic mechanisms. Those differing pathways could offer attractive therapeutic targets, allowing limitation of pathological thrombosis without unwanted side effects on normal hemostasis. Mechanisms unique to immunothrombosis and absent in normal hemostasis include delivery of active TF by innate immune cells to sites of microbial invasion, degradation of endogenous anticoagulants and provision of pro-coagulant surfaces, such as extracellular nucleosomes[43]. cfDNA, provided by NETs, acts as a pro-coagulant cofactor, similar to other by augmenting activation of proteases of the contact pathway including factors (F) XII and XI. cfDNA was shown to bind to FXII and high molecular weight kininogen. Histone H4 can generate thrombin independently of the coagulation cascade by directly binding to thrombin[44]. In addition, the polyanionic surface of the DNA enhances the ability of neutrophil elastase to inactivate TFPI and potentially thrombomodulin[43]. NETs intercalate within a fibrin clot forming a network resistant to lysis[45]. cfDNA has a concentration-dependent effect on fibrinolysis, accelerating the process at low and inhibiting it at high concentrations. At physiological concentrations, cfDNA potentiates the activation of plasminogen by tissue-type plasminogen activator (tPA). At higher concentrations, cfDNA inhibits fibrinolysis by competing with fibrin for plasmin[5].

Over time, pathogens have responded to the concept of immune-thrombosis by developing sophisticated evasion strategies. Group A, C and G streptococci express streptokinase, a plasminogen activator that encourages plasmin formation and degradation of fibrin. Group A streptococci use streptokinase to invade and spread into host tissues[43]. Several pathogens have developed strategies to escape NETs. *Streptococcus pyogenes* and *pneumoniae* express DNase that degrades NETs and inhibits NET-mediated killing. Staphylocoagulase and von Willebrand factor-binding protein expressed by *S. aureus* stimulate fibrin generation through activation of prothrombin. Fibrin is then used to build a multilayered barrier which forms part of an abscess pseudo-capsule[43]. Coagulation disorders, including subclinical and clinical DIC, are commonly observed in systemically ill horses of all ages[46-51]. Microvascular fibrin deposits were detected in a large number of horses with ischemic and inflammatory intestinal disease and critically ill foals, suggesting that immunothrombosis, albeit not under this name, is a major issue in critical equine patients contributing to morbidity and mortality[52-54].

REFERENCES

[1]Brinkmann V, et al. Science 2004;303(5663):1532.[2]Branzk, N, et al. Nat Immunol 2014;15(11):1017.[3]Rao AN, et al. World J Cardiol 2015;7(12):829.[4]Yipp BG, et al. Blood 2013;122(16):2784.[5]Gould TJ, et al. Thromb Haemost 2015;13 Suppl 1:S82.[6]Alghamdi AS, et al. Biol Reprod, 2005;73(6):1174.[7]Rebordao MR, et al. J Reprod Immunol 2014;106:41.[8]Cote O, et al. PLoS One 2014;9(4):e96217.[9]Weyrich AS, Set al. J Thromb Haemost 2003;1(9):1897.[10]Gawaz M, Cardiovasc Res 2004;61(3):498.[11]Pitchford SC, et al., Am J Respir Crit Care Med 2008;177(6):604.[12]Weyrich AS, et al. Trends Immunol 2004;25(9):489.[13]Czapiga M, et al. Exp Hematol 2005;33(1):73.[14]Yaguchi A, et al. J Thromb Haemost 2004;2(12):2096.[15]Levi M, Crit Care Med 2005;33(12 Suppl):S523.[16]Pitchford SC, et al. J Allergy Clin Immunol 2003;112(1):109.[17]Tweardy DJ, et al. Shock 2006;26(4):386.[18]Yoshida A, et al. Br J Pharmacol 2002;137(2):146.[19]Andres O, et al. Thromb Haemost 2015;113(1):3.[20]Tang YQ, et al. Infect Immun 2002;70(12):6524.[21]Dunkel B, et al. Equine Vet J 2007;39(6):557.[22]Dunkel B, et al. Equine Vet J 2012;44(3):346.[23]Segura D, et al. J Vet Intern Med 2006;20(3):581.[24]Holmsen H, et al. Biochem J 1972;129(1):67.[25]McCarrel T, et al. J Orthop Res 2009;27(8):1033.[26]Aktan I, et al. Vet Immunol Immunopathol 2013;152(3-4):209.[27]Watts AE, et al. Am J Vet Res 2014;75(8):760.[28]Aktan, I, et al. Vet Immunol Immunopathol 2013;152(3-4):209.[29]Bailey SR, et al., Inflamm Res 2000;49(5):224.[30]Jarvis GE, et al. Res Vet Sci 1994;57(3):317.[31]Delesalle C, et al. Equine Vet J 2008;40(4):326.[32]Torfs SC, et al. Can Vet J 2015;56(2):178.[33]Khakpour S, et al. Innate

Immun 2015;21(8):827.[34]Chelazzi C, et al., Crit Care 2015;19:26.[35]Opal SM. et al. J Intern Med 2015;277(3):277.[36]Carestia A, et al. J Leukoc Biol 2016;99(1):153.[37]Brooks AC, et al. Vet Immunol Immunopathol 2009;129(3-4):174.[38]Hedges JF, et al. Immunology 2001;103(4):498.[39]Menzies-Gow NJ. et al. Inflamm Res 2014;63(8):637.[40]Baxter G, Am J Vet Res 1994;55(4):551.[41]Menzies-Gow, N.J., et al. Equine Vet J 2014;46(3):317.[42]Suagee JK, et al. J Vet Intern Med 2013;27(1):157.[43]Engelmann B. et al. Nat Rev Immunol 2013;13(1):34.[44]Barranco-Medina S, et al. J Biol Chem 2013;288(50):35749.[45]Fuchs TA, et al. Proc Natl Acad Sci USA 2010;107(36):15880.[46]Dallap BL, et al. Journal of Veterinary Emergency and Critical Care 2003;13(4):215.[47]Dolente BA, et al. J Am Vet Med Assoc 2002;220(7):1034.[48]Dallap Schaer BL., et al. J Vet Emerg Crit Care (San Antonio) 2009;19(1):88.[49]Dunkel, B., et al., J Vet Intern Med 2010;24(6):1467.[50]Epstein KL, et al. J Vet Emerg Crit Care (San Antonio) 2013;23(5):504.[51]Epstein KL. et al. J Vet Intern Med 2011;25(2):307.[52]Cesarini C, et al. J Vet Intern Med 2015.[53]Cotovio M, et al. J Vet Intern Med 2008;22(6):1403.[54]Cotovio M, et al. J Vet Intern Med 2007;21(2):308.

CHARACTERIZATION OF THE FECAL BACTERIAL MICROBIOTA OF HEALTHY AND DIARRHEIC CALVES

Diego Gomez[1], Luis Arroyo[2], Laurent Viel[2], J. Scott Weese[1]

[1]Department of Pathobiology, Ontario Veterinary College, University of Guelph., Guelph, Ontario, Canada, [2]Department of Clinical Studies, Ontario Veterinary College, University of Guelph., Guelph, Ontario, Canada

The intestinal microbiota plays a key role in health and disease, and microbiota disruption (dysbiosis) is increasingly be associated with diarrhea and other clinical abnormalities. This study aimed to characterize the fecal bacterial microbiota of healthy and diarrheic calves.

The fecal microbiota of 20 healthy and 20 diarrheic calves (<4 wk) from 2 farms (*F*1 and *F*2) was characterized using the Illumina MiSeq platform targeting the V4 region of the 16S rRNA gene.

Fecal bacterial microbiota of healthy and diarrheic calves within each farm was significantly different in community membership (Jaccard index) and structure (Yu and Clayton index) (Parsimony and AMOVA; P<0.05). Healthy calves from F1 and F2 also differed in bacterial community membership and structure (Parsimony and AMOVA; P<0.05).

Healthy and diarrheic calves from F1 had higher bacterial richness than those from F2 (Chao-1; P=0.0008). Diarrheic calves from F1 had higher bacterial diversity (Inverse-Simpson; P=0.039) and evenness (Shannoneven; P<0.044) than those from F2. In F1, healthy calves had higher relative abundance of Actinobacteria (median: 22%) than diarrheic (8%) (P=0.0006), whereas in F2 healthy calves had lower abundance of Firmicutes (44%) than diarrheic (55%) (P=0.04).

LefSe analysis showed that the *Bifidobacterium* genus was enriched in healthy calves, while *Oligella,* Lachnospiracea incertae sedis, unclassified Deinococcales and unclassified Lachnospiracea were enriched in diarrheic calves. PICRUSt analysis revealed overrepresentations of genes for tetracycline biosynthesis, metabolism of energy, porphyrins, and lipids in diarrheic calves and overrepresentation of genes for protein digestion and absorption, vitamin B6 metabolism and biosynthesis of vancomycin group antimicrobials in healthy calves.

This study provides the basis for generation of hypothesis to investigate the roles that different bacterial taxa and dysbiosis play in the development of diarrhea in calves.

HIGH PRESSURE PROCESSING OF BOVINE COLOSTRUM: IMPACT ON QUALITY, PATHOGENS, AND TRANSFER OF PASSIVE IMMUNITY

Derek Foster[1], Keith Poulsen[4], Kaitlyn Casulli[3], Brian Farkas[2]

[1]*NC State University, Raleigh, NC, USA,* [2]*Purdue University, W. Lafayette, IN, USA,* [3]*Michigan State University, East Lansing, MI, USA,* [4]*University of Wisconsin, Madison, WI, USA*

The objectives of this study were to 1) measure the effects of high pressure processing (HPP) on microbial load, viscosity, and immunoglobulin concentration when applied to bovine colostrum as an alternative to thermal pasteurization and 2) determine the impact of HPP on the transfer of passive immunity to calves.

First milking colostrum from 6 Holstein-Friesian cows were inoculated with known concentration of microorganisms and pressure processed at 300 MPa for up to 60 minutes and at 400 MPa for up to 30 minutes. The extent of inactivation of native bacteria, *Escherichia coli*, *Salmonella enterica* subsp. *enterica* serovar Dublin, *Mycobacterium avium* subsp. *paratuberculosis* (MAP), bovine herpesvirus type 1, and feline calicivirus (as model for non-enveloped viruses) were determined after processing. Colostrum immunoglobulin G (IgG) content was measured before and after pressure processing. Shear stress and viscosity for each treatment was determined over shear rates encompassing those found during calf feeding and at normal bovine body temperature (38.6˚C). Colostrum was then collected from 15 cows and processed at 400MPa for 15 min. Serum IgG concentration was measured and apparent efficiency of absorption was calculated in 15 calves fed 3.8 L of pressure processed colostrum and compared to 15 calves fed heat treated colostrum.

High pressure processing at 300 MPa (30, 45, 60min) and 400MPa (10, 15, 20min) was shown to significantly decrease native bacteria, *E. coli*, and *S.* Dublin populations in bovine colostrum, however, no decrease in MAP was observed. Both viral pathogens were reduced below the limit of detection. These treatments resulted in significant increases in viscosity with added time and pressure. The impact high pressure on IgG content of colostrum appeared to be variable and somewhat related to the initial IgG concentration. Calves fed colostrum processed at 400 MPa for 15min had similar serum IgG concentrations and rates of failure of transfer of passive immunity but a lower efficiency of absorption than calves fed heat-treated colostrum.

The results of this study suggest that high pressure processing of bovine colostrum generally maintains an acceptable feeding viscosity and IgG concentration while decreasing bacterial and viral load. Additional studies to assess the impact of the combination of high pressure and heat on the viability of MAP and to understand the effect of high pressure processing on the efficiency of IgG absorption are needed.

ANTIMICROBIAL DRUG USE AND RISK FACTORS FOR INCREASED TREATMENT INCIDENCE AND MORTALITY IN VEAL CALVES

Matteo Lava[1], Getraud Schüpbach[2], Adrian Steiner[1], Mireille Meylan[1]

[1]Clinic for Ruminants, Vetsuisse Faculty, University of Berne, Berne, Switzerland, [2]Veterinary Public Health Institute, Vetsuisse Faculty, University of Berne, Berne, Switzerland

The present study aimed at describing and quantifying antimicrobial use, and at identifying the major risk factors for increased antimicrobial drug use and mortality in veal farms rearing own and purchased veal calves in Switzerland.

Ninety-one Swiss veal farms producing under a label with improved welfare standards were visited between August and December 2014 to investigate risk factors for increased antimicrobial drug use and mortality. A questionnaire on management factors was filled with the herd managers, and data on antimicrobial use were retrieved from the veterinary bills and the treatment records of the farms. The animal defined daily dose (ADD) methodology was used to quantify treatment incidence on each farm (TI_{ADD}, in average number of daily doses of antimicrobial drugs per animal year). Two multivariable models were constructed for risk factor analysis, i.e. for increased antimicrobial drug treatment incidence (53 farms) and increased mortality (91 farms).

All farms fattened own and purchased calves, with a median of 77.4% purchased calves. The calves' mean age was 29±15 days at purchasing and the fattening period lasted a mean of 120±28 days. The mean carcass weight was 125±12 kg. Mean herd size was 58±33 fattened calves per year, and purchased calves were bought from a mean of 20±17 farms of origin. The mean TI_{ADD} was 21±15 daily doses per calf and year. The mean mortality risk was 4.1%, calves died at a mean age of 94±50 days, and the main causes of death were bovine respiratory disease (BRD, 50%) and gastro-intestinal disease (33%). Risk factors associated with increased antimicrobial treatment incidence included the lack of clinical examination of purchased calves and lack of quarantine measures upon arrival at the farm, and shared air space for several groups of calves. Maximal group size and weight differences >100 kg within a group were associated with an increased mortality risk, while vaccination and beef breed were associated with a decreased mortality risk. The majority of antimicrobial treatments (84.6%) were given as group treatments with oral powder fed through an automatic milk feeding system. Combination products containing chlortetracycline combined with tylosin and sulfadimidine or with spiramycin were used for 54.9%, and amoxicillin was used for 43.7% of the oral group treatments. The main indication for individual treatment was BRD (73% of the cases). The mean age at the time of treatment was 51 days, corresponding to an estimated weight of 80-100 kg. Individual treatments were applied through injections in 88.5% of the cases, and included administration of fluoroquinolones in 38.3%, penicillines (amoxicillin or benzylpenicillin) in 25.6%, macrolides in 13.1%, tetracyclines in 12.0%, 3th and 4th generation cephalosporines in 4.7%, and florfenicol in 3.9% of the cases.

The present study allowed for identifying risk factors for increased antimicrobial drug treatment and mortality. This is an important basis for future studies aimed at reducing treatment intensity and mortality in veal farms, as most of these factors are fairly easily amenable to improvement (physical examination and quarantine upon arrival, group size and age distribution within groups, vaccination). These results indicate that improvement is needed in the selection of drugs for the treatment of veal calves according to the principles of prudent use of antibiotics.

IDENTIFICTION OF A HAPLOTYPE ASSOCIATED WITH HYPOCHOLESTEROLEMIA AND INCREASED JUVENILE MORTALITY IN HOLSTEIN CATTLE

Sandra Kipp[1], Dierck Segelke[1], Sven Schierenbeck[1], Friedrich Reinhardt[1], Reinhard Reents[1], Christine Wurmser[2], Hubert Pausch[2], Ruedi Fries[2], Georg Thaller[3], Jens Tetens[3], Josef Pott[4], Dorothea Haas[5], Barbara Raddatz[6], Marion Hewicker-Trautwein[6], Ioannis Proios[7], Marion Schmicke[7], Walter Gruenberg[7]

[1]*Vereinigte Informationssysteme Tierhaltung w.V. (vit), Verden, Germany,* [2]*Chair of Animal Breeding, Technische Universitaet Muenchen, Freising, Germany,* [3]*Chair of Animal Breeding, Christian-Albrechts-Universitaet zu Kiel, Kiel, Germany,* [4]*Masterrind GmbH, Verden, Germany,* [5]*University Children's Hospital Heidelberg, Division of Neuropediatrics and Metabolic diseases, Heidelberg, Germany,* [6]*Department of Pathology, University of Veterinary Medicine, Hannover, Foundation, Hanover, Germany,* [7]*Clinic for Cattle, University of Veterinary Medicine Hannover, Foundation, Hanover, Germany*

Over the last decades, several genetic disorders affecting animal health and wellbeing have been discovered in cattle. Recently, German cattle farmers reported calves with chronic diarrhea and retarded growth. The etiology of the condition remained obscure despite of considerable efforts from field veterinarians to elucidate the underlying cause. Affected calves were unresponsive to symptomatic treatment and died within the first months of life. Examination of pedigree and genotype information revealed that affected calves could be traced back to a popular sire of the Holstein population. The purpose of the study presented here was to determine the etiology and pathophysiology of this novel condition affecting health and survival of Holstein calves.

A total of 5 calves displaying the suspect phenotype characterized by growth retardation and recurrent or chronic diarrhea of unknown etiology and having a specific Holstein sire as ancestor were referred to the teaching hospital of the University of Veterinary Medicine Hannover, Foundation for complete diagnostic work-up. Furthermore a genome-wide association study was conducted as case-control study including 9 clinically affected and 21,077 control animals.

Affected calves were underdeveloped in weight and showed progressive and severe emaciation despite of normal feed intake. Hallmark findings of the blood biochemical analysis were pronounced hypocholesterolemia and deficiency of fat soluble vitamins. Results of the clinical and blood biochemical examination had striking similarities with findings reported in human abetalipoproteinemia. Post mortem examination revealed near complete atrophy of the body fat reserves including the spinal canal and bone marrow. In order to identify the causal region, we performed a genome-wide association study with 9 affected and 21,077 control animals obtained with the Illumina bovineSNP50 Bead chip revealing a strong association signal on BTA 11 ($P = 3.9 \times 10^{-66}$). Subsequent autozygosity mapping identified a disease-associated haplotype encompassing 1.01 Mb. The segment of extended homozygosity contains six transcripts, among them the gene *APOB* which is causal for cholesterol disorders in humans. However, results from multi-sample variant calling of one affected and 37 unaffected animals did not detect any putative causal mutation. The disease-associated haplotype has an important negative effect on calf mortality in the homozygous state when comparing survival rates of risk-matings vs. non-risk matings. Blood cholesterol values of animals are significantly associated with the carrier status indicating a codominant inheritance. The frequency of the haplotype in the current Holstein population was estimated to be 4.2 %.

This study describes the identification and phenotypic manifestation of a new Holstein haplotype characterized by pronounced hypocholesterolemia, chronic emaciation, growth retardation and increased mortality in young cattle, denominated as Cholesterol Deficiency Haplotype (CDH). Our genomic investigations and phenotypic examinations provide additional evidence for a mutation within the *APOB* gene causing cholesterol deficiency in Holstein cattle.

EFFECT OF INTRAVENOUS PLASMA TRANSFUSION ON LEUKOCYTE ACTIVITY IN CALVES WITH FAILURE OF PASSIVE IMMUNITY

Victoria Yang, Maire Rayburn, Munashe Chigerwe

UC Davis, Davis, CA, USA

Plasma administration has been recommended in calves older than 48 hours of age with failure of passive immunity (FPI) to provide immunity consistent with adequate colostral ingestion. The protective serum immunoglobulin G (IgG) concentrations (≥ 1000 mg/dL) of plasma derived IgG only lasts up to 12 hours thereby rendering supplementation of IgG through plasma administration likely unjustifiable in clinical practice.

In septic foals, plasma administration was beneficial by enhancing neutrophil function. This study hypothesized that intravenous bovine plasma administration would improve oxidative and phagocytic activity of granulocytes and monocytes in calves with FPI.

A non-probability, stratified sampling clinical trial was performed. Twenty- seven Jersey calves were assigned into three groups. The colostral (CL, N = 9) group received 3 L of colostrum. Two other groups of calves received 1 L of colostrum and were assigned based on their health status (sick or non-sick) at 4 days of age, as the sick-group (SG, N = 7) or the non-sick (NG, N = 11) groups. At 4 days of age, the SG and NG groups were administered plasma intravenously administration at 30 mL/kg. Granulocyte and monocyte oxidative and phagocytic activity was determined by flow cytometry.

There was no significant difference in the granulocyte and monocyte oxidative or phagocytic activity among the 3 groups ($P > 0.05$). Plasma administration had no significant effect on the oxidative or phagocytic activity of granulocytes or monocytes.

In clinical practice, intravenous plasma administration for enhancing activity of granulocytes or monocytes may not be justified in calves with FPI.

RISK FACTORS INFLUENCING LISTERIA MONOCYTOGENES PREVALENCE IN MIDDLE-SIZE DAIRY FARMS

Petra Bandelj[1], Urska Jamnikar Ciglenecki[1], Matjaz Ocepek[1], Rok Blagus[2], Modest Vengust[1]

[1]*University of Ljubljana, Veterinary Faculty, Ljubljana, Slovenia,* [2]*University of Ljubljana, Institute for biostatistics and Medical informatics, Ljubljana, Slovenia*

Listeria monocytogenes (LM) is a significant food-borne pathogen affecting animals and humans. Family farming is the most common farming model in the European Union. The purpose of this study was to determine 1) the prevalence of LM and 2) risk factors for LM perpetuation in middle-size dairy farms.

Fecal samples were collected from cows and calves on 20 mid-size family dairy farms in two week intervals for a period of one year. Environmental samples were collected once in May. LM was detected using qPCR after one day of incubation in the enrichment broth. The univariate assessment of the association between each farms' management/environmental related risk factors and LM prevalence was performed by means of logistic regression. P-values were adjusted with Benjamini-Hochberg method (P.bh ≤ 0.05).

LM was isolated from all farms on at least one sampling day. The average prevalence between farms was 18.2% (98/540) and 8.4% (43/511) in cows and calves, respectively. LM was isolated from manure, dirt, maize silage and grass hay. The only risk factor associated with LM prevalence was the meteorological season, with highest prevalence during spring (OR: 4.6; 95% CI: 2.2-9.7; P.bh=0.0005).

The longitudinally of this study indicated that LM prevalence is variable and associated with the meteorological season. Diet was not directly associated with LM prevalence in this study (OR: 1.1; 95% CI: 0.5-2.6; P.bh=0.9). However, high LM prevalence during spring is most probably associated with the degrading quality of conserved feed and favorable environmental conditions for LM proliferation. Farmers and veterinarians should use this information when implementing strategies to reduce risks for LM dissemination.

IODIDE SUPPLEMENTATION AS A STRATEGY FOR ENHANCING BOVINE INNATE AIRWAY DEFENSES

Brian Shoemake[2], Brian VanderLey[2], Benjamin Newcomer[3], Kristin Clothier[1], P. K. Galik[3], Mallory Klingenberg[2], Rachel Nolan[2], Allison Meyer[2], Loren Schultz[2], Meera Heller[1]

[1]University of California Davis, Davis, CA, USA, [2]Univeristy of Missouri, Columbia, MO, USA, [3]Auburn Univeristy, Auburn, AL, USA

Bovine respiratory disease is an important cause of economic loss in the beef industry accounting for 31% of mortality in calves over three weeks of age. The airway epithelium has several innate defense mechanisms to protect the health of bovine lungs; one of these is lactoperoxidases (LPO) and hydrogen peroxide (H_2O_2) which react with excreted halide ions to form hypohalous acids on the epithelial surface. The objective of this study was to determine if important bovine bacterial and viral respiratory pathogens, parainfluenza 3 virus (PI3), bovine herpesvirus-1 (BoHV1), and bovine viral diarrhea virus (BVDV), *Manheimia hemolytica* and *Bibersteinia trehalosi (*formerly *M. hemolytica* biotype T*)*, are susceptible to inactivation by the LPO/H_2O_2/halide system that is present in bovine airways. A second objective was to determine if a single oral bolus of sodium iodide would increase the iodine content of upper respiratory secretions in calves.

Bacterial isolates from bovine pneumonia cases were identified via traditional biochemical testing and confirmed via Maldi-tof. Experiments were carried out on three different strains of each bacteria. The susceptibility of these pathogens to inactivation by hypoiodous acid was determined by generating HOI in vitro using LPO, sodium iodide (NaI) and H_2O_2. Concentrations of NaI from 25 µM to 250 µM were tested, and controls were run in parallel. Subsamples were removed prior to addition of H_2O_2 (catalyst) and at 5 and 30 minutes. Bacterial killing was assessed via standard plate counts. Stock cultures of each virus were incubated with LPO, H_2O_2, and sodium iodide at a concentration of 0, 10, 100, or 250 µM for 5, 15 or 60 minutes. Each assay was repeated in triplicate. Virus titration was performed by direct visualization of cytopathic effect in cultured cells. At all concentrations of NaI, both *M. hemolytica* and *B. trehalosi* were killed by the complete reaction. Surprisingly, NaI alone at higher concentrations (above 250uM) inactivated both pathogens *in vitro*. A dose-dependent decrease in viral concentration was observed for PI3, with decreased concentrations seen in all samples at even the lowest concentration of iodine included. For BoHV1, significant reduction in viral titer was seen at 100 µM iodine concentration. Of the 3 viruses, BVDV showed the least susceptibility to the system with no decrease in titer observed.

For the second objective, sixteen crossbred beef calves were divided into two equal groups by random allocation to either a treatment or control group. Blood and nasal secretions were collected immediately prior to the beginning of the study, then every 12 hours for 72 hours after the administration of a 70 mg/kg bolus of sodium iodide (treatment) or water (control). The calves that received sodium iodide had a marked increase of nasal fluid iodine concentration with a peak concentration at 12 hours and Cmax = 181.40 µg/mL (1.2mM).

These data indicate that iodine can be supplemented for incorporation into the LPO/H2O2/iodine system and that NaI shows exciting promise as a preventative strategy for bovine respiratory disease which may also reduce antimicrobial use in food producing animals.

CHARACTERIZATION OF AN EMERGING NEUROLOGICAL ENTITY WITHIN THE OVINE FLOCK OF QUEBEC

Hélène Ruel, Gilles Fecteau, Joane Parent

University of Montreal, Saint-Hyacinthe, QC, Canada

This study aims to characterize "crampage" [cramping] or "spring leg," a relatively new condition in the sheep industry in Quebec. This condition is easily recognized by the particular gait displayed by affected sheep, characterized by a hyperflexion (hip, stifle, hock) of one or both pelvic limbs while walking. Five naturally-affected and 5 clinically normal lambs, matched by age and weight, were subjected to physical, neurological and orthopedic examinations followed by advanced imaging (CT, MRI), and electrodiagnostic testing (electromyogram, motor and sensory nerve conduction velocities). Complete blood count, serum biochemistries and cerebrospinal fluid analysis were also performed. At the study's completion, the lambs were humanely euthanazied and necropsied. The cross-sectional area of the vertebral canal on CT images, at the level of the second lumbar vertebra, was smaller in the affected group compared to clinically normal lambs ($p = 0.045$). No secondary change was observed in the corresponding spinal cord segment. In addition, the S2 nerve root was more narrow in the affected group ($p = 0.01$). This study excludes a musculoskeletal origin for this problem. A spinal cord lesion was also excluded as an etiology for this condition. A sensory impairment of the S2 nerve root, altering the sensation of the affected limb, may serve as the underlying cause. A cerebral or a functional abnormality in the lumbar intumescence, however, could also be considered. Without an obvious muscular abnormality, the lay term "crampage" is inaccurate. We propose more descriptive terminology, such as "hyperflexion syndrome" or "high stepping gait".

Camelid Medical Emergencies

Christopher K. Cebra, VMD, MA, MS, DACVIM (LAIM)
Corvallis, OR

INTRODUCTION

The last 30 years have seen a tremendous increase in the populations of New World Camelids in the United States and certain other countries outside of South America. This increase has led to corresponding increases in the calls for veterinary care of these species. Even though the body of knowledge about them is increasing also, veterinarians frequently find themselves frustrated in their dealings with these new camelid patients. Among other issues, I believe that major causes for this frustration are 1) that veterinarians are often unfamiliar with common camelid disorders, 2) a lack of comprehensive references, 3) lack of convincing or specific physical examination findings, 4) lack of convincing or specific clinical pathology findings, 5) lack of success with treatment modalities that work in other species, and 6) that even necropsies frequently offer little to no enlightenment. The purpose of this talk is to give some overview of common camelid emergencies, their recognition and treatment.

Camelids generally have a reputation for being healthy 99% of the time, and on the verge of death the other 1%. Thus, beyond routine herd health and reproductive work, it seems almost all the veterinary work we do on camelids is emergencies. Some of these are bona fide acute disorders, but many are the final or critical stages of chronic diseases such as anemia, endoparasitism, neoplasia, starvation, and mycobacterial diseases. Owner education and monitoring strategies can potentially lead to earlier disease recognition and less urgent care, but those are another topic. Neonates also represent another large and distinct source of critical care cases, which cannot be handled adequately in this lecture.

AIRWAY AND RESPIRATORY EMERGENCIES

As with other species, the most urgent emergencies concern a compromised airway. This is relatively common in camelids, because sick and stressed camelids frequently dorsally displace their soft palate. This leads to pharyngeal collapse, dyspnea, stertorous, open-mouthed breathing, and potentially death. Aggressive intervention frequently worsens the condition, and common interventions, such as administration of nasal oxygen, do not help. Frequently, the best response is to leave the camelid alone for a period of time. This is easiest to do if the response is acute, and the clinician is convinced there is no more permanent lesion.

If the handling stress increases or if the camelid is simply that sick, the camelid may progress to a characteristic behavior that I have dubbed "Dying Llama/Alpaca Syndrome [DLS/DAS]." The major features of this are retroflexion of the head over the back and a gradual slide from sternal to lateral recumbency. The eyes frequently roll back into the head, and some twitching or struggling may be seen. Spontaneous regurgitation of gastric contents may occur, especially if the forestomach is full. Some relief can be obtained by maintaining the camelid in sternal recumbency with the head and neck gently extended forward.

Pharyngeal and possibly pulmonary edema secondary to severe hypoproteinemia represent another way that camelids with various disorders often present as respiratory emergencies. Once again, the initial disease is often missed for a long while, until a major, visible complication develops. Too rapid administration of intravenous fluids (pulmonary edema) and tight neck bandages (pharyngeal edema) may exacerbate or create these conditions.

Common primary causes of an obstructed airway include congenital stenosis or atresia,[1] nasal bots, snake bite,[2] or insect bite. With nasal bots, there may be serosanguinous nasal discharge. With insect or snake bite, there is usually severe nasal or facial swelling, and fang marks (for snakes) may be visible. For these more severe physical obstructions, tracheostomy should be considered to provide rapid relief.

Pneumonia is rare beyond the neonatal period, except as a manifestation of sepsis. Pulmonary edema is far more common, and is usually iatrogenic. As with the stress-induced upper respiratory issues, evidence of lower respiratory diseases usually warrants a systemic search for disease. There are a few organisms that primarily invade the lower respiratory tract. These include *Streptococcus zooepidemicus,*[3] *Cryptococcus gatti*, *Coccidioides immitis*, *Aspergillus*, and *Histoplasma*. The fungal diseases are regional in occurrence. Streptococcus frequently involves direct or indirect exposure to horses. These diseases result in dyspnea and tachypnea. Abnormal lung sounds can be difficult to detect, so imaging studies, and follow-up thoracocentesis or tracheal or bronchoalveolar lavage may be necessary to diagnose pneumonia.

As a relatively unique camelid disease, diaphragmatic paralysis has been reported in a number of camelids.[4] Thoracic-abdominal asynchrony, hypercapnea, dyspnea, and evidence of ineffective airflow area present. In the eastern half of the U.S., this is sometimes blamed on damage to the phrenic nerve caused by migrating *Parelaphostrongylus tenuis*. It is also occasionally blamed on vertebral instability, or vitamin E deficiency. The condition may be bilateral or unilateral, with the bilateral form presenting with more severe signs and having a poorer response. Treatment response to mechanical ventilation is usually unsatisfactory, due to lack of recovery.

SHOCK

Many of the camelid emergencies that we see can simply be classified as shock. The major signs are anorexia, weakness, lethargy, recumbency, dyspnea, and the major laboratory findings are dehydration, mild anemia, hypoproteinemia, potentially acidosis and hyperlactemia, and potentially evidence of a disorder in energy metabolism. The causes are multiple and can include sepsis, endoparasitism,[5] heat stress,[6] cold stress, tick paralysis, trauma, ruptured gastric or intestinal ulcers, clostridiosis (myonecrosis/hepatitis/nephritis), metabolic disorders, and various intoxications. Infectious diseases can be exacerbated by underlying immunodeficiency, as with Juvenile Llama Immunodeficiency Syndrome[7] or persistent infection with Bovine Viral Diarrhea Virus.[8] The worst intoxications leading to shock and death include Oleander,[9] Red Maple,[10] Jimsonweed, Water Hemlock, Death Camas, and nitrates. Oleander is common in California. Red Maple and nitrates each cause hemolytic anemia.

Metabolic disorders are common, and may by primary or secondary to another condition. Regardless, the metabolic condition often becomes life-threatening and must be addressed. The more common disorders are those of fat metabolism, including Fat Mobilization Syndrome or Ketosis (high non-esterified fatty acids and ketones), hyperlipemia,[11] and fatty liver.[12] Based on our research, evidence of fat mobilization usually precedes increases in liver enzymes by 2 to 5 days,[13] so we treat these disorders aggressively, even in the early stages. Hyperlipemia appears to represent a subset of this disorder, affecting about a third of camelids with fat mobilization. Disorders of fat metabolism can occur in camelids of all ages and genders. Disorders of mineral metabolism are relatively uncommon. On lush pasture, Grass Tetany has been seen, and in lactating, anorexic camelids, hypocalcemia may occur.

Organ specific signs help localize the lesion.

SPECIFIC EMERGENCIES NOT ASSOCIATED WITH SHOCK

Camelids are susceptible to a number of disorders which have specific presentations that aid in their identification. One common disorder, particularly in male alpacas, is urolithiasis, with the accompanying stranguria.[14] In females of any camelids species, dystocia is a major cause for veterinary intervention. It is important to remember that uterine torsion is common in camelids,[15] as well as the more conventional fetal malposition, size mismatch, and failure of the cervix to dilate. Regurgitation may be a sign of choke or megaesophagus,[16] whereas vomiting is characteristic of ingestion of laurel, azalea, pieris, or rhododendron leaves. Colic is a common sign with a variety of abdominal disorders including various entrapments, obstructions, volvulus, peritonitis, pancreatitis, hepatic disease, and other maladies.[17] Diarrhea may be see with parasitic and viral enteritis, but may also be absent.

Seizures may be seen with polioencephalomalacia, hyperosmolar disorder, infection, hydrocephalus, cerebral edema (overhydration), fungi, or idiopathic (familial?) epilepsy. Polioencephalomalacia has been seen in camelids of all ages, and does not appear to have the same risk factors as in ruminants. Cerebrospinal fluid analysis is helpful in distinguishing some of these diseases. This fluid is fairly easy to obtain from the lumbosacral space in camelids. Other neurologic conditions include Rabies, EHV-1, meningeal worm, Cryptococcus, Tetanus, West Nile Virus/EEE. Grass Staggers, Otitis, vertebral malformations, and tick paralysis.

Trauma to the eyes, dog bites, fight wounds, fractures, gunshots, and burns can occur and should be handled routinely.

IATROGENIC LESIONS

Sometimes, our actions can lead to emergency situations. These mimic similar spontaneous emergencies. The most common situation is the stress response caused by overhandling, as outlined above. The next most common situation is overhydration with intravenous fluids in the face of hypoproteinemia. Generally, unless blood protein analysis suggests that higher rates will be tolerated, giving more than 50 ml/kg/day, including short "shock" boluses, is potentially deleterious. Plasma or synthetic colloid administration may be necessary, even at that low rate. Clinical signs of overhydration and pulmonary edema include respiratory distress, obtundation, lethargy, and in the worst cases, seizures. Treatment options include cessation or reduction of fluid administration, administration of colloids, and use of diuretic agents such as mannitol. In some cases, fluids can cause metabolic disturbances such as hyperglycemia and hypernatremia. These should be identified and treated appropriately. Bleeding disorders are relatively rare, but a small percentage of camelids develop significant intrabdominal hemorrhage after liver biopsy. Weakness, pale membranes, and tachycardia are characteristic signs. Fluids and blood transfusions are the best treatments. Carotid catherization or puncture can lead to bleeding in the neck and mediastinum. The volume of blood lost is usually not large, but some associated factor, possibly pressure on the trachea or on the vagus leads to severe respiratory signs, weakness, lethargy, and nystagmus. Tracheostomy is tempting, but rarely beneficial. In fact, I have found minimal intervention on such shocky camelids to be more helpful than invasive interaction. Most recover within 15 to 30 minutes.

Other induced lesions include postanesthetic hepatic necrosis, and acute renal failure due to aminoglycosides or non-steroidal anti-inflammatory medications. These are best addressed through standard supportive care. Nephrotoxicosis is potentially avoidable through correct dosing and proper maintenance of the patient's hydration during treatment. Rectal or uterine tears or rupture can occur during manipulation. They are best assessed endoscopically or surgically. Full-thickness tears into the peritoneal cavity require emergency surgery and intensive antibiotic treatments for successful resolution.

ULCERS?

I mention this as a separate category, because of the concern many veterinarians and owners have about this disorder. In my experience, gastric ulcers are rarely clinical before they rupture, and then peritonitis is more the problem than the ulcer. I do not typically use ulcer prophylaxis in hospitalized camelids, in part because most medications have been demonstrated to be ineffective in camelids, and in part, because other diagnoses predominate in this clinic. Specific treatment of the primary condition appears to decrease the clinical importance of any possible ulcers in most camelids.

REFERENCES

[1]Bertin FR, et al. J Vet Intern Med 2015;29(1):368. [2]Sonis et al. J Vet Intern Med 2013;27(5):1238. [3]Jones M, et al. J Vet Intern Med 2009;23(1):220. [4]Byers S, et al. J Vet Intern Med 2011;25(2):380. [5]Cebra CK, et al. J Am Vet Med Assoc 2007;230(1):94. [6]Norton PL, et al. Can Vet J 2014;55(10):992. [7]Hutchison JM, et al. Vet Immunol Immunopathol 1995;49(3):209. [8]Bedenice D, et al. J Vet Intern Med 2011;25(3):605. [9]Kozikowski TA, et al. J Am Vet Med Assoc 2009;235(3):305. [10]Dewitt SF, et al. J Am Vet Med Assoc 2004;225(4):578. [11]Waitt LH, et al. J Am Vet Med Assoc 2008;232(9):1362. [12]Tornquist SJ, et al. J Am Vet Med Assoc 1999;214(9):1368. [13]Tornquist SJ, et al. Am J Vet Res 2001;62(7):1081. [14]Duesterdieck-Zellmer KF, et al. Aust Vet J 2014;92(8):313. [15]Cebra CK, et al. J Am Vet Med Assoc 1997;211(5):600. [16]Watrous BJ, et al. J Vet Intern Med 1995;9(2):92. [17]Cebra CK, et al. Vet Surg 1998;27(2):112.

Managing Colic in Camelids

Christopher Cebra, VMD, MA, MS, DACVIM (LAIM)
Corvallis, OR

INTRODUCTION

The subject of abdominal pain in camelids has been addressed a number of times over the years, with the discussion becoming progressively more detailed as we gain experience with these species. We were truly learning from scratch, as the South American approach (hands off and deal with the consequences) was generally unsuitable to North American tastes. The tone has varied over the years, from extremes such as "we had better cut it, because it's worth so much," to, "why cut it, because they all die." The bases for these extreme views were the high proportion of underlying lesions that were a major morphologic defect, such as a ruptured ulcer, and they general uncertainty on how to proper assess the likely severity and nature of the disease in a sick camelid. As we have become more knowledgeable about disease states and more proficient and comfortable in dealing with them, we have improved in both those regards.

CAUSES OF COLIC

Camelids develop abdominal pain for the same reasons as other animals. The major mechanisms are activation of stretch-pain receptors in the mesentery, various ligaments, or organ capsules, muscle spasms, inflammation, and ischemia. Similar to cattle, the capacity and location of the gastric compartments combined with the ability to regurgitate excess fluid decrease the likelihood of gastric distention or is tug on the mesentery contributing to pain signs. Additionally, relatively complete forestomach fermentation in adult camelids decreases the likelihood of aberrant intestinal fermentation, and hence gas colics are uncommon. However, aberrant fermentation can occur in younger camelids, or adults with depletion of their normal gastric microbial population. In crias, this "milk colic" may be severe, but usually only lasts a few hours. As in other species, poorly digestible milk replacers or viral damage to the intestinal brush border may contribute to aberrant fermentation. In adults, it is downright rare. In all ages, infection by gas-producing bacteria may contribute to signs with some forms of enteritis.

Capsular stretch is a more important source of pain when it involves non-digestive abdominal organs. Colic signs may be severe with distention of the kidney, ureters, or to a lesser degree, bladder. Milder signs are evident with enlargement of the liver due to lipidosis, hepatitis, or fluke infection.

Anatomically, the camelid stomach and intestines bear more similarity to cattle than horses. Aside from thrombosis due to systemic events, ischemia is most likely in regions with long mesenteric attachments and relative mobility. In camelids, this includes the distal phlange of the small intestine and the whole of the ascending colon, including its spiral loop. Rotation or displacement of the third compartment has not been described. Described ischemic events occur as a results of strangulation involving the epiploic foramen (rare), body wall defects (intermediate frequency), diaphragmatic defects (rare), mesenteric defects (rare), rotation around the mesenteric root (most common), intraabdominal adhesions (intermediate), or intussusception (rare). Non-strangulating entrapments can occur with less severe involvement of most of these structures as well. Intussusception has been described involving the jejunum, cecum, or ascending colon. Uterine torsion is among the more common causes of colic. Liver lobe or splenic torsion is extremely rare. Risk factors for entrapments obviously include the presences of body wall, mesenteric, or diaphragmatic defects. Enteritis or coccidial infection may promote mesenteric volvulus or intussusception, particularly when they involve the colon.

Intraluminal obstruction is usually caused by hair or plant fiber balls. In most cases, these appear to form in the gastric compartment, where multiple may be found. Some, smaller balls may form in the proximal loop of the ascending colon. Hairballs are most common in animals less than one year old. Ectoparasites, boredom, and post-weaning suckling behavior are thought to be risk factors. Phytobezoars may develop in animals with poor chewing or on excessively fibrous feed. Long-stem spring pasture appears to increase risk in the Pacific Northwest. Obstruction tends to occur near the pylorus or in the beginning of the duodenum in alpacas and smaller llamas, and in the centrifugal loops of the spiral colon in adult llamas. With time, bezoars lodged in the duodenum may migrate further along the intestine, but it is rare for them to make it into the colon or pass entirely. Tapeworm impaction with ulceration throughout the jejunum has been described in one alpaca. Foreign body obstruction is very rare.

Over recent years, a number of important causes of non-surgical colic have been identified. The earliest work on this identified that many camelids with colic signs that abated with time or medical treatment had serum biochemical abnormalities that reflected either obstruction (hypochloremic metabolic alkalosis) or enteritis (hyponatremia, hypochloremia, and often hypoalbuminemia). Our subsequent experiences suggest that "obstruction-like" episodes are rare. Small bezoars may pass, but that is uncommon and may result in mucosal abrasion. Hypochloremic metabolic alkalosis appears uncommon with ileus in sick camelids, which generally trend toward hyperchloremia. Whether ruptured ulcers with omental bursitis can cause this biochemical abnormality is unclear.

Camelids with "enteritis-like" blood work and colic signs may have a variety of conditions. The single most common in many different geographic areas appears to be *Eimeria macusaniensis* infection. These large coccidia invade the gut epithelium and lamina propia, in contrast to most GI parasites, which remain in the epithelium. It is possible that this deeper invasion has a more profound effect on GI motility or is simply more inflammatory, and hence more likely to cause colic. Other clinical evidence, including gut fluid distention and ileus detected ultrasonographically, by succussion, or at surgery, suggest that pooling of intraluminal fluid resulting in mesenteric tug is the main contributor to colic signs. This ileus can confound diagnosis, because it inhibits or delays the onset of diarrhea. Supportive evidence for a diagnosis of *Eimeria macusaniensis* infection includes hypoalbuminemia disproportionately severe for the level of anemia, occasionally hyponatremia, and identification of oocysts on fecal examination, if the infection has achieved patency. *Trichuris* (whipworms), which mainly invades the cecum and spiral colon, can also cause significant colic, usually with the continued passage of loose feces.

Other common causes of enteritis without severe inflammation include coronavirus and rotavirus, *Giardia*, and *Cryptosporidia*. Salmonellosis is fairly uncommonly reported in camelids, and may or may not cause severe inflammation. Other bacterial enteritides include *Clostridium perfringens* and occasionally coliform organisms or *Bacteroides*-like organisms. In North America, the bacterial enteritides, except for Salmonellosis, are rarely considered primary diseases, and usually occur secondary to parasitic or viral infections, physical or chemical damage to the epithelium, or gut thrombosis. The less inflammatory enteritides are rarely associated with colic, and it usually resolves rapidly with the onset of diarrhea. Affected camelids may have hyponatremia and hypochloremia, relatively specific indicators of intestinal diseases in camelids, and hypoalbuminemia. Acidosis is generally less common or milder than with comparable disorders in cattle. Camelids with bacterial enteritis may additionally have fever, tachycardia, abnormal mucus membranes, high liver enzymes, lactic acidosis, and an abnormal leukogram. All abnormalities need not be present simultaneously to support this diagnosis. Diarrhea tends to be scant and less aqueous with bacterial enteritis or Eimeria macusaniensis infection, as opposed to profuse and watery with viral or other parasitic enteritides.

Camelids with colic due to bacterial enteritis, whipworms, or *Eimeria macusaniensis* infection can be very difficult to differentiate from camelids with gut strangulations, entrapments, or intraluminal obstructions, especially if the surgical lesion has been allowed to progress. Both types of disorder result in a camelid with low fecal output and also in declining systemic condition. For this reason, combined with a historical lack of rapid identification of surgical lesions, and the reluctance to watch hospital patients deteriorate without trying every possible treatment modality, many camelids with inflammatory conditions in the abdomen were surgically explored in the past. Exploration is helpful in confirming the lesion and establishing the prognosis in camelids with inflammatory diseases, and gives the opportunity to obtain biopsy specimens, which is especially helpful in diagnosing prepatent *Eimeria macusaniensis* infection, but is unlikely to lead to clinical improvement.

Slaframine poisoning is an uncommon, but important cause of non-surgical colic. It presumably has the same pathogenesis as in other species, and is associated with ingestion of the indolizidine alkaloid produced by fungus *Rhizoctonia leguminicola*. This fungus colonizes legumes, particularly red clover in damp areas and subterranean clover in dry areas, causing discoloration of the clover known as 'black patch'. Camelid exposure appears to be most common in the damp seasons, particularly Spring and Fall, which is conducive to both the clover and the fungus. Multiple animals may be affected simultaneously, which is uncommon with all other causes of colic. Signs occur within hours of ingestion in other species. In camelids, they last from 1 to 3 days, usually with a steady clinical improvement after the first 24 hours. In addition to colic, affected camelids may have profuse salivation ('Slobbers'), other SLUD signs, tremors, and a stiff gait. Treatment is not usually necessary, and the specific combination of signs, especially if multiple animals are affected or the discolored clover can be found, helps give the attending veterinarian confidence of the diagnosis. Clinical pathologic abnormalities have not been reported.

CLINICAL SIGNS AND PHYSICAL EXAMINATION

To the practitioner most acquainted with colic in equids, tachycardia, abdominal distension, absence of appetite, borborygmal sounds, and fecal passage as well as activities including pawing, restlessness, and stretching the abdomen are frequently taken as signs of colic. Although all of these signs may be present in some colicky camelids, most have neither tachycardia nor active signs of pain. Camelids have been described as being stoic animals that try to hide signs of illness and pain. While it is true that signs in camelids are less visible and violent than in horses, their behavior is restive and abnormal enough to be apparent to the careful observer. Specific signs of colic include lying in an abnormal position: sternal recumbency with one or more legs extended to the side or lateral recumbency; repeatedly getting up and lying down; kicking at the abdomen; and frequently changing position while recumbent. Lying quietly in sternal or lateral recumbency is not specific for colic.

Level of pain and presence or absence of other signs can be used to estimate the likelihood of the various categories of lesions. Small intestinal obstruction often leads to more subtle signs than other types of lesions, because fluid is refluxed into the gastric compartments and inflammation is minimal. This is most true for obstructions near the pylorus; the further into the jejunum the bezoar is pushed, the more tension there is on the mesentery and the more painful the camelid appears. Thus, in addition to the importance of recognizing pain signs, it is important to remember that camelids with mild to subtle signs still may have life-threatening obstructive lesions.

Very cranial obstructions can cause several other abnormalities that help differentiate them from other lesions. Fluid accumulation in the forestomach leads to bilateral, ventral abdominal distension, and a pot-bellied appearance. Gas may be eructated until the gastric fluid level reaches the cardiac sphincter, causing muscular spasms. After that time, bloat may be noticed. Other signs include dehydration, depression, and in rare cases, vomiting. Abdominal distension is difficult to assess in full-fleeced camelids; direct external palpation ("bear-hug") may be necessary. Other conditions, including advanced pregnancy, effusive peritonitis, venous congestion (as with strangulation), and enteritis can lead to abdominal distension without gastric engorgement. These conditions may be distinguished by other characteristics. It is also appropriate to mention here that subjective assessment of cranial abdominal pain by palpation of the flanks is extremely difficult to interpret; most camelids do not like being touched in that area and it is natural for them to flinch, jump, or lie down.

Dehydration and depression with small intestinal obstruction are due to sequestration of ingested and secreted fluids, with minimal functional intestinal absorptive surface area. Signs worsen with proximity to the pylorus, completeness of the obstruction, and time. Thus, a camelid with duodenal phytobezoar obstruction is likely to have subtle colic signs that abate after a few days due to progressive dehydration and depression. In contrast, a camelid with a colonic bezoar obstruction that leaves most of its intestinal absorptive epithelium functional is likely to have more obvious pain signs and develop minimal dehydration.

Camelids with strangulating obstructions or sepsis also develop signs of dehydration and depression (and also distension, as previously mentioned) due to venous pooling, peritoneal effusion, and the inflammatory response. Such camelids usually appear more painful and differ in other aspects of the diagnostic evaluation, but may be extremely difficult to differentiate from camelids with jejunal obstruction or any small intestinal obstruction with inflammatory complications.

In addition to the more obvious colic signs and possible dehydration and depression, camelids with inflammatory or cecocolic diseases may also show tenesmus. Camelids usually defecate and urinate at a communal dung pile several times a day, but rarely spend more than a minute or two at the pile or assume the posture when away from the pile. Tenesmus with obstruction is usually non-productive, whereas camelids with enteritis or peritonitis may have diarrhea. Thus, tenesmus with diarrhea and colic signs suggests a severe inflammatory condition. Tenesmus is also a common sign with urogenital disorders.

Anorexia, lack of feces, and gastric hypomotility are common findings in camelids with gastrointestinal disease, but lack specificity because they are also seen with diseases of other organ systems. However, finding appetence, normal fecal passage, and normal ruminations in a camelid with colic strongly supports a mild, transient problem, a non-inflammatory, non-gastrointestinal condition (i.e. uterine torsion), or the early stages of more severe disease. These signs should be monitored for resolution or progression.

Fecal passage may provide other insights into the nature of the disease. Camelids with obstructions may pass normal feces for 1 to 3 days (longer with small intestinal than cecocolic obstruction), but eventually change to dry, mucus-covered feces, and later stop defecating altogether: camelids with entrapments may continue to pass small quantities of normal or blood-tinged feces, whereas camelids with strangulations and necrotic enteritis and peritonitis may pass diarrhea in decreasing quantities. Consistent passage of small amounts of diarrhea in spite of ongoing pain signs for up to 4 days is common with *E. macusaniensis* infection. Continued, profuse diarrhea or normal feces are rare in camelids with colic due to gastrointestinal lesions requiring surgical repair.

Vital signs have been of little value in determining severity of colic and the presence of surgical lesions. Camelids with all types of lesions have been recorded as having low, normal, and high pulse and respiratory rates. It is our general bias that the more normal the vital signs, the better the prognosis, but this is not always true. High rectal temperatures are most common with peritonitis, but are not specific to the cause of inflammation. Low rectal temperatures can occur with most disorders and are usually a sign of circulatory compromise and more severe disease. When interpreting heart rate, it is imperative to remember that most healthy adult camelids have heart rates of 72 beats/minute or less, and that values between 72 and the published high end of 90 beats/minute may reflect pain or shock.

FREQUENCY OF COLIC

The types of lesions described here have been blamed for up to 19% of camelid deaths in camelids. It is possible that the percentages are even higher because colic signs may not have been recognized in all camelids, even if they were present. In our practice, colic is responsible for approximately 5% of camelid admissions. The percentage is likely to be lower in general practice, though poor recognition of signs may lead to underdiagnosis. The prevalence of individual diseases is likely to vary based on geographic, dietary, breed prevalence, and other factors. With many of these diseases, colic signs are transient, subtle, and may not be noticed without careful, constant observation.

RECTAL PALPATION

Although juvenile llamas and most alpacas may be too small, rectal examination of most adult llamas is possible with good restraint (preferably in a camelid chute), adequate lubrication and if necessary, chemical sedation. We find that adding 20 ml of a local anesthetic agent (2% lidocaine HCl) to 100ml of lubricant (for adults) and injecting the mixture into the rectum by plastic dose syringe or insemination pipette relaxes the anal sphincters and caudal rectum sufficiently to perform a good rectal examination. Individuals with large hands may not be able to do this and should stop if they encounter too much resistance. Rectal injuries and bladder rupture have been reported.

Rectal examination of the llama is similar to that of the cow; the bladder, female reproductive tract, left kidney and caudal first gastric compartment usually can be palpated. The right abdomen usually has no palpable viscera. Rectal palpation is most useful for abnormalities in the caudal abdomen, especially urogenital lesions. Obstruction, inflammation, entrapment, or strangulation of the terminal small intestine and ascending colon are likely to be detected by the presence of distended bowel or tight mesenteric roots, whereas generalized inflammatory conditions and small intestinal obstructions generally cause no palpable abnormalities.

Except for diagnosing uterine torsion, rectal examination of colicky camelids has been largely supplanted by flank ultrasonography in my practice.

CLINICAL PATHOLOGY

A thorough clinicopathologic evaluation helps in identification of the lesions and selection of treatment. In general, the more normal the evaluation, the better the health of the bowel and the prognosis for survival. Camelids with hematologic abnormalities are likely to have inflammatory lesions and camelids with small intestinal obstruction are likely to have metabolic abnormalties.

Complete Blood Count- Because of the frequency of inflammatory lesions in camelids with colic, differential blood counts are frequently abnormal. These changes can be obscured by stress neutrophilia, a common finding in sick camelids. More specific evidence of inflammation includes immature granulocytes, toxic changes in granulocytes, and hyperfibrinogenemia. Inflammatory changes are to be expected in camelids with peritonitis or enteritis; in camelids with obstructive bowel diseases they are an undesirable sign of advanced disease.

Acid-base balance- Camelids with small intestinal obstruction frequently develop metabolic alkalosis due to the sequestration of gastric HCl within the gut lumen. Small intestinal entrapments and strangulations and long-standing cecocolic obstruction occasionally cause mild metabolic alkalosis, but such lesions usually cause so much pain that veterinary attention precedes development of this abnormality. Camelids with inflammatory or ischemic conditions have a tendency towards metabolic acidosis. This is more likely to be due to lactate production than intestinal bicarbonate loss, and is often accompanied by hyperlactemia and a high anion gap. Metabolic acidosis may also be seen with long-standing intestinal obstructive lesions as they become more inflammatory; hence, it is a sign of more severe disease.

Electrolytes- Serum chloride is the most important electrolyte for diagnosing gastrointestinal causes of colic. Chloride ion is sequestered in the gastric compartments with cranial obstructions, but to a much lesser degree with more aboral obstructions. Thus, hypochloremia (< 106 mEq/L) is usually present under similar conditions as metabolic alkalosis.[2,6] Hypochloremia may also be found with metabolic acidosis in camelids with enteritis; hyponatremia and hypoproteinemia usually accompany it in these cases. *E. macusaniensis* infection is the most common cause associated with colic.

Sequestered chloride can also be detected by analyzing gastric fluid. This fluid may be retrieved by stomach tube or paracentesis. When analyzed correctly, forestomach fluid supernatant appears to have a similar chloride concentration as cattle (normal: < 20 mEq/L; camelids with anorexia: < 30 mEq/L). Higher values can be expected in camelids with obstructions that are more cranial, complete, and long-standing. Values >40 mEq/L almost certainly reflect an obstruction. Smaller increases are seen with strangulating lesions due to the rapid decline in clinical parameters.

Hypokalemia is almost ubiquitous in camelids with gastrointestinal disease, and worsens with duration and severity of anorexia and tendency towards metabolic alkalosis. Thus camelids with small intestinal obstructions have some of the lowest serum potassium concentrations, but this finding should be expected in camelids with other gastrointestinal disorders.

Other blood constituents- Measures of dehydration and renal function are also likely to be most severe in camelids with small intestinal obstruction and are also in camelids in the shock phase which occurs with many of the inflammatory conditions. Creatinine is of more value than urea nitrogen in some cases, because urea nitrogen is affected by the rate of ruminal utilization. Albumin and serum total protein also tend to increase with dehydration, but camelids appear to lose or catabolize protein with many gastrointestinal disorders. Therefore, high and low albumin concentrations are associated with more severe disease. Additionally, protein is lost due to enteritis. Moderate to severe hypoproteinemia combined with continued passage of soft feces is suggestive of *E. macusaniensis* infection.

Abdominal paracentesis- Using the right paracostal approach, abdominal fluid can be obtained from most sick camelids. Ultrasound guidance may improve yield but is usually not necessary. Protein concentrations and nucleated cell counts of peritoneal fluid are lowest in camelids with the least damage to the bowel and without peritonitis. Thus, they are most normal with early intraluminal obstructions and non-strangulating entrapments. Abnormal peritoneal fluid with these lesions (often accompanied by abnormal CBC results) indicates progression of the lesion and inflammatory complications. Strangulating obstructions with venous occlusion may increase the volume and red cell count of the peritoneal fluid before they increase protein and white cell counts. Very high (< 4.0 g/dL) protein concentrations and nucleated cell counts (< 10,000 cells/μL) are indicative of inflammatory diseases, as are low cell counts with toxic changes to neutrophils. Bacteria and plant material may be seen with sepsis and visceral rupture. Peritoneal fluid activities of amylase and lipase higher than serum activities may be associated with pancreatitis.

Peritoneal lactate higher than systemic lactate tends to reflect compromise of bowel in general, strangulation, or effusive cellular peritonitis. The greater the divergence, the more severe the lesion. Uroperitoneum is similar in camelids to other species. In general, the more normal the peritoneal fluid, the less severe the lesion. Exceptions to this rule include camelids with mature adhesions, which may have uncorrectable lesions but normal fluid and camelids with certain inflammatory conditions (pancreatitis, streptococcal peritonitis), which may have correctable lesions in spite of very abnormal fluid.

ABDOMINAL IMAGING

Radiographs- Plain abdominal imaging studies are of limited usefulness in diagnosing the cause of colic. In adult camelids, most abdominal detail is obscured by feed material in the gastric compartments, and few lesions lead to gas accumulation within the bowel. Neonatal llamas have a higher incidence of abnormal intestinal fermentative disorders and smaller gastric compartments; thus imaging studies of colicky crias may reveal elaborate intestinal gas patterns, or potentially impactions of the forestomach or colon.

Contrast studies have been performed on few occasions. They may be diagnostic with some obstructive lesions, and are useful in distinguishing whether a distended viscus is first compartment or not. Unfortunately, intestinal motility is often impaired, making the interpretation of contrast studies difficult.

Ultrasound- Ultrasonographic imaging of the abdomen and its viscera may be performed transrectally or transcutaneously. It is ideal for detecting fluid accumulation within the bowel or abdomen. Transrectal imaging is best for the same lesions as are palpated per rectum, namely those in the caudal abdomen. This technique has the advantage of being possible in many camelids that are too small for palpation. Transcutaneous imaging through the right paralumbar fossa shows the same structures, as well as revealing the size and contents of the gastric compartments, and the amount, location, and nature of free fluid.

Ultrasonography allows assessment of bowel contents and motility. Changes in wall thickness are usually subtle. Normal intestines should have intermittent pockets of hypoechoic fluid and good evidence of peristalsis. Obstructed or damaged bowel has larger, stable pockets of fluid with minimal evidence of progressive motility: this is easier to find with lesions of the colon or jejunum due to their accessibility. Cranial obstruction may be inferred on finding fluid distension or atony of the third gastric compartment. Distinguishing obstructed from severely inflamed bowel is difficult; on occasion, the lumen of the inflamed bowel may have gas shadowing and that of obstructed bowel may contain a hyperechoic mass of fiber or intussusception. Inflamed, distended bowel may also show evidence of progressive motility.

Large amount of abdominal fluid are present with effusive peritonitis, severely inflamed bowel, uroperitoneum, venous occlusion with strangulation, and other ascites-causing conditions. The more inflammatory the cause, the more hyperechoic flecks are likely to be present. As mentioned previously, finding large pockets of fluid ultrasonographically aids in successful paracentesis. Ultrasonography may also be used to detect uterine torsion, urolithiasis, renal disease, liver disease, mass lesions, and rarely pancreatitis.

Advanced imaging techniques, such as computed tomography, are seeing increasing use in camelids, and have been used to identify gut distention and mass lesions, including intraluminal phytobezoars, in camelids.

TREATMENT

Surgery vs medical therapy- Making the decision to surgically explore a camelid with an acute abdomen can be difficult. Historically, there has been some reluctance to perform surgery, mainly because of the perception that most camelids with colic (including many with intraluminal obstructions) eventually have uncorrectable lesions. This perception has been reinforced by poor surgical survival rates. However, improvements in diagnostic modalities, anesthetic management and clinical decision-making have led to many successful surgeries.

Our experiences with camelids are far less extensive than those with horses and cattle, but using similar criteria to those used in these species appears to give us reasonable guidelines for when to do surgery in camelids with colic. This includes clinical signs that worsen or do not improve with medical treatment. Deciding how long to attempt medical treatment before resorting to surgery is difficult. Although there is evidence that some intraluminal obstructions resolve with medical treatment, there is also ample evidence that delaying surgery in camelids with intraluminal obstruction can be detrimental; many intestinal ruptures found at surgery or necropsy occur at the site of a chronic intraluminal obstruction. Based on our best estimates, gut friability becomes life-threatening within 4 days of obstruction. Because cranial obstructions usually cause less overt signs than caudal obstructions, damage to the bowel wall may be more advanced on first examination of camelids with cranial obstructions. Abnormalities in the leukogram or peritoneal fluid and signs of sepsis often accompany bowel compromise.

Rising awareness of *E. macusaniensis* infection as a cause of colic has complicated the use of pain as the major determinant for surgical exploration. Camelids with parasitic enteritis tend to pass some abnormal feces, have blood work suggestive of enteritis, and have ultrasonographic evidence of progressive motility.

MEDICAL TREATMENT

Medical treatment of colic focuses on the principles of patient stabilization and maintenance, pain control, treatment of infectious and inflammatory conditions, and occasionally attempts to resolve the obstruction.

Patient stabilization involves shock treatment and restoration of hydration, electrolyte, acid-base, and protein norms. Shock treatment is most vital with strangulation and inflammatory lesions. Intravenous fluids are key. An initial bolus of up to 5% of body weight (10% in neonates) may be administered, if the patient will tolerate it, and is not too hypoproteinemic. With alkalosis, 0.9% sodium chloride may be used, but with most disorders, a polyionic, buffered fluid is adequate. Sodium bicarbonate may be added to correct severe acidosis (pH <7.25), but this treatment is not particularly effective in the face of lactic acidosis. Supplemental potassium chloride, administered at <0.5 mEq/kg/h may also be helpful. Because of the small patient size and rarity of severe hyponatremia or hypochloremia, hypertonic saline is rarely necessary. Plasma transfusion or administration of a synthetic colloid may be necessary to treat hypoproteinemia. I typically administer plasma if the blood albumin concentration is <2.0 g/dL or total protein is <4.0 g/dL.

After the initial fluid bolus, I typically continue intravenous fluids at 2 ml/kg/h in adults and up to 4 mg/kg/h in neonates until the camelid is able to obtain its own water. Stabilization also involved common sense procedures such as protection from the elements, removal from stressful circumstances, provision of heat, water, and possibly feed, and protection from other animals. A bedded, sheltered stall with a controlled temperature serves nicely.

Pain control can be difficult to achieve and has the temporary danger of masking a surgical lesion. Flunixin meglumine (1.1 mg/kg, IV, q12h) or butorphanol tartrate (up to 0.05 mg/kg, IV or SC, q 4 to 6h) offer relief from mild pain, but are usually inadequate to control colic. In one case under my care, butorphanol appeared to allow sufficient patient relaxation for an alpaca to pass an enormous (>5cm in diameter) bezoar, but that is rare. More aggressive pain management includes Lidocaine (1.3 mg/kg IV over 15 minutes as a loading dose, followed by 0.05 mg/kg/min), ketamine (40mcg/kg/min), or fentanyl as constant rate infusions. Other options include epidural morphine (0.1 to 0.3 mg/kg of preservative-free morphine diluted to 12 ml), Buprenorphine (3 mcg/kg IM q4-6h), or transdermal fentanyl (300 mcg/h in an adult llama). Alpha-2 adrenergic agonists generally should be avoided due to their contribution to cardiovascular compromise and displacement of the soft palate.

Flunixin meglumine is also useful as an anti-inflammatory. I typically use 0.5 mg/kg, IV, q 12h, unless circumstances justify a higher dose. Antimicrobials round out the medical treatment plan. Broad spectrum antibiotics, such as pencillin-aminoglycoside combinations may be augmented with metronidazole to increase the anerobic coverage. Ceftiofur sodium may not have sufficient distribution to the peritoneal cavity to be effective there. If coccidia are suspected, the appropriate medication should be used.

Oral and parenteral agents to help move on obstruction have shown little promise in my hands. Oral agents tend to sit in the atonic, distended stomach, and parenteral agents cannot move the mass.

Determining how long to attempt medical treatment requires careful assessment of the patient. If I cannot control the pain, I am reluctant to delay surgery more than 24 hours in any camelid showing evidence of obstruction and diminishing fecal passage. To treat even that long, the camelid must show no evidence of bowel degradation on physical, blood, or peritoneal fluid examination. If the camelid continues to pass diarrhea or soft feces, possibly has a fever, and ultrasonography reveals evidence of progressive motility, the high prevalence of *E. macusaniensis* infection in our area often warrants extending medical treatment for 2 to 3 days.

Criteria for surgery include persistent pain and lack of feces, finding evidence strongly suggestive of an obstructive lesion, entrapment, or strangulation, and sufficient confidence in ruling out inflammatory conditions such as peritonitis, pancreatitis, enteritis, and hepatitis. Evidence for a small intestinal obstruction includes hypochloremic metabolic alkalosis, a high forestomach chloride concentration, abdominal distension, and possibly regurgitation or vomiting. Evidence for a distal small intestinal or colonic obstruction includes rectal or ultrasonographic identification of fluid-distended loops of bowel without good progressive motility in camelid showing overt pain signs. Characteristics of volvulus include initial high levels of pain, seroanguinous peritoneal fluid, and high concentrations of lactate in blood and especially peritoneal fluid.

SURGICAL APPROACHES

Ventral midline, paracostal, and paralumbar celiotomy all have been performed successfully to diagnose and treat gastrointestinal disorders in camelids. Left paralumbar laparotomy offers the best exposure to the first gastric compartment, but poor exposure to the remaining abdominal viscera, and thus is not recommended unless the lesion is thought to affect those compartments. The choice between ventral midline and right paralumbar exploratory laparotomy often is based on the preference of the surgeon, as both approaches allow for palpation, visualization, and exteriorization of most abdominal viscera, including the spiral colon, cecum, most of the small intestine, and the pyloric antrum of the third gastric compartment. The pyloric antrum may be difficult to exteriorize through a ventral midline incision when the third gastric compartment is distended with fluid. Therefore, a right paracostal or paralumbar incision may be preferable when a gastric or duodenal lesion is suspected. Herniation of gastrointestinal viscera through the body wall incision may occur after ventral midline laparotomy, although we have noted this to occur more frequently after caesarean section in large female llamas than after general surgical exploration of the abdomen of camelids.

SELECTED REFERENCES

[1]Garry F. Proceedings. 11th Annu ACVIM Forum 1993;817. [2]Cebra CK, et al. Vet Surg 1998;27:112. [3]Cebra CK. Proceedings. 15th Annu ACVIM Forum 1997;447. [4]Navarre CB, et al. J Am Vet Med Assoc 1999;214:812. [5]Pearson EB. Proceedings. 13th Annu ACVIM Forum 1995;671. [6]Cebra CK, et al. J Am Vet Med Assoc 1997;211:600. [7]Sullivan EK, et al. Vet Surg. 2005;34(5):524. [8]Cebra CK, et al. J Am Vet Med Assoc. 2008;232(9):1357.

Herd Health Planning for Camelid Herds – A Valuable Opportunity

Claire E. Whitehead, BVM&S, MS, DACVIM (Large Animal Internal Medicine), FHEA, MRCVS
Oxford, UK

INTRODUCTION

Herd health planning is considered a standard practice for traditional livestock farms. Camelid owners follow often-untraditional pathways to livestock farming and have varied levels of husbandry knowledge. This offers opportunities for large animal vets to interact with their clients, providing tools for prevention of disease and offering improved animal welfare.

The goal of any herd health management programme is to minimise the potential detrimental effects of predictable diseases and events and to ensure good standards of health that will protect against exposure to unexpected ones. This is achieved by having appropriate processes in place that involve veterinary prophylaxis, nutritional planning and farm infrastructure that allows for ease of management and care of animals whether conditions are good or bad.

Camelids are fleece-producers primarily although in South America meat production is a viable part of the industry. As populations of camelids grow in the developed regions of the world, meat production is likely to become more important and this will have an influence on veterinary drug use. In the US, an EPD programme has been running for a number of years and this will help to clarify fibre production targets within the industry and in turn help direct management systems that optimise production accordingly.

INFLUENCES ON HERD HEALTH & PRODUCTION

Geographical location and climate

Geographical location and climate can result in varied conditions. This may be between countries or regions within countries. Within Europe there are very differing climates with the UK being generally quite temperate with lush pasture; in Norway in the winter it may be very cold (some areas average -15°C) and pasture may be covered in snow for months at a time; Italy tends to be hot and dry with sparse grazing and mild winters. Within the US, there is very similar variation with some areas having lush pasture, others been sparse, mountainous and cold/snowy in winter, with Florida being hot or hotter year-round! Each of these areas will have different considerations affecting management. Regardless of these external influences, a customised herd health plan can be designed that takes these into consideration. Latitude and altitude has a particular influence on camelid health in that camelids are prone to vitamin D deficiency in non-equatorial latitudes and at low altitudes due to reduced ultraviolet radiation: supplementation is therefore required to prevent rickets in growing animals and to optimise immunity, foetal growth and lactation in adults.[1-3]

Herd health planning should take account of expected weather variations over the course of the year but also consider contingency planning for extreme conditions that may occur in either summer or winter. Although camelids are adapted to the Andes mountain ranges of South America, weather conditions tend not to be extreme, and when winters are harsh, mortality rates increase dramatically due to unavailability of shelter. Generally speaking, camelids can live outdoors throughout the year, but they do need some sort of shelter from inclement weather, especially rain and excessive cold or snowfall in winter but also from heat in the summer. Camelids in fleece are well-insulated from snow in winter provided that they were not shorn late in the summer, but they can get soaked through and chilled in rain and need to have some way to get out of the rain. A three-sided shelter facing away from the prevailing wind may be adequate in the UK but this sort of shelter needs to be large enough to allow even the recessive animals space to get inside. Barns can be made use of when there are larger numbers of animals and may be more appropriate in some climates. Finally, alpacas can be housed indoors completely provided that they have sufficient space to exhibit normal behaviour. At the opposite end of the spectrum heat stress is a major concern for camelids, particularly in the southern US. Shearing must be done early enough in the year to minimise this concern: llamas may be especially at risk if only shorn around the chest and abdomen in a barrel-cut. When the Heat Stress Index (humidity % + temperature in °F) is over 160, camelids are considered to be at high risk of heat stress. Handling should be avoided where possible or completed during the cooler parts of the day. Shade provision with fans, misting devices or paddling pools may be available. Trees can provide excellent canopy shade but depending on the species of tree, seed cases in autumn can cause concern due to toxicity and/or fleece damage. Maintainance of good airflow will result in improved cooling.

Pasture

Pasture availability will directly influence management of camelid herds. It is closely related to stocking density with more pasture required for greater stocking numbers and impacting the need for additional feed provision and parasite control measures. The quality of the available pasture will influence the stocking density with lower quality pasture able to support fewer animals. If no pasture is available at all and animals are confined to dry lots, emphasis shifts to provided forages and additional supplements but also social considerations will become more important as the animals' behaviour will become altered in confinement. Camelids are very well adapted to poor quality forages but provision of only poor quality forage will not yield optimum production in terms of fetal growth, lactational performance and subsequent cria growth. Oversupplementation also has a negative impact on fibre quality and can result in health problems such as an increased incidence of dystocia and poor lactation. Therefore, nutritional management plays a key role in maintaining optimal health and production.

Pasture quality will affect its nutritional value and therefore determine whether supplementation of either forages and/or concentrates are required. It should be remembered that there are groups of animals with differing energy requirements within the herd and these different groups should be managed separately if at all possible. In small herds, this can represent additional challenges but management systems can be designed to enable feeding of certain individuals with increased (or decreased) requirements separately. This should help to avoid the situation where lactating females are fed too little and lose weight while non-pregnant and non-lactating females gain weight through troughing the additional feed aimed at the nursing dams. Ideally pasture should be seeded with a mixture of grasses and legumes and also species that thrive in both the cool and dry/hot seasons. For breeders that make their own hay, they will need to be aware of the factors that affect hay quality in terms of the grass maturity and weather at time of cutting. Whether home-grown or bought-in hay is used, hay analysis will help determine how much additional supplementation will be required in order to meet energy demands of the various groups.

Farm layout affects grazing efficiency. Long thin paddocks tend only to be grazed at the front while shelter location within the paddock may also influence grazing distribution. With static grazing (as opposed to rotation grazing), there is greater pressure to remove manure because camelids will ignore lush grazing around dunging areas resulting in effectively less grazing space. Regular manure removal is also an invaluable tool for parasite control.

Stress

Stress is a big factor in camelid operations. Camelids are extremely sensitive to the effects of stress from transportation, changes in social structure, changes in environmental conditions including the weather and of course presence of disease.[4,5] Even short periods of transportation result in elevated cortisol concentrations while longer periods of transportation have resulted in extended periods of ongoing disease susceptibility. When alpacas are being grouped together for various management reasons, it's important to consider social interactions and try to keep certain animals together if they have a tight bond if possible as this will reduce social stresses. Stress can also result in pregnancy losses, particularly if that stress occurs in the first or last 60 days of pregnancy: therefore ideally handling should be minimized during these periods to optimize breeding success. This may be challenging for certain husbandry procedures that must take place during these times such as shearing.

Gastrointestinal Parasites & Coccidia

Parasitism is one of the biggest health problems affecting camelids worldwide. This is particularly the case under the more intensive grazing conditions found outside South America where conditions for parasite survival are optimal and there is increased exposure. Intensive use of anthelmintics for prevention of meningeal worm (location-dependent) or mange, and also inappropriate use associated with poorly considered parasite control programmes have resulted in problems of drug resistant parasites. The emergence of multi-drug resistant parasites have been reported among alpaca farms in the South-eastern United States.[6] Therefore, it is important that a more holistic approach is taken to parasite control that makes use of body condition scoring, faecal screening, judicious use of anthelmintics and anticoccidials where required, pasture management and biosecurity measures.

Co-grazing Species

Occasionally camelids are pastured alongside other livestock but neighbouring livestock can potentially also cause health issues in camelids. In the UK, bovine tuberculosis is an enormous problem in parts of the country and poor control has resulted in spillover of infection into camelids. In the US and elsewhere, BVD has been known to cause health issues including premature births, abortions, and persistent infections in cria that in turn will affect production of affected herds. Co-grazing with sheep can result in increased problems with parasite control while contact with horses has been associated with EHV and *Strep zooepidemicus* infections.

Biosecurity

Given the recent emergence of several infectious diseases and the unknown effects of these diseases on South American Camelids, it seems prudent that camelid breeders take stock of their approach to biosecurity in order to protect these animals from disease. Biosecurity is essential in herd health planning because the aim of biosecurity is to prevent the introduction and spread of disease in susceptible populations or herds of animals. Farms should be assessed from the point of view of biosecurity in any herd health plan with changes made to existing practices where appropriate.

HERD HEALTH PLANNING

Development of a herd health plan should be done in close consultation with individual herds, taking into account local and regional conditions and considering the key areas discussed above. This must include an on-site visit to observe existing practices or opportunities may very easily be missed. For example, owners may verbalise that they are doing something that theoretically sounds adequate when in practice the way that they are doing it is completely ineffective. It is thus also important to ensure that owners are competent in performing the various tasks that are set for them in their herd health plan.

Body Condition Scoring

Body condition scoring is the most important tool available for monitoring the health of animals in the herd. It allows assessment of the adequacy of nutrition provided but is also useful for picking up on herd level and individual animal health issues. This should be performed on a regular basis and records kept so that trends can easily be spotted. Herd management software is ideal for this purpose. It is ideal to also have weigh scales available to enable proper dosing of veterinary drugs and to monitor the growth rates of young animals.

Parasite Control Strategy

This is another essential component of any herd health plan. A holistic approach should be taken that considers body condition scoring, faecal screening, judicious use of anthelmintics and anticoccidials where required, pasture management and biosecurity measures.

Vaccinations

In most parts of the world, camelids require vaccinating against clostridial diseases. It may be necessary in certain areas to vaccinate for other diseases based on local disease prevalence, for example leptospirosis, rabies, bluetongue virus or West Nile Virus. There is very little data available about the best vaccination protocol to use and no products are licensed for use in camelids, so strategies are largely extrapolated from use in other species. Clostridial vaccine use is often advised pre-partum to optimise passive immunity in crias while others have suggested neonatal vaccination to minimise pregnancy loss that may be associated with handling in late pregnancy. The relative advantages and disadvantages of each strategy should be discussed with owners and fully-informed decisions made on the basis of the most appropriate strategy for an individual farm.

Pasture & Nutritional Management

For pasture management, the following factors should be considered: stocking density, availability of spare pasture, pasture layout, appropriateness of fencing and pasture cleanlinless (including whether or not manure removal is practiced and how frequently). The herd health plan will need to be adapted to the conditions found on each farm and take into consideration what the owners may be able to realistically achieve.

Body condition scoring will assess the adequacy of nutritional management. In addition, the herd health planning visit should assess adequacy of trough and hay feeding space, hay quality if applicable and also the availability of fresh clean water.

Inspection of the animals may suggest whether trace element nutrition is good enough, but it may be desirable to assess this by blood sampling and make changes to the nutritional plan accordingly.

Vitamin D Supplementation

A strategy should be provided for supplementation during the winter to all growing animals and adults. Crias generally require regular dosing during the winter depending on the method of supplementation chosen (oral or injectable) while adults generally only require one or two doses.

Biosecurity

A risk analysis should be performed for each farm taking into consideration animal movements, quarantine provision, farm borders, feed and water sources and storage, and arrangements for visitors. Recommendations can be made that are appropriate to the individual farm.

Other Husbandry Procedures

Owners should be aware of and have plans available for shearing, foot and dental care and also have protocols for handling animals that will minimise stress.

Contingency Planning

Although owners may follow the best laid plans as outlined above, it is inevitable that unexpected events may occur. This includes extremes of weather or unexpected disease outbreaks. It is important that the management systems in place will be able to cope with those types of events that may have *potential* to occur. Space should be available for accommodating sick animals and the amount of space should be appropriate to the number of animals in the herd. These pens should ideally be located on concrete for ease of disinfection.

REFERENCES

[1] Van Saun RJ, et al. J Am Vet Med Assoc 1996;209:1128. [2] Judson GJ, et al. Aust Vet J 1999;77:310. [3] Smith BB, et al. Am J Vet Res 2001;62:1187. [4] Bravo PW, et al. Anim Reprod Sci 2001;67:125. [5] Anderson DE, et al. Vet J 1999;157:189. [6] Gillespie, R.A., et al. VetParasitol 2010;172:168

Bovine TB in Camelids

Claire E. Whitehead, BVM&S, MS, DACVIM (Large Animal Internal Medicine), FHEA, MRCVS
Oxford, UK

INTRODUCTION

Tuberculosis has been reported in South American Camelids (SACs) due to *M bovis, M tuberculosis, M microti*, and *M kansasii*. Infections with *M avium* subsp *avium* and *M avium* subsp *paratuberculosis*, the causative agent of Johne's Disease, have also been reported in SACs.

Historically, SACs had been thought not to be particularly susceptible to tuberculosis. Only rare reports exist of TB in SACs in South America despite it being common in cattle, sheep and humans in the same areas. Outside South America, TB due to *M bovis* has been reported in the US and Canada, the Netherlands, Spain, Ireland, and UK, while *M microti* has been reported in Switzerland and has also been documented in the UK. The population of alpacas and llamas in the UK being kept as pets and also as commercial livestock has risen dramatically over the last 10-15 years. The number of registered alpacas doubled from just under 15,000 to over 30,000 in the 6 year period between 2006 and 2012. The vast majority of these alpacas are distributed around Central and Southern areas with nearly one third of the population in the South West. Over the same time period, the incidence of bovine TB in the UK has risen dramatically, particularly in the South West of England and Wales. This has clearly brought concentrations of camelids into areas that are endemically infected with *M bovis*.

The first reported case of TB due to infection with *M bovis* in the UK was in January 1999 when a female llama died following a period of weight loss. This female was necropsied: gross lesions included abscesses in the lungs, some in the liver, enlarged bronchomediastinal lymph nodes, and grape-like lesions over the thoracic parietal pleura, lungs and pericardium. *M bovis* was cultured and this matched the spoligotype found in cattle and badgers in the area. Curiously, another female llama died on the same day that had originated from this herd, but that llama had been sold the previous autumn to another farm: that female had similar lesions and *M bovis* was also cultured. Since 1999, the number of infected SAC herds has been increasing in line with cattle breakdowns, particularly since 2008 when there have been between 7 and 14 new infected herds per year. By far the majority of cases of bovine TB in camelids have been associated with "spillover" of infection from local animal reservoirs, either cattle or wildlife.[1] This has been confirmed by spoligotyping of cultured *M bovis* isolates.

Clinical Presentation of TB in Camelids

As prey species, camelids are particularly good at masking clinical signs of disease. Some animals have been reported to have shown no clinical signs of disease at all prior to death, even with severe lung lesions found at post-mortem: camelids are physiologically exceptionally tolerant to hypoxia. In general terms, TB should be considered a differential diagnosis for any camelid presenting with weight loss, with or without respiratory difficulty, particularly in areas endemically affected with TB.

Problems reported by owners with animals diagnosed with TB (regardless of causal agent) include weight loss sometimes to the point of emaciation, lethargy, anorexia, recumbency, and respiratory signs varying from increased respiratory rate to dyspnoea. Clinical identification of disease appeared to be the most consistent indicator of whether or not an animal turned out to have visible lesions at PM in one large llama outbreak.[2] It should be noted that most of these clinical signs are non-specific but may have value in identifying infected animals in breakdown scenarios.

In six llamas that were experimentally infected with *M bovis* via intratracheal challenge, progressive disease was observed in some animals with the first clinical signs noted at 62 days post-infection (PI) (coughing up blood followed by weight loss at day 66 PI) and with the first mortality occurring 68 days post-infection.[3] Interestingly, the single intradermal skin test (bovine PPD only) was positive in 4 out of 5 infected llamas at day 80 PI. Two further deaths occurred at days 136 and 143 PI. All 5 llamas had exhibited respiratory signs in the form of coughing (in some cases producing bloody sputum) by day 80 PI.

In one case series of 10 llamas and one alpaca affected with *M microti* TB, clinical signs included weight loss, recumbency, and anorexia in the late stages of disease: 6 of the 11 animals exhibited respiratory problems manifested by dyspnoea or tachypnoea (difficulty breathing or increased respiratory rate). Three of these animals also had enlarged submandibular lymph nodes.

Several papers have detailed ultrasonographic findings in camelids infected with *M microti* and *M kansasii*.[4,5] Findings included multifocal hyperechoic structures measuring up to 4cm diameter throughout the livers of all 8 animals examined by ultrasound in one study (*M microti*): two animals also had a large mineralised mass in the retropharyngeal area.[4] In one alpaca diagnosed with *M kansasii* tuberculosis, ultrasonographic evaluation was again able to show multiple hyperechoic lesions measuring 1-3cm in diameter. Given the frequent occurrence of liver lesions in camelids affected with TB and the ease of ultrasonographic evaluation of the liver in camelids, this may be a particularly useful diagnostic tool for clinical use.

Radiographic evaluation of camelids affected with TB has only been performed infrequently, despite the obvious involvement of the respiratory system. Two out of 3 alpacas radiographed in a Spanish outbreak of *M bovis* exhibited cavitary lesions of heterogeneous density with diffuse consolidation while the third alpaca had diffuse consolidation with micronodular and interstitial infiltrate.[6] Two alpacas in Ireland showed diffuse or patchy consolidation consistent with granulomatous, broncho-interstitial pneumonia.[7] Radiographic evaluation of the thorax of camelids infected with *M microti* revealed variable-sized thoracic masses in two animals while all 5 of those that were radiographed had increased interstitial and bronchial patterns over the entire lung fields.[4]

Bovine TB in Camelids - Pathology

By far the majority of bovine TB herd incidents in camelids are disclosed at post mortem examination. This is because there has been no routine surveillance for bovine TB in camelids to date. In the majority of cases, lesions are found in the lungs, pleura and associated lymph nodes. However, occasionally abdominal lesions have been more pronounced than those found in the lungs.[7] Lesions are also found in the liver and hepatic and mesenteric lymph nodes relatively commonly. They may be found virtually anywhere in the body but the superficial lymph nodes have only infrequently been affected. A single discharging skin lesion was reported in a solitary case with acid-fast bacilli identified in the exudate.[2] No mammary lesions have been reported in camelids and transmammary infection has not yet been documented.

Lesions in the lungs are typically multifocal granulomatous caseonecrotic lesions and these vary in size. The lung lesions may be coalescing and affected areas of lung are consolidated. Lesions are variably calcified. The pleura are often affected when the lungs are involved. Cavitating lesions have sometimes been described.[1,2] In humans, the presence of cavitation is recognised as a significant feature associated with the generation of further cases of TB by increasing the risk of transmission of infection.[2,8] In some reported camelid cases, caseonecrotic lesions have been observed to be communicating directly with the bronchial tree ("open TB") which would clearly increase the risk of transmission to other animals and potentially their owners.[6,7] This has led to concern amongst regulatory authorities in dealing with camelids. It has been suggested that in animal populations without active surveillance such as wildlife and camelids, cavitation may contribute to a potentially higher risk of onward transmission in these species. **(Cassidy, VetMicro 2006)** Having said this, it seems that camelids develop lesions more rapidly such that even with routine surveillance, these lesions may still be found.

At histopathology, intralesional acid-fast bacteria are usually seen and this may provide early stage support for a diagnosis of mycobacterial infection following identification of suspect visible lesions. Where visible lesions exist at PM, this tissue will be submitted for culture confirmation of mycobacteria and spoligotyping to enable epidemiological investigations. Where there are no visible lesions (NVL) in animals that have been culled due to a positive antemortem test, pooled lymph node samples are cultured.

Based on the available literature, it appears to be extremely uncommon for any camelid with visible lesions to culture negative: in other words, where lesions exist, camelids normally culture positive for *M bovis*. It also appears to be rare for *M bovis* to be cultured from NVL animals. However, sometimes environmental bacteria have been cultured from NVL llamas.[2] *M microti* has also been found in camelids in the UK. *M microti* can be just as clinically significant to the individual Camelid and to herd health so is just as important to identify as *M bovis*.

TB Testing

The sensitivity of intradermal skin testing for detecting antemortem TB infection in Camelids is fairly well recognised as being poor (14-20% based on published studies). In general, it appears that in practice where there is a positive skin test, usually there are visible lesions and *M bovis* is cultured. However, inconsistencies and poor sensitivity mean that any routine surveillance for bovine TB in these species cannot rely on skin testing alone. Therefore, a camelid-specific gamma interferon test and a number of serological tests were evaluated by the Animal Health & Veterinary Laboratory Agency (AHVLA).[9] From these studies in the UK, new recommendations have been formulated by the regulatory authorities that attempt to maximise sensitivity of testing in breakdowns and these will be presented later in these notes.

The sensitivity of the individual serological tests as assessed by the above-mentioned study is shown in the table below (Table 1).

Table 1. Serological test sensitivity and specificity, as shown by the AHVLA study.[9]

Serological Test	n	Sensitivity % [95%CI]	n	Specificity % [95%CI]
Stat-pak[a]	35/52	67.3 [54.5 – 80.8]	8/306	97.4 [95.6 – 99.2]
DPP[a]	30/52	57.7 [44.3 – 71.1]	10/306	96.7 [94.1 – 98.4]
IDEXX[b]	36/52	69.2 [56.7 – 81.7]	8/306	97.4 [95.6 – 99.2]
ENFERplex[c]	32/48	66.7 [53.4 – 80.0]	8/257	96.9 [94.8 – 99.0]

[a] Chembio Diagnostic Systems Inc., Medford, NY, USA; [b] IDEXX Laboratories Inc., Maine, USA; [c] Enfer Scientific, Co. Kildare, Ireland

No significant differences were found between the tests, and sample sizes are relatively small such that real differences may be masked. [One potential limitation of this trial was that the Enferplex test was a late addition to the trial such that samples that didn't test positive to Statpak were not subjected to Enferplex testing. It is therefore possible that the Enferplex test may in fact be more sensitive than the trial suggests since the Enferplex test uses 7 antigens compared with 3 for Stat-pak and 2 for the Idexx and DPP tests. Further research would be needed to show whether in fact this was the case and to provide improved estimates of sensitivities and specificities for all of these tests.]

When combinations of pairs of serological tests were used, this resulted in increased sensitivity to between 73 and 81% depending on the combination used (Table 2). Again, no significant differences were observed. When all 4 serological tests were used together, sensitivity only increased further to 83% while specificity reduced to 90% such that around 17% of infected animals in this study were not detected by any of the serological methods used.

Table 2. Serological test combinations showing sensitivity and specificity (AHVLA study).[9]

Serological test combination	n	Sensitivity % [95%CI]	n	Specificity % [95%CI]
Stat-pak/IDEXX	39/48	81.3 [71 – 91.6]	11/257	95.8 [93.3 – 98.2]
Stat-pak/Enferplex	37/48	77.1 [65.2 – 89.0]	14/257	94.6 [91.8 – 97.4]
Stat-pak/DPP	36/48	75.0 [63.6 – 86.4]	14/257	94.6 [91.8 – 97.4]
IDEXX/DPP	35/48	72.9 [61.1 – 84.6]	13/257	94.9 [92.2 – 97.6]
IDEXX/ENFERplex	37/48	77.1 [65.2 – 89.0]	12/257	95.3 [92.7 – 97.9]
Enferplex/DPP	35/48	72.9 [61.1 – 84.6]	15/257	94.2 [91.3 – 97.1]

Since this study was published, the Stat-pak and DPP tests have become unavailable. A Cervid-DPP test was proposed as a replacement for the DPP or Statpak tests to be used in breakdown testing: unfortunately only 15 test-positive samples remained from the original trial on which to test the new version. When the Cervid-DPP was used with the IDEXX test and the results interpreted in parallel (positive to either test returns a seropositive outcome), sensitivity was found to be 66.7% with specificity of 93% [95% CI 89.3 – 96.7].[10] No confidence intervals were provided for the sensitivity data because the data set is too small. Although again not statistically significant, the quoted specificity is worse for the paired Cervid-DPP/IDEXX and this will cause concern to breeders. In the author's opinion, the Cervid-DPP test has not been adequately validated to be offered as a viable alternative.

In the original study, the specificity of the camelid-specific γIFN test appeared to be compromised by its detection of *M microti*. Sensitivity could be increased to 80%, but specificity was only 80.9%. [The specificity of the cattle γIFN test is much better at 96.7% (GB Cattle Specificity Trial).] Using a different interpretation gave better specificity at 98.1% but sensitivity dropped to 34.6%. By maximising specificity to 100% for use in surveillance, sensitivity dropped off to only 16.4%. Therefore it was thought unlikely to be useful alone for routine surveillance in camelids. If the γIFN test was combined with single or dual serological testing, sensitivity could be maximised to greater than 90% but with corresponding specificity of only 75-80%. The authors of the study noted that since alpacas with visible lesions were identified originally only after γIFN and Stat-pak testing in breakdown herds that these combined sensitivity figures should be viewed with caution.

A different study in several herds of South American Camelids, both llamas and alpacas, evaluated the Stat-pak test alongside a multiantigen print immunoassay (MAPIA) to investigate serological responses in healthy animals (n=79) and those that were infected with *M microti* (n=8).[11] The MAPIA test used a panel of 14 mycobacterial antigens as compared with the Stat-pak's 3 (all of which were included in the MAPIA panel): Stat-pak exhibited a sensitivity of 63% and specificity of 90% while the MAPIA had sensitivity of 88% and specificity of 97%. Thus, the sensitivity of the Stat-pak in detection of *M microti* in camelids is similar to that for *M bovis* suggesting that culture confirmation of test-positive camelids is important for differentiation of the two types of infection. Furthermore, the inclusion of additional antigens was shown to enhance the sensitivity of the MAPIA as compared with the Stat-pak test suggesting that inclusion of further antigens in any test used for camelids would be advantageous. This is where the Enferplex test could carry potential since it is possible for this test to use up to 9 different mycobacterial antigens (it currently uses 7): further research could therefore define the most suitable antigens for detection of TB in camelids and allow potential differentiation of *M bovis* from *M microti* or indeed other environmental mycobacteria.

It is reported in cattle that sensitivity of serological testing can be boosted by prior skin testing (SCITT), known as "priming". This boosting effect was observed in 2 llamas (out of 14 VL llamas) that were found to be Stat-pak positive only 3 weeks after single comparative intradermal tuberculin testing, having previously tested negative at the time of skin testing.[12] [Unfortunately only 6 out of the 14 llamas were blood tested 3 weeks after skin testing: therefore this boosting effect was not statistically significant, presumably due to the low numbers of animals included.] In another large herd outbreak in Spain during which 41/67 alpacas that were necropsied were found to be infected with *M bovis* (out of a herd of 115 alpacas), the highest proportion of test-positive alpacas that proved to be culture-positive for *M bovis* was found when the single intradermal test (SIT) was used instead of the SICTT and when serological testing (using a novel ELISA based on a single mycobacterial antigen) was performed 15-30 days after skin testing with the tests being interpreted in parallel.[13] Based on their serological test alone, detection of culture-positive alpacas prior to tuberculin injection was only around 35% while detection using combination testing (in parallel) of SIT and serology at time of tuberculin injection increased to 68.7 to 90%: using the serological test on blood samples collected 15-30 days after tuberculin injection increased detection rate to between 76.9 and 100%. The effect of priming has not been evaluated in non-breakdown herds.

As a result of these findings, in the UK the Animal & Plant Health Agency (formerly AHVLA) is recommending that in confirmed TB breakdowns in camelid herds, skin testing (SIT) is employed in combination with paired serological testing (owners have a choice of two tests out of Enferplex, Cervid-DPP and IDEXX). The results of these tests will be interpreted in parallel such that a camelid is declared a reactor if it tests positive to either of the serological tests (and/or the skin test). Blood testing should take place 10-30 days after skin testing.

Surveillance Testing

In the UK, owners currently have their choice of the privately-available Enferplex test or Cervid-DPP/IDEXX combination test available at APHA. The Enferplex test is interpreted at 4-antigen level instead of the 2-antigen level used in the AHVLA study: this optimises the specificity at 99.99%, giving a reduced sensitivity of 55.1% (personal communication, MV Diagnostics). The test combination offered by APHA for voluntary testing uses the *quantitative* Cervid-DPP test interpreted in serial with the IDEXX (both tests must be positive for the alpaca to be reported as seropositive): the corresponding sensitivity and specificity of this combination test are 40% and 99.5% respectively.

Surveillance testing can be done at the individual animal level for pre-movement or pre/post-purchase testing, for pre-export testing or at the herd level to clarify freedom from infection. There is a Camelid Voluntary Health Scheme for Bovine Tuberculosis in the UK, operated by Sure Farm Ltd and this scheme uses the Enferplex test. For larger herds over 50 animals, a statistically-derived proportion of the herd is tested that allows for a threshold number of positive test results (at 2-antigen level) to be interpreted as inconclusive and retested. To date, voluntary participation in this Scheme has been minimal due to the expenses involved and the fear of false positive test results due to the poor level of compensation available. Unfortunately, there are other issues surrounding the control of bovine TB in general in the UK, including wildlife control and market-value compensation for cattle that are hampering voluntary involvement in this scheme.

REFERENCES

[1] Twomey DF, et al. Vet Rec 2009;165:121. [2] Twomey DF, et al. Vet Rec 2010;167:475. [3] Stevens JB, et al. Can J Vet Res 1998;62:102. [4] Zanolari P, et al. J Vet Intern Med 2009;23:1266. [5] Braun U, et al. Schweiz Arch Tierheilkd 2009;151:287. [6] Garcia-Bocanegra I, et al. J Clin Microbiol 2010;48:1960. [7] Ryan E, et al. Ir Vet J 2008;61:527. [8] Rodrigo T, et al. Int J Tuberc Lung Dis 1997;1:352. [9] Rhodes S, et al. AHVLA; 2012. [10] Rhodes S. APHA; 2014. [11] Lyashchenko KP, et al. Vet Microbiol 2007;125:265. [12] Waters WR, et al. Clin Vaccine Immunol 2011;18:1882. [13] Bezos J, et al. Prev Vet Med 2013;111:304.

Current Status of Anthelmintic Resistance in Gastrointestinal Nematodes of Cattle and Recommendations for Sustainable Control

Ray M. Kaplan, DVM, PhD, DACVM, DEVPC
Athens, Georgia

Overview of resistance situation in gastrointestinal nematodes of cattle

Beginning with phenothiazine in the 1950s, followed by the benzimidazoles in the 1960s, the imidazothiazole/tetrahydropyrimidines in the 1970s and the avermectin/milbemycins (AM) in the 1980s, a new class of anthelmintics was introduced into the marketplace each decade. This arsenal of highly effective and relatively inexpensive drugs led to recommendations for parasite control that were based almost solely on the frequent and or strategic use of anthelmintics, the goals of which were to maximize livestock health, productivity, and profitability. Though this approach was highly successful for a number of decades, we are now experiencing ever-increasing levels of anthelmintic resistance in all drug classes, involving virtually all of the most economically important parasites of all livestock species. Resistance in parasites of cattle was slower to develop than in the small ruminant and equine sectors, but over the past decade we have seen a rapid escalation in the levels and distribution of anthelmintic resistance in parasites of cattle worldwide.

Though there are some published case reports of resistance in parasites of cattle in the US[1,2], no studies have been performed to establish the national prevalence of resistance. Thus, we do not know how severe and widespread the problem is nationally. However, studies performed by my laboratory on a number of cow-calf farms in Georgia and on stocker cattle purchased at various stockyards in the southern region suggest that AM resistance in cattle is both common and widespread. In fact, we have not tested a farm in the last 5 years that did not have AM-resistant *Cooperia*, other than a single organic farm with a long history of no anthelmintic use. Resistance in *Cooperia* spp. and *Haemonchus* spp. are the most common, but we also have seen evidence in other species as well. Outside the US, there is a large amount of published data indicating that resistance is becoming a very serious problem; a study in New Zealand reported that ivermectin resistance was evident on 92% of cattle farms and resistance to both ivermectin and albendazole was evident on 74% of farms.[3] More recently resistance in *Ostertagia ostertagi* has been found on numerous New Zealand beef farms (Dr. Dave Leathwick, AgResearch, NZ, 2015). Very high prevalences of resistance have also been reported in studies performed in Brazil, Argentina and Australia.[4-6] *Cooperia* is consistently the species with the most resistance, but resistance in *Haemonchus* is also common. Resistance in *Oesophagostomum* and *Ostertagia* are reported less commonly, but recent evidence suggests increases in these species as well.

Historically *Cooperia* was not considered a very important pathogen. However over the past few decades, as a consequence of heavy use of AM drugs, the relative intensity of *Cooperia* compared to other species has risen substantially. Though *Cooperia* does not impact animal health and productivity to the degree that *Ostertagia* does, a recent study confirmed that *Cooperia* infections do have a significant negative effect on growing cattle.[7] So, although clinical disease in cattle due to *Cooperia* may be uncommon, there is little doubt that significant production losses can result from high levels of infection. Consequently, there is little evidence to support the opinion of some that AM resistance in *Cooperia* is not a major concern.

Luckily, resistance in *Ostertagia* is not yet a major problem in most of the world. However, recent evidence suggests that this might be in the process of changing. Should avermectin/milbemycin resistance emerge at high levels in *Ostertagia*, the problem of resistance in cattle parasites will reach a new level of importance and concern, as *Ostertagia* is a highly pathogenic species that can produce not only production loss, but also severe clinical disease and occasional deaths. The problem of anthelmintic resistance also needs to viewed with an eye to the future. No new classes of anthelmintic have been introduced for use in cattle since ivermectin in 1981, 35 years ago. Other second generation AM drugs have provided some improvements since then, but AM resistance demonstrates a class effect; resistance to any one AM drug tends to confer resistance to all AM drugs. Additionally, no new novel classes of drugs have become available in the US over this time. The new drug monepantel (Zolvix®; Elanco) is sold throughout much of the world for sheep, but this drug has not yet been approved for use in the US, and it is unknown when or even if it ever will be. Currently there are no other new anthelmintic prospects in the late phase pipeline, thus, we are left in a situation where it could be a long while before a new anthelmintic class is sold for cattle. This makes it important that the efficacies of currently available products are protected as much as is reasonably possible.

Given this situation, the problem of anthelmintic resistance in parasites of cattle should not be ignored. Clearly, there is a great need for new research to address this issue, but waiting for this research before acting is not advisable. It is recommended that anthelmintic resistance in parasites of cattle be considered a major threat to cattle productivity, and that steps be taken to mitigate the potential problems resistance can cause. Because almost no research has been done in this area, no one can say for sure what are the best approaches to reduce the rate with which anthelmintic resistance evolves in cattle. However, it would seem logical to follow some of the recommendations for sheep, which are based on sound research. To do nothing seems irrational and short-sighted.

Strategies for mitigating the problem of anthelmintic resistance in cattle

There are several approaches that have proven effective in reducing the rate with which resistance develops in sheep nematodes: (1) using drug combinations (two or more active compounds from different drug classes administered at the same time), (2) leaving the heaviest 10% of the flock untreated, (3) treating selectively based on some measure of parasitism or growth rate, (4) not treating the ewes and only treating the lambs.[8-10] However, cattle are not sheep – thus some will be more difficult to implement in cattle, and some may be less effective in cattle. Still, some of these are easily adapted for cattle and are likely to be effective. For instance, not treating cows will not provide as much benefit as not treating ewes. This is due to differences between the species of importance and host-parasite interaction. In sheep, both ewes and lambs share the same parasite species in similar proportions, and periparturient ewes have relatively high EPG. Thus not treating the ewes provides a source of refugia of the same species that infect the lambs. However, there is no perparturient rise in EPG in cattle, thus the relative level of egg shedding in cows remains low so long as nutrition is adequate. Furthermore, cows usually have very good immunity to *Cooperia*, and are predominantly infected with *Ostertagia*, whereas calves are infected primarily with *Cooperia*. Consequently, not treating cows will not have a major impact on the resistance levels in *Cooperia*. Still, this practice may be quite beneficial for slowing the development of resistance in *Ostertagia,* which is of growing concern. Quite simply – if *Ostertagia* develop resistance at high levels, the cattle industry will suffer great economic losses, and the health and welfare of cattle will suffer. Thus, implementation of any practices that can reduce this likelihood is a good idea, so long as it does not cause significant productivity losses now.

I have often heard people say that cattlemen will not leave some animals untreated since this goes against what they have been told for years, and goes against their common sense of what is best for maximizing productivity. This may sound reasonable at first, but how many of these same cattlemen are currently using anthelmintics that are poorly effective without knowing it? Based on my experience in testing cattle farms for resistance, this may be the majority of farms. So – yes they are treating all animals with the full associated costs, but they are not getting a highly effective result, and in some cases getting almost no benefit. I have tested several farms that had 0% reduction in FEC, and the cattleman had not suspected resistance at all prior to the test. Studies in sheep have clearly demonstrated that the production cost of subclinical parasitism as a result of using an anthelmintic product that is less than fully effective due to resistance can greatly exceed the cost of routine testing of anthelmintic efficacy.[11] Cattle farmers would thus be much better off in the present, and have greater sustainability for the long term, if they used effective anthelmintic treatments and left some animals untreated. Testing the efficacy of drugs with a FEC reduction test (FECRT) is the best way to make sure they are using effective drugs. The reluctance of most cattlemen to test for anthelmintic resistance is not rational from an economic perspective.

Recent research has also demonstrated quite clearly that the use of anthelmintics in combination is a beneficial practice. In fact, in Australia and New Zealand there are few products sold as single actives; most products contain 3, 4, or 5 different anthelmintic classes (note that they have some anthelmintic classes that are not available in the US). There are 3 major benefits to using drugs in combination: (1) one gets an additive effect with each drug used, thus the efficacy of the treatment increases, sometimes dramatically (Table 1). (2) Provides broad-spectrum efficacy; resistance is species and drug specific, thus a second (or third) drug may kill any species resistant to the first drug. This will then return the broad-spectrum result that one aims to achieve (and that is specified on the product label). (3) By achieving a higher efficacy, there are fewer resistant survivors, thus there is a greater dilution of resistant worms by the susceptible portion of the population. For example, if 2 drugs each with 90% efficacy are used in rotation, then each time cattle are treated 10% of the worms (resistant) survive. In contrast if the 2 drugs are used in combination then the efficacy would be 99%; this yields10X fewer resistant survivors (first drug kills 90%, second drug kills 90% of the remaining 10%).

Thus, cattleman should perform a FECRT to determine which drugs are effective and then knowing this, they should optimally use 2 drugs in combination. Additionally, they should leave 10% untreated (selected from best looking animals; upper quartile) to provide un-treated refugia. Using this new approach, they will be getting a highly effective treatment in most of the herd, which will greatly diminish egg shedding thus reducing subsequent pasture contamination and re-infection. And by leaving some animals untreated they will be sustaining a drug-susceptible refugia, which will dilute out the small number of resistant worms that survive the treatment, thus maintaining a predominantly drug-susceptible worm population. The production loss in the 10% that are untreated is likely to be small because these were in the upper quartile of animals before the treatment, and so their growth was apparently not being heavily impaired by parasites. Studies in sheep comparing productivity of groups where both traditional and targeted selective treatment programs were used demonstrated no significant differences in growth of lambs.[12,13] By using effective drugs (and preferably combinations of drugs) and managing refugia by leaving some animals untreated is highly likely to improve overall herd productivity, and the susceptibility of the worms to the drugs will be sustained much longer into the future. This approach is likely to be even more effective when using new long-acting formulations, though if long-acting formulations are used without managing refugia, resistance might be accelerated. Research is needed to investigate this issue more fully.

In summary, given the current situation, it is recommended that cattle farms not assume high efficacy of their treatments and rather they should test for drug resistance using the fecal egg count reduction test (FECRT). Doing so will allow farms to make future treatment decisions based on the knowledge of which drugs are effective and which are not. Failing this, they should assume they have resistance and move immediately to the use of anthelmintic combinations. It is also important to appreciate that resistance develops slowly over many years and is undetectable during this time, but then suddenly reaches clinically detectable levels rapidly. Because the last phase of resistance development can happen quite quickly the FECRT should be repeated every few (2-3) years. Currently there are no published standards for performing the FECRT in cattle, so there are different things being recommended by different veterinarians and parasitologists. This is a problem because an improperly designed or analyzed FECRT can result in erroneous conclusions. Thus it is important that clear guidelines for the FECRT in cattle be established; such guidelines are currently in preparation by the World Association for the Advancement of Veterinary Parasitology (WAAVP). Nevertheless, the guidelines below should closely resemble the final published protocol, thus I recommend the following procedures be used until such guidelines are published.

Table 1: Impact of using anthelmintics in combination on the efficacy of treatments. Note the increases in efficacy is due to an additive effect.

Drug 1 (%)	Drug 2 (%)	Drug 3 (%)	Combination (%)
80	80		96
80	80	80	99.2
90	90		99
90	90	90	99.9
60	95		98
60	60	95	99.2
99	99		99.99

Recommended (RMK) Procedures For Performing A FECRT In Cattle:

- 15 animals per treatment group should be sampled
  - a minimum of 10 cattle should be used, but 20 is preferred when EPG are low
  - Each animal selected for testing should be uniquely identified
  - The same cattle must be sampled for FEC both pre and post treatment
- Cattle selected should be relatively uniform in age, breed, grazing history, anthelmintic exposure, other management considerations, etc.
- Weaned animals <16 months of age are preferable
  - FEC of adult cows are generally too low to perform a FECRT (see note below)
- If goal is to unequivocally detect drug resistance then only use injectable or oral formulations
  - pour-ons have high animal-to-animal variability and many sources of application error that can lead to reduced efficacy not due to resistance
- If goal is determine if the currently used pour-on product is effective then by all means it should be tested. However, border-line results should be interpreted carefully with regards to resistance.
  - If testing a pour-on it is critical that all cattle in the pasture be treated with the pour-on
    - Normal licking behavior by cattle will cause drug transfer to any untreated animals, decreasing the dose absorbed by treated cattle
- Weigh each animal so proper dose is given
  - If each animal is not weighed and dosed individually, then all cattle should be dosed to the heaviest in the group.
  - Never dose to the average weight, as this results in ½ the animals being under-dosed
- Use label recommendations for dosage and application
- Ensure dosing equipment is calibrated and operating properly
- Fecal samples should be collected per rectum at time of Tx and 14–21 days post-Tx depending on the drugs tested.
  - Non-AM drugs (e.g. benzimidazoles or levamisole): 10-14 days
  - Ivermectin and other avermectin drugs: 14-17 days
  - Moxidectin: 17-21 days
  - If Non-AM and AM tested at same time then use: 14 days
  - Inserting two fingers into the rectum and gently massaging the dorsal rectum will stimulate most cattle to deposit feces into your hand; active or forceful collection is rarely needed.
- Data can be analyzed one of two ways
  1. Compare pre-Tx and post-Tx FEC of individual cattle
     - [(PreTx FEC – PostTx FEC)/PreTx FEC] *100

- Then calculate the mean % reduction for the group

2. Compare mean pre-Tx FEC and post-Tx FEC for the entire group of cattle
    - [(Mean PreTx FEC – Mean PostTx FEC)/Mean PreTx FEC] *100
    - This method is easier and is recommended unless a full statistical analysis is planned.

- Interpretation of egg count reduction
    - NOTE: due to many sources of variability the accuracy of results of FECRT are highly dependent on the number of animals tested and the number of eggs counted in the pre-treatment FEC. With few animals and low EPG there is a possibility of the observed results being different from the true situation.
    - To properly interpret results of a FECRT it is necessary to calculate 95% confidence intervals, however, to simplify interpretation the following guidelines can be used
        - >95% = effective, no evidence of resistance
        - 90-95% = reduced efficacy, suspected resistance in early stages
        - 80-90% = reduced efficacy, highly suspicious of resistance
        - <80% = ineffective, resistance is highly likely
    - Of course, efficacy data can only be correctly interpreted if the following are true: (1) cattle were treated with the proper dose, (2) cattle were treated using proper administration technique, (3) drug used was within the expiration date and was stored properly, (4) cattle were positively identified, (5) the same cattle were sampled both pre and post treatment, (6) fecal samples were labeled and stored correctly, (7) proper lab technique was used when performing FEC
- Use a FEC technique with sensitivity of 5 EPG or less
    - Mini-Flotac (and Fill Flotac) is my recommended procedure.
        - New method that resembles McMaster but provides a higher level of detection sensitivity and better precision
    - Modified Wisconsin can also be used, but testing in our lab shows it is both more time consuming and less accurate than the mini-Flotac
    - NOTE: FEC sensitivity that is required will depend on the magnitude of the EPG in the group being tested.
        - For the most accurate result it is recommended that at least 200 eggs be counted pre-treatment. However, so long as at least 100 eggs are counted, results will be accurate more often than not.
        - Thus if using a technique with 5 EPG detection, then the mean EPG will need to be 100 if 10 cattle are tested and 65 is 15 cattle are tested. If mean FEC is <50 EPG, then there are 3 options:
            a. Use a procedure with a more sensitive detection level.
            b. Repeat the FEC or read an additional chamber for each animal
            c. Test more animals
        - If mean FEC are >500 EPG (group size treated = 10 cattle) or >250 EPG (group size treated = 20 cattle), then a McMaster can used.
        - Adult Cattle: adult cattle tend to have very low EPG, often averaging less than 2 EPG. Thus, it is extremely difficult to perform a FECRT in adult cattle with accuracy unless many cows are tested and a very sensitive FEC method (1 EPG or less) are used.

Recommendation for control of gastrointestinal nematodes in cattle

There is no such thing as a one size fits all parasite control program. Parasite control programs must be tailored to the farm – differences in climate, weather, type of cattle, grazing management, cattle/milk prices, etc. will impact decisions for what constitutes the optimal control program for a farm. Selection of drugs may be based on many factors, the most important of which are: label efficacy against the targeted species/stages, what the expected prevalence of anthelmintic resistance is in your region, whether drug efficacy testing (using FECRT) has been performed on the farm, type of cattle, time of year, cost and convenience. Gastrointestinal helminths usually do not cause clinically apparent disease in cattle, but rather produce subclinical disease that is measured as lowered productivity. When clinical disease does occur, it is most likely seen in young animals less than 2 years old. This is true for both the GI nematodes and liver flukes.

Though anthelmintics can produce important improvements in the productivity of cattle, total reliance on anthelmintics for parasite control is no longer recommended, and will not be sustainable into the future, as drug resistance becomes a more serious problem. There are numerous anthelmintic products currently available for use in cattle. Many of these drugs are both highly effective and convenient to administer, making treatment of parasites easier than ever. However, many of these drugs are expensive and may not be working at the levels they did in the past; money can be lost to unnecessary or ineffective drug treatment just as easily as to parasites. And there is always the risk of losing the ability to effectively control parasites in the future should anthelmintic resistance develop due to heavy use of these drugs. With this is mind it is important to distinguish treatment from control. Treatment with an effective drug kills those parasites infecting animals at the time of treatment (and for a short period after treatment if the drug has residual activity). However, depending upon many factors, removal of these parasites may or may not have a significant health and/or economic benefit. On the other hand, control of parasites implies developing strategies to prevent cattle from becoming infected with economically and/or clinically important levels of infection. This most often relies on the strategic use of anthelmintics at times of the year when treatments will have the greatest benefit. This brings parasite control into the realm of cost-benefit decision-making and requires knowledge of the epidemiology of the important parasites. However, strategic control when used optimally, places strong selective pressures for the development of anthelmintic resistance. Thus, strategic control should be used together with resistance-mitigating strategies outlined above.

The clinical disease syndrome produced by gastrointestinal nematodes is referred to as parasitic gastroenteritis. This can be a very serious disease that is characterized by weight loss, diarrhea, submandibular edema, and anemia. It is caused by a variety of nematodes infecting the abomasum and intestine (primarily of the family trichostrongylidae), but *Ostertagia ostertagi* (abomasum) historically has been the most important. Other common species include *Haemonchus placei* (abomasum), *Trichostrongylus axei* (abomasum), and *Cooperia* spp. (small intestine). Less common and important genera are *Oesophagostomum* and *Nematodirus*, but these too can cause diseases if present in high enough numbers.

Historically, *Ostertagia ostertagi* was the most common, pathogenic, and economically important parasite of cattle throughout most of the temperate world. In contrast, *Cooperia* spp. were not viewed as being a very important parasite of cattle. However, this view is changing; emerging avermectin/milbemycin (macrocyclic lactone) resistance in *Cooperia* spp., particularly *C. punctata*, is raising concerns regarding the pathogenic potential and economic impact of this parasite species. And *Haemonchus placei* is known to be pathogenic if infections reach certain thresholds, and resistance to this species is also on the rise.

Immunity to GI nematodes is slow to develop and requires several years of grazing to reach adequate levels. Therefore, when developing appropriate parasite control programs it is important to consider the age and use of the cattle. Healthy cows (older than 3 years) that are in good body condition and on a good nutritional plane tend to have fairly strong immunity to GI nematodes. In this class of cattle, when otherwise healthy, clinical disease from GI parasites will almost never be seen. However, subclinical reductions in production are still likely. On the other hand, young recently weaned calves in their first year at grass are highly susceptible to parasitism. The potential for high fecal egg counts (>500 epg) combined with the production of a large amount of feces due to their increasing body size, make this group of animals at greatest risk for production loss and clinical disease.

Favorable climatic conditions for hatching of nematode eggs and development of trichostrongylid larvae include moisture and moderate temperatures ranging from 40 to 85 degrees F. Temperatures beyond either extreme are often fatal (too high) or inhibitory (too low). During hot times of the year moisture then will be the limiting factor that determines the magnitude of infective larvae development. In contrast, once the infective L3 stage is reached, a different set of climatic conditions are favorable for continued survival. L3's persist longest under moist, cool conditions and are rapidly killed by high temperatures, freeze-thaw cycles, and desiccation. Thus, optimal periods of transmission will vary among regions of the USA. This will then impact the optimal times for administering anthelmintics.

Anthelmintic Therapy in Cattle

There are 3 primary classes of anthelmintics available for use in treatment of helminth infections in cattle: (1) benzimidazoles (BZ), (2) imidazothiazoles/tetrahydropyrimidines (I/T) also referred to as membrane depolarizers, and (3) avermectin/milbemycins (AM) (also referred to as macrocyclic lactones and macrolide endectocides) All 3 major anthelmintic classes are broad spectrum nematocides, but effectiveness against other groups of parasites varies widely. Cattle are typically infected with multiple species of helminth parasites that differ in their susceptibilities to the different anthelmintic drugs, but most of the commonly used drugs are effective against the most economically important species (if not resistant). This is important because some species of parasites are much more pathogenic than others and therefore are much more important to target with treatment. It can generally be assumed that virtually all cattle will benefit from an effective anthelmintic treatment; but the magnitude of the health and cost benefit is related to the timing of treatment in relation to the parasite transmission dynamics and the management system. Appropriate meat and milk withdrawal times must be followed for all anthelmintics administered to food producing animals and anthelmintics not labeled for use in dairy cattle should not be administered to dairy cattle of breeding age or older.

Classes of Anthelmintics

Benzimidazoles (BZ): Benzimidazoles are broad spectrum nematocides, with albendazole also having activity against tapeworms and flukes, and fenbendazole being effective against tapeworms and Giardia. Many BZ kill larval stages of nematodes as well as the adults, but often larval efficacy is lower than for adults. In some cases increasing the dose and/or duration of treatment (e.g. treat for multiple days) will increase efficacy against larval stages. BZ also have moderate to good activity against hypobiotic (arrested) larvae of Ostertagia - but this efficacy is considerably lower than that seen with the avermectin/milbemycin drugs. BZ are very safe; the toxic dose is 10 to 100 times the therapeutic dose depending upon the drug. Fenbendazole is approved for use in lactating dairy cattle. BZ must be administered orally only, but are available in numerous formulations such as feed premix, bolus, drench, paste, mineral/protein block, and pellets. Albendazole is teratogenic in mice but little evidence of teratogenocity exists in ruminants. Nevertheless, there is a label warning to not use in cows during the first 45 days of gestation. Though drug resistance is reported, the serious problem of BZ resistance seen in parasites of small ruminant and horses has not yet emerged as a problem in cattle. This may to due largely to the fact that BZ have been used much less frequently in cattle than the avermectin/milbemycin drugs over the past 30 years. BZ drugs are non-persistent, rapidly metabolized; however, a rumen reservoir and gut recycling leads to prolonged gut levels of BZ in ruminants (but still short duration, <24 hr). Because of this, efficacy can be improved by restricting feed intake for 24 hr before treatment. This increases digesta residence time by decreasing digesta transit rate, which in turn increases drug availability and contact time of drug and parasite.

Membrane Depolarizers: This group is comprised of two chemically unrelated groups of drugs having very similar mechanisms of action. These drugs act as cholinergic (nicotinic) agonists causing paralysis of worms. Levamisole (imidazothiazole) is a broad spectrum nematocide, but has no effect on any other groups of parasites, and is not effective against hypobiotic (arrested) Ostertagia larvae in cattle. Similar to the BZ, this drug is very short acting, but may have added benefits as an immune stimulator by potentiating T-cells, and stimulating phagocytosis in monocytes. Safety can be an issue with levamisole; toxicity appears at 3-5 X the therapeutic dose. Overdose of levamisole resembles organophosphate poisoning, with transient ataxia, salivation/muzzle foaming, and muscle fasciculations as common symptoms. Levamisole also may potentiate organophosphates, especially in Brahman cattle. Because of this it is best to not use levamisole in debilitated animals or in animals also being treated with organophosphate insecticides, but levamisole is considered safe in pregnant cattle so long as it is dosed appropriately. Morantel (tetrahydropyrimidines) is a broad spectrum nematocide that is approved for use in lactating animals. In contrast to levamisole, morantel is very safe & gentle with no contraindications for debility, pregnancy, or age. Presently, morantel is available only as feed premix for cattle making accurate and uniform dosing a little more challenging.

Avermectins/Milbemycins (AM, Macrocyclic lactones, Macrolide endectocides): Avermectin/Milbemycins are derivatives of naturally occurring antibiotic like compounds secreted by soil dwelling bacteria of the genus Streptomyces; presumably as a defense against bacterial feeding soil nematodes. Drugs in this group are broad spectrum, being effective against most nematodes and arthropod parasites. They have excellent efficacy (>99%) against both adult and larval stages of virtually all trichostrongyle parasites including arrested (hypobiotic) larvae (if not resistant). Efficacy against *Nematodirus* spp. and *Trichuris* is often less than that for most other worm species, but still should be >90% if the parasites have not developed drug resistance. AM resistance has been reported in all major species of parasites infecting cattle, but is most prevalent in *Cooperia* and *Haemonchus*. AM drugs have excellent efficacy against most ectoparasites and fly larvae (myiasis), although drugs in the avermectin group (ivermectin, eprinomectin, doramectin) are more potent against arthropod parasites than those in the milbemycin group (moxidectin). AM Drugs are highly lipophilic and provide residual activity against reinfection of many parasites for periods of time that vary depending upon the worm species.

Though commonly administered as pour-ons with a dose of 0.5 mg/kg, the bio-availability is superior when administered orally or by injection at a dose of 0.2 mg/kg. Though very convenient, and thus popular with cattlemen, pour-ons are a very poor way to administer an anthelmintic. It is well established that pour-on formulations produce high variability in plasma drug concentrations between animals, due to both allo- and self-licking.[14,15] In fact, drug exchange can be quite high; in one study untreated animals ingested up to 27.4% of a pour-on dose.[13] In another study, 4 calves were treated with ivermectin pour-on and housed together with 6 untreated calves. Percent reduction in FEC of the untreated calves ranged from 0% (one heifer) to approximately 95% (one heifer), with intermediary values from approximately 30% to 80% in the four other untreated animals.[16] In addition, haircoat type and length, soiling of haircoat, and poor application technique can all lead to reduced efficacies not related to drug resistance. Because of these delivery and pharmacokinetic (PK) issues, the pour-on route is less forgiving to dosing errors and this often leads to reduced efficacy, even when resistance is not present. For example, poor efficacies were observed following administration of a pour-on formulation of doramectin in Highland calves, whereas subsequent treatment with injectable doramectin yielded 100% reduction in FEC.[17] Furthermore, a recent study in New Zealand demonstrated superior efficacy when AM drugs were administered orally as compared to injectable and pour-on routes.[18] Though oral administration is the least convenient, it generally provides the best efficacy because that is where the worms are. There is little metabolism of AM drugs - most of the drug is excreted unchanged in the feces.

Considerations for Treatment and Control of Gastrointestinal Nematodes in Cattle

As mentioned above, there is no such thing as a one size fits all parasite control program, and parasite control programs must be tailored to the farm. Nevertheless, there are some general recommendations that will apply in most situations. Beef cows in poor body condition due to sub-optimal winter nutrition should be given a treatment in the late winter. This treatment is usually best given just before calving, but of course optimal timing will vary depending on the time frame of calving. Once new grass growth begins to occur, a return to improved nutrition will allow the cow's immune system to take back control, and the cow will limit its own worm burden. In dairy cows there is no consensus on whether treatment is economically justified. The more time dairy cows spend grazing, the more likely a treatment at calving will be of economic benefit. However, cows housed in confinement systems where they have little access to pasture may not gain much benefit from treatment. Due to the high cost of deworming 1500 lb dairy cattle, confinement operations may optimally do an on farm study to determine if deworming is beneficial. This is easily accomplished by matching pairs of cows and milking heifers based on age, production and calving date, and randomly deworming one within the pair, and leaving the other untreated. Then after a year milk production data can be compared to see if deworming led to a significant increase in milk production.

Stockers and replacement heifers are at the greatest risk for production loss from parasites, and use of anthelmintics will be important to their health and production. However, optimal strategies for applying those anthelmintic treatments will vary greatly, depending on the management and grazing system. Thus, it is not possible to provide a general recommendation. In general, strategic treatments, resistance-mitigating approaches, and sound pasture management must be used together.

Finally, it should be understood that any recommendation given will not be uniformly accepted by all parasitologists and veterinarians who work with cattle. There is no single best worm control program, and there is plenty of room for disagreement among experts; the best and most cost effective strategy will differ from region to region and farm to farm depending on many factors. And what is optimal today will not be optimal tomorrow. Constant vigilance to changes in the host-parasite-environment dynamic (including the emergence of drug resistance) is required. Finally, as noted above, the risk/problem of anthelmintic resistance must be considered. Drug resistance is not something that only happens to ones neighbors; it is extremely common and worsening all the time.

REFERENCES

1. Edmonds MD, Johnson EG, Edmonds JD. Anthelmintic resistance of Ostertagia ostertagi and Cooperia oncophora to macrocyclic lactones in cattle from the western United States. *Veterinary Parasitology* 2010;170:224-229.
2. Gasbarre LC, Smith LL, Hoberg E, et al. Further characterization of a cattle nematode population with demonstrated resistance to current anthelmintics. *Veterinary Parasitology* 2009;166:275-280.
3. Waghorn TS, Leathwick DM, Rhodes AP, et al. Prevalence of anthelmintic resistance on 62 beef cattle farms in the North Island of New Zealand. *New Zealand Veterinary Journal* 2006;54:278-282.
4. Suarez VH, Cristel SL. Anthelmintic resistance in cattle nematode in the western Pampeana Region of Argentina. *Veterinary Parasitology* 2007;144:111-117.
5. Soutello RGV, Seno MCZ, Amarante AFT. Anthelmintic resistance in cattle nematodes in northwestern Sao Paulo state, Brazil. *Veterinary Parasitology* 2007;148:360-364.
6. Rendell DK. Anthelmintic resistance in cattle nematodes on 13 south-west Victorian properties. *Australian Veterinary Journal* 2010;88:504-509.
7. Stromberg BE, Gasbarre LC, Waite A, et al. Cooperia punctata: Effect on cattle productivity? *Veterinary Parasitology* 2012;183:284-291.
8. Leathwick DM, Waghorn TS, Miller CM, et al. Managing anthelmintic resistance – Use of a combination anthelmintic and leaving some lambs untreated to slow the development of resistance to ivermectin. *Veterinary Parasitology* 2012;187:285-294.
9. Bartram DJ, Leathwick DM, Taylor MA, et al. The role of combination anthelmintic formulations in the sustainable control of sheep nematodes. *Veterinary Parasitology* 2012;186:151-158.
10. Leathwick DM, Miller CM, Atkinson DS, et al. Managing anthelmintic resistance: Untreated adult ewes as a source of unselected parasites, and their role in reducing parasite populations. *New Zealand Veterinary Journal* 2008;56:184-195.
11. Miller CM, Waghorn TS, Leathwick DM, et al. The production cost of anthelmintic resistance in lambs. *Veterinary Parasitology* 2012;186:376-381.
12. Kenyon F, McBean D, Greer AW, et al. A comparative study of the effects of four treatment regimes on ivermectin efficacy, body weight and pasture contamination in lambs naturally infected with gastrointestinal nematodes in Scotland. *International Journal for Parasitology: Drugs and Drug Resistance* 2013;3:77-84.
13. Busin V, Kenyon F, Laing N, et al. Addressing sustainable sheep farming: Application of a targeted selective treatment approach for anthelmintic use on a commercial farm. *Small Ruminant Research* 2013;110:100-103.
14. Laffont CM, Alvinerie M, Bousquet-Melou A, et al. Licking behaviour and environmental contamination arising from pour-on ivermectin for cattle. *International Journal for Parasitology* 2001;31:1687-1692.
15. Bousquet-Melou A, Mercadier S, Alvinerie M, et al. Endectocide exchanges between grazing cattle after pour-on administration of doramectin, ivermectin and moxidectin. *International Journal for Parasitology* 2004;34:1299-1307.
16. Bousquet-Melou A, Jacquiet P, Hoste H, et al. Licking behaviour induces partial anthelmintic efficacy of ivermectin pour-on formulation in untreated cattle. *International Journal for Parasitology* 2011;41:563-569.
17. Sargison N, Wilson D, Scott P. Relative inefficacy of pour-on macrocyclic lactone anthelmintic treatments against Cooperia species in Highland calves. *Veterinary Record* 2009;164:603-604.
18. Leathwick DM, Miller CM. Efficacy of oral, injectable and pour-on formulations of moxidectin against gastrointestinal nematodes in cattle in New Zealand. *Veterinary Parasitology* 2013;191:293-300.

The Science Behind Alternative Endoparasite Control Strategies for Small Ruminants

Lisa H. Williamson, DVM, MS, DACVIM
Athens, GA

INTRODUCTION

Gastrointestinal nematode parasites are the primary health threat on many small ruminant farms, especially in temperate parts of the world. Management of extremely pathogenic nematodes such as *Haemonchus contortus* has become increasingly difficult over the past 15 years as a result of multi-drug anthelmintic resistance, which continues to increase in both prevalence and magnitude. Total anthelmintic resistance is already a reality on many sheep and goat farms, and ever more farms are down to their last working anthelmintic. Consequently, non-pharmaceutical management practices, and use of alternative treatments that minimize parasitic burdens in sheep and goats have become an important area of research. The science supporting use of alternative control methods such as copper oxide wire particles, the condensed tannins found in *Sericea lespedeza*, and *Haemonchus contortus* vaccination will be reviewed.

COPPER OXIDE WIRE PARTICLES (COWP)

After administration, copper oxide wire particles mix with ingesta in the rumen, and subsequently adhere to the abomasal mucosa. Copper slowly dissolves from the particles in the acidic abomasal environment. Ionic copper is absorbed in the intestinal tract, and any excess absorbed copper is stored in the liver. Copper oxide wire particles are currently marketed as Copasure® for cattle (15 gram boluses), and small ruminants (2 and 4 gram boluses) to alleviate copper deficiency. Over 15 years ago, Bang et al. reported that COWP also had an anthelmintic effect.[1] Administration of 5 grams of COWP to 10 week old lambs five days prior to inoculation with either *H. contortus*, *O. circumcincta*, and *T. colubriformis* reduced parasitic establishment by 95, 56, and zero percent, respectively.[1] Lambs parasitized by *O. circumcincta* had significantly increased abomasal pH, which most likely interfered with copper metabolism.[2] Bang also noted that ionic copper concentrations peaked at day 4 in the proximal duodenum post treatment, and after 10 days, steadily declined.

In 2000, Chartier et al. noted that administration of 4 grams (g) of COWP to adult dairy goats with established *H. contortus* infections reduced worm burdens by 75%, and lowed fecal egg counts by 37-95%.[3] Treatment variably limited establishment of new infections for several weeks. However, COWP treatment did little to prevent established infections of *Teladodorsagia* sp, *Trichostronylus* sp and *Oesophagostomum* sp.[3] Several years later, Knox et al demonstrated that COWP treatment (either 2.5 or 5g) was highly effective at reducing established *H. contortus* infections in yearling sheep.[4] When COWP was administered, followed by weekly administration of *H. contortus* larvae, no difference was noted in fecal egg counts between treated and control hoggets.[4] After 8 weeks, worm counts were reduced by only 37% in the treated group, which indicated the benefit of COWP administration was transient.[4] These results also indicated that COWP treatment was more effective at eliminating established *H. contortus* infections that in preventing establishment of new ones.[4]

In 2004, Burke et al. published a study that investigated various dosages of COWP in lambs.[5] Her goal was to determine an optimal dose that would maximize anti-parasitic effect, and at the same time, minimize risk factors such as hepatic copper accumulation. Copper toxicity is a potential consequence of copper administration to sheep and goats. Sheep in particular are less able to excrete excess hepatic copper stores, and are therefore more prone to accumulating toxic hepatic levels than goats. Six month-old male lambs received 0, 2, 4, or 6g COWP 28 days after administration of *H. contortus* third stage larvae; blood and feces were collected weekly.[5] Two months later, lambs were necropsied, and abomasal worms were counted and liver copper concentrations were measured. All doses of COWP resulted in much lower fecal egg counts (FEC) and much higher packed cell volumes (PCV) compared to untreated controls. Liver copper concentrations at the end of experiment were 62.2, 135.7, 161.1, and 208.4 ppm (wet matter basis) in the 0, 2, 4, and 6 g groups respectively.[5] None of lambs showed signs of copper toxicity. This study determined that 2 g COWP was the optimal dose in lambs of the ones studied in this experiment.[5] Since the 2 gram treatment practically doubled hepatic copper concentrations compared to controls, recommendations were made to assess the copper, molybdenum, iron, sulfur, and zinc status in flocks prior to using COWP treatment, and to not repeat treatment for at least 12 months in sheep.[5]

Several other researchers investigated the 2g COWP dose in lambs and goat kids. Soli et al. showed that the 2g COWP treatment reduced FEC and increased PCV in weaned, naturally *H. contortus* - infected lambs and kids for up to 6 weeks, compared to untreated lambs and kids grazing the same pasture.[6] Research conducted in South Africa by Vatta et al. evaluated 2 and 4g COWP dosages in young Zulu goats with established *H. contortus* infections.[7] At necropsy (study week 10-11), worm burdens were reduced 95% and 93% in the 2 and 4g COWP treatment groups, respectively.[7] In contrast to what was noted in Burke's lamb study, liver copper concentrations were not higher than control hepatic concentrations in either 2 or 4g COWP treatment group.[7] Whether this finding reflects a species difference, a difference in sample timing, or other factors, is unclear. Copper concentration in muscle tissue was not significantly different in treated goats than in controls, suggesting a prolonged meat withdrawal period in not necessary in treated goats.[7] Similar to what had been reported previously, Vatta's study showed that neither the 2 or 4g COWP treatment prevented establishment of new *H. contortus* infections, further demonstrating COWP treatment has minimal prophylactic benefit. In a subsequent study, Vatta et al. determined that the efficacy of COWP treatment persisted in young (8-11 month old) goats treated with 4g COWP for only 28 days.[8] Hepatic copper concentrations were determined at 3 different time points post-administration. At days 35, 56, and 84, mean hepatic copper concentration (wet matter basis) in treated goats was 191 ppm (control 120 ppm), 148 ppm (control 121 ppm) and 163 (control 150 ppm), respectively.[8]

Given concerns over the potential to induce unsafe hepatic copper concentrations following COWP treatment at the 2g (and higher) doses, several studies evaluated lower doses of COWP to determine if they had anti-parasitic benefit. Burke et al. showed that both the 0.5 and 1g COWP treatments had anti-parasitic benefit in lambs.[9] Further, these lower doses were safely administered 3 times, at 6 week intervals, to weaned lambs within the same grazing season without increasing hepatic copper into a toxic range (although they were higher than control lambs), or increasing plasma aspartate aminotransferase concentrations.[9] Similarly, lower COWP doses (0.5 and 1g) improved FEC and PCV in goat kids compared to untreated controls.[10] The 0.5g dose was considered optimal in kids, because it was effective, and was less likely to induce copper toxicity than higher doses.[10]

Less work has been done to establish optimal doses of COWP in mature does and ewes, but several papers have provided some insights. Mature goats grazing fall pasture were treated with either 0 or 4g of COWP; treated goats had lower PCV and FEC than untreated goats, but the results were somewhat ambiguous, possibly due to time of the year and low infection intensity.[10] In a second experiment, mature goats that were treated with either 5 or 10g COWP 28 days after an *H. contortus* infection was established had a 75% decrease in adult *H. contortus* recovered at necropsy at day 7. However, differences were not apparent on days 21 and 35 when compared with untreated controls.[10] Since this research was also conducted in the late fall and early winter, researchers wondered if the decline in *H. contortus* in the control groups was a natural event related to time of year.[10] Hepatic copper concentrations were higher in COWP-treated goats (5g: 153; 10g: 149 □□34 mg/kg of dry matter) than untreated goats (56 □□34 mg/kg of dry matter; *P* □□0.001); no signs of copper toxicity were noted.[10] Unfortunately, lower doses were not evaluated in adult goats to determine the lowest dose that would provide benefit. However, Burke et al. evaluated a variety of COWP doses in mature Polypay ewes.[11] Ewes with natural parasite infections (70% *H. contortus*) received 0, 0.5, 1, or 2g doses of COWP 60 days after lambing.[11] Seven days after administration, ewes treated with 1 or 2 g COWP had lower FEC than ewes treated with either 0 or 0.5 g COWP.[11] Similarly, PCV declined in all groups but much less so in ewes treated with 1 and 2 g COWP dose. Aspartate aminotransferase activity did not differ among groups; hepatic copper concentrations were not measured. The 2g dose appeared to be give slightly better results than the 1g dose, and that the 0.5 dose had little anti-parasitic effect. The researcher concluded that the 2g dose was the optimal dose in mature ewes in this study.[11]

All previous studies with COWP in sheep and goats had been conducted by administering the wire particles within a gelatin capsule using a balling gun. Since this method of administration involves individual handling, a simpler delivery method would provide a practical advantage. Burke et al. demonstrated that administration of 2 g COWP in feed was as effective at reducing fecal egg counts in goats as 2g COWP administered in a gelatin capsule.[12] The researcher advised that treatment be given on an individual basis rather than free choice to a group of animals to avoid inadvertent over or under-dosing. This more convenient route of administration could be particularly important to handlers of exotic hoof stock, as administration of a COWP capsule involves more inherent danger to people and animals. Previous research conducted at a zoo in Florida showed that treatment of oryx, roan antelope, blackbuck, and blesbok with 12.5g COWP reduced FEC by over 90% for 28 days.[13]

In summary, COWP treatment effectively reduces established *H. contortus* infections in sheep and goats, based on decrease of FEC of up to 97%. However, it does not have much activity against other abomasal or intestinal parasites. Benefit appears to persist only about 28 days, and treatment does not effectively limit re-establishment of new infections.[4,7,8] Treatment with COWP appears to have greater anti-parasitic benefit in lambs and kids than in mature sheep and goats, for reasons that are unclear.[10] Dosing recommendations need further refinement, as little attention was given to dosage related to body weight in various studies. Different breeds of sheep and goats vary substantially in size and weight. Further, breed differences possibly exist with regards to tolerance to copper administration, and geographic factors will influence the amount of copper, molybdenum and other minerals already in the diet. Even though copper is released slowly and incompletely released from copper oxide wire particles, several studies clearly indicated that treatment can elevate hepatic copper concentration.[5,8] Even though clinical copper toxicity was not induced in test subjects by COWP administration in any of the published studies, use of COWP in client-owned animals should be undertaken with caution. Assessment of copper and other minerals in hepatic tissue submitted from healthy animals, or animals that die suddenly on farms, is recommended to determine if benefits of COWP outweigh potential risks. If used more than once within a grazing season, liver enzymes should be monitored.[9] Researchers currently recommend selective (rather than whole herd or flock) use of COWP. Since some worms survive exposure, it is possible that resistance to COWP treatment (as has been seen with conventional anthelmintic treatment) could develop.[10] Use of COWP can be combined with an anthelmintic to make treatment more broad-spectrum when infections are mixed.[14] The FAMACHA System can be used to detect small ruminants with clinically significant *Haemonchus contortus* burdens based on their degree of anemia.[10] Further, treatment efficacy can be assessed using pre- and 10-14 day post-treatment fecal egg counts.

CONDENSED TANNINS

Many studies have recently focused on the use of plants such as sulla, sainfoin, trefoil, chicory and Sericea lespedeza (SL) for their nutraceutical advantages, including anti-parasitic activity[15]. Benefit is derived through direct activity against parasites, improved protein utilization, and also indirectly through enhancement of immunological processes.[15] Plant polyphenols, specifically condensed tannins, are credited for many of the beneficial effects. From a nutritional standpoint, condensed tannin benefits ruminants by binding dietary protein in the rumen, and allowing it to bypass rumen degradation, thereby improving nutrient utilization. When condensed tannins are consumed in low or moderate concentrations, increases in growth and production (fiber, milk) are reported. However, high consumption of condensed tannins can be detrimental to production, as it can have a negative impact on rumen microbiota, and cause a decline in appetite. Anti-parasitic benefits include reduction in nematode numbers, reduced fecundity, and significant decreased in fecal egg output.[15] It is likely that this action occurs through binding of condensed tannins to the protein rich outer surface of parasites. Disruption of the parasitic cuticle is visible on scanning electron microscopy after in vitro incubation with condensed tannins (CT).[15]

Sericea lespedeza is a warm-season, non-bloating perennial legume that is rich in condensed tannins. It has been used for soil conservation and as a livestock forage in the United States for over 80 years. It grows well in a variety of soils, including acidic, otherwise infertile soils, and is also drought tolerant and insect resistant. Sericea lespedeza's anti-parasitic properties have been the focus of many research projects in small ruminants in the past 10-15 years. Min et al reported that naturally parasitized goats grazing SL-rich pasture had a 71.7% reduction in FEC, and a reduction in fecal egg hatchability of 41%, compared to goats on crabgrass/rye pasture.[16] Based on copraculture, 91% of larvae were *Haemonchus contortus*, with the balance being *Trichostrongylus* and Teledorsagia spp.[16] Ingestion of SL pasture by naturally parasitized Angora does and kids resulted in reduced fecal egg counts and reduced egg hatchability, compared to does and kids on crabgrass pasture.[17] Further, milk somatic cell counts were lower, and intradermal response to phytohemagglutinin testing more robust in the does that grazed SL.[17]

A recent report provided valuable insight into the anti-parasitic impact of feeding SL. [18] Twenty Boer bucks experimentally infected with *H. contortus* infections (trickle infection) were housed indoors, and fed either Bermuda grass (control) or SL ration for 7 weeks. The ration was balanced between groups for protein and calories. Fecal egg counts in the SL group dropped 79.7% by week 1, and maintained at 88% reduction the remainder of the experiment. In contrast, the control group's FEC continued to increase over baseline. The SL group maintained higher PCV than the control group. Also, ova collected from the SL group were less likely to develop into infective larvae. At slaughter, significantly less abomasal worms (*H. contortus* and *Teladorsagia circumcincta*) and intestinal worms (*Trichostrongylus colubriformis*) were recovered from the SL group compared to the control group. This study showed that SL has a direct toxic effect on abomasal and intestinal worms, especially females, and that the lethality was most pronounced in *H. contortus*. It is likely that it also reduced parasite egg viability and larval development in feces. This study demonstrated that SL maintains it's anti-parasitic properties after being dried and made into hay, and that it was highly palatable to the goats. [18] Another study done using SL hay provided further insights. [19] Lambs fed SL hay had 67-98% lower FEC during the feeding trial, and the FEC remained significantly lower than controls on Bermuda grass hay throughout the feeding period. [19] The FEC increased quickly once the lambs being fed SL were switch to Bermuda grass hay. This finding indicates that SL works partly by suppressing egg production by female worms. [19] Feeding SL hay was associated with a 67.2% reduction in established worm burdens overall, but was less effective at limiting establishment of incoming larvae. [19]

Research efforts were directed at evaluating the anti-parasitic effect of different SL preparations. Pelleted preparations would be more convenient to store and ship than hay, but the heat and pressure involved in pelleting feed could reduce biologic activity of the CT fraction. An experiment evaluated the anti-parasitic effect of SL hay and pellets in goats, using a Bermuda grass hay control group.[20] Both LS hay and pellets effectively lowered FEC compared to controls, and the effect was greatest in the goats receiving the SL pellets. Worm burdens, especially *H. contortus*, were significantly reduced in SL fed groups, compared to controls.[20] Pastured young goats supplemented with SL pellets (75% or 95% SL leaf meal) had lower FEC than controls that received a commercial pellet supplement. [21] The group receiving the 95% SL pellet gained more weight than goats supplemented with 75% SL pellet, and both groups gained more weight than the controls. [21] One question that needed to be addresses was, how much SL must animals consume to gain an anti-parasitic benefit? A recent project showed that the anti-parasitic effect of a SL enriched diet is optimized when it comprises about 70% of the total diet, but feeding it as low as 50% of the diet will also reduce the fecal egg count compared to controls.[22]

A recently completed study that compared SL hay, ensiled SL, and Bermuda grass hay diets in goats (as 70% of the diet, which was balanced for protein and energy) showed that both SL preparations reduced fecal egg counts and *Eimeria* oocyst counts by 75.5% and 88.6%, respectively, compared with controls that received a Bermuda grass diet.[23] Lambs with naturally infected with nematodes and coccidia that received SL pellets 30 days before, and 21 days after weaning shed lower numbers of oocysts than controls (conventional creep feed). In addition, none of the lambs on the SL pelleted diet required treatment for clinical coccidiosis, whereas 33% of controls had symptoms significant enough to warrant treatment.[24] Feeding SL leaf meal pellets to recently weaned goats was also shown to be beneficial at reducing parasitic ova and oocysts.[25] One group of kids received SL leaf meal pellets that had been stored for 3 years, another received SL leaf meal pellets that had been stored less than 6 months, and a control group was fed commercial pellets. Within 7 days, and for the remainder of the trial period, lambs that receiving the pellets had 66.2% to 79.2% lower fecal egg counts, and 92.2% to 91.2% lower oocyst per gram counts than controls.[25] These results show that SL has great potential for reducing coccidian burdens in lambs and kids. Current recommendations are to feed SL pellets as a creep, at about 50% or more of the diet, to youngsters 2 weeks prior to weaning. Feeding can continue for 6 weeks after weaning, but not indefinitely. Trace mineral deficiencies and a negative impact on weight gain have been noted in youngsters, but not in adults on long-term supplementation.[23] As an aid to management of H contortus, SL can be fed at approximately 50% of the diet in any of it's various forms (fresh, hay, pellets, silage) to animals during periods of high risk. Once SL feeding is discontinued, animals should be closely monitored for signs of parasitism.

HAEMONCHUS CONTORTUS VACCINE

Many studies have been conducted in sheep to investigate the immunoprophylactic potential of *Haemonchus contortus*-derived hidden gut antigens (H-gal-GP and H11),[26,27,28] cysteine proteinases,[29,30] and excretory-secretory products.[31,32] All these experimental vaccines caused a reduction in FEC and *H. contortus* worms in vaccinated animals. Use of "hidden" somatic antigens in vaccines potentially offer an advantage over soluble and exposed antigens, because the hidden antigens are not recognized by the immune system during natural infection.[32] As a result, no natural selection pressure has been imposed, so blood-feeding worms should remain highly susceptible to antibodies present in the vaccinated host's blood.[32] However, repeated vaccination within the same grazing season is necessary to maintain the protective effects using hidden gut antigens.[28,33]

Despite the remarkable success achieved by the experimental vaccine, commercialization was only thought feasible if a recombinant version of the antigen proved to be effective. Researchers did not think it would be possible to mass-produce enough native antigen to meet market demand. Unfortunately, recombinant antigens failed to achieve the reductions in FEC and worm counts attained by native antigen vaccines.[34,35] The recent discovery that a much lower dose than what was used in the initial field trials also induced a protective response in sheep renewed interest in commercial vaccine development. Material obtained from 1 infected sheep could provide sufficient worm antigen to make several thousand doses.[36] In addition, a worm harvesting machine was developed that could efficiently harvest worms from slaughtered lambs without reducing the carcass value.

A field trial in Western Australia was conducted from October 2010 to January 2011 using the vaccine produced by Moredun Research Institute scientists to determine the optimal dosage and injection regimen in cross bred lambs.[36] The lambs were dosed with *H. contortus* larvae twice a week during the first half of the trial, then at weekly intervals for the remainder of the trial. The mean FEC of control lambs was 7,000 epg despite the fact 7 of the 20 controls received anthelmintic treatment for haemonchosis. Treated sheep received a 2, 5, 10, or 50 microgram vaccine dose. A second dose was given 3 weeks later, and a third dose was given either 6 or 8 weeks after the second dose. The sheep that received the "early" 3rd dose at 6 weeks, received a 4th dose 6 weeks after the 3rd dose. A significant reduction in FEC was seen in sheep that received the 3rd vaccine at 6 weeks compared to at 8 weeks after the 2nd dose. The researchers concluded that a dose as little as 5 microgram of *H. contortus* native gut antigen vaccine was sufficient to provide a high level of protection during periods of intense larval challenge; mean FEC were reduced by over 85% in vaccinates compared to controls.[37] Further, vaccinated animals gained more weight, had higher PCVs, and fewer required anthelmintic treatment compared to controls.[37] The study results indicate that 2 priming doses should be given 3 weeks apart, and that vaccine boosters should be given at 6-week intervals to maintain protective effect during periods of high parasitic challenge.[37]

As a result of the successful collaboration between Moredun Research Institute scientists in Scotland and the Department of Agriculture and Food in Western Australia, Barbervax® was released in October 2014 for use in sheep. The first batch of Barbervax® consisted of 30,000 doses, which were sold in 1 week by word of mouth. Barbervax® is marketed in 250 ml containers, and has a shelf life in the refrigerator of 2 years. Current recommendations are to administer 1 ml subcutaneously, using the dosing interval described above. Applications have also been made to South African regulatory agencies to allow use of the vaccine, which will be marketed as Wirevax®.[37] Currently, the manufacturer is not interested in marketing the vaccine in the United States (David Smith, personal communication 2015). However, if there is sufficient demand, this attitude could change. The vaccine is currently undergoing 33 field trials in various species around the world.

REFERENCES

[1]Bang KS, et al. Res Vet Sci 1990;49(6):132-7. [2]Bang KS, et al. Res Vet Sci 1990;49(3):306-14. [3]Chartier C et al. Vet Res Commun 2000; 24:389–399. [4]Knox MR Aust Vet J 2002;80:224–227. [5]Burke JM et al. Vet Parasitol 2004;123:235–243. [6]Soli F et al. Vet Parasitol 2010;168:93–96. [7]Vatta AF et al. Vet Parasitol 2009;162:306–313. [8]Vatta AF et al. Vet Parasitol 2012;190:159-166. [9]Burke JM et al. Vet Parasitol 2006;139:145-149. [10]Burke JM et al. J Anim Sci 2007;85:2753–2761. [11]Burke JM et al. 2007 Vet Parasitol 2007;146:372-375. [12]Burke JM et al. Vet Parasitol 2010;168:346-350. [13]Fonttenot DK et al. J Zoo Wildlife Med 2007;39:642-645. [14]Burke JM Vet Parasitol 2016;215;1-4. [15]Hoste H Trends Parasitol 2006;22:253-261. [16]Min BR et al. Small Ruminant Res 2004;51:279-283. [17]Min BR et al. Vet Parasitol 2005;130:105-113. [18]Shaik SA et al. Vet Parasitol 2006;139:150-157. [19]Lange KC et al. Vet Parasitol 2006;141:273-278. [20]Terrill TH et al. Vet Parasitol 2007;146:117-122. [21]Gujja S et al. Vet Parasitol 2013;191:1-2. [22]Terrill TH et al. Vet Parasitol 2009;163:52-56. [23]Terrill TH W4 Congress 2015. [24]Burke JM et al. Vet Parasitol 2013;193:39-46. [25]Kommuru DS et al. Vet Parasitol 2014;204:191-198. [26]Munn EA et al. Int J Parasitol 1993;23:261-269. [27]Smith WD Res Vet Sci 1993;54:94-101. [28]Smith WD et al. Vet Parasitol 2001;98:285-297. [29]Redmond DL Vaccine 2004;22:4252-4261. [30]Ruiz A et al. Vet Res 2004:565-572. [31]Bakker N et al. Vaccine;22:618-628. [32]De Vries E et al. Vet Res 2008;40:41. [32]Newton SE et al. Parasitol Today 1999;15:116-122. [33]LeJambre LF et al. Vet Parasitol 2008;153:3012-312. [34]Newton SE et al. Parasite Immunol 2003;25:283-296. [35]Cachat E et al. Parasite Immunol 2010;32;414-419. [36]Beiser B et al. Proceedings Aust Sheep Vet Conference 2012:14-18. [37]Bredenkamp H et al W4 Congress 2015.

The Strong Ion Approach to Acid Base - How to and Why

Angelika Schoster, DVM, DVSc, PhD, DACVIM (Large Animal Internal Medicine), DECEIM and
Henry Staempfli, DMV, DACVIM (Large Animal Internal Medicine)
Zurich, Switzerland

INTRODUCTION

Acid base and electrolyte disturbances are frequently encountered in animals with different diseases. The signs of the underlying disease often overshadow clinical manifestation of acid base and electrolyte abnormalities. Only very severe disturbances are fatal, but mild to moderate disturbances can also lead to complications and prolonged recovery times from the underlying disease(s). Correcting acid base and electrolyte abnormalities has a clinical and physiological benefit and should therefore be undertaken. To correct disturbances, these have to be diagnosed correctly. There are several approaches to determining the acid base and electrolyte status of a patient. Initial traditional approaches mainly focused on bicarbonate and hydrogen ions as the major parameters deranged, while quantitative approaches, such as the strong ion approach yield more accurate assessments and allow understanding of underlying pathologic causes.

ACID BASE AND ELECTROLYTE BALANCE

Bronsted defined a base is a substance which can accept H^+, and acid as a substance which can donate H^+. Unfortunately these definitions have led to H^+ being synonymous with acid and being considered as independent. More correct would be Stewart's definition, which suggests that **acids are substances that result in a H^+ increase, and bases are compounds that result in a H^+ decrease,** which then indicates that Cl^- is an acid and Na^+ is a base. This is based on the chemical laws of aqueous solutions, including blood and plasma to fulfill electroneutrality, mass conservation and dissociation equilibria. Acids are byproduct of cellular metabolisms. There is one volatile acid (CO_2) and several non-volatile acids such as lactate, acetate, pyruvate and citrate. Strictly speaking CO2 is a gas, not an acid, however it forms into carbonic acid (sometimes termed respiratory acid) when it comes in contact with water, making it an effective acid. There are two main components of the acid base homeostasis, the respiratory and metabolic component (Table 1). These two components are interdependent and need to be considered together. The different approaches to acid base all differ in their assessment of the metabolic component (Table 2). There are benefits to using the strong ion approach to understand underlying derangements and to assist more accurately treatment decisions .

Table 1. Components of acid base and electrolyte homeostasis

Respiratory	Metabolic	
Co_2 (volatile acid)	Strong ions	Na, Cl, K, ketoacids, uremic acids, d/l-lactate
	Organic weak acids	Blood Proteins
	Inorganic weak acids	Sulfate, phosphate
	Volatile weak base	Bicarbonate

Table 2. Conceptual differences between the traditional and quantitative strong ion approach

Traditional approach	Quantitative strong ions approach
Centered on pH, bicarbonate, hydrogen ions	Centered on strong ions, weak acids and bases
HCO_3 and H^+ are independent	HCO_3 and H^+ are dependent on strong ions and weak acids and bases
pH is maintained within narrow margins in health	Concentrations of strong ions are maintained within narrow margins in health
Buffering of acids and bases	Change in the dissociative state of weak acids to satisfy rule of electroneutrality, conservation of mass and dissociation equilibrium

TRADITIONAL APPROACH

Henderson focused on the unique role the bicarbonate-carbonic acid buffer system played in stabilizing the acid-base equilibrium and his knowledge led to the development of the Henderson-Hasselbach equation [1]. This approach to acid base centers on ph, bicarbonate and H^+ as the main determinant of the metabolic component of acid base homeostasis. Non-bicarbonate buffers, such as proteins and serum electrolytes are not measured or taken into account using this approach. H^+ is a strong acid and considered the most important acid in the human body. As H^+ is present in the body in very small amounts (0.0004mEq/L), the log function is used to make the 'pH' better comprehensible ($pH = -logH^+$). pH and pCO2 are nowadays easily and very accurately measured on ion specific electrodes and bicarbonate calculated by using the Henderson-Hasselbach equation. According to the traditional approach pH is generally maintained within a very narrow range, this is important for protein configuration, enzyme activity, trans-cellular transport mechanisms and general cellular function. The traditional approach recognizes 4 abnormalities:

- Respiratory alkalosis and acidosis
- Metabolic alkalosis and acidosis

TRADITIONAL APPROACH PLUS

To take into account the non-bicarbonate buffers, the buffer base was introduced and added to the traditional approach. The change in buffer base was termed base excess (BE), and defined as the amount of strong base or acid needed to return one liter of blood to pH 7.4. In a later step the equation was normalized for hemoglobin and the resulting variable called standard base excess, or extracellular fluid BE (BE_{ecf}). This approach helped separate respiratory from metabolic components of acid base, but is still not able to differentiate between the different components of the metabolic side and physically does not really exist. Base excess has been shown to be normal in complex acid base derangements.

PITFALLS OF THE TRADTIONAL APPROACH

Focusing on pH (or indirectly bicarbonate and H^+) as the sole determinant of acid base homeostasis is not sufficient, as pH can be normal, or the severity of the abnormality masked in complex acid base derangements e.g. pH can be normal with hyponatremia and concurrent hypoproteinemia. Therefore the underlying abnormality might be missed or at least its severity under- or overestimated. Especially the role of electrolytes is largely ignored by the traditional approach and this could lead to selection of inadequate types of fluids for therapy e.g. sodium bicarbonate in cases of lactic acidosis. Several clinical anomalies have been identified when using these traditional approaches. Using the Henderson-Hasselbach approach it cannot be explained why the administration of $CaCl_2$ in cattle, increased amounts of total protein and administration of certain fluid types are acidifying.

QUANTITATIVE STRONG ION APPROACH

Peter Stewart, a chemist at brown University challenged the traditional approach to acid base using fundamental biochemical and mathematical models in 1978 [2]. This theory did not attract much attention until the 1990s when it was simplified by several investigators to allow its use in clinical settings [3]. The strong ion approach focuses on the role of independent and dependent variables, on the role of strong ions and weak electrolytes (acids and bases) in the changes of the metabolic component of acid base which have been ignored by the previous approaches. Complex mixed abnormalities, which cancel each other out can so be detected and diagnosed. Independent variables influence the acid base system based on their concentration while dependent variables are influenced directly and predictably by changes in independent variables. This approach describes the whole acid base and electrolyte balance based on three independent variables, and as a result 6 major acid base disturbances are known (Table 3).

- Carbon dioxide (respiratory component, pCO_2)
- Strong ion difference (SID = Na + K - Cl)
- Total non volatile weak acids (A_{tot})

In the strong ion approach HCO_3 and H^+ are dependent variables and change their dissociation state due to changes in independent variables pCo2, total mass of non volatile weak acids (A_{tot}) and concentrations of strong electrolytes (depicted as SID). Weak acids and bases are present in a dissociation system (undissociated and dissociated) and the amount of dissociation changes depending on the concentrations of strong ions to maintain electroneutrality. This represents what is known as 'buffers' in the traditional approach. Buffering is a term discouraged by the quantitative strong ion approach. Electroneutrality can be visualized using **Gamblegrams.** The mathematical concept behind electroneutrality is: sum of cations (measured and unmeasured) = sum of anions (measured and unmeasured). Anions include Cl^-, negative charge of proteins (A-), lactate, ketoacids, uremic acids, phosphate, sulfate, fatty acids and others. Cations include Na, K, Ca and Magnesium. The term unmeasured anion or cation can refer to different substances, depending on what can be measured and included in calculations.

Strong ions

Strong ions exist completely dissociated in solutions. Strong cations are sodium, potassium, magnesium and calcium. Strong anions are chloride, lactate mainly but also to be complete sulfate, ketoacids, non-esterified fatty acids and uremic acids, which are not clinically relevant in large animals. **Clinically mainly Cl, K and Na are present in high concentrations, therefore SID is generally calculated as = (Na + K) – (Cl) equaling approx. 38-42 mEq/L.** Lactate, also considered a strong ion, is usually not included in this calculation when not measured and is therefore usually included in the 'unmeasured anion' group. **Lactate is a strong anion and has major acid base implications and if measured should be included in the SID = (Na + K) – (Cl+ Lactate).** As lactate concentration is 1 -2 mmol/L in healthy individuals, the normal SID of 38-42 remains unchanged when including lactate. An increase in SID results in metabolic alkalosis, a decrease results in metabolic acidosis (Table 3).

A_{tot}, A^-

A_{tot} is the total concentration of nonvolatile weak acids, mainly comprised of proteins and in ruminants to probably phosphate as well. **Proteins play the most important role in A_{tot}.** In cases where A_{tot} (total protein concentration) is increased significantly, metabolic acidosis results (often called hyperproteinemic acidosis). In cases where A_{tot} (total protein concentration) is decreased significantly, metabolic alkalosis results (often called hypoproteinemic alkalosis). Total proteins are weak acids and therefore present in both forms undissociated and dissociated. A- is the dissociated form and carries a net negative charge. Once total protein concentration is known, a **useful estimate of net negative charge (A^-) of proteins in mEq/L (or mmol/L) = 0.2 x total protein (g/L).**

CLINICAL APPLICATION OF QUANTITATIVE ACID BASE

The above concepts should be integrated into acid base evaluation, as suggested by the simplified strong ion approach (Table 3). Several studies have shown the applicability and also advantages of using the quantitative approach to acid base [4-8]. A venous blood sample is sufficient in most cases to determine acid base status. If a strong respiratory component is assumed, an arterial blood sample might useful for specific evaluation of pCo2.

1. **Evaluate pCO2**
   - ➢ If abnormal take respiratory component into account when formulating a treatment plan
   - **Respiratory adaption due to metabolic acid base abnormalities (termed compensation in the traditional approach) is negligible** i.e. there is no or very little (about max 1 mEq/)L adaption
2. **Calculate SID (Na+K) – (Cl+Lactate) = 38-42**
   - ➢ If abnormal, determine which electrolyte abnormalities are present and cause the derangement and take into account when formulating a fluid plan, particularly when choosing the fluid type
   - Common causes of low SID are hyponatremia, hypokalemia, hyperchloremia, hyperlactatemia
   - Common causes of high SID are hypochloremia, hypernatremia
3. Determine A_{tot} (clinically: **measure total protein concentration)**
   - ➢ If abnormal, recognize the acid base consequences as outlined above
   - Note that in severe renal failure significantly increased phosphate concentrations also influence severity of acid base problem present

Table 3. Clinical Interpretation of acid base and electrolyte abnormalities

	PCO_2	**SID = (Na+K) – (Cl + Lactate)**	**A_{tot}**
Low	Respiratory alkalosis	Strong ion acidosis (SID <38)	Non volatile buffer acidosis (Hyperproteinemia)
High	Respiratory acidosis	Strong ion alkalosis (SID>42)	Non volatile buffer alkalosis (Hypoproteinemia)

Note that Lactate is not always measured in which case it must be considered an unmeasured anion and removed from the SID equation. The normal value of the SID is still 38-42 as lactate concentrations in health are usually less than 2mmol/L. D-Lactate cannot be measured at all in clinical settings

If clinical signs (dehydration, weakness, loss of suckle) cannot be explained by changes in the SID, pCO2 or total protein concentration in the blood, the reason for acidosis must be assumed with an unmeasured strong ion – most often lactate (d/L).

Determination of unmeasured anions

Clinically it is of interest to determine the presence of unmeasured anions. These can include L-lactate, D-lactate, ketoacids, uremic acids and sulfate. **In the majority of cases D-lactate or L-lactate are increased.** There are 2 methods of determining the concentrations of unmeasured anions, the anion gap and strong ion gap. The strong ion gap hold is based on the SID theory and holds advantages over the anion gap.

Anion Gap (AG)

The AG is based on the concept of electroneutrality. The AG can be calculated as (Na + K) – (Cl + HCO3) of HCO3 / TCO2 is available . It is the difference between unmeasured anions and cations, however electroneutrality between cations and ions has to be preserved in the body, therefore there is no real anion gap. The net negative of total proteins usually causes up to 90% of the AG, the remainder is caused by the unmeasured anions and cations. Usually it is assumed that unmeasured anions and cations cancel each other out, however if there is a significant increase in one of the anions (common) or cations (uncommon) a 'gap' can be calculated as the A_{tot} is not taken into account in this calculation. **Clinically the anion gap often stands for the increase in unmeasured anions, such as lactate, uremic acids and ketoacids and proteins. An increase in lactate is the most common clinical cause for an increased AG.** The anion gap was a useful determinant in times when lactate could not be measured and it was trying to be determined whether lactic acidosis was present. Nowadays that lactate is routinely measured it has become less useful. If HCO_3 is not available, totalCO2 can be used as a proxy in practice.

Strong ion gap

The strong ion gap includes the net anionic charge of proteins in the calculation of unmeasured anions and uses electroneutrality, and is therefore more useful in determining the concentration of mostly strong unmeasured anions in in our sample analyzed. **The strong ion gap is defined as (Na + K)-(Cl + HCO3) - A^-** and so effectively represent lactate and other unmeasured anions such as ketoacids (mainly small animals) and very rarely uremic acids. It can also be viewed as SIG= AG - A-. **The main player in the SIG is lactate and in the bovine particularly D- and L-lactate.** Several calculations have been devolved to calculate SIG based on total protein or albumin, where the former is preferred. An online calculator has also been developed to assist in this calculation (http://www.uoguelph.ca/~pjf/1.4.html)- As described above, A^- (mEq/L) can be estimated as: Total Protein (g/L) x 0.2.

When all calculations have been performed, it can be judged which variables have part in the acid base and electrolyte abnormalities. Particularly in complex cases these might cancel out each other's effect on pH or at least 'mask' the severity of the other problem. Based on pCO_2, SID and total protein concentrations (A_{tot}) and potentially assessment of unmeasured anions (lactate) an integrative fluid plan can be developed to correct electrolyte abnormalities and acid base abnormalities. In case the respiratory component is affect appropriate treatment has to be selected.

SUMMARY

While the quantitative analysis of acid base appears more complicated, it actually provides better mechanistic insights into acid base abnormalities. Particularly when total protein concentrations are abnormal this approach offers a great advantage over the Henderson-Hasselbach approach. A further advantage is a better way of choosing fluid type using SID and A_{tot}.

REFERENCES

[1] Henderson, L.J. (1908) The theory of neutrality regulation in the animal organism. *Am J Physiol* **21**, 427-448.

[2] Stewart, P.A. (1978) Independent and dependent variables of acid-base control. *Respiration physiology* **33**, 9-26.

[3] Constable, P.D. (1997) A simplified strong ion model for acid-base equilibria: application to horse plasma. *Journal of applied physiology* **83**, 297-311.

[4] Gomez, D.E., Lofstedt, J., Stampfli, H.R., Wichtel, M., Muirhead, T. and McClure, J.T. (2013) Contribution of unmeasured anions to acid-base disorders and its association with altered demeanor in 264 calves with neonatal diarrhea. *Journal of veterinary internal medicine / American College of Veterinary Internal Medicine* **27**, 1604-1612.

[5] Gomez, D.E., Biermann, N.M. and Sanchez, L.C. (2015) Physicochemical Approach to Determine the Mechanism for Acid-Base Disorders in 793 Hospitalized Foals. *Journal of veterinary internal medicine / American College of Veterinary Internal Medicine* **29**, 1395-1402.

[6] Constable, P.D. (2002) Calculation of variables describing plasma nonvolatile weak acids for use in the strong ion approach to acid-base balance in cattle. *American journal of veterinary research* **63**, 482-490.

[7] Stampfli, H.R., Schoster, A. and Constable, P.D. (2014) Clinical utility of serum biochemical variables for predicting acid-base balance in critically ill horses. *Veterinary clinical pathology / American Society for Veterinary Clinical Pathology* **43**, 547-556.

[8] Trefz, F.M., Constable, P.D. and Lorenz, I. (2015) Quantitative physicochemical analysis of acid-base balance and clinical utility of anion gap and strong ion gap in 806 neonatal calves with diarrhea. *Journal of veterinary internal medicine / American College of Veterinary Internal Medicine* **29**, 678-687.

Update on Fluid Resuscitation in Calves with Diarrhea

Geof W. Smith, DVM, PhD, DACVIM (Large Animal Internal Medicine)
Raleigh, NC

INTRODUCTION

Neonatal diarrhea remains the most common cause of death in both beef and dairy calves. Despite significant progress in understanding the pathophysiology of neonatal diarrhea, it continues to be a major cause of economic loss to the cattle industry. A complete review of the pathophysiology of diarrhea is beyond the scope of this presentation but has been covered in a recent review.[7] Some pathogens cause secretory diarrhea, causing the small intestine to move from a net absorption of fluid to a net secretion of chloride, sodium, and water into the intestinal lumen. This increase in secretion overwhelms the absorptive capacity of the large intestine resulting in diarrhea. Other pathogens damage the small intestinal villi which results in failure to absorb electrolytes and water (malabsorptive diarrhea). However, regardless of the pathogen or the mechanism involved, diarrhea increases the loss of electrolytes and water in the feces of calves and often decreases milk intake. This results in dehydration, strong ion acidosis, electrolyte abnormalities (usually decreased sodium and increased **OR** decreased potassium), increased D-lactate concentrations, and a negative energy balance (from anorexia and malabsorption of nutrients). Therefore diarrhea is by far the most common indication for fluid therapy in neonatal calves. The primary goals of treating calf diarrhea are to 1) correct free water and electrolyte abnormalities; 2) correct acid-base deficits (acidemia); 3) provide nutritional support; and 4) eliminate and/or prevent E. coli bacteremia. Three of these 4 goals can be met with fluid therapy. The purpose of this article is to provide an overview of fluid therapy in calves with particular emphasis on treating diarrhea. Practical options for fluid therapy that can be performed on the farm will be emphasized during the seminar.

ORAL ELECTROLTYE THERAPY

Considerable variability exists in the quality of commercial oral electrolyte solutions available today and the practitioner must put some thought into the product they choose to use in practice. There are several important factors to consider when deciding on a product. Current knowledge indicates that an OES must satisfy the following four requirements: 1) supply sufficient sodium to normalize the ECF volume; 2) provide agents (glucose, citrate, acetate, propionate, or glycine) that facilitate absorption of sodium and water from the intestine; 3) provide an alkalinizing agent (acetate, propionate, or bicarbonate) to correct the acidosis usually present in calves with diarrhea; and 4) provide energy, as most calves with diarrhea are in a state of negative energy balance.[13] Factors to consider when choosing an oral electrolyte solution include the following.

1) **Sodium concentration**- Sodium is the osmotic skeleton of the extracellular fluid and therefore of plasma. Because sodium is the principal determinant of the volume of the ECF volume, it must be present in an oral electrolyte solution to rapidly correct the losses that have occurred with dehydration and diarrhea. The ideal sodium concentration for oral rehydration therapy in calves is not completely known, however most research would suggest it should be between 90 and 130 mM/L. Products containing sodium at lower concentrations are not able to adequately correct dehydration. Oral electrolyte products with very high sodium concentrations might be expected to cause hypernatremia and have also been shown to delay abomasal emptying rates because of increased osmolality.[11]

2) **Chloride concentration**- Although calves lose chloride during diarrhea, this loss does not occur nearly to the same degree as sodium.[5] A general guideline has been that oral electrolyte products should contain chloride in concentrations between 40 and 80 mEq/L.

3) **Potassium concentration**- Like sodium and chloride, potassium is lost in the feces of calves with diarrhea. Therefore all calves with diarrhea have a total body deficit of potassium. However, in acute cases of diarrhea calves may have elevated blood potassium concentrations (hyperkalemia). With dehydration, aldosterone is released from the pituitary gland. Aldosterone acts on the kidney to conserve sodium and water at the expense of increased potassium losses. Therefore in chronic cases of diarrhea, calves can have profound depletion of body potassium stores and generally have low serum concentrations of potassium. Clinical signs of hypokalemia include profound muscular weakness which is often present in calves with chronic diarrhea. General recommendations are that oral electrolyte products used in calves with diarrhea contain potassium concentrations between 10 and 30 mM/L.

4) **Sodium absorption** - Sodium absorption by the small intestine is a passive process, and is linked to the movement of actively absorbed or secreted solutes. If sodium is present in the lumen of the small intestine without either glucose or amino acid, there is either a small net absorption or no net sodium movement across the jejunum. Glucose can be co-transported with sodium from the intestinal lumen to the inside of the enterocyte at the brush border membrane. Neutral amino acids such as glycine, alanine or glutamine can also facilitate sodium absorption in the small intestine by a mechanism similar to glucose. Whether amino acids are needed in addition to glucose in oral electrolyte solutions is not well understood, however the addition of glycine does appear to further improve water absorption in the intestine. In addition, volatile fatty acids such as acetate or propionate have been shown to

facilitate sodium absorption in the gut. The mechanism by which volatile fatty acids stimulate sodium absorption in the intestine appears to be different from that of glucose or amino acids. Therefore acetate seems to have an additive effect to glucose and amino acids, meaning you can expect a significant increase in intestinal sodium absorption in electrolyte products containing volatile fatty acids, even when they already contain high concentrations of glucose and/or glycine.

5) **Osmolality** - Commercially available oral electrolyte products in North America can range from roughly isotonic (280-300 mOsm/L) to extremely hypertonic (700-800 mOsm/L).[15] The primary difference in most of these products is the amount of glucose that is added. High osmolality solutions provide greater nutritional support to calves relative to lower osmolality products. However research has demonstrated that milk replacer is better able to maintain normal serum glucose concentration than either hypertonic or isotonic oral electrolyte solutions. However, as expected, oral electrolyte solutions rehydrated calves and prevented the development of metabolic acidosis more effectively than did milk replacer because they have a much higher sodium concentration. Oral electrolyte solutions with extremely high osmolalities have been shown to slow abomasal emptying rates as compared to isotonic products.[10,12] This suggests that electrolyte products with a very high osmolality (or high glucose concentrations) would be likely to induce abomasal ileus, thus increasing the risk of bloat and/or abomasitis. So although the ideal osmolality of an oral electrolyte solution for calves is not completely understood, a moderate osmolality solution (400-600 mOsm/L) would be ideal in dairy calves or in beef calves that have been separated from the dam. Certainly if milk is to be withheld for any length of time, a hypertonic oral electrolyte solution would be indicated to provide energy to the calf. However, lower osmolality solutions might be appropriate for beef calves that were still suckling or in conjunction with milk replacer in dairy calves that maintain a very good appetite. I recommend avoiding extremely high osmolality oral electrolyte products (>700 mOsm/L).

6) **Alkalinizing agents** - Acidemia and metabolic acidosis occur in almost all cases of calf diarrhea. This was originally attributed to bicarbonate loss in the feces along with a decreased glomerular filtration rate in response to severe dehydration. However more recent data in calves with diarrhea have indicated that the acidosis in calves with diarrhea actually results from differences in strong ion balance.[5] This is discussed in significant detail in a recent review article.[9] It is imperative that either oral or intravenous fluid therapy protocols be able to increase blood pH. Classically this has been done by adding alkalinizing agents (ie. bicarbonate, acetate, or propionate) to oral electrolyte mixtures. More recently, there has been growing interest in looking at the strong ion difference (SID) of electrolytes as they relate to the efficacy of a different product to promote alkalinization. In reality, both (having an alkalinizing agent and a high SID) are likely important.

Acetate, propionate, bicarbonate and citrate are all considered alkalinizing agents and are frequently present in commercial oral electrolyte solutions. Bicarbonate-containing fluids are very effective at correcting a severe acidosis, since bicarbonate reacts directly with H^+ ions to form CO_2 and H_2O. Acetate and propionate are also alkalinizing agents and have been shown to have alkalinizing effects similar to bicarbonate. Acetate and propionate are only effective alkalinizing agents when they are metabolized; a process which forms water and creates bicarbonate ions (bicarbonate precursors). This metabolic process appears to still function efficiently in calves with severe diarrhea as the alkalinizing ability of the acetate has been shown to be as effective as bicarbonate. Acetate and propionate have several advantages over bicarbonate:

a. As discussed above, acetate and propionate facilitate sodium and water absorption in the calf small intestine whereas bicarbonate does not.
b. Acetate and propionate produce energy when metabolized, whereas bicarbonate does not.
c. Acetate and propionate do not alkalinize the abomasum whereas bicarbonate does – low abomasal pH is a natural defense mechanism against bacterial proliferation.

The feeding of oral electrolyte products containing bicarbonate has been shown to alkalinize the abomasum in calves.[14] Suckling of bicarbonate-containing oral electrolyte solutions can cause a large and sustained increase in abomasal pH. A similar effect is not seen with acetate based products. Abomasal acidity provides a natural barrier to ingested bacteria, and maintaining a low abomasal pH will decrease the number of viable coliform bacteria that reach the small intestine. This increases nonspecific resistance to intestinal colonization. Therefore the increase in abomasal pH seen with electrolyte products that contain high concentrations of bicarbonate may facilitate growth of bacterial diarrheal pathogens and thus increase the severity, duration, and mortality rate associated with diarrhea in calves. The ideal electrolyte solution for use in calves with diarrhea should contain at least 50 mM/L of a alkalinizing agent (preferably acetate and/or propionate), and have a SID of at least 60-80. Unfortunately, products without alkalinizing agents and with very low SIDs are commonly available in North America and should be avoided in calves with diarrhea.

In summary, oral electrolytes continue to be the hallmark of routine therapy for treating neonatal calf diarrhea. It is important that practitioners are able to assess dehydration accurately and understand how and when to use oral electrolyte products. There are tremendous differences in the formulation of commercially available electrolyte products found both in North America and around the world. Certainly all products are not "created equally" and choosing which of these products to use in practice should be an important decision. Practitioners should focus on selecting oral electrolyte solutions that satisfy the following four requirements: 1) supply sufficient sodium to normalize the ECF volume; 2) provide agents that facilitate absorption of sodium and water from the intestine; 3) correct the acidosis usually present in calves with diarrhea; and 4) provide energy. Additionally, an oral electrolyte should not cause any deleterious effects (such as abomasal bloat or abomasal alkalinization). Since veterinarians are often not directly involved with the administration of oral electrolytes to calves, it is important that they examine the electrolyte product being used in their clients' herds and make recommendations when appropriate.

INTRAVENOUS FLUID THERAPY

Although oral electrolyte therapy is generally much easier to perform on the farm, there are times where the use of intravenous fluids is critical when trying to resuscitate calves with diarrhea. Like oral electrolytes, intravenous fluids are given primarily to correct dehydration, electrolyte imbalances, acidosis and reduce the elevated D-lactate concentrations often seen with diarrhea. The key for ruminant practitioners is to be able to decide if IV fluid therapy is necessary in sick calves based on clinical examination rather than on laboratory values. Important clinical parameters to guide decision making on fluid therapy are obtained from the evaluation of hydration status and central-nervous-system (CNS) function. Degree of enopthalmus is the best predictor of dehydration in calves, followed by skin elasticity determined on the neck and thorax. Detailed information on the clinical assessment of hydration status in calves was previously presented.[13] In clinically sick calves, it is important to evaluate hydration status along with other clinical signs. These include the ability of the calf to suckle, severity of CNS depression, and whether or not the calf can stand (degree of weakness). These factors in combination are used to determine whether or not IV fluid therapy is indicated. Calves with diarrhea and dehydration during their first week of life are less acidotic than older calves, and will require less sodium bicarbonate to correct their acidemia. Calves that are unable to stand or have a weak or absent suckle reflex have a more severe acidosis and require intravenous sodium bicarbonate to correct their acidemia. D-lactic acidosis may be present in sick calves with or without diarrhea and dehydration that are recumbent or wobbly, tired, listless or comatose, and with a delayed, incomplete, or absent palpebral reflex.[16] Also if the suckle reflex is absent, weak, or the calf is chewing irregularly instead of suckling normal, D-lactic acidosis may be the underlying disease state.

Solutions for Intravenous Administration

Thorough reviews of fluid therapy in ruminants and options for fluid therapy in calves with diarrhea have been published previously.[2,4] Because acidemia is very common in calves with diarrhea, generally these animals require an alkalinizing fluid type to increase their blood pH. These can include:

1) **Lactated Ringer's (LRS)** - A traditional isotonic fluid that is sometimes be used to correct dehydration and electrolyte abnormalities in neonatal ruminants. Lactate is a metabolizable base and therefore LRS is considered an alkalinizing fluid (that can increase blood pH). However since the lactate must be metabolized to produce an alkalinizing effect, this fluid type is considered to have weak or slow alkalinizing ability and is not recommended for neonates with severe acidemia. Although LRS can be successfully used to treat dehydration and electrolyte abnormalities in neonates, it is difficult and expensive to administer in the field requiring intravenous catheterization, delivery equipment, animal restraint, large fluid volumes (3 to 5 liters in a calf depending on size and degree of dehydration), and monitoring. A theoretical disadvantage of commercially available LRS is that the lactate is a racemic equimolar mixture of L-lactate and D-lactate and its use should be avoided in severely acidemic calves since D-lactate concentrations may already be elevated (Lorenz)

2) **Acetated Ringer's** – Similar to LRS but contains acetate instead of lactate as the metabolizable base. Would also be considered a weak or slow alkalinizing fluid type and would have to be given in large volumes to correct dehydration in diarrheic calves. Acetated Ringer's solution is theoretically superior to lactated Ringer's solution because acetate is metabolized faster, therefore alkalinization is more rapid. In addition, acetate would not exacerbate the D-lactic acidosis present in the majority of calves with diarrhea.

3) **Isotonic sodium bicarbonate** – Sodium bicarbonate is often referred to as a "strong" alkalinizing fluid since bicarbonate does not have to be metabolized by the liver to have an alkalinizing effect on the blood. Sodium bicarbonate has proven to be more effective than other metabolizable bases (such as lactate or acetate), bicarbonate precursors, or synthetic bases. Isotonic sodium bicarbonate is often given as a 1.3% solution and can be easily prepared by adding baking soda ($NaHCO_3^-$) to sterile (or distilled) water at 13 grams per liter (155 mEq/L HCO_3^-) and administered via an intravenous catheter. The amount of isotonic bicarbonate required to correct an acidemia is usually calculated based on either blood total CO_2 or bicarbonate concentrations, or base excess values but usually ranges between 2 and 5 liters depending on the calf's weight and severity of acidosis (which can be estimated using depression scores). Base excess values calculated from blood gas analysis or estimated from depression scoring charts are multiplied by body weight, and with a factor that considers the volume of distribution for bicarbonate ions in the body (generally 0.6) according to the following formula.

Bicarbonate requirement [mEq] = body weight [kg] x base deficit [mEq/L] x 0.6 [L/kg]

Another simple but successful rule of thumb is to administer isotonic sodium bicarbonate solution at approximately 10% body weight over a period of several hours (ie. 4 liters to a 40 kg calf). The disadvantage of isotonic sodium bicarbonate is that it still requires an intravenous catheter and the administration of a relatively large volume of fluids which can sometimes be difficult to accomplish under field conditions.

4) **Hypertonic saline** - Over the past 10 years we have discovered that hypertonic saline (2400 mOsm/L) can be used to rapidly expand plasma volume in a severely dehydrated calf. When combined with oral electrolyte solutions, this therapy can be as effective in resuscitating severely dehydrated calves as large volume lactated Ringer's administration and is less expensive and much easier to administer. Hypertonic saline solutions can be purchased commercially in 1000 ml containers and should be given at a rate of 4 to 5 ml/kg administered slowly over a 4 minute period. Keep in mind however that hypertonic saline does not correct an acidemia. Hypertonic saline is also indicated for the treatment of hyperkalemia in calves. It is very effective in rapidly decreasing serum potassium concentrations and reversing electrocardiographic abnormalities associated with hyperkalemia.

5) **Hypertonic sodium bicarbonate** – In recent years, the use of hypertonic sodium bicarbonate (HSB) combined with oral electrolytes has gained popularity for the correction of acidosis and dehydration in neonatal ruminants with diarrhea. Generally HSB is commercially available as an 8.4% solution which contains sodium bicarbonate at 1 mEq/mL of solution. The total osmolality is approximately 2000 mOsm/L. Therefore the product should theoretically generate an osmotic movement of water and electrolytes from the gastrointestinal tract to the extracellular fluid space similar to hypertonic saline. However an added benefit would be that it is an alkalinizing fluid and should significantly increase blood pH at the same time. Although there are some theoretical disadvantages to HSB administration, a study in anesthetized calves demonstrated that rapid administration of HSB was safe when administered to anesthetized calves.[1] It was effective in reversing an experimentally-induced acidemia and did not cause cerebrospinal fluid pH to decrease (paradoxical CSF acidosis) as has long been hypothesized. Other recent studies have also been done using HSB in calves with diarrhea. In one study, 28 calves with naturally occurring diarrhea were divided into 2 groups.[8] One group received hypertonic saline (5 ml/kg of body weight over 4 minutes) and the other group received hypertonic (8.4%) sodium bicarbonate (10 ml/kg of body weight over 10 minutes). Intravenous fluids were followed by 3 liters of an isotonic oral electrolyte solution. During the 72-hour period following treatment, more calves recovered that had received the HSB as compared to hypertonic saline (many calves that received only hypertonic saline required additional fluid therapy). Another study done in Turkey involved 50 calves with diarrhea, dehydration, and strong ion acidosis.[3] Thirty calves in this study received isotonic sodium bicarbonate (65 ml/kg of body weight over 3 hours) while 20 calves received 8.4% HSB (10 ml/kg of body weight over 20 minutes). No oral electrolytes were given to either group. Although isotonic sodium bicarbonate was able to increase plasma volume to a greater extent than HSB, both fluid types were effective in rapidly increasing venous pH and resuscitating calves. Although more studies are needed to determine the proper dose of HSB for use in calves, this can be an effective fluid type for correcting metabolic acidosis. It can be used in calves with diarrhea when an intravenous catheter or long-term fluid therapy is not practical. However, HSB should probably not be used in diarrheic calves that have concurrent respiratory disease, since these calves may not be able to effectively exhale the excess CO_2 generated in buffer reactions.

In summary, alkalinizing fluids are the appropriate choice for the intravenous rehydration of calves with diarrhea and dehydration). At present, sodium bicarbonate is the recommended solution for intravenous treatment for diarrheic calves that have lost their suckle reflex, palpebral reflex and are unwilling to stand. Sodium bicarbonate (either hypertonic or isotonic) rapidly corrects both acidosis and dehydration, and will restore normal cellular function. When the calf's suckle reflex is reestablished, further treatment can be given orally.

REFERENCES

1. Berchtold, JP, et al. J Vet Int Med 2005; 19;240.
2. Berchtold J. Vet Clin Food Anim 2009; 25:73.
3. Coskun A, et al. J Am Vet Med Assoc 2010; 236:1098.
4. Constable PD. Vet Clin Food Anim 2003; 19:557.
5. Constable PD, et al. J Vet Int Med 2005; 19:581.
6. Constable PD, et al. Vet J 2001; 162:129.
7. Foster DM, et al. Vet Clin Food Anim 2009; 25:1.
8. Koch A, et al. J Vet Intern Med 2008; 22:202.
9. Lorenz I, et al. Vet Clin Food Anim 2014; 30:317.
10. Nouri M, et al. J Vet Intern Med 2006; 20:620.
11. Sen I, et al. Am J Vet Res 2006; 67:1377.
12. Sen I, et al.. J Am Vet Med Assoc 2009; 234:926.
13. Smith GW. Vet Clin Food Anim 2009;25:55-72.
14. Smith GW, et al. J Am Vet Med Assoc 2012; 241:1075.
15. Smith GW, et al. Vet Clin Food Anim 2014; 30:409-427.
16. Trefz FM, et al.. J Vet Intern Med 2012; 26:162.

Fluid Therapy in Adult Cattle

Geof W. Smith, DVM, PhD, DACVIM (LAIM)
Raleigh, NC

INTRODUCTION

Fluid therapy in adult ruminants is often difficult to accomplish because large volumes are needed, animals must be restrained, proper therapy can be very time consuming, and monitoring is often impossible. For these reasons, fluid therapy is often avoided in adult ruminants, however there are clinical situations where either oral or intravenous fluids are necessary and cannot be avoided.

Indications for and Principles of Fluid Therapy in Adult Ruminants

Obviously the most common indication for fluid therapy in adult ruminants is to correct dehydration that can occur as a sequella to any number of primary conditions. However accurately predicting the degree of dehydration in adult ruminants is often difficult. A well designed study has been published demonstrating the eyeball recession and skin tent duration are the most accurate indicators of dehydration in neonatal calves and provides validation for estimating the percent dehydration based on extent of eyeball recession or duration of skin tent.[2] Unfortunately this type of study does not exist in adult ruminants and therefore we can only roughly estimate the extent of dehydration as follows:

Mild dehydration (6-8%)	slight eyeball recession, skin tent slightly prolonged (2-4 seconds), mucous membranes moist
Moderate dehydration (8-10%)	eyes obviously sunken, skin tent obviously prolonged (4- 8 seconds), mucous membranes tacky
Severe dehydration (10-12%)	eyes severely sunken into orbits, skin remains tented indefinitely, mucous membranes dry

When treating dehydrated cattle, the practitioner must not focus solely on correcting the degree of dehydration, but must also consider the animal's maintenance fluid requirements. Non-lactating cattle require at least 50 ml/kg of water per day for maintenance and lactation adds to the requirement depending on the cow's production level.[6]

We must also remember that adult ruminants rarely develop metabolic acidosis. In a study of over 500 cattle older than 1 month of age, blood gas and serum electrolyte analyses determined that 60% of dehydrated animals had a normal blood pH with most of the remaining cattle exhibiting a metabolic alkalosis.[5] Therefore alkalinizing fluids (such as lactated Ringer's or sodium bicarbonate) or oral electrolyte solutions containing alkalinizing agents (bicarbonate) are rarely indicated in adult ruminants except in cases of grain overload, hepatic lipidosis, and in occasional cases of choke (excessive salivation leads to bicarbonate loss) and severe gastrointestinal disease (small intestinal strangulation, mesenteric torsion, etc).

ORAL FLUID THERAPY

Oral electrolyte solutions have classically been used to replace fluid losses and correct electrolyte abnormalities in adult ruminants because they are cheap and easy to administer on-farm. Since most dehydrated cattle have a metabolic alkalosis, it is important to use a non-alkalinizing oral electrolyte solution that does not contain bicarbonate, acetate, or propionate. We generally don't attempt to correct an alkalosis by administering acid, instead our goal is to provide extracellular anions in relative excess to cations.[6] In practice, this is accomplished with chloride-rich, high potassium solutions. By simply adding NaCl (7 grams/L), KCl (1.25 grams/L) and $CaCl_2$ (0.5 grams/L) to a liter of water (or 140 grams NaCl, 25 grams KCl, and 10 grams $CaCl_2$ in 20 liters or roughly 5 gallons of water), a non-alkalinizing oral electrolyte solution for adult ruminants can be created that will effectively rehydrate animals without alkalinizing the blood pH. Feed-grade salts are weighed out on a small laboratory scale and combined into individual bags (such as Whirl-Pak® or Zip-Lock). Once prepared, these bags can be carried on a truck and have a long shelf life. Most of the commercial oral electrolyte solutions made for cattle are designed for dehydrated calves which typically have a metabolic acidosis. Therefore the majority of these products contain bicarbonate and are not indicated for use in adult ruminants. Some products such as Bovine Bluelite® (Techmix Inc., Stewart, MN) or Magna-Lyte (First Priority Inc., Elgin, IL) contain no bicarbonate and could be used to rehydrate adult cattle. However the homemade solutions are more economical and likely more effective. Oral fluids typically can be used successfully in cows with mild to moderate dehydration, however in cows with severe dehydration or in those that are truly "toxic," intravenous fluids are generally preferred.

INTRAVENOUS FLUID THERAPY

In cases of moderate to severe dehydration or with specific electrolyte abnormalities, intravenous fluid therapy is indicated. Although it is frequently avoided in adult ruminants due to reasons listed above, there are certain clinical situations for which there is no substitute. Fluid types commonly used in adult ruminants include:

Saline or Ringer's solution

Administration of iostonic fluids is difficult and expensive to accomplish in the field requiring intravenous catheterization, delivery equipment, and monitoring. However for animals in severe shock, large volumes of isotonic fluids are preferred to restore plasma volume, improve blood flow, and replenish electrolytes. For large volume replacement in adult ruminants, non-alkalinizing isotonic fluid types such as saline or Ringer's would be indicated as a good base solution. In cases of severe dehydration, these isotonic, non-alkalinizing solutions are generally recommended for replacement of large fluid volumes in adult ruminants.[6] If mild to moderate hypokalemia is present, potassium chloride can be added at a rate of 20 to 40 mEq/L during routine fluid administration (1 gram of KCl contains 14 mEq of K^+). If mild to moderate hypocalcemia is suspected, a 500 ml bottle of calcium gluconate can be added to 20 liters of fluids intended for intravenous administration. In dairy cattle it is also a good idea to add a 500 ml bottle of calcium gluconate to the first 20 liters of fluids intended for intravenous administration. Studies in cattle with both experimentally induced and naturally occurring endotoxemia have shown significant decreases in serum calcium concentrations. Although the reason for the decrease in calcium concentrations is not fully understood, toxic cattle that are already at risk for hypocalcemia should be supplemented with calcium in addition to what is present in the isotonic fluids. Potassium chloride can be added at a rate of 20 to 40 mEq/L during routine fluid administration (1 gram of KCl contains 14 mEq of K^+).

It is generally assumed that 40 ml/kg/hr is the maximum safe flow rate for administration of intravenous fluids in animals, which is equivalent to approximately 20 liters (5 gallons) per hour in a 500 kg cow. However these maximal flow rates are rarely achieved in practice with conventional catheters and gravity flow delivery systems. For rapid administration of large fluid volumes, a jugular catheter is required (14 gauge or larger) and at least 40 to 60 liters of fluids is recommended over the first 12 to 24 hours in a toxic cow. Large volumes of fluids can often be cost prohibitive in cattle when using commercially available products, however homemade fluids can be made quite economically. By adding reagent-grade salts to filtered de-ionized water, either Ringer's, sodium chloride, or a ruminant electrolyte solution containing additional potassium and calcium can be prepared quite economically. Although cumbersome or impossible to use on farm, fluids prepared in autoclavable 20 liter containers (Nalgene Labware, Rochester, NY, catalog number 2319-0050; available from Fisher Scientific as product number 02-963-2B) are quite practical in clinics that are able to hospitalize cattle. Pieces of old simplex tubing can be used to connect fluid delivery lines to the spigot on the containers. Rapid administration of large volumes of non-alkalinizing fluids in the severely toxic cow will produce better clinical results than oral fluids alone or even hypertonic saline and oral fluids.

Hypertonic Saline

Over the past 10 years we have discovered that hypertonic saline (2400 mOsm/L) can be used to rapidly expand plasma volume in a severely dehydrated animals. When combined with oral electrolyte solutions or oral water (without added electrolytes), this approach can be extremely effective in rehydrating dehydrated cattle and is much easier to administer than volumes of isotonic fluids. Adult ruminants have a large water reservoir (rumen) that enables the animal to go without water for days and then rapidly rehydrate without any negative effects.[3] The main force for water movement across the rumen wall is the gradient of osmolality between ruminal fluid (which is normally isotonic to plasma) and blood perfusing ruminal epithelium. This is the basic principle of hypertonic saline. Basically we increase plasma osmolality in adult ruminants with the administration of hypertonic saline while simultaneously decreasing rumen osmolality through oral administration of water or a hypotonic oral electrolyte solution containing sodium. The increase in osmolar gradient across the rumen wall following this combined treatment will cause a large movement of water from the rumen into the extracellular space, thereby expanding the plasma volume and correcting dehydration.[3]

Hypertonic saline solutions can be purchased commercially in 1000 ml containers and should be given to ruminants at 4 to 5 ml/kg administered slowly over a 4 minute period (approximately 2 liters for an adult cow). Cattle should be immediately given a supply of fresh water after treatment and most animals will drink 5-10 gallons over the next 10 minutes. Cattle that do not drink water within 10 minutes of hypertonic saline should have 5 gallons of water pumped into their rumen. Hypertonic saline should never be given alone without providing the animal fresh water to drink or ororuminal administration of water. This protocol has been shown to be safe and effective for the correction of dehydration in adult cattle.[4,8,9]

Dextrose

Often indicated for cattle in early lactation with severe ketosis, hepatic lipidosis, or hypoglycemia. Glucose as a 5% solution can be administered at a slow rate for several days, however this delivers free water and can cause dilution of serum electrolytes. In general, it is preferable to add 2.5 to 5% glucose to a non-alkalinizing fluid type (ie. Ringer's) and administer a slightly hypertonic solution than to administer isotonic dextrose by itself.

Isotonic Potassium Chloride

Only used for severe hypokalemia in cattle (serum K^+ concentration <2.3 mEq/L) with severe muscle weakness or recumbency. This is usually seen in dairy cattle following the administration of isoflupredone acetate (Predef®), a corticosteroid used to treat ketosis in lactating cows. Isoflupredone acetate has been shown to have mineralocorticoid activity and has been associated with severe hypokalemia in cattle following multiple or higher than recommended doses.[7] Recommended treatments involve the intravenous administration of isotonic potassium chloride (11.5 grams of potassium chloride per liter of sterile water) at a rate of 4 ml/kg/hour. Combined with large doses of oral potassium salts (ie. 200 grams of KCl per day), the hypokalemia can be rapidly corrected.

REFERENCES

1. Carter RR, et al. J Anim Sci 1990; 68:2811.
2. Constable PD, et al. J Am Vet Med Assoc 1998; 212:991.
3. Constable PD. Vet Clin Food Anim 1999; 15:559.
4. Roeder BL, et al. Am J Vet Res 1997; 58:549.
5. Roussel AJ, et al. J Am Vet Med Assoc. 1998; 212:1769.
6. Roussel AJ. Vet Clin Food Anim 2014; 30:429.
7. Sielman ES, et al. J Am Vet Med Assoc 1996: 210:240.
8. Tyler JW, et al. J Am Vet Med Assoc 1994; 204:1949.
9. Tyler JW, et al. Am J Vet Res 1994; 55:278.

Multiomic Approaches to Study Host Pathogen Interactions

Paul J. Plummer, DVM DACVIM (LAIM), DECSRHM
Ames, IA

INTRODUCTION

Over the last several decades, research into the causes and pathophysiology of infectious disease has become a major focus in order to allow for the development of novel treatment and prevention modalities. During that period of time, the development of novel research tools has allowed for the continual improvement in our understanding of these disease processes. Improvements in our ability to cultivate both bacterial and viral organisms played a critical role in the initial identification of a significant number of etiologic agents and improved our ability to diagnose disease. Later, the development of the polymerase chain reaction (PCR) methodology revolutionized diagnostics and allowed for the basis of our modern diagnostic systems that rely heavily on that technology. Likewise, PCR opened the door for the development of real-time reverse transcriptase PCR and allowed, for the first time, the ability to monitor mRNA levels and gene expression of individual gene targets. The development of Sanger based DNA sequencing gave us the ability to sequence portions of both the bacterial and host genomes, however the complexity of the genomes along with the lack of good bioinformatics support (both computer power and bioinformatics tools) maintained the research at a relatively slow pace. During the last decade the development of microarray technology allowed for the simultaneous comparison of gene expression across a larger subset of know gene targets. While all of these technologies have contributed greatly to improving our understanding of the pathophysiology of disease, they have in large part not allowed for the large-scale systems level approach that would allow the global picture of how the host and pathogen interact and impact one another. Also lacking was an appreciation of the fact that the pathogens are present, not as a monoinfectious process, but instead are part of a much more complex microbial community whose members also influence pathogen colonization, growth and replication. The recent development and rapid advancement of next generation sequencing methodologies have for the first time opened to door to a more comprehensive "systems-based" approach to studying the host-pathogen interaction at a molecular scale. These technologies have allowed for whole genome scale studies, on both the pathogen and host side, as well as providing a means by which to look at the bacterial and viral microbial communities on a global scale. As such, the simultaneous application of several next generation based sequencing technologies to models of infectious disease is opening new avenues of research and holds to potential to impact animal health at levels never reached before. The goal of this presentation is to provide the audience with an appreciation of the new technologies and how they can be applied to study pathogen-host interactions on a systems level. While the technical aspects of the process are far beyond the scope of this presentation, we will focus on the basic principles and approaches used in this type of research.

WHAT IS MULTIOMICS?

The suffix *–ome* is widely used in molecular biology to refer to the study of something in its totality. Examples would include genomics, transcriptomics, metabolomics, methylomics and proteomics, which refer to the study of the complete genome, transcriptional profile, metabolic profile, methylome profile and proteome of a given organism respectively. It is important to note however, that the data derived from an –omics approach is highly dependent on the methodology used and may not always represent the "total" view. For instance, the recent discovery of small regulatory RNA molecules in bacteria and eukaryotes has challenged the central dogma of molecular biology (this issue will be further addressed in the next session) but has also challenged RNA purification techniques. In many purification protocols "small" RNA is intentionally removed to prevent contamination with degraded mRNA. As a consequence however, legitimate sRNAs are removed and would not be included in a transcriptomics study that used that methodology.

For the purposes of this discussion, however, we are interested in "multiomics". This term is used to apply to research designs that incorporate the concurrent and integrated evaluation of multiple "–omics" based technologies to study a process at the systems level with a more global perspective. In the specific example of studying pathogen-host interactions this approach would utilize one or more technologies applied to both the pathogen and the host in order to look at the global impact they have on the other.

WHAT TYPES OF "-OMICS" TECHNOLOGIES ARE USED?

While it seems that new –omics technologies are being developed on a regular basis, we will focus on the major technologies that have been developed, validated and are being widely deployed for these types of host-pathogen studies. In the subsections below we will cover the basics of each of these technologies, what they are used for and what types of outputs they each provide. In the context of the "multiomics" paradigm it is important for the reader to be thinking about how the concurrent use of two or more of these technologies would augment and synergize our ability to address a research hypothesis.

Genomics

As the name would imply, genomics is the whole scale evaluation of the genetic sequence of a given organism. With the advent of the next generation sequencing platforms, whole genome sequencing has become commonplace and affordable on both the pathogen and host side of the equation. Current cost estimates for full genome sequencing of the human genome are in the range of $1000[1] and bacterial genomes can be sequenced for as little as $12 (based on success sequencing 100 *C. jejuni* bacterial genomes on a single lane of HiSeq in our laboratory) per genome depending on the scale, genomic size, desired coverage and previous DNA sequences to be used as a backbone for assembly of the organism. By using genomics, the researcher can start to address the questions of how differences in the pathogen genetic sequence impact the disease causing ability of the pathogen. It is likely intuitive that the addition of specific virulence genes to a pathogen might increase virulence, and such changes are readily identified by PCR or genomics. This type of approach is commonly used in veterinary diagnostic medicine where we might test an *Escherichia coli* isolate for the *stx-1* gene. Perhaps, less intuitive is the idea that individual single nucleotide polymorphism (SNPs) present in the gene of a pathogen could result in significant changes in virulence. These types of minor changes (only a single nucleotide change) can only be identified with full genome sequencing and functional genomic comparisons. Comparative genomics relies on the bioinformatics analysis of multiple isolates know to have varying virulence and the identification of specific sequences associated with virulence. Genomics can be applied to multiomic approaches to both better understand the development of pathogen virulence as well as to better understand the role of the host or microbiome on the microevolution of the pathogen.

Metagenomics is a methodology that relies on the use of next generation sequencing to perform large scale genomic sequencing of total DNA collected from a given environment. Using this tool we are able to study bacterial populations at the community level by sequencing the 16S ribosomal sequences (molecular tag used for phyologenetic classification of bacteria) of a representative sample of the organisms present in the total DNA collected from an environment. This approach is termed phylogenomics, and allows for the evaluation of polymicrobial disease processes as well as the co-occurrence of particular bacteria or bacterial communities. In many cases the presence of other bacterial organisms allows for production of metabolites or the development of a favorable environment for pathogen colonization. Metagenomics can also be utilized to sequence the total DNA of a sample focused on what genes are present and what their likely source is. In this approach we can extend the genomic data to evaluate for the presence of target genes of interest (regardless of host), metabolic capacity of the community or functional categories of genes. Both technologies are widely utilized in multiomic approaches.

Transcriptomics

Transcriptomics is the study of the products of RNA transcription at the organisms level. These products could include mRNA, rRNA, tRNA and sRNA (small regulatory non-coding RNAs) depending on the methodology used to purify the RNA from the sample. Evaluation of the mRNA and/or sRNA are the most likely to be useful for improving our understanding of host pathogen interactions and are the typical focus of transcriptomics. Given that upwards of 90% of the RNA in most organisms is rRNA and tRNA, sample processing to remove or minimize the level of these types of RNA is often beneficial.[2,3] By evaluating the expression level (counted as relative number of RNA sequences of a gene or target of interest), we can evaluate how exposure to different environmental conditions, host or pathogen cells or drugs impact the expression of genes. This information provides unique insights into how both the pathogen and host adapt to each other, and how gene expression regulates virulence or immunity. For example by monitoring the transcriptome of both the host and the pathogen over time we can gain important insights into how the host initiates and focuses an immune response while the pathogen regulates expression of its virulence factors.

Metatranscriptomics is an emerging field in which total RNA obtained from an environment is sequenced to determine the community level gene expression in the environment.[4] The benefits of this approach is that the experiment to host and pathogen independent. This means that we don't have to know what organisms are present in the environment (for instance a guttural pouch or a mammary gland) but can still look and see what the community level transcriptional response is doing. When combined from metagenomic techniques and host transcriptomics we can start to get a systems level picture of what genes are present in the environment (from metagenomics) and how are they being expressed (from metatranscriptomics). Furthermore, comparative studies over time or during pertubations start to provide insights into the underlying molecular mechanisms that drive host-pathogen interaction.

Proteomics

Proteomics combines modern mass spectrometry with bioinformatics to evaluate the repertoire of proteins present in a biological sample of interest.[5] The proteins can be either host or pathogen in origin and can be quantitated. Proteomics provides the researcher will the ability to evaluate if changes in gene expression are resulting in actual changes in protein concentration. Such experiments are critical to improving our understanding of post-transcriptional regulation by bacterial and eukaryotic cells. Furthermore, by identifying unique proteins present in specific pathogen-host responses, we have the potential to identify unique biomarkers of disease that can be utilized for early disease detection and diagnostics.

Metabolomics

Advances in mass spectrometry over the last decade have allowed for the development of high throughput systems that can allow for analysis of the presence and quantitation of thousands of metabolites, chemicals produced during cellular growth and activity, in biological samples. [6] These metabolites can be derived from the host or the pathogen of interest and can be measured simultaneously in tissue, urine, blood, milk or any other biologic sample desired. When coupled with metatranscriptomics and proteomics, one can now validate that the changes observed in gene expression via transcription and protein expression via proteomics are resulting in changes in down stream metabolites as expected. In such a case the techniques complement and validate the results of each other by demonstrating that multiple methods of analysis provide the same outcome. Similar to proteomics, and unlike genomics and transcriptomics, this methodology does not rely on next generation sequencing.

Methylomics

DNA methylation is widely recognized as a major driver of gene expression via epigenetic modification of the DNA.[7] While low throughput methods of methylation identification have been utilized for the last decade, more recent development of next generation sequencing platforms that can simultaneously identify methylated nucleotides have revolutionized the study of this epigenetic regulatory mechanism for both bacterial and host DNA. One example of such a platform is the Pacific Biosystems SMRT sequencing technology. Since this approach relies on next generation sequencing it can be easily integrated as part of or along side a genomic sequencing approach.

BIOINFORMATICS

In many cases, the collection of samples and sequencing or mass spectrometry is the easiest component of the process. The "bottle-neck" often occurs at the bioinformatics level. While bioinformatics processes are well developed for each of the individual platforms alone, very few bioinformatics tools have been developed to integrate and evaluate the various platforms together. Hence, at present, most datasets are analyzed first by the platform specific bioinformatics tools, then manual evaluation and basic comparisons are performed across platforms to evaluate for common themes and outcomes. This process is tedious at best and lacks robust methods of statistical analysis. Computing power can also hinder analysis. Many of these processes require significant resources for storage of the data and for analysis of the data. In many cases high performance computing clusters are required and command-line programming skills are essential using predominantly R, PERL and PYTHON. As these cross platform analyses become more common new bioinformatics tools are sure to emerge and become important in furthering this research approach.

CONCLUSIONS

The development and use of multiomic approaches is likely to accelerate our understanding of host-pathogen interactions in veterinary medicine. The technologies continue to be developed as a robust rate and provide new methodologies on a regular basis. The rapidly dropping cost of applying these technologies combined with the high data output provides an excellent cost-benefit ratio for using these experimental approaches. However, the current bioinformatics limitations require the involvement of considerable bioinformatics resources and expertise and slow adoption.

REFERENCES

[1] National Institutes for Health Human Genome Research Institute. "DNA sequencing cost." Accessed at https://www.genome.gov/sequencingcosts/ . Accessed February 2, 2016. [2] Peano C. et al. Microbial Informatics and Experimentation 2013 3:1. [3] He S. et al. Nature Methods, 2010, 7, 807–812. [4] Moran M. Microbe. 2009. Volume 4, Number 7. [5] Jensen O. Nature Reviews Molecular Cell Biology. 2006 7, 391-403. [6] Patti G. et al. Nature Reviews Molecular Cell Biology. 2012 13, 263-269. [7] Smith Z. et al. Nature Reviews Genetics. 2013 14, 204-220.

Beyond the Central Dogma of Molecular Biology in Bacterial Pathogens

Paul J. Plummer, DVM DACVIM (LAIM), DECSRHM
Ames, IA

INTRODUCTION

In 1956 Francis Crick coined the term "the central dogma of molecular biology". He stated it as follows, "The central dogma of molecular biology deals with the detailed residue-by-residue transfer of sequential information. It states that such information cannot be transferred back from protein to either protein or nucleic acid." [1] Underlying this concept was the idea that genetic information flows sequentially from DNA to RNA in a process known as transcription, then from RNA to protein in a process called translation. While the dogma recognizes that DNA can replicate itself to produce more DNA, it generally argues (with a couple notable exceptions) that RNA cannot become or "transfer information" back to DNA and that proteins cannot become or "transfer information" back to nucleic acids. Notable exceptions to this process that are well described and validated are the use of reverse transcription by some RNA viruses and the identification (albeit still somewhat controversial in some circles) of prions or self reproducing proteins. More recently, advances in our understanding of bacterial gene regulation and epigenetics have identified additional mechanisms that fall outside of the dogma described above. The purpose of this session is to cover the basic principles of how these new discoveries influence bacterial pathogenesis.

POST-TRANSCRIPTIONAL REGULATION OF GENE EXPRESSION

Gene expression in bacterial pathogens can be regulated at several different levels. Traditionally, the central dogma suggested that the main regulatory control occurred at the level of transcription and was regulated by a variety of molecular mechanisms including gene promoter activity, sigma factors (proteins that regulate initiation of RNA synthesis) and RNA polymerase. Changes in gene expression were consequently identified by changes in mRNA quantities, measured by RT-PCR, microarray, or more recently transcriptomics. Given that both RT-PCR and microarray technology required one to know what target they were looking for a priori, the principle of the central dogma of molecular biology forced us to focus on traditionally recognized and annotated open reading frames (ORFs). However, with the development of next generation sequencing based transcriptomic techniques, the research community was now able to get a full transcriptome without a priori knowledge of what was there. It didn't take long for people to realize that there was a significant portion of the total RNA that was being transcribed from the opposite strand of ORFs or from intergenic regions with no known gene function. Based on the central dogma, neither of these sites should have transcription occurring at them, thus leading to exploration of the function of these RNA transcripts. These transcripts soon became known as small non-coding regulatory RNAs (sRNAs) due to their relatively short length, their demonstrated role in post-transcriptional regulation of gene expression and their similarity to previously described sRNAs of eukaryotes. The field of bacterial sRNA is still in its formative years and many questions still remain about how these RNA transcripts evolve, function and are regulated. It is clear that many of these sRNAs function through direct base-pairing with mRNA leading to either increased or decreased translation and protein expression. Additional mechanisms identified to date include, direct binding of the sRNA to the DNA to control transcription as well as direct binding of the sRNA to proteins and resultant impacts on protein function. Current estimates suggest that model bacterial organisms likely encode 200-600 sRNAs, most with unknown function at the present time. [2,3]

Identify and Predicting sRNAs

Bioinformatic based in silico prediction of sRNAs has proven to be difficult. While several tools are available, their accuracy is questionable particularly when applied to non-model pathogens. There are several reasons for this. Their short length (25-400 nucleotides) combined with even shorter areas of target nucleotide complementary sequence (as short as 7 nucleotides) makes in silico identification challenging. [4] Consequently, identification still largely relies on next generation sequencing based transcriptomics and co-precipitation of RNA targets with known sRNA chaperones in some species.[5]

TYPES OF BACTERIAL RIBOREGULATORS

sRNAs are often subclassified by where in the genome they are located in reference to their gene targets. *Cis*-encoded sRNAs are encoded on the opposite strand of the target RNA in either the 5'-UTR or a region of the mRNA ORF. By having direct homology to these regions these sRNAs can impact gene regulation with direct binding to the DNA or mRNA, thus impacting transcription or translation. Alternatively, *trans*-encoded sRNAs are encoded on a portion of the genome distant to their target(s). A third classification is known as *cis*-acting mRNA elements. These elements are located within the mRNA where they induce secondary RNA structural changes associated with either the presence of a small molecule ligand (riboswitchs) or in some cases in response to temperature (thermosensors). Where changes occur in the secondary structure, these allow for changes in ribosomal binding and consequently protein expression. Finally, regulation is sometimes associated with extended transcription of untranslated regions (either 5' or 3') that overlap with divergent operons. These are known as extended UTR elements.

ACTIVITIES OF sRNAs

In general, sRNAs can function to either activate or repress gene expression. There are two major pathways by which sRNAs function to activate genes. In the first scenario, the sRNA binds to a target mRNA in such a manner that it exposes a previously sequestered (through secondary folding of the mRNA) ribosomal binding site. By exposing the ribosomal binding site it allows for increased translation of the mRNA and increased protein expression. Alternatively, some sRNAs activate expression through stabilizing the mRNA and preventing its degradation by RNases. The stabilized mRNA can then accumulate to a higher abundance resulting in increased ability for translation and expression. Repression is also accomplished through two major mechanisms. In some targets the sRNA binds to the 5' UTR or ribosomal binding site and hence prevents translation by blocking nucleotide interaction. Alternatively, some RNases recognize and degrade double stranded RNA, so when the sRNA binds to its target this can lead to increased degradation of the mRNA target and repression of gene expression.

One additional form of regulation involves the function of sRNAs as RNA sponges. In this case the sRNA functions to sequester one of several targets, including other sRNAs or protein regulators. In many cases the alteration in gene expression is mediated by the sequestration of the regulatory element, however in some cases the elements may be degraded or altered in their modulating properties.

Riboswitches

Riboswitches occur in some mRNAs with extended structured 5' UTR. In these cases the longer than normal untranslated region develops secondary structure (stem loops or similar structures) that allows binding to small molecules associated with regulation of that gene. Riboswtiches most commonly occur in genes associated with metabolic pathways and the ligands for binding to riboswitches are often the intermediates or the end products of the associated pathway. Thiamine, amino acids, purines, charged tRNAs and magenesium have all been shown to function in riboswitch modulation of some bacterial species. Binding of the aptamer results in a conformational change of the secondary folding structure that can expose the ribosomal binding site or expose a RNase clevage site.

Thermosensors

Similar to riboswitches, secondary structure located in an extended 5' UTR are the key element of thermosensors. However, unlike riboswitches, the stimulus for the change in secondary structure is temperature as opposed to small molecules. In thermosensors a rise in environmental temperature results in denaturation of the double stranded interaction of the RNA resulting in exposure of either the RBS or a RNase cleavage site. Thermoswitches are most commonly found in heat-stress genes of a variety of bacterial species, however they also have been identified in pathogen associated islands and virulence genes. In such cases, the bacteria utilizes the change in temperature to trigger expression of the virulence factor in response of infecting a host. Variability in the length and nucleotide composition of the thermosensor impact the melting temperature and consequently the temperature at which the thermosensor is triggered.

DNA METHYLATION

Similar to sRNAs, DNA methylation has been demonstrated to impart additional information in the DNA sequence beyond the simply order of the nucleotides described by Crick in the central dogma. In bacteria one major role for DNA methylation is the differentiation of the template DNA strand from the nascent newly synthesized strand. This allows DNA repair mechanisms to know which strand is the template and which should be repaired in instances where an incorrect nucleotide is incorporated. DNA methylation mediated by restriction enzyme-modification systems is also important in bacterial gene transfer as it allows the recognition of foreign DNA (either viral DNA from phages or transposons). Finally, it is now well recognized that DNA methylation can also impact gene regulation.[6] Bacterial DNA can be methylated at three different sites. Cytosine can be methylated at the C-5 or N-4 positions while adenosine can only be methylated at the N-6 position. DNA methylation is mediated by methyltransferases using s-adenosylmethionine as substrate.

Recent advances in next generation sequencing have resulted in the ability to obtain methylation data concurrent to sequence data. Particularly the Pacific Biosystems SMRT based sequencing uses minute differences in the length of nucleotide incorporation between DNA template with non-methylated versus methylated nucleotides to recognize methylation sites. This development has allowed for the study of the full methylome and the impact of this modification on bacterial physiology and pathogenicity. It is likely that continued work in this area facilitated by the new sequencing technologies will result in major changes to the paradigm moving forward.

CONCLUSIONS

Sixty years ago when Francis Crick first verbalized the central dogma of molecular biology it seemed pretty simple. DNA was used to make RNA, which was then used to make protein. All one needed to know was the nucleotide sequence of the DNA and one could predict the RNA and protein sequence, and that remains true. However, the idea that "all of the knowledge" contained in DNA, RNA or a protein was contained in their basic sequence is now clearly known to be an oversimplification. In this session we have discussed how recent developments in sequencing technology have allowed for the discovery of several new mechanisms of bacterial gene and virulence regulation. These mechanisms challenge the central dogma by demonstrating that multiple forms of knowledge carried by the nucleotides can be involved in the regulatory process of bacteria. Moving forward, our improved understanding on these mechanisms and the role that they play in bacterial pathogenesis will play an important role in improved diagnostics and treatment interventions for bacterial diseases of interest to the veterinary community.

REFERENCES

[1] Crick F. Nature. 1970. 227 (5258): 561–3. [2]Zhang Y et. al. Bioinformatics. 2004. 20: 599-603. [3]Tran T. et. al. Bioinformatics. 2009 25:2897-2905. [4] Backofen R et. al. RNA Biol 2010, 7. [5] Zhang A. et. al. Mol Microbiol 2003, 50:1111. [6]Low D. et al. Infect Immun. 2001 Dec; 69(12): 7197–7204.

AN UPDATE ON THE INFLUENCE OF HUSBANDRY AND MANAGEMENT ON CALF HEALTH

Ingrid Lorenz, DMV, DMV habil., DECBHM
Poing, Germany

INTRODUCTION

Calf health was prioritized as one of the most important animal health issues facing the Irish livestock industry in an expert -Delphi study conducted on behalf of Animal Health Ireland (AHI)[1]. AHI was founded in 2009 as an industry-led, not-for-profit partnership between livestock farmers, processors, service providers and government, with the goal to improve the profitability, sustainability and competitiveness of Irish livestock farmers and related industries through superior animal health. In 2010 the development of a national program to improve calf health and survival in Ireland was initiated.

In 2010 a group of experts (Technical Working Group, TWG) was commissioned to provide evidence-based advice on calf health and disease management to Irish farmers, agricultural advisers and veterinary practitioners. As an initial step, a three-part review series on calf health from birth to weaning has been generated[2,3,4], specifically to provide a scientific evidence base for the development of advisory tools on calf health, and to identify gaps in current knowledge to be filled with targeted research. Even though the envisaged output was specific for Irish husbandry systems, the scope of the reviews make them useful for the same purpose worldwide.

To provide an update on the most important topics covered in the above cited reviews, a search of the PubMed database was performed in November 2015 using the search term: (diarrh* [tiab] OR pneumon* [tiab] OR nutrition [tiab] OR feeding [tiab] OR husbandry [tiab] OR management [tiab] OR colostrum [tiab] OR housing [tiab]) AND (calf [ti] OR calves [ti]), taking into account peer-reviewed articles published since 2012. Review papers and original research articles that are most likely to have a practical impact on calf husbandry and management are included. Information without citation is summarized from the original review series[2,3,4].

HOUSING

After birth, the dam and the environment in the calving area are the most important source of infection for the dairy calf. For that reason it is recommended to remove the dairy calf from the dam immediately after calving, and to house it in a clean pen without contact with cows or older calves.

Calves are born with functional thermoregulatory mechanisms. Therefore, healthy calves are readily able to deal with outdoor temperatures as long as they receive adequate amounts of energy and are provided with a dry, well-bedded and draft-free shelter.

In general, outdoor housing is superior to indoor housing, and the risk for transmission of infection is lower in individually housed calves than in groups. If calves are housed in groups the risk is lowest in small, stable groups of calves of similar age.

Recently, the behavioral and performance advantages of group housing or housing of calves in pairs have gained attention. Studies have shown that pair housing or social housing of dairy calves reduces behavioral responses to weaning and improves performance during and after weaning[5,6]. It has also been shown that pair-housing early in life improves early starter intake and average daily weight gains throughout the milk feeding period[7]. Mandel et al. concluded that pair-housing could be considered a viable compromise between the disadvantages of group housing, concerning the spread of infectious diseases on one side, and the behavioral and performance implications of individual housing on the other[8]. One study carried out in a research facility did not find a difference in health and spread of infection between individually and pair-housed calves[9], however, the situation could be more unfavorable in a less controlled environment.

It also appears that the advantages of social housing are only evident if calves are fed a higher than traditional nutritional level (see below)[8,10,11].

COLOSTRUM MANAGEMENT

It is commonly acknowledged that successful colostrum management is the single most important factor influencing the future health of the calf.

In dairy cows, the quality of colostrum has been found to be problematic to begin with, decreasing substantially with every hour that passes between calving and first milking. Also, the ability of the calf to absorb immunoglobulins decreases rapidly so that any delay in harvesting the colostrum and feeding the calf increases the risk of failure of passive transfer.

Normal sized Holstein calves should drink at least 3 L of good quality colostrum (> 50 g/L IgG) within 4 hours, and a total of 4 L within 12 hours from birth if nipple fed. Calves that are not ingesting the crucial 3 L should be stomach-tubed. Alternatively, calves can be given 3 L of good quality colostrum by stomach tube within 2 hours.

Feeding colostrum by a nipple feeder is more natural and the absorption of antibodies is slightly better, but to ensure adequate colostrum is fed, it may be necessary to feed colostrum by stomach tube. Leaving the calf with the cow to drink is not a reliable method for colostrum intake in dairy calves.

It has been shown that bacterial counts and IgE levels in colostrum are affected by heat-treatment. Godden et al.[12] compared serum IgG concentration and health in calves fed fresh colostrum with calves fed colostrum that had been heat-treated in a commercial batch pasteurizer at 60 degrees C for 60 min. They found significantly higher serum IgG concentrations, as well as lower risk to be treated for scour, or to be treated for any cause, in calves that had ingested heat-treated colostrum. A further study comparing heat-treated (at 60 degrees C for 30 min) and fresh colostrum with either high or low bacteria count showed that the beneficial effect on IgG absorption when feeding heat-treated colostrum is indeed due to lowering of the bacterial count[13].

Failure of Passive Transfer (FPT) is traditionally defined as serum IgG concentrations below 10 g/l. Since direct measurement of IgG is expensive and not possible as on farm assessment, evaluation of serum total solids by refractometry has been used as a proxy. Values of 52 and 55 g/l have been suggested as cut-off points indicating FPT.

Recent findings indicate that there may be a need to revise these traditional recommendations. Windeyer et al.[14] confirmed that a cut-point of 52 g/l was most predictive of death before 5 weeks of age, however, a higher cut-point of 57 g/l was the most predictive of respiratory disease before 5 weeks of age. In this study total serum solids were a poor indicator for the occurrence of neonatal calf diarrhea. A further study including 1290 heifers from one single calf raising facility, using only mortality as indicator of FPT, suggested serum IgG and TP concentrations of 20-25 g/l and 58-63 g/l, respectively, as best indicators of adequate passive transfer of colostral immunity in dairy calves[15].

Hydrometers for on-farm testing of colostrum quality are commercially available, but fragile and temperature sensitive, and therefore not widely used. Recent studies suggest Brix refractometry as a cheap, easy and reliable alternative for the assessment of colostrum quality on farm. Cut-points for good colostrum quality of 21[16], and 22[17] Brix percent were reported.

POST-COLOSTRAL NUTRITION AND WEANING

Traditionally, dairy calves have been fed milk or milk replacer to an amount of approximately 10% of the calf's body weight (BW) per day (i.e. 4 litres for Holstein calves). This level of nutrition (restricted feeding) allows only for maintenance requirements, and was introduced to encourage calves to eat concentrates as early as possible, and thus minimize costs for relatively expensive liquid feeds. However, starter concentrate intake is negligible in the first 3 weeks of life. Therefore, calves on restricted feeding regimes are, at most, only able to achieve 20 to 30% of their biologically normal growth, with all its associated negative implications for health, future performance and animal welfare. After the first 3 weeks of life, starter concentrate intake increases and the calves start to grow rapidly. It has been known for a long time that calves can grow a lot faster if they are supplied with more nutrients. Calves suckling their dam or otherwise fed ad libitum ingest about 20% of BW per day and reach up to 1 kg of daily weight gain.

Research interest in the benefits of enhanced levels of nutrition in young calves didn't really materialize until the start of this century. The available literature has been reviewed[18]. It is clear from several studies that calves can double their weight gain if fed ad-libitum, or amounts that are close to ad-libitum intake, when compared with traditional restricted feeding. Very high milk rations or ad libitum feeding, on the other hand, can have a negative impact on concentrate intake, which will either delay weaning or cause a growth check after weaning. Khan et al.[18] came to the conclusion that the optimal milk ration is likely greater than that used in conventional practice, but less than ad libitum intake.

An intermediate volume of milk, or an equivalent amount of good quality milk replacer (approximately 15% of the calf's birth weight per day), is sufficient to allow calves to reach over 50% of their growth capacity under moderate weather conditions and is likely to avoid the negative impact on the calf's immune system that is associated with lower feeding levels. This intermediate volume of milk does not decrease starter intake and therefore weaning is not delayed. However, calves will still be hungry as evidenced by increased numbers of unrewarded visits to the automatic feeder, which indicates suboptimal animal welfare[19].

It is also well established by now that enhanced feeding of the young calf increases performance of the cow later in life. Soberon et al.[20] found that pre-weaning average daily weight gain accounted for 22% of the variation in first lactation milk yield in two dairy herds. However, it is not clear if a similar effect can be expected after intermediate feeding.

The choice of liquid feed usually depends on availability and producer preferences. If non-salable "waste" milk is used for feeding dairy calves, pasteurization is strongly recommended. Milk from cows under withdrawal period from antibiotic treatment should not be fed to calves. Milk replacers are lower in energy content than whole milk and vary widely in composition and quality. Products containing non-milk proteins may not be suitable for young calves. Since the protein requirements increase rapidly with increased growth rates, products with increased concentration of crude protein (25 – 27 %) should be used for the recommended higher feeding volumes.

The benefits of feeding forage, in addition to calf starter concentrates, to the pre-weaned dairy calf has been another area of heightened scientific interest. During the past two decades forage feeding has not been recommended for the pre-weaned calf since it was assumed that it would decrease starter intake, and thus delay ruminal development. It is well understood that the presence of fermentation end products (i.e. volatile fatty acids) triggers the development of ruminal papillae, therefore the emphasis has been on the feeding of grain-based starter feed to pre-weaned calves to improve rumen development.

A recent review of the available literature has called the sole reliance on starter concentrate for ruminal development into question[21]. Specifically, the additional offering of forage increases rumen weight, expansion, and motility, and promotes cellulolytic bacteria and protozoa. It also stimulates rumination and the development of the salivary glands, therefore increasing the buffering capacity. Exclusive feeding of calf starter on the other hand increases the risk of lactate production, low ruminal pH and ruminal parakeratosis.

MANAGEMENT OF THE SCOURING CALF

Preferable good management and husbandry would prevent calves from getting diarrhea. However, if diarrhea occurs, appropriate management of the diseased calves is important to avoid poor performance and an additional negative impact on the immune system.

Not much new information is available concerning the management of the diarrheic calf.

Calves with diarrhea lose an average of about 10 % (but up to 20 %) of their body weight in fluid in addition to their normal losses. For this reason oral rehydration therapy is the single most important treatment to be provided to a diarrheic calf.

Good quality oral rehydration solutions (ORS) should be given as one or more additional feedings to counteract the fluid and electrolyte losses caused by the diarrhea.

Not all commercially available ORS are adequate for the treatment of diarrheic calves. Since diarrheic calves more commonly develop severe metabolic acidosis than non-ruminant species and humans with diarrhea, some ORS especially lack in buffering capacity.

Oral rehydration solutions should contain[22]:

- 90 to 130 mmol/L sodium
- glucose and/or other facilitators of the absorption of sodium and water
- an alkalinizing capacity of 60 to 80 mmol/L from bicarbonate or bicarbonate precursors (the strong ion difference can be calculated to assess the alkalinizing capacity: SID = alkalinizing capacity = [Na+]+[K+]-[Cl-])
- 10 to 30 mmol/l potassium

Continued milk feeding

It is recognized that milk feeding does not worsen or prolong the course of diarrhea, despite a somewhat lowered digestive capacity. Rather, withdrawal of milk rapidly results in malnourishment and weight loss. Continued milk feeding not only provides the energy required for weight gain and growth throughout the period of diarrhea, but also provides the nutrients that are necessary for the recovery of the intestinal mucosa.

Antibiotic treatment

The benefits of antibiotic treatment of calves with diarrhea have been reviewed[23,24].

Since the infectious agents commonly causing calf diarrhea are viruses and parasite,s reasons for the use of antibiotics are not self-evident. However, there is evidence that the risk of bacteraemia or septicaemia, as well as bacterial overgrowth of the small intestine, is increased in diarrheic calves. Also, antibiotic treatment has been shown to reduce mortality in diarrheic calves. It is generally accepted that calves with severe systemic disease can benefit from antibiotic treatment. The treatment of systemically unaffected calves, however, is probably not in accordance with recommendations on the prudent use of antimicrobials in veterinary medicine.

If treatment is indicated it should be targeted towards coliform bacteria. Antimicrobials most likely to be beneficial are amoxicillin, ampicillin, cephalosporins, and fluoroquinolones.

However, the European Commission recommends that antibacterials used as last resort in human medicine (e.g. fluoroquinolones, 3rd/4th generation cephalosporins) should only be used in single animals for a limited number of strict indications where other antibiotics fail. Also, the use of fluoroquinolones in food-producing animals is prohibited in the US.

REFERENCES

[1]More S, et al. Prev Vet Med 2010; 95: 198. [2]Lorenz I, et al. Ir Vet J 2011; 64: 10. [3]Lorenz I, et al. Ir Vet J 2011; 64: 9. [4]Lorenz I, et al. Ir Vet J 2011; 64: 14. [5]De Paula Vieira A, et al. J Dairy Sci 2010; 93: 3079. [6]De Paula Vieira A, et al. J Dairy Sci 2012; 95: 3218. [7]Costa JH, et al. J Dairy Sci 2015; 98: 6381. [8]Mandel R, et al. J Dairy Sci 2016; 99: 1695. [9]Jensen MB and Larsen LE. J Dairy Sci 2014; 97: 5035. [10]Jensen MB, et al. J Dairy Sci 2015; 98: 2568. [11]Pempek JA, et al. J Dairy Sci 2016; 99: 569. [12]Godden SM, et al. J Dairy Sci 2012; 95: 4029. [13]Gelsinger SL, et al. J Dairy Sci 2015; 98: 4640. [14]Windeyer MC, et al. Prev Vet Med 2014; 113: 231. [15]Chigerwe M, et al. J Dairy Res 2015; 82: 400. [16]Quigley JD, et al. J Dairy Sci 2013; 96: 1148. [17]Bielmann V, et al. J Dairy Sci 2010; 93: 3713. [18]Khan MA, et al. J Dairy Sci 2011; 94 :1071. [19]de Passillé AM, et al. J Dairy Sci 2011; 94: 1401. [20]Soberon F, et al. J Dairy Sci 2012; 95: 783. [21]Khan MA, et al. J Dairy Sci 2016; 99: 1. [22]Smith GW. Vet Clin North Am – Food An Pract 2009; 25: 55. [23]Constable PD. J Vet Intern Med 2004; 18: 8. [24]Smith GW. Vet Clin North Am – Food An Pract 2015; 31: 47.

Neonatal Calf Diarrhea: Intestinal Bacterial Overgrowth and Systemic Health

Brent Credille, DVM, PhD, DACVIM (Large Animal Internal Medicine)
Athens, Georgia

INTRODUCTION

Neonatal calf diarrhea (NCD) remains a common occurrence throughout both the beef and dairy cattle industries. NCD is responsible for approximately 60% of all mortality in preweaned dairy heifers, 14% of all mortality in beef calves less than 21 days of age, and 23% of all mortality in beef calves older than 21 days of age.[1-3] While dehydration and acidosis contribute to the sometimes unfavorable outcome of calves with diarrhea, overgrowth of intestinal bacteria and translocation of these organisms to the bloodstream play an important role in the disease process. The purpose of this manuscript is review the microbiology of the intestinal tract in the clinically normal calf and contrast that to the microbiology of the gastrointestinal tract in the calf with diarrhea. This manuscript will also review how changes in the intestinal microflora, precipitated by changes in the environment and management, contribute to the morbidity and mortality associated with NCD.

MICROBIAL COLONIZATION OF THE GASTROINTESTINAL TRACT IN HEALTHY NEWBORN CALVES

Much of our current knowledge regarding the microbiota of the mammalian gastrointestinal tract (GIT) comes from species other than cattle. During and after parturition, the mammalian GIT is colonized by a wide variety of microbes. Generally, the process of microbial colonization is a cooperative interaction between host and microbe. Host factors that affect GIT colonization include GIT luminal pH, local immune defenses, and GIT transit time. Microbial factors affecting GIT colonization include adhesion capability and other survival mechanisms. Other external factors affecting microbial colonization of the neonatal GIT include maternal fecal microbiota, method of delivery (vaginal, caesarean), and antimicrobial use.

In most mammals, *Staphylococcus* spp., *Streptococcus* spp., *Enterococcus* spp., and Enterobacteriaceae are the initial microbial colonizers and utilize the available oxygen in the GIT to create an environment suitable for anaerobic GIT flora.[4] Anaerobic microflora, particularly *Bifidobacterium* spp. and *Bacteroides* spp., are the main bacteria present in human infants and play a considerable role in the development of local and systemic immune function.[4] The earliest studies specifically evaluating bacterial colonization of the GIT in the neonatal calf using culture-dependent methods found that *E. coli* and *Streptococcus* spp. were the predominant microbes cultured from all regions of the GIT within 8 hours of birth.[5] However, within 1 week, strict anaerobes such as *Lactobacillus* spp., *Clostridium perfringens*, and *Bacteroides* spp. predominated, a pattern that closely mimics the dynamics of other mammalian species.[5] Studies using culture-independent methods (rRNA cleaveage analysis, 16S rRNA amplification) have found increased numbers of *Lactobacillus* spp. and *Bifidobacterium* spp. throughout the GIT of newborn calves than previous studies.[6] Also, it has been shown that increased numbers of *Faecalibacterium* spp. in the neonatal GIT have a positive effect on calf health and weight gain. Recently, pyrosequencing and quantitative PCR (qPCR) analysis have been used to evaluate the differences in microbial composition at different locations in the neonatal GIT. This work revealed distinct populations of microbes in the content of the large and small intestines of clinically normal calves with the small intestinal contents consisting of > 95% *Firmucutes* and the large intestinal contents harboring a mixed population of *Bacteroidetes* and *Firmicutes.*[7] In addition, it was found that populations of bacteria associated with the small intestinal mucosa were much different than the bacteria associated with the small intestinal contents, with *Bacteroidetes*, *Firmicutes*, and *Proteobacteria* being found in relatively high abundance at the mucosal surface.[7] Thus, from this work, it's clear that fecal-sample based studies may not represent the GIT microbiome as a whole. It is also clear that the composition and diversity of the intestinal microbiome in neonatal calves is unique and new work is beginning to shed light on the importance of the composition of the microbiome to the health of the calf.

MICROBIAL COLONIZATION AND ENTERIC INFECTIONS

While viral, bacterial, and protozoal pathogens have been implicated in the development of clinical disease, NCD is invariably associated with overgrowth of coliform organisms in the small intestine. Indeed, culture-based studies completed in the late 1920's revealed increased numbers of *E. coli* in the abomasum, duodenum, and jejunum of calves with diarrhea.[8] Similarly, more recent studies using culture-based techniques have shown that numbers of *E. coli* increase 5-10,000-fold in the duodenum, jejunum and ileum of diarrheic calves, regardless of causative agent.[9] The largest increase in coliform numbers is seen in the distal jejunum and ileum. Furthermore, this overgrowth persists long after the primary pathogen has gone.[10] Unfortunately, studies using culture-independent methods to evaluate the role of changes in the GIT microbiota are few and far between. One study found a negative association with the abundance of *Faecalibacterium* spp. in the feces of calves and risk of diarrhea.[11] Another study found that calves fed colostrum (fresh and heat-treated) had greater bacterial numbers (as represented by copies of 16S rRNA) than colostrum deprived calves.[12] The aforementioned study also found that calves fed colostrum had lower numbers of mucosa-attached *E. coli* and greater numbers of *Bifidobacterium* spp.[12] Calves deprived of colostrum had significantly greater populations of *E. coli* in the small intestinal contents and attached to the small intestinal mucosa. Of note is that the difference in bacterial populations was most pronounced in calves fed heat-treated colostrum immediately after birth. Calves fed heat-treated colostrum immediately after birth had significantly greater numbers of *Bifidobacterium* in the small intestine than calves in the other study groups.[12] These findings suggests that the type and number of bacteria present in the environment and colostrum, as well as timing of colostrum feeding relative to birth, impact the development of the microbial community in the neonatal calf's GIT. Unfortunately, this study did not evaluate the health of calves in the different treatment groups. Nevertheless, other work evaluating health outcomes in calves fed either heat-treated or fresh colostrum has shown that, when calves are fed heat-treated colostrum, the risk of diarrhea and subsequent antimicrobial treatment is significantly reduced.[13] While the exact mechanism for these findings was not investigated, path analysis suggested a reduction in colostrum coliform content and improved serum IgG levels were responsible for the differences in calf health.[13]

While it is essential to have a basic understanding of the dynamics of microbial colonization of the GIT in the neonatal calf, an understanding of the clinical consequences of the overgrowth of an abnormal flora is just as important. Increased numbers of coliform organisms within the lumen of the gastrointestinal tract lead to altered GIT function, morphologic damage, and increased susceptibility to bacteremia.[10,14] In addition, it has been shown that co-infections with viruses and *E. coli* increase the amount of damage done to the intestinal mucosa and primary viral infection facilitates systemic invasion with normal flora.[14] Studies have demonstrated that calves with diarrhea had significantly shorter intestinal villi with lower activities of intestinal lactase. Other work has shown that intestinal colonization with coliforms, particularly, *E. coli*, is associated with impaired glucose and fat absorption, factors that place affected calves at risk for negative energy balance.[14]

It is likely that commensal GIT microbes have many roles in calf health but the mechanisms by which they prevent disease is likely through manipulation of local immune responses (differentiation and proliferation of mucosal T cell populations, inhibition of local inflammatory responses) and production of compounds (acetate, butyrate) that inhibit the growth of pathogenic microorganisms. For example, work in gnotobiotic mice has shown that, when inoculated with *Bifidobacterium longum* prior to challenge with *E. coli* O157:H7, these mice were significantly less likely to die than mice not inoculated.[15] In addition, inoculated mice had significantly lower levels of expression of *E. coli* virulence genes in GIT contents. Furthermore, less shiga toxin translocated from the GIT to the bloodstream in inoculated mice. In this study, production of acetate by *Bifidobacterium longum* enhanced the immune effector functions of the host epithelial cells, reducing the risk of bacterial translocation and death.[15] Thus, it's clear that colonization of the GIT with abnormal flora has significant consequences on the subsequent health of the animal in question.

INTESTINAL BACTERIAL OVERGROWTH AND SYSTEMIC HEALTH

Numerous studies have shown that calves with diarrhea are at risk for bacteremia. Early work performed in the 1960's showed that colostrum deprived calves that subsequently developed diarrhea were significantly more likely to be bacteremic (14/17 calves, 82%) than colostrum fed calves that developed diarrhea (0/26, 0%).[16] More recent work evaluating dairy calves on a calf rearing facility found that, of the 169 calves with severe diarrhea, 47 (28%) were bacteremic with Gram-negative enteric pathogens (75% of all isolates), particularly *E. coli*, being the predominant organisms isolated (51% of all isolates).[17] In this study, calves with failure of passive transfer (FPT, 44/129, 34%) were more significantly more likely to be bacteremic than calves with adequate passive transfer (3/40, 8%).[17] Furthermore, calves that were blood culture positive had a mean age of approximately 8 days.[17] Another study characterizing the clinical and clinicopathologic features of septicemic calves found that 7/10 (70%) of calves from which blood cultures were submitted were positive for bacterial or fungal growth.[18] Affected calves ranged from 3-14 days of age with a median of 5 days. Again, *E. coli* accounted for most of the organisms isolated (4/8, 50%).[18] Additionally, work from Prince Edward Island found that 103 of 252 (31%) calves with diarrhea that were sampled were blood culture positive.[19] Again, this study showed that bacteremia was significantly more likely to occur in calves with FPT (47/103, 46%) than calves with adequate passive transfer (21/116, 18%).[19] Here, calves less than 5 days of age were approximately 2.5 times more likely to be bacteremic than calves older than 5 days of age. Like previous studies, *E. coli* was the predominant organism isolated and represented 55% of all positive cultures.[19] From this data it's easy to surmise that bacterial invasion of the bloodstream is a common occurrence in calves with diarrhea. One of the most significant risk factors for the development of bacteremia is FPT. Nevertheless, it must be remembered that 8 to 18% of calves with adequate passive transfer can develop bacteremia.[10,17] Therefore, bacteremia should be expected to occur in at least 30% of calves with diarrhea. Furthermore, it is important to note that Gram-negative enteric organisms, particularly *E. coli*, are the organisms most commonly isolated from ill calves. However, it is important to note that, although Gram-negative pathogens predominate, Gram-positive organisms can be isolated in up to 25% of cases.[17,19] As mentioned previously, calves with diarrhea experience an overgrowth of coliform organisms, particularly *E. coli*, within the lumen of the intestinal tract. Given that this bacterial overgrowth is associated with damage to the small intestine and *E. coli* is the predominant organism isolated from the bloodstream of most blood culture positive calves, it is safe to assume that the overgrowth of bacteria in the intestinal lumen is the source of bacteria that invade the bloodstream in calves with diarrhea. Furthermore, if a viral pathogen is present concurrently the amount of damage done to the intestinal epithelium can be infinitely more severe and the risk of bacterial translocation significantly greater.

While it's clear that calves with diarrhea develop bacteremia and that FPT is a significant risk factor for the development of both diarrhea and bacteremia, the effects of bacterial invasion of the bloodstream on calf survival have also been investigated. In one study, blood culture positive calves were significantly more likely to die (57.4% morality risk) than blood culture negative calves (15.4% mortality risk).[17] Another study found that only 3 of 25 (12%) of calves diagnosed with sepsis based on clinical signs survived to discharge.[18] Additional work found that while 81% of blood culture negative calves survived to discharge, only 23 of 78 (29.5%) of bacteremic calves survived to discharge.[19] Thus, it's clear that bacterial invasion of the bloodstream has a significant impact on survival and calf well-being.

Since blood cultures can take several days to yield any results and delays in the implementation of therapy can lead to poorer short and long-term outcomes, reliable tools that are based on easily obtainable data have been evaluated to help predict the risk of bacteremia in bovine neonates. One clinical sepsis scoring system based on calf signalment and presenting clinical signs was 75% sensitive and 71% specific in predicting bacteremia.[20] In this study, bacteremic calves were usually > 7 days of age and were more likely to have a focal site of infection than blood culture negative calves. Another model developed to predict risk of bacteremia was 40% sensitive and > 90% specific and found that calves < 5 days of age, calves that were unresponsive or comatose, calves with a weak or absent suckle reflex, and calve with abnormal mucous membranes were significantly more likely to be bacteremic than calves without these findings.[19] Other studies have found that signs of dehydration (recession of the eyes in the orbit and prolonged skin tent) and hypovolemia (cold extremities, prolonged jugular fill) are also consistently found in blood culture positive calves.[18] In all studies mentioned thus far a diagnosis of FPT places a calf with diarrhea at significantly greater risk of bacteremia. As a result, it is safe to assume that calves that have a reduced or absent suckle reflex, > 6% dehydration or hypovolemia, or calves that are obtunded or comatose are at a greater risk of bacteremia than calves that appear systemically healthy, regardless of passive transfer status. It is also safe to assume that calves with diarrhea and failure of passive transfer are at risk of bacteremia as well.[10]

CONCLUSIONS

The establishment of a normal intestinal flora has significant implications for both the health and performance of neonatal calves. It is likely that a combination of factors influences the establishment of a normal intestinal flora and changes in the population of GIT microbes can be brought about through alterations in management practices, farm environment, or method of delivery, to name a few. Bacteremia is a common occurrence in calves with diarrhea and it is safe to assume that roughly 30% of calves with diarrhea will be bacteremia. The presence of bacteremia has a significant impact on calf survival and failure of passive transfer is the single most significant factor contributing to an increased risk of bacteremia in calves with diarrhea. Clinical signs can be used to predict the likelihood of bacteremia and assist the practitioner in developing a treatment plan.

REFERENCES

1. NAHMS. Dairy 2007 Part II: Changes in the U.S. Dairy Cattle Industry, 1991-2007. 2. NAHMS. Dairy 2007 Part III: Reference of Dairy CAttle Health and Management Practices in the United States, 2007. 3. NAHMS. Mortality of Calves and Cattle on U.S. Beef Cow-calf Operations. 4. Malmuthuge N, et al. Front Vet Sci 2015;2:36. 5. Smith HW. J Pathol Bacteriol 1965;90:495. 6. Uyeno Y, et al. Lett Appl Microbiol 2010;51:570. 7. Malmuthuge N, et al. Appl Environ Microbiol 2014;80:2021. 8. Smith T, et al. J Exp Med 1925;41:89. 9. Isaacson RE, et al. Am J Vet Res 1978;39:1750. 10. Constable PD. J Vet Intern Med 2004;18:8. 11. Oikonomou G, et al. PLoS One 2013;8:e63157. 12. Malmuthuge N, et al. FEMS Microbiol Ecol 2012;79:337. 13. Godden SM, et al. J Dairy Sci 2012;95:4029. 14. Youanes YD, et al. Am J Vet Res 1987;48:719. 15. Fukuda S, et al. Nature 2011;469:543. 16. Smith HW. J Pathol Bacteriol 1962;84:147. 17. Fecteau G, et al. Can Vet J 1997;38:95. 18. Aldridge BM, et al. J Am Vet Med Assoc 1993;203:1324. 19. Lofstedt J, et al. J Vet Intern Med 1999;13:81. 20. Fecteau G, et al. Can Vet J 1997;38:101.

Antimicrobial Use in Calves with Diarrhea: Considerations for Clinical Decision Making

Brent Credille, DVM, PhD, DACVIM (Large Animal Internal Medicine)
Athens, Georgia

INTRODUCTION

Neonatal calf diarrhea (NCD) remains a common disease throughout both the beef and dairy cattle industries. NCD is responsible for approximately 60% of all mortality in preweaned dairy heifers, 14% of all mortality in beef calves less than 21 days of age, and 23% of all mortality in beef calves older than 21 days of age.[1-3] While dehydration and acidosis contribute to the sometimes unfavorable outcome of calves with diarrhea, overgrowth of intestinal bacteria and translocation of these organisms to the bloodstream play an important role in the disease process.[4,5] The use of antimicrobials in calves with diarrhea is controversial with strong opinions on both sides of the argument. Nevertheless, evidence does exist to suggest that antimicrobials do have a place in treating calves with diarrhea and, when used appropriately, can improve outcome. The purpose of this manuscript is to review the clinical rationale for the use of antimicrobials in calves with diarrhea and provide recommendations to help assist the practitioner with decision making when faced with a critically ill calf.

ANTIMICROBIAL USE IN CALVES – CURRENT PRACTICES AND FUTURE DIRECTIONS

Antimicrobials are frequently added to calf milk replacer to prevent disease, improve feed efficiency, and improve growth. A recent survey of 113 dairy farms in Pennsylvania found that calves on 70% of the dairies were fed milk replacers containing a combination of neomycin and tetracycline.[6] On the national level, 57.5% of dairy producers feed medicated milk replacers to preweaned heifers with neomycin and tetracycline being the antimicrobials most frequently used.[7] In addition, this same study found that 62% of all dairy operations used antimicrobials to treat calves with diarrhea and approximately 20% of all preweaned heifers in the United States were treated specifically for diarrhea.[7] In the case of treatment, β-lactams, sulfonamides, and tetracyclines are the compounds most frequently used.[6] Less data on patterns of antimicrobial use in individual beef calves is available but it is clear that fewer operations use antimicrobials in milk replacer to prevent diarrhea or improve growth and feed efficiency. Approximately 70% of all cow-calf operations use oral or injectable antimicrobials to treat disease with approximately 30% of operations using antimicrobials specifically for digestive disorders in preweaned calves.[8] At the individual calf level, approximately 4% of all calves in the United States are affected with diarrhea and 60% of these calves are treated with either an oral or parenteral antimicrobial. In these calves, sulfonamides and tetracyclines are the antimicrobials most frequently administered orally, while the tetracyclines are the antimicrobials most frequently administered parenterally.[8] Of some concern is the fact that 4% of cow-calf operations in the United States use parenteral fluoroquinolones to treat calves with diarrhea, a practice that is strictly prohibited by the United States Food and Drug Administration (FDA).[8]

New federal regulations will likely have a significant impact on the amount and type of antimicrobials used in calves. First, the practice of feeding medicated milk replacers from birth to weaning was restricted in 2010 and these products may only be fed for a period of up to 14 days. Second, the ratio of neomycin to tetracycline found in combination products was changed from 2:1 to 1:1, a regulation that significantly increases the cost of medicated milk replacers and reduces the overall cost-effectiveness of the herd-level use of these products. Lastly, as of January 1, 2017, using medically important antimicrobials in feed or feed products to improve feed efficiency or promote weight gain will be strictly prohibited. In addition, feed products containing medically important antimicrobials will no longer be available over the counter and the use of these products will require a Veterinary Feed Directive (VFD) to be written by a veterinarian that has a valid veterinary-client-patient relationship established with the operation in question. Internationally, the European Union (EU) has prohibited the sale of milk replacers and animal feeds containing antimicrobials for almost 20 years. In the EU, feeds are sold as non-medicated and antimicrobials are added only if needed for a therapeutic purpose. Thus, it appears that the routine use of antimicrobials in milk replacers and feeds for calves is being reevaluated worldwide and is falling out of favor.

ANTIMICROBIAL USE IN CALVES WITH DIARRHEA – GOALS AND CONSIDERATIONS

The goal of choosing antimicrobials for use in calves with diarrhea is to protect animal health while reducing the unnecessary use of ineffective or inappropriate medications. Also, practitioners should be focused on two important factors when treating calves with diarrhea. First, antimicrobials must decrease the number of coliforms in the intestinal lumen. Second, bacterial translocation from the GIT to the bloodstream must be addressed. In addition, due to concerns regarding the development of antimicrobial resistance in pathogens of human importance, potential impacts of unnecessary or inappropriate antimicrobial use on human health should be considered. Therefore, the following considerations are of utmost importance when using antimicrobials in calves with diarrhea:[9]

1. Administering medications per label recommendations or with the input of a veterinarian if medications are to be used in a legal extra-label manner
2. Selecting an antimicrobial agent with an appropriate spectrum of activity
3. Selecting a dose that maintains therapeutic concentrations at the site of infection (blood and intestinal tract)
4. Treating for an appropriate duration
5. Avoid side effects and violative tissue residues

6. Minimizing the transfer of resistant bacteria from animal to animal or animal to humans

As mentioned previously, calves with diarrhea experience an overgrowth of coliform bacteria, primarily *E. coli*, in the intestinal tract. In addition, approximately 30% of calves with diarrhea will be bacteremic and most of the pathogens (75%) isolated from these calves are Gram negative enteric bacteria with *E. coli* again being the predominant isolate (50%) in most cases.[4] Therefore, antimicrobials chosen for use in calves with diarrhea should achieve concentrations effective against Gram negative enteric pathogens in both plasma and the intestinal lumen.[5,9] Furthermore, antimicrobials should be targeted to calves that are most likely to benefit from their use and not given to every affected calf. Thus, calves that are systemically ill (weak suckle, > 6% dehydrated or hypovolemic, recumbent, comatose) are most likely to benefit from the use of antimicrobials. In addition, calves with blood in the feces and calves that have failure of passive transfer concurrent with diarrhea will likely benefit from antimicrobials because of the risk of bacteremia.[5,9]

ANTIMICROBIAL USE IN CALVES WITH DIARRHEA – THE EVIDENCE

Currently, parenteral oxytetracycline and sulfachloropyridazine and oral chlortetracycline, neomycin, oxytetracycline, sulfachloropyridazine, sulfamethazine, and tetracycline are the only antimicrobials labeled for the treatment of calf diarrhea in the United States. Unfortunately, little evidence exists to support the routine use of these compounds for this purpose. Generally, there are no published studies that include the appropriate control groups or insufficient information exists to make definitive conclusions. However, there is some evidence to suggest that certain products may produce deleterious effects when used routinely. Several studies have shown that the routine inclusion of neomycin and tetracycline in the diet of milk fed calves slows growth rate and tends to increase the proportion of calves developing diarrhea, number of days with diarrhea, and decreases glucose absorption compared to untreated controls.[5] When looking specifically at treatment of calves with naturally occurring diarrhea, one study showed no benefit to twice daily oral administration of neomycin compared to untreated controls.[5] Another study found no benefit to the oral administration of a combination of sulfamethazine and neomycin when compared to untreated controls.[5] Other work has shown that blanket therapy of calves with diarrhea using a combination of parenteral penicillin benzathine, oral trimethoprim-sulfamethoxazole, and parenteral spectinomycin increases number of days with diarrhea and tends to reduce growth rate and grain consumption when compared to targeted therapy of only calves with signs of systemic illness such as fever and obtundation.[7] Furthermore, it was found that targeted therapy could save dairy producers an estimated $10 per calf.[2] Recently, studies have shown that the routine use of neomycin/tetracycline combinations in milk replacer can transiently increase the proportion of resistant *E. coli* isolates in the feces of treated calves.[10] These studies also found that even therapeutic antimicrobial administration increased the proportion of resistant *E. coli* in feces of treated calves.[10] More recently, studies have shown that reducing the use of medicated milk replacer can reduce the prevalence of antimicrobial resistance without negatively impacting calf health.[11] Thus, it's clear that the use of antimicrobials in calves with diarrhea is not benign and carries potentially significant consequences for both the calf and people if appropriate treatment regimens are not followed.

Based on the information currently available it is unlikely that the antimicrobials labeled for use in calves with diarrhea will be effective for their intended purpose. As a result, the extra-label use of drugs approved for use in cattle is justified. Unfortunately little data from controlled studies are available to help guide clinical decision making. Nevertheless, data from experimental infections and pharmacokinetic studies have the potential to help guide practitioners.

It is likely that oral antimicrobials are most appropriate for calves with diarrhea and mild signs of systemic illness (obtundatation and depressed suckle with normal hydration and rectal temperature). Studies evaluating amoxicillin trihydrate (10 mg/kg PO q 12 hrs for 3 days) have found it to be effective in reducing duration of diarrhea and mortality risk in calves experimentally infected with enterotoxigenic *E. coli*.[5] Work comparing amoxicillin trihydrate (400 mg PO q 12 hrs) and ampicillin trihydrate (400 mg q 12 hrs) found similar clinical responses with 80 % of calves in each group responding to therapy.[4] Amoxicillin-clavulanic acid (12.5 mg/kg PO q 12 hrs) has been recommended due to its increased spectrum of activity relative to amoxicillin alone.[6] It is important to remember that the use of the aforementioned antimicrobials and dosing regimens are extra-label and appropriate sources should be consulted regarding appropriate withdrawal times.

In more severely affected calves (recumbency, dehydration, hypovolemia, blood in feces), parenteral antimicrobials are most appropriate due to the ability to achieve higher drug concentrations in plasma.[11] Of the parenteral compounds available, ceftiofur is the most logical choice due to its broad spectrum of bactericidal activity and relatively wide safety margin in compromised patients. One study evaluating the efficacy of ceftiofur in treating calves with experimentally induced salmonellosis found that, when given at a dose of 5 mg/kg IM q 24 hrs, ceftiofur reduced rectal temperature, diarrhea severity, and reduced fecal *Salmonella* spp. shedding when compared to non-medicated calves.[12] Unfortunately, no effect on risk of mortality was seen. While little work evaluating the efficacy of ceftiofur containing products in calves with diarrhea is available, studies in pigs have been done and ceftiofur hydrochloride (2 mg/kg IM once) was shown to reduce the severity of diarrhea and mortality risk in pigs with experimentally induced colibacillosis.[13] Work in mice has shown ceftiofur sodium (5 mg/kg PO q 24 hrs) to be effective in improving outcome in an experimental model of colibacillosis.[14] Previous reviews on the use of antimicrobials in calves with diarrhea have suggested that a dose of 2.2 mg/kg IM or SQ q 12 hrs should be used.[5,9] Unfortunately, as of April 2012, the use of cephalosporins at doses, routes, or frequencies of administration that deviate from those listed on the approved label is illegal. As a result, the previously evaluated or recommended dosing regimens would violate federal law. Recently, a study comparing the plasma disposition of ceftiofur sodium and ceftiofur crystalline free acid (CCFA) when given at their label dose and route was published. In this study healthy neonatal calves were given ceftiofur sodium (2.2 mg/kg SQ once) or CCFA (6.6 mg/kg SQ at the base of the ear once) and the disposition in plasma evaluated.[15] This study found that, when dosed according to the current label guidelines, concentrations of ceftiofur and its related metabolites remained above the MIC_{90} for *E. coli* isolated from the feces of calves with diarrhea (2.0 µg/ml) for 21 hours in the calves treated with ceftiofur sodium and 84 hours in calves given CCFA.[15] The most significant and clinically relevant difference found between groups was the time it took to reach maximum plasma concentrations (T_{max}) with ceftiofur sodium reaching maximum concentrations within 1 hour and CCFA reaching maximum concentrations within 12 hours.[15] This delay in reaching therapeutic plasma concentrations in animals given CCFA represents a relative delay in the initiation of therapy and could potentially impact outcome in certain scenarios, particularly when therapeutic concentrations of antibiotic are needed rapidly. It is important to note that, based on the results of this study, clinicians can have confidence that they are administering a dose at a frequency that should be effective and not violate federal law. A study evaluating the impact of ceftiofur on resistance in enteric bacteria found that while concentrations of ceftiofur in the intestinal lumen following a label dose were sufficient to reduce concentrations of intestinal bacteria, short exposure times (< 4 hours) had the potential to select for resistant organisms, particularly enterococci.[16] While not necessarily important in and of themselves, enterococci are known to be prolific carriers and transmitters of resistance determinants and may play a role in the dissemination of genes that impact therapy in the future.[16] Therefore, it is important to remember that while potentially effective, ceftiofur should be used judiciously and only in calves where it is warranted to reduce the risk of contributing to the emergence of resistant strains of enteric pathogens.

CONCLUSIONS

Antimicrobials are frequently used in dairy and beef operations. The use of antimicrobials is not without risk and unnecessary use impacts not only the health of treated cattle but also has the potential to increase the prevalence of antimicrobial resistance in pathogens of both veterinary and human importance. There are certain situations in which antimicrobials are warranted in calves with diarrhea and there is some evidence that exits to support the use of antimicrobials in an extra-label manner. More data are needed to evaluate the impact of antimicrobial use on the development of resistance in pathogens of veterinary and human importance. As time goes on, the use of strategies that prevent the occurrence of diarrhea will become infinitely more important as it is anticipated that the availability of certain antimicrobials will be limited further limited by governmental regulations.

REFERENCES

1. NAHMS. Dairy 2007 Part II. 2. NAHMS. Dairy 2007 Part III 7. 3. NAHMS. 4. Fecteau G, et al. Can Vet J 1997;38:95. 5. Constable PD. J Vet Intern Med 2004;18:8. 6. Sawant AA, et al. J Dairy Sci 2005;88:2991. 7. Berge AC, et al. J Dairy Sci 2009;92:4707. 8. NAHMS. Antimicrobial Drug Use and Antimicrobial Resistance on U.S. Cow-Calf Operations, 2007-2008. 9. Constable PD. Vet Clin North Am Food Anim Pract 2009;25:101. 10. Berge AC, et al. Appl Environ Microbiol 2006;72:3872. 11. Kaneene JB, et al. J Clin Microbiol 2008;46:1968. 12. Fecteau ME, et al. Am J Vet Res 2003;64:918. 13. Yancey RJ, Jr., et al. Am J Vet Res 1990;51:349. 14. Yancey RJ, Jr., et al. Am J Vet Res 1987;48:1050. 15. Woodrow JS, et al. J Vet Pharmacol Ther 2015. 16. Foster DM, et al. J Vet Pharmacol Ther 2016;39:62.

Methods of Measuring Antibiotic Concentrations in the Airway
Derek Foster, DVM, PhD, DACVIM (Large Animal Internal Medicine)
Raleigh, NC

INTRODUCTION

It is now accepted that drug concentration at the site of infection is a critical piece of information to predict antimicrobial efficacy. In respiratory disease of cattle, the initial site of infection is the airways of the lower respiratory tract as bacterial pathogens move from the upper airways with the assistance of superimposed viral infections and the stress of transport and comingling. Therefore sampling the bronchi and bronchioles would be ideal, though difficult. Furthermore, once respiratory disease is established, bacterial pathogens may invade across the epithelium and interstitial fluid moves into the airways across the damaged epithelium. This suggests that measuring drug concentrations in the interstitial fluid of the lungs may also be useful in understanding the efficacy of antimicrobial therapies. There are multiple methods to determine antibiotic concentrations in the airways of cattle in order to predict clinical efficacy. These approaches range from measurement of drug concentrations in the plasma or lung homogenates to direct sampling of the airways. Understanding the biases and limitations in these techniques is important in applying research findings to clinical disease.

Other considerations include the impact of protein binding on the measured drug concentrations. The unbound drug concentration is widely considered the microbiologically active component. Some sampling techniques will inherently collect only the unbound drug, while most will collect both. In these cases, assessment of measured drug concentrations must be done in light of the drug's protein binding in that particular fluid. As alluded to above, infection can alter the barrier of the respiratory epithelium which would be expected to alter the movement of drugs into the airways. Additionally, it is now recognized that systemic disease of any type could alter the pharmacokinetics of many drugs. Yet most research in this field is done in healthy animals, so one should use caution when interpreting the findings.

INDIRECT MEASUREMENTS

Due to the invasive nature of sampling the airways, drug concentrations in the lungs has often been extrapolated from indirect measurements. Samples may be collected from the plasma, a transudate or exudate in a tissue cage or interstitial fluid. There are inherent advantages of these techniques as the sampling can be humanely repeated overtime in the same animals without requiring euthanasia. This reduces both the cost and animal use associated with the study, and reduces the inter-animal variability. Limitations of these techniques come from the observation that these fluids may not accurately reflect drug concentrations in the airways or lung tissue. Unfortunately, the correlation of drug concentrations between these fluids and the airways can vary dramatically between drug classes, and even within drugs, making extrapolation difficult.

Plasma Concentrations

Historically pharmacokinetic modeling was based primarily on plasma concentrations due to the ease of sample collection. There is still value in assessing plasma concentrations to assess drug absorption, distribution, and elimination. Yet it is now clear that use of plasma concentrations to predict drug concentrations at the site of infection is problematic. This is particularly true of the macrolides. For this class of drugs, antimicrobial concentrations in plasma are low and rapidly eliminated while samples collected from the lung demonstrate significantly higher drug levels for much longer periods of time.[1-3] Most other classes of antimicrobials, on the other hand, typically achieve higher concentrations in the plasma than any other measured areas as demonstrated by the combined enrofloxacin and ciprofloxacin concentrations in steers treated with enrofloxacin (Figure 1).

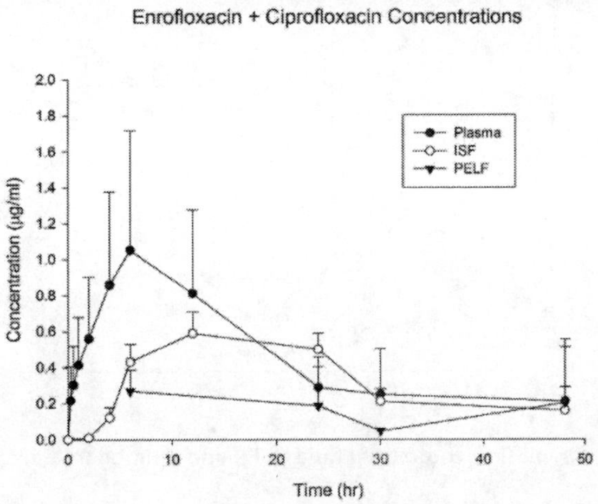

Figure 1. Combined enrofloxacin and ciprofloxacin concentrations in plasma, interstitial fluid (ISF), and pulmonary epithelial lining fluid (PELF) from steers treated with enrofloxacin.

Tissue Chambers

Tissue chambers are perforated, inert chambers that can implanted subcutaneously in a wide variety of species. Initially, acute inflammation leads to fibrin deposition around the cage. Over time, this resolves and a mild, chronic inflammatory process persists, but stabilizes by 40 days post implantation. By this time the chamber is encapsulated by fibrous tissue.[4] Samples can be repeatedly collected by aspirating fluid with a needle and syringe from the chamber. The chambers can be injected with an inflammatory agent to create a sterile inflammatory process or inoculated with bacteria. A non-inoculated chamber in the same animal can be sampled to assess the differences due to inflammation or infection. Due to the large size of cattle and the ease of standing flank anesthesia, insertion of tissue cages in the flank of steers has been routinely used to evaluate the distribution of antimicrobials into these inflammatory transudates and exudates.

Advantages of this approach include the relative ease of placement and sampling. The cages can be as simple as plastic, practice golf balls.[5] Inoculation of the cage with a clinically relevant pathogens such as *Mannheimia hemolytica* allows for determination of pharmacokinetic and pharmacodynamic parameters by repeatedly culturing the fluid to assess drug concentrations and bacterial counts.

Disadvantages of this approach include the long delay from implantation to sampling, and the invasive nature of implanting the chambers. The inflammatory conditions associated with implantation are milder and more chronic than would be expected in most cases of respiratory disease, and the drug concentrations measured will include both bound and unbound drug. Both of these confound interpretation of the pharmacokinetic data derived from these studies. Most studies demonstrate that drug diffusion into the chamber and subsequent elimination lags behind what would be expected, and is driven by the ratio of the surface area to volume of the chamber.[6] In studies in which both are measured, it does not appear that the tissue cage drug concentration reflects the airway concentration[7-8], while correlations between tissue cage concentrations and infected lung tissue are unclear.

Ultrafiltration

Measurement of active drug concentrations in the extracellular fluid may be the preferred method to correlate PK-PD indices to clinical efficacy with some classes of antibiotics as unbound antibiotic concentrations in the interstitial fluid at the target site are responsible for the antibacterial effect in extracellular infections. In order to sample interstitial fluid without implantation of a tissue chamber, ultrafiltration probes have been placed subcutaneously in cattle. The ultrafiltration probes contain three semi-permeable loops connected to a nonpermeable tube extending to the exterior of the animal and attached to a serum collection (red top) vacutainer tube. Samples are collected by changing the vacutainer tube. The vacutainer tube provides the negative pressure for fluid collection through the small pores in the loop membrane. The membrane in the probe consists of pores allowing water, electrolytes, and low molecular weight molecules (<30 000 Da) to pass and excludes the passage of protein, protein-bound drugs, cells, and other large molecular weight compounds. Studies in cattle have demonstrated that sampling of interstitial fluid from the subcutaneous space, pleural cavity, and intramuscularly yields similar drug concentrations.[9]

Due to the small size of the pores in the ultrafiltration probes, protein bound drug is excluded from the sample allowing for direct measurement of the active drug concentration. Further, implantation and removal of the probe is simpler than a tissue chamber and is assumed to create less inflammation, allowing for sampling immediately after implantation. Studies comparing drug concentrations in interstitial fluid with pulmonary epithelial lining fluid demonstrate that they do not directly correspond and interstitial fluid can both over (Figure 1) and underestimate (Figure 2) airway concentrations. Yet it is assumed that these peripheral interstitial fluid drug concentrations would reflect the interstitial fluid concentrations in the lung and would be relevant in cases of respiratory disease, but this has not been proven.

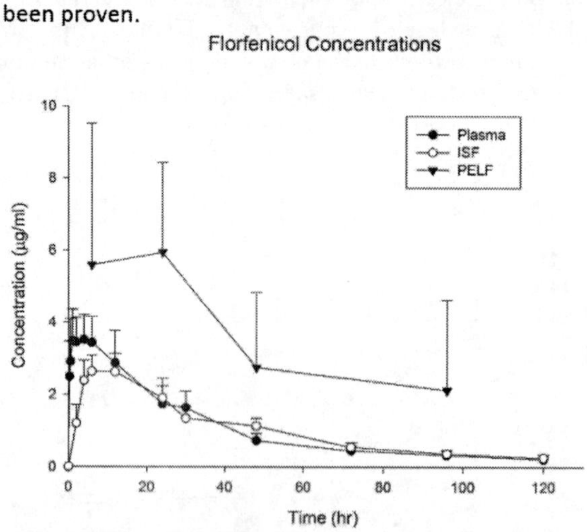

Figure 2. Florfenicol concentrations in plasma, interstitial fluid (ISF), and pulmonary epithelial lining fluid (PELF) from steers.

DIRECT MEASUREMENTS

In order to overcome the inaccuracies of extrapolation from the indirect measurement techniques mentioned above, various methods of directly sampling the lungs of cattle have been used. Due to the morbidity (and mortality in the case of lung homogenates) associated with these techniques, the samples obtained are typically obtained less frequently than plasma or interstitial fluid sampling.

Lung Homogenates

Originally, lung sampling was done by collecting a portion of the lung from calves euthanized at specific time points in the study. The tissue is then homogenized and the drug extracted and measured. The obvious immediate limitation to this technique is the large number of animals required to have adequate replicates at each time point for statistical relevance. The cost, animal welfare concerns, and inter-animal variability associated with this technique has made it largely fall out of favor. Furthermore, the data obtained is likely less clinically relevant than we would like to believe. Certainly any drug in the airway and interstitial fluid of the lung would be measured, but this is confounded by the blood contamination, intracellular drug concentrations and protein binding. These clinically irrelevant concentrations lead to overestimation of lipophilic drugs and underestimation of more hydrophilic drugs.

Bronchoalveolar Lavage

Originally developed as a diagnostic test for various respiratory diseases, bronchoalveolar lavage (BAL) has more recently been used to sample the airways in order to determine drug concentrations. To collect a sample, a catheter is passed through the nose, into the trachea, and lodged into place in a bronchi. There a balloon is inflated, and sterile saline is infused. For animals greater than 200kg, typically 120-240ml of saline is used. After infusion, the saline is rapidly aspirated in an attempt to recover as much as possible. Recovery volumes vary from 10-50% of the infused saline. The recovered saline can be directly analyzed or can be centrifuged to pellet the leukocytes suspended in the saline. Drug concentrations in the recovered fluid versus the supernatant and the cell pellet can differ greatly.[2,10] In order to control for the dilution due to the lavage, a urea correction must be performed. As urea should be equivalent between the pulmonary epithelial lining fluid and the plasma, urea concentrations in the recovered saline can be compared to plasma samples obtained at the same time and control for the dilution.

With this approach a relatively large area of the lung will be sampled as fluid is flooded into the airways distal to the bronchus where the catheter is wedged. The sample will also collect fluid from both larger bronchi and smaller airways as the fluid is lavaged in and aspirated. Disadvantages of this approach include the limited frequency of sampling due to the volume of saline infused. Additionally, the calculated drug concentrations can be extremely variable due to the urea correction method. Increasing the dwell time of the saline by as little as 60s can increase the measured urea concentration in the lavage fluid by 200%.[11]

Swab of Pulmonary Epithelial Lining Fluid

An alternative to BAL is the collection of pulmonary epithelial lining fluid (PELF) via a guarded swab. The guarded swab is passed through either the nose or mouth, down the trachea to a bronchus. The swab is then passed into the bronchus, and PELF is absorbed in the swab. PELF is then extracted from the swab allowing for measurement of drug concentration. Advantages of the swab technique include direct measurement of drug concentrations without the need to correct for dilution and the ability to repeatedly sample the same animal without concern for the volume of fluid required for sampling. Collection of BAL and PELF swabs from the same animals have demonstrated that there is significantly less variability with this technique. Correlation of the drug concentrations between the two collection methods appear to depend on the drug of interest. With enrofloxacin, BAL concentrations exceed the PELF, plasma and ISF, while with tilmicosin, the BAL and PELF concentrations are relatively similar. These reasons for this are unclear at this time. The limitations of this technique include the small area of the lung sampled, which is not as deep as a BAL. The samples cannot be partitioned to assess drug concentrations in the resident leukocytes, and it would be expected that fewer leukocytes would be obtained with the swab compared to a BAL.

REFERENCES

[1]Villarino N, et al. J Vet Pharmacol Therap 2013;37:211. [2]Giguere S, et al. Am J Vet Res 2011;72(3):326. [3]Menge M, et al. J Vet Pharmacol Therap 2011;35:550. [4]Clarke C, et al. Res Vet Sci 1989;47:195. [5]Alibadi F, et al. Res Vet Sci 2003;74:247. [6]Bentgsson B, et al. J Vet Pharmacol Therap 1992;15:202. [7]Halstead S, et al. Can J Vet Res 1992;56:269. [8]McKellar Q, et al. Antimicrob Agents Chemotherap 1999;43(8):1988. [9]Davis J, et al. J Vet Pharmacol Therap 2007;30:564. [10]TerHune T, et al. Am J Vet Res 2005;66(2):342. [11]Rennard S, et al. J Appl Physiol 1985;60(2):532.

Antibiotic Concentrations in the Airways Of Cattle

Derek Foster, DVM, PhD, DACVIM (Large Animal Internal Medicine)
Raleigh, NC

INTRODUCTION

Antimicrobial concentration at the site of infection is a critical piece of information to predict clinical efficacy. In respiratory disease of cattle, the initial site of infection is the airways of the lower respiratory tract as bacterial pathogens move from the upper airways with the assistance of superimposed viral infections and the stress of transport and comingling. Therefore sampling the bronchi and bronchioles has become the gold standard. Furthermore, once respiratory disease is established, bacterial pathogens may invade across the epithelium and interstitial fluid moves into the airways across the damaged epithelium. This suggests that measuring drug concentrations in the interstitial fluid of the lungs may also be useful in understanding the efficacy of antimicrobial therapies. As discussed in the previous presentation, the approach to sampling the airways and interstitial fluid to measure drug concentration can influence the results that are obtained. Nonetheless, multiple studies have been performed to attempt to determine the drug concentrations at these sites.

Antibiotic concentrations in interstitial fluid (ISF) can be predictive of the active concentration necessary for treating most infections. However, the respiratory tract presents another challenge: the diffusion of antibiotics across the blood-alveolar barrier (also referred to as the blood-bronchus barrier in some publications). The concentration of drug that penetrates the blood-alveolar barrier can be assessed by collecting the pulmonary epithelial lining fluid (PELF). The PELF may be an important site of infection in pneumonia, but there are limitations on the interpretation of these data. The importance of adequate antibiotic concentrations in PELF notwithstanding, we also recognize that lung infection can disrupt the alveolar wall and invade the interstitial space. Therefore, a healthy pulmonary epithelium may not represent the actual environment during clinical infection. In addition, during established pneumonia which may occur in BRD, the area of consolidation may not resemble PELF. In reality, both ISF and PELF concentrations may be important to evaluate: ISF drug concentrations may be predictive of respiratory concentrations during infection when disruption of the blood alveolar barrier occurs; but drug concentrations in PELF may be more predictive of drug concentrations in the airway secretions and may be helpful for infecting agents that colonize the airways. It is possible that drugs that attain adequate concentrations in PELF (adequate to meet PK-PD targets) are appropriate to prevent colonization in the airway. When the airway is inflamed and the blood-alveolar barrier disrupted, the infection can spread to the interstitium where the ISF fluid concentration is relevant. Currently prediction of these concentrations based on drug properties is difficult. Protein binding is believed to largely drive ISF concentrations, but the influence of protein binding or lipophilicity on PELF concentrations is unclear.

CEFTIOFUR

Ceftiofur protein binding ranges from 52% and 63%, representing an average fraction unbound (*fu*) of 0.42. Terminal plasma half-life (T½) for the slow-release formulation, as expected, was over 103 hours. The ISF concentrations were lower than anticipated based on the protein of ceftiofur but correspond closely with protein binding of the metabolite, with penetration to the ISF only 5% of the plasma concentration. This is in contrast to a study by Halstead[1] in which they demonstrated much higher penetration into tissue cages after a ceftiofur sodium dose of either 2.2 or 4.4 mg/kg. The measured penetration in that study (calculated from a ratio of the AUC values) ranged from a mean of 47% to 79%, depending on the dose and frequency of administration. This difference can be explained by the method of collection. Tissue cages collect fluid containing protein and other transudate components. Drug concentrations in transudate represent both the bound and unbound form. Our study collected protein-free ultrafiltrate, which represents the biologically active form.

Penetration of ceftiofur and associated metabolites to the PELF was higher and represented 40% of the plasma concentration. This value is close to the degree of penetration reported in the study by Halstead[1] who used a similar method of collecting PELF. They reported penetration of 30.5% and 42.2% at the high and low dose, respectively. These values (from both studies) agree with the fraction of drug unbound (*fu*) described earlier of 0.42. In a review[2] only three comparisons between drug concentration in PELF and plasma were available for β-lactam antibiotics. For each study, the ratios for these β-lactams were < 0.5.

PELF concentrations exceeded the MIC_{90} cited by the manufacturer for BRD pathogens (0.2 μg/mL) for 192 hours. Concentrations in PELF exceeded the MIC breakpoint listed by CLSI for susceptible BRD pathogens (≤ 2 μg/mL) for only a short time at approximately 48 hours after administration. Because of low penetration, concentrations in the ISF never reached a level above the MIC_{90} cited by the manufacturer for BRD pathogens (0.2 μg/mL) at any time point.

ENROFLOXACIN

Terminal T½ for enrofloxacin and its metabolite ciprofloxacin in plasma was 9.23 hr and 14.7 hrs, respectively. This was slightly longer than the half-life from a previous study in other calves using a higher dose.[3] Protein binding was approximately 46% for enrofloxacin and 19% for ciprofloxacin, representing a *fu* of 0.54 and 0.81, respectively. The ISF concentration for enrofloxacin and ciprofloxacin was 52% and 78% of the plasma concentration, respectively, agreeing closely with the expected concentrations predicted from the fraction unbound (*fu*), as one expects the unbound plasma concentration in plasma to be in equilibrium with the interstitial fluid concentration. PELF concentrations were 24% and 40% of the plasma concentration for enrofloxacin and ciprofloxacin, respectively. The PELF concentrations (enrofloxacin + ciprofloxacin combined) exceeded the reported MIC_{90} of 0.06

μg/mL at 48 hours after administration. The PELF area-under-the-curve (AUC) concentrations were 5.72 μg hr/mL. Using a ratio of AUC/MIC > 100, this indicates that AUC in PELF is close to achieving the target for MIC values < 0.06 μg/mL. Using the same target AUC/MIC ratio of 100, the AUC for the ISF of 12.85 μg hr/mL exceeds the target for MIC values of 0.06 μg/mL by a factor of more than 2-fold.

The enrofloxacin and ciprofloxacin penetration from plasma to the tissue fluid (ISF) was similar to predictions and consistent with other studies. But it is unclear what factors caused lower concentrations in PELF. The concentrations of enrofloxacin and ciprofloxacin in PELF were 24% and 40%, respectively of the plasma concentration. However, this appears to be in agreement with a previous study in which enrofloxacin was administered to calves at a dose of 2.5 mg/kg.[4] In that study the authors point out that collection of bronchial secretions in calves with their technique was difficult and there were not enough samples for pharmacokinetic analysis. However, estimating the bronchial fluid concentrations from their figures it appears that peak (Cmax) PELF concentrations were 24% and 45% of the enrofloxacin and ciprofloxacin concentrations, respectively. Their study did not report AUC for the bronchial fluids. In a follow-up study we examined enrofloxacin and ciprofloxacin concentrations in PELF after administration of 12.5 mg/kg to calves and the same technique used in this study. At the higher dose, we observed the drug concentration in the PELF to be 124% of the plasma concentration. Therefore, there are obviously differences among studies, the cause of which are undetermined, but it may be related to differences among groups of calves or because of the higher dose administered in our most recent study.

FLORFENICOL

Florfenicol protein binding was only 5% at the high concentration and was negligible at the low concentrations, representing a *fu* of essentially 1.0. The terminal T½ was approximately 28 hours, agreeing with label information. The ISF concentration was almost 98% of the plasma concentration, as expected from the low protein binding (high *fu*). The PELF concentrations exceeded both plasma and ISF concentrations and were over 200% of the plasma concentrations. The PELF concentrations exceeded the breakpoint of ≤ 2 μg/mL, and the MIC90 for *Mannheimia haemolytica* (1.0 μg/mL) for BRD pathogens for the duration of the study.

TULATHROMYCIN

Tulathromycin concentrations in PELF greatly exceeded the plasma drug concentrations throughout the collection period. PELF exposure was over 9 x higher than plasma. In other studies, the terminal half-life was (mean) 90 and 64 hours in beef calves and Holstein calves, respectively. In this study we calculated a half-life of 81 hours (CV 38%). The reported plasma area-under-the-curve from zero to infinity (AUC) in Villarino's paper was listed as 18,700, 14,100, and 14,000 ng hr/mL, depending on the study.[5] In our study we found 14,478.84 (CV 28.61%). Thus, we believe based on these comparisons that our study was consistent with others with respect to plasma concentrations.

The PELF AUC to plasma AUC ratio in the study by Cox et al.[6] was 53 and in our study was 9.1. The PELF half-life for tulathromycin in Holstein calves was 330 hours[6] and in our study it was 153 hours (CV 52.5%). The peak (Cmax) and AUC in PELF were reported as 3,730 ng/mL and 492 μg hr/mL[6] and in our study was 867 ng/mL (CV 29.5%), and 87.6 μg hr/mL (CV 21.74%), respectively. We calculated the AUC from samples out to 288 hours and the Cox study calculated to 360 hours.[6] But this is probably not enough to account for the large differences in PELF concentrations between the two studies.

The differences in PELF concentrations between our study and the Cox study[6] can be attributed to several factors. PELF concentration measurements are inherently highly variable and affected by the method of collection. The Cox et al study, as well as others listed in the Villarino review[5] collected samples for PELF measurement by bronchoalveolar lavage (BAL). This method flushes the entire airway, including alveolar space. The direct sampling method used in our study samples the bronchial fluid directly. The alveolar tissue is more richly perfused with blood and there is faster equilibrium between blood and the PELF. The bronchial mucosa receives less blood and is slower to equilibrate. This may partially explain the lower concentrations measured in our study.

The discrepancies in sampling techniques notwithstanding, these concentrations measured in this study can be viewed in relation to the reported MIC values for cattle pathogens. As reported by the drug sponsor (Zoetis Animal Health, Florham Park, NJ, USA) BRD pathogens have MIC90 values that range from 1 to 4 μg/mL, with *Histophilus somni* the highest at 4 μg/mL and *Pasteurella multocida* the lowest at 1 μg/mL. At no point, either in plasma or PELF as measured in our study, was the mean concentration above the lowest MIC90 (1.0 μg/mL). In fact, only 1 out of the 6 calves had PELF concentrations slightly above this value. This raises questions about the property of tulathromycin that produces clinical benefits in treated animals. Perhaps, as suggested in other reviews, the anti-inflammatory effects of macrolides are responsible for clinical effects.[7] Tulathromycin was shown to have significant effects on neutrophils and inflammatory cytokines.[8] Obviously, further study is needed to characterize the effects of tulathromycin and other macrolides in animals with pneumonia as there is significant evidence demonstrating their efficacy.

IMPACT OF DRUG PROPERTIES ON ISF AND PELF CONCENTRATIONS

There are differences among drugs in the penetration of antimicrobials into the ISF and PELF that may affect the therapeutic use. It appears that some antimicrobials are best for control of respiratory disease based on high PELF concentrations while others may be more effective for treatment of pneumonia due to high ISF concentrations. In Table 1, we present the penetration values for ISF and PELF in relation to the drug's protein binding (shown as fraction unbound) and lipophilicity (shown as

LogD). For all drugs (or drug metabolite for ceftiofur) there is a relationship between protein binding and the penetration to the ISF; but, a similar relationship is not present for PELF. Likewise, lipophilicity (the higher the LogD value, the higher the lipophilicity) does not appear to influence the penetration into either ISF or PELF. Thus, the properties of protein binding and lipophilicity – often cited as determinants of drug penetration into the PELF – did not have an influence on the penetration of the antimicrobials administered to the calves of this study.

Antimicrobials from different classes have distinct differences in the penetration into both the ISF and PELF that do not always directly correlate with protein binding or lipophilicity. Drugs including florfenicol and ceftiofur with high PELF concentrations are expected to be effective in the control of respiratory disease while those with high ISF concentrations including enrofloxacin and florfenicol may be more effective in treatment of active respiratory infections.

Table 1. Effect of Lipophilicity and Protein Binding on Penetration into Interstitial Fluid (ISF) and Pulmonary Epithelial Lining Fluid (PELF): LogD = log of the partition coefficient at pH 7.4. A higher LogD value indicates a greater lipophilicity (Source: ChemSpider, www.chemspider.com). Fraction unbound is the mean from three replicates. * Data for ceftiofur metabolite was derived from incurred samples (after calves converted ceftiofur to main metabolite).

			Percent Penetration	
	LogD	Fraction Unbound (*fu*)	ISF	PELF
Florfenicol	-0.12	0.95-0.99	98%	261%
Enrofloxacin	0.66	0.54	52%	24%
Ciprofloxacin	-1.26	0.81	78%	40%
Ceftiofur	-1.72	0.42	-	-
Ceftiofur metabolite *	-	0.067	5%	40%
Tulathromycin	-1.06	0.53-0.68	44%	905%

REFERENCES

[1]Halstead S, et al. Can J Vet Res 1992;56:269. [2]Kiem S, et al. Antimicrob Agents Chemotherap 2008;52(1):24. [3]Davis J, et al. J Vet Pharmacol Therap 2007;30:564. [4]McKellar Q, et al. Antimicrob Agents Chemotherap 1999;43(8):1988. [5]Villarino N, et al. J Vet Pharmacol Therap 2013;37:211. [6]Cox S, et al. Intern J Appl Res Vet Med 2010;8(3):129. [7]Giamarellos-Bourboulis E. Int J Antimicrob Agents 2008;31(1):12. [8]Fischer C, et al. Antimicrob Agents Chemotherap 2011;55(1):338.

Methods of Determining Antibiotic Concentrations in the Gastrointestinal Tract of Cattle

Derek Foster, DVM, PhD, DACVIM (Large Animal Internal Medicine)
Raleigh, NC

INTRODUCTION

Fluid from the lumen of the gastrointestinal tract (GIT) is commonly collected in research projects ranging from nutrition to pharmacology in a wide variety of laboratory animals from rodents to cattle. Substances of interest have included nutritional metabolites including electrolytes and volatile fatty acids, drugs, toxic substances, and pH measurement. Yet, several anatomic and physiologic characteristics of the GIT make frequent sampling challenging. Repeated aspiration of a portion of the GIT carries significant risk of peritonitis due to the bacterial flora throughout most of the GIT and significant welfare concerns. Further, repeated sampling of the same area is difficult in the small and large intestine due to the continuous motility of these segments. This motility also makes implantation of a permanent cannula difficult to impossible in most of the GIT. Therefore, these are most commonly only implanted in the rumen or proximal duodenum.

A variety of approaches to collection of luminal GIT fluid have been used. These include euthanasia of multiple animals in order to collect GIT contents at multiple time points, requiring a significant number of experimental animals and yielding a relatively low number of animals per time point. Currently, this approach has only been used in pigs,[1-2] and has not been reported in cattle. Several cannulas and tubes have been used for similar purposes including rumen and duodenal cannulas, and gastrotomy tubes. Implanting these devices requires significant surgical skill and carries inherent risks of leakage, dislodgement, and peritonitis. Additionally, we are somewhat limited in the sections of the GIT into which these devices can be placed. They can also disrupt normal intestinal motility and flow of ingesta as they must be anchored to the abdominal wall.

Measurement of drug concentrations in the feces of cattle has been reported in environmental studies, but the use of these concentrations in pharmacokinetic and pharmacodynamics studies is problematic. These include both protein-bound and unbound drug, and may not reflect the actual concentration of drug higher in the GIT. Drugs may be concentrated by extraction of water in the colon or degraded or transformed to inactive forms by intestinal bacteria.[3-4]

Ultrafiltration probes have been used to collect interstitial fluid (ISF) in a number of species to measure many of these same substances. Typically, the probes are placed subcutaneously, though they can be placed in the muscle, bone, or pleural space. The ultrafiltration probe typically contains 3 semi-permeable loops connected to a non-permeable tube extending to the exterior of the animal and attached to a Vacutainer® tube. The membrane in the loop consists of pores allowing water, electrolytes, and low molecular weight molecules (less than 30,000 Daltons) to pass while excluding the passage of protein, protein-bound drugs, and other large molecular weight compounds. The Vacutainer® provides the negative pressure for fluid collection through the small pores in the loop membrane.

SURGICAL METHOD

The ultrafiltration probes can be placed in the GIT using either general anesthesia or in a standing animal with regional anesthesia of the flank. The right paralumbar fossa is clipped and scrubbed for sterile surgery. A vertical 20 cm skin incision is made 6 cm below the transverse processes of the lumbar vertebrae and approximately equidistant from the last rib to the tuber coxae in the paralumbar fossa. Sharp dissection continues through the external and internal abdominal oblique muscles, transverse abdominal muscle, and peritoneum to enter the peritoneal cavity. The cecum is identified and exteriorized caudally to expose the ileocecal fold. The ileocecal fold is used to identify the ileum, the location of the first probe placement.

The site for probe insertion is on the antimesenteric border of the ileum, approximately 20cm orad to the ileocecal orifice. Intestinal contents are gently milked orad and aborad from the chosen site. A pursestring is pre-placed in the ileal wall using 3-0 polydioxanone (PDS II; Ethicon). A stab incision is made with a number 11 scalpel blade in the center of the pursestring. An introducer needle is placed through the stab incision into the lumen of the ileum. The collecting end of the ultrafiltration probe is inserted through the introducer needle aborally towards the cecum so the entire collecting apparatus is within the lumen. The introducer needle is fed off the probe, and the pursestring is tightened and tied to secure the probe in place and produce a leak-free seal. The remaining 3-0 PDS suture is used to fingertrap the free end of the probe for added security.

The cecum is then reflected cranially to locate and exteriorize a portion of the spiral colon. A probe is introduced into the lumen of the spiral colon in the same manner as the ileum in an aborad direction. The free ends of the ileal probe and spiral colon probe are identified and kept separate. The abdominal contents are replaced in the body cavity, while allowing the free ends of the probes to lightly settle in the abdominal cavity between the omental sling and abdominal wall. The free ends of the probes are exteriorized cranial to the dorsal third of the skin incision by using the introducer needle to create a tunnel from the abdominal cavity through the muscle layers. A skin incision with a number 10 blade further exteriorizes the introducer needle so that the free end of the probe could be fed through to the outside of the body. Mild slack is left in the tubing inside the abdominal cavity to allow normal intestinal motility, and avoiding intestinal entrapment. Fingertrap sutures using 2-0 nylon (Ethilon; Ethicon) are placed to secure the probe tubing to the skin.

SURGICAL OUTCOMES

Surgical implantation of ultrafiltration probes is easily accomplished from a right flank celiotomy using the cecum as the landmark for identification of the ileum and spiral colon. There were no complications postoperatively, and the steers tolerated the probes well. At 3-4 days post-surgery, there were no clinically significant findings at necropsy in any of the 12 steers. The suppurative inflammation found in the peritoneal fluid was consistent with that found after routine abdominal surgery in cattle. Using the probes, we were able to obtain an adequate number of samples for a pharmacokinetic study.

We believe there are several key benefits to use of ultrafiltration probes to collect GIT fluid. The first is the ability to continuously sample the same animals over time without euthanasia, allowing for significant reduction in the number of experimental animals. Additionally, continuous sampling will reduce inter-animal variation in the data. As implantation is relatively quick with a short postoperative period, several animals can be instrumented immediately prior to a study; whereas animals with rumen and duodenal cannulas are typically surgically prepared well in advance and must be maintained between studies.

For pharmacokinetic studies, the collection of a protein-free ultrafiltrate is ideal. This ensures that drug measurements only include the active, non-protein bound drug. Other fluid collection techniques will necessarily include both free and protein-bound drug and, therefore, will overestimate the amount of active drug present in the fluid.

Because of the small size of the ultrafiltration probes, they could potentially be implanted into many different laboratory animals. The locations are not limited to those used in this study, as implantation would be feasible in any surgically accessible portion of the GIT. As there is no need to anchor the loop of bowel being sampled, multiple probes can be inserted in one animal with no impact on intestinal motility. The surgery is relatively atraumatic with rare complications, and we feel this may be a more humane alternative to other approaches of collecting intestinal fluid.

Potential drawbacks to this technique include the longevity of the probes in the GIT as most will cease collecting between two to three weeks after implantation. Further, this may vary from one location to another. In this study, there was no observed decrease in volume of fluid collected over 5 days, in spite of significant accumulation of ingesta on the membrane at necropsy. Further, the fragile membrane can easily be damaged by drier, firmer ingesta. When the probes were implanted in a canine colon, the membranes were rapidly damaged and the tubing clogged.

The size selectivity of the ultrafiltration probes is a distinct advantage for pharmacokinetic studies, but this will hinder the use of these probes in other studies. Nothing larger than 30,000 Da can be collected which will exclude many proteins and all fiber and bacteria. Therefore, clear knowledge of the protein of interest is critical before using these probes for that particular purpose. A common outcome measure of ruminant nutrition trials is a measure of digestibility based on the fiber content of the duodenum which would not be possible with this technique. Finally, it would be ideal to collect bacteria from the GIT over time to measure changes to the flora in response to drug administration or diet changes, but these organisms would be excluded from the fluid collected by this technique. We have attempted to cut open the loops in order to collect bacterial samples. We have been able to isolate bacteria from these modified probes, but they rapidly clog. One final drawback is that the probes are not easily removed when implanted in the GIT by this method, as an additional surgery would be required to remove them. If the animal was to be slaughtered after completing its withdrawal time, the abattoir should be notified of the foreign material as it would not be obvious externally that the tubing and probes were in place. With appropriate precautions though, the probes and tubing could be removed at time of slaughter without contaminating the food supply with foreign material.

Future suggestions for this procedure include a few minor changes to ensure more consistent sample collection. We recommend only leaving enough slack in the impermeable tubing to allow for normal GIT motility, but no more as longer lengths could become entangled in the intestine. Additionally, the probes should be inserted in an aboral direction so that they are less likely to become folded and obstructed as intestinal contents flow by the area. This can be readily achieved in the ileum as the probe can be inserted towards the cecum, but determining aboral direction in the spiral colon is difficult unless centrifugal loops can be distinguished from centripetal loops. Finally, we believe that this surgical procedure could easily be performed in a standing animal. The cecum could be similarly identified via an incision in the right flank, and then used to identify the ileum and spiral colon.
We believe that the use of surgically implanted ultrafiltration probes provides a novel and more humane method of collecting intestinal contents over time from young steers. While not suitable for all applications, this technique has the potential to replace other surgically implanted cannulas and tubes for many sampling procedures.

ANTIBIOTIC CONCENTRATIONS

Antimicrobial concentrations in the GIT of steers administered a single dose of either 7.5 mg/kg of enrofloxacin or 2.2 mg/kg of ceftiofur sodium subcutaneously. Figure 1 demonstrates that the ileal concentrations of the fluoroquinolones appear to be closely associated with the ISF concentrations, potentially suggesting passive diffusion across the epithelium as might be expected for a relatively lipophilic drug with little protein binding. On the other hand, colonic concentrations peaked later and higher than plasma concentrations, which we believe was caused by transit from the ileum and subsequent concentration due to water extraction rather than active transport. In both the ileum and colon, enrofloxacin contributed greater than ciprofloxacin to the total combined concentration.

Ceftiofur, on the other hand, (Figure 2) peaks earlier at 6 hours simultaneously in the ileum and colon. As it is a poorly lipophilic drug with moderate to high protein binding, little ceftiofur diffuses into the ISF. The low ISF concentrations suggest that there is no passive diffusion from the ISF to the intestine, but rather, there may be an active efflux transport mechanism as the driving force from the plasma to the intestinal lumen. These transporters are known to exist in other animals, but have not been studied in ruminants.[5] Ceftiofur is seen to be rapidly eliminated from the GIT which is likely due to degradation by GI flora.[3-4]

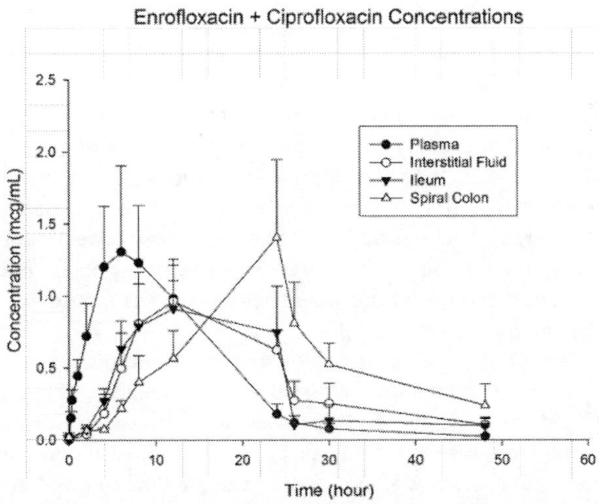

Figure 1. Combined enrofloxacin and ciprofloxacin concentrations in the plasma, interstitial fluid, and GIT of steers administered 7.5 mg/kg of enrofloxacin subcutaneously.

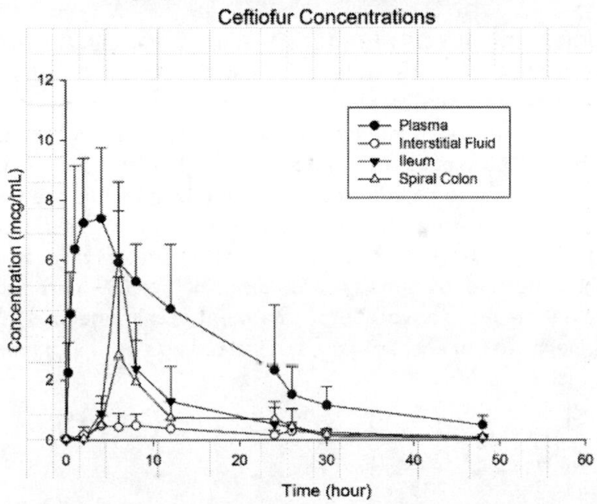

Figure 2. Combined concentrations of ceftiofur and its metabolites in the plasma, interstitial fluid, and GIT of steers administered 2.2 mg/kg of ceftiofur sodium subcutaneously.

REFERENCES

[1]Ferran A, et al. Int J Antimicrob Agents 2013;42:429. [2]Lindecrona R, et al. Res Vet Sci 2000;68:211. [3] Erickson B, et al. Vet Microbiol 2014;172:499. [4]Wagner R, et al. Antimicrob Agents Chemotherap 2011;55:4990. [5]Shugarts S, et al. Pharm Res 2009;26(9):2039.

Impact of Gastrointestinal Antibiotic Concentrations on Antimicrobial Resistance

Derek Foster, DVM, PhD, DACVIM (Large Animal Internal Medicine)
Raleigh, NC

INTRODUCTION

The National Action Plan for Combating Antibiotic-Resistant Bacteria released on September 18, 2014 calls "for agricultural stakeholders to facilitate identification and implementation of interventions to reduce the spread of antibiotic-resistance" and to "identify validated interventions to reduce the spread of antibiotic resistance and work with public and private sector partners to incorporate them into veterinary practice."[1] Antibiotics administered in feed and water are being phased out, but there will always be a need for parenteral antibiotics to treat specific diseases. Parenteral antibiotics administered to treat infections in animals diffuse into the intestine and influence the population and susceptibility of the intestinal bacteria. However, there is little is known about the duration of this effect, particularly on bacteria that might become foodborne pathogens. Earlier attempts have not measured drug concentrations in the animal's gastrointestinal tract (GIT), and rarely assess microbial changes over time. Measurement of active, unbound drug concentrations at the site of action (e.g. GIT lumen) is critical to correlate pharmacokinetic-pharmacodynamic (PK-PD) indices with microbial changes.

Use of therapeutic antimicrobials is critical to cattle health, and many antimicrobials used in cattle are prescription drugs deemed "critically important" for human use. Epidemiological associations between systemic antimicrobial administration in cattle and resistance in enteric organisms,[2-3] and studies reporting the association between antimicrobial use and subsequent presence of antimicrobial resistant bacteria within the feces are of limited value as they usually evaluate only a single bacterial species or isolate.[4] Changes in the microbiome have been observed,[5] but only at a single time point after treatment. To our knowledge, no study has correlated the intestinal pharmacokinetics (PK) of the antimicrobials, shedding of foodborne pathogens, and changes in the microbiome over time after treatment. This is primarily due to an inability to measure active drug concentrations in the GIT.

IN VITRO ASSESSMENT MICROBIOLOGICAL IMPACT OF GI TRACT ANTIBIOTIC CONCENTRATIONS

S. enterica was largely eliminated by the peak enrofloxacin concentrations in 4 hours and by the tail concentrations in 24 hours (Figure 1A). Based on the relatively broad peaks created by the slow rise and decline after the C_{MAX} in both the ileum and colon, we believe this accurately represents the *in vivo* impact. Similarly, the tail concentrations appear to persist for more than 24 hours *in vivo* suggesting that these concentrations could eliminate wild-type *S. enterica* in the GIT. Exposure of *E. faecalis* to the C_{MAX} of enrofloxacin in the ileum or colon, though, led to an initial reduction, but by 24 hours the reduction had stabilized, suggesting the bacteria became tolerant to the remaining active drug (Figure 1B); interestingly, this was not tied to an obvious increase in the MIC. In the United States, the only approval of enrofloxacin for cattle is to treat respiratory disease in beef cattle or dairy heifers less than 20 months of age. Extralabel use of this drug in food animals is prohibited by the FDA in order to reduce the amount used in an attempt to mitigate AMR. This work is not meant to advocate for treatment of salmonellosis in cattle with enrofloxacin as this would constitute an off-label, and therefore illegal, use of enrofloxacin, but instead was used as a model for other similar enteric organisms with a potential for foodborne transmission.

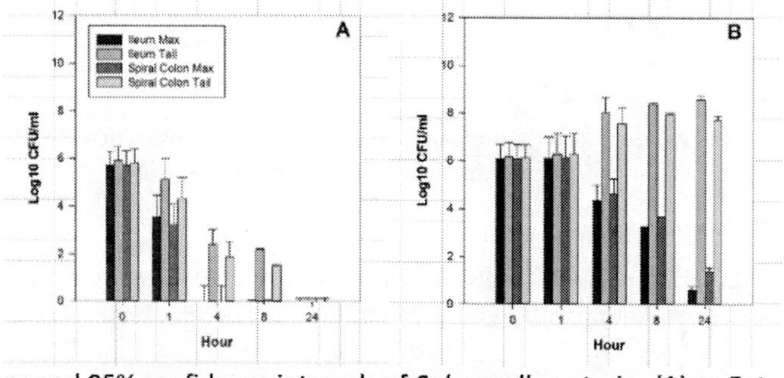

Figure 1. The least square means and 95% confidence intervals of *Salmonella enterica* (A) or *Enterococcus faecalis* (B) recovered from *in vitro* treatments of enrofloxacin at maximum and tail concentrations achieved in the ileum (C_{MAX}=0.7 ug/ml, tail= 0.07 ug/ml) and spiral colon (C_{MAX}=0.85 ug/ml, tail=0.17 ug/ml) of steers.

There was a significant reduction of *S. enterica* by the peak concentrations of ceftiofur, which was only statistically significant after 4 hours of exposure (Figure 2A). Based on the pharmacokinetic data, these peak concentrations persisted for less than 4 hours, suggesting that *in vivo* these concentrations would not to be bactericidal and may be more likely to select for resistant organisms. As expected, ceftiofur did not inhibit the growth of *E. faecalis* (Figure 2B), as this *Enterococcus* species is inherently resistant to cephalosporins. Although we did not expect ceftiofur to inhibit the *E. faecalis*, determining its growth potential at a time of antibiotic selection pressure is significant, especially as enterococci are known to carry and transfer AMR elements. These results do not imply that ceftiofur resistance will be shared, as these are inherent resistance mechanisms, but instead suggest that if additional AMR genes or elements already exist in these populations, they may thrive under this selection pressure.

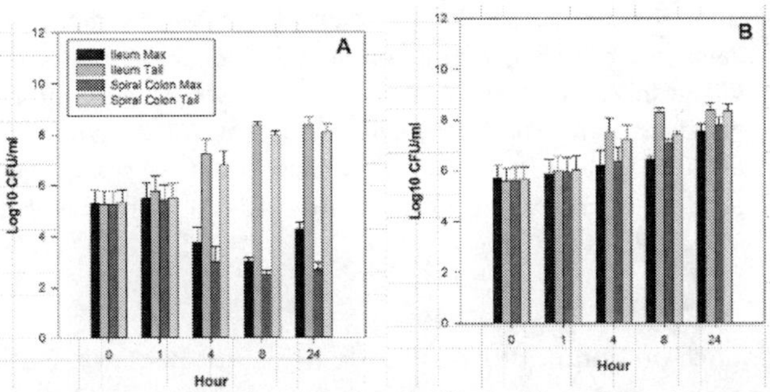

Figure 2. The least square means and 95% confidence intervals of *Salmonella enterica* (A) or *Enterococcus faecalis* (B) recovered from *in vitro* treatments of ceftiofur at maximum and tail concentrations achieved in the ileum (C_{MAX}=4.53 ug/ml, tail=0.05 ug/ml) and spiral colon (C_{MAX}=2.96 ug/ml, tail=0.1 ug/ml) of steers.

IN VIVO ASSESSMENT MICROBIOLOGICAL IMPACT OF GI TRACT ANTIBIOTIC CONCENTRATIONS

Steers were administered enrofloxacin either once (12.5 mg/kg subcutaneously) or once a day (5 mg/kg subcutaneously) for 3 days. After the sampling period and a 1 week washout, steers were reassigned to the other treatment group. Fecal samples were collected every 12 to 24 hours during treatment and for 7 days after the last dose. Fecal samples were collected twice a week from a subset of the calves for 1 month after the last treatment. Samples were cultured on selective media to quantify *E. coli* and *Campylobacter* spp. Ten isolates from each time point were further characterized by microboth dilution to determine the minimum inhibitory content (MIC) of enrofloxacin.

Fecal *E. coli* concentrations decreased significantly immediately after treatment in both groups, but returned to baseline by 72 hr in the high dose group and 144 hr in the low dose group (Figure 3). The impact of dosing on *E. coli* MIC was dependent on treatment order. In the steers that received the high dose first, the median *E. coli* MIC did not change after the high or low dose. In the steers administered the low dose first, the median *E. coli* MIC increased significantly in the low dose group from time 0 to day 8, but returned to baseline by day 12. At that time, they received the high dose, and the MIC again increased significantly at day 7.

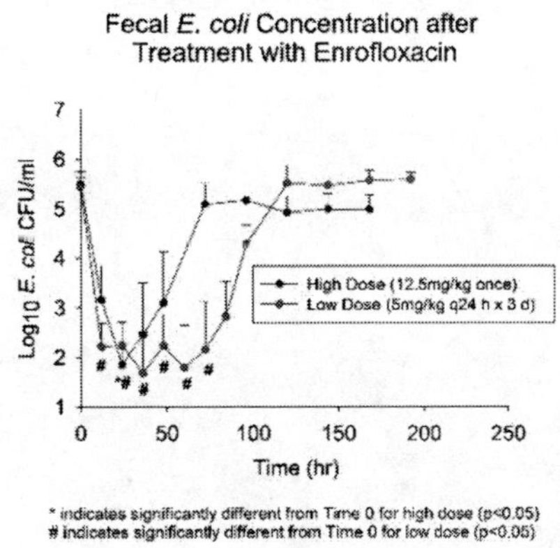

Figure 3. E. coli concentration in feces of steers treated with enrofloxacin at two different doses.

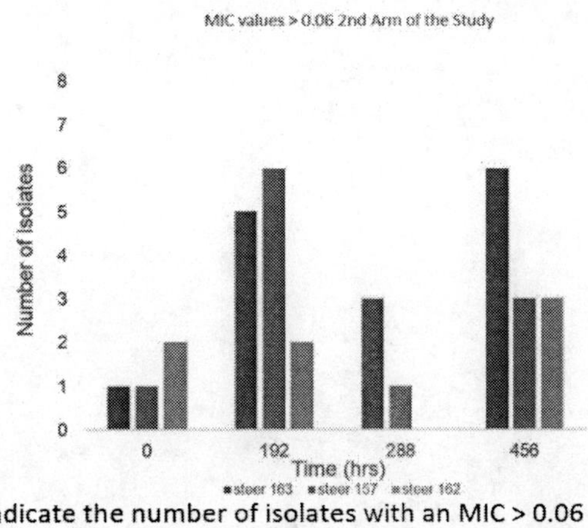

Figure 3. The bars indicate the number of isolates with an MIC > 0.06 for each steer. n = 3

All animals were positive for enrofloxacin-susceptible *Campylobacter jejuni* either by direct plating or by enrichment (or both). *Campylobacter* could not be recovered for the first 24-120h after treatment, but samples from 144-196h post-treatment were again *Campylobacter*-positive, suggesting that the populations of the organism were temporarily reduced below detection levels in response to treatment. Resistance prevalence differed among sampling periods with isolates prior to treatment being *C. jejuni* susceptible to nalidixic acid and ciprofloxacin, while most samples during the washout period and the second treatment period (100 and 75%, respectively) were nalidixic acid-resistant (Nal^R). However, these Nal^R isolates were susceptible to ciprofloxacin and appeared to be *C. fetus*, reported previously to be frequently Nal^R but susceptible to ciprofloxacin. These findings suggest a shift in *Campylobacter* spp. populations in response to enrofloxacin treatment, with Nal^R campylobacters, primarily *C. fetus*, predominating subsequent to the administration of the antibiotic.

REFERENCES

[1]Presidential Advisory Council on Combating Antibiotic-Resistant Bacteria. National action plan for combating antibiotic-resistant bacteria. 2014. [2]Berge A, et al. J Am Vet Med Assoc 2010;236:1338. [3]Tragesser LA, et al. Am J Vet Res 2006;67:1696. [4]Mann S, et al. Foodborne Pathog Dis 2011;8:861. [5] Chambers L, et al. PLoS One 2015;10(8):e0133764.

Phosphorus Balance Disorders in Cattle: What We Know and What We Don't

Walter Grünberg, DMV, MS, PhD, DECAR, DECBHM, assoc. DACVIM (Food Animal)
Hanover, Germany

INTRODUCTION

Phosphorus (P) homeostasis in dairy cattle has received increased attention in the last decades because of increasing concerns about the environmental impact of excessive amounts of P in manure but also because of the possible association between hypophosphatemia and the downer cow syndrome. Legal incentives to reduce environmental pollution with P through manure have led to a critical reassessment of the dietary P requirements of ruminants. Current estimates of daily dietary P requirements for cattle are lower than estimates used in the 70ies and 80ies of the last millennium but have been found to be adequate for most of the lactation cycle.[1] Concerns remain however because of the high incidence of hypophosphatemia in dairy cows during the first days of lactation, a period characterized by the sudden increase of P requirements for milk production in combination with low dietary P intake due the periparturient feed intake depression.

PHOSPHORUS AND PHOSPHATE

Although the terms phosphorus (P) and phosphate (or inorganic phosphorus, Pi) are often used synonymously it is essential to understand that P refers to phosphorus present in any organic or inorganic form in the body (i.e. total P), while the Pi fraction only comprises the immediately biologically available fraction of P dissolved in body fluids. While P bound in organic molecules (e.g. phospholipids, DNA, carbohydrate metabolites) is critically important for the organism, only P present as dissolved Pi is available for biochemical reactions that depend on the availability of P such as the phosphorylation reaction or the synthesis of ATP- or creatine phosphate. In other words the "currency for trade and traffic" of P in the body is Pi. Pi can thus be considered as an index of "cash flow" while total P would be comparable to the total capital.

PHOSPHORUS HOMEOSTASIS IN DAIRY CATTLE

The organism's P balance is dictated by the equilibrium between uptake and excretion of P to which the exchange of P between bone and extracellular space as well as between intra- and extracellular space have to be added. Uptake of P exclusively occurs through the gastrointestinal tract from where P is absorbed and then transferred to the extracellular space for further distribution throughout the body. Absorbed P may be retained and transferred into bone or the intracellular space of different tissues or excreted from the organism.

Feces and (in lactating cows) milk present the most important routes of P excretion in ruminants.[13] Fecal phosphorus consists of unabsorbed dietary P, endogenously lost P as well as unreabsorbed salivary P. Excretion of P in saliva is a peculiarity of ruminants that assures adequate P supply to ruminal microorganisms and provides a buffer to rumen fluid. In contrast to monogastric species in ruminants under physiological conditions, renal excretion of P in quantitatively relevant amounts only occurs in preruminating young animals as well as in non-lactating ruminants fed a high concentrate diet with high P content.

In high yielding dairy cows the mammary gland contributes considerably to daily P losses, with the amount of P excreted in milk easily exceeding 40 g in high yielding dairy cows. This P drain presents a considerable challenge to the mechanisms regulating P homeostasis particularly at the onset of lactation.

Exchanges of P between bone and the P pool are balanced and quantitatively near nil in adult animals with adequate P supply. Growing animals have a net flux of P towards bone to assure adequate mineralization of the bone matrix while older and P depleted animals tend to have a net flux of P from bone to the P pool. Acidogenic diets trigger the demineralization of the bone matrix, thereby releasing calcium (Ca) and Pi. Longstanding metabolic acidosis therefore results in a net flux of P together with Ca from bone to the P pool.[2]

The equilibrium between intracellular and extracellular space is a major factor affecting the extracellular and thus the serum Pi concentration ([Pi]). The regulation and pathway of P transport in and out of the cells are still poorly understood. Changes in the equilibrium of Pi between intra- and extracellular space can occur suddenly and have a considerable impact on the extracellular [Pi]. Factors affecting the compartmental redistribution of Pi are among others the insulin and catecholamine secretion of the body or the acid-base status of the organism. An increase in plasma insulin concentration as observed after feeding or parenteral dextrose administration increases the uptake of Pi by insulin responsive cells, thereby resulting in a decrease of the extracellular [Pi].[3,4]

In vertebrates 80 to 85% of the body's P is located in the skeleton primarily in the form of insoluble salts which present the biologically inert storage form of P. Approximately 15 to 20% of the body P is distributed in fluids and soft tissues of the body, thus forming the body P pool. The bulk of the body P pool is located in the intracellular space, while less than 1% of the total body P content is found in the extracellular space, which includes blood.

PHOSPHORUS DEFICIENCY AND HYPOPHOSPHATEMIA

Hypophosphatemia refers to a decline of the serum [Pi] below the established reference range but does not necessarily need to be associated with P depletion of the body as a subnormal serum [Pi] can be the result of inadequate dietary supply or excessive loss, a compartmental shift of P from the extracellular into the intracellular space or a combination of these mechanisms.

Only the former mechanism is consistent with P depletion of the organism, while a compartmental shift of P to the intracellular space results in hypophosphatemia without a concomitant reduction of the total P pool size.

Hypophosphatemia occurring in cattle in early lactation is of multifactorial etiology.[5] The sudden onset of milk production is without doubt an important contributing factor to the deregulation of the P homeostasis around calving and the decline in dry matter intake observed in the last week before calving is likely to contribute to hypophosphatemia around parturition or at least to hamper counter regulation. It is noteworthy though that a marked drop in serum [Pi] around calving is also observed in mastectomized cows not producing any milk.[6,7] This observation underscores that factors other than milk production must be involved in the etiology of periparturient hypophosphatemia of dairy cows. It ensues that the plasma [Pi] is neither representative of the intracellular Pi nor the total body P status.

CLINICAL SIGNS ASSOCIATED WITH HYPOPHOSPHATEMIA

A plethora of clinical signs and conditions such as unthriftiness, anorexia, pica, impaired growth and fertility, muscle weakness and recumbency, intravascular hemolysis, osteomalacia and many other have been associated to P deficiency in ruminants. Chronic P depletion caused by long term ingestion of P deficient diets or ingestion of compounds binding P in the gastrointestinal tract is characterized by anorexia, pica, unthriftiness, poor growth or weight loss, impaired fertility and milk production and most of all by bone demineralization leading to rickets or osteomalacia. Weight loss, decreased fertility or productivity are likely the result of energy and nutrient deficiency ensuing from decreased feed intake.[1,8]

Acute hypophosphatemia in fresh cows is of particular importance to the dairy industry. Apart from feed intake depression, hypophosphatemia in early lactating cows has been associated with clinical signs and syndromes such as the downer cow syndrome or periparturient hemoglobinuria.[4] In particular the empirical observation that hypophosphatemia appears to be more common or more pronounced in recumbent periparturient cows unresponsive to intravenous Ca infusion than in cows responding to intravenous Ca administration has led to the widely held assumption that hypophosphatemia may be a causative or at least contributing factor to the so called downer cow syndrome.[9,10] Nonetheless experimentally decreasing plasma [Pi] to values as low as 0.3 mmol/L (0.9 mg/dL) by limiting dietary P intake did not result in recumbency or signs of muscle weakness or disturbed muscle function, nor do recumbent cows with hypophosphatemia consistently respond to oral or parenteral Pi supplementation. It must be emphasized though that models inducing hypophosphatemia by reducing dietary P uptake result in a gradually developing P depletion and therefore are poorly suited to mimic the sudden decline of plasma [Pi] at the onset of lactation where a dramatic increase of the P drain through the udder results in a sudden disequilibrium of P homeostasis, potentially overwhelming regulatory mechanisms.

Postparturient hemoglobinuria (PPH) characterized by acute Intravascular hemolysis, that is occasionally observed in dairy cows in early lactation is another condition believed to be caused by hypophosphatemia.[11] A decrease in intracellular ATP content in erythrocytes can increase osmotic fragility of the erythrocytes and may result in intravascular hemolysis in severe cases.[12] However, after inducing hypophosphatemia in cows by limiting the dietary P supply, self-limiting hemoglobinemia but not hemoglobinuria was observed.[12] The understanding of the etiology of this condition is further obscured by the fact that animals suffering from PPH do not consistently respond to parenteral supplementation of phosphate salts. Furthermore a recent study conducted on past-peak lactating dairy cows where pronounced hypophosphatemia was experimentally induced by feeding a low P diet over several weeks revealed that red blood cells were able to maintain a constant intracellular Pi concentration despite of sustained severe hypophosphatemia.[13] Other etiologies such as copper deficiency, increased oxidative stress, ketosis, acidosis or unidentified hemolysing substances contained in plants such as the sugar beet have been discussed as alternative or contributing causes. The discrepancy between the high incidence of periparturient hypophosphatemia and the negligible incidence of PPH, make it unlikely that P deficiency is anything more than a contributing factor to PPH.

NEW FRONTIERS

While in dairy production medicine concerns with hypophosphatemia in early lactation remained focused on recumbency and PPH over the last 30 years, hypophosphatemia in other species has also been associated with disturbed liver and immune cell function.[14] Impaired phagocytic activity of leukocytes in states of hypophosphatemia is well recognized in human medicine and has been experimentally induced in rats. The underlying mechanism has been associated with intracellular ATP depletion.[15,16] Granuloyctes had diminished chemotactic and phagocytic function in individuals with severe hypophosphatemia, abnormalities that resolved when patients were treated with P. One recent study was able to reproduce these findings in lactating dairy cows, where experimentally induced hypophosphatemia resulted in a significant decrease of the granulocyte count and decreased survival of granulocytes after phagocytosis.[17] The clinical relevance of this finding for the periparturient dairy cow remains to be determined.

Phosphorus is widely accepted to be relevant for liver function and most of all liver cell regeneration in human patients diagnosed with acute liver failure, and supplementation of Pi forms part of a state of the art conservative treatment in these patients.[18] In cattle only one study investigated liver tissue P contents of periparturient cows and reported a decline of the liver tissue P content in the range of 20% in the first two weeks of lactation, as well as a decline of the P content in liver tissue with increasing liver lipid content.[19] The clinical relevance of these results again needs to be studied in further clinical trials.

TREATMENT OF HYPOPHOSPHATEMIA

There is general consensus that chronic P deficiency requires correction of the negative P balance ideally through long term supplementation of the diet with phosphate salts. In contrast the clinical relevance of acute hypophosphatemia in fresh cows and thus the necessity of therapeutic intervention is still under contentious debate.[4] In practice the decision if and how acute hypophosphatemia is treated will be based on the personal experience and preference of the attending veterinarian. With Pi being the only form of P the organism can use metabolically it is essential to supplement P either as Pi or a form that can be converted to Pi in the body. Parenteral treatment in cattle often consists of infusion or injection of solutions containing organic P compounds such butafosfan or toldimfos or phosphite, none of which is converted to Pi. These compounds must therefore be considered as unsuitable for P supplementation therapy. Currently there are no pharmaceutical products containing phosphate salts approved for intravenous treatment in food animal species, making the use of phosphate salts an extralabel treatment in food producing animals.

Oral Pi supplementation in cattle was shown to be effective to correct hypophosphatemia and P depletion within hours .[4] For the treatment of acute hypophosphatemia where rapid absorption of phosphate is essential, the solubility characteristics of the specific phosphate salt are important. Monosodium phosphate (NaH_2PO_4) is the most commonly used compound for this purpose with well documented efficacy to rapidly restore serum [Pi].[4] Monopotassium or monocalciumphosphate salts may be considered alternatively in animals with concomitant hypocalcemia or hypokalemia but are slightly less effective than monosodium phosphate due to lower solubility.[19] For all compounds oral administration of the equivalent to 50-60 g of elemental P per individual animal dose has been suggested.[4]

REFERENCES

[1]National Research Council 2001. 7th rev. ed. Natl. Acad. Sci., Washington, DC; [2] Grünberg et al. J Dairy Sci 2011;94:727-745; [3] Grünberg et al. J Am Vet Med Assoc 2006;229:413-420; [4] Grünberg et al. J Vet Intern Med 2006;20:1471-1478; [5]Grünberg Vet Clin Food Anim 2014;30:383-408; [6]Robertson et al. Vet Rec 1956;68:173-180; [7]Goff et al. J Dairy Sci 2002;85:1427-1436; [8]Call et al. Am J Vet Res 1987;48:133-136; [9]Menard et al. Can Vet J 2007;48:487-491; [10]Gerloff and Swenson. J Am Vet Med Assoc 1996;208:716-719; [11]Jubb et al. Aust Vet J 1990;67:86-89. [12]Ogawa et al. Am J Vet Res 1989;50:388-392; [13]Grünberg et al. J Vet Intern Med 2015;29:395-399; [14]Knochel. Arch. Intern. Med. 1977;137:203-220; [15]Craddock et al. New Engl J Med 1974;290:1403-1407; [16]Kiersztejn et al. J Am Soc Nephrol 1992;2:1484-1489; [17]Eisenberg et al. J Dairy Sci 2014;97:5176-5184; [18] Baquerizo et al. Transplantation 2003;75:2007-2014; [18] Grünberg et al. J Dairy Sci 2009;92:2106-2117; [19] Idink and Grünberg Vet Rec 2015; 176:494.

Strategies for Prevention of Hypocalcemia Revisited: Is There a New Perspective to an Old Problem?

Walter Grünberg, DMV, MS, PhD, DECAR, DECBHM, assoc. DACVIM (Food Animal)
Hanover, Germany

INTRODUCTION

Periparturient hypocalcemia remains one of the most important metabolic disorders of the periparturient dairy cow despite of intensive research activity over the past decades. Hypocalcemia associated with clinical signs such as muscle weakness and recumbency occurs at a rate of around 5% in most industrialized countries and costs associated with a clinical case have been estimated with approximately 300 $.[1] Thus far the focus of producers and veterinarians when dealing with hypocalcemia on a herd level was to reduce the occurrence of clinical cases as these are perceived as the cause of largest part of economic losses. In recent years however subclinical hypocalcemia that is not associated with typical signs of milk fever has been recognized as gateway disturbance increasing the predisposition of early lactating dairy cows to other periparturient disorders such as abomasal displacement, ketosis, metritis or mastitis.[2] On average around 40% of periparturient cows are estimated to undergo a transient period with subnormal serum calcium concentration, with over 50% of multiparous cows and 25% of primiparous cows affected.[2] The economic damage per case of subclinical hypocalcemia resulting from decreased productivity and increased disease occurrence has been estimated with 125 $ in the recent literature.[1]

With the awareness that not only clinical milk fever but also subclinical hypocalcemia present a metabolic disturbance with an impact on animal health a paradigm shift when formulating the objectives of prevention of periparturient hypocalcemia is required. While thus far the primary objective of milk fever prevention was to reduce the incidence rate of clinical hypocalcemia, modern day milk fever prevention should also and foremost target at limiting the severity and the duration of the decline of the serum calcium concentration in periparturient dairy cows. The objective of the present paper is to revisit current strategies for milk fever prevention with an emphasis on their suitability not only to prevent clinical cases of milk fever but also to alleviate duration and severity of subclinical hypocalcemia.

REDUCTION OF DIETARY CALCIUM AVAILABLE FOR INTESTINAL ABSORPTION

One of the oldest strategies for milk fever prevention is based on the concept of depriving the bovine organism of dietary calcium for at least 2 weeks before the expected parturition in order initiate a negative calcium balance and trigger counter-regulatory circuits before calving.[3] Initiation of bone mobilization and increasing intestinal calcium absorption shortly before rather than after the occurrence of the periparturient dip in serum calcium is indeed suitable to accelerate the response to hypocalcemia and thereby to shorten and soften the decline of serum calcium around calving. The major limitation of this method is the difficulty of formulating a balanced dry cow ration with calcium content sufficiently low to induce a negative calcium balance.[4] With a dietary calcium requirement of a dry cow in the close-up period of below 40g /d and an estimated dry matter feed intake of between 20 and 30 lbs per day the dietary calcium content of an adequate dry cow ration must not exceed 0.3% of the dry matter to induce a negative calcium balance. Because a calcium content this low is hardly achievable with standard ration ingredient, this approach has little practical relevance.

An alternative approach following the same concept is the use of compounds binding dietary calcium that are added to the close-up ration. Feeding rations containing calcium binders reduces the bioavailability of the dietary calcium thereby facilitating the creation of a negative calcium balance. Zeolites are used for many years in the detergent industry as water softeners and descalers; a compound modified for the use as feed additive, zeolite A has been proposed and tested as feed additive for close-up rations.[5] Doses of up to 2 lbs /d/head have been tested for their suitability to increase periparturient calcium concentration in dairy cows. Treated cows indeed showed significantly higher serum calcium levels compared to untreated animals, but this effect was accompanied by pronounced feed intake depression and a concomitant decline of the serum concentrations of phosphorus and magnesium. Lower doses in the range of 250 g /d/head were effective in increasing the periparturient calcium concentration in older cows (third lactation and higher) and did not alter the serum magnesium concentration.[6] Treated animals however still showed a 10% decrease in dry matter feed intake and lower serum phosphate concentrations when compared to untreated controls.[6] While feed intake depression is a major side effect precluding the use of doses sufficiently high to induce a pronounced effect on the calcium homeostasis of the periparturient cow, it has been suggested that the combination of zeolite A with a ration containing the lowest possible amount of calcium may be an adequate alternative for milk fever prevention.[7]

More recently the use of formaldehyde treated rice bran has been proposed and tested as feed ingredient to improve the calcium balance of periparturient dairy cows.[8] Rice bran is rich in phytate that tends to bind calcium and thereby to reduce the availability of dietary calcium. To protect phytate from bacterial degradation in the rumen formaldehyde treatment is currently used. The calcium binding capacity of rumen protected rice bran is estimated with 4-5g per Lb of rice bran. Currently the addition of up to 6 lbs to the close-up dry cow ration are recommended.[8] Recent studies indicate that this compound that is sufficiently energy dense not to induce or exacerbate negative energy balance, does not depress dry matter feed intake and increases the serum calcium concentration in periparturient cattle compared to untreated controls.[9] Wanting to reduce the amount of available amount of calcium to below 30g/day however requires the combination of the use of this calcium binding compound with a close-up diet containing less than 0.6% calcium.

ORAL CALCIUM SUPPLEMENTATION AROUND CALVING

The strategy of supplementing oral calcium as drench, gel or bolus to dairy cows at the time of and shortly after calving is based on the intention to facilitate active and passive intestinal calcium absorption at a time when counter regulatory mechanisms to hypocalcemia are already upregulated, while feed- and thereby calcium intake is at its nadir.[7] Supplementing calcium at single doses containing the equivalent to 50 -100g of elemental calcium was successful in increasing the serum calcium concentration of periparturient cattle for at least 6 h.[10] The efficacy of this treatment not only depends on the amount of calcium administered but also the solubility of the used calcium containing compound. While inorganic salts such as calcium chloride are highly soluble and rich in calcium, organic calcium salts such as calcium gluconate or fumarate contain less calcium, thus requiring a large volume per dose and tend to be less soluble, thereby resulting in a slightly delayed effect. Advantages of organic compounds are that they are not caustic and are palatable.[1] Administration of oral calcium is certainly among the most labor intensive milk fever prevention strategies but has been recommended not only as prophylactic treatment but also as follow-up treatment for previously recumbent milk fever cows.[1]

PARENTERAL ADMINISTRATION OF VITAMIN D

The parenteral administration of vitamin D at a dose of 1 Mio IU / 90 lbs between 3 and 10 days before the expected parturition is a widely used prophylactic treatment with which the incidence of hypocalcemic recumbency can be reduced efficiently.[11] Large supraphysiologic doses of vitamin D not only increase intestinal calcium absorption beyond requirements thereby elevating the serum calcium concentration, but also downregulate the release of parathyroid hormone (PTH), a key hormone for the regulation of the calcium homeostasis.[12] The effect on intestinal calcium absorption subsides after 2 weeks of parenteral vitamin D administration while the negative feedback mechanism on PTH is more sustained. The result is a declining serum calcium concentration in treated animals with a concomitantly hampered regulation of the calcium homeostasis in the weeks after treatment.[12] Clinical studies have confirmed that dairy cows treated with parenteral vitamin D in the recommended time frame have higher serum calcium concentrations at the time of calving, but have significantly lower serum calcium concentrations than untreated animals for up to two weeks post-partum.[14] The injection of vitamin D is thus reducing the risk of clinical hypocalcemia around calving at the cost of increased incidence of subclinical hypocalcemia in early lactation and can therefore not be recommended as first choice strategy for prevention of periparturient hypocalcemia.

DCAD DIETS

Feeding acidogenic diets with a pronounced negative dietary cation-anion balance (DCAD) for two to three weeks before calving results in enhanced mobilization of calcium from bone.[15] Enhanced calcium release from bone is accompanied by a concomitant increase in renal calcium excretion ante-partum, which results in unchanged serum calcium concentrations but increased calcium turnover ante partum.[7] At the onset of lactation the sudden drag of calcium through the mammary gland triggers a switch of calcium excretion from the kidney to the mammary gland.[16] The increase of the calcium flux triggered by feeding a DCAD diet thus alleviates the transient drop of the plasma calcium concentration occurring around parturition. The poor palatability of commonly used anionic salts limits the amount of salts that can be fed without negative effects on feed intake. In particular rations high in potassium, a cation antagonizing the effect of anionic salts, would require doses of anionic salts that would negatively affect dry matter intake.[7]

PARENTERAL CALCIUM ADMINSITRATION AROUND CALVING

Intravenous administration of calcium salt solutions at a doses also used for treatment of clinical milk fever, i.e. 10-15 g/ cow is commonly used under field conditions as a prophylactic measure in clinically healthy cows at or shortly after calving. While this treatment is the undisputed state of the art treatment for clinically affected cows, concerns have been raised with the preventive intravenous calcium administration.[1,9] A rapid bolus infusion transiently converts a presumable subclinical hypocalcemia into a pronounced hypercalcemia, with PTH as key regulatory hormone responding to this change within minutes.[17] Counter regulation to hypocalcemia that at the point of treatment was still efficient in prophylactically treated animals is thus interrupted and will only be upregulated with serum calcium levels reaching subnormal values.[1.9] A recent study comparing the prophylactic intravenous treatment to untreated control animals documented markedly increased serum calcium concentrations for hours after treatment that was followed by serum calcium concentrations that were significantly lower than in untreated animals for at least 48 h post treatment.[18] Although this approach may prevent recumbency in a few animals that would have become recumbent within hours without treatment it exacerbates subclinical hypocalcemia in the majority of treated animals for at least 48 h.

Subcutaneous administration of calcium salt solutions, containing the equivalent of 10 g of elemental calcium presents an alternative to the intravenous approach and is considered to be less disruptive to regulatory mechanisms of the calcium homeostasis as peak calcium concentrations to not exceed the reference range.[1] Treatment effect and duration appear to be comparable to oral treatment. Concerns with this treatment are associated with tissue irritation when administering these mildly acidic solutions subcutaneously.[1]

REFERENCES

[1]Oetzel et al. Vet Clin Food Anim 2013;29:447-455; [2] Reinhardt et al. Vet J 2011;188:122-124; [3]Goings et al. J Dairy Sci 1974;57:1184-1188. [4]Thilsing-Hansen et al. Acta Vet Scand 2002;43:1-19. [5]Thilsing-Hansen et al. J Dairy Sci 2002;85:1855-1862; [6]Grabherr et al. J Anim Physiol Anim Nutr 2009;93:221-236; [7]Goff. Vet Clin Food Anim 2014;30:359-381; [8]Martin-Tereso et al. J Anim Physiol Anim Nutr 2014;98:775-784; [9]Martin-Tereso and Martens Vet Clin Food Anim2014;30:643-670; [10] Goff and Horst. J Dairy Sci 1993;76:101-108. [11]Littledike and Horst. J Dairy Sci 1979;62:189; [12] Littledike and Horst. J Dairy Sci 1980;63:89; [13]Horst et al. Acta Vet Scand Suppl. 2003;97:35-50; [14] Hove and Kristiansen. Acta Vet Scand 1984;25:510-525; [15]Block. J Dairy Sci 1984;67:2939-2948; [16]Grünberg et al. J Dairy Sci 2011;94:727-745; [17]Braun et al. Vet Rec 2009;164:296-299; [18] Blanc et al. J Dairy Sci 2014 ;97:6901-6906.

Methods of Teaching Clinical Reasoning

Amy L. Warren, DVM, PhD, DACVP (Anatomic Pathology)(Clinical Pathology)
Calgary, AB, Canada

INTRODUCTION

Clinical (and diagnostic) reasoning ability is one of the most critical skills a clinician develops through their education and career[1]. It is the differences in reasoning capability that often distinguish clinicians' excellence. Reasoning can be both diagnostic (the thought processes that drive the process from presentation to diagnosis) or therapeutic (the thought processes that drive the treatment plan, choices and alterations). Despite the clear importance of this skill, the cognitive processes underlying clinical reasoning is still poorly understood and as a consequence, effective methodologies to teach clinical reasoning are even less studied. To understand how best to teach clinical reasoning, it is necessary to understand how reasoning occurs in experts and novices. This presentation will first focus on the current theories of how clinical reasoning occurs in novices and experts, and will then focus on teaching and learning in clinical reasoning.

THEORIES OF CLINICAL REASONING

Clinical (and diagnostic) reasoning, a critical skill clinicians develop through their education and career, relies on the integration and application of thinking and learning skills to make decisions in familiar and unfamiliar clinical experiences.[1,2] It relies on both theoretical (basic and clinical science knowledge) and experiential knowledge (exposure to and handling of cases). Experiential knowledge is built as new experiences (cases) force the revision of past knowledge and skill development. Several theories about the cognitive process of diagnostic reasoning have been proposed over the years: hypothetico-deductive, knowledge structures, script theory, pattern recognition and dual process theory. Despite multiple attempts to ascertain a single reasoning process, it is more probable that experts' reasoning ability is the result of an extensive and multidimensional knowledge base and that multiple reasoning processes are employed, based on the context of the case.[3,4] Each of the theories of clinical reasoning will be briefly discussed and points related to teaching and learning of clinical reasoning will be highlighted.

Hypotheticodeductive Model

Hypothetico-deductive reasoning, proposed in the 1970s and 1980s,[5] is similar to the reasoning used in scientific investigation where knowledge grows through a succession of tested hypotheses for which experimental evidence is sought to prove or disprove.[6,7] It was thought that clinical reasoning utilized this generic problem-solving strategy. Backward reasoning underpins hypothetico-deductive reasoning and is characterized as reasoning from the diagnosis (hypothesis) back to the data.[8] Forward reasoning, used more frequently by experts, starts with observations and data to generate a diagnostic hypothesis.[8] Hypothetico-deductive, forward and backward reasoning are all analytical reasoning processes. The generalizability of hypothetico-deductive reasoning is questionable, however, when applied to experts as much of expert reasoning is highly reliant on an extensive knowledge base and is pattern based. From a teaching and learning perspective however, it is often thought that novices start in a very analytic frame of reasoning and hypotheticodeductive reasoning is taught to novices to bridge early gaps between their current knowledge base and forming a diagnosis.

Knowledge Structures

Boshuizen and Schmidt[9] (1992) demonstrated through think-aloud and post-hoc explanation, that experts had more in-depth biomedical knowledge than novices and intermediates, and suggested that experts' more superior reasoning ability was related to their underlying knowledge structures. They proposed that knowledge expertise developed in three stages—acquisition, practical experience and encapsulation (integrating theoretical and experiential knowledge). Experts made more inferences from fewer data points in the text as experts were better able to distinguish diagnostically useful data from the case.[10] From a teaching and learning perspective, this highlights the importance of pre-clinical biomedical science and clinical science learning in clinical reasoning. A solid understanding of the pathophysiology and biology that underpins diagnosis and treatment is related to clinical reasoning success.

Script Theory

In 1990, Schmidt et al.[11] proposed that as important as the knowledge clinicians possess was the *knowledge structures* they used to organize the information. Building on a cognitive science theory of "schemes", they proposed that similar cognitive structures, "illness scripts" were possessed by experienced clinicians.[11-14] Illness scripts are knowledge structures constructed during repeated experiences that organize medical knowledge into readily mobilized schemes that are goal-directed and increase the efficiency of task performance.[11] Illness scripts are activated when a practitioner first perceives a patient's symptoms and interprets characteristics and features of a case to infer a differential diagnosis.[13] Each generated hypothesis (i.e., differential diagnosis) is an activated illness script. Illness scripts have hierarchical knowledge structures "slots" containing relevant medical information pertaining to the clinical signs, including pre-stored knowledge about different predisposing conditions, pathophysiology and diagnostic testing, and acceptable or not acceptable values and default values.[12,13,15] Once a selection of scripts (or script) are identified as working clinical hypotheses, data collected is used to ascertain if the clinical signs fall within the pre-stored acceptable values for the script.[13] If unacceptable values are identified for a clinical sign or test, the script is rejected. Clinical signs and test results within a script are also weighted in that some clinical signs are more indicative of a diagnosis than others.[13]

Based on this, Schmidt et al.[11] (1990) proposed there were four stages to developing illness scripts in clinical reasoning learning: 1) elaborated causal networks 2) abridged networks 3) illness scripts and 4) instance scripts. Elaborated causal networks link concepts based on cause and consequence, including linking biomedical aberrations to presenting clinical signs or pathophysiologic consequences.[16] As students are exposed to extensive and repeated application of causal networks, abridging of the networks occur. From this high-level, simplified causal models develop amalgamating signs and symptoms into a diagnosis. As students' experiences with cases grow, they develop illness scripts and later, as experienced clinicians, instance scripts. It is thought that the repertoire of illness scripts that clinicians acquire with experience allow problem solving by recognizing the presenting problem as being similar to ones previously solved (i.e., pattern recognition).[17]

Pattern Recognition

Emerging simultaneously with work on knowledge structures was the recognition that while analytical styles of reasoning explained some of the clinical reasoning process, particularly in novices, experts used a different form of reasoning that was intuitive and pattern based.[15,18] Analytical reasoning analyzes a case on a feature-by-feature basis whereas in non-analytical processing is holistic (i.e., the global impression of the case is compared to prior experienced exemplars).[19] In pattern recognition models of reasoning, case presentations resemble previously seen disease examples and trigger disease categorization and automatic knowledge retrieval from well-structured knowledge patterns.[35] Similar theories of clinical reasoning, including scripts[16], use a pattern-based approach to reason in an intuitive manner. The two seemingly dissonant groups of reasoning, those reliant on analytic skills and the other pattern based, can be explained by a psychological theory of decision making: dual processing theory.

Dual Processing Theory

A prominent emerging theory of human judgment is dual process theory where two systems of cognition are proposed to exist: System 1) heuristic and intuitive, and System 2) systematic, deliberate and analytical.[20,21] Numerous attributes have been ascribed to the systems, reflecting levels of consciousness, evolution, functional characteristics and individual differences (Table 2.1).[22] System 1 cognition is generally fast, high capacity, unconscious (first impression), perceptive, contextualized, associative and highly reliant on pattern recognition. It is also prone to error and due to its context-bound nature, prone to subconscious influences of emotion and prior experiences.[23] In contrast, System 2 cognition is slow, deliberate, analytical and conscious, and is less prone to error.[23] As System 2 is relatively slow and effortful, it is susceptible to interference associated with time pressure, fatigue, cognitive difficulty of a task, and competing thought processes.[1,24]

While two delineated systems have been proposed, cognitive theorists suggest that a continuum (also referred to as cognitive continuum theory) rather than a dichotomy exists and that these systems likely act in parallel during the reasoning process.[21,24-26] While System 1 (intuitive) is a default response, it can be over-ridden or altered by analytical reasoning (System 2).[27] Triggering of analytical reasoning occurs with strong deductive reasoning instructions, dissonant patterns and tends to occur more often in individuals with high cognitive ability.[27,28] System 2 reasoning is believed to be a gate-keeper, monitoring System 1 reasoning processes.[29]

Table 2.1. Attributes of System 1 and System 2 thinking (adapted from Evans et al., 2008)[22]

Attribute cluster	System 1	System 2
Consciousness	Unconscious	Conscious
	Implicit	Explicit
	Automatic	Controlled
	Low effort	High effort
	Rapid	Slow
	High capacity	Low capacity
	Holistic, perceptual	Analytic, reflective
Evolution	Evolutionally old, shared with animals	Evolutionally recent, uniquely human
Functional characteristics	Associated	Rule-based
	Contextualized	Abstract
	Pragmatic	Logical
	Parallel	Sequential
Individual differences	Independent of intelligence and working memory	Linked to intelligence and limited to working memory capacity

In the context of diagnostic reasoning, System 1 (intuitive) is highly related to experience.[23] System 2 (analytic) tends to be used by novices, relying on hypothetico-deductive, backward reasoning,[2] aborization (algorithmic), and critical and logical thought.[23] It is thought that experts revert to System 2 (analytic) thinking when faced with unknown or ambiguous presentations,[23,24,30,31] relying more heavily on pathophysiological causality and biomedical knowledge.[1] In the acquisition of highly skilled performance, there is a transformation from System 2 (effortful) to System 1 (effortless) activity[24] with repetition, highly relevant to novice-expert learning.[22,23] System 1 reasoning is thought to be based on visual similarity and to a lesser degree verbal descriptions,[32] pertinent to fields where visual reasoning predominates (e.g., dermatology, radiology and pathology).[1]

Context of Clinical Reasoning

Recently, theories of ecological or situated cognition have been explored in the clinical reasoning processes of medical experts.[33,34] The central tenet of these theories is that knowledge is situated in the "experience", referring to not only the participants in the experience but also the physical environment. In a clinical situation, it has been shown that the clinical reasoning of experts was influenced by the context of the presenting case.[33] Further, context appeared to affect diagnostic reasoning more than therapeutic reasoning.[34]

Despite multiple attempts to ascertain a single reasoning process used by clinicians, it is more probable that experts' reasoning ability is the result of an extensive and multidimensional knowledge base and that multiple reasoning processes are employed, based on the context of the case.[3,4] It is likely that experts in clinical reasoning in general employ a combination of system 1 (pattern recognition) and system 2 (analytic) reasoning in parallel during their diagnostic reasoning process. It is also likely that using system 1 reasoning, they activate a subset of illness scripts, activating knowledge structures. Subsequent to this, an analytical process, using knowledge structures, rules in or out diagnostic hypotheses and employs checks and balances in a metacognitive manner to verify the diagnosis.

TEACHING AND LEARNING CLINICAL REASONING

DVM students and veterinary specialist residents, as adults, have particular learning traits that are important in optimizing clinical reasoning learning. In adult learning, two key factors drive the learning experience: the learner and the context in which learning takes place.[35] Important to adult learning is that learners:

1) organize knowledge in structures described as "schemes",[35]
2) have a great deal of prior knowledge and new learning experiences should ideally build on this knowledge,[35,36]
3) make sense of knowledge through active manipulation of information,[35]
4) have optimal learning when it is experiential and contextually and socially bound,[35]
5) are intrinsically rather than extrinsically motivated to learn,[35,36] and
6) knowledge acquisition is hampered by excessive cognitive load.[37,38]

Knowledge Structures

Key to adult learning is not only the knowledge that is acquired, but also the manner by which that knowledge is structured and subsequently accessed (recalled). Adult learners, as compared to children, have a wealth of prior knowledge gained through experiences and problem solving which plays an important role in their learning. It is believed that the prior knowledge possessed by adults is organized in mental schemes (categorical memory structures).[35] During learning, schemes accumulate new knowledge that is tuned (gradually modified) or restructured (reorganized or new schemes are created to better represent a new understanding).[35] As well as adding, fine-tuning and re-structuring schemes, the ability of adults to meta-cognitively reflect on the strengths and weakness of their mental schemes is important in the highest levels of evaluating and synthesizing knowledge. There is some suggestion that the structure in which students learn new material is also the structure used to recall material.[39,40] This has important instructional design implications for teaching biomedical and clinical science in that students taught basic sciences in an organ-based manner will tend to recall their knowledge in the context of organs. This is disconnected from the way cases are presented to veterinarians and may hamper biomedical knowledge transference.[40]

Organization of knowledge is key to clinical veterinary expertise in that experts have more organized and detailed knowledge structures than novices.[12,16,41] It stands to reason that a goal of biomedical and clinical education is for novices to create or learn knowledge schemes that are similar to those used by experts.[42] During the process of learning, rather than ignore students' existing knowledge schemes, instructors can build on these schemes by correcting their structure or adding new material.[42] An ideal way to achieve this is through classroom discussions with feedback on the understanding of the concepts from both the student and instructor. Another role of instructors or experts is to model their own knowledge structures. Many experts have abbreviated ways of understanding a concept or performing a task.[42] To transfer knowledge structures or understanding to a student, instructors need to be self-aware of which steps they skip and make explicit the links between the biomedical and clinical science and clinical signs in a case.[43] The use of concept maps, algorithms and schemes are ways to provide students with cognitive structures and encourage deep learning.[42]

A comprehensive biomedical and clinical knowledge base is essential in illness script development and clinical problem solving.[39] Teaching biomedical and clinical science by emphasizing and making the causal links to clinical signs[44] explicit improves retention of biomedical and clinical knowledge[45] and diagnostic success in difficult cases[46] (both achieved by developing the knowledge schemes in students). Teaching biomedical science in schemes that represent expert knowledge structures has been shown to enhance students' knowledge acquisition and problem solving capabilities.[39]

Prior Knowledge, Experiential and Active Learning

Fundamental to knowledge schemes is the prior knowledge that adults possess. In essence, the understanding of new knowledge and concepts cannot be isolated from prior experience.[35] Prior knowledge and experience is also identified as one of the key internal motivators for adult learning.[36] For experiential learning to be effective, new learning experiences need to be interactive and build on prior knowledge gained from earlier experiences (i.e., connect what is already known to the new experience).[35] These two features have important implications for clinical reasoning education in that optimal adult learning experiences are contextual and active.

Key to experiential learning is that learning is an interactive process between the learner and the content. Knowledge can be viewed as a tool; it is only fully understood through its use.[47] Active learning in biomedical and clinical science is often discovery-based, uncovering new knowledge in the context of how it will be used (in clinical medicine), and building on what the student already knows.[48] By engaging in a discussion about a biomedical problem, students relate the material to their prior experiences, re-assess their understanding based on others' opinions, and ultimately better understand and retain the knowledge.[48]

Models of experiential learning represent a cyclical process of abstract conceptualization, planning for implementation, active experimentation, concrete experience and reflective observation.[35] In the context of biomedical science education, the learner is introduced to a concept (e.g., abstract conceptualization in a lecture, text book or e-learning), provided with a concrete example of the concept (e.g., a clinical case illustrating the biomedical principle in small group problem or case-based learning session) and encouraged to engage in the problem (e.g., active experimentation).[49]

For most learning situations, particularly concepts that are difficult to understand, learning occurs not in a single cycle but in a spiral, where the concept is experienced multiple times with modifications and gains in comprehension with each passing stage of the learning process.[35] This has implications for instructional design and how it needs to be built on prior knowledge structures and understanding. Presenting similar concepts (disease processes as an example) and concrete experiences (cases) in increasing complexity allows the learner to build on their prior knowledge in an experiential manner.[49,50] The role of the instructor in experiential learning is to relate new material to the prior knowledge of the learner, provide concrete experiences that students can actively engage in, and provide conceptual experiences in the form of narration for experiences that are difficult to deliver to a classroom setting.[35] An additional role of the instructor is to create cognitive dissonance (in a safe environment) in students' understanding. This highlights the limitations of the students' current understanding, forcing students to re-examine and potentially restructure their knowledge schemes.[42]

Contextual Learning

Cognitive scientists argue that learning and cognition are fundamentally situated (i.e., contextually bound) and that ignoring this results in learning that is less robust and usable.[47] To create an analogy from learning languages, learning basic and clinical science as isolated facts out of context to their medical story is akin to learning words from a dictionary with no reference to sentence construction and use. Biomedical and clinical knowledge is situated within a clinical context and ignoring this context jeopardizes the effectiveness of the learning environment and successful transference of the knowledge.

Cognitive Load Theory

Cognitive load theory is an emerging theory that argues learning is largely restricted by the limitations of the working memory. The capacity of the working memory to recall and process new information is limited in that it is able to hold only 7 plus or minus 2 new pieces of information at any one time.[35] Working memory, however, has no known limit when processing information retrieved from the long-term memory. This has important implications for instructional design, in that the instructor is left with the task of identifying strategies to optimize the learner's memory available to structure and recall new information.

Three types of cognitive load have an impact on learning: 1) intrinsic load (the complexity of the learning task), 2) extraneous load (the manner which the learning tasks are presented) and 3) germane load (the load associated with actual learning).[37,38] It is believed that these three forms are additive and, hence, educational strategies should be employed that maximize the germane load while minimizing the intrinsic and extraneous loads.[38] The intrinsic load (complexity) can be reduced by introducing multiple steps to problem solving, sequencing clinical cases from low to high levels of complexity, and by gradually increasing the fidelity of the case.[38] There are several approaches to reducing the extraneous load. These include providing a non-specific goal (e.g., to not to initially strive for a diagnosis, but to identify the biomedical principles at play in a case), introducing a worked through example of a problem, partially completing a problem, putting multiple sources of information in one location (e.g., temporally and spatially), replacing written explanations with visual aids (e.g., diagrams, images), and replacing multiple sources of the same information within one location.[37,38] Germane load can be optimized by varying the cases (e.g., encouraging comparison between cases and identifying unifying principles), varying the types of cases encountered (e.g., mixed practice instead of blocked practice), and self-explanation of concepts to increase prior-knowledge elements.[38,49]

Teaching Clinical and Diagnostic Reasoning

Several themes emerge in the clinical reasoning education literature that builds on the cognitive underpinnings of illness scripts, and System 1 and System 2 reasoning processes. It is generally agreed that the most effective method for acquiring illness scripts and converting System 2 to System 1 reasoning is the repeated and directed exposure of students to real cases.[3,4,28,32,41,51]

Rather than case exposure being a passive process, several educational principles are recognized: 1) learners need to be actively engaged, 2) new information is articulated on student's prior (biomedical) knowledge, 3) intermediate stages of clinical reasoning are explained, 4) learners use their clinical knowledge to judge clinical information to support or reject a hypothesis, and 5) immediate feedback from peers and teachers validates newly acquired knowledge.[3,12,40] Deliberate application and integration of biomedical knowledge is important in script building, placing biomedical knowledge in an accessible clinical-presentation based location.[13,16] Teaching the basic and clinical sciences in a clinical context facilitates the encapsulation of biomedical knowledge in illness scripts.[16] Both increased case reasoning exposure and practice, as well as the teaching of biomedical and clinical knowledge in the context of clinical cases, are supported by newer medical curricula including case-based and clinical-presentation-based curricula.[30,52,53] The teaching of knowledge "schemes" has been suggested to enhance memory organization and diagnostic success.[17] Finally, use of multiple reasoning processes (providing redundancy) lead to the most success (i.e., accuracy) in diagnostic reasoning.[41,51,54] Analytic and non-analytic reasoning processes likely occur interchangeably and additively in a reasoning process and use of both should be encouraged in an educational context.

REFERENCES

1. Pelaccia T, Tardif J, Triby E, et al. An analysis of clinical reasoning through a recent and comprehensive approach: the dual-process theory. Med Educ Online 2011;16.
2. Higgs J, Jones MA, Loftus S, et al. Clinical reasoning in the health professions. In, 3rd ed. Philadelphia, PA: Elsevier Butterworth Heinemann; 2008:504.
3. Norman G. Research in clinical reasoning: past history and current trends. Med Educ 2005;39:418-427.
4. Eva KW. What every teacher needs to know about clinical reasoning. Medical Education 2004;39:98-106.
5. Elstein AS, Shulman LS, Sprafka SA. Medical problem solving: an analysis of clinical reasoning. Cambridge, MA: Harvard University Press; 1978.
6. Hopayian K. Why medicine still needs a scientific foundation: restating the hypotheticodeductive model - part two. Br J Gen Pract 2004;54:402-403; discussion 404-405.
7. Hopayian K. Why medicine still needs a scientific foundation: restating the hypotheticodeductive model - part one. Br J Gen Pract 2004;54:400-401; discussion 404-405.
8. Norman G, Trott AL, Brooks L, et al. Cognitive differences in clinical reasoning related to postgraduate training. Teaching and Learning in Medicine 1994;6:114-120.
9. Boshuizen HC, Schmidt HG. On the role of biomedical knowledge in clinical reasoning by experts, intermediates and novices. Cognitive Science 1992;16:153-184.
10. Schmidt HG, Boshuizen HP. On the origin of intermediate effects in clinical case recall. Mem Cognit 1993;21:338-351.
11. Schmidt HG, Norman GR, Boshuizen HP. A cognitive perspective on medical expertise: theory and implication. Acad Med 1990;65:611-621.
12. Charlin B, Tardif J, Boshuizen HP. Scripts and medical diagnostic knowledge: theory and applications for clinical reasoning instruction and research. Acad Med 2000;75:182-190.
13. Charlin B, Boshuizen HP, Custers EJ, et al. Scripts and clinical reasoning. Med Educ 2007;41:1178-1184.
14. Barrows HS, Feltovich PJ. The clinical reasoning process. Med Educ 1987;21:86-91.
15. Elstein AS, Schwartz A. Clinical problem solving and diagnostic decision making: selective review of the cognitive literature. BMJ 2002;324:729-732.
16. Schmidt HG, Rikers RM. How expertise develops in medicine: knowledge encapsulation and illness script formation. Med Educ 2007;41:1133-1139.
17. Coderre S, Mandin H, Harasym PH, et al. Diagnostic reasoning strategies and diagnostic success. Med Educ 2003;37:695-703.
18. Papa FJ, Shores JH, Meyer S. Effects of pattern matching, pattern discrimination, and experience in the development of diagnostic expertise. Academic medicine : journal of the Association of American Medical Colleges 1990;65:S21-22.
19. Regehr G, Cline J, Norman GR, et al. Effect of processing strategy on diagnostic skill in dermatology. Academic medicine : journal of the Association of American Medical Colleges 1994;69:S34-36.
20. Sloman SA. Two systems of reasoning. In: Gilovich T, Griffin D, Kahneman D, eds. Heuristics and biases: The psychology of intuitive judgment. New York: Cambridge University Press; 2002:379-396.
21. Kahneman D. A perspective on judgment and choice: mapping bounded rationality. Am Psychol 2003;58:697-720.
22. Evans JS. Dual-processing accounts of reasoning, judgment, and social cognition. Annu Rev Psychol 2008;59:255-278.
23. Croskerry P. A universal model of diagnostic reasoning. Acad Med 2009;84:1022-1028.
24. Kahneman D, Frederick S. A model of heuristic judgment. In: Holyoak K, Morrison RG, eds. The Cambridge Handbook of Thinking and Reasoning. Cambridge, UK: Cambridge University Press; 2005:267-294.
25. Norman G, Monteiro S, Sherbino J. Is clinical cognition binary or continuous? Academic medicine : journal of the Association of American Medical Colleges 2013;88:1058-1060.
26. Custers EJ. Medical education and cognitive continuum theory: an alternative perspective on medical problem solving and clinical reasoning. Academic medicine : journal of the Association of American Medical Colleges 2013;88:1074-1080.
27. Evans JS. The heuristic-analytic theory of reasoning: extension and evaluation. Psychon Bull Rev 2006;13:378-395.

28. Lee A, Joynt GM, Lee AK, et al. Using illness scripts to teach clinical reasoning skills to medical students. Fam Med 2010;42:255-261.
29. Kahneman D. Thinking Fast and Slow, 1st ed. New York, NY: Farrar, Straus and Giroux; 2011.
30. Norman GR, Trott AL, Brooks LR, et al. Cognitive differences in clinical reasoning related to postgraduate training. Teaching and Learning in Medicine 1994;6:114-120.
31. Moulton CA, Regehr G, Lingard L, et al. Slowing down to stay out of trouble in the operating room: remaining attentive in automaticity. Academic medicine : journal of the Association of American Medical Colleges 2010;85:1571-1577.
32. Norman G, Young M, Brooks L. Non-analytical models of clinical reasoning: the role of experience. Med Educ 2007;41:1140-1145.
33. Durning S, Artino AR, Jr., Pangaro L, et al. Context and clinical reasoning: understanding the perspective of the expert's voice. Medical Education 2011;45:927-938.
34. Durning SJ, Artino AR, Boulet JR, et al. The impact of selected contextual factors on experts' clinical reasoning performance (does context impact clinical reasoning performance in experts?). Advances in health sciences education : theory and practice 2011.
35. Merriam SB, Caffarella RS. The learning process. In: Merriam SB, Caffarella RS, eds. Learning in adulthood: a comprehensive guide., 2nd ed. San Franscisco, CA: Jossey-Bass; 1999:193-220.
36. Dale VH, Sullivan M, May SA. Adult learning in veterinary education: theory to practice. Journal of Veterinary Medical Education 2008;35:581-588.
37. Khalil MK, Paas F, Johnson TE, et al. Interactive and dynamic visualizations in teaching and learning of anatomy: a cognitive load perspective. Anat Rec B New Anat 2005;286:8-14.
38. van Merrienboer JJ, Sweller J. Cognitive load theory in health professional education: design principles and strategies. Medical Education 2010;44:85-93.
39. Mandin H, Jones A, Woloschuk W, et al. Helping students learn to think like experts when solving clinical problems. Acad Med 1997;72:173-179.
40. Bowen JL. Educational strategies to promote clinical diagnostic reasoning. N Engl J Med 2006;355:2217-2225.
41. Kassirer JP. Teaching clinical reasoning: case-based and coached. Acad Med 2010;85:1118-1124.
42. Horii CV. Teaching insights from adult learning theory. Journal of Veterinary Medical Education 2007;34:369-376.
43. Woods NN. Science is fundamental: the role of biomedical knowledge in clinical reasoning. Med Educ 2007;41:1173-1177.
44. Woods NN, Brooks LR, Norman GR. The value of basic science in clinical diagnosis: creating coherence among signs and symptoms. Medical Education 2005;39:107-112.
45. Woods NN, Neville AJ, Levinson AJ, et al. The value of basic science in clinical diagnosis. Academic medicine : journal of the Association of American Medical Colleges 2006;81:S124-127.
46. Woods NN, Brooks LR, Norman GR. The role of biomedical knowledge in diagnosis of difficult clinical cases. Adv Health Sci Educ Theory Pract 2007;12:417-426.
47. Brown JS, Collins A, Duguid P. Situated cognition and the culture of learning. Educational Researcher 1989;18:32-42.
48. Hines SA, Collins PL, Quitadamo IJ, et al. ATLes: the strategic application of Web-based technology to address learning objectives and enhance classroom discussion in a veterinary pathology course. Journal of Veterinary Medical Education 2005;32:103-112.
49. Norman G. Teaching basic science to optimize transfer. Medical Teacher 2009;31:807-811.
50. Bender HS, Lockee BB, Danielson JA, et al. Mechanism-based diagnostic reasoning: thoughts on teaching introductory clinical pathology. Vet Clin Pathol 2000;29:77-83.
51. Bloch RF, Hofer D, Feller S, et al. The role of strategy and redundancy in diagnostic reasoning. BMC Med Educ 2003;3:1.
52. Mandin H, Harasym P, Eagle C, et al. Developing a "clinical presentation" curriculum at the University of Calgary. Acad Med 1995;70:186-193.
53. Papa FJ, Harasym PH. Medical curriculum reform in North America, 1765 to the present: a cognitive science perspective. Acad Med 1999;74:154-164.
54. Eva KW, Hatala RM, Leblanc VR, et al. Teaching from the clinical reasoning literature: combined reasoning strategies help novice diagnosticians overcome misleading information. Med Educ 2007;41:1152-1158.

*** Formatting Issue - Best Available Paper/Graphic ***

Evaluation of Interventions to Improve the Health of Working and Food Animals

Andy Stringer, BVSc, PhD, MRCVS; Paul Lunn, BVSc, MS, PhD MRCVS, DACVIM
Raleigh, NC

INTRODUCTION

Despite considerable efforts by the international community, inequalities still exist between high-income and low-income countries globally. By 2050, the human population is estimated to increase to over nine billion. Millions of households depend on agriculture as part of their livelihoods and about 70 percent of the world's -extreme scale farmers, with an estimated 2.6 billion people surviving on less than $2 a day in the developing world [1]. In many developing countries, up to 80% of the population is employed in agriculture. The United Nations Sustainable Development Goals (SDGs), agreed in 2015, are a set of 17 goals to respond to the world's main development challenges [2]. Livestock and working equids are central to achieving the SDGs and are directly relevant to the majority of the 17 goals. Animals are vital to the livelihoods of individuals, families and communities through livelihood strategies [1,3].

One approach to understanding how animals and livelihoods are linked, is the United Kingdom Department for International Development's sustainable livelihoods framework [4]. At the core of the livelihoods framework (Figure 1) are five asset categories – human capital, natural capital, financial capital, social capital and physical capital – on which livelihoods are built. These assets are utilised in a number of processes that determine the livelihood status of the household and the wellbeing of its members. Increasing access to a range of assets, through ownership or a right of use to assets, and the utilisation of these assets through processes (livelihoods strategies), is the primary mechanism to a sustainable livelihood and human wellbeing.

Animals (working equids and livestock) contribute to several asset types and livelihoods through processes including the direct production of produce for consumption (for example, meat, eggs and milk), providing services for other households that can be exchanged for goods or cash, and, if necessary, such as in times of immediate need, animal assets can be liquidated. Analysis of those living in poverty have identified that people's ability to escape from poverty is critically dependent upon their access to assets [4].

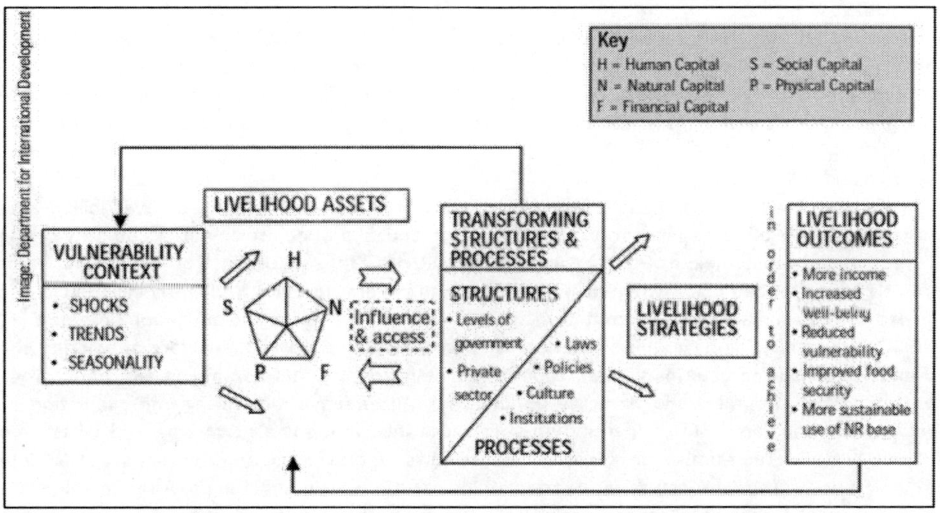

Figure 1: The Department for International Development's Sustainable Livelihoods Framework.

Non-governmental organizations (NGOs) and government agencies located in high-income countries have invested considerably in working equid and livestock development programs as a means to alleviate poverty and improve livelihoods. The complex relationships between livestock keeping and the poor provide a multitude of entry points for livestock based interventions, which is reflected in the variability of livestock programs [5]. Organizations across all sectors assume their programmatic interventions make a positive difference, although this is not always the case. The process of formalized monitoring and evaluations (M&E) has gained increasing importance within the development agenda over recent years as programs that are not evaluated waste time, money and have little impact on future development [6].

Livestock development programs have a long history as a means of poverty alleviation for the global poor; however the relationships between livestock ownership and livelihood strategies are very complex, with both positive and negative impacts on outcome measures for household assets. Evaluating the effects of such interventions is difficult and requires a carefully designed study intervention as well as local, contextual knowledge of the livestock/human/poverty interface for the targeted population [7]. Where resources are finite, robust program evaluations enable organizations to identify the most effective interventions to allocate resources for maximum benefit.

LIVESTOCK

It is estimated that there are 1.43 billion cattle, 1.87 billion sheep and goats, 0.98 billion pigs, and 19.60 billion chickens globally with a growing demand for livestock products. Livestock accounts for at least 40% of agricultural Gross Domestic

Product (GDP) in a number developing countries and this is increasing. The growing demand for livestock products (particularly in low-income countries) is driven by population growth, higher incomes and urbanization, and presents an important opportunity for millions of smallholders to meet this increased market demand and rise out poverty. Four of the five highest value agricultural products are livestock products (milk, pork, beef, chicken) with demand for milk and meat predicted to triple in Africa by 2050. Globally, infectious disease is estimated to reduce livestock productivity by 25% (valued at $300 billion per year).

One of the most widely recognized forms of livestock interventions is the rotational lending scheme. In this intervention the donor makes an initial investment for livestock purchase, and the recipient is expected to repay this loan by passing on offspring from the livestock to other community members. This form of intervention allows the poor to overcome a lack of financial capital and/or credit acquisition for the initial livestock purchase [8]. Other interventions focus on increasing animal production as a means to improve beneficiary nutritional status [9]. These interventions include for example, National Dairy Development Programs to increase local dairy production by smallholder farmers through zero-grazing based interventions with crossbred cows and cultivated fodder. Programs have been shown to increase household income, food purchases, school fee payments and book purchases; beneficiary households also reported increased intake of animal protein through increased household milk consumption [10].

Programs may also focus on improving livestock health directly. Livestock kept by the poor are extremely vulnerable to disease and poor health due to increased exposure to infectious diseases, suboptimal management, and lack of access to animal health services [11]. Frequently, poor livestock owners have relatively few animals, and therefore the loss of an individual animal has greater significance compared to farmers with fewer resource limitations. One study into the impact of a Newcastle disease vaccination program for poultry on household food intake and food insecurity in rural Tanzania found that households that participated in the vaccination program had increased ownership of chickens and increased household egg consumption compared to control households not participating in the program [12]. Improving the access of poor farmers to animal services has the potential to improve their livelihoods. For improvements to be made to animal health in low-income countries by primary animal healthcare services certain important criteria must be considered: the affordability, accessibility, availability, acceptability and quality of potential animal health services [13,14] are essential for sustainability

WORKING EQUIDS

It is estimated that the total world equine population is around 112 million (approximately 58.5 million horses, 43 million donkeys and 10.5 million mules). Working equids (working horses, mules and donkeys) have an essential role in the livelihoods of millions of individuals worldwide [15,16]. These equids perform numerous vital tasks, including the transportation of goods, people and construction materials, as well as being used in agricultural and tourism activities [17,18]. Working equids are often used in the mo landless, women and children.

The welfare of working equids is affected by numerous problems including the availability, affordability and owner knowledge constraints relating to feed and nutrition; water and hydration; sustainable equine working capacity; effective disease prevention and equine healthcare management, and the need for services such as farriery and harness making. Working equids currently have a limited place within many animal health systems (are not frequently included in disease eradication and vaccination campaigns), and are largely absent from agricultural policy, research and education programs [17, 19].

NGOs working globally have utilised numerous approaches to improve the welfare of working equids including, direct veterinary interventions (free veterinary care); capacity building; training and continuing education programmes, community engagement; education programs and advocacy (influencing). An effective monitoring and evaluation system is required to ensure that organisations are able to identify whether these interventions are working and whether objectives are being achieved. There are a limited number of published peer-reviews reports focused on the impact of NGO programs on working equids and their owners. These reports cover a variety of interventions including, the provision of animal health services, owner training, distribution of anthelmintic prophylaxis and the provision of improved equipment. A summary of these interventions and their impact is provided in one review article [6].

The design of the six published studies was generally robust, with three of the studies utilizing control groups. A number of the studies incorporated baseline measurements although the information provided lacked detail. A number of the studies reported reasonably large sample sizes but only one provided details of a sample size calculation to enable evaluation of the power of the study. Although several studies involved elements of community participation in the evaluation of the intervention, it was not clear to what extent there had been community involvement in defining the criteria against which the intervention would be evaluated [6]. Monitoring and evaluation of programs undertaken by NGOs operating in the working equid sector are still in their infancy and there is considerable need for a stronger evidence base.

CONCLUSIONS

There are considerable challenges to evaluating interventions within resource-poor contexts, including the complex nature of multi-level partnerships, the extended timescales of interventions and their effect, and the frequent lack of applicability of conventional statistical methodologies. Evaluations should focus on the use of outcome-based (e.g. disease prevalence) indicators rather than output-based indicators (e.g. number of animals treated), and need to distinguish between the questions 'did the intervention work?' and 'did the intervention make a difference?' [6].

Potential recommendations to improve programmatic evaluations of interventions in the animal health sector include, defining the objectives of an intervention clearly; establishing an objective measure of baseline prior to starting an intervention; ensuring that the community of owners, users and service providers participate in the evaluation process from the start, and

*** Formatting Issue - Best Available Paper/Graphic ***

establishing a clear statement of the theory of change that underpins the design of the intervention. Key priorities for organisations working to improve the health and productivity of working equids and livestock globally should be the use of evidence-based interventions, demonstrating the impact of these interventions, and the utilisation of this data to influence key decision makers and policy agendas. Global networks of organisations and researchers can play a critical role in building this robust evidence base through interdisciplinary programmes and increased partnerships.

REFERENCES

1. FAO. Livestock's Many Virtues. 2012
2. UN. https://sustainabledevelopment.un.org
3. Perry B, et al. Phil. Trans. R. Soc. B. 2009. 364:2643
4. UK DFID. http://www.livelihoods.org/info/guidance_sheets_pdfs/section1.pdf
5. Carlson-Bremer D, et al. Cochrane Database Sys Reviews.
2013. http://www.cochranelibrary.com/enhanced/doi/10.1002/14651858.CD010624
6. Upjohn MM, et al. Vet J. 2014. 199:210
7. Randolph TF, et al. J Anim Sci. 2007. 85:2788
8. Heffernan C, et al. Livestock and Wealth Creation. 2005.
9. Leroy JL, J Nutr. 2007. 137:2311
10. Mullins G, et al. Hum Ecol. 1996. 24:231
11. Perry B, et al. Livestock and Wealth Creation. 2005
12. Kneuppal D, et al. Food Nutr Bull. 2010. 31:436
13. Catley A, et al. Prev Vet Med. 2001. 49:95
14. Catley A, et al. Community-based Animal Healthcare. 2002
15. Perry B, et al. Investing in animal research to alleviate poverty. 2002
16. Thornton P, et al. Mapping poverty and livestock in developing countries. 2002
17. Pritchard JC. Vet J. 2010. 186:271
18. Stringer AP, et al. Prev Vet Med. 2011. 100:90
19. Stringer AP, et al. Equine Vet J. 2015. 47:6

Insights Gained for Large Animal Referral Veterinarians fFrom an Evidenced-Based Study of Equine Practitioner Expectations

Colleen Best, DVM, PhD
Guelph, ON, Canada

The presentation will describe research findings and insights from two studies - one conducted with equine referring veterinarians (rDVMs) and the other with equine referring veterinarians and specialists. The abstracts for these studies are provided below.

Equine Referring Veterinarians' Expectations of Equine Veterinary Specialists and Referral Centers

The objective of this study was to explore rDVMs' expectations of equine veterinary specialists and equine referral centers. Six focus groups with equine rDVMs (n=48) were conducted and the conversations were analyzed using thematic analysis, a recognized form of qualitative research.

The overarching theme of participants' discussions was that the specialist and referral center should act as an extension of the care the rDVM provides to their clients and patients. The relationship that participants had with their clients was a significant driver of their expectations of the specialist and referral care. Participants spoke about wanting to remain involved in the horse's care when the horse was under the specialist's care. Participants spoke about desiring mutual respect, collegiality, professionalism, and trust in the relationships they had with specialists. Communication played a central role in many of the positive experiences participants shared, and it was perceived to be an essential component of the referral process. Displeasure and frustration were voiced about referral centers that competed with the rDVM's practices or solicited their clients. The results indicated that participants expected to have the care they provide equine patients and clients integrated with that provided by the specialist and referral center through frequent, two-way communication and collegial relationships, such that the outcome is the product of a combined effort. In conclusion, strengthening the relationship of and communication between rDVMs and specialists has the potential to improve the referral process, as well as patient and client outcomes.

Survey of Equine Referring Veterinarians' Satisfaction with their Most Recent Equine Referral Experiences

The objective of this study was to investigate equine rDVM satisfaction with their most recent referral experience, and factors considered by rDVMs when referring. An online survey was administered to both rDVM and specialists. Respondents included 184 referring veterinarians and 87 specialists (referral care providers). Referring veterinarian satisfaction with their most recent referral experience was evaluated by rDVMs. Both rDVMs and specialists were asked to identify factors that influence a rDVM's decision where to refer, and the top 3 factors they perceive are barriers to referral care.

Median rDVM satisfaction with their most recent referral care experience was 80 out of 100 (mean 75, range 8-100). Referring veterinarians were least satisfied with "The competition the referral hospital poses to your practice" (mean 56.96, median 62, range 0-100). The top factor that rDVMs identified as influencing their decision where to refer was "quality of care", while specialists identified "quality of communication and updates from the clinician". The top barrier to referral care for rDVMs was "high cost of referral care", and for specialists was "poor service provided to the client by the referral hospital." Areas exist where rDVMs and specialists differ in what they view as important to the referral process. Exploring opportunities and resources to overcome these differences and their potential impacts on the outcome of referral care is likely to support referral hospitals' efforts to improve caseload and ultimately to support high quality patient care.

Opportunities for Residents to Become Clinician Scientists: What the NIH is Looking for in Veterinary Trained Clinician Scientists

Franziska B. Grieder, DVM, PhD
Bethesda, MD

INTRODUCTION

Completing a veterinary residency program provides in-depth specialized graduate medical training. This training is specific to a clinical branch or area in veterinary medicine and encourages advanced analytical thinking and reasoning. However, a different aspect of graduate training is the scientific research training that involves the systemic investigation and study to either establish facts, reach a conclusion, or discover new knowledge, and is different from, and not always part of, a residency program. Advanced analytical thinking and reasoning is guided toward biomedical questions linked to a broad range of topics extending from molecular pathways and cell-receptor-ligand interactions, to drug discovery and therapeutic interventions; some of these topics may not be immediately linked to a clinical setting or patient, but all will enhance, in the long term, clinical outcomes and patients well-being. The National Institutes of Health (NIH) supports scientific research investigations and training that lead to creating fundamental knowledge about the nature of living systems as well as the application of such knowledge to improve health and reduce illness and disability in humans. As a veterinary resident trainee or graduated resident/fellow, you may be interested in how to transition from being a clinician to a clinician scientist or research scientist who focuses on behavioral, applied, translational, or basic science questions.

The following paragraphs will provide some guidelines on how to approach the challenging evolution from clinician to clinician scientist and what steps to consider in forging a successful path as a clinician scientist.

DEFINING YOUR RESEARCH OBJECTIVES AND GOALS

As indicated in the introduction, it is imperative to define what you are trying to accomplish. Is your goal to become a biomedical bench scientist and your clinical training was an important step towards achieving that goal? Or are you a clinician at heart, but realize that in your desired academic position at a teaching hospital, active and engaged participation in research will advance your career as a clinician and academician? If either one of these settings is the case (the truth may also be somewhere in between), you need to keep in mind that research training, like clinical training, takes a long time and is acquired over many years. If a successful research career is your goal, appropriate training is an essential step that should not be underestimated. Optimal mentorship, a productive environment, and hard work are some of the key factors that you need to seek and commit to; there is no short-cut or fast track solution. A degree-awarding program may not always be necessary, but avoiding exposure to and immersion in meticulous strategic thinking, systematic analysis, and methodical discussion of problems and solutions to test hypotheses will be an open invitation to failure in the search for new knowledge.

Finding an area of biomedical research that both fascinates and challenges you is a very good starting point. Identifying a mentor who is able and willing to take you under his or her wing to teach you the art of research in a setting that is conductive to learning is critical. For you to find the time to conduct research (and sometimes the associated patience and courage) is an essential ingredient that you will learn to make happen. You may already be trained in research – this training may have taken place during a research project which you conducted as an undergraduate, a basic laboratory research experience or rotation you completed as a veterinary student, or a research project you finished as part of a Master's program during your residency. These experiences can all be excellent building blocks on which to continue assembling your research career. As a veterinary clinician scientist, you are in a very strong position to contribute as a member of a multi-disciplinary research project overseen by a team of scientists. Alternatively, as a smart and innovative scientist with an inquisitive mind, strong observation skills, and sharp analytical intellect, you may be leading a research investigation in your own laboratory setting. Whether you want to be a bench researcher or a clinician scientist, you will need to use similar approaches and tactics to be successful. .

APPROACHES TO THE CHALLENGE OF SCIENTIFIC RESEARCH SUCCESS

If your goal is to become an NIH-funded scientist, there are many aspects of your planning that are independent of whether you call yourself a clinician, clinician scientist or bench scientist. Some of the fundamental pieces that you need to consider include: available resources, collaborations and team approach, and of course, funding support through research proposals or grant applications.

Resources: Resources can have many flavors ranging from research training programs to the institutional office of sponsored projects to infrastructure elements such as core facilities for specific technologies or centers providing research materials and expertise (*e.g.*, animals, antibodies, probes). Your goal should be to align all required components that will become parts of your project so you can secure support or request contributions during the process of strategizing and planning for the project; this could include having individuals from these resources writing part of the application or letters of support.

Collaborations: As a young scientist, you may wonder if independence is better than being the second in command attached to a well-established research investigator. Your experience, expertise, the specific research project under evaluation, the research environment including resources and infrastructure at the performance site, your past relationship with the collaborating members of the research team, and the description of the responsibilities of the partners in the project, all influence how the younger, less experienced member of the team will be judged. Everyone needs to start at somewhere; mentorship is essential. If your proposal lacks required expertise or demonstrates insufficient rationale or alternative approaches, senior partners would have been a significant benefit. Describing every team members' role and responsibility carefully and providing clear justification of what is expected will make any application strong.

The Research Proposal or Grant Application: You will need financial support for your research project(s) or for your research training. The earlier you start gaining insight and expertise into the grant writing process, the better you are positioned for ultimate success. The technical aspects of grant writing have been addressed many places and there are courses you can take (added advice: don't forget your senior colleagues). What young, less experienced scientists frequently overlook is the need to understand the mission and goals of the funding agency. These last paragraphs will focus on three aspects of securing grant support: funding opportunity announcements (FOA), funding mechanisms, and review criteria.

The NIH accepts research applications addressing topics of relevance to the NIH mission; such research proposals need to be in response to a FOA, a written notice published in the NIH Guide for Grants and Contracts [https://grants.nih.gov]. Such FOAs communicate the NIH's intent to accept medical and behavioral research proposals in a specific area of research and to award funds on a competitive basis. FOAs can be found at Grants.gov [https://grants.nih.gov/] and proposals must be submitted electronically by a specified deadline. Reading the FOA text very carefully and many times is going to help every potential applicant avoid errors that can be difficult to correct or in certain cases make applications non-responsive or ineligible. Discussing FOAs with your collaborating research team and potentially your institution's office of sponsored projects will help set timelines required to complete tasks and meet specific deadlines as extensions are seldom granted. The FOA will also list the name of the NIH program official who can and should be contacted with any questions that you might have.

Each FOA will provide mountains of information that need to be understood and digested. In addition to deadlines and requested funding amounts, the FOA also lists the funding mechanism. At first glance, this may be of low importance to you, but it carries significant weight. If you seek support for career development, you are likely applying for a K or F award; please note, letters from your institution/department or mentor will be required. If you are part of a core or resource project, your application could be a P mechanism; please note, support letters from core/resource users and collaborators will be critical parts of the application. If you write a research project grant under an R21 or R01 mechanism, your understanding of the key research question targeted and your experience with the topic and associated research field will contribute to the success of your application. If you don't know the specific mechanism that is listed in an announcement, ask questions and obtain clarification.

Last but not least, your proposal will be reviewed by a peer panel of experts [http://grants.nih.gov/grants/peer_review_process.htm]. The FOA will list the standard review criteria as well as any specific criteria [http://grants.nih.gov/grants/peer/guidelines_general/Review_Criteria_at_a_glance.pdf], if applicable. You need to pay close attention to the scored criteria (*e.g.*, Significance, Investigator(s), Innovation, Approach, and Environment). Reviewers will address these five aspects. All five are highly important, but do not necessarily carry the same weight for each application type (*e.g.*, a resource application may not have to be exceptionally innovative). For certain criteria, you have opportunities to influence how your application will be judged. For example, if your institution lacks a key technical element, under approach and environment, your description of an established collaboration with an expert at a different institution may remedy that situation. Novel ideas and approaches are very desirable; out of the box thinking drives research discoveries. However, a lack of understanding of accepted standards without providing substantive rationale why the new approach will succeed partnered with a description of alternatives may be viewed as naïve or overly ambitious. Identifying the path that convinces the reviewers is a critical target to aim for and hit.

HOW TO LEARN MORE AND WHERE TO GET HELP

This last paragraph provides some additional thoughts that may help your future research career: 1) Keep in mind that the internet and specific web sites can be invaluable sources of information. The NIH has extensive and well-developed guidelines and examples of the application process [http://www.hsph.harvard.edu/research-strategy-and-development/faculty-resources/general-nih-resources/]; the peer review system is also covered (see above). 2) If you have an opportunity to participate in a peer review panel, a time-consuming and sometimes intimidating process, you will greatly benefit from that experience. If it can't be an NIH peer review venue, look for participation in another setting, another agency, organization, or your local institution. 3) Make sure you write applications. Most importantly, have your draft proposal reviewed and critiqued by your senior colleagues. Their insight and input is invaluable, not just for this individual application, but for you to learn about writing style, approach (proposing alternatives and suggesting remedies), and presentation (font size, cramming). 4) Finally, you should contact the funding organization that you are targeting for support, ideally prior to writing, but definitely prior to submitting your application. This is the only way you will find out if the focus of your research fits with the agency's goals. Reading the announcement and convincing yourself that there is a match is an easy mistake that has been committed by others many times. Having a dialogue with the program staff who know what the agency is looking for is a critical step. Keep in mind that if it is a good fit, you did not lose anything by asking.

Others have encountered similar career questions and challenges that you may face at this junction of your professional development. Careful planning, a solid and informed network of colleagues and mentors, and the courage to reach out and ask questions will be quintessential elements in your journey to become a productive NIH-supported veterinary clinician scientist.

Overview of the Pathophysiology and Pharmacology of Pain

Johann F. Coetzee, BVSc, Cert CHP, PhD, DACVCP, DACAW
Ames, IA

INTRODUCTION

Societal concern about the moral and ethical treatment of animals is increasing. In particular, negative public perception of pain associated with routine animal management practices such as dehorning and castration is mounting, with increasing call for the development of practices to relieve pain and suffering in livestock. Preemptive analgesia can be applied in advance of the painful stimulus thereby reducing sensitization of the nervous system to subsequent stimuli that could amplify pain. Agents that could be used to provide preemptive analgesia include local anesthetics, nonsteroidal anti-inflammatory drugs (NSAIDs), opioids, α2-agonists, and *N*-methyl *D*-aspartate receptor antagonists. However, less than 20% of U.S. veterinarians currently report using analgesia routinely at the time of dehorning and castration.[1] The capacity to experience pain is considered to have a protective role by eliciting behavioral responses aimed at reducing further tissue damage and enhance wound healing.[2] However, persistent pain syndromes offer no biological advantage and are associated with suffering and distress.[2] Pathological pain states in cattle occur as a result of tissue damage, nerve damage and inflammation and are frequently associated with pain hypersensitivity. Pain hypersensitivity manifests as hyperalgesia (exaggerated responses to painful stimuli) and allodynia (pain resulting from normally innocuous stimuli). Hyperalgesia has been reported to persist in dairy cattle and lame sheep for at least 28 days after the causal lesion has resolved.[3,4] Consequently, chronic pain associated with lameness is considered one of the most significant welfare concerns in dairy cows. Inflammatory pain associated with lameness responds modestly to treatment with non-steroidal anti-inflammatory drugs (NSAIDs)[5,6] but neuropathic pain (due to nerve damage or neuronal dysfunction) is considered refractory to the effects of NSAIDs and many opioid analgesics.[5] Therefore, there is a need to identify novel drugs and drug targets for alleviating chronic pain of neuropathic origin in animals. In this session we will review the challenges associated with providing analgesia in food animals in the U.S. and the salient pharmacokinetic and pharmacodynamics features of the analgesic compounds that are commonly used in livestock. These proceedings contain excerpts that were previously published in a recent issue of Veterinary Clinics of North America: Food Animal Practice (March 2013) and readers are urged to consult this text for more detailed information on this topic.[7]

CHALLENGES ASSOCIATED WITH PROVIDING ANALGESIA IN FOOD ANIMALS

There are several challenges associated with providing effective analgesia in food animals in the United States. Firstly, there are currently no analgesic drugs specifically approved for the alleviation of pain in livestock. Therefore, use of any drug for pain relief constitutes extra-label drug use (ELDU). Under the Animal Medicinal Drug Use Clarification Act of 1994 (AMDUCA), ELDU is permitted for relief of suffering in cattle provided specific conditions are met. These conditions include that (1) ELDU is allowed only by or under the supervision of a veterinarian, (2) ELDU is allowed only for FDA approved animal and human drugs; (3) ELDU is only permitted when the health of the animal is threatened and not for production purposes; (4) ELDU in feed is prohibited and (5) ELDU is not permitted if it results in a violative drug residue in food intended for human consumption. Therefore, use of an analgesic to alleviate pain associated with castration in calves in the United States would be required by law to comply with these regulations.

A second challenge to providing effective analgesia in cattle is that there is often a delay between the time of drug administration and the onset of analgesic activity. For example, local anesthetics require 2 - 5 minutes before a maximal effect is achieved. This may slow animal processing as producers must wait for local anesthesia to take effect. This delay may serve as a disincentive for them to provide routine preemptive analgesia. Furthermore, the requirement for large numbers of animals to be processed quickly may result in procedures being initiated before optimal analgesia is achieved. A third challenge is that the route or method of analgesic drug administration may require specialized training and expertise or may be hazardous to the operator. For example, the NSAID flunixin meglumine is only approved for IV administration in the United States. Therefore, administration requires the animal to be adequately restrained and the operator to be proficient in IV administration. Similar issues are encountered with epidural analgesic drug administration and administration of local anesthesia into the scrotum. The latter procedure is also considered especially hazardous by many livestock handlers. In addition, the majority of analgesic drugs that are available in the U.S. have a short elimination half-life necessitating frequent administration in order to be effective. This increases the stress on the individual animal and increases labor and drug costs.

In addition to the regulatory considerations discussed previously, certain drug classes such as the opioid and NMDA receptor antagonists are designated as Schedule 3 drugs and are subject to regulation by the U.S. Drug Enforcement Administration (DEA). Therefore administration of these compounds to provide pre-emptive analgesia is restricted to use by licensed veterinarians. Finally, the cost associated with providing preemptive analgesia contributes to the reluctance of producers to adopt these measures especially since there is no perceived economic benefit for doing so. It may also be difficult for producers and veterinarians to determine if analgesic compounds are effective because cattle may not show overt signs of pain and distress. Thus determining the need for analgesia and the dose, route, duration and frequency of drug administration in cattle can be especially challenging.

PATHOPHYSIOLOGY OF PAIN IN FOOD ANIMALS

Pain perception involves the transduction of chemical signals at the site of injury into electrical energy (**Figure 1**).[7] This is followed by transmission of the electrical signal via nerve fibers up the spinothalamic tracts where modulation may occur in the dorsal horn.[8] Finally the impulse is projected to the brain where pain perception occurs.[9] The initial response to a noxious stimulus is typically brief, well-localized and somewhat proportional to the intensity of the insult. The second phase of the response is prolonged, diffuse and often associated with hypersensitivity around the point where the initial stimulus was applied.[9] This effect may lead to persistent post-injury changes in the central nervous system resulting in pain hypersensitivity or central sensitization ("wind-up").[9] These effects lead to hyperalgesia (increased pain from previously painful stimuli) and allodynia (a previously non-painful stimulus now produces pain).[11]

Surgery-induced pain and central sensitization consist of two phases: an immediate incisional phase and a prolonged inflammatory phase that arises primarily due to tissue damage.[10] The goal of administering analgesic compounds prior to castration is to mitigate both the incisional and inflammatory phase of the pain response.[11] Effective analgesia therefore requires a multimodal approach using compounds that act on different receptor targets along the nociceptive pathway (**Figure 1**).[7] This can be achieved through a combination of local anesthesia, nonsteroidal anti-inflammatory drugs (NSAIDs) and sedative-analgesic combinations of opioids, α2-agonists, and *N*-methyl *D*-aspartate receptor antagonists.

Figure 1

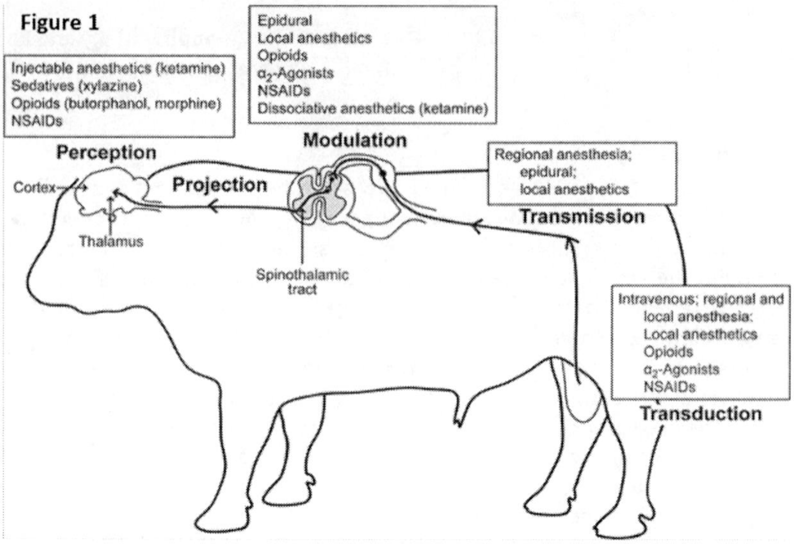

ANALGESIC COMPOUNDS AND THEIR EFFECT IN FOOD ANIMALS

LOCAL ANESTHESIA

Local anesthetics are the most commonly prescribed pre-emptive analgesic drugs used in food animal practice. These compounds produce reversible loss of sensation in a localized area without causing loss of consciousness. Local anesthetics enter and block open sodium channels of nerve cells and prevent generation and propagation of nerve impulses. Repeatedly stimulated nerve cells are therefore more susceptible to the effects of local anesthetics. Furthermore, unmyelinated nerve fibers that transmit pain signals are preferentially blocked by local anesthetics compared with myelinated fibers that are responsible for pressure sensation and motor activity. The quality of local anesthesia in an acidic environment, such as infected tissues, is often poor because these compounds are weak bases that must dissociate in an alkaline environment to exert their effect. Lidocaine has a fairly rapid onset of activity (2 to 5 minutes) and an intermediate duration of action (90 minutes). Local anesthetic administration into the epidural space has also been shown to provide regional analgesia of the perineal region commencing 5 minutes after administration of 0.2 mg/kg lidocaine and lasting 10 – 115 minutes.[12]

NON-STEROIDAL ANTI-INFLAMMATORY DRUGS (NSAIDS)

NSAIDs produce analgesia and anti-inflammatory effects by reducing prostaglandin (PG) synthesis through inhibition of the enzyme cyclo-oxygenase (COX) in the peripheral tissues and central nervous system. COX exists in two isoforms. COX-1 is constitutively expressed in both peripheral and central nervous systems although expression is enhanced by pain and inflammatory mediators. COX-2 is ubiquitous in the CNS but only becomes the major enzyme for PG synthesis after induction by factors released during cell damage and death.[11] It takes 2 – 8 hours for maximal COX-2 mRNA expression to occur in the peripheral tissues therefore initial release of PG is primarily due to COX-1.[13] PG in the peripheral tissues lowers the activation threshold of sensory neurons and may initiate nociceptive activity. PG also works in concert with substance P, histamine, calcitonin gene-related peptide (CGRP) and bradykinin to lower the firing threshold of sensory nerves and produce inflammation. Therefore NSAIDs that inhibit COX-1 may have a more immediate impact on pain by inhibiting PG production in the periphery than COX-2 selective compounds. However, NSAIDs that inhibit COX-1 may be associated with increased risk for adverse gastrointestinal and renal effects.

Spinal PG, notably PGE_2, is responsible for increased excitability of the dorsal root ganglia leading to centrally mediated hyperalgesia. Given that COX-2 is constitutively expressed in the CNS, inhibition of spinal PGE_2 production by NSAIDs that inhibit COX-2 may be an important mechanism in preventing the establishment of hyperalgesia. The effect of NSAIDs on both central and peripheral PG synthesis suggests that these compounds have an important role in multimodal analgesic protocols.

The dose and pharmacokinetic parameters of the commonly used NSAIDs in the United States are summarized in Table 1.

Table 1

Drug	Approved Species	Indications	Dose (Cattle)	T ½ in cattle	Withhold period
Flunixin meglumine (Merck®)	Cattle, horses and pigs	Antipyretic, Anti-inflammatory. BRD and mastitis	2.2 mg/kg IV ONLY! Oral F =60%	3-8 h	Meat- 4 days Milk- 36 hours
Phenylbutazone	Horses and dogs	Anti-inflammatory	4 mg/kg IV ONLY!	40 – 55 h	Not approved in cattle in the USA
Ketoprofen (Merial ®)	Horses and dogs	Anti-inflammatory	1.5 mg/kg IV, IM	0.42 h	Not approved in cattle in the USA
Aspirin	No FDA approval Horses and Cattle	Reduction of fever Relief of minor muscle aches and joint pain	50 – 100 mg/kg PO Oral F < 20%	0.5 h (IV salicylate)	No formal FDA approval Not for use in lactating cattle
Carprofen (Zoeitis®)	EU approval in cattle Dogs	Adjunctive therapy of acute respiratory disease and mastitis	1.4 mg/kg bodyweight IV or SC Oral Tablets	Age Dependent <10 weeks: 49.7 hours	Not approved in cattle in the USA EU- 21 days (Meat), 0 days (Milk)
Meloxicam (Boehringer Ingelheim®)	EU and Canadian approval in cattle Dogs and cats	Adjunctive for BRD; diarrhea and acute mastitis(EU). Analgesia after disbudding (Can)	0.5 IV, SC 0.5 - 1 mg/kg PO Oral F = 100%	27 hours (Range: 19.97 – 43.29 hours)	Not approved in cattle in the USA. 15 days EU and 20 d Canada. FARAD 21 days (Meat)
Firocoxib (Merial ®)	Dogs and horses	Anti-inflammatory	0.5 mg/kg (PO) Oral F =98.4%	18.8 hours (Range: 14.2 – 25.5 hours)	Not approved in cattle in the USA or EU.

SEDATIVE-ANALGESIC DRUGS

Opioids, α2-agonists, and *N*-methyl *D*-aspartate (NMDA) receptor antagonists are the most commonly used sedative analgesic compounds in veterinary medicine. These compounds may act synergistically and are therefore increasingly co-administered. A recent survey of Canadian veterinarians found that respondents that did use an analgesic at the time of castration used xylazine (>50% of respondents) more frequently than lidocaine (< 30% of respondents).[14] Administration of local anesthetics into the testicles is considered by some to be dangerous and time consuming with unpredictable efficacy, especially when circumstances do not allow sufficient time for maximal anesthesia to take effect.[12] Sedative-analgesic compounds may replace the need for intra-testicular anesthetic injection and thus enhance animal well-being and operator safety. A sub-anesthetic combination of xylazine, administered at 0.02 - 0.05 mg/kg and ketamine at 0.04 - 0.1 mg/kg given IV or IM ("Ketamine Stun") is reported to provide mild sedation without recumbency in cattle. Butorphanol (0.01 mg/kg) or morphine (0.05 mg/kg) may be included for enhanced analgesic effects.[15]

FUTURE PROSPECTS FOR TREATING CHRONIC PAIN AND CENTRAL SENSITIZATION IN CATTLE.

GABAPENTIN

Gabapentin (1-(aminomethyl) cyclohexane acetic acid) is a γ-aminobutyric acid (GABA) analogue originally developed for the treatment of spastic disorders and epilepsy. Studies have reported that gabapentin is also effective for the management of chronic pain of inflammatory of neuropathic origin. Although the mechanism of action of gabapentin is poorly understood, it is thought to bind to the α2-δ subunit of voltage gated calcium channels acting pre-synaptically to decrease the release of excitatory neurotransmitters. Efficacy of gabapentin in humans is associated with 2 μg/mL plasma drug concentrations. It has also been reported that gabapentin can interact synergistically with NSAIDs to produce antihyperalgesic effects. In a recent study we report a mean peak plasma gabapentin concentration (Cmax) of 3.40 μg/mL (Range: 1.70 – 4.60 μg/mL) at 7.20 hours (Range: 6 – 10 hours) after oral gabapentin administration at 15 mg/ kg. An elimination half-life (T ½ λz) of 7.9 hours (Range: 6.9 – 12.4 hours) was recorded.[16] Oral administration of gabapentin at 15 mg/kg may be associated with plasma concentrations of > 2 μg/mL for up to 15 hours. The pharmacokinetics of gabapentin suggests that this compound may be useful in mitigating chronic neuropathic and inflammatory pain in ruminant cattle.

REFERENCES

1. Coetzee, J.F. et al. 2010. BMC Vet. Res.; 6 :12. doi:10.1186/1746-6148-6-12.
2. Woolf, C.J., Mannion, R.J., 1999. Lancet 353, 1959–1964.
3. Whay, H.R. et al. 1998. The Veterinary Journal 156, 23–29.
4. Whay, H.R., et al. 2003. The Veterinary Record 153,197-202.
5. Whay, H.R. et al. 2005. The Veterinary Record 157, 729 – 733.
6. Flower, F.C. et al. 2008. Journal of Dairy Science 91, 3010-14.
7. Coetzee J.F. 2013. *Vet Clin North Am Food Anim Pract*. 29(1):75-101.
8. Muir, W.W et al. 2001. J AmVet Med Assoc. 219 (10): 1346 – 1356.
9. Gottschalk, A., Smith, D.S. 2001. Am. Fam. Phys, 63(10): 1979-84.
10. Kissin, I. 2000. Anesthesiology 93:1138 – 43.
11.Ochroch, E.A., et al. 2003. Drugs. 63(24) 2709- 2723.
12. Muir, W.W. et al. 1995. In: Muir WM, Hubbell JAE, Skarda R, and Bednarski RM (eds), Handbook of Veterinary Anesthesia (2nd ed). Mosby, St-Louis. 1995:53-77.
13. Svensson, C.I., Yaksh, T.L. 2002. Annu Rev Pharmacol Toxicol; 42: 553- 583
14. Hewson, C.J. et al. 2007. Can Vet J 48(2), 155 - 164.
15. Abrahamsen, E.J., 2009. In: Anderson, D.E., Rings, D.M. (Eds.) Current Veterinary Therapy Food Animal Practice (5th Ed). Saunders Elsevier., St Louis, pp. 546 – 549.
16. Coetzee, J.F et al.. 2010. Vet J. doi:10.1016/j.tvjl.2010.08.008.

Cross-Species Techniques in Large Animal Pain Assessment
Johann F. Coetzee, BVSc, Cert CHP, PhD, DACVCP, DACAW
Ames, IA

INTRODUCTION

The development of robust biomarkers for the objective measurement of pain is necessary for evaluating the efficacy of analgesic treatment regimens during routine animal husbandry procedures such as castration, dehorning and lameness. This process is especially complex in a prey species, such as cattle, that inherently conceal pain. Pain is defined as "An aversive feeling or sensation associated with actual or potential tissue damage resulting in physiological, neuroendocrine, and behavioral changes that indicate a stress response". In previous research, markers for the evaluation of pain and distress associated with noxious animal husbandry procedures have focused on assessing behavioral, physiological and neuroendocrine changes. A change in animal behavior has been assessed using visual pen scoring, videography, vocalization, chute exit speed measurement, pedometers and accelerometers. Physiological changes have been assessed using serum cortisol measurement, heart rate determination, feed intake and average daily gain. Neuroendocrine changes have been assessed through measurement of the neuropeptide substance P, infrared thermography, heart rate variability (HRV), skin electrical impedence (electrodermal activity) and electroencephalography (EEG). Several of these tools have also been used to assess the efficacy of analgesic compounds. In this session we will review the cross-species techniques in large animal pain assessment. These proceedings contain excerpts that were previously published in a recent issue of Veterinary Clinics of North America: Food Animal Practice (March 2013) and readers are urged to consult this text for more detailed information on this topic.[1]

ASSESSMENT TOOLS USED TO DETERMINE THE NEED FOR AND EFFICACY OF ANALGESIC DRUGS

Assessment of behavioral changes

Assessment tools that have been used to quantify changes in animal behavior following castration include visual scoring systems, videography, vocalization assessment, chute behavior, pedometers and accelerometers. The literature pertaining to behavioral responses associated with castration has been summarized in an excellent review by Stafford and Mellor (2005).[2] The authors conclude that assessments of individual animal behavioral changes in response to pain are highly subjective. Escape behaviors demonstrated at castration but not seen afterwards may reflect a pain response or a desire to escape confinement. Fell et al. (1986) reported that surgically castrated calves struggle and kick during the procedure but calves castrated with rubber rings are quieter.[3] Macauley et al. (1986) found that calves castrated surgically were less active than control calves or calves castrated using a Burdizzo.[4] Robertson et al. (1994) found that rubber-ring, Burdizzo and surgical castration caused significant behavioral responses indicative of pain during the first 3 hours after castration.[5] Fisher et al. (2001) found that 14 month old bulls castrated surgically stamped their hind feet, swished their tails and grazed less following castration than control bulls and bulls castrated using bands. Behaviors indicative of a painful sensation such as turning the head towards the hindquarters, alternate lifting of the hindlegs, abnormal postures and slow movement of the tail has been reported weeks after rubber-ring castration).[6] Currah et al (2009)[7] and González et al (2010)[8] used videography to determine stride length of calves before and after surgical and band castration respectively. Both studies report that stride length was significantly shortened after castration. Furthermore, Currah et al. (2009)[7] concluded that calves took significantly fewer steps after surgical castration in a study that used pedometers to compare step count before and after the procedure. In the same study only 10 out of 71 calves (14%) vocalized during castration and did not report a difference in chute behavior assessed using load cells. White et al. (2008)[9] used accelerometers to evaluate standing and laying behavior in calves before and after surgical castration. The study concluded that calves spent significantly more time standing after castration.

Assessment of physiological changes

Physiological changes after a painful event such as castration have been assessed using serum cortisol, heart rate, feed intake and average daily gain measurements.

Several studies have evaluated acute cortisol response as a method of determining the extent and duration of distress associated with castration in cattle. Studies reviewed by Stafford and Mellor (2005)[2] indicate that surgical and latex band castrations, especially when performed in older cattle, appear to elicit higher plasma cortisol responses that remain above pre-treatment levels for longer. Peak cortisol concentration following surgical castration occurs within the first 30 minutes after castration and range from 45 nmol/L following rubber ring castration to 129 nmol/L following surgical castration. The duration of plasma cortisol response above pre-treatment levels typically ranges from 60 minutes following Burdizzo castration to 180 minutes following surgical castration.

Cortisol has been widely used as a measurement of distress since its response magnitude, as indicated by peak height, response duration and/or integrated response usually accords with the predicted noxiousness of different procedures. At each end of the cortisol response range, however, interpretation is less straightforward. At the lower end, for example, studies have shown that tail docking with a ring and tail docking with a docking iron cause similar cortisol responses to control handling in older lambs. At the upper end of the range, there are several examples where cortisol responses do not increase proportionally to the severity of different treatments as might be expected. This may suggest a "ceiling effect" on plasma cortisol responses. Other studies have

Figure 1

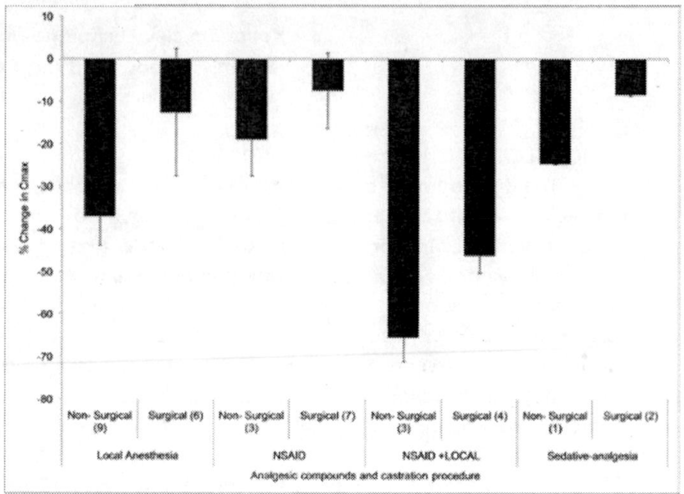

Figure 2

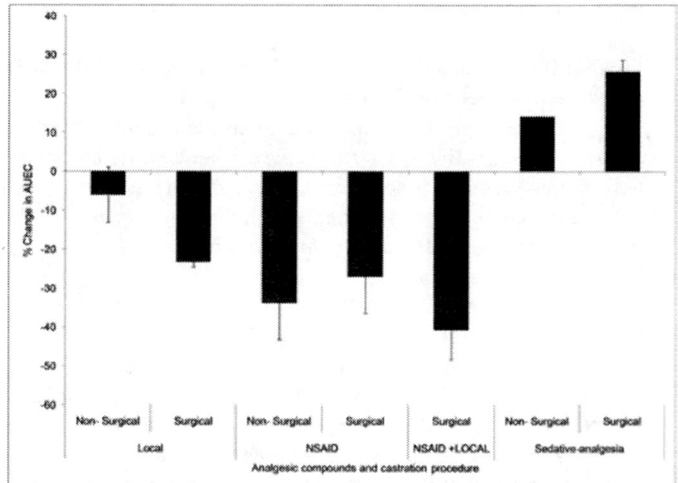

shown that plasma cortisol concentrations following surgical castrations vary greatly between animals. Based on these data, it has been hypothesized that low responses may be due to individuals having high pain thresholds. Variations may also come about due to differences in the way in which a particular castration method is performed by different operators. These data suggest that plasma cortisol levels may not always accurately reflect the extent of the pain response in animals.

A meta-analysis of studies that evaluated cortisol as a primary outcome to determine analgesic drug efficacy revealed that the average percent reduction in peak plasma cortisol concentration (Cmax) in calves receiving local anesthesia alone prior to castration when compared with castrated controls was 25.8% (95% CI: 2.46– 49.1%) (**Figure 1**).[1] However, the average area under the effect curve (AUEC) for cortisol was only reduced by 16. 3% (95% CI: 2.91% – 29.7%) (**Figure 2**). This suggests that local anesthetics alone are effective in reducing acute distress associated with castration. However, the overall AUEC is only modestly reduced in calves receiving local anesthesia prior to castration likely due to the absence of analgesic and anti-inflammatory effects extending into the post-operative period. The average percent reduction in peak plasma cortisol concentration (Cmax) in calves receiving only an NSAID prior to castration when compared with castrated control calves was 10.8 % (95% CI: 4.2 % increase – 25.9% decrease in cortisol). However, the area under the effect curve (AUEC) for cortisol was reduced by an average of 29% (95% CI: 13.2 – 44.8% reduction) (**Figure 2**). This suggests that NSAIDs alone are not effective in reducing acute distress associated with castration. However, the reduction in overall AUEC was greater in calves receiving an NSAID compared with calves administered only local anesthesia prior to castration. This is likely due to the analgesic and anti-inflammatory effects of NSAIDs extending into the post-operative period.

Schwartzkopf-Genswein et al. (2005)[10] evaluated the heart rate in 15 calves before and after surgical castration. It is noteworthy that heart rates were significantly lower at 15 and 30 minutes after castration compared with pre-castration rates. However there was a significant increase in heart rate at 120 minutes after castration. The authors concluded that castration had little effect on heart rate. However these results may have been confounded because calves had been dehorned 21 days previously and the overall heart rates were higher before and after castration than they were prior to the dehorning period.

Production parameters are often too imprecise to reflect the pain experienced by animals following castration. Furthermore, weight gain following castration may be negatively influenced by a decrease in testosterone following removal of the testes. However, assessment of production parameters is critical if animal well-being research is to have relevance to livestock producers. In studies reviewed by Stafford and Mellor (2005)[2], Burdizzo or surgical castration were found to have no effect on average daily gain (ADG) over a three month period following castration. However, the ADG of 7 week-old calves during the 5 weeks following castration using rubber rings, clamp or surgery was found to be lower than non-castrated calves but similar between the different castration methods. Rubber ring and surgical castration were reported to cause a decrease in ADG of 50% and 70% respectively in cattle aged 8 to 9 months. When 8, 9 and 14 month old cattle were castrated surgically or using latex bands, cattle castrated later had poorer growth rates than those castrated at weaning. Cattle castrated with latex bands also had lower growth rates than those castrated surgically during the following 4 – 8 weeks.

In a study conducted by Oklahoma State University, 162 bull calves were used to determine the effects of latex banding of the scrotum or surgical castration on growth rate. Bulls that were banded at weaning gained less weight than bulls that were banded or surgically castrated at 2 to 3 months of age. In a second study, 368 bull calves were used in two separate experiments to examine the effect of method of castration on receiving health and performance. In the first experiment latex banding intact males shortly after arrival was found to decrease daily gain by 19% compared with purchasing steers, and by 14.9% compared with surgically castrating intact males shortly after arrival. In the second experiment purchased, castrated males gained 0.58 lbs more and consumed 1.26 lbs more feed per day than intact males surgically castrated shortly after arrival.

Assessment of neuroendocrine changes

Neuroendocrine changes have been assessed through measurement of the neuropeptide substance P, infrared thermography, heart rate variability (HRV), electrodermal activity and electroencephalography (EEG).

Substance P is an 11-amino acid prototypic neuropeptide that regulates the excitability of dorsal horn nociceptive neurons and is present in areas of the neuroaxis involved in the integration of pain, stress, and anxiety. One study found that plasma SP concentrations are up to 27-fold greater in human patients with soft tissue injury than healthy controls. In a study to evaluate plasma substance P (SP) and cortisol response following castration, no significant difference in plasma cortisol response between castrated and uncastrated calves was observed over time (p=0.644). In contrast, mean plasma SP concentrations were significantly higher in castrated calves compared to uncastrated controls over the course of the study (p=0.042). Significant increases in plasma SP concentration post-castration suggest that this measurement may be associated with nociception although further investigation is necessary.[11]

Figure 3

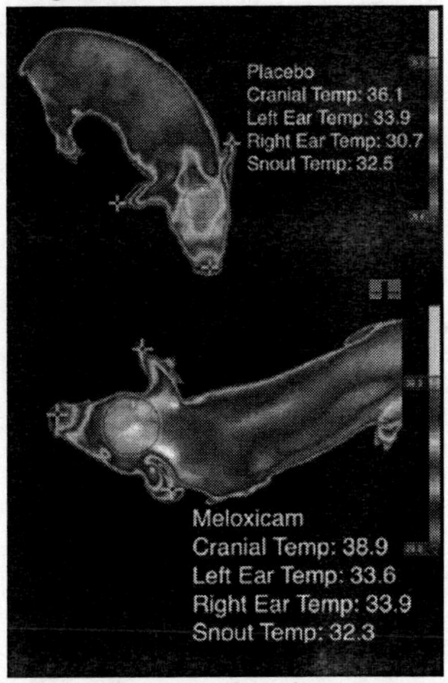

Infrared thermography evaluates changes in surface temperature. Epinephrine release associated with castration causes changes in sympathetic tone. The adrenergic effects on cutaneous blood flow cause changes in skin temperature that can be quantified with a thermography camera (**Figure 3**). A decrease in cranial skin temperature observed following castration of piglets without meloxicam has been attributed to sympathetically mediated alterations in blood flow in capillary beds.[12]

Heart rate variability (HRV) measurement assesses variation in the intervals separating consecutive heart beats. HRV is used to investigate the functioning of the autonomic nervous system, especially the balance between sympathetic and vagal activity. It has been hypothesized that HRV measurement provides a more detailed measure of a stress response than heart rate alone (Stewart et al, 2010).[13] This is because HRV makes it possible to measure the balance between sympathetic and parasympathetic tone, therefore providing a more detailed interpretation of autonomic activity. HRV data are analyzed using frequency domain measures including high frequency power (HF; 0.30 to 0.80 Hz), low frequency power (LF; 0.04 to 0.30 Hz), and the LF:HF ratio. These outcomes are calculated by fast Fourier transformation.

Stewart et al. observed a significant increase in high frequency power from baseline in calves castrated surgically without local anesthesia. In contrast, a significant decrease in low frequency power compared to baseline was observed in calves castrated surgically with local anesthesia.[13] The authors conclude that an increase in HF power indicates an increase in parasympathetic activity that may be associated with deep visceral pain as might occur when the spermatic cords are torn.

Electrodermal activity (EDA) is the measurement of the electrical resistance between 2 electrodes applied to the skin. EDA can be influenced by changes in resistance as a result of changes in sympathetic outflow. The Pain Gauge® (Public Health Information Systems, Inc, Dublin, OH) is purported to be a device capable of measuring EDA although there is a paucity of data to support this use in livestock species. A study that used the Pain Gauge® in rats found it ineffective for accurately assessing postoperative pain because pain scores did not decrease with increasing dosages of analgesic regimens.[14] Similar results were reported by Kotschwar et al. (2009) in calves subjected to an amphoterecin B lameness model.[15] Baldridge et al. (2010) observed a significant decrease in EDA measurement coinciding with the presence of quantifiable plasma xylazine, ketamine and butorphanol concentrations.[16] After 90 minutes, EDA increased and was not significantly different from other treatment groups. It is noteworthy that a difference in EDA between a sham and actual castration and dehorning period both with analgesia was not observed. Therefore, it was concluded that EDA measurement was not a reliable indicator of pain associated with dehorning and castration in calves and that EDA effects were likely associated with other physiological changes associated with drug exposure.

EEG responses of calves to noxious stimulus associated with scoop dehorning using a minimal anesthesia model have been studied. However, the use of general anesthesia may have confounded these results. Further studies are needed to determine the relevance of this research to understanding pain in the conscious calves. Our research group is currently investigating the effect of age and method of castration on EEG response in conscious calves.[17] Prior to castration, a distinct, low frequency wave pattern was found to predominate. Immediately following castration there is a significant shift towards high-frequency, low-amplitude brain activity (beta frequency). Relative power of the low and high frequency waves decreased and increased respectively between baseline and castration periods. This activity is known as desynchronization of waves and is associated with nociception. Delta bands showed a tendency toward an increase during the first recovery period suggesting attempted synchronization within 5-10 minutes after castration. The results of this study suggest that EEG may be a sensitive and specific measure of changes in brain electrical activity associated with castration.

REFERENCES

1. Coetzee JF. (2013). *Vet Clin North Am Food Anim Pract.* 29(1):75-101.
2. Stafford, K.J., Mellor, D.J. 2005. N Z Vet J. 53:271–278.
3. Fell, L.R. et al. 1986. Aust. Vet J. 63:16–18.

4. Macauley, A.S. et al. 1986. J Anim Sci. 63:166.
5. Robertson, I.S. et al. 1994. Res Vet Sci. 56: 8–17.
6. Fisher, A.D. et al. 2001. Aust. Vet J. 79(4):279-284.
7. Currah, J,M. et al. 2009. Can Vet J, 50(4), 375–382.
8. González, L.A. et al. 2010. J. Anim, Sci. 88, 802–810.
9. White, B. J. et al. 2008. Amer J Vet Res. 69 (8): 1005-1012.
10. Schwartzkopf-Genswein, K. S. et al. 2005. Can. J. Anim. Sci. 85:131–138.
11. Coetzee, J.F. et al. 2008. Amer J Vet Res. 69(6): 751 – 762.
12. Bates, J.L. et al. 2014. PLoS ONE 9(12): e113678. doi:10.1371/journal.pone.0113678
13. Stewart, M. et al. 2010. J. Dairy Sci. 93 :3602–3609.
14. Richardson, C.A. et al. 2007. Lab Anim 41: 46 – 54.
15. Kotschwar, J.L. et al. 2009. J Dairy Sci J 92(8): 3731-43.
16. Baldridge SL. Et al. 2011. *American Journal of Veterinary Research*. 72(10):1305-17.
17. Bergamasco L, et al. 2011. *Journal of Veterinary Pharmacology and Therapeutics.* 34(6). 565 -576.

What is the Risk: Chagas Disease?

H. Edward Durham Jr., CVT, LATG, VTS (Cardiology),
St Kitts and Nevis, W.I.

INTRODUCTION

Chagas Disease is the common name of infection with *Trypanosoma cruzi,* a protzoan organism seen in South America, Central America and the southern United States. It is also known as American trypanosomiasis . The disease is transmitted by a vector; a species of reduviidae insect, more commonly known as "kissing bug", "conenose bugs" or "assassin bugs". The disease is named for Carlos Chagas who first describe the infection in 1909 and identified the *T. cruzi* organism which he named for his mentor, Oswaldo Cruz. Infection with *T. Cruzi* is endemic to South America, and parts of Central America. In the United States, the greatest concentration of *T. cruzi* has been detected in south-east Texas. It presents a serious health risk in human, will infect dogs, and some other insect eating mammals. Several mammal species have been identified as reservoirs, including raccoons, opossum, rabbits and armadillos. In recent years Chagas disease has been diagnosed in the United States, mostly in Texas. Virtually every state in the contiguous United States has had a case diagnosed in humans. This can be a big public health concern, especially in the blood bank community, as latent infections of *T. cruzi* could be passed to unsuspecting recipients of blood transfusions. The ABB (formerly the American Blood Bank- now simply ABB) recommends testing for *T. cruzi* in donated samples especially those donors who are from or traveled to endemic areas.

PATHOGENESIS

Trypanosoma cruzi is a parasitic euglenoid protozoan which occurs in three morphologic forms; the **trypomastigote**, the **amastigote**, and the **epimastigote**. The amastigote and epimastigote forms are the reproductive stages. The life cycle of *T. cruzi* begins with the vector taking a blood meal from an infected host or reservoir. The trypomastigote is the form that invades the host by use of the vector or directly crossing the mucus membranes such as the eye mucosa or ingestion. The trypomastigote is a flagellated single cell organism that is travels freely in the circulation system of the host and invades either macrophages or myocytes. Once intracellular infection occurs, the trypomastigote quickly transforms to the amastigote. In the amastigote form the parasite quickly reproduces overwhelming the host immune response to allow great numbers of *T. cruzi* to be created. In the myocyte the reproduced amastigotes transform back into trypomastigotes and are released into the circulation again when the myocyte ruptures. The circulating trypomastigotes are then available to be picked up during the feeding of the vector. In the gut of the vector bug, the trypomastigote transform into the epimastigote, reproduces through binary fission, reverts to the trypomastigote, and is ready to be deposited into another host.

Infection with T Cruzi in humans begins with being bitten by the vector. (Figure 1) A sub-family of the reduviidae family of insects, *Triatoma*, is the vector for infection Although often wrongly identified, this sub-family does not actually include "assassin bugs" since assassin bugs feed exclusively on other insects. The Triatome sub-family of Reduviidae are blood sucking insects. The species of these insects in South America that most typically bites humans will defecate shortly after or during the feeding process. The host will then usually spread the *T. cruzi* organism to the open bite wound by scratching the area. The trypomastigote then enters the host circulation through the bite wound. Any subspecies of *Triatoma* potentially can potentially transmit the parasite. The five that are most reported are *Triatoma infestans, Rhodnius prolixus, Triatoma dimidiate, Triatoma brasiliensis, Panstrongylus megistus* with the first three being most common in the United States. In dogs, transmission does not seem to be as common through the traditional vector system, but generally requires the dog to eat the vector, or potentially meat containing the parasite, such as eating the carcus of a reservoir animal.

Chagas has three phases during the course of the disease. Shortly after exposure to *T. cruzi*, an acute phase begins about 3 days post infection. During this phase, the parasite has travel to the myocytes, transformed to amastigotes, and begun to multiply within the myocyte. During the acute phase there age of the patient impacts the appearance of the disease in dogs. Puppies under 6 months may show lethargy, fever, anorexia and lymphadenopathy; or may be asymptomatic. Dogs greater than 6 months generally do not generally show signs at all. If they do it is typically some transient mild lethargy. The acute phase is followed by a latent phase in which the parasite has fully reproduced, transformed back into a trypomastigote and begun circulating again. Here again sugns are not common The trypomastigotes not ingested by a vector, will then be eliminated by the immune system. The chronic phase occurs after the parasite has been removed from the host, and is a result of the residual damage to the myocardium and gut tissue of the host. This is the phase in which most of the clinical signs are noted.

DIAGNOSIS

Diagnosis of Chagas disease is not always easy. The first critical step is having an index of suspicion. Since young puppies are typically the only group that shows clinical signs during the acute phase, most veterinarian may not suspect *T. cruzi* infection for the average transiently lethargic dog. In puppies less than 6 months, a definitive diagnosis during the acute phase may never be made unless the clinician suspects Chagas as a differential; and may treat the puppy symptomatically. Trypomastigotes can possible be seen on a blood smear during the acute phase, but there number in the circulation are low and easily missed. Microscopic evaluation of the buffy coat may concentrate the trypomastigotes allowing for detection by noting the flagellated motility. A process of plasma centrifugation has been described[1] that allows for further concentration of the trypomastigotes, which can then be sent for polymerase chain reaction (PCR) analysis or used to grow epimastigotes in a liver infusion tryptose growth media. Cardiac troponin-I will greatly elevate over time with infiltration of the myocardium during the acute and early latent phase, but will return to normal as the parasitism clears the host system. Serology and PCR can demonstrated antibodies specific to the organism but if indirect fluorescent antibody, enzyme-linked immunosorbent or radioimmunoprecipitation assays are used one needs to be aware that these tests will be cross reactive with leishmaniosis. Clear historical evidence can help differentiate the two, but serology analysis can be useful in the latent and chronic phases, when trypomastigote levels are low. Serology titers generally become positive at about 21 days post infection and will persist for the rest of the dog's life.

PATHOPHYSIOLOGY

The real pathology of trypanosomiasis occurs during the chronic phase, when the parasite load is at it's lowest or absent. *Trypanosoma cruzi* has is greatest effect on the heart and gastrointestinal tract, although the nervous system and skeletal muscle may also be damaged. Megacolon and megaesophagus are seen in cases with the gastrointestinal form, accompanied with vomiting and diarrhea. These patients will often lose considerable weight. The more common manifestation is cardiac complications. After infiltration of the myocytes, and subsequent reproduction of the amastigotes, the myocyte ruptures due to the mass of parasites in it. This damages the myocytes within the heart. Clinically this is notes as arrhythmias, dilatation of the ventricles eventually leading to depressed contractility and poor cardiac output. During an evaluation by echocardiogram, these dogs will appear to have dilated cardiomyopathy, demonstrating left ventricular enlargement and systolic dysfunction. This may be accompanied by atrial enlargement. In severe cases, right atrial and ventricular enlargement is also noted. Histologically, the ruptured myocytes are replaced by fibro-fatty infiltrates. A host of cardiac arrhythmias may be present including; ventricular arrhythmias, atrioventricular blocks, atrial fibrillation, supraventricular tachycardia, and possible even atrial standstill. Due to the cardiac conditions described above, the dog may develop congestive heart failure, or succumb to sudden death from a fatal arrhythmia.

TREATMENT

Treatment of Chagas disease is directed at management of the clinical signs. Antiparasitic treatment has been attempted in dogs with benznidazole and nifurtimox, which are used in human medicine. These medications need to be administered during the acute phase, which is not often diagnosed. They both can cause significant vomiting, but do lower serum titers. Unfortunately, neither drug is available in the United States at this time. Drugs such as ketoconazole, gossypol, allopurinol, imidazole and verapamil have been tried in dogs, based on promising results in human, but were ineffective.

The best therapy in chronic disease is management of the cardiac signs. Antiarrhythmics are used to treat the specific arrhythmia that may be present. Amiodarone is an antiarrhythmic that may also provide some anti-inflammatory effect. In dogs with clinically significant atrioventricular blocks or atrial standstill pacemakers therapy may be warranted. Decreased contractility may be helped by administration of pimobendan or digoxin. Dogs with congestive heart failure usually receive "triple therapy" of furosemide, enalapril and pimobendan. Outcomes are directly related to the severity of the cardiac damage. Some as yet unpublished data suggests a median survival of 232 days (range 6-1280 days) once cardiac manifestations are diagnosed and treated.

PUBLIC HEALTH ISSUES

Chagas Disease or American trypanosomiasis is considered a zoonotic disease by the Centers for Disease Control. Dogs presenting to the veterinarian from regions that are known to have *T.cruzi* and are diagnosed with dilated cardiomyopathy should be serotested to the organism. This is particularly true of the dog is an atypical breed for dilated cardiomyopathy. Chagas disease is considered a reportable disease by the state of Texas and the Centers for Disease Control. To date this disease is still uncommon in the United States, but could proliferate. Human exposure in the United States has mostly been related to laboratory accidents, or inadvertent needle puncture in the clinic. When working with dogs that are suspected of having *T. cruzi* infection great care must be used not to allow any blood contamination.

Because of its location within the epicenter of American trypanosomiasis in the United States, Texas A&M University in conjunction with the TAMU College of Veterinary Medicine have the most experience with the disease in the US. Their website: kissingbug.tamu.edu is an excellent resource for information. This group also serves as the center of research and identification of Chagas disease in Texas, and the greater USA for veterinary medicine.

The following web address is to the Texas Department of State Health Services. Any confirmed case of Chagas should be reported. Instructions for the reporting process are available on the website.
https://www.dshs.state.tx.us/idcu/disease/chagas/

Suggested Reading

Centers for Disease Control. Parasite – American Trypanosomiasis. http://www.cdc.gov/parasites/chagas/

Texas A&M Kissing Bug and Chagas Disease website. http://kissingbug.tamu.edu/

Barr, S. J., Simpson, R. M., et. al. 1989 Chronic dilitative myocarditis caused by *Trypanosoma cruzi* in two dogs. *Journal of the American Veterinary Medical Association* 195:9, 1237-1241.

Beard, C.B., Pye, G., et al. 2003. Chagas disease in a domestic transmission cycle in southern Texas, USA. *Emerging Infectious Diseases*. 9:1, 103-105

REFERENCES

1. Barr, Stephen C. 2009. Canine Chagas Disease (American Trypanosomiasis) in North America. *Veterinary Clinics in Small Animals* 39: 1055-1064

Figure 1

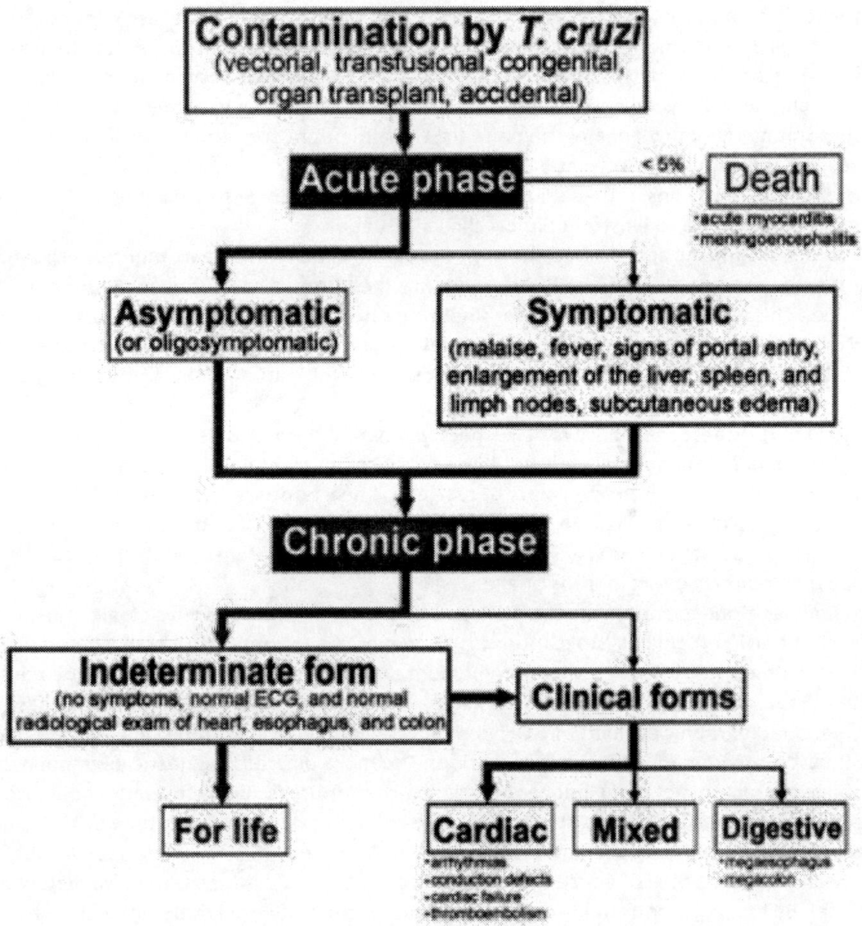

http://web.stanford.edu/group/parasites/ParaSites2008/Raquel_Cristin/AmericanTrypanosomiasis_C_Weekley_R_Kronen.htm

Diagnostic Approach to Asymptomatic Cats with a Heart Murmur - A Primer for Veterinary Technicians

William Rausch, DVM, DACVIM (Cardiology)
Portland, OR

The discovery of a heart murmur in an asymptomatic feline patient is a common clinical scenario in veterinary practice. Although a heart murmur is frequently our first clinical indication of heart disease in the cat, the finding of a murmur is not specific for any particular cardiac disorder nor necessarily indicative of cardiac disease at all. A heart murmur merely indicates abnormal turbulent blood flow within the heart or great vessels. Such flow disturbances can be due to various congenital or acquired primary cardiac changes, non-cardiac disease affecting the heart or benign variants. Moreover, in the cats who do have underlying occult heart disease, the finding of a new heart murmur gives us no specific information as to the severity of any underlying condition.

Ideally, in this situation there would be a single diagnostic test that was entirely specific (no false positive results), sensitive (no false negative results), non-invasive, globally available and equally diagnostic in the hands of all users. Such a test does not exist and the many diagnostic tools at our disposal each suffer from the lack of one or more of these conditions. Although there is no one best diagnostic approach to all asymptomatic cats with newfound murmurs, logical approaches can be developed based upon the particulars of the case. The relative sensitivity and specificity of our diagnostic tools vary, depending upon the underlying cause for the murmur. The positive and negative predictive values of the results of a test also vary, depending on the likelihood that a given disorder exists. By incorporating our knowledge of the patients' signalment, history and physical exam findings with the natural history of feline diseases, we can carefully construct a logical diagnostic approach based upon our index of suspicion for a given disease, thereby minimizing the limitations of the various tests and getting us closer to a diagnosis. This lecture will discuss such an approach in detail, using case examples to illustrate typical clinical situations.

The formulation of a diagnostic approach to the asymptomatic feline with a heart murmur relies heavily on the patient's age when the murmur is first apparent. It is generally assumed that the murmur has not been present, but undetected, for any significant amount of time. This assumption will not always hold true however, as it may have been many years since the owner sought veterinary care or this may be a new client with incomplete past medical records. When the age of murmur onset cannot be reasonably approximated or the age of the patient is in question, some combination of the approaches discussed below must be employed.

While congenital heart disease must always be suspected when a murmur is detected in a kitten, some kittens will have non-pathologic innocent murmurs that abate by early adulthood. These innocent physiologic murmurs have certain characteristics that can be useful to distinguish them from the murmurs of congenital heart disease. Innocent developmental murmurs tend to be soft (usually grade I-II/VI), early to mid-systolic, high frequency murmurs. The point of maximum intensity is generally at the left heart base and they tend not to radiate extensively. Such murmurs may vary in intensity with changes in body position and heart rate and tend to disappear completely by six months of age.

Physical exam findings alone cannot make the diagnosis of a juvenile innocent physiologic murmur as many congenital heart diseases can also have a soft left-sided systolic murmur. However, if a murmur in a kitten is particularly loud (greater than III/VI), associated with a palpable thrill, is loudest on the right, contains a diastolic component, is associated with an arrhythmia, or is present after six months of age, congenital heart disease is likely. The best diagnostic approach in this case would include echocardiographic, ECG and radiographic evaluation. Congenital echocardiographic studies in kittens can be especially challenging and referral to a cardiologist is recommended if this is an option. The more difficult diagnostic dilemma occurs when there is a murmur that fits the above characteristics of an innocent physiologic murmur, as such a murmur could also be due to congenital heart disease. The options in this case include waiting to see if the murmur goes away with time or pursuing cardiac evaluation as described. The cardiac evaluation is not invasive and thus the only detriment to this proactive approach is financial. The downside of a wait-and-see approach is the potential of worsening heart disease that could have been forestalled with prior intervention. An intermediate approach can be taken, by performing only radiographs and/or electrocardiography. However, there is risk of false assurance with this approach, given the relative insensitivity of these modalities for detecting heart disease in the asymptomatic patient. If radiographic or ECG suggestions of cardiac disease were found, this would present a clear indication for echocardiographic evaluation. A PCV/TS should be considered as part of the workup when a murmur is detected in an asymptomatic kitten, as the anemia that accompanies many diseases of young felines can be associated with a murmur due to a decrease in blood viscosity. Murmurs secondary to anemia generally share the physical exam characteristics described above for innocent murmurs.

The various feline primary cardiomyopathies represent the most common heart diseases seen in adult cats. They generally first appear between young adulthood and middle age. While radiographs and electrocardiography are invaluable in dyspneic and arrhythmic cats, the results of these tests will often be unremarkable in asymptomatic young adult cats with a new murmur. Echocardiography is a more sensitive and specific diagnostic test in this population and is better suited to determine severity of disease and necessity and choice of any therapy. Referral to a cardiologist for an echocardiograph would be an appropriate first line diagnostic in this situation. Minimum database bloodwork will occasionally identify anemia secondary to systemic occult disease as a cause for a murmur in this population. We will also discuss the use of biomarkers, including serum NT-proBNP, as screening tools for cardiomyopathies in adult cats. If we are using a blood test as a screening tool we would like it to have a very high sensitivity (very few false negatives). Like any test, the positive predictive value (the likelihood that a positive result is a true positive) of NT-proBNP in this setting is related to the incidence of disease in the tested population. That is to say that the positive predictive value will be relatively higher in cats who are more likely to have underlying heart disease - such as cats with heart murmurs - and relatively lower in the general population of cats. This has implications for the use of cardiac biomarkers as screening tools for occult cardiomyopathies. In this lecture we will review the evidence from the recent literature and discuss possible uses for these tests as well as their limitations.

In middle-aged cats, hyperthyroidism and systemic hypertension are potential causes for a new murmur, but are generally less likely than a primary cardiomyopathy. The median age of diagnosis for HCM has been reported to be from 4.8 to 7 years, while the median age at diagnosis for hyperthyroidism has been reported to be 13 years. A minimum laboratory database in a murmur workup for cats six years or older should include a T4 level, but the incidence will rise with advancing age. A chemistry panel and urinalysis can sometimes be useful in identifying early chronic renal failure, which is often associated with systemic hypertension. If these are normal, echocardiography still maintains the advantages of sensitivity and specificity in middle-aged asymptomatic cats as described above.

A new murmur in an asymptomatic geriatric cat raises different prioritizations of clinical suspicions. This population is more likely than their middle-aged counterparts to have a new murmur as a result of hyperthyroidism, systemic hypertension, anemia or neoplasia. As a result, acquisition of a minimum laboratory database (CBC/Chemistry Panel/UA/T4) and blood pressure measurements are appropriate first line tests. Systemic hypertension can be difficult to diagnose, as the labile nature of feline blood pressure in the clinical setting can make it difficult to interpret elevated measurements. Certain procedural steps can help limit this problem such as gentle handling, performing the measurement with the owner present after a period of acclimation, repeating the test if there are elevated measurements and obtaining the blood pressure prior to other procedures. Despite such precautions, high measurements will invariably be obtained in cats without systemic hypertension. A retinal exam and the laboratory results can be helpful diagnostics in these cases. Retinal hemorrhage, retinal detachment or vessel tortuosity is highly suggestive of systemic hypertension in this setting. Lab findings indicating chronic renal failure or glomerulonephropathy should raise the index of suspicion for hypertension. Echocardiography is often a second tier diagnostic in these patients, based upon the results of initial tests, as echocardiography loses some of its diagnostic specificity in this population. Case examples will be provided to illustrate these principles.

Management of Arrhythmias: Advanced Concepts
A Primer for Veterinary Technicians

William Rausch, DVM, DACVIM (Cardiology)
Portland, OR

AIMS OF THIS TALK

This lecture is intended for veterinary technicians who play an active role (or desire to play a more active role) in monitoring the ECG. Technicians are often tasked with being the primary guardian of the cardiac rhythm - both in surgery and in unstable hospitalized patients. As such, it is critical not only to recognize variances from a normal sinus rhythm, but also to understand that not all arrhythmias are created equal. It takes quick and accurate judgement to recognize which arrhythmias need to be acted upon and why. This lecture is designed to bring you closer to that point.

First, we will briefly review ECG basics. Second, we will discuss a wide selection of common arrhythmias with a focus on when and why they occur and the hemodynamic consequences that often determine how and if we intervene. Lastly, we will discuss our therapeutic options why some treatment choices are better in particular situations and how we should measure success/failure of therapy.

ECG BASICS

This section will be covered briefly in the lecture and is provided here for review purposes. The ECG is a graphic representation of the depolarization and repolarization of cardiac myocytes as recorded from the body surface. Time is expressed in millimeters/second and is plotted on the horizontal axis versus voltage, which is expressed in millivolts on the vertical axis. The ECG is a superb diagnostic tool for determination of heart rate and rhythm. Many electrocardiographic recording devices will automatically count the heart rate, but these automatic counts can lead to inaccuracies - especially in veterinary medicine. The relatively wide breadth of morphologic normality for T waves and QRS complexes in our patients can lead machines to undercount or overcount the true heart rate. Likewise, ECG monitors that purport to automatically yield a rhythm diagnosis can often prove misleading. These machines have a particularly difficult time interpreting the variations in sinus rate that are often a normal variant in our patients. Thus, there is still no substitute for careful assessment of cardiac rhythm and an accurate manual heart rate. Acquisition of an accurate heart rate is especially important when managing bradycardic and tachycardic patients. Accurate manual heart rate assessment simply involves counting the number of QRS complexes in three or more seconds and multiplying appropriately in order to yield a per minute rate. If the cardiac rhythm is highly irregular or slow, then using at least six seconds of an ECG strip to extrapolate a heart rate will yield a more accurate count.

INDICATIONS AND LIMITATIONS

While the ECG is uniquely poised to give information regarding cardiac rate and rhythm, it is important to realize its limitations. A major limitation is that the ECG is a relatively insensitive test for detecting cardiac disease and is generally non-specific as to the underlying disease in the presence of abnormal ECG findings. There are other tests that are better suited to determine if primary heart disease is present, make a cardiac diagnosis, evaluate hemodynamic status and characterize cardiac function. The ECG is but one indispensable part of the cardiac evaluation. Another important limitation is that the transient nature of some rhythm disorders can lead to false negative ECG findings if the arrhythmia does not occur during the recording period. This is a limitation that can often be overcome with ambulatory ECG monitoring (such as Holter) and typical case examples utilizing these technologies will be discussed.

Accurate ECG assessment can be limited by the quality of the recording. Proper recording technique can help to avoid many of the pitfalls associated with artifact. The standard position for recording a small animal ECG is right lateral recumbency. If the patient is too dyspneic for lateral recumbency a standing or sternal ECG can be recorded. Gentle limb restraint will often help to minimize tremble artifact. It is important to avoid contact between the leads and to limit lead and wire movement with respiration. Such movement can be avoided by proper placement of the limb leads away from the chest (just above the olecranon and in the area of the patellar ligaments. Case examples will be presented that focus on methods to determine if abnormal 'complexes' represent arrhythmia or artifact. A helpful guideline to keep in mind is that an actual arrhythmia has to make physiologic sense in the context in which it is seen. Thus, the complexes surrounding the suspect deflection can often yield important information that alters the likelihood that the deflection represents an arrhythmia. For example, a deflection that may represent a VPC is likely to be so if it is immediately followed by a non-conducted P wave, but more likely to be artifact if it appears to fall in the QT interval of a sinus beat without altering the T wave of the sinus complex. An ECG device that is capable of recording multiple leads simultaneously can also be helpful when trying to distinguish artifact from arrhythmia, as certain motion artifact will be detected in some leads but not others. Such a finding would suggest artifact, rather than arrhythmia, as at least some portion of an arrhythmic complex would be expected to be seen in each lead tracing.

PHYSIOLOGIC CORRELATES AND ECG ASSESSMENT

A regular sinus rhythm emanating from the sinus node is the normal rhythm of the heart. The sinus node is a highly organized cluster of specialized cells located in the right atrium near the junction with the cranial vena cava. The cells within the sinus node have the ability to spontaneously depolarize during phase 4 of the cardiac cycle. While this is not a trait exclusive to cells of the sinus node, the sinus node assumes the role of the dominant cardiac pacemaker under normal physiologic conditions because of two basic physiologic properties of sinus node pacemaker cells. These cells have both a low level of resting diastolic membrane potential and a relatively rapid rate of rise of phase 4 diastolic depolarization, as compared to other cells in the heart that have the ability to act as pacemakers. Thus, the sinus node normally controls the heart rate and rhythm, and by doing so suppresses the activation of subsidiary pacemakers. The importance of these phenomena in understanding bradyarrhythmia and tachyarrhythmia will be discussed in this lecture.

The initial depolarization of the cells of the sinus node is a relatively small electrical event and is not significant enough to be detected on a surface ECG. After emerging from the sinus node the cardiac impulse traverses both atria in its path to the atrioventricular (AV) node. The normal P wave represents the depolarization of the atria that results from sinus node firing. The sum of the wavelets depolarizing the atria is an electrical wave that predominantly moves caudally and from right to left, and thus the normal P wave is a positive deflection when viewed from lead II. Disease processes that structurally alter the atria can affect the height or width of the P wave. A tall P wave (P pulmonale) is classically associated with an enlarged right atrium and wide P waves (P mitrale) are often associated with an enlarged left atrium. These are not entirely specific findings however, and these ECG changes are relatively insensitive markers of chamber enlargement. The AV node is the only electrical pathway between the atria and the ventricles under normal circumstances. Electrical activation of the AV node begins somewhere during the period of the P wave on the ECG. Conduction proceeds relatively slowly through the AV node, which has implications in the medical management of arrhythmias as will be discussed in this lecture. Because it cannot be known precisely when AV nodal conduction begins, the PR interval (the time from the beginning of the P wave to the beginning of the QRS complex) is used as a surrogate for measuring AV conduction velocity. Normal ventricular activation occurs via the rapidly conducting specialized Purkinje system. Cardiac or pulmonary diseases that alter the relative distribution of myocardial mass can affect QRS morphology and the mean electrical axis of the ECG. Cardiac diseases that lead to eccentric or concentric myocardial hypertrophy can lead to changes in complex size and sometimes cause conduction disturbances.

With this information as background, we will evaluate a collection of challenging arrhythmias. Once we make a diagnosis we will consider how it is that we decide if a particular arrhythmia warrants therapy. Many arrhythmias that we see in practice have severe manifestations which require intervention and less severe forms that may not. We will use case examples to explore in detail the following questions - the answers to which will determine if and how we choose therapy:

1. Is the arrhythmia hemodynamically significant?
2. Does it appear to be causing clinical signs?
3. Can a correctable, or self-limiting, underlying cause for the arrhythmia be identified?
4. Does the arrhythmia appear likely to degenerate into a less stable rhythm?
5. Is there a consideration for interactions between anti-arrhythmic therapy and current therapy?
6. Is the patient under or soon to be under general anesthesia?
7. What are the risks of beginning therapy?
8. What measures will I use to determine if my therapy is effective?

Cardiac Pacemakers in Small Animals

H. Edward Durham Jr., CVT, LATG, VTS (Cardiology)
St. Kitts and Nevis, W.I.

INTRODUCTION

Cardiac pacemaker implantation is increasingly available for treatment of symptomatic bradyarrhythmia in small animals The goals of this article are to: 1) familiarize the veterinary technician with the indications for cardiac pacing, 2) introduce in simple terms the types of equipment and systems available, 3) review implantation of a pacemaker and 4) discuss proper handling techniques and care for patients that have received cardiac pacemakers. Although veterinary cardiologists perform most pacemaker implantations, increasingly veterinary technicians in private practice manage dogs and even cats with a pacemaker for follow up care; and usually these patients return to their primary veterinarian for veterinary care unrelated to the pacemaker. Awareness of appropriate care for pacemaker patients can alleviate potential problem with the pacemaker system.

Pacemaker implantation is more commonly performed in dogs than cats since dogs are more apt to develop bradyarrhythmias and have a slower ventricular escape rate. It is this slow heart rate that leads to clinical signs.[1] The first reported use of cardiac pacing in veterinary medicine was in a horse in 1967.[2] The first use of a cardiac pacemaker in a dog was just a year later.

INDICATIONS

The most common indications for cardiac pacing in veterinary medicine are symptomatic bradycardias.[1] The predominant bradyarrhythmias that necessitate pacing are 3rd Degree atrioventricular block (AVB), 2nd Degree AVB, Sick Sinus Syndrome (SSS), and Persistent Atrial Standstill. Emergency pacemaker implantation may be required for 3rd degree AVB, which can result in sudden death. Familiarity with the normal electrocardiogram (ECG) will aid the technician in identifying these arrhythmias.

Third degree AVB is a complete disassociation of atrial to ventricular impulses. It is characterized clinically by profound bradycardia, weakness, and/or syncope. Electrocardiographically 3rd degree AVB appears as non-conducted P waves with a junctional or ventricular escape rhythm. The patient's hemodynamic status is maintained by the ventricular escape rhythm. Heart rates between 30 and 60 beats per minute are not uncommon in dogs with 3rd Degree AVB[1], although escape rates can be faster in cats (100-110 bpm).

Sick Sinus Syndrome is an arrest of impulses from the sinoatrial node causing asystole for a period of time sufficient to cause syncope and is typically unresponsive to atropine or related drugs. The asystole can be interrupted by another sinus impulse or an escape impulse that "restarts" the heart and the patient recovers form the syncope. This condition is not generally considered an emergency. High-grade 2nd degree AVB is not usually considered an emergency, but is an arrhythmia that should be promptly addressed. It is defined as a normal to prolonged P to R interval with many non-conducted P waves. High-grade second-degree AVB can progress to third degree AVB and is therefore an indication for permanent cardiac pacemaker implantation.[1] Persistent atrial standstill is rare and associated with atrioventricular muscle disease and a lack of atrial contractions. This differs from hyperkalemia-associated atrial standstill.

EQUIPMENT

The pacemaker system consists of two main components: the pulse generator and the pacing lead. The pulse generator delivers the electrical impulse through an electrode in the pacing lead to stimulate (depolarize) the heart. Either the ventricle alone (single chamber pacing) or the ventricle and the atrium (dual chamber pacing) can be paced. In recent years dual chamber pacing has become more common. The pulse generator is powered by a lithium iodide battery that has a life expectancy of 8-10 years when new. The pulse generators used in veterinary medicine are usually human units that have exceeded their shelf life for human implantation. These pulse generators have 4-6 years of battery life when acquired for veterinary use, but this will meet the needs of most veterinary patients.

There are two types of generators available, unipolar and bipolar. A bipolar system uses a lead with two electrodes at the tip. Current is delivered through one electrode and returns to the generator via the other electrode and the pacing lead. Unipolar systems use the patient's body as the return conduction system. The bipolar system is preferred because it is less likely to cause muscle fasciculations as the impulse returns to the pulse generator.

There are also two basic lead placement systems available for cardiac pacing: endocardial and epicardial. Epicardial pacing systems consist of a generator with a pacing lead that is attached to the outer surface (epicardium) of the left ventricle. This system is typically implanted using an abdominal incision with a transdiaphragmatic approach to the pericardium or by a lateral thoracotomy. The lead is literally screwed into the epicardium with a corkscrew-type tip on the lead. The lead is then brought through the diaphragm and attached to the generator, which is in the abdomen, or in the subcutaneous tissue outside of the abdominal cavity. Epicardial pacing is most commonly used in very small patients (e.g. cats), when vascular access is a problem, when there is another reason for a celiotomy (e.g. abdominal tumor), or when training or equipment for transvenous pacing is lacking. Epicardial leads are typically unipolar.

Endocardial transvenous pacing is less invasive than epicardial pacing and potentially can be implanted in a shorter period of time. Endocardial leads can be either unipolar or bipolar. In the endocardial transvenous pacing system, the pacing lead is passed via the jugular vein (transvenous) through the right atrium into the apex of the right ventricle. Either jugular vein may be used. This procedure must be done with fluoroscopic guidance to insure proper placement of the lead. Once in place, the lead is either actively screwed into the endocardium with a corkscrew-like tip or is held in place passively by grappling hook-type projections tines that snag the trabeculae of the right ventricle. These systems hold the lead tip in place until scar tissue surrounds the lead tip, permanently securing it. The jugular vein is ligated distal to the lead insertion point and the lead is secured with the proximal portion of the vessel (ligation of the jugular vein causes no problems in dogs because collateral vascularization takes over venous return). The remaining proximal portion of the lead is then tunneled under the superficial muscles of the neck and connected to the pulse generator. The lead is secured to the pulse generator with setscrews. These tiny screws are essential as they establish electrical contact and secure the attachment between the lead and the generator. The pulse generator is sutured into a pocket under the superficial muscles on the dorsolateral aspect of the neck. Lead placement can be done under heavy sedation, local anesthetic, and neuromuscular blockade, thereby reducing the risk of general anesthesia for patients until the cardiac rhythm can be controlled. The endocardial lead can also be attached to temporary pulse generator while the generator pocket is dissected and the permanent system connected. Currently, the endocardial pacing system is most commonly used due to equipment availability, ease of placement, lower morbidity, and decreased anesthetic complication rate compared to epicardial pacing.[3]

The pacemaker batteries, though relatively low voltage (about 3 volts), are capable of producing sufficient energy (measured in volts known as *amplitude*) to depolarize the heart. This voltage delivered as a "pulse" of brief duration (measured in mSecs, known as *pulse width*). Current is the amplitude multiplied by the pulse width. The lead and heart tissue constitute the resistance (measured in ohms) that must be overcome to stimulate the heart.

The relationship (Ohm's Law) is: **current = voltage / resistance**

Current needs to be controlled such that sufficient energy is delivered to depolarize the heart, but battery strength is not wasted. The pacemaker discharge rate (i.e., the heart rate) also must be controlled. These and other pacemaker settings can be manipulated with a programmer that controls the pulse generator non-invasively with a magnetic field guided by a computer. A programmer is available for all pacemakers, although they are <u>not</u> interchangeable. Each manufacturer of pacing systems also makes programmers for their systems. Modern programmers can program a variety of models from one manufacturer. The available parameters that are programmable vary with the model of pulse generator, but nearly all pacemakers allow control of the pulse (heart) rate, voltage amplitude, pulse width, and level of sensing native heartbeats. Newer generators are highly programmable allowing for programming of sleep modes, variable rates, automatic threshold tests, and arrhythmia recording and other parameters.

Pacing modes are identified by a 4 letter code system. This system designates the first letter to be the chamber paced, the second letter is the chamber sensed, and the third letter indicates the pulse generator function when a native beat is sensed, the final letter indicates if a pacemaker increases pace rate in response to activity. The abbreviations used are: V= ventricle, A= atria, O= nothing, I= inhibit, R = rate responsive. Most systems are programmed to fire only when the patient's rate drops below a preset value (known as VVI or demand mode). In VVI mode the designation indicates that the ventricle is paced, the ventricle sensed, and the pulse generator is inhibited from firing when a native beat is detected. This inhibition keeps the pacemaker from firing at inappropriate times. Other pacing modes including dual chamber pacing to achieve a more physiologic contraction are now available. Pacing modes and variables can be changed non-invasively using the programmer.

{PRIVATE }PRE-IMPLANTATION EVALUATION{TC \L 2 "PRE-IMPLANTATION EVALUATION"}

In the case of emergency pacing, the patient evaluation may be abbreviated. The suspected diagnosis is confirmed with an ECG. Other diagnostic tests such as radiographs are performed as indicated to determine the anesthetic risk and identify concurrent disease. The presence of concurrent heart disease that could alter the long-term prognosis is also evaluated by echocardiogram. In the situation of a non-emergency pacing, time permits a full workup, including blood serum chemistries, complete blood count, urinalysis, thoracic radiographs, ECG, echocardiogram and non-invasive blood pressure. In some cases a 24 hour ambulatory ECG (Holter Monitor) may be required to confirm the diagnosis. These data will be used to rule out other causes of the arrhythmia that may obviate the need for cardiac pacing and give indication of prognosis. In the face of severe dilated cardiomyopathy for example, the long-term benefits of pacemaker therapy are keenly diminished. ***Venepuncture should not be performed using either jugular vein prior to referring a patient for pacemaker implantation!*** A hematoma around the jugular may preclude using that vein for the pacing lead. The lateral saphenous vein is preferred for blood collection so the cephalic veins can be preserved for IV catheter placement.

Patients with signs of congestive heart failure should be stabilized appropriately with medications such as furosemide, nitroglycerin and sometimes enalapril with or without temporary pacing. In some cases, a constant rate infusion of isoproterenol or dopamine may be administered to increase heart rate of escape rhythms and augment cardiac contractility. Lidocaine, digoxin and β-blockers, or calcium channel blockers are **contraindicated**, as these drugs can worsen the bradycardia or suppress the ventricular escape beats that are sustaining the patient.

Patients with severe bradycardia or escape activity that changes rate erratically may require temporary transvenous pacing with a special catheter to reduce the risk of sudden death from anesthesia. A combined anesthetic protocol of heavy intravenous sedation, neuromuscular blockade, intubation and ventilation, and local anesthesia can be used to gain vascular access and place a pacing lead for this function. An external temporary pulse generator is then attached using sterile connectors. This allows control of the heart rate during the procedure and allows safe administration of inhalation anesthesia. An alternative to temporary catheter pacing is transthoracic pacing. In transthoracic pacing, two wide electrodes are adhered to each side of the shaved thorax of the patient through which just enough current is applied to depolarize the heart with an external pacing system.

{PRIVATE }POSTOPERATIVE CARE

Transvenous endocardial pacing is the primary method of pacing today. The surgical site is small and post-operative pain is less than with epicardial pacing. Post-implantation care involves analgesia, sedation, and continuous electrocardiographic monitoring. Antibiotics are given peri-operativly and post-operatively. Opiate sedation/ analgesia also keep the patient quiet after surgery to avoid dislodging the lead thereby necessitating surgical repositioning. A sterile bandage is placed around the neck to protect the surgical site and reduce seroma formation. Radiographs are taken in the right lateral and dorsoventral views the following day to confirm the lead placement. Avoid positioning the patient on its back during the first two weeks following implantation as this may dislodge the lead. Neck leads or choke chain collars should **NEVER** be used on a dog with a transvenous pacemaker system; they should always wear a harness. An ECG is used to verify pacemaker function and rate. Pulse generator parameters are programmed to a base setting and the patient is released from the hospital 1-3 days after implantation. At discharge, the client is instructed how to take the patient's heart rate and asked to check it daily until the next visit. If the patient's heart rate goes below the pacing rate, the client is instructed to call the clinic immediately.

The clients are asked to return for suture removal 10 to 14 days from discharge. At this visit an ECG and thoracic radiographs are performed. The pulse generator is also evaluated with the programmer. As the endocardium develops a reaction around the pacemaker lead tip where it contacts the heart, increased resistance to the pacing impulse can develop. The minimum current for the pulse generator needed to "capture" the ventricle, causing it to depolarize and contract is determined. To verify capture, a continuous ECG is performed while programming the pacemaker while the current is gradually reduced until capture is lost. When this "threshold" is determined, the pacemaker is programmed at a sufficiently high output to assure capture in the event that resistance increases. Typically, an output setting of two times the threshold is programmed to allow a sufficient margin of safety.

A similar follow up visit is scheduled about one month post implant and again at 3 months post implant at which time radiographs and pulse generator evaluation are performed. Unless complications arise, then the client is asked to see their local veterinarian for biannual radiographs and to visit the cardiologist annually for pacemaker system evaluation. At all follow-up visits an ECG, echocardiogram and pacemaker programming are repeated. Other tests may be added as indicated.

COMPLICATIONS

Complications of pacemakers have been well documented [1, 4] and can range from mild to serious. The most common complications seen are seroma, or hematoma around the pulse generator, dislodgment of the pacing lead from the heart, congestive heart failure from concurrent heart disease or subsequent myocardial failure, tissue reaction sufficient to prevent pacing capture (known as exit block), and malfunctions of the pacing system. Infections occur infrequently, but can be months after the implantation, from hematogenous spread. System malfunctions include lead fracture or dislodgement, battery depletion, or damage to the pulse generator circuitry from electrical impulses (i.e. electrocautery or defibrillation).

CAVEATS FOR OTHER VETERINARY CARE

- Keep conspicuous records of pacemaker details, including rate settings, in the patient's chart along with the pacemaker system's location in the patient. Label cage ID with "Pacemaker" sign.
- **ALWAYS** use a harness for pacemaker patients. **NEVER** put leashes or leads around the patient's neck
- **NEVER** attempt venipuncture in the vein that carries the pacing lead.
- Avoid placing the patient on its back or over extending the neck.

Although the stable pacemaker patient can resume a substantially normal life there are details of patient care that must be remembered. In epicardial pacing, the pacemaker generator is usually located in the abdomen or in between abdominal muscle layers. It must be remembered where the pacemaker is located during palpation so as not to mistake it for a mass. For endocardial transvenous systems the generator and leads are more superficial and greater care must be exercised. During the physical exam the generator is usually palpable and should not be moved, rotated, or mistaken for a mass. Likewise the lead in the jugular vein is easy to feel. In obtaining blood for laboratory analysis, it is crucial to know where the pacemaker and lead are located. ***One should NEVER attempt venipuncture in the jugular vein that carries the pacing lead.*** Striking the lead with a hypodermic needle could cause damage to the insulation around the lead or the actual metallic core of the lead resulting in immediate malfunction of the system and potential life threatening complications for the patient.

The veterinary technician should become acquainted with the electrocardiogram of pacemaker patients.[5] Patients with pacemakers have a distinctive electrocardiogram pattern that shows a "pacing spike" and typically demonstrate wide QRS complexes. Knowing the pacemaker rate setting can also alert the technician to depletion of battery life in a pulse generator. Pulse generators will automatically reduce their rate when the battery weakens, indicating it is time to replace the pacemaker. Modern pacemakers are not susceptible to interference from microwave ovens, garage door openers or pet identification microchip readers. Patients with pacemakers should remain clear of strong magnetic fields such as magnetic resonance imaging or children playing with magnets. Patients that have pacemakers implanted cannot wear electric shock collars like those used with invisible fencing systems.

REFERENCES

[1]Moïse, N. ***Textbook of Can and Fel Cardio*** 1999, WB Saunders pp400. [2]Taylor, D. et al. **JAVMA** 1967; 151(9):pp1172. [3]Bonagura, J. et al. **JAVMA** 1983; 182: pp149. [4]Kittleson, M. ***Small Ani Cardiovasc Med*** 1999, Mosby pp525. [5] Durham, H.E. **Vet Tech** 2000, 21(11): pp642

Technician's Role in Equine Sports Medicine

Emmanuelle van Erck, DVM, ECEIM
Waterloo, Belgium

INTRODUCTION

Equine sports medicine is a particular field of veterinary medicine. The aim is to accompany working equine athletes throughout a career so that they can reach and maintain peak performance during the successive competitive seasons. Horses can be involved in many different equestrian disciplines and it is important as preamble to have a good knowledge of the demands of each sport. Each sport has its particular physiological demands, rules and pitfalls that will eventually determine the performance outcome. Equine sports medicine plays an important role in insuring the horse's best health and welfare at periods that will be physically demanding. An important part of sports medicine is aimed at preventing disease and injuries; the veterinarian's scope goes beyond the horse's basic physical assessment, it needs to encompass the environment, diet and training. Performance is the result of the work and professionalism of an entire team: the horse, the rider, the groom, the owner, the coach, the farrier and, of course, the veterinary team. During the competitive season, good communication and a constructive dynamic spirit will lead the team to success.

ASSESSING HEALTH

Working equine athletes are valuable animals that support the work and life of an entire stable and, possibly later, that of a stud farm when the sporting career comes to its term. It is important to keep these athletes in peak shape and regular veterinary assessment is important to check for any underlying subclinical problems. Regular check-ups need to be programmed according to the horse's competitive schedule. These should initially include at least a general examination, a basic lameness examination and a blood sample analysis. As with human athletes, horses should be given a body condition score and their weight should be regularly assessed. In some disciplines, the examination of videos can be helpful to check the horse's behavior and stamina during competition; gathering of the footage of a race is usually accessible on the Internet and in English sport horses, riders usually have their class recorded for subsequent analysis (showjumping rounds, dressage,...). The veterinary technician plays a central role in coordinating the required examinations and samples according to the horse's working schedule and time of the season. Careful planning requires to take into consideration the areas where the horse is going to show, distances travelled, and making sure that boosters and necessary travelling and veterinary forms have been appropriately filled and sent to the authorities organizing the competitions and regulating the transport of horses.

ASSESSING FITNESS

Similarly to common human sports medicine practice, the investigation into equine exercise physiology has been initially and extensively based on the evaluation of horses on a treadmill. Investigation of a horse on a treadmill requires transport to a specialized center and the intervention of a qualified team of several persons. Treadmill exercise enables to perform numerous and comprehensive measurements in a controlled environment. It also ensures standardization and repeatability of the procedures. Studies have shown the pertinence of measuring specific parameters to evaluate the athletic capacity and metabolic response of horses during exercise and their relation to performance. Among these parameters, maximal oxygen consumption ($VO2max$), arterial blood gases ($PaO2$, $PaCO2$), heart rate (HR) or lactate (LA) against speed have served as reference values in both physiological and pathological conditions. Each parameter brings forth specific information, complementary to others, and there is not to this day a universal predictive parameter of athletic capacity or performance.

On a practical level, referring a horse to a facility that has a treadmill can be time-consuming and costly for riders that already have a busy schedule. Furthermore, exercising a horse on a treadmill requires acclimation and does not necessarily reproduce the same effort as when the horse is exercised under a saddle and it has sometimes been wrongly perceived as unsafe (1). Because of these issues, the implementation of treadmill exercise tests have never become routinely integrated to fitness or clinical evaluations of equine athletes.

This has encouraged the development of alternative techniques that allow performing similar evaluations but in the field. These different techniques have benefited from the miniaturization of measuring and recording equipment and techniques as well as telemetric transmission of the data. Many interesting measurements can now be done with the horse exercising in normal conditions or conditions simulating competition, with the effect of its rider, tack etc... taken into account.

Field exercise tests

Amongst the techniques, which are now commonly used by riders and trainers, HR-monitors coupled to GPS or accelerometric speed recordings are useful to evaluate the horse's physical condition on a regular basis and monitor the horse's progress with training (2). Evaluations may be integrated to routine work sessions (Kingston et al., 2006,). They may also be useful to detect early-on subclinical problems (nevertheless without identification of the disease) as these will directly or indirectly influence the horse's cardiac and metabolic response to exertion (3,4). Portable lactate analyzers are also easy to use in the field; however results are less reliable than laboratory measurements. From the coupling of LA, HR and speed measurements, functional parameters such as VLA4 (speed at which a lactatemia of 4mmol/l is reached) or V200 (speed at which a HR of 200 bpm is reached) may be derived. These are useful to compare the horse's condition from one session to the other with the pursuit of training, or even for inter-individual comparisons (5).

More sophisticated than HR-meters but very useful for the veterinary clinician are the ECG monitors. Small recording units are now available to record and eventually transmit ECG signals throughout an exercise session, recovery included. The recordings can be afterwards downloaded and analyzed with special software. This not only enables to check the disappearance of physiological arrhythmias detected at rest but also to detect cardiac arrhythmias occurring only during exercise or paroxystic episodes that may be responsible for poor performance.

A critical and experienced eye will be necessary in the field environment where more factors of bias are present in comparison to a well-controlled laboratory setting. In order to obtain the most reliable parameters, the veterinary technician needs to communicate very clearly what will be required from the rider during these field tests. Standardized exercise tests yield very useful information only when they are appropriately standardized and conducted according to clear and strict instructions. Organizational skills are important as they will determine the proper acquisition of precious data during the work session. In some cases, such as a test at maximal exertion or a run to fatigue, it will be impossible to re-iterate measurements on the same day: if anything goes wrong, tests will have to be conducted at least a week later, meaning the loss of precious time. Good knowledge of the functioning of the equipment is paramount to obtaining reliable and reproducible measurements. The veterinary technician familiarized with these systems is responsible to download all the acquired data and check its validity, identify it to the horse and extract or calculate the parameters allowing the assessment of the horse's level of fitness. Some blood samples may need to be processed during the field tests.

Seasonal training data

Field exercise tests are most valuable when they are repeated several times over the season. A log-book needs to be kept to compare the horse's given test to the ones previously recorded. This will enable to verify if the horse' level of fitness is progressing or if its training regimen needs to be adapted. An individual training program can be established on the basis of the horse's test results.

It is also important to assess the horse's diet and body condition score as the requirements will change throughout the season either due to changes in physical requirements (intensification of training and competing) or to changes in the quality of forage (depending on its storage or time of the year). Scoring and estimation of body weight needs to be monitored regularly during the season. Regular weighing and analysis of the food is also recommended and adaptations can be made based on the results of the blood sample analysis.

LOOKING OUT FOR SUBCLINICAL DISEASES

Early detection of subclinical disease enables veterinarians to take immediate action and prevent the occurrence of poor performance. Regular exercise testing will indicate if the horse maintains his physiological parameters within his own normal range or the range of horses involved in the same sport and at the same level. As tests are usually very reproducible, any abnormality will indicate a possible subclinical cause. Increases in heart rate for a given exercise can indicate the occurrence of one or several problems including pain (locomotor, digestive, genital,...), weight gain, dehydration or cardiac disease . Lameness or more generally locomotor issues are recognized as the first cause of poor performance, closely followed by respiratory disease. Cardiac and muscular diseases are less prevalent.

Horses are naturally limited by their respiratory system, even in healthy and performing equine athletes. Maintaining respiratory health in working equine athletes should be a priority, as any form of respiratory disease will lead to reduced performances. A critical evaluation of the horse's environment should be done (levels of ventilation, dust, contamination,...) and measures to improve it, such as disinfection should be routinely programmed. In case of suspected respiratory disease, ancillary exams can be performed. Airway endoscopy and sampling can be safely and regularly performed in competition horses, even in the field. These are sensitive techniques that help determine the degree of severity of the disease, chronicity and causative factors. Exercise testing can be an indirect way of estimating the functional consequences of respiratory disease.

Probably one of the most remarkable recent advances for clinical investigations in the field is overground endoscopy. Video recordings of the dynamic behavior of the upper airways enable the diagnosis of obstructive disorders that would be absent at rest. They are useful to determine the origin of a respiratory noise during exercise, assign a severity score and determine appropriate treatment (medical or surgical). Their use is reported in horses of all disciplines, from Thoroughbred racehorses to dressage Warmbloods (6). In contrast to treadmill evaluations, the effect of the rider or driver may be assessed and riding has been shown to influence upper airway stability. Poll flexion induces increased upper airway resistance and promotes dorso-ventral collapse of the pharynx (6). Some conditions are therefore preferably diagnosed when the horse is ridden, especially in dressage horses.

Other problems that can only appear when the horses are exercising are cardiac arrhythmias. Recording ECGs during an exercise, preferably one that simulates levels of exertion similar to competition, can reveal abnormal changes in electrical conduction and rhythm. These recordings can now be made overground and monitored by telemetric devices.

TREATMENTS

Use of treatments in sports either during training or during competition can be submitted to some strict antidoping rule. In equestrian sports ruled by the FEI (Fédération Equestre Internationale), the administration of most medication prior to or during competition is forbidden. It is essential to check that the applied treatment regimen is compatible with the horse's participation to a competition and if not, that proper withdrawal times are respected.

A veterinary technician with a thorough knowledge of the requirements of equestrians sports as well as its rules is a precious asset to a sports medicine practice. Being part of a team that leads top level equine athletes to peak performance is an incredibly fulfilling and exhilarating experience !

REFERENCES

1. Franklin et al. 2010. Equine Vet. J., 42 (Suppl. 38):70-75.
2. Vermeulen and Evans. 2006. Equine Vet. J. Suppl., 36:113-117.
3. Hamlin et al. 2002, Equine Vet. J., 34(4):383-388.
4. Richard et al. 2010, Vet. J., 184(3):282-289.
5. Couroucé et al., 2002, Equine Vet. J., 34(1):91-97.
6. Van Erck, 2011, Equine Vet J, 40(40):18-24.

Update on Equine Endocrine Disease

Janice Kritchevsky VMD, MS, DACVIM (LAIM)
West Lafayette, IN

INTRODUCTION

The discovery that high insulin concentrations are associated with laminitis has led to a large increase in interest and research on disorders of equine carbohydrate metabolism. There are new findings reported every month about endocrine diseases that challenge what we once thought. While exciting, new information is often contradictory to long-held beliefs, leading to confusion. The goal of this session is to present some of the more recent discoveries in a way that is clinically relevant and useful. This will be done by using three specific situations to introduce larger topics.

"LABOR DAY LAMINITIS"

Autumn can be a particularly difficult time for a horse with pituitary pars intermedia dysfunction (PPID). Changes in hormone concentrations and pasture composition both play a role in making the time between mid-August and mid-October an extremely high-risk period for the PPID horse prone to laminitis. [Note: The changes observed are linked to decreased day length. Thus, in the southern hemisphere, the crucial time would be 6-months opposite, from mid-February to mid- April]. During times of the year when the day length is shortening, the concentrations of all the hormones which originate in the pars intermedia increase.[1] The increased hormonal concentrations can cause an exacerbation of the clinical signs associated with PPID in the fall. Insulin also increases in the autumn, and laminitis is closely associated with high blood insulin concentrations.[2]. Additionally, warm days and cold nights produce conditions that are most conducive to increased fructan levels in pasture grasses.[3] Fructan is a form of soluble carbohydrate that is stored in plants, and ingesting too much can cause laminitis.

The recognition that every horse's ACTH increases in the fall has led to some controversy about the best time to perform diagnostic testing in a horse suspected of PPID. There is a possibility of a false positive if a single ACTH measurement happens to be taken at a point when a horse is experiencing the normal fall rise. Evocative or dynamic tests, including the dexamethasone suppression test, have a higher percent of false positives in the fall as well. For this reason, some clinicians suggest that diagnostic testing not be performed between July and November.

Recently it has been suggested, however, that September is actually the best time to perform diagnostic testing to confirm PPID. While a normal horse experiences some increase in ACTH in the fall, horses with PPID often have extremely exaggerated increases. For this reason, the difference in ACTH concentration between normal horses and horses with PPID is most profound in the autumn, making it easier to separate these two populations.[4] If a horse's ACTH concentration is in the normal range in September, it is likely to be normal all 12 months of the year. On the other hand, if the concentration is extremely elevated, 2X the upper limit of the normal range or more, it is likely that an adenoma is present. A horse that is in the "grey zone" in September (i.e. close to the upper limit of normal in either direction), can either be retested during a different time of the year or have a thyroid releasing hormone (TRH) stimulation test performed to help confirm the diagnosis.

Endogenous ACTH concentration can be determined on blood that has been collected into a plastic EDTA tube. It should be refrigerated within 3 hours of collection. Once the plasma is separated, it can either be frozen or kept refrigerated. If collected and handled in this manner, equine ACTH is quite stable. It does not need to be rushed to a laboratory, shipped on dry ice, or have a preservative added to the sample.[5]

FAT VS. EQUINE METABOLIC SYNDROME – DOES IT MATTER?

Equine metabolic syndrome (EMS) is a cluster of clinical signs and laboratory findings that includes obesity with abnormal adipose tissue deposition and a cresty neck, laminitis, and an inability to metabolize glucose and other carbohydrates correctly.[6] In the past, EMS has been termed hypothyroidism, peripheral Cushings disease, Syndrome X, and pre-laminitic syndrome; none of those terms are used presently. Additionally, there are some horses that react abnormally to an oral glucose load although they have normal responses after intravenous glucose or insulin is given. These horses are said to suffer from insulin dysregulation (ID). Obesity is the single largest risk factor for EMS. Additionally, obese, sedentary horses that have been diagnosed as insulin resistant (IR) have dramatically improved values after a weight loss and exercise plan has been followed.[6] These observations lead one to believe that any horse with a high condition score should have its diet and exercise program improved. It is hoped that once the horse has lost weight, the problems with carbohydrate metabolism will take care of themselves. While it is true that over condition is associated with health problems in horses,[7] horses that cannot metabolize carbohydrate correctly due to EMS or ID are not like other fat horses. Furthermore, many horses with EMS do not grade as extremely obese on standard body condition scales because their fat is distributed differently from horses that are just too fat. Animals with insulin dysregulation often suffer from abnormal regional adiposity, and the first step towards differentiating abnormal from merely overweight is to specifically look at the neck, topline around the tail head, and over the ribs for fat deposits. Increased amounts of fat in the neck can be quantified and assigned a cresty neck score.[8] There is a definite genetic contribution, as the tendency towards ID and laminitis runs in lines of ponies.[9] Breeds that are most at risk include English island breed ponies, donkeys, Arabian, Morgan, Mustang, Saddlebred, and Saddlebred-type horses. No matter what the overall condition score, topline score, or neck score, any equid with elevated resting blood insulin concentrations above 20 µIU/ml when fasting or eating only low-carbohydrate hay is at risk of laminitis and needs to be managed through a combination of diet, exercise, and medication.

The most important way to control ID and EMS is through dietary management. The carbohydrate content of the diet should be carefully monitored. If the horse is overweight, the absolute caloric intake should be reduced as well. A horse or pony on pasture will consume 100% of its daily caloric needs fairly rapidly, and the first step towards regulating intake is to slow down the rate of feed ingestion. For an animal on pasture, a grazing muzzle may be sufficient. In stabled or dry lot horses, using slow feeders to decrease the rate of hay ingestion is useful. Housing on a dirt lot may be required initially as the content of pasture is quite variable and horses tend to consume more pasture than dry hay in the same time span. If slow feeders are not an option, soaking hay can remove some of the soluble carbohydrate content. Owners should be aware that minerals and other macronutrients are lost with soaking, in addition to soluble carbohydrates, so a vitamin/mineral ration balancer should always be included if soaked hay is fed.

Working with owners to feed a low carbohydrate diet correctly can be challenging. The primary feeding practice most horse owners have experienced consists of prolonged times on pasture, hay that has been purchased and fed based on its appearance, and sweet feed or some other supplement whether the horse needs it or not. Feeding an ID horse requires a radical departure from these standard practices. The term "low starch" is not regulated by the FDA or any other body, so any feed manufacturer can make that claim on its label no matter what the content of the feed inside the bag. It is up to the owner to sort through the feed company claims and innumerable bits of advice off the internet. Working to develop an individual horse's dietary plan is a valuable role that a technician can play within a practice. The first step entails taking a careful dietary history. In many instances one finds that the amounts fed are way more than is needed. If an ID horse is overweight, it does not need any supplement other than a vitamin/mineral ration balancer. Hay should be fed at a rate of no more than 1.5 % of a horse's body weight per day and should contain no more than 10 - 12% soluble or nonstructural carbohydrate.[10] The amount of carbohydrates in feed can be further reduced by soaking the hay in water. One of the more challenging animals to treat is the horse that has ID but is also thin. In these instances, adding non-carbohydrate calories in the form of additional fat or a low-carbohydrate feed, such as beet pulp, may be helpful. Unless there is severe lameness present, horse with EMD or ID should be exercise at a trot or canter for 30 minutes at least three time a week. Lame horses can be hand walked or turned out in an arena or dry lot.

"BUT HE LOOKED SO NORMAL" – TROUBLE SHOOTING EARLY ENDOCRINE DISEASE

Far too often, the first time an owner considers the possibility of endocrine disease is after their horse has developed acute laminitis. In many instances, some of the warning signs of metabolic disease or PPID were present, yet they were either too subtle to be noticed by a layperson or thought to be a normal aging phenomenon. Screening for metabolic disease should be done as part of every adult horse's annual physical examination. If a horse is deemed at high risk due to advanced age, body condition score, or being a susceptible breed, diagnostic testing should be considered as a part of the examination. Because you are testing for conditions in their earliest stages, screening blood tests will often be normal. Instead, one should consider performing dynamic or challenge testing.[6] The goal is to use dynamic testing to identify the horses with sub-clinical disease so that they can be managed before clinical signs appear.

Screening for metabolic disease in the horse up to eight years of age would consist primarily of body condition scoring and neck/topline scoring to specifically look for the presence of an increased crest in the neck and regional adiposity. The hooves should be examined carefully for rings or other signs of low-grade laminitis. One should also get in the habit of including a photograph of the horse(side view, front, and back) as part of the physical examination medical record. In addition to a physical examination, a careful dietary history should be taken. More than identifying specific problems, the goal in the younger adult horse is to establish baseline values. This is also a good time to provide sound nutritional advice and to discuss the importance of feeding for an animal's use and body condition score.

Problems with insulin regulation are the main issue in the "middle aged" horse between the ages of 8 and 18 years. Cresty neck or topline scoring is particularly important in a horse of this age. Diagnostic testing should be performed on any horse with a cresty neck, abnormal regional adiposity, and on equids that are prone to EMS. This includes donkeys, ponies, and "easy keeper" horse breeds such as Morgan, Mustang, and Saddlehorse types). The annual photographs can prove extremely useful when comparing the horse's neck and overall appearance to earlier years. The hooves should also be examined closely for rings or any patterns of abnormal growth. Many EMS horses have radiographic and visual evidence of past laminitis episodes despite having no history of pain. While a high fasting blood insulin concentration might be present in a horse with advance EMS, in the earliest stages the blood insulin concentration will probably be normal. The oral sugar test and the insulin challenge test should be considered as they will reveal problems with glucose handling.[11,12] Ideally, testing will take place when the horse is not stressed, recently trailered, or fasted for a prolonged period of time. If the suspicion of an insulin problem has come up because of a recent bout of laminitis, testing should be put off until the feet have stabilized. Horses that are very painful with acute laminitis may have high blood concentrations of the stress hormones cortisol and epinephrine, which can affect the results.

A horse that is over conditioned should be placed on a calorie and carbohydrate controlled diet. Taking the dietary history is a great time to discuss how to best feed a diet that has restricted soluble sugars. As the horse crosses above 15 years, PPID should also be screened for. Compare past years' pictures to determine whether muscle tone and overall musculature has decreased. Before PPID horses have long hair coats all year long, they often shed out at a later date than other horses or retain long guard hairs, particularly on their legs, belly, and under their chin. It is important to debunk the notion that the dexamethasone suppression test is the gold standard to diagnose PPID. It is an acceptable test, but can produce false positive and negative results like any other. At this point in time, the TRH response test is probably the most consistent and reliable means to identify horses with PPID. A horse with "early" PPID is unlikely to have a high resting ACTH in the winter, spring or summer. There are two strategies to identify horses with pituitary tumors before they grow the characteristic long hair coat. The first is to perform a dynamic test, in this case the thyroid-releasing hormone stimulation test. Alternatively, one can collect blood for a resting ACTH determination in the autumn. While all horses experience some increase in their ACTH levels between mid-August and mid-October, horses with PPID have an extremely elevated increase at this time. A horse may have normal ACTH most of the year, and a modest elevation to a "grey zone" value in the fall. In those instances, a TRH stimulation test might help settle the issue. Any horse over 18 years of age should be tested for PPID and insulin dysregulation annually. Additionally, the thyroid should be palpated to screen for thyroid adenoma, which is an extremely common benign tumor in aged horses.

REFERENCES

[1]Beech, J et. al *JAVMA* 2009;235:715-722 [2]Schreiber, C et. al *JAVMA* 2012;241:241-248 [3]Hoffmann, RM, et. al *J Anim Sci* 2004;79:500-506 [4]Copas, V et. al *Eq Vet J* 2012;44:440-443 [5]Durham, A et. al *Eq Vet Edu* 2014;26:216-223 [6]Morgan, R. et. al *Veterinary Record* 2015;177:173-179 [7] Johnson, P et. al J *Diabetes Sci Technol* 2009;3:163-174 [8]Carter, R et. al *Vet J* 2009;179:204-210 [9]Treiber, K et. al *JAVMA* 2006;228:1538-1545 [10]McGregor-Argo, C *Vet J* 2009;179:158-160 [11]Frank, N et. al *Eq Vet J* 2014;46:103-112 [12]Bertin, F et. al *Dom Animal Endocrinology* 2013;44:19-25

Technical Tips for Treadmill Training of Horses

Anna M Firshman, BVSc, PhD, DACVIM (LAIM), DACVSMR
Saint Paul, MN
Erica McKenzie BVSc, PhD, DACVIM (LAIM), DACVSMR
Corvallis, OR

INTRODUCTION

Treadmills were originally used to study equine exercise physiology in the 1960s by the pioneer of equine exercise science, Professor Sune Persson. At the present time most academic veterinary institutions and several private practices in the United States have high speed above ground equine treadmills for use in research studies, clinical investigation of cardiorespiratory and musculoskeletal systems, and for rehabilitation and training purposes. Investigations that can be performed on a treadmill include gait analysis, endoscopic evaluation of upper airway function, and tests of exercise performance. Being able to standardize and control the amount of exercise and the exercise surface, and the ease of measuring and evaluating the horse (e.g. endoscopy and telemetry) offer distinct advantages to using treadmills. However there are significant differences that occur when a horse is exercised on a treadmill compared to exercising on land, including differences in exercise intensity at similar speeds, recruitment of different muscles and different biomechanical function. A handful of choices of high speed equine treadmill are available in the United States. Most of them are of the brands 'Mustang', 'EquiGym' and 'Sato'. All of these treadmills are capable of speeds in the region of 16 meters/second (~36 mph, "high speed") and can usually incline up to a 10% grade.

EQUIPMENT

When performing treadmill exercise sessions with horses, additional supplies and considerations include:

- A nylon (non-breakaway) halter and lead rope(s) with a bull snap
- Overreach boots (bell boots) for the front limbs
- Splint boots for all four limbs
- A whip
- Gloves for the handler
- A sturdy rope to place behind horse (if desired/needed)
- A timer (if not using display screen or other treadmill time source)
- A cart to place other necessary items within reach of the treadmill operator
- Floor mats or wall mats for the horse's protection
- Portable or permanent fans for cooling during exercise
- +/- miscellaneous equipment: heart rate monitor, jugular catheter, equine safety belt, endoscopy equipment
- An appropriate number of informed personnel with horse handling experience.

TREADMILL ROOM SETUP

Treadmill room dimensions are somewhat dependent on the amount of available space. The room should be large enough to have generous space around the treadmill to allow the horse to move around it safely. In addition, planning of the room should include appropriate entry points and space for delivery and installation of the treadmill. Treadmills are noisy and so the choice of location should be away from other animals or areas that would be disturbed by noise. The room should be set up with an aim to allow the horse to concentrate on the task at hand, without distraction that may cause it to startle. For example if windows overlooking the room are required, they should be installed at a height that is above the horse's visual line of sight. Alternatively blinds may be installed over windows to provide a privacy screen so that there are no sudden appearances of personnel or other horses at the windows that may cause distraction. Doors without windows are recommended for the same reason. Rubber mats should be placed over any cement surfaces behind and in front of the treadmill to prevent the horse from slipping when entering or exiting the treadmill. The wall behind the horse should also be padded and all unnecessary portable objects should be removed from around the treadmill area. Before the horse enters the treadmill room, the operator should check that the treadmill will turn on and that any display screens are functioning normally. The fans should also be turned on if needed to reduce the chance of startling the horse by turning them on during exercise. The lubricant reservoir should also be checked.

PREPARATION OF THE HORSE

Acclimation of inexperienced horses to the treadmill requires organization and experience. In general, horses with limited or no experience will require at least one full day of training to prepare them for an exercise test involving high speed and/or endoscopy. This acclimation period is important as it allows the operator to ascertain that the horse is adequately comfortable with the treadmill and capable of accomplishing the appropriate gaits to make the test successful. Athletic horses need to demonstrate a reasonable degree of comfort with high speeds on the treadmill, and dressage or gaited horses may need to adapt to moving comfortably in a collected frame or in the required gait. The goal is to have the horse moving comfortably at the desired pace while wearing any equipment it usually wears when it displays the clinical signs for which it is under examination.

Most horses perform adequately if given at least two or three training sessions before an exercise test is attempted. Ideally, the owner should be prepared to leave their horse at the facility for at least one night to provide enough time for appropriate training. However every horse may respond to the treadmill differently and developing a training protocol and timeframe for performing an exercise test depends on the following factors:

1. The horse's previous experience with treadmill exercise
2. The horse's temperament and initial reaction to the treadmill experience
3. The horse's willingness and safety to move at faster or specific gaits on the treadmill.
4. The availability of appropriately qualified staff to assist with training and testing and accessory personnel as needed (e.g. trainers).

At our clinic, every horse presenting for a high speed treadmill evaluation is evaluated by a lameness clinician before training begins. The horse is evaluated at walk and trot gaits on a firm surface. Ideally, the horse should have its shoes removed to avoid damage to the treadmill belt and to reduce the possibility of injury to the horse (e.g. in case the horse falls or catches its shoes at the plate at the back of the belt) or to personnel (from thrown shoes). If front shoes are absolutely necessary for soundness reasons, they should be newly set, so that they are tightly fitted, and toe clips should never be present. All four feet should be picked clean, and the limbs inspected, splint boots and over reach boots should be placed. The halter and lead rope should be checked. We use a nylon halter and a strong lead rope with a bull snap and avoid halters and ropes that will break easily. A chain shank or bridle should never be used for restraint due to potential injury to the horse. Heart rate monitors or safety belts can also be applied outside or inside the room before the horse is on the treadmill. The treadmill room doors should be fully closed after entry of the horse and a sign placed outside prohibiting entry. This is to prevent the horse becoming distracted or startling as a result of a person suddenly entering the room during a training session. All personnel in the room should be fully informed in advance about their expected role. Personnel should be placed in appropriate locations depending on their level of comfort and experience. Observers should be warned not to make loud noises or sudden movements. They should also be instructed never to walk behind the treadmill while the horse is working. No one standing near to the treadmill belt should be wearing loose clothing. In addition any personnel close to the treadmill belt should be instructed to keep feet and rope ends clear of the horse and moving belt. For safety reasons the horse handler should always wear gloves and never wrap the rope around their hands or over their shoulders or necks. The treadmill operator should be comfortable and experienced with the treadmill and should be familiar with the expected speeds for each equine gait - walk: 1.6-1.8 meters/second; trot: 3.0-4.5 m/s; canter: 7-8 m/s; gallop: 9-14 m/s. Whenever the treadmill belt is moving one hand should be positioned over the emergency stop.

On occasion we use judicious doses of sedation (such as acepromazine or xylazine) to help with treadmill training of particularly nervous or difficult horses, or if a potential cause of anxiety for the horse is anticipated, such as large numbers of people in the room, flash cameras, etc. Often times once the horse becomes comfortable with the treadmill, it will no longer require sedation after the introductory period. Sedation is contraindicated during resting or high speed endoscopy of the respiratory tract in case it alters respiratory function and endoscopic findings.

TREADMILL TRAINING OF HORSES

First Training Session

When introducing a new horse to the treadmill, four personnel are useful. One person is responsible for holding the horse, another for operating the treadmill and whip, a third can stand across from the operator to encourage the horse to move if needed, and the fourth person can serve as a runner or extra hand if needed. The goal of the initial session is to get the horse used to the noise and motion of the treadmill, and to teach it to walk and stop with a reasonable degree of comfort and obedience. This can be accomplished in a short series of steps:

1. Bring the horse into the room and allow it to walk around and to settle.
2. Leave the chest bar down and have a gloved handler walk the horse over and FORWARD off the treadmill 3-5 times in a row while the treadmill operator encourages it with verbal commands and by slapping the rump if needed. Ideally use verbal commands the owner routinely uses. The handler should walk to one side of the treadmill, not over the belt, to avoid injury should the horse rush through.
3. Stand the horse to the side of the treadmill and allow it to look at and listen to the treadmill as the operator turns the treadmill belt on for the first time and the speed is increased.
4. Once the horse is comfortable with these steps, the chest bar is raised and the handler walks the horse up to the bar and loops the rope around the side bar at least one time to create a source of friction between the horse and their hands if the horse pulls back. The same commands used to encourage the horse shortly before should be applied and the treadmill immediately started. Most horses will need strong encouragement to take their first steps. They may lean on one side bar which is acceptable as long as they stay close to the chest bar. If the horse's hind feet get very close to the back plate the treadmill should be stopped, the horse re-positioned, and another attempt made. In some circumstances it may be helpful to place a sturdy rope behind the horse looped around the side bars with a non-slip knot (such as a clove-hitch), to provide

encouragement to the horse when it is first learning. The horse should never be shown or allowed to back off the treadmill as it may learn that this is its preferred method of exit. Backing off is dangerous when the horse is moving at a speed higher than a trot.

5. In some circumstance (in particular jumping horses) it may be prudent to leave the chest bar down (or remove it) so that it is not mistaken as a jumping obstacle.
6. The position of the horse on the treadmill is important. It's chest should not be pushing up against the chest bar but its head should be positioned forward of the chest bar so that it is not too far back on the treadmill belt. This is important at higher speed when slight hesitation of the horse can result in the horse moving too far backwards on the treadmill belt.
7. Once the horse is walking and not leaning on the rear rope or side bars it should be allowed to walk for several minutes until it becomes more comfortable and centers itself on the belt. At this point the horse should be verbally commanded to stop, immediately before the treadmill is stopped and the horse is then rewarded with praise and treats. Practice stop-go-stop transitions several times over 15 minutes. If the horse appears to adjust quickly, the initial session may even end with an episode of 1 to 2 minutes of trotting.
8. The horse should be cooled down appropriately with at least 4 minutes of walking on or off the treadmill and hosing or blanketing if needed.

Second Training Session

The second session can occur after as little as a one hour of rest and often occurs on the same day as the first. The second session usually starts by leading the horse directly onto the treadmill with the chest bar up. The same verbal commands should be given and the horse encouraged to walk. The horse should be warmed up at a walk for at least 4 minutes at 1.8 – 1.9 m/s before one or two episodes of trotting are attempted for 2 minutes at a time, followed by 2 minutes of walking each time. It is typical to change gaits at 2 minutes intervals in order to allow the horse to develop a routine. In order to trot the horse is verbally asked to trot and the speed of the treadmill is immediately increased to trot speed. Encouragement may be needed. It is important to attain the trot speed comfortably and not to over shoot the speed and have the horse trotting too fast. Slight and careful adjustments can be made to the speed to find a rate that seems comfortable for the horse. Usually if a horse is comfortable at a walk on the treadmill it will learn to trot relatively easily. If the horse shows reasonable comfort moving from walk to trot, an attempt can be made at cantering. This can also be left until the third session if a horse is not learning to use the treadmill very well. Do not attempt to canter the horse until it is comfortable with the walk and trot progression, and is moving forward willingly.

A horse is taught to canter on the treadmill by increasing speed from the trot gait. Horses that are new to the treadmill will commonly try to trot with a progressively longer stride as speed is increased. It can be challenging to obtain the canter gait and treadmill speed may need to be increased to as much as 9.5 or 10 m/s while encouraging the horse to canter. At the appropriate time it is helpful to verbally command the horse to canter while simultaneously slapping the horse on the rump so that it learns this signal for the canter transition. Once the horse breaks into a canter the speed can be gradually reduced down to a more sustainable level for the individual horse. Generally try to hold this gait for 1-2 minutes at a speed that is comfortable for the horse. Once a comfortable speed is found, keep this speed steady so that the horse can find its balance without having to continuously readjust its stride. Cantering can be very difficult to achieve in gaited horses but in general is easier to achieve the longer the horse is trained on the treadmill prior to attempting canter. Horses that have walked and trotted on the treadmill multiple times over days or weeks will usually transition into canter very easily when asked to do so for the first time. In general, the whip is used for visual encouragement when necessary and depends on the forward going nature of the horse. If a horse is reluctant to move forward the whip can be used carefully for tactile encouragement to prevent the horse from exiting the back of the treadmill.

Third Training Session

A third training session may or may not be performed prior to a maximum exercise/endoscopy session. However, before testing, a horse should be comfortable breaking into a canter and be in balance at that pace on the treadmill. If incline is to be used at any point during the examination it is best to apply it when the horse is in a walking stage to avoid unexpected reactions from the horse due to the changes in noise and motion of the treadmill. The third session is conducted similarly to the second session, but if upper respiratory endoscopy is to be performed at this time, the horse should be fitted with an endoscopy halter (which usually has Velcro attached to the side straps and nose band) prior to commencing. This halter can be placed over the nylon halter and two lead ropes can be attached on either side if desired. The horse should be accustomed to the new equipment in the room. The session should start with walking and at least two bouts of trotting before stopping the treadmill to insert the endoscope (for which a twitch may be necessary). Handlers should be prepared to insert the scope, fasten it securely and then immediately start the treadmill to distract the horse. In some individuals it may be easier to insert the endoscope with the horse walking. The horse can then enter two minute bouts of trot, walk, canter, and if needed, gallop then walk. In cases requiring special investigation (such as dressage horses in collection) a surcingle to fasten long reins can be applied before the horse is on the treadmill. An appropriate personnel member or the horse's trainer can then stand to one side of the treadmill and slowly bring the horse into collection during the walk or trot, then maintain that position at the canter if needed.

TREADMILL MAINTENANCE

This should be performed on a routine basis by designated personnel. Specific tasks include:

- Checking for sufficient lubrication of the belt: Achieved by examination of the underside of the belt or lubricant gauge or reservoir which should be marked with a line at the lubricant level and the date of inspection using a dry erase pen. Ensure the belt is not slippery.
- The manure pit should be regularly cleaned
- The treadmill belt should be cleaned intermittently after exercise by sweeping off debris and manure followed by hosing. Ideally cleaning should occur on Friday afternoon to allow sufficient time for the belt and area to dry over the weekend.
- Clean the belt using a detergent that leaves no residue. Residue causes the belt to be slippery, which may cause injury to the horse or personnel or may damage the equipment.

ALTERNATIVES TO HIGH-SPEED TREADMILL EXAMINATION

Dynamic over ground endoscopy systems are now available for horses, and should be considered when a land treadmill is not available, a specific horse cannot be safely examined on the treadmill, or where the particular exercise activity cannot be adequately duplicated on a treadmill.

UNDERWATER TREADMILL USE IN HORSES

The use of underwater treadmills exercise for training and rehabilitation of horses has become increasingly popular in recent years. Many claims are made as to its effects in horses; although relatively little study has been done to substantiate these assertions. The training of horses to the underwater treadmill follows a similar protocol to that described above, although in many cases walk is the fastest gait that can be achieved, except for below ground underwater treadmills were trotting may also performed.

REFERENCES

Hodgson, David, Catherine McGowan, Kenneth McKeever. The Athletic Horse: Principles and Practice of Equine Sports Medicine, 2nd Edition. W.B. Saunders Company, 2014.

Hinchcliff, Kenneth. Equine Sports Medicine and Surgery: Basic and Clinical Sciences of the Equine Athlete, 2nd Edition. Elsevier Health Sciences, 2014.

Nutritional Support for Hospitalized Horses
Imogen Johns, BVSc, DACVIM (Large Animal Internal Medicine)
North Mymms, United Kingdom

Nutritional support plays an important role in stabilising patients and providing the body with a means of recovery from illness. In certain conditions a negative energy balance can be life threatening.

IMPORTANCE OF NUTRITIONAL SUPPORT

Many hospitalized patients will have a decreased voluntary food intake resulting in a negative energy balance. This in turn is associated with fatigue, an excessive loss of lean body mass, immuno-suppression, delayed wound healing and mobilisation of fatty acids.

PATIENTS REQUIRING NUTRITIONAL SUPPORT

All hospitalised animals with a decreased appetite will benefit from support, but certain cases will present more of a challenge, in particular those at high risk of developing hyperlipemia and conversely, thin horses. Patients with gastro-intestinal disease, including colic and colitis, will also warrant special attention and can be very difficult to manage. Those predisposed to hyperlipemia; including ponies, miniature horses, donkeys and pregnant and lactating mares, must be closely monitored and treated as necessary. Finally, long term residents, such as limb fractures, lose weight gradually that can be hard to track as they are seen every day but often cannot be weighed accurately on scales.

MONITORING TECHNIQUES

It is imperative to know what and how much an individual horse has eaten over any 24 hour period; this can be particularly problematic in a hospital setting where the animal may be fed, and stable cleared, several times a day by multiple personnel. To combat this, every hospitalised horse should have a feeding chart kept if possible by the stable door for easy access and it should be impressed upon all staff that this must be kept up to date. This chart should also form part of the animal's medical record. Feed, both being given and discarded, should be weighed accurately, not recorded in number of 'scoops,' and the weight should be recorded.

Animals should be weighed on a regular basis, preferably on a walk-on weigh scale if in a hospital setting. Weigh tapes can be useful for weight to be tracked over a longer period and at home. Remember that changes in hydration will markedly affect body weight and this must be taken into account in acutely ill patients.

METHODS OF NUTRITIONAL SUPPORT

Some basic rules still apply when feeding horses with decreased appetites such as feeding little and often, keeping a regular routine and not making any sudden feed transitions. Forages should still form the majority of the diet and, for horses that can be walked out in hand, grazing is a useful appetite stimulant. Alternatively, grass can be handpicked and offered. During inappetent periods horses can be offered a 'smorgasbord' of different feeds and adding a little molasses or warm water will help release aromas.

Enteral nutrition

"If the gut works, use it" – many human studies have shown that enteral nutrition is more effective than parenteral nutrition in supporting organ function and blood flow, patient weight gain and immune function principally due to effects on the gastrointestinal mucosa. Enteral feeding helps maintain the role of the intestinal epithelium as a principal barrier to pathogenic microorganisms.

Enteral nutrition can encompass everything from normal feedstuffs, to slurry diets made from normal feeds, to specially formulated component liquid feeds. Horses with a decreased appetite can be fed initially via a nasogastric tube but placement of an indwelling esophagostomy feeding tube may be necessary if prolonged feeding is expected in horses with dysphagia, oral or head and neck trauma, or prolonged anorexia with a functional GI tract. Long term nasogastric intubation requires the use of a small bore flexible soft tube which usually precludes the use of slurry feeds and necessitates the use of equine liquid formulations. Traditionally, these formulations have been associated with complications such as diarrhea and laminitis, although newer formulations with higher fiber contents exist and can be used successfully in the short to medium term.

Parenteral nutrition

This is required when the use of enteral nutrition is contraindicated because of a lack of gastrointestinal function, such as in the presence of gastric reflux. Its aim is to minimize catabolism and malnutrition but parenteral nutrition has several disadvantages, such as an increased risk of thrombophlebitis, added metabolic complications and the loss of the positive effects of enteral nutrition on GIT motility and function. Patients on parenteral nutrition also require intensive nursing. Components used in parenteral solutions include proteins, lipids, carbohydrates, vitamins, electrolytes, minerals and trace elements. A dedicated catheter, or a double lumen catheter with a port dedicated to parenteral administration, is recommended. Formulations can be made up in house using aseptic technique with the required ratios of dextrose, amino acids and lipids, or commercially available, pre-mixed solutions containing dextrose and amino acids can be used.

An example of an adult feeding program utilizing TPN is:

Adult (500kg):		
Glucose:	5g/kg/d → 2500g	→5000ml of 50% dextrose
Amino acids*:	1g/kg/d → 500g	→5000ml of 10% AA solution
Lipids:	0.5g/kg/d→ 250g	→2500ml of 10% lipid solution
(Use of an 11% amino acid solution provides 1.1g/kg/d of protein)		
Total volume:		7.5L/d
Hourly rate:		313ml/h
Calories (non-protein) :		26kcal/kg/d (13000kcal/d)
Calories (total):		30kcal/kg/d (15000kcal/d)

TIMING OF NUTRITIONAL SUPPORT

Certain groups of patients require immediate support such as foals, patients with or at risk of hyperlipemia and lactating or pregnant mares (in the last 3 months of gestation in particular). In others the goal should be to address nutritional support within 48 hours of hospitalization, including patients withheld from feed for >24-48h, malnourished or inappetent patients and those who have oral / esophageal disease. Other patients may improve once their level of pain is better controlled and can be watched for a few days to see if this occurs.

NUTRITIONAL GUIDELINES FOR SPECIFIC DISORDERS

Colic and colitis

There are minimal evidence based recommendations for feeding horses with colic, and as such, most feeding guidelines are based on personal experience and anecdotal recommendations. In general, most horses with a 'simple' colic episode (non-surgical large colon displacement, colonic impaction) can be re-fed relatively rapidly starting as soon as the colic episode has resolved and when normal GI function is anticipated. The forage element of the diet is usually introduced first, and grain can be withheld for several days to minimize colonic gas production. Similar guidelines are probably appropriate for horses following surgery for large intestinal disease, starting with feeding small amounts of grass or soft hay q2-3 hours as early as 12 hours after surgery. More conservative recommendations exist for horses following surgery or with small intestinal disease, especially if a resection has been performed. Enteral feeding should not be introduced until ileus is no longer a concern (the horse is no longer refluxing), and when it is introduced, a low bulk feed is usually recommended to minimize distention at the anastomosis site. Long stem roughage can then be introduced after 3-4 days of soft feeding.

Many horses with colitis are inappetent, so offering a variety of palatable feeds to encourage eating is essential. Once a good appetite is established, using complete pelleted feeds may be beneficial to minimize the requirement for large intestinal digestion. Dietary modification appears to be a key part of the management of horses with right dorsal colitis, where a complete pelleted feed and grazing should be the predominant part of the diet.[1]

Hepatic Disease

Traditionally, diets for horses with hepatic insufficiency have contained low protein concentrations with the emphasis on soluble carbohydrates as an energy source. This was based on currently outdated recommendations in people suggesting that hepatoencephalopathy (HE) was worsened when people ate high protein meals. This is no longer the recommendation, and malnutrition (in people) is associated with a poorer prognosis. Based on this information and considering the relatively low protein diet of most horses, protein restriction is not recommended. Instead the emphasis should be on providing sufficient protein to prevent endogenous protein catabolism.[2] In horses with a poor appetite, encouraging them to eat is the most important factor, so that hepatic gluconeogenesis and muscle catabolism is minimised. Maximising grazing and free choice access to quality forage is recommended, along with supplemental grain meals fed 4-6 times a day. These feeds can be made up with commercially available diets of 8-10% protein and a moderate digestible energy (DE) content, or a mix of sugar beet pulp/chopped alfalfa (to provide highly fermentable fibre) and micronized maize or molasses (to provide glucose and improve palatability). Because of the possible role of an imbalance between aromatic and branched chain amino acids (BCAA) in the development of HE, feeding a diet with protein containing increased BCAA (such as hays, wheat bran, maize and sorghum) may be considered. In hospitalized horses which have a poor appetite, supplementation with intravenous glucose at 1-2mg/kg/minute may be beneficial.

Emaciated horses

The refeeding syndrome is a somewhat poorly defined set of biochemical changes and clinical signs which can occur during the adaptation process following starvation or malnutrition. Although most horses appear to adapt well to the introduction of feed after a period of starvation, biochemical changes and clinical signs consistent with this refeeding syndrome have been seen in chronically starved horses. Up to 20% of horses in poor body condition may die during the refeeding period, suggesting that careful attention to the nutrition of these horses is required, in particular for horses in very poor condition where feed deprivation is known to have occurred.[3] Although not always feasible, close monitoring of electrolyte levels is also recommended with supplementation made as necessary, based on measured levels. A multivitamin and mineral supplement, and thiamine supplementation is also recommended.

Varying guidelines for introducing feed to horses in poor/very poor body condition exist, but in general, most begin with provision of 50% of DE requirements based on the horse's anticipated normal weight. Although some practitioners recommend feeding of relatively poor quality hay or straw and avoiding diets higher in protein such as alfalfa, most current dietary recommendations suggest that alfalfa based diets may help prevent some of the biochemical disturbances of the refeeding syndrome because of their relatively low glycaemic index, and high dietary phosphorous and magnesium content.[4,5]

As a guide, the following slow adaptation to feeding can be utilised, with a combination of alfalfa based feed and hay.[6] Energy density of hays is variable, and thus forage analysis may be beneficial.

Days 1 to 3: Feed 50% of D.E. requirement in 6 feedings with four hour intervals.
Days 4 to 5: Feed 75% of D.E. requirement in 6 feedings with four hour intervals.
Days 6 to 10: Feed 100% of D.E. requirement in 3 feedings every 8 hours.
Days 10 to 120: Continue feeding 2 to 3 times per day, increasing quantity if consumed. It is recommended not to feed grain until body condition score of 3/9 or higher is achieved, usually after 2 months of re-feeding the emaciated horse.

DE can be calculated as: (for resting horses; in kcal)
=(kg BW X 0.03) + 1.4 (<600kg horse)
=1.82 + (-.0383 X kg BW) – [0.000015 X (kg BW)^2] (>600kg horse)

DE for idle horses (may be most appropriate calculation for hospitalised horses):
= (BW X 0.021) + 0.975

Hyperlipemia

Treatment of this condition has historically been unrewarding although an awareness of animals at risk (in particular donkeys, ponies, obese animals), prompt treatment and, potentially, prophylactic partial parenteral nutrition can result in a positive outcome for these cases. In our hospital all animals considered at risk of developing hyperlipemia have their appetite and feed intake closely monitored, and triglyceride concentrations measured if the appetite is considered decreased. If a prolonged period of decreased feed intake or starvation is anticipated, provision of intravenous glucose at 1-2mg/kg/minute can prevent the development of significant fat mobilisation and can be used to treat those animals which are already hyperlipemic.

REFERENCES

1.Cohen N, et al. JVIM 1995;9(4):272. 2. Geor R, et al (eds). Saunders 2013. 3. Whiting T, et al Can Vet J 2006;46:320. 4. Witham C, et al. JAVMA 1998;212:691. 5. Stull C, et al Prof Anim Sci 2003;19:47. 6. Johns I. In Practice 2014;36:144. *Additional Reading:* 7. Carr, E. et al. *Vet. Clin. N. Am.: Equine Pract* 2003;25: 93. 8. Hardy, J. *Clin. Tech. Equine Pract.* 2003;2:193

Cancer Cachexia: Pathophysiology and Treatment in Dogs and Cats

Kara M. Burns, MS, MEd, LVT, VTS (Nutrition)
Wamego, KS

In companion animals, cancer is one of the most common causes of nonaccidental death.[1] Pet owners are very aware of cancer in humans; therefore the healthcare team needs to approach pets with cancer and their owners in a positive, compassionate, and knowledgeable fashion. Cancer cachexia is a complex neoplastic syndrome that adversely alters the patient's functional status.[2,3] As a result of the cancer disease process, cancer cachexia manifests as weight loss, reduced food intake, and systemic inflammation.[4] This discussion will take a look at cancer cachexia and the role nutrition plays in managing cancer cachexia.

STARVATION VS. CACHEXIA

Cancer associated weight loss differs from that seen with simple starvation. In cachexia there is an equal loss of muscle and fat characterized by increased catabolism of skeletal muscle. During starvation, fat is mobilized first sparing muscle proteins, resting energy expenditure is decreased and glucose metabolism is reduced. In contrast patients with cancer cachexia have normal or elevated resting energy expenditures and glucose turnover. Adequate nutrition will halt and reverse the metabolic alterations that accompany simple starvation but will not completely reverse the metabolic disturbances associated with cancer cachexia.

PREVALENCE AND DIAGNOSIS

Cancer cachexia has been reported to affect 30-85% of human cancer patients. Cachexia is reported as the cause of death in approximately 20% of human patients with cancer.[5] Similar data are not available for veterinary patients. However, one study did document clinically relevant muscle wasting in 15% of dogs presented to a veterinary oncology service. In addition, 4% exhibited cachexia as defined by a body condition score of ≤ 3/9 and 68% had documented weight loss of 5% to > 10%.[2,4,6] There are no specific criteria for diagnosis of cachexia in humans but the clinical features which characterize the syndrome include progressive involuntary weight loss with depletion of lean body mass, muscle wasting and weakness, edema, and declines in motor and mental function. There are no specific criteria for diagnosis of cancer cachexia in veterinary patients except weight loss. The diagnosis of cachexia in both human and veterinary patients is complicated by the fact that malnutrition occurs long before clinical signs are evident. Veterinary technicians should suspect cancer associated malnutrition in all tumor bearing animals even in the absence of documented weight loss.

PATHOPHYSIOLOGY

Cancer associated starvation occurs as a consequence of an imbalance between the nutritional needs of the patient, the demands of the tumor and the availability of nutrients in the body.[3,4] The competition for nutrients between the tumor and the host promotes a variety of metabolic disturbances including alterations in carbohydrate, lipid and protein metabolism. Cytokines play a key role as the main humoral and tumor derived factors involved in cancer cachexia and may be responsible for the majority of metabolic changes associated with cancer cachexia. Table 1 summarizes the effects of cytokines on nutrient metabolism in patients with cancer cachexia.[3]

Table 1.

CHO	Increased resistance to insulin Increased glucose synthesis Increased Cori cycle activity
Protein	Increased protein breakdown (catabolism) Increased liver (acute phase proteins) and tumor protein synthesis
Lipid	Increased lipid mobilization Elevated levels of triglycerides

THERAPEUTIC STRATEGIES

Despite years of research in human medicine, practice guidelines for the prevention and treatment of cancer cachexia are lacking. Even less information is available for veterinary patients. The optimal therapy is to cure the underlying cancer; obviously this is not always possible. In veterinary medicine therapeutic strategies generally include management of anorexia, nutritional support / supplementation, and omega-3 fatty acids.

MANAGING ANOREXIA / HYPOREXIA

Nutritional counseling should be a part of the pretreatment plan for all patients. Veterinary technicians should help owners understand the importance of measuring intake and should have a sequential plan for maintaining nutritional support throughout the treatment process. This may include simple strategies to increase consumption initially (see table) which may progress to assisted feeding if necessary. In a patient with a functional GI tract enteral feeding is preferred. Strategies for managing anorexia in dogs and cats have been thoroughly reviewed.[7] Table 2 is a brief review of those recommendations.

Table 2.

Create ambiance /improve service	Create a quiet comfortable feeding area away from disturbances.
Increase Palatability	
Increase moisture	Switch to canned food / add water to kibble. Caution: some cats may have texture preferences and actually prefer dry food
Increase fat	Higher fat diets are more energy dense so less food may be required to meet needs.
Increase protein	Dogs may select foods with higher protein level
Sweet and salty	Adding sweet flavor (No artificial sweeteners) as top dressing may increase palatability for dogs but not cats. Adding salt (if appropriate) may increase palatability for dogs but not cats.
Freshness, aroma, and food temperature	Provide 'fresh' canned / dry foods or home prepared foods. Warming foods to no greater than body temperature may also release aromas.
Rarity	Uncommon or rare food may entice some dogs and cats to eat. Best to offer foods that are uncommon but not completely novel.
Variety	If a variety of options are provided each should be carefully measured to ensure accurate accounting of food intake /preferences. This approach should not be used if food aversions are likely, so that the patient will not develop a learned aversion to all of the therapeutic options.
Drug administration	Never top dress food with medication or use the desired diet to administer medication. Avoid giving medications immediately before/ after a meal if possible.
Eliminate physical barriers	Be sure food bowls are accessible and account for any limitations in movement (raise bowls, remove E collars etc)
Appetite Stimulating drugs	No recommended as effects are unpredictable, intermittent and short lasting. If above approaches are ineffective assisted feeding is recommended.

FUNCTIONAL FOODS FOR PATIENTS WITH CANCER

Most dogs and cats do not have a requirement for soluble carbohydrates and fiber in their diet. Soluble carbohydrates may be poorly used by animals with cancer and can contribute to increased lactate production. Consequently, soluble carbohydrates should comprise less than 25 percent of the food's dry matter content. Dietary protein should be highly digestible and go above levels normally used for maintenance of adult animals. This is due to patients with cancer having altered protein metabolism which may result in loss of lean muscle mass (cachexia). Protein levels should be 30 to 45 percent dry matter in foods for dogs with cancer and 40 to 50 percent dry matter in foods for cats with cancer. Some tumor cells have difficulty using lipids as a fuel source, while host tissues continue to oxidize lipids for energy. As discussed previously, this finding led to the hypothesis that foods relatively high in fat may benefit animals with cancer compared with foods relatively high in carbohydrates. Pets in North America receive most of their nutrient intake from commercial dry pet foods. These foods are usually high in soluble carbohydrate (25 to 60%) and relatively low in fat (7 to 25%). These characteristics make most commercial dry foods inappropriate for nutritional management of animals with cancer. Omega-3 fatty acids, especially those found in certain types of fish and fish oil (eicosapentaenoic acid [EPA], docosahexaenoic acid [DHA]), are probably the most important nutraceuticals to consider for animals with cancer.[1,3]

Alternative Feeding Strategies

It is recommended to feed a food which contains high levels of dietary protein and fat to help meet nutritional demands of dogs undergoing chemotherapy or radiation treatment for cancer to increase survival time and improve quality of life. Studies indicate that approximately 10% of dogs may develop soft stools when initially placed on a canned therapeutic neoplasia dog food. Of these dogs, about half will adjust to the product and develop normal stool consistency. The other half (about 5% of total cases) continue to have GI or stool consistency issues that pet owners find unacceptable. Several strategies are outlined below to help reduce potential discomfort for the pet and the owner.

GI Intolerance

1. Transition from the previous food to therapeutic food over a minimum 10 to 14 day period. The longer transition period helps most patients adjust to the higher fat intake.
2. If unacceptable stools or discomfort persists despite a longer transition period, additional fiber can be added to the overall diet of the dog.

Additional fiber options include:

- Add psyllium husk fiber (Metamucil®); 1 teaspoon per 5 to 10 kg body weight with each meal.
- Replace approximately 10% of the total daily calories with higher fiber diet.
- Add a high-fiber cereal to therapeutic food by thoroughly mixing ½ to 1 cup of cereal per can of food (e.g., Post® 100% Bran™ cereal [28% fiber], Kellogg® All-Bran™ cereal [50% fiber], or General Mills® Fiber One™ cereal [43% fiber]).

Feline Cancer Feeding Strategies

The metabolic abnormalities associated with various forms of cancer in cats are poorly understood. However, the metabolic changes cats with cancer experience may be similar to those found in dogs with cancer and therefore similar feeding strategies should be employed.

Food Recommendations for Cats with Cancer					
Cat Size lbs.	5	8	10	12	15
Cat Size kg	2.3	3.6	4.5	5.5	6.8
Recommended kcal/day	170	215	250	280	330
Key Nutritional Factors for Cats					
Protein = 45.7% DMB					
Fat = 28.7% DMB					
NFE = 16.5% DMB					
Arginine, mg/d	631	798	928	1,039	1,224
Total n-3, mg/d	806	1,019	1,185	1,327	1,564

SUMMARY

It is imperative that veterinary technicians recognize and understand cachexia in cancer patients. It is also necessary for healthcare team members to employ nutritional strategies to help manage anorexia and the imbalance between the nutritional needs of the patient, the demands of the tumor and the availability of nutrients in the body. Feeding strategies should be aimed at alleviating the competition for nutrients between the tumor and the host which result in a variety of metabolic disturbances including alterations in carbohydrate, lipid and protein metabolism.

REFERENCES

1. Saker, KE, Selting, KA. Cancer. In: Hand MS, Thatcher CD, Remillard RL, Roudebush P, Novotny B. eds. *Small Animal Clinical Nutrition, 5th ed.* 2010:587-607.
2. Forrester SD, Roudebush P, Davenport DJ. Nutritional Management of the Cancer Patient. In Cancer Management in Small Animal Practice. C. Henry CJ, Higginbotham ML, eds. Elsevier, 2010: 170-176. St. Louis, MO
3. Wortinger A, Burns KM. Nutrition and Disease Management for Veterinary Technicians and Nurses. Wiley Blackwell. 2015: 202-207.
4.Mauldin GE. Nutritional Management of Oncological Diseases. In *Applied Veterinary Clinical Nutrition*. Fascetti AJ, Delaney SJ, eds. 2012. Wiley-Blackwell, Ames, IA.
5. Hemming L, Maher D. Understanding cachexia and excessive weight loss in cancer. *Brit J Comm Nursing* . (2005) 10 (11); 492-495.
6. Michel KE, Sorenmo K, Shofer FS. Evaluation of body condition and weight loss in dogs presented to a veterinary oncology service. *JVIM* (2004) 18; 692-695.
7. Delaney SJ. Management of anorexia in Dogs and Cats. *Vet Clin Sm An Pract* (2006) 36; 1243-1249.

Using Nuclear Medicine to Diagnose and Treat Cancer in Cats and Dogs

Kim A. Selting, DVM, MS, DACVIM (Oncology), DACVR (Radiation Oncology)
Columbia, MO

INTRODUCTION

The use of radioactive iodine to treat hyperthyroid cats has become common over the past 20 years. Administration is minimally invasive, the therapy spares the parathyroid glands, and licensing to handle and administer such small doses is readily obtained. Many private specialty practices provide this service. The concept of using radiation (energy) to treat abnormal tissue has many other applications, and identification of appropriate patients for referral can increase options for diagnosis and treatment.

A rapidly expanding armamentarium of isotopes will change veterinary cancer care. Radioisotopes such as ^{18}F-fluorodeoxy glucose (^{18}F-FDG) can exploit the metabolism of the cancer cell, allowing improved and minimally invasive staging, as well as the ability to locate an elusive pathologic process. Additionally, by using inherent and manufactured targeting characteristics, and by selecting the desired energy for a biologic effect, radioisotopes can be used to target the cancer cell while sparing normal tissue. In veterinary medicine, facilities that work with radioactive diagnostic and therapeutic isotopes are sparse, and the goal of this talk is to inform the listener of the appropriate applications for nuclear medicine for veterinary patients. This discussion will help identify cases for which referral to a facility with nuclear medicine will be beneficial.

External beam radiation therapy has been the mainstay of radiation therapy for many years. This entails fractionated treatments over the course of approximately 3-6 weeks. The use of cross sectional imaging (CT scans) for computer-guided planning allows relative sparing of normal tissue and targeting of tumor tissue. Brachytherapy involves implanting a radioactive source into the tumor, and the energy that is emitted is highly concentrated around the point sources, resulting in sometimes complete sparing of the surrounding normal tissue. However, proper orientation of sources can be challenging. Brachytherapy can also be accomplished using targeted pharmaceuticals such as radioiodine and samarium. The University of Missouri houses the largest research reactor on a university campus in North America, and the active radiopharmaceutical program offers unique opportunities for research and clinical treatment of animals with cancer.

NUCLEAR MEDICINE

Diagnostic

Radioisotopes will undergo decay (emit energy) until they are energetically stable. Scintigraphy using Tc-99m-MDP identifies osteoblastic activity and is a common application of diagnostic nuclear medicine for assessing bone pathology in dogs, cats, and horses. Technetium-based scintigraphy has been in use in veterinary medicine for many years, but positron emission tomography (PET) scanning is a more recent addition to our repertoire, and is available in a small number of centers.

Positron emission tomography (PET) scanning can assess the function of a population of cells. When glucose is labeled with a radioisotope of fluorine, F-18 labeled FDG is produced and will detect any population of cells in the body with an unusually high metabolism of glucose, such as cancer cells. The positron emitted by the F-18 travels a very short distance then undergoes an annihilation reaction with its counterpart, an electron, thus resulting in emission of two gamma rays traveling in opposite directions. These are detected by the PET scanner which can then display the origin. In veterinary medicine, PET has been used to image various tumors including sarcomas, head and neck tumors (nasal and oral), mast cell tumors, and prostatic carcinomas. PET scans are most often performed using ^{18}F-FDG and a growing body of literature has characterized uptake in many canine tumors and normal tissues.

Other agents are being explored to image hypoxia and proliferation. Staging of canine lymphoma using ^{18}F-FDG is currently being investigated at the University of Missouri, as is the use of ^{18}F-FDG based PET to determine optimal area of soft tissue sarcoma for the most accurate histologic prediction of tumor behavior. Eventually, it is likely that PET will be used to assess response to cancer therapy as is done in humans. This can offer prognostic information as well as allow tailored therapy. For example, a mass that has been treated with radiation but did not regress may be inactive on PET scan, suggesting a dormant tumor, whereas PET avidity would suggest that additional therapy is needed immediately. PET has been anecdotally used to distinguish between late effects of radiation (osteopenia) and radiation-induced osteosarcoma 3 years after radiation therapy for soft tissue sarcoma.

Technetium (^{99m}Tc) is used commonly for nuclear scintigraphy. In cats, dogs, and horses ^{99m}Tc-Methylenebis(phosphonic acid), also called ^{99m}Tc-medronic acid (^{99m}Tc-MDP) is used on an almost daily basis for skeletal imaging. The 140 keV gamma provides very good resolution of images. The 6 hour half life allows timely and efficient management of patients and radiation safety. Scintigraphy is very sensitive and as such is an excellent screening tool for skeletal pathology (such as primary and metastatic neoplasia). Increased uptake in potential secondary bone lesions in dogs with primary osteosarcoma has been variable but is estimated to be 8%. An additional 10% of dogs can show non-specific uptake due to possible osteoarthritis and other non-neoplastic conditions, and increased intensity of uptake in dogs with osteosarcoma has also been correlated with earlier time to metastasis.

In addition to skeletal imaging, thyroid tissue will uptake sodium pertechnetate via the sodium iodide symporter, thus allowing scintigraphic imaging of benign and malignant thyroid masses. This can predict uptake of radioactive iodine (^{131}I) for therapy, regardless of thyroid function.

Therapeutic

Therapeutic applications of nuclear medicine include the well-known use of radioactive iodine for the treatment of hyperfunctional thyroid disorders. In addition, dogs with non-functional malignant thyroid tumors have been successfully treated. The ability to handle the amount of radioiodine to treat dogs and cats with malignant thyroid cancer is limited to only a few centers in the United States. Dogs with thyroid carcinoma treated with radioiodine have median survival times of 2 ½ to 3 years with locoregional disease even if nonresectable, and 1 year with metastatic disease.

Samarium-153-EDTMP (Quadramet™) was developed at the University of Missouri and has been used to treat both primary and metastatic bone tumors. Samarium-153 has a decay half-life of 46 hours, primarily by beta emission. Using the same premise as the ^{99m}Tc-MDP, Sm-153-EDTMP is injected intravenously and targets areas of high bone turnover; diagnostic bone scans are used to determine optimal candidates for this treatment. The transient myelosuppression that ensues typically occurs 2-4 weeks after treatment and can be associated with a transient anemia and increased ALP. This has been used primarily for palliation of bone pain. A new formulation to deliver radioactive samarium to sites of bone pathology is being developed at the University of Missouri. The new conjugate appears to be less myelosuppressive, which may facilitate concurrent or earlier use of chemotherapy. Samarium has been used to treat nonresectable tumors of the skull (osteosarcoma and multilobular osteochondrosarcoma) with a median survival of 144 days and a 20% clinical response rate. When used to treat osteosarcoma, the intent is primarily palliative to decrease pain, with over 60% of dogs exhibiting less pain 2 weeks after injection. Samarium has also been used for synovectomy in horses, and intra-arterially for osteosarcoma in dogs.

The use of physiologic processes to target radiation to tumor cells aims to deliver radiation to each cell. However, traditional brachytherapy involves implanting physical sources of radiation (such as seeds, ribbons, or needles) directly into the tumor tissue. Sources include radioactive iridium and cesium among others. At the University of Missouri, the injection of radioisotopes suspended in a liquid or colloid has more recently been investigated. This may result in more homogenous dose distribution as the liquid diffuses throughout the tumor tissue. However, determination of the resulting dosimetry is challenging. This technique has been used with bone and soft tissue tumors at the University of Missouri, with radioisotopes such as gold-198 and yttrium-90. Prostate cancer has been treated in men with brachtherapy for many years. This approach has been mimicked in dogs, and we have explored the use of liquid brachytherapy with radioactive gold nanoparticles injected directly into prostatic tumors in dogs.

RADIATION SAFETY

Radioactive substances naturally exist in our bodies and in our environment (such as radon and potassium-40). Naturally occurring isotopes are a source of radiation exposure, however as caregivers we aim to minimize exposure to both personnel and pet owners. Time, distance and shielding are the most important aspects of keeping radiation exposure as low as is reasonably achievable (ALARA). Radiation exposure decreases exponentially with distance (law of inverse square) so even minor increases in distance from the treated patient can significantly decrease exposure. After diagnostic or therapeutic procedures with radioisotopes, dogs and cats are confined to specially designed facilities for some length of time depending on local, regional, and national regulations. Pets are discharged to their owners with additional instructions to limit close exposure for some period of time based upon the decay half-life of the radioisotope. When a minimum of 10 half-lives have been surpassed, the amount of radioactivity remaining in the body is considered to be equivalent to or below background radiation.

Specialized training, facilities, and licensing are needed for the use of radioisotopes. With proper patient identification, animals can be referred to appropriate facilities for advanced care and more treatment options. When these patients return for follow up with their primary care veterinarians or specialists, it is important to understand what additional precautions might be needed. Radioisotopes such as radioactive iodine and samarium, have long enough half-lives that there is still excess radiation in the body when dogs or cats are presented for monitoring bloodwork such as complete blood counts to watch for expected myelosuppression. Veterinarians should request records from the treating facility that explain the duration of time during which additional precautions should be exercised. Pregnant or immunocompromised technical staff should not have prolonged (or perhaps any) direct contact with treated animals for some length of time. For radioactive iodine, 10 half -ives is almost three months at which time the patient has long since been considered to be free of radiation. However, with proper precautions patients can be adequately and safely managed and exposure to personnel kept very low such that the risk above background radiation is considered minimal.

REFERENCES

1. Lattimer JC et al. Clinical and clinicopathologic response of canine bone tumor patients to treatment with samarium-153-EDTMP. J Nucl Med. August 1990;31(8):1316-25.
2. Vancil JM et al. Use of samarium Sm 153 lexidronam for the treatment of dogs with primary tumors of the skull: 20 cases (1986-2006). J Am Vet Med Assoc. June 2012;240(11):1310-5.
3. Worth AJ et al. Radioiodide (131I) therapy for the treatment of canine thyroid carcinoma. Aust Vet J. April 2005;83(4):208-14.
4. Turrel JM et al. Sodium iodide I 131 treatment of dogs with nonresectable thyroid tumors: 39 cases (1990-2003). J Am Vet Med Assoc. August 2006;229(4):542-8.

5. Lamb CR et al. Preoperative measurement of canine primary bone tumors, using radiography and bone scintigraphy. J Am Vet Med Assoc 1990;196(9):1474-8.
6. Statham-Ringen KA et al. Evaluation of a B-cell leukemia-lymphoma 2-specific radiolabeled peptide nucleic acid-peptide conjugate for scintigraphic detection of neoplastic lymphocytes in dogs with B-cell lymphoma. Am J Vet Res 2012;73(5):681-8
7. Jankowski MK et al. Nuclear scanning with 99mTc-HDP for the initial evaluation of osseous metastasis in canine osteosarcoma. Vet Comp Oncol 2003;1(3):152-8
8. Robben JH et al. In vitro and in vivo detection of functional somatostatin receptors in canine insulinomas. J Nucl Med 1997;38(7):1036-42

Top Tips for Managing Cancer Patients in Your Practice

Sue Ettinger, DVM, DACVIM (Oncology)
Wappinger Falls, NY

Cancer is not a death sentence in pets. Chemotherapy is well tolerated in the majority of dogs and cats undergoing treatment. With treatment, many cancer patients are not only living longer, but living well.

IMPORTANCE OF ONCOLOGY

The pet is a family member, and owners often want same standard of care for their dogs and cats as they do themselves. Sadly, cancer is leading cause of death in pets. "Cancer" is a scary word that is often equated with death. There is often a visceral fear of cancer, and people think cancer equals pain and suffering. Owners think cancer treatment will just make the patient sicker. With cancer, there is no hope. I disagree. Cancer is not a death sentence. While we all want a cure for cancer, I encourage thinking about many cancers as chronic conditions that may require chronic therapy, such as kidney or heart disease. As an oncologist, I recommend treatment when the pet is likely to live longer with it than without. Thankfully, most dogs and cats feel good, if not great, during treatment. I believe it is important to approach the topic of cancer with knowledge, compassion, and a positive attitude.

CHEMOTHERAPY

Conventional Chemotherapy

Conventional chemotherapy is typically given at high dosages, known as maximum tolerated dose, or MTD. The goal is to kill the rapidly dividing cancer cells. But some normal cells that also have high turnover often can be temporarily damaged by MTD chemo. Most commonly, it is the GI tract cells and the neutrophils that are temporarily damaged. As a result there is a break period to allow these cell populations to recover. MTD is typically given weekly to every 3 weeks.

Chemotherapy drugs attack rapidly dividing cells. The normal tissues that typically are most sensitive to chemotherapy are the bone marrow, hair follicles (alopecia), and the gastrointestinal lining. This is often referred to as "BAG".

Bone marrow suppression most commonly results in a neutropenia but cats seem to be more tolerant than dogs. Neutrophils and platelets are at greatest risk due to the shorter circulating lifespan, and shorter bone marrow transit times. Neutropenia is the dose-limiting toxicity in veterinary oncology.

When giving a potentially myelosuppressive drug like doxorubicin, carboplatin and Lomustine, I personally like to check the expected nadir (low neutrophil count) to see if antibiotics and/or a dose reduction are needed. The nadir typically occurs 7 days after chemo administration. Pay attention to the *neutrophil* count, not the total white blood cell count. For some chemotherapy drugs the nadir is more variable such as carboplatin and Lomustine. For cats, the nadir is can occur 7 to 28 days after treatment. In dogs the nadir for carboplatin in day 10-14. Chlorambucil tends to cause delayed neutropenias and thrombocytopenias after chronic use.

Alopecia (hair loss) is due to damaging the rapidly dividing hair follicle. In dogs, potentially affected breeds have continuously growing coats and include Poodles, Scottish Terriers, and Westies. In cats, alopecia is rare, but shaved areas tend to grow back more slowly (limb catheters, abdominal ultrasounds). Cats more commonly lose their whiskers. The good news is that hair and whiskers will re-grow once the treatments have completed. Occasionally, hair will grow back a different texture or color. In cats it is typically softer, aka the "chemo coat". It is important to remember pets do not care about this cosmetic side effect, and it does not impact the quality of life. However, pet owners like to be advised about the whiskers and coat so they are not surprised.

Gastrointestinal (GI) toxicity includes vomiting, diarrhea, decreased appetite, nausea. It typically occurs 1 to 5 days after chemotherapy and is self-limiting – lasting on average 2-3 days. These side effects are less common in feline chemotherapy patients than dogs. I often will use Cerenia or mirtazapine as needed.

How toxic is chemotherapy?

The overall toxicity rate is very low in veterinary chemotherapy patients. In my experience, only 15-20% experience side effects, and this is even less common in cats than dogs. The primary goal is to provide the best quality of life possible for as long as possible. As I say, **live longer, live well.** Most side effects are mild and medically manageable.

If there are side effects, I also typically will add prophylactic medications to prevent side effects like nausea, vomiting or diarrhea as indicated. I recommend Cerenia SQ with the following drugs: vincas, doxorubicin, mitoxantrone, carboplatin, and the MOPP protocol. I always recommend oral Cerenia for 4 days after doxorubicin in dogs to prevent nausea and vomiting. If the GI side effects are more severe in a patient, the drug type or dosage may be adjusted at subsequent treatments to minimize the chance of side effects recurring.

When a chemotherapy drug is used that is known to have a high potential for bone marrow suppression, a complete blood count (CBC) is often checked after the treatment to see if the WBC are low. Antibiotics may be prescribed as a preventive measure. Subsequent doses of chemotherapy are adjusted based on the results of the CBC. Unlike dogs, I do not routinely use prophylactic antibiotics or GI medications unless the cat had issues with a prior treatment.

In my experience, there is less than a 5% chance that a patient will need hospitalization. If this does occur, these patients are usually hospitalized for typically 24-48 hours with supportive care including IV fluids and antibiotics. In my experience most chemotherapy patients can successfully receive that drug again with a dose reduction.

Metronomic chemotherapy

In contrast to MTD chemotherapy, metronomic chemotherapy is pulse or low-dose continuous chemotherapy. This is typically administered daily or every other day. The target is endothelial cells in that line tumor blood vessel. The goal may be that the tumor is stabilized, but this prevents further growth and spread. Common chemotherapy drugs include Palladia, cyclophosphamide, and chlorambucil and also NSAIDS. There is still much to be learned including best drugs, dose, schedule, tumor types, and toxicity. This can be considered for some dogs and cats with advanced metastatic disease.

What to do at the nadir visit?

In addition to running a CBC, it is important to get a good history, TPR (that temperature is so important in chemotherapy patients), and a complete physical examination. I am always interested in knowing how the patient handled chemo –did she eat well, any vomiting/diarrhea, did the owner use any nausea or diarrheal medications? For the exam, did he lose weight, was she febrile? The nadir CBC should not be a technician appointment to just pull the blood sample. The history and exam are very important.

Pay attention to the *neutrophil* count, not the total white blood cell count. The nadir typically occurs 7 days after chemo administration, but can vary (see above). I recommend antibiotics if the neutrophil count is <1500. If the patient has <1500 neutrophils and is afebrile and feeling well, I recommend managing as an outpatient. However, if the patient has <1500 neutrophils and is febrile and sick, I recommend admitting for supportive care. Remember a febrile neutropenic is an oncologic emergency.

Also, I prefer that we get blood samples from the jugular veins for patients getting IV chemotherapy (unless thrombocytopenic). Save those peripheral veins for treatment please. Finally many times the oncologist has run a recent chemistry panel, so check with the oncologist, and try not to repeat unneeded blood work to keep costs down.

In addition to the chemotherapy targeting rapidly dividing bone marrow stem cells, other mechanisms for neutropenia includes bone marrow infiltration with neoplastic cells (leukemia, advanced stage lymphoma, multiple myeloma) and increased consumption due to infection.

Neutrophil count (per uL)	Fever, systemic signs	Plan
1500-2500	No	Monitor +/- treatment delay 2 to 4 days
<1500	No	Oral antibiotics treatment delay Consider dose change
<1500	Yes	ATH for IVF & IV antibiotics treatment delay Dose change

Sepsis

Sepsis in chemotherapy patients is typically due to patient's own flora - Gram negative from GI bacteria: *E. coli, Klebsiella, Pseudomonas;* Gram positive from skin bacteria: *Staphylococcus epidermitidis and aureus*; anaerobes from oral bacteria. Predisposing factors include neutropenia (infection risk well correlated with degree and duration), cellular immune dysfunction, humoral immune dysfunction, prolonged hospitalizations, indwelling catheters, and poor nutrition.

History and clinical signs are typically straightforward - cytotoxic chemotherapy was administered typically 5 to 7 days ago. Remember, the **febrile neutropenic patient is an oncologic emergency!!!** In addition the patient may have an inability to mount an inflammatory response, so the lack of fever, pyuria, or radiographic changes of pneumonia does not rule out sepsis. Signs of illness are unrelated to absolute neutrophil count, but are related to an increased susceptibility to local and systemic infections when neutropenic. Gastrointestinal, urogenital, and respiratory infections are most common. Shock is also possible

The sepsis work up includes: CBC, Chemistry panel, UA & UCS (if >50,000 platelets). If respiratory signs are present, chest radiographs are recommended, and TTW should be considered. Blood cultures may be needed, but uncommon in my experience. Culture any catheters suspected as the infection source.

Treatment for sepsis includes: IVF and broad-spectrum IV antibiotics. Neupogen is human recombinant G-CSF. The MOA is stimulation of proliferation & maturation of neutrophil precursors, and monocyte precursors to a lesser extent. It also primes neutrophil for cell killing & neutrophil migration. The benefit for the febrile & febrile neutropenic patient is contradictory, and in my experience, Neupogen is rarely needed. The recommended dose is 5 ug/kg SQ SID until neutrophil >1000.

When should I lower chemotherapy dose?

Dose Intensity is chemotherapy given at MTD & shortest possible interval. It is important to remember than small dose changes can have significant impact on cancer control. Dose reductions as small as 20% can decrease drug efficacy up to 50%. ***Dose reductions should not be considered lightly.***

VOMITING AND DIARRHEA

Acute vomiting is typically associated with cisplatin, doxorubicin (Adriamycin), dacarbazine (DTIC), cyclophosphamide, actinomycin, 5-FU streptozotocin. This can typically be prevented with pre-treatment

Delayed vomiting is more common in our patients. This is due to direct damage to rapidly dividing GIT cells (crypt cells) or via the centrally mediated CRTZ stimulated via gut vagal efferents. Delayed vomiting is most commonly 2 to 5 days post-chemo and seen with doxorubicin and the vinca alkaloids. Clinical signs include vomiting, diarrhea, anorexia, lethargy, weakness, $\pm$ dehydration.

For work up, I recommend CBC, chemistry panel, UA, +/- fecal floatations and cultures. If abdominal pain is present, consider AXR or AUS to rule out foreign body, obstruction, and intussusception. For patients with GI neoplasia, it can be challenging to differentiate chemotherapy side effects vs. disease, and a good history can be key.

For outpatient treatment, I recommend NPO, food & water trial, bland diet, anti-emetics, antibiotics with severe diarrhea and a probiotic. Do not forget to discontinue oral chemotherapy or delay chemotherapy treatment. In addition, I recommend prophylactic therapy with the next chemotherapy

For inpatient, I add injectable antiemetics, IV fluid therapy, and IV antibiotics. An important note, I strongly encourage owners to NOT EUTHANIZE at this time. It is amazing with 1 to 2 days of good supportive care how quickly these patients improve. And with prophylactic therapy and a dose reduction, these patients can tolerate the same chemotherapy drug.

DON'T TREAT CATS LIKE SMALL DOGS WHEN IT COMES TO CHEMOTHERAPY

Some chemotherapy drugs are dosed differently in cats. In dogs, weight and body surface area are used to determine the carboplatin dose. In cats there is now a more accurate method to dose carboplatin in cats based on glomerular filtration rate, which is determined with an Iohexol clearance test.

Side effects in cats are also different. Cardiotoxicity is a well-described adverse effect in dogs treated with doxorubicin, but it has not been reported in cats. Sterile hemorrhagic cystitis (SHC) is a relatively uncommon complication of cyclophosphamide in dogs and ifosfamide therapy in dogs and cats. SHC is typically associated with long-term use, but possible after single dose, and can progress to bladder fibrosis. The incidence with cyclophosphamide has been reported to be 9% in dogs (7-24%), 3% in cats, and 24% in humans. Unlike dogs, concurrent administration of furosemide with cyclophosphamide is not recommended in cats. Mesna, which binds the SHC-inducing acrolein, is recommended for cats and dogs when administering ifosfamide.

DON'T MONITOR THE BUMP OR LUMP. DO GET AN ASPIRATE OR BIOPSY. Why Wait? Aspirate. ®

Visual monitoring of skin and subcutaneous masses is not enough. Even the most experienced veterinarian or oncologist cannot look at or palpate a mass and know whether it is malignant or not. Cancer is a cellular diagnosis! It is always recommended to evaluate masses that are growing, changing in appearance, or irritating to the patient. But these guidelines are not enough.

"See Something Do Something. Why Wait? Aspirate. Dr. Sue Cancer Vet®" (SSDS) provides guidelines for evaluating superficial masses in dogs and cats. These guidelines will increase client awareness and will promote early cancer detection, diagnosis, and early surgical intervention. In veterinary medicine, most skin and subcutaneous tumors can be cured with surgery alone if diagnosed early when tumors are small.

See Something: When a skin mass is the size of a pea (1 cm) and has been present for at least 1 month,

Do Something: Aspirate or biopsy, and treat appropriately!

Why diagnose early? Obtaining a definitive diagnosis with cytology or biopsy *early and before excision* <u>will</u> lead to improved patient outcomes for superficial masses. When smaller, superficial tumors are detected early, surgery is likely curative – this is especially true for benign lesions and tumors that are only locally invasive with a low probability of metastasis. If tumors are removed with complete surgical margins, the prognosis is often good with no additional treatments needed.

- Pet owners need to be aware of the "pea" size requirement to have masses evaluated.
- Veterinarians must measure and document the size of the mass in order to compare growth.
- If > 1 cm (or size of large pea) and present for a month, the mass should be aspirated or biopsied.
- Knowing the tumor type prior to the FIRST surgery will increase success of a curative-intent surgery

REFERENCES

Chretin JD, et al, JVIM, 2007; 21(1):141-8. Rau SE, et al, JVIM 2010; 24(6):1452-7. Vaughan MA, et al, JVIM 2007; 1(6):1332-9

Electrochemotherapy in Horses, Dogs & Cats

Chelsea Tripp, DVM, MS, DACVIM (Oncology)
Seattle, WA

What is electrochemotherapy?

Electrochemotherapy is an anti-cancer treatment that couples local administration of chemotherapy drugs to the delivery of pulses of a specific waveform. The application of these pulses allow for alterations in the cell membrane through an electrically mediated reorganization of the plasma membrane of cells such that there is increased uptake of the chemotherapy drugs. Ultimately this leads to cancer apoptotic death.

What is the difference between electropulsation and electrochemotherapy?

- **Electropulsation** – *the direct delivery of electric pulses to cells*
- **Electropermeabilization/Electroporation/Electrochemotherapy** – *using exogenous chemotherapy to enter cells after electropulsation*

Electrochemotherapy uses a substance (in this case a chemotherapy agent) in conjunction with the pulses to help facilitate higher delivery of the drug to the cytosol of the cell. The two drugs of choice are bleomycin and cisplatin; as they both are relatively large molecules and poorly permeable on their own. Electroporation does not enhance the therapeutic potential of lipophilic agents. Gehl, et al observed that, for the lipophilic anticancer drugs daunorubicin, doxorubicin, etoposide and paclitaxel, there was no increase in cytotoxicity when combined with electropermeabilization. Investigators are looking at carboplatin, 5-FU and using concomitant peri-lesional administration of cytokines, such as IL-2.

Electrochemotherapy for our veterinary patients:

Electrochemotherapy, in veterinary oncology (dogs, cats, exotics and horses), has been utilized more extensively by centers in Brazil, France, Italy, Ireland, Slovenia and the UK. We are seeing growth in the application within the United States and Canada with facilities that have trained specialists in this technique. This may be coming to a hospital near you!

It is important from a technician standpoint, to be able to manage these patients for both the process of electrochemotherapy and the side effects that you may have to counsel clients through. The process does require heavy sedation but is overall well tolerated and many tumors can be treated in 1-3 treatment sessions. Larger tumors may require more sessions and more time to regress.

There are two types of probes that are typically used to administer the pulsation for electrochemotherapy; planar plate electrodes and parallel needle based electrodes. These probes offer different applications and abilities to treat gross disease or deeper tissue segments. The electrical field generated by these probes will differ somewhat and there is variability within tumor characteristics that will further alter the electrical field. Electrode systems have recently been developed for intraluminal delivery by endoscopes, laparoscopic systems, or intravascular catheters.

What electrical concerns are present with electrochemotherapy?

The actual pulse is an electric field that shocks the patient as well as anything surrounding. Having the patient on a non-kinetic surface like a pad on top of metal is advised. All metal should be removed from the patient and no internal probes used that have metal associated with them. Staples within the surgical field should be removed prior to electrochemotherapy or sutures should be pre-emptively used.

Many times patient position is also affected by whether the machine senses electrical activity. Tumors along the axillary region or thoracic wall are slightly more difficult to treat due to the electrical activity of the heart underlying these tissues. Many of these patients can be successfully treated by putting the patients on their back and leaning within a V-tray or propped with towels to have the heart fall away from the field of desired treatment.

The electrical pulse will cause a muscular contraction and/or muscle fasciculation. Depending on the muscle group that is being treated, this can cause the patient to jump or lurch. With horses, this is why we recommend heavy sedation in stocks or full anesthesia. These patients can be slightly sore after the procedure if this is over a major muscle group like the flank or thigh.

The sedation necessary for small animals to accomplish electrochemotherapy is actually very brief. It is a balance of having them in a plane of anesthesia but also very short in nature. The actual treatment of electrochemotherapy takes anywhere from 30 seconds to three minutes at most and then the patient can wake up. The sedation protocols used for dogs and cats include standard doses of midazolam, torbugesic (dog), oxymorphone (cats) and propofol to effect. We intubate most small animal patients (not all exotics) to protect their airway but many do not need to be maintained on actual gas. They are maintained on oxygen during the procedure.

All chemotherapy administered should be performed with personal protective equipment. Closed delivery devices for all injections is ideal. Intralesional chemotherapy carries inherent higher risk and should be only performed by qualified staff and professionals comfortable with the administration of chemotherapy in this manner.

The following are examples of tumor types which are amenable to electrochemotherapy. In general, tumors need to be superficial or easy to access the tumor bed to be successful with treatment:

- Fibrosarcoma in the mouth or eyelid.
- Melanoma in the mouth, eyelid or paw
- Squamous cell carcinoma in the mouth, eyelid, nasal planum or paw
- Soft tissue sarcomas post debulking surgery where aggressive surgery would necessitate amputation
- Feline vaccine-associated fibrosarcoma after surgery
- Localized cutaneous lymphoma in dogs or cats
- Plasmacytic tumors of the footpad, lip or skin
- Low to intermediate grade mast cell tumors on the distal limb, trunk and face
- Perianal & rectal tumors
- Operation site margins where there is a significant risk of remaining tumor tissue
- Sarcoids and squamous cell carcinoma in horses
- Superficial tumors on exotics (small mammals, birds & zoo animals)
- Some internal tumors – tracheal and esophageal (on a case by case and location basis)

Side effects of electrochemotherapy?

Electrochemotherapy is generally very well tolerated with most side effects being localized to the area of treatment and related to the type of tumor being treated. Mast cell tumors contain a variety of vasoactive substances and when treated with electrochemotherapy there is an initial release of histamines as the first wave of cells rupture. This continues over the next 7-10 days after electrochemotherapy has been performed to a mast cell tumor. We recommend having all mast cell tumor patients on Benadryl prior to electrochemotherapy and for up to two weeks after treatment to minimize this side effect. Sensitive areas like the eyelid may also require prednisone to help contain the post-treatment swelling and degranulation.

Large tumors or high grade tumors may die off rather quickly and the body has to process this. We usually see this happen about 7-10 days after a session with high grade tumors. Counseling clients appropriately is paramount so they know what to expect. This is a good sign that the tumor responded but can be somewhat disconcerting if they are not expecting it. A second session of electrochemotherapy may be delayed due to this side effect as the necrosis resolves and remaining tumor is assessed.

Quality of Life in Dogs and Cats: How to Measure It and Why It Matters

Lisa M. Freeman, DVM, PhD, DACVN
North Grafton, MA

INTRODUCTION

Quality of life (QoL) in people depends on factors such as health, social relationships, mental stimulation, health, food consumption, stress, and degree of control over one's life. Measures of QoL in people rely on patient self-report [patient-reported outcome measures (PROs)], where patients are available and able to report their own health status or on clinician or caregiver reports [observer-reported outcome measures (OROs)] when patients are unable to self-report their health status (e.g., children, cognitively impaired individuals). When QoL is viewed from a health perspective, it is referred to as health-related QoL (HRQoL). Since QoL/HRQoL in companion animals is defined by the owner's perception, instruments assessing an animal's QoL/HRQoL are classified as OROs.

Quality of life instruments can be either general (i.e., not disease-specific) or specific to an individual disease, such as heart failure, arthritis, or asthma. Quality of life instruments in people, whether general or disease-specific, can be important as endpoints for clinical trials, for early diagnosis of disease, for assessment of changes in health status or clinical response to medical or nutritional interventions, or to monitor the progression of disease in individual patients. The importance of QoL assessment is people is well-established, and while the study of QoL/HRQoL in companion animals has lagged, recent studies have focused primarily on measures for specific diseases, such as heart disease, chronic kidney disease, cancer, osteoarthritis, diabetes mellitus, spinal cord injury, dermatologic disease, and epilepsy. Much of the work on HRQoL has focused on dogs, with fewer publications on cats. Fewer instruments assessing generic HRQoL or QoL are available.

There are a number of excellent reviews of the importance of QoL assessment and methodologies in companion animals. The QoL/HRQoL instruments developed thus far to study dogs and cats, range from a single-answer Likert or visual analog scales to more complicated instruments, and the methods used to demonstrate validity and reliability vary considerably. Whether for specific diseases or general health status, it is important to rigorously demonstrate the reliability and validity of a new instrument to ensure that it measures the intended constructs and that any differences observed between animals are reliable and valid.

HUMAN VS VETERINARY HRQoL INSTRUMENTS: LIMITATIONS

Quality of life is an important assessment in people and companion animals with a variety of diseases, but heart failure is a condition in which the use of HRQoL instruments is well-established, both for clinical and research purposes because HRQoL has been shown to be an independent predictor of survival. In people and companion animals with heart failure, using a health-related QoL questionnaire is likely important to address common clinical signs such as exercise intolerance, coughing, or adverse effects of medication. The clinical signs of congestive heart failure (CHF) in dogs and cats, such as weakness, dyspnea, anorexia, exercise intolerance, and syncope, are thought to adversely affect HRQoL. Whether the animals' QoL or the owners' *perception* of their animals' QoL is adversely affected, these changes are important because QoL can affect survival time. A study of owners of dogs with CHF that had been euthanatized found that 79% of dogs were considered by their owners to have had a fair or poor QoL in the 2 weeks prior to euthanasia, and 13% of owners reported that poor QoL as the single most important factor in their decision to have their dogs euthanized (Mallery et al, 1999).

In addition to its role in survival time, HRQoL can be an important end-point for success or failure of medical or surgical treatment in patients, and measures of HRQoL may complement typical clinical measures of disease status. Health-related QoL has been used as a primary endpoint in human clinical trials or to optimize treatment regimens for individual patients. Studies also have shown that QoL is extremely important to dog and cat owners: 86% of dog owners and 93% of cat owners responded that they would be willing to trade survival time for QoL in their pet with heart disease (Oyama et al, 2008; Reynolds et al, 2010). This information can have an important influence in optimizing treatment for individual patients.

Of the many questionnaires designed specifically to assess health-related QoL in people with heart failure, the Minnesota Living with Heart Failure Questionnaire is one of the most widely used. This questionnaire has been well validated in people but has low applicability to dogs because of the nature of the questions (e.g., swelling in the legs and loss of self-control). As a result, the FETCH and CATCH Questionnaires were developed and validated for assessing HRQoL in dogs and cats, respectively, with heart disease (Freeman et al, 2005). Both were shown to have good validity and reliability, and have been used in clinical trials of dogs and cats with heart disease.

Individual questions on the FETCH and CATCH Questionnaires (e.g., appetite, difficulty administering medications, etc) can be used as items for discussion with owners to identify areas of difficulty that could affect adherence with treatment recommendations and, potentially, survival. For example, some owners have great difficulty administering medications to their cats and respond with a high score on the item assessing how much giving medications affects their cats' QoL. However, other owners, even some of those administering multiple medications, respond that medication administration has no effect on their cats' QoL. This may be related to individual owner or cat personalities or owner skill at administering medications. However, the owner's perception of the effect of these factors on his or her cat's QoL is likely still important to assess as it may influence the decision of euthanasia or medication adherence. Assessment of individual items on the CATCH Questionnaire also may provide support for utilization of other medications, techniques to reduce the frequency of medication administration, nutritional strategies that could improve appetite, or other factors that could enhance a dog or cat's QoL.

One challenge with all QoL questionnaires for companion animals is that they provide a proxy measure of the animal's true QoL, since it is actually providing the owner's *perception* of the animal's QoL (i.e., an ORO). This is a similar issue in human pediatric medicine where parents of infants and very young children typically provide information used to assess QoL. Another issue is the responsiveness of QoL questionnaires, whether for heart failure, osteoarthritis, or chronic kidney disease. The responsiveness refers to the instrument's sensitivity to changes associated with treatment. Information on responsiveness is important to establish a questionnaire's usefulness as an endpoint in clinical trials. Recently, the responsiveness of the CATCH questionnaire to changes associated with medical treatment in cats with cardiomyopathy and acute CHF was evaluated (Rush et al, 2015). Health-related QoL scores were highly variable and scores did not correlate with either NT-proBNP concentrations or International Small Animal Cardiac Health Council (ISACHC) stage. Therefore, the CATCH Questionnaire may not be ideal for use in a cat with acute CHF and its use should be reserved for cats after the disease has been stabilized and under chronic management.

CONCLUSIONS AND FUTURE DIRECTIONS

Assessment of HRQoL is an important issue for dogs and cats with a variety of diseases, both for studying more effective methods of treatment in clinical trials and ideally for determining optimal therapy for an individual patient. Further evaluation of the currently available HRQoL instruments for companion animals is needed to be able to better understand this important issue for veterinarians and pet owners. In addition, well-validated generic HRQoL instruments could play an important role in veterinary medicine to screen for medical issues, monitor progress over time, and facilitate patient-centered care. Although much additional research is needed, well-validated instruments have the potential to provide tools for veterinarians and owners to monitor the ongoing QoL of dogs and cats for earlier diagnosis of disease or decline in health status, progression of disease, or the change in overall health status with interventions, such as diet, medication, or surgery.

REFERENCES

1. Benito J, Gruen ME, Thomson A, et al. Owner-assessed indices of quality of life in cats and the relationship to the presence of degenerative joint disease. J Fel Med Surg 2012;14:863-870.
2. Bijsmans ES, Jepson RE, Syme HM, et al. Psychometric validation of a general health quality of life tool for cats used to compare healthy cats and cats with chronic kidney disease. J Vet Intern Med 2015;30:183-191.
3. Budke CM, Levine JM, Kerwin SC, et al. Evaluation of a questionnaire for obtaining owner-perceived, weighted quality-of-life assessments for dogs with spinal cord injuries. J Am Vet Med Assoc 2008;233:925-930.
4. Favrot C, Linek M, Mueller R, Zini E. Development of a questionnaire to assess the impact of atopic dermatitis on health-related quality of life of affected dogs and their owners. Vet Dermatol 2010;21:63-69.
5. Freeman LM, Rush JE, Farabaugh AE, Must A. Development and evaluation of a questionnaire for assessing health-related quality of life in dogs with cardiac disease. J Am Vet Med Assoc 2005;226:1864-1868.
6. Freeman LM, Rush JE, Oyama MA, et al. Development and evaluation of a questionnaire for assessment of health-related quality of life in cats with cardiac disease. J Am Vet Med Assoc 2012;240:1188-1193.
7. Giuffrida MA, Kerrigan SM. Quality of life measurement in prospective studies of cancer treatments in dogs and cats. J Vet Intern Med 2014;28:1824-1829.
8. Green TC, Mellor DJ. Extending ideas about animal welfare assessment to include 'quality of life' and related concepts. New Zealand Vet J 2011;59:263-271.
9. Iliopoulou MA, Kitchell BE, Yuzbasiyan-Gurkan V. Development of a survey instrument to assess health-related quality of life in small animal cancer patients treated with chemotherapy. J Am Vet Med Assoc 2013;242:1679-1687.
10. Kiddie JL, Collins LM. Development and validation of a quality of life assessment tool for use in kennelled dogs (*Canis familiaris*). Appl Anim Behav Sci 2014;158:57-68.
11. Lascelles BDX, Hansen BD, Roe S, et al. Evaluation of client-specific outcome measures and activity monitoring to measure pain relief in cats with osteoarthritis. J Vet Intern Med 2007;21:410-416.
12. Lavan RP. Development and validation of a survey for quality of life assessment by owners of healthy dogs. Vet J 2013:97:578-582.
13. Linek M, Favrot C. Impact of canine atopic dermatitis on the health-related quality of life of affected dogs and quality of life of their owners. Vet Dermatol 2010;21:456-462.
14. Lynch S, Savary-Bataille K, Leeuw B, Argyle DJ. Development of a questionnaire assessing health-related quality-of-life in dogs and cats with cancer. Vet Comp Oncol 2010;9:172-182.
15. Mallery KF, Freeman LM, Harpster NK, Rush JE. Factors contributing to the euthanasia decision in dogs with congestive heart failure. J Am Vet Med Assoc 1999;214:1201-1204.
16. Niessen SJ, Powney S, Guitian J, et al. Evaluation of a quality-of-life tool for dogs with diabetes. J Vet Intern Med 2012;26:953-961.
17. Noli C, Minafo G, Galzerano M. Quality of life of dogs with skin diseases and their owners. Part 1: development and validation of a questionnaire. Vet Dermatol 2011;22:335-343.25.
18. Noli C, Colombo S, Cornegliani L, et al. Quality of life of dogs with skin disease and of their owners. Part 2: Administration of a questionnaire in various skin diseases and correlation to efficacy of therapy. Vet Dermatol 2011;22:344-351.

19. Oyama MA, Rush JE, O'Sullivan ML, et al. Perceptions and priorities of owners of dogs with heart disease regarding quality versus quantity of life for their pets. J Am Vet Med Assoc 2008;233:104-108.
20. Peddle GD, Singletary GE, Reynolds CA, et al. Effect of torsemide and furosemide on clinical, laboratory, radiographic and quality of life variables in dogs with heart failure secondary to mitral valve disease. J Vet Cardiol 2012;14:253-259.
21. Reid J, Wiseman-Orr ML, Scott EM, Nolan AM. Development, validation and reliability of a web-based questionnaire to measure health-related quality of life in dogs. J Small Anim Pract 2013; 54:227-233.
22. Reid J , Wiseman-Orr ML, Nolan AM, Scott EM. Health-related quality of life measurement. In: Gaynor JS, Muir WW III (eds). Handbook of veterinary pain management, 3rd ed, 2014. Philadelphia: Elsevier: 98-110.
23. Reynolds CA, Oyama MA, Rush JE, et al. Perceptions of quality of life and priorities of owners of cats with heart disease. J Vet Intern Med 2010;24:1421-1426.
24. Rush JE, Freeman LM, Cunningham SM, Yang VK, Bulmer BJ, Antoon KN. Assessment of the responsiveness of the Cats' Assessment Tool for Cardiac Health (CATCH) questionnaire. J Vet Cardiol 2016;17:S341-S348.
25. Scott EM, Nolan AM, Reid J, Wiseman-Orr ML. Can we really measure animal quality of life? Methodologies for measuring quality of life in people and other animals. Animal Welfare 2007;16 (Suppl 1):17–24.
26. Spofford N, Lefebvre SL, McCune S, Niel L. Should the veterinary profession invest in developing methods to assess quality of life in healthy dogs and cats? J Am Vet Med Assoc 2013;243:952-956.
27. Wessman A, Volk HA, Parkin T, et al. Evaluation of quality of life in dogs with idiopathic epilepsy. J Vet Intern Med 2014;28:510-514.
28. Wiseman-Orr ML, Nolan AM, Reid J, Scott EM. Development of a questionnaire to measure the effects of chronic pain on health-related quality of life in dogs. Am J Vet Res 2004;65:1077-1084.
29. Wiseman-Orr ML, Scott EM, Reid J, Nolan AM. Validation of a structured questionnaire as an instrument to measure chronic pain in dogs on the basis of effects on health-related quality of life. Am J Vet Res 2006;67:1826-1836.
30. Wojciechowska JI, Hewson CJ, Stryhn H, et al. Development of a questionnaire to assess nonphysical aspects of quality of life in dogs. Am J Vet Res 2005;66:1453-1460.
31. Wojciechowska JI, Hewson CJ, Stryhn H, et al. Evaluation of a questionnaire regarding nonphysical aspects of quality of life in sick and healthy dogs. Am J Vet Res 2005;66:1461-1467.
32. World Health Organization Quality of Life Assessment Group. The World Health Organization Quality of Life Assessment (WHOQOL). Development and psychometric properties. Soc Sci Med 1998;46:1569-1585.
33. Yazbek KV, Fantoni DT. Validity of a health-related quality-of-life scale for dogs with signs of pain secondary to cancer. J Am Vet Med Assoc 2005;226:1354-1358.
34. Yeates J, Main D. Assessment of companion animal quality of life in veterinary practice and research. J Small Anim Pract 2009;50: 274-281.
35. Zamprogno H, Hansen BD, Bondell HD, et al. Item generation and design testing of a questionnaire to assess degenerative joint disease–associated pain in cats. Am J Vet Res 2010; 71:1417-1424.

Beating the Bugs: How Veterinary Technicians Prevent Hospital Acquired Infections

Dennis Spann, DVM, DACVIM, (Small Animal Internal Medicine)
Roseville, CA

INTRODUCTION

Hospital or healthcare acquired infections (HAI's) have emerged as a major health care risk in human and veterinary medicine. Antibiotic resistance is an increasingly prevalent aspect of HAI's. Multiple governmental bodies have sounded and alarm to increase awareness and mobilize medical doctors, veterinarians and, perhaps most importantly, nursing staff to adapt strategies that minimize the occurrence of HAI's and the development of antibiotic resistance. HAI's are frequently associated with microbial resistance because the organisms in the hospital setting have a strong selective pressure from frequent use of antibiotics in sick patients. The magnitude of the problem is unknown in veterinary medicine, but it is clearly important. We are all seeing bacteria cultured with extensive resistance patterns. The World Health Organization has called bacterial resistance a serious enough problem to" reverse the advances of modern human medicine". Prevention of hospital acquired infections and prevention of resistant infections are complimentary tasks. Multipronged strategies need to be employed to prevent occurrence and spread of HAI's. There are several key components of an infection control program: establishment of a policy, education of clients and staff (medical and reception), patient risk assessment, infection surveillance (passive or active depending on facility size and resources), minimization of patient and staff exposure, environmental contamination control, and antibiotic stewardship.

ESTABLISHMENT OF AN INFECTION CONTROL POLICY

The first step to establishing an infection control policy is to establish an infection control committee and infection control officer. The committee should be composed of nursing, veterinary and reception staff. This committee will develop written policy appropriate to the hospital. The infection control officer will be the "go-to" person for case management and troubleshooting policy. The infection control officer can be an advanced veterinary technician or a veterinarian. This person may also take the lead in coordinating the development of an infection control policy. This policy should be a written document that is readily available and disseminated to all staff. It will be most useful if it is practical, terse and readable. A bulleted reference version can also help make the policy more useful to post and implement. The protocol should address specific procedures ranging from handling patient inquires on the phone through patient discharge, case transfer and even autopsy. Simply put, it should address all the contacts and procedures between admission and discharge.

EDUCATION

There are many resources available to educate clients, veterinary medical and lay staff about hospital infection control. A few of these resources are listed in the references below. Important topics include responsible use of antibiotics to minimize resistance. This topic is generally antibiotic stewardship. Antibiotic education of clients and veterinary staff should occur at the appropriate level for their involvement. Reception staff should be educated to identify at risk patients. Anticipating which appointments should be approached with a heightened awareness gives a" leg-up" on preventing hospital acquired infection. These cases can be managed with a variety of strategies to minimize the spread of pathogens through the hospital and among patients. The actual steps to minimize infection spread vary by clinic but are generally outlined below:

- Identify at risk patients. These are critically ill patients, patients that have gastroenteritis, immune compromised patients (from immune suppressants, chemotherapy or sepsis), and patients with difficult to heal wounds or wounds that produce copious discharge. Animals with coughing and or oculonasal discharge. Animals with these problems should have hygiene practiced aggressively to prevent infection and to prevent spread of that infection should it occur. Immobile, debilitated, heavily instrumented animals are all at risk for developing and potentially spreading ordinary and resistant infections. Known infectious animals should be isolated to prevent exposure to other patients. Suspected infectious patients should be separated from healthy and at-risk patients.

- Tracking infections is an important aspect of infection control. Records of patients' infections and particularly records of resistant infections should be maintained. Ideally, the hospital location of every patient should be recorded where possible. A binder of positive cultures can be maintained and resistance patterns monitored. Maintaining and reviewing records avoids conjecture and vague anecdotal assessments. One of the significant limitations in veterinary medicine is that lack of centralized collection of data from multiple hospitals. This handicaps the recognition of evolving resistance patterns in our patients and hospitals. Reference labs could serve as a valuable resource for recognition of local and regional resistance patterns. Knowledge of local susceptibility patterns would allow for educated empirical choices of antibiotics while waiting for culture results. A collaborative effort between academia and reference labs could be very productive in this effort.

- Hand hygiene is the act of disinfecting hands with soap and water or utilizing an antiseptic sanitizer to decrease the bacterial load on health-care workers' hands. Pathogens can survive an hour or more on clinicians' hands and the degree of contamination is proportional to the time spent with clinical cases without effective washing. Despite this knowledge, clinicians routinely have compliance less than 50 % in observational studies with reports as low as 17 %. Washing should last 20 seconds and contact all surfaces and up the forearms. The time to sing "happy birthday to you" twice approximates 20 seconds. Washing should occur before an after contact with a patient, before and after contact with the patients environment (cages, blankets, etc.), bodily fluids, and before putting on or taking off gloves. Wearing gloves does not replace hand washing, and wearing the same pair of gloves through multiple cases is not helpful.

- Personal protective equipment also prevent us from becoming fomites. Street clothes should be covered by smocks or lab coats in the clinic. Scrubs and lab coats should not be worn outside of the clinic. Lab coats, gowns, scrubs and smocks should be laundered frequently (daily is best), and changed with first gross contamination or after handling any patient with know or suspected infectious disease. Gloves should be worn when handling bodily fluids, when performing laundry or when cleaning environmental surfaces. Face protection should be used if splashes or other aerosols are expected particularly during dentistry. Foot covers should be worn when in isolation or when cleaning grossly contaminated laundry or working with patients with infections. Several current recommendations suggest that clinicians should be bare below the elbows, free of jewelry and watches, and ties. Stethoscopes and other shared equipment should be cleaned and disinfected between patients. Cell phones are ubiquitous and should also be cleaned frequently.

- Environmental risk reduction means assuring high levels of patient cleanliness, identification of patients at risk of soiling themselves or the environment, removal of organic debris and sterilization from the environment (cages, floors, exam tables, sinks, stethoscopes, thermometers, clippers and other commonly used equipment), and appropriate use of patient spacing or isolation. Soiling with any bodily discharges should be rapidly cleaned and disinfected. Linens (towels, blankets, etc.) should be changed at least daily. Cleaning and disinfection need to be separate steps. Cleaning to remove gross organic material mechanically usually with soap and water, while disinfection helps remove microscopic pathogens with disinfectant solutions. Mop buckets should be changed frequently. A review of disinfectants is beyond the scope of this talk, but the choice of disinfectant depends on the surface, degree of penetration required, the pathogens being destroyed and toxicity considerations. Choices include: alcohols, biguanides, halides, peroxides, phenolics and quaternary ammoniums. Care should be given to assure hygienic handling of all bodily fluids/discharge. These are biological waste should be treated as such. Bandages should be kept clean and dry and free of strike though. Hospital or lab areas that contact biologics should be cleaned and disinfected. Waste should be categorized as routine versus biological waste. Biomedical waste includes: Sharps, blood or bodily fluid soaked material including bandages, tissues and dead animals. Elimination areas should be separated for patients shedding infectious organisms. Ideally a cleanable surface should be chosen for elimination for those patients. Clients should generally be restricted from visiting animals with infectious diseases. Raw food should never be fed to hospitalized patients.

- Surgical patients are at risk for hospital-acquired infection. Surgery creates a breach in the skin, the natural barrier for infectious agents. Surgery creates some degree of vascular and lymphatic compromise and tissue damage. Environmental cleanliness, surgical scrub and gown cleanliness and hand hygiene are paramount. Sterilization of equipment is an obvious necessity and indicators of sterilization are essential. Anesthetic equipment should cleaned regularly since hands touch patients and then the equipment. Endotracheal tubes should be meticulously clean and ideally sterilized. Preoperative clipping should be as close to the surgical time as possible to avoid microtrauma to the skin barrier. Dirty patients may require a bath prior to surgery. Different clipper blades should be used for contaminated and clean procedures and both should be regularly cleaned and replaced. Perioperative antibiotics when indicated should be given so that they are at therapeutic levels at the time of first incision. Post operatively; incisions should not be touched with bare hands to avoid introducing bacteria. Other important possible sources of infection could be failure to change disinfection scrub pots, or inappropriate cleaning of multidose vials.

- Wounds are important avenues of infection. When handling wounds, sterile gloves should be used. Glove changes should occur as necessary as the wound prep becomes cleaner. Bandages must be carefully monitored as specific timed patient orders and any strike-through warrants a bandage change. Soiled bandages should be treated as infectious material. Personnel protection is needed if the wound has known multiresistant bacteria and these patients should have some form of isolation.

- Antibiotic stewardship is integral to the preventing the increase of antibiotic resistance and hospital acquired infections. Antibiotics should be delayed long enough to obtain appropriate cultures, and then administered aggressively. Antibiotic choice should be based on culture results where available and de-escalation should occur as indicated once cultures are final. We should have a decision process and create a criterion for use of higher tier antibiotics. Where empirical choices are required, antibiotics should be chosen based on the likely organisms for the organ system affected. Antibiotics should be chosen on pharmacologically sound principles. They should be able to reach therapeutic levels in the affected organ. They should be used for an appropriate length of time for the disease process. Owners should be educated about the importance of antibiotic compliance to avoid treatment failure and contributing to antibiotic resistance.

Summary

Prevention and limitation of hospital acquired infections and prevention or antibiotic resistance requires an organized multifaceted approach. Veterinary hospitals of every size should have formalized infection control policies. We currently lack some of the centralized information collection that we need to track infection and resistance patterns. Despite this limitation, institution of well organized infection control policies can provide guidelines for optimal management of our routine and at-risk patients. Improved infectious disease control is going to require the development of centralized information about infection patterns and resistance. These will require greater efforts individually to obtain appropriate cultures, coupled with academic and industry collaboration to develop a clearing house for larger patterns of infection and resistance than can be appreciated at by individual clinicians or practices.

REFERENCES

1. Infection Prevention and Control Best Practices for Small Animal Veterinary Clinics – Canadian Committee on Antibiotic Resistance. 2. Healthcare Personnel Attire, Recommendations to help prevent healthcare-associated infections transmitted through clothing. Society for Healthcare Epidemiology of AmericaStudy contact: Gonzalo Bearman MD, MPH. Ther Adv Drug Saf 2014, Vol 5(6) 229 – 241. B. Spellberg et al. Trends in Antimicrobial Drug Development: Implications for the Future Clin Infect Dis. (2004) 38 (9): 1279-1286. Chang-Ro Lee et al. Strategies to Minimize Antibiotic Resistance Int. J. Environ. Res. Public Health 2013, 10, 4274-4305, Sabino et al. Can Vet J 2006;47:779–782. Jumaa et al. Hand hygiene: Simple and complex. International Journal of Infectious Diseases (2005) 9, 3—14

Tricks of the Trade: How to be an Exceptional Technician

Amy M. Oliver, LVT, VTS (ECC)
Auburn, AL

INTRODUCTION

Most veterinary technicians know how to handle, care, and treat multiple species in various practices. Nursing staff choose careers ranging from animal shelters, research, general practice, industry, and academia. They all have their elite skills that help them excel at their job daily. But, as most people grow to learn, with strength comes weakness. Recognizing weaknesses and learning to strengthen them paves the way to becoming the best veterinary technician and patient advocate possible.

COMMUNICATE

When you are dealing with other people day in and day out, communication skills are imperative. Being able to effectively and efficiently communicate a change in a patient's status with a doctor could mean the difference between life and death of that patient. Patients rely on veterinary technicians to be vigilant in patient care. Be an advocate and speak up when your patient cannot speak for itself. That is one skill they never teach you in school, yet every technician should utilize from day one. Your job is treating animals. Part of the concept of treatment is learning how to communicate with all parties involved in each patient so that they receive the best care achievable.

Inversely, be open and receptive to comments, suggestions, and critiques. Whether you have been in the field for ten years or ten days, you can always learn something new; a new piece of equipment, a new skill, a new way of taping in an intravenous catheter, or a new outlook on your daily duties. Be respectful of other individuals and their opinions. The best favor you can do for yourself is to understand that there are **always** more ways than your way of doing anything in the medical field.

STRIVE

No matter where you are working, there will always be that superstar technician; the go-to when every one else fails at getting a blood draw or IV catheter in; the one that seems to just have everything together. Strive to be that technician. That technician's secret is experience and willingness to learn. They found ways of doing things that worked best for them and picked up new skills as they progressed in their career. Look at each new colleague or coworker you meet as a potential opportunity to learn something new rather than the new person that you will have to train. Keep an open mind.

As veterinary technicians, our primary collective goal is patient care. Our patients rely on us heavily while they are being treated in the hospital, whether it be for a weight loss consult in an obese diabetic cat or intensive nursing care on a mechanically ventilated ivermectin toxicity dog. Anticipate what your patients will need and have it ready to go. Your ability to anticipate patient demands will develop with experience and education. While doctors are in exam rooms talking to clients or on the phone with drug representatives, technicians are spending their time providing medical care and TLC. With this comes experience. Utilize this time to grow comfortable with your patients so that you notice the slight changes.

Pay attention to how other technicians do things such as record a history or secure an esophagostomy tube in a bandage. You may find that a small change in the way you perform your every day tasks can make you more efficient and comfortable in your job. Understand that you did not graduate from technician school knowing all there is to know about veterinary medicine. Strive to learn as much as you can and be the best advocate for your patients that you can be. Ask questions and educate yourself.

LEARN

As mentioned before, there are many ways to do the many aspects of our jobs as veterinary technicians. This lecture will cover techniques that may make your jobs a little less stressful. We will talk about intravenous catheter placement and stabilization, central line placement, blood donor collection techniques, female urinary catheter placement, Doppler blood pressure measurements, feeding tube management, bandaging and wound management, and various other duties that veterinary technicians perform daily. The goal is to help each of you become the best technician you can be so that you can become the best advocate for your patient, big or small.

Commonly Encountered Errors in Collecting and Reporting Laboratory Tests

Jean-Sébastien Palerme, DVM, DACVIM (SAIM)
Ames, IA

INTRODUCTION

Laboratory testing is divided in to three phases: (1) the pre-analytical phase, which comprises the collection, storage and transportation of samples, (2) the analytical phase, which involves the specific steps that occur during the actual performance of the test and (3) the post-analytic phase, which involves the reporting of test results. Given the centralized and specialized nature of many laboratory test facilities and the improvements in patient sample identification, most cases of "laboratory errors" occur before samples are even processed by laboratory staff. Incorrect or erroneous test results can result in incorrect diagnoses, unnecessary ordering of additional tests and potentially life threatening interventions. Consequently, it is vital for veterinary staff to have a good knowledge of the various pre-analytical factors that can influence test results. This talk will address some commonly encountered errors in sample collection as well as some species and breed peculiarities that can influence results of minimum database tests.

PATIENT-RELATED ABNORMALITIES

Breed

One of the most important, and commonly overlooked, factors to consider when interpreting blood work is the breed or species it was taken from. Greyhounds are often cited for their multiple hematologic and biochemical differences from other breeds.[1] In fact most sighthounds (e.g. whippets, Afghans, Borzois) share these peculiarities. Reported abnormalities in these breeds include increased hematocrit and mean corpuscular volumes as well as lower platelet counts. Biochemical differences include increased creatinine values and lower blood protein concentrations. Macrothrombocytopenia is a platelet anomaly that is frequently described in Cavalier King Charles Spaniels and has been sporadically described in other breeds. This autosomally inherited trait results in a lower number of circulating platelets (50 – 100, 000/ul) that are also increased in mean size. Though their platelet counts are substantially lower than most dogs, dogs with macrothrombocytopenia do not have increased bleeding tendencies. A genetic test is now available.[2] It is important to recognize this condition since failure to do so can result in erroneous suspicion of an acquired cause of thrombocytopenia. Lastly, variations in red blood cell sizes have been encountered with some dog breeds: macrocytosis or microcytosis has been encountered in miniature poodles and Akitas, respectively.

Concurrent laboratory abnormalities

Apart from hematological and biochemical abnormalities that are intrinsic to certain breeds, concurrent abnormalities such as increased albumin or increased platelet counts can influence other readings on complete blood counts or serum chemistries. Pseudohyperkalemia refers to the incorrect measurement of increased potassium levels that are not reflective of *in vivo* levels. It is most commonly secondary to the release of intracellular stores from platelets[3] when the blood clot is formed but it can occasionally occur from lysis of red blood cells (in some Asian dog breeds). If pseudohyperkalemia secondary to thrombocytosis is suspected, a platelet count should be performed. If thrombocytosis is confirmed, measurement of potassium from a heparinized plasma sample is recommended to eliminate the effect of platelet-released potassium on the measurement.

The total calcium measured in routine chemistries is composed of three components: the active ionized form, the complexed form and the protein-bound form. Given that more than a third of total calcium is bound to albumin, it is important to appreciate that variations in serum albumin will have a direct effect on reported total calcium levels. Historically, an equation to correct total calcium values to take into account variations in serum albumin has been used:

$$\text{Corrected } T_{calcium} = \text{Measured } T_{calcium}\ (\text{mg/dl}) - \text{albumin}\ (\text{g/dl}) + 3.5$$

However, one should remember that this equation is an estimate. More recent studies comparing the corrected calcium value to measured ionized calcium values show that calculated calcium values often fail to correctly categorize the calcium status of cats and dogs.[4,5] This is especially true for patients with chronic kidney disease where over 50% of dogs had incorrect calcium estimates using the equation. Given the potentially serious consequences of hypo- or hypercalcemia as well as the repercussions of incorrectly treating these conditions, ionized calcium measurement should be performed on any case where the clinician is concerned about the calcium status of the patient.

SAMPLING ERRORS

Hemolysis

The most common cause of sample rejection in human medicine is hemolysis.[6] When hemolysis is noted on blood collection or reported on the serum chemistry report, the first thing that needs to be determined is whether the hemolysis is true (*in vivo*) or spurious (*in vitro*). An *in vivo* cause of the hemolysis should be suspected if other bloodwork abnormalities are present such as increased bilirubin, the presence of spherocytes or schistocytes, or a decrease in hematocrit. Some factors associated with increased rates of spurious hemolysis include prolonged tourniquet times, difficult blood draws, use of a peripheral catheter instead of a direct venipuncture, use of smaller gauge (<22 Ga) needles, and prolonged times between sampling and centrifugation.[7] Hemolysis interferes with sample analysis by two main mechanisms. The first is biological: lysis of the red blood cells can increase the concentrations of certain analytes that have high intracellular concentrations (magnesium, calcium, AST, cholesterol, lactate, ammonia, phosphorus, +/- potassium) but also causes a release of intracellular water, resulting in lower measured values of some analytes. Hemolysis also results in spectrophotometric interference: by increasing the optical absorbance of the serum which affects measurements of bilirubin, iron, lipase and GGT.[6]

EDTA contamination

Ethylenediaminetetraacetic acid (EDTA) is a commonly used anticoagulant in medicine. Its anticoagulant properties are due to its capacity to bind divalent cations, such as calcium, and inhibit the coagulation cascade. Contamination of serum samples is thought to occur when blood is put into an EDTA container prior to being put into the serum tube using the same needle. Since EDTA is most commonly formulated as a salt complexed with potassium, the changes on serum chemistries are usually predictable. The finding of low divalent cation (calcium and magnesium) concentrations with concurrent high potassium concentrations should warrant suspicion for EDTA contamination.[8] Though prolonged incubation (>5 hours) of blood samples with EDTA can cause important changes to red blood cell morphology, white blood cells can be stored up to 12 hours at 4^0C with minimal changes in morphology.

REFERENCES

1. Sambasivarao S V. *Vet Clin Pathol.* 2011;40:1199. 2. Gelain ME, et al. *Vet Clin Pathol.* 2014;43:317. 3. Reimann KA, et al. *J Vet Intern Med.* 1989;3:47. 4. Schenck P et al. *Am J Vet Res.* 2005;66:1330. 5. Schenck P et al. *Can J Vet Res.* 2010;74:209. 6. Lippi G et al. *Clin Chem Lab Med.* 2008;46(6):764. 7. Lippi G et al. *Crit Rev Clin Lab Sci.* 2011;48(3):143. 8. Davidson DF. *Ann Clin Biochem.* 2002;39:273-280.

The Importance of a Balanced Gut Microbiome for Intestinal Health of Dogs and Cats

Jan S. Suchodolski, DrMedVet, PhD, AGAF, DACVM (Immunology)
College Station, TX

INTRODUCTION

We are just now beginning to understand how intestinal bacteria interact with the host, thereby exhibiting an influence that reaches beyond the gastrointestinal tract. The intestinal microbiota is defined as the collection of all living microorganisms (i.e., bacteria, fungi, protozoa, and viruses) that inhabit the gastrointestinal (GI) tract. The use of novel molecular analysis tools, based most commonly on sequencing of the bacterial 16S rRNA gene, has revealed that the GI microbiota of mammals is highly diverse, comprising several hundred bacterial species.[1,2] It is estimated that the intestine harbors a total of 10^{10}-10^{14} microbial cells, which is approximately 10 times more than the number of cells in the host body. It is, therefore, obvious that this highly complex microbial ecosystem will have a crucial effect on host health and disease. Gut microbes are useful to the host by fending off transient pathogens, aiding in digestion and energy harvest from the diet, providing nutrition for enterocytes, and playing a very important role in the development and stimulation of the host's immune system. On the other side, the intestinal microbiota can also have a detrimental influence on GI health. Various studies have associated alterations in the composition of the intestinal microbiota (dysbiosis) with chronic GI diseases of humans, dogs, and cats.[1,3] Furthermore, the gut microbiota is also strongly linked to disorders outside of the gut, as dysbiosis has been associated with obesity, diabetes mellitus, autism, and others.

CHARACTERIZATION OF GASTROINTESTINAL MICROBIOTA

Until a decade ago, traditional bacterial culture was the recommended method for identification of the bacteria in the GI tract. Bacterial culture is still a useful method for the detection of specific enteropathogens in a targeted approach, such as culture for *Salmonella* or *Campylobacter jejuni*. But it is now well established that bacterial culture has not enough resolution to characterize the many different bacteria within the complex GI ecosystem. Therefore, a culture based method underestimates total bacterial numbers, and also does not allow identifying the majority of bacterial groups present in the GI tract. This is because we lack the knowledge regarding optimal growth requirements for most GI bacteria, especially anaerobic bacteria that are predominant in the GI tract and which are prone to sampling and handling damage. Furthermore, many selective culture media lack sufficient specificity and often other organisms than the targets are enumerated. The current standard for characterization of the intestinal microbiota is the use of DNA based methods. Molecular high-throughput sequencing techniques (i.e., 16S rRNA gene based 454-pyrosequencing and/or Illumina sequencing) allow identification of most microbes. For detection of specific pathogens, species specific PCR assays can be employed. However, these techniques are currently available only for research studies, but in the near future may also become available to analyse the microbiota of individual patients in routine clinical settings.

Many ingested nutrients, but also drugs and other xenobiotics are metabolized by intestinal microbes and then absorbed by the host. Therefore, it is obvious that a balanced composition of the intestinal microbiota will play an important part in intestinal homeostasis, and any changes in the microbiota may directly or indirectly influence metabolic host pathways. More recent studies are attempting to study the functional properties of the intestinal microbiota. This is important because it remains challenging to correlate the presence of specific bacterial groups with GI health and disease. For example, administration of antibiotics to healthy dogs leads to decreases in some of the beneficial bacterial groups, but this change leads to gastrointestinal problems only in a subset of animals. It is believed that a functional redundancy exists in the GI tract, with several members of the bacterial community performing similar functions, and if one group is displaced because of perturbations (e.g., antibiotic therapy), other members of the community appear to maintain a stable ecosystem function. Therefore, it is important to evaluate the intestinal microbiome as an entity, including phylogenetic relationships and metabolic functions. Novel metabolomics approaches using multiple mass spectrometry platforms allow assessing changes in metabolite profiles, either produced by the host or microbiota, and therefore both together yield a better understanding of the pathophysiology of GI disease.

GUT BACTERIA IN HEALTHY DOGS AND CATS

Due to differences in anatomical and physiological factors along the GI tract (i.e., differences in pH, bile composition, intestinal motility), the microbial composition differs along the length of the GI tract. Also, differences exist between luminal and mucosa-adherent microbiota. Of special note is that each dog and cat harbors a very unique and individual microbial profile. For example, in one study, it was shown that almost all evaluated cats harbored *Bifidobacterium* spp. in their GI tract, but only a small percentage of cats harbored the same species of *Bifidobacteria*.[4] These differences in bacterial composition between individual animals are important, as they may explain why there is a highly individualized response to therapeutic approaches that are designed to modulate intestinal microbiota, and not every animal will respond similarly to dietary changes, or administration of antibiotics or nutraceuticals (i.e., probiotics).

The phyla Firmicutes, Bacteroidetes, Proteobacteria, Actinobacteria, Spirochaetes, and Fusobacteria constitute almost 99% of all gut microbiota in dogs and cats. A few less abundant phyla are Tenericutes, Verrucomicrobia, Cyanobacteria, and Chloroflexi. It is also likely that low abundant bacterial lineages have not yet been identified. Aerobic bacteria occur in higher abundance in the small intestine, whereas anaerobic bacteria predominate in the large intestine. In the stomach, mucosa-adherent *Helicobacter* spp. are the major group, followed by various lactic acid bacteria (i.e., *Lactobacillus* and *Streptococcus* spp.) and *Clostridia* spp. The most abundant groups in the small intestine are Clostridia, Lactobacillales, and Proteobacteria, whereas Firmicutes, Bacteroidetes, and Fusobacteria are the predominant bacterial phyla in the large intestine. The phylum Firmicutes comprises many phylogenetically distinct bacterial groups, the so called *Clostridium* clusters. Clusters XIVa and IV encompass many important short-chain fatty acid producing bacteria (i.e., *Ruminococcus* spp., *Faecalibacterium* spp., *Dorea* spp.) and are the major groups in the ileum and colon of cats and dogs.[4,5] In contrast, species like *C. perfringens* are a very minor component of the intestinal microbiota.

Differences also exist in the number of total bacteria in the different compartments of the GI tract. The stomach harbors between 10^1 and 10^6 cfu/g of contents. Bacterial counts in the duodenum and jejunum of dogs and cats can range from 10^2 to 10^9 cfu/ mL of contents .[6] The distal small intestine (i.e., ileum) contains a more diverse microbiota and higher bacterial numbers (10^7 cfu/ mL of contents) than the proximal small intestine. Bacterial counts in the colon range from 10^9 and 10^{11} cfu/ml of intestinal content.

In addition to bacteria, the mammalian GI tract harbors various fungi, archaea, protozoa, and viruses. While recent molecular studies have provided more information about the diversity of these microorganisms in healthy animals, their role in health and disease remain unclear.[7]

MICROBIOTA IN IMMUNITY AND HOST HEALTH

The commensal microbiota contributes to the development of the physiological gut structure. For example, germ-free raised animals exhibit an altered epithelial architecture (e.g., decreased number of lymphoid follicles) and an underdeveloped immune system. The microbiota in early life is important for establishing oral tolerance in order to prevent onset of inappropriate immune responses against bacteria and food antigens, which have been associated with chronic gastrointestinal inflammation. This importance is emphasized by recent epidemiological studies in humans that are linking dysbiosis due to early childhood exposure to antibiotics with increased risk for asthma, obesity, and Ulcerative Colitis.[8]

There is an important communication between intestinal bacteria and the host immune system. This cross-talk is mediated through a combination of microbiota-derived metabolites and microbial surface molecules that activate receptors of the innate immune system (e.g., Toll-like receptors). The resident microbiota is also a crucial part of the gut barrier system which protects the host from negative impacts of invading pathogens as well as deleterious microbial toxins. Examples include the competition for nutrients, mucosal adhesion sites, and the creation of a physiologically restrictive environment for non-resident bacterial species (e.g., secretion of antimicrobials, alterations in the pH in the gut, hydrogen sulfide production). Some of the major nutrient sources of bacteria are complex carbohydrates, including starch and dietary fiber such as cellulose, pectin, and inulin. The fermentation of these substrates results in the production of short chain fatty acids (SCFA; e.g., acetate, propionate, and butyrate) and other metabolites. SCFA in general are important energy sources for the host, as it has been shown that up to 7% of the metabolic energy of dogs is produced by microbial fermentation.[9] Furthermore, SCFA are important growth factors for epithelial cells. They are also highly immunomodulatory and influence intestinal motility. Butyrate has, for example, the ability to reduce oxidative damage to DNA. Acetate can modulate intestinal permeability, thereby decreasing the systemic translocation of gut microbiota derived endotoxins.[10] Furthermore, the intestinal microbiota produces various other immunomodulatory metabolites that are yet to be explored. For example, indole, a metabolite produced by many gut bacteria is anti-inflammatory, decreases IL-8 expression, induces expression of mucin genes, and also increases gene expression that strengthens tight junction resistance.[11]

DYSBIOSIS AND DISEASE

Several studies have recently demonstrated alterations in microbiota in animals with acute and chronic GI disease. Typically there is a decrease in normal anaerobic bacteria, and an increase in aerobic bacteria. Due to the important role of microbial-derived metabolites for host health, these alterations in bacterial groups are also associated with metabolic changes within the intestinal lumen. Some of these major metabolic pathways that are altered include bile acid, lactate, and tryptophan metabolism. These changes have an effect on the entire host metabolism. For example, an increase in serum D-lactate in cats with gastrointestinal disease is possibly due to disarrangements of the intestinal microbiota and increased bacterial production of D-lactate, and D-lactate has been shown to lead to neurological signs in some cats.[12] Bile acid dysmetabolism, due to a dysbiosis and improper conversion from primary to secondary bile acids, leads to bile acid diarrhea.

CONCLUSIONS

The above studies indicate that acute and chronic GI disease is associated with microbial dysbiosis and altered systemic metabolic states. Better understanding of the interactions of microbial-derived metabolites and the host will yield insights into the pathophysiology of GI diseases. Furthermore, some of these metabolic pathways can be targeted to develop better treatment approaches.

Table 1. Disorders that have been associated with changes in the intestinal microbiome

Acute hemorrhagic diarrhea	dogs
Atopic dermatitis	humans, mouse models, dogs
Autism	humans
Calcium oxalate (CaOx) urolithiasis	dogs
Diabetes Mellitus Type II	humans, rodent models
Inflammatory bowel disease	humans, rodent models, dogs
Irritable Bowel Syndrome	humans
Metabolic syndrome	rodent models
Obesity	mouse models
Stress diarrhea	humans, rodent models, dogs
Stress, anxiety, depression-related behavior	mouse models

REFERENCES

1. Frank DN et al. Molecular-phylogenetic characterization of microbial community imbalances in human inflammatory bowel diseases. *Proceedings of the National Academy of Sciences of the United States of America* 2007;104:13780-13785.
2. Suchodolski JS et al. The effect of the macrolide antibiotic tylosin on microbial diversity in the canine small intestine as demonstrated by massive parallel 16S rRNA gene sequencing. *BMC Microbiol* 2009;9:210.
3. Suchodolski JS. Microbes and gastrointestinal health of dogs and cats. *J Anim Sci* 2010:1-34.
4. Ritchie LE et al. Characterization of fecal microbiota in cats using universal 16S rRNA gene and group-specific primers for Lactobacillus and Bifidobacterium spp. *Veterinary Microbiology* 2010;144:140-146.
5. Suchodolski JS et al. Analysis of bacterial diversity in the canine duodenum, jejunum, ileum, and colon by comparative 16S rRNA gene analysis. *FEMS Microbiology Ecology* 2008;66:567-578.
6. Johnston KL. Small intestinal bacterial overgrowth. *Veterinary Clinics of North AmericaSmall Animal Practice* 1999;29:523-550.
7. Foster ML et al. Characterization of the fungal microbiome (mycobiome) in fecal samples from dogs. *Vet Med Int* 2013;2013:658373.
8. Ungaro R et al. Antibiotics associated with increased risk of new-onset Crohn's disease but not ulcerative colitis: a meta-analysis. *Am J Gastroenterol* 2014;109:1728-1738.
9. Stevens CE et al. Contributions of microbes in vertebrate gastrointestinal tract to production and conservation of nutrients. *Physiological Reviews* 1998;78:393-427.
10. Fukuda S et al. Acetate-producing bifidobacteria protect the host from enteropathogenic infection via carbohydrate transporters. *Gut Microbes* 2012;3.
11. Bansal T et al. The bacterial signal indole increases epithelial-cell tight-junction resistance and attenuates indicators of inflammation. *Proc Natl Acad Sci U S A* 2010;107:228-233.
12. Packer RA et al. D-lactic acidosis secondary to exocrine pancreatic insufficiency in a cat. *Journal of Veterinary Internal Medicine* 2005;19:106-110.

Management of Canine Lower Urinary Tract Disease
Vicky L. Ograin, MBA, RVT, VTS (Nutrition)
Topeka, KS

INTRODUCTION

Veterinary technicians play a crucial role in aiding the veterinarian with diagnosis and treatment of canine lower urinary tract disease (LUTD). The information discerned from the owner aids in the diagnosis of LUTD and the discussion of the diagnostic and treatment plan with the client is crucial to the client's understanding and compliance with the veterinarian's recommendation and ultimately the health of the pet. Canine lower urinary tract disease (LUTD) signs are seen in 2 to 3% of dogs per year. Urinary tract infections and uroliths are seen most commonly[1]. The most common uroliths in dogs are struvites (magnesium, ammonium, phosphate) and calcium oxalates making up 90+% of uroliths seen in dogs[1].

PREVALENCE

In 1981, 78% of canine uroliths were composed of struvite and only 5% were calcium oxalate. In the mid- to late-1980s, the occurrence of calcium oxalate uroliths began to increase. In 2011 the Minnesota Urolith Center analyzed 57,152 uroliths, with calcium oxalates making up 42%, struvites 39%, "other" (mixed and compound) uroliths 12.4%, purines 4.2%, cystine 1.3%, silica and calcium phosphate both 1% and magnesium hydrogen phosphate and xanthine each less than 0.1%[2].

CLINICAL SIGNS

Regardless of underlying cause, LUTD is characterized by dysuria, pollakiuria, stranguria, hematuria, and/or periuria (urination in inappropriate places). It is important that veterinary technicians are aware of signs of LUTD when talking with clients[3].

DIAGNOSTIC EVALUATION

Diagnostic evaluation of dogs with recurrent or persistent lower urinary tract signs should include a urinalysis, culture (if inflammatory cells are evident or if struvites are suspected) and diagnostic imaging, hematology and serum chemistry. If the urolith is removed, urolith analysis is imperative for proper management of the patient[3].It is very important to perform complete urinalysis, including urine specific gravity and sediment evaluation. Survey radiographs are helpful for identifying radiopaque uroliths. Normal survey radiographs do not exclude radiolucent uroliths (urate/purine, cystine), small uroliths (< 2 mm), neoplasia, blood clots, or anatomic defects. Abdominal ultrasonography and/or contrast urethrocystography is helpful in these cases.

STRUVITES UROLITHS

RISK FACTORS

Females (~85%) are affected more often than males (~15%). The median age is 6-years old, but the range is 1 month to 24 years[4]. They are usually found in the lower urinary tract, 99% compared to 1% found in the upper urinary tract. Recurrence is common, especially if the bladder infection is not taken care of; in one study 21% recurred[5]. Struvites can form within a few days to a few weeks when a dog has a urease-producing microbial urinary tract infection[6]. More than 95% of dogs with struvite urolithiasis have infection as an underlying cause for the urolith growth[5]. It has been demonstrated many times in a variety of studies that the formation of struvite uroliths in dogs are caused by urease-producing bacteria- eg. *Staphylococcus* and *Proteus spp*. That is why struvites are sometimes identified as infection stones. The urease producing bacteria release ammonia which causes the urine to be more alkaline or basic. Sturvite crystals are precipitated in a basic pH. Also, bacterial infection produces inflammation and protein debris which may serve as a nidus for urolith formation. *E.coli* and other non-urease-producing microbes have not been linked with struvites in dogs[4].

SURGERY VS. DISSOLUTION

Treatment options for dogs with struvite uroliths include physical removal of uroliths or dissolution via nutritional management. For dogs with suspected struvite uroliths, it is appropriate to transition to feeding a dry or canned calculolytic food over a 7-day period. The appropriate antibiotics should be given to treat the UTI. Dogs should be re-evaluated every 2-4 weeks (urinalysis and abdominal radiographs). Urine pH should remain < 6.1 and specific gravity should be < 1.040, if canned food is being fed exclusively. Nutritional management (dissolution) and antibiotics should be continued 1 month beyond radiographic resolution of the urolith. As the stone dissolves, viable bacteria released from the urolith could allow the infection to persist and cause a urolith to reform, so it is important to keep the patient on antibiotics while the dog is on the dissolution food.

NUTRITIONAL MANAGEMENT

After dissolution or removal of struvite uroliths nutritional management can reduce the building blocks of struvite crystals. Food can decrease the risk factors or building blocks of a struvite urolith. Monitoring for future urinary tract infection is also important for managing urolith recurrence.

STRUVITE UROLITH KEY NUTRITIONAL FACTORS

Water: Water should be increased in both dissolution and management of struvite uroliths. It is recommended to feed canned food or if feeding dry, adding water to dry food should be considered. Increased water intake will help achieve a specific gravity of 1.020 or less. In addition increased water intake will increase urination, reducing crystal retention time and minimize crystal growth. Adequate water intake may be the single most important factor in determining whether or not uroliths recur[4].

Protein: To dissolve struvite uroliths, protein should be more restricted than when only management is the objective. For dissolution, protein should be 8% DMB (dry matter basis) or less. As an aid to the management of struvite uroliths where there is high risk, protein content of food should be less than 25%DMB to decrease urinary concentration of urea. Also high protein foods may be high in phosphorous another nutrient that should be controlled[4].

Phosphorous: For dissolution, phosphorous should be restricted to 0.1% DMB or less and for management of struvite uroliths, phosphorous should be less than 0.6% DMB[4].

Magnesium: Struvite crystals will form when there is excess magnesium in the urine. Avoiding excess magnesium can decrease the concentration of magnesium in the urine. Magnesium should be less than 0.02% DMB for dissolution; for management magnesium should be 0.06% DMB[4].

Urinary pH: Struvite management foods may help notify the veterinarian that there is a possible infection even if the dog is not showing signs. Struvite urolith management foods produce an acid urine pH of 6.2-6.4 or below. If a dog is on a struvite management food and has urine a pH of 8 or above, it warrants further testing for a urinary infection. For dissolution, the pH target is 5.9-6.1 and for management 6.2-6.4[4].

CALCIUM OXALATE UROLITHS
RISK FACTORS

Breed: Schnauzers have been shown to have a higher urinary excretion of calcium; this may explain why we see a higher number of calcium oxalate uroliths in schnauzers. The most common breeds affected were miniature and standard schnauzers, Lhasa apso, Yorkshire terrier, bichon frise, shih tzu, and miniature and toy poodle. The golden retriever, German shepherd dog and cocker spaniel are at lowest risk for calcium oxalate uroliths. Age and Sex: The median age of calcium oxalate urolith removal is 8.7 years. Less than 1% of uroliths are retrieved from dogs less than 1 year of age. Males were affected 74% more than females. Neutered males, overweight and pet dogs were affected the most Recurrence rate for calcium oxalates is common: 25-48% will recur[5].

SURGERY

Calcium oxalate stones cannot be nutritionally dissolved as of yet. Uroliths can be surgically removed. Clinicians are choosing to perform nonsurgical retrieval of the stones. Voiding urohydropulsion is gaining in popularity.

NUTRITIONAL MANAGEMENT

After stones are removed, it is important to manage or eliminate the risk factors for reforming. Calcium oxalate uroliths are considered metabolic stones and can recur. With nutrition we can hopefully minimize risk factors for the uroliths reforming and slow the rate of recurrence.

CALCIUM OXALATE UROLITH KEY NUTRITIONAL FACTORS

Water: Managing water intake is the same as already discussed for struvite managment[7].

Protein: Feeding foods high in protein could contribute to calcium oxalate uroliths by increasing calcium excretion. Hypercalciuria can occur in healthy dogs fed high protein foods. Excessive protein should be avoided in a dog's food with a history of calcium oxalate uroliths. The recommended protein should be 10-18% DMB[7].

Calcium and oxalic acid: Reduction of both calcium and oxalate in the food are important for the management of calcium oxalates. Calcium should be reduced to 0.4-0.7% DMB[7].

Phosphorous: Phosphorous should be 0.3-0.6% DMB and the calcium/ phosphorous ratio should be 1.1:1 to 2:1[7].

Sodium: Sodium chloride can increase thirst and urine volume. However, excess sodium increases urine calcium excretion and therefore is a risk factor for calcium oxalate and calcium phosphate urolithiasis, especially if the urinary pH is highly basic. High sodium level in foods can also contribute to hypertension in salt- sensitive dogs. For active calcium oxalate urolith formers, <0.3% DMB is advised for sodium[7].

Magnesium: There has been some indication for supplementing magnesium in humans, but more studies need to be completed for dogs. Currently it is not recommended to supplement calcium oxalate dogs with magnesium. Magnesium should be 0.15 -0.4% DMB[7].

Vitamin C: Acidifying urine is not recommended for dogs with calcium oxalate uroliths, so currently it is not recommended to give ascorbic acid. Ascorbic acid is also a precursor of oxalates[7].

Urinary pH: Acid load is reflected in a healthy dog's urinary pH. A pH of not less than 7 is recommended to manage dogs with calcium oxalates[7].

Urine alkalinizing agents: Potassium citrate can be used to obtain the desired pH. A desirable pH is 7.1-7.57.The potassium citrate also gets excreted into the urine and is a stone inhibitor for calcium oxalates.

Vitamin B6: A deficiency of vitamin B6 should be avoided, because it promotes endogenous production of oxalic acid. Healthy dogs should get 1.5mg/kg DMB[7].

Vitamin D: Vitamin D excesses should be avoided for dogs with a history of calcium oxalate uroliths. Excessive vitamin D promotes intestinal absorption of calcium. For dogs with a history of calcium oxalate uroliths 500 to 1500 IU/kg DMB is recommended[7].

FOLLOW UP

All dogs should be monitored for recurrence, including urinalysis, potentially with a culture, every 3 months to detect struvites or calcium oxalate crystalluria, with diagnostic imaging every 6 months to detect uroliths.

SUMMARY

Increased understanding of specific causes of LUTD has allowed diagnostic and therapeutic efforts to be directed toward identification, removal and management. The most common types of uroliths is struvites and calcium oxalates.Veterinary technicians play a critical role in the treatment of LUTD. The history obtained from discussions with the owner aids in the diagnosis of LUTD. The technicians' discussion of the treatment plan with the client is crucial to the client's understanding and compliance with the veterinarian's recommendation and ultimately the health of the pet.

REFERENCES/SUGGESTED READING

1. Bartges J, et al., Applied Veterinary Clinical Nutrition, Wiley- Blackwell, West Sussex, UK, 2012; 269 2. Osborne C.A., et al., DVM Magazine, Advanstar Communications, May 1, 2012 3. Osborne C.A, et al. Small Animal Clinical Nutrition, 5th edition, Mark Morris Institute, Topeka, Kansas, 2010: 813- 832. 4. Osborne C.A, et al., Small Animal Clinical Nutrition, 5th edition, Mark Morris Institute, Topeka, Kansas, 2010: 891-914. 5. Chew D.J., et al., Elsevier Saunders, St. Louis MO, 2011: 272-305. 6. Osborne C.A., DVM360, Advanstar Communications, April 1, 2004. 7. Lulich J.P, et al., Small Animal Clinical Nutrition, 5th edition, Mark Morris Institute, Topeka, Kansas, 2010: 854- 870.

Hemolytic Transfusion Reactions in Dogs: Are You Really My Type?

Marie-Claude Blais, DMV, DACVIM
Montréal, QC, Canada

INTRODUCTION

What can be done when a transfusion reaction occurs? From a meticulous blood donor selection to the blood product administration itself, every step is important to offer safe and effective transfusion therapy. This session will offer practical recommendations on how to recognize, prevent and investigate transfusion reactions, emphasizing on in-house blood compatibility testing and appropriate blood bank management.

CLASSIFICATION OF TRANSFUSION REACTIONS

Hemolytic transfusion reactions mostly occur during or shortly after a transfusion (acute hemolytic transfusion reaction (AHTR); within ≤ 24 hours), but later onset may also occur (delayed hemolytic transfusion reaction). These reactions can be immunologic or non-immunologic. It may not be easy to differentiate between the two, but the cause can be ascertained by careful evaluation of blood bank management, patient's clinical history and signs, and laboratory investigations.

Non-immune hemolytic transfusion reactions occur when red blood cells (RBCs) are damaged before the transfusion or during its administration. They may result in hemoglobinemia and hemoglobinuria. Although the clinical signs are usually milder, acute, life-threatening or fatal, non-immune hemolytic transfusion reactions have recently been reported by *Patterson et al* in 4 dogs most likely secondary to improper RBCs storage[17].

Acute immune hemolytic transfusion reactions are generally the most severe transfusion reactions and are caused by blood type incompatibilities. Because dogs do not possess clinically significant naturally-occurring alloantibodies, such reactions generally are only observed in previously transfused dogs (≥4 days) receiving RBC products of mismatched blood type.

Delayed immune hemolytic transfusion reactions are associated with the production of RBC antibodies within days of a transfusion, which result in a significant decline in the patient's post-transfusion hematocrit. Because they are not associated with acute clinical signs, they may be clinically unnoticed.

RECOGNITION OF TRANSFUSION REACTIONS

It is critical to monitor patients throughout the administration of blood products. The recognition of transfusion reactions may be improved by using a standardized transfusion monitoring form. In addition to containing information concerning the blood unit administered (unit identification, blood type of donor and recipient, volume administered, crossmatch status), baseline clinical data should be recorded including: temperature, heart rate, pulse quality, respiration rate/effort, mucous membrane color/capillary refill time, blood pressure, mentation as well as any other pertinent observations concerning the patient. It is necessary to monitor the patient closely and to treat any manifestations (e.g. skin rashes, tachycardia, tachypnea, sudden aggressive behavior) that develop during a transfusion as a potential transfusion reaction.

Different acute transfusion reactions share common manifestations (Table 1). In addition, some features can be obscured by the patient's underlying condition or treatments. For instance, fever is one of the most common presentations in an acute hemolytic transfusion reaction, but can also be caused by a febrile non-hemolytic transfusion reaction (patient's reaction to donor white blood cells or to biological response modifiers in the component) or a septic reaction (bacterial contamination).

In dogs sensitized by previous blood transfusions, the binding of antibodies (anti-DEA 1 or others) to donor RBCs leads to RBC lysis, activation of complement and coagulation pathways and cytokine release. Complement activation induces hemolysis, while activation of the coagulation cascade and thrombin generation predisposes the patient to DIC. Consequences of intravascular hemolysis are hemoglobinemia associated with a rapid decline in PCV and a positive Coombs test, and hemoglobinuria. Hypotension may also be present. Severe anxiety is often described in human medicine ("feeling of impending doom"). If this reaction is not recognized, the patient can progress to renal ischemia, acute renal failure, DIC, and death.

Table 1: Clinical signs of acute transfusion reaction in dogs

Clinical signs	Type of transfusion reaction	Comments
Fever	FNHTR, AHTR, TRALI, bacterial contamination, or unrelated to the blood transfusion.	Defined as a temperature >39.5°C (103.1°F) or a rise of 1-2 °C (2-4°F) from baseline. Can coexist with other symptoms such as chills, rigors, myalgia, nausea or vomiting, dyspnea, hypotension (≥30mmHg below baseline) and tachycardia (HR > 40 bpm above baseline)
Urticaria, hives or pruritus Angioedema	Allergic transfusion reaction or anaphylaxis	Can be mild and localized or more severe with generalized urticarial
Dyspnea or hypoxia	TRALI, TACO or severe allergic transfusion reaction	
Pulmonary edema	TRALI or TACO	
Hypotension	AHTR, severe allergic transfusion reaction, anaphylaxis, bacterial contamination, TRALI, TACO or unrelated to the transfusion	
Severe anxiety	AHTR or bacterial contamination	
Acute onset of bleeding	AHTR or bacterial contamination	Caused by disseminated intravascular coagulation

FNHTR: febrile non-hemolytic transfusion reaction; AHTR: acute hemolytic transfusion reaction; TRALI: transfusion-related acute lung injury; TACO: transfusion-associated circulatory overload.
Table adapted from Al-Riyami AZ et al. SQU Medical Journal. August 2014[1]

INITIAL CLINICAL ASSESSMENT AND MANAGEMENT

The first step to approaching any acute transfusion reaction is to stop the transfusion. While maintaining the venous access with an isotonic solution such as 0.9% NaCl, the patient's airway, breathing and circulation (ABC) need to be assessed next. The blood unit should be examined for the presence of any unusual clumps, particulate matter or discoloration suggestive of bacterial contamination. Finally, although it is unlikely in veterinary medicine that a clerical error leads to the wrong patient receiving the wrong blood product, it remains essential to verify that the identification of the blood donor is available for eventual transfusion reaction investigations.

Following the successful management of mild FNHTR or a mild allergic transfusion reaction (e.g. antihistamines and/or steroids), the transfusion may be resumed in patients with no or limited changes in their vital signs and who are lacking other clinical signs, but at a slower rate and under direct observation. In patients with all other types of transfusion reactions, the transfusion should be discontinued. Any suspicion of an acute hemolytic transfusion reaction or bacterial contamination upon visual assessment of the unit or by the patient's clinical signs, should prompt further investigation. If an AHTR is suspected, any hypotension should be promptly addressed in order to maintain an adequate renal blood flow.

LABORATORY INVESTIGATIONS

Reports of moderate or severe transfusion reactions should prompt testing to rule out AHTR and bacterial contamination. Post-transfusion blood (EDTA and serum) and urine samples should be collected, and the implicated unit should be put aside. In addition to performing/repeating compatibility testing (see below), a complete blood count, Coombs test, coagulation panel, and renal and liver function tests should be considered. The urine should be inspected for the presence of hemoglobinuria. If there is any suspicion of bacterial contamination, blood cultures should be obtained from the patient and the implicated unit.

If the cause of the acute hemolytic reaction is not obvious (ie immunologic or non-immunologic), a thorough evaluation of the hospital transfusion practices should be done including inspection of in-house transfusion material (collection bags, pumps, transfusion sets, fluids), blood collection techniques (ex.: negative pressure used during collection if suction is used), blood storage conditions and administration techniques. As a preventive measure, the color and presence of gross hemolysis (centrifuged blood sample of the bag) should be evaluated for all blood bags before its administration.

As previously mentioned, compatibility testing between the donor and the recipient should be performed/repeated. Pre and post-transfusion crossmatch should be performed to assess the serologic compatibility or incompatibility between donor and recipient. An incompatible major crossmatch result supports an acute immune hemolytic anemia. A standardized tube crossmatching is an appropriate technique, which requires very little material but it requires some expertise. Even with experience, the grading of the agglutination reactions (+/- hemolysis) is subjective. Gel column techniques (DiaMed and DMS gel tube assay) have been evaluated and found to be simple, sensitive, and standardized methods to crossmatch dogs. Following the discontinuation of DiaMed's veterinary products in North America, our clinical laboratory (University of Montreal) has continued to perform all crossmatch tests using the gel column technology but with human saline cards provided by Ortho Clinical. The sensitivity of the gel column technique and the objectivity of the results make it an ideal tool when investigating potential blood incompatibility cases.

Obviously, a pre-transfusion compatible crossmatch does not prevent sensitization of the patient against donor cells within the following weeks. In the same regard, it does not give information about the blood type of the dog. Therefore, previously transfused dogs should always be crossmatched. A minimum of 4 days following a transfusion is generally required for a dog to produce alloantibodies against a blood type it does not possess, which may last in circulation for months to years; this is supported by our preliminary data obtained following the sensitization of two *Dal*-negative dogs.

Because of the strong antigenicity of DEA 1 and its prevalence in the canine population (grossly 50% of dogs are DEA 1 positive, with geographical and breed variation), typing of donors and recipients for DEA 1 is strongly recommended. Indeed, DEA 1 invariably elicits a strong alloantibody response after sensitization of a DEA 1- dog with DEA 1+ blood, and may be responsible for an acute hemolytic transfusion reaction in previously transfused dogs. Standardized DEA 1 blood typing is readily available and may be performed using monoclonal antibodies based on agglutination reactions (DMS/RapidVet-H) or on chromatographic techniques (Alvedia).

Transfusion reactions against DEA 4, *Dal*, and another unspecified common RBC antigen have been described, and other clinically important blood types may be found in the future. If DEA 1 incompatibility does not explain a blood incompatibility, extended blood typing is recommended. Unfortunately, typing for all recognized DEA might be difficult as some reagents are no longer commercially available (polyclonal antibodies from sensitized dogs are used for typing). Little has been published on the prevalence of blood type according to dog breed, which may eventually help narrow the search for compatible blood. A limited survey revealed that Doberman pinschers and Dalmatians are at increased risk of blood incompatibility, as >1/3 of Doberman pinschers and >1/4 of Dalmatians are *Dal*-negative while all non-Doberman blood donors tested to date (>130 dogs) are *Dal*-positive.[11] *Dal* blood typing is now available at the Blood Bank of the University of Montreal and at the Transfusion Laboratory at the University of Pennsylvania's School of Veterinary Medicine (PennGen). The administration of "test volumes" of donor blood to assess blood-type compatibilities is highly inappropriate, as small amounts (< 1 ml) of incompatible blood can result in life-threatening reactions.

CONCLUSION

The prompt recognition, management and reporting of transfusion reactions are important aspects in ensuring patient safety during and after transfusions. Most transfusion reactions can be avoided by careful selection of healthy donors, appropriate collection, storage, and administration techniques, adequate pre-transfusion compatibility testing (blood typing +/- crossmatching), and the transfusion of only required blood components. As transfusion reactions may have overlapping manifestations, only thorough investigation can identify the cause, which may be crucial to improve hospital transfusion practice and the safety of the patients. Finally, blood incompatibility investigations require time and expertise, but remains essential to improve our knowledge in the constantly growing field of transfusion medicine.

REFERENCES

[1]Acierno MM et al. J Vet Intern Med 2014; 28: 592-598. [2]Al-Riyami AZ et al. SQU Medical Journal 2014; 14(3): 306-318. [3]Blais et al. J Vet Intern Med 2007; 21: 281-286. [4]Blais et al. J Vet Intern Med 2009; 23: 462-465. [5]Blois SL et al. J Vet Emerg Crit Care 2013; 23: 340–343. [6]Conversy B et al. J Am Anim Hosp Asso 2013; 49 (5) : 301-307. [7]Ergul Ekiz E et al. Vet Clin Pathol 2011: 40: 518–523. [8]Ferreira RR et al, Vet Clin Pathol 40: 198–201. [9]Giger U et al. Am J Vet Res 2005; 66:1386–1390. [10]Giger U.: In: Kirk's CVT 2014. [11]Goulet S et al. Research Abstract . J Vet Intern Med 2014; 28(3); 976-1134. [13]Hohenhaus AE Transfus Med Rev 2004; 18:117–122. [14]Iazbik MC et al. Vet Clin Patho 2010; 39:433-435. [15]Kessler RJ et al. Vet Clin Patho 2010; 39:306–316. [16]Kohn B et al. J Vet Diagn Invest 2012; 24:539–545. [17]Patterson J et al. J Vet Intern Med 2011; 25:927–933. [18]Seth M et al. Am J Vet Res 2012; 73:213–219. [19]Tinegate H et al. British Journal of Haematology 2012; 159: 143-153. [20]Wardrop JK et al. J Vet Intern Med 2016; 30:15-35, [21]Zimring JC and Spitalnik SL. Annu Rev Pathol Mech Dis 2015; 10: 83-110.

Alternative Transfusion Methods in Dogs and Cats

Kenichiro Yagi, BS, RVT, VTS (ECC, SAIM)
Los Altos, CA

INTRODUCTION

Our ability to perform blood component transfusions has become a vital therapeutic option in treating many ailments affecting our patients. One of the most common uses of blood components involve replacement of red blood cells (RBC) in the case of clinical anemia leading to tissue hypoxia. The amount of oxygen delivered (DO_2) being reduced to the point critical oxygen exchange ratio is exceeded (signifying inadequate DO2 to meet oxygen demand, VO_2) leads to reliance on anaerobic respiration for energy production, and is ultimately unsustainable. Supplementation of oxygen carrying capacity is warranted in these cases, and the most accessible method in providing this is through RBC transfusions. Thus, the demand for RBC products has historically been high, and the struggle for blood banks to meet demands still continues. The unavailability of RBC product is exacerbated by the presence of RBC antigens (blood types) and the potential for immunologic complications without careful consideration of compatibility risks. Alternative methods of transfusions are sought out in the effort to minimize our reliance on homologous transfusions to meet oxygen carrying capacity demands.

XENOTRANSFUSIONS

A recent news report of a feline patient who was determined to require a RBC transfusion with the alternative being death has brought to the forefront of our minds of the challenge obtaining feline blood matching our patient can be. A practice, for example, with an in-house bank has two units of type B blood banked regularly, with a type B donor in the donor program ready to donate at beck-and-call would be considered very well prepared for a type B patient requiring transfusions. The practice still would be out of options for RBCs of this rare feline blood type if a patient requires multiple transfusions, or multiple patients come in and happen to be of the same type. Many practices face shortage of RBC products in general, regardless of the type. In this particular news report, the feline patient received canine blood as a part of his treatment; an example of xenotransfusion in practice.

Xenotransfusions, or transfusion of blood products between individuals from different species, has been recorded in literature dating back to 1667 involving transfusion of blood from lambs, calves, and dogs to human for various reasons. Xenotransfusion has been employed in the 1800's with a good degree of success, though the discovery of blood types and improved knowledge on blood compatibility from the 1900's encouraged the practice of intraspecies, or homologous, transfusions. The practice of intraspecies transfusions has significantly improved the safety and effectiveness of RBC transfusions since varying degree of hemolysis and shortened life span of RBCs were seen in xenotransfusions, with the development of high antibody titer against foreign RBCs leading to immunologic consequences. Veterinary transfusion largely is practiced in this manner, with adequate feline blood supply being challenging to sustain with the small donor pool and more complicated collection process. Being able to use, for example, canine donors of a typical donation volume of 450mL to supply our feline patients with a typical (whole blood) unit volume of 50mL would allow a single donation to give us enough blood for nine transfusions. This seems like a great opportunity for a solution to the blood shortage. Let's discuss this further.

In order to implement xenotransfusion into our practices, we would first need to ensure the effectiveness and safety of xenotransfusions. There are several published studies specifically evaluating the effects of canine to feline transfusions, evaluating pre-transfusion predictors of immunologic complications, immunologic complication signs seen during and after transfusions, and the same on repeat xenotransfusions at varying timings.

The studies that have been conducted heavily focused on compatibility testing and immunologic complications, and less information on efficacy of the transfusion is available. Literature giving insight to the efficacy of canine to feline transfusion indicates a rapid improvement of clinical symptoms, leading to the conclusion that a positive effect of the transfusion is expected. In a case report involving a type B cat receiving blood from a Labrador retriever resulted in an increase in PCV taken after the transfusion performed over 48 hours. The effect is relatively short lived, however, since the transfused RBCs seem to have an average lifespan of 4 days. In comparison, a typical life span of RBCs of homologous feline transfusion is 30 days. The loss of transfused RBC is attributed to delayed hemolytic transfusion reaction resulting from antibody production against RBC antigens introduced. The production of antibodies by the immune system when exposed to foreign antigens (sensitization) occurs in a delayed manner reaching significant titer levels 4-7 days post transfusion, allowing the transfused RBC to persist for 4 days on average. The recipients have been observed to develop icterus from increased intravascular hemolysis and bilirubin load.

Statements regarding safety of canine to feline transfusions can be separated into three specific situations. In almost all transfusions performed in these studies, there were no acute clinical consequences to performing canine to feline transfusions as long as there was no history of previous canine blood transfusions. Subsequent transfusions performed within 4 days of the initial exposure also were performed without acute symptoms of immunologic complications. Subsequent transfusions performed more than 46days past initial exposure resulted in anaphylaxis, which is often fatal.

Cats seem to be tolerant of the first canine blood transfusion they receive without signs of acute hemolytic or allergic reactions, indicating a lack of naturally occurring antibodies against canine RBC antigens. These observations correlate with the majority of compatibility testing methods (slide agglutination and cross match hemolysis testing) showing no signs of incompatibility. Occasional positive reactions were seen on the minor (recipient RBC mixed with donor plasma) cross match, which upon transfusion would cause some hemolysis of recipient RBC. This effect is minimized from the use of pRBC, containing very little plasma, and dilution by recipient blood volume upon transfusion. Cats in the studies rarely exhibited signs of complication, and when they did they were mild, involving tachypnea and pyrexia during or within 24 hours of the transfusion. Delayed hemolysis occurring after 4-7 days of transfusion seems inevitable, with presence of anti-canine RBC antibodies made evident through positive slide-agglutination and cross match results when performed after this timing.

Subsequent transfusions after the initial exposure within the 4-7 day period did not result in clinical signs. Slide agglutination and cross match tests performed during this time period did not result in a positive reaction as well. This indicates a "grace period" in which antibodies are being produced to incite an anamnestic response when the next exposure occurs. Felines in the study that were given doses of canine RBC 1 and 2 days after initial exposure did not exhibit signs of immunologic complications. Test subjects having subsequent transfusions performed later than 6 days after the initial exposure showed signs of anaphylaxis with more than 66% resulting in death. Some of these cats were treated with cyclophosphamide as an immune suppressant prior to second exposure with the hypothesis of immune suppression reducing chances or severity of immunologic complications, with no positive effect seen. This serves as further evidence of the ineffectiveness of immune suppressive agents as premedication for transfusion (Human studies show premedication to be ineffective).

Transfusion of canine blood, whether whole blood or packed RBC, can be performed without immediate consequences if it is the first transfusion of this kind. However, the benefit is short lasting, and should be reserved for patients are in dire need of blood. The basic considerations in employing this therapeutic option should include situations where all of the following are conditions are met. (1) The patient has no source of RBCs with compatible blood type (type B cat with no stocked blood, donor, or nearby hospital with stock, for example) or hemoglobin-based oxygen carrier solution. (2) The patient is imminently going to pass away or is thought to be in danger of sustaining irreversible hypoxic damage (certainly up to clinical judgment) without the ability to obtain compatible blood in a timely manner. (3) The patient is expected to benefit from a short term oxygen carrying capacity gain. In this case, the patient may have a condition where the cause of anemia can be controlled swiftly enough to allow regenerative response to take over, or allow for time to obtain sources of compatible RBCs in the meantime. (4) The patient has never received canine blood products. (5) The owner understands the risks and consequences of performing a canine to feline transfusion. As the method is non-traditional, the owner should know our exact state of knowledge and potential consequences to give fully informed consent to the procedure.

If these conditions are met, use of transfusions of canine RBC to feline patients can be a life-saving therapeutic option in true emergency situations. If a xenotransfusion is decided on, a major and minor cross match should be performed to screen donor-recipient matches resulting in any signs of immunologic complications, and a therapeutic plan formulated with the benefit lasting 4-7 days in mind. Clear understanding of outcomes of future canine to feline xenotransfusion by the owners is important in preventing a second exposure beyond the 4-7 day mark. As responsible veterinary professionals, xenotransfusion should not become common practice in its current state. Effort invested in maintaining a good source and stock of feline RBC products should not decrease simply because "we can always turn to dog blood if we really need to".

"In its current state" is used as a qualifier in the previous passage because there is research ongoing for the human medical field involving "immunocamouflaging", or biochemical alteration of RBC antigens to prevent detection of the antigen as foreign by the immune system. The technique is still under development and its application to veterinary medicine would have an unknown timeline. Implementation of such techniques in xenotransfusions is theoretically beneficial in preventing immunologic consequences. A couple of case reports have been published in 2014 documenting similar experiences in efficacy, short term gain, relative safety, and recovery by the feline patient.

AUTOLOGOUS TRANSFUSIONS

Several forms of autologous transfusions, or the act of infusing blood products derived from the patient themselves, are available in providing the patient with components necessary. These techniques can be useful in specific situations.

Autotransfusion is the act of recycling blood lost by the patient by collection and reinfusion. The term autotransfusion is specific to the form of autologous transfusion involving blood that is lost from the patient's condition or injury. In veterinary medicine, autotransfusion is most commonly performed as unwashed red cells transfused through a filter. This is accomplished through transfer of suctioned blood into an intravenous fluid bag and administered with a blood administration set. An alternative method involves the use of a 3-way stopcock attached to a syringe and extension sets. The stopcock is first opened to the extension leading to the pooled blood and syringe, and blood pulled into the syringe. The stopcock is then opened to the syringe and another extension leading to an intravenous catheter, and the blood pushed through an in-line blood filter into the patient. The syringe method can also be employed prior to surgical intervention, through percutaneous insertion of a catheter into a body cavity hemorrhaging is occurring in. Autotransfusion is observed to be effective in alleviating compensatory signs from anemia and improvement in consciousness, whereas replacement of the volume with lactated ringers did not result in the same effect.

An autologous RBC salvage system or cell-saver device is a piece of equipment with the capability of collecting blood lost into the patient, separating blood components through centrifugation, washing the RBC, and suspending them back into a suspension for administration. The salvage process can be discontinuous (batched) or continuous (uninterrupted reinfusion) depending on the device, and has the advantage of having the ability to provide higher concentration of RBC than collected blood containing fluid, as well as removing contaminants through the separation and wash process.

Autotransfusion has advantages, including the alleviation of blood bank demands in performing transfusions. Blood bank supply is often strained, with upkeep being a significant financial and staff training investment. In addition, by administering autologous blood, the concerns of immunologic complications are eliminated. Blood typing and cross-matching can also be omitted, since there should be no better match to a recipient than their own blood. Storage lesions, which include accumulation of hazardous levels of electrolytes, metabolites, and inflammatory mediators as well as RBC changes reducing their efficacy as oxygen carriers lead to blood stored longer than 14 days to be less effective in treatment of anemia, have shorter survival time, and incite negative effects. These storage lesions can be virtually completely avoided through employment of autotransfusion, as no storage time is involved.

One of the main concerns with autologous transfusions is the theoretical potential for iatrogenic metastases of neoplastic cells that may be contained in the blood suspension. While no veterinary evidence is available, human studies indicate the lack of such consequences upon autotransfusion. This may be a combination of several factors, such as the thought of the cell salvage process and leukoreduction of the blood leading to effective removal of neoplastic cells. In addition, neoplastic cells may already be in circulation in many cases. Furthermore, allogenic transfusions are known to cause transfusion related immunomodulation resulting in down-regulation of the immune system, associated with increased incidences of cancer recurrence. There is potential for autologous transfusions to be more beneficial for cancer patients.

Another concern with autologous transfusions lies in the case of bacterial contamination of the blood being salvaged. In the case of blood salvaged by cell-saver devices undergoing leukoreduction, 99% of the bacterial load is seen to be removed prior to transfusion, eliminating this concern. Blood collected in clean procedures not involving obvious sources of bacterial contamination such as bowel perforation or penetrating trauma is seen to contain bacteria as well, questioning the notion of all bacterial contamination being avoidable as well. While the effect of bacterial load in blood prepared for autotransfusion without cell-saver devices and leukoreduction is uncertain, the general approach to these patients involve antibiotic therapy, and there is no evidence supporting increased incidences of bacteremia as a result of autotransfusion. The use of cell-salvage devices may also lead to cell-salvage syndrome resulting in disseminated intravascular coagulation, acute respiratory distress syndrome, acute kidney injury, or death, and is a potential consequence.

Autologous transfusions may be performed as *preoperative donations*, an intentional form of blood collection prior to procedures with reasonable chances of it resulting in hemorrhaging and subsequent demand for a transfusion. Preemptively collecting blood from the patient, typically a week prior to the procedure will allow for regenerative response to have helped replace the RBC lost through collection. The collected blood is stored in preparation for the anticipated blood loss, serving as a source of autologous RBC product if need arises. Advantages similar to autotransfusion techniques exist with this form of autologous transfusion, including the elimination of need for compatibility testing and alleviation of allogenic blood demand. A differing severity of storage lesion will be seen depending on the timing of blood collection and can lead to increase in complication related to storage lesions. The risk of bacterial contamination upon blood collection should be at a rate similar to allogenic units, dependent on quality of blood collection procedures and storage.

The detrimental effects of storage anticipated by performing autologous transfusions through pre-operative donation can be circumvented through a technique called *acute euvolemic hemodilution*. The technique involves blood collection from the patient undergoing a surgical procedure with likelihood of hemorrhaging just prior to the procedure, with replacement of lost blood volume with crystalloids to prevent hypovolemia. While the patient will have reduced circulating RBC mass as the procedure is started, the amount of blood collected is limited to tolerable amounts to meet oxygen demands. The blood lost during the procedure will be diluted in concentration of RBC (PCV), reducing the RBC mass lost when compared to the same volume of blood with the original concentration. The collected blood can be transfused if the need for a RBC transfusion arises, with the same benefits of any autologous transfusion technique, with the additional benefit of the blood being fresh. The technique serves as a method in alleviating allogenic blood transfusion need, though is often not sufficient to eliminate needs.

The quest for providing a sufficient supply of safe RBC product is an area of ongoing improvement, as the demand for blood products is ever-present, while the negative effects of blood transfusions in current practice are increasingly understood. Ensuring supply and safety in transfusions rely on methods in reducing the need for RBC transfusions (conscientious blood sampling, assertive treatment of underlying cause, proper assessment of indication), use of alternatives to RBC transfusions, proper risk assessment and compatibility testing, proper patient monitoring, and swift response to complications. Xenotransfusions and autologous transfusions are a couple of strategies increasing our supply in RBC products, each with their situational effectiveness and appropriate considerations.

REFERENCES

Bovens C, et al. J Fel Med Surg 2012;15(2):62-67. Hirst C, et al. J Vet Emerg Crit Care 2012;22(3):355-360. Kellett-Gregory LM, et al. J Vet Emerg Crit Care 2013;23(1):82-86. Kisielewicz C, et al. Vet Anaesth Analg 2014;41:233-242. Prittie JE. J Vet Emerg Crit Care 2010;20(2):167-176.Safaei M, et al. J Res Med Sci 2011;16(10):1332-1339. Sarpataki O, et al. Bulletin UASVM Vet Med 2014;71(2):513-514. Weingram T. Isr J of Vet Med 2014;69(1):50-52.

Update on Blood Banking for Dogs and Cats

Sarah E. Musulin, DVM, DACVECC
Raleigh, NC

INTRODUCTION

Veterinary blood banking is an expanding field in response to the advances in veterinary medicine and surgery. Currently, there are neither official guidelines nor governing bodies overseeing the quality of veterinary blood products. There is value in careful extrapolation of human blood banking and transfusion research to the veterinary field. Additionally, new research is being generated in veterinary medicine allowing for evidence-based practices in blood banking and transfusion medicine. In order to ensure safe and efficacious blood products, responsible practices in veterinary blood banks are paramount.

BLOOD DONOR SELECTION AND CONSIDERATIONS

When selecting blood donors, the health and safety of both the donor and the recipient must be equally considered. Emphasis must be placed on choosing healthy blood donors that will not be compromised by the donation process and that the transfusion product is safe for the recipient. Refer to Table 1 for a summary of donor criteria and considerations. Donors should be young adults that have no history of prior transfusion, infectious disease or chronic/active health conditions. The screening process should begin with temperament testing in animals that have historically demonstrated comfort in unfamiliar environments and minimal stress with handling. Temperament testing should include mock donations that evaluate ease with restraint and venipuncture. Donor selection can be divided into multiple visits to allow for continued temperament evaluation. Initial health screening may include physical examination, blood work (complete blood count, chemistry panel, +/- coagulation panel, +/- coagulation factor quantification, +/- plasma vWF concentration), urinalysis and fecal. Blood type must be determined in all donors and may be complemented with plasma antibody screening for anti-DEA antibodies in donors with unknown histories. Infectious disease testing should be guided by knowledge of diseases endemic to the donor's geographical area. The American College of Veterinary Internal Medicine published updated consensus guidelines in 2016 for infectious disease testing for canine and feline blood donors[1]. Health screening and infectious disease testing should be completed annually or more frequently in donors that become ill, have high vector exposure and/or live in blood-borne pathogen endemic areas. Optimal preventive healthcare should be ensured in all donors including vaccinations, deworming, and flea, tick and heartworm prevention. Some blood donor programs assume the costs of screening and preventive healthcare as an incentive for healthy donor participation. As with human blood banking standards, vaccinations and medications are discouraged in the pre-donation period.

Canine

Multi-bag collection sets designed for human blood donation may be used for canine blood donations. The advantage of these collection sets are that they are closed systems containing anticoagulant and red blood cell preservative solutions, thus allowing for sterile component processing. Canine donors must weigh at least 50 pounds to safely donate the 450 mL +/- 45mL required in these collection sets for balanced anticoagulant-blood ratio. Previous laboratory animal research found that canine donors can donate ~18 mL/kg (~20 % of total blood volume) every 3-4 weeks for 1 year without evidence of anemia[2]. More recent canine studies have focused on measuring iron status parameters to evaluate if iron deficiency is a consequence of frequent or chronic blood donation. Mild iron deficiency has been documented in experienced canine donors (~14 consecutive blood donations every ~8 weeks), although in this study the indicators of iron status remained within normal canine reference internvals[3]. A 2015 study measuring RBC and reticulocyte indices in canine donors, supports the presence of iron-deficient erythropoiesis in dogs donating ≥ 6 times annually[4]. Although further research investigating the consequences of chronic donation on RBC regeneration in dogs and cats is warranted, current considerations may include selecting for larger donors, increasing donation interval, providing iron supplementation and monitoring donors for iron deficiency.

Feline

Feline donors should weigh a minimum of 5 kg (lean body mass) to safely donate a standard unit of blood (~ 50 mL). Commercial feline blood donation collection sets are available that are equipped for processing components. These collection sets require the addition of anti-coagulant and red blood cell preservative solution and are therefore considered a semi-closed system. Cats may donate as frequently as every four weeks, but iron supplementation is recommended at this frequency. Research investigating optimal donation frequency or necessity for iron supplementation in donor cats is lacking.

BLOOD COLLECTION

Affirmation of continued good health status prior to each donation may include a physical examination, temperature, heart rate, respiratory rate, packed cell volume, total solids, blood urea nitrogen and blood glucose. Baseline blood pressure may also be included. Packed cell volume should be >40% in canine donors, and >35% in feline donors prior to donation. For community donors, it is recommended to have a short questionnaire completed prior to each donation confirming continued health, current medications, recent travel history, etc. When appropriately selected, canine donors rarely require sedation for blood donation. The vast majority of cats will require chemical restraint for blood donation. Considerations when choosing a sedation or anesthetic protocol include hemodynamic and thermoregulatory consequences, administration route, length of restraint, ease and time of recovery , and cost. Monitoring feline donors for hypotension and body temperature changes during the donation procedure is suggested.

Strict aseptic technique is imperative for blood donation to prevent contamination of the product. Fur is shaved at the jugular venipuncture site, followed by a thorough scrub. Sterile gloves and equipment should be used. The collection of blood is most commonly performed via gravity or suction methods. A prospective study comparing the two collection methods in dogs revealed no significant difference in clinical parameters (heart rate, respiratory rate, systolic blood pressure, temperature), hematoma formation, frequency of needle repositioning, hemolysis index or hematocrit between methods[5]. Although the suction method was noisier, it was significantly faster than gravity flow and was associated with less donor repositioning[5]. The use of vascular access ports (VAPs) for feline blood donors has been studied to eliminate the need for sedation during donations[6,7]. Other advantages of VAPs include less hypotension, decreased personnel required for donations, and shorter collection times. Drawbacks include port-related complications, contamination of blood products and increased osmotic fragility.

BLOOD PROCESSING, COMPONENTS, STORAGE

Whole blood collected during donation is immediately mixed with an anticoagulant or anticoagulant-preservative solution. Whole blood can then be immediately transfused (fresh whole blood; FWB), stored in the refrigerator (stored whole blood; SWB) or processed into components. Fresh whole blood may be indicated in patients with anemia secondary to significant blood loss, thrombocytopenia or a coagulopathy. Depending on the anticoagulant-preservative solution, stored red blood cell (RBC) products may be refrigerated for up to 42 days. Refrigerated storage of whole blood leads to loss of labile coagulation factors and platelet function limiting the utility of this whole blood product. When centrifuge and separation equipment are available, processing of FWB into RBC and plasma components allows for the provision of only those blood components necessary without excess volume. Packed RBCs (pRBCs) are a concentrated source of RBCs appropriate for anemic patients. Plasma provides coagulation factors, albumin, fibrinogen, antithrombin and immunoglobulins.

Stored RBC transfusion products, such as SWB and pRBCs, must be kept refrigerated at 1-6°C. Human FDA standards for RBC storage require ≤1% hemolysis at the end of storage and that at least 75% of red cells remain in circulation at 24 hours[8]. Improper storage of transfusion products not only negatively impacts effectiveness of blood components, but also recipient safety due to contamination or cell damage. The importance of proper blood product storage is highlighted in a 2011 case series of four dogs that experienced acute, life-threatening non-immune mediated hemolytic transfusion reactions associated with improper storage of RBC products[9]. RBC products should be stored in a dedicated blood bank refrigerator with temperature regulation and an alarm system equipped to alert during temperature fluctuations. Blood bags should be stored upright or hung in separate compartments and rotated every other day to ensure RBC-anticoagulant-preservative admixture. Blood product bags should be inspected daily for evidence of discoloration and darkening that may be seen with bacterial contamination.

Traditional processing of plasma components involves separation from RBCs and freezing within eight hours of collection to maintain coagulation factor concentrations. More recent investigation has proven that coagulation factors remain stable in whole blood stored at ambient temperature for up to 24 hours[10]. Fresh frozen plasma stored for up to 1 year at -30°C or 6 months at -20°C maintains acceptable therapeutic levels of coagulation factors and vWF[11]. Because thawing FFP can cause an unfavorable therapeutic delay (>30 min) in treatment of critically ill patients, refrigerated storage of canine plasma has been examined. Refrigerated storage of plasma resulted in only minor loss of coagulation factors with no evidence of bacterial contamination[12].

INVENTORY MANAGEMENT

Veterinary transfusion products are a valuable resource warranting practical and shrewd inventory management. Inventory should be sufficient to supply patient demand, but with minimal to no waste due to product expiration. This delicate balance is compounded by concern over biomechanical, biochemical and immunomodulating changes that occur during the storage of RBC products, coined RBC storage lesion. Computer-based blood donor and inventory management systems can be created to meet individual blood bank and/or hospital needs.

Table 1. Blood Donor Selection Criteria and Donation Specifications

	Canine	Feline	Comments
Temperament	Easy-going, minimally stressed with shaving, venipuncture and laying still for prolonged periods of time	Easy-going, minimally stressed with injections, sedation, shaving and venipuncture	Canine donors often do not require sedation Feline donors classically require sedation
Age	1 – 6 years old	1-8 years old	Young adult; age ranges may vary based on health of donors
Weight	At least 23 kg	At least 5 kg	Body weights based on lean body mass
Health Screening	CBC, chemistry, urinalysis, fecal, +/- coagulation panel, +/- vWF concentration RBC and reticulocyte indices	CBC, chemistry, urinalysis, fecal, +/- coagulation panel RBC and reticulocyte indices	Health screening should be performed at least annually Consider monitoring RBC and reticulocyte indices more frequently if donation interval is ≤ 8 weeks
Infectious Disease Screening	2016 ACVIM Consensus Statement[1]	2016 ACVIM Consensus Statement[1]	
Maximum Donation Volume	15-19 mL/kg	11-15 mL/kg	
Maximum Donation Frequency	Every 8 weeks	Every 4 weeks	Donations may be more frequent if donation volumes are below maximum and/or donor is receiving iron supplementation

REFERENCES

1. Wardrop KJ, et al. J Vet Intern Med 2016;30:15.
2. Potkay S, et al. Lab Anim Care 1969;19(2):192.
3. Zaldivar-Lopez S, et al. J Vet Intern Med 2014;28:211.
4. Foy DS, et al. J Vet Intern Med 2015;29:1376.
5. Conversy B, et al. J Am Anim Hosp Assoc 2013;49:301.
6. Morrison JA, et al. J Am Vet Med Assoc 2007;230(6):855.
7. Aubert I, et al. Can J Vet Res 2011;75(1):25.
8. FDA. Workshop on red cells stored in additive solutions systems. 1985
9. Patterson J, et al. J Vet Intern Med 2011;25:927.
10. Walton JE, et al. J Vet Intern Med 2014;28:571.
11. Wardrop KJ, et al. Vet Clin Pathol 2001;30:91.
12. Grochowsky AR, et al. J Vet Emerg Crit Care 2014;24(4):388.

Clinical Use of Neuroimaging in Small Animals: CT and MRI

Laurie Cook, DVM, DACVIM (Neurology)
Columbus, OH

INTRODUCTION

Cross sectional imaging is critical to diagnosis of neurologic diseases affecting the brain and spinal column. Computed Tomography (CT) and Magnetic Resonance Imaging (MRI) both have utility for diagnosis of many conditions with advantages and disadvantages associated with each modality. Much is still being learned about how to maximize the utility of each. As we further our utilization of CT and MRI for diagnoses of neurologic diseases, we are able to learn more about these diseases and how to best treat our patients and serve our clients.

MRI

MRI manipulates hydrogen protons in biologic tissues to produce images of pathologic changes. Disease processes often alter content, distribution and environment of these hydrogen protons in tissues. The magnetic field applied by the MRI scanner initially aligns all hydrogen protons in a patient's body. A radiofrequency pulse is applied which "knocks" the protons out of alignment. They gradually return to their original alignment with variable "relaxation times" depending on the tissue type, pathology, etc. These differences in relaxation times create different signal intensities. Different MRI sequence s exploit the changes in hydrogen protons in different magnetic fields. [1] The term intensity is used to describe the brightness of certain tissues on MRI. Tissues with high signal intensities appear brighter while tissues with low signal intensities appear darker (**Table 1**).

Table 1. MRI sequences and tissue intensities

Sequence	Fat	Cerebrospinal fluid	High protein fluid
T1*	High signal intensity	Low signal intensity	Low signal intensity
T2	High signal intensity	High signal intensity	High signal intensity
FLAIR	High signal intensity	Low signal intensity	High signal intensity
STIR	Low signal intensity	High signal intensity	High signal intensity
Gradient echo (GRE)**	High signal intensity	High signal intensity	Low signal intensity

**Contrast enhancement may be demonstrated with diseases that interrupt the blood-brain barrier (inflammation, neoplasia).*
***GRE generally used to show hemorrhage as a low signal intensity/signal void.*

MRI is the preferred imaging modality for many brain and spinal diseases due to its ability to image soft tissue changes with superior contrast resolution. [1] Bone pathology may be imaged as well with MRI but in some cases Computed Tomography (CT) can provide better bone detail.

COMPUTED TOMOGRAPHY (CT)

CT images are produce by x-rays similar to conventional radiography. Images come from computer processing of x-ray attenuation data providing improved spatial resolution. In general, CT is better for imaging bone than soft tissues. It also is less costly and takes less time than MRI, and in some cases patients can be imaged under sedation v.s. general anesthesia which is needed with MRI. [2]

SPINAL IMAGING

Intervertebral disc disease

MRI allows imaging of acute disc extrusions as well as chronic disc protrusion. The normal intervertebral disc has a hyperintense nucleus pulposus (NP) on T2 weighted sequences due to high proteoglycan content in the NP. The annulus fibrosis is a hypointense fibrocartilaginous ring on T2 weighted images that surrounds the NP. With disc degeneration, the NP becomes iso- to hypointense NP on T2 weighted images. When a disc extrudes or protrudes into the spinal canal, compression of the spinal cord, subarachnoid space and epidural fat can be seen on T1 and T2 weighted sequences. Extruded disc material usually appears hypointense within the spinal canal with the exception of hydrated nucleus pulposus extrusion which is hyperintense. Hyperintensity within the spinal cord on T2 weighted images may be appreciated secondary to edema, gliosis or spinal cord atrophy and in some cases may indicate a poorer prognosis for recovery. Hemorrhage may also be noted secondary to disc extrusions and may show a signal void on GRE sequences. Contrast enhancement on T1 weighted images with intervertebral disc extrusions or protrusions is common. [2]

Non-contrast CT is sensitive for mineralized disc extrusions and epidural hemorrhage. Mineralized disc extrusions appear as hyperdense material within the vertebral canal, loss of epidural fat, and distortion of the spinal cord. Hemorrhage often appears as irregular areas of hyperdensity surrounding the extruded disc material. If extruded disc material is not mineralized, identification on non-contrast CT is more difficult and may require myelography or MRI for diagnosis. [2]

Infectious/inflammatory spinal diseases

CT and MRI are more sensitive for diagnosis of discospondylitis than plain radiography. CT may demonstrate bone lysis, contrast enhancement, or instability in the spine secondary to discospondylitis. MRI exploits difference in tissue signal intensities of inflamed and infected tissues which enhances diagnostic accuracy over radiography.

Meningomyelitis may be apparent on MRI as hyperintensity within the spinal cord and/or meninges on T2 weighted images. Contrast enhancement of the spinal cord and meninges may be see on T1 weighted images.

Cervical Spondylomyelopathy (CSM, Wobbler Syndrome)

MRI is generally preferred for diagnosis of CSM. Soft tissue changes such as chronic disc protrusions, ligamentous hypertrophy, and spinal cord hyperintensity can be appreciated better with MRI. CT/myelography has historically been used for this purpose and may have some clinical utility. Bone changes including osteoarthritis of the articular facet joints and changes in the shape of the vertebral canal and intervertebral foramen can be appreciated on CT. CT myelography is necessary to visualize areas of spinal cord compression.

Degenerative Lumbosacral Stenosis (DLS)

MRI and CT may be useful for diagnosis of DLS in dogs. Disc protrusion, osteoarthritis of the articular facets, and ligamentous hypertrophy all contribute to stenosis of the spinal canal in these patients. MRI shows soft tissue changes better than CT, but bone changes may be superior on CT.

Spinal Trauma

CT and MRI may both be useful for imaging patients with spinal trauma. CT is gold standard for spinal trauma imaging in humans. Both CT and MRI are more sensitive than radiographs for diagnosis of fractures/luxations. Therefore, if fracture/luxation is not noted on radiographs, it should not be ruled out if clinical signs are suggestive of a spinal injury. In addition, cross sectional imaging allows for recognition of bone fragments or hemorrhage within the spinal canal that might need to be addressed surgically. MRI may be useful for identification of changes in spinal cord intensity.

BRAIN IMAGING

Brain tumors

Many types of primary brain tumors exist in domestic animals including meningiomas, glial tumors, pituitary tumors, choroid plexus tumors, ependymomas, metastatic neoplasia and others. In general, MRI is superior for diagnosis of most tumors. Most appear hyperintense on T2 weighted images, Iso- to hypointense on T1 weighted images, have varying degrees of contrast enhancement, and may appear hyperintense on FLAIR images. GRE sequences may be used to demonstrate hemorrhage within or around tumors. They have varying degrees of border definition with meningiomas and pituitary tumors being more well defined and glial tumors often appearing less well defined. Some tumors are extra-axial arising from outside the brain such as meningiomas and pituitary tumors. Others are intra-axial arising from within the brain such as glial tumors and metastatic neoplasias. Metastatic neoplasia usually is apparent at the grey-white matter junction and often is multifocal. Choroid plexus tumors and ependymomas both usually are apparent within a ventricle and both tend to be strongly contrast enhancing. While certain imaging findings are suggestive of tumor type, definitive diagnosis of tumor type cannot be confirmed based on MRI alone. Histopathology is necessary to confirm the diagnosis suspected based on imaging findings. Peri-tumoral edema usually appears as a poorly defined area of hyperintensity on T2 weighted images surrounding the mass. Mass effects such as shift of the midline, compression of ventricles or signs of brain herniation may be present. Obstructive hydrocephalus may be present in particular with some caudal fossa masses.

CT in general is less sensitive for some tumors, but meningiomas often are apparent on CT and may readily contrast enhance. Nasal tumors's extending through the cribriform plate often are apparent on CT. Other large brain masses that cause mass effect may also be apparent with CT.

Inflammatory brain diseases

Meningoencephalitis of unknown etiology (MUE) is a common neurologic disease in dogs. Cross sectional imaging often plays an important role in the diagnosis of this disease. MRI may be normal or may demonstrate focal or multi-focal lesions. Lesions may appear hyperintense on T2 weighted images, iso- to hypointense on T1 weighted images, and may show mild to marked contrast enhancement. FLAIR images often confirm presence of hyperintense lesions noted on T2 weighted images. Edema and mass effects may be noted as well.

CT is generally less effective for diagnosis of inflammatory brain diseases unless pronounced contrast enhancement is present or mass effects are noted.

Congenital Hydrocephalus

MRI and CT may both be useful for diagnosis of congenital hydrocephalus. Dilation of the lateral and 3rd ventricles is often apparent on MRI or CT. Enlarged fontanelles may be apparent in particular with CT.

Cerebrovascular disease

Cerebrovascular disease refers to disorders that impair blood supply to a region of the brain. The most commonly of these recognized in domestic animals are ischemic and hemorrhagic infarcts. With increasing availability of MRI, these conditions are being recognized more frequently in dogs and cats. MRI is superior to CT in diagnosis of cerebrovascular diseases. Ischemic infarcts usually appear as a sharply demarcated area of hyperintensity on T2 weighted images and iso- to hypointensity on T1 weighted images. They are usually confined to grey matter, but may affect white matter in severe cases. Ring enhancement is common >1 day post-onset of neurologic signs due to development of neovascularization. Often the lesion appears in a wedge-shaped manner representing the blood supply to that region. Diffusion weighted imaging (DWI) may show regions of restricted diffusion (impaired Brownian motion) in areas of secondary cytotoxic edema. This appears as a hyperintensity on DWI and a hypointensity on apparent diffusion coefficient (ADC) map. These changes generally become apparent 12-24 hours from onset. [3]

Hemorrhagic stroke occurs when a bleed occurs secondary to an infarct resulting in hematoma formation. The hematoma may appear as a mass lesion of variable size and intensity and brain edema may be present in the subacute stages. Appearance of hemorrhage on MRI changes over time due to changes in the state of hemoglobin, and this change may assist in staging the timing of a hematoma (**Table 2**).

Table 2. MRI appearance of hemorrhage over time

Stage	Time Frame	Intensity T1 weighted	Intensity T2 weighted
Peracute	<12-24 hrs	Iso- to hypointense	Hyperintense
Acute	1-3 days	Iso- to hypointense	Hypointense
Early subacute	>3 days	Hyperintense	Hypointense
Late subacute	>7 days	Hyperintense	Hyperintense
Chronic	>14 days	Hypointense	Hypointense

REFERENCES

1. Hecht, S. and W.H. Adams, *MRI of brain disease in veterinary patients part 1: Basic principles and congenital brain disorders.* Vet Clin North Am Small Anim Pract, 2010. **40**(1): p. 21-38.
2. da Costa, R.C. and V.F. Samii, *Advanced imaging of the spine in small animals.* Vet Clin North Am Small Anim Pract, 2010. **40**(5): p. 765-90.
3. Hecht, S. and W.H. Adams, *MRI of brain disease in veterinary patients part 2: Acquired brain disorders.* Vet Clin North Am Small Anim Pract, 2010. **40**(1): p. 39-63.

Neurologic Manifestations of Endocrine Diseases in Dogs and Cats

Laurie Cook, DVM, DACVIM (Neurology)
Columbus, OH

INTRODUCTION

Endocrinopathies are commonly diagnosed in dogs and cats and may present with varied clinical signs. These metabolic diseases can affect the central and peripheral nervous systems. Patients may present with neurologic signs as their first indication of illness. Therefore, it is important to recognize the various neurologic sequelae that may be seen with some endocrinopathies.

HYPERADRENOCORTICISM

Myopathy/Myotonia

Myotonia is a sign of prolonged muscle contraction with delayed relaxation after movement or stimulation. It occurs as a result of hyperexcitability of the skeletal muscle membrane resulting in repetitive action potential generation in the muscle fiber after stimulation. Myopathy or myotonia may be seen in dogs with hyperadrenocorticism. The physiologic mechanism is not completely understood, but elevated glucocorticoid levels may interfere with muscle fiber mitochondrial function. [1] Often it appears as muscle stiffness without cramping. Gait appears stiff, stilted and weak. Proximal limb muscles may be enlarged with atrophy of other muscles. Percussion myotonia, dimpling of a muscle with percussion by a reflex hammer, may be noted as well. Affected dogs may have delayed proprioception. Bloodwork may show mild elevations in creatine kinase and aspartate aminotransferase. Diagnosis is based on clinical signs of myopathy in a dog with hyperadrenocorticism. Electromyographic testing may show myotonic potentials. Muscle biopsies show Type II muscle fiber atrophy. Treatment of hyperadrenocorticism may improve signs of myotonia but improvement is often slow and incomplete. There is no known specific treatment for the myotonia itself, though procainamide given at 13 mg/kg every hours was reported to be effective in 1 dog. [2]

Central Nervous System Disorders

Pituitary dependent hyperadrenocorticism may be caused by pituitary microadenoma or macroadenoma. Macroadenomas may be large enough to compress the hypothalamus, thalamus and in severe cases the brainstem. These tumors are often slow growing and may cause no neurologic signs. However neurologic signs including mental obtundation, inappetence, circling, blindness, pain and seizures may occur due to compression of brain structures and optic pathways. [1]

Ischemic or vascular encephalopathies may occur secondary to hyperadrenocorticism in dogs. Hyperadrenocorticism is one of the most common underlying causes of thromboembolic disease in dogs. Systemic hypertension is commonly associated with hyperadrenocorticism and is likely one precipitating factor to thromboembolic events in addition to increased hepatic synthesis of coagulation factors and elevated platelet synthesis. [1, 3]

HYPOTHYROIDISM

Peripheral Neuropathy

Motor neuropathies affecting cranial nerves and peripheral nerves have been reported in dogs with hypothyroidism. Physiologic cause is not known for sure but is thought to be due to reduced mitochondrial activity or accumulation of acid mucopolysaccharides in peripheral nerves. Signs may include dysfunction of Cranial nerves V, VII, VIII, laryngeal paralysis, and megaesophagus. Additionally lower motor neuron paresis to the limbs can occur. Diagnosis is based on clinical signs of a neuropathy in a hypothyroid dog. Histopathology of peripheral nerves may show wallerian like degeneration and myelin irregularities. Signs typically improve with thyroid supplementation over a few weeks to months.

Myopathy

Humans and dogs with hypothyroidism may develops signs of myopathy. Physiologic mechanism is not known but abnormal carbohydrate metabolism or myofiber mitochondrial activity have been proposed. Clinical signs are often subclinical, but may be partially responsible for the lethargy and exercise intolerance sometimes seen in affected dogs. Diagnosis may be made in a known hypothyroid dog. Muscle biopsies may show type II muscle fiber atrophy. Resolution of signs may occur with thyroid supplementation. [1]

Myasthenia gravis

Hypothyroidism has been seen in dogs with myasthenia gravis. The link is not known definitively, but in humans a link between hypothyroidism and myasthenia gravis exists. A hypothesis is that acetylcholine receptor antibodies responsible for myasthenia gravis may cross react with self antigens present on the thyroid gland. [4]

Central Nervous System Disorders

Atherosclerosis is thought to lead to vascular encephalopathies in dogs. Affected dogs present with acute onset of focal encephalopathic signs and corresponding imaging findings. Central vestibular disease has also been reported to occur in relation to hypothyroidism. Cause is not definitively known, but cerebellar or brainstem infarction may be the cause in some affected dogs.

Myxedema coma is a rare, life-threatening syndrome that occurs in hypothyroid dogs presenting with stupor or coma. Physiologic mechanism is thought to be altered neurotransmitter release and reuptake or lack of thyroid hormone transport within the brain. Myxedema is the accumulation of acidic and neutral mucopolysaccharides and hyaluronic acids which leads to binding of water development of tissue edema. This edema in the brain is what leads to depressed mentation or coma. Myxedema coma requires a precipitating event which overwhelms normal homeostatic mechanisms. In humans, usually this event is an infection. Other signs of myxedema include non-pitting edema of the face and jowls, altered thermoregulation, and bradycardia. Signs can be reversible with intravenous thyroid supplementation. [4, 5]

DIABETES MELLITUS

Peripheral Neuropathy

A polyneuropathy caused by diabetes mellitus has been reported in dogs and cats. Physiologic mechanism is thought to be abnormal Schwann cell and myelin function. The cause of this dysfunction is hypothesized to be due to 1) microvascular compromise, 2) accumulation of sorbitol with subsequent free-radical formation and membrane damage or 3) immune mediated axonal/myelin damage. Clinical signs in dogs vary but may include lower motor neuron tetraparesis or paraparesis with proprioceptive deficits, decreased reflexes and muscle atrophy. In cats, the typical presentation is symmetric lower motor neuron paraparesis with a plantigrade stance. Bilateral Horner's syndrome has also been reported in dogs and cats. Diagnosis depends on documentation of clinical signs of a peripheral neuropathy in a diabetic animal. Electrodiagnostic testing and nerve and muscle biopsy results may be abnormal supporting the diagnosis. Treatment is focused on maximizing diabetic control. Signs may recover function slowly over time but may take months, and many patients have incomplete recoveries. [1]

HYPERTHYROIDISM

Hypokalemic myopathy

While hypokalemic myopathy is more commonly recognized in cats with chronic renal dysfunction, it is also reported in cats with hyperthyroidism. Hypokalemia leads to hyperpolarization of muscle membranes making it refractory to depolarization and contraction. Clinical signs include myalgia, reluctance to walk, a stiff stilted gait, and neck ventroflexion. Diagnosis is based on clinical evidence of a myopathy in a cat with hypokalemia, usually <3.5 mEq/l. Treatment involves oral or IV potassium supplementation as well as correction of the hypothyroidism. [1]

REFERENCES

1. Dewey, C. 2 ed. A Practical Guide to Canine and Feline Neurology, ed. C. Dewey. 2008, Ames, IA: Wiley Blackwell.
2. Vite, C.H., *Myotonia and disorders of altered muscle cell membrane excitability.* Vet Clin North Am Small Anim Pract, 2002. **32**(1): p. 169-87, vii.
3. Nichols, R., *Complications and concurrent disease associated with canine hyperadrenocorticism.* Vet Clin North Am Small Anim Pract, 1997. **27**(2): p. 309-20.
4. Bertalan, A., M. Kent, and E. Glass, *Neurologic manifestations of hypothyroidism in dogs.* Compend Contin Educ Vet, 2013. **35**(3): p. E2.
5. Finora, K. and D. Greco, *Hypothyroidism and myxedema coma.* Compend Contin Educ Vet, 2007. **29**(1): p. 19-31; quiz 31-2.

Urinary System Care nd Maintenance In Neurologically Impaired Dogs nd Cats

Jessica M. Weber, BS, LVT, CVT, VTS (Neurology)
East Greenwich, RI

MICTURATION

The voluntary and involuntary control of urination is called micturition. This process is a regulated by a complex series of pathways in the central and peripheral nervous systems. In the normal adult canine or feline patient there are two phases of micturition, the filling and emptying phases. During the filling phase signals from stretch receptors in the bladder are sent along afferent (sensory) nerves to the spinal cord and then on to the brain with information that the bladder is filling with urine. If it is an inappropriate time or place to urinate signals are sent to the bladder to loosen the bladder wall muscles to allow for more urine to enter. Signals are also sent along efferent (motor) neurons to the pelvic floor muscles to contract as well as to the internal and external pseudo-sphincters of the urethra to continue to contract to prevent voiding of urine. As the bladder fills, signals continue to travel to the brain advising the brain that the bladder is nearly full and urination should commence. Once a suitable place or time for urination is found, or if the bladder becomes too distended, the emptying phase of micturition commences. During the emptying phase, signals from the brain travel to the detrusor muscle of the bladder to signal it to contract. At the same time inhibitory signals are sent to the internal and external urethral sphincters allowing them to relax. This synergistic effort allows urine to flow from the body. Once the bladder is empty the filling phase begins again as the muscles of the pelvic floor and urethra contract once more.

NEUROGENIC BLADDER

Damage to any part of this complex system can lead to myriad of problems voiding the bladder and are often lumped into the general term "neurogenic bladder". There are two main types of neurogenic bladder disorders, the upper motor and lower motor neuron bladders, though some patients may have a combination of the two.

Upper Motor Neuron Bladder:

With upper motor neuron (UMN) bladder disease there is often a problem along the spinal cord from the area of T3 to L3, otherwise known as the thoracic spine or mid-spine, or possibly the cervical spine or brain. In veterinary medicine this is most commonly seen as a consequence of intervertebral disk disease (IVDD) disk herniation. Trauma to the spinal cord from external forces, like being hit by a car, or internal forces like a Fibrocartilaginous Embolism (FCE) in area of the mid spine or higher may also cause UMN disease. The spinal cord damage may prevent the afferent information coming from the bladder to reach the brain and/or prevent the efferent information coming from the brain to the bladder. This leads the bladder to be in a constant state of filling with the urethral sphincters and pelvic floor muscles contracted. The patient also has an inability to relax those muscles, thereby preventing the emptying phase to occur.

Patients with UMN bladders often have tense bladders due to increased bladder and urethral tone. They may also have tense abdominal muscles and the bladder may be difficult or impossible to express due to the constriction of the urethra. With mild UMN disease of the bladder the patient may have an overactive bladder, the need to urinate frequently, and may have an inconsistent stream of urine during the emptying phase. With moderate to severe disease the bladder often fills over the standard capacity, stretching the detrusor muscle and may damage the bladder wall. When stretched that far the pressure on the urine in the bladder is often high enough to leak through the constricted urethra causing a form of incontinence called overflow. Overflow may also happen when a mostly full bladder is pushed from the outside of the abdomen, such as when picking up an animal or sling walking. Since there is no voluntary action to initiate the urine flow, only pressure changes, these patients are considered unable to urinate voluntarily. Since the patient is unable to empty the bladder, a large residual urine volume may be present leading to urinary tract infections. Damage to the bladder wall may also predispose these patients to bladder rupture, especially with aggressive bladder expression. Given time and proper treatment UMN bladder dysfunction often resolves once the patient regains motor function of the rear limbs. Though, if the patient has severe paralysis, normal bladder function may not return. In this situation, many patients begin revert to pediatric micturition reflexes, which are more dependent on bladder and urethral pressure to urinate. These patients are often much easier to manually express than at the time of the original trauma.

Treatment of the UMN bladder first involves reducing or removing the disk compressing the spinal cord (if present) and waiting for the body to repair the pathways. While waiting for the body to heal, which may take weeks to months, the bladder must be managed so as not to overfill, or to remain full for more that 12-24 hours, this may help prevent bladder damage and/or urinary tract infections. Bladder management often includes a combination of bladder expression, urinary catheterization and medications to decrease urethral tone. If left untreated, the UMN bladder patient may develop pyelonephritis, bladder wall damage – Atony or rupture, and/or urethral damage. Patients with chronic incontinence may also develop skin lesions and infections due to urine scald.

Lower Motor Neuron Bladder

With lower motor neuron (LMN) bladder disease there is often a problem in the L4-S3, or lower lumbar, spinal cord; or a problem with the peripheral nervous system such as a polyneuropathy. Lower lumbar problems from IVDD, FCE, or trauma including tail-pull injuries damage the nerves in the cauda equina (an area where the spinal cord separates out into nerves going to the rear legs and tail). These nerves are crucial for transmitting signals to the bladder and urethra allowing the filling phase of micturition to occur, and also disrupt the emptying phase.

LMN bladders are often "loose" with decreased bladder and urethral sphincter tone. The bladder is easily expressed and the patient will often dribble urine constantly. Due to the laxity of the bladder wall though, the patient is often unable to fully empty the bladder. The complications of chronic urinary incontinence include urinary tract infections and skin problems due to urine scalding. Unlike UMN bladders, the LMN bladder does not tend to overfill enough to cause bladder wall or urethral damage.

Treatment of the LMN bladder, like the UMN bladder, first involves removing any spinal cord or peripheral nerve compression that may be present. Unlike UMN bladder disease though, motor function may still be present in the rear legs, even though the bladder is not functional, and even after any compression is removed. The incontinence of LMN bladders may take weeks to months to resolve, but may be permanent. While waiting for the body to heal, management of the LMN bladder revolves around keeping the bladder as empty as possible to reduce incontinence and the associated urinary tract infections and urine scald. This is usually accomplished with manual bladder expression, though urinary catheterization and/or medications to increase urethral and bladder tone may be used.

BLADDER MAINTENANCE

There are three general ways to help a neurologically impaired patient to empty the bladder. Manual bladder expression, Urinary catheterization, and pharmacologic aids can be used alone or in conjunction to facilitate bladder management.

Manual Bladder Expression

The management of both UMN and LMN bladders in veterinary medicine often involves manual bladder expression. It is used to initiate a urine stream and to empty bladder every six to eight hours while the patient is in the hospital. Bladder expression can be taught to owners and used to manage these patients at home.

To manually express the bladder of a canine or feline patient the patient can be in either lateral recumbency or in a supported stand. Hands are placed on either side of the mid to caudal abdomen, approximately halfway between the last rib and the pelvis. The bladder is palpated and using steady firm pressure the bladder is squeezed until urine is voided. This action simulates the pressure changes in a patient that induce overflow incontinence, but since there is steady pressure, the urine stream may be maintained long enough to empty the bladder. Once the bladder is small, the patient should be allowed to rest for about one minute, this allows the bladder shape to reform if any urine is remaining in the bladder. The bladder should be palpated once more and if still full, the remaining urine should be expressed. For videos about how to express a bladder see: http://www.dodgerslist.com/literature/Expressing.htm and https://www.youtube.com/watch?v=CAaLdeJClkw .

Bladder expression is not without risks though. It should not be attempted on patients with abdominal trauma unless it is known that the bladder wall and/or is intact. Even then, it may be too painful to press on a bruised abdomen and should be avoided. Patients with unstable fractures or luxations of the spine, pelvis or legs should also not be expressed due to the chance of worsening the trauma and pain that could be caused. Patients with bladder or urethral disease such as bladder stones or cancer should only be expressed with extreme caution, as the pressure changes may lead to bladder or urethral rupture. In all these cases it may be better to place an indwelling urinary catheter for bladder maintenance.

It may also be more difficult or even impossible to express the bladder of an obese or very large patient. In such a case, you may only be able to put general pressure on the abdomen and not directly on the bladder. This may or may not elicit a voiding response. If possible monitoring the bladder size with an ultrasound may be the only way to know if the bladder was reduced in size significantly. Large non-ambulatory patients may do better with an indwelling urinary catheter for cleanliness and proper bladder management. Fractious and very anxious patients also present a challenge to manual bladder expression. These patients often contract the abdominal wall as a response to fear or pain, making it difficult to palpate the bladder. The patient must also be carefully restrained so as to not hurt itself or the caregiver. Medications to reduce anxiety and or pain should be used if these patients are to be expressed. Again an indwelling urinary catheter may also be used in these patients, at the doctor's discretion.

Always **stop and ask a doctor** if any of the below is true:

- Painful or fractious patient
- Abdomen is tense/unable to palpate bladder
- Abdomen is bruised
- The bladder difficult to express
- The patient has motor function in the rear limbs

In most of the above cases it is prudent to measure the bladder on ultrasound and note bladder size and any urine voiding in the patient record. If the bladder is still large after eight to twelve hours it may be necessary to start medications to facilitate urination and/or to place a urinary catheter. If the bladder remains small, but no urine voiding is noted, the patient may be dehydrated. If the patient has motor function (with UMN bladder disease) bladder function may have returned and the patient is refusing to urinate due to hospital conditions or the presence of a sling. If possible give these patients an extra-long walk with and without sling support if able, a visit/walk with the owner, or a trial at home to see if voluntary urination occurs. Signs of voluntary urination include lifting the tail or attempting to squat before urinating, sniffing an area then urinating, or urinating in the kennel and immediately moving away from the urine. If the actual urination is not witnessed it may be difficult to distinguish from overflow incontinence without ultrasound guided measurement of bladder size.

Urinary Catheterization

In cases where manual bladder expression is contra-indicated, difficult, painful, or at the doctor's discretion, placing a urinary catheter may be used to manage both UMN and LMN bladders. As with manual bladder expression, urinary catheterization aims to empty the bladder to prevent over-distension of the bladder as well as to maintain patient cleanliness. A catheter may be placed in cases or bladder or urethral rupture to aid in healing after surgical repair. Urinary catheterization may also be indicated if the patient has a communicable disease such as leptospirosis, to prevent infection of caregivers.

The placement of the catheter may introduce bacteria into the urinary system and/or bacteria may migrate up the catheter into the urinary system over time. Because of these complications, urinary catheters should only be placed as sterilely as possible and kept as clean as possible while in place. Intermittent catheterization may be used once to twice daily to avoid bacterial migration, but each placement increases the risk of bacterial contamination. Indwelling catheters do increase the chance of bacterial migration, but lessen the chance of bladder or urethral damage due to multiple catheterizations. The method and frequency of urinary catheterization is doctor and patient specific as no one method is suitable to all patients. In one study of urinary system maintenance of dogs with IVDD, the duration of management was more predictive of urinary tract infection than of which type of management was used. Dogs were more likely to have an infection the longer they needed management regardless if the patient was manually expressed or had intermittent or indwelling cathererization.[2]

Urinary catheters are more difficult to place in feline patients and female dogs. Patients with known urethral or bladder ruptures should only be catheterized with extreme care. Other complications of urethral catheterization include over-distention of the bladder with air or saline during placement or with urine if the system becomes obstructed in some way. Forceful catheterization may cause urethral or bladder trauma, as can using too large or too long of a catheter. Patients can damage themselves attempting to remove a catheter so should be monitored closely and/or wear and Elizabethan collar. Cleaning of indwelling catheters and lines or repeated placement of intermittent catheters may cause the mucous membranes of the penis or vulva to become dry or ulcerated, use mild cleansers and rinse well to avoid this.

Video demonstration of urinary catheterization can be found here: https://www.youtube.com/watch?v=8Au5ZP1mRuE https://www.youtube.com/watch?v=RaoF9yOg8Yo

Pictorial demonstration can be here: http://www.vetmed.wsu.edu/resources/Techniques/urinary.aspx

Pharmacologic Aids

Medications to aid in decreasing or increasing urethral tone or increasing bladder tone may be helpful in patients to urinate more easily or to facilitate bladder expression or catheterization. In painful animals opioids are used to decrease pain, but may lead to urine retention. Some sedatives like acepromazine may be useful for fractious animals but may cause incontinence. Medications with a direct effect on the bladder or urethra that are used with urinary dysfunction include bethanechol, phenoxybenzamine, prazosin, diazepam and phenylpropanolamine.

Bethanechol: This medication increases bladder tone for cases of UMN and LMN bladders. Since bladder tone is already increased in many patients with UMN disease, this drug should be used with caution as it may induce bladder rupture if another agent is not used to decrease urethral tone. It can be helpful in patients with bladder decreased bladder tone due to overdistension or LMN bladder disease.

Phenoxybenzamine: This medication decreases tone of the internal urethral sphincter. It may also be used to decrease blood pressure, and may cause syncope in patients that are normotensive or hypotensive before treatment. Treatment also takes about three days before it is effective. It is often used in cases of UMN disease to loosen the sphincter and allow for urine voiding, either voluntary or by bladder expression.

Prazosin: This medication works similarly to phenoxybenzamine to reduce internal sphincter tone. It may also reduce blood pressure. Treatment is much faster though, with only as little as one hour, but sometimes upwards of a day to be effective. It is often used in cases of UMN disease to loosen the sphincter and allow for urine voiding, either voluntary or by bladder expression.

Diazepam: This medication is most often used as a sedative, but it also works to relax the external urethral sphincter. It can be used in combination with prazosin or phenoxybenzamine to relax both the urethral sphincters in UMN bladder patients. It is generally regarded as a safe medication in dogs but owner misuse could be an issue. It can be given orally or intravenously in dogs, but should only be given intravenously in cats.

Phenylpropanolamine: This medication increases tone of the internal urethral sphincter and can be used in cases of LMN disease. It is more commonly used for hormone related incontinence.

CLIENT CARE

Since UMN and LMN bladder dysfunction can take weeks to months to resolve, and can be permanent, client communication and education is paramount to providing these patients with a good quality of life. Manual bladder expression is the most common method of home care, along with pharmacologic aids, though urinary catheterization of the male dog can be taught as well. While many veterinary technicians have months or years or medical background before ever attempting these procedures, the clients may have little to no background knowledge. Be patient with them! Schedule multiple sessions on how to properly express the bladder. Show them multiple methods as well. If possible have the client watch you express the bladder first and then watch videos at home before practicing in the office.

REFERENCES

1. A Practical Guide to Canine and Feline Neurology
   Curtis Dewey DVM, 2nd Edition, 2008
2. Urinary tract infection in dogs with thoracolumbar intervertebral disc herniation and urinary bladder dysfunction managed by manual expression, indwelling catheterization or intermittent catheterization.
   Vet Surg. 2008 Dec;37(8):791-800, Bubenik L[1], Hosgood G.

Physical Rehabilitation for Spinal Cord Disease in Dogs and Cats

Stephanie N. Gilliam, RVT, CCRP, VTS (Neurology)
Columbia, MO

INTRODUCTION

Neurologic disease can cause loss of motor and autonomic function as well as sensory abnormalities such as analgesia, paresthesia or hyperesthesia. The secondary complications from these problems such as muscle atrophy or contracture, decubital ulcers, urinary tract infections and respiratory infections can be debilitating.[1] Typical rehabilitation goals for a patient with neurologic disease are to minimize pain, reestablish neural pathways, prevent secondary complications and return the animal to independent function.[2] The rehabilitation program should be designed along with other appropriate treatments for the underlying problem.

ACUTE SPINAL CORD DISEASE

The most common conditions that cause acute spinal cord injury are Hansen type I intervertebral disc disease (IVDD), traumatic injuries and fibrocartilaginous emboli (FCE). The ability to distinguish what types of neural tissue have been damaged is critical in order to predict the expected recovery and to design a rehabilitation program. The goals of rehabilitation for acute spinal cord disease are to minimize postoperative pain, maintain joint range of motion, minimize muscle atrophy and restore neuromuscular function.[1]

Before beginning any course of rehabilitation for a neurologic patient it is important to consider the disease process and the stage of healing for that particular patient. Many post-operative neurologic patients will need to be strictly rested for 6-8 weeks. If this is the case it is important to allow time for adequate healing before beginning many exercises. For example, with a post-operative intervertebral disc disease patient it would be appropriate to start passive range of motion and assisted standing exercises the day following surgery, as long as adequate pain control is obtained. Water therapy may be appropriate with the correct modalities and appropriate monitoring. Other exercises should be restricted until following the 6-8 week period of cage rest.

Thermal Modalities

Indications for the use of cryotherapy include the acute phase of tissue healing after injury and after exercise to minimize an inflammatory response. It can help reduce pain and can also be effective at decreasing edema when combined with compression and elevation. The physiologic effects of cryotherapy include vasoconstriction, cellular metabolism and permeability, decreased nerve conduction velocity, analgesia, edema reduction, decreased muscle spasm and temporary reduction of spasticity. The easiest method of application is to wrap a freezer bag filled with crushed ice in a damp cloth and place on affected area. A compression wrap can be used to secure the bag to the body. The treatment should be applied for 15 minutes at a time. The area should be monitored closely for frostbite.[2]

Heat therapy is indicated after the acute inflammatory phase of healing is over. Premature application of heat can exacerbate swelling and pain, therefore if in doubt apply cold until the dog is well into the subacute phase of healing. Superficial heat therapy results in vasodilation, muscle relaxation, increased pain threshold and superficial connective tissue extensibility. It may be applied to a joint to reduce stiffness in order to facilitate exercise. Treatment duration and frequency depends on the injury and desired outcome. The treatment time can be cycled at 30 minutes on, 30 minutes off if needed. Superficial heat is contraindicated during acute inflammation, over an area of hemorrhage or malignant tissue. It should be used with caution in patients with edema, impaired circulation or over open wounds.[2]

Range of Motion Exercises

Passive range of motion is motion of a joint that is performed without muscle contraction. It is used when a patient is unable to move joints on its own, or when active movement may be deleterious to the patient. Some of the common indications for PROM are immediately after surgery, before active weight-bearing, to help prevent joint contracture, maintain mobility of soft tissue, reduce pain, enhance blood and lymphatic flow and improve synovial fluid production. It will help maintain joint health; however, it will not improve strength or prevent muscle atrophy. Proper technique for PROM is important. The patient should be relaxed in lateral recumbency and the limb should be supported. The upper limbs are put through a comfortable flexion and extension for 15-20 cycles. The limb is then moved through a "bicycle" pattern another 15-20 times. This is repeated on each limb and is performed 3-4 times a day until the patient is ambulatory.

Active assisted ROM is motion of the joint that is guided by the rehabilitation practitioner with assistance from the patient's muscle activity. This is most beneficial for patients that are weak but do have motor function. This can be performed with the patient in lateral recumbency, with the patient in a sling, while the patient is on a treadmill or while swimming. Active assisted ROM will combat the negative effects of immobilization just as PROM does, while also promoting muscle strengthening. It may also aid with neuromuscular reeducation and proprioceptive training.[2]

Active ROM is the motion of the joint by active muscle contraction. This active ROM can be performed during a normal gait cycle or under special conditions meant to promote use of the full ROM. These activities can include swimming, walking in tall grass, water, or sand, climbing stairs, crawling and walking over cavaletti rails. As the patient improves it is important to continue to perform PROM to achieve a complete ROM, and then perform an active ROM. More strength is required for the active ROM so it is important to remember that a transition between active assisted and active ROM may be necessary.[2]

If the patient has upper motor neuron disease, the withdrawal reflex will cause active flexion of the elbow and carpus or stifle and tarsus, thereby improving muscle tone. This is done by putting the patient in lateral recumbency and pinching in between the toes to elicit a withdrawal reflex. When the patient pulls the leg back the therapist creates a gentle tug of war. This can be performed for 3-5 repetitions per limb, 3-4 times a day.[1]

Therapeutic Exercise

Therapeutic exercises are beneficial to help improve muscle strength, balance and coordination in patients that have voluntary motor function. Assisted standing exercises are beneficial for patients that are not able to stand and support their own body weight. Standing, even if partially assisted, will aid in strengthening and proprioceptive training as well as improve circulation and respiration. The purposes of these exercises are to reeducate muscles, encourage neuromuscular function, develop strength and enhance proprioception.[2] If necessary, standing exercises may be performed with the aid of a sling or inflatable exercise roll.

Weight shifting exercises may be performed when the animal is able to stand unassisted. A treat or ball may be used to encourage weight shifting. As the treat is moved the dog will follow it up, down and side to side. This movement causes the dog's center of gravity to shift, which makes the dog shift its weight to maintain balance. The rehabilitation practitioner may also gently push the animal at the hips or shoulders or lift the unaffected limb to encourage weight shifting. This may also be performed as the animal is walking.[2] Sit-to-stand exercises may be performed if the patient has enough strength to stand up without assistance and after the cage rest period is over. This exercise strengthens stifle and hip extensor muscles. It should be repeated 3-5 times and done 2-3 times a day.

Assisted walking is beneficial when the patient has some voluntary motor function. Several short walks a day will help improve strength and coordination. A sling may be used for support if needed. A land or underwater treadmill may also be used if available. The goal is to walk as slowly as possible to encourage weight bearing on all limbs for 2-5 minutes depending on the patient's ability.[1] Multiple short walks per day are more beneficial than 1-2 longer ones.

Once the patient is able to walk unassisted and has had adequate healing time, incline walks may be added to the program. Pelvic limb strength is required for a dog to walk uphill while thoracic limb strength is required for walking downhill. Walking downhill also requires the pelvic limbs to reach under the body, requiring flexion of the hock, stifle and hip. When beginning, start with gentle slopes and progress to steeper ones as the patient is able.[2]

Balance and Proprioceptive Exercises

There are many exercises that may be performed to improve balance and coordination in patients that have motor function but decreased proprioception. Weight shifting as previously discussed with also aid in proprioceptive training. Balance boards, wobble boards, balance balls, cavaletti rails and air mattresses can be used to help improve balance and coordination. These exercises should be tailored to the patient and should continue until they have a normal gait.

To challenge balance a dog may stand with all limbs on a physioroll. Standing in a static position is challenging to the core muscles. The ball may be rocked to provide an additional challenge. A balance dome (half therapy ball, half flat surface) is less challenging so can be used for dogs just beginning a rehabilitation program.

A wobble or balance board may also be used for balance and proprioceptive training. The dog may stand on the board while being rocked back and forth or side to side. Initially the movements should be small but can be increased as the patient improves.

Cavaletti rails are poles that are spaced apart for the dog to walk over. This encourages a greater active range of motion and lengthened strides. They are also used for proprioceptive and balance training. The poles should be placed about the dog's body length apart. The distance between poles should only allow for a single step.[2]

Weaving is used to encourage weight shifting on to specific limbs as well as flexion of the spine. Cones should be placed about the dog's body length apart to encourage the dog to bend the trunk in between poles.

Aquatic Therapy

Aquatic therapy may be beneficial to improve strength and joint range of motion. Swimming may need to be delayed in postoperative patients for 4-6 weeks due to the forceful muscle contraction that is involved, however, underwater treadmill may be performed sooner.[1] Underwater treadmill may be instituted as soon as the surgeon is comfortable with the incision going in the water. Petroleum jelly may be placed over the incision to help protect it from the water.

An underwater treadmill can facilitate active movements while supporting the dog's body weight through buoyancy. For example, a dog bears 91% of its body weight with the water at the level of the hock. Increasing the water level to the stifle the dog bears 85%, and to the level of the hip the dog bears 38% of its body weight. The resistance of the water allows for an increase in muscle strength. Water is much more resistant than air, making it a better strengthening exercise than land walks. Hydrostatic pressure of the water has been shown to reduce edema and swelling which may be of benefit in nonambulatory patients. Walking or swimming in water has also been shown to improve general circulation. The water temperature should be kept warm because of the beneficial effects of heat when applied to tissues. Heat provides an increase in elasticity and blood flow of the tissue. It also helps to relax the patient.

There are many advantages to an underwater treadmill as compared to swimming. Joint extension is more complete here than with swimming. The speed of the treadmill can be adjusted as needed depending on the condition and fitness level of the patient. The level of the water may be adjusted to reduce the risk of injury. It also allows for earlier intervention than swimming.

Pool therapy allows for active or passive range of motion while also providing the heat, buoyancy and resistance of water. It is used to increase muscle and cardiovascular fitness with minimal risk of injury, or when weight bearing and balance are problems. It can also be used to decrease muscle spasms or peripheral edema.

Pool therapy has some advantages over an underwater treadmill. Since the patient is non-weight bearing and the thoracic and lumbar spines are in a neutral position there may be a reduced risk of injury. It promotes a better active range of motion, especially flexion.[1] It is also a great exercise for overweight patients who may be unable to perform vigorous exercise on land. Some of the techniques used in pool therapy include interval swimming, free swimming, swimming against a current, swimming while wearing an adaptive device, or passive movements performed by the therapist.[1]

Neuromuscular Electrical Stimulation (NMES)

Neuromuscular electrical stimulation (NMES) may be beneficial to increase tissue perfusion, decrease pain and delay the onset of muscle atrophy. It can be used to delay the onset of neurogenic muscle atrophy in patients with lower motor neuron disease. It is contraindicated over the incision following a hemilaminectomy. NMES should be applied to affected muscle groups once daily for 15 minutes until the patient is ambulatory.[1]

CHRONIC SPINAL CORD DISEASE

Chronic neurologic conditions are common in geriatric large and small breed dogs. They are usually degenerative conditions such as Hansen type II IVDD, spinal malformations, degenerative myelopathy, neoplasia and degenerative lumbosacral disease. The expectations and goals for recovery are different for chronic disease versus acute disease. The spinal cord lesion in these cases often results from an underlying, poorly understood, structural abnormality. They can also be difficult to treat because of the insidious onset of signs. It is preferable to begin treatment and rehabilitation when the animal is still ambulatory. The goals of a rehabilitation program include decreasing pain, improving joint range of motion, correcting muscle atrophy and restoring neuromuscular function.[1]

Passive range of motion exercises should be performed if the patient has proprioceptive deficits or decreased strength. Joint range of motion should be measured in these patients to determine a baseline and establish which joints are the most compromised in order to develop a plan. If range of motion is decreased, stretching exercises may be added to help restore function. The joint and adjacent muscles may be warmed packed, then passive range of motion is performed. When the endpoint of flexion or extension is reached the therapist exerts gentle traction. A "bouncing" motion may also be used to assist in breaking down fibrous tissue. If the patient is uncomfortable following treatment, cryotherapy may be applied.[1]

Withdrawal reflex stimulation may be applied as with acute disease. It should be performed 20 times, 2-3 sessions a day.[1] Muscle atrophy may be as important as loss of neuromuscular function in patients with chronic disease, and the rehabilitation program must address this. The time period needed for rehabilitation of chronic disease may be much longer, sometimes even indefinite. Therapeutic exercises such as sit-to-stands, assisted walking, balance, and coordination exercises as well as aquatic therapy are similar to those described for patients with acute disease.

SUMMARY

Neurologic rehabilitation involves a combination of nursing care, passive and active exercises and therapeutic modalities. Many of these exercises do not require special equipment and can even be taught to the client to be performed at home. This requires cooperation from the patient, client and therapist.

REFERENCES

1. Olby,N, Halling,KB, Glick, TR. Rehabilitation for the neurologic patient, *Veterinary Clinics of North America Small Animal Practice:* 35 (2005), 1389-1409.
2. Millis, DL and Levine, D. Canine rehabilitation and physical therapy (2nd ed.). Elsevier, 2014.
3. Platt,SR, Olby,NJ. BSAVA manual of canine and feline neurology (3rd ed.). 2004, BSAVA.

Canine Osteosarcoma

Sandra M. Nguyen, BVSc (Hons I), DACVIM (Oncology)
Sydney, Australia

INTRODUCTION

Osteosarcoma is the most common bone tumour in dogs and accounts for 85% of all bone tumours in that species (1). It more commonly affects the appendicular skeleton in large breed dogs, but this cancer can also affect the axial skeleton, and small dogs.

In the appendicular skeleton the metaphyseal regions of long bones are affected with the distal radius, proximal humerus and distal femur the most common primary tumour locations. The proximal and distal tibia are equally affected when this long bone is involved.This has led to the phrase 'away from the elbow and towards the knee' as a means of remembering the locations to be concerned about. The forelimb is more commonly affected than the hindlimb.

Osteosarcoma is highly metastatic. With amputation of an appendicular lesion alone, the 1-year survival time is only 10%, with death usually due to pulmonary metastasis. At the time of diagnosis, dogs with radiographically appreciable pulmonary metastasisis about 10% (1).

ETIOLOGY

Large breed dogs, such as Rottweilers, Irish wolfhounds and Newfoundlands, are more commonly affected. Other breeds with a predisposition to osteosarcoma include greyhounds, Great Danes and St. Bernards. This then points towards a genetic etiology for this cancer type, or some reason where size would lead to the development of the cancer.

In Scottish deerhounds, about 15% die from osteosarcoma. It has been found that nearly 70% of risk in this breed is directly attributable to heritable traits (2). Thirty three inherited risk loci have been identified in greyhounds, Rottweilers, and Irish wolfhounds which accounts for 55 to 85% of phenotype variance in each breed (3).

Host factors for development of osteosarcoma include large breed, and it has been postulated that it is the microfractures caused by rapid growth in the first year predispose these bones to osteosarcoma. This has led to the development of 'large breed' dog food, allowing for the appropriate development of the long bones. Early neutering , previous fracture (both with and without implants), and external beam radiation have also been associated with the development of osteosarcoma(1).

DIAGNOSIS

The patient usually presents to the veterinarian with lameness, pain and swelling. The lameness may be acute or chronic, as well as acute on chronic. Survey radiographs are usually adequate to appreciate an aggressive bone lesion. Occasionally the lesion can be subtle and advanced imaging, such as computed tomography (CT) and magnetic resonance imaging (MRI) may be required if the survey radiographs are not diagnostic. Radiographs of the distal radius in a Great Dane with confirmed osteosarcoma are below.

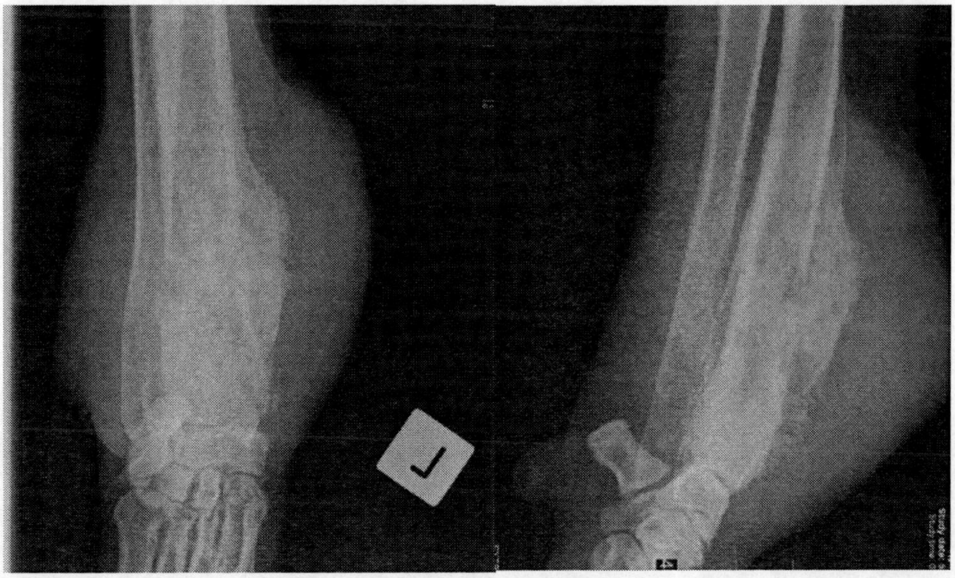

Due to the highly metastatic nature of this cancer, complete staging is recommended. CT is more sensitive than radiographs to assess for metastatic pulmonary disease (4), but survival times quoted are from publications where screening is done with radiography. While abdominal ultrasound usually does not reveal abdominal metastasis, about 5% of patients have concurrent neoplasia (5), which can change treatment plans. Nuclear scintigraphy is sensitive when screening for bony metastatasis, but not very specific.

If a biopsy for definitive diagnosis is performed, it must be taken from the center of the lesion in order to increase the diagnostic yield. Personally, I have also had good success with aspirates from the center of the lesion, and if either (biopsy or fine needle aspirate) is to be undertaken the client needs to be counseled regarding the chance for pathologic fracture.

PROGNOSIS

Histologic grade and mitotic index are tumor associated factors that have been shown to influence survival, with high grade tumours having a worse prognosis. Patient factors include age, with young dogs having a shorter survival time than older dogs; serum ALP, those with high serum ALP have a poorer outcome; and a lymphocytosis and monocytosis at the time of diagnosis, also having shorter disease free intervals (1). In another study of dogs with osteosarcoma living greater than one year after diagnosis, 48% of those dogs had an elevated serum ALP at the time of diagnosis (6).

From a clinical perspective, it is difficult to know what to do with some of this data, as I still recommend definitive treatment in dogs with an elevated ALP and monocytosis. When I am recommending amputation I rarely have a grade of tumour. Most studies give us the median survival times and disease free intervals for a group of dogs with osteosarcoma, and at this point we still cannot predict with certainty what will happen in the individual. In terms of the conversation I have with the owner, I am inclined to ask whether or not they want to know what haematologic factors for survival their dog has and whether it will help or hinder them in their treatment making decisions.

TREATMENT

The treatment of choice remains amputation and chemotherapy (platinum based protocol) resulting in a median survival of 10-12 months. In an older study of 107 dogs, 28% were alive at 2 years and 16% alive at 3 years (7). In the more recent study of 90 dogs that lived to greater than one year post diagnosis, only 19 (21%) survived to 2 years. (6).

Amputation can be a difficult decision for owners; and the anthropomorphic tendencies of owners and some veterinarians can be a significant psychological barrier. Thankfully, I have never had a patient not rise from an amputation; and some of those patients have had significant concurrent orthopaedic disease. A single obese patient that did not rise within 24 hours of amputation was discovered to have a mis-placed nerve block, which resolved with time.

In a survey of owners after amputation, 91% perceived no change in their dog's attitude after amputation with 73% reporting no change in the dog's recreational activities. Only 6% said they would not do it again with most saying that the recovery was much better than they had anticipated (8).

Metastatic disease

When dogs develop metastatic disease there are additional therapies available. In a study looking at non-mast cell tumor response to tocerinib, dogs with metastatic osteosarcoma were enrolled and 48% of them experienced clinical benefit; some of which also received metronomic cyclophosphamide and a non steroidal anti inflammatory (9). I have personally had 3 patients now live more than 12 months with pulmonary metastatic disease from osteosarcoma which, a few years ago would have been unheard of. Unfortunately adding this therapy after carboplatin chemotherapy has not resulted in an increase in survival time (10).

COMPARATIVE ASPECTS

Osteosarcoma is the most common form of bone cancer in children, and because of this the dog has been, and is, used as a spontaneous model for cancer. The intact immune system of the dog, along with the similar histologic and molecular aspects of the neoplasia, make it an effective large animal model for drug development (11).

REFERENCES

(1)Withrow et.al Small Animal Clinical Oncology 5th ed. (2). Phillips JC1 Genomics. 2007 Sep;90 (3):354-63.3. Karlsson EK, Genome Biol. 2013 Dec 12;14(12):R132 (4) Eberle et al Vet Comp Oncol. 2011 Jun;9(2):131-40 (5) Wallace et al J Am Anim Hosp Assoc. 2013 Jul-Aug;49(4):243-5 (6) Culp WT, et al J Am Vet Med Assoc. 2014 Nov 15;245(10):1141-6. (7) Berg J et al: Effect of timing of postoperative chemotherapy on survival of dogs with osteosarcoma. Cancer 1997;79:1343. (8) Dickerson VM J Am Vet Med Assoc. 2015 Oct 1;247(7):786-92. (9) London C, Vet Comp Oncol. 2012 Sep;10(3):194-205
(10) London et. Al PLoS One. 2015 Apr 29;10(4):e0124889. (11) Fenger ILAR J. 2014;55(1):69-85.

Feline Inflammatory Bowel Disease Vs. Lymphoma: Nebulous Cat Cases

Sandra (Axiak) Bechtel, DVM, DACVIM (Oncology) and Anne Corrigan, DVM, DACVIM (SAIM)
Columbia, MO and Grenada, West Indies

INTRODUCTION

The presenting complaints, clinical signs and diagnostic tests necessary to identify feline inflammatory bowel disease (IBD) and alimentary tract small cell lymphoma are exceedingly similar and, usually, identical.[1] There are a variety of complex interactions with both diseases including but not limited to: inflammatory mediators, intestinal microbiome, diet, immune system function or dysfunction, genetic predispositions and environmental factors. Differentiating between the two requires a complete workup, diagnostic imaging and ultimately biopsy and histopathology.[1-4]

Feline IBD is an exceedingly common diagnosis in small animal practice and feline small cell (lymphocytic) lymphoma is the most common malignancy in cats. Lymphocytic infiltration of the alimentary tract in part or whole is possible with both diseases, and clinical signs are indistinguishable. Small cell lymphoma is considered low grade or well differentiated, whereas large cell (lymphoblastic) lymphoma is high grade. Large cell (lymphoblastic) lymphoma is readily differentiated from other disease of the intestines on cytology, therefore, for purposes of this discussion, we will focus on differentiating low grade/small cell lymphoma from IBD in cats.[4,6]

PHYSICAL EXAM FINDINGS AND CLINICAL SIGNS

Common presenting complaints and clinical signs associated with both IBD and small cell lymphoma include any of the following: diarrhea, vomiting, weight loss, appetite alterations (i.e. anorexia/hyporexia/pica/polyphagia), melena or hematochezia, thickened bowel loops, borborygmus, flatulence, and protein loss, and ascites.[1-6]

DIAGNOSTIC TESTING AND IMAGING

Many organ specific/systemic diseases can present with diarrhea and weight loss (examples include hepatic, renal, pancreatic and endocrine diseases). Therefore, a basic minimum database - complete blood count, chemistry panel, and urinalysis as well as feline leukemia virus/ feline immunodeficiency virus testing should be performed. This will help us begin to understand the particular disease manifestations that are occurring, and to help us plan out supportive care options while other diagnostic tests are pending. Further testing will include: fecal examination and specific testing for parasites or infectious agents; evaluation of intestinal function (i.e. cobalamin, folate, and fecal α1-PI), and a total T4 for hyperthyroidism .[3,5]

Diagnostic imaging is the next step in evaluation of cats with the above clinical signs and physical examination findings; imaging is invaluable for these patients. Abdominal radiographs can help identify anatomic abnormalities and can be used to screen for the many diseases that can present with the clinical signs discussed above. Abdominal ultrasound allows for more advanced evaluation, including intestinal wall thickness, layer anatomy and echogenicity, presence of free abdominal fluid, enlargement of mesenteric lymph nodes, and GI motility. Computed tomography and MRI have been also used, but advantages over abdominal ultrasound for IBD or small cell lymphoma are questionable given the additional expense.[3-5,7]

DEFINITIVE DIAGNOSIS

The above diagnostics can generally help narrow the differential diagnoses to IBD or small cell lymphoma. However, due to the remarkable similarities between these two diseases, a cellular sample is required to obtain a definitive diagnosis. Options include fine needle aspirates of thickened intestine, endoscopically obtained mucosal biopsies or surgically obtained full thickness biopsies. Fine needle aspirate and cytology is inexpensive, the least invasive method, and results are available fairly rapidly. However, while large cell (lymphoblastic) lymphoma is readily diagnosed on cytology in most cases, IBD and small cell lymphoma cannot be differentiated from each other based on cytology alone.

Endoscopy allows for visualization of the mucosa of the stomach and duodenum and pinch biopsies can be obtained; it is less invasive than exploratory laparotomy, but disadvantages are the ability to evaluate only the mucosal surface of the stomach and intestines, so if the disease is located deeper within the intestinal wall the diagnosis can be missed. Exploratory laparotomy is the most invasive technique for obtaining samples, but is also the most likely to provide the samples needed for diagnosis. With full thickness biopsies you can evaluate the layers of the intestines as well as multiple portions of the intestines, lymph nodes and other abdominal organs.[4-6] When considering biopsy for diagnosis, thought must be given to the number of biopsy samples obtained—more is better, as well as the grading scheme used by the pathologists, as there can be large variations in the diagnosis. Care must be given to ensure appropriate evaluation of the pathological description and diagnosis.[8-9]

TREATMENT OPTIONS

IBD has been treated with a variety of options, usually dependent upon the clinical signs and supportive care necessary for any individual patient. The mainstay of IBD treatment usually requires corticosteroid administration for its anti-inflammatory and immunosuppressive effects, the most common choice being prednisolone, however budesonide has been successful in treating cats that cannot handle prednisolone or exhibit untoward side effects. In cats refractory to corticosteroids, or without a complete response, chlorambucil given orally is well tolerated and can induce remission in cats with IBD.[1,2]

Antiparasitics or antibiotics (usually metronidazole or tylosin) can be used if diagnostic tests or the clinical picture suggests treatment, but often these will be used in a therapeutic treatment trial to see if any improvement will occur. Dietary changes, considering that food allergies increase inflammation and could promote clinical signs, is an excellent first line therapeutic. Several diet types have been utilized in cats with chronic diarrhea including: hypoallergenic/low allergenic diets, hydrolyzed highly digestible diets with low fat as well as low carbohydrate high fat diets; all have shown promise in cats with chronic diarrhea, and diet trials may be necessary .[1,2] Cobalamin levels can drop due to malabsorption in diseased intestines, and hypocobalaminemia can in itself cause weight loss and anorexia in cats. Therefore levels should be measured and supplementation considered. Nutritional supplements including but not limited to prebiotics and probiotics are treatment options. Currently, there is little evidence based data for evaluation, however, adjusting the intestinal microbiome shows promise and treatment can be considered. [1, 2] New therapies are under evaluation, but all in all the clinical response is quite good. A prolonged remission or at least some improvement of the clinical signs is achieved in most cats with IBD, with or without concurrent changes in the histopathology.

Small Cell (Lymphocytic) Lymphoma

Small cell lymphoma is most commonly treated with a combination of glucocorticoids (prednisolone or budesonide) and chlorambucil. Response, defined as improvement in clinical signs, is reported to be as high as 96% with a median duration of over two years.[10]

CONCLUSION

The clinical signs and imaging findings for IBD vs small cell (lymphocytic) lymphoma are exceedingly similar. Since the clinical signs for these diseases are the same as those seen in many diseases of older cats, a thorough physical examination, minimum database, and imaging (radiographs and ultrasound) of the abdomen will aid in guiding further diagnostics. Biopsy with histopathology is required to differentiate between IBD and small cell lymphoma; often, full thickness surgical biopsies are essential. IBD responds well to supportive care and glucocorticoids, with chlorambucil utilized for refractory cases. Cats with small cell lymphoma respond well to a combination of glucocorticoids and chlorambucil, with long durations of remissions reported.

REFERENCES

[1]Ettinger and Feldman eds. Textbook of Veterinary Internal Medicine. 7th ed. Chp. 270; [2]August, John R. ed. Consultations in Feline Internal Medicine. Vol. 6 Chpt 17, 22, 11; [3]Berghoff and Steiner. "Laboratory Tests for the Diagnosis and Management of Chronic Canine and Feline Enteropathies" in VCNA SAP Chronic Intestinal Diseases of Dogs and Cats. Vol 41. 2, March 2011; [4]Norsworthy, Estep, Hollinger et al. JAVMA Vol 247. 6, Sept 15, 2015; [5]Norsworthy, Estep, Kiupel et al JAVMA, Vol 243, 10, Nov 15, 2013. [6]Gieger VCNA SAP Chronic Intestinal Diseases of Dogs and Cats. Vol 41. 2, March 2011;[7] Gaschen, LorrieVCNA SAP Chronic Intestinal Diseases of Dogs and Cats. Vol 41. 2, March 2011; [8]Willard and Mansell "Correlating Clinical Activity and Histopathologic Assessment of Gastrointestinal Lesion Severity: Current Challenges" in VCNA SAP Chronic Intestinal Diseases of Dogs and Cats. Vol 41. 2, March 2011;[9]Willard, Jergens et al JAVMA Vol 220, 8, April 15, 2002; [10] Stein TJ et al. J Am Ani Hosp Assoc 2010, 46, 6.

Tumor or Fungus: Which is Among Us?

Anne Corrigan, DVM, DACVIM (SAIM) and Sandra Axiak-Bechtel, DVM, DACVIM (Oncology)
Grenada, West Indies and Columbia, MO

INTRODUCTION

When dogs and cats are presented with clinical signs of disease, the signs are often non-specific and may be consistent with multiple disorders. In particular, both localized and systemic fungal and neoplastic diseases may have very similar presenting complaints and imaging characteristics. While signalment and history may aid in the construction of a list of differentials and in prioritizing them, definitive diagnosis requires microscopic evaluation of tissue using cytology and/ or histopathology, and in some cases fungal culture. Given the contrasting therapies for fungal and neoplastic diseases, it is very important to methodically perform diagnostics that will differentiate between the two diseases and educate pet owners on therapy and prognosis.

CLINICAL SIGNS

Clinical signs of fungal disease and neoplasia will vary depending on location of localized disease and areas of dissemination of systemic disease. Fungal infection usually commences through the respiratory tract, so clinical signs of cough, dyspnea, and exercise intolerance may occur. Infection may also begin in the gastrointestinal tract, with clinical signs of diarrhea and weight loss, or in the skin leading to cutaneous nodules. Typical clinical signs of systemic fungal infection are fever, lymphadenopathy, and weight loss. Additional differentials for fever, lymphadenopathy and weight loss include rickettsial diseases, brucellosis, mycobacteriosis, protothecosis, leishmaniasis, and immune mediated disease. These clinical signs are also common in dogs and cats with lymphoma and can occur with diffuse metastatic cancer of other cell types.[1-4]

Dogs and cats with nasal fungal infections and cancer typically present with nasal congestion, decreased air flow, nasal discharge, epistaxis, facial deformity, sneezing, stertorous breathing, and exophthalmus. Other differentials for these clinical sings are bacterial rhinitis, lymphoplasmacytic rhinitis, nasal parasites, foreign body, trauma, and developmental abnormalities (age dependent). If epistaxis is the main presenting complaint, clotting disorders and hypertension should also be considered.[1-4]

Cutaneous manifestations of fungal disease and cancer include raised, proliferative cutaneous or subcutaneous nodules potentially with ulceration, alopecia, and erythema.[1-4]

Bone is an additional primary fungal infection/ cancer site and dogs may present with a bone mass, soft tissue swelling, lameness, and pain due to an osteolytic and proliferative bone lesion, or may present with signs consistent with systemic fungal infection or metastatic osteosarcoma (cough, lethargy, pain at the site of the bone lesion, etc).[1-4]

CYTOLOGIC DIAGNOSIS

Fine needle aspiration and cytology of a mass lesion may differentiate between fungal and neoplastic diseases. For example, macrophages are numerous, or predominate, in lesions caused by fungal infections. It is important to note that macrophages can vary in morphology, from the appearance of blood monocytes shortly after tissue migration (similar in size to an eosinophil, elongated and U or muiltilobulated nuclei, lacy to ropey chromatin pattern, grainy gray-blue cytoplasm) to very large cells with cytoplasm that may be foamy with bi- or multi-nucleated cells. Macrophages may also appear as epithelioid cells and multinucleated giant cells in granulomatous inflammation. Epithelioid macrophages are characterized by blue cytoplasm with minimal vacuolation, contain little phagocytic debris, and may contain organisms. Due to the large and sometimes bizarre appearance of tissue macrophages, they can mistaken for neoplastic cells. Neutrophils may be present and in some cases the predominant cell type; eosinophils are also commonly found, especially with hyphating fungi. Low numbers of lymphocytes, plasma cells, and fibroblasts may also be seen. While most bacterial infections are diagnosed by a neutrophilic response, *Mycobacterium* spp. may cause granulomatous inflammation, and protozoal infections (cytauxzoonosis, toxoplasmosis, leishmaniasis) and algal infections (protothecosis) may elicit a granulomatous or pyogranulomatous response. Therefore, macrophages indicating granulomatous inflammation are not pathognomonic for fungal infection.[1,2]

In addition to inflammation, the fungal agent may be detected on cytologic samples and provide a diagnosis. Below are the common culprits of fungal infection, the organs systems they tend to infect, and their cytologic appearance.

Sporothrix schenckii:

- Infects: Skin primarily - raised, proliferative lesions, ulceration common. Three forms – cutaneous, cutaneolymphatic, and disseminated disease. Zoonotic (especially feline disease). Dissemination more frequently in cats than dogs.
- Cytology: Pyogranulomatous inflammation. In cats, relatively easy to diagnose as many organisms are produced and found on cytology. In dogs, however, organisms may be scarce and therefore more difficult to diagnose.
- Organism Appearance: Round, oval, or fusiform, staining pale to medium blue with pink or purple nucleus (Romanowsky stain).[1,2]

Blastomyces dermatitidis

- Infects: Usually originates in lungs and disseminates to lymphatics, skin, eyes, bone, and other organs. Pulmonary signs are common as is diffuse lymphadenopathy. 10-15% of dogs will have osteomyelitis.
- Cytology: Pyogranulomatous inflammation with few to many organisms.
- Organism Appearance: Blue, spherical, thick-walled, broad based budding with a single bud (Romanowsky stain). Lack of capsule differentiates from Cryptococcus.
- Antigen enzyme immunoassay – serum and urine (cross reactivity with histoplasmosis, coccidioidomycosis, penicilliosis, paracoccidioidomycosis). Useful for monitoring response to treatment.
- Serology of antibodies – low sensitivity and specificity. [1,2]

Histoplasma capsulatum:

- Infects: Lungs, other internal organs, skin. Usually originates in lungs, sometimes GI tract, and disseminates.
- Cytology: Pyogranulomatous inflammation. Many organisms within macrophages and some organisms found extracellularly, occasionally may find organism in neutrophils.
- Organism Appearance: Round to slightly ovoid, ¼- ½ size of RBC, pale purple to blue, eccentric crescent shaped pink to purple nucleus, clear halo around the yeast (Romanowsky stain).[1,2]

Cryptococcus neoformans:

- Infects: Infects through respiratory system – lungs, nasal cavity, or paranasal tissue. Can also infect subcutaneous tissue, central nervous system..
- Cytology: Granulomatous inflammation with epithelioid and/ or inflammatory giant cells, many organisms present.
- Organism Appearance: Pleomorphic - ranges from spherical to fusiform (spindle) shaped; thick mucoid capsule, stains pink to blue purple and may appear slightly granular. Capsule usually clear and homogenous, may stain light pink. Narrow based budding (Romanowsky stain).[1,2]

Coccidiodes immitis:

- Infects: Lung, bones. Cutaneous lesions and draining tracts from bone lesions may occur.
- Cytology: Pyogranulomatous to granulomatous inflammation.
- Organism Appearance: Organisms scarce in cytologic preparation. Variable size – large blue to clear spheres, finely granular protoplasm that may appear folded; round endospores seen in larger organisms. Differentiated from blastomycosis by the variation in size and presence of endospores (Romanowsky stain).[1,5]

Aspergillus fumigatus (and other *Aspergillus* spp) and *Penicllium* spp:

- Infects: Nasal cavity, lungs,
- Cytologic Appearance: Neutrophilic inflammation; often free and phagocytized bacteria are found.
- Organism appearance: Negatively staining or large dark blue branching septate hyphae (Romanowsky stain).[1,2,6]

Lymph node cytology of fungal disease reveals lymphadenitis, or infection/inflammation of the lymph node. White blood cells - neutrophils, eosinophils, or macrophages- migrate into the lymph node. Neutrophilic/pyogenic or granulomatous lymphadenitis can occur with certain infections including *Neorickettsia* sp, *Mycobacterium* sp, and some fungal agents.[1] Lymphadenomegaly due to neoplasia can be due to primary lymphoid neoplasia (lymphoma) or secondary neoplasia from metastatic disease. Lymphoma in the dog most commonly results an overabundance of lymphoblasts, greater than 50% and usually 80-90% of the lymph node population. Metastatic disease to the lymph nodes is also quite common from tumors that metastasize via lymphatics, including carcinomas, melanomas, and round cell tumors (particularly mast cell tumors). The presence of cells not normally found in the lymph node (such as epithelial cells) or an increase in numbers (such as large aggregates of mast cells) suggest metastatic neoplasia.[3]

Cytology of masses caused by neoplasia may reveal a population of epithelial, mesenchymal, or round cells with criteria of malignancy. Determination of sample quality, expected population of cells or abnormal population of cells, determination of inflammation (inflammation may cause normal cells in the area to become reactive, and reactive cells may look similar to neoplastic cells), and determination of criteria of malignancy. In the skin, multiple tumor types including epithelial, mesenchymal, and round cell tumors may occur. In the nasal cavity, adenocarcinoma most common in dogs and lymphoma is the most common tumor diagnosed in cats. Sarcomas, other carcinomas, and other round cell tumors may also occur. The most common bone tumor diagnosed in the dog and cat is osteosarcoma, characterized by osteoblasts that have indistinct cell borders, multiple nucleoli, eosinophilic staining cytoplasm with granulation (~50%), and eccentric nuclei; cytoplasmic staining with alkaline phosphatase may help differentiated osteosarcoma from other bone tumors.[3]

IMAGING FINDINGS

Radiographs are commonly performed to detect the source of lameness in the limbs of both dogs and cats. Often, bone lysis is seen in addition to new bone formation, soft tissue swelling, and periosteal reaction. These radiographic findings are seen with primary bone neoplasia and with fungal osteomyelitis.[4] CT scan is a most helpful imaging tool for imaging of the nasal cavity. Findings of fungal infection and of nasal cancer in both the dog and cat are quite similar and include: bone destruction, destruction of the sphenoid sinus, abnormal soft tissue in the retrobulbar space, nasopharyngeal invasion. Detection of a mass in the nasal cavity is not specific for a diagnosis of neoplasia as idiopathic inflammatory disease and fungal infections can also present with these findings.[5] For pulmonary signs of fungal disease, radiographs may reveal a nodular interstitial, diffuse interstitial, or bronchointerstitial pattern; other findings may include an alveolar pattern, tracheobronchial lymphadenopathy, or less commonly a mediastinal mass, solitary pulmonary mass, or pleural effusion.[2] Dogs and cats with lymphoma may have similar pulmonary parenchymal changes including an interstitial or alveolar lung pattern, mediastinal mass, pleural effusion, and thoracic lymph node enlargement. Furthermore, metastatic sarcomas and carcinomas to the lungs generally have a nodular interstitial pattern, and solitary pulmonary masses may also represent primary lung cancer.[3]

While diagnostic imaging is vital to define disease location and progression, there are no pathognomonic changes that will differentiate fungal disease from neoplasia. Therefore, further investigation using cytology or histopathology is necessary for definitive diagnosis.

TREATMENT

Therapy for fungal infections is dependent on the patient's health status, location of infection, and organism. Antifungal agents include amphotericin B, azole drugs (ketoconazole, itraconazole, fluconazole), flucytosine, terbinafine, and capsofungin. Similarly, therapy for neoplasia is dependent on the patient's health status, type of cancer, and location of disease.[2] While nasal adenocarcinomas tend to be locally aggressive and are successfully treated with radiation therapy (as are most nasal lymphomas in the cat), multicentric lymphoma in the dog and cat are treated with multi-agent chemotherapy protocols (prednisone, doxorubicin, vincristine, cyclophosphamide). Bone tumors are treated locally with amputation, radiation therapy, or limb-sparing procedures, and, since osteosarcomas also have a high risk of metastasis, chemotherapy is recommended in addition to localized therapy.[3-5]

CONCLUSION

The clinical signs and imaging findings in fungal infections and cancer in both the dog and cat can be very similar. While signalment may help increase the likelihood of infection or neoplasia, a diagnosis requires visualization of fungal organs or cancer cells, respectively, using cytology or histopathology. Fungal cultures or antigen titers may also be used in some cases to help differentiate between infection and neoplasia. Therapy of these diseases is dependent on the definitive diagnosis and health status of the patient.

REFERENCES

[1]Tyler RD et al in Cowell, Tyler, Meinkoth, and DeNicola eds. Diagnostic Cytology and Hematology of the Dog and Cat 3rd ed. 83-88;[2]Taboada J and Grooters AM in Ettinger and Feldman, eds. Textbook of Veterinary Internal Medicine 7th ed. 971-988; [3]Vail DM et al in Withrow, Vail Page eds. Small Animal Clinical Oncology 5th ed. 608-678; [4]Ehrhart NP et al. in Withrow, Vail Page eds. Small Animal Clinical Oncology 5th ed 463-503; [5]Turek M and Lana S. in Withrow, Vail Page eds. Small Animal Clinical Oncology 5th ed 435-437; [6]De Lorenzi et. al. J Small Anim Prac 2006; 47: 316-19; [7]Reinhardt S et. al. 46, 65-70, 2005, J Sm Anim 2005; 46: 65-70.

New Therapeutics Impacting Small Animal Oncology Practice
Chad M. Johannes, DVM, DACVIM (SAIM, Oncology)
Ames, IA

It is an exciting time in veterinary oncology as many new therapeutics tailored for canine and feline patients are on the horizon. As treatment options continue to expand in human oncology, those advancements are finding their way to veterinary medicine. Technicians are often the frontline in receiving questions regarding new and/or novel treatment alternatives for patients, so it is critical to stay up to date on the new technology entering our ever-evolving specialty. Although many of these therapeutics are in the earlier stages of development, they hold strong potential and are beginning to become clinically available for patient care.

IMMUNO-ONCOLOGY DEVELOPMENTS

Anti-Cancer Monoclonal Antibodies

1. **Monoclonal Antibody for Aid in Treatment of Canine B-cell Lymphoma (Aratana Therapeutics, Inc.)**
   This caninized MAb (AT-004) was fully licensed by the USDA in January 2015 as an aid in the treatment of canine B-cell lymphoma. Initial clinical data presented in abstract form showed signs of biologic activity[1-3] however no peer-reviewed data is available to date. Additional clinical studies are reportedly underway to further evaluate the best use and timing of this MAb in conjunction with cytotoxic chemotherapy, however it is not widely commercially available at this time. Aratana has stated that scientific studies suggest that AT-004 is not as specific to the CD20 target as expected.[4]
2. **Monoclonal Antibody for Aid in Treatment of Canine T-cell Lymphoma (Aratana Therapeutics, Inc.)**
   This caninized MAb (AT-005) received conditional licensure by the USDA in January 2014 as an aid in the treatment of canine T-cell lymphoma. The stated target of this MAb, CD52, has not yet been described in canine lymphoma cells. Aratana is currently conducting two prospective studies to further evaluate AT-005 in conjunction with cytotoxic chemotherapy (CHOP-based protocol and lomustine) for treatment of canine T-cell lymphoma. Unfortunately, the MAb does not seem to be adding significant progression free survival in these two studies. Aratana has also stated that scientific studies suggest that AT-005 is not as specific to the CD52 target as expected.[4] AT-005 remains available to many veterinary oncologists however no peer-reviewed clinical evidence of efficacy has been published at this time.
3. **Anti-CD20 Monoclonal Antibody for Canine Lymphoma (Elanco)**
   The generation and characterization of a rituximab-like anti-CD20 antibody intended as a candidate treatment for canine B-cell lymphoma has been described.[5] Monoclonal antibody 1E4 binds to approximately the same location in the extracellular domain of CD20 as rituximab, and 1E4-based chimeric antibodies co-stain canine B cells in flow cytometric analysis of canine leukocytes using an anti-canine CD21 antibody. Chimeric monoclonal antibodies were assembled. Both 1E4-cIgGB and 1E4-cIgGC were observed to significantly deplete B-cell levels in healthy beagle dogs. The *in vivo* half-life of 1E4-cIgGB in a healthy dog was ~14 days. The antibody 1E4-cIgGB has been selected for further development as an agent for the treatment of canine B-cell lymphoma.[5]

Therapeutic Anti-Cancer Vaccines

1. **Canine Lymphoma Vaccine, DNA (Merial)**
   This xenogeneic murine CD20 DNA therapeutic vaccine for use in dogs with B-cell lymphoma was conditionally licensed by the USDA in September 2015. Utilizing a similar technology platform as Oncept® for canine melanoma, the anti-CD20 vaccine is administered using a needle-free transdermal device. A prospective, multi-center clinical trial in dogs with B-cell lymphoma receiving CHOP-based chemotherapy followed by 4 doses of anti-CD20 vaccine has been described but results have not yet been published.[6]
2. **Canine Osteosarcoma Vaccine (Aratana Therapeutics, Inc.)**
   A recombinant HER2/neu expressing Listeria therapeutic vaccine (AT-014) is being studied as an aid in the treatment of canine osteosarcoma. In a study conducted at the University of Pennsylvania, Dr. Nicola Mason administered AT-014 to 18 dogs with appendicular osteosarcoma following amputation and chemotherapy (4 doses of carboplatin). The median survival time (MST) of 11 historical control dogs was 316 days; the MST for the treated group had not yet been reached ($p<0.0001$). However, eleven of the 18 treated dogs had surpassed the MST of the control group. Adverse events were mild to moderate and primarily consisted of fever, lethargy, and nausea/vomiting.[7]
3. **Feline IL-2 Immunomodulator (Merial)**
   Feline IL-2 Immunomodulator is a recombinant canarypox virus (ALVAC) expressing feline interleukin-2 (IL-2). This therapeutic was approved by EMA in 2013 (Oncept IL-2). Conditional license was granted by USDA in March 2015 (indicated to delay postsurgical recurrence of feline fibrosarcoma in adult cats with stage 1 disease).[8] A recent report outlines adjuvant treatment of feline injection-site sarcomas with this immunotherapy in complement to surgery and brachytherapy.[9]
4. **Telomerase Vaccine (Takis)**[10]
   A DNA telomerase therapeutic vaccine has been described. As observed in humans, telomerase reverse transcriptase (TERT) activity is largely confined to tumor tissues and absent in the majority of normal dog tissues. Therefore, dog TERT (dTERT) may constitute a valid target for translational cancer immunotherapy. The ability of adenovirus serotype 6 (Ad6) and DNA

electroporation to induce immune responses against dTERT in dogs affected by lymphoma was evaluated in combination with standard chemotherapy [cyclophosphamide, vincristine, prednisone (COP)].[11] Current studies are evaluating this immunotherapeutic in treatment of several other types of cancer.[12]

5. **ImmuneFx™ Cancer Vaccine (Morphogenesis Inc)**
   A multi-indication cancer vaccine, IFx-VET, is available. This product reportedly combines cell and gene therapy to promote cross-presentation and priming of the dog's immune system. It can be delivered as direct DNA or autologous vaccine.[13] Clinical benefit of this therapeutic is not well defined as no peer-reviewed data is available to date. As an autologous vaccine, this therapeutic is not currently approved or licensed by a regulatory agency.

SMALL MOLECULE INHIBITOR DEVELOPMENTS

1. **Verdinexor/KPT-335 for Canine Lymphoma (Karyopharm Therapeutics)**
   KPT-335 is a novel orally bioavailable XPO1 inhibitor. Exportin 1 (XPO1) is the sole nuclear exporter of several major tumor suppressor and growth regulatory proteins (TSPs and GRPs, including p53, p75, Rb, p21, p27, STAT3, FOXO and IkB among others). Expression of XPO1 is known to be upregulated in a variety of both hematologic malignancies and solid tumors and this correlates with a poor prognosis, indicating that changes in nuclear-cytoplasmic trafficking resulting in aberrant localization of key proteins can contribute to tumorigenesis and potentially resistance to therapy.[14] A Phase I clinical trial of KPT-335 was performed in 17 dogs with non-Hodgkin lymphoma (NHL, naive or relapsed), mast cell tumor or osteosarcoma. The maximum tolerated dose was 1.75 mg/kg given orally twice/week. Clinical benefit including partial response (PR, n = 2) and stable disease (SD, n = 7) was observed in 9/14 dogs with NHL with a median time to progression (TTP) for responders of 66 days (range 35–256 days). Toxicities were primarily gastrointestinal and were manageable; hepatotoxicity, anorexia and weight loss were the dose limiting toxicities.[14] A Phase 2b clinical trial of Verdinexor in dogs with lymphoma has been conducted. Verdinexor has received a Minor Use / Minor Species, or MUMS, designation from the Center for Veterinary Medicine (CVM) of the FDA for the treatment of lymphomas in dogs.[15]

2. **RV-1001 for Canine Lymphoma (Rhizen Pharmaceuticals)**
   PI3 kinase (PI3K) is a protein involved in cell signaling. It has been shown that for lymphomas and leukemias, PI3K isoforms γ and δ are particularly important in maintaining tumor growth. RV-1001 is an orally bioavailable inhibitor of PI3K family members, having a strong binding affinity towards PI3Kδ. A phase I clinical trial performed in dogs with naïve or relapsed B or T-cell lymphoma was conducted with oral dosing beginning at 10mg/kg PO q24h and escalating to 25mg/kg PO q24h. Grade 3 and 4 hepatotoxicity was noted in all dosing groups, prompting a change in dosing to once daily on Monday through Friday rather than 7 days/week. This Monday-Friday dosing regimen markedly reduced toxicity while maintaining response to therapy.[16] Nine dogs have been entered into the Phase I clinical trial beginning at a dose of 10 mg/kg q24h using a standard 3x3 design. One dog with T cell NHL has had a durable complete response to therapy (10 mg/kg), and 3 additional dogs have had partial responses (15 mg/kg [n=2] and 25 mg/kg [n=1]) for an objective response rate of 44%.[17]

3. **Toceranib phosphate/PALLADIA® (Zoetis)**
   Although first approved by the FDA in 2009, there are a number of recent publications to review that help to further define the best clinical use of this powerful therapeutic.
   A. **Pulse-dosed Palladia + lomustine for nonresectable canine mast cell tumor (MCT)**
      Lomustine at a dosage of 50 mg/m^2 once every 3 weeks combined with pulse-dosed Palladia was well-tolerated. The objective response rate of 46% for this protocol is comparable to what previously has been reported for single-agent protocols, but considerably higher than that reported with single-agent lomustine. Combining pulse-administered Palladia with lomustine may be a reasonable treatment option for dogs with unresectable or metastatic MCT.[18]
   B. **Maintenance Palladia following doxorubicin-based chemotherapy for canine splenic hemangiosarcoma (HSA)**
      Results of this study indicate that use of Palladia following doxorubicin chemotherapy does not improve either DFI or ST in dogs with stage I or II HSA compared to previously published reports of adjuvant chemotherapy for splenic HSA.[19]
   C. **Impact of Palladia/piroxicam/cyclophosphamide maintenance therapy on canine osteosarcoma (OSA)**
      The addition of Palladia to metronomic piroxicam/cyclophosphamide therapy following amputation and carboplatin chemotherapy did not improve median DFI, OS or the 1-year survival rate in dogs with OSA.[20]
   D. **Evaluation of Palladia administered to dogs with solid tumors at doses below the maximum tolerated dose**
      Palladia administered at doses between 2.4-2.9 mg/kg PO q48h resulted in an average 6–8 hr plasma concentration ranging from 100–120 ng/ml, well above the 40 ng/ml concentration associated with target inhibition. The lower doses of Palladia used in this study were associated with a substantially reduced adverse event profile compared to the established label dose of 3.25 mg/kg q48h.[21]
   E. **Continuous Palladia dosing in combination with cytotoxic chemotherapeutics**
      Studies evaluating the use of continuous Palladia dosing in combination with vinblastine,[22] CCNU,[23] and doxorubicin[24] have been described. These reports provide initial information on how to potentially utilize Palladia concurrently with other anti-cancer agents.

CYTOTOXIC CHEMOTHERAPEUTIC DEVELOPMENTS

1. **Paccal Vet®-CA1 (Oasmia Pharmaceutical)**
   The active ingredient in Paccal Vet-CA1 is paclitaxel, an antimicrotubule agent that works by stabilizing microtubules. Paccal Vet-CA1 has conditional approval and is labeled to treat: 1) Nonresectable stage III, IV, or V mammary carcinoma in dogs that have not received previous chemotherapy or radiotherapy; and 2) Resectable and nonresectable squamous cell carcinoma in dogs that have not received previous chemotherapy or radiotherapy. Federal law prohibits extra-label use of conditionally approved drugs meaning that Paccal Vet-CA1 can only be utilized for the two labeled indiations.[25]

2. **Doxophos® Vet (Oasmia Pharmaceutical)**
   Doxophos Vet is a patented formulation of doxorubicin which is being developed for treatment of canine lymphoma. Oasmia is reportedly conducting a Phase I study of Doxophos Vet which will include approximately 15 dogs. The FDA has granted MUMS Designation for Doxophos Vet for the indication of canine lymphoma.[26]

3. **Rabacfosadine/VDC-1101/Tanovea™ (VetDC)**
   Tanovea is a prodrug of the nucleotide analogue 9-(2-phosphonylmethoxyethyl) guanine (PMEG). A phase I/II trial was conducted in 38 dogs with non-Hodgkin lymphoma (NHL) using different dose schedules. Thirty (79%) dogs receiving Tanovea monotherapy achieved clinical remission [23 (61%) CR and 7 (18%) PR] and responses were noted in all eight treatment cohorts. Dose-limiting toxicities were generally manageable and reversible and included dermatopathy, neutropenia, and gastrointestinal signs. Two dogs developed signs of pulmonary dysfunction after they had been withdrawn from study after completing all Tanovea treatments.[27] In June 2013, the FDA granted Tanovea MUMS designation for use in canine lymphoma, allowing VetDC to move toward filing for full regulatory approval. Tanovea is also currently being evaluated for use in cats with lymphoma.[28]

ONCOLOGY SUPPORTIVE THERAPEUTIC DEVELOPMENTS

1. **Canalevia™ (Jaguar Animal Health)**
   Canalevia is a canine-specific formulation of crofelemer, an active pharmaceutical ingredient isolated and purified from the *Croton lechleri* tree, that contains anti-secretory properties. As an oral, enteric-coated, twice daily formulation of crofelemer, Canalevia is being developed for the treatment of chemotherapy-induced diarrhea, or CID, in dogs. The product is not absorbed systemically at the therapeutic dose, but acts locally in the gastrointestinal tract. Jaguar is pursuing a Minor Use/Minor Species (MUMS) approval from the CVM for Canalevia for CID in dogs. If conditional approval is received, Jaguar expects to launch Canalevia for CID in dogs in 2016.[29,30]

2. **Capromorelin/AT-002 (Aratana Therapeutics, Inc.)**
   Capromorelin is an orally active small molecule that mimics the action of ghrelin which causes growth hormone secretion and appetite stimulation. Capromorelin has been shown to cause increased food intake and weight gain in both laboratory and client-owned dogs.[31,32] Additionally, capromorelin has been demonstrated to cause increases in IGF-1 and increased food intake and body weight in cats.[33] Based on current timelines, Aratana anticipates FDA approval and commercialization of capromorelin for treatment of inappetence in dogs in 2016.[34] This therapeutic has significant potential to positively impact the treatment of inappetence related to chemotherapy and/or underlying cancer in our canine and feline patients.

3. **Grapiprant/AT-001 (Aratana Therapeutics, Inc.)**
   Grapiprant is a selective antagonist of the EP4 receptor, one of four receptors that mediates the action of prostaglandin E_2 (PGE_2). This receptor is believed to play a key role in mediating pain associated with osteoarthritis. Long-term safety in laboratory dogs has been demonstrated.[35] FDA approval and commencement of commercialization is anticipated in 2016.[36] This may provide an alternate pain management therapeutic for cancer patients experiencing chronic pain.

REFERENCES

[1]Bulman-Fleming J, et al. Abstract, 2014 Veterinary Cancer Society; St. Louis, MO. [2]Ogilvie GK, et al. Abstract, 2014 Veterinary Cancer Society; St. Louis, MO. [3]Rodriguez CR, et al. Abstract, 2015 ACVIM Forum; Indianapolis, IN. [4]http://aratana.investorroom.com/2015-09-24-Aratana-Therapeutics-Provides-Product-Updates. [5]Rue SM, et al. Vet Immunol Immunopathol 2015;164(3-4):148-159. [6]http://c.ymcdn.com/sites/www.nysvms.org/resource/collection/824A08A0-F7AF-4D95-A850-C088F60C9C8E/NYSMVS_SeptOct_proof_06.pdf. [7]Mason N, et al. Abstract, 2014 ACVIM Forum; Nashville, TN. [8]http://todaysveterinarypractice.epubxp.com/i/504690-may-jun-2015/18. [9]Jas D, et al. Trials in Vaccinology 2015;4:1–8. [10]http://www.takisbiotech.it/index.php/news/30-takis-veterinary-cancer. [11]Peruzzi D, et al. Mol Ther 2010;18(8):1559-67. [12]http://www.vetcancertrials.org/studies/vschv-telomerase-vaccine-for-canine-cancer. [13]http://www.morphogenesis-inc.com/veterinary-oncology/veterinarians/. [14]London CA, et al. PLOS One 2014;9(2):e87585. [15]http://karyopharm.com/drug-candidate/oral_verdinexor_kpt-335/. [16]https://iom.nationalacademies.org/~/media/Files/Activity%20Files/Disease/NCPF/2015-JUN-8/London.pdf. [17]London CA, et al. Abstract, 2015 American Association for Cancer Research; Philadelphia, PA. [18]Burton JH, et al. J Vet Intern Med 2015;29:1098–1104. [19]Gardner HL, et al. BMC Veterinary Research 2015;11:131. [20]London CA, et al. PLOS One

2015:DOI:10.1371/journal.pone.0124889. [21]Bernabe LF, et al. BMC Veterinary Research 2013;9:190. [22]Robat C, et al. Vet Comp Oncol 2012;10(3):174-183. [23]Pan X, et al. Vet Comp Oncol 2014. [24]Pellin M, et al. Abstract, 2015 Veterinary Cancer Society; Vienna, VA. [25]http://www.fda.gov/AnimalVeterinary/ResourcesforYou/ucm402476.htm [23] http://www.oasmia.com/news.asp?c_id=373. [26]http://www.oasmia.com/pages.asp?c_id=313 [27]Vail DM, et al. Clin Cancer Res 2009;15(10):3503-10. [28]http://vet-dc.com/products/tanovea-for-lymphoma/. [29] http://jaguaranimalhealth.com/animal-health/. [30] http://www.businesswire.com/news/home/20150701005278/en/Jaguar-Completes-Study-Dogs-Chemotherapy-Induced-Diarrhea-Commercial#.Vc4B8ZNViko. [31] Zollers B, et al. Abstract, 2014 American College of Veterinary Internal Medicine; Nashville, TN. [32]Zollers B, et al. Abstract, 2014 ACVIM Forum; Nashville, TN. [33]Zollers B, et al. Abstract, 2015 ACVIM Forum; Indianapolis, IN.[34]http://www.aratana.com/aratana-announces-positive-results-in-a-pivotal-field-effectiveness-study/. [35]Rausch-Derra L, et al. Abstract, 2014 ACVIM Forum; Nashville, TN. [36]http://www.aratana.com/aratana-reports-positive-results-from-its-pivotal-field-study-of-at-001-grapiprant/.

Gastrointestinal Parasites of Ruminants: Treatments, Management Techniques, and Resistance

Melissa G. Ackerman, DVM, MS
Pullman, WA

INTRODUCTION

Livestock around the world are affected by external and internal parasites. Parasites cause economic losses due to decreased production, illness, death, and cost of treatment and prevention. As with many pathogens, interactions between host factors, such as genetics and nutrition status, environmental factors, such as temperature, moisture, and climate, and infectious agent factors, such as drug resistance, determine the severity of parasitism in an animal or herd. This discussion focuses on internal parasites that infect the gastrointestinal (GI) tract of ruminant species in the United States (US). The common GI parasites include nematodes, cestodes, and protozoa.

Table 1. Categories, genera, and anatomical distribution of GI parasites in ruminants

Category of Parasite	Genera	Anatomical Distribution
Nematodes	*Haemonchus*	Abomasum
	Ostertagia	Abomasum
	Teledorsagia	Abomasum
	Trichostrongylus	Abomasum
	Marshallagia	Abomasum
	Nematodirus	Small Intestine
	Trichostrongylus	Small Intestine
	Cooperia	Small Intestine
	Oesophagostomum	Cecum/Large Intestine
	Trichuris	Cecum/Large Intestine
Cestodes	*Taenia*	Small Intestine
	Monezia	Small Intestine
Protozoa	*Eimeria*	
	Isospora	
	Cryptosporidia	
	Giardia	

NEMATODE DISEASES OF THE GI TRACT

Parasitic Gastroenteritis

The nematode genera *Trichostrongylus, Ostertagia, Teladorsagia, Cooperia, Oesophagostomum,* and *Nematodirus* cause parasitic gastroenteritis (PGE). Nematodes are members of the trichostrongylid worm family, and different species have both host and anatomical preferences. Species prevalence and behavior varies depending on region, season, weather, and management factors but the general pathogenesis of PGE is similar.[2]

As the ruminant host grazes it consumes the infective larvae, which enter either the gastric glands of the abomasum or the crypts of the small intestine, depending on the anatomical preferences outlined in Table 1 above. *Ostertagia* develop and molt in the gastric glands, and after 18-21 days emerge into the lumen for further development to the adult stage. It is this emergence that triggers a cascade of events resulting in a spectrum of clinical signs that include diarrhea, dehydration, colic, weight loss, poor wool or hair growth, reduced or no weight gain, weakness, lethargy, reduced appetite or anorexia, and tissue edema. The parasite residence within the gastric glands make them become distended and nodular, causing local hyperplasia and gland dysfunction. When the glands no longer secrete acid (their normal function), the pH rises within the abomasum so that incoming ruminal bacteria and protozoa are no longer killed. Additionally, pepsinogen cannot be converted to pepsin at this high pH so that protein digestion is compromised. Trichostrongylid infections of the small intestine cause inflammation-associated mucosal thickening with stunted villi. This results in reduced epithelial enzyme activity, but because the lesions are focused proximally, malabsorption is not a hallmark of this disease. However, with excessive mucus production, protein losses from sloughed epithelia, and ongoing protein-losing enteropathy, hypoalbuminemia and edema can be severe.[2]

The trichostrongylid worms have a direct lifecycle. Eggs are passed in feces and hatch as the first larval stage when environmental conditions are favorable. On pasture, they molt into the second and then the infective third larval stage. *Nematodirus* species differ in that the infective larvae develop within the egg. Once in the host's gastric glands or small intestinal crypts as previously described, the parasites molt, emerge from the mucosa, and return to the lumen to molt again into a mature adult. Once egg-laying females are present, the prepatent period (PPP) ends and the host passes eggs into the environment through the feces. *Ostertagia/Teladorsagia* species cause two types of disease processes, Type I and Type II. Type I is characterized by completion of a normal life cycle in one season. Type II is characterized by hypobiosis, or arrested development, of the parasite, which occurs in cold and/or unfavorable weather conditions. In this case, the larvae remain dormant in the gastric glands for 4-5 months before emerging. This accumulation of parasites and simultaneous emergence can result in significant GI damage and severe clinical signs.[2]

Haemonchosis

Haemonchus species affect all ruminants (*H. contortus* in sheep and goats and *H. placei* in cattle) and are widespread throughout the US. Also called the "barberpole worm", it has a characteristic red and white intertwining appearance. Fecundity of *Haemonchus* is such that one female can lay up to 10,000 eggs daily for several months. This worm resides in the abomasum and sucks the blood of its host; therefore the clinical signs are similar to other trichostrongylid infections but are related to the severity of anemia and hypoproteinemia due to blood loss. Pale mucous membranes and severe edema, especially submandibular, are present in severely affected animals. Acutely infected animals may die with exertion, others may be found suddenly dead.[2]

Like other Trichostrongylids, the eggs of *Haemonchus* are passed in the host feces to the environment, where the non-infective larvae hatch and molt to the infective third larval stage. The host grazes and ingests the infective larvae, which have the ability to become hypobiotic in the glands of the abomasum. Upon emergence, the adults lay eggs about 18 days post-infection (*H. placei* has a PPP of 26 days). The length of time of development varies depending on region and environmental conditions like weather, temperature, and moisture. A warm, humid climate allows for fast development in the environment to the infective larval stage, making the southeast US a favorable region for this parasite. This coupled with the high fecundity of the female worms allow for very high numbers of infective larvae to accumulate quickly in pastures. [2]

Oesophagostomosis

Oesophagostomum species may cause enough damage to the ruminant host to cause colitis and anemia. Hypoproteinemia may occur if the cecum and colon become edematous. Larvae may become encased in the mucosa leading to peritonitis, adhesions, intussusception, or stenosis. These nodules can be quite painful, and affected sheep may have an arched back and stilted gait. The lifecycle is similar to that of the other common nematodes.[2]

Whipworms

Trichuris infections in livestock typically are not clinically problematic. The worms embed their anterior end into the cecum, which may enable gut bacteria to escape the lumen. It has a direct lifecycle, and its eggs are very hardy in the environment and may survive for years. The infective larvae develop inside the egg. After hatching, they are ingested by the host and mature into adults and lay eggs that are passed to the environment in feces.[2]

CESTODE INFECTIONS OF THE GI TRACT

Tapeworms

Taenia and *Moniezia* infections are often not significant in ruminants, but in susceptible animals heavy worm burdens can cause unthrifty appearance, poor hair coat, constipation and diarrhea. Tapeworm eggs are distinctive in appearance and may be visible around the tail base or in feces. Eggs are passed in feces into the environment and are infective. Pasture mites ingest them and mature worms develop when the host swallows infected mites while grazing.[2]

PROTOZOAL INFECTIONS OF THE GI TRACT

Coccidiosis

Infection with *Eimeria* and *Isospora* species, collectively called coccidia, may lead to clinical coccidiosis, especially in crowded conditions where hygiene and biosecurity are poor. Young stock are the most commonly affected when oocyst levels in the environment are high and the animals are brought together in smaller spaces, especially for weaning or in the winter months. There are many species of coccidia and they are all host specific. Clinical signs include diarrhea, which may contain varying degrees of mucus and blood. Straining with rectal prolapse may occur, and inappetence is common.[2]

Unsporulated oocysts are passed in feces and develop into infective stages in the environment. When ingested, sporocysts are released and enter epithelial cells of the intestinal mucosa. There, a cycle of asexual and sexual reproduction occurs, rupturing and reinvading epithelial cells during the process. The resulting damage to the mucosal epithelium results in the clinical signs of the disease.[2]

Cryptosporidiosis

Young ruminants are commonly infected with *Cryptosporidium parvum*. Clinical disease is common in calves; outbreaks in lambs and kids may be associated with severe diarrhea and death. Only a small number of oocysts are necessary to cause infection, and young animals will shed high numbers of oocysts once infected. The PPP is 2-12 days, depending on the species. Oocysts are shed in feces and ingested from the environment. Once ingested, the parasites reproduce and can sporulate inside cells. The parasites attach to intestinal epithelial cells and cause damage, blunting the intestinal villi and resulting in impaired digestion and malabsorption with diarrhea.[2]

Giardiasis

Most infections are not symptomatic and the species is commonly found incidentally on fecal exams. However, it is more recently thought to be a potential primary cause of diarrhea in ruminants. Infection often occurs through contaminated water. Ingested cysts release life stages that attach to intestinal epithelial cells and cause inflammation and consequent damage to the intestinal villi. As with *Cryptosporidium,* the infective dose is small and significant fecal shedding occurs especially with young animals. Affected animals may have fluid or pasty diarrhea with mucus, along with growth depression.[2]

CONTROL METHODS

Drugs are an important part of parasite control as they can save lives and restore animals to health. In most livestock populations 80% of the parasite burden is concentrated in 20% of the animals.[3] Recommended parasite control protocols incorporate this idea into the management system, and reflect the complex interactions of the host, environment, and pathogen. Before designing a parasite control protocol, it is important to examine animals exhibiting clinical signs and use diagnostic tests such as fecal egg counts to determine species presence and the extent of parasitism in the herd. Once a protocol is selected, ongoing monitoring of the herd is essential to gauge the success of the protocol. Animal response to therapy, as well as other diagnostic tests such as the fecal egg count reduction test can help identify whether anthelmintic resistance is significantly prevalent in the parasite population.

There are many factors to consider when designing a control method. Young, infected animals may be shedding high amounts of eggs. Additionally, preparturient animals often shed eggs and oocysts at high levels (pre-parturient rise). Host nutrition status is closely linked to the immune system, so animals in a healthy body condition are more able to fight parasite infection.[3,4] Heritability of resistance and tolerance of parasite infection exists in ruminants, although it varies by species, breed, and is incompletely understood. The goals of successful parasite control include reducing the rate and amount of contamination into the environment in order to lessen parasite exposure, identifying and protecting the most vulnerable animals, reducing the selection pressure for anthelminthic resistance to preserve the efficacy of anthelmintics, monitoring worm infections, and selecting within the herd for animals that are resistant to parasite infections. [4,8]

Drug Therapy for Protozoal Infections

Drugs used to treat coccidiosis in ruminants include sulfadimethoxine, amprolium, and ponazuril. Lasalocid and monensin are often used to prevent clinical disease but not infection.[3] Unfortunately, there are no approved drugs to treat giardiasis, as metronidazole use is banned in food animals. Fortunately, drug resistance in protozoa is not problematic.

Anthelmintic Drugs and Parasite Resistance

There are three classes of anthelmintic drugs: benzimidazoles (including albendazole and fenbendazole), cholinergic agonists (including levamisole and morantel), and macrocyclic lactones or avermectins and milbemycins (including ivermectin and moxidectin).[1] There have been no new anthelmintic drug developments for at least 25 years. The efficacy of these drugs varies depending on the nematode species and the stage of development present in the host. Also, the residual activity of the drug can vary depending on host species and individual factors.[2] For example, goats metabolize these drugs quite differently than other ruminants and therefore need different dosages for the drug to be effective.[10] There are many formulary sources from which to choose an appropriate drug and dosage according to species. Some are listed in the references below and therefore specific dosage and therapy will not be discussed here.

Parasite resistance appeared soon after these drugs became widely available and persist today. GI nematode populations are huge because of the high fecundity, which means each population has high inherent genetic diversity. Additionally, the genes are easily disseminated because of host movement. This combination primes nematodes for rapid, genetically successful resistance to drugs. Once present in a population, these drug resistance genes are rapidly spread with host movement. Use of the drug increases selection pressure, resulting in a drug resistant population.[1] Parasite resistance to anthelmintics is often controlled by interactions of more than one gene. Nematode resistance to each class of anthelmintics is widespread. In the US, *H. contortus* populations may be resistant to all three classes.[4]

Some management practices inadvertently result in anthelmintic resistance. Underestimating weight results in underdosing and increased selection pressure for drug resistant parasite populations.[4] Improper use of a drenching gun can stimulate the esophageal groove reflex and cause the drug to be deposited largely in the abomasum, rather than the rumen. This can reduce the residual effect of some anthelmintics.[2] The practice of rotating the use of multiple anthelmintics within a grazing season selects for parasite populations where resistance to all drugs are present.[4] Treating the entire herd at regular intervals will also allow for resistant parasites to survive in populations.[4]

MANAGEMENT TECHNIQUES

FAMACHA or Smart Drenching Method

This parasite control method involves targeted treatment of the most heavily burdened animals. FAMACHA was designed to detect *Haemonchus* blood sucking nematode clinical infections. This system uses a color-coded chart to score adult animals with pale conjunctival mucous membranes. The palest animals with the most severe clinical signs are treated. This helps reduce cost of drug treatment, reduces the pasture load by treating the highest shedders, and allows for parasites that are not drug resistant to stay in the population and continue to contribute to the gene pool, diluting out the number of drug resistant parasites and maintaining the efficacy of anthelmintics in that population.[3,4] Additionally, animals that have high fecal egg counts and low packed cell volumes may be removed from the breeding pool, selecting for a herd with genetic predisposition to parasite resistance.[4]

Five Point Check

This method is designed to detect animals burdened with parasites that do not only include *Haemonchus contortus.* The system involves monitoring regularly for conjunctival mucous membrane color, nasal discharge, submandibular edema, back body condition scoring, and fecal staining around the perineum. As with the FAMACHA system, those with the lowest scores would be treated or removed.

Pasture Grazing Strategies

Because each parasite species differs in its persistence in the environment, this strategy will depend on which parasites are causing clinical problems in the herd. The goal is to provide "clean" pastures for grazing by reducing numbers of infective stage larvae. Because young animals are highly susceptible to infection and disease, rotating the young through the pastures before placing the adults on the same pasture is effective in protecting them. Also, any immunocompromised animals should be included in this group. Additionally, dividing pastures into smaller lots will allow for long enough rest periods that the infective stage larvae and/or eggs will die. However, stocking rate should not be increased to the point of increased risk of infection. Pasture areas that are persistently wet, like ponds and marshes will build up "hot spots" of high worm burdens and should be eliminated. [8] Obviously, weather, season, and climate will differ with region and will affect the survivability of the parasite in the environment. Overgrown pastures prevent ultraviolet light penetration and drying out of the larvae and can prolong parasite survival. Strategic deworming to eliminate arrested larvae in infected animals or recently emerged nematodes before they develop into adults and contaminate the pasture will greatly reduce pasture burden. Also, larvae will often hatch after rains or recent moisture, and strategic deworming at those times will accomplish the same thing.[1,4,8]

Data shows that grazing plants high in condensed tannins such as *Lespedeza cuneate* can reduce fecal egg count, larval development in feces, and adult worm numbers in the GI tract. These plants may be added to the pasture or to the feed.[1] Nematode-trapping fungi occur in the soil and feed on free-living nematodes. Incorporating this into the daily feed is a management strategy that has been experimentally demonstrated to help control worms.[1]

Alternating grazing with other livestock species may help reduce *Haemonchus* load in pastures. If adequate parasite control is present in cattle, small ruminants can graze after them because of the host specificity present in GI parasites. These species may even graze together; ingesting the parasites of the other helps reduce the potential infective larvae available to the preferred host species.[1]

Copper Oxide Wire Particles

Copper oxide needles are available in bolus form. These are given orally and sit in the reticulum, and are slowly broken down over time. Traditionally, these are used to supplement dietary copper to livestock. Copper has anthelmintic activity against abomasal worms, but not other GI parasites. This treatment method is effectively used in areas where *H. contortus* is problematic, and does not contribute to parasite resistance. Due to the increased sensitivity of sheep to copper toxicosis, it is recommended not to exceed 4 grams (in one dose) per year.[1]

REFERENCES

1. Fleming, S.A. et al. 2006. *Anthelmintic resistance of gastrointestinal parasites in small ruminants.* Journal of Veterinary Internal Medicine, 20:435-444.
2. Radostits, O.M. et al. 2007. Veterinary medicine: A textbook of the diseases of cattle, horses, sheep, pigs, and goats, 10th ed. Saunders, Elsevier.
3. Smith, B.P. 2015. Large Animal Internal Medicine, 5th ed. Elsevier, St. Louis, MO.
4. Craig, T.M. 2006. *Anthelmintic resistance and alternative control methods.* Veterinary Clinics of North America: Food Animal Practice, 22:567-581.
5. Saddiqi, H.A. 2011. *Small ruminant resistance against gastrointestinal nematodes: a case of Haemonchus contortus.* Parasitology Research, 109: 1483-1500.
6. Pugh, D.G. 2002. Sheep and Goat Medicine, Saunders, Philadelphia, PA.
7. Yazwinski, T.A. et al. 2006. *A sampling of factors relative to the epidemiology of gastrointestinal nematode parasites of cattle in the United States.* Veterinary Clinics of North America: Food Animal Practice, 22: 501-527.
8. Bath, G. 2011. *Non-pharmaceutical control of endoparasitic infections in sheep.* Veterinary Clinics of North America: Food Animal Practice, 27: 157-162.
9. Ballweber, L. 2006. *Diagnostic methods for parasitic infections in livestock.* Veterinary Clinics of North America: Food Animal Practice, 22: 695-705.

10. Hoste, H. et al. 2011. *Control of endoparasitic nematode infections in goats.* Veterinary Clinics of North America: Food Animal Practice, 27: 163-173.
11. Miller, J.E. et al. 2006. *Immunological aspects of nematode parasite control in sheep.* Journal of Animal Science, 84 (E

Critical Care of Neonatal Camelids

Claire E. Whitehead, BVM&S, MS, DACVIM (Large Animal Internal Medicine), FHEA, MRCVS
Oxford, UK

INTRODUCTION

Crias are extremely rewarding to treat as large animal patients: treat them like foals and they respond like calves! This presentation will discuss the differences of camelids as neonatal patients, including recognition of some camelid-specific congenital conditions, prevention of failure of passive transfer (FPT) and how to deal with the crashing neonatal cria.

THE CAMELID NEONATE 101

Management of the sick newborn alpaca or llama requires familiarity with the cria's normal state in order to correctly identify abnormalities and to prevent or treat potential problems.

Gestation lengths may vary considerably: although the average is quoted as being 340-346 days[1], they can be between 335 and 375 days. Season has a significant effect on gestation length with spring-birthing alpacas having longer gestation lengths than autumn-birthing alpacas[2].: this effect becomes more pronounced at increasing latitudes. Any cria born at <335 days' gestation should be considered premature. Recognition of the clinical signs of prematurity is vital in newborn crias as they may deteriorate rapidly andearly intervention is key to survivals. One report showed a 75% survival rate and this was attributed to the fact that 87.5% of cases were presented within 24 hours of birth.[3] Clinical indications of prematurity include low birth-weight (<7kg for alpacas, <8kg for llamas), unerupted incisors, floppy ears or ears that are bent backwards, poor or absent suckle reflex, and silky haircoats. Tendon laxity may result in overextension of the carpal and fetlock joints so that crias may be seen walking on their fetlocks – occasionally support bandaging may be required. Crias may also have a rubbery covering on the nails of the feet and a thicker epidermal membrane that persists for longer than normal. The incisors normally erupt two to three weeks prior to birth, so unerupted incisors suggest that a cria is at least two to three weeks premature. Premature crias tend to be weak and may have difficulty standing or holding their heads up to nurse due to poor muscle development. Furthermore, premature crias may have incomplete structural and functional maturation of vital body systems and this leaves them susceptible to a range of potential neonatal complications. These crias are at risk for hypothermia as their thermoregulatory control is poor and they are prone to hypoglycemia. Respiratory system immaturity may result in poor lung expansion due to lack of pulmonary surfactant and this also reduces oxygen transfer across the alveoli. This, together with reduced respiratory rate results in hypoxia. It is unknown when surfactant is produced in the lungs of camelids, but those with unerupted incisors, indicating at least 2-3 weeks' prematurity, are especially at risk. The intestinal mucosa may also be inadequately mature in premature crias resulting in inefficient absorption of colostral antibodies. More than half of the premature crias presenting to one referral hospital had FPT.[3]

Newborn llamas and alpacas should be standing within 30-60 minutes and nursing within two to four hrs but often both events will occur within 30-60 minutes. Meconium should be passed within 18 hours. Normal alpaca crias should weigh at least 7 kg and llama crias 8 kg; those weighing less should be monitored more closely for the first few days of life. The mean birth weight of 424 alpaca crias born at altitude in Peru was 7.7kg[4], compared with 8.1kg for 186 alpaca crias in a UK-based study (Davison, unpublished). Crias will normally lose 0.25 kg or so during the first 24hours but then should gain 0.1-0.45 kg in body weight daily (average 0.25kg).

The newborn cria is covered in a thin, translucent epidermal membrane that is attached at the mucocutaneous junctions. The function of this membrane is unknown but it is thought that it may aid lubrication in the birth canal and also act as a windbreaker in the immediate post-partum period. It usually rubs off during the first few hours of life. Crias normally nurse frequently, two to three times per hour in the neonatal period, for one to three minutes. They will spend only a short amount of time on each teat before moving on to the next. A cria that is seen to be nursing frequently and for long periods should alert the observer to the possibility that the dam has insufficient milk to cope with the demands of the cria.

Crias are born severely hypogammaglobulinemic due to the diffuse epitheliochorial nature of placentation in camelids, like that of the horse.[5,6] Ensuring adequate colostrum intake (10-20% of body weight in the first 24 hours) is vital in order to acquire passive immunity. Failure of passive of transfer of immunoglobulin (FPT) has been demonstrated to be a major cause of neonatal mortality in alpacas.[7] In this study, 9% of 82 crias were found to have FPT. Camelid colostrum contains a predominance of immunoglobulin G (IgG). Absorption of IgG declines in efficacy over the first 24 hours of the cria's life so there is a narrow window of opportunity.[8] If a female has insufficient milk, then frozen camelid colostrum can be used, or goat, cow or sheep colostrum as alternatives. Llama milk has been shown to have higher sugar content (6.5%) and less fat (2.7%) than the milk of other domestic ruminants and it appears that either goat or cow milk is the most suitable alternative.[9] The milk can be either fresh or frozen withfrozen colostrum retaining immunoglobulin levels for up to 1 year in the freezer.

Congenital Defects

Camelids do tend to have a higher incidence of defects than other domestic species. Regardless of outcome, animals with defects should not be bred. Of particular importance in newborn crias are the following congenital defects:

- CHOANAL ATRESIA: This causes respiratory difficulty. The defect is an embryogenic failure of the openings at the choanae (back of the nostrils) such that a membranous and sometimes bony defect prevents air from passing through the nostrils to the back of the mouth. Because of this , they cannot breathe normally and have to mouth-breathe. Some crias compensate for this relatively well while the majority show respiratory difficulty shortly after birth. Often crias will puff out cheeks while breathing and may struggle to nurse because they get out of breath. Making the diagnosis is fairly simple:

a red rubber catheter or other soft tube, should be passed up the nostril and a repeatable obstruction will be noted at the level of the medial canthus of the eye. It is not usually necessary to perform further diagnostic procedures such as contrast radiography and endoscopy. Secondary aspiration pneumonia is common. Surgery is possible but the prognosis is not great since most affected crias already have FPT and sepsis by the time of diagnosis in addition to hypoxaemia (low blood oxygen concentrations), which causes them to be further compromised. Surgery should only be considered in animals that are stable clinically or in cases with unilateral atresia. These latter cases may not present with difficulties until they are weaned since they are able to breathe normally through one side.

- IMPERFORATE VULVA: The cria strains to urinate with an imperforate vulva. Sometimes the vulva is completely imperforate (no hole) or may be incomplete (small hole). With this condition the vulva pouches out below the anus and appears to be fluid-filled. This is an easy problem to correct by making an incision in the midline to make a normal sized vulva opening. Topical local anaesthetic is ideal but not necessary since the procedure is very quick. Post-operatively, twice daily cleaning ensures that the new opening remains patent and topical application of antibiotic ointment (especially one containing an anti-inflammatory) assists in keeping the wound clean. No suturing is required, but a pre-op epidural would facilitate this if desired.
- CLEFT PALATE: Affected crias normally have nasal discharge and may develop aspiration pneumonia easily. The defect is usually severe involving the entire hard and soft palates. This condition carries a poor prognosis.
- UMBILICAL HERNIA: If the umbilical hernia is small and reducible, then an abdominal wrap may be used – elasticated bandage wrap with 10x10cm gauze swabs folded over the defect to push the contents of the hernial sac back into the abdomen. The wrap may be left on for 1-2 weeks but it is important to make sure it doesn't get too tight as the cria is growing. If it is still not reduced after this time, and the defect is more than one finger-tip in diameter, surgical correction can be done to help prevent this from being a problem in later life. Animals with fair-sized hernias requiring surgical correction should not be used for breeding.
- HEART DEFECTS: These may cause a cria to be weak in the first few days of life if the defect prevents the heart from being able to pump blood efficiently. The cria may be weak and floppy, stand with its head down, or become easily out of breath: it is important to differentiate these presentations from problems caused by hypoglycemia, hypothermia or sepsis as a result of FPT. A thorough physical examination should detect a significant heart murmur. An echocardiogram is required to determine the cause of the murmur. Crias that have murmurs do not necessarily have defective hearts since other problems may cause murmurs, such as hemodynamic alterations in acutely ill cases; however, the murmurs of non-cardiac origin are typically not loud (grade III or less out of VI).
- ATRESIA ANI: An occasional defect in crias. This must be differentiated from a more extensive atresia coli since only the former type of defect is surgically correctable.

ASSESSMENT OF CAMELID NEONATES

In the US, camelid owners are generally very efficient at monitoring their newborn crias. Crias will often be weighed daily allowing owners to spot the first indication of illness. At this time, routine physical examination findings (Table 1) may be within normal limits. However, evaluation of laboratory data at this time may prove critical: within 24 hours crias may become systemically very unwell, particularly from complications due to FPT and subsequent sepsis. Routine hematology and biochemistry should be performed along with assessment of passive transfer using camelid-specific IgG tests. Evaluation of serum total protein and globulin concentration may give a good estimate of whether a cria has ingested insufficient colostrum, but these tests are highly dependent on hydration status.. A dehydrated cria is likely to have falsely increased protein concentrations, mainly due to albumin, but if they are low they may become lower following rehydration. Glucose concentration can also be rapidly evaluated by the use of handheld devices.

Table 1. Normal physical examination findings for neonatal crias.

Variable	Range
Birth weight (kg)	Alpaca: 7 – 11 Llama: 8 - 12
Body temperature	37.8-38.9°C (100-102°F)
Heart rate (bpm)	70-100
Respiratory rate (bpm)	20-30

The "crashing" cria may present unable to stand or even hold up its head. Weakness of the long neck muscles means that they can appear clinically worse than other neonates but may still respond to appropriate emergency care. There may also be quite significant biochemical and /or hematologic derangements. The most common reasons for a cria to present in such a critical state are hypothermia, hypoglycemia (usually in crias that have not yet nursed), prematurity, complications arising from FPT, and congenital defects. Hypothermia and most congenital defects can be ruled out on initial physical examination. It is important that the initial physical assessment is done as quickly as possible in order to assess the need for immediate intervention and stabilisation measures such as oxygen therapy or fluid therapy to address severe hypoglycemia.

The eyes should be checked with a bright light source for hypopyon (pus in the anterior chamber) that may occur secondary to sepsis. This requires immediate treatment to prevent r glaucoma secondary to adhesions that form between the lens and cornea as a result of fibrin formation. Eyes should be checked daily because this may occur at any time during hospitalisation.

Arterial blood gas analysis can be very useful in evaluating the sick neonate. One of the easiest sites from which to collect this sample is from the saphenous artery as it passes over the inside of the hind limb just caudal to the stifle. A 25g needle is recommended for this technique in order to minimise the extent of any hematoma.

Table 2. Normal arterial blood gas variable in healthy neonatal alpaca crias aged 2-7 days.[10]

Variable	Mean	95% CI
pH	7.47	7.46-7.48
Bicarbonate (mEq/L)	20.4	19.6-21.3
PaO_2	92	87.7-96.4
$PaCO_2$	30.5	27.2-33.8

MANAGEMENT OF HOSPITALISED NEONATES

Despite sometimes fairly dramatic presentations of unresponsive crias with major biochemical and/or hematologic derangements, crias do seem to be remarkably responsive to intensive care, which is in contrast to foals, and those that respond well in the first 24 hours of treatment will often proceed to make a full recovery. The exceptions are those crias with congenital defects that are not amenable to treatment, such as choanal atresia or major cardiac abnormalities. For this reason it is especially important for crias to be fully evaluated on presentation for any defects that will lead to costly treatment failure and cause further distress to the cria, its dam and the owner.

Intravenous catheter placement may be done early during evaluation to permit blood sample collection and minimise the number of venipuncture sites. Venous access is likely to be required for fluids (plasma and/or crystalloids) and medications. Intraosseous access may be a suitable alternative if hypovolemia is severe.

Oxygen can be administered via face-mask or nasal insufflation tubes. Those used for humans can be secured around the back of the head and are generally better tolerated than face-masks. Alternatively, a small diameter flexible catheter can be inserted up one nostril to the level of the medial canthus and secured in position for oxygen supplementation. If intubation is required for a cria that is not breathing by itself, an endotracheal tube (size 4-6) can be placed with the aid of a long-bladed laryngoscope and a dog urinary catheter can be used as a stylet to guide placement.

In the case of severe hypoglycemia, a slow infusion of 50ml of dextrose-containing fluid (ideally about 10% dextrose) can be administered over about 5-10 minutes. This provides a safe volume of resuscitation fluid and can be repeated 20-30 minutes later if required. Large volumes of fluids can very easily induce pulmonary edema in crias. Boluses of hypertonic dextrose solutions (e.g. 50% dextrose) may worsen any pre-existing CNS problems and should generally be avoided.[11] Further fluid therapy decisions can be made once the results of biochemistry are available. In general, maintenance fluid therapy requirements are about 50% higher than those of adults. The author uses a guide of 3.3 ml/kg/hr for maintenance in crias and this can be administered by continuous rate infusion using an infusion pump. When crias are more active and able to get up and nurse for themselves, bolus administration of fluids every 2-4 hours can be easier to manage since the cria does not need to be separated from its dam. When administering glucose or dextrose to sick crias, it is important to monitor blood glucose concentrations serially as crias are prone to a hyperglycemic hyperosmolar syndrome.[12] For this reason, checking blood glucose concentration is warranted from the perspective of assessing and managing hyperglycemic crias.

Many sick neonates have FPT and plasma will be required to address this. Plasma should be administered at 20-25ml/kg but higher doses than this are regularly given owing to the average size of plasma bags available commercially (about 300ml). Plasma is much more effective than crystalloid fluids at expanding volume in very ill neonates, since these crias often have increased endothelial permeability due to inflammatory changes as a result of sepsis and endotoxemia. Clinically, the observed response during a plasma transfusion can be quite rewarding. The safest method of plasma administration is intravenously using a blood-products administration set.

As part of ongoing monitoring, crias should be weighed twice daily to assess response to treatment and adequacy of nursing. If crias are well enough to stand, they should be assisted to nurse every couple of hours. If bottle-feeds are required, aim to feed 10-15% of the cria's bodyweight over 24 hours, divided into feedings every 2 hours ideally initially. Feeding small amounts frequently more closely simulates nursing behaviour and reduces the potential for overfeeding resulting in milk entering the first stomach compartment (C1). Larger volume feeds that are safe in calves appear to result in a higher occurrence of C1 acidosis in crias. This could be due to anatomical differences between species. The esophageal groove is poorly developed in camelids and consists of only a single lip rather than two.[13] It is also possible that the functional and/or anatomical barriers to backflow of ingesta may be less well developed in camelids. Bottle-feeding is preferred, although tube-feeding may be necessary initially. Repeated tube-feedings may cause esophagitis in the author's experience.

Recumbent crias may need assistance to maintain sternal recumbency. If this is not possible, crias should be turned regularly. Warming aids should be refreshed as necessary and ensured that they are not too hot. Infectious causes of diarrhea may result in crias needing to be kept in an isolation unit. Although salmonellosis is rare in camelids, some of the causative agents are zoonotic such as *Cryptosporidia* spp., so that attention to personal hygiene is essential when caring for these cases.

REFERENCES

[1] Sumar JB. J Reprod Fertil Suppl 1999;54:169. [2] Davis GH, et al. Anim Reprod Sci 1997;46:297. [3] Gerspach C. International Camelid Health Conference for Veterinarians, OSU 2006;90. [4] Bravo PW, et al. Anim Reprod Sci 2008. [5] Weaver DM,

et al. Am J Vet Res 2000;61:738. [6] Bravo PW, et al. Small Ruminant Research 1997;26:145. [7] Garmendia AE, et al. Am J Vet Res 1987;48:1472. [8] Wernery U. J Vet Med B Infect Dis Vet Public Health 2001;48:561. [9] Morin DE, et al. J Dairy Sci 1995;78:1713. [10] Bedenice D. Vet Clin North Am Food Anim Pract 2009;25:407. [11] Vaala WE, et al. In: Smith BB, ed. Large An Int Med, 2002:294. [12] Cebra CK. J Am Vet Med Assoc 2000;217:1701. [13] Fowler ME. In: Med & Surg of SACs. 1998:305.

Emergency Medical Management of Camelids

Christopher K. Cebra, VMD, MA, MS, ACVIM (LAIM)
Corvallis, OR

INTRODUCTION

New World Camelids can be very frustrating cases. They often present in critical condition, the severity of their condition is often underappreciated, and they often die in spite of the best-intentioned treatments. Understanding them better can help in all these areas. Strategies to recognize diseases sooner allow for earlier, and often more successful, veterinary interventions. Knowledgeable assessments allow for better appreciation of all that is wrong, and hence allow for more accurate and comprehensive treatments. Knowing more about camelids also allows for more accurate treatments and also hopefully allows the health care team to avoid serious mistakes – extrapolation from other species is helpful in some ways, but disastrous in others.

Camelids' recent economic devaluation is adding to the older frustrations. Previously, veterinarians had considerable financial leeway to evaluate and treat camelids. Now the budget is more likely to be restrictive and some owners are more likely to delay the call for veterinary attention until the condition becomes truly grave, even if they recognize it in the early stages.

GETTING STARTED

The veterinary technician can play many key roles in the assessment and treatment of sick camelids. These may begin before the patient even arrives, and include getting a phone history and securing a quiet, warm, protected place for initial assessment and interventions. The history can be critical in knowing what to expect and set up for. Size, age, and mobility or lack thereof are all valuable pieces of information. Pads, warming equipment (blankets, heat lamps, hot water bottle, and forced-air blankets), cooling equipment (fans and water or ice baths), support systems (slings, pads, or flotation tanks), IV fluid set-ups, and oxygen lines can be set up in advance, if indicated.

The next task is getting the patient into the clinic. Recumbency is often the first sign of illness noticed by the owners, and ambulatory camelids often resist leading or gentle driving by lying down. Smaller youngsters and even some adult alpacas may be physically carried, but for the larger individuals, gurneys, hay carts, and rescue glides may be necessary.

A quiet, protected area is often all that is needed for examination. More vigorous llamas may require restraint chutes or sedatives, but the need for restraint of any type decreases in critical cases, and overhandling or over-sedation may lead to struggling, collapse, and airway compromise.

INITIAL CRITICAL ASSESSMENT

The two most critical initial signs are copious bleeding and an obstructed or labored breathing pattern. Hemorrhage is luckily one of the areas where extrapolation from other species can go a long way. Weakness, tachycardia, and mucus membrane pallor all are suggestive, and determination of PCV, blood hemoglobin, or red cell numbers will provide supportive evidence in all but peracute cases. Immediate interventions such as applying compression are pretty much the same in camelids as in other species. The most common cause of acute severe hemorrhage is a bite wound, usually from a dog or predator. Dominance fighting, gun or arrow shots, motor vehicle accidents, obstetric or neonatal bleeding, and impalement or laceration represent less common possibilities. If a clinic has a high camelid case load, having access to a blood donor camelid is very useful.

Obstructed or labored breathing is far more common and frequently life-endangering. Camelids are often described as obligate nasal breathers; open-mouthed breathing is therefore a sign of serious disease or distress. Tachycardia, tachypnea, and mucus membrane cyanosis are all suggestive; arterial blood gas analysis can by confirmatory, but unless that sample can be obtained quickly and easily, it is often better to start corrective interventions based on suspicion. It should be noted that nostril flare is a reflex, and does not signify air movement. Congenital airway malformations,[1] snake bites,[2] nasal bots, tumors, abscesses, and stress- or sedative-induced dorsal displacement of the soft palate are all possibilities for obstructed nasal breathing.

Hypoxic camelids, as well as camelids sick for other reasons, may exhibit a characteristic behavior that I have dubbed "Dying Llama/Alpaca Syndrome [DLS/DAS]." The major features of this are a bending back of the head and neck over the back and a gradual slide from sternal to lateral recumbency. The eyes frequently roll back into the head, and some twitching or struggling may be seen. Spontaneous regurgitation of gastric contents may occur, especially if the forestomach is full. Some relief can be obtained by maintaining the camelid in sternal recumbency with the head and neck gently extended forward.

If the obstruction is severe enough and likely to be long-lasting, tracheostomy may be indicated. Camelids have long, straight necks, which affords many possible sites. With stress- or sedative-induced palate displacement, time and minimal handling may allow the camelid to self-correct. If it progresses to DLS/DAS, gentle correction of the head position may be helpful. With any respiratory distress without complete airway obstruction, supplemental oxygen delivered by nasal catheter is usually helpful.

Pneumonia, pulmonary edema, neoplasia, heat stress,[3] and diaphragmatic paralysis[4] are all relatively common causes of non-obstructive dyspnea in camelids. Hernias are rarer. Diaphragmatic paralysis may be recognized by observing the asynchronous movement of the rib cage and the abdominal muscles. Normally, the abdominal muscles contract as the ribcage relaxes, pushing the abdominal viscera against the lungs to complete expulsion of air; they then relax during inspiration, allowing the diaphragm to push against the viscera and facilitate lung expansion. Thus, the abdomen should expand during inspiration and contract during expiration. The opposite occurs with diaphragmatic paralysis. The other pulmonary conditions typically cause tachypnea and exercise intolerance, with varying degrees of obtundation, dyspnea, and cyanosis. In they are severe, nasal oxygen should be administered as soon as possible, and further procedures should only be done if the patient can tolerate them. It is important to remember that rapid breathing can also be a sign of stress, hyperthermia, or metabolic acidosis.

EVALUATION, CATHETERIZATION, AND SAMPLE COLLECTION

Once immediate hemorrhagic and hypoxic conditions are addressed, everyone can relax a bit and progress methodically into the next phases of evaluation and stabilization. Camelids are prey species, and heavy handling can lead initially to struggling and later to a freezing or death reflex. Struggling can exhaust the critical patient, alter their blood profile, and complicate medical procedures. The death reflex is essentially DLS/DAS. Fighting these reflexes leads to escalation, and many sedatives can contribute by decreasing cardiac output and loosening the epiglottis' hold on the soft palate. Generally, it is important at this time to limit interventions to the most critical ones, perform those procedures quickly and skillfully, use light sedatives, such as butorphanol, if necessary, and allow the initial treatments to affect some improvement before moving on to the more stressful or less essential procedures. The "Anthill" approach is often counterproductive and might even kill the patient.

The most important procedures are the physical examination and usually placement of a venous line and collection of blood or other fluid samples. Important components of the physical examination are similar as in other species, and in the emergency case are often more directed to establishing the presence and severity of shock than to identifying the primary disease. Rectal temperature determination is helpful for diagnosing hypo- and hyperthermia, with hypothermia (< 37.5C or 99.5F) suggestive of shock and associated with poor outcomes.[5] Rapid determination of body temperature is especially important in neonates and in the colder months, when hypothermia is more common. Heart rates are quite variable in normal camelids, some sources list fairly high upper values, but generally rates >72 bpm suggest stress or shock. Respiratory rates are also variable, often influenced by fleece cover and environmental temperatures.

Camelids rarely develop gas bloat, so gastric intubation is not usually on the emergency list. The procedure can be useful to obtain a fluid sample for pH (forestomach acidosis), toxin (andromeda toxin and others), or chloride (obstructions) analysis, but in those cases, collection can be delayed a short time until the patient has higher stress tolerance. Assessing body condition, abdominal contour, colic or stranguria, neurologic deficits, and other organ specific signs should be accomplished during the examination with minimal stress.

The jugular vein is the most common site for venous access, and also may be hard to find, due to thick overlying fleece and skin. Shock makes it more difficult, and the blind nature of jugular sticks makes the likelihood of accidental carotid puncture higher than in horses or domestic ruminants. Bleeding after venipuncture or artery puncture may make subsequent attempts difficult, so it is common to place a jugular catheter as early as possible, and to draw all subsequent venous samples through the catheter. The best location for jugular venipuncture is usually fairly high in the neck on the right side, allowing enough room for the hub of the catheter and fluid line connector to remain below the mandible. The jugular is reasonably straight and superficial there, and runs just lateral to the trachea. The carotid is further lateral and deeper, so a forward approach to the ideal site is less risky than a lateral one. A quick look with the ultrasound may be helpful in identifying the depth and location of the jugular vein and also the location of the carotid, especially in shocky camelids or those with thick skin. The left jugular may be used, especially if the right is damaged or inaccessible, but the risk of esophageal trauma is greater.

The animal is restrained in sternal or lateral recumbency with the neck held straight. The site is best clipped and prepared. Some owners are reluctant to authorize clipping for show purposes, so that conversation may need to occur. Others appreciate it, if the clipped fleece is collected and returned to them. Subcutaneous local anesthetic is used sparingly (0.5 cc), because larger amounts compress the jugular without forming a visible bleb. A full-thickness stab incision with a #15 blade or large needle oriented in the eventual direction of the catheter (about 30 degrees off the skin) will facilitate passage of the catheter. Lifting the skin as much as possible while making the stab incision reduces the chance of accidentally lacerating a vessel. In shocky camelids, the jugular should be held off for at least 30 seconds to allow the vein to fill. Slowly introducing smooth-sliding Teflon catheters work well, as will others that minimize tissue drag. The shockier the camelid and more upright the neck, the poorer the hub will fill with blood after successful venipuncture, so having a second person apply negative pressure with a 20cc syringe to an attached and heparinized saline-filled extension set during the catheterization procedure may allow better recognition of success before the catheter passes through the far wall of the vein. If used, neck bandages should be placed loosely, so that at least two fingers comfortably fit under the bandage. If the bandage is too tight, the camelid may develop pharyngeal edema, dysphagia, and respiratory distress.

The major complications of catheter placement relate to accidental vessel puncture and too-tight neck bandaging. Hitting the vein on the first try and quickly placing the catheter is the best way to avoid excess bleeding. If the jugular or especially the carotid is accidentally encountered, applying pressure to the area for up to five minutes is helpful in preventing a large hematoma. These hematomas or vascular spasms confound future venipuncture attempts at the same site or lower in the neck, and may induce DLS/DAS. If the hematoma is bad enough, it may be best to wait an hour or more for the next venipuncture attempt.

If deemed necessary, arterial blood gas samples are best collected from the medial saphenous vein, which is usually easily palpable coursing upward from the stifle on the medial thigh. The median artery, which runs proximal from the elbow along the cranial edge of the medial humerus, is another possible site, mainly in neonates. It should be noted that camelids are adapted to high altitude conditions, including that their hemoglobin has a high affinity for oxygen. Even their venous blood can appear reasonably bright red at times, and they often look clinical brighter than horses or ruminants with the same level of hypoxemia. This bright blood can sometimes be confusing during catheterization attempts; removing the syringe to see if the blood is under pressure or running a sample through a blood gas analyzer will help confirm whether the catheter is in the artery or the vein.

Blood work and other fluid assessments play a myriad of important roles in the diagnosis of sick camelids. From a critical care standpoint, the assessments of dehydration and shock play the earliest roles. Camelids are desert-adapted, have some pigmented membranes, alter the oral membrane's tackiness through open-mouthed breathing, and have a wide range of reported heart rates, so physical assessments of dehydration and shock are sometimes lacking or difficult to interpret. Blood indicators can also be equivocal: hemoconcentration and increased protein concentrations are often used to diagnosis dehydration, but sick camelids frequently have low-grade anemia and moderate to severe hypoproteinemia regardless of hydration status. Blood urea nitrogen (BUN) and creatinine are also frequently used as indirect indicators of hydration and renal blood flow. As in ruminants, BUN can drop with acute anorexia due to gastric microbial consumption of urea. Creatinine comes from muscle mass, and hence may be low in emaciated camelids. Therefore, with all these blood components, high values support a diagnosis of dehydration in most sick camelids, but normal or low values do not exclude it.

Camelids have diabetic-like qualities, which means they develop stress hyperglycemia easily, and mainly clear their excess glucose in the urine.[6] Under normal conditions, they will replace the lost water by drinking, but sick camelids are often unable to do this. Sodium climbs with the water loss, and thus hypernatremia is a common and important indicator of dehydration. Lactate is also a good indicator of dehydration and shock, and is becoming more widely available as a test.

TREATMENT AND IATROGENIC COMPLICATIONS

Confirming the presence of dehydration, shock, or blood loss usually justifies fluid therapy. This is an important intervention in critical camelids, but to be most effective, it requires good understanding of a number of camelid-specific features. Aspects of missing the indications for fluid therapy have already been addressed. Equally important is the need not to create harm. Camelids' trend towards hypoproteinemia when sick dictates conservative crystalloid resuscitation or the use of colloids. Slow correction of dehydration, through limiting cumulative volumes of crystalloids to no more than 50 ml/kg/d and avoiding shock doses, may be safer than rapid dosing. Of course, camelids with high to normal blood protein concentrations will tolerate boluses and high fluid volumes better, so decisions should be made on a case-by-case basis.

Fluid overloading is probably the most common iatrogenic complications in treated camelids in North America. It results in ascites and pulmonary, pharyngeal, or cerebral edema, leading to progressive abdominal distention, obtundation leading eventually to seizures, and respiratory distress. Monitoring blood protein, using conservative fluid volumes, administering colloids, and monitoring physical exam parameters are all helpful to avoid this condition. Decreasing fluid rates, using colloids, giving intranasal oxygen, and administering mannitol or other diuretics are the most useful treatments. Camelids that succumb to this frequently have evidence of pulmonary and cerebral edema.

Due to the high prevalence of hyperglycemia and hypernatremia in sick camelids, and the shifts in body water related to them (from tissue to blood to urine), hypertonic glucose or sodium should only be given to camelids after the need is established. Nephrotoxic drugs, such as aminoglycoside antibiotics and non-steroidal anti-inflammatories should be used with caution in camelids that might be dehydrated. Delaying those treatments for an hour or two while IV fluids run is rarely harmful.

Camelids' unusual energy metabolism is thought to contribute to the hyperglycemia and hypoproteinemia seen in many sick individuals.[6] These appear to signify that camelids' use glucose poorly, and tend to consume body protein when they go off feed. To improve glucose utilization and potentially spare body protein, it is not uncommon for sick camelids to be treated with partial parenteral nutrition solutions and insulin.

Avoiding the major iatrogenic complications - the stress response to overhandling, over-sedation, carotid punctures and cervical hematomas, underappreciation of dehydration, and overhydration – will increase the success of camelid critical care. Combining this with effective management and treatment methods will increase it even further.

SELECTED REFERENCES

[1]Fenwick et al. J Am Vet Med Assoc 1982;181(11):1409. [2]Sonis et al. J Vet Intern Med 2013;27(5):1238. [3]Norton PL, et al. Can Vet J 2014;55(10):992. [4]Byers S, et al. J Vet Intern Med 2011;25(2):380. [5]Cebra et al. J Am Vet Med Assoc 2007;230(1):94. [6]Cebra et al. Am J Vet Res. 2001;62(5):682.

Cannabinoids in the Neurologic Patient

Christine Weaver, LVT, VTS (Neurology)
Leesburg, VA

The use of medicinal cannabis has a long history in human civilization. First evidence of its use dates to 5000 years ago in China. It was also reportedly used throughout history in India, the Middle East, South American, South Africa and Southeast Asia. It was used for various ailments including insomnia, fevers, malaria and surgical analgesia. It wasn't until the 19th century that medicinal cannabis reached the western world where it was used to treat hysteria, delirium, epilepsy and migraine, among other ailments.

THE ENDOCANNABINOID SYSTEM

One of the kind of compounds within cannabis, cannabinoids, work on the body through the endocannabinoid system. The endocannabinoid system is ubiquitous is all mammals. It is a group of endogenous cannabinoid receptors located in the mammalian central and peripheral nervous systems, consisting of neuromodulatory lipids and their receptors, located throughout the body. It is made up of endocannabinoids, the G-protein coupled cannabinoid receptors, the endogenous ligands that activate them, and the enzymes that are responsible for endocannabinoid biosynthesis, and well as their inactivation and degradation. The endocannabinoid system has important physiological functions in the CNS and PNS. Research has shown that it is involved the regulation of feeding behavior and modulating energy metabolism and control metabolic function[1]. It is also involved in neuromodulation and immunmodulation[2]. Two G-Protein coupled cannabinoid receptors, CB_1 and CB_2, have been identified.

CB_1 receptors are expressed in the central and peripheral nervous systems, but are found to have higher expression in the basal ganglia nucleus and moderately high expression in the hippocampus, cerebellum and neocortex in the brain. It is found in much lower levels in the periphery but has been detected in a variety of circulating immune cells. The function of CB_1 receptors is thought to be related to neuromodulation and may be anti-inflammatory effects[2].

CB_2 receptors are expressed primarily in the periphery and on various circulating and resident immune cells. Found in particular abundance in the peripheral organs with immune function (macrophages, spleen, tonsils, thymus, leukocytes, as well as the lungs and testes). The agonism of these receptors is typically associated with a dampening of their pro-inflammatory activities; inhibition of the release of inflammatory mediators (nitric oxide, interleukin 2, TNF-α), inhibition of activation of the cell mediated immune processes and inhibition of proliferation and chemotaxis.

Cannabinoids are the group of compounds that act on the cannabinoid receptor cells in the body. Cannabinoids can be plant based (phytocannabinoids), endogenous (endocannabinoids) or synthetic. Endocannabinoids are the endogenous lipids that are the physiologic ligands for the cannabinoid receptors. Endocannabinoids, as the prefix suggests, are cannabinoids that our bodies make for ourselves, to act of the cannabinoid receptors. Five endocannabinoids have been identified, but there are two main endcocannabinoids – 2-arachidono glycerol (2-AG), and N-arachidonoyl ethanolamide (anandamine, AEA). These two endocannabinoids have been the most extensively studied and understood. Phytocannabinoids are cannabinoids that occur naturally in the cannabis plant and are found in abundance in the resin of the plant. Of over 480 different compounds in the cannabis plant, 66 are cannabinoids. Not all phytocannabinoids are derived from the cannabis plant.

CANNABINOIDS IN NEUROLOGIC DISEASE

Neurodegenerative Disorders

Cannabinoids may impact the neurodegenerative process by two main pathways; one, the regulation in the activity of neurons, and two, brain immune function[2]. CB_2 is primarily associated with immune modulation and is expressed on circulating and resident immune cells. As such, the agonism of CB_2 is associated with the dampening of the pro-inflammatory activities of these cells and inhibits the release of inflammatory mediators, and inhibition of the activation of the cell mediated immune processes.

There is evidence to suggest that cannabinoids inhibit the excitotoxic neurotransmitter glutamate and counteract oxidative damage to dopaminergic neurons and as such, could be useful in the treatment of Parkinson's disease in humans. The antioxidant, anti-inflammatory and anti-excitotoxic properties of cannabinoids make them neuroprotective and can be used in a variety of neurodegenerative disorders including Parkinson's disease and multiple sclerosis[3]. These protective effects exerted by CB1 and CB2 have also been seen in amyotrophic lateral sclerosis, a disease most closely mimicked in the veterinary world as Degenerative Myelopathy.

The current understanding is that cannabinoids impact the neurodegeneration process through neuromodulation and immunomodulation. The thought is that the dysregulation of the endocannabinoid systems may be responsible for the symptoms of neurodegenerative diseases. Studies have highlighted the potential role of endocannabinoids may play in protecting the cells in the disease process and/or symptoms[2].

Studies have shown that the activation of CB1 has been effective in limiting cell death following excitotoxic lesions. Activation of CB2 receptors dampens inflammatory immune response to diseases[2]. This could hold promise in treatment of chronic disease states such as degenerative myelopathy, acute injury such as traumatic brain injury[4], or inflammatory conditions. The endocannabinoid system also helps regulate energy balance and metabolism, immune function and neuromodulation. This leaves a great deal of potential for treatment of various diseases.

Seizures

The CB^1 receptor has been implicated in the regulation of neuronal excitability. Studies have shown that the endogenous cannabinoid system is rapidly activated after seizure activity[5], which provides neuroprotection against excessive neuronal activity by reducing excitability of pyramidal neurons and activating intracellular cascades. Exogenous cannabinoids have been shown to effectively control seizures in a rat seizure model[5,6]. As such, it is thought that CB_1 activation with exogenously applied cannabinoids can decrease neuronal excitability and inhibit glutamatergic transmission[4]. In some animals, the antagonism of the CB^1 receptor resulted in protracted seizures that were consistent with status epilepticus[5].

There has been some controversy regarding the potential anti-seizure properties of Cannabis sativa. However, as far back at 1981, the anticonvulsant nature of the phytocannabinoid cannabidiol showed potential in controlling grand mal (now generalized tonic-clonic seizure), cortical focal and complex partial seizures[7].

More recently cannabidiol has been shown in rodent models to reduce the percentage of animals experiencing the most severe seizures and significantly decreased the mortality as a result of seizure[8]. Cannabidiol has the added benefit of not possessing the psychotropic effects.

Cannabidivarin is another non-psychoactive cannabinoid that has demonstrated anti-seizure effects in a number of experimentally induced seizure models and had the added benefit of minimal effects on normal motor function[7].

Pain

The anecdotal reports of the benefits of medicinal marijuana for pain has been well documented. Specifically, the cannabinoid CBD has been shown to suppress chronic inflammatory and neuropathic pain[10]. The use of cannabinoids has been shown to suppress allodynia (painful response to non-painful stimulus) and hyperalgesia forms of neuropathic pain[11]. Both CB1 and CB2 receptors are involved in the mediation of inflammatory pain[10].

There are experimental studies that have shown that cannabinoids work synergistically with opioids, allowing for lower doses of opioids to be used and decreasing the side effects of short term and or chronic opioid therapy[10].

Cannabinoids can also decrease inflammation it has been shown that CB2 is upregulated in response to inflammatory cues[2].

Studies have shown that the activation of CB1 has been effective in limiting cell death following excitotoxic lesions[2]. Activation of CB2 receptors dampens inflammatory immune response to diseases[2]. This could hold promise in treatment of chronic disease states such as degenerative myelopathy, acute injury such as traumatic brain injury[4], or inflammatory conditions. The endocannabinoid system also helps regulate energy balance and metabolism, immune function and neuromodulation. This leaves a great deal of potential for treatment of various diseases particularly in the field of neurology.

REFERENCES

[1]Pagotto U et al., Endocr Rev. 2006; 27(1):73. [2]Scotter EL et al., Br J Pharmacol. 2010; 160(3):480. [3]De Lago E, et al., CNS Neurol Disord Drug Targets 2007; 6(6):337. [4] Croxford JL et al., CNS Drugs 2003; 17(3):179. [5] Wallace MJ et al., J Pharmacol Exp Ther 2003; 307:129. [6] Maricano G et al., Science 2003; 302:84. [7] Karler R, et al., J Clin Pharmacol 1981;8-9 suppl; 437S. [8] Jones NA et al., Seizure, 2012; 21(5): 344. [9] Hill AJ, et al., British Journal of Pharmacology 2012; 167(8): 1629. [10] Xiong, W, et al., J Exp Med. 2012; 209(6): 1121. [11] Elikottil J, et al., J Opioid Manag. 2009;5(6):341.

Canine Degenerative Myelopathy: Perspectives on Therapy Approaches

Joan R. Coates, DVM, MS, DACVIM (Neurology)
Columbia, MO

REVIEW CANINE DEGENERATIVE MYELOPATHY

Canine degenerative myelopathy (DM) is a progressive adult-onset neurodegenerative disease that has many similarities to human amyotrophic lateral sclerosis (ALS – Lou Gehrig's disease) and may serve as an important, novel disease model for therapy development. Mutations in the superoxide dismutase 1 gene (*SOD1*) alter the conformation of the SOD1 protein and underlie most cases of canine DM and some forms of human ALS. Canine DM was first described by Averill in 1973 as an insidious, progressive, general proprioceptive (GP) ataxia and upper motor neuron (brain and spinal cord = UMN) spastic paresis of the pelvic limbs ultimately leading to paraplegia and necessitating euthanasia.[1] The disease was termed "degenerative myelopathy" because of its histopathologic nature as a nonspecific degeneration of spinal cord tissue of undetermined cause. In 1975, Griffiths and Duncan also reported hyporeflexia and nerve root involvement, and they termed the condition chronic degenerative radiculomyelopathy.[2] Similarities between the canine and human nervous systems and the homogeneity in onset and clinical progression of canine DM will facilitate translation of therapies into human applications.

Signalment

There is no sex predilection. Degenerative myelopathy affects older, adult dogs. Ages at onset range between 8 and 14 years. Mean age at onset is 9 years in large dogs.[1] In the Pembroke Welsh Corgis (PWC) reported mean age of onset is 11 years; presumably due to breed size differences.[3] Although first described in German Shepherd dogs, DM is now being recognized as a common neurologic problem in many pure breeds and mixed.[4]

Clinical Spectrum

Dogs with DM exhibit a predictable pattern of clinical signs that begins with UMN paresis and GP ataxia in the pelvic limbs, progress to LMN paraparesis, and then spreads to involve the thoracic limbs and brainstem.[3,5-8] The clinical course of DM is rather uniform after the presumptive diagnosis but ambulatory (paresis/ataxia) to nonambulatory status occurs within a median time of 10 months (95% CI, 9-12; unpublished data, Kanazono and Coates, 2013 ACVIM Abstract) from onset of signs. Pet owners usually elect euthanasia when their dogs can no longer support weight in their pelvic limbs. The pet owner can care for smaller dogs over a longer time. The median disease duration in the Pembroke Welsh Corgi was 19 months.[3]

Four disease stages in the progression of DM-affected dogs have been defined (Figure 1).[5] Asymmetric spastic weakness and general proprioceptive ataxia in the pelvic limbs (*stage 1*) progresses to paraplegia (*stage 2*) within 1 year from onset of signs. At disease onset, spinal reflexes are consistent with UMN loss. Dog owners often elect euthanasia when their dog becomes paraplegic. When euthanasia is delayed, weakness spreads to the thoracic limbs (*stage 3*) and lower motor neuron (nerve and muscle = LMN) signs emerge as flaccid quadriplegia, widespread muscle atrophy, difficulty swallowing and inability to bark (*stage 4*).

Early Disease – The earliest clinical signs of DM are GP ataxia and mild spastic paresis in the pelvic limbs. Worn nails and asymmetric pelvic limb lameness are apparent upon physical examination. Asymmetry of signs at disease onset is frequently reported.[1] At disease onset, spinal reflex abnormalities are consistent with UMN paresis localized in the T3 to L3 spinal cord segments. Patellar reflexes may be normal or exaggerated to clonic; however, hyporeflexia of the patellar reflex has also been described in dogs at similar disease stage. Flexor (withdrawal) reflexes may also be normal or show crossed extension (suggestive of chronic UMN dysfunction). Often dogs progress to nonambulatory paraparesis and are euthanized during this disease stage.

Stage	Neurologic Signs
1 Early	**UMN Paraparesis** • Progressive general proprioceptive ataxia • Asymmetric spastic paraparesis • Intact spinal reflexes
2 Early	**Nonambulatory Paraparesis to Paraplegia** • Mild to moderate loss of muscle mass • Reduced to absent spinal reflexes in pelvic limbs • +/- urinary and fecal incontinence
3 Late	**LMN Paraplegia to Thoracic Limb Paresis** • Signs of thoracic limb paresis • Flaccid paraplegia • Severe loss of muscle mass in pelvic limbs • Urinary and fecal incontinence
4 Late	**LMN Tetraplegia and Brain Stem Signs** • Flaccid tetraplegia • Difficulty with swallowing and tongue movements • Reduced to absent cutaneous trunci reflex • Generalized and severe loss of muscle mass • Urinary and fecal incontinence

Figure 1: Disease grading system for *SOD1* associated canine DM with clinical severity scores ranging from 1-4; **1-2** being early disease stages and **3-4** being late disease stages.

Late Disease – If the dog is not euthanized early, clinical signs will progress to LMN paraplegia and ascend to affect the thoracic limbs. Flaccid tetraplegia occurs in dogs with advanced disease. The paresis becomes more symmetrical as the disease progresses. Lower motor neuron signs emerge as hyporeflexia of the patellar and flexor reflexes, flaccid paralysis, and widespread muscle atrophy beginning in the pelvic limbs as the dogs become nonambulatory. Widespread and severe loss of muscle mass occurs in the appendicular muscles in the late stage of DM. Most reports attributed loss of muscle mass to disuse but flaccidity in dogs with protracted disease suggests a peripheral neuropathy component.[6,9] Cranial nerve signs include swallowing difficulties and inability to bark. Onset of urinary and fecal continence can vary during disease course.

Genetic Basis /Mode of Inheritance:

Awano *et al.* identified a *c.118G>A* transition in Exon 2 of the *SOD1* gene that predicted an E40K missense mutation, which underlies most cases of canine DM.[6] Not all *SOD1:c.118A* homozygotes develop clinical signs. Initially DM appeared to be an autosomal recessive disease with incomplete penetrance; whereas most human *SOD1* mutations in human ALS are autosomal dominant. Thus, homozygosity for the E40K mutation in SOD1 is a major risk factor for canine DM. More recently, DM has also been histopathologically confirmed in few heterozygous dogs.[4] The occurrence of DM in a heterozygote seems plausible since most human *SOD1* mutations cause dominant ALS. Nonetheless, the age at onset for many dogs has exceeded the mean life expectancy of dogs indicating that DM has an age-related, incomplete penetrant mode of inheritance.

We have identified a second *SOD1* missense mutation (*c.52 A> T*) that was homozygous in an affected Bernese Mountain dog.[10] This mutant allele *c.52T* allele appears to be restricted to the Bernese Mountain Dog breed where it is less common than the *c.118A* allele.[4] We have identified a few *c.52T + c.118A* compound heterozygous Bernese Mountain dogs with DM.[4] This finding serves as a reminder that direct DNA tests indicate the presence or absence of disease-causing alleles but cannot be used to rule-out a diagnosis because other sequence variants in the same gene or in a different gene might produce a similar disease phenotype.

Diagnostic Approach

Clinical Examination – Accurate antemortem diagnosis is based on recognition of the pattern progression of clinical signs supported by inclusionary and exclusionary diagnostic testing. A careful neurologic examination is fundamental for developing a diagnostic approach. Paw replacement (proprioceptive positioning) is a very useful test that distinguishes between orthopedic and neurologic diseases because it does not require weight-bearing. Animals with orthopedic disease will not have paw replacement deficits. Lack of paraspinal hyperesthesia is a key clinical feature of DM that distinguishes it from differentials of compressive myelopathy or inflammatory disorders.

Neuroimaging – An antemortem diagnosis of canine DM is based on ruling-out spinal cord compressive diseases. A presumptive diagnosis of DM often is based on lack of clinically relevant compressive myelopathy as determined by magnetic resonance imaging (MRI). If MRI is unavailable, computed tomography (CT)/myelography also can be performed. Often imaging reveals disk protrusions that can confound a diagnosis of DM. Survey spinal radiography will show no changes that reflect DM, although concurrent degenerative vertebral changes (e.g. spondylosis deformans, lumbosacral spondylosis, narrowed intervertebral disc spaces, spinal osteoarthritis) may be visualized. Ultimately the clinician must be guided by clinical experience to take into account the rate of disease progression, presence of paraspinal hyperesthesia and amount of spinal cord compression to estimate significance of the compressive myelopathy.

Electrodiagnostic – Electrodiagnostic testing is useful for detecting evidence of neuromuscular disease in DM. Early in the progression of DM, when UMN signs predominate, electromyography (EMG) and nerve conduction studies are within normal limits. Later in the disease with emergence of LMN signs, EMG reveals multifocal spontaneous activity, fibrillation potentials and positive sharp waves, in the appendicular musculature. Recordings of compound muscle action potentials (M waves) from stimulation of mixed nerves have shown decreases in amplitudes consistent with axonopathy, and temporal dispersion and decreased motor nerve conduction velocities that also signify demyelination.[6,7]

Genetic Testing – Genetic testing will complement other antemortem diagnostics for DM. Most DM-affected dogs are homozygous for the *SOD1:c.118G>A* allele.[4,6]

Neuropathology

Canine DM is definitively diagnosed by histopathology of the spinal cord and nerves. Axonal degeneration of the upper and lower motor neurons is one of the pathologic hallmarks shared by ALS and DM. The pathologic features of DM include axonal degeneration with secondary demyelination and astroglial proliferation (sclerosis) in all spinal cord funiculi, but consistently most severe in the dorsal portion of the lateral funiculus and in the dorsal columns of the middle to lower thoracic region.[1,3,6,8,11] In addition, cytoplasmic aggregates that bind anti-SOD1 antibodies are usually present in spinal cords of DM-affected dogs.[4,6] Histopathologic studies of dogs in the late disease stage with LMN signs have documented denervation atrophy in muscle, nerve fiber loss with axonal degeneration and secondary myelin loss in myelinated fibers of peripheral nerves.[7]

Because owners of companion dogs with DM have their pets euthanized at different stages of disease progression, by examining tissues from these dogs collected at the time of euthanasia, it is possible to study the disease at all stages of progression and thereby gain a better understanding of the chronological relationships between the pathologies that develop in the spinal cord, peripheral nerves, and muscles. Studies we conducted on the thoracic intercostal muscles of dogs with DM and the associated thoracic motor and sensory neurons indicate that there are significant atrophic changes in these muscles at stages of the disease in which there is no apparent degeneration of the associated motor neurons.[12,13] Neuronal cell body degeneration or loss in the ventral horn of the spinal cord is not a prominent histopathologic finding until terminal disease.[14] Significant sensory neuron degeneration also preceded any evidence of motor neuron pathology.[13] Thus, the distribution of lesions and clinical disease progression in DM are similar to that reported for the UMN-onset ALS[15]; whereby, UMN signs in DM-affected dogs later progress to

LMN signs.[5,6] The clinical spectrum of DM has now been broadened to involve both the UMN and LMN systems and is considered a multi-system disease involving both central and peripheral sensory and motor axons.[5]

Treatment

Pharmacotherapies including drugs and nutritional supplements for canine DM have been advocated; however, the efficacies of these therapies have not been established. A study involving 12 dogs administered epsilon-aminocaproic acid, N-acetylcysteine, prednisone, vitamins B, C, and E, and exercise therapy indicated that there was no benefit over physical therapy alone.[16]

Although well-controlled studies are still needed to establish treatment efficacy, physical rehabilitation may have positive effects in early DM. Kathmann et al. reported survival data from 22 DM affected dogs that received varying degrees of physiotherapy.[17] Dogs that received intensive physiotherapy had significantly longer survival times (mean = 255 days) compared to dogs that received moderate (mean = 130 days) or no physiotherapy (mean = 55 days). The physiotherapy regimen consisted of active and passive exercises without taking into account disease stage or UMN/LMN signs. Study limitations included lack of randomization and definitive diagnosis, small group size, and bias from owner perception; still results warrant further investigation into the efficacy of rehabilitation in DM-affected dogs. Cautions must be taken when rehabilitation, especially therapeutic exercise, is considered as these dogs can be easily exhausted. Inducing fatigue in an already diseased muscle can potentially hasten the ongoing disease process.

CANINE DM AS A DISEASE MODEL FOR THERAPEUTIC APPROACHES

When translating therapies for ALS from animals to people or vice versa, sensitive and objective disease markers shared across species are necessary to diagnose and monitor disease progression so that therapeutic efficacy can be accurately evaluated.[18,19] Reliable methods for measuring progression of damage to upper and lower motor neurons will provide early diagnosis of DM, determine the contribution of the lesion to the clinical syndrome and serve as objective functional measures or biomarkers of disease progression in treatment trials. As we learn more about the clinical and neuropathological spectrums of DM, functional scales, cross sectional and functional imaging, electrodiagnostic testing and CSF disease markers are being investigated as potential longitudinal measures to follow disease progress and therapeutic efficacy.

We propose that canine DM as a disease model will be beneficial in the preclinical development of treatments to more accurately predict therapies that will halt or slow progression of ALS. Furthermore, dogs with DM offer a ready clinical population on which therapies can be evaluated in an environment closely mimicking human clinical trials. This approach has proven successful in developing cancer chemotherapies in canine patients which have then been applied to humans.[20] Additionally, rodent (mouse and rat) mutants have been established as useful ALS models, as similar CNS pathology manifestations and observables are found. Canine DM represents an ideal disease model for this work on ALS because (i) a naturally-occurring *SOD1* mutation (E40K) is causative for canine DM, a common disease of older aged companion dogs, (ii) dogs have a relatively large spinal cord/brain, (iii) DM-affected dogs have a homogeneous pattern of disease progression and (iv) histopathologic findings and functional deficits in DM share similarities to some forms of human ALS. In summary DM, a *naturally-occurring* disease model, will serve as a potentially useful intermediate sized model of ALS, which could be valuable for temporal studies of disease progression and evaluation of ALS and DM targeted new therapeutic/diagnostic regimes.

One of the major challenges regarding therapies to treat the CNS is delivery of the therapy to target cells at levels that results in a therapy that is both safe and effective. Administration of therapeutic agent into the CSF, intrathecal (IT) or directly into the nervous tissue has been evaluated for several neurodegenerative diseases. Specifically, IT administration is a delivery approach for some therapies that has shown good distribution of therapeutic agents for treatments in the CNS. We will further discuss potential therapy approaches with relation to repression of the SOD1 protein.

REFERENCES

[1]Averill DR. J Vet Med Assoc 1973;162(12):1045. [2]Griffiths IR, et al. J Sm Anim Pract 1975;39(8):1309. [3]Coates JR, et al. J Vet Int Med 2007;21:1323. [4]Zeng R, et al. J Vet Intern Med 2014;28:515. [5]Coates JR et al. Vet Clin N Am Sm Anim Pract 2010;40(5):929. [6]Awano T, et al. Proceed Nat Acad Sci USA 2009;106(8):2794. [7]Shelton GD, et al. J Neurol Sci 2012;318:55. [8]March PA, et al. Vet Pathol 2009;46:241. [9]Matthews NS, et al. J Am Vet Med Assoc 1985;186(11):1213. [10]Wininger FA, et al. J Vet Int Med 2011;25(5):1166. [11]Johnston PEJ, et al. Vet Rec 2000;146(22):629. [12]Morgan BR, et al. J Neurosci Res 2013;91:1639. [13]Morgan BR, et al. J Neurosci Res 2014;92:531. [14]Ogawa M, et al. Vet Pathol 2014;51(3):591. [15]Gordon PH, et al. Neurology 2009;72(22):1948. [16]Polizopoulou ZS, et al. Acta Vet Hungarica 2008;56:293. [17]Kathmann I, et al. J Vet Int Med 2006;20:927. [18]Turner MR, et al. Lancet Neurol 2009;8:94. [19]Turner MR, et al. Amytroph Lateral Scler Frontotemporal Degener 2013;14:19. [20]Paoloni M, et al. Nat Rev Cancer 2008;8:147-156.

Nutritional Management of Cognitive Dysfunction in Dogs

Kara M. Burns, MS, MEd, LVT, VTS (Nutrition)
Wamego, KS

Human and veterinary medicine have seen tremendous advances in medical care and nutrition resulting in humans and their pets living longer. In the United States alone, it is estimated that 20 to 30 million dogs are over 7 years of age and are considered seniors.[1-3] Consequently, veterinary healthcare team members are seeing many age related changes and disease conditions such as cancer, renal disease, and cognitive decline, in their canine patients that have been associated with aging in human patients. Cognitive decline in dogs is typically manifested as behavioral changes in one or more of the following areas: disorientation (D), altered interactions with family members (I), disruptions in sleep patterns (S), loss of housetraining (H), and decreased levels of activity(A). Many pet owners attribute behavior changes in their dogs to simply be part of the aging process and believe nothing can be done to help their beloved pet. Unfortunately, some of these behavioral changes lead to owners placing their senior pet in a shelter or electing euthanasia. The combination of these behaviors, known by the acronym DISHA, is diagnosed as Cognitive Dysfunction Syndrome (CDS)[2-4], and nutrition plays a key role in helping combat this syndrome.

The veterinary healthcare team plays an important role in diagnosing behavioral changes in geriatric patients. These behavioral changes may be noticed by owners but not considered severe enough by the family to warrant a visit to the veterinarian. Oftentimes the owners think the changes are "just due to old age". The effects of aging on the brain are oftentimes subtle and progress slowly; therefore it is crucial that the healthcare team, specifically the veterinary technician, educate owners about the signs and symptoms associated with cognitive decline so these symptoms can be addressed as soon as possible. In fact, when owners were educated about the behavioral signs of DISHA, 75% stated that they had seen at least one sign. Furthermore, 37% admitted having observed one or more signs several times a week and 32% reported three of more signs. The alarming fact from this study is that only 12% actually talked to their veterinarian about a sign observed in their dog.[5] Behavioral changes may be the first indication of a physical or health problem. If a medical condition is the cause of the changes, early reporting of these changes in the pet by the owners can help lead to early diagnosis and potential medical intervention.

The acronym DISHA, as described above, is used to describe signs often seen in cognitive dysfunction. However, this is not all inclusive as there are other clinical signs and owner observations that may be associated with brain aging. The healthcare team, especially when presented with a senior pet, should institute a cognitive impairment screening checklist as part of the routine physical examination for senior pets. A checklist, given to the pet owner upon check-in to the hospital, will allow for that owner to consider whether or not their pet is exhibiting certain behaviors and is a great way to get clients thinking about their pets' behavior and imply that it may not be simply 'old age behavior', but rather a disease condition that can be treated. This behavior checklist will also allow the healthcare team to ask more in depth questions regarding the observations of the client. Oftentimes the client is observing signs in their pet relating to confusion, social relationships, activity level changes, depression/apathy, increased anxiety, sleep-wake cycle alterations, and learning and memory changes and deficits.[6] It is important to note if the patient is going to the wrong side of the door, vocalizing at night, not responding to previously learned commands, searching for an increase or exhibiting a decrease in the amount of affection sought with their owner, increasingly irritable, not wanting to play, having no interest in eating, pacing, and exhibiting compulsive behaviors (increased licking) not previously seen. (Table 1) This information will assist healthcare team members in a proper diagnosis of CDS. In conjunction with the patient history, a complete physical exam, blood work, urinalysis, and imaging, will help to rule out any medical conditions or disease processes that may affect the patient's mental attitude or behavior.

AGING AND ITS EFFECT ON THE BRAIN

Aging in most species studied to date is accompanied by the progressive accumulation of oxidative damage in many tissues. The brains of older dogs and the brains of older humans have several key features that have been observed. Both have a number of morphologic changes including cortical atrophy, ventricle widening, white matter myelin degeneration, accumulation of degraded proteins, damage to the DNA, and reduction of endogenous antioxidants.[7] The brain in older dogs also accumulates proteins within and around neurons and these proteins may be toxic to the neurons. The result is mitochondrial dysfunction and impaired energy metabolism in cortical neurons. Oxidative damage in older animals' brains may contribute to neuronal dysfunction and accumulation of neuropathologic lesions. The mitochondria use oxygen for normal, aerobic, energy production and one side reaction that may occur is conversion to free radicals within the mitochondria. As mitochondria age, the production of free radicals is increased and subsequently, the mitochondria are less efficient and produce less energy.[6] The brain is especially susceptible to the effects of free radicals due to its' high lipid concentration, demand for oxygen, and limited ability for antioxidant defense and repair. If neurodegenerative processes leading to cognitive dysfunction are influenced by oxidative damage then it is expected that antioxidants would improve learning and memory.

NUTRITIONAL MANAGEMENT OF COGNITIVE DYSFUNCTION SYNDROME

Physiologic antioxidants act to decrease the effects of free radicals in a biologic system. These antioxidants differ from those that are well known and used as preservatives to keep food fresh. Antioxidants such as vitamins E and C help to neutralize free radicals thus preventing cellular damage. Lipoic acid and l-carnitine increases the efficiency of energy conversion while decreasing the production of free radicals resulting in mitochondrial health.[7] Carotenoids and flavonoids, typically found in fruits and vegetables also aid in reducing cellular damage by inactivating free radicals. Together these create a strong antioxidant bundle which helps to counter cellular alterations resulting from oxidative stress and the neuropathology that comes from aging. Additionally, fatty acids docosahexaenoic acid (DHA) and eicosapentaeoic acid (EPA) contribute to neuronal cell health and are found in the brain in high concentrations

The addition of DHA and EPA along with the antioxidant package to a food should improve mitochondrial efficiency, decrease free radical production, and absorb free radicals thus resulting in decreased cell damage in the brains of dogs. A series of studies looked at a test food (fortified with antioxidants and mitochondrial cofactors) vs. a control food (non-fortified food) to see if the test food would decrease signs of brain aging as manifested by the following: 1) cognitive abilities and 2) improved age related changes in dogs as observed by owners in their homes. Oddity discrimination tasks (where the dogs had to chose which object was different), discrimination learning (where the dogs had to discriminate between objects, size of objects, and black and white), landmark discrimination tasks (where the dog had to use spatial information to locate objects), and in home behavioral testing (where owners had to assess changes in their dog's behavior) were studied.[6,7] In addition, behavioral enrichment was studied in conjunction with a fortified food in the landmark discrimination study.[8,9] In all studies, it was found that a food fortified with the antioxidant package described above (the test food) will delay or partially reverse age-related deteriorations in learning. Environmental enrichment (exercise, novel play toys, and continued testing) was also found to be an important factor in improving the effects of brain aging in dogs and those dogs who participated in the combined group which had fortified food and environmental enrichment both as treatments found their learning ability to be the best preserved.[2,9]

CONCLUSION

It is necessary that the veterinary healthcare team begin to identify behavioral changes in senior dogs. Dog owners are recognizing changes in their senior dogs, but are not discussing these changes with their pet's healthcare team. Utilizing a behavior checklist will open the conversation between the healthcare team and the pet owner and will help to discover certain behaviors that the owner may be seeing in their pet, but attribute to old age. These behaviors should not be dismissed simply as 'old age' especially given the fact that nutritional management has been clinically shown to increase learning ability and attentiveness to solving tasks. Also, antioxidant fortified nutrition has led to reports of a 74% reduction in house-soiling accidents and a 61% increase in enthusiasm in greeting family members by senior dog owners feeding the antioxidant fortified food. By pinpointing behaviors consistent with CDS, the healthcare team can provide a nutritional solution to alleviate Cognitive Dysfunction Syndrome and strengthen the human-animal bond by improving the quality of life for senior dogs.

REFERENCES

1. AVMA. Total pet ownership and pet population. US pet ownership and demographics sourcebook. Schaumburg, Ill: AVMA Membership & Field Services, 2002; pp 6-43.
2. Burns KM. Teaching an Old Dog New Tricks: Nutritional Management of Cognitive Dysfunction. *The NAVTA Journal*. 2010 Convention Issue. Pp: 17 – 22.
3. Christie LA, Pop V, Landsberg GM, et al. Cognitive dysfunction in Dogs. In In: Hand MS, Thatcher CD, Remillard RL, Roudebush P, Novotny B. eds. *Small Animal Clinical Nutrition, 5th ed.* 2010:715-730.
4. Ruehl WW, Bruyette DS, DePaoli A, Cotman CW, Head E, et al. Canine cognitive dysfunction as a model for human age-related cognitive decline, dementia and Alzheimer's disease: clinical presentation, cognitive testing, pathology, and response to l-deprenyl therapy. Prog. Brain Res. 1995; 106; pp 217-225.
5. US Marketing Research summary. Omnibus Study on Aging Pets. November, 2000. Data on File. Hill's Pet Nutrition, Inc.
6. Landsberg G, Hunthausen W, Ackerman L. 2003. Handbook of behavior problems of the dog and cat, 2nd ed. Elsevier, St. Louis, pp 269-304.
7. Roudebush P, Zicker SC, Cotman CW et al. Nutritional Management of Brain Aging in Dogs. Journal of the American Veterinary Medical Association 2005; 227: 722-728.
8. Dodd CE, Zicker SC, Jewell DE, Fritsch DA, Lowry SR, Allen TA. Can a fortified food affect behavioral manifestations of age-related cognitive decline in dogs? Veterinary Medicine 2003;98:396-408.
9. Milgram NW, Head E, Zicker SC, et al. Learning ability in aged beagle dogs is preserved by behavioral enrichment and dietary fortification: a two-year longitudinal studt. Neurobiology of Aging 2005; 26: 77-90.

Table 1 [2]

Clinical Signs of Brain Aging
▪ Going to wrong side of door
▪ Diminished awareness
▪ House-soiling
▪ Depression or apathy
▪ Vocalizing at night
▪ Changes in sleep/wake cycle
▪ Not responding to previously learned commands
▪ Looking for increased/decreased affection from owner
▪ Increasing irritability
▪ Not interested in playing
▪ Change in appetite
▪ Pacing/Restless
▪ Compulsive behaviors (i.e., increased licking)
▪ Altered relationships with people and other pets

Compassion Fatigue in the Veterinary Practice

Carol Fitzpatrick, CVT, VTS (Cardiology), ACC
Philadelphia, PA

"Veterinarians deal with death at five times the rate of any other healthcare profession, but we don't offer five times the training to deal with death and morbidity". (Brandt, DVM 360, 2014)

Compassion fatigue has become an important topic of discussion in veterinary medicine but what exactly is it? Compassion fatigue is defined as a secondary traumatic stress (Figley 1983); while not personally traumatized, the trauma was witnessed or affected people who cared for the trauma victim. In short, compassion fatigue occurs in people who are in caregiving positions such as doctors, nurses, and first responders. While similarities exist between burn out and compassion fatigue, they are distinguished because burn out is related to the workplace and is relieved if the workplace is changed but compassion fatigue occurs in people who are in caregiving positions, who identify as caregivers, and so can follow an affected person to a new position. Compassion fatigue can affect people on all levels: mental, emotional, spiritual, and physical. In short, a person who is suffering from compassion fatigue has run out of resilience.

Symptoms of compassion fatigue include, but are not limited to, irritability, sadness, apathy, persistent physical ailments, disturbed sleep, substance abuse, and difficulty concentrating (Compassion Fatigue Awareness Project, compassionfatigue.org). Virbac Animal Health conducted a Compassionate Care survey (2010) which revealed 90% of veterinarians and 66% of technicians reported being mentally and physically tired. In addition, 67% of veterinarians and 43% of technicians reported feelings of sadness and apathy while in the clinic.

Unrecognized compassion fatigue at work leads to increased injuries, decreased morale (increased negativity and complaints), decreased job satisfaction, and increased turnover. All of these can affect patient and client care.

Four core needs of employees have been identified (Schwartz and Porath, 2014): mental, emotional, physical, and spiritual. Research shows that improvements in one area results in improvements in productivity and job satisfaction. The more effectively the four core needs are met, the more likely it is that employees will show loyalty, engagement, and positive energy at work and their perceived level of stress will decrease.

Individuals suffering from compassion fatigue disconnect from or isolate themselves from previously supportive habits and relationships. Therefore, individual recovery from compassion fatigue includes reconnecting with the impulse to do the work, cultivating nourishing relationships and habits, developing a resiliency plan, and stress reduction. In addition, supportive, healthy workplaces can improve outcomes. Promoting continuing education for all staff, encouraging "open door" communication policies and good communication between workers, and implmenting an in-house de-brief process after traumatic events, and ensuring staff members take breaks.

RESOURCES

Compassion Fatigue Awareness Project, compassionfatigue.org; "When Helping Hurts, Compassion Fatigue in the Veterinary Profession" Kathleen Ayl PsyD, AAHA Press 2013; "Why You Hate Your Job" Tony Schwartz and Christine Porath, New York Times, May 2014.

Nursing Care of the Patient Hospitalized in Congestive Heart Failure

Kristie Garcia, LVT, VTS (Cardiology)
Hillsborough, NJ

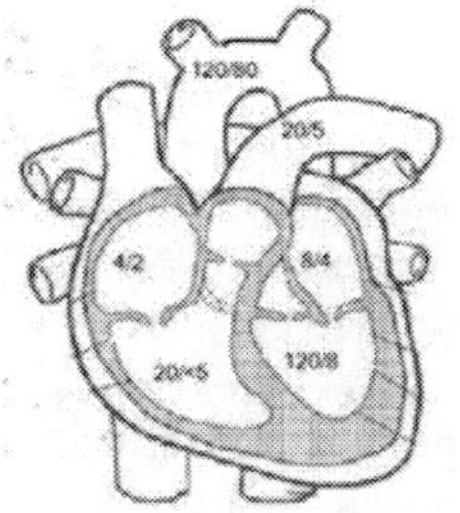

Heart failure as defined in Small Animal Cardiovascular Medicine –Kittleson/ Kienle "The end result of severe heart disease and is a clinical syndrome seen as congestion or edema, poor peripheral perfusion and /or systemic hypotension." (p.136)

Clinical Signs of Heart Failure

Left sided:

Cough, dyspnea, orthopnea, tachycardia. Fluid build up in the lungs.

Right sided:

Potentially jugular distention, Hepatic vein congestion, ascites, tachycardia

Treatment protocols

Chronic vs Acute heart failure:

Patients with chronic episodes of heart failure require and respond to less aggressive therapies.

Diuretics- help body remove the "extra" fluid building up and causing congestion.

Positive inotropes – add power to the heart.

Afterload reducers- drop blood pressure and decrease the amount of pressure the heart has to pump against.

Patients with an acute onset of heart failure are more unstable.

Diuretics, positive inotropes and afterload reducers are still the standard, however therapy is much more aggressive. These are the patients that warrant our concern and require our best nursing skills.

Heart failure medications

Diuretics - help body remove the "extra" fluid building up and causing congestion.

Angiotensin converting enzyme inhibitors (ACE I) – help to drop peripheral blood pressure to decrease the amount of pressure the heart has to pump against.

Positive inotropes – add power to the heart.

Vasodilators – also works to decrease the blood pressure and therefore decrease the amount of work the heart has to do to push blood forward.

Recognizing patient responses to treatment

Patients being treated for acute heart failure require intense monitoring as non invasively as possible. Respiratory exhaustion is real and kills.

Nursing care of the patient in congestive heart failure demands that we use all of our abilities to help each patient. While heart failure is a common illness that we need to recognize, the care of each patient is unique and should be tailored to their severity and how they are responding to therapy.

Tips for handling fragile / dyspnic patients

1. Less is more
2. Placing an IV in a rear leg requires little to no restraint in the most dyspnic patient.
3. Put Nitro in the groin / abdomen-ears are not really perfusing in these patients and therefore its not an ideal place for nitro ointment.
4. These patients temperatures will read lower- don't increase the temp in the oxygen cages as this will only add to their discomfort.
5. In the cases of saddle thrombus, pain control before heart meds!

Amplatz® Canine Ductal Occluder for the Treatment of Patent Ductus Arteriosus

Hally E. Niehaus, BS, RVT, VTS (Cardiology)
Akron, OH

INTRODUCTION

The Ductus Arteriosus (DA) is a muscular blood vessel that connects the pulmonary artery and descending aorta in utero. This communication originates from the sixth aortic arch and helps with fetal circulation when the fetus's lungs are collapsed and filled with amniotic fluid. It acts as a bypass around the collapsed lungs when oxygenated blood is supplied via the placenta. The DA normally closes on its own by the time the animal is one week of age. A Patent Ductus Arteriosus (PDA) is a congenital defect in which this communication persists post-natally. PDA is hereditary in dogs and there are many breeds that have been found to be predisposed to the defect.

PDAs most commonly result in left-to-right shunting due to the resistance of blood flow through right sided circulation being less than that of the left. This shunting results in a continuous murmur often described as a washing machine murmur. Correction of the defect is recommended for a left-to-right shunting PDA. A good prognosis, with clinical normality, is seen with correction. If left untreated, many dogs will die from left sided heart failure in the first year of life.

Right-to-left shunting can occur when the defect is so large it provides no resistance to blood flow. Because of this, pulmonary hypertension develops and pressures can equalize. A clinically significant shunt will cause deoxygenated blood to be circulated to the organs. Polycythemia can occur when the kidneys release erythropoietin in response to exposure to blood with low oxygen content. A hematocrit of 55 to 65% can be valuable in helping to compensate for the decrease in tissue oxygen. If the pet is symptomatic for polycythemia, routine phlebotomy may be needed to maintain the hematocrit around 60%. Surgical correction of the right-to-left PDA is not recommended because of the potential of catastrophic right heart failure due to loss of the pop-off valve of the PDA.

PDAs can be grouped into four classifications based on angiographic appearance (I, IIA, IIB, III). The morphology of the PDA can help determine the route of correction best suited for the patient. Options for PDA correction include surgical ligation and less invasive catheter based procedures such as coils, vascular plugs and the Amplatz® Canine Ductal Occluder (ACDO).

HISTORY

The ACDO is a device designed and developed at the Veterinary Clinical Sciences Department, College of Veterinary Medicine, University of Minnesota, St. Paul in collaboration with AGA Medical Corporation from March 2005 to December 2006. In January 2007, the device became commercially available through Infiniti Medical, LLC. The device was developed to address the differences between human and canine morphology and minimal ductal diameter that human devices did not take into account.

DESCRIPTION

The ACDO is a self-expanding device with 2-3 layers of nitinol wire mesh, each containing approximately 140 tightly woven wires. A 1-1.5mm waist separates a flat distal disc from a cupped proximal disc. Sizes of the device are typically described by the waist's diameter. Available sizes are 3mm to 10mm, in 1mm increments, along with a 12mm and a 14mm size. The diameters of the waist are 4-6mm less than the discs and the height of the cupped disc in anywhere from 2mm to 4mm. The device comes attached to a delivery cable within a hoop dispenser with a hub for flushing with heparinized saline.

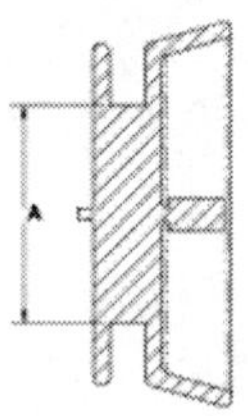

Figure 1. *Schematic of device indicating measurement of the waist diameter (A). Waist diameter determines the device size.*

Device Size	Minimum Required Inner Diameter of Delivery System (in)	Recommended Sheath (Fr)	Recommended Guiding Catheter (Fr)
3mm	.056	4 or 5	5
4mm	.060	4 or 5	6
5mm	.060	4 or 5	6
6mm	.060	4 or 5	6
7mm	.073	5	7
8mm	.073	5	7
9mm	.086	6	8
10mm	.099	7	9
12mm	.099	7	9
14mm	.099	7	9

Source: http://infinitimedical.com/products/canine-ductal-occluder/

CONTRAINDICATIONS/COMPLICATIONS

An ACDO is becoming the preferred method for PDA occlusion among veterinary cardiologists but that's not to say it is always the best. Patients that are too small may not have vessels large enough to accommodate the guiding catheter needed to place the device. In our catheterization lab, our current recommended weight for the ACDO is 3kg or larger. Any dog weighing less than this is sent for surgical ligation through our soft tissue surgeon. A new low profile ACDO prototype has been developed utilizing a smaller 4 French guiding catheter which may help rectify this limitation in the future.

Morphology of the PDA can be a deciding factor in the placement of the ACDO. Sizing of the Minimal Ductal Diameter (MDD) is essential for proper device selection. The manufacturer recommends upsizing close to 200% the measurement of the MDD on angiogram. Measurements can be taken by echocardiogram but these are two-dimensional and are not as accurate as angiographic measurements. Nevertheless, the echo can be a good tool in estimating the size of the MDD to aid in sheath selection. Given that the largest ACDO available is 14mm, a PDA with a MDD of greater than 7mm would not be a good candidate for the ACDO and surgical ligation would be recommended.

Presence of a thrombus or active infection would both be contraindications for implantation of the ACDO device. An implant of any kind would be strongly contraindicated in the presence of an infection due to the possibility of bacteria seeding within the device.

Complications with the ACDO are similar to those of coils and vascular plugs and can include pulmonary embolization, persistent ductal flow, femoral artery tear, and hemorrhage/bruising/bleeding. There is one documented case of delayed embolization of an ACDO two days post-op after the owners, against medical advice, allowed their dog to be unsupervised outdoors. The pet presented on emergency with coughing, severe tachypnea, dyspnea and lethargy. Radiographs revealed the ACDO within the main pulmonary artery at the origin of the left pulmonary artery along with a severe alveolar pattern in the right lung fields. Shortly after radiographs were taken, the dog went into cardiac arrest and cardiopulmonary resuscitation was unsuccessful. There have also been anecdotal reports of sudden death following ACDO implantation (from immediately post-op to a few weeks). An MRI revealed a cerebral infarct in one patient that developed severe neurological symptoms one hour post-recovery. Two other patients developed neurologic signs, and the reasons are unknown. Several other dogs died following surgery that had ventricular arrhythmias documented at their initial presentation.

PROCEDURE IN OUR FACILITY

The procedure for ACDO implantation is considered a minimally invasive one. As compared to surgical ligation requiring a thoracotomy and open chest surgery, the ACDO implantation is a catheter based procedure with access obtained via the femoral artery. The patient is placed in right lateral recumbency and the right inguinal area is shaved and aseptically prepped. The right femoral artery is surgically isolated and lidocaine is used to prevent constriction of the vessel when manipulated. Using a modified Seldinger technique, the needle and guidewire are advanced into the vessel, the needle is removed and a long introducer sheath is advanced over the guidewire. The size of the introducer is based on the size of the patient and estimated size of the ductus (Figure 1). The tip of the introducer sheath is positioned in the descending aorta just cranial to the left atrium with the use of fluoroscopy and the guidewire is removed. Once in place, a test injection with Ultravist, an iodinated contrast agent, is performed followed by a full angiogram using 1ml/kg Ultravist. A pigtail catheter is ideal for an angiogram, given its pigtail curl and side holes to prevent muscle staining; however, we have had poor angiographic resolution of the MDD using the pigtail with hand injection. We have found that we can obtain excellent angiographic resolution with hand injection when the angiogram is performed using the introducer sheath. We are comfortable with the safety of this technique for hand injection, particularly as the tip of the sheath is not positioned in the heart so the contrast cannot be injected directly into the heart muscle. The angiogram is played back, the PDA anatomy is evaluated, and the MDD is measured. Due to magnification from fluoroscopy, a calibration device must be used to calculate an accurate measurement. We have used a dime in the past because it has a consistent diameter of 18mm. More recently, we have been using a catheter marked with 1cm reference markers passed down the patient's esophagus. We are able to accurately measure the size of MDD by using the following formula: actual MDD in mm = MDD on angiogram in mm x 10 mm/reference marker in mm. Once the ductal anatomy type is determined, and the MDD is measured, an appropriate ACDO is selected. The sheath is repositioned through the ductus into the main pulmonary artery by passing a Judkins right catheter and guidewire through the sheath. The Judkins right catheter is used due to the angle of the tip allowing easy directing into the ductus. Once in the pulmonary artery, the guidewire and Judkins right catheter are removed and the ACDO, attached to its delivery cable, is advanced into position, after being examined and flushed with heparinized saline. The flat distal disc is deployed within the main pulmonary artery and slight tension is placed, on both the delivery cable and sheath to avoid further deployment, until the disc fits snuggly against the pulmonary ostium, or opening, of the PDA. The introducer sheath is retracted further, while the delivery cable is held stationary, to allow the remainder of the ACDO to be deployed resulting in the waist of the device spanning the MDD and the proximal cupped disc forming within the body of the ductus. Then, slight pressure is placed on the delivery cable, until slight bowing is seen, to ensure a snug fit of the device. This is necessary to form the appropriate shape of the proximal disc. After five minutes, a test injection is repeated to assess for residual ductal flow. If no flow is seen, the ACDO is released from the delivery cable and a post-release angiogram is performed to confirm occlusion of the ductus. If residual flow is seen, a test injection can be repeated every 5 minutes (after the first 5 minutes) for a total of 15 minutes in hopes that the flow will cease. If significant flow persists, a larger device can be attempted with the thought that the original device size may have been under estimated.

Once occlusion is confirmed, the delivery cable and introducer sheath are removed and the artery is double ligated above and below the entry site along with routine closure of the incision site. Thoracic radiographs are taken post-op to document placement of the device. An echocardiogram is performed one day following the procedure to assess for any residual flow and mitral regurgitation and to reassess cardiac dimensions and contractility. Three months post procedure, an echocardiogram is repeated. Mild heart muscle dysfunction is common to see post occlusion, but only those animals with moderate to severe changes are followed out further in our practice.

CONCLUSION

Over the past decade, since the development of the ACDO, it has been found to be superior to other transvascular PDA occlusion devices. The ease of use, rate of closure and the low risk of complications can be attributed to the fact that the device was specifically designed for the canine patient. When the conditions are right, in our canine patients, the ACDO is our choice for correction of the PDA. When the patient is too small or the PDA too large, we refer them to our surgery department.

REFERENCES

Bonagura, JD, Lehmkuhl, LB; *Textbook of Canine and Feline Cardiology* 1999: 505. Carlson JA et.al; *JVC* 2013; 15: 271. Gorgon SG et.al; *JVC* 2010; 12: 85. Kittleson, MD, Kienle RD; *Small Animal Cardiovascular Medicine* 1998: 218. Miller MW et.al; *JVC* 2006; 8: 109. Nguyenba TP et.al; *JVC* 2007; 9: 109. Singh MK et.al; *JVIM* 2012; 26: 85. Stauthammer CD et.al; *JVC* 2015; 17(3): 203.

Areteriosus occlusion333333333333333333333333333333333333333333333333333

Technician's Role in Equine Electrocardiogram (ECG): Resting and Exercising

Gayle D. Hallowell, MA, VetMB, PhD, CertVA, DACVIM, DACVECC, PFHEA, MRCVS
Nottingham, UK

INTRODUCTION

Electrocardiography is the diagnostic test of choice for the assessment of unexplained tachycardia, for abnormal or irregular cardiac rhythms at rest and to assess safety for ridden exercise in horses with cardiac murmurs and/or rhythm abnormalities at rest. It provides information on cardiac rate and rhythm, but due to the extensive Purkinje fibre network compared with other species, does not provide any information on cardiac size. Electrode placement and length of recording is going to significantly vary depending on the presenting signs and envisaged outcome of the results of the test. In order to form any valid judgements from the electrocardiogram recorded, a good quality electrocardiogram and knowledge of normal, variations of normal and abnormal is required. Good quality recordings are relatively straightforward to obtain at rest, but more challenging at exercise.

Usually three or four electrodes are used to obtain recordings in multiple leads. A lead is the electrical recording between two electrodes, where one is positive and one negative. Whether the deflections are positive or negative is dependent upon whether the electrical depolarization is towards (positive) or away (negative) from the positive electrode.

Electrical activity in the heart originates in the right atrium at the sino-atrial (SA) node. Electrical depolarization then spreads across both atria leading to a bifid, positive or biphasic p wave on the ECG. The atria and ventricles are electrically isolated due to their separation with fibrous tissue except for the atrio-ventricular (AV) node. Conduction through the AV node is slow and is affected by parasympathetic tone. As such conduction is even slower through this region at low heart rates. This period corresponds to PR interval. Once the electrical activity contacts the ventricular myocardium and Bundle of His, it spreads rapidly leading to the large QRS deflection. Following ventricular depolarization, the ventricular has to repolarize, which manifests as the T wave. It can be of great practical value to remember this fact when evaluating ECG's before diagnosing abnormalities. T waves are variable in appearance, especially at different heart rates; they also appear different following abnormal beats.

RESTING ELECTROCARDIOGRAMS

The resting electrocardiogram can be recorded in a variety of ways. Although 6 and sometimes 12 lead ECG are routinely used in small animal practice, due to the size and anatomy of a horse, these are not always practical [1]. Six or 12 lead ECGs result in a poor quality trace due to frequent shifts in limb position and muscle tremors (Robertson 1990). It is widely documented that the base-apex lead is sufficient to detect rhythm disturbances in the horse[2-5]. The base-apex lead configuration consists of the positive electrode located over the left apex of the heart and the negative electrode located in the right jugular groove[5]. This results in a positive P wave, large, negative QRS (although the Q wave is often not visible) and a positive or negative T when examined in lead 1. Multiple leads may allow more accurate differentiation between normal and abnormal complexes[6], which can be very valuable in some cases. In addition to the surface base-apex ECG, intracardiac or transesophageal leads[7] can be used for individual cases. Although there are stringent guidelines regarding how to record a base-apex ECG, in fact as we are only interested in acquiring information regarding rate and rhythm, electrodes can be placed anywhere that allow for all aspects of the ECG to be visualized..i.e. if the P, QRS and T waves are visible then that will be adequate. Recent technology (AliveCor ECG, http://www.alivecorvet.com) has questioned the need to have electrodes so far apart when different algorithms are used. The two electrodes in this device sit side by side on a piece of plastic that attaches to a mobile phone. The human app seems to work better for horses than the veterinary in the author's opinion and short, good quality ECG's can be obtained using this technology.

For long-term ECG recordings, using a modified base-apex ECG system can be advantageous. There are various places described to place the electrodes, but the key is that they will ideally run from the apex to the heart dorsally. Some people feel that traces are improved when they are recorded with electrodes on both the left to right sides, but there is no data in the peer-reviewed literature to support this. Practically this works best when the electrodes are placed under a roller or surcingle to help hold the electrodes in place. There are now some very nice commercial girths on the market to facilitate these recordings. The author prefers using a 'traffic light' system beneath the girth area where the electrodes are placed just dorsal to the epaxial muscles, midway between the shoulder and elbow, and then placed on the ventral midline, but many other combinations are used and work very effectively. The advantage to multiple leads is that it allows diagnostic ECG's to be recorded even if one electrode falls off.

Electrodes and patient preparation

In order to obtain the best quality ECG's self-adhesive electrodes rather than crocodile clips should be used. These clips elicit pain and therefore likely change the autonomic tone as well as causing unnecessary artefacts. In addition, not every self-adhesive electrode is the same and using good quality, equine specific (which contain more gel and are more adhesive) pads will also make a difference to the quality of the ECG that you record. Ideally patients should not be clipped prior to placement of the self-adhesive pads, although very thick coats may need trimming to allow the center of the pad making contact with the skin. Alcohol-based hand gels or ultrasound coupling gel may be applied in small quantities in the center of the pad to improve electrical contact. If too much is applied, the pad will not adequately stick. Clipping also results in poor adhesion, particularly if the horse sweats. Some people use superglue or tissue glue to improve adhesiveness of the pads to the skin. However, in a small proportion of horses, growth of circular white hairs in these regions has been reported and thus should be used with caution. The author has also had issues with surgical spirit resulting in severe areas of dermatitis. These self-adhesive pads are not required with the AliveCor system, although sometimes alcohol hand gel or ultrasound coupling gel may be needed to acquire appropriate contact with the metal electrodes.

Recording devices

There are many recording devices available now that are affordable. These include mobile phone apps, anesthetic monitoring machines, small, hardy portable devices and small, battery operated devices. The latter device can easily be attached to the horse to allow ECG's not only to be recorded at exercise, but also resting ECG's to be obtained when the horse is relaxed and alone in its stall. If you are recording paper trace ECG's, it is convention to record ECG's at 25mm/second unless the horse's heart rate is greater than 80-100 beats per minute, when 50mm/second may provide a more diagnostic trace, with an initial sensitivity of 10mm/mV. The ECG should be two-thirds of the width of the paper. If larger than this, you can turn down the sensitivity or move the electrodes further apart and vice versa if they are too small, which is the most common problem. In the author's opinion, simply recording heart rate using Polar heart rate monitors or equivalent is not sufficient.

Artefacts

Try and avoid all forms of artefact if possible. The main annoyances identified on the ECG are movement, particularly skin twitching, which can be minimized by using self-adhesive pads rather than crocodile clips, reducing skin contact with the horse and providing a fly free environment. Other artefacts commonly encountered are electrical in origin and include mobile phones, clippers and tumble dryers. Try and switch off if at all possible. Another potential artefact results from poor contact between the skin and recording device or due to poor connections or movement at the point where the electrode leads join – it can be advantageous to hold this connection still.

Length of ECG recording

The length of ECG required depends upon the aim of the ECG. If you wish to confirm a rhythm abnormality present permanently or ascertain the cause of a tachycardia, a short 30 second to 2 minute strip should be adequate. If you are trying to identify an intermittent abnormal rhythm or problem then a 24 hour or longer Holter recording may be required. Ensuring an accurate history will allow you to most accurately choose the type and length of ECG recording required- if you choose the wrong type, then you will significantly reduce your chances of obtaining a definitive diagnosis.

ECG evaluation at rest

Horses obviously should have a low resting rate (28-44 beats per minute) and will frequently have a regularly irregular rhythm due to the presence of second degree atrioventricular blockade. As horses have a large degree of cardiac reserve, they control their cardiac output at rest by reducing heart rate using a variety of strategies including second degree atrioventricular blockade, but rarely may also have sinoatrial pause or arrest. These events are vagally mediated. Sinus dysrhythmia is quite unusual in the horse at rest when compared with dogs, although is seen concurrently with respiratory disease. Rarely these vagally mediated rhythms do not disappear at exercise and thus may result in poor performance, weakness or collapse. Other differential diagnoses for an irregular rhythm at rest would include: atrial fibrillation, where there is an irregular rhythm, no p waves, f or fibrillation waves with normal qrs complexes; supraventricular premature depolarisations (SVPD's) where there is a largely normal complex that is premature - there is a p wave, but this sometimes fuses with the preceding t wave so it is important to look for a change in t wave morphology; and then ventricular premature depolarisations (VPD's), where there is a premature beat with no p wave, an abnormal qrs complex and then usually a compensatory pause. Although VPD's are described as being 'wide and bizarre', this is not always true in the horse and it can be challenging to differentiate from SVPD's. However, changes in t wave morphology are likely to be seen with VPD's than with SVPD's.

The role of the technician

As stated above, irregular rhythms at rest are common and if found on routine or initial physical differentiating physiological from pathological is important, not only for the safety of the animal, but also to prevent wasting client money. This is of particular importance prior to anesthesia. For example a horse that has multiple SVPD's at rest is at risk of developing atrial fibrillation under anesthesia with the administration of multiple pharmacological agents that reduce heart rate. Acute onset atrial fibrillation under anesthesia results in a very unstable, challenging and often stressful anesthetic! In addition, horses with rhythm abnormalities at rest may not be safe to be ridden and as such a routine procedure for an orthopedic condition may be inappropriate and a waste of client money.

EXERCISING ELECTROCARDIOGRAMS

As with resting electrocardiograms, these are also recorded in a variety of ways. The reason for this is to try and obtain the best quality trace for the given rider and circumstance. This can be challenging for a number of reasons. The main reason for a poor quality ECG is the rider and the movement caused by their seat and legs. This 'artifact' is often less with professional riders and when jockeys are largely off the horses' back. Other causes of artifact include muscle activity, fat movement or movement of any part of the recording device. Data by the author has shown that most artifact regarding recording devices comes from movement of the electrodes and then movement at the electrode-wire connection. This is the main limitation of this technique and makes recording consistently diagnostic ECG's in all horse types significantly challenging. Electrode position needs to be such that all sections of the ECG are present and diagnostic and ideally different leads can be assessed. So in an ideal world this would result in electrodes being placed in a base-apex or modified base-apex position with electrodes being on both the left and right sides of the horse. However these positions can be problematic regarding muscle, limb and rider movement. Alternative methods may include positioning electrodes under the saddle, although different algorithms may be required for ECG analysis.

The aims of the recording ECG include whether the horse has appropriate heart rates for a given speed, evaluate recovery from exercise to help decide levels of fitness and to establish if normal rhythms are present during peak exercise in order to assess safety for ridden work and also to establish any potential causes of poor performance. For these to be diagnostic, they need to represent the type of the exercise that the horse normally undertakes and be performed in a horse that is fit enough to undertake the exercise that is required of it; many of these horses present de-trained after identification of a problem. For example it is inappropriate to do an exercising ECG'S on a lunge in an event horse or racehorse. The ECG needs to be recorded for long enough to obtain heart rates representative of what the horse does. The horse needs to not be lame as pain as excessive activation of the sympathetic nervous system can lead to rhythm abnormalities that are not usually there. The same is true if ECG's are assessed when the horse is unduly stressed and as such there are times where this test is better undertaken at home or the usual place that the horse is exercised rather than in the clinic. It should be noted that normal horses commonly have aberrant beats at exercise[8,9], thus making it challenging sometimes to discern significant rhythm abnormalities from insignificant ones.

Exercising ECG's should contain a section recorded at rest, the warm-up phase, a period of peak exercise and then the recovery phase with the ECG ideally recorded until the horse's heart rate has returned to normal or nearly normal. Interpretation of these phases should however be undertaken with the knowledge that horses are more likely to develop aberrant beats when there is an imbalance between the sympathetic and parasympathetic nervous system – i.e in the warm-up and warm-down periods, and their significance in these phases is controversial.

The role of the technician

I think that the recording of a diagnostic exercising ECG trace should be one of the roles of the technician. This role begins with obtaining an accurate history of what the horse does, how much exercise it is currently doing and the problem list for that case. For example, a significant number of horses that have dysrhythmias also have concurrent airway disease[8] and as such if they are known to make a noise at exercise, the exercising ECG recorded should be recorded alongside dynamic exercising endoscopy. The link between these two conditions is thought to relate to hypoxia. If the horse seems unduly stressed in the clinic environment, the exercising ECG may be more reflective if it performed elsewhere. The next decision-making step is how to best mimic the type of work that should be undertaken for the exercise test and whether it is to be performed on a lunge or under saddle. As a general rule, horses do not work as hard on a lunge as they do when they are ridden, but this has to be balanced with the liability factor that the horse may collapse. School work is generally appropriate for dressage and showjumping horses, whereas event horses and racehorses should be exercised in a larger space at gallop. Prior to the exercise test, establishing that the horse is not lame, with examinations in a straight line and on circles to rule-out bilateral lameness is essential. Electrode placement should then be chosen based on shape of horse, how fat it is and rider position if this is relevant. Examples will be provided in the session to help with this decision making. A brief electrographical interpretation should be performed prior to the examination concluding in order to ensure a diagnostic ECG trace has been obtained. Excessive baseline movement or where there is a significant signal-to-noise ratio, particularly at peak exercise, will make interpretation impossible.

SUMMARY

In summary, to ensure the diagnostic recording of ECG's at rest and exercise, a detailed and targeted history for a particular horse needs to be obtained in order that the most appropriate ECG is recorded and combined with any other appropriate diagnostic tests. Placement of electrode position will be dependent upon the type and length of ECG required and also other factors including shape and body condition score of the horse, type of rider and saddle and position in the saddle. Being flexible with electrode placement and adapting to different situations, with experience, will improve the ECG recordings you obtain.

REFERENCES

1. Morgan R. *Companion Animal* 2012;17:10-13; 2. Marr CM, Bowen M. *Cardiology of the Horse*: Saunders Elsevier, 1999; 3. Robertson SA. *In Practice* 1990;12:59-67; 4.Reef VB, Bonagura J et al. *J Vet Intern Med* 2014;28:749-761; 5. Reed SM, Bayly WM, Sellon DC. *Equine Internal Medicine*. 2nd ed: Saunders, 2004; 6. Reimer JM et al . *J Am Vet Med Assoc* 1992;201:1237-1243; 7. Verheyen T, et al. Oesophageal electrocardiography in healthy horses. *Equine Vet J* 2012;44:640-645; 8.Martin BB, Jr., et al. *J Am Vet Med Assoc* 2000;216:554-558; 9. Ryan N et al *Equine Vet J* 2005;37:265-268.

Influenza

Meri Hall, RVT, LVT, CVT, LATG, VTS (SAIM)
Palm Beach Gardens, FL

Influenza viruses are caused by RNA viruses of the Orthomyxoviridae family.<B7> </C>There are three genera that contain the influenza viruses: influenza virus A, influenza virus B, and influenza virus C. The structure is similar for all influenza viruses, with enveloped virus particles that are either filamentous or spherical in shape and range in size from 80 to 120 nm in diameter.<B8> Viruses can only replicate in living cells. They affect all species of birds and mammals but are usually host specific. Transmission of the influenza virus interspecies is a concern due to the emergence of novel endemic influenza strains and reverse transmission.<B9></C>

The subtypes of the virus are labeled for the HA and NA proteins of the virus. HA (H) codes the hemagglutinin glycoprotein and NA (N) codes the neuraminidase enzyme. There are 16 different H antigens and nine different N antigens at this time.<B7></C> Each subtype has mutated into a variety of strains that are pathogenic to either one or multiple species.

There is only one species of the influenza virus A, but there are many strains of influenza A. Wild aquatic birds are the natural hosts for a great many of the influenza A viruses. They can be the most virulent of the influenza types. The H5N1 strain, also known as avian flu, has been transmitted to domestic and wild felids that were fed uncooked chickens with the virus. The H3N8 strain was originally an equine influenza, which crossed species in the last decade to create what we know as canine influenza.

There is only one species of the influenza virus B. Humans are the host for the vast majority of the influenza B viruses. It has been shown that ferrets and seals can be infected with the influenza B virus. <B10></C>Influenza virus C also has only one species, but unlike influenza virus B, it does infect humans, dogs and pigs. Influenza C is less common but can cause both severe illness and local epidemics.<B13></C>

Pathogenesis and transmission

Transmission of the virus is by direct contact, aerosolization, and fomites, and the duration of transmission is 5–7 days postinfection. In seasonal human flu outbreaks, the strains of influenza that are seen change from year to year. Influenza viruses are capable of surviving cold weather. They can survive in the environment for 24–48 h on hard, nonporous material, 15 min on paper and 5 min on the skin. In mucous, the virus has been shown to survive for 17 days on paper.<B14></C>

Clinical signs

Clinical signs for all of the influenza viruses can appear 1–2 days postinfection and may include fever, cough, nasal discharge, lethargy, dehydration, pain, reddened eyes, and, occasionally, diarrhea.

Diagnostic testing

Hematology may reveal moderate lymphopenia, normal leukogram, and hemoconcentration. Radiographs may reveal a bilateral, caudodorsal alveolar pattern.

There are PCR tests that can be utilized to diagnose influenza, but typically the animal has already past the infectious stage when diagnostics are completed.

Pharmacology and treatment

Treatment options are primarily supportive care, rest, and nutrition support. Antiviral drugs such as oseltamivir (Tamiflu) or zanamivir (Relenza) may be beneficial to halt the spread of the disease. Many owners do not realize an animal has been affected until secondary bacterial infections, such as pneumonia, occur.

Effective reduction of the transmission is through minimizing human/animal contact when owners are ill and includes frequent hand washing and covering coughs and sneezes. Influenza viruses can be inactivated by sunlight, disinfectants, and detergents. <B15></C>Thoroughly sanitizing surfaces with quaternary ammonium compounds or diluted bleach is indicated.

Leptospirosis

Rachel Poulin, RVT, VTS (SAIM)
Capital Veterinary Specialists
Tallahassee, Fl

INTRODUCTION

Leptospirosis, a zoonotic disease of worldwide significance, is caused by spirochetes of the genus Leptospira. Leptospires are 0.1 to 0.2 µm wide by 6-12 µm long, motile, filamentous spirochetes that infect animals and humans[1]. These bacteria can cause renal and/or hepatic disease. While the bacteria is typically found in free standing water it is also associated with alkaline soil in rural and suburban areas. Indirect contact with contaminated water, soil, bedding or food can also lead to infection. The bacteria are transmitted by direct contact with infected urine, placental transfer, bite wounds or ingestion of infected tissue.

There are multiple serovars that infect dogs and cats[2].

Dogs	Cats
L. australis	L. bratislava
L. autumnalis,	L. canicola
L. ballum	L. grippotyphosa
L. bratislava	L. Pomona
L. bataviae	
L. canicola	
L. grippotyphosa	
L harjo	
L. icterohaemorrhagiae	
L. Pomona	
L. tarassovi	

While most mammals are susceptible to leptospirosis, natural hosts do not show clinical signs naturally shed the infection for months to years. It is the infection of the incidental host that results in clinical signs.

CLINICAL SIGNS

Leptospirosis can develop in any breed at any age or gender. Most dogs will be subclinical for infection [3]. Clinical signs are usually non specific but can include vomiting, diarrhea, anorexia and severe weakness and depression. Severe clinical signs can include fever, pale mucous membranes, and tachycardia. Thrombocytopenia and disseminated intravascular coagulation (DIC) results in petechiae, ecchymosis, melena and epistaxis. It is possible for a peracute infection to result in death before renal or hepatic signs are noted. These animals usually present in the sub-acute phase, after the survival of the sub-acute phase some animals may develop renal or hepatic signs characterized by clinical signs of polyuria/polydypsia, anorexia, weight loss, and ascites.

DIAGNOSIS

Because leptospirosis suspects pose a zoonotic risk and can serve as a reservoir it is important to establish a true diagnosis as quickly as possible. Recent advances in technology allow us access to serology and polymerase chain reaction (PCR) that will grant us a definitive diagnosis. PCR tests for the presence of bacterial RNA and DNA. These tests can still present challenges for clear diagnosis due to cross reactions that can occur and not all serovars being present in testing. However, if clinical signs are suggestive of leptospirosis then specific testing with serology and PCR is indicated. It is important to point out that sample acquisition of blood and/or urine should be done prior to the administering of antibiotics for PCR testing.

TREATMENT

Supportive care is indicated in patients presented with kidney injury or other clinical signs such as vomiting, lethargy and/or diarrhea which can quickly result in dehydration. IV fluid therapy should be administered to replace losses and dieresis. IV fluid therapy choices should be bases on biochemical profile in order to treat any acid-base imbalances.

Gastro-Intestinal supportive care is indicated for stabilization of vomiting and diarrhea. Antiemetic, pro-motility agents and antacids are indicated when vomiting, diarrhea, and hyporexia/anorexia is present. Doses may need to be adjusted when hepatic and/or renal injuries are present due to decreased function of these organs. Antibiotic therapy is the key to treating leptospirosis. Due to the risk of zoonotic risk and the spread disease, antibiotic therapy should be initiated as soon as possible even if diagnosis is not yet confirmed. Doxycycline (administered orally) or penicillin and its derivatives (i.e, ampicillin (IV) or amoxicillin (orally) are the antibiotics of choice for treatment. These drugs terminate letpospiremia within 24 hours, which in turn prevents urinary shedding and transmission of the organism and significantly decreases the risk of zoonotic transfer[4]. While termination of leptospiremia occurs after 24 hours, leptospiruria can continue to be present for 7-10 days after starting antibiotic therapy[5]. Therefore, aseptic technique and zoonotic protocol are still essential in preventing human exposure. Duration of antibiotic therapy is dependent on severity of disease and tolerance and response to therapy.

VACCINE

There is a vaccine for Leptospirosis but it is not considered a core vaccine. Dogs may be vaccinated for Leptospirosis if they are considered to be at risk for the disease. Risk factors include lifestyle, geographic location, and season. The vaccine available is a quadrivalent leptospirosis bacterin vaccine containing serovars Grippotyphosa, Canicola, Icterohemorrhagica, and Pomona. The most commonly reported serovars infecting dogs in North America in recent years include Grippotyphosa, Pomona, Autumnalis, Bratislava[6]. Vaccine should be given annually while patient is considered to be at risk, however dogs that have an exceptionally high risk should have repeat vaccines every 6-9 months due to lack of duration of all serovars[7].

Leptospira Spp. Serovars and hosts[8]

Genomospecies	Serovar	Known primary reservoir host	Incidental domestic host	Wild animal hosts	Serovar included in vaccine
Leptospira interrogans	Autumnalis	Mouse	Cow, dog, human	Opossum, rat, raccoon	No
L. interrogans	Bataviae	Dog, rat, mouse	Cat, cow, human	Armadillo, hedgehog, shrew, vole	No
L. interrogans	Bratislava	Rat, pig, horse	Cow, dog, horse, human	Fox, hedgehog, mouse, opossum, raccon, skunk, vole, weasel	No
L. interrogans	Canicola	Dog	Cat, cow, horse, human, pig	Armadillo, hedgehog, raccoon, rat, skunk, vole	Yes – bivalent and quadrivalent product
Leptospira Kirschneri	Grippotyphosa	Vole, opossum, skunk, raccoon	Cat, cow, dog, goat, human, pig, rabbit, sheep	Bobcat, fox, hedgehog, mouse, muskrat, rat, shrew, squirrel, weasel	Yes – quadrivalent product only
L. interrogans	Icterohemorrhagica	Rat	Cat, cow, dog, horse, human, pig	Ape, civet, fox, hedgehog, mouse, muskrat, opossum, skunk, woodchuck	Yes – bivalent and quadrivalent product
L. interrogans	Hardjo	Cow	Dog, horse, human, pig, sheep		No
L. interrogans	Pomona	Skunk, opossum, pig, cow	Cat, cow, dog, gerbil, goat, humans, pig, rabbit, sheep	Civet, fox, deer hedgehog, mouse, sea lion, vole, wolf, woodchuck	Yes – quadrivalent product only

ZOONOTIC RISK AND PREVENTION

Leptospirosis can be transmitted to humans and other animals, therefore protective measures must be taken with these patients to ensure safety to staff and other patients. These Patients should be walked should have a special area where they are walked and other patients will not have access to there urine. In addition, bedding and any samples taken should be separated and labeled appropriately. Isolation protocols should be in place including gowns, gloves and goggles to protect mucous membranes from transmission. The organism is susceptible to bleach as well as cleaning detergents. The veterinary team needs to spend time educating the owner about the zoonotic potential of this disease and advised to seek medical attention if symptoms occur.

PROGNOSIS

With early recognition and appropriate treatment, the survival rate for dogs is approximately 80%[9]. Without diagnosis and appropriate treatment this disease is progressive and likely to be fatal.

REFERENCES

[1,2,3] Nelson, et.al; Small Animal Internal Medicine, Mosby 2003, [4] Idexx diagnostic update 2014, [5] J Vet Intern Med. 2011 Jan; 25(1) 1–13, [6,7,8] Merrill, Small Animal Internal Medicine, Wiley-Blackwell 2014

Fanconi Syndrome

Rachel V. Poulin, RVT, VTS (SAIM)
Tallahassee, Fl

INTRODCUTION

Fanconi syndrome is a proximal tubular defect resulting in mismanagement of glucose, electrolytes and aminoacids. It is characterized by excessive urinary losses of the aforementioned substances of leading to glucosuria, bicarbonaturia, phosphaturia, kaliuresis, proteinuria, ,and aminoaciduria. These losses leads to hypoglycemia, hypophosphatemia, hypokalemia and a metabolic acidosis. Ketonuria can also be present. Glomerular filtration rate (GFR) may be normal initially but decreased later in the course of the disease. This can lead to azotemia. Defective urinary concentration ability can also be seen. This can lead to isosthenuria or hyposthenuria.

There are two different types of Fanconi syndrome: inherited and acquired. Inherited Fanconi syndrome is most common in Basenjis. It affects 10% to 30% of the breed (Ettinger and Feldman 2010). This defect in the Basenji is thought to be the result of a metabolic or membrane defect (DiBartola 2006). Inherited Fanconi syndrome has rarely been reported predominantly in other breeds. Acquired Fanconi syndrome as an acquired condition has been linked to jerky treat ingestion, gentamicin/amoxicillin nephrotoxicosis, or potentially toxic substances such as lead or copper. It also exists as an idiopathic form (Brown 2013).

CLINICAL SIGNS

Clinical signs of Fanconi syndrome include but are not limited to lethargy, vomiting, anorexia, diarrhea, polyuria, polydypsia, dehydration, weight loss, and a poor hair coat. The most common clinical sign and complaints by clients are polyuria and polidypsia (Brown 2013). Diagnosis is made by testing the levels of solutes such as glucose, sodium, phosphorus, , potassium, bicarbonate and amino acids in the urine and the blood. Dogs with Fanconi syndrome will have low levels of these solutes in the blood and elevated levels in the urine This is due to the proximal tubular inability to reabsorb these substances. ,. This eventually leads to dehydration, disruption in acid base balance and electrolyte deficiencies. As the disease progresses hyperchloremia, metabolic acidosis and renal failure (azotemia) occurs. Progression is variable; some dogs may develop renal failure within a few months of the onset of clinical signs, whereas others may remain stable for years (Ettinger and Feldman 2010). Although diagnosis is made by blood and urine testing other disease processes should still be ruled out. Patients with glucosuria should be screened for Leptospirosis as well as for pyelonephritis. Ideally imaging to evaluate the patient for metal toxicities is recommended. Abodminal ultrasonography to evaluate the renal architecture is ideal.

TREATMENT

Treatment for the underlying cause may lead to resolution of the syndrome. Patient's with pyelonephritis, Leptospirosis and toxin exposure can have resolution of the syndrome within 2-3 months of therapy, Otherwise, the treatment is supportive to address protein and electrolyte deficiencies as well as the patient's azotemia. Supportive treatment can be determined by level of clinical signs, presence of glucosuria and ketonura and presence of renal disease. If diagnostics reveal severe azotemia, electrolyte deficiencies and/or metabolic acidosis then hospitalization with aggressive fluid therapy is indicated. For dogs that are less severely affected or that have stabilized after aggressive treatment an out patient regimen is appropriate. Fresh water should be readily available at all times. As already stated, these patients usually have moderate to severe polyuria and polydypsia. This makes the likelihood of dehydration relevant. Even though the patient may drink excessively and urinate quite often, water consumption should not be restricted at any time. Subcutaneous fluids are usually indicated for outpatient treatment for stable dogs. The fluid choice is based on the severity of acid/base imbalance. Because of the loss of bicarbonate through the urine an alkalizing fluid is necessary such as Lactated Ringer's®, Normosol-R® or Plasmalyte-156®. For the patient who is azotemic but has a normal appetite and is otherwise stable, subcutaneous fluids administered at home is appropriate at half of the maintenance rate for 24 hours (Morales 2014). For nauseous and hyporexic patients the recommended subcutaneous dose is the entire maintenance rate for a 24 hours period (Morales 2014). Additives to the fluids such as KCL and KP04 may be necessary, doses should be based on blood gas analysis but should not exceed 30-40mEq/L KCL per day because of localized irritation (Plumb 2005). Slowly weaning off IV fluids while in the ICU and quantifying water consumption in the hospital can help determine the amount of subcutaneous fluids that will allow the patient to do well at home. Subcutaneous fluids may be necessary for weeks to months.

Gastrointestinal support plays a role in treatment as well. Antacids, motility agents and antiemetics may be necessary when vomiting, lethargy, anorexia and/or hyporexia are present. The necessity of these agents may be greater in the presence of ketonuria, uremia and azotemia as these conditions can cause the patient to feel nauseous and therefore may be reluctant to eat.

Gastrointestinal Support Agents

Medication:	Function:	Canine Dose:	Feline Dose:	Warnings:
Famotdine (pepcid®)	H_2 receptor antagonist used to reduce GI acid production	0.5 -1 mg/kg PO SID-BID	0.5mg/kg PO once daily	H_2 have been demonstrated to be relatively safe and exhibit minimal adverse effects.
Metoclopramide (reglan®)	Stimulates upper GI motility and has antiemetic properties	0.2-0.4mg/kg PO BID-TID	0.2-0.4mg/kg PO BID-TID	Behavior changes; Changes in mentition, behavior, disorientation and constipation.
Maropitant (cerenia®)	neurokinin (NK1) receptor antagonist that blocks the pharmacological action of substance P in the central nervous system (CNS)[35]	2mg/kg PO once daily for four days	Off label for cats	has not been evaluated in dogs used for breeding, pregnant or lactating bitches, dogs with gastrointestinal obstruction, or dogs that have ingested toxins. Use with caution in dogs with hepatic dysfunction.
Ondansetron (zofran®)	5-HT_3 receptor antagonist for severe vomiting	0.1-1mg/kg PO SID-BID	0.22mg/kg BID-TID	Potential of constipation, extrapyramidal symptoms, arrhythmias and hypotension
Sulcralfate (carafate®)	Locally acting treatment for GI ulcers	0.5-1 gram PO BID-QID	0.25-0.5 gram PO BID-TID	Constipation possible Should be administered on an empty stomach
Omemprazole (prilosec®)	Proton pump inhibitor used for GI ulcers and erosion	0.5-1 mg/kg PO SID	0.7 mg/kg PO SID	Hypersensitivity is possible Caution should be taken in patient with hepatic or renal disease

Nutritional support is equally important in the treatment of Fanconi syndrome. Dr. Steve Gonto established a treatment protocol in 2003, Fanconi Disease Management Protocol for Veterinarians. While we do not have any evidence-based medicine that these recommendations are beneficial, they have become popular guidelines for therapy and management of Fanconi Syndrome. This protocol recommends a good quality food without protein restriction be fed. A hybrid protocol is recommended for patients with uremia (BUN > 60) and/or stage 2 renal disease or above. The hybrid protocol calls for a low protein diet.

In addition to subcutaneous fluid administration, supplements are frequently used to help replace the lost solutes. These supplements and gastroprotectants tend to be well tolerated in patients that are eating and drinking well.

PET-TAB PLUS®	Vitamin & mineral supplement tablet	One tablet, divided ½ tab BID for asymptomatic dogs	Two tabs daily, given one tab BID for symptomatic dogs
PET-CAL® or CALCIUM PHOSPHORUS (By PET-AG) **CENTRUM VITAMIN**	Vitamin D-phosphorus replacement tablet Complete vitamin/mineral tablet (high potency)	Given ½ tab BID in asymptomatic dogs One tab/week in dogs with PU/PD	One tab BID in symptomatic dogs no dose needed in fully asymptomatic dogs
AMINO FUEL® ("Stack" by Twinlabs®)	Complete amino acid preparation.	One tab/week if asymptomatic	Titrate up as high as one tab, every-other-day, in cases of extreme muscle wasting, poor coat or unresolved skin problems
SODIUM BICARBONATE ANTACID TABLETS **TUMIL-K® (2 MEQ/Tab) or UroCit-K ®(5 MEQ/Tab)**	Sodium Bicarbonate 10-grain antacid tablets POTASSIUM SUPPLEMENTATION	Dose dependant on blood gas analysis Dosage is based on blood chemistry	**THE MOST IMPORTANT OF ALL THE SUPPLMENTS** These tablets must also be given intact, especially UROCIT-K®, which is a "timed release matrix" delivery vehicle. **Crushing some potassium tablets can result in catastrophic overdose for the dog.**

(Gonto 2003)

PROGNOSIS

Prognosis for Fanconi's patients tends to be very good if the patient has transitory Fanconi syndrome and if the patient is not uremic. A controlled Fanconi dog can have a normal life span compared to an unaffected dog. Diligence and maintenance is key along with frequent veterinary evaluation and constant supply of fresh water at all times. Acquired Fanconi syndrome has a good prognosis with aggressive treatment and management. Inherited Fanconi syndrome has a variable prognosis that is dependent on progressive renal disease. Death usually results from acute renal failure and papillary necrosis or acute pyelonephritis (DiBartola 2006).

A technician plays a large role in the management of a patient with Fanconi syndrome. An excellent and detailed patient history is vital in developing a plan and possibly finding a cause.. Client education is equally vital and a role that can be taken by the technician. Client education such as subcutaneous fluid administration and nutritional support is another role that can be taken on by tan educated technician and will add to the probability of an excellent patient outcome.

REFERENCES

Ettinger S, Felmdan E, Textbook of Veterinary Internal Medicine 7th ed. St. Louis, MI: Saunders Elsevier, 2010
Brown A, The Merk Veterinary Manual 18th ed. Whitehouse Station, NJ: Merck Research Laboratories
Gonto S. Fanconi Disease Management Protocol for Veterinarians. http://www.basenji.org/fanconiprotocol2003.pdf
Morales S, Personal Communication, Coral Springs, Fl: 2014
DiBartola S, Fluid, Electrolyte, and Acid-Base Disorders in Small Animal Practice 3rd ed. St. Louis, MI: Saunders Elsevier 2006

The Five Key Questions in Diagnostic Testing: What, When, Where, Why and How

Julie Allen, BVMS, MS, MRCVS, DACVIM (SAIM)
Raleigh, NC

INTRODUCTION

The number and complexity of diagnostic tests available to veterinarians is ever increasing. Veterinary technicians are often tasked with discovering where a particular test is performed as well as the specimen requirements, any specific protocols and the associated costs. Unfortunately there is often not time to explain why the test is being performed and what the results could mean for the patient. The purpose of this talk is to cover the five key questions in diagnostic testing: what, when, where, why and how. Hopefully this will then not only provide a resource to refer to, but also greater understanding of the indications, methodologies and interpretation of the many tests available in veterinary medicine today. The table below is obviously not all inclusive and we will highlight the most commonly requested tests in the lecture. Additionally, it is always best to contact the specific laboratory if you ever have questions regarding a particular assay.

What?	When?	Where?	Why?	How?
Inhibin	To aid in the diagnosis of pheochromocytoma	UC Davis http://www.vetmed.ucdavis.edu/phr/labs/endolab/canine_inhibin_testing.cfm	Dogs with pheochromocytomas have similar inhibin levels to normal neutered animals (dogs with pituitary dependent hyperadrenocorticism and adrenocortical carcinoma have higher levels)	Fasted sample, minimum 1 ml serum, no lipemia, ship overnight on ice.
Metanephrine	To aid in the diagnosis of pheochromocytoma	Marshfield Laboratories https://www.marshfieldlabs.org/sites/ltrm/Vet/Pages/25191.aspx	Urinary normetanephrine-to-creatinine ratios are significantly increased in dogs with pheochromocytoma compared with dogs with hyperadrenocorticism and normal dogs.	Min. 5 ml urine in sterile plastic tube – add 0.15 gram sulfamic acid. Stable for 2 weeks. Performed Mon and Thurs.
Insulin-like growth factor-1	For the diagnosis of growth hormone deficiency and acromegaly	MSU Diagnostic Center for Population and Animal Health https://www.animalhealth.msu.edu/Bin/Catalog.exe?Action=Test&NAME=igf&Id=1401	In pituitary dwarfs, IGF-1 levels are typically low but diagnosis is not definitive. IGF-1 values > than 1000ng/mL are expected in cats with acromegaly.	Min. 0.5ml serum or plasma. Refrigerate or freeze prior to shipment with icepack. Run on Wed.
Anti-Müllerian hormone (AMH)	To distinguish between spayed and ovarian intact females or intact or neutered males (ovarian remnant / cryptorchid)	UC Davis http://www.vetmed.ucdavis.edu/phr/local_resources/docs/2015_Laboratory_Instructions,_reference_ranges_and_submission_form.pdf Cornell University Animal Health Diagnostic Center https://ahdc.vet.cornell.edu/test/list.aspx?Species=&Test_Name=amh&TstTyp=&WebDisc=7	Negative test = spayed female. If positive, highly accurate in cats for detecting ovarian remnant syndrome. Some false negatives in dogs. Recommend concurrent progesterone level if clinical signs indicate a remnant, or if the animal has been previously spayed. AMH test has not been fully evaluated for detecting cryptorchidism, but a positive test is consistent with presence of testicular tissue.	Fasted sample, no lipemia or hemolysis, minimum 0.2ml, ship overnight on ice. Min 30 days post-surgery. Performed on Tues only (Davis).
Iohexol clearance	To determine glomerular filtration rate (GFR)	MSU Diagnostic Center for Population and Animal Health https://www.animalhealth.msu.edu/Bin/Catalog.exe?Action=Test&NAME=iohexol&Id=1578	This is a kidney function test using iohexol clearance. A computerized model is utilized to calculate GFR in cats and dogs to assess global renal function, prior to the development of azotemia, allowing for earlier diagnosis and intervention.	Protocol available online. Serum chilled or frozen shipped overnight with gel packs.

Symmetric dimethylarginine (SDMA)	For the early diagnosis of kidney disease	Idexx Laboratories https://www.idexx.com/smallanimal/reference-laboratories/directory-tests-services.html	SDMA is almost exclusively eliminated by renal filtration so it is a good estimate of GFR. It increases on average with 40% loss of kidney function versus creatinine, which does not increase until 75% of kidney function is lost. SDMA is unaffected by lean body mass, which can lead to a decreased creatinine in old or cachexic animals.	1 ml serum (preferred) or 1 ml EDTA or heparinized plasma (spun and separated). Stable 14 days at 2-8° C; 4 days room temp.; frozen longer.
DGGR lipase	To aid in the diagnosis of pancreatitis	Antech Diagnostics http://precisionpsl.com/index.html	This test is believed to be more sensitive and specific than general lipase assays and some studies suggest substantial agreement with the pancreatic lipase immunoreactivity.	Min. 0.5 ml serum. Refrigerated. Gross lipemia may affect results.
Canine and feline Pancreatic lipase immunoreactivity (cPLI and fPLI)	To aid in the diagnosis of pancreatitis	Texas A&M GI Laboratory http://vetmed.tamu.edu/gilab/service/assays/pli	Dogs: levels greater than 400 μg/L are consistent with pancreatitis (201 to 399 μg/L questionable range). Cats: levels ≥5.4 μg/L consistent with pancreatitis (3.6 - 5.3 μg/L questionable range).	0.5 ml fasting (12-18 hours) non-hemolyzed serum.
Cobalamin (vitamin B12) /folate	Test of intestinal function/small intestinal dysbiosis	Texas A&M GI Laboratory http://vetmed.tamu.edu/gilab/service/assays/b12folate	Decreased cobalamin levels suggests distal small intestinal disease, exocrine pancreatic insufficiency or small intestinal dysbiosis. Increased cobalamin levels have no known significance. Increased folate levels suggests small intestinal dysbiosis; decreased levels suggest proximal small intestinal disease.	Minimum 400 μl of serum. Serum samples for folate analysis should be kept at 4ºC for not longer than 48 hours, freezing preferred.
Trypsin-like immunoreactivity	Used to diagnose exocrine pancreatic insufficiency (EPI)	Texas A&M GI Laboratory http://vetmed.tamu.edu/gilab/service/assays/tli	In dogs, values < 2.5 μg/L are diagnostic for EPI. In cats, values equal to or below 8.0 μg/L are diagnostic for EPI.	Dogs: 0.5 ml fasting (12-18 hours) non-hemolyzed serum. Cats: 0.2 ml. Extremely stable; can be shipped at ambient temps.
C reactive protein	Marker of inflammatory disease	Texas A&M GI Laboratory http://vetmed.tamu.edu/gilab/service/assays/canine-c-reactive-protein	Increased levels suggest the presence of inflammatory disease such as inflammatory bowel disease (IBD). Correlates well with the IBD clinical disease activity index, so can be used to monitor response to therapy of IBD.	0.5 ml fasting (12 hours) serum; DOGS ONLY
Fluorescent in situ hybridization (FISH)	Used to identify bacteria within formalin fixed tissues	Cornell University http://vet.cornell.edu/labs/simpson/FISHFAQS.cfm	Fluorescent probe binds to bacterial ribosomes in tissue sections, visualized using fluorescent microscope. Especially important to determine if there is bacterial component in certain diseases e.g. chronic granulomatous diseases, cholangiohepatitis etc. Particularly important in boxers with colitis to assess for evidence of intramucosal and intracellular bacterial invasion. If bacterial invasion is present, specific	Min. 5 unstained paraffinized sections, 4-5 μm thick. Room temp. Request special kit for boxers. Tues or Wed.

			treatment is required.	
Influenza Virus Matrix PCR	Can aid in diagnosis of canine influenza	Cornell University Animal Health Diagnostic Center https://ahdc.vet.cornell.edu/test/list.aspx?Species=&Test_Name=canine%20influenza&TstTyp=&WebDisc=	Detects most influenza viruses including H1N1 and H3N8. Likely to be of highest yield early in the course of illness (e.g., first 1- 3 days), or in exposed dogs that have not yet developed clinical signs. Conversely, serology requires 2 separate samples 2 weeks apart and may be negative in first 10 days of illness.	Submit nasal swab in sterile, sealed vial with several drops of saline. Refrigerate, ship on ice overnight. Mon – Fri.
Protein C	To help differentiate portosystemic shunt from portal vein hypoplasia	Cornell University Animal Health Diagnostic Center https://ahdc.vet.cornell.edu/sects/coag/test/proteinC.cfm	Non-invasive measure of portal blood flow. Dogs with congenital portosystemic vascular anomalies (PSVA) typically have low protein C activity, with values < 70% of normal.	0.5 to 1.0 ml of citrate plasma, shipped overnight on a cold pack Mon-Fri.
PARR	To aid in the diagnosis of lymphoid neoplasia	Colorado State University http://csu-cvmbs.colostate.edu/academics/mip/ci-lab/Pages/PARR-Instructions.aspx NC State University https://cvm.ncsu.edu/research/labs/population-health-pathobiology/clinical-immunology/	The test evaluates DNA to distinguish reactive from neoplastic lymphocyte expansions.	Stained slides are OK and can be kept indefinitely; no special shipping. Samples in suspension (blood, cavity fluid, CSF) sent with cold pack using express mail.
Flow cytometry	Used to diagnose and phenotype lymphoma or leukemia	Colorado State University http://csu-cvmbs.colostate.edu/academics/mip/ci-lab/Pages/Flow-Cytometry-Instructions.aspx NC State University https://cvm.ncsu.edu/research/labs/population-health-pathobiology/clinical-immunology/	Different types of lymphocytes express different proteins (e.g. T cells express protein CD3, B cells express protein CD21). Flow cytometry evaluates how many cells of each type are present, allowing determination of lineage of cells present and whether they are more consistent with neoplasia or a reactive process. The phenotype e.g. B cell vs T cell is also prognostic and may alter treatment options. Can submit blood, bone marrow, aspirates of lymph nodes, masses or organs as well as body fluids.	See website. Send overnight with cold pack for delivery Mon-Fri. Blood and bone marrow only Saturday (10 am, CSU). Do NOT freeze!

REFERENCES

1. Brömel C et al. J Vet Intern Med 2013; 27: 76.
2. Salesov E et al J Vet Intern Med 2015; 29 (2):597.
3. Niessen SJ et al. Vet Clin North Am Small Anim Pract. 2013; 43(2):319.
4. Place NJ et al. J Vet Diagn Invest. 2011; 23(3):524.
5. Collignon CM et al. Am J Vet Res. 2012; 73(11):1841.
6. Nabity NB et al. J Vet Intern Med 2015; 29(4):1036.
7. Kook PH et al. J Vet Intern Med 2014; 28(3):863.
8. Xenoulis PG. J Small Anim Pract. 2015; 56(1):13.
9. Berghoff N et al. Vet Clin North Am Small Anim Pract. 2011; 41(2):311.
10. Simpson KW et al. Infect Immun. 2006; 74(8):4778.
11. Toulza O et al. J Am Vet Med Assoc. 2006; 229(11):1761.
12. Thalheim L et al. J Vet Intern Med. 2013; 27(6):1509.

FELINE CARDIOMYOPATHY UPDATE

Virginia Luis Fuentes, VetMB, PhD, CertVR, DVC, MRCVS, DACVIM (Cardiology), DCVIM-CA (Cardiology)
Hatfield, UK

CARDIOMYOPATHY CLASSIFICATION

Traditionally, feline cardiomyopathies have been based on human classifications.[1,2] Human classifications have been revised periodically, but subdivision according to phenotype (hypertrophic, restrictive, dilated, arrhythmogenic) has remained even though etiology (genetic/ familial versus acquired/ non-familial) plays a greater role in more recent classifications.[3-5] The etiology is unknown in most cats with myocardial disease, so classification based on phenotype makes the most sense for cats. However, a recent online survey of board-certified cardiologists showed a lack of consensus on the criteria for each feline cardiomyopathy phenotype with poor inter-observer and intra-observer agreement on echocardiographic classification.[6] This suggests either our criteria are not well-defined, or the criteria are not applied consistently, or the criteria do not match the clinical features. Although pathology is often thought of as the gold standard, inter-pathologist agreement is also poor. Longitudinal studies of cats with cardiomyopathy are lacking, although 'end-stage' forms of hypertrophic cardiomyopathy (HCM) are recognized with phenotypes that differ markedly from 'classic HCM'.[7] As typical HCM features such as myofiber disarray and interstitial fibrosis are also recognized in restrictive cardiomyopathy (RCM)[8] and some human sarcomeric mutations are associated with both HCM and RCM phenotypes, there is considerable blurring of boundaries between phenotypes. HCM remains the most important category of cardiomyopathy in cats.

PREVALENCE

Epidemiologic studies are very difficult to conduct in small animal practice, and most prevalence studies are conducted from referral centers. One prospective HCM prevalence study of cats belonging to Veterinary School staff[9] and two prospective prevalence studies conducted at cat shelters[10,11] all found a very similar HCM prevalence of around 1 in 7 cats (15%). Other cardiomyopathies were rare in these studies, in contrast to referral populations.[2] This is considerably higher than the prevalence in people, which is around 1 in 500,[12] and also probably much higher than the prevalence of heart failure in cats. Cardiovascular mortality in cats has been reported at around 30 deaths per 10,000 cat years in Sweden[13] and aortic thromboembolism was reported in 0.3% of cats over an 8 year period in first opinion practice.[14] This suggests that many cats with HCM have a reasonable prognosis, but cardiac complications are still unacceptably high in the general cat population.

AETIOLOGY

Current human classifications stress the importance of genetic mutations in the development of HCM, but other causes are also recognized. While left ventricular (LV) hypertrophy due to hypertension is usually excluded from definitions of HCM, other causes of apparent LV hypertrophy in man include glycogen and lysosomal storage diseases (which can be familial) and obesity. The importance of storage diseases in feline myocardial disease is unclear, as it is not common to screen for these diseases in cats. The role of obesity in the development of LV hypertrophy in cats is intriguing. One study of Maine coon cats found that cats with HCM were heavier and had a higher body condition score than Maine coons without HCM.[15] Another study of 780 shelter cats found that male sex, increasing age, louder murmur, and higher body condition score were all associated with a greater likelihood of HCM. It is not clear whether LV hypertrophy may be linked to obesity because of alteration in cardiac loading or whether there is some other association. Similarly, the reason why HCM is more prevalent in males (even when inheritance appears to be autosomal dominant transmission) is unclear.

Familial HCM is often recognized in cats, both in pedigree breeds[16] and non-pedigree cats[17], but no further genetic progress has been made since the discovery of mutations in myosin binding protein C in Maine coons and Ragdolls. Around 60% of human HCM patients have a sarcomeric mutation, but up to 30% of people with HCM have no obvious underlying etiology, despite intensive searching with whole genome sequencing.[18]

DIAGNOSIS

Echocardiography remains the mainstay of HCM diagnosis, despite the challenges of echoing cats. A diagnosis of HCM is still based on maximum cut-off values of LV wall thickness, despite more sophisticated methods available to assess diastolic[19] and systolic[20] function. The interpretation of echocardiographic findings may vary in terms of type of cardiomyopathy,[6] but a clearer picture is emerging of echocardiographic risk factors.[21] Left atrial indices in general appear to have consistently high prognostic value[21-23] and may be more applicable for use by non-cardiologists that subtle interpretation of mild LV wall abnormalities. For the general practitioner, diagnosis of HCM is less important than diagnosis of the 'high risk' cat with HCM, whereas cardiologists are often faced with differentiating the cat with mild HCM from a normal cat. Cardiac biomarkers can be a very useful tool for identifying high risk cats.[24] In cats with respiratory distress, a point-of-care test can be used to identify cats with NT-proBNP <100 pmol/l (ie. unlikely to have heart failure). In asymptomatic cats, NT-proBNP can be used to pre-screen which cats may benefit from echocardiography. Radiography is more difficult to use, and left atrial enlargement may be difficult to identify even in cats with congestive heart failure.[25]

HCM SCREENING

Many pedigree cat breeders are anxious to screen for HCM in their breeding stock, but difficulties abound in differentiating normal cats from cats with mild HCM using echocardiography.[26] There is little standardisation between schemes. In Maine coons and Ragdolls, there is an option to use genetic testing. This may help to reduce the number of genotype positive cats in the breeding pool, and in particular, reduce the proportion of homozygous affected cats t(which have an increased likelihood of cardiac death).[27,28] There is a need for consensus guidelines on the echocardiographic approach to screening for HCM. Issues such as how to adjust allowable LV wall thickness according to size[29] and age-related penetrance have not been resolved. Many breeders want their young breeding cats screened, but a normal scan at this age can be meaningless. The phenotype of parents and grandparents may be more helpful.

PROGNOSIS

Echocardiography remains the most valuable tool for calculating prognosis in HCM, with increasing left atrial (LA) size, presence of extreme LV hypertrophy and LV systolic dysfunction rating highly as prognostic indicators. Physical exam is less helpful, but presence of a gallop sound or arrhythmia is associated with increased cardiac mortality. Presence of a murmur is associated with HCM, but is not associated with risk of cardiac death. NT-proBNP has prognostic value and can help general practitioners evaluate feline patients.[30,31] Elevated troponin-I has prognostic value independent of echocardiographic findings.[31] Risk factors for congestive heart failure (CHF) in cats with HCM include a history of CHF and LV systolic dysfunction.[32] Independent predictors of aortic thromboembolism (ATE)) within 2 years of diagnosis include prior history of ATE and reduced LA fractional shortening.[33] The risk factors for sudden death are still poorly understood, but risk factors in univariable analysis included a history of syncope.[33] Sudden cardiac death probably often goes unrecognized in the general cat population, although heart disease is commonly present in cats with sudden unexplained death.[34]

TREATMENT

The biggest breakthrough in management of feline cardiomyopathy has been the FatCat trial, which showed that clopidogrel is better than aspirin for extending the time for recurrence of ATE in cats with a prior ATE history.[35] This is likely to be extrapolated to primary prevention of ATE. No other prospective studies to date have shown any extension of survival in cats with HCM, although a number of studies are underway. A case control study of pimobendan suggested an improvement in survival in cats with cardiomyopathy.[36] Conversely, atenolol had no effect on 5 year survival rates in asymptomatic cats.[37]

FUTURE DIRECTIONS

There are many, many areas in feline cardiomyopathy where more work is required. Longitudinal studies and better standardization of phenotypic assessment are needed, as genotypic characterization will be difficult without accurate phenotyping. Diagnosis needs to be earlier if we are to implement therapy that can alter natural history. For many cats presenting with ATE, owner and veterinarian were ignorant of pre-existing heart disease. While clopidogrel may not prevent all ATE events, it may reduce their frequency of occurrence. Better understanding of the pathophysiology of HCM may be needed before we can prevent progression to sudden death, CHF and ATE

REFERENCES

1. Fox PR. *Canine and Feline Cardiology.* New York: Churchill Livingstone, 1988.
2. Ferasin L, Sturgess CP, Cannon MJ, et al. Feline idiopathic cardiomyopathy: a retrospective study of 106 cats (1994-2001). *Journal of Feline Medicine & Surgery* 2003;5:151-159.
3. Maron BJ, Towbin JA, Thiene G, et al. Contemporary definitions and classification of the cardiomyopathies: an American Heart Association Scientific Statement from the Council on Clinical Cardiology, Heart Failure and Transplantation Committee; Quality of Care and Outcomes Research and Functional Genomics and Translational Biology Interdisciplinary Working Groups; and Council on Epidemiology and Prevention. *Circulation* 2006;113:1807-1816.
4. Elliott P, Andersson B, Arbustini E, et al. Classification of the cardiomyopathies: a position statement from the European Society Of Cardiology Working Group on Myocardial and Pericardial Diseases. *Eur Heart J* 2008;29:270-276.
5. Arbustini E, Narula N, Dec GW, et al. The MOGE(S) classification for a phenotype-genotype nomenclature of cardiomyopathy: endorsed by the World Heart Federation. *J Am Coll Cardiol* 2013;62:2046-2072.
6. Wilkie LJ, Luis Fuentes V, Rishniw M. Online survey to asses inter- and intra-observer agreement on echocardiographic classification of cardiomyopathy in cats. ACVIM Forum 2015.
7. Baty CJ, Malarkey DE, Atkins CE, et al. Natural history of hypertrophic cardiomyopathy and aortic thromboembolism in a family of domestic shorthair cats. *Journal of Veterinary Internal Medicine* 2001;15:595-599.
8. Fox PR, Basso C, Thiene G, et al. Spontaneously occurring restrictive nonhypertrophied cardiomyopathy in domestic cats: a new animal model of human disease. *Cardiovasc Pathol* 2014;23:28-34.
9. Paige CF, Abbott JA, Elvinger Fo, et al. Prevalence of cardiomyopathy in apparently healthy cats. *Journal of the American Veterinary Medical Association* 2009;234:1398-1403.

10. Wagner T, Fuentes VL, Payne JR, et al. Comparison of auscultatory and echocardiographic findings in healthy adult cats. *Journal of Veterinary Cardiology* 2010;12:171-182.

11. Payne JR, Brodbelt DC, Luis Fuentes V. Cardiomyopathy prevalence in 780 apparently healthy cats in rehoming centres (the CatScan study). *Journal of Veterinary Cardiology* 2015;17, Supplement 1:S244-S257.

12. Maron BJ, Gardin JM, Flack JM, et al. Prevalence of Hypertrophic Cardiomyopathy in a General Population of Young Adults : Echocardiographic Analysis of 4111 Subjects in the CARDIA Study. *Circulation* 1995;92:785-789.

13. Egenvall A, Bonnett BN, Häggström J, et al. Morbidity of insured Swedish cats during 1999-2006 by age, breed, sex, and diagnosis. *Journal of Feline Medicine & Surgery* 2010;12:948-959.

14. Borgeat K, Wright J, Garrod O, et al. Arterial thromboembolism in 250 cats in general practice: 2004-2012. *J Vet Intern Med* 2014;28:102-108.

15. Freeman LM, Rush JE, Cunningham SM, et al. A randomized study assessing the effect of diet in cats with hypertrophic cardiomyopathy. *J Vet Intern Med* 2014;28:847-856.

16. Marz I, Wilkie LJ, Harrington N, et al. Familial cardiomyopathy in Norwegian Forest cats. *Journal of Feline Medicine & Surgery* 2015;17:681-691.

17. Kraus MS, Calvert CA, Jacobs GJ. Hypertrophic cardiomyopathy in a litter of five mixed-breed cats. *J Am Anim Hosp Assoc* 1999;35:293-296.

18. Elliott PM, Anastasakis A, Borger MA, et al. 2014 ESC Guidelines on diagnosis and management of hypertrophic cardiomyopathy: The Task Force for the Diagnosis and Management of Hypertrophic Cardiomyopathy of the European Society of Cardiology (ESC). *Eur Heart J* 2014;35:2733-2779.

19. Schober KE, Chetboul V. Echocardiographic evaluation of left ventricular diastolic function in cats: Hemodynamic determinants and pattern recognition. *J Vet Cardiol* 2015;17 Suppl 1:S102-133.

20. Wess G, Sarkar R, Hartmann K. Assessment of Left Ventricular Systolic Function by Strain Imaging Echocardiography in Various Stages of Feline Hypertrophic Cardiomyopathy. *Journal of Veterinary Internal Medicine* 2010:no-no.

21. Payne JR, Borgeat K, Connolly DJ, et al. Prognostic Indicators in Cats with Hypertrophic Cardiomyopathy. *Journal of Veterinary Internal Medicine* 2013;27:1427-1436.

22. Smith S, Dukes-McEwan J. Clinical signs and left atrial size in cats with cardiovascular disease in general practice. *Journal of Small Animal Practice* 2012;53:27-33.

23. Linney CJ, Dukes-McEwan J, Stephenson HM, et al. Left atrial size, atrial function and left ventricular diastolic function in cats with hypertrophic cardiomyopathy. *J Small Anim Pract* 2014;55:198-206.

24. Borgeat K, Connolly DJ, Luis Fuentes V. Cardiac biomarkers in cats. *Journal of Veterinary Cardiology* 2015;17, Supplement 1:S74-S86.

25. Schober KE, Wetli E, Drost WT. Radiographic and echocardiographic assessment of left atrial size in 100 cats with acute left-sided congestive heart failure. *Veterinary Radiology & Ultrasound* 2014;55:359-367.

26. Haggstrom J, Luis Fuentes V, Wess G. Screening for hypertrophic cardiomyopathy in cats. *J Vet Cardiol* 2015;17 Suppl 1:S134-149.

27. Borgeat K, Casamian-Sorrosal D, Helps C, et al. Association of the myosin binding protein C3 mutation (MYBPC3 R820W) with cardiac death in a survey of 236 Ragdoll cats. *Journal of Veterinary Cardiology* 2014;16:73-80.

28. Granstrom S, Godiksen MT, Christiansen M, et al. Genotype-phenotype correlation between the cardiac myosin binding protein C mutation A31P and hypertrophic cardiomyopathy in a cohort of Maine Coon cats: a longitudinal study. *J Vet Cardiol* 2015;17 Suppl 1:S268-281.

29. Scansen BA, Morgan KL. Reference intervals and allometric scaling of echocardiographic measurements in Bengal cats. *J Vet Cardiol* 2015;17 Suppl 1:S282-295.

30. Singletary GE, Rush JE, Fox PR, et al. Effect of NT-pro-BNP Assay on Accuracy and Confidence of General Practitioners in Diagnosing Heart Failure or Respiratory Disease in Cats with Respiratory Signs. *Journal of Veterinary Internal Medicine* 2012;26:542-546.

31. Borgeat K, Sherwood K, Payne JR, et al. Plasma Cardiac Troponin I Concentration and Cardiac Death in Cats with Hypertrophic Cardiomyopathy. *J Vet Intern Med* 2014;28:17-31-1737.

32. Payne JR, Borgeat K, Brodbelt DC, et al. Risk factors associated with sudden death vs. congestive heart failure or arterial thromboembolism in cats with hypertrophic cardiomyopathy. *Journal of Veterinary Cardiology* 2015;17, Supplement 1:S318-S328.

33. Antony R, Daghem M, McCann GP, et al. Cardiovascular magnetic resonance activity in the United Kingdom: a survey on behalf of the british society of cardiovascular magnetic resonance. *J Cardiovasc Magn Reson* 2011;13:57.

34. Wilkie LJ, Smith K, Luis Fuentes V. Cardiac pathology findings in 252 cats presented for necropsy; a comparison of cats with unexpected death versus other deaths. *J Vet Cardiol* 2015;17 Suppl 1:S329-340.

35. Hogan DF, Fox PR, Jacob K, et al. Secondary prevention of cardiogenic arterial thromboembolism in the cat: The double-blind, randomized, positive-controlled feline arterial thromboembolism; clopidogrel vs. aspirin trial (FAT CAT). *J Vet Cardiol* 2015;17 Suppl 1:S306-317.

36. Reina-Doreste Y, Stern JA, Keene BW, et al. Case-control study of the effects of pimobendan on survival time in cats with hypertrophic cardiomyopathy and congestive heart failure. *Journal of the American Veterinary Medical Association* 2014;245:534-539.

37. Schober KE, Zientek J, Li X, et al. Effect of treatment with atenolol on 5-year survival in cats with preclinical (asymptomatic) hypertrophic cardiomyopathy. *Journal of Veterinary Cardiology* 2013;15:93-104.

AUTHOR INDEX

AUTHOR INDEX

AUTHOR INDEX

AUTHOR INDEX

AUTHOR INDEX

American College of Veterinary Internal Medicine (ACVIM)
1997 Wadsworth Blvd.
Lakewood, CO 80214-5293
www.acvim.org

ISBN 978-1-5108-2540-6

9 781510 825406

CURRAN ASSOCIATES INC.
proceedings.com